MICROBIOLOGY

BERNARD D. DAVIS, M.D.

Adele Lehman Professor of Bacterial Physiology,
Harvard Medical School, Boston, Massachusetts

RENATO DULBECCO, M.D.

Distinguished Research Professor, The Salk
Institute, San Diego; Senior Clayton
Foundation Investigator; Professor, Medical
School Departments of Pathology and Medicine,
University of California, San Diego, LaJolla,
California

HERMAN N. EISEN, M.D.

Professor of Immunology, Department of Biology
and Center for Cancer Research, Massachusetts
Institute of Technology, Harvard–M.I.T. Division
of Health Sciences and Technology, Cambridge,
Massachusetts

HAROLD S. GINSBERG, M.D.

John Borne Professor and Chairman, Department
of Microbiology, College of Physicians and
Surgeons, Columbia University, New York, New York

With 20 additional contributors

MICROBIOLOGY

Including Immunology and Molecular Genetics

THIRD EDITION

HARPER & ROW, PUBLISHERS

PHILADELPHIA

Cambridge
New York
Hagerstown
San Francisco

London
Mexico City
São Paulo
Sydney

1817

84–85—10-9-8-7-6-5-4

Library of Congress Cataloging in Publication Data
Main entry under title:

Microbiology, including immunology and molecular genetics.

  Includes bibliographies and index.
  1. Medical microbiology. 2. Immunology. 3. Molecular genetics. I. Davis, Bernard David, 1916–
[DNLM: 1. Microbiology. 2. Immunity. QW4 M626]
QR46.M5393 1980   616′.01   80-13433
ISBN 0-06-140691-0

Cover illustration: Recombination between two DNA molecules, visualized at the stage where a strand of each has been opened and its ends have been connected with those of the other opened strand. (H. Potter and D. Dressler, Proc. Natl. Acad. Sci. USA 74:4168, 1976) For further details see Figure 11-11.

To the memory of W. Barry Wood, Jr.
(1910–1971),
who assembled the group of authors
for the first edition of this text.
His perspective and judgment as a coauthor
contributed a great deal to its style,
and his spirit still pervades the present volume.

CONTENTS

† Deceased

† Deceased

LIST OF
CONTRIBUTORS

ROBERT AUSTRIAN, M.D.
CHAPTERS 25, 27
John Herr Musser Professor and Chairman, Department of Research Medicine,
The University of Pennsylvania School of Medicine, Philadelphia, Pennsylvania

JOEL B. BASEMAN, Ph.D.
CHAPTER 39
Associate Professor, Department of Bacteriology and Immunology, University of
North Carolina School of Medicine, Chapel Hill, North Carolina

ROBERT M. CHANOCK, M.D.
CHAPTER 42
Chief, Laboratory of Infectious Diseases, National Institute of Allergy and
Infectious Diseases, National Institutes of Health, Bethesda, Maryland; Professor
of Child Health and Development, George Washington University School of
Medicine, Washington, D.C.

BERNARD D. DAVIS, M.D.
CHAPTERS 1–9, 12–14, 24, 26, 43, 67
Adele Lehman Professor of Bacterial Physiology, Harvard Medical School,
Boston, Massachusetts

RENATO DULBECCO, M.D.
CHAPTERS 10, 11, 46–52, 67
Distinguished Research Professor, The Salk Institute, San Diego; Professor,
Medical School Departments of Pathology and Medicine, University of California,
San Diego, LaJolla, California

HERMAN N. EISEN, M.D.
CHAPTERS 15–23
Professor of Immunology, Department of Biology and Center for Cancer
Research, Massachusetts Institute of Technology, Cambridge, Massachusetts

RONALD J. GIBBONS, Ph.D.
CHAPTER 44
Senior Staff Member, Department of Microbiology, Forsyth Dental Center;
Clinical Professor or Oral Biology, Departments of Oral Biology and
Pathophysiology, Harvard School of Dental Medicine, Boston, Massachusetts.

HAROLD S. GINSBERG, M.D.
CHAPTERS 50, 53–65
John Borne Professor and Chairman, Department of Microbiology, College of
Physicians and Surgeons, Columbia University, New York, New York

EMIL C. GOTSCHLICH, M.D.
CHAPTER 30
Professor and Senior Physician, Departments of Bacteriology and Immunology, The Rockefeller University, New York, New York

GEORGE S. KOBAYASHI, Ph.D.
CHAPTERS 38, 45
Professor, Departments of Medicine and of Microbiology and Immunology, Washington University School of Medicine, St. Louis, Missouri

LORETTA LEIVE, Ph.D.
CHAPTER 6
Research Biologist, Laboratory of Biochemical Pharmacology, National Institute of Arthritis, Metabolism, and Digestive Diseases, National Institutes of Health, Bethesda, Maryland

G. PHILLIP MANIRE, Ph.D.
CHAPTER 41
Kenan Professor and Chairman, Department of Bacteriology and Immunology, University of North Carolina, Chapel Hill, North Carolina

MACLYN McCARTY, M.D.
CHAPTER 28
John D. Rockefeller, Jr. Professor, The Rockefeller University, New York, New York

ZELL A. McGEE, M.D.
CHAPTER 43
Professor, Department of Medicine, Vanderbilt University School of Medicine, Nashville, Tennessee

STEPHEN I. MORSE, M.D., Ph.D.
CHAPTERS 29, 34
Professor and Chairman, Department of Microbiology and Immunology, State University of New York, Downstate Medical Center, Brooklyn, New York

EDWARD S. MURRAY, Ph.D. (deceased)
CHAPTER 40
Professor, Department of Microbiology, Harvard School of Public Health, Boston, Massachusetts

ROGER L. NICHOLS, M.D.
CHAPTER 41
Irene Heinz Given Professor of Microbiology, Department of Microbiology, Harvard School of Public Health, Boston, Massachusetts

ELIORA Z. RON, Ph.D.
CHAPTER 14
Associate Professor, Department of Microbiology, Faculty of Life Sciences, Tel-Aviv University, Tel-Aviv, Israel

SIGMUND S. SOCRANSKY, D.D.S.
CHAPTER 44
Head, Department of Periodontology, Forsyth Dental Center, Boston,
Massachusetts

ALEX C. SONNENWIRTH, Ph.D.
CHAPTERS 31, 32, 33, 43, 44
Professor, Departments of Microbiology and Immunology, and Pathology,
Washington University School of Medicine; Director, Division of Microbiology,
The Jewish Hospital of St. Louis, St. Louis, Missouri

P. FREDERICK SPARLING, M.D.
CHAPTER 30
Professor, Departments of Medicine, and of Bacteriology and Immunology,
University of North Carolina School of Medicine, Chapel Hill, North Carolina

MORTON N. SWARTZ, M.D.
CHAPTERS 33, 35, 36
Chief, Infectious Disease Unit, Department of Medicine, Massachusetts General
Hospital; Professor, Department of Medicine, Harvard Medical School, Boston,
Massachusetts

JOSEPH G. TULLY, Ph.D.
CHAPTER 42
Chief, Mycoplasma Section, Laboratory of Infectious Diseases, National Institute
of Allergy and Infectious Diseases, National Institutes of Health, Bethesda,
Maryland

EMANUEL WOLINSKY, M.D.
CHAPTER 37
Professor, Department of Medicine, Case Western Reserve University School of
Medicine; Director of Microbiology and Chief, Division of Infectious Diseases,
Department of Medicine, Cleveland Metropolitan General Hospital, Cleveland,
Ohio

PREFACE TO THE THIRD EDITION

This textbook continues to be designed for the student who seeks to understand microbiology and immunology in some depth, and as a growing science. Hence despite the pressures for space we continue to try to indicate the nature of the evidence underlying each major conclusion, rather than only summarizing existing knowledge.

To make room for the new material nearly every paragraph has been rewritten. In addition, we have eliminated some detail in areas that are also covered in biochemistry courses (energy production, biosynthesis, and protein synthesis). However, we have expanded another biochemical topic, metabolic regulation, because it integrates molecular genetics with metabolism and with microbial ecology, and because studies in bacteria and in bacteriophages are continuing to provide profound new insights. Moreover, we have retained the chapters that provide a background in molecular genetics, since advances in bacterial physiology and virology, and now also in immunology, depend heavily on the concepts and the technics of this subject.

The most dramatic recent advances have been the discovery of several mechanisms for rearranging nucleic acid sequences (both in nature and in the laboratory), and the development of simple technics for isolating and sequencing short segments of DNA. This material will appear in many places—in the chapters on Organization of the Genetic Material, on Animal Cell Cultures, and on Multiplication of Animal Viruses; in a new chapter on Plasmids and Gene Manipulation; and in a reorganized chapter called Immunoglobulin Molecules and Genes. Cellular and molecular aspects of immunology have also advanced explosively, requiring a new chapter on the Cellular Basis of the Immune Response, and major revision of most of the chapters in immunology. A third area of rapid growth has been the molecular analysis of cell surfaces, including the attachment sites of bacteria and of host cells, and the various cell receptors involved in immune responses. Indeed, from these studies, and from molecular studies on bacterial toxins, complement, and inflammation, a true molecular biology of bacterial infections is emerging.

Authorship of the various sections and chapters is noted in appropriate places. In the section on virology, Dulbecco was responsible for general virology and the tumor viruses, and Ginsberg for multiplication of animal viruses, viral pathogenesis and the specific agents. Overall editing for uniformity of style was provided, except for certain chapters on specific viruses, by Davis.

Since immunology and virology are now often taught in separate courses, the sections on these subjects will also be made available as separate volumes.

The late W. Barry Wood, Jr. initiated this text, and his contributions are still present throughout the section on Bacterial and Mycotic Infections. He would have been pleased to see how his often expressed vision of a molecular biology of infectious diseases is now being realized.

B. D. Davis
R. Dulbecco
H. N. Eisen
H. S. Ginsberg

PREFACE TO THE FIRST EDITION

"What is new and significant must always be connected with old roots, the truly vital roots that are chosen with great care from the ones that merely survive."

This principle, professed by the composer Bela Bartok, is as applicable to science as it is to music. Indeed, it highlights the most difficult aspect of writing a modern textbook of microbiology, for few branches of natural science have been so rapidly altered by recent advances. Only a few years ago microbiology was largely an applied field, concerned with controlling those microbes that affect man's health or his economic welfare, but with the recent development of molecular genetics, stemming largely from the study of microbial mutants, microbiology has rapidly been drawn to the center of the biological stage.

As a result, infectious disease no longer constitutes the sole bridge between microbiology and medicine. An additional, rapidly broadening span is provided by the use of microbes as model cells in the study of molecular genetics and cell physiology, for the principles and the successful approaches developed in such studies will surely prove widely applicable to human cells, which can now be cultured much like bacteria. In addition, studies at a molecular level are also rapidly providing a deeper insight into problems directly related to infectious disease, including the action of chemotherapeutic drugs, the structure of antibodies and cellular antigens, and the nature of viruses. Hence to prepare the student for the scientific medicine of his future it has seemed to us desirable to increase emphasis on the molecular and genetic aspects of microbiology. At the same time, the authors, having all had clinical experience, are vividly aware of the importance of providing a thorough understanding of host-parasite relationships and mechanisms of pathogenicity, even though many aspects cannot yet be explained in molecular terms.

In short, we have tried to identify the "truly vital roots" of classical bacteriology, immunology, and virology, and to engraft upon them the recent molecular advances. To keep the volume to a reasonable size we have eliminated much traditional information that did not seem to have either theoretical or practical importance for the student of medicine. Moreover, the clinical and epidemiological aspects of infectious diseases have been largely left for later courses in the medical curriculum, and we have provided only a small number of selected references, primarily for access to the original literature and not for documentation. In the hope of making the book more useful and versatile, we have included in smaller type a good deal of material that seemed not essential for an introduction to the subject, but still likely to interest many readers.

The demands of the medical curriculum frequently lead to a condensed memorizing of conclusions; yet courses in the basic medical sciences should surely illustrate the scientific method as well as transmit a body of information. We have therefore briefly reviewed the history of many major discoveries in order to show how scientific advances may depend on new concepts or technics, or on ingenious experiments, or an alertness to the significance of unexpected observations. Moreover, we have endeavored throughout to indicate the nature of the evidence underlying the conclusions presented—for otherwise the student sees only the shadow and not the substance of science.

This book is designed primarily as a text for students and investigators of medicine and the allied professions: hence the exposition proceeds from general principles to specific pathogenic microorganisms. However, we hope that the discussion of general principles will also prove useful to graduate students and investigators in the biological sciences.

The preparation of this volume has been a truly cooperative effort: the chapters drafted by each author have been critically reviewed by most or all of the others. We are deeply grateful for the education and for the warm friendships that have resulted.

A new book of this size will inevitably contain errors and weaknesses. We shall welcome corrections and suggestions for future editions.

B. D. Davis
R. Dulbecco
H. N. Eisen
H. S. Ginsberg
W. B. Wood, Jr.

ACKNOWLEDGMENTS

We are grateful to the contributors who revised various chapters or portions of chapters. These authors should not be held accountable for any defects in expression, since their contributions were extensively modified to fit the style of the rest of the book.

We are also deeply indebted to the many colleagues who have critically reviewed various chapters. These consultants are listed on the following page. Special thanks go to A. M. Pappenheimer, Jr. for his extensive contributions to Chapter 26, and to Thomas J. Yang for the monumental task of checking the entire book for errors.

It is a pleasure to acknowledge the skillful help, pleasant cooperation, and patience of the publisher's staff and of our several secretaries, during the three years of preparation of this revision. In addition, we are indebted to the Marine Biological Laboratory at Woods Hole, Massachusetts; to the many investigators who provided illustrations; and to the publishers who granted permission to reprint figures. Sources are acknowledged in the legends. Finally, we gratefully note the skill of Stephen Shaw in preparing the new illustrations and the careful preparation of the index, from the perspective of a student, by Jonathan H. Davis.

LIST OF CONSULTANTS

K. FRANK AUSTEN, M.D.
THOMAS L. BENJAMIN, Ph.D.
MICHAEL J. BEVAN, Ph.D.
BARRY R. BLOOM, Ph.D.
ZANVIL A. COHN, M.D.
MALCOLM L. GEFTER, Ph.D.
E. PETER GEIDUSCHEK, Ph.D.
EDWARD J. GOETZL, M.D.
ALFRED L. GOLDBERG, Ph.D.
RICHARD N. GOLDSTEIN, Ph.D.
DONALD R. HELINSKI, Ph.D.
MORRIS J. KARNOVSKY, M.D., M.B.B. Ch.
FRED KARUSH, Ph.D.
EUGENE P. KENNEDY, Ph.D.
EDWIN D. KILBOURNE, M.D.
LAWRENCE KUNZ, Ph.D.
WERNER K. MAAS, Ph.D.
A. M. PAPPENHEIMER, JR., Ph.D.
LEWIS I. PIZER, Ph.D.
EFRAIM RACKER, M.D.
R. WALTER SCHLESINGER, M.D.
PHILIP A. SHARP, Ph.D.
SAUL J. SILVERSTEIN, Ph.D.
LISA STEINER, M.D.
JACK L. STROMINGER, M.D.
MICHAEL SYVANEN, Ph.D.
P.-C. TAI, Ph.D.
TSU-JU (THOMAS) YANG, D.V.M., Ph.D.
CHARLES YANOFSKY, Ph.D.

MICROBIOLOGY

chapter I

EVOLUTION OF MICROBIOLOGY AND OF MICROBES

BERNARD D. DAVIS

There are similarities between the diseases of animals or man and the diseases of beer and wine. . . . If fermentations were diseases one could speak of epidemics of fermentation.

L. PASTEUR

EVOLUTION OF MICROBIOLOGY

THE FIRST MICROSCOPIC OBSERVATIONS

The spread of certain diseases within populations long suggested the existence of invisible, transmissible agents of infection. In the poem *De rerum natura* Lucretius (96?–55 B.C.) presciently recognized not only the atomistic nature of matter but also the existence of "seeds" of disease. But however logical this inference, the existence of living microscopic organisms (microbes) was not established, and was not generally accepted, until they were seen under a simple (one-lens) microscope (Fig. 1-1 in 1677, by Antony van Leeuwenhoek. Even then the development of microbiology as an advancing science, and recognition of its relation to medicine, had to wait for nearly 150 years.

Leeuwenhoek's contributions are all the more remarkable in the light of his isolation from the learned world and his lack of formal education. A cloth merchant in Delft, Holland, with a political sinecure as custodian of the Town Hall, he spent much of his time in grinding tiny lenses of high magnification (probably up to 300×). With these this patient and curious man discovered a whole new world, including the major classes of bacteria (spheres, rods, and spirals; Fig. 1-2), protozoa, algae, yeasts, erythrocytes, spermatozoa, and the capillary circulation. Leeuwenhoek's discoveries were described in a flow of letters to the Royal Society of London.* Moreover, by keeping his methods secret he remained throughout his long lifetime the sole occupant of the field he had created; no other observers succeeded in using single lenses so effectively.

---

* The following observations on a decayed tooth illustrate Leeuwenhoek's charming, colloquial style, and his reliance on motility as the only criterion of life then available for a microscopic object. "I took this stuff out of the hollows in the roots, and mixed it with clean rainwater, and set it before the magnifying-glass. . . . I must confess that the whole stuff seemed to me to be alive. But notwithstanding the number of these animalcules was so extraordinarily great that 'twould take a thousand million of some of 'em to make up the bulk of a coarse sand-grain, . . . the animalcules, with their strong swimming through the water, put many little particles which had no life in them into a like motion, so that many people might well have taken these particles for living creatures too."

The compound microscope had already been invented, but it had little use in microbiology until its serious optical aberrations were eliminated, early in the 19th century. Linnaeus in 1767 distinguished only 6 species in assigning microbes to the class "Chaos," but 600 types were figured in Ehrenberg's *Atlas* in 1838.

The advance from a descriptive to an experimental science of microbiology proceeded only slowly, with the development of a special methodology. The key was the use of sterilized materials and aseptic techniques; for while purity is defined in chemistry in terms of percentage of contaminating material, in microbiology a single contaminating cell can ruin an experiment. Only after learning to avoid such contamination could investigators recognize and characterize a great variety of microbes. The development of reliable methods was stimulated by a prolonged controversy over an issue with religious overtones: the spontaneous generation of life.

SPONTANEOUS GENERATION

Until the last century it seemed self-evident that living organisms can arise spontaneously in decomposing organic matter. In the seventeenth century Redi demonstrated that the appearance of maggots in decomposing meat depended on the deposition of eggs by flies, but the idea of spontaneous generation persisted for the new world of microbes, delaying recognition of their relation to the rest of biology.

The question would appear to have been settled in the eighteenth century when Spallanzani (1729–1799) introduced the use of sterile culture media: he showed that a "putrescible fluid," such as an infusion of meat, would remain clear indefinitely if boiled and properly sealed.†

---

† Indeed, the soundness of this discovery was confirmed in the early nineteenth century, when a Parisian confectioner, Appert, competing for a prize offered by Napoleon, developed the art of preserving food by canning.

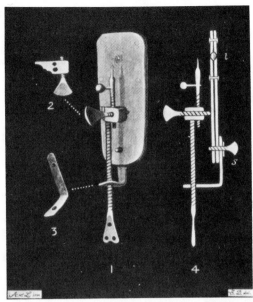

FIG. 1-1. A Leeuwenhoek microscope, viewed from the back (**1**) and in diagrammatic section (**4**). The specimen, on a movable pin, is examined through a minute biconvex lens (**I**), held between two metal plates. (Dobell C: Anton van Leeuwenhoek and His "Little Animals." New York, Dover, 1960)

Moreover, in 1837 Schwann obtained similar results even when air was allowed to reenter the cooling flask before sealing, provided the air passed through a heated tube. Skeptics could claim, however, that the absence of decomposition in these sealed vessels was due to a limitation in the supply of air rather than to the exclusion of dustborne living contaminants. To answer this objection Schroeder and von Dusch introduced the use of the cotton plug, which is still used today to exclude airborne contaminants.

Nevertheless, the controversy continued, for some investigators were unable to reproduce the alleged stability of dust-free sterilized organic infusions. Louis Pasteur (1822–1895) then entered the lists. He showed that boiled medium could remain clear in a "swan-neck" flask, open to the air through a sinuous horizontal tube in which dust particles would settle as air reentered the cooling vessel (Fig. 1-3). Pasteur also demonstrated that in the relatively dust-free atmosphere of a quiet cellar, or of a mountain top, sealed flasks could be opened and then resealed with a good chance of escaping contamination.

Pasteur's experiments were a public sensation. Though his contributions were in principle no more decisive than those of his predecessors, his zeal and skill as a polemicist were largely responsible for laying the ghost of spontaneous generation. Because of his crusading spirit, as well as his experimental skill and intuitive genius, Pasteur was for the nineteenth century, in Dubos' words, "not only

the arm but also the voice, and finally the symbol, of triumphant science." For example, from a lecture delivered at the Sorbonne in 1864:

"I have taken my drop of water from the immensity of creation, and I have taken it full of the elements appropriate to the development of microscopic organisms. And I wait, I watch, I question it!—begging it to recommence for me the beautiful spectacle of the first creation. But it is dumb, dumb since these experiments were begun several years ago; it is dumb because I have kept it sheltered from the only thing man does not know how to produce, from the germs which float in the air, from Life, for Life is a germ and a germ is Life. Never will the doctrine of spontaneous generation recover from the mortal blow of this simple experiment!"

The Problem of Spores. Despite Pasteur's dramatic success it turned out that his accusations of technical incompetence did not really explain why his opponents' boiled infusions stubbornly refused to remain clear. The key difference was their use of infusions of hay (while Pasteur used yeast extract). The decisive experiments were provided by the British physicist John Tyndall, whose interest in atmospheric dust led him into this biologic problem. After bringing a bale of hay into his laboratory Tyndall could no longer achieve sterility in the same room by boiling, and he showed that the hay had contaminated his laboratory with *an incredible kind of living organism: one that could survive boiling.* In the same year (1877) Ferdinand Cohn demonstrated the resistant forms as small, refractile **endospores** (see Spores, Ch. 6), a special stage in the life cycle of the hay bacillus (*Bacillus subtilis*). Since even the most resistant spores are readily sterilized in the presence of moisture at 120°C,

FIG. 1-2. Leeuwenhoek's figures of bacteria from the human mouth, from letter of 17 September, 1683. Dotted line between B and D indicates motility.

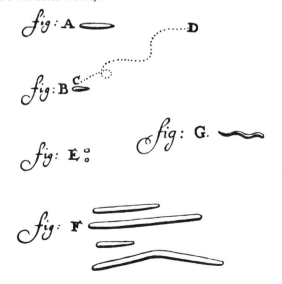

FIG. 1-3. Pasteur's "swan-neck" flasks. After their use in his studies on spontaneous generation these flasks were sealed, and they have since been preserved, with their original contents, in the Pasteur Museum. (Courtesy of Institut Pasteur, Paris)

the autoclave, which uses steam under pressure, became the hallmark of the bacteriology laboratory.

Though a rich variety of microbes, adapted to different ecologic niches, were soon demonstrated, the tools needed to analyze what went on in such tiny cells were not available, and microbiology quickly became primarily an applied science, concerned with isolating various organisms and identifying and controlling their beneficial or harmful activities. The areas of greatest interest were useful fermentations, soil microbiology, and infectious diseases of plants, animals, and man—the last dominating the field. However, along with the emphasis on applications there were fundamental advances: the study of fermentations contributed much to the development of biochemistry, while soil microbiologists developed profound insights into the geochemical cycle.

THE ROLE OF MICROBES IN FERMENTATIONS

We have seen that practical men proceeded to preserve foods while the savants continued to dispute spontaneous generation. Useful fermentations* have an even longer history of achievement without

* In general usage the term **fermentation** refers to the microbial decomposition of vegetable matter, which contains mostly carbohydrates, while **putrefaction** refers to the formation of more unpleasant products by the decomposition of high-protein materials, such as meat or eggs. (For a more rigorous definition of fermentation see Ch. 3.) The etymology of the word ferment (L. *fervere,* to boil), and its figurative use today, reflect the heating and bubbling that are generated in a vat of fermenting grape juice.

a theoretic foundation: lost in antiquity are the origins of leavened bread, wine, or the fermentations that preserve food through the accumulation of lactic acid (soured milk, cheese, sauerkraut, ensilage).

In the 1830s, with the development of microscopes of sufficient resolving power, Schwann, Cagniard-Latour, and Kutzing concluded independently that the sediment of microscopic globules accumulating in an alcoholic fermentation consists of tiny growing plants whose metabolic activities are responsible for the fermentation. However, the leading chemists of the day considered that fermentation was a chemical process, due to a self-perpetuating instability of the grape juice initiated by its exposure to the air. The amorphous sediment would thus be a byproduct of the fermentation, analogous to the frequent crystalline precipitates of tartaric acid. In particular, the distinguished Liebig, father of biochemistry, advanced this view with such vehemence that Schwann's excellent evidence was essentially discarded for two decades.† Liebig's authority was eventually refuted by Pasteur.

Educated as a chemist, Pasteur became interested in fermentations through discovering optical isomerism as a property of certain fermentation products (isoamyl alcohol, tartaric acid). In 1857, in his first paper on fermentation, Pasteur showed that *different kinds of microbes are associated with different kinds of fermentation:* spheres of variable size (now known as yeast cells) in the

† Indeed, in 1839 Liebig and Wohler published an anonymous, scatologic hoax in which they reported the microscopic visualization of an animal, shaped somewhat like a distilling apparatus, that took in sugar at one end and excreted alcohol at the other, while its large gonads released bubbles of CO_2.

alcoholic fermentation, and smaller rods (lactobacilli) in the lactic fermentation.*

In the course of establishing the nature of fermentation Pasteur founded the study of **microbial metabolism** and developed profound insight into many of its problems. Indeed, our knowledge of microbial physiology hardly advanced any further until the development of sophisticated biochemical technics many decades later. In particular, Pasteur showed that *life is possible without air,* that some organisms (obligatory anaerobes) are even inhibited by air, and that fermentation is much less efficient than respiration in terms of the growth yield per unit of substrate consumed.

Selective Cultivation. Since the nature of a specific fermentation depends on the organism responsible, how can a given kind of substrate, when not deliberately inoculated with a particular microbe, regularly undergo a given kind of fermentation? Pasteur recognized that the explanation lay in the principle of selective cultivation: various organisms are ubiquitous, and the one best adapted to a given environment eventually predominates. For example, in grape juice the high sugar concentration and the low protein content (i.e., low buffering power) lead to a low pH, which favors the outgrowth of acid-resistant yeasts and thus yields an alcoholic fermentation. In milk, in contrast, the much higher protein and lower sugar content favor the outgrowth of fast-growing but more acid-sensitive bacteria, which cause a lactic fermentation.

An excess of the wrong organism in the starting inoculum, however, can grow out sufficiently to cause formation of a product with a poor flavor; hence specific microbes play a role in "diseases" of beer and wine, as well as in different normal fermentations. This finding led Pasteur to a most fruitful suggestion: *that specific microbes might also be causes of specific diseases in man.*

With an eye always to practical as well as to theoretic problems, Pasteur developed the procedure of gentle heating (**pasteurization**) to prevent the spoilage of beer and wine by undesired contaminating microbes. This process was later used to prevent milkborne diseases of man.

Of great economic importance was the extension of industrial fermentations from the production of foods and beverages to that of valuable chemicals, such as glycerol, acetone, and later vitamins, antibiotics, and alkaloids. Successful developments have depended

---

* As in many scientific disputes, the losing argument possessed at least a grain of truth. The view of fermentation as a chemical rather than a vital process was ultimately vindicated in 1897, though in a profoundly modified form, when Edouard Buchner accidentally discovered that a cell-free extract, made by grinding yeast cells with sand, fermented a concentrated sugar solution added to preserve the extract.

largely on the selection of microbial mutants with increased production of the desired compounds.

SOIL MICROBIOLOGY: GEOCHEMICAL CYCLES

Though the similarity of the pathways of carbohydrate metabolism in some microbes and in mammals led eventually to the concept of the **unity of biology at a molecular level,** toward the end of the Pasteurian era investigators exploring the microbiology of the soil (notably Winogradsky in Russia and Beijerinck in Holland) discovered an astonishing **variety of metabolic patterns** by which different kinds of bacteria are adapted to different ecologic niches. These organisms were isolated by an extension of Pasteur's principle of selective cultivation: **enrichment cultures,** in which only a particular energy source is provided, and growth is thus restricted to those organisms that can use that source. Some of the unusual patterns of bacterial energy production, ranging from the oxidation of sulfur to the formation of CH_4 from CO_2 and H_2, are described in Chapter 3.

Soil microbiology revealed that *the major role of microbes in nature is geochemical:* mineralization of organic carbon, nitrogen, and sulfur (i.e., conversion to CO_2, NH_3 or NO_3^-, SO_4^{2-} or S^{2-}), so that these elements can be used cyclically for growth of higher plants and animals, rather than being tied up in dead organic matter. (Unlike all natural compounds, however, many synthetic organic compounds, such as polystyrene or fluorocarbons, cannot be recycled because no microbes have evolved to attack them.) In the synthesizing half of the cycle algae and photosynthetic bacteria, as well as higher plants, reduce CO_2 to organic compounds by photosynthesis. In addition, **nitrogen-fixing** bacteria reduce atmospheric N_2 to NH_3, and the **nitrifying** bacteria enhance soil fertility by converting the volatile NH_3 into a nonvolatile form, nitrate.

These beneficent geochemical roles deserve emphasis, since the prominent historical connection of microbiology with infectious disease has given rise to the popular image of a malignant and hostile microbial world. In fact, human pathogens constitute only a small fraction of the recognized bacterial species, and an infinitesimal fraction of the total mass of microbes on the earth: most microbes attack organic matter only after it is dead and buried.

MICROBIOLOGY AND MEDICINE

THE GERM THEORY OF DISEASE

Among the major classes of disease infections have undoubtedly presented the greatest burden to mankind: not only were they a leading cause of death, but these deaths were often especially heartbreaking because they were so frequent among the young. Moreover, by their epidemic nature infections have disabled and terrorized communities and have determined the fate of armies and nations: thus smallpox permitted a few dozen Spaniards to take over a flourishing Mexican civilization. Clearly the control over infectious diseases, and over microbial pollution of the environment, has been the greatest achievement of medical science. Today it is easy to take these advances for granted while focusing on problems created by technology; but an earlier public enshrined Pasteur and Robert Koch (1843–1910) as national heroes.

Epidemiologic Evidence. The discovery of infectious agents was long preceded by the concept of **contagious** disease, i.e., one initiated by **contact** with a diseased person or with objects contaminated by him. Thus even though the ancient Hebrews viewed pestilences as punishments visited on peoples by the Lord, the Mosaic code also includes numerous public health regulations, including the isolation of lepers, the discarding of various unclean materials, and the avoidance of shellfish and pork as foods. Later shrewd observers of epidemics, such as Lucretius and Boccaccio, explicitly recognized their contagious nature.

In 1546 Fracastorius of Verona presented an impressive body of evidence in *De Contagione*. This book founded the science of **epidemiology,** which *analyzes the distribution in human populations of events affecting health.* After carefully studying several epidemic diseases, including plague and syphilis, Fracastorius concluded that they were spread by **seminaria** ("seeds"), transmitted from one person to another either directly or via inanimate objects. (But while recommending avoidance of exposure to patients, or to plagued communities, he perpetuated the view that the initial seeds in an epidemic were generated by supernatural or telluric forces.) Despite the impressive evidence and reasoning of Fracastorius, even a century later such leading physicians as William Harvey continued, like Hippocrates and Galen, to ascribe epidemics to **miasmas,** i.e., poisonous vapors created by the influence of planetary conjunction or by disturbances arising within the earth.

Part of the difficulty arose from the existence of many **communicable** diseases that are not contagious in a strict sense. We now know that these are transmitted by less obvious routes, such as air, water, food, and insects; and it is easy to see how an airborne disease could logically be ascribed to poisoned air until the particulate nature of the agent was demonstrated. Moreover, before the development of chemistry the idea that living organisms too small to be seen could exist, and could mortally harm large animals, was clearly contrary to common sense. Thus even the recognition of transmissible "seeds of disease" did not quite hit the mark: the notion of seed was taken literally, the actual *contagium vivum* within the diseased body being considered something more complex derived from the seed. Experimental evidence was required, and it accumulated slowly, from several directions: transmission of infection, its prevention, and finally identification of the agents.

Transmission of infection was demonstrated boldly in the eighteenth century by the renowned surgeon John Hunter, who inoculated himself with purulent material from a patient with gonorrhea; unfortunately, both for him and for the study, he transmitted at the same time a much more serious disease, syphilis. The use of experimental animals was introduced later: for example, in 1865 Villemin so transmitted tuberculosis, though the nature of the responsible agent was not established until 20 years later.

The role of **indirect transmission** was not recognized until the 1840s, when Semmelweis in Vienna, and Oliver Wendell Holmes (perhaps better known as a poet than as a physician) in Boston, shockingly (but unsuccessfully) blamed obstetricians, moving with unwashed hands from one patient to the next, for the prevalence in hospitals of puerperal sepsis, a frequent cause of maternal death. Communicability was demonstrated more convincingly in 1854, for waterborne enteric infections, when John Snow terminated a localized epidemic of cholera in London by closing the Broad Street pump, whose supply of water for the neighborhood had been contaminated by sewage.

Preventive measures also lent support to the germ theory. In 1796, Edward Jenner had introduced vaccination (L. *vacca,* cow) against smallpox, using material from lesions of a similar disease of cattle (cowpox), though the practice then had no theoretic foundation. In the 1860s Joseph Lister introduced antiseptic surgery, on the basis of Pasteur's evidence for the ubiquity of airborne microbes. He reasoned that such organisms might also be responsible for the frequent development of pus in surgical and in traumatic wounds, and he found that application of a disinfectant, phenol, markedly reduced the incidence of serious infections. As with the canning of food, practice outran theory: this major advance was achieved before any specific agent was identified.

RECOGNITION OF AGENTS OF INFECTION

The epidemiologic evidence for communicability, though logically convincing, did not carry the weight of a direct demonstration of the agents of infection. The first to be recognized were **fungi,** which are larger than bacteria: in 1836 Agostino Bassi demonstrated experimentally that a fungus was the cause of a disease (of silkworms), and 3 years later Schönlein discovered the association of a fungus with a human skin disease (favus). In 1865 Pasteur entered the field of pathogenic microbiology with the discovery of a **protozoon** that was threatening to ruin the European silkworm industry.

The etiologic role of **bacteria** was unequivocally established by Robert Koch in 1876 for anthrax, which offered the investigator several advantages: the organism is unusually large and is readily identified morphologically; the disease, primarily one of cattle and sheep, may be conveniently transmitted to small animals; and dense bacterial populations may appear in the blood.

Indeed, Davaine in 1850 had already seen rod-shaped bodies in the blood of sheep dying of anthrax, and he later transmitted the disease by inoculating as little as 10^{-6} ml of blood. However, this evidence did not prove whether these bodies were the cause or a result of the disease. Koch solved the problem by isolating the anthrax bacillus in pure culture, with which he transmitted the disease to mice. In addition, the cultures developed spores (recognized from their refractility: Fig. 1-4) whose resistance to sterilization explained why fields once inhabited by anthrax-infected animals could infect fresh herds years later.

Pure Cultures. The key to the identification of bacteria as pathogens was the isolation of pure cultures. Lister obtained such cultures by the method of **limiting dilutions,**

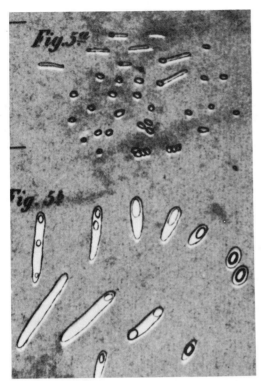

FIG. 1-4. Spore formation in *Bacillus anthracis*, as independently described and simultaneously published by Robert Koch ("Fig. 5a") and Ferdinand Cohn ("Fig. 5b"). (Courtesy of Koch-Institut, Berlin)

in which the source material is diluted until the individual inocula each contain either one infectious particle or none; but this method is awkward. Koch meticulously perfected the technics of identification that are used today, including the use of **solid media,** on which individual cells give rise to separate colonies, and the use of **stains.** Koch's genius is perhaps best reflected in his patient modifications of his own earlier methods, which finally led to the identification of the tubercle bacillus in 1882: because this organism grows very slowly the usual 1–2 days of cultivation had to be extended to several weeks, and because it is so impervious the usual few minutes of staining had to be extended to 12 h.

After identifying the tubercle bacillus Koch formalized the criteria, introduced by Henle in 1840 but known as **Koch's postulates,** for distinguishing a pathogenic from an adventitious microbe: 1) the organism is regularly found in the lesions of the disease, 2) it can be isolated in pure culture on artificial media, 3) inoculation of this culture produces a similar disease in experimental animals, and 4) the organism can be recovered from the lesions in these animals. These criteria have proved invaluable in identifying pathogens, but they cannot always be met: some organisms (including all viruses) cannot be grown on artificial media, and some are pathogenic only for man.

The powerful methodology developed by Koch introduced the "Golden Era" of medical bacteriology. Between 1879 and 1889 various members of the German school isolated (in addition to the tubercle bacillus) the cholera vibrio, typhoid bacillus, diphtheria bacillus, pneumococcus, staphylococcus, streptococcus, meningococcus, gonococcus, and tetanus bacillus. Studies naturally followed on the mechanisms of pathogenicity of these organisms, the host responses, and the methods of prevention and treatment.

Curiously, Pasteur, despite his early start with anthrax, did not enter the race to identify pathogens; as a chemist he was uninterested in the problems of isolating and classifying organisms. Instead, he devoted his later years to the development of **vaccines.** By accident he found that chickens injected with an old culture of the bacterium of chicken cholera were subsequently resistant to a fresh, virulent culture. He quickly recognized that this observation suggested an important variability in the properties of the same organism, supplementing his earlier emphasis on the constancy of the organisms responsible for different fermentations. Within the incredible space of 4 years, and without any understanding of the immune response, Pasteur discovered four methods of "**attenuating**" organisms and thus converting them to useful vaccines: aging of the culture (chicken cholera), cultivation at high temperature (anthrax), passage through another host species (swine erysipelas), and drying (rabies). As we shall see in later chapters, this "attenuation" comprises two distinct processes: selection of less virulent mutants, and killing of virulent organisms with retention of their immunizing activity.

VIRUSES

The term virus (L., poison) was long used as a synonym for "infectious agent," but it became restricted some decades ago to agents smaller than bacteria (hence separable by filtration) and unable to multiply outside a living host. These characteristics, however, overlap with those of some especially small bacteria (rickettsiae, chlamydiae). Viruses are now sharply defined in terms of their characteristic structure and their **mode of replication,** i.e., dissociation into components within a host cell, use of host machinery to synthesize the components coded for by the viral genes, and formation of new units by reassembly rather than by cellular enlargement and division.

The first virus to be recognized as filterable was a plant pathogen, tobacco mosaic virus, discovered independently by Ivanovski in Russia in 1892 and by Beijerinck in Holland in 1899. Filterable animal viruses* were first demonstrated for foot-and-mouth disease of cattle by

---

* It is curious that though the most dramatic part of Pasteur's work in the 1880s on the development of vaccines was performed with tissues containing rabies virus, he did not recognize the filterability of the agent.

Löffler and Frosch in 1898, and for a human disease, yellow fever, by the US Army Commission under Walter Reed in 1900. Viruses that infect bacteria (**bacteriophages**) were discovered by Twort in England and by d'Herelle in France in 1916–1917.

For the first third of this century viruses could be detected only by their pathogenic effects on living hosts, and progress was slow. Eventually sophisticated physical and chemical methods were developed for purifying and characterizing viruses, while the development of the electron microscope and the advance of molecular genetics made it possible to analyze their mechanism of reproduction. Precise quantitative studies became possible with the development of monolayer cultures of host cells, in which viruses can form discrete plaques analogous to the bacterial colonies formed on solid media.

With the resulting dramatic expansion of virology it has been recognized that the agents of acute viral diseases are only the most conspicuous members, but not the bulk, of the viral kingdom: more and more viruses are being identified that have delayed effects or no apparent effect at all. In addition, the genetic discontinuity between virus and cell has become blurred, with the finding that viruses can become integrated into host chromosomes and can also pick up host genes. These developments suggest that viruses have evolved from cellular chromosomes, and that they have evolved not primarily as parasites but as agents for transmitting blocks of genetic material from one organism to another.

The first crystallization of a virus (tobacco mosaic virus), by Stanley in 1935, was widely hailed as bridging the gap between the living and the nonliving. In reality, however, crystallinity simply reflects structural uniformity and surface complementarity of the particles, leading to orderly aggregation.

THE HOST RESPONSE: IMMUNOLOGY

Vertebrates infected by a microbial parasite exhibit a specific "immune" response, which contributes to recovery and also protects against reinfection. Analysis of this response has given rise to the major field of immunology. But long before the development of any immunologic theory practical successes were achieved: vaccination against smallpox in 1798, and Pasteur's several vaccines in the 1880s.

Resistance to infectious diseases involves not only specific induced immune responses but also other factors, including **phagocytic cells** that engulf the parasites, **enzymes** that attack them, and differences in **cell surface receptors** for attachment and entry of viruses and even of larger parasites. The science of immunology is thus more restricted than immunity in its broadest sense, being concerned with only its specific, induced aspects. But in another sense immunology is a broader branch of pathophysiology, extending beyond infectious disease, for the same responses are elicited when foreign substances of nonmicrobial origin gain access to the tissues (e.g., pollens, insect venoms, drugs, foreign serum or other proteins, transplanted tissues), and sometimes when cancer cells arise in the host. In addition, **autoimmune** responses to an individual's own tissue components are being recognized in an increasing variety of diseases; and since the immune response involves selective gene activation it is also of broad biologic interest as a model for **cell differentiation.** Finally, immunologic methods are now widely used in biochemistry, both in the study of protein structure and in assays for trace amounts of innumerable compounds.

Immunology can thus no longer be considered a branch of microbiology. Nevertheless, the two areas are still inextricably connected: immunologic reactions are of great importance in identifying and classifying various microbes, in understanding their pathogenicity, and in identifying individuals who have been infected. Moreover, infectious disease has been a major selective force in the evolution of the immune response; and this response has doubtless contributed to the evolution of diversity among pathogenic microbes.

CONTROL OF INFECTIOUS DISEASES

Identification of the agents of various infectious diseases soon led to several remarkably effective methods of control. 1) In technologically advanced countries environmental **sanitation*** and improved personal **hygiene** have strikingly reduced the incidence of certain diseases, and sometimes even eliminated them, particularly those spread by water or food (e.g., typhoid, cholera) or by insects (e.g., typhus, yellow fever). However, a knowledge of these diseases is still essential for the physician, since they may at any time be reintroduced by travelers. 2) **Vaccination** has drastically reduced the incidence of several serious epidemic diseases (e.g., smallpox, diphtheria, whooping cough, poliomyelitis); but for many organisms vaccination is not effective or feasible. It has been especially valuable for diseases transmitted by respiratory droplets, whose distribution is difficult to control. 3) In the most striking advance in medical bacteriology since the 1880s, the development of **antibacterial chemotherapy** has dramatically reduced the seriousness of many infectious diseases and the incidence of some.

In principle it should ultimately be possible to **eradicate,** by one or more of these methods, those organisms that are **obligatory** human pathogens. However, this

---

* Sanitary measures began to develop, largely on aesthetic grounds, long before the germ theory of disease. A major advance was the popularization of indoor toilets through the invention, by Thomas Crapper in Victorian England, of the trap and air vent that block the return of sewage gas.

hope does not exist for those occasional pathogens that can also be widely carried by man without causing disease, or for those that have reservoirs in lower animals or in the soil.

THE IMPACT OF MICROBIOLOGY ON THE CONCEPT OF DISEASE

The discovery of specific agents of infection represented a tremendous theoretic advance for medicine. But the limits of the new principle, as usual, were not promptly recognized: the success of the Pasteurian approach led to unwarranted confidence that a single cause was waiting to be discovered for every disease. Such an oversimplified view still appears today, as in the irrelevant argument that tobacco smoking cannot be **the** cause of bronchogenic cancer since this disease occasionally arises in nonsmokers.

The principle of **multifactorial causation** is applicable even to many infectious diseases. Thus, though the concept of the tubercle bacillus as the etiologic agent of tuberculosis proved much more fruitful than the preceding concept of the "phthisical **diathesis**" (i.e., a constitutional tendency to develop this disease), tuberculin testing has shown that many more people are **infected** with tubercle bacilli than have the **disease.** Hence, the presence of the tubercle bacillus is a necessary but not a sufficient condition for the disease tuberculosis: the genetic constitution and the physiologic state of the host can be decisive.

THE DEVELOPMENT OF MICROBIAL AND MOLECULAR GENETICS

We have noted that for many years microbiology was largely separated from the rest of biology. The gap narrowed as the advance of biochemistry, in the 1930s, revealed the **unity of biology at a molecular level,** i.e., the close resemblance between microbial cells and the cells of higher organisms in their building blocks, enzymes, and metabolic pathways. In the 1940s, with the development of bacterial genetics, single-celled organisms became models for studying universal problems in cell physiology.

In such studies bacteria have several advantages. These include a relatively simple structure, homogeneous cell populations, and extremely rapid growth. But by far the greatest advantage lies in the possibility of easily cultivating billions of individuals and selecting, from these huge populations, rare mutants and rare genetic recombinants between these mutants. Mutants with a single biochemical defect have proved to be exceptionally sharp tools for dissecting complex intracellular processes.

As the use of bacterial mutants moved beyond the analysis of biosynthetic pathways (**biochemical genetics**) to the study of gene action and its regulation, it converged with the biochemistry of nucleic acids and proteins and with electron microscopy, giving rise to the vigorous interdisciplinary activity called **molecular genetics.** We shall review this field in some detail, not only because it is crucial for understanding many aspects of microbiology, but even more because it is being rapidly extended to human cells.

MICROBIOLOGY AND EVOLUTION

Darwin and Pasteur showed no interest in each other's work; and for nearly a century biologic thought continued to flow in two separate main streams: **evolutionary** (genetic), concerned with the origin of various organisms, and **mechanistic** (physiologic), concerned with their structure and function. The two streams were finally brought together by the development of molecular genetics. But while this convergence has explained in remarkable detail hereditary **variation,** which provides the substrate for evolution, we must recognize that the direction of evolution is largely determined by **natural selection**—a process that cannot be reduced to molecular terms but must be studied at the level of population dynamics.

Indeed, exclusive attention to the role of variation in evolution, ignoring the stabilizing role of natural selection, contributed to widespread anxiety over the conceivable distortion of evolution by novel bacterial genotypes produced by molecular recombination of DNA in vitro (see Ch. 12).

Microbial Evolution. The Darwinian process of genetic adaptation of organisms to various ecologic niches occurs very rapidly (and is easy to demonstrate) with microbes, both because they grow so rapidly and because they are subject to very strong selection pressures (for example, selection for drug-resistant mutants in the presence of a drug). Accordingly, a full understanding of infectious processes must take evolutionary principles into account. For example, natural selection explains the existence of variants of a pathogen that differ from each other only in their surface antigens: a strain with an altered surface is better fitted to spread in a host population that has developed immune responses to the original surface antigens. Conversely, within the host species diseases with significant mortality will tend to select for the progeny of survivors with genetic resistance to those diseases.

For example, the virus of myxoma, a highly lethal disease of rabbits, was introduced into Australia to reduce crop destruction by a plague of these animals; and at first this biologic warfare was dramatically effective. Within a few years, however, the rabbit population recovered its initial density through the outgrowth of strains with increased resistance to myxoma. In addition, a less virulent mutant strain of the virus outgrew the original form, evidently because it did not kill off its hosts so rapidly, and hence had more opportunity to spread.

Since hosts and parasites thus have a marked reciprocal effect on each other's evolution, it seems appropriate to dwell briefly on the impact of microbes (and of microbiology) on the human gene pool.

ROLE OF MICROBES IN HUMAN EVOLUTION

Though the 3 million years of hominid evolution have been associated with an extraordinarily rapid increase in intelligence and manual dexterity—the features most responsible for man's dominating position—the selection pressures changed markedly 10,000 years ago, when the development of agriculture and then urbanization permitted the formation of larger, more freely communicating groups. As J. B. S. Haldane pointed out, under these changed conditions epidemic diseases must have become an increasingly prominent cause of death; and the resulting selection for increased resistance to these diseases would blunt the selection for more interesting traits.*

Selection pressure from infections was strong: a century ago in the United States (and still today in some parts of the world) 25%–50% of children died of these diseases before reaching puberty; today this figure is below 2%.

Molecular Individuality in Man. Since a species occupies varied, and often fluctuating, environments, the total population (i.e., the gene pool) will be selected for diversity in the traits that help to fit it for various environments, rather than being selected for the single type best fitted for a single environment. One important source of variation in the environment is the encounter with various infectious agents. Since no one human genotype is maximally resistant to all possible agents, selective pressures from a **variety** of infectious agents will promote **genetic heterogeneity (polymorphism)** in the genes that influence resistance to these agents. The results of such selection can be clearly seen with sickle-cell hemoglobin; this variant, which protects against falciparum malaria, is prevalent in African tribes constantly exposed to this disease, even though the gene is strongly selected against in other environments because its homozygous state causes a serious anemia. Other infectious agents must exert similar selective pressures, though the specific gene products that they select for are as yet obscure.

---

* It is a well-established principle that selection for one trait will interfere with the efficiency of selection for another. For example, a school may select its student body on the basis of both ability to pay tuition and personal qualities; but if scholarships are used to eliminate the first selection the second will become more effective because of the enlarged pool of applicants.

This consideration may explain why the cells of a vertebrate species possess an extraordinary variety of specific surface antigens (Ags), many of them differing from one individual to another. We recognize these Ags because they prevent successful organ transplantation between individuals, but they can hardly have evolved for this function. A possible alternative explanation is that a specific Ag promotes resistance to a specific disease, just as different surfaces on microbial cells influence their virulence for different hosts. There is some evidence that different allelic forms of the best-studied human Ags, the ABO blood group substances (Ch. 23), are associated with differences in resistance to major killers of the past, such as smallpox. Moreover, the ethnic distribution of these Ags is very uneven and shows some correlation with the historical distribution of certain major infectious diseases.

EVOLUTION AND TELEONOMY

The convergence of the evolutionary and the mechanistic approaches to biology has also legitimized teleologic considerations in biology. **Teleology** (Gr. *tele,* goal), as originally formulated by Aristotle, proposed that structures or mechanisms are found in a living organism because they have value or purpose for that organism, and because some agency—a "final cause"—had foreseen this value. After Darwin biologists could profitably employ the same concept in a modified form, substituting the hindsight of natural selection for divine foresight. However, the term teleology continued to have supernaturalistic connotations. The term **teleonomy** has therefore been introduced: it simply implies that an organism's genetic characteristics reflect evolutionary adaptation to its environment.

Teleonomic reasoning has long been held suspect in biochemistry, for characterization of the components of a reaction was much more solid than speculation about its "purpose." However, the "purpose," or function, of an enzyme can now be unequivocally determined, by observing the metabolic effects of mutations that alter it, and by identifying the endproducts that regulate its formation and its activity. Indeed, the chapter on regulatory mechanisms (Ch. 14) would have little meaning without the concept of a purpose for these mechanisms, namely, the increased efficiency of growth for which natural selection inexorably presses.

We can now flesh out the concept implicit in the etymology of the word **organism:** a living entity with unique properties derived from the interactions of its component parts. In bacteria these interactions permit a cell $1\mu m$ in diameter to double in 20 min. Our growing insights into this process are opening up for us as new a world as did Leeuwenhoek's discovery of invisible animalcules.

PLANTS, ANIMALS, AND PROTISTS

When the microbial world began to be systematically explored, in the early nineteenth century, the newly discovered organisms were fitted into the familiar pattern of two kingdoms. Those single-celled organisms that have a flexible cell integument, as do animal cells, were considered the most primitive animals, or **Protozoa.** The **Algae,** on the other hand, were considered plants, since they are photosynthetic and have a rigid cell wall, like plant cells. On vaguer grounds the **Fungi,** or **Eumycetes** (molds and yeasts), were also lumped with the plants, as were the

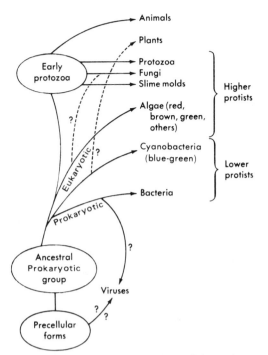

FIG. 1-5. Probable evolutionary relations of the major groups. Postulated ancestral, extinct groups are encircled.

Bacteria (also called **Schizomycetes,** or fission-fungi, because they divide by transverse fission).

Nevertheless, this Procrustean classification into plants and animals presented too many inconsistencies. Fungi and most bacteria are nonphotosynthetic; many bacteria are motile; some fungi and algae have motile spores (zoospores) which by themselves could be taken for protozoa; and the small group of slime molds were claimed by both zoologists and botanists. Moreover, with the recognition of evolution by natural selection, and the later evidence that both plants and animals have a common ancestry (see Precellular Evolution, below), it became more reasonable to regard the microbes of today as descendants, perhaps with little change, from the primitive common ancestors of the plant and the animal kingdoms. The logical solution was proposed in 1866 by Haeckel but was long ignored: the establishment of a **third** biologic kingdom, the **Protista** or **Monera** (Fig. 1-5), which are distinguishable from animals and plants by their **relatively simple,** fundamentally single-celled organization.*

Some protists form large, multicellular, superficially plantlike structures, such as seaweeds (many marine algae) and mushrooms

---

* The earlier classification into two kingdoms is reflected in the present volume by the omission of the protozoa, which are traditionally taken up in the medical curriculum along with the higher animal (metazoan) parasites.

(the basidiomycete group of fungi). However, these organisms are aggregates of similar cells, with only very primitive differentiation, whereas true plants and animals have highly differentiated multicellular adult forms, reproducing via transient unicellular gametes.

PROKARYOTES AND EUKARYOTES†

The electron microscope revealed a fundamental division among the protists, based on complexity of organization. The cells of higher protists (protozoa, fungi, and algae) are **eukaryotic** (Gr., true nucleus), like those of plants and animals: they contain a nucleus with a nuclear membrane, multiple chromosomes, and a mitotic apparatus to ensure equipartition of the products of chromosomal replication to the two daughter cells. The lower protists (bacteria) are **prokaryotic,** i.e., the "nucleus" is a single, circular chromosome, without a nuclear membrane and without histones.‡ Prokaryotic cells also differ from eukaryotes in lacking other membrane-bound organelles (mitochondria, lysosomes), in lacking endocytosis and intracellular digestive vacuoles, in lacking (with few exceptions) steroids, and in having unique components (muramate and often diaminopimelate) in the cell wall. Specialized and unique properties of certain groups of the prokaryotes include N_2 fixation, obligate **anaerobiosis** (a mode of photosynthesis that does not release O_2), derivation of energy from the oxidation of inorganic compounds, and carbon storage as poly-βhydroxybutyrate.

The prokaryotes thus appear to be a distinct evolutionary group, stabilized at a primitive stage in the evolution of the cell. A wide evolutionary gap (the real "missing link" in evolution) separates them from the eukaryotes. A long period of prokaryotic evolution yielded the blue-green bacteria, whose plantlike photosynthesis released O_2 from water, providing the atmospheric conditions required for the respiration found in most eukaryotes.

The further evolution of higher plants and animals depended on aggregation, differentiation, and specialization of cells, but not on any radical change in cell design. Hence yeasts, which offer the experimental advantages of single-celled organisms, are widely used as model cells for studying the molecular genetics of eukaryotes.

The prokaryotes have been involved at additional levels in the evolution of eukaryotes. Typical bacteria, capable of free living, are found as essential **endosymbionts** within some protozoa and in cells of cellulose-digesting insects and nitrogen-fixing plants. The **mitochondria** of eukaryotes that supply respiratory energy in animal cells, and the **chloroplasts** that supply photosynthetic energy in plant cells, appear to be derived from bacteria: they have a similar simple circular chromosome, and ribosomes that are similar in size and in antibiotic sensitivity (see Ch. 13).

---

† The spelling originally introduced was procaryote and eucaryote, but for consistency with such terms as karyotype many prefer to transliterate the Greek kappa as k.

‡ One group of prokaryotes (*Cyanobacteria,* blue-green bacteria) were formerly classified as **blue-green algae** because of their plantlike photosynthesis (see Ch. 3).

The classification of bacteria is considered in Chapter 2, and fungi are considered in Chapter 45. Algae do not cause infectious disease (although some are extremely poisonous when ingested). For a broader introductory survey of bacterial and other microbial groups we recommend *The Microbial World* by Stanier et al. (see Selected References).

PRECELLULAR EVOLUTION AND THE ORIGIN OF LIFE

As we have seen, the work of Pasteur and of Tyndall dispelled earlier claims of spontaneous generation of life from nonliving matter under experimental conditions. However, this result does not exclude the possibility that it could arise under other conditions, given eons of time. Indeed, Darwin's theory of evolution (published within 2 years of Pasteur's first paper on spontaneous generation) logically required such an initial evolution of life, preceding the evolution of the contemporary living world.

A satisfactory general explanation, now widely accepted, was proposed independently in the 1920s by Oparin in the Soviet Union and by Haldane in England. They suggested that the development of living organisms was preceded by a period of **chemical, prebiotic evolution,** occupying perhaps the first 2 billion of the earth's 4 to 5 billion years. During this period bodies of water accumulated an increasing variety of organic compounds, formed with the aid of such agencies as ultraviolet light, lightning, volcanic heat, inorganic surface catalysis, and concentration by freezing. These substances could accumulate because the primitive earth lacked the two agents that make them so unstable under present terrestrial conditions: microbial cells and molecular oxygen. The former would be absent by definition; and there is geologic evidence that free oxygen appeared in the earth's atmosphere rather late, arising as a consequence of the biologic evolution of photosynthesis and/or the inorganic splitting of water followed by loss of the light H_2 molecules from the gravitational field.

The thin "soups" resulting from this organic accumulation are believed to have developed systems that slowly catalyzed their own formation from simpler substrates. With the selection of improved catalysts, which permitted an increasingly complex system to be condensed in a smaller volume, chemical evolution would merge into precellular biologic evolution. This process would eventually yield the minimal unit of life that we can recognize: a genome-containing, membrane-bounded cell, within which a concentrated and efficient set of catalysts brings about both replication of the genetic material and synthesis of further catalysts.

This hypothesis eliminates the concept of a sharp division between the living and nonliving: in early evolution there would be no moment at which the first living being suddenly began to stir, or at least to grow. Viewed in these terms, **life** cannot be defined in terms of a cellular pattern of organization, or in terms of a given kind of molecule, but is defined in terms of self-replication from simpler substrates. And though nucleic acids provide the basis for this genetic continuity in the organisms that we know today, the same function may originally have been provided, in a primitive form, by a system of catalytic molecules whose overall capacity for self-replication was not concentrated in any special informational macromolecules.

The postulated precellular living systems have not been demonstrated. They would be difficult to recognize; and though multiple primitive systems may well have arisen independently, we would expect them to be displaced by spread of the first efficient cellular system, thus yielding the evidently **monophyletic** origin (i.e., from a single ultimate ancestor) of the present biologic kingdom. The presence of only L-amino acids in proteins, and not the equally suitable D-amino acids, suggested a single origin, now confirmed by the universality of the genetic code.

Fossils of bacterial cells, morphologically like various present species, have been dated back 3.5 billion years of the estimated 4.5 billion years of existence of the earth's crust. The rates of divergence of DNA sequences yield almost as early a date for the ultimate common ancestor.

Experimental Approaches. The Haldane–Oparin hypothesis received considerable support when Miller and Urey showed that various amino acids can be formed, in detectable amounts, by the action of an electric spark on a gas designed to simulate the atmosphere of the primitive earth (a mixture of water vapor, ammonia, methane, and hydrogen). Similar nonspecific reactions have yielded products as complex as nucleic acid bases, and have formed polynucleotides and polypeptides from their monomers. The problem of the origin of life has thus become a respectable area of experimentation, spurred on by the prospect of exploring for **extraterrestrial life** (in perhaps quite unfamiliar form). However, since an indispensable condition for life is the presence of a liquid in which molecules can diffuse, it is hardly realistic to fear contamination of the earth from exploration of an arid satellite or planet.

SELECTED READING

BARGHOORN ES: The oldest fossils. Sci Am: May 1971, p. 30

BERNAL JD: The Origin of Life. Cleveland, World Publishing, 1967

BROCK TD (ed and trans): Milestones in Microbiology. Englewood Cliffs, Prentice-Hall, 1961. Reprinted by American Society for Microbiology, 1975. Paperback. An excellent selection of classic papers, with helpful annotations; probably the best introduction to the history of the field.

BULLOCH W: The History of Bacteriology. London, Oxford University Press, 1960, Dover PB: A detailed account, with emphasis on medical bacteriology, 1979

CAIRNS J, STENT GS, WATSON JD (eds): Phage and the Origins of Molecular Biology. New York, Cold Spring Harbor Lab, 1966

CALVIN M: Chemical Evolution. London, Oxford University Press, 1969

CARLILE MJ, SKEHEL JJ (eds): Evolution in the microbial world. 24th Symposium, Society for General Microbiology. Cambridge University Press, 1974

COHEN SS: Are/were mitochondria and chloroplasts microorganisms? Am Sci 58:281, 1970

COLLARD P: The Development of Microbiology. Cambridge University Press, 1976. An excellent, short history.

DELEY J, KERSTERS K: Biochemical evolution in bacteria. In Florkin M, Stotz EH (eds): Comprehensive Biochemistry, Vol 29B. New York, Elsevier, 1975, p 1

DOBELL C: Antony van Leeuwenhoek and His "Little Animals." London, Constable, 1932. Reprinted in paperback by Dover, New York, 1960

DOUGHERTY EO: Neologisms needed for the structure of primitive organisms. J Protozool [Supp] 4:14, 1957

DUBOS RJ: Louis Pasteur: Free Lance of Science. Boston, Little, Brown, 1950

DUBOS RJ: The Professor, the Institute, and DNA. New York, Rockefeller University Press, 1976. On the life and scientific achievements of OT Avery

HOROWITZ NH, HUBBARD JS: The origin of life. Annu Rev Genet 8:393, 1974

JACOB F: The Logic of Life: A History of Heredity. New York, Pantheon, 1974

LARGE EC: Advance of the Fungi. London, Cope, 1940. Reprinted in paperback by Dover, New York, 1962. An entertaining account of the development of the infectious microbiology of higher plants

LECHEVALIER HA, SOLOTOROVSKY M: Three Centuries of Microbiology. New York, McGraw-Hill, 1965; Dover (PB), 1974

LONDON J: Evolution of proteins in prokaryotes and bacterial physiology. Trends Biochem Sci: Nov 1977, p. 256

MARGULIS L: Origin of Eukaryotic Cells. New Haven, Yale University Press, 1970

MESELSON M: Chemical and biological weapons. Sci Am: May 1970. p. 15

Microbial classification. 12th Symposium, Society for General Microbiology. London, Cambridge University Press, 1962

OPARIN AI: The origin of Life On Earth, 3rd ed. New York, Academic Press, 1957; or 2nd ed, 1938, reprinted in paperback by Dover, New York, 1953

PORTER JR: Antony van Leeuwenhoek: tercentenary of his discovery of bacteria. Bacteriol Rev 40:260, 1976

SCHOPF JW: Precambrian micro-organisms and evolutionary events prior to the origin of vascular plants. Biol Rev 45:319, 1970

SCHWARTZ RM, DAYHOFF MO: Origins of prokaryotes, eukaryotes, mitochondria, and chloroplasts. Science 199:395, 1978

STANIER RY: Some aspects of the biology of cells and their possible evolutionary significance. In Charles HP, Knight BC (eds): Organization and Control in Procaryotic and Eucaryotic Cells. Cambridge University Press, 1970

STANIER RY, ADELBERG EA, INGRAHAM J: The Microbial World, 4th ed. Englewood Cliffs, Prentice-Hall, 1976. An excellent introduction to general microbiology.

VALLERY-RADOT R: The Life of Pasteur. London, Constable, 1901 Reprinted as paperback by Dover, New York, 1960. This biography is detailed and chronological; that of Dubos (see above) is more interpretive.

VAN NIEL CB: Natural selection in the microbial world. J Gen Microbiol 13:201, 1955

WILSON GS, MILES AA: Topley and Wilson's Principles of Bacteriology and Immunology, 3rd ed, 2 vols. Baltimore, Williams & Wilkins, 1964. An excellent reference work on medical bacteriology, including its history.

part I

BACTERIAL PHYSIOLOGY

BERNARD D. DAVIS

chapter 2

BACTERIAL STRUCTURE AND CLASSIFICATION

For naught so vile that on the earth doth live,
But to the earth some special good doth give.

W SHAKESPEARE, ROMEO AND JULIET

BACTERIAL STRUCTURE

GENERAL

Bacteria (prokaryotes: Ch. 1) are the smallest organisms that contain all the machinery required for growth and self-replication at the expense of food stuffs; their diameter is usually about 1μm (10^{-3} mm). The limited resolving power of the light microscope (0.2μm) can reveal little detail in such small cells, and bacteria were long regarded as essentially bags of enzymes. However, the electron microscope revealed a distinctive architecture, represented schematically in Figure 2-1. Inside, prokaryotes are simpler than animal cells, lacking a membrane-bounded nucleus, an extensive endoplasmic reticulum, and mitochondria. However, they have a more complex surface structure, with a **rigid cell wall** surrounding the **cytoplasmic membrane** (cell membrane, plasma membrane).

The membrane provides the osmotic barrier and the active transport required to maintain an appropriate intracellular concentration of specific ions and metabolites, and the wall protects the cell against osmotic rupture in dilute media and against mechanical damage, In addition, the wall is responsible for many of the taxonomically significant features of bacteria: their shapes, their major division into gram-positive and gram-negative organisms (see Staining, below), and antigenic specificities that are important in classification and in the interactions of pathogens with their hosts. The cell wall and membrane are often referred to together as the **cell envelope**, while the more optional capsule, flagella, and pili are considered appendages.

This chapter will introduce bacterial structure in general; the wall, membrane, and flagella will be discussed in detail in Chapter 6, and intracellular structures in various other chapters.

GROSS FORMS OF BACTERIA

The light microscope reveals two principal forms of eubacteria: more or less spherical organisms known as **cocci** (Gr. and L., berry), and cylindrical ones called **bacilli** (L., stick).* Incompletely separated cocci may appear in a number of different patterns (Fig. 2-2), depending upon the planes in which they divide: when predominantly in pairs they are known as **diplococci;** in chains, as **streptococci** (Gr. *streptos,* twisted); and in clusters, as **staphylococci** (Gr. *staphyle,* bunch of grapes). Cocci that remain adherent after splitting successively in two or three perpendicular directions, yielding square tetrads or cubical packets, are known as **sarcinae** (L., bundles). Bacilli when unusually short are referred to as **coccobacilli;** when tapered at both ends, as **fusiform bacilli;** when growing in long threads, as **filamentous forms;** and when curved, as **vibrios** or **spirilla.**

METHODS OF ANALYSIS

Staining. Since classification has depended heavily on size, shape, and staining properties, the light microscope was long the hallmark of the bacteriologist. Unstained preparations may be used, especially with the phase-contrast microscope, but medical bacteriologists have generally studied heat-fixed, stained preparations.

Special stains used for certain organisms (e.g., mycobacteria, corynebacteria) will be described in the chapters on these groups. For most organisms the **Gram stain** is preferred. This method was developed empirically by

---

* The term "bacillus" is unfortunately used both as a general name for rod-shaped bacteria and as the name of a particular genus (capitalized and italicized).

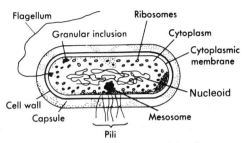

FIG. 2-1. Diagram of prototype eubacterial cell.

the Danish bacteriologist Christian Gram in 1884. The cells are first fixed to the slide by heat and stained with a basic dye (e.g., crystal violet), which is taken up in similar amounts by all bacteria. The slides are then treated with an iodine-KI mixture to fix (mordant) the stain, washed with acetone or alcohol, and finally counterstained with a paler dye of different color (e.g., safranin). Gram-positive organisms retain the initial violet stain, while gram-negative organisms are decolorized by the organic solvent and hence show the counterstain.

Nearly 75 years after the introduction of this very useful procedure gram-positive and -negative bacteria were found to differ fundamentally in their wall structure. Salton then showed that the gram-positive wall is not stained itself but presents a permeability barrier to elution of the dye–I_2 complex by alcohol. This mechanism accounts for the observation that gram-positive cells often become gram-negative in aging cultures (in which autolytic enzymes attack the wall).

Darkfield microscopy is a special procedure that recognizes objects through their reflection of light admitted from the side of the field of vision, rather than by virtue of their opacity or refractility. It is useful for identifying objects that are too thin for resolution in the light microscope but are long enough to have a characteristic shape (e.g., certain spirochetes, Ch. 39).

Electron microscopy has increased the available resolving power at least 200-fold—i.e., to about 1 nm (10^{-6} mm). Bacteria are commonly prepared for electron mi-

croscopic examination by using one or more of the following procedures. For studying external surfaces and shapes **shadowcasting** with metal vapor reveals appendages of intact bacteria (Figs. 2-5 and 2-6, below), as well as shapes of viruses (see Fig. 46-2, Ch. 46). **Negative staining** is accomplished by drying in the presence of a solution of an electron-dense material, which forms thicker deposits in crevices (see Fig. 46-3, Ch. 46). (Electron density depends on the electron-scattering power of heavy atoms, such as tungsten in phosphotungstate or uranium in uranyl acetate.) This procedure resolves features that are obscured in metal-shadowed preparations, e.g., the fine structure of flagella, or the surface of the cell wall or the ribosome.

For revealing internal organization the use of **thin sections** (about 0.02 μm) is the most important method. In this procedure the cells are pelleted, fixed (e.g., with glutaraldehyde or OsO_4), dehydrated, and embedded with a nonshrinking plastic resin. After thin-sectioning with an ultramicrotome the sections are stained with a heavy metal. Different treatments are optimal for demonstrating different structures.

Another approach eliminates artefacts due to the fixation and drying ordinarily required in preparing specimens: **freeze-fracturing,** in which the cells are frozen at a very low temperature ($-150°$ C) in a block of ice, which is then cleaved with a knife, and a replica of the surface, with shadowing, is prepared. The fracture often passes through cells, and the new surfaces thus formed may separate along natural lines of cleavage, e.g., between wall and membrane, and also between the inner and outer faces of a membrane. Both the inner and outer surfaces of a layer (concave and convex) can be studied. In **freeze-etching** some cell water is sublimed from the fresh surface after cleavage, to reveal underlying structures (e.g., proteins protruding from a membrane surface, see Fig. 6-1).

Fractionation of bacteria (by mechanical disintegration, differential centrifugation, and selective digestion or solubilization) permits study of the chemical composition and molecular organization of specific components;

FIG. 2-2. Bacterial forms: **A,** diplococci; **B,** streptococci; **C,** staphylococci; **D,** bacilli; **E,** coccobacilli; **F,** fusiform bacilli; **G,** filamentous bacillary forms; **H,** vibrios; **I,** spirilla; **J,** sarcinae.

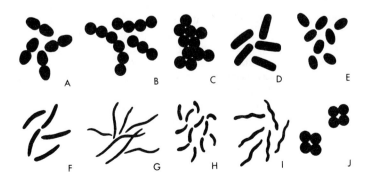

electron microscopy has been invaluable in establishing the identity and homogeneity of the fractions. Because the cells are small and have tough walls it has been necessary to develop special methods of lysis.

Thus by violent agitation (e.g., in a Waring Blendor) it is possible to separate flagella and pili from bacteria. Shaking with glass beads disrupts the envelope as well, releasing the cell contents; the envelope fraction can be recovered by differential centrifugation. Bacteria can also be disrupted by grinding in a wet paste with abrasives (e.g., alumina), by sonic or supersonic oscillation (**sonication**), or by explosive release of high pressure (e.g., release past a needle valve in a French press); the products are often more useful for enzymatic than for morphologic studies.

For gentle disruption, with minimal fragmentation of fragile structures (e.g., long DNA or RNA chains), the wall is digested by lysozyme (often assisted by freeze-thawing), and the resulting fragile spheres (see Protoplasts and Spheroplasts, below) may be lysed by transfer to a hypotonic medium or by adding reagents that dissolve the membrane (lipase, deoxycholate). Because the DNA threads are relatively unfragmented such lysates are very viscous, in contrast to those produced by grinding or sonication; hence treatment with DNase is helpful in further fractionation.

Other Methods. Radioautographic studies have occasionally been valuable, e.g., for localizing DNA by means of radioactive thymidine, or for following wall synthesis. Cytochemical procedures have had limited use because of the small dimensions involved, but they have been of great value in animal virology, as will be seen in later chapters: they include the use of specific antibodies tagged with a fluorescent dye (see Fig. 16-2) or coupled to a substance, such as ferritin, that forms a sharp image in electron micrographs (see Fig. 27-1).

We shall consider here the characteristic structures of only the typical or **"true" bacteria (eubacteria)**; later chapters will deal with the less typical bacteria (including actinomycetes, spirochetes, mycoplasmas, rickettsiae, and chlamydiae), as well as with fungi and viruses.

FINE STRUCTURE OF BACTERIA

CELL WALL

The presence of a rigid cell wall, outside the cytoplasmic membrane, may be demonstrated by **plasmolysis,** i.e., exposure to a hypertonic solution, which causes the protoplast to contract and shrink away from the wall (Fig. 2-3). Moreover, in bacterial preparations subjected to mechanical damage (e.g., crushing) many of the cells lose refractility, from loss of their intracellular contents, but leave a pale, empty ghost which retains the characteristic shape of the original cell (see Fig. 6-3; Ch. 6).

Electron microscopy of thin sections reveals the walls of gram-positive organisms as a relatively thick (15–80 nm), uniform, dense layer, with the cytoplasmic membrane closely apposed to its inner surface (Fig. 2-4). This dense mterial was eventually found to be a **peptidoglycan:** a network of polysaccharide chains cross-linked by an oligopeptide. Gram-negative cells have a much thin-

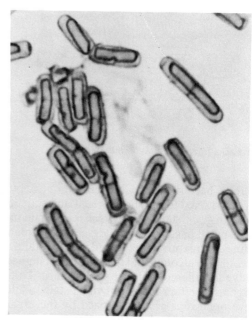

FIG. 2-3. Plasmolysis of *Bacillus megaterium*. The cells on a slide were successively treated with ether vapor (which loosens the attachment of membrane to wall), air-dried and postfixed with Bouin's fluid (which cause contraction of the protoplast), and stained with Victoria blue (which enhances the visibility of the membrane enclosing the protoplast). (Courtesy of C. Robinow)

ner peptidoglycan layer, covered by a closely apposed outer membrane. The two layers are not usually distinct in electron micrographs (see Fig. 6-2). The details of wall and membrane structure, function, and biosynthesis will be presented in Chapter 6.

Protoplasts and Spheroplasts. When the peptidoglycan layer of the cell wall is digested by lysozyme (Ch. 6), or when its synthesis is specifically blocked, the cell ordinarily lyses. However, in a hypertonic medium (e.g., 20% sucrose or 0.5 M KCl) the cell survives as an osmotically sensitive sphere. With gram-positive organisms this product is free of wall constituents and is called a **protoplast.** With gram-negative organisms these osmotically sensitive spheres retain much of the outer membrane; they are called **spheroplasts.**

CAPSULES

Capsules are optional bacterial structures, of particular importance in pathogens since they protect the bacteria from phagocytosis. They are loose, gel-like structures, most easily demonstrated by negative staining (e.g., suspension in India ink), where they form a clear zone between the opaque medium and the more refractile (or stained) cell body. Exposure to specific antibodies increases the refractility of capsules and prevents their

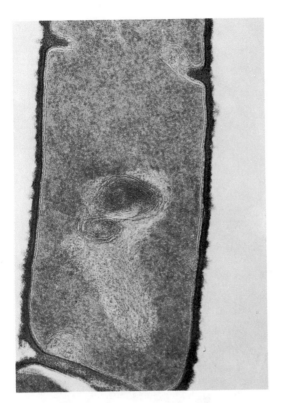

FIG. 2-4. Section of portion of gram-positive *Bacillus subtilis*, including a completed but not yet cleaved **septum** at one end of the cell and a beginning septum forming an equatorial ring in the midzone of the cell. Note the well-defined **plasma membrane** ("double track" lining the inner surface of the dense wall), continuous with the small **vesicular mesosome** at one portion of the growing septum. Also note the two large, concentric **lamellar mesosomes** in the nuclear region; their connection to the plasma membrane is not seen in this section. (Courtesy of A. Ryter)

contraction on drying, thus permitting their visualization in stained, dried specimens ("quellung" reaction) and in electron micrographs (see Fig. 27-1).

The capsular layer usually does not reveal any structural detail in electron micrographs, and because the gel is so loose its mode of attachment is difficult to determine. If it is not stabilized by antibody the fixation procedure may convert it artificially to contracted ropes and knobs. Soluble capsular substance can often be detected immunologically in culture filtrates, being released by partial hydrolysis (but possibly also in some cases by diffusion of a loosely attached "slime" layer). The thickness of the capsule varies with growth conditions. In addition, some mutants make an invisible but immunologically detectable **microcapsule,** with unchanged specificity.

A given bacterial species may include strains of different antigenic types, with immunologically distinct capsules. Most capsules consist of relatively simple polysaccharides, containing repeating sequences of two or three sugars and often of uronic acids. The structures of some representative capsules of the pneumococcus are

described in Chapter 27, and others are noted briefly in chapters on other organisms.

FLAGELLA

Flagella (*L. flagellum,* whip), when present, are responsible for the motility of eubacteria. Motility can be recognized under the microscope in liquid medium (in a hanging drop or under a cover slip), where it must be carefully distinguished from brownian movement; it can also be revealed by the spread of visible growth in semisolid medium (e.g., 0.3% agar).

Flagella are long ($3\mu m$–$12\mu m$) filamentous appendages, readily visualized by darkfield microscopy; they are too thin (12–$25nm$) to be seen by ordinary microscopy unless heavily coated with special stains containing a precipitating agent, such as tannic acid (Fig. 2-5A). Their waviness, long thought to be a product of a whipping motion (hence the name flagellum), actually reflects the fixed helical shape of the filament, which propels the cell by rotation (see Flagellar Rotation, Ch. 6). The wave length of the helix, easily seen in two dimensions in fixed preparations, varies from one bacterial strain to another.

Some species exhibit **peritrichous** (pron. trīkus) **flagellation** (Gr. *trichos,* hair), with flagella distributed at random over the cell surface (Fig. 2-5B); in others one or a few flagella are found only at one or both poles **(polar flagellation).** The pattern is genetically stable: strains that mutate to a loss of flagellation revert to the original pattern. On this basis the common bacteria are separated into two orders, Eubacteriales (with peritrichous flagellation) and Pseudomonadales (with polar flagellation).

The vigor of bacterial movement depends on the number of flagella, which varies widely. Certain species (e.g., *Proteus*) may produce a huge number (Fig. 2-5C), especially when cell division is slowed; such cells spread in a thin film ("swarm") on the surface of the usual moist agar media.

Other Types of Locomotion. In spirochetes (Ch. 39) the cell proper forms a helix around axial filaments; motility is associated with marked bending of the cell, possibly due to contraction of actin tubules. Certain myxobacteria (slime bacteria), mycoplasmas, and cyanobacteria exhibit a slow gliding movement on solid surfaces, also involving bending of the cell.

PILI (FIMBRIAE)

With many gram-negative bacilli the electron microscope revealed a group of still finer filamentous appendages, called **pili** (L., hairs) or **fimbriae** (L., fringe). They are shorter, thinner, and straighter than flagella (Fig. 2-6); and on the same cell they vary in thickness (about 7.5–10 nm) and length (up to several micrometers). They consist of a protein (pilin); they are not motile.

Somatic pili, up to hundreds per cell, function in bacte-

FIG. 2-5. Flagella. **A.** Flagellated bacillus stained with tannic acid–basic fuchsin, a flagellar stain. **B.** Electron micrograph of palladium-shadowed bacillus, showing peritrichous flagellation. (>13,000) **C.** Highly motile form of *Proteus mirabilis* ("swarmer"), with innumerable peritrichous flagella. (**A,** Leifson E et al: J Bacteriol 69:73, 1955; **B,** Labaw LW, Mosley VM: Biochim Biophys Acta 17:322, 1955; **C,** Hoeniger JFM: J Gen Microbiol 40:29, 1965)

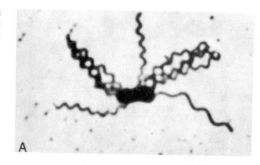

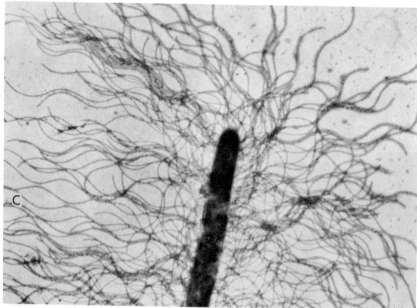

rial adherence to specific surfaces (Ch. 24), as shown by the effect of their loss by mutation. In addition, as we shall see in Chapter 9, conjugation depends on one or two special **conjugal pili** on the male cell. When pili are mechanically removed cells rapidly form them again.

CYTOPLASMIC CONTENTS

Mesosomes. Thin sections of bacteria often reveal one or more large, irregular invaginations of the plasma membrane called **mesosomes**; additional membrane within these pockets may be **lamellar** or **vesicular** (Fig. 2-4). These structures evidently compartmentalize the cell, somewhat like the endoplasmic reticulum in eukaryotes, but they are not as well understood. Their function will be discussed in Chapter 6.

Ribosomes. The cytoplasm of bacteria is thickly populated with ribosomes: roughly spherical, densely stained objects about 18 nm in diameter (Fig. 2-4). Their struc-

ture and function will be considered in Chapter 13. The ribosomes are mostly grouped in chains called **polysomes,** but these cannot be recognized in a cell section because of the close packing. In gently lysed protoplasts polysomes are found not only free but also attached to the membrane (see Protein Secretion, Ch. 6).

Granular Inclusions. Many kinds of bacteria contain large granules of reserve materials (e.g., poly-β-hydroxybutyrate, glycogen), in amounts that vary with nutritional conditions; their high molecular weights permit nutrients to be stored without increasing the osmolarity of the cytoplasm. Iodine reveals the characteristic red of **glycogen** or the blue of a starchlike **granulose.** Polymetaphosphate granules $[(PO_3-)_n]$ are stained **metachromatically** (i.e., with a change in color) by certain dyes (methylene blue or toluidine blue).

NUCLEAR BODY (NUCLEOID)

In bacteria, unlike eukaryotic cells, the cytoplasm has a high concentration of RNA and hence is as basophilic as the nuclear region; accordingly, the latter cannot be re-

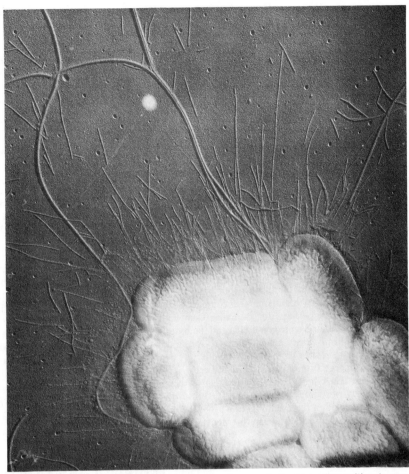

FIG. 2-6. Pili. Piliated strain of *Escherichia coli*, grown in liquid medium without aeration. Each cell possesses hundreds of pili (diameter 7 nm), and their presence promotes aggregation. Many isolated, broken pili are also seen. A few flagella, much longer and of larger diameter (14 nm), extend from the cells to the edge of the photograph. (Platinum shadowed, >45,000, reduced) (Courtesy of Charles C. Brinton, Jr)

vealed by basic dyes. However, discrete nuclear bodies in bacteria can be recognized under the light microscope through special procedures, such as the DNA-specific Feulgen (fuchsin sulfite stain) method, or selective hydrolysis of the RNA (in fixed cells) by HCl or RNase, followed by a basic stain (Fig. 2-7). Most bacilli contain two or more such bodies per cell, since cell division lags behind nuclear division. More detailed study of these bodies with the electron microscope established the concept of bacteria as prokaryotic cells, lacking the discrete chromosomes, mitotic apparatus, nucleolus, and nuclear membrane of eukaryotic cells. Hence the **nuclear region** in bacteria is referred to as the **nuclear body** or **nucleoid.**

With most methods of preparing sections the nuclear region in bacteria appears as a compact, centrally located mass, less dense than the cytoplasm, with parallel strands in some areas, and excluding ribosomes (Fig. 2-8). This compact body can also be seen without fixation, by fluorescence microscopy, or by phase contrast in cells growing in a medium of high refractive index. Moreover, if cells are gently lysed in the presence of a high salt concentration (especially divalent cations), to neutralize electrostatic repulsion of the nucleic acid chains, the nucleoid can be recovered intact: when freed of membrane by detergents it contains about 60% DNA, 30% RNA, and 10% protein. It is less dense than the cytoplasm, for though it contains about 10% of the cell volume, DNA is only 2%–3% of the cell's dry weight.

As we shall see later, the bacterial **chromosome** is a closed circle (see Fig. 10-26); because of its length (about 1100 μm) it must be greatly folded. Its organization in the nucleoid is not well under-

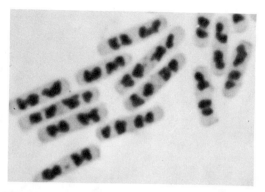

FIG. 2-7. Bacterial nuclear bodies. Demonstration of "chromatin bodies" in *Bacillus cereus* by a basic stain (Giemsa) after hydrolysis of the RNA by treatment with HCl. (>3600) (Courtesy of C. Robinow)

stood, but the effects of enzymatic treatment, and of inhibition of RNA or protein synthesis, suggest that both nascent RNA and protein contribute to maintaining a compact structure. In this structure there are perhaps 100 domains, each with supercoiled DNA (i.e., the chain is coiled on itself); and the bases of these loops must be held firmly, since a nick in the DNA of a domain uncoils only that region of the DNA. These domains may be related to the loops that emerge from a partly disrupted cell (Fig. 2-9).

Since the RNA being transcribed on DNA is simultaneously translated by ribosomes (see Fig. 14-14), yet no ribosomes penetrate the nucleoid, it has been suggested that the regions of DNA being actively transcribed at any moment are looped, invisibly,

outside the nucleoid. This model would suggest a rapid churning of the DNA regions in the nucleoid.

CYTOPLASMIC SOLUTES

Bacterial cell volumes can be determined from measurements of the volume of packed wet cells, corrected for intercellular space (about 20%) by the use of a macromolecule that does not penetrate or adsorb to the bacteria (e.g., hemoglobin, or isotopically labeled dextrin or albumin). Dry weight determinations then show that an average bacterium contains about 75% water. The wall is dense and contributes a substantial fraction of the cell weight. The cytoplasm averages about 80% water. It contains a great variety of small molecules (the metabolite pool) and inorganic ions, as well as ribosomes and enzymes.

The **metabolite pool** can be recovered for chemical estimation by transferring cells to boiling water or cold 5% trichloroacetic acid, which disrupts the cytoplasmic membrane and precipitates the macromolecules. When radioactively labeled substances are used for assay the cells must first be washed free of extracellular fluid. Washing on a nitrocellulose membrane filter is much more rapid than centrifugation and thus minimizes metabolic alteration of the pool during its isolation.

The high osmotic pressure of the bacterial cell shows that the metabolites in the pool are largely in a free state. This **osmotic pressure** can be estimated by determining at what point increasing external osmolarity causes an abrupt decrease in cell volume (i.e., plasmolysis). The values observed are about 20 atmospheres (atm) in some gram-positive organisms and about 5 atm in thinner-walled gram-negative organisms. (It will be recalled that a 1-M solution of a nonelectrolyte ideally has an osmotic pressure of 22.4 atm.)

BACTERIAL CLASSIFICATION

PHYLOGENY VERSUS DETERMINATIVE KEY

The classification of organisms (**taxonomy**) has two purposes. One is purely descriptive: to group organisms of the same kind, and to describe the basis for that grouping, so that the information collected about a given kind can be pooled and compared. The ordering of the entities in such a scheme could then be simply a matter of convenience, providing a determinative key for applying a succession of criteria until the correct description for an unknown specimen is reached. The second purpose is to provide a "natural," phylogenetic classification, in which the successive **taxons** (levels of ordering) in a **hierarchical family tree** (species, genus, family, tribe, order) aim at reflecting lines of evolutionary descent. In higher organisms the fossil record, supported by morphologic, physiologic, and developmental homologies, yields such an ordering.

Evolution by natural selection implies a historical continuity between all organisms. Nevertheless, animals and plants can be sharply divided, with few exceptions, into

several million **species:** populations of organisms that are capable of interbreeding only among members of the same group.

Such limitations to the range of gene flow (species boundaries) had to evolve with sexual reproduction, since crosses between individuals with excessive genetic differences would result in progeny with unworkable combinations of genes. Accordingly, in the formation of new species two populations, reproductively separated (usually by a geographic barrier), accumulate increasing differences in their total gene pool, through the random appearance of new mutations, chance differences in the multiplication of alleles, and differences in selection pressures in different environments; and these shifts first yield races (subspecies), which are still interfertile, and eventually lead to separate species. The transitional forms generally are soon lost, as stable new species emerge.

With bacteria, in contrast, multiplication is asexual, and so there are no "natural" species boundaries. Accordingly, in bacteria the definition of a species is arbitrary, and taxonomists can themselves be classified into "lumpers" and "splitters."

macromolecules (usually detected immunologically); and habitat (including the ability to parasitize higher organisms and to cause disease). Taxonomists add other fundamental properties.

These include flagellar pattern, energy-yielding pathways, and chemical composition (especially of the peptidoglycan, and total lipids analyzed by gas–liquid chromatography); nucleic acid sequences will be discussed below. Biosynthetic pathways vary little; but differences in their regulatory mechanisms (see Table 14-1) furnish another variable.

By consulting published descriptions of such features, by comparing **"type cultures"** (i.e., standard strains) that are maintained in several countries, and by exchanging strains, bacteriologists can communicate taxonomically with each other. Following the Linnean* tradition of zoology and botany, each "species" of bacterium is assigned an official Latin **binomial** (in italics), with a capitalized genus followed by an uncapitalized species designation. Unitalicized vernacular names may be derived from, or may be identical with, the official name (e.g., pneumococcus = *Diplococcus pneumoniae*, now *Streptococcus pneumoniae;* salmonella = *Salmonella* spp.)

Some of the major "families" of bacteria are described in Table 2-1. The major guide to classification is a large cooperative effort: *Bergey's Manual of Determinative Bacteriology.* The species **nomenclature** is based on relatively stable descriptions; the less important ordering into higher taxa fluctuates markedly in the periodic revisions.

Because international codes of nomenclature were developed late, have not always been adhered to, and have undergone frequent change, the same organism can be encountered in the literature under several names: e.g., *Bacillus typhosus, Bacterium typhosum, Eberthella typhosa,* and *Salmonella typhi.* In virology, which developed later and emphasized molecular properties, the connection with systematic botany and zoology has been weak, and a proposed binomial system has been controversial.

The Meaning of Bacterial Species. Because the asexual reproduction of bacteria does not require species barriers one might conceivably find a truly continuous spectrum in this kingdom, reflecting the continuity of past evolution. In fact, however, evolution does not select a mutation in isolation but selects for a **balanced genome,** well adapted to a particular range of environments. A mutant gene that improves one genome may decrease the fitness of another. Hence genes do not vary independently among the successful organisms that have survived in nature. Bacterial speciation is based on recognition of these partial discontinuities; and the "art" of the systematic bacteriologist lies in distinguishing secondary details

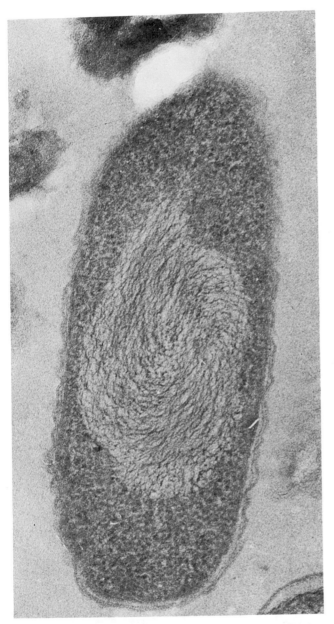

FIG. 2-8. Section of *Agrobacterium tumefaciens*. Special fixation to bring out the fibrillar structure of the nucleoid. (Courtesy of A. Ryter)

For **identifying an unknown organism** and relating it to previously described organisms, diagnostic bacteriologists must rely on properties that are relatively easy to recognize: visible features (including shape, size, color, staining, motility, capsule, and colonial morphology); formation of characteristic chemical products; nutrition (including both growth requirements and ability to utilize various foods); presence of characteristic surface

* Carolus Linnaeus (Carl von Linné) (1707–1778), Swedish botanist and professor of medicine.

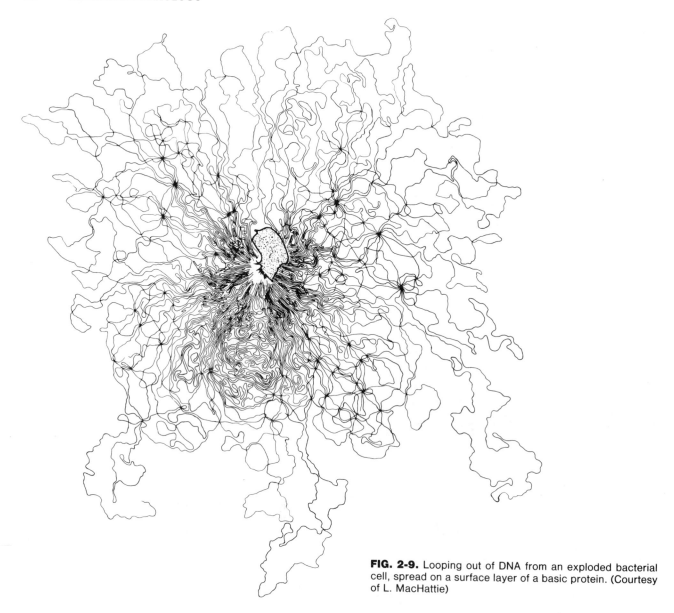

FIG. 2-9. Looping out of DNA from an exploded bacterial cell, spread on a surface layer of a basic protein. (Courtesy of L. MacHattie)

from more stable and fundamental properties. Complicating the problem, organisms isolated from nature often change rapidly (on the basis of rapid growth and strong selection pressures) when propagated in the new environment of a laboratory medium.

Accordingly, in bacteria, perhaps even more than in higher organisms, a species usually includes a spectrum of organisms with a relatively wide range of properties. In the Enterobacteriaceae in particular, which appear to undergo much genetic recombination in nature, a large fraction of the strains encountered are intermediate in properties rather than identical with one or another type culture; each species thus represents a **cluster of biotypes**, more or less resembling a strain that is maintained as the type culture of the species.

The immunologic characterization of surface macromolecules plays a large role in medical bacteriology and virology (though not in general microbiology), since these molecules strongly influence host–parasite interactions.* The deeper layers of the wall, and the cell mem-

---

* However, since polysaccharides, based on repeating units, exist in a finite variety (in contrast to the virtually unlimited variety of proteins), the same antigenic determinant occasionally appears in widely separated organisms. For example, the surface polysaccharide of *Escherichia coli* type O-86 crossreacts with blood group substance B of human red cells.

TABLE 2-1. Main Groups of Bacteria*

I. GRAM-POSITIVE EUBACTERIA

Cell shape	Motility	Other distinguishing characteristics		Genera	Families
Cocci	Nearly all permanently immotile	Cells in cubical packets Cells irregularly arranged		*Sarcina* *Micrococcus* *Staphylococcus*	Micrococcaceae
		Cells in chains Lactic fermentation of sugars		*Streptococcus* *Leuconostoc*	Streptococcaceae
Straight rods	Nearly all permenently immotile	Lactic fermentation of sugars Propionic fermentation of sugars Oxidative, weakly fermentative		*Lactobacillus* *Propionibacterium* *Corynebacterium* *Listeria* *Erysipelothrix*	Lactobacillaceae Propionibacteriaceae
	Motile with peritrichous flagella, and related immotile forms	Endospores produced	Aerobic	*Bacillus*	Bacillaceae
			Anaerobic	*Clostridium*	

II. GRAM-NEGATIVE BACTERIA, EXCLUDING PHOTOSYNTHETIC FORMS

Cell shape	Motility	Other distinguishing characteristics		Genera	Families
Cocci	Permanently immotile	Aerobic Anaerobic		*Neisseria* *Veillonella*	Neisseriaceae
Straight rods	Motile with peritrichous flagella, and related immotile forms	Facultative anaerobic	Mixed acid fermentation of sugars	*Brucella* *Bordetella* *Pasteurella* *Hemophilus* *Escherichia* *Erwinia* *Shigella* *Salmonella* *Proteus* *Yersinia*	Brucellaceae Enterobacteriaceae
		Aerobic	Butylene glycol fermentation	*Enterobacter* *Serratia*	
			Free-living nitrogen fixers	*Azotobacter*	Azotobacteraceae
			Symbiotic nitrogen fixers	*Rhizobium*	Rhizobiaceae
	Motile with polar flagella	Aerobic	Oxidize inorganic compounds	*Nitrosomonas* *Nitrobacter* *Thiobacillus*	Nitrobacteraceae
			Oxidize organic compounds	*Pseudomonas* *Acetobacter*	Pseudomonadaceae
		Facultative anaerobic		*Photobacterium* *Zymomonas* *Aeromonas*	
Curved rods	Motile with polar flagella	Comma-shaped Spiral	Aerobic Anaerobic	*Vibrio* *Desulfovibrio* *Spirillum*	Spirillaceae

(continued)

TABLE 2-1. Main Groups of Bacteria (cont'd)

III. OTHER MAJOR GROUPS

Characteristics	Genera	Orders (-ales) or families (-aceae)
Acid-fast rods	*Mycobacterium*	Actinomycetales
Ray-forming rods (actinomycetes)	*Actinomyces*	
	Nocardia	
	Streptomyces	
Spiral organisms, motile	*Treponema*	Spirochetales
	Borrelia	
	Leptospira	
	Spirocheta	
Small, pleomorphic; lack rigid wall	*Mycoplasma*	Mycoplasmataceae
Small intracellular parasites	*Rickettsia*	Rickettsiaceae
	Coxiella	
Small intracellular parasites, readily filtrable	*Chlamydia*	Chlamydiaceae
Intracellular parasites; borderline with protozoa	*Bartonella*	Bartonellaceae

*Some of these traditional names have been officially replaced.

(Modified from Stanier RY et al: The Microbial World. Englewood Cliffs, N.J., Prentice-Hall, 1963)

brane, vary much less among related organisms, for they are subject to less selection pressure from agents that attack the surface (Abs, enzymes, bacteriophages). Hence immunologic or chemical characterization of these deeper components is a frequent basis for grouping species of bacteria or viruses into a genus.

Numerical Taxonomy. To avoid the arbitrary weighting of characters some investigators, with the aid of computers, have revived an approach to taxonomy introduced in the eighteenth century by the French biologist Adanson. In this system a large number of characters are determined for each strain, and strains are grouped on the basis of the proportion of characters shared, without giving any characters more weight than others.

TAXONOMY BASED ON MACROMOLECULAR SEQUENCE HOMOLOGY

The approaches described above have ordered bacteria, rather tentatively, in terms of phenotypic traits. The development of molecular genetics, however, has now revolutionized taxonomy by providing a precise, quantitative basis for defining biologic relatedness.* For as organisms

* The earlier attitude of many bacterial physiologists toward problems of classification is reflected in Duclaux' story of the microscopist who pointed out to Pasteur that an organism which he had taken for a coccus was in reality a small bacillus. The reply was, "If you only knew how little difference that makes to me!" However, as the study of visible structure has merged with that of molecular structure, the gap between the taxonomist and the physiologist has narrowed a good deal.

drift apart in evolution, through the accumulation of mutations, their genes differ not only in their products but also in their base sequences.

Composition. The simplest comparison is the base composition of the DNA, estimated easily from melting temperature or buoyant density (see Physical Properties, Ch. 10). The composition, though essentially the same in all vertebrates (about 40 moles % guanine ˏcytosine [G + C] and 60% adenine + thymine [A + T]), varies remarkably among bacteria, from about 30 to 70 moles % G + C (Table 2-2).† Moreover, the chromosome of a bacterium is strikingly homogeneous in this respect: in a given strain the fragments produced by harsh methods of lysis have a narrow range of densities. The base compositions presented in Table 2-2 confirm most previously accepted ideas concerning taxonomic relations, but have also revealed some interlopers.

For example, a similar composition (38%–40% G + C) is observed for various streptococci, pneumococci, and lactobacilli, which have traditionally been grouped together as lactic acid bacteria because of their characteristic fermentation. On the other hand, *Lactobacillus bifidus* is far removed (56% G + C), and is now renamed *Bifidobacterium*. Considerable spread is also seen within other large groups that have been considered a single genus primarily on morphologic grounds, e.g., *Proteus*, *Bacillus*, *Coryne-*

† Chapter 13 will show how such variation can be reconciled with a universal genetic code and a relatively uniform protein composition.

TABLE 2-2. DNA Base Compositions of Representative Bacteria

G + C (moles %)	Organism
28–30	*Spirillum linum*
30–32	*Clostridium perfringens, C. tetani*
32–34	*C. bifermentans,* Leptospira pomona, Staphylococcus aureus
34–36	*Bacillus anthracis,* other bacilli, *Clostridium kluyveri,* Leptospira pomona, *Mycoplasma gallisepticum, Pasteurella aviseptica, Staphylococcus albus, Streptococcus faecalis, Treponema pallidum*
38–40	*Bacillus megaterium, Hemophilus influenzae, Streptococcus pneumoniae, Lactobacillus acidophilus, Leuconostoc mesenteroides, Streptococcus pyogenes,* other streptococci, *Proteus vulgaris, Sporosarcina ureae*
40–42	*Bacillus laterosporus, Leptospira biflexa, Neisseria catarrhalis*
42–44	*Bacillus subtilis, B. stearothermophilus, Bacteroides insolitus, Coxiella burneti*
46–48	*Bacillus licheniformis, Clostridium nigrificans, Corynebacterium acnes, Yersinia pestis, Vibrio cholerae*
48–50	*Neisseria gonorrhoeae,* other neisseriae
50–52	*Bacillus macerans, Escherichia coli,* other escherichiae, *Erwinia* spp., *Neisseria meningitidis, proteus morgani, Salmonella* spp., *Shigella* spp.
52–54	*Enterobacter* spp., *Corynebacterium diptheriae, Erwinia* spp.
54–56	*Enterobacter* spp., *Alcaligenes faecalis, Azotobacter agile, Brucella abortus*
56–58	*Corynebacterium* spp., *Lactobacillus bifidus*
58–60	*Agrobacterium tumefaciens, Corynebacterium* spp., *Serratia marcescens*
60–62	*Azotobacter vinelandii, Pseudomonas fluorescens, Rhodospirillum rubrum, Vibrio* spp.
64–66	*Desulfovibrio desulfuricans, Pseudomonas* spp.
66–68	*Pseudomonas aeruginosa, Mycobacterium tuberculosis*
68–70	*Pseudomonas saccharophila, Sarcina flava*
70–80	Micrococcus lysoideikticus, *Mycobacterium smegmatis, Nocardia* spp., *Sarcina lutea, Streptomyces* spp.

G + C = Guanine + cytosine.

(Modified from Marmur J et al: Annu Rev Microbiol 17:329, 1963)

bacterium, Pseudomonas. Finally, the weakness of gross morphology as a major taxonomic criterion (cf. whales and fishes) is well illustrated by the anomalous position of *Sporosarcina ureae,* the only coccus that sporulates: this organism has a base composition (38%–40% G + C) close to that of the sporulating rods (bacilli and clostridia), and very far from that of other sarcinae (70%–80% G + C).

Sequence. Similarity of base composition respresents only a minimal basis for close genetic relatedness, since even distant organisms can by chance have a similar composition. Moreover, as groups diverge in evolution

the DNA sequence changes long before the base composition. Homology of sequence (**DNA–DNA homology**) can be measured quantitatively in terms of the ability of DNA strands from two different sources to form molecular hybrids in vitro (see Homology; Hybridization; Ch. 10). Surveys among higher organisms have revealed close parallelism between the results of such hybridization and the phylogenetic relations inferred on other grounds.

Among bacteria **DNA–DNA hybridization** is useful only within closely related groups, because it quickly vanishes in the broad range of variation between more distant organisms. **Ribosomal RNA hybridization to DNA** is useful for estimating more distant kinship among bacteria, since these sequences are conserved in evolution much more than most DNA sequences, and since this RNA is easily isolated. Nucleic acid hybridizations have not yet provided a phylogenetic bacterial taxonomy to replace the classical determinative key.

Since base sequences are translated into amino acid sequences, evolutionary relations can also be illuminated by analysis of the amino acid sequences of **homologous proteins** (e.g., cytochrome *c*, ferredoxin, various enzymes). This approach has been used widely with higher organisms (see Fig. 13-21, Ch. 13). Much simpler is the measurement of **immunologic crossreactions,** which offer a rough index of sequence homologies above 60%; they have been widely used to study relations within large groups (e.g., Enterobacteriaceae; lactic acid bacteria). Proteins vary widely in their rates of evolutionary change; some (e.g., ferredoxins) are partly conserved in sequence even across the gap between prokaryotes and eukaryotes.

THE RANGE OF BACTERIA

The weight of bacteria found in soil, decomposing organic matter, and bodies of water is sufficient to be a substantial fraction of the total terrestrial biomass. It is estimated that each person carries some 10^{14} bacteria (more than the number of his own cells!), and that the total human population excretes from its collective gut 10^{22}–10^{23} bacteria per day.

Bacteria have evolved to fill an enormous variety of ecologic niches. Their physical environments range from hot springs at 80° C to refrigerated foods, and from distilled water with trace contaminants to the Dead Sea; and their energy-yielding mechanisms range from the oxidation of Fe^{2+} to Fe^{3+}, or of H_2S to H_2SO_4, to a special form of photosynthesis that cannot yield O_2 (see Ch. 3). Though the era of explosive discovery of new bacteria is over, the 3000 species described in *Bergey's Manual* probably represent quite an incomplete set: these are almost all organisms capable of growth in pure culture on known media, and the range of the yet undiscovered population with symbiotic or otherwise unfamiliar requirements is unknown. The natural historian's approach of Leeuwenhoek is still revealing new "animalcules."

For example, a bacterial parasite on bacteria, *Bdellovibrio* (Lat. *bdellus,* leech), was discovered only in 1962, although the organism is widespread and easily isolated. This very small bacterium bores a hole in the wall of a specific host bacterium (enzymatically and mechanically), occupies the space between protoplast and wall, grows to several times its initial length at the expense of the protoplast, and then divides into cells of the original length. Another remarkable organism, a **magnetotactic** spirillum, was recently isolated from swamp waters by exposure to a magnetic field. It contains crystals of a magnetic iron oxide, and it swims toward the north pole.

Archaebacteria. The methanogenic bacteria (e.g. *Methanobacterium*) derive energy from a process with a very low yield, but one that could have been supported by the presumed primitive atmosphere of the earth.

$$CO_2 + 4H_2 \rightarrow CH_4 + 2H_2O.$$

Moreover, this group (and certain halophiles) lack muramic acid (in their walls) and glyceryl esters (in their lipids); and in the sequences of their ribosomal rRNA they differ from other bacteria about as much as the latter differ from eukaryotes. These organisms therefore appear to be relics of a primitive group that transformed the initial environment by accumulating organic substances before the evolution of the complex apparatus of photosynthesis. It has been proposed that they be called archaebacteria.

The earliest fossils appear to be bacteria, observed in rocks over 3×10^9 years old. However, the fossil record is not detailed enough to help trace bacterial evolution.

SELECTED READING

STRUCTURE

BOOKS AND REVIEW ARTICLES

BAYER ME, THUROW H: Polysaccharide capsule of *E. coli:* microscopic study of its size, structure, and sites of synthesis. J Bacteriol 130:911, 1977

COSTERTON JW: The role of electron microscopy in the elucidation of bacterial structure and function. Annu Rev Microbiol 33: 459, 1979

FULLER R, LOVELOCK DW (eds.): Microbial Ultrastructure. New York, Academic Press, 1976

GUNSALUS IC, STANIER RY (eds): The Bacteria, Vol I, Structure, New York, Academic Press, 1960

LEIBOWITZ PJ, SCHAECHTER M: The attachment of the bacterial chromosome to the cell membrane. Int Rev Cytol 41:1, 1975

PETTIJOHN DE: Prokaryotic DNA in nucleoid structure. CRC Crit Rev Biochem 4:175, 1976

REMSEN CC, WATSON SW: Freeze-etching of bacteria. Int Rev Cytol 33:253, 1972

RYTER A: Association of the nucleus and the membrane of bacteria: a morphological study. Bacteriol Rev 32:39, 1968

SHIVELY JM: Inclusion bodies of procaryotes. Annu Rev Microbiol 28:167, 1974

SMITH RW, KOFFLER H: Bacterial flagella. Adv Micro Physiol 6:219, 1971

SPECIFIC ARTICLES

BRINTON CC JR: Contributions of pili to the specificity of the bacterial surface. In Davis BD, Warren L (eds): The Specificity of Cell Surfaces. Englewood Cliffs, Prentice-Hall, 1966

DUGUID JP, WILKINSON JF: Environmentally induced changes in bacterial morphology. In Microbial Reaction to Environment, 11th Symposium, Society for General Microbiology. London, Cambridge University Press, 1961

GRIFFITH JD: Visualization of prokaryotic DNA in a regularly condensed chromatin-like fiber. Proc Natl Acad Sci USA 73:563, 1976

WORCEL A, BURGI E: On the structure of the folded chromosome of E. coli. J Mol Biol 81:127, 1972

CLASSIFICATION

AINSWORTH GG, SNEATH PHA (eds): Microbial Classification, 12th Symposium, Society for General Microbiology. London, Cambridge University Press, 1962

BRENNER RJ, FALKOW S: Molecular relationships among members of the Enterobacteriaceae. Adv Genet 16:81, 1971

BUCHANAN RE, GIBBONS NE (eds): Bergey's Manual of Determinative Bacteriology, 8th ed. Baltimore, Williams & Wilkins, 1974

COWAN ST: Sense and nonsense in bacterial taxonomy. J Gen Microbiol 67:1, 1971

COWAN ST: Manual for the Identification of Medical Bacteria. London, Cambridge University Press, 1974

LONDON J: Evolution of proteins in prokaryotes and bacterial phylogeny. Trends Biochem Sci, November, 1977, p 256

MANDEL M: New approaches to bacterial taxonomy: perspective and prospects. Annu Rev Microbiol 23:239, 1969

MAYR E: Populations, Species, and Evolution. Cambridge, Harvard University Press, 1970

WOESE CR, FOX GE: Phylogenetic structure of the prokaryotic domain: the primary kingdoms. Proc Natl Acad Sci USA 74:5088, 1977

ENERGY PRODUCTION

METABOLIC VERSATILITY

Growing cells require a constant supply of metabolic energy, in a form that can be used for biosynthesis. The microbial world has evolved extraordinarily diverse energy-yielding patterns, filling every possible ecologic niche. Some of these use inorganic electron donors (H_2, H_2S, Fe^{2+}, CO) and various electron acceptors (O_2, NO_3^-, SO_4^{2-}) to provide energy for forming their organic constituents from CO_2. In evolution these early, **autotrophic** (Gr., self-feeding) patterns, later expanded by photosynthesis, led to a rich accumulation of organic compounds, and this accumulation made possible the evolution of the great variety of **heterotrophic** patterns (using organic fuels) now found in most bacteria (and in animals).

Heterotrophic metabolism not only vastly expanded the variety of microbes but also gave them a new role: the **mineralization** of organic matter to CO_2 and H_2O, which constitutes, together with the reduction of CO_2 by photosynthesis, the geochemical cycle. Indeed, the atmospheric reservoir of CO_2, if not constantly replenished, would support the current rate of photosynthesis for only 20 years. It is therefore axiomatic that every naturally occurring organic compound can be metabolized by some microbe.*

This chapter will survey the energy-yielding patterns of bacteria. For details the reader is referred to textbooks of biochemistry.

FERMENTATIONS

The evolution of gas from grape juice revealed one kind of fermentation (L. *fervere,* to boil) long before the development of microbiology. Because fermentations yield their major products in large amounts, these products were the first microbial compounds to be identified, initiating the study of microbial metabolism.

Pasteur defined fermentation as **life without air;** and he recognized that metabolic energy is derived, in this remarkable process, by the organism's "property of performing its respiratory functions, somehow or other, with the oxygen existing combined in sugar." Later studies on the "somehow or other" led to a redefinition of fermentation as *metabolism in which organic compounds serve as both the electron donors and the electron acceptors.*

A fermentation must balance: the average level of oxidation, i.e., the number of moles of C, H, and O, must be the same in the products as in the substrates. For example,

$$C_6H_{12}O_6 \rightarrow 2\ C_3H_6O_3\ (CH_3CHOHCOOH)$$
Glucose   Lactic acid
$$C_6H_{12}O_6 \rightarrow 2\ CH_3CH_2OH + 2\ CO_2$$
Ethanol

The anaerobic conditions required for fermentation are easily achieved with liquid media. In the laboratory it is convenient to use completely filled bottles, provided with loose glass stoppers to allow venting of any gas that may be produced. In industry deep vats are sufficient: oxygen at the surface does not penetrate far, and immediately above the surface it is displaced by the heavier CO_2 evolved. Accordingly bacteria, which require an aqueous environment, have mostly retained a fermentative metabolism, though it is much less energy-efficient than respiration.

In many contexts the term fermentation refers to the anaerobic metabolism of carbohydrates, yielding various pleasant-smelling products. **Putrefaction** refers to a fermentation, primarily of proteins (as in infected tissues), that yields ill-smelling products. In industrial jargon the term fermentation is loosely extended to any large-scale microbial synthesis, even when the process involves strong aeration, as in antibiotic production.

---

* Since this reciprocal relation between the variety of substrates and the variety of microbes is the product of a long evolution, it is not surprising that some modern synthetic organic compounds are not attacked. Accordingly, **biodegradable** substitutes are being sought for toxic compounds that are persistent and are concentrated by plants and animals.

SOURCE OF ENERGY IN FERMENTATION

The energetics of fermentation may be considered from several points of view. In **thermodynamic** terms fermentations proceed because the products have a lower energy content than the substrates. Thus the molar free energy difference (ΔF) between 1 mole of glucose and 2 moles of lactate* (at neutral pH) is 58 kcal. This difference arises essentially from the much lower energy level of the carboxyl group and H_2O, compared with the same atoms in carbonyl or hydroxyl groups. Fermentation may thus be viewed, in terms of **bond energy,** as a regrouping of atoms in organic molecules to yield low-energy products. Similar considerations explain the much higher energy yield (688 kcal) obtained on oxidizing all the C and H of glucose to CO_2 nd H_2O.

COUPLINGS OF FERMENTATION WITH PHOSPHORYLATION

In terms of **intermediate metabolism,** to generate ATP a sequence of spontaneous, energy-losing reactions in a fermentation must lead up to an organic phosphate whose energy of hydrolysis ($-\Delta F$) exceeds the 8-kcal difference between ATP and ADP + P_i. An enzyme can then transfer the phosphate group from the donor to ADP.†

Such "substrate-level" phosphorylation provides all the useful energy from fermentation, and also a small part of the energy of respiratory metabolism (see below). The reader is referred to a biochemistry text for a full exposition. We may note briefly that three classes of reactions yield such phosphorylation. These reactions, all present in the alcoholic fermentation, are 1) dehydration of a phosphate ester to a high-energy enolphosphate; 2) oxidation of an aldehyde to a carboxy-P anhydride (in which the aldehyde and the –SH of an enzyme form a thiohemiacetal; this compound is then oxidized to a high-energy thioester, and the enzyme –SH is then displaced by phospate); and 3) activation of an α-keto acid by combination with thiamine pyrophosphate, making possible the release of formic acid (HCOOH); the rest of the molecule can then be converted to the high-energy acyl CoA and thence to acyl-P.

COENZYMES

A variety of coenzymes (cofactors), of relatively low molecular weight and high thermostability, were initially distinguished from

* For organic acids terms such as lactic acid and lactate will generally be used interchangeably in this volume; the formula will be generally written as the free acid, though at the usual pHs organic acids are largely ionized.

† Abbreviations: The phosphate group (-PO_3H_2) is symbolized by the letter P in the names of compounds, and by a circled P in structural formulas; P_i = inorganic phosphate; ~P = high-energy phosphate. ADP and ATP = adenosine 5′-di- and triphosphate; NAD^+ and NADH = the oxidized and the reduced forms of nicotinamide adenine dinucleotide (NAD = DPN [diphosphopyridine nucleotide]); $NADP^+$ and NADPH = the comparable forms of nicotinamide adenine dinucleotide phosphate (NADP = TPN); CoA = Coenzyme A.

enzymes because they could be removed from extracts by dialysis, with a consequent loss of enzymatic activity; and activity could be restored by supplying the supernate from an extract in which the enzymes had been denatured by boiling. Such cofactors include 1) CoA for acyl transfer; 2) thiamine pyrophosphate (TPP) for transferring groups derived from a ketone (e.g., decarboxylation of α-keto acids, or the transketolase reaction of ketoses described in Ch. 4); and 3) biotin for CO_2 transfer. Various biosynthetic reactions employ pyridoxal phosphate for amino acid transamination, decarboxylation, and racemization; tetrahydrofolate for 1-carbon group transfer and reduction; and cobamide (from vitamin B_{12} [cobalamin]) in methyl group transfers and certain reductions.

Redox cofactors play a major role in directing metabolic flow: in order to ensure a smooth flow of energy and to promote economy, the cell contains cofactors of different redox potential. In increasing order these are ferredoxin, lipoic acid, NAD and NADP, flavins, and hemes. Flavins in flavoproteins and hemes in cytochromes are tightly bound to proteins and therefore function as prosthetic groups rather than free cofactors; the different proteins endow them with a wide range of redox potentials.

An example of the direction of metabolic flow by cofactors is the reduction of pyruvate to lactate and its reversal. The NAD-linked reaction in the lactic fermentation is reversible in vitro, but the equilibrium strongly favors lactate. To form lactate cells employ this reaction, but to oxidize lactate they employ a quite different, flavin-linked, lactic oxidase, thus ensuring flow in the desired direction.

The necessity of ensuring a given direction of flow also explains the presence in the same cell of two pyridine nucleotides that are very similar in structure and in redox potential. In a well-aerated culture in the steady state NADP is present predominantly as NADPH, generated primarily by the pentose phosphate pathway and by isocitrate dehydrogenase, and oxidized only by certain biosynthetic reactions. NAD, in contrast, is present predominantly in the oxidized form, because it is directly linked to the powerful oxidizing system of electron transport. Hence the cell can simultaneously carry out, in the same pool, NAD-linked oxidative biosynthetic reactions (e.g., histidinol → histidine) and NADP-linked reductive reaction (e.g., α-ketoglutarate → glutamate; acetate → fatty acid).

Ferredoxins. Most of the coenzymes noted above are equally important in all cells. Only some bacteria and higher plants, however, have iron-containing cofactors, ferredoxins, with a remarkably low standard redox potential (-417 mV), similar to that of the H_2 electrode:

$$2\,H^+ + 2\,e^- \rightleftharpoons H_2$$

This factor makes possible the release of H_2 gas by certain anaerobes, and also the utilization of H_2 as a fuel for respiration (see Autotrophic Metabolism, below). After their discovery in bacteria ferredoxins were found to have a much wider role in bacteria and in plants, as an essential component in photosynthesis and in N_2 fixation. The ferredoxins of various bacteria are small proteins of mol wt 6000–10,000, containing 8 atoms of Fe^{2+} (or Fe^{3+}) complexed with an equivalent amount of acid-labile S that readily yields H_2S.

Flavodoxins have a similar low potential and can replace ferredoxins in certain reactions. They lack Fe; the redox cofactor is flavin mononucleotide (FMN). Flavodoxins are synthesized by many bacteria, especially in low-Fe media.

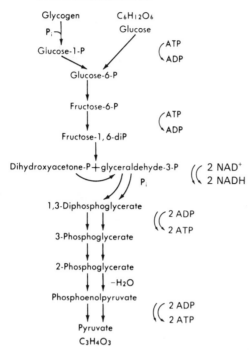

FIG. 3-1. The glycolytic formation of pyruvate (Embden–Meyerhof pathway). Sum: Glucose + 2 ADP + 2P$_i$ + 2NAD$^+$ → 2 Pyruvate + 2 ATP + 2 NADH + 2 H$^+$. **Double arrows** signify 2 moles reacting per mole of glucose.

Thioredoxin is a widely distributed small protein that appears to function in the reduction of ribose to deoxyribose, of disulfides in proteins, and of sulfate; it is also, curiously, a component of certain phage-induced DNA polymerases.

ALTERNATIVE FATES OF PYRUVATE

A number of fermentations are based on the **glycolytic (Embden–Meyerhof) pathway** (Fig. 3-1). This pathway generates ATP twice: first through the oxidation of glyceraldehyde-3-P (triose-P) by NAD$^+$, and again through the conversion of P-enolpyruvate to pyruvate. To balance the fermentation, and allow the NAD to recycle, the NADH must be reoxidized by pyruvate or a derivative. Microbes have evolved a variety of pathways for this purpose, illustrated in Figure 3-2. The relative advantages of these pathways, adapted to different nutritional conditions, will be considered below (see Adaptive Value of Different Fermentations).

Lactic Fermentation. This is the simplest fermentation: a one-step reaction, catalyzed by NAD-linked lactic dehydrogenase (really a pyruvate reductase), reduces pyruvate to lactate. No gas is formed. Since two ATP molecules are consumed in the formation of hexose diphosphate from glucose, and since four ATP molecules are subsequently produced, the net yield is two ATPs per

hexose. This fermentation is the first stage in cheese manufacture; it is also identical with the glycolysis in mammalian cells.

The **homolactic fermentation**, which forms only lactate, is characteristic of many of the lactic bacteria: e.g., *Lactobacillus casei, Streptococcus cremoris,* and pathogenic streptococci. Others carry out a rather different, **heterolactic, fermentation** (see Phosphogluconate Pathways, below), which converts only half of each glucose molecule to lactate. Both these fermentations are responsible for the souring (acidification) of milk and certain other foods (sauerkraut, pickles), which preserves them as long as they are kept anaerobic; these processes also provide interesting flavors (especially highly developed in the secondary fermentations of lactic acid and amino acids in cheeses).

Alcoholic Fermentation. Pyruvate is converted to CO$_2$ (retained in some beverages) plus acetaldehyde, which is then reduced to ethanol in an NAD-linked reaction. This fermentation is characteristic of yeasts; as a major pathway it is uncommon in bacteria.

Propionic Fermentation. This pathway extracts additional energy from the substrate. Pyruvate is carboxylated to yield oxaloacetate, which is reduced to yield succinate and then is decarboxylated to yield propionate.*

Organisms possessing this pathway can eke out a living by fermenting lactate, ordinarily an endproduct. The lactate is first oxidized to pyruvate; part is then reduced to propionate and the rest oxidized to acetate and CO$_2$:

$$3\ CH_3CHOHCOOH \xrightarrow{-6[H]} 3\ CH_3COCOOH \xrightarrow{+6[H]}$$
$$\text{Lactate} \qquad\qquad \text{Pyruvate}$$
$$2\ CH_3CH_2COOH + CH_3COOH + CO_2 + H_2O$$
$$\text{Propionate} \qquad \text{Acetate}$$

This arduous process of extracting energy from lactate, by the overall shift of oxygen from three hydroxyls to one carboxyl, yields only one ATP per nine Cs fermented. Hence propionic acid bacteria grow slowly. In Swiss cheese their late formation of CO$_2$, after the completion of the lactic fermentation of milk by other organisms, is responsible for the formation of the holes, while the propionic acid contributes to the flavor.

The **mixed acid (formic) fermentation** is characteristic of most Enterobacteriaceae. These organisms dispose of their substrate partly through a lactic fermentation, but mostly through a fermentation characterized by the splitting of pyruvate (without net oxidation or reduction) to formate and acetyl CoA; the latter in turn generates an ATP (Fig. 3-2).

Since this reaction does not absorb the 2[H] released in forming pyruvate, fermentation balance requires that an equal amount of

---

* The decarboxylation is accomplished via a cobamide-linked reaction in which succinyl CoA is rearranged to form the less stable methylmalonyl CoA [CH$_3$CH(COOH)CO–CoA]. This compound is then split to yield propionyl CoA and CO$_2$: a reaction that is necessary, in a closed system, to provide CO$_2$ for further cycles.

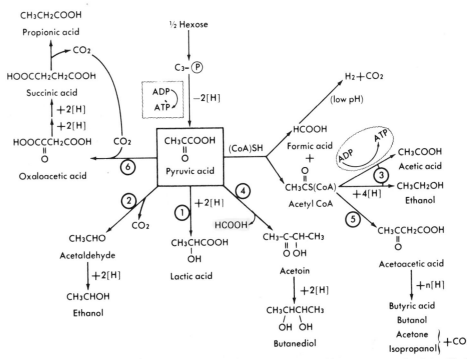

FIG. 3-2. Key role of pyruvate in principal fermentations. 1. Lactic (*Streptococcus, Lactobacillus*). 2. Alcoholic (many yeasts, few bacteria). 3. Mixed acid (most Enterobacteriaceae). 4. Butanediol (*Enterobacter*). 5. Butyric (*Clostridium*). 6. Propionic (*Propionibacterium*). (For abbreviations, see text footnote.)

pyruvate be reduced by pathways absorbing more [H]: a) formation of ethanol and formate, and b) reduction (after carboxylation) to succinate (as in the propionic fermentation).

(a) Glucose $\xrightarrow{-[4H]}$ 2 $CH_3COCOOH$ $\xrightarrow{(CoA)SH}$ 2 HCOOH +

$\quad$ 2 $CH_3CO-S(CoA)$ $\xrightarrow{+[4H]}$ CH_3CH_2OH + (CoA)—SH

$\qquad\qquad\qquad\searrow^{P_i}$

$\qquad\qquad\qquad\quad CH_3COO\,\textcircled{P}$ + (CoA)—SH

(b) $CH_3COCOOH$ $\xrightarrow[+[4H]]{CO_2}$ $HOOCCH_2CH_2COOH$ + H_2O

$\quad$ Pyruvate $\qquad\qquad\qquad$ Succinate

Either process consumes 4[H] per pyruvate reduced. The formic fermentation thus yields three ATP per glucose fermented (compared with two in the lactic fermentation).

Gas Formation. The formate from the mixed acid fermentation may remain as such if the pH is kept alkaline, but most fermentations become acidic; and at a pH of 6 or less "gas-formers" (e.g., *Escherichia coli*) form an enzyme, formic hydrogenlyase, that converts formic acid ($HCOOH$) to CO_2 and H_2. Enterobacteriaceae that do

not form this enzyme (e.g., *Shigella*) produce acid but no gas, like homolactic fermenters. Gas formation is an important diagnostic test (see Table 31-1).

Intestinal gas accumulates when bacteria in the colon produce H_2 and CO_2, and in some people CH_4, faster than these compounds can be absorbed. The required substrates increase in the colon when disorders of peristalsis, or of pancreatic or biliary secretion, impair their absorption in the small intestine; in addition, some foods contain carbohydrates that can be absorbed by bacteria but not by the intestine. Impaired colonic peristalsis also promotes the formation of free gas, since it hinders the convection of fluid containing dissolved gases to the mucosal surface, where the compounds are transferred to the blood.

The methane fermentation, a major process in ruminants, is carried out by *Methanobacterium*, which oxidizes fatty acids to carbohydrates at the expense of the reduction of CO_2 to CH_4. Curiously, the distribution of this fermentation among humans appears to be a familial trait, established early in life: it is highly correlated among siblings but not between spouses.

Butanediol (Acetoin) Fermentation. This pattern, observed in *Enterobacter* (formerly *Aerobacter;* Ch. 31), and also in certain other Enterobacteriaceae and some species of *Bacillus,* also releases formate. However, the remaining "active acetaldehyde" is not oxidized but condenses with a second pyruvate. The product is converted to butylene glycol (butanediol) in the following reactions, which balance the 4[H] created in forming the two pyruvates.

$$CH-CO-COOH + TPP \xrightarrow{[2H]} [CH_3-CHO]-TPP + HCOOH \xrightarrow{H_2 + CO_2}$$

"Active acetaldehyde"

$$[CH_3-CHO]-TPP + CH_3-CO-COOH \xrightarrow{TPP} CH_3-CO-\underset{\underset{OH}{|}}{\overset{\overset{COOH}{|}}{C}}-CH_3$$

Acetolactate

$$\xrightarrow{-CO_2} CH_3-CO-CHOH-CH_3 \xrightarrow{[2H]} CH_3-CHOH-CHOH-CH_3$$

Acetoin (acetylmethyl carbinol)   2,3–Butylene glycol

This fermentation, like the alcoholic fermentation, yields only neutral products and produces two ATPs per glucose. It is often called the **acetoin fermentation** because exposure to air oxidizes some of the butylene glycol to acetoin, which is readily recognized by a specific color test **(Voges–Proskauer)**. In sanitary engineering this test is of considerable diagnostic value (Ch. 31) in discriminating between *E. coli,* which reaches bodies of water primarily from the mammalian gut, and *Enterobacter,* originating primarily from vegetation.

The formation of neutral rather than acidic products permits the fermentation of larger amounts of carbohydrate without self-inhibition, and therefore the production of more gas; hence the old name *Aerobacter aerogenes.*

Butyric–butylic fermentation. This pattern of pyruvate reduction is seen in certain strict anaerobes. The initial scission yields H_2, CO_2, and 2-C fragments at the acetate level of oxidation. Two such fragments are then condensed, not head to head as in acetoin but head to tail as in fatty acid synthesis. The resulting acetoacetyl CoA undergoes several reactions, in varying proportions: decarboxylation to acetone, and reduction by the H_2 (activated by ferredoxin) to yield isopropanol, butyric acid, and *n*-butanol.* These patterns are useful for classification (e.g., *Bacteroides,* Ch. 31).

Mixed amino acid fermentations. These fermentations occur where there is considerable proteolysis; they are prominent in putrefactive processes, including the gangrene associated with anaerobic wound infections. Certain amino acids (or their deamination products) serve as electron donors and others as acceptors. For example:

---

* The use of this fermentation for the industrial production of acetone was developed by Chaim Weizmann in England in 1915. This scientific contribution, which solved an urgent problem in explosives manufacture in World War I, is said to have promoted the Balfour Declaration, and thus contributed eventually to the founding of the state of Israel, with Weizmann as its first president.

$$CH_3-CHNH_2-COOH + 2\ CH_2NH_2-COOH$$
Alanine   Glycine
$$+\ 2\ H_2O \rightarrow 3\ CH_3-COOH + CO_2 + 3\ NH_3$$
Acetate

In addition, decarboxylation of various amino acids, and further reactions, yield products that are pharmacologically active (e.g., histamine) or malodorous (e.g., indole from the breakdown of tryptophan, and even more mephitic –SH compounds derived from cysteine and methionine). More pleasant, empirically selected fermentations of minor constituents are responsible for the characteristic flavors of various wines and cheeses, while the toxic effect of poorly fermented wines is largely due to longer-chain aldehydes (fusel oil) derived from amino acids.

PHOSPHOGLUCONATE PATHWAYS

Not all metabolism of glucose proceeds via the Embden–Meyerhof pathway. For example, the **heterolactic** fermenters (e.g., *Leuconostoc mesenteroides*) yield approximately equimolar quantities of lactate, ethanol, and CO_2, via P-gluconate decarboxylation (pathway B, Fig. 3-3); and in pseudomonads the major route of hexose metabolism is the **ketodeoxygluconate (Entner–Doudoroff) pathway** (C, Fig. 3-3). The latter is also used in *E. coli* to metabolize gluconate.

ALTERNATIVE PATHWAYS WITHIN AN ORGANISM

Glucose is metabolized in *E. coli* mainly via the Embden–Meyerhof pathway, but the so-called **hexose monophosphate shunt** also converts it via P-gluconate, to pentose-P (Fig. 3-3). This compound can be oxidized (via acetyl-P and triose-P) or fermented by a complex series of 2-C and 3-C transfers that convert it to triose-P.

The shunt is not essential: its block by mutation does not slow growth on glucose. On the other hand, a mutant blocked in the Embden–Meyerhof pathway can grow at one-third the normal rate by using the shunt to capacity. Isotopic studies show that in wild-type *E. coli* the shunt functions simultaneously with the glycolytic pathway. The explanation is that while both pathways yield triose-P they also have different functions: glycolysis generates NADH (which functions primarily in energy metabolism), while the shunt generates pentose-P and also NADPH (which is required for reductive steps in biosynthesis). The alternative pathways allow reducing power to be distributed as needed between the two factors; and a block in the shunt pathway has no perceptible effect because the cell has alternative routes both to pentose-P and to NADPH.

It is clear that cells do not carry alternative pathways simply as options or reserves: apparent alternatives have different functions, though they may replace each other when necessary.

ADAPTIVE VALUE OF DIFFERENT FERMENTATIONS

The central role of the glycolytic pathway in many different fermentations presumably reflects evolutionary selection for the most effective mechanism. The diversity of final products, on the other hand, reflects the variety of

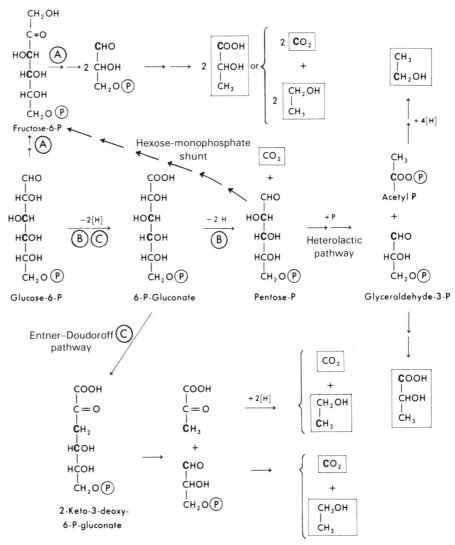

FIG. 3-3. Alternative ways of metabolizing glucose to lactate and/or ethanol. Glucose-3, 4-^{14}C yields alcohol labeled (*heavy letter*) either **A**) in neither C atom (Embden–Meyerhof pathway), **B**) in the CH_2OH (heterolactic pathway), or **C**) in the CH_3 (ketodeoxygluconate pathway).

ecologic opportunities available, which select for different variants of the basic mechanism. Thus, as we have already noted in Chapter 1, Pasteur recognized that grape juice is much richer in sugar than milk but poorer in protein (i.e., buffering power). Hence in the fermentation of grape juice lactic acid bacteria soon produce a self-inhibitory pH, while the yeasts, though slower-growing, produce a neutral product and outgrow the bacteria. (In addition, grape juice fails to meet the requirement of these bacteria for various amino acids.) Milk, in contrast, supports the growth of either organism, and the faster-growing bacteria soon predominate.

Similarly, the vertebrate colon, with a meager supply of unabsorbed carbohydrates and growth factors, selects for enteric bacteria (e.g., *E. coli*) that have no specific growth requirements and that extract as much energy as possible from their fermentation. In contrast *Enterobacter*, which grows primarily on vegetation, can ferment larger concentrations of carbohydrate (to butanediol), but with a less efficient energy yield.

It is evident that nature selects for a variety of fermentative talents, which might be compared to sprinters and long-distance runners.

Substrate–Specific Pathways. Glucose is considered a typical substrate for fermentation, for it is much the most widely distributed sugar (being the monomer of starch and cellulose), and most bacteria can utilize it. The first steps in the metabolism of most other substrates convert them to some intermediate in a central pathway (glycolysis or tricarboxylic acid cycle). The specific connecting pathway is often quite short.

For example, a single step will convert fructose to fructose-6-P; a one-step hydrolysis will yield glucose from sucrose or from lactose; three steps are necessary to convert galactose to glucose-1-P (reversing the reactions of Fig. 4-10); and two will convert glycerol to glyceraldehyde-P.

Under aerobic conditions facultative organisms will eventually oxidize, and not simply ferment, a substrate that they can use; but this response is still referred to as "fermentation-positive," since it is most conveniently recognized through the initial accumulation of acid.

Though a facultative organism can make both fermentative and respiratory enzymes, the existence of obligate aerobes and obligate anaerobes suggests that there is selective advantage in dispensing with the genes for one set when the customary habitat employs only the other. For example, yeasts, which grow both on the surfaces of living fruits and within the juices of crushed or decaying fruit, are generally facultative, while the closely related molds, which form fluffy aerial growths (mycelia) on surfaces (Ch. 45), tend to be strictly aerobes.

RESPIRATION

AEROBIC AND ANAEROBIC METABOLISM

In fermentation the electron acceptor is an organic compound, while in respiration it is usually O_2 (aerobic respiration). Hence for most bacteria fermentation and respiration are virtually synonymous with anaerobic and aerobic metabolism, respectively. (The exception of anaerobic respiration, using an inorganic electron acceptor other than O_2, will be discussed below.)

Bacteria fall into several groups with respect to the effect of O_2 on their growth and metabolism. These properties are important for pathogenesis, and also for the isolation and identification of the organism.

1) **Obligate aerobes** (e.g., the tubercle bacillus and some spore-forming bacilli) require O_2 and lack the capacity for substantial fermentation.

2) **Obligate anaerobes** (e.g., clostridia, propionibacter) can grow only in the absence of O_2. A subgroup, called **microaerophilic,** can tolerate, or even prefer, O_2 at low tension, but not at that of air.

3) **Facultative organisms** (e.g., many yeasts, enterobacteria) can grow without air but shift in its presence to a respiratory metabolism.

4) **Aerotolerant anaerobes** (e.g., most lactic acid bacteria) resemble facultative organisms in growing either with or without O_2, but their metabolism remains fermentative.

In contrast to the relatively simple set of enzymes required for the "substrate-level" phosphorylation occurring in fermentation, respiration requires an elaborate, membrane-bound system for deriving energy by electron transport from substrate to O_2 (see below). Strictly fermentative organisms lack this system, and facultative organisms cease making its components when growing anaerobically.

Some aerotolerant anaerobes, however, can partly oxidize glucose via a flavoprotein that is directly reoxidized by O_2. This short respiratory chain yields little or no ATP but may serve to scavenge dissolved O_2.

Superoxide. In anaerobes the presence of O_2 can be shown to inhibit some enzymes required for fermentative growth. In addition, *most obligate anaerobes not only are inhibited but are rapidly killed by air.* The poison is a highly reactive free-radical form of O_2, superoxide (O_2^-), formed by flavoenzymes: this product is destroyed by **superoxide dismutase,** an enzyme present in aerobes and aerotolerant organisms but not in strict anaerobes. The requirement of this enzyme for growth in O_2 is further shown by mutations that inactivate it. The dismutase reaction is:

$$2\ O_2^- + 2H^+ \rightarrow H_2O_2 + O_2$$

The peroxide (H_2O_2) formed by this reaction (and others) is also toxic, and it is either destroyed by catalase ($2\ H_2O_2 \rightarrow 2\ H_2O + O_2$) or used for oxidation by a peroxidase.

Superoxide is formed, in both anaerobes and aerobes, as an intermediate in the flavoprotein-linked reduction of O_2, either to H_2O_2 or to H_2O.

Oxygenases carry out a small class of oxidative reactions in which molecular oxygen is added across a double bond in the substrate. These reactions do not yield ATP, but they do convert a refractory compound to one that is useful, either for structural purposes (e.g., saturated → unsaturated fatty acids) or for further metabolism. For example, some pseudomonads use an oxygenase to open benzenoid rings, thus initiating the conversion of various aromatic compounds into useful intermediates.

Bioluminescence is a curious accompaniment of respiration in certain bacteria, as in some higher forms. It is caused by the oxidation of a flavoprotein (luciferin) by O_2 in the presence of the enzyme luciferase. Though its value in the mating of a fish or a firefly is not hard to imagine, in bacteria it may be an adventitious concomitant of a primitive respiratory system. Luminescent bacteria have proved useful as delicate detectors of O_2 in solution, because their light production responds to even very low concentrations of O_2.

TABLE 3-1. Energetics of Metabolism

Reaction	No. of ATP generated	$-\Delta F$ (kcal)	Efficiency* (%)
Glucose $\longrightarrow$ 2 Lactic acid	2	58	28
Glucose $\longrightarrow$ 2 Ethanol + 2 CO_2	2	57	28
Glucose + 6 $O_2 \longrightarrow$ 6 CO_2 + H_2O Stages:			
Glucose $\xrightarrow{-4[H]}$ 2 Pyruvate (substrate) (2 NAD)	2 6		
$\xrightarrow{-4[H]}$ 2 AcCoA + 2 CO_2 (2 NAD)	6		
$\xrightarrow{-16[H]}$ 4 CO_2 + 4 H_2O (8 NAD or equivalent)	24		
Total	38	688	44

ΔF = Free energy difference; $-\Delta F$ = energy of hydrolysis.

*Assuming ΔF of ATP formation = 8 kcal.

TERMINAL RESPIRATION: INCOMPLETE OXIDATIONS

Ultimately respiration converts organic compounds to CO_2 and H_2O. In the commonest pattern (see Table 3-1 above) pyruvate, from the glycolytic pathway, is oxidized to acetyl CoA and CO_2, and the acetyl CoA is oxidized via the tricarboxylic acid (TCA) cycle (Fig. 4-2, presented in Ch. 4 because of its intimate connection with biosynthesis). This function of the TCA cycle is called **terminal respiration.**

Respiration not only increases the energy yield from fermentable fuels; it also permits the use of fuels too highly reduced to be fermented. These are converted, by short pathways, into acetate and TCA cycle intermediates. The degradative pathways have been worked out for a variety of amino acids and bases, and for many other compounds.

Incomplete Oxidations. Respiration does not always lead to complete oxidation of each molecule attacked. For example, wine exposed to air is converted to vinegar, through the conversion of ethanol to acetic acid (incomplete oxidation) by *Acetobacter*. Some members of this genus lack the genes required for the further oxidation of acetate, while others have these genes but do not use them until after the ethanol is consumed. This preferential partial oxidation of a readily available substrate is not rare: indeed, in most organisms positive fermentation tests (for acid) appear under aerobic as well as under anaerobic conditions.

Because of such incomplete oxidations microbes yield many valuable products of respiration as well as of fermentation. For example, oxidation of sorbitol to the previously rare sugar sorbose has made possible the inexpensive synthesis of ascorbic acid; similarly microbial oxidation of C-11 of the steroid nucleus has facilitated the synthesis of steroid hormones. Apart from such specific conversions, the commercial production of various metabolic products (e.g., antibiotics, amino acids, citric acid) uses respiring organisms.

The conversion of wood or petroleum to palatable microbial protein is a major area of current exploration.

ELECTRON TRANSPORT

The electrons from various donors, transferred to NADH or a flavoprotein, are funneled into the **electron transport system:** a set of reversible electron carriers (Fig. 3-4), embedded in the cell membrane. These carriers have successively higher redox potentials (oxidizing power), in the sequence from NADH to O_2, and the terminal one transfers electrons to O_2 (or another acceptor). In three of the steps in this sequence the drop in energy is used to extrude a pair of protons (H^+) from the cell, setting up an **electrochemical gradient (protonmotive force)** across the electrically insulating (and proton-impermeable) membrane. The protons can return through a channel in an ATPase that traverses the membrane, with a drop in potential that allows each pair of protons to provide the enzyme with the large amount of energy required to reverse its ATPase activity (**oxidative phosphorylation**):

$$ATP \underset{O_x.\ Phos.}{\overset{ATPase}{\rightleftharpoons}} ADP + P_i$$

The overall free energy of oxidizing 2[H] from NADH, about equivalent to burning H_2 gas, is about 50% recovered as ATP energy. Since the protonmotive force is also used directly for energy-requiring functions of the membrane it is described in greater detail in Chapter 6 (see Fig. 6-27).

Electron transport has been studied most intensively in mitochondria; bacteria have been especially helpful in providing mutants blocked in single steps. In *E. coli* at least 5 "uncoupled" (=*unc*) genes are found, in one cluster (operon). The oxidation of NADH by O_2 is fundamentally the same in all organisms. However, those bacteria that use donors with a higher redox potential than NADH, or terminal acceptors with a lower potential than O_2,

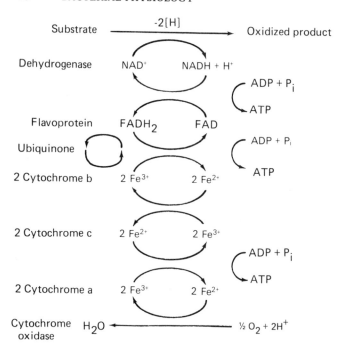

FIG. 3-4. Electron transport and associated oxidative phosphorylation (P/O < 3). The bacterial system also contains nonheme iron proteins, whose position in the scheme is uncertain. Some bacteria appear to use naphthoquinones rather than a benzoquinone (ubiquinone); they presumably go through a reversible cycle of reduction to hydroquinones.

use a shorter electron transport chain and hence derive less energy per mole.

Cytochromes. The three cytochromes (heme-proteins) of an electron transport chain (Fig. 3-4), which differ in their heme or in the nature of its attachment to protein, each have a characteristic sharp absorption band in their reduced form, and all three, when present, can thus be detected even in intact bacteria. In different organisms, the cytochrome of each kind has a slightly different position for its main absorption band, depending on the protein. It has thus been possible to identify a great variety of cytochromes in the bacterial kingdom, in contrast to the relative uniformity seen in mitochondria.

Many facultative organisms (e.g., *E. coli*) lack cytochrome *c*. Its presence or absence can be detected in a diagnostically useful, empiric **oxidase test,** in which cells containing the enzyme rapidly catalyze oxidation of N,N,dimethyl-*p*-phenylenediamine into a colored product.

Other constituents. Some bacteria have **ubiquinone** (coenzyme Q), a benzenoid quinone, as do mitochondria; others have a **naphthoquinone** (e.g., menaquinone [vitamin K]); and some have both. A long alkyl side chain localizes these small molecules in the membrane.

The **flavoproteins,** which cover the whole range of biologic redox potentials, were discovered by Warburg on the basis of simple observations on lactobacilli, which lack the red cytochromes: on exposure to air the intact cells become yellow (*L. flavus,* yellow). The pigments, derived from riboflavin, function as prosthetic groups tightly bound to their enzymes rather than as diffusible hydrogen carriers.

ENERGETICS AND GROWTH

Table 3-1 shows that respiration is considerably more efficient than glycolysis: it yields about 10 times as much free energy (ΔF) and 19 times as much ATP per mole of glucose metabolized. Accordingly, a facultative organism grown on a limited amount of glucose will exhibit a larger **growth yield** (dry weight of bacteria/weight of substrate metabolized) under aerobic than under anaerobic conditions.

The energy not used in ATP formation, plus part of the ATP energy subsequently used in biosynthesis, appears as heat. Indeed, dissipation of heat is one of the major problems of large-scale fermentation; if inadequate, a fermentation may sterilize itself.

Storage Materials. With cells that are not growing (for lack of a nitrogen source) oxidative assimilation may convert much substrate into carbohydrates or, in many bacteria, into a unique storage product, poly-β-hydroxybutyric acid, an incompletely reduced, highly polymerized equivalent of fat. This substance is intermediate, in composition $(C_2H_3O_2)_n$ and in energy content, between fat and carbohydrate $(CH_2O)_n$; it is presumably more readily available than fat as a reserve material.

$$2n\ CH_3\!-\!\overset{\overset{\displaystyle O}{\|}}{C}\!-\!S(CoA) \rightarrow (\!-\!OCH\!-\!CH_2\!-\!\overset{\overset{\displaystyle O}{\|}}{C}\!-\!)_n$$

Acetyl CoA   Poly-β-hydroxybutyrate

The Pasteur Effect. In the presence of air a facultative anaerobe, such as the yeast of alcoholic fermentation, not only forms more CO_2 per mole of glucose consumed but also yields more growth and grows faster. Yet as Pasteur discovered, the **rate** of CO_2 evolution and of glucose utilization decreases. A reasonable explanation for this old paradox is that respiration, producing much more ATP per glucose molecule, soon saturates the cell's capacity to utilize the ATP; i.e., the ADP/ATP ratio (**energy ratio**) is critical.

In more specific terms, glycolysis requires a supply of ADP for accepting phosphate from 1,3-di-P-glycerate (Fig. 3-1), and respiration will keep the ADP level low and the ATP high. However, the Pasteur effect may well involve various of the many regulatory mechanisms in which metabolite levels influence the level and the activity of specific enzymes (Ch. 14).

AUTOTROPHIC METABOLISM

CHEMOAUTOTROPHY AND ANAEROBIC RESPIRATION

In autotrophic metabolism microbes tap various sources of energy and reducing power, which they use to reduce CO_2 to organic compounds. This feature of photosynthesis was recognized early, in plants. Bacteria also have a unique form, chemoautotrophy (chemosynthesis), discovered before 1900 by the Russian soil microbiologist Winogradsky.* This process derives energy from the respiration of various inorganic electron donors (Table 3-2).

Anaerobic respiration is another form of unusual use of inorganic substrates, also confined to bacteria. Thus, while some chemoautotrophs are aerobic and use O_2 as the ultimate electron acceptor, others engage in anaerobic respiration, using an inorganic terminal electron acceptor other than O_2 (Table 3-2, class II). Moreover, these acceptors can be used by some bacteria for the oxidation of organic rather than inorganic compounds.

*Winogradsky (1856–1953) had a remarkable history. After studying with Pasteur he returned to Russia and established the fundamentals of soil microbiology; then, tiring of struggles with bureaucracy, he retired at about 1900 to his family estate. Driven out by the Revolution, he returned to laboratory work, in France, at the age of 65, and began a long period of further important contributions.

The use of **nitrate** as terminal electron acceptor, in a short, flavoprotein-dependent electron transport system, is common among heterotrophs (including *E. coli*), probably because nitrate is universally present as the principal form of storage of N in soil.

PHOTOSYNTHESIS

In photosynthesis energy from light is used to provide both energy and reducing power for a succession of reactions that convert CO_2 to triose-P and hence to other cell constituents. In this process a photon of absorbed visible light is converted to chemical energy by causing a charge separation at the photocenter of a chlorophyll molecule (a protein-bound Mg^{++}–tetrapyrrole pigment) embedded in a membrane. The energized chlorophyll ejects an electron, which is accepted by an adjacent ferredoxin molecule; and the positive charge on the resulting chlorophyll free radical is then neutralized by donation of an electron from a cytochrome on the other side of the insulating membrane. These reactions occur very rapidly, and the large potential difference between the oxidized cytochrome and the reduced ferredoxin is then used in either of two ways, whose proportions vary with the needs of the cell (Fig. 3-5).

First, in **cyclic photophosphorylation** the electron from reduced ferredoxin passes through an electron transport system, with associated phosphorylation of ADP to ATP. This process is essentially the same as oxidative phosphorylation, except that the terminal ac-

---

TABLE 3-2. Autotrophic Modes of Metabolism

Organism or group	Source of energy	Remarks
I. Aerobic lithotrophs (chemoautotrophs)		Use inorganic (litho-) electron donors
Hydrogen bacteria	$H_2 + 1/2\ O_2 \longrightarrow H_2O$	
Sulfur bacteria (colorless)	$\begin{cases} H_2S + 1/2\ O_2 \longrightarrow H_2O + S \\ S + 1.5\ O_2 + H_2O \longrightarrow H_2SO_4 \end{cases}$	Can produce H_2SO_4 to pH as low as 0
Iron bacteria	$2\ Fe^{2+} + 1/2\ O_2 + H_2O \longrightarrow 2\ Fe^{3+} + 2\ OH^-$	
Nitrifying bacteria		
Nitrosomonas	$NH_3 + 1.5\ O_2 \longrightarrow HNO_2 + H_2O$	Convert soil N to nonvolatile form, used
Nitrobacter	$HNO_2 + 1/2\ O_2 \longrightarrow HNO_3$	by plants
II. Anaerobic respirers	Use inorganic electron acceptors	Most can also use organic electron donors
Denitrifiers	$H_2 + NO_3^- \longrightarrow N_2O, N_2, NH_3$	Cause N loss from anaerobic soil
Desulfovibrio	$H_2 + SO_4 = \longrightarrow S\ or\ H_2S$	Odor of polluted streams, mud flats
Methane bacteria	$4\ H_2 + CO_2 \longrightarrow CH_4 + 2\ CH_2O$	Sewage disposal plants; prevent accumulation of free H_2 in nature
Clostridium aceticum	$4\ H_2 + 2\ CO_2 + \longrightarrow CH_3COOH + 2\ H_2O$	
III. Photosynthesizers	Light	"Bacterial" photosynthesis; $H_2(A) =$ various electron donors
Purple sulfur bacteria	$4\ CO_2 + 2\ H_2S + 4\ H_2O \longrightarrow 4\ (CH_2O) + 2\ H_2SO_4$	
Nonsulfur purple bacteria	Light $CO_2 + 2\ H_2(A) \longrightarrow (CH_2O) + H_2O + 2\ (A)$	
Algae; higher plants	Light $CO_2 + 2\ H_2O \longrightarrow (CH_2O) + 1/2\ O_2$	"Plant" photosynthesis

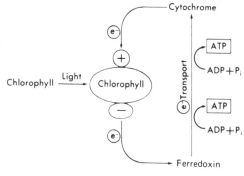

Cyclic photophosphorylation

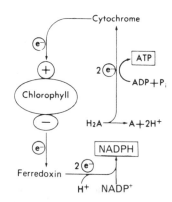

Noncyclic photophosphorylation

FIG. 3-5. Electron flow in photosynthesis. In photosynthetic bacteria H_2A is H_2, H_2S, or various organic compounds; two electrons must be derived by photosynthesis per molecule oxidized. In plant photosynthesis H_2A is H_2O; a second reaction center chlorophyll (not shown), with a higher oxidized potential, must be coupled to the first.

ceptor is the positively charged chlorophyll; the system is thus closed or cyclic.

In the second process, **noncyclic photophosphorylation**, the reduced ferredoxin does not return its electrons to chlorophyll but instead transfers them to NADPH, which is then used in biosynthetic reductions. The positive charge, transferred from chlorophyll to a cytochrome, must therefore be neutralized in another way, and here bacteria and plants differ. In the familiar photosynthesis of higher plants (which permits them to grow simply on CO_2, H_2O, and some inorganic ions) the oxidized chlorophyll is reduced at the expense of oxidizing water:

$$H_2O \rightarrow \frac{1}{2} O_2 + 2H^+ + 2e^-$$

But a very high redox potential, exceeding that of O_2, is necessary for this reaction; and a single quantum of light does not provide enough energy to create the difference between the potential ($+0.82$ V) and the very low potential of reduced ferredoxin (-0.42 V). Accordingly, plants evolved their remarkable capacity to metabo-

lize water, and thus to eliminate their dependence on other fuels, by coupling two photoactivated chlorophyll reaction centers in series, much like a pair of electrical batteries. Bacteria, however, retain a more primitive type of photosynthesis, employing only one reaction center. They therefore require a more easily oxidized electron donor than water, such as H_2 or H_2S (Fig. 3-5).

Bacterial photosynthesis thus does not release O_2, and photosynthetic bacteria are obligate anaerobes. The process is very restricted in location today, but in the primitive earth, with H_2 and no O_2 in the atmosphere, it no doubt contributed to the initial accumulation of organic compounds. The next stage was the evolution of oxygenic photosynthesis, in the **cyanobacteria** (blue-green bacteria). These prokaryotes were formerly called **blue-green algae** because they evolve O_2) in photosynthesis, like the eukaryotic **true algae** (unicellular plants). The latter subsequently became the major agent of photosynthesis in bodies of water, while higher plants spread on land. Organisms of these three classes presumably generated the atmospheric O_2 that then permitted the electron transport system of photosynthesis to evolve into the powerful energy-yielding system of respiration. Blue-green bacteria may also have given rise in evolution, by symbiosis, to chloroplasts (chlorophyll-containing organelles) in eukaryotic plants.

In plants noncyclic photophosphorylation is essential, since no other source of reducing power is generally available. In bacteria, however, some of the reducing compounds used (e.g., H_2) can also reduce NADP directly, and so most photophosphorylation would then be cyclic.

Some photosynthetic bacteria show the property of **phototaxis**: a gradient of light intensity elicits movement toward the light.

AUTOTROPHIC ASSIMILATION OF CO_2

With the use of the energy and the reducing power supplied by photosynthesis, or by chemosynthesis, an autotrophic cell derives its carbon from CO_2 (the "dark reactions" of photosynthesis, which take place after a pulse of light).

This pathway involves surprisingly few novel enzymes, in addition to the enzymes present in universal pathways of carbohydrate metabolism (including pentose synthesis). The basic mechanism for the initial fixation of CO_2 is:

$$
\begin{array}{ccc}
CH_2O\,\textcircled{P} & \left[CH_2O\,\textcircled{P} \right. & CH_2O\,\textcircled{P} \\
| & | & | \\
C{=}O & C{-}OH & CHOH \\
| & \| & | \\
CHOH \longrightarrow & C{-}OH & \xrightarrow[\text{Carboxy-}]{CO_2} COOH \\
| & | & \text{dismutase} + \\
CHOH & CHOH & COOH \\
| & | & | \\
CH_2O\,\textcircled{P} & \left. CH_2O\,\textcircled{P} \right] & CHOH \\
& & | \\
& & CH_2O\,\textcircled{P}
\end{array}
$$

Ribulose di-P   3-P-glycerate

At this stage the assimilated C is still at the fully oxidized, carboxy level. To serve as a general source of C 3-P-glycerate is reduced to triose-P by NADPH, with additional energy provided by ATP. In a complex cycle (the Calvin cycle) six triose-P (18C) regenerate 3 pentose-P (15C), and one triose-P represents the net gain of fixed C (Fig. 3-6).

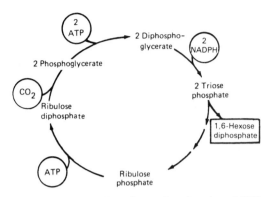

FIG. 3-6. The ribulose-diP cycle requires 3 moles of ATP and 2 moles of NADPH per mole of CO_2 converted to carbohydrate.

NITROGEN CYCLE; NITROGEN FIXATION

Microbes play an essential role in the geochemical **nitrogen cycle,** in which some of the chemoautotrophic bacteria listed in Table 3-2 participate. The N in decomposing organic matter is at first largely converted to NH_3. This volatile compound is then stabilized in the soil by oxidation, through the action of **nitrifying** bacteria, to non-volatile nitrate, which can be reduced by plants to organic amino compounds. However, some NH_3 is lost immediately to the atmosphere, and additional amounts are lost, from anaerobic regions of the soil, through the action of denitrifying bacteria, which reduce nitrate to N_2 and NH_3. Maintenance of the biosphere therefore requires constant **fixation of atmospheric N_2.**

Biologic N_2 fixation is accomplished only by bacteria and algae. The numerically most important group appears to be the bacterial genus *Rhizobium* (Gr. *rhizo,* root), which infects the roots of leguminous plants, leading to the formation of symbiotic N_2-fixing nodules. N_2 can also be fixed, however, by many other bacteria, including the photosynthetic purple sulfur bacteria (e.g., *Rhodospirillum rubrum*). These remarkably self-sufficient organisms can, therefore, derive all their major atoms (C, H, N, and O) from the atmosphere and water.

N_2 fixation, nitrification, and denitrification involve several different valence levels of N. Enzymes in two of these processes have been shown to contain molybdenum, which can also occupy several valence levels. The reduction of N_2 to NH_3, which involves ferredoxin, can now be accomplished in vitro.

ALTERNATIVE SUBSTRATES

GENOTYPIC ADAPTATION (SELECTION OF ORGANISMS)

A small sample of soil may contain many different kinds of organisms, which play different roles in the process of mineralization; but different nutritional conditions select for very different populations. In the initial stages of decay of animal and plant tissues, generally in anaerobic environments, most organic constituents are first hydrolyzed and fermented; the products then diffuse to locations where they are ultimately oxidized, by a different group of organisms. Fatty compounds are too highly reduced to be readily fermented, and so their insoluble globules tend to be attacked by aerobes with a lipophilic surface, generally mycobacteria. Petroleum is formed when large amounts of organic matter decay below water; the hydrocarbons that remain after all the oxygen has been eliminated from the molecules represent the ultimate in the natural reduction of organic matter.

Many organisms have only a narrow range of foodstuffs, but some pseudomonads (a major group of soil scavengers) can use more than 100 carbon sources.

PHENOTYPIC ADAPTATION (SELECTION OF ENZYMES)

Not only do organisms differ genetically in their capacity to metabolize a given substrate, but the cells of a competent strain may or may not possess the enzymes required, depending on their immediate past history. This **adaptive enzyme formation** was first recognized by Karström in Finland in 1931; when a strain of lactic bacteria was grown on various sugars and then tested as "resting" cells (i.e., without a N source, to prevent further protein synthesis) most sugars could be fermented only by cells that had grown in their presence. However, the enzymes of glucose metabolism were **constitutive,** i.e. glucose could be fermented regardless of the C source present during growth.

The regulatory mechanisms of enzyme adaptation will be discussed in detail in Chapter 14. Meanwhile, as a background for the next few chapters, we should note

FIG. 3-7. Diauxic growth of *E. coli* on glucose and sorbitol in the proportions 1/3 (**A**), 2/2 (**B**), and 3/1 (**C**). Minimal medium; inoculum grown on glucose. (Monod J: *La croissance des cultures bactériennes.* Paris, Hermann, 1942)

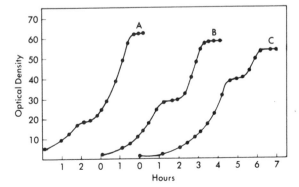

that this response not only is **stimulated** by the presence of the substrate but also may be **blocked** by the presence of an alternative food that supports faster growth. Thus with *Micrococcus denitrificans,* which can use either O_2 or nitrate as electron acceptor, Kluyver's group in Holland discovered in 1940 that nitrate induces, but O_2 blocks, formation of the enzyme connecting nitrate with electron transport. In 1942 Monod in France found a similar interference between alternative C sources, called **diauxie:** *E. coli* growing on glucose plus almost any other C source consumes all the glucose first and then, after a lag, resumes growth at the expense of the second source (Fig. 3-7). Further work on this phenomenon established *E. coli* as the prototype species, and induction of β-galactosidase by lactose as the prototype system, for studying gene regulation.

Such competition between alternative substrates also occurs within a pathway, where it accounts for the incomplete oxidations noted above. Thus *E. coli* growing aerobically on glucose will at first oxidize this substance only as far as acetic acid and CO_2, yielding about 40% of the possible ATP (Table 3-1); only after acetic acid is heavily accumulated will it induce the formation of enzymes required for its metabolism via the TCA cycle.

SELECTED READING

BOOKS AND REVIEW ARTICLES

BARKER HA: Bacterial Fermentations. New York, Wiley, 1957

BARTSCH RG: Bacterial cytochromes. Annu Rev Microbiol 22:181, 1968

BENEMANN JR, VALENTINE RC: High-energy electrons in bacteria. Adv Microbiol Physiol 5:135, 1971

BUCHANAN BB, ARNON DI: Ferredoxins: chemistry and function in photosynthesis, nitrogen fixation, and fermentative metabolism. Adv Enzymol 33:119, 1970

DOWNIE JA, GIBSON F, COX GB: Membrane adenosine triphosphatases of prokaryotic cells. Annu Rev Biochem 48:103, 1979

FRAENKEL DD, VINOPAL RT: Carbohydrate metabolism in bacteria. Annu Rev Microbiol 27:69, 1973

FRIDOVICH I: Oxygen: boon and bane, Am Sci 63(1):54, 1975

GIBSON F, COX GB: The use of mutants of *E. coli* K12 in studying electron transport and oxidative phosphorylation. Essays in Biochemistry 1974, p 1

GUNSALUS IC, STANIER RY (eds): The Bacteria, Vol II New York, Academic Press, 1961 Especially Cyclic Mechanisms of Terminal Oxidation (Krampitz), Survey of Microbial Electron Transport Mechanisms (Dolin), Cytochrome Systems in Aerobic Electron Transport (Smith), and Fermentation of Carbohydrates and Related Compounds (Wood)

HADDOCK BA, HAMILTON WA: Microbial Energetics. Society for General Microbiology Symposium 27. Cambridge University Press, 1977

HADDOCK BA, JONES CW: Bacterial respiration. Bacteriol Rev 41:47, 1977

KELLY DP: Autotrophy: concepts of lithotrophic bacteria and their organic metabolism. Annu Rev Microbiol 25:177, 1971

KLUYVER AJ, VAN NIEL CB: The Microbe's Contribution to Biology. Cambridge, Harvard University Press, 1956. A thought-provoking set of lectures, providing perspective on the unity of biology and the evolutionary aspects of microbiology

MANDELSTAM J, MCQUILLEN K (eds): Biochemistry of Bacterial Growth. New York, Wiley, 1973

MORRIS JG: The physiology of obligate anaerobiosis. Adv Microbial Physiol 12:169, 1975

MORTENSON LE: Nitrogen fixation: role of ferredoxin in anaerobic metabolism. Annu Rev Microbiol 17:115, 1963

ORNSTON LN, SOKATCH JR (eds): The Bacteria, Vol VI, Bacterial Diversity. New York, Academic Press, 1978 (Includes comprehensive articles on pathways for utilization of organic growth substrates, energy-yielding pathways, and bacterial photosynthesis)

YOCH DC, CARITHERS RP: Bacterial iron-sulfur proteins. Microbiol Rev 43:384, 1979

YOCH DC, VALENTINE RC: Ferredoxins and flavodoxins of bacteria. Annu Rev Microbiol 26:139, 1972

SPECIFIC ARTICLES

DICKERSON RE, TIMKOVICH R, ALMASSY RJ: The cytochrome fold and the evolution of bacterial energy metabolism. J Mol Biol 100:473, 1976

LEVITT MD, INGELFINGER FJ: Hydrogen and methane production in man. Ann NY Acad Sci 150:75, 1968

chapter

BIOSYNTHESIS

In contrast to the energy-yielding pathways, the biosynthetic pathways involve many endergonic steps; these could not be reproduced outside the cell until the role of ATP* as an energy coupler was understood. In addition, the large flow of material through the central pathways is subdivided into many much narrower biosynthetic pathways, branching from many different origins. These hundreds of reactions could not be dissected until two powerful new tools were developed: microbial mutants defective in specific biosynthetic enzymes, and radioactively labeled precursors.

This chapter will review the biosynthesis of small molecules very briefly, emphasizing fundamental principles and methodology. Most intermediates will be identified only by name; the reader is referred to biochemistry texts for their structural formulas. Later chapters will consider the synthesis of lipids (Ch. 6) and the polymerization of building blocks into polysaccharides (Ch. 6), nucleic acids (Ch. 10), and proteins (Ch. 13).

Studies in these areas, and in energy metabolism, have by now revealed a considerable fraction of the enzymes of *Escherichia coli*. The known enzymes number about 1000, while the total DNA would code for about 3000 proteins of ordinary size, including regulatory proteins as well as enzymes.

PRINCIPLES

MICROBIAL COMPOSITION AND THE UNITY OF BIOCHEMISTRY

The enormous diversity in the nutritional requirements of bacteria (Ch. 5), ranging from autotrophic growth on CO_2 to organic requirements even more complex than those of mammals, initially suggested a corre-

sponding diversity in metabolic complexity. However, improved analytic methods showed that the proteins, nucleic acids, and enzyme cofactors of the most varied cells are made of the same building blocks. Accordingly, cells differ not in their **essential metabolites** but in their ability to make these metabolites, i.e., *nutritional complexity reflects biosynthetic deficiencies*.

When the pathways of biosynthesis of the major building blocks were worked out they also were found to be uniform throughout the living world, with few exceptions. This finding not only reflects a common ancestry: it also suggests that the pathways that evolved early could not be further improved.

METHODS OF ANALYZING BIOSYNTHETIC PATHWAYS

Auxotrophic Mutants. Microbes offer several advantages in studies of biosynthesis: they use nearly all their energy for growth, rather than for maintenance and motion; the nutritionally simple strains have a full complement of biosynthetic pathways; and, above all, it is possible to isolate a great variety of **auxotrophic mutants,** i.e., strains that require a particular metabolite for growth because they have acquired a genetic defect in its synthesis. Beadle and Tatum first systematically isolated such mutants, in 1943, from the bread mold *Neurospora crassa*. With bacteria similar mutants have proved even easier to isolate and to study quantitatively (see Fig. 8-6).

Originally isolated for genetic studies, auxotrophic mutants proved useful also for the identification of biosynthetic intermediates: these are normally present in cells only in trace amounts, but most auxotrophic mutants accumulate and excrete the substrate of the blocked reaction in large amounts, often exceeding the weight of the cells. In addition, the intermediate in question can often support the growth of other mutants blocked earlier in the same pathway. Hence two mutants that are auxotrophic for the same endproduct can often exhibit unidirectional **crossfeeding** (Fig. 4-1). Once some intermediates in a pathway are available the enzymes that connect them can be characterized; additional enzymes and intermediates can then be identified by **enzyme fractionation.**

Radioactive Isotopes. These have also been extremely useful, in several applications:

*In this chapter the same abbreviations as in Chapter 3 are used. In addition, DAP = diaminopimelate; Me = methyl; P-enolpyruvate, or PEP = phosphoenolpyruvate; PP = pyrophosphate; PRPP = 5-phosphoribosyl-1-pyrophosphate; TCA = tricarboxylic acid; THF = tetrahydrofolate; UMP, CMP, GMP, IMP, and AMP = the ribonucleotide (monophosphate) of uracil, cytosine, guanine, inosine, and adenine, respectively; dUMP, etc. = deoxyuridine monophosphate, etc.; dTTP, etc. = deoxythymidine triphosphate, etc.; UDP, etc. = uridine diphosphate, etc.; UTP, etc. = uridine triphosphate, etc.; UDPG = uridine diphosphate glucose.

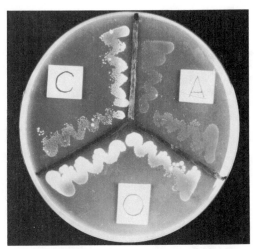

FIG. 4-1. Cross feeding between different bacterial mutants blocked in the same pathway.

Note that a streak of mutant A, growing slightly on medium containing a trace of arginine, stimulates the growth of adjacent streaks of mutants C and O, while C similarly stimulates O but not A. Such growth responses can be used for bioassays (e.g., disc assays on a plate heavily seeded with the responding mutant); these have been very useful in the isolation of unknown intermediates. However, some accumulated intermediates cannot be detected by nutritional tests because they cannot penetrate the cells. (Davis BD: Experientia 6:41, 1950)

In **precursor competition** a labeled possible precursor, such as glycine, is supplied along with an unlabeled general carbon source (e.g., glucose); when compounds subsequently recovered from the cells by hydrolysis are heavily labeled (e.g., purines), glycine has served as a direct precursor. Another approach is **pulse-labeling.** For example, the pentose cycle for the autotrophic fixation of CO_2 was largely revealed by growing cells for a few seconds on highly radioactive CO_2, followed by analysis, at intervals of a few seconds, of the flow of the radioactivity into successive compounds.

The organism can also be supplied with a single carbon source (e.g., glucose) **selectively labeled in a specific atom;** various end-products are then isolated and are degraded in a way that permits determination of the isotope concentration (specific activity) in individual atoms. The results indicate where a given biosynthetic pathway branches off from a known central route. The coherence of the several kinds of evidence (mutant, enzymatic, and isotopic) has firmly established the pathways.

Finally, in the extension of the study of biosynthesis to complex macromolecules the amounts formed in vitro have generally been too small to detect by ordinary analytic methods. To measure these reactions radioactive precursors are incorporated into polymers, which are then precipitated by reagents (e.g., trichloroacetic acid) that do not precipitate the substrates.

Criteria for a Biosynthetic Intermediate. By definition, a substance that is incorporated serves as a precursor. But it does not necessarily follow that the compound serves in the cell as an intermediate in growth on a general nutri-

ent. The distinction is significant for cell physiology: while most precursors have also turned out to be intermediates, there are exceptions.

For example, histidine provided in the medium induces *Enterobacter* to form degradative enzymes that convert its 5-C chain to α-ketoglutarate; hence it can serve as a source of glutamate. However, the level of histidine formed endogenously in the cell is not sufficient to induce its own degradation, and histidine biosynthesis has no connection with normal glutamate biosynthesis.

These considerations have led to the formulation of the following criteria: A and B are obligatory intermediates in the biosynthesis of product X

$$\rightarrow \ \rightarrow A \rightarrow B \rightarrow \ \rightarrow X$$

if 1) compounds A and B can each give rise to X (in the cell or in extracts), 2) a single enzyme converts A to B, and 3) loss of that enzyme (e.g., by mutation) results in a requirement for X.

These criteria exclude, for example, free purines and pyrimidines as normal intermediates, even though they may be excreted by certain mutants and used by others: the bases are synthesized at the level of nucleotides (see below).

AMPHIBOLIC PATHWAYS

As the various biosynthetic pathways were worked out they were found to branch off from various intermediates in the major energy-yielding pathways: glycolysis, pyruvate oxidation, and the TCA (Krebs) cycle (Fig. 4-2). Accordingly, these three pathways, though initially considered purely catabolic, are just as directly involved in biosynthesis, and so they are now designated as **central** or **amphibolic** (Gr. *amphi,* either) pathways.

Conversion of Amphibolic to Biosynthetic Pathways. Under various nutritional conditions, which are no less "normal" than aerobic growth on glucose, parts of the amphibolic pathways become purely biosynthetic. In particular, any C source that enters metabolism via the TCA pathway will require net flow into a reversed glycolytic pathway, to keep pace with the biosynthetic drains on various glycolytic intermediates. The reactions that connect these two pathways will be discussed below (Fig. 4-5). In mammals, of course, metabolic flow patterns are similarly affected by the use of different fuels (e.g., gluconeogenesis).

For example, a cell growing aerobically on succinate (Fig. 4-3) oxidizes it to oxaloacetate, part of which enters biosynthesis, both directly and via the TCA cycle; the rest is converted into pyruvate or P-enolpyruvate, which are used for biosynthesis (partly via reversed glycolysis) and for terminal respiration. Similarly, when cells grow on glycerol, which is readily converted to triose-P, the lower part of the glycolytic pathway is used amphibolically and the upper part, in reverse, biosynthetically. Finally, when facultative organisms shift from aerobic to anaerobic conditions the TCA cycle no longer functions as a cycle. It is then converted into a pair of biosynthetic pathways, branching from oxaloacetate: a portion of the original cycle continues to form α-ketoglutarate, while the

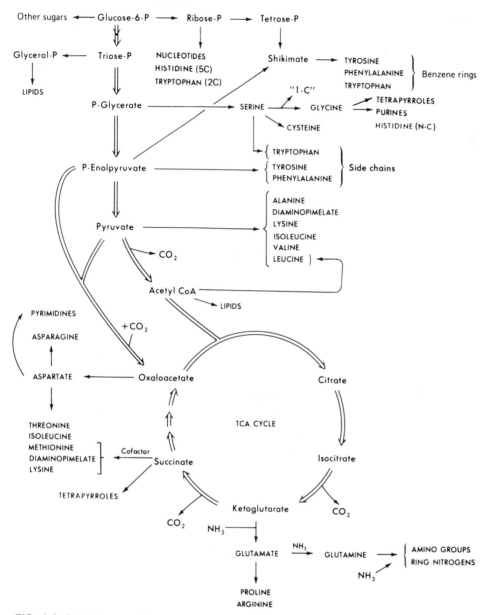

FIG. 4-2. Relation of amphibolic pathways (**heavy arrows**) to main anabolic pathways. End products are in capitals. "1-C" is a 1-carbon fragment.

usual pathway from succinate to oxaloacetate is reversed (by the induction of new enzymes) to form a reductive branch (Fig. 4-4). α-Ketoglutarate oxidase, which would be a useless enzyme, disappears.

Flow Between Glycolytic and TCA Pathways. The 3-C and the 4-C central intermediates are interconverted by carboxylation or decarboxylation reactions that differ in different organisms and under different conditions (Fig. 4-5). For example, malate can be oxidatively decarboxylated to pyruvate either directly (generating NADPH) or via oxaloacetate (generating NADH): the proportion

depends on the cell's need for NADPH (for biosynthetic reductions) or for NADH (for electron transport). Similarly, oxaloacetate can be converted either to P-enolpyruvate, to initiate reversed glycolysis, or to pyruvate, leading to terminal respiration.

BIOSYNTHESIS FROM 2-C COMPOUNDS: THE GLYOXYLATE CYCLE

The central pathways reviewed above can account for the ability of organisms to grow on many compounds.

FIG. 4-3. Pathway of metabolic flow in aerobic growth on succinate. The glycolytic intermediates would be included in the **box**, but are not specified.

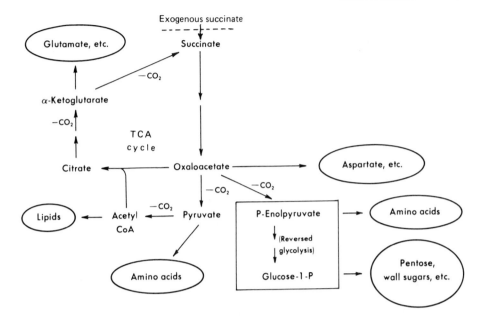

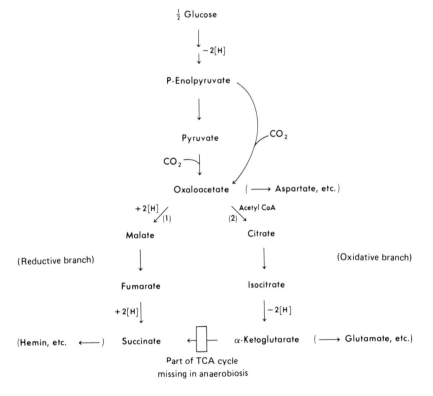

FIG. 4-4. Biosynthesis of TCA intermediates in anaerobic growth on glucose in *E. coli*. The TCA cycle becomes split into a reversed reductive branch and a normal oxidative branch.

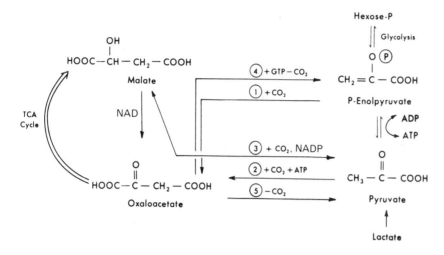

FIG. 4-5. Carboxylation and decarboxylation reactions connecting the glycolytic and the TCA pathways under various conditions and in different organisms. When the carbon source is glucose carboxylation is by reaction 1 in *E. coli* (using energy of PEP [P-enolpyruvate]); by reaction 2 in *Pseudomonas* (using energy of ATP). When it is lactate carboxylation is by reaction 2. When the carbon source is a TCA component (such as succinate) decarboxylation to high-energy PEP occurs by reaction 4 (using energy of GTP), for biosynthetic needs, and also by reactions 3 and 5 to pyruvate, for amphibolic purposes and to generate NADPH and NADH.

However, the aerobic organisms that can oxidize acetate do not have a system with enough reducing power to reverse the oxidative decarboxylation of pyruvate, yet they must replenish the biosynthetic drain on the central pathways. Indeed, depletion of TCA cycle intermediates is the cause of the acidosis of a diabetic mammal forced to consume fat (via acetyl CoA) without carbohydrate. Nevertheless, many aerobic bacteria can thrive on acetate (or on lipids that are metabolized via acetyl CoA). The mechanism, discovered by H. Kornberg, is a bypass or epicycle on the TCA cycle, involving reactions 1 and 2:

$$\underset{\underset{COOH}{|}}{CH_2}\text{---}\underset{\underset{COOH}{|}}{CH}\text{---}\underset{\underset{COOH}{|}}{CHOH} \xrightarrow[\text{(Isocitrate lyase)}]{\text{Isocitratase}}$$

$$\underset{\underset{COOH}{|}}{CH_2}\text{---}\underset{\underset{COOH}{|}}{CH_2} + \underset{\underset{COOH}{|}}{CHO} \quad (1)$$
$$\qquad\qquad\text{Succinate}\qquad\text{Glyoxylate}$$

$$\underset{\underset{COOH}{|}}{CHO} + \underset{\underset{COOH}{|}}{CH_3} \xrightarrow[\text{synthetase}]{\text{Malic}} \underset{\underset{COOH}{|}}{HOCH}\text{---}\underset{\underset{COOH}{|}}{CH_2} \quad (2)$$
$$\text{Glyoxylate}\quad\text{Acetate}\qquad\qquad\text{Malate}$$

$$\text{Succinate + Acetate} \xrightarrow[\substack{\text{(TCA cycle} \\ \text{reactions)}}]{-4(H)} \text{Isocitrate} \quad (3)$$

$$\text{Net: 2 Acetate} \xrightarrow{-4(H)} \text{Malate}$$

The succinate and malate formed in reactions 1 and 2 can be used to regenerate isocitrate through part of the TCA cycle (summed as reaction 3), yielding a glyoxylate cycle that uses acetate for net synthesis of 4-C compounds (Fig. 4-6). An organism growing on acetate will funnel part of the supply through this cycle and part through the TCA cycle.

In organisms that can use the glyoxylate bypass its formation is repressed by the simultaneous supply of a more rapidly used substrate, such as glucose or succinate.

Ferredoxin and Reductive Carboxylation. Though aerobes and facultative organisms cannot reductively carboxylate acetate to pyruvate, photosynthetic bacteria and certain obligate anaerobes have a ferredoxin (Ch. 3) with a low enough redox potential to catalyze this reaction; hence they do not employ the glyoxylate cycle. Similarly, some anaerobes (including bacteroides of the human gut) have no TCA cycle reactions at all but proceed from pyruvate to oxalacetate, succinate, and α-ketoglutarate by reductive carboxylation.

PROMOTION OF UNDIRECTIONAL FLOW

Reversible and Irreversible Reactions. As we have noted, in the cell some sequences (e.g., glycolysis) may exhibit net flow in either direction, depending on circumstances. For some steps in this reversal the same enzyme appears to be used reversibly. For others, however, cells have developed a somewhat different reverse reaction, presumably because the equilibrium of the first reaction is unfavorable.

For example, aldolase catalyzes either the conversion of triose-P to hexose-diP or the reverse, depending on whether the C source is glycerol or glucose. On the other hand, the condensation of acetate and oxaloacetate to citrate, though reversible, has an equilibrium that is not useful in both directions: the "clockwise" flow of the TCA cycle is favored by the release of acyl CoA energy in the reaction of a citrate synthase:

$$\text{Acetyl CoA + Oxaloacetate} \rightleftharpoons \text{Citrate + CoA}$$

while organisms fermenting citrate start with its direct hydrolysis:

$$\underset{\underset{CH_2\text{---}COOH}{|}}{\overset{\overset{CH_2\text{---}COOH}{|}}{HO\text{---}C\text{---}COOH}} \xrightarrow{\text{Citratase}} \begin{array}{l} CH_3\text{---}COOH \\ \text{Acetate} \\ + \\ O{=}C\text{---}COOH \\ \quad | \\ \quad CH_2\text{---}COOH \end{array}$$
$$\text{Citrate}\qquad\qquad\qquad\text{Oxaloacetate}$$

FIG. 4-6. Krebs TCA cycle, and within it the "glyoxylate bypass" (**heavy arrows**). The glyoxylate cycle substitutes these reactions, which conserve carbon, for the two decarboxylative reactions of the TCA cycle (**dashed arrows**), which release carbon as CO_2. In an organism growing on acetate alone the glyoxylate cycle provides net 4-C (and thus 3-C) synthesis to replenish the biosynthetic drain (**circled compounds**). Since the glyoxylate cycle involves two oxidative steps, linked to electron transport, it also provides some energy; but most of the cell's energy is derived from the simultaneous oxidation of other acetate molecules via the TCA cycle.

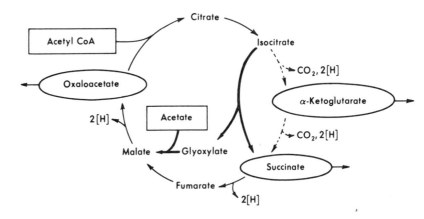

The oxaloacetate can then be fermented by dismutation, some molecules being oxidized via pyruvate and others reduced to fumarate.

Though many reactions in biosynthetic pathways are reversible, most of the sequences include a virtually irreversible reaction, usually early and usually ATP-linked. This reaction functions rather like the initial power-driven climb in a rollercoaster.

Flow in the desired direction is also promoted by the **release of pyrophosphate** (PP) rather than inorganic phosphate (P_i) from ATP, or from an intermediate such as PRPP. The standard free-energy drop (i.e., at unit concentration of all components) is essentially the same for either of these hydrolyses of a P–O–P bond; but since the cell has a substantial P concentration, whereas PP is rapidly removed by a pyrophosphatase, the actual free-energy drop in the overall reaction is greater for PP release: with PP hydrolysis the overall energy expenditure is two high-energy bonds rather than one. On the other hand, some PP bond energy may be conserved by the reversible transfer of P from PP into polyphosphate.

AMINO ACIDS

The amino acids are conveniently grouped, on the basis of their points of origin from central pathways, into a few biosynthetic families (Fig. 4-2). Though these pathways were worked out largely in bacteria, the reactions are, with few exceptions, universal. The amino acids of proteins all have the L configuration, which will not be specified in the following descriptions.

D-Amino acids, which are present in bacterial cell wall polypeptides, capsules, and antibiotics, are synthesized by a racemase that acts on alanine, plus a special transaminase that can transfer the D-amino group from D-alanine to various keto acids.

THE GLUTAMATE FAMILY

Glutamate and glutamine are formed by reductive amination of α-ketoglutarate, by two different pathways.

These are depicted, along with their regulation, in Chapter 14, under Regulation of Nitrogen Assimilation.

The NH_3 may be supplied in the medium as such, or it may be derived by degradation of nitrogenous organic compounds or, with certain organisms, by reduction of NO_3^- or fixation of N_2.

Glutamate and glutamine not only are constituents of proteins; glutamate serves as the source, by transamination via pyridoxal-P, of the α-amino group of all other amino acids; the amide group of glutamine provides N, at a higher energy level, in the biosynthesis of many other compounds; and glutamate provide the C skeleton of **proline** and **arginine,** as follows:

$$\text{Glutamate} \xrightarrow[\text{ATP}]{\text{NADPH}} \text{Glutamic } \gamma\text{-semialdehyde} \xrightarrow{\text{Spont.}}$$

$$\Delta'\text{-Pyrroline-5-carboxylate} \xrightarrow{\text{NADPH}} \text{Proline.}$$

$$\text{Glutamate} \rightarrow \text{N-Acetylglutamate} \xrightarrow[\text{ATP}]{\text{NADPH}} \text{N-Acetylglutamate}$$

$$\gamma\text{-semialdehyde} \rightarrow \text{N}^\alpha\text{-Acetylornithine} \rightarrow \text{Ornithine} \xrightarrow{\text{Carbamyl-P}}$$

$$\text{Citrulline} \xrightarrow{\text{Aspartate}} \text{Argininosuccinate} \xrightarrow{-\text{Fumarate}} \text{Arginine.}$$

Arginine in turn gives rise to **polyamines** (see below).

In **proline biosynthesis** a γ-aldehyde spontaneously condenses with the NH_2 group, much as in its reaction with an OH group in the formation of the furanose ring in sugars. In arginine biosynthesis this cyclization would interfere with the reduction of the open chain, and it is prevented by first blocking the α-NH_2 group of glutamate with an acetyl group, just as one might do in a laboratory synthesis.

The **carbamyl-P** (NH_2COOP) used for building up the guanidino group in arginine is formed from the amide of glutamine, CO_2, and the terminal P of ATP. The same reactions, plus arginase to release urea and regenerate ornithine, are found in the urea cycle in mammalian liver.

THE ASPARTATE FAMILY

Aspartate is synthesized by transamination of oxaloacetate. It gives rise to several other amino acids, by the se-

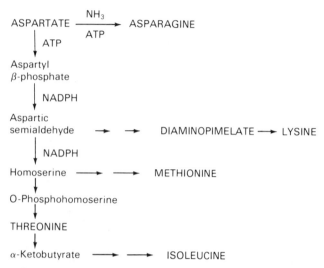

FIG. 4-7. Biosynthesis of the aspartate family. Building blocks (endproducts) are in capital letters.

quences in Figure 4-7. Aspartate is also a direct precursor of the pyrimidine ring (see Nucleotides, below), nicotinamide (a component of NAD), and N atoms in the purines and in the guanidino group of arginine.

Methionine is synthesized via cystathionine, a thioether formed by substituting the thiol group of cysteine for an acyl group (acetyl or succinyl) in *O*-acylhomoserine. Cleavage of this compound transfers the S from the 3-C to the 4-C moiety; the resulting homocysteine is then methylated to yield methionine.

The methyl group is transferred via cobamide in *Enterobacter,* but the closely related *Escherichia coli* cannot make cobamide and uses 5-methyltetrahydrofolate; curiously, when this path is blocked *E. coli* can use cobamide (which it can form if provided with vitamin B_{12}).

Methionine links many processes in metabolism via **S-adenosyl methionine,** a derivative in which the energy of the $S–CH_3$ bond is increased by conversion to the sulfonium (R_3S^+) ion. This intermediate provides the methyl group for various methylations, which yield modified bases in DNA and RNA and the cyclopropane ring in some fatty acids. It also provides the $–(CH_2)_3NH_2$ portion of **spermidine.**

Synthesis of **diaminopimelate** (DAP; $HOOC–CHNH_2–(CH_2)_3–CHNH_2–COOH$) begins with condensation of aspartic semialdehyde with pyruvate; the product spontaneously cyclizes to form dihydrodipicolinate:

$$HOOC \underset{\text{N}}{\bigcirc} COOH$$

Dihydrodipicolinic
acid

After reduction to tetrahydrodipicolinate the N = C is spontaneously hydrolyzed, and the open-chain form is stabilized by a covering of the amino group with a succinyl or acetyl group, which is

removed after transamination has yielded the L,L-DAP chain. Isomerization then yields *meso*-DAP. Some organisms incorporate L,L-DAP and others *meso*-DAP in their cell wall (Ch. 6).

L-Lysine is synthesized by decarboxylation of *meso*-DAP. A mutant with a block preceding DAP synthesis cannot grow for lack of both DAP and lysine; but if given lysine it will lyse, because it continues to synthesize protein but not wall. This property has been used in constructing enfeebled host strains for research with recombinant DNA (Ch. 12).

Lysine is the only amino acid for which **quite different biosynthetic routes** have been found in different organisms. In fungi, whose walls do not contain DAP, it is synthesized by a route analogous to the citric acid pathway to glutamate, but with chains one C longer. The route starts with the condensation of acetyl CoA with α-ketoglutarate and yields α-aminoadipate ($HOOC–CHNH_2–(CH_2)_3–COOH$). The terminal COOH of this compound is reduced and then transaminated, yielding lysine.

THE PYRUVATE FAMILY: ALIPHATIC AMINO ACIDS

Alanine is derived in most bacteria by transamination of pyruvate (but in some bacilli by direct reductive amination). It also serves as the starting point for formation of the longer-chain aliphatic amino acids: **valine, isoleucine,** and **leucine.** These pathways are unusual in having a single set of enzymes that catalyze two parallel sequences, whose substrates differ by a $–CH_2–$ group (Fig. 4-8).

Leucine formation involves elongation of the chain of valine by one C, through addition of an acetyl group to "ketovaline," followed by rearrangement of the C skeleton and then oxidative decarboxylation.

The C skeleton of valine also takes up a hydroxymethyl group, via tetrahydrofolate, to yield pantoate, a component of CoA. Acetolactate, a precursor of valine, is also an intermediate preceding acetoin in the fermentation of *Enterobacter* (see Fig. 3-2, Ch. 3). The regulation of the biosynthetic and the degradative acetolactate synthetase will be discussed in Chapter 14 (Branched Pathways).

THE SERINE FAMILY: 1-C FRAGMENTS

Serine is formed by the reactions:

3-P-Glycerate → 3-P-Hydroxypyruvate → 3-P-Serine → Serine

Serine in turn can transfer CH_2OH to tetrahydrofolate (THF), giving rise to glycine plus a 1-C fragment (hydroxymethyl-THF). In addition, the glycine can react with THF to yield a second molecule of $HOCH_2–THF$, plus CO_2 and NH_3.

All these reactions are reversible: hence the cell can make serine from glycine, as well as the reverse; and it can make the two compounds (and 1-C fragments) in whatever proportions are needed.

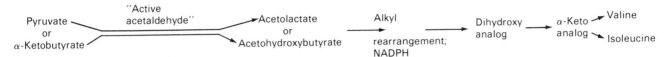

FIG. 4-8. Pathway to valine and isoleucine.

Cysteine is formed from serine by the reactions:

$$\text{Serine} \xrightarrow{\quad} \textit{O}\text{-Acetylserine} \xrightarrow{\text{H}_2\text{S}} \text{Cysteine} + \text{Acetate}$$

In the usual media the H_2S is formed by **reduction of sulfate,** which is first activated by the use of two ATP molecules to yield 3'-phosphoadenosine-5'-phosphosulfate:

$$\text{Adenine-ribose-3'-P'5'-P-OSO}_3\text{H}$$

1-C Transfer. The hydroxymethyl-THF derived from serine or glycine is used in a variety of biosynthetic reactions: it can be isomerized to 5,10-methylene-THF, reduced to 5-Me-THF, or oxidized to 5,10-methenyl-THF and its isomer 1-formyl-THF. At these various levels of oxidation it provides the methyl group of methionine or of thymine, the formyl group of formyl-Met-tRNA (see Initiation, Ch. 13), or closure of the two rings in purines. Inhibition of THF synthesis (by sulfonamides), or of its function (by trimethoprim), are important chemotherapeutic mechanisms.

HISTIDINE

This amino acid derives its 5-C backbone from PRPP. This compound first condenses at C-1 with N-1 of ATP, which donates that N and the adjacent C to the chain. (The residue of the ATP, after the N–C donation, is a normal intermediate in purine biosynthesis and hence is recycled.) After addition of an NH_2 group from glutamine the ring is closed with C-2 of the 5-C backbone to yield imidazole-glycerol-P. The two OH groups in the side chain are rearranged to form a keto group, which is transaminated; dephosphorylation then yields the aminoalcohol L-histidinol; and oxidation of the terminal CH_2OH to COOH yields histidine. Histidine is unique among amino acids in having its COOH formed late, rather than present in its initiating precursors.

AROMATIC COMPOUNDS

The key to this pathway was provided by the finding that certain mutants of *E. coli* require multiple aromatic compounds and can satisfy this requirement with **shikimic acid,*** while other similar mutants accumulate this hydroaromatic compound:

*This compound was first isolated from fruits of the shikimi tree in Japan, like malic acid from apples (L. *malus,* apple) and citric acid from citrus fruits. It is not clear why certain plants accumulate large amounts of various biosynthetic intermediates.

In this pathway erythrose-4-P condenses with P-enolpyruvate to form the 7-C compound 3-deoxyarabinoheptulosonic acid-7-P, which is cyclized, dephosphorylated, and dehydrogenated to yield **shikimate,** with one double bond in the ring. Another double bond is introduced by dehydration, and an enolpyruvyl group is attached, to form the main branch compound, **chorismic acid** (Gr., fork):

Shikimic Acid

Chorismic Acid

This compound replaces its enolether linkage with an attachment of its CH_2 group to C-1 of the ring, with a shift of the double bonds, to yield an intermediate whose decarboxylation provides the ketoacid precursors of **tyrosine** and **phenylalanine.** Other reactions, in which chorismic acid releases the enolpyruvyl group, yield **anthranilate** (→ tryptophan), **p-aminobenzoate, p-hydroxybenzoate** (→ **benzoquinones**), and 3,4-**dihydroxybenzoate** (→ **naphthoquinones; enterochelin:** cf. Ch. 5).

In the synthesis of **tryptophan,** anthranilate (*o*-aminobenzoate) adds C-1 of a 5-P-ribosyl group to its NH_2, and displacement of the COOH by C-2 of the ribose forms an imidazole ring, yielding indole-glycerol-P. The residual 3-C side chain exchanges with serine to yield tryptophan.

POLYAMINES

Two polyamines constitute half the total product of the arginine pathway in *E. coli.* The diamine **putrescine** (1,-4-diaminobutane) is made by the decarboxylation of ornithine, and also by the decarboxylation of arginine followed by conversion of the guanidino to an amino group. The triamine **spermidine** is made by addition to putrescine of a $-CH_2-CH_2-CH_2-NH_2$ group from S-adenosylmethionine. The tetramine **spermine,** present in eukaryotic cell nuclei, has not been found in bacteria, but it can be taken up from the medium, replacing endogenous spermidine.

Auxotrophic mutants, so useful for the study of other metabolites, could not be obtained for polyamines until the presence of a double pathway was recognized. This unusual feature promotes economical synthesis, either from exogenous arginine, when available, or from a simpler intermediate when endogenous synthesis is required.

Polyamines bind readily to DNA and to RNA: their dimensions are suitable for forming ionic crosslinks between opposite phosphates in a double helix, and they increase resistance to strand-separating agents. Moreover, polyamines can be found in preparations of DNA, ribosomes, and tRNA. Their main location in bacterial cells is probably in ribosomes, since their total concentration roughly parallels the ribosome content under different growth conditions. However, the distribution and the functions of polyamines in the cell have been difficult to determine in detail, for these compounds are readily replaced by Mg^{2+}, in various functions as well as in binding.

In particular, protein synthesis in extracts (and the integrity of the ribosomes, Ch. 13) is highly dependent on polyvalent cation concentrations. But while this requirement can be satisfied by polyamines or by a somewhat higher concentration of Mg^{2+}, optimal accuracy requires a mixture of both. Moreover, the polyamines evidently have a specific effect on the conformation of tRNA, since its methylation in vitro does not achieve the correct pattern in the presence of Mg^{2+} alone but requires polyamines.

The best-defined function of polyamines is not in cells but in viruses. A large fraction of the charges on viral nucleic acid (DNA or RNA) are neutralized by polyamines, whose crosslinking evidently promotes the required tight packing of the nucleic acid in the viral coat.

Polyamines are required for, or promote, the growth of some bacteria (*Pasteurella, Mycoplasma*). However, this effect has not thrown light on their function in the cell, for they can be replaced by other, inorganic polycations, which evidently function from the outside to preserve the integrity of the cell membrane.

It seems likely that in cells polyamines serve to optimize and regulate a number of functions of nucleic acids. Their different affinities in reversibly crosslinking strands could provide fine control over the opening and closing of double-stranded regions.

NUCLEOTIDES

Purine and pyrimidine nucleotides serve several functions: They are 1) building blocks of nucleic acids; 2) components of many coenzymes; 3) activators for the transfer and transformations of sugars, wall peptides, and complex lipids; 4) covalently added modifiers of enzymes; and 5) constituents of some antibiotics. The pathways to the nucleotides were largely elucidated in

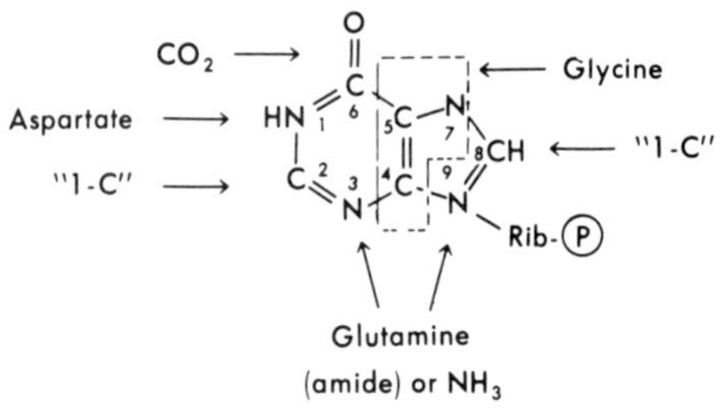

FIG. 4-9. Origin of purine ring atoms.

animal tissues, by the use of isotopic and enzymatic methods.

Purine nucleotides are built up on the ribose-P chain from PRPP, glycine, and single C and N additions in a long, highly endergonic sequence requiring at least three ~P. The origin of the various ring atoms is summarized in Figure 4-9.

In this pathway the ring is completed as hypoxanthine ribonucleotide (inosinic acid, IMP), which has a 6-keto group. This intermediate is converted to adenylate (AMP) by replacing the keto by an amino group (from aspartate), and to guanylate (GMP) by retaining the 6-keto group and adding a 2-amino group (from NH_3). Interestingly, these reactions do not depend for energy on their own products: AMP synthesis requires GTP, and GMP synthesis requires ATP.

Pyrimidine nucleotides are made via a series of carboxyl-containing intermediates; unlike purines, the ribose-P is added late.

The pathway starts with carbamyl-P (also used in arginine synthesis: see The Glutamate Family, above), adding the carbamyl group to aspartate; ring closure and dehydrogenation form a pyrimidine, **orotic acid**:

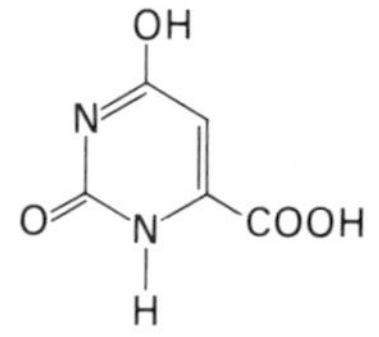

Addition of ribose-P and decarboxylation then yield uridylate (UMP), which forms cytidylate (CMP) by adding an NH_2 group.

Pentoses. Ribose-5-P can be formed in *E. coli* oxidatively, from glucose-6-P via 6-P-gluconate, or nonoxidatively. In the later process a sequence of transfers from sugar phosphates forms a heptose-P, from which transfer of a 2-C fragment to triose-P yields two pentose-P. Since the oxidative pathway generates NADPH from NADP, while the second does not, the presence of both permits the supply of this biosynthetic reductant to be adjusted to the cell's needs under various conditions.

As noted above, ribose-5-P enters nucleotide biosynthesis from its 1-pyrophosphate derivative, PRPP.

Deoxyribose residues are formed by NADPH-linked reduction of the ribose of ribonucleotides.

The substrates for the reduction are the nucleoside **triphosphates** in *Lactobacillus* and the **diphosphates** in *E. coli*. The reduction involves a derivative of vitamin B_{12} in the former organism, whose normal habitat (milk) contains that vitamin, while *E. coli*, which often grows in a simple environment, has evolved quite a different pathway that uses instead a novel small-protein factor, **thioredoxin.** In this compound a pair of –SH groups on adjacent cysteines undergo reversible oxidation to an –S–S group.

The presence of thymine (5-Me-uracil) instead of uracil in DNA presents a problem. A route like that to the other deoxyribonucleotides would convert UDP to dUDP; but this compound would easily generate dUTP and thus incorporate U into DNA. Instead, dTTP is formed from dCMP, with deamination and methylation of the ring, by a sequence that avoids dUDP:

$$dCDP \rightarrow dCMP \rightarrow dUMP \rightarrow dTMP \rightarrow dTDP \rightarrow dTTP$$

As further protection against error, *E. coli* has a specific dephosphorylase for dUTP, and also an N-glycosidase that cleaves any dU that may be inserted in DNA.

Salvage Pathways. Many bacteria, like mammalian cells, can use exogenously supplied free purines and pyrimidines, or their nucleosides or deoxyribonucleosides, as sources of nucleotides. The wide distribution of these "salvage" pathways (which convert added bases into nucleotide intermediates) probably reflects the presence of nucleosides and free bases wherever cells are degraded, and the ready penetration of these products (unlike the corresponding nucleotides) into cells.

OTHER PATHWAYS

In contrast to the informational molecules (nucleic acids, proteins), which are built of the same building blocks in all organisms, *the components of the structural lipids and carbohydrates vary widely.* Since lipids in bacteria are present only in membranes their synthesis will be described in Chapter 6.

SUGARS

Free sugars are not metabolic intermediates within the bacterial (or the mammalian) cell: they are phosphorylated during or immediately after entry. Conversion of glucose into some other sugars (e.g., pentoses) takes place at this phosphorylated level. However, most conversions of sugars occur at the level of **sugar nucleotides (nucleoside diphosphate sugars)**. These are formed by specific pyrophosphorylases (named after the reverse of the biosynthetic reaction):

$$\text{Nucleoside-P-P-P} + \text{Hexose-1-P} \rightleftharpoons$$
$$\text{Nucleoside-P-P-hexose} + \text{P-P}$$

The utilization of such a nucleoside diphosphate "handle" in the conversion of glucose to galactose (and vice versa) is depicted in Figure 4-10. Other conversions include amination, reduction to deoxysugars, or oxidation to uronic acids. Nucleotide sugars are also the donors in transglycosylation reactions in the formation of polysaccharides.

Intermediates in these pathways, as in amino acid biosynthesis, have often been discovered through their accumulation by mutants blocked in the synthesis or transfer of a particular sugar. Indeed, such mutants often accumulate not only the precursor of the blocked reaction but also nucleotides of various other sugars, for the absence of one component prevents the incorporation of any others into a repeating polysaccharide. Figure 4-11 outlines some pathways that have been worked out in this way. The use of different nucleotide carriers may decrease the interference between pathways to different sugars.

POLYSACCHARIDES

Capsules. Most bacterial capsules are repeating heteropolymers (i.e., they contain more than one kind of monomer), formed from sugar nucleotides by transglycosylation reactions. Examples will be given in Chapter 27. In their synthesis each repeating subunit is built upon a membrane lipid, which then transfers it across the cytoplasmic membrane, as will be described in Chapter 6 for wall peptidoglycan and for outer membrane lipopolysaccharides.

Certain lactic bacteria (e.g., *Leuconostoc*) form a homopolymer from exogenous sucrose (glucose-1,2-fructoside) by a simple reaction in which the energy of the glycoside bond is used to polymerize one of the two residues. Some species release and absorb the fructose moiety and polymerize the glucose (= dextrose) to a high-molecular-weight **dextran**; others free the glucose and polymerize the fructose (levulose) to a **levan**.

$$\text{n Sucrose} \longrightarrow (\text{Glucose})_n + \text{n Fructose}$$
$$or$$
$$\text{n Glucose} + (\text{Fructose})_n$$

In these polymers the predominant linkage is α-1,6, in contrast to the α-1,4 of starch or glycogen.

The polymerization is catalyzed by an enzyme on the outer surface of the cell. The polymer promotes adhesion of bacteria to surfaces—a major mechanism by which sucrose promotes dental caries (Ch. 44).

Cellulose (poly-β-1,4-glucose), indistinguishable from that of higher plants, is formed by some acetobacters, obligate aerobes that oxidize ethanol to acetate. It appears to be excreted as separate macromolecules; on reaching the exterior it crystallizes into a mat. By trapping both cells and bubbles of CO_2 the mat floats the cells to the surface, which is obviously advantageous for obligate aerobes.

Glycogen (poly-α-1,4-glucose) is stored by many bacteria. It is formed, as in mammals, from UDPG (Fig. 4-10A), and is utilized via phosphorolysis:

$$\text{Glycogen} + nP_i \rightarrow \text{n Glucose-1-P}$$

A   Biosynthetic formation of galactose from glucose

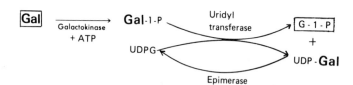

D-Glucose-1-P (G-1-P)   Uridine diphosphate   Uridine diphosphate
                                          glucose (UDPG)   galactose (UDP Gal)

B   Utilization of galactose (Gal) as carbon source

Gal →(Galactokinase + ATP)→ **Gal**-1-P →(Uridyl transferase)→ G-1-P + UDP-**Gal**, with UDPG and Epimerase

FIG. 4-10. Epimerization of galactose and glucose. The anabolic sequence from glucose to galactose (**A**) evidently employs the same isomerase as the pathway for the amphibolic utilization of galactose (**B**), for a *gal⁻* mutant of *E. coli* that lacks this enzyme has lost the ability both to use galactose as a carbon source and to make a galactose-containing wall polysaccharide from glucose. (These two sequences are also found in mammalian cells.)

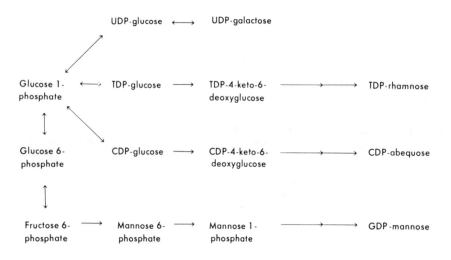

FIG. 4-11. Biosynthesis of the precursors of a specific wall polysaccharide (O ag) of group B *Salmonella*. (Nikaido H et al: J Bacteriol 91:1126, 1966)

The highly varied **wall polysaccharides** will be considered in Chapter 6.

POLYPEPTIDES

Homopolymeric Polypeptides. The synthesis of amino acid homopolymers, and of short peptides (e.g., the tripeptide glutathione; wall peptides [Ch. 6]), employs a simple mechanism, in which each successive monomer is recognized by the appropriate enzyme. Each addition costs only one ~P, from ATP, whereas protein synthesis, with an elaborate template mechanism for synthesizing an unlimited variety of sequences and lengths, costs more than four ~P per residue.

Some species of *Bacillus* form a capsule of poly-D-glutamic acid, or a mixture of poly-D- and poly-L-glutamic acid; in the anthrax bacillus this capsule is essential for virulence. The residues are linked by a γ-peptide bond (in contrast to the α-peptide of proteins), and the molecular weight may reach 250,000.

Peptide antibiotics (10–20 residues) are synthesized by a mechanism intermediate in complexity between protein synthesis and homopolymer synthesis. The sequence of additions is determined by a sequence of sites on a complex, **multiheaded enzyme.** An example is the synthesis by *Bacillus brevis* of **gramicidin S**, a head-to-tail ring of two pentapeptides:

L-Leu-D-Phe-L-Pro-L-Val-L-Orn
| |
L-Orn-L-Val--L-Pro-D-Phe-L-Leu

In this process the amino acids are activated as aminoacyladenylates (as in protein synthesis), and each is then transferred to a high-energy thioester link with an –SH on one of two enzymes. Enzyme I racemizes and transfers the phenylalanine, to provide the initiating D-Phe residue. Enzyme II, a multi-enzyme complex, can form thioester bonds with the other four amino acids of the pentapeptide, each in a specific site, but it does not polymerize them until **initiated** by the transfer of D-Phe from enzyme I. Moreover, if any of the four amino acid residues is absent enzyme II retains multiple growing chains, up to the empty site. The multienzyme complex (mol wt 280,000) has a single, long prosthetic group, pantetheine-P, covalently attached in a position from which its free –SH end can reach the more peripheral sites of attachment of the four amino acids (Fig. 4-12).

This model is based on a very similar succession of transthiolation reactions by pantetheine in fatty acid biosynthesis (Ch. 6). The multiple sites in the latter process specify the reaction sequence undergone by each acetate residue, while in peptide synthesis the multiple sites serve as a kind of template for specifying amino acid sequence.

As Lipmann pointed out, before the genetic code could be evolved it was probably necessary that the primitive precursors of cells discover the extraordinary functional pliability of proteins. The current mechanism of synthesis of peptide antibiotics may be a metabolic "fossil" of such a process. On the other hand, some have speculated that nucleic acids, with their capacity for self-replication as well as for tertiary structure, preceded proteins in evolution; peptide synthesis on multiheaded enzymes could then have evolved as a special, economic device for creating short peptides.

SELECTED FEATURES OF SMALL-MOLECULE BIOSYNTHESIS

Certain interesting generalizations emerge from the pathways reviewed above.

HETEROTROPHIC CO_2 FIXATION

All heterotrophs assimilate CO_2 (though on a much smaller scale than autotrophs). This property was long masked by their large production of CO_2, in metabolizing organic fuels. The first demonstration of CO_2 uptake, in 1936, involved the special case of the fermentation of glycerol, which yielded more organic matter than that supplied because CO_2 was fixed in making succinate as a major product. The later use of radioactive CO_2 demonstrated its fixation by organisms in general. When biosynthetic pathways became known this CO_2 was found to be assimilated into a few specific C atoms of purines, pyrimidines, and several amino acids, and not to be generally distributed as in autotrophic metabolism.

The following reactions have been demonstrated:
1) Carboxylation of pyruvate or of P-enolpyruvate (Fig. 4-5), the largest route of heterotrophic CO_2 fixation. In these reactions CO_2 yields one of the carbons of oxaloacetate (or malate), and thus of aspartate, glutamate, and their numerous derivatives.

FIG. 4-12. Peptidyl transfer by transthiolation in the synthesis of a peptide antibiotic on a multienzyme complex. The pantetheine (**Pant-SH**) is attached at one end to some central position in the protein. Its free, reactive–SH end can accept an amino acid (**A**) or peptide from a thioester and transfer it to the amino group of the next amino acid in the sequence.

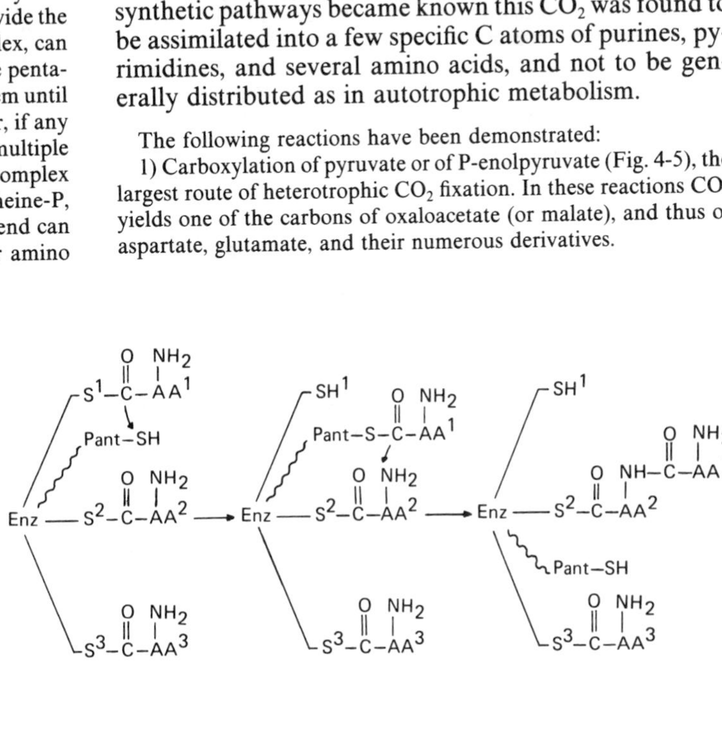

TABLE 4-1. The Number of Enzymes Involved in the Biosynthesis of the Amino Acids

Synthesized by mammals		Required by mammals		
Amino acid	No. of enzymes	Amino acid	No. of enzymes	Plus shared enzymes
Alanine	1	Arginine	8	1
Aspartate	1	Histidine	9	
Asparagine (from aspartate)	1	Threonine	5	1
Glutamate	1	Methionine	5	9
Glutamine (from glutamate)	1	Lysine	7	
Proline (from glutamate)	2	Isoleucine	4	6
Serine	3	Valine	1	9
Glycine (from serine)	1	Leucine	3	9
Cysteine (from serine and S^{2-})	2	Tyrosine	10	
		Phenylalanine	2	8
Total	13	Tryptophan	6	7
		Total	60	

The enzymes counted are those of the anabolic pathways, after they have branched off from amphibolic pathways. For branched anabolic pathways the enzymes of a shared portion are counted for the first endproduct cited, and are listed as "shared" for subsequently cited products.

(Modified from Davis BD: Cold Spring Harbor Symp Quant Biol 26:1, 1961

2) Formation of carbamyl-P, which contributes both the guanidino C or arginine and C-2 of the pyrimidine ring.

3) Contribution of C-6 of the purine ring.

In addition, CO_2 participates cyclically, without being assimilated, in fatty acid biosynthesis.

THE IMPORTANCE OF BEING IONIZED

Although various nonionized compounds are excreted by cells as endproducts (e.g., ethanol), or circulate between cells in higher organisms (e.g., glucose), all known intermediates in bacteria contain one or more groups that are largely ionized at physiologic pH: generally a phosphate (as in glycolysis) or a carboxyl (as in the TCA cycle). The regular presence of such a group is illustrated by comparison of the purine and the pyrimidine pathways. From the first reaction of purine biosynthesis the intermediates are attached to ribose-P; otherwise several of them would be un-ionized. In contrast, in the first reaction of pyrimidine biosynthesis aspartate provides a carboxyl, which is eliminated only after a phosphate has been added. In glycolysis, similarly, two phosphates are added to

hexose from ATP and are then recovered from two P-enolpyruvates; there is no gain or loss of energy, but without this addition many intermediates would be un-ionized.

The function of these ubiquitous dissociable groups is not certain; they may promote 1) the retention of a compound by a cell, and 2) the efficiency and specificity of enzyme action.

ECONOMY IN MAMMALIAN BIOSYNTHESIS

Man can synthesize about half of his amino acids but is an auxotroph for the other half. Analysis of the pathways in *E. coli* has shed light on this curious evolution: the 9 amino acids synthesized in man arise by pathways of 1–3 enzymes each, while the 11 required amino acids have individual path lengths of 6–13 enzymes (Table 4-1). The human species has thus spared itself genes for about 60 enzymes by losing the pathways to 11 amino acids, while the retained pathways to the other 9 amino acids are relatively inexpensive, involving only 13 enzymes in all.

SELECTED READING

BOOKS AND REVIEW ARTICLES

COHEN SS: Some roles of polyamines in microbial physiology. Adv Enzyme Regul 10:207, 1972

FLAVIN M: Methionine biosynthesis. In Greenberg DM (ed): Metabolic Pathways, Vol VII. New York, Academic Press 1975, p 457

GINSBURG V: Sugar nucleotides and the synthesis of carbohydrates. Adv Enzymol 26:35, 1964

GUNSALUS IC, STANIER RY (eds): The Bacteria, Vol III, Biosynthesis. New York, Academic Press, 1962 (Especially Chpts 2, 4, 6)

KORNBERG HL: Anaplerotic sequences in microbial metabolism. Angew Chem [Eng] 4:558 1965

LIPMANN F, WIELAND G, KLEINKAUF H, ROSKOSKI RJ: Polypeptide synthesis on protein templates: the enzymatic synthesis of gramicidin S and tyrocidine. Adv Enzymol 35:1, 1971

MANDELSTAM J, MCQUILLEN K: Biochemistry of Bacterial Growth, 2nd ed. New York, Wiley-Interscience, 1973

TABOR CW, TABOR H: 1,4-Diaminobutane (putrescine), spermidine, and spermine. Annu Rev Biochem 45:285, 1976

THELANDER C, REICHARD, P: Reduction of ribonucleotides. Annu Rev Biochem 48:133,1979

TONN SJ, GANDER JE: Biosynthesis of polysaccharides by prokaryotes. Annu Rev Microbiol 33:169, 1979

TROY FA II: The chemistry and biosynthesis of selected bacterial capsular polymers. Annu Rev Microbiol 33:519, 1979

UMBARGER HE: Amino acid biosynthesis and its regulation. Annu Rev Biochem 47:533, 1978

chapter 5

BACTERIAL NUTRITION AND GROWTH

NUTRITION

ORGANIC GROWTH FACTORS

Microbiologists learned early that they could grow a wide variety of bacteria in "broths," obtained by cooking animal or vegetable tissues. In 1923 Mueller, undertaking to redefine this vague requirement in terms of specific compounds (see quotation above), discovered the previously unknown amino acid methionine. Bacterial nutrition became a lively field a dozen years later. With recognition of the unity of biochemistry (Ch. 3) the many different nutritional patterns turned out to be simply minor variations on a central theme, and the study of microbial nutrition lost much of its theoretic interest. Nevertheless, this field retains its practical importance, and it still presents challenges; for example, the leprosy bacillus, the treponeme of syphilis, and rickettsiae still cannot be cultivated in artificial media.

An important consequence of the study of natural growth requirements was the development of **quantitative microbial (bio)assays** for amino acids and vitamins, using principles described in Chapter 4 (see Auxotrophic Mutants). The simplicity of these assays, compared with those in animals, greatly facilitated the isolation of novel factors; hence the majority of vitamins were first identified in this way, and were only later found to be also essential for mammals.

Nutritional investigations with animals and with bacteria have sometimes converged. Thus vitamin B_{12} was independently isolated as a hematopoietic factor in patients with pernicious anemia and as a growth factor for *Lactobacillus leichmannii.*

Bacteria that are adapted to growth in animal tissues or milk, or on mucous surfaces, often require various amino acids, nucleic acid bases, and vitamins. The requirements of many pathogens (e.g., streptococci) are more complex than those of mammals. Yeasts and molds usually exhibit requirements only for vitamins.

Other compounds required by various microbes include inositol and choline (as components of phospholipids or cell walls), vitamin K, hemin (or occasionally porphyrin), unsaturated fatty acids, mevalonic acid (a precursor of isoprenoid compounds), and polyamines.

INORGANIC REQUIREMENTS

Oxygen. The metabolic reasons for aerobic, anaerobic, and facultative requirements have been discussed in Chapter 3. O_2 has a low solubility in water: a solution in equilibrium with air at $34°C$ contains about $5\mu g/ml$, which would be consumed in <10 sec by a fully grown culture of an aerobe. The diffusion of O_2 across the air–water interface therefore limits the density attained by an aerated, well-nourished culture: for example, with aeration by swirling in a flask growth is often limited to 1–2 mg dry weight per ml. Moreover, during the last portion of such growth the culture becomes anaerobic.

Methods that increase the area of the liquid–air interface, such as rapid bubbling of air through a porous sparger, or recycled dripping, will support heavier growth. The problem of adequately aerating dense cultures makes large-scale production more difficult with bacteria than with yeasts, which respire more slowly (surface-to-volume ratio about 1/100 as great).

Anaerobiosis. The establishment of a strictly anaerobic atmosphere for the cultivation of obligate anaerobes (Ch. 3) on plates is difficult, since oxygen tensions as low as 10^{-5} atm can be inhibitory. However, supplementation of the medium with a sulfhydryl compound, such as sodium **thioglycollate** ($HSCH_2COONa$), permits some strict anaerobes, such as *Clostridium tetani,* to be grown in tubes exposed to air. It is helpful to add a layer of oil or paraffin to slow the diffusion of oxygen, plus semisolid agar (0.2%–0.3%) to prevent convection. Practical methods are further described in Chapter 31 (Bacteroides).

In nature mixed cultures are the rule, and the strict anaerobes may depend on neighboring facultative organisms to scavenge oxygen.

Some organisms are **microaerophilic,** initiating growth well at reduced but not at fully aerobic O_2 tensions. Some facultative organisms (e.g., *E. coli*) require cystine to permit initiation of anaerobic growth in a minimal medium. Presumably strict anaerobiosis reduces, and cystine restores, some S–S bonds required for growth.

Carbon Dioxide. The role of CO_2 as a universally essential nutrient has been discussed in Chapter 4 (Heterotrophic CO_2 Fixation). Some organisms (e.g., meningococci, gonococci), especially when first isolated, initiate growth better at a pCO_2 higher than that found in air (about 0.03% outdoors); they presumably have some enzyme with a low affinity for CO_2. Elevated pCO_2 is conveniently provided in a **candle jar,** a closed vessel in which a candle is allowed to burn until it extinguishes itself. The accompanying lowering of pO_2, however, does not provide strict anaerobiosis.

With a small inoculum there may be a long lag in the initiation of growth in a minimal medium because the low pCO_2 limits the required flow of carbon from glucose into the tricarboxylate (TCA) cycle. Added TCA cycle intermediates may then exert a "sparking" effect in overcoming this lag, until the culture becomes dense enough to build up its pCO_2. However, even with these additions growth can be prevented by measures that further reduce the CO_2 tension in the vessel. The explanation was provided by the discovery that fatty acid synthesis requires CO_2 for the conversion of acetyl CoA to malonyl CoA (Ch. 6). Hence even though this CO_2 is recycled rather than assimilated, *all growing cells have an absolute requirement for an adequate* pCO_2, for their lipids cannot all be supplied from without.

The **inorganic ions** required in substantial quantity are PO_4^{3-}, K^+, and Mg^{2+}. In the absence of organic sources of N and S, NH_3 and SO_4^{2-} (or a reduced product) are also required. Unlike mammalian cells most bacteria can thrive in a broad range of concentrations of the required ions. This flexibility long prevented the recognition of their need for constant internal ion concentrations (e.g., K^+, Mg^{2+}, and PO_4^{3-}), which depend on active membrane transport systems (Ch. 6).

K^+ is required by many enzymes, and is especially important for ribosomal function: when the K^+ content of bacterial cells is progressively lowered (in a mutant with a defective K^+ transport system) protein synthesis ceases, while glycolysis continues. Mg^{2+} is also essential for the integrity of ribosomes (Ch. 13) and for the function of many enzymes.

Trace Elements. Iron is required not only for heme proteins in aerobes but also for certain nonheme enzymes. Its concentration markedly influences diphtheria toxin formation (Ch. 26). Other requirements include Zn^{2+} and Mn^{2+} for certain enzymes, and Mo^{2+} for N_2 fixation and nitrate reduction. Co^{2+} is required by those bacteria that make vitamin B_{12}; since plants do not contain this vitamin bacteria may be its ultimate source for man. Cu^{2+} has

not been found in any bacterial enzymes, though it is widely used in higher organisms.

Ca^{2+} does not appear to be required by gram-negative organisms. However, it is a major constituent of the wall of gram-positive bacilli and of their spores.

Siderophores and Fe Transport. Iron transport presents a special problem, because at neutral pH Fe^{3+} forms very insoluble colloidal hydroxides. Accordingly, many bacteria and fungi form and excrete compounds that chelate Fe, i.e., form tight, soluble coordination complexes. These complexes are taken up via specific membrane receptors (See Outer Membrane, Ch. 6), and the Fe^{3+} is released within the cell by hydrolysis of the chelator. Three classes of natural Fe-binding compounds (siderophores) have been recognized: hydroxamic acids ($-CONH_2OH$), known as **sideramines;** catechols (2,3-dihydroxybenzene derivatives); and **citrate.**

Enteric organisms (including *E. coli*) form a cyclic trimer of 2,3-dihydroxybenzoylserine known as **enterobactin** or **enterochelin.** One species of *Mycobacterium* requires for growth a sideramine, **mycobactin,** formed by other mycobacteria. Conversely, some organisms form **sideromycins** (e.g., albomycin): nonchelating, **antibiotic** analogs that interfere with the uptake of sideramine–Fe^{3+} complexes. The supply of available Fe^{3+} may play an important role in limiting the multiplication of pathogens in tissues.

PHYSICAL AND IONIC REQUIREMENTS

Most bacteria can withstand a rather wide range of temperature, osmotic pressure, and pH. Moreover, in filling all possible ecologic niches the bacterial world has evolved members that can grow under conditions too extreme for any other group of organisms: temperatures up to 90° C, pH below 1.0, salinity up to 30% NaCl. These organisms are of interest as experiments of nature that can help to correlate macromolecular structure and function.

Temperature. Most bacteria can grow over a **temperature range** of 30° or more but have quite a narrow range for optimal growth. Below the optimum the decline in growth rate with decreasing temperature at first has a slope typical of enzyme reactions, but then it becomes very steep, giving rise to a fairly well defined **minimal growth temperature** (Fig. 5-1). Above the optimum the growth rate decreases steeply with increasing temperature, giving rise to a sharply defined **maximum growth temperature.**

The lower temperature limit of growth may depend on solidification of membrane lipids (Ch. 6), or on the marked sensitivity of the initiation process in protein synthesis (Ch. 13) to cooling. Slightly above the upper temperature limit, in contrast, many enzymes are denatured and the cell dies: i.e., in general a cell evidently does not build enzymes with more stability than is useful. At an intermediate temperature, which slows growth without killing, the new growth rate is set by a reversible decrease in the activity of a particularly sensitive biosynthetic enzyme.

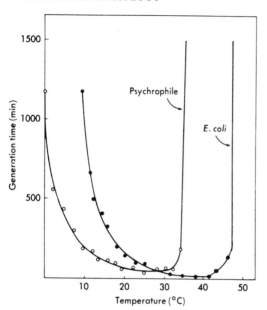

FIG. 5-1. Effect of temperature on the generation time of a typical mesophile (*Escherichia coli*) and a psychrophilic pseudomonad. (After Ingraham JL: J Bacteriol 76:75, 1958; modified according to data of Ron EZ, Davis BD: J Bacteriol 107:391, 1971)

The **temperature range** for growth of an organism is a stable characteristic, of considerable taxonomic value. It is customary to divide bacteria into mesophiles, psychrophiles (or cryophiles), and thermophiles. Most bacteria are **mesophiles**. Those found in the mammalian body have a temperature optimum of 37°–44°C, but many others found in nature (e.g., *Bacillus megaterium*) grow better at 30°.

Psychrophiles (predominantly pseudomonads) can grow at low temperatures, many down to 0°C. These organisms are important in spoilage of refrigerated foods and are also found in naturally cold waters and soils. **Thermophiles** (predominantly bacilli), in contrast, may have temperature optima as high as 50°–55°, with tolerance to 90°. They are found especially in hot springs and compost heaps. Thermophiles (e.g. *Bacillus stearothermophilus*) are useful as sources of stable forms of those enzymes that are unstable when extracted from most cells.

The existence of thermophiles shows that nature can evolve proteins, for all essential cellular functions, with stability far beyond the usual range; the extra stability can be provided by very few additional weak bonds within the folded protein (Ch. 13, Genetic Determination of Protein Structure). Conversely, many temperature-sensitive mutants isolated in the laboratory (Ch. 8) form specific altered enzymes that denature at ordinary temperatures.

Cold Shock. Though bacteria are often preserved successfully in the refrigerator, the sudden chilling of exponentially growing cells of some species (*E. coli, Pseudomonas*) results in substantial killing (>90%). This curious phenomenon is not observed with gradual cooling or with stationary-phase cells; it may be related to the

changing composition of the cell membrane lipids with changes in temperature and in growth phase (Ch. 6).

pH. The pH range tolerated by most microorganisms extends over 3–4 units, but rapid growth may be confined to 1 unit or less. *E. coli* cannot withstand a pH much above 8 or below 4.5, while pathogens adapted to tissues (*Pneumococcus, Neisseria, Brucella*) have a narrower range. Vinegar-forming *Acetobacter* and sulfur-oxidizing bacteria can tolerate the acid that they produce up to 1 N (pH ∼ 0 for sulfuric acid). In contrast, a few bacterial species (urea splitters, *Alcaligenes faecalis,* the cholera vibrio) thrive at a pH of 9.0 or more. Most yeasts and molds are highly acid-tolerant, and this feature is exploited in selective media for their cultivation.

The lower pH cutoff point depends in part on the concentrations of organic acids in the medium; a lower pH increases the proportion of an acid in the undissociated (and hence more permeable) form, thus making it more inhibitory. Hence a lactate-producing fermenter inhibits itself when it reaches a certain concentration of free lactic acid rather than a given pH.

A low external pH need not correspond to the internal pH; in the absence of accurate measurements of intracellular pH we can only be sure that the cell wall and membrane are exposed to the external conditions.

In a culture growing aerobically on a limiting amount of sugar the pH often falls and then rises as acid accumulates and then is utilized. To restrict pH changes during growth media are often heavily buffered; and for fine control automatic continual titration is sometimes employed. Incidentally, the concept of pH, which first clearly distinguished **extent** and **intensity** of acidity, was originally formulated by Sorensen in the course of determining what limits the growth of microbes in various media.

Halophiles. Na^+ and Cl^- are not widely required by bacteria, though moderate concentrations are generally tolerated. Most bacteria isolated from the ocean, however, are slightly halophilic, requiring NaCl in a concentration approaching that of their natural habitat (3.5%). In addition, moderate and extreme halophiles, with NaCl requirements up to 20%, and with optima approaching saturation (slightly above 30%), are found in flats and lakes where salt water is evaporated, and in pickling fluids.

The high Na^+ concentration of the medium functions osmotically to permit the intracellular accumulation of a high concentration of K^+, required by the ribosomes and many enzymes of these organisms. However, in some halophiles the integrity of the cell wall specifically requires a high Na^+ level.

Water. In contrast to higher organisms, whose specialized integument retains water, the metabolism of bacteria is dependent on ambient water.

PRACTICAL BACTERIAL NUTRITION

Though media of chemically defined compositions are valuable for special purposes, the traditional rich "soups" are still generally employed in diagnostic bacteriology: they are less expensive, and initiation of growth by small inocula is often more reliable. These media are based primarily on **meat digest** (tryptic digest, peptone, nutrient broth), the soluble product of enzymatic hydrolysis of meat or fish. Many types are marketed, differing in source material or in method of preparation, and often in suitability for cultivating specific organisms.

To provide vitamins and coenzymes media are often further enriched with **meat extract (meat infusion)** or **yeast extract,** containing the stable small molecules released from the cells and concentrated by boiling. Yeast extract is rich in nucleotides. **Casein hydrolysate** is often used as an inexpensive source of amino acids in chemically defined, relatively rich media. In the usual acid hydrolysate (e.g., Casamino acids) tryptophan and glutamine have been destroyed. Enzymatic hydrolysates contain all the amino acids, but mostly as small peptides.

Blood, in **blood agar,** provides not only nutrients but also a diagnostically useful index of hemolysis.

The genus *Hemophilus* requires heme, NAD, or both (Ch. 34). In heated blood agar (called **chocolate agar** because of its color) heat has released the heme from hemoglobin and has inactivated an enzyme that hydrolyzes NAD.

Some organisms thrive best in media containing **serum** (e.g., 20%), which provides not only nutrients but also a **protective, nonnutrient growth factor,** albumin: a protein whose versatile affinity protects cells from such toxic compounds as fatty acids (soaps) and heavy metal ions. Starch also binds fatty acids. Protective growth factors are especially valuable in promoting the initiation of growth by small inocula, for with inhibitors that bind tightly a few molecules contaminating the glassware may be sufficient to inhibit a few cells.

The compositions of some representative media are given in Table 5-1.

ATTACK ON NONPENETRATING NUTRIENTS: EXOENZYMES

Microbes can take up foods of low molecular weight, including oligopeptides, nucleosides, and small organic phosphates, e.g., glycerol phosphate (Ch. 6); nucleotides generally cannot penetrate at a substantial rate. But the organic matter initially returned to the soil in dead plants and animals is predominantly in macromolecules, which must be hydrolyzed before they can be taken up by bacteria. For this purpose various bacteria and fungi elaborate a variety of **exoenzymes.** Many of these are secreted into the medium (especially by gram-positive rods) as **extracellular enzymes;** in pathogens these often play an important role by attacking tissue constituents. In gram-negative organisms exoenzymes are often **periplasmic,** i.e., are retained between the plasma membrane and the wall. The mechanism of enzyme secretion will be discussed in Chapter 6.

Extracellular enzymes include proteases and peptidases; polysaccharidases (amylase, cellulase, pectinase); mucopolysaccharidases (hyaluronidase, chitinase, lysozyme, neuraminidase); nucleases; lipases; and phospholipases. Some proteases are released from the cells as inactive zymogens, which catalyze their own activation. Protease formation can be repressed by a high concentration of amino acids in the medium.

TABLE 5-1. Composition of Representative Bacteriologic Media

Minimal medium for *Escherichia coli*

	G/liter
K_2HPO_4	7.0
KH_2PO_4	3.0
Na_3citrate—$3H_2O$	0.5
$MgSO_4$—$7H_2O$	0.1
$FeSO_4$	0.01
$(NH_4)_2SO_4$	1.0
Glucose*	2.0

Penassay broth (typical rich medium)

Peptone	5.0
Beef extract	1.5
Yeast extract	1.5
NaCl	3.5
Dipotassium phosphate	3.7
Monopotassium phosphate	1.3
Glucose	1.0

B_{12} Assay medium for *Lactobacillus leichmannii*

Vitamin-free casein hydrolysate	15.0
Tomato juice	10.0
Glucose	40.0
Asparagine	0.2
Sodium acetate	20.0
Ascorbic acid	4.0
Monopotassium phosphate	1.0
Dipotassium phosphate	1.0
Sorbitan monooleate	2.0
$MgSO_4$	0.4
NaCl	0.02
$FeSO_4$	0.02
$MnSO_4$	0.02
L-Cystine	0.4
DL-Tryptophan	0.4
Adenine sulfate	0.02
Guanine hydrochloride	0.02
Xanthine	0.02
Uracil	0.02
	Mg/liter
Riboflavin	1.0
Thiamine	1.0
Niacin	2.0
p-Aminobenzoate	2.0
Ca pantothenate	1.0
Pyridoxine	4.0
Folic acid	0.2
Biotin	0.008

*Glucose is autoclaved separately, for when autoclaved in the presence of phosphate it produces discoloration and some toxicity.

Some Macromolecular Substrates. Recognition of various hydrolases is useful in diagnostic work: extracellular protease is generally detected by the liquefaction of gelatin (denatured collagen) around a colony, and lecithinase by the formation of an opaque product from egg yolk (Ch. 36). Plasma is clotted by coagulase (Ch. 29) and clots are lysed by streptokinase and fibrinolysin (Ch. 28).

Some microbial hydrolases cleave only a particular site and hence have been helpful in the analysis of protein structure.

Starch is hydrolyzed by many bacteria and molds but by only a few yeasts. Hence in the fermentation of grain to yield beer germinating barley is used to convert the starch to the disaccharide maltose, which is then fermented by brewer's yeast (*Saccharomyces cerevisiae*). The mold *Aspergillus oryzae* is used in Japan in a one-step process that both splits and ferments starch.

GROWTH IN LIQUID MEDIUM

METHODS OF MEASUREMENT

For biochemical studies bacterial growth is usually defined in terms of **mass** of cellular material, while for studies of genetics or of infection **cell number** is more pertinent. The ratio of number to mass is fixed under conditions of steady-state growth, but can vary with growth conditions.

Cell Mass. This can be measured in terms of dry weight, packed cell volume, or nitrogen content. A convenient index is **turbidity,** whose rapid measurement in a photoelectric colorimeter or spectrophotometer allows the density of a culture to be followed during growth. The absorption of light by colored cell constituents is negligible; most of the decrease in transmission is due to light scattering, and is dependent on the high refractive index of the bacteria (dry wt. about 25% compared with 1%–2% in the medium).

Wave lengths between 490 and 550 nm are generally used: the lower the wave length the greater the light scattering. However, below 490 nm absorption by yellow products of autoclaving may become significant. Turbidity is linear with bacterial density between 0.01 mg dry weight (about 10^7 cells) and 0.5 mg/ml. Increased osmotic pressure of the medium, by shrinking the cells, increases the refractive index and hence the light scattering.

Cell Number. To determine the **viable number** of a culture a series of 10-fold or 100-fold dilutions is plated in or on a solid medium; the number of colonies is counted in those Petri plates that are not too crowded (<400 colonies). This method is useful down to extremely low bacterial densities. Precision is limited by the statistical sampling error: the **standard deviation** (SD) is the square root of the number counted (e.g., in a count of 400 colonies SD = 20, or 5%). It is therefore customary in precise work to make several replicate plates.

It is sometimes desirable, especially in studying antimicrobial action, to measure the **total cell count,** which is ordinarily identical with the viable count. Cells can be counted under the microscope in specially designed chambers, but it is more convenient to use an electronic particle analyzer (e.g., the Coulter counter), in which a pair of electrodes detect the effect of a passing particle on the impedance.

Cell number cannot be determined accurately when cells adhere to each other after division (e.g., streptococci) or when they aggregate. Freedom from aggregation is one of the properties that makes *E. coli* especially convenient for physiologic investigation.

GROWTH CYCLE

Growth of bacteria is characteristically exponential (see next section). However, such a culture, called log phase, eventually slows down and ceases growth, either because a required nutrient (often O_2) becomes limiting, or because inhibitory metabolic products (often organic acids or alcohol) accumulate. In this transition to the **stationary phase** (Fig. 5-2) the **cells become smaller,** as a result of dividing faster than they grow; moreover, the macromolecular composition changes (Ch. 14).

Stationary-phase cells transferred to fresh medium exhibit a **lag phase,** which varies with the recent history of the organisms and with the medium; it is much more pronounced in a minimal than in a rich medium. Cell number lags more than cell mass (Fig. 5-2), for the small stationary-phase cells increase in size before they begin to divide.

Even when cells are transferred from an exponentially growing culture to fresh, identical medium there may be a lag in resuming growth, especially in minimal medium and with small inocula. The causes include slowing of growth by trace contaminants in the medium (e.g., soap or heavy metal ions), and the need to accumulate the CO_2 required for biosynthesis.

Cells in the stationary phase develop adaptive **chemical changes** (Chs. 6, 14) that increase their stability. However, on prolonged incubation cells do die and lyse, through membrane damage and resulting activation of autolytic enzymes. The released products support **cryptic growth** of surviving cells (i.e., survival without net growth); hence the progeny of mutants more resistant to the limiting conditions may accumulate.

EXPONENTIAL KINETICS

In the exponential phase of growth the rate of increase of bacterial mass at any time is proportional to the mass present:

$$dB/dt = \alpha B \tag{1}$$

where B is bacterial mass, t is time, and α is the **instantaneous growth rate constant** for that culture (i.e., the relative increase per unit time). Hence

$$dB/B = \alpha dt \text{ (or } d \ln B = \alpha dt) \tag{2}$$

Integrating,

$$B_t = B_0 e^{\alpha t}, \text{ and} \tag{3}$$

$$\ln B_t/B_0 = \alpha t, \text{ or } \ln B_t = \ln B_0 + \alpha t \tag{4}$$

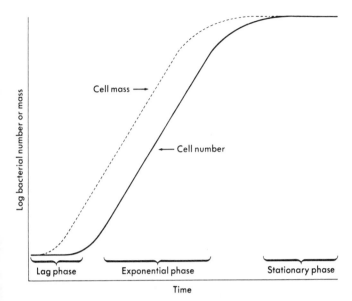

FIG. 5-2. Phases of bacterial growth, starting with an inoculum of stationary-phase cells. Note that the classic phases, defined in terms of cell number, do not precisely coincide with the phases of changing growth in terms of protoplasmic mass.

Hence in this phase a plot of the logarithm of B against time gives a straight line (Fig. 5-2). This semilogarithmic plot is generally used for bacterial growth curves. The exponential phase is also often called the **log phase,** because the logarithm of the mass increases linearly with time.

It is sometimes convenient to convert the instantaneous growth rate constant α, which has the dimensions of time^{-1}, into the more familiar dimensions of time: $\tau = 1/\alpha$. This **instantaneous generation time,** τ (Gr. *tau*), represents the time that would be required for a doubling of mass **if** the growth rate at zero time, αB_0, continued unchanged. In exponential growth, however, the value of B, and hence of αB, is constantly increasing, and at the end of a doubling the rate of cell synthesis is twice what it was at the beginning.

Hence the actual **doubling time,** t_D, is shorter than τ (Fig. 5-3). The relation between the two is derivable by setting B_t at $2B_0$ (i.e., one doubling) in equation 4:

$$\ln B_t/B_0 = \ln 2 = \alpha t_D$$
$$t_D = (1/\alpha) \ln 2 = 0.69 \, (1/\alpha) = 0.69 \, \tau$$
$$\text{or} \quad \tau = 1.45 t_D$$

The doubling time is also called the **mean generation time (MGT).** Growth rate is usually expressed in terms of t_D or its reciprocal, μ ($= 1/t_D$), which is the **exponential growth rate constant,** expressed as generations per hour.

Figure 5-3 demonstrates the curvature of exponential growth when plotted linearly, rather than logarithmically, against time. Precisely the same curve would

be obtained if cell number were measured instead of cell mass, because ordinary bacterial cultures are **asynchronous:** since the cells at any moment are randomly distributed with respect to stage in the division cycle, the rate of cell formation rises continuously rather than discontinuously. The growth of a hypothetic perfectly synchronized culture is compared in Figure 5-4.

The linear relation between logarithm of number (or mass) and time is obtained regardless of the base of the logarithm. Logarithms to the base 10 are conventional, but the base 2 is more pertinent since the unit of growth then represents a doubling (one generation). The conversion is made through the relation:

$$\log_2 x = \log_{10} x \, / \, \log_{10} 2 = 3.3 \log_{10} x$$

It is convenient to remember that $2^{10} = 1024$: i.e., 10 generations equal a thousandfold increase and 20 generations a millionfold. In exponential growth plotted in terms of the conventional $\log_{10}$, an increase of 0.3 units = 1 generation.

From these considerations it is evident that exponential growth must be more the exception than the rule in the life of bacteria. A bacterium that doubles in 20 min will yield 10 generations (10^3 cells) in 3.3 h, and 10^9 cells in 10 h; while a bacterium doubling every 60 min would require three times as long. (These are approximately the values for many species on a rich and on a minimal medium.) Hence single cells yield grossly visible colonies (about 10^6–10^7 cells) in overnight cultures. Furthermore, since the volume of an average bacterium is $1\mu^3$, or 10^{-12} cm^3, the volume of the earth (about 4×10^{27} cm^3) is equivalent to 4×10^{39} bacteria; and the progeny of our rapidly growing cell would reach this volume in only 45 hours if growth remained exponential. Fortunately, something becomes limiting earlier.

RELATION OF GROWTH TO SUBSTRATE CONCENTRATION

Bioassays with microbes are performed by measuring the **total growth** after incubation with limited amounts of a required factor. In a satisfactory assay the response is strictly proportional to the amount of the factor provided (Fig. 5-5).

With limiting building blocks or fuel the transition in **growth rate** from exponential growth to plateau is rapid (about 1% of the range of visible growth; Fig. 5-5). With limiting cofactors, however, the transition is more gradual (Fig. 5-5), as their concentration in the cell decreases.

THE CHEMOSTAT

In physiologic studies reasonably reproducible conditions are achieved by harvesting cells that are in exponential growth. However, the cells are growing in an ever-changing environment. To obtain bacteria that have grown in a medium of truly constant composition Novick and Szilard, and Monod, devised the chemostat, which permits steady-state growth in a continuous-flow culture (Fig. 5-6).

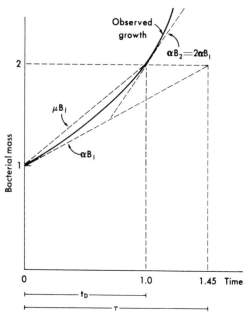

FIG. 5-3. Kinetics of exponential growth. B_1 = initial bacterial mass, B_2 = doubled mass, t_D = doubling time, τ = instantaneous generation time, α = instantaneous growth rate constant, μ = exponential growth rate constant.

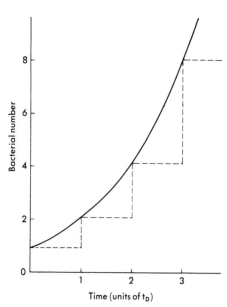

FIG. 5-4. Arithmetical plot of the increase in **cell number** in asynchronous (**solid line**) and hypothetic synchronous (**broken line**) exponential growth. In either type of growth **mass** would follow the solid line.

In this instrument fresh medium flows into a filled, stirred growth chamber at a constant, carefully controlled rate; each drop causes a drop of culture to overflow, and so the volume of growing culture is constant. A specific required growth factor, provided in the medium at a **growth-limiting concentration,** determines the **cell density** in the steady-state culture. Moreover, the **rate of flow of the medium** determines the **rate of growth,** at any level below the potential growth rate of the cells (i.e., the rate observed with an excess of the required nutrient). In the steady state the culture fluid will contain the limiting nutrient at a low concentration, which just permits the membrane transport units (Ch. 6) to take up the nutrient as the rate corresponding to the growth rate.

The chemostat thus permits the indefinite growth of bacteria in a constant medium, with independent control of the growth rate and population density. It has made possible very precise analysis of mutation rates (Ch. 8) and regulation of enzyme formation (Ch. 14). Since the mass of growing bacteria remains constant, growth of the culture is **linear** rather than exponential. The observed doubling time is therefore identical to the instantaneous generation time (τ), as defined above.

SYNCHRONIZED GROWTH

In a growing bacterial culture the cells are distributed among all stages in their division cycle.* Hence chemical

---

* Bacteriologists frequently speak of "old" cells in referring to cells from old cultures, which may mean either cultures in the stationary phase or those stored for a long time. However, the concept of age in a physiologic sense is not useful for bacteria, since they undergo binary fission rather than a life cycle that includes senescence and death.

analyses yield only average values. But there are a number of things we would like to know about the properties of the individual cell in the course of its division cycle: the kinetics of any cyclic change (or lack of change) in the synthesis of its major components (particularly DNA synthesis in relation to cell division: Ch. 14); variations in mutability or stability of the genetic material; susceptibility to lethal agents; competence for transfer of genetic material. Since it is not possible to study such problems with isolated cells, technics have been sought for

FIG. 5-5. Growth curves (semilogarithmic plot) of auxotrophs given limiting amounts of a required building block (tryptophan) or a required cofactor (biotin).

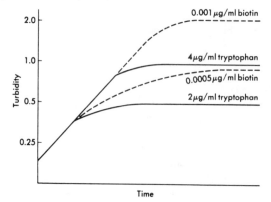

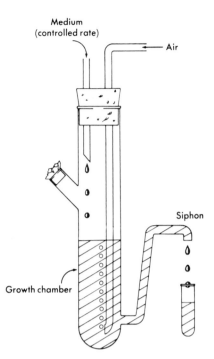

FIG. 5-6. Simplified diagram of the chemostat. (For a detailed description see Novick A, Szilard L: Proc Natl Acad Sci USA 36:708, 1950)

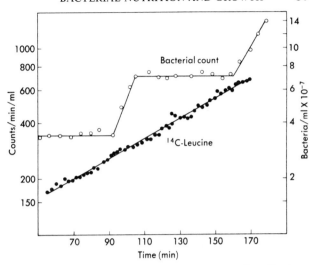

FIG. 5-7. Exponential synthesis of protein in an *E. coli* culture synchronized by filtration. Protein synthesis was measured as incorporation of radioactive leucine from the medium into cell components precipitated with trichloroacetic acid. (Adapted from Abbo FD, Pardee AB: Biochim Biophys Acta 39:478, 1960)

synchronizing the growth of bacteria, i.e., for producing cultures in which all the cells would be in approximately the same part of the division cycle.

Synchronization of bacterial growth was first achieved by various procedures that delayed the initiation of DNA replication and then restored the conditions required (e.g., repeated alternation between short periods of incubation at 37° C and at 25°, or transient deprival of a required amino acid). However, these procedures distort the composition of the cells and hence may change the regulatory patterns from those of balanced growth.

Physiologic synchronization has been achieved by **mechanical separation** of the smallest cells, just produced by cell division, from those in later stages of the cell cycle. One approach is to filter exponentially growing cultures through a number of layers of filter paper, which slows down the larger cells. In a more effective procedure cells are absorbed on or wedged in the **pores of a membrane filter;** subsequent reverse flow of fresh medium through the filter provides a continuous supply of newborn cells, derived by fission from the permanently trapped cells. The released cells, accumulated in a chilled tube, exhibit two or three generations of synchronized growth when incubated in fresh medium.

During a synchronized doubling of cell size the synthesis of the major constituents proceeds exponentially (Fig. 5-7), except that DNA synthesis proceeds linearly until the chromosome is repli-

cated (Ch. 10). This finding leads to the conclusion that *exponential growth, $dB/dt = \alpha B$, is a property of the growing cell,* and not simply a statistical property of the culture: a cell on the verge of dividing has twice as many enzyme molecules and ribosomes as a cell just formed, and it grows twice as fast.

The loss of synchronization after two to four generations is in harmony with early microscopic evidence for a rather large dispersion in the division times of single cells on solid media. The cause may be the absence of a mechanism for ensuring precise equipartition of cytoplasm in cell division.

GROWTH ON SOLID MEDIUM

SOLIDIFYING AGENTS

The earliest method available for enumerating bacteria or for isolating pure cultures (clones)* was **extinction dilution,** i.e., dilution up to loss of infectivity. An important advance was the introduction by Koch of **solid media.** The initial materials used were unsatisfactory: the cut surface of a potato might permit confluent growth, while gelatin is solid only at relatively low temperatures and is liquefied by many organisms. The world is therefore much indebted to Frau Hesse, the wife of a German physician and an amateur bacteriologist, for introducing agar, a Japanese product used for thickening soups.

Agar (Malay *agar-agar*) is an acidic polysaccharide derived from certain seaweeds (various red algae); it is primarily a polymer of

---

* A **clone** is defined as a group of organisms derived by vegetative reproduction from a single parental organism. Thus any pure culture of bacteria is a clone, and the progeny of a mutant arising in it are a subclone.

galactose, with one sulfate per 10–50 residues. Agar proved to be an ideal substance for making solid media. It is nontoxic to bacteria, and very few attack it. At 1.5%–2% agar the surface is wet enough to support growth but dry enough to keep colonies separate. (Exceptionally motile organisms, such as *Proteus,* require about 5% agar to prevent swarming.) After melting at 80°–100° C agar solutions remain fluid at temperatures down to 45°–50°, which most bacteria can withstand briefly; and after solidifying at room temperature they remain solid far above 37°. The fibrous structure of the gel is fine enough to prevent motility of bacteria within it, but coarse enough to permit diffusion even of macromolecular nutrients.

Being a material of natural origin, agar contains traces of various metabolites; hence many auxotrophs will yield microscopic colonies on minimal agar. Being an anionic macromolecule, agar binds cations such as Mg^{2+} and Ca^{2+}. A sulfate-free fraction, **agarose,** is available for special purposes. Purer solid media can be made, though inconveniently, through the use of **silica gel,** formed by neutralizing dilute Na_4SiO_4 with HCl.

USES OF SOLID MEDIA

For **enumerating viable cells** dilute suspensions are spread on the surface of a prepared agar plate or are added, at 45° C, to melted agar medium which is then immediately poured. After incubation each cell capable of growth on the medium yields a colony.

In using a selective solid medium to **isolate pure cultures** from a naturally occurring mixture an important precaution must not be overlooked: if the selected cells are only a minor proportion of the population the colony may overlap, and even feed, viable cells of other species that cannot grow on the medium used. These contaminants may then become predominant if the colony is transferred to a nonselective medium for storage. Indeed, the literature is replete with results obtained with allegedly pure cultures that were actually mixtures isolated from a single plating. To isolate clones reliably a colony from the first plate should be streaked again to yield single colonies on **a second plate of the same medium;** one of these colonies can then be used to furnish the stock culture.*

Solid media are used most widely for the **identification** of microorganisms, including pathogens. A practiced eye can recognize a variety of species in a single throat culture, simply on the basis of size, color, shape, opacity, hemolytic activity, and surface texture (see below) of the colonies; the presumptive diagnosis can then be confirmed by microscopic examination and by subculture onto special diagnostic media.

Because molecules can diffuse through agar while bacteria cannot, agar media have lent themselves to a variety of special applications.

---

* Two successive platings are sufficient because of the diluting effect of each: if the originally predominant contaminant has become a minor component of a colony in the first plating, it will rarely contaminate colonies produced from dispersed cells on the second plate.

1) **Nutritional activity** of spots of added substances, known and unknown, can be detected on plates of inadequate medium containing a test organism (10^2–10^4 cells/ml); moreover, with known amounts of material added on a disc of thick filter paper the sizes of the zones of response yield reasonably accurate bioassays. Similarly, paper chromatograms can be incubated on large seeded plates to detect the positions of growth factors.

2) Conversely, with nutritionally adequate, heavily seeded plates a similar procedure provides a qualitative test or a quantitative bioassay for **antimicrobial agents** (See Fig. 7-2, Ch. 7).

3) Streaking two strains close to each other on a medium that supports poor growth can reveal **crossfeeding** of one by a growth factor released by the other **(syntrophism).** This phenomenon, originally recognized as **satellite** growth on diagnostic plates, has proved invaluable in the recognition of biosynthetic intermediates excreted by auxotrophic mutants (See Fig. 4-1, Ch. 4).

At the edge of a zone of inhibition (e.g., around an antibiotic test disc) growth may be heavier than elsewhere, owing to extra nutrient diffusing from the zone of inhibition.

COLONIAL MORPHOLOGY

Surface Texture. One of the most important diagnostic features of a colony is the texture of its surface, ranging from rough (R) to smooth (S) to mucoid (M) (Fig. 5-8). **Smooth** colonies reflect the presence of a capsule or other surface component that promotes a compact cellular orientation. **Rough** colonies have a dry and sometimes wrinkled surface; they are formed by cells that lack such a component or by cells growing in a filamentous manner. Different degrees of roughness are possible within a species, and are often correlated with differences in virulence.

Thus with the tubercle bacillus (Ch. 37) **virulence** is associated with a lipophilic surface component that causes the cells to adhere to each other in serpentine **cords,** while mutations to decreased virulence are often associated with a decreased tendency to form cords and hence with a smoother surface of the colonies. With many organisms, in contrast, rough strains are avirulent, and **virulent** strains form a surface component associated with a **smooth** colony: polysaccharide side chains on the outer membrane in gram-negative organisms (e.g., Enterobacteriaceae, Ch. 31), and a capsule in gram-positive organisms (e.g., pneumococci, Fig. 5-8). In the former a capsule may also be formed, yielding a **mucoid** or **glossy** colony. The structure of these surface molecules will be further considered in Chapter 6 and their role in virulence in Chapter 24.

Capsule formation, and therefore colonial morphology, sometimes depends on environmental as well as on genetic factors. Some nonmucoid Enterobacteriaceae become mucoid if grown at a low temperature or in a medium with an excess of carbon and limited nitrogen or phosphorus. Mucoid colonies may be huge, and even liquid, because of the volume of capsular material produced.

Size. Colony size provides a more sensitive comparison of growth rate than does the most careful direct measurement of growth rate in liquid medium. Thus, the latter cannot detect directly a 1% difference in the growth rate of two organisms. However, growth from

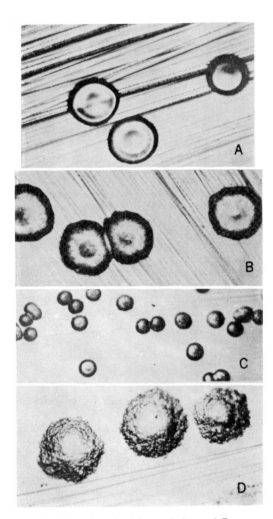

FIG. 5-8. Variations in colonial morphology of *Pneumococcus* strains. **A.** Smooth colonies of capsulated, nonfilamentous cells. **B.** Rougher colonies of capsulated but filamentous variant. **C.** Nonfilamentous noncapsulated variant. **D.** Roughest variant, filamentous and noncapsulated. All photographs ×18, after 24 h incubation at 36°C on blood agar. (Austrian R: J Exp Med 98:21, 1953)

1 cell to the 10^7 cells of a colony involves about 23 generations; and if the differential of 1% is maintained through this period the faster grower multiplies through $(1.01/1)^{23} = 1.3$ times as many generations, and, hence, yields a proportionately larger colony.

Differential media may reveal specific characteristics without being selective, and are very useful in bacterial identification. **Blood agar,** containing 5% sheep or horse blood, can reveal production of a hemolysin. In **fermentation plates** utilization of the particular sugar provided is revealed by indicator dyes, such as an eosin–methylene blue (EMB) mixture, which not only changes color but precipitates in the presence of acid; hence the colony itself is stained, and precipitation is sufficiently localized to demarcate stained **sectors** in a colony that contains a mutant subclone (Fig. 5-9). Fermentation plates are used largely for facultative orga-

nisms; they are effective even when incubated in air, because such organisms convert sugars to organic acids faster than they can burn the latter to carbon dioxide.

A fermentation plate must contain other nutrients besides the test sugar, to permit the growth of fermentation-negative organisms. When such negative colonies are fully grown they may give rise, on prolonged incubation, to positive **papillae** (Fig. 5-9) derived from mutant cells arising late in the growth of the colony. Indeed, even on nondifferential media prolonged cultivation frequently gives rise to papillae that can grow on nutrients not adequate for the mother colony; hence colonies on old plates are frequently warty and irregular.

In **diagnostic bacteriology** it is important to deal with well-separated colonies: **crowded colonies** are too small

FIG. 5-9. A. Lac− (unstained) sectors in colony derived from an ultraviolet-irradiated lac+ *E. coli* cell, plated on EMB-lactose medium (which stains only lac+ cells). Note sharp demarcation of sectors, and adjacent lac+ colony without lac− mutants. **B.** Lac+ papillae, arising late in lac− colony incubated for several days on EMB-lactose medium. (**A,** courtesy of H. B. Newcombe; **B,** courtesy of V. Bryson)

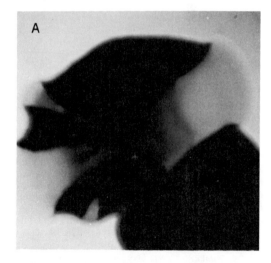

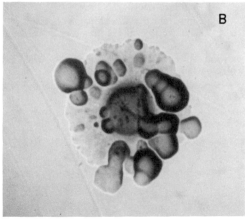

to reveal characteristic morphology, and their extensive production of acid can obscure the staining of fermenting colonies and can cause diffuse nonspecific hemolysis. It is also important to inspect plates after a standard period of incubation (usually 18–24 h).

In diagnostic work the bacteria in a colony are often further identified by overnight growth in tubes of various media **(fermentation tubes),** which grow out larger quantities of cells and at the same time test for their ability to carry out biochemical reactions on components of the media. Some biochemical tests can be carried out directly, and hence more rapidly, on the initial colonies, e.g., catalase or oxidase production. Additional tests for enzyme activities, divorced from simultaneous growth, will no doubt be developed.

SELECTIVE MEDIA

Cell types that predominate in a mixed population can be readily cloned on a plate by simple dilution. Minor components, however, require selective media. Selective liquid media **enrich** the population with respect to the desired organisms, while solid media permit direct **isolation.**

Organisms that can utilize a given sugar are easily screened for by making that compound the only carbon source. Similarly, the use of a minimal medium will exclude fastidious organisms. Selection in the opposite direction (nonutilizers; requirement for specific factors) is not always possible and is indirect, i.e., it is based on **selective inhibition** of the unwanted organisms (e.g., by unfavorable pH, salts, specific inhibitors).

Useful selective inhibitors, empirically discovered, include tellurite for growth of diphtheria bacilli in throat cultures, bismuth for pathogenic *Salmonella* and *Shigella* in stool cultures, and various dyes (e.g., malachite green), as well as preliminary treatment with strong acid or alkali, to permit recovery of the slow-growing but hardy tubercle bacillus.

Newer experimental methods for isolating rare cell types depend on their surface Ags, which permit adsorption of cells by specific Abs fixed on a column, or separation by a cell sorter that recognizes immunofluorescent staining.

SELECTED READING

BOOKS AND REVIEW ARTICLES

ALEXANDER M: Microbial Ecology. New York, Wiley, 1971

BROCK TD: Microbial growth under extreme conditions. 19th Symp Soc Gen Microbiol. Cambridge Univ Press, 1969, p 15

EMERY T: Hydroxamic acids of natural origin. Adv Enzymol 35:135, 1971

FARRELL J, CAMPBELL LL: Thermophilic bacteria and bacteriophages. Adv Microb Physiol 3:83, 1969

GRAY TRG, POSTGATE JR (eds): The Survival of Vegetative Microbes. 26th Symp, Soc Gen Microbiol, Cambridge Univ Press, 1976

GUIRARD BM, SNELL EE: Nutritional requirements of microorganisms. In Gunsalus IC, Stanier RY (eds): The Bacteria, Vol IV. New York, Academic Press, 1962, p 33

HUGHES DE, WIMPENNY JWT: Oxygen metabolism by microorganisms. Adv Microb Physiol 3:197, 1969

INGRAHAM JL: Temperature relationships. In Gunsalus IC, Stanier RY (eds): The Bacteria, Vol IV. New York, Academic Press, 1962, p 265

LANYI JK: Salt-dependent properties of proteins from extremely halophilic bacteria. Bacteriol Rev 38:272, 1974

LARSEN H: Biochemical aspects of extreme halophilism. Adv Microb Physiol 1:97, 1967

NEILANDS JB (ed): Microbial Iron Metabolism. New York, Academic Press, 1974

SNOW GA: Myobactins: iron-chelating growth factors from bacteria. Bacteriol Rev 34:99, 1970

Symposium: Continuous culture methods and their application. In Recent Progress in Microbiology, 7th International Congress of Microbiology, Stockholm, 1958. Springfield, Ill, Thomas, 1959

TEMPEST DW: The place of continuous culture in microbiological research. Adv Microb Physiol 4:223, 1970

THAYER DW: Microbial interactions with the physical environment. Benchmark Papers in Microbiology, Vol 9. Stroudsburg, PA, Dowden, Hutchinson & Ross, 1975

SPECIFIC ARTICLES

BROCK TD: Life at high temperatures. Science 158:1012, 1967

HITCHENS AP, LEIKIND MC: The introduction of agar-agar into bacteriology. J Bacteriol 37:485, 1939

MONOD J: La technique de culture continue: théorie et applications. Ann Inst Pasteur Lille 79:390, 1950

SPUDICH JL, KOSHLAND DE, JR: Nongenetic individuality: chance in the single cell. Nature 262:467, 1976

ZOBELL CE, KIM J: Effects of deep-sea pressures on microbial enzyme systems. Symp Soc Exp Biol 26:125, 1972

CELL ENVELOPE; SPORES

LORETTA L. LEIVE
BERNARD D. DAVIS

CELL WALL

In this chapter we shall consider the molecular organization, biosynthesis, and functional properties of the bacterial envelope, which consists of a cytoplasmic membrane surrounded by a rigid wall. We shall also discuss spores, whose envelope is very different from that of the parent cell.

The bacterial cell wall is of particular interest in medical bacteriology. Its unique and varied surface antigens (Ags) dominate the interactions of bacteria with host defense mechanisms; the serologic identification of these Ags is a major diagnostic tool; and wall biosynthesis is the site of action of many antibiotics (which will also be reviewed in this chapter). The membrane, in contrast, is of interest more as a model system, with properties shared by all cells.

GRAM-POSITIVE AND GRAM-NEGATIVE WALLS

The Gram stain (Ch. 2) divides bacteria into two classes, which differ in their ability to retain a basic dye after fixation by iodine. This difference is based on major differences in the structure of the cell wall, demonstrated by electron microscopy and by chemical analysis.

Morphology. In sections fixed by OsO_4 **gram-positive** cells exhibit a rather thick (20 to 80 nm) electron-dense wall, composed of peptidoglycan (see below), surrounding a cytoplasmic (plasma) membrane (7.5 nm) (see Fig. 2-4). In the envelope of **gram-negative** cells, in contrast, the similar cytoplasmic membrane is surrounded by two layers, so tightly apposed that they often cannot be distinguished in sections: a very thin peptidoglycan layer (about 1 nm), surrounded by an outer membrane (OM). Both membranes have the trilaminar ("railroad track") cross section typical of biologic membranes, in which the charged groups on the two surfaces become stained while the lipid interior forms a clear band. Freeze-etching, which in effect peels apart the layers (Ch. 2), confirms the presence of two layers in gram-positive organisms and three in gram-negatives (Fig. 6-1), and digestion of the peptidoglycan with lysozyme removes the middle layer from the latter (Fig. 6-2).

As was noted in Chapter 2, if the peptidoglycan is dissolved by lysozyme in a hypertonic medium, which prevents osmotic lysis, the cells are converted into osmotically sensitive spheres: **protoplasts** (bounded by the cytoplasmic membrane) from gram-positive cells, and **spheroplasts** (which also retain the outer membrane) from gram-negative cells. In the latter cells the access of lysozyme to the peptidoglycan requires prior damage to the OM (e.g., by EDTA* or by freezing and thawing).

Composition. The rigid layer of the cell envelope (the basal wall) can be purified by mechanically disrupting the cells (e.g., by ultrasound, by an abrupt decrease in pressure, or by shaking with glass beads) and then separating the insoluble envelope from the soluble contents and dissolving the membrane with detergents. The residual wall, which retains the shape of the cell (Fig. 6-3), is the layer responsible for the rigidity of the cell and for its resistance to osmotic lysis.† (This peptidoglycan layer is also called glycopeptide, mucopeptide, or murein.)

Membranes are obtained by osmotic lysis of protoplasts or spheroplasts: the membranes break up into small fragments, which close into vesicles (Fig. 6-4). These fragments are larger than the other cell constituents, and they can be recovered by differential centrifugation. The inner and outer membrane fragments from spheroplasts can be separated by density gradient centrifugation, because the lipopolysaccharide in the OM increases its density.

In an alternative method for preparing the OM the cells are lysed mechanically and then fractionated by centrifugation. The OM adheres to the peptidoglycan, which can be digested away by lysozyme.

Surface Topography. Negative staining of some gram-positive cells, or of their isolated wall preparations, reveals a surface layer of regular, hexagonally packed spherical subunits of 8–12 nm (Fig.

---

* EDTA (ethylenediamine tetraacetic acid; sodium edetate; Versene) chelates polyvalent metal cations, such as Mg^{2+}, and thereby weakens ionic bonds.

† Though the peptidoglycan is rigid enough to give various bacteria characteristic shapes, it is not inflexible: in spirochetes (which contain a peptidoglycan) motility involves extensive bending of the long, helical cells.

FIG. 6-1. Surface of *Escherichia coli* revealed by freeze-etching. **W** = Surface of cell wall. At the **arrow** the outer membrane has been peeled away, revealing an underlying shelf of peptidoglycan. Both these layers have been removed over most of the surface, revealing the pebbly surface (due to embedded proteins) of the fracture face between the two leaflets of the plasma membrane (**PM**). (Bayer ME, Remsen CC: J Bacteriol 101:304, 1970)

6-5), analogous to a tile wall. The subunits have been identified as proteins; in some pathogens an outer protein layer is important in virulence and in classification (e.g., the M protein of streptococci). The walls of some other bacteria and of fungi, in contrast, have a matted, irregular fibrous texture, analogous to a thatched wall, but this appearance can also be produced by the drying of a slime layer.

PEPTIDOGLYCAN

STRUCTURE

Acid hydrolysis of purified walls yields a small number of amino acids (some in the D configuration) and two sugars, *N*-acetylglucosamine (shown as GlcNAc in the figures) and its 3-O-D-lactyl ether, *N*-acetylmuramate (MurNAc; L. *murus,* wall). Additional covalently attached residues vary from one organism to another. The

enzyme lysozyme (from egg white) was found to hydrolyze a specific glycoside bond in the polymer, yielding a substituted disaccharide, GlcNAc-MurNAc. Later the use of some 20 additional bacteriolytic enzymes from various sources, each specific for a different linkage, yielded a variety of fragments. Their identification, together with that of accumulated biosynthetic intermediates (see Biosynthesis, below), revealed for *Staphylococcus aureus* the structure shown in Figure 6-6.

The essential features of this structure, found in virtually all bacteria,* are a **backbone** of alternating MurNAc and GlcNAc residues in β-1,4 linkage, a **tetrapeptide** substituent (with alternating L and D residues), and a

* Exceptions are 1) the mycoplasmas (Ch. 42), which lack rigid wall and characteristic shape, and 2) certain halophilic bacteria, whose evolutionary adaptation to a high external salt concentration has evidently eliminated the need for a peptidoglycan layer to prevent osmotic lysis.

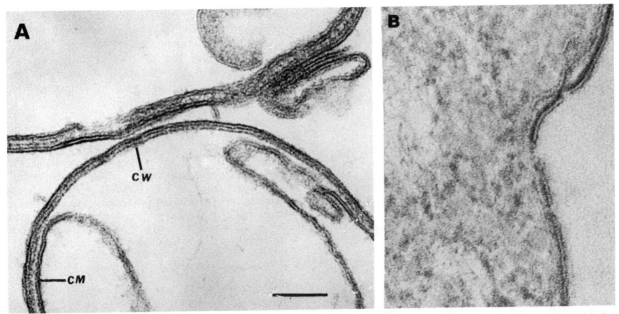

FIG. 6-2. A. Separation of the cytoplasmic membrane (**CM**) from the cell wall (**CW**) in purified envelope of *E. coli.* **B.** Partial dissolution of peptidoglycan layer of the cell wall of *E. coli* on brief treatment with lysozyme. The upper half of the section of wall has a thick, dense band outside the plasma membrane and separated by a clear layer of constant thickness from an outer, thin, dense band; untreated cells have the same structure. In the lower half lysozyme has removed much of the thick layer but a thinner residue remains. Thus in this gram-negative wall the peptidoglycan is fused with the outer layer, which is left after its removal. (**A,** Schnaitman CA: J Bacteriol 108:545, 1971; **B,** courtesy of R. G. E. Murray. See Can J Microbiol 11:547, 1965)

FIG. 6-3. Shadow-cast electron micrograph of purified wall preparation from *Bacillus megaterium.* Note flattened structure compared with intact cell. Latex balls: diam. 0.25μ. (Salton MRJ, Williams RC: Biochim Biophys Acta 14:455, 1954)

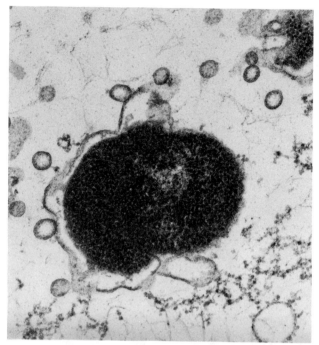

FIG. 6-4. Spheroplasts of *E. coli* being lysed by dilution. The cell is in the process of disruption, and the outer membrane is peeling away, fragmenting and resealing to form vesicles. This cell has apparently not yet lost its inner membrane, for the cell contents have not been released. However, ribosomes released from burst cells can be seen in the background. (Kulpa CF, Leive L: J Bacteriol 126:467, 1976)

FIG. 6-5. Regular hexagonal array of granules on the outer surface of isolated cell wall of a spirillum. Similarly regular rectangular arrays are found on many gram-positive bacteria, but the layer bearing the pattern is easily lost in preparation. Negatively stained. (Courtesy of R. G. E. Murray) (×204,000, reduced)

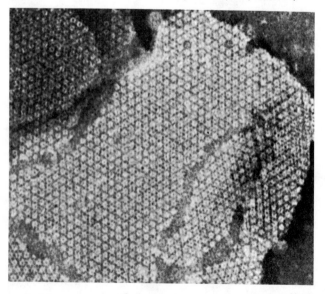

peptide bridge from the terminal COOH of one tetrapeptide to an available group (most often the free NH_2 of lysine or of diaminopimelate) of a neighboring tetrapeptide. Within this framework many variations are found: in amino acids 2 and 3 of the tetrapeptide, in the structure of the crosslinking bridge, and in the frequency of crosslinks (Table 6-1).

Variation is especially marked in gram-positive organisms, and is of value in taxonomy. Gram-negative bacteria often have a direct peptide bond between two tetrapeptides. The frequency of crosslinking varies, from about 75% (in *S. aureus*) to about 25% (in *Escherichia coli*). In most species all the MurNAc residues are substituted with tetrapeptides, even though not all are crosslinked. However, in *Micrococcus lysodeikticus* some of the tetrapeptide units have been transferred from muramate to form long crossbridges, consisting of several such units linked end to end, between the remaining MurNAc-linked tetrapeptides. The resulting very loose structure may account for the exceptional sensitivity of this organism to lysozyme.

The crosslinked structure not only joins the "backbone" polysaccharide chains into indefinite two-dimensional sheets but evidently also forms bridges **between** sheets, since the concentric layers fail to peel apart in isolated walls. The entire peptidoglycan of a cell is thus **one giant, bag-shaped, covalently linked molecule** (murein sacculus).

The amount of peptidoglycan in *Bacillus subtilis* corresponds to 40 layers; that in *E. coli* to only one layer, 1 nm thick.

Additional features of the peptidoglycan contribute to its toughness. The β-1,4 link provides a particularly compact, strong polysaccharide chain: it is also found in chitin (a poly-*N*-acetylglucosamine in fungal cell walls and crustacean exoskeletons) and in cellulose. Moreover, synthetic polymers of alternating D and L amino acids (as in the peptidoglycan tetrapeptide) are stronger than those of either kind of monomer alone. Finally, extensive hydrogen bonding may be expected between the peptide groups, as in proteins.

BIOSYNTHESIS

Peptidoglycan biosynthesis occurs in four stages: 1) synthesis of water-soluble complex precursors; 2) attachment to a membrane lipid, followed by further additions; 3) formation of linear polymers outside the membrane; and 4) crosslinking of these polymers, which may occur simultaneously with step 3.

Soluble Intermediates. The key to peptidoglycan biosynthesis came not from a direct attack on this problem but from Park's observation that penicillin-treated staphylococci accumulate a novel type of compound: a short peptide attached to a nucleotide. When the basal wall was later isolated it was found to contain the same few amino acids; thus began a long mutual interaction of

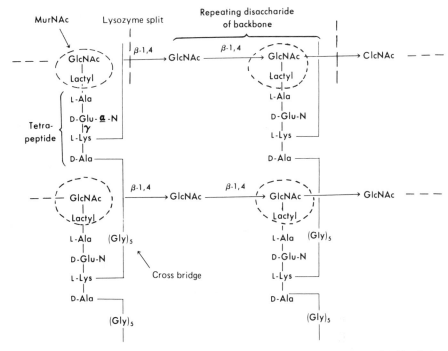

FIG. 6-6. Structure of peptidoglycan of *Staphylococcus aureus*. The polysaccharide chains (backbone) are β-1,4-linked polymers of alternating residues of *N*-acetylglucosamine (GlcNAc) and its 3-D-lactic ether (*N*-acetylmuramic acid = MurNAc; structure in Fig. 6-7). (The chain can also be considered a substituted homopolymer of GlcNAc, but because the lactic ether bond is very stable the muramic acid is ordinarily recovered as such in hydrolysates.) The COOH of the lactic group is attached to a tetrapeptide, which is in turn linked by a peptide cross-bridge to a tetrapeptide on a neighboring polysaccharide chain (which may be above, below, or in the depicted plane). Teichoic acid chains are attached to occasional MurNAc residues.

studies on wall synthesis and on penicillin action.

The accumulated compounds were found to be uridine nucleotides of MurNAc, with the carboxyl of the lactic acid moiety attached to a **pentapeptide** (Fig. 6-7) or to shorter precursors. The biosynthetic sequence (Fig. 6-8) was then established by Strominger's in vitro enzymatic studies. The synthesis of the complete UDP-MurNAc-pentapeptide precursor, from ordinary metabolites, requires 15 enzymes, present in the cytoplasm.

Unlike the synthesis of protein (Ch. 13) and of peptide antibiotics (Ch. 4), synthesis of this wall precursor utilizes a separate enzyme for each successive addition of an amino acid (except that the terminal D-Ala-D-Ala is made as a dipeptide and then added). Moreover, the amino acids are activated by the split of P_i from ATP, rather than by the more expensive split of PP used in these other processes.

Lipid-Attached Intermediates. Further studies with radioactive precursors showed that particulate, lipid-containing enzyme preparations, presumably derived from the cell membrane, can form a high-molecular-weight,

acid-precipitable peptidoglycan from appropriate precursors. In this process P-MurNAc-pentapeptide is first transferred from its nucleotide to the phosphate of a **membrane carrier lipid**, identified as **undecaprenol-P**:

$$H - (CH_2 - \underset{\underset{CH_3}{|}}{C} = CH - CH_2)_{11} - OP$$

GlcNAc is then added from its UDP derivative, and any future polypeptide bridge is built up at this stage on the pentapeptide (Fig. 6-9).

The activated amino acids required for synthesis of bridge peptides are provided by the corresponding aminoacyl-tRNAs (Ch. 13). In some cases (e.g., glycyl-tRNA of *S. aureus*) the cell has a special tRNA that functions only in peptidoglycan synthesis.

Polymerization. In the polymerization reaction a completed disaccharide subunit on a lipid carrier accepts the growing end of a polysaccharide chain from another molecule of carrier, releasing the latter as undecaprenol-PP. This compound then loses a P (Fig. 6-9), re-

TABLE 6-1. Peptidoglycan Variations in Gram-positive Bacteria

Type structure

```
                                         Bridge
            ①    ②  γ ③      ④          |
MurNAc-L-Ala-D-Glu-L-Lys-D-Ala—C=O
                    | α        | ε
                  HOOC       NH-bridge
```

Variations in tetrapeptide

```
                       α
Position 2: D-Glu-COOH
                       α
            D-Glu-CONH₂
                       α
            D-Glu-Gly
Position 3: L-Lys
            meso-DAP
            L,L-DAP
            L-Ornithine
            L-α,γ-Diaminobutyric acid
            L-Homoserine
            L-Glu
            L-Ala
```

Variations in cross-bridge from D-Ala (position 4) of one tetrapeptide to NH₂ or COOH in another tetrapeptide:

```
To NH₂ at position 3 via
   Direct peptide bond (i.e., COOH of D-Ala to NH₂ of Lys)
   (Gly)₅
   (L-Ala)₄-L-Thr
   (Gly)₃-(L-Ser)₂
   L-Ser-L-Ala
   L-Ala-L-Ala
   D-Asp-NH₂
   D-Asp-L-Ala
   Standard tetrapeptide
To COOH at position 2 (requiring a diamino acid in the cross-
bridge):
   D-Ornithine
   D-Diaminobutyric acid
   Gly-L-Lys
```

generating the form that can again accept a precursor, and possibly providing energy for reorientation toward the inner face of the membrane.

Crosslinkage to Existing Wall. The growing glycan chain is eventually released from its lipid carrier, at a length of 10–50 disaccharides. Meanwhile, it is incorporated covalently into existing wall by crosslinking of its peptides to those of other chains. The temporal relation between chain growth, termination, and crosslinking is not clear and may be variable; an oligomeric intermediate with 12 disaccharide units accumulates, without crosslinking, in *Bacillus megaterium.*

Since these incorporating reactions take place outside the cell membrane the energy source must be built into the substrate. One step, the transglycosylation from a phosphorylated intermediate in chain growth, is a familiar mechanism. The peptide crosslink, however, is formed by a novel reaction: an energetically neutral

transpeptidation (Fig. 6-10), in which a free NH₂ group displaces a terminal D-alanine, releasing that residue and forming a bridge to the subterminal D-alanine. This reaction was detected through its specific block by penicillin (see Antibiotics, below).

This interference with crosslinking was demonstrated not only in extracts but also in cells. Thus in *S. aureus* in the presence of penicillin 1) NH₂-terminal glycine accumulates, 2) an excess of labeled D-alanine is retained in the polymer, and 3) the product is much more soluble. Moreover, cells accumulate large amounts of amorphous material between the membrane and the wall, presumably in zones of growth (Fig. 6-11); sometimes this material also is extruded.

ENZYMES ATTACKING PEPTIDOGLYCAN

Bacteria contain several kinds of **autolysins** (peptidoglycan hydrolases): **glycosidases** (specific for one or the other glycoside bond between the alternating sugars in the backbone), an **amidase** (which releases the tetrapeptide from MurNAc), and **endopeptidases** (which attack various bonds in the bridge). Though the autolysis of dead cells first revealed the existence of these enzymes their main function is probably morphogenetic (i.e., splitting and reshaping the wall in the region of a dividing septum), since mutants lacking one or another autolysin may be unable to separate daughter cells or may have aberrant shapes.

Enzymes that attack various bonds in peptidoglycans are also widely distributed outside bacteria, where they supply food for animals that ingest bacteria (e.g., many protozoa), break down organic matter in the soil, and protect against infections (in plants as well as in animals). Though the chemotherapeutic use of such enzymes initially seemed promising, immune reactions proved to interfere. Lysozyme (which splits the MurNAc-GlcNAc bond) is important in human resistance to infection.

TEICHOIC ACIDS

Teichoic acids (Gr. *teichos,* wall) are found only in gram-positive bacteria. They consist primarily of chains of up to 30 **glycerol** or **ribitol** residues with phosphodiester links. They were discovered by Baddiley in cell walls, where they are attached to the 6-OH of occasional Mur-NAc residues. In some species they are major surface Ags (e.g., group D carbohydrate of streptococcus), carrying various substituents (sugars, choline, D-alanine) and present in large amounts. In addition, though some gram-positive organisms lack wall teichoic acid, all appear to have a second kind, **lipoteichoic acid** (membrane teichoic acid), in which the chain is linked not to the wall but to a glycolipid embedded in the membrane.

Teichoic acids are formed from CDP-ribitol or CDP-glycerol by a membrane-bound enzyme, as follows:

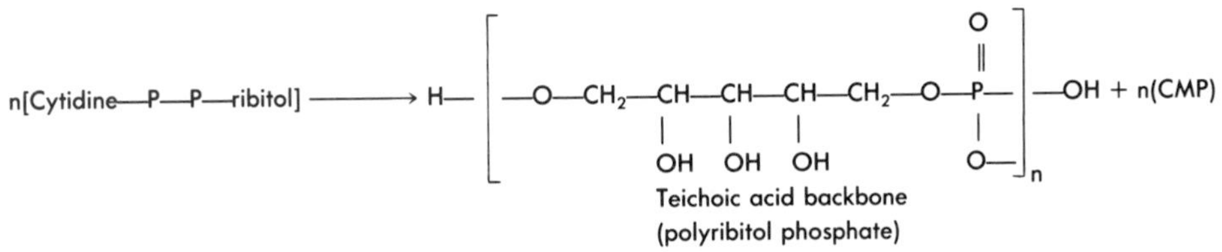

$$n[\text{Cytidine—P—P—ribitol}] \longrightarrow H—\left[—O—CH_2—CH—CH—CH—CH_2—O—\overset{\displaystyle O}{\underset{\displaystyle O^-}{\overset{\|}{P}}}—\right]_n OH + n(\text{CMP})$$

Teichoic acid backbone
(polyribitol phosphate)

The polymers are assembled on a membrane carrier, but it is not certain whether the glycolipid serves that function, or whether the membrane teichoic acid is a precursor of wall teichoic acid.

Environmental conditions can influence the amount and nature of the teichoic acid, with multiple physiologic (and presumably pathogenic) effects. Thus when the normal choline of the teichoic acid in pneumococci is replaced by the analog ethanolamine the peptidoglycan becomes resistant to autolysis, the daughter cells remain linked in long chains, and the cells lose competence to take up transforming DNA (Ch. 9). Moreover, in a me-

FIG. 6-7. Structure of UDP-Mur-NAc-pentapeptide, a uridine nucleotide from penicillin-treated *S. aureus*.

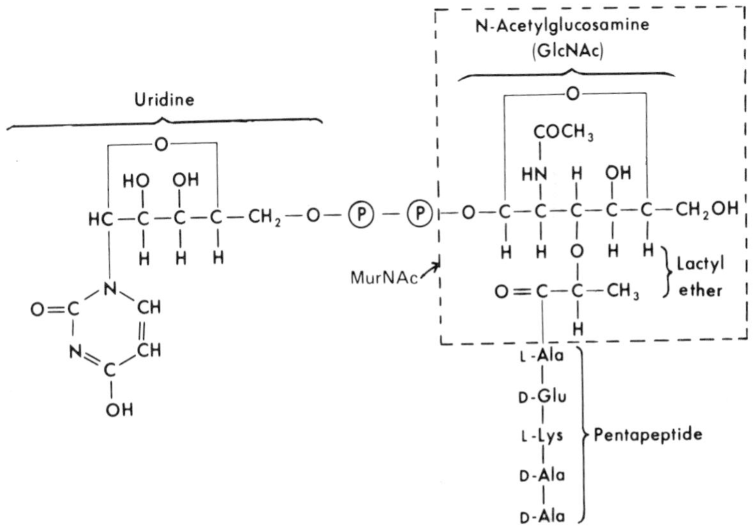

FIG. 6-8. Pathway of biosynthesis of the UDP-muramic-pentapeptide wall precursor in *S. aureus*. The individual amino acids are added with the use of energy from ATP. The sites of action of various inhibitors, which have been useful in working out this pathway, are indicated. (After Strominger JL, Tipper DJ: Am J Med 39:708, 1965)

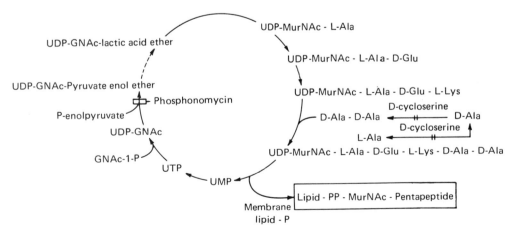

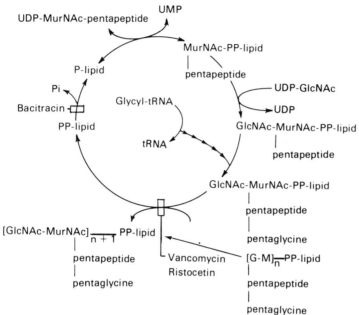

FIG. 6-9. Lipid carrier cycle in peptidoglycan synthesis in *S. aureus*. The carrier lipid is undecaprenol, which enters the cycle as a monophosphate and builds up a substituted disaccharide unit attached as a pyrophosphate. That unit, on the carrier, serves as recipient for transfer of a growing chain of *n* units from an identical carrier. The latter is then recycled. (Modified from Matsuhashi M et al: Proc Natl Acad Sci USA 54:587, 1965, Ghuysen JM: Bacteriol Rev 32:425, 1968)

FIG. 6-10. The penicillin-sensitive transpeptidation reaction in *S. aureus*, which completes the crosslink between different peptide side chains: the D-ala → D-ala peptide bond (CO → NH) is replaced by a similar D-ala → gly bond. Some species use other crosslinks than the pentaglycine bridge (Table 6–1).

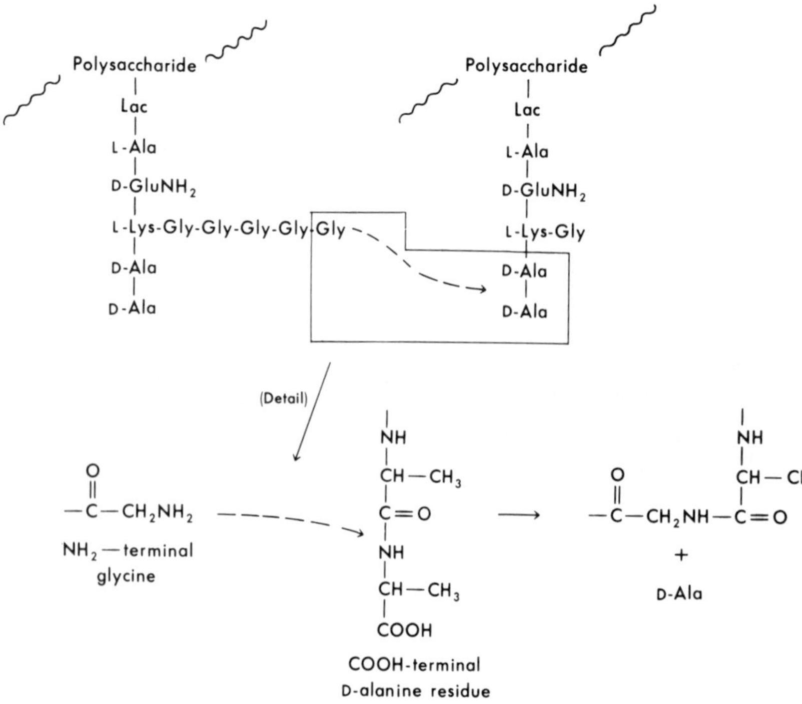

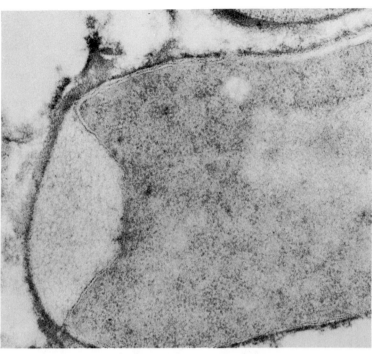

FIG. 6-11. Disorganized cell wall formation in *B. megaterium* grown in the presence of penicillin. Note the pocket of fibrous material between cell membrane and cell wall. Such accumulations are usually seen earliest in the region of the growing septum. (Fitz-James P, Hancock R: J Cell Biol 26:657, 1965)

dium limited in phosphate *B. subtilis* no longer produces teichoic acid but makes instead another acidic polymer, teichuronic acid, containing glycosidically linked uronic acid residues instead of polyol phosphates.

ANTIBIOTICS ACTING ON PEPTIDOGLYCAN SYNTHESIS

Several specific sites of inhibition of cell wall biosynthesis have been identified as targets of antibiotic action. (Other aspects of these antibiotics are discussed in Chapter 7.)

Penicillin, as we have seen, blocks crosslinking in peptidoglycan synthesis (Fig. 6-10, 6-11).

A basis for this inhibition is suggested by the structural analogy (shown in atomic models) between the regions surrounding the peptide bond in the four-membered β-lactam ring of penicillin (see Fig. 7-7, Ch. 7) and the bond in D-Ala-D-Ala that undergoes transpeptidation. The energy stored in the strained β-lactam ring of penicillin makes it an active chemical reagent, which can form a stable, covalently linked penicilloyl derivative of the enzyme that it inhibits. Radioactive penicillin forms such derivatives in bacterial cells with multiple proteins, which may well carry out the same crosslinking reaction in different regions of the wall, but with morphogenetically important differences. The filamentous growth

observed at low penicillin concentrations may be due to special sensitivity of a septal crosslinking enzyme.

In gram-negative cells the outer membrane may limit the access of penicillin to its sites of action. Thus ampicillin and penicillin G (Ch. 7) are equally effective in extracts of *E. coli* but the former is more effective with intact cells, presumably because its additional positive charge promotes transfer across the negatively charged lipopolysaccharide (LPS).

Lysis by penicillin (or spheroplast formation in a hypertonic medium) does not result simply from expansion of the growing protoplast through a fixed peptidoglycan layer; rather, it depends on an autolytic enzyme (a murein hydrolase), which splits the tetrapeptide from the backbone. When the activity of this enzyme is impaired (in growth at low pH, or in certain mutants, or in pneumococci with ethanolamine instead of the normal choline in their teichoic acid) penicillin (and other inhibitors of peptidoglycan synthesis) become **bacteriostatic** rather than bactericidal and lytic.

These studies revealed an unexpected freedom of exchange between the cell envelope and the exterior: growing cells of an autolysis-defective pneumococcal mutant can take up the normal autolysin, which then mediates lysis by penicillin.

Cycloserine, formed by a streptomycete, is a structural analog of D-alanine (Fig. 6-12). It blocks two successive

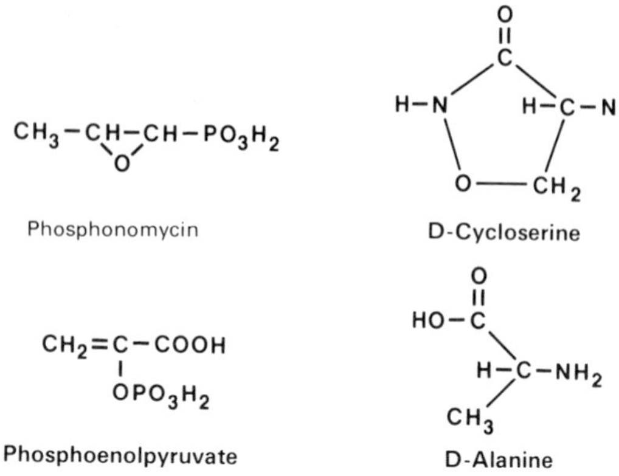

FIG. 6-12. Antibiotic metabolite analogs that inhibit peptidoglycan synthesis, and the corresponding metabolites.

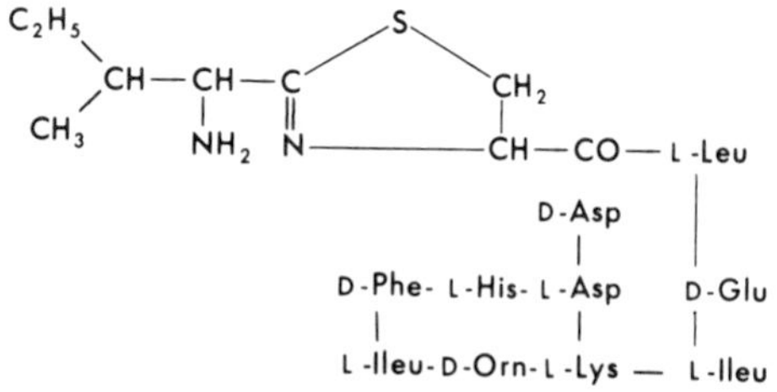

FIG. 6-13. Bacitracin A. The portion presented in detail consists of an isoleucine and a cysteine residue, condensed to form a thiazoline ring rather than the usual peptide linkage of a polypeptide.

reactions that involve this metabolite (Fig. 6-8), and D-alanine competitively reverses this inhibition, both in vitro and in cells.

The affinity of the analog for the dipeptide synthetase in *S. aureus* is about 20-fold greater than that of the natural substrate. Evidently the cyclic structure of the analog fixes it in the conformation required for binding, whereas the freely rotating bonds of D-alanine permit many conformations.

Fosfomycin (phosphonomycin: Fig. 6-12) is an antibiotic analog of P-enolpyruvate that inhibits its conversion to the ether-linked lactyl group of UDP-MurNAc (Fig. 6-8). In this biosynthesis the P-enolpyruvate attaches transiently to an SH in the active site of the enzyme, but the analog, through its highly reactive epoxide group, forms a stable derivative, irreversibly blocking the enzyme. The penetration of fosfomycin via known transport systems will be described in Chapter 7.

Vancomycin and the very similar but more toxic **ristocetin** are polyphenolic glycopeptides (mol wt about 3500), from an actinomycete of the genus *Nocardia.* They

block elongation of the peptidoglycan backbone (Fig. 6-9), in cells or in extracts.

Bacitracin, a cyclic peptide (Fig. 6-13), is produced by a strain of *B. subtilis,* from an oligopeptide cleaved from a vegetative protein during sporulation. This antibiotic forms a 1:1 complex with undecaprenol-PP and blocks its hydrolysis to the active carrier, undecaprenol-P (Fig. 6-8). (Indeed, the PP derivative was discovered as a result of its accumulation in the presence of the antibiotic.)

All these blocks in wall synthesis are bactericidal. Since vancomycin and bacitracin block the undecaprenol cycle, which is also essential in synthesis of the outer membrane LPS, these antibiotics (unlike penicillin) also inhibit the growth of spheroplasts.

OUTER MEMBRANE OF GRAM-NEGATIVE ORGANISMS

ORGANIZATION AND FUNCTIONS

Structure. The outer membrane (OM) has the fundamental features of a biologic membrane: it consists of a bilayer of amphipathic molecules, with interspersed protein molecules. On the other hand, it is so closely attached to the underlying peptidoglycan that it can also be considered the outer layer of the cell wall. Thus when the cell is fragmented mechanically the OM remains adherent to the pieces of peptidoglycan; but when these are digested with lysozyme the released OM fragments form vesicles. Functionally the OM is much more permeable than other membranes, serving as a coarse **molecular sieve.** It also serves as a **carrier of surface Ags and receptors.** In accord with this limited variety of functions, it contains only a few different proteins, whereas the cytoplasmic membrane has many different enzymes and transport systems (see Cytoplasmic Membrane, below).

The LPS is unique to the OM, its lipid terminus providing a large part (and perhaps all) of the lipid phase of the outer leaflet of the bilayer, while its long polysaccharide side chains provide surface Ags. The inner leaflet of the OM bilayer, in contrast, contains phospholipids, like other membranes.

The OM is firmly bound to the underlying peptidoglycan by two kinds of protein. A small **lipoprotein** (mol wt 7200) has its lipid, at one end of the protein, embedded in the OM, while the other end, as shown by Braun, provides a covalent link to the peptidoglycan in some species (e.g., *E. coli*) (Fig. 6-14). In addition, matrix proteins (mol wt about 36,000), present in the OM as trimers, remain attached firmly to the peptidoglycan after phospholipid is dissolved (Fig. 6-15): they detach only if heated above 60° C in a strong detergent, or if ionic bonds are weakened by a high salt concentration. Thus the OM is tightly organized: many of its proteins do not

LAYERS   COMPONENTS

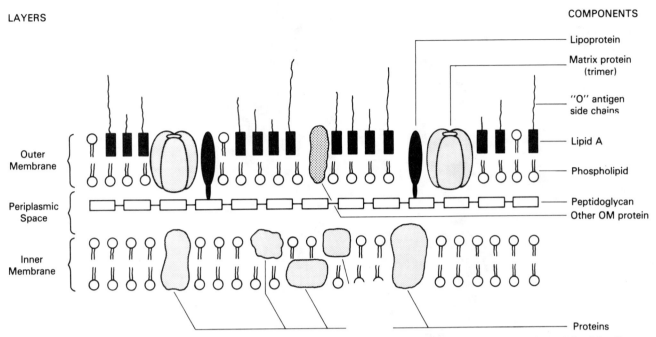

FIG. 6-14. Diagram of a gram-negative cell envelope. Components are listed on the right. The trimers of matrix protein of the Om are associated with lipoprotein and with LPS (of variable polysaccharide length), and lipoprotein is covalently bound to peptidoglycan. Diagram also illustrates some general properties of membranes (see Cytoplasmic Membrane). Phospholipid molecules are illustrated with a **circle** for the polar groups, and a **line** for each fatty acid acyl moiety.

move about freely in a two-dimensional fluid layer of lipid.

Mutations in the LPS core can result in an OM deficient in matrix proteins, suggesting an interdependence of these components during OM assembly. Moreover, in reconstitution experiments matrix proteins will interact with phospholipid and LPS to form vesicles with OM permeability properties (see below). However, formation of a normal periodic array (Fig. 6-15) also requires peptidoglycan.

Of the approximately 250,000 molecules of lipoprotein per *E. coli* cell about one-third form bridges between membrane and peptidoglycan. The covalent link of the other terminus of the protein to peptidoglycan has been demonstrated by enzymatic treatment of an envelope preparation: proteases leave a short remnant of the protein on one-tenth of the DAP residues of the peptidoglycan, while digestion with lysozyme leaves short peptidoglycan residues attached, via the lipoprotein, to the OM fragments. The link is between the ϵ-amino group of the C-terminal lysine and the COOH of DAP; the energy for forming this bond, outside the cytoplasm, is provided by displacement of the terminal D-Ala of the peptidoglycan tetrapeptide. The lipid of the lipoprotein has an unusual structure, with three fatty acid chains on an N-terminal glyceryl-S-cysteine (Fig. 6-16). The chains are presumably embedded in the inner leaflet of the OM.

Like many other membrane and secreted proteins, the lipoprotein and matrix proteins are synthesized as larger precursors, with a predominantly hydrophobic additional leading segment that is cleaved in the course of incorporation.

Permeation. In its function as a molecular sieve the OM allows diffusion of molecules up to about 800 mol wt (which can also then pass the loose underlying peptidoglycan), whereas the cytoplasmic membrane, like most membranes, excludes almost all hydrophilic molecules except water itself. The pores in the OM are provided by matrix proteins, in association with lipid and LPS. The exact molecular structure of the pore is unknown, but it may be related to the arrangement of the matrix protein molecules as trimers, demonstrated by electron microscopy (Fig. 6-15) and by isolation from membranes treated to crosslink neighboring protein molecules.

This model for the molecular sieve is supported by functional studies. Thus mutations in the matrix protein change the permeability to various substances of moderate molecular weight. Moreover, if a radioactive sugar solution is enclosed in synthetic vesicles, made of phospholipid plus LPS with or without various OM proteins, only vesicles containing matrix protein allow the sugar to escape.

The OM also contains **specific transport proteins** for certain nutrients that are too large to diffuse through its pores, e.g., vitamin B_{12}, nucleosides, maltose oligosaccharides, or Fe^{3+} chelates. Most of these proteins also appear to serve as **receptors** for specific phages or colicins (Ch. 12), since both properties can be eliminated by a single mutation. The double function suggests that phages and colicins may have evolved the capacity to parasitize surface structures that are already present to serve other, essential functions.

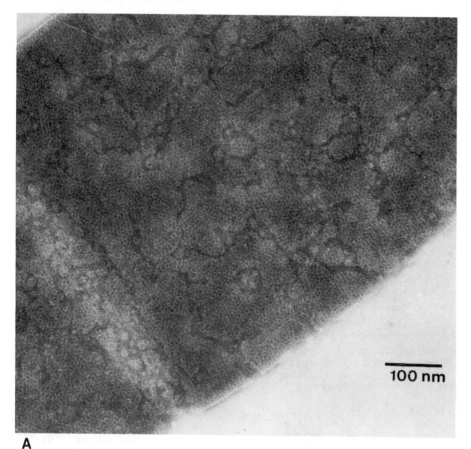

A

FIG. 6-15. Matrix protein of outer membrane (OM) of *E. coli*. **A.** Periodic array on the surface of the peptidoglycan, after extraction of other OM constituents by 2% SDS at 60° C. Negatively stained. Portions of borders of cell and septal region are shown. **B.** Same treatment of spheroplasts. Note that fragments of the matrix protein layer retain the same periodic arrangement without supporting peptidoglycan. (Steven AC et al: J Cell Biol 72:292, 1977)

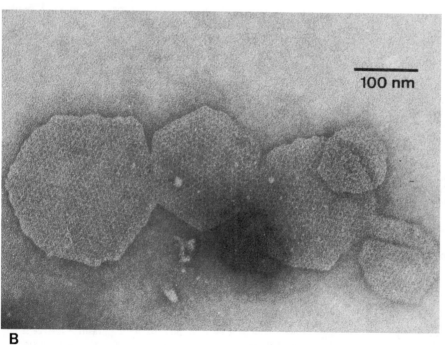

B

FIG. 6-16. Lipid-substituted *N*-terminal cysteine of lipoprotein (mol wt 7200) of outer membrane. **FA** = Long-chain fatty acid.

The composition of the OM can vary quite widely. For example, the presence of maltose, or a limitation of Fe^{2+}, can induce increased formation of the corresponding receptors; and even major proteins can be individually eliminated by mutation without loss of viability.

Despite its permeability to small molecules, the OM is less permeable than other membranes to hydrophobic or amphipathic molecules. For this reason, gram-negative organisms are generally less sensitive to antibiotics, and so they have become more prominent in human infections since the widespread use of antibiotics. Permeability to these molecules can be increased both by mutations in LPS and by loss of matrix proteins, presumably by affecting the tight OM organization described above. The OM is also resistant to detergents; and it shields the peptidoglycan from lysozyme.

LIPOPOLYSACCHARIDE STRUCTURE

LPS is also known as **endotoxin.** It can be extracted from intact cells (e.g., by incubation with 45% phenol at 90°C) and then split by mild acid hydrolysis into **lipid A** (which retains the toxicity) and a **polysaccharide** (which is responsible for the antigenic specificity).* In aqueous solution LPS forms aggregates with mol wt $>10^6$, but hot detergent separates these into the true units with six fatty acids each.

The polysaccharide portion can be lost by mutation without injury to the organism. In contrast, lipid A (including the first sugar, KDO) plays an essential role in the cell envelope (Fig. 6-14), as shown by conditionally lethal mutants (Ch. 8).

* When the phenol–water mixture separates into two phases, on cooling, LPS is found in the aqueous phase. In contrast, lipid A contains a higher proportion of lipid to carbohydrate and is soluble in lipid solvents rather than in water; hence it is called a glycolipid.

CELL ENVELOPE; SPORES **85**

Among Enterobacteriaceae lipid A is virtually constant (Fig. 6-17). The polysaccharide of LPS, however, provides hundreds of different O Ags, which are of great diagnostic importance (Ch. 31).

Lipid A. The lipid moiety of LPS from salmonellae is a glycophospholipid, consisting of β-1,6-D-glucosamine disaccharide units with all the hydroxyl and amino groups substituted (Fig. 6-17). One of the substituents in LPS is the O polysaccharide, two are phosphate or pyrophosphate, and six are long-chain fatty acids. The phosphates, and the COOH of KDO, give the outer surface of the cell a strong negative charge, which may be reduced by variable substitution with ethanolamine.

The Core. The first step in analyzing the isolated polysaccharides of various strains consisted of identifying the monomers released by complete acid hydrolysis. Westphal and Lüderitz discovered that O Ags collectively contain a wide variety of sugars, often novel (e.g., deoxyhexoses, dideoxyhexoses; Ch. 31). In addition, five sugars are common to all wild-type, smooth (S) *Salmonella* LPSs, and these are also present in certain rough (R) mutants which have lost their specific O Ags. The common sugars evidently form a constant **core** polysaccharide, to which the **O-specific** residues of any given S strain are added.

Various R mutants were found to be blocked either in the **formation** (Ch. 4) or in the specific **transfer** of various individual components of the polysaccharide. Since the core polysaccharide chain is built up sequentially (from

FIG 6-17. Structure of unit of lipid A from *Salmonella* lipopolysaccharide. Some of the 1-phosphate groups are replaced by a pyrophosphate group. Brackets enclose the three units of ketose-linked ketodeoxyoctonate (**KDO**, 3-deoxy-D-mannooctulosonic acid), which link the lipid to a variable polysaccharide and may be considered part of lipid **A**. **HM** = β-hydroxymyristic acid (a C_{14}-saturated fatty acid characteristic of lipid A); **FA** = other long-chain fatty acids. One of the HM residues has its hydroxyl group esterified with a myristic acid. The fatty acids are very similar in all enterobacteria studied but are quite different in some other gram-negative organisms (e.g., *Pseudomonas*, *Brucella*). (After Rietschel T et al: Eur J Biochem 28:166, 1972)

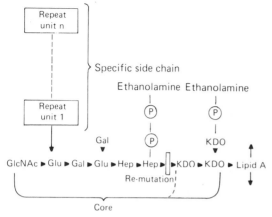

FIG 6-18. Structure of the core of *Salmonella* lipopolysaccharide (**Hep** = L-glycero-D-mannoheptose). The sugars are attached with their reducing group toward the lipid A (1→4, etc.); hence there are no free reducing groups.

In the core biosynthesis, from right to left, a specific enzyme adds each residue. The genes for the core transferases map in a cluster (the *Rfa* locus), while those for the synthesis or transfer of a component of a specific, repeating chain (Fig. 6-19) map in another cluster (*Rfb* locus; see Ch. 11 for definition of locus). Re ("extreme rough") mutants have the most incomplete LPS compatible with viability.

right to left in Fig. 6-18) these mutants yield polysaccharides with different degrees of incompleteness, and the full set revealed the sequence of the *Salmonella* core.

As Figure 6-18 shows, this sequence, proceeding inward, consists of five hexose (or glucosamine) residues, followed by two heptoses and then three residues of the 8-C sugar acid KDO. Other major groups of enterobacteria (e.g., *Shigella*) have similar but not identical cores.

Specific Side Chains. In a given LPS every side chain contains the same repeating sequence, in which the subunit is a linear trisaccharide (Fig. 6-19) or a branched tetra- or pentasaccharide. The chains vary in length (even in the same organism), ranging up to 40 repeat units. The O Ag is thus a mat of "whiskers" of variable length, each attached, via a core chain, to lipid A embedded in the outer membrane.

Since the polysaccharide chains are not charged they are not stained in the usual electron microscope preparations. Their length and location explain the presence of an electron-transparent zone of 2–4 nm between adjacent smooth cells but not between R cells.

The structures of various O Ags were recognized not only by the usual chemical methods (analysis of partial hydrolysis products and derivatives), but also by immunologic tests. As later chapters

FIG. 6-19. Biosynthesis of repeating chain of LPS of *Salmonella newington*. **A.** Inner face of plasma membrane: unit synthesis. **B.** Outer face of plasma membrane: polymerization. Und = Undecaprenol (a C_{55} polyisoprenol) carrier lipid. The nucleoside diP sugars (UDP-galactose, TDP-rhamnose, GDP-mannose) are synthesized as described in Ch. 4. The chemically reactive reducing group (C-1) of a sugar, by which it is linked to the nucleoside diP, is involved in its transfer to a growing repeat unit (**A**), in transfer of the "head" end of the growing chain to each additional repeat unit (**B**), and in final transfer (not shown) of the chain to the free end of a core chain (Fig. 6-18). (After Robbins PW et al: Science 158:1536, 1967. Copyright 1967 by the American Association for the Advancement of Science)

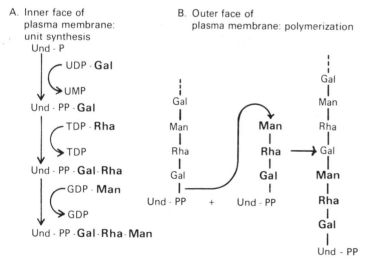

will show (Hapten Inhibition, Ch. 16), a particular Ab recognizes not only a particular sugar but also its immediate neighbor and the position and configuration of the linkage between them. Hence an oligosaccharide or glycoside of known structure competitively inhibits the reaction of specific Abs with the corresponding group in the macromolecule being analyzed.

LPS structures are further discussed in Chapter 31.

LIPOPOLYSACCHARIDE SYNTHESIS

All components of the LPS are assembled on the cytoplasmic membrane: it contains the required enzymes, and radioactive pulse–chase experiments reveal growing chains there before they are incorporated into the outer membrane.

Lipid A is synthesized by addition of the fatty acids and KDO to the glucosamine disaccharide; it is initially embedded in the inner membrane. There the **core polysaccharide** is built up on lipid A, which serves as the **carrier** and also the primer. Each successive sugar is added, by a specific transferase, from a nucleoside-diP-sugar (Ch. 4). As we have noted, various R mutants accumulate various incomplete chains, and these have provided the substrates for studying the successive enzymatic steps.

The **repeating side chains** are made by successive transfers from nucleoside-diP-sugars on another **carrier lipid**, identified by Robbins as **undecaprenol-P** (see Peptidoglycan, Biosynthesis, above). Each short repeat unit is made by **tail addition,** and when it is complete it is incorporated, as radioactive pulses show, by **head addition,** i.e., transfer of the growing polymer from its carrier to the next, carrier-attached unit (Fig. 6-19). [This process, like the synthesis of protein (Ch. 13) or of fatty acid (see below), permits synthesis of a long chain, whose free end may wander far from the site of growth.] Finally, when the chain reaches an appropriate length (or happens to encounter the transfer enzyme) it is transferred to the free end of a **core** side chain, already built up on lipid A.

Undecaprenol-P is also employed as a carrier in the synthesis of the peptidoglycan (see above), capsular polysaccharides, mannan, and perhaps teichoic acid. Capsular polysaccharides are built up on it by essentially the same mechanism described for the side chains of LPS.

TRANSFER TO THE OUTER MEMBRANE

LPS, completed on the cytoplasmic surface of the inner membrane, must finally be transferred to the outer

FIG. 6-20. Areas of adhesion between wall and inner membrane of *E. coli*. After plasmolysis (shrinkage in 20% sucrose for 2 min) the cells were fixed in formaldehyde and OsO₄, embedded, and sectioned. Numerous duct-like extensions from membrane to wall are seen; these are the sites of attachment of various phages to the outer surface of the wall, and the probable sites of synthesis of outer membrane. (Bayer ME: J Gen Microbiol 53:395, 1968)

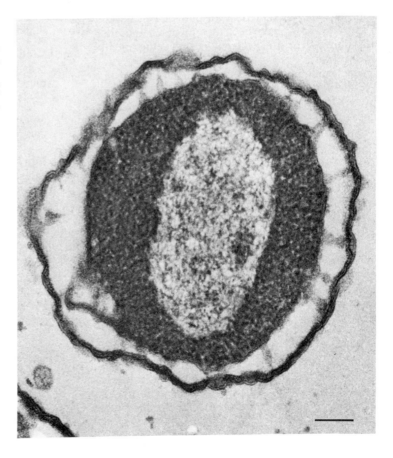

membrane. This process apparently occurs at **zones of adhesion** between the two: when gram-negative cells are fixed in a hypertonic medium the retracted cytoplasmic membrane adheres to the outer membrane at several hundred sites (Fig. 6-20). Moreover, osmotic shock leads to protrusion of the cytoplasmic membrane through holes in the peptidoglycan (about 10 nm in diameter), presumably at the same sites; and these are also sites of adsorption of some phages (reflecting a special composition). At these sites the inner leaflet of the inner membrane should be continuous with the outer leaflet of the outer membrane, and the LPS formed on the former evidently diffuses laterally to the latter.

Thus a *gal⁻* mutant forms LPS without a side chain when grown without galactose; but when galactose is supplied side chains are formed, and they may be located with labeled Ab. Such studies revealed that newly inserted LPS molecules are located directly above the adhesion zones, but in 2–3 min they spread over the entire membrane, by diffusion within the lipid bilayer. Similar evidence suggests that the proteins of the OM are also incorporated through the adhesion sites.

The attachment of LPS to the outer membrane is not uniform, for about half of the LPS of *E. coli* can be released by treatment with EDTA, while the entire LPS can be solubilized by phenol. These results suggest that the attachment depends both on hydrophobic bonds and on ionic bonds mediated by polyvalent cations (e.g., Mg^{2+}).

CYTOPLASMIC MEMBRANE

COMPOSITION

Reasonably pure preparations of fragmented cytoplasmic membrane can be obtained by methods described above. The protein content is slightly higher than in mammalian cell membranes (60%–70%, with 20%–30% lipid and small amounts of carbohydrate). The membrane (including invaginations) may contain as much as 20% of the total protein of a bacterium. Many proteins with specific functions in transport or in catalysis (see Enzymes, below) have been identified; whether there are also purely structural proteins, as in the outer membrane, is uncertain.

The lipids of the cytoplasmic membrane of bacteria are mainly phosphatides, with traces (<1%) of undecaprenol-P. Virtually no bacteria (except for certain mycoplasmas) contain sterols.

ORGANIZATION AND FUNCTIONS

The cytoplasmic membrane provides an **osmotic barrier,** traversed at intervals by **specific transport systems** (though they cannot be seen in the deceptively uniform cross section of the membrane in electron micrographs). The barrier retains metabolites and excludes external compounds: ions, and nonionized molecules larger than glycerol, penetrate very slowly except by specific transport systems. In addition, this membrane contains a wide variety of enzymes of biosynthesis and of respiration; it provides an insulating barrier across which energy can be built up in the form of a membrane potential; and it energizes flagellar movement.

Because phospholipids are **amphipathic,** with a hydrophilic polar region and long, hydrophobic lipid chains, they aggregate from dispersions in aqueous solution, in vitro as well as in the cell, to form a highly oriented, thin **bilayer,** with the polar regions exposed at the surface (Fig. 6-14). This film forms the osmotic barrier. The proteins in natural membranes possess large lipophilic surface regions, which account for their stable incorporation into the membrane and their insolubility in water unless coated with detergent. As is diagrammed in Fig. 6-14, membrane proteins may span the membrane (transmembrane proteins), or they may be exposed on only one face. The latter type are easier to remove; they may be bound mostly to the exposed surface of transmembrane proteins. The two types have been demonstrated by radioactive labeling of both surfaces, in isolated membrane, and the outer surface only, in intact cells, followed by solubilizing and fractionating the membrane proteins.

Some halophilic bacteria form gas bubbles by secreting N_2 **between** the two lipid layers of a specialized intracellular membrane, thus regulating their depth in a liquid.

Lipid Interactions. The mutual attractions of lipid molecules, unlike those of proteins and nucleic acids, are relatively nonspecific: hence lipids readily exchange their contacts within either leaflet of a membrane. (They rarely flip-flop between the two leaflets, except during synthesis.) Membranes are consequently two-dimensional fluids. Since bilayers with only saturated fatty acids are rigid at physiologic temperature, some of the fatty acids must be unsaturated (see Fatty Acids, below): mutants unable to make these compounds require them, and on deprival they die when the membrane reaches a composition incompatible with fluidity.

Since different membranes in the same cell, and the inner and outer leaflets of any membrane, differ in lipid and protein composition, it seems necessary to infer a good deal of specificity for the lipid–lipid and lipid–protein interactions that presumably guide the insertion of various components. At least some membrane proteins are incorporated with the help of hydrophobic leading segments, while being synthesized by ribosomes on the inner surface; but proteins that do not traverse the membrane may be adsorbed from the cytoplasm. The proteins of the OM are probably incorporated, like LPS, at the zones of adhesion between IM and OM (see above, Fig. 6-20).

Enzymes. The cytoplasmic membrane contains enzymes involved in 1) electron transport and oxidative phosphorylation, 2) complex lipid synthesis, 3) synthesis of wall constituents, and 4) DNA replication. It also contains transport systems and chemotactic systems (see below: Membrane Transport; Motility and Chemotaxis), with enzymelike specificity. Finally, the membrane is involved in protein secretion, and, much like the endoplasmic reticulum in eukaryotic cells, it binds a considerable fraction of the ribosomes in the cell (see Protein Secretion, below).

Membrane proteins vary widely in their ease of solubilization: the sequence of increasingly drastic reagents is buffers, EDTA, osmotic shock, nonionic detergents, deoxycholate, dodecylsulfate, phenol, and finally hot phenol. Membrane enzymes often are inactive when purified but regain an active conformation when the appropriate phosphatide is provided.

MESOSOMES

Electron micrographs of gram-positive bacteria show large, irregular convoluted invaginations of the cytoplasmic membrane; in gram-negative cells they are smaller. These bodies appear to be connected with the cytoplasmic membrane, and cells in electron-dense media reveal connections between mesosomes and the periplasmic space (Fig. 6-21). The exact form of mesosomes in vivo, however, is uncertain, since freeze-fracture studies reveal such shapes in fixed but not in unfixed cells.

Nevertheless, mesosomes appear to be entities with at least two identifiable functions. **Septal** mesosomes participate in DNA replication and cell division, often remaining as polar mesosomes after division. **Lateral** mesosomes, attached to nonseptal regions, function in **secretion.** Thus during induced formation of penicillinase in a *Bacillus* such mesosomes increase in number; in addition, they appear to give rise to periplasmic vesicles (Fig. 6-21).

These vesicles are a specialized apparatus for penicillinase excretion in such cells: they may be recovered by conversion of the cells to protoplasts. Lampen has shown that they contain the bulk of the cell-bound penicillinase, which they release on lysis in vitro (and presumably at the cell surface). The remainder of the cellular penicillinase is attached to the cell membrane via a short peptide "leash" with a terminal phosphatide; this bound precursor of the exoenzyme is in turn formed from a still larger precursor with an additional N-terminal leading sequence.

Although mesosomes in most bacteria are not enriched for components of the electron transport chain, photosynthetic bacteria have a similar internal membranous structure, the chromatophore, which contains the photosynthetic pigments and all other factors required for photosynthesis.

FIG. 6-21. Mesosomes and periplasmic vesicles in *Bacillus licheniformis.* Two parallel cells, with negative stain penetrating into large septal mesosomes (**S1**) and into smaller lateral mesosomes; **arrow** shows vesicles being released from a lateral mesosome into periplasmic space. These cells are excreting penicillinase; in comparable cells not engaged in enzyme secretion lateral mesosomes and periplasmic vesicles are rare or absent. (Ghosh, BK et al: J Bacteriol 100:1002, 1969)

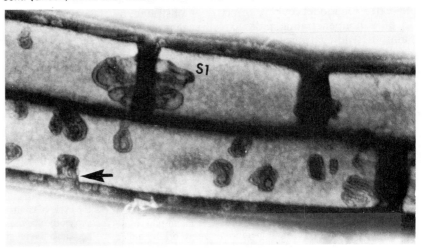

PERIPLASMIC PROTEINS AND EXOENZYMES

Gram-positive bacteria excrete various exoenzymes into the medium, sometimes via vesicles (see preceding section). Most of these enzymes (e.g., proteases, nucleases) convert impermeable substances into permeable foodstuffs; others destroy harmful substances (e.g., penicillinase) or serve as virulence factors (exotoxins) of pathogens.

Gram-negative bacteria may release exotoxins to the medium, but most of their excreted proteins are retained in the **periplasmic space,** between the inner and the outer membrane. The periplasm thus sequesters potentially destructive enzymes in a useful location, much like the lysosomes of animal cells. The periplasm also contains specific binding proteins, whose function in membrane transport and in chemotaxis will be discussed in later sections. The periplasmic location of a protein* is demonstrated by its release when the cell is converted to a spheroplast or is subjected to osmotic shock.†

PROTEIN SECRETION

Animal cells that secrete proteins are rich in membrane-bound ribosomes, and this apposition suggested that the proteins cross the membrane (into the endoplasmic reticulum) as growing chains. Moreover, in these cells (e.g., those forming immunoglobulins, Ch. 19), and then in bacteria, many secreted proteins were found to be synthesized as larger precursors with an additional leading (N-terminal) segment, of about 20 predominantly hydrophobic amino acid residues. This "signal" sequence evidently directs the ribosome to the membrane, initiates secretion, and then is cleaved by an enzyme located in the membrane.

With bacterial spheroplasts or protoplasts, which secrete directly to the outside, this simultaneous translation–extrusion model could be directly demonstrated, by use of a nonpenetrating radioactive reagent to label chains protruding from the surface. After gentle fractionation labeled chains were found still attached to polysomes in the cell.

* When a cell is broken up the apparent location of some enzymes may be distorted by artefacts of adsorption. For example, in lysates of *E. coli* cells RNase I, a highly basic protein, appeared to be a component of the ribosome, but it is now known to be a periplasmic enzyme: when the cells were converted to spheroplasts it was released, and subsequent lysis yielded ribosomes without the enzyme.

† In **osmotic shock** cells are exposed to EDTA in a hypertonic medium and then are rapidly diluted in a chilled solution of low osmotic strength. All layers may be damaged: the EDTA releases a large fraction of the LPS and makes the outer membrane more permeable; the cold makes both membranes more brittle; and a sudden outflow of osmolytes evidently places stress on the peptidoglycan.

Up to half the polysomes in a gentle bacterial lysate are recovered in membrane–polysome complexes. These synthesize membrane proteins and secretory proteins, while the free polysomes make cytoplasmic proteins. The membrane-bound ribosomes may be attached solely by their nascent chains, since they are freed when the chains are released (Ch. 13) by puromycin.

STRUCTURE AND BIOSYNTHESIS OF LIPIDS

Aspects of these biochemical topics with special features in bacteria will be briefly discussed here.

FATTY ACIDS

The apolar components of bacterial lipids are predominantly long (C_{14} to C_{18}) fatty acid chains, which may be saturated or monounsaturated, but not polyunsaturated. Unusual fatty acids are also found, containing methyl branches, cyclopropane rings, or extremely long chains (e.g., mycolic acids with two long branches totaling C_{83} in tubercle bacilli). The **biosynthesis of fatty acids** in bacteria, as in higher organisms, proceeds by successive addition of acetyl residues to the chain at its carboxyl end.

The reactivity of the acetyl group (initially on CoA) is first increased by adding CO_2 to form malonyl CoA (Fig. 6-22), and the malonyl group is then transferred to the -SH of a long-chain prosthetic group (phosphopantetheine; see Fig. 4-12) on a small acyl carrier protein (ACP). (This "head addition" resembles the growth of specific wall polysaccharides; Fig. 6-19.)

The growing chain, on an -SH of the condensing enzyme, is transferred to the active CH_2 of the malonyl group, displacing its free carboxyl and creating a β-keto acid attached to the ACP. (The added CO_2 is thus not incorporated, though it is essential for growth; Ch. 5.) The β-keto acyl group on the flexible pantetheine "leash" of the ACP is successively exposed to several enzymes linked in the fatty acid-synthesizing complex, and the β-keto group is thereby reduced to a CH_2 group. The resulting longer fatty acid is then transferred back to the condensing enzyme. This cycle is repeated until the fatty acid reaches the proper length for transfer to an acceptor to form a complex lipid.

The common even-numbered, straight-chain products are formed by initiation by acetyl-CoA. Fatty acids with an odd-numbered chain, or with a subterminal methyl branch, may be formed respectively by initiation with propionyl-CoA or with the acyl-CoA derived from a branched-chain amino acid. In mycobacteria a partly methylated oligosaccharide solubilizes the acyl-CoA as it is extended beyond the usual limit of 18 C atoms.

Unsaturation. Some aerobic bacteria, like plants and animals, make unsaturated fatty acids by O_2-linked oxidation of a saturated fatty acid. Bacteria growing anaerobically, however, cannot use this route. Instead, they modify the formation of a double bond in the cycle of chain elongation. Usually the dehydration steps in this cycle produce *trans* double bonds, which are then reduced; but at a certain chain length a *cis* double bond is produced instead in some molecules, by a special enzyme, and this bond is not a substrate for reduction.

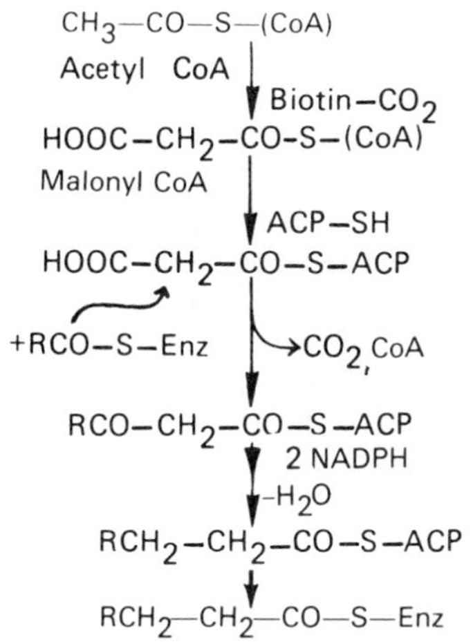

$$CH_3—CO—S—(CoA)$$

Acetyl CoA

$$\downarrow \text{Biotin—}CO_2$$

$$HOOC—CH_2—CO—S—(CoA)$$

Malonyl CoA

$$\downarrow \text{ACP—SH}$$

$$HOOC—CH_2—CO—S—ACP$$

$$+RCO—S—Enz \qquad \searrow CO_2, \text{CoA}$$

$$RCO—CH_2—CO—S—ACP$$

$$\downarrow \text{2 NADPH}$$

$$\downarrow \text{—}H_2O$$

$$RCH_2—CH_2—CO—S—ACP$$

$$\downarrow$$

$$RCH_2—CH_2—CO—S—Enz$$

FIG. 6-22. Chain elongation in fatty acid biosynthesis. ACP-SH = Acyl carrier protein; RCO- = growing chain. The elongated acyl group is transferred from -S-ACP to -S-Enz, recycling the ACP-SH.

COMPLEX LIPIDS

In most of the complex lipids of bacteria (other than lipid A, see above) two of the hydroxyls of glycerol are attached to hydrocarbon chains, through either ester, vinyl ether (plasmalogen), or saturated ether linkages; the third hydroxyl is attached either to phosphate (in phosphatides) or to a sugar (in some glycolipids). The phosphate in turn is linked either to a nitrogenous compound (serine, ethanolamine) or to a polyol (glycerol).

In enterobacteria most of the phosphatide in the outer membrane is the neutral phosphatidylethanolamine, whereas the cytoplasmic membrane contains in addition the acidic phosphatidylglycerol and diphosphatidylglycerol (cardiolipin). Phos-

phatidylcholine (lecithin) is rare in bacteria. The cytoplasmic membrane of *B. subtilis* contains predominantly derivatives of ethanolamine in the inner layer and of glycerol in the outer layer. Some bacteria form **lipoamino acids,** with an amino acid (derived from its aa-RNA) on one of the free hydroxyls of phosphatidylglycerol.

Mevalonic Acid. As noted above, most bacteria do not contain steroids, which are synthesized in yeasts and higher organisms from acetyl CoA via mevalonic acid (β,γ-dihydroxy-β-methylvaleric acid). However, this intermediate also gives rise to the polyisoprenoid membrane carrier lipid, to carotenoids (present especially in photosynthetic bacteria), and to side chains on other constituents (e.g., quinones). Indeed, mevalonic acid was initially discovered as a growth factor for *Lactobacillus acidophilus*.

VARIATIONS IN COMPOSITION

The composition of bacterial lipids varies markedly with conditions of growth, including temperature, pH, and composition of the medium. At lower temperatures, for example, cells synthesize a larger proportion of unsaturated fatty acids (which have a lower melting point). Moreover, as cells approach the **stationary phase** of growth (Ch. 5) their chances of survival through a period of starvation are increased by changes in their lipids that "toughen" the membrane (e.g., conversion of unsaturated fatty acids to cyclopropane fatty acids, by addition of a methylene group from S-adenosyl methionine across the double bond).

The mechanism of the regulation by temperature is twofold: in direct response to temperature the enzymes of fatty acid biosynthesis change the ratio of saturated to unsaturated residues that they produce, and the enzymes that transfer long-chain acyl residues from ACP or AcCoA to glycerol-P shift their selection from the mixture.

ANTIBIOTICS ACTING ON THE MEMBRANE

DAMAGE TO THE MEMBRANE

Polymyxins, a group of cyclic polypeptides produced by *Bacillus polymyxa*, resemble cationic detergents in having basic groups plus a fatty acid side chain (Fig. 6-23). They cause direct membrane damage by a detergentlike action (Ch. 67) but at much lower concentrations.

The membrane damage can be recognized by the decrease of turbidity, leakage of soluble constituents (including nucleotides and inorganic ions), penetration of normally excluded substrates into the cell, and staining of the cell by a dye that fluoresces when bound to pro-

teins. Furthermore, a fluorescent derivative of polymyxin becomes concentrated at the cell membrane (Fig. 6-24). The lysis caused by polymyxin differs from that caused by penicillin or lysozyme: it leaves the intact wall as a relatively nonrefractile ghost, and it is not prevented by hypertonic media.

Polymyxin is unique among useful chemotherapeutic agents (Ch. 7) in being bactericidal even in the absence of cell growth. Complexing occurs with several phosphatides but not with phosphatidyl choline (which is present in animal cells but not in bacteria): this difference may be related to the chemotherapeutic selectivity of the antibiotic.

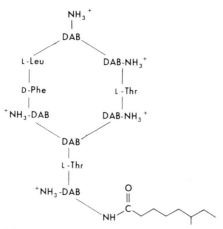

FIG. 6-23. Polymyxin B. DAB = α,γ-diaminobutyrate [$NH_2CH_2CH_2CH(NH_2) COOH$]. Aliphatic residue is 6-methyloctanoic acid.

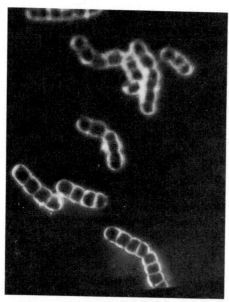

FIG. 6-24. Fluorescence photomicrograph of *B. megaterium* treated with a fluorescent derivative of polymyxin. (Newton, BA: J Gen Microbiol 12:226, 1955)

Polymyxin also binds to LPS, which may explain why it is more effective against gram-negative than positive organisms: mutations in LPS that reduce binding also reduce sensitivity.

Polyene antibiotics (Ch. 45) similarly impair membrane integrity, but by complexing with **sterols.** They are therefore effective against certain fungi (Ch. 45) but not against bacteria (except those mycoplasmas that have incorporated sterols in their membrane, Ch. 42).

IONOPHORES

These molecules greatly increase the permeability of membranes to specific ions: they form rings that have a nonpolar periphery and therefore spontaneously insert themselves ("dissolve") in membranes, while the interior of the ring closely fits a particular inorganic cation and also provides carbonyl groups that coordinate with it, replacing its normal hydration shell. (Such caged ions are called **clathrates.**) These antibiotics are bactericidal but they are not selective enough to be useful in therapy; they are of interest chiefly as tools for studying membrane physiology.

TABLE 6-2. Classes of Antibiotic Ionophores

Enniatins	Cyclic depsipeptides with ring of 18 atoms
Gramicidins	Cyclic (gramicidin S) and open-chain (A,B,C) peptides
Actins (e.g., Nonactin)	Tetralactone, ring of 32 atoms, 4 methyl or ethyl substituents
Nigericin, monensin	Open-chain: form ring by hydrogen-bonding the terminal COOH to 2 OH groups at other end

Several chemical classes of antibiotics have been found to act as ionophores. Some have a covalently closed ring, while others are linear but fold into a ring; some are neutral and others have a carboxyl group (Table 6-2). For example, **valinomycin,** which is highly specific for transporting K, is a cyclic **depsipeptide** (i.e., a chain of alternating α-amino and α-hydroxy acids connected by peptide and ester bonds):

$$[D - hydroxyvalerate - D - valine - L - lactate - L - valine]_3$$
Valinomycin

As atomic models show, the aliphatic side chains readily face outward from the ring and the carboxyl groups face inward, which explains how these compounds form selective pores across membranes.

The carrier function depends on diffusion or on conformational change of the ionophore within its lipid matrix; the function is lost at temperatures low enough to solidify the lipids. Neutral ionophores permit the charged ion to move across the membrane in the direction of the electrical potential, thus eliminating that potential. (The simultaneous block in oxidative phosphorylation supports the view that in electron transport energy is transferred in the form of a membrane potential.) Acidic ionophores can exchange their metal ion for H^+; hence they eliminate pH differences rather than potential differences across the membrane.

THE ENVELOPE IN CELL DIVISION

WALL GROWTH AND SEPARATION

The division of bacteria (by binary fission) produces two daughter cells of approximately equal size, though the equality may be quite rough. Yeasts generally divide by the budding of a small daughter cell from the larger mother cell (Ch. 45).

Division of bacteria starts with ingrowth of cytoplasmic membrane, sometimes in association with a mesosome (see Fig. 2-4, Ch. 2). The accompanying ingrowth of wall eventually forms a complete transverse septum (cross-wall), which is thicker than the peripheral cell wall. Septal cleavage, progressing inward from the periphery, leads to cell separation.

Differences in the details of septum formation and separation create characteristic differences in bacterial shape and arrangement. For example, linked cocci are formed by incomplete cleavage of septa: streptococci form successive septa in parallel orientation, yielding linear chains, while staphylococci begin a new septum perpendicular to a still unseparated old one, yielding three-dimensional clumps. Prolonged delay in the cleavage of bacilli, often seen under conditions of impaired growth, yields long **filaments.**

Mutants with many different temperature-sensitive alterations in cell division have been isolated. For example, when grown at a nonpermissive temperature, which causes expression of the genetic defect, mutants that cannot make cross-walls yield **filaments,** while others frequently make a cross-wall near one end of the cell and thus form **minicells** (Fig. 6-25) without a chromosome. Some mutants with impaired peptidoglycan synthesis form **fragile** cells, and still others develop highly **aberrant shapes.** Clearly, many genes contribute to the orderly process of cell division. The pleomorphic spontaneous L-phase variants (Ch. 43) of various organisms, which lack a rigid wall, evidently have a genetic defect in wall synthesis.

Lysozyme-induced protoplasts cannot divide, though they can increase their protoplasm severalfold. Mycoplasmas (Ch. 42) and spheroplasts can divide slowly (especially in the depths of agar, whose fibers may help to pinch off daughter cells); the daughter cells vary widely in size.

SITES OF WALL GROWTH

The **equatorial zone** around a growing septum is a major region of elongation and reshaping of the peripheral wall. One might expect such a region to be weak because of an abundance of "loose ends" of the growing peptidoglycan chain and the presumed presence of morphogenetic enzymes. Indeed, on limited autolysis staphylococci release the wall as hemispheres.

In more direct tests, with **gram-positive** streptococci, the surface is labeled by staining with fluorescent Abs or by incorporation of a tritiated component, and growth is then continued without further labeling. The unlabeled new wall is laid down in an equatorial zone, where cell division finally takes place (Fig. 6-26). **Gram-negative** enterobacteria also have an equatorial zone of rapid peptidoglycan assembly, as shown by the initial bulging of this region in cells growing in the presence of penicillin, and by radioautography after labeling peptidoglycan with a short pulse of ^{3}H-DAP.

After more prolonged growth the radioactivity is found more diffusely distributed in enterobacteria, possibly through formation of additional zones of elongation, or through formation and resealing of the sites of adhesion between inner and outer membrane.

MEMBRANE TRANSPORT

EXISTENCE OF TRANSPORT SYSTEMS IN BACTERIA

That bacteria admit and concentrate substances selectively was suggested early by two findings: certain enzymes are **cryptic** (i.e., they can attack added substrate in a lysate but not in intact cells); and amino acids and mineral ions are often found at concentrations

FIG. 6-25. *E. coli* mutant producing a minicell, without chromosome. (From Adler HI et al: Proc Natl Acad Sci USA 57:321, 1967)

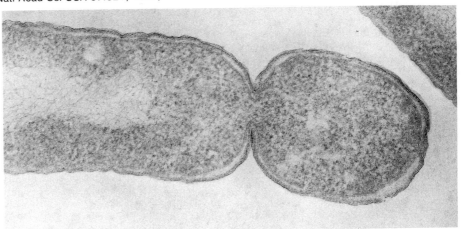

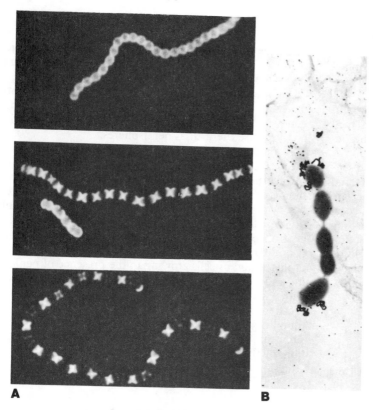

FIG. 6-26. Equatorial growth of gram-positive cell wall. **A.** Top: Streptococci, with surface Ag stained with fluorescent Ab (Ch. 16). The next two samples were taken after 15 and 30 min of growth without Ab: fresh, unstained wall has been deposited in each cell in an equatorial ring, which separates the older, stained parts of the cell. Each stained portion remains attached to the similar portion of the adjacent cell, forming an ''X.'' **B.** Pneumococcus with wall of original cell labeled in its teichoic acid with ^{3}H-choline, and then grown for several generations with ethanolamine (which promotes formation of chains of cells rather than separation into diplococci). Electron microscope radioautographs reveal localization of silver grains (from disintegration of ^{3}H) only at ends of chain, indicating formation of new wall between conserved halves of original, labeled wall. (**A,** Cole RM: Bacteriol Rev 29:326, 1965; **B,** Briles EB, Tomasz A: J Cell Biol 47:786, 1970)

within the cell many times that outside. Nevertheless, bacteria were long regarded as semipermeable bags of enzymes.

Specific transport systems in bacteria, capable of active transport against a concentration gradient, were unequivocally demonstrated by the induction of some systems.* Thus exogenous citrate can be metabolized by *Enterobacter* cells grown in its presence, but not by cells grown on glucose. Since the enzymes for its metabolism must be active in both kinds of cells, the adaptive response to citrate has evidently increased the access of exogenous substrate. Cohen and Monod similarly showed that the adaptive response of *E. coli* to β-galactosides (Ch. 3, Diauxie) involves formation not only of the enzyme β-galactosidase but also of machinery permitting entry of its substrate into the cell, and also active transport of nonmetabolizable analogs of the substrate (e.g., methyl β-thiogalactoside, MTG; see Table 14-1, in Ch. 14). The "downhill" transport of substrate and the "uphill" transport of the analog evidently employ the same system: the two kinds of compounds compete for entry, and the two activities respond in parallel to induction (and to mutation).

Active transport (or active concentration) can be defined as transport from a given external concentration (or, more precisely, chemical potential) to a higher level within the cell. It is usually studied by incubating cells with a radioactively labeled permeant

---

* Specific inducible transport systems in bacteria were initially called **permeases,** but this term no longer seems useful, since these systems involve multiple proteins and they do not alter the substrate enzymatically.

and measuring the amount retained after rapid filtration. The radioactive compound recovered from the cells must be the same as that supplied, and not a derivative; hence to demonstrate active transport of a sugar in bacteria it is usually necessary to block subsequent metabolism (e.g., by a mutation) or to use a nonmetabolizable analog.

Carrier-Mediated Transport. Active transport requires the expenditure of energy. However, under conditions of **downhill flow** (i.e., entry followed by rapid metabolism), or in the absence of an energy supply (e.g., cells poisoned by azide), some systems capable of active transport become **uncoupled** from the energy supply but continue to shuttle back and forth, functioning as carrier-mediated (but not energy-linked) transport systems. Carrier-mediated transport is also called **facilitated** or **passive transport, or facilitated diffusion.** However, it has saturable, Michaelis kinetics, while true diffusion increases in rate indefinitely with increasing concentration of the permeant.

Systems capable of carrier-mediated transport, but incapable of active transport, are found widely for sugars in RBCs and in yeasts, for which sugars are rarely limiting in their natural habitat. The presence of active transport systems in bacteria evidently reflects the evolutionary adaptation of these cells to a poor supply of the substrate.

Because an increase in the concentration of a permeant above the saturating level cannot further raise the intracellular concentration, spheroplasts can be protected osmotically by hypertonic solutions, even of compounds (e.g., sugars, KCl) that can enter the cell.

KINETICS OF ACTIVE TRANSPORT

The kinetics of active transport can be illustrated by the β-galactoside system in *E. coli* and may be summarized as follows. In the presence of an energy source cells possessing this system take up a corresponding radioactive nonmetabolized permeant (e.g., MTG) at an **initially constant** velocity (V_{in} = **influx** rate). (The rate of **net** uptake soon slows because the efflux begins to balance the influx.) V_{in} varies with the external concentration of permeant (C_{ex}) according to the mass-law kinetics of a saturable system, with a characteristic dissociation constant (K_m). The maximal uptake rate (V_{in}^{max}) is the value obtained at a saturating concentration of permeant; and K_m, like the Michaelis constant of an enzyme, is the concentration yielding half the maximal rate. Thus:

$$V_{in} = V_{in}^{max} \cdot C_{ex}/(K_m + C_{ex}). \qquad (1)$$

In contrast to the Michaelis kinetics of influx, the rate of efflux at any time is proportional to the internal concentration (C_{in}) of permeant, and to the exit rate constant (K). The overall kinetics is thus:

$$\text{Net uptake} = \text{Influx} - \text{Efflux}$$
$$dC_{in}/dt = V_{in}^{max} C_{ex}/(K_m + C_{ex}) - KC_{in} \qquad (2)$$

At equilibrium (reached after 2–3 min at 37° C) the influx and the efflux rates are equal; hence

$$V_{in}^{max} \cdot C_{ex}/(K_m + C_{ex}) = KC_{in} \qquad (3)$$

Linear exit kinetics can be explained both by nonspecific diffusion through the membrane and by transport via systems for which the permeant has low affinity (and hence remains on the early, linear part of the Michaelis curve). Diffusion probably becomes important only at extremely high internal concentrations of permeant or when the membrane is injured. Thus, in the commercial production of amino acids by bacterial mutants the yield may be increased by controlled damage to the membrane.

Table 6-3 shows that different permeants using the same systems differ from each other independently in various parameters: affinity for the entry system, maximal rate of entry, exit rate constant, and the resultant maximal concentration ratio. Ratios of 100 or more are common.

SPECIFIC TRANSPORT SYSTEMS

Many **sugar transport systems** have been described, with properties similar to the β-galactoside system of *E. coli*. In addition to their high affinity for one sugar (K_m about 10^{-4} M) most also have a far lower affinity for other sugars, which may account for apparently nonspecific transport (see Nonspecific Entry, below). (An additional

TABLE 6-3. Parameters of β-Galactoside Transport System

Parameter	Thiomethyl-galactoside	Thiodiga-lactoside	Lactose
Michaelis constant, K_m (mole/liter)	5×10^{-4}	2×10^{-5}	7×10^{-5}
Capacity Y (μmole/g) at saturation	300 (14°) 160 (26°) 52 (34°)	40	550 (4°) — 125 (34°)
Maximal rate of uptake, V_{in}^{max} (μmole/g/min)	148	20.4	158
Exit rate constant, K_{ex} (ml/g/min)	0.82	0.59	—
Maximal concentration ratio, C_{in}/C_{ex}	65	400	1950

(Kepes A, Cohen GN: In The Bacteria, vol. IV, p. 179. New York, Academic Press, 1962.)

mechanism of sugar uptake, group translocation, will be discussed below.)

Amino acids are actively concentrated by a similar set of transport systems, with even higher affinities, the K_m values ranging mostly from 10^{-6} to 2×10^{-8} M. These low values may reflect the lower concentrations found in the microbial environment.

A rather complex set of specificities has been found in *E. coli* and related organisms: amino acids of similar structure are often transported by the same carrier (e.g., aromatic; aliphatic; basic; Gly-Ser-Ala; Cys-DAP); but there are also individual transport systems for some amino acids in the same group. This array of carriers permits the simultaneous uptake of many different amino acids, over a wide range of concentrations, without mutual interference. Amino acid transport systems, unlike those for sugars, transport their substrates faster than they can be utilized; hence active concentration can be seen even without a block in protein synthesis. These systems are generally constitutive (i.e. are present under all growth conditions), but some high-affinity "scavenger" systems are repressed by high concentrations of substrate.

Oligopeptides of varying length and varying composition, and even covalently attached to other compounds, can be transported by the same carrier. Substances that lack a transport system can thus enter in this form and be released by hydrolysis of the peptide.

Most **phosphorylated compounds** (e.g., nucleotides) are excluded from bacteria. However, glycerol-P and glucose-P induce systems that transport the intact molecule. Since these compounds are normal metabolites within the cell the induction by exogenous molecules evidently involves some undefined external interaction with the membrane.

Inhibitors that act inside a cell presumably enter via a hospitable system that ordinarily transports a normal metabolite (e.g., sulfonamides compete with *p*-aminobenzoate for entry; Ch. 7). Moreover, **competition** for a shared transport system is an occasional mechanism of **growth inhibition**. For example, arginine blocks the entry of lysine, thus inhibiting the growth of lysine auxotrophs (in response to added lysine) but not growth of the wild type (which makes its own lysine). In human cells such interference with transport may be significant in diseases in which metabolites (such as phenylalanine) accumulate.

Inorganic ions (e.g., K^+, Mg^{2+}, PO_4^{3-}) normally have a relatively constant intracellular level as free ions (in addition to that bound to nucleic acids and proteins), even when the external concentration is very low. This **homeostasis** is maintained by membrane transport systems. Thus mutations that impair the transport of a specific ion (K^+, Mg^{2+}) markedly increase the extracellular concentration required for growth. *E. coli* has developed a transport system specific for **extrusion** of the toxic Ca^{2+} ion.

Nonspecific Entry. If cells (including mutants) possess the appropriate enzymes but not the transport system for a small substrate they may metabolize it at a very low rate (<1% of the rate with a transport system). This entry is linearly proportional to the substrate concentration, and it is probably primarily due to low-affinity transport by systems specific for other, similar compounds. However, injury to the cell membrane may allow true diffusion (which is also linear with concentration). Indeed, such injury by toluene is useful in quantitating intracellular enzymes: these are still retained in the cells, but they are now freely accessible to substrates.

MECHANISM OF ACTIVE TRANSPORT

Membrane Carrier Proteins. The molecular mechanism of active transport has been difficult to analyze by the usual biochemical approach of isolating the components, for this procedure destroys the characteristic asymmetric (vectorial) orientation of the process and its linkage to an energy supply. Moreover, isolation destroys the binding activity of most binding components (presumably through a conformational change), just as with many membrane enzymes.

The first membrane transport protein, for β-galactosides, was isolated by Fox and Kennedy: an -SH group in its binding site was masked with substrate during a preincubation of the cells with *N*-ethylmaleimide; the remaining -SH groups were labeled with radioactive reagent in the absence of substrate; and the preferentially labeled protein was isolated. Its specificity was confirmed by its absence from cells lacking the transport system. The isolated protein had lost its binding activity, but in later work several similar isolated carrier proteins were reactivated by incorporation into vesicles (see Membrane Vesicles, below).

Specific transport proteins appear to constitute much of the cell membrane: in a fully induced cell the β-galactoside-binding protein (about 8000 molecules) constitutes 0.35% of the cell protein. The specific binding proteins for various permeants may be linked in the membrane to a common protein that couples them to an energy source, since some one-step mutants have lost active transport for multiple permeants.

Periplasmic Binding Proteins. In gram-negative bacteria certain permeants are bound by specific soluble proteins in the periplasmic space, and not by fixed membrane proteins. (In *E. coli* these permeants include glutamine, sulfate, galactose, and maltose.) The periplasmic proteins are involved in transport: their elimination by osmotic shock, or by conversion of cells to spheroplasts, inactivates the corresponding transport systems, and a mutation that eliminates one of these proteins (or that makes it temperature-sensitive) has the same effect on the corresponding transport.

It is not clear how these proteins interact with the membrane in active transport, nor why the same cell has some systems with fixed and others with periplasmic binding proteins. The periplasmic systems are reported to differ from the others in requiring ATP or acetyl-P, rather than a membrane potential (see Membrane Energetics, below), as the source of energy. They also appear to have a greater capacity for concentrating their permeant from a dilute solution.

Membrane Vesicles. The analysis of active transport has been considerably advanced by the development, by Kaback, of a system that retains vectorial orientation but is considerably simpler than the cell: the closed vesicles that form spontaneously when bacterial membranes are fragmented. Using energy derived from oxidation, and in the absence of ATP or cytoplasmic proteins, these vesicles can concentrate the substrates of those transport systems that have fixed binding proteins in the membrane. Moreover, it has been possible to incorporate additional transport systems into the membrane of such vesicles by inserting appropriate proteins (with the aid of detergents and sonication); and such effective insertions have also been accomplished with vesicles reconstituted from phospholipids and proteins (liposomes).

Cell-derived vesicles can metabolize many compounds, but only a few can support active transport. This paradox was resolved when the use of Abs to an enzyme on the inner surfce (ATPase) showed that under some conditions membrane fragments may form inside-out (inverted) as well as right-side-out vesicles. The former can oxidize nonpenetrating substrates, but they cannot carry out active transport; while only penetrating substrates can fuel active transport by the latter.

Conformational K_m Model for Active Transport. Active transport converts metabolic energy into osmotic work. Coupling of the two by transfer of a high-energy group to the permeant, based on the model of such transfer to substrates of endergonic reactions in biosynthesis, can be ruled out: covalent attachments would not be chemically possible for inorganic ions (e.g., K^+); and a covalent reaction with organic substrates would be difficult to reconcile with the reversible binding observed with the β-galactoside transport protein in the membrane. It therefore seemed likely that energy is being used to **modify the carrier system,** instead of the substrate. More specifically, an **energy-dependent conformational change** could account for active transport if it gives the binding site a high affinity (low K_m) for its ligand when facing outward and a low affinity when facing inward.

Since rotation of a transmembrane protein between the two faces of a phospholipid bilayer is thermodynamically excluded, the binding site presumably faces in either direction within a protein

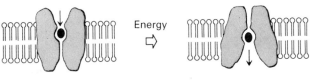

Outside

Energy

Cytoplasm

FIG. 6-27. Model for conformational change in membrane binding protein, accepting permeant on one side of membrane and discharging it on the other side. In active transport (not shown) application of energy changes the affinity for the permeant on one side. (After Singer SJ: Annu Rev Biochem 43:805, 1974)

channel (Fig. 6-27), taking up permeant at a low external concentration and unloading it in the cell in the face of a high internal concentration. This mechanism is supported by evidence that if cells preloaded with some permeants are abruptly deprived of energy the efflux exhibits a marked lowering of K_m, to the level characteristic of normal influx.

MEMBRANE ENERGETICS AND ACTIVE TRANSPORT

As we have seen, active transport does not depend directly on ATP but results from an **energized state of the membrane,** which is produced under aerobic conditions by electron transport. This state also generates ATP from ADP (oxidative phosphorylation), by transfer of energy to the reversible ATPase on the inner surface of the membrane. Conversely, under anaerobic conditions, where there is no respiration, the ATPase reacts (in the opposite direction) with cytoplasmic ATP, formed by fermentation, to generate the energized state of the membrane that is required for active transport.

These reactions are demonstrated by mutations that inactivate the ATPase. Such "uncoupled" cells can still carry out electron transport, and can use the energy for active transport. However, they cannot use electron transport for ATP synthesis; neither can they use ATP, synthesized anaerobically, for active transport.

The problem of how energy is stored in a state of the membrane, and is connected reversibly with ATP synthesis or is transduced into active transport, has been illuminated by the **chemiosmotic theory** of Mitchell. Departing from the classic focus of biochemistry on the directionless transformations of compounds in solution, this approach emphasizes oriented, vectorial reactions, in which electrons and protons (H^+) are **translocated** across the osmotic and electrically insulating barrier of a membrane by enzymes and carriers traversing it.

The key experimental finding (obtained first with mitochondria and then with bacteria and with membrane vesicles) is that the outward transport of a pair of electrons is associated with the extrusion of a pair of protons (H^+) at each of three (or in some bacteria each of two) energy-yielding steps (Fig. 6-28). As a result the interior of the cell develops a **negative electrical potential** and also an **alkaline pH,** relative to the outside medium. These together determine the **electrochemical potential of** H^+ (also called the **protonmotive force**) across the membrane. With the energy provided by this potential the return of the extruded H^+ to the ATPase would generate

ATP; return through other channels would drive active transport (see below).

The theory is supported by much experimental evidence. Perhaps the most decisive is the finding that membrane fragments (or reconstituted systems) cannot generate ATP unless they form closed vesicles with high electrical resistance. Moreover, phosphorylation is "uncoupled" by lipid-soluble proton-conducting agents (e.g., dinitrophenol), which short-circuit the potential; and in artificial vesicles incorporation of the hydrophobic component of ATPase (F_o) evidently provides a channel for return of protons, since it markedly increases electron transport even though the absence of the remainder of the ATPase (F_1) prevents oxidative phosphorylation or ATP hydrolysis. Finally, in the absence of metabolism a brief wave of active transport or of ATP synthesis can be induced in bacteria (or in mitochondria) by elevating the external H^+ concentration, and also by artificially inducing a positive external electrical potential (by using valinomycin (see above) to permit the outflow of K^+ from a high intracellular concentration to a low external one).

Though the chemiosmotic model has been presented as an alternative to a conformational model for energy transfer, the two may focus on different aspects of the same phenomenon. For in oxidative phosphorylation the energy provided by the protonmotive force must ultimately affect the ATPase, by changes in its conformation or charge, in a way that results in formation of a high-energy bond in ATP. Similarly, **active transport** cannot be explained simply by the suggestion that the protonmotive force drives the carrier in: the carrier would still have to unload its ligand in the face of a high internal concentration. Conformational (or charge distribution) changes, in response to proton flow, thus seem logically inescapable.

GROUP TRANSLOCATION

Bacteria have evolved an additional transport mechanism, group translocation, in which a compound does undergo a covalent change in the course of entering the cell. However, in this process, discovered by Roseman, the covalent reaction is not a step specific for active transport but is the first step in the metabolism of the compound. The substrates of such a system in *E. coli* include glucose and mannose, their reduction products (sorbitol and mannitol), and fructose. Each of these is linked by a specific membrane enzyme to a common **phosphotransferase system,** which converts each to its 6-phosphate derivative before it enters the cytoplasm.

In this system three protein components have been separated (Fig. 6-29). Cytoplasmic enzyme I transfers P from P-enolpyruvate to HPr, a small cytoplasmic protein of mol wt 9000; the HPr transfers the high-energy P to various sugar-specific enzymes, either soluble or membrane-bound (enzyme III). These, in conjunction with other sugar-specific membrane-bound enzymes (enzyme II complex), transfer the P to a specific substrate on the outer surface, transport the resulting phosphorylated derivative to the inner surface, and release it into the cytoplasm. This compound is ordinarily further metabolized,

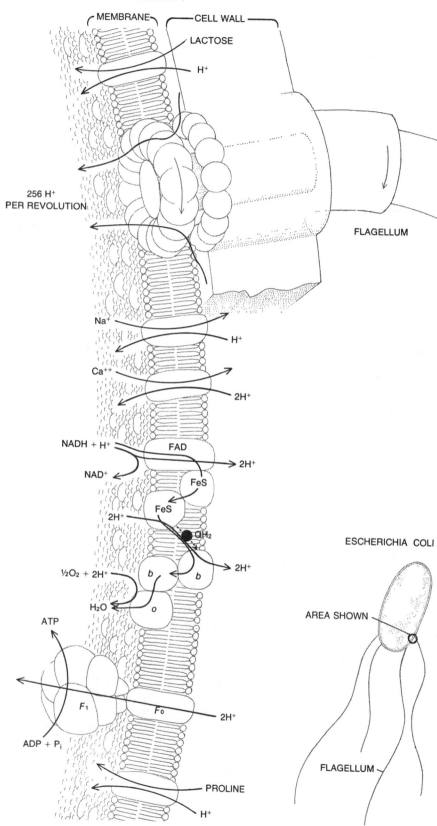

FIG. 6-28. Proton transfer associated with various activities of the cytoplasmic membrane of *E. coli*, according to chemiosmotic hypothesis. FAD is reduced by NADH, thus transferring both electrons and protons to the outer surface. The two H^+ protons are extruded and the electrons are returned to the inner surface by iron–sulfur proteins (FeS), which can transfer only electrons. These electrons, together with protons from the cytoplasm, reduce ubiquinone (QH_2), which similarly extrudes two H^+ protons and transfers the electrons to the cytochrome system. The final transfer to oxygen is linked in some organisms to extrusion of a third pair of H^+. The H^+ potential is used to provide energy for phosphorylation by ATPase (designated as F_0 and F_1 components), for active transport of sugars, amino acids, and (not shown) K^+ and Mg^{2+} ions, for extrusion of Na^+ and Ca^{2+}, and for flagellar rotation. (Hinkle PC, McCarty RE: Copyright © 1978 by Scientific American, Inc. All rights reserved.)

FIG. 6-29. The phosphotransferase transport system. Each sugar uses a specific enzyme II (which accepts the sugar) and a specific enzyme III (which accepts -P from P-HPr).

Enzyme I
(Cytoplasm)
$$\text{P-enolpyruvate + HPr} \rightleftharpoons \text{Pyruvate + P-HPr} \quad (\text{Reaction 1})$$

Enzyme II, III
(Membrane)
$$\text{P-HPr + Sugar} \longrightarrow \text{Sugar-P + HPr} \quad (\text{Reaction 2})$$

$$\text{P-enolpyruvate + Sugar} \longrightarrow \text{Sugar-P + Pyruvate} \quad (\text{Over-all reaction})$$
(outside)   (inside)

but it may also be dephosphorylated, which leads to the same overall result as active transport.

A mutant defective in any particular enzyme II cannot take up the corresponding substrate, whereas defects in enzyme I or in HPr affect uptake of all the substrates of the system. Indeed, this pleiotropic effect of a single mutation first revealed group translocation.

Group translocation is not limited to the phosphotransferase system. Analogous systems convert fatty acids to internal acyl-CoA, or thiamine to its pyrophosphate.

Organisms differ in their choice of the two mechanisms. For example, lactose is taken up by a carrier system in *E. coli* but by a phosphotransferase in *Staphylococcus*. Phosphotransferases have been found only in anaerobic or in facultative organisms, where the resulting economic uptake of sugar from dilute solution is especially valuable since fermentations yield only 2–2.5 ~P per hexose.

MOTILITY AND CHEMOTAXIS

The cytoplasmic membrane is also intimately involved in the **motility** exhibited by many bacteria, and in the associated **chemotaxis,** i.e., net oriented movement in a concentration gradient of certain compounds. While these processes have been known since the early days of microbiology, only in recent years have the basic mechanisms been recognized. They have aroused considerable interest as the simplest models for a sensory system linked to a motor system.

FLAGELLA

Structure. We have seen that the motility of eubacteria depends on the action of flagella: thin filaments several times the length of a cell (see Fig. 2-5). In fixed preparations these structures appear as regular sinusoidal curves (now recognized as flattened helices), whose wavelength is characteristic of the species.

Purified filaments can be obtained by mechanical agitation followed by differential centrifugation. They dissociate in mild acid to yield a globular protein, **flagellin*** (mol wt 40,000 in some species), which can reassemble in neutral solution. The overlapping arrangement of subunits mimics a helix composed of several strands (Fig. 6-30).

When the wall is digested away the protoplasts retain the flagella, which in turn can be isolated by dissolving away the membrane. The filament is then found to end in a thicker **basal body.** In this structure the filament is attached to a short, hooklike **sheath,** which evidently gives flexibility to the angle between filament and cell axis; and at the very base of the flagellum are thin parallel **rings** (Fig. 6-30). In gram-negative organisms there are four rings, while in gram-positive organisms, with fewer envelope layers, there are only two.

Biosynthesis. In *E. coli* 12 genes for flagellar components have been identified from mutants (*fla*) with absent or defective flagella, and several basal body proteins (as well as flagellin) have been isolated. Incomplete flagella formed by various of these mutants show that synthesis proceeds outward, from the inner ring to the filament. Flagellin, which adds to the distal tip in vitro, also does so in the cell; it evidently passes through a hollow core in the filament. Its rate of addition decreases with filament length, thus self-regulating that length.

The location of the new segment at the tip of a growing filament was demonstrated not only by radioactive labeling but, even earlier, by pulsing with *p*-fluorophenylalanine, whose incorporation yields filaments with a different wavelength.

Flagellar Rotation. The wavy shape of flagella in fixed preparations (see Fig. 2-5) was long believed to be due to fixation of the whiplike motion of a flexible filament. However, in 1973 Berg showed, by attaching polystyrene beads, that the filament rotates; hence it must have an intrinsic helical structure, serving as a propeller.

---

* Some flagellins contain a novel amino acid, ϵ-*N*-methyl lysine, formed by modification of lysine after its incorporation (like the conversion of proline to hydroxyproline in collagen). The capacity for such modification of residues in structural proteins thus evolved in prokaryotes.

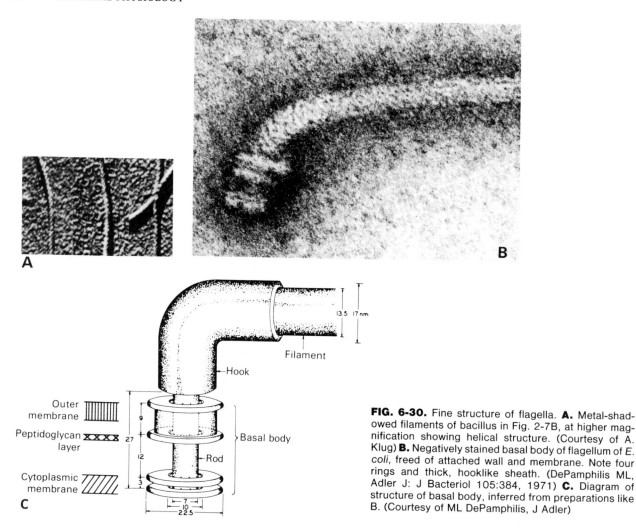

FIG. 6-30. Fine structure of flagella. **A.** Metal-shadowed filaments of bacillus in Fig. 2-7B, at higher magnification showing helical structure. (Courtesy of A. Klug) **B.** Negatively stained basal body of flagellum of *E. coli*, freed of attached wall and membrane. Note four rings and thick, hooklike sheath. (DePamphilis ML, Adler J: J Bacteriol 105:384, 1971) **C.** Diagram of structure of basal body, inferred from preparations like B. (Courtesy of ML DePamphilis, J Adler)

Rotation was also demonstrated by using Abs to adsorb a flagellum to a glass surface: the cells then rotated around this fixed attachment. Moreover, mutants with straight flagella were no longer motile, though the cells rotated on them. The multiple flagella present in many bacteria evidently rotate in bundles, with the filaments sliding past each other: motion ceases when the flagella are crosslinked by bivalent Abs, but not when they are coated with univalent Ab fragments.

The energy for rotation is believed to be derived from the protonmotive force in the cytoplasmic membrane, applied to the flagellar ring embedded in the membrane (Fig. 6-28). The outer rings evidently serve as bearings, minimizing friction and leakage.

CHEMOTAXIS

It has been known since the 1880s that motile bacteria exhibit chemotaxis. Various sugars and amino acids serve as attractants **(positive chemotaxis,** Fig. 6-31), while some substances (e.g., phenol, acid, alkali) serve as repellants **(negative chemotaxis).** The basis was long assumed to be a gradient of energy supply in the cell, reflecting the external gradient. However, over the 2-μm length of a bacterial cell the differences in external concentration would be infinitesimal.

In 1969 a radically different mechanism was discovered: Adler found that chemotaxis can be evoked by nonmetabolizable analogs of various foods, which implies that the cell responds to a gradient of **information** rather than a gradient of energy supply. Since the information must be transmitted from **sensory receptors** to **effectors,** bacteria evidently possess the elements of a **nervous system.**

Sensors. Bacteria "taste" their environment by means of **specific chemoreceptors.** They evidently rely on a limited

FIG. 6-31. Positive chemotaxis of *E. coli* cells toward the tip of a capillary containing a concentrated solution of glucose. (Courtesy of J. Adler)

sample, since many nutrients are not chemotactic; only about 20 different chemoreceptors have been found in *E. coli*. These turned out to be *the same binding proteins* (either fixed or periplasmic) *that are used in active transport*.

Thus chemotaxis and transport show the same relative affinities for various analogs of galactose, and mutations in the binding proteins affect the two processes in parallel. The processes may also be eliminated separately, by mutations in other, unshared components. With nonchemotactic nutrients (e.g., gluconate, which supports as rapid growth of *E. coli* as glucose) the similar binding proteins are evidently not connected to the chemotactic system.

Memory. The shift of focus from an energy to an information gradient still left the problem of how so short a cell can sense a gradient. The solution is that the sensing mechanism recognizes **temporal** rather than **spatial** differences. Motion pictures with a rapid tracking microscope showed that in a uniform medium bacteria move in a straight line for a fixed time (a second or so), then "twiddle" or "tumble" aimlessly for another fixed time, then set off again in a random direction. However, the linear run is longer in the direction of an attractant (and shorter in the direction of a repellant). The cell thus compares the concentration at the start of a run with later concentrations.

For quantitative analysis of these relations the effects of abrupt changes in concentration proved more suitable than a gradient. The results suggested that a set of receptors with a characteristic K_m bind a given chemotactic agent, and the fraction of sites occupied determines the change in the length of the run. The molecular mechanism appears to involve a rapidly reversible methylation: methionine limitation impairs chemotaxis, and the **methylation of a specific glutamate** γ-COOH in a membrane protein increases during a run and decreases during twiddling.*

The **direction of rotation of the flagella** is the key to the two alternative kinds of cell motion. When the flagella rotate counterclockwise they move in parallel intertwined groups and support linear movement of the cell, while in the reverse direction they somehow cause **twiddling** (or backup in cells with polar flagellation). There is evidence that under the stress of reversed rotation the subunits snap into an altered conformation, resulting in a different curvature of the flagellum.

The simultaneous reversal in direction of all the flagella suggests that triggered stimuli to shift gears, in either direction, are rapidly conducted through the cell membrane. The chemotactic response somehow changes the timing of the trigger event.

* A requirement for S-adenosyl methionine, the methyl donor, has also now been demonstrated for leukocyte chemotaxis.

The analysis of these phenomena is being aided by the isolation of chemotaxis (*che*) mutants. Some have lost rotation (in one direction or in the other), and others have lost coupling to the sensory system.

Adaptive Value. The expected evolutionary value of these mechanisms has been confirmed by experiments on reconstruction of the intestinal flora in animals: nonmotile mutants of *E. coli* are at a disadvantage. It is of interest that the related pathogen *Salmonella,* which grows primarily in the lumen, is also motile, while the nonmotile pathogen *Shigella* invades the intestinal epithelium.

SPORES

Under conditions of a limitation in the supply of C, N, or P certain gram-positive rods (aerobic *Bacilli* and anaerobic *Clostridia*), and a few sarcinae and actinomycetes, form highly resistant, dehydrated forms, called **endospores** or **spores** (Gr., seed). The surrounding mother cell, from which they are eventually released, is called the **sporangium.** Like the exospores of fungi (Ch. 45) and of certain actinomycetes (Ch. 38) and the seeds of higher plants, which are also adapted for dissemination, bacterial endospores have no metabolic activity. However, bacterial spores are particularly adapted for prolonged survival under adverse conditions: they are relatively resistant to killing by heat, as well as by drying, freezing, toxic chemicals, and radiation.

Spores are medically important in the dissemination of a few diseases, and in the preparation of sterile materials. They also have aroused wide interest as a simple unicellular model for cell differentiation. The regulatory aspects of this process, studied mostly in *B. subtilis* and *B. megaterium,* will be described here, but they depend on principles of molecular genetics presented in later chapters.

Though resistance to heat has received most attention, the main ecologic role of spores is probably **survival in the dry state** (or in a nonnutrient medium). In soil preserved with plant specimens in herbaria the numbers of viable sporulating *Bacilli* decreased with the duration of storage, but organisms could be cultured from samples over 300 years old.

FORMATION AND STRUCTURE

Spores are unusually dehydrated, impervious, highly refractile cells. They do not take ordinary stains (Gram's, methylene blue). In the **light microscope** the first visible stage in **sporulation** is the formation of an area of increased refractility, the **forespore,** at one end of a cell. The refractility gradually increases and the mature spore is completed in 6–8 h and then is freed by autolysis of the sporangial wall. In a well-sporulating culture most cells form a spore.

Spores are usually smooth-walled and ovoid, but in some species they are spherical or have characteristic ridges. In bacilli the spores usually fit within the normal cell diameter, but in the slender clostridia they cause a bulge, which may be either terminal ("drumstick") or more central (Fig. 6-32).

Blocks in DNA synthesis show that sporulation starts, like normal cell division (Ch. 14), at a critical stage in a round of DNA replication; and in **electron micrographs** the first detectable change is conversion of the compact nucleoid into an **axial filament** (Fig. 6-33). The resulting

FIG. 6-32. Visualization of spores: light microscope. **A.** *Bacillus cereus:* elongated subterminal spores, nigrosin stain. **B.** *Clostridium pectinovorum:* large, terminal spores, and spores freed from parent cell (sporangium); cells stained with I₂ (which stains granulose). (Courtesy of C. Robinow) (×3600)

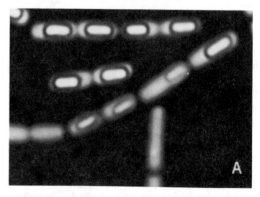

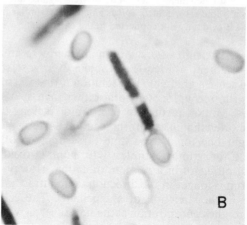

FIG. 6-33. Stages in sporulation.

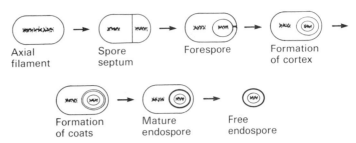

Axial filament → Spore septum → Forespore → Formation of cortex →

Formation of coats → Mature endospore → Free endospore

movement of one chromosome to the pole of the cell triggers a specialized, asymmetric cell division, with the ingrowth of a double layer of cytoplasmic membrane to form a subpolar **spore septum** (without peptidoglycan). By migration of its peripheral zone of attachment toward the pole of the cell this septum finally engulfs the chromosome and surrounding cytoplasm in a double membrane, forming the forespore (Fig. 6-34).

Spore Integument. The specialized spore integument (envelope) is laid down between the two membrane layers of the forespore, which are initially extensions of the mother cell membrane but become differentiated in composition and function. Both facing surfaces corre-

spond to the wall-synthesizing surface of the parental cell membrane, and in the maturation of the spore a large amount of material is laid down between the two membranes. The resulting thick envelope eventually occupies over half the spore volume, surrounding the **protoplast (core).** Several layers, successively initiated, can be distinguished (Fig. 6-35).

1) The innermost layer is the **germ cell membrane.** 2) Next is the thickest layer, the **cortex,** which contains a concentric laminated structure. 3) Outside the cortex is the densely stained **coat.** 4) Spores of some species are further loosely shrouded in a delicate **exosporium.**

The **cortex** contains many layers of spore peptidoglycan, much more loosely crosslinked than that in the vege-

FIG. 6-34. Early stage in sporulation. **A.** The protoplasm at one end of the cell, containing a chromosome (**CHR**), is cut off from the rest of the cell by a transverse spore septum (**SPS**), formed by ingrowth of a double membrane and mesosome (**M**) from the protoplasmic membrane of the other cell. **B.** The periphery of this double membrane moves toward the tip of the cell, ultimately enclosing the whole forespore. (Ohye DF, Murrell WG: J Cell Biol 14:111, 1962)

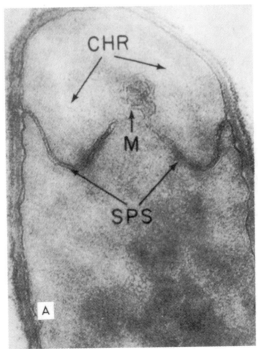

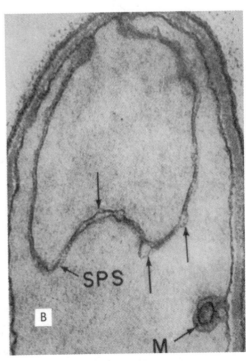

tative cell. Electron micrographs, and studies with radioactively labeled DAP, show that during germination (see below) this part of the peptidoglycan is rapidly autolyzed and released to the medium, while an inner layer (20% of the total peptidoglycan in *B. subtilis*) is more tightly crosslinked and is retained. This stable peptidoglycan layer plus the underlying membrane are sometimes called the **spore cell wall**: it provides osmotic stability for the germinating protoplast, and also a primer for restoration of the vegetative cell wall. The wall peptidoglycan is probably laid down by the inner forespore membrane, and the loose cortical peptidoglycan by the outer membrane. In *B. subtilis* only 6% of the muramate residues in the cortex are crosslinked.

This loose structure is accompanied by substitutions, on the -OH of many muramate residues, that block crosslinking: L-Ala instead of a tetrapeptide, or a lactam ring formed within the muramate.

The **coat** is made of a keratinlike protein, rich in S-S, which constitutes as much as 80% of the total protein of a spore; it is laid down by the mother cell from a larger precursor. *The impervious protein coat is responsible for the resistance of spores to chemicals.*

Muramic lactam

Thus chemical resistance is lacking until the stage of coat formation, and it is permanently absent in mutants with a defective coat. The coat is composed largely of one to four kinds of proteins in various species. These are laid down in one to three layers, seen in freeze-etched preparations as a close-packed hexagonal array. The protein cannot be solubilized unless its S-S groups are reduced. The moderately increased resistance of spores to killing by ultraviolet or ionizing radiation may also depend on the coat, impeding entry of radiochemical products formed in the medium. In addition, S-S groups can combine with free radicals.

The **exosporium** is a lipid–protein membrane, with 20% carbohydrate. It is not essential for survival, and its function is unknown.

Small Molecules. A striking feature of spores is their huge content of Ca^{2+}, for which active transport units appear in the mother cell early in sporulation. Normally the Ca^{2+} is accompanied by a roughly equivalent amount of **dipicolinic acid,** which can chelate Ca^{2+}. Dipicolinate is almost unique to bacterial spores and may constitute as

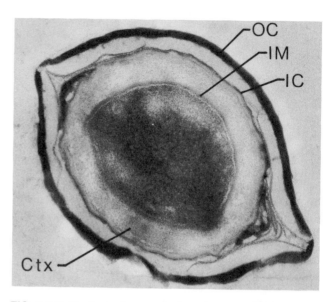

FIG. 6-35. Electron micrograph of sectioned spore of *B. megaterium*. The core is surrounded successively by the inner membrane (**IM**), cortex (**Ctx**), inner coat (**IC**), and outer coat (**OC**). With other methods of fixation the cortex is much denser and is seen to be laminated. (Courtesy of P. Fitz-James) (×11,200)

much as 15% of their weight. It is formed by a one-step reaction from DAP (a precursor of lysine) in the mother cell, from which it is rapidly accumulated by the maturing core.

Dipicolinic acid

(Pyridine-2,6-dicarboxylic acid; for biosynthesis see bacterial pathway to lysine, Ch. 4)

The location of dipicolinate primarily in the core was established by determining the attenuation of the soft β emission of ^{3}H-labeled molecules by the surrounding envelope layers. ^{3}H-Dipicolinate is attenuated like ^{3}H-uracil (known to be in the core), while more peripherally located materials (^{3}H-diaminopimelate in the cortex, ^{3}H-lysine in the coat) are less attenuated.

Dehydration. The mechanism for accomplishing the large thermodynamic work of eliminating the water from the maturing spore is a mystery, but the cortex seems to play a key role. Thus impairment of cortex formation (by mutation or by penicillin) prevents effective dehydration, though the spore develops a normal coat. The **contractile cortex theory** suggests that water may be expelled by contraction of the loose polyanionic cortical peptidoglycan, resulting from ionic crosslinking by Ca^{2+}. Conversely, it has been suggested that loss of ionic crosslinks causes the cortex to expand within the fixed volume of the coat.

COMPOSITION OF THE CORE

The spore core contains much less material than the corresponding vegetative cell, but it must contain everything necessary for resuming growth. It lacks the components of the vegetative cell that are particularly unstable or are readily replaced early in germination. Thus the DNA of one chromosome (3×10^9 daltons) is present, together with small amounts of all the stable components of the protein-synthesizing machinery (including ribosomes, tRNAs, and accessory factors and enzymes), but there is no detectable messenger RNA. Amino acids (and their biosynthetic enzymes) are virtually absent; they are supplied early in germination by hydrolysis of a storage protein of low mol wt, constituting about 20% of the total protein.

Deoxyribonucleotides (which are not needed early) and ribonucleoside triphosphates (which are unstable) are absent, but a pool of ribonucleoside monophosphates and diphosphates is available for building messenger RNA (Table 6-4). Energy for initiating germination is stored as the stable **3-P-glycerate**, which is readily converted to the ~P donor P-enolpyruvate. Further energy is derived by glycolysis; the cytochrome content is low and TCA cycle enzymes are absent.

Proteins and Heat Resistance. A few enzymes (e.g., aldolase) are cleaved by a protease to yield a smaller, more stable enzyme in the spore than that in the vegetative cell; and some enzymes are unique to spores. However, many enzymes purified from spores are identical with those in the parental cells and are not especially thermostable in solution. Hence the striking stability of enzymes in a spore must depend largely on their environment, including dehydration and ionic conditions; there is some evidence for binding of stabilizing factors.

In some species a reduction in Ca dipicolinate content is paralleled by a reduction in resistance to killing by heat.

TABLE 6-4. Levels of Metabolites in Dormant Spores and in Growing Cells of *B. megaterium*.

Compound	Amount (μmol/g)	
	Spores	Growing cells
ATP	3	725
ADP + AMP	544	195
GTP + CTP + UTP	<10	680
G + C + U nucleotides	530	860
Deoxyribonucleotides	<1.5	181
Phosphoglyceric acid	6,800	—
Amino acids*	150	1,400

(After Setlow P: In *Spores VI*, Amer Soc Microbiol, 1975, p. 443)

*Setlow P, Primus G, ibid., p. 451; germinating instead of growing cells.

REGULATORY CHANGES IN SPORULATION

Triggers. Innumerable metabolites change concentration under the conditions of deprivation that trigger the commitment to sporulate. However, a key regulatory role for some purine nucleotide seems likely, since blocks in the purine pathway are effective triggers. A novel purine metabolite, **adenosine bis-triphosphate** ($5',3'$-p_3Ap_3), may play a key role: it appears very early in sporulating cells, and mutations blocking its synthesis prevent sporulation. (Closely related compounds, $3',5'$-cyclic AMP and p_2Gp_2, play key roles in the response of vegetative cells to a limitation in the supply of energy, or of amino acids: Ch. 14.)

Another early event in sporulation is the appearance of one or more intracellular **protease(s)**. These enzymes cause increased protein turnover, providing starved sporulating cells with the amino acids needed for synthesis of new proteins; they may also be involved in specific cleavages that contribute to regulation.

Shifts of Gene Activity. Genomes of both the forespore and the sporangium are active in transcription during spore formation, as shown by radioautography and also by biochemical studies on isolated forespores. Studies of mRNA specificity (by hybridization), as well as of enzymatic composition, show that some of the genes used in vegetative growth continue to be active during sporulation, others are turned off, and still other genes unique to sporulation (about 20% of the genome) are turned on.

A major regulatory mechanism is a change in the **specificity of the RNA polymerase**, first revealed as a loss of ability of sporulating cells to transcribe the DNA of a phage that can multiply in vegetative cells. The change was traced to a decreased affinity of the **initiating factor** (σ) for the polymerase (Ch. 11), possibly due to an inhibitor. Sporulating cells also form novel, sigma-like proteins that bind to the enzyme and may influence its specificity.

Further evidence for a role of the polymerase is the finding that mutations in one of its subunits (β), selected by rifampin resistance, can prevent sporulation. (This subunit was also found in altered form in extracts of sporulating cells, but the change turned out to be an artefact of proteolysis during extraction.)

In addition to changes in the polymerase initiator, there must be an intricate **cascade** of successive changes in the **activation or repression of many operons** (Ch. 14), like that analyzed in detail in the process of phage synthesis (Ch. 47). Not only can a regular sequence of stages be observed, from formation of the spore septum to completion of the coat and dehydration of the core; in addition, the numerous **asporogenous mutants** (*spo*) that have been isolated exhibit blocks in at least 12 morphologically or biochemically distinguishable stages.

The mutants thus far isolated occupy 30 known operons. From the ratio of new to old mutants encountered it is estimated that about 200 genes are involved specifically in sporulation. Their pleiotropic effects, and their failure to accumulate compounds that relieve the block in other mutants, have prevented identification of the products of these genes (except for a serine protease required at the start). Cloning of *spo* genes by molecular recombination (Ch. 12) will surely help in the identification of their products.

Antibiotics are made by a few bacteria (and by many streptomycetes) during sporulation (Ch. 7), and many asporogenous mutations halt antibiotic production (depending on the stage of the block). A regulatory role of an antibiotic is suggested by a mutant that has ceased to make gramicidin and that forms defective, heat-sensitive spores: addition of the antibiotic restores the formation of heat-resistant spores.

GERMINATION

The overall process of converting a spore into a vegetative cell is often called germination. It is much faster than sporulation (about 90 min from onset to cell division in rich medium). Three stages can be distinguished: **activation, germination proper** (initiation), and **outgrowth.**

Activation. Though some bacterial spores will germinate spontaneously in a favorable medium, others (especially if freshly formed) remain dormant unless they are activated by some traumatic agent, such as heat, low pH, or an SH compound. **Aging,** with its multiple, undefined consequences, is probably the most important natural cause. Activation presumably damages the impermeable coat, since grinding with glass powder is also effective.

Dormancy (i.e., unresponsiveness to a germination medium) has the same biologic function in spores as in plant seeds. In many wild-type higher plants the seeds will not germinate until an outer coat has been damaged by some agent provided in the ecology of that plant: abrasion, chemical or bacterial attack, heat, freezing, light, maceration. Germination is thereby spread in time and space, promoting survival of the species by preventing uniform germina-

FIG. 6-36. Electron micrograph of germinating spore of *B. megaterium.* Cortex has lost its laminated appearance, and the coats have begun to disintegrate. Note deposition of vegetative cell wall (**CW**) outside membrane. Appearance of the cytoplasm already closely resembles that of a vegetative cell, with mesosomes, many ribosomes, and well-demarcated nuclear body. (Courtesy of C. Robinow and J. Marak)

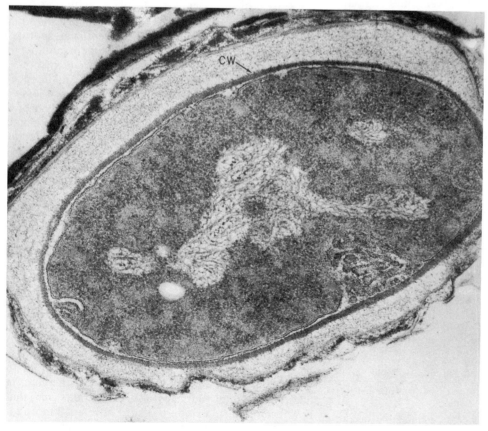

tion in response to conditions that are only temporarily favorable. Domesticated plants, in contrast, have been selected for uniformity of germination.

Germination. Unlike activation, germination requires water and a triggering **germination agent.** Various species respond to various metabolites (e.g., alanine, dipicolinate) or inorganic ions (e.g., Mn^{2+}), which penetrate the damaged coat. Germination can also be triggered by nonmetabolizable analogs, which suggests an allosteric (Ch. 14) action on the spore membrane, resulting in activation of a spore-lytic enzyme and release of Ca dipicolinate.

The resulting hydrolysis of the cortical peptidoglycan (Fig. 6-36) proceeds to completion within minutes: up to 30% of the dry weight of the cell is released, mostly as peptidoglycan fragments ("spore peptide") and Ca dipicolinate. With the breakdown of the cortical barrier the cell rapidly takes up water, accompanied by K^+ and Mg^{2+}, and so the cell loses its refractility (the usual test for germination), along with its resistance to heat and to staining. Most of the energy stored in phosphoglyceric acid is converted to ATP within 5 min; and the storage protein noted above is hydrolyzed within 10 min, providing amino acids and further energy.

FIG. 6-37. Sequence of germination and outgrowth in *B. megaterium*, observed with the light microscope. **A.** 5 min: all cells highly refractile. **B.** 50 min: most cells have lost refractility and begun to grow out of spore integument. **C.** 135 min: elongated cells still attached to refractile spore coat. **D.** 165 min: cells longer, some freed from spore coat. (Courtesy of C. Robinow) (×3600, reduced)

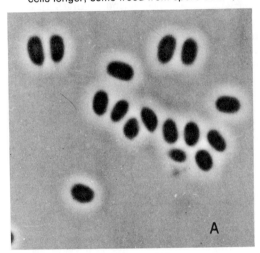
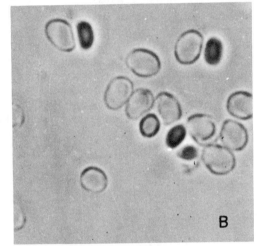
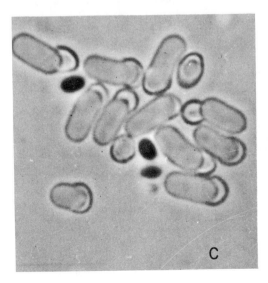
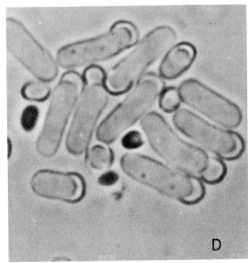

The **loose crosslinking** of the cortical peptidoglycan of spores undoubtedly promotes its rapid digestion during germination. Two observed changes may activate the process: the shift of the digestive enzyme from a particle-bound to a soluble state, and loss of Ca^{2+} ions bound to the peptidoglycan.

Outgrowth. In a nutrient medium germination leads to immediate outgrowth. However, in a starvation medium, or in the presence of an inhibitor of protein synthesis, the cell becomes rehydrated but it cannot become a vegetative cell.

Outgrowth is a gradual resumption of vegetative growth (Fig. 6-37): protein synthesis increases progressively, as its initially scanty machinery is expanded by protein and RNA synthesis; the spore wall becomes a thicker vegetative cell wall; and after about an hour DNA synthesis begins and the cell, twice its initial volume, begins to burst out of the spore coat (Fig. 6-37). The σ factor activity of vegetative RNA polymerase is restored early.

SUMMARY

We have seen that bacterial endospores are differentiated cells formed within a vegetative cell; they encase a genome in an insulating, dehydrated vehicle that makes the cell ametabolic and resistant to various lethal agents, but permits subsequent germination in an appropriate medium. Spores are formed by the invagination of a double layer of cell membrane, which closes off to surround a chromosome and a small amount of cytoplasm. A thin spore wall, and a thicker cortex with a much looser peptidoglycan, are synthesized between the two layers; outside the cortex is a protein coat, rich in disulfide crosslinks. Selective synthesis, hydrolysis, and uptake of metabolites yield a core containing the minimal complement of the stable constituents necessary for the resumption of growth. The stages of sporulation are presented diagrammatically in Figure 6-33.

The characteristic features of spore physiology are beginning to be understood. The keratinlike properties of the coat account for the resistance to staining and to attack by deleterious chemicals, while the dehydration contributes to the heat resistance. We do not know the mechanism by which these cells become essentially completely dehydrated, but the cortex appears to play a major role. In germination, following activation by mechanical or chemical damage to the surface coat, the attack of a lytic enzyme on the peptidoglycan of the cortex permits uptake of water and loss of Ca dipicolinate by the core.

Sporulation involves an extensive shift in the pattern of gene transcription, brought about in part by a change in the specificity of RNA polymerase. This change probably results from the attack of a specific protease. Other proteases are responsible for the extensive protein turnover that accompanies sporulation. The sequential regulatory processes in this microbial differentiation are being further analyzed through the use of mutants blocked at various stages in sporulation.

SELECTED READING

BOOKS AND REVIEW ARTICLES

CELL ENVELOPE

ADLER J: Chemoreceptors in bacteria. Science 166:1588, 1969

ADLER J: Chemotaxis in bacteria. Annu Rev Biochem 44:341, 1975

ARCHIBALD AR: The structure, biosynthesis, and function of teichoic acid. Adv Microb Physiol 11:53, 1974

BAYER ME: Role of adhesion zones in bacterial cell surface function and biogenesis. In Tzagoloff A (ed): Membrane Biogenesis: Mitochondria, Chloroplasts, and Bacteria. New York, Plenum, 1975, p 493

BERG HC: Chemotaxis in bacteria. Annu Rev Biophys Bioeng 4:119, 1975

BERG HC: Bacterial behaviour. Nature 254:389, 1975

BLOCH K: Control mechanisms for fatty acid synthesis in *Mycobacterium smegmatis*. Adv Enzymol 45:1, 1977

BLUMBERG PM, STROMINGER JL: Interaction of penicillin with the bacterial cell: penicillin-binding proteins and penicillin-sensitive enzymes. Bacteriol Rev 38:291, 1974

BRAUN V, HANTKE K: Biochemistry of bacterial cell envelopes. Annu Rev Biochem 43:89, 1974

COHEN GN, MONOD J: Bacterial permeases. Bacteriol Rev 21:169, 1957

COX GB, GIBSON F: Studies on electron transport and energy-linked reactions using mutants of *Escherichia coli*. Biochim Biophys Acta 346:1, 1974

CRONAN JE JR: Molecular biology of bacterial membrane lipids. Annu Rev Biochem 47:163, 1978

DEMAIN AL, BIRNBAUM J: Alteration of permeability for the release of metabolites from microbial cell. Curr Top Microbiol Immunol 46:1, 1968

DI RIENZO JM, NAKAMURA K, INOUYE M: The outer membrane proteins of gram-negative bacteria. Annu Rev Biochem 47:481, 1978

DOWNIE JA, GIBSON F, COX GB: Membrane adenosine triphosphatases of prokaryotic cells. Annu Rev Biochem 48:103, 1979

GHUYSEN JM: Use of bacteriolytic enzymes in determination of wall structure, and their role in cell metabolism. Bacteriol Rev 32:425, 1968

GOLDFINE H: Comparative aspects of bacterial lipids. Adv Microb Physiol 8:1, 1972

GREENAWALT JW, WHITESIDE TL: Mesosomes: membranous bacterial organelles. Bacteriol Rev 39:405, 1975

HADDOCK BA, HAMILTON WA (eds): Microbial energetics. Soc Gen Microbiol Symposium 27, Cambridge University Press, 1977

HAROLD FM: Antimicrobial agents and membrane function. Adv Microb Physiol 4:46, 1970

HAROLD FM: Membranes and energy transduction in bacteria. Curr Top Bioenergetics 6:84, 1977

HENNING U: Determination of cell shape in bacteria. Annu Rev Microbiol 29:45, 1975

HEPPEL LA: The effect of osmotic shock on release of bacterial proteins and on active transport. J Gen Physiol 54:953, 1969

IINO T: Genetics of structure and function of bacterial flagella. Annu Rev Genet 11:161, 1977

ITO K, SAITO T, YURA T: Synthesis and assembly of the membrane proteins in E. coli. Cell 11:511, 1977

KABACK HR: Transport across isolated bacterial cytoplasmic membranes. Biochim Biophys Acta 265:367, 1972

KABACK HR, NEURATH H, RADDA GK, SHWYZER R, WILEY WR (eds): Molecular Aspects of Membrane Phenomena. Berlin, Springer, 1975

KNOX KW, WICKEN AJ: Immunological properties of teichoic acids. Bacteriol Rev 37:215, 1973

LECHEVALIER MP: Lipids in bacterial taxonomy—a taxonomist's view. CRC Crit Rev Microbiol 5:109, 1977

LEIVE L (ed): Membranes and Walls of Bacteria. New York, Dekker, 1973

LENNARZ WJ, SCHER MG: Metabolism and function of polyisoprenol sugar intermediates in membrane-associated reactions. Biochim Biophys Acta 265:417, 1972

LODISH HF, ROTHMAN JE: The assembly of cell membranes. Sci Am: 481, 1979

MITCHELL P: Membranes of cells and organelles: morphology, transport, and metabolism. In Charles HP, Knight BCJG (eds): Organization and Control in Prokaryotic and Eukaryotic Cells. Society Gen Microbiol Symposium, Cambridge University Press, 1970, p 121

NEILANDS JB: Transport functions of the outer membrane of enteric bacteria. Horizons Biochem Biophys 5:65, 1978

PURCELL EM: Life at low Reynolds number. Am J Physics 45:3, 1977

RAETZ CRN: Enzymology, genetics, and regulation of membrane phospholipid synthesis in E. coli. Microbiol Rev 42:614, 1978

REISSIG JL: Decoding of regulatory signals at the microbial surface. Curr Top Microbiol Immunol 67:44, 1974

ROSEN BP (ed): Bacterial Transport. New York, Dekker, 1978

ROTHFIELD LI, ROMEO D: Role of lipids in the biosynthesis of the bacterial cell envelope. Bacteriol Rev 35:14, 1971

RYTER A: Structure and functions of mesosomes of gram-positive bacteria. Curr Top Microbiol Immunol 49:151, 1969

SALTON MRJ: Membrane-associated enzymes in bacteria. Adv Microb Physiol 11:213, 1974

SCHLEIFER KH, HAMMES WP, KANDLER O: Effect of endogenous and exogenous factors on the primary structures of bacterial peptidoglycans. Adv Microb Physiol 13:245, 1976

SCHLEIFER KH, KANDLER O: Peptidoglycan types of bacterial cell walls and their taxonomic implications. Bacteriol Rev 36:407, 1972

SILVERMAN M, SIMON MI: Bacterial flagella. Annu Rev Microbiol 31:397, 1977

SINGER SJ: Thermodynamics, the structure of integral membrane proteins, and transport. J Supramol Struc 6:313, 1977

TANFORD C: The hydrophobic effect and the organization of living matter. Science 200:1012, 1978

WEINBAUM G, KADIS S, AJL SJ (eds): Microbial Toxins, Vol IV, Bacterial Endotoxins. New York, Academic Press, 1971

WILSON DB: Cellular transport mechanisms. Annu Rev Biochem 47:933, 1978

SPECIFIC ARTICLES

BRAUN V, WOLFF H: The murein-lipoprotein linkage in the cell wall of Escherichia coli. Eur J Biochem 14:387, 1970

GOODELL EW, LOPEZ R, TOMASZ A: Suppression of lytic effect of beta lactams on Escherichia coli and other bacteria. Proc Natl Acad Sci USA 73:3293, 1976

INOUYE H, BECKWITH J: Synthesis and processing of an Escherichia coli alkaline phosphatase precursor in vitro. Proc Natl Acad Sci USA 74:1440, 1977

KOSHLAND DE Jr: A response regulator model in a simple sensory system. Science 196:1055, 1977

MUHLRADT PF, MENZEL J, GOLECKI Jr: Sites of export of newly synthesized lipopolysaccharide on the bacterial surface. Eur J Biochem 35:471, 1973

NAKAE T, NIKAIDO H: Outer membrane as a diffusion barrier in Salmonella typhimurium. J Biol Chem 250:7359, 1975

OSBORN MJ, GANDER FE, PARISI E: Mechanism of assembly of the outer membrane of Salmonella typhimurium. Site of synthesis of lipopolysaccharide. J Biol Chem 247:3973, 1972

ROGERS HJ, WARD JB, BURDETT IDJ: Structure and growth of the walls of gram-positive bacilli. In Stanier R, Rogers HJ, Ward JB (eds): Relations between Structure and Function in the Prokaryotic Cell. Society Gen Microbiol Symposium 20, Cambridge University Press, 1978, p 139

SCHWARZ U, RYTER A, RAMBACH A, HELLIO R, HIROTA Y: Process of cell division in E. coli: differentiation of growth zones in the sacculus. J Mol Biol 98:749, 1975

SMITH WJ, TAI PC, THOMPSON RJ, DAVIS BD: Extracellular labeling of nascent polypeptides traversing the membrane of Escherichia coli. Proc Natl Acad Sci USA 74:2830, 1977

STEVEN AC, TEN HEGGELER B, MULLER R, KISTLER J, ROSENBUSCH JP: Ultrastructure of a periodic protein layer in the outer membrane of E. coli. J Cell Biol 72:292, 1977

WINKLER HH, WILSON TH: The role of energy coupling in the transport of β-galactosides by E. coli. J Biol Chem 241:2200, 1966

SPORES

ARONSON AI, FITZ-JAMES P: Structure and morphogenesis of the bacterial spore coat. Bacteriol Rev 40:360, 1976 (Rather detailed and speculative, but beautifully illustrated)

DOI RH: Genetic control of sporulation. Annu Rev Genet 11:29, 1977

GERHARDT P, COSTILOW RN, SADOFF HL (eds): Spores, VI. Washington DC, American Society for Microbiology, 1975

GOULD GW, HURST A (eds): The Bacterial Spore. New York, Academic Press, 1969

GOULD GW, DRING GJ: Mechanisms of spore heat resistance. Adv Microb Physiol 11:137, 1974

HALDENWANG WG, LOSICK R: A modified RNA polymerase transcribes a cloned gene under sporulation control in Bacillus subtilis. Nature 285:256, 1979

HOCH JA: Genetics of bacterial sporulation. Adv Genet 18:69, 1976

KEILIN D: The problem of anabiosis or latent life: history and current concept. Proc R Soc Lond [Biol] 150:149, 1959

KEYNAN A: The transformation of bacterial endospores into vegetative cells. In Microbial Differentiation. 23rd Symposium on Society for General Microbiology, Cambridge University Press, 1973, p 85

KEYNAN A: Spore structure and its relations to resistance, dormancy, and germination. In Spores, VII. Washington DC, American Society for Microbiology, 1978

KORNBERG A, SPUDICH J, NELSON DL, DEUTSCHER MP: Origin of proteins in sporulation. Annu Rev Biochem 37:51, 1968

LOSICK R, PERO J: *Bacillus subtilis* RNA polymerase and its modification in sporulating and phage-infected bacteria. Adv Enzymol 44:165, 1976

MANDELSTAM J: The Leeuwenhoek lecture. Bacterial sporulation: a problem in the biochemistry and genetics of a primitive developmental system. Proc R Soc Lond [Biol] 193:89, 1976

PIGGOT PJ, COOTE JG: Genetic aspects of bacterial endospore formation. Bacteriol Rev 40:900, 1976

RHAESE HJ, GROSCURTH R, RUMPF G: Molecular mechanism of initiation of differentiation in B. subtilis. In Spores, VII. Washington DC, American Society for Microbiology, 1978

SONENSHEIN AL, CAMPBELL KM: Control of gene expression during sporulation. In Spores VIII. Washington DC, American Society for Microbiology, 1978

SNEATH PHA: Longevity of microorganisms. Nature 195:643, 1962

TIPPER DJ, GAUTHIER JJ: Structure of the bacterial endospores. In Halvorson HO, Hanson R, Campbell LL (eds): Spores, V. Washington DC, American Society for Microbiology, 1972, p 3

WARTH AD: Molecular structure of the bacterial spore. Adv Microb Physiol 17:1, 1978

chapter 7

THE BASIS OF CHEMOTHERAPY

GENERAL PRINCIPLES

The impact of antibacterial chemotherapy on medicine can hardly be exaggerated: the sulfonamides in 1936, followed shortly by a series of antibiotics, introduced a second golden age in the field of infectious disease, comparable to the earlier recognition of individual etiologic agents. In addition, many antibiotics have been valuable tools in advancing our understanding of bacterial physiology; we have noted the contribution of penicillin to the study of wall synthesis, and we shall see further examples when we take up protein synthesis (Ch. 13).

Of the innumerable compounds that are inhibitory in cultures, only a much smaller group are selective enough to be **chemotherapeutic agents,** i.e., compounds that can inhibit the proliferation of an infecting organism at drug concentrations that are tolerated by the host. Some of these agents are **bacteriostatic** (i.e., they reversibly inhibit growth), while others are **bactericidal** (i.e., they have an irreversible lethal action).

In this chapter we shall consider the development of chemotherapeutic agents, their actions on bacterial cultures, some molecular mechanisms, and the causes of variations in bacterial sensitivity. Other chapters will take up actions on nucleic acids (Ch. 10) or on the ribosome (Ch. 13), antibacterial action in the animal host (Ch. 25), antifungal agents (Ch. 45), viral chemotherapy (Ch. 51), and nonselective antimicrobial action (disinfection: Ch. 67).

DEVELOPMENT OF CHEMOTHERAPY

Origins. The idea of a direct, selective action of a drug on infecting microbes was forcefully advanced by Paul Ehrlich, in Germany, at a time when organic chemistry was blossoming. While a medical student in 1870 he introduced dyes that are still used in histology for the selective staining of basophilic and acidophilic cell components, and he later formulated the "side-chain" theory to account for the extraordinarily specific interactions of Abs with Ags. In 1904, drawing on these experiences, he began to seek synthetic chemicals that would exhibit a greater affinity for parasites than for host cells (*nihil agit nisi fixatur*); and for this selective action he coined the word **chemotherapy.** Ehrlich discovered dyes that were useful against trypanosomes, and arsenicals against spirochetes, but, disappointingly, no "magic bullets" against eubacteria.

Since these drugs were inactive in cultures, Ehrlich's concept met great resistance, and for 20 years after his death it was generally believed that in the struggle between host and parasite one could help only by stimulating host defenses. However, we now know the explanation for the lack of activity in vitro: these early drugs became active only after metabolic conversion by the host.

Sulfonamides. Antibacterial chemotherapy was launched in 1935, when Domagk, in Germany, developed a dye, Prontosil, that dramatically cured streptococcal infections. It proved inactive in vitro, but Tréfoüel, in France, showed that patients receiving it excreted a simpler product, sulfanilamide (see Figs. 7-5 and 7-6), which was active in vitro as well as in vivo. The effect of this development was dramatic, for it finally established Ehrlich's principle of direct chemotherapeutic action. More potent derivatives, called the sulfonamide drugs, were soon developed.

Antibiotics. The success of sulfonamides also renewed interest in antibiotics: **antimicrobial agents of microbial origin.*** Fleming reported in 1929 that a contaminating colony of the mold *Penicillium notatum* lysed adjacent colonies of staphylococci (Fig. 7-1); but the lytic agent seemed too unstable to be useful. However, when 10 years later Chain at Oxford purified the active material, called penicillin, it proved remarkably effective in certain infections.

---

* Inhibition of some microbes by others was first recorded in 1877 by Pasteur, who observed the sterilization of anthrax bacilli in a contaminated culture. In the next half-century a number of similar accidental observations were reported, together with a few abortive attempts at putting the inhibitory substances to therapeutic use.

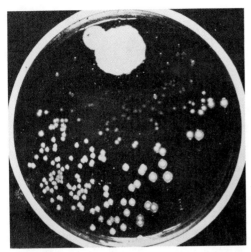

FIG. 7-1. The discovery of penicillin. Note lysis of colonies of *Staphylococcus aureus* surrounding a large contaminating colony of *Penicillium notatum*. (Fleming A: Br J Exp Pathol 10:226, 1929)

This success encouraged the search for additional antibiotics. In 1944 Waksman, in a laboratory devoted to soil microbiology, discovered streptomycin (from a soil actinomycete). This development extended chemotherapy to the tubercle bacillus and to many gram-negative organisms; it also established the value of a systematic, trial-and-error search for antibiotics. Massive screening, largely by pharmaceutical industries, has since yielded many useful products. Antibiotics soon became the largest class of prescription drugs: the production of penicillins alone, worldwide, is over 10^8 pounds/year.

The term **chemotherapeutic** has often been used to distinguish synthetic compounds from antibiotics. However, Ehrlich's original definition, emphasizing the distinction between selective and nonselective toxicity, is more useful. Moreover, some drugs discovered as antibiotics are now produced by chemical synthesis.

EFFECTS OF ANTIMICROBIALS ON GROWTH AND VIABILITY

The study of any new antimicrobial agent begins with a determination of its **antimicrobial spectrum** (i.e., the range of sensitive organisms). The sensitivity of an organism may be determined by the **tube dilution method** (in which identical inocula are incubated in tubes of medium containing different concentrations of the drug), by the more convenient **agar diffusion** method (Fig. 7-2), or by **automated** measurement of the early development of **turbidity**. These tests, of great clinical importance, will be discussed in Chapter 25.

Studies on cultures can also reveal 1) the kinetics of inhibition of growth, 2) a bactericidal versus a bacteriostatic effect, 3) lysis when present, and 4) effects of environmental conditions on antimicrobial action. More-

over, while all biosynthetic incorporation soon ceases when growth is inhibited, the primary area of inhibition can be detected by kinetic studies on the incorporation of specific radioactive precursors, to see whether DNA, RNA, protein, lipid, or wall synthesis is inhibited first.

The **kinetics of inhibition,** and the presence of **lysis,** may be observed by **turbidimetric** measurements following the addition of the drug to a growing culture. Several patterns are illustrated in Figure 7-3A: thus sulfonamide inhibition of growth is delayed for several generations; chloramphenicol and streptomycin cause an almost immediate leveling off of turbidity; and penicillin causes lysis, indicated by a sharp fall in turbidity.

Bactericidal action is, of course, evident when there is gross lysis (e.g., with penicillin). To quantitate the rate of killing, however, and also to recognize killing without lysis, **viability counts** are required (Fig. 7-3B). For this purpose samples are taken at intervals, diluted appropriately, and plated on an adequate medium. The dilutions must be great enough to eliminate further action of the drug.

Indirect indices of cell death, such as inhibition of respiration or of dye reduction, or intracellular staining, are less reliable than viability counts, especially for drugs that do not damage the cell membrane.

At borderline concentrations some bactericidal agents appear to be bacteriostatic; conversely, bacteriostasis from any cause is eventually followed by a decrease in viable number. Nevertheless, the distinction is real and useful: with a bacteriostatic agent the viable count remains essentially constant for several hours, while with a bactericide it declines by decimal orders of magnitude.

FIG. 7-2. Original diffusion plate assay for an antibiotic. Plate was heavily seeded on the surface with the standard test strain of *S. aureus*. To the open cylinders were added equal volumes of penicillin solutions containing 4, 2, 1, 0.5, and 0.25 U/ml, and the plates were incubated overnight. Today commercially prepared, stable, dry filter paper discs are used instead of cylinders. (Chain E, Florey H: Endeavour, Jan. 1944, p 3)

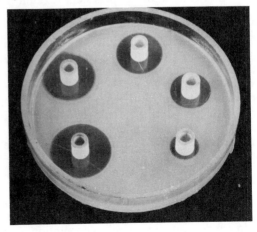

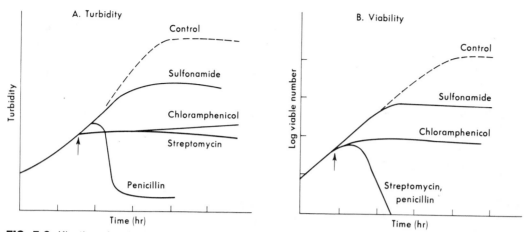

FIG. 7-3. Kinetics of antimicrobial action of representative drugs. A. Turbidity. B. Viability. Drug added at **arrow** to exponentially growing culture.

MECHANISMS OF BACTERICIDAL ACTION

Bactericidal action depends not on irreversible interaction between an agent and its receptor, but on irreversible damage to an element of the cell that cannot be replaced, such as **DNA** (whose damage and repair will be discussed in Ch. 11) or the **cell envelope** (Ch. 6). Most species of enzymes, after irreversible inactivation, can be regenerated by protein synthesis if the genome remains intact. However, one exceptional group is also a site of bactericidal action: **any protein required for RNA or protein synthesis.** Hence ribosomes are a frequent site of bactericidal action.

Persisters (Phenotypic Resistance). When a bactericidal agent is added to a growing culture killing is initially more or less exponential, but after some time it levels off: successive samples continue, for hours, to contain a small number of cells that generate colonies when transferred to fresh solid medium. Such "persisters" exhibit **phenotypic** rather than **genotypic** resistance: their progeny show no increase in resistance.

Phenotypic resistance is difficult to study, since it is not perpetuated in the progeny. Evidently some cells, perhaps damaged by abnormal cell division, have a block in the **metabolism required for a bactericidal response** (e.g., wall synthesis, protein synthesis), and the block is spontaneously overcome during the subsequent test for viability. This phenomenon is undoubtedly even more extensive in the body, where nutritional conditions are less uniform. The presence of persisters explains why the cure of an infection, even with a bactericidal agent, usually requires relatively prolonged chemotherapy.

Fortunately, immune mechanisms complement chemotherapy, for they attack bacterial surfaces and hence do not depend on metabolic cooperation. Accord-

ingly, persisters, like cells subject to bacteriostatic action, are finally cleaned up by immune and phagocytic responses.

Antagonism. Since most bactericidal actions depend on a particular metabolic activity they are antagonized by bacteriostatic agents, which directly or indirectly inhibit most metabolic activities. These interactions are seen not only in vitro (Fig. 7-4) but also in experimental animals; extrapolation to patients seems reasonable (see Combined Therapy, Ch. 25), though differences in the distribution of two drugs, in time and space, may decrease their antagonism.

FIG. 7-4. Rapid protection by chloramphenicol **(CA)** against killing by streptomycin **(SM)** added to exponentially growing *Escherichia coli* at 0 time. At intervals samples were transferred **(arrows)** to flasks containing chloramphenicol. These flasks were similarly incubated. Samples were removed for viable counts as noted. (Plotz P, Davis BD: J Bacteriol 83:802, 1962)

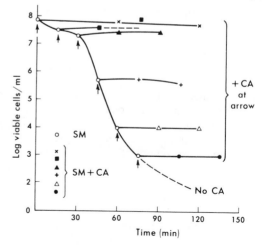

Synergism. The bactericidal actions of penicillin and of an aminoglycoside are synergistic: in mixtures both the rate of killing and the minimal effective concentrations reflect a greater than additive action. Damage to the wall by penicillin evidently facilitates the entry of streptomycin, and/or the loss of antagonistic ions, since sublethal pretreatment with penicillin increases sensitivity to streptomycin and accelerates its uptake, whereas pretreatment with streptomycin does not influence the subsequent interaction with penicillin.

Another mechanism, synergism between bacteriostatic agents acting on the same pathway, is mentioned under sulfonamides.

METABOLITE ANALOGS

COMPETITIVE INHIBITION

Yeast extract was found to antagonize the inhibition of bacterial growth by a sulfonamide, and Woods, in England, identified the antagonist as p-aminobenzoate (PAB, Fig. 7-5), a structural analog of sulfonamides. He further showed that the antagonism is **competitive;** i.e., with a doubled concentration of the analog one can restore growth by also doubling the concentration of substrate. This development dramatized a principle that now pervades pharmacology: analogs compete with a normal metabolite for an enzymatic or a receptor site.

With enzymes the characteristics of competitive inhibition follow from the **mass law.** If a substrate (S) forms a reversible, active complex (ES) with an enzyme (E), and if an analog (A) forms a reversible, inactive complex (EA), the ratio of the two complexes (which determines the degree of inhibition) will be proportional to the ratio of the free concentrations of the two ligands, S and A. Competitive inhibition of growth by analogs of an essential metabolite is thus readily explained, whether the metabolite arises endogenously or is an essential nutrient.

Most analogs compete reversibly with the substrate, but some react irreversibly with the corresponding enzyme (e.g., penicillin, Ch. 6).

Noncompetitive Reversal of Growth Inhibition. Inhibition of growth can be reversed not only competitively, by the substrate of the blocked enzyme, but also noncompetitively, by a later product in the same sequence: i.e., an adequate quantity of product restores growth, regardless of how tightly a preceding reaction is shut off. This principle will be illustrated by the sulfonamides, below.

SYNTHETIC ANTIMETABOLITES

The explanation of sulfonamide action generated the hope that a rational synthesis of **antimetabolites** (i.e., analogs of known essential metabolites) could replace the purely empiric search for antimicrobial agents. However, the thousands of analogs of vitamins, bases, and amino acids that were synthesized failed to exhibit the selectivity required for antibacterial chemotherapy, for they were modeled on metabolites that are, unlike PAB, essential for the host as well as for the bacteria.

FIG. 7-5. Metabolism blocked by sulfonamides. These analogs of p-aminobenzoate (PAB) block (and also replace) its condensation with a pteridine to form dihydropteroate (not shown), which condenses with glutamate to form folic acid. This compound and its reduced forms, dihydrofolic acid (DHF) and tetrahydrofolic acid (THF), go through a cycle of reductions and oxidations in the reduction and transfer of 1-C fragments derived from serine; only the products are shown. Trimethoprim, an analog of DHF, blocks its reduction to THF; hence trimethoprim and sulfonamides interfere with different steps in the same pathway.

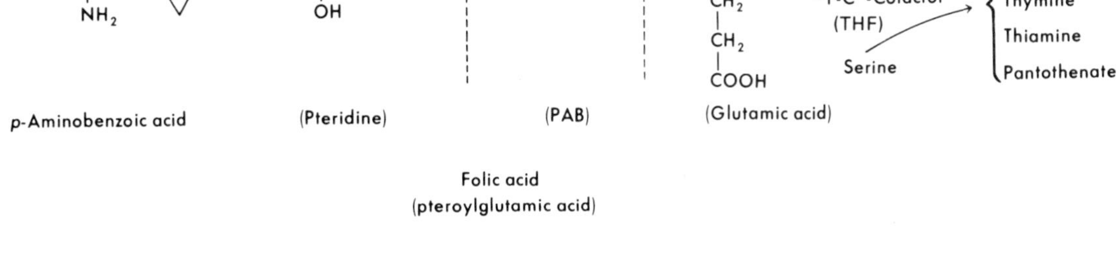

Antimetabolites have proved clinically more useful in cancer chemotherapy (where a higher level of toxicity is acceptable) and in viral chemotherapy (e.g., arabinosyl nucleosides). Moreover, the concept of structural analogy has explained the action of many empirically discovered drugs, including some antibiotics.

OTHER SITES OF COMPETITION

Though inhibition of the enzyme acting on the competitive metabolite was long assumed to be the only basis for competitive inhibition, other mechanisms were discovered later.

Biosynthetic Incorporation. Radioactive labeling showed that many analogs (including sulfonamides) not only can **block** but also can **deceive** enzymes, i.e., can serve as **competitive substrates,** which are incorporated into coenzymes, proteins, RNA, or DNA.

Some **amino acid analogs** (e.g., *p*-fluorophenylalanine) can replace as much as 50% of the corresponding amino acid in the protein being formed, whereas others (e.g., 5-methyltryptophan) inhibit growth without being substantially incorporated. The incorporation may or may not impair the function of the product.

Base analog incorporation has been extensively studied in relation to cancer, inhibition of viral synthesis, and mutagenesis (Ch. 11). A few key findings will be summarized here.

To affect growth, bases must first be converted to the corresponding nucleotides, by the enzymes (one set for purines and one for pyrimidines) that convert normal bases and nucleosides to nucleotides. Some base analogs are incorporated primarily into DNA and others into RNA. Thus 5-bromouracil substitutes well for thymine and can largely replace it in bacteria or bacteriophages, with little loss of viability but with mutagenic consequences. 5-Fluorouracil, in contrast, resembles uracil more than thymine and is incorporated into RNA but not detectably into DNA. However, it also becomes converted to a deoxyribonucleotide, 5-fluorodeoxyuridylate, which inhibits the conversion of the normal deoxyuridylate, by thymidylate synthetase, to deoxythymidylate.

Competition for Transport. Specific membrane transport systems are an additional site of competition: metabolites can restore growth by blocking the entry of the analog; and analogs, conversely, can inhibit growth by inhibiting the entry of an essential nutrient. Hence the sensitivity of a cell to an inhibitor may be determined by the affinity of its transport system rather than by the affinity of an intracellular receptor. Indeed, even the classic competition between sulfonamides and PAB within intact cells appears to reflect competition at a transport system rather than at an enzyme.

In **pseudofeedback inhibition** an analog mimics the endproduct of a biosynthetic pathway in inhibiting the first enzyme of the pathway. This process will be described in Chapter 14.

SPECIFIC AGENTS

SULFONAMIDES

PAB is a biosynthetic precursor of folic acid, a required vitamin for mammals that becomes a cofactor, cycling between dihydrofolate and tetrahydrofolate, in the synthesis of several cell constituents (Fig 7-5). Hence even though PAB is not a metabolite of mammals, one might expect the products of these pathways to reverse sulfonamide action noncompetitively. Fortunately, two aspects of these products make sulfonamide chemotherapy possible. First, most bacteria (unlike animal cells) cannot utilize folic acid. Second, of the several compounds whose biosynthesis depends on folate only methionine and the vitamins are present in our body fluids: purines and pyrimidines are made in our cells as nucleotides, and they do not circulate (at least not as the free bases or nucleosides that can be utilized by bacteria).*

Noncompetitive antagonism explains a serious limitation of sulfonamide therapy: **ineffectiveness in sites of extensive tissue destruction** (such as purulent exudates, burns, and wounds), in which host cell autolysis releases nucleic acid bases (as well as other antagonists). Moreover, the gradual dilution of a coenzyme after sulfonamide inhibition explains why the bacteriostatic effect is delayed rather than immediate (see Fig. 5-5).

Among the derivatives of sulfanilamide potency was increased by various substitutions on the sulfonamide group (N^1) that increased the negativity of its sulfone component ($-SO_2-$) and hence its resemblance to the COO group of PAB. Improvements in pharmacologic properties (distribution, inactivation, toxicity) were also achieved.

Figure 7-6 shows two such N^1 derivatives. Compounds substituted on N^4 (e.g., sulfasuxidine) are inactive in vitro, but they have been used in intestinal antisepsis: they are poorly absorbed and are slowly hydrolyzed to yield an active compound.

Ionization of the sulfonamide group shifts the center of negativity away from the sulfone. Hence the sulfonamide drugs selected for maximal activity at pH 7.4 are less potent at the usual pH of urine (5–6) than those with a lower pK, such as sulfacetamide.

Except for the few bacterial species that require exogenous folic acid, all others can be inhibited in culture by sulfonamides. However, the required concentrations vary widely, and the effective antibacterial spectrum (at concentrations attainable in the host) is quite limited. Since this spectrum is essentially the same for all

---

* Thus mutations of *Salmonella* to various amino acid requirements do not affect virulence in mice; but a requirement for PAB or for purines eliminates virulence unless the required factor is injected along with the organisms.

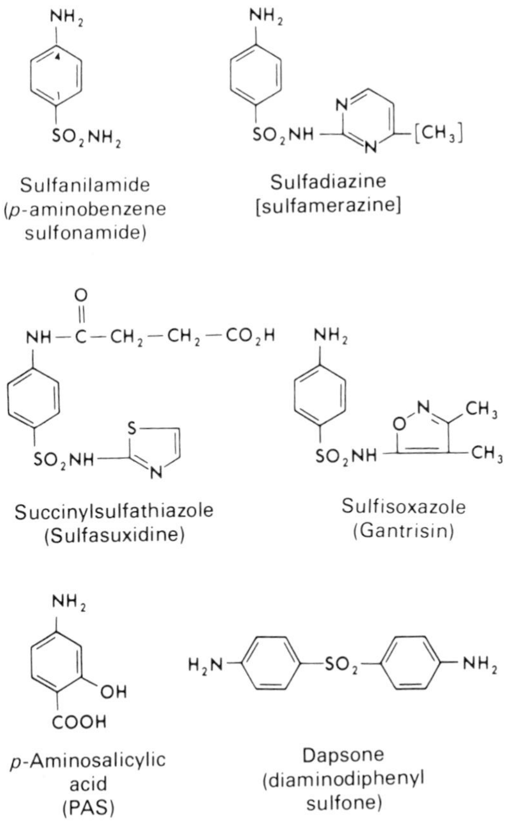

FIG. 7-6. Sulfonamides and other PAB analogs.

fonamides, suggests that the receptor for PAB in *Mycobacterium tuberculosis* differs significantly from that in many other bacteria.

ANTIBIOTICS

The mechanism of action of most antibiotics remained a mystery until macromolecule synthesis became accessible to analysis. It is now possible to classify almost all these agents in terms of mechanism, as well as origin, structure, or antimicrobial spectrum.

PENICILLIN: PRODUCTION AND CHEMISTRY

We have noted above the discovery of penicillin, and in Chapter 6 we described its interference with a reaction unique to bacteria (crosslinking in the peptidoglycan). Here we shall consider its production and its modifications.

This valuable antibiotic, discovered in England in 1929, was developed during World War II, chiefly in the United States, as one of the first large government-supported medical research projects. The procedures devised for its large-scale production have provided a model for subsequent antibiotics, and their development also provided an early model for large-scale government support of medical research.

Structure. The substance originally designated as penicillin was found to be a mixture of compounds, with different acyl side chains attached to a binucleate structure, 6-aminopenicillanic acid. This structure is a cyclized dipeptide, formed by condensation of L-cysteine and D-valine (Fig. 7-7).

The four-membered (β-lactam) ring of penicillin has a strained configuration, and its CO–N bond is therefore readily hydrolyzed, yielding an inactive product, penicilloic acid (Fig. 7-7). Hence penicillin is unstable in solution and is relatively rapidly destroyed by the acid of the stomach. The lability to acid varies with different penicillins; with G it permits roughly 1/5 of an orally administered dose to be absorbed.

Modified Biosynthesis. The nature of the penicillin produced can be influenced by the medium: **benzylpenicillin (penicillin G),** the most satisfactory of the original penicillins, can be obtained in almost pure yield by providing an excess of the corresponding acyl donor (phenylacetic acid). It is still the most potent and inexpensive penicillin, but it was not patented; hence more profitable derivatives are vigorously promoted.

SEMISYNTHETIC PENICILLINS

Though penicillin G is an extremely effective and nontoxic drug, it has several limitations: a relatively narrow

sulfonamides of similar potency, a single member of the group suffices in sensitivity tests.

Synergistic Combined Therapy. Simultaneous partial inhibition of two steps in a pathway is synergistic, since the two degrees of inhibition are multiplied. Hence the effectiveness of sulfonamides is increased by adding an inhibitor of dihydrofolate reductase, **trimethoprim.**

Dapsone (Diaminodiphenyl Sulfone). This drug (Fig. 7-6) was synthesized as a congener of sulfanilamide. It does not have a sulfonamide group, however, and unlike the sulfonamides it exhibits marked specificity: it is valuable in the treatment of only one bacterial infection, **leprosy.**

p-**Aminosalicylic Acid (PAS).** Analogs of PAB can also be created by substitutions on the ring. Among these *p*-aminosalicylate (PAS; Fig. 7-6) is useful in tuberculosis. It has the advantage over streptomycin of penetrating mammalian cells, but it is only bacteriostatic. PAS is antagonized competitively by PAB.

The striking selectivity of PAS, quite different from that of sul-

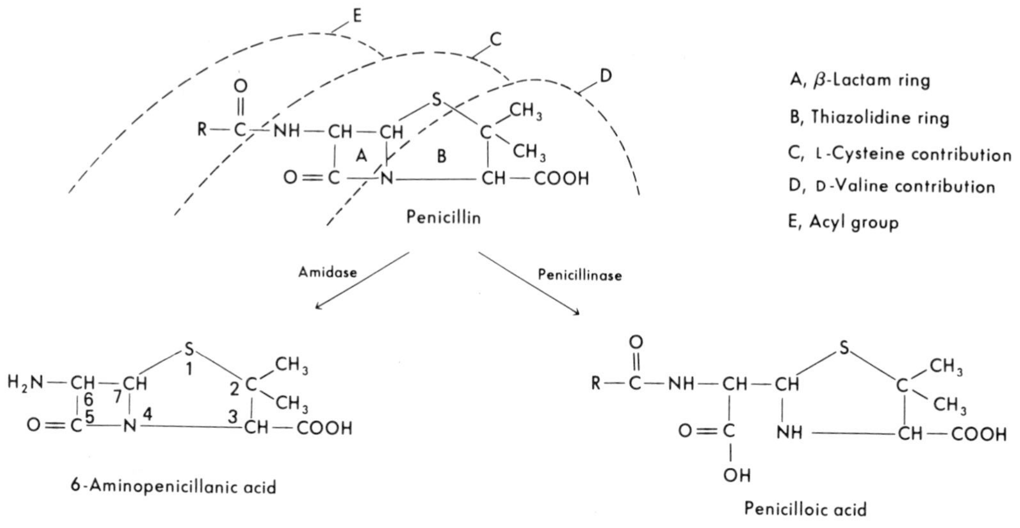
A, β-Lactam ring
B, Thiazolidine ring
C, L-Cysteine contribution
D, D-Valine contribution
E, Acyl group
Penicillin
Amidase
Penicillinase
6-Aminopenicillanic acid
Penicilloic acid

Side chain (R)	*Penicillin*
$C_6H_5CH_2-$	Benzyl (G)
$C_6H_5OCH_2-$	Phenoxymethyl (V)

FIG. 7-7. Structure of some penicillins.

antimicrobial spectrum (see below), destruction by acid, destruction by penicillinase, and elicitation of allergic responses. The molecule is too complex for commercial synthesis, but modifications became possible with the discovery that in a medium lacking any acyl side chain donor (or with certain mutants) the mold produces an inactive precursor, **6-aminopenicillanic acid (6-APA).** This compound can be condensed chemically with any carboxylic acid, yielding a virtually unlimited number of possible penicillins. Two major classes of new drugs emerged: penicillins resistant to penicillinase, and those with a broad antimicrobial spectrum. Several are illustrated in Figure 7-8.

β-Lactamase and Drug Resistance. Many staphylococci, and certain other species, are resistant to penicillin G because they form **β-lactamase (penicillinase),** which hydrolyzes the β-lactam bond (see Drug Resistance, below). However, those semisynthetic penicillins with bulky groups near the site of hydrolysis (Fig. 7-8) are essentially inert to the enzyme. Moreover, some of these derivatives (oxacillin, nafcillin) have the further advantage that they can be given orally; but with organisms that do not form β-lactamase they are only about one-tenth as potent as penicillin G.

The penicillinase-resistant penicillins not only fail to be hydrolyzed but they stabilize penicillinase in an inactive conformation, as tested by the subsequent addition of a sensitive substrate. However, on simultaneous exposure this effect is not seen, since the enzyme has a much greater affinity for the sensitive than for the resistant penicillin.

Penicillins that withstand penicillinase are nevertheless limited, like other penicillins, by another mechanism of bacterial resistance: mutations with decreased permeability.

Broadening of Antimicrobial Spectrum. The spectrum of the original penicillins is largely limited to various gram-positive organisms, treponemes, and neisseriae. However, the presence of an amino group on the side chain makes **ampicillin** (α-aminobenzyl penicillin; Fig. 7-8) much more active against many gram-negative bacilli, perhaps because the positive charge enhances penetration through the negatively charged lipopolysaccharide. Ampicillin is half as active as penicillin G against gram-positive organisms, and it is sensitive to penicillinase. It can be given orally.

The **amidinopenicillins** (e.g., mecillinam) are semisynthetic products with an amidino group (–NH–CH = N–) instead of a peptide (–CONH–) between a substituent and 6-APA. They are particularly effective against gram-negative organisms.

Penicillinlike antibiotics. The crosslinking reaction in **wall synthesis** is the site of action of all β-lactam antibiotics. Several antibiotics that act at other stages in this synthesis have been noted in Chapter 6: they are more toxic than the β-lactams and have limited chemotherapeutic use. These include **bacitracin** (a polypeptide), **vancomy-**

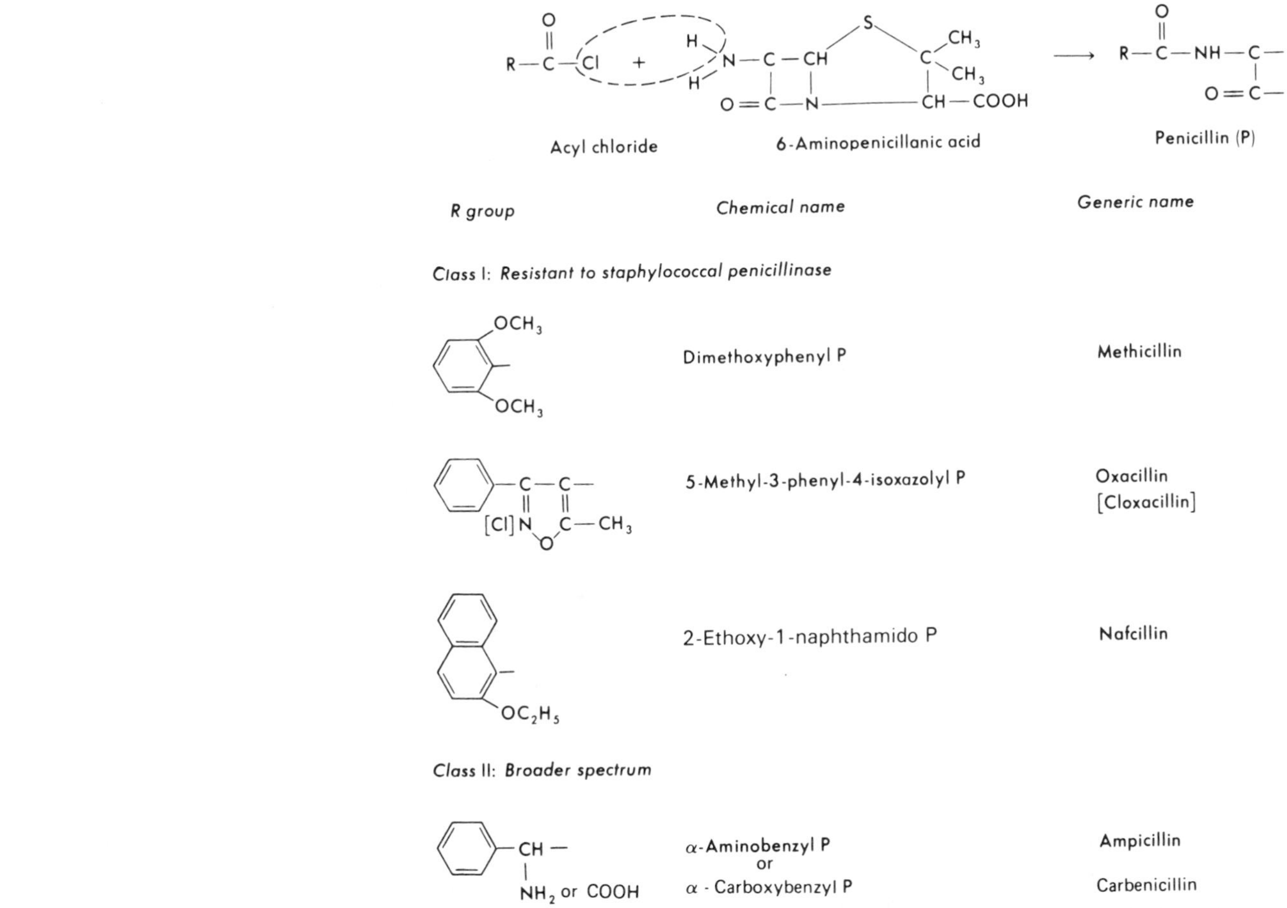

| Acyl chloride | 6-Aminopenicillanic acid | | Penicillin (P) |

R group	Chemical name	Generic name

Class I: Resistant to staphylococcal penicillinase

	Dimethoxyphenyl P	Methicillin
	5-Methyl-3-phenyl-4-isoxazolyl P	Oxacillin [Cloxacillin]
	2-Ethoxy-1-naphthamido P	Nafcillin

Class II: Broader spectrum

| | α-Aminobenzyl P *or* α-Carboxybenzyl P | Ampicillin Carbenicillin |

FIG. 7-8. Some semisynthetic penicillins.

cin, and **cycloserine.** Antibiotics with a spectrum similar to that of penicillin G but with very different, bacteriostatic, actions include erythromycin, lincomycin, and novobiocin (see below).

CEPHALOSPORINS

Cephalosporin C, a β-lactam antibiotic similar to the penicillins in structure and action, was isolated from a mold of the genus *Cephalosporium.* Though the original cephalosporin had low potency, the nucleus (7-amino-

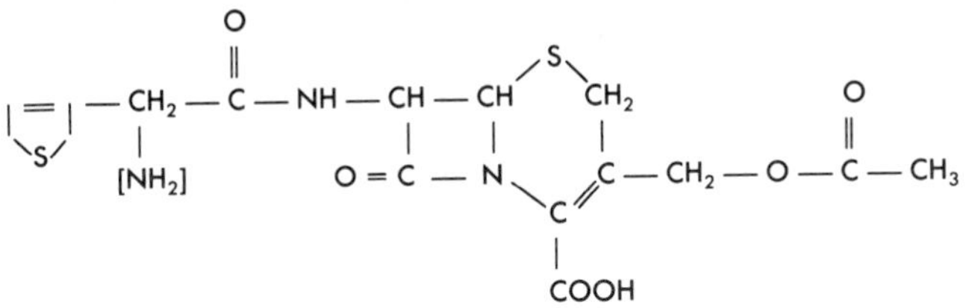

Cephalothin
[Cephaloglycin]

cephalosporanic acid) was used as the basis for a series of useful semisynthetic derivatives, including cephalothin, cephaloridine, and cefoxitin.

The initial advantage of the cephalosporins arose from the prevalence of allergy to penicillin. In this reaction the actual allergen is usually a derivative, formed in the body, with the β-lactam ring opened up (Ch. 22). Since the major immunologic determinant is thus the S-containing ring, in which cephalosporins differ from penicillins, *cephalosporins do not cause allergic reactions in most patients who are allergic to penicillin.* However, allergy to cephalosporins can also develop.

In addition, the cephalosporins in use, unlike any penicillins, both have a broad spectrum and are resistant to β-lactamase. They are therefore widely used where the sensitivities of the organism are not known. However, with sensitive organisms penicillins are considerably less expensive.

Cephamycins are a similar group of compounds produced by certain streptomycetes.

119

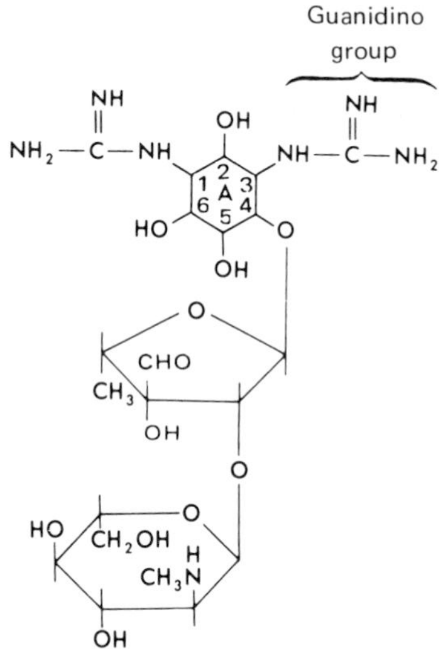

FIG. 7-9. Streptomycin (an aminoglycoside antibiotic) and spectinomycin. The aminoglycosides all contain an aminocyclitol ring and one or more aminosugars; spectinomycin also contains an aminocyclitol but no aminosugar. In streptomycin the aminocyclitol is streptidine (1,3-diguanidino inositol); in some aminoglycosides it is streptamine (1,3-diaminoinositol); and in many it is 2-deoxystreptamine. Neomycin and paromomycin have aminosugar substituents on positions 4 and 5 of deoxystreptamine, while in the kanamycin group (including amikacin, gentamicins, and tobramycin) they are on positions 4 and 6.

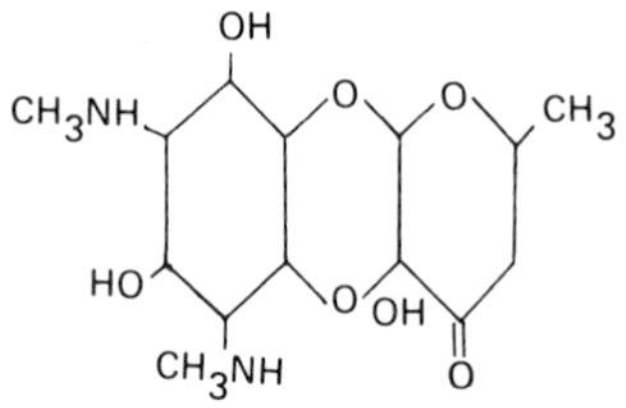

AMINOGLYCOSIDES

We have noted the discovery of streptomycin above. Other clinically useful aminoglycosides, with structures similarly involving an aminocyclitol ring and one or more amino sugars (Fig. 7-9), have a similar action and antimicrobial spectrum, and variable crossresistance. The aminoglycosides act on the ribosome, which is an even more frequent site of antibiotic action than wall synthesis; fortunately, many antibiotics act selectively on bacterial and not on eukaryotic ribosomes.

The aminoglycosides in clinical use include **streptomycin, kanamycin, gentamicin,** * and the semisynthetic

---

* The aberrant spelling of gentamicin results from the ruling of a committee that the suffix "mycin" should be restricted to products of *Streptomyces.* A particularly confusing alternative was chosen

tobramycin and **amikacin; neomycin** and **paromomycin** are more toxic and are used topically.

New members of this group are being developed by **mutasynthesis:** mutants blocked at certain stages in the synthesis of an aminoglycoside can form novel, active products when provided with replacements for the missing sugar or aminocyclitol.

Here we shall note the features of aminoglycoside action on bacterial cells that are directly relevant to their therapeutic use. The interactions with the ribosome will be discussed in Chapter 13.

1) The **bactericidal** action of aminoglycosides requires a small amount of protein synthesis: it can be prevented by starvation, or by reversible inhibition of protein synthesis (see Fig. 7-4). Penicillin, in contrast, requires substantial growth before the cell wall is irreversibly damaged.

2) Unlike the case with penicillin, the rate of killing increases with the drug concentration up to a very high level. Evidently **penetration** into bacteria, rather than growth rate, determines the rate of killing.

3) **Anaerobiosis** antagonizes aminoglycosides, possibly by an effect on penetration. These drugs are not effective with any obligate anaerobes, and with facultative organisms their potency decreases about ten times on a shift from aerobic to anaerobic conditions.

4) **Acidity** and **salts** (especially polyvalent ions such as Mg^+ and HPO_4^{2-}) strikingly decrease their effectiveness, perhaps by impairing penetration into the bacterial cell.

5) Aminoglycosides are relatively ineffective against **intracellular bacteria.** It is not certain whether the drugs fail to penetrate host cells or whether the cells contain too high a concentration of antagonistic ions.

6) Aminoglycosides are more **toxic** than β-lactams, and systemic administration requires injection. They are therefore used systemically only for serious infections.

Spectinomycin (Fig. 7-9) is an aminocyclitol but not an aminoglycoside. It also acts on the ribosome, but it is bacteriostatic rather than bactericidal (Ch. 13). It is used in treating gonorrhea.

OTHER ANTIBIOTICS

Chloramphenicol and **tetracycline** (Fig. 7-10), though differing markedly in structure, have an essentially identical antimicrobial spectrum: they were the first broad-spectrum antibiotics, effective against many gram-negative as well as gram-positive organisms. Both reversibly block the ribosome and hence are bacteriostatic.

---

for gentamicin, which was isolated originally from another genus of actinomycetes (though subsequently also from *Streptomyces*). After two editions of using "gentamycin" this text is abandoning the struggle for a rational nomenclature.

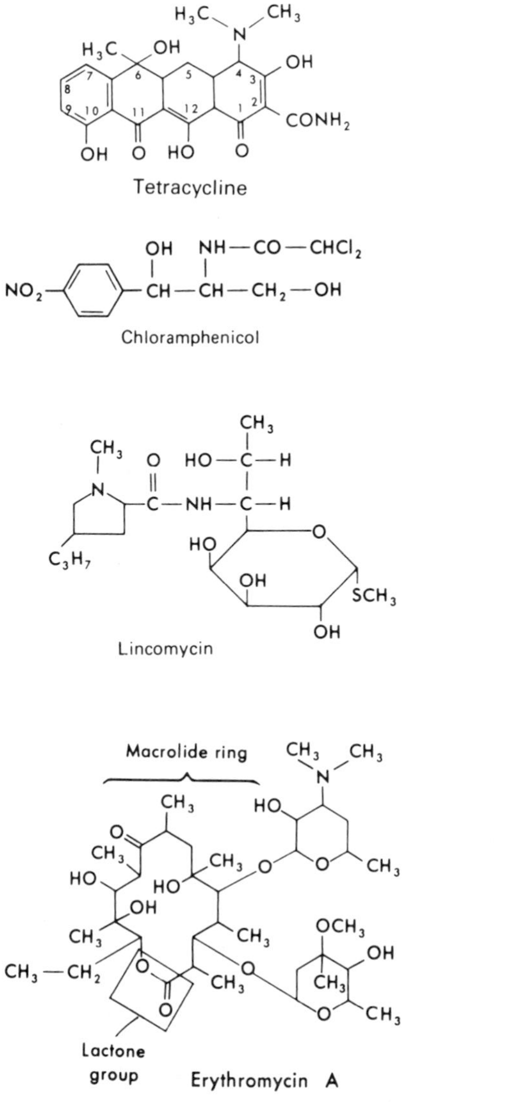

Tetracycline

Chloramphenicol

Lincomycin

Macrolide ring

Lactone group   Erythromycin A

FIG. 7-10. Antibiotics that act reversibly on the ribosome. Chlortetracycline is 7-chlorotetracycline; oxytetracycline is 5-hydroxytetracycline; demethylchlortetracycline is chlortetracycline without the 6-methyl group. Erythromycin is one of several macrolides with similar structure and action. Lincomycin has been largely replaced by clindamycin (7-chloro-7-deoxylincomycin). Chloramphenicol has no congeners in clinical use.

Chloramphenicol is produced by *Streptomyces venezuelae.* It contains a nitro and a dichloroacetyl group, both unusual in a natural product. The **tetracyclines** have four fused rings; hence the name. **Chlortetracycline** and the chlorine-free compound **tetracycline** are both produced by *Streptomyces aureofaciens,* depending on the chloride content of the fermentation medium.

The several tetracyclines have identical antimicrobial spectrums and complete crossresistance, but as naturally occurring compounds possessing some chemical difference they are independently patentable "compositions of matter."

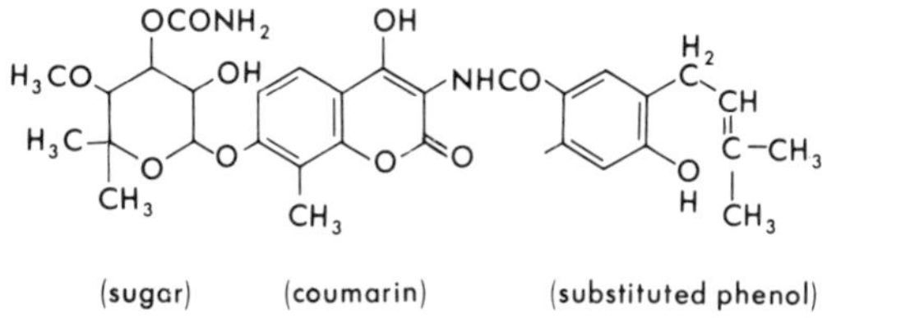

(sugar)   (coumarin)   (substituted phenol)

FIG. 7-11. Novobiocin. Nalidixic acid, with a very similar action but different structure, is shown in Fig. 7-13.

Lincomycin (Fig. 7-10) has a narrower spectrum, resembling that of penicillin, but its action is very similar to that of chloramphenicol. These actions on the ribosome will be discussed in Chapter 13.

Novobiocin (Fig. 7-11), produced by *Streptomyces niveus,* blocks DNA synthesis: it competes with ATP for **DNA gyrase,** an ATP-requiring enzyme that catalyzes the negative coiling of closed duplex DNA (Ch. 10). The synthetic compound **nalidixic acid** (see Fig. 7-13) blocks the same enzyme, binding to a different subunit. Resistant mutants, altered in either subunit, have been useful in studying this enzyme.

Macrolides are antibiotics with a large ring, usually formed by lactone closure of an aliphatic chain with frequent keto and hydroxyl groups. The most widely used macrolide is the reversible ribosome inhibitor **erythromycin** (Fig. 7-10), produced by *Streptomyces erythreus;* it has a 14-member ring.

A classification based on a large ring is too general, for it also includes antibiotics with quite different actions: certain **ionophores** (Table 6-2) and **polyenes** (Ch. 45), which both affect membrane permeability. In the **rifamycins** (e.g. **rifampin**), which block transcription (see Fig. 11-35, Ch. 11), a large chain forms a ring by two attachments (nonlactone) to an aromatic nucleus.

Polymyxin B (Fig. 6-23) and colistin, very similar polypeptides, are unique among antibiotics in attacking a cell constituent (the membrane) without requiring metabolic activity. The action of polymyxins is described in Chapter 6 (Fig. 6-24).

Other antibiotics, mostly not of therapeutic use, act on still other metabolic processes and have been useful experimental tools. For example, **cerulenin** blocks fatty acid synthesis; **citrinin (Antimycin)** blocks electron transport between cytochromes b and c; **oligomycin** uncouples phosphorylation from electron transport; and **tunicamycin** blocks glycosylation of proteins. Antimycin is fungicidal.

ADDITIONAL SYNTHETIC ANTIMICROBIALS

ISONIAZID AND ETHIONAMIDE

Isoniazid (isonicotinic acid hydrazide = INH: Fig. 7-12) is the only major synthetic agent discovered since the sulfonamides. Clinically it is useful only against *M. tu-*

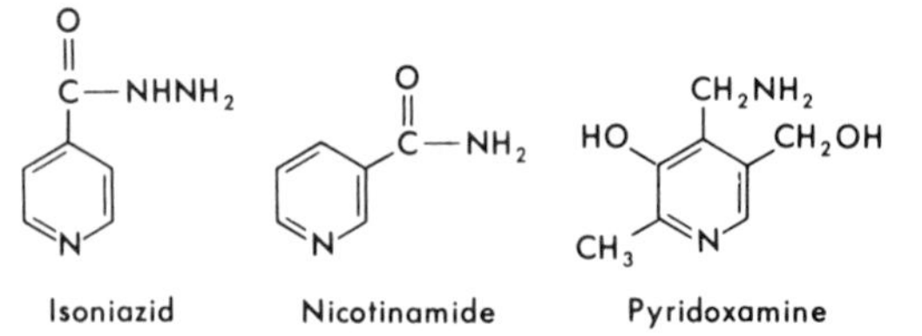

FIG. 7-12. Isoniazid (isonicotinic acid hydrazide; pyridine-4-hydrazoic acid) and two metabolites that it resembles.

FIG. 7-13. Synthetic compounds used to treat urinary tract infections.

berculosis. Because it is bactericidal even at very low concentration (<1 μg/ml), and it acts (unlike streptomycin) on intracellular as well as extracellular organisms, its chemotherapeutic effect is dramatic. Other agents used in treating tuberculosis will be noted in Chapter 37.

INH is active in cultures only against certain myobacteria, and the closely related nocardiae and corynebacteria. The explanation is that it inhibits (in cells and in extracts) synthesis of a class of compounds unique to these organisms, the **mycolic acids;** the inhibited reaction is not known. These very long-chain fatty acids (see Ch. 37) form part of the waxy outer layer of the wall of these organisms, and they are linked covalently to the underlying peptidoglycan; hence INH is bactericidal by interfering with wall synthesis. As with other agents with this locus of action (in contrast to streptomycin), mutations cause only small-step increments of resistance; moreover, INH-resistant mutants are often less virulent (at least in guinea pigs) than the sensitive parent strains.

Isoniazid is an analog of nicotinamide (Fig. 7-12), and it can be incorporated enzymatically in place of nicotinamide, forming an analog of NAD. NAD participates in several reactions besides electron transfer (e.g., DNA ligation), but their significance for isoniazid action is not known. Isoniazid also resembles pyridoxamine, and it can inhibit a number of enzymes that require pyridoxal (or pyridoxamine) phosphate as cofactor. The clinical toxicity of isoniazid is reported to be antagonized by large doses of pyridoxine.

Ethionamide (α-ethyl isonicotinyl thioamide) resembles INH in structure and in action. It is also useful in treating tuberculosis.

URINARY ANTIBACTERIAL THERAPY

Several synthetic bacteriostatic compounds reach effective concentrations in the urine, though the concentrations attained in the body are too low to provide any systemic chemotherapy. Among the **nitrofurans,** which inhibit a variety of gram-positive and gram-negative bacteria, **nitrofurantoin** (Fig. 7-13), is widely used for chronic urinary tract infections. **Nalidixic acid** (Fig. 7-13) is also used in gram-negative infections of the urinary tract.

Methenamine, a cyclic product of the condensation of formaldehyde and ammonia (Fig. 7-13), is not active in vitro, but on reaching the urine it is split (if the urine is acidic), and the formaldehyde is bactericidal. A mixture with mandelic acid (methenamine mandelate) is often used to promote acidity of the urine. *Proteus* is re-

sistant because its urease splits urea to CO_2 and NH_3, and hence makes the urine alkaline.

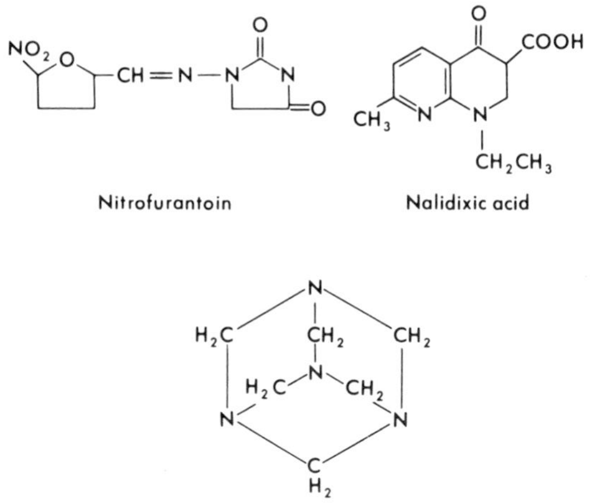

Nalidixic acid, noted above, is used in gram-negative infections of the urinary tract.

ANTIBIOTIC PRODUCTION

Nearly 3000 antibiotics have been isolated; of these only several dozen are selective enough to be useful in chemotherapy. Over 50% are from actinomycetes (mostly of the genus *Streptomyces*): most polypeptide antibiotics are produced by *Bacilli*, and the penicillins and cephalosporin by fungi. Genes for antibiotic synthesis, like genes for other products not essential for growth, are generally present on plasmids (Ch. 12), which sometimes also carry genes that make the producing organism insensitive to its own antibiotic.

After penicillin, the various antibiotics were discovered by routine screening of random isolates of soil microbes. The initial cultures have only a low concentration, but the yield is greatly increased (as much as 5000 times) by two empiric procedures: large-scale testing for higher yield mutants, and modifications of the culture medium. The organisms that have produced antibiotics are obligate aerobes, which are conventionally cultivated in shallow layers of medium; but on a large scale submerged growth has proved more economic in well aerated fermenters of 50,000 gallons.

BIOSYNTHESIS

Isotopic studies, as well as structural relations, show that all the major biosynthetic pathways contribute precursors to the various antibiotics, and often a single antibiotic incorporates the products of several pathways. The polypeptides are derived from amino acids (by a nonribosomal mechanism described in Ch. 4); the varied sugar residues of many antibiotics arise from the carbon chain of glucose.

The long chains and rings of macrolides, polyenes, tetracyclines, and portions of other antibiotics are derived from lipid precursors (acetyl CoA and propionyl CoA). These undergo sequential condensation, just as in fatty acid biosynthesis (Ch. 6), but without the reductive steps between successive condensations. The resulting poly-β-ketones (**polyketides**)

$$RCO-CH_2CO-CH_2CO \ldots$$

can readily condense internally to form rings, both large and small. Some aromatic rings are derived via shikimic acid (Ch. 4).

ECOLOGIC ROLE; SECONDARY METABOLITES

Antibiotics were initially assumed to participate in an antagonistic ecologic relation between competing organisms (i.e., antibiosis, the opposite of symbiosis). However, several considerations oppose this view. 1) Antibiotic producers are only a tiny fraction of the microbial population in soil samples, and they thus do not appear to have a striking advantage. 2) Antibiotics generally appear only after growth has ceased, rather than during competition for growth. 3) The strains found in nature excrete only small amounts of antibiotic; heavy excretion is an artifact developed in the laboratory.

If the inhibitory property of antibiotics is adventitious, from an ecologic perspective, the problem of their evolution becomes that of the broader class to which they belong: **secondary metabolites.** These compounds, which are formed after growth ceases, exist in great variety, and they exhibit less biosynthetic specificity than primary (essential) metabolites: many organisms excrete a **group** of closely related compounds rather than a single one (e.g., several penicillins, actinomycins A to D, etc). This property suggests an almost playful activity of enzymes, condensing accumulated metabolites when they cannot engage in the serious business of growth.

It seems more than coincidence, however, that *antibiotics appear only in sporulating organisms;* moreover, they appear generally at the time of sporulation, and their production is eliminated by mutations that block any early stage in this process.

Since sporulation is associated with synthesis of a new kind of wall and degradation of old wall, and since walls, like many antibiotics, contain D-amino acids and novel sugars, the altered wall metabolism of sporulating organisms might conceivably provide the precursors that are forged into some antibiotics. Alternatively, a possible regulatory role of antibiotics in sporulation is suggested by the observation that the addition of gramicidin (a polypeptide antibiotic) in a mutant blocked in its synthesis will alter the spores that are formed.

Since plasmids play an evolutionary role in spreading blocks of genes, as well as a role in antibiotic production, it seems possible that these compounds (and other secondary metabolites) may reflect evolutionary experiments in the formation of new compounds.

Another intriguing question is how antibiotic-producing bacteria (including streptomycetes) escape damage by their products. Some plasmids bearing genes for aminoglycoside production, in streptomycetes, also have resistance genes for inactivating the compounds.

While antibiotic activity is the most easily screened property of secondary metabolites, microbial filtrates are proving to be promising sources of compounds with various other potentially useful pharmacologic activities, e.g., protease inhibitors, antitumor agents.

DRUG RESISTANCE

That microbes can become resistant to a drug during treatment was discovered by Ehrlich with protozoa. The rediscovery of this phenomenon with antibacterial drugs led to the fear that in our race with the adapting microbial population we would be required, like the Red Queen, to keep running faster and faster merely to stand still. Indeed, the problem is clinically very important (see Drug Resistance, Ch. 25), especially in infections with staphylococci, enterobacteria, and tubercle bacilli, and it continues to increase.

The term **drug resistance** ordinarily refers not to the natural resistance of a species but to **acquired genotypic changes,** which persist during cultivation in the absence of the drug. The change may be brought about either by **mutation,** which alters a cell constituent, or by **infection by a plasmid,** which brings in genes for new enzymes; the drug plays only a selective and not a directive role. A few antibiotics, notably streptomycin, select not only for resistant but also for dependent mutants. The mechanism of dependence will be discussed in Chapter 13.

PHYSIOLOGIC (PHENOTYPIC) MECHANISMS OF DRUG RESISTANCE

Several phenotypic mechanisms have been observed. Resistance to the same drug may depend on different mechanisms in different strains.

1) **Increased destruction of the drug** is the usual mechanism of plasmid-borne resistance, to be discussed below.

2) **Decreased activation of the drug** is seen in mutants resistant to purine or pyrimidine analogs, which must be

converted to nucleotides before they can interfere with essential reactions. Since the enzymes involved in the conversion are not essential for the cell their deletion does not impair viability and hence may yield resistant mutants. Compounds of this type are used in cancer chemotherapy, but among antimicrobial agents those in current clinical use are all active without further chemical alteration.

3) **Formation of an altered receptor** is an important mechanism: for example, the alteration in specific **ribosomal proteins** (Ch. 13) by one-step mutations that increase many hundredfold the level of resistance to streptomycin, erythromycin, or spectinomycin. Similarly, with some metabolite analogs (e.g., sulfonamides, p-fluorophenylalanine) resistant mutants have been found to form **altered enzymes,** which discriminate better between the normal substrate and the analog.

Mutations to streptomycin resistance occur at a **low frequency,** compared with auxotrophic mutations, because they depend on a substitution in a small region of a ribosomal protein, while mutations that inactivate an enzyme can occur virtually anywhere in its gene.

4) **Decreased permeability,** in mutants resistant to amino acid analogs, involves the loss or alteration of a membrane transport system that also transports the corresponding normal amino acid. In addition, normal permeability barriers are responsible for much "natural resistance," the outer membrane thus contributing to the generally lower sensitivity of gram-negative organisms.

For example, actinomycin inhibits RNA synthesis in many gram-positive bacteria but not in gram-negative bacilli; yet the extracts of the latter organisms are equally sensitive. Furthermore, gram-negative cells become sensitive following chemical treatment or mutation that increases the permeability of the outer membrane.

In another example, fosfomycin, an analog of p-enolpyruvate in muramate synthesis (see Antibiotics Acting on Peptidoglycan Synthesis, Ch. 6), penetrates via both a transport system for L-glycerophosphate and one for hexose-6-phosphate. Mutational loss of the former causes resistance, which is overcome by induction of the latter. The mechanism of entry of most agents is not so well known.

Decreased uptake of the drug by **intact cells** in often used as an index of decreased permeability but is not reliable: it can also be due to reduction in the number or the affinity of intracellular binding sites. Moreover, much of the binding may involve receptors that are irrelevant to drug action. However, studies on binding to **isolated cell components** (enzymes, ribosomes, membranes) have yielded significant results.

5) **Increased level of an enzyme** can increase resistance. This mechanism may involve either increased formation of the competitive metabolite (observed in some sulfonamide-resistant bacterial mutants) or an increased number of copies of the inhibited enzyme (resulting from gene amplification in plasmids; Ch. 12).

Induced Phenotypic Resistance. The level of resistance of a strain is usually a fixed property of the cells (or the population). However, growing some strains with a subinhibitory concentration of erythromycin or lincomycin induces a rapid increase in resistance on subsequent exposure to a higher concentration. There is no genotypic change, for on growth without the antibiotic the effect is reversed in a few generations. The mechanism depends on an unusual feature of this resistance: it is due to methylation of specific nucleotides in the ribosome, which can be stimulated by interaction of the ribosome with the antibiotic.

PLASMID-MEDIATED RESISTANCE (R FACTORS)

Since drug resistance had been decisively shown to arise in pure cultures by spontaneous mutations, each specific for and selected by a single drug (see Fig. 8-1), the rapid spread of **multiple resistance,** first observed in Japan, came as a surprise. Specifically, after a few years of widespread chemotherapy of bacillary dysentery, strains of the organism, and of other enteric bacteria, began to appear with multiple resistance to various widely used drugs: sulfonamides, streptomycin, chloramphenicol, and tetracyclines, and later also neomycin, trimethoprim, and ampicillin. In 1959 this phenomenon was found, also in Japan, to be due not to mutation but to **infectious heredity** (Ch. 9): the multiple resistance could be transferred in one step by cell contact (conjugation) between resistant and sensitive strains, in mixed cultures and also in the gastrointestinal tract.

The agents responsible, called **R factors (RF),** vary widely in their content of genes that convey resistance to various inhibitors. They spread rapidly as a result of the wide use of antibiotics: in one hospital in Japan the frequency of multiple resistance in *Shigella* isolates rose from 0.2% in 1954 to 52% in 1964. Only some years after the initial discovery were such factors sought in Western countries, and they were found to be not at all peculiar to the ecology of Japan. (This development illustrates how fragmentary is our knowledge of microbial variation and distribution.)

R factors are members of the broad group of **plasmids:** nonchromosomal genetic elements. The R plasmids code for their transmission by conjugation, but many other plasmids do not: the familiar resistance to penicillin in many staphylococci is due to production of a β-lactamase by such a nonconjugative plasmid. With this mechanism the rapid spread of resistant strains in hospitals has apparently depended entirely on their direct selection, rather than on transfer of the plasmid.

Resistance plasmids are of great clinical and theoretic importance. Their molecular properties will be described in Chapter 12.

ENZYMES DIRECTED BY RESISTANCE FACTORS

The predominant mechanism of resistance mediated by R factors was revealed by the finding that extracts of such strains could inactivate the drugs to which the cells were

TABLE 7-1. Mechanisms of Plasmid-Mediated Resistance

Agent	Reaction
β-Lactam antibiotics	Hydrolysis of β-lactam ring
Aminoglycosides	O-phosphorylation
	O-adenylylation
	N-acetylation
Spectinomycin	O-adenylylation
Chloramphenicol	O-acetylation
	Reduction of nitro group (? plasmid)
Erythromycin, lincomycin	Methylation of ribosomal RNA
Tetracycline	Modification of membrane permeability
Sulfonamides	Resistant dihydropteroate synthetase
Trimethoprim	Resistant dihydrofolate reductase
Hg^{2+}	Reduction to Hg (volatile)

resistant. A number of such reactions were then defined; these are presented in Table 7-1. The β-lactamase, like that in staphylococci, is a hydrolytic exoenzyme, but the other RF-directed enzymes are intracellular and require intracellular metabolites (ATP, acetyl CoA); the resulting levels of resistance are modest.

Because of the increasing distribution of genes for these enzymes the development of many aminoglycosides has provided an advantage, similar to the advantage of having various penicillins resistant to inactivation by β-lactamases: these aminoglycosides differ markedly in their susceptibility to various inactivating enzymes.

For example, kanamycin is a substrate for inactivation by substitution, by different enzymes, at six different positions. Some of these enzymes can also act on certain other aminoglycosides, but not on all.

Most bacteria make small amounts of a chromosomally coded, low-affinity β-lactamase, located in the periplasmic space; this enzyme may function in wall morphogenesis, and its contribution to resistance is not certain. Its amount, and its specificity for various β-lactams, varies widely. In addition, plasmids may code for a much more active cell-bound enzyme. The formation of some β-lactamases is increased by growth in the presence of a subinhibitory concentration of penicillin.

Some plasmid-mediated resistance involves mechanisms other than drug inactivation: a resistant dihydropteroate synthetase in sulfonamide resistance, and increased excretion of tetracycline (shown by increased concentration by inverted vesicles) in tetracycline resistance.

Population Density. In resistant staphylococci the β-lactamase is largely excreted rather than intracellular, and resistance depends on destroying the surrounding reservoir of drug before the cells grow (and commit suicide) in its presence. With such strains the observed level of resistance can vary 1000-fold with population density; resistance is recognized in the usual disc assay because a heavy inoculum is used. Clinically, such strains are also completely resistant, since staphylococci produce local-

ized infections with a high bacterial density. Indeed, in mixed infections resistant staphylococci may protect other sensitive bacteria by destroying the drug.

In contrast to this **social mechanism of resistance,** the resistance mediated by intracellular, metabolite-dependent inactivating enzymes (Table 7-1) is relatively independent of bacterial density. Here the level of resistance is evidently determined by the rate of penetration of the antibiotic and the rate of its inactivation in the individual cell.

PREVENTION OF RESISTANCE

Since no method is available for effectively decreasing spontaneous mutation rates one cannot prevent **formation** of resistant mutants, but one can prevent their **selection.** Where resistance develops through the accumulation of small increments, by successive mutations, therapeutic "escape" can be avoided by continuously maintaining drug concentrations high enough to inhibit the first-step mutant.

A more general approach, applicable even to one-step mutations to high-level resistance (e.g., with streptomycin), is the simultaneous use of two non-crossresistant drugs. Such **combination therapy,** originally suggested by Ehrlich, has a simple genetic rationale: if 1 cell in 10^6 mutates to resistance to drug A and 1 in 10^7 to drug B, only 1 in 10^{13} will develop both mutations. This approach has been especially successful in the treatment of tuberculosis (see Combined Therapy, Ch. 25, and Therapy, Ch. 37).

The spread of resistance plasmids can be decreased by avoiding indiscriminate use of the drugs. However, the problem cannot be eliminated in this way, since necessary as well as indiscriminate therapy exerts this selective effect. Attempts to prevent or cure the infection of bacteria by plasmids have not been successful.

The use of low levels of antibiotics in **animal feeds** presents a special problem: this procedure has real economic value in increasing meat production, but it has been shown to increase the frequency of resistance plasmids in gut flora, not only of the animals but also of the farm workers. Accordingly, antibiotics commonly used in man have been banned from feeds in Great Britain since 1971. Unfortunately, subsequent tests for a resistance plasmid (tetracycline) in pigs showed little decrease during the next 4 years.

GENETIC DOMINANCE OF SENSITIVITY OR RESISTANCE

Some genes for resistance are dominant to sensitivity and others are recessive. Analysis of the mechanisms of resistance has provided an explanation. If a drug acts on a soluble enzyme (e.g., sulfonamides) it would inhibit only the sensitive fraction of the mixture of sensitive and resistant enzyme molecules present in a heterozygote: hence resistance is dominant. Similarly, when resistance is due to the production of an inactivating enzyme by an RF it is "dominant." (More precisely, it is epistatic, for one is dealing with interactions of products of different genes, rather than the products of a sensitive and a resistant allele of the same gene.) On the other hand, with synthesis on a template the sensitive unit inhibited by the antibiotic blocks access or movement of the resistant unit on the template, and sensitivity is dominant. Examples include blockade of the movement of ribosomes by streptomycin (Ch. 13), of DNA gyrase by novobiocin, and of RNA polymerase by rifampin.

SELECTED READING

BOOKS AND REVIEW ARTICLES

ALBERT A: Selective Toxicity, 4th ed. London, Methuen, 1968

CORCORAN JW, HAHN FE (ed): Antibiotics III. New York, Springer-Verlag, 1975

DAVIES J, SMITH DH: Plasmid-determined resistance to antimicrobial agents. Annu Rev Microbiol 32:469, 1978

DEMAIN AL: Cellular and environmental factors affecting the synthesis and excretion of metabolites. J Appl Chem Biotechnol 22:345, 1972

DEMAIN AL: How do antibiotic-producing microorganisms avoid suicide? Ann NY Acad Sci 235:601, 1974

EHRLICH P: Collected Papers, Vol III, Chemotherapy. New York, Pergamon Press, 1960

FALKOW S: Infectious Multiple Drug Resistance. London, Pion, 1975

FRANKLIN TJ: Antibiotic transport in bacteria. CRC Crit Rev Microbiol 2:253, 1972–73

GALE EF, CUNDLIFFE E: The Molecular Basis of Antibiotic Action. New York, Wiley, 1972

HAHN FE (ed): Antibiotics: Mechanism of Action. New York, Springer-Verlag, 1979

HOPWOOD DA: Extrachromosomally determined antibiotic production. Annu Rev Microbiol 32:373, 1978

HOPWOOD DA, MERRICK MJ: Genetics of antibiotic production. Bacteriol Rev 41:595, 1977

KORZYBSKI T, KOWSZYK-GINDIFER Z, KURYLOWICZ W: Antibiotics: Origin, Nature, and Properties. Washington, DC, Am Soc for Microbiol, 1979

LACEY RW: Antibiotic resistance plasmids of *Staphylococcus aureus* and their clinical importance. Bacteriol Rev 39:1, 1975

MEYNELL GG: Drug Resistance and Other Bacterial Plasmids. London, Macmillan, 1972

MITSUHASHI S, HASHIMOTO H: Microbial Drug Resistance. Baltimore, University Park Press, 1975

NEWTON BA: The properties and mode of action of the polymyxins. Bacterial Rev 20:14, 1956

RICHMOND MH, SYKES RB: The β-lactamases of gram-negative bacteria and their possible physiological role. Adv Microb Physiol 9:31, 1973

ROLINSON GN: 6-APA and the development of the β-lactam antibiotics. J Antimicrob Chemother 5:7, 1979

SCHAEFFER P: Sporulation and the production of antibiotics, exoenzymes, and exotoxins. Bacteriol Rev 33:48, 1969

SYKES RB, MATTHEW M: The β-lactamases of Gram-negative bacteria and their role in resistance to β-lactam antibiotics. J Antimicrob Chemother 2:115, 1976

TAKAYAMA K, DAVIDSON LA: Antimycobacterial drugs that inhibit mycolic acid synthesis. Trends Biochem Sci, Dec 1979, p 280

TOMASZ A: The mechanism of the irreversible antimicrobial effects of penicillins. Annu Rev Microbiol 33:113, 1979

UMEZAWA H (ed): Index of Antibiotics from Actinomycetes. Univ of Tokyo Press, Vol I, 1967; Vol II, 1979

VANDAMINE EJ: Enzymes involved in β-lactam antibiotic biosynthesis. Adv Applied Microbiol 21:89, 1977

WATANABE T: Infectious drug resistance in bacteria. Curr Top Microbiol Immunol 56:43, 1971

ZÄHNER H, MAAS WK: Biology of Antibiotics. New York, Springer, 1972. (A brief, clear presentation, designed for interested medical students, paperback)

SPECIFIC ARTICLES

BROWN GM: Biosynthesis of folic acid. II. Inhibition by sulfonamides. J Biol Chem 237:536, 1962

MUKHERJEE PK, PAULUS H: Biological function of gramicidin: Studies on gramicidin-negative mutants. Proc Natl Acad Sci USA 74:780, 1977

ROBERTS M, ELWELL LP, FALKOW S: Molecular characterization of two beta-lactamase-specifying plasmids isolated from *Neisseria gonorrheae*. J Bacteriol 131:557, 1977

SUGINO A, et al: Energy coupling in DNA gyrase and the mechanism of action of novobiocin. Proc Natl Acad Sci USA 75:4838, 1978

WINDER FG, COLLINS PB: Inhibition by isoniazid of synthesis of mycoloic acids in *Mycobacterium tuberculosis*. J Gen Microbiol 63:41, 1970

part II

BACTERIAL AND MOLECULAR GENETICS

BERNARD D. DAVIS
RENATO DULBECCO

chapter 8

BACTERIAL VARIATION AND POPULATION DYNAMICS

Bacterial genetics and phage genetics have provided much of the foundation for molecular genetics—a subject of great importance for microbiology and for immunology, among other areas. We shall take up bacterial genetics in the next two chapters and then proceed to molecular genetics. Phage genetics, however, will be postponed until the section on viruses (Ch. 47, 48).

GENOTYPIC AND PHENOTYPIC VARIATION

Early studies in bacteriology were plagued by continual changes in the properties of cultures repeatedly transferred in the laboratory. With the development of reliable pure culture technics in the 1870s (Ch. 1) much of this variation disappeared. However, residual changes were observed, and after many decades they were found to be of two types: **genotypic** and **phenotypic.**

These two kinds of adaptation, fundamental to all organisms, can be distinguished in bacteria by a simple criterion: in phenotypic variation **all the cells** of a culture adapt physiologically, within the range of potential of the genotype, to changes in the environment; while in genotypic adaptation a **rare mutant** is **selected** (i.e., its progeny outgrow the parental strain) because it is better adapted to the new environment.

Each bacterial cell normally gives rise to a **clone,** i.e., genetically identical progeny, produced by vegetative multiplication. Hence with parental and mutant cells that produce detectably different colonies on the same solid medium (Ch. 5) their changing proportion, in samples of a culture undergoing genetic adaptation, is readily measured. But though a colony is initially a pure clone, during its prolonged growth additional mutants of various kinds may appear among the newly formed cells on the plate. Such mutant subclones may yield visibly sectored colonies (see Fig. 5-9, Ch. 5), e.g., for S–R (smooth–rough surface) variation or for color mutants.

Several factors contributed to the long delay in the recognition of these elementary principles. 1) Because neither a nucleus nor genetic recombination could be demonstrated in bacteria, inheritance in these organisms was long believed to reside in some vague plastic properties of the entire cell. 2) Because bacteria grow so much more rapidly than higher organisms, and because their rare spontaneous mutants are often subject to strong (but not obvious) selection

pressures, their populations can change at a rate that seemed far too rapid to fit into the familiar evolutionary mechanism of mutation and selection. 3) Genotypic and phenotypic variation (or adaptation) can often bring about the same phenotypic change, such as capsule formation; the quantitative technics and the concepts necessary to distinguish between the two processes were not developed until late.

However, in the early 1940s it was recognized that the emergence of some readily quantifiable variants in bacteria (e.g., phage- or drug-resistant) had the same range of frequencies per unit population as mutations in higher organisms (10^{-5}–10^{-9}), and the rate was increased by known mutagenic agents (e.g., ultraviolet radiation). Hence inheritable changes in bacteria began to be referred to as mutations.*

RANDOM OR DIRECTED MUTATIONS?

Definitive evidence of spontaneous, undirected bacterial mutation was furnished by a statistical approach called **fluctuation analysis,** designed by Luria and Delbruck in 1943. This experiment was based on the idea that in liquid medium spontaneous mutations, occurring before exposure to a drug, would arise in different tubes at different times and hence would differ widely in their number of cell divisions before the culture ceased growing and was plated with the inhibiting drug. But if the mutants appeared only after exposure to the drug, the numbers in various tubes should fluctuate much less, i.e., to the same degree as in samples plated from a single flask. The results (Fig. 8-1) showed conclusively that the resistant mutants had appeared spontaneously, i.e., were only selected, and not directed, by the inhibitor. This proof did much to overcome a widespread anthropomorphic reluctance to accept the key role of chance events in an adaptive (and hence apparently intelligent) process in bacteria.

Lederberg subsequently used separation on solid medium to demonstrate the same effect more directly: a heavy inoculum of sensitive bacteria (about 10^4 cells) was grown on plates without drug to form a lawn, which was then **replica-plated** onto several plates of drug-containing medium. The results showed that the

---

* In 1900, within a few months after de Vries in Holland discovered mutations in higher plants, his countryman Beijerinck proposed that the same mechanism must be responsible for the heritable changes observed in bacteria. But this idea was premature by nearly half a century.

FIG. 8-1. Fluctuation analysis of mutation to streptomycin **(Sm)** resistance. A small number of sensitive cells were inoculated in a flask containing 100 ml of broth, and also in 100 tubes each containing 1 ml of the same medium. After full growth was reached, 1-ml samples were inoculated in plates of medium containing the drug, and the number of Sm-resistant colonies appearing after overnight incubation was determined. The fluctuation in their number was much greater among the samples that had grown out in separate tubes than among those from the same flask. (Based on Luria SE, Delbruck, M: Genetics 28:491, 1943)

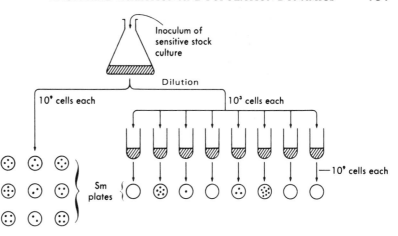

lawn contained localized clusters of resistant cells, i.e., clones that had been formed **before** exposure to the drug (Fig. 8-2).

It was thus gradually recognized that mutable unit factors, indistinguishable from the genes of higher organisms, govern inheritance in bacteria. However, bacterial genes could not be studied in depth until the discovery, described in the next chapter, of methods for recombining these units between different individuals.

EXTRAGENIC INHERITANCE

Certain "inheritable" changes in the cell envelope do not appear to involve changes in DNA. Thus penicillin converts bacteria uniformly to wall-deficient, spherical L forms (Ch. 43) by interfering

with wall synthesis (Ch. 6); and in some species these variants are stable and can grow indefinitely without reversion after penicillin is removed. The cell apparently needs a wall material as a **primer** before more can be synthesized.

SELECTION PRESSURES AND GENETIC ADAPTATION

Many of the inheritable changes observed in bacteria during cultivation have obvious **adaptive value** (i.e., they increase fitness for a new environment). For example, on first isolation a pathogen often grows slowly in a laboratory medium (because it has been adapted to conditions in the animal host), but on repeated transfer it will adapt to that medium through selection of rare faster-growing

FIG. 8-2. Use of replica plating to demonstrate undirected, spontaneous appearance of streptomycin-resistant mutants. About 10^5 sensitive cells were spread on a plate of drug-free solid medium and allowed to reach full growth (10^{10}–10^{11} cells). Sterile velveteen, covering the flat end of a cylindrical block, was pressed lightly on this continuous heavy lawn ("master plate") and was then pressed successively on two plates of medium containing streptomycin at a concentration that killed sensitive cells. A few colonies of resistant cells appeared on each plate, usually in coincident positions, and cells harvested from the corresponding positions on the master plate yielded a much larger proportion of resistant colonies than cells harvested from other parts of the plate. Evidently resistant clones, arising in the absence of drug on the master plate, were the source of most of the resistant colonies on the replica plates.

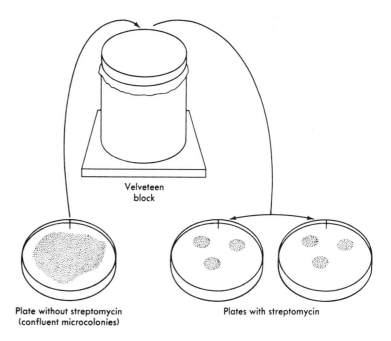

Velveteen block

Plate without streptomycin
(confluent microcolonies)

Plates with streptomycin

mutants that appear among the progeny. This adaptation is often accompanied by a decreased adaptation to the animal host, i.e., a **loss of virulence (attenuation).** Conversely, virulence may be restored by passage of a large inoculum through an animal host, which selects rare virulent mutants.

Repeated transfer under unusual conditions is widely used, as an empiric practice, to produce attenuated strains of bacteria (or of viruses) for use as vaccines.

S–R Variation and Phase Variation. With some kinds of variation in bacterial populations the adaptive advantage is not obvious and even seems to be belied by spontaneous reversal during further cultivation. These phenomena were long considered to reflect some kind of life cycle (analogous to sporulation or gametogenesis) and were given special names. In **S–R variation** pathogens isolated from patients plate out homogeneously as **smooth** (S) colonies, but after repeated passage in liquid medium they plate out mostly or entirely as **rough** (R) colonies (Fig. 8-3); and the transition may later be reversed (i.e., $R \rightleftharpoons S$).

Analysis of kinetics showed that these population shifts are due to shifts in factors exerting selection pressure (e.g., lower pO_2 and pH; accumulation of metabolic products) as the medium is altered by growth, especially in cultures approaching the stationary phase (see Growth Cycle, Ch. 5). Hence the patterns of change during culture transfers vary not only with different media but also with the inoculum size and the duration of incubation before transfer.

Thus with a mixture of R and S *Brucella* cells the S cells grow faster in fresh medium. However, as the culture matures the S cells excrete D-alanine and the pO_2 decreases; and under these conditions S cells grow more slowly than R cells.

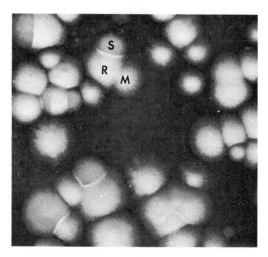

FIG. 8-3. Rough **(R)** and smooth **(S)** colonies of *Brucella abortus*. Because of the difference in the reflection of light in the photograph the R colonies appear considerably lighter than the S colonies, as well as more stippled. (Courtesy of W. Braun)

In **phase variation** an enteric organism with one antigenic specificity of its flagella rapidly shifts during cultivation from this type (phase 1) to another (phase 2); and on further transfers the cultures may shift back and forth between the two "phases." The alternative genotypes are less subject to differential selection pressures than with S–R variation, and the population can reach a stable equilibrium (Fig. 8-4) that depends on the mutation rates in the two directions.

Genetic studies showed that the two alternative proteins are coded for by two genes, H1 and H2; H2 can somehow be switched on or off; and in the "on" phase it forms not only the H2 protein but also a product that represses gene H1. Recent molecular studies

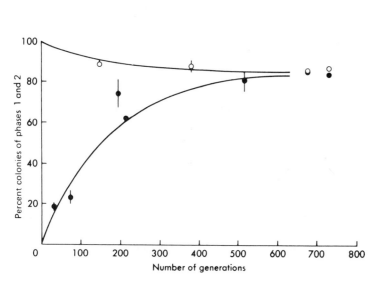

FIG. 8-4. Attainment of equilibrium between two phases of *Salmonella typhimurium* after prolonged exponential growth. Each initial culture contained cells of phase 1 (•) only or 2 (o) only. Growth was kept exponential in a series of successive subcultures by transferring from each a small inoculum taken before the population density approached saturation. From these kinetics it can be calculated that the mutation rates were 5.2×10^{-3} per division for phase $1 \rightarrow 2$ and 8.8×10^{-4} for $2 \rightarrow 1$. (After Stocker BAD: J Hyg 47:398, 1949)

have shown that the switching on of H2 depends on a reversible inversion of a segment of its DNA (see Phase Variation, Ch. 12).

Population genetics is clearly pertinent to the properties of pathogenic bacteria in the host, for during an infection selection pressures are exerted by the changing bacterial density, nutritional conditions, and host defenses. The most frequent example is the emergence of drug-resistant mutants (or recombinants) in treated individuals. In addition, other examples of population shifts have been observed (e.g., the periodic replacement of one antigenic type by another in relapsing fever: Ch. 39).

DETECTION AND SELECTION OF MUTANTS

Most studies in genetics are based on phenotypic differences produced by alternative (**allelic**) forms of a gene. Some alleles are found in nature (**polymorphism**), but their variety has been enormously expanded by the experimental isolation of mutants. We shall discuss here some of the selective methods used; mutagenic agents will be discussed in Chapter 11.

Bacteria are **haploid**, i.e., lack the paired homologous chromosomes present in higher, **diploid** organisms. Accordingly, **heterozygosity** and **dominance**, so prominent in classic genetics, are not present to influence the expression of mutations (except under the special circumstances that produce partial diploids: see Ch. 9).

Wild-Type and Mutant Alleles. It is customary to refer to strains found in nature as the **wild type**, and then to distinguish as **mutants** any recognizable altered strains derived from them, usually in the laboratory. However, the concept of wild type is not applicable to genes that are **polymorphic** (i.e., that have two or more prevalent alleles), such as those determining surface Ags.

The relative meaning of wild type is illustrated by the classic K12 strain of *Escherichia coli*. It is the "wild-type" parent of most mutants used in genetic studies in this species, but after 50 years in the laboratory K12 can no longer survive in the human gut, and it would probably not be recognized in a diagnostic laboratory as *E. coli*.

Auxotrophic Mutants. The concept of a mutant allele is useful in considering those that cannot survive in nature, e.g., **auxotrophic** mutations, causing defects in the synthesis of an essential metabolite. (Strains reversed from auxotrophy to restoration of synthesis are called **prototrophs**.) The introduction of auxotrophs revolutionized microbial genetics. It greatly increased the variety of available genetic markers; it brought genetics closer to biochemistry by providing a powerful tool for exploring biosynthetic pathways (see Ch. 4); it laid a foundation for molecular genetics by suggesting that each gene makes a corresponding enzyme; and recombinations between al-

lelic auxotrophs, to yield easily selected prototrophs, have been invaluable in studies of fine-structure genetics (Ch. 11).

Some geneticists had recognized decades earlier that a simple enzymatic defect probably underlies certain visible mutations in higher organisms, including color in plants, eye color in *Drosophila*, and albinism in man (Garrod). However, such isolated instances had little impact until Beadle and Tatum, in 1941, systematically isolated microbial mutants (of the mold *Neurospora*) with various specific enzymatic defects. The resulting **one gene–one enzyme hypothesis** established proteins as concrete intermediates between the genes and the characters of formal genetics. The exceptions, which evoked much resistance, were later explained when studies in molecular genetics showed that 1) most genes are translated into a corresponding **polypeptide**, and many enzymes contain more than one kind of polypeptide; and 2) some genes **regulate** the function of multiple other genes.

Screening and Scoring. The enormous size of microbial populations proved to be a great asset for genetic studies, but only because it is possible to **select** certain kinds of rare genotypes with great efficiency. Thus even one lactose-positive mutant in 10^9 lactose-negative cells can be **screened** by inoculation on a medium in which lactose is the sole carbon source. On the other hand one cannot similarly screen for mutations from positive to negative, though they can be readily detected (**scored**) on an appropriate plate containing a dye that will stain only fermentation-positive colonies (see Fig. 5-9, Ch. 5).

Quantitative selection is possible for several classes of mutants: fermentation-positive (or able to use a new source of nitrogen, sulfur, etc.), prototrophic, drug-resistant, and phage-resistant. Mutations in the opposite direction, or to alterations in appearance or in the excretion of various metabolites, are readily scored but are not readily selected.

Replica plating (Fig. 8-2) is useful in scoring clones, in a mixed population, for their ability to grow in different media (Fig. 8-5). Ten replica platings onto different selective media, from a plate containing 100 colonies, can provide the same information as 1000 repetitive transfers of cells.

Selection of Auxotrophs. The isolation of auxotrophic mutants has been aided by an enrichment procedure based on the use of penicillin (or of other agents that kill only growing cells): in a minimal medium auxotrophs cannot grow and therefore largely survive, while the parental cells, which can grow, are selectively killed (Fig. 8-6).

Conditionally Lethal Mutations. Some mutants are auxotrophic at $35°–40°$ C but not at $20°$. These strains form a **temperature-sensitive** enzyme which is rapidly denatured even at moderate temperatures.

Temperature sensitivity marks one type of a broader group: conditionally lethal mutations, whose lethal effects may be presented under special conditions. The

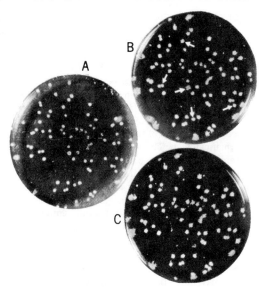

FIG. 8-5. Detection of auxotrophic mutants by replica plating. The master plate **(A)**, containing enriched medium, was replicated by a velvet press (see Fig. 8-2) onto a plate of enriched medium **(B)** and one of minimal medium **(C)**. **Arrows** indicate colonies of auxotrophic mutants, which grow on B but not on C. (Lederberg J, Lederberg EM: J Bacteriol 63:399, 1952)

E. coli culture

↓ Mutagen

Mixture of killed parental, live parental, and a few live mutant genotypes

↓ Intermediate cultivation in enriched medium

Phenotypic expression of mutant genotype (Fig. 8-8)

↓ Penicillin in minimal medium

Selective killing of parental cells

↓ Plate on enriched medium

Colonies from surviving cells, with parental type much decreased in frequency

↓ Parallel inoculation of each colony on minimal and enriched medium

Growth of parental clones on both media, auxotrophic mutants only on enriched medium

↓ Inoculation of mutant colony in minimal pour plate

Identification of growth requirement(s) of mutant by "spot tests" with growth factors

FIG. 8-6. Penicillin method for selecting auxotrophic mutants of bacteria. Similar results may be obtained with other agents that also kill only growing cells, e.g., metabolite analogs such as 8-azaguanine, or radioactive metabolites that are allowed to disintegrate during prolonged storage after their incorporation.

mutants can thus be grown under **permissive** conditions and their defects expressed under **nonpermissive** conditions. (Auxotrophs are, in a sense, conditional lethals, except that their deficit can be supplied exogenously.) This principle has permitted the isolation of mutations in many genes whose alterations would ordinarily be lethal because their products cannot be supplied exogenously. Such mutations have been especially valuable in mapping viral genomes (see Chs. 47, 48, and 50).

DELAYED EXPRESSION OF MUTATIONS

After a mutation has occurred it is still not phenotypically expressed until the cells have undergone some growth. Thus if a suspension of auxotrophic cells is mutagenized the prototrophic reversions that have been induced yield few colonies unless a trace of the previously required growth factor is provided in the selective minimal medium (Fig. 8-7), to initiate protein synthesis and thus permit expression of the mutant gene. Similarly, in the selection of auxotrophs intermediate cultivation after mutagenesis is essential, to allow phenotypic expression before exposure to penicillin.

The latter delay in phenotypic expression has two main components (Fig. 8-8): 1) **nuclear segregation,** separating a recessive mutation (e.g., an inactive biosynthetic gene) from the dominant parental allele in a companion chromosome, and 2) a **phenotypic lag,** which lasts until a change in the genome is reflected in the phenotype. Thus an auxotrophic mutation is not expressed until the enzyme molecules previously formed by the gene have been diluted out by further growth.

MUTATION RATES

With bacterial mutations that are easy to select quantitatively from large populations (e.g., mutations to resistance or to prototrophy) it is possible to measure even low mutation rates with precision, i.e., to detect small increments to the background "spontaneous" mutation rate. Many weakly mutagenic substances have thus been detected, expanding enormously the class of recognized mutagens (see Ch. 11).

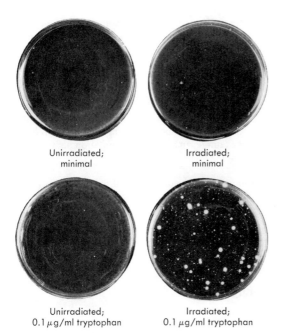

Unirradiated;
minimal

Irradiated;
minimal

Unirradiated;
0.1 μg/ml tryptophan

Irradiated;
0.1 μg/ml tryptophan

FIG. 8-7. Lag in phenotypic expression of induced mutations. Part of a culture of a tryptophan auxotroph was irradiated with ultaviolet to 0.1% survival, and equal numbers of cells, taken before and after irradiation, were plated on minimal medium to select for prototrophic mutants. Identical inocula were also plated on medium supplemented with 0.1μg/ml of tryptophan, which permitted a few generations of growth of the large number of auxotrophic cells plated. It is seen that many back-mutants were induced, but very few of these initiated colony formation unless provided with a trace of the required growth factor.

Definition of Mutation Rate. In growth at different rates the **spontaneous mutation rate** (i.e., the probability of appearance of a given type of mutant) remains relatively constant **per cell division** rather than per cell per unit of time. Accordingly, it is customary to define mutation rate, α, as

$$\alpha = m/d,$$

where m is the number of **mutations** and d the number of cell **divisions**.

If one grows a population from a small inoculum the value of d essentially equals the final number of cells present, since the initial number is negligible and each division produces one additional cell. The value of m, on the other hand, cannot be determined simply by screening for the total number of **mutants** of a given kind, i.e., the **mutant frequency** in that population. That number represents not only the mutants produced in the most recent generation but also the accumulated progeny of mutations that occurred in earlier generations; hence the mutant frequency ordinarily becomes higher the later a culture is harvested. The mutation rate, in contrast, is constant for a given class of mutations under constant conditions.

Determination of Mutation Rate. 1) One reliable method for determining the true mutation rate is to measure the slope of the increase in mutant frequency with continued growth. Especially smooth curves are obtained in the chemostat (see Fig. 5–6, Ch. 5), an apparatus which provides prolonged steady-state growth with a constant population size. This method is applicable only when the mutant multiplies at the same rate as the wild type.

2) Another method utilizes mutation on solid medium, in which the progeny of an early mutant can be held together in a colony. For example, 10^5 streptomycin-sensitive cells may be spread on a

FIG. 8-8. Delay in phenotypic expression of recessive mutations. o = Wild-type allele; • = mutant allele; x = product of the wild-type allele.

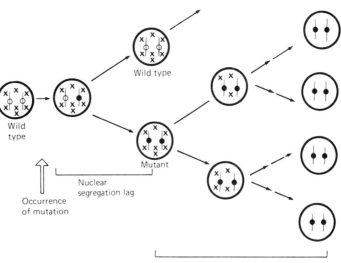

Wild type

Wild type

Mutant

Occurrence of mutation

Nuclear segregation lag

Phenotypic lag

porous membrane on nonselective medium and allowed to grow to 10^9 cells (determined by enumerating the cells washed off a plate grown in parallel). When the membrane is then transferred to a drug-containing plate the progeny of each mutation to resistance, whether a single cell or a clone, will give rise to one colony.

3) A widely used statistical method is based on the Poisson distribution of random, chance mutations in a series of identically inoculated and incubated tubes of culture medium. The probability that X mutations occur in a tube ($P_{(x)}$) depends on m, the average number of mutations per tube, averaged over all the tubes, and is

$$P_{(x)} = (m^x / x!)e^{-m}$$

where e is 2.718, the base of natural logarithms. The value of m cannot be measured directly, since, as we have seen, the number of mutants in any tube does not measure the number of mutational events in that tube. However, the number of tubes containing **no** mutants corresponds to the Poisson prediction, which reduces in this case to

$$P_{(0)} = e^{-m}$$
$$\text{or} \quad \ln P_{(0)} = -m$$

Hence from a determination of the proportion of mutant-free tubes, $P_{(0)}$, one can calculate the average mutation frequency per tube, m. For example, an average of 1 mutation per tube yields $P_{(0)} = e^{-1} = 1/e = 0.37$.*

* For a more extensive discussion of the use of the Poisson distribution see Appendix, Chapter 46.

SELECTED READING

BOOKS AND REVIEW ARTICLES

ADELBERG EA (ed): Papers on Bacterial Genetics. Boston, Little, Brown, 1966. A collection of reprints of outstanding papers, with a valuable historical introduction and bibliography.

BEADLE GW: Genetics and metabolism in *Neurospora.* Physiol Rev *25:*643, 1945

SAGER R: Cytoplasmic Genes. New York, Academic Press, 1972

SPECIFIC ARTICLES

LANDMAN OE, HALLE S: Enzymically and physically induced inheritance changes in *Bacillus subtilis.* J Mol Biol *7:*721, 1963

LEA DE, COULSON CA: The distribution of the mutants in bacterial populations. J Genet *49:*264, 1949. In Adelberg collection.

LEDERBERG J, IINO T: Phase variation in *Salmonella.* Genetics *41:*743, 1956

LEDERBERG J, LEDERBERG EM: Replica plating and indirect selection of bacterial mutants. J Bacteriol *63:*399, 1952

LURIA S, DELBRUCK M: Mutations of bacteria from virus sensitivity to virus resistance. Genetics *28:*491, 1943. In Adelberg collection.

NOVICK A: Experiments with the chemostat on spontaneous mutations of bacteria. Proc Natl Acad Sci USA *36:*708, 1950

chapter

GENE TRANSFER IN BACTERIA

EVOLUTIONARY SIGNIFICANCE OF GENE TRANSFER

Evolution depends equally on variation and selection; and it would be very slow if the variation arose only through the accumulation of successive mutations in a line of descent. Accordingly, evolution was accelerated enormously by the development of a powerful additional mechanism of genetic diversification: gene transfer between organisms.

In eukaryotes gene transfer is accomplished by sexual reproduction, in which the organisms are **diploid** (possess pairs of homologous chromosomes), and the process of **meiosis** (reviewed in Ch. 45) more or less randomly assorts one member from each pair of homologous genes into each **haploid gamete**. A male and a female gamete then fuse to form a diploid **zygote**. It was long believed that such **genetic recombination** evolved only in eukaryotes, and that bacteria multiply only **clonally (by binary fission)**. However, between 1944 and 1952 three different mechanisms were discovered for transferring genes from one bacterial cell to another: 1) uptake of naked DNA **(transformation,** 2) mating between cells in contact **(conjugation),** and 3) infection by a nonlethal virus **(transduction)**. It is therefore clear that gene transfer first evolved in prokaryotes, and the resulting increase in the rate of evolution then led to the emergence of eukaryotes.

Transformation was discovered accidentally in 1928, and its genetic significance was not appreciated until it was shown, in 1944, to be due to the transfer of DNA. This discovery led to a search for sexual reproduction in bacteria, but the mechanism discovered, conjugation, turned out to have major differences from sex. Transduction was also discovered accidentally, in studies designed to extend conjugation from *E. coli* to *Salmonella.*

The bacterial mechanisms for gene transfer are all primitive, producing not zygotes but partial diploids called **merozygotes** (Gr. *meros,* part): part of a genome from a donor cell **(exogenote)** is transferred into an intact recipient cell. The two blocs of DNA may then undergo recombination,* in which genes from the exogenote exchange with or add to the recipient genome, by means of molecular crossover.

Infectious Heredity. The discovery of mutable units of inheritance (genes) in bacteria (Ch. 8) brought the previously vague field of bacterial variation into the fold of genetics. Moreover, these genes were found to be linked in a chromosomal chain, and to be able to recombine, as in higher organisms. But gene transfer in bacteria also has unique features, summed up in the phrase infectious heredity. Thus transduction and conjugation often transfer blocs of DNA that can engage in **autonomous replication** in the cell: viruses, and plasmids (small autonomous circles of DNA; Ch. 12). The transferred genes may then be added to the genome (by persistence of the autonomous unit in the progeny), or they may be substituted for chromosomal genes (by recombination) (Fig. 9-1).

Transformation, which ordinarily transfers only pieces of DNA, might be simply an adventitious property of bacterial cells. However, since it involves special enzymes for DNA uptake it probably evolved as a specific, primitive mechanism of gene transfer. Conjugation protects the DNA better, and the plasmids that mediate it probably evolved later. Finally, autonomous blocs of DNA, being primitive forms of life, inevitably undergo evolutionary selection for maximizing their reproduction. Hence it is hardly surprising that evolution produced not only plasmids but also viruses, at the other end of the spectrum of infectious heredity: units that function both as infectious parasites and as agents of transfer of host genes.

Spread of Genes among Prokaryotes. Genes are most efficiently transferred in bacteria between members of the same species, but they may also cross even distant boundaries. With transformation or transduction of chromosomal fragments, which depends on recombination between regions of DNA homology, interspecific re-

* The term **recombination** is now used in two senses: the overall process of assembling new combinations of genes from two parents, or the molecular process of crossing-over between two DNA chains. The latter usage would include rearrangements within a genome (Ch. 11).

Gene substitution

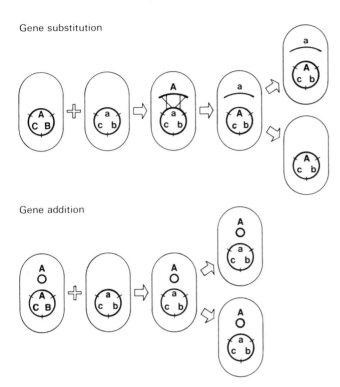

Gene addition

FIG. 9-1. Two types of gene transfer in bacteria, gene substitution and gene addition.

combination falls off rapidly with evolutionary distance. (It also varies with the evolutionary lability of the marker, being most effective for highly conserved gene sequences such as those for ribosomes.) Plasmids, in contrast, do not depend on homology with the host chromosome for the persistence of their genes, or for picking up host genes. Some plasmids are highly promiscuous, moving between organisms as distant as *Neisseria, Escherichia coli,* and *Pseudomonas;* others have a narrow host range.

It is evident that prokaryotic genes move widely, at low, variable frequencies (Fig. 9-2). *The entire prokaryotic gene pool may thus be linked by interspecific circuits.* But despite this flow, the evolutionary requirement for balanced genomes maintains clusters of similar genotypes, recognizable as stable bacterial species (Ch. 2).

TRANSFORMATION

Transformation was discovered adventitiously in the course of a study of virulence in pneumococci.

Virulent strains have one or another type-specific capsule, and they form smooth (S) colonies. On repeated transfer in culture they often yield avirulent variants, without a capsule: these form rough (R) colonies. Conversely, when a large number of R cells are inoculated in a mouse virulent S cells emerge and kill the animal.

These shifts (Ch. 8) were not yet recognized as muta-

FIG. 9-2. Genetic interconnections demonstrated between bacterial groups, either by transformation or by conjugation. (Courtesy of F. E. Young)

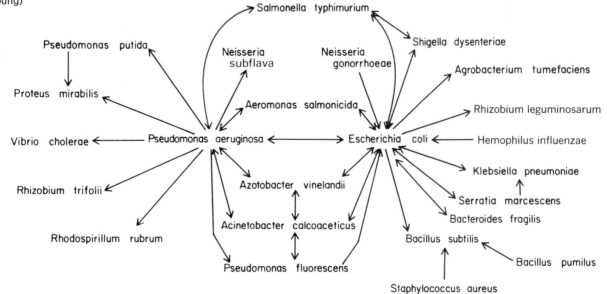

tions when Griffith in England, in 1928, discovered transformation while testing an erroneous hypothesis: that R cells retain traces of S antigen and release it on disintegration, and that this material from many cells, accumulated by a single surviving R cell, somehow converts it into an S cell. Indeed, heat-killed S cells injected in mice along with live R cells did promote the appearance of virulent S cells. However, killed S cells not only could enhance an apparent reversion of R cells to their original S type: they also could transform live R cells derived from another S type into the S type of the killed cells (Fig. 9-3). Neither Griffith nor his readers recognized the profound implication: that a substance conferring a new heritable property on an organism must itself be replicated.

IDENTIFICATION OF THE TRANSFORMING SUBSTANCE

Transformation was then taken up by Avery (who had discovered the role of capsular polysaccharide in pneu-

FIG. 9-3. The Griffith experiment. Rough (R) cells not only were apparently reverted to smooth (S) cells of their parental type by heat-killed S cells of the same type, but also were transformed to a different S type by heat-killed cells of that type.

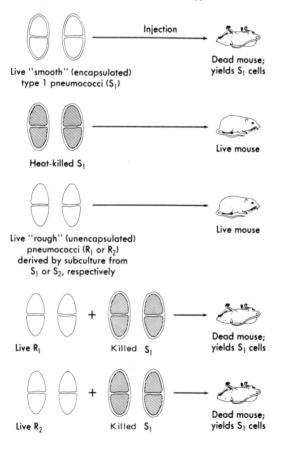

Live "smooth" (encapsulated) type 1 pneumococci (S_1)

Injection → Dead mouse; yields S_1 cells

Heat-killed S_1 → Live mouse

Live "rough" (unencapsulated) pneumococci (R_1 or R_2) derived by subculture from S_1 or S_2, respectively → Live mouse

Live R_1 + Killed S_1 → Dead mouse; yields S_1 cells

Live R_2 + Killed S_1 → Dead mouse; yields S_1 cells

mococcal virulence). It was extended, over the next decade, to increasingly simplified systems: from mice to the test tube, then from heat-killed cells to extracts, and finally to fractions of extracts. In vitro transformation was facilitated by the use of antiserum to agglutinate R cells: these then sedimented in the culture, while any S cells that appeared grew diffusely in the medium. After years of patient fractionation, in 1944 Avery, MacLeod, and McCarty succeeded in isolating the "transforming principle" and in identifying it as deoxyribonucleic acid (DNA; Fig. 9-4).

The biologic significance of transformation was not rapidly appreciated. Indeed, even identification of the active substance was not rewarded with a Nobel Prize, though it founded molecular genetics. The delay in recognition may have been due in part to the fact that virulence, in an organism with no known genetics, was not readily seen by geneticists and biochemists as a genetic trait. Moreover, the pinpointing of DNA as the active material was too revolutionary to be readily accepted. Cytologists had identified DNA, as well as protein, in the chromosomes of animal and plant cells; but the enormous variety of genetic specificity was universally assumed to reside in the protein: the DNA contained only four different monomers, and their presence in roughly equal proportions in the early samples suggested a monotonous repeating tetranucleotide sequence, consistent with a structural role. And even having accepted DNA as the active substance, investigators could not be sure that it was the stuff of heredity, rather than a mutagenic or an inducing agent. Finally, members of the phage school, investigating the simplest living entities in a rational approach toward characterizing the substance of the gene, were skeptical about the solution to this problem as a byproduct of studies on pathogenicity. However, phage studies provided confirmation, when Hershey and Chase showed in 1952 that the DNA of infecting, radioactively labeled phage penetrates into bacteria while most of the protein remains outside (see Fig. 47-4).

More quantitative studies of the process of transformation became possible through the use of genes suitable for sharp selection: e.g., those conferring drug resistance or fermentative ability. In such studies the frequency of transformants at low DNA/cell ratios measures the effectiveness of a DNA preparation, while at high ratios it measures the state of **competence of the cells** to accept DNA.

Free DNA is highly susceptible to mechanical and to enzymatic damage. Moreover, the nutritionally fastidious, fragile pneumococcus is one of the most difficult organisms to work with. It is therefore remarkable that bacterial gene transfer was first discovered as transformation. The key was Avery's lifelong concentration on the then leading cause of death in Western countries.

MECHANISM OF PHYSIOLOGIC TRANSFORMATION

Physiologic transformation has been observed in cultures of a fairly wide variety of genera: including *Hemophilus, Neisseria, Streptococcus, Staphylococcus, Bacillus,* and *Acinetobacter.* Artificial (induced) transformation, in other organisms, has a rather different mechanism, to be described below; it has been extended to yeasts.

FIG. 9-4. Transformation of R to S pneumococci by DNA from S cells. Left (1): Colonies on blood agar of an R variant derived from type 2 pneumococci. (2): Colonies from cells of the same strain that had grown in the presence of DNA from type 3 pneumococcus, plus antiserum to R cells (see text). (Type 3 forms especially glistening mucoid colonies.) (Avery OT et al: J Exp Med 79:137, 1944)

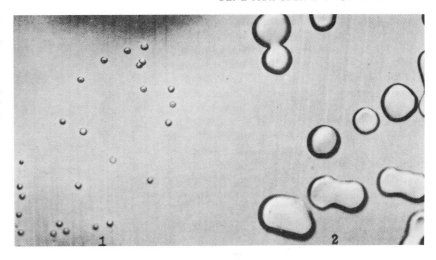

Adsorption of DNA. Physiologic transformation, studied mostly in the pneumococcus, is observed with DNA fragments of mol wt from 3×10^5 to 10^7 or more, and its effectiveness increases with the fragment length. With pneumococcal uptake of DNA the sites of entry appear to recognize any double-stranded DNA, for DNA from distant species can competitively inhibit the penetration of transforming DNA. Single-stranded DNA, however, neither penetrates nor inhibits transformation.

Competence. The uptake of DNA requires cellular energy, and in synchronized cultures (Ch. 8) the proportion of cells competent to accept DNA varies regularly during the cycle of cell division. These findings, and radioautography with tritiated DNA, suggest that DNA enters in zones of wall synthesis. In addition, in an unsynchronized culture competence rises and then falls during the growth cycle; and during the brief optimal phase the supernatant contains a protein, competence factor, whose presence increases competence.

With improved technics it has become possible to transform about 5% of the cells for a given marker, compared with Avery's fraction of 10^{-4}. When pains are taken to keep the DNA relatively undegraded (i.e., free of single-strand "nicks") it is highly efficient in transformation: nearly one transformant for a given marker can be produced per genome-equivalent of DNA taken up by the recipient culture.

Entry and Integration. Transformation in pneumococci (or in bacilli) involves a number of cell components that appear to have evolved for this function. Thus the molecules of double-stranded DNA are cut, by an endonuclease in the membrane, to a more or less constant length (7–10 kilobases), of which *only one strand enters.* In consequence, there is an **eclipse** period, of about 5 min, during which the transforming DNA is protected within the cell from external DNase but cannot be recovered in active form.

Studies with heavy-isotope-labeled DNA show that after entry the single-stranded DNA recombines in a special way, first forming a **heteroduplex** (see Fig. 10-15) by hybridizing with one strand of homologous recipient DNA (and displacing the other strand), and eventually becoming covalently integrated (and recovering transforming activity).

Transformation in the **gram-negative *Hemophilus*** differs from that in gram-positive organisms: the DNA enters as a double strand, and heterologous DNA does not compete. Uptake depends on a specific sequence, of about 10 nucleotides, that appears with much higher frequency in *Hemophilus* (about 600 copies per genome) than in other species; a membrane protein binds fragments containing this sequence. In the gram-negative *Neisseriae* transformation occurs only in piliated strains.

ARTIFICIAL TRANSFORMATION: TRANSFECTION

With enteric bacteria DNA uptake requires **modification of the cell envelope:** it was first accomplished with spheroplasts, but is more convenient with cells whose envelope has been made more permeable to DNA by brief exposure to $CaCl_2$. This artificially induced competence permits the cells to take up, very inefficiently, **double-stranded chromosomal fragments.**

Such modified cells can also take up the *intact* double-strained DNA extracted from **viruses,** or that of **plasmids** extracted from cells. This process is essentially the same as transformation, but it is often designated as transfection, because it results in infection by an abnormal route of entry.

Infection with a helper phage causes a transient change in cell permeability that allows inefficient entry of other DNA, in both enteric bacteria and *S. aureus*.

SIGNIFICANCE OF TRANSFORMATION

Transformation can be used for genetic mapping, by conventional linkage measurements (over short distances), or by measuring the time of doubling of the transforming DNA for various genes in a synchronized culture. However, its main value lies in its use for incorporating into bacteria DNA that has been synthesized or modified in vitro, including DNA recombined with a plasmid or a phage vector (Ch. 12). Transformation has also been extended to eukaryotes (yeast, animal cells).

Transformation probably occurs in nature. It has been achieved with live cultures mixed in the test tube. Moreover, in the peritoneal cavity of a mouse injected with pairs of pneumococcal strains of low virulence recombinants with increased virulence can be selected.* Recombination by transformation may therefore have epidemiologic significance. However, in interspecific transfers transforming DNA is much less efficient than plasmids.

With enteric bacteria transformation does not occur at a detectable rate unless the cell envelope is damaged. However, because of the enormous scale of production of these organisms in nature (about 10^{22} cells are excreted per day by the human species), and the inevitable appearance of rare phenotypic variants with an altered envelope, it seems very possible that enteric bacteria do occasionally take up DNA from each other (and from lysing host cells).

TRANSDUCTION

Transduction of bacterial genes by a bacteriophage, like transformation, introduces only a small fraction of a bacterial chromosome. Because the DNA is protected from damage by the surrounding phage coat, transduction is much easier to perform and more reproducible than transformation. Originally discovered by Lederberg and Zinder in *Salmonella,* it has been observed in a wide range of bacteria.

Two mechanisms are known. In **generalized transduction** the phage may pick up fragments of host DNA at random; hence it may transfer any genes. In **specialized transduction** phage DNA that has been integrated into the host chromosome is excised along with a few adjacent genes, which the phage can then transfer. Detailed consideration of these processes will be deferred to Chapter 48. We shall consider here only the use of generalized transduction for obtaining desired recombinants, and for mapping the bacterial chromosome. (Genetic mapping is further discussed in Ch. 11.)

In generalized transduction the phage coat occasionally encloses a fragment of the disintegrating host chromosome, instead of the phage genome (Fig. 9-5). Such a **transducing particle,** released from the cell along with normal phage, can then inject its DNA into a recipient cell, just as in phage infection. After the DNA penetrates the cell it frequently recombines with a homologous region of the chromosome.

* The alteration of virulence by recombination is not surprising, for virulence (Ch. 24) is a quantitative, **polygenic** character, i.e., one influenced by many genes.

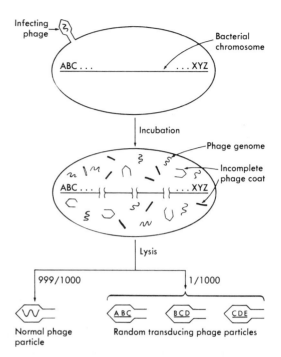

FIG. 9-5. Formation of transducing phage particles by phage-infected bacterium. The entry of the phage genome leads to its own replication and also to formation of all the other components that are finally assembled into the mature phage (Ch. 48). In this assembly the coat occasionally encloses a fragment of bacterial DNA instead of the usual phage DNA, thus forming a transducing particle.

Because the donor lysate contains only rare transducing particles, it is usually added to recipient cells at densities that produce multiple infection. Hence transduction is ordinarily carried out with temperate phages (see Lysogeny, Ch. 48), whose superinfection of a transduced cell usually does not cause lysis. Some virulent phages can also carry out transduction, if superinfection is avoided by low multiplicity of infection.

MAPPING BY TRANSDUCTION

Mapping by transduction is based on the fact that the population of phage particles includes an assortment of segments, of relatively uniform length, derived at random from the donor chromosome (Fig. 9-6). Each bacterial gene should then be represented with more or less equal frequency. The distance between two mutational sites that can **cotransduce** in the same particle can be estimated in two ways: from the frequency of **joint transduction** of distinct genes, or from the frequency of **recombination** between sites within a gene.

Joint Transduction. In classic genetic mapping the degree of linkage in the transmission of two markers is assumed to decrease in proportion to the distance between

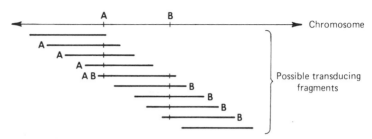

FIG. 9-6. Relation of the distance between two markers to the frequency of their joint transduction. The various segments indicate possible fragments of cellular DNA of the proper length to be transferred in phage particles. The probability of including any marker in a fragment is proportional to its length, but the probability of including two markers is proportional to the difference between that length and the intermarker distance. Hence the frequency of cotransduction falls off sharply with increasing distance between A and B. As a first approximation, the breaks that form transducing fragments are randomly located, and the lengths of the fragments are uniform. Refined studies, however, have revealed some preference in sites, and some variation in length.

them. In mapping by transduction the frequency of cotransduction falls off much more rapidly with distance between the markers, because a cotransducing fragment must be formed by a pair of breaks outside the pair of linked markers but within about 1% of the length of the chromosome (Fig. 9-6); cotransduction reaches zero when the distance exceeds the length that can be transduced.

Map distances and sequences can be determined in **two-factor crosses,** selecting for one marker and comparing the frequency of cotransduction of various others. Sequences are more reliably determined, however, in **three-factor crosses.** In this procedure pairs of genes are selected, and a third, unselected, marker is scored for its frequency of cotransduction (see Fig. 11-16).

In the example provided in Table 9-1 the donor was wild-type *E. coli* and the recipient was a mutant negative for leucine synthesis (*leu⁻*), threonine synthesis (*thr⁻*), and arabinose utilization (*ara⁻*). In two-factor crosses transductants were selected separately for the wild-type allele of each marker. For example, *leu⁺* was selected by plating on a medium lacking leucine but containing threonine and a carbon source other than arabinose. The selected colonies were then scored for each of the unselected markers. It is seen that thr is relatively distant from *ara* (6.7% and 4.3% linkage when one or the other was selected) and even more distant from *leu* (4.1% and 1.9% linkage), whereas *leu* and *ara* are closely linked (55% and 72% linkage). These results suggest the sequence diagrammed at the top of the table.

This sequence was verified by selection of various pairs in a three-factor cross. As Table 9-1 shows, selection for the two extremes (*thr⁺* and *leu⁺*) from the donor yielded a very high frequency of unselected *ara⁺* (80%) also from the donor, as would be expected if *ara* were the "inside" marker.

The **frequency of transductional recombination** between similar mutations in donor and recipient has been extremely valuable in **fine-structure genetics,** i.e., map-

ping of sites at very close intervals (see Units of Information, Ch. 11). In this procedure the strains tested all have the same phenotypic deficiency, due to different mutations in the same gene (or in adjacent genes of similar function). One of these strains is transduced with phage from each of the others, and the wild-type recombinants are selected on an appropriate medium. Their frequency **increases** with the distance between the sites of mutation in donor and recipient (Fig. 9-7). Frequencies as low as 10^{-10}, corresponding to mutations in adjacent nucleotides, can be detected.

TABLE 9-1. Joint Transduction in *E. coli* by Phage P1

	thr⁺	*ara⁺*	*leu⁺*	
Donor: Wild type				(Fragment)
Recipient: Multiple mutant	*thr⁻*	*ara⁻*	*leu⁻*	(Chromosome)

Selected marker	Number of transductants per phage P1 plated	Percentage of selected colonies that also have the following unselected marker		
		leu⁺	*thr⁺*	*ara⁺*
Two-factor				
thr⁺	2.5×10^{-5}	4.1	—	6.7
leu⁺	5.0×10^{-5}	—	1.9	55.4
ara⁺	3.5×10^{-5}	72.6	4.3	—
Three-factor				
thr⁺leu⁺	1.0×10^{-7}	—	—	80.0

The dashed line in the diagram depicts the double crossover necessary to exchange the chromosomal segment containing all three markers. Crossovers in various other regions would result in transduction of one or two markers. In transduction, unlike conjugation, the donor need not be "counterselected" since it is introduced as a lysate rather than as intact cells.

(After Gross J, Englesberg E: Virology 9:314, 1959)

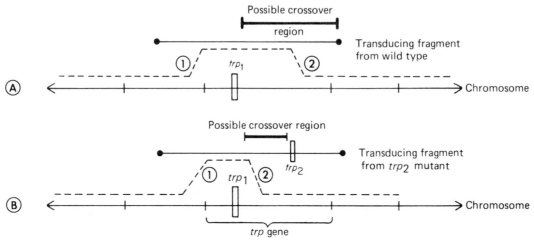

FIG. 9-7. Recombination between a transducing and a recipient marker in same gene. **A.** A trp^- recipient is transduced with phage from a wild-type donor. Of the set of transducing fragments (Fig. 9-6) carrying a normal replacement for the defective trp_1 site one typical member is shown. Given a crossover **(1)** to the left of the trp_1 site, the second crossover **(2)** required for recombination may occur anywhere to the right of that site, up to the end of the fragment. **B.** The same experiment is performed with transducing phage from another trp^- mutant (trp_2). To produce a prototrophic recombinant the second crossover must now occur between trp_1 and trp_2; otherwise the correction of one defect in the cell will be accompanied by the introduction of another.

With various donors, the shorter the distance between their trp^- site and trp_1 the lower the frequency of prototrophic recombinants. In "self-transduction" of $trp_1 \times trp_1$ none are obtained.

The sequence of trp_1 and trp_2 relative to an outside marker (in another gene) can be determined unequivocally by a three-factor cross with two different alleles of the marker in the two parents. In a reciprocal transduction, with either strain serving as donor in the $trp_1 \times trp_2$ cross, one donor will bring in a nearby unselected marker to the right of the cross and the other will bring in one to the left.

CONJUGATION

The universality of sexual reproduction in higher organisms led to early efforts to visualize mating in bacteria, but without success. The discovery of transformation in the pneumococcus renewed interest in the subject, and the discovery of auxotrophic mutants of bacteria (Ch. 8) offered a new approach. Lederberg, then a medical student, undertook a search for genetic recombination in *E. coli,* and in 1946 he demonstrated the formation of prototrophic recombinants in mixed cultures of two auxotrophs (Fig. 9-8).

Recombination turned out to be so infrequent (about 10^{-6} per cell) that it was obscured by the similar rate of reversion of singly auxotrophic parents. Success depended on crossing double auxotrophs ($A^-B^- \times C^-D^-$): these could not produce prototrophs ($A^+B^+C^+D^+$) at a detectable rate by mutation, since two independent, rare mutations in the same cell would be required.*

Conjugation requires cell contact (unlike transformation and transduction); and the simultaneous transfer of multiple markers clearly indicated extensive genetic linkage. These findings suggested that in bacteria, as in higher organisms, two haploid cells fuse to form a diploid zygote. But unlike higher organisms, in

* Success also depended on the fortunate use of *E. coli* K12, a strain that had been carried in the laboratory since 1922, with resultant extensive changes in the cell surface that promote mating. Most subsequently tested strains proved to be infertile.

which the degree of linkage between markers in various crosses yields a consistent map of the chromosomes, crosses between various sets of mutations in bacteria did not yield such consistency. Eventually the search for unity was replaced by recognition that bacterial mating forms merozygotes (partial zygotes), rather than the expected true zygotes.

MECHANISM OF GENE TRANSFER

Polarity. All the early mutant cultures of *E. coli* K12 were interfertile, and so conjugation did not appear to involve two mating types. However, later purified subcultures revealed not only two types, but also a novel polarity: some cultures behaved as donors and others as recipients.

Hayes in London discovered this polarity serendipitously. In a study of the physiology of mating, employing streptomycin resistance (str^R) as a selective marker, a control test included this lethal antibiotic during mating, rather than as usual in the subsequent selective plates. Recombinants were formed when a particular parent was str^R (resistant) and the other str^S (sensitive), but not when these alleles were reversed in the two parents. Hence one of the parents could apparently donate genetic material even though killed by streptomycin, while the other parent had to be viable.

These findings showed that cells of *E. coli* K12 exist in either of two inheritable mating types, called F^+ and F^-. The F^+ cells serve as **genetic donors (males),** and the F^- cells as **recipients (females);** only the latter must be vi-

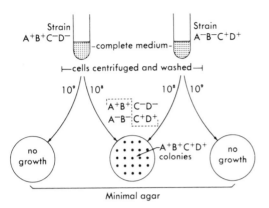

FIG. 9-8. Diagrammatic representation of the initial experiment of Lederberg (1946). The mutants were cultivated in a medium that included growth factors A, B, C, and D; the test for recombination, and the control tests for reversion, were carried out in a minimal medium lacking these factors.

able. Evidently genetic material is transferred from male to female through some kind of **conjugation bridge.** $F^- \times F^-$ crosses are uniformly sterile, and $F^+ \times F^+$ have low fertility.

The F Plasmid. The donor property of F^+ cells is due to presence of a **sex factor** (fertility factor), called the **F agent.** This agent was later discovered to be a plasmid that leads with high frequency to transfer of itself, and much less frequently to transfer of the chromosome. Thus when F^+ and F^- cells, labeled with different genetic markers, are grown together (at population densities sufficient for frequent cell contact), within an hour most of the F^- cells become F^+. Moreover, the F plasmid was later observed physically in lysates of F^+ but not of F^- cells. It was the first of the class of **conjugative plasmids,** to be discussed in Chapter 12: extrachromosomal genetic elements that code for a conjugation apparatus.

The F plasmid is easily lost from an F^+ cell (especially during prolonged incubation of cultures in the stationary phase): it does not appear to have a chromosome-like mechanism ensuring its regular distribution to daughter cells, and cells lacking it may have a slight advantage in growth. The interfertility of the various mutant stocks in earlier studies is thus explained by the presence of a mixture of F^+ and F^- cells in each culture.

Asymmetric Genetic Contributions. The anomalies of the genetic linkage map in *E. coli* were further clarified by the discovery that the two parents make **unequal genetic contributions:** unselected markers are derived much more frequently from the F^- than from the F^+ parent. This finding suggested that the donor transfers only part of a chromosome, yielding a **merozygote** that is a transient stage in the formation of stable haploid cells.

High-Frequency Recombination (Hfr). The mechanism of conjugation became much clearer when Cavalli in Italy by chance isolated a subclone, from an F^+ strain, with a 1000-fold increased rate of recombination with F^- strains. Such Hfr (high frequency of recombination) organisms are evidently derived from F^+ by a change in the state of the F plasmid, since Hfr can revert to F^+, while F^- cells (which lack the plasmid) never mutate to Hfr. Jacob and Wollman in Paris recognized (on grounds described below) that in Hfr strains the F plasmid has become integrated in the chromosome (Fig. 9-9), so that in conjugation transfer of its initiating segment is followed by the chromosome.

Hfr mutants can be isolated (about 10^{-5}/cell) by **indirect selection.** When a lawn of F^+ cells is replica-plated (see Fig. 8-2, Ch. 8) onto a thin lawn of a complementary F^- strain, on a medium that would select for recombinants, a few colonies appear; and from the corresponding regions on the stored F^+ plate one can obtain Hfr clones. Thus chromosome transfer by F^+ cultures involves rare mutation to, and then crosses by, Hfr cells.

KINETICS OF MATING AND TRANSFER

With cultures conjugating at low frequency one could study only progeny many generations after the mating. However, with Hfr conjugation one can study the actual conjugants and thus analyze the component steps of the mating process: cell pairing, DNA transfer, and integration.

In the first experiments the F^- strain was present in excess, and time was allowed for every Hfr cell to mate. The frequency of various alleles in the resulting recombinants was then determined by plating on appropriate selective media. It was found that the various markers could be ordered in a sequence based on a **gradient of decreasing transmission** from the Hfr to the F^- cells. This finding could be explained if the Hfr strain starts transferring its chromosome at a characteristic locus (the **origin of transfer**), and if the various markers, following in a sequence, have a decreasing chance of transmission with increasing distance from the origin.

The sequence was confirmed by experiments in which mating was **artificially interrupted,** at various times after mixing, by mechanical agitation (by means of a Waring Blendor, rapid pipetting back and forth, or strong vibration). Entry was found to start at about 8 min for a gene very close to the origin of transfer; and for other markers the increasing **time of initial entry** (Fig. 9-10) was found to correspond to their order in the gradient of decreasing transmission (Fig. 9-11).

Mapping by Interrupted Mating. For a given marker the time of initial entry represents the time required to form the earliest mating pairs (which is the same for all genes), plus that required for transfer of the necessary length of

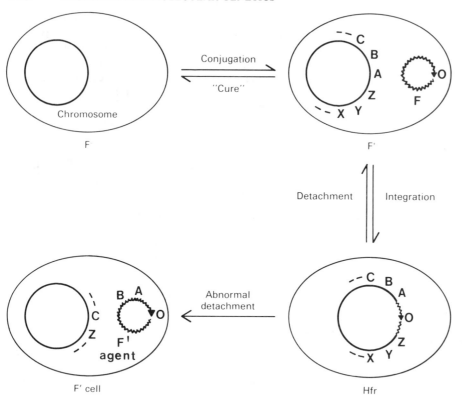

FIG. 9-9. Transitions between F⁻, F⁺, Hfr, and F′ cells. The reasons for inferring cyclic structures arose later, and will be presented below. **O** = origin of transfer. Though the chromosome giving rise to an F′ plasmid is defective in the primary F′ cell, as shown, the F′ plasmid soon becomes stabilized, by transfer, in cells with a normal chromosome.

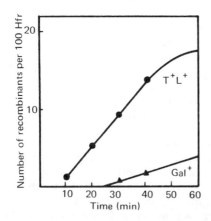

FIG. 9-10. Kinetics of conjugation, studied by interrupting further chromosomal transfer, as well as further pair formation, at various times. Cross: HfrH str^S thr^+ leu^+ gal^+ × F⁻ str^R thr^- leu^- gal^-. Exponential broth cultures were mixed at time 0 (10^7 Hfr + excess of F⁻ cells/ml) and aerated in broth. At intervals samples were diluted, agitated briefly in a Waring Blendor, and plated 1) on glucose minimal medium plus streptomycin to select thr^+ leu^+ str^R recombinants; and 2) on galactose minimal medium plus threonine, leucine, and streptomycin, to select gal^+ str^R recombinants. (After Wollman EL, Jacob F, Hayes W: Cold Spring Harbor Symp Quant Biol 21:141, 1956)

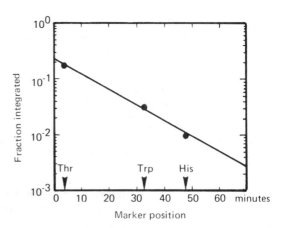

FIG. 9-11. Gradient of transfer of markers from *Escherichia coli* K12 strain HfrH. The ordinate represents the **total entry** of each marker (fraction of input Hfr, with excess of F⁻), ranging from 20% to about 0.1%. Abscissa represents the **time of initial entry** for each marker; the distal end of the chromosome requires 90 min at 37°. (After Wood TH: J Bacteriol 96:2077, 1968)

chromosome (which proceeds at a constant rate). Accordingly, *map distances on a chromosome can be directly measured, in units of time.* In all previous genetic systems these distances could only be inferred, from the relative frequency of crossover between loci in recombination tests.

At 37°C about 1% of the chromosome (15μm, or 50,000 bases per strand) is transferred per min. This genetic length contains about 20 recombination units (i.e., 20% crossing over) in *E. coli.* Mapping by interrupted mating is thus ideal for long distances in the chromosome (> 1 min), complementing the use of cotransduction, and recombination frequencies in conjugation, for short distances.

INTERACTIONS OF F PLASMID AND CHROMOSOME

The Cyclic Chromosome. Independently isolated Hfr mutants were found to differ in their sequence of gene transfer. The patterns could be fitted together by the model illustrated in Fig. 9-12, in which *the chromosome is circular,* it is *mobilized for transfer by integration of the F plasmid,* and the *variable site and orientation of integra-*tion determines both the origin and the direction of transfer of the chromosome. Circularity of the chromosome was later confirmed by radioautography following gentle lysis (see Fig. 10-26). The F plasmid (discussed further in Ch. 12) is also circular and is about 2% as long as the chromosome; F+ cells contain one to two copies per chromosome.

The discovery of circularity provided a simple explanation for the mechanism of **integration:** a **single crossover** between two rings of DNA, yielding an enlarged single ring (Fig. 9-12; for details see Fig. 48-13). This process, ordinarily viewed as the integration of F into the chromosome, can also be regarded as an integration of the chromosome into F. Hence the mechanism of chromosome transfer is the same as that of F itself: the integrated and the free plasmid initiate at the same site and differ only in the sequence that follows.

Though F can integrate into various sites on the chromosome, with varying frequency, the number of sites is limited. The mechanism of integration will be considered further in Chapter 12.

F+ donors usually transfer the entire F plasmid, be-

FIG. 9-12. Formation of different Hfr strains by integration of F agent at different locations on the bacterial chromosome. Site of initial entry (origin) is depicted as an **arrowhead** in the middle of F.

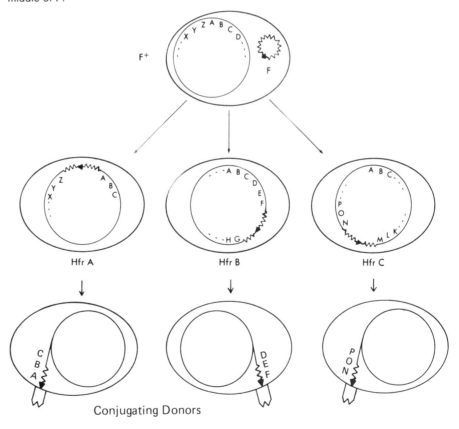

Conjugating Donors

cause its short DNA chain is rarely broken during conjugation; hence the product is F⁺. With Hfr donors, however, the much longer chain is usually broken, as we have seen, before complete transfer, and the product is F⁻.

F′ Plasmids. The integration in an Hfr strain is reversible at low frequency, usually restoring the original F⁺. However, the integrated F plasmid occasionally is excised imprecisely and carries a segment (of variable length) from either adjacent region of the host chromosome; the resulting **hybrid plasmid** is called F′ (Fig. 9-9). Since these hybrids are transmitted to F⁻ cells with high efficiency (as in a normal F⁺ × F⁻ cross), and they are easily separated from the chromosome in lysates, they are very useful for the transfer and the isolation of selected genes.

F′ transfer has been called **F-duction** (or **sexduction**); it is the equivalent of the specialized transduction of a limited segment of host chromosome attached to phage DNA, except that F′ can vary much more in size. The abnormal cuts in the chromosome that create F′ plasmids do not occur entirely at random but involve special insertion sequences, described in Chapter 11.

The F′ plasmid can have several fates in a recipient cell. 1) It can replicate autonomously. 2) Its segment from the donor chromosome can recombine by a double crossover with the homologous recipient segment (as in transformation). 3) A single crossover can integrate the plasmid in the region of homology. Because of the extensive homology this last integration (and its reversal) is much more frequent than the normal integration of a plasmid into variable sites; hence the F′ will oscillate frequently between the two locations, and the culture will transfer the F′ from some cells and the chromosome from others (**intermediate donor cultures**).

Ingenious methods have been developed for selecting a desired F′. For example, in crosses with *recA⁻* (recombination-deficient) recipients the donor DNA cannot enter the recipient chromosome, and so transferred genes can be perpetuated only in an autonomously replicating plasmid (yielding a **transconjugant**). Since the plasmids from an Hfr donor often contain adjacent genes from either side of the integration site, selection with *recA⁻* recipients for such a gene readily yields an F′ plasmid carrying it.

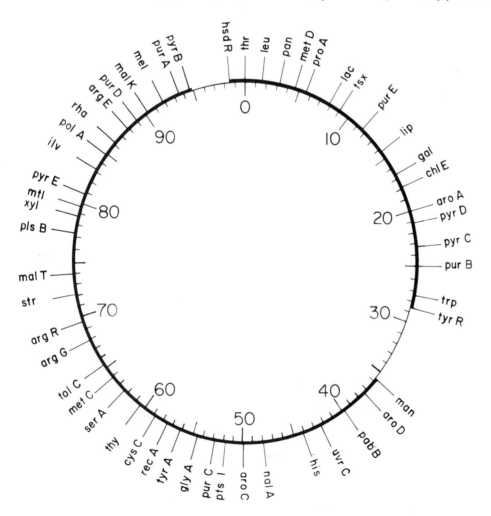

FIG. 9-13. Map of a few genes in the chromosome of *E. coli*. Numbers represent minutes required for transfer (broth, 37°) from the origin (zero) of the first Hfr strain isolated, HfrH. The parts of the circle in heavy lines contain markers whose distances have been confirmed by cotransduction. In the earlier literature the map was calibrated to a total of 90 rather than 100 min, and so the positions of genes distant from the origin were given lower numbers. (Modified from Bachmann BJ, Low KB and Taylor AL: Bacteriol Rev 40:116, 1976)

THE GENETIC MAP

With the several unique features of bacterial conjugation thus elucidated, a consistent, detailed chromosome map of *E. coli* could be constructed. Figure 9-13 presents some of the known markers. The distances are expressed in terms of time of entry: total 90–100 min.

A reliable map is important for understanding the organization of the genome. In addition, in the experimental creation of specific genotypes nonselectable genes can be inserted by taking advantage of their known linkage to selectable markers.

PHYSIOLOGY OF CONJUGATION

F Pili. The mechanism of initiation of mating was revealed by the study of bacteriophages that can infect only "male" (F^+) cells of *E. coli* (see Viral Sites for Adsorption, Ch. 47). These phages were found to adsorb onto the sides of special pili (Ch. 2) that are present only on the male cells, in one to three copies; they do not adsorb onto the many other pili found on both male and female cells. The tips of these special **sex pili** were then shown to attach firmly to the surface of F^- cells, in mixed cultures (Fig. 9-14). Mutations showed that the **receptor** in the F^- cells is an outer membrane protein (which also serves as receptor for certain female-specific phages). In F^+ cells this receptor becomes ineffective (see Entry Exclusion, Ch. 12), explaining the low $F^+ \times F^+$ fertility, for otherwise identical F^+ cells would uselessly exchange plasmids with each other.

The Bridge. The F pilus is made of an 11,800-dalton phosphoprotein, **pilin,** and it appears in electron micrographs to be a hollow tube. It could therefore provide the bridge through which the DNA passes; however, no decisive evidence (such as isolation of pili containing DNA) is available. Moreover, most sets of mating cells are found to be aggregated, in wall-to-wall contact; and exposure to sodium dodecyl sulfate, at concentrations that normally dissolve pili, does not cause disaggregation, nor does it block further DNA transfer. It has therefore been suggested that the pilus may serve as only a recognition and triggering device, followed perhaps by its retraction, and by formation of some kind of cytoplasmic bridge between the cell envelopes.

Whichever the mechanism, it is clear from genetic studies that **triparental mating** can occur in the mating aggregates (i.e., simultaneous mating of two different Hfr cells with a single F^-). Mating aggregates break up more rapidly after F^+ than after Hfr matings, suggesting that completion of DNA transfer to the recipient initiates a process of active disaggregation.

DNA Transfer. The replication of any autonomous bloc of DNA is initiated at a specific site and moves contin-

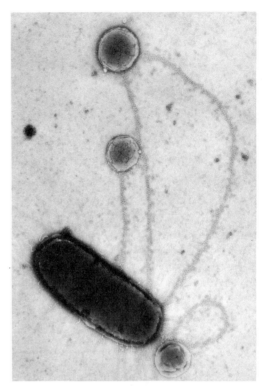

FIG. 9-14. Attachment of an F^+ *E. coli* cell, by means of F pili, to three F^- minicells (products of mutants that occasionally separate a small cell without a nucleus). The few long F pili are covered with F-specific MS2 phage particles. These phages are icosahedral (Ch. 46); other, filamentous, male-specific phages have been found to adsorb only to the tips of F pili. (After Curtiss R III et al: J Bacteriol 100:1091, 1969)

uously along the DNA molecule (see DNA Replication, Ch. 10). Normal, symmetric replication, whether of a free F agent or of a chromosome, simultaneously synthesizes a new, complementary strand along each old, template strand. In conjugation, however, the plasmid (whether free or integrated) is somehow triggered into initiating an **asymmetric** mode of replication (see Symmetry of Replication, Ch. 10): a particular one of the original strands is cut at a specific site and enters the bridge, while the other remains in the donor cell (Fig. 9-15). Moreover, a single strand is transferred, as was shown by transfer to minicells (which lack other DNA: see Fig. 6-25) and by use of physical methods for distinguishing the two strands. Normally the separated strands are replicated immediately, one in the donor, and the other as it enters the recipient.

It is clear, from the experiment in minicells just noted, that replication in the recipient is not necessary for DNA transfer. Evidence is not conclusive as to whether DNA synthesis in the donor is required, possibly providing the energy for transfer.

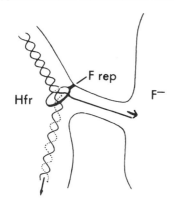

FIG. 9-15. Model for the mechanism of DNA transfer during conjugation. Attachment of the tip of a pilus of an F⁻ cell somehow activates a machinery **(F rep),** attached to the cell membrane, that begins DNA replication at the transfer origin of an F agent and directs an unreplicated strand into the conjugation bridge. (Modified from Jacob F, Brenner S and Cuzin F: Cold Spring Harbor Symp Quant Biol 28:329, 1963)

This mechanism of replication and transfer can explain several features of conjugation: why the chromosome requires 100 min to transfer its whole length but only 45 min for its normal replication (which proceeds in both directions from the origin: see DNA Replication, Ch. 10); how an F^+ cell infects an F^- cell in conjugation without itself becoming F^-; and why a cell can serve as donor even after being killed by streptomycin [which irreversibly inhibits the ribosomes (Ch. 13) but permits continued energy production and DNA synthesis].

The molecular details and genetic control of conjugation will be considered in Chapter 12.

PROTOPLAST FUSION

Zygote formation with bacterial protoplasts has recently been achieved experimentally, with a procedure developed earlier with plant cell protoplasts: brief exposure to 40% polyethylene glycol. Under these conditions protoplasts of *Bacillus subtilis* undergo occasional fusion, and on resynthesis of the cell wall they segregate stable haploid recombinants, with a total frequency of about 10^{-4}. Much rarer fusion of protoplasts also occurs without the fusing agent, which suggests that this form of gene transfer may occur in nature.

SELECTED READING

BOOKS AND REVIEW ARTICLES

ADELBERG EA (ed): Papers on Bacterial Genetics. Boston, Little, Brown, 1966 (A paperback reprinting of key articles, with a useful historical introduction and bibliography)

DUBOS RJ: The Professor, the Institute, and DNA. New York, Rockefeller University Press, 1976 (An affectionate life of Avery, including the circuitous route to the identification of the transforming substance)

HAYES W: The Genetics of Bacteria and Their Viruses, 2nd ed. New York, Wiley, 1968

JACOB F, WOLLMAN EL: Sexuality and the Genetics of Bacteria. New York, Academic Press, 1961

JONES D, SNEATH P: Genetic transfer and bacterial taxonomy. Bacteriol Rev 34:40, 1970

LOW KB, PORTER RD: Modes of gene transfer and recombination in bacteria. Annu Rev Genet 12:249, 1978

SANDERSON KE, HARTMAN PE: Linkage map of *Salmonella typhimurium,* V ed. Microbiol Rev 42:471, 1978

STANISICH VA, RICHMOND MH: Gene transfer in the genus *Pseudomonas.* In Clarke PH, Richmond MH (eds): Genetics and Biochemistry of Pseudomonas. New York, Wiley, 1975, p 163

SPECIFIC ARTICLES

TRANSFORMATION AND TRANSDUCTION

AVERY OT, MACLEOD C, MCCARTY M: Studies on the chemical nature of the substance inducing transformation of pneumococcal types. J Exp Med 79:137, 1944

COHEN SN, CHANG ACY, HSU L: Genetic transformation of *E. coli* by R factor DNA. Proc Natl Acad Sci USA 69:2110, 1972

DUBNAU D: Genetic transformation of *B. subtilis.* In Microbiology 1976. Washington DC, American Society for Microbiology, 1976, p 14

HARTMAN PE, LOPER JC, SERMAN D: Fine structure mapping by complete transduction between histidine-requiring *Salmonella* mutants. J Gen Microbiol 22:323, 1960

HOTCHKISS RD: Gene, transforming principle, and DNA. In Cairns J, Stent GS, Watson JD (eds): Phage and the Origins of Molecular Biology. Cold Spring Harbor, NY, Cold Spring Harbor Laboratory, 1966, p 180

HOTCHKISS RD, and GABOR M: Bacterial transformation. Annu Rev Genet 4:193, 1970

LACKS SA: Binding and entry of DNA in bacterial transformation. In Reissig J (ed): Microbiol Interactions. New York, Halsted-Wiley, 1977

NOTANI NK, SETLOW JK: Mechanisms of bacterial transformation and transfection. Prog Nucl Acid Res Mol Biol 14:39, 1974

SISCO KL, SMITH H: Sequence-specific DNA uptake in *Haemophilus* transformation. Proc Natl Acad Sci USA 76:972, 1979

TOMASZ A: Some aspects of the competent state in genetic transformation. Annu Rev Genet 3:217, 1969

CONJUGATION

ACHTMAN M, MORELLI G, SCHWUCHOW S: Cell-cell interactions in conjugating *E. coli:* role of F pili and fate of mating aggregates. J Bacteriol 135:1053, 1978

CAVALLI-SFORZA LL, LEDERBERG J, LEDERBERG EM: An infective factor controlling sex compatibility in *Bacterium coli.* J Gen Microbiol 8:89, 1953

IHLER G, RUPP W: Strand-specific transfer of donor DNA during conjugation in *E. coli*. Proc Natl Acad Sci USA 63:138, 1969

LEDERBERG J: Gene recombination and linked segregation in *Escherichia coli*. Genetics 32:505, 1947 (In Adelberg collection)

LOW KB: Formation of merodiploids in matings with a class of recipient strains of *Escherichia coli* K12. Proc Natl Acad Sci USA 60:160, 1968

LOW KB: *E. coli* K12 F-prime factors, old and new. Bacteriol Rev 36:587, 1972

OU J, ANDERSON TF: Role of pili in bacterial conjugation. J Bacteriol 102:648, 1970

SANDERSON KE, ROSS H, ZIEGLER L, MAKELA PH: F, Hfr, and F strains of *Salmonella*. Bacteriol Rev 36:608, 1972

SCHAEFFER P, CAMI B, HOTCHKISS RD: Fusion of bacterial protoplasts. Proc Natl Acad Sci USA 73:2151, 1976

VAPNEK D, RUPP WD: Asymmetric segregation of the complementary sex-factor DNA strands during conjugation in *Escherichia coli*. J Mol Biol 53:287, 1970

chapter 10

STRUCTURE AND REPLICATION OF NUCLEIC ACIDS

We have seen that DNA carries the **genetic information** of a cell, i.e., the instructions for its own replication and also for the structure of the other macromolecules, whose activities are in turn responsible for the total structure, function, and growth of the cell. In most viruses the genetic information similarly resides in DNA, while in others it is in RNA. In this and the following chapters we shall examine the main molecular aspects of genetic information: 1) how it is encoded in nucleic acids and other macromolecules, 2) how it is transmitted from one macromolecule to another, 3) how its expression is regulated, and 4) how it is altered.

The molecular events connected with storage and transmission of information are governed by two general principles. 1) The informational molecules are **linear** and have **periodic** backbones (the phosphate–sugar chain in DNA, the peptides in proteins), which allow stepwise chain synthesis and also formation of regular secondary structures. The information itself is carried, in vast amounts, in an **aperiodic sequence** of side chains (bases in nucleic acids, amino acid side chains in proteins).* 2) Nucleic acids and proteins are synthesized on a **template,** which serves neither as a catalyst nor as a substrate, but as a source of information determining the sequences being synthesized. Thus a DNA strand serves as a template for the complementary strand in **replication,** or for an RNA strand (the messenger RNA) in **transcription,** while the latter in turn serves as template for a polypeptide chain in **translation.**

The idea of a template found its first significant support when Watson and Crick, in 1953, proposed the double-stranded structure of DNA: the complementarity of the two strands immediately suggested that each serves as a direct template for the synthesis of the other. This prediction was confirmed in 1958, when Kornberg discovered a DNA polymerase, in extracts of *Escherichia*

coli, that requires a template DNA and faithfully replicates its composition.

The study of nucleic acid structure and synthesis, coordinated with bacterial genetics, gave rise to the interdisciplinary field of molecular genetics (sometimes called molecular biology). When this field was extended from gene replication to gene expression, it was found to have converged with the pursuit of the mechanism of protein synthesis by biochemists. Even though molecular biologists were accused of "practicing biochemistry without a license," a dialogue between the two groups led to rapid advances in analyzing many fundamental problems at a molecular level. These results will be reviewed in this and the following chapters.

Protein Synthesis. The critical step was the development of a system for synthesizing proteins in vitro, achieved in 1954 by Zamecnik with extracts of mammalian cells and subsequently extended to bacteria. The involvement of an RNA template was suggested by the requirement for RNA but not DNA. Two classes of RNA were separated: ribonucleoprotein particles (now called **ribosomes**), which were pelleted at 100,000 g, and smaller, supernatant "soluble" RNA molecules (the **transfer RNA [tRNA]**), to which, as Hoagland showed, specific amino acids were covalently attached by specific activating enzymes. On the basis of studies of the regulation of gene expression (see Ch. 14) Jacob and Monod, in 1961, discovered an unstable additional fraction of cellular RNA, the **messenger RNA (mRNA),** into which the information for protein sequences is **transcribed** from DNA. Biochemical studies with synthetic messengers then rapidly deciphered the **genetic code,** which governs the **translation** of the nucleotide sequence of mRNA into a polypeptide sequence.

As background for this and the following chapter, we may further note that the bacterial ribosome contains many different proteins and three molecules of **ribosomal RNA (rRNA),** designated from their sedimentation constants as 23S, 16S, and 5S. As the ribosome moves along the mRNA, each coding unit (**codon**) that it encounters directs the binding of the corresponding **aminoacyl-tRNA (aa-tRNA),** whose amino acid is then incorporated into the growing polypeptide chain. These developments will be described in Chapter 13.

---

* Chapter 14 will consider the processing of another kind of information that is also important for the cell: the concentrations of its metabolites.

PROPERTIES OF NUCLEIC ACIDS

MOLECULAR ORGANIZATION OF NUCLEIC ACIDS

Extraction. Nucleic acids are extracted with the aid of substances that denature the proteins with which they are associated, e.g., sodium dodecyl sulfate (SDS) or water-saturated phenol. The conditions for extraction (temperature, ionic strength, and special additives) depend on the type of nucleic acid (RNA or DNA) and on the nature of the contaminants (e.g., whether lipid-containing or not). Analyses are facilitated by the fact that RNA is completely hydrolyzed by concentrations of alkali that do not hydrolyze DNA.

PRIMARY STRUCTURE

All nucleic acids contain four main bases: guanine, adenine, cytosine, and either thymine (in DNA) or uracil (in RNA), commonly symbolized as G, A, C, T, and U. The presence of thymine in DNA reduces mutation frequency: uracil produced from frequent deamination of cytosine is a mutation in RNA, but in DNA it is removed by the action of uracil–DNA glycosidase and subsequent repair (see Repair of DNA Damage, Ch. 11). The DNA of certain bacteriophages contains 5-hydroxymethylcytosine instead of cytosine, or 5-hydroxymethyluracil instead of thymine. Each nucleic acid strand has a **polarity,** because phosphodiester bonds connect the 3′ position* of one nucleotide residue to the 5′ position of the next; polynucleotide chains terminate with a free 3′ position at one end (the 3′ end) and a free 5′ position at the other end (the 5′ end) (Fig. 10-1).

The free rotation of the phosphodiester bonds between adjacent nucleosides tends to make the chains very flexible, but this tendency is compensated for by the reciprocal attraction of the planar aromatic rings of the bases (especially purines), which tend to **stack** neatly on top of each other, stiffening the chain. Nucleic acids have multiple negative charges owing to the primary phosphoryl groups; in vivo these are neutralized by inorganic cations (especially Mg^{2+}) and by basic organic molecules, such as polyamines or, in cells of higher organisms, histones.

Various **modifications** are found in nucleic acid bases occupying specific positions in polynucleotide chains; some of them protect DNA from nucleases (see Modifications of DNA, below), whereas others affect secondary structure and interaction with cellular components, as in tRNA (see tRNA, Ch. 13).

Most common in DNA are **methylations** that yield 5-methylcytosine and 6-methyladenine. In addition, in some bacteriophages the DNA is **glucosylated** on the hydroxyl group of hydroxymethylcytosine or hydroxymethyluracil. Transfer RNAs contain many methylated derivatives, including 5-methyluracil (ribothymine), 5,6-dihydrouracil, and pseudouridine (with the ribose attached to ring C-5 rather than N-1). Ribosomal RNA contains 2-methyl substituted ribose as well as methylated bases. Modifications are carried out by specific enzymes after polymerization; at least eight different tRNA methylases have been identified. Eukaryotic mRNA contains special sequences at the 5′ end, which will be considered in Chapter 49.

SECONDARY STRUCTURE

Chargaff showed that in DNA the proportion of A equals that of T and the proportion of G equals that of C. This finding, and the x-ray crystallographic studies of Franklin and Wilkins, led Watson and Crick to recognize in 1953 that the structure of the usual DNA is a **double-stranded helix,** or **duplex,** with a diameter of about 2 nm.* A turn of the helix encompasses about ten nucleotide pairs, but this number varies slightly, depending on conditions, with important consequences for cyclic DNAs (see Cyclic DNA, below). In the helix the two strands are **complementary** and **antiparallel,** i.e., with inverse polarity (Fig. 10-1). Complementarity is deter-

FIG. 10-1. Diagram of double-stranded DNA. The two strands are connected by hydrogen bonds (vertical parallel lines) between complementary bases. Each strand has a polarity determined by the direction of the phosphodiester bonds. The 5′ → 3′ direction of each strand is indicated by an **arrow.** The two strands constituting the same molecule have opposite polarities.

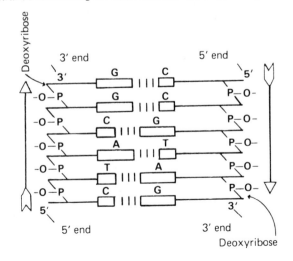

---

* Primed numbers in nucleotides refer to positions on the ribose or deoxyribose residue, and other numbers of positions in the purine or pyrimidine rings.

* 1 nm (nanometer) = 10 A.

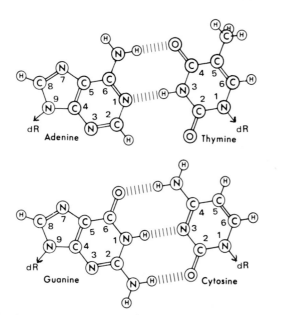

FIG. 10-2. Pairing of adenine with thymine (two hydrogen bonds) and of guanine with cytosine (three hydrogen bonds) in double-stranded DNA. **Arrows** indicate bonds to 1'-C of deoxyribose **(dR)** in the sugar–phosphate backbone of DNA (Fig. 10-1). Similar pairing occurs in helical RNA with U substituting for T and ribose for deoxyribose.

mined by the steric relations of the bases: A of one strand is always paired with T of the other, by two hydrogen bonds, as is G with C, by three hydrogen bonds (Fig. 10-2). The **base ratio** of a given DNA is thus $(A + T)/(G + C)$. The same relations apply to **double-stranded RNA,** with U substituting for T. In **single-stranded DNA and RNA,** in contrast, the proportions of the four bases can vary independently.

In helical nucleic acids the genetic information, i.e., the sequence of the bases, is inside, near the axis of the helix, and it cannot be read unless the strands separate. The information is **redundant** since it is duplicated in each strand. This feature is essential not only for replication, where each strand determines its complement, but also for recombination and repair of damage, where one strand is partly demolished and is then rebuilt as a complement of the intact strand (see Repair of DNA Damage, Ch. 11).

The hydrogen bonds between complementary bases and the stacking of the bases in each strand confer on helical molecules the properties of **elastic rods** (as if they were made of rubber). Their substantial **rigidity** is recognized from their shape in electron micrographs and the high viscosity of their solutions. Yet these molecules undergo considerable **dynamic changes** (especially near single-strand breaks, also called **nicks**), which are of great significance in replication, transcription, and recombination. Thus intact duplexes can form short unpaired segments that move rapidly along the molecule ("breathing"); and at a nick a molecule can exchange a strand with an identical strand of another molecule (**strand displacement;** see Recombination, Ch. 11).

In contrast, *molecules of single-stranded nucleic acids are very flexible* and in solution form **random coils.** However, they tend to give rise to internal base pairs and are therefore **folded** to some degree (Fig. 10-3). The proportion of paired bases (also called the **helical content**) varies with the composition of the nucleic acid and the nature of the solvent; if it is high, as in tRNA (see Fig. 13-4), definite **secondary** and **tertiary** structures may result. Differences in the secondary structure of diverse RNAs are probably important for their specific interactions with certain proteins, as in the complexing of tRNA with activating enzyme and with ribosomes, in the assembly of ribosomes and in viruses.

PHYSICAL PROPERTIES OF DNA MOLECULES

SIZE AND CONFIGURATION

The lengths of nucleic acids are generally given as the number of bases per strand, with 1000 bases as a unit (**kilobase [Kb],** corresponding to a mol wt of 3.3×10^5 for single-stranded molecules). Because nucleic acids are extremely long and thin a constant concern in determining their size is to avoid breaks caused by shear. With large

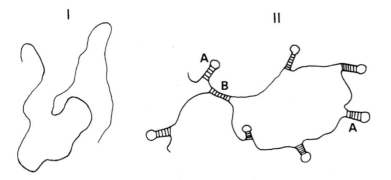

FIG. 10-3. Schematic representation of single-stranded nucleic acids in solution. I. Unfolded random coil present either at high temperature or in the presence of denaturing agents. II. Folded random coil present at low temperature in salt solutions; parallel lines represent hydrogen-bonded bases. Short-range folds, i.e., those between adjacent sequences ("hairpins") **(A),** are thought to be much more common than long-range folds **(B).** (Modified from Studier FW: J Mol Biol 41:189, 1969. Copyright by Academic Press, Inc. (London) Ltd.)

DNAs even pipetting causes extensive shearing, and so most preparations consist of fragments.

The following methods are generally used for examining nucleic acids of moderate length.

Electron Microscopy. Elegant electron micrographs of DNA are obtained by Kleinschmidt's method, in which the molecules are spread on a solution of a basic protein and then are collected on a membrane. The basic molecules adsorb to the nucleic acid, increasing its thickness and therefore its visibility. Single strands are coiled and must be extended before spreading, for instance by binding a helix-destabilizing protein, such as the gene 32 product of phage T4 (see Ch. 47).

Radioautography. Tritium-labeled thymidine of high specific activity is incorporated into replicating DNA, and the molecules of the DNA, extracted by mild methods, are collected on membrane filters or glass surfaces. These are overlaid with a photographic emulsion and kept in the dark for several weeks or months, during which β-radiations from disintegrating tritium atoms cause the formation of silver grains in the adjacent emulsion. After photographic development the radioautographs reveal the shape of the DNA molecules and their lengths (the distance between nucleotide pairs is 0.34 nm along the helix axis).

Sedimentation. The sedimentation coefficient ($S^o_{20,w}$) is based on the velocity of sedimentation in a centrifugal field in water at 20°C and at very low nucleic acid concentration. Absolute values are determined in the analytic ultracentrifuge. Relative values are often obtained in the preparative ultracentrifuge by **zonal centrifugation** through a **density gradient,** commonly of sucrose or CsCl, formed in a centrifuge tube by the appropriate mixing of two solutions of different specific gravity (Fig. 10-4). This technic is also invaluable for separating macromolecules in a mixture.

Sedimentation coefficients of macromolecules and viruses are usually given in Svedberg units (S). The sedimentation coefficient of nucleic acids is related to the molecular weight by empiric relations, which differ for single-stranded and double-stranded molecules. For single-stranded molecules the S values are strongly affected by the degree of helical content, which varies with ionic strength and temperature; molecular weights are best determined in solutions that prevent formation of secondary structures, such as low concentrations of formaldehyde or methylmercuric hydroxide or a high concentration of dimethylsulfoxide.

Agarose or Polyacrylamide Gel Electrophoresis. The sample is applied to one end of a glass cylinder containing gelled agarose or polymerized acrylamide; then a constant voltage is applied across the length of the cylinder. The negatively charged nucleic acid molecules move toward the anode, the larger molecules moving more slowly. At the end of the run molecules of different sizes are distributed as thin bands, at various distances from the origin, which may be visualized by staining. The rates of migration of the molecules depend on their molecular weights and on the concentration of the gel.

Very Large DNA. Chromosomal DNAs from bacteria and other cells are so long (millimeters or centimeters) that their size cannot be determined by the foregoing procedures, because they are easily broken. Special methods must be used, such as the following: 1) **Sedimentation**—The preparation is centrifuged at an extremely low speed and for a long time in order to minimize a shear-dependent aggregation. 2) **Viscoelastic method**—A cylinder is rotated in the DNA solution. When the driving force is discontinued the cylinder slows down to a stop and then rotates backwards, as the stretched DNA molecules retract to their normal random coiled state. The rate of exponential decay of the back rotation reflects the size of the longest DNA molecules; hence the results are not affected by breakage of some molecules.

Such determinations of molecular size, combined with measurements of the amount of nucleic acid per cell or particle, show that bacterial and many viral **genomes** consist of a *single copy of a nucleic acid;* in some viruses it is in several pieces. These molecules must be tightly coiled, because in their extended form they are extraordinarily long in comparison with the dimensions of the microbe to which they belong (see Table 10-1). For instance, the DNA of *E. coli* is about 400 times longer than the long axis of the cell, and that of bacteriophage T2 is 500 times longer than the entire viral particle.

Several other properties of nucleic acids are of particular biologic interest for understanding the structural basis of their activities and as tools for purification and fractionation.

Buoyant Density. Buoyant density is determined by **equilibrium density gradient centrifugation** in the ultracentrifuge (Fig. 10-4). Homogeneous nucleic acid accumulates as a symmetric **band** whose width is related to the molecular weight; smaller molecules diffuse more rapidly and so their bands are wider.

Several characteristics of nucleic acids determine their buoyant density. Thus at neutral pH 1) RNA has a higher density than DNA of the same strandedness; 2) single-stranded DNA is denser than double-stranded DNA of the same average base composition; and 3) the density of double-stranded DNA increases linearly with its proportion of G+C (Fig. 10-5,I). Buoyant density is very useful for determining properties of DNA that is available only in small quantities. However, the presence of substitutions on the bases, especially glucosylation (which is found in some viruses), alters the relation between base ratio and density.

Chromatographic Properties. An especially useful column for fractionating DNA is hydroxyapatite, which selectively retains helical nucleic acids since their phosphates are more accessible than those of random coils.

Enzymatic Digestion. Nucleases are widely used for identifying nucleic acids, for determining strandedness and details of structure, and for sequencing. The properties of some of the most widely used enzymes are given in Table 10-2.

DENATURATION; MELTING TEMPERATURE (Tm)

Transient breakage of individual bonds between the paired bases occurs continuously in DNA at physiologic temperatures, but only at high temperatures are enough bonds disrupted simultaneously to cause the collapse or **melting** of the helical structure. A sudden **helix–coil** transition results. This **denaturation,** and its reversal, have provided the powerful analytic tool of nucleic acid

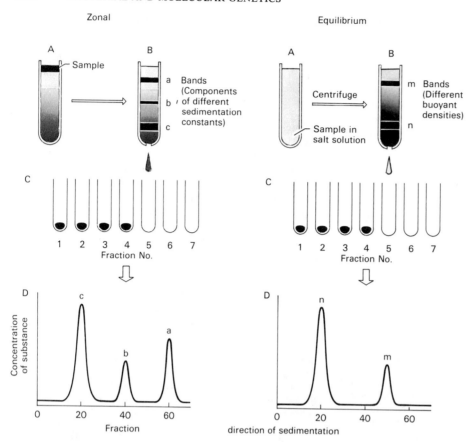

FIG. 10-4. Zonal and equilibrium density gradient centrifugation. In **zonal centrifugation** a linear density gradient is prepared with an inert solute in a plastic centrifuge tube (A): its purpose is simply to prevent convection. The sample is layered at the top of the gradient. After centrifugation the various components of the sample have moved different distances, depending on their sedimentation coefficients, and thus form **bands** (B). Components are separated on the basis of differences in their sedimentation coefficient which depend on particle size, shape, and density.

In **equilibrium density gradient centrifugation** the sample is mixed in the centrifuge tube with a solution of a salt of high molecular weight (e.g., CsCl or Cs_2SO_4) to obtain a mixture of uniform specific gravity similar to that of the nucleic acid (A). Centrifugation causes the salt to form a concentration—and therefore density—gradient; each component of the sample collects in a band at a level where its density equals that of the gradient (B). The band width is inversely proportional to the square root of the molecular weight, because larger molecules have less tendency to diffuse away from the band against the pull of the gravitational field.

In both technics the bottom of the tube is punctured with a fine needle and fractions are collected, in the form of drops, into a series of tubes (C); during this operation the bands maintain their relative positions because the density gradient prevents mixing. Different bands are therefore collected in different groups of tubes. When their contents are analyzed a diagram similar to that of D is obtained.

TABLE 10-1. Lengths of Different Kinds of Nucleic Acid Molecules

	Kilobases per strand	Length (μm)
Transfer RNAs	0.07–0.08	~0.03
5S Ribosomal RNA	0.120	~0.04
16S Ribosomal RNA	1.6	0.5
23S Ribosomal RNA	3.1	1.0
Phage MS2 RNA	4	1.3
Polyoma DNA (cyclic)	5	1.7
Paramyxovirus RNA	21	7.6
Phage λ DNA	47	17.0
Phage T2 DNA	200	68.0
Escherichia coli DNA	4×10^3	1,350 (1.35 mm)
Mammalian, largest strand observed	5×10^3	1,800 (1.8 mm)
Drosophila (total haploid genome)	2×10^5	68,000 (68 mm)
Human (total haploid genome)	3×10^6	10^6 (1 meter)

hybridization (see below), which has been of immense importance.

Denaturation increases the **optical density** of nucleic acids at 260 nm (OD_{260}), the peak of their absorption spectrum. The extent of this **hyperchromic shift** is proportional to the change in the helical content; a complete helix–coil transition causes an increase of about 40%. A plot of OD_{260} as a function of temperature yields an S-shaped curve (Fig. 10-6). The slope is steep for helical nucleic acids, in which the shift from helix to random coil occurs rapidly in the whole molecule; the slope is shallow and variable for single-stranded nucleic acids, in which individual helical segments melt independently at different temperatures, depending on their length and composition.

The **melting temperature (Tm)** is defined as that temperature at which the increase in OD_{260} is 50% of the maximum. For double-stranded DNA the Tm is a strict function of the base ratio (Fig. 10-5,II), since the triple hydrogen bonds of GC pairs require a higher temperature for rupture than the double hydrogen bonds of AT.

TABLE 10-2. Properties of Some Widely Used Nucleases

Enzyme*	Substrate	Point of attack	Products after exhaustive digestion
DNases—endonucleases Pancreatic DNase I	ds or ss		5'-Nucleoside monophosphates and oligonucleotides with 5'-P and 3'-OH
S1 endonuclease	ss		
Restriction nucleases†	ds	Specific sequences	Fragments terminated with characteristic sequences
DNases—exonucleases *E. coli* exonuclease I	ss	3'-OH end	5'-Nucleoside monophosphates
E. coli exonuclease III	ds	3'-OH end	5'-Nucleoside monophosphates and single strands of high molecular weight
Snake venom exonuclease	ds or ss	3'-OH end	5'-Nucleoside monophosphates
Spleen exonuclease	ds or ss	5'-P end	3'-Nucleoside monophosphates
RNases—endonucleases Pancreatic RNase A	ss	3'-End of pyrimidines	3'-Pyrimidine nucleoside monophosphates, purine oligonucleotides terminated by a 3'-OH pyrimidine nucleoside mono-phosphate
RNase T1	ss	3'-End of guanines	Oligonucleotides terminated by a 3'-guanosine phosphate
RNase H	DNA-RNA hybrids	RNA	5'-Nucleoside monophosphates and oligonucleotides
RNases—exonucleases Snake venom exonuclease Spleen exonuclease	Same as for DNase action		

ds = Double-stranded nucleic acid; ss = single-stranded nucleic acid.

*Endonucleases hydrolyze phosphodiester bonds within a polynucleotide chain; exonucleases hydrolyze strands from one end.

†See text: Modifications of DNA

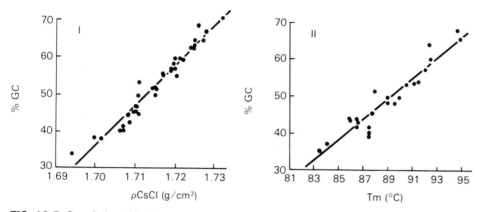

FIG. 10-5. Correlation of the base ratios (% G+C) of different DNAs with the buoyant densities (ρ) in CsCl (I) and with Tm (melting temperature) (II). (Data from Kropinski AM: J Virol 13:753, 1974)

Small cyclic DNA molecules, however, behave atypically (see Cyclic DNA, below).

Denaturation can also be brought about at low temperature by extremely low or high pHs, which disrupt hydrogen bonds (Fig. 10-7); indeed, sedimentation in an alkaline sucrose gradient (pH 12.5) is a standard procedure for isolating single strands of DNA. Denaturation is favored by substances (such as formamide, dimethylsulfoxide, or certain salts, e.g., sodium perchlorate or trifluoroacetate) that interact with the bases, disrupting the bonds between them.

FIG. 10-6. Melting curves of two nucleic acids of the same length but of different structure. Double-stranded DNA of bacteriophage φX174 gives a sharp transition (curve **A**). Single-stranded DNA of the same phage gives a very flat curve **(B)**. **OD$_{260}$** = Optical density at 260nm. **TM** = Melting temperature, i.e., the temperature at which 50% of the hyperchromic effect of heating has appeared. For sample A, Tm is 79° C. (Chamberlin M, Berg P: Cold Spring Harbor Symp Quant Biol 28:67, 1963)

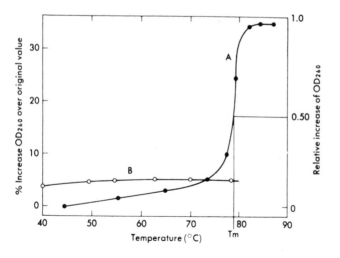

RENATURATION

Heat-denatured DNA tends to re-form duplexes **(renaturation)** on cooling, especially when it is made up of relatively small, homogeneous molecules, as is viral DNA. Renaturation is maximized by **annealing**—i.e., maintaining the denatured DNA at a temperature somewhat below Tm, to melt out the folds—and at a high ionic strength (e.g., 0.3 M NaCl), to decrease electrostatic repulsion between the strands; once **nucleation** is initiated between two short complementary sequences, the helical region extends rapidly in a zipperlike fashion by a process of one-dimensional crystallization. Renaturation of heat-denatured DNA is minimized by rapid cooling **(quenching)**, which permits intrastrand folding to be stabilized before complementary strands have time to pair. Cyclic structures, and the covalent crosslinks between complementary strands produced by some cytocidal agents (see Ch. 11), prevent the melted strands from separating completely and thus promote renaturation.

Renaturation is promoted, at low temperature, by formamide or dimethylsulfoxide, which melt out the folds. In nature, renaturation at physiologic temperature (e.g., in recombination) is facilitated by **helix-destabilizing proteins** which bind to single-stranded DNA.

CYCLIC DNA

Covalently closed, ring-shaped DNA molecules are the genetic material of some viruses and plasmids, as well as of mitochondria and plastids of higher organisms. In these molecules the number of helical turns is fixed, because the strands cannot wind or unwind. This fixation results in several special properties which afford useful methods for separating cyclic DNA from linear DNA.

First, molecules of cyclic DNA are usually twisted **(superhelical)** (Fig. 10-8). This is because the number of

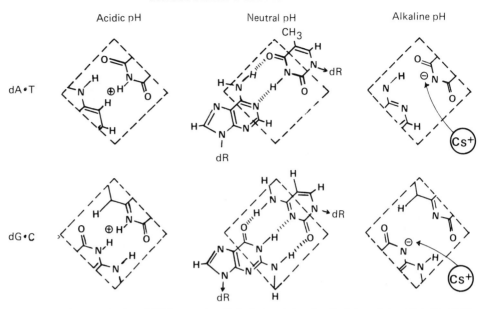

FIG. 10-7. Effect of pH on DNA base pairs. Only the parts involved in interpair bonding are represented. Both acidic (<3) and alkaline (>12) pHs cause ionization (of different kinds at the two pHs) of ring nitrogens which, by abolishing the H bonds and creating a steric hindrance, disrupt the DNA helix. At alkaline pH the negatively charged thymine and guanine bind Cs^+, which therefore increases the buoyant density; hence equilibrium sedimentation in alkaline CsCl yields two bands with DNAs in which one strand is rich in T + G. Short parallel lines = hydrogen bonds. (Modified from McConnell B, von Hippel P: J Mol Biol 50:297, 1970. Copyright by Academic Press Inc. (London) Ltd.)

helical turns per molecule required for stability under the artificial conditions of analysis is greater than during intracellular synthesis. Since the total number cannot be changed in a closed circle, the molecules develop internal stresses on extraction, like rubber rods whose two ends are twisted in opposite directions. These stresses are partially relieved by the twisting, since each twist compensates for the deficiency of one helical turn. The stress is also partly relieved by **local denaturation** of AT rich segments, where the interstrand bonds are weakest. Another property of cyclic molecules is their **faster sedimentation** because they are more compact than linear molecules of the same length. The internal stresses are entirely relieved if a single phosphodiester bond is broken **(nick)**, allowing the free ends of the broken strand to rotate around the intact strand; then a cyclic molecule loses its superhelical twists **(relaxes)**. Nicking is frequent during handling of large cyclic molecules, and prevents recognition of the superhelix.

For cyclic molecules the *degree of twisting varies with the composition of the solution.* For instance, an alkaline pH lowers the required number of helical turns; at the pH at which this number coincides with that of built-in turns, the molecules become untwisted, without strand breakage. Twisting is also strongly affected by acridine dyes (e.g., ethidium bromide) (Fig. 10-9): **intercalation**

of one such planar molecule between adjacent nucleotide pairs almost doubles their normal distances, causing the helix to unwind. Maximum intercalation is about one dye molecule for every two nucleotides in linear molecules, which can freely unwind, but is much less in cyclic molecules, whose unwinding is limited. Since intercalation lowers the buoyant density of the DNA, in the presence of the dye **cyclic molecules have a higher buoyant density** than linear DNA.

Finally, cyclic molecules *do not denature at the temperature expected* from their base ratio because the hindered unwinding prevents extension of the initially melted areas. At much higher temperatures there is a sudden collapse of all the interbase bonds, generating a compact coil with the two strands still irregularly wound around each other. This structure sediments much faster than the native molecule (see Fig. 10-8).

ORGANIZATION OF SEQUENCES IN NUCLEIC ACIDS

NUCLEOTIDE SEQUENCES

The biologic functions of macromolecules depend on minute details of their structure. In DNA the details of particular interest center on the sequence of the bases.

The determination of the sequences has met with for-

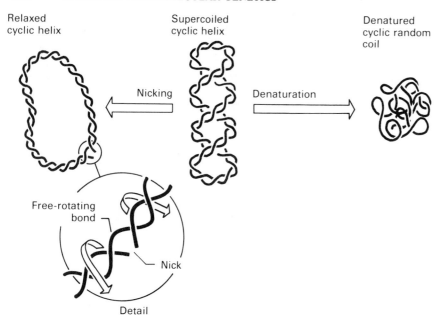

Relaxed cyclic helix

Supercoiled cyclic helix

Denatured cyclic random coil

Nicking

Denaturation

Free-rotating bond

Nick

Detail

FIG. 10-8. Characteristics of cyclic helical DNA. Cyclic DNA molecules in solution usually have fewer helical turns per unit length than linear molecules under the same conditions. For instance, linear DNA of the length of polyoma DNA (Ch. 66) would have 500 helical turns in solution, but native polyoma DNA has 485 built-in turns. Whereas in linear DNA the number of helical turns per unit length can easily be changed by a rotation of one strand around the other, in cyclic molecules this rotation is prevented by the covalent continuity of the two strands. However, cyclic DNAs can change the required helical turns per unit length by developing superhelical turns (or twists). Polyoma DNA acquires 15 twists.

The introduction of a single-strand break **(nick)** in a cyclic molecule restores free rotation of one strand with respect to the other. The result is a disappearance of the superhelical turns. Cyclic molecules become denatured when the bonds stabilizing the helix completely collapse; although the two strands retain the same number of helical turns (which is a topologic invariant), they are locked out of register in the random coil. The entangled strands cannot become renatured when the denaturing conditions are removed.

midable methodologic difficulties. These have recently been overcome, both by methods of subdividing DNA into manageable, defined fragments, and by improvements in the analytic technics. This subdivision is accomplished by 1) **incorporation into a carrier** phage or plasmid, either by natural or by artificial recombination; 2) formation of enormously amplified **clones** by the multiplication of the carrier; and 3) fragmentation of the cloned DNA into characteristic pieces by using **restriction endonucleases** (see Modifications of DNA, below).

Cloning of DNA Segments. Clones containing bacterial sequences can be obtained in vivo by **specialized transduction** (which will be reviewed in Ch. 48). In a powerful, more general approach DNA fragments from any source can be cloned after molecular recombination in vitro with a suitable carrier (a phage or a plasmid). Two methods for splicing the fragment into the carrier are portrayed in Figure 10-10.

With a phage carrier the recombinant DNA is used to infect bacteria in such a way that a single molecule produces a plaque (see The Plaque Method, Ch. 46). The phage recovered from the plaque is a clone, i.e., a genetically homogeneous population derived from a single progenitor. This phage is now grown to a high titer in a large bacterial culture. One can then recover the cloned fragment by extracting the DNA from the phage, cutting the links between inserted segment and carrier by a suitable nuclease, and separating the molecules on the basis of size.

The cloning of molecular recombinants will be further described in Chapter 12.

Analytic Methods. Sequencing of DNA was long possible only through its transcription into RNA, whose secondary structures allow nucleases to attack only certain sites. These pieces are then cut further into different sets of small fragments by the use of RNases of different specificities, such as A and T1 (see Table 10-2). After chromatographic or electrophoretic separation of the frag-

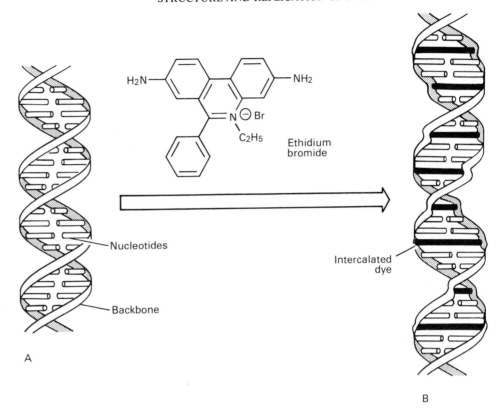

FIG. 10-9. Sketches representing the DNA helix seen from the side (A), and the helix of DNA containing ethidium bromide (B). Both the nucleotides **(white bands)** and the intercalated dye **(dark bands)** are seen edgewise. The phosphate–deoxyribose backbones appear as smooth coils. The intercalation increases the length of the DNA and the stability of the helix; the pitch of the helix and the number of helical turns per nucleotide are decreased.

ments, and chemical determination of their sequences, the whole sequence is deduced from overlaps between the sets.

DNA sequencing is now considerably simpler, through several procedures, such as the methodology presented in Figure 10-11. These advances allow the direct sequencing of a 100-base DNA segment in a few days. Recently similar methods have been developed for RNA segments.

A number of nucleic acids have been completely sequenced, including many tRNAs, some viral RNAs and DNAs, and many fragments of cellular DNAs. The many important conclusions deriving from this work will be reviewed in the appropriate sections of this book.

HOMOLOGY: HYBRIDIZATION

Sequence homology in different nucleic acids can be measured, without actually determining sequences, by the powerful technic of nucleic acid hybridization. In this procedure a nucleic acid of unknown sequence and a reference nucleic acid, denatured and annealed together in solution, form hybrid helices between homologous regions (Fig. 10-12). Even quite short sequences of complementary bases (between 30 and 50) form hybrid helices, allowing the recognition of **partial homology.**

The hybrid helices are usually purified either by enzymatic digestion with nuclease S1 (specific for single-stranded DNA), which removes unhybridized sequences, or by chromatography through hydroxyapatite columns, which retain only double-stranded DNA at 60° C (in 0.12 M phosphate buffer, pH 6.8) and then release it in 0.5 M buffer.

The helices can be tested for **accuracy of pairing** by measuring their Tm. If it is the same as that of native DNA the sequences are accurately matched, while a decrease of 1° C corresponds to approximately 1.5% of unpaired bases.

Filter Hybridization. Hybridization between an unknown labeled RNA and a reference DNA has been very important in studies of transcription. Filter hybridization is widely used: the denatured DNA binds to nitrocellulose filters, which are then dried and immersed in the RNA solution under annealing conditions. When the filters are washed they retain only RNA that has hybridized with the DNA.

In an important application the same principle is combined with agarose gel electrophoresis (Southern technic). DNA fragments of

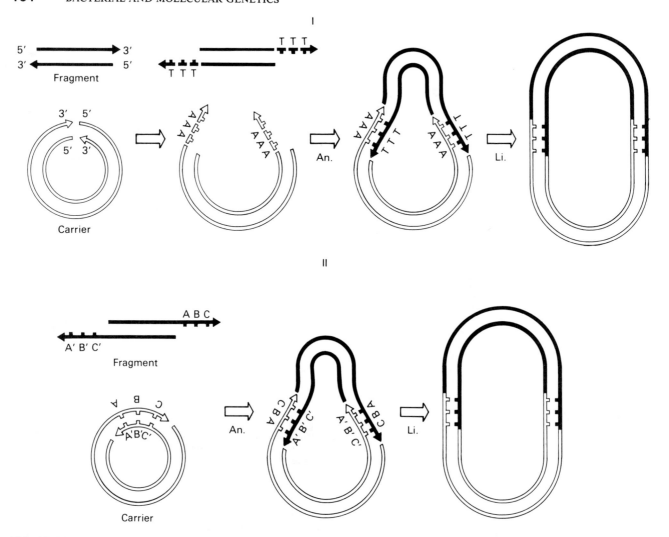

FIG. 10-10. Incorporation of a DNA fragment into a self-replicating carrier, represented as a cyclic DNA. I: Incorporation mediated by the addition of poly-A and poly-T by a terminal transferase. II: Incorporation through cohesive ends formed by restriction endonucleases. A, B, C = bases complementary to A', B', C', respectively; **An** = annealing; **Li** = ligase.

different characteristic lengths are produced by a restriction endonuclease (see Modification of DNA, below), separated by electrophoresis, denatured by immersing the gel in alkali, and "blotted" onto a sheet of nitrocellulose. Hybridization to saturating amounts of a probe of labeled DNA or RNA, followed by quantitative radioautography of the sheet, gives the sizes and amounts of the DNA fragments homologous to the probe. A modification of this technic is applicable to RNAs.

Measurement of the Degree of Homology. The proportion of sequences shared by two nucleic acids with **complementary** sequences (such as two DNAs, or a DNA and a RNA) is determined by **saturation** experiments, with one component labeled and the other unlabeled and in much greater molar proportion (in excess). After an-

nealing, the unhybridized label is removed by appropriate nucleases. The proportion of label protected from hydrolysis measures the proportion of the labeled sequences homologous to the other component (Fig. 10-13,I). Hybridization of RNA to labeled separate strands of isolated DNA fragments produced by restriction endonucleases (Fig. 10-14) greatly improves the significance of these measurements, because it also positions the RNA with respect to the map of restriction sites on the DNA (see Modifications of DNA, below).

The degree of homology between nucleic acids containing stretches of **identical sequences** (such as different RNAs) may also be determined by **competition experiments**, which measure the proportion of a labeled RNA hybridized to DNA in the presence of

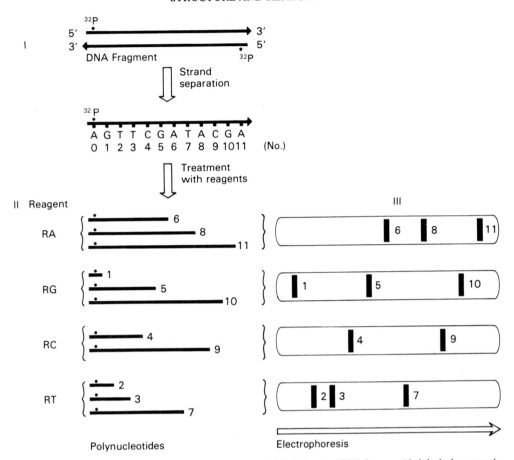

FIG. 10-11. Analytic procedure for sequencing DNA. I. Purified DNA fragment is labeled enzymatically with ^{32}P at both 5′ ends and the strands are then separated. II. Selected strand is subdivided into four batches which are treated with chemical reagents (indicated as **RA, RG, RC,** and **RT**) each capable of reacting with a given base, breaking the polynucleotide chain at the 5′ side of that base. The treatment applied is weak, so that on the average only one break is introduced at random in each fragment, generating in each batch a collection of polynucleotides of all possible lengths, all labeled at the 5′ end and terminated at the 5′ side of the reacted base. Number at the end of each polynucleotide indicates the reacted base. III. Polynucleotides are separated by gel electrophoresis and revealed by autoradiography, yielding the somewhat idealized pattern shown. Numbers on the bands correspond to those on the polynucleotides in I and II. The relative distances of the bands from the bottom give directly the sequence in the 5′ → 3′ direction. The first base is recognized by its label after complete digestion of the DNA to 5′ mononucleotides. (Maxam A, Gilbert W: Proc Natl Acad Sci USA 74:560, 1977)

an excess of another, unlabeled, RNA (Fig. 10-13,II). A third method, under certain circumstances, is to transcribe one RNA into labeled **complementary DNA (cDNA)**, using reverse transcriptase (see Ch. 11). Then the homology of the cDNA to other RNAs can be determined by saturation experiments.

Heteroduplex Analysis. An approach of enormous power is to examine the annealed nucleic acids by **electron microscopy**, without any enzyme treatment. Molecules formed by two heterologous strands **(heteroduplexes)** show double-stranded parts (corresponding to homologous segments) separated by loops, or terminated by tails, of single-stranded DNA (corresponding to nonhomologous segments). Since specific markers are often recognizable in the electron micrograph, it is also possible to position the homologous regions on the maps of the two genomes (Fig. 10-15).

An important offshoot of this technic is the hybridization of RNA to double-stranded DNA in the presence of formamide at a partially denaturing concentration. The RNA binds tightly enough to complementary DNA sequence to displace the other DNA strand, generating a loop (**R loop**) that identifies the region of homology.

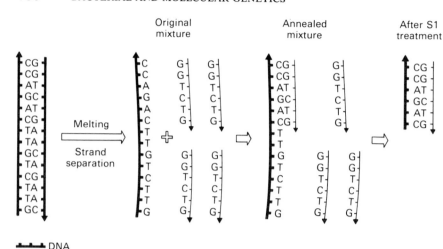

Original mixture — Annealed mixture — After S1 treatment

FIG. 10-12. DNA–RNA hybridization experiment between double-stranded labeled DNA **(heavy lines)** and single-stranded RNA (in excess). After annealing, the mixture is exhaustively digested with nuclease S1 specific for single-stranded DNA. The only label not converted to mononucleotides corresponds to the homologous sequences.

DNA
RNA

Separation of the Strands of Helical DNA. After denaturation of DNA complementary strands can sometimes be separated by gel electrophoresis if they fold differently, or by equilibrium centrifugation if they differ in base composition. Strands differing in the proportion of G and T can be separated in alkaline CsCl gradients, where Cs^+ binds to these two bases (see Fig. 10-7). Another method uses a neutral CsCl gradient containing a **synthetic homopolyribonucleotide,** whose binding to runs of the complementary base increases the buoyant density of the DNA (since RNA is much denser). For instance, polyguanylic acid increases the density of strands containing runs of cytosine. If these approaches fail, strands can still be separated by hybridization to RNA complementary to the DNA **(cRNA)**; the cRNA is made in vitro through the use of RNA polymerase, which, under suitable conditions, transcribes only one strand.

Renaturation Kinetics. The rate of renaturation of a given DNA is proportional to the square of the concentration of DNA molecules (second-order kinetics), since it requires the encounter of two different strands. According to these kinetics the proportion of unrenatured DNA decreases as a function of the product of initial concentration of DNA molecules (Do) × time of reannealing (t), or Dot. In many cases the concentration of DNA molecules is unknown, but concentration of nucleotides (Co) is known. Since Co = concentration of DNA molecules (Do) × length of a molecule (L), or DoL, unrenatured DNA decreases as a function of Cot/L. L is defined as **complexity.**

In the analysis of data the proportion of unhybridized DNA is plotted versus Cot; these "Cot curves" are similar in shape for different DNAs but more complex DNAs require proportionally higher Cot values for the same degree of renaturation (Fig. 10-16). Hence the complexity can be determined from renaturation kinetics.

Renaturation kinetics are also used to determine the **number of copies** of a reference DNA (e.g., of a gene or a virus), available in pure form and of known complexity, in a cell. This number is deduced from the shortening of the time required for 50% renaturation of a constant amount of labeled reference DNA upon addition of DNA extracted from cells. Since the Cot product for the reference DNA must remain constant, the shortening of t is compensated for by an identical increase of Co (Fig. 10-17).

DISTRIBUTION OF SEQUENCES OVER THE GENOME

A general feature of nucleic acid sequences is that in all organisms their *base ratios are not uniform over the genome;* thus, when the DNA is broken into fragments of 2000 to 3000 nucleotides, these range from about 35% to 55% G + C in *E. coli,* and similarly in phage λ and in man. Moreover, heterogeneity on a smaller scale is revealed by local melting, recognizable by electron microscopy, of AT-rich segments at temperatures below the Tm of the whole DNA.

An important feature is the **repetition and clustering** of similar genes. This form of differentiation along the chromosome is seen, for instance, in ribosomal RNAs, which are manufactured in the cell at much higher rates than other RNAs. Hybridization of cellular DNA with either of the main ribosomal RNAs reveals that the total genome contains about 7 copies of these genes in bacteria, 14 in yeast, and 400–700 in amphibians.

There are about 25,000 genes for 5S RNA in the genome of a higher organism, and they are repeated in tandem in several clusters. Clustered also are genes for tRNAs in bacteria and eukaryotes, and for histones in eukaryotes.

Gene clusters have probably evolved by **duplications** through unequal crossing over, according to the scheme of Figure 10-18. Thus *E. coli* genes for 5S RNA are constituted of two quite similar portions, clearly arisen by duplication (Fig. 10-19); and many proteins contain repetitions of very similar sequences (e.g., the immunoglobulins). Duplications can also be large; in fact, the order of related genes of the *E. coli* circular chromosome reveals repetitions

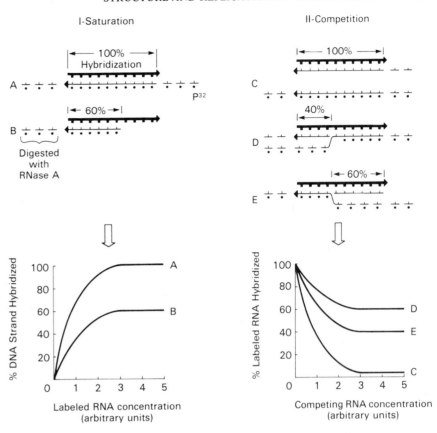

FIG. 10-13. Results of hybridization experiments between denatured DNA immobilized on nitrocellulose membrane filters and labeled RNA. I. In the **saturation experiment** in RNA excess each curve results from a number of determinations with the same amount of DNA on a filter and increasing concentrations of RNA in the hybridization mixture. In each case, annealng is carried out until the number of counts hybridized approaches a maximum (usually within hours). The filters are then treated with RNase A specific for single-stranded RNA (see Table 10-2), to remove the unhybridized RNA. Radioactivity is then measured and the proportion of DNA hybridized is calculated from the bound radioactivity and the known specific activity of the RNA. Horizontal part of the curves gives the saturation level of hybridization. Curves **A** and **B,** obtained with the same DNA and two different RNAs, show that RNA A can completely saturate one of the DNA strands (i.e., the maximum, since only one DNA strand is transcribed into RNA), whereas RNA B saturates only 60% of a strand. Thus, RNA A transcribes all sequences of the DNA, and RNA B 60% of them. II. In the **competition experiment,** constant amounts of DNA on filters are hybridized, first to increasing amounts of an unlabeled unknown RNA and then to saturating amounts of a labeled known RNA. If both RNAs share sequences with the DNA, the unlabeled RNA will compete with the hybridization of the labeled one. The amount of labeled RNA hybridized without competing RNA is taken as 100%. Curve **C** shows complete competition caused by an RNA identical to the labeled RNA (or perhaps containing additional sequences). Curves **D** and **E** show the results obtained with use of partially homologous competing RNAs, which allow partial hybridization by the labeled RNA; the competing RNA D has 40% of the sequences of the labeled RNA and RNA E has 60%.

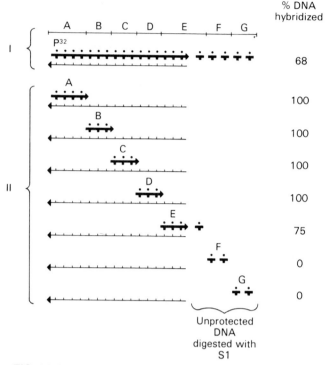

% DNA hybridized

I: 68

II:
A: 100
B: 100
C: 100
D: 100
E: 75
F: 0
G: 0

Unprotected DNA digested with S1

FIG. 10-14. Determination of the degree of homology between a labeled DNA and an unlabeled RNA (RNA in exesss) using separate DNA strands from restriction endonuclease fragments. The percent DNA hybridized is the proportion of the DNA protected against S1 digestion. In the hybridization to whole DNA (I) the total extent of hybridization (about 70%) is known, but the position of the homologous part is undetermined, whereas it is determined very precisely using fragments (II). Also, the accuracy is greater using fragments because for most of them protection is either 100% or nil.

at 90° and 180° intervals, suggesting that large parts of the genome have been duplicated.

Renaturation experiments have shown that, in contrast to prokaryotic DNA, eukaryotic DNA contains extensive repeat sequences. The resulting kinetics are complex, because they derive from the interactions of sequences of different lengths, degrees of homology, and numbers of copies. Britten has shown that with proper analysis these curves define the lengths (complexities) of certain classes of genes and their number of copies (Co/L, where Co is the amount of DNA and L the complexity in a given class) (Fig. 10-20). In the mouse, for instance, 10% of the whole genome contains about 10^6 repeats of sequences each 200–300 nucleotides long; these renature at low Cot values. About 30% is present in about 100 repeats; and 60% has essentially unique sequences, which hybridize at very high Cot values.

In some animal species the repeat sequences consist of **tandem repetitions** (i.e., one after another in succession and in the same direction) with base ratios different from those of the bulk DNA. On equilibrium centrifugation in CsCl they give rise to one or more **satellite** bands which are rather sharp (since their DNA is homogeneous) and distinct from the broad band of the rest of the DNA (which is heterogeneous in base ratios). Tandem repeats are best studied by using restriction endonucleases (see Modifications of DNA, below), which cut each repeat identically; the fragments so generated afford precise information. For instance, mouse satellite DNA has a basic repeat 240 base pairs long; in other species repeats of 10–15 base pairs have been observed.

Electron microscopy has provided **direct visual evidence** for repetitive DNA in animal cells. On renaturing partially or totally denatured DNA fragments Thomas obtained a high proportion of circular structures (Fig.

FIG. 10-15. I, II. Heteroduplex DNA molecule produced by annealing a strand of colicinogen E1 DNA (see Ch. 12) to the complementary strand of an in vitro DNA recombinant formed by colicinogen E1 DNA with a fragment of *Drosophila melanogaster* DNA which contains three genes for 5S RNA. One of the genes has been hybridized to 5S RNA covalently bound to ferritin, a large iron-containing molecule, opaque to electrons. III. Heteroduplex shows an **insertion loop** of single-stranded DNA corresponding to the *Drosophila* DNA. The A:T stem derives from the annealing of the poly-A and poly-T linkers used in constructing the recombinant DNA (see Fig. 10-10). Marker = 1 Kb. (Hershey ND, et al: Cell, 11:585, 1977. Copyright © MIT, published by the MIT Press)

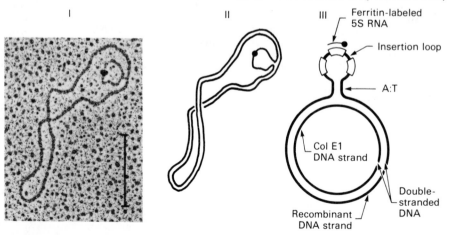

I   II   III   Ferritin-labeled 5S RNA

Insertion loop

A:T

Col E1 DNA strand

Double-stranded DNA

Recombinant DNA strand

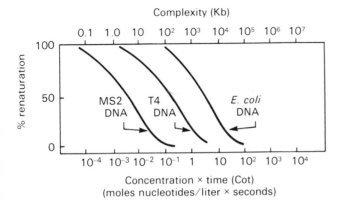

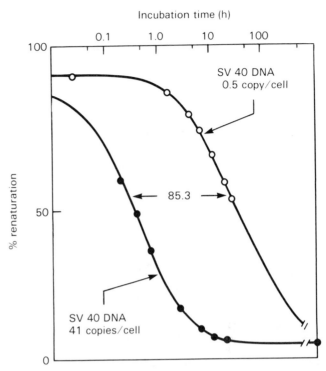

FIG. 10-16. Renaturation kinetics of double-stranded nucleic acids. The nucleic acids were fragmented to a constant length of about 500 nucleotides, denatured, and reannealed, using different concentrations and reannealing times. The proportions renatured were determined by chromatography through hydroxyapatite columns, which selectively retain helical nucleic acids.

The curves give the proportion of renatured nucleic acid versus the product of concentration of denatured nucleotides × time (*Cot*). The curves were obtained with nucleic acids that do not contain appreciable proportions of extensively repeated sequences: double-stranded RNA of phage MS2, DNA of phage T4, and *E. coli* DNA. These curves have similar shapes, reflecting the second-order kinetics of renaturation (unrenatured fraction = $1/(1 + KCot)$, where K is a constant), and are displaced in proportion to the molecular lengths (complexities). The lengths, determined from the *Cot* value at 50% renaturation, are similar to those determined by other means (see Table 10-1). Therefore, the *Cot* value of 50% renaturation can be used to determine the length of unique sequences in any double-stranded nucleic acid. (Modified from Britten RJ, Kohne DE: Science 161:529, 1968. Copyright 1968 by the American Association for the Advancement of Science)

FIG. 10-17. Kinetics of renaturation of SV40 DNA at various concentrations. In this reconstruction experiment different amounts of sheared ^{32}P-labeled SV40 DNA were added to a constant amount of sheared unlabeled salmon sperm DNA to reach the concentrations indicated in the figure. An 82-fold increase in SV40 DNA concentration caused an 85.3-fold decrease in the time required for reaching 50% renaturation of the labeled DNA, showing good agreement between observed and expected results. (Modified from Gelb LD et al: J Mol Biol 57:129, 1971. Copyright by Academic Press Inc. (London) Ltd.)

10-21), which are produced by hybridization between repeat sequences, a few hundred bases long, **interspersed among unique sequences** every three kilobases or so.

Another special feature of essentially all DNAs is the presence of **inverse** repeated sequences (**palindromes**), interspersed with other DNA. On denaturation and reannealing the complementary sequences on the same strand often pair, generating double-stranded hairpins with or without a single-stranded loop (see Fig. 11-20).

Meaning of Unique and Repeat Sequences. Unique sequences probably are the structural genes for cellular proteins. Highly repetitious sequences, either tandem or interspersed, must have other functions, since they are not present in RNA. They may have been produced by **genetic drift,** i.e., by mutations, and also by unequal crossovers between short homologous sequences in DNA, not subject to evolutionary pressure. (In fact they can be generated from random base sequences by a computer simulation based on these assumptions.) Some

repeat sequences may perform **specific functions:** for instance, tandem repeats around the centromere may be involved in binding of the spindle in mitosis or meiosis; and palindromes may be targets for regulatory proteins, since their rotational symmetry may match the symmetry of dimeric proteins. However, some repeat sequences appear to move around on the genome, like the transposable elements (Ch. 11) of bacteria and plasmids.

EVOLUTIONARY CONSIDERATIONS

The various kinds of duplication discussed above show that **tandem reduplication of the genetic material** (see Fig. 10-18) is a basic evolutionary mechanism. Moreover, duplications underlie the expansion and divergence of the genome that is characteristic of evolution. Thus with one gene of an initial duplicate pair continuing to provide an essential function, the other can change rap-

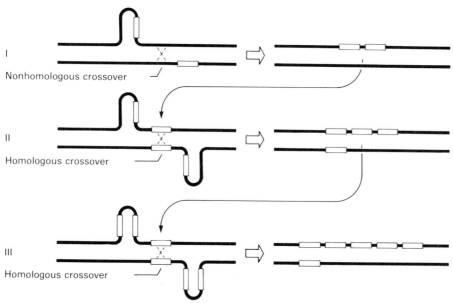

FIG. 10-18. Origin of gene clusters. The clusters are thought to arise by a rare initial nonhomologous recombination (I; see Genetic Recombination, Ch. 11) and to increase in size by subsequent unequal homologous recombinations (II, III), which are more frequent. The latter process can also lead to elimination of some of the replicas of the gene. Whether the clusters grow or shrink during evolution depends on their selective value.

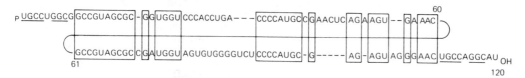

FIG. 10-19. Sequence of *E. coli* 5S ribosomal RNA, showing the large homologous segments of the two halves **(boxes)**. **Dashes** indicate gaps introduced in the sequence to maximize homology. Two short sequences of the two ends **(underlined)** are also equal. The numbers indicate the order of nucleotides from the 5′ end. (Brownlee GG et al: J Mol Biol 34:379, 1968. Copyright by Academic Press Inc. (London) Ltd.)

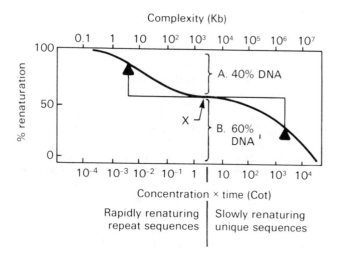

FIG. 10-20. Renaturation kinetics of calf thymus DNA. The experimental procedures were as in Fig. 10-16. The renaturation curve splits up into two curves separated by the inflexion at X. Curve **A,** consisting of 40% of the DNA, contains sequences corresponding to a length of 10 Kb, as shown by the position of its midpoint **(arrow)**; curve **B,** consisting of 60% of the DNA, contains sequences of the length of 3×10^6 Kb **(arrow).** Since there are 6×10^6 Kb of DNA in a calf thymus cell, component B must be composed of independent sequences all different from each other, present in different genes or different chromosomes. In contrast, component A, which corresponds to 2.4×10^6 (i.e., 40% of 6×10^6) Kb per cell, contains highly repeated sequences ($2.4 \times 10^6/10 = 2.4 \times 10^5$ per cell). (Modified from Britten RJ, Kohne DE: Science 161:529, 1968. Copyright 1968 by the American Association for the Advancement of Science)

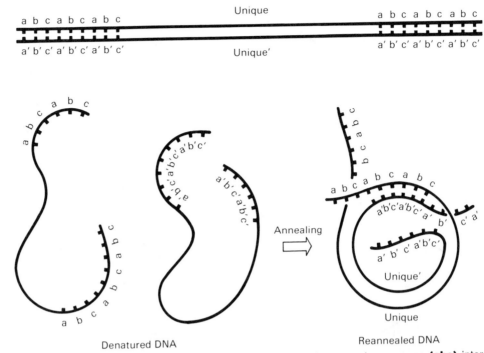

FIG. 10-21. Circularization of amphibian DNA due to tandem repeat sequences **(abc)** interspersed with unique sequences. Primed symbols are complementary to the unprimed ones. The DNA was denatured and reannealed. The rings, recognizable by electron microscopy, had circumferences between 0.6 ad 15 K long and their sum was equal to up to ⅓ of the total DNA. Therefore the interspersed repeat sequences are present throughout the genome. (Modified from Thomas CA Jr et al: J Mol Biol 51:621, 1970. Copyright by Academic Press Inc. (London) Ltd.)

idly in sequence and in function. The rate at which alterations accumulate depends on mutation rate, genetic drift, and selection.

Curiously enough, among eukaryotes *evolutionary divergence in the sequences of structural genes is slow:* indeed, ribosomal 5S RNA and cytochrome C show marked resemblance even between bacteria and man. It may be that the ability of these essential proteins and RNAs to change is severely limited by functional requirements. In contrast, *the highly repetitive DNA sequences in eukaryotes are different even in closely related species.* It is not clear whether the correlation between their divergence and speciation is causal or accidental.

The *uniformity of certain highly reiterated genes* (such as those for ribosomal RNA) *within an organism* is most remarkable. This uniformity implies either that some unknown mechanism counteracts mutations within such a set of multiple genes in an organism, or that there is strong selection against organisms carrying such mutations.

DNA REPLICATION

SEMICONSERVATIVE REPLICATION

The Watson–Crick structure of DNA suggests a model for its reproduction called **semiconservative,** whereby each strand (i.e., half-molecule) serves as a template for, and combines with, a new strand (Fig. 10-22), thereby being conserved in the new duplex. By taking advantage of the wonderful symmetry built into the DNA structure, this model accounts in a natural way for the transmission of genetic information from parent to progeny.

The semiconservative model was verified by Meselson and Stahl in a classic experiment, presented in Figure 10-23. The most notable feature is the total conversion of the DNA, after one generation, into a form with hybrid density, i.e., containing **half old and half new DNA.** Since the density remained unchanged after the DNA was fragmented into short pieces, the old and new strands were complementary rather than joined end to end.

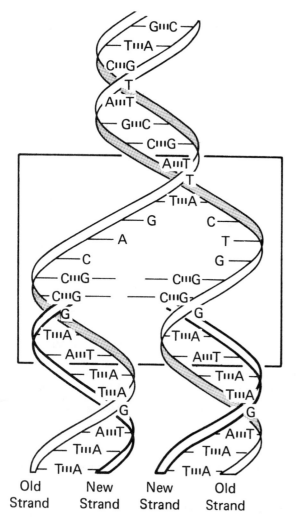

Old Strand   New Strand   New Strand   Old Strand

FIG. 10-22. Replication of DNA. **Box** indicates the growing fork. (Modified from Watson JD: Molecular Biology of the Gene, 2nd edition. Menlo Park, California, Benjamin/Cummings, 1970)

ORGANIZATION OF THE REPLICATING CHROMOSOME

Sequential Bidirectional Replication. Replication of DNA occurs at a **growing point (fork)** that moves linearly from an **origin** to a **terminus**, usually in both directions **(divergent replication).**

Genetic evidence for sequential bidirectional replication was obtained in an *E. coli* culture synchronized by blocking DNA synthesis for a time with nalidixic acid. Upon removal of the inhibitor a new wave of DNA replication began at the origin in all cells. Samples of the cultures were then briefly exposed at various times to the mutagen nitrosoguanidine, which is especially effective on replicating genes. At each time peaks of mutagenesis were found for genes at equal distance, on either side,

from a point at about 80 min on the standard genetic map (see Ch. 9) (Fig. 10-24). **Autoradiography** supplies additional evidence for bidirectional replication in bacteria (Fig. 10-25) and in animal cells, and electron microscopy does the same in bacteriophages (see Fig. 47-13).

Unidirectional replication occurs more rarely, e.g., in the cyclic DNAs of mitochondria and of some plasmids.

Replicative Intermediate. The *E. coli* genome in the process of replication was visualized by Cairns, using radioautography. The observations clearly showed, as was suggested earlier by the variation of Hfr strains in conjugation (Ch. 9), that *the* E. coli *chromosome is cyclic throughout replication* (Fig. 10-26), and that the ring is doubled between the origin and a growing point.

Symmetry of Replication. The scheme of semiconservative replication is symmetric because the two strands have equal general roles. A different, **asymmetric** mode of replication, called the **rolling circle,** has been proposed by Gilbert and Dressler (Fig. 10-27). The two schemes postulate two different types of replicative intermediate: the **symmetric intermediate** corresponds to Cairns' radioautograph; the **asymmetric intermediate** is observed in electron micrographs of certain replicating viral DNAs. The asymmetric model is mostly observed during the conversion of cyclic into linear molecules, which takes place in these viruses.

Units of Replication (Replicons). The above observations show that DNA replication begins at special DNA sites which are identically located in related bacterial strains. Topographically, therefore, it is possible to distinguish between **chain initiation** and **elongation.** The two functions appear to require **different proteins,** because when the temperature is raised some *E. coli* mutants with a temperature-sensitive block in DNA synthesis cannot initiate replication but can complete already initiated chains, while others cannot complete chain growth.

Some of the proteins involved appear to be specific for bacterial DNA synthesis, since mutations in the corresponding genes do not prevent synthesis of bacteriophage DNA in the same cell. Conversely, some temperature-sensitive mutations of the F plasmid (Ch. 9) can prevent its replication at 40° C without affecting that of cellular DNA. These findings define, in both anatomic and functional terms, units of DNA replication (**replicons**), each with a **unique origin, initiating protein, and termination.**

Characteristics of Replicons. Replicons vary in number and size in different organisms. Thus, in *E. coli* there is one replicon per chromosome; it contains 4.5×10^6 nucleotide pairs and is completely replicated in about 45 min. Mammalian cells, containing 30,000–40,000 replicons with about 10^5 necleotide pairs each, have

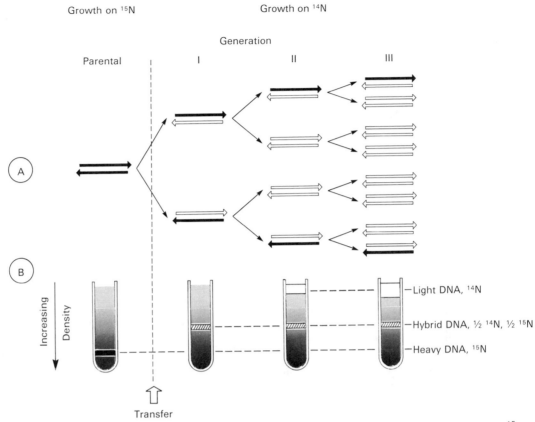

Growth on ^{15}N Growth on ^{14}N

Generation

FIG. 10-23. Verification of semiconservative replication in Meselson and Stahl's experiment. A culture of *E. coli* is grown with $^{15}NH_3$ (heavy isotope) and then transferred to $^{14}NH_3$ (light isotope). A. Expected changes in the constitution of the DNA; **heavy lines** = the heavy parental DNA strands; **thin lines** = the light newly synthesized DNA strands. B. Results obtained from DNA extracted from samples taken at various times after transfer and banded by equilibrium density gradient centrifugation in CsCl. The results confirmed the predictions: before transfer the DNA is heavy **(HH);** one generation after transfer it is hybrid **(HL,** i.e., one strand heavy, the other light); and at subsequent generations a constant amount of DNA per culture is hybrid and an increasing proportion is light **(LL).**

a much slower replication rate; these replicons are **linearly arranged** in the same DNA molecule, and their radioautographs after ^3H-thymidine incorporation show that they are *not replicated at the same time.* Thus in animal cells the time required for replicating a chromosome depends not only on the rate of DNA chain elongation but also on the number of active replicons. In rapidly multiplying bacteria new initiations can occur on a chromosome before the previous fork has reached the terminus (see Fig. 14-23, Ch. 14).

Role of Membrane in Replication. Electron microscopy shows that the bacterial DNA is always attached to the cell membrane, sometimes via a **mesosome** (Ch. 6). The attachment probably occurs at the **growing point,** because after a short exposure to a radioactive precursor the DNA just synthesized is found preferentially associated with the membrane fraction when the cell is lysed. Furthermore, the membrane fraction also contains the DNA polymerase III (see Enzymes Involved in Replication, below), which is directly involved in replication. The

origin or the **terminus** may also be permanently attached to the membrane. The cell membrane thus has an essential role in DNA replication.

The cell membrane also determines the regular **segregation** of the daughter molecules to the daughter cells at the end of replication. Electron microscopy has revealed that this occurs when the mesosome is split by localized growth of the membrane. Some plasmids (Ch. 12) segregate together with the chromosome, suggesting that they are attached to the same mesosome. The cell membrane, therefore, appears to have a role **like that of the spindle** that segregates the chromosomes of higher cells at mitosis.

MECHANISMS OF REPLICATION

Initiation and chain growth must be considered separately. The **initiation complex** appears to include the cell

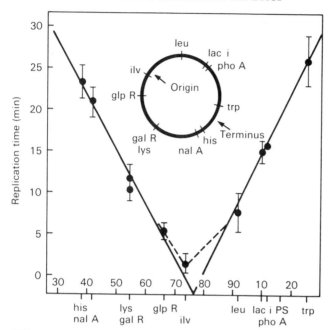

FIG. 10-24. Replication of various *E. coli* mutant markers determined as the moment of highest frequency of induction of mutations by nitrosoguanidine, as a function of the marker position on the genetic map. Time is measured from the initiation of replication by the removal of nalidixic acid. The bar attached to each point is the standard deviation resulting from a series of measurements. (Modified from Hohlfeld R, Vielmetter W: Nature New Biol 242:130, 1973)

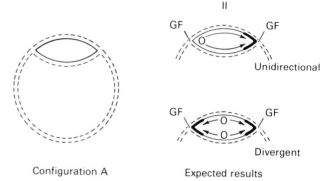

FIG. 10-25. Radioautographic evidence for divergent replication of the bacterial chromosome. Synchronous cultures of germinating *Bacillus subtilis* spores at the start of the second round of replication were first exposed to a low concentration of ³H-thymidine, and then for a brief time, before extraction of the DNA, to a high concentration. I. When extracted the chromosome was in configuration A, where the **continuous lines** indicate radioactive atoms (irrespective of grain density). II. The different positions of the parts with high grain density under the two models of unidirectional or divergent replication are shown; **heavy lines** represent the heavily labeled DNA. Experimental results clearly conformed to the expectations of the divergent replication model. **O** = Origin; **GF** = growing fork; **T** = terminus. (Modified from Gyurasitis EB, Wake RG: J Mol Biol 73:55, 1973. Copyright by Academic Press Inc. (London) Ltd.)

membrane, the DNA template, and new proteins. RNA polymerase and ribonucleoside triphosphates generate a temporary **primer** for DNA synthesis. Initiation is therefore blocked by inhibitors of either RNA synthesis or protein synthesis. A **requirement for RNA synthesis** arises because DNA-synthesizing enzymes (see Enzymes Involved in Replication, below) can add nucleotides only to a preexisting primer, whereas RNA polymerase can initiate chains without primer (see Transcription of DNA, Ch. 11). The RNA appears to be transiently incorporated in the DNA chain. The **proteins** specifically involved in initiation may be defined by initiation mutants of *E. coli*. A special DNA-binding protein, whose synthesis precedes initiation, may be required for binding the growing point to a cell membrane site.

Important in **chain growth** are the nature of the **synthesizing enzymes** and the **structural changes** of the DNA. The discovery by Kornberg of a DNA polymerase seemed to solve the first problem, but later studies showed that replication is carried out by other enzymes discovered more recently. Moreover, it became apparent that these enzymes could not account by themselves for semiconservative replication, for this process requires the essentially simultaneous replication of both antiparallel strands, while the enzymes synthesize DNA only in the $5' \rightarrow 3'$ direction. After a vain search for an enzyme able to synthesize DNA in the opposite direction, the solution came from the study of the **structures present at the growing point.**

THE REPLICATING DNA

The essential events of DNA replication take place at or near the **replicating fork,** where the unreplicated molecule and the two daughter molecules meet (see Fig 10-22). The newly synthesized DNA is recognized by its radioactive labeling upon very brief exposure of cells (5–10 sec) to ³H-thymidine. In many bacteria and phages this DNA contains single-stranded segments (**Okazaki segments**) a few thousand nucleotides long, which are recognized by their low sedimentation value (10S) in alkali gradients (Fig. 10-28). If the short labeling is followed by a chase (i.e., further synthesis without a label) the originally incorporated label becomes covalently bound to the large, double-stranded DNA.

The Okazaki segments are evidently *synthesized in the $5' \rightarrow 3'$ direction,* since ³H-thymidine incorporated during a short pulse is rapidly released by exonuclease I, which attacks single-stranded DNA from the 3′-OH end. Okazaki segments start with a short polyribonucleotide **primer,** which is removed before synthesis is completed. After completion, the segments are connected by **ligase,**

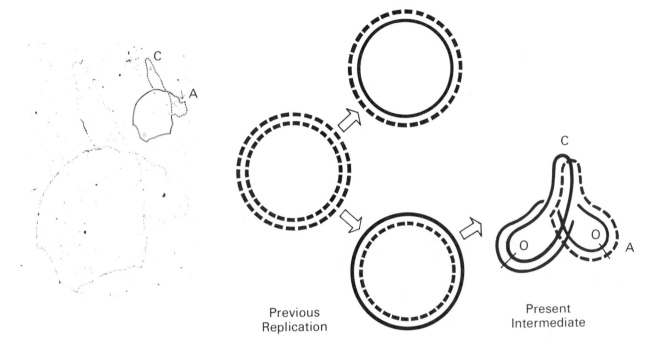

Previous
Replication

Present
Intermediate

FIG. 10-26. Radioautograph of *E. coli* DNA after labeling for about 1.8 generations. It shows a replicating chromosome with three branches **(A, B,** and **C).** Of these B is probably newly replicated, because it has the highest grain density. A possible interpretation is given to the **Dashed lines** = unlabeled strands; **continuous lines** = labeled strands; **O** = possible origin of replication. This type of replicative intermediate is referred to as **theta** (*θ*) or **Cairns intermediate.** For comparison see Fig. 10-27. (Cairns J: Cold Spring Harbor Symp Quant Biol 28:43, 1964)

since they accumulate unjoined in ligase-deficient mutants. These findings support the replication scheme of Figure 10-29, in which the 5'-ended **lagging strand** is synthesized in an overall 3' → 5' direction by discontinuous synthesis of Okazaki segments in the opposite direction.

The 3'-ended chain (the **leading strand**) is evidently synthesized continuously (i.e., without segments), since in *E. coli* carrying an integrated λ prophage, which is replicated as a part of the cell chromosome (see Lysogeny, Ch. 48), Okazaki segments extracted from the cells were found to hybridize to **only one** of the λ DNA strands.

ENZYMES INVOLVED IN REPLICATION

The complex enzymology of DNA replication has been considerably clarified by Kornberg and by Hurwitz, who studied the replication of the DNA of small bacteriophages (Ch. 47) in vitro, using as the source of enzymes extracts from wild-type *E. coli* cells or from mutants defective in DNA replication. **More than ten different proteins** are involved, of which some are well characterized.

Three DNA **polymerases** (I, II, and III) have been isolated from *E. coli;* enzymes with similar properties exist in other microorganisms and in animal cells. In vitro these enzymes catalyze the reaction:

$$n(\text{dATP, dGTP, dTTP, dCTP}) \xrightarrow[\text{DNA template}]{\text{Enzyme}}$$

deoxynucleoside triphosphates

$$(\text{dAMP, dGMP, dTMP, dCMP})n + n(\text{P·P})$$

polynucleotide $\quad$ pyrophosphate

All three enzymes extend a **primer** polynucleotide (DNA or RNA, paired to template DNA strand) by linking nucleotides to its free 3' end; a hairpin of the template serves as primer in some cases (see Fig. 49-14, Ch. 49). The polymerases also have a **3' → 5' exonuclease** activity, i.e., they remove nucleotides from the same end at which they elongate. This is thought to be an **editing function,** able to remove a nucleotide that has been incorrectly paired with the template. In addition *polymerase I is a 5' → 3' exonuclease*, active on double-stranded DNA or on an RNA–DNA hybrid helix. This activity allows the enzyme to carry out **nick translation,** i.e., to elongate the free 3' end at a nick while removing nucleotides from the 5' end (Fig. 10-30). The exonuclease

Symmetric Replication

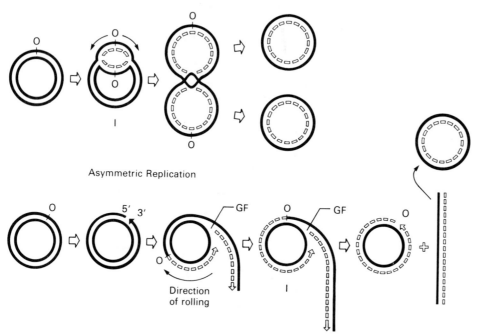

Asymmetric Replication

FIG. 10-27. Symmetric and asymmetric replication of cyclic DNA. In the symmetric replication the newly synthesized strands are not covalently connected to the preexisting strands. The replicative intermediate (I) is theta-shaped (θ). The product is two identical cyclic molecules. In the asymmetric replication, one of the old strands is broken at the origin and its 3'-end is continued by new synthesis, while an unconnected second new strand is synthesized as the complement of the 5'-ended old strand, which in the process is detached from its old complement. The replicative intermediate is sigma-shaped(σ). The elongation of the 3'-ended strand causes the circle to roll if the fork is held fixed. When the origin has completed a revolution, the extended strand is set free by another single-strand break at the origin, releasing a cyclic and a linear molecule. The latter may subsequently be cyclized. **O** = Origin; **GF** = growing fork.

can be separated by proteolytic cleavage from the polymerase.

Of these enzymes *polymerase III is the main replicating enzyme* for bacterial DNA, since mutations in its gene (*E. coli Pol C*) are lethal. *Mutations eliminating the 5' → 3' exonuclease activity of polymerase I are also lethal,* whereas those affecting the synthesizing activity are not. In the former case the Okazaki segments are not connected. It therefore appears that in bacteria the main function of polymerase I is to remove the RNA primer from the segments and to **fill the gaps,** by nick translation. Since this enzyme dissociates from the template after each catalytic step, as soon as the RNA is removed and the gap is filled **ligase** can seal the residual nick, stopping the process. Nick translation by polymerase I is also essential for **repairing damages in DNA** produced by ultraviolet light or by alkylating agents (Ch. 11).

Primases are required in many systems for synthesizing a primer RNA; in addition to the regular transcriptase (see Transcription of DNA, Ch. 11), other RNA polymerases are used in the replication of viral DNAs. **RNase H** (see Reverse Transcription, Ch. 11), which breaks down RNA in RNA–DNA hybrids, is needed in some systems to degrade the primers.

Polynucleotide ligases, which join polynucleotide chains, are present in all cells and some are specified by certain phages. They are essential in DNA replication, which is prevented by mutations that abolish ligase activity. *E. coli* ligase restores a phosphodiester bond break (a **nick**) in double-stranded DNA between a 3'-OH and 5'-P end. In fact, this is the discontinuity left when a gap in one strand is filled by extending the 3' end with polymerase; ligase then reconstitutes the continuity of the strand. The restoration is perfect, as shown by the *return of biologic function in transforming DNA.* The ligase of phage T4 can join two evenly terminated DNA duplexes **(blunt end ligation).**

Endonucleases may participate in DNA replication, because mutations in phage genes specifying endonucleases inhibit replication. A specific **swivel enzyme** may be responsible for unwinding the helix during replica-

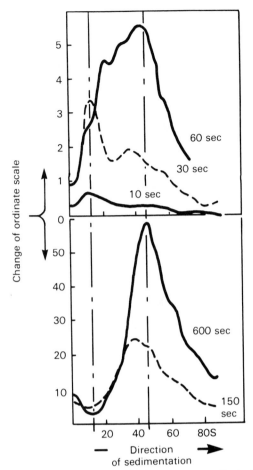

FIG. 10-28. Sedimentation in alkaline sucrose gradient of *E. coli* DNA labeled by progressively longer puses of ³H-thymidine. The length of the pulse, at 20° C, is indicated near each curve. After very short pulses a large proportion of the label has a sedimentation constant of 10S (Okazaki segments). The amount of radioactivity in that fraction increases until 30 sec, then remains essentially constant (considering the change of scale), as would be expected of an intermediate. In contrast, the 45S peak, which under the conditions of extraction contains the bulk of the DNA, continues to increase. The label, therefore, flows from the nucleotide pool into the Okazaki segments and from them into long DNA. (Modified from Okazaki R et al: Cold Spring Harbor Symp Quant Biol 33:129, 1968)

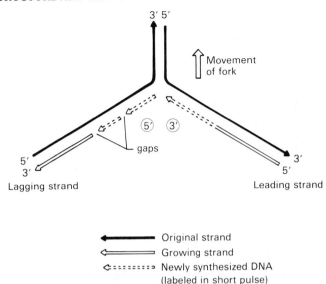

FIG. 10-29. General characteristics of synthesis at a DNA fork. Synthesis always occurs in the 5′ → 3′ direction. The 5′-ended growing strand (**circled,** left branch) is replicated by discontinuous synthesis, whereas the 3′-ended growing strand (**circled,** right branch) is elongated by continuous synthesis. Gaps are later closed by ligase. Synthesis of the lagging strand is usually behind because it must wait until that of the leading strand has progressed sufficiently for the synthesis of a new Okazaki segment.

FIG. 10-30. Nick translation of a cyclic DNA by polymerase I. The combination of exonuclease action attacking the 5′ end and polymerase action elongating the 3′ end leads to a complete replacement of the old with a new strand. By using ³²P-labeled nucleoside triphosphates as precursors, radioactive strands of high specific activity can be prepared by this method; they are essential analytic tools (**probes**) for tracing the DNA by hybridization in cells or after fractionation in gels.

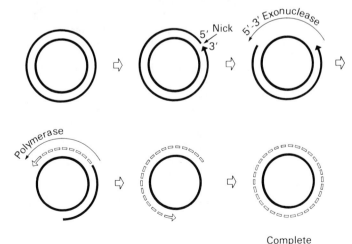

tion, since enzymes converting cyclic helical DNA from the twisted to the relaxed state by a series of nicking–closing reactions are present both in bacteria and in the nuclei of animal cells. Twists are removed one by one by breaking one of the strands, allowing unwinding, and then resealing. Conversely, a **gyrase,** by introducing negative supercoils ahead of the fork, makes the DNA helix easier to open. The essential role of gyrase is shown by the sensitivity of bacterial DNA replication to **nalidixic acid** and other antibiotics that inactivate the enzyme.

I

II

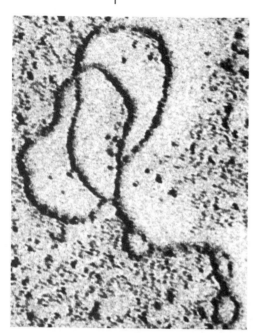

FIG. 10-31. Replication of cyclic SV40 (Ch. 66) DNA without permanent breaks in the parental strands, suggesting the participation of a "swivel" enzyme. The molecule seen in the electron micrograph (Kleinschmidt's technic) (I) is about half replicated; its constitution is shown in II, where **continuous lines** represent the parental strands and **dashed lines** the progeny strands. Continuity of the parental strands is shown by the tight supercoiling of the unreplicated part: had a strand been nicked, the whole molecule would have relaxed. (Sebring ED et al: J Virol 8:478, 1971)

Helix-destabilizing proteins (see Renaturation, above) are required for DNA replication, in order to keep locally denatured (single-stranded) DNA from folding at random, and also to protect the DNA from nucleases.

The requirement for so many different proteins in DNA replication is probably related to **changes of the template** during replication. At first the helical DNA is opened locally by the regular, rifampin-sensitive RNA polymerase (see Transcription of DNA, Ch. 11), which probably recognizes specific sequences where it builds the initial RNA primer. DNA synthesis in the hydrophobic environment of the membrane is then initiated, in a series of steps and in cooperation with other proteins, by DNA polymerase III, which is itself rather hydrophobic. As synthesis progresses the DNA strands are dissociated ahead by gyrase and helix-destabilizing protein; the latter, as the leading strand is elongated, continues to protect the lagging strand. RNA primers are laid down on this strand by another, rifampin-resistant primase (specified by E. coli

gene *dan G*), perhaps able to recognize the template strand covered by the unfolding protein; Okazaki segments are then synthesized. The process is finally completed by the nick translation function of polymerase I and by ligase.

Unwinding of the helix must occur as the fork proceeds, and the mechanism has long been a puzzle. If it occurred through rotation of the whole molecule, which is often of great length, formidable hydrodynamic problems would arise. The difficulty is even more serious in the replication of cyclic DNAs in which the parental strands remain intact during replication (Fig. 10-31). These difficulties were solved by the discovery of single-stranded nicks in replicating DNA, and later of the swivel enzyme. The nicks act as swivels, since the two broken ends can freely rotate with respect to each other. Both mechanisms allow local unwinding near the growing fork without rotation of the whole molecule.

MODIFICATIONS OF DNA; RESTRICTION ENDONUCLEASES

Modifications of DNA (methylation or glycosylation of bases) have great significance as **species-specific** (and strain-specific) **markers,** which affect the survival of a

DNA when it enters another organism. They were discovered through the study of the selective destruction of certain bacteriophage DNAs (see Fig. 47-19B) in some

TABLE 10-3. Examples of Some Target Sites and of Various Cutting Modes of Restriction Endonucleases

Enzyme and bacterial species	Target site*	Type of ends produced		

Eco RI
Escherichia coli†
	5'-G⌄A A T T C-3'
	    | | | | |
	3'-C T T A A⌃G-5'
		-G-3'
		 |
		-CTTAA-5'
			5'-AATTC-
			     |
			3'-G-
				5'-ended
				tails

Hae I
Haemophilus aegyptius I
	5'-Pu G C G C⌄Py-3'
	    | | | | | |
	3'-Py⌃C G C G Pu-5'
		-PuGCGC-3'
		 |
		-Py-5'
			5'-Py-
			  |
			3'-CGCGPu-
				3'-ended
				tails

Hpa I *Haemophilus parainfluenzae*	5'-G T T⌄A A C-3' | | | | | | 3'-C A A⌃T T G-5'	-GTT-5' | | | -CAA-3'	3'-AAC- | | | 5'-TTG-	flush ends

*Arrows are sites of cuts.

†Carrying an *fi*⁺ R factor (see Ch. 12)

Pu = Purine; Py = pyrimidine

---

special hosts (**restriction**), which was traced to special **restriction endonucleases;** but they also can cause recombination at the cleaved sites. The purification of several such enzymes, by Arber and by H. Smith, revealed that they *recognize specific short sequences in DNA;* the same sequences in the cell's DNA are not attacked because some of their bases are **modified by methylation** shortly after incorporation.

These nucleases were also shown by Nathans to be extremely useful **tools for mapping DNA,** and for studying its functions, because they cut DNA into precise, moderately short fragments. Soon many similar enzymes were discovered in various bacteria. Some have been shown to carry out **both restriction and modification** of DNA, and perhaps most of them do so.

The basis for this dual function in the *E. coli* K and B enzymes is the presence of **three different subunits.** The effects of mutations in the respective genes show that *each subunit performs a different function:* one is responsible for recognizing the DNA sequence (**target**), the second for the restricting activity, and the third for the modifying activity.

Target Sequences. The outstanding feature of restriction endonucleases is their recognition of fixed **target sequences.** In most cases these are **palindromes,** three to six base pairs long (Table 10-3). The symmetry of the target is probably related to the symmetry of the enzyme (Fig. 10-32). Recognition of the target is abolished by mutations that change a base pair within the target, or by methylation of target bases. Most enzymes (called **class 2**) produce **two symmetric cuts** within the target, either coincidental or staggered, one in each strand; the symmetry of the cuts is probably related to that of the target–enzyme complex (Fig. 10-32). Staggered cuts leave ends that are terminated by short complementary single

FIG. 10-32. Postulated structure and mode of action of the Hpal enzyme of *Hemophilus parainfluenzae*. I. Target site. II. Postulated enzyme–target complex. (Modified from Kelly TJ, Smith HO: J Mol Biol 51:393, 1970. Copyright by Academic Press, Inc. (London) Ltd.)

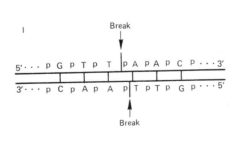

strands **(cohesive ends);** at physiologic temperatures these ends are too short for stable pairing, and a pair of staggered cuts is in fact a complete scission of the duplex. However, cohesive ends are useful for certain applications (see below). Some enzymes (called **class 1**) cut at some constant distance from the target (which is not a palindrome). The ends of the DNA fragments generated by these enzymes are variable, because they are not re-lated to the sequence of the target, in contrast to the constant ends generated by class 2 enzymes.

Large DNAs contain many targets for any restriction enzyme. The *frequency of a given target is mostly related to its length,* since the sequence of bases in DNA is often essentially random. Thus the expected frequency of a four-base target is about 1:200 bases, and of a six-base target 1:4000 bases. The observed frequencies are usually in agreement with the expectations, but *some targets are much rarer,* and have apparently been selected against.

Restriction endonucleases find important **applications** in sequencing of DNA (see Organization of Sequences in Nucleic Acids, above), in mapping of genes in viruses (Fig. 10-33) (see also Genetic Mapping, Ch. 11 and Fig. 66-5), and in the formation of DNA recombinants in vitro (Fig. 10-34; Ch. 12). The last-named use is based on the ability of cohesive ends to form fairly stable duplexes at low temperature; the duplexes can then be sealed co-valently by a ligase.

EVOLUTIONARY CONSIDERATIONS

The biologic consequences of restriction are very evident in the destruction of DNA of bacteriophages or plasmids attempting to infect a new host. Hence restriction may have evolved as a **defense of bacteria** against these parasites. Moreover, not only bacteria but some plasmids and phages specify restriction endonucleases; hence the enzymes may also be involved in the **competition among parasites,** and the rarity of certain restriction endonuclease targets may be the result of selection against an enzyme produced by a wide-spread parasite.

Recombinations at restriction sites have been observed in bacterial cells, which suggests that restriction enzymes contribute to evolution by promoting the spread of genes by plasmids and viruses. On the other hand, these enzymes also interfere with bacterial conjugation (e.g., *E. coli* B and K have enzymes that attack each other's DNA).

Modification must have evolved as a **defense against autorestriction.** This concept is supported by observations on the interaction of purified *E. coli* B restriction enzymes. With modified (MM) and unmodified (00) phage fl DNA, and artificial heteroduplexes modified in one strand only (0M), the enzyme cleaved 00 DNA rapidly but left 0M and MM uncleaved. In addition, 0M was rapidly modified to MM, whereas 00 was modified only slowly. Thus the enzyme respects its own DNA (MM), rapidly destroys foreign DNA (00), and rapidly modifies its own replicating DNA (0M, since the newly made strand is at first unmodified).

SIGNIFICANCE OF NUCLEASES FOR NUCLEIC ACID DYNAMICS

Nucleases were first discovered as enzymes that destroy nucleic acids and hence appeared to play a nutritional role. More recently, however, many far more specialized functions have emerged, often associated with the ability to **recognize selected sites** on a nucleic acid. The restric-

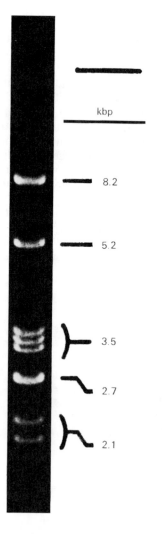

FIG. 10-33. DNA fragments generated by the action of the restriction endonuclease Hind III (from *Hemophilus influenzae*) on Adenovirus type 2 DNA. Because the targets are at fixed locations each DNA molecule produces an identical set of fragments. These have different lengths and so they can be separated as bands by gel electrophoresis, the shorter fragments migrating more rapidly. The bands are visualized here by soaking the gel with ethidium bromide, which intercalates between base pairs in double-stranded DNA; the bound dye is then revealed by its flourescence when illuminated with ultraviolet light. Kbp=Kilobase pairs. (Courtesy of I Verma)

kbp

8.2

5.2

3.5

2.7

2.1

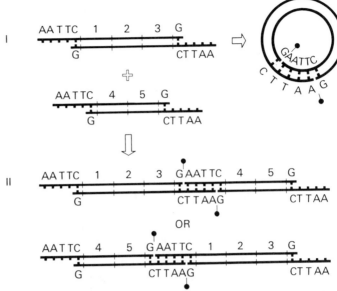

FIG. 10-34. Use of cohesive ends produced by endonuclease Eco R1 of *E. coli* to join DNA fragments by pairing of complementary bases. I. A single fragment can cyclize. II. Two or more fragments can join together in all possible arrangements (only two of the four possible ones are shown). ⌐ = Missing phosphodiester bond after pairing; ligase can form this bond.

tion endonucleases are one example. Another is found in the **nucleases that process RNAs.** Contrary to DNA, whose final sequences are entirely determined at synthesis, RNA is often (perhaps regularly) synthesized as a precursor of the final molecule, containing extra sequences that must be removed **(processing).**

Two processing endoribonucleases are known in bacteria, RNase III and RNase P. **RNase III** has been identified as the enzyme that cuts the primary T7 mRNA, and the *E. coli* rRNA precursor, at specific sites, generating functional segments. In vitro this enzyme completely hydrolyzes double-stranded RNA, and in processing it may recognize and cut local folds. **RNase P** has been identified in both bacteria and animal cells as the enzyme that cleaves dimeric tRNA precursors to produce the two functional tRNAs. Since different precursors are cleaved at different sequences, the enzyme, which contains an RNA component essential for function, probably recognizes the secondary and tertiary structure of the precursors.

Highly specific nucleases are involved in **repair of DNA damages** (Ch. 11). Thus phage T4 specifies an endonuclease that cuts only at the 5′ end of lesions induced by ultraviolet light, and also two exonucleases that excise the altered DNA beginning at that point. Other nucleases recognize **damages by x-rays;** others, **apurinic sites.**

Nucleases of little site-specificity are essential in fundamental processes such as **DNA replication, recombination,** and **mRNA degradation.** It seems that they are precisely **regulated by other proteins.** An example is *E. coli* **exonuclease V,** which is formed by the combined products of the *rec B* and *rec C* genes and is essential for recombination. Its activity is regulated by the product of the *rec A* gene and by the helix-destabilizing protein: if the functions of these proteins are prevented by mutations the exonuclease destroys the replicating phage DNA.

SELECTED READING

BOOKS AND REVIEW ARTICLES

ALBERTS B, STERNGLANZ R: Recent excitement in the DNA replication problem. Nature 269:655, 1977

ARBER W: DNA modification and restriction. Prog Nucleic Acid Res Mol Biol 14:1, 1974

DRESSLER D: The recent excitement in the DNA growing point problem. Annu Rev Microbiol 29:525, 1975

GEIDER K: Molecular aspects of DNA replication in *Escherichia coli* systems. Curr Top Microbiol Immunol 74:21, 1976

KORNBERG A: DNA Synthesis. San Francisco, Freeman, 1974

LEHMAN IR, UYEMURA DG: DNA polymerase I: essential replication enzyme. Science 193:963, 1976

PERRY RP: Processing of RNA. Annu Rev Biochem 45:605, 1976

ROBERTS RJ: Restriction endonucleases. Crit Rev Biochem 4:123, 1976

SZYBALSKI W, SZYBALSKI EH: Equilibrium density gradient centrifugation. Prog Nucleic Acid Res Mol Biol 2:311, 1971

TOMIZAWA J, SELZER G: Initiation of DNA synthesis in *E. coli*. Ann Rev Biochem 48:999, 1979

WATSON JD: Molecular Biology of the Gene, 3rd ed. New York, Benjamin, 1975

SPECIFIC ARTICLES

BAUER W, VINOGRAD J: Interaction of closed circular DNA with intercalative dyes. I. The superhelix density of SV40 DNA in the presence and absence of dyes. II. The free energy of superhelix formation in SV40 DNA. J Mol Biol 33:141, 1968; 47:419, 1970

BOTHWELL LM, STARK BC, ALTMAN S: Ribonuclease P substrate specificity: cleavage of bacteriophage φ80-induced RNA. Proc Natl Acad Sci USA 73:1912, 1976

BOYER HW, CHOW LT, DUGAICZYK A, HEDGPETH J, GOODMAN HM: DNA substrate site for the *Eco RII* restriction endonuclease and modification methylase. Nature 244:40, 1973

CECH TR, HEARST JE: Organization of highly repeated sequences in mouse main-band DNA. J Mol Biol 100:227, 1976

COHN RH, LOWRY JC, KEDES LH: Histone genes of the sea urchin (*S. purpuratus*) cloned in *E. coli*: order, polarity and strandedness of the five histone-coding and spacer regions. Cell 9:147, 1976

GEFTER ML, HIROTA Y, KORNBERG T, WECHSLER JA, BARNOUX C: Analysis of DNA polymerases II and III in mutants of *E. coli* thermo-sensitive for DNA synthesis. Proc Natl Acad Sci USA 68:3150, 1971

GELLERT M, MIZUUCHI K, O'DEA MH, NASH HA: DNA gyrase: an enzyme that introduces superhelical turns into DNA. Proc Natl Acad Sci USA 73:3872, 1976

KLEID D, HUMAYUN Z, JEFFREY A, PTASHNE M: Novel properties of a restriction endonuclease isolated from *Haemophilus parahaemolyticus*. Proc Natl Acad Sci USA 73:293, 1976

KUROSAWA Y, OGAWA T, HIROSE S, OKAZAKI T, OKAZAKI R: Mechanism of DNA chain growth, XV RNA–linked DNA pieces in *Escherichia coli* strains assayed with spleen exonuclease. J Mol Biol 96:653, 1975

LEE CS, THOMAS CA JR: Formation of rings from *Drosophila* DNA fragments. J Mol Biol 77:25, 1973

LOUARN JM, BIRD RE: Size distribution and molecular polarity of newly replicated DNA in *Escherichia coli*. Proc Natl Acad Sci USA 71:329, 1974

MESELSON M, STAHL FW: The replication of DNA in *Escherichia coli*. Proc Natl Acad Sci USA 44:671, 1958

MOISE H, HOSODA J: T_4 gene 32 protein model for control of activity at replication fork. Nature 259:455, 1976

OLIVEIRA BM, HALL ZW, ANRAKU Y, CHIEN JR, LEHMAN IR: On the mechanism of the polynucleotide joining reaction. Cold Spring Harbor Symp Quant Biol 33:27, 1968

SCHEKMAN R, WEINER JH, WEINER A, KORNBERG A: Ten proteins required for conversion of φX 174 single-stranded DNA to duplex form *in vitro*. J Biol Chem 250:5859, 1975

SMITH GP: Evolution of repeated DNA sequences by unequal crossover. Science 191:528, 1976

SOUTHERN EM: Base sequence and evolution of guinea pig α-satellite DNA. Nature 227:794, 1970

STARK BC, KOLE R, BOWMAN EJ, ALTMAN J: Ribonuclease P, an enzyme with an essential RNA component. Proc Natl Acad Sci USA 75:3717, 1978

TAIT RC, SMITH DW: Roles for *E. coli* polymerases I, II, and III in DNA replication. Nature 249:116, 1974

VOVIS GF, HORIUCHI K, ZINDER ND: Kinetics of methylation of DNA by a restriction endonuclease from *Eschichia coli* B. Proc Natl Acad Sci USA 71:3810, 1974

WETMUR JG, DAVIDSON N: Kinetics of renaturation of DNA. J Mol Biol 31:349, 1968

WICKNER W, BRUTLAG D, SCHEKMAN R, KORNBERG A: RNA synthesis initiates in vitro conversion of M13 DNA to its replicative form. Proc Natl Acad Sci USA 69:965, 1972

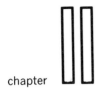

chapter **11**

ORGANIZATION, ALTERATION, AND EXPRESSION OF THE GENETIC INFORMATION

ORGANIZATION OF GENETIC INFORMATION

UNITS OF INFORMATION

The existence of units of genetic information (**genes**) was first inferred from observations in pea plants, on **phenotypic traits** or characters, such as color or height, that differ in different strains. These traits were shown to segregate in breeding experiments through **recombination.** The alternative traits are due to alternative (**allelic**) forms of a gene, and cytologic evidence in higher organisms showed that each gene has a specific locus in a chromosome. Combined genetic and biochemical studies of Beadle and Tatum in the mold *Neurospora* later led to the concept of **one gene—one enzyme,** which provided an invaluable foundation for molecular genetics.

A new momentum was given by studies of **fine-structure genetics,** initiated by Benzer in phages and by Yanofsky in bacteria. Recombinants occurring even at extremely low frequencies could be selected in these organisms, and they showed that recombination could occur not only between genes but between many mutational sites within a gene. The definition of a gene was then given a more precise, structural basis: **one gene—one polypeptide chain** (Fig. 11-1). (Genes also specify discrete RNA molecules that are not translated into polypeptides: rRNA and tRNA.) Moreover, the boundaries of a gene could be identified operationally by a **complementation test** (Fig. 11-2), which is applicable even when the gene product is unknown. In this test, two mutant genomes of similar phenotype are brought into the same cell. When the two mutations eliminate different gene products the wild-type allele of each gene, present in the other chromosome, still functions, and so the heterozygous cell is prototrophic (**intergenic complementation**). Moreover, since many oligomeric enzymes are made up of different polypeptide chains (monomers) which assemble spontaneously, extracts of two mutants lacking the same enzyme can regain activity when mixed, if the mutations alter different polypeptides. In general, mutants that are altered at different sites within the same gene do not complement each other; exceptions will be discussed in Chapter 13, under intragenic complementation.

The gene is now known to consist of a **chain of nucleotides,** whose transcripts are read without interruption in polypeptide synthesis, each group of three successive nucleotides (**codon**) specifying an amino acid. Since this reading can begin at any base, three different **reading frames** are possible. In bacteria each locus seems to have only one correct reading frame (which is specified by the mechanism of initiation), and gene and polypeptide are **colinear.**

Recent developments have revealed additional, more complex relations of a polypeptide sequence to a genetic locus. In viruses, which evidently maximize the use of their small genome, **overlapping genes** (Fig. 11-3, I) have been recognized. In another refinement, genes in eukaryotic cells often contain **intervening sequences** that are not utilized for specifying the polypeptide chain, because they are removed (spliced) from the mRNA (see Fig. 49-18) before it is translated (Fig. 11-3, II). Hence genes may be made up of **noncontiguous pieces** (at least eight in an ovalbumin gene); and some pieces might be read in different frames. In a third variant mechanism, found in some viruses, a long polypeptide chain is cleaved to yield separate fragments, each with a separate function (Fig. 11-3, III). *Each fragment then is a unit of information* at the polypeptide level, and the corresponding gene is *the collection of codons specifying that fragment.* Such a gene would not be bounded by an initiator and a terminator codon.

The definition of a gene in terms of its functional polypeptide product is only minimal, because additional sequences also contribute to its action. Thus with polypeptides that are inserted into or secreted through a membrane the functional peptide is often carved out of a larger **preprotein** (Fig. 11-3, III). More generally, the mRNA sequences that specify peptides are flanked by sequences that ensure a reproducible initiation and termination of translation; and the gene DNA contains still other flanking sequences that act as signals for initiation and termination of transcription. This complexity may explain why only one frame is ordinarily read at any one locus: overlapping genes must interweave not only the coding sequences but also the punctuation signals, and these cannot be readily varied.

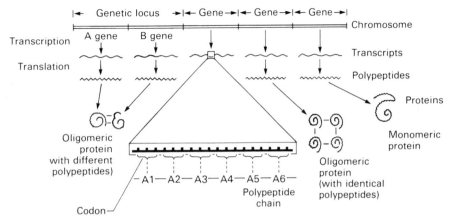

FIG. 11-1. Units of genetic information. This figure represents the simple forms of genes and their one-to-one relationship to polypeptide chains. A_1, A_2, etc. = aminoacids.

FIG. 11-2. Complementation test showing that the r_{II} locus of bacteriophage T4 contains two genes. *E. coli* K_{12} (λ) cells (represented as rectangles), which will not support multiplication of an r_{II} mutant, were infected simultaneously by two r_{II} mutants in three different experiments. The phage genomes are represented by straight lines and the mutations by dots. The two mutations were previously localized by recombination tests either in the A segment (**I**), in the B segment (**II**), or one in each segment (**III**). In **I** only B protein presumably is made, and in **II** only A protein, since in neither is there viral multiplication (except for a small proportion of cells in which wild-type phage has been produced by genetic recombination). In **III**, however, viral multiplication takes place in all cells, so presumably both A and B proteins are made. The A and B segments, therefore, are two separate genes.

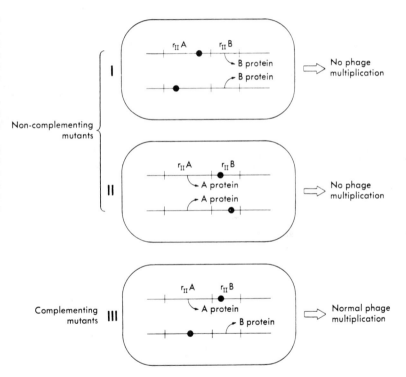

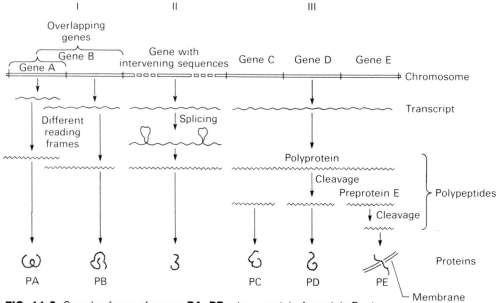

FIG. 11-3. Complex forms of genes. **PA, PB,** etc = protein A, protein B, etc.

DNA DAMAGES AND THEIR REPAIR

Various physical or chemical agents in the environment cause structural changes in DNA. If these are not repaired they may be recognized as a surviving mutation or as lethal damage. The latter may be due to a lethal mutation or to an interference with transcription or replication.

FIG. 11-4. Thymine dimer resulting from ultraviolet irradiation of DNA. Thymine-cytosine and cytosine-cytosine dimers are also formed.

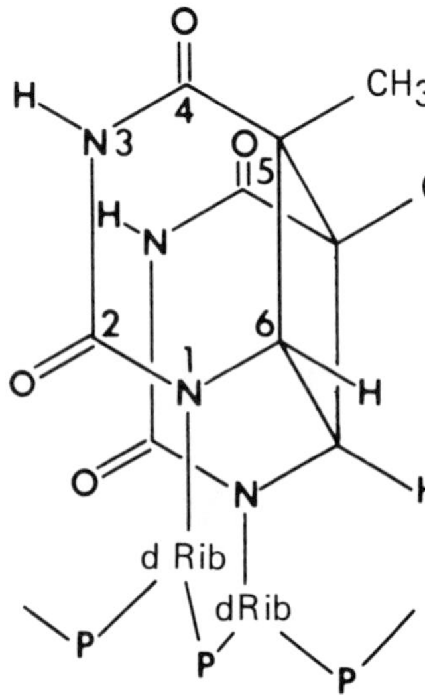

The DNA alterations are of four main types:

1) **Dimerization of two adjacent pyrimidines** on the same strand under the action of ultraviolet (UV) light. The pyrimidines become connected by a cyclobutane ring after activation of the 5:6 double bond (Fig. 11-4). UV light also causes other, less frequent pyrimidine changes. The photoproducts are chemically stable and are readily recovered by chromatography of hydrolysates of the altered nucleic acid. Dimers block replication, which can later resume beyond the dimer, leaving a gap in the newly synthesized strand. Transcription is also arrested, causing the **functional inactivation of genes** situated beyond the dimer in the same transcription unit.

2) **Chemical alterations of bases** by deaminating or monofunctional alkylating agents (see below).

3) **Crosslinks** between the two strands, mainly by UV and by bifunctional alkylating agents (e.g., mechlorethamine [nitrogen mustard], psoralen in the presence of light, mitomycin). These are revealed by the immediate renaturation, on cooling, of heat-denatured DNA, whose complementary strands are held together and can snap back in register.

4) **Breaks** of one or both strands, mostly by ionizing radiations, through the formation of highly reactive radicals in the water surrounding the DNA, or by the decay of radioactive atoms (e.g., ^{32}P, ^{3}H) incorporated in the DNA. **Single-strand breaks** are revealed by a decreased sedimentation rate of the DNA in an alkaline gradient

(which separates the strands) and **double-strand breaks** by a similar decrease in a neutral gradient.

REPAIR OF DNA DAMAGE

Although the mutations induced by some of the above agents are important for evolution, too big a rate is a burden, and several specialized repair mechanisms have been evolved. These mechanisms are also involved in the important phenomena of **recombination** and **mutation.**

Direct repair causes a return to the original structure, and is only applicable to pyrimidine dimers. A more general form, **excision repair,** takes advantage of the double-strandedness of DNA (Fig. 11-5): a segment of the strand, containing the alteration, is excised; the gap is then filled by new error-free synthesis (copying the intact strand) and is finally closed by ligase. If excision repair fails the altered DNA is replicated; then the new DNA strand may contain an incorrect base, generating a mutation. Alternatively, the new strand may have a gap at the level of the original alteration; a third mechanism may then come into action to eliminate the gap (**post-replication repair).**

Single-strand breaks may be repaired directly by ligase or by excision repair. Double-strand breaks (i.e., two single-strand breaks close to each other on different strands) cannot be repaired since the continuity of the duplex is not maintained.

Direct Repair: Photoreactivation. This phenomenon was discovered through the curious observation that parallel platings of a bacterial suspension exposed to UV light yield highly variable numbers of colonies, depending on how long the inoculated plates remain on the laboratory bench before being placed in a dark incubator. The cause is an enzyme that combines specifically with pyrimidine dimers: on irradiation with light of the long UV or short visible region the enzyme cleaves the dimers, restoring the original pyrimidine residues. Photoreactivation occurs with UV-damaged RNA as well as DNA. UV-irradiated virus particles do not contain the photoreactivating enzyme, but in infected cells the cellular enzyme can act on the viral nucleic acid.

Photoreactivation increases, sometimes enormously, the viable fraction of UV-treated bacteria or viruses, as shown by mutants defective in this repair. After maximal photoreactivation other, less frequent, damages still remain.

Excision Repair (Fig. 11-5). This type of repair, also known as **dark reactivation,** is observed if an UV-irradiated suspension of bacteria is stored for a few hours in the cold, or in an inadequate medium, before being allowed to resume growth; the fraction of colony-forming cells increases appreciably during this preincubation. The first step is **incision,** which is recognized from a reduction in the sedimentation velocity of denatured DNA in an alkaline gradient; it is carried out by specific endonucleases that recognize specific abnormalities in the DNA. The second step is **excision:** the altered bases are released as acid-soluble material, usually by the action of DNA polymerase I, leaving a short gap (about 20 nucleotides long). This gap is filled by the same enzyme and then ligated.

Crosslinks between complementary strands are also removed by excision repair, but the process must be repeated twice, first to remove the altered base on one strand, then that on the other strand (Fig. 11-6).

Different endonucleases recognize pyrimidine dimers, or pairing irregularities (**mismatches;** for instance, two purines at the same position on opposite strands), or the **apurinic acid residues** that result from the elimination of certain abnormal bases (such as alkylated purines or U deriving from the deamination of C) by specific DNA glycosidases.

Complete repair of a dimer takes about 2 min at 32° C; in a cell containing many dimers they are repaired one by one, often sequentially along the chromosome. This minimizes the possibility of entanglement or recombination between various gaps.

Excision repair is hindered by several chemicals, including caffeine (whose mutagenic effect may thereby be explained), and by the incorporation of 5-bromouracil instead of thymine in DNA.

Postreplication Repair. In mutant bacteria incapable of excision repair or photoreactivation the presence of even hundreds of pyrimidine dimers per cell may still not abolish viability. In these cells the newly replicated DNA

FIG. 11-5. Steps in excision repair of pyrimidine dimers (**PD**).

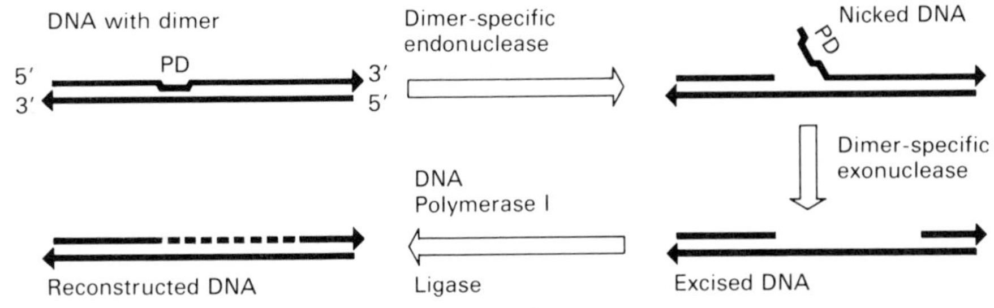

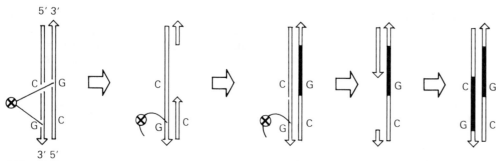

FIG. 11-6. Removal of a crosslink produced by a bifunctional alkyl residue (**X**) between two guanine residues (**G**) in opposite strands. The process consists of a sequence of degradation and resynthesis, as in repair of UV damage. **Heavy lines** = new synthesis.

strands are initially in relatively short pieces (corresponding to the intervals between dimers), as shown by slow sedimentation in an alkaline gradient. Upon incubation of the cells for an hour or so, the pieces become longer and reach the normal length, i.e., the strands are reconstructed. The mechanism, recognizable by electron microscopy of DNA, is **crossovers** between the two daughter molecules, whose gaps do not usually coincide (Fig. 11-7).

The SOS System. Another mechanism is the **gap-filling error-prone repair,** which, although lethal or detrimental in some cells, rescues others from an otherwise lethal damage. The gaps in the new strands are filled by synthesis that does not copy the template, thereby causing many mutations. Error-prone repair becomes active only some time after DNA damage has occurred, because its enzymes must be **induced** by the DNA that was altered by incomplete replication; therefore it will not operate on gaps produced during excision repair, which are rapidly filled.

DNA damages leading to error-prone repair can be induced not only by UV but also by thymine starvation or by some chemicals affecting DNA—for instance, the most potent bacterial mutagen, nitrosoguanidine (see Action of Chemical Mutagens, below). Error-prone repair is part of a system of reactions to persistent DNA damage, inducible as a last resort to save the cells (hence named the SOS system); it also causes induction of prophages and inhibition of cell division, with resulting filamentous growth. Once the SOS system is induced in a cell, it will act also on DNA introduced in that cell; thus a virus exposed to UV light will display a higher survival (and enhanced mutation frequency) after infecting UV-treated cells.

The activation of the SOS system seems to depend on **proteolytic cleavage** of the repressors of genes for the SOS functions: protease inhibitors prevent the activation of the system (see cI Repressor Inactivation, Ch. 48). Apparently the protease is specified by the *rec A* gene (also involved in normal recombination), which usually is repressed by the product of the *lex A* gene, and therefore there is expressed at a low level. Some effector formed from damaged DNA (perhaps a DNA breakdown product) enhances the activity of the *rec A* protease; increased breakdown of the *lex A* product causes, by positive feedback, hyperproduction of the protease, and consequently increased cleavage of the SOS repressors. Mutations, probably in the *rec A* operator (*tif⁻*), or in *lex A,* make *rec A* constitutive, and thus elicit the SOS response without DNA damage.

FIG. 11-7. Repair of pyrimidine dimers by recombination. When DNA-containing dimers (**PD**) replicate, the new strands (**heavy lines**) have corresponding gaps. Multiple crossovers (**Co**) between the two daughter helices restore an intact molecule. Evidence for this mechanism was obtained by growing *E. coli* cells in heavy isotopes (^{13}C, ^{15}N). The cells were irradiated and transferred to light isotopes (^{12}C, ^{14}N) in the presence of ^{3}H-thymidine. Shortly after irradiation the label was in short strands of light density; after repair it was in long strands of density intermediate between light and heavy. Sonication to break the strands produced short pieces, some of light, some of heavy density. (Rupp WD et al: J Mol Biol 61:25, 1971. Copyright by Academic Press Inc. (London) Ltd.)

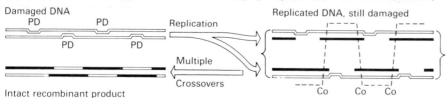

FACTORS DETERMINING SENSITIVITY TO DNA-DAMAGING AGENTS

Sensitivity to the lethal effects of UV light is measured by the slope of the survival curve (see Appendix, below). Organisms with double-stranded DNA are usually very resistant, and repair is very efficient owing to the duplication of the genetic information in the complementary strands: as many as 2500 dimers may be needed for a single lethal hit. In contrast, single-stranded DNA is very sensitive in the dark: it cannot be repaired by dimer excision or by recombination, but only by gap-filling error-prone repair after replication (Fig. 11-8). However, organisms with double-stranded DNA may become much more sensitive as a result of mutations affecting repair.

The *phr⁻* mutants lack photoreactivation, which normally eliminates up to 90% of UV damages. Mutations in genes *uvrA,B,C,* which together specify the dimer incision enzyme, affect excision repair. Double mutants *polA⁻ polC⁻* (which lack both DNA polymerases I and III), and ligase-deficient mutants, cannot carry out reconstruction: they show extensive DNA degradation after UV irradiation, but not if the cells lack excision repair (i.e., are also *uvr⁻*).

REPAIR OF DNA DAMAGES IN HIGHER ORGANISMS

Repair mechanisms similar to those described above, and similar repair enzymes, have been recognized in higher organisms. Thus exicision repair removes damages produced by UV light or by certain carcinogenic agents (such as *N*-acetyl-2-acetylaminofluorene). Error-prone repair is probably induced in cells exposed to carcinogens or UV light, since they exhibit increased survival of UV-irradiated viruses; it may contribute to the induction of cancer-inducing mutations.

The significance of DNA repairs in humans is dramatically demonstrated in patients with the rare fatal hereditary (autosomal recessive) disease **xeroderma pigmentosum.** Homozygotes for this gene have defects of various types in repair of UV damages. Complementation tests, carried out by measuring repair in heterokaryons (obtained by fusing cells of two individuals: see Cell Hybridization, Ch. 49), identify at least five complementation

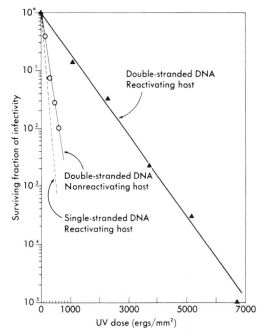

FIG. 11-8. UV inactivation of the single-stranded and double-stranded DNA of bacteriophage φX174 (Ch. 47). Each DNA in solution was exposed to increasing doses of UV light and then was assayed for infectivity in the dark on protoplasts of two *E. coli* strains, one capable and the other incapable of carrying out repair. The proportion of the residual titer is plotted versus the dose of irradiation. (Data from Yarus M, Sinsheimer RL: J Mol Biol 8:614, 1964. Copyright by Academic Press Inc. (London) Ltd.)

groups, affecting incision, recognition of apurinic sites, photoreactivation, postreplication repair, and perhaps also recognition of UV lesions in DNA complexed with proteins in chromatin. Repair defects have also been recognized in homozygotes of other recessive hereditary diseases: ataxia telangiectasica, Fanconi's anemia, and Bloom's disease. Individuals affected by all these diseases frequently develop cancers, probably from somatic mutations during error-prone repair. They also tend to have neurologic disorders, perhaps due to cell death from accumulation of DNA damages, since nerve cells are not renewable.

GENETIC RECOMBINATION

Genetic recombination is recognized by analysis of the distribution of markers (identifiable alleles) among the progeny derived from a cross between two parental genomes (Ch. 8). We shall examine here the molecular events.

Early genetic studies with eukaryotes indicated that the pairing of homologous chromosomes **(synapsis)** must be extremely precise, since a cross between two different mutants regularly yielded the wild-type and the reciprocal recombinant (double mutant), with no other genetic changes in either. Studies in fine-structure genetics localized the site of recombination precisely: between **two adjacent nucleotides.** Thus, if one of the glycines in tryptophan synthetase is replaced by glutamic acid, argi-

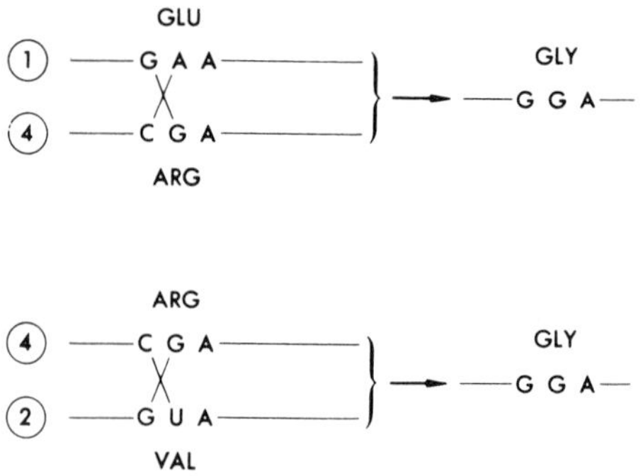

FIG. 11-9. Recombination within a codon. The crosses involved three mutants of the A protein of tryptophan synthetase (identified by numbers on the left), with different amino acids replacing the glycine. The nucleotide composition of the codons is inferred from the amino acid changes in the mutants and the restoration of glycine in the recombinants (see Genetic Code, Fig. 13-8, Ch. 13). As would be expected from the known codons, the cross of mutants 1 and 2 failed to restore glycine. Recombination is indicated as a reciprocal crossover but may have occurred by gene conversion (see Figs. 11-14 and 11-15). (Data from Yanofsky C: Cold Spring Harbor Symp Quant Biol 28:581, 1963)

nine, or valine, crosses between pairs of these mutants yield a very low frequency of recombinants in which the glycine has been restored (Fig. 11-9). These results would have appeared extraordinary before the genetic code was discovered, but they are now easily understood as the consequence of rare recombination **within a codon.**

The **molecular aspects of recombination** have been revealed primarily by a combination of formal genetic analysis in fungi and molecular studies with phages.

CROSSING OVER

This process is studied by the recombination of **distant markers:** it occurs by **breakage and reunion of DNA molecules.**

The evidence derives from a classic experiment of Meselson and Weigle, in which cells were mixedly infected with **heavy DNA** (labeled with the isotopes ^{13}C and ^{15}N) of two strains of bacteriophage λ, while the **new DNA** formed during replication was light, containing only the normal isotopes ^{12}C and ^{14}N. The **progeny phage particles** had different densities, depending on the DNA they contained, and accordingly formed different bands in density gradient equilibrium centrifugation in CsCl (bands 1 and 3 in Fig. 11-10).

Particles with nonreplicated DNA were the heaviest and formed a separate band (band 1), from which they could be isolated. Some of these particles were found genetically to be recombinant and therefore must have arisen by union of fragments of unreplicated DNA molecules present in the infecting particles. These special recombinants had another important property—they were **slightly lighter** than the original parental particles (band 2), indicating that a small proportion of the heavy parental DNA had been replaced by new synthesis. This finding suggests that limited DNA synthesis of repair type is required to seal the fragments together.

This conclusion appears to be valid for all organisms, because in eukaryotes recombination occurs at the pachytene stage of meiosis, **after** the DNA has replicated, and it is also accompanied by a small amount of repair synthesis.

DIFFERENT TYPES OF RECOMBINATION

The perfect conservation of the original base sequence in crossover is explained by two basic features of recombination: 1) only one strand of each parental molecule is broken at any time, and the sequence is maintained by the continuous strand; 2) the broken strands are aligned by base pairing with the continuous complementary strands. In some cases, specialized proteins may contribute to holding the loose components together. The characteristics of the single-strand cuts, and of the mechanism of sequence conservation, distinguish different types of recombination.

In **generalized recombination** the first parental strand cut can occur anywhere in the DNA, sometimes as an expression of DNA damage or repair (e.g., breaks produced by ionizing radiations, strand incision during excision repair). Single-stranded tails are thus generated from one parental molecule and presumably pair with complementary sequences of the other partly unwound molecule.

In bacteria and phages this type of recombination (*recA*-dependent) requires the products of many genes; some of them are also required for DNA replication (e.g., DNA polymerases, ligase, DNA helix-destabilizing protein). Principally involved in the recombination are exo- and endonucleases, such as *E. coli* exonuclease V (specified by the *rec B and C* genes, and controlled by *rec A*), which in vitro, in conjunction with helix-destabilizing protein, unwinds the DNA helix.

Site-specific recombination is *recA*-independent; the two broken parental strands pair at sites of very limited homology. This process is used in the integration of certain genetic units into the bacterial chromosome, as well as in exchange of portions of either. With some units (see below, Transposable Elements; Insertion Sequences) the crossover site is specific in the unit but more or less random in the chromosome. With phage λ both sites are specific (i.e., λ integrates at a definite receptor site: Ch. 48).

Finally, an **illegitimate recombination,** of obscure mechanism, occurs at a low frequency in the bacterial chromosome, using neither extensive homology nor specific sites; it causes gene duplication and some types of deletions.

Reciprocal Recombination. With cells a crossing-over must produce two reciprocal recombinants for distant markers, in order to maintain the integrity of the genome. With phage λ the bacterial *rec* system acts in the same way, but in addition the phage *red* system (see Ch. 48, the Genetics of λ) can also cause **nonreciprocal recombination.** This process can rescue smaller DNA fragments.

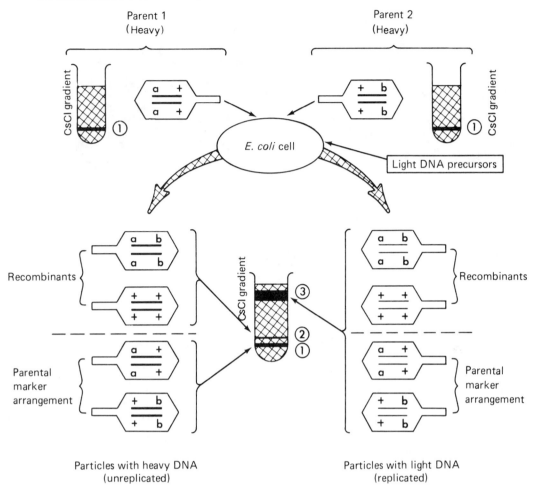

FIG. 11-10. Scheme of the Meselson and Weigle experiment, showing recombinant formation by breakage and reunion of molecules of phage λ DNA. The differences between particle densities in CsCl density gradients have been exaggerated for clarity. (Meselson M, Weigle J: Proc Natl Acad Sci USA 47:857, 1961)

MODEL OF GENERALIZED RECOMBINATION

A detailed model of reciprocal crossover is based on the electron microscopic observation of **recombination intermediates** as molecules with an X or H shape (Fig. 11-11); these occur in DNA extracted from phage-infected bacteria in which recombination is initiated, but without DNA replication (owing to mutations or metabolic inhibitors). These structures contain two DNA molecules joined at homologous sites; if introduced into cells by transfection they generate a high proportion of recombinants in the phage progeny.

The model given in Fig. 11-12 makes two important predictions that have been verified experimentally: 1) heteroduplex DNA (with one strand from each parent) is present at the site of crossing over; and 2) the intermediate can segregate into molecules that are recombinant, and others that are nonrecombinant, for distant markers.

Heteroduplex DNA. In cells mixedly infected by a phage mutant and wild type, 1% of the progeny phage are heterozygous for the mutation: i.e., on subsequent replication, half of the progeny of such a particle are mutant, and half wild type. The heterozygous particles are caused by heteroduplex DNA, and not by diploid heterozygosity as found in eukaryotic cells (Fig. 11-13): the **intramolecular heterozygosity** of phages is lost upon replication, whereas heterozygosity of diploid cells is intermolecular, and therefore is maintained indefinitely, by mitosis, through cell duplication. The location of the heteroduplex segment, at the junction of the DNA fragments, is supported by recombination of phage DNA differing at three not closely linked markers (e.g., ABC, abc); molecules heterozygous at the middle marker (Bb) are frequently recombinant for the outside markers (Ac or aC). The probability of heterozygosity, which is about 1% for any marker, is equal to the ratio between the lengths of

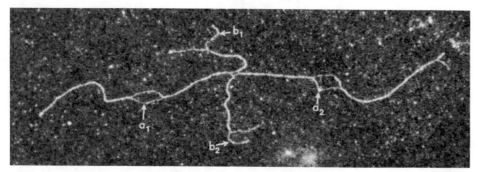

FIG. 11-11. Electron micrograph of a recombination intermediate of colicin E1 DNA. After the two partners were linearized by cutting with restriction endonuclease RI (Ch. 10) the DNA was spread in the presence of a high concentration of formamide, which denatures AT-rich regions. The local denaturations (**arrows**) allow the identification of homologous regions in the two molecules taking part in recombination. In this intermediate the homologous arms are characterized, one by an internal (**a**), the other by a terminal (**b**) denaturation; clearly the homologous arms are in *trans* configuration. This intermediate corresponds to the X figure predicted by the model of Fig. 11-12, configuration 6. (Potter H, Dressler D: Proc Natl Acad Sci USA 73:3000, 1976)

FIG. 11-12. Model of reciprocal recombination. Panel I: 1. A DNA duplex carrying markers **a** and **B** is nicked and partially unwound. 2. The free strand pairs to a complementary sequence of another intact duplex which carries complementary markers **A** and **b.** Such **assimilation of the free strand** is facilitated by a superhelical configuration of the duplex (which appears to be common in cells) since the helix tends to denature locally (see Ch. 10). After assimilation the recipient duplex is nicked by an endonuclease. 3. Elongation of the assimilated strand by a polymerase causes a **strand displacement** in the upper duplex and creates a **heteroduplex segment (boxed)** in the lower duplex. 4. Covalent attachment of the ends generates an **integrated molecule.** Cutting the bridge by an endonuclease (**arrow**) and resealing releases (5) two recombinants in which the flanking markers are still in parental arrangement (**aB** and **Ab**). Panel II. Intermediate 4 undergoes a series of isomerizations generating a figure X (6). This generates a different intermediate (8) which, after bridge cutting, produces two molecules (9) that are **recombinants for the flanking markers (AB** and **ab)** and again contain heteroduplex tracts (**boxed**). Figures H can be generated from 6. (Modified from Meselson MS, Radding CM: Proc Natl Acad Sci USA 72:358, 1975)

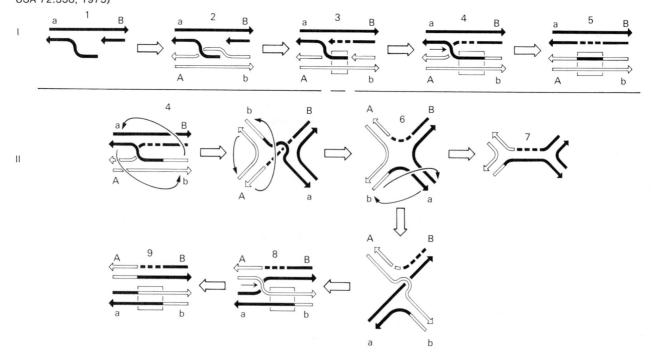

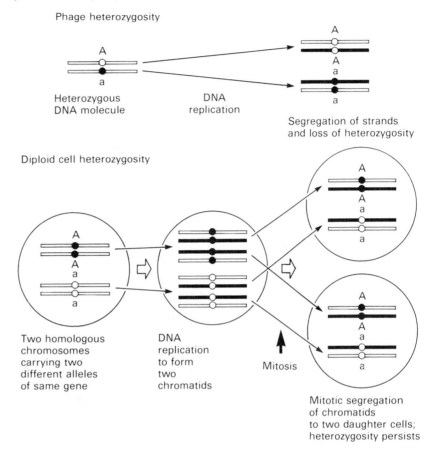

FIG. 11-13. Phage and cell heterozygosity.

the heteroduplex region and of the whole genome. From this probability it can be estimated that the heterozygous segment is about 2000 base pairs long; hence molecules can be heterozygous for several markers only when these are closely linked.

Formation of the heteroduplex DNA in crossing-over is probably common to all organisms, although not directly demonstrable with longer chromosomes. In eukaryotes its existence is indicated by the aberrant marker ratios in the progeny of certain crosses, described in the next section (Gene Conversion).

The second feature of the model of Fig. 11-12, possible *segregation of a recombination intermediate into two nonrecombinant molecules,* is verified in fungi, in which gene conversion (which must occur in a recombination intermediate; see below) is frequently observed without crossover of flanking markers.

GENE CONVERSION

With phages or cells distinguished by several markers more than one recombination may be observed in the same DNA molecule. If these occurred independently the frequency of multiple recombinants should equal, as a first approximation, the product of the frequencies of each recombination by itself. With distant markers this is observed. With closely linked markers, however, multiple recombinations are considerably **more frequent** than expected. These multiple recombinations occur between closely linked markers for which phage DNA molecules are heterozygous. The recombination cannot be due to conventional crossovers, because distant markers flanking the group of close markers often do not show enhanced recombination.

The phenomenon was clarified by studies in fungi, in which the four (or eight) spores in each ascus define the four products resulting from recombination among four chromatids (each a helical DNA molecule) in one meiosis (Fig. 11-14). In a cross of mutant (m) and wild type (wt), the ratio of mutant to wild type in each ascus is normally 2:2, but occasionally it is 3:1 or 1:3. This exceptional result, called gene conversion, implies that a part of one chromatid, instead of being exchanged, becomes **changed** to match a chromatid deriving from the

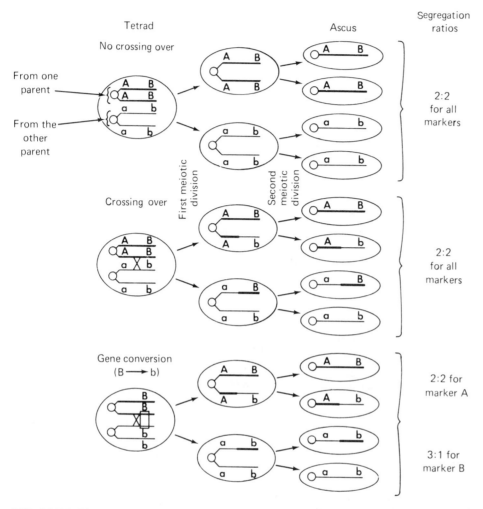

FIG. 11-14. Chromatid segregation during meiosis in ascomycetes, showing the composition of asci and the effect of gene conversion. The **boxed area** indicates where gene conversion occurs. It is represented as associated with crossing over, but the association is not required (see Fig. 11-12). Asci are shown with four spores, but often they contain eight, each spore being doubled by a subsequent mitotic division. The events presumably occurring in the boxed area are shown in Fig. 11-15.

other parent. *One gene conversion is formally equivalent to two crossovers,* since it replaces one allele by another without recombination of flanking markers.

Gene conversion is responsible for almost all recombinations between very closely linked markers in fungi, and possibly in phage and higher organisms as well. It is not important, however, in classic recombination between more distant markers (which is due to crossing over), except insofar as it throws light on the underlying molecular mechanism.

Mechanism. Gene conversion is caused by **mismatch correction** (see Repair of DNA Damage, above) **in heteroduplex DNA** (Fig. 11-15): it is absent in mutants de-

fective for an endonuclease that initiates correction. A correction initiated by a mismatched site may extend to other mismatched sites in nearby hybrid DNA, causing the simultaneous conversion of two or more markers.

Mismatch correction can be reproduced experimentally with heteroduplex phage DNA made in vitro by annealing the DNAs of two strains differing at several alleles: the heterozygosity progressively disappears after transfection into cells. Either allele can be corrected, but not at random: the direction and frequency of correction depends on the markers and the neighboring nucleotides.

Mismatch correction is a mechanism for ensuring the stability of genetic information: thus in a heteroduplex

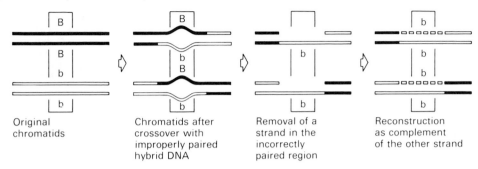

Original chromatids | Chromatids after crossover with improperly paired hybrid DNA | Removal of a strand in the incorrectly paired region | Reconstruction as complement of the other strand

FIG. 11-15. Hypothetic molecular events in gene conversion, representing the segments of only two of the four participating chromatids (i.e., half tetrad) of Fig. 11-14 (**boxed area**). Incorrect pairing in hybrid DNA is thought to activate the error-correction mechanism to form a correctly paired DNA, thus eliminating the mutant (or wild-type) information in one of the strands.

formed by an old strand, in which bases are methylated, and a new one, in which they are not yet, the unmethylated strand is preferentially corrected. Mismatch correction thus minimizes errors that might have occurred during replication.

GENETIC MAPPING

Genetic mapping aims at determining the order of markers in DNA and the distances between them. It can be achieved by recombinational, topologic, and physical methods.

Determining recombination frequencies is the classic method, intermarker distances being calculated from the proportion of recombinants between markers. The standard procedure uses a **three-factor cross,** with parents differing in three markers as shown in Figure 11-16. By crossing various overlapping sets of three markers in this way all the available markers in a chromosome can be ordered in a unique sequence: **the genetic map.** Phenotypically markers are assigned to the same or to different genes by means of complementation tests.

Distances are evaluated from the proportion of recombinants between fairly close markers, for with more distant markers mul-

tiple crossovers tend to decrease the recombination values, instead of remaining additive, the values asymptotically approach a maximum, with increasing distance, of 50%. For intance, if we take into account only single and double crossovers the recombination frequencies expected between three markers, with the order A, B, and C, are given by

$$P_{ac} = P_{ab} + P_{bc} - 2P_{ab} \cdot P_{bc}$$

where P_{ac}, P_{ab}, and P_{bc} are the proportions of the respective recombinants.

Recombination mapping is generally valid down to fairly short distances. However, it is less accurate for very close markers (e.g., within a gene) because of the role of gene conversion, whose frequency does not depend simply on distance (Fig. 11-17).

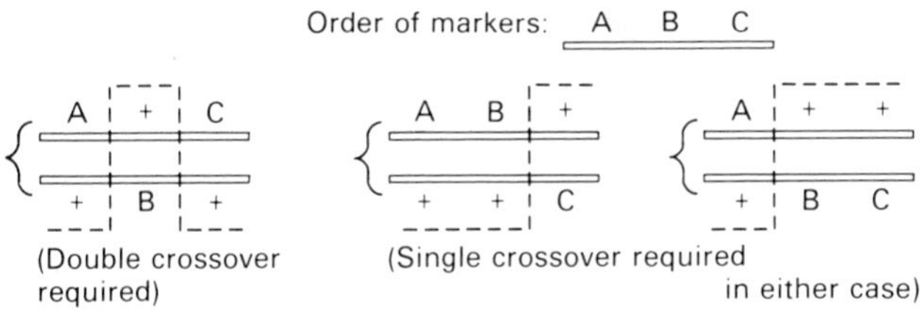

Deletion mapping is based on the inability of a deletion, in a pairwise cross, to produce wild-type recombinants with an overlapping deletion or with any point mutation within the deleted region. *Since the crosses give a simple yes or no answer, the deletion ends and their relation to point mutations can be ordered unambiguously* (Fig. 11-18). This approach gives marker order over any distance, whether minute or large; it is not affected by gene conversions and it is applicable to many biologic systems. Deletion mapping, however, does not provide distances. These can sometimes be obtained by physical mapping of the product as in Figure 11-17.

Transduction mapping (Ch. 9) takes advantage of the transfer, by generalized transducing phages (Ch. 48), of DNA segments of **nearly constant length,** taken at random from the DNA of the donor cell. In **abortive transduction** (see Fig. 48-17) the probability of cotransduction of two genes depends only on the ratio of their physical distances to the known length of the transduced DNA fragments; hence the results afford an unambiguous way to determine both order and distances. In **stable transduction** the distances are less reliable because the factors that complicate recombination similarly influence integration of the transduced segment.

Physical mapping is the most direct. Protein mapping

FIG. 11-16. Three-factor cross. **A, B,** and **C** = the mutant alleles; + = the corresponding wild-type (**wt**) alleles. The test consists of measuring the proportion of wild-type genomes in the progeny of double-mutant × single-mutant crosses. When the single mutant is the central marker the cross yields many fewer wt recombinants because two crossovers are required.

Order of markers: A   B   C

A | + | C
+ | B | +
(Double crossover required)

A   B | +
+   + | C
(Single crossover required

A | + | +
+ | B | C
in either case)

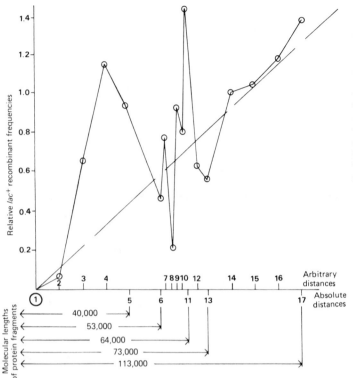

FIG. 11-17. Relative recombination frequencies in the *E. coli* β-galactosidase gene when various chain-terminating (nonsense) mutants are crossed to mutant 1. The **order** of the markers 1–17 was independently determined by deletion mapping, which affords an unbiased order (see text). Moreover, the **distances** between some of the markers and marker 1 are given by the molecular weight of the protein fragments produced by the mutants. The distances between the other markers are arbitrary. By conventional mapping, mutants 6–10 would incorrectly be considered closer to 1 than mutant 4. The classic expectation of a linear dependence of recombination frequency on distance (**dashed line**) is thus verified for only some of the markers. (Modified from Norkin LC: J Mol Biol 51:633, 1970. Copyright by Academic Press Inc. (London) Ltd.)

FIG. 11-18. Properties of deletions. Deletions (**heavy lines**) are identified by genetic crosses with selected point mutants. Cross **1** between the deletion mutant and point mutant **F** and cross **2** between the deletion mutant and point mutant **S** yield no wild-type recombinants; however, cross **3** between the two point mutants yields wild-type recombinants, showing that mutations F and S occur at different sites. The deletion mutant gives wild-type recombinants in cross **4** with point mutant **Z,** whose site is outside the deletion.

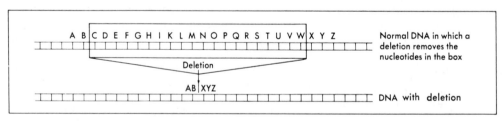

IDENTIFICATION OF DELETION BY GENETIC CROSSES

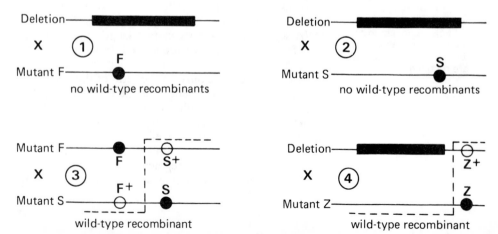

is useful for sites within a gene; it can be carried out by measuring the length of the polypeptide chains resulting from the introduction of terminator (nonsense) mutations (see Fig. 11-17). Nucleic acid mapping is best performed by hybridizing DNA from strains differing by two or more deletions or insertions, to form heteroduplexes. As Figure 11-19 shows, where two strands lack homology the unpaired DNA forms a "bush," and the length of the paired region between bushes can be unambiguously determined.

Restriction enzyme mapping uses specific fragments produced by restriction endonucleases (Ch. 10). The order of cleavage sites for a variety of endonucleases is determined from the overlaps of fragments produced by different enzymes. The location of a gene on this restriction site map is then established by various procedures, such as marker rescue (see Fig. 66-5) or hybridization of fragments to mRNA or to genetically altered DNA (e.g., with deletions).

A **comparison of genetic maps** obtained by physical means or by generalized recombination shows a general agreement over long distances, or in selected DNA segments. However, there are marked discrepancies over short distances, and "hot spots," where recombination frequencies are abnormally high. These discrepancies show that the local frequency of recombination depends on the DNA sequences.

TRANSPOSABLE ELEMENTS; INSERTION SEQUENCES

Cells contain a variety of transposable elements that can become inserted into various specific sites in the bacterial chromosome by a *recA*- and homology-independent mechanism; and upon excision (by a similar mechanism) these elements occasionally carry with them bacterial genes. The biological significance of site-specific recombination differs from that of generalized recombination, but may be as great—i.e., as an evolutionary mechanism

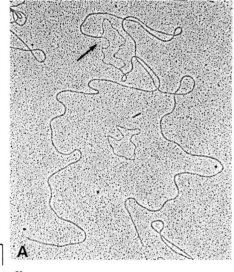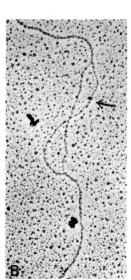

FIG. 11-19. Physical mapping of deletions in the DNA of a transducing derivative of phage ϕ80, which has exchanged 0.7μm of its DNA for 1μm of *E. coli* DNA containing two tRNA genes. **A.** Heteroduplex DNA was made by hybridizing complementary DNA strands deriving from the mutant and from the wild-type ϕ80. The left end of the exchanged piece corresponds to the ϕ80 attachment site (**att^{80}**), which is therefore precisely mapped. The DNA was prepared under conditions that prevented random folding of the single-stranded loops. **B.** The bacterial tRNA site is recognizable by hybridizing it with tRNA labeled with the iron-containing (and electron-opaque) protein ferritin (**arrow** in B). (Courtesy of M. Wu and N. Davidson)

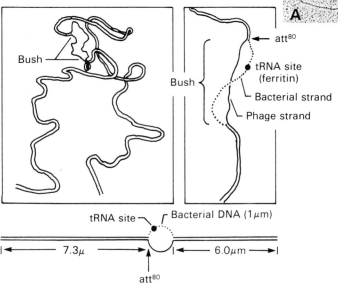

for expanding the genome through the incorporation of sections from the same or from a different organism. Lack of dependence on DNA homology is essential for this role.

Of great medical importance is the transfer, among bacteria, of genes for resistance to antibiotics (see Resistance Factors, Ch. 12). These are found in a type of transposable element (i.e., DNA sequences) apparently incapable of autonomous existence, the **transposons** (or **Tn elements**), which transpose blocks of genetic material back and forth between the cell chromosome and smaller replicons (plasmids: Ch. 12); these can then transfer the blocks to other cells. Many transposons are known, containing different sets of bacterial genes.

All have **terminal repeat sequences,** 80–1700 base pairs long, either in direct or in inverted orientation. Transposons with long inverted repeat sequences are recognizable in electron micrographs, after denaturation and self-annealing, as "lollipops," in which the stalk derives from pairing of the inverted sequences on the same strand, and the loop contains the bacterial genes (Fig. 11-20).

Another class of transposable elements, much smaller, was revealed by their inactivation of the gene in which they are inserted (see Polar Mutation, Ch. 14), rather than by coding for an additional product. These **insertion sequences (IS elements)** resemble Tn elements in their mechanism of recombination.

Several different IS elements are known (IS1, IS2, etc.), each with a distinctive length (700–1400 base pairs)

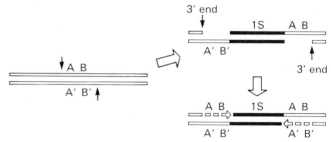

FIG. 11-21. Possible mechanism of insertion of an IS element (**heavy lines**), causing a direct repeat of the host sequences AB/A′B′ by formation of two staggered nicks, followed by insertion and filling of the gaps.

and sequence; they are present in bacterial and plasmidial DNAs. IS1, which has been entirely sequenced, has an imperfect inverted terminal repetition of 30 bases.

It seems that the repetitious ends of Tn and IS elements are important for two kinds of transposition events. In one, IS elements provide homologous sequences recognized by the *rec* bacterial system for the integration and excision of plasmids such as the F factor (see Conjugation, Ch. 9, Ch. 12). In addition, *recA*-independent recombination between nonhomologous segments causes deletions and inversions; these terminate at one end near the transposable element, which is generally not altered; the other end is at a variable distance. The details of the recombination event are not known, but it is probably carried out by an enzyme (provided by the transposon) that associates with the repeated ends.

Inserted IS elements are regularly flanked by short (5-9 bases), direct repeats of host sequences, possibly resulting from staggered cuts in the chromosome when the element is inserted (Fig. 11-21). The repeats are different for different insertions of the same element, showing that the recombinase does not recognize a specific site of insertion in the host DNA, but like certain restriction endonucleases, it might recognize a host site at some distance from the insertion. Indeed, there are host sequences near IS1 insertions that are similar to the common sequences of promoters (see Transcription below), hinting at a possible role for RNA polymerase in this recombination.

Many IS elements contain strong **translational terminators,** which explains their induction of polar mutations (see Ch. 14). One element contains both a strong **transcriptional terminator** and a **promoter,** and may therefore act as a **switch** for turning genes on or off, depending on its orientation; it can be inverted by a crossover between the terminal repetitions (Fig. 11-22). Such a mechanism may generate the antigenic **phase shifts** in bacteria (Ch. 12).

IS or Tn elements are probably also present in eukaryotes: the **transposable mutators** of maize and *Drosophila* may be examples.

FIG. 11-20. Transposons. Selfannealing of the inverted repeat sequences (heavy lines) on one strand produces the "lollipop" figure visible in electron micrographs, with the bacterial genes (G1-G3) in the single-stranded loop. AB . . . Z = nucleotide sequence, A′B′ . . . Z′ = its complement.

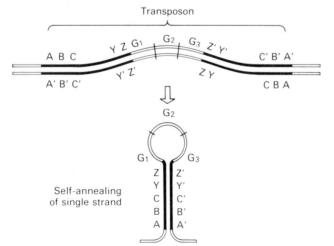

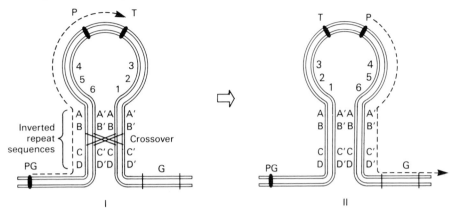

FIG. 11-22. Inversion of IS2 acting as a switch controlling an adjacent gene. A crossover between the inverted repeat sequences causes an inversion of the loop between the sequences. The original direction can be similarly restored by a second crossover. The loop contains a transcription terminator (**T**) and a promoter (**P**). In orientation **I,** transcription from the promoter **PG,** which normally initiates transcription of gene **G,** is terminated at T, inactivating the function of gene G; in the opposite orientation (**II**), promoter P initiates transcription, and G is transcribed even if there is no initiation at PG. **Dashed lines** = transcription; **A,B,1,2,** etc. = base sequences; primed letters = sequences complementary to unprimed ones.

MUTATIONS

GENERAL PROPERTIES

The term **mutation** applies broadly to all heritable changes in nucleotide sequence arising within an organism; they may be **spontaneous,** or **induced** by mutagenic agents. Though mutations are ordinarily recognized through their effect on the phenotype, many have no phenotypic consequences (i.e., are **silent**): in these the new codon may specify the same amino acid, or may cause an amino acid replacement that does not alter the function of the protein, or may be compensated for by suppression. Because of silent mutations, overall measurements of mutation rates give minimal values.

Reversion and Suppression. Many mutations are revertible. However, genotypic or true reversion must be clearly distinguished from suppression—a change at a different site in the genome that phenotypically corrects the mutation. The distinction is made by crossing an apparent revertant to the wild type: as Figure 11-23 indicates, with suppressed mutants the original mutated gene can be segregated from its suppressor, again producing the mutant phenotype. The mechanisms of suppression will be considered in Chapter 13.

Classes of Mutations. The DNA changes in various kinds of mutations are precisely inferred from studies of genes whose base sequences in DNA and amino acid sequences in the protein are completely known. Mutational changes include **nucleotide replacements, deletions,** and **insertions** (Table 11-1). Among replacements, in **transitions** a purine is replaced by a purine and a pyrimidine by a pyrimidine (e.g., AT→ CG), and in **transversions** a purine is replaced by a pyrimidine and vice versa (e.g., AT → CG) (Fig. 11-24).

Microdeletions and **microinsertions,** usually of a single nucleotide, generate **shifts of the reading frame;** these completely change the amino acid sequence of the protein downstream from the mutation, but the effect can be limited by another frame shift, opposite in direction, a short distance away (Fig. 11-25). The longer deletions and insertions, produced by other mechanisms, involve sequences of many nucleotides (up to thousands); they may or may not shift the reading frame.

Nucleotide replacements, and **single nucleotide** deletions and insertions, are **point mutations,** i.e., they all give wild-type recombinants with each other, and they undergo true reversion through appropriate replacements, insertions, or deletions. However, **larger deletions** give wild-type recombinants only with other strains that can replace the entire deleted region, and they do not revert. Because of this genetic stability they are especially useful for metabolic and genetic studies where reversions would interfere; they are also extremely useful in genetic mapping (see above).

These properties afford criteria for the recognition of various kinds of mutations. Most difficult to recognize

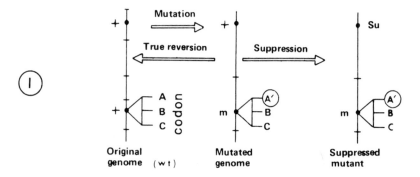

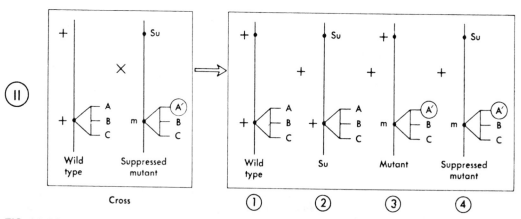

FIG. 11-23. Phenotypic reversions. I. Phenotype reversions caused either by true reversion of a mutation or by suppression. II. Genetic cross between a suppressed mutant (with wild-type phenotype) and a true wild-type yields four types of progeny, in one of which (3) the primary mutation is expressed. (The recombinant carrying only the suppressor mutation [2] may or may not be phenotypically distinguishable from the wild type.) A true reversion would not produce mutant progeny in a similar cross. + = Wild-type allele; **Su** = suppressor mutation; **m** = suppressible mutation; **ABC** = wild-type codon; **A′BC** = mutated codon.

TABLE 11-1. Main Properties of Different Kinds of Mutations

Nature of nucleic acid change	Effect on coding properties	Recombinational behavior	Production of reversions	Consequences for protein		Other properties
				Structure	Function	
Nucleotide replacement	Missense	Point mutation	Yes	Amino acid substitution	1) None 2) Temperature-sensitive 3) Lost	CRM may be present
	Nonsense	Point mutation	Yes	Premature termination of polypeptide chain	Usually lost	Extragenic suppression
Microdeletion; microinsertion		Point mutation	Yes	Frame shift	Usually lost	Intragenic suppression
Insertion		Point mutation	Yes	Altered	Usually lost	May introduce terminator codons
Deletion		Segment	No	Altered	Usually lost	
Silent				1) No amino substitution 2) Amino acid substitution with little effect on over-all structure	Conserved	

CRM = Crossreacting material.

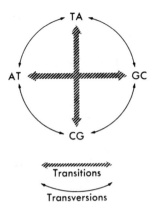

FIG. 11-24. Classes of nucleotide replacement. When a mutation substitutes one nucleotide for another in a strand, at the next replication the new nucleotide is paired to its regular partner. The resulting pair is represented. Example: If in the upper TA pair T is replaced by C, the result is the lower CG pair. Note that each base pair can undergo one kind of transition and two kinds of transversions.

are insertions, which behave formally like point mutations; they can be distinguished readily only if the inserted DNA has special properties (e.g., contains recognizable genes), or, with small DNAs, by electron microscopy, heteroduplexes formed by a deletion mutant and a wild-type DNA strand show unpaired areas (Fig. 11-19).

CONSEQUENCES OF MUTATIONS FOR PROTEIN STRUCTURE

These are outlined in Table 11-1. Among the replacements, **missense mutations** cause one amino acid to substitute for another. The protein may remain functional (though often with quantitative alterations) if the substitution does not markedly affect its tertiary structure or its active site. Some altered enzymes exhibit increased sensitivity to heat denaturation; if this effect is detectable in the whole organism, the mutation is called **temperature-sensitive (ts).** Other altered enzymes have lost function,

FIG. 11-25. Effect of frame shifts on a sequence of bases which are read three at a time, from the left. The sequences are arbitrary, but in order to emphasize the consequences of frame shifts the three letters in each codon (underlined) are the first letters of amino acids (glycine, leucine, valine, arginine, lysine). **Plus frame shift** means the insertion, and **minus** the deletion, of a nucleotide. The scheme shows that two mutations of opposite signs can correct each other (except for the region between them), two of the same sign cannot, but three of the same sign can. The heavy lines indicate the abnormal region. If these stretches are short, the function of the gene may be conserved.

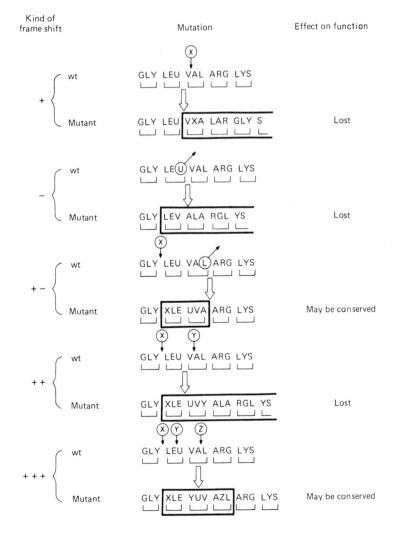

but the mutant protein remains recognizable immunologically as a **crossreacting material** (CRM). **Nonsense (terminator) mutations** create a codon that prematurely terminates the growing peptide chain and almost always destroys the function of the protein.

A **shift of the reading frame** causes the production of a jumbled protein with loss of function (see Fig. 11-25). **Larger deletions** destroy function except in rare cases when they remove small, inessential parts of proteins. Deletions at the boundary of two genes may cause the two polypeptide chains to be synthesized as a single chain, often with partial retention of one or both functions. **Insertions,** usually caused by transposons or insertion sequences (see above), and by prophages (Ch. 48), frequently destroy the function of the protein; moreover, they may introduce terminator codons, which affect the expression of distal genes (**polarity;** Ch. 14).

INDUCTION OF MUTATIONS

The induction of mutations by a **mutagenic agent** was first shown in 1927 by Muller, applying x-rays to the fruit fly *Drosophila*. With bacteria, which do not require the penetrating power of x-rays, the more convenient UV irradiation is equally effective, and powerful chemical mutagens were subsequently discovered. In addition, through the selection of rare mutants from large populations of bacteria (Ch. 8), a slight mutagenic action was later demonstrated for many chemicals (e.g., formaldehyde, caffeine, Mn^{2+}) and even for elevated temperature. Bacteria thus provide sensitive test systems for detecting potential environmental mutagens for man.

ACTION OF CHEMICAL MUTAGENS

Studies on the effects of chemical mutagens on DNA have revealed several molecular mechanisms of mutagenesis. The chemical mutagens fall into several main groups: 1) base analogs, 2) deaminating agents, 3) alkylating agents, and 4) acridine derivatives. The alkylating agents may be monofunctional (attacking one DNA strand) or bifunctional (and hence able to cross-link two-strands). Additional agents, of less significance for present purposes, include Mn^{2+} and formaldehyde. Some agents (base analogs, acridines, Mn^{2+}) *require replication for their action,* because they affect only the product of replication, whereas the chemically reactive alkylating and deaminating agents cause mutations *even in a nonreplicating template,* including transforming DNA or phage DNA in vitro. However, a very reactive, strongly mutagenic alkylating agent, **nitrosoguanidine,** *acts only on cells;* it produces mutations in closely linked clusters, at the growing fork of replicating DNA, by inducing error-prone repair (see above).

We shall now consider these agents in terms of the types of mutations they induce.

Point Mutations. Transitions (substitutions of AT for GC, or vice versa) are produced in DNA of cells or viruses grown in the presence of certain base analogs, such as **5-bromouracil** (5-BU) or its deoxynucleoside 5-bromo-deoxyuridine (5-BUdR). In cells these are converted to the corresponding nucleoside triphosphates and thus can be incorporated into new DNA in place of a normal base. Indeed, 5-BU so closely resembles thymine (T) that its substitution for as much as 90% of the T in bacteriophages is compatible with subsequent normal replication, provided the organisms are not exposed to blue or UV light, which cause strand breaks at the sites of incorporation. However, 5-BU tends to undergo a transient internal rearrangement (**tautomerization**) from the keto to the enol state, in which it pairs with G instead of with A (Fig. 11-26).

Mostly 5-BU is incorporated correctly (i.e., in place of T), but its subsequent tautomerization occasionally leads to a **replication error,** causing the positioning of a G instead of A in the new strand (a mispair); the result is genetic instability, with **delayed mutation.** Less frequently, tautomerization of 5-BU in the triphosphate precursor can lead to an **incorporation error,** since it is recognized

FIG. 11-26. Regular and unusual base pairing of 5-BU. I. Regular base pairing (in the common keto form) with adenine. II. Base pairing (in rare enol form) with guanine. The **heavy arrow** in II indicates the displacement of the proton in the tautomerization of 5-BU.

I

Adenine   5–Bromouracil (normal keto state)

II

Guanine   5–Bromouracil (rare enol state)

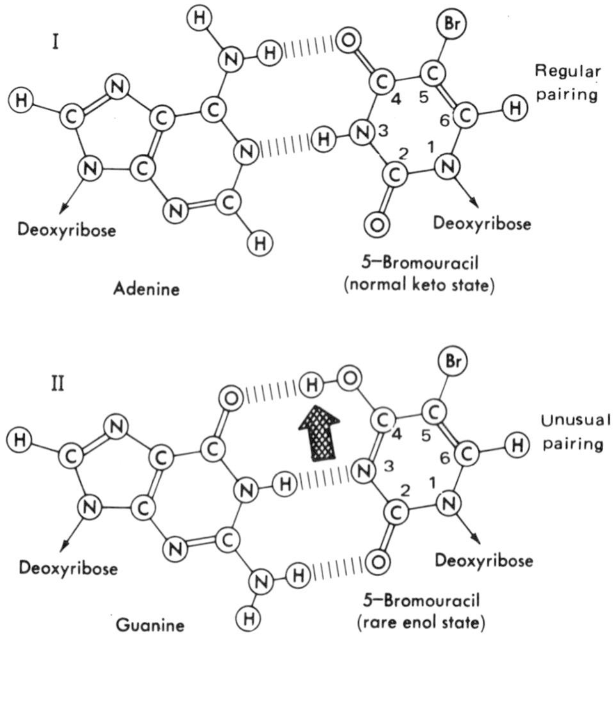

FIG. 11-27. Induction of mutations and their reversions by 5-BU: an incorporation error corrects a previous replication error. The sequence can be inverted, a replication error correcting a previous incorporation error. BU* = Transient enol form.

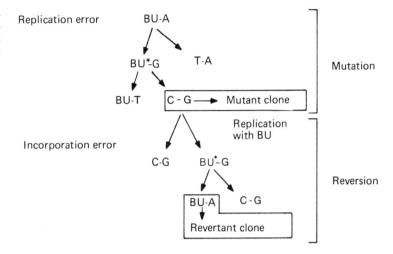

as though it were dCTP; the subsequent correct copying of this 5-BU immediately yields a mutant clone.

Because of this double action 5-BU can induce **reversion** of the mutations that it has produced (Fig. 11-27). Hence mutations of unknown origin can be defined operationally as transitions if 5-BU induces them to undergo true reversion. Only a fraction of the tautomerizations yield stable mutations, because many mispairs are removed by the **editing function of DNA polymerase** (see Enzymes Involved in Replication, Ch. 10) before polymerization moves to the next step; they are also corrected by the repair mechanisms noted above.

2-Aminopurine, a purine analog, causes transitions about as effectively as 5-BU, its tautomerization causing it to be read as either A or G.

Selective Transitions. Several **chemically reactive mutagens** induce more selective transitions. **Nitrous acid** (Fig. 11-28) deaminates A to hypoxanthine (which resembles G), and C to U. **Hydroxylamine** specifically converts C to a derivative that pairs with A, producing a GC → AT transition, which is not reversed by hydroxylamine. **Monofunctional alkylating agents,** such as ethyl ethanesulfonate (EES), also produce a similar transition, primarily through alkylation of the oxygen at the 6 position in G; the altered base then mispairs with T instead of C. These observed chemical interactions are evidently responsible for most of the mutagenic effects of these several agents in vivo, since the amino acid substitutions observed are those predicted from these chemical changes according to the genetic code (see Ch. 13). **Heat** produces GC → AT transitions by deaminating C to U.

Transversions. These are recognized operationally as point mutations that are not reverted by the agents that induce transitions or those that induce frame shifts. Though frequent among spontaneous mutants, transversions are produced by only a few mutagens, for instance the potent carcinogen 4-nitroquinoline 1-oxide (which also induces other types of mutation). Heat also induces GC → CG transversions. Some transversions are produced by **malfunctions of the DNA polymerase:** they are increased by certain polymerase mutations (see Mutations Affecting Mutability, below) and by Mn^{2+}, which also increases the effect of the mutant polymerase. Other transversions are the result of error-prone repair.

Reading Frame Shifts. These mutations, already discussed above, are induced by **acridine** derivatives. Usually only one base of the three-base **codon** is removed or inserted (though additions of up to four bases have been observed). Unknown mutations can be identified as frame shifts if they can be reversed by acridine (or by a second frame shift, Fig. 11-25) but not by mutagens of the preceding groups.

Acridines shift the reading frame by intercalating between successive base pairs in DNA (see Fig. 10-9, Ch. 10).

These mutations evidently require single-strand breaks, for they preferentially affect replicating DNA, and their frequency is increased when the closing of a gap is slowed by mutational impairment of DNA ligase. They also seem to require the presence of *short runs of the same base pair,* which promote **illegitimate** (i.e., shifted) **pairing** of a free end and the uninterrupted strand (Fig. 11-29). As is suggested in the model of this figure, the intercalated acridine may act by stabilizing the temporary illegitimate pairing until the gap is closed.

The considerable recombinational activity of certain phages (e.g., coliphage T4) explains why they have a high frequency of

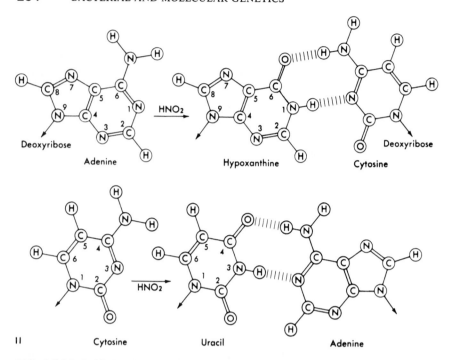

FIG. 11-28. Oxidative deamination of DNA by nitrous acid and its effects on subsequent base pairing. I. Adenine is deaminated to hypoxanthine, which pairs with cytosine instead of thymine. II. Cytosine is deaminated to uracil, which pairs with adenine instead of guanine.

FIG. 11-29. Possible mechanism of frame shift by acridines at sites of single-base reiterations near a single-strand break. The intercalated acridine molecules (shown as thick lines between base pairs) would stabilize the illegitimate pairing long enough to allow repair, leading to base insertion (I) or deletion (II).

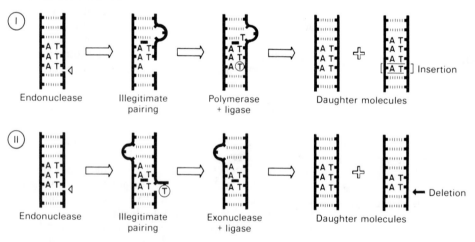

frame shifts. In bacteria, frame shifts are induced only by special aklylating acridine derivatives (e.g., ICR 191), which not only intercalate but also covalently bind to DNA.

Epoxides of carcinogenic polycyclic hydrocarbons (which represent the active forms of the carcinogens) cause frame shifts in bacteria, suggesting a correlation between the mutagenic and the carcinogenic activity of these compounds.

Deletions. Deletions in DNA are induced by agents that can cause **crosslinks** between the complementary strands: nitrous acid, bifunctional alkylating agents, and irradiation. Apparently a segment of DNA around the crosslink is not replicated, while the segments on either side replicate and join with each other. Excision of transposable elements (see above) is also an important cause of deletions.

INDUCTION OF MUTATIONS BY RADIATIONS

The main source of UV-induced mutations is the error-prone postreplication repair, which is part of the inducible SOS system (see above). In fact, *recA* mutants, which lack error-prone repair, show no induction of mutations by UV, although they are more sensitive to its lethal action. Conversely, mutant strains unable to excise dimers (*uvr*) show a marked increase in both induction of mutations and lethality; and caffeine, which interferes with dimer excision, also increases both effects. The mutations induced by UV light are mostly base substitutions (both transitions and transversions), sometimes affecting two consecutive bases (i.e., both the bases opposite to a dimer); frame shifts and deletions also occur.

Mutations induced by **x-rays** or by ^{32}**P decay**, as well as by some chemical carcinogens and crosslinking agents, also appear to result from imperfections of postreplication repair, since they are rare or absent in recombination-deficient strains.

SPONTANEOUS MUTATIONS

These represent mostly frame shifts, deletions, and transversions. Thus frame shifts constitute about 80% of all spontaneous mutations in phage T4; large deletions or frame shifts predominate in the I gene of the *E. coli lac* operon and transversions constitute more than 80% of spontaneous reversions of amber mutations in phage T4, and almost 50% of the mutations in electrophoretically altered human hemoglobins. This pattern shows that *spontaneous mutations are not induced by the usual chemical mutagens or by background radiation.* Rather, they may arise primarily from enzymatic imperfections during DNA replication or recombination; in bacteria, de-

letions may possibly arise from transient insertions of transposable elements.

INFLUENCE OF SITE ON THE FREQUENCY OF MUTATIONS

Fig. 11-30 shows that the frequencies of the GC → AT transitions at various sites (yielding amber or ochre termination mutations, Ch. 13) in the I gene of the *E. coli lac* operon are far from random; furthermore, the spectrum

FIG. 11-30. Distribution of GC → AT transition mutations to either an amber or an ochre terminator codon at the different normal codons of the I gene protein of the *E. coli lac* operon. The normal codons are numbered 1–320 in the abscissa. The number of mutations at each site in which the transition can generate a terminator codon is given on the ordinate for different mutagens and for spontaneous mutations. Numbers exceeding 60 are written on the bar. 2AP = 2-Aminopurine; UV = ultraviolet light; EMS = ethyl methanesulfonate (a monofunctional alkylating agent). (Coulondre C, Miller JH: J Mol Biol 117:577, 1977. Copyright by Academic Press Inc. (London) Ltd.)

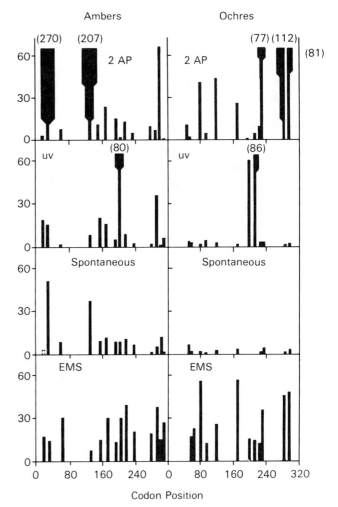

is different for mutations induced by various agents or arising spontaneously. Certain sites, with different locations for different mutagens, act as mutational **hot spots.** Hence the frequency of mutation at a nucleotide pair is strongly influenced by the local features, such as the base sequence or sites of methylation.

Thus hot spots of spontaneous mutations occur at the second CG pair of the sequence

CCAGG
GGTCC

(the target for the Eco II restriction endonuclease, Ch. 10), whose C is methylated in most *E. coli* strains to prevent restriction. Methylation of C determines the hot spots, which are absent in nonmethylating *E. coli* strains. A possible mechanism is the frequent deamination of 5-methyl C to generate 5-methyl U—i.e., T.

The mutagenic replacement of G by 2-aminopurine is strongly influenced by the base at the 5′-side of C in a GC pair: G > C > A > T. UV-induced mutations, as expected, occur mostly at pyrimidine–pyrimidine sequences; and intercalating agents induce frame-shifts mostly at reiterative sequences, as noted above.

MUTATIONS AFFECTING MUTABILITY

Mutations that increase or that decrease the spontaneous mutation rate have been observed: the former in bacteria and in phage, and the latter in phage. These **mutator** and **anti-mutator** mutations alter various proteins that participate in DNA replication or repair. For example, the editing function of the DNA polymerase (i.e., the $3' \rightarrow 5'$ exonuclease activity: see Enzymes Involved in Replication, Ch. 10) may be decreased or increased, in a way that affects the **correction of specific mispairings.**

Thus the mutator mutations in *E. coli* favor TA $\rightarrow$ GC transversions; those in phage T4 favor the opposite transversions; and the observed antimutator mutations decrease the rate of transitions.

Evolutionary Consequences of Mutations Affecting Mutation Rates. These findings could explain the large variations in the proportion of G + C observed among different organisms. Indeed, after the introduction of a mutator gene with TA $\rightarrow$ GC bias in an *E. coli* strain the proportion of G + C increased 0.5% in about 1500 generations. In order to be tolerated, the substitutions must occur where they have acceptable consequences for the proteins, i.e., mostly in the third bases of codons (see The Genetic Code, Ch. 13).

From the ease of selection of mutants with an increased or a decreased mutation rate it is evident that *natural selection favors some balanced value,* rather than maximal stability. This value is correlated, in various organisms, with the size of the genome (Table 11-2), so that the **rate per genome, rather than per nucleotide,** is relatively constant.

An advantage of higher mutation rates under some circumstances has been demonstrated by chemostat experiments (Ch. 5), in which an *E. coli* mutator strain outgrew the wild type.

DIRECTED MUTAGENESIS

After a long search, selective mutagenesis in a given gene can now be attained by several procedures, based on the ability of restriction endonucleases to cause a cut at precise points in DNA. The DNA molecules thus altered are then used 1) to prepare small deletions by imperfect resealing of the cut, 2) to produce a single-stranded gap by additional enzymatic action and then mutagenize the accessible persisting strand; and 3) to produce segments that can be treated with mutagenic agents and then inserted into homologous DNA (Fig. 11-31).

CHEMICAL MUTAGENESIS IN HIGHER ORGANISMS

Most work on mutagenesis in higher organisms has been carried out with ionizing radiations because of their ability to penetrate. Some mutagenic chemicals, such as alkylating agents, are also active, but others may not reach

TABLE 11-2. Spontaneous Mutation Rates in Different Organisms

Organism	AT base pairs per genome	Mutation rate per AT base pair replication	Total mutation rate per genome
Bacteriophage	4.8×10^4	2.0×10^{-8}	1.2×10^{-3}
Bacteriophage T4	1.8×10^5	1.7×10^{-8}	3.0×10^{-3}
Salmonella typhimurium	4.5×10^6	2.0×10^{-10}	0.9×10^{-3}
Escherichia coli	4.5×10^6	2.0×10^{-10}	0.9×10^{-3}
Neurospora crassa	4.5×10^7	0.7×10^{-11}	2.9×10^{-4}
Drosophila melanogaster	2.0×10^8	7.0×10^{-11}	1.4×10^{-2}
E. coli with a mutator mutation	2.0×10^6	3.5×10^{-6}	

(Data from Drake JW: Nature 221: 1132 1969)

the DNA in germ cells. Moreover, most mutagens effective on animal cells are not mutagenic by themselves but need **metabolic activation** to generate the mutagenic intermediate.

In the Ames test the mutagenic activity of such compounds is measured by using auxotrophic bacterial strains with different kinds of mutations, which can be reversed by different kinds of mutagenesis. The bacteria are incubated with the substance to be tested in the presence of a liver extract capable of carrying out the metabolic activation; in this way even labile nascent mutagenic products can be recognized. With this system about 90% of carcinogens also score as mutagens, and vice versa. The result suggests that **somatic mutations** are the mechanism by which carcinogenic agents cause cell transformation (Ch. 49) in vitro, or cancers in animals.

The same is probably true of radiations, as suggested by the high incidence of skin cancers in *xeroderma pigmentosum* patients (see Repair of DNA Damages in Higher Organisms, above), and by the effects of the atomic bomb explosion in Hiroshima: in the following 12 years the incidence of leukemia among the heavily irradiated survivors was about 50 times higher than in a comparable nonirradiated population.

The widespread use of chemicals with potential mutagenic action, and the relation between mutagenesis and carcinogenesis, are serious problems for human health.

TRANSCRIPTION OF DNA

The transfer of genetic information from DNA to RNA was inferred from the function of mRNA and from the requirement of DNA for RNA synthesis in vitro (see below); it was later demonstrated directly by the formation of a **hybrid RNA–DNA helix** when a template and its transcript are annealed together (see Homology: Hybridization, Ch. 10). The term **transcription** emphasizes retention of the language of base pairs.

With the powerful technic of hybridization a given RNA can be used to identify and quantitate its template, and vice versa (see. Fig. 10-13). Physiologic transcription (in the cell, and with unbroken DNA in vitro) is found to be generally asymmetric, i.e., the RNA transcribes only one DNA strand. (However, both strands of mitochondrial and certain viral DNA are transcribed at first and one transcript is then destroyed.) In the same DNA chain different segments may be transcribed on opposite strands (and consequently, because of the polarity of DNA, in opposite directions).

RNA POLYMERASE AND SIGMA (σ) FACTOR

Transcription can be achieved in vitro with suitable enzymes extracted from cells. The product is similar to native RNA, for coupling of its formation with protein synthesis yields small amounts of functional products.

RNA is synthesized by a **DNA-dependent RNA polymerase**, also called **transcriptase**, which catalyzes the reactions:

$$n(\text{ATP, GTP, UTP, CTP}) \xrightarrow[\text{Enzyme}]{\text{DNA template}}$$

Nucleoside
triphosphates

$$\cdot(\text{AMP, GMP, UMP, CMP})n + n(\text{P·P})$$

Polynucleotide   Pyrophosphate

As in DNA synthesis, the chains grow at the 3′-end (see Fig. 10-27), as shown by digestion with suitable nucleases after a short labeling pulse. The initiating, 5′-terminal nucleotide is always a purine, and it retains its triphosphate.

Purification of RNA polymerase of *E. coli* yields both **core enzyme** and **holoenzyme.** The core polymerase (mol wt about 400,000) is composed of two α chains (mol wt 41,000), one β chain (mol wt 155,000), and one chain whose similar size (mol wt 165,000) led it to be designated at β′. The holoenzyme contains an additional protein called **sigma (σ) factor** (mol wt 86,000). This **initiation factor** promotes attachment of the enzyme to specific initiation sites and then is released. It has provided the key to dramatic shifts in specificity of transcription observed in sporulation and in phage infection (see Alterations of RNA Polymerase, Ch. 14).

Eukaryotic transcriptases are even more complex, but other transcriptases are simpler: one specified by phage T7 is a single polypeptide of mol wt 100,000, while *Neurospora* mitochondria have one of mol wt 64,000.

The interaction of transcriptase with DNA involves three main steps: initiation, chain elongation, and termination (Fig. 11-32).

INITIATION

Physiologic initiation requires the complete enzyme (with σ-factor), the proper nucleoside triphosphate (which is always a purine), and a special DNA sequence called a **promoter.** Promoters are identified by mutations that prevent them from initiating, in vivo and in vitro. Binding of the complete enzyme to a promoter goes through a loose, reversible stage and then becomes tighter; the second step is much more temperature-dependent, since it involves local melting of the DNA. The

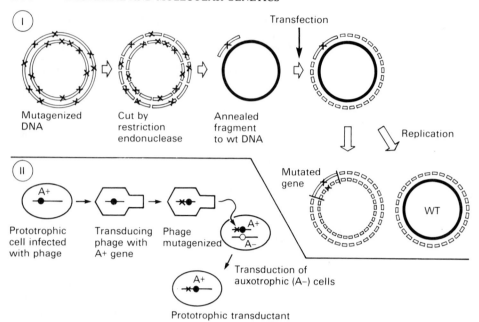

FIG. 11-31. Directed mutagenesis. I. Viral or plasmidial DNA is highly mutagenized and cut by restriction endonucleases. A selected fragment is denatured and annealed with single-strand circular wild-type DNA (obtained by denaturation after nicking double-stranded DNA) and is then introduced into cells where the fragment is completed by synthesis incorporating the mutation into the wt DNA. II. A similar approach is taken with bacteria, in which DNA adjacent to a selected gene (A⁺) and isolated into a transducing phage is highly mutagenized and then transferred to an A recipient bacterium in which it becomes incorporated by recombination. Both methods are suitable for the isolation of conditionally lethal mutations, allowing the propagation of the mutated organism under permissive conditions. (II, Hong JS, Ames BN: Proc Natl Acad Sci USA 68:3158, 1971)

initial nucleoside triphosphate is then bound and the σ factor is released, to engage in another cycle (Fig. 11-33); the 5'-terminal nucleotide retains its triphosphate. Many RNA chains are initiated at the same promoter before the first complete molecule is released (Fig. 11-34).

A **binding site** in the promoter, of about 40 base pairs, is protected by RNA polymerase from DNAse digestion (Fig. 14-18). However, mutations located as far as 20 base pairs from the binding site (as well as mutations within the site) can change the efficiency of initiation; they identify a recognition site, to which the enzyme initially binds. Except for a relatively constant heptanucleotide, *promoters differ in their sequences;* these differences account for the varied initiation efficiencies of different promoters, which fit them to specific biologic goals. Moreover, the affinity of certain promoters for the transcriptase can be markedly changed by their binding of various regulatory factors (Regulation of Initiation of Transcription, Ch. 14).

CHAIN ELONGATION

In this phase mononucleotides complementary to the template are incorporated. The growing end of the RNA is attached noncovalently to the DNA by the enzyme, and perhaps by a short stretch of RNA-DNA base pairs, in a region of local DNA strand separation. Thus if RNA is labeled by a brief radioactive pulse the most recently synthesized molecules are found connected to the DNA, and the complex can be broken by mild heating or by proteolysis. As the complex threads along the template the double helix apparently opens ahead of it and closes behind it, i.e., the DNA simultaneously unwinds and rewinds. At the end of transcription the template DNA is unaffected.

TERMINATION

Transcription terminates at specific DNA sites (terminators), where the stability of the DNA helix undergoes a sudden change (e.g., a short GC-rich stretch followed by an AT run). Terminators, like promoters, vary in sequence and hence in efficiency; a proportion of the chains grow beyond weak terminators. Since in bacteria and phage many clusters of adjacent genes are transcribed as a continuous unit, variation in termination is important, along with variation in initiation, in regulating gene function (see Polarity; Attenuation, Ch. 14).

FIG. 11-32. Model of transcription.

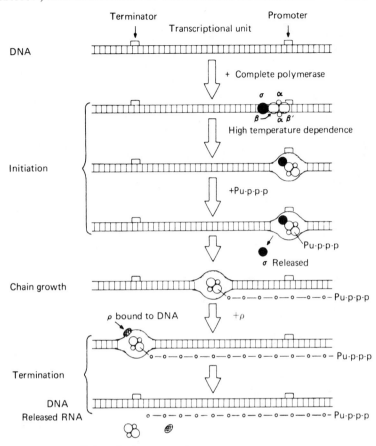

FIG. 11-33. Recycling of RNA polymerase core enzyme and of σ factor.

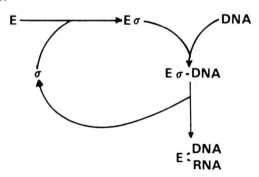

Rho (ρ) Factor. In transcription of DNA in vitro many bacterial or phage terminators are effective only along with a cytoplasmic factor called *rho*. This tetrameric protein (mol wt 200,000) appears to move along the DNA at the expense of ATP hydrolysis and to interact with the transcriptase. It evidently contributes to regulating gene expression, because its terminating efficiency varies for different messengers and can be neutralized by other factors. Moreover, mutations either in *rho* or in the polymerase (as well as in a terminator site) can increase or decrease the efficiency of termination, and some polymerase mutations make *rho* dispensable.

ANTIBIOTICS AFFECTING TRANSCRIPTION

Initiation is blocked by a group of antibiotics with an aromatic chromophore spanned by a long aliphatic bridge (Fig. 11-35): the naturally occurring **rifamycins,** the more effective semisynthetic derivative **rifampin,** and the **streptovaricins.** These antibiotics inhibit bacterial RNA polymerase and the small RNA polymerase of *Neurospora* mitochondria, but not the cytoplasmic enzyme of eukaryotes. They do not interfere with chain extension, nor do they prevent binding of the enzyme to DNA, but they prevent completion of the process of initiation.

The binding site is on the β subunit: rifampin binds tightly, in a 1:1 molar ratio, to this isolated subunit; and in resistant mutants the subunit has decreased affinity for the inhibitor. Mutants selected

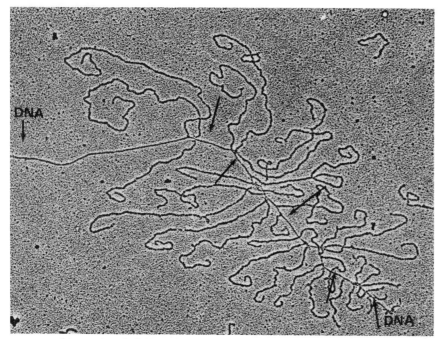

FIG. 11-34. Transcription of the DNA of bacteriophage T7 (**arrows**) in vitro by *E. coli* DNA polymerase. The transcriptase recognizes a promoter at the right end of the DNA (in the figure), and transcribes about 15% of the DNA. The remainder (not shown) is not transcribed owing to absence of the required T7-specified transcriptase. The RNA molecules are held in the stretched state by addition of the antifolding protein specified by phage T4 (see Ch. 10). The molecules are shortest near the promoter and grow in length as they approach the terminator at the left. This picture shows that transcription can reinitiate many times before the first complete molecule is released. (Courtesy of H. Delius and N. Axelrod)

for resistance to one of these antibiotics are also resistant to the others.

Chain extension can be blocked in two ways. **Streptolydigin** (Fig. 11-35) binds to the polymerase. In contrast, **actinomycin D (dactinomycin)** binds to helical DNA at GC pairs; the chromophore (Fig. 11-35) intercalates into the helix, as demonstrated earlier for acridines, and the attached cyclic peptides appear to bind to the external surface. Like all antibiotics that bind to DNA, actinomycin inhibits both transcription and replication, but the former is much more sensitive. This difference suggests that the unwinding of the helix in replication, behind a swivel point, can dislodge actinomycin, whereas the local, transient unwinding in transcription is probably less forceful.

Many other antibiotics that bind to DNA inhibit transcription and replication to different degrees. Some of them have intercalating polycyclic chromophores with attached sugars.

The use of rifampin to block RNA chain initiation, and actinomycin to block all RNA synthesis, has proved most valuable in studying the kinetics of RNA synthesis and breakdown. With many intact organisms, including *E. coli,* the effective use of some

of these inhibitors requires the elimination of a permeability barrier, either by chemical treatment or by mutation.

The selectivity of rifamycins has made them useful in antibacterial chemotherapy (especially against tuberculosis). Agents that act on DNA, of course, act equally on the DNA of animal and bacterial cells; actinomycin has thus been useful in tumor chemotherapy but not in antimicrobial chemotherapy.

REVERSE TRANSCRIPTION

The enzyme **RNA-dependent DNA polymerase** (or **reverse transcriptase**), present in retroviruses (Ch. 66), inverts the usual flow of genetic information by synthesizing DNA on an RNA template, thus building an intermediate for the replication of the viral RNA. In experimental work this method for synthesizing complementary DNA copies (cDNAs) of eukaryotic mRNAs or viral RNAs has been very useful, as will be detailed in later chapters.

The reverse transcriptase also has an RNase activity specific for DNA–RNA hybrids (**RNase H**), probably essential for its physiologic activity. Like DNA-depen-

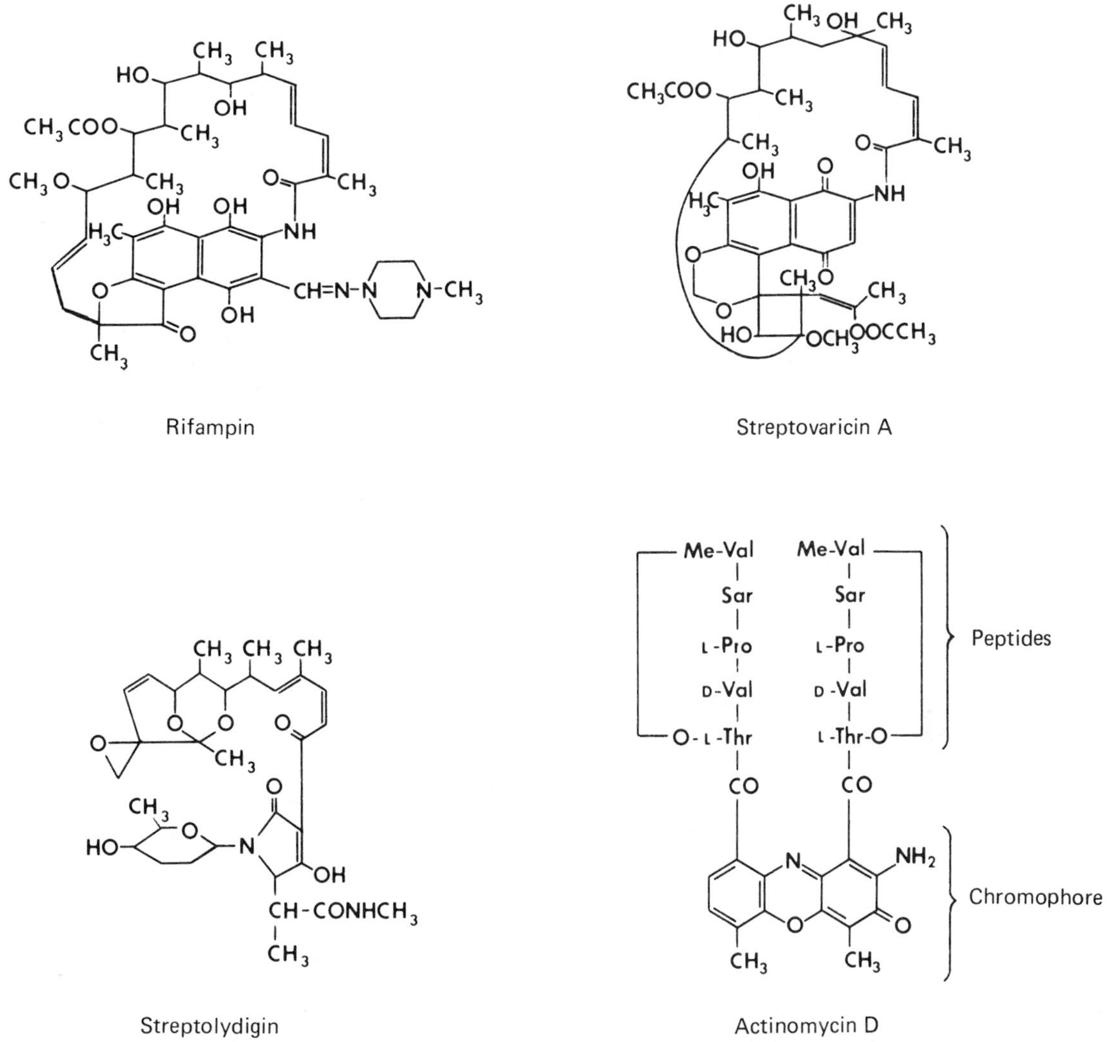

Rifampin

Streptovaricin A

Streptolydigin

Actinomycin D

Me-Val   Me-Val
Sar   Sar
L-Pro   L-Pro
D-Val   D-Val
O-L-Thr   L-Thr-O
CO   CO

Peptides

Chromophore

FIG. 11-35. Antibiotics that interfere with initiation of transcription (rifampin, streptovaricin) or chain extension (streptolydigin, actinomycin D.) **Sar** = Sarcosine (N-Me-glycine); **Me-Val** = N-Me-L-valine.

dent DNA polymerases, it synthesizes $5' \rightarrow 3'$, adding to a primer on a template. However, the template can be either DNA or RNA. With viral RNA as template the normal primer is a particular tRNA bound to it by base pairs; with mRNAs it is a short polythymidylic acid, paired with a polyadenylic acid tail at the $3'$ end of the RNA. The enzyme produces first a DNA–RNA hybrid, and later, perhaps after hydrolyzing the RNA strand with its RNase, single- and double-stranded DNA. The activity is inhibited by certain rifamycin derivatives.

APPENDIX

QUANTITATIVE ASPECTS OF KILLING BY IRRADIATION

We shall consider here certain mathematical relations between the dose of radiation and its lethal effect, which will be useful later for understanding problems involving uses of radiation in research and for sterilization.

The **inactivation** or **death** of a microorganism is defined as the loss of its ability to initiate a clone: this effect is the consequence of a certain number of **chemical events,** each consisting, for instance, in the unrepaired change of a chemical group or the breaking of a chemical bond. In a population, those individuals that have experienced enough of these events are inactivated. The relation between the dose of radiation and the proportion of surviving organisms can be calculated as follows:

It is assumed that the events occur randomly and independently in the susceptible chemical groups in various individual organisms, with a probability P per group, proportional to the dose. If there are n susceptible groups per organism, there will be, **on the average,** Pn inactivating events per organism. If a single such event is sufficient for inactivation, the pro-

211

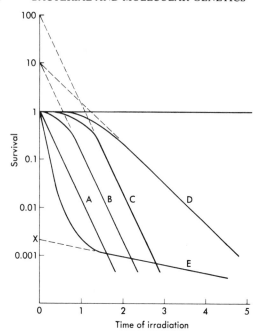

FIG. 11-36. Survival curves of microorganisms irradiated by UV light. **A,** a single-hit curve; **B,** a 10-hit curve; **C,** a 100-hit curve, all with the same target size; **D,** a 10-hit curve for organisms with a smaller target size than that of B; **E,** a multicomponent curve produced by a population containing organisms with different target sizes. X = proportion of organisms with the smallest target size. The dose rate is assumed to be constant.

portion of surviving organisms (S) which have experienced no such event is, from the Poisson distribution (see Appendix, Ch. 46), $S = e^{-Pn}$. In turn, $P = kD = krt$, where k is a constant that measures the probability of unrepaired damage of the chemical group, D is the dose of radiation, r is the dose rate (i.e., the dose of radiation per unit time), and t is the time. Thus the basic equation of inactivation is $S = e^{-krtn}$. The equation is used in its logarithmic form: $\log S = -krtn \log e = -Krtn$, where $K = k \log e$.

When the inactivation of bacteria or viruses by UV light, x-rays, or ^{32}P decay is followed by plotting the surviving fraction (S) versus time on semilogarithmic paper, the **survival curve** is often seen to be a **straight line passing through the origin:** curve A in Figure 11-36. Such a curve is called a **single-hit** curve, since it is generated by a process in which a **single event** in an organism destroys its viability.

The slope of a one-hit survival curve, $-Krn$, is the basis for radiation analysis of **target size;** as a first approximation this slope is proportional to the size of the genome (e.g., n) if the proportion of repaired damages is constant. If K is the same for two different organisms, the ratio of the values of Kn, or

relative target size, can be directly determined from the slopes of their survival curves. However, differences in efficiency of repair influence the value of K and hence the observed target size.

Other important types of survival curves, called **multiple-hit curves,** are represented by curves B, C, and D in Figure 11-36; they have a shoulder near the origin before becoming linear, because several events (hits) must accumulate in a viable unit before it is inactivated. To determine the number of events required for inactivation, the straight part of the survival curve is extrapolated back to meet the ordinate axis. The position of the intersection measures the number of hits in logarithmic units. Curve B corresponds to 10 hits and curve C to 100 hits; curve D also corresponds to 10 hits but has a different slope. The slope of the straight part of a multiple-hit curve has the same meaning as for a single-hit curve, i.e., it is proportional to the target size of the organism. Thus, curves A, B, and C reflect the inactivation of organisms with the same target size.

Multiple-hit curves are obtained whenever one susceptible component can replace another in the reproduction of the unit whose survival is being measured. For instance, when bacteria are infected by several UV-irradiated virus particles, curves of this type describe the survival of the ability to yield active virus (see Multiplicity Reactivation, Ch. 47). Similarly, in clumps of bacteria, the survival of a single organism is sufficient to maintain the colony-forming ability of a clump. Since the straight part of the curve is reached when the surviving clumps have only one surviving organism, its slope measures the inactivation (and the target size) of the last surviving individual in the clump. Curve A would then describe the inactivation of single organisms, curve B clumps of 10 organisms, and curve C clumps of 100 organisms.

Another mechanism generating multiple-hit curves is a decrease of repair efficiency at high radiation doses. Thus postreplication repair becomes inefficient when the distances between dimers are short and recombination is infrequent.

A third type of survival curve is **multicomponent** (E in Fig. 11-36), generated by a population composed of organisms with different target sizes. Those with the larger target, i.e., more sensitive, are inactivated first with a steep slope, while at the end only the most resistant ones survive, with a shallower slope for their survival curve. The proportion of this group can be estimated by extrapolating back the final straight part of the curve to the ordinate axis (about 2×10^{-3} in Fig. 11-36). Curves of this type are commonly obtained when viruses are exposed to chemical inactivating agents, where the differences in sensitivity depend primarily on differences in penetration. Recognition of these curves is important in the preparation of safe vaccines.

SELECTED READING

BOOKS AND REVIEW ARTICLES

ADHYA S, GOTTESMAN M: Control of transcription termination. Annu Rev Biochem 47:967, 1978

ANDERSON RP, ROTH JR: Tandem gene duplications in phage and bacteria. Annu Rev Microbiol 31:473, 1977

CLEAVER JE, BOOTSMA D: Xeroderma pigmentosum: biochemical and genetic characteristics. Annu Rev Genet 9:19, 1975

COX EC: Bacterial mutator genes and the control of spontaneous mutations. Annu Rev Genet 10:135, 1976

GREEN M, GERARD GF: RNA-directed DNA polymerase. Properties and functions in oncogenic RNA viruses and cells. Prog Nucleic Acid Res Mol Biol 14:1, 1974

FREESE E: Molecular mechanisms of mutation. Chem Mutagen 1:1, 1969

HOTCHKISS R: Models of genetic recombination. Annu Rev Microbiol 28:445, 1974

LOSICK R, CHAMBERLIN M (eds): RNA Polymerase. New York, Cold Spring Harbor Laboratory, 1976

MILLER RW: Delayed radiation effects in atomic-bomb survivors. Science 166:569, 1971

ROTH JR: Frameshift mutations. Annu Rev Genet 8:319, 1974

SETLOW RB: Repair deficient human disorders and cancer. Nature 271:713, 1978

TRAVERS A: RNA polymerase specificity and the control of growth. Nature 263:641, 1976

TROSKO JE, CHU EHY: The role of DNA repair and somatic mutation in carcinogenesis. Adv Cancer Res 21:391, 1975

WEISBERG RA, ADHYA S: Illegitimate recombination in bacteria and bacteriophages. Annu Rev Genetics 11:451, 1977

WITKIN EM: Ultraviolet metagenesis and inducible DNA repair in *Escherichia coli*. Bacteriol Rev 40:869, 1976

SPECIFIC ARTICLES

AHMAD A, HOLLOMAN WK, HOLLIDAY R: Nuclease that preferentially inactivates DNA containing mismatched bases. Nature 258:54, 1975

BRENNER S, BARNETT L, CRICK FHC, ORGEL L: The theory of mutagenesis. J Mol Biol 3:121, 1961

BURGESS RR, TRAVERS AA, DUNN JJ, BAUTZ EKF: Factor stimulating transcription by RNA polymerase. Nature 221:43, 1969

CASSUTO E, MURSALIM J, HOWARD-FLANDERS P: Homology-dependent cutting in trans: an approach to the enzymology of genetic recombination. Proc Natl Acad Sci USA 75:620, 1978

CLEAVER JE: Defective repair replication of DNA in xeroderma pigmentosum. Nature 218:652, 1968

COLE RS, LEVITAN D, SINDEN RR: Removal of psoralen interstrand cross-links from DNA of *Escherichia coli:* mechanism and genetic control. J Mol Biol 103:39, 1976

COULONDRE C, MILLER JH, FARABAUGH PJ, GILBERT W: Molecular basis of base substitution hot spots in *Escherichia coli*. Nature 274:775, 1978

COX EG, YANOFSKY C: Altered base ratios in the DNA of an *Escherichia coli* mutator strain. Proc Natl Acad Sci USA 58:1895, 1967

DAS GUPTA UB, SUMMERS WC: Ultraviolet reactivation of herpes simplex virus is mutagenic and inducible in mammalian cells. Proc Natl Acad Sci USA 75:2378, 1978

DAVIS RW, DAVIDSON N: Electron-microscopic visualization of deletion mutations. Proc Natl Acad Sci USA 60:243, 1968

DONIGER J: DNA replication in ultraviolet light irradiated Chinese hamster cells: the nature of replicon inhibition and post-replication repair. J Mol Biol 120:433, 1978

HOWARD-FLANDERS P, BOYCE RP, THERIOT L: Three loci in *Escherichia coli* K-12 that control the excision of pyrimidine dimers and certain other mutagen products from DNA. Genetics 53:1119, 1966

LEE F, YANOFSKY C: Transcription termination at the *trp* operon attenuator of *Escherichia coli* and *Salmonella typhimurium:* RNA secondary structure and regulation of termination. Proc Natl Acad Sci USA 74:4365, 1977

MCENTEE K: Protein X is the product of the *recA* gene of *Escherichia coli*. Proc Natl Acad Sci USA 74:5275, 1977

MESELSON M, RADDING CM: A general model for genetic recombination. Proc Natl Acad Sci USA 72:358, 1975

OKADA Y, AMAGASE S, TSUGITA A: Frameshift mutation in the lysozyme gene of bacteriophage T4: demonstration of the insertion of five bases, and a summary of *in vivo* codons and lysozyme activities. J Mol Biol 54:219, 1970

PIECZENIK G, HORIUCHI K, MODEL P, MCGILL C, MAZUR BJ, VOVIS GF, ZINDER ND: Is mRNA transcribed from the strand complementary to it in a DNA duplex? Nature 253:131, 1975

PRINBROW D: Bacteriophage T7 early promoters: nucleotide sequences of two RNA polymerase binding sites. J Mol Biol 99:419, 1975

SARTHY PV, MESELSON M: Single burst study of rec- and red-mediated recombination in bacteriophage lambda. Proc Natl Acad Sci USA 73:4613, 1976

SEDGWICK SG: Inducible error-prone repair in *Escherichia coli*. Proc Natl Acad Sci USA 72:2753, 1975

SINDEN RR, COLE RJ: Topography and kinetics of genetic recombination in *Escherichia coli* treated with psoralen and light. Proc Natl Acad Sci USA 75:2373, 1978

TOPAL MD, FRESCO JR: Complementary base pairing and the origin of substitution mutation. Nature 263:285, 1976

TSUJIMOTO Y, OGAWA H: Intermediates in genetic recombination of bacteriophage T7 DNA. J Mol Biol 109:423, 1977

WAGNER R JR, MESELSON M: Repair tracts in mismatched DNA heteroduplexes. Proc Natl Acad Sci USA 73:4135, 1976

WITKIN EM: Thermal enhancement of ultraviolet mutability in a *tif-1 uvr* derivative of *Escherichia coli* B/r: evidence that ultraviolet mutagenesis depends upon an inducible function. Proc Natl Acad Sci USA 71:1930, 1974

chapter 12

PLASMIDS; GENE MANIPULATION

PLASMIDS

GENERAL PROPERTIES AND CLASSIFICATION

The F agent turned out to be the first recognized member of a broad group of **extrachromosomal genetic elements,** called **plasmids:** autonomously replicating, cyclic, double-stranded DNA molecules distinct from the cellular chromosome. It was discovered through its mediation of fertility (i.e. transfer of chromosomal genes, Ch. 9). The other major types of plasmids were also discovered by virtue of some recognizable function: resistance to various antibiotics (R factors), or synthesis of bactericidal proteins (bacteriocinogens). In addition, physical technics have now demonstrated plasmids in cell lysates of virtually all bacterial groups. In general, *plasmids carry genes that are not essential for host cell growth,* while the chromosome carries all the necessary genes.

Plasmids fall into two main classes. The **large plasmids** (e.g., F, R, and certain bacteriocinogens), 60–120 Kb long,* are mostly **conjugative** (also called self-transmissible); i.e., they code, within a bacterial cell, for an apparatus for their own transfer, by contact, to another cell. The **small plasmids** (e.g., some bacteriocinogens, some resistance determinants), 1.5–15 Kb long, are **nonconjugative,** but they can often be **mobilized** for transfer by a conjugative plasmid in the same cell. Plasmids can also be transferred, without cell contact, by **transfection** (uptake of infecting DNA from solution: Ch. 9).

The study of plasmids has built heavily on earlier knowledge of another class of autonomously replicating units, bacterial viruses (bacteriophages); but we shall discuss these later, as a foundation for animal virology. Though plasmids and temperate phages share many properties (Table 12-1), an important difference, besides their mode of transfer, is that plasmids may vary widely in size, as they gain or lose genes that are not essential for their replication or transfer, whereas viral DNA is fixed in size by the capacity of the viral coat.

The term **episome** was initially proposed for those plasmids and viruses that not only can replicate autonomously but also can be integrated into the bacterial chromosome. However, the term has lost its usefulness: it has no sharp boundary, since the ability to integrate into a chromosome varies quantitatively; and it breaks up two groups that do have sharp boundaries. The adjective **episomal** is still used to designate those replicons (plasmids or viruses) that are observed to integrate.

Classification. Plasmids differ widely in the number of their copies in a cell. The **specificity of the control system for copy number** provides a basis for classification into **incompatibility groups:** different plasmids in the same group are **incompatible,** i.e., they cannot persist for many generations within the same cells, for reasons discussed below. In addition, conjugative plasmids differ in the **specificity of the sex pilus** that initiates conjugation, and also in the **system that regulates its formation** (fertility inhibition; see Repression, below). Incompatibility groups can therefore be collected into overgroups, each specifying the same kind of pilus. In *Escherichia coli* most pili are either F (like that of F agent) or I (named for colicinogen I; see Bacteriocinogens and Bacteriocins, below).

Classification on the basis of **phenotypic effects** (e.g., the spectrum of antibiotic resistance, for the R factors) has decreased in importance, with the finding that plasmids can carry various combinations of genes for diverse properties, ranging from the formation of toxins to the use of special carbon sources (Table 12-2). (Antibiotic resistance has been particularly conspicuous, because of its easy recognition and widespread testing.) Some small plasmids are **cryptic** (i.e., have no known phenotypic effect).

The taxonomy of plasmids has epidemiologic applications in tracing the spread of resistance genes, both from man to man in hospitals and from agricultural sources to man.

Because plasmids exist in enormous variety, and because they continually exchange genetic elements, their **nomenclature** presents problems. Each new isolate, whether obtained from nature or modified in the laboratory, is now identified by the initials of the investigator and a serial number (e.g., pSC101). Early isolates of resistance plasmids were designated by R followed by a number.

---

* Plasmid size is also often expressed in terms of length or molecular weight: $1\mu m$ length = 2 megadaltons (1 Mdal = 10^6 daltons) = 3 kilobase pairs (Kbp); 1 Kb is also about the length of an average gene.

TABLE 12-1. Comparison of Plasmids (P) and Temperate Bacterial Viruses (V)

Shared properties
Autonomously replicate in host cell*
Host shows specificity
Some detectably affect host phenotype
Code for apparatus for own transfer (not all P)
May repress formation of that apparatus
Reversibly integrate into chromosome
Transfer of chromosomal genes
Transfection by free DNA

Differences
Transfer by conjugation (P) vs. infection (V)
Highly variable (P) vs. relatively fixed (V) size of DNA, and genetic composition, within a class
Viruses but not P found free in nature, in bacteria-free filtrates
Viruses can cause cell lysis

*But with use of cell machinery.

REPLICATION; COMPATIBILITY

Replication of plasmids is symmetric (see Fig. 10-27). With **large plasmids** it is **bidirectional,** and it is essentially synchronous with replication of the host chromosome. It also uses the same enzymatic machinery (including DNA polymerase III), but initiation requires, and is controlled by, products of plasmidial genes (see Genetics, below). With **small plasmids** replication is more primitive: it usually proceeds in **one direction,** terminating back at the origin; it uses polymerase I instead of III; it may not re-

TABLE 12-2. Observed Phenotypic Effects of Plasmids on Bacteria

Trait	Bacterial species
Fertility	Most gram-negative organisms; *Streptococcus faecalis; Streptomyces*
Resistance to:	
Various antibiotics	Widespread
Various metal ions	Widespread
Ultraviolet irradiation	*Escherichia coli*
Phages	*E. coli*
Serum bactericidal activity	*E. coli*
Ethidium bromide	*Staphylococcus aureus*
Production of:	
Bacteriocins	Widespread
Proteases (cheese)	*Streptococcus lactis*
Exotoxin	*Clostridium botulinum*
Enterotoxin	*E. coli, S. aureus*
Exfoliatin	*S. aureus*
Surface antigens	*E. coli*
Hemolysins	*E. coli, S. faecalis*
H_2S	*E. coli*
Chloramphenicol	*Streptomyces*
Metabolism of:	
Various sugars	Widespread
Hydrocarbons (toluene, xylene, camphor, salicylate, etc.)	*Pseudomonas*
Tumorigenesis in plants	*Agrobacterium tumefaciens*

quire any plasmid-coded protein; and it is not synchronous with replication of the chromosome.

In lysates plasmids are predominantly found free, but they are also found associated with membranes and with chromosomes. Nevertheless, small plasmids in particular do not appear to segregate with the chromosomes, since they are found in minicells (see Fig. 6-25), which lack chromosomes. The role of membrane attachment in plasmid replication is not certain.

Copy Number Regulation. In a growing culture of a given host each plasmid has a characteristic number of copies: 1 per chromosome for large plasmids, but often 10–20 for small ones. Regulation appears to depend on a **replication repressor** coded for a plasmid gene: plasmid copy number may be increased by a mutation in that gene. Evidently the concentration of the repressor depends on the number of copies of its gene (and hence the number of copies of the plasmid). Unlike the bacterial chromosome, many plasmids do not require synthesis of protein for initiation of their replication, and so when the synthesis of protein is blocked by chloramphenicol the extra metabolic energy is diverted to DNA synthesis, and the number of copies may reach 1000 per cell.

Incompatibility. When two different plasmids are present in the same cell random distribution of the small numbers to the two daughter cells causes unequal segregation at each cell division. If the plasmids are regulated by different replication repressors this fluctuation is separately corrected by each repressor in the next replication, and so each plasmid can maintain its characteristic average number within a cell line; these plasmids are **compatible.** However, if the plasmids have a common repressor, which regulates their **total number,** the fluctuation is not corrected, and most cells will eventually end up with one or the other kind of plasmid. Hence *plasmids with the same repressor specificity* (and usually extensive DNA homology) *form an incompatibility group:* its members cannot coexist for many cell generations in the absence of strong selection for each.

CONJUGATIVE PLASMIDS

MECHANISM OF TRANSFER

As Chapter 9 noted, in conjugation the attachment between the tip of a sex pilus and a receptor on another cell initiates transfer of a single strand of a plasmid. This process is normally associated with asymmetric DNA replication (possibly of the rolling circle type: see Fig. 10-27): One strand is nicked (at a constant site), transferred (5′ end first), and released when the circle gets back to the origin. Some process then initiates separation of the cells. The recipient cell synthesizes a complementary strand on

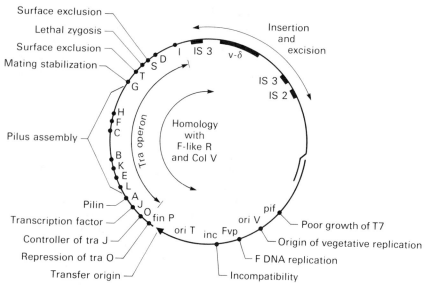

FIG. 12-1. Map of the F plasmid. **Letters** (A to T) indicate genes of the transfer operon (*tra*). F-like R plasmids have an identical *tra* operon, but also an additional gene, *finO*, which represses its function (Fig. 12-2). Insertion elements IS2, IS3, γ–δ provide the sites of integration in various Hfr strains. During conjugation the order of penetration is *oriT–inc–frp*, etc. The *tra* operon is thus last, and so in transfers from Hfr donors it enters only after the whole inserted bacterial chromosome. The separation of the *tra* and the *inc* operons accounts for the finding of different *inc* groups among plasmids of the same pilus type. Col V = colicinogen V.

the transferred strand, and it then seals the ends to form a closed circle.

Initiation of Transfer. A cut at the origin of transfer (*oriT*) of the plasmid (Fig. 12-1) initiates transfer; it appears to be mediated by a plasmid-specified endonuclease, specific for a particular DNA sequence. Moreover, deletion of the nick site abolishes transferability without impairing vegetative replication (which initiates at another site). This mechanism permits a given conjugative plasmid to mobilize certain nonconjugative plasmids in the same cell, i.e., those with the same *oriT* sequence.

The mechanism of transfer has been further illuminated by the isolation, after gentle cell lysis, of **"relaxation complexes"**: supercoiled plasmid DNA complexed with an endonuclease and other proteins. Exposure to detergents artificially triggers the endonuclease to produce a specific single-strand cut, which relaxes the supercoil. A **"pilot"** protein remains covalently bound to the 5' end of the nicked strand; it presumably directs and attaches the DNA to the membrane at the site of transfer.

GENETICS, REGULATION, AND RANGE OF TRANSFER

The Transfer (*tra*) Operon. In conjugative plasmids physical mapping of mutations (see Ch. 11, Genetic Mapping) has localized genes for the machinery of transfer and also for vegetative functions such as DNA replication (*rep*) and incompatibility (*inc*). The large size (95

Kb) of the F plasmid suggests that many genes are still unrecognized. The genes for conjugative transfer all belong to the transfer (*tra*) operon (Fig. 12-1), in which at least eight genes are concerned with formation of the pilus.

Repression. *With most conjugative plasmids (such as R) only rare cells form sex pili:* the transfer operon is ordinarily repressed by **fertility-inhibition** (*fin*) genes (Figs. 12-1 and 12-2), but derepressed (*fin*$^-$) mutants form sex pili in most or all cells. The F plasmid is such a high-fertility mutant, found in a bacterial strain that had been isolated from nature decades earlier.

R factors were initially classified as *fi*$^+$ (for fertility inhibition) or *fi*$^-$, on the basis of their ability or inability to inhibit conjugation induced by F plasmids in the same cell. We now know that a *fin* gene in *fi*$^+$ strains (Fig. 12-2) complements the missing gene in F, while that in *fi*$^-$ strains has a different specificity.

Transient Derepression. When donor cells with repressed pilus formation encounter a population of recipient cells a process occurs that markedly enhances spread of the plasmid. The initial recipients of the rare transfer lack fertility repressor, and they accumulate it only slowly; hence the transfer operon of the incoming plasmid is active, resulting in **transient high-frequency transfer (HFT)** from those cells. Accordingly, *spread is exponential.* After a few generations the progeny of an

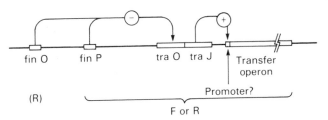

FIG. 12-2. Repression of the transfer operon in F-like plasmids. The transcription of the operon requires a product specified by gene *tra J*, which is in turn controlled by *traO*. The combined *finO* and *finP* products form a *traO* repressor. Gene *finO* is absent in the F plasmid, which therefore is derepressed; but the *finO* product of fi⁺ R plasmids in combination with the *finP* product of F causes repression of the transfer operon of the F factor.

infected cell accumulate repressor and acquire the usual state of **low-frequency transfer (LFT).**

This repression of sex pilus production, between episodes of plasmidial spread to virgin cell populations, evidently has survival value for the plasmids: it provides economy for the host cells and protects them from infection by the widespread male-specific phages.

Entry exclusion is another form of regulation under the control of the transfer operon: cells that form the same type of pilus rarely mate. Otherwise identical donor cells would uselessly exchange genes with each other. Mutations show that several *tra* genes are involved, some interfering with pair formation and others with DNA transfer.

Interspecific Transfer. The F plasmid can be transferred to various (but not all) strains of *E. coli*, and also to *Salmonella, Shigella,* and even *Proteus.* Some other plasmids (especially P, from *Pseudomonas*) exhibit even broader spread (see Fig. 9-2). The efficiency of transfer varies greatly; it may be limited by poor cell pairing, repression of plasmid replication, or enzymatic destruction of foreign DNA. Since identical genes are found on plasmids in distantly related bacteria it is clear that *interspecific transfer, though infrequent, extends over a broad range in nature.*

The spread of plasmids is balanced by their continual loss, observed both during culture and on storage, as discussed in Chapter 9; otherwise more and more would accumulate.

Conjugative plasmids have been found in about 1/3 of freshly isolated *E. coli* strains, and so far in about 30 other genera (mostly gram-negative). Small plasmids are even more common. Some streptococci conjugate, even though this genus lacks demonstrable pili; and they excrete sex pheromones, which promote aggregation with cells of opposite mating type.

INTEGRATION

Integration of a plasmid in the host chromosome is achieved by a single reciprocal crossover between the two circles; in detachment this crossover is reversed (see Fig. 48-13).

Integrated plasmids do not normally contribute to replication of the host chromosome. However, when the chromosomal origin is inactivated, by a temperature-sensitive (*ts*) mutation, replication can initiate at the origin of an integrated F plasmid.

Most integrations occur by crossovers between insertion (IS) elements (see Transposable Elements, Ch. 11) in the plasmid (Fig. 12-1) and specific regions in the chromosome. F may be integrated either by a *rec*-independent, transposon-like mechanism, or by a *rec*-dependent mechanism involving an IS element (see Ch. 11, Site-Specific Recombination). F′ plasmids also use the *rec* system for much more frequent crossovers, between their bacterial genes and homologous host genes (see F′ Plasmids, Ch. 9).

Stable Merodiploids. F′ merodiploid cultures are unstable even under conditions that select for plasmid genes, since these genes can exchange with the homologous chromosomal region, and the plasmid is then easily lost without impairing growth. However, in a *recA⁻* host, which cannot exchange genes, a merodiploid can be maintained by growth in a medium in which some plasmid genes are required. Studies of genetic dominance in such stable, heterozygous strains have shed light on mechanisms of drug resistance and regulation.

Chromosome Mobilization. Any conjugative plasmid can transfer a bacterial chromosome with it which has become integrated, but plasmids other than F do so very rarely. The main reason may be their repression of pilus formation.

RESISTANCE PLASMIDS (R FACTORS)

R factors, a group of conjugative plasmids of unusual medical interest, and much more widely distributed than F, were discovered in 1959 in Japan as the cause of multiple drug resistance in enteric bacteria. The rapid recent spread of R factors, and their mechanisms of resistance (mostly the formation of drug-inactivating enzymes), have been described (Ch. 7). These factors were later found to carry other kinds of genes as well (Table 12-2).

Origin. Antibiotics have not induced, but have selected for, resistance plasmids: these have been found even in strains of bacteria that were stored before the use of antibiotics. It seems likely that the enzymes formed by resistance genes evolved to carry out reactions with other, unknown substrates.

Medical Significance. The selection for R plasmids by antibiotic therapy, together with their interspecific transfer, creates a growing, cumulative threat to the treatment of bacterial diseases. These plasmids have been encountered not only in Enterobacteriaceae but also in the most

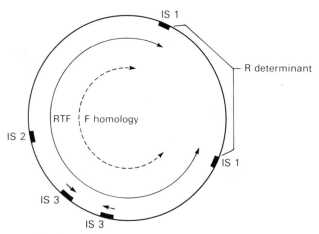

FIG. 12-3. Map of an R factor (R6), 97 Kb long. The resistance transfer factor (RTF), which comprises most of the molecule, is largely homologous to the F factor and contains similar genes (see Fig. 12-1); however, it also contains a finO gene, which is absent in F. The R determinant and the RTF are separated by IS1 elements, which promote the exchange of the R determinant. Two inverted IS3 elements generate a "lollipop" figure when the DNA is denatured and then reannealed, by intrastrand duplex formation (see Fig. 11-20, Ch. 11). They probably delimit a tetracycline transposon.

prevalent enteric organism, the anaerobic *Bacteroides*. Moreover, a penicillinase-producing R factor appeared in gonococci after two decades of spread in enteric bacteria. The nonconjugative plasmids coding for penicillinase in staphylococci do not appear to be closely related to enteric R factors.

Structure. Classic R factors are large plasmids with two functionally distinct parts (Fig. 12-3). One is the **resistance transfer factor (RTF)**; about 80 Kb long; it contains the genes for autonomous replication and for conjugation. The other part is a smaller **resistance determinant (R determinant)**; it varies widely in size and in its content of genes for drug resistance (**R genes**). Plasmids are genetically labile: many R genes are located in highly mobile transposons (see Transposable Elements, Ch. 11) 3–15 Kb long.

For example, the widely distributed transposon Tn 1 carries ampicillin resistance (Ap), and Tn 10 carries tetracycline resistance (Tc). Such translocatable elements migrate not only between R plasmids but also into other replicons (the chromosome, a cryptic plasmid, or a virus).

MODIFICATIONS OF R PLASMIDS

Dissociation. The RTF and the R determinant together usually form one unit. However, IS elements at the boundaries between them promote crossover, and in some R plasmids (especially when transferred from *E.*

coli to *Proteus*) these components dissociate (Fig. 12-4).

The dissociated components can be readily separated in lysates by virtue of a difference in buoyant density; their lengths add up to that of the parent plasmid. They can recombine in the cell, but they can also segregate into daughter cells. Cells have been found in nature that carry an R determinant on defective, non-conjugating plasmids; when mixed with cells carrying RTF the latter spreads to the other cells and causes transfer, with or without fusion, of the R determinant. Reversible dissociation suggests an explanation for the wide distribution of nonconjugative plasmids, and for the wide size range of plasmids.

Gene Amplification. With cells containing a gene for resistance to a given drug, in an autonomous R determinant, growth with a subinhibitory concentration of the drug often causes the emergence of cells with high resistance due to selection of **oligomeric plasmids.** These contain tandem duplications of the determinant (or of an R gene within it), formed by recombination between daughter strands during replication. This selectively amplified resistance gene can then be integrated into an RTF, yielding an R plasmid (cointegrate) that carries increased resistance. Conversely, tandem duplications of a determinant in a plasmid may be excised by recombination with each other (looping out), yielding a high copy number of the dissociated determinant.

Restriction Enzymes. Certain R plasmids from *E. coli* specify restriction endonucleases (Ch. 10), e.g., Eco RI and Eco RII. However, most R plasmids do not appear to have this property.

BACTERIOCINOGENS AND BACTERIOCINS

In 1925 Gratia in Belgium discovered that a protein released by a strain of *E. coli* inhibited the growth of a limited number of other strains. Subsequently some 20 such **colicins,** of different specificity and action, were recognized and designated by letters. They are widespread: 20% of tested enteric bacteria yielded colicins against a single test strain of *E. coli.* Similar substances have been isolated from *Pseudomonas, Bacillus,* and ultimately many other genera. The group as a whole are now called **bacteriocins.**

The formation of a given bacteriocin is due to a corresponding plasmid, called a **bacteriocinogen** (and identified by the same letter). Most bacteriocinogens repress their bacteriocin formation, and the escape of an occasional cell from repression, followed by lysis, appears to be responsible for the small amount of bacteriocin normally present in culture filtrates. Moreover, ultraviolet irradiation can induce some bacteriocinogenic strains to form and release bacteriocin, just as it induces certain

FIG. 12-4. Diagram of hypothetic formation of resistance factors and their observed occasional segregation of an RTF and a resistance-determinant plasmid. These two forms may be found in the same cell or in different cells; when in the same cell they can reversibly associate and dissociate.

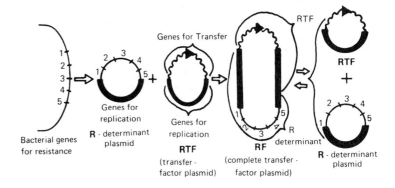

strains carrying a temperate phage to form infectious phage (Ch. 48). Most bacteriocinogens, therefore, behave like **highly defective prophages,** which can produce only one protein rather than the several proteins of a phage.

Some bacteriocinogens are conjugative. As noted above for R factors, pilus formation—F pili from colicinogen (Col) V, I from Col I—is normally repressed, and it is transiently derepressed on spread to a new population. Small bacteriocinogens (e.g., Col E1, E2, E3, K) are not conjugative.

MECHANISMS OF ACTION OF BACTERIOCINS

Bacteriocins, being proteins, are much larger than antibiotics (mol wt 50,000–80,000). They also differ in having a narrower antimicrobial spectrum, and they are much more potent: one molecule is often sufficient to kill a bacterium. Like phages (Ch. 47), bacteriocins attach to specific outer membrane (OM) receptors, and they can select for **resistant** bacterial mutants, which lack effective receptors.

Indeed, some bacteriocins morphologically resemble components of a phage tail (Fig. 12-5). Furthermore, some share a receptor with a phage (e.g., Col K and phage T6), since a mutational alteration in a receptor can cause resistance to both.

Some colicins are nucleases and act after entering the cell: Col E2 attacks DNA, and E3 stops protein synthesis by cleaving a specific fragment from the 16S RNA in a ribosome (see Fig. 13-15). Some colicins (E1, I, K) act on the cytoplasmic membrane, increasing permeability to

FIG. 12-5. Electron micrographs of a *Rhizobium* bacteriocin. **A.** The purified bacteriocin, which is similar to a phage tail, with a tube and a contractile sheath; the two components often separate during preparation; most sheaths are contracted. **B.** The bacteriocin perpendicularly adsorbed to sensitive cells, much like phage tails (see Fig. 47-3, Ch. 47). (Lotz W, Mayer F: J Virol 9:160, 1972)

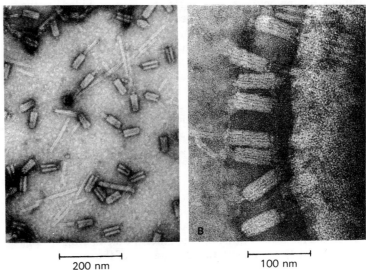

200 nm   100 nm

ions or inhibiting active transport. With this group bacterial mutations to **tolerance** to the bacteriocin alter the cytoplasmic membrane, without affecting binding of the colicin to the OM receptor.

Strains that produce a bacteriocin are **immune** to its effects, and some have been shown to contain a small **immunity protein,** which reversibly binds and inactivates the bacteriocin. Colicin E3 is found in the medium mostly as such a complex, which absorbs to the receptor; there the enzymatically active colicin is freed and enters the cell.

SIGNIFICANCE OF PLASMIDS

Plasmids and viruses were probably both derived ancestrally from cell chromosomes by cyclization of a small excised segment of DNA, including the site for initiating autonomous replication. This primordial DNA could then have evolved by gene duplication and mutation, plus incorporation of additional host DNA. Some of these replicons developed structures for their transfer: capsids for infection, or pili for conjugation. Viruses thereby gained the ability to spread without cell contact, at the expense of killing the host; plasmids developed a more symbiotic relation, advancing their own survival by assimilating genes useful for the host.

Plasmids have played a key role in the recent recognition that genetic recombination is not restricted to the classic homology-dependent mechanism, which regularly generates variation within higher species. Mechanisms that do not require extensive DNA homology bring about less regular recombinational events, such as gene transpositions and gene amplification. It is now clear that these events also occur in higher organisms (Ch. 49).

The evolutionary value of plasmids, in producing novel genotypes, depends on their ability to recombine with unrelated genes in this way. Accordingly, the evolution of plasmids may have proceeded hand in hand with the evolution of the elements that promote nonhomologous recombination: insertion sequences, trans-posons, and the special enzymes that recognize them.

The **practical effects** of plasmids are very considerable. The spread of resistance to antibiotics is best known; but plasmids also play a valuable role in the geochemical cycle by spreading genes for the degradation of complex organic compounds.

GENE MANIPULATION

For several decades the technics of gene transfer in bacteria have been used to produce recombinants between different mutations in a species: the classic foundation of genetic studies. More recently new technics have vastly expanded the possibilities for creating genetic novelty. First, ingenious methods for **rearranging genes within the cell** have made possible a wide variety of **genetic fusions** between normally unlinked segments of DNA. Second, methods for **manipulating DNA in vitro** have provided the simpler and much broader procedure of **molecular recombination (gene splicing),** in which a segment of DNA from **any source** can be inserted into a bacterial replicon (phage or plasmid). Moreover, these hybrid replicons can be multiplied **(cloned)** in a bacterial host and then recovered. Such cloning of genes within a species, through the use of specialized transducing phages, has been of immense value in studying the sequence, expression, and regulation of bacterial genes. The extension of cloning to DNA of any source promises to have an extraordinary range of uses, both fundamental and practical.

GENETIC FUSIONS IN THE CELL

Operon Fusion. In 1965 Jacob described a mutation that deleted the regulatory initiating region of the *lac* operon (see Fig. 14-22), along with the termination of the adjacent *purE* operon; the fused operons were now transcribed as a single unit, and the *lac* genes were controlled by purines and not by lactose. The broad application of this finding depended on Beckwith's development of technics for placing desired regions on either side of a segment in which deletions were not lethal and could be easily selected.

Genes can be fused by several methods. For example, transpositions to the neighborhood of *tonB*, the gene for the surface receptor for virulent phage T1, are favorable, because deletions in this gene cause resistance to T1 and hence are readily selected (Fig. 12-6).

Various tricks can be used to transpose genes from distant regions to this neighborhood. For example, phage φ80 normally integrates at an attachment site near *tonB*. In a host lacking that site it will integrate in many other locations, where it can pick up adjacent host genes (see Specialized Transduction, Ch. 48); and it can then insert these genes, in a normal host, by integration at its usual attachment site.

Since the products of the *lac* operon are particularly easy to quantify, fusions that yield these products have been valuable in studying the regulation of other operons whose products are less accessible. In another application, transfer of a gene from a low-activity regulatory system to a high-activity one can provide large amounts of an otherwise scarce desired product.

Protein Fusions. Frequently a fusion deletion will end within two genes, even in different operons; when the two fused genes are in phase (see Fig. 13-10, Ch. 13) they can be translated as a single unit,

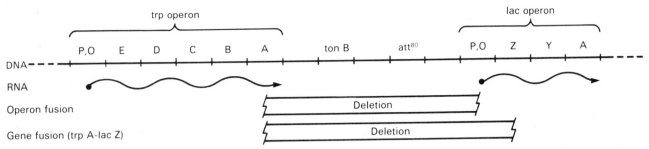

FIG. 12-6. Operon and gene fusion. The several genes for the enzymes of tryptophan synthesis (**A-E**) are transcribed into a single mRNA molecule, starting in a regulatory region (**p, o**) that initiates and controls that transcription: the whole constitutes the *trp* operon. The genes for lactose utilization similarly constitute the *lac* operon. When (as indicated) the *trp* operon is transposed to be near the *lac* operon and the region between them is deleted the *lac* mRNA is formed as a continuation of *trp* mRNA, subject to *trp* regulation. Moreover, when the ends of the deletion lie within a gene for a *trp* enzyme and one for a *lac* enzyme a hybrid gene is formed, and its mRNA may yield a hybrid protein, with a *trp* NH_2-terminal portion and a *lac* COOH-terminal portion. Gene *tonB* is for receptor for phase T1; its deletion is readily selected for. Gene att^{80} is the attachment site for phage $\phi80$.

yielding a fused protein (Fig. 12-6). These manipulations can help to define the region of the protein responsible for a particular function—for example, enzymatic activity, or insertion into a membrane.

In another application, the unknown product of a gene can be identified by fusion of that gene with another. Antibodies to the known product of the second gene will precipitate the fused protein, which is then used to raise Abs to both its moieties. These Abs can precipitate the intact unknown protein.

MOLECULAR CLONING (RECOMBINANT DNA)

The in vitro splicing of DNA into a bacterial replicon (e.g., a plasmid or a phage DNA), followed by cloning in cells, was first accomplished by Berg and by Kaiser in 1972 with defined segments of prokaryotic DNA, then by Cohen and Boyer with eukaryotic DNA, and in its most flexible form by Hogness with random fragments of DNA. This development, accomplished by the integration of a number of earlier, esoteric technics, illustrates vividly the unforeseeable consequences of basic scientific knowledge.

A key finding was that certain restriction endonucleases make staggered cuts in double-stranded DNA, with overlapping ends, at regions of rotational symmetry (palindromes) of 4–8 base pairs (see Fig. 10-33, Ch. 10). After separation these complementary ends can be reannealed, and they can then be covalently closed by polynucleotide ligase. A given endonuclease produces the same self-complementary ends in all its products; hence in a mixture of a purified replicon and some other DNA this treatment not only will reconstitute the original, cyclic replicon but will occasionally also form a **recombinant (chimeric) replicon,** with a foreign segment spliced into it (see Fig. 10-10, Ch. 10). The DNA can then be introduced into *E. coli* cells rendered competent by warming with $CaCl_2$ solution; any replicon infecting a cell will be perpetuated **(cloned)** in the cell's progeny.

After lysis the circular replicons are easily separated from the fragmented chromosomal DNA; in addition, a chimeric replicon can be cleaved, at the specific sequences joining its components, by the same enzyme that created the cohesive ends, and the components can be separated. Moreover, with high-copy-number plasmids, or with a temperate phage that can be induced to multiply vegetatively, many copies can be obtained per cell. In this way *large amounts of a specific segment of DNA, of any origin, can be readily obtained in pure form.*

Chimeric replicons can also be made by creating flush double-stranded ends, by mechanical fragmentation or by certain restriction endonucleases, and using an enzyme to add single-stranded **complementary homopolymer tails** (see Fig. 10-10). For example, an opened plasmid with a poly(dA) tail on the 3′ end of each strand will cohere with fragments of another DNA prepared with poly(dT) tails. Since the two ends of the opened plasmid are identical, rather than complementary, this procedure has the advantage that only recombinant plasmids are cyclized and therefore able to replicate. However, unlike the joints that are made and unmade by endonucleases, there is no enzyme that will cleanly cleave the two moieties of the recovered chimera.

Vectors. Two kinds of bacterial replicon have been used for cloning foreign DNA: plasmids and temperate phages (Ch. 48). Both have been improved as vectors by a severe reduction in size (through fragmentation of a naturally occurring replicon), so that their circle of DNA will be cut at only one site by certain endonucleases. (These derivatives must, of course, retain the genes required for autonomous replication.)

Suitable vectors have further extended molecular cloning to **eukaryotic host cells.** A hybrid of a bacterial and a yeast plasmid can transfect either host; and DNA from tumor viruses can transduce spliced DNA into cultured animal cells. Genes from a variety of eukaryotic sources can thus be expressed in a standard cell culture.

With phage λ *the DNA can be packaged within a phage coat in vitro* (efficiency ca. 10^{-3}), and so it can be introduced into bacteria by infection (efficiency ca. 10^6 times that of transfection of naked DNA). Moreover, small, highly defective derivatives cannot fill a phage coat unless enlarged by recombination; hence only recombinants of appropriate size, and not the unchanged vector, are recovered in phage particles (see Phage λ as DNA-cloning Vector, Ch. 48).

To minimize the possibility of inadvertent spread of chimeric DNA from the laboratory, via transmission from weakened hosts (see below) to normal hosts, the plasmids in use lack genes required for conjugation, and they have also been selected for a low frequency of mobilization by other, conjugative plasmids. Similarly, the phages used have been selected for defects that prevent their transmission to organisms encountered in nature.

Selection of Recombinants. We have seen that with phage highly efficient infection by recombinants is possible. With plasmids, taken up as free DNA, transfection is inefficient (about 10^{-6} of the cells), and with endonuclease splicing only about 10^{-3} of the transfectants may be recombinant. Accordingly, efficient selection of the recombinant cells is essential.

One procedure uses as vector a plasmid carrying two different antibiotic resistance genes, one containing an endonuclease site suitable for splicing. After exposure to splicing conditions, followed by transfection, three classes of cells are readily distinguished: those without a plasmid are not resistant to either antibiotic; those with the unchanged plasmid are resistant to both antibiotics; and those with a recombinant plasmid are resistant only to one, since any insert has inactivated the other resistance gene.

Identification of Specific Cloned Genes. With the procedures described the entire genome of a prokaryote or a yeast can be transferred into a set of several thousand clones within a few weeks. Individual segments are then identified: occasionally by complementation of an auxotrophic mutation, but primarily by **hybridization with radioactive** complementary RNA or DNA, called a probe.

Sensitivity is increased if the radioactive probe is not pure DNA but that DNA inserted in a much larger vector. The amount of radioactivity bound is thereby amplified many-fold.

In a convenient blotting technic, developed by Southern, DNA fragments separated electrophoretically in a gel adsorb spontaneously to a nitrocellulose film placed on the surface of the gel, and the film is then flooded with a radioactive probe and rinsed. Complementary fragments can then be revealed by autoradiography.

Colony hybridization is especially useful: "shotgun" recombinant cultures (i.e. containing random fragments from the total DNA of an organism) are grown on a nitrocellulose filter and then exposed to a reagent that lyses the cells and denatures their DNA; the DNA of each colony adheres to the adjacent filter, which is then exposed to a radioactive probe. Several thousand colonies can be tested per Petri dish. Even greater resolution

can be attained with recombinant phage, released in small plaques on a bacterial lawn (Ch. 48) and then lysed by alkali. With these technics a desired mammalian gene, constituting only about 10^{-6} of the mammalian genome, can be readily isolated.

For example, the gene of β-globin has been cloned by Leder and by Maniatis with a λ phage vector, using as probe ^{32}P-labeled globin complementary DNA, enzymatically prepared as a complement to the mRNA of erythroid cells. Such an isolation can be facilitated by preliminary fractionation of the DNA, using the labeled cDNA for assay, and then cloning a fraction enriched for the desired gene.

APPLICATIONS OF MOLECULAR CLONING

Eukaryote–Prokaryote Transfers. The broadest promise of molecular cloning for advancing our understanding would appear to lie in the insertion of eukaryotic genes into prokaryotes or suitable animal viruses. The resulting availability of any segment of DNA from any organism (including mammals and their viruses), together with simple methods for determining DNA sequences, revolutionizes the level of genetic analysis. For example, this approach has shown that immunoglobulin genes undergo **somatic rearrangement** during development (Ch. 17). Moreover, the determination of neighboring sequences can identify the **chromosomal locations** of various human genes and also of integrated viruses. In a medical application, the diagnosis of **hereditary disease** in a fetus, by studies on epithelial cells recovered by amniocentesis, has hitherto been restricted to defects expressed or visible in these cells, but it has now been extended to certain defects recognizable by DNA analysis (e.g., in the gene for a β-globin).

In the long run, the greatest promise may lie in the use of molecular cloning, in bacteria or a yeast, to study **gene regulation** in mammals and its changes in differentiation and in cancer. This approach provides means, for example, of detecting rearrangements in the DNA and changes in the quantity of specific genes. It also provides a simpler environment for the study of gene function, and though *differences in regulatory elements generally prevent the expression of eukaryotic genes in prokaryotes,* molecular cloning should help to identify the differences.

This barrier to gene expression presents an obstacle to another promising application: the use of bacteria to **manufacture desired proteins,** such as human hormones or interferon. One way to overcome this barrier has been to fuse the mammalian sequence with bacterial regulatory elements before splicing. Moreover, since untranslated intervening sequences are present in mammalian genes, but not in the corresponding mRNA, where an mRNA is available (e.g., that of globin) it has been useful to clone the corresponding cDNA, prepared by reverse transcription. Alternatively, chemical synthesis has provided a DNA sequence for a small protein (e.g., the 14-residue polypeptide hormone somatostatin), inferred from its amino acid sequence. Finally, to increase bacterial synthesis of insulin, by bringing about its secretion from the cell, the gene for this mammalian protein has been fused with the leader sequence of the gene for a secreted bacterial protein (penicillinase).

Prokaryote–Prokaryote Transfers. The unlimited transfer of DNA segments between bacteria also has many uses, in research and in technology. Molecular recombination can be used to create operon fusions and gene fusions, whose value has been noted above. The chemically synthesized gene for a prokaryotic tRNA has been cloned, opening prospects for an unlimited variety of synthetic "mutants" of various genes: these should be useful for correlating structure and function of gene products, and for creating new enzymes with useful properties. Cells containing novel mixtures of genes from different bacteria may improve the production of valuable chemicals (e.g., antibiotics). Finally, in the study of phase variation in bacteria transducing phages created by in vitro splicing have revealed a **rapidly reversible inversion of DNA**—a mechanism that undoubtedly has broad significance in regulation.

In **phase variation** (Fig. 8-3) a strain of *Salmonella* shifts, rapidly and reversibly, between making one and then the other of two immunologically distinct types of flagella. The process is directed by genes H1 and H2. In the active state gene H2 not only makes the H2 protein but also makes a repressor of gene H1; and when H2 is

FIG. 12-7. Reversible inversion of DNA in phase variation. A phage carrying the H2 region was recovered from a culture in which the H2 gene was expressed, and from another in which this gene was repressed (i.e., in the H1 phase). An endonuclease fragment carrying a regulatory sequence adjacent to the H2 gene was isolated from the phage DNA, denatured, and renatured. Mixed H1 and H2 preparations yielded heteroduplexes of two forms: (A) the ends were perfectly paired, but an unpaired segment of 800 nucleotides formed a bubble in the middle, as shown; or (B) this segment was paired but the ends were free. Evidently the reversible shift between the two phases is due to inversion of this 800-nucleotide segment. (Zieg J et al: Science 196:170, 1977. Copyright 1977 by the American Association for the Advancement of Science)

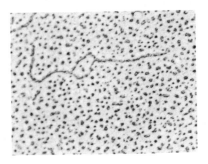

somehow inactivated H1 becomes active. Insertion of the H2 region into a phage made it possible to identify the mechanism of this rapidly reversible inactivation, for when the phage was harvested from cells in the H1 or in the H2 phase it yielded a particular endonuclease fragment in two different forms, a segment in one being inverted relative to its sequence in the other (Fig. 12-7). It would be surprising if such a readily reversible mechanism of regulation were not extended to other systems—for example, similar metastable states of gene activity that are seen in higher organisms.

REGULATION OF RESEARCH ON RECOMBINANT DNA

Concerned by the theoretic possibility that molecular cloning might inadvertently create dangerous new organisms, in 1973 a group of investigators called for a moratorium on certain experiments (e.g., incorporation of tumor virus genes into bacteria). The resulting extensive discussion yielded the valuable principle of **biologic containment**, i.e., the use of a host strain (EK2) with multiple mutational defects which ensure rapid self-destruction of the cells outside the laboratory. For example, in media that lack diaminopimelate (including the human gut contents) mutant cells requiring this compound will soon burst for lack of muramate synthesis (Ch. 4), leaving fewer than 10^{-8} survivors in 24 h. EK2 strains of vectors are also available that not only lack the means of transmission to other cells (as described above), but also have additional requirements for multiplication (e.g., a low temperature, or a particular suppressor mutation in the host).

Various governments subsequently set up guidelines, providing a sliding scale of restrictions for different kinds of experiments: physical containment ranged from P1 (ordinary medical bacteriology practice) to P4 (a maximum security facility). The usual *E. coli* K12 strain plus a non-conjugative plasmid is host-vector system 1 (HV1), while presumably more dangerous experiments require the more restrictive HV2 or HV3. The restrictions were later relaxed substantially, but administrative requirements for the approval of almost all experiments have persisted.

It seems appropriate to note here some of the scientific considerations that contributed to this partial relaxation. 1) After several years of experience, in dozens of laboratories, no illness or environmental damage could be traced to a recombinant. 2) *E. coli* K12, adapted for 50 years to laboratory media, cannot colonize the human bowel: when given to volunteers in large doses (10^9 cells) it regularly disappeared from the feces within 3–6 days. Moreover, an EK2 derivative showed, as predicted, rapid disappearance and lack of transmission of its plasmid to viable strains. 3) The promiscuous interspecific transfer of genes (Fig. 9-2) suggests that the incorporation of mammalian genes into *E. coli* is not altogether novel but occurs naturally. Indeed, *E. coli* cells carrying a restriction enzyme were found to insert added DNA fragments into a plasmid at the very same sites used by that enzyme in vitro. 4) Evolutionary principles emphasize that natural populational shifts in genotype depend not only on the creation of genetic novelty but also on selection pressure for adaptedness; and adaptedness requires a balanced genome. Insertion of DNA from a distant source is likely to impair rather than to improve that balance. 5) Epidemiologic considerations speak against a significant menace to the community. Thus it is highly unlikely that a recombinant of *E. coli* could be as virulent as many natural pathogens. Moreover, in work with such pathogens the occasional infection of laboratory personnel has been inevitable; but spread outside the laboratory has been rare and very limited in scope. Finally, 6) a recombinant *E. coli* with 0.1% foreign DNA would have the same habitat in nature as common *E. coli* (the vertebrate gut), and would share its mode of spread; and the spread of enteric pathogens is limited by sanitation.

Ironically, the concern that initiated the alarm has been reversed. It is now generally agreed that the DNA of an animal virus is more safely prepared from recombinant bacteria than from conventional animal cell cultures: spread by transfection of DNA is 10^6 times less hazardous than infection by viral particles, and a viral genome in *E. coli* has proved even less infectious than the viral DNA.

SELECTED READING

BOOKS AND REVIEW ARTICLES

BASSFORD J JR, et al: Genetic fusions of the *lac* operon: a new approach to the study of biological processes. In Miller JH, Reznikoff WS (eds): The Operon. Cold Spring Harbor Lab, 1978, p 245

BEERS R (ed): Genetic Alteration: Impact of Recombinant Molecules on Genetic Research. New York, Raven, 1977

BENNETT PM, RICHMOND MH. Plasmids and their possible influence on bacterial evolution. In Ornston LN, Sokatch JR (eds): The Bacteria, Vol VI. New York, Academic Press, 1978, p 1

BENZINGER R. Transfection of *Enterobacteriaceae* and its applications. Microbiol Rev 42:194, 1978

BUKHARI AI, SHAPIRO JA, ADHYA SL: DNA Insertion Elements, Plasmids, and Episomes. Cold Spring Harbor Lab, 1977

CLARK AJ, WARREN GJ: Conjugal transmission of plasmids. Annu Rev Genetics 13:99, 1979

COHEN SN: Transposable genetic elements and plasmid evolution. Nature 263:731, 1976

DNA Recombinant Molecule Research: Supplemental Report II. Congressional Research Service, Library of Congress. Washington, US Government Printing Office, 1976

FALKOW S: Infectious Multiple Drug Resistance. London, Pion, 1975

FRANKLIN NC: Genetic fusions for operon analysis. Annu Rev Genet 12:193, 1978

HELINSKI DR: Plasmid-determined resistance to antibiotics: molecular properties of R factors. Annu Rev Microbiol 27:437, 1973

KLECKNER N: Translocatable elements in procaryotes. Cell 11:11, 1977

KONISKY J: The bacteriocins. In Ornston LN, Sokatch JR (eds): The Bacteria, Vol VI. New York, Academic Press, 1978, p 71

LEWIN B: Gene Expression, Vol 3. Plasmids and Phages. New York, John Wiley & Sons, 1977

NEVERS P, SAEDLER H: Transposable genetic elements as agents of gene instability and chromosomal rearrangements. Nature 268:109, 1977

NOVICK RP, CLOWES RC, COHEN SN, et al: Uniform nomenclature for bacterial plasmids: a proposal. Bacteriol Rev 40:168, 1976

Recombinant DNA Research, Proposed Revised Guidelines. Fed Reg 43:33042, 1978; 44:69210, 1979

ROWBURY RJ: Bacterial plasmids with particular reference to their replication and transfer properties. Prog Biophys Molec Biol 31:271, 1977

STARLINGER P: DNA rearrangements in procaryotes. Annu Rev Genet 11:103, 1977

STARLINGER P, SAEDLER H: IS-elements in microorganisms. Curr Top Microbiol Immunol 75:111, 1976

SPECIFIC ARTICLES

BENTON WD, DAVIS RW: Screening of λgt recombinant clones by hybridization to single plaques in situ. Science 196:180, 1977

BERG PD, et al: Potential biohazards of recombinant DNA molecules. Science 185:303, 1974

BLATTNER FR, et al: Charon phages: safer derivatives of bacteriophage lambda for DNA cloning. Science 196:161, 1977

BLATTNER FR, et al: Cloning human fetal γ globin and mouse α-type globin DNA: preparation and screening of shotgun collections. Science 202:1279, 1978

CASADABAN MJ: Fusion of the *E. coli lac* genes to the *ara* promoter: a general technique using bacteriophage Mu-1 insertions. Proc Natl Acad Sci USA 72:809, 1975

CHAKRABARTY AM, FRIELLO DA, BOPP LH: Transposition of plasmid DNA segments specifying hydrocarbon degradation and their expression in various microorganisms. Proc Natl Acad Sci USA 75:3109, 1978

CLARKE L, CARBON J: A colony bank containing synthetic ColEl hybrid plasmids representative of the entire *E. coli* genome. Cell 9:91, 1976

COHEN SN, CHANG ACY, BOYER HW, HELLING RB: Construction of biologically functional bacterial plasmids *in vitro*. Proc Natl Acad Sci USA 70:3240, 1973

DAVIS BD: The recombinant DNA scenarios: Andromeda strain, Chimera, and Golem. Am Sci 65:547, 1977

GRUNSTEIN M, HOGNESS DS: Colony hybridization: a method for the isolation of cloned DNAs that contain a specific gene. Proc Natl Acad Sci USA 72:3961, 1975

HAMER DH, THOMAS CA JR: Molecular cloning of DNA fragments produced by restriction endonucleases Sal I and Bam I. Proc Natl Acad Sci USA 73:1537, 1976

ITAKURA K, HIROSE T, CREA R, RIGGS AD, et al: Expression in *E. coli* of a chemically synthesized gene for the hormone somatostatin. Science 198:1056, 1977

MANIATIS T, et al: The isolation of structural genes from libraries of eucaryotic DNA. Cell 15:687, 1978

MORROW JF, et al: Replication and transcription of eukaryotic DNA in *E. coli*. Proc Natl Acad Sci USA 71:1743, 1974

SHARP P, COHEN S, DAVIDSON NJ: Electron microscope heteroduplex studies of sequence relations among plasmids of *E. coli*. II. Structure of drug resistance (R) and F factors. J Mol Biol 92:529, 1973

SILHAVY TJ, CASADABAN MJ, SHUMAN HA, BECKWITH JR: Conversion of β-galactosidase to a membrane-bound state by gene fusion. Proc Natl Acad Sci USA 73:3423, 1976

SMITH HW: Survival of orally administered *E. coli* K12 in the alimentary tract of man. Nature 255:500, 1975

TILGHMANN SM, LEDER P, TREMEIR DC, POLSKY F, et al: Cloning specific segments of the mammalian genome: bacteriophage λ containing mouse globin and surrounding gene sequences. Proc Natl Acad Sci USA 74:4406, 1977

VILLA-KAMAROFF L, GILBERT W, et al: A bacterial clone synthesizing proinsulin. Proc Natl Acad Sci USA 75:3727, 1978

WARREN GJ, TWIGG AJ, SHERRATT DJ: ColEl plasmid mobility and relaxation complex. Nature 274:259, 1978

WATANABE T: Infective heredity of multiple drug resistance in bacteria. Bacteriol Rev 27:87, 1963

YAGI J, CLEWELL DB: Plasmid-determined tetracycline resistance in *Streptococcus fecalis:* tandem repeat resistance determinants in amplified form of pAMα 1 DNA. J Mol Biol 102:583, 1976

ZIEG J, SILVERMAN M, HILMEN M, SIMON M: Recombinational switch for gene expression. Science 196:170, 1977

chapter 13

PROTEIN SYNTHESIS

Biochemical analysis of the complex process of protein synthesis has been aided by repeated convergence with genetic studies in bacteria, and with studies of antibiotic action. Because the resulting pattern is so elaborate we shall first present a general survey and then review additional features.

OVERALL SURVEY OF PROTEIN SYNTHESIS

The first reproducible incorporation of radioactively labeled amino acids into acid-precipitable protein by cell-free systems, achieved by Zamecnik, Hoagland, and their colleagues in the early 1950s, led to recognition of many components of the system. The existence of messenger RNA (mRNA), however, was revealed by genetic studies on the regulation of gene expression, as will be described in Chapter 14. The use of synthetic ribonucleotide polymers as messengers then revealed the genetic code, and also most features of polypeptide chain elongation, under conditions that allowed messengers to bind without going through physiologic initiation. When the discovery of RNA phages provided viral RNA as a uniform, unfragmented natural messenger, initiation also became accessible to study. Finally, in 1967, Gesteland and Zubay achieved the synthesis of complete proteins in vitro, using a coupled transcription–translation system to form a phage enzyme directed by phage DNA. This approach can now be extended to bacterial genes, since the required high concentration of the particular gene can be provided by cloning DNA segments in a phage or a plasmid. Such systems permit very direct studies on the regulation of transcription and translation.

In describing the translation of the nucleotide sequence of an mRNA molecule into a corresponding polypeptide sequence it is convenient to postpone consideration of the process of chain initiation at the beginning of a gene (Fig. 13-1), and to start with chain elongation, i.e., the successive microcycles of amino acid polymerization.

CHAIN ELONGATION

In translation between a language of four nucleotides and one of 20 amino acids the specific **transfer RNA** (tRNA) molecules and their charging enzymes (**aminoacyl-tRNA synthetases**) are the units that read both languages: each enzyme attaches the correct amino acid to the 3′ terminus of a corresponding tRNA, by an ester bond (which leaves the amino group of the amino acid free); and the tRNA in turn, recognizing a particular trinucleotide **codon** in **messenger RNA (mRNA)**, serves as an "adapter" between codon and cognate amino acid, on the ribosome. *Accurate translation requires high fidelity in both the charging of tRNA and its pairing with the proper codon;* the ribosome promotes accurate pairing, as well as causing successive codons to be read and the corresponding amino acids to be covalently linked.

The **bacterial ribosome** is a particle with a sedimentation constant of 70S, mol wt of about 3 million, and diameter of 18 nm; it contains about 2/3 RNA (in 3 molecules) and 1/3 protein (in 54 molecules). In chain elongation the growing polypeptide shuttles back and forth between two sites that noncovalently bind tRNA: sites **A (aminoacyl or acceptor)** and **P (peptidyl or donor)**. This microcycle has three major steps (Fig. 13-2):

1) **Recognition:** A **trinucleotide codon** in the mRNA, positioned in the recognition region of the ribosome, specifies the binding, in the A site, of an **aminoacyl-tRNA (aa-tRNA)** that has a complementary anticodon (**c** in Fig. 13-2).

2) **Peptidyl transfer:** The nascent polypeptide, residing as **peptidyl-tRNA (pp-tRNA)** in the P site, replaces its ester bond to the tRNA with a peptide bond to the α-amino group of the aa-tRNA in the A site. The chain is thus transferred from the P to the A site and becomes one residue longer.

3) **Translocation:** The pp-tRNA moves back from the A to the P site, displacing the free tRNA (**b** in Fig. 13-2) left in the P site by the preceding peptidyl transfer. The mRNA moves with the pp-tRNA by the length of one codon, bringing the next codon (**d**) into the A site. The ribosome is now ready for another cycle.

In the **recognition process** the aa-tRNA first complexes with **elongation factor EFTu** (a soluble protein), and also with guanosine triphosphate (GTP), in the cytosol; and it is this ternary complex that binds to the ri-

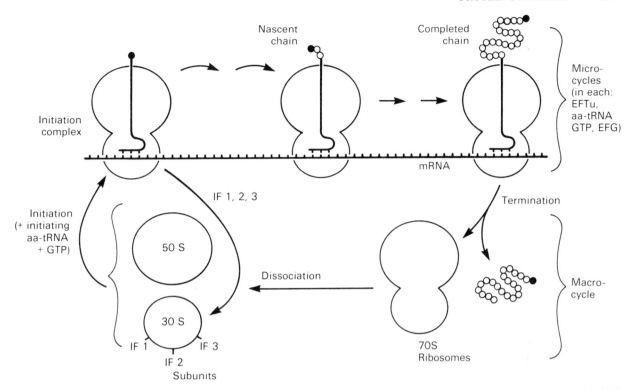

FIG. 13-1. Overall scheme of protein synthesis, portrayed as successive microcycles of chain elongation within a macrocycle of initiation, termination, dissociation, and reassociation during initiation.

bosome in recognizing a codon. Recognition has two steps. The initial, **reversible** binding excludes most incorrect aa-tRNA. Hydrolysis of the GTP (to GDP + P_i) then provides energy for the second step, which makes the binding of the aa-tRNA **irreversible,** thus permitting secure retention before acceptance of the expensive nascent chain. This reaction also serves as an **editing step (proofreading,** see also Fidelity of Recognition, below), for the shift in the position of the ligand provides a second opportunity for the ribosome·codon complex to discriminate between the correct aa-tRNA (which is retained) and occasional nearly correct ones (which are released along with the EFTu·GDP). After release the EFTu·GDP complex is dissociated (in the cytosol) by another **elongation factor, EFTs,** freeing EFTu for another cycle.

The **peptidyl transfer step,** unlike recognition (and translocation), does not require external factors: the enzyme is part of the ribosome, and the energy for peptide bond formation is provided by the somewhat higher energy of the ester bond to tRNA.

The **translocation step** involves another **elongation factor, EFG,** which hydrolyzes GTP on the ribosome and thus provides the energy for the movement in translocation. The EFG·GDP complex is then released (and spon-

taneously dissociates). The ribosome is now ready for another microcycle.

Cost. We should note that much energy is expended in chain elongation, to ensure flexibility, accuracy, and security against premature release. With each incorporation of an amino acid a pyrophosphate is released in the charging of tRNA (see below), and two GTPs are hydrolyzed on the ribosome.

THE MACROCYCLE

The microcycles of chain elongation occur within a macrocycle of initiation, termination, and reinitiation (Fig. 13-1). **Initiation** at the beginning of a gene is signaled by an AUG (or rarely a GUG) codon, which codes for the binding of a ribosome along with a special **initiating tRNA, N-formyl-Met-tRNA$_F$** (fMet-tRNA). This tRNA, unlike aa-tRNA, binds to the P site. Surprisingly, the AUG codon is not unique for initiation: it also codes for Met within a gene, where it binds a different tRNA, Met-tRNA$_M$, to the A site.

Initiation involves a macrocycle of **ribosome dissocia-**

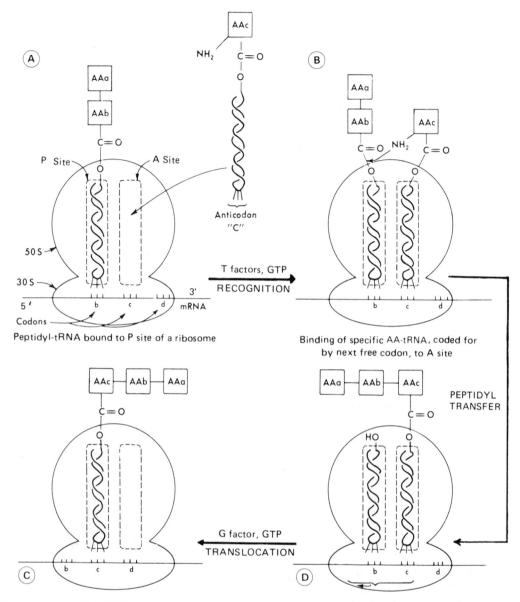

Peptidyl-tRNA bound to P site of a ribosome

Binding of specific AA-tRNA, coded for by next free codon, to A site

Movement of peptidyl-tRNA, along with corresponding region of mRNA, from A to P site, ejecting free tRNA

Peptide chain transfer to A site by formation of peptide bond with A A-tRNA bound to that site

FIG. 13-2. A. Peptidyl-tRNA bound to peptidyl (P) site of a ribosome. **B.** Binding of specific aa-tRNA, coded for by next free codon, to aminoacyl (A) site. **C.** Peptide chain transfer to A site by formation of peptide bond with aa-tRNA bound to that site. **D.** Movement of peptidyl-tRNA, along with corresponding region of mRNA, from A to P site, ejecting free tRNA.

tion and reassociation (Fig. 13-3), and three **initiation factors:** IF1, 2, and 3. The 70S ribosome released at termination dissociates spontaneously and reversibly into a **large (50S)** and a **small (30S) subunit,** but under physiologic conditions the equilibrium is unfavorable for dissociation; IF3 shifts the equilibrium by complexing with the small subunit, which then also binds IF1 and IF2, providing a reservoir of initiating small subunits. These complex with mRNA, fMet-tRNA, and GTP; and in subsequent steps the 50S subunit is added, GTP is hydrolyzed and released, and the three IFs are released. The product is a completed 70S **initiation complex,** carrying fMet-tRNA in the P site and ready for recognition of aa-tRNA in the A site.

Termination is considerably simpler than initiation. Any of three termination codons (UAG, UAA, UGA) cause the ribosome to bind a protein **release factor** (RF), which hydrolyzes the bond between polypeptide and tRNA and releases the polypeptide.

TRANSFER RNA

The charging of each tRNA by its specific aa-tRNA synthetase derives the required energy from the pyrophosphorolysis of ATP:

$$\text{Amino acid} + \text{ATP} + \text{Enz}$$
$$\downarrow$$
$$5'\text{-Aminoacyl} - \text{AMP} - \text{Enz} + \text{PP}_i$$
$$\downarrow \text{tRNA}$$
$$\text{Aminoacyl—tRNA} + \text{AMP} + \text{Enz}$$

The two-step process, linking the amino acid first to the enzyme and then to tRNA, increases the accuracy of the reaction, since the second step provides a correction mechanism that can recognize and excise an incorrect amino acid attached in the first step. In this way amino acids as similar as leucine and isoleucine are distinguished with an error frequency of $=1{:}10{,}000$.

In the cell the tRNA released from the ribosome is rap-

FIG. 13-3. Ribosome–polysome macrocycle. S = small subunit; L = large subunit; Term = termination triplet. On reaching a termination codon the polypeptide is released by a release factor (**RF**); the ribosome and the tRNA are then released from mRNA by the ribosome release factor (not shown). The free, 70S ribosome dissociates slightly, and initiation factor **IF3** increases the dissociation by stabilizing S in a conformation that does not couple with L. At some stages in the formation of the 70S initiation complex the initiation factors IF1, IF2, and IF3 are released, and they then reattach to a small subunit from a free, runoff ribosome. (After Subramanian AR, Davis BD: Proc Natl Acad Sci USA 61:761, 1968)

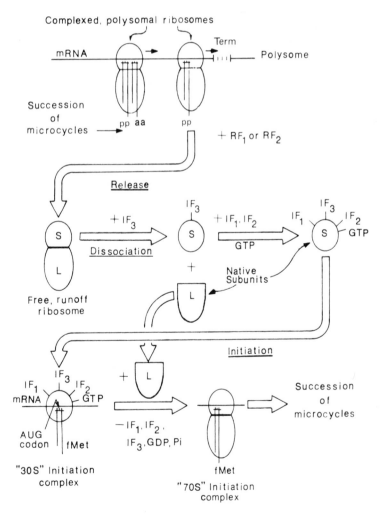

idly charged and then complexed with EFTu and GTP. Hence in steady-state growth most of the tRNA is in the form of the **ternary aa-tRNA·EFTu·GTP complex,** waiting to bind to the A site when the proper codon appears.

The affinity of each charging enzyme for its amino acid normally fits the steady-state endogenous level of that compound. Mutations that reduce that affinity can make a strain **pseudoauxotrophic:** i.e., it requires an added amino acid (which raises the intracellular level), though the biosynthetic pathway is unimpaired.

Structure. The tRNAs are 75–85 nucleotides long (mol wt about 25,000). Sequencing, first achieved by Holley, suggested a cloverleaf structure, with four base-paired arms and three unpaired loops, plus a variable extra arm (Fig. 13-4). Chemical studies on the accessibility and crosslinking of groups, and x-ray crystallography and other physical studies, have confirmed the predicted base pairing and have revealed additional folding into a compact, L-shaped structure, with specific binding sites for polyamines and Mg^{2+} ions.

Various tRNAs differ widely in sequence, which permits the charging enzymes to recognize each tRNA and match it with the proper amino acid. At the same time, the uniformity in general structure permits interchangeable binding of all tRNAs to the ribosome; and the uniform 3′-terminal -C-C-A, to which the amino acid is attached, permits the same ribosomal enzyme to incorporate the various amino acids, regardless of side chain, into the growing polypeptide.

Loop I (starting from the 5′ end) is called the **dihydroU (DHU)** loop (or arm) because in most tRNAs it contains this residue. The **anticodon** loop (II) was identified by the presence of the predicted anticodon sequence in various specific tRNAs. Arm III (the "extra arm") is highly variable in length. Loop IV (**pseudoU** loop) always contains a GTψC sequence (ψ = pseudoU).

The anticodon is always flanked at its 5′ end by two successive pyrimidines, which provide a flexible region with little base stacking, while at the 3′ end two purines provide a more rigid region with strong stacking. Accordingly, helical pairing of anticodon to codon allows a certain degree of ambiguity at the 5′ end of the anticodon (see Genetic Code, below).

Over 20% of the bases in tRNAs carry methyl or occasionally other groups. By permitting additional weak bonds at variable angles and distances, these modifications make possible a folding, and a conformational flexibility, more complex than that of a double-helical nucleic acid. Thus tRNA is an evolutionary product that has retained the base pairing of a nucleic acid while gaining some of the characteristics of proteins.

Indeed, as in proteins, mutational replacement of a single residue can yield a temperature-sensitive product. Moreover, aa-tRNA and uncharged tRNA exhibit conformational differences, which undoubtedly contribute to the cycling of tRNA on and off the ribosome.

RIBOSOMES

An *Escherichia coli* cell contains up to 15,000 ribosomes per genome, the number depending on the growth rate (see rRNA/DNA in Table 14-5). Their integrity and their function require suitable concentrations of Mg^{2+} (2–20 mM) and K^+ (50–100 mM); the intact ribosome has a sedimentation constant of 70S (mol wt about 3

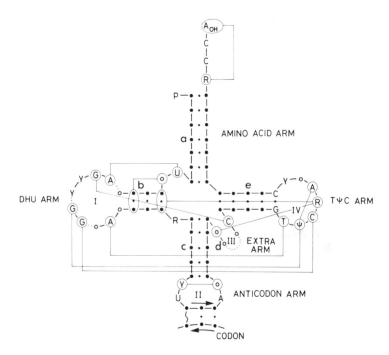

FIG. 13-4. Generalized cloverleaf model of tRNA. **I–IV** = unpaired regions (loops); **a–e** = base-paired regions; **solid small circles** with centered **dots** = base pairs; **R** = purine; **Y** = pyrimidine; **T** = ribothymidine; ψ = pseudouridine; **arrow** = 5′ → 3′ direction. **Letters** indicate nucleotides common to all sequences (but often with substituents, not indicated). **Circled** nucleotides, joined by light lines, are known to be paired or adjacent in the tertiary structure. (Levitt M: Nature 224:759, 1969)

million), and at low Mg^{2+} levels it dissociates artificially (and reversibly) into a **large (50S)** and a **small (30S) subunit.** When growing cells are disrupted gently (e.g., by enzymatic digestion of the wall), and the membrane fragments (with many attached ribosomes) are removed by low-speed centrifugation, the ribosomal particles remaining in the supernate are distributed (Fig. 13-5A), on zonal sedimentation, about 10% as **free 70S ribosomes,** 80% as **polysomes** (in which varying numbers of ribosomes are connected by a strand of mRNA; Fig. 13-6A), and 10% as "native subunits" (whose significance will be discussed below: see Initiation). When protein synthesis is blocked by means that prevent ribosomal movement on mRNA (e.g., by various antibiotics) the same pattern is seen; when initiation is specifically blocked the "runoff" ribosomes accumulate as 70S particles (Fig. 13-6B).

Ribosomal RNA (rRNA). Each subunit is about 2/3 RNA and 1/3 protein. The small subunit contains one 16S RNA molecule (mol wt 0.6×10^6), and the large subunit contains one each of 23S (mol wt 1.1×10^6) and 5S (mol wt 40,000). The RNA molecules, in the ribosomes or in solution, have much secondary structure; the sequences suggest many possible ways of folding.

The **ribosomal proteins** are tightly associated with each other and with the RNA, and rather drastic means (e.g., 2 M LiCl-urea) are required for their dissociation and solubilization. However, the separated proteins (mostly of mol wt 10,000–20,000) renature readily under normal ionic conditions, are soluble, and elicit specific Abs. The small subunit contains one each of 21 different proteins, named S1 to S21 according to their positions in gel electrophoresis. The large subunit contains proteins L1 to L34.

Protein S20 is identical with L26: the one copy per ribosome may be recovered in either subunit. This protein presumably provides contact between the subunits.

Electron microscopy (Fig. 13-6) shows that each ribosomal subunit has a highly irregular, characteristic shape, and that the 70S ribosome has a deep groove between the subunits, suggesting the possibility of movement at a limited interface.

Reconstitution (Assembly). In 1969 Nomura reconstituted active subunits of bacterial ribosomes from a solution of their RNA and protein molecules. This achievement was a milestone in the study of an exceedingly complex organelle: it emphasized the role of spontaneous, specific aggregation in organelle morphogenesis (i.e., in the organization of inert components into active structures), and it made possible a wide range of manipulations that shed light on ribosome structure and function. Thus a single protein from one kind of ribosome (e.g., mutant or labeled) can be inserted into a ribosome

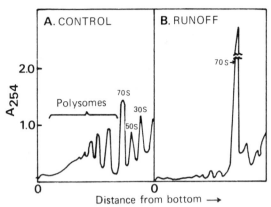

FIG. 13-5. Production of 70S ribosomes by runoff in *E. coli.* Control growing cells (**A**) were rapidly chilled by pouring the culture onto ice. Runoff cells (**B**) were slowly cooled, which allowed the ribosomes to complete their chains but not to reinitiate. (A smaller sample was analyzed because of the high 70S peak.) The cells were gently disrupted by treatment with lysozyme (to lyse the wall) and deoxycholate (to lyse the membrane). The lysates were analyzed in a sucrose gradient (see Fig. 10-4, Ch. 10). Similar results were obtained when net runoff was promoted 1) by starvation for a required amino acid or a required nucleic acid base, 2) by treatment with puromycin to release ribosomes prematurely, or 3) by treatment with actinomycin to cause depletion of mRNA. (After Subramanian AR et al: Cold Spring Harbor Symp Quant Biol 34:233, 1969)

with all the other components from another source. Such studies demonstrated early that streptomycin-resistant (str^r) mutations alter protein S12, since only this protein from str^r ribosomes conferred resistance on reconstituted ribosomes.

Success in ribosome reconstitution depended on the use of an "annealing" temperature (about 40° C). At 0° C (previously used to minimize denaturation and enzymatic degradation) improper contacts are evidently "frozen," as in the renaturation of DNA.

In the cell, ribosome assembly evidently occurs in a specific sequence: cold-sensitive subunit-assembly-deficient (*sad*) mutants form active ribosomes at 40° but accumulate specific incomplete precursor particles at 20°. Moreover, in vitro some proteins bind directly to rRNA, while others bind only after various members of the first group (**assembly mapping**). Those that bind directly bind to the RNA region that is synthesized first, suggesting that in the cell the ribosomal proteins assemble on RNA while it is still growing.

THE GENETIC CODE

DEGENERATE TRIPLET CODE

The four different kinds of bases in mRNA yield 64 possible sequences of 3 bases, and to specify the 20 different standard amino acids in proteins evolution has used 61 of the 64. Accordingly, most amino acids have two or more codons. The code is thus **degenerate;** but it is **not**

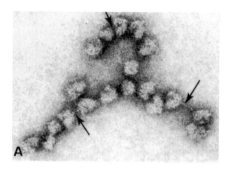

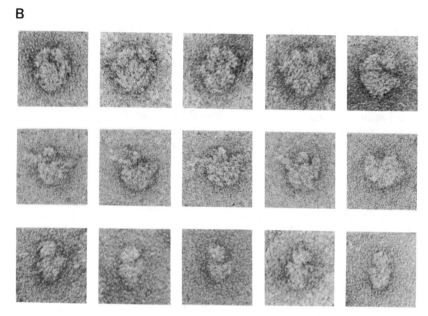

FIG. 13-6. Electron micrograph of negatively stained polysomes and ribosomes. **A.** Polysomes; arrows show mRNA strand connecting the ribosomes. (From Nonomura Y, Blobel G, and Sabatini D: J Mol Biol 60:303, 1971. Copyright by Academic Press Inc. (London) Ltd.) **B.** Various views of *E. coli* 70S ribosomes (top row), 50S subunits (middle row), and 30S subunits (bottom row). (Courtesy M. Boublik)

ambiguous, i.e., no codon specifies more than one amino acid.

Genetic studies of Crick et al provided formal evidence for a triplet code by showing that three frame shifts in the same direction could restore the reading frame (see Fig. 11-24). However, specific codons could be identified only by in vitro translation of known sequences. A breakthrough occurred in 1961, when Nirenberg discovered that polyU, a synthetic homopolymer of uridylic acid, directs the formation of polyphenylalanine: hence UpUpU (designated also as UUU) is the coding triplet for phenylalanine. Similar studies with additional synthetic polynucleotides of known composition and random sequence rapidly revealed the **composition** of the other codons.

The nucleotide **sequences** of the codons were revealed by the use of known trinucleotides to code for the binding of corresponding aa-tRNAs to ribosomes. However, some of the results were ambiguous, since additional variables influence the binding. A definitive approach came when Khorana prepared polynucleotides with known repeating sequences and showed that their translation generated polypeptides with repeat sequences of amino acids, as illustrated in Figure 13-7. Such studies reliably established the entire genetic code (Fig. 13-8).

Origin of Degeneracy. The translation of different codons into the same amino acid results only in part from the presence of different **isoaccepting tRNAs,** which accept the same amino acid but read different codons. In an additional, more economic mechanism, **anticodon degeneracy,** one tRNA can read two or three codons for the

same amino acid. This multiplicity depends on the loose positioning of the base at the 5′ end of the anticodon (i.e., the 3′ end of the codon), noted above; the resulting **"wobble"** (changes in angle relative to the mRNA) allows nonstandard pairing. Indeed, the four normal bases can form pairs, with either two or three hydrogen bonds, in 29 theoretically possible ways; and from a consideration of their angles and distances Crick accounted for the known patterns of degeneracy by predicting the specific wobbles depicted in Figure 13-9. The predicted anticodon sequences were then found in those tRNAs that bind to multiple codons.

Reading Frame and Direction. The bases are read in a continuous sequence, without punctuation between codons; hence initiation establishes the reading frame for the subsequent sequence. Indeed, this feature provided one kind of evidence that the code observed in vitro applies also in the cell: when two frame-shift mutations, of opposite sign, were introduced into a gene the shifted sequence between the two mutations yielded precisely the predicted amino acid substitutions (Fig. 13-10). In addition, these and other findings showed that mRNA is translated 5′ → 3′. Since this direction is the same as that of transcription (Ch. 11), translation of a chain can proceed simultaneously with transcription; i.e., the ribosomes join the growing mRNA soon after the 5′ end is released from DNA, and they move toward the growing end (see Fig. 14-14).

As we shall see later, in some viruses the same sequence may contribute information to more than one product, as a result of initiation in more than one reading frame. In addition, the splicing of

FIG. 13-7. Use of repeat polynucleotides for determining the genetic code. The polynucleotides were used as messengers in vitro in conjunction with a protein-synthesizing system from *E. coli*. The polypeptide chains produced were isolated and analyzed, and their composition defined the coding properties of the contributing triplets. (Adapted from Khorana G: Harvey Lect 62:79, 1968)

Repeat unit	Coding properties of polynucleotides	Expected product
UC	UCU CUC UCU CUC...... ser - leu - ser - leu -------	Single sequence irrespective of phase
AAG	AAG AAG AAG AAG...... lys - lys - lys - lys ------ A AGA AGA AGA AG..... arg - arg - arg - arg ----- AA GAA GAA GAA G....... glu - glu - glu ------	Different sequences in different phases
UAUC	UAU CUA UCU AUC UAUC..... tyr - leu - ser - ileu - tyr ----	Single sequence irrespective of phase
GAUA	GAU AGA UAG AUA GAU AGA UAG AUA asp - arg - - ileu - asp - arg - - ileu nonsense nonsense	Single response irrespective of phase but no long peptide, owing to periodic termination

FIG. 13-8. The genetic code. **Glu, Asp** = glutamic and aspartic acid; **Gln, Asn** = glutamine and asparagine; **ochre, amber, opal** = terminator codons. **Shaded triplets** code for polar amino acids. This diagram also summarizes the pattern of degeneracy (Fig. 13-9). **Brackets** at the right of amino acids indicate codons that are recognized by the same tRNA, in which wobbling of the 5'-end base of the anticodon leads to ambiguous reading of the 3' end of the codon. Some codons are recognized by more than one anticodon, as indicated by overlapping brackets.

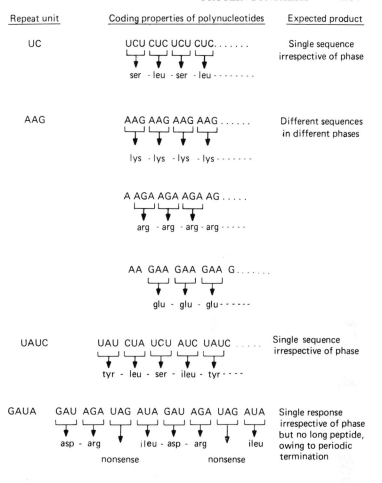

First letter of triplet (5' end)	Second letter of triplet				Third letter of triplet (3' end)
	U	C	A	G	
U	Phe] Phe] Leu Leu]	Ser] Ser] Ser Ser]	Tyr] Tyr] Ochre Amber	Cys] Cys] Opal Try]	U C A G
C	Leu] Leu] Leu Leu]	Pro Pro Pro Pro	His] His] Gln Gln	Arg] Arg] Arg]] Arg]	U C A G
A	Ileu]] Ileu Ileu] Met	Thr Thr Thr] Thr]	Asn] Asn Lys] Lys]	Ser] Ser] Arg] Arg]	U C A G
G	Val] Val] Val Val] or Met	Ala] Ala Ala] Ala]	Asp Asp Glu] Glu]	Gly] Gly] Gly] Gly]	U C A G

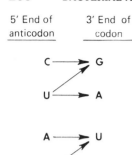

5' End of anticodon	3' End of codon

FIG. 13-9. Rules for anticodon degeneracy determined by wobble of the 5'-end base of the anticodon corresponding to the 3' end of the codon. **Arrows** indicate possible stable pairing, from anticodon to codon. I = inosinic acid (hypoxanthine nucleotide). (Adapted from Crick FHC: J Mol Biol 19:548, 1966)

eukaryotic mRNA, by the excision of intervening nontranslated sequences, may result in useful frame shifts.

Punctuation. As we have noted, AUG codes for initiation, and UAG, UAA, or UGA for termination. The terminating triplets were identified by producing a **nonsense mutation** (which interrupts chain completion) and selecting for point mutants that restored chain completion by changing the nonsense codon to one coding for an amino acid. The set of replacements was identified by sequencing the altered protein from these revertants;

with the known code, only **UAG** could provide the full set by single base substitutions (Fig. 13-11). This terminator codon was named **amber** (a translation of the name of a contributing student, Bernstein).

The other terminator codons were similarly identified as UAA (ochre) and UGA (opal). More direct studies later showed that the three nonsense trinucleotides do not bind any normal aa-tRNA, but they do bind a termination factor.

EVOLUTION OF THE CODE

Universality. The same code is found in man, bacteria, and a plant virus; hence all present terrestrial life probably had a common (monophyletic) origin. Moreover, the code could not change once life had achieved a certain degree of complexity, for a mutation that shifted the meaning of any codon would alter the sequence of practically every protein. However, organisms vary in their use of the different codons. Thus in different bacteria the proportion of G + C in DNA ranges from about 30% to 70%. These differences fall within the limits of degeneracy of the code; they may be accounted for by selection for the use of triplets with C or G rather than U or A (or vice versa) at the less specific 3' end of codons (see Mutations Affecting Mutation Frequency, Ch. 11).* Moreover, one

---

* DNA rich in G + C would have codons primarily of the type XXG or XXC; that poor in G + C would have codons of the type XXA or XXU. With equal nucleotide frequencies in the first two positions, in codons of the first type the proportion of G + C would be $(\frac{1}{2} + \frac{1}{2} + 1)/3 = 2/3$; in those of the second type it would be $(\frac{1}{2} + \frac{1}{2} + 0)/3 = 1/3$.

FIG. 13-10. Consequences of a shift of the reading frame in *E. coli* tryptophan synthetase. **I.** The normal correspondence between mRNA and polypeptide chains; the vertical lines indicate the reading frame. **II.** Acridine-induced **deletion** of a nucleotide, indicated as a **circled minus sign,** shifts the reading frame and causes production of a jumbled, inactive protein (indicated by **heavy box,** open on right). **III.** A second mutation, causing the **insertion** of a nucleotide at the **circled plus sign,** restores the reading frame beyond that point. The result is a polypeptide chain with only a short jumbled segment, compatible with function. (Data from Brammar WJ et al: Proc Natl Acad Sci USA 58:1499, 1967)

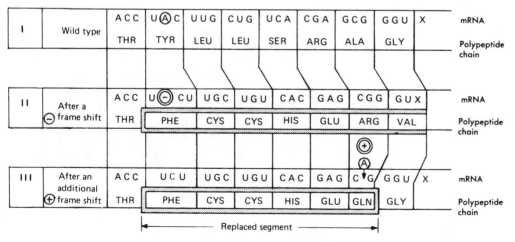

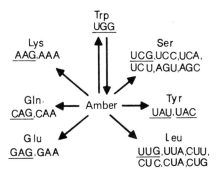

FIG. 13-11. Identification of an amber nonsense triplet. The diagram summarizes the amino acid substitutions found in the alkaline phosphatases produced by various revertants of an amber mutant, in a position originally occupied by a tryptophan residue. The standard codons for the substituted amino acids are listed: the **underlined** codons are those related to UAG by a single base change. UAG is the only triplet that has this relation to at least one codon for each of the substituted amino acids. (Garen A: Science 160:149, 1968. Copyright 1968 by the American Association for the Advancement of Science)

exception to the universality has been found: UGA is normally a terminator codon, but in mitochondria it codes for tryptophan (usually UGG).

In contrast to the code itself, the specificity of the machinery for its translation is not universal: mutations in one component can be balanced by mutations in another. Thus tRNAs of lower eukaryotes (yeasts) show some reactivity with aa-tRNA synthetases, but none with ribosomes, from *E. coli*.

Evolutionary Advantages. Degeneracy is not a sloppy consequence of the use of 64 codons to specify 20 amino acids plus some punctuation: it is a necessity for the evolution of new genes by the duplication of preexisting DNA, followed by sequential selection for missense mutations that yield an increasingly useful protein. If only 20 of the 64 codons specified an amino acid most nucleotide replacements would give rise to a nonsense triplet, thus halting further improvement of the evolving protein; but with a degenerate code they yield an altered protein. In addition, with the present code many mutations (and errors in base pairing in transcription) have slight or no effect on the protein. Thus in most codons a transition (purine-purine or pyrimidine-pyrimidine replacement) in the 3′ position would not change the amino acid specified. Moreover, with most other single nucleotide shifts the substituted amino acid is functionally similar to the original (e.g., polar or nonpolar; see shading in Fig. 13-8). If terrestrial life had to start again it might well develop nearly the same genetic code.

FURTHER ANALYSIS OF PROTEIN SYNTHESIS

RIBOSOME STRUCTURE

The intricate shape of the ribosomal surface is shown in Figure 13-6. The complexity and plasticity of this organelle present a great challenge to analysis at a molecular level. Problems include the topographic relations of its 54 macromolecules, the binding sites of its ligands, and its variations in conformation during various steps in its cycles. Unfortunately, unlike enzymes, the ribosome is too large and complex for detailed structural analysis by x-ray crystallography. A variety of other methods, however, are contributing useful information.

These include crosslinking of protein–protein, RNA–protein, and RNA–RNA neighbors; covalent attachment of analogs of normal ligands; variations in the accessibility of groups to chemical modification, as an index of conformational change; energy transfer between fluorescent dyes attached to various pairs of proteins in reconstituted ribosomes; neutron scattering in deuterated proteins incorporated into reconstituted ribosomes; identification of ribonucleoprotein fragments released by partial hydrolysis of RNA; and competition for binding between an Ab and a ligand. Perhaps the most powerful tool for topographic studies has been **immunoelectron microscopy**, in which Abs to a specific ribosomal protein (or to a ligand) link two particles, and the Ab attachment site can be localized in relation to the recognizable subunit shape.

Key Features. A number of features of the ribosome have emerged. First, the RNA and the proteins are interdigitated in a complex manner: parts of all the proteins, and many regions of the rRNAs, are accessible at the surface; and many of the proteins have a highly extended shape and reach the surface at multiple, distant positions (Fig. 13-12). (A single ribosomal protein, S4, has been crosslinked to as many as seven others.) Moreover, the RNA clearly contributes specificity, like the proteins, to the ribosomal structure: a single protein can bind specific loops from regions far apart on an RNA chain, and certain rRNA sequences pair, presumably functionally, with specific mRNA or tRNA sequences. Finally, because the ribosomal components are so intricately interwoven, a change at one site (mutation, drug binding) may alter or restrict the conformation at a distant region.

RIBOSOMAL GENES

Genes for Ribosomal RNAs. Hybridization to total DNA showed that the genes for rRNAs, unlike all other known bacterial genes, are present in **several copies per genome.** More refined studies with transducing phages or plasmids, carrying these genes associated with various neighboring genes, revealed seven copies, scattered on the chromosome. This multiplicity allows formation of

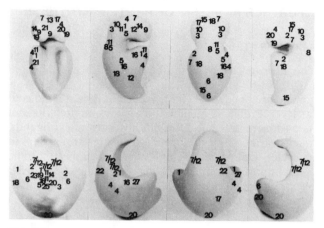

FIG. 13-12. Three-dimensional model of *E. coli* ribosomal subunits, with approximate locations of antibody-binding sites for individual proteins. **Above:** model for 30S subunit, in successive 90° rotations. **Below:** same for 50S subunit. The orientation of the two subunits in the 70S ribosome forms a channel between them. (Courtesy H. G. Wittmann.)

the required large amounts of the rRNAs. Moreover, at each site the genes for the three rRNAs are linked in an operon, providing the required equimolar quantities by including all three (as well as one or more tRNAs) in a **single large precursor transcript.** Specific nucleases, recognizing specific secondary structures, cleave and trim these precursors to yield the final products.

The precursors were recognized through their accumulation in a mutant defective in one of these nucleases (RNase III). They have the sequence 16S-tRNA-23S-5S, with an additional terminal tRNA in some. Though the seven rRNA operons differ markedly in their spacers (including three different tRNAs), only very slight sequence heterogeneity has been observed in the rRNAs; there must be strong selection for conserving optimal fit with the ribosomal proteins.

Genes for Ribosomal Proteins.

These proteins (r-proteins) form a very small free pool in the cell: they aggregate with rRNA as it is formed, and the synthesis of the two sets of components is closely coordinated.

Because the r-proteins are all essential their genes could not be mapped by the classic use of defective mutants. However, drug-resistant mutants with an altered r-protein identified the first few of these genes; and since then mutations generating additional phenotypes have been recognized for most r-proteins.

These include temperature-sensitive assembly, suppression of drug dependence, altered electrophoretic mobility, and altered fidelity of translation; they have been invaluable not only in gene mapping but also in correlating structure and function. Finally, newer technics have made it possible to bypass mutations and to recognize many more r-protein genes by identifying their normal

products, i.e., by using cloned DNA segments for coupled transcription–translation, either in vitro or in cells (infected after ultraviolet treatment has inactivated the cell genome).

About half the r-protein genes in *E. coli* are found (along with genes for other components of protein synthesis: i.e., elongation factors, RNA polymerase) in a cluster, the *str-spc* region, at 72 min on the 100-min map. The sequence of genes in this cluster could be identified by correlating overlapping restriction nuclease fragments of the DNA with the proteins produced by these fragments (including those responsible for resistance to streptomycin or to spectinomycin). Moreover, by interrupting the transcription of an operon by a polarity insertion (see Operon Polarity, Ch. 14) the genes in the cluster could be shown to be linked in four operons.

Designations of ribosomal genes in terms of initially observed phenotypes are now being replaced by more systematic designations based on the protein products. Thus *strA*, the gene for mutations to streptomycin resistance in protein S12, is now named *rpsL* (for ribosomal protein, small, 12th letter).

Evolution.

It is not surprising that the ribosome, with its intricate interdependence of many components, has changed very slowly in evolution: even bacteria with gross differences in DNA composition retain strikingly similar rRNA sequences. However, eukaryotic cytoplasmic ribosomes have evolved quite different RNA and protein sequences. They retain the same general structure as bacterial ribosomes but have somewhat larger RNA molecules (28S and 18S, as well as 5S), and an even greater increase in protein content. In consequence, they have increased in size by about one-third (80S, made of a 60S and a 40S subunit).

The ribosomes in chloroplasts are as small as those of bacteria, while those of animal cell mitochondria are even smaller; both types resemble bacterial ribosomes in their patterns of antibiotic sensitivity. It has accordingly been suggested that mitochondria have evolved from prokaryotic cytoplasmic symbionts. However, their mRNAs, unlike those of prokaryotes, undergo splicing.

CHAIN ELONGATION

A high Mg^{2+} level permits bypassing of physiologic initiation. This happy accident, unrecognized at the time, made it possible to dissect the genetic code with synthetic messengers, as described above. In addition, with these messengers Gilbert discovered the key features of chain elongation: attachment of the growing polypeptide to tRNA (pp-tRNA), by an ester linkage as in aa-tRNA; simultaneous binding of pp-tRNA and aa-tRNA to the ribosome; and the noncovalent nature of this binding (since the ligands can be released by a low Mg^{2+} level). These findings suggested a cycle in which pp-tRNA alternately occupies the two binding sites (P and A) on the ribosome (Fig. 13-2). Its presence in the P site can be demonstrated through release of the peptide by puromycin, a weakly held analog of aa-tRNA that binds to part of the A site (see Fig. 13-19).

Supernatant Factors. In chain elongation the "soluble" portion of the cell lysate supplies not only amino acids, tRNAs, and charging enzymes, but also three **elongation factors.** These proteins, separated by Lipmann, are designated EFTu, EFTs, and EFG (on the basis of the early mistaken assignment of the T factors to peptidyl transfer and the G factor to GTP hydrolysis). As noted above, the T factors function in recognition and EFG functions in translocation.

As fits its binding in a ternary complex with aa-tRNA and GTP, EFTu has a molar concentration in the cell about equal to that of total tRNA; it is specified by two genes. Only one-tenth as much EFTs (which does not bind to ribosomes) is present.

Ligand Binding. Transfer RNA straddles both subunits of the ribosome: at high Mg^{2+} levels aa-tRNA can bind to a 30S·mRNA complex, or nonspecifically to a 50S subunit; its physiologic binding can be impaired by Abs to various proteins of either subunit; and it can be cross-linked to both subunits. This double contact suggests a speculative two-step model for the complex movement in translocation, with each half of the pp-tRNA, in turn, binding firmly to one subunit while the other half moves relative to the other subunit.

Ribosomal RNA as well as proteins probably participates in tRNA binding: the 5S RNA of the large subunit has an accessible sequence complementary to the GTψC loop common to all tRNAs, and the 23S RNA has a sequence complementary to the DHU loop.

Both mRNA and the nascent polypeptide have a considerable length (about 30 residues) buried in the ribosome, i.e., protected from digestion by a nuclease or a protease, respectively. *The binding of EFTu and that of EFG* (both to the large subunit) *are mutually exclusive,* as can be shown by blocking their release with antibiotics.

Fidelity of Recognition. We have noted that recognition occurs in two steps, thus providing two successive opportunities for discrimination between correct and incorrect aa-tRNA (Fig. 13-13). The first binding, of the ternary complex, is **reversible:** the retained aa-tRNA is not yet

FIG. 13-13. Proofreading in the two-step recognition of aa-tRNA. The initial, reversible binding of a ternary complex (aa-tRNA•EFTu•GTP) probably involves a site that overlaps with the A site, since the aa-tRNA recognizes a codon that must already be in position in the A site. However, the aa-tRNA is not fully in the A site (i.e., is not able to receive peptidyl transfer) until after GTP hydrolysis and ejection of EFTu•GDP. This second step not only makes the binding of a correct aa-tRNA irreversible, but it somehow also allows recognition and ejection of an incorrect aa-tRNA (i.e., one whose anticodon pairs with only two nucleotides, instead of three, in the codon). Grossly incorrect aa-tRNAs are rejected in the first step.

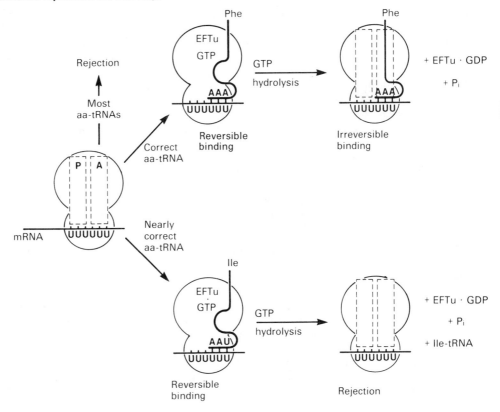

fully in the A site, and it cannot accept peptidyl transfer. This stage can be demonstrated with a ternary complex containing a nonhydrolyzable analog instead of GTP (GMPPCP, with a methylene instead of an oxygen between two phosphates). The second step, which requires GTP hydrolysis, locks the aa-tRNA **irreversibly** into the A site.

As noted earlier, the first step rejects most incorrect aa-tRNAs: i.e., binding is reasonably specific even when it is limited to the first step by the use of GMPPCP instead of GTP, or when it is carried out without EFTu at a high Mg^{2+} level. However, mistakes are made, and their elimination by a proofreading process was demonstrated by comparing binding and GTP hydrolysis of various ternary complexes (with ribosomes freed of EFG to prevent translocation and hence additional cycles of GTP hydrolysis). With polyU (see Degenerate Triplet Code, above) stable binding of the cognate aa-tRNA (Phe) from a ternary complex was associated with one equivalent of GTP hydrolysis; a nearly cognate aa-tRNA (Leu, normally specified by codon UUA) stimulated slight GTP hydrolysis but no stable binding; and a distant aa-tRNA stimulated neither.

INITIATION

The Initiating tRNA. With one of the two bacterial tRNAs for Met ($tRNA_F^{Met}$) a specific enzyme attaches a formyl group, after charging, to yield *N*-formyl-Met-$tRNA_F$ (fMet-$tRNA_F$); the other ($tRNA_M^{Met}$) does not undergo this reaction. The fMet-$tRNA_F$ serves as the initiating tRNA in bacteria, fixing the reading frame at AUG (or occasionally GUG) at the beginning of a gene, while Met-$tRNA_M$ recognizes internal AUG (i.e., within a gene).

The difference in binding of the two Met-tRNAs appears to depend on two structural differences. In $tRNA_F^{Met}$, unlike all other tRNAs (Fig. 13-4), the 5′-terminal nucleotide is not paired. This feature prevents formation of a complex with EFTu and allows complex formation with the initiation factors, which results in binding in the P site rather than in the A site. In addition, the formyl group eliminates the charge on the Met and thus increases the resemblance to pp-tRNA, the usual inhabitant of the P site; unformylated Met-$tRNA_F$ binds less well to that site.

In mammalian cell cytoplasm the initiating tRNA (Met-$tRNA_i$) is not formylated. Curiously, it can be formylated by the *E. coli* enzyme; and simpler eukaryotes (yeast) use an fMet-tRNA.

The Ribosomal Macrocycle. The finding that ribosomes dissociate into two subunits began to acquire significance when studies with heavy isotopes showed that the subunits exchange during protein synthesis (Fig. 13-14); hence at some stage ribosomes must go through a pool of subunits. Analysis of this process was delayed until the recognition of two kinds of 70S ribosomes ("complexed"

with mRNA and pp-tRNA, and free), and two kinds of 30S subunits ("native" complexed with initiation factors, and free). These findings revealed the macrocycle depicted earlier in Figure 13-3.

Dissociation. Complexed ribosomes are stabilized by their ligands: they are less readily dissociated by low Mg^{2+} levels, or by the high hydrostatic pressure in an ultracentrifuge.

Initiation is more sensitive than chain elongation to low temperature or abnormal ion concentrations; reversible dissociation may be the limiting factor in growth under these conditions. In bacteria evolved for life at unusual temperatures or salinity the dissociability of the ribosomes is suitably modified.

The **initiation factors** (IF1, 2, and 3) were discovered adventitiously, in the course of efforts to purify the system for protein synthesis in vitro. When ribosomes were freed of adsorbed proteins by washing with 1 M NH_4Cl they retained full activity in translating polyU, but their activity with viral RNA as messenger, which involves physiologic initiation, was lost: it could be restored by proteins in the wash fluid. The three initiation factors were separated by Ochoa and by Revel and Gros, and they were then shown to be present only on the native subunits and not on free or polysomal ribosomes.

In the course of initiation IF1 and IF3 are released as the 30S subunit forms a 30S initiation complex with

FIG. 13-14. Exchange of subunits by density-labeled ribosomes in *E. coli* cells. After lysis the "light" ribosomes sediment in a sucrose gradient at 70S and the "heavy" ribosomes at above 76S. Hybrid ribosomes (one heavy, one light subunit) were detected as a large peak of intermediate S value; they could also be distinguished from HH and LL ribosomes by equilibrium density gradient centrifugation. (After Kaempfer R et al: J Mol Biol 31:277, 1968)

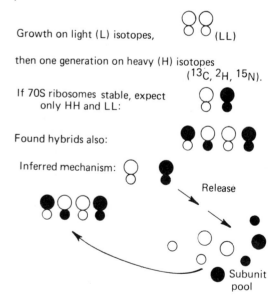

Growth on light (L) isotopes, (LL)

then one generation on heavy (H) isotopes (^{13}C, ^2H, ^{15}N).

If 70S ribosomes stable, expect only HH and LL:

Found hybrids also:

Inferred mechanism:

Release

Subunit pool

mRNA, fMet-tRNA, and GTP. A 50S subunit is added, and GTP hydrolysis then places the fMet in the P site and releases the IF2. (This mechanism resembles the release of EFTu in the final GTP-dependent positioning of aa-tRNA in the A site in recognition.)

Ribosome Binding Sites in mRNA. In binding at an initiating AUG (and not at an internal AUG) the ribosome recognizes additional features of the mRNA. The region that is bound can be isolated by allowing ribosomes to initiate at a specific viral gene but not to proceed with chain elongation: the ribosome protects an mRNA segment of 30 nucleotides from digestion by RNase, and the segment recovered from a given gene proved to be sufficiently uniform for analysis of its sequence. Such initiating segments from various genes have two common features: 1) all have an AUG at the middle (i.e., the groove binding mRNA on the ribosome extends on either side of the site of codon–anticodon pairing); and 2) about 10 nucleotides before the AUG all have the purine-rich sequence AGGAGGU (totally or in part). This sequence is complementary to a sequence found in the 3′ terminal region of 16S rRNA (Fig. 13-15); and this complementarity seems likely to be significant for initiation, since this region of the 16S RNA can also be cross-linked to initiation factors in an initiation complex.

Since colicin E3 (see Bacteriocinogens and Bacteriocins, Ch. 12) specifically cleaves the 49-nucleotide 3′-terminal segment of 16S RNA in the ribosome, it could be shown that the postulated pairing actually occurs on the ribosome: with ribosomes initiated on viral RNA, treated with RNase to leave only the bound region of the messenger, and treated with colicin E3, a complex between the cleaved 3′-terminal fragment of 16S RNA and the intergenic viral binding region of the viral RNA could be recovered (Fig. 13-15). In addition, the same purine-rich sequence has also been found at an appropriate position in bacterial mRNA, e.g., near the end of the *trp B* gene, which precedes the *trpA* gene (without an intergenic sequence).

Differences in the initiating (binding) sequences of various genes probably influence their efficiency of translation. **Phage RNAs** have an additional regulatory mechanism: **secondary structures** mask sequences that otherwise could be initiating.

CHAIN TERMINATION: RELEASE FACTORS

We have noted that the termination triplets recognize a release factor, which releases the polypeptide by hydrolyzing its bond to tRNA (i.e., by catalyzing peptidyl transfer to water). This activity could be assayed, in a purified protein-synthesizing system, by the release of a short polypeptide halted at a termination mutation early in a phage gene, or by the use of separate trinucleotide molecules as codons (i.e., the release of fMet from fMet-tRNA bound in the presence of the ApUpG trinucleotide plus a termination trinucleotide). Two **release factors** were isolated from *E. coli* lysates: RF1 recognizes either UAG or UAA, and RF2 either UAA or UGA.

An additional protein from the supernate, **ribosome release factor (RRF)**, is required to release the ribosome from the mRNA after polypeptide release.

FIG. 13-15. Base pairing between the binding region of mRNA and the 16S RNA of *E. coli*, as proposed by Shine and Dalgarno and supported by physical evidence. **A.** Probable base pairing within the 3′-terminal fragment of 16S RNA, cleaved by colicin E3. **B.** Altered secondary structure of that fragment on pairing with the initiation region for the A protein of RNA phage R17; the sequence on the 3′ side of the underlined AUG normally is translated, while that on the 5′ side appears to play a role only in positioning the RNA on the ribosome. Initiation regions of other genes show a greater or shorter length of pairing with the 16S RNA terminus. (After Steitz JA, Jakes K: Proc Natl Acad Sci USA 72:4734, 1975)

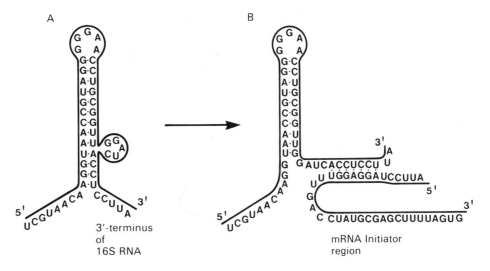

Some viral RNAs have a long untranslated sequence between successive genes, and it seems clear that the ribosome attaches at the beginning of each viral gene and is released at its end. The polygenic mRNA produced by a bacterial operon may lack such sequences, and it is not clear whether the ribosome drops off at the end of a gene or "reads through" to the next one. Indeed, sequencing in the *trp* operon has shown that two contiguous genes even overlap; i.e., the same A residue serves in a terminal UA*A* and an initiating *A*UG. Since the ribosomes translating the proximal gene might then be expected to hinder initiation by another ribosome at the distal one, it has been suggested that the terminating ribosome (or its 30S subunit) may slip back one nucleotide and then initiate.

General Features. Several features of protein synthesis deserve comment from the perspective of their selective value in evolution. First, this elaborate mechanism provides unlimited variation in polypeptide sequence, high fidelity, reliable retention of the growing chain, and reasonable speed (about 15 amino acids per second per engaged ribosome). These features exact a high economic price (four ~P per residue, plus rapid mRNA turnover): nonribosomal synthesis of the same peptide bond in short peptides (peptidoglycan, certain antibiotics) costs only 1 ~P. Second, the two-stage binding of aa-tRNA or fMet-tRNA, first reversible and then irreversible, is an important part of the mechanism for using energy to promote accuracy. Finally, the orderly succession of attachment and detachment of various factors is evidently a necessity: they impose on the ribosome a series of conformations that eventually return it, after each cycle, to its original state; and GTP hydrolyses ensure the unidirectional flow of these cycles.

MODIFICATIONS OF PROTEIN SYNTHESIS

SUPPRESSION

While reversal of the phenotypic effect of a mutation can be caused by a **true reversion,** which restores the original gene, more often it is caused by a mutation elsewhere. In classic genetics this effect is called **genotypic suppression:** the original mutation is still present, as can be demonstrated by its segregation, in further crosses, from the second mutation. Studies in bacteria have revealed several molecular mechanisms: some are **extragenic** (i.e, in another gene), as in the classic cases, but others are **intragenic** (i.e., within the originally mutated gene). In addition, environmental factors that decrease the accuracy of gene expression result in occasional misreading of a mutation, thus producing **phenotypic suppression.**

GENOTYPIC SUPPRESSION

The following mechanisms of genotypic suppression have been identified.

Intragenic Suppressors:

1) With intracodon suppressors a second mutation in the same codon calls for an amino acid that differs from the original amino acid but that nevertheless restores function to the protein.

2) **Reading-frame mutations** add a shift in the frame opposite in direction to the one already introduced in the gene. This compensatory shift restores normal reading, except for the segment between the two mutations (Fig. 13-10).

3) A distant **amino acid substitution,** in the same polypeptide as the primary mutation, can sometimes restore a folding required for function.

For example, tryptophan synthetase A of *E. coli* is inactivated by a particular Gly → Glu substitution, or by a distant Tyr → Cys: but with both mutations the protein formed is active.

Extragenic Suppressors:

4) **Codon-specific suppressors** each cause a particular error in the translation of a particular codon, and this effect can correct a mutational defect involving that codon. Both nonsense and missense mutations can be suppressed in this way.

Bacterial strains exhibiting such suppression are designated as su^+ and those lacking it as su^-, just as trp^+ and trp^- represent the ability or inability to make tryptophan. However, we should note that the mutant allele is designated + for *su* and − for most other bacterial genes.

5) **Generalized translational suppressors,** which are not codon-specific, will be discussed in a later section (Genotypic Ribosmal Ambiguity).

6) **Metabolic suppressors** of various types, which are **gene-specific** rather than codon-specific, have also been observed. These mutations may supply an enzyme that carries out or bypasses a blocked reaction, or they may restore activity to a mutant enzyme by altering the concentration of a substrate, cofactor, or inhibitor. They have been useful primarily in the study of metabolism and will not be considered further here.

7) **Polarity suppressors** will be discussed in Chapter 14 (Polarity Mutations).

MECHANISM OF CODON-SPECIFIC TRANSLATIONAL SUPPRESSION

A large majority of suppressor mutations are nonsense suppressors, which reverse the effect of a mutation that creates a terminator (nonsense) codon within a gene. Thus a phage carrying a nonsense mutation (amber) in the gene for a major protein produces only a prematurely terminated fragment of that protein when infecting normal cells, but in host cells carrying an amber suppressor the mutant gene produces both the fragment and a protein of normal length. It is thus clear that the suppressor causes **occasional but not regular insertion of an amino acid** at the site of the amber codon (Fig. 13-16).

Analysis of such products showed that *each su muta-*

FIG. 13-16. Genetic suppression of nonsense mutation. In the translation of an amber mutant gene (in a cell
or in extracts) an su^+ tRNA reads the amber codon
(UAG) in competition with an **R** (release) protein. A particular suppression causes incorporation of a particular
amino acid at the amber site, corresponding to the su
tRNA present; the various amber su tRNAs are derived
from those tRNAs in which a replacement of one nucleotide can yield the required anticodon (CUA⃗), which
reads UAG⃗ in antipolar fashion. The **efficiency** of suppression is the fraction of the readings yielding complete protein. The rate of cell growth restored by an su^+
mutation depends not only on its efficiency but also on
the frequency of harmful interference of the su tRNA
with normal termination elsewhere in the genome.

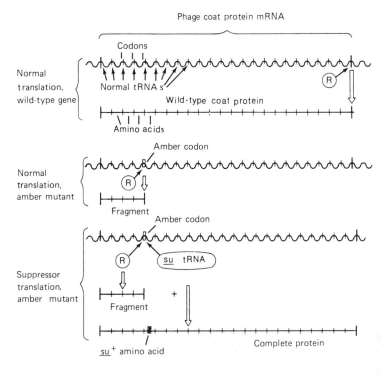

tion causes the insertion of a particular amino acid. This
result suggested that each su mutation might have
changed some tRNA, so that it would now read the
amber codon; the resulting translation of that codon as
an amino acid, rather than as termination, would allow
continuation of the polypeptide; and the competition
between an su tRNA and a release factor for the codon
would account for the formation of both a complete and
a short polypeptide. This hypothesis was readily confirmed by studies of protein synthesis in vitro.

Thus extracts of an su^+ bacterial strain (but not those of an su^-,
wild-type strain) allowed readthrough of an amber mutation, in the
mRNA of a phage gene; they also allowed the trinucleotide of the
amber codon to direct ribosomal binding of an aa-tRNA. The active component in the su^+ extract was a tRNA, and these properties
could be used to guide its isolation.

Structural Changes in Suppressor tRNAs. The set of
amber suppressors insert the same set of amino acids as
the set of reversions from the amber codon: i.e., with
codons that differ from the amber codon in only one base
(Fig. 13-11). Moreover, similar 1-nucleotide shifts are
found in the suppressors for other terminator mutations
(see Genetic Code, above). As these findings predicted,
*su tRNAs generally have a base substitution in the antico-
don* (Fig. 13-17). However, some substitutions in other
locations have also been found.

Frame-shift suppressors have also been observed. Where a
frame-shift mutation has expanded a sequence of three Cs in the
mRNA to a sequence of four the suppressor tRNA reads these as
though they were CCC. It follows that in translocation the tRNA is
moved on the ribosome, and the mRNA moves with it.

Translational suppressor mutations have made it possible to recover and maintain mutations (otherwise lethal) in essential genes
(see Conditional Lethal Mutants, Ch. 11), to distinguish the three
termination codons, to correlate structure and function in mutant
tRNAs, and to map genetically (and hence to manipulate the
quantity of) various tRNAs.

EFFICIENCY OF SUPPRESSION

Every nonsense suppressor has a characteristic efficiency, shown by **the fraction of the mutant chains that
are completed.** This fraction depends on the concentration of the su tRNA in the cell, its affinity for its charging
enzyme, and its affinity for the ribosome in competition
with release factors (which read the same codon). The su
mutations themselves usually slow down growth, because the misreading that corrects the mutational error at
one site creates many errors by misreading the same
codon elsewhere in the genome. In addition, the cell may
be deprived of a valuable tRNA.

In cells synthesizing predominantly a single protein (e.g., phage
coat protein) the efficiency of suppression can be measured precisely, as the ratio of completed protein to mutant fragment. A

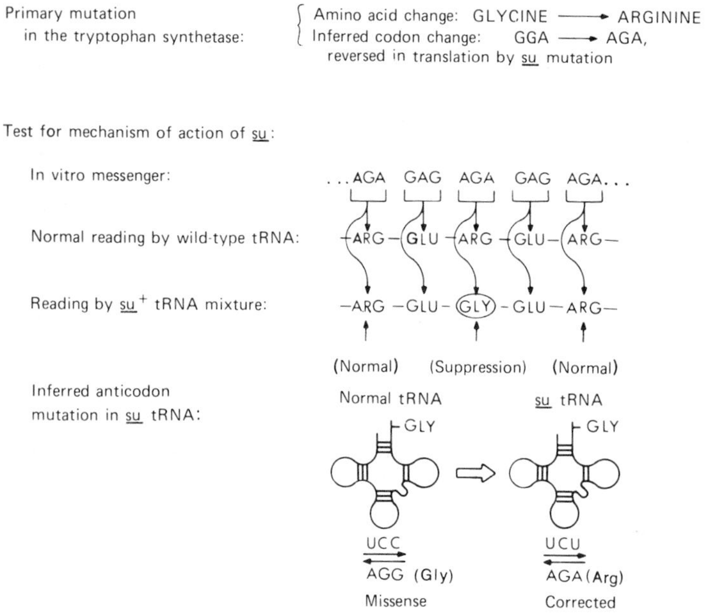

Primary mutation
  in the tryptophan synthetase:

{ Amino acid change: GLYCINE ⟶ ARGININE
Inferred codon change: GGA ⟶ AGA,
reversed in translation by <u>su</u> mutation

Test for mechanism of action of <u>su</u>:

In vitro messenger:

...AGA GAG AGA GAG AGA...

Normal reading by wild-type tRNA:

—ARG—GLU—ARG—GLU—ARG—

Reading by <u>su</u>⁺ tRNA mixture:

—ARG —GLU —(GLY)— GLU —ARG—

(Normal)   (Suppression)   (Normal)

Inferred anticodon
  mutation in <u>su</u> tRNA:

Normal tRNA   <u>su</u> tRNA

GLY   GLY

UCC   UCU
AGG (Gly)   AGA (Arg)
Missense   Corrected

FIG. 13-17. Evidence for altered codon recognition by a missense suppressor tRNA. Unfractionated tRNA was obtained from an *E. coli* strain carrying a mutation (*su*⁺) that suppressed a Gly → Arg mutation in the tryptophan synthetase gene. This tRNA was tested for altered codon recognition with a repeating AG polymer, together with ribosomes and protein factors from wild-type *E. coli*. In addition to the expected incorporation of Glu (GAG) and Arg (AGA), some Gly was incorporated by a suppressor tRNA$_{Gly}$ that recognized AGA instead of GGA, the normal Gly codon. (Data from Carbon J et al: Proc Natl Acad Sci USA 56:764, 1966)

more general but less precise measure is the degree to which the activity of a mutant enzyme is restored: suppressors with even very low efficiency (2%–15%) can be isolated.

Paradoxically, with a primary mutation in a codon widely distributed in the genome growth would be optimal with a low-efficiency (weak) suppressor. Conversely, if the suppressed codon is rare the corrections outweigh the errors and a more efficient suppressor will support faster growth. In fact, amber suppressors (anticodon CUA, with antipolar complementarity [Ch. 10] to UAG) are highly efficient (30%–75% restoration) and do not themselves slow growth. In contrast, the known ochre suppressors (anticodon UUA) are all weak (4%–12% efficiency) and do slow growth, which suggests that strong ochre suppressors would be lethal, presumably by interfering too often with normal termination. It therefore seems likely that the ochre codon (UAA) is a more frequent normal terminator than amber (UAG).

Missense suppressors are theoretically possible in great variety, but they are isolated much less frequently than nonsense suppressors. Unlike nonsense mutations (which prevent the formation of an active enzyme wherever they occur), missense mutations will be isolated only when located in certain critical positions in a polypeptide chain; and these mutations will often not be suppressible, since in these locations most shifts to other amino acids will also be deleterious. The most frequently encountered missense suppressors replace a larger amino acid by glycine, whose unobtrusive presence is apparently more easily tolerated than most other substitutions.

tRNA Gene Duplication. The altered specificity of a tRNA is compatible with viability only when the cell also possesses a nonmutated gene for the same or for an isoaccepting tRNA; otherwise one codon could no longer be translated. Some codons seem to be served by only one gene, since corresponding *su* mutations are

found only in cells that have become diploid for the appropriate region of the genome.

GENOTYPIC RIBOSOMAL AMBIGUITY

The fidelity of translation can also be influenced by mutations that alter the ribosome. Mutations in ribosomal protein S12, selected as streptomycin-resistant (*str*ʳ: see Streptomycin, etc., below), **restrict misreading,** while certain mutations in S4 or S5, selected to reverse this effect, cause **increased ribosomal ambiguity** (*ram*). Unlike the codon-specific suppression observed with *su* mutations in tRNA, *ram* suppresses a wide variety of mutations, including nonsense and frame shifts.

In addition, *str*ʳ mutations restrict, and *ram* mutations increase, amber suppression (*su*). Hence these genetic distortions of the ribosome evidently influence its relative affinity for a release factor and a *su* tRNA, competing for a termination codon.

Normal Ambiguity. Clearly some error is inevitable in transcription and translation in the cell, just as in DNA replication. The increased accuracy (and slower growth) seen in *str*ʳ mutants indicates that nature has not selected, in the wild type, for minimal ambiguity. Excessive precision must involve, as in any manufacture, too high a cost (e.g., in catalytic rate); and there might also be a biologic value to low-level ambiguity.

While a specific error in DNA replication (mutation) can be identified, as a result of its amplification in a clone of mutant progeny, errors in translation are harder to study since their products can be recognized only as a mixed population and not as specific altered protein molecules.

PHENOTYPIC SUPPRESSION

The background ambiguity in transcription and in translation can be increased by the addition of certain substances. These effects are not codon-specific, and they are most easily recognized through their suppression of various growth-limiting mutations.

Stimulation of Ribosomal Ambiguity. Gorini observed that streptomycin and related aminoglycoside antibiotics, at sublethal concentrations, can cause phenotypic suppression: in certain auxotrophs of E. coli they permit slow growth, by restoring the synthesis of a small amount of active enzyme by the mutant gene. The inferred mechanism, an increased frequency of errors in codon–anticodon pairing, was confirmed in vitro (see Streptomycin, etc., below).

Various other agents can also increase ribosomal ambiguity in vitro, including elevated Mg^{2+}, elevated temperature, and various organic solvents. (Indeed, 5% methanol can also suppress auxotrophy in cells.) Evidently many environmental effects on conformation can influence the noise level in the process of information transfer.

Ambiguous Messengers. Fluorouracil (Fig. 13-18) is incorporated extensively into RNA in place of U but then is sometimes misread as C. Its incorporation into RNA viruses is mutagenic, while in mRNA it causes phenotypic suppression of certain mutations.

FIG. 13-18. Fluorouracil (5-FU). The highly electronegative fluoro group makes the adjacent 4-carbonyl more negative and hence increases tautomerization to the enol form, which is an analog of cytosine rather than of uracil.

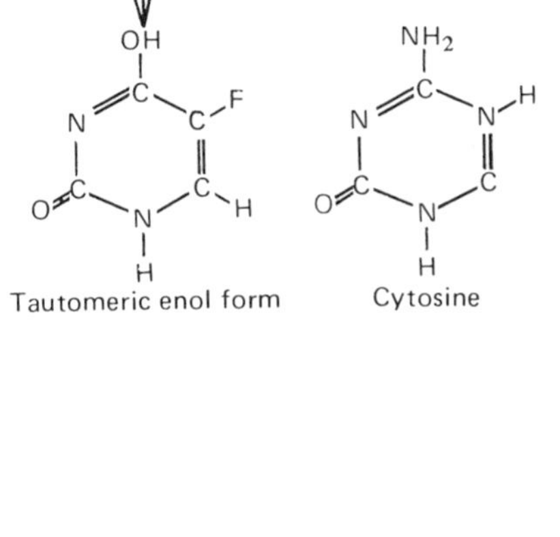

Fluorouracil

Uracil

Tautomeric enol form

Cytosine

INHIBITORS OF PROTEIN SYNTHESIS

Protein synthesis is the most frequent site of antibiotic action. Some of these inhibitors are useful in chemotherapy (Ch. 7); an even larger number have been helpful in analyzing ribosomal action.

METHODS OF ANALYSIS

Since *the ribosome is a body of conformationally interacting macromolecules,* and not a group of discrete enzymes, binding of an antibiotic at one site may influence other sites (and more than one function). Accordingly, while some antibiotics clearly block a particular step, others have pleiotropic effects. Moreover, the binding of different antibiotics may require different conformations that appear in the course of the ribosome cycle. This requirement divides ribosomal antibiotics into two general classes: those that can act on **chain-elongating ribosomes** (whose conformational mobility is restricted by their bound ligands), and those that can act only on the more flexible **free ribosomes,** at initiation.

The first class (e.g., puromycin, tetracycline, chloramphenicol), interfering with the microcycles of chain elongation, could be studied effectively with synthetic messengers, and they were shown early to act variously on recognition, peptidyl transfer, or translocation. The second class (e.g., erythromycin, spectinomycin) could not be understood until viral RNA made a good initiating system available. The aminoglycosides are unusual: they act, with different effects, on both free and complexed ribosomes.

To distinguish these actions required comparison between preparations that could initiate synthesis and others that could carry out only chain elongation, without reinitiation. The first was provided by viral RNA as messenger, and the second by purified polysomes from cell lysates, freed of native subunits and hence of initiation factors.

Antibiotics can also be classified in terms of their **locus of action** on the ribosome: e.g., the **subunit** that binds a radioactively labeled antibiotic, or that carries resistance (for those antibiotics that select for resistant ribosomes). For more refined localization the altered protein or RNA in resistant ribosomes can be identified by reconstitution; but because of the conformational interactions noted above, a protein responsible for resistance is not necessarily in the antibiotic-binding site. Affinity labeling of the ribosome with a reactive analog of the antibiotic better defines that site.

Some antibiotics are highly specific for ribosomes of bacteria, and others act equally on those of bacteria or of eukaryotes. Ribosomes of mitochondria generally resemble those of bacteria in antibiotic sensitivity.

ANTIBIOTICS ACTING ON RECOGNITION

Puromycin serves as an analog of the terminal amino-acyl-adenosine of tRNA (Fig. 13-19); binding to the part of the A site on the large subunit, it can accept the nascent polypeptide (or fMet) from the P site; but its binding is freely reversible (in contrast to that of aa-tRNA, locked into the A site). Accordingly, puromycin rapidly causes premature release of the nascent polypeptide chain, on either prokaryotic or eukaryotic ribosomes.

The release can be demonstrated in several ways: loss of the nascent polypeptide from the ribosomes, the accompanying separation of free ribosomes from mRNA (seen as a breakdown of polysomes), and the identification of free peptidyl-puromycin.

The puromycin reaction has been very useful in studying peptidyl transfer: it showed that peptidyl transferase is part of the large subunit, and that its action does not require GTP or soluble factors.

Tetracycline (Fig. 7-10, Ch. 7) directly inhibits recognition: it binds to the small subunit of bacterial ribosomes and blocks the binding of aa-tRNA (or pp-tRNA) to the A site. The pp-tRNA therefore remains in the P site, i.e., reactive with puromycin.

Though tetracycline prevents *stable* binding in the A site, it does not inhibit the GTP hydrolysis associated with aa-tRNA recognition; hence it evidently permits binding but makes the complex very unstable.

Pulvomycin interacts with EFTu and blocks the formation of the ternary complex (with aa-tRNA and GTP) required for recognition. **Kirromycin** blocks EFTu·GDP release from the ribosome, thus preventing the completion of recognition. Aminoglycosides also interfere with recognition, in a complex manner that will be discussed below.

ANTIBIOTICS ACTING ON PEPTIDYL TRANSFER

Chloramphenicol and **lincomycin** (see Fig. 7-10) bind to the large subunit of the bacterial ribosome and block peptidyl transfer. Chloramphenicol is the prototype inhibitor of this reaction. It should be noted that a block in the peptidyl transfer reaction does not imply action on the catalytic center of the enzyme: distortion of the ribosomal surface might keep either the donor or the acceptor group away from that center.

Lincomycin and several other inhibitors of peptidyl transfer compete with chloramphenicol binding, though they vary markedly in structure. The competition may well not be at the same site but may involved mutually incompatible conformations, like the competition between ligands at distant sites on an allosteric protein (Ch. 14).

A block in polypeptide transfer to puromycin could reflect inhibition either of peptidyl transfer or of translocation, since the latter block would keep the pp-tRNA from reaching the P site. However, the binding of fMet-tRNA in the P site does not involve translocation; hence inhibition of the release of fMet by puromycin has identified inhibitors of peptidyl transfer.

ANTIBIOTICS ACTING ON TRANSLOCATION

Erythromycin (Fig. 7-10) binds, reversibly, to free ribosomes but not to polysomal ribosomes. It remains during initiation, allows the formation of a very short polypeptide, and blocks further synthesis. The blocked complex is unstable: the ribosomes are released from the mRNA, recycle onto new mRNA, and are blocked again, as noted for streptomycin below. The block appears to be in translocation: the blocked polypeptide does not react

FIG. 13-19. Puromycin and its metabolic analog, the aminoacyl end of tRNA. (After Yarmolinsky MB, de la Haba GL: Proc Natl Acad Sci USA 45:1721, 1959)

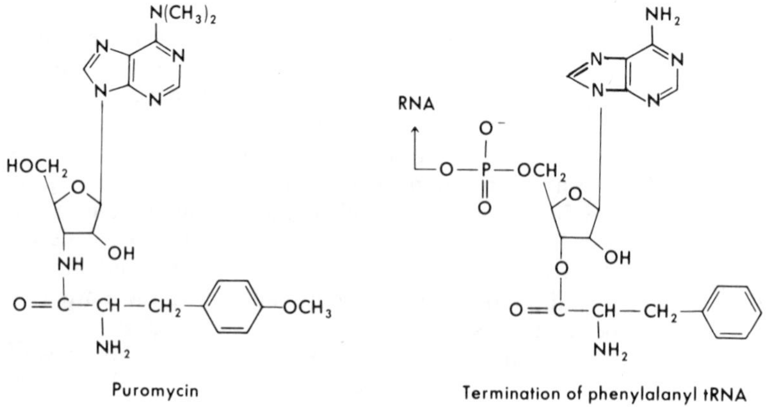

Puromycin

Termination of phenylalanyl tRNA

with puromycin, and erythromycin also fails to block the puromycin reaction with fMet, which rules out a simple block in peptidyl transfer.

However, the dependence of the inhibition on peptide length (and composition) suggests that erythromycin may well interfere only with a part of translocation that is virtually part of peptidyl transfer, i.e., placement of the peptide in the site required for peptidyl transfer. A relation to the peptidyl transfer center is also suggested by the finding that erythromycin, like many inhibitors of this transfer, blocks the binding of chloramphenicol.

Resistance to erythromycin can be caused by a change either in a protein of the large subunit or in the methylation of 23S RNA.

Viomycin, a polycationic peptide, blocks translocation (possibly because it tightens the association between the subunits), and also formation of the initiation complex. It binds to either subunit, and resistant mutants can be altered in either.

Fusidic acid (Fig. 13-20), a steroid, allows one round of translocation but prevents further cycles by inhibiting release of the EFG • GDP complex, on prokaryotic or eukaryotic ribosomes. It binds to EFG, and fusidic-resistant mutants are altered in EFG. As we noted above, this block shows that EFTu and EFG cannot be bound simultaneously. Moreover, **thiostrepton** (Fig. 13-20) and the structurally related **siomycin,** which bind to the large subunit, each block the binding of both factors. These findings reveal a strong interaction between the binding sites for EFTu and for EFG.

FIG. 13-20. Antibiotics that interfere with translocation.

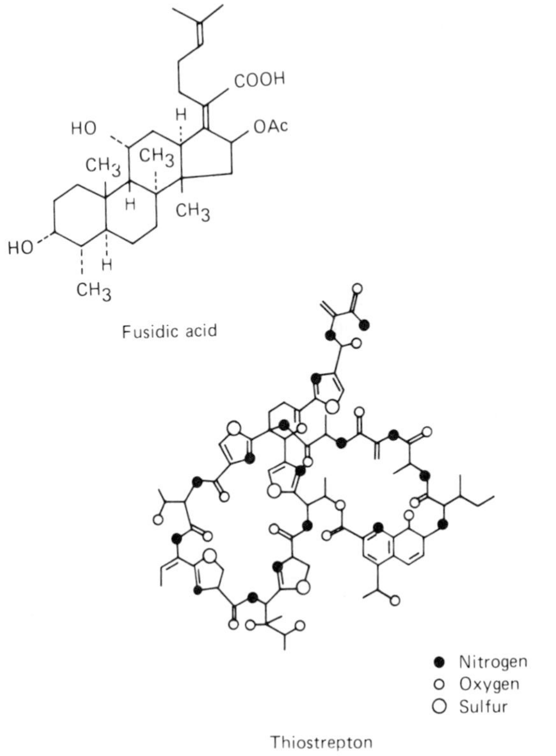

Fusidic acid

● Nitrogen
○ Oxygen
◯ Sulfur

Thiostrepton

STREPTOMYCIN AND RELATED AMINOGLYCOSIDES

The aminoglycosides are highly polar compounds with several cationic substituents, i.e., amino or guanidino groups on an inositol ring (aminocyclitol), and one or more aminosugars (see Fig. 7-9). The actions of this group have been studied mostly with streptomycin (Str).

In different concentration ranges aminoglycosides have two different, mutually exclusive effects on sensitive cells: at chemotherapeutic concentrations they inhibit protein synthesis completely, but at subinhibitory concentrations they stimulate misreading. Since only one molecule of Str binds to the ribosome these findings presented a paradox, which was explained when studies with purified systems showed that the effect of this binding depends on the state of the ribosome: with free ribosomes Str fixes the initiation complex, while with ribosomes already engaged in chain elongation it causes a less drastic distortion, which permits continued synthesis but impairs its accuracy. In the cell adequate concentrations evidently reach all the initiating ribosomes and hence block all protein synthesis; with lower concentrations there will be some initiation, and Str may then encounter the ribosome in chain elongation.

Effects on Initiation. Studies with an initiating system showed that Str allows initiation but blocks the initiation complex from becoming a chain-elongating ribosome: fMet is the only amino acid bound on the ribosomes. In addition, the blocked initiation complex is unstable: the ribosomes are released from mRNA, with a half-life of about 5 min; and though the released ribosome is no longer active in protein synthesis it is not inert: it recycles onto mRNA, forming a blocked initiation complex again. This **cyclic blockade** explains why **sensitivity to killing is dominant** in *str*[s]/*str*[r] heterozygotes, despite the presence of resistant ribosomes as well as sensitive ones.

Thus if the *str*[s] ribosomes were permanently held in the blocked initiation complexes they would soon all be engaged (since mRNA continues to be made in the presence of Str), and the new mRNA would then be read by the *str*[r] ribosomes. However, the turnover of the blocked *str*[s] ribosomes (and the mRNA) provides a supply of released Str-ribosome complexes that can continue to block new mRNA.

The cyclic blockade was revealed by the unexpected finding that cells blocked at initiation by Str nevertheless contain polysomes. But unlike the polysomes "frozen" by agents that block chain elongation (e.g., chloramphenicol), the Str polysomes turned out to be unstable **blocked polyinitiation complexes:** they carried fMet but not other amino acids, and the polysomol mRNA turned over rapidly.

Misreading Effect on Chain Elongation. Studies with synthetic messengers confirmed the misreading effect of

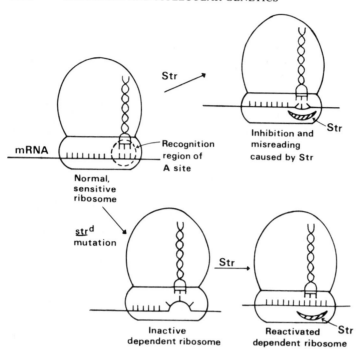

mRNA

Recognition region of A site

Normal, sensitive ribosome

Str

Inhibition and misreading caused by Str

Str

str^d mutation

Inactive dependent ribosome

Str

Reactivated dependent ribosome

Str

FIG. 13-21. Diagram to show how the same distorting effect of streptomycin (Str) is believed to impair the activity of sensitive ribosomes but reactivate dependent ribosomes.

aminoglycosides, predicted from their phenotypic suppression (see above). The incorrectly incorporated amino acids all had an anticodon that differed from the correct one in only one nucleotide.

More physiologic studies with natural messenger confirmed the misreading effect of Str on ribosomes engaged in chain elongation. Thus when purified polysomes (able to complete their chains but not to reinitiate) were halted in protein synthesis by deprival of an aa-tRNA (e.g., by using a supernate with a *ts* synthetase), Str could restore activity—i.e., by stimulating misreading it permitted the chains to fill an otherwise empty position. The distorting effect of Str on chain-elongating ribosomes also results in slowing of synthesis.

The misreading effect identified the recognition region of the ribosome as a major site of action of Str. This discovery also initiated an interest in the more general problem of variations in the fidelity of translation.

Dependence. Str can select not only for one-step high-level resistant mutants (*str^r*) but also for dependent (*str^d*) mutants, which require Str for growth. Both have various amino acid substitutions in a small region of protein S12, which explains their low frequency (ca. 10^{-9}). The explanation for dependence is diagrammed in Figure 13-21: **loosening** of codon-anticodon interaction by the antibiotic compensates for a mutational distortion of the ribosomes that **restricts** this interaction.

This interpretation is supported by the finding that a *ram* (ribosomal ambiguity) mutation (see Genotypic Ribosomal Ambiguity,

above), which promotes misreading much like Str, can replace Str in supporting the growth of an *str^d* mutant; conversely, *str^r* and *str^d* mutations restrict the misreading elicited by Str (or by *ram* mutations). *Str^r* mutations thus appear to distort codon–anticodon interaction in the same direction as *str^d*, but less drastically.

These distortions not only influence the competition of aa-tRNAs for a codon: they also strikingly influence the competition between an *su* tRNA and a release factor, both correctly reading a termination codon. Thus Str increases the efficiency of nonsense-suppressing mutations; conversely, many *str^r* mutations decrease that efficiency.

Variation in Aminoglycosides. Although the characteristic misreading action of the various aminoglycosides suggests that they bind to closely related sites, these cannot be identical, for *str^r* or *str^d* mutations have little effect on the interaction of the ribosome with most aminoglycosides other than Str. Moreover, Str-resistant mutants are altered in protein S12, but neamine-kanamycin-resistant mutants are altered in other S proteins.

Not all aminoglycosides select for high-level resistance or for dependence. Gentamicin binds to multiple sites: the first binding primarily causes inhibition, and the second paradoxically decreases the inhibition but causes marked misreading. Gentamicin and kanamycin, unlike Str, also block translocation.

Bactericidal Action. Chloramphenicol, which fixes the ribosomes in polysomes, prevents killing of bacteria by Str (see Fig. 7-4). In contrast, puromycin, which equally prevents protein synthesis but allows the ribosomes to recycle, even accelerates this killing. It thus appears that killing by Str does not require protein synthesis but does require an opportunity for Str to interact with all the ribosomes as free ribosomes—an interaction that is evidently irreversible in the cell.

The production of abnormal proteins by misreading does not appear to be responsible for the lethal action of Str, for *ram* mutations cause marked misreading without causing killing. The lethal action is accompanied by early damage to the cell membrane (without gross lysis); the mechanism, and the possible role in killing, are unknown.

Spectinomycin (see Fig. 7-9) is an aminocyclitol but not an aminoglycoside. Like Str, it specifically blocks initiating ribosomes as unstable initiation complexes and it selects for one-step mutations to high-level resistance (altered in protein S5). However, unlike the aminoglycoside group, it is bacteriostatic, and it does not cause misreading.

OTHER ACTIONS ON PROTEIN SYNTHESIS

Initiation complex formation appears to be blocked selectively by kasugamycin, edeine (a polypeptide), and aurintricarboxylate (a synthetic dye). It also is inhibited by agents (e.g., trimethoprim) that interfere with synthesis of the formyl group and hence of fMet-tRNA.

Amino acid activation is a minor area of antibiotic action: borrelidin specifically inhibits threonyl-tRNA synthetase. Some **synthetic analogs of amino acids** block charging of the corresponding tRNAs, while others can become attached and then incorporated.

SUMMARY

Table 13-1 summarizes the actions of various antibiotics, in terms of the major effect on a particular ribosomal function. However, there may also be additional effects, since each antibiotic restricts the conformational mobility of the ribosome to a greater or lesser degree. The aminoglycosides are especially pleiotropic.

Some antibiotics act, as noted in Table 13-1, on eukaryotic as well as on prokaryotic ribosomes, while others are specific for the latter. **Cycloheximide** blocks peptidyl transfer only on eukaryotic ribosomes.

THREE-DIMENSIONAL PROTEIN STRUCTURE

GENETIC DETERMINATION OF PROTEIN STRUCTURE

The **phenotypic effects of changes in genes** depend ultimately on how the corresponding polypeptide sequences are related to three-dimensional structure and to function. A principle of the greatest importance has emerged from studies of these problems: *information encoded in a one-dimensional sequence can specify a three-dimensional protein.* In addition, many proteins undergo some post-

translational processing: cleavage, addition of substituents, S-S crosslinking.

In tRNA, as we have already seen, interactions between various residues within the chain of a macromolecule can lead to specific folding. Proteins, which probably followed nucleic acids in evolution, have an infinitely greater range of three-dimensional structures, for their greater variety of residues permits a number of attractive and repulsive interactions, leading to intramolecular contacts over a wide range of angles, and to finely graded interactions with hydrophilic, hydrophobic, and charged components of the environment.

Hierarchy of Structures. In proteins **primary structure** refers to the sequence of amino acid residues, **secondary structure** to a helical or other orderly arrangement in a segment of the primary chain, and **tertiary structure** to the more or less globular product of further folding. Many proteins also form a **quaternary structure** by assembling identical or different **monomers** (polypeptide chains) to yield an **oligomer.** Monomers held together only by multiple weak bonds can be separated by detergents, acid, or concentrated salts; S-S bonds must be reduced.

In soluble proteins the residues exposed on the surface are mostly polar, and those in the core are mostly nonpolar. Proteins that are incorporated in membranes, however, have nonpolar surface regions in contact with lipids.

Specification of Higher-Order Structures. After denaturation (unfolding of specific conformation) many small proteins may be renatured under appropriate ionic conditions and temperature. However, with large proteins energy barriers often prevent renaturation. The initial selection of their unique native conformation may therefore depend on progressive folding during chain synthesis. In addition, binding with a **natural ligand** (i.e., a small molecule that interacts physiologically with the protein) can also direct or select for the native conformation of the released polypeptide.

Some functional proteins are monomeric but most are oligomeric. **Oligomer assembly** in vitro is often accelerated and stabilized, like protein folding, by specific ligands. Thus aggregation of the components of a viral coat (capsomers) may require the viral nucleic acid (see Virus Capsid, Ch. 46); and assembly of monomers of the long bacterial flagellum is initiated by a membrane-bound structure that forms its base.

STRUCTURAL FEATURES RELATED TO FUNCTION

Another important feature of proteins is **conformational flexibility,** dependent largely on the central core. Flexibility is essential both for catalytic activity and for its regulation: substrates and regulatory effectors, in forming weak bonds with a protein, alter its conformation (**induced fit**).

Such changes may be recognized by x-ray crystallography, by altered accessibility of side chains to specific chemical reagents, and by changes in spectral or antigenic properties. The size of cores

TABLE 13-1. Antibiotics Inhibiting Protein Synthesis

Antibiotic	Cell type	Site of action Subunit	Step	Altered in resistant mutants	Specific effects
Puromycin	Eu, Pro	L	R,P	—	Releases peptidyl-puromycin
Tetracyclines	Pro	S	R	—	Block binding in A site
Streptomycin	Pro	S	R	Protein S12	1. Irreversible: blocks movement of initiation complex
Neomycin	Pro	S	I	Proteins S5, S12	2. Causes misreading by elongating ribosome
Kasugamycin	Pro	S	I	Protein S2; 16S RNA	Blocks formation of initiation complex
Spectinomycin	Pro	S	T	Protein S5	Blocks movement after initiation
Chloramphenicol, lincomycin	Pro	L	P	—	Blocks fragment binding to P site, puromycin reaction with pp-tRNA
Sparsomycin	Eu, Pro	L	P	—	Promotes fragment binding in unreactive position; blocks puromycin reaction
Erythromycin	Pro	L	P,T	23S RNA, Proteins L4, L26	Blocks translocation soon after initiation
Siomycin, thiostrepton	Pro	L	T,R	23S RNA	Irreversible; blocks binding of EFG + GTP, and of EFT · aa-tRNA · GTP
Micrococcin	Pro	L	T	—	Blocks translocation
Kirromycin	Pro	S	R	EFTu	Blocks release of EFTu · GDP
Pulvomycin	Pro	S	R	—	Blocks formation of aa-tRNA · EFTu · GDP
Fusidic acid	Eu, Pro	L	T	EFG	Blocks release of EFG and GDP
Viomycin	Pro	S,L	I,T	S,L	
Rifamycins	Pro	Transcription		β-Subunit	Blocks initiation
Streptolydigin	Pro	Transcription		Polymerase	Inhibits extension
Dactinomycin	Eu, Pro	Transcription		—	Binds to DNA

Eu, Pro = eukaryotic, prokaryotic cells; L, S = large, small ribosomal subunit; R = recognition; P = peptidyl transfer; T = translocation, I = initiation.

seems to be limited: a very large chain, such as β-galactosidase (mol wt 135,000) or an immunoglobulin, forms two or more globular **domains** connected by a flexible hinge.

Interactions with specific macromolecules can also increase the activity of some enzymes, evidently by inducing the necessary conformation. For example, with many oligomeric enzymes made of identical chains these are inactive until aggregated. Moreover, though Abs generally inhibit normal enzymes (by covering a site or reducing flexibility), Abs to normal β-galactosidase can **activate** certain inactive mutant forms of the enzyme.

Conformational changes in the allosteric regulation of enzyme activity will be discussed in Chapter 14.

Significance of Quaternary Structure. In oligomers the monomers do not simply aggregate: they strongly influence each other's conformation, yielding a protein with important new features. (The same principle is extended in the interaction of proteins with RNA in ribosomes, or of proteins with lipids in membranes.) Some of the effects include altered activity or substrate specificity, interaction of catalytic and regulatory monomers, greater stability to denaturing agents (through the mutual reinforcement of interacting monomers), and formation of multienzyme complexes that bring together enzymes catalyzing successive reactions.

INTRAGENIC COMPLEMENTATION

The complementation test (see Fig. 11-1) has been useful to distinguish whether phenotypically similar mutations reside in the same gene (alleles) or in different genes: ordinarily only the latter exhibit complementation in a heterozygote, since the two genes each have one normal allele, which yields a normal product. However, certain pairs of allelic mutations exhibit intragenic (interallelic) complementation. This effect was puzzling until the enzymes involved were found to contain multiple copies of a monomer, which suggested that different mutant monomers might interact to yield a functional hybrid molecule.

This possibility was verified in vitro: with dimeric alkaline phosphatase, or with tetrameric β-galactosidase, those pairs of mutants that displayed complementation in vivo also yielded extracts that became active when mixed. This effect is restricted to certain monomers whose abnormality in tertiary structure can be corrected by the mutual stresses generated by interaction between subunits (Fig. 13-22). As a rule the correction is imperfect

FIG. 13-22. Intragenic complementation with an enzyme that normally contains two identical chains. Neither mutant alone produces functional dimers. The hybrid dimers, produced in the cell or in extracts, are functional because they mutually correct each other.

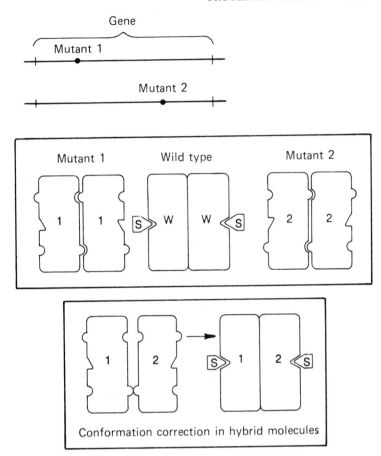

Conformation correction in hybrid molecules

and *the level of enzyme activity is low.* Since intergenic complementation, in contrast, yields a normal level, the complementation test for defining a gene is still useful.

Intragenic complementation usually involves missense mutations. However, it can also occur between deletion mutants in large genes, which yield complementary globular domains.

EVOLUTION OF PROTEINS

Hybridization of the denatured DNA from one species with that from another (Ch. 10) provides a quantitative measure of their degree of evolutionary relatedness, especially if attention is focused on the part of the DNA whose kinetics of hybridization (see Fig. 10-17) reflect the presence of single copies, rather than highly reiterated sequences. Evolutionary distances can also be estimated by comparing the products of a single homologous gene in different organisms—an approach initiated many decades ago with determinations of immunologic crossreactivity, and immensely improved with determinations of protein sequence. Studies of specific proteins,

however, test only a small sample of the genome, and they miss mutations to synonymous codons (about one-fourth of all possible base substitutions). Moreover, since bacteria interchange genes promiscuously a comparison of their single-gene products may be less reliable than the results of DNA hybridization.

The **phylogenetic tree** inferred from one extensively studied protein, cytochrome *c* (Fig. 13-23), closely parallels that inferred by earlier biologists from the total phenotypic differences between species; and consistent results have been obtained with several other proteins.

Among different proteins the rate of evolution varies widely, ranging from histone IV, with only 2 of its 101 residues differing between peas and cattle, to a short fibrinopeptide released on the activation of fibrinogen, with 10 of its 16 residues differing between rabbits and cattle. This variation depends primarily not on differences in mutation rate but on differences in the fraction of possible amino acid substitutions that are compatible with satisfactory function. Thus globular proteins evolve much more slowly than the terminal fibrinopeptide, whose structural requirements are obviously much less stringent. Such comparative studies thus also contribute, by identifying the invariant residues, to the correlation of structure with function.

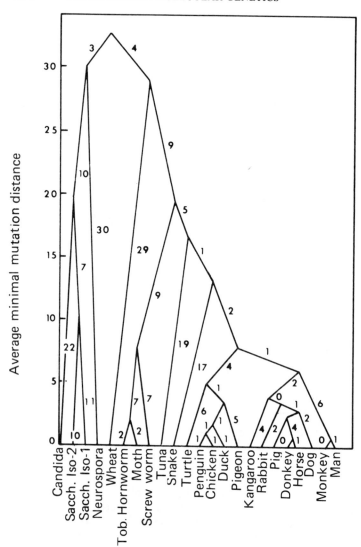

FIG. 13-23. Phylogenetic tree derived from amino acid sequences of cytochrome *c* of different species. The total of the numbers along the shortest path between any two species (''mutation distance'') represents the minimal number of nucleotide substitutions required to account for the observed differences in their sequences. The nodes, representing hypothetic common progenitors, connect those species, or subsets of species, which are closer in sequence to each other than to any other members of the total set. The long evolution and rapid reproduction of microbial species has generated large mutation distances, as illustrated by the large differences between yeasts (the four species at the left) compared with the differences between animals. (Margoliash E et al: Brookhaven Symp 21:259, 1968; see also Fitch WM, Margoliash E: Science 155:279, 1967. Copyright 1967 by the American Association for the Advancement of Science)

The relative constancy of the rate of evolution in a given protein has suggested that most of the observed substitutions are not selected for but are **neutral mutations,** which, after appearing in a progenitor, become fixed in a new species through statistical chance (**genetic drift**). Of course, such unselected variations may later be stabilized by selection if a second mutation, in the same codon or elsewhere, has a favorable phenotypic effect when combined with the first.

As with nucleic acids, divergence is much wider in microbes, with their long history and short generation time, than in higher organisms (see Fig. 13-23). Thus in *B. subtilis* a protease (subtilisin) in two strains differs in 83 of 275 residues, though it has essentially the same specificity.

Though accumulation of neutral mutations appears to be prominent at the level of macromolecular sequence, the more important selective, Darwinian evolution depends on those mutations that produce altered proteins with useful phenotypic effects. Data on protein sequences support the assumption that innovation proceeds by duplication of DNA segments, followed by mutational modification of the redundant genes. Thus some large polypeptides (such as immunoglobulin heavy chains) contain two or more segments with much homology, suggesting derivation by tandem gene duplication followed by fusion and cumulative mutation.

Mutations that change the specificity of an enzyme are especially important for evolution. This change does not necessarily require a long sequence of mutations. For example, wild-type *Enterobacter* can grow on ribitol but not on its isomer, xylitol; but selection with xylitol as the sole carbon source can yield, in one step, a mutant whose ribitol dehydrogenase has gained the ability to act on xylitol.

SELECTED READING

BOOKS AND REVIEW ARTICLES

BRIMACOMBE R, et al: The ribosome of *E. coli*. Prog Nucl Acid Res 18:1, 1976

BRIMACOMBE R: The structure of the bacterial ribosome. In Stanier RY, Rogers HJ, Ward JB (eds): Relations between Structure and Function in the Prokaryotic Cell. 28th Symp Soc Gen Microbiol, Cambridge University Press, 1978, p 1

BRIMACOMBE R, STOFFLER G, WITTMANN HG: Ribosome structure. Annu Rev Biochem 47:163, 1978

DAVIES J: Errors in translation. Prog Mol Subcell Biol 1:47, 1969

DAVIS BD: Role of subunits in the ribosome cycle. Nature 231:153, 1971

FITCH WM, MARGOLIASH E: The usefulness of amino acid and nucleotide sequences in evolutionary studies. Evol Biol 4:67, 1971

GORINI L: Informational suppression. Annu Rev Genet 4:107, 1970

HAHN FE (ed): Mechanisms of Antibiotic Action, Vol V. New York, Springer-Verlag, 1979

JUKES TH: The amino acid code. Adv Enzymol 47:375, 1978

NOMURA M, MORGAN EA, JASKUNAS SR: Genetics of bacterial ribosomes. Annu Rev Genet 11:297, 1977

NOMURA M, TISSIERES A, LENGYEL P (eds): Ribosomes. Cold Spring Harbor Lab, 1974. (A comprehensive, authoritative set of reviews)

WEISSBACH H, PESTKA S (eds): Molecular Mechanisms of Protein Synthesis. New York, Academic Press, 1977. (Includes detailed review by Pestka on antibiotic action)

WITTMANN HG: Structure and function of *E. coli* ribosomes. Fed Proc 36:2075, 1977

WOESE CR: The Genetic Code. New York, Harper & Row, 1967

ZABIN I, VILLAREJO MR: Protein complementation. Annu Rev Biochem 44:295, 1975

SPECIFIC ARTICLES

ABELSON JN, et al: Mutant tyrosine transfer ribonucleic acids. J Mol Biol 47:15, 1970

BODLEY JW, ZIEVE FJ, LIN L: The hydrolysis of a single round of guanosine triphosphate in the presence of fusidic acid. J Biol Chem 245:5662, 1970

CHINALI G, WOLF H, PARMEGGIANI A: Effect of kirromycin on elongation factor Tu. Location of the catalytic center for the ribosome. Eur J Biochem 75:55, 1977

HELD WA, BALLOU B, MIZUSHIMA S, NOMURA M: Assembly mapping of 30S ribosomal proteins from *E. coli*. J Biol Chem 249:3103, 1974

LAKE JA: Ribosome structure determined by electron microscopy of *E. coli* small subunits, large subunits, and monomeric ribosomes. J Mol Biol 105:131, 1976

LINDAHL L, et al: Mapping of a cluster of genes for components of the transcriptional and translational machineries of *E. coli*. J Mol Biol 109:23, 1977

MONRO RE: Catalysis of peptide bond formation by 50S ribosomal subunits from *Escherichia coli*. J Mol Biol 26:147, 1967

PLATT T, YANOFSKY C: An intercistronic region and ribosome-binding site in bacterial mRNA. Proc Natl Acad Sci USA 72:2399, 1975

SHINE J, DALGARNO L: The 3′-terminal sequence of *E. coli* 16S RNA: complementarity to nonsense triplets and ribosome binding sites. Proc Natl Acad Sci USA 71:1342, 1974

STEITZ JA, JAKES K: How ribosomes select initiator regions in mRNA-. Proc Natl Acad Sci USA 72:4734, 1975

TAI P-C, WALLACE BJ, DAVIS BD: Streptomycin causes misreading of natural messenger interacting with ribosomes after initiation. Proc Natl Acad Sci USA 75:275, 1978

THOMPSON RC, STONE PJ: Proofreading of the codon: anticodon interaction on ribosomes. Proc Natl Acad Sci USA 74:198, 1977

TISCHENDORF GW, ZEICHHARDT H, STOFFLER G: Architecture of the *E. coli* ribosome as determined by immune electron microscopy. Proc Natl Acad Sci USA 72:4820, 1975

WALLACE BJ, DAVIS BD: Cyclic blockade of initiation sites by streptomycin-damaged ribosomes in *Escherichia coli*: an explanation for dominance of sensitivity. J Mol Biol 75:377, 1973

YOURNO J, KOHNO T: Externally suppressible proline quadruplet. Science 175:650, 1972

chapter 14

METABOLIC REGULATION

BERNARD D. DAVIS
ELIORA Z. RON

INTRACELLULAR REGULATION

Once the intermediates and the enzymes of metabolism had been identified its dynamic aspects could be studied in detail. The results revealed several elaborate mechanisms for regulating the distribution of the metabolic flow into various pathways, and for adapting this flow to changing circumstances. Mutants have played a large role in the analysis of these mechanisms, which are much better understood in bacteria than in animals.

Several fundamental features of metabolic regulation have emerged. 1) In the formation of aperiodic ("informational") macromolecules (DNA, RNA, and protein), by an enzyme or a catalytic particle that moves along a template **(processive synthesis),** the rate of synthesis is ordinarily governed by the **frequency of initiation** and not by the rate of chain growth. 2) The **signal** for this regulation is the concentration of some small molecule in the cell. 3) This **effector** molecule interacts with the initiation site indirectly, by reversibly altering the confor-

mation (and hence the activity) of a **regulatory protein (allostery;** see Allosteric Transitions, below). 4) The formation of building blocks is governed by a similar allosteric mechanism, in which the level of an endproduct molecule regulates the rate of initiation of each biosynthetic sequence (i.e., the activity, per molecule, of the first enzyme). 5) Functionally related genes are often coordinately regulated as clusters in the chromosome, and 6) the same cluster may be subject to multiple regulatory mechanisms.

The discovery of these principles has had a large impact on many areas of biology. These include selective gene expression in differentiation in higher organisms; the allosteric basis of responses to pharmacologic, hormonal, and sensory stimuli (as well as to metabolite levels); and alterations of allosteric regulation in human disease. Moreover, teleonomy (Ch. 1) has now been extended from the level of organ physiology to the molecular level, where questions of "why" were long considered outside the range of science.

ALLOSTERIC REGULATION OF ENZYME ACTION

ENDPRODUCT (FEEDBACK) INHIBITION

Two observations independently suggested that endproducts of metabolic pathways somehow influence their own biosynthesis. First, an unlabeled amino acid added to a growing culture rapidly becomes the source of that component for protein synthesis, cutting off its endogenous synthesis (e.g., from radioactive glucose). Second, precursor accumulation by auxotrophic mutants (Ch. 4) was found to occur only after the cells had exhausted the endproduct of the blocked pathway and had ceased to grow. Moreover, readdition of the endproduct inhibits further accumulation within a few seconds (Fig. 14-1).*

*Though blocks in purely biosynthetic pathways cause accumulation of intermediates only after growth has ceased, blocks in central or catabolic pathways may lead to accumulations during growth that affect growth. For example, when galactose is fed to a mutant that can convert it only as far as the corresponding phosphate this product accumulates in toxic concentrations.

Though it was already known that endproducts can **repress formation** of their own biosynthetic enzymes (see Enzyme Repression, below), the effect of this mechanism on the formation of endproducts is too gradual to explain the immediate responses observed. In 1957 Umbarger, studying isoleucine biosynthesis, and Pardee, studying pyrimidine biosynthesis, discovered the underlying mechanism, **endproduct inhibition (feedback inhibition),** in which *the endproduct of a pathway directly inhibits the first enzyme of that pathway.* The action is direct and immediate, as shown by the effect of the endproduct on the activity of the extracted and purified enzyme. The action is also very specific for one metabolite (the endproduct) and one enzyme (the first) in the pathway. Such endproduct inhibition of the initial enzyme has now been observed in many biosynthetic pathways, and also in higher organisms.

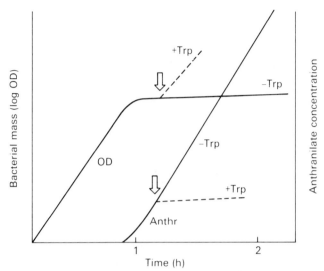

FIG. 14-1. Feedback inhibition by an endproduct. A trytophan auxotroph, growing on limiting Trp, begins to accumulate a precursor (anthranilate) in the medium at the time when growth (OD = optical density) ceases. Addition of Trp to part of the culture stops the accumulation, with simultaneous resumption of growth. In extracts the inhibition of accumulation can be shown to occur in seconds.

ENZYME ACTIVATION

Some pathways are regulated by **positive effectors.** Thus aspartate transcarbamylase not only is inhibited by its endproduct, cytidine triphosphate (CTP), but is stimulated by ATP (see Fig. 14-6, below). Purine accumulation thus promotes pyrimidine synthesis. Similarly, va-

line stimulates the first enzyme of isoleucine synthesis. In general, however, enzyme inhibition seems to be more important than activation in biosynthesis.

In an example involving regulation of the energy supply in mammals, ATP utilization leads to the accumulation of AMP, which stimulates the action of muscle phosphorylase; and this enzyme initiates glycogen breakdown, thus providing a source of energy for regenerating ATP from AMP.

BRANCHED PATHWAYS

Branched pathways present a special problem in regulation, for if feedback from one branch blocked the common pathway any other branch would be starved.

One way of solving the problem is the formation of **more than one enzyme (isozymes)** for the same reaction. For example, the biosynthesis of the aspartate family involves a common pathway that later branches to threonine and isoleucine, to methionine, and to diaminopimelate and lysine (Fig. 14-2). *Escherichia coli* makes three species of the first enzyme of this common pathway (aspartokinase); and these are subject to inhibition by different endproducts, though they all contribute to a common pool of the precursor, aspartyl phosphate. In addition, at each subsequent **fork** in the pathway the **initial enzyme of each branch** is also subject to inhibition by the appropriate endproduct (Fig. 14-2). Flow is thus regulated by the presence of a valve at each branch point, just as in a proper hydraulic system.

The formation of such regulatory isozymes was discovered with acetolactate synthetase (Fig. 14-3), which contributes to both an anabolic pathway (the biosynthesis of valine) and a catabolic

FIG. 14-2. Sites of endproduct inhibition in *Escherichia coli* in the family of amino acids derived from aspartate. **Thick lines** = biosynthetic pathway; **thin lines** = inhibition; **AK₁, AK₂, AK₃** = isozymic aspartokinases. AK₁ and AK₃ are subject to endproduct repression of enzyme formation (see below), as well as inhibition of enzyme actions; AK₂ is not inhibitable, but it is repressed by methionine.

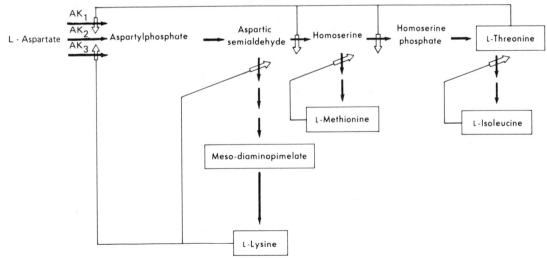

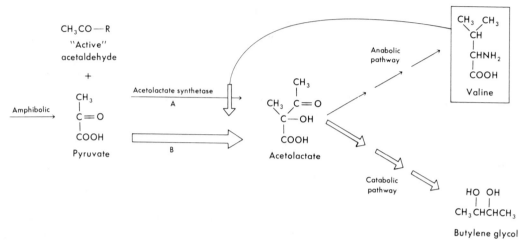

FIG. 14-3. Anabolic and catabolic metabolism of acetolactate. The enzymes of the catabolic pathway, including acetolactate synthetase B, are induced in large amount by fermentative conditions but are not influenced by valine. Acetolactate synthetase A, in contrast, is both repressed and inhibited by valine.

pathway in *Enterobacter*. In aerobic growth the level of this enzyme is only that required for valine biosynthesis, while under anaerobic, acidic conditions the level of activity increases many-fold, as the cells engage in the butylene glycol fermentation (Ch. 3). However, the additional activity involves a second enzyme that catalyzes the same reaction. The "catabolic" enzyme and the "anabolic" enzyme differ in ways that fit their functions: the former has a much lower pH optimum, and only the latter is inhibited by valine. In this way a common pool of acetolactate can meet the cell's needs for biosynthesis or fermentation or both, without mutual interference.

Several additional mechanisms for regulating branched pathways have been discovered in bacteria (Fig. 14-4). 1) **Cumulative inhibition** is seen with glutamine synthetase (in *E. coli*), which plays a key role in forming nitrogenous compounds from carbohydrates (see Regulation of Nitrogen Assimilation, below): many endproducts each contribute only partial inhibition, even at high concentration, and their effects are additive. 2) In **concerted feedback** a single enzyme responds to a mix-

ture of two or more endproducts but not to individual ones (at physiologic concentrations). 3) In **sequential feedback** each endproduct governs only the initiation of its branch; the resulting accumulation of the branch-point intermediate, in the common pathway, then inhibits the first enzyme of that pathway.

It is not clear why different groups of organisms employ different patterns to regulate the same branched pathway, e.g., the aspartate pathway (Table 14-1) or the aromatic pathway.

In some common pathways in the mold *Neurospora* parallel enzymes are compartmented: i.e., the intermediates in a sequence do not enter a common pool but remain enzyme-bound. In mammalian cells (Ch. 49) compartmentation extends to membrane-bound organelles, but isozymes are also prominent: their different kinetic and regulatory properties, often dependent on different proportions of two alternative subunits, are adapted to a variety of physiologic circumstances.

Regulation by **chemical modification of enzymes** (e.g., phosphorylation, adenylylation) is prominent in animal systems and is

TABLE 14-1. Patterns of Control of Aspartokinase Activity in Some Bacterial Species

Species	Mode of regulation	Feedback inhibitors
Escherichia coli	Isoenzymes	Lysine (enzyme 3)
Salmonella typhimurium		Threonine (enzyme 1)
Enterobacter aerogenes		None (enzyme 2)
Pseudomonas aeruginosa	Concerted feedback inhibition	Lysine plus threonine
Rhodopseudomonas capsulata		
Brevibacterium flavum		
Rhodopseudomonas spheroides	Sequential feedback inhibition	Aspartic semialdehyde
Bacillus subtilis	Isoenzymes and concerted feedback inhibition	Meso-diaminopimelate (enzyme 1)
Bacillus stearothermophilus		Lysine plus threonine (enzyme 2)

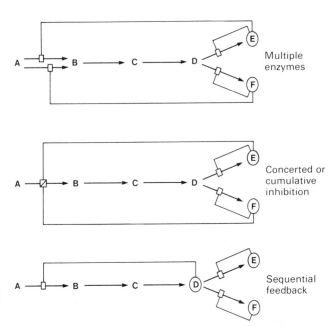

FIG. 14-4. Alternative patterns of regulation of branched pathways. Concerted and cumulative inhibition are similar in their response to the mixture of endproducts, but only the latter shows a partial response to the individual products.

also found in bacteria. One example, the adenylylation of glutamine synthetase, will be discussed below under Regulation of Nitrogen Assimilation.

ALLOSTERIC TRANSITIONS

Endproduct inhibition usually exhibits **competitive antagonism** between substrate and endproduct, even though the two compounds differ markedly in structure

(e.g., aspartate and CTP). Unlike the competition between antimetabolites and metabolites at the same site (Ch. 7), competition between substrate and endproduct is based on binding at different sites, and on the relation of this binding to the special ability of these enzymes to assume two **alternative stable conformations.** An **inhibitory effector,** complexing with an **effector site,** stabilizes (or induces) the inactive conformation, which has very low affinity for the substrate; conversely, the **substrate,** complexing with a **catalytic site,** stabilizes (or induces) the active conformation, which has less affinity for the inhibitory effector (Fig. 14-5). Both reactions are rapidly reversible. Enzymes with such alternative stable shapes, and able to interact with molecules also of quite different shapes, are designated as **allosteric** (Gr. *allos* and *stereos*, other shape).

Positive effectors may stabilize the more effective conformation, or they may block the binding site for a negative effector.

Though allosteric proteins, with two distinct binding sites, came as a great surprise, they are now seen to be essential for regulatory flexibility. If regulatory effectors had to bind directly to the catalytic site they would be limited to compounds with structural similarity to the substrate.

Molecular Mechanism. Allostery has become a major topic in protein chemistry and can be reviewed here only very briefly. A special feature of many allosteric enzymes, of profound physiologic significance, is the presence of symmetrically arranged polypeptide subunits, which provide multiple catalytic sites and multiple effector sites, and which also influence each other's conformations through the complementarity of their contacting surfaces. Accordingly, the binding of substrate (or of effector) to a subunit stabilizes a conformation that

FIG. 14-5. Regulatory changes in an allosteric model that assumes spontaneous transitions in tertiary structure, which can be stabilized by binding various ligands (indicated by **arrow**). The inhibitory effector stabilizes the inactive state of the enzyme; the substrate (and for some enzymes the stimulatory effector) stabilizes the active state. (After Changeux JP: Sci Amer April 1965, p 36. Copyright 1965 by Scientific American, Inc. All rights reserved.)

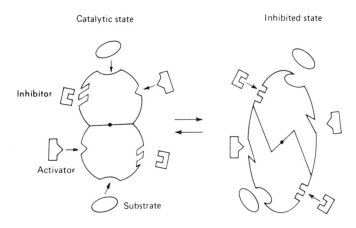

Oligomeric enzyme

Catalytic state   Inhibited state

Inhibitor

Activator

Substrate

favors the same conformation, and hence high affinity for the same ligand (and low affinity for the other), in the other subunits. As a result of these cooperative interactions the enzyme exhibits **cooperative kinetics** rather than first-order mass-law kinetics, i.e., the variation of activity with substrate concentration, and the variation of inhibition with effector concentration, are both **sigmoidal** (Fig. 14-6), with a steep rise in activity over a narrow concentration range. These subunit interactions illustrate the importance of quaternary structure (Ch. 13).

Usually a catalytic site and an effector site are paired on the same polypeptide subunit. However, the first enzyme of pyrimidine biosynthesis, aspartate transcarbamylase (see Fig. 14-6), has separate catalytic and regulatory subunits: when separated one kind can catalyze the reaction (with first-order kinetics), and the other can bind the effector.

FIG. 14-6. Kinetics of an allosteric enzyme, aspartate transcarbamylase from *E. coli*. This initial enzyme of the pathway to pyrimidine nucleotides converts aspartate plus carbamyl phosphate to carbamyl aspartate. In these experiments enzyme activity in an extract is measured at various concentrations of aspartate, with the other substrate constant.

Solid lines = native enzyme. The sigmoid middle curve **(a)**, obtained without effectors, indicates cooperative interaction of multiple substrate molecules with an enzyme molecule. **(b)** The endproduct of the reaction, cytidine triphosphate (CTP), causes feedback inhibition, which is overcome competitively by increased concentrations of aspartate. **(c)** Conversely, ATP is stimulatory, and since it was added at a high enough concentration to stabilize all the enzyme molecules in a fully active conformation the kinetics are of the Michaelis–Menten type, with a hyperbolic curve (i.e., binding of the first molecule of substrate by such a stabilized enzyme does not influence the binding of additional molecules). ATP and CTP can antagonize each other's effects.

Dashed line = enzyme "desensitized" by heating at 60° C or by treatment with 10^{-6} M Hg^{2+}, which separates the catalytic from the regulatory subunits. The sigmoidal shape is lost, as is inhibitability by CTP. (After Gerhart JC, Pardee AB: J Biol Chem 237:891, 1962)

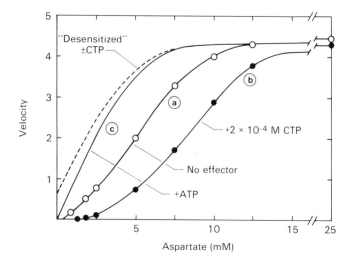

The most stable states of the enzyme are those in which all identical subunits are in the same conformation, and the enzyme is symmetric. Monod, Wyman, and Changeux have been able to interpret many enzyme kinetics in terms of varying proportions of the enzyme molecules in these two states. However, Koshland has shown that at subsaturating ligand concentrations the intermediate states can sometimes be detected; moreover, a high concentration of both ligands may strain some enzymes enough to bind both. It thus appears that effectors (and substrates) are not restricted to **stabilizing** a spontaneously formed conformation but may also use their binding energy to **induce** a conformation (**induced fit:** see Structural Features Related to Function, Ch. 13).

Whatever the detailed mechanism, it is clear that an excess either of substrate or of effector can fix an allosteric enzyme in one or the other stable conformation. This conclusion from kinetic studies is now supported by additional lines of evidence. The fixation is confirmed by the decreased susceptibility of the enzyme to proteases, denaturing agents, or chemical reagents, for susceptibility to these agents is promoted by flexibility, which results in transient accessibility of certain ordinarily buried groups. Morever, after partial denaturation of an enzyme either ligand will promote restoration of the native state (by induced fit); hence with such denatured enzymes, which lag in their catalytic response to substrates, a **negative** effector can paradoxically **stimulate** the initial activity by promoting the transition from a denatured, inactive form to a native but inhibited form, which can then be activated by substrate.

Competition between substrate and effector, as we have seen, derives from a conformational interaction in which the binding of either **decreases the affinity** for the other; i.e., the effector increases the K_m of the enzyme. With some enzymes, however, such as the first enzyme of histidine biosynthesis in *E. coli*, binding of the effector **decreases the turnover number** (V_{max}) instead of increasing the K_m. This kind of conformative interaction is not associated with competition between substrate and effector.

Desensitization. The **connection** between the catalytic and the effector site is evidently **more labile** than is either site, for mild denaturation, by heat or by chemicals (e.g., Hg^{2+}, high pH), destroys the allosteric interaction between these sites without destroying either the catalytic activity or the ability to bind the effector. Desensitizing treatments also eliminate cooperation between catalytic sites, yielding classic, first-order Michaelis–Menten kinetics, i.e., a hyperbolic curve instead of the sigmoid curve of the native enzyme (Fig. 14-6). Many allosteric enzymes have not been recognized as such because of loss of these properties during purification.

PHYSIOLOGIC SIGNIFICANCE OF THE COOPERATIVE RESPONSE

The sigmoid allosteric curves of Figure 14-6 resemble the oxygen dissociation curve of hemoglobin, which permits a large fraction of the bound O_2 to be transferred between lungs and tissues with only a moderate pO_2 difference. In allosteric enzymes, similarly, the higher-order kinetics cause a large response of enzyme activity to small changes in the concentration of a substrate (or of a regulatory endproduct). Allosteric enzymes are thus designed, like hemoglobin, for efficient **homeostasis**, i.e., for *maintaining the concentration of their ligands in the organism within a narrow range*. Moreover, the molecular mechanism similarly involves a multimeric protein: when hemoglobin is dissociated into its monomers the equilibrium shifts from sigmoidal to first-order.

The marvelous homeostatic properties of hemoglobin thus derive from a property of proteins that had evolved already in bacte-

ria. The allosteric transition inferred for enzymes has been demonstrated with hemoglobin in detail, by x-ray crystallography, as a large difference in conformation between the oxygenated and the reduced forms, largely stabilized by electrostatic bonds (salt bridges) within the folded polypeptide.

REGULATION OF GENE FUNCTION

Bacteria regulate the amount as well as the activity of various enzymes: many substrates induce, and many endproducts repress, the formation of enzymes specifically involved in their metabolism. In classic studies these two actions were found to have the same fundamental mechanism, in which the regulating compound affects the **initiation of transcription** of a small group of adjacent genes (an **operon**). However, later studies revealed several additional mechanisms, some acting on much larger classes of genes.

ENZYME INDUCTION

Many enzymes are formed in bacteria only when the cells are grown on the corresponding substrates (see Pheno-

typic Adaptation, Ch. 3). Detailed study of this **adaptive enzyme formation** was begun in 1940 by Monod, at the Pasteur Institute, with β-**galactosidase,** the enzyme that hydrolyzes lactose (glucose-4-β-D-galactoside) and other β-galactosides. Activity of various compounds as a substrate did not parallel activity as an **inducer:** in fact, the sulfur analog methyl-β-D-thiogalactoside was an effective inducer even though it cannot be metabolized at all (Table 14-2). Accordingly, the increase in enzyme concentration cannot be simply a response to increased activity. With induction thus separated from adaptive function, adaptive enzyme formation was renamed **induced enzyme formation.**

Further studies showed that within a few minutes after the addition of an excess of inducer the new enzyme be-

TABLE 14-2. Induction of β-Galactosidase by Various Compounds

Compound		Substituent	Relative induction*	Relative hydrolysis rate†	Relative affinity‡
β-D-galactosides		Glucose (product: lactose)	17	30	14
		Phenyl	15	100	100
β-D-thiogalactosides		Isopropyl	100	0	140
		Methyl	78	0	7
		Phenyl	0	0	100

*Relative induction is the maximal enzyme concentration attained, after growth with a saturating concentration of inducer, relative to the value obtained with isopropyl-β-thiogalactoside (IPTG), designated as 100.

†The **hydrolysis rate**, obtained at a saturating concentration, is given relative to that of phenyl-β-galactoside (100).

‡The **affinity** is given relative to that of phenyl-β-galactoside (100); for substrates it is determined from the concentration giving half-maximal hydrolysis, and for nonsubstrates it is determined from the inhibitory effect of a given concentration on the hydrolysis of a substrate of known affinity.

(Jacob F, Monod J: J Mol Biol 3:318, 1961. Copyright by Academic Press Inc (London) Ltd)

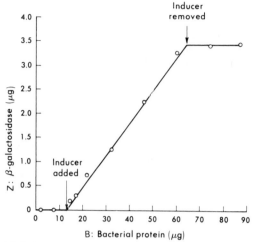

FIG. 14-7. Kinetics of induced enzyme synthesis in a growing culture of *E. coli*. Accumulation of β-galactosidase is plotted as a function of total protein synthesis (dZ/dB = differential rate of synthesis). At a saturating concentration of inducer the value for this enzyme, as for other enzymes under conditions of maximal induction, turned out to be surprisingly high (about 5% of total protein). (Jacob F, Monod J: J Mol Biol 3:318, 1961. Copyright by Academic Press Inc. (London) Ltd.)

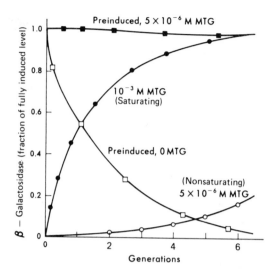

FIG. 14-8. Slow induction and full maintenance of β-galactosidase by a low concentration of inducer. At zero time methyl thiogalactoside (MTG), a nonmetabolized inducer, was added at the indicated concentration to both a preinduced and an uninduced culture of *E. coli*, in steady-state growth in the chemostat (Ch. 5), and the amount of β-galactosidase per cell was determined at intervals. It is seen that uninduced cells **(circles)** responded rapidly to a saturating concentration of inducer, and very slowly to a nonsaturating concentration. However, with preinduced cells **(squares)** the same low concentration can maintain the induced state, though in the absence of inducer the enzyme is no longer formed and its concentration is exponentially diluted by cell growth and division.

The response of a **cell** to inducer is all or none. The slow response of the **culture** to low concentrations represents a slow increase in the fraction of induced cells; each cell that begins to respond forms transport units that ensure its further rapid induction. (Adapted from Novick A, Weiner M: Proc Natl Acad Sci USA 43:553, 1957).

comes a constant fraction of the total new protein synthesized (Fig. 14-7). Moreover, the inducer stimulates a previously inactive gene to form a new protein, rather than activating a preexisting zymogen. Enzyme induction thus became a key to a very general problem: gene regulation.

Persistent Induction: Inducible Transport Systems. In contrast to the rapidly maximal response of all cells to excess inducer, at nonsaturating concentrations of inducer the formation of β-galactosidase increases gradually over a long period of time (Fig. 14-8), and it involves only a fraction of the cell population. The explanation is the induction of an **active transport system** (Ch. 6), which concentrates the inducer in the cell. Thus at a low concentration of inducer only occasionally will a cell form a transport protein, but once that process is initiated the increased uptake of inducer becomes **autocatalytic,** and the cell soon becomes fully induced. Moreover, because of this active transport the induced state can be **maintained** by concentrations of inducer too low to **initiate** a uniform response in uninduced cells (Fig. 14-8).

ENZYME REPRESSION

In the converse of the induction of catabolic enzymes, an endproduct (e.g., an amino acid) added to the medium **decreases** the formation of the enzymes of its own **biosynthetic** pathway. This regulatory mechanism is known as **endproduct repression.**

Repression has been observed in various kinds of metabolic pathways. For example, alkaline phosphatase is formed in *E. coli* cells when one of its substrates (e.g., glycerophosphate) is used as an

obligatory source of phosphate, but not when inorganic phosphate (i.e., the endproduct) is supplied. Similarly, in the shift of facultative organisms from aerobic to anaerobic conditions (and vice versa) many enzymes are induced and others are repressed.

Repression is not confined to biosynthetic endproducts; in fact, many inducible enzymes are repressed by glucose, as shown in the early observation of diauxie (Fig. 3-7). The mechanism will be discussed below (see Catabolite Repression).

The term **constitutive** was initially applied to those enzymes (mostly biosynthetic) that do not require induction. However, since these enzymes are also regulated in their formation, the term is now used primarily to characterize mutations that eliminate regulatory responses.

Derepression. The regulation of biosynthetic enzyme formation responds not only to exogenous supplies of an endproduct but also to its endogenous level. That is, in steady-state growth in minimal medium the genes for the enzymes are not working up to capacity but are partly repressed by feedback from the endogenous endproduct.

These genes can be **derepressed** (i.e., the level of the enzymes can be increased above the normal) by conditions that decrease the level of the endproduct (see Fig. 14-25, below), or by a mutation that eliminates the regulatory system.

Branched Pathways. Branched (i.e., multifunctional) pathways have special arrangements for regulating enzyme formation, just as for endproduct inhibition (see above), in order to avoid interference between the branches. Two mechanisms have been observed. To form carbamyl phosphate, a precursor both of arginine and of pyrimidine nucleotides, *E. coli* has two parallel carbamyl phosphokinases, each repressible by one of the endproducts. The common pathway to isoleucine, leucine, and valine, in contrast, has multivalent control of a single set of enzymes, requiring all three endproducts (like the concerted feedback inhibition illustrated in Fig. 14-4).

The branched aromatic pathway has parallel initial enzymes in enteric bacteria but multivalent control in gram-positive bacilli. The relative advantages of these two mechanisms are not clear.

REGULATOR GENES; NEGATIVE CONTROL

Genetic studies of Jacob and Monod on the induction of β-galactosidase distinguished two genes that affect its synthesis: **structural** and **regulatory**. Gene *z* determines its structure, since some *z*− mutants form an **altered protein,** detected in some strains as an immunologically **cross-reactive material** and in others as a **temperature-sensitive** (*ts*) enzyme. Mutations in the R gene (I-lac)*, in contrast, affect only the **quantity** of the enzyme: R− mutants are **constitutive** for its formation, no longer requiring an inducer.

* For historical reasons the regulator locus for the lactose system is still designated as **I** (for inducible), but other regulator loci are now designated as **R**.

FIG. 14-9. Dominance of R⁺*lac* (inducible) over R⁻*lac* (constitutive) allele. **Dashed arrows** indicate that the product of the R gene influences the activity of a *z* gene in a separate chromosome (*trans* effect) as well as in its own chromosome (*cis* effect). The donor may be either an Hfr or an F′-*lac* cell.

The R gene acts by producing a diffusible cytoplasmic product, designated as the **repressor** or **aporepressor;** and the function of this regulatory protein is influenced by a specific small-molecule effector, a β-galactoside. The key evidence for a diffusible repressor is summarized in Figure 14-9: in a merozygote heterozygous for R⁺/R⁻ the R⁺ allele is dominant and exerts a *trans* effect, i.e., it makes the *z* gene inducible on the other chromosome (as well as on its own).

These findings led to a model of **negative control,** in which combination with an active repressor turns off a group of genes. *In inducible systems* the repressor is active *in its natural conformation,* and complexing with the inducer stabilizes a reversible allosteric transition to an inactive form. In **endproduct repression** the mechanism is the same, except that *the repressor (R gene product) is inactive in its natural conformation*: complexing with the endproduct effector (also called **corepressor**) makes it active.

Mutations not only can inactivate R proteins but can alter them in other ways, e.g., to temperature-sensitivity; and in superrepressed (R^S) mutants the repressor remains active even in the presence of inducer.

We shall now consider in some detail the negative control of gene expression. This mechanism appeared for a decade to be the only one, but as we shall see below, several kinds of positive control have also turned out to be important in bacteria.

REGULATION OF INITIATION OF TRANSCRIPTION

OPERONS AND THE OPERATOR

The evidence for a cytoplasmic repressor led to several important predictions. 1) A specific repressor must act on a corresponding specific, genetically determined recep-

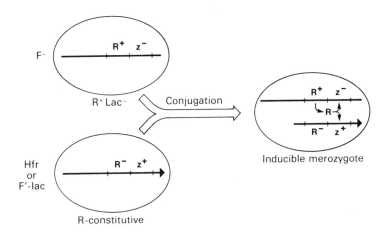

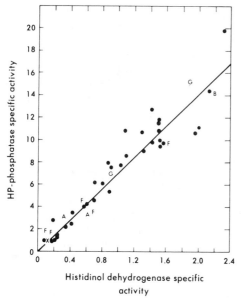

FIG. 14-10. Coordinate synthesis of enzymes of a biosynthetic pathway. Cells of several different histidine auxotrophs were harvested in different stages of derepression, caused by limitations in the supply of histidine. The levels of several enzymes were determined, including histidinol phosphate **(HP)** phosphatase and histidinol dehydrogenase. Throughout the range of derepression these enzymes (and others of the histidine pathway) retained a constant ratio of concentrations. Glutamic dehydrogenase, an enzyme not on the histidine pathway, did not vary in parallel. (Ames B, Garry B: Proc Natl Acad Sci USA 45:1453, 1959)

tor, called the **operator** (O). 2) Since the genes of a given metabolic pathway are often adjacent to each other in bacteria, a single operator might be able to regulate the whole cluster, thus explaining the **coordinate regulation** observed when the enzymes of a biosynthetic pathway were repressed to varying degrees (Fig. 14-10). 3) The

postulated operator should be subject to mutational alterations that would destroy its response to the repressor, rendering the cell **operator-constitutive** (O^C). 4) If the operator regulates an adjacent gene sequence then O^C mutations should be dominant (in contrast to the recessive R^- constitutivity); and 5) they should be **cis-dominant,** i.e., affecting the z gene on the same chromosome only (in contrast to the *trans* dominance of R^+).

These predictions were confirmed by the isolation of *cis*-dominant constitutive *lac* mutants, using the ingenious selective conditions shown in Figure 14-11. Thus in heterozygotes, formed with plasmids, $O^C z^+ R^+/O^+ z^- R^+$ is constitutive, but $O^C z^- R^+/O^+ z^+ R^+$ is inducible (see also Table 14-3).

O^C *lac* mutants were found to be constitutive not only for β-galactosidase but also for two additional proteins of β-galactoside metabolism. Such a group of linked genes and regulatory elements, which functions as a **unit of transcription,** is called an **operon** (Fig. 14-12). An operon starts with a **promoter (P),** which binds RNA polymerase and initiates transcription of the operon (Transcription of DNA, Ch. 11); it was initially defined by P^- mutants, which prevent expression of the whole operon. The **operator,** defined by O^C mutations, was found to lie between P and the sequence of structural genes. However, later studies on the interaction of these regions with RNA polymerase and repressor showed that O and P overlap (see Fig. 14-18).

The regulator (R) gene, which synthesizes a soluble repressor, is not part of the operon that it regulates. (In the *lac* system it is next to the operon, but in many systems the two are separated by unrelated genes.) The effects of mutations in P, O, and R are illustrated in Table 14-3.

Not all biosynthetic pathways have all their genes linked in a single operon. However, in some pathways (e.g., arginine) the several separate operons still constitute a functional unit **(regulon),** since they are regulated by the same R gene. Operons vary widely

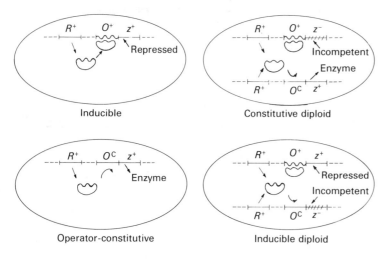

Inducible

Constitutive diploid

Operator-constitutive

Inducible diploid

FIG. 14-11. Dominance of operator constitutivity (O^C). Repressor **(R)** attaches to the wild-type operator **(O$^+$)** (blocking expression of the *cis* **z$^+$** gene) but not to the **O^C** operator. In isolating O^C mutants merodiploid cells were used in order to avoid the more frequent (R$^-$) constitutive mutants. Thus in an R$^+$O$^+$z$^+$/R$^+$O$^+$z$^+$ cell, mutation of either R$^+$ gene to R$^-$ would usually be recessive and hence not expressed, whereas mutation of O$^+$ to O^C would be dominant and hence would be expressed.

TABLE 14-3. Effects of Regulatory Mutations on β-Galactosidase Synthesis.

Genotype	Approximate enzyme level	
	Uninduced	Induced
R mutants		
R^+Z^+ (wild type)	1	1000
$R^+Z^+/F'—R^+Z^+$	2	2000
R^-Z^+	1000	1000
$R^-Z^+/F'—R^+Z^-$	1	1000
R^SZ^+	1	1
$R^SZ^+/F'—R^+Z^-$	1	1
O mutants		
$R^+O^+Z^+$ (wild type)	1	1000
$R^+O^CZ^+$	500	1000
$R^-O^CZ^+$	1000	1000
$R^+O^CZ^+/F'—R^+O^+Z^-$	500	1000
$R^+O^CZ^-/F'—R^+O^+Z^+$	1	1000
P mutants		
$R^+O^+P^+Z^+$ (wild type)	1	1000
$R^+O^+P^-Z^+$	0.01	10
$R^+O^+P^-Z^+/F'—R^+O^+P^+Z^-$	0.01	10
$R^+O^+P^-Z^-/F'—R^+O^+P^+Z^+$	1	1000

F' is a plasmid carrying the designated genes in a merozygote.

R^- fails to make repressor; Z^- fails to make β-galactosidase. R^S makes a "superrepressor" that fails to be altered by the inducer. O^C is a constitutive operator (seen to be still slightly responsive to repressor). P^- is a nearly inactive promoter. The **basal level** of enzyme in uninduced cells will be discussed in the text under Mechanism of Action of Repressors and Activators.

The data show that R^+ is *cis*- and *trans*-dominant to R^-, and R^S is *cis*- and *trans*-dominant to R^+. O and P, in any allele, are *cis*-dominant.

in size; one of the largest contains 10 genes of histidine biosynthesis. Some operons (e.g., *mal* for maltose utilization) are **divergent**; i.e., a single O–P region controls genes on both sides.

Before considering further how regulatory proteins act we shall review some properties of messenger RNA (mRNA), including its remarkable discovery as part of the operon story. The interaction of mRNA with the ribosome has already been reviewed (Ch. 13).

MESSENGER RNA

When the operon model of regulation was proposed in 1961 the process of information transfer from DNA to protein was still obscure. Protein could be synthesized in vitro without DNA, and so there must be **intermediate templates;** but chemical analyses revealed no well-defined RNA fractions except tRNAs and rRNAs. The initial assumption of a diversity of ribosomal templates proved unsatisfactory, since experiments with heavy isotopes showed that the ribosomes do not turn over, and yet the introduction of a z^+ gene into R^-z^- cells led to a full rate of synthesis of the corresponding enzyme within a few minutes. Moreover, analogs of ribonucleotides, in-

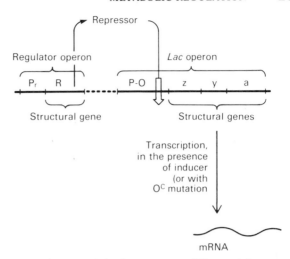

FIG. 14-12. Map of the *lac* operon and its regulator operon. Structural genes, **z, y,** and **a** code respectively for β-galactosidase, the β-galactoside transport protein (Ch. 6), and β-galactoside transacetylase (an enzyme of unknown function in the cell's economy). Transcription is initiated at the promoter **(P),** and its frequency is regulated by the overlapping operator **(O)** locus. The regulator operon is a primitive operon: it has a single structural gene R (for a repressor protein) and must have its own promoter (P_r), but it does not have an operator.

corporated into RNA, rapidly blocked specific enzyme synthesis. Jacob and Monod therefore concluded that the intermediate templates of protein synthesis must be a novel RNA species that turns over very rapidly.

Chemical Evidence. Once this novel "messenger" RNA fraction became logically necessary it was directly demonstrated within a year. The first demonstration was with phage infection, which blocks further synthesis of host RNA and protein but allows incorporation of radioactive precursors into new RNA, coded for by the phage DNA and shown by Brenner, Jacob, and Meselson to be complexed with the ribosomes (formed before infection).

In uninfected cells labeling of mRNA is harder to detect, because it is masked by the accumulation of labeled stable forms of RNA. However, at any moment *more than half the RNA being synthesized in steady-state growth in bacteria is mRNA*. Accordingly, this RNA can be heavily labeled (compared with the stable RNA) by exposing cells to a radioactive precursor (phosphate or base) for a brief period (<1 min). With this **pulse-labeling** technic growing cells were shown to have a rapidly labeled, labile RNA fraction transiently complexed with the ribosomes (Fig. 14-13). Moreover, development of the hybridization technic showed that the rapidly labeled mRNA was complementary to phage DNA in infected cells, and to cellular DNA in uninfected cells.

Studies with Specific mRNAs In Vitro. Subsequent studies were extended to specific messengers: first synthetic polynucleotides, then natural viral RNA (see Ribosomal

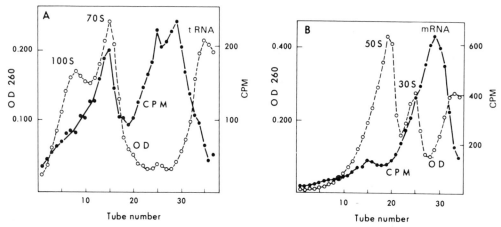

FIG. 14-13. Zonal sedimentation of **pulse-labeled** RNA. RNA was labeled in *E. coli* by growth for 20 sec with ^{14}C-uracil, and further metabolism was stopped rapidly by pouring on ice and adding azide. The cells were lysed by grinding with alumina, the lysates were sedimented in a sucrose gradient with Mg^{2+}-tris buffer; and fractions were collected through the bottom of the tube and analyzed for RNA (optical density at 260 nm; open circles) and for radioactivity **(CPM). A.** Extracted and sedimented in 10 mM Mg^{2+}, which preserves the integrity of the ribosomes and their association with mRNA. The pulse-labeled RNA **(closed circles)** sediments partly with the 70S (ribosome) and 100S (disome) peaks, and partly as a broad slower peak of free mRNA and nascent rRNA. **B.** Sedimented in 0.1 mM mg^{2+}, which dissociates the ribosomes into their 50S and 30S subunits and releases the labeled RNA, now all seen as a broad, slower peak. The lysis by grinding in this early work produced much fragmentation; gentler methods later yielded much longer polysomes, binding a larger fraction of the pulse-labeled RNA. (Gros F et al: Nature 190:581, 1961)

Macrocycle, Ch. 13). Finally, in "coupled" systems, carrying out both transcription and translation, active enzymes were synthesized in vitro by using a rich source of a particular gene (e.g., a plasmid carrying that gene) plus all the components needed.

SIMULTANEOUS SYNTHESIS, TRANSLATION, AND BREAKDOWN OF mRNA

We saw in Chapter 13 that transcription and translation take place in the same direction along RNA ($5' \rightarrow 3'$). Electron micrographs showed that ribosomes move along the mRNA as it is still growing, and that many chains grow simultaneously on the same operon. Thus gently prepared and spread out bacterial lysates (Fig. 14-14) reveal "trees" with parallel branches, of regularly increasing length, on either side of the DNA. Each branch is an mRNA chain, connected to the DNA by an RNA polymerase molecule, and carrying ribosomes at regular intervals. Each tree obviously corresponds to an operon, beginning at the peak and ending where the mRNA is completed and released. Figure 14-14 further shows that translation keeps pace with transcription, since the ribosomes on the growing polysomes press close to the attachment to DNA.

The same conclusions were reached from hybridization studies with the DNA of selected parts of the tryptophan operon (Fig. 14-15): after abrupt initiation of transcription many copies of mRNA from the first gene were made before any mRNA from the last gene. In ad-

dition, a brief pulse of derepression is seen to cause a wave of transcription followed by degradation, with the early part of the messenger (at the $5'$ end) disappearing before transcription reaches the last part. Hence *mRNA breakdown proceeds in the same direction as synthesis and translation* (i.e., $5' \rightarrow 3'$), and simultaneously. The enzyme (or enzymes) responsible for breakdown has not been identified.

The rate of these processes in *E. coli* at 37° C is about 45 nucleotides (= 15 amino acids per ribosome) per second. (Protein synthesis in vitro, in much more dilute solutions, reaches only about one-tenth this rate.)

The kinetics of breakdown of mRNA can be studied by pulse-labeling the mRNA in growing bacteria and then abruptly stopping its synthesis by the addition of actinomycin. In cells continuously exposed to a labeled precursor (e.g., ^{32}P$_i$) only 5% of the label is destroyed after addition of actinomycin (i.e., most of the label has accumulated in the stable rRNA and tRNA); in cells only briefly exposed (45 sec) over 50% is destroyed after the same addition, with a half-life of about 2 min, at 37° C. Measurements of a specific mRNA, by hybridization, have also shown strictly exponential decay (Fig. 14-16), suggesting that breakdown initiates (at the $5'$ end) at random times. However, there is evidence that different mRNAs have different half-lives.

After the onset of repression the capacity for residual β-galactosidase synthesis declines in strict proportion to the amount of its residual mRNA (Fig. 14-16). This close coordination of physical

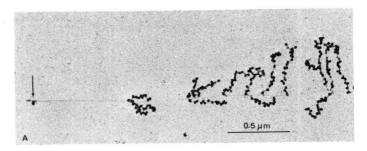

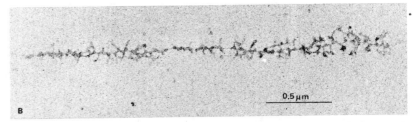

FIG. 14-14. Electron micrographs of active operons recovered in gentle lysate of *E. coli*. The rectilinear thin fibers can be destroyed by treatment with DNase, while the attached chains can be destoyed by RNase. (They are also released by protease, which suggests that RNA polymerase is essential for their attachment.) **A.** Formation of messenger. The growing mRNA chains are essentially fully loaded with ribosomes, and they exhibit an irregular gradient of increasing length along the gene. Most regions of DNA are free of such appended polysomes, which is consistent with other evidence that only a small fraction of the genome is being transcribed at any time. An RNA polymerase molecule (mol wt 4×10^5) is visible at the presumptive site of initiation **(arrow)** and at the site of attachment of most chains. **B.** Formation of rRNA. Since ribosomes do not attach to this RNA it is seen as coils, probably with aggregated protein, rather than as extended chains. The products exhibit two adjacent gradients of size, whose lengths of DNA correspond reasonably to those expected for the formation of 16S and 23S RNA. This operon for rRNA is much more crowded with growing chains than is the typical mRNA-synthesizing operon shown in **A.** (Miller OL Jr et al: Science 169:392, 1970. Copyright 1970 by the American Association for the Advancement of Science)

and functional loss suggests that the degradative enzyme normally moves closely behind or with the last ribosome. This conclusion fits other evidence that mRNA breakdown occurs only in regions that are not protected by close packing of ribosomes: when polysomes are "frozen" in the cell (e.g., by chloramphenicol) the mRNA is much more stable; conversely, removing ribosomes has the opposite effect, as will be described below (see Polarity Mutations). The other kinds of RNA in the cell are evidently stabilized in other ways: secondary structure and methylation in tRNA and ribosomes, and also bound proteins in the latter.

From measurements of the amount of *trp* mRNA in a cell (determined by hybridization), compared with the rate of synthesis of the corresponding enzymes, this mRNA appears to be translated about 30 times before destruction. This value should equal the average number of ribosomes on that species of polysome in midphase, with one end of the mRNA growing and the other being degraded.

Special features of the mRNA in mammalian cells are discussed in Chapters 11 and 49. One of these features, a poly(A) tail added to the 3′ end after transcription (see Cytoplasmic mRNA, Ch. 49), also appears to be present, but less regularly, in bacterial mRNA.

REPRESSORS

The detailed understanding of the action of regulator proteins required their isolation and study of their interactions in vitro.

Isolation. The first repressors, of *lac* and of phage λ, were isolated from *E. coli* in 1966 and were found to be proteins that recognize specific DNA sequences. These isolations required considerable ingenuity, because the repressor is normally present only in very small amounts, and there was no direct assay for its activity.

The **lac repressor** was isolated by Gilbert and Müller-Hill by assaying fractions of the cell extract for their capacity to bind an inducer, isopropyl-β-D-thiogalactoside, or IPTG (Table 14-2). The final product was a tetrameric protein of mol wt 160,000, constituting only about 0.002% of the normal cell protein (about 10 molecules per cell).

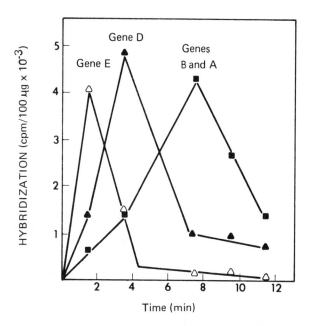

FIG. 14-15. Sequential synthesis and degradation of the mRNA of the tryptophan operon in *E. coli.* Cells were derepressed at 0 time to initiate transcription, with ^{3}H-uridine present to label the RNA formed. After 1.5 min tryptophan was added to restore repression, i.e., to prevent new initiations. At various times RNA was extracted and portions were hybridized to the DNA of different transducing ϕ80 phages carrying selected parts of the operon, as indicated in the upper part of the figure. The curves show that such a pulse of derepression causes a wave of transcription that moves from one end of the operon to the other in about 6 min. (After Morse DS et al: Cold Spring Harbor Symp Quant Biol 34:725, 1969)

The concentration in the cell has now been raised 1000-fold by selection for mutations that increase the quantity of the R gene product, and by multiple infection with transducing phage carrying this mutant operon.

The inducer-binding protein was identified as repressor by its absence or alteration in appropriate mutants. Moreover, *lac* DNA (in a transducing phage) specifically binds the protein, and O^c mutations weaken that binding.

Phage λ repressor (see The Immunity Repressor, Ch. 48) was isolated by Ptashne, without benefit of even a binding assay, by isolating a chromatographic peak of protein that appeared in cells after infection with this phage, but not with a phage from which the repressor gene had been deleted.

Repressors are now conveniently assayed by their effect on a **coupled transcription–translation** system in vitro (see Table 14-4, below), using DNA from a transducing phage carrying the desired operon. Moreover, purification is now greatly facilitated by **affinity chromatography:** passage through a column of agarose with covalently linked inducer, which specifically retains the binding protein. Repressors of a number of operons have now been isolated.

With the identification of repressor proteins the classic terms, structural genes and regulatory genes, become imprecise: the latter also form structures.

POSITIVE CONTROL BY AN R GENE PRODUCT

The model of regulation by negative control has been extended to many operons. However, it is not universal: positive regulation (i.e., **gene activation** by the R gene product) has been observed with the arabinose (*ara*) and maltose (*mal*) genes, and also with some genes of phage λ lysogenization (Ch. 48). Hence the "R" of the R gene now stands for regulator, rather than repressor.

The *ara* system has a regulator locus (C) that influences the *ara* genes in a *trans* chromosome as well as in the *cis* chromosome; hence it must form a cytoplasmic product, like the usual R loci. However, studies with mutants and with merodiploids have shown that one allosteric form of the product is required to activate the system, while the other represses it, and mutations can make the product a classic repressor or a constitutive activator.

PROMOTER VARIATION

The affinity for RNA polymerase determines the frequency of initiation, and hence the maximal rate of transcription (in the absence of repression). Promoters

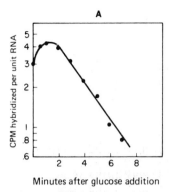

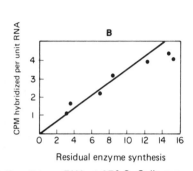

FIG. 14-16. Kinetics of decay of *E. coli lac* mRNA at 37° C. Cells synthesizing β-galactosidase were grown with ³H-uridine long enough for steady-state labeling of the mRNA. The *lac* operon was then repressed by adding glucose (see text, Catabolite Repression), which blocks further initiation of transcription of this operon. **A.** Cells were harvested at intervals and lysed, and the *lac* mRNA was measured by hybridization to the DNA of a transducing phage (Ch. 48) carrying the *z* gene of the *lac* operon. It is seen that the *lac* mRNA level is maintained for about 2 min (completion of initiated chains) and then decays exponentially (half-life 2.2 min). **B.** Following repression the declining further synthesis of β-galactosidase was measured at short intervals. The residual enzyme synthesis at a given time is the amount of enzyme still synthesized between that time and the cessation of further synthesis. This measure of mRNA activity is seen to be proportional, at various times, to the amount of mRNA measured by hybridization. (Adesnik M, Levinthal C: Cold Spring Harbor Symp Quant Biol 33:451, 1970)

vary widely in this affinity, the yield of various operons ranging from 10 molecules per cell (e.g., certain regulator proteins) to 10^4–10^5 molecules (various proteins involved in protein synthesis or in membrane structure). The variation in affinity is due to a wide variation in DNA sequence.

This primitive, inflexible method of regulation seems to be adequate for the R genes, for which a low, constant rate of production meets the cell's needs. Otherwise an infinite series of regulators would be required.

Curiously, other low-activity genes, such as those coding for the enzymes for synthesizing trace metabolites such as biotin, have operators and repressors.

Promoter mutations can increase or decrease the maximal rate of transcription. In addition, inefficient **secondary promoters** are found in the middle of some operons (e.g., *trp*), and they may be created by mutation; they permit weak expression of distal genes despite repression of the proximal genes.

CATABOLITE REPRESSION: LOSS OF CYCLIC AMP ACTIVATION

In the mechanisms discussed so far a metabolite affects a single operon. However, glucose has a much broader effect (see Phenotypic Adaptation, Ch. 3), preventing the induction of many operons concerned with the utilization of various other carbon sources. This action spares the synthesis of enzymes that would not be useful, since glucose alone evidently supplies carbon and energy as rapidly as they can be used.

Mechanism: cAMP. The glucose effect involves an interaction with the promoter, rather than with the operator or the repressor, since it is not eliminated by mutations to R⁻ constitutivity. The effect was named **catabolite repression** on the assumption that some intermediate of catabolism served as corepressor. However, the actual signal turned out to be not a catabolite but adenosine-3′,5′-monophosphate (**cyclic AMP, cAMP**). Thus Sutherland, after discovering this regulator in mammalian cells, showed that it is also present in bacteria, and its level varies inversely with the efficiency of the carbon source. Moreover, added cAMP was later found to overcome catabolite repression (Fig. 14-17). Cyclic AMP is made from adenosine triphosphate (ATP), with loss of pyrophosphate (PP), by a membrane-bound enzyme, adenyl cyclase.

The cAMP system acts positively on transcription, by interaction with a 45,000-dalton **cAMP receptor protein** (**CRP;** also called **CAP,** for catabolite activator protein). In catabolite repression the decrease in cAMP level fails to provide the cAMP·CRP complex needed to permit induction of certain operons. Thus mutants blocked in cAMP synthesis (adenyl cyclase–deficient = *cya*), or

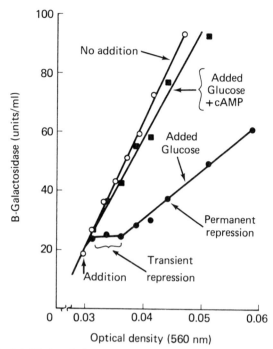

Cyclic AMP

FIG. 14-17. Catabolite repression of β-galactosidase synthesis and its reversal by cyclic AMP. Glucose, with or without cAMP, was added to a culture of *E. coli* growing on succinate in the presence of an inducer of the enzyme, isopropyl-β-D-thiogalactoside (IPTG). The cAMP is seen to overcome both the transient (complete) and the permanent (partial) repression by glucose. The brief lag before repression reflects the completion and the translation of already initiated messenger. (After Pastan I, Periman R: Science 169:339, 1970. Copyright 1970 by the American Association for the Advancement of Science)

mutants with a defective CRP, can no longer grow on (and be induced by) many different carbon sources; and addition of cAMP can correct these defects in the *cya* group. Finally, in extracts the transcription of *lac* DNA requires both CRP and cAMP, as well as an inducer (Table 14-4).

Compounds below glucose in the rate of energy supply can repress the response to carbon sources still lower in the hierarchy: different carbon sources yield different levels of cAMP, and operons differ in their sensitivity to it. Moreover, when protein synthesis is partly blocked by a borderline concentration of an antibiotic the synthesis of induced enzymes is preferentially inhibited because the decreased expenditure of energy results in an elevated level of ATP and a concomitant drop in cAMP.

Catabolite repression can also be exerted by a nonmetabolizable analog of glucose that is transported by the phosphotransferase system (see Group Translocation, Ch. 6). Cyclic AMP formation may therefore be controlled, at least in part, by the level of phosphorylation of the membrane protein involved in that transport system, and not directly by energy supply. Nevertheless, since cAMP is formed directly from ATP its usual increase in association with a lower energy supply is paradoxic.

Cyclic AMP promotes a variety of activities, including the formation of flagella (which aid a starved cell in the search for food). Its rate of destruction (by phosphodiesterase) offers another possible control mechanism. *Bacilli* employ cGMP instead of cAMP.

MECHANISM OF ACTION OF REPRESSORS AND ACTIVATORS

Kinetics of Binding to DNA. The *lac* repressor is a 150,-000-dalton molecule of four identical subunits. It binds to double-stranded (but not to single-stranded) DNA of the operator, with an extremely high affinity (dissocia-

tion constant about 10^{-12} M); hence very few molecules of repressor per cell suffice for regulation.

Its allosteric nature was confirmed in vitro: binding of an inducer (IPTG) to the *lac* repressor markedly decreases its affinity for DNA. Moreover, the inducer actively causes a conformational change, for it markedly accelerates the otherwise very slow release of already bound repressor. This finding explains the striking speed of induction in the cell.

The multivalent binding of the tetrameric repressor contributes to its very tight binding to DNA; and the allosteric cooperativeness of the subunits, like that noted for allosteric enzymes (Fig. 14-6), can account for its release by even a low concentration of inducer.

Uninduced cells have a **basal level of enzyme,** about 1/1000 of the fully induced level (Table 14-3). The explanation may be the release of bound repressor as the strands separate in DNA replication, yielding daughter loci transiently free of repressor. For similar reasons a **transduction escape synthesis** is seen when the limited supply of repressor in a cell is exceeded by a large number of copies of an operon, resulting from infection with a transducing phage.

TABLE 14-4. Effect of Cyclic AMP and its Receptor (CRP) on DNA-directed In Vitro Synthesis of β-Galactosidase

Source of bacterial extract	Cyclic AMP (5×10^{-4} M)	CRP	β-Galactosidase (relative values)
CRP⁻ strain	–	–	1
CRP⁻ strain	+	–	1
CRP⁻ strain	–	+	1
CRP⁻ strain	+	+	5
CRP⁺ strain	–	–	1
CRP⁺ strain	+	–	20
CRP⁺ strain	–	+	1
CRP⁺ strain	+	+	24
CRP⁺ strain without IPTG	+	+	1

CRP = cAMP receptor protein.

The incubation system contained DNA from a transducing phage ($\phi80dlac$) carrying the *lac* operon, as well as IPTG (Table 14-2) to overcome repression, a crude bacterial extract (S-30), and all the additional components required for transcribing the DNA and then translating the product. The enzyme formed was assayed after incubation for 1 h, which yielded maximal formation.

The control mixtures, lacking some component required for normal initiation, are seen to have about 1/20 the full activity. This background synthesis represents imperfect regulation in the in vitro system, i.e., false initiation distal to the regulatory loci.

(Modified from Zubay, G et al: Proc Natl Acad Sci USA 66:104, 1970)

Promoter–Operator Sequences. The promoter region that binds RNA polymerase is about 44 base pairs long, as shown by protection of the DNA from digestion by DNase. The binding unwinds the DNA 1–1½ turns.

Promoters vary widely in sequence, as befits their variation in efficiency and their specificity for different repressors. The greatest similarity is an AT-rich region of seven nucleotides, located shortly before the site where transcription initiates.

The **operator (O),** now defined as the region protected from DNase by binding the repressor, extends for 35 base pairs in the *lac* operon. It overlaps extensively with the promoter (P) site (Fig. 14-18), as had been suggested by the observed binding competition between repressor and polymerase. As Figure 14-18 shows, the *lac* O region has extensive dyadic (-twofold rotational) symmetry (14 of 17 base pairs on either side of the axis), which fits the binding of a symmetric, tetrameric repressor molecule. O^C mutations cluster near the center of the O region.

Gene Activation. The **cAMP·CRP protein complex** (which is required for transcription of many operons) binds immediately upstream (in the direction of transcription) from the RNA polymerase site, as shown by the sites of mutations that affect its binding to the *lac* operon. This **effector protein** evidently provides an adjacent surface that increases binding of the polymerase to an otherwise weak promoter. Two point mutations in the promoter are sufficient to eliminate dependence on cAMP.

Presumably the same kind of compensation for a low-affinity

promoter underlies other mechanisms of activation, by operon-specific R gene products (e.g., *ara*, noted above), and by ppGpp stimulating the operons of amino acid biosynthesis (see Mechanism Linking rRNA Synthesis to Protein Synthesis, below).

Regulation of Structural Proteins. Proteins that function as components of complex structures (e.g., ribosomes, membrane), rather than as enzymes, do not yield specific endproducts and hence must employ some other regulatory mechanism. For example, the 53 ribosomal proteins are synthesized in strict proportion to each other, and to rRNA, even though they are formed by several different operons (see Ch. 13, Ribosome Structure). It appears that one or more proteins made by each operon serves as **direct repressor** for that operon. Thus when genes for part of a ribosomal operon are fused to the *lac* regulator in a plasmid, induction of those genes by lactose causes overproduction of the corresponding proteins, and repression of the whole chromosomal operon. (See Regulation at the Level of Translation, p. 275.)

ALTERATIONS OF RNA POLYMERASE: PHAGE INFECTION AND SPORULATION

Initiation factor sigma (σ) governs the recognition of promoters by RNA polymerase (Ch. 11), and mutations in σ can compensate for promoter mutations that decrease af-

FIG. 14-18. The *lac* operator sequence of *E. coli,* i.e., the region protected by *lac* repressor from digestion by DNase. The sequence has extensive dyadic symmetry (see text), with its axis at base pair 18; the bases involved are indicated by lines above and below. The identified point mutations that have yielded an O^C phenotype are indicated; their locations show that the central region of the operator plays an especially large role in binding repressor.

Bases in close contact with the repressor have been identified by several additional approaches. These include 1) influence of repressor binding on methylation of A or G by Me_2SO_4; 2) ultraviolet-induced crosslinking of the protein to DNA containing 5-bromouracil (5-BU, a photoreactive analog) in place of T; and 3) altered binding of repressor to synthetic sequences in which various analogs have replaced specific single bases. The results indicate that the repressor binds to one side of the operator, alternately contacting a few bases from one strand and then a few from the other, in the next turn of the helix.

Transcription starts and proceeds, to the right from base pair 8 (denoted by an **asterisk**). The promoter (i.e., the region protected by RNA polymerase from digestion) extends 38 base pairs "upstream" from this point and 6 "downstream"; hence roughly half of the promoter (P) and of the operator (O) overlap. (After Ogata R, Gilbert W: Proc Natl Acad Sci USA 74:4973, 1977)

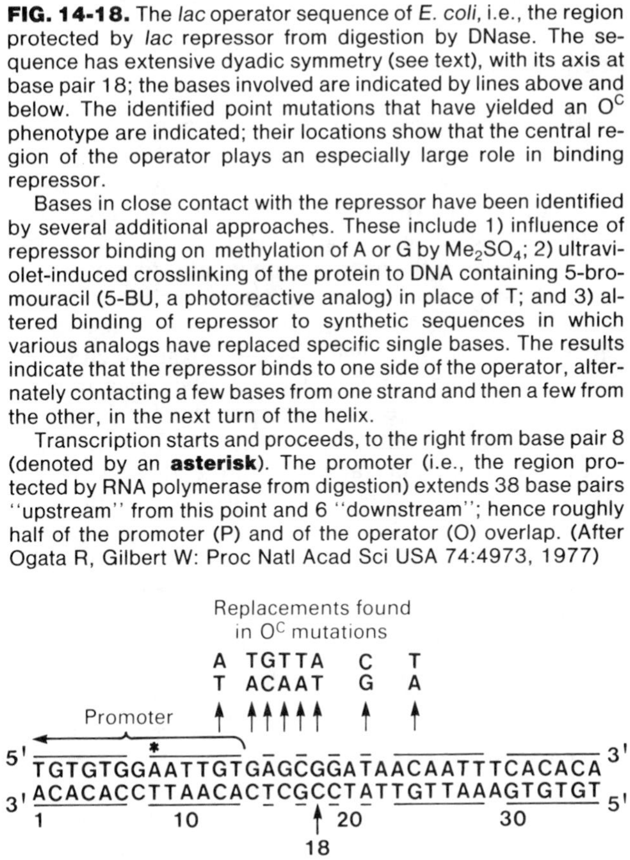

```
                    Replacements found
                    in O^C mutations
                    A TGTTA   C   T
                    T ACAAT   G   A
   Promoter         ↑ ↑↑↑↑↑   ↑   ↑
5' ←──────────────────────────────────────── 3'
       *
   TGTGTGGAATTGTGAGCGGATAACAATTTCACACA    3'
   ACACACCTTAACACTCGCCTATTGTTAAAGTGTGT    5'
3' 1        10        ↑ 20        30
                      18
```

finity for the enzyme. In the drastic reprogramming of transcription that occurs in **sporulation** (Ch. 6) or in **phage infection** (Ch. 47) modifications in the structure or the availability of σ alter the specificity of promoter recognition. However, this mechanism does not appear to play a role in vegetatively growing cells. Additional regulatory mechanisms in sporulation, involving protease action and a novel regulatory nucleotide, pppAppp, are described in Chapter 6.

REGULATION OF TRANSCRIPTION TERMINATION AND OF TRANSLATION

In the intricate network of regulation the general inhibition of translation (e.g., by amino acid deprival) decreases transcription, in cells and also in extracts. Much of this response is indirect: the accumulation of unusable endproducts, and of energy, represses many operons. However, within an operon impaired translation may also affect transcription directly, by two processes, polarity and attenuation. For some operons these mechanisms, which prematurely terminate transcription, may be as important as the regulation of initiation.

OPERON POLARITY

Polarity Mutations. Nonsense mutations, yielding translation termination codons UGA, UAA, or UAG, not only directly block distal expression of the gene in which they occur (by releasing ribosomes); they sometimes also decrease the expression of distal ("downstream") genes of the operon, without affecting proximal ones. The degree of inhibition increases with the length of the "dead space" between the termination site and the beginning of the next gene (Fig. 14-19), at which translation is initiated again. A lack of ribosomes moving through this space is evidently responsible for this **polarity effect,** for the inhibition is eliminated by suppressor mutations that permit translation of the nonsense codon (Ch. 13) and thus restore access of ribosomes to the dead space.

The key to polarity was provided by the selection of a novel class of mutants that **suppress the polarity effect,** i.e., that restore the activity of the distal genes but not of the gene containing the nonsense mutation. These mutations (initially called SuA) were found to reside in the gene for the **transcription termination factor, rho** (ρ; see Transcription of DNA, Ch. 11). Rho apparently binds to mRNA in regions free of ribosomes, moves along the mRNA (at the expense of ATP hydrolysis) until it reaches the RNA polymerase, and promotes release when the polymerase reaches a weak termination site

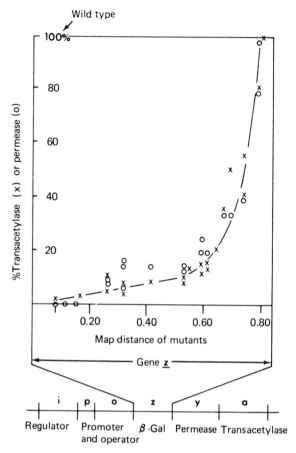

FIG. 14-19. Gradient of polarity effect in relation to position of mutation. *E. coli* strains with a nonsense (translation-terminating) mutation at various positions in the β-galactosidase gene (z) were grown under conditions of full induction and were then assayed for the products of the *lac* operon. None of the mutants yielded any z gene product. The amounts of the y and the a gene products are seen to increase, in parallel, with decreasing length of the "dead space" between the mutation and the start of the next gene. Studies with genetic deletions, on either side of the nonsense mutation, have shown that the important variable is the length of this dead space and not the length of the translated fragment of the mutant gene. (Newton WA et al: J Mol Biol 14:290, 1965. Copyright by Academic Press Inc. (London) Ltd.)

(which would normally be masked by ribosomes). The longer the mRNA dead space without ribosomes, the more frequently ρ could bind to it and cause release in this way.

The mutant form of ρ in polarity suppressors evidently cannot terminate transcription at weak sites, but can at the strong sites at the ends of operons; hence the cells are viable.

Polarity mutants also exhibit **hyperlability of the ribosome-free mRNA** of the dead space. The subsequent mRNA is protected by ribosomes initiating at the next gene.

Normal polarity. Some operons make their several polypeptide products in equal molar quantities, but in others the proportion decreases with distance from the promoter. In the *lac* operon, for example, the molar ratio of *z*:*y*:*a* products (Fig. 14-12) is about 5:2:1. This regulatory mechanism depends on transcription termination signals of varying strength: in addition to those that regularly terminate operons (beyond the end of translation), and those just described that depend on interruption of translation within a gene, a third class causes occasional termination within a normal operon.

Some insertion sequences (IS) of DNA, described in Chapter 11, contain sequences that terminate transcription. These elements cause polarity in the operon in which they are inserted.

ATTENUATION; THE LEADER SEQUENCE

In the classic regulation of amino acid operons the endproduct acts as a corepressor of the initiation of transcription. However, in already derepressed R$^-$ mutants synthesis could be further stimulated by deprival of the endproduct, and also by internal deletions of the region between the operator and the first structural gene. These findings led to recognition of a second mechanism of regulation, involving variable, premature termination of transcription in this region. This process is called **attenuation.** It is a major mechanism; in growth in minimal medium it releases 85% of the messengers initiated in the *trp* operon and in the presence of excess tryptophan the fraction is even larger.

Yanofsky elucidated the mechanism by analysis of the properties of the early mRNA (the **leader sequence**) of the *trp* operon (Fig. 14-20). A part of this sequence was found to code for a short **leader peptide;** and variations in this translation, dependent on the supply of the endproduct, influence the frequency of termination of transcription at the **attenuator** (*a*) site still farther ahead.

The main findings (see Fig. 14-20), and a reasonable interpretation, are the following. 1) Both cells and extracts synthesize a leader sequence of 162 ribonucleotides, starting at the site for initiating transcription of the operon, in several-fold higher molar yield than the transcript of the subsequent regions. 2) An early part of this sequence codes for the leader peptide of 14 amino acids. 3) This peptide has two successive Trp residues, several units before its end. 4) The end of the leader sequence, just before the attenuator site, resembles other known regions of transcription termination in having a string of seven or eight Us preceded by a region rich in G and C. 5) The region of the mRNA between the termination of translation and that of transcription can form two alternative, stable stem-and-loop structures, which are mutually exclusive because the same short region forms a strand in either stem. 6) Whether the ribosomes are held up at the Trp codons, or pass on to complete the peptide, depends on the supply of Trp-tRNA; and this behavior of the ribosomes apparently determines which loop is formed. The choice of secondary structures in turn influences the frequency of attenuation, as outlined in the legend to Figure 14-20.

In the Phe (and in the His) operon the leader peptide contains seven residues of the endproduct amino acid; and in the operon for aliphatic amino acid synthesis (*ilv*) the multivalent control (p. 265), requiring Ileu, Leu, and Val together, is explained by a leader peptide with multiple copies of each. Moreover, the role of aa-tRNA (Ch. 13) in attenuation explains the earlier observation that various amino acid operons are derepressed not only by deprival of the amino acid, but also by mutations in the corresponding tRNA or in its aa-tRNA synthetase.

The release of RNA polymerase in attenuation also involves the transcription termination protein ρ, since mutations in this factor can decrease the effect. It is noteworthy that in attenuation ribosomal holdup **prevents** this release, whereas in polarity (see above) such holdup, within a gene, **promotes** release by providing a long stretch of untranslated mRNA.

It is not known whether attenuation occurs in bacterial operons other than those for amino acids. The regulation of temperate phage λ (Ch. 48) involves a phenomenon resembling attenuation, and also involves an antiterminator protein as well as ρ.

REGULATION AT THE LEVEL OF TRANSLATION

In an important new mechanism, impaired rRNA synthesis leads to accumulation of free ribosomal proteins and certain of these directly inhibit translation (rather than transcription) of their mRNA. Efficiency of translation can also be varied in other ways, of less certain of regulatory significance: messengers have been reported to differ in their rate of initiation of translation, their rate of chain elongation (possibly limited by the supply of tRNA for a particular codon), and their rate of degradation.

In addition, infection with some phages (e.g., T4) appears to turn off translation of host mRNA, thus accelerating the shift to phage mRNA within the brief period between infection and lysis. Modifications of the initiation factor IF3 (Ch. 13) have been reported to be involved.

REGULATION OF STABLE RNA SYNTHESIS; ppGpp

CELL COMPOSITION AND GROWTH RATE

In bacteria, as in animal cells, the RNA content parallels the rate of protein synthesis. While this phenomenon was long thought to reflect some central control over all RNA synthesis, it turned out to reflect specific regulation of the operons that synthesize stable RNA (rRNA and tRNA).

The underlying mechanisms were revealed by Maaloe, starting with a comparison of cells at various rates of **steady-state, balanced growth** (i.e., exponential growth, with unchanging composition), controlled by the carbon source or by use of the chemostat (Ch. 5). As Table 14-5 shows, *the ratio of protein to DNA varies little,* but *the ratio of rRNA to DNA is essentially proportional to the rate of growth,* over a tenfold range. Since the rRNA re-

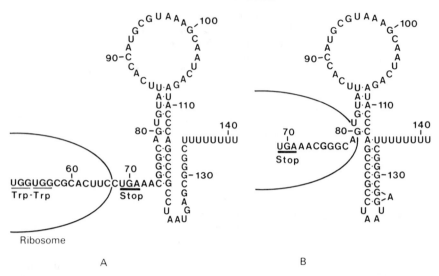

FIG. 14-20. Possible secondary structures of mRNA involved in attenuation in the *trp* operon of *E. coli*. The leader sequence starts being transcribed at nucleotide 1 (not shown), and it codes for a 14-residue leader peptide from nucleotide 27 through 68, followed by termination codon UGA **(underlined).** That peptide contains two Trp residues (codons UGG, **underlined).** If Trp is available the leader peptide is completed, and the leader mRNA sequence is terminated at the attenuator site (141); if Trp is not available the ribosome is held up and the leader sequence continues to be transcribed, reaching the first structural gene of the operon at nucleotide 163 (not shown).

The connection between leader peptide synthesis and attenuation depends on an effect of the ribosome on the alternative secondary structures of this mRNA. Ribosomes mask a dozen nucleotides on either side of the codon being recognized: hence a ribosome held up at UGG should not interfere with the formation of **form I** (A) of the leader sequence. However, if Trp-tRNA is present the ribosomes proceed to the end of the peptide, where they interfere with the first stem. The second stem, in **form II** (B), can then form. This secondary structure, extending close to the site of potential premature termination of transcription, evidently somehow promotes this termination, thus linking it to the supply of Trp. (After Lee F, Yanofsky C: Proc Natl Acad Sci USA 74:4365, 1977)

flects the number of ribosomes, it appears that each engaged ribosome moves along mRNA at the same rate in fast and in slow steady-state growth. The economic advantage is obvious: in steady-state growth the *ribosomes evidently work to capacity, and they are formed only in the amount needed.*

The **maximal growth rate** of a bacterium may be set by the minimal amount of cytoplasm required to synthesize the constituents that must still be made endogenously in a rich medium. (However, slow-growing bacteria, and eukaryotic cells, are presumably limited by other factors.) In *E. coli* growing at the maximal rate (doubling time about 20 min at 37° C) the average distance between

TABLE 14-5. RNA Distribution and Protein Synthesis in *S. typhimurium* **at Various Growth Rates**

| Carbon source | Growth rate (generations/h) | DNA (μg/mg bact. dry wt) | rRNA/ DNA | tRNA/ DNA | Protein/ DNA | Protein synthesis per h | |
						Per unit RNA	Per unit rRNA
Broth	2.4	30	8.3	2.0	22	3.7	4.5
Glucose	1.2	35	3.9	2.4	21	2.8	4.6
Glycerol	0.6	37	2.4	2.4	21	1.8	3.6
Glutamate	0.2	40	0.9	2.1	21	1.0	3.3

The growth rate was controlled by providing carbon sources that differ in their maximum rate of utilization. The cells were harvested at low densities, below 0.15 mg dry weight/m. Above this value alterations of the medium by the metabolizing cells began to cause unbalanced growth, resulting in progressive changes in cell composition. Conventionally plotted growth is much less sensitive to such changes and appears to remain exponential until about half-way to the saturation level of 1–2 mg/ml.

(Kjeldgaard NO: Dynamics of Bacterial Growth. Thesis, Univ. Copenhagen, 1963.)

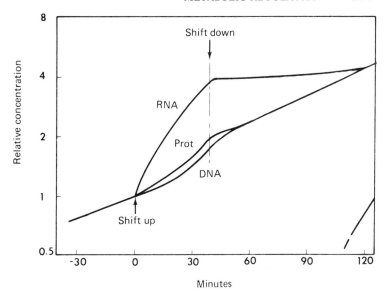

FIG. 14-21. Response of DNA, net RNA, and protein synthesis to shift up and shift down of the medium. The curves are idealized, and the amount of each component per milliliter of culture is normalized to 1.0 at the time of the initial shift. The shift down depicted does not involve a diauxic lag. When there is such a lag, as in adapting to a new carbon source, the absence of an energy supply temporarily halts net synthesis of all three classes until turnover of mRNA and protein (see text, Protein Breakdown) has built up the required new enzymes.

ribosomes in the nonnuclear region is approximately the diameter of a ribosome. Such cells contain about 25,000 ribosomes per genome.

Response to Shifts of Growth Rate. After regulatory mechanisms were shown to influence steady-state levels of various macromolecules, studies of the dynamics of the transition from one level to another showed that an abrupt shift in medium causes a rapid, efficient transition to the appropriate new composition. Thus in a **shift up** to a richer medium the **rate of net RNA synthesis** (largely rRNA) increases **abruptly** (Fig. 14-21), while the **rate of protein synthesis** increases only **gradually,** as the cells build up their ribosome concentration. The ratio of the two components (and rates) becomes constant when the cells reach the ribosome concentration required for the rate of steady-state growth supported by the new medium. In a **shift down** to a poorer medium, where the cell initially has more ribosomes than it needs, *protein synthesis slows immediately* because of the diminished supply of building blocks and energy; but *net RNA (and ribosomal protein) synthesis ceases abruptly,* and it resumes only when further growth has diluted the ribosomes to the level appropriate for the new growth rate.

MECHANISM LINKING rRNA SYNTHESIS TO PROTEIN SYNTHESIS: ppGpp

The adjustment of ribosome synthesis to growth rate involves a response to the supply of aa-tRNAs: when protein synthesis is stopped by a block **before** the charging of any tRNA (either by deprival of a required amino acid or by a defect in aminoacylation) rRNA synthesis stops immediately; but a block **after** this step (e.g., inhibition of the ribosomes by chloramphenicol) even stimulates RNA synthesis. The key was provided by mutants that had lost the response of their rRNA regulation to amino acid deprival. These strains were designated as having **relaxed control** of rRNA synthesis (RC^{rel}, or *rel A⁻*), in contrast to the wild-type **stringent control** (RC^{str}, or *rel A⁺*). More than a decade after the isolation of these strains Cashel and Gallant discovered that during deprival of any amino acid the wild-type cells, but not the relaxed mutants, rapidly accumulate a novel nucleotide, **guanosine 3′-diphosphate 5′-diphosphate (ppGpp);*** and in cells in steady-state growth the rate of synthesis of rRNA is inversely related to the level of ppGpp. In vitro studies showed that this compound is synthesized on "idling" ribosomes (i.e., ribosomes encountering a codon for an unavailable aa-tRNA).

The nucleotide ppGpp is formed by an enzyme (called **stringent factor**) that is found on *rel A⁺* ribosomes but is absent from *rel A⁻* cells. This enzyme has little activity unless on ribosomes complexed with mRNA. When such ribosomes (with stringent factor) encounter a codon for which the cognate aa-tRNA is missing they bind the uncharged tRNA instead (in the acceptor site), and this interaction stimulates the stringent enzyme to transfer PP from ATP to the 3′ group of GDP. The tRNA is simultaneously released, so that the reaction is constantly dependent on the ratio of charged to uncharged tRNA.

In this elaborate mechanism a limitation in the supply of any aa-tRNA is equally effective. Moreover, mutations in the ribosome

---

* Also called guanosine tetraphosphate, or MS-I (for magic spot I). The accumulation of ppGpp is generally accompanied by that of ppGppp (MS-II).

that impair its interaction with uncharged tRNA, or with stringent factor, also have the relaxed phenotype.

During starvation for an amino acid the ppGpp level in the cell is about ten times the basal level. The response to change is rapid: replacement of the missing amino acid halts further ppGpp synthesis at once, and the accumulated compound is destroyed with a half-life of only 30 sec.

Bacilli accumulate similar polyphosphates of A as well as of G, and *rel*⁻ mutations block formation of both.

A decreasing production of rRNA leads to parallel decrease in ribosomal protein production (see Regulation at the Level of Translation, p. 275).

OTHER ASPECTS OF ppGpp REGULATION

The accumulation of ppGpp is promoted not only by amino acid limitation but also by a decrease in energy supply. The mechanism involves decreased destruction of the compound (which can also be brought about by mutations called *spo T*). The regulation of both formation and destruction clearly fits the compound for the task of informing operons about variations in the supply of both aa-tRNA and energy.

In addition, ppGpp mediates many other useful responses to these variations. It decreases synthesis of phospholipids and polyamines; as shown in vitro, it stimulates directly the activity of operons of amino acid synthesis (just as cAMP activates operons for the utilization of sugars); and it activates a "starvation-dependent" protease that can attack normal proteins and thus provide amino acids (see Protein Breakdown, below).

A very similar compound, 3′,5′ppApp, appears to play a key role in the shifts in gene regulation in sporulation (Ch. 6).

tRNA, ELONGATION FACTOR, AND POLYAMINE SYNTHESIS

Though net (i.e., stable) RNA synthesis (Fig. 14-21) includes tRNA as well as rRNA, the two are not regulated identically. As Table 14-5 shows, at various growth rates the tRNA concentration remains at about twice the DNA and one-tenth the protein concentration. Since the frequency of contact of any ribosome with a required tRNA is proportional to the concentration of the latter, this constancy fits the constant efficiency of ribosomal function.

Polyamines are largely bound in the cell to ribosomes and tRNAs. When RNA synthesis is blocked polyamine synthesis continues, but the normal ratio to RNA is maintained by excretion of the excess as acetylated derivatives.

REGULATION OF OTHER COMPONENTS

Cell Envelope. Regulation of the formation of the cell wall and membrane is not well understood. As we saw in Chapter 6 (see Antibiotics Acting on Peptidoglycan Synthesis) a block in basal wall synthesis (e.g., by penicillin) causes intracellular accumulation of nucleotide-linked precursors, which implies lack of tight feedback inhibition. An adaptive change in the stationary phase increases the stability of the already present lipids by converting the double bond in unsaturated fatty acids into a **cyclopropane ring** (Ch. 6).

REGULATION OF NITROGEN ASSIMILATION: GLUTAMINE SYNTHETASE

The classic regulatory mechanisms described above employ one allosteric protein to govern the concentration of the enzymes of a biosynthetic pathway, and another to govern the rate of the first step of the pathway. However, a single protein, **glutamine synthetase**, performs both these functions in the elaborate regulation of nitrogen assimilation, worked out by Stadtman and by Magasanik. The underlying principle is one widely employed in mammalian cells: **reversible chemical modification of enzymes,** governed by a cascade of regulatory enzymes. Such a cascade enormously amplifies both the range of the response and its sensitivity to the intensity of the signal.

Nitrogen is assimilated in enteric bacteria by two alternative pathways, both of which yield glutamate (which provides the amino groups of most amino acids by transamination) and glutamine (which provides other nitrogen atoms). When NH_3, the preferred source of N, is supplied in adequate concentration it is assimilated into glutamate by reaction (1), and part of the glutamate is then converted to glutamine by reaction (2):

$$NH_3 + \alpha\text{-ketoglutarate} + NADPH + H^+ \xrightarrow[\text{dehydrogenase}]{\text{Glutamic}} \text{glutamate} + NADP^+ + H_2O \quad (1)$$

$$\text{Glutamate} + NH_3 + ATP \xrightarrow[\text{synthetase}]{\text{Glutamine}} \text{glutamine} + ADP + P \quad (2)$$

However, when NH_3 is supplied at a low concentration (<1 mM NH_4^+), or when it is derived from a slowly degraded source, the equilibrium of reaction (1) is unfavorable. The NH_3 is then all taken up, at the expense of ATP hydrolysis, via glutamine in reaction (2), and the glutamine in turn yields glutamate by reaction (3):

$$\text{Glutamine} + \alpha\text{-ketoglutarate} + NADPH + H^+ \xrightarrow[\text{synthetase}]{\text{Glutamate}} 2 \text{ glutamate} + NADP^+ \quad (3)$$

The sum of reactions (2) and (3) is thus the same as reaction (1), plus the hydrolysis of ATP.

Regulation of Enzyme Activity. The shifts between the two modes of assimilation involve a cycle of changes in the structure (as well as in the concentration) of glutamine synthetase (GS). A high NH_3 level increases the ratio of glutamine to α-ketoglutarate, which triggers a sequence that causes the specific activity of GS (i.e., the activity per molecule present) to be decreased by covalent addition of adenylyl (AMP) groups (reaction 4); and when the NH_3 level drops the enzyme is reactivated by reaction (5):

$$GS \text{ (active)} + ATP \rightarrow GS\text{-}AMP \text{ (inactive)} + PP \qquad (4)$$
$$GS\text{-}AMP + P_i \rightarrow GS + ADP \qquad (5)$$

The enzyme has 12 identical subunits, which can each attach one AMP group; and the activity of a GS molecule is proportional to the number of its unmodified subunits.

The adenylylation is controlled by two levels of regulation (Fig. 14-22). First, an unmodified enzyme complex carries out reaction (4), and its uridylylated derivative carries out reaction (5). Second, two additional enzymes add or remove, respectively, this UMP group; and it is these enzymes that are responsive to the ultimate effector in this cascade: the ratio of α-ketoglutarate to glutamine. Thus low NH_3 results in accumulation of α-ketoglutarate and a short supply of glutamine, which leads to high uridylylation and low adenylylation, and therefore to high activity of GS. High NH_3 has the opposite effects.

This **indirect feedback inhibition** of GS by glutamine, through regulation of its adenylylation, is further supplemented by **direct feedback inhibition,** by endproducts of pathways to which glutamine contributes nitrogen: e.g., tryptophan, histidine, glycine, glucosamine-6-P, and several nucleotides.

Regulation of Enzyme Concentration. Adenylylation plays a role in regulating the formation as well as the activity of GS. In addition, elevation of free GS is associated with repression of the formation of glutamic dehydrogenase (of reaction 1)—also an economically valuable response.

Utilization of Alternative Nitrogen Sources. When a single foodstuff, such as histidine, can provide both carbon and nitrogen its enzymes must respond to either need; hence they are regulated by a rather elaborate mechanism, worked out by Magasanik. In the usual minimal medium, with NH_3 and glucose, the genes for histidine utilization (*hut*)* cannot be induced by histidine because of catabolite repression (see above). If carbon becomes limiting cAMP activates this induction; while if nitrogen becomes limiting (in the presence of glucose) the accumulating nonadenylylated form of GS performs the same role (as shown by its effect on in vitro transcription).

Not surprisingly, genes for enzymes that supply nitrogen but not carbon (e.g., urease) respond similarly to GS, but not to cAMP.

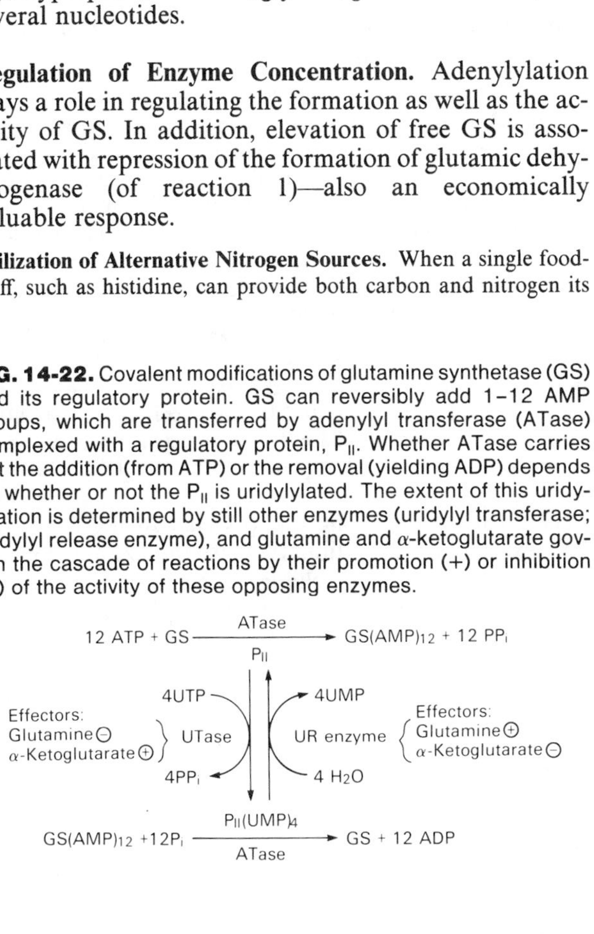

PROTEIN BREAKDOWN

The rate of degradation of cell proteins is determined both by structural features of the individual molecules and by the nutritional status of the cell. Goldberg has shown that normal and abnormal proteins are broken down by different systems.

Steady-State Breakdown. Metabolism in bacteria is normally linked to growth, and not turnover: when an enzyme is no longer formed its concentration per cell decreases primarily by exponential dilution. However, there is slow turnover during steady-state growth, shown by both incorporation and release of labeled amino acids; the average rate in *E. coli* is about 1% of the protein per hour at 37°C.

This breakdown seems to function mainly to cleanse the cell of proteins with abnormal structures. Thus the rate of breakdown is markedly accelerated when errors are introduced (by growth in the presence of amino acid analogs, by missense mutations, or by misreading antibiotics), or when proteins are prematurely released (by puromycin or by a nonsense mutation), or when denaturation is accelerated (by temperatures above the optimum). Incorrect proteins are undoubtedly also made during normal growth, at a low rate, and they are presumably removed by this mechanism.

Starvation-Induced Breakdown. When *E. coli* is starved for a carbon source (or a nitrogen source or a required amino acid) the rate of protein breakdown jumps from 1% to 5%/h. The adaptive value of the resulting turnover is clear, since it provides amino acids for the synthesis of enzymes that become useful under these conditions. This mechanism degrades some normally stable proteins, such as those of ribosomes, in preference to others.

The key to this response to starvation is ppGpp. As we have seen, limitations of either an amino acid or energy elevate this compound; and when its synthesis is blocked

FIG. 14-22. Covalent modifications of glutamine synthetase (GS) and its regulatory protein. GS can reversibly add 1–12 AMP groups, which are transferred by adenylyl transferase (ATase) complexed with a regulatory protein, P_{II}. Whether ATase carries out the addition (from ATP) or the removal (yielding ADP) depends on whether or not the P_{II} is uridylylated. The extent of this uridylylation is determined by still other enzymes (uridylyl transferase; uridylyl release enzyme), and glutamine and α-ketoglutarate govern the cascade of reactions by their promotion (+) or inhibition (−) of the activity of these opposing enzymes.

---

* These genes are not found in *E. coli*, which would rarely encounter a supply of histidine in its natural habitat, but they are found in the closely related free-living organism *Enterobacter aerogenes* (Ch. 31).

(by *rel⁻* mutations) these limitations no longer stimulate protein breakdown. It appears that ppGpp activates an enzyme already present, rather than inducing a new one, since the stimulation occurs within seconds.

A membrane-bound endopeptidase, stimulated by ATP, initiates the degradation of abnormal proteins. Moreover, when such proteins are experimentally caused to increase in quantity they can be seen to aggregate into dense granules that bind to the membrane. Degradation of normal proteins is initiated by another enzyme, which differs in not being stimulated by ATP and in being inhibited by serine protease inhibitors. The peptides produced by either system (and also those taken up from the outside) are probably cleaved to amino acids by the same soluble peptidases.

REGULATION OF DNA SYNTHESIS AND CELL DIVISION*

GROWING POINTS; THE REPLICON

As with other informational macromolecules, the overall rate of synthesis of DNA in steady-state growth depends on the frequency of initiation of a growing point, which then moves along the chain. (Actually, since the replicating machinery of the growing point is attached to the cell membrane, it is the bacterial chromosome that moves processively.) The rate of synthesis at each growing point is independent of the growth rate, and in *E. coli* about 45 min are required between the initiation and the completion of replication of a chromosome at 37°C. Thus radioautographic studies with ³H-thymidine show that in growth with a doubling time of 50 min about 90% of the cells are engaged in DNA synthesis, and in a slower growth the gap between rounds of replication is greater.

Studies of chromosome replication and cell division in bacteria have benefited from the use of **synchronized cultures** (Ch. 5), in which all the cells are at the same stage in their growth cycle. As described in Chapter 10 (see DNA Replication), in this cycle DNA replication is initiated at a specific region of the chromosome (the **origin**), from which it proceeds **bidirectionally** until the two **replication forks** reach the site of termination, at a position on the circular chromosome about equidistant on either side from the origin. In *E. coli* the origin is at about 80 min on the 100-min genetic map (see Fig. 9-13, Ch. 9) and the terminus is at about 30 min. The terminus appears to provide information for special functions in the completion and separation of the daughter chromosomes.

There is evidence that the origin, the terminus, and the growing point of a chromosome may all be attached to the membrane at the same time, moving in a coordinated way at cell division.

The Replicon. Maaloe showed that when protein synthesis is blocked any replication of DNA that is already under way is completed but no new rounds are initiated. Hence initiation evidently depends on the formation of one or more **replication proteins,** which can act only once (perhaps becoming a permanent part of the membrane attachment site of a new chromosome). Studies with *ts* mutants blocked in various aspects of initiation led Jacob, Brenner, and Cuzin to infer a functional unit of regulation of DNA replication, the **replicon,** in which one genetic element (analogous to an R gene) produces a specific cytoplasmic **initiator,** and another element, the **replicator** locus (analogous to an operator), interacts with this product to initiate replication. *Different replicators* (e.g., on the bacterial chromosome, the phage genome, or the F plasmid) *respond to different initiators.*

DNA REPLICATION AND CELL DIVISION

Cell division is normally triggered by completion of replication of the chromosome.

Some key evidence is that blockage of DNA synthesis prevents initiation of septum formation while allowing further growth. Conversely, after starvation for a required amino acid has resulted in the completion of replication of all the chromosomes, in some strains restoration of the missing amino acid will trigger synchronized septum formation and cell division, independent of cell mass.

With the recognition of the triggering role of completion of replication, the interval between this event and the completion of cell **division** (i.e., physical cell separation) was defined as period **D** of the cell cycle. We have seen that in *E. coli* at 37°C the time required for replication to traverse the length of a **chromosome**, period **C**, is about 50 min (Fig. 14-23). The D period occupies 20–25 min, creating a **constant phase difference** between the rhythm of chromosome completion and that of cell division (Fig. 14-24).

DNA Synthesis in Nongrowing Cells. In the transition from exponential to stationary phase the size and composition of the cells may change even after cessation of overall growth. Thus when growing cells of *E. coli* are abruptly made stationary (by transfer to a medium lacking a nitrogen or a carbon source) the DNA increases until the current rounds of DNA are completed, and the cells then divide.

CONTROL OF INITIATION

Multiple Growing Points. The finding of a fixed chromosomal replication time (C) at different growth rates pre-

* Contributed by Eliora Z. Ron, Ph.D., Department of Microbiology, Tel-Aviv University.

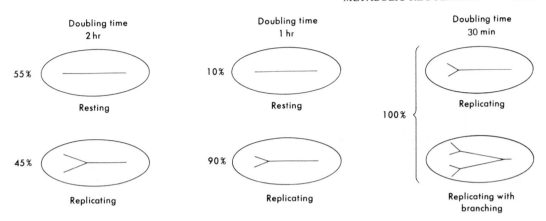

FIG. 14-23. States of chromosomal replication in *E. coli* cultures growing at different rates. Numbers denote percent of the population in each state. For convenience the cyclic structure of the chromosome is ignored.

FIG. 14-24. DNA replication and cell division at various rates of initiator synthesis. For convenience the cyclic chromosome is diagrammed as though it were linear. **I** = doubling time = time required for initiating a new round of replication after last initiation. **C** = time required for travel of a growing point through the length of a chromosome. At completion of a chromosome pair (time **X**) the attachment site is doubled, the two chromosomes begin to separate, and a septum is initiated **(dashed line)** between them. **D** = time between X and physical separation of the daughter cells.

When the growth rate is slow (e.g., I = C + D) the chromosome duplicates with a single fork, there is a gap without DNA synthesis before cell division, and the daughter cells start with one unbranched chromosome. When I = C the gap is eliminated: there is still only a single fork but a new fork is formed just when the old one ends, and the daughter cells have a branched chromosome. During faster growth (I = C) there is multiforked replication, yielding daughter cells with a branched chromosome. (Modified from Helmstetter C et al: Cold Spring Harbor Symp Quant Biol 33:809, 1968)

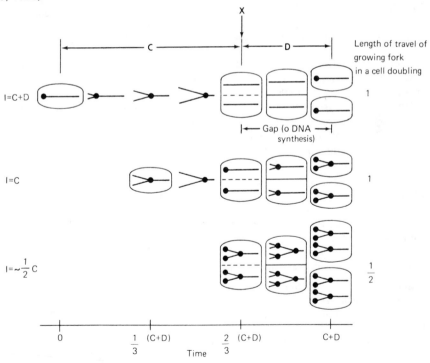

sented a problem: how can a cell have a **doubling time** (τ) less than C? The solution is that **additional growing points (forks)** are formed before completion of the first round of replication (Fig. 14-23): i.e., the already replicated initiator region on each incomplete daughter chromosome (each branch of the first fork), which has passed through the membrane-attached site of replication, reattaches to the membrane and forms a new fork. The presence of multiple branches can be demonstrated by quantitating the transforming DNA in cell lysates: with a doubling time of 50 min the ratio of the DNA of genes near the origin to that near the termination of the sequence of replication is about 2:1, but in fast growth it is almost 4:1.

The time period between two successive initiations has been defined as another parameter, **I**, which is the time for genes at one fork to reach the position of the next fork (Fig. 14-24). In steady-state growth I is identical with τ.

In general, at each growth rate the number of origins has a characteristic ratio to cell mass (and to cell protein); i.e., an initiation occurs after a fixed increment of cell mass is reached.

The molecular mechanism for relating initiation to cell growth in this way is still not clear. It might involve accumulation of an initiator protein synthesized at a rate proportional to overall protein synthesis, or dilution of a "pulse" of repressor synthesized at each initiation event.

FUNCTIONS OF VARIOUS REGULATORY MECHANISMS

We have seen that expression of a gene is regulated in bacteria by various mechanisms, in which **transcription is central.** Regulation at this level, rather than at translation, maximizes economy in macromolecule synthesis.

Multiple mechanisms for influencing the same operon increase not only flexibility but also the range of the response. Thus in the *trp* operon the 70-fold range of variable initiation of transcription is further multiplied 10-fold by variable attenuation.

RESPONSES TO ENDOGENOUS STIMULI

Feedback inhibition and repression were both discovered as mechanisms that spare synthesis in response to **exogenous** supply of a metabolite. However, their effects on the **endogenous** or **domestic** economy may be even more important for the efficiency of growth.

The key role of **endproduct feedback inhibition** in this regulation of **endogenous metabolite levels** is shown by mutants with an altered initial enzyme that is insensitive to feedback by the endproduct of the pathway: these mutants synthesize more of the compound than the cell can use, and it accumulates to such levels that it is excreted despite the normal impermeability of the membrane. In contrast, derepressed mutants, with elevated levels of all the enzymes of a pathway, do not excrete the endproduct if they possess normal feedback inhibition.

Mutants defective in endproduct inhibition are readily recognized as **feeder colonies** when plated on media heavily seeded with an auxotroph requiring the excreted product. Indeed, loss of such inhibition underlies the empiric selection of mutants for the commercial production of various microbial products.

Reserve Capacity for Enzyme Formation. In steady-state growth repression by the endogenous endproducts causes most operons to act at only a fraction of their capacity.

Full capacity is normally used only temporarily, in adapting to exhaustion of an exogenous endproduct. It can also be demonstrated in **derepressed mutants** (O, R, or attenuator), or in auxotrophic mutants grown on a limiting supply of the endproduct.

Thus prolonged growth in the presence of arginine severely represses the enzymes of arginine biosynthesis, and growth ceases when the exogenous supply is exhausted. Within 10 min, however, growth becomes normal again. During that interval, as Figure 14-25 shows, there is striking preferential synthesis of the required enzymes of the arginine pathway; these enzymes are thus restored quite quickly (and then level off), even though total protein synthesis is severely limited by the trickle of arginine from protein turnover. Figure 14-25 also shows that when arginine is continuously growth-limiting, in an auxotroph in the chemostat, the preferential synthesis of these enzymes continues and their level becomes high.

In contrast, genes subject to inflexible regulation, by a low-activity promoter, provide no reserve capacity. When such a gene (e.g., R^+ for the *lac* repressor) is introduced into an R^- cell the conversion from constitutive to inducible requires over 60 min.

Repression of Enzyme Formation versus Inhibition of Activity. Repression has often been viewed as a coarse and sluggish control over the synthesis of metabolic building blocks, and inhibition as a fine and immediate control. However, the properties just reviewed make it clear that repression is concerned primarily with economy in macromolecule synthesis, and endproduct inhibition with economy in metabolite synthesis.

Gene Dosage Effects. When enzymes are regulated by endogenous feedback their levels are not ordinarily affected by the number of copies of the corresponding gene per cell (e.g., the doubling in merodiploids). However, with a fully induced or derepressed gene, expressed to capacity, synthesis is then proportional to the number of copies present. In man, curiously, biosynthetic enzymes do not seem to be regulated by such endogenous feedback, for in **carriers** heterozygous for an enzymatic defect (e.g., galactosemia)

FIG. 14-25. Preferential synthesis of a derepressed enzyme of arginine biosynthesis, ornithine transcarbamylase. **A.** An *arg⁻his⁻* mutant, grown with **excess Arg** and His and transferred to a chemostat with limiting His and excess Arg. **Complete repression** of the enzyme continues. **B.** The same cells, in a chemostat with **limiting Arg** and excess His, show **completely derepressed** synthesis of the Arg enzyme: i.e., because of continued Arg limitation this enzyme continues indefinitely to be synthesized up to the capacity of the gene. Only the early part of the curve of exponential increase, which finally levels off at 45 arbitrary units per cell, is shown. Thus, 50% of the final level is reached after one generation, 75% after two, etc., just as with a fully induced enzyme. **C.** An *arg⁺his⁻* mutant, **grown with excess Arg** and His and transferred to a chemostat with limiting His and **no Arg.** The enzyme is initially synthesized at the fully derepressed rate, relative to total protein synthesis, as in **B.** However, within about 0.05 generations the restoration of Arg synthesis in this *arg⁺* strain causes formation of the enzyme to level off at the steady-state, **partly repressed** rate (maintaining 2.5 units per cell) characteristic of the wild type in minimal medium.

We thus see that normal synthesis uses only about 1/20 of the capacity of the gene, while the excess capacity is used for rapid adjustment to deprival of exogenous arginine. (Gorini L, Maas WK: Biochim Biophys Acta 25:208, 1957)

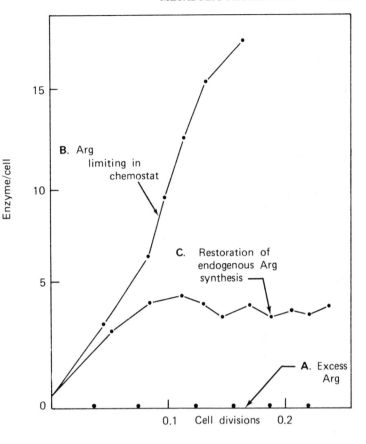

the cells produce only half as much of the enzyme as the cells of normal persons.

Levels of Intermediates. Regulation can also be studied by measuring the level of various constituents in the **intracellular metabolite pool** (released by pouring the culture into boiling water or into trichloroacetic acid solution). Like operons, the enzymes of intermediate metabolism provide flexibility by working at less than capacity in steady-state growth: that is, the concentration of each intermediate is such that it does not saturate the next enzyme but is in a range not far from the K_m.

PSEUDOFEEDBACK INHIBITION BY METABOLITE ANALOGS

Regulatory mechanisms have obviously had to be selected in evolution for great accuracy in discriminating between those normal metabolites with similar structures (e.g., leucine versus isoleucine). However, *synthetic analogs, which have not been exerting evolutionary selection pressure for specificity, can often deceive regulatory mechanisms,* just as they can deceive enzymes. For example, 5-methyltryptophan inhibits growth by mimicking the feedback effect of tryptophan on the first enzyme of its pathway, thus preventing further synthesis of tryptophan. Added tryptophan restores growth noncompetitively.

Mutants resistant to such pseudofeedback inhibition are readily selected by growth in the presence of the analog. Their altered enzyme is often resistant to normal feedback as well; hence analogs provide a means for isolating feedback-resistant mutants.

TELEONOMIC VALUE OF VARIOUS REGULATORY MECHANISMS

The intricate set of feedback circuits that evolved in the primitive bacterial cell contribute to **adaptedness** (i.e., faster growth): harmonious steady-state synthesis of metabolites, in proportion to need, promotes **efficient conversion of foodstuffs;** and the **faster growth on a rich medium** is due to the exogenous sparing of certain pathways and the resulting channeling of the endogenous material and energy into the remaining pathways.

Evolution also selects for **adaptability** to altered circumstances; we have seen several elaborate mechanisms for rapid recovery from the exhaustion of an exogenous amino acid, and for shifting to a new energy source. Without pump-priming by protein turnover and by the use of energy stores, and without preferential synthesis of the needed enzymes (and repression of synthesis of already adequate cell constituents), the cell might never recover in such a transition.

In contrast to RNA and protein synthesis, chromosome replication proceeds to completion even under conditions of extreme starvation. Since a growing point in DNA, with single-stranded segments, is less stable than the one-dimensional crystal of the double helix, the completion of replication evidently promotes the subsequent survival of the chromosome. Cell division also tends to be completed during starvation, which may further promote survival since the zone of septum formation appears to be a region of weakness in the cell wall.

DIFFERENTIATION

The systems of gene regulation revealed in bacteria provide a partial model of the even more complex regulatory processes in animal cells. However, these homeostatic bacterial mechanisms, whose **negative-feedback loops** promote **constancy** of the cell in **different** environments, do not provide an adequate model for the epigenetic changes of gene regulation in higher organisms, in which differentiated cells maintain their **differences** in an **identical** environment. Differentiation may well involve **positive-feedback loops,** in which activation of a group of genes results in the formation of a product (or a circuit of products) that ensures perpetuation of the activation.

INFORMATION, CYBERNETICS, AND EVOLUTION

The discovery of universal molecular regulatory mechanisms in cells has had a large impact on basic concepts in biology. We have already noted the increased acceptability of **teleonomic** reasoning. In addition, the theoretic concepts of cybernetics (self-regulating feedback loops), initially developed for mechanical and electronic devices, have now been brought into biology. Finally, the concept of **molecular information transfer** has now been extended. The demonstration of the transcription and translation of nucleic acid templates had already established this concept for genetic information. Since the concentration of effector molecules is also information, the response of allosteric proteins to these concentrations is the molecular basis for acquiring information from the environment. Allosteric proteins, determined by genes and in many cases acting on genes, thus provide the molecular basis for the gene–environment interactions that determine the phenotype, whether in bacteria or in man.

SELECTED READING

BOOKS AND REVIEW ARTICLES

ADHYA S, GOTTESMAN M: Control of transcription termination. Annu Rev Biochem 47:967, 1978

ANDREWS KJ, LIN ECC: Selective advantages of various bacterial transport mechanisms. Fed Proc 35:2185, 1976

BAUTZ EKF: Regulation of RNA Synthesis. Prog Nucl Acid Res Mol Biol 12:129, 1972

BECKWITH J, ZIPSER D (eds): The Lactose Operon. Cold Spring Harbor, NY, Cold Spring Harbor Laboratory, 1970

BOURGEOIS S, PFAHL M: Repressors. Adv Protein Chem 30:1, 1976

CASHEL M: Regulation of bacterial ppGpp and pppGpp. Annu Rev Microbiol 29:301, 1975

CHAPMAN AG, ATKINSON DE: Adenine nucleotide concentrations and turnover rates. Their correlation with biological activity in bacteria and yeast. Adv Microb Physiol 15:254, 1977

Cold Spring Harbor Symposia: Cellular Regulatory Mechanisms, Vol 26, 1961; Synthesis and Structure of Macromolecules, Vol 28, 1963; Replication of DNA in Microorganisms, Vol 33, 1968; Transcription of Genetic Material, Vol. 35, 1970

DICKSON RC, ABELSON J, BARNES WM, REZNIKOFF WS: Genetic regulation: the lac control region. Science 187:27, 1975

DOI RH: Role of RNA polymerase in gene selection in bacteria. Bacteriol Rev 41:568, 1977

ENGLESBERG E, WILCOX G: Regulation: positive control. Annu Rev Genet 8:219, 1974

GERHART JC: A discussion of the regulatory properties of aspartate transcarbamylase from *Escherichia coli.* Curr Top Cell Regul 2:275, 1970

GILBERT W, MULLER-HILL B: The lactose repressor. In Beckwith J, Zipser D (eds): The Lactose Operon. Cold Spring Harbor, NY,

Cold Spring Harbor Laboratory, 1970, p 93. (This volume contains many additional relevant articles)

GOLDBERG AL, DICE JF: Intracellular protein degradation in mammalian and bacterial cells. Annu Rev Biochem 43:835, 1974

GOLDBERG AL, ST. JOHN AC: Annu Rev Biochem 45:747, 1976

GOLDBERG AL, HOWELL EM, LI JB, et al: Physiological significance of protein degradation in animal and bacterial cells. Fed Proc 33:1112, 1974

GOLDBERGER RF (ed): Biological Regulation and Development, Vol I, Gene Expression. New York, Plenum, 1979

HELMSTETTER CE, COOPER S, PIERUCCI O, REVELAS E: On the bacterial life sequence. Cold Spring Harbor Symp Quant Biol 33:809, 1968

HELMSTETTER CE, PIERUCCI O, WEINBERGER M, HOLMES MR, TANG MS: Control of cell division in *Escherichia coli.* In Ornston LN, Sokatch JR (eds): The Bacteria, Vol VII. New York, Academic Press, 1978

KJELDGAARD NO, GAUSING K: Regulation of biosynthesis of ribosomes. In Nomura M, Tissieres A, Lengyel P (eds): Ribosomes. Cold Spring Harbor, NY, Cold Spring Harbor Laboratory, 1974, p 369

KJELDGAARD NO, MAALOE O (eds): Control of Ribosome Synthesis. New York, Academic Press, 1976

KOLTER R, HELINSKI DR: Regulation of initiation of DNA replication. Annu Rev Genetics 13:355, 1979

MAGASANIK B: Classical and postclassical modes of regulation of the synthesis of degradative bacterial enzymes. Prog Nucl Acid Res 17:99, 1976

MATSUSHITA T, KUBITSCHEK HE: DNA replication in bacteria. Adv Microb Physiol 12:247, 1975

MILLER JH, REZNIKOFF WS (eds): The Operon. Cold Spring Harbor, NY, Cold Spring Harbor Laboratory, 1978

NIERLICH DP: Regulation of bacterial growth, RNA, and protein synthesis. Annu Rev Microbiol 32:393, 1978

NOMURA M, MORGAN EA, JASKUNAS SR: Genetics of bacterial ribosomes. Annu Rev Genet 11:297, 1977

PARDEE AB, WU PC, ZUSSMAN DR: Bacterial division and the cell envelope. In Leive L (ed): Bacterial Membranes and Walls. New York, Marcel Dekker, 1973

PASTAN I, ADHYA S: Cyclic adenosine 3′,5′-monophosphate in *E. coli*. Bacteriol Rev 40:527, 1976

PASTAN I, PERLMAN R: Cyclic adenosine monophosphate in bacteria. Science 169:339, 1970

PERUTZ MF: Electrostatic effects in proteins. Science 201:1187, 1978

PITTARD J, GIBSON F: The regulation of biosynthesis of aromatic amino acids and vitamins. Curr Top Cell Regul 2:29, 1970

ROBERTS JW: Transcription termination and its control in *E. coli*. In Losick R, Chamberlin M (eds): RNA Polymerase. Cold Spring Harbor, NY, Cold Spring Harbor Laboratory, 1976, p 247

SANWAL BD: Allosteric controls of amphibolic pathways in bacteria. Bacteriol Rev 34:20, 1970

SCHIMKE RT, KATANUMA N (eds): Intracellular Protein Turnover. New York, Academic Press, 1975

SLATER M, SCHAECHTER M: Control of cell division in bacteria. Bacteriol Rev 38:199, 1974

STADTMAN ER: The role of multiple molecular forms of glutamine synthesis in the regulation of glutamine metabolism in *Escherichia coli*. Harvey Lect 65:97, 1969–1970

STADTMAN ER, CHOCK PB: Interconvertible enzyme cascades in metabolic regulation. Curr Cell Regul 13:53, 1978

STADTMAN ER, GINSBURG A: The glutamine synthetase of *E. coli*: structure and control. In Boyer P (ed): The Enzymes, Vol 10, 3rd ed. New York, Academic Press, 1974, p 755

UMBARGER HE: Amino acid biosynthesis and its regulation. Annu Rev Biochem 47:533, 1978

YAMADA K, KINOSHITA G, TSUNODA T, AIOTA K (eds): The Microbial Production of Amino Acids. New York, Wiley (Halsted), 1972

ZUBAY G: In vitro synthesis of proteins in microbial systems. Annu Rev Genet 7:267, 1973

SPECIFIC ARTICLES

BERTRAND K, KORN L, LEE F, PLATT T, SQUIRES CL, SQUIRES C, YANOFSKY C: New features of the regulation of the tryptophan operon. Science 189:22, 1975

DE CROMBRUGGHE B, CHEN B, GOTTESMAN M, PASTAN I, VARMUS HE, EMMER O, PERLMAN RL: Regulation of *lac* mRNA synthesis in a soluble cell-free system. Nature 230:37, 1971

DENNIS PP, NOMURA M: Stringent control of the transcriptional activities of ribosomal protein genes in *E. coli*. Nature 255:460, 1975

GILBERT W, MAIZELS N, MAXAM A: Sequences of controlling regions of the lactose operon. Cold Spring Harbor Symp Quant Biol 38:845, 1974

GILBERT W, MULLER-HILL B: Isolation of the lac repressor. Proc Natl Acad Sci USA 56:1891, 1966

GOEDEL DV, YANSURA DG, CARUTHERS MH: How lac repressor recognizes lac operator. Proc Natl Acad Sci USA 75:3578, 1978

HASELTINE WA, BLOCK R, GILBERT W, WEBER K: MSI and MSII made on ribosome in idling step of protein synthesis. Nature 238:381, 1972

HIRAGA S, YANOFSKY C: Hyper-labile messenger RNA in polar mutants of the tryptophan operon of *E. coli*. J Mol Biol 72:103, 1972

JACOB F, MONOD J: Genetic regulatory mechanisms in the synthesis of protein. J Mol Biol 3:318, 1961

LEE F, YANOFSKY C: Transcription termination at the *trp* operon attenuators of *E. coli* and *S. typhimurium*: RNA secondary structure and regulation of termination. Proc Natl Acad Sci USA 74:4365, 1977

MONOD J, CHANGEUX J-P, JACOB F: Allosteric proteins and cellular control systems. J Mol Biol 6:306, 1963

NARGANG FE, SUBRAHMANYAM CS, UMBARGER HE: Nucleotide sequence of *ilv*GEDA operon attenuator region of *E. coli*. Proc Natl Acad Sci USA 77:1823, 1980

NOMURA M, ENGBAECK F: Expression of ribosomal protein genes as analyzed by bacteriophage Mu-induced mutations. Proc Natl Acad Sci USA 69:1526, 1972

PARDEE AB, JACOB F, MONOD J: The genetic control and cytoplasmic expression of "inducibility" in the synthesis of β-galactosidase by *E. coli*. J Mol Biol 1:165, 1959

POUWELS PH, VAN ROTTERDAM J: In vitro synthesis of enzymes of the tryptophan operon of *E. coli*. Evidence for positive control of transcription. Mol Gen Genet 136:215, 1975

RICHARDSON JP, GRIMLEY C, LOWERY C: Transcription termination factor rho activity is altered in *E. coli* with *suA* gene mutations. Proc Natl Acad Sci USA 72:1725, 1975

RON EZ, GROSSMAN N, HELMSTETTER CE: Control of cell division in *E. coli*: effect of amino acid starvation. J Bacteriol 129:569, 1977

RYTER A, CHANG A: Localization of transcribing genes in the bacterial cell by means of high resolution autoradiography. J Mol Biol 98:797, 1975

ST. JOHN AC, GOLDBERG AL: Effects of reduced energy production on protein degradation, guanosine tetraphosphate, and RNA synthesis in *E. coli*. J Biol Chem 253:2705, 1978

STEPHENS JC, ARTZ SW, AMES BN: Guanosine 5′-diphosphate-3′-diphosphate (ppGpp): positive effector for histidine operon transcription and general signal for amino acid deficiency. Proc Natl Acad Sci USA 72:4389, 1975

YATES JL, ARFSTEN AE, NOMURA M: In vitro expression of *E. coli* ribosomal protein genes: autogenous inhibition of translation. Proc Natl Acad Sci USA 77:1837, 1980

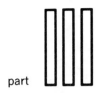

part III

IMMUNOLOGY

HERMAN N. EISEN

chapter 15

INTRODUCTION TO IMMUNE RESPONSES

In a remarkably versatile set of adaptive processes, animals form specifically reactive proteins and cells in response to an immense variety of organic molecules. These **immune responses** are encountered only in vertebrates, for whose survival they are of vast importance: they constitute the principal means of defense against infection by pathogenic microorganisms and viruses, and probably also against host cells that undergo transformation into cancer cells.

THE ORIGINS OF IMMUNOLOGY

Almost the earliest written records reveal awareness that persons who recover from certain diseases cannot contract them again: in today's terminology, they become immune. Thucydides, for example, pointed out 2500 years ago, in a remarkable description of an epidemic in Athens (possibly typhus fever or plague), that whatever attention the sick and dying received was "tended by the pitying care of those who had recovered, because they . . . were themselves free of apprehension. For no one was ever attacked a second time, or with a fatal result." This awareness led to deliberate attempts, beginning in the Middle Ages, to induce immunity by inoculating well persons with material scraped from skin lesions of persons suffering from smallpox (**variolation**). The procedure was hazardous, but in the late eighteenth century a safe, related procedure was established by the English physician Jenner.

It was widely believed that those individuals who had had cowpox (a benign disease acquired from cows infected with a mild form of smallpox) were spared in subsequent smallpox epidemics (hence the famous complexion of milkmaids). To test this belief Jenner inoculated a boy with pus from a lesion of a dairymaid who had cowpox; some weeks later reinoculation from a patient in the active stage of smallpox failed to cause illness. Repetition of the experiment many times led to Jenner's classic report, establishing that **vaccination** (L. *vacca,* cow) leads to immunity against smallpox.

Jenner's conception was not generalized until nearly 100 years later. While studying chicken cholera Pasteur happened to use an old culture of the causative agent (*Pasteurella aviseptica*) to inoculate some chickens, which failed to become ill, and proved to be immune when reinoculated with a fresh virulent culture. This finding may have prompted Pasteur's epigram that "chance favors only the prepared mind." His observations were soon applied to many other infectious diseases.

Various procedures have been used to destroy the viability or attenuate the virulence of pathogenic organisms for purposes of vaccination; examples are aging of cultures or passing the microorganisms through "unnatural hosts" (e.g., the agent of rabies passed through the rabbit). In nature the passage of smallpox virus through cows has probably selected for a variant virus that multiplies unusually well in cows but poorly in humans, while retaining the ability to induce immunity against the smallpox virus that is virulent in man.

Immunity to infectious diseases, it was found, could also be induced with products or fractions of the causative microorganism. Following the demonstration of a powerful toxin in culture filtrates of diphtheria bacilli in 1888, von Behring showed that nonlethal doses of the filtrates could induce immunity to diphtheria. Ehrlich and Calmette similarly established immunity to toxins of nonmicrobial origin, e.g., snake venoms, and ricin from castor beans.

The basis for these immune responses was revealed in 1890, when von Behring and Kitasato demonstrated that induced immunity to tetanus was due to the appearance in the serum of a capacity to neutralize the toxin; this activity was so stable it could be transferred to normal animals by infusions of blood or serum. Moreover, Ehrlich showed that protection against the toxic effects of ricin on red blood cells in vitro involved the combination of specifically reactive components of the serum with the toxin; a similar combination presumably accounted for the effects of immune serum on infectious agents. These observations opened the way to analyses of substances responsible for immunity, and also to the practical treatment of many infectious diseases by injections of serum from immune animals.

Within the ensuing 10 years most of the now known serologic* reactions were discovered. For example, serum from immunized animals caused: 1) **bacteriolysis** (first observed as disintegration of cholera vibrios, 1894), 2) **precipitation** (reported with cell-free culture filtrates of plague bacilli, 1897), and 3) **agglutination** (clumping of bacteria, reported in 1898).

These reactions were all specific: an immune serum

---

* The specific reactions of immune sera are usually referred to as **serologic** reactions.

reacted only with the substance that had induced the immune response, or, as we shall see later, with substances of similar chemical structure. By about 1900 it was realized that immune responses can also be elicited by injection of nontoxic agents, such as proteins of milk or egg white.

DEFINITIONS

The inoculated agents and the substances whose appearance in serum they evoked were called **antigens (Ags)** and **antibodies (Abs),** respectively.

An almost limitless variety of macromolecules can behave as Ags—virtually all proteins, many polysaccharides, nucleoproteins, lipoproteins, numerous synthetic polypeptides, and many small molecules if they are suitably linked to proteins or to synthetic polypeptides.

An **antigen** has two properties: 1) **immunogenicity,**[*] i.e., the capacity to stimulate the formation of the corresponding Abs, and 2) the **ability to react specifically** with these Abs. The distinction is important: substances known as **haptens** (described more fully under Antigenic Determinants, below) are not immunogenic but do react specifically with the appropriate Abs. **Specific** means that the Ag will react, in a highly selective fashion, with its corresponding Ab and not with the multitude of other Abs evoked by other Ags.

It is important to recognize the operational nature of the definition of an Ag. For example, when rabbit serum albumin (RSA) is isolated from a rabbit and then injected back into the same animal, Abs specific for RSA are **not** formed. Yet the same preparation of RSA can elicit copious amounts of anti-RSA Abs in virtually any other species of vertebrate. Moreover, the formation of these Abs depends not only on the injection of RSA into an appropriate species, but also on other conditions, namely, the quantity of RSA injected and the route and frequency of injection. It is clear, therefore, that *immunogenicity is not an inherent property of a macromolecule*, as is, for example, its molecular weight or absorption spectrum; rather, immunogenicity is dependent on the system and the conditions employed. *One cardinal condition is that the putative immunogen be somehow recognized as alien (i.e., not self) by the responding organism.*

The term **antibody** refers to the proteins that are formed in response to an Ag and that react specifically with that Ag. All Abs belong to a special group of serum proteins, the **immunoglobulins (Igs),** whose properties are considered in Chapter 17.

Though the definition states that Abs are formed in response to Ag, sera may contain Igs that react specifically with certain Ags to which the individuals concerned

have had no known exposure. These Igs are called **natural antibodies.** When present in serum they are usually in low titer, but they probably exercise a significant role in conferring resistance to certain infections. It is not clear whether natural Abs are formed without an immunogenic stimulus or in response to unknown exposure to naturally occurring Ags, e.g., in inapparent infections or in food.

CELLULAR AND HUMORAL IMMUNITY

Coincident with early studies on the role of serum Abs, Metchnikov discovered phagocytosis as a mechanism for the destruction of microbes. The resulting controversy between advocates of "humoral" immunity, due to Abs, and "cellular" immunity, due to cells, was reconciled by the finding, in 1903, of cooperation between the two components: the coating of particles by Abs (**opsonins;** Gr., to prepare food) was shown to increase their susceptibility to phagocytosis.

However, a very different kind of cellular immunity was recognized much later as a result of studies of **allergy (hypersensitivity)**—a state induced by an Ag, in which a subsequent response to the Ag causes inflammation or acute shock and death. Though certain allergic states can be transferred by serum Abs, Landsteiner and Chase showed (1942) that others can be transferred only with living leukocytes (actually lymphocytes). Various other immune responses were subsequently also found to be mediated by lymphocytes rather than by Ab molecules, e.g., immunity to tubercle bacilli and many other infectious agents, the capacity for accelerated rejection of grafted cells from genetically different individuals, and resistance to many experimental cancers.

These findings focused the attention of immunologists on **lymphocytes,** which are now recognized as the key **specificity-determining cells** in all immune responses, humoral as well as cellular. Dispersed throughout the body, these cells migrate through tissues, circulate in blood and lymph, and accumulate (transiently) in lymphatic organs (spleen, lymph nodes). In accord with the **clonal selection hypothesis,** proposed principally by Burnet in the 1950s, each lymphocyte expresses on its surface one of a large library of recognition molecules. When an Ag is encountered and is bound with sufficient affinity by these recognition molecules, and when certain regulatory signals are received, the cell is triggered to differentiate into one of several forms.

There are two main types of lymphocytes, named for the sites where they develop into immunologically competent cells. **B lymphocytes** mature in the bone marrow (mammals) or the bursa of Fabricius (birds). **T lymphocytes** mature in the thymus. Upon being triggered by an Ag, B cells divide and the daughter cells synthesize and secrete **Ab molecules** with the same Ag-binding site as

---

[*] The term **immunogen** is often used for the substance that stimulates the formation of the corresponding Abs.

the parental B cell's recognition molecule (itself an Ab molecule). T cells, however, do not secrete their recognition molecules (whose structure is still unknown); instead, these cells proliferate and differentiate into a variety of **effector T cells.** Some T effectors are **"killer" cells** (cytotoxic T lymphocytes) that specifically destroy any target cell with the appropriate surface Ag, e.g., virus-infected cells. Others produce substances that cause local inflammation **(delayed-type hypersensitivity).** Still others control the responses of various other Ag-triggered lymphocytes, either as **helpers** that promote the maturation of Ag-stimulated B and T cells, or as **suppressors** that block the enhancing activity of the T helpers.

The only recognition molecules of the immune system whose structure is well understood are Abs. Each individual in the vertebrate kingdom (fish, birds, mammals) produces an enormous number of these molecules (nearly all estimates are upward of one million), each with a distinctive combining site and specificity for an Ag. Each Ab molecule is a glycoprotein (i.e., an Ig) made up of **light (L)** and **heavy (H) polypeptide chains;** and each combining (Ag-binding) site on the molecule is formed by the pairing of one L and one H chain. The great diversity of an individual's Abs derive from 1) the diversity of the amino acid sequence in particular regions of the L and H chains, and 2) the capacity of each L chain to associate stably with many different H chains (and vice versa), each pair probably forming a distinctive combining site. The three-dimensional shapes of the molecules and their combining sites have been established through x-ray diffraction of crystallized Ab fragments.

Though each individual produces many different L and H chains, any particular B cell (or clone of B cells) makes one kind of L chain and essentially one kind of H chain, restricting its recognition capability to one (or a few) Ags out of the millions that the individual can respond to. However, the number of genes for L and H chains that are transmitted from parent to progeny seems to be far smaller than the number of chains the immunologically mature individual can produce; hence it appears that some modifications of the inherited genes take place during the somatic development of lymphocytes or their precursors.

The analysis of Abs has also progressed to where the polydeoxynucleotide sequences in some genes coding for Ab combining sites have been determined. From all of these studies several novel concepts have emerged, with implications that extend far beyond immunology: e.g., that several genes can code for different parts of a single polypeptide (Ig) chain, and that mutation–selection probably operates in the somatic development of cells of the immune system, as a remarkable form of Darwinian microevolution within each individual.

The Abs elicited by an Ag, though they are all defined by specificity for that Ag, are nevertheless highly **diverse:** Ags are large molecules and usually have several (in the case of proteins many) different antigenic determinants (see below), and even a single determinant usually evokes the formation of a family of Ab molecules of overlapping specificity and different affinity. The study of many aspects of immunology has therefore been greatly aided by the recent development of a general method for producing stable clones of cells that will each produce only one molecular species of Ab **(monoclonal Abs).**

ANTIGENIC DETERMINANTS

The reaction between an Ag and the corresponding Ab involves an actual combination of the two. We shall consider in Chapter 16 the nature of this combination whose characteristics are fundamental to almost all immunologic phenomena. For the present, however, it is useful to distinguish between the Ag molecule as a whole and its antigenic determinants, i.e., those restricted portions of the Ag that determine the specificity of Ab–Ag reactions. Attempts to evaluate the size and conformation of an antigenic determinant involve indirect approximations (see Ch. 17) which indicate that these determinants are much smaller than a typical macromolecule, being equivalent in volume to perhaps six or seven amino acid residues.

The great diversity of antigenic substances was first emphasized by Obermayer and Pick (1903), who attached NO_2 groups to rabbit serum proteins, injected these proteins into rabbits, and found that the resulting serum reacted with the nitrated proteins of rabbit, horse, or chicken serum, but not with the corresponding unmodified proteins. They inferred, therefore, that the Abs formed were capable of specifically recognizing the nitro groups or other uniquely altered structures in the nitrated proteins.

To explore the chemical basis of antigenic specificity Landsteiner then used diazonium salts to couple a wide variety of aromatic amines to proteins, as indicated by the representative reactions in Figure 15-1. Rabbits injected with *p*-azobenzenearsenate globulin form Abs that react with other proteins containing *p*-azobenzenearsenate substituents, but not with the latter proteins in unsubstituted form. Later, it was found that *p*-aminobenzenearsenate itself can competitively inhibit the reaction with *p*-azobenzenearsenate proteins, whereas other aromatic amines (with a few important exceptions) do not.

Thus this simple compound combines, in a highly selective manner, with the Abs formed in response to injection of the corresponding azoprotein, though injections of benzenearsonate itself do not evoke the formation of Abs. Substances of this type are defined as **haptens:** *they react selectively with appropriate Abs but*

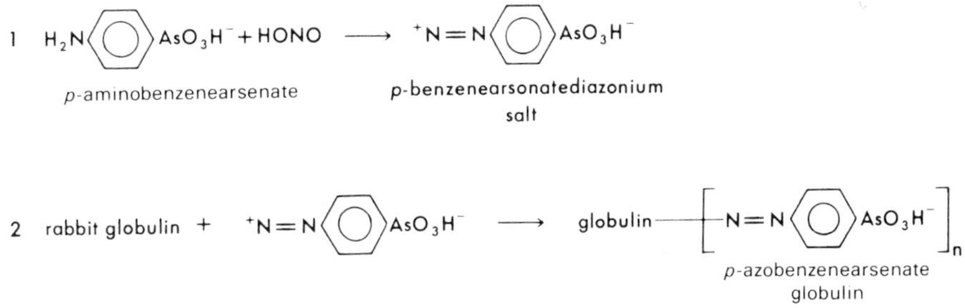

1 H_2N–C_6H_4–AsO_3H^- + HONO ⟶ $^+N{=}N$–C_6H_4–AsO_3H^-
p-aminobenzenearsenate
p-benzenearsonatediazonium salt
2 rabbit globulin + $^+N{=}N$–C_6H_4–AsO_3H^- ⟶ globulin–[–N=N–C_6H_4–AsO_3H^-]$_n$
p-azobenzenearsenate globulin

Fig. 15-1. Attachment of an aromatic amine to a protein via azo linkage (-N=N-) formed by reaction with diazonium ions.

they are not immunogenic. In the example cited the hapten is a small molecule, but some macromolecules can also function as haptens.

While the formal difference between Ag and hapten is clear, in practice it may be difficult to decide whether a substance is weakly immunogenic or completely nonimmunogenic. Generally, however, small molecules (mol wt <1000) are not immunogenic, unless covalently linked to proteins in vitro (Fig. 15-2) or in vivo (see Contact Skin Sensitivity, Ch. 22). With certain macromolecules the binding to proteins is also necessary but it need not be covalent, presumably because enough weak bonds form to establish stable (nondissociating) complexes.

The diazo reaction introduces azo groups as substituents in tyrosine, tryptophan, histidine, and lysine residues (Fig. 15-2). Other methods for coupling haptens to proteins will be referred to in subsequent chapters.

Fig. 15-2. Some azo substituents on tyrosine, lysine, and histidine residues of a representative azoprotein.

N=N–C_6H_4–AsO_3H^-
H_2C
OH
N=N–C_6H_4–AsO_3H^-
N=N–C_6H_4–AsO_3H^-
$(CH_2)_4$–N
N=N–C_6H_4–AsO_3H^-
CH_2–C=C–N=N–C_6H_4–AsO_3H^-
N NH
C
N=N–C_6H_4–AsO_3H^-

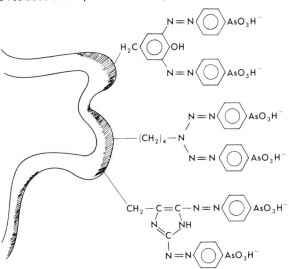

Proteins with substituents covalently linked to their side chains are referred to as **conjugated proteins;** and the substituents, sometimes including the amino acid residues to which they are linked are called **haptenic groups.** These distinctions are shown in Figure 15-3. While a haptenic group is thus part of an antigenic determinant, it is usually not clear just how much of the complete antigenic determinant it represents.

IMMUNOLOGY AND IMMUNITY

It has long been evident that immune responses can be induced against an almost limitless variety of substances, of which microbial Ags are a small minority: hence immunity to infection represents only a small facet of a more general adaptive response. Furthermore, diverse pathologic effects have been traced to immune responses to noninfectious and nontoxic foreign Ags (e.g., allografts, red blood cells, pollens, drugs), and even to an individual's own constituents (autoimmunity); hence immunologic considerations have become important for an extraordinary variety of disorders.

In addition, interest has grown in 1) Ab formation as a model for cellular differentiation; 2) the use of Ab molecules as highly specific and sensitive analytic reagents for exploring the structure of complex macromolecules (e.g., enzymes, blood group glycopeptides, cell surface macromolecules), and for measuring trace amounts of many physiologically important substances (e.g., hormones, vasoactive peptides, cyclic AMP, prostaglandins); 3) the Ab–Ag reaction as a model for specific noncovalent interactions in general (e.g., those between viruses and their host cell membranes, or those causing cell aggregation in such phenomena as fertilization and morphogenesis; and 4) the organization of structural genes for Igs as a model for the chromosomal organization of the multigene clusters in higher forms that code for a large family of similar proteins. Finally, the immune response to distinctive Ags of tumor cells has generated hope that immunology may ultimately contribute to the control of cancer.

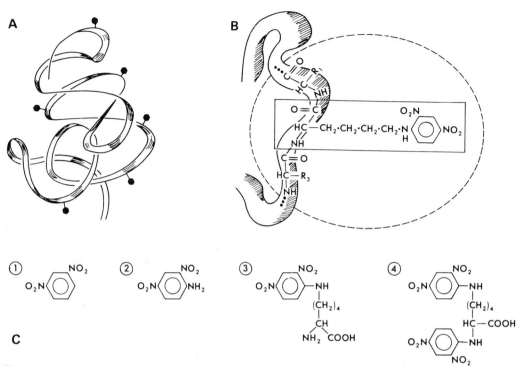

Fig. 15-3. Distinctions between conjugated proteins, antigenic determinants, haptenic groups, and haptens. **A. Conjugated protein** with substituents represented as solid hexagons. **B.** A representative **haptenic group:** a 2, 4-dinitrophenyl (Dnp) group substituted in the ε-**NH₂** group of a lysine residue. The haptenic group is outlined by the solid line; the **antigenic determinant** by the broken line. Amino acid residues contributing to the antigenic determinant need not be the nearest covalently linked neighbors of the ε-Dnp-lysine residue, as shown; they could be parts of distant segments of the polypeptide chain looped back to come into contiguity with the Dnp-lysyl residue. **C.** Some **haptens** that correspond to the haptenic group in **B:** 1) m-dinitrobenzene; 2) 2, 4–dinitroanilline; 3) ε-Dnp-lysine; 4) α, ε-bis-Dnp-lysine. With respect to Abs specific for the Dnp group, haptens 1, 2, and 3 are univalent (one combining group per molecule), and hapten 4 is bivalent.

Though there is, thus, far more to immunology than immunity in the literal sense, the historic role of infectious diseases has left an indelible imprint on nomenclature, which is not always appropriate: for example, the induction of Ab formation and of specific cellular responses is referred to as **immunization** even when no infectious agents are involved. **Vaccination** is the term reserved for immunization in which a suspension of infectious agents (or some part of them), called a **vaccine**, is given to establish resistance to an infectious disease.

In the chapters that follow we shall consider the molecular properties of Abs and their interactions with Ags, and then proceed to the more complex reactions of cellular immunity. Since the present view of Ab structure is extensively based on the use of Ab molecules themselves as analytic reagents, we shall begin with a consideration of specific Ab–Ag reactions. Some properties that will be emphasized are summarized in Table 15-1, where Abs are compared with enzymes.

TABLE 15-1. Comparison of Enzymes and Antibodies

Property	Enzymes	Antibodies
Phylogenetic distribution	Ubiquitous; made by all cells	A late evolutionary acquisition; made only in certain lymphocytes of vertebrates
Structure	Proteins with variable chemical and physical properties; an enzyme of a given specificity and from any particular organism is homogeneous; many have been crystalized	A group of closely related proteins having a common multichain structure with the chains held together by —SS—bonds. Molecules of a given specificity are heterogeneous in structure and function
Function	Specific reversible binding of ligands* *with* breaking and forming covalent bonds	Specific reversible binding of ligands* *without* breaking or forming covalent bonds
Reaction with ligands*	Wide range of affinities; populations of enzyme molecules of a given specificity are uniform in affinity for their ligand	Wide range of affinities; populations of Ab molecules of the same specificity are usually heterogeneous in affinity for their ligand
Affinity	Usually measured kinetically	Usually measured at "steady-state" (equilibrium, because the reactions are so fast)
Number of specific ligand-binding sites per molecule	Different in different enzymes, depending on number of polypeptide chains per molecule; usually one site per chain	2 per molecule of the most prevalent type (mol wt ~ 150,000); *each site is formed by a pair of chains (a light plus a heavy chain)*

*Ligand = substrate or coenzyme in case of enzymes, and antigen or hapten in case of antibodies.

SELECTED READING

BOOKS AND REVIEW ARTICLES

AMOS BD (ed): Progress in Immunology. Proceedings of the First International Congress of Immunology. New York, Academic Press, 1971

ARRHENIUS S: Immunochemistry. New York, Macmillan, 1907

BENACERRAF B, UNANUE ER: Textbook of Immunology. Baltimore, Williams and Wilkins, 1979

BRENT L, HOLBOROW J (eds): Progress in Immunology. Proceedings of the Second International Congress of Immunology. Amsterdam, North Holland, 1974

EHRLICH P: Studies in Immunity. New York, Wiley, 1910

GOLUB ES: The Cellular Basis of the Immune Response. Sunderland, MA, Sinauer, 1977

HAUROWITZ F: Immunochemistry and the Biosynthesis of Antibodies. New York, Interscience, 1968

HOOD LE, WEISSMAN IL, WOOD WB: Immunology. Menlo Park, CA, Benjamin/Cummings, 1978

HUMPHREY JH, WHITE RG: Immunology for Students of Medicine, 3rd ed. Philadelphia, FA Davis, 1970

KABAT EA: Structural Concepts in Immunology and Immunochemistry, 2nd ed. New York, Holt, Rinehart & Winston, 1976

LANDSTEINER K: The Specificity of Serological Reactions, rev ed. Cambridge, Harvard University Press, 1945; reprinted by Dover Publications, New York, 1962 (paperback)

MANDEL TE (ed): Progress in Immunology. Proceedings of the Third International Congress of Immunology. Australian Academy of Science, 1977

METCHNIKOFF E: Lectures on the Comparative Pathology of Inflammation. Kegan Paul, Trench, Trübner and Co, 1893; reprinted Dover Publications, New York, 1968 (paperback)

ROITT IM: Essential Immunology. 2nd ed. Oxford, Blackwell Scientific, 1974

TOPLEY WWC, WILSON GD: The Principles of Bacteriology and Immunity, 2nd ed. Baltimore, Williams and Wilkins, 1936

ZINSSER H, ENDERS JF, FOTHERGILL LD: Immunity: Principles and Applications in Medicine and Public Health, 5th ed. New York, Macmillan, 1939

chapter 16

ANTIBODY–ANTIGEN REACTIONS

The combination of antibody (Ab) with antigen (Ag) is the fundamental reaction of immunology. However, most Ags in wide use are macromolecules, especially proteins, and even when their covalent structure is completely established we rarely know the identity and conformation of the groups that react with Abs, or even the number and variety per Ag molecule. We shall therefore first consider specific Ab reactions with simple haptens, on which most of our understanding of the Ab–Ag reaction is based. Subsequently the more complicated reactions with macromolecular Ags will be taken up. Since the distinction between haptens and Ags, based on immunogenicity (see Definitions, Ch. 15), is largely irrelevant for the present discussion, we shall frequently use the generic term **ligand** to include both.

REACTIONS WITH SIMPLE HAPTENS

Abs to simple haptens are usually obtained by immunizing animals with the hapten attached covalently to a protein. In addition, Abs formed against high-molecular-weight polysaccharides may react with small oligosaccharides corresponding to repeated short sequences in the immunogen, e.g., the cellobiuronic acid of type 3 pneumococcus (see Fig. 16-17). Since Abs that react with haptens are easily isolated (Appendix, this chapter), the formation of specific Ab–ligand complexes can be examined in detail with relatively simple systems.

VALENCE AND AFFINITY OF ANTIBODIES

In the simplest reaction a small **univalent ligand,** with one combining group per molecule, binds reversibly to a specific site on the Ab. At low concentrations of ligand only a small proportion of the Ab's combining sites are occupied by ligand molecules; as the ligand concentration increases the number of occupied Ab sites rises until all are filled. At saturation the number of univalent ligand molecules bound per Ab molecule is the **antibody valence** (or n; see Equilibrium Dialysis, below).

If we assume that the binding sites on a population of Ab molecules are equivalent and act independently of each other the **intrinsic binding reaction** can be represented as

$$S + L \underset{k'}{\overset{k}{\rightleftharpoons}} SL \tag{1}$$

where S is a representative binding site on Ab, L is ligand, and k and k' are the rate constants for association and dissociation, respectively. The ratio k/k' is the **intrinsic association constant** K (also called **equilibrium constant**) which measures the tendency of site and ligand to form a stable complex: K represents the intrinsic affinity of the representative Ab binding site for the ligand, or

$$K = \frac{k}{k'} = \frac{[SL]}{[S][L]} \tag{2}$$

the terms in brackets referring to the concentrations, at equilibrium, of the occupied Ab sites (SL), vacant Ab sites (S), and free (unbound) ligand molecules (L). If the total concentration of binding sites ($S + SL$) and of ligand molecules ($L + SL$) are known, a measurement at equilibrium of any one term on the right side of equation 2 can lead to the association constant. It is usually most convenient to measure the free ligand concentration, and several methods are available to distinguish free (L) from bound (SL) ligand. The most generally applicable is equilibrium dialysis; it is also the most satisfactory method because it avoids perturbing the equilibrium (unlike some more rapid methods that physically separate free and Ab-bound ligands).*

Equilibrium Dialysis. The principles are shown in Figure 16-1. A solution containing Ab molecules specific for a

---

* Hapten inhibition of precipitation, described later in this chapter, is widely used to obtain *relative* values of association constants for a series of homologous haptens in relation to a reference hapten and a reference Ag. Other methods are also useful with special systems, e.g., ultracentrifugation, fluorescence quenching (see below).

A

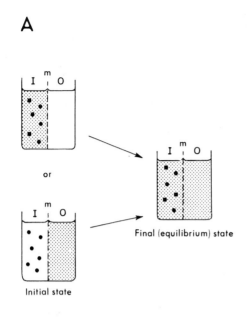

or

Initial state

Final (equilibrium) state

B

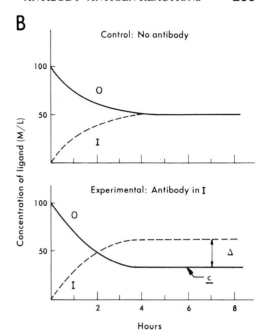

Control: No antibody

Experimental: Antibody in I

Hours

Fig. 16-1. Equilibrium dialysis. **A.** Small univalent haptens **(small dots)** can diffuse freely between the compartments **(I, O)**, but Ab molecules **(large dots)** cannot. At equilibrium the greater concentration of hapten in **I** is due to binding by Ab. **B.** Change in hapten concentration with time. Equilibrium is reached in about 4 h (the time varies with temperature, volumes of the compartments, nature and surface area of the membrane **(m)**, etc.). In **B** the hapten was initially in compartment **O**; *c* is the concentration of free (unbound) hapten at equilibrium in **I** and **O**; Δ is the concentration of Ab-bound ligand in compartment **I** at equilibrium.

simple haptenic group, such as 2,4-dinitrophenyl (Dnp), is placed in a compartment (I, for inside) which is separated by a membrane (m) from another compartment (O, for outside) that contains a solution of an appropriate univalent ligand (e.g., 2,4-dinitroaniline; see Fig. 15-3, Ch. 15). The membranes used are permeable to small molecules (mol wt < 1000), but not to Abs. If the concentration of ligand is measured periodically, it is observed to decline in O and to rise in I until equilibrium is reached; thereafter the concentrations in the two compartments remain unchanged. If compartment I simply contained the solvent, or a protein that was incapable of binding, the ligand's concentration would ultimately become the same in both compartments. But if I contains Ab molecules that can bind dinitroaniline, the final (equilibrium) concentration of total ligand in I will exceed its concentration in O.* The difference represents ligand molecules bound to Ab molecules. Because the reaction is **readily reversible** the same final state is attained

* Even without specific binding by Ab, charged ligands could be unevenly distributed across the membrane at equilibrium because of the net charge on the protein. This inequality, or **Donnan effect,** is avoided by carrying out the dialysis in the presence of a sufficiently high salt concentration, e.g., 0.15 M NaCl. Donnan effects are irrelevant with uncharged ligands.

regardless of whether the Ab and ligand are placed initially in the same or in adjacent compartments (Fig. 16-1A).

By dividing the numerator [SL] and the denominator [S] by the Ab concentration, equation 2 may be expressed as:

$$K = \frac{r}{(n-r)c} \quad (3)$$

or, more conveniently, as:

$$\frac{r}{c} = Kn - Kr^{\dagger} \quad (4)$$

where, at equilibrium, r represents the number of ligand molecules bound per Ab molecule at c free concentration of ligand, and n is the maximum number of ligand molecules that can be bound per Ab molecule (i.e., the **Ab valence).** A set of values for r and c is obtained by examining a series of dialysis chambers at equilibrium, each with the same amount of Ab but a different amount of ligand.

† This equation, called the **Scatchard equation,** is derived more fully in the Appendix, this Chapter.

According to equation 4 a plot of r/c versus r should give a straight line of slope $-K$, providing all Ab sites are identical and independent. Linearity or nonlinearity of this plot can therefore provide information on the uniformity of the ligand-binding sites in the sample of Ab molecules. The number of binding sites per Ab molecule can also be determined by the same plot: when the concentration of unbound ligand (c) becomes very large r/c approaches 0 and r approaches n; i.e., the number of ligand molecules bound per Ab molecule approaches the number of binding sites, or **Ab valence.**

Antibody Valence; Heterogeneity with Respect to Affinity. Figure 16-2 shows representative data for the binding of univalent ligands by Abs of the most prevalent type (mol wt 150,000, IgG class, see Ch. 17). Two main points are apparent.

1) At saturation (c becomes very large) the limiting value of r is 2: thus these Abs have two binding sites per molecule. Bivalence of Abs has been repeatedly observed, even in the more complex reactions with macromolecular Ags (see Fig. 16-10). (Less common Abs with 5 times higher molecular weights seem to have 5 times more binding sites, or ten per molecule; see IgM Abs, Ch. 17.)

2) The relation between r/c and r is not linear for most Ab–ligand systems (Fig. 16-2A–C). Moreover, when bivalent Ab molecules are cleaved by proteolytic enzymes into univalent fragments (Ch. 17), the fragments also exhibit nonlinearity in their reaction with the ligand. These findings and several independent lines of evidence demonstrate that the combining sites of Ab molecules of any particular specificity are usually highly diverse in their affinity for the ligand, even when the Abs are derived from a single serum. (The structural and cellular bases for this diversity are considered at length in Chapters 17, 18, and 19.) However, certain unusual Abs are homogeneous and have uniform combining sites (see Myeloma proteins, Fig. 16-2E and Ch. 17, and Hybridomas and Monoclonal Abs, Ch. 19).

For an Ab–ligand pair with diverse binding constants it is useful to determine an average value, K_0, which is defined by the free ligand concentration required for occupancy of half the Ab binding sites. Thus, substitution of $r = n/2$ in equation 4 leads to

$$K_o = \frac{1}{c} \qquad (5)$$

K_0 is designated the **average intrinsic association constant;** it is a measure of **average affinity,** and is usually referred to simply as "the affinity." Its unit is the reciprocal of concentration, i.e., liters/mole when c is in moles/liter. The higher the affinity of a given population of Ab molecules for the ligand, the lower the concentration of free ligand required for the binding sites to become occupied to any

specified extent. (The analogy to the equilibrium constant for ionization reactions should be evident; by convention, however, the ionization constant is expressed as the dissociation constant [reciprocal of the association constant], which for the reaction $HA \rightleftharpoons H^+ + A^-$ is the H^+ concentration at which half of A binds a hydrogen ion.)

The heterogeneity of Abs with respect to affinity is often evaluated by the **Sips distribution function,** which is similar to the normal distribution function commonly used in statistics. The Sips function leads to the explicit statement:

$$log \frac{r}{n-r} = a \; log \; K + a \; log \; c \qquad (6)$$

where r, n, K, and c have the same meanings as before, and a is an index of the dispersion of equilibrium constants about the average constant, K_0. The term a is similar to the standard deviation in the more familiar normal distribution.

Some representative average association constants are shown in Table 16-1. With different conditions of immunization substantial differences are observed among Ab molecules of the same specificity (Ch. 19); for example, different populations of anti-Dnp Abs differ as much as 100,000-fold in their affinity for ϵ-Dnp-lysine.

Thus *the term* **antibody,** *which is used in the singular, is in most instances a collective noun referring to a population or set of molecules defined by the capacity to bind a given functional group.* With simple ligands it is relatively easy to specify the functional group and thereby the corresponding set of Ab molecules. For example, the set that binds lactose is the "antilactose Ab." With protein Ags, however, the functional groups (**antigenic determinants**) have only rarely been identified; but each of these groups also probably specifies a set of heterogeneous Ab molecules.

Effects of Temperature, pH, and Ionic Strength; Rates of Reaction. Because Abs are relatively stable proteins, their reactions can be studied over a wide range of conditions. The resulting changes in association constants provide insight into the forces that stabilize the Ab–ligand complex. For example, the binding of an ionic ligand, such as p-aminobenzoate by antibenzoate, decreases as the pH is dropped from 7 to 4 and as the salt concentration is raised from 0.1 to 1 M NaCl. But similar changes do not affect the binding of nonionic ligands, such as 2,4-dinitroaniline by anti-Dnp Abs. This means that the COO^- group of benzoate probably interacts with a positively charged group of the Ab combining site, but ionic interactions are not important for the binding of the neutral Dnp group.

In interactions of Abs with simple ligands an increase in temperature may decrease the association constant or may have no effect; it does not increase affinity. Nevertheless, the general practice of incubating mixtures of antiserum and Ag at 37 °C is often helpful, because it speeds up some of the secondary reactions of Ab–Ag complexes, e.g., precipitation and agglutination (see below) and complement fixation (see Ch. 20).

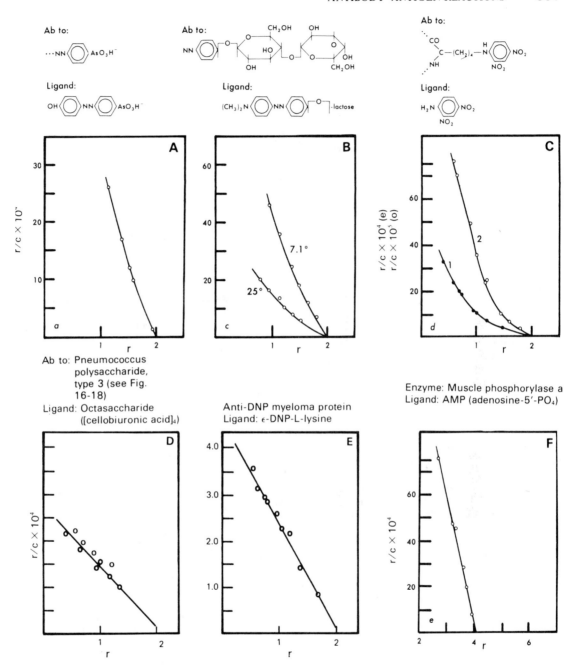

Fig. 16-2. Specific binding of ligands plotted according to equation 4. For all Ab–ligand systems **(A–E)** the extrapolation shows two combining sites per Ab molecule (mol wt 150,000, IgG class; see Ch. 17). In **B** affinity is higher at 7°C than at 25°. In **C** two purified anti-2,4-dinitrophenyl (anti-DNP) Abs differ about 30-fold in affinity for dinitroaniline. In **E** the anti-DNP protein was produced by an Ab-producing tumor (Myeloma tumor; see Ch. 17). Nonlinearity, showing heterogeneity with respect to affinity of binding sites, is pronounced in **B** and **C**, but slight in **A**; linearity, showing uniformity of binding sites, is evident with the antipolysaccharide Abs **(D)**, the anti-DNP myeloma protein **(E)**, and the enzyme muscle phosphorylase a (used as a model for a protein with uniform combining sites), which has four binding sites per molecule for adenosine monophosphate **(F)**. (Data based on : **A** , Eisen HN, Karush F: J Am Chem Soc 71:363, 1949; **B,** Karush F: J Am Chem Soc 79:3380, 1957; **C,** Eisen HN, Siskind G: Biochemistry 3:996, 1964; **D** Pappenheimer AM et al: J. Immunol 100:1237, 1968; **E,** Eisen HN et al: Cold Spring Harbor Symp Quant Biol. 32:75, 1967; **F,** Madsen NB, Cori CF: J Biol Chem 224:899, 1957).

TABLE 16-1. Average Intrinsic Association Constants for Representative Antibody–Ligand Interactions

Antibody specific for	Ligand	Average intrinsic association constants (K_0) in liters/mole
p-Azobenzenearsonate	OH⟨O⟩NN⟨O⟩AsO$_3$H$^-$	3×10^5
p-Azobenzoate	I⟨O⟩COO	4×10^4
ε-Dnp-lysyl	OOC(NH$_2$)CH-(CH$_2$)$_4$NH⟨O⟩-NO$_2$ (NO$_2$)	1×10^7
p-Azophenyl-β-lactoside	(CH$_3$)$_2$N⟨O⟩NN⟨O⟩-O-lactose	2×10^5
Mono-Dnp-ribonuclease	Mono-Dnp-ribonuclease	1×10^6

Dnp = 2,4-dinitrophenyl.

Thermodynamics. The formation of the Ab–ligand complex results in a change in free energy, ΔF, which is exponentially related to the average association constant by

$$\Delta F = -RT \ln K_0 \qquad (7)$$

where R is the gas constant (1.987 cal/mole-deg.), T the absolute temperature, and $\ln K_0$ the natural logarithm of the average intrinsic association constant. The standard free energy change, $\Delta F°$, is the gain or loss of free energy in calories, as 1 mole of Ab sites and 1 mole of free ligand combine to form 1 mole of bound ligand; when the units of K_0 are liters per mole (i.e, reactant concentrations are in moles per liter) ΔF in equation 7 is $\Delta F°$.

Values of $\Delta F°$ for various Ab–hapten pairs range from about −6,000 to −11,000 calories (per mole of hapten bound), corresponding to association constants of 10^3–10^9 liters/mole at 30° (Table 16-1).

It is sometimes useful to determine whether the free energy change comes about from a change in the heat content (**enthalpy**) or in the entropy of the system. This determination is based on

$$\Delta F = \Delta H° - T\Delta S° \qquad (8)$$

where $\Delta H°$ is the change in enthalpy (measured in calories), T is absolute temperature, and $\Delta S°$ is the entropy change. $\Delta H°$ is determined experimentally in sensitive calorimeters or by measuring the average intrinsic association constant (K_0) at two or more temperatures:

$$\Delta H° = \frac{R \ln \dfrac{K_2}{K_1}}{\dfrac{1}{T_1} - \dfrac{1}{T_2}} \qquad (9)$$

where K_1 and K_2 are average intrinsic association constants (K_0) at temperatures T_1 and T_2. $\Delta H°$ values range from 0 (no change in affinity with temperature), in which case the driving force for complex formation is the $T\Delta S$ term of equation 8, to −30,000 cal/mole ligand bound, in which case the decrease in heat content drives the reaction. The formation of apolar or **hydrophobic bonds** is essentially athermal ($\Delta H° \cong 0$), whereas the formation of hy-

drogen bonds is exothermal ($\Delta H° \cong -1000$ cal/hydrogen bond).

Rates of Reaction. Abs combine with ligands at various rates. The forward (association) reaction for binding small haptens is one of the fastest biochemical reactions known: the rate constant (k in equations 1 and 2) is only about 10 times less than the theoretical limit of 10^9 liters/mole/sec for diffusion-limited reactions. This implies that a high proportion of collisions between ligand and Ab are fruitful; perhaps they lead so often to specific binding because of a "cage effect" of the water solvent, which holds collided ligand and Ab molecules together until the ligand slips into the Ab's binding site. The association rate constants with protein Ags, however, are about 1000 times slower (about 1×10^5 liters/mole/sec), probably because incorrect molecular orientations cause many collisions to be fruitless.

With any type of ligand, a series of structural analogs (e.g., Figs. 16-5, and 16-6 below), having different equilibrium binding constants, will usually bind with virtually the same forward rate constant but dissociate with markedly different backward rate constants: *thus differences in affinity of an Ab for various ligands are determined largely by differences in the rates at which bound ligands leave the Ab combining site.*

SPECIFICITY

The forces responsible for the stability of the Ab–ligand complex have been studied by comparing the binding of structurally related ligands. The illuminating early studies by Landsteiner were based on the ability of antisera to a conjugated protein, such as sulfanilate-azo-globulin, to form specific insoluble complexes (precipitates) with other proteins conjugated with the same azo substituent, e.g., sulfanilate-azo-albumin, but not with the albumin itself. Thus the reaction was specific for the sulfanilate azo group. The Abs' reactivity with diverse groups could be evaluated by comparing the precipitating effectiveness of various conjugates, each substituted with a different azo group. Representative results are shown in Figures 16-3 and 16-4.

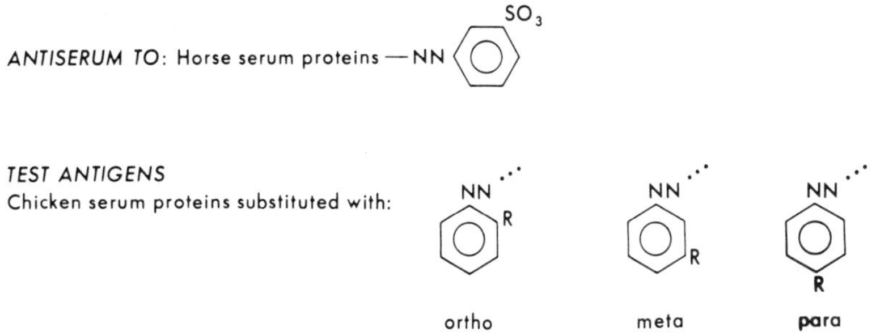

Fig. 16-3. Prominent effect of position and nature of acidic substituents of haptenic groups on the reaction between Abs to *m*-azobenzenesulfonate and various test Ags. **R** in the test Ag refers to the acidic substituents SO_3^-, AsO_3H^-, and COO^-. The homologous reaction is most intense (largest amount of precipitation) and is shown in heavy type. (Landsteiner K, van der Scheer J: J Exp Med 63:325, 1936).

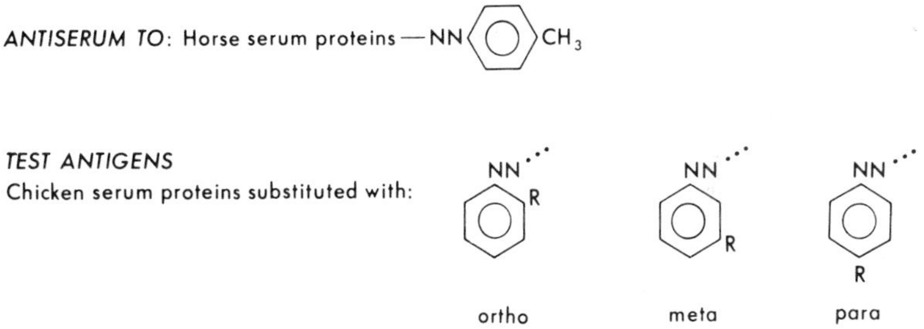

ANTISERUM TO: Horse serum proteins — NN

TEST ANTIGENS
Chicken serum proteins substituted with:

	ortho	meta	para
$R = SO_3^-$	$+\pm$	$++$	$\pm$
$R = AsO_3H^-$	0	$+$	0
$R = COO^-$	0	$\pm$	0

Fig. 16-4. Effect of nature and position of uncharged substituents of haptenic groups on the reactions between Abs to the *p*-azotoluidine group and various test Ags. The homologous reaction is shown in heavy type. (Landsteiner K, van der Scheer J: J Exp Med 45:1045, 1927)

ANTISERUM TO: Horse serum proteins — NN CH$_3$

TEST ANTIGENS
Chicken serum proteins substituted with:

	ortho	meta	para
$R = CH_3$	$+\pm$	$+\pm$	$++$
$R = Cl$	$+$	$+\pm$	$++$
$R = NO_2$	$\pm$	$+$	$+\pm$

Some more recent examples of the dependence of affinity on the structure of ligands are given in Figures 16-5 and 16-6. The following generalizations have been drawn from these and many other examples.

1) The ligands bound most strongly by a given set of Ab molecules are those that most closely simulate the structure of the determinant groups of the original immunogen. This generalization is part of the broader rule that Abs react more effectively with the Ag that stimulated their formation than with other Ags; within this context the former is generally designated **homologous Ag** and the latter **heterologous Ags**. Similarly, haptens that resemble most closely the haptenic groups of the immunogen are the **homologous haptens**.*

* Rare **heteroclitic** Abs have a higher affinity for crossreacting than for homologous ligands. The best-studied example (Abs elicited in certain strains of mice with 3-nitro-4-hydroxyphenylacetyl [NP] bind the 5-iodoanalogue, 3-nitro-5-iodo-4-hydroxyphenylacetyl [NIP], better than NP) is discussed under Antigen-Binding Receptors on T Cells in Chapter 18.

2) Those structural elements of the determinant group that **project distally** from the central mass of the immunizing Ag are **immunodominant**; i.e., they are especially influential in determining the Ab's specificity. Thus, Abs to *p*-azophenyl-β-lactoside and to 2,4-dinitrophenyl (Dnp) bind the **terminal residues** almost as well as they bind the larger haptenic structures which have these residues as their end groups: for example, compare lactose with a phenyl-β-lactoside (Fig. 16-5), and dinitroaniline with ε-Dnp-lysine (Fig. 16-6). A particularly striking example is provided by Abs to human blood group substances: anti-A is specific for terminal *N*-acetyl galactosamine residues, and anti-B is specific for terminal galactose residues (see the ABO System, Ch. 23).

However, nonterminal residues also contribute to specific binding, sometimes decisively. For example, in the cell wall lipopolysaccharide that determines the serologic specificity of various groups of *Salmonella* the sugars that react specifically with Abs to group E organisms are

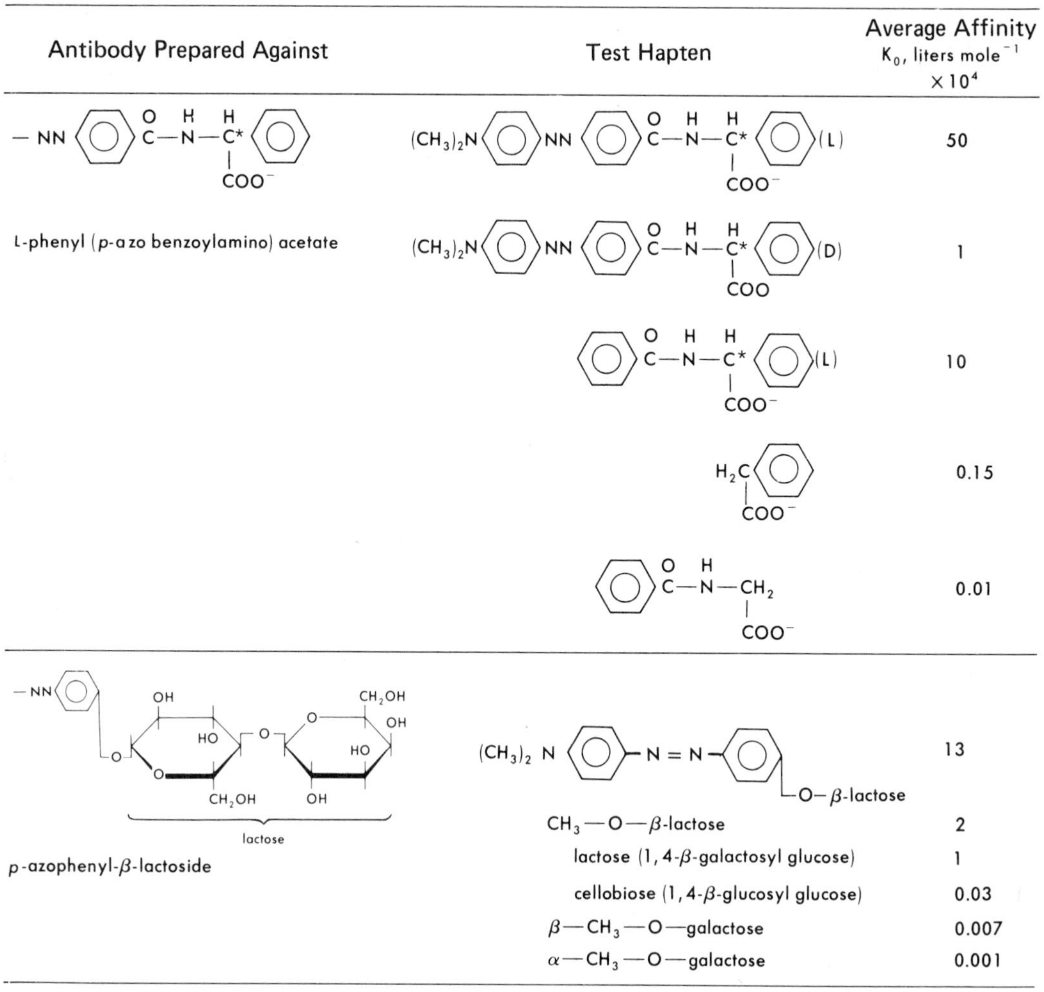

Antibody Prepared Against	Test Hapten	Average Affinity K_0, liters mole^{-1} $\times 10^4$

Fig. 16-5. Specificity of Ab–hapten reactions: Dependence of affinity on the structure of the hapten. (Karush F: J Am Chem Soc 78:5519, 1956; 79:3380, 1957).

nonterminal mannosyl rhamnose residues (see Fig. 16-19).

3) Abs are generally as discriminating as enzymes. For instance, some Abs readily distinguish between two molecules that differ only in the configuration about one carbon (e.g., glucose versus galactose, or D- versus L-tartrate; see also Fig. 16-5).

4) The specific binding of a ligand by an Ab molecule may be regarded as a competitive **partition** of the ligand between water and Ab binding sites, which are relatively hydrophobic. Hence ligands that are **sparingly soluble** in water, such as dinitrophenyl haptens, tend to form particularly **stable complexes** with Ab, whereas ligands that are highly soluble in water, such as sugars and organic ions (e.g. benzoate), tend to form more dissociable complexes.

Many observations on the interactions between Abs and their ligands have made clear that the strength of the overall bond between an Ab and a ligand reflects the sum of many noncovalent interactions among atomic groups of the ligand and side chains of amino acid residues in the binding site of the Ab. The greater the number and strength of the interactions the more stable (i.e., the less dissociable) is the Ab–ligand complex. It is apparent intuitively that the more closely the three-dimensional surface of the ligand matches up with the three-dimensional contour of the Ab site the greater the number of bonds that can be formed. However, binding strength depends not only on geometric complementarity but on chemical features of the paired groups: for example, it is greater when the groups attract, as when an anionic group of the ligand is next to a cationic group of the Ab, or a hydrogen-bond acceptor of the ligand is close to a hydrogen-bond donor of the Ab.

Antibody Prepared Against	Test Hapten	Average Affinity K_0, liters mole^{-1} $\times 10^5$

	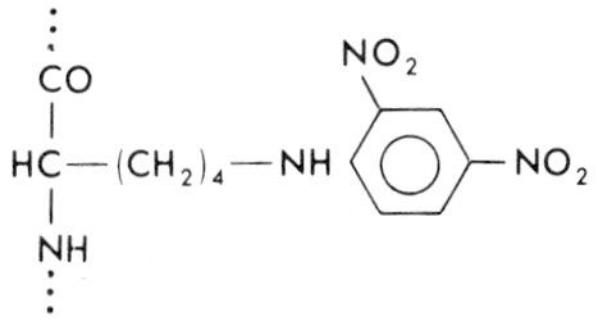 2,4-dinitrophenyl-L-lysyl group of Dnp protein	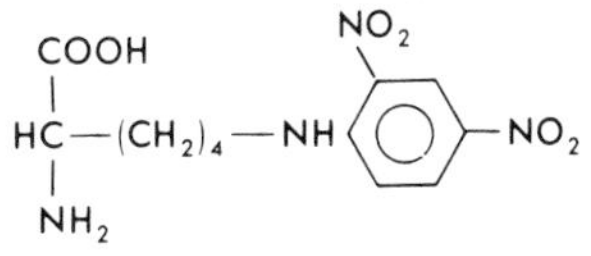 ε-Dnp-L-lysine	200
		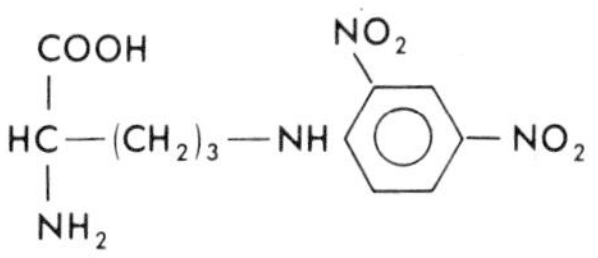 δ-Dnp-L-ornithine	80
		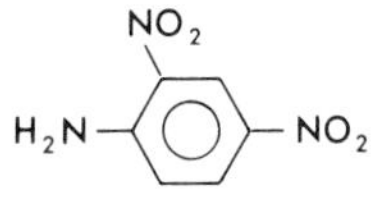 2,4-dinitroaniline	20
		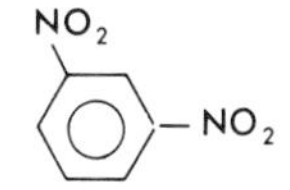 m-dinitrobenzene	8
		 p-mononitroaniline	0.5

Fig. 16-6. Specificity of Ab–hapten reactions: Dependence of affinity on the structure of the hapten. The haptens that approximate the haptenic group of the immunogen are bound more strongly. (Eisen HN, Siskind GW: Biochemistry 3:996, 1964).

Virtually all the known noncovalent bonds appear to participate in various Ab–ligand interactions: ionic, hydrogen, apolar (hydrophobic), charge-transfer. For some bonds the strength is inversely proportional to distance to the sixth or seventh power. Hence the stability of immune complexes is critically dependent on the closeness of approach of ligand groups to Ab groups. Bulky substituents on ligands can hinder close approach and thereby diminish the binding (**steric hindrance;** for example, see Fig. 16-16).

Ab binding sites have long been thought to be shallow depressions on the molecule's surface, rather than deep clefts, because they are accessible to determinant groups on macromolecules, including those on cell surfaces. This view has been confirmed by x-ray diffraction of crystallized Abs (see Fig. 17-27B, Ch. 17).

Specificity and Affinity. The **specificity** of an Ab refers to its capacity to discriminate between ligands of similar structure by combining with them to different extents: *the greater the difference in affinity for two closely related structures, the more specific the Ab.*

Specificity can thus be defined as $S = k_h/k_x$, where k_h and k_x are intrinsic affinities for the homologous (corresponding to the immunogen) and the crossreacting ligands, respectively. In practice, differences in true specificity are often obscured by an affinity threshold in most Ab-Ag assays: below the threshold reactions do not take place (see precipitation and agglutination, below). Thus, if k_h/k_x were 1000 for antiserum A and 100 for antiserum b, elicited by the same immunogen, A would be more specific than B; yet B could appear to be exquisitely specific (no crossreaction) if k_h were low but just above the threshold and k_x just below it, whereas antaserum A could appear to be much less specific (strong crossreaction) if k_h and k_x were both well above the threshold. *Abs for carbohydrates generally have low affinity for their ligands, which could account for their extraordinary ability to discriminate, in many assays, between closely related Ags, and their resulting great practical value in diagnostic typing of blood groups* (Ch. 23), *salmonellae* (Chs. 6 and 31), etc.

Similar considerations could explain why antisera to ionized groups generally appear, in practice, to be more discriminating than antisera to uncharged ligands. For instance, antisera to *m*-azobenzenesulfonate usually distinguish sharply between variously substituted analogs (sulfonate, arsonate, carboxylate) at *o, p,* and *m* positions (see Fig. 16-3), whereas antisera to *p*-azotoluidine hardly differentiate methyl, chloro, or nitro in the *p* position (Fig. 16-4).

This difference probably arises because the latter, apolar groups tend to react with their Abs with higher affinity.

Carrier Specificity. Some of the Abs made against a hapten–protein conjugate seem to react exclusively with the haptenic group: they combine no better with the immunogen than with other conjugates in which the same hapten is attached to unrelated proteins. However, many other Abs exhibit **carrier specificity:** for maximal reactiv-ity (highest affinity) they require not only the haptenic group and the amino acid residue to which it is attached but also (in varying degree) neighboring residues of the immunogen. Other carrier-specific Abs react only with the protein moiety of the conjugate. Carrier specificity is particularly significant in some of the reactions of Ags with specific receptors on lymphocytes, as we shall see in connection with Ab formation (Ch. 19) and cell-mediated immunity (Ch. 22).

REACTIONS WITH SOLUBLE MACROMOLECULES

The complexes formed by Abs and small univalent ligands, considered in previous sections, are soluble. With macromolecular Ags, however, the complexes frequently become insoluble and precipitate from solution. Though the Abs responsible for this **precipitin reaction** used to be regarded as members of a unique class, called "precipitins," it is now clear that most Abs are capable of precipitating with their Ags.

PRECIPITIN REACTION IN LIQUID MEDIA

THE QUANTITATIVE PRECIPITIN REACTION

Since its discovery in 1897 the precipitin reaction has been used extensively as a qualitative or semiquantitative assay for estimating Ab titers in serum. Attempts to measure precipitates quantitatively were of limited value until Heidelberger and Avery discovered, in 1923, that an important Ag of the pneumococcus, capsular, was a polysaccharide. This discovery had several important consequences: 1) it established that some macromolecules besides proteins could be immunogenic; 2) because the antigenic determinants of polysaccharides, unlike those of proteins, are not markedly influenced by the macromolecule's conformation, small oligosaccharides could be used as simple haptens to explore the structural basis for the specificity of some Ags; 3) the Abs precipitated from serum were identified as proteins; and 4) quantitative procedures for measuring proteins could be applied to the analysis of precipitated Abs, because the included Ag did not interfere.

The quantitative procedure is no longer commonly used experimentally. However, it is necessary to understand it in order to appreciate the general characteristics of Ab reactions with high-molecular-weight Ags. As illustration consider an antiserum prepared by immunizing a rabbit with type 3 pneumococci. When the purified capsular polysaccharide is added to the antiserum a precipitate appears. The reaction is specific: it does not occur with serum obtained before immunization or from rab-bits immunized with other Ags. Analysis of the precipitate after thorough washing reveals only protein and the type 3 polysaccharide. Moreover, when the precipitated protein is separated from the polysaccharide (see Appendix, this chapter) it can be precipitated completely and specifically by the type 3 polysaccharide. Thus, the precipitated protein is evidently Ab, which can be measured with precision. (Trace amounts of the proteins known as complement are also precipitated; they form the subject of Ch. 20.)

As is shown in Figure 16-7 and in Table 16-2, in a series of tubes with the same volume of antiserum the amount of protein precipitated increases with the amount of polysaccharide added, up to a maximum beyond which larger amounts of the Ag lead to progressively less precipitation. The precipitation of a **maximum** amount of Ab by an **optimal** amount of Ag may appear inconsistent with the binding reaction discussed earlier, in which the number of Ab sites occupied by ligand increases steadily without going through a maximum. This apparent discrepancy is due to special features of precipitation, which are discussed below, under Lattice Theory.

When the Ag is a protein instead of a polysaccharide (Table 16-3) the precipitated Ag protein must be deducted from the total precipitated protein, and so one must know the Ag content of the precipitate. Fortunately, in certain regions of the precipitin curve (Ab-excess and equivalence zones in Fig. 16-7) the precipitated Ag is essentially equivalent to the total amount of Ag added. (This conclusion was inferred initially from the absence of Ag in the corresponding supernatants [see next paragraph], and it was later demonstrated directly by the use of labeled Ags.)

Zones of the Precipitin Curve. Useful information can be obtained from qualitative tests of the supernatants to detect unreacted Ab and unreacted Ag. For this purpose each supernatant is divided into aliquots: to one is added a small amount of fresh Ag (to detect excess Ab), and to the other a small amount of fresh antiserum (to detect excess Ag). If the Ag is homogeneous (i.e., consists of a

Fig. 16-7. Precipitation curve for a monospecific system: one Ag and the corresponding Abs.

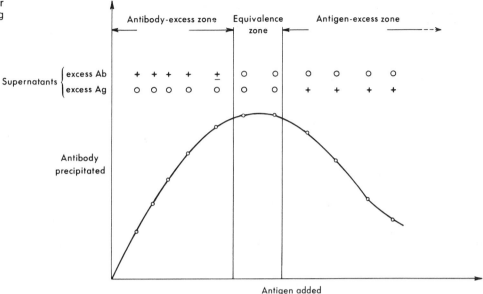

Antigen added

single uniform group of molecules), or if the Ag preparation is heterogeneous but the antiserum is capable of reacting with only one of the components, then **none** of the supernatants contains **both** unreacted Abs and unreacted Ag in detectable amounts. Instead, the residual soluble reactants are distributed as shown in the precipitin curve of Figure 16-8: on the ascending limb, or

TABLE 16-2. Precipitin Reaction with a Polysaccharide as Antigen

Tube no.	Antigen (S3) added (mg)	Total protein (or antibody) precipitated (mg)	Supernatant test
1	0.02	1.82	Excess Ab
2	0.06	4.79	Excess Ab
3	0.08	5.41	Excess Ab
4	0.10	5.79	Excess Ab
5	0.15	6.13	No Ab, no S3
6	0.20	6.23	Slight excess S3
7	0.50	5.87	Excess S3
8	1.00	3.76	Excess S3
9	2.00	2.10	Excess S3

The Ag (S3) is purified capsular polysaccharide of type 3 pneumococcus. Each tube contained 0.7 ml of antiserum obtained by injecting rabbits repeatedly with formalin-killed, encapsulated type 3 pneumococci. The supernatant of tube 6, which had the maximum amount of precipitated Ab, contained a slight excess of Ag; this is often observed and reflects the presence of some nonprecipitable or poorly precipitable Ab (see text).

(Based on Heidelberger M, Kendall FE: J Exp Med 65:647, 1937)

Ab-excess zone, the supernatants contain free Ab; on the descending limb, or **Ag-excess zone,** the supernatants contain free Ag. In the **equivalence zone** or **equivalence point,** the supernatants are usually devoid of both detectable Ab and detectable Ag, and the amount of Ab in the corresponding precipitate is taken to represent the weight of Ab in the volume of serum tested. (Sometimes, as in Tables 16-2 and 16-3, the maximum amount of Ab is precipitated when there is a slight excess of free Ag in the supernatant; this is commented on below—see Nonprecipitating antibodies.)

Up to this point we have been considering a **monospecific** system, i.e., one in which only one Ag and the corresponding Abs form the precipitates. However, most Ags, including those that satisfy the usual physical and chemical criteria of purity, are actually contaminated by small amounts of unrelated Ags, which may provoke independent immune responses. The precipitin reaction is then the sum of two or more independent monospecific precipitin reactions.

This complex (but commonplace) situation is shown schematically in Figure 16-8. Contrary to what was observed with a monospecific system, some supernatants contain **both** unreacted Abs and unreacted Ag, because the Ag-excess zone of one system overlaps the Ab-excess zone of another. Qualitative testing of supernatants in the precipitin reaction thus provides a simple means for detecting the existence of multiple systems. However, as we shall see later, the precipitin reaction in agar gel provides a more sensitive test, and it also reveals the *number* of monospecific systems.

TABLE 16-3. Precipitin Reaction with a Protein as Antigen

Tube no.	Antigen (EAc) added (mg)	Total protein precipitated (mg)	Antibody precipitated by difference (mg)	Supernatant test	Ab/Ag in precipitates	
					Weight ratio	Mole ratio
1	0.057	0.975	0.918	Excess Ab	16.1	4.0
2	0.250	3.29	3.04	Excess Ab	12.1	3.0
3	0.312	3.95	3.64	Excess Ab	11.7	2.9
4	0.463	4.96	4.50	No Ab, no EAc	9.7	2.4
5	0.513	5.19	4.68	No Ab, trace EAc	9.1	2.3
6	0.562	5.16	(4.60)	Excess EAc	(8.2)	(2.1)
7	0.775	4.56	(3.79)	Excess EAc	(4.9)	(1.2)
8	1.22	2.58	—	Excess EAc	—	—
9	3.06	0.262	—	Excess EAc	—	—

Each tube contained 1.0 ml of antiserum obtained by injecting rabbits repeatedly with alum-precipitated crystallized chicken ovalbumin (EAc).

The Ab content of precipitates in tubes 6–9 could not be determined by difference because too much EAc remained in the supernatants. The latter was measured independently in the supernatants of tubes 6 and 7, allowing an estimate to be made of EAc and Ab in the corresponding precipitates (values in parentheses).

Mole ratio Ab/Ag was estimated by assuming mol wt for EAc and Ab of 40,000 and 160,000, respectively.

(Based on Heidelberger M, Kendall FE: J Exp Med 62:697, 1935)

Antibody/Antigen Ratios in the Precipitin Reaction. With most monospecific systems the Ab/Ag ratio in precipitates varies nearly linearly over the Ab-excess zone with the amount of Ag added (Fig. 16-9). With Ab in large excess the mole ratio greatly exceeds 1, showing that many Ab molecules can be bound by one molecule of Ag; i.e., **Ag is multivalent.** In the Ag-excess zone the mole ratio of Ab/Ag tends to plateau with a limiting value of slightly over 1 (Table 16-3, Fig. 16-9). These variations are explicable in terms of the lattice theory.

LATTICE THEORY

To account for precipitation and for varying ratios, Marrack, and Heidelberger and Kendall, suggested that precipitation could be a consequence of the growth of Ab–Ag aggregates in such a way that each Ag molecule is linked to more than one Ab molecule and each Ab molecule is linked to more than one Ag molecule. When the aggregates exceed some critical volume they settle out of solution spontaneously (because sedimentation rate is

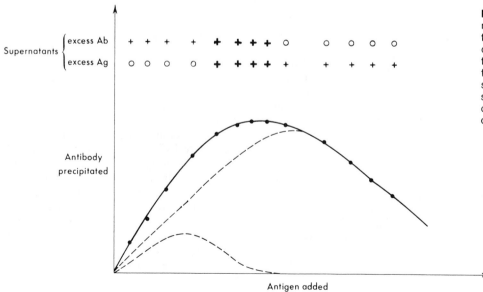

Fig. 16-8. Precipitin curve for a multispecific system. The precipitation observed (——•——) is the sum of two or more precipitation reactions (– – –). The significant difference from the monospecific system shown in Fig. 16-7 is that some supernatants have **both** excess Ag and excess Abs (indicated by pluses in heavy type).

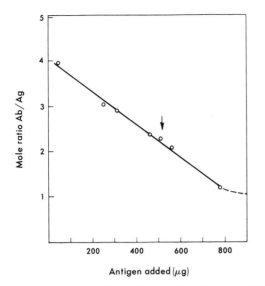

Fig. 16-9. Continuous decline in the Ab/Ag ratio of precipitates with increasing amount of Ag added to a fixed volume of antiserum. Chicken ovalbumin (EAc) is the Ag and the serum is rabbit anti-EAc. The limiting mole ratio and slope vary (about 40%) in different anti-EAc sera. **Arrow** marks the equivalence zone. (Data are those of Table 16-3.)

proportional to $V(\rho - \rho_0)g$, where V is the volume of a particle, ρ and ρ_0 are the densities of the particle and solvent, respectively, and g is the gravitational field). The assumption that Abs are multivalent was validated many years later by equilibrium dialysis with univalent haptens, as noted above (Fig. 16-2).

As Fig. 16-10 shows, **alternation** of multivalent Ag and Ab molecules (i.e., the **lattice theory**) also accounts for the wide and continuous variations in Ab/Ag ratios (Fig. 16-9). With systems in which the Ag is distinctively labeled, and can thus be measured directly, precipitates formed in the presence of excess Ag are found to have Ab/Ag mole ratios that approach 1 as a limiting value, suggesting a large linear aggregate with alternating Ab and Ag molecules (... Ab•Ag•Ab•Ag•Ab•Ag ...). In the region of Ag-excess, complexes of even lower Ab/Ag ratios are found, but they remain in the supernatant because they are small (Fig. 16-10D–F); they account for the descending limb of the precipitin curve (Fig. 16-8). The soluble complexes have been demonstrated by ultracentrifugation and electrophoresis, and their mole ratios vary considerably; e.g., 0.75 (Ab_3Ag_4) in slight Ag excess, 0.67 (Ab_2Ag_3) in substantial Ag excess. In extreme excess the ratio approaches 0.5 ($AbAg_2$), as expected from the bivalence of Ab molecules.

VALENCE AND COMPLEXITY OF PROTEIN ANTIGENS

The limiting mole ratio of Ab/Ag in extreme Ab excess is often taken as a measure of the Ag molecule's valence

(e.g., 4 in Fig. 16-9). Because Ab molecules are bivalent, however, the actual number of binding sites on the Ag can approach twice the limiting mole ratio. And the limiting ratio provides only a **minimal** estimate of the Ag valence: a larger number of reactive sites could fail to react, either because of spatial limitations in the packing of more Ab molecules about one Ag molecule, or because the particular antiserum used lacks Abs for some antigenic determinants.

As is noted in Table 16-4, an Ag molecule of high molecular weight can generally bind more Ab molecules (simultaneously) than an Ag of low molecular weight, and *all Ags capable of giving a precipitin reaction have a valence of at least two.* Even some small bivalent haptens form specific precipitates; for example, α,ϵ-bis-Dnp-lysine (see Fig. 15-3, Ch. 15, and Fig. 16-11).

The multivalency of many **polysaccharides** results from their repeating residues (e.g., Fig. 16-17). With **proteins,** however, the chemical basis for their multivalency is less obvious. Groups of amino acid residues almost never recur as repetitive sequences in a given polypeptide chain; and so each antigenic determinant, consisting of several residues, is likely to occur only once per chain. Nevertheless, protein molecules made up of a single chain behave as though multivalent, because the corresponding antiserum usually contains Abs to the chain's many different determinants.

Thus, proteolytic cleavage of bovine serum albumin (BSA, one chain of 70,000 daltons) yields several large fragments, each of which gives a precipitin reaction with different sets of Abs in antiserum prepared against the intact BSA molecule. Further evidence is supplied by the model system illustrated in Figure 16-11: the small ligand R-X, in which the functional groups R and X each occur once per molecule, does not precipitate with an anti-R serum or with an anti-X serum, but does precipitate with a mixture of the two. These considerations lead to the schematic view of the precipitin reaction shown in Figure 16-12: protein precipitation usually depends upon a mosaic of Ag determinants and the cooperative effects of Abs to different determinants.

The validity of the scheme shown in Figure 16-13 is illustrated with bovine pancreatic ribonuclease (one chain per molecule), to which one Dnp group has been added (mono-Dnp-RNase). In the antiserum made against this immunogen some Abs are anti-Dnp and others are specific for other determinant groups (X, Y, Z in Fig. 16-13). Mono-Dnp-RNase behaves as a multivalent molecule with this antiserum, giving a classic precipitin reaction. But if the anti-Dnp molecules are isolated from the antiserum and then mixed with mono-Dnp-RNase only soluble complexes are formed, because the same Ag is univalent with respect to this particular set of Ab molecules. These observations emphasize the operational nature of the definition of valence: *a given molecule of Ag can be univalent with respect to some Ab molecules and multivalent with respect to others.*

The number of sets of Abs in an antiserum to a globular protein

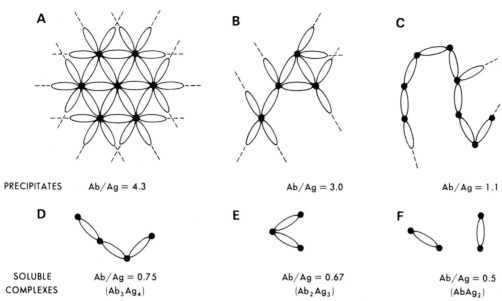

Fig. 16-10. Hypothetic structure of immune precipitates and soluble complexes according to the lattice theory. Numbers refer to mole ratios of Ab to Ag. **Dotted lines** with precipitates indicate that the complexes continue to extend as shown. The precipitates may be visualized as those found in the Ab-excess zone **(A)**, the equivalence zone **(B)**, and the Ag-excess zone **(C)**. The soluble complexes correspond to those in supernatants in moderate **(D)**, far **(E)**, or extreme **(F)** Ag excess. **Black circles,** Ag molecules; **open ellipses,** Ab molecules.

is at least equal to the number of binding sites per Ag molecule (Ag valence; i.e., the limiting mole ratio of Ab/Ag in the far Ab-excess region of the precipitin curve). Because each set is probably made up of Ab molecules of different affinities, and even of different immunoglobulin classes (Ch. 17), terms such as "the anti-BSA Ab" or "tetanus antitoxin" are deceptively simple.

TABLE 16-4. Correlation Between Molecular Weight and Valence of Antigens

Antigen	Molecular weight	Approximate mole ratio Ab/Ag of precipitates in extreme antibody excess*
Bovine pancreatic ribonuclease	13,600	3
Chicken ovalbumin	42,000	5
Horse serum albumin	69,000	6
Human γ-globulin	160,000	7
Horse apoferritin	465,000	26
Thyroglobulin	700,000	40
Tomato bushy stunt virus	8,000,000	90
Tobacco mosaic virus	40,000,000	650

*The mole ratios are representative, and tend to be higher with antiserum obtained late in the course of immunization. Since the Ab molecules involved are at least bivalent (cf. IgG and IgM, Ch. 17), the Ag valences must be somewhat higher than the ratios listed.

(Based on Kabat EA: In Kabat and Mayer's Experimental Immunochemistry, 2nd ed. Springfield, Ill., Thomas, 1961)

Homospecificity of Antibodies. Since most Ags have diverse functional groups it may be asked whether the binding sites of a given bivalent Ab molecule are specific for two different ligand groups. All existing evidence shows that this does not occur: *bivalent Abs are homospecific.* For example, in antiserum prepared against dinitrophenylated bovine γ-globulin (Dnp-BγG) the removal of all the Abs that can react with BγG does not reduce the capacity of the antiserum to bind Dnp ligands. This finding is explained by the symmetric structure of Ab molecules (Ch. 17) and the one cell–one Ab rule (Chs. 18, 19).

NONSPECIFIC FACTORS IN THE PRECIPITIN REACTION

The precipitin reaction involves **two distinct stages: rapid** formation of soluble Ab–Ag complexes and **slow** aggregation of these complexes to form visible precipitates. By measuring free Ag concentrations it has been found that the specific interactions are completed within a few minutes, but precipitate formation usually requires several days to reach completion.

While the lattice theory is clearly relevant for the first stage, the second, slow stage probably results from the close packing of Ab molecules in Ab–Ag aggregates, allowing neighboring Ab molecules to react nonspecifically with each other, probably by way (mostly) of ionic bonds between oppositely charged groups. The aggregates, which are predominantly made up of Ab molecules (Figs. 16-9, 10), thus become poorly charged, and this contributes to their insolubility. This view is supported by the effects on precipitation of modifying the charge on Ab molecules by changing the

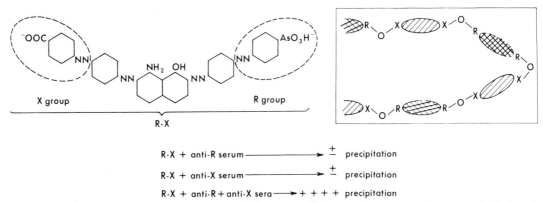

R-X + anti-R serum ⟶ $\pm$ precipitation

R-X + anti-X serum ⟶ $\pm$ precipitation

R-X + anti-R + anti-X sera ⟶ + + + + precipitation

Fig. 16-11. Cooperation between Abs of different specificities (anti-R and anti-X) in the precipitation reaction with a synthetic ligand, R-X. (The small amount of precipitate ($\pm$) formed by R-X with anti-R alone or with anti-X alone is probably due to some aggregation of R-X.) **Inset** shows a hypothetic segment of the precipitate with alternation of anti-R and anti-X molecules in a linear aggregate. (Based on Pauling L, Pressman D, Campbell DH, J Am Chem Soc 66:330, 1944).

salt concentration (e.g., lowering the salt increases the number of charged groups per molecule and decreases precipitability) or by chemical substitution. For instance, when the negative charge on Ab molecules is increased by acetylation of free amino groups the ability to precipitate can be lost without impairing the Ab's ability to form specific, soluble complexes with Ag.

Fig. 16-12. Diversity of Abs formed against a pure Ag and their cooperation in the precipitation reaction with that Ag. Further complexities arise because each set of Abs (**anti-1, anti-2**, etc.) is probably heterogeneous with respect to affinity for the corresponding antigenic determinant; in addition, Abs of the same specificity may differ considerably in structure (Ch. 17). Antigenic determinants are labeled **1, 2, 3, 4.**

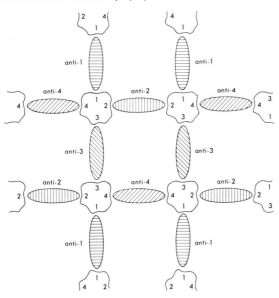

NONPRECIPITATING ANTIBODIES

Nonprecipitating Abs were first recognized because of discrepancies in the amounts of Ab precipitated after the addition of Ag by two different procedures. In **procedure A** each of a series of tubes containing the same amount of antiserum receives a different amount of Ag, as in Table 16-2. By analysis of the precipitates the quantity of Ag necessary to precipitate the maximal amount of Ab is established. In **procedure B** about 1/10 of this optional amount of Ag is added repeatedly to a single volume of antiserum, from which the precipitate formed after each addition is removed before the next addition. The sum of all the Ab precipitated is usually considerably less than that precipitated at the equivalence point in procedure A. The difference represents **"nonprecipitating" antibodies,** i.e., molecules that are not precipitable unless they bind to insoluble Ab–Ag aggregates having unoccupied binding sites on the Ag.

Under the influence of the lattice theory, immunologists assumed that **nonprecipitating Ab molecules** have a single combining site. However, careful studies have shown that native Ab molecules are always multivalent, and the inability of some Abs to precipitate with Ags arises from other mechanisms. One cause is the low affinity of some Abs: the Ag must then be added at relatively high concentrations in order to occupy a significant proportion of the Ab's combining sites, and if this Ag level is in stoichiometric excess only small soluble Ab–Ag complexes are formed.

MULTIVALENT BINDING

Monogamous Bivalency. Another mechanism could arise from this phenomenon: some bivalent Ab molecules with

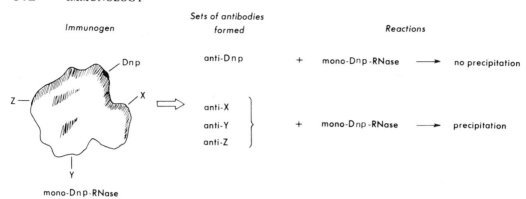

Fig. 16-13. Antigen valence illustrated with mono-DNP-RNase, which induces the formation of anti-DNP and other Abs, arbitrarily called anti-X, anti-Y, and anti-Z. The immunogen is **univalent** with respect to the anti-DNP Abs, with which it forms soluble complexes. The same immunogen, however, is **multivalent** with respect to the mixture of diverse sets of Abs (anti-X, anti-Y, etc.) that are formed against its various nonhaptenic determinants (X, Y, DNP, etc.). (Based on Eisen HN, Simms ES, Little JR, Steiner LA: Fed Proc 23:559, 1964).

a high affinity for Ag preferentially combine with two determinant groups of a **single** Ag particle, forming cyclic complexes (Ab:Ag) rather than crosslinked ones (Ag·Ab·Ag; Fig. 16-14). This type of "multivalent" binding, called **monogamous bivalency** by Karush, requires that a given antigenic group occur repetitively on the Ag surface (as occurs in hapten–protein conjugates, polysaccharides, and multichain proteins). This situation is commonly found with surface determinants of bacteria and other cells: Abs that combine specifically with these cells, but do not agglutinate them, are called blocking Abs (or incomplete Abs; see below).

Abs in general would be expected to engage preferentially in monogamous bivalent binding (because it is energetically advantageous to form binary Ab:Ag rather than ternary Ag·Ab·Ag complexes). However, most Abs seem unable to do so, perhaps because they are not flexible enough to fit their two combining sites to neighboring sites on a given Ag particle (see Overall Structure of Abs, Ch. 17).

For Abs that can form monogamous complexes the equilibrium constant with multivalent ligands can be many orders of magnitude higher than with the corresponding univalent hapten. For instance, the reaction between anti-Dnp Abs and Dnp-bacteriophage, with many Dnp groups per phage particle, has an association constant about 100,000-fold greater than the reaction between the same Abs and univalent Dnp-lysine (e.g., 10^{12} versus 10^7 liters/mole): evidently the two bonds in the monogamous complex greatly decrease the probability that the Ab:Ag pair will dissociate. Because unmodified viruses and bacterial surfaces usually have repeating,

Fig. 16-14. Monogamous bivalent binding **(A),** leading to cyclic Ab:Ag complexes, is contrasted with conventional binding **(B),** which leads to cross linking of Ag particles (. . . Ag • Ab • Ag . . .).

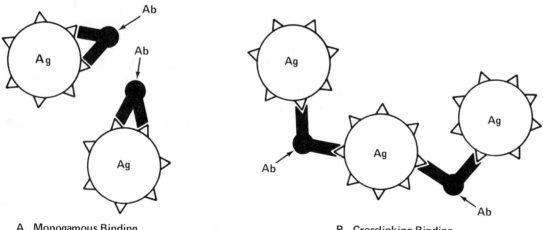

A. Monogamous Binding

B. Crosslinking Binding

identical antigenic groups, their formation of similar cyclic complexes can explain why exceedingly low concentrations of some Abs (e.g., 10^{-12} moles/liter or about $1.5 \times 10^{-4} \mu g$/ml) can neutralize pathogenic microbes under in vivo conditions.

REVERSIBILITY OF PRECIPITATION

After a precipitate has formed, its complexes can dissociate and reequilibrate with a fresh charge of Ag. When the latter is in sufficient excess small soluble complexes are formed and the precipitate dissolves. This reversibility provides the basis for many of the procedures used to isolate purified Abs (see Appendix, this chapter). In practice, however, it may be difficult to observe dissociation: hence the formation of Ab–Ag aggregates was formerly believed to be irreversible.

A striking example is the **Danysz phenomenon.** When an equivalent amount of diphtheria toxin is added all at once to an antitoxin serum the mixture is nontoxic; but if the same quantity of toxin is added in portions at intervals of about 30 min. the mixture is toxic. This phenomenon was interpreted to mean that, in the latter procedure, the first addition of toxin led to the formation of irreversible complexes with a high Ab/Ag ratio, leaving insufficient free Ab to neutralize all the toxin subsequently added. However, the complexes that appear to be nondissociable no doubt really have unusually slow dissociation rates, requiring days or even months, rather than minutes, to reestablish equilibrium when the system is perturbed.

Apparent irreversibility is especially striking with many Ab–virus complexes: some are so stable that they do not perceptibly dissociate even when a mixture of virus and antiserum is diluted many thousandfold, with a corresponding reduction in the concentrations of free virus and free Ab (Ch. 52). The extraordinary stability of such complexes probably derives from monogamous multivalent binding (see above). If the intrinsic association constant is sufficiently high ($>10^8$ liters/mole) even ordinary single-bonded complexes of Ab with small univalent haptens can appear to be irreversible; e.g., Ab cannot be completely freed of the hapten by dialysis. Nevertheless, reversibility can always be demonstrated by "exchange": the addition of free ligand, in great excess, will replace the molecules that appeared to be irreversibly bound.

HAPTEN INHIBITION OF PRECIPITATION

Qualitative Hapten Inhibition. Long before its quantitative features were characterized the precipitin reaction was widely used as a visual qualitative test to detect Ab–Ag reactions. It was, in fact, in just this simple fashion that Landsteiner had exploited it, by means of hapten inhibition, in his classic investigations of immunologic specificity. In general terms, in the precipitin system that he used the antiserum was prepared against one conjugated protein, which

may be designed X-azoprotein A, and to eliminate reactions with the protein part of the conjugate the precipitating agent was a conjugate with the same azo substituent attached to a different protein, X-azoprotein B (with many X-azo groups per protein molecule). Simple haptens with one X group per molecule formed soluble complexes with anti-X and could competitively inhibit the precipitin reaction (Fig. 16-15). The greater the Ab's affinity for the hapten, relative to its affinity for the precipitating azoprotein, the more effectively was precipitation inhibited.

Quantitative Hapten Inhibition. By combining hapten inhibition with quantitative measurements of precipitates, Pauling and Pressman were able to obtain more insight into the specificity of Ab reactions. They added equal volumes of an anti-X serum to a series of tubes with varying amounts of univalent X haptens, and after a few minutes they added a multivalent precipitating agent (e.g., an X-azoprotein) in that amount which, in the absence of hapten, would give roughly maximal precipitation of anti-X Abs. They found that the amount of Ab precipitated decreases as the concentration of added hapten increases. By convention, the concentration of hapten required for 50% inhibition, compared for different haptens with a particular antiserum and Ag, provides an index of the **relative affinity** for that set of precipitating Abs (Fig. 16-16.)

Conformational versus Sequential Determinants. Hapten inhibition of precipitation has been used to identify determinant groups of complex Ags, such as proteins and blood group substances (Ch. 23). The reactions of globular protein Ags are usually inhibitable by some of the large fragments (mol wt well above 1000) derived from the Ag by proteolytic enzymes, but not by small peptides or by the denatured protein. Such results suggest that most determinant groups of proteins are **conformational,** representing a particular three-dimensional spatial arrangement of a cluster of amino acids, rather than simply their sequence. However, **sequential** determinants—small dialyzable oligopeptides derived from the Ag—are conspicuous in fibrous proteins (e.g., silk fibroin, synthetic poly-

Fig. 16-15. Competitive inhibition of the precipitation of a multivalent antigen **(A)**.

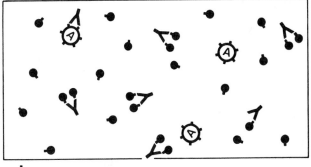

(A) Antigen

● Univalent ligand

< Antibody

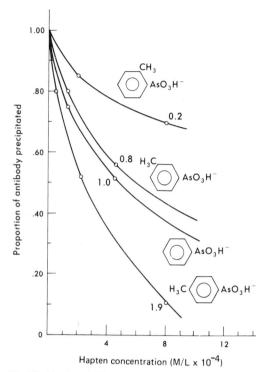

Fig. 16-16. Hapten inhibition of the precipitin reaction, illustrated with various univalent benzenearsonates. The number on each curve gives the affinity for the corresponding hapten relative to benzenearsonate (= 1.0). The antiserum was prepared against *p*-azobenzenearsonate; hence the hapten with $-CH_3$ in the para position is bound best, i.e., it is most inhibitory ($K' = 1.9$). With the methyl group in the meta or ortho position steric hindrance reduces affinity for Ab. (Pressman D. Adv Biol Med Phys 3:99, 1953).

peptides); they are conspicuous only in certain globular proteins (e.g., myoglobin).

CROSS REACTIONS

Besides reacting with its immunogen, an antiserum almost invariably reacts with certain other Ags (heterologous Ags) that are sufficiently similar to the immunogen; some crossreactions are illustrated in Figures 16-17 and 16-18.

Crossreactions Due to Impurities. Most immunogens are complex mixtures of diverse antigenic molecules. This is obviously true when the immunogen is a cell. It is also usually true, though less obvious, with purified proteins, because these are nearly always contaminated with other proteins. Even at trace levels (e.g., 1%) the contaminants can elicit the formation of detectable amounts of Ab. Hence, antisera usually consist of several Ab populations, each reactive with one Ag.

If, for example, crystallized chicken ovalbumin (EAc) contaminated with trace amounts of chicken serum proteins were used as immunogen, the anti-EAc serum would probably contain low levels of Abs to the contaminants and would probably form precipitates (i.e., crossreact) with some chicken serum proteins. Supernatant tests would probably reveal that the precipitin reaction was not monospecific; and this would be even easier to establish by reactions in agar gel (see Fig. 16-26).

Crossreactions Due to Common or Similar Functional Groups. We have already noted that a single protein contains many different antigenic determinants, each of which can evoke the formation of a corresponding set of Abs. If two different Ag molecules happen to have one or more groups in common they usually crossreact. This type of crossreaction is frequently observed with polysaccharides, and it provides the basis for **classifying** many groups of closely related bacteria.

The 1500 or more varieties of salmonellae, for example, have been arranged into several serologic groups, each identified by mutual crossreactions: an antiserum to any strain of a particular group reacts with the other strains of that group. The common antigenic determinants responsible for these **group-specific reactions** have been shown to be short sequences of particular sugar residues. For example, the determinant that defines group O *Salmonella* has as its principle residue colitose (3,6-dideoxy-L-galactose), attached as a terminal sugar on a branch of the cell envelope lipopolysaccharide. Other terminally attached dideoxyhexoses are responsible for the crossreactions that characterize other groups of salmonellae. The determinants responsible for crossreactions need not, however, be terminal residues. The strains of group E *Salmonella*, for example, crossreact because their lipopolysaccharides all possess nonterminal mannosylrhamnose residues as repeating units (Fig. 16-19).

Another example is provided by the pneumococcus. The capsular polysaccharide of type 3 is a linear polymer of repeating cellobiuronic acid residues (β-1,4-glucuronidoglucose), whereas in type 8 these residues alternate with glucosyl galactose residues. Hence antisera to either Ag crossreact extensively with the other (Fig. 16-17).

A similar principle accounts for the crossreactions that are commonplace with conjugated proteins; for example, antiserum to Dnp-bovine γ-globulin reacts with Dnp-human serum albumin because Dnp groups are present in both conjugates.

Crossreacting groups need not be identical; they need only be sufficiently similar. For example, Abs to *m*-azobenzenesulfonate crossreact with *m*-azobenzenearsonate (Fig. 16-3), and Abs to the Dnp- (2,4-dinitrophenyl-) lysyl group crossreact with 2,4,6-trinitrobenzene. *All Abs exhibit some crossreactions: i.e., they bind some Ags with determinant groups that are not identical with those in the immunogen.*

General Characteristics of Crossreactions. The following generalizations are drawn from the study of many crossreactions.

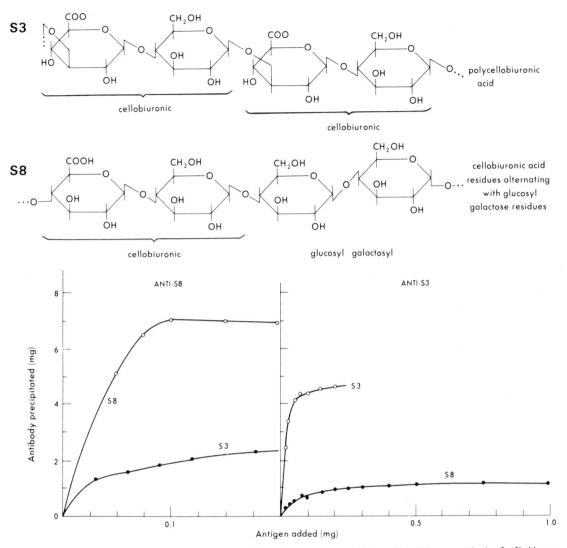

Fig. 16-17. Cross reactions between S3 and S8, capsular polysaccharides of pneumococci of types 3 and 8, respectively. **Left:** Horse antiserum to type 8 pneumococcus, reacted with purified S8 and S3 polysaccharides. **Right:** Horse antiserum to type 3 pneumococcus, reacted with S3 and S8 polysaccharide. (Based on Heidelberger M J Exp Med 65:487, 1937; Heidelberger, M et al: J Exp Med 75:35, 1942).

1) An antiserum precipitates **more copiously** with its immunogen than with crossreacting Ags (Figs. 16-17 and 16-18), because a) a heterologous ligand usually reacts with only a fraction of the total Ab to the immunogen, and b) Abs of a given specificity will almost always have **greater affinity** for the homologous ligand than for a cross-reacting ligand (Fig. 16-5 and 16-6).

2) The mutual crossreactions are usually not quantitatively equivalent; Abs to the first Ag may react more extensively with the second Ag than Abs to the second react with the first (Fig. 16-17).

3) Different antisera to a given immunogen are likely to vary in the extent of their crossreactions with diverse heterologous Ags; this is true even when the antisera are obtained from the same animal (at different times after immunization; Ch. 19)

4) Crossreacting ligands tend to be bound more strongly by Abs with high affinity for the homologous ligand than by those with low affinity. Thus, fewer crossreactions are exhibited by low- than by high-affinity Abs (see Specificity and Affinity, above). Because polysaccharide–antipolysaccharide systems are, in general, characterized by low affinities, this rule accounts for the great specificity of the sera used to type bacteria and red blood cells according to their surface saccharides (Ch. 23).

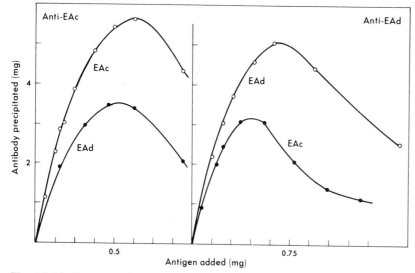

Fig. 16-18. Cross reactions between chicken and duck ovalbumin, EAc and EAd, respectively. **Left:** Rabbit antiserum to EAc, reacted with EAc and EAd. **Right:** Rabbit antiserum to EAd, reacted with EAd and EAc. (Based on Osler AG, Heidelberger M: J Immunol 60:327, 1948).

Fig. 16-19. Portion of the cell wall lipopolysaccharide of three strains of salmonellae. The strains are assigned to the same group (O) because of their serologic cross reaction, due to their common mannosyl rhamnose residues (boxed in). They are also distinguishable serologically because each has some unique structural feature, e.g., the terminal glucose **(G)** residue in *S. minneapolis*, and the α or β glycosidic bond linking galactose **(Gal)** to mannose **(M)**. (Based on Robbins PW, Uchida T: Fed Proc 21:702, 1962).

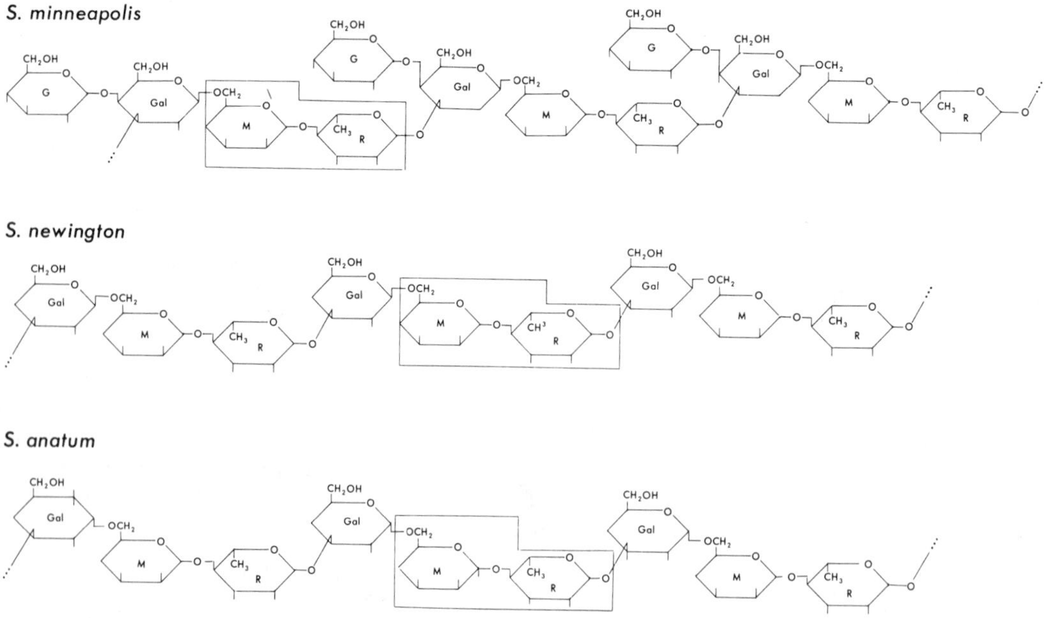

S. minneapolis

S. newington

S. anatum

Comparison of Various Types of Crossreactions. The crossreactions discussed above may be contrasted as follows. 1) The presence of antigenic impurities is trivial conceptually but important in practice, as in serologic reactions. 2) **Identical** functional groups in different Ags are found with groups that are relatively insensitive to the fine structure of macromolecules, such as lactose, Dnp, etc. Groups of this type are usually found in polysaccharides, nucleic acids, and conjugated proteins. 3) Crossreactions based on **similarity** rather than identity of determinants are probably of particular importance for proteins, whose determinant groups are mainly conformational and not likely to be precisely duplicated in different proteins (Fig. 16-20).

Removal of Crossreacting Antibodies (Adsorption and Absorption). Before an antiserum is sufficiently **monospecific** for use as an analytic reagent (e.g., in typing bacteria), it is usually necessary to remove certain crossreacting Abs by allowing the antiserum to react with the appropriate crossreacting Ags. Large complexes (e.g., precipitates or Abs bound to cells) are easily removed along with the crossreacting Abs. When, however, the complexes are soluble and difficult to remove, crossreacting ligand may be added in large excess to saturate the crossreacting Abs, eliminating them functionally

Fig. 16-20. Cross reactions of variant forms of a determinant group of a protein Ag. A synthetic peptide, with an S–S bonded loop, duplicates the loop of residues 64–82 of hen's egg lysozyme. Conjugated to a protein or high-molecular-weight polypeptide carrier, the synthetic loop elicits Abs (in rabbits and goats) that react with the loop peptide and with native lysozyme. Various synthetic analogs, each with a D-alanine in place of an indicated residue **(arrows)**, differ in their ability to inhibit the reaction between these Abs and lysozyme (lower half). Disruption of the loop, by cleaving the S–S bond ("oxidized S–S bond") virtually eliminates reactivity. (Teicher E et al: Immunochemistry 10:265, 1973; Arnon R et al: Proc Nat Acad Sci USA 68:1450, 1971).

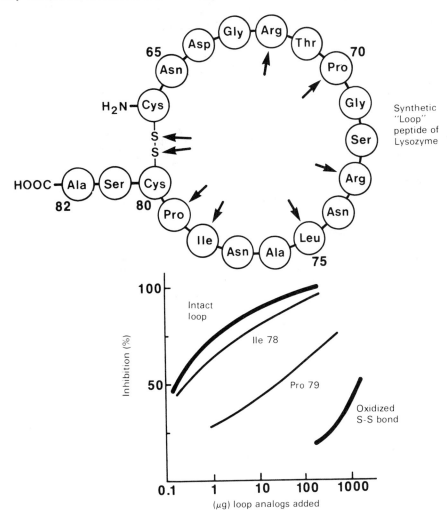

though not physically. In a frequently used method the antiserum is passed over a column of agarose beads to which the crossreacting Ag is attached: the crossreacting Abs stick to the column and other Abs emerge with the "pass-through." We shall use the terms **adsorption** when Abs are removed by particulate Ags and **absorption** when they are merely neutralized by reaction with soluble Ags.

AVIDITY

Early in this chapter we saw that an Ab's affinity for univalent ligands is measured by the intrinsic association constant, *K,* which is a property of a representative **single site** (see also Appendix). However, Ab molecules and most Ag particles are multivalent, and their tendency to pair depends not only on intrinsic affinities (per site), but also upon the **number** of sites involved. For instance, the equilibrium constant in the formation of multivalent monogamous Ab:Ag complexes can exceed by perhaps 100,000-fold the intrinsic equilibrium constant of the same Ab for the corresponding univalent ligand. In most reactions the number of functional groups on Ag particles is obscure and different ones are likely to be engaged by different sets of Abs (see Fig. 16-12); in addition, the pairing of Ab and Ag can also be influenced by nonspecific factors involved in the aggregation or closepacking of macromolecules.

Because of these complexities the term **avidity** is often used to denote the overall tendency of Abs to combine with Ag particles, the term **affinity** being reserved for the intrinsic association constant that characterizes the binding of a univalent ligand (see Appendix).

Differences in the avidity of different antisera for the same Ag can sometimes be demonstrated simply by diluting antiserum–Ag mixtures. The concentrations of free Ab and free Ag are reduced, causing dissociation of the immune complexes: *dissociation is less evident with more avid antisera.* This procedure is particularly useful when the unbound Ag can be measured with high sensitivity (as toxin or infectious virus).

Differences in avidity can arise because Abs differ 1) in intrinsic affinity for a given determinant (Fig. 16-21), 2) in the number of sets of Abs for diverse determinants of the Ag, 3) in the proportion of Ab molecules that can engage in monogamous binding, or 4) in the number of combining sites per molecule (e.g., IgG has two and IgM has ten.) (Abs with more combining sites have a greater tendency to bind ligands than Abs with fewer sites, when both have the same intrinsic association constant: see Appendix, especially equation 16.) Avidity differences among antisera are commonplace and easily recognized, but the basis for the differences is usually obscure.

MULTISPECIFICITY

The great diversity of Abs to a given Ag, combined with the immense number of different Ags that elicit Ab responses, suggests that the number of different Abs produced by one individual might be virtually limitless. Alternatively, it is possible that the individual Ab molecule is **multispecific**: i.e., that it can bind a variety of structurally different ligands. According to this hypothesis, all of the many Ab subsets that characterize an antiserum would, of course, be capable of binding the Ag that elicited the antiserum, but each subset would also be capable of binding (and being elicited by) other, structurally unrelated Ags. Any one of the latter **"strange" crossreactions** would ordinarily escape attention, because it would be displayed by only a very small fraction of the total population of Ab molecules in the antiserum. Direct evidence for this attractive idea is still limited.

Fig. 16-21. Precipitin curves showing differences in avidity of four antisera for the same Ag. The sera were prepared against DNP–bovine γ-globulin and tested with human serum albumin substituted with about 30 DNP groups per molecule (DNP–HSA). The order of avidity of the sera is A > B > C > D. A similar order in affinity for ε-DNP–lysine was found for anti-DNP molecules isolated from each of the sera: K_0 in liters/mole were: (A) > 10^8, (B) 1×10^7, (C) 5×10^6, and (D) 1×10^6. (Steiner LA, Eisen HN: In Samter M, ed.: Immunological Diseases, Boston, Little, Brown, 1965)

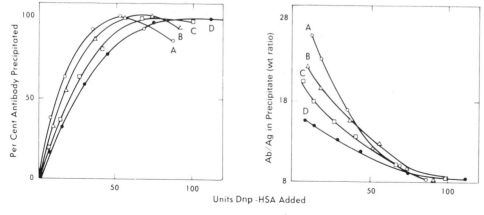

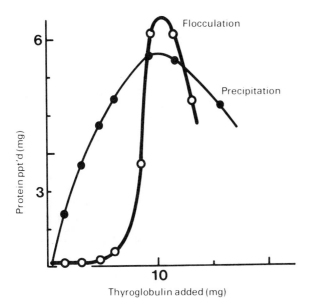

Fig. 16-22. Contrast between precipitation and flocculation reactions. Serum from some humans with autoimmune (Hashimoto's) thyroiditis (Ch. 22) gives a flocculation reaction with human thyroglobulin, whereas serum from other patients with the same disease gives a typical precipitin reaction. Value at each point is expressed as though each reaction was carried out with 1 ml serum. (Based on Roitt IM et al: Biochem 69:248, 1958).

THE FLOCCULATION REACTION

In anomalous precipitin reactions, called **flocculation,** precipitation is observed only over a narrow range of Ab/Ag ratios: insoluble aggregates are not formed until a relatively large amount of Ag has been added (Fig. 16-22). These reactions are given only by certain antisera (e.g., horse antisera to diphtheria toxin and to some streptococcal toxins, some human antisera to thyroglobulin). It is possible that the flocculating antisera contain some high-affinity, nonprecipitating Abs (see Monogamous Binding, above), which must be saturated with Ag before the remaining, conventional Abs can bind and precipitate with the additional Ag.

PRECIPITIN REACTION IN GELS

When Abs and Ag are introduced into different regions of an agar gel they diffuse toward each other and form opaque bands of precipitate at the junction of their diffusion fronts. Simple and ingenious applications of this principle provide powerful methods for analyzing the multiplicity of Ab–Ag reactions within a system. The most widely used methods are illustrated in Figure 16-23.

Single Diffusion in One Dimension (Fig. 16-23A). This procedure, developed by Oudin in France, is generally performed by placing a solution of Ag over an antiserum that has been incorporated in a column of agar gel. By diffusion, a concentration gradient of Ag advances down the agar column (provided the concentration of Ag in the upper reservoir is high relative to the concentration of Ab in the agar phase), and a precipitate forms in the agar at the advancing diffusion front. This precipitate extends upward to that level in the gradient at which Ag excess is sufficient to prevent precipitation. With continuing diffusion of Ag from the reservoir, the leading edge of the precipitate advances downward. The trailing edge of the precipitate also advances, because the additional Ag migrating into the specific precipitate dissolves it by forming soluble complexes, as expected from the precipitin reaction in liquid. Thus, a band of precipitate migrates down the column of agar. The distance traveled is proportional to the square root of time, in accord with the laws of diffusion.

The rate of migration depends on the diffusion coefficient of the Ag (which varies with molecular weight and shape) and on its concentration in the upper reservoir (Fig. 16-23A). Accordingly, when several Ags diffuse into an antiserum that can react with each of them, several bands of precipitation are observed and each migrates at a distinctive rate. Because different Ab–Ag systems can form overlapping bands that migrate at indistinguishable rates the observed bands represent the minimum number of Ab–Ag systems in the substances being analyzed.

The rate of band migration varies with the concentration of Ag introduced, and the optical density of a precipitate (which can be measured by photometers) depends on the concentration of Ab in the antiserum. Hence with the use of appropriate standards (solutions of known Ag concentration and antisera of known Ab con-

Fig. 16-23. Some arrangements for gel diffusion precipitin reactions. **A.** Single diffusion in one dimension. **B.** Double diffusion in one dimension. **C.** Double diffusion in two dimensions. **Arrows** show direction of diffusion. **Stippled areas** are opaque precipitation bands.

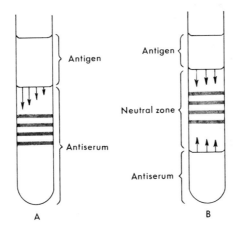

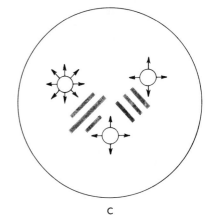

Identity

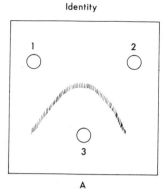

Nonidentity

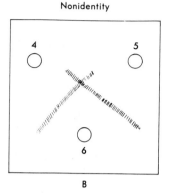

Partial identity

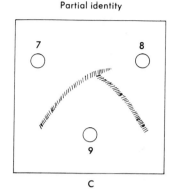

Fig. 16-24. Double diffusion precipitin reactions in agar gel illustrating reactions of identity **(A)**, nonidentity **(B)**, and partial identity **(C)**. In **A** the same Ag was placed in wells **1** and **2**, and the antiserum was in well **3**. In **B** different Ags were placed in wells **4** and **5**, and antisera to both were placed in well **6**. In **C** an Ag and its antiserum were placed in wells **7** and **9**, respectively, and a crossreacting Ag was placed in well **8**.

centration), it is possible to measure the concentration of Ag in unknown solutions and the concentration of precipitating Abs in antisera.

Double Diffusion in One Dimension (Fig. 16-23B). In this procedure the antiserum in agar is overlaid by a column of clear agar which in turn is overlaid by Ag, either added as liquid or incorporated into agar. Ag and Ab molecules diffuse across the respective interfaces into the clear neutral zone and advance toward each other. At the junction of their diffusion fronts where Ab and Ag are in equivalent proportions a precipitation band forms, and it increases in density with time as more Ab and Ag molecules continue to enter this zone in optimal (i.e., equivalent) proportions. If the Ag and Ab are added to their respective reservoirs in proportions that correspond to the equivalence zone in solution, the precipitation band has maximal sharpness and is stationary. If, however, either is added in relative excess, the band migrates slowly away from the excess.

This method is particularly valuable for determining the number of Ab–Ag systems in complex reagents. For example, with highly purified diphtheria toxin and a human antiserum to diphtheria toxin as many as six bands of precipitation have been observed. These same reactants in liquid solution would very likely have displayed a precipitin curve with a single zone of maximum precipitation; with careful supernatant analysis it might have been possible to recognize that more than one Ab–Ag system was involved, but not how many.

Double Diffusion in Two Dimensions. Considerable insight can be obtained by the simple but elegant procedure, developed mainly by Ouchterlony in Sweden, of placing Ag and Ab solutions in separate wells cut in an agar plate (Fig. 16-23). A large number of geometric arrangements are possible; some of the simpler ones are shown in Figures 16-24 to 16-26. The reactants diffuse from the wells, and precipitation bands form where they meet at equivalent proportions. If the concentration of

introduced Ab or Ag is in relative excess, the band forms closer to the other well.

The arrangements shown in Figures 16-24 to 16-26 are particularly useful for comparing Ags for the presence of **identical or crossreacting components.** If a solution of Ag is placed in two adjacent wells and the corresponding Ab is placed in the center well, the two precipitin bands eventually join at their contiguous ends and fuse (Fig. 16-24A). This pattern, known as the **reaction of identity,** is seen whenever indistinguishable Ab–Ag systems react

Fig. 16-25. Use of purified Ags to identify components of a complex series of precipitin bands. Center well **(AS)** contains rabbit antiserum prepared against unfractionated human serum. Wells **a** and **c** contain human serum; well **b** has purified human serum albumin **(HSA)**; and well **d** has a purified human immunoglobulin **HIg)**. Thus, bands **4** and **6** correspond to HIg and HSA, respectively.

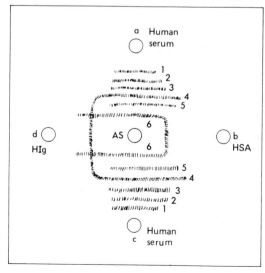

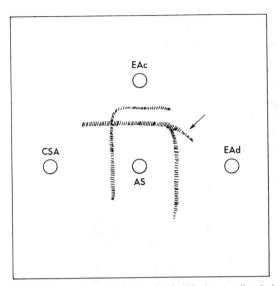

Fig. 16-26. Example of the use of gel diffusion to discriminate between two types of cross reactions. In liquid medium an antiserum **(AS)** prepared against crystallized chicken ovalbumin **(EAc)** precipitates copiously with EAc and less well with crystallized duck ovalbumin **(EAd;** see Fig. 16-18); it also precipitates slightly with crystallized chicken serum albumin **(CSA)**. From the gel precipitin pattern shown it is concluded that 1) EAc is contaminated with CSA, hence the antiserum contains some Abs to CSA; 2) CSA and the main component of EAc are unrelated antigenically (reaction of nonidentity); and 3) some antigenic determinants of EAd are similar to or identical with some determinants of EAc (reaction of partial identity). The Abs that cause the spur (arrow) are those that precipitate with EAc but not with EAd in the quantitative precipitin reaction of Fig. 16-18.

In the event that neither of the Ags is homologous with respect to the antiserum (i.e., neither is the immunogen) a pattern of partial identity might be observed, wtih the spur projecting toward the less reactive Ag; or there might be partial fusion with two crossing spurs, indicating that some Abs react only with with one of the crossreacting Ags and some only with the other. Unless inspected carefully for attenuation and curvative of the spurs, double spur formation can be mistakenly interpreted to mean that the adjacent systems are unrelated.

As in diffusion in one dimension, if one well contains a mixture of different Ags and the facing well contains Abs to several of them, the number of bands formed between the wells represents the minimum number of Ab–Ag systems involved. These bands may be identified if the appropriate pure Ags are placed in adjacent wells, as illustrated in Figure 16-25.

Double diffusion in two dimensions provides a simple means for evaluating the basis for crossreactions observed in liquid media. Thus crossreactions that arise from a common impurity can usually be recognized unambiguously, as is shown in Figure 16-26. On the other hand, when two purified Ags, such as chicken and duck ovalbumins (Fig. 16-18), give rise to a crossreaction, gel diffusion analysis cannot decide whether their common determinant groups are identical or only similar; in both cases partial fusion would be observed, with projection of the spur toward the crossreacting Ag (Fig. 16-26).

Because rates of diffusion vary inversely with molecular weight, the curvature of the precipitin band also provides a clue to the molecular weight of the Ag (provided Ag and Ab are present in roughly equivalent amounts). If the Ag and Ab have about the same molecular weight the precipitation band appears as a straight line; if not, the band is concave toward the reactant of higher molecule weight.

Radial Immunodiffusion (Fig. 16-27). This procedure is carried out in a layer of agar (e.g., on a glass slide) containing a uniformly dispersed monospecific antiserum. Diffusion of Ag from a well cut in the agar leads to the formation of a ring of precipitation whose area is proportional to the initial concentration of the Ag. From the results obtained with known concentrations a calibration curve is constructed, permitting quantitation of the Ag concentration. This simple procedure has been widely adopted from measurement of many Ags (e.g., immunoglobulins in diverse biologic fluids).

in adjacent fields. If, on the other hand, unrelated Ags are placed in adjacent wells and diffuse toward a central well that contains Abs for each, the two precipitin bands form independently of each other and cross (**reaction of nonidentity,** Fig. 16-24B). If, however, the Ag in one of the wells and the antiserum in the central well constitute a homologous pair, and the Ag in an adjoining well is a crossreacting Ag, the precipitation bands fuse, but in addition form a spurlike projection that extends toward the crossreacting Ag (**reaction of partial identity,** or **crossreaction,** Fig. 16-24C).

From what is known of precipitin reactions in liquid the **spur** can be readily interpreted: it represents the reaction between homologous Ag and those Ab molecules that do **not** combine with the crossreacting Ag and hence diffuse past its precipitation band. Because these noncrossreacting Abs represent only a fraction of the total Ab involved in the homologous precipitin reaction (see Figs. 16-17 and 16-18, for example), the spur is usually less dense than the band from which it projects and it tends to have increased curvature toward the antiserum well.

IMMUNOELECTROPHORESIS

By combining electrophoresis with precipitation in agar, Grabar and Williams developed a simple but powerful method for identifying Ags in complex mixtures (Fig. 16-28). The mixture of Ags is introduced into a small well in agar that has been cast on a plate, say an ordinary

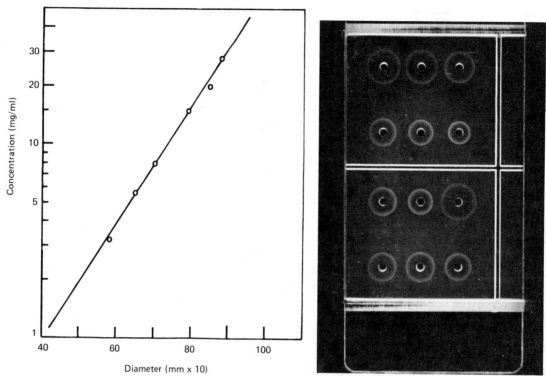

Fig. 16-27. Radial immunodiffusion measurement of Ag concentrations. A standard preparation of Ag, human immunoglobulin G (HIgG), was added at six different concentrations to wells in agar containing goat antiserum to HIgG (**upper right panel**). The diameters of the resulting circular precipitates are plotted on the **left.** After a longer incubation, when each precipitate has attained its maximum size, linearity is obtained by plotting log Ag concentration versus area or (diameter)^2. The **lower right panel** shows six human sera with different concentrations of HIgG. (Courtesy of Dr. C. Kirk Osterland)

microscope slide. An electric field applied across the plate for 1–2 h causes the proteins to migrate, each with its distinctive electrophoretic mobility. The electric gradient is then discontinued, and antiserum is introduced into a trough whose long axis parallels the axis of electrophoretic migration. The Abs and Ags now diffuse toward each other, the precipitin bands form at the intersection of their diffusion fronts. The principles involved in the precipitation stage are those described above: precipitation reactions of identity, nonidentity, and partial identity may be seen.

Immunoelectrophoresis has revealed as many as 30 different Ags in human serum. Many applications of this method appear in subsequent chapters.

Rocket Immunoelectrophoresis (Fig. 16-29). Radial immunodiffusion and electrophoresis have been combined to provide a rapid method for measuring Ags. By electrophoresing the Ag (from solution in a small well) into agar that contains the antiserum (at excess), the time required to reach maximum precipitation is shortened to a few hours (from several days in the radial diffusion method shown in Fig. 16-27). The height of the resulting **rocket-shaped zone of precipitation** is proportional to Ag concentration.

REACTIONS WITH PARTICULATE ANTIGENS

THE AGGLUTINATION REACTION

Bacterial and other cells in suspension are usually clumped (agglutinated) when mixed with their antisera. The principles involved are fundamentally the same as those described above for reactions with soluble Ags. Ag-

glutination provides a simple and rapid method for identifying various bacteria, fungi, and types of red blood cells; conversely, with the use of known cells it provides a simple test to detect and roughly quantitate Abs in sera.

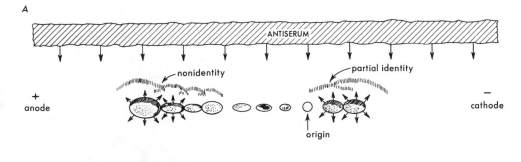

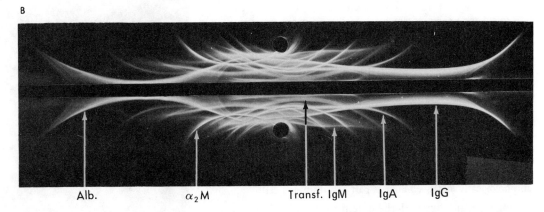

Fig. 16-28. Immunoelectrophoresis. **A.** A thin layer of agar gel (about 1–2 mm) covers a glass slide, and a small well near the center (**origin**) receives a solution containing various Ags. After electrophoresis of the Ags the current is discontinued and antiserum is added to the trough. Precipitation bands form as in double diffusion in two dimensions. The apex of each precipitin band corresponds to the center of the corresponding Ag. **B.** Human serum, placed at the origin, was analyzed with an antiserum prepared against unfractionated human serum. Alb, γ₂M, Transf, etc., refer to various serum proteins (Courtesy of Dr. Curtis Williams)

Mechanisms of Agglutination. Agglutination is carried out in physiologic salt solution (0.15 M NaCl). The ionic strength is important, for at neutral pH bacteria ordinarily bear a net negative surface charge which must be adequately damped by counter-ions before cells can approach each other closely enough for Ab molecules to form specific bridges between them. Hence, even with Abs bound specifically to bacteria, agglutination may not occur if the salt concentration is too low (e.g., $<10^{-3}$ M NaCl). Conversely, the addition of excess salt can lead to agglutination even in the absence of Abs.

When a mixture of readily distinguishable particles, such as nucleated avian RBCs and nonnucleated mammalian RBCs, is added to a mixture of the respective antisera, each clump that forms consists of cells of one or the other type (see Fig. 17-10, Ch. 17). Thus, as expected from the lattice theory (above), *each cell–Ab system agglutinates independently of the others in the same mixture.* The basic similarity between agglutination and precipitation is also brought out by the quantitative agglutination reaction, which measures the amount of Ab adsorbed by bacteria: the maximum amount bound by encapsulated pneumococci is identical with the maximum amount precipitated from the same volume of serum by the cells' isolated, soluble capsular polysaccharide (see Unitarian Hypothesis).

Titration of Sera. The agglutination reaction is widely used as a semiquantitative assay. A given number of cells is added to a series of tubes, each with the same volume of antiserum at a different dilution, usually increasing in twofold steps. The reaction is speeded up by shaking and warming to 37 °C. (sometimes to 56 °); then, after the cells have settled or been centrifuged lightly, the clumping is detected by direct inspection. The relative strength of an antiserum is expressed as the reciprocal of the highest dilution that causes agglutination. If, for example, a 1:512 dilution causes agglutination but a 1:1024 dilution does not, the titer is 512.

Agglutination titers are not precise (± 100%), but they are easily obtained and provide valid indications of the **relative** Ab concentrations of various sera with respect to a particular strain of bacteria. Hence agglutination titrations are useful for following changes in Ab titer during the course of acute bacterial infections (see Diagnostic Applications of Serologic Tests below).

However, agglutination titers obtained with **different** systems are not necessarily comparable: for example, an antiserum to type 1 pneumococcus with 1.5 mg Ab/ml agglutinated these organisms at a dilution of 1:800, while an antiserum to type 1R pneumococcus with 9.6 mg Ab/ml agglutinated the 1R organisms to a titer of only 1:80. Apparently the number and distribution of determinant groups on the bacterial cell surface can markedly influence the titer.

Agglutination reactions are, in principle, neither more nor less specific than other serologic reactions. However, they present difficulties in practice because a cell surface has a great diversity of antigenic determinants, of which some will usually crossreact with different but related cells. In order to achieve a high level of specificity it is nearly always necessary to adsorb antisera with sufficient amounts of crossreacting cells.

Surface versus Internal Antigens of Cells. When a bacterial, fungal or alien animal cell is introduced as an immunogen it is broken up in the host animal, and many of its surface and internal components (of cytoplasm and nucleus) are immunogenic. However, the Abs that cause agglutination are specific for surface determinants, which are often called **agglutinogens** (i.e., they induce formation of **agglutinins, the agglutinating Abs**). Surface Ags are usually much more potent immunogenically when administered as part of a morphologically intact cell than as part of a disrupted cell.

Prozone. By analogy with the precipitin reaction it would be expected that agglutinating activity would decline with progressive dilution of an antiserum. Curiously, however, some sera give effective agglutination reactions only when diluted several hundred- or thousandfold: when undiluted or slightly diluted they do not visibly react with the Ag. The latter region of the titration is called the **prozone,** as in tubes 1–3.

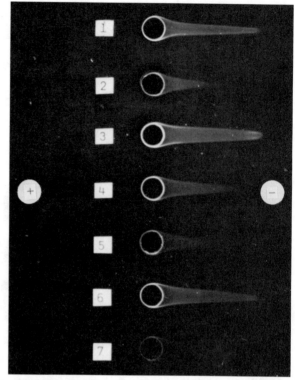

Fig. 16-29. Rocket immunoelectrophoresis measurement of Ag concentrations. A. The rocket-shaped areas of precipitation result from electrophoresing solutions with various concentrations of an Ag into agar that contains the corresponding antiserum. Rocket heights (or areas of precipitates) are linearly proportional to Ag concentration. (Based on Claman, H.N. et al. J Lab Clin Med 69:151, 1967)

Tube No.	1	2	3	4	5	6	7	8	9
Serum dilution	1:8	1:16	1:32	1:64	1:128	1:256	1:512	1:1024	1:2048
Clumping	0	0	0	+	+	+	+	+	0

By means of labeled Abs, or the antiglobulin test described below, it can be shown that unagglutinated cells in the prozone actually have Abs adsorbed on their surface. Indeed, it might be expected, on statistical grounds, that when Ab-molecules are in great excess relative to the number of functional groups on the cells, the simultaneous attachment of both sites of individual bivalent Ab molecules to different cells would be improbable. Nevertheless, the prozone phenomenon is not due simply to Ab excess, but often involves special blocking or incomplete Abs.

BLOCKING OR "INCOMPLETE" ANTIBODIES

In certain sera with a pronounced prozone some of the Abs appear to react with Ag particles in an anomalous manner: the bound Ab not ony fails to elicit agglutination but actively inhibits it, as shown by subsequently mixing the particles with antiserum at a dilution that would otherwise evoke a brisk clumping reaction. Some sera contain only the inhibitory Abs. These Abs are referred to as **blocking** or **incomplete Abs;** they are particularly evident in certain human antierythrocyte sera (anti-Rh; see Rh Antigens, Ch. 23).

Blocking Abs were once thought to be univalent. However, they are probably bivalent molecules that engage in monogamous bivalent binding (to repeating antigenic groups of the cell surface), like some nonprecipitating Abs in the precipitation reaction noted above.*

* Other Abs are called "blocking" Abs for a different reason: they seem to protect the cells on which they are adsorbed against

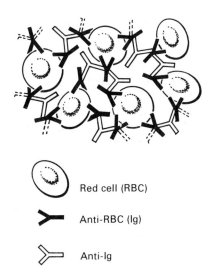

Red cell (RBC)

Anti-RBC (Ig)

Anti-Ig

Fig. 16-30. Antiglobulin (Coombs) test for incomplete Abs. Cells coated with specifically bound, nonagglutinating "incomplete" Abs are clumped by other Ab molecules (from another species), which react specifically with antigenic determinants of the non-agglutinating Abs.

The Antiglobulin (Coombs) Test. An ingenious method for detecting incomplete or blocking Abs, called the **Coombs** or **antiglobulin test,** is important for recognizing certain hemolytic diseases. The test exploits the fact that Ab molecules can be highly immunogenic, especially in a foreign species, and thus can bind simultaneously to Ags on the RBC surface and to Abs to themselves. For example, rabbits injected with human immunoglobulins (HIg) form anti-HIg that reacts with most HIg molecules, regardless of their specificities as Abs. Hence RBCs with nonagglutinating human Abs bound to their surface can be specifically clumped by rabbit anti-HIg serum (Fig. 16-30).

PASSIVE AGGLUTINATION

In the agglutination reactions described above the functional ligand groups are components of the cell wall or cell surface membrane. The same reaction can be extended to a wide variety of soluble Ags by attaching them to the surface of particles. In these **passive agglutination** reactions the most widely used particles are RBCs (**passive hemagglutination),** or a synthetic polymer such as polystyrene, or a mineral colloid such as bentonite. Attachment ordinarily depends on **adsorption** by noncovalent bonds.

destructive attack by specifically reactive lymphocytes; see Interference by Antibodies, Chapter 22, and Enhancement (under the Allograft Reaction) and Blocking Antibodies (under Tumor Escape Mechanisms), in Chapter 23.

RBCs readily adsorb many polysaccharides. For the attachment of proteins, however, it is usually necessary first to treat the cells with tannic acid or chromic chloride (whose mechanisms of action are not established). **Covalent linkage** of proteins to the red cell surface has also been achieved with bifunctional crosslinking reagents (Fig. 16-31).

As with conventional agglutination tests for estimating Ab levels, passive agglutination is highly sensitive (see below) but not very precise (at best ± 100%); and Abs are measured only in relative terms, not in weight units.

DIFFERENCES IN SENSITIVITY OF PRECIPITATION AND AGGLUTINATION REACTIONS

As a means of detecting Abs, agglutination is considerably more sensitive than precipitation: large particles, coated with Ag, serve to **amplify** the reaction. Thus precipitin reactions in liquid media or gels are usually not observed when antisera are diluted more than tenfold, whereas many antisera to bacteria or RBCs retain their agglutinating activity after being diluted many thousandfold. Passive hemagglutination is particularly sensitive: some mouse antisera to hemocyanin have titers of 10^5.

In one example, an antiserum to hen ovalbumin (EA) lost the ability to precipitate EA when diluted 1:5, but at 1:10,000 dilution it

Fig. 16-31. Passive hemagglutination. Bis-diazotized benzidine is used to attach protein X to the RBC surface. RBCs coated with X are specifically clumped by anti-X Abs. The antiglobulin test shown in Fig. 16-30 may be regarded as another form of passive hemagglutination, the Ag attached to the RBC surface being a specifically adsorbed immunoglobulin (nonagglutinating ["incomplete"] Ab). Various other bifunctional reagents have also been used, e.g., toluene-2, 4-diisocyanate (Fig. 16-40). $CrCl_3$ is most widely used currently to bind proteins and polypeptides (noncovalently) to RBCs.

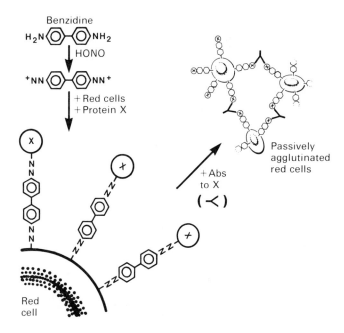

TABLE 16-5. Minimal Concentrations of Antibody Detectable by Various Methods

Method	Antibody concentration (μg/ml)*
Precipitin reactions in liquid media	20
Precipitin reactions in agar gel	60
Bacterial agglutination	0.1
Passive hemagglutination	0.01
Complement fixation†	1
Passive cutaneous anaphylaxis‡	0.02
Phage neutralization	0.001–0.0001

*Approximate Values

†See Chapter 20.

‡See Chapter 21.

could still agglutinate collodion particles coated with EA. These large differences are probably related to the difference in the number of aggregated particles needed for a visible reaction; e.g., about 10^7–10^8 for agglutination versus about 6×10^{12} molecules of soluble Ag (1μg of a protein of mol wt 100,000) for precipitation.

The sensitivities of some routine assay methods for Abs are compared in Table 16-5.

RADIOIMMUNOASSAYS

Any substance that can serve as Ag or hapten can be measured by radioimmunoassays (Table 16-6). These assays, introduced by Berson and Yalow to measure serum insulin concentrations, are usually based on competition for Ab between a radioactive indicator ligand (L*) and its unlabeled counterpart (L) at unknown concentration in the test sample: the higher the level of L, the less L* is bound. The concentration of L is readily de-

TABLE 16-6. Some of the Many Substances Measured by Radioimmunoassay

Insulin	Testosterone
Growth hormone	Estradiol
Parathyroid hormone	Aldosterone
Adrenocorticotropic hormone	Intrinsic factor (see Ch. 22)
Glucagon	Digitalis
Secretin	Cyclic AMP (cAMP), cGMP,
Vasopressin	cIMP, cUMP
Bradykinin (see Ch. 21)	Prostaglandins
Thyroglobulin	Australia antigen (HB$_s$Ag; see
Gastrin	Ch. 65)
Calcitonin	Carcinoembryonic antigen
	(see Ch. 23)
	Human IgE (see Ch. 21)
	Morphine

Each substance corresponds to ligand X in Figure 16-32. The same principle has been applied in nonimmune assays, in which Abs are replaced by other specific binding proteins; for example, Vitamin B$_{12}$ is measured by its binding to intrinsic factor.

termined by comparison with a calibration curve prepared with purified L at known concentrations (Figs. 16–32 and 16–33).

These assays can be exceedingly discriminating because antisera can be chosen with great specificity for the ligand. The sensitivity is limited primarily by the amount of radioactivity that can be introduced into L*: with carrier-free radioactive iodine (^{125}I) as extrinsic label, ligands can usually be detected at the level of a nanogram (1×10^{-9}g).

Because of the extremely low concentrations used, specific precipitates are not evident even if the ligand is multivalent. Nevertheless, various procedures easily permit free and bound indicator ligand (L* and Ab–L*) to be separated (and counted). Some may take advantage of special chromatographic or electrophoretic properties of particular ligands. In a widely used, more general method the Ab–L* complexes are precipitated with antiserum to the Ab moiety of the complex, leaving unbound L* in the supernate (Fig. 16-32). The latter approach is feasible because, as noted before, the Abs of one species (donor) are usually immunogenic in other species: the resulting antisera react with essentially all Abs of the donor species, regardless of their ligand-binding specificities.

In the double Ab reaction the first Ab (antiligand) is added at a low level to enhance competition between labeled and unlabeled ligand, i.e., the Ab/L ratio corresponds to the Ag-excess zone in the precipitin reaction (see Fig. 16-7). The second Ab (anti-Ab) is introduced in excess (i.e., Ab-excess zone in Fig. 16-7) to ensure complete precipitation of the first Ab and its complexes.

Various **solid-phase assays** eliminate the need for the second antiserum (anti-Ab). In one popular version, Abs to the ligand (anti-L) are simply adsorbed to the walls of plastic (e.g., polystyrene) tubes (which have the property of adsorbing a monomolecular layer of protein). Though the adsorbed molecules are probably denatured, many of them evidently retain at least one effective combining site, for when L* is introduced it sticks specifically to the adsorbed Abs and is easily counted. In the presence of competing L, proportionately less L* sticks, and the remainder is simply decanted. The method is rapid, economic, and sensitive: e.g., a tube prepared with 1μg of Ab can measure <0.001μg of Ag.**

Like other assays based upon inhibition, radioimmunoassays may have difficulty in determining whether the inhibitor in the unknown is identical to the indicator L*, or only similar enough to crossreact.

** The need for a second antiserum (anti-Ab) can also be circumvented with certain staphylococci (Cowen strain): a special protein on these bacteria (protein A) binds many immunoglobulins (Igs) and soluble Ab–Ag complexes. (However, certain Ig classes [Ch. 17] are not bound [IgM] or are bound only weakly [IgA].) Instead of staphylococci, purified protein A can be used to bind or precipitate soluble Ab–Ag complexes.

Fig. 16-32. **A.** Radioimmunoassay. The Ab–ligand complexes formed in step I are separated from unbound ligand by specific precipitation in step II with antiserum prepared against the Abs used in step I. The effect of competition between labeled (●) and unlabeled (○) ligand molecules is illustrated in Fig. 16-33. **B.** "Solid Phase" radioimmunoassay. In contrast to the "double Ab" assay illustrated in **A** in this assay the Ab is bound (noncovalently) to a plastic surface, which is than coated with an unrelated protein, such as serum albumin, to prevent the subsequently added ligands from adhering to plastic. Then the labeled ligand and an unlabeled competitive inhibitor are added. The difference between the amount of labeled ligand added and the amount remaining in solution is the amount bound by the immobilized Ab.

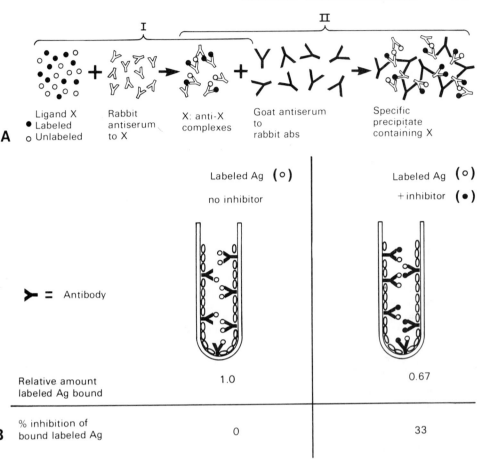

	Labeled Ag (○) no inhibitor	Labeled Ag (○) +inhibitor (●)
Relative amount labeled Ag bound	1.0	0.67
% inhibition of bound labeled Ag	0	33

SOME OTHER ASSAYS

Enzyme-Linked Immunosorbent Assays (ELISA) (Fig. 16-34). These assays are based on the use of covalently linked enzyme–Ab complexes (E–Ab) in which the catalytic and immunologic activities are preserved. Substrates that yield easily detected colored or fluorescent products measure the enzyme, and thereby its linked Ab. The enzymes used (e.g., alkaline phosphatase, β-galactosidase) are stable, unlike radioactive labels, and they are cheaper and simpler (and just as sensitive) to measure. Hence these assays are especially useful under conditions (as in underdeveloped societies) where radioimmunoassays are prohibitively costly and complex.

Many variations of ELISA are possible. For instance, to estimate the level of serum Abs, the corresponding Ag, adsorbed to plastic (polystyrene) plates, is used to bind Abs from the test serum (Ab no. 1). The specifically adsorbed Abs are then measured with an E–Ab complex whose Ab is specific for Ab no.1 (actually, for immunoglobulins (Igs) from the species in which Ab no. 1 was produced; e.g., E-linked goat antihuman Ig can be used to measure almost any human Ab).

Essentially the same procedure is used to measure Ags, which are specifically taken up from the test solution by Abs that were previously adsorbed to the polystyrene plate. As is suggested by Figure 16-34, the same E–Ab preparation can be used to measure a great variety of Abs and Ags (e.g., human Abs to cytomegalovirus or measles, herpes, rubella, or mumps viruses).

Agglutination-inhibition. Like radioimmunoassays, this assay also uses the inhibition principle to measure ligand concentration. For example, polystyrene particles or RBCs with an adsorbed hormone (corresponding to L*) can be agglutinated by the appropriate antiserum; and, with antiserum in limiting amount, any added L, in proportion to its concentration in test samples of serum or urine, will inhibit the agglutination. Assays of this type are widely used to measure some hormones, e.g., urinary chorionic gonadotropin in a rapid test for pregnancy.

Phage Neutralization. Bacteriophage with covalently attached haptens or Ags remains infectious for bacteria, and infectivity can be specifically blocked ("neutralized") by Abs to the attached ligands. Because phage can be measured at exceedingly low levels (about 6000 virions/ml, or 1×10^{-17} M, assuming each virion is a single molecule), neutralization assays are among the

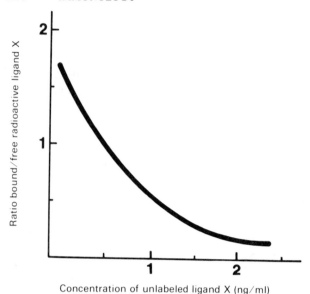

Fig. 16-33. Calibration curve for a radioimmunoassay. A fixed amount of radioiodine-labeled ligand X competes with various amounts of unlabeled ligand X for a limiting amount of anti-X Abs.

most sensitive known (e.g., they can measure Ab production by single cells). These assays have been used primarily to measure Abs: the initial rate of neutralization, reflecting the forward rate constant (Ab + phage → Ab-phage complex), is directly proportional to Ab concentration (and independent of the Ab's affinity, as emphasized before—see Rates of Reaction, above).

Neutralization assays can also be used to measure Ags and haptens, because unbound ligand can compete with ligand attached to phage and inhibit the neutralization of infectivity caused by a standard amount of a standard antiserum. From the extent of inhibition it is possible to measure trace amounts of unbound ligand in serum, lymph, urine, etc., following the same principle and with about the same sensitivity as in radioimmunoassays (e.g., see Fig. 16-34, below).

Ammonium Sulfate Precipitation (Farr Technic). Relative levels of Abs in sera can be estimated with another use of radiolabeled ligands. Ammonium sulfate at high concentration (40% of saturation) precipitates Abs and Ab–ligand complexes but not most unbound ligands. Hence addition of the salt at this concentration to a mixture of antiserum and labeled ligand can, if the specific binding reaction is not affected by high ionic strength, separate free and Ab-bound ligand. The **ABC (Ag-binding capacity)** of an antiserum is determined, by serial dilution, as the volume required to bind an arbitrarily selected proportion (e.g., 33%) of the ligand, at any arbitrarily chosen total level. Estimates of Ab activity are obtained rapidly and with small volumes (e.g., a few microliters) of antiserum; and the method is more precise than is, say, agglutination. Though widely used experimentally (e.g., to test blood samples from individual mice), the method is limited because an unknown proportion of Abs may not have sufficient affinity to bind the ligand at the level selected, and because the proportion of these low-affinity Abs may vary widely from one serum to another.

Ammonium sulfate precipitation has also been used with radioactive univalent ligands to estimate intrinsic association constants. Even though the high salt concentration can modify the over-all Ab structure (e.g., cause the molecule to become highly insoluble) the ligand-binding site seems to be largely unperturbed: under these conditions the affinity for some ligands has been close to that measured in solution by equilibrium dialysis.

Fig. 16-34. Enzyme-linked immunosorbent assay (ELISA test). This technic can measure serum Abs **(A)** or Ag **(B)**. It depends upon the ability of certain plastics (polystyrene, polyvinyl) to adsorb proteins (Ags, Abs) strongly. Only the component added in Step I is bound to the plastic, which is then coated with unrelated proteins to make sure that the substances added subsequently will stick only specifically to reactants, rather than nonspecifically to the plastic.

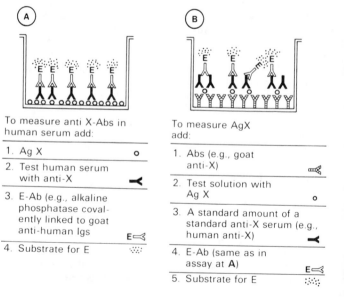

To measure anti X-Abs in human serum add:

1. Ag X   o
2. Test human serum with anti-X   ◄
3. E-Ab (e.g., alkaline phosphatase covalently linked to goat anti-human Igs)   E⊲
4. Substrate for E   ⋮⋮

To measure AgX add:

1. Abs (e.g., goat anti-X)   ▱
2. Test solution with Ag X   o
3. A standard amount of a standard anti-X serum (e.g., human anti-X)   ◄
4. E-Ab (same as in assay at **A**)   E⊲
5. Substrate for E   ⋮⋮

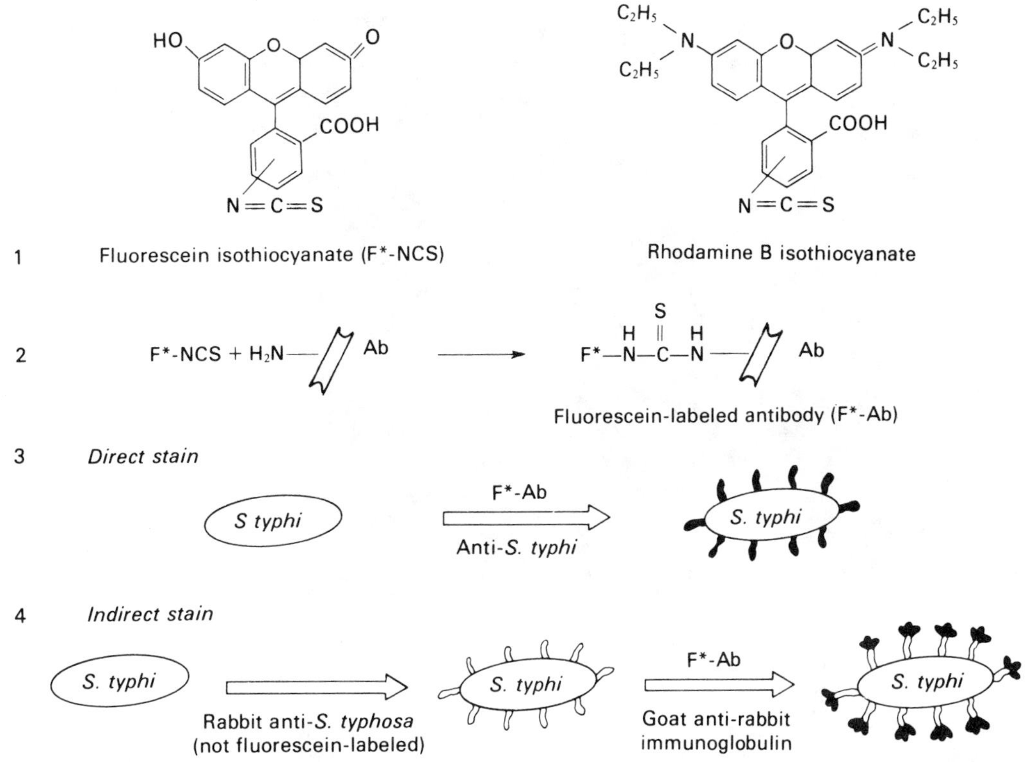

Fig. 16-35. Immunofluorescence. The fluorescent isothiocyanates (1) form thioureas, substituted with the fluorescent group and ϵ-NH$_2$ groups of lysine residues of Abs or other proteins (2). In the **direct** staining reaction (3) the labeled Abs are specific for the Ag of interest. In the **indirect** reaction (4) the labeled Abs are specific for the Abs of another species (e.g., goat anti-rabbit immunoglobulin).

FLUORESCENCE AND IMMUNE REACTIONS

The fluorescent* properties of Ab molecules and of certain organic residues that can be attached to them provide the basis for a number of analytic methods, the most important of which is **immunofluorescence.** Introduced by Coons, this method is widely used to identify and localize Ags in cells and on cell surfaces (Ch. 18) and to identify bacteria rapidly in infected materials.

IMMUNOFLUORESCENCE

Of the reagents that introduce fluorescent groups into proteins, the most widely used is fluorescein isothiocyanate (Fig. 16-35). With one or two fluorescein residues per molecule, Abs are intensely fluorescent and retain their specific reactivity. After a specimen on a slide

---

* When molecules absorb light they subsequently dispose of their increased energy by various means, one of which is the emission of light of longer wave length. When the emission is of short duration (10^{-8}–10^{-9} sec for return of the excited molecule to the ground state) the process is called **fluorescence.** (When the emission is of long duration the process is called **phosphorescence.**)

has been covered for several minutes with a solution of fluorescein-labeled Abs (usually as the globulin fraction of an antiserum), the slide is rinsed to remove unbound fluorescent protein and then is examined in the light microscope, with a suitable light source and filters to provide the proper incident light. Since the emitted light (about 530 nm) is of longer wave length than the background incident light (about 490 nm) the Ab-coated cell stands out as a sharply visible yellow-green object (Fig. 16-36).

As with all immunologic reactions it is necessary to establish specificity with suitable controls. Thus staining is blocked by preliminary treatment of the smear or tissue section with unlabeled Abs (to saturate antigenic sites) and by addition of excess soluble Ag (to saturate the labeled Abs).

Fluorescein-labeled Abs can be used in both direct and indirect reaction sequences (outlined in Fig. 16-35) to detect and localize Ags. The **indirect reaction** amplifies the response and permits one labeled preparation to be used in a wide variety of serologically specific reactions.

Different chromophores can be introduced with other reagents, e.g., orange-red fluorescence with rhodamine B isothiocyanate

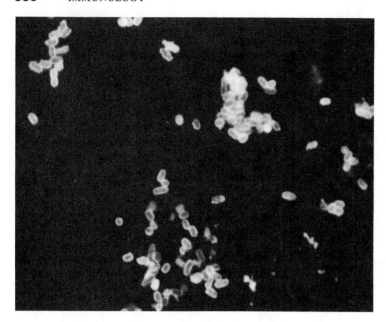

Fig. 16-36. Immunofluorescence staining. Stained with a fluorescein-labeled globulin fraction of a specific antiserum, *E. Coli* 0127:B8 is clearly visible in a fecal smear from a patient with infantile diarrhea. Many other types of bacteria are abundant in the smear but are not stained. (Courtesy of Center for Disease Control, Atlanta) (About ×1000)

(Fig. 16-35.). Immunofluorescence can thus be applied to a number of special problems, e.g., to determine if two Ags on a cell surface are coupled (see Major Histocompatibility Complex, Ch. 18).

In all these applications a recurrent problem arises from the nonspecific staining of tissues, especially by proteins with more than two or three fluorescein residues per molecule. To minimize this difficulty ion-exchange chromatography and adsorption with acetone-dried tissue powders are used to remove the more highly substituted proteins. In addition, the use of high-titer antisera (e.g., >2 mg of Ab/ml) at high dilution reduces nonspecific staining.

OTHER FLUORESCENCE TECHNICS

Fluorescence Quenching. Like other proteins, unlabeled Ab molecules fluoresce in the upper ultraviolet region by virtue of their aromatic amino acid residues. This fluorescence is excited maximally at 290 nm and the emitted light has maximal intensity at about 345 nm (characteristic of the absorption and emission spectrum of tryptophan). When the binding sites in the Ab are specifically occupied by certain kinds of haptens (or Ags), the energy absorbed by the irradiated tryptophan residues is transferred to the bound ligand, which emits light at some still longer wave length or does not fluoresce at all. The net result is quenching or damping of the Ab's characteristic fluorescence. Thus the binding sites of a preparation of purified Abs can be titrated as shown in Figure 16-37, and the titrations can be used to calculate average intrinsic association constants, which agree with those determined by equilibrium dialysis.

For quenching to occur, the ligand's absorption spectrum must overlap the Ab's emission spectrum to some extent (Fig. 16-37A). A nonquenching ligand, whose absorption spectrum does not overlap, can sometimes be converted into a quenching one by linking it to a chromophore group with the correct absorption spectrum. For

instance, the fluorescence of antilactose Abs is not quenched by lactose alone but is quenched by lactose to which a Dnp group is covalently attached as a "sensor" (Fig. 16-38); similarly, the fluorescence of antidextran Abs is not quenched by isomaltose (its natural ligand) but is quenched by a nitrophenyl derivative of isomaltose.

Fluorescence quenching is not as generally useful or rigorous a method as equilibrium dialysis but it offers several advantages: it is rapid, requires only small amounts of Ab, and can be carried out with ligands that are too large to be dialyzable (for example, cytochrome c quenches Ab to cytochrome c).

Fluorescence Polarization. When fluorescent molecules are excited by a beam of polarized light the extent of polarization of the emitted light varies with the molecule's size: emission from small molecules, with much rotational motion, exhibits little polarization, but binding of the fluorescent molecule to a relatively large particle, such as an Ab molecule, restricts its rotational movement and increases the polarization of its fluorescence. Hence the proportion of ligand molecules bound by Abs can be determined from the polarization of fluorescent emission. This method is potentially useful with those small Ags and haptens whose fluorescence spectrum (natural or introduced by fluorescent substituents) is distinctly different from that of Abs.

Fluorescence Enhancement. Certain small organic molecules, such as anilinonaphthalenesulfonates, have strikingly different fluorescence spectra when they exist as free molecules in water and when in the binding sites of Abs, which have a much lower dielectric constant than water. It is thus possible to measure the specific binding of such ligands to the corresponding Abs by following the appearance of fluorescence at the appropriate wave length. The method is exquisitely sensitive and can be used with Abs in complex media, such as serum.

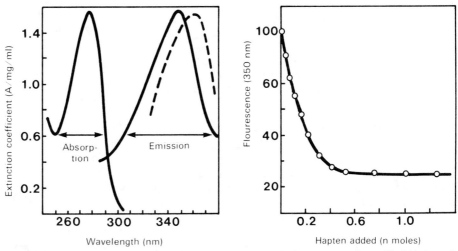

Fig. 16-37. Fluorescence quenching. **A.** Absorption and emission spectra of a purified Ab (———) and the overlapping absorption spectrum of a representative ligand (– – – –, ε-DNP-L-lysine) of the type that, when bound specifically, quenches the fluorescence of the Ab (see B). (The height of the Ab emission band is shown as equal to the height of its absorption maximum, though only about 20% of the energy absorbed by Abs appears as fluorescence emission.) **B.** Decline in fluorescence emission of anti-DNP Ab as it binds ε-DNP-L-lysine; about 75% of the fluorescence is quenched when the ligand-binding sites of the Ab are saturated. (Velick SF et al: Proc Natl Acad Sci USA 46:1470, 1960).

Fig. 16-38. Fluorescence of antilactose Ab is quenched by lactose carrying a covalently linked DNP-lysyl group **(B)**, but not by lactose itself **(A)**, though the Ab binds both ligands approximately to the same extent. The reason is that the absorption spectrum of DNP (maximum absorption is at 360 nm), but not that of lactose, overlaps the tryptophan emission spectrum of the Ab (see Fig. 16-37). (Based on Gopalakrishnan PV, Karush F: J. Immunol 114:1359, 1975).

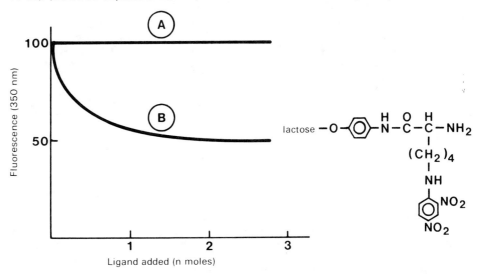

OTHER SPECTRAL METHODS

Ligand binding has also been measured by other methods that depend upon differences in spectroscopic properties of free and Ab-bound ligands; e.g., binding of haptens with covalently attached nitroxide **spin-label** substituents has been measured by **electron-spin resonance (ESR) spectroscopy.**

DIAGNOSTIC APPLICATIONS OF SEROLOGIC TESTS

The etiologic diagnosis of infectious diseases is usually established both by isolating the putative causative microorganism and by demonstrating Abs to it. Several principles are involved in the diagnostic use of serologic tests.

Changing Titer. Ab formation initiated by an infectious agent may continue for months or even years after the clinically apparent infection has subsided. Hence the presence of an elevated Ab titer to a given microbe indicates only that infection (or vaccination or exposure to a crossreacting Ag) has occurred at some time in the past. To establish that an acute illness is due to a particular agent it is desirable to show a **change** in titer during the illness: the absence (or low level) of Abs during the first week, their appearance and progressive increase in titer during the second and subsequent weeks, and perhaps their eventual decline. For practical purposes a **pair** of serum samples is compared, one drawn in the **acute** phase of the illness, the other during **convalescence.** If only a single serum sample is available, from late in the illness, a high titer can sometimes be accepted as provisional diagnostic evidence; the critical level depends upon the disease and the assay, and it is necessary to be sure that the high titer is not the result of prior vaccination.

Identification of Etiologic Agent. Even when no likely agent has been isolated, a provisional etiologic diagnosis can be established by testing the serum with Ags from microbial stains under suspicion: the Ag that elicits the highest titer reaction is assumed to be from the etiologic agent. This procedure, however, can be misleading: other Ags, not tested, might have elicited a still higher titer; and crossreactions may occur between very different microorganisms. In a classic example persons with rickettsial infections form Abs that react in high titer with certain *Proteus* strains (Weil–Felix Reaction; see Laboratory Diagnosis, Ch. 40).

When mixed flora are recovered from the site of infection the reactivity of particular isolates with convalescent antisera has been helpful for identifying new agents, e.g., Legionnaires' disease (Ch. 43).

Time Course of Immune Response in Infectious Disease. In the course of a given infectious disease the appearance and the persistence of Abs to the etiologic agent follow different time courses when measured with different antigenic preparations and different assays. In brucellosis, for example, agglutination titers appear early in infection and persist at low levels for years afterward, whereas precipitin levels appear later and disappear much sooner; and in many rickettsial and viral diseases the Ab titers measured by complement fixation (Ch. 20) appear later and subside sooner than those measured by neutralization of infectivity (Ch. 30). It is important for practical purposes to be aware that such differences exist even though the basis for these differences is usually not clear.

A number of possibilities can account for these variations. 1) The many Ags in a microbe differ in amount, immunogenicity, and stability. 2) There are structural and functional differences in the Abs specific for a given antigenic determinant, and different types of Abs are formed at different stages of the immune response; e.g., IgM Abs (Ch. 17) are formed early in the response and are especially effective agglutinators. 3) Serologic methods differ in sensitivity (Table 16-5) and, other things being equal, a method that requires less Ab will detect Ab earlier and more persistently.

THE UNITARIAN HYPOTHESIS

In the first decades of immunology it became apparent that most antisera display diverse activities. A serum prepared, for example, against the cholera vibrio could 1) protect guinea pigs against an otherwise lethal infection with virulent cholera vibrios, 2) agglutinate a suspension of these organisms, 3) form a precipitate with a filtrate of the broth in which they had grown, 4) specifically enhance their phagocytosis by leukocytes, etc. This versatility led to a **pluralistic view,** according to which each of these diverse activities was ascribed to a qualitatively different molecular form of Ab, named agglutinins, precipitins, opsonins (promoters of phagocytosis), protective Abs, etc.

With improved methods of analysis in the 1930s, however, this view became untenable, although the different names are still used. For example, Heidelberger found that all the Ab activities in antiserum to type 3 pneumococci were removed by precipitation with the organism's purified capsular polysaccharide; and the protein subsequently isolated from the precipitate duplicated all the diverse activities of the antiserum. Moreover, this capsular polysaccharide is a polymer of cellobiuronic acid (Fig. 16-18): and it was found that a conjugated protein substituted with cellobiuronic acid elicited the same kind of protective and precipitating antiserum as the organisms.

These and similar observations led to the **unitarian hypothesis,** according to which a given population of Abs will, on uniting with the corresponding ligand, produce any of the diverse consequences of Ab–Ag combination, depending on the state of the ligand and the milieu in which combination takes place: if the ligand is soluble and multivalent, precipitation; if the ligand is a natural constituent of a particle's surface or artificially attached to the surface, agglutination; if the ligand is part of the surface of a virulent bacterium, protection against infection.

Thus the early pluralistic view maintained that a given Ab molecule is **unipotent,** whereas the unitarian view held that the Ab molecule is **totipotent.** It is now clear that neither view is entirely correct. In subsequent chapters we shall see that some of the Ab molecules for a given ligand can elicit certain reactions, such as anaphylaxis (Ch. 21), and others cannot. **Abs are multipotent, but not totipotent.**

APPENDIX

Purification of Antibodies

Methods of purifying Abs are based on the dissociability of Ab–ligand complexes. Two stages are usually involved. 1) Abs are precipitated from serum with soluble Ags or (more commonly) are bound to Ag or hapten covalently attached to an insoluble adsorbent. In the most widely used procedure (**affinity chromatography**) antiserum is passed slowly through a column of agarose beads (an **immunoadsorbent**) to which the Ag or hapten has been coupled, as in Fig. 16-39. 2) After all extraneous serum proteins are washed away adsorbed Abs are eluted by specific or nonspecific procedures.

Specific Procedures. With aggregates whose stability depends largely on specific ionic interactions, strong salt solutions (e.g., 1.8 M NaCl) elute purified Abs effectively.

When the specific antigenic determinants are simple haptenic groups, such as 2,4-dinitrophenyl, small univalent haptens that encompass the crucial part of the determinant (e.g., 2,4-dinitrophenol) displace the insoluble Ag, yielding soluble Ab–hapten complexes. Diverse procedures are then used to separate the hapten from the Ab (e.g., ion-exchange resins, dialysis, gel filtration).

To elute Abs it is desirable to use those haptens that are both 1) **weakly bound** by the Ab and 2) **highly soluble.** Highly concentrated solutions of hapten can then elute the Ab in high yield, and the weakly bound hapten is easily removed by dialysis or gel filtration, leaving purified Abs.

Nonspecific Procedures. To elute Abs from protein Ags on the immunoadsorbant it is usually necessary to expose the adsorbed Abs (and the Ag) to conditions that cause reversible denaturation of many proteins, allowing Ab to dissociate from the immobilized Ag. Acids (HCl-glycine or acetic acid) at pH 2–3 are often effective. Because Abs usually regain their native structure on being restored to physiologic conditions, neutralization of the acidified eluate yields purified Ab with most of its activity retained.

Yield and Purity. Abs can be isolated from serum in high yield (usually about 50%). Though usually of high purity, in the sense that nearly all ($>90\%$) of the recovered protein reacts specifically with the Ag, the purified Ab molecules are usually heterogeneous with respect to affinity (see Fig. 16-2) and in many of the physical and chemical properties described in Chapter 17.

Ferritin-Labeled Antibodies

As noted earlier, fluorescein-labeled Abs provide specific stains for the detection and localization of Ags in the light microscope. For use in the electron microscope Abs must be rendered much more highly electron-scattering than proteins in general: for example, by attaching ferritin (Fig. 16-40). This protein, of about the same size as an Ab molecule, has high electron-scattering capacity because of its uniquely high iron content (about 20%).

Ferritin and Hybrid Antibodies. Unmodified ferritin is also useful in conjunction with artificial "hybrid" Ab molecules in which one combining site is obtained from anti-ferritin Abs and the other from Abs to an Ag of interest, e.g., a particular cell surface Ag. The hybrid Ab can thus link a ferritin mole-

Fig. 16-39. Purification of Abs by affinity chromatography. Agarose beads activated by CNBr bind any ligand with free NH_2 groups, as shown. Packed into a column, the beads with attached ligand bind Abs specifically. After other proteins are washed away the Abs can be eluted by specific or nonspecific means (see text). Similar immunoadsorbent columns with covalently attached cross reacting ligands are used for specifically removing undesired cross reacting Abs from antisera (see Cross reactions, this chapter).

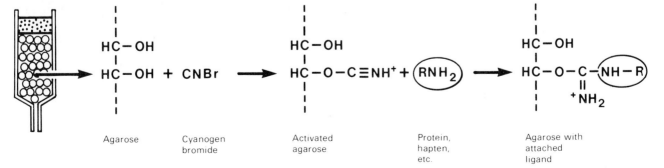

| Agarose | Cyanogen bromide | Activated agarose | Protein, hapten, etc. | Agarose with attached ligand |

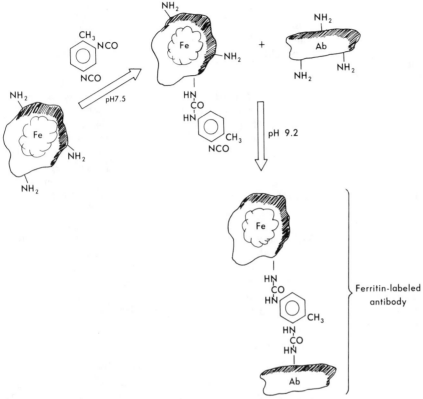

Fig. 16-40. Preparation of ferritin **(Fe)** -labeled Ab by reaction with toluene-2,4-diso-cyanate. (Based on Singer SJ, Schick AF: J Biophys Biochem Cytol 9:519, 1961).

cule (noncovalently) to the Ag target. The preparation of hybrid Abs is described in Chapter 17 (Fig. 17-11) and their use in electron microscopy is illustrated in Figure 21-3, Chapter 21.

Intrinsic and Actual Affinities

The reversible binding of a univalent ligand by Ab can be described by the **Scatchard equation** (eq. 4, above: see Equilibrium Dialysis), whose derivation follows.

The successive steps in the binding of a univalent ligand (L) by a multivalent antibody (B) with n combining sites can be represented by the reactions on the left and their equilibrium constants on the right.

$$B + L \rightleftarrows BL \qquad k_1 = \frac{[BL]}{[B][L]}$$

$$BL + L \rightleftarrows BL_2 \qquad k_2 = \frac{[BL_2]}{[BL][L]}$$

$$\cdots \qquad \cdots$$

$$BL_{i-1} + L \rightleftarrows BL_i \qquad k_i = \frac{[BL_i]}{[BL_{i-1}][L]} \qquad (10)$$

$$\cdots \qquad \cdots$$

$$BL_{n-1} + L \rightleftarrows BL_n \qquad k_n = \frac{[BL_n]}{[BL_{n-1}][L]}$$

We wish to determine the **average** number of ligand molecules bound per Ab molecule, i.e., the ratio of all bound ligand/all Ab molecules, or L_b/B_t.

$$L_b = BL + 2BL_2 + 3BL_3 \cdots + iBL_i \cdots + nBL_n, \text{ or}$$
$$= B[k_1 L + 2k_1 k_2 (L)^2 \cdots + ik_1 k_2 \cdots k_i (L)^i \cdots$$
$$+ nk_1 k_2 \cdots k_n (L)^n] \qquad (11)$$

and

$$B_t = B + BL + BL_2 + BL_3 \cdots + BL_i \cdots + BL_n, \text{ or}$$
$$= B [1 + k_1 L + k_1 k_2 (L)^2 \cdots + k_1 k_2 \cdots k_i (L)^i \cdots$$
$$\cdots + k_1 k_2 \cdots k_n (L)^n] \qquad (12)$$

Assume all Ab combining sites are equivalent and independent. Then for the representative step in which the ith site becomes occupied (third step in equation 10), the concentration of vacant Ab sites (S) is:

$$[S] = [n - (i - 1)] [BL_{i-1}] \qquad (13)$$

and the concentration of occupied sites (SL) is:

$$[SL] = i [BL_i] \qquad (14)$$

When equations 13 and 14 are combined, the equilibrium constant for the ith reaction becomes

$$k_i = \frac{n-i+1}{i} \cdot \frac{[SL]}{[S][L]} \tag{15}$$

$[SL]/[S][L]$ is defined as K, the **intrinsic association constant,** or the **intrinsic affinity,** for the general reaction in which a representative site binds a ligand molecule ($S + L \rightleftharpoons SL$); i.e.,

$$k_i = \frac{n-i+1}{i} K \tag{16}$$

The constants for the individual steps in equation 10 can thus be expressed in terms of the intrinsic constant (e.g., $k_1 = nK; k_2 = (n-1)K/2; k_n = K/n$); this makes it possible to reduce equations 11 and 12, with the aid of the binomial theorem, to*:

$$L_b = nK[B][L](1 + K[L])^{n-1} \tag{17}$$

and

$$B_t = [B](1 + K[L])^n \tag{18}$$

or,

$$\frac{L_b}{B_t} = \frac{nK[L]}{1 + K[L]}$$

Since L_b/B_t is r (moles ligand bound per mole Ab) and $[L]$ is c (equilibrium concentration of free ligand) we have

$$r = \frac{nKc}{1 + Kc} \tag{19}$$

* This derivation is due to Dr. B. Altschuler, New York University.

which is the same as equation 4:

$$\frac{r}{c} = Kn - Kr \tag{20}$$

Intrinsic affinity, K, provides a convenient and rigorous basis for analyzing Ab combining sites and for comparing sites of different Abs with each other and with those of other proteins. However, the multiplicity of binding sites on Ab molecules and on most Ag particles means that for biologically significant Ab–Ag reactions in vivo the actual equilibrium constants can differ greatly from the intrinsic constants. Because surface Ags on cells and virions occur as repeated copies, their complexes with Abs probably conform in most instances to Karush's concept of the monogamous complex (see Monogamous, Multivalency, above); the equilibrium constants for formation of these complexes can exceed by perhaps 10^5 the Ab's intrinsic affinity. The great difference represents a form of statistical cooperativity between sites on the same molecule. This cooperativity is widespread in nature; for example, the bond between two polynucleotide strands with many complementary pairs is enormously greater (Ch. 10) than the bond between a pair of complementary nucleotides (e.g., A–T).

Even without monogamous binding, actual and intrinsic binding constants need not be the same. For instance, in a reaction that is limited (say by a low level of Ag) to only one site of an Ab molecule that has two or ten sites (IgG or IgM, respectively; see Ch. 17), the observed equilibrium constant will be 2 and 10 times the respective intrinsic constants (see equation 16). Other things being equal, Ab molecules with more combining sites will form more stable complexes with ligands.

SELECTED READING

BOOKS AND REVIEW ARTICLES

CAMPBELL DH, GARVEY JS, CREMER NE, SUSSDORF DH: Methods in Immunology. New York, WA Benjamin, 1970

DAY ED: Advanced Immunochemistry. Baltimore, Williams & Wilkins, 1972

DELISI C: Antigen antibody interactions. In Levin S (ed): Lecture Notes in Biomathematics, Vol 8. Berlin, Spring-Verlag, 1976

KABAT EA: Structural Concepts in Immunology and Immunochemistry, 2nd ed. New York, Holt, Rinehart, & Winston, 1976

KARUSH F: The affinity of antibodies: range, variability and the role of multivalence. In Immunoglobulins, Vol 3 of Comprehensive Immunology. Plenum Press, N.Y., 1977

LANDSTEINER K: The Specificity of Serological Reactions, rev ed. Cambridge, Harvard University Press, 1945; reprinted by Dover, New York, 1962

LUFT R, YALOW R (eds): Radioimmunoassay: Methodology and Applications in Physiology and in Clinical Studies. George Thieme Verlag, Stuttgart, 1974

PARKER CW: Radioimmunoassay of biologically active compounds. Englewood Cliffs. Prentice-Hall, Inc., 1976

PRESSMAN D, GROSSBERG AL: The Structural Basis of Antibody Specificity. New York, Benjamin, 1968

ROSE NR, FRIEDMAN H (eds): Manual of Clinical Immunology. (Washington DC), American Society for Microbiology, 1976

WEIR DM (ed): Handbook of Experimental Immunology, 3rd ed. (London,) Blackwell Scientific, 1976

WILLIAMS CA, CHASE MW (eds): Methods in Immunology and Immunochemistry, Vols I, II, III, IV, V New York, Academic Press, 1967, 1968, 1971

SPECIFIC ARTICLES

CROTHERS DM, METZGER H: The influence of polyvalency on the binding properties of antibodies. Immunochemistry 9:341, 1972

VON SCHULTHESS GK, COHEN RJ, SAKATO N, BENEDEK GB: Laser light scattering spectroscopic immunoassay for mouse IgA. Immunochemistry 13:955, 1976

FRACKELTON AR JR., WELTMAN JK: Diffusion control of the binding of carcinoembryonic antigen (CEA) with insoluble anti-CEA antibody (Use of enzyme-linked immunoassay). J. Immunol, 1979, in press

HASELKORN D, FRIEDMAN S, GIVOL D, PECHT I: Kinetic mapping of the antibody combining site by chemical relaxation spectrometry. Biochemistry 13:2201, 1974

HORNICK CF, KARUSH F: Antibody affinity- III The role of multivalence. Immunochemistry 9:325, 1972

IMMUNOGLOBULIN MOLECULES AND GENES

For many years antibodies (Abs) were regarded simply as γ-**globulins,** the class of plasma proteins with the least electrophoretic mobility. The γ-globulin level was increased by intensive immunization and was lowered when Abs were removed from the serum by specific precipitation with antigen (Ag) (Fig. 17-1). Moreover, purified Abs, isolated from specific precipitates, seemed to be indistinguishable from γ-globulins in their electrophoretic mobility, solubility, amino acid composition, and reactivity as Ags; for instance, when goats were immunized with γ-globulins from a nonimmune rabbit the resulting antiserum reacted not only with the immunogen but with rabbit Abs of diverse specificities.

However, some of the electrophoretically faster-moving serum proteins also exhibit Ab activity; and many γ-globulins appear to lack Ab activity, probably because they are specific for unidentified ligands.

Accordingly, all proteins that function as Abs, or that have antigenic determinants in common with Abs, are now called **immunoglobulins** (abbreviated **Igs**).

As will be seen in this chapter, all Igs have a similar structural organization, but they are an immensely diversified family that can be arranged into groups and subgroups on the basis of variations in antigenic properties and amino acid sequences. In this chapter we emphasize the relations between these properties and the **two functions** that are **characteristic** of every Ab molecule: 1) specific binding of one or a few ligands, and 2) participation in certain general effector reactions, e.g., activating of complement proteins (Ch. 20), stimulating macrophages to phagocyte and destroy microbes, triggering mast cells to release vasoactive amines (Chs. 18, 19, and 21). As we shall see, the two functions are carried out by different parts of the Ab molecule.

HETEROGENEOUS AND HOMOGENEOUS IMMUNOGLOBULINS

HETEROGENEOUS NORMAL IMMUNOGLOBULINS

Abs are strikingly heterogeneous, not only in the range of affinities for a given ligand (Ch. 16), but also in electrical charge (reflecting differences in amino acid sequences: Amino Acid Sequences section). For instance, in serum electrophoresis, albumin and other proteins travel as compact bands, each with a characteristic mobility, because the molecules of each protein have the same net charge at a given pH. Igs, however, migrate as a broad band, because they vary considerably in electrical charge (see Fig. 17-2).

HOMOGENEOUS IMMUNOGLOBULINS

Myeloma Proteins. The cells that normally produce Igs (Chs. 18 and 19) sometimes become cancerous and grow as **myeloma tumors** (also called plasma cell tumors or plasmacytomas). The cells in each tumor apparently constitute a single clone (i.e., they derive from a single progenitor cell); and each tumor usually produces a homogeneous Ig, called a **myeloma protein,** sometimes in

huge amounts (serum levels of 50 mg/ml are not unusual). Insight into Ig heterogeneity was initially derived from analyses of myeloma proteins.

Myeloma proteins are also called **monoclonal** or **M proteins,** whereas normal Igs and Abs are sometimes referred to as **polyclonal,** implying an origin from many clones. Figure 17-2 shows the electrophoretic distinction between the diffuse migration of normal (polyclonal) Igs and the compact migration of some myeloma (monoclonal) Igs. When their corresponding Ags can be found, myeloma proteins react like typical Abs, i.e., with similar affinity and specificity, and with two binding sites per molecule of 150,000 daltons, located in the proper parts of the molecule (i.e., in the Fab domains; see Fragmentation below). It is clear that *normal Igs and Abs consist of a great many homogeneous subsets,* each with a characteristic net charge and mobility (reflecting, as we shall see, individually distinctive amino acid sequences).

Myeloma tumors arise in other mammals (dogs, cats, etc.) besides humans. They have been studied particularly intensively in mice: they appear with high frequency (in about 50%) in certain inbred strains (BALB/c, NZB) 3–9 months after intraperitoneal

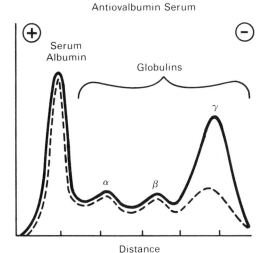

Fig. 17-1. Electrophoresis of serum from a rabbit intensively immunized with ovalbumin, before (————) and after (– – –) removal of Ab molecules by specific precipitation with ovalbumin. (Tiselius A, Kabat EA: J Exp Med 69:119, 1939)

FIG. 17-2. Electrophoresis of human sera at pH 8.4. **A.** Normal serum; note diffuse Igs, proteins with least mobility to anode (+). **B.** Serum with elevated heterogeneous **(polyclonal)** Igs. **C-E.** Sera from three patients with myeloma tumors. Each myeloma protein **(arrow)** is a compact band **(monoclonal** Ig); note their different mobilities. Serum albumin is the compact dark band at left. (Courtesy of Dr. C. Kirk Osterland)

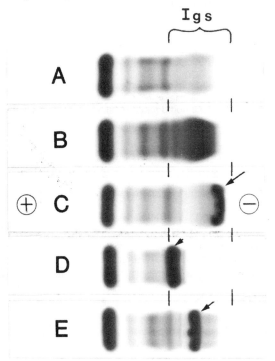

injections of mineral oil or certain plastics. The tumors can be maintained indefinitely by serial passage through mice of the same inbred strain (or even grown in culture), providing large amounts of homogeneous Ig for analysis. The tumor cells are also a rich source of mRNA and DNA for analyzing Ig genes.

Homogeneous Antibodies. Homogeneous Abs can also be elicited against special Ags. After intensive immunization with certain streptococcal and pneumococcal vaccines some rabbits and mice produce Abs (to the bacterial polysaccharides) that may match myeloma proteins in homogeneity and high serum levels (Fig. 17-3). In many of these animals Ab levels reach 20–50 mg/ml of serum and consist of just two or three homogeneous sets. An occasional animal produces a massive amount of a single Ab, as homogeneous as a myeloma protein; but unlike the case with myeloma tumors, production depends on continued administration of the Ag.

Hybridomas. An ingenious approach to the deliberate production of homogeneous Abs against almost any Ag is based upon **cell fusion.** When two different ("parental") cell populations are incubated with certain surface-active agents—e.g., Sendai virus (inactivated by ultraviolet light to eliminate infectiousness) or polyethylene glycol—cells fuse to form **heterokaryons:** single cells with a nucleus from each parent. Some heterokaryons become **hybrids:** i.e., the nuclei fuse. If one of the parents is "immortal" (i.e., from a stable cultured cell line) and the other is a normal "mortal" cell (i.e., not capable of growing in culture), many hybrids can be maintained indefinitely as immortal cells that have both parental phenotypes. Ab-producing immortal hybrids are prepared by fusing a cultured mouse myeloma cell

FIG. 17-3. Homogeneous Ab **(arrow)** in a rabbit. Serum samples were analyzed by electrophoresis on a cellulose acetate membrane. **A.** Before immunization. **B.** After 16 injections (4 weeks) of streptococcal vaccine (group C); the sample had about 40 mg Ig **(arrow)**/ml of serum. **C.** Absorption of the sample in **B** with purified envelope carbohydrate of group C streptococci removes the Ab. Compact, dark bands of serum albumin are at the right. (Eichmann K et al: J Exp Med 131:207,1970)

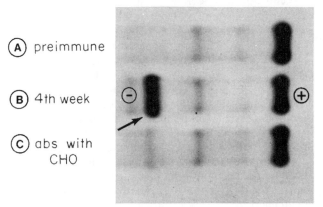

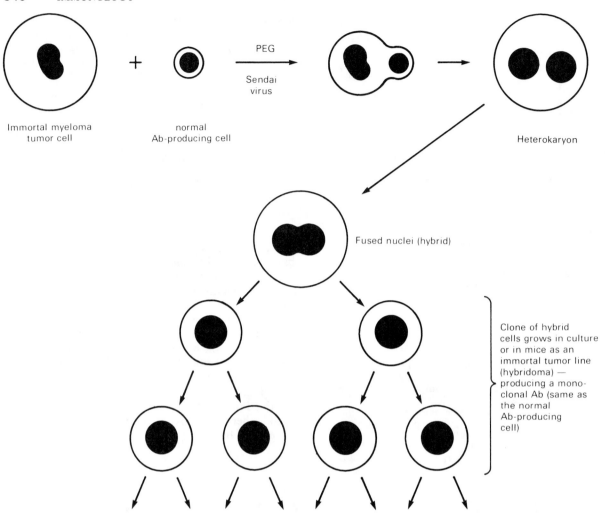

FIG. 17-4. Development of an Ab-producing **hybridoma**. An Ab-producing cell from the spleen of a normal, immunized mouse is fused to a "immortal" myeloma tumor cell (with the aid of polyethylene glycol (PEG) or inactivated Sendai virus). The resulting hybrid clones grow in vitro as a cultured line or in vivo as a tumor (hybridoma) that produces large amounts of a homogeneous (monoclonal) Ab. (Based on Kohler G, Milstein C: Nature 256:495, 1975).

line with spleen cells from a suitably immunized mouse that serves as a source of Ab-producing normal parent cells (Fig. 17-4).

Because a small proportion of spleen cells make Abs to any particular Ag, only a small proportion of the hybrid cells produce Abs to the Ag of interest. These few can be detected by testing culture media from many individual clones (usually thousands) of hybrid cells. When a clone that produces an Ab of interest is detected, the Ab can be mass-produced by passage of the clone as a tumor (**hybridoma**) through mice with compatible transplantation Ags (see Major Histocompatibility Complex, Ch. 18, 23). Serum or ascites fluid from mice carrying a transplanted hybridoma can have extraordinarily high levels of the **monoclonal Ab** (e.g., 5–15 mg/ml).

ANTIGENIC CLASSIFICATION OF IMMUNOGLOBULINS: ISOTYPES, ALLOTYPES, AND IDIOTYPES

Igs are highly immunogenic when injected into heterologous species, and they can be classified on the basis of differences in their antigenic determinants. The principal tools in classification are antisera raised against selected Igs and rendered specific by absorption with other Igs. Reactions with these absorbed antisera divide Igs into various of sets and subsets.

1) **Isotypic determinants** (Gr. *iso,* same). These are shared by all Ig molecules of a given class (e.g., IgG or IgA); hence they are present in all normal individuals of

a given species. The Ig classes, called **isotypes,** are recognized by antiisotypic sera: properly absorbed heterologous antisera that are produced in one species against the Igs of another species **(heteroantisera).** The classes differ in effector functions (See Appendix, Table 17-6), but not necessarily in their ligand-binding specificities: Abs of different classes can be specific for the same Ag.

2) **Allotypic determinants** (Gr. *allos,* different). The Igs within certain isotypes can be separated into sets, called **allotypes,** that are distinguished by small antigenic differences. These differences reflect alternative amino acid substitutions within otherwise similar amino acid sequences, and they do not affect ligand binding or effector functions. Antiallotypic Abs are generally produced within a species by immunizing an individual who lacks a particular allotype with Igs from those who possess it (yielding **alloantisera).** Because allotypic determinants reflect differences between allelic genes, they are valuable probes for exploring Ig genes.

3) **Idiotypic determinants** (Gr. *idios,* individual). This term refers to the individually distinctive antigenic determinants in or close to the specific ligand-binding sites of Ig molecules. Some of these determinants are limited to a single monoclonal Ig **(unique or "private" idiotypes);** others are shared by a small number of Igs **(shared or "public" idiotypes).** Idiotypes are recognized by heteroantisera or alloantisera, or by isologous antiidiotypes—i.e., Abs to the idiotype of another Ig made in the same or in a genetically identical individual.

The ability of one Ab to elicit other Abs against the ligand-binding site of the first Ab has profound implications for the regulation of immune responses (see Idiotype–Antiidiotype Network, Ch. 19)

ISOTYPES: CLASSES OF HEAVY CHAINS AND TYPES OF LIGHT CHAINS

Electrophoresis initially identified three classes of human Igs (G, A, and M: Fig. 17-5); more refined antisera subsequently distinguished four IgG classes (IgG1 to IgG4), and two IgA classes. Two additional classes, IgD and IgE, present in low concentrations in normal serum, were detected by more sensitive methods, e.g., radioimmunoassays.

Heavy-Chain Isotypes. Chemical studies described below showed that Ig molecules are made up of heavy (H) and light (L) polypeptide chains. Each of the nine classes of Igs has antigenically distinctive **heavy chains** (named with the corresponding Greek letter—i.e., γ, α, μ, δ, and ϵ), but the light chains are similar in all classes. In a heteroantiserum to a human Ig of one class, say IgG, some Abs are specific for the H (γ) chains and others for the L chains. After absorption with γ chains the residual Abs are specific for the L chains, and they react with human Igs of all classes. If, however, the original heteroantiserum is absorbed with L chains, or with Igs of other classes, the remaining Abs are specific for γ chains and recognize only IgG proteins (Fig. 17-6).

FIG. 17-5. Immunoelectrophoresis showing IgG, IgA, and IgM. Human serum in the center well **(c)** was subjected to electrophoresis in agar in 0.1M barbital, pH 8.6; then the corresponding rabbit antiserum, prepared against unfractionated human serum, was placed in the long trough, parallel to the axis of migration. (See Fig. 16–28; Courtesy of Dr. C. Kirk Osterland)

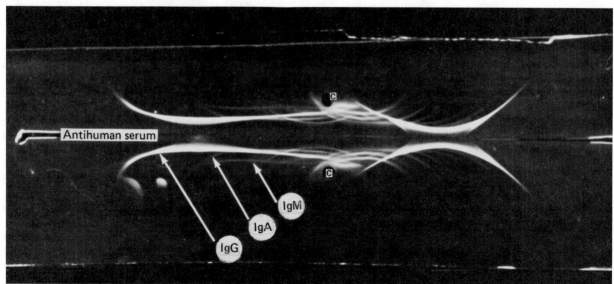

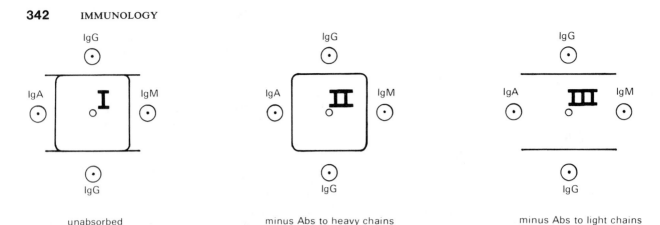

FIG. 17-6. Gel diffusion precipitin reactions of human IgG, IgM, and IgA with rabbit antiserum to human IgG. The antiserum (center wells) was used without absorption **(I)** or after absorption with heavy (H) chains **(II)** or light (L) chains **(III)** from IgG proteins.

Kappa and Lamda Light Chains. There are also two main **L-chain types.** Because they are both present in all Ig classes, they were overlooked until it was discovered that many myeloma tumors produce L chains that are unassociated with H chains. The free L chains are excreted in the urine. Named **Bence Jones proteins** after the 19th century physician who first recognized them, the free L chains have the unusual property of precipitating on being heated to 45°–60° C, redissolving on boiling, and precipitating again on cooling. (In contrast, complete Ig molecules [H + L chains] are, like most other proteins, coagulated irreversibly on boiling.) Some myeloma tumors (in about 10% of patients) produce only Bence Jones proteins, and others (about 40%) produce only a whole myeloma protein; in those who produce both, the Bence Jones protein is nearly always identical with the L chain of the myeloma protein. Because patients often excrete Bence Jones protein in huge amounts (e.g., 3–4 gm per day), these proteins have been intensively analyzed for amino acid sequences and antigenic properties.

Heteroantisera to individual Bence Jones proteins distinguish two types of L-chain, kappa (κ) and lambda (λ). These differ considerably in amino acid sequences (see below); *each normal Ig molecule, regardless of its H-chain class, has one type or the other, not both.* About 60% of human Ig molecules have κ chains; 40% have λ chains. The ratio varies somewhat in different Ig classes (see Appendix, Table 17-6). Igs in other vertebrate species also have κ and λ chains, as defined by homologous amino acid sequences. Most species have predominantly one type (about 95% of all L chains in mice are κ; almost 100% in horses and in birds are λ), while a few (e.g., guinea pigs) have, like humans, nearly equal amounts.

In the following parts of this chapter we shall consider the main classes of Igs, then genetic variants, and indi-

vidual Igs. The final sections consider the structural basis for the specificity of Ab combining sites, the genetic basis for diversity among Igs, and the organization of Ig genes.

THE IgG IMMUNOGLOBULINS

These proteins, originally referred to as 7S γ-globulins because of their sedimentation coefficient, have a mol wt of about 150,000. They constitute over 85% of the Igs in the serum of most individuals.

LIGHT AND HEAVY CHAINS

The IgG proteins have many disulfide (S–S) bonds per molecule, mostly within chains but a few between chains. The chain structure was established in 1959, when Edelman discovered that after cleaving the S–S bonds the molecular weight dropped and electrophoresis (in 8 M urea to prevent the grossly denatured protein from aggregating) revealed two components. Shortly thereafter Porter et al. selectively reduced the few interchain S–S bridges. Soluble chains separated when the reduced molecules were exposed to an organic acid (acetic, propionic), which broke noncovalent hydrophobic bonds and imposed many positive charges on the chains, leading to their mutual repulsion. By gel filtration two components were then recovered, with essentially 100% yield (Fig. 17-7A); the mol wt of the heavier (H) was about 50,000 and of the lighter (L) about 25,000. On the assumption that each fraction was made up of a single kind of chain, *the original molecule of 150,000 daltons consists of two H chains (2 × 50,000) plus two L chains (2 × 25,000).*

Though analysis of antigenic differences and of amino acid sequences have revealed a great variety of H chains and of L chains, *any particular Ig molecule has identical H*

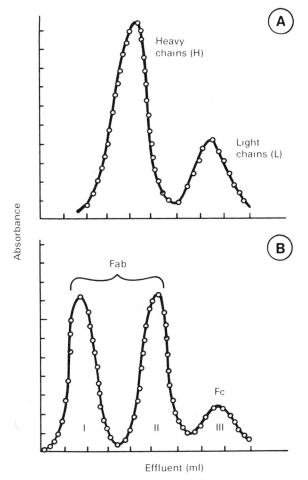

FIG. 17-7. A. Separation of H and L chains of IgGs. The protein was reduced and then alkylated to prevent the liberated –SH groups from again forming S–S bridges:

$$...S{-}S{\rightarrow}...SH + I{\bullet}CH_2{\bullet}COOH{\rightarrow}...S{\bullet}CH_2{\bullet}COOH$$

Gel filtration in an organic acid (1M propionic) yielded H and L chains. **B.** Separation of Fab and Fc fragments of Igs. The protein was digested with papain; after dialysis (during which some of the Fc crystallized) the digest was chromatographed on carboxymethylcellulose. Two Fab peaks **(I, II)** were seen because of the heterogeneity of the IgG sample: when the fragments are produced from a monoclonal Ig (e.g., a myeloma IgG) the Fab fraction emerges as one large peak, the sum of I + II. (**A** based on Fleischman JB et al: Arch Biochem Suppl 1: 1974, 1962. **B** based on Porter RR: Biochem J 73: 119, 1959)

chains and identical L chains (e.g., see Rabbit allotypes; below).

FRAGMENTATION WITH PROTEOLYTIC ENZYMES: Fab AND Fc FRAGMENTS

Attempts to analyze the architecture of Ab molecules by enzymatic fragmentation were not illuminating until certain S–S bonds were also cleaved.

Papain Digestion. In 1959 Porter showed that digestion of rabbit IgG with papain split the molecule into fragments with loss of only about 10% of the protein as small dialyzable peptides. Ion-exchange chromatography separated the digest into three fractions. It was subsequently recognized that fractions I and II (Fig. 17-7B) were the same except for small differences in charge, corresponding to charge differences among the IgG molecules from which the fragments were derived (Fig. 17-2, above). When obtained from purified Abs, these fragments retained the specific combining sites of the parent molecules but were **univalent**, i.e., they contained a single ligand-binding site per fragment; and their total yield accounted fully for the ligand-binding activity of the original molecules. These fragments are now called **Fab (antigen-binding fragments).**

The third fragment (III) did not combine with Ags. It was so uniform that it could be crystallized, in contrast to the heterogeneous Abs from which it was derived. This fraction is therefore called **Fc (crystallizable fragment).**

Of the total digest prepared with papain about two-thirds is Fab (mol wt about 45,000) and one-third is Fc (mol wt 50,000). Hence *an intact, bivalent IgG molecule is made up of two univalent Fab fragments joined through peptide bonds to one Fc fragment.*

Other Proteases. Other proteolytic enzymes also split Igs, and **pepsin** has been especially useful: it cleaves from IgG Abs a bivalent fragment, of mol wt about 100,000, which can be split by reduction of one or a few S–S bonds to yield univalent fragments (Fig. 17-8). The latter are indistinguishable from the Fab fragments prepared with papain with respect to ligand-binding activity, and antigenicity, but the pepsin fragment has a molecular weight about 10% higher. To indicate the small difference the univalent fragment obtained with pepsin is called **Fab′**; the corresponding bivalent fragment is **F(ab′)₂**. Fc is not recovered in peptic digests, but is fragmented.

The variety of fragments obtained with other proteases from Igs of many species fit into a coherent pattern: *all IgG Abs have two compact globular domains (corresponding to Fab fragments) joined to a third compact domain (corresponding to the Fc fragment) by connecting regions that are highly susceptible to attack by proteases* (Fig. 17-8).

OVERALL STRUCTURE

Relations Between Chains and Fragments. The immunogenicity of the proteolytic fragments made it possible to match fragments with dissociated chains. Thus, goat antiserum to the isolated Fc fragment of rabbit IgG forms specific precipitates with H chains, as well as with Fc, but not with L chains. In contrast, antiserum to Fab fragments reacts with both L and H chains (Table 17-1). These findings led Porter to the schematic structure for

TABLE 17-1. Goat Antibodies to Rabbit Fab and Fc Fragments: Precipitation Reaction with Chains Isolated from Rabbit IgG

	Antiserum to	
	Fab	Fc
Heavy chains	+	+
Light chains	+	−

IgG shown in Figure 17-8, which fits all the information now available.

Shape. Various analytic approaches (based on hydrodynamic and fluorescence properties, electron microscopy, and x-ray diffraction) suggest that IgG molecules are **T- or Y-shaped,** with a "hinge" at about the middle of the H chain connecting the two Fab domains with the Fc domain (Fig. 17-9). There may also be segmental flexibility: by movement about the hinge, the angle between Fab segments might vary from a few degrees to almost 180°.

DISULFIDE BRIDGES

Stability. IgG proteins are unusually stable. In serum or as purified Abs they can remain unaltered for years at 0° C. Moreover, after exposure to denaturing conditions

FIG. 17-8. Four-chain structure of IgG, showing some interchain S–S bonds and regions susceptible to proteolytic cleavage (papain, pepsin, at **arrows**). The piece of heavy **(H)** chain within the Fab fragment is the **Fd piece.** The spherical Ag represents an Ag molecule in a ligand-binding site. [Four-chain model from Fleischman JB et al: Biochem J 88:220, 1963; Y shape at top based on x-ray diffraction studies; scheme for pepsin digestion based on Nisonoff A et al: Arch Biochem 89:230, 1960]

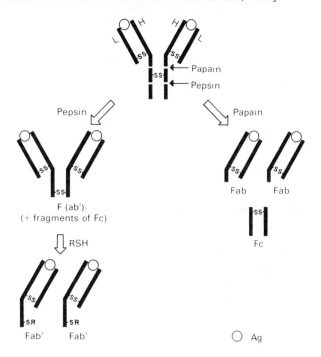

(e.g., 70° C; pH 11 or pH 2; 8 M urea) they largely recover their native structure when returned to dilute salt solution at physiologic conditions of pH and temperature; a major factor is the presence of a large number of disulfide (S–S) bonds (16–24 per molecule). Four to twelve of these bonds link the four chains together; the remainder are distributed within the chains, stabilizing their respective conformations (see Figs. 17-14 and 17-15, below).

Under mild conditions that cleave H-H interchain S–S bonds the IgG molecule can be split into symmetric halves (each with one H and one L chain), which separate at low pH; they reassociate to form native molecules when returned to neutral pH. Even if their SH groups are blocked chemically, e.g., by treatment with iodoacetic acid (see Fig. 17-7A), the halves can still associate (though without S–S bridges) to form molecules that are essentially indistinguishable from the original e.g., in mol wt and antigenicity.

Hybrid Antibodies. By combining appropriate fragments from different Abs it is possible to produce bivalent molecules with a different specificity at each ligand-binding site. These "hybrids" are useful experimentally; e.g., they can localize easily detected marker molecules (ferritin or virions) at particular cell surface Ags (see Appendix, Ch. 16, and Fig. 21-3, Ch. 21). Hybrids are generally prepared with the F(ab')₂ fragments obtained by pepsin digestion (of rabbit Abs): mild reduction yields Fab' fragments, each with one ligand-binding site and one free SH group. If these fragments are prepared from two purified Abs of different specificity, mixed, and allowed to reoxidize, **hybrid bivalents** appear with a different ligand-binding specificity at each of the two sites. The large proportion of hybrids indicates that the recombination of Fab' fragments is random (Fig. 17-10).

Symmetry. As we shall see later, an Ig-producing cell usually makes, at any one time, only one type of L chain and one type of H chain (Ch. 18, One cell–One Ig rule). Hence *natural IgGs are symmetric: in any particular molecule the two light chains are identical, as are the two heavy chains and the two ligand-binding sites* (Ch. 16, Ab Affinity and Valence).

IgG CLASSES

Antigenic analysis with heterologous antisera revealed that proteins of the human IgG group can be differentiated into four classes (IgG1 through IgG4), each with a distinctive H chain (γ1 through γ4). IgG1, 2, 3, and 4 make up about 70%, 19%, 8%, and 3%, respectively, of human IgG proteins. Their characteristic antigenic determinants are localized in their Fc segments.

Antisera vary greatly in capacity to differentiate among the classes. Rabbit antiserum to the total IgG fraction of human serum does not distinguish among them, while monkey antiserum can distinguish three of the four. The sharpest distinctions are made with antisera against individual myeloma proteins. Absorption of

FIG. 17-9. Three IgG anti-Dnp Ab molecules joined in a triangular complex by a ligand with two Dnp groups per molecule [Dnp-NH-$(CH_2)_8$-NH-Dnp]. The ligand is too small to be seen. The complex in **I** forms the basis for the diagram in **III**. The picture in **II** was obtained after treatment of the complex in **I** with pepsin, removing the corner projections, which correspond to Fc. (Valentine RC, Green NM: J Mol Biol 27:615, 1967) (I and II, electron micrographs, ×500,000)

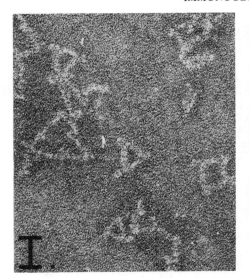

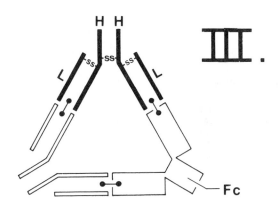

Dnp ligand

25 Å

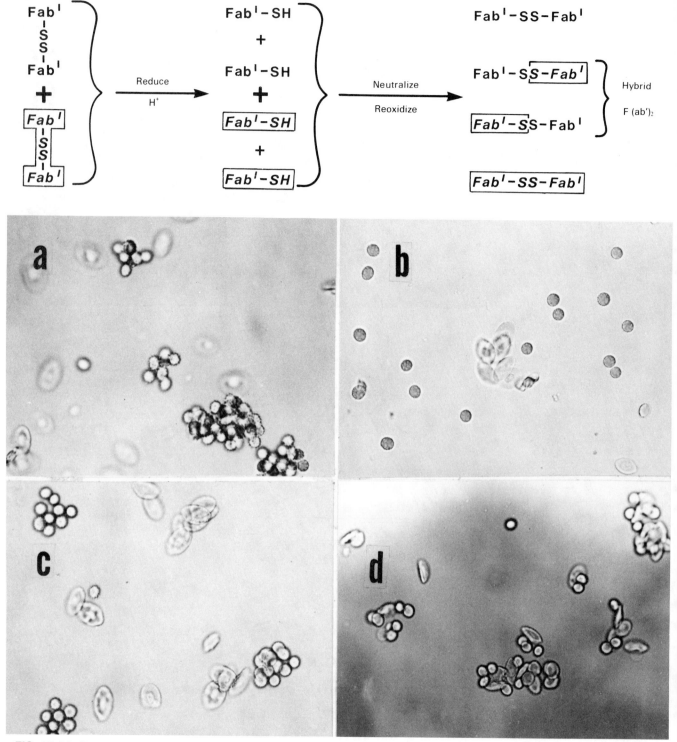

FIG. 17-10. Formation of hybrid bivalent Ab fragments. **Top.** F(ab′)$_2$ fragments from Abs of different specificities were mixed, reduced to Fab′ fragments, and then allowed to reoxidize. **Bottom.** Dual specificity of hybrids revealed by mixed hemagglutination. The Fab′ fragments were obtained from rabbit Abs specific for chicken ovalbumin (EA) or for bovine γ-globulin (BγG). Human red blood cells (RBCs) (small, spherical, nonnucleated) were coated with EA, and duck RBCs (large, oval, nucleated) were coated with BγG. Reoxidized F(ab′)$_2$ fragments prepared exclusively from anti-EA clumped only the human cells **(a)**. Reoxidized fragments prepared exclusively from anti-BγG clumped only the duck cells **(b)**. A mixture of the two F(ab′)$_2$ fragments of **a** and **b** formed separate clumps of human and of duck cells **(c)**. In **d** hybrid F(ab′)$_2$ fragments formed mixed clumps, with both human and duck cells. (Fudenberg H et al: J Exp Med 119:151, 1964)

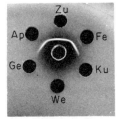
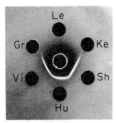

FIG. 17-11. IgG classes. Twelve IgG myeloma proteins (each from a human with a myeloma tumor) were tested with monkey antiserum (center well in each panel) to the protein from patient **Zu**, after the antiserum had been absorbed with Igs of other classes. Five additional proteins were precipitated (**Ap, Fe, Vi, Hu, Sh**); hence they belong to the same class as **Zu** (IgG3). The six nonreacting IgGs belong to other classes. (Grey HM, Kunkel HG: J Exp Med 120:253, 1964)

such an antiserum by myeloma proteins of other IgG classes leaves monospecific Abs that react only with Igs whose heavy chains belong to the same class as the immunogen (Fig. 17-11).

Four classes of IgG have also been identified in the mouse, the only other species in which many myeloma proteins have been studied. Fewer IgG classes have been recognized in other species (e.g., IgG1 and IgG2 in the guinea pig), probably because homogeneous Igs have not yet been obtained from them and the only IgG classes recognized are those that happen to be relatively abundant and to differ substantially in electrophoretic mobility.

Antigenic differences among the IgGs are less pronounced than those between IgGs and the other Ig classes. For example, after Abs to L chains are removed, antiserum prepared against an IgG1 myeloma protein crossreacts with IgG2, IgG3, and IgG4, but not with IgM, IgA, IgE, or IgD. Amino acid sequence data (below) confirm these relations by revealing greater homologies among γ1, γ2, γ3, and γ4 chains than among H chains of other classes (Fig. 17-38).

AMINO ACID SEQUENCES: VARIABLE (V) AND CONSTANT (C) SEGMENTS

The basis for antigenic and other differences among Igs has been greatly illuminated by amino acid sequences of chains from myeloma proteins. (Except for unusually homogeneous Abs, only partial sequences have been de-

FIG. 17-12. Variable (**V**) and constant (**C**) segments illustrated by selected amino acid sequences from some human light chains (named **SH**, etc., for the patient from whom the myeloma protein was derived). Sequences are aligned for maximal homology. From positions marked **C_L** to the COOH terminus (to right of **vertical dashed line**) the κ chains are all alike, as are the λ chains; κ-λ differences are especially easy to discern in C_L. Amplitude of the **jagged line** above the sequences represents extent of variability. (Based on references in Dayhoff M (ed): Atlas of Amino Acid Sequence and Structure, vol. 5. Silver Spring, Md, National Biomedical Research Foundation, 1972). Note the one-letter code for amino acids.

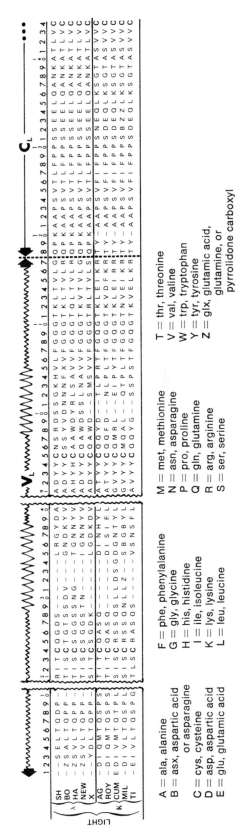

A = ala, alanine
B = asx, aspartic acid or asparagine
C = cys, cysteine
D = asp, aspartic acid
E = glu, glutamic acid

F = phe, phenylalanine
G = gly, glycine
H = his, histidine
I = ile, isoleucine
K = lys, lysine
L = leu, leucine

M = met, methionine
N = asn, asparagine
P = pro, proline
Q = gln, glutamine
R = arg, arginine
S = ser, serine

T = thr, threonine
V = val, valine
W = trp, tryptophan
Y = tyr, tyrosine
Z = glx, glutamic acid, glutamine, or pyrrolidone carboxyl

termined for chains of normal Igs, and even then only with great difficulty.) The first extensive sequences were obtained with two human L chains (κ Bence Jones proteins) by Hilschmann and Craig and by Putnam et al. These chains had different amino acid residues at many positions in the amino-terminal half (the **variable or V segment**) while the carboxyl half (**constant or C segment**) was the same in these and in all other human κ chains (except at two positions: see Allotypes, below). The remarkable distinction between the V and C segments, illustrated in Fig. 17-12, has since been found to be characteristic of all Ig chains, H and L.

The IgG γ chains are about twice as long as the L chains. The first completed H-chain sequence (a γ1 chain, by Edelman and colleagues) revealed linear repeats of similar segments: each segment had about 110 amino acids with an S–S bond that established a loop of approximately 60 residues. Moreover, similarities in sequence suggested that each of these **domains** has an approximately similar shape, and this was later confirmed by x-ray diffraction of Ig crystals (see Three-Dimensional Structure, below). Comparison of various γ chains showed that, as with κ chains, variations are localized in the amino-terminal segment (ca. 115 residues); the remaining sequence is the same in all chains of a particular class (except for small differences in allotypes, discussed below). Thus six domains are recognized in IgG molecules: two in L chains (V_L and C_L) and four in H chains (V_H, C_H1, C_H2, C_H3; Fig. 17-13).

The strong serologic cross-reactions among the four IgG classes is matched by amino acid sequence homology among their respective C_H domains. Human γ1 and γ2, for instance are over 90% homologous in C_H1 and 60%–70% homologous in C_H2 and C_H3. In some species homology is somewhat less; generally, homology is greater in C_H1 than in C_H2 or C_H3 domains.*

* In aligning chains or domains to determine homology, gaps are introduced in some chains in order to obtain maximal homology.

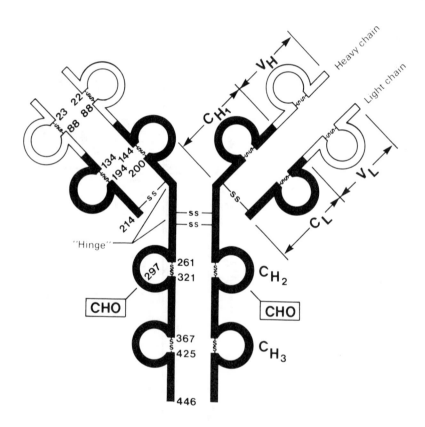

FIG. 17-13. Linear periodicity in amino-acid sequences of an IgG molecule shows that light (L) and heavy (H) chains have repeating domains, each with about 110 amino-acid residues and an approximately 60-membered S–S bonded loop. Domains with variable sequences are represented by open bars (V_L, V_H); those with constant sequences in a given class of H or type of L are represented by solid lines (C_L, C_H1, C_H2, C_H3; see Fig. 17-12). Numbered positions refer to cysteinyl residues that form S–S bonds, or to the point of attachment of an oligosaccharide (CHO). Disulfide (S–S) bonds joining distant parts of the same chain are **intrachain bonds;** those linking L to H or H to H chains are **interchain bonds.** For other arrangements of interchain S–S bonds see Fig. 17-14. (Based on Edelman GM: Biochemistry 9:3197, 1970)

—ss— Disulfide bonds

☐ Variable region

▬ Constant region

Hinge Region. A stretch of about 15 amino acid residues between C_H1 and C_H2 corresponds to the flexible junction between Fab and Fc regions in $\gamma1$, $\gamma2$, and $\gamma4$ chains (see Three-Dimensional Structure, below). This "hinge" region has no obvious homology with other domains. It contains the S–S bonds that link the two H chains of each molecule and the $\gamma1$ and L chains of IgG1 (Fig. 17-14). The hinge is about four times longer and has about four times more S–S bonds between the H chains in $\gamma3$ than in the other γ chains, owing to duplications of the normal hinge sequence (Table 17-2; see also Fig. 17-16 for a duplicated IgA hinge).

The few cleavages caused by proteolytic enzymes in *native* Ig molecules are localized in the hinge region (Fig. 17-8). The abundance of hinge residues that confer rigidity (proline and cysteine; see Table 17-2) evidently exposes this part of the backbone to proteolytic enzymes. When Igs are *denatured* (following reduction of intrachain S–S bonds) most proteases cause extensive degradation of the entire molecule into many small peptides.

OTHER IMMUNOGLOBULIN CLASSES

IgM

The first Abs studied in the ultracentrifuge, from horses immunized with pneumococci, had high sedimentation values (predominantly 19S), but most Abs subsequently proved to be 7S IgG molecules. However, sensitive assays later revealed that the 19S Abs, now assigned to the IgM class (M for macroglobulin), are formed early in every Ab response, though usually only at low levels so that they are soon overshadowed by larger amounts of IgG Abs to the same Ag (see Chs. 18 and 19).

In man and other mammals IgM normally accounts for about 5%–10% of the serum Igs. Likewise, few myeloma proteins are of the IgM class. (The monoclonal IgMs used to establish the detailed structure outlined in Figure 17-15 are more frequently obtained from humans with Waldenström's macroglobulinemia, a syndrome due to a less malignant lymphocytic tumor.)

The human IgM molecule (about 900,000 daltons) is a pentamer made up of five four-chain subunits (Fig. 17-15), each with a pair of H(μ) chains and a pair of κ or λ L chains. Its distinctive features were established by identifying the complete amino acid sequence of the μ chains from two monoclonal IgM proteins (determined by Hilschmann et al in West Germany and Putnam et al in the USA): 1) The μ chains lack a hinge (Table 17-2). 2) Instead, μ chains have about 130 more amino acid residues, or one more domain, than γ chains. 3) The extra domain is located (like the hinge in other Ig molecules) between domains that are homologous, by amino acid sequence, to C_H1 and C_H2 of γ chains. 4) As expected, different μ chains differ in the amino-terminal (V) domain, and not in other domains. 5) The four-chain subunits are linked by two S–S bonds between μ chains (one

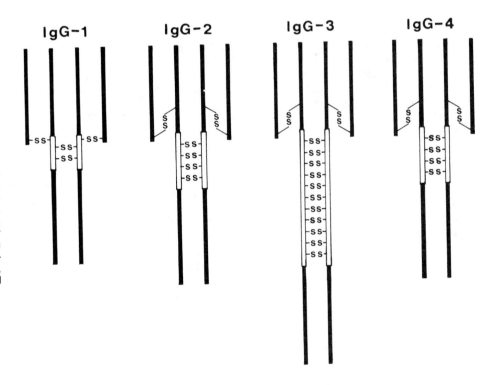

FIG. 17-14. Interchain S–S bridges in different human IgG classes. The diversity of these bonds contrasts with the virtual constancy of intrachain S–S bonds (Fig. 17-13). Note that the hinge in $\gamma3$ is fourfold longer than in the other γ chains. All H–H S–S bridges are in the hinge (**open rectangles**); variations in their number and arrangement are probably related to effector and other differences between these classes (e.g., susceptibility to papain digestion). The H chain cysteinyl residue forming the H–L S–S bridge in IgG classes 2, 3, and 4 is about 100 residues closer to the amino terminus than in IgG1. (Based on data from Frangione B et al: Nature 211:145, 1969; Michaelson TE et al: J Biol Chem 252:883, 1977)

TABLE 17-2. Comparison of Hinge Regions of Human Heavy Chains

Heavy chain class (isotype)	No. amino acids in hinge*	No. cysteine residues	No. proline residues	Site of hinge or of flexible interface between domains
$\gamma 1$	18	3†	5	$C_H 1/C_H 2$
$\gamma 2$	15	4	4	$C_H 1/C_H 2$
$\gamma 3$	65	11	22	$C_H 1/C_H 2$
$\gamma 4$	15	2	5	$C_H 1/C_H 2$
$\alpha 1$	26‡	3	10	$C_H 1/C_H 2$
$\alpha 2$	13‡	3	7	$C_H 1/C_H 2$
μ	0§	—		$C_H 2/C_H 3$¶
ϵ	0§	—		$C_H 2/C_H 3$¶

*Position 216 from the amino-terminus is the initial residue of the hinge.

†One of these contributes to the S–S bond between L and H chains (see Fig. 17-14).

‡Based on Plaut AG et al: Science 190:1103, 1975 (see Fig 17-17)

§μ and ϵ chains have no amino acid sequence corresponding to the hinge of γ or α chains.

¶Some evidence suggests that segmental flexibility may also be present at the $C_H 1/C_H 2$ junction in μ and ϵ chains.

(Based largely on Metzger H: Contemp. Top Mol Immunol, in press)

is in $C_H 3$, while the other, in $C_H 4$, is the penultimate residue of the chain).

Under mild reducing conditions, IgM is cleaved into subunits, each with two H (μ) and two L chains. Resembling an IgG molecule, the subunit, called **IgM$_s$**, has two ligand-binding sites per molecule (shown by equilibrium dialysis; Ch. 16), but loses its ability to participate in agglutination and many other immune reactions. The dependence of these latter activities on more than two binding sites per molecule is probably due to the fact that IgM Abs usually have very low intrinsic affinity (per site) for Ag. Accordingly, multivalent binding to Ag (Ch. 16) is usually necessary for IgM molecules to manifest biologic activity (see Fig. 17-15C and Appendix, Ch. 16).

Low-molecular-weight IgM, corresponding to the 7S IgM$_s$ subunit produced by mild reductive cleavage of the 19S pentamer, also occurs naturally. The surface Igs of most B lymphocytes (precursors of Ab-secreting cells; see Chs. 18 and 19) are modified IgM$_s$ molecules (see also IgD, p. 351). IgM$_s$ also occurs at trace levels in normal human serum, and at high levels (about 1 mg/ml) in Waldenström's macroglobulinemia and in certain immunologic disorders (e.g., rheumatoid arthritis, systemic lupus erythematosus).

J Chains. In addition to the ten L and ten H chains per molecule, pentameric IgM has one extra chain, called J.

This chain (15,000 daltons) does not share antigenic determinants or amino acid sequences with Igs. However, J chains are S–S bonded to μ chains, and they are synthesized in the same cells as the Ig molecules to which they are linked. Because they are present only in IgM and in multimeric forms of IgA (see below), J chains are thought to join the subunits of multimeric Igs.

IgA

Immunoelectrophoresis of human serum led to the identification of IgA (Fig. 17-5), whose basic structural unit corresponds to the four-chain molecule shown in Figure 17-13 (see also Fig. 17-16). Unlike IgM, which is a pentamer, and IgGs, which are uniformly monomers, IgA occurs in various polymeric forms: monomers (i.e., H_2L_2), dimers (H_4L_4), and even higher multimers. The multimers are S–S- linked monomers: mild reduction dissociates them into the basic 4-chain Ig molecule (H_2L_2).

IgA normally accounts for about 15% of the total Igs in human serum. In addition, in man and in other mammals IgA is the principal Ig in exocrine secretions (colostrum, respiratory and intestinal mucin, saliva, tears, genitourinary tract mucin, etc.).

Antisera to human IgA myeloma proteins distinguish two classes, IgA1 and IgA2, with the respective H chains, $\alpha 1$ and $\alpha 2$. IgA1 amounts to 80%–90% of serum IgA, but in secretions of the exocrine glands the two classes are approximately equally abundant.

Both $\alpha 1$ and $\alpha 2$ chains have four domains (one V_H, three C_H). A duplicated proline-rich sequence in the hinge region of $\alpha 1$ is largely lacking in $\alpha 2$ chains (Fig. 17-16). Bacterial proteases from certain streptococci and from gonococci cleave particular bonds in this region: IgA1 molecules are readily split by these enzymes into Fab$_\alpha$ and Fc$_\alpha$ fragments, whereas IgA2 molecules are entirely resistant. Thus IgA2 Abs might confer greater protection than IgA1 against such bacteria (e.g., gonococci in the genitourinary tract).

IgA occurs mostly as a monomer (H_2L_2) in human serum but as a dimer (H_4L_4; about 400,000 daltons) in secretions (**sIgA**). The dimer sIgA also has a single J chain (see IgM, above) and still another chain, called **secretory piece (SP)**, which is also (like the J chain) not homologous to Igs, antigenically or in amino acid sequences.

Secretory piece is a glycoprotein (9% carbohydrate). Made in mucosal epithelial cells of exocrine glands, it becomes joined via noncovalent and S–S bonds to α chains of IgA molecules that are produced, together with J chains, in nearby plasma cells (Ch. 19) of the gland's submucosal layer. Secretory piece can also be found as a free chain unattached to IgA in colostrum, saliva, etc., especially in individuals who are unable to synthesize IgA (see Immunodeficiency Diseases, Ch. 19). Isolated SP can bind to IgA molecules ($\alpha 1$ and $\alpha 2$). In patients with IgA deficiency some IgM molecules have SP.

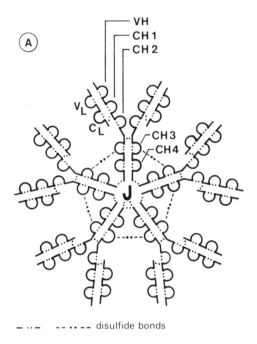

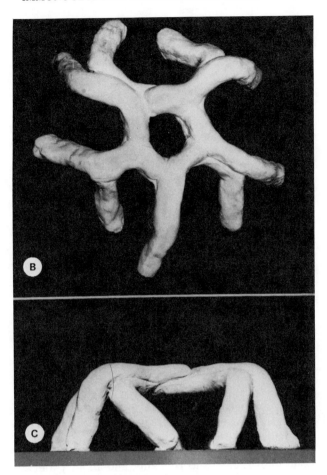

FIG. 17-15. Structure of human IgM. **A.** Five four-chain subunits are joined by S–S bridges and linked by other S–S bonds to the J chain. "Hot" trypsin (60°C) cleaves between C_H2 and C_H3, yielding ten Fab fragments and one Fc pentamer per molecule. **B** and **C.** Models of the IgM molecule, based on electron microscopy. Top view **(B)** displays flexible subunits. Profile view **(C)** show the molecule bound at multiple sites to the surface of an Ag. (**A** based on Putnam FW et al: Science 182:287, 1973); **B** and **C**, Feinstein A, Munn EA: Nature 224:1307, 1969); Svehag SE et al: J. Exp Med 130:691, 1969)

Cells that produce IgA are particularly abundant around the mucin-secreting glands of the intestinal, respiratory, and genitourinary mucosa. In response to Ags that enter locally (by ingestion, inhalation, via the conjunctivae, etc.) the mucosa secretes IgA Abs, which are probably important in protecting mucosal surfaces from invasion by microbes: e.g., in colostrum these Abs probably protect the suckling newborn from infection. Secretory piece also enhances the resistance of secretory IgA to proteolysis, and thereby adds to the protective function of these Ig molecules.

Though IgA accounts for only a small proportion of normal serum Igs (see above), it makes up about 65% of the myeloma proteins in mice (of the BALB/c strain). This probably comes

about because in mice myeloma tumors usually arise in the chronic inflammatory tissues of the peritoneal cavity, elicited by intraperitoneal injection of mineral oil, and IgA-producing cells are particularly abundant in lymphoid tissue of the peritoneum and intestine (see Lymphatic Organs, Ch. 18).

IgD

These proteins were discovered not as Abs but as rare monoclonal Igs that failed to react with monospecific antisera to the other Ig classes known at the time (IgG, IgA, and IgM). Produced by rare myeloma tumors (about 1%) or other malignant lymphomas, these proteins and the corresponding normal Igs are called IgD.

IgD levels vary considerably in normal human sera

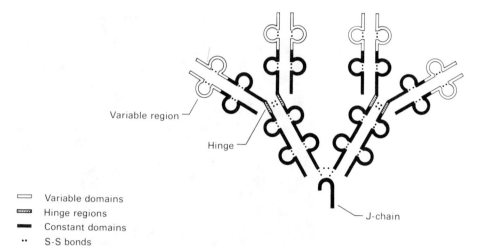

FIG. 17-16. IgA dimer shown with S–S bonds linking the $H_2 L_2$ monomers to each other and to a J chain. The duplicated hinge of the $\alpha 1$ chain is shown in the rectangle; note deletion within the $\alpha 2$ hinge. **Arrows** point to the threonine-proline bonds cleaved in IgA1 molecules by a streptococcal protease (1) and a gonococcal protease (2), yielding in each case Fabα and Fcα fragments (IgA2 molecules are resistant to these proteases). Secretory piece is not shown. For amino acid abbreviations, see Fig. 17-12. (Based on Plaut AG et al: Science 190:1103, 1975)

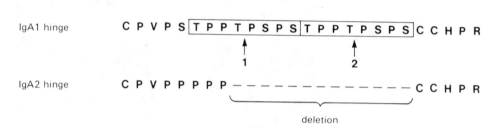

but, on average, they are about $30\mu g/ml$ or 0.2% of the total Igs. The overall structure is the same as that of IgG. The molecular weight of the isolated δ chain is 70,000, suggesting five domains (one V_H, four C_H). A single S–S bond joins L chains to H chains, and another joins the two H chains. An unexplained high proportion of IgD myeloma proteins (about 80%) have λ light chains.

Compared to other Igs, IgD molecules are relatively heat-labile (antigenic properties are lost after heating at $56°$ C for 1 h). They are also extremely sensitive to proteases: in serum stored in the cold for a few days IgD is broken down into Fab and Fc fragments, probably through attack by serum enzymes. Though present at only trace levels in serum, *IgD is surprisingly abundant on the surface of B lymphocytes,* the precursors of Ab-secreting cells (Antigen-Recognition Molecules on B Cells, Ch. 18).

IgE

IgE proteins were discovered by K. and T. Ishizaka while studying the special Abs (called **reagins;** Ch. 21) that mediate certain acute allergic reactions in man. These Abs were precipitated by rabbit antiserum to reagin-rich Ig fractions but not by monospecific antisera to the then known Ig classes. The inference that reagins belonged to still another Ig class was verified with the aid of the independent finding by Johannson and Bennich in Sweden

of an unusual human myeloma protein that also did not react with antisera to the known Ig classes. Rabbit antiserum to the Swedish myeloma protein was then found to react specifically with reagins, and antireagin serum reacted with the Swedish protein (Fig. 17-17).

Since then, a few other human IgE myeloma proteins have been identified. They have, per molecule (188,000–196,000 daltons), a pair of ϵ H chains and a pair of κ or λ L chains. The carbohydrate content is high (11.5%). One S–S bond joins L to H chains and four appear to join the pair of ϵ chains. The mol wt of ϵ chains is about 75,000, or 61,000–65,000 daltons if carbohydrate is deducted. This makes ϵ chains slightly larger (by about 3,000 daltons) than μ chains. Amino acid sequence homology suggests that, as with μ chains, the extra ϵ domain lies between domains that correspond to $C_H 1$ and $C_H 2$ of γ chains.

IgE levels in normal human serum are, on average, about $0.3\mu g/ml$: i.e., about 1 in 50,000 serum Ig molecules is IgE. Levels are elevated slightly in some patients with severe reagin-mediated disease (e.g., hay fever), and even more in those with chronic parasitic infestations of the intestinal tract (e.g., African children with ascariasis, hookworm, or schistosomiasis), where levels up to $140\mu g/ml$ have been described. Perhaps because of frequent chronic intestinal parasitism, IgE levels are also elevated markedly in laboratory rats, in which a high proportion of myeloma tumors produce IgE myeloma proteins. However, in man only five or six out of many

Fig. 17-17. Identification of reagins as IgE protein. A reagin-rich fraction of human atopic serum was placed in the center well and antisera for each of the human Ig classes were placed in peripheral wells. Antisera to reagin and to the IgE myeloma gave reactions of identity. (Ishizaka K: In Merler E (ed): Immunoglobulins: Biologic Aspects and Clinical Uses. Washington, DC, Natl Acad Sciences, 1970)

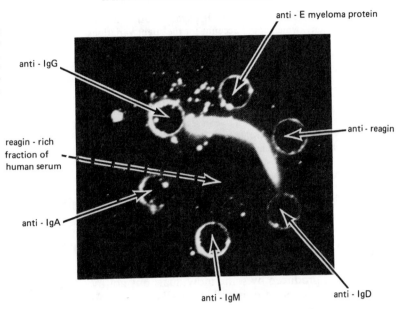

thousands of myeloma proteins have so far been identified, worldwide, as IgE proteins; but these few have had an enormous impact on the management of acute allergic disorders of man (discussed in Ch. 21). The reason is that serum from an individual with an IgE-producing myeloma tumor can have huge IgE levels (e.g., 40 mg/ml), and it takes only a few milligrams to elicit liters of high-titered monospecific goat antiserum to IgE. With the antiserum thus available, radioimmunoassays have been developed on a practical clinical scale for rapid, sensitive measurement of serum levels of total IgE Abs to particular Ags (Ch. 21).

As we shall see later in detail, many acute allergic reactions depend upon the extremely high affinity of the Fc region of IgE molecules for special cell surface receptors (Fc$_\epsilon$ receptors) on tissue mast cells and on basophilic leukocytes. These cells contain powerful vasoactive amines (histamine, etc.) which are released by the cross linking of IgE molecules and Fc$_\epsilon$ receptors on the cell surface (by Ag or, experimentally, by anti-IgE) (Ch. 21).

LIGHT CHAIN ISOTYPES

Amino acid sequences distinguish between κ and λ L chains (Fig. 17-12), and within each type minor sequence differences in the C domains reveal additional variant forms, which are also isotypes (because they are present in all normal individuals of the species; see Antigenic Classification, above).

Rabbits (but not humans or mice) have two κ isotypes, κ_A and κ_B, differing in the number of cysteine residues. In mice, there are three

λ isotypes ($\lambda1$, $\lambda2$ and $\lambda3$): they differ antigenically and at multiple positions in their C domains. In man there are several types of λ chains differing from each other at about ten scattered positions in the C domain.

CARBOHYDRATE SIDE CHAINS*

Igs are glycoproteins, with carbohydrate making up 3%–12% of the molecular weight. Oligosaccharide units are attached to one or more positions in the C regions of all Ig H chains, and of rare L chains (see Appendix, Fig. 17-40).

Except for some of the carbohydrate on $\alpha1$ chains, the oligosaccharides are attached via *N*-glycosidic linkage of glucosamine to asparagine in the sequence *Asn-X-Ser/Thr* of the Ig chains, where Asn is asparagine, *X* is any amino acid, and the third position is serine or threonine. Where this triplet is exposed on the surface of native Ig it apparently can serve as acceptor for a glycosylating enzyme (*N*-acetylglucosamine-asparagine transglycosylase). Almost all of these oligosaccharides have a common core containing mannose and glucosamine, and branches with *N*-acetylglucosamine, mannose, galactose, and sometimes sialic acid. In addition, μ and ϵ chains also have high-mannose oligosaccharides in which multiple residues of mannose are attached to the same core.

The $\alpha1$ chain also contains five oligosaccharides that are in O-glycosidic linkage (galactosamine attached to the hydroxyl of serine). All five units are located in the duplicated proline-rich hinge region of $\alpha1$; they are not found in $\alpha2$ chains, which lack this hinge segment (Fig. 17-16).

---

* With assistance of Lisa A. Steiner, Massachusetts Institute of Technology.

An interesting feature of many of the Ig oligosaccharides is their microheterogeneity, which could account for the slight variations in electrophoretic motility seen even with monoclonal Igs. The heterogeneity may be due to "ragged ends" arising from irregular losses of sialic acid end groups, due to serum glycosidases. (It could also arise in part from irregular losses of amides from asparagine and glutamine residues, due to serum amidases.)

The function of the oligosaccharides in Igs is not understood. It has been suggested that they play a role in secretion but this seems unlikely since free L chains, containing no carbohydrate, can be secreted. Interestingly, carbohydrate tends to occur at homologous positions in various human and animal Ig classes (Appendix, Fig. 17-40), indicating conservation in the evolution of the acceptor sequences. Some properties of various classes of human Igs are summarized in the Appendix (Table 17-6).

ALLOTYPES: GENETIC VARIANTS

An Ag that is produced by some individuals but not by others of the same species is called an **alloantigen (alloAg)**; the corresponding alloantisera are readily elicited by injecting the alloAg into individuals (of the same species) that lack it. The most prominent alloAgs are on cell surfaces: they determine host responses to blood transfusions and organ transplants (Ch. 23), they mark various differentiation stages of lymphocytes (Ch. 18), and, as Ags of the major histocompatibility complex, they play a key role in the cell–cell interactions of diverse immune responses (Chs. 18, 19, 22, and 23). Alloantigenic forms of **Igs**, called **allotypes**, were the first soluble alloAgs discovered; they have contributed greatly to our understanding of Ig structure and of the organization of Ig genes.

HUMAN ALLOTYPES

A chance observation led to the discovery of human Ig allotypes. Serum from persons with rheumatoid arthritis often contains **rheumatoid factors**, which are Igs (generally IgM) that react with pooled human IgG; for example, they specifically agglutinate human red blood cells (RBC) coated with human anti-RBC Abs of the incomplete type (see Blocking Abs, Ch. 16, and Rh Antigens, Ch. 23). Grubb noticed that rheumatoid factor from some patients reacted only with RBC that had been coated with Abs from certain other persons. Moreover, any particular agglutination reaction could be specifically inhibited only by serum from particular individuals, whose Igs presumably carried the same antigenic determinants as the anti-RBC Abs and therefore competed for the same rheumatoid factor (Fig. 17-18).

Some normal sera also contain Igs that agglutinate particular Ab-coated RBCs. These SNaggs (serum normal agglutinators), discriminate more sharply among related IgG variants than do most

rheumatoid agglutinators (Raggs), but both are valuable reagents for typing human Igs.

Some of the agglutinating activities in normal serum are probably due to maternal–fetal incompatibilities. In about 10% of human pregnancies the mothers have Ig allotypes that their babies lack. With transfer of maternal Ig across the placenta about half of these babies are stimulated to make antiallotype Abs. Detected with the sensitive hemagglutination assay (Fig. 17-18), these Abs generally reach peak titers at about 6 months of age (after the large amount of maternal Ig present in the infant's blood at birth has been eliminated; see Serum Igs in the Newborn, Ch. 19. Antiallotypes are also detected in about 1% of normal adults, and they are frequently formed in persons who receive multiple blood transfusions (for chronic anemia, open-heart surgery, etc.), because the donors often have allotypes foreign to the recipients.*

Through agglutination reactions with sera from thousands of persons, many human allotypes have been identified and classified into several groups, of which the principal ones are *Gm* on γ chains, *Am* on α chains, and *Km* on κ L chains.

Gm Allotypes. A numerical system, resembling those used for Rh blood factors and for Salmonella Ags, has recently been adopted for Gm allotypes. Agglutinating sera of a given specificity are designated by number. Igs that react with a particular agglutinator receive the corresponding positive number; those not reacting when tested receive a number with a minus sign. For instance, Igs with allotype Gm(1, 23−) react with anti-Gm(1) but not with anti-Gm(23); the absence of other numbers in this example would mean that the Ig was not tested with other agglutinators. As is conventional with other systems, alleles are italicized, their products are not: e.g., the allele *Gm⁵* specifies a γ chain of allotype Gm(5).

Over 20 Gm allotypic determinants have been recognized. Each of the four IgG classes has its own group of allotypes (Table 17-3), specified by codominant† alleles at an **autosomal** (i.e., not sex-linked) **locus**. Hence a given allotype is associated with all, or with approximately half, of the molecules of the appropriate Ig class, depending on whether the individual is homozygous or heterozygous at the appropriate locus. Fragments and isolated chains from myeloma proteins have been particularly useful in localizing the allotypes to particular domains and, in some instances, to particular amino acid substitutions (Table 17-3). *The chain specified by one al-*

---

* Most human Abs to IgG allotypes appear to have low affinity, and transfusions of blood with foreign allotypes of this class rarely cause reactions (Chs. 21 and 23), even in recipients with high titers of the corresponding antiallotype. Severe reactions have occurred, however, when the discordant allotype is IgA, some of which is polymeric and may form more stable complexes with its antiallotype (see Am Allotypes, this Ch., and Multivalent Binding, Ch. 16).

† Alleles in a heterozygous diploid individual are **codominant** when both are expressed more or less equally.

+ test Ig (same allotype
as the indicator)
+ human Abs to the
indicator's allotype

no agglutination (inhibited
by test Ig)

Human
red blood
cells

indicator
human
Ig

Sensitized red
cells, coated
with indicator

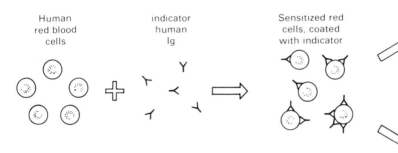

+ test Ig (different allotype than the indicator)
+ human Abs to the
indicator's allotype

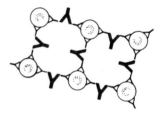

agglutination

FIG. 17-18. Agglutination-inhibition assay for human allotypes. In the assay for Gm and Km allotypes the indicator Ig is nonagglutinating Ab specifically bound to RBCs (anti-Rh on Rh⁺ RBCs; Ch. 23). In the assay for some other allotypes the indicator Ig is adsorbed non-specifically on $CrCl_3$-treated RBCs. (Based on Grubb R, Laurell AB: Acta Pathol Microbiol Scand 39:195,390, 1956)

lele can have several allotypic determinants, as expected from the multiplicity of antigenic determinants on individual protein chains (Ch. 16). Thus some IgG1 molecules are Gm(1, 17): their γ1 chains have the Gm(1) marker in the C_H3 domain (Fc fragment position 355–358) and the Gm(17) marker in the C_H1 domain (Fd piece; position 214; see Table 17-3).

Allotypes are inherited as mendelian alleles. For example, a $Gm^{1,17}$ homozygous mother [whose γ1 chains and IgG1 molecules are all Gm(1,17)] and a Gm^4 homozygous father [whose γ1 chains and IgG1 molecules are all Gm(4)] invariably have heterozygous children, $Gm^{1,17}/Gm^4$, in whom about half the IgG1 molecules are Gm(4) and the others are Gm(1,17). The Gm allotypes are of considerable value in studying population genetics and in anthropology. In cases of disputed paternity they

are accepted as legal evidence in some countries (Norway, France).

Nonmarkers. Some H-chain markers that occur as allelic variants in one γ class are present in all H chains of some other γ classes. These so-called **"nonmarkers"** are sometimes designated by a minus sign to the **left** of the number. For instance, Gm(−1) is an allelic form of Gm(1) in γ1 chains (see Table 17-3 for the corresponding amino acid substitutions), but *all* γ2 and γ3 chains are Gm(−1). On the other hand, γ4 chains lack both Gm(1) and Gm(−1); instead, they come in two allelic variants, Gm(4a) and Gm(4b), which are also nonmarkers in the sense that all γ1 and γ3 chains are Gm(4a) and all γ2 chains are Gm(4b). Because normal serum contains all Ig classes, it is necessary in testing for nonmarker allotypes to assay purified Igs of the appropriate class.

It has been speculated that nonmarkers trace the evolution of γ chains. Thus an ancient γ4a allele might have given rise by dupli-

TABLE 17-3. Some Representative Human Gm Allotypes

Class	Chain	Gm allotype*	Amino acids responsible for serologic specificity of allotype†	Comments
IgG1	γ1	1	355 358 ...Arg-Asp-Glu-Leu...	Gm(1) and Gm(17) are usually on the
		−1	...Arg-Glu-Glu-Met...	same γ1 allele. Gm(−1) and Gm(4) are encoded by another γ1 allele.
		17	Lys(214)‡	Gm(1) and Gm(−1) are on Fc
		4	Arg(214)	fragments; Gm(17) and Gm(4) are on Fab fragments.
IgG2	γ2	23	Fc	Antiserum for the allelic product [Gm(23−)] is not available
IgG3	γ3	11	Phe(436)	
		−11	Tyr(436)	
		21	Tyr(296)	
		−21	Phe(296)	
IgG4	γ4	4a	308 ...Val-Leu-His...	Like Gm(−1), Gm(−11), and Gm(−21),
		4b	...Val——-His...	Gm(4a) and Gm(4b) are nonmarkers

*A minus sign to the left of a number means a nonmarker, i.e., present on all chains of some other classes (to the right it would mean the marker is absent).

†Numbers above amino acids, or in parentheses, refer to positions in the sequence, counting from the chain's amino-terminus.

‡A conformation-dependent determinant in the Fd region; to be recognized in serologic reactions, the Fd piece or H chain must be associated with the L chain.

(Based on Grubb R: Genetic Markers of Human Immunoglobulins, NY, Springer, 1970; Natvig JB, Kunkel HG: Advan Immunol 16:1 1973; Steinberg AG: Ann Rev Genetics 3:25, 1969)

cation to genes that evolved into those of present-day γ1 and γ3, and an ancient γ4b allele might have similarly given rise to γ2; through subsequent mutation-selection the duplicated genes would have eventually evolved their own allelic variants while retaining the imprint of the γ4 allele from which they originated.

The alleles for Gm markers are linked in complexes (**allogroups**) that differ conspicuously in diverse racial groups (Table 17-4). Each of these complexes behaves as

TABLE 17-4. Common Gene Complexes in Different Human Populations

Racial group	Allogroups*	IgG1	IgG2	IgG3	IgG4
Caucasoid	1	1,17	23−	21	4a
	2	−1,4	23	5,−21	4b
	3	−1,4	23−	5,−21	4a
	4	1,2,17	23−	21,−5	4a
Negroid	5	1,17	23−	5,11	
Mongoloid	6	1,4	23	5,11	4a
	7	1,17	23−	21	
	8	1,17	23−	11	

*The alleles of each allogroup are aligned in horizontal rows; e.g., allogroup 1 has allotypic specificities 1,17(γ1), 21(γ3), and 4a (γ4) and lacks allotype specificity 23 (i.e., it is 23− for γ2). For the significance of minus signs to the left or right of a number, see Table 17-3.

(Based on Steinberg AG: Ann Rev Genet 3:25, 1969; Grubb R: Genetic Markers of Human Immunoglobulins, NY Springer, 1970; Natvig JB, Kunkel HG: Adv Immunol 16:1, 1973)

a stable heritable unit. For example, a person who is heterozygous for γ1 and γ3 genes and homozygous for γ2, with, say, γ1:1, 17, 4; γ2:23; γ3: 21, 5, would have inherited $Gm^{1,17}$, Gm^{23}, and Gm^{21} as linked alleles (for γ1, γ2, and γ3, respectively) from one parent and Gm^4, Gm^{23}, and Gm^5 from the other parent (Table 17-4).

Surveys of large populations suggest that the crossover frequency between genes for γ1, γ2, and γ3 is extremely low; indeed, these genes probably represent the most closely linked human genes recognized so far. However, a few pedigrees and rare "hybrid" chains provide evidence for recombination and suggest the gene order shown in Figure 17-19.

One of these chains was detected in IgG molecules with a γ3 allotype in the Fab fragment (C_H1 domain) and a γ1 allotype in the Fc fragment (C_H2 or C_H3 domain). Another was found to have a γ4 allotype in the Fab fragment and a γ2 allotype in the Fc fragment. These chains probably resulted from recombination; they suggest that the gene for γ3 is next to the gene for γ1 and that the genes for γ4 and γ2 are also near neighbors. The order γ4-γ2-γ3-γ1 (Fig. 17-19) has been proposed because of the especially pronounced homology between γ2 and γ3, which suggests that one was derived from the other by gene duplication (see Evolution of Immunoglobulins, below). The recombinant chains are called **Lepore-type hybrids**, after a rare hemoglobin chain that presumably arose from a similar intragenic recombination.

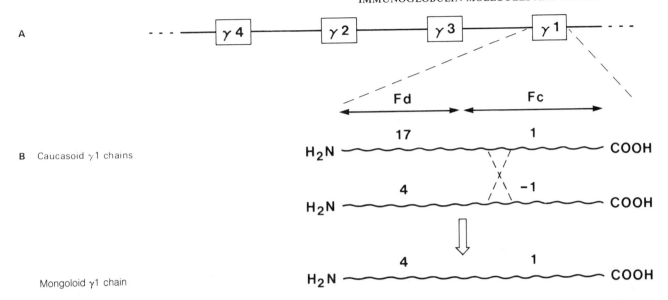

FIG. 17-19. A. Possible order of linked genes for γ chains of human Ig. **B.** Alignment of allotypic markers in common γ1 chains. Recombination **(dashed lines)** between alleles for the caucasoid chains could account for the combination of markers in some mongoloid γ1 chains. (Based on Natvig JB et al: Cold Spring Harbor Symp Quant Biol 32:173, 1967)

Am Allotypes. Genetic variants of IgA were discovered through reactions to blood transfusions: recipients who lack an IgA variant can suffer severe reactions when transfused more than once with blood that contains it. (The first transfusion induces formation of the antiallotype, which reacts with that allotype in subsequent transfusions.) With these human antiallotypes and the hemagglutination assay described above (Fig. 17-18), two heritable variants of α2 chains have been detected in IgA2; by analogy with Gm allotypes of IgG they are called A2m(1) and A2m(2).

Km Allotypes. Three allelic variants of L (κ) chains are distinguished by agglutinating human sera. The differences are due to amino acid substitutions at two positions, 153 and 191, in the C domain (Table 17-5). In the three-dimensional Ig structure (see below) these positions are close together; they probably constitute a single antigenic site for anti-Km Abs.

TABLE 17-5. Amino Acid Substitutions Corresponding to Km Allotypes in Human κ Chains

Serologic marker*	Amino acid at position	
	153	191
Km(1)	Val	Leu
Km(1,2)	Ala	Leu
Km(3)	Ala	Val

*Km was previously called InV (because Igs with these allotypes inhibited the reactions of a typing serum from patient V). Km(2) is always associated with Km(1), but some individuals are Km(1^+2^-).

Serologic recognition of Km markers depends upon Ig conformation, and not only on sequence. When separated from H chains, L chains lose their reactivity with anti-Km Abs; and the same Km markers are more reactive with anti-Km serum when the L chains are combined with γ1 or γ3 than with γ2 or γ4 chains.

Rare individuals are Km(−1,−2,−3), suggesting additional Km alleles. The geographic and ethnic distribution is as uneven for Km as for Gm alleles; e.g., Km(1,2) is present in 10%–20% of Europeans and in >90% of Venezuelan Indians.

Allotypic variants of human λ light chains have not yet been definitely identified.

RABBIT ALLOTYPES

Oudin in France discovered genetic variants of rabbit Ig chains at the same time that human allotypes were discovered in Sweden by Grubb's chance observation. Though the principles are the same, studies of rabbit allotypes contribute additional important insights.

Rabbit allotypes were first recognized by precipitin reactions with alloantiserum from rabbits immunized with Abs of other rabbits.* Examination of rabbit populations with such antisera eventually disclosed many sets of allotypes. In addition to allotypic variants of the C segments of all H and L chains, an important set, the *a* allotypes, are localized in the V_H domains. Thus anti-*a* serum reacts with Fab fragments, with isolated H chains, and even

* Unlike the sensitive hemagglutination assay for human allotypes (Fig. 17-18), precipitin reactions require high titers of Ab (see Table 16-5, Ch. 16). Hence the rabbit alloantisera are elicited by intensive immunization (e.g., injecting rabbit Igs in aggregated form and in adjurants; see Adjuvants, Ch. 19).

with an unusual fragment that appears to consist of just the V_H domain of an IgG protein.

The *a* allotypes were, surprisingly, found by Todd on H chains of *all* classes (γ, α, μ, etc.), and this finding first prompted serious consideration for the then radical view that two genes (one for V and the other for C segments) join to specify a single Ig chain. As we shall see later, there is now clear chemical evidence that separate sequences in DNA encode V and C domains of Igs (Ig genes, below).

About 75% of rabbit IgG molecules have *a* allotypes; by immunizing rabbits with *a*-deficient IgGs from other rabbits, antisera to two other V_H domain markers have been found (*x*32, *y*33). The *x* and *y* markers also are associated with all H chain classes, but appear not to be allelic with each other or with the *a* variants: *a*, *x*, and *y* are probably separate genetic loci for what are defined below as V_H groups (see Variable Domains, below). Breeding studies show close linkage of *all* the genes that specify allotypic markers on H chains—for C_H domains (γ, α1, α2, μ) as well as for V_H domains (*a*, *x* and *y*).

As with the H chain linkage groups in man, alleles for diverse rabbit H chains are inherited as genetically stable complexes or **allogroups**, which are only rarely separated by recombination. However, *gene loci for the H, κ, and λ chains all segregate independently, as though located on three different chromosomes.*

Rabbit allotypes were instrumental in establishing the identity of the H chains and of the L chains in any given molecule. For example, successive additions of antisera to κ chain allotypes, say *b*4 and then *b*5, precipitate 50% and 30%, respectively, of the Ig molecules in serum from a heterozygous *b*4,*b*5 rabbit. When the order of additions is reversed (anti-*b*5, then anti-*b*4) the amounts precipitated by each antiallotype are unaltered. Thus anti-*b*4 does not precipitate molecules with the *b*5 marker and anti-*b*5 does not precipitate those carrying *b*4 (Fig. 17-20).

Similar findings with pairs of antisera to *a* allotypes, for H chains, and to *c* allotypes, for λ chains, established that despite the great diversity of Ig chains synthesized in an individual, any particular Ig molecule has identical H chains and identical L chains (Fig. 17-20). The reason is that Ig-producing cells synthesize, at any one time, only one kind of H chain and one kind of L chain, even in individuals who are heterozygous at Ig loci. This means that an Ig-producing cell expresses only one of the two alleles at a given locus for an Ig chain (**allelic exclusion**). This exclusivity contrasts with the general rule that the diploid cell with two allelic variants expresses both, e.g.,

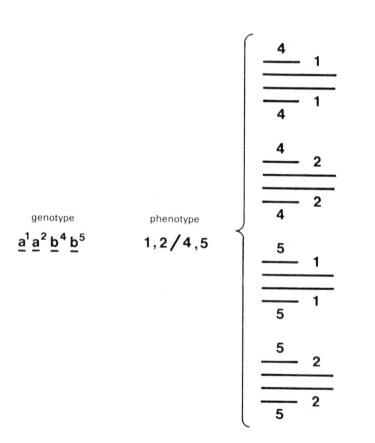

genotype

$\underline{a}^1\underline{a}^2\underline{b}^4\underline{b}^5$

phenotype

1,2 / 4,5

Heterozygous at *a* and *b* loci

FIG. 17-20. Some allotypes of rabbit IgG showing the diversity of Ig molecules in a rabbit who is heterozygous at H and L chain loci. In spite of the diversity, in any individual Ig molecule the L chains **(short bars)** are identical and so are the H chains **(long bars)**. (Based on Dray S, Nisonoff A: Proc Soc Exp Biol Med 113:20, 1963)

individual red blood cells with sickle and normal hemo-globin chains, or with blood groups A and B (Ch. 23).

Simple versus Complex Allotypes. Simple allotypes are associated with a few (one or two) amino acid substitutions per domain of about 110 residues, e.g., *d*12 and *d*11, with threonine and methionine, respectively, at position 225 in rabbit γ chains. However, some rabbit allotypes are **complex**: they differ from each other by many substitutions, e.g., 15 out of 100 in *a* allotypes on H chains and as many or more in the *b* and *c* allotypes of κ and λ chains.

While simple allotypes are inherited as dominant alleles, some observations suggest that complex allotypes may be **pseudoalleles**; e.g., a rare rabbit can have more than two complex allotypic variants of a particular L chain. Hence it is probable that complex allotypes are actually **isotypes** (i.e., they are specified by structural genes that are present in *all* individuals of the species) and that their expression may be controlled by different allelic variants of a regulatory gene.

MOUSE ALLOTYPES

Inbred mouse strains simplify the recognition of allotypes. All members of a given strain are virtually identical genetically, and many strains differ in Ig allotypes. Hence monospecific Abs are readily produced by immunizing mice of one strain with allotype-different Igs from another strain (**alloantisera**). The assignment of allotypic determinants to Ig classes, and their localization within chains, has been greatly aided by the availability of many myeloma proteins from certain strains (particularly BALB/c). Some mouse allotypes are listed in Table 17-7 (see Appendix, this Ch.).

A uniform set of terms for mouse Ig classes and their genetic loci has not yet been adopted. Usually, however, the 7S Ig classes are called IgG1, IgG2a, and IgG2b (a fourth class, present at trace levels, is IgG3), without any intent to imply homology with human Ig classes having the same or similar designations.

Nearly all of the alloantisera against mouse Igs are specific for determinants in the Fc fragments (C_H2 or C_H3 domains), but a few have been raised against Fab fragments and appeared to be specific for determinants in C_H1. So far, in mouse Igs, unlike the situation in rabbits, no allotypic determinants have been recognized in V_H or in L chains. (However, heritable κ L chain variants have been detected electrophoretically and as recurrent small differences in amino acid sequence near the amino-terminus; see Variable segments: Framework and Hypervariable Regions, below.)

Inbred strains of mice are homozygous at all loci (see Congenic Mice, Ch. 18) and each strain has only one kind of allele for the C segment of each H chain class. The set of linked H chain alleles in a given chromosome is an **allogroup** (or **phenogroup**, or **heavy-chain linkage group**).

No recombinants have been detected in thousands of progeny of crosses and backcrosses between inbred strains with different H chain allotypes. Thus genes for Cγ1, Cγ2a, Cγ2b and Cα segments appear to be as closely linked in the mouse as are the various H chain alleles in man and in rabbit.

IDIOTYPES: UNIQUE DETERMINANTS OF INDIVIDUAL IMMUNOGLOBULINS

Unlike any particular isotype or allotype, which are found on many Igs, a particular idiotype (Id) is restricted to the Ig molecules made by one or a few clones of Ig-producing cells. Ids were discovered through studies of rabbit Abs and human myeloma proteins. Oudin injected anti-*Salmonella* Abs from one rabbit (donor) into recipient rabbits with the same allotypes. Some recipients formed Abs (antiidiotypes) that reacted specifically with the immunogen, but not with other Igs from the donor or with anti-*Salmonella* Abs from other rabbits. Basically similar observations were made in man by Kunkel and colleagues, but with heterologous antisera: a rabbit antiserum to one human myeloma protein was absorbed with other myeloma proteins having the same H- and L-chain isotypes and allotypes, and the residual Abs were then specific for the "individually unique" antigenic determinants (i.e., the Id) of the immunogen.

Ids of some relatively homogeneous human Abs (e.g., to dextrans) have also been identified by rabbit antisera formed against the purified Abs from one person and absorbed with other Igs from the same person.

Each Ig molecule has a **unique (or "private") Id**, shared by molecules with identical amino acid sequences in the V segment and produced by cells of the same clone. Some Abs also have **shared (or "public") Ids**, which are shared by many of the Abs elicited by a given Ag in all individuals of the same genetic background (e.g., an inbred mouse strain). One shared Id, for instance, is present on about 15% of the Abs made against benzenearsonate in A/J mice; another is present on virtually all Abs made in BALB/c mice to phosphorylcholine. About one-third of the 20–30 Ags tested in inbred mice elicit Abs with shared Ids.

All Ids are located in V segments. Many Ids, especially private ones, must be in or extremely close to the ligand-binding site, for haptens can specifically and competitively block their binding of antiidiotypic Abs (Fig. 17-21): the higher the hapten's affinity the more effectively it inhibits.

The occasional failure of haptens to block private Ids can have several explanations. 1) Hapten and antiidiotype could bind to different zones within a large combining site of the antihapten Ab. 2) The antiidiotype could have a much greater affinity for the site than the hapten has.

Because ligand-binding sites are formed by a precise combination of V segments of H and L chains (see Combining Sites, below), isolated chains do not react with Abs to private Ids. Recombined

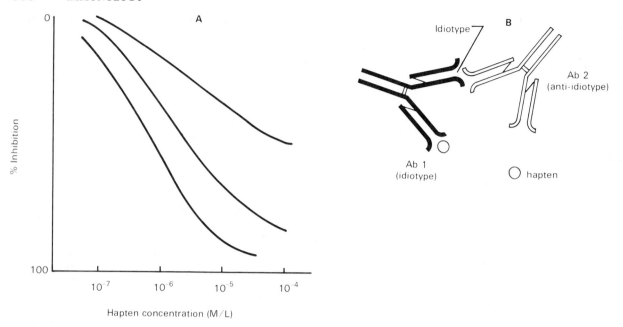

FIG. 17-21. A. Specific inhibition of the idiotype/antiidiotype reaction. The reaction of an antiidiotypic Ab can be blocked specifically by ligands (such as haptens) that bind to the Ab with the idiotype. The 3 inhibition curves were obtained with 3 haptens having different affinity for the Id-bearing Ab: The hapten with highest affinity elicited the curve at the left, and the hapten with lowest affinity, the curve at the right. **B.** A hapten competing with an antiidiotypic Ab **(Ab no. 2)** for the binding site of the antihapten Ab **(Ab no. 1)**. (Based on Brient BW, Nisonoff A: J Exp Med 132:951, 1970)

chains (see Reconstruction of Combining Sites, below) do react, but only if they correctly reconstitute molecules in which the proper ligand-binding activity is regained.

Relation between Private and Public Idiotypes. This relation has been illuminated by sets of monoclonal anti-hapten Abs (see Hybridomas, above), each elicited in a different mouse of the same genetically uniform strain (e.g., against benzenearsonate in A/J mice). These monoclonal Abs have a shared (public) Id, and so they crossreact extensively with anti-Id antisera elicited against any one of them (or elicited against conventional polyclonal anti-benzenearsonate Abs of A/J mice). The amino acid sequences of their variable domains are also remarkably similar, though few, if any, have entirely identical sequences. Each of these Abs can be shown to be serologically unique, i.e., to have a private Id. Thus *Abs that share a particular public Id constitute a family of closely related but still unique Igs.*

VARIABLE SEGMENTS: FRAMEWORK AND HYPERVARIABLE (COMPLEMENTARITY-DETERMINING) REGIONS

Comparison of Ig chains of the same class reveals that they usually differ from each other at many positions in the 110-115 residues at the amino end, hence the term **V segment (or V domain).** The diversity is especially extreme in several **hypervariable regions** (Fig. 17-22). Because hypervariable regions determine, or actually form, the combining sites for ligands (see below), they are also referred to as **complementarity-determining regions (CDRs).** There are three of these regions in L chains, and

three in H chains; each consists of five to ten residues centered at positions 30–35, 50–55, and 95–100. (H chains may have an additional hypervariable region at 85–90.) When the V segments of various chains are aligned for comparison it is necessary to introduce gaps into some of them to achieve maximal sequence homology, and the gaps also tend to cluster in the hypervariable regions.

The less variable sequences on either side of the hypervariable regions (CDRs) are called **framework regions (FRs).** Thus each V region consists of four FRs and three CDRs (in the following order: FR1, CDR1, FR2, CDR2, FR3, CDR3, FR4). The relative invariance of certain framework positions distinguishes **three types of V sequences : V_H, associated with H chains; V_κ, associated with κ chains; and V_λ, associated with λ chains.**

V Groups. Variations within the framework regions of the three types further distinguish **groups**, each characterized by distinctive amino acids at positions dispersed throughout almost the entire length of the V region (Fig. 17-23).

Just how many framework differences between two V sequences are required to assign them to different groups is not precisely defined. The prevailing view is that V domains with more than three differences in the amino-terminal 1–25 positions belong to different groups. In mice, where about 95% of L chains are κ, there appear to be about 50 V_κ groups. For human Ig chains, fewer of which have

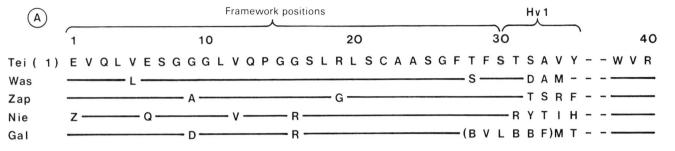

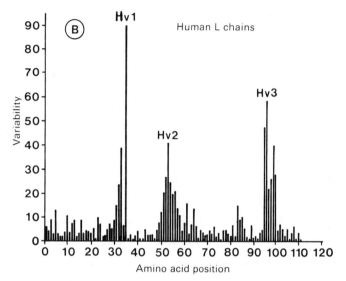

FIG. 17-22. Framework and hypervariable (complementarity determining) regions of V domains of human H chains. Sections with extreme variability are **hypervariable (Hv)**; other sections are **framework** regions. **A.** Sequences from parts of some human H chains. (Based on Capra D, Kehoe M: Adv Immunol 1975) (See Fig 17-12 for abbreviations). **B.** Variability (v) at different positions in human L chains. v is defined as the number (n) of different amino acids at a given position divided by the frequency (f) of the most common amino acid at that position ($v=n/f$). v can range from 1.0 (no variation) to 400 (20 amino acids). (Based on data compiled by "Prophet Information Handling System", National Institutes of Health, by H. Bilofsky, T.T. Wu, and E. A. Kabat, 1978)

been sequenced, fewer V groups have so far been discerned. Despite the lack of precision in their definition, the V groups figure prominently in theories on the genetic basis for the origin of Ig diversity (below).

Combining Sites of Antibodies. *These are formed by hypervariable regions* (Fig. 17-24). Thus, several Abs with the same ligand-binding specificities have virtually identical residues in hypervariable regions of their V_L domains; the same is true of their V_H domains. In contrast, some of these residues are rarely present in the corresponding positions of chains from randomly selected Igs.

Chemical studies by affinity labeling (below) established independently that ligand-binding sites are formed by hypervariable regions, and this conclusion has been elegantly confirmed and extended by x-ray diffraction studies of the three-dimensional structure of Fab fragments of myeloma proteins (see Three-Dimensional Structure, below).

AFFINITY LABELING OF COMBINING SITES

A number of chemical methods have been devised to identify the amino acid residues that form the Ab active site. In the method of **affinity labeling**, ligands with chemically reactive substituents serve as univalent haptens. In reacting with Ab, these reagents are first bound specifically in the active site by noncovalent bonds, and then form stable covalent bonds with susceptible amino acids in or close to the site. For example, Abs to the 2,4-dinitrophenyl (Dnp) group bind *m*-dinitrobenzenediazonium salts, which can form azo derivatives of tyrosine, histidine, or lysine residues. The derivatives are stable during subsequent separation and fragmentation of the chains, making it possible to establish the position of the labeled residue.

Because of their homogeneity, myeloma proteins are especially useful for affinity labeling, furnishing labeled

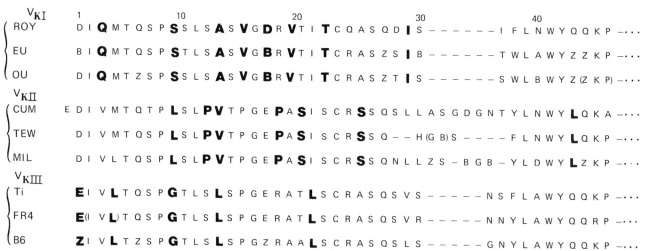

FIG. 17-23. Some V region groups (e.g., V$_{KI}$, V$_{KII}$, V$_{KIII}$, illustrated with the amino terminal sequences of nine human κ chains. The framework residues that characterize each group are in **boldface**. Alignment maximizes homology; **dashes** (gaps) represent deletions (or insertions). Residues in parentheses are known by peptide composition, and not actually sequenced. See Fig. 17-12 for amino acid abbreviations. (Data are from several laboratories, reviewed by Hood L, Prahl J: Adv Immunol 14:291, 1971; see Fig. 17-34 for probable genetic significance of a V group.)

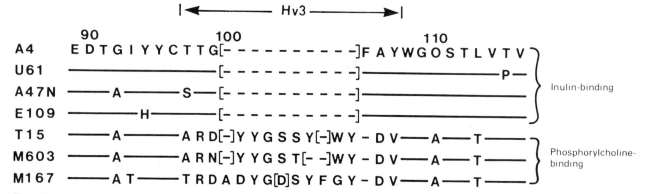

Fig. 17-24. Evidence that hypervariable regions determine the specificity of combining sites. Comparison of sequences in and around the third hypervariable (Hv) region of H chains from four inulin-binding and three phosphorylcholine-binding myeloma proteins. **Bracketed hyphens** represent deletions (compared to a prototype sequence). Horizontal lines refer to sequences that are the same as the topmost protein (A4). For amino acid abbreviations, see Fig. 17-12. (Vrana M et al: Proc Nat Acad Sci USA 75: 1957, 1978)

peptides in high yield. From studies of these proteins, and also of conventional, heterogeneous Abs, it is clear that amino acid residues of both heavy and light chains of a single Ig can be specifically labeled, and that the *labeled residues fall within the hypervariable regions of the V$_L$ and V$_H$ segments* (Fig. 17-25).

RECONSTRUCTION OF COMBINING SITES: CHAIN RECOMBINATION

Light Chain–Heavy Chain Interaction. Though interchain S–S bonds stabilize the Ig molecule, they are not indispensable for its characteristic structure. When H and L chains are separated, and the SH groups that constituted their S–S links are chemically blocked (e.g., with iodoacetate), the chains can reassociate through noncovalent bonds, spontaneously forming four-chain molecules that closely resemble native (H$_2$L$_2$) Abs in physical, chemical, and antigenic properties; they bind ligands with the same specificity and affinity and even yield Fab fragments when treated with papain.* Chains derived from different Igs, and even from different species, e.g., human L and rabbit H chain, also pair and

---

* In the commonplace IgA myeloma proteins of mice (BALB/c strain) L and H chains are normally associated noncovalently in stable H$_2$L$_2$ molecules (and also in dimers [H$_4$L$_4$] and higher multimers) of IgA.

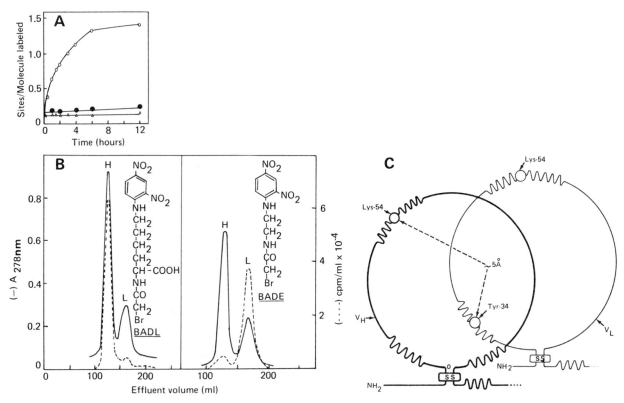

FIG. 17-25. Affinity-labeling of a mouse myeloma protein with antihapten activity (anti-Dnp). **A.** Rate of labeling the combining sites of the myeloma protein (○) with a bromoacetyl-Dnp reagent (BADE; structure shown in right panel of **B**). The presence of excess ϵ-Dnp-L-lysine specifically blocked the binding and reaction of BADE (●) ("hapten protection"). In a control test BADE did not label nonspecific mouse Ig (△). **B.** Separation of H and L chains of the myeloma protein after reaction with ^{14}C-BADL, which labeled only the H chain, or with ^{14}C-BADE, which labeled only the L chain. **C** Localization of labeled residues in hypervariable regions **(jagged lines)** of V_L **(light lines)** and V_H **(heavy lines)** domains. A bifunctional Dnp reagent with bromoacetyl groups at two positions (about 0.5 nm apart), one resembling its position in BADL and the other its position in BADE, labeled **both** Lys-54 on V_H and Tyr-34 on V_L, crosslinking H and L chains **(dashed lines)**. (Based largely on Haimovich J et al: Biochemistry 11:2389, 1972)

form H_2L_2 molecules. Nonetheless when a mixture of L chains from diverse Igs compete for a limiting amount of H chain from one Ig, the homologous pair (derived from the same Ig molecule) usually recombine more readily with each other than with heterologous chains (from different Igs). The mutual affinity of L and H chains from diverse Igs probably derives from invariant sequences in C domains, whereas the preferential binding of homologous chains probably is due to additional interactions between residues in V_H and V_L domains.

Indeed, an Ig cleaved by pepsin between V_L and C_L and between V_H and C_H domains yields a stable fragment (called **Fv**) that consists of V_L linked to V_H entirely by noncovalent bonds; this small fragment retains all the ligand-binding activity of the intact Ig molecule from which it is derived. Nevertheless, it is clear that many L chains pair equally well with that many H chains (and vice versa).* Hence the number of different Ig molecules that can be

* In some hybridoma clones (Fig. 17-4), made by fusing a κ-producing myeloma cell with a normal cell that produces a λ and a

assembled from *l* light and *h* heavy chains could be $l \times h$, if all chain combinations formed stable molecules (see Origin of Diversity, below).

Activity of Recombined Chains. The H chains isolated from an Ab can bind the corresponding ligand, though much less strongly than the original molecule. The isolated L chains bind far less well. Nevertheless, the L chains must make a profound and specific contribution to the combining site, because 7S molecules reconstituted from the chains of a given Ab have much greater affinity for the ligand than those formed from the same H chain and other L chains.

The H_2L_2 molecules reconstituted from the chains of a homogeneous Ab are virtually as active as the native

γ (heavy) chain, some of the secreted IgG have, in the same molecule, one κ-chain and one λ-chain, both associated with a pair of identical γ-chains. These "mixed" κ-λ molecules are entirely stable.

molecule. However, when reconstitution is carried out with chains from more heterogeneous, conventional Ab preparations the reconstituted molecules have far less activity for the test Ag than the native ones, evidently because many novel H–L pairs are formed. Though these lack appreciable affinity for the original ligand they doubtless possess their own unique combining sites, specific for other (unknown) ligands.

THREE-DIMENSIONAL STRUCTURE

The view of the Ab binding site drawn from amino acid sequences and from affinity-labeling studies has been greatly extended by x-ray diffraction analyses of Ig crystals (Fab fragments, L-chain dimers, and a whole IgG molecule). In addition, amino acid sequence homologies between different domains have turned out, as predicted, to reflect a common three-dimensional structure in all V and C domains.

Each domain is essentially a compact globular unit, approximately 3.5 nm in diameter, made up of seven extended polypeptide stretches (in so-called β-structure), three in one plane and four in another, forming a characteristic sandwichlike, or barrellike, structure, with adjacent chains of the same plane in antiparallel orientation (Fig. 17-26). A key difference between V and C domains consists of the hypervariable regions, which are brought together from different parts of the V domain.

In Fab fragments the hypervariable sections of V_L and V_H domains are clustered to form a pocketlike or cleftlike ligand-binding site at the tip of the Fab (Fig. 17-26). In two of the crystals subjected to high-resolution x-ray diffraction analysis the bound ligand has actually been visualized within the combining site.

As predicted by hapten-binding studies, the binding sites differ in size and shape in different molecules. In one crystal the site is a deep cleft, 2 nm × 1.5 nm across and 1.2 nm deep; in another it is a shallow groove, 1.6 nm × 1.7 nm at the surface and 0.6 nm deep; in a third it is shaped like a truncated cone, 0.7 nm deep, 1.5 nm wide at the surface, and 1 nm wide at the base. Hydrophobic amino acids are abundant in the binding sites, especially at their depths.

The angle between V_H and C_H1 domains (Fig. 17-14) is 135° in the crystals of a Fab fragment and 170° in those of the intact IgG molecule from which the fragment was derived. This result demonstrates **flexibility** at the V_H–C_H1 junction. Other differences between Fab and

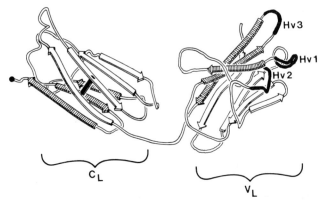

FIG. 17-26. Three-dimensional structure of V and C domains of a human λ chain (McG). **Thick arrows** indicate direction of chain polarity (NH$_2$→COOH). **White arrows** represent twisted "β-stretches" in one plane, **hatched arrows** in the other plane. Each short **black bar** is the intradomain S–S bond (cf. Fig. 17-13). Hypervariable (Hv) regions are brought together at one end of the molecule. In intact Ig molecules and in Fab fragments apposition of Hv sections of V_L and V_H form the ligand-binding site. (Based on Schiffer M et al: Biochemistry 12:4620, 1973)

the intact molecule demonstrate flexibility at the **hinge**, between C_H1 and C_H2. With these flexible interdomain joints, a "domino effect" could account for conformational changes in C_H2 or C_H3 domains induced by binding of ligand in V domains. Such induced conformational changes could explain the formation in Ab–Ag complexes of Ig sites for binding complement (Ch. 20) or to receptors for Fc domains on lymphocytes and macrophages (Ch. 18).

With the formation of combining sites by approximately 15 hypervariable positions in V_L and a like number in V_H, an enormous number of variations in the shape and specificity of the site becomes possible. For instance, if 12 different amino acids could occupy each of the 30 positions ($15V_H + 15V_L$) there would be 12^{30}, or about 10^{36}, variations. *It is evident how an almost limitless variety of Ag-binding specificities and affinities and idiotypes can exist in Ig molecules that are otherwise monotonously alike in overall structure.* Whether the genetic information for all these variants is carried in germ-line DNA (i.e., in gametes), or is generated during development of somatic cells is considered below (Origin of Diversity of V Genes). First, however, we shall consider what is known about Ig genes.

IMMUNOGLOBULIN GENES

TWO GENES—ONE CHAIN

In different κ chains the same C segment is linked to diverse V segment amino acid sequences. Hence, Dreyer and Bennett suggested that various V genes can each join the same C gene to form a composite gene for a complete L chain. Many observations and arguments, a few outlined in the following paragraphs, have provided indirect support for this hypothesis.

1) For a single inherited $V\kappa$ gene to account for the many $V\kappa$ groups—about 50 in inbred mouse strains, based on amino acid sequences (Fig. 17-23)—a highly improbable number of parallel and independent somatic mutations would be required; hence it appears that an individual inherits many $V\kappa$ genes (probably several hundred; see below). In contrast, the C segments of κ chains of the mouse (and similarly of man) have a uniform amino acid sequence (except for minor allotypic variation), indicating that there is a single C gene in each species.

2) Certain myelomas synthesize two Igs, e.g., an IgM and IgA protein. In one such double-producer tumor an identical V_H amino acid sequence was associated with $C\mu$ in the μ chains and with $C\alpha$ in the α chains.

IMMUNOGLOBULIN GENE CLUSTERS

There are stringent limitations to the associations between classes of V and C amino acid sequences: V_κ sequences are always associated with C_κ, V_λ sequences with C_λ, and V_H sequences with the various C_H classes ($\gamma1$, $\gamma2$, μ, α, etc.).

Breeding experiments, with allotypes as markers for C genes, show that the *genetic loci for C segments of H, κ, and λ chains are unlinked* (see Rabbit Allotypes, above). However, each C locus is linked to a set of V genes (see Mapping V_H Genes, below). Thus *there are three independently segregating Ig gene clusters: one V-C set for H chains, another for λ chains, and a third for κ chains.*

GENE REARRANGEMENT

Nucleic acid analyses have now shown directly that there are many V genes and only one (or a few) C genes for most classes and that certain V and C genes become closely linked as a cell differentiates into an Ig producer. Several technical developments made this direct demonstration possible. 1) **Purified mRNA** for Ig chains can be isolated from myeloma tumor cells. 2) The use of radiolabeled mRNA as **probes**, specifically binding (**hybridizing**) to complementary deoxynucleotide sequences, permits detection of the corresponding genes in DNA fragments. 3) Bacterial restriction endonucleases cleave cellular DNA at specific sites, reproducibly yielding precisely defined **"restriction" fragments.** 4) Recombinant DNA technology permits the isolation, mass production and sequencing of genes.

Applying these technics, Tonegawa et al. discovered that the deoxynucleotide sequences encoding the V and C regions of a λ L chain were much closer to each other in DNA from the cells (of a mouse myeloma tumor) that were making that chain than in DNA from other cells (e.g., mouse embryo). Thus, with an intact mRNA molecule for this λ chain (corresponding to both the V and C domains) and with a fragment of that mRNA that corresponded just to the C domain it was shown the encoding sequences for V λ and C λ were present in separate fragments of DNA (cleaved with restriction enzymes) from the embryo cells but in the same fragment of DNA from the λ-producing myeloma cells

FIG. 17-27. The deoxynucleotide sequences that encode the V and C regions of a $\lambda1$ L chain are closer together in DNA of the cells that produce this chain than in DNA of other cells. The $\lambda1$ chain was made by mouse myeloma tumor A. DNA from this tumor **(A)**, from another myeloma tumor **(B)**, producing a different Ig, and from a 13-day mouse embryo **(C)** were each digested with endonuclease and electrophoresed in agar. The separated fragments, blotted onto a nitrocellulose filter (**Southern blots**) were allowed to bind ("hybridize") a radioactive cDNA "probe" that corresponded to the mRNA for the entire $\lambda1$ chain ($V_{\lambda1} + C_{\lambda1}$). In DNA from the embryo **(C)** and from the myeloma producing another Ig **(B)**, three binding fragments were recognized: on the basis of deoxynucleotide sequences, one was shown to encode $C_{\lambda1}$, another to encode $V_{\lambda1}$, and a third to encode $V_{\lambda2}$, the V domain of another L chain, called $\lambda2$, that closely resembles (and cross-hybridizes with) $\lambda1$. However, *in the myeloma that produced the $\lambda1$ chain there was also a fourth fragment that encodes two sequences: $V_{\lambda1}$ and $C_{\lambda1}$.* Though close together in the same fragment of DNA from the cell that makes the $\lambda1$ chain, the $V_{\lambda1}$ and $C_{\lambda1}$ encoding sequences are still separated in this fragment by an intervening sequence (intron) of 1200 bases. (Brack C et al: Cell 15:1, 1978)

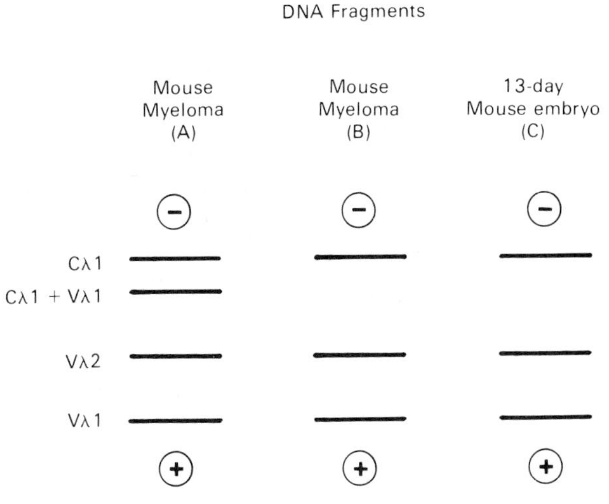

DNA Fragments

(Fig. 17-27). Similar evidence for a rearrangement of V and C genes has been observed in myeloma cells that make κ chains and H chains.

This discovery has consequences that extend far beyond immunology. Previously, it was thought that all cells in the body, regardless of type, have exactly the same genome (except for rare mutations) and that different cells expressed different genes. Instead, it is now evident that there is genomic diversity of an individual's cells: his B cells differ from each other, in accord with the DNA sequences that are eliminated in the rearrangements of different V and C encoding sequences. It is likely that similar rearrangements control the expression of other genes and that the genomic diversity of B lymphocytes, which make up about 5% of the cells in the body, applies to some other cell types as well.

In DNA from cells that make a particular Ig chain another copy of the encoding sequence for that chain's C region can occupy the same position as is found in embryo DNA, or this sequence may undergo another (abortive) rearrangement in which it is shifted from its position in embryo DNA but is still not associated with a V sequence. These observations suggest an explanation for the principle of allelic exclusion (see Rabbit Allotypes, this chapter): *V and C genes are expressed only in the chromosome in which they have been brought together but not in the homologous chromosome, in which they remain widely separated.*

IMMUNOGLOBULIN GENES ARE SPLIT

Deoxynucleotide sequences, determined in cloned DNA, reveal how the encoding sequences for Ig chains are distributed and suggest how these sequences become rearranged. The cloned DNA is prepared by the usual procedures of recombinant DNA technology. For instance, restriction fragments of cellular DNA are inserted at random into the DNA of bacteriophage, which then infects *E. coli.* Alternatively, cDNA copies of mRNA are inserted into the phage. Of the resulting hundreds of thousands of phage plaques, many containing one bit of mouse DNA, the few plaques that contain the sequences of interest are identified by their ability to bind radioactive mRNA (or its DNA copy) for the appropriate Ig chain. These phages are propagated in mass culture, and the mouse DNA segment is excised and its deoxynucleotide sequence determined.

Sequences for several cloned fragments show that genes for λ, κ, and H chains have essentially the same arrangement. The first established sequences, for a λ L chain, are illustrative (Fig. 17-28). In cells *not* making that chain (e.g., embryo cells) there are three separate encoding sequences for it: one for most of the Vλ region (amino acid positions 1 to about 97), another for the Jλ (J for junction) segment that encodes the rest of the V region's amino acids (98–110), and a third that encodes the entire C region (Cλ, 111–214). Jλ and Cλ are separated by an untranslated intervening sequence of 1200 nucleotide (1.2 kilobases or Kb). The distance between Jλ and Vλ is not defined but is probably much greater.

However, *in the cells that produce that λ chain the Vλ and Jλ sequences are joined to form a continuous sequence for the entire V domain;* but Jλ and Cλ remain separated by the same intervening sequence.

Electron microscopy demonstrates vividly the intervening sequences that interrupt Ig genes. As is shown in Fig. 17-29, when mRNA for a κ chain is annealed to a fragment of DNA from cells that produce that chain the mRNA binds to the V and C encoding sequences and the intervening sequence is revealed by a prominent loop of DNA: the length of the loop corresponds exactly to the number of deoxynucleotides found, by sequencing, in the intervening sequence.

Thus, *in the development of an Ig-producing cell (B lymphocyte, Ch. 19) from its precursor, several rearrangements take place:* J_L joins V_L to form an active L chain gene (κ or λ) and J_H joins V_H to form an active H chain gene. However, intervening sequences still interrupt the otherwise continuous, mature Ig gene, e.g., *a large intervening sequence lies between the J and C encoding sequences* (another is in the "leader" sequence, see Figs. 17-28 and 17-30).

Exons and Introns. In split genes the deoxynucleotide sequences that are translated into amino acid sequences are called **exons**; the intervening untranslated sequences are called **introns**.

Transcription of such a split gene does not immediately yield the messenger RNA (mRNA) that is ultimately translated into the protein chain. Instead, *the primary RNA transcript must be converted, by excision of introns and splicing together of exons, into a mature mRNA molecule* whose ribonucleotide sequence is precisely colinear with, and translated into, the complete Ig chain (Fig. 17-30). It is likely that special nucleotide sequences at intron-exon boundaries provide signals for splicing enzymes.

Exons Correspond to Domains. There is a single encoding sequence for the C segment of each type of L chain. However, for the longer C segment of each class of H chain the encoding sequence is split into exons by several introns; and deoxynucleotide sequences show that *each of these exons corresponds to a particular C_H domain* (as defined by amino acid sequences (Fig. 17-13) and by x-ray crystallographic studies (Fig. 17-26)). For instance, *the C segment of γ1 heavy chains is encoded by four exons, one for each of the three $C_\gamma1$ domains (C_H1, C_H2, C_H3), and one for the hinge region* (Fig. 17-31).

J Segment. Each type of chain has its own J segments. There is one J for each Cλ, but for the one Cκ there are five different J_κ segments. The H family has four J_H

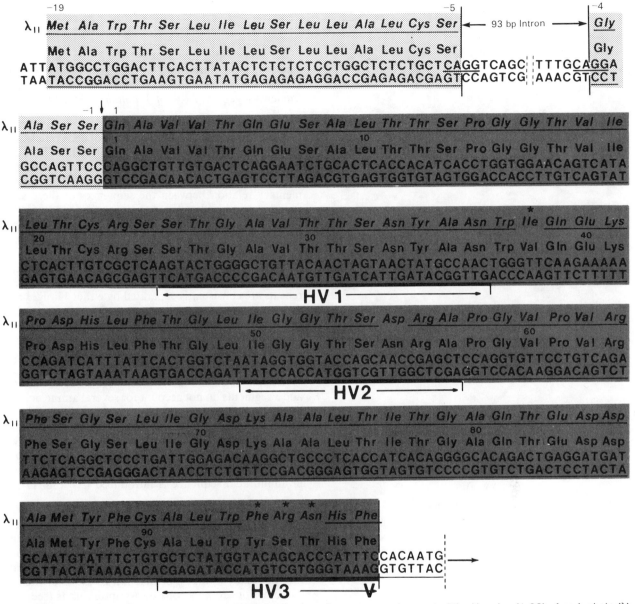

FIG. 17-28. Cloned fragment of mouse embryo DNA with the encoding sequence for most of the V region (1-98) of an L chain (λ). Encoding sequences are underlined (note the **bold line** beneath hypervariable regions); above them, in **regular type**, is the corresponding amino-acid sequence; and above that, in **italics**, is the amino-acid sequence of a λ L chain produced by a myeloma tumor. Note that there are only four differences (**asterisks**), indicating (because the V sequence shown occurs only once in embryonic [or germ line] DNA) that this germ line sequence and the corresponding sequence in a somatic (myeloma) cell differ at four codons.

The amino terminus of the mature λ chain is marked by a vertical arrow (at 1, Gln). The encoding sequence for the 19 additional ("leader") amino acids (found in newly synthesized chains), extending from the Met (−19) to Ser (−1) is interrupted by a 93 base pair untranslated intervening sequence or "intron" (most of which is not shown to save space) that falls between −5 and −4. Unlike V_λ, there are hundreds of encoding sequences for V_κ and also for V_H: so far, all of those examined have the same arrangement as is shown here for a Vλ. (Based on Tonegawa S et al: Proc. Nat Acad Sci 74:3518, 1978)

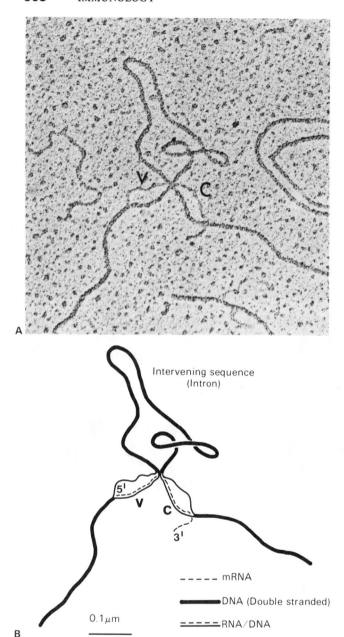

A

Intervening sequence
(Intron)

5'

V   C

3'

- - - - - mRNA

———— DNA (Double stranded)

═════ RNA/DNA

0.1 μm

B

FIG. 17-29. Electron micrograph **(upper)** showing the RNA-displacement loops (R-loops) formed by annealing mRNA for a κ chain to the corresponding encoding sequences in a piece of DNA derived from the myeloma cells that produced that chain. The diagram **(lower)** shows mRNA displacement loops for the V and C encoding sequences and the long (approximately 3000 base pair) double-stranded intervening DNA sequence between them. (Seidman JB & Leder P: Nature 276:790, 1978)

segments. Comparison of amino acid and deoxynucleotide sequences shows, for κ and H chains, that *various J segments can combine with various V encoding sequences of the same family.* Independent determinations of amino acid sequences show that *J segments correspond to part of the third hypervariable region and to the entire fourth framework region (FR4) of V domains* (Fig. 17-22).

V-J Recombination. Close inspection of deoxynucleotide sequences revealed certain unusual stretches, 7- and 9-residues long, in the intervening sequences between V and J segments of the λ, κ and H families. The complementary relation between the sequences of these short stretches would allow them to become aligned, via "inverted repeats" (see Fig. 17-35), in such a way as to bring together widely separated V and J segments, which can then be joined after excision of the length of DNA (>4,000 bases in κ chains) that previously separated them. These curious 7- and 9- residue complementary repeats are consistently separated by either 12 or 23 residue "spacers", of varying sequence. Since there are 10.4 residues per turn in the DNA helix *the spacers evidently correspond to 1 or 2 turns, suggesting that a "recombinase" enzyme probably bridges them in bringing together a V and J segment for the joining step.*

In some H chains the encoding sequences of their V and J segments do not account for several amino acids in the 3rd hypervariable region, suggesting still other separate encoding sequences, between V and J. Several short exons have been found in the predicted region: they are called *D segments* because it is likely that, in joining at one end to V and at the other to J, diversity in codons and in amino acid sequence is introduced (see Fig. 17-35).

Evolutionary Significance of Split Genes. The prevalence of split genes in higher forms (animals, plants, fungi) indicates that they make an important contribution to evolution. It has been suggested that by permitting the ready addition or deletion of exons or their recombination from various transcription units (see Short Ig Chains and Heavy Chain Disease, below) split genes allow much more rapid and extensive variation of proteins during evolution than would be possible if the variations were confined to single amino acid residues. Hence split genes increase the pool of variants that can be acted upon by natural selection, thereby permitting relatively rapid evolution of species with long generation times.

SHORT IMMUNOGLOBULIN CHAINS

The separation and rearrangements of exons, and the need to splice their copies in RNA, suggests explanation for a group of previously puzzling variant Ig chains that

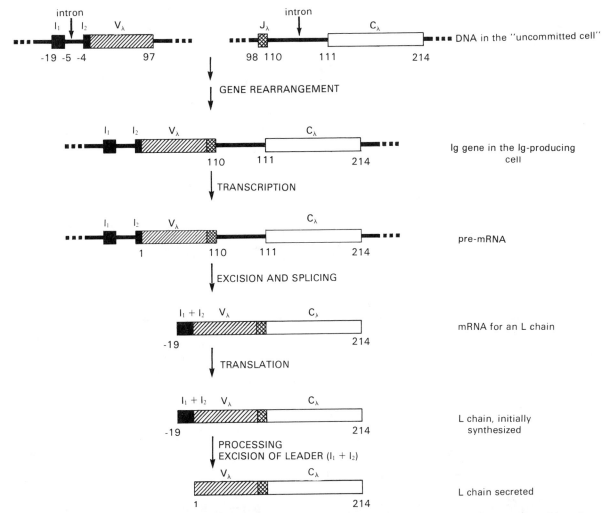

FIG. 17-30. Summary of the transitions in DNA, RNA, and protein leading to the production of a finished λ chain. Based largely on references in Figs. 17-27, 17-28, and 17-29. In the κ family there are several hundred V_κ sequences (for 1-97) followed, in the uncommitted cell, by five clustered J_κ sequences (for 98-110), which are separated by an intron, about 4 Kb long, from the C_κ gene. The H chain family also has several hundred V sequences (each for 1-97) followed by four clustered J_H sequences, which are separated by a long intron from the series of C_H genes, each organized as in Fig. 17-31. The order of C_H genes (in the mouse) is: μ, $\gamma1$, $\gamma2a$, $\gamma2b$, α.

are produced by rare myeloma or other lymphatic tumors. For instance, a variant line of one mouse myeloma tumor produces γ chains that lack the C_H1 domain (residues 121–214): *the missing amino acids correspond exactly to those encoded by exon 1* (Fig. 17-31). Another mouse myeloma secretes a short κ chain, consisting of just the C_κ domain.

Three types of **short human H chains** have been described.

Type 1: **Deletion of part of the V segment and all of C_H1.** These H chains are initiated normally and have normal C_H2 and C_H3 domains. But a gap extends from a position within the V region to residue 216, the beginning of the hinge in γ chains.

Type 2: **Deletion of part of the V segment, all of C_H1, and the hinge.** With this extreme deletion the beginning of V appears normal, but a gap that begins in the V region extends all the way to the end of the hinge (position 232). C_H2 and C_H3 are normal.

Shortened H chains of types 1 and 2 are not associated with L chains and they are excreted in the urine. Patients with these anomalous chains are said to have **heavy chain disease.**

Type 3: **Deletion of the hinge.** This deletion complements type 1. The V segment and C_H1 domains (and also C_H2 and C_H3) are normal: only the hinge (216–232) is missing. In contrast to types 1 and 2, H chains with this defect are associated with L chains in four-chain (H_2L_2) molecules. But because the hinge, with its cysteines, is missing, the usual S–S bridges that link L and H chains cannot form; instead the two L chains are S–S linked to each other. Some of these hingeless human myeloma proteins crystallize exceptionally well, perhaps because they have decreased molecular flexibility.

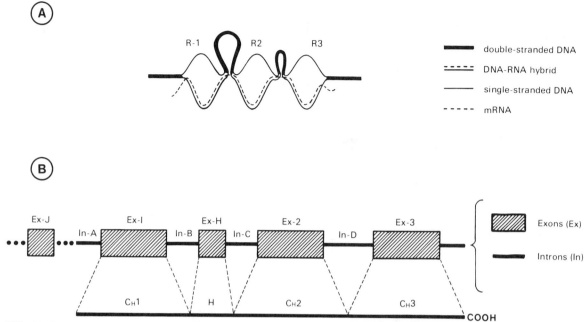

FIG. 17-31. Alternation of exons (encoding for constant domains and the hinge region) and introns (I-1, etc.) in the segment of mouse DNA that specifies an H chain. **A.** Idealized tracings of electron micrograph showing RNA-displacement loops formed by hybridizing mRNA for a mouse γ1 chain to DNA. The single-stranded loops (R1, R2, R3) correspond to exons for C_H1, C_H2, and C_H3. **B.** The order of exons and introns for the C segment of a γ chain. The small exon for the hinge region (Ex-H) was found by sequencing cloned DNA. Ex-J is the exon for the J segment that links a V_H to a C_H1. (Sakano H et al: Nature 277:627, 1979)

The large gaps in the shortened Ig chains can be explained by defective joining of exons or by defective splicing of primary mRNA transcripts: either anomaly could account for the deletion of one or more exons. Whether splicing defects are the result of abnormal splicing enzymes, or of abnormalities in DNA (i.e., mutations) that cause normal enzymes to act abnormally, is not known.

Mapping of V_H Genes. The capacity to produce Abs with shared idiotypes (see Idiotypes, above) is inherited as dominant trait linked to alleles for C_H allotypes. Thus mice of some strains (here called Id⁺ strains) can make Abs with a shared idiotype to Ag X, while in mice of other strains (here called Id⁻) the anti-X Abs lack that idiotype. F1 hybrids (Id⁺ × Id⁻) are always Id⁺, and the genetic loci for the idiotype and C_H allotypes are clearly linked, because backcrosses of the hybrids to the Id⁻ parental strain yield 50% Id⁺ progeny, which have the C_H allotypes of the Id⁺ parent, and 50% Id⁻ progeny, which have the C_H allotypes of the Id⁻ parent.

Hence the V_H genes that are marked by shared, heritable idiotypes are linked to C_H genes. The assumption that V_L and C_L genes are similarly linked has not been tested because heritable serologic markers for C_L regions are not known for mouse Igs.

While in various crosses the appropriate C_H and idiotype markers nearly always segregate together, they are occasionally separated by recombination. From the recombination frequencies a provisional map of some

mouse V_H genes has been constructed. Because about ten V_H genes have already been identified by the shared idiotypes of the Abs elicited against only a small number of conventional Ags (about 30), there must be many V_H genes.

Cis-**Trans Rearrangement.** As mentioned above, the three V–C complexes (V_H-C_H, $V_κ$-$C_κ$, $V_λ$-$C_λ$) are unlinked and each is probably located in a different chromosome. Does this mean that only nucleotide sequences of the same chromosome (the *cis* chromosome) can give rise to the RNA transcript that, after excision and splicing, forms an Ig mRNA molecule (Fig. 17-31)?

Breeding experiments provide evidence that V and C segments of any particular Ig chain are encoded by exons from the same chromosome. For example, when rabbits homozygous for one allotype in the V_H domain (*a*1) and for another allotype in the hinge region of γ chains (*d*11) were crossed with rabbits homozygous for allelic allotypes (*a*3 in V_H and *d*12 in the γ chain's hinge) the F1 progeny were doubly heterozygous and had the expected variety of markers (*a*1/*a*3, *d*11/*d*12) in their total pool of serum Ig, but *their individual γ chains were almost exclusively of parental type: i.e., either a1, d11 or a3, d12.* However, 1% of their γ chains (and up to 5%–10% of α chains in parallel experiments with α-chain allotypes) had *trans* combinations, with a V_H allotype from one parent on the same chain as a C_H allotype from the other parent. The origin of the "scrambled" chains is obscure; perhaps during the rearrangement of V and J segment the V of one chromosome is occasionally joined to the C gene of the homologous chromosome.

THE PRE–IMMUNOGLOBULIN LEADER

Ig chains produced in vitro by translation of myeloma cell mRNA contain an N-terminal segment of about 20 amino acids that is missing in the chains that are secreted by intact cells. The extra piece, the **leader**, is made up predominantly of hydrophobic amino acids: it can contain up to 25% leucine, some in repetitive . . . Leu-Leu-Leu . . . stretches. The leader is synthesized in the cell as part of the nascent Ig chain and is later cleaved off to yield the definitive, and ultimately secreted, chain. Many other secreted proteins (in bacteria [Ch. 6] as well as in animal cells) are also initially synthesized with an extra N-terminal hydrophobic leader (e.g., preproinsulin, pre-proparathyroid hormone) that is removed before the protein molecule is secreted.

With proteins destined for secretion, the leader evidently guides the growing chain to an appropriate receptor in the membrane of the endoplasmic reticulum and initiates transfer across the membrane. A membrane enzyme then cleaves off the leader. This process is followed, for Igs, by assembly of H and L chains into H_2L_2 molecules within the vesicles of the endoplasmic reticulum, and then by extrusion from the cell (Fig. 19-9, Ch. 19). The mechanism of chain transfer across the membrane is unknown.

Different L chains of the same V group have the same leader; those from different groups have different leaders, as though the leader were part of the subgroup's characteristic framework. In the DNA sequence of Fig. 17-28 the codons for the first 15 leader amino acids are separated by an intervening sequence (of 93 bases) from the codons of remaining four leader amino acids (see also Fig. 17-31).

ORIGIN OF DIVERSITY OF V GENES

As noted above, each C is coded for by a single structural gene, transmitted from generation to generation in the germ line. In V domains, however, the number of distinctive amino acid sequences is enormously greater: in producing an immense variety of Abs any individual must be able to synthesize many thousands of V_L and V_H sequences that differ, especially in hypervariable (HV) regions. Figure 17-32 summarizes the main theories that try to account for the origin of this great multitude of V sequences.

Germ Line Theory. According to this theory germ cells carry structural genes for all the Ig sequences an individual can synthesize. These genes would have arisen during evolution through conventional mechanisms for gene duplication, mutation, and selection. Though a great many V genes are required (perhaps 10^4 or more) they could probably be accommodated in the vertebrate genome.

A typical diploid mammalian cell has about 10^{-11} g of DNA, which could code for about 4×10^8 amino acid residues in proteins.* Hence as many as 10^4 V genes (each coding for 110 residues) would occupy about 0.25% of the genome (e.g., $10^4 \times 110/4 \times 10^8$). That number of V_L and of V_H genes, pairing at random, could generate 10^8 different Abs—or perhaps 10^7 if only 10% of the V_L-V_H pairs resulted in stable Ig molecules (see Reconstruction of Combining Sites, above). This number is probably sufficient to account for the apparently limitless diversity of Abs that an individual can make.

Somatic Mutation Theory. This theory maintains that a small number of germ line V genes (e.g., one for each V subgroup) become diversified through mutation in somatic cells, yielding differentiated clones of immunologically competent lymphocytes. Though the number of V genes per somatic cell would be the same as in germ cells, their variety in the individual animal would be enormously greater, because there are an immense number of Ab-forming lymphocytes (about 10^{12} per human), highly diversified in their specificity for Ags (see Ch. 18).

Special mechanisms have been proposed for increasing mutation rates in cells destined to make Abs, but they seem unnecessary: even with ordinary mutation rates (about 10^{-7}/base pair/cell division) the mutation rate per hypervariable region might well be about $1/10^5$ cell divisions, because there are about 300 nucleotide base pairs per V gene, with about 1/6 in hypervariable regions. Hence about 1000 immunologically competent cells with V gene mutations could arise each day during the development of even so small an animal as the mouse, which is estimated to have about 10^8 precursors of Ab-forming lymphocytes (Chs. 18 and 19). However, to account for the observed diversity, cells with hypervariable region mutations would have to be selected for, relative to the vastly greater number of cells with nonmutated V genes.

If a cell expressing a particular mutant V gene should happen to encounter the appropriate Ag the cell would be stimulated to proliferate (**positive selection**) (B Cell Responses to Antigenic Signals, Ch. 19). On the other hand, the need for a large number and variety of selective Ags during ontogeny is bypassed in Jerne's proposal for **negative selection**: germ line V genes are assumed to code for V domains that are specific for certain self-Ags (histocompatibility Ags on cell surfaces; Chs. 18 and 23), leading to suppression of the cells that express these antiself genes (selftolerance; see Ch. 19); but cells with V gene mutations, coding for Igs that are specific for **anything** other than self-Ags, would be preferentially preserved. The resulting selection would thus operate in favor of mutations in hypervariable regions, without requiring the fortuitous presence of the corresponding Ags. Random mutations in C genes and in relatively invariant regions of V genes probably also occur but would not accumulate, either because they are selected against (if, for instance, they are incompatible with a functional Ig molecule) or because they are neutral and not subject to any selection pressure.

---

* On the basis of the triplet code and a weight of 600 daltons for each nucleotide pair, 1×10^{-11} g DNA/$3 \times 600 = 6 \times 10^{-15}$ moles amino acid equivalents; and $6 \times 10^{-15} \times 6 \times 10^{23}$ (Avogadro's number) $= 36 \times 10^8$ amino acid residues in proteins; we assume, however, that only 10% of the genome is devoted to structural genes (i.e., about 4×10^8 amino acids in proteins).

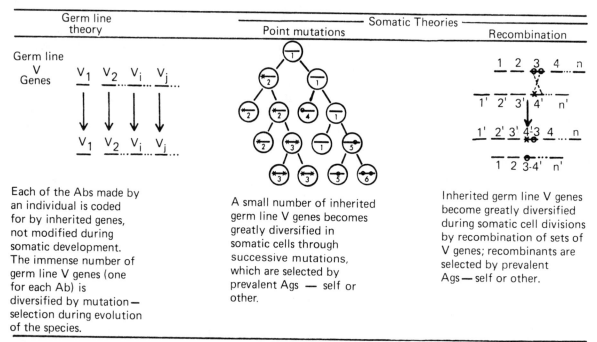

Germ line theory	Somatic Theories	
	Point mutations	Recombination

Germ line V Genes

V_1 V_2 ... V_i ... V_j ...

↓ ↓ ↓ ↓

V_1 V_2 ... V_i ... V_j ...

Each of the Abs made by an individual is coded for by inherited genes, not modified during somatic development. The immense number of germ line V genes (one for each Ab) is diversified by mutation—selection during evolution of the species.

A small number of inherited germ line V genes becomes greatly diversified in somatic cells through successive mutations, which are selected by prevalent Ags — self or other.

Inherited germ line V genes become greatly diversified during somatic cell divisions by recombination of sets of V genes; recombinants are selected by prevalent Ags — self or other.

FIG. 17-32. Theories of the origin of diversity of immunoglobulin V genes. (Based on Edelman GM: In Schmitt FO (ed): Neurosciences; Second Study Program N.Y. Rockefeller Univ Press, 1970)

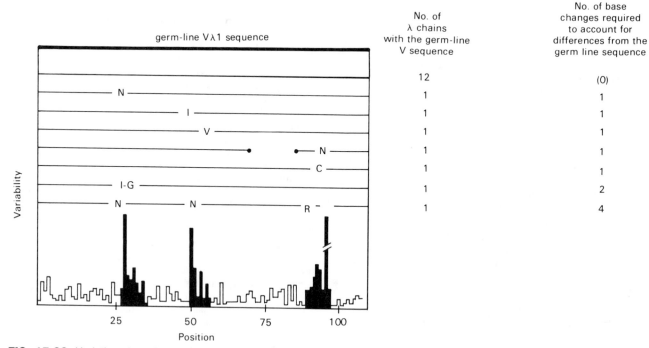

	No. of λ chains with the germ-line V sequence	No. of base changes required to account for differences from the germ line sequence
N	12	(0)
I	1	1
V	1	1
N	1	1
C	1	1
I-G	1	2
N N R	1	4

germ-line Vλ1 sequence

Variability / Position

FIG. 17-33. Variations in amino-acid sequences of V domains of mouse λ1L chains (each from a different myeloma tumor). The amino-acid sequence corresponding to the germ line λ1 gene was based on the deoxynucleotide sequence of DNA cloned from mouse embryo (e.g., Fig. 17-28). Differences from the germ line sequence are clustered in hypervariable regions (for amino acid abbreviations see Fig. 17-12), represented by the Kabat-Wu variability plot (see Fig. 17-22B). (Based on Cohn M et al: Sercarz EE et al (eds.), The Immune System, Genes, Receptors, Signals. NY, Academic Press, 1974; germ line sequence is from references in Figs. 17-28, 17-29, and 17-31)

Somatic Recombination of Germ Line Genes. According to this theory, diversity arises in somatic cells through recombination among a limited number of germ line V genes for each group. If there are *a* different codons at each of *n* positions, unlimited recombination could yield a^n sequences. If, for example, among the several germ line V genes for a given V group there were 8 variable positions and 3 different codons represented at each of these positions, unlimited recombination could result in 3^8, or about 10^4 different sequences in this one V group alone (see Fig. 17-34).

The hypothetic recombination in somatic cells is postulated to occur among V genes of the same chromosome (see Immunoglobulin Gene Clusters: Translocons, above). Recombination between homologous chromosomes would require their pairing, which is exceedingly rare in mitosis (unlike its regular occurrence during meiosis in the formation of gametes).

Striking similarities in the deoxynucleotide sequences of the introns that flank V_κ genes (exons) of the same group supports the possibility that genetic recombination is an important source of variation. The similarities suggest that these segments of DNA can readily pair and undergo recombination during cell division, either in somatic or in germ cells. The plausibility of this suggestion is reinforced by the parallel analyses of a pair of hemoglobin genes that also have similar amino acid sequences but vary only slowly in evolution: nucleotide sequences in the introns that flank the hemoglobin exons show virtually no homology, indicating that, unlike the highly variable Ig V genes, these evolutionarily fixed genes have a limited target for pairing and recombination (see below).

How Many V Genes in the Germ Line? If an individual has fewer germ line V genes than V amino acid sequences in his Ig molecules diversity must arise from somatic mechanisms during the individual's lifetime. However, if there are a great many germ line V genes (over 10^4) all Ig diversity could already exist in the germ line, indicating an origin over evolutionary time. The number of germ line V genes is not known, but estimates based on nucleic acid hybridization and other approaches suggest that there are probably too few to account for the great diversity of V domains.

There seems, for instance, to be a single germ-line gene for the V domain of λ1 L chains, and it differs at several hypervariable positions from the V domains of λ1 chains (Fig. 17-33).

In accord with the much greater frequency of κ than λ chains in the mouse (about 20:1), there appear to be many more V_κ germ-line genes. Thus a radiolabeled probe for the V domain of a particular κ chain hybridized to six different restriction DNA fragments in the genome, from two of which the corresponding genes were cloned. Their deoxynucleotide sequences were very similar but neither matched precisely the original V_κ probe. Hence it is probable that the set of six V_κ genes corresponds to one V_κ group (defined by marked amino acid sequence homology, see Figs. 17-23 and 17-34). Approximately 50 V_κ groups have been recognized in the mouse; if each also has six germ-line genes, there are in all about 300 V_κ germ-line genes. While the number of germ-line V_K genes is far greater than had previously been expected the number still seems far too low to account for the many V_K sequences synthesized in each mouse. Thus V_K and other V genes must somehow become

FIG. 17-34. A possible arrangement of genes at the locus for κ L chains. A radioactive mRNA or cDNA "probe" for a given V_κ or V_H region usually detects (e.g., by methods illustrated in Fig. 17-27) a *group* of 5–10 similar ("cross-hybridizing") sequences in the genome, represented here as V_κ groups 1, 2, etc. Each member of a group is probably a germ line gene and probably represents a V subgroup: i.e., a subset of V sequences that differ among themselves by only a few (e.g., 3 or 4) amino acids, especially in hypervariable regions. The members of a subgroup are thought to derive from its germ line representative by somatic mutation. This *one germ line gene–one subgroup* concept is illustrated by the λ chains shown in Fig. 17-33. In any particular group the sequences flanking the coding region of each gene are very similar, suggesting that, due to the high level of sequence homology, they can readily pair and undergo recombination, generating additional diversity of coding sequences. ● =suggested site for joining V and J encoding sequences during gene rearrangement. ○ =suggested site for splicing RNA transcripts. L =leader sequence (see Figs. 17-28, 17-30). (Gottlieb PD: Molec Immunol, in press)

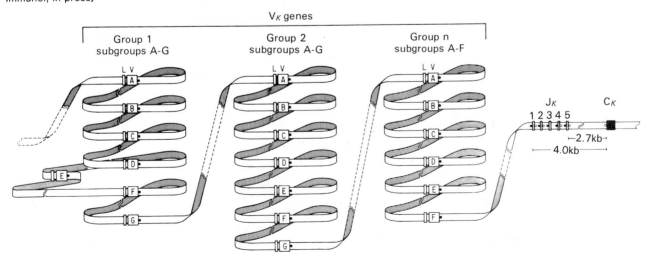

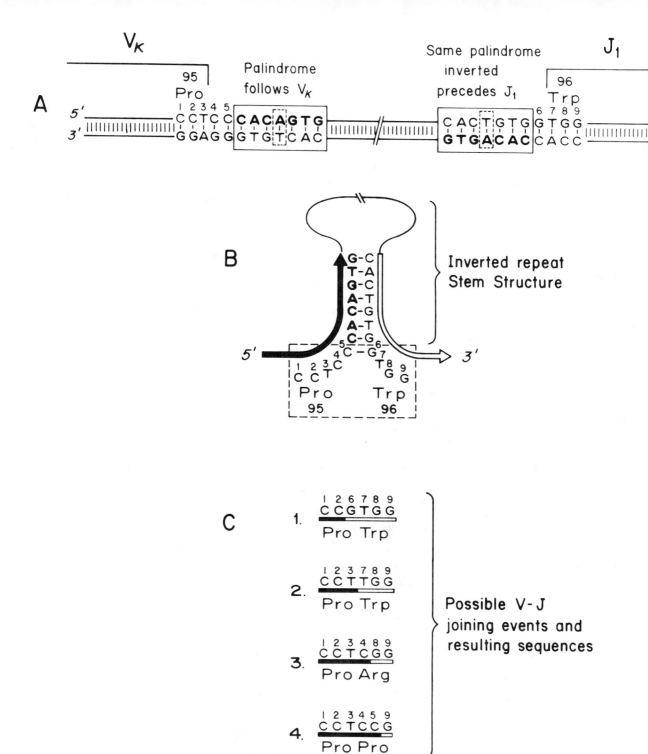

FIG. 17-35. Gene rearrangement (V-J joining) generates codon variation, resulting in amino-acid sequence variation in the third hypervariable (complementarity determining) region of a V domain. The key is the DNA **palindrome** sequence (it reads the same way from each end, as in MADAM IM ADAM), which occurs both at the end of a V_κ encoding sequence and at the beginning of a J_κ sequence (J_1). As is shown with the top strand, the palindrome sequences form an inverted stem structure, bringing bases 1-5 next to bases 6-9. Breaking and rejoining the DNA strand at various points between 1-9 lead to different codons for position 96 of the complete L chain. (Gottlieb PD: Molec Immunol (in press); based on Sakano H, et al. Nature 280:288, 1979 and Max EE, et al: Proc Nat Acad Sci USA 76:3450, 1979

during somatic development. The key mechanisms are obscure, but several mechanisms are already plausible:

1) The combination of different V sequences with different J sequences (of the same family) can yield a large number of **combinatorial variants.**

2) The individual members of a given V group have virtually identical framework sequences and are flanked by untranslated sequences that are also markedly similar. Hence, pairing and **recombination among members of a V group** should be facilitated. If individual members of the group, in the germ-line, differ even slightly from each other in their respective hypervariable regions, recombination among them could yield a rich crop of variants (Fig. 17-34).

3) In the gene rearrangement that links the amino-end of a J segment to the carboxy-end of a V encoding sequence small **variations at the site for joining V to J** can generate new codons, leading to especially pronounced variability in th 3rd hypervariable region at the V-J junction (Fig. 17-35).

EVOLUTION OF IMMUNOGLOBULINS

Comparison of Igs from various species provides the raw material for reconstructing the evolution of Igs. Invertebrates lack Igs; vertebrates possess them. All vertebrate species have high molecular-weight Igs resembling human IgM, but with varying numbers of four-chain subunits per molecule: five in sharks (as in mammals), four in bony fish, six in toads. Low-molecular-weight Igs are also present in sharks and other lower vertebrates: they are probably subunits of the IgM-like molecules, which they resemble antigenically.

Abs have been elicited in cyclostomes (hagfish, lampreys), the most primitive vertebrates, but their molecular structure is still poorly characterized. The lungfish is the most primitive species with a second, distinctive non-IgM class, and in amphibia, which generally represent a watershed of diversification in vertebrate evolution, there is a striking divergence: the salamander (a urodele) has only the IgM-like class, but frogs (anurans) have IgM and two low-molecular-weight classes of Igs. Reptiles resemble frogs in having IgM and two other antigenically distinct classes. IgA has been recognized (so far) only in mammals and in birds; Igs in secretions of reptiles and lower forms are probably IgM. In general, the relationships between Igs of different species are poorly defined so long as they are compared only by determining H-chain molecular weights in gel electrophoresis (see Homologies Based on Amino Acid Sequences, below).

Light chains. The ratio of κ to λ L chains varies widely. Sharks have only κ chains, birds only λ; as noted earlier most mammalian species have both, but often in greatly divergent proportions (e.g., in the mouse, 95% are κ).

J chains. These chains are present in birds, mammals, toads, and catfish, and possibly in the shark; they probably coevolved with IgM.

Complement. These serum proteins (Ch. 20) are present in all vertebrates as important amplifiers of Ab activity; some invertebrates may have related proteins. Cobra venom factor, which activates a major complement component in mammals (Ch. 20), can also activate some invertebrate serum proteins, which become capable, like activated vertebrate complement, of lysing red blood cells.

HOMOLOGIES BASED ON AMINO ACID SEQUENCES

Because serologic analyses are sensitive to small antigenic variations they readily reveal **differences** among Ig chains. Amino acid sequences, in contrast, are especially useful for revealing **similarities** among chains, including those that do not cross-react serologically. The relatedness of two chains is apparent from the frequency with which corresponding positions are occupied by the same amino acid. A more searching comparison determines the minimum number of mutations necessary to change from one chain to the other. Both approaches give essentially the same result, and for both it is usually necessary to introduce occasional gaps in one or another chain,

FIG. 17-36. Representative C-terminal amino-acid sequences illustrating differences between heavy-chain classes and between κ and λ light chains. The cysteine that is C-terminal in κ chains and penultimate in λ chains contributes to the S–S bonds that link light and heavy chains (see Fig. 17-14). For abbreviations see Fig. 17-12. (Based largely on references in Dayhoff MD, ed: Atlas of Amino Acid Sequence and Structure, vol. 5, National Biomedical Research Foundation, Silver Spring, Md, 1972)

COOH ENDS →

Human γ1 ...	Z	K	S	L	S	L	S	P	G
Human γ2 ...	Z	K	S	L	S	L	S	P	G
Human γ3 ...	Z	K	S	L	S	L	S	P	G
Human γ4 ...	Z	K	S	L	S	L	S	L	G

Human μ ...	M	S	B	T	A	G	T	C	Y
Human α ...	M	A	E	V	D	G	T	C	Y
Mouse α ...	M	S	E	G	D	G	T	C	Y

Human κ ...	K	S	F	N	R	G	E	C
Human λ ...	K	T	V	A	P	T	E	C	S

Mouse κ ...	K	S	F	N	R	N	E	C
Mouse λ1 ...	K	S	L	S	R	A	D	C	S
Mouse λ2 ...	K	S	L	S	P	A	E	C	L

		Light Chains			Heavy Chains				
		Human		Mouse		Human γ1			Rabbit γ	
		C_κ	C_λ	C_κ	C_λ	C_{H1}	C_{H2}	C_{H3}	C_{H2}	C_{H3}
Human	C_κ	100	38	(60)	35	29	30	34	23	23
	C_λ		100	43	(65)	26	21	26	18	23
Mouse	C_κ			100	40	23	22	25	21	26
	C_λ				100	23	20	29	15	25
Human γ1	C_{H1}					100	31	29	10	25
	C_{H2}						100	29	(63)	28
	C_{H3}							100	22	(66)
Rabbit γ	C_{H2}								100	35
	C_{H3}									100

FIG. 17-37. Matrix comparison of frequencies of identical amino acid residues in homologous positions of the constant **(C)** regions of various Ig chains, expressed as identities per 100 residues (%). The most striking homologies (60%) are **circled**. Because there are 20 amino acids, 5% identity is the background level expected when randomly chosen chains are compared. (In fact, when the first 110 residues of the human hemoglobin α chain are compared with the above Ig chains identities range from 4%–8%.) (Based on sequences and references in Dayhoff, MD, ed: Atlas of Protein Sequence and Structure, vol. 5. National Biomedical Research Foundation, Silver Spring, Md, 1972)

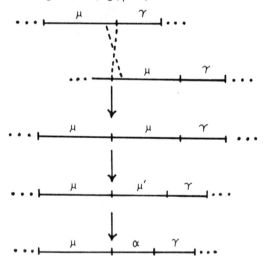

FIG. 17-38. Diagram showing that illegitimate crossing over could generate additional genes, whose occasional missense mutations could be selected for, leading eventually to the appearance of new Ig classes (e.g., $\mu \rightarrow \alpha$).

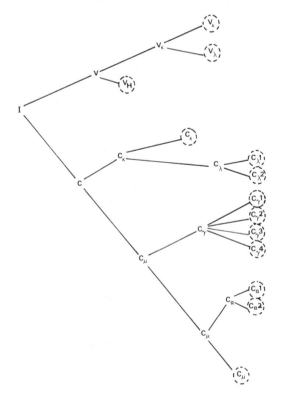

FIG. 17-39. Possible order of evolution of contemporary Ig genes (in **dashed circles**) from a primitive precursor (I) over approximately 4×10^8 years. Chains δ and ϵ are not included.

corresponding to genetic deletions or insertions, to bring out the full extent of their homology.

The sequences shown in Fig. 17-36 illustrate variations in homology, and the matrix of Fig. 17-37 summarizes homology relations among a number of C segments. Extensive analyses of this kind lead to the following generalizations.

1) Homology is greater between the C_κ sequences of different species (e.g., man and mouse), and between the C_λ sequences of different species, than between C_κ and C_λ of the same species.

2) Successive domains in the constant part of heavy chains of a particular class (C_H1, C_H2, or C_H3 of γ chains) have extensive homology with each other, with the corresponding domains of other species (e.g., in human and rabbit γ), and also with C_κ and C_λ sequences.

3) Only slight homology is detectable between V and C sequences, even when those of the same chain are compared.

4) In spite of variations among different molecules, there is more homology **within** each of the V_κ, V_λ, and V_H sets than **among** these sets. Nevertheless, discernible low-level homology among all V sequences sets them apart from all the C sequences.

5) The V_H and V_L sequences of the same molecule have no more homology than those from different molecules.

Order of Evolution. From these observations it appears that structural genes for present-day Igs evolved from a primordial gene coding for a polypeptide with the general characteristics of V and C segments, namely, about 110 amino acids with two cysteine residues that form an approximately 60-membered S–S loop. The small degree of homology between V and C sequences suggests that, as one of the earliest events, duplication of the primordial gene probably gave rise to the primitive precursors of V and C genes. Genes for constant segments of diverse heavy chains could have arisen from unequal ("illegitimate") crossovers, yielding duplicated C genes (Fig. 17-38).

The L chains in Igs of sharks (among the most primitive existing vertebrates) appear to be only of the κ type. This finding and the extensive homology between avian and mammalian λ chains, suggest that C_κ is the more primitive C_L gene and that it probably gave rise by duplication to a C_λ gene some time before birds and mammals diverged, about 250 million years ago. Present day μ chains are assumed to correspond most closely to primordial heavy chains, because sharks and teleost fish have only one class, resembling μ of higher forms (Fig. 17-38). Because chains resembling γ are present only in lungfish, in some amphibia, and in higher vertebrates, C_γ may have arisen from C_μ (with a deletion). Recognized clearly so far only in mammals and in birds, α chains probably arose by a still later duplication, most likely of the μ gene, since μ and α are more alike than either is like γ, at least near the C-terminus (see Fig. 17-39). This order of evolution (Fig. 17-39) is in accord with the successive appearance of Igs during embryonic development and infancy (IgM first, then IgG, finally IgA; Ch. 19), and with the principle that ontogeny recapitulates phylogeny.

APPENDIX

TABLE 17-6. Some Properties of Classes of Human Immunoglobulins*

Property	IgG				IgA		IgM	IgD	IgE
	IgG1	IgG2	IgG3	IgG4	IgA1	IgA2			
Sedimentation coefficient (S)	7	7	7	7	7–13	7–13	19	7	8
Molecular weight ($\times 10^{-3}$)	150	150	150	150	150–600	150–600	900	?	190
Heavy chains	$\gamma1$	$\gamma2$	$\gamma3$	$\gamma4$	$\alpha1$	$\alpha2$	μ	δ	ϵ
Light chains: κ/λ ratio	2.4	1.1	1.4	8.0	1.4	1.6	3.2	0.3	?
Carbohydrate (%, approx.)	3	3	3	3	7	7	12	13	11
Average conc. in normal serum (mg/ml)	8	4	1	0.4	3.5	0.4	1	0.03	0.0001
Half-life in serum (days, in vivo)	23	23	8	23	6	(6?)	5	3	2.5
Heavy chain allotypes	Gm	Gm	Gm	Gm		Am			
Earliest Ab in primary immune responses[†]							+		
Most abundant Ab in most late immune responses[†]	←———————		+	———————→					
Conspicuous in mucinous exocrine secretions					+	+			
Transmitted across placenta[†]	+	±	+	+	0	0	0	?	–
Effector functions									
Principle surface Ig on B cells[†]							+	+	
Binds to macrophages	++	+	++	±					
Complement fixation[‡]	++	+	++	0§	0§	0§	++	?	0§
Sensitizes human mast cells for anaphylaxis (homocytotropic)[¶]									+
Sensitizes guinea pig mast cells for passive anaphylaxis (heterocytotropic)[¶]	+		+	+					

*For hinge region differences see Fig. 17-15 and Table 17-2.

†Chapters 18 and 19.

‡Chapter 20.

§Can activate complement via bypass mechanism (Ch. 20).

¶Chapter 21; IgE molecules are reagins, responsible for atopic allergy.

TABLE 17-7. Allotypes of Mouse H Chains

Heavy chain class	Locus	Alleles or serologic specificity	Representative mouse strains
$\gamma2a$(IgG2a)	Ig-1	Ig-1[a]	BALB/c
		Ig-1[b]	C57BL/10J
		Ig-1[c]	DBA/2J
		Ig-1[d]	AKR/J
		Ig-1[e]	A/J
		Ig-1[f]	CE/J
α(IgA)	Ig-2	Ig-2[a,h]	BALB/c
		Ig-2[b]	C575BL/10J
		Ig-2[c,g]	DBA/2J, RIII/J
		Ig-2[d,e]	AKR/J, A/J
		Ig-2[f]	CE/J
$\gamma2b$(IgG2b)	Ig-3	Ig-3[a,c,h]	BALB/c, DBA/2J
		Ig-3[b]	C57BL/10J
		Ig-3[d]	AKR/J
		Ig-3[e]	A/J
		Ig-3[f]	CE/J
		Ig-3[g]	RIII/J
$\gamma1$(IgG1)	Ig-4	Ig-4[a,c,d,e,f,g,h]	BALB/c, DBA/2J, AKR/J, A/J, CE/J, RII/J,
		Ig-4[b]	C57BL/10J

(Based on Herzenberg LA et al: Ann Rev Genet 2:209, 1968; Mage RG et al: in Sela M (ed): The Antigens, Vol 1, p. 299. NY, Academic Press, 1973)

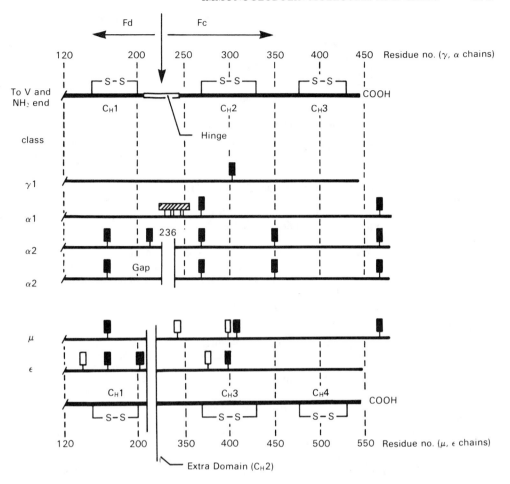

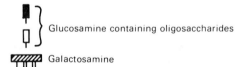

FIG. 17-40. Locations of carbohydrate in human $\gamma1$, $\alpha1$, $\alpha2$, μ and ϵ chains. **Vertical** rectangles are glucosamine-containing oligosaccharides and the **horizontal** rectangle is a galactosamine-containing oligosaccharide; **shaded** rectangles are those with homologous positions in two or more chains. The upper scale gives residue positions for γ and α chains; those in the lower scale for μ and ϵ chains. The extra domain (C_H2) in μ and ϵ has been omitted. (Torano A et al: Proc Nat Acad Sci 74:2304, 1977)

SUGGESTED READING

BOOKS AND REVIEW ARTICLES

AMZEL LM, POLJAK RJ: Three-dimensional structure of immunoglobulins. Annu Rev Biochem 48:961, 1979

EDELMAN GM, GALLY JA: The genetic control of immunoglobulin synthesis. Annu Rev Genet 6:1, 1972

FRANKLIN EC, FRANGIONE B: Structural variants of human and murine immunoglobulins. (An extensive review of "heavy chain disease".) Contemp Topics in Molec Immunol 4:89, 1975

GOTTLIEB P: Immunoglobulin genes. Molec Immunol (in press), 1980

GRUBB R: The Genetic Markers of Human Immunoglobulins. NY, Springer, 1970

HOOD L, CAMPBELL J, ELGIN S: The organization, expression, and evolution of antibody genes and other multigene families. Annu Rev Genet 9:305, 1975

KOSHLAND ME: Structure and function of J chain. Adv Immunol 20:41, 1975

MELCHERS F, POTTER M, WARNER N (eds): Lymphocyte Hybridomas. Curr Top Microbiol Immunol 81: 1978

DIAMOND BA, YELTON DE, SCHARFF MD: Monoclonal antibodies: a new technology for producing serological reagents. N Engl J Med (in press), 1980

MAGE R, LIEBERMAN R, POTTER M, TERRY WD: Immunoglogulin allotypes. In *The Antigens* Vol. I (Sela M editor) Academic Press, NY pp. 299 1973.

NISONOFF A, HOPPER JE, SPRING SR: The Antibody Molecule. Academic Press, NY 1975. (A classic monograph)

ORIGIN OF LYMPHOCYTE DIVERSITY. *Cold Spring Harbor Symp. in Quant. Biol.* vol. 41 1977

POTTER M: Immunoglobulin-producing tumors and myeloma proteins of mice. Physiol Rev 52:631, 1972

SEIDMAN JG, LEDER P, MAU M, NORMA B, LEDER P: Antibody diversity. Science 202:11, 1978

WEIGERT M, RIBLET R: The genetic control of antibody variable regions in the mouse. Springer Seminars in Immunopathology 1:133, 1978

SPECIFIC ARTICLES

BERNARD O, HOZUMI N, TONEGAWA S: Sequences of mouse immunoglobulin light chains before and after somatic changes. Cell 15:1133, 1978

BIRSHSTEIN B, PREUD'HOMME JL, SCHARFF MD: Variants of mouse myeloma cells that produce short immunoglobulin chains. Proc Nat Acad Sci USA 71:3478, 1974

BRACK C, HIRAMA M, LENHARD-SCHULLER R, TONEGAWA S: A complete immunoglobulin gene is created by somatic recombination. Cell 15:1, 1978

BURSTEIN Y, SCHECHTER I: Amino-acid sequence of the NH₂-terminal extra piece segments of the precursors of mouse immunoglobulin λ-type and κ-type light chains. Proc Natl Acad Sci USA 74:716, 1977

EARLY P, HUANG H, DAVIS M, CALAME K, HOOD L: An Ig H Chain V region gene is generated from three segments of DNA: V_H, D, and J. Cell (in press), 1980

HONJO T, KATAOKA T: Organization of immunoglobulin heavy chain genes and allelic deletion model. Proc Natl Acad Sci USA 75:2140, 1978

HOESSLI D, OLANDER J, LITTLE JR: Heterologous recombination between mouse myeloma protein 315 heavy chains and rabbit antibody light chains. J Immunol 113:1024, 1974

JERNE NK: The somatic generation of immune recognition. Eur J Immunol 1:1, 1971

KOHLER G, MILSTEIN C: Continuous cultures of fused cells secreting antibody of predefined specificity. Nature 256:495, 1975

KABAT EA, WU TT, BILOFSKY H: Variable region genes for the immunoglobulin framework are assembled from small segments of DNA- a hypothesis. Proc Natl Acad Sci USA 75:2429, 1978

KNIGHT KL, MALEK TR, HANLY WC: Recombinant rabbit secretory immunoglobulin molecules: alpha chains with maternal (paternal) variable-region allotypes and paternal (maternal) constant-region allotypes. Proc Natl Acad Sci USA 71:1169, 1974

MAX EE, SEIDMAN JG, LEDER P: Sequences of five potential recombination sites encoded close to an immunoglobulin κ constant region gene. Proc Natl Acad Sci USA 76:3450; 1979

MCKEAN DJ, BELL M, POTTER M: Mechanisms of antibody diversity: multiple genes encode structurally related mouse κ variable regions. Proc Natl Acad Sci USA 75:3913, 1978

OI VT, BRYAN VM, HERZENBERG LA, HERZENBERG LA: Lymphocyte membrane IgG and secreted IgG, are structurally and allotypically distinct. J Exp Med 151:1260, 1980

POLJAK RJ, AMZEL LM, CHEN BL, PHIZACKERLEY RP, SAUL F: Three dimensional structure of the Fab' fragment of a human myeloma immunoglobulin at 2.0 A resolution. Proc Natl Acad Sci USA 71:3440, 1974

PUTNAM FW, FLORENT G, PAUL C, SHINODA T, SHIMIZU A: Complete amino acid sequence of the mu heavy chain of a human IgM immunoglobulin. Science 182:287, 1973

RODKEY LS: Production and characterization of autoantidiotypic antisera. J Exp Med 139:712, 1974

RODWELL J, KARUSH F: A general method for the isolation of the V_H domain from IgM and other immunoglobulins. J Immunol 121:1528, 1978

SAKANO H, ROGERS JH, HÜPPI K, BRACK C, TRAUNECKER A, MAKI R, WALL R, TONEGAWA S: Domains and the hinge region of an immunoglobulin heavy chain are encoded in separate DNA segments. Nature 277:627, 1979

SAKANO H, MAKI R, KURASAWA Y, ROEDER W, TONEGAWA S: The two types of somatic recombination necessary for generation of complete H chain genes. Nature (in press), 1980

SCHILLING J, CLEVINGER B, DAVIE J, HOOD L: Amino-acid sequence analyses of homogeneous antibodies to dextran: diversity patterns suggest that DNA rearrangements between V and J segments occur in heavy chains. Nature (in press)

SEARS DW, MOHRER J, BEYCHOK S: A kinetic study in vitro of the reoxidation of interchain disulfide bonds in a human IgGk. Correlation between sulfhydryl disappearance and H₂L₂ assembly. Proc Natl Acad Sci USA 72:353, 1975

SEIDMAN JG, LEDER A, EDGELL MH, POLSKY F, TILGHMAN SM, TIEMEIER D, LEDER P: Multiple related immunoglobulin variable-region genes identified by cloning and sequence analysis. Proc Natl Acad Sci USA 75:3881, 1978

WEIGERT M, GATMAITAN L, LOH E, SCHILLING J, HOOD L: Rearrangement of genetic information may produce immunoglobulin diversity. Nature 276:785, 1978

WU TT, KABAT EA, BILOFSKY H: Some sequences similarities among cloned mouse DNA segments that code for λ and κ light chains of immunoglobulins. Proc Natl Acad Sci USA 76:4617, 1979

chapter 18

THE CELLULAR BASIS
FOR IMMUNE RESPONSES

GENERAL FRAMEWORK AND THE CLONAL SELECTION HYPOTHESIS

The cells that recognize antigens (Ags) and initiate immune responses are **lymphocytes.** These cells are dispersed throughout the body and, next to erythrocytes and fibroblasts, they are probably the most numerous cell type in the body: an adult human has about 10^{12} lymphocytes—or about 10% of all cells in the body*—weighing in toto about 1 kg, or approximately as much as the human brain.

Lymphocytes vary only moderately in appearance (Fig. 18-1), but they are greatly diversified in function. Individual cells seem to be virtually **monospecific,** i.e., to be able to react with only one or a few Ags. Binding of Ag causes some lymphocytes to differentiate into **effector cells:** antibody (Ab)-secretors (Ch. 19) or effectors of cell-mediated immunity (Ch. 22). Other Ag-stimulated lymphocytes develop into **regulatory cells (helpers or suppressors)** that control the differentiation of the effector lymphocytes. Still other lymphocytes seem to regulate the maturation or activity of the regulatory lymphocytes (T Cell Regulatory Circuits, Ch. 19).

The theoretical framework for viewing the responses of these cells has undergone several radical transitions since it was recognized, about 75 years ago, that an individual can respond selectively to each of an immense number of different Ags (probably on the order of millions). Earlier theories focused entirely on Abs, because cell-mediated and humoral (i.e., Ab-mediated) immune responses were not clearly distinguished until about 20 years ago. However, the early theories can also be applied to cell-mediated immunity by substituting Ag-binding receptors of the appropriate lymphocytes (thymus-derived; see T Cell Differentiation, below) for Ab molecules. Two kinds of theories are prominent. According to **selective theories** the immunogen† stimulates the synthesis of an Ab by cells that already make the Ab

at a low level in advance of immunization: i.e., all of the necessary genetic information is present before the cell encounters the Ag.‡ **Instructive theories,** in contrast, assume that the formation of an Ab cannot occur until the Ag's presence provides the necessary information (Fig. 18-2).

The first detailed theory, proposed by Ehrlich in 1900, was selective. Supposedly, cell surfaces were covered by a variety of Ab-like receptors; and in combining with matching ones the immunogen is somehow selectively stimulated their further production and secretion by the cells. This theory was abandoned by the 1920s, because of the enormous variety of immunogenic substances, including newly synthesized organic chemicals: it seemed beyond belief that Abs to an almost limitless diversity of Ags could all preexist.

In its place an instructive **antigen template theory** (suggested by Breinl and Haurowitz, Pauling, and others) came to be widely favored in the 1940s and 1950s. According to this view the specificity of an Ab molecule was determined not by its amino acid sequence but by the process of molding the nascent molecule around the immunogenic determinant; the immunogen acted as a template at the site of protein synthesis, creating a complementary conformation at the combining sites of the Ab.

Clonal Selection. The template theory was abandoned when it became evident that the shape of a protein, and therefore its activity, is determined by its amino acid sequence: this sequence is specified by the encoding sequence in DNA and cannot be dictated by the presence of an Ag. In support of this argument, Haber and Tanford showed that when an Ab's three-dimensional structure is reversibly destroyed (denaturation) the molecule can refold **in the absence of the Ag** and regain its original specificity. There is now general acceptance of the **clonal selection theory,** proposed by Jerne, by Talmage, and especially by Burnet in 1959.

According to this theory, which was largely inspired by the success of mutation–selection in explaining microbial adaptations, an individual's immunologically responsive lymphocytes are an immensely diversified pool of cells, each capable of responding to only one Ag (or a few re-

---

* The estimate of approximately 10^{13} cells in the adult human is extrapolated from measurement of total body DNA in the rat and the assumptions that 1) the total amount of DNA per nucleus is essentially constant, and 2) there is one nucleus per cell.

† In this and in subsequent chapters **immunogen** is used as a synonym for Ag when the emphasis is on its function in inducing immune responses.

‡ A modification of the selective theory maintains that additional variation follows the introduction of Ag.

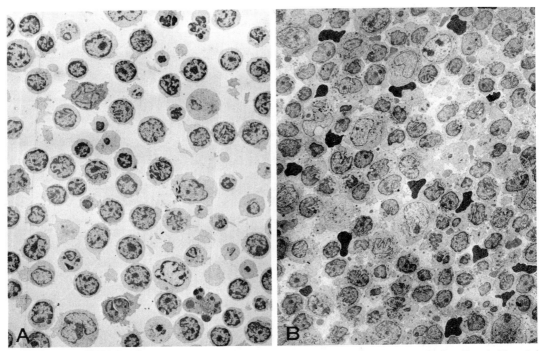

FIG. 18-1. Lymphocytes in lymph from the thoracic duct. A. From a rat. (Zucker-Franklin D: Seminars in Hematol 6:4, 1969) (×2400) B. From a human. (Zucker-Franklin D: J Ultrastruct Res 325, 1963) (×1200)

lated ones), and this capacity is somehow acquired before the Ag is encountered (see Origin of Diversity of V Genes, Ch. 17). When an immunogen penetrates the body, it binds to a recognition molecule (an Ab or an Ab-like receptor) on the surface of the corresponding lymphocyte. This stimulates the cell to proliferate and generate a clone of differentiated effector cells **(primary response)**. Some Ag-stimulated lymphocytes circulate through blood, lymph, and tissues as an expanded reservoir of Ag-sensitive **memory cells:** the same immunogen encountering these cells months or years later evokes a more rapid and copious **secondary response.**

Older theories are contrasted with the clonal selection theory in Figure 18-2. Several lines of evidence support the clonal hypothesis. 1) A given Ag is bound specifically by only a small proportion of an individual's lymphocytes, and inactivation of these cells eliminates the ability

to form the corresponding Abs. 2) An Ab-forming cell makes immunoglobulin (Ig) molecules with only one kind of combining site; and in an animal stimulated with two Ags, A and B, or even with one Ag that has two distinguishable determinants, some cells make anti-A, others make anti-B, and no cell makes both.

Later developments also support clonal selection, but in a more complex form, for there are not only two classes of lymphocytes, B and T cells, but a diversity of types within each class, especially among T cells (see T Cell Subsets, below). Before considering the main differences between the various subsets of lymphocytes, however, we shall review the principal lymphatic organs, as their experimental removal first paved the way to the recognition of the two main cell types and their fundamentally different contributions to the immune system.

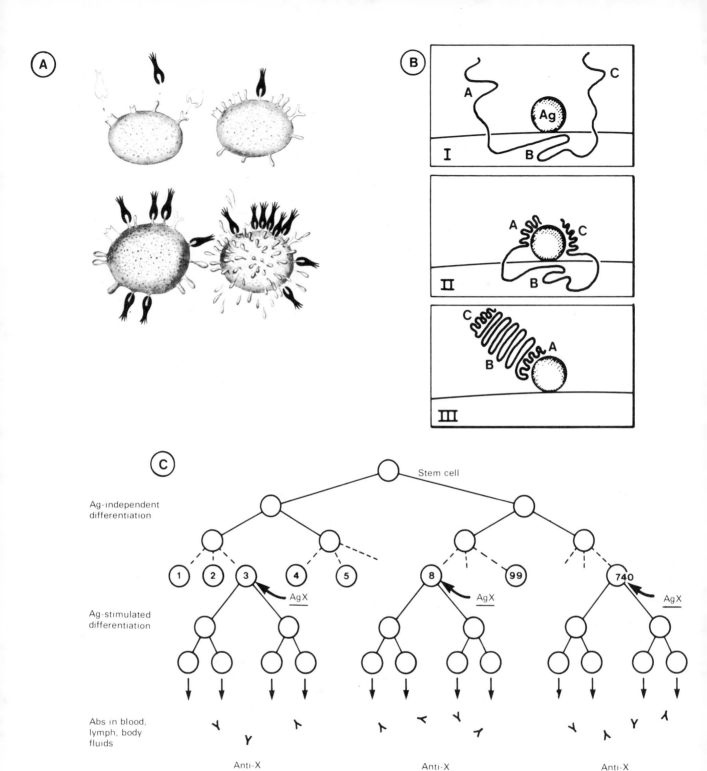

FIG. 18-2. Comparison of selective and instructive theories of Ab formation. A. In Ehrlich's selection hypothesis, multiplication and shedding of a cell receptor follows its combination with an Ag; each cell was assumed to be able to combine with any Ag. (Ehrlich P: Proc R. Soc Lond [Biol] 66:424, 1900) B. In the antigen-template theory an Ag reacts with the nascent Ig chain, **A-B-C,** causing it to acquire sites that are complementary to determinants of the Ag. (Pauling L: J Am Chem Soc 62:2643, 1940) C. Clonal selection hypothesis resembles A, but with the important difference that each reactive cell (3, 8, 740, etc.) can respond to only one or a few Ags. (Based on Burnet FM: Clonal Selection Theory of Acquired Immunity. Nashville, Vanderbilt Univ. Press, 1959)

LYMPHATIC ORGANS

DIFFERENTIATION OF HEMATOPOIETIC CELLS

Like other cells that circulate in the blood, lymphocytes are derived from self-renewing, pluripotential hematopoietic (Gr. *haima, haimatos,* blood; *poiesis,* forming) **stem cells.** These cells first appear in the yolk sac of the developing embryo and then in the fetal liver. After birth they are located in the bone marrow, where they persist during adult life.*

Stem cells are self-renewing. Some of their daughter cells remain stem cells; others become more differentiated progenitor cells, programmed to yield progeny that differentiate further into particular types of blood cells. For instance, some progenitors develop into immature

* When the bone marrow is irreversibly destroyed in an adult by disease, stem cells and active hematopoiesis become evident in the spleen and sometimes even in the liver.

erythroid and myeloid cells that eventually differentiate into erythrocytes and granulocytes, respectively. The latter forms are terminally differentiated "end cells": they have a limited life span and undergo few if any additional divisions (Fig. 18-3).

Other progenitors develop into lymphocytes in the special environments provided by the main lymphatic organs. T lymphocytes mature in the thymus. B lymphocytes develop in the bursa of Fabricius of birds and probably in the bone marrow of mammals, which have no bursa. In these so-called **primary lymphatic organs** lymphocytes become **committed** 1) to behave as B cells or as particular types of T cells, and probably 2) to react with particular Ags. In secondary lymphatic organs (spleen and lymph nodes), and also more diffusely in all tissues, the committed lymphocytes encounter Ags and complete their differentiation into fully mature

FIG. 18-3. Hematopoiesis: Self-renewing pluripotential stem cells give rise to more stem cells and to various progenitors **(A, B, C, . . . F)** of diverse lineages (erythrocytes, platelets, various granulocytes, macrophages, and diverse lymphocytes). The terminally differentiated "end cells" (underlined) vary in life expectancy; e.g., erythrocytes, 120 days; granulocytes, 12 h; plasma cells (Ab-secreting form of B lymphocytes), 3–4 days.

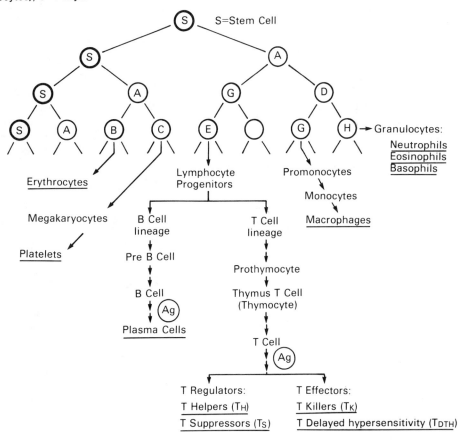

forms: regulator T cells and effector B and T cells (Fig. 18-3).

PRIMARY LYMPHATIC ORGANS

THYMUS

The thymus is a **lymphoepithelial** organ: it consists of a mass of lymphocytes (about 10^8 in a young mouse) and a much smaller number of epithelial cells (Fig. 18-4).

The thymus develops in the embryo from branchial pouches of the pharynx: epithelial buds grow out, pinch off, and migrate (in higher vertebrates) to the midline of the upper thorax, where they eventually become populated by thymic lymphocytes (called **thymocytes**). Through the transfer of cells from various tissues of mice with a distinctive chromosomal marker to mice that lack the marker (but are otherwise the same genetically) it has been shown that **thymocytes derive from migratory hematopoietic stem cells.** Frequent mitotic figures in the thymus, and rapid incorporation of

radioactive thymidine into DNA, show that the immigrant cells divide rapidly; but only a small proportion of their progeny survive and become mature T cells. A polypeptide hormone produced by thymic epithelial cells appears to promote the differentiation of primitive thymocytes into T cells (see T Cell Differentiation, below). About 90% of thymocytes are immature and are readily destroyed by corticosteroids. The cortisone-resistant 10%, located mostly in the medulla (Fig. 18-4), have surface Ags like those of mature T cells of peripheral tissues (Table 18-2, below), which they also resemble functionally (e.g., in reactions with foreign cells; see Graft-vs-Host Reactions, Ch. 23).

A role for the thymus in the immune response was long suspected (because it is essentially a mass of lymphocytes), but decades of study showed that this organ did not form Abs, trap Ags, or contain more than a rare Ab-forming cell. Beginning, however, in 1960, largely with J. Miller's work in England, mice that were thymectomized at birth were observed to develop a coherent variety of immune defects: a drastically reduced number of blood lymphocytes; a reduced ability to reject tissue

FIG. 18-4. Anatomy of the thymus. A. The darkly stippled cortex consists of a dense mantle of immature T cells (thymocytes) surrounding the lightly stippled medulla, which is made up of epithelial cells plus T cells that are virtually as mature as T cells in peripheral tissues and blood (see T Cell Differentiation, this Ch.). B. Portion of a thymus lobule (area enclosed in A) in greater detail. (Based on Weiss L: Cells and Tissues of the Immune System. Englewood Cliffs, N.J., Prentice-Hall, 1972)

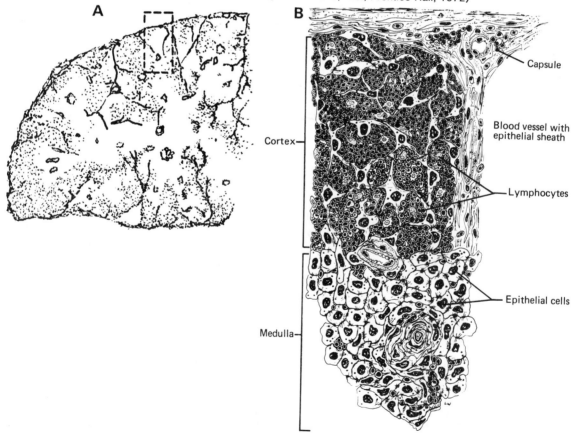

grafts from genetically different (histoincompatible) mice (see Major Histocompatibility Complex, this Ch. and Histocompatibility Genes, Ch. 23); moreover, in response to many Ags they produce little Ab, and the Abs that are formed are predominantly of the IgM class.

The thymus is larger at birth than in the adult. It begins to atrophy around puberty or in occasional rapid bursts after severe stress, probably in response to high levels of adrenal corticosteroids (see Nonspecific Immunosuppression, Ch. 19). Removal of the thymus from an adult usually has minimal effects, because peripheral tissues are already populated with diversified T cells. However, the adult thymus is functional: e.g., some of its lymphocytes serve as competent T cells and T cell precursors, especially of suppressor T cells (see below).

The effects of neonatal thymectomy are duplicated in genetically defective **nude mice** that lack a thymus. The primary defect, in epidermal function, is inherited as a recessive trait associated with hairlessness. Mature, functional T cells are almost totally missing, and so these mice do not make cell-mediated immune responses: they do not reject skin grafts from other mouse strains or even from other species (e.g., chicken skin grafts grow well and even form feathers). Most lymphocytes in nude mice are B cells. Some of the serum Ig levels are normal (IgM, IgG3); others are greatly depressed (IgG1, IgA). What Abs they make are largely IgM molecules.

Nude mice are valuable for analyzing T cell functions. (They are also of practical use for maintaining human cancer cells, some of which can be propagated indefinitely by serial transfer among these mice.) However, nude mice are difficult to produce and maintain: because they make no cell-mediated responses, and poor Ab responses to most Ags, they are vulnerable to infection and do not generally survive more than a few weeks after birth unless raised in a protected environment that reduces exposure to pathogens.

BURSA OF FABRICIUS

One of the earliest clues to the distinction between B and T cells derived from Glick's serendipitous discovery that chicks deprived at birth of the bursa of Fabricius are unable to produce Abs. The bursa resembles the thymus anatomically, and it also develops by lymphocytic infiltration of an epithelial outpouching from the intestine (but from the cloaca, instead of the foregut). However, lymph follicles of the bursa, unlike those of the thymus, are packed with Ig-producing B cells (plasma cells; see Ch. 19).

Extirpation experiments showed that *the bursa and thymus regulate complementary functions* (Table 18-1). In contrast to animals without a thymus, neonatally bursectomized chicks have normal cell-mediated immune responses (e.g., they reject skin allografts [Ch. 23]) but they have few Ab-producing cells, and their severe deficit

TABLE 18-1. Effects of Extirpation of Thymus or Bursa of Fabricius*

Property	Thymectomy	Bursectomy
Cellular changes		
Circulating lymphocytes	↓↓	0
Spleen		
Germinal centers	0(↓)	↓↓
B zones†	0	↓↓
T zones†	↓↓	0
Plasma cells	0(↓)	↓↓
Functional changes		
Serum Ig level	0(↓)	↓↓
Antibody formation	0(↓)	↓↓
Cell-mediated immunity (e.g., rejection of skin allografts)	↓↓	0

*Composite effects of bursectomy and thymectomy in chicks at hatching and of thymectomy in newborn mice.

†Lymph nodes in birds are poorly organized clusters of lymphocytes alongside lymphatic channels; B and T zones are discernible in the spleen.

0 = No change; ↓↓ = marked decrease; ↓ = modest decrease; 0(↓) = variable decrease, probably secondary to role of T cells as helpers in Ab responses to T-dependent Ags (see Ch. 19).

(Based on Szenberg A, Warner NL: Nature 194:146, 1962, Warner NL, Szenberg A: Nature 196:784 1962)

in serum Ig levels and in Ab-forming ability is greater the earlier in life the bursa is removed.

Like the thymus, the bursa also undergoes atrophy at puberty (about 4 months of age in chickens), probably as a result of rising levels of steroid sex hormones. In fact, the bursa can be eliminated by simply dipping embryonated eggs for a few minutes in a solution of testosterone (**hormonal extirpation**).

Mammals lack a bursa of Fabricius, and the mammalian structure with an equivalent function has not been identified. Possibilities include the bone marrow itself, various gut-associated lymphoepithelial structures (appendix, tonsils), the lymphocyte-infiltrated epithelium that lines the intestinal tract, and lymph nodes, which are well-developed structures only in mammals.

SECONDARY LYMPHATIC STRUCTURES: SPLEEN AND LYMPH NODES

The spleen, many lymph nodes, and smaller, less organized clusters of lymphoid cells in many tissues and organs constitute the secondary structures in which committed lymphocytes undergo further differentiation in response to Ag stimulation. These structures are highly effective Ag-trapping filters in which lymphocytes pack the interstices of a dense three-dimensional fibrous network in which many sticky macrophagelike **reticular cells** are attached to reticulum fibers. Injected lymphocytes accumulate in the spleen and lymph nodes, as

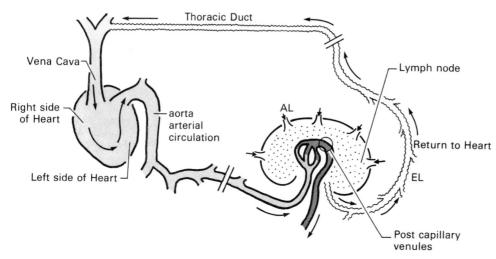

FIG. 18-5. Circulation of lymphocytes. Lymphocytes leave the blood stream at postcapillary venules (small veins) in a lymph node, exiting from the node in efferent lymph channels which merge with the lymph channels from hundreds of other lymph nodes to form, eventually, the large collecting thoracic duct. This empties into the great vein (vena cava) returning blood from the venous circulation to the right side of the heart. A typical lymph node is normally much less than 1/1000 the size of the heart. **AL** = afferent lymph channel carrying lymph *to* the lymph node from nearby tissue spaces. **EL** = efferent lymph channel, carrying lymph *from* the lymph node toward the thoracic duct. (Redrawn from Gowans JL: Hosp Pract 3 (3):34, 1968)

though expressing the **homing instinct** that normally causes circulating lymphocytes to leave blood vessels in these structures (via postcapillary venules or sinusoids) and to reside for a time in appropriate areas (B and T areas; see below) before recirculating.

Lymph Nodes. These small ovoid structures (normally <1 cm diam. in man) are widespread. They generally are situated at major

junctions of the network of lymphatic channels, which pass, tree like, from twigs in the more superficial tissues to a large central collecting trunk, the **thoracic duct**; this vessel pours lymph and lymphocytes into the great vein that returns blood to the heart (Fig. 18-5). Ags deposited in the tissues are carried via lymphatic channels through successive lymph nodes, in which they are likely to be trapped by macrophages. (The brain, spinal cord, and eye lack lymphatic drainage, and Ags carried away in the blood from these

FIG. 18-6. Mammalian lymph node. **A.** Schematic view. ''Naive'' lymph nodes—i.e., before antigenic stimulation—have multiple primary nodules (**PC,** primary cortex) separated by a tertiary, or diffuse, cortex (**DC**). After intense antigenic stimulation primary nodules become secondary nodules (**SN**) with prominent germinal centers (**GC**). The lymphocytes of primary and secondary nodules are mainly B cells; those of the diffuse cortex are mainly T cells. **AL** and **EL** are afferent and efferent lymphatic channels, respectively (see Fig. 18-5). **SS,** the subcapsular sinus, has crisscrossing reticulum fibers that extend throughout the node. The central portion (M) is less packed with cells (see Medulla in **B**). **B.** Human lymph node. The **secondary nodule** label points to a germinal center with its surrounding mantle of small B lymphocytes. The **tertiary cortex** is a mass of T cells. (×45) (B, Weiss L: Cells and Tissues of the Immune System. Englewood Cliffs, N.J., Prentice-Hall, 1972)

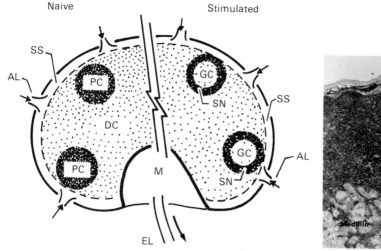

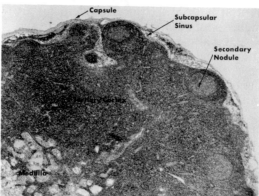

A   B

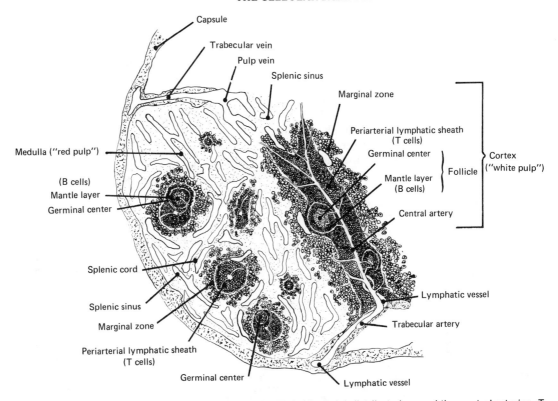

FIG. 18-7. Representative area of the spleen showing masses of lymphoid cells (white pulp) distributed around the central arteries. T cells make up the periarterial lymphatic sheath; B cells form the mantle layer around germinal centers. (Weiss L: Cells and Tissues of the Immune System. Englewood Cliffs, N.J., Prentice-Hall, 1972)

areas are likely to be trapped in the spleen or distant lymph nodes.)

B and T cells are concentrated in different areas of lymph nodes. B cells are accumulated in subcapsular nodules (**B cell areas;** Fig. 18-6) and T cells occupy the spaces between nodules (**T cell areas**). The latter are greatly depleted in nude and in neonatally thymectomized mice.

At the center of each subcapsular nodule is a **germinal center** which is made up of a mass of rapidly dividing large B lymphocytes (generation time about 12 h). Macrophages with ingested pyknotic nuclei and DNA debris are also abundant, as in many other areas with rapidly dividing cells (as though many of the rapidly formed progeny are defective and are disposed of by phagocytosis). *The size of germinal centers is proportional to the intensity of antigenic stimulation:* they are greatly enlarged in secondary Ab responses (Ch. 19) and are essentially absent in germ-free mice. These centers become similarly hyperplastic in other secondary lymphatic structures when intensely engaged in producing Abs.

Spleen. This large encapsulated vascular organ (about 200 g in adult man) traps Ags carried in the blood. Like other lymphatic organs, its cortex has densely packed lymphocytes (with germinal centers), and its loose medulla has wide vascular spaces with a variety of lymphocytes, macrophages, and other cells (Fig. 18-7). In the spleen, however, the medulla (called "red pulp"), encloses cortical areas (called "white pulp"), rather than the reverse as in other lymphatic structures. Cortical areas surround small central arteries with leaky walls, through which Ags are possibly driven by the arterial blood pressure. Separate areas enriched for T and B cells are also discernible in the white pulp (Fig. 18-7); the T area is relatively smaller in the spleen than in lymph nodes, and after intense antigenic stimulation T cells are virtually replaced by sheets of Ab-producing plasma cells (in contrast to lymph node T areas, where plasma cells are almost never seen).

Gut-associated Lymphoid Cells. Lymphocytes (many producing IgA) are spread throughout the inner layers of the intestinal wall as isolated cells or as small cell clusters. Larger clusters form distinct follicles with germinal centers; some follicles become confluent and interdigitated with the overlying epithelium, forming small **lymphoepithelial** structures which (unlike the thymus and bursa) remain associated with the intestinal wall throughout life. The main ones in man are 1) tonsils (in the pharyngeal wall), 2) appendix (at the junction of small and large intestine), and 3) Peyer's patches (oblong lymphoid aggregates found mostly in the wall of the terminal part of the small intestine). As in other secondary lymphatic structures, follicular (B) areas around germinal centers are well developed and interfollicular (T) areas are atrophic in individuals without a thymus.

LYMPHOCYTES

B AND T CELLS

Lymphocytes are ubiquitous in blood, lymph, and connective tissues. The two fundamentally different classes, B and T, look alike: small, motile, nonphagocytic cells of varying size (about 6–15 μm diameter) (Fig. 18-8). In the smallest ones (6–8 μm) mitochondria are scanty and no endoplasmic reticulum is discernible; in the light microscope the cytoplasm forms a barely perceptible rim around the dense nucleus (Fig. 18-8A). Specific binding of Ag by membrane-bound receptors on the cell surface stimulate the **transformation** of small lymphocytes into large ones (up to 15 μm diameter), whose more abundant cytoplasm is richer in mitochondria and in polysomes; some large B lymphocytes also contain a secretory system (endoplasmic reticulum, Golgi apparatus) and secrete Abs (Fig. 18-8B; also see Fig. 19-7, Ch. 19). Some large lymphocytes of B lineage differentiate into mature plasma cells (Fig. 18-8C), the most active of all B cells in the synthesis and secretion of Igs. Other large lymphocytes are thought to revert back to small ones, which probably function as memory cells (see Primary and Secondary Responses, Ch. 19). The majority of small lymphocytes in the periphery are small quiescent cells until stimulated by Ag, when they become not only larger but also highly motile, and may divide (generation time, about 8–48 hrs).

Some lymphocytes ("long-lived") can persist without division for long periods. For instance, even after 40 days' continuous infusion of ^{3}H-thymidine into mice or rats the DNA in most of the small lymphocytes in the blood remains unlabeled, showing that these cells synthesized no DNA during this interval. The longevity of nondividing blood lymphocytes has also been demonstrated by exposing them in intensive x-ray therapy and then, after varying intervals, to mitogens (see below), which force the cells to undergo mitosis; cytologic examination reveals some cells with such pronounced chromosomal aberrations that they are unlikely to have

FIG. 18-8. Principal cells involved in immune responses. A. Small ("resting") lymphocyte. B. Large lymphocyte or lymphoblast. C. Plasma cell. D. Macrophage. The macrophage contains a phagocytized red cell (**arrow,** inset D), corresponding to the larger of the dark cytoplasmic inclusions in the associated electron micrograph. The plasma cell (inset C) has its secretory vesicles distended with Ig molecules. The small and large lymphocytes in panels A and B could be of B or T lineage. The plasma cell (C) is the Ab-secreting, terminally differentiated cell of B lineage. The macrophage (Mφ, D) is an accessory cell that binds Ag nonspecifically and "presents" it to lymphocytes (especially T cells); Mφ are derived from a separate lineage (see Fig. 18-3). Electron micrograph magnifications are A, ×7500; B, ×6600; C, ×6600; D, ×5500. Upper left insets are light microscope photographs of the corresponding cells, stained with toluidine blue; magnifications are approximately A, ×4000; B, ×3000; C, ×3000; D, ×2000. (Courtesy of Dr. R. G. Lynch)

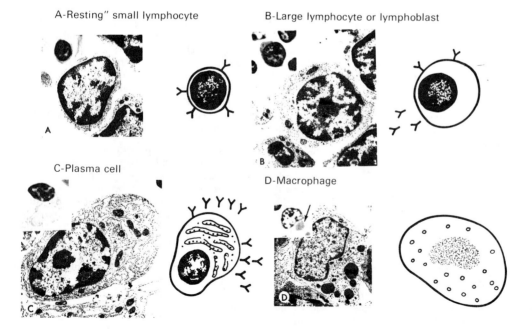

A-Resting" small lymphocyte

B-Large lymphocyte or lymphoblast

C-Plasma cell

D-Macrophage

divided once since incurring their x-ray damage. From the decreasing incidence of such cells after irradiation it has been calculated that *the mean survival time of circulating small lymphocytes in man is about 5 years,* which might account for the long persistence of immunologic memory (Ch. 19).

CELL SURFACE ANTIGENS THAT DISTINGUISH B AND T CELLS

B and T cells differ in many surface molecules. These are sometimes called **differentiation Ags** (because they change as cells progress from one stage of differentiation to another), and many of them are **alloantigens** (alloAgs): like Ig allotypes (Ch. 17), variants of these Ags distribute in a breeding population as though encoded by allelic genes. Consequently they are conveniently recognized by reciprocal crossimmunization of inbred mouse strains, which are homozygous at essentially all genetic loci (see Major Histocompatibility Complex, this chapter). Hence alloAbs are readily elicited by injecting mice of a strain that lacks an alloAg with cells from a strain that possesses it (and vice versa).

Congenic Strains. To restrict the Ab responses of the mutually crossimmunized strains of mice to the alloAgs of interest it is advantageous to use **congenic pairs of strains:** e.g., strain A.B is derived from A by special breeding programs, developed by Snell (one is illustrated in Fig. 18-9), that substitutes a bit of chromosomal material from B, bearing the allele for the alloAg of interest, for the homologous chromosomal material of A. Ideally, because Abs are not elicited by self Ags (Tolerance, Ch. 19), the only Abs made by A mice against A.B cells are directed against the product of the substituted B gene. Usually, however, the B chromosomal fragment introduced into the A genome probably consists of several linked genes; moreover, over the several years of breeding (at least 10 generations) required to establish a congenic strain, random mutations can cause the strains to differ in other genes. Nevertheless, *congenic strains represent a powerful means for isolating one or a few genes, and determining their effect, by transfering them from one genome to another.*

Assays for Cell Surface Antigens. Cell surface Ags of lymphocytes are often detected by **cytotoxicity assays.** When Abs of an appropriate class (e.g., IgM) bind to cell surface Ags, the addition of serum complement (Ch. 20) damages the cell membrane, causing cell death (**"lysis"**). Permeability to anionic dyes, such as trypan blue, is commonly used to count dead cells: the dye penetrates into and stains dead cells while living cells, with intact surface membranes, remain unstained.

Lymphocytes are unusually susceptible to damage by Abs plus complement. To determine whether less susceptible cells share

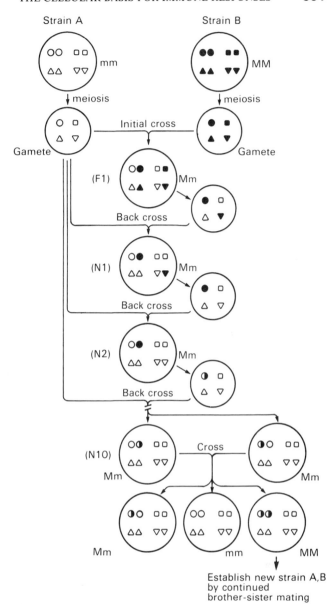

FIG. 18-9. Production of congenic strain A.B. Following an initial cross between an inbred A mouse and an inbred B mouse, and a back-cross of the F1 generation to strain A, a series of successive back-crosses to strain A is carried out with progeny that possess the strain B marker of interest (M; e.g., a particular alloantigen). Offspring of the tenth back-cross generation **(N10)** are intercrossed, and progeny that are homozygous for M (e.g., they have the alloantigen of the B strain, not of the A strain) are then inbred by successive brother–sister matings to establish the new congenic A.B strain. This strain is identical genetically with inbred strain A except for the transferred chromosomal segment of strain B that bears the allele of interest. The principle is a general one: a congenic strain can be established for any gene. (If the transferred allele codes for a histocompatibility Ag and leads to graft rejection, the congenic A.B line is sometimes called a **congenic resistant (CR) line,** because it resists being killed by a grafted A strain tumor.) (Courtesy of Dr. R. J. Graff)

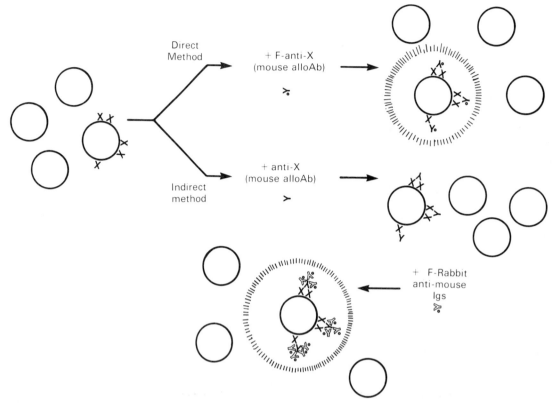

FIG. 18-10. Visualization of a cell surface Ag by immunofluorescence (IF). In the **direct method,** the surface Ag **(X)** is stained with fluorescein-labeled **anti-X** (e.g., a mouse alloantibody, F-anti-X). The same Ag is detected with greater sensitivity in the **indirect ("sandwich") method:** mouse anti-X binds to X and then fluorescein-labeled rabbit Abs to mouse Igs combine with the mouse anti-X Abs. Because several rabbit Ab molecules can bind to one mouse Ig molecule, more fluorescence is localized on X in the sandwich technic than in the direct method.

surface Ags with lymphocytes, the suspected cells are usually tested for their ability to adsorb the Abs that lyse the appropriate lymphocytes **(adsorption assay).**

Immunofluorescence with fluorescein-labeled Abs to a membrane Ag provides another generally useful method to detect cell surface Ags, since Abs do not penetrate into intact cells (Fig. 18-10).

The following surface Ags have been particularly useful for distinguishing B and T cells and some of their subsets (Table 18-2) in the mouse. Parallel findings in human T cells are described at the end of this Chapter (see Differentiation Ags and Human T Cell Subsets).

Thy-1 and Immunoglobulins. The first lymphocyte surface alloAg discovered by cross-alloimmunization has been especially useful. Originally called **theta** (θ) and now known as **Thy-1,** it is present on all T lymphocytes but not on B cells. In contrast, B cells have abundant surface Ig, readily detected by immunofluorescence; T cells lack conventional Igs, (though they have surface receptors with V domains of Ig heavy chains; see Antigen-Binding Receptors on T Cells, below). Accordingly, *surface Thy-1 and Ig distinguish the two major lymphocyte classes: B cells are Ig^+ $Thy\text{-}1^-$ and T cells are Ig^- $Thy\text{-}1^+$.*

There are two allelic variants of Thy-1 in mice (Thy-1.1 and Thy-1.2); each inbred mouse strain has one or the other. Besides being present on thymocytes and peripheral T cells, Thy-1 (a glycoprotein) is also found in brain and on cultured fibroblasts.

Ly Antigens. These Ags are found only on lymphocytes: **Lyt on T cells, Lyb on B cells.** The Lyt Ags have been shown by Boyse and Cantor to distinguish T cell subsets that differ in function. Almost all thymocytes and about 50% of peripheral T cells are **Ly-123 cells;** i.e., they are lysed by alloantisera (plus complement) to three Lyt Ags: Lyt-1, Lyt-2, and Lyt-3. About 35% of peripheral T cells are **Ly-1 cells** (lysed only by anti–Lyt-1) and about 5%–10% are **Ly-23 cells** (lysed only by anti-Lyt-2 or

TABLE 18-2. Some Cell Surface Antigens That Distinguish Between Mouse B and T Lymphocytes

Surface antigen	B cells	T cells
Immunoglobulins	+	(0)*
Thy-1	0	+†
Lyt-1, 2, 3	0	+‡
Ia (encoded by IA and IE loci of the H-2 complex)	+	0
Ia (encoded by IJ locus of the H-2 complex)	0	+
TLa (thymus-leukemia Ag)	0	+§
GIX	0	+
Fc receptors	+	(0)
Complement receptors	+	0
Lyb-4	+	0
Ly-5	0¶	0¶
Qa	0	+§§
PC-1	+**	0

*The Ag-recognition molecules on some T cells seem also to contain V_H and possibly V_L segments of Igs.

†The allelic variant Thy-1.1 is present in AKR mice, and the variant Thy-1.2 is present in most of the other inbred mouse strains. Thy-1 is also present in brain.

‡There are two allelic forms of each Lyt gene: 1.1 and 1.2, 2.1 and 2.2, 3.1 and 3.2. Genes for Lyt-2 and Lyt-3 are linked to each other and to the structural genes for V_k (on chromosome 6); the Lyt-1 gene is on chromosome 12.

§Normally only on thymocytes of certain strains of mice (TL⁻); also on leukemic T cells of all strains (TL⁻ as well as TL⁺).

¶On null cells, which are lymphocytes that lack surface Ig and Thy-1.

§§Qa-1,2,3,4,5 Ags are found predominantly on T cells, though Qa-2 and Qa-3 are also on Thy-1⁻ cells. For the function of Qa-1 + T cells see T Cell Regulatory Circuits, Ch. 19.

**PC-1 is found on the most mature B cells (plasma cells, see Ch. 19), and also on liver cells.

anti–Lyt-3).* *The T helper (T_h) cells that enhance Ag-stimulated maturation of B cells into Ab-secreting plasma cells (Ch. 19), and the T cells (T_{DH}) that mediate delayed hypersensitivity (Ch. 22), are Ly-1 cells, while the T suppressor (T_s) cells, which block the activity of T_h cells, and T "killer" (T_k) cells, which specifically lyse diverse target cells, are Ly-23 cells* (Table 18-4, below).

Ly-123 cells account for nearly all T cells in early postnatal life and are precursors of Ly-1 and Ly-23 cells. Preliminary observations suggest that Ly-123 cells serve as the main regulators of immune responses by controlling the balance of activity between T_h and T_s cells (see T Cell Regulatory Circuits, Ch. 19). Though these Ags are now recognized only in mice, because the diagnostic alloantisera are made by crossimmunizing inbred strains,

* Immunofluorescence shows that Ly-23 cells also have some Lyt-1 Ag, but evidently not enough (or not appropriately distributed) to render them susceptible to elimination by anti-Lyt-1 Abs (and complement).

preliminary findings indicate that similar surface Ags distinguish comparable T cell subsets in man (see p 417).

Ia Antigens of the Major Histocompatibility Complex. Like all nucleated cells, B and T cells have surface Ags coded for by certain loci of the major histocompatibility complex (see that section below). However, the surface Ags (called Ia, for I-associated) specified by the I locus of this complex in mice are confined to lymphocytes (and macrophages), and are distributed differently on B and T cells. Ia Ags specified by certain loci of this region (IA, IB, IE, and IC) are on B cells and the Ia Ag of another I locus (IJ) is on certain T cells (suppressor T cells, see below).

TL Antigen. The "thymus-leukemia" Ag (TLa) has two distinctive features: 1) Some mouse strains (TL⁺ express it only on thymocytes; other strains (TL⁻) do not express it on any normal cells. In F1 hybrids (TL⁻ × TL⁺) and in back-crosses of the F1 hybrids to TL⁻ strains the TL Ag behaves as though specified by an autosomal dominant allele. However, 2) TL is also present on the malignant neoplastic T cells of leukemic mice of TL⁻ strains. Hence TL⁻ strains possess the structural gene for TL; the apparent allelism (TL⁺/TL⁻) is probably due to a regulatory gene.

GIX. This surface alloAg, present on thymocytes and other T cells of many mouse strains, is absent from B cells. The same molecule is also present in the envelope of murine leukemia viruses (MuLV), where it is called **gp 70** (glycoprotein of 70,000 daltons). Like other MuLV genes, the one that encodes GIX (gp 70) is integrated into the genome of many mouse strains, but the GIX gene appears to be unique in that it is expressed regularly, as a differentiation alloAg, during normal development of T cells in the thymus of some strains. GIX is also produced by many other normal cells besides T lymphocytes; it is present, for example, at an especially high level in seminal fluid.

Fc Receptors. Present on nearly all B cells, on T_s cells, and on macrophages, these receptors bind Igs (aggregated specifically by Ag or nonspecifically by heat denaturation) via the Fc domain; i.e., they do not bind Fab, Fab', or F(ab')₂ fragments (even if aggregated). Whether the receptors on B and T_s cells and on macrophages are identical is not known.

Complement Receptor. These cell surface structures bind special proteolytic fragments of the third complement (C) component (C3b and C3d; see Ch. 20) in Ab-Ag-C complexes. Immature B cells lack the C receptor (CR⁻) and can be distinguished from mature CR⁺ B cells by **rosette formation**: red blood cells with Ab and C proteins on their surface bind to CR⁺ B cells, forming rosettes (Fig. 18-11).

PC-1. In mice, plasma cells (the Ab-secreting, terminally differentiated form of the B cell; Ch. 19) have a distinctive cell surface alloAg, PC-1, recognized by alloantiserum elicited against the plasma cells of malignant myeloma tumors. PC-1 is present in some strains (PC-1⁺) and absent in others (PC-1⁻). It is expressed late in differentiation: the only cells in the B lineage with PC-1 are plasma cells.

Lyb-5. This differentiation Ag marks a subset of B cells that can be stimulated to proliferate and secrete Igs by a special set of "T-in-

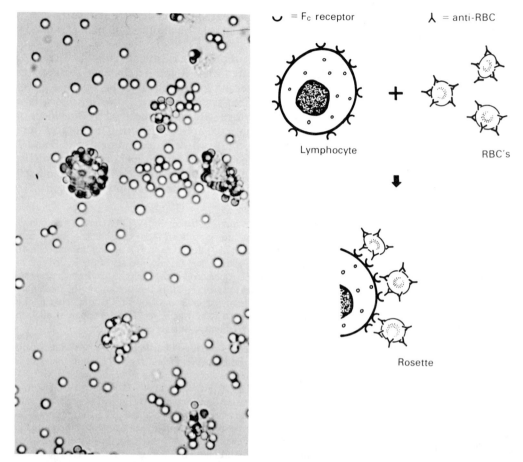

FIG. 18-11. Identification of cell surface receptors by rosette formation, i.e., by the adherence of red blood cells (RBCs) that are coated with appropriate ligands. For example, to detect lymphocyte receptors for Fc domains of Igs (FcR), the RBCs are coated with Abs (anti-RBC). To detect receptors for complement (C), the RBCs are coated with anti-RBC Abs and C components (C3 is essential; see Ch. 20). Rosetted lymphocytes sediment more rapidly than nonrosetted ones; the two can thus be separated physically, yielding sets of cells with different surface Ags. More efficient automated isolation of cells with particular surface Ags is achieved in the fluorescence-activated cell sorter (FACS): cells with the surface Ag bind the corrresponding fluorescent Ab and are separated from nonfluorescent cells, which lack the Ag. (Courtesy of Drs. K. Hannestad, M. S. Kao, and R. G. Lynch)

dependent" Ags, such as trinitrophenyl-ficoll (see mitogens, below, and Ch. 19). Mice of the CBA/N strain lack B cells with Lyb-5, do not respond to these particular T-independent Ags, and can be used to produce alloantiserum to Lyb-5. F1 hybrids made by crossing CBA/N with mice of a normal strain, such as BALB/c, have shown that expression of the Lyb-5 marker is linked to the X (sex) chromosome.

As is discussed below (see Null Cells), Lyb-5 is present on "natural killer" lymphocytes, which lack surface features of B and T cells.

MITOGENS AND BLAST TRANSFORMATION

Various substances of plant and bacterial origin stimulate lymphocytes to proliferate and undergo **blast trans-**

formation: the cells enlarge, the nucleolus swells, polysomes, rough endoplasmic reticulum, and microtubules develop, and the rates of protein and nucleic acid synthesis increase markedly (Fig. 18-12). An increase in the rate of ^{3}H-thymidine incorporation into DNA provides a convenient measure of the effects of these mitosis-promoting agents, or **mitogens.** Among the mitogens are plant proteins called **lectins** (L. *legere*, to pick and choose), which react specifically with diverse oligosaccharides of cell surface glycoproteins. Some lectins, such as phytohemagglutinin A (PHA) and concanavalin A (con A), are mitogenic for T cells, not for B cells.

Other mitogens act only on B cells, causing them to secrete Igs (mostly IgM). Because the resulting Igs have various allotypes and types of L chains these mitogens

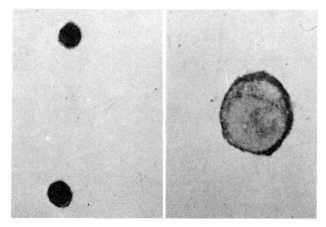

FIG. 18-12. Blast transformation. A. Small lymphocytes. B. The same after treatment with an appropriate lectin or Ag.

are **polyclonal B activators** (PBA). Usually they are high-molecular-weight substances with repeating units. Lipopolysaccharide (LPS) of *E. coli* is one of the most potent. Different subsets of B cells are stimulated by different PBAs—e.g., dextran acts on B cells from embryonic mice, LPS on mature B cells. The T cell mitogens, PHA and con A, are polyclonal activators of T cells.

In contrast to a polyclonal activator, an Ag, when presented to lymphocytes in the appropriate manner (see Macrophages, below, and Mechanisms of Ir Gene Control, Ch. 19), induces mitogenesis and blast transformation of only one or a few clones of B or T cells.

ANTIGEN-BINDING RECEPTORS (RECOGNITION MOLECULES) OF B LYMPHOCYTES

Ags are bound to B cells by cell surface Igs. These receptors were originally thought to be the same as the Ab molecules that are secreted by mature B cells after antigenic stimulation. However, the receptor Igs and the secreted Igs differ in C_H domains. The μ chains of the membrane IgM and of the secreted IgM differ at the COOH-end. This end, in the membrane form, is extended by about 15 amino acids; the latter are highly hydrophobic, suggesting that they anchor the surface IgM molecule in the cell membrane's lipid bilayer (Fig. 18-13). Similarly, the γ chain in membrane IgG is slightly longer than in secreted IgG.

Most secreted Ig molecules are IgGs, but relatively few B cells have surface IgG (less than 10% of mouse spleen cells). Instead, the surface Ig on most B cells is monomeric (7S) IgM and many B cells also have surface IgD, which is normally present at only trace levels in serum (Ch. 17).

The proportions of B cells with various surface Ig classes, and the change in these proportions with time after antigenic stimulation, are considered in detail in Ch. 19.

One Cell–One Idiotype. When a B cell has IgM and IgD on its surface, the molecules of both classes have the same type of L chain (λ *or* κ) and the same idiotype (revealed by immunofluorescence with antiidiotypic Abs). Indirect evidence (see Increase in Affinity with Time, Ch. 19) also suggests that surface and secreted Igs of a given B clone have the same specificity and affinity for ligands (implying the same V_L and V_H domains and the same idiotype). Thus what was once regarded as the **one cell–one Ab rule** has become, with more information, the **one cell–one idiotype rule. The significance is that each B cell produces Igs with a single Ag-binding specificity, even though the cell may produce Igs of several classes** (e.g., two on the surface, another secreted).

ANTIGEN-BINDING RECEPTORS (RECOGNITION MOLECULES) ON T CELLS

The nature of the Ag-recognition molecules on T cells is not yet established. Some Ags elicit T cells whose Ag-recognition molecules have idiotypes (Id) that are the same as, or similar to, those of the Ab molecules elicited by the same Ag, suggesting that the V domain of Igs are involved. However, antisera to C domains of Igs react poorly, if at all, with T cells (hence, these cells, in contrast to B cells, are Ig⁻ by immunofluorescence). Moreover, Ag-recognition by many T cells is strongly dependent upon surface histocompatibility (H) Ags (see Major Histocompatibility Complex, MHC, below). It is believed that Ag-recognition by T cells depends upon IgV domains that may perhaps be associated with other

FIG. 18-13. Ag-binding receptors on B and T lymphocytes. B cells have surface Ig molecules (usually 7S IgM and IgD); some B cells also have Igs of other classes. T cells lack conventional Igs, but some of them share idiotypes (representing V domains of Igs) with some Abs. For instance when mice of certain strains are immunized with Ags containing the NP group, the resulting Abs, B cells, and some T cells share a particular idiotype (recognized by an anti-idiotype Ab). (See Fig. 17-21, Ch. 17). (Based on Krawinkel U, et al: Eur J Immunol 8:566 1977 and Wall KA, et al: in press, 1980)

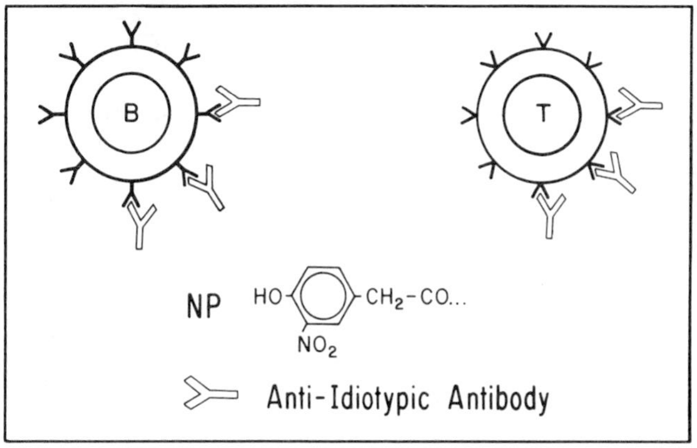

domains, encoded by genes of the MHC, rather than with conventional C_H or C_L domains. Once considered an outlandish idea, this possibility has become more plausible with the discovery that encoding sequences from widely separated parts of a chromosome can join to specify different domains of the same polypeptide chain (Ch. 17, Ig genes). (Of course, it has long been clear that different chains of the same molecule [e.g. L and H chains of Ig] can be specified by genes on different chromosomes.)

CIRCULATION OF B AND T CELLS

As noted before, as lymphocytes circulate through the blood, lymph, and peripheral lymphatic tissues (lymph nodes, spleen, Peyer's patches of intestinal wall), T cells tend to congregate in special interfollicular areas, while B cells tend to cluster in follicles (see Figs. 18-6 and 18-7). Lymphocytes subsequently leave these areas of lymph nodes via efferent lymph channels and return to the blood via the thoracic duct (Fig. 18-5). Normally, the majority of circulating lymphocytes are T cells: e.g., about 80% in the thoracic duct lymph.*

If a circulating lymphocyte specific for a particular Ag encounters that Ag in tissues, the lymphocyte adheres to it and temporarily drops out of the circulation. The bound lymphocytes subsequently proliferate, become lymphoblasts, dissociate from the tissue Ags, and reenter the circulation. The lag between binding and release (about 2 days) provides the basis for use of the living animal as a filter to fractionate lymphocytes by a form of "affinity chromatography" (see Appendix, Ch. 16): a population of diverse lymphocytes, some of which are specific for a particular tissue Ag, is injected intravenously and thoracic duct lymph is collected. For the first 1–2 days, the collected thoracic duct lymphocytes are depleted of the Ag-specific cells, but a few days later these cells emerge as lymphoblasts and make up a large proportion of the thoracic duct cells (Fig. 18-14).

CAP FORMATION

When the living B cell reacts with fluorescein-labeled anti-Ig the cell's surface is initially stained diffusely. After a few minutes, the fluorescent material aggregates into patches, and then coalesces into a polar cap (Fig. 18-15). *Patch formation represents a two-dimensional pre-*

---

* In adult humans with chronic lymphatic leukemia, the greatly increased number of circulating lymphocytes (e.g., 100 times normal) are usually neoplastic B cells (one tumor clone per patient). In children, lymphocytic leukemia is a more acute disease, with neoplastic T cells in some patients, neoplastic B cells or null cells (lymphocytes without T or B markers) in others. Leukemia in mice usually arises in the thymus and the affected animal has an enormous clone of neoplastic T cells that circulate in the blood and infiltrate the tissues.

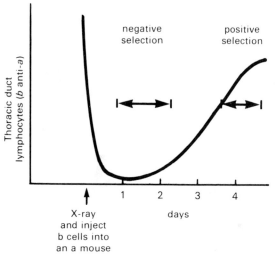

FIG. 18-14. Use of the living animal to remove Ag-specific lymphocytes from the circulation and then to return them in large numbers. A recipient mouse of strain *a* is heavily x-irradiated (to eliminate most of its lymphocytes) and injected **(arrow)** with spleen cells from a mouse of strain *b*. During the following 1–2 days (period of **negative selection**) the specifically reactive lymphocytes (*b* anti-*a*) leave the circulation and react with *a* Ags in the tissue, proliferating and differentiating into blasts. After 3 days the lymphoblasts reenter the circulation in large numbers (period of **positive selection**). Based on Sprent, J. et al: Cell Immunol 2:171 (1971).

cipitation reaction, crosslinking surface molecules that are diffusible in the cell's fluid membrane: it does not occur if the fluorescein-labeled Fab fragment (univalent) of the anti-Ig Ab is used. Caps seem to arise from patches that are swept together in the cell's moving membrane; hence the formation of caps, but not patches, requires cellular metabolic activity.

Surface Ig molecules can also be aggregated into patches and caps by multivalent Ag. This aggregation could be a key step in the Ag-stimulated proliferation and differentiation of B cells (Ch. 19).

Though capping of B cell surface Igs occurs rapidly and dramatically, the same process can occur on other cells with other surface Ags crosslinked by Abs (or even by multivalent lectins). For instance, capping can also be brought about on T cells, and on many other cells, by Abs to surface histocompatibility Ags.

Capping can serve to determine whether various macromolecules are normally linked on the cell surface: when linked, they are capped together (**cocapped**) by an Ab to one of them. For instance, the reaction with a mixture of two Abs having different fluorescent chromophores shows the independence of IgM and IgD on the same B cell: flourescein-labeled anti-μ (green) and rhodamine-labeled anti-δ (red) yield separate red and green caps.

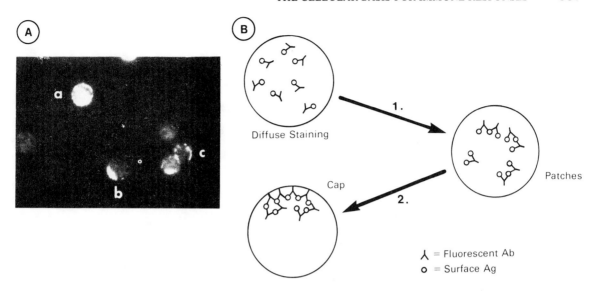

FIG. 18-15. Aggregated Igs in the surface membrane of lymphocytes. A. Living white blood cells (of rabbit) were stained with fluorescein-labeled Abs to a rabbit Ig allotype. The speckled appearance of cells **a** and **c** represents patches of membrane Igs crosslinked by the anti-Ig; the patches are accumulated into a polar cap in cell **b.** The reaction was at 4° and the cells were examined at room temperature. At 37° most of the stained cells would have caps as in **b.** B. Schematic representation of the separate requirements for patching and capping. (Courtesy of Dr. J. M. Davie)

The caps formed by aggregated surface Igs are eventually shed or ingested by the cell and degraded into fragments. B cells thus lose their receptors, which cannot be stained again until regenerated. The time required (6–8 h in mature B cells) agrees with other evidence for a slow dynamic turnover of surface Igs.

Membrane movement is required for capping, and conditions that block motility inhibit cap formation; e.g., lowering the temperature to 4°C, blocking ATP generation with azide or cyanide, or adding demecolcine (Colcemid), an agent that disaggregates cytoplasmic microtubules and disorganizes the cell's locomotory system. Coalescence of patches into a polar cap can also occur in motile fibroblasts but not in nonmotile epithelial cells.

T CELL SUBSETS

Broadly speaking, there are two kinds of T cells: **effectors** and **regulators** (Table 18-3). The effectors cause various cell-mediated immune reactions: e.g., T_{DTH} cells cause slowly progressive inflammatory reactions called delayed-type hypersensitivity (DTH; see Ch. 22), and are responsible for increased resistance to many infectious agents. Cytotoxic T lymphocytes (CTLs, or T_K, for T killer cells) cause lysis of particular target cells. In contrast, regulator T cells control the maturation of effector lymphocytes—both the various types of effector T cells and effector B cells (Ab producers). T helpers (T_h cells) enhance, and T suppressors (T_s cells) block, the development of effectors but most T cells are regulators (T_r) that control the extent to which T_h or T_s cells dominate the response to an Ag. In mice, the T_r cells belong to the

Ly-123 set. They are considered later in connection with Ab formation (under T Cell Circuits, Ch. 19). The various T cell sets differ not only in function but in Lyt surface Ags (Table 18-4). The conditions that lead to production of the effector T cells, and their mechanisms of action, will be considered in Ch. 22, while the production and activity of B cells (Ab production) are considered in

TABLE 18-3. T Cell Subsets* and Their Functions

T Cell Type	Function
T_h†, T helpers	Aid Ag-stimulated development of B cells and of T effector cells (T_{DTH} and T_k)
T_s, T suppressors	Block induction and/or activity of T_h cells
T_r, T regulators	Develop into T_h or into T_s cells and control the balance between enhancement and suppression of responses to Ag
T_{DTH}	Mediate the inflammation and the nonspecific increased resistance to many infectious agents associated with activated macrophages in delayed-type hypersensitivity
T_k, T killer cells; cytotoxic T lymphocytes, CTLs	Lyse target cells

*Various sets differ in surface Ags (e.g., Ly; see table 18-4).

†It is not clear whether the T_h cells that cooperate with B cells are the same as the T cells (sometimes called T amplifiers or T inducers) that aid the development of various effector T cells (T_{DTH}, T_k).

TABLE 18-4. Correlation Between Lyt Surface Antigens and T Cell Function†

Lyt Group			
Designation	Lysed by*	Function	Prevalence
Ly-123	Anti-Lyt-1 or Anti-Lyt-2 or Anti-Lyt-3	Precursors of Ly1 and Ly2,3 cells. Regulate the development of T_h and T_s	50% of peripheral T cells >95% of thymocytes >95% of peripheral T cells in newborn (1–2 wk)
Ly-23	Anti-Lyt-2 or Anti-Lyt-3	T_k cells (cytotoxic T cells) T_s cells (T suppressors)	5%–10% of peripheral T cells
Ly-1	Anti-Lyt-1	T_h cells: help Ag-driven differentiation of B cells and T_k cells T_{DTH} cells: mediate delayed-type hypersensitivity	30%–35% of peripheral T cells

*With complement (see Ch. 20).

(Based on Cantor H, Boyse EA: J Exp Med 141:1376, 1390; 1975)

†For mouse lymphocytes. For comparable sets of human T cells, see below, this chapter.

Ch. 19. The regulatory T cells are considered here because they profoundly influence virtually all immune responses, Ab production as well as cell-mediated reactions.

T HELPER CELLS

The discovery that immune responses to most Ags depend upon interactions among diverse lymphocytes came about through **reconstruction experiments:** mice deprived of their own lymphocytes by whole-body x- or γ-irradiation can be restored to immunologic function by an injection of living lymphocytes from other mice (**adoptive transfer;** e.g. Fig. 19-3, Ch. 19). In carrying out such experiments, Claman and coworkers discovered that the amount of Ab formed against sheep red blood cells (RBCs) by irradiated recipients of both normal bone marrow cells and thymus cells greatly exceeded the sum of the amounts formed by recipients of either one alone (Table 18-5): *the cooperation of both cell populations was clearly necessary to restore an optimal Ab response.*

Subsequent analyses showed that *the Ab-producing cells are derived from the marrow.* Thus, when marrow and thymus cells were obtained from congenic donors (Fig. 18-9) that differed only in Ig heavy chain allotype, the allotype of the resulting Abs was always that of the bone marrow donor. Thymus cells, whose contribution can be eliminated with anti–Thy-1 Abs (plus complement), are T helper cells or their immediate precursors. We shall see later that cooperation between T helper and B cells requires the compatibility of cell surface glycoproteins encoded in the major histocompatibility complex (see H-2 Restriction, below).

T cells (T_h) that help B effectors belong to the Ly-1 set (in mice; see Table 18-4). Some Ly-1 cells have a similar amplifying effect on

the Ag-stimulated development of T effectors (T_{DTH} and T_k cells). Whether these T amplifiers and the T_h cells are the same is not clear. T_h cells are discussed further in relation to Ab production (see The Role of T Helper Cells, Ch. 19).

T SUPPRESSOR CELLS

Unlike T helpers, T suppressors (T_s cells) block the development of effector B and T cells. T_s cells were discovered by Gershon and colleagues while studying immune tolerance, an unresponsive condition in which an Ag that ordinarily elicits immune responses fails to do so (see Tolerance, Ch. 19). They discovered that tolerance to sheep RBCs in mice can be "infectious": i.e., it could be

TABLE 18-5. Antibody Responses are Promoted by Cooperation Between Thymus and Bone Marrow Cells*

	Recipients	
Donor cells	% Spleen fragments secreting antibody	Serum antibody activity†
None	0.8	None detected
Bone marrow (1×10^7)	1.3	None detected
Thymus (5×10^7)	12.3	0.2
Bone marrow (1×10^7) Thymus† (5×10^7)	53.7	2.1

*X-irradiated (800 r) mice were injected with the indicated numbers of thymus or bone marrow cells (or both, or neither) from normal donors, and the recipients were then given sheep red cells (SRBCs) as Ag. Eleven days later Ab production was determined by measuring the serum Ab titer and the percentage of spleen fragments that secreted hemolytic anti-SRBC Abs in culture.

†Log₂ hemolysin titer.

(Based on Claman HN et al: J Immunol 97:828, 1966)

transferred by spleen cells from such tolerant donors to normal recipients, who became tolerant to the sheep RBCs, but not to other Ags (including horse RBCs). The responsible cells are T cells of the Ly-23 type, and, as mentioned above (see Ia Antigens of the Major Histocompatibility Complex), they have surface Ags encoded by the IJ region of the major histocompatibility complex. Specific T_s cells have now been demonstrated to play a large role in tolerance to many proteins, both in Ab- and in cell-mediated immune responses. In addition, genetic unresponsiveness to some Ags is due to the greater stimulation of T_s than of T_h cells by these Ags (see Mechanisms of Ir Gene Control, Ch. 19).

There appear to be two types of T_s cells, **T_s-1 cells** produce, in response to Ag, a soluble factor with unusual properties **(suppressor factor)**: it binds the Ag specifically, but is not an Ig molecule. It is small (about 50,000 daltons) and lacks determinants of constant domains of known Ig chains, but it apparently has V_H domains, for it has some idiotypes in common with Abs that are specific for the same Ag, and it also has structural features in common with products of the IJ region of the major histocompatibility complex. Hence the two major gene complexes of the immune system—the major histocompatibility complex and the linked genes for Ig H chains—encode parts of the suppressor molecule.

The soluble factor from T_s-1 cells seems to act on other T cells, causing their development into T_s-2 cells, which block T_h activity; hence B cells, that would otherwise be stimulated, fail to mature into Ab-secreting cells.

An effect of T_s on T_h cells is indicated by the following analysis of allotype suppression (Ch. 19). A cross between *a* and *b* strains of mice with different Ig heavy chain allotypes yields hybrid progeny (*a* × *b*) that normally produce roughly equal amounts of the *a* and *b* allotypes, called Ig-1^a and Ig-1^b (see Table 17-7, Ch. 17). If, however, the hybrids are injected at birth with Abs to the Ig-1^b allotype, the mice are unable to produce Igs of this allotype, but they produce other Igs, including normal (or elevated) amounts of Ig-1^a. Spleen cells from these allotype-suppressed hybrids transfer the suppression to adoptive (irradiated) hosts: i.e., these spleen cells block the production of Ig-1^b molecules, but not of Ig-1^a or other Igs, by simultaneously transferred spleen cells from a normal *a* × *b* hybrid. When suppressor cells are eliminated from the spleen cell population by treatment with anti-Lyt-2 serum (plus complement) T_h cells are unaffected, because they are Ly-1 cells. However, the remaining T_h cells still cannot enhance the production of Ig-1^b molecules, though they can enhance the formation of Ig-1^a and other Igs. These observations suggest that the T_s cells eliminated the corresponding T_h cells (specific for Ig-1^b) before they themselves were eliminated by the anti-Lyt-2 Abs; in the absence of these T_h cells, B cells are unable to produce Ig heavy chains with the corresponding (Ig-1^b) allotype.

As is generally found with cells in other differentiation pathways, precursors of T_s cells (pre-T_s cells) are readily inactivated by agents that destroy rapidly dividing cells, such as low doses of x-irradiation and of cyclophosphamide (see Nonspecific Immunosuppression, Ch. 19). Mature T_s cells are resistant to these agents. T_s cells, especially T_s-1 cells, are short lived. Hence in systems where Ab production and cell-mediated immunity are strongly suppressed, because the Ag elicits more T_s than T_h cells, elimination of T_s-1 cells (by low doses of cyclophosphamide or x-ray), at about the time the Ag is administered, can result in *enhancement of the immune response*. If the blocking agents are given after the Ag has already elicited T_s cells, no enhancement results.

NULL CELLS

A small proportion of mouse lymphocytes lack Thy-1 and Igs. These null cells are easily isolated from complex mixtures, such as spleen cells: they survive treatment with anti–Thy-1 and complement, and are not bound by anti-Ig Abs on solid adsorbents, which eliminate T and B cells, respectively. In culture, some null cells acquire surface Igs or surface markers of T cells; some others develop into hemoglobin-forming cells. Null cells are probably a heterogeneous mixture of various immature hematopoietic cells.

K Cells. These null cells, called K (for killer) cells, have receptors for Fc domains of aggregated Igs, such as Ag-crosslinked Abs. By binding to Abs that are attached to surface Ags of diverse target cells, the K cells cause target cells to undergo lysis. K cells appear to provide a host defense against some transplanted lymphomas (i.e., lymphocyte tumors), but their role in natural resistance to spontaneous tumors is not clear (see Ab-Dependent Cell-Mediated Cytotoxicity, Ch. 23).

Natural Killer (NK) Cells. Some null cells in the spleen can destroy various tumor cells, particularly those that are persistently infected with various enveloped viruses. Present in unimmunized mice (and at high levels in nude mice), these "natural killer" (NK) cells lack Fc receptors for Igs. They possess the surface alloantigen Ly-5; and alloantiserum to this Ag blocks their cytotoxic activity, suggesting that Ly-5 may function in a binding or cytolytic step. These cells may have slight amounts of surface Thy-1, and may be pre-T cells.

MACROPHAGES

Macrophages (Mϕs) bind, ingest, and degrade Ags, and are thought to "process" and "present" Ags to lymphocytes. Unlike B and T cells, which are Ag-specific, Mϕs seem to be able to bind any Ag. Large (15–25 μm diameter) and motile, Mϕs are distinguished by intense phagocytic activity, abundant lysosomes (cytoplasmic vesicles filled with proteases, nucleases, lipases, phosphatases, lysozyme, etc.), tight adhesion to glass surfaces, and secretion of proteolytic enzymes and various soluble factors that act on T cells (Fig. 18-8D). Mϕs derive from bone marrow promonocytes (Fig. 18-3), which give rise to circulating blood monocytes (less phagocytic and fewer lysosomes than mature Mϕs). Monocytes penetrate into tissues, where they differentiate into Mϕs.

Because Mϕs ingest Ags, these cells were long suspected of being essential for Ab formation. However, convincing evidence for their role in this process emerged only when it became possible to elicit Ab production in cultures containing distinct cell populations (obtained from spleen or lymph nodes).

Suspensions of spleen cells consist of adherent cells (essentially Mϕ), which stick to glass surfaces, and nonadherent cells, which are primarily B and T cells. *For an Ag to induce the development of Ab-secreting cells in culture, both adherent and nonadherent cells are necessary.* (The procedures for stimulating cells in culture to produce Abs and for counting the resulting Ab-secreting cells by the hemolytic plaque technic are described in Chapter 19).

Though Ab-secreting cells can be elicited by adding soluble Ag to a mixture of adherent and nonadherent cells, 100 to 1000 times less Ag is needed if it is first bound to Mϕ and these **Ag-pulsed Mϕs** are then added to the cultured spleen cells (or just to the nonadherent lymphocytes). Ag-pulsed Mϕs are also highly effective stimulators of proliferation by T cells (probably T helpers) from individuals that were previously primed with that Ag.

Mϕs do more than bind Ag and present it on an inert surface to Ag-specific lymphocytes. Thus certain **responder (R)** strains of mice and guinea pigs can make Abs to a particular Ag and **nonresponder (NR)** strains cannot; hybrids (R × NR) are also responders (see Genetic Controls, Ch. 19). These differences are reflected in their respective Mϕs: in culture, T cells from Ag-primed (R × NR) hybrids can be stimulated to proliferate by Ag-pulsed Mϕs if the Mϕs are from responders (R or the R × NR hybrid), but not if they are from nonresponders.

The explanation for these differences probably lies in the Mϕ surface **Ia molecule** encoded by the major histocompatibility complex (see Ia Antigens, below). Since Ia differs in responders and nonresponders, and anti-Ia serum blocks the stimulating effect of Ag-pulsed Mϕs, it has been suggested that Ia and Ag molecules form complexes on the Mϕ surface and that it is the "proper" complexes that are recognized by and stimulate T$_h$ cells,

rather than the adsorbed Ag alone (see H-2 Restrictions, below).

As will be described later, differences in immune responses elicited by Ags in various physical states, and administered by different routes, suggest that Ag presentation is critical. If the Ag is present on Mϕs that have appropriate surface Ia molecules, T$_h$ cells are stimulated rather than T$_s$ cells, and an active immune response ensues. If, however, the Ag bypasses Mϕs and interacts directly with T cells, T$_s$ cells tend to be stimulated, rather than T$_h$, and unresponsiveness results (see Tolerance, and Genetic Controls, Ch. 19; also Ch. 22).

Ags, especially when aggregated, readily adhere to Mϕs. When complexed with Abs of certain classes ("opsonins", IgG) and with activated complement (Ch. 20), Ags are bound to special receptors on Mϕs. Fc receptors **(FcR)** are specific for Fc domains of Igs: complement receptors **(CR)** are specific for certain complement fragments (C3b, C5b, Ch. 20). The binding of Ag-Ab-C complexes to these receptors stimulate phagocytosis: the Ag is ingested into a phagocytic vacuole (phagosome) that fuses with lysosomes to form the phagolysosome, in which degradative enzymes break down the ingested material.

Macrophages wander through tissues. Together with reticular cells that are immobilized on reticulum fibers and other phagocytic cells in the endothelial lining of vascular, sinuslike spaces (of spleen, lymph nodes, liver, etc.), they constitute the scavenger **reticuloendothelial system** which traps, ingests, and degrades foreign particulate matter (bacteria, many viruses, aggregated macromolecules).

Mϕs can divide, but they probably survive for only a short time in acute inflammatory lesions (caused by infection, irritants, or adjuvant mixtures with Ag), where, as noted above, they arise by differentiation of monocytes carried to tissues via the blood. The natural life span of mature tissue Mϕs is not known, but they can survive many weeks in tissue culture. In some chronic inflammatory lesions Mϕs differentiate further into epithelioid cells and into multinucleated giant cells.

An additional role of Mϕs in cell-mediated immunity will be discussed in Chapter 22: Ags react specifically with effector T lymphocytes (T$_{DTH}$ of delayed-type hypersensitivity) and cause the release of factors that stimulate Mϕs to differentiate into **activated, or "angry," cells,** which have greatly increased numbers of lysosomes and are more actively phagocytic. Activated Mϕs are especially well adapted to destroy ingested microbes (bacteria, fungi, etc.), and they are responsible for the nonspecific increased resistance to many infectious agents that is seen in some cell-mediated immune responses (see Delayed-Type Hypersensitivity, Ch. 22).

THE MAJOR HISTOCOMPATIBILITY COMPLEX (MHC)

Certain cell surface glycoproteins on B and T cells and on Mϕs have a profound influence on the reactions of these cells with Ags and with each other. The critical glycoproteins are encoded by a large cluster of genes called the

major histocompatibility complex (MHC). Our understanding of the MHC derives largely from studies on inbred mice.

Inbred mouse strains are produced by successive

brother–sister matings. Except for rare, mutations these matings lead after about 20 generations to 1) genetic uniformity of all members of the inbred strain (except for the sex chromosome), and 2) homozygosity at all genetic loci.* Mice of the same strain are thus **isogenic** or **syngeneic**† (Gr. *iso* and *syn*, same); those of different strains are **allogeneic** (Gr. *allos*, different). Normal tissue grafts or tumor grafts from a donor of one strain are accepted by recipients of the same strain (**isografts succeed**), but are rejected by recipients of another strain (**allografts fail**), because of the recipient's immune response to histocompatibility alloantigens (alloAgs) on the surface of the grafted cells.

There are many histocompatibility Ags (on the order of perhaps 100 in the mouse), encoded by many unlinked sets of genes. However, one set—the MHC—specifies the cell surface molecules that elicit the most rapid allograft rejection (i.e., after 7–10 days). Under normal conditions, as we shall see, the function of the MHC glycoproteins is to govern the interactions among an individual's lymphocytes and Mϕs.

H-2 COMPLEX IN THE MOUSE

The MHC was discovered by Gorer in England in the 1930s while he was studying alloAgs on RBCs of inbred mouse strains (Ch. 23) and resistance to tumor allografts exchanged between the strains. He realized the identity of a gene for tissue "histocompatibility" Ags that elicited rapid rejection of the test graft with the gene for a red cell Ag called II. Designated as H-2, this "gene" was shown by Snell to be a cluster of several genes, now called the H-2 complex.

When recipient mice of one strain reject an allograft, they generally develop at the same time Abs that agglutinate RBCs of the donor strain and of several other strains. The graft rejection is actually due to cell-mediated, not humoral, immunity (i.e., it can be transferred with T cells, not with Abs; Chs. 22 and 23), but the Abs are extremely useful for antigenic analysis of some H-2 products. Tested on cells from various strains,‡ these

* The proportion of loci that are heterozygous in the progeny of successive brother–sister matings is approximated, after five to six generations, by the following expression:

$$H_g \simeq (0.809)^g \, H_o$$

where H_o is the proportion of heterozygous loci in the original parents and H_g is the proportion after g generations. Only 1% of the originally heterozygous loci are still heterozygous at generation no. 22, and only about 0.01% at generation no. 50.

† In common usage, **syngeneic** refers to the absence of discernible tissue incompatibility; i.e., to identity of the genes controlling histocompatibility Ags. It approximates the situation in **isogenic** individuals who, as in monozygotic twins, are identical with respect to all their genes (except for rare somatic muations).

‡ By agglutination of RBCs or lysis of lymphocytes.

Abs, or the antiserum resulting simply from repeated injections of allogeneic cells, reveal many antigenic specificities: some (**"private specificities"**) are unique for the product of a particular H-2 allele, but others (**"public specificities"**) are shared by products of different alleles. Each specificity is numbered (see footnote to Table 18-6) and is identified by the alloantiserum resulting from crossimmunization of particular mouse strains. Each strain has a characteristic set of H-2 specificities (private and public), and a few strains have the same set, i.e., the same H-2 complex.

Genetic crosses showed that the H-2 Ags are specified by a single autosomal (i.e., not sex-linked) H-2 locus each designated by a superscript (H-2^a, etc.; Table 18-6). The loci are dominant: all progeny resulting from a cross between an H-2^a and an H-2^b strain are H-2$^{a/b}$ heterozygotes in which each cell expresses the specificities characteristic of both H-2^a and H-2^b.

TABLE 18-6. H-2 Haplotypes, Alleles, and Specificities of Some Commonly Used Inbred Mouse Strains

Haplotype	Alleles* at loci				Representative strains†
	K	I	S	D	
H-2^b	b	b	b	b	C57BL/10 (B10), C57BL/6 (B6), BALB.*b*
H-2^d	d	d	d	d	BALB/c, DBA/2, NZB, B10.D2
H-2^k	k	k	k	k	AKR, C3H, CBA, B10.BR, BALB.k
H-2^a‡	k	k-d	d	d	A, B10.A

*Alleles at K and D are defined by reactions of alloantisera (agglutinate red blood cells or lyse lymphocytes), as in the following examples (the specificities of the alloantisera and the corresponding Ags are numbered):

K^b (*b* allele at K locus) has private specificity 33 and public specificities 5 and 36

D^b (*b* allele at D locus) has private specificity 32 and public specificities 6

K^d (*d* allele at D locus) has private specificity 31 and public specificities 3 and 8

D^d (*d* allele at D locus) has private specificity 4 and public specificities 3, 6, 13, 36, 41, 42 and 43

K^k (*k* allele at K locus) has private specificity 23 and public specificities 3, 5, 8, 11 and 25

D^k (*k* allele at D locus) has private specificity 32 and public specificities 3 and 5

†Notice the congenic strains: BALB.*b* and BALB.k, with H-2^n and H-2^k haplotypes respectively, were derived by breeding chromosomal segments with these haplotypes onto the BALB/c (H-2^d) strain (see Fig. 18-9). Similarly, the congenic lines B10.D2 (H-2^d), B10.BR (H-2^k), and B10.A (H-2^a) were all derived from the B10 (H-2^b) strain by a breeding program like that shown in Fig. 18-9.

‡The H-2^a haplotype is a recombinant: its alleles at the left side of the I region have serologic and functional properties of the H-2^k haplotype and at the right side of this region the properties of the H-2^d haplotype (see H2-K and H2-D Loci, and I Locus, this chapter and Fig. 18-16).

H-2K and H-2D LOCI

The existence of more than one locus within H-2 was disclosed when H-2$^{k/d}$ heterozygotes were found to accept a tumor graft from an H-2^a strain. This puzzling observation was explained by the assumption that H-2 consists of two linked loci, designated **H-2K** and **H-2D**, or simply **K** and **D**, with H-2^a mice having at one locus the allele of the H-2^k strain and at the other the allele of the H-2^d strain. The H-2^a tumor graft was not rejected because its K and D alleles were both present in the H-2$^{k/d}$ recipient and, as noted below, allograft rejection depends largely upon the Ags specified by the K and D loci.

Alleles at the K and D loci can be clearly identified serologically (Table 18-6 footnote), but terminology can be confusing because the alleles are designated with the same small letters that are used for the entire complex. Thus in the H-2^k complex the alleles at K and D are called H-2K^k and H-2D^k, and in the H-2^d complex they are H-2K^d and H-2D^d. In A-strain mice, with H-2^a, they are H-2K^k and H-2D^d, suggesting that this strain arose from a mouse with a genetic recombination between K and D loci. Systematic breeding studies have since revealed that about 1% of the F1 hybrids derived from crosses between H-2–differing parents have a recombination between these loci (Fig. 18-16).

The view that the H-2 complex includes separate K and D loci was later supported by the finding that K and D are linked to two other loci: Ss, which specifies some complement components (C3 and C4, Ch. 20), and the I locus, which controls immune responses to many Ags and specifies certain cell surface glycoproteins (Ia Antigens; see below and Genetic Controls, Ch. 19). Rare recombinants that appeared in breeding experiments showed that I and Ss are located between K and D (Fig. 18-16). *Except for rare recombinants, all of the alleles of a given H-2 complex are inherited as a stable linked cluster or* **haplotype** (Table 18-6, Fig. 18-16).

The order of loci shown in Fig. 18-16 has been determined by typing recombinant hybrid mice in which crossover between two chromosomes (one from each parent) is evident within the H-2 complex. If, for instance, the H-2 alleles on the chromosome from one parent were *bbbbbbbbb* for the 9 loci, K–D, shown in Fig. 18-16, and on the other were *ddddddddd,* the recombinant might be typed as *bb/ddddddd,* indicating a crossover between the IA and IB loci.

Once detected in a rare offspring, the new recombinant chromosomal segment can be preserved by genetic crosses that establish it in a new **congenic mouse strain** (according to the breeding scheme in Fig. 18-9). In this way, about 50 extremely valuable H-2 recombinant mouse strains have been established.

I LOCUS

This part of the H-2 complex was recognized through its control over immune responses to many Ags. As described later (see Genetic Controls, Ch. 19), the ability to produce high Ab titers to certain Ags is associated with various H-2 haplotypes. Through breeding experiments

FIG. 18-16. Genetic map of the H-2 complex. Numbers refer to map distances (recombination frequency) between the indicated loci on mouse chromosome 17. **Class I loci** specify integral membrane glycoproteins: the 45,000-dalton *h* chains of major histocompatibility Ags (each associated noncovalently on the cell surface with a chain of β2-microglobulin (12,000 daltons), which is encoded by a gene in another chromosome; see Fig. 18-20). **Class II loci** specify Ia (I-associated) Ags (noncovalently associated 35,000- and 28,000-dalton chains), control immune responses to T-dependent Ags (Ch. 19), and elicit proliferation of T cells in mixed lymphocyte reactions (see H-2 Complex, this Ch.). The **class III locus** codes for the expression of some complement components (e.g., C4 and C3 in the mouse, C2 in man, Bf in guinea pigs; Ch. 20). **Qa-1 and Qa-2** are cell surface glycoproteins that resemble those specified by K and D loci; they also elicit strong allograft reactions, but they are less polymorphic than K and D. The **TL locus** is defined by the thymus-leukemia Ag (TLa) on thymocytes and (in leukemia) on neoplastic T cells. By convention, the centromere **(O)** is placed at the left. Mutants at the T locus affect the embryonic development of the neuroectoderm and the recombination frequency between T and H-2 loci. **G** specifies a red cell surface Ag.

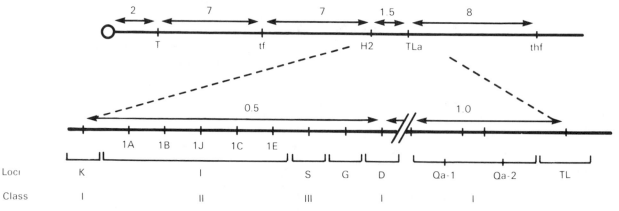

with strains that make high or low responses to various Ags, and by examining the responses of mice with recombinant H-2 complexes, it has been established that the controlling genetic locus, called **Ir-1** (immune response-1), lies **between the K and D loci in the region called I or H-2 I** (Fig. 18-16). Genes governing the responses to various Ags, and genes specifying various Ia Ags, have been separated by recombinations that divide the I region into several subregions (Fig. 18-16).

Some strains that respond poorly to a particular Ag (X) make strong Ab responses to X when X is coupled with a suitable carrier molecule; e.g., methylated bovine serum albumin (highly cationic) forms stable electrostatic complexes with, and serves as a potent carrier for, a glutamic acid–rich polypeptide Ag (highly anionic). This finding suggests that there may be defective recognition of carrier determinants by T_h cells in low-responder mice. With other low-responder strains, however, poor Ab production seems to be due to excessive T suppressor cell activity. Indeed, the ability to produce T_s cells in response to certain Ags is influenced by a genetic locus that also maps in the I region (see Is Genes, Ch. 19). In later chapters, dealing with Ab formation (Ch. 19) and cell-mediated immunity (Ch. 22), genetic regulation by H-2 linked and other Ir genes (not H-2 linked) will be considered further.

Ia-ANTIGENS

Cell surface Ags coded for by I region genes are called **Ia (I-associated) Ags.** Intense interest focuses on these molecules because they may represent the means by which Ir genes regulate immune responses. Unlike the products of K and D loci, which are present on virtually all cells, the Ia Ags are found only on B lymphocytes, some Mφs, some sets of T cells (e.g., T suppressors), and perhaps a few other cells (sperm cells; Langerhans cells, which may be a special type of Mφ in epidermis). Because anti-Ia serum blocks the stimulation of T_h cells by Ag on Mφs, Ia molecules may have a special role in Mφ presentation of Ags.

The Ia Ags are identified by alloantisera produced in various ways. The most effective is reciprocal immunization of pairs of congenic mouse strains that have the same genome (including H-2K and H-2D) except for the I region. Repeated injections of spleen cells from mice of one strain into recipients of the other elicit Abs to I region products (Ia Ags) on the donor cells.

Ia Ags can also be identified by alloantiserum that contains both anti–H-2 and anti-Ia Abs, provided the serum is tested on cells that lack the corresponding H-2 specificities; and it is also possible to remove the anti–H-2 Abs selectively by adsorption on the many cell types that lack Ia (e.g., fibroblasts and erythrocytes). The resulting anti-Ia sera (plus complement) lyse cells carrying the corresponding surface Ia Ags. To establish that an antiserum or the assay conditions are specific for Ia, the gene coding for the reactive Ag must be localized to the I region (by testing spleen cells derived from mice with diverse recombinant haplotypes). In this way *over 20 Ia allelic products have been identified within a short time. There may be as many different Ia alleles as there are H-2K and H-2D alleles* (see Polymorphism, below).

Anti-Ia sera cause capping of Ia Ags on intact cells. When B cells are treated sequentially with anti-H2 and anti-Ia sera the Ia molecules do not form joint caps (cocap) with H-2K or H-2D molecules; thus these molecules are not linked in the cell membrane.

POLYMORPHISM

Genes of the H-2 complex are highly diversified in the commonly used inbred strains, especially at K and D loci. Among wild mice the diversity is far greater. By breeding trapped wild mice with inbred strains about 30 congenic strains have been established, each with the H-2 complex of a wild mouse trapped in a different locality. The H-2 haplotypes of these wild mice differ serologically not only from all the inbred strains but from each other. The actual number of K and D alleles is unknown but must be immense: altogether among the inbred strains used in the United States and in Europe, and in the few wild mice tested, there are about 50–60 H-2K and 45 H-2D alleles; about 2500 different K–D combinations are possible and most of these probably exist. In contrast to this extraordinary polymorphism of H-2 there is only one gene (no allelic variants) for about 60% of all other known loci (monomorphic) and polymorphism at the rest is limited to a few (two or three) alleles. The large number of K and D alleles suggests that the mutation rate at these loci must be unusually high.

Mutations. Mutations in histocompatibility (H) genes are detected by exchange of skin grafts among mice of the same genotype, each mouse serving as the donor of grafts to, and the recipient from, two other mice (Fig. 18-17). Mutations detected by graft rejection can be preserved in the genome of an inbred line by establishing a new congenic strain (see Fig. 18-9, page 391), with the mutant gene in homozygous form.

The mutation rate in H genes is extremely high: in one study there were 32 mutants among the 2752 mice tested, or $12/10^3$ zygotes and $6/10^3$ gametes. Most of these are probably H-2 mutations, but even if they were distributed over 100 H loci, this frequency would be at least 10 times greater than the mutation rate of most genes. With an increasing interest in environmental mutagens (many of them carcinogens as well) there is a growing recognition of the need to study mutations in mammalian genes. With its many loci, and its readily detectable mutations, the system of H genes in the mouse could become the standard for analyzing mutagens that are active in intact mammals.

Because of the high mutation rate and extensive polymorphism, variability at H-2K and H-2D loci probably confers important selective advantages in evolution. Nonetheless, several mouse strains that have the same haplotype ($H-2^k$) and that have been separated by 50–60 years of inbreeding (CBA, AKR, C3H) have serologically indistinguishable H-2 products: e.g., cells from each strain absorb completely the anti-$H-2^k$ Abs elicited by the others. Hence there must also be stabilizing pressures that limit the diversity of these highly variable loci. The possible advantages of extensive polymorphism of H Ags (K,D,I), and some reasons for limits to their diversity, are discussed below in connection with MHC-restriction of Ag-recognition by T cells (H-2 Restriction, below).

MIXED LYMPHOCYTE REACTION (MLR)

Cell surface alloAgs due to the H-2 complex can be recognized in vivo by allograft rejection, or in vitro either by mixed lymphocyte reactions (MLRs) or serologically (by reactions with antisera to H-2K, H-2D, and Ia Ags).

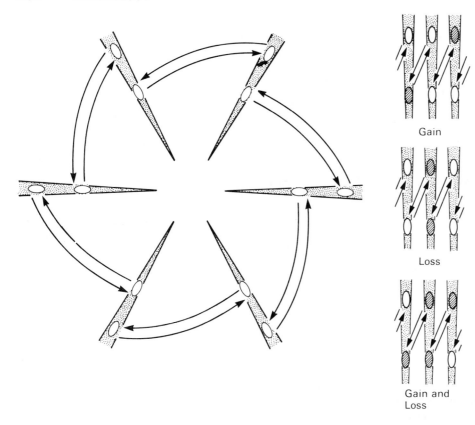

Gain

Loss

Gain and
Loss

FIG. 18-17. Detection of mutations in histocompatibility (H) genes by exchange of grafts of tail skin. The mice are F1 hybrids (H-$2^{a/b}$) of H-2–differing strains: because they have one copy of each H-2 haplotype they permit the detection of mutations that result in the loss of an antigenic determinant. As is shown in the right panel, a mutant that gains an H antigenic determinant accepts grafts but its donated grafts are rejected; a mutant that loses a determinant rejects grafts and its donated grafts are accepted; a mutant with both changes (gain and loss) rejects grafts, and its donated grafts are also rejected. **Open ovals,** accepted grafts; **shaded ovals,** rejected grafts. (Based on Bailey DW, Kohn HI: Genet Res 6:330, 1965)

With congenic strains of mice that differ in restricted regions of the H-2 complex it has been established that *products of the K and D loci elicit the most rapid and intense allograft reactions, while products of the I locus (probably Ia Ags) elicit the strongest MLR.*

The MLR is carried out by coincubating, for about 4–5 days, lymphocytes from two strains that differ in H-2. ^{3}H-thymidine is then added, and the extent of its incorporation into DNA measures the proliferative response of T cells (mainly T_h) of one lymphocyte population to H-2 alloAgs of the other.

The mutual stimulation **(two-way reaction)** can be made **one-way** by treating one lymphocyte population with mitomycin C or irradiating it with x- or γ-rays. Cells of the treated population cannot divide but they can still serve as stimulators. Some cells of the untreated **(responder)** population proliferate and undergo blast transformation. Most of the blast cells become T helpers that promote the proliferation and differentiation of precursors of cytotoxic T lymphocytes (T_k cells), specific for H-2K and H-2D products of the stimulators: these killer cells can lyse any cell with the same K or D alleles as the stimulator cells (Cytotoxic T Lymphocytes, Ch. 22).

When responders and stimulators differ in K, I, and D loci the proliferative and cytotoxic responses are most vigorous. When they differ only in K and D the prolif-

erative (and cytotoxic) responses are much reduced. Hence products of the I locus, probably Ia Ags, account for most of the proliferative stimulus.

Taken together, all these observations mean that the MLR consists of two reactions. In the first, some responder Ly-1 T cells react against Ia Ags on stimulator B cells and Mφs by proliferating and differentiating into T helpers. In the second reaction, the T helpers enhance the differentiation of precursor cells in the responder population into mature cytotoxic T lymphocytes, specific primarily for products of the H-2K or H-2D alleles of the stimulator population.

Graft-vs.-Host Reaction. A reaction analogous to the MLR occurs in vivo when lymphocytes from an allogeneic donor are injected into an **immunologically defective recipient** (e.g., a newborn or mature individual whose immune system has been inactivated). In the resulting **graft-vs.-host (GvH) reaction** the injected T cells react against H-2 products on the host's cells (primarily Ia in the **stimulator step,** and K and D in the **effector step**). The reaction, which can be lethal, is characterized by skin rash, diarrhea, loss of weight, widespread inflammation, and focal proliferation of diverse host cells (see Ch. 23).*

---

* The GvH reaction does not occur in a normally competent recipient mouse. If donor and recipient differ at H-2 the transplanted allogeneic T cells are rejected, as would be any other allograft (Ch. 23), and the G$_v$H reaction cannot occur.

THE HUMAN HLA COMPLEX

Like mice, humans and other mammals have a tightly linked cluster of genes for the cell surface glycoproteins that regulate immune cell interactions and evoke intense reactions to allografts. In man this complex, called HLA (human leukocyte antigens), also consists of three types of loci (Fig. 18-18A). **Class I loci** (HLA-A, -B, and -C, analogous to K and D of mouse H-2) specify Ags that are present on virtually all cells and that elicit strong allograft reactions. **Class II loci** (HLA-D, analogous to I of mouse H-2) specify Ags on B lymphocytes and probably on Mɸs: they stimulate responses of T cells in the mixed lymphocyte reaction (see above) and probably modify the responses to Ags on Mɸs (see Macrophages, above, and H-2 Restriction, below). **Class III loci** (analogous to S in the mouse H-2) contain genes for complement components. In the HLA complex the class II and III loci

FIG. 18-18. A. Diagram of human chromosome 6 with expanded HLA complex. Numbers refer to map distances in recombination units (as frequency [%] of recombinants among progeny). For definition of classes I, II, and III see text and map of the H-2 complex (Fig. 18-16). The complement locus (II) includes structural and/or regulator genes for the second and fourth complement components (C2, C4) and for factor Bf of the alternate pathway for complement activation (Ch. 20). Centromere **(O)** is at the left, by convention. Also shown are loci for phosphoglucomutase (PGM3), glyoxalase (Glo), and urinary pepsinogen 5 (PG5). B. Comparison of the major histocompatibility complexes of humans (HLA), the mouse (H-2), and the Rhesus monkey (RhLA). (Based largely on Snell GD, Dausset J, Nathenson S: Histocompatibility. N.Y., Academic Press, 1976)

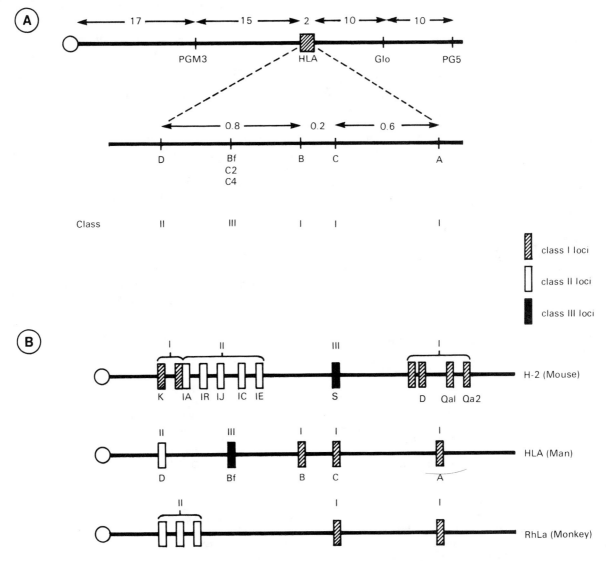

lie outside class I loci, rather than between them as in the mouse H-2 (Fig. 18-18B).

The breeding programs that contribute so much to the analysis of the mouse H-2 complex are obviously inapplicable to human populations, and entirely different approaches have to be used in man. HLA alleles were initially recognized by the human alloAbs that appear following 1) multiple blood transfusions in which donor white blood cells bear Ags that the recipient lacks, and 2) pregnancies involving parental disparities in HLA: the pregnant mother is exposed to the father's HLA Ags on fetal leukocytes that cross the placenta. Multiple pregnancies (same father) lead to especially high-titered antisera. The Abs in these sera were initially recognized by white blood cell agglutination (leukoagglutination), but they are currently measured by cytotoxicity assays: the Abs (plus complement) damage leukocytes with the corresponding surface Ags.

HLA-A, -B, AND -C LOCI (CLASS I)

The many Ags revealed by human antileukocyte sera have been arranged into simple groups by family studies. The sera divide the children of any particular family into at most four sets, suggesting that the corresponding Ags are products of single genetic locus: if each parent is heterozygous (i.e., has two HLA haplotypes; see below) and the parents have no HLA haplotypes in common, an $a/b \times c/d$ mating should yield four types of progeny—$a/c, a/d, b/c,$ and b/d. The children of homozygous parents, or of parents with a shared haplotype will fall into fewer than four sets. (If recombination takes place within one parent's HLA complex there may be more than four sets among the children, and this has been observed in rare families.)

Family studies are especially valuable for identifying individual loci and their alleles; and those families in which one parent has two HLA specificities that are lacking in the other parent are particularly illuminating. The probability is high that the two specificities are encoded by alleles of the same gene if, in many families, each child receives one, and only one, of them. Many alleles have been identified through such laborious pairwise analyses: e.g., alleles A1 and A2, then A2 and A3 (Table 18-7). In this way **three allelic series,** corresponding to the HLA-A, -B, and -C loci, were established. As additional antisera have accumulated and panels of leukocyte donors have been better defined, the number of allelic variants has grown. Currently recognized are 20 alleles at the A locus, 20 at B, and 8 at C; many more will doubtless be identified.

Within families the Ags specified by the A, B, and C loci segregate in two groups. For instance, a parent heterozygous at the three loci, with specificities A1/A2, B1/B2, and C1/C2, will have children who inherit, say, either A1, B1, and C1, or A2, B2, and C2; each group represents an *HLA haplotype, i.e., the set of HLA alleles on one chromosome. Each person inherits two HLA haplotypes, one from each parent, and transmits one haplotype to each of his progeny.*

The tight linkage of the multiple HLA alleles that are transmitted en bloc from one generation to the next thus resembles other linkage groups in the immune system, e.g., the Ig heavy chain linkage group for which Ig allotypes represent the phenotypic expression (Ch. 17). However, in about 0.8% of children (corresponding to about 1/200 gametes) the HLA-A and HLA-B alleles represent a recombination between the A and B loci in one parent; recombinants between B and C are much rarer (Fig. 18-18).

THE HLA-D LOCUS (CLASS II)

The fourth HLA locus, D, like I of the mouse H-2 complex, specifies the Ags on B cells that elicit T cell proliferation in mixed lymphocyte reactions (MLR). In man

TABLE 18-7. Some Common Alleles at HLA Loci and the Frequencies of the Corresponding Antigens*

Locus			
HLA-A	**HLA-B**	**HLA-C**	**HLA-D**
HLA-A1 (25.1)	HLA-B5 (15.2)	HLA-Cw1 (10.0)	HLA-Dw1 (19.3)
HLA-A2 (44.8)	HLA-B7 (18.2)	HLA-Cw2 (14.9)	HLA-Dw2 (15.2)
HLA-A3 (22.6)	HLA-B8 (16.7)	HLA-Cw3 (26.0)	HLA-Dw3 (16.4)
HLA-Aw24 (18.2)	HLA-B12 (32.5)	HLA-Cw4 (19.7)	HLA-Dw4 (15.6)
HLA-A11 (11.8)	HLA-B14 (8.8)	HLA-Cw5 (14.0)	HLA-Dw5 (14.6)
HLA-A28 (9.8)	HLA-B18 (11.3)		HLA-Dw6 (10.5)
HLA-A29 (10.3)	HLA-Bw35 (15.2)		
HLA-Aw32 (9.8)	HLA-B40 (13.7)		
	HLA-B15 (12.3)		

*The frequencies of the HLA Ags are in parentheses (as percent); at any locus the percentages add up to >100% because most individuals have two alleles at each locus (heterozygous). Values are for the caucasoid population of France. Provisional assignments are indicated by w.

(Based on Snell GD et al: Histocompatibility, New York, Academic Press, 1976)

these reactions are evoked by incubating together peripheral blood leukocytes (PBLs) from genetically different individuals: one person's cells are stimulated to proliferate (measured by increased ^{3}H-thymidine incorporation into DNA) by D-differing Ags on the other's cells. As with mouse cells, the human MLR can be made one-way by treating one person's cells with mitomycin C or with irradiation.

The MLR is negative (^{3}H-thymidine incorporation is not increased above background) between siblings with the same serologic HLA specificities, and it is positive between those with different HLA specificities. Hence the MLR-specifying locus is linked to the HLA-A, -B, and -C loci (whose alleles are identified by antisera). Moreover, platelets and fibroblasts cannot stimulate the MLR, though they can adsorb the Abs that identify the HLA-A, -B, and -C loci; hence other Ags, specified by another locus (D), trigger the MLR. The finding of rare individuals whose HLA Ags are serologically the same but whose leukocytes are mutually stimulating in the MLR also showed that a different Ag, encoded by a separate locus, elicits the MLR.

In intra–H-2 recombinants in mice the I locus maps closer to H-2K than to H-2D (Fig. 18-16). In man, also, rare recombinants within HLA show that the D locus is closer to B than to A and establish the order as D-B-C-A (Fig. 18-18). Negative MLR reactions between blood leukocytes of unrelated individuals are extremely rare (about 1/10^4). Hence the D locus either contains many MLR-specifying genes or has fewer extremely polymorphic ones.

Linkage Disequilibrium. If the frequency in the population at large is 0.1 for allele A1 at the A locus and 0.1 for allele B2 at the B locus, the frequency of individuals with A1 and B2 should be 0.01 (0.1 × 0.1). With most HLA alleles the observed and expected frequencies agree, but in some instances the observed is many times greater than the expected. Such preferential associations, called **linkage disequilibrium**, have been found with certain D, B, and A alleles. Whether it is due to evolutionary selection for some immunologic advantage or to the reproductive separation of populations (races) with different proportions in their gene pool, is not clear.

Asymmetric Mixed Lymphocyte Reactions. In some MLRs the white blood cells of person X stimulate those from person Y but not vice versa. The following example suggests an explanation: if two D-locus heterozygous parents share a D allele (e.g., D1/D2 × D1/D3) some of their progeny will be homozygous (D1/D1) and will react, in the MLR, against either parent, but neither parent will react against the homozygous child's cells.

Cells that are homozygous for different D alleles are valuable for typing purposes. These reference cells can be kept frozen in liquid N$_2$ and after thawing can be used in MLRs to help establish the D phenotype of other cells: e.g., the homozygous reference cells will only stimulate D-differing cells. Individuals with D-homozygous cells are common in reproductively isolated communities with frequent first-cousin marriages, e.g., the Amish communities in the United States.

Antisera to Ia-like Antigens. D alleles can also be identified with human alloantisera that block MLRs. These Abs are occasionally found in multiply transfused patients and in multiparous women; but unlike Abs to Ags of HLA-A, -B, and -C loci, the anti-D Abs are not adsorbed on platelets or fibroblasts. Rather, like Ia Ags of

the mouse, human D Ags (also called **DR**, or D-related) are restricted to B lymphocytes (and probably to Mφs): these cells adsorb the alloAbs. So far eight D alleles have been identified. (For use as target cells and adsorbents, B cells are readily obtained in large numbers from the peripheral blood of patients with chronic lymphocytic leukemia; B cells can also be cultured from almost any person's blood after the cells are transformed into "immortal" cell lines by infection in vitro with Epstein–Barr [EB] virus.)

The antisera to products of D and of the other HLA loci are valuable not only for establishing an individual's phenotype but also for monitoring the isolation of the corresponding membrane Ags for structural studies.

STRUCTURE OF SOME MHC PRODUCTS

Human B lymphocytes infected with EB virus can be grown as cell lines and in mass culture. They have 20–40 times more HLA Ag per cell than normal lymphocytes and constitute an unusually rich source for isolation of these Ags. The corresponding molecules of mouse cells are less available and have been less extensively analyzed, but so far they are similar in overall structure.

The number of chains per molecule, and their molecular weight, and some other structural features of the MHC products have been determined by a series of procedures that are applicable to almost any cell surface Ag.

1) The surface proteins are labeled in the living cell in various ways: e.g., by iodinating accessible tyrosines with lactoperoxidase, peroxide, and ^{125}I, or by labeling some galactose residues of oligosaccharides in glycoproteins by successive treatment with galactose oxidase and ^{3}H-sodium borohydride.

2) Proteins are extracted from the labeled cell's membrane (and cytoplasm) with a nonionic detergent, in which protein antigenicity is retained.

3) Antiserum is added, forming complexes with the corresponding radiolabeled Ag (*Ag).

4) The Ab–*Ag complexes are specifically isolated from the other proteins by adsorption on a special strain of staphylococci that has a protein (**protein A**) that happens to bind specifically the Fc domain of Igs of many classes.

5) *Ag is separated from the washed staphylococci with a dissociating solvent, such as the anionic detergent sodium dodecylsulfate (SDS), and subjected to electrophoresis in a polyacrylamide gel (see Fig. 18-19).

Because the SDS binds to all proteins to about the same extent (in proportion to the number of peptide bonds) all proteins in SDS have approximately the same charge density and a protein's electrophoretic migration is inversely proportional to the log of its molecular weight. Judicious use of conditions that reduce S–S bonds (to SH) and block SH groups, before and after labeling and extracting surface proteins, provides information on chain linkage, whether by noncovalent or by S–S bonds.

Molecules Specified by Class I Loci (A, B, and C of HLA and K and D of Mouse H-2). Each of these loci codes for a polypeptide chain of about 45,000 daltons. On the cell membrane these chains are associated noncovalently with a lighter chain, β$_2$**-microglobulin** (12,000 daltons),

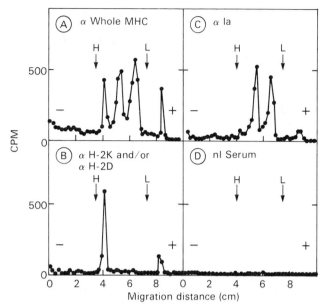

FIG. 18-19. Electrophoresis of cell surface proteins encoded by genes of the major histocompatibility complex. The surface proteins of intact cells were labeled with ^{125}I, extracted from cell membranes, immunoprecipitated with specific alloantisera, redissolved in a dissociating solvent (sodium dodecyl sulfate, SDS), and subjected to electrophoresis in SDS–polyacrylamide gels. All proteins bind SDS and are thus negatively charged (i.e., they migrate to anode [+]).

The alloantisera (α) were elicited against products of: A. the entire H-2 complex; B. H-2K and/or H-2D loci; C. the I locus; D. nothing (normal serum). Migration distance is proportional to the log of the molecular weight. **Arrows** point to mol wt markers (heavy [H] and light [L] Ig chains, 55,000 and 23,000 daltons respectively). In B the main peak is the heavy (*h*) chain (45,000 daltons) and the small peak is probably the light (*l*) chain, β2-microglobulin (12,000 daltons; see Fig. 18-20). In C the two peaks are α and β chains of Ia Ags (see Fig. 18-20). In A all four peaks are evident. (Sachs DH: In Contemp Top Mol Immunol 5:1, 1976)

which is specified by a gene on another chromosome. The resulting cell surface molecule appears to have one heavy (*h*) chain, the class I polypeptide, and one light (*l*) chain, β2-microglobulin (Fig. 18-20).

The *l* chain is the same in all of these molecules (within a particular species), but the *h* chains specified by different alleles differ in amino acid sequence. Oligosaccharides are found on the *h* chain (carbohydrate constitutes about 10% by weight) but not on the *l* chain; their removal with deglycosylating enzymes leaves the serologic specificity unchanged. Hence *the multitude of serologic specificities of these molecules within a given species* (Tables 18-6, and 18-7) *is due to polymorphism at class I loci and the resulting amino acid sequence variations in the* h *chains.*

The *h* chain (45,000 daltons) is cleaved (Fig. 18-21) by treating intact cells with papain into a large N-terminal water-soluble

fragment (35,000 daltons) and a C-terminal fragment (10,000 daltons) whose hydrophobic segment (demonstrated by amino acid sequence) probably interacts with lipids in the cell membrane's lipid bilayer. The C-terminal fragment traverses the cell membrane, suggesting that the *h* chain can transmit signals from the cell surface to cytoplasmic organelles, such as microfilaments and microtubules. *Parts of the* h *chains resemble Ig chains*: some S–S bonds form approximately 60-membered peptide loops (Ig domains; see Fig. 17-13, Ch. 17) and amino acid sequences reveal some homology of the 3RD domain (counting from the amino end) to C domains of Igs (Fig. 18-21). The N-terminal domains, however, lack sequence homology to Igs. Thus an Ig-like and non-Ig exons (see Ig Genes, ch. 17) probably combine to specify a single polypeptide chain.

β2-Microglobulin, the *l* chain, is also Ig-like. Its amino acid sequence is homologous with that of C_H domains of Ig chains (Fig. 18-22); it also resembles an Ig domain in mol wt (12,000) and in containing a peptide loop with about 60 amino acids joined by one disulfide (S–S) bond. β2-Microglobulin was discovered (in the 1960s) in the urine of individuals with impaired kidney reabsorp-

FIG. 18-20. Schematic representation of cell surface molecules determined by some genes of the major histocompatibility complex. A. The *h* chain is encoded by class I loci (K and D of mouse H-2 and B, C, and A of human HLA). The same I chain (β2-microglobulin) is associated noncovalently with each of the many different *h* chains in a given species. The *h•l* molecule elicits intense allograft rejection and restricts the specificity of T killer cells (cytotoxic T lymphocytes). B. The α and β chains of Ia (I-associated) Ags are encoded by IA and IE regions of class II loci (I region of the mouse H-2, D of human HLA). The α•β molecule stimulates proliferation of T helper (T_h) cells in the mixed lymphocyte reaction and restricts Ag recognition by T_h cells and by the T cells (T_{DTH}) that mediate delayed-type hypersensitivity. Oligosaccharides are indicated by branched units (. . . .). Arrows show where cleavage occurs at the cell surface when intact cells are treated with papain. (Based on Strominger J et al: Fed Proc 37:1467 1978)

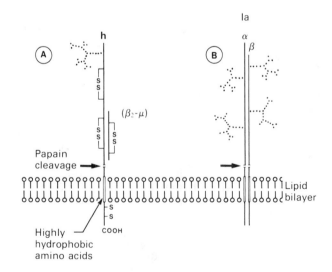

Heavy Chain of a Histocompatibility Antigen (HLA-B7)

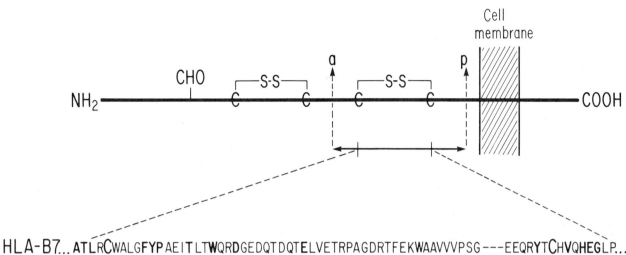

HLA-B7... ATLR**C**WALG**FYP**AEITLTWQRDGEDQTDQT**E**LVETRPAGDRTF**E**KWAAVVVPSG---EEQRYT**C**HVQHEGLP...

C$_\lambda$... AT**L**V**CL**ISD**FYP**GAVTVA**W**KADSSPVKAGV**E**TTTPSKQSNNKYAASSYLSLTPEQWKSHRSYS**C**QVTHEG--...

FIG. 18-21. Amino acid sequence of a part of a domain of the heavy (h) chain of a human histocompatibility Ag (HLA-B7). The same amino acid is found at 23% of the positions (bold face) in the HLA h chain and in the C domain of human light (λ) Ig chains. This degree of homology is about the same as that between C domains of heavy and light Ig chains (see Fig. 17-38); it indicates the Ig nature of this part of histocompatibility Ags. The N-terminal section of the HLA h chain has no sequence homology with Igs; different histocompatibility Ags have different amino acids at various positions of the N-terminal section of h chains. For amino acid abbreviations, see Fig. 17-12, p. 347. (Based on Orr HT, et al: Nature 282:266, 1979)

FIG. 18-22. Amino acid sequence homology between human β_2 microglobulin, the light (l) chain of histocompatibility Ags, and a constant domain (C$_H$3) of a human heavy (γ) Ig chain. The degree of homology (23%) is about the same as between C domains of light and heavy Ig chains (see Fig. 17-28 and Fig. 18-21). The entire β_2 microglobulin (12,000 daltons) corresponds to a single Ig domain. (Based on Smithies O, Poulik MD: Science 175:187, 1972 and Peterson PA, et al: Proc Natl Acad Sci USA 69:1697, 1972)

Light Chain (β_2 Microglobulin) of Histocompatibility Antigens

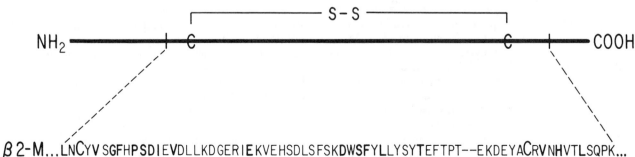

β2-M... L**NC**YVSGFH**PSD**IEVDLLKDGERIEKVEHSDLSFSK**DW**S**FYLLYS**YTEFTPT--EKDEYA**C**RVNHVTLSQPK...

Cγ3... LT**C**LVKGFY**PS**DIAVE**W**ESNDGEPENYKTTPPVLDSDGSFFLYSKLTVDKSRWQQGNYFS**C**SVMHEALHNHY...

tive function (e.g., due to cadmium poisoning). It is currently obtained in gram quantities from urine collected within the first 24 h following transplantation of human kidney allografts.

Though information on the amino acid sequences of *h* chains is still limited, it is clear that 1) there is marked homology, with conservative differences (e.g., leu–val interchanges), among the *h* chains specified by various alleles at the same locus or at different loci (HLA-A versus HLA-B), and even between human and mouse *h* chains. Hence all of the alleles at the A, B, and C loci of human HLA and at K and D of mouse H-2 probably evolved from the same primordial gene. The exon that encodes the 3RD domain of the *h* chain probably evolved from an Ig gene.

The same subunit structure (one 45,000-dalton *h* chain and one 12,000-dalton β2-microglobulin *l* chain) has been found for the thymus-leukemia or TL Ag of mouse thymocytes (TL Ag; see above). The gene for the TL *h* chain is also located on the same mouse chromosome (17) as the H-2 complex, near the H-2D locus (Fig. 18-16). It is thus possible that much of this large chromosomal segment in mammals and in other vertebrates codes for various Ig-like cell surface glycoproteins. A possible role of these proteins in distinguishing normal ("self") cells from either abnormal ones ("altered self") or foreign cells ("nonself") is considered below, under H-2 Restrictions).

Ia Antigens. Anti-Ia sera precipitate these Ags from detergent extracts of mouse spleen cell membranes; homologous molecules, isolated from cultured human B lymphocytes, are similarly precipitated by antisera to Ags specified by the HLA-D locus. *These Ags are also glycoproteins (mol wt about 60,000) with two chains per molecule* (Fig. 18-20B): α *(33,000) and* β *(28,000).* (In some molecules α and β are disulfide [S–S] linked, but in others they are associated by noncovalent bonds.)

Separation of Genes for Multichain Proteins. Fusion of cultured human and mouse cells (see Fig. 17-5, Ch. 17) yields "hybrid" cell lines which express human and mouse genes. However, the human chromosomes are gradually lost. By matching the retained human chromosomes with persistent cell surface HLA Ags it has been shown that the HLA class I loci, specifying the *h* chains, are located on chromosome no. 6, and that the *l* chain is specified by a gene on a different chromosome. The chains of some other proteins with multiple structurally homologous chains are also specified by genes on different chromosomes (e.g., for α and β chains of hemoglobin and for heavy, κ, and λ Ig chains [Ch. 17]). The location of genes with homologous sequences on different chromosomes probably preserves their individuality by minimizing opportunities for their genetic recombination. But for some other multichain protein molecules the genes for the different chains are located on the *same* chromosome, as in Ia Ags (Fig. 18-20) and complement component C4 (Ch. 20). Perhaps all of the latter proteins, like insulin, are synthesized as a single, large precursor polypeptide from which one or more internal peptides is excised during biosynthesis, yielding a molecule with two or more chains.

Is There Variability Among the Molecules Specified by an Individual's MHC? As stated above the *h* chains determined by class I loci of the MHC (human A, B, and C, and mouse K and D) bear some structural similarity to Ig heavy chains. In addition, there are hints that Ia molecules on Mϕs of different H-2 haplotypes interact differently with various Ags (see Macrophages, above). Do the *h* and Ia chains produced by a given individual have the extensive variability expected of molecules that resemble Igs, either in structure or in the ability to discriminate between Ags? So far amino acid sequences and electrophoretic mobilities (isoelectric focusing) of class I and class II molecules from individual humans and mice show little or no variability. Thus *while there is an extraordinary degree of polymorphism of these MHC genes within a species, there seems to be no variability of the products of these genes within the individual.*

CELL INTERACTIONS

H-2 RESTRICTIONS

The immune responses of most lymphocytes depend upon both specific binding of particular Ags and interactions with regulatory lymphocytes and Mϕs. Nude mice provided the earliest suggestive evidence that MHC products are critically involved in the cellular interactions: the inability of these T-deficient mice to make Abs to many Ags can be corrected by an injection of normal T cells, but only if the T cells and the nude recipients share the same H-2 haplotype. This provocative observation has been extensively analyzed in several systems: 1) T cell–B cell interaction in Ab production (see Role of T Helper Cells, Ch. 19), 2) T cell–Mϕ interactions associated with T cell proliferation and delayed-type hypersensitivity (Ch. 22) and, especially, 3) the lysis of diverse target cells by cytotoxic T lymphocytes (CTLs, or T_k cells; see T Killer Cell Reactions, below, and Cytotoxic T Lymphocytes, Ch. 22).

As the following sections show, in all of these systems a specific T cell response to Ags on cell surfaces depends not on recognition of the Ag alone but on recognition of the Ag **plus** certain products of the MHC on the same cell surface. Inbred mouse strains with recombinant H-2 haplotypes have been especially illuminating: they show that *CTLs (T_k cells), which are Ly-23 cells, are specific for Ag plus products of H-2K and H-2D loci, whereas T helpers and T_{DTH} cells (effectors of delayed-type hypersensitivity), both of which are Ly-1 cells, are specific for Ag plus products of the H-2 I locus. The dependence of the T cell's specific reactivity on Ag plus H-2 products, rather than on the Ag alone, is called* **H-2, or MHC, restriction.**

T KILLER CELL REACTIONS WITH TARGET CELLS

In the MLR T_k cells (anti–H-2K and anti–H-2D), are easily elicited in the responder population by allogeneic (i.e., H-2–differing) stimulator cells (see Mixed Lympho-cyte Reaction, above). Under normal in vivo conditions, where allogenic stimulators are, of course, absent, T_k cells can also be readily elicited, by "self" cells that carry new surface Ags; e.g., Ags specified by virus in virus-infected cells. However, the specificity of these antiviral T_k cells is also determined by H-2: Zinkernagel and Doherty made the important discovery that the only (or best) virus-infected target cells are those with the same H-2K and H-2D alleles as the original virus-infected stimula-tors (Table 18-8).

H-2 restriction is also evident in model systems where the anti-genic moieties are **haptenic** groups, such as Tnp (2,4,6-trinitro-phenyl groups) attached covalently to cell surface proteins.

In another valuable model system, **minor histocompatibility Ags ("minors")** serve as natural cell surface Ags: mice of some strains (e.g., DBA/2 and BALB/c) have the same H-2 complex (H-2^d) but differ in many minors, causing DBA/2 tissue grafts to be rejected by BALB/c mice and vice versa (see Non–H-2 Loci, Ch. 23). T_k cells elicited by injecting BALB/c cells into DBA/2 mice (anti-minors) will lyse BALB/c cells but not cells from the congenic BALB.b strain, which have the same minors as BALB/c but a dif-ferent H-2 (H-2^b).

In the H-$2^{a/b}$ hybrids resulting from crosses between homozygous H-2^a and H-2^b mice there are two sets of T_k cells for any Ag (call it X), whether due to virus, hapten, or minors: one set is specific for X+H-2^a and other for X + H-2^b. Moreover, in either set some T_k cells are specific for X + H-2K and others for X + H-2D of the same haplotype. Hence in the H-$2^{a/b}$ individual there are **four sets** of anti-X T_k cells, specific for X + H-2K^a, X + H-2D^a, X + H-2K^b, and X +H-2D^b. In accord with the clonal selection hypothesis the corresponding four sets of precursors appear to exist before X is encountered.

The specificity of some of these precursors has been demon-strated by "suicide" experiments, based on the finding that cells proliferating in the presence of the thymidine analog 5-bromo-deoxyuridine (BUdR) incorporate this base into DNA and are thereby rendered susceptible to destruction by light (at the wave length where BUdR absorbs maximally). The experiment is out-lined in Figure 18-23. When H-$2^{a/b}$ hybrid spleen cells were stim-ulated with X-containing H-2^a parental cells in the presence of BUdR, the precursors that recognized X + H-2^a proliferated, in-corporated BUdR, and were killed on exposure of the culture to light: when the surviving cells were restimulated with X + H-2^a and X + H-2^b cells, the resulting T killer cells specifically lysed only X + H-2^b targets.

Unlike the K and D loci, the I locus has no influence on the speci-ficity of the mature T_k cells. This indifference can be demonstrated with congenic intra–H-2 recombinant strains: T_k cells do not dis-tinguish between Ag-bearing target cells prepared from a congenic pair of mice that differ in I but have the same K and D loci.

H-2 RESTRICTION OF OTHER T CELLS

As noted before, the H-2K and H-2D restricted T killers are Ly-23 cells. Ly-1 T cells are also H-2 restricted, but with these cells the restricting locus is I, not K or D. Since I alleles are expressed predominantly on B cells and Mφs, it is in the reaction of these cells that H-2I restriction is evident. The following paragraphs summarize some of the evidence for H-2I restriction.

T Cell Proliferation. Because the Ag-pulsed Mφ (see Macro-phages, above) is much more effective than soluble Ag in stimu-lating the proliferation and development of T_h cells, it probably represents the normal mode for stimulating these cells in vivo. The stimulating effect of Ag-pulsed Mφs is especially striking in the guinea pig, where two inbred strains, 2 and 13, differ at the locus that corresponds to I in the mouse H-2 complex but are the same in loci corresponding to K and D. As is shown in Fig. 18-24, T cells from Ag-primed 2×13 hybrids can be restimulated in culture by Ag-pulsed Mφs from either parental strain, but in the subsequent assay a proliferative response can only be elicited with Ag-pulsed Mφs from the same parent (i.e., with the same I allele) that was used to restimulate the T cells. The proliferating lymphocytes in this assay almost certainly correspond to Ly-1 T helpers in the mouse.

T Cells That Cause Delayed-Type Hypersensitivity (DTH). This type of allergic inflammatory reaction is thought to result from in-teraction in vivo between special effector T cells (T_{DTH}) and Ag on Mφs (Ch. 22). H-$2^{a/b}$ heterozygous mice (made by crossing homo-zygous H-2^a and H-2^b strains) can be immunized with Ag-pulsed, homozygous Mφs from either parent. When the resulting "primed" T cells are then transferred to mice of the H-2^a or H-2^b strains, a DTH response is elicited by Ag only in the recipient whose H-2 haplotype is the same as that of the Ag-pulsed Mφs used to prime the hybrid mouse (Fig. 18-25). Intra–H-2 recombinant strains show that correspondence in the I locus, not in K or D, is essential. Thus the T_{DTH} cell (an Ly-1 cell in the mouse) is probably specific not for

TABLE 18-8. Schematic Representation of H-2 Restriction of Target Cell Lysis by Cytotoxic T Lymphocytes (CTLs).

Responder cells	Stimulator cells	Target cells	Target cell lysis
A (H-2^a)	X-A (H-2^a)	X-A (H-2^a)	+
		A (H-2^a)	−
		X-B (H-2^b)	−
		X-B.A (H-2^a)	+
		B.A (H-2^a)	−
		X-A.B (H-2^b)	−

X = a particular surface Ag (e.g., virus or haptenic group, such as trinitrophenyl) on stimulator cells and on some target cells, but absent on cells of the responder population. A and B = cells from inbred strains with H-2 haplotypes H-2^a and H-2^b, respectively (see H-2 Complex, Ch. 18, and Table 18-6). A.B = a congenic strain whose H-2 is H-2^b; otherwise the genome is A. B.A = a congenic strain whose H-2 is H-2^a; otherwise the genome is B. See text and Fig. 18-9; for Histo-compatibility Genes, see Ch. 23; and for more on CTL see Ch. 22.

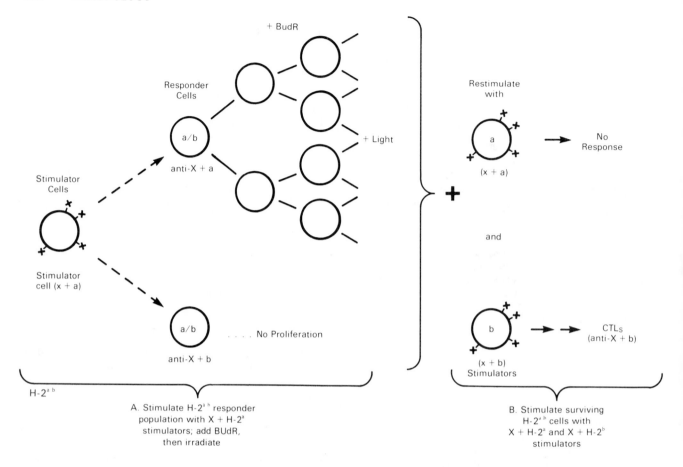

FIG. 18-23. Clonal distribution of the H-2 restricted specificity of T_k cells in H-2 heterozygous (H-2$^{a/b}$) mice, demonstrated by a BudR (bromodeoxyuridine) "suicide" experiment. X on stimulator cells is a neoantigen (due to infecting virus, attached hapten, or normal minor histocompatibility Ags). (Based on Janeway CA, Jr et al: J Exp Med 147:1065 1978)

the Ag itself, but for the Ag plus a product of the I locus (probably Ia) on the Mφs that initially stimulated the development of the T_{DTH} cells.

T Helper Cells and Antibody Formation. As noted elsewhere, the response of B cells to Ag is enhanced by T_h cells that are activated by that Ag presented on Mφs. In the activation step, the T_h cell (or its precursor) recognizes not the Ag alone but the Ag and MHC products (probably Ia molecules) on the Mφs. Thus Ag-primed T_h cells are stimulated to proliferate by the Ag on Mφs, if the Mφs have the same I-encoded surface molecules as the Ag-pulsed Mφs that initially primed the T_h cell (see The Role of T Helper Cells, Ch. 19).

Correlations between Lyt surface Ags, T cell function, and the H-2 loci that restrict T cell specificity are summarized in Table 18-9.

Dual Receptor and Altered-Self Hypotheses. Two
models have been suggested to account for the dual re-

quirement by T cells for Ag plus H-2 products. The dual receptor model postulates that the T cell has two receptors, one for Ag and the other for products of self H-2K, D, or I, and both receptors must be occupied for the T cell to respond. In contrast, the altered-self model proposes that interaction of Ag and H-2–encoded molecules, on the target cell's surface, forms a complex ligand (**altered-self**) that is recognized by a single receptor on the T cell (Fig. 18-26).

The existing observations do not eliminate either hypothesis, but they help sharpen the issues. The lysis of target cells bearing an Ag, call it X, by anti-X T_k cells can be specifically inhibited by other cells if X and the correct H-2 are on the same cell, but not if they are on different cells, suggesting **the need for proximity of Ag and H-2 products.**

However, in their reaction with target cells, T cells, as a group, can recognize an immense number of different Ags and haptens,

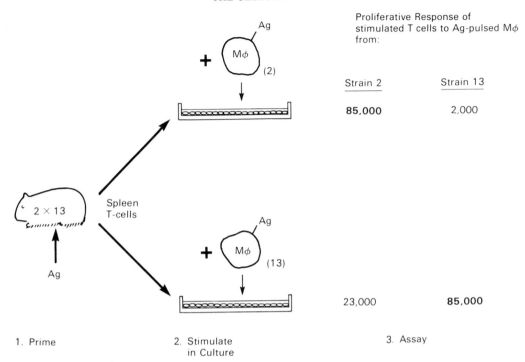

Proliferative Response of
stimulated T cells to Ag-pulsed Mϕ
from:

	Strain 2	Strain 13
	85,000	2,000
	23,000	**85,000**

1. Prime   2. Stimulate in Culture   3. Assay

FIG. 18-24. Evidence for restriction of T cell–macrophage interaction by the major histocompatibility complex (MHC). Inbred guinea pigs of strains 2 and 13 differ in that part of the MHC that corresponds to the I region of the mouse H-2 complex. Spleen cells from an Ag-primed (2 × 13) hybrid guinea pig were stimulated in culture with Ag-pulsed Mϕs from one parent (2 or 13); in a subsequent test, the T cells that responded were found to proliferate more actively (i.e., to incorporate more ^{3}H-thymidine into DNA) in response to a second exposure to Ag-pulsed Mϕs when the Mϕs were derived from the same parent that provided the stimulators in culture than when derived from the other parent. Numbers on the right refer to counts per minute of ^{3}H-thymidine incorporated into T-cell DNA. (Based on Paul WE et al: Cold Spring Harbor Symp Quant Biol 41:571, 1976)

and it is difficult to see how H-2 molecules can bind so many different structures, forming in each instance a singular complex ligand. According to one suggestion, the extremely small volume in which the cell's surface proteins are free to diffuse means that these proteins exist on the surface membrane at very high concentrations; hence they may be able to form many reversible complexes even though their mutual affinities (equilibrium association constants) are extremely low.

T Cell Recognition of Self H-2 is Not Programmed by the Cell's H-2 Genes. MHC restriction means that an individual's T cells have receptors for cell surface molecules that are encoded by his own MHC genes, perhaps modified by viral or other surface Ags (altered-self model; see above). To determine whether the T cell receptors are also specified by the T cell's own MHC genes, specially reconstructed mice have been used: for instance, lethally irradiated mice of one H-2 haplotype rescued by the injection of hematopoietic stem cells (from fetal liver or bone marrow) of another H-2 haplotype. The survivors, called **chimeras** (after the mythical Greek monster with a lion's head and goat's body), are **mosaics**: lymphocytes, and other blood cells have one H-2 type and the rest of the animal has another H-2 type. (As conventionally denoted, $a \rightarrow (a \times b)$F1 means that stem cells of type a were injected into an irradiated $(a \times b)$F1 recipient, where a and b refer to different H-2 haplotypes.)

Other reconstructed mice are lethally irradiated and thymectomized, then restored with stem cells of one type and a thymus graft of a different type. As the following examples indicate, *the specificity of H-2 restriction is determined by the thymic environment in which stem cells differentiate into mature T cells, rather than by the T cells' own H-2 haplotype.*

1) Chimeras of $a \rightarrow a \times b$ type, prepared by injecting stem cells from an H-2^a parent into an irradiated H-2$^{a/b}$ hybrid recipient, were immunized by infection with virus X. The resulting T killer cells, though genetically H-2^a, lysed X-infected target cells from either the H-2^a or the H-2^b parent, but not from the unrelated H-2^c strain (Fig. 18-27).

2) In the reverse $a \times b \rightarrow a$ chimeras (i.e., H-2$^{a/b}$ hybrid stem cells

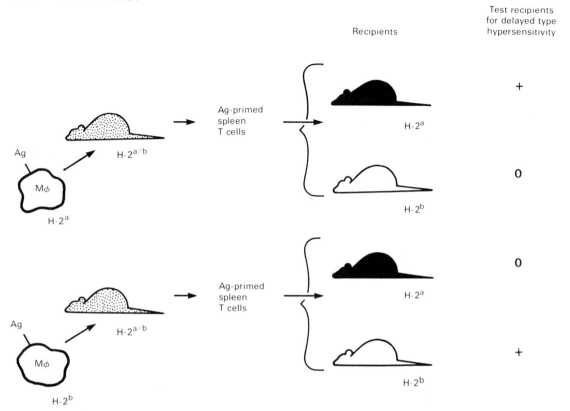

FIG. 18-25. H-2 restriction of T cell specificity in delayed-type hypersensitivity (DTH). In the assay for DTH, an inflammatory skin response is elicited by injecting a test dose of Ag into the skin (Ch. 22). Here the tested mice were recipients of Ag-primed T cells (Ly-1 type) from H-2$^{a/b}$ hybrids (H-2^a × H-2^b) that had been primed in vivo with Ag-pulsed Mφs from the H-2^a or H-2^b parental lines. (Based on Miller JFAP, Vadas MA: Cold Spring Harbor Symp Quant Biol 41:579, 1976)

injected into the irradiated H-2^a parental recipient), the T$_k$ cells elicited with Ag X have both H-2^a and H-2^b haplotypes, but they can only lyse X-containing target cells from the H-2^a parent (Fig. 18-27).

3) Lethally irradiated and thymectomized H-2$^{a/b}$ hybrids were restored to immunologic function by an injection of H-2$^{a/b}$stem cells plus a thymus graft (required for differentiation of pre-T stem cells), and were then immunized with Ag X. If the thymus graft was

from the H-2^a parent, the resulting T$_k$ cells, though H-2$^{a/b}$ hybrids and developing in an H-2$^{a/b}$ mouse, lysed only X+H-2^a and not X+H-2^b target cells. Hence *it is in the thymus during the differentiation of pre-T cells that precursors of T$_k$ cells learn which H-2-encoded molecules to regard as self* (Table 18-10). The thymic cells that provide instruction resist high doses of x-ray (hence need not proliferate); they may be epithelial cells or special Mφs (perhaps analogous to Langerhans cells in the epidermis).

4) T helper cells also learn during differentiation in the thymus which H-2 to regard as self, but for these T cells what matters is not the K and D loci but the I locus (probably Ia Ags). Normally, H-2$^{a/b}$ hybrid T$_h$ cells can aid B cells of either parental type, H-2^a or H-2^b, to develop into Ab-secreting cells (**"semisyngeneic cells cooperate"**). However, in $a \times b \rightarrow a$ chimeras, the hybrid T$_h$ cells elicited by Ag X cooperate well with B cells of the H-2^a parent (like the environment in which the hybrid stem cells matured) but not with B cells of the other parent. In addition, T$_h$ cells elicited in the reverse chimeras, $a \rightarrow a \times b$, can help B cells of either parent, H-2^a or H-2^b. And T$_h$ cells elicited in adult thymectomized H-2$^{a/b}$ hybrid mice, reconstituted with H-2$^{a/b}$ stem cells and an H-2^a (or H-2^b) thymus graft also display **haplotype preference (or best) with B cells that have the same H-2I locus as the thymus in which the pre-T cells differentiated from stem cells.**

TABLE 18-9. Correlation Between Lyt Differentiation Antigens, T Cell Functions, and the H-2 Loci that Restrict T Cell Specificity

Lyt subset	Functional subset*	H-2 locus that restricts the T cells' "recognition" of Ag
Ly-1	T$_h$	I
	T$_{DTH}$	I
Ly-23	T$_k$ (CTL)	K,D
	T$_s$	?

*For explanation of abbreviations and functions, see Table 18-3.

TABLE 18-10. The Thymic Environment in which T Cells Undergo Differentiation Determines What H-2 Encoded Cell Surface Molecules the Mature T Cell Recognizes as "Self".*

H-2 haplotype of donor of thymus graft	Effect of T_k cells from recipient mice on target cells of haplotype	
	a	*b*
a	lysis	0
b	0	lysis

a and *b* = H-2 haplotypes

*Thymus grafts taken from donor mice were placed under the skin or kidney capsule in lethally irradiated, thymectomized $(a \times b)$ hybrid recipient mice reconstituted with $(a \times b)$ stem cells. The stem cells were fetal liver cells or adult bone marrow cells treated with antiserum to Thy-1 (and complement) to eliminate any mature T cells. The thymus graft was able to function even after irradiation destroyed its own lymphocytes, presumably because its epithelial cells remained active. The results show that $(a \times b)$ T_k cells, with genome *a* and *b*, will recognize as self either *a* or *b*, depending on the genome of the thymus in which the T_k cell matured.

(Based on Zinkernagel RM et al: J Exp Med 147:882 1978; and Fink PJ, Bevan MJ: Exp Med 148:766 1978)

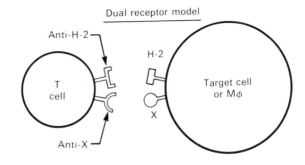

FIG. 18-26. Models for H-2 restriction of T cell specificity.

FIG. 18-27. The H-2 identified as self by cytotoxic T lymphocytes (T_k cells) is determined by the chimeric host in which pre-T stem cells differentiate, rather than by the T_k cells' own H-2 haplotype. The letters *a, b,* and *s* refer to H-2 haplotypes (H-2^a, H-2^b, and H-2^k in the actual experiments), while the Ag was a virus-specified neoantigen on infected lymphocytes and Mϕs. (Based on Zinkernagel R. M et al: J Exp Med 147:882 1978)

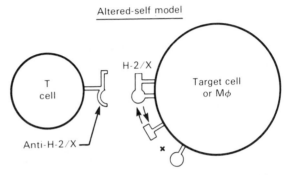

T CELL DIFFERENTIATION

As discussed in earlier sections of this chapter, precursors of T cells arise in the yolk sac and liver of the fetus and in the bone marrow of the normal adult. They migrate to the thymus, where their maturation involves 1) extensive cell proliferation, 2) expression of new surface molecules, 3) acquisition of responsiveness to various Ags and to lectins, and 4) commitment to function as a particular type of T cell (T_h, T_s, T_{DTH}, T_r, or T_k; see Table 18-3).

Arriving in the thymus, the immigrant cells lodge in the outer cortex (subcapsular cells), divide rapidly, and migrate deeper into the cortex; they then complete their maturation either in the medulla or in peripheral tissues. The average transit time through the thymus is about 3–4 days. The average generation time of the dividing cells is about 9 h, and more thymocytes are produced in the thymus (about 1000 per immigrant precursor) than leave it: most thymocytes die in the thymus.

To follow the time course of maturation, thymocytes have been analyzed for

1) cell surface Ags (Thy-1, H-2, TLa, Lyt, and Qa; see Table 18-2);

2) susceptibility to agglutination by a peanut agglutinin (which binds terminal galactosyl-N-acetyl-galactosyl groups and agglutinates many thymocytes and bone marrow stem cells but not mature, peripheral T cells); and

3) susceptibility to destruction by corticosteroids.

In order of increasing maturity thymocytes can be classified as subcapsular, cortical, and medullary (see corresponding layers of the thymus in Fig. 18-4). When cells just beneath the capsule are selectively labeled with ^{3}H-thymidine (which is carefully injected into the intact organ), the label subsequently appears in cells of the cortex and medulla, suggesting that the latter are derived from the subcapsular cells. The properties of thymocytes at various positions within the thymus are summarized in Table 18-11.

Prothymocyte Induction. Bone marrow and spleen contain some cells that lack the characteristic T cell Ags but acquire them when appropriately stimulated. When these isolated **pre-T cells (prothymocytes)** are cultured with various agents that raise intracellular cyclic AMP levels they promptly form Thy-1, TLa, and GIX (see Table 18-2).

Over ten factors isolated from thymus or blood cause this transformation of prothymocytes; each has been reputed to be the thymus hormone responsible for maturation in vivo. Some of the better known factors are listed here:

1) **Facteur thymique serique** (serum thymus factor) is a nonapeptide from serum. It causes pre-T cells to express the surface molecule that is responsible for T cell binding of sheep RBCs (rosette formation; see Fig. 18-11).

2) **Thymopoietin,** a small protein (49 amino acid residues), has been sequenced and synthesized. It causes pre-T cells to express Thy-1, TLa, and GIX and to become somewhat sensitive to mitogens, but not to become immunologically competent.

3) **Thymosin,** a mixture of many active low-molecular-weight polypeptides (1000–15,000 daltons), probably comes closer than any others to behaving like the natural hormone. Besides inducing expression of Thy-1 and other T cell Ags, it restores immune function to neonatally thymectomized mice, duplicating the restorative effect of a thymic graft that is enclosed in a chamber with walls that are permeable to macromolecules but not to cells.

Thymic hormones are probably produced by thymic epithelial cells and promote T cell maturation locally and also in peripheral tissues. However it is also probable that maturing thymocytes have to make **direct contact with thymic epithelium** in order to learn to recognize as self the cell surface molecules encoded by the MHC (see H-2 Restrictions, above).

Acquisition of T Cell Specificity. A high proportion (>50%) of T cells respond to immunization with foreign MHC-encoded molecules on allogeneic cells (see Transplantation Immunity, Ch. 23). Moreover, as noted before (see H-2 Restrictions, above), in recognizing other Ags on cells, T cells recognize not the Ag alone but the Ag

TABLE 18-11. T Cell Development: Some Properties of Thymocytes at Various Levels within the Mouse Thymus*

Location in the thymus	Abundance (% of total thymocytes)	Cellular properties
Subcapsule	~ 5–10	Large blasts; high mitotic rate; cortisone-sensitive; large amounts of Thy-1; agglutinated by peanut lectin.
Cortex	80	Small, dense cells; high levels of Thy-1, TLa, Lyt-1,2,3 (i.e., all are Ly-123 cells) but low levels of H-2 and no Qa-1; cortisone-sensitive; agglutinated by peanut lectin; respond poorly to con A and not at all to PHA†
Medulla	10–15	Larger and less dense than cortical cells, with more H-2 encoded Ag (per cell) but less Thy-1 and no TLa; some are Ly-1 cells but most are Ly-123; relatively resistant to cortisone; not agglutinated by peanut lectin; immunologically competent (e.g., cause graft-vs-host reaction; Ch. 23).

*For anatomy of the thymus, see Fig. 18-4.

†con A = concanavalin A; PHA = phytohemagglutinin.

and MHC-encoded molecules on the same cell. It is thus probable that in acquiring immunologic specificity T cells develop two receptors, or one receptor with two distinct sites: one directed to MHC-encoded molecules and the second to other Ags. These other Ags seem to be as diverse as the Ags recognized by B cells and Ab molecules (e.g., any protein, including Ig idiotypes and many small haptenic groups), except that, so far as is known, few polysaccharides are recognized by T cells.

As noted above, the receptors for these Ags share idiotypes with Abs that are specific for the same Ags, and the shared idiotypes are probably encoded by V_H genes (see Antigen-Binding Receptors on T Cells, above). Hence the diversity of T cell receptors for non-MHC molecules probably originates by the same mechanisms that generate the diversity of Abs (see Origin of Diversity of V Genes, Ch. 17).

The nature of the receptors for MHC molecules (here referred to as anti-histocompatibility or anti-H) are more obscure. Each individual must produce a sufficiently diverse set of them to recognize virtually all histocompatibility (H) Ags of the species, including his own. It may be speculated, following an early hypothesis of Jerne and the clonal selection theory, that of the many anti-H molecules produced by the individual each prothymocyte expresses one kind. On being exposed to self-H in the thymus, those developing prothymocytes with anti-self-H molecules will proliferate selectively and populate peripheral tissues.

If the self-H is a new mutation, or is produced by an H-2-different thymus cell, as in the chimera experiments described above, the selection would still result in mature T cells that recognize these unusual H molecules as self. Given enough heterogeneity and cross-reactivity, the set of anti-self-H receptors could still be sufficiently diverse to include subsets that cross-react with other H Ags of the species, thus accounting for the high frequency of alloreactive anti-H T cells.

Because recognition of self-H is necessary for T cell regulation of many immune responses, it is probably crucial for resistance to infections and even for survival. Hence the need for anti-H molecules that are complementary to self-H could exercise an important restraining influence on the proliferation of H polymorphism. Any mutation of H for which a complementary anti-H molecule is not produced in the same individual would be lethal. Thus, though the mutation rate and polymorphism of MHC genes are extraordinarily high, there are limits to their variability: as noted before, three inbred mouse strains (AKR, CBA, C3H), separated by over 50 years of intensive inbreeding seem still to have identical H Ags (H-2^k haplotype).

DIFFERENTIATION ANTIGENS AND HUMAN T CELL SUBSETS

Human T cells are divisible into sets resembling those of mice. By immunofluorescence, surface Ig is readily detected on human B cells but not on T cells. However, **human T cells fortuitously bind sheep red blood cells** and the resulting rosettes (Fig. 18-11) permit T cells to be counted and separated from other lymphocytes. Subsets of T cells, analogous to mouse Ly-1 helpers and Ly-23 suppressors and killers are distinguished by some heteroantisera (e.g., rabbit antiserum elicited with one person's T cells and adsorbed with the same individual's B cells [made immortal by Epstein-Barr virus infection and maintained as a lymphoblastoid B cell line]). Of particular value are the Abs produced by fusing mouse myeloma cells with spleen cells from mice immunized against human T cells. The resulting cell lines (Hybridomas, Fig. 17-4, ch. 17) are immortal and produce extremely high levels of monoclonal Abs. These Abs specifically precipitate surface glycoproteins from human T cells that resemble those of mouse T cells. (For instance, monoclonal anti-"T4" precipitates molecules that resemble Lyt-2 and Lyt-3 [33,000 and 30,000 daltons, respectively], anti-

"T5" precipitates molecules that resemble Lyt-1 [60,000 daltons], and anti-"T6" precipitates molecules that resemble the thymus-leukemia Ag TL.) With these Abs several functionally distinctive T cell subsets have been identified.

T Inducers. This subset (T4$^+$, T5$^-$), representing about 70% of human T cells, resembles mouse Ly-1 helper cells. These cells have to be present for the generation of cytotoxic T lymphocytes (CTLs) in mixed lymphocyte reactions (MLRs), but they do not themselves become CTLs. They are also necessary for the response of B cells (cell proliferation and secretion of Igs) to pokeweed mitogen.

T Suppressor and T Killers. This subset (about 10% of all human T cells) is analogous to mouse Ly-23. It includes precursors of the CTLs that are formed in the MLRs; and it also includes cells that, in response to concanavalin A, produce a "suppressor factor" that blocks the development of CTLs. Cells of this subset are T4$^-$ T5$^+$.

DIFFERENTIATION OF HUMAN T CELLS

The differentiation of lymphocytes in the human thymus can also be followed with monoclonal Abs to human T cells. Most human thymocytes (about 70%) are T4[+], T5[+], T6[+] and resemble mouse Ly-123 cells. One small subset (about 10%) resemble immigrant prothymocytes (T9[+], T10[+]), whereas another 10% are like mature cortical T cells of the mouse: some are T4[+] T5[−] (inducers), others are T4[−] T5[+] (suppressors/killers), and all lack T6, the thymus-leukemia like antigen.

Acute Lymphatic Leukemia (ALL). The restricted distribution of surface T Ags on these cells is consistent with the view that the *ALL cells in any particular patient represent a single clone of immature T cells, arrested at some stage in development.* In about 70% of individuals with ALL, the leukemic cells are T9[+]T10[+] (resembling immature prothymocytes, see above); in other patients (about 20%) the ALL cells are T4[+] T5[+] T6[+], like most immature thymus cells; in only rare patients are ALL cells T3[+] T5[+], like mature cytotoxic/suppressor cells.

SUGGESTED READING

BOOKS AND REVIEWS

CANTOR H, BOYSE EA: Lymphocytes as models for the study of mammalian cellular differentiation. Immunol Rev 33:105, 1977

FINK PJ, BEVAN MJ: The influence of thymic H-2 antigens on the specificity of maturing killer and helper cells. Immunol Rev 42:3, 1978

KLEIN J: Biology of the Mouse Histocompatibility-2 Complex. New York, Springer-Verlag, 1975

MARCHALONIS JJ, COHEN N (eds): Self/Non-self discrimination. Contemp Topics in Immunobiol 9: (in press), 1980

MILLER JFAP: Influence of the major histocompatibility complex on T-cell activation. Adv Cancer Res 29:1, 1979

Origens of Lymphocyte Diversity. Cold Spring Harbor Symp Quant Biol, Vol 41, 1977

PAUL WE, BENACERRAF B: Functional specificity of thymus-dependent lymphocytes. Science 195:1293, 1977

RAJEWSKY K, EICHMANN K: Antigen receptors of T helper cells. Contemp Topics in Immunobiol 7:69, 1977

REINHERZ EL, SCHLOSSMAN SF: The differentiation and function of human T lymphocytes: A review. Cell (in press, 1980)

ROBERTSON M: The life of a B lymphocyte. Nature 149:332, 1980

SHREFFLER DC, DAVID CS: The H-2 major histocompatibility complex and the immune response region: genetic variation, function and organization. Adv Immunol 20:125, 1975

SNELL GD, DAUSSET J, NATHENSON S: Histocompatibility. New York, Academic Press, 1976

STUTMAN O: Intra-thymic and extra-thymic T cell maturation. Immunol Rev 42:138, 1978

ZINKERNAGEL RM, DOHERTY PC: MHC-restricted cytotoxic T cells: studies on the biological role of polymorphic major transplantation antigens determining T-cell restriction, specificity, function, and responsiveness. Adv Immunol 27:51, 1979

SPECIFIC ARTICLES

ABRAMSON S, MILLER RG, PHILLIPS RA: The identification in adult bone marrow of pluripotent and restricted stem cells of the myeloid and lymphoid systems. J Exp Med 145:1567, 1977

ADORINI L, HARVEY MA, MILLER A, SERCARZ EE: Fine specificity of regulatory T cells. II. Suppressor and helper T cells are induced by different regions of her egg-white lysozome in a genetically nonresponder mouse strain. J Exp Med 150:293, 1979

CAMPBELL DG, WILLIAMS AF, BAYLEY PM, REID KBM: Structural similarities between Thy-1 antigen from rat brain and immunoglobulin. Nature 282:341, 1979

FRELINGER JG, HOOD L, HILL S, FRELINGER JA: Mouse epidermal Ia molecules have a bone marrow origen. Nature 282:321, 1979

HÜNIG T, BEVAN MJ: Self H-2 antigens influence the specificity of alloreactive cells. J Exp Med 151:1288, 1980

KATZ DH: Lymphocyte differentiation, recognition, and regulation, Academic Press, NY, 1977

KATZ SI, TAMAKI K, SACHS DH: Epidermal Langerhans' cells are derived from cells originating in bone marrow. Nature 282:324, 1979

LINDAHL KF, WILSON DB: Histocompatibility antigen–activated cytotoxic T lymphocytes. I. Estimates of the absolute frequency of killer cells generated in vivo. J Exp Med 145:500, 1977

ORR HT, LANCET D, ROBB RJ, LOPEZ DE CASTRO JA, STROMINGER JL: The heavy chains of human histocompatibility antigen HLA-B7 contains an immunoglobulin-like region. Nature 282:266, 1979

ROTHENBERG E, BOYSE EA: Synthesis and processing of molecules bearing thymus leukemia antigen. J Exp Med 150:777, 1979

SANDERSON RJ, RULON K, GROENEBOER EG, TALMAGE DW: The response of murine splenic lymphocytes to concanavalin A and to co-stimulator. J Immunol 124:207, 1980

SCOLLAY R, KOCHEN M, MUTCHER E, WEISSMAN I: Lyt markers on thymus cell migrants. Nature 276:79, 1978

TADA T, TANIGUCHI M, DAVID CS: Suppressive and enhancing T-cell factors as I-region gene product: properties and subregion assignment. Cold Spring Harbor Symp Quant Biol 41:119, 1977

TANIGUCHI M, TAKEI I, TADA T: Functional and molecular organization of an antigen-specific suppressor factor from a T-cell hybridoma. Nature 283:227, 1980

WILLIAMSON AR: Three-receptor, clonal expansion model for selection of self-recognition in the thymus. Nature 283:527, 1980

chapter 19

ANTIBODY FORMATION

The central problem of antibody (Ab) formation is to understand how an individual can form Ab molecules to an almost limitless variety of foreign antigens (Ags) but not to his own substance (self-Ags). Previous chapters noted that there are a great number of Ab-forming B lymphocytes (close to 10^{12} in an adult human); these are divided up into a great many clones, each derived from one B cell. Each clone can synthesize immunoglobulin (Ig) molecules with a singular Ag-combining region made up of the variable domains of heavy (H) and light (L) chains ($V_H + V_L$). Because the Ig is an integral membrane protein with its combining site exposed on the B cell surface, each clone is specific for a particular Ag and is marked by its Ig's unique V-domain antigenic individuality, or **idiotype.** When, under appropriate conditions, Ag is encountered and combines with the B cell's surface Ig, the cell is triggered to proliferate and to form some daughter cells that secrete Ab molecules with the same combining sites as the B cell's surface Ig. Under other conditions, the Ags (particularly self-Ags) have the opposite effect: they induce B cells to become tolerant, i.e., not to yield Ab-secreting daughter cells, because the B cells are either eliminated or otherwise rendered unable to function.

In this chapter we will focus largely on the various "appropriate conditions" that lead to one or the other of these responses: **immunogenic or tolerogenic.** We will also review the development of B cells, their synthesis and assembly of Ig chains, and some B cell deficiency diseases of man.

ANTIBODY RESPONSES TO T-DEPENDENT ANTIGENS

With most Ags the Abs made at increasing intervals after immunization show a switch in their type of H chain and an increase in affinity for the Ag. The changes depend upon T cells: hence these Ags are **T-dependent.** In contrast, with a small group of **T-independent** Ags (mostly a group of highly polymerized molecules of bacterial origin) the Ab responses do not change with time and are not helped by T cells. The shifting Ab responses elicited by T-dependent Ags are especially evident when the initial **(primary)** and subsequent **(secondary)** responses to the Ag are compared.

After initial exposure to the Ag the Abs detected earliest (usually after several days) are predominantly IgM molecules that combine weakly with Ag. After a few weeks Ab production virtually ceases, but the immune system does not return to its preimmune status. Instead, it remains **primed,** as though the Ag is remembered, to give an enhanced secondary response on a subsequent encounter with the same Ag (Fig. 19-1). The secondary response is thus also called a **memory,** or **anamnestic, response** (Gr. *anamnesis,* recall).

By comparison with the primary response, the secondary response is characterized by

1. a shorter lag period between the encounter with Ag and the appearance of Abs,
2. Ab production that has a higher rate and is more persistent,
3. higher Ab titers at the peak of the response,
4. a predominance of IgG molecules, and
5. Abs with a higher affinity for Ag than those produced in the primary response.

These differences are partly accounted for by the number of responding B cells. In the primary response only a few of the B cells that are present have the specificity that permits a particular Ag to trigger their development into Ab-secreting B cells (plasma cells; see below). The initial antigenic stimulus also results in the proliferation and formation of a large number of **secondary** or **memory B cells,** capable of responding to the same Ag. Thus at the time of the secondary stimulus many more B cells can respond to the Ag.

In the absence of overt antigenic stimulation, memory cells seem to persist for prolonged periods. Thus tetanus toxoid can evoke a powerful secondary response in man as long as 20 years after primary immunization. Enduring memory and the corresponding capacity for a vigorous secondary response can thus provide for long-lasting immunity against infection: many years after the primary response, and long after Abs cease to be detectable in serum, reintroduction of the immunogen stimulates memory cells to undergo rapid division, resulting in early and vigorous production of Abs. Moreover, these are

FIG. 19-1. Comparison of serum Ab concentrations following the first (priming) and second (booster) injections of immunogen **(arrowheads).** Note the logarithmic scale for Ab concentration. Time units are left unspecified to indicate the great variability encountered with different immunogens under different conditions. Because secondary (memory) B cells at the time of the second injection usually outnumber the primary B cells at the time of the first injection, the secondary response has a shorter lag and results in higher and more sustained Ab levels.

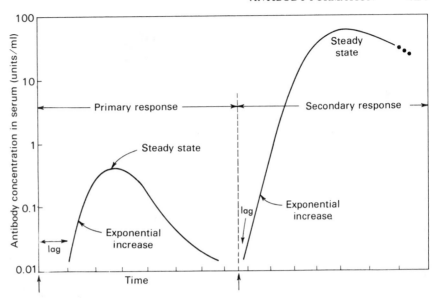

high-affinity IgG molecules, in contrast to the IgM Abs of the primary response.

The duration of the primary and secondary responses varies with the dose and mode of administration of Ag. If the Ag is administered in solution, and is rapidly eliminated, the primary response is short-lived and another injection of Ag is required to elicit the secondary response. If, however, the Ag is retained for a long period, as when given in a water-in-oil emulsion (Freund's complete adjuvant), the resulting Ab production continues at a high level for a prolonged period, and the transition from the production of IgM to IgG molecules, and from low- to high-affinity Ab molecules, occurs gradually and without the need for another injection. When the prolonged response finally subsides, after many months, a subsequent injection of the same Ag promptly elicits the formation of IgG Ab molecules of the same high affinity as those made at the end of the first response.

INCREASE IN AFFINITY WITH TIME

Because all the Ig molecules made by a B cell have the same combining site, the increase in affinity with time means that cells making low-affinity Abs predominate early in the response and those making high-affinity Abs predominate later. The shift probably arises for the following reasons:

1) When Ag is first encountered B cells with low-affinity Ig outnumber those with high-affinity Ig, because affinity reflects the fit of the Ig combining site for the Ag and there is a greater probability that randomly arising Ig molecules, if they fit at all, will fit poorly and have low affinity.

2) High Ag levels are required to bind low-affinity cell surface Abs and to trigger the clones that produce these molecules, whereas cells capable of producing high-af-

finity molecules can be triggered by low (and high) Ag concentration.

3) Initially, Ag levels are relatively high and most of the stimulated cells are secretors of low-affinity Abs. With time, the Ag level falls, high-affinity producers are selectively stimulated, and the average affinity of secreted Abs rises. A larger initial dose of Ag can, accordingly, prolong the production of low-affinity Abs (Table 19-1).

Analysis of individual Ab-secreting cells, by Jerne's hemolytic plaque assay, confirms that the increase in average affinity of serum Ab is due to a shift in the population of Ab-secreting cells. Because the hemolytic plaque assay is a powerful general tool for quantitating and characterizing Ab responses, it is described here in some detail.

The Hemolytic Plaque Assay. In a typical test spleen cells from a mouse immunized with sheep erythrocytes

TABLE 19-1. Sequential Changes in Affinity of Anti-Dnp Antibodies (IgG Immunoglobulins) Made with Increasing Time After Immunization*

Group	Ag injected per rabbit (mg)	Average intrinsic association constants for binding ϵ-Dnp-L-lysine at		
		2 weeks	5 weeks	8 weeks
I	5	0.86	14	120
II	250	0.18	0.13	0.15

*Immunogen was 2,4-dinitrophenyl bovine γ-globulin (Dnp-BγG). Values are averages for five animals per group, and are given in liters/mole $\times 10^6$ (30°C).

(Eisen HN, Siskind GW: Biochemistry 3:966, 1964)

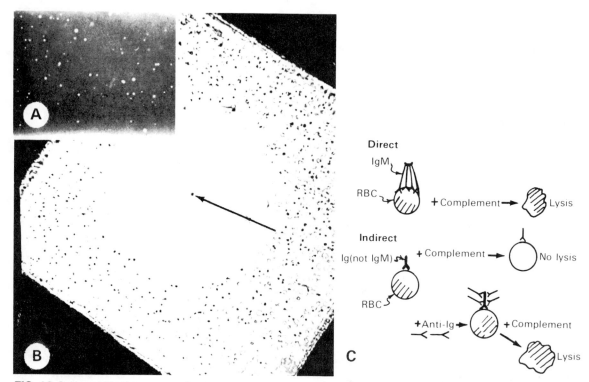

FIG. 19-2. Hemolytic plaque assay for Ab-producing cells. **A.** Multiple plaques in a Petri dish. (Courtesy of Dr. L. Claflin) (about ×15) **B.** Single plaque with its central Ab-secreting cell **(arrow).** Representative plaque-forming cells (PFCs) are shown in Fig. 19-7. (Harris TN et al: J Exp Med 123:161, 1966) (about ×100). **C.** Contrast between **direct plaques,** which reveal cells that secrete Abs of the IgM class, and **indirect plaques,** due to cells that secrete Abs of other classes and require appropriate anti-Igs to obtain complement-dependent lysis of RBCs.

(SRBCs) are plated in agar with the SRBCs. During the following incubation some of the cells synthesize and secrete Abs that lyse surrounding SRBCs when complement is added. This leaves clear plaques with the Ab-secreting cell in the center (Fig. 19-2), resembling the plaques produced by lytic phage on a lawn of susceptible bacteria. By using RBCs with various covalently attached haptens and Ags this method can be extended to detect cells forming a wide variety of Abs.

The unmodified ("direct") plaque technic counts cells that produce IgM Abs but not those that produce IgG Abs. The reason is that lysis of an Ab-coated RBC by complement (Ch. 20) requires only a single IgM molecule on the erythrocyte's surface, whereas a pair of adjacent IgG molecules is necessary: with relatively few Ab molecules secreted by a single cell, the close packing of bound Ab molecules is likely to be infrequent, especially if spacing of the RBCs antigenic determinants is unfavorable. However, IgG-secreting cells can be counted by the "indirect" technic: addition of an antiserum to IgG results in clusters of anti-IgG Abs on each molecule of bound anti-RBC Ab, and the RBC can then be lysed by complement (Fig. 19-2). If the added antiserum is specific for a particular allotype it is possible to recognize cells that produce that allotype. Antiserum to IgA, which does not fix complement, simi-

larly allows "indirect" enumeration of cells that produce Abs of the IgA class.

Moreover, an added soluble ligand with the same structure as the antigenic determinant on the RBCs combines with the secreted Abs, reducing their effective concentration and diminishing the plaques to the point of invisibility. Low levels of the soluble ligand eliminate only plaques due to cells that secrete high-affinity Abs; high ligand levels also eliminate plaques due to low-affinity producers. Hence a titration curve with ligands at various concentrations provides a rough view of the distribution of plaque-forming cells (PFCs) with respect to the affinity of the Ab molecules they secrete.* The results of this approach also support the idea that PFCs early in the response predominantly produce low-affinity Abs, while later most PFCs produce high-affinity molecules.

IgM–IgG SWITCH

In the transition from IgM to IgG Abs, in response to T-dependent Ags, analysis of single B cells and B cell clones

---

* The underlying assumption is that the rate of Ab secretion is about the same in PFCs producing high- or low-affinity Abs. However, if cells vary widely in the amount of Ab secreted, this variation, as well as variation in the affinity of the Ab, would influence the result.

suggests that *individual cells first produce IgM and then IgG or IgA molecules.*

Thus, using antisera to μ, γ, or α chains to type the Ab secreted by PFCs, it was shown that 3 days after primary immunization the PFCs were all IgM producers: after 4–5 days many PFCs produced IgM and IgG or IgM and IgA; and after 7 days all PFCs secreted Ig of one class (IgM or IgG or IgA).

The **splenic focus technic** shows definitively that one B cell clone can make Abs of various Ig classes. In this technic, as is described later (see B Cell Repertoire, below), the Abs secreted by individual clones can be analyzed. Though all the Ab molecules made by a given clone appear to have the same combining site ($V_H + V_L$), some are IgM and others are IgG or IgA. Nevertheless, all of the diverse H chains made by any particular clone seem to be encoded in the same chromosome. Thus, when the clones are derived from a hybrid mouse, made by crossing parental strains that differ in H chain allotypes (Ch. 17), the allotypes of all of the various H chains made by a given clone appear to be derived from the same parental chromosome.

The shift from production of an IgM to an Ig of a different class probably results from a recombination event like the V-J joining event described earlier (see Rearrangement of Ig Genes, Ch. 17). To visualize the "switching-rearrangement", imagine that the encoding sequences for all the C_H classes (μ, $\gamma 1$, ... etc.) are aligned on the same chromosome, separated from each other by intervening sequences, with $C\mu$ being closest to the J region (but still separated from J by an intervening sequence several thousand bases long [e.g., see Fig. 17-29]). If a switching site in the J-$C\mu$ intervening sequence becomes joined to a switching site between $C\mu$ and, say, $C\gamma 1$, then, by looping out and excising the $C\mu$ gene, the V-J sequence previously next to $C\mu$ will be next to $C\gamma 1$.

This view is supported by analyses of restriction enzyme fragments of DNA from different myeloma tumors, producing Igs of different classes: cells making IgM have a full complement of C_H genes (μ, α, $\gamma 1$, etc.); however, cells making an IgG1 ($\gamma 1$ chain) were missing $C\mu$ genes, and cells making an IgG2b were missing $C\mu$ and $C_{\gamma 1}$ genes, while cells making an IgG2a were missing $C\mu$, $C_{\gamma 1}$, and $C_{\gamma 2b}$ genes. These findings suggest that switching is unidirectional, and that the order of C_H genes is: μ-$\gamma 1$-$\gamma 2b$-$\gamma 2a$-α. It must be emphasized that *despite the C_H switches, all cells of a given clone are thought to express the same V_H-J_H sequence and the same L chain ($V_L \cdot J_L \cdot C_L$); hence their Igs have the same idiotype and the same Ag-binding specificity and affinity.*

Abs with high intrinsic affinity for their ligands are IgG molecules; IgM Abs nearly always have low intrinsic affinity (but relatively high avidity, due to increased possibilities for multivalent binding; see Avidity and Appendix Ch. 16). This relation is puzzling, since the V_H (and V_L) domains determine affinity, and considerable evidence suggests that any particular V_H can be linked to any C_H class (Ig genes Ch. 17). Thus if cells forming IgM differentiate to form IgG, the affinity of their products must be the same. Therefore, *the ultimate formation of high-affinity IgG must involve the selection of new clones, as well as a switch in H chain class.*

THE ROLE OF T HELPER CELLS

That the switch from production of IgM Abs of low affinity to IgG Abs of high affinity depends upon T helper (T_h) cells can be shown with specially prepared **T-deficient mice,** from whom essentially all lymphocytes are initially eliminated by thymectomy plus intense wholebody x- or γ-irradiation. To prevent death, syngeneic bone marrow stem cells are injected (freed of T cells by treatment with antiserum to Thy-1 [Ch. 18] together with complement [Ch. 20]): the injected cells lodge in the recipient's bone marrow and generate new B cells (as well as RBCs, etc.). The antihapten Abs elicited in these T-deficient **B-mice** by a hapten–protein conjugate consist only of IgM molecules with low affinity (e.g., anti-2, 4-dinitrophenyl [anti-Dnp] Abs elicited by dinitrophenylated bovine γ-globulin [Dnp-BγG]). If, however, syngeneic T cells are also injected, before the Ag, IgG Abs of high affinity are eventually produced.

How do T Helper Cells Enhance B Cell Responses? Two steps can be distinguished.

1) Precursor T_h cell + Ag–macrophage → activated T_h cell. In this process receptors on T_h cells react with both Ag and products of the major histocompatibility complex (MHC), probably Ia molecules, on macrophages (see H-2 Restrictions Ch. 18). Some evidence suggests that soluble factors released by the macrophages stimulate the proliferation and activation of the T_h cells.

2) Activated T_h cell + Ag–B cell → Ab-secreting B cells (plasma cells) + memory cells. In this second step activated T_h cells stimulate the proliferation of Ag-bearing B cells, some of whose daughter cells become Ab-secreting plasma cells and others become memory cells. Whether the T_h cells recognize Ag alone on the B cell or Ag in conjunction with MHC products (? Ia molecules) on the B cells is not clear. The activated T_h cell appears to stimulate B cells by releasing a **helper factor,** which could be a soluble form of the T_h cell's Ag-recognition molecule.

The mitogenic stimulus provided by activated T_h cells is probably necessary for vigorous B cell proliferation when the triggering Ag is T-dependent. Because the B cells whose surface Ig molecules have high affinity for Ag are probably much rarer to begin with than B cells with low-affinity surface Igs, the former cells probably require extensive proliferation merely to produce enough Ab to elevate the average affinity of serum Abs. Hence *the increasing average affinity of the Abs made with increasing time requires activated T_h cells and the resulting marked B cell proliferation.* Whether the activated T_h cell also causes the Ig-secreting B cell to switch its Ig H chain from μ to γ or to α is not known.

T_h–B Interactions. One basis for T_h–B cooperation is illuminated by the **carrier effect,** observed with hapten–protein conjugates. For instance, when an animal is primed with Dnp-BγG a second injection of the same Ag several weeks later elicits a vigorous secondary response (high-affinity IgG anti-Dnp molecules). However, if the second injection is made with Dnp attached to another carrier protein, say ovalbumin (OA), the secondary response hardly occurs unless the animal has also been previously primed with OA itself.

Moreover, when an animal is irradiated and then given spleen cells from a mouse primed with hapten–protein X it does not make an optimal secondary antihapten Ab response to hapten–protein Y unless it is also given T_h cells from a second donor primed with protein Y (Fig. 19-3). Thus a vigorous secondary response requires T_h cells and also recognition of the carrier (protein Y), not just the haptenic group.

These results suggest that cooperation between T_h and B cells depends upon **associative recognition**: the T_h cells are specific for determinants on the carrier molecule and the B cells are specific for haptenic groups (Fig. 19-4). This view explains why *haptens are not immunogenic unless attached to a carrier*. Proteins are especially effective as carriers (and as immunogens) because their great variety of potential carrier determinants per molecule can engage many T_h cells.

The distinction between hapten and carrier determinants also applies to natural immunogens (proteins, RBCs), some of whose determinants (haptenlike) bind the elicited Abs (formed by B cells), while others (carrierlike) react with helper T cells in the induction of the B cells. For instance, a brisk secondary response in formation of noncrossreacting Abs to a and to b subunits of tetrameric lactic acid dehydrogenase (porcine) could be elicited in rabbits primed with the a_2b_2 form of the enzyme and then boosted with a_2b_2, but no Abs to either subunit were formed in animals that were boosted with b_4, as though b were a haptenlike subunit requiring the carrier effect of the a subunit.

Though the functional distinction between carrier and haptenic group is clear (carrier groups are recognized by T_h cells and haptenic groups by B cells) there is no systematic structural difference between them: T_h cells can also recognize small structures, such as Dnp. This result is in accord with the finding that T and B cells specific for the same Ag can have the same idiotype on their respective surface Ag-binding receptors. However, hapten–protein conjugates have many more nonhaptenic than haptenic determinants, and so it is probable that the various T_h cells specific for the former can collectively have a greater helper effect.

The presence of the same idiotype on T_h and B cells further suggests another model for T_h-B cell interactions, in which antiidiotype Abs, acting without any Ag, can specifically crosslink T_h and B cells and thereby stimu-

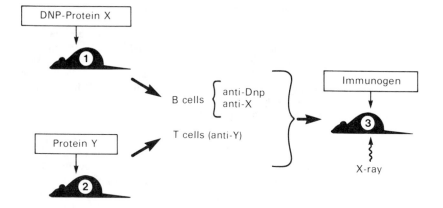

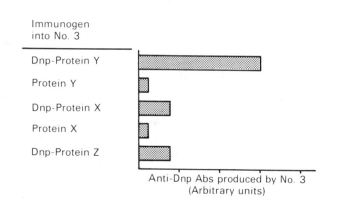

FIG. 19-3. Carrier effect, revealed by adoptive transfer. Mouse no. 1, primed against protein X, provided B cells. Mouse no. 2, primed against protein Y, provided T cells. Mouse no. 3 was given, in order, a) sufficient X- or γ-irradiation (e.g., 600 r) to inactivate its own lymphatic system, b) a mixture of B cells from mouse no. 1 and T cells from mouse no. 2, and c) various immunogens as noted. One week later the anti-Dnp Ab response of mouse no. 3 was measured by analyzing either its serum Ab concentration or the number of spleen cells secreting anti-Dnp Abs in a hemolytic plaque assay. The most intense response is elicited with Dnp–protein Y.

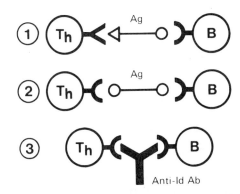

FIG. 19-4. Models for T$_h$–B cell interaction. In two models T$_h$ and B cells are crosslinked by Ag **(associative recognition):** the cells are specific for different determinants in model no. **1** (the usual situation), and for the same determinants in no. **2**. In model no. **3** the cells are assumed to have the same (or similar) idiotypes on their respective Ag-recognition molecules, and they are crosslinked by an antiidiotypic (anti-Id) Ab.

late the B cells to produce Ab molecules (Fig. 19-4). This possibility is of particular interest in connection with considerable evidence for an idiotype–antiidiotype network in the control of Ab formation (see Regulation of B Cell Responses, p. 431).

Multiple T Helpers. Though the highest Ab yields are obtained when the cooperating T and B cells are specific for different determinants on the **same immunogenic particle (associative recognition;** see T$_h$–B Interactions, above), helper effects are also evident when the cooperating T$_h$ and B cells react at the same time and in the same locale with **separate immunogens;** e.g., Dnp-primed B cells and ovalbumin (OA) primed T cells cooperatively respond to a mixture of Dnp-BγG and OA. The effect is probably due to the release of locally acting soluble factors.

Some of the soluble factors probably act locally on diverse B cells, specific for unrelated Ags, causing the latter to produce their Igs. This nonspecific effect on **"bystander" B cells** could explain why an antigenic stimulus, in addition to eliciting production of the corresponding Abs, usually also evokes considerable production of other Igs (e.g., often only half, or less, of the incremental increase in serum Ig that follows intense immunization is due to Abs that react specifically with the immunogen).

It also appears that certain T$_h$ cells are required to help the B cells that produce particular Ig classes, such as IgE, or particular Ig allotypes or idiotypes. These T$_h$ cells probably recognize determinants on the surface Ig molecules of B cells. Whether these helpers are also Ag (carrier)-specific is not known; and whether all B cells require two kinds of T helpers, one to recognize Ag and the other to recognize Ig on the B cell surface, is also not clear.

SUPPRESSOR T CELLS

Ags can also stimulate T suppressor (T$_s$) cells, which hinder Ab production by blocking T$_h$ cells or, perhaps, by acting directly on B cells. Whereas Ag bound to macrophages stimulates T$_h$ cells, it has been suggested (see Macrophages, Ch. 18) that Ag that is not bound to macrophages stimulates T$_s$, not T$_h$, cells. This distinction could explain the frequent observation that high doses of Ag, especially when the Ag is monomeric and injected intravenously (conditions that minimize Ag binding by macrophages), elicit more T$_s$ than T$_h$ activity and, correspondingly, little or no Ab production. T$_s$ cells are discussed further below, under Tolerance and under Regulation of B Cell Responses. (T$_s$ cells, including T$_s$-1 and T$_s$-2 cells, Ch. 18).

GENETIC CONTROL

The responses of B cells to T-dependent Ags are under genetic control. It has long been known that selective breeding can generate strains of animals able to make high titers, or only low titers, of Abs to complex Ags (RBCs, etc.). However, insight into genetic control accumulated rapidly only after use was made of inbred strains, together with Ags of limited immunogenicity (synthetic polypeptides, alloantigens, and unusually small doses of conventional Ags).

Ir Genes. Benacerraf et al. found that one inbred strain of guinea pigs (R, for responders) made high Ab responses against Dnp or other haptenic groups carried on the synthetic polypeptide poly-L-lysine (PLL), whereas another strain (NR, for nonresponders) made little or no Abs to the same Ag. F1 hybrids (R × NR) were also high responders, as were 50% of the progeny of a back-cross between the F1 hybrids and the NR strain (R/NR × NR). Evidently the antihapten response to hapten–PLL was controlled by a single dominant autosomal (non-sex-linked) locus, the **PLL gene.**

McDevitt and Sela obtained similar results in inbred mice, using various synthetic polypeptides as immunogens (Table 19-2). For instance, mice of the C57BL strain make about 10 times higher Ab titers than mice of the CBA strain to the tyrosine–glutamate (T,G) determinant of the (T,G)-A—L Ag, whereas the relative responses are reversed with (H,G)-A—L, a similar Ag in which histidine replaces tyrosine (Fig. 19-5).

Of particular interest was McDevitt's finding that *the controlling locus, called Ir (immune response), is linked to the major histocompatibility complex* (H-2 in the mouse; see Ch. 18). By use of congenic mouse strains with genetic recombination between the K and D ends of the H-2 complex the Ir locus was mapped between the K and S loci (see Figs. 18-16 and 18-18, Ch. 18); and tests with diverse Ags eventually distinguished three immune response–controlling regions within this locus: IA, IB, and IC.

Rather than specifying V segments of Abs, Ir genes seem to determine some structures involved in the recognition of carrier determinants. Thus animals of a strain that is a

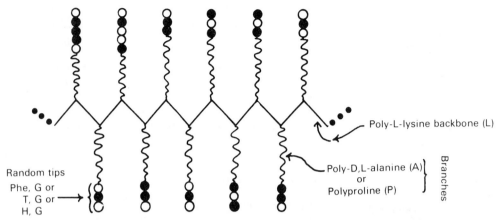

Random tips
Phe, G or
T, G or →
H, G

Poly-L-lysine backbone (L)

Poly-D,L-alanine (A) }
or   } Branches
Polyproline (P) }

FIG. 19-5. Synthetic branched polypeptides used to study genetic control of Ab formation. The most useful ones have a poly-L-lysine (L) backbone and poly-D,L-alanine (A) (or polyproline (P)) branches, with tips carrying short random amino acid sequences of either tyrosine and glutamate (T,G), histidine and glutamate (H,G), or phenylalanine and glutamate (Phe,G). The corresponding polymers are designated (T,G)-A—L, (H,G)-A—L, and (Phe,G)-A—L. The antigenic determinants are usually the short terminal sequences (Phe,G or T,G, or H,G), but in a few mouse strains the branched backbone (poly-pro-polylysine) is immunogenic. (Based on Sela M: Harvey Lect 67:213, 1973)

nonresponder to Ag X can make anti-X Abs when immunized with X linked to potent carrier molecules. For example, strain 13 guinea pigs are nonresponders to Dnp-PLL, but they respond to Dnp-PLL when it is administered in a complex with bovine serum albumin; moreover, the anti-Dnp Abs made under these circumstances cannot be distinguished from those elicited by Dnp-PLL in responder animals.

Immune Suppressor (Is) Genes. Though nonresponders fail to make Abs to Ag X, they can produce X-specific T cells. For instance, mice of certain strains are nonre-

TABLE 19-2. Ir Gene-Controlled Responses to Synthetic Polypeptides in Some Inbred Mouse Strains*

H-2 haplotype	Polypeptides			
	(T,G)-A—L	(H,G)-A—L	(Phe,G)-A—L	GLφ
a	l	h	h	l
b	h	l	h	n
d	m	m	h	h
k	l	h	h	h
p	l	l	h	l
q	l	l	h	h
s	l	l	l	l

l = low response; h = high response; m = moderate response; n = no response

*For (T,G)-A—L, (H,G)-A—L, and Phe, G)-A—L see Fig. 19-5; GLφ is a random linear polymer with about 55% L-glutamate, 35% L-lysine, and 10% L-phenylalanine.

(Based on Benacerraf B, Dorf ME: Cold Spring Harbor Symp Quant Biol 41:465, 1976, and Klein J: Biology of the Mouse Histocompatibility-2 Complex. Springer, N.Y., 1975)

sponders to the glutamate–tyrosine random copolymer, GT, but make anti-GT Abs in response to GT complexed with methylated bovine serum albumin (GT/MBSA); and immunization with GT elicits T cells (T_s) that specifically block a response to subsequent immunization with GT/MBSA. Moreover, the T_s cells are elicited by GT only in certain nonresponder (to GT) strains; and genetic crosses show that production of these cells is also a dominant trait controlled by a locus (Is, for immune suppressor) in the I region of the H-2 complex.

Indeed, careful mapping has revealed two Is loci, one on either side of the **IJ region;** and this region itself encodes a surface Ag characteristic of T_s cells and of the soluble suppressor factor produced by them (Ch. 18). The two Is loci are complementary. Thus, in a cross between two different Is⁻ strains, each unable to produce anti-GT T_s cells, the hybrid progeny are able to form these cells. This finding suggests that the specific suppressor molecule has two chains, one encoded by each Is locus.

To search for the I-associated (Ia) structures that are specified by Ir (or by Is) genes, congenic mice that differ only in the I region were crossimmunized with each other's lymphocytes. As noted in Chapter 18, the resulting alloantisera distinguish many allelic variants of the Ia Ags on cell surfaces. However, unlike the H-2K and H-2D glycoproteins, which are present on virtually all nucleated cells in the body, the Ia glycoproteins are present only on B cells, macrophages, and T_s cells (and their soluble suppressor factor). The Ia Ags on B cells and macrophages are determined by genes in the IA and IC loci, while those on T_s cells (and on the soluble suppressor factor) are determined by the IJ locus.

Mechanisms of Ir Gene Control. Broadly speaking, Ir genes appear to determine whether antigenic stimulation results mainly in T helper cell activity (in responders) or mainly in T suppressor activity (in nonresponders). Though the phenomena are not well understood, many observations support two mechanisms.

1. T_h Cell–Macrophage Interactions. The key to this mechanism is **macrophage presentation of Ag.** The basic observation was described in Chapter 18 (Fig. 18-24): T cells from Ag-primed hybrid mice (R × NR), made by crossing a strain (R) that can respond to Ag X with a strain (NR) that cannot, are stimulated by X on macrophages when the macrophages are derived from the R parent but not when they are derived from the NR parent. The critical R–NR difference is determined by the I region of the major histocompatibility complex, and probably by Ia molecules. Thus, T_h cells are probably specifically activated by macrophages that bear X plus the proper allelic (R) variant of Ia.

2. Soluble T Suppressor and T Helper Factors. Judging from analyses on certain strains, this mechanism suggests that Ag-specific helper factor (from T_h cells) is defective in nonresponders and Ag-specific suppressor factor (from T_s cells; Ch. 18) is defective in some responders. Information on the helper factor is fragmentary, but it appears, grossly, to resemble suppressor factor.

Both factors bind Ag specifically; they have idiotypes that are serologically indistinguishable from those of Abs that are specific for the same Ag; the idiotypes are encoded by V_H genes or genes that are linked to V_H genes. Whether V_L genes contribute is not known. Both factors also have structural elements that are encoded by genes in the major histocompatibility complex: helper factor by a gene in the IA locus, suppressor factor by a gene in the IJ locus.

ANTIBODY RESPONSES TO T-INDEPENDENT ANTIGENS

In contrast to the responses elicited by T-dependent Ags, those evoked by T-independent Ags (TIAs) are the same in mice that are normal, or T-deficient (athymic nude mice or neonatally thymectomized mice), or supplemented with T cells. These Ags elicit, almost entirely, IgM molecules with low intrinsic affinity. In addition, memory effects are lacking: the response to a second injection of these Ags is no different than to the first.

Most TIAs are distinguished by two other properties. 1) Many are high-molecular-weight polysaccharides (lipolysaccharide [LPS] of gram-negative bacteria, dextran sulfate, Ficoll [polymer of sucrose], etc.), with **multiple copies of the antigenic determinant** per molecule; hence they are capable of multivalent binding to B cells. 2) At high concentrations these Ags (especially LPS) are **poly-**clonal activators of B cells; i.e., they stimulate diverse B cell clones to proliferate and to secrete their Ig molecules, which are not specific for the TIAs.

A mitogenic effect (see Mitogens and Blast Transformation, Ch. 18) is probably an essential feature of the T-independence of these Ags. Consider, for instance, LPS. At high concentrations it is a polyclonal B cell activator; at low concentrations it is bound and its mitogenic effect is concentrated on just those B cells that can make anti-LPS Abs. These properties suggest that the LPS molecule has separate antigenic and mitogenic sites, which bind to separate receptors on B cells.

B CELL RESPONSES TO ANTIGENIC SIGNALS

ONE VERSUS TWO SIGNALS

B cell proliferation and development into Ab-secreting cells (see B Cell Development, below) probably requires two signals: one due to Ag binding to the cell's surface Ig, and the second due either to T_h cells, when the Ag is T-dependent, or to the Ag itself, via a mitogenic site, when it is T-independent.

The ability of LPS and some other T-independent Ags, at high concentrations, to stimulate many different B cell clones that are specific for diverse noncrossreacting Ags suggests, however, that the mitogenic signal itself, without the Ag signal, can also be sufficient: Ag binding to Ig on the B cell membrane serves, in this view, primarily to focus mitogens, including T_h factors, on the B cell.

Experiments using the **"hapten-sandwich" technic** also suggest that one signal can be sufficient, and that occupancy of the cell's surface Ig receptor by Ag is not essential. This general technic for labeling any cell surface Ag (Fig. 19-6) uses two Abs: Ab no. 1,

FIG. 19-6. "Hapten-sandwich" technic, used to evaluate the effects of T_h cells on B cells without binding Ag to the B cell's receptor. X, an integral cell membrane protein, reacts with Ab no. 1, labeled with a hapten (e.g., **R** = benzenearsenate), and then with anti-R Abs (Ab no. 2) linked covalently with another ligand (*****). The ligand ***** is usually radioactive or fluorescent (to count or to visualize X-bearing cells), but it can be another protein, Y, and can then serve to evaluate the effect of concentrating anti-Y T cells on the X-bearing B cell (see text). (Based on Wofsy L et al: Contemp Top Mol Immunol 7:215, 1978)

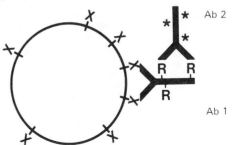

directed to a cell surface macromolecule, is covalently linked to a hapten; Ab no. 2 is antihapten and is covalently linked to another ligand (usually fluorescent, radioactive, etc.). With these reagents it can be shown that the *accumulation of T_h cells on the B cell surface, without involving an Ag that is recognized by the B cells, can deliver a signal that is sufficient to trigger the B cell's development into Ab-producing cells (Fig. 19-6).*

Thus B cells primed to make anti-X were treated successively with 1) hapten-labeled Abs to, say, Ia molecules (on the B cell), then with 2) antihapten Abs linked covalently to another protein, Y, and finally with 3) T cells from Y-primed mice. The resulting interaction of Y-specific T cells with anti-X B cells caused the production of anti-X Abs, even though Ag X and the B cell's surface Ig were not at all involved in the reaction.

Once the proper signal is delivered, B cells divide rapidly (doubling time about 20 h) and form two kinds of daughter cells: Ab-secreting plasma cells and secondary (memory) B cells. *The primary and secondary B cells look alike but differ in some subtle ways:* e.g., the secondary cells circulate more actively from blood to lymph, and they survive for longer intervals between mitoses (many months in the rat).

PLASMA CELLS

The morphology of Ab-secreting cells is evident on examining the cell at the center of the hemolytic plaque (Fig. 19-2). Some of these cells, called **immature plasma cells** or **large lymphocytes,** are probably transitional forms between the small B cell and the fully developed plasma cell.

All of the Ig molecules secreted by any particular plasma cell have the same L chain (V_L + C_L) and the same V_H

FIG. 19-7. Electron micrographs of representative Ab-secreting cells (hemolytic plaque-forming cells, PFCs, as in Fig. 19-2). Cells in **A, B,** and **C** are from lymph nodes and spleen; the one in D is characteristic of PFCs in efferent lymph emerging from an antigenically stimulated lymph node. Most PFCs from within lymph nodes and spleen are plasma cells (C), except during the first few days after immunization when lymphocytes (A) and transitional cells (B) constitute the majority. (A and B, Gudat FG et al: J Exp Med 134:1155, 1971; C, Gudat FG et al J Exp Med 132:448, 1970; D, Hummeler K et al: J Exp Med 135:491, 1972)

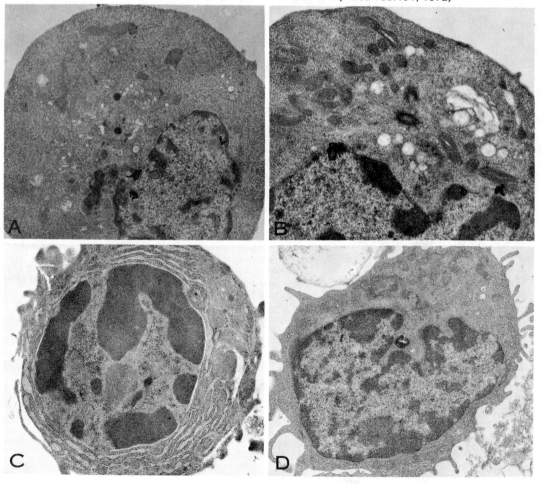

domain: hence they all have the same idiotype and the same specificity for Ag. This finding forms the basis for the **1 cell- 1 Ab rule.** However, a single cell or clone can express various C_H domains and thus produce Abs of more than one class. As noted above (IgM-IgG switch), and previously (Gene Clusters, Ch. 17), cells usually make IgM and then perhaps IgG or IgA; and some cells simultaneously produce IgM and IgD (see Igs on B cells, below).

In its fully differentiated form the plasma cell is anatomically distinctive: the nucleus is eccentric and contains coarse, radially arranged chromatin (in a cartwheel formation), the cytoplasm has a conspicuous Golgi apparatus, and abundant rough and smooth endoplasmic reticulum is usually packed into thin onionskinlike lamellae or, sometimes, distended with Igs (as shown by immunofluorescence; see Fig. 18-8, in Ch. 18, and Fig. 19-7).

The mature plasma cell is believed to survive only a few weeks and to die after a few (or no) cell divisions; in contrast, **neoplastic plasma cells (myeloma cells),** like other neoplastic differentiated cells, are capable of an unlimited number of cell divisions.

Immunoglobulins on B Cells and Plasma Cells. Nearly all of the Ig molecules synthesized by small B cells become integral membrane proteins at the cell surface. In primary B cells the surface Ig is monomeric IgM and/or IgD; some secondary B cells have IgG or IgA in place of IgM. The surface Ig of B cells is released and replaced very slowly (turnover time about 20 h); this release might account for the low levels of IgD and monomeric ($\mu_2 L_2$) IgM in normal serum.

In plasma cells, in contrast, Ig is synthesized rapidly (about 2000 molecules/cell/min), and nearly all of it is rapidly secreted, either as pentameric (19S) IgM or as IgG or IgA; only small amounts, detected with difficulty, are present in the surface membranes of these cells. The transitional cells (immature plasma cells) have intermediate properties: they secrete Ig and also insert much of it into the cell membrane. (See Ag-binding receptors of B Lymphocytes, Ch. 18, for differences at the COOH-end of μ chains of membrane-bound and secreted IgM molecules.)

SYNTHESIS, ASSEMBLY, AND SECRETION OF IMMUNOGLOBULINS

In plasma cells (provided with radioactive amino acids) Igs, like secreted proteins in general, are synthesized on polyribosomes attached to the rough endoplasmic reticulum. Synthesis by membrane-free polysomes from myeloma cells has shown that the chains are formed as larger precursors, with an additional N-terminal sequence of about 20 amino acids (the **"leader"**) that is cleaved in the course of transfer across the membrane into the lumen of the endoplasmic reticulum (see Pre-Ig Leader, Ch. 17 and Fig. 17-28). The newly synthesized Ig molecules are secreted from the cells after a 20–30 min lag (Fig. 19-8), during which they pass through cisternae of the endoplasmic reticulum, traverse the Golgi apparatus, and then move toward the cell surface in secretory vesicles, which fuse with the surface membrane and release their Ig molecules. During this migration the completed chains are assembled into molecules, the interchain S-S bonds are formed, and sugars are added successively by hexosetransferases to form the oligosaccharide groups of complete Ig molecules (Fig. 19-9).

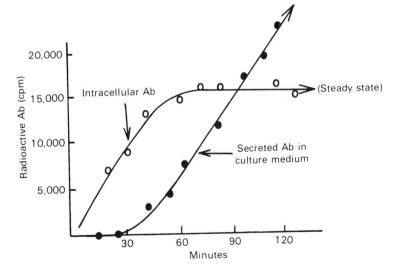

FIG. 19-8. Lag in secretion of Abs. Labeled amino acid was added at zero time to a suspension of lymph node cells from an immunized rabbit, and cells and supernate were then assayed at intervals for newly synthesized Abs. (Based on Helmreich E et al: J Biol Chem 236:464, 1961)

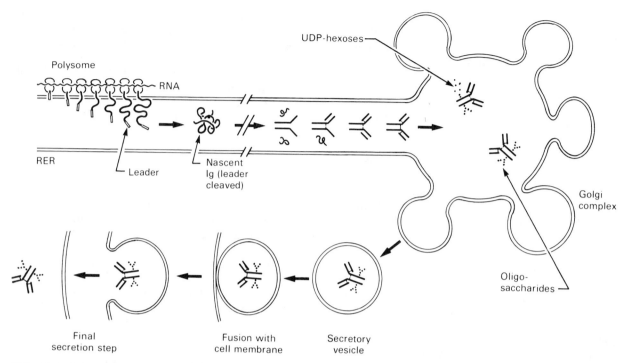

FIG. 19-9. Scheme for cleavage of leader, chain assembly, intracellular transport, and secretion of Igs. **RER** = rough endoplasmic reticulum, i.e., with bound polyribosomes. In B cells that produce IgM there are two types of mRNA for μ chains. At the COOH-terminus of μ chains encoded by one type there is a highly hydrophobic amino acid sequence, which is probably responsible for holding the corresponding IgM molecules in the cell's surface membrane (m IgM). The other μ-mRNA encodes μ chains that, lacking the hydrophobic "tail," are in secreted IgM molecules (s IgM). Secreted and membrane forms of other Igs are probably also encoded by separate mRNA molecules. (See Fig. 17-27 for the hydrophobic "leader" at the amino-terminus of an Ig chain.)

The order of chain assembly has been studied primarily in myeloma tumors, in which various patterns have been deduced from the variety of incomplete molecules found intracellularly and from pulse–chase experiments. Thus when cells are harvested at various times after a 30-sec pulse of radioactive amino acid and labeled nascent light (L) or heavy (H) chain can be followed as it joins with other chains to establish complete four-chain Ig molecules ($H_2 \cdot L_2$) by various assembly patterns: complete molecules are made by joining two H·L half-molecules; or H-chain dimers (H_2) add one L chain at a time. Regardless of the pattern, assembly and secretion are orderly: interchain S-S bonds form slowly (2–20 min after the constituent chains are completed); the rate of secretion equals the rate of synthesis and, accordingly, the intracellular level of Ig molecules is constant; the more recently synthesized molecules leave the cell only after previously synthesized ones have been secreted (Fig. 19-8).

In myeloma cells that make IgA and IgM molecules the intracellular Ig seems not to progress beyond the four-chain monomer, though polymers are found in the culture medium. The monomers probably associate with each other and with J chains (which are made in the same cells; see Ch. 17) as Ig molecules are secreted.

The order of incorporation of radioactive sugar precursors (hexoses and hexosamines) is in accord with the sequence of sugar residues in Ig oligosaccharides (see Appendix, Ch. 17). Glucosamine, which links oligosaccharides to Ig chains, is incorporated first, followed by mannose, galactose, and sialic acid. Fucose appears to become attached last, as Ig molecules exit from the cell.

Tunicamycin, an antibiotic that prevents glycosylation of newly synthesized proteins, blocks the secretion of the most highly glycosylated Igs (IgM, IgA, IgE; see Appendix, Ch. 17), which suggests that with these Abs the oligosaccharides impose a conformation that is necessary for secretion. However, the secretion of IgG, which is less glycosylated than the other Igs, is not blocked; and most Bence Jones proteins (L chains) are not glycosylated at all and yet are secreted with the same time course as whole Ig molecules (Fig. 19-8).

B CELL REPERTOIRE

How many different B cell clones does an individual have? Because the Ig molecules made by each clone have a distinctive combining site an answer is suggested by the individual's Ab repertoire. The trace levels of some "natural" Abs to common Ags have long suggested that the total pool of serum Ig consists of more than 10^6 different Igs.

A more direct approach is provided by the **splenic focus assay.** In this procedure, developed by Klinman, a highly dilute suspension of spleen cells, enriched for B lymphocytes by the elimination of T cells with anti-Thy-1 plus complement, is injected intravenously into a lethally irradiated recipient. About 1 in 20 of the injected cells lodges in the recipient's spleen, which is removed the next day and cut up into 50 fragments, each of which is then cultured in a separate well along with the Ag. (Because the recipients were irradiated their own lymphocytes cannot respond to the Ag, but these cells can provide considerable T helper activity if the recipient had been primed with the Ag before irradiation).

A single Ag-stimulated B cell can, with T cell help, yield in 10 days almost 1000 daughter cells, each producing about 2000 Ig molecules per minute (see Plasma Cells, above). Hence a fragment that has one responsive B cell to start with can secrete enough Ab in 10 days to allow the determination, by radioimmunoassay, of the Ab's characteristics (Ag binding, H and L chain type, etc.). Allotype differences between donor and recipient show that all Ab-forming clones are of donor origin; and all the Ab molecules produced by an active fragment appear to have the same Ag-binding sites (idiotypes). Hence the Ab-secreting cells in each fragment probably represent one clone.

The number of fragments (clones) that produce Abs to Ab X, divided by the number of B cells that lodge in the spleen (estimated with ^{51}Cr-labeled cells), provides a measure of the frequency of anti-X B cells. The number of different B cells seems to be on the order of 10^7 in the adult mouse.

For instance in an adult mouse about 1 B cell in 10,000 can make anti-Dnp Abs. Since independent estimates suggest that there are about 1000 different anti-Dnp Abs (distinguishable by isoelectric focusing electrophoresis), and each is presumably the produce of a distinctive clone, the frequency of any particular anti-Dnp B cell is about 1 in 10^7 ($1/10^4 \times 1/10^3 = 1/10^7$). Roughly similar values have been found for B cells to some other Ags.

REGULATION OF B CELL RESPONSES:
IDIOTYPE–ANTIIDIOTYPE NETWORK AND T CELL CIRCUITS

Besides Ab production, Ags elicit other responses, of a self-regulatory nature, that have the effect of both facilitating the Ab response and preventing it from developing to an excessive level. Two broad regulatory responses are recognized: one involves mainly Ig combining sites (**idiotype–antiidiotype network**) and the other involves various regulatory and effector T cells (**T cell circuits**).

Idiotype–Antiidiotype Network. An individual normally does not make an immune response to his own Ags (see Tolerance, below). However, before it increases markedly in response to an Ag, the idiotype of an Ab may be present at such an extremely low level that it is not recognized by the individual's immune system as a self-Ag. Accordingly, with appropriate immunization conditions, an animal produces Abs against idiotypes of some of its own Igs. Because an antiidiotypic Ab is itself an Ig, with its own idiotype, it can probably elicit still other antiidiotypes, also of autologous (self) origin, against its own V domains.

These observations support Jerne's proposal that the immune system is a network of idiotypes and antiidiotypes: the combining region of an Ig has both a distinctive idiotype and an antiidiotypic activity directed toward the idiotype of another Ig, made by another B cell clone within the same individual (Fig. 19-10). It has not yet been established that antiidiotypes (Abs or T cells) normally regulate the production of the corresponding idiotypes, but this possibility is supported by the time course of Ab responses to certain Ags.

Thus a pneumococcal vaccine elicits Abs (Ab no. 1) to phosphorylcholine (PC); as the level of anti-PC Abs decreases with time, other Abs (Ab no. 2) appear and react specifically with the idiotype of the anti-PC molecules. Presumably, Ab no. 2 was elicited by the idiotype of Ab no. 1 and suppresses further production of this Ab. However, other antiidiotypic Abs, elicited in other species or strains, can sometimes stimulate production (or suppression) of the corresponding idiotype. Because various antiidiotypic Abs can stimulate as well as suppress, quantitative models of the proposed network are complicated. Other evidence for the regulation of Ig production by antiidiotypic Abs is considered (see Allotype Suppression, Idiotype Suppression, and Fig. 19-17).

FIG. 19-10. Idiotype–antiidiotype network. The Ag elicits Ab no. 1, which elicits Ab no. 2, directed to the idiotype (Id) of Ab no. 1. Besides stimulating or suppressing the clone that produces Ab no. 1, Ab no. 2 (anti-Id no. 1) elicits Ab no. 3, which is specific for the Id of Ab no. 2 and which can stimutate (or suppress) the clone producing Ab no. 2, as well as elicit Ab no. 4, specific for the Id of Ab no. 3 (etc.).

The network can involve T as well as B cells. Thus T_h cells can recognize Ids of Abs, and the Ag-recognition molecules of T cells also have Ids (indistinguishable in some instances from those of Ab molecules). It is clear that 1) idiotypes can elicit antiidiotypes, even in autologous (same-individual) responses, and 2) antiidiotypic Abs can stimulate or suppress responses. However, nearly all experiments involve extreme or unnatural conditions. Whether the network regularly operates under physiologic conditions, in many natural responses, is not yet established. (Based on Jerne NK: Ann Immunol (Paris) 125:373, 1974)

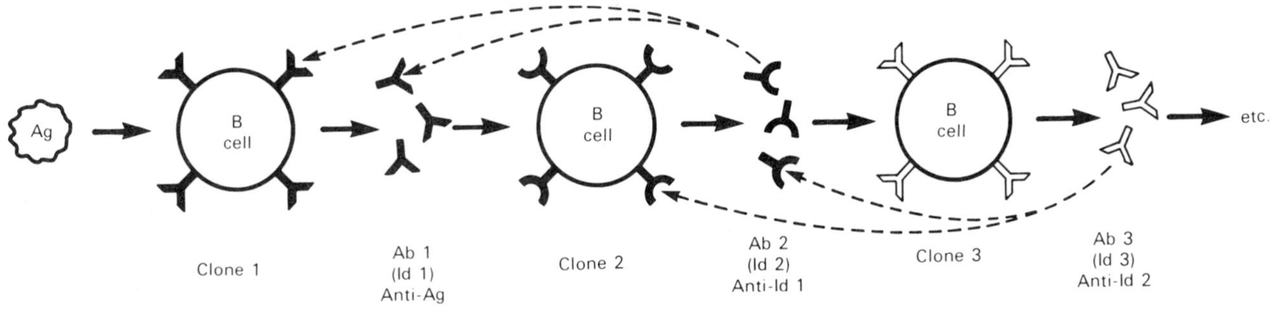

| Ag → | B cell (Clone 1) → | Ab 1 (Id 1) Anti-Ag → | B cell (Clone 2) → | Ab 2 (Id 2) Anti-Id 1 → | B cell (Clone 3) → | Ab 3 (Id 3) Anti-Id 2 → etc. |

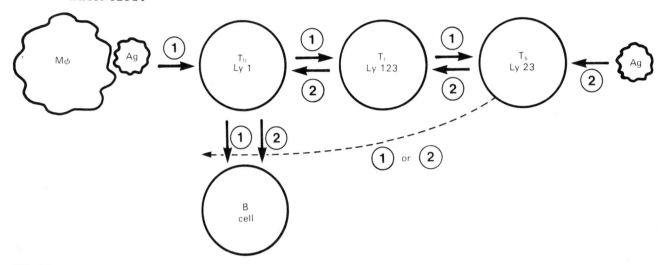

FIG. 19-11. Proposed T cell feedback regulatory circuits. In pathway no. 1 **Ags** on macrophages **(Mφ)** stimulate T helpers **(T$_h$),** which stimulate T regulators **(T$_r$)** to develop into T suppressors **(T$_s$).** The latter block the activity of the T$_h$ cells on effector lymphocytes **(B cells** in this case, but T effectors of cell-mediated immunity in others [Ch. 22]). In pathway no. 2 free **Ag** (not on Mφ) elicits **T$_s$** cells, which stimulate **T$_r$** cells to develop into **T$_h$** cells that promote the development of effector cells (B cells, or T effectors of cell-mediated immunity). (Based on Gershon R, Cantor H: Fed Proc, 1979).

T Cell Regulatory Circuits. The amount of Ab produced in response to an Ag depends not only on the number of responsive B cells but also on the number and types of T cells that are specific for that Ag and on the regulatory interactions among these cells. The T cells are drawn from various Ly sets (see Ly Antigens, Ch. 18). Ly-1 cells (about 30% of all T cells) provide the helpers (T$_h$) that enhance, and Ly-23 cells (about 10% of all T cells) provide the suppressors (T$_s$) that block Ab responses. Ly-123 cells, the most abundant of all T cells (about 50%) seem to be bifunctional regulators (T$_r$).

Preliminary observations suggest that Ag-stimulated T helpers induce Ly-123 T$_r$ cells to develop into T$_s$ cells that block, in a feedback circuit, the T helpers' stimulation of B cells. Conversely, Ag-activated T$_s$ cells stimulate Ly-123 T$_r$ cells to become T$_h$ cells. Thus Ly-123 cells seem to function as "buffers," developing rapidly into either Ly-1 helpers or Ly-23 suppressors (hence their somewhat facetious designation as **hermaphrocytes**) and preventing excessive help or suppression (Fig. 19-11).

The effects of aberrations in these feedback circuits is becoming evident in certain defective inbred mouse strains and in certain human disorders. For instance, a deficiency of T$_s$ cells (possibly from a defect in the T regulatory circuit) in NZB mice probably accounts for the many autoimmune diseases in these animals (Ch. 22). The effect of deficient T$_s$ cells on autoimmunelike reactions elicited by human bone marrow grafts is discussed in Chapter 23 (Bone Marrow Allografts).

All T helpers are Ly-1 cells, but those that induce T regulators to become T suppressors also have the Qa-1 surface Ag, whereas those activated T helpers that act on B cells lack Qa-1. Hence the ratio of Ly-1$^+$Qa-1$^+$ T$_h$ cells to Ly-1$^+$Qa-1$^-$ T$_h$ cells may determine whether an Ag elicits Ab production or a suppressor response that specifically prevents production of that Ab.

How do T cells communicate with each other? Ag-activated lymphocytes crawl on the surface of other cells, but whether proximity suffices, or actual contact is necessary, for T$_h$ and T$_s$ cells to interact with each other or with T$_r$ cells is not clear. The Ag-recognition sites of some T cells and of their soluble factors share heritable idiotypes with Abs that are specific for the same Ags, and these idiotypes are encoded by genes for H-chain V domains, or by genes that are closely linked to these V$_H$ genes. In culture, Ag-stimulated T$_h$ cells induce T$_r$ cells to generate T suppressor activity only when the T$_h$ and T$_r$ cells are from strains of mice with the same V$_H$ genes. It is possible that interactions between receptors on these T cells basically resemble the idiotype–antiidiotype reactions described above.

TOLERANCE

The ability of an individual's immune responses to distinguish between his own Ags and foreign Ags was recognized in 1900, in Ehrlich's doctrine of **horror autotoxicus:** one can form Abs to almost any substance except components of one's own tissues. This discrimination is clearly evident in the crossimmunizations, described in preceding chapters, between mice of different congenic strains: *a mouse makes Abs to alloantigens that it lacks, but not to those it possesses.*

The occasional breakdown of this **self-tolerance** can

result in serious autoimmune diseases (Ch. 22). Conversely, under certain conditions a foreign Ag acts as a **tolerogen,** establishing a state in which the animal fails to form Abs (or to develop cell-mediated immunity) to that particular Ag, even when the Ag is later given in what would ordinarily be an optimal immunogenic form. These conditions have made it possible to analyze the mechanisms responsible for self-tolerance. Unlike genetic unresponsiveness due to Ir genes the unresponsiveness of tolerance requires the Ag's presence for its development: i.e., tolerance is induced by an Ag.

Diversity and Specificity. Any of the substances that induce Ab formation can establish tolerance: proteins, polysaccharides, haptenic groups, etc. As with other immune responses *tolerance is specific and is directed to particular determinants.* For instance, rabbits made unresponsive to the Fc fragment of human IgG can still respond to immunization with intact IgG, but they then form Abs only to Fab domains.

Tolerance and immunity are alternative responses. The circumstances under which Ags tend to induce tolerance are summarized below.

CONDITIONS THAT FAVOR DEVELOPMENT OF TOLERANCE

Newborn versus Adult. In the fetus, and also shortly after birth in species where the newborn is relatively immature, tolerance is more easily established than in the adult. One important reason seems to be that immature B cells are more readily made tolerant than mature B cells (due to **"Modulation,"** see Clonal Deletion, below). The effect of immaturity was first recognized by studies of foreign tissue grafts (see Acceptance of Allografts, Ch. 23; and Tolerance, Ch. 22).

Dose and Form of the Antigen. Every Ag has an optimal immunogenic dose range. Much larger amounts elicit **high-zone tolerance.** With T-dependent Ags, but not with T-independent ones, lower amounts can also cause tolerance (**low-zone tolerance,** Fig. 19-12).

The effects of high doses were first recognized with pneumococcal capsular polysaccharide, a T-independent Ag. Mice injected with 0.01μg–1.0μg of a pneumococcal polysaccharide, say of type 2, become resistant to infection with type 2 pneumococci and produce Abs to the polysaccharide; but if given 1000μg of the same substance they fail to become resistant or to form detectable Abs. The unresponsiveness is specific: though no response to type 2 can be elicited, the mice react normally to immunogenic doses of other Ags, including other pneumococcal polysaccharides. Similar effects can be elicited with bovine serum albumin, a typical T-dependent Ag, at doses that are much above or much below the immunogenic range (Fig. 19-12).

Another factor with many Ags is their **physical state:** with bovine gamma globulin, for instance, aggregated

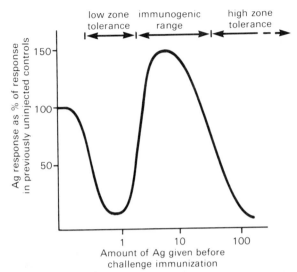

FIG. 19-12. High-zone and low-zone tolerance. Effect of daily injections of different amounts of flagellin from *Salmonella adelaide* on the Ab response of rats to a later challenge with 10 μg of flagellin. Ab levels are given as the percent of values in control rats (no pretreatment). Low-zone tolerance is probably due to stimulation of T_s cells. High-zone tolerance could be due to the deletion of antiflagellin T or B cells. The responses that exceed 100% are due to the priming effects of the initial injections. (Based on Allison AC: Clin Immunobiol 1:113, 1972, and derived from data of Shellam and Nossal)

molecules are immunogenic, while monomers are tolerogenic. It is thus difficult or impossible to establish tolerance to many particulate Ags (viruses, bacteria, etc.), which are usually highly immunogenic.

Route of administration is another determinant: soluble Ags tend to be immunogenic when injected into tissues, but to be tolerogenic when given intravenously.

CELLULAR MECHANISMS

B Cells versus T Cells. Tolerance to **T-independent Ags** derives from unresponsiveness of the corresponding B cells. T cells seem not to be involved; thus T-deficient (e.g., athymic) mice are as easily made tolerant as normal mice by these Ags in excessive doses.

In contrast, tolerance to **T-dependent Ags** can be due to unresponsiveness of either T or B cells. Thus in irradiated mice given B and T cells an Ag fails to stimulate Ab formation if either set of cells is drawn from a donor previously rendered unresponsive to that Ag. However, *T cells are made tolerant much more readily than B cells,* as shown by treating prospective donors with various doses of Ag at varying times before T and B cells are separately transferred (Fig. 19-13): compared to B cells, unresponsiveness of T cells is established sooner (1 versus 10 days), lasts longer (100 versus 50 days), and can be

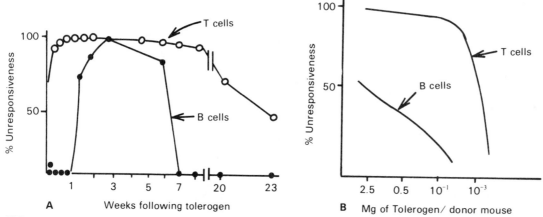

FIG. 19-13. Induction and persistence of tolerance in B and T cell populations. Thymus (T) and bone marrow (B) cells were removed at various times from mice rendered tolerant with various amounts of bovine γ-globulin (BγG, the tolerogen) and were tested, with complementary cells from normal donors, for the ability to cooperate in Ab formation when transferred to irradiated syngeneic mice. Results are given as percent of values in controls (untreated donors). In A tolerance was induced with 2.5 mg of BγG. In B the cells were removed 15–20 days after the doses of Ag shown (abscissa). Tolerance appeared sooner, lasted longer, and was established with lower Ag doses, in T than in B cells. Tolerance was established sooner with B cells from lymph nodes than with those from bone marrow (3 vs. 10 days). (Chiller JM et al: Science 171:813, 1971)

initiated by exceedingly low Ag levels. At these levels, much below the immunogenic range (Fig. 19-12), T cells can become unresponsive while B cells remain competent.

Many self-Ags are probably T-dependent and present at low levels. Hence *natural self-tolerance is probably often due to low-zone tolerance;* i.e., to lack of T_h activity, while the corresponding B cells are present and potentially responsive. The implications of this situation for the breakdown of self-tolerance are discussed below (also in Ch. 22).

Two **general cellular mechanisms** appear to account for tolerance: **clonal deletion,** in which particular clones of T or B cells are eliminated or inactivated, and **suppression,** in which the clones are present but unresponsive because they are blocked by excessive activity of T_s cells. Either mechanism can account for tolerance to T-dependent Ags, but only deletion (or inactivation) of the appropriate B cells can cause tolerance to T-independent Ags.

Clonal Deletion. Ag is required to induce tolerance, but how contact with an Ag eliminates the corresponding lymphocytes (clonal deletion) is not clear. Some observations suggest that with B cells the loss of cell surface Ig is the critical event.

As noted before, the binding of multivalent Ag by a B cell aggregates the cell's surface Ig in the plane of the membrane (cap formation; see Fig. 18-15, Ch. 18). The aggregates are then ingested by the cell and degraded, or are shed, leaving the surface denuded of Ig: until this Ig is regenerated the Ag can elicit no response and the cell is

"tolerant." For convenience, this stripping or **modulation** of surface Ig has usually been studied with anti-Ig Abs, rather than with multivalent Ag.* The ease, completeness, and persistence of the effect differ with B cells at different stages of development (see B Cell Development, below).

In **mature B cells** (from adult spleen and lymph nodes) the modulation of surface Ig is slow (requiring about 1 h) and is seldom complete, and regeneration occurs in a few days. However, in **immature B cells** (from fetal or newborn mouse liver or spleen) the disappearance is more rapid and is complete, and regeneration seems not to occur at all. The difference suggests why tolerance is much more easily established in the fetus and newborn than in the adult. The inactivation of B cells by contact with Ag at a critical early stage in their differentiation into immunologically competent cells has been termed **clonal abortion.**

Though mature B cells are relatively resistant to modulation they also can be inactivated by contact with Ag, provided the dose is high enough and T_h activity is lacking. For instance, B cells exposed to high doses of Dnp conjugated to pneumococcal polysaccharide or to autologous IgG molecules (i.e., to carriers that elicit little or no T_h cell response) become unable to produce anti-Dnp Abs when subsequently challenged with optimal doses of Dnp on potent carriers. How mature B cells are inactivated ("deleted") is not known.

---

* In the analogous process of **antigenic modulation** the disappearance of a cell's surface Ag is caused by binding the corresponding Ab (see TL Ag, Ch. 18).

Suppressor T Cells. Unresponsiveness of T cells may be due to deletion, or inactivation, of T helpers, perhaps caused by a modulation mechanism like the one that inactivates B cells. However, considerable evidence indicates that another mechanism, excessive T_s cell activity, is usually responsible. The existence of T_s cells, in fact, was first recognized when Gershon discovered that, in mice, tolerance of sheep RBCs can be transferred from tolerant donors to normal mice (**infectious tolerance**). The responsible cells were T lymphocytes; it was later shown that these suppressor T cells belong to the Ly-23 set of T cells, and that they have surface molecules encoded by the IJ region of the major histocompatibility complex (Ch. 18). Since then tolerance to several T-dependent Ags, especially low-zone tolerance, has been transferred with T_s cells. Hence these cells are now believed to be largely responsible for natural tolerance to many self-Ags.

As noted before (T Suppressors, Ch. 18), there are at least two types of T suppressors: T_s-1 cells produce a soluble factor that activates T_s-2 cells and the latter block the activity of T_h cells (and also the activities of the effector T cells responsible for various forms of cell-mediated immunity [Ch. 22]); whether the T_s-2 cells can also act directly on B cells is not clear. Why Ags sometimes elicit T_s cells, rather than T_h cells, is also not clear. Macrophages may be the key. As noted before, T_h cells are stimulated by macrophages that present Ag to the T cells (in conjunction with Ia molecules on the macrophage surface). However, under circumstances that allow the Ag to bypass macrophages and to act directly on T cells (such as intravenous injection of low doses of monomeric proteins) T_s cells seem to be preferentially elicited.

Because macrophages are rare in the thymus, Ag penetration into the thymus may be a particularly effective way to elicit T_s cells

from thymocytes. In fact, it was among thymocytes that T_s cells were first found and the T suppressor factor was first isolated.

Though T_s cells have been demonstrated in tolerance to several model proteins it is clear that these cells are not the only cause of tolerance to T-dependent Ags. For instance, persistent tolerance in mice to human γ-globulin (HGG) cannot be transferred to normal recipients with T cells. Moreover, agents that interfere with the production or activity of T_s cells, such as colchicine (see Adjuvants, below), do not prevent the induction of tolerance to this Ag, and do not abrogate tolerance even when HGG-specific T_s cells are demonstrable in the tolerant animal's spleen. In these instances deletion of anti-HGG B cell clones may be responsible.

Termination of Tolerance. Natural tolerance of self-Ags normally persists for the individual's lifetime, but occasional breakdowns result in autoimmune responses and disease (Ch. 22). The termination of tolerance to model foreign Ags suggests the following mechanisms.

1) **Persistence of Ag** is necessary for the maintenance of tolerance. New B and T cells continue to be generated from stem cells, and once the tolerance–eliciting Ag is gone responsiveness to it returns. Hence the duration of high-zone tolerance depends upon the amount of Ag injected, its rate of breakdown in tissues, and the rate at which new B and T cells of the appropriate specificity arise. Recovery is retarded by thymectomy (in the case of T-dependent Ags), and it is accelerated by the injection of lymphocytes from normally responsive syngeneic donors.

The duration of experimentally induced tolerance varies widely. For instance levan (polyfructose) and many other polysaccharides are not broken down in mammalian tissues (which lack the appropriate enzymes), and a single high dose of these Ags establishes lifelong tolerance. Proteins, in contrast, are broken down readily; a few weeks after high-zone tolerance to bovine serum albumin is estab-

FIG. 19-14. Abrogation of tolerance to an Ag with an altered form of the Ag. **Tg** = thyroglobulin; **dTg** = denatured thyroglobulin, with a new carrier determinant, **(Z)**, and retained determinants of the native protein **(X,Y)**.

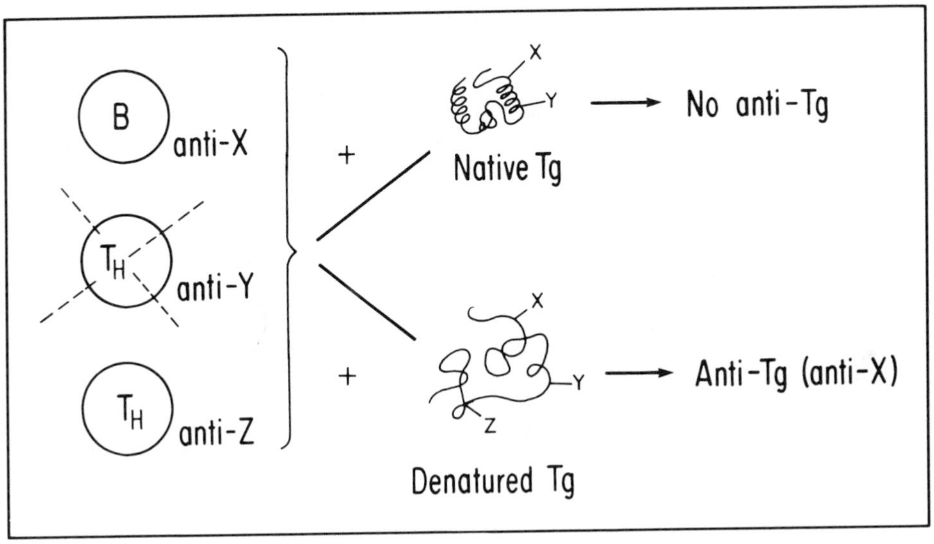

lished in mice, the declining level of Ag reaches an immunogenic range, causing a short burst of Ab (anti-BSA) synthesis.

2) **Abrogation by a Crossreacting Ag.** After tolerance to a T-dependent Ag, X, has been established by means of deleted or suppressed T_h cells a crossreacting Ag, X', can terminate the tolerance. The effect is due to other T_h cells, specific for determinants that are lacking on X but present on X'.

As noted earlier, T_h cells for some determinants on a protein can help Ab-forming B cells for other determinants on the same protein (Fig. 19-14). Hence T_h cells for

determinants that are unique to X' interact with B cells for determinants that are shared by X and X'.

Thus mice that are unable to make anti-Dnp Abs in response to Dnp-ovalbumin, because they were rendered tolerant to this Ag, should make anti-Dnp Abs in response to Dnp-BγG. In another example, closer to human autoimmune disease, animals with natural tolerance to their own thyroglobulin (Tg) have been induced by immunization with denatured Tg to make Abs that react with native Tg: denaturation probably exposes new carrier determinants (which can engage new T_h cells), while preserving many determinants of native Tg (Fig. 19-14). Immunity to self-Ags (autoimmunity) is further discussed in Chapter 22.

FACTORS INFLUENCING ANTIBODY PRODUCTION IN THE WHOLE ANIMAL

The amounts and types of Abs formed vary widely with the conditions of immunization, some of which are reviewed in this section.

ROUTE OF ADMINISTRATION OF ANTIGEN

Natural Immunization. Lymphatic tissues are probably bombarded almost constantly with Ags from transiently invasive or indigenous microbes (normal flora of skin, intestines, etc.), and by those that enter the body by inhalation (e.g., plant pollens), by ingestion (e.g., foods, drugs), and by penetration of the skin (e.g., catechols of poison ivy plants). The resulting stimulation is probably responsible for the familiar histologic appearance of lymph nodes and spleen, for the normal concentration of Igs in serum (about 15 mg/ml), and for **natural Abs**—those Igs that react or crossreact (one cannot be sure which) with Ags that have not been known to serve as immunogens in the individual under test. Animals reared under **germ-free conditions** synthesize Igs at about 1/500 the normal rate, have **exceedingly low serum Ig levels** (especially of IgG), and have small, poorly developed lymph nodes and spleen.

Deliberate Immunization. For this purpose immunogens are usually injected into skin (intradermally or subcutaneously) or muscle, depending upon the volume injected and the irritancy of the immunogen. Intraperitoneal and intravenous injections are also used in experimental work, especially with particulate Ags. Regardless of the route, most Ags eventually become distributed widely throughout the body via lymphatic and vascular channels.

Because most Ags are degraded in the intestines, **feeding** is effective only under special circumstances; e.g., with attenuated poliomyelitis vaccine (Ch. 57), which can invade the intestinal wall. Allergic responses to food are probably due to Ags that resist degradation by intestinal enzymes. **Inhalation** can also be used: e.g., aerosol administration of attenuated strains of *Pasteurella tularensis*. **Preferential synthesis of IgA Abs** occurs when immunogens are introduced into the respiratory or intestinal tract, many of whose B cells are committed to produce Igs of this class.

ADJUVANTS

The immunogenicity of soluble proteins is enhanced if they persist in tissues; for example, repeated small injec-

tions of diphtheria toxoid evoke a greater Ab response than the same total amount of toxoid given as a single injection. Accordingly, a widely used procedure involves the administration of inorganic gels (e.g., alum, aluminum hydroxide, or aluminum phosphate) with adsorbed immunogens so that they are released slowly for a prolonged period. The term **adjuvant** is applied to any substance whose admixture with an injected immunogen increases the response.

The most effective adjuvants are the water-in-oil emulsions developed by Freund, particularly those in which living or dead mycobacteria are suspended (**complete Freund's adjuvant**). After a single subcutaneous or intramuscular injection (e.g., 0.5 ml in a rabbit) droplets of emulsion metastasize widely from the site of injection; Ab formation, detected as early as 4 or 5 days later, may continue for 8 or 9 months or longer (Fig. 19-15).

The intense, chronic inflammation around the deposits of emulsion precludes their use in man. However, emulsions without my-

FIG. 19-15. Influence of adjuvants: Schematic view of amounts of Ab produced by rabbits in response to one injection (**arrow**) of a soluble protein such as bovine γ-globulin, in dilute salt solution (**A**), adsorbed on precipitated alum (**B**), or incorporated in a water-in-oil emulsion containing mycobacteria (Freund's complete adjuvant, **C**).

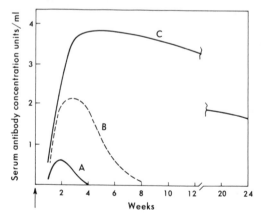

cobacteria (**incomplete Freund's adjuvant**) are less irritating and have been used clinically; their enhancing effect is less than that of complete Freund's adjuvant. The adjuvant activity of the mycobacteria is due largely to a complex glycolipid, whose activity has been duplicated by a small, synthesized glycopeptide, **muramyl dipeptide:** N-acetyl-muramyl-L-alanyl-D-isoglutamine.

Most adjuvants act not only by increasing Ag persistence, but by somehow **increasing macrophage and T_h activities.** Some other adjuvants are killed *Bordetella pertussis,* the lipopolysaccharide (LPS) of gram-negative bacteria, and large polymeric anions (e.g., dextran sulfate). A few adjuvants (LPS, dextran sulfate) are **polyclonal B cell activators** and possibly act by promoting B cell proliferation.

Colchicine is unusual: at doses that evidently block proliferation of T_s cells preferentially it enhances Ab production, presumably by permitting unchecked T_h cell activity.

DOSE

Dose effects of Ag were discussed above, under Tolerance. It should be noted, however, that dose effects vary with conditions of immunization. For instance, with a typical protein injected in solution in a rabbit the smallest effective dose might be about $100\mu g$, whereas injected in complete Freund's adjuvant it might be $1\mu g$–$10\mu g$. Similarly, the threshold dose is usually much lower in previously primed than in immunologically "virgin" ("naive") animals. Moreover, a subthreshold dose can sometimes prime animals for a pronounced secondary response without actually eliciting Abs in a detectable primary response.

Aggregated versus Soluble Form. Protein Ags are more immunogenic when administered in aggregated than in soluble form. Thus chemically cross-linked protein molecules (e.g., by gluteraldehyde) and Ag-Ab complexes, prepared in slight Ag-excess (Fig. 16-7) are usually highly immunogenic. (When, however, the complexes are prepared in Ab-excess their immunogenicity is greatly reduced, probably in part because the antigenic determinants are blocked; see Suppression by Ab as Antiimmunogen, below.)

ANTIBODY TURNOVER AND DISTRIBUTION

The level of Ab in the serum reflects the balance between rates of synthesis and degradation. When the rates are equal, the serum Ab concentration is constant (**steady state**). The rate of synthesis depends upon the total number of Ab-producing cells, which varies enormously with conditions of immunization. By contrast, the rate of degradation (expressed as half-time, or $t_{1/2}$) is determined by the H chain class (Table 19-3): *IgM and IgA are normally broken down much more rapidly than IgG molecules.*

However, infusion of a trace amount of ^{125}I-labeled IgG into individuals with widely varying IgG levels (from agammaglobulinemia or multiple myeloma) revealed an inverse relation between the half-time for IgG degradation and the total concentration of this Ig class: at high and low levels of IgG the $t_{1/2}$ was about 11 and 70 days, respectively, as compared with 23 days at normal levels. Injected Fab fragments and light chains disappear rapidly ($t_{1/2}<1$ day), but the Fc fragment has the same half-life as intact IgG. Analogy with some other serum proteins suggests that shortening of oligosaccharide branches, by random removal of terminal sialic acid or other residues by blood glycosidases, makes Ig molecules

TABLE 19-3. Some Metabolic Properties of Human Immunoglobulins

Properties	IgG*	IgA	IgM	IgD	IgE
Serum concentration (mg/ml) (average, normal individuals)	12.1	2.5	0.93	0.023	0.0003
Half-life (days)†	23	5.8	5.1	2.8	2.5
Rate of synthesis (mg/kg body weight per day)	33	24	6.7	0.4	0.016
Catabolic rate (% of intravascular pool broken down per day)	6.7	25	18	37	89

*IgG-1, IgG-2, and IgG-4 have the same half-life (~ 23 days), but that of IgG-3 is shorter ($t_{1/2} = 8$–9 days), perhaps because the unusually large hinge region of $\gamma 3$ chains (Fig. 17-15) increases susceptibility to proteolysis. The half-life of IgG in some other species is (in days): rabbit (6), rat (7), guinea pig (7–9).

†Half-life ($t_{1/2}$) is the time required for the concentration (at any particular moment) to drop to $\frac{1}{2}$ the value; it is related to the first-order rate constant for degradation, k (in days^{-1}):
$$t_{1/2} \text{ (in days)} = 0.693/k.$$
(Based on Waldmann TA et al: In Immunoglobulins. Merler E. (ed): Washington, D.C., National Academy of Sciences, 1970)

susceptible to uptake and degradation in liver macrophages (Kupffer cells).

The actual serum concentration of Ig also depends upon the volume in which the molecules are distributed. The total mass of IgG is about the same in blood and in extravascular fluids and about 25% exchanges between the two compartments each day.

CROSS-STIMULATION

A secondary response can sometimes be elicited with an immunogen that is not quite identical to the primary Ag. Surprisingly, most of the Abs made will then react more strongly with the first than with the second immunogen. This phenomenon, called **original antigenic sin,** was initially recognized in epidemiologic studies with cross-reacting strains of influenza virus.

For example, in rabbits that had been primed with Dnp-proteins, the secondary response evoked with 2,4,6-trinitrophenyl (Tnp)-proteins months later consisted primarily of Abs with the properties of anti-Dnp, rather than of anti-Tnp, molecules (Table 19-4). The effect is probably due to specific cross-stimulation, by Tnp-proteins, of an enlarged population of anti-Dnp memory B cells remaining after the primary response to the Dnp immunogen.

Once a clone has been triggered by one Ag, X, it can evidently be restimulated by a crossreacting Ag, X', that itself is unable to stimulate a primary response of that clone. This principle has been exploited in **serologic archeology,** e.g., in testing human sera during an influenza epidemic with diverse strains of the virus. A given patient's serum tends to react less strongly with the strain

TABLE 19-4. Cross-Stimulation (Original Antigenic Sin) Illustrated in the Secondary Response to 2,4-dinitrophenyl (Dnp)-protein and to 2,4,6-trinitrophenyl (Tnp)-protein

Rabbit no.	Primary stimulus	Secondary stimulus	Affinity of antibodies formed 7–8 days after secondary stimulus		Ratio of affinities DNT/TNT
			For 2,4-dinitrotoluene (DNT) (liters/mole $\times 10^{-7}$)	For 2,4,6-trinitrotoluene (TNT) (liters/mole $\times 10^{-7}$)	
1	Dnp-BγG	Tnp-BγG	25.	0.49	50.
2	Dnp-BγG	Tnp-BγG	17.	0.85	20.
3	Dnp-BγG	Tnp-BγG	34.	0.84	40.
4	Tnp-BγG	Dnp-BγG	0.20	0.30	0.7
5	Tnp-BγG	Dnp-BγG	0.62	1.0	0.6

The primary stimulus was 1 mg Dnp-bovine γ-globulin (Dnp-BγG) or 1 mg Tnp-BγG; 8 months later animals with no detectable serum Ab were reinjected with the immunogen shown, and they produced Abs in abundance. In control rabbits given Dnp-BγG in both injections the ratio of affinities for DNT/TNT ranged from 2 to 100 (i.e., they formed **anti-Dnp** Abs). In other controls given Tnp-BγG in both injections this ratio ranged from 0.2 to 0.9 (i.e., they made **anti-Tnp** molecules). Most of the Abs produced within 1 week of the secondary stimulus had binding properties that correspond to the primary rather than the secondary immunogen.

(Based on Eisen HN et al: Isr J Med Sci 5:338, 1969)

causing his current illness than with the strain that caused his first attack of influenza in some previous epidemic. From the study of sera from very elderly patients it has thus been possible to identify strains that probably caused major epidemics in the past, e.g., in 1918, long before the influenza virus was discovered.

FATE OF INJECTED ANTIGEN

Following intravenous injection of a soluble Ag the decline in its concentration in serum exhibits three sharply distinguishable phases: 1) a brief **equilibration phase** due to rapid diffusion into the extravascular space, 2) slow **metabolic decay** during which the Ag is degraded, 3) rapid **immune elimination,** which identifies the onset of Ab formation; during this phase the Ag exists largely as soluble Ab–Ag complexes, which are taken up and degraded by macrophages. Free Ab appears at the end of the immune elimination stage.

Extensively phagocytized particulate Ags, such as bacteria and red cells, do not diffuse into extravascular spaces, and hence do not exhibit the initial equilibration phase of rapid decrease in serum concentration after intravenous injection. Trace antigenic fragments can persist in lymphoid tissues long after the Ag is no longer detectable in blood.

Do Ag-recognizing lymphocytes react with native proteins or with denatured proteins? Abs elicited by globular proteins react with native, not denatured, protein; hence B cells, which have Ab on the cell surface probably recognize native proteins, not protein that is denatured or fragmented by macrophages. However, some of the T cells that are elicited by globular proteins seem not to distinguish native and drastically denatured forms of the protein, suggesting that perhaps it is the primary amino acid sequence of the protein, bound and denatured and perhaps fragmented on macrophages, that is recognized by T cells. This unusual behavior of T cells is difficult to reconcile with other evidence that T cells and Abs (i.e., B cells) have similar idiotypes in their Ag-recognition sites (see Antigen-Binding Receptors on T Cells, Ch. 18) when they react with the

same Ag, implying that T and B cells have similar Ag-binding activity and specificity.

A later chapter describes the extensive alterations undergone in vivo by some small molecules as they combine with tissue proteins to form immunogenic conjugates (e.g., penicillin, Ch. 22).

Antigenic Competition. The response to an Ag may be diminished if an unrelated Ag is injected at the same time or shortly before. For instance, rabbits injected with a foreign serum (say from the horse) produce Abs to serum globulin but not to serum albumin, although serum albumin alone elicits anti-albumin Abs. Similarly, poly-L-alanyl-protein readily elicits Abs to the L-polypeptide, but on coimmunization with poly-D-alanyl-protein only the latter elicits Abs to its polypeptide. This **intermolecular antigenic competition** is important in practical immunization programs (see Vaccination against Microbial Antigens, below), which often involve giving several different Ags in one vaccine. Adjusting the amounts of the several components ("balancing") can overcome the competitive effect.

Different determinants on the same molecule can also compete (**intramolecular competition**): thus a protein with both D- and L-polyalanyl peptides substituted on the same molecule evokes synthesis only of Abs to the D-polyalanyl groups. Presumably, the available cell surface receptors for the two determinants differ in affinity, and one determinant becomes dominant because the corresponding cells bind the limited supply of Ag. The mechanism for intermolecular competition is more obscure: it evidently involves a suppressive effect of T cells on B cells. Thus *the phenomenon is not observed in thymus-deprived mice unless they are given thymus cells.* However, the suppression is nonspecific in the sense that T_s cells elicited by Ag X can block anti-Y B cells.

INTERFERENCE WITH ANTIBODY FORMATION

SUPPRESSION BY ANTIBODY AS ANTIIMMUNOGEN

It was noted earlier (Aggregated versus Soluble Form of Ag) that an Ag is sometimes more immunogenic when administered as an

Ab–Ag complex, in slight Ag excess, than as the Ag alone. Usually, however, the addition of Abs to the immunogen is likely to block the induction of Ab synthesis, particularly when the added Ab is in relative excess, largely because the resulting Ab–Ag complexes are then degraded rapidly (see Fate of Injected Antigen, on page 438).

Inhibition by excess Ab is important clinically. For instance, severe hemolytic disease of the newborn, due to maternal Abs against fetal Rh^+ red cells (RBCs, see Rh Antigens, Ch. 23), has been greatly reduced in incidence by routine injections of anti-Rh Abs into Rh^- mothers at the time they deliver Rh^+ babies: the baby's RBCs, entering the maternal circulation in profusion as the placenta separates, are eliminated by the anti-Rh Abs. This prevents the mother from becoming immunized and reduces the risk of an anamnestic anti-Rh response during a subsequent pregnancy with another Rh^+ baby. Another example arises from placental transfer of maternal Abs to measles, polio, etc.: these can block the induction of Ab synthesis by the corresponding viral Ags in the young infant. Hence active immunization of the newborn is postponed until 6–9 months of age, by which time all maternal Abs have been eliminated (see Fig. 19-23).

SUPPRESSION BY ANTIBODY AS ANTIRECEPTOR

Abs to Ag-binding receptors on B or T cells can block the production of Igs of certain classes (isotypes) or allotypes or idiotypes, depending upon the specificity of the anti-receptor Abs and the conditions under which they are administered. The mechanisms are not well understood, but **immunologic intervention** of this type is a potentially powerful tool for manipulating immune responses.

Isotype Suppression. Injecting newborn mice with antiserum to μ chains leads to the absence of serum IgM, IgG, and IgA in the growing animal. Similar injections of anti-α or anti-γ antiserum block only the production of Igs with the corresponding heavy chains (IgA and IgG, respectively). The effects persist for several weeks, and recovery ensues gradually. Similar suppression cannot be established in adults, probably because their high levels of serum Igs neutralize the injected Abs.

Isotype suppression can also be shown in primary spleen cell cultures, in which the addition of both sheep red blood cells (SRBCs) and certain anti-Igs block the production of various classes of Abs to SRBCs: *antisera to μ chains inhibit the formation of Abs of all classes*, while antisera to γ1 or to γ2 chains block only the formation of those anti-SRBCs of the corresponding class.

These findings support the idea that *B cells with IgM on their surface are precursors of cells that secrete Igs of various classes* (see IgM–IgG switch, above, and B Cell Development, below).

Allotype Suppression. The first discovered case of suppression by anti-Ig was found by Dray among hybrid offspring of crosses between rabbits with different Ig allotypes (see Ch. 17, rabbit *a* and *b* allotypes of heavy chains and of κ light chains, respectively). If the newborn receives antiserum to the paternal allotype, it produces hardly any Ig of that allotype for many months; however, a compensating overproduction of the maternal allotype maintains the total Ig at a normal level (Fig. 19-16). Antiserum to the mother's allotype is not effective in the newborn, whose high levels of maternal Ig, acquired transplacentally or by suckling, neutralize the injected Ab. (Antiserum to the father's allotype is suppressive only if it is given before a neutralizing level of this allotype has been actively produced, i.e., before the third week after birth in rabbits.)

Newborn rabbits that produce no κ-containing Igs at all can also be elicited by Abs to κ-chain allotypes: e.g., by injecting into a foster mother who has one kind of κ allotype, and who carries implanted fetuses with a different κ allotype, Abs to the fetus's allotype. The total serum Ig level in the resulting κ-suppressed newborn is normal, because all of its L chains are λ type (which is ordinarily present in only about 10% of rabbit Ig molecules).

Like all Ig molecules that circulate in vivo, the suppressing Abs

FIG. 19-16. Allotype suppression in rabbits. The parents were homozygous for different light-chain allotypes (father, b4/b4; mother, b5/b5). At birth half the offspring (b4/b5) were given 2.2 mg anti-b4 Abs (– – –); control littermates did not receive Abs (———). For over 1 year, the target allotype (b4) was virtually lacking in the treated animals' Igs while the other allotype (b5) was overproduced; the level of total Ig in serum was normal as were the heavy-chain allotypes. The changes in control rabbits at 5–10 weeks are due to onset of synthesis of the paternal allotype (b4☐) and loss of passively acquired maternal Igs (b5,■). Relative concentrations of b4 and b5 in controls are not unusual for products of codominant alleles. Levels of individual allotypes are expressed as percent of the sum of both allotypes (b4 + b5). (Mage RG: Cold Spring Harbor Symp Quant Biol 32:203, 1967)

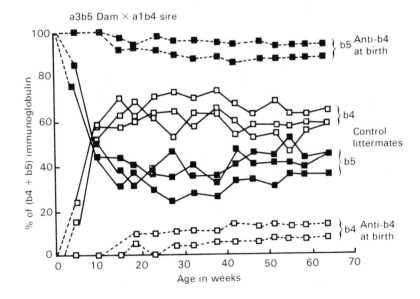

a3b5 Dam × a1b4 sire

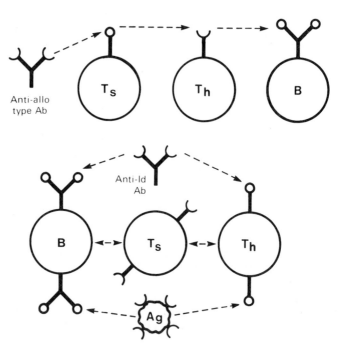

A Sequence: Injected anti-allotype Abs elicit T$_s$ cells, which block allotype-specific T$_h$. Result: B cells making that allotype do not become Ig-secreting cells.

B Sequence: Inject anti-Id Abs and Ag; the resulting activated B and T$_h$ cells elicit T$_s$ cells: the anti-Id Abs provide short-term, and the anti-Id T$_s$ cells long-term, suppression of Id⁺ T$_h$ and B cells.

FIG. 19-17. Hypothetical schemes for Ab-triggered suppression of the production (by inbred mice) of particular Igs. **A.** Allotype suppression. Injected antiallotype Abs elicit T$_s$ cells, which block allotype-specific T$_h$. As a result, B cells making that allotype do not become Ig-secreting cells. **B.** Idiotype suppression. Anti-Id Abs and Ag are injected; the resulting activated B and T$_h$ cells elicit T$_s$ cells: the anti-Id Abs provide short-term, and the anti-Id T$_s$ cells long-term, suppression of Id⁺ T$_h$ and B cells. In A the antiallotype Abs were from allogenic mice. In B the antiidiotype Abs were from rabbits. (A, based on Herzenberg et al: Cold Spring Harbor Symp Quant Biol 41:33, 1976; B, based on the Ju S-T et al: Cold Spring Harbor Symp Quant Biol 41:699, 1976)

T$_s$ cells seem to be a complex family of cells. In some responses, Ag stimulates T$_s$-1 cells to release a soluble factor, T$_s$-1F, that binds Ag and, from its serologic properties, appears to have polypeptide sequences encoded both by Ig V genes (V$_H$) and by genes of the major histocompatibility complex (IJ of the mouse H-2 complex). T$_s$-1F stimulates T$_s$-2 cells to produce another factor, T$_s$-2F, which causes still other T$_s$ cells, T$_s$-3, to act with broad specificity on diverse effector lymphocytes.

are broken down in a few weeks (Table 19-3). However, the suppressive effect persists for many months, because the Abs somehow trigger the development of specific T$_s$ cells. The mechanisms are not known but one possibility is considered in Fig. 19-17A.

Idiotype Suppression. Abs to an idiotype (anti-Id) can suppress production of Igs with that idiotype. Thus, rabbit Abs to the shared idiotype that is present on a high proportion of the anti-arsenate (anti-ARS) Abs made by A-strain mice can block the production of these anti-ARS Abs and not of other anti-ARS Abs made by the same mice. The suppression is induced by injecting mice with the anti-Id Abs (from rabbits) and then immunizing them with an ARS–protein conjugate: anti-ARS Abs are produced, but they all *lack* the shared idiotype. The suppressed mice develop T$_s$ cells that bind Id⁺ anti-ARS Ab molecules but not Id⁻ anti-ARS molecules. The suppressive activities of these T$_s$ cells are, perhaps, exerted directly on B cells with surface Id⁺ anti-ARS, or on T helpers with the same Id on their surface receptors (Fig. 19-17B). T$_h$ and B cells that share the same Id are probably specific for the same determinant (in this instance the benzenearsonate group).

NONSPECIFIC IMMUNOSUPPRESSION

Whole-body x- (or γ-) irradiation and a variety of cytotoxic drugs (Fig. 19-18) can prevent the initiation of Ab

formation by any Ag. But once Ab synthesis is under way they cannot interrupt it, unless given in doses that are severely cytotoxic for cells in general. *All these agents are much more effective in blocking the primary than the secondary response:* either memory cells are more resistant than virgin Ag-sensitive cells, or the increased number of cells that can respond increases the probablity that some will initiate Ab formation before they can be blocked. Some immunosuppressants act primarily as inhibitors of cell division (**antiproliferative**, e.g., the antimetabolites); others act primarily by destroying lymphocytes (**lympholytic**, e.g., 11-oxycorticosteroids); and others are both lympholytic and antiproliferative (x-rays, radiomimetic alkylating agents).

Most of the **antimetabolites** (purine, pyrimidine, and folate analogs; Chs. 7 and 11) were developed as byproducts of cancer chemotherapy screening programs. Their principal action is to **block cell proliferation**, and they are especially toxic for rapidly dividing cells—in tumors, bone marrow, and intestinal and skin epithelium, as well as lymphocytes that proliferate in response to stimulation by Ag. Accordingly, the margin of clinical safety between therapeutic and toxic doses is small, but for some of these

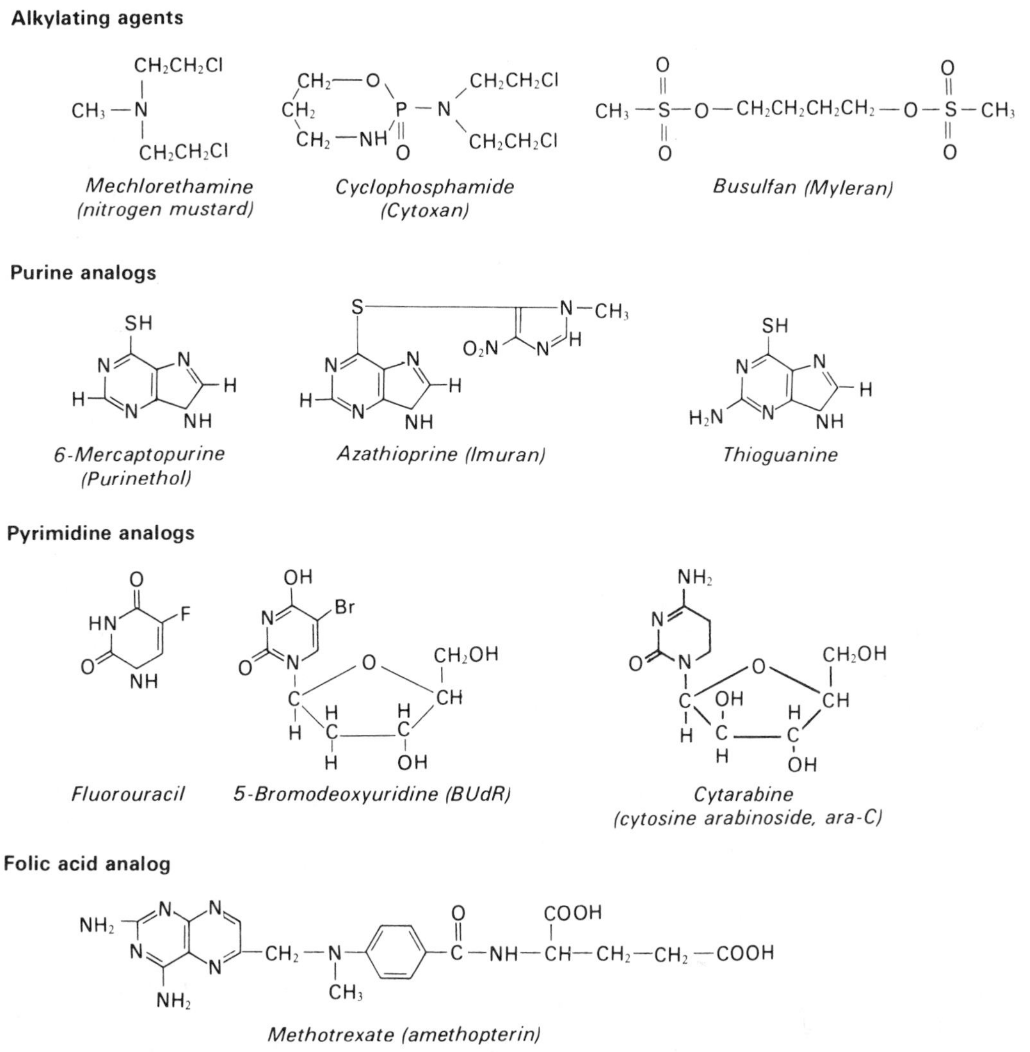

Alkylating agents

Mechlorethamine
(nitrogen mustard)   Cyclophosphamide
(Cytoxan)   Busulfan (Myleran)

Purine analogs

6-Mercaptopurine
(Purinethol)   Azathioprine (Imuran)   Thioguanine

Pyrimidine analogs

Fluorouracil   5-Bromodeoxyuridine (BUdR)   Cytarabine
(cytosine arabinoside, ara-C)

Folic acid analog

Methotrexate (amethopterin)

FIG. 19-18. Some immunosuppressive drugs. Azathioprine (Imuran) is converted in vivo to 6-mercaptopurine.

drugs it is great enough for routine clinical use (e.g., azathioprine and cyclophosphamide).

The frequency of successful kidney transplants in man is now largely due to skillful use of immunosuppressive agents: currently favored are combinations of azathioprine (an antimetabolite), prednisone (a corticosteroid), and antilymphocytic serum (ALS) or its IgG fraction (ALG), which selectively destroys circulating T cells (see Chs. 22 and 23).

Lympholytic agents (x-rays, radiomimetic alkylating drugs, corticosteroids) cause prompt and massive destruction of lymphocytes. The chromosomal damage caused by x-rays and alkylating agents also impairs the capacity of surviving lymphocytes to undergo mitosis, blocking their normal response to subsequent antigenic stimulation.

In the following comparison of various agents it is useful to contrast their activity in the three phases of the Ab response: 1) the preinductive phase, before immunogen is administered; 2) the inductive phase, between immunogen administration and rise in titer of the corresponding serum Ab; and 3) the productive phase, when Ab is being synthesized vigorously. *The lympholytic agents are effective immunosuppressors when given in the preinductive phase, and the antiproliferative agents are most effective in the inductive phase.* All known agents appear to be ineffective in the productive phase, unless given in highly toxic doses. Though this division into phases is convenient, it is an oversimplification, for lymphoid cells do not respond synchronously: the immunogen can remain active for months, and cells in all three phases may be present simultaneously.

X-rays. Whole-body irradiation with sublethal doses of x-rays (400–500 r) can suppress the response to most immunogens for many weeks, but usually not permanently. Suppression is greatest when the irradiation is given 24–48 h before the Ag, because mas-

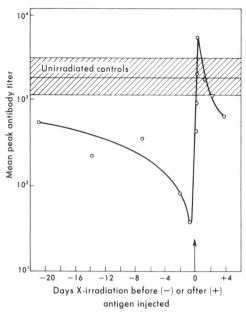

FIG. 19-19. Effect of 500-r x-irradiation of rabbits at various times before and after immunization **(arrow)** with sheep red cells. Ab titers were measured about two weeks after the injection at time zero **(arrow).** (Based on Taliaferro WH, Taliaferro LG: J Infec Dis 95:134, 1954)

sive disintegration of lymphocytes occurs promptly after whole-body irradiation. After a large dose (e.g., 500 r) active proliferation of lymphoid cells is resumed after about 3–4 weeks, and lymph nodes appear normal shortly thereafter. However, damage to chromosomes (the main targets of x-ray) can be "stored": i.e., it may not be expressed until the cells attempt to divide, perhaps months or years later. In the meantime some *nondividing cells (e.g., T helper cells) can apparently function normally.*

When an immunogen is given **after** massive irradiation it may be eliminated before the capacity to initiate Ab formation is restored. However, if the immunogen is given just **before** irradiation the stimulated cells can apparently continue their differentiation and eventually forms Abs, while already differentiated cells continue to synthesize Abs. The responding cells may even yield more Ab in animals irradiated 1–2 days after immunization (Fig. 19-19), perhaps because T_s cells are more readily inactivated than T_h or B cells by small doses of x-rays.

The **secondary response** is relatively **resistant to irradiation**: the appearance of Abs may be delayed, but peak titers are usually normal. As in the primary response, Ab formation may be increased by irradiation given shortly after the booster injection.

Alkylating Agents. Compounds such as the nitrogen mustards (Fig. 19-18) block cell division by cross linking strands of DNA. Often called **radiomimetic drugs**, their biologic effects (such as massive destruction of lymphocytes) resemble those of x-irradiation. However, recovery is more rapid than after x-rays, and these drugs are

therefore usually given at frequent intervals to bring about sustained immunosuppression.*

Corticosteroids. Large doses of 11-oxycorticosteroids cause extensive destruction of small lymphocytes. However, the surviving small T cells appear to be unusually active in some reactions, e.g., graft-versus-host (Ch. 23). Nonetheless, if the steroids are given just before the Ag they inhibit Ab formation in some species (rats, mice, rabbits); but at the doses used clinically a significant suppression has not been observed in man. In therapeutic doses these drugs inhibit inflammation, whether due to allergic reactions (Chs. 21 and 22) or to nonspecific irritants such as turpentine. Accordingly, they are widely used clinically to suppress allergic inflammation, especially if it is persistent (as in some types of bronchial asthma).

Antimetabolites. In contrast to x-rays, which are most inhibitory when given just before the immunogen, the antiproliferative metabolite analogs, such as 6-mercaptopurine, usually suppress Ab formation best when their administration is begun 2 days afterward, when the Ag-induced proliferation of lymphocytes is particularly active (see Selectivity, below). The difference is readily understood: x-rays (and alkylating agents) can damage DNA whether or not it is replicating, while the *immunosuppressive antimetabolites damage only cells with replicating DNA.*

If **dactinomycin** is added together with the immunogen the formation of Ab is substantially inhibited, because the synthesis of new mRNA is required for initiation of Ab synthesis. However, once Ab formation has started it can persist for many days in cell culture in the presence of this drug, suggesting that the corresponding mRNA is relatively stable.

Selectivity. Though immunosuppressants generally affect immune responses as a whole, rather than particular manifestations, some selective suppression has been observed. For instance, x-rays, corticosteroids, and antiproliferative drugs suppress IgG production more than IgM, suggesting that fewer cell divisions may be needed to initiate production of Abs of the IgM class. As will be noted later (Ch. 22), antisera to lymphocytes (ALS, ALG) seem to block cell-mediated more than humoral immunity.

Selectivity has also been brought out by combining the administration of Ag and antiproliferative immunosuppressant therapy. The clones whose proliferation has been stimulated are selectively eliminated by the drugs, as in the selective destruction of growing bacterial cells by penicillin.

---

* At low doses these agents (e.g., cyclophosphamide) are particularly damaging to T_s cells and they can, therefore, **enhance** immune responses.

Complications of Immunosuppression. Prolonged immunosuppression is dangerous: not only are the agents highly toxic for various cells, but they can activate latent infections and greatly increase susceptibility to serious infection with prevalent opportunistic fungi, bacteria, and viruses that ordinarily have little pathogenicity (e.g., *Candida, Nocardia,* cytomegalovirus, herpesvirus). In addition, chronically immunosuppressed individuals, like those with congenital immunodeficiencies (see Immunodeficiency Diseases, below), have an increased incidence of certain cancers. Suppression of cell-mediated immunity rather than of Ab formation is probably responsible, because effective immune responses to most tumors seem to be cell-mediated (Ch. 23).

B CELL DEVELOPMENT

PRE-B TO MATURE B CELLS

B lymphocytes, like other blood cells, arise from pluripotential hematopoietic stem cells. In embryogenesis, the first recognizable cells of the B lineage make μ chains but not L chains. More differentiated cells, called **pre-B cells,** are detected in the fetal liver about half way through gestation. They appear as large and small lymphocytes with cytoplasmic IgM but no surface Ig molecules.* Without further maturation in vivo or in vitro they are not responsive to mitogens or to Ags. Similar cells are seen in small numbers in adult bone marrow, indicating that B cells continue to arise from stem cells throughout life.

Pre-B cells give rise to **immature B cells,** which have **surface IgM (sIgM$^+$)** but no cytoplasmic Ig (Fig. 19-20). They lack several of the surface molecules that are characteristic of mature B cells (Ia, FcR [receptors for the Fc domains of aggregated Igs; Ch. 17], and C3R [receptors for the C3b fragment of complement; Ch. 20]); and, as noted before, they are highly susceptible to the removal of surface Ig by multivalent Ags or by Abs to surface Ig molecules (see Modulation, under Clonal Deletion, above).

Mature B cells arise from the immature ones. Besides sIgM, the mature cells usually have one or two other surface Ig isotypes: most have sIgM + sIgD, and some have sIgM + sIgG or sIgM + sIgA.

From the percentage of cells that bind fluorescent Abs to various isotypes it appears, but has not been demonstrated directly, that some cells with sIgM + sIgG, or with sIgM + sIgA, also have sIgD. Only infrequent mature B cells have a single surface Ig isotype (usually IgM, sometimes IgD or IgA). Mature B cells also have surface Ia molecules and receptors for Igs (FcR) and for a complement fragment (C3R). As noted before (see Tolerance, above) these cells, unlike immature B cells, are relatively resistant to removal of sIg molecules by multi-valent Ags and by anti-Ig Abs, and they are not easily made tolerant.

As noted above, the **terminal differentiation steps** in which mature B cells are stimulated by T-dependent Ag to develop into **plasma cells** (which secrete IgG or IgA), and into **memory cells,** are greatly influenced by T_h cells and macrophages. In contrast, the maturation step in which immature B cells begin to express surface Ig isotypes other than IgM seems not to depend upon either T cells or Ags, for it occurs normally in athymic mice raised under pathogen-free conditions.

In human fetuses lymphocytes with IgM or IgG on the cell surface are detected by immunofluorescence by the 10th week of gestation, and cells with surface IgA are detected slightly later. By the 15th week the proportion of fetal spleen and blood cells with each Ig class is essentially the same as in the normal adult. However, secretion of Igs by fetal cells occurs much later, perhaps because T helpers are absent or because restricted diversity of Ag-sensitive cells and of foreign Ags in the sheltered fetus reduces the chances for Ag-triggering of the fetus's lymphocytes. IgM and lesser amounts of IgG normally begin to be synthesized and secreted by fetal spleen cells in about the 20th week, and production of IgA seems to start only some weeks after birth. However, with severe fetal infection, as in congenital syphilis, Abs (of the IgM class) are formed vigorously, and plasma cells are abundant in the infected 6-month fetus (Fig. 19-21).

The central role of IgM-bearing cells in the maturation sequence is also evident from inhibition experiments. Abs to μ chains, administered to newborn mice, block the development of all Ig-producing cells and result in agammaglobulinemic mice, whereas Abs to γ or to α chains block the production of just IgG or IgA molecules, respectively.

Taken together, all these results suggest that the order in which Ig H chains are produced in embryogenesis repeats the apparent evolutionary sequence (**"ontogeny recapitulates phylogeny"**): *IgM, then IgG and IgA* (see Ch. 17, Fig. 17-40).

Acquisition of Specificity. The initial synthesis of Ig marks the transition from stem cells to pre-B cells. The transition probably requires gene rearrangements that

---

* Immunofluorescence is used to detect Igs in or on cells. Extracellular Abs (labeled with fluorescein or rhodamine) can penetrate into dead cells, but not into living ones. Accordingly, cytoplasmic Ig is revealed by staining "fixed" (i.e., killed) cells with fluorescent Abs, and surface Ig is revealed by staining living cells (see Fig. 18-10, Ch. 18).

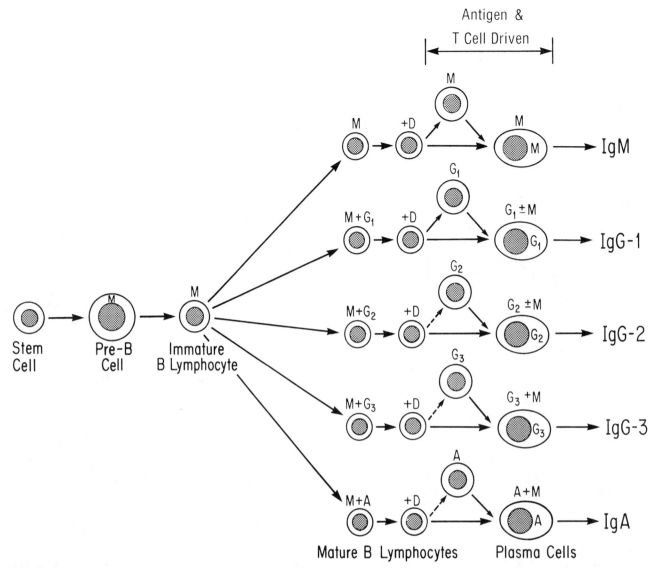

FIG. 19-20. Scheme for B cell development, showing isotype diversification in a single clone, each of whose cells presumably has the same specificity for Ag and expresses the same V_H and V_L genes. Surface IgE (sIgE) also occurs on some IgM-bearing cells (not shown). Letters (M, G, D, etc.) above cells refer to surface Ig; those inside cells refer to readily demonstrable cytoplasmic Ig. (Abney ER et al: J Immunol 120:2041, 1978)

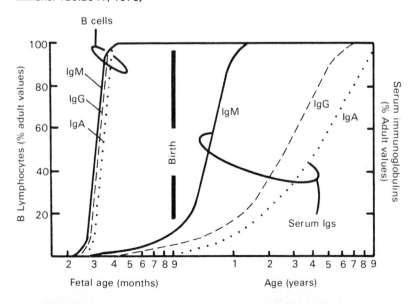

FIG. 19-21. Maturation of B cells and serum Ig levels in man. With both cells and their secreted products the ontogenetic order (IgM, IgG, IgA) recapitulates the presumptive evolutionary sequence of their appearance (see Fig. 17-39). (Based on Cooper MD, Lawton AR: Am J Pathol 69:513, 1972, from data of Lawton AR et al: Clin Immunol Immunopathol 1:104, 1972)

bring encoding sequences for C and V segments together: V_L close to C_L and V_H close to C_H (Translocation, Fig. 17-30). Each B cell clone expresses only one V_H and one V_L gene, out of the many available in the genome, and it has been generally assumed that the V genes translocated in any particular precursor cell are chosen at random, implying random appearance of responsiveness to particular Ags during development. Surprisingly, however, it seems that responsiveness to various Ags appears in a fixed and orderly sequence: for instance, the developing fetal lamb can respond to ferritin at the end of one month's gestation and to diphtheria toxin a few months later, at the time of birth; and in the mouse anti-Dnp B cells appear at the time of birth and antiphosphorylcholine B cells about 7 days later (Fig. 19-22).

The increase in size of the B cell repertoire, from about 10^4 different clones in the newborn mouse to about 10^7 shortly afterward (Fig. 19-22), may also be the result of some type of programmed developmental process, rather than a response to stimulation by haphazardly encountered environmental Ags.

SERUM IMMUNOGLOBULINS IN THE NEWBORN

Though synthesis of Igs can be detected in mammalian fetuses, protective levels of Abs to common pathogens are not produced until some time after birth. The newborn would thus be vulnerable to many infections were it not for the maternal Abs it receives passively before or shortly after birth. In some species the offspring receives maternal Igs only from colostrum, which is usually rich in IgA and IgG, while in others the Igs are transferred both in utero and by suckling (Table 19-5).

In man and in higher primates absorption from colostrum is probably of minor importance compared with transfer across the placenta. Maternal IgG, but not IgM, IgA, or IgE, is transmitted freely to the human fetus in utero, suggesting selective transport. Special sites on the Fc domain are evidently required: in rabbits, Fc frag-

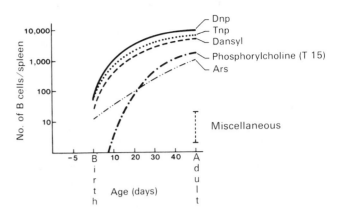

FIG. 19-22. Number of some hapten-specific B cells in the spleen of mice (BALB/c strain) at various ages. The numbers, determined by the splenic focus assay, provide the basis for determining the frequency of each of the B cell types and for estimating the size of the total repertoire (about 10^4 different clones in the newborn and about 10^7 3 weeks later), assuming that most clones are about the same size (see B Cell Repertoire, this Ch.). (Based on Klinman NR et al: Cold Spring Harbor Symp Quant Biol 41:165, 1977).

ments (of γ chains) are transferred as readily as intact IgG molecules and much more rapidly than Fab fragments.

Because IgG molecules are readily transferred from maternal to fetal blood and the other Igs are not, the newborn infant's blood contains high levels of IgG (with the Gm allotype of the mother; Ch. 17), traces of IgM (of fetal origin), and essentially no IgA. The newborn's total IgG (and total Ig) declines until about 8–10 weeks of age, when it starts to rise as the neonate's biosynthesis becomes sufficiently active. Adult serum levels of IgM are reached at about 10 months of age, of IgG at about 4 years, of IgA at about 9–10 years (Figs. 19-21 and 19-23), and of IgE at about 10–15 years.

IMMUNODEFICIENCY DISEASES INVOLVING B CELLS AND SERUM IMMUNOGLOBULINS

Genetic and congenitally acquired defects in the ability to make immune responses have been recognized since the 1950s, when it became possible to prolong the survival of affected individuals through the use of antibiotics and of Igs transferred from normal donors. These immunodeficient states help to illuminate mechanisms in normal immune responses, just as mutations have clarified many other pathways. Several clinical patterns are recognized (Table 19-6). Here we summarize some of the syndromes that derive from defective B cells; those that

derive from defective T cells or defective T and B cells, as a consequence of abnormal stem cells, are discussed in Chapter 22.

Suspected subjects with recurrent severe infections or with close kinship to affected persons are **evaluated** by procedures listed in Table 19-7. **It is important to avoid the use of live vaccines**—e.g., attenuated viruses (polio, vaccinia, measles, rubella, mumps) or bacteria (the bacillus Calmette-Guerin [BCG] variant of the tubercle bacillus)—which can cause overwhelming infections in

TABLE 19-5. Placental Structure and Mode of Passive Transfer of Immunoglobulins to the Fetus*

Species	No. of tissue layers between maternal and fetal circulation at term	Placental or amniotic transmission	Importance of transmission via colostrum
Pig	6	−	+++
Ruminants	5	−	+++
Carnivores	3	±	+
Rodents	2	+ (yolk sac)	+
Man	2	+++ (placenta)	−

*In chickens, and presumably in other birds, β-globulins containing Abs are transmitted from hen to egg via follicular epithelium and are stored in the yolk sac, from which the proteins are absorbed into the fetal circulation shortly before hatching.

(Good RA, Papermaster BW: Adv Immunol 4:1 1964; based on Vahlquist B: Adv Pediat 10:305, 1958)

severely affected persons. By staining blood lymphocytes with fluorescein-labeled, class-specific anti-Igs, it is possible to determine the proportion of cells with cytoplasmic or surface Igs, and thus to determine whether defects occur before or after the differentiation of pre-B cells into B cells.

Infantile X-Linked Agammaglobulinemia. This X-linked genetic defect occurs in male infants, who begin to suffer from recurrent bacterial (pyogenic) infections at about 9–12 months of age, when maternal Igs received transplacentally have disappeared (Fig. 19-23). These patients form exceedingly little if any Ab, and no plasma cells, following deliberate immunization, and their serum Igs of all classes are greatly depressed (IgG to less than 10% and IgA and IgM to about 1% or less of the normal level, Fig. 19-24). Correspondingly, their blood lymphocytes lack surface Igs, as though the defect occurred in stem cells or in pre-B to immature B cell differentiation. However, differentiation into T cells seems to be unaffected, as all their cell-mediated immune responses are normal. *Since these children generally recover without difficulty from measles, mumps, and other viral diseases of childhood, it appears that resistance to most virus infections is due to cell-mediated immunity (T cells) rather than to Abs.*

Selective IgA Deficiency. This is the commonest abnormality, occurring in about 0.1% of all persons, often without any clinical manifestations. In some families the defect is inherited as an autosomal dominant trait and in others as autosomal recessive. IgA deficiency is also associated with two other autosomal genetic abnormalities—ataxia–telangiectasia and partial deletion of chromosome 18—and it is also frequently associated with congenital infections (rubella, toxoplasmosis), suggesting that in some cases the defect may be acquired. Despite the diversity of associated genetic and congenital lesions, patients with selective IgA deficiency may have a **normal number of blood lymphocytes with surface IgA.** Moreover, the addition of a polyclonal B activator (e.g., the mitogenic protein from pokeweed plants) that stimulates blast transformation of normal B cells can also stimulate this change in IgA-bearing cells from these patients. The defect seems, therefore, to lie in Ag triggering of the terminal differentiation of the IgA-bearing B lymphocytes into IgA-secreting plasma cells.

A **block in terminal differentiation of B cells** can also occur in other Ig deficiencies; for example, in rare children who are deficient in serum Igs of all classes but who have a normal number of lymphocytes with surface Igs of diverse classes.

However, **concordant deficiencies** can also occur: the absence of serum Igs of a given class and of lymphocytes carrying surface Igs of that class suggests a block in differentiation of B cells with one type of Ig into B cells with another type (see IgM–IgG Switch and B

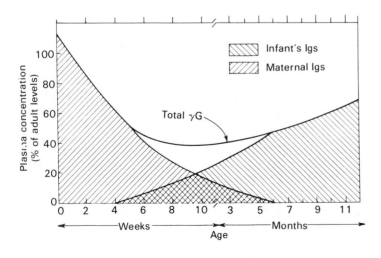

FIG. 19-23. Changing plasma levels of Igs in the human infant during the first year. Maternal IgG, which accounts for almost all the infant's Igs at birth, has essentially disappeared by about 6 months. (Based on Gitlin D: Pediatrics 34:198, 1964)

Cell Development, above, and Fig. 19-20). For instance, some individuals with **X-linked agammaglobulinemia with hyper-IgM** have elevated levels of serum IgM and of lymphocytes carrying IgM but lack IgA and IgG, both as serum molecules and as surface receptors on B cells: the defect here may be in the mechanism for switching from μ to γ or α chains (Gene Rearrangement, Ch. 17).

Common Variable Hypogammaglobulinemia. A mixed group of patients with intermittent or persistent marked depression of serum Igs suffer from a variety of disorders. The B cells in some of these individuals seem to be missing; in others they appear normal by immunofluorescence staining for Ig, but fail to respond to polyclonal B activators; and in still others they appear normal, and they synthesize Igs, but they do not secrete them. Of particular interest is a group whose B cells are inhibited by excessive nonspecific suppressor T cell activity: the isolated B cells from these individuals behave normally, while their isolated T cells suppress mitogen-induced Ig synthesis by B cells from normal individuals. The mechanism and significance of this nonspecific suppression by T cells are not understood.

VACCINATION AGAINST MICROBIAL ANTIGENS

The many immunization procedures that increase resistance to infectious agents (Table 19-8) justify Edsall's statement: "Never in the history of human progress has a better and cheaper method of preventing illness been developed than immunization at its best." Unfortunately, the development of the best procedure for a given microbe is a laborious and almost entirely empiric process. Useful generalizations are meager; some of them follow.

Number of Injections. Multiple injections of immunogen (commonly at 1- to 6-month intervals) are usually necessary to establish a long-lasting ability to give an effective secondary response, either to natural infection or to a prophylactic (booster) injection.

Soluble versus Insoluble Antigens. Immunogens that are aggregated, or are adsorbed on alum or other gels, are usually more effective than soluble immunogens. The aggregated Ags, by binding more effectively to macrophages, are more stimulatory than dispersed Ags for T_h cells; and the slow desorption of Ags from gels or emulsions maintains the Ag in tissues for long periods.

Systemic versus Local Immunization. The choice of site for injection of Ag is usually determined by convenience, because the ensuing immunity is generally due to systemically disseminated Ab molecules (or T lymphocytes in cell-mediated immunity). However, preferential activation of local groups of Ag-sensitive cells can sometimes be valuable, by concentrating the immune response at sites where target microbes invade, e.g., by achieving high IgA Ab levels in secretions of respiratory or intestinal tracts for agents that primarily infect these regions. This may in part account for the effectiveness of attenuated polio vaccine, which is administered orally, and for the contention that influenza vaccine elicits more protection when given by aerosol inhalation than by injection into skin or muscle. Trachoma vaccines instilled in the conjunctival sac may similarly generate more protection against trachoma eye infections than does injected vaccine.

Though current immunization practices stimulate immune responses effectively, it is still not possible to stimulate them selectively: e.g., to elicit Ab formation without cell-mediated immunity or vice versa, or to avoid producing certain types of Abs, such as those of the IgE class, that can cause serious allergic reactions (Ch. 21).

TABLE 19-6. Some Primary Immunodeficiency Disorders*

Probable defect in	Disease
I. B cells	Infantile X-linked agammaglobulinemia
	Selective Ig deficiency (usually IgA)
	Transient hypogammaglobulinemia of infancy
	X-linked Ig deficiency with hyper-IgM
II. T cells	Thymic hypoplasia (pharyngeal pouch syndrome or DiGeorge's syndrome)
	Episodic lymphopenia with lymphocytotoxin
III. T cells, B cells, and stem cells	Immunodeficiency with ataxia-telangiectasia†
	Immunodeficiency with thrombocytopenia and eczema (Wiskott–Aldrich syndrome)‡
	Immunodeficiency with thymoma
	Immunodeficiency with dwarfism
	Immunodeficiency with generalized hematopoietic hypoplasia
	Severe combined immunodeficiency
	Autosomal recessive
	X-linked
	Sporadic
	Variable immunodeficiency (commonest type, largely unclassified)

*Excludes immunodeficiencies secondary to x-irradiation, cytotoxic drugs, lymphomas with replacement of normal by neoplastic lymphoid cells (e.g., multiple myeloma), or excessive loss of Igs through leaky lesions in intestine (exudative enteropathy) or kidneys (nephritis, nephrosis). Severe, recurrent infections are also seen in rare children with deficiencies in complement components C3 or C5 (see Ch. 20) or in bactericidal activity of granulocytes (due to defective metabolic production of H_2O_2). Cell-mediated immunity is conspicuously deficient in categories II and III, but normal in I.

†Inherited as autosomal recessive character; 80% lack serum and secretory IgA; IgM and IgG are usually normal.

‡Inherited as X-linked character. IgM levels are usually low; IgG and IgA are usually normal or elevated. Pronounced inability to make Abs to polysaccharides.

(Based on Bull WHO 45:125, 1971)

TABLE 19-7. Some Tests Used for Clinical Evaluation of Immune Status

Tests for B cell functions	Serum Ig levels*
	Serum Ab levels†
	Biopsies examined for plasma cells by histologic methods and immunofluorescence‡
	Viable blood lymphocytes stained for surface Igs (also stained to test for regeneration of surface Igs after these are removed by treating the cells with trypsin)
	Susceptibility to infection by Epstein–Barr (EB) virus, tested with fluorescent Abs to EB virus Ag
Tests for T cell functions	Skin tests for delayed-type hypersensitivity§
	Rosette formation with sheep red blood cells
	Lymphocyte transformation induced by plant mitogens (phyto-hemagglutinin or concanavalin A) or by incubation with allogeneic lymphocytes¶
	Release of macrophage migration inhibitory factor (MIF) upon incubation of lymphocytes with common Ags#

*Radial immunodiffusion (Ch. 16) is preferred. It requires little serum (about 10 μl), and it is accurate ($\pm10\%$), sensitive (≥10 μg/ml), and not too slow (24 h): with radiolabeled anti-Igs the sensitivity can be increased 1000-fold to where, for instance, 0.001μg/ml of IgE can be detected. The distribution of Ig levels in the normal population is essentially bell-shaped and lacks the discontinuities that permit sharp definition of an Ig deficiency. By common agreement, deficient levels are <2 mg/ml for IgG and $<10\mu$g/ml for IgA.

†Commonly measured are "natural" Abs to blood group substances A and B (Ch. 23) or to sheep RBCs or to *E. coli;* or induced Abs after immunization with potent, harmless, and potentially helpful Ags, such as diphtheria and tetanus toxoids and *Bordetella pertussis* ("triple vaccine"), or polysaccharides from pneumococci, or *Hemophilus influenzae*, or *Neisseria meningitidis.*

‡Plasma cells are sought in biopsies of lymph nodes that drain intracutaneous sites of injection of Ags or in biopsy of the rectal mucosa, whose lamina propria layer normally contains many plasma cells (IgA-containing).

§The Ags injected intradermally are derived from prevalent microbes; e.g., mumps virus, tuberculin (from *Mycobacterium tuberculosis*), streptokinase–streptodornase from hemolytic streptococci, and culture media supernatants from various fungi (e.g., *Candida, Trichophyton, Coccidioides* [useful in California], or *Histoplasma* [useful in Mississippi Valley]). Skin patch tests are performed after deliberate skin sensitization with 2,4-dintrichlorobenzene (Ch. 22).

¶Blast transformation can be evaluated by changes in cell morphology (see Fig. 18–12, Ch. 18), by radioautography, or by measuring the incorporation of ^{3}H-thymidine into DNA. For the mixed lymphocyte reaction, see Ch. 18.

#See Lymphokines, Ch. 22.

TABLE 19-8. Vaccines Preventing Infectious Disease in Man

Disease	Immunogen
Diptheria	Purified diphtheria toxoid
Tetanus	Purified tetanus toxoid
Small pox	Infectious (attenuated) virus
Yellow fever	Infectious (attenuated) virus
Measles	Infectious (attenuated) virus
Mumps	Infectious (attenuated) virus
Rubella	Infectious (attenuated) virus
Poliomyelitis	Infectious (attenuated) virus or inactivated virus
Influenza	Inactivated virus
Rabies	Inactivated virus
Typhus fever	Killed rickettsiae (*Rickettsia prowazekii*)
Typhoid and paratyphoid fever	Killed bacteria (*Salmonella typhi,* *S. schottmülleri*, and *S. paratyphi*)
Pertussis	Killed bacteria (*Bordetella pertussis*)
Cholera	Crude fraction of cholera vibrios
Plague	Crude fraction of plague bacilli
Tuberculosis	Infectious (attenuated) mycobacteria (bacille Calmette-Guérin, or BCG)
Meningitis	Purified polysaccharide from *Neisseria meningitidis*
Pneumonia	Purified polysaccharides from *Streptococcus pneumoniae*

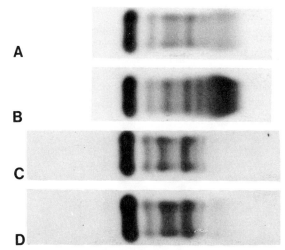

FIG. 19-24. Agammaglobulinemia, revealed by electrophoresis of human sera on a cellular acetate film. **A.** Normal serum. **B.** Serum with elevated heterogeneous Igs (polyclonal hypergammaglobulinemia). **C,D.** Sera for children with virtual absence of Igs. Serum albumin in the compact dark band at the left. (Courtesy of Dr. C. K. Osterland)

SELECTED READING

BOOKS AND REVIEW ARTICLES

B Cell Differentiation and Commitment. In Origins of Lymphocyte Diversity. Cold Spring Harbor Symposia on Quant Biol: 41:129–251, 1977

Immunodeficiency. Report of a World Health Organization Scientific Group. WHO Technical Report Series 630, 1978

Immunological tolerance. In British Medical Bulletin: 32[2], 1976

ROSENTHAL AS: Determinant selection and macrophage function in genetic control of the immune response. Immunol Rev 40:135, 1978

SCHWARTZ RH, YANO A, PAUL WE: Interaction between antigen-presenting cells and primed T lymphocytes. Immunol Rev 40:153, 1978

SPRENT J: Role of H-2 gene products in the function of T helper cells from normal and chimeric mice measured in viro. Immunol Rev 42:108, 1978

UHR JW, CAPRA JD, VITETTA ES, COOK RG: Organization of the immune response genes. Science 206:292, 1979

SPECIFIC ARTICLES

ABNEY ER, COOPER MD, KEARNEY JF, LAWTON AR, PARKHOUSE RME: Sequential expression of immunoglobulins on developing mouse B lymphocytes: a systematic survey that suggests a model for the generation of immunoglobulin isotype diversity. J Immunol 120:2041, 1978

ALT FW, ENEA V, BOTHWELL ALM, BALTIMORE D: Activity of multiple light chain genes in murine myeloma cells producing a single functional light chain. Cell 21:1980 (in press)

ANDERSSON J, LERNHARDT W, MELCHER F: The purified protein derivative of tuberculin, a B-cell mitogen that distinguishes in its action resting small B cells from activated B-cell blasts. J Exp Med 150:1339, 1979

ARANEO BA, YOWELL RL, SERCARZ EE: Ir gene defects may reflect a regulatory imbalance. I. Helper T cell activity revealed in a strain whose lack of response is controlled by suppression. J Immunol 121:961, 1979

BARCINSKI MA, ROSENTHAL AS: Immune response gene control of determinant selection. I. Intramolecular mapping of the immunogenic sites on insulin recognized by guinea pig T and B cells. J Exp Med 145:726, 1977

BLACK SJ, LOU VDW, LOKEN MR, HERZENBERG LA: Expression of IgD by murine lymphocytes. Loss of surface. IgD indicates maturation of memory B cells. J Exp Med 147:984, 1978

BROWN JC, RODKEY LS: Autoregulation of an antibody response via network-induced auto-anti-idiotype. J Exp Med 150:67, 1979

EARDLEY DD, HUGENBERGER J, MCKAY-BOUDREAU L, SHEN FW, GERSHON RK, CANTOR H: Immunoregulatory circuits among T-cell sets. I. T-helper cells induce other T-cell sets to exert feedback inhibition. J Exp Med 147:1106, 1978

GODING JW, LAYTON JE: Antigen induced co-capping of IgM and IgD on mouse B cells. J Exp Med 144:852, 1976

HIRAI Y, NISONOFF A: Selective suppression of the major idiotypic component of an antihapten response by soluble T-cell derived factors with idiotypic or anti-idiotypic receptors. J Exp Med 151:1213, 1980

JERNE NK: Towards a network theory of the immune system. Ann Immunol (Inst. Pasteur) 125C:373, 1971

MURPHY DB, HERZENBERG LA, OKAMURA LA, HERZENBERG LA, MCDEVITT HO: A new subregion (I-J) marked by a locus (Ia-4) controlling surface determinants on suppressor T lymphocytes. J Exp Med 144:699, 1976

PARKS DE, DOYLE MV, WEIGLE WO: Effect of lipopolysaccharide on immunogenicity and tolerogenicity of HGG in C57Bl/6 nude mice: evidence for a possible B cell deficiency. J Immunol 119:1923, 1977

PERELSON AS, GOLDSTEIN B: Antigen modulation of antibody forming cells: the relationships between direct plaque size, antibody secretion rate and antibody affinity. J Immunol 118:1649, 1977

ROSENWASSER LJ, ROSENTHAL AS: Adherent cell function in murine T lymphocyte antigen recognition III. A macrophage-mediated immune response gene function in the mouse. J Immunol 123:1141, 1979

PIERCE S, CANCRO M, KLINMAN N: Individual antigen-specific T lymphocytes. Helper function in enabling the expression of multiple antibody isotypes. J Exp Med 148:759, 1978

SWIERKOSZ JE, MARRACK P, KAPPLER JW: The role of H-2-linked genes in helper T cell function. V. I-region control of helper T cell interaction with antigen-presenting macrophages. J Immunol 123:654, 1979

TADA T, TAN/GUCHI M, DAVID CS: Properties of the antigen-specific suppressive factor in the regulation of antibody response of the mouse. IV. Special subregion of the gene(s) that code for the suppressive T cell factor in the H-2 histocompatibility complex. J Exp Med 144:713, 1976

TEALE JM, LAYTON JE, NOSSAL GJV: In vitro model for natural tolerance to self-antigens. J Exp Med 150:205, 1979

THEZE J, WALTENBAUGH C, DORF ME, BENACERRAF B: Immunosuppressive factor(s) specific for L-glutamic acid^{50}-L-tyrosine^{50} (GT). II. Presence of I-J determinants on the GT-suppressive

factor. J Exp Med 146:287, 1977

WEITZMAN S, SCHARFF MD: Mouse myeloma mutants blocked in assembly, glycosylation, and secretion of immunoglobulin. J Mol Biol 102:237, 1976

WOODLAND RT, CANTOR H: Idiotype specific T helper cells are required to induce idiotype-positive B memory cells to secrete antibody. Eur J Immunol 8:600, 1978

chapter 20

COMPLEMENT

It was recognized in the 1890s that a normal serum component contributes to host defenses through modifying the behavior of diverse antibody–antigen (Ab–Ag) complexes. Thus cholera vibrios were lysed within a few minutes when added to serum from immunized animals. However, if the serum had been previously heated to 56° C for a few minutes, or simply allowed to age for a few weeks, it lost its lytic activity, though its Abs were retained; and the addition of fresh **normal** serum to the inactivated antiserum restored its bacteriolytic capacity. Hence lysis required **both** specific Ab and a complementary, labile factor present in normal (as well as in immune) serum.

Originally called **alexin** (Gr. *alexein,* to ward off), the unstable serum factor was subsequently named **complement** (at first referred to as **C'** and now as **C**). Complement is now known to consist of 14 proteins that act in an ordered sequence (like the many proteins of the blood clotting system). The C proteins make up about 10% of the globulins in the normal serum of vertebrates. These proteins are not immunoglobulins (Igs), and they are not increased in concentration by immunization.

They exert their effects primarily on cell membranes, causing lysis of some cells and functional aberrations in others.

Of the many effects, lysis of red blood cells (RBCs hemolysis) is especially easy to measure in vitro and has been widely adopted as a model for analyzing C proteins and their reaction mechanisms.

COMPLEMENT (C) PROTEINS

Complement was shown to consist of more than one substance by the finding that activity could be restored when preparations inactivated in various ways were recombined. From four classic fractions (C'1, C'2, C'3, C'4), defined in Table 20-1, eleven active proteins have been isolated; their properties and those of three additional proteins of the alternate C pathway (discussed below) and of four inhibitory proteins are summarized in the table. The C proteins are diverse high-molecular-weight glycoproteins that vary greatly in physical properties and in the number of chains per molecule.

THE REACTION SEQUENCE LEADING TO CELL LYSIS

As noted above the C proteins react in an ordered sequence: in several of the steps one protein cleaves the next reacting protein, and the largest fragment behaves as an activated proteolytic enzyme, cleaving and thereby activating the next protein. Unlike most proteases *the C proteases are remarkably specific: each apparently confines its attack to a particular bond in a particular C protein.*

Some of the smaller proteolytic fragments have **"phlogistic"** (Gr. *phlogistos,* burnt) activity: i.e., they cause inflammatory tissue changes, such as increased vascular permeability, and attraction of polymorphonuclear leukocytes (**chemotaxis**). Activated C proteins with enzymatic activity are designated by an overbar (e.g., C1 is the enzymatically active form of C1). The reaction sequence is illustrated in Figure 20-1, and individual steps are discussed below.

An overview of the reaction sequence. It is useful to distinguish the early from the late stages. In the early steps

(up to cleavage of C3) there are two different pathways, the classic and the alternate.

In the **classic pathway** three proteins (C1q, C1r, C1s), making up the C1 complex, are triggered by appropriate Ab–Ag complexes (or by aggregated Igs) to become active proteases that cleave and thereby activate C4 and C2, which form an enzyme complex (**C3 convertase**) that specifically splits C3. (The proteins C4, C2, and C3 are numbered in the order of their discovery, which preceded an understanding of the reaction sequence.) The resulting fragments have pronounced phlogistic effects, and one of them is critical for the activation of the remaining proteins in the sequence (C5–C9).

The **alternate pathway** does not require Ab–Ag complexes or aggregated Abs. Instead, certain polysaccharides (mainly from bacterial and fungal cell walls) activate a special set of serum proteins, which cleave C3 into the same fragments as the C3-splitting complex of the classic pathway.

TABLE 20-1. Purified Proteins of the Human Serum Complement System*

Classic serum fraction*	Purified C proteins	Mol wt ($\times 10^3$)				Approximate serum concentration (μg/ml)	Comments	
		Unactivated forms		**Activated forms**				
		Intact molecule	Chains†	Intact molecule	Chains†			
C'1	C1‡ { C1q	410	24 (6A) 23 (6B) 22 (6C)	no change		180	Classic initiation pathway	Early components
	C1r§	83	83 (1)	83	56 (1a) 27 (1b)	50		
	C1s§	83	83 (1)	83	56 (1a) 27 (1b)	50		
C'4	C4	206	93 (1α) 78 (1β) 33 (1γ)	200	87 (1α') 78 (1β) 33 (1γ)	400		
C'2	C2	110	110 (1)	110	70 (1) 30 (1)	25		
	Properdin	220	? (4)				Alternate initiation pathway	
	Factor B	100	100 (1)					
	Factor D	24	24 (1)			2		
C'3	{ C3§	180	105 (1α) 75 (1β)	170	95 (1α') 75 (1β)	1600		
	C5§	180	105 (1α) 75 (1β)	170	95 (1α') 75 (1β)	80		Late components
	C6	95	95 (1)	no change		75		
	C7	110	110 (1)	no change		75		
	C8	165	? (3)	no change		80		
	C9	79	? (?)	no change		230		
	C1 Inhibitor	105	105 (1)					
	C3 Inactivator	93	? (2)					Inhibitors and inactivators
	βIH	150	150 (1)					
	Anaphylatoxin inhibitor (A1)	310	36 (8)					

*The symbol for complement, now C, was originally C'. The four classic crude fractions (C'1–C'4) were originally characterized by conditions that inactivated or removed them from fresh serum: C'1 and C'2 are extremely heat-labile (inactivated in a few minutes at 56°C, whereas inactivation of C'3 and C'4 at this temperature requires 20–30 min); C'1 is precipitated with "euglobulins" by dialysis against H_2O at pH 5, whereas C'2, a "pseudoglobulin," is soluble under these conditions; C'3 is inactivated by zymosan (see Alternate Pathway, in text) and C'4 by exposure to ammonia or hydrazine. C'1 to C'4 were numbered in the order of their discovery, which, unfortunately, is not the same as the order in which they react in the C cascade.

‡C1 is a readily dissociable tetramer: one molecule of C1q, two of C1r (which also exists as a free dimer in serum), and one of C1s.

§Note the structural similarity of C1r and C1s (see also Fig. 20–5), and of C3 and C5 (Fig. 20–6).

†For each type of molecule, its polypeptide chains (A,B,C, α, β, etc.), and the number of chains per molecule, are given in parentheses; e.g., C1r consists of one 83,000 dalton chain, whereas C3 (also C5) consists of one α chain of 105,000 daltons and one β chain of 75,000 daltons.

(Based largely on Porter RR, Reid KBM: Nature 125:699, 1978, and Muller-Eberhard HJ: Ann Rev Biochem 44:697, 1975)

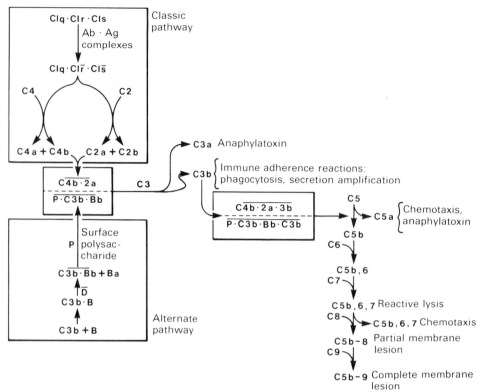

FIG. 20-1. Reaction sequence of the complement (C) system. The leaky membrane lesions that develop after C7–C9 react probably correspond to the ''holes'' (Fig. 20-7) and local membrane swellings seen in electron micrographs. **P** = properdin, **B** = factor B, **D** = factor D. **Bars** over components = active form. The catalytically active components and complexes, and their substrates (in parentheses) are: $\overline{C1r}(C1s)$; $\overline{C1s}(C4,2)$; $\overline{C4b\cdot2a}(C3)$; $\overline{D}(C3b\cdot Bb)$; $\overline{P\cdot C3b\cdot Bb}$ (C3); $\overline{C4b\cdot2a\cdot3b}$ (C5); $\overline{P\cdot C3b_2\cdot Bb}$ (C5). (Based in part on Austen KF: Harvey Lectures, 73:93, 1979

In the late stages of the C sequence large multimolecular complexes formed by the sequential reactions of the five "late" proteins (C5, C6, C7, C8, and C9) cause cell membrane lesions that result in cell death.

EARLY STEPS

THE CLASSIC PATHWAY

Competent and Incompetent Ab–Ag Complexes. The classic pathway is initiated by large multimolecular Ab–Ag aggregates, if the Abs are of the appropriate class: the most effective classes of human Ig are IgM, IgG1, and IgG3. The great variety of effective Ags suggested long ago that activation of C is due to the aggregate's Ab moiety; indeed, when Igs of the appropriate class are simply aggregated by mild heat or by crosslinking chemicals (e.g., bisdiazotized benzidine) they can activate the classic C pathway in the same way as Abs that are crosslinked by Ags. Because close packing of IgG molecules is necessary it is understandable that the small complexes

formed by these Abs with univalent haptens (Ab·H or AbH$_2$), or with multivalent Ags in great excess (Ag·Ab·Ag), are ineffective (see Appendix, this Ch.).

Conformational changes in Ig molecules, induced by aggregation, expose a critical site in the C_H2 domain of γ1 and γ3 chains. Thus the classic C pathway can be activated by aggregated IgG1 or IgG3 if the Ig molecules are intact, or if they are missing just the H chain's C_H3 domain, but not if the H chains lack both C_H2 and C_H3 domains (as in F(ab)$_2$ fragments). In addition, a fragment corresponding to just C_H2 of γ1 or γ3 can specifically block the binding of IgG1 to C1, whereas Fab fragments (and a fragment corresponding to C_H3) are ineffectual.

Comparison of amino acid sequences in C_H2 domains of several human IgG classes, only some of which bind C1q, does not reveal any single amino acid or sequence of amino acids that correlate with C1q-binding activity. It is possible that there is a potential binding site for C1q in the C_H2 domain of *all* γ chains, and that in some IgG classes this site is sterically blocked by Fab regions. Thus, C1q is bound by the Fc fragment of human IgG4 but not by the intact IgG4 molecule. Synthetic peptides corresponding to a short sequence of amino acids in C_H2 of human IgG1 can activate the complement system.

Sequential steps leading to cleavage of C3. These are described chiefly in relation to RBC lysis.

1) **S + A ⇌ SA.** Lysis of RBCs is initiated by the specific binding of one IgM Ab molecule or one pair of IgG molecules (**A**) (see One-Hit Theory, below) to an antigenic site (**S**) on the cell surface, forming a "sensitized" cell. The surface Ag need not be a natural constituent of the cell membrane: in **passive hemolysis** soluble Ags or small haptens are attached artificially to the red cell membrane, and the cells are then sensitized with the corresponding Abs.

The requirement for a pair of adjacent IgG molecules implies that the effective antigenic determinants on the cell surface must be closely spaced. This requirement also suggests why many more IgG than IgM molecules must be bound per cell for C-activated lysis: e.g., an estimated minimum of 800 IgG Ab molecules per RBC. If the antigenic determinants on the cell are too far apart C will not be activated, no matter how many IgG Ab molecules are bound: hence Abs to many cell surface Ags are unable to cause C-mediated lysis (for example, Abs to Rh and to many other Ags on RBCs; Ch. 23).

2) **SA + C1 + Ca^{2+} ⇌ SAC$\overline{1}$.** With the attachment of one IgM or a pair of closely spaced IgG molecules on the cell membrane, sites in the H chains' C_H2 domains become exposed and interact specifically with C1. C1 exists in serum as an aggregate of three proteins—C1q, C1r, and C1s—which are loosely associated noncovalently in a complex with Ca^{2+}

C1q, the recognition unit of C1, is an unusual macromolecule (Figs. 20-2 and 20-3), combining collagenlike fibrils with globular domains. It has, per molecule, six triple-stranded subunits, each composed of one A, one B, and one C chain, linked by S-S bonds. For 78 residues near the amino-terminus the chains have a typical collagenlike sequence: glycine occupies every third position; hydroxyproline and hydroxylysine are frequent; and a galactose–glucose disaccharide is attached to the hydroxyl of many hydroxylysine residues. Electron microscopy shows that each of the six subunits has a globular domain at the end of its collagen like fibril (Fig. 20-4).

After the fibrils are digested with collagenase the globular heads remain and the ability to bind aggregated Ig is retained. The binding constant of the individual C1q globular fragments for Ig appears to be about 100–1000 times less than that of intact C1q, suggesting that the stability of normal C1q–Ab–Ag aggregates is due to multivalent binding: the several globular domains of an intact C1q molecule bind to several of the Fc domains of an IgM molecule or of two or more IgG molecules in Ab–Ag complexes. C1q can, in fact, form a specific precipitate with many soluble Ab–Ag complexes, and this reaction has been used to detect such complexes in the serum, in diseases where they seem to be responsible for severe inflammatory lesions (e.g., rheumatoid arthritis; see Evidence for Persistent Soluble Immune Complexes, Ch. 21).

Unlike most other C proteins, C1q has stable combining sites and requires no activation. However, on binding to Ig it probably undergoes a conformational change that

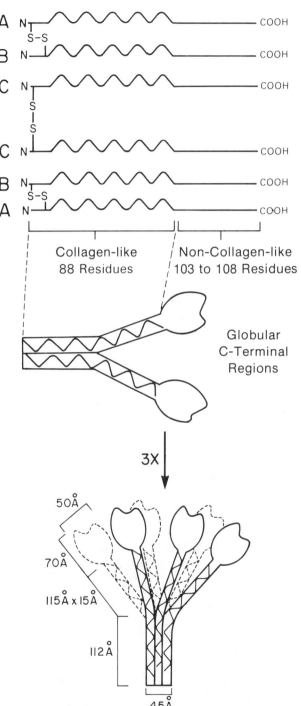

Equimolar amounts of A, B and C chains
(Each ~ 23,000 mol. wt.)

Collagen-like 88 Residues / Non-Collagen-like 103 to 108 Residues

Globular C-Terminal Regions

3X

50 Å
70 Å
115 Å x 15 Å
112 Å
45 Å

FIG. 20-2. Proposed peptide chain structure of human C1q. **Wavy lines** = proposed triple helix sections (collagenlike fibrils). Dimensions are from electron microscope studies. (Porter RR, Reid KBM: Nature 275:699, 1978)

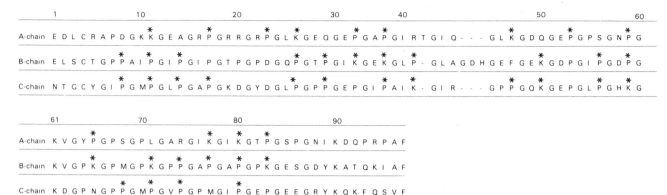

FIG. 20-3. N-terminal amino acid sequence revealing the collagenlike structure of the chains of human C1q. The sequences are given in the single-letter code (see Fig. 17-12, Ch. 17). The numbering is based on the B-chain sequence. Each chain is approximately 200 residues long. **Asterisk** = hydroxylated residue. The resemblance to collagen is strengthened by the glucosylgalactosyl disaccharides attached to the OH of the hydroxylysine residues (K). Based on Porter RR, Reid KBM: Nature 275:699, 1978; and Reid KBM: Unpublished data)

results in activation of C1r and then C1s. In serum the C1q·C1r·C1s complex is readily dissociable and has no proteolytic activity. With the binding of C1q to the proper Ab–Ag complexes, C1r is cleaved and becomes a proteolytic enzyme (C̄1r) that splits C1s, its only known substrate, to C̄1s.

C1r and C1s have the same mol wt (83,000), and in the cleavage–activation steps each is similarly split into two fragments (56,000 and 27,000 daltons; see Table 20-1) that remain linked by an S-S bond. The similarity between C1r and C1s is also evident from 1) inhibition of

FIG. 20-4. Electron micrograph of human C1q. Each molecule looks like a bunch of flowers with the lower halves of the stems close together and the upper halves diverging to display six separate stems, each terminating in a globular head. The latter contain the binding sites for C_H2 domains of γ chains in human IgG1 and IgG3 and of μ chains in IgM. The structure shown here agrees with the peptide chain structure of Fig. 20-2. (Knobel HR, Villeger W, Isliker H: Eur J Immunol 5:78, 1975)

their proteolytic activity by diisopropylphosphorofluoridate, DFP, which reacts with a serine residue in the catalytic site of each molecule, and 2) their amino acid sequence homology with each other and with other "serine-type" proteases, such as trypsin (Fig. 20-5). These similarities suggest that C1r and C1s are encoded by a duplicated gene. C4 and C2, the next C proteins that react in the C cascade, are substrates for activated C̄1s.

The stability of the C1 complex (C1q·C1r·C1s) depends on *calcium ions (Ca²⁺), which are thus essential for immune lysis of cells.* Chelating agents that bind Ca^{2+} cause C1 complexes to come apart and to dissociate from Ab–Ag complexes; the dissociated C1 can transfer to other SA sites.

3) **SAC̄1 + C4 → SAC̄1,4b̄ + C4bᵢ + C4a.** The next protein to react, C4, has three chains per molecule (α, β, γ, see Table 20-1). C̄1s cleaves a small N-terminal fragment of about 6000 daltons (C4a) from the α chain. The remaining molecule, C4b, has a short-lived active site: some C4b molecules bind to the membrane-bound SAC̄1 complex, others bind to the membrane itself, and, the rest deteriorate rapidly into an inactive form (C4bᵢ).

$$4)\ \mathrm{SAC\overline{1,4b}} + \mathrm{C2} \xrightarrow[\mathrm{C2_a^d}]{\mathrm{Mg^{2+}}} \mathrm{SAC\overline{1,4b,2a}} + \mathrm{C2a_i} + \mathrm{C2b}.$$

In the presence of Mg^{2+}, C2 binds to cell-bound C4b and is split by the C̄1s moiety of SAC̄1,4b̄. One fragment, C2a (mol wt 70,000), remains bound, forming the next activated complex, SAC̄1,4b,2a: the **C4b, 2a component** is called **C3 convertase,** because through an active site in its C2a moiety it attacks C3.

The binding of C2 by the C4b̄ moiety of SAC̄1, 4b̄ before C2 is cleaved (by C̄1) minimizes the loss of C2a, which quickly deteriorates in solution to C2aᵢ. The C2a moiety of the SAC̄1,4b,2a complex is also unstable (half-life at 37° C is 10 min): in dissociating

Residue No					(5)				(10)					(15)				(20)
Human C1r 'b' chain	Ile	Ile	Gly	Gly	Gln	Lys	Ala	Lys	Met	Gly	Asn	Phe	Pro	Trp	Gln	Val	Phe	Thr	Asn Glx
Human C1s 'b' chain	Ile	Ile	Gly	Gly	Ser	Asp	Ala	Asp	Ile	Lys	Asn	Phe	Pro	Trp	Gln	Val	Phe	Phe	Asp Asn
Bovine chymotrypsin A	Ile	Val	Asn	Gly	Glu	Glu	Ala	Val	Pro	Gly	Ser	Trp	Pro	Trp	Gln	Val	Ser	Leu	Gln Asp
Bovine trypsin		Ile	Val	Gly	Gly	Tyr	Thr	Cys	Gly	Ala	Asn	Thr	Val	Pro	Tyr	Gln	Val	Ser	Leu	Asn Ser
Bovine thrombin B chain	Ile	Val	Glu	Gly	Gln	Asp	Ala	Glu	Val	Gly	Leu	Ser	Pro	Trp	Gln	Val	Met	Leu	Phe Arg
Human plasmin B chain	Val	Val	Gly	Gly	Cys	Val	Ala	His	Pro	His	Ser	Trp	Pro	Trp	Gln	Val	Val	Leu	Leu Arg

FIG. 20-5. Homology in the N-terminal sequences of C1r and C1s b chains and of other "serine" proteases. Residues that are identical in the b chains of both C1r and C1s are **underlined twice;** those that are identical in either the C1r or C1s b chain and in one or more of the proteases are **underlined once.** (Porter RR: Fed Proc 36:2191, 1977)

from the complex (as inactive $C2_a^d$) it leaves $SA\overline{C1,4b}$, which can bind and cleave additional C2 to form more $SA\overline{C1,4b,2a}$.

C3 Cleavage in the Classic Pathway: $SA\overline{C1,4b2a}$ + C3→$SA\overline{C1,4b,2a,3b}$ + C3b$_i$ + C3a.

In this critical step each $SA\overline{C1,4b,2a}$ unit cleaves hundreds of C3 molecules. A small fragment, C3a (mol wt 7000), has pronounced phlogistic activity (see Anaphylatoxins, below). The large fragment, C3b, decays in solution (to C3b$_i$) or binds to the cell membrane; some C3b fragments join membrane-bound $\overline{C4b,2a}$, forming the next catalytic unit, $\overline{C4b,2a,3b}$, which cleaves C5 (below). Other C3b fragments are scattered over the cell membrane; they are not catalytically active and do not promote cytolysis, but they contribute to immune adherence (below) and greatly increase the cell's susceptibility to phagocytosis. Thus, *C3b on the cell surface is a powerful opsonic agent* (below).

THE ALTERNATE PATHWAY

In another mechanism opsonic C3b fragments are produced and the late steps (C5–C9) of C-mediated cell lysis are activated without involving Ab–Ag complexes or the early C components (C1, C4, and C2).

This alternate pathway was discovered by Pillemer et al (in the 1950s), who found that, on addition of zymosan (a yeast cell-wall polysaccharide) to fresh serum, C3 was lost (with negligible reduction in levels of C1, C4, and C2), and the serum also acquired some antimicrobial activities (e.g., destroying *Shigella dysenteriae*). They regarded the reaction as a manifestation of a powerful, immunologically nonspecific, antimicrobial system, due to activation of a novel protein, which they named **properdin** (L. *perdere*, to destroy). At first, since properdin migrated electrophoretically with γ-globulins most immunologists dismissed it as a natural Ab to zymosan, and attributed the loss of C3 to conventional activation of the C

sequence (C1, C4, etc.). However, the zymosan effect was later elicited in serum from a C4-deficient strain of guinea pigs (see Genetic Deficiencies, below), and properdin was eventually purified and shown not to be an Ig.

C3 Cleavage in the Alternate Pathway.

This sequence requires Mg^{2+} and depends upon the C3b fragment and three serum proteins that do not participate in the classic sequence: **properdin (P), factor B,** and **factor D** (Fig. 20-1).

Factor B binds to the trace amounts of C3b that exist in normal serum, forming a C3b,B complex from which factor D (a 23,000-dalton active protease in serum) splits off the small Ba fragment (about 20,000 daltons), leaving $\overline{C3b,Bb}$, which splits more C3b from C3. *$\overline{C3b,Bb}$ readily dissociates and loses activity; but by binding serum properdin it forms P,$\overline{C3b,Bb}$, a relatively stable C3-splitting proteolytic complex.*

P,$\overline{C3b,Bb}$ is usually inactivated by two regulatory proteins: **C3b inactivator (C3bINA)** splits C3b into inactive C3c and C3d fragments, while βIH enhances the activity of C3bINA and also promotes the dissociation of P,$\overline{C3b,Bb}$ into its individually inactive components.* However, the complex is shielded from attack by the inactivators when it is adsorbed by certain polysaccharides (zymosan, inulin, etc.) or by many microbial cells. The resulting stable complex cleaves C3 in the same way that $\overline{C1,4b,2a}$ does in the classic pathway; i.e., it generates C3a (anaphylatoxin) and C3b fragments.

The C3b has two actions: by binding to bacteria and viruses it promotes their destruction by phagocytosis (see

* An analog of C3b in cobra venom is resistant to C3bINA. When purified, this factor is nontoxic and, substituting for C3b, it can form, with P and Bb, a stable, C3-splitting enzyme (see Cobra Venom Factor, below in this chapter, for its effect in vivo in eliminating C3 and thereby inactivating the complement system).

Opsonization, below); and it combines with $\overline{P,C3b,Bb}$ to form $\overline{P,C3b_2,Bb}$, which specifically cleaves C5 and activates the late-acting C components (below).

Initiation of the Alternate Pathway. In the first step of this pathway C3b is both a reactant (in the C3b,Bb complex) and a product. What is the source of the C3b in the initiating $\overline{C3b,Bb}$ complex? Factor B in serum is probably cleaved slowly but continuously by serum factor $\overline{D}$. The resulting traces of fragment Bb bind C3, and the complex, C3,Bb, can also cleave C3b slowly from C3. Though the C3b is largely inactivated by C3bINA, all of these reactions combine to maintain a low concentration of C3b. The level is evidently sufficient to trigger the full pathway simply by the introduction of certain special particles whose surfaces can provide C3b with a microenvironment in which it is protected from its inactivators (C3bINA and βIH).

The particles having such **triggering surfaces** include 1) diverse polysaccharides, especially of microbial origin (zymosan, inulin, lipopolysaccharide of gram-negative bacteria, teichoic acid of gram-positive bacteria, etc.); 2) various gram-negative and gram-positive bacteria; 3) some parasites (e.g., *Schistosoma mansoni* larvae); 4) some mammalian RBCs (rabbit, certain mouse strains). Sialic acid–poor carbohydrate surfaces are especially effective; treatment of some RBCs with a sialic acid–cleaving enzyme (sialidase) converts the cells from inactive to active triggering particles.

The alternate pathway can also be initiated by aggregated Igs of classes that are ineffectual in the classic pathway because they do not bind C1q: e.g., aggregated human IgG4 and IgA. It can also be triggered by aggregated F(ab)$_2$ fragments, which, as was noted above, cannot initiate the classic pathway because they lack a C_H2 domain.

LATE STEPS

C5 CLEAVAGE

The C3-splitting enzymes of the classic and alternate pathways can each combine with fragment C3b to form C5-splitting enzymes or **convertases**: $\overline{C4b,2a,3b}$ from the classic pathway and $\overline{P,C3b_2,Bb}$ from the alternate pathway. C5 resembles C3 in molecular weight and subunit composition (one α and one β chain; see Table 20-1). Like C3, C5 is split by its convertase near the amino-terminus of the α chain (Fig. 20-6). The products are C5a and C5b.

C5a (about 12,000 daltons) constitutes the NH$_2$-terminal 75 amino acids of the α chain. This fragment and C3a (the corresponding fragment of the α chain of C3) have about 40% amino acid sequence homology (Fig. 20-6) and, as noted below, both have anaphylatoxin activity; C5a is, in addition, an attractant for polymorphonuclear leukocytes and macrophages. **C5b**, the remainder of C5, initiates assembly of the terminal C5b–9 complex that can cause cell lysis.

THE FINAL MEMBRANE-ATTACK COMPLEX

Unlike the steps that lead up to cleavage of C5 the remaining reactions involve the formation of progressively larger complexes, without breaking peptide bonds. On the cell membrane the C5b fragment cumulatively binds, in succession, C6, C7, C8, and finally C9. (In the fluid phase C5b alone rapidly becomes inactive C5b$_i$, and the C5b–C9 complex quickly loses cytotoxic activity.)

MEMBRANE LESIONS

When assembled on a cell the C5b–C9 complex impairs surface membrane permeability: water enters RBCs, which swell and then burst, while nucleated cells become leaky, losing first low and then high-molecular-weight substances (e.g., initially K$^+$, later nucleotides, finally proteins). In addition, ionic dyes that are excluded by intact cells penetrate into the cytoplasm: hence staining by trypan blue or eosin is often used to detect C-damaged nucleated cells (see Assays for Cell Surface Ags, Ch. 18).

FIG. 20-6. Homologous structures for C3 and C5. The two proteins have the same molecular weight and chain structure (Table 20-1), and each is split by its convertase at a susceptible arginine bond 75 residues from the NH$_2$-terminus of the α chain. Carbohydrate (CHO) is present on this region of human C5 (and in C5a). (Hughli TE: Contemp Top Mol Immunol 7:181, 1978)

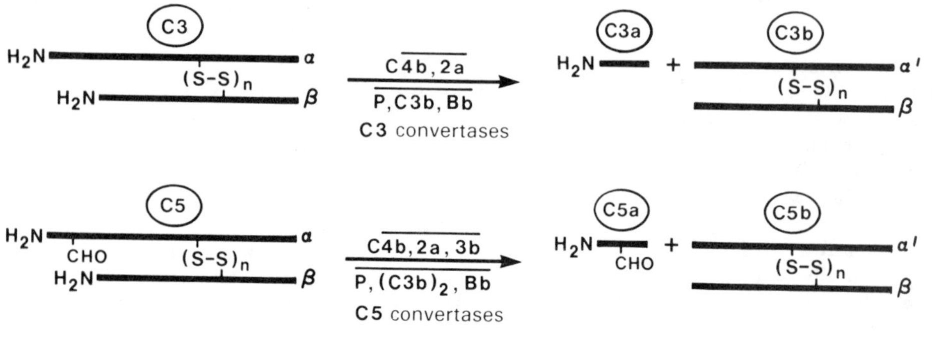

The degree of leakiness depends upon the number of C5b–9 complexes per cell (see One-Hit Theory, below), suggesting that the terminally acting complexes produce discrete lesions rather than a diffuse change in the cell membrane.

How are the lesions formed? The possibility of an enzymatic attack by C5b–9 on the membrane has been virtually eliminated by the finding that activated C causes the same impaired permeability in **liposomes** (synthetic lipid vesicles) as in cells, and no covalent changes have been found in any of the liposome's chemically well characterized constituents.

A detergentlike action is also unlikely, because with authentic detergents the number of molecules required for lysis is about 10^8–10^{10} per RBC, while only one or a few C5b–9 units per cell is sufficient (see One-Hit Theory, below).

The most attractive possibility, advanced by Mayer, is that in the course of assembling C5b–9, the subunits undergo conformational changes, exposing hydrophobic segments that interact with lipids of the cell membrane. As a result, the complexes presumably form transmembrane channels which enlarge with stepwise additions to the growing C5b–9 complex, allowing first small and then large molecules to escape. The proposed channels could well correspond to the "holes" seen in electron micrographs of C-damaged cells (Fig. 20-7).

Because the proposed conformational changes are probably unstable, the exposed hydrophobic sites would be expected to slip back promptly into the protein interior, unless they interact with lipids. This would explain why the C5b–9 complexes that form in the fluid phase quickly lose their cytotoxic activity. This also explains why *lysis due to C is sharply restricted to just those cells on which the initiating events occur; neighboring "innocent bystander" cells escape damage.*

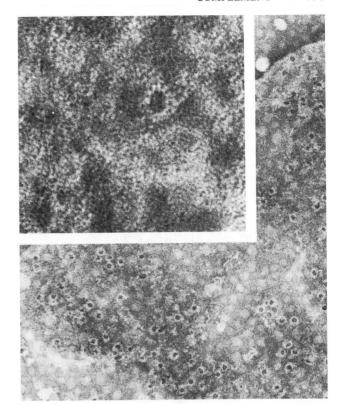

FIG. 20-7. Electron micrograph of the membrane of a sensitized sheep RBC lysed by C. Many defects ("holes") are evident. (× 187,000, reduced). **Inset** shows a representative lesion at greater magnification (× 720,000, reduced). (Humphrey JH, Dourmashkin RR: In Wolstenholme GEW, Knight J (eds): Complement. Ciba Foundation Symposium, Boston, Little, Brown, 1965)

THE ONE-HIT THEORY

The sigmoidal shape of the curve that relates the extent of red cell lysis to the amount of C (e.g., Fig. 20-10, Appendix) was previously interpreted to mean that lysis depends on the accumulation of many damaged sites (S*) per cell. However, with a limiting amount of C, the extent and velocity of hemolysis are not reduced by increasing the total number of Ab-sensitized RBCs in the reaction mixture. This finding suggests that lysis depends not on the accumulation of many damaged S* lesions per cell, but rather on one S* per cell.

Mayer, who advanced this view, has, with his colleagues, provided much evidence to support it. For instance, with RBCs that are coated with Abs (SA), with Abs and C$\overline{1}$ (SAC$\overline{1}$), or with Abs, C$\overline{1}$, and C4 (SAC$\overline{1}$,4), the average number of lytic lesions formed per cell is linearly related to the concentration of C1, C4, or C2,

respectively. (The average number of lytic lesions per RBC is calculated from the proportion of cells that are *not* lysed, by applying the Poisson distribution.†) Similarly, the activation and binding of C1 is linearly related to the concentration of IgM anti-RBC Abs, or to the square of the concentration of the corresponding IgG Abs.

These linear relations indicate that *the S site is established by the binding of one IgM or one pair of IgG Ab molecules, followed by one molecule each of C1, C4, and C2.* This is sufficient to activate and bind many molecules of C3 and the remaining C components and to cause lysis.

† $P(k) = e^{-m} m^k / k!$, where $P(k)$ is the proportion of cells with k lytic lesions per cell, m is the average number of lytic lesions per cell, and e is the base of natural logarithms. Unlysed cells have no lytic lesions ($k = 0$); their frequency is e^{-m} (because $0! = 1$), and $m = -2.303 \log_{10} P(0)$. For more on the Poisson distribution, see Ch. 46, Appendix.

COMPLEMENT AND INFLAMMATION

Though cell lysis has dominated the in vitro study of C, the most noticeable effects of activated C in vivo are related to cellular and tissue changes associated with inflammation. As is shown in Table 20-2, particular effects are due to particular C components and fragments. Because these phlogistic activities are generated in the immediate vicinity of the complexes that activate C, they lead to increased local concentrations of serum proteins (including Igs and C) and blood leukocytes (including "activated" phagocytes). The resulting destruction or containment of pathogens is probably more important to host defenses than is cell lysis itself, especially since many bacteria and viruses are not susceptible to the lytic action of C (see Bactericidal Reactions, below).

Anaphylatoxins. This term is derived from the old observation that guinea pigs can undergo fatal shock, resembling anaphylaxis (Ch. 21), when injected with normal serum that has been incubated briefly with various substances (e.g., inulin, Ab–Ag complexes, talc) that are now known to activate C through either the classic or the alternate pathway. The effect was attributed to the appearance of "anaphylatoxins," which are now known to be the polypeptides cleaved from C3 and C5 during C activation (C3a and C5a; Fig. 20-6).

The effects of C3a and C5a arise largely from their stimulating mast cells to secrete histamine, causing (among other effects) a marked increase in capillary permeability (see Pharmacologically Active Mediators, Ch. 21). Injection of C3a into human skin promptly elicits a wheal, and the effect can be specifically blocked by antihistamine drugs (see Fig. 21-8B, Ch. 21).

Both C3a and C5a have carboxy-terminal arginine, indicating their origin from a restricted trypsinlike attack of C3 convertase

and C5 convertase on C3 and C5, respectively (see Fig. 20-6). The carboxy-terminal arginine is required for anaphylatoxin activity, which is lost when just this residue is removed by a specific serum enzyme (carboxypeptidase), called **anaphylatoxin inhibitor.**

Though similar in origin and structure, C3a and C5a seem to act on different mast cell receptors: isolated guinea pig ileum treated repeatedly in vitro with C3a loses its responsiveness to this substance (**tachyphylaxis**), but still responds to C5a, and vice versa. Both peptides are active at extremely low concentrations (1×10^{-8} to 5×10^{-10}M).

Chemotaxis. With diffusion chambers divided by porous membranes it has been shown that certain C derivatives in one compartment attract polymorphonuclear leukocytes from the adjacent compartment. This **chemotactic** activity is exhibited by C5a and the ternary complex C5b,6,7.

Immune Adherence and Opsonization. The binding of C3b to Ab–Ag aggregates and to Ab-sensitized cells and viruses causes them to adhere to polymorphonuclear leukocytes, macrophages, and certain other cells (e.g., platelets, primate RBCs). This **immune adherence** is probably responsible for the **opsonic activity** of C3b: i.e., the increased susceptibility to phagocytosis of bacteria, viruses, and other particles with Ab and C3b on their surface. C3b inactivator (C3bINA), a serum protein, can specifically inactivate and reverse the opsonic activity of C3b in solution and on many cells. (Other cells, on which C3bINA is ineffective, are initiators of the alternate pathway.)

Bactericidal Reactions. Gram-negative bacteria coated with specific Ab can be lysed by C through the same reaction sequence as in RBC lysis. Gram-positive bacteria and mycobacteria, however, are not susceptible to the lytic action of C. Their resistance is probably due to peculiarities of their cell wall, because protoplasts of gram-positive bacteria are readily lysed by C.

Tissue Destruction. Though activation of C potentiates the defensive function of Abs, it can also cause severe damage to normal tissues. In some allergic disorders (especially those involving small blood vessels and glomeruli) the binding and activation of C by otherwise innocuous Ab–Ag complexes attracts leukocytes, whose subsequent degeneration releases lysosomal enzymes, causing local necrosis of host tissues (Arthus reaction, Ch. 21).

Immune Conglutination. A variety of Ab–Ag–C complexes are specifically precipitated by **immune conglutinin**, an IgM Ab specific for activated C3 and C4. Present in low titer in most normal sera, the level of immune conglutinin increases after various infections

TABLE 20-2. Principal Activities of Activated Complement (C) Proteins and Their Fragments

Activity	C protein or fragment
Anaphylatoxin: Histamine release from mast cells and increased permeability of capillaries	3a; 5a
Chemotaxis: Attract polymorphonuclear leukocytes and macrophages	5a; 5b,6,7
Chemokinesis: Increased locomotion of leukocytes (polymorphs, macrophages), but not lymphocytes	5c
Immune adherence and opsonization; Adherence of Ab–Ag–C complexes to leukocytes, platelets, etc., increasing susceptibility to phagocytosis by granulocytes and macrophages	3b; 5b
Membrane damage: Lysis of red cells; leakiness of plasma membrane of nucleated cells; lysis of gram-negative bacteria	8; 9

and after immunization with many Ags. This anti-C Ab is directed against hidden determinants in native C3 and C4, which become exposed when these proteins are activated. Since a given individual's activated C reacts with his own immune conglutinin, the latter could be regarded as an autoantibody (Ch. 22).

COMPLEMENT GENES

Genetic variants of human C3, C4, C6 and factor B have been identified by their altered electrophoretic mobilities. Perhaps because of the high serum concentration of C3 (see Table 20-1), more variants of C3 (over 20) than of other C proteins have been recognized. C proteins are specified by autosomal codominant genes. Like Ig allotypes (Ch. 17) and blood group substances (Ch. 23), C3 variants are useful in paternity tests. Genes for several polymorphic C proteins have been localized at the *B* locus of the human HLA complex (Factor B, C2, C4) and at the *S* locus of the mouse H-2 complex (C4, ?C3) (see Figs. 18-16 and 18-18, Ch. 18).

In the mouse two closely linked genes at the S locus specify two C4 proteins that are distinguishable by slight differences in molecular weight and in antigenicity. C4 appears to be synthesized as a single large polypeptide (**pro-C4**) and is then split into the three chains of the mature C4 molecule. A difference in cleavage sites for the products of the two genes could account for the small structural and serologic differences between them. The mode of inheritance of polymorphic forms of human C4 suggests that there are also duplicated, closely linked genes for this protein in the human HLA complex.

GENETIC DEFICIENCIES IN COMPLEMENT

DEFICIENCIES IN MAN

Genetic defects have been recognized in many human C proteins: C1r, C4, C2, C3, C5, C6, C7, C8 (Table 20-3). That a serum defect is due to the loss of a particular C protein, rather than to the gain of an inhibitor, has been demonstrated in several instances by the recovery of full activity following the addition of the missing component in purified form.

Heterozygotes with a defective C allele have about half the normal serum level of the corresponding C protein (measured by radioimmunoassay or by hemolytic tests), and they usually have no associated illness. About 1% of the population in Boston, for instance, is reported to have a heterozygous C2 defect. However, many who are homozygous for a defective C allele are subject to recurrent infections or other illnesses: e.g., recurrent infections with pyogenic bacteria in the rare individuals with homozygous C3 deficiency; prolonged systemic gonococcal infections in the few with homozygous C6 or C8 deficiency. Curiously, skin rashes and joint pains simulating lupus erythematosus (a probable autoimmune disease, see Ch. 22) have been observed in individuals with homozygous defects in C1r, C4, or C5, and in a few with a C2 defect, though others with virtually no C2 are apparently healthy, perhaps because the alternate pathway for generating C3b is sufficiently protective.

TABLE 20-3. Genetic Deficiencies In Complement Components and Inhibitors

Species	Deficient component	Abnormalities in some individuals with homozygosity of the defective alleles
Man	C1r	Lupus erythematosus (auto-immune disease)
	C4	Lupus erythematosus
	C2	Lupus erythematosus
	C3	Recurrent severe bacterial infections
	C5	Lupus erythematosus
	C6	Disseminated gonococcal infection and recurrent meningococcal meningitis
	C7	
	C8	Disseminated gonococcal infection
	C3bINA	Marked C3 deficiency with recurrent bacterial infections
	C1INH	Hereditary angioneurotic edema (recurrent episodes of localized edema)
Guinea pig	C4	No obvious illness
Mouse	C5	No obvious illness
Rabbit	C6	No obvious illness

(Based largely on Alper CA, Rosen FS: In Mechanisms of Immunopathology. S Cohen, PA Ward, RT McCluskey (Eds), John Wiley & Sons, N.Y., 1979)

DEFICIENCIES IN LABORATORY ANIMALS

C4-Deficient Guinea Pigs. Guinea pigs without C4 were recognized through the observation that they formed Abs (anti-C4) in response to injections of guinea pig C4 (self-tolerance; Ch. 19). Since these guinea pigs lack the classic pathway (C1,4,2) for activating the C cascade, they also provide valuable evidence for the importance of the alternate pathway: e.g., their serum bactericidal and opsonic activities, and their ability to eliminate injected bacteria, are essentially normal. Moreover, these animals have normal

C-dependent Ab-mediated allergic inflammatory reactions (Arthus reactions, Ch. 21). Isolated macrophages from normal guinea pigs synthesize C4 but macrophages from the C4-deficient strain do not.

C5-Deficient Mice. Many inbred mouse strains lack C5* but appear healthy, perhaps because they are usually maintained in relatively clean animal quarters. However, when deliberately exposed to *Corynebacterium kutscheri*, a bacterium pathogenic for mice, they exhibit increased susceptibility. The defect behaves as an autosomal codominant trait: hybrids of C5$^-$ and C5$^+$ mouse strains (C5$^+$ × C5$^-$) have half the serum C5 level of the C5$^+$ strains. However, other evidence suggests that the defect resides in a regulatory rather than in a structural gene: macrophages from C5$^-$ mice do not synthesize C5, but somatic cell hybrids, made by fusing these macrophages with chicken cells, produced mouse C5.

C6-Deficient Rabbits. The heritable absence of C6 in some rabbits has been confirmed by their ability to produce anti-C6 Abs in response to injections of purified rabbit C6. The deficient animals have impaired serum bactericidal activity against salmonellae.

DEFICIENCIES IN REGULATORY PROTEINS

The potential dangers of uncontrolled C activity are evident in the disorders associated with rare genetic defects in C3b inactivator (C3bINA) and C1 inhibitor (C$\overline{1}$INH), two normal serum proteins that inhibit or degrade active C proteins (Table 20-1).

C3bINA Deficiency. An individual without C3bINA would have virtually unchecked activity of C3b, leading to excessive breakdown of C3 by C$\overline{3b,Bb}$. In one person with a homozygous defect in C3bINA, the serum C3 was reduced to 1/20 of the normal level; also reduced were factor B and C5, bactericidal activity for gram-negative bacteria, opsonic activity for Ab-coated pneumococci, and chemotactic activity. The affected individual suffered from severe, recurrent infections with β-hemolytic streptococci and other pyogenic bacteria, much like those rare individuals with a homozygous defect in C3 itself (see above).

Deficiency of C$\overline{1}$ Inhibitor. Individuals with an inherited deficiency in C$\overline{1}$INH have **hereditary angioneurotic edema**: they suffer periodically from acute, transitory, local accumulations of edema fluid, which can become life-threatening when localized in the larynx, obstructing the tracheal airway.

The disease is transmitted in an autosomal dominant pattern: i.e., heterozygotes are affected. One would expect their serum level

of C$\overline{1}$INH (normally about 160μ/ml) to be reduced by about half, but it is actually much lower (17% of the normal value, on average), suggesting that the inherited defect is expressed as blocked synthesis. Moreover, in immunofluorescence of liver biopsies about 5% of hepatocytes are stained by fluorescein-labeled Abs to C$\overline{1}$INH in normal individuals but none are stained in affected individuals (heterozygotes mostly). Of all the serum glycoproteins, C$\overline{1}$INH is the most highly glycosylated (it has 40% carbohydrate, about half as neuraminic acid); whether the impaired biosynthesis (if it exists) is due to a gene defect for the protein moiety of C$\overline{1}$INH or for glycosylating enzymes is not clear, but some affected persons produce a serum protein that crossreacts with Abs to C$\overline{1}$INH. The clinical course is the same in those with or without the crossreactive material.

Serum obtained from patients during attacks has increased C$\overline{1s}$ activity, leading to decreased C4 and C2. Injected into skin, the serum causes increased permeability of cutaneous blood vessels (Hives; see Cutaneous Anaphylaxis in Man, Ch. 21). The responsible factor is a small polypeptide, **C-kinin**, which is split from C2 by C1s; its activity resembles that of bradykinin (a nonapeptide that increases capillary permeability; see Kinins, Ch. 21), but the two are distinguishable by radioimmunoassay.

Administration of ϵ-aminocaproate, an inhibitor of plasmin activation of C1, reduces the frequency of attacks, and antihistamines reduce their intensity. Curiously, about half the individuals with this disease are completely freed of symptoms by treatment with methyltestosterone; the reason is not known, but one possibility is that it relieves blocked synthesis of C$\overline{1}$INH.

Cobra Venom Factor. Because animals with genetic or congenital C deficiencies are not always available for the study of C function in vivo, soluble Ab–Ag complexes, formed in vitro, have been injected to deplete C levels (with simultaneous administration of antihistamines to prevent anaphylaxis due to the production of anaphylatoxin and massive release of histamine; see Aggregate Anaphylaxis, Ch. 21). However, the most selective and prolonged depletion (of C3) is brought about with venom from the cobra (*Naja naja* or *N. haja*).

As noted above, the active component, cobra venom factor or CoF (mol wt 140,000), is homologous to C3b but is entirely resistant to C3bINA; hence it forms a stable $\overline{P,CoF,Bb}$ complex that cleaves C3. The released C3b in turn forms more C3-cleaving $\overline{P,C3b,Bb}$. Through this autocatalytic sequence, intact C3 is eliminated and remains undetectable for 4–96 h after CoF is administered; and without C3 the entire C system is effectively eliminated.

COMPLEMENT SYNTHESIS

C proteins are produced by cultured cells and also by tissue slices. Macrophages synthesize C1q, C2, C4, and C5; blood monocytes also produce C2; intestinal epithelial cells synthesize C1, and liver cells (hepatocytes) C3, C6, and C9. The liver appears to account for most or all C3. Thus, after a human recipient's diseased liver was replaced by a liver allograft from a donor with a different C3 allo-

type, the recipient's serum C3 soon became replaced by C3 of the donor's type. As noted above, (see Complement Genes), C4 is synthesized as a single chain, pro-C4, and is then split to yield the three chains per molecule of serum C4. C3 is likewise synthesized as one large chain (pro-C3), which is then cleaved to form the α and β chains. In view of the collagen like structure of C1q it is of interest that cultured human fibroblasts synthesize and secrete C1q (and also C1r and C1s).

* Some C5-deficient strains: DBA/2, A.

SUMMARY

The complement (C) system consists of 14 proteins that, acting in order, can cause lysis of many microbial and animal cells and also tissue inflammation. Nearly all of the C proteins exist in normal serum as inactive precursors; when activated some of them are highly specific proteolytic enzymes whose substrate is the next protein in a sequential chain reaction. The entire C sequence can be triggered by either of two initiation pathways. In one (the classic pathway) appropriate Ab–Ag complexes (with Abs of the correct class and Ab/Ag in proper mol ratio) bind and activate C1, C4, and C2 to form a C3-splitting enzyme. In the second (the alternate pathway), various polysaccharides, common on the surface of many bacteria and fungi, bind with trace amounts of a C3 fragment and then with two other proteins (factor B and properdin) to form another C3-splitting enzyme. Once C3 is split by either pathway the way is open for the remaining sequence of steps.

Many of the inflammatory (phlogistic) effects of the C sequence can be accounted for by individual components or their activated products: C3b on the surface of a cell (or virus) is a powerful opsonic agent, increasing susceptibility to phagocytosis by polymorphonuclear leukocytes and macrophages; C3a and C5a cause vasodilation and increase capillary permeability (anaphylatoxin activity); C5a, C5b•6•7, and probably C3a, attract leukocytes (chemotactic activity); C8 and especially C9 increase the permeability of cell membranes, resulting in lysis of some cells (cytotoxic activity).

The cytotoxic effects of activated C are almost entirely confined to those cells that initiate the sequence. Neighboring cells, without the triggering immune aggregates or the initiating complexes of the alternate pathway, are spared because the activated C proteins are extremely unstable and rapidly lose activity in the fluid phase.

The importance of the C system to host defenses against microbial pathogens is brought out by the increased susceptibility to infections shown by rare persons with genetic defects in C3 formation. Though largely beneficial, the C system can also cause damage in tissues around otherwise benign Ab–Ag aggregates, by causing local accumulation of leukocytes whose degeneration releases destructive lysosomal enzymes. Thus C amplifies both the protective effects of the immune system and its capacity to cause tissue damage (hypersensitivity).

APPENDIX

The Complement-Fixation Assay

Because RBC lysis is simple to measure, it provides the basis for the C fixation assay. This assay has been used to detect and measure many different Ags and Abs. Though largely replaced now by radioimmunoassays, the C fixation test sheds light on the C system and on Ab–Ag reactions generally.

In addition to the test Ag and antiserum, the assay requires several standard reagents (referred to as the "immunologic zoo"): **sheep** RBCs, **rabbit** Abs to sheep RBCs (these Abs are also often called **hemolysins**),* and fresh **guinea pig** serum (more active as a source of C than serum from other species). Because C proteins are labile, the serum is used promptly after collection, or it may be stored at $-70°$ C or lyophilized.

If sheep RBCs are optimally coated with nonagglutinating amounts of Abs to the cells, the addition of C in the presence of adequate concentrations of Mg^{2+} and Ca^{2+} promptly causes the cells to lyse. The extent of lysis is evaluated by inspection, or quantitatively by determining the concentration of supernatant hemoglobin after sedimentation of intact cells and stroma. Ab-coated (**sensitized**) RBCs thus become indicators to detect active C: when Abs combine with Ags in the presence of C some components of C are bound and inactivated. As a result, C activity, i.e., the ability to lyse sensitized RBCs, is lost.

C fixation assays are therefore performed in two stages. In **stage 1** antiserum and Ag are mixed in the presence of a carefully measured amount of C and then incubated, usually for 30 min at $37°$ or overnight at about $4°$. If the appropriate Ab–Ag complexes are formed C is inactivated or **fixed**. In **stage 2** a suspension of sensitized RBCs is added to determine whether active C has survived. **Hemolysis** indicates that C persists and, therefore, that an effective Ab–Ag reaction **has not** occurred in stage 1. Conversely, **absence of hemolysis** indicates that C has been fixed and, therefore, that an Ab–Ag reaction **has** occurred in stage 1 (a positive C fixation reaction). These steps are outlined in Figure 20-8. With a known Ag the assay can be used to detect (and measure) Abs in unknown sera; and with a standard antiserum it can be used to detect a known Ag in complex biologic materials.

---

* Rabbits immunized with intact sheep RBCs form two kinds of hemolysins: **isophile** Abs are species-specific (for determinants on RBCs of sheep), and **heterophil** Abs are specific for the Forssman Ag, a glycolipid found in many species and even in some bacteria (see Ch. 23 for a discussion of this Ag and human colon cancer). To obtain more uniformly effective antiserum it is preferable to immunize rabbits with boiled stromas of sheep RBCs, which evoke only the anti-Forssman (heterophil) Abs. The latter are almost entirely IgM, whereas the isophil Abs belong to both IgG and IgM classes. As noted repeatedly in this chapter, IgM Abs are more efficient in the reaction with C than IgG Abs.

STAGE 1 | STAGE 2

1. Ab + Ag + C' $\longrightarrow$ Ab-Ag-C' | + EA $\longrightarrow$ No lysis

2. Ab + C' $\longrightarrow$ Ab + C' | + EA $\longrightarrow$ Lysis

3. Ag + C' $\longrightarrow$ Ag + C' | + EA $\longrightarrow$ Lysis

FIG. 20-8. Complement fixation assay. **Ab** = antibody; **Ag** = antigen; **C** = guinea pig complement; **EA** = sheep RBCs complexed with rabbit Abs to the cells (sensitized RBCs) as indicator for active C. Complement fixation has occurred in reaction 1 but not in reaction 2 or 3.

Conditions of Assay. If the amount of C used in the assay is excessive, some active C may persist and cause hemolysis even though an Ab–Ag reaction has taken place in stage 1. But if the amount of C added is insufficient its deterioration may lead to the absence of lysis in stage 2 even if an Ab–Ag reaction has not occurred in stage 1. Accordingly, just the right amount of C is required. Usually 5 units is an effective compromise in conventional assays (10^8 sensitized cells in a total reaction volume of 1.5 ml). The unit of C is discussed below (Table 20-4).*

 * Interpretation of the assay depends on the outcome of **controls**. For example, the Ag and the antiserum must be individually tested to determine that they are not "anticomplementary," i.e., that each does not inactivate C without the other. Anticomplementarity is not unusual, especially with Ags prepared from tissue homogenates. Hence this difficulty is especially troublesome with C fixation

Dependence on Mass and Ratio of Antibody–Antigen Complexes. The amount of C fixed depends on the mass of Ab–Ag complexes formed and their Ab/Ag ratio. When the ratio corresponds to that in the Ab-excess or the equivalence regions of the precipitin curve (see Fig. 16-8, Ch. 16), C is fixed most effectively.

When Ag is present in excess, fixation is less; and with Ag in extreme excess C is not fixed at all (see Fig. 20-9). Since information about optimal proportions is not generally available, C fixation tests are carried out in a "checkerboard titration" with both Ag and antiserum concentrations varied as shown in Table 20-4.

Measurement of Complement

The proportion of sensitized RBCs that are lysed increases with the amount of C added (Fig. 20-9). Because 100% lysis is approached asymptotically it is convenient to *define the hemolytic unit of complement (the CH_{50} unit) as that amount which lyses 50% of sensitized RBCs* under conditions that are arbitrarily standardized with respect to the concentration of sensitized RBCs, the concentration and type of sensitizing Ab, the ionic strength and pH of the solvent, the concentrations of Mg^{2+} and Ca^{2+}, and the temperature.

assays in the serologic diagnosis of diseases due to viruses, rickettsiae, and chlamydia, in which antigenic material is obtained from infected tissue. Undiluted or slightly diluted antisera are also frequently anticomplementary, usually owing to some denatured and aggregated Igs (see Competent and Incompetent Ab–Ag Complexes, above).

TABLE 20-4. Complement Fixation Test for Human Serum Antibodies to Poliovirus*

| Virus dilution | Serum dilution | | | | | | | | Control with | |
	1:10	1:20	1:40	1:80	1:160	1:320	1:640	1:1280	5 CH_{50}	3 CH_{50}
1:20	0	0	0	½	4	4	4	4	4	4
1:40	0	0	0	0	2	4	4	4	4	4
1:80	0	0	0	0	0	3½	4	4	4	4
1:160	0	0	0	0	0	0	4	4	4	4
1:320	0	0	0	0	0	0	2	4	4	4
1:640	0	0	0	0	0	0	1	4	4	4
1:1280	½	½	½	½	½	½	3	4	4	4
1:3200	4	3	3	3	3	3	4	4	4	4
1:6400	4	4	4	4	4	4	4	4	4	4
Control with										
5 CH_{50}	4	4	4	4	4	4	4	4		
3 CH_{50}	4	4	4	4	4	4	4	4		

 *Assay with 5 CH_{50} units of complement; fixation at 4° for 20 h before addition of sensitized indicator RBCs. 0 = no lysis (positive C fixation); 4 = complete lysis (negative C fixation). The highest dilution of Ag (virus) that gives positive C fixation is 1:1280. A 1:640 dilution of serum gives positive C fixation with dilute virus (1:640), but with more concentrated virus the system is then in Ag excess and it is necessary to use less dilute serum. The Ag alone and the serum alone are not anticomplementary at any dilution examination, even when tested in controls with only 3 CH_{50} units of complement.

(Mayer M et al, J Immunol 78:435, 1957)

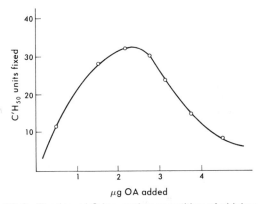

FIG. 20-9. Fixation of C by varying quantities of chicken ovalbumin **(OA)** and 12.5μg Ab from rabbit antiserum to OA. Note resemblance to precipitin curves of Chapter 16, with decreasing C fixation in the region of Ag excess. (Osler AG, Heidelberger M: J Immunol 60:327, 1948)

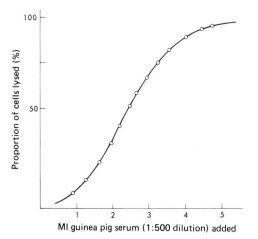

FIG. 20-10. Dose–response curve of immune hemolysis. The curve follows the empiric von Krogh equation (see text).

The dose–response curve of Figure 20-10 follows the von Krogh equation,

$$x = K\left(\frac{y}{1-y}\right)^{1/n}$$

in which x is the amount of C added (i.e., milliliters of guinea pig serum), y is the proportion of cells lysed, and n and K are constants. The curve described by this equation (which was arrived at empirically) is sigmoidal when $1/n < 1$; for fresh normal guinea pig serum $1/n$ is usually about 0.2. In estimating the number of CH_{50} units per milliliter of guinea pig serum it is convenient to plot $\log x$ vs. $\log (y/1-y)$; the data fall on a straight line,

$$\log x = K + \left(\frac{1}{n}\right)\log\frac{y}{1-y}$$

in which the intercept at 50% lysis ($y = 1-y$; $\log y/1-y = 0$) gives the volume of guinea pig serum that corresponds to one CH_{50} unit.

The Quantitative Complement Fixation Assay

The amount of C fixed in a reaction between soluble Ag and Ab can be determined as the difference between the amount added and the amount remaining after the reaction has gone to completion. When the reaction of antiserum with increasing amounts of Ag is followed (Fig. 20-9), the amount of C fixed varies like the amount of precipitate in the precipitin reaction, increasing over the Ab-excess region to a maximum at the equivalence zone and then decreasing in the Ag-excess zone. The C reaction can thus be used as an alternative to the precipitin reaction, e.g., to measure quantitatively the concentrations of Ab or of Ag, or to compare closely related Ags for their reactivity with a standard antiserum. C fixation offers the advantage that it can be used with high precision to measure small amounts of Ab or Ag, detecting as little as 0.5 μg of Ab. However, it measures the reactivity of only certain classes of Abs (see Competent and Incompetent Ab–Ag Complexes, above), and it measures concentrations in relative rather than absolute weight units.

SELECTED READING

BOOKS AND REVIEW ARTICLES

ALPER CA, ROSEN FS: Human Complement Deficiencies. In Mechanisms of Immunopathology. S Cohen, PH Ward, RT McCluskey (eds) John Wiley & Sons, N.Y. pp. 289-305, 1979

BIANCO C, NUSSENZWEIG V: Complement receptors. Contemp Topics in Mol Immunol 6:145, 1977

COOPER NR, ZICCARDI RJ: The nature and reactions of complement enzymes. In Proteolysis and Physiological Regulations, Miami Winter Symposia 11. Ribbons DW, Breed K (eds) New York, Academic Press, 1976, p 167

FEARON DT: Activation of the alternative complement pathway. In Critical Reviews in Immunology, Vol 1, CRC Press, Boca Raton, Fla, 1980

MAYER MM: Complement, Past and Present. Harvey Lect. New York, Academic Press, 1978, p 139

MÜLLER-EBERHARD HJ: Complement. Annu Rev Biochem 44:697, 1975

OHANIAN SH, SCHLAGER SI, BORSOS T: Molecular interactions of cells with antibody and complement: influence of metabolic and physical properties of the target on the outcome of humoral immune attack. Contemp Top in Mol Immunol 7:153, 1978

OSLER AG: Complement: Mechanisms and Functions. Foundations

of Immunology Series. Englewood Cliffs, NJ, Prentice-Hall, 1976

PORTER RR, REID KBM: The biochemistry of complement. Nature 275:699, 1978

SHREFFLER DC: The S region of the mouse major histocompatibility complex (H-2): genetic variation and functional role in the complement system. Transplant Rev 32:140, 1976

SPECIFIC ARTICLES

BHAKDI S, TRANUM-JENSEN J: Molecular nature of the complement lesion. Proc Natl Acad Sci USA 75:5655, 1978

EINSTEIN LP, ALPER CA, BLOCH KJ, HERRIN JT, ROSEN FS, DAVID JR, COLTON HR: Biosynthetic defect in monocytes from human beings with genetic deficiency of the second component of complement. N Engl J Med 292:1169, 1975

FEARON DT: Regulation by membrane sialic acid of βIH-dependent decay-dissociation of amplification C3 convertase of the alternative complement pathway. Proc Natl Acad Sci USA 75:1971, 1978

FEARON DT, AUSTEN KF: Activation of the alternative complement pathway due to resistance of zymosan-bound amplification convertase to endogenous regulatory mechanisms. Proc Natl Acad Sci USA 74:1683, 1977

HALL RE, COLTON HR: Cell-free synthesis of the fourth component of complement (C4): identification of a precursor of serum C4 (Pro-C4). Proc Natl Acad Sci USA 74:1717, 1977

O'NEILL GJ, YANG SY, QAND-DUPONT B: Two HLA-linked loci controlling the fourth component of human complement. Proc Natl Acad Sci USA 75:5165, 1978

REID KBM, SOLOMON E: Biosynthesis of the first component of complement by human fibroblasts. Biochem J 167:647, 1977

ROOS MH, ATKINSON JP, SHREFFLER DC: Molecular characterization of the Ss and Slp (C4) proteins of the mouse H-2 complex: subunit composition, chain size polymorphism, and an intracellular (Pro-Ss) precursor. J Immunol 121:1106, 1978

SCHREIBER RD, MÜLLER-EBERHARD HJ: Assembly of the cytolytic alternative pathway of complement from 11 isolated plasma proteins. J Exp Med 148:1722, 1978

SHREFFLER DC, OWEN RD: A serologically detected variant in mouse serum: inheritance in association with the histocompatibility locus-2. Genetics 48:9, 1963

chapter **21**

ANTIBODY-MEDIATED (IMMEDIATE-TYPE) HYPERSENSITIVITY

With the discovery of antitoxins and antimicrobial antibodies (Abs) the immune response appeared at first to be purely protective. However, it was soon found, surprisingly, that immune responses also possess dangerous potentialities. In 1902 Portier and Richet, studying the toxicity of extracts of sea anemones, observed that dogs given a second injection several weeks after the first often became acutely ill and died within a few minutes. The response was called **anaphylaxis** (Gr. *ana,* against, and *phylaxis,* protection),* and almost simultaneously observers in the United States and in Germany noted that guinea pigs responded similarly to various nontoxic antigens (Ags). Later, when injected horse and rabbit antisera were used to treat various infectious diseases in man, diverse pathologic consequences of the immune response to the foreign proteins became commonplace.

To develop a coherent terminology, von Pirquet introduced the term **allergy** (Gr. *allos* and *ergon,* altered action) to cover any altered response to a substance induced by previous exposure to it. Increased resistance, called **immunity,** and increased susceptibility, called **hypersensitivity,** were regarded as opposite forms of allergy. Usage has modified these definitions. **Allergy** and **hypersensitivity** are now synonymous: *both refer to the altered state, induced by an Ag, in which pathologic reactions can be subsequently elicited by that Ag, or by a structurally similar substance.*

In previous chapters the administration of an **immunogen** (i.e., an Ag) to stimulate Ab formation was called **immunization.** In discussions of the allergic response, however, the immunogen is often referred to as the **allergen** or **sensitizer,** immunization as **sensitization,** and the immunized individual as **sensitive, hypersensitive,** or **allergic.**

Two Basic Types. Various allergic responses have different time courses, which reflect fundamentally

* Magendie reported about 60 years earlier the sudden death of dogs repeatedly injected with egg white, and Flexnor noted shortly afterward that "animals that had withstood one dose of a foreign serum would succumb to a second dose given after the lapse of some days or weeks, even when this dose was not lethal for a control animal." However, as often happens, these valid observations were ignored until they could be accommodated within a conceptual framework.

different mechanisms. **Immediate reactions,** involving IgE Abs, begin within minutes and subside after about ½ h. **Subacute reactions,** involving IgG or IgM Abs, begin after about 1–3 h and last for about 10–15 h. **Delayed reactions,** involving a special set of T lymphocytes (T_{DTH}), become evident only after 1–2 days and persist from several days to a few weeks. To emphasize these differences the Ab-mediated reactions (immediate and subacute) are lumped and called **immediate-type hypersensitivity** ("type" means that immediate is not to be taken literally). The **delayed-type hypersensitivity** reactions are also often referred to as **cell-mediated hypersensitivity** (i.e., T-cell–mediated): they are part of a larger group of reactions, called **cell-mediated immunity,** in which similar mechanisms are also involved in resistance to many infectious agents and to neoplastic cells.

In this chapter we consider the allergic reactions due to soluble Ab molecules, and in the next those due to T lymphocytes. Both reactions can be involved in allergy to drugs and in autoimmunity, which will be discussed at the end of the next chapter.

Antibody-Mediated Responses. The most important Ab-mediated responses are grouped in Table 21-1 on the basis of underlying mechanisms. The arrangement reflects the principle that a *combination of Ab and Ag is seldom damaging unless the immune complexes trigger certain cells to release various mediators,* which serve as the immediate causes of pathologic change. However, aberrations may follow directly from the combination of Abs (and complement) with certain cell surface Ags, e.g., destruction of transfused red cells (Ch. 23).

ANAPHYLAXIS

Anaphylactic reactions are due to special Abs that bind with exceptionally high affinity to receptors on tissue mast cells and blood basophils. In man these Abs are of the IgE class; in other species they are IgE or, less effectively, a special class of IgG (see Homocytotropic Abs, below). Unless otherwise indicated, it should be assumed that IgE Abs are responsible.

Injection of an Ag into a hypersensitive individual can cause an explosive response within 3–4 min. If the Ag is

TABLE 21-1. Antibody-Mediated ("Immediate-Type") Allergic Reactions

Prototype*	Examples	Mechanism	
		Activated cells	Mediators released
Anaphylaxis	Anaphylactic shock Wheal-and-erythema responses Hay fever Asthma (some forms) Hives	Mast cells; Basophils† (platelets in some species)	Low molecular weight, e.g., histamine (see Fig 21-8)
Serum sickness	Arthus reaction Serum sickness syndrome Immune-complex diseases (glo- merulonephritis, ?rheumatoid arthritis, etc.)	Neutrophils†	High molecular weight, (lyso- somal en- zymes)
Reactions to transfused blood	Red cell incompatibilities (e.g., maternal–fetal, as in Rh disease; Ch. 23) Autoantibodies to some self- Ags (Ch. 22); e.g., to platelets, or to antihemophilic globulin, causing bleeding and purpura	None	None

*Hypersensitivity reactions are sometimes classified into four types:

Type I—Anaphylaxis (i.e., due to Abs bound to mast cells and basophils)

Type II—Cytotoxic reactions due to Abs against autologous (self) Ags, as in blood transfusions, Rh incompatibility, etc. (Ch. 23).

Type III—Due to Ab–Ag (immune) complexes, as in serum sickness, Arthus reactions, etc.

Type IV—Delayed-type reactions (i.e., mediated by T cells rather than by Abs).

†The principal leukocytes are polymorphonuclear granulocytes, monocytes (i.e., not fully differentiated macrophages), and lymphocytes. On the basis of affinity for various dyes the granulocytes are classified as neutrophils (>95%), basophils (about 1%), or eosinophils (about 1%).

injected intravenously the response, called **systemic** or **generalized anaphylaxis,** can lead to shock, vascular engorgement, and asphyxia due to bronchial and laryngeal constriction; if death does not follow promptly recovery is complete within about 1 h. If the Ag is injected into the skin the same type of reaction occurs in miniature form at the local site: called **cutaneous anaphylaxis,** it is characterized by transient redness and swelling, with complete return to normal appearance in about 30 min. Both reactions can occur not only in actively immunized individuals, but also in those who are **passively sensitized** with antiserum containing IgE Abs.

The basic mechanisms have been largely illuminated by studies of **passive cutaneous anaphylaxis,** which can be elicited simply and safely at multiple sites in the same individual, providing opportunities for controlled observations on the mediating Abs and Ags.

CUTANEOUS ANAPHYLAXIS IN HUMANS

The response begins 2 or 3 min after Ag is injected into the skin of a sensitive person: itching at the injected site is followed within a few minutes by a pale, elevated, irregu-

lar wheal surrounded by a zone of erythema (**hive** or **urticarium**). This **wheal-and-erythema** response (Fig. 21-1) reaches maximal intensity about 10 min after the injection, persists for an additional 10–20 min, and then gradually subsides.

Atopy. Certain persons, constituting about 10% of the population in the United States, are especially prone to hypersensitive responses of the anaphylactic (IgE) type. These individuals readily become sensitive "spontaneously" (i.e., without deliberate immunization) to a variety of **environmental allergens:** e.g., allergens in airborne pollens of ragweed, grasses, or trees, or those in fungi, animal danders, house dust, or foods. As a class these Ags are relatively resistant to proteolytic destruction. Nonetheless, the basic difficulty derives from the tendency of affected individuals to produce IgE Abs. As a result, when they inhale or ingest the appropriate allergen they promptly develop hives or the manifestations of hay fever and asthma. The tendency to develop this form of allergy, called **atopy** (Gr. *a* and *topos,* out of place), is heritable (see under Genetic Control of IgE Production).

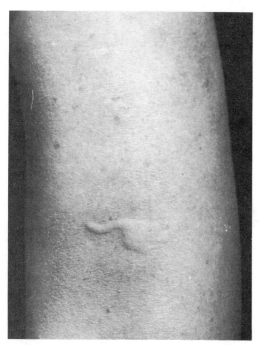

FIG. 21-1. Cutaneous anaphylaxis (wheal-and-erythema response) in man. At 15 min before the photograph was taken the subject was injected intradermally with 0.02 ml containing about 0.1 μg protein extracted from guinea pig hair. Note the irregularly shaped wheal, with striking pseudopodia; the surrounding erythema is not easily visible. No reaction is seen at the control site where 0.02 ml of buffer alone was injected.

Passive Transfer. The serum of an atopic person, even after extensive dilution (1000-fold or more), can passively sensitize the skin of normal persons. Passive sensitization is performed by injecting about 0.05 ml of dilute serum from the sensitive donor into the skin of a nonsensitive recipient. After 1 day, and up to as long as 4 weeks, injection of the corresponding Ag into the same skin site elicits the wheal-and-erythema response. To elicit the reaction it is necessary to allow a **latent period** of at least 10–20 h after the injection of serum.

This transfer response is called the **Prausnitz–Küstner (P-K) reaction** after those who first described it.* Patients are commonly tested for wheal-and-erythema responses to intradermal injections of extracts of plant pollens, fungi, food, animal danders, etc. to identify etiologic Ags. P-K tests were sometimes used to avoid direct skin tests (on young children or on adults with disse-

* As described in 1921, Küstner was extremely sensitive to certain fish, but his serum gave no detectable reaction with extracts of these fish. Prausnitz injected a small amount of Küstner's serum into a normal person's skin, and injected fish extract into the same site 24 h later: the immediate wheal-and-erythema response provided the basis for much of the clinical and experimental work on allergy of succeeding decades.

minated skin disease), but they no longer are, because of the danger of serum hepatitis. Removal of IgE, but not other immunoglobulin (Ig) classes, from human serum eliminates activity in P-K tests. For decades before IgE was discovered, the distinctive properties of these **skin-sensitizing Abs** were recognized: they were named **reagins** to distinguish them from other Abs.

REAGINS (IgE) AND BLOCKING ANTIBODIES

If ragweed extract is injected repeatedly into nonatopic human volunteers antiragweed Abs (predominantly of IgG class) appear in the serum and may be detected by conventional assays. However, these Abs are incapable of sensitizing human skin for wheal-and-erythema responses; instead, they combine with Ag and specifically **block** its ability to evoke this response in a sensitive person's skin (or in a normal person's skin at a P-K site). These **blocking antibodies** differ substantially from the skin-sensitizing Abs (reagins) that cause the wheal-and-erythema reaction (Table 21-2).

Reagins, like IgE molecules in general, are heat-labile and do not cross the human placenta, whereas blocking Abs (like IgGs in general) are heat-stable and readily cross the placenta. Most important, *IgE Abs persist at passively prepared (P-K) skin sites for several weeks, whereas blocking Abs diffuse away completely within 1–2 days.* In addition to containing reagins, serum from atopic individuals usually contains some blocking Abs of the same specificity. Until the blocking Abs diffuse away they can competitively inhibit the reaction of injected Ag with reagin at P-K sites. The latent period in the P-K reaction probably also reflects the time required for reagins to bind to tissue receptors (see IgE Receptors on Mast Cells and Basophils).

CUTANEOUS ANAPHYLAXIS IN GUINEA PIGS

Cutaneous anaphylaxis can also be elicited in actively or passively sensitized guinea pigs. **Passive cutaneous anaphylaxis (PCA)** has been especially well developed by Ovary into a powerful model system for evaluating the ability of various Abs and Ags to elicit anaphylactic responses. PCA and the human P-K reaction are fundamentally the same; but special measures are necessary to increase the visibility of the response in animal skin.

In PCA an antiserum (or purified Ab) is injected intradermally. After a latent period of several hours the corresponding Ag is injected intravenously along with a dye, such as Evans blue, that is strongly bound to serum albumin. Hence, as serum proteins rapidly leak into the dermis at the site of the reaction the response appears as an irregular circle of stained skin; the area is an index of the reaction's intensity (Fig. 21-2).

Two kinds of guinea pig Abs produce PCA reactions in guinea pigs: 1) reagins and 2) IgG molecules of the γ1 class. These reagins resemble human IgE: present at trace levels in serum (nanograms per milliliter), they persist at injected skin sites for weeks, and their

TABLE 21-2. Comparison of Human Reagins (IgE) and Blocking Antibodies (Chiefly IgG) To Pollen Antigens*

	Reagins	Blocking antibodies
Immunoglobulin class	IgE	IgG (predominantly)
Activity in Prausnitz–Kustner (P-K) test	Yes	No (inhibits)
Persistence in human skin (P-K test)	Up to about 4 weeks	Up to 2 days
Stability		
To heat (56°C, 4 h)	Labile	Stable
To sulfhydryl reagents (0.1 M 2-mercaptoethanol)	Labile	Stable
Transfer to fetus across human placenta	No	Yes
In vitro assays	Radioimmunoassay	Hemagglutination and others
Molecular weight†	~185,000	~150,000
Sedimentation coefficient ($S^\circ_{20,w}$)	8.2	6.6
Carbohydrate†	12%	3%
Amino acid residues per heavy chain†	~550	~440
Heavy chains	ϵ	γ (predominantly)
Light chains	κ, λ	κ, λ

*Highly purified protein Ags have been isolated from ragweed and grass pollen extracts by ion-exchange chromatography. Several active fractions have been obtained from each extract, and different fractions are active in different persons. As little as 10^{-4} μg of some ragweed fractions evoke specific wheal-and-erythema responses. Fatal anaphylaxis has been known to occur in response to skin tests with small amounts of crude extracts. Indeed, pollen antigen may be as potentially lethal (on a weight basis) for a pollen-sensitive person as botulinus toxin is for humans in general.

†Values for reagins are those for human IgE myeloma protein.

skin-sensitizing activity is destroyed by sulfhydryl compounds and by heat (56°C for 4 h). In contrast, IgG1 molecules are present at much higher serum levels (several milligrams per milliliter), are stable to heat and S-S–splitting reagents, and persist at injected skin sites for only 1-2 days.

Homocytotropic Antibodies. The Abs that mediate anaphylaxis are called **cytotropic** because they bind to mast cells (and also to circulating basophils). Since guinea pig IgE and IgG1 molecules bind to mast cells of the guinea pig, but not of other species, they are **homocytotropic.** (The other major class of guinea pig IgG, called IgG2, is **heterocytotropic**: these Igs can sensitize mouse but not guinea pig skin, because they fortuitously bind with sufficient affinity to receptors on mouse mast cells.)

Many other species (rat, rabbit, dog, mouse) also have two classes of homocytotropic Abs: IgE, which binds persistently to mast cells, and a class of IgG, which binds much less persistently (see IgE Receptors on Mast cells and Basophils, below). In man, only IgE molecules are homocytotropic. (However, human blocking Abs are heterocytotropic: they can bind, fortuitously, to mast cells of phylogenetically distant species—e.g., guinea pig—but not to human cells.) Unlike IgE and IgGs, *Abs of the IgM and IgA classes do not sensitize animals of the same or other species for anaphylactic responses.*

GENERALIZED ANAPHYLAXIS

Systemic anaphylaxis in man is a rare event brought on occasionally in hypersensitive individuals by insect stings (especially bees, wasps, or hornets), or by injection of an Ag, e.g., horse serum or (more commonly at present) penicillin. Because of this hazard patients who receive foreign proteins (e.g., horse antitoxin) or penicillin are first questioned about previous allergic reactions, and sometimes tested for wheal-and-erythema responses: when horse serum is used, greatly diluted serum (1000-fold or more) is injected, for even minute amounts of undiluted serum (e.g., 0.05 ml) can precipitate systemic anaphylaxis. (For penicillin hypersensitivity and skin tests, see Ch. 22.)

The fundamental mechanisms in systemic anaphylaxis are the same as in the cutaneous wheal-and-erythema response, but certain features of the generalized reaction, which has been studied mostly in the guinea pig, are illuminating.

Mode of Administration of Antigen. Anaphylaxis depends not only on the **amount** but also on the **rate** of Ag–Ab complex formation, for the complexes act by causing the release of pharmacologically active mediators that are rapidly degraded (below). Hence intravenous injection of Ag, or its inhalation in aerosols, can provoke fatal shock, whereas responses elicited by subcutaneous and intraperitoneal injections come on more slowly and are less often fatal.

Fixation of Antibodies. Only those Abs that bind to mast cells can mediate passive anaphylaxis. Less Ab is needed if a latent period intervenes between injection of antiserum and of Ag: for example, 180μg of anti–egg albumin rendered guinea pigs uniformly susceptible to fatal shock when they were challenged 48 h later, whereas about 100 times more was required if the Ag was injected immedi-

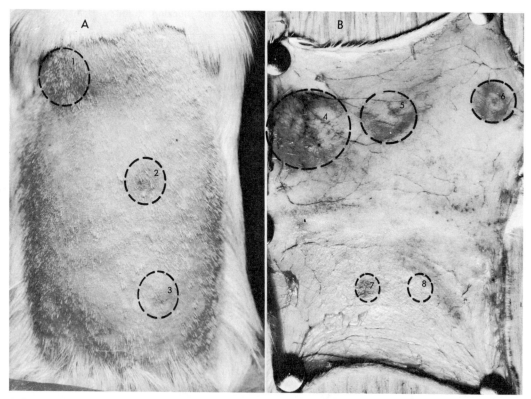

FIG. 21-2. Passive cutaneous anaphylaxis in the guinea pig. In **A** the guinea pig was injected intradermally at three sites with 0.1 ml saline containing **(1)** 100μg rabbit antichicken ovalbumin (EA),**(2)** 10μg anti-EA, and **(3)** buffered saline. Four hours later 1ml containing 2 mg EA and 5 mg Evans blue was injected intravenously; the photo was taken 30 min later. Note blueing at **1** and at **2,** and absence of blueing at the control site **(3)**. In **B** a similar sequence was followed but at 30 min after the intravenous injection of EA the animal was sacrificed and skinned; the photo is of the skin's undersurface. The amount of rabbit anti-EA injected initially was 100μg (at **4**), 10μg (at **5**), 1μg (at **6**), and 0.1μg (at **7**). The control site, which did not turn blue **(8),** had been injected with buffered saline. Another site (not shown) had been injected with 0.01μg anti-EA; it also failed to react.

ately after the antiserum. *During the latent period cytotropic Abs are bound to mast cells; in addition, the circulating level of unbound Ab, which competes with cell-bound Ab for the Ag, is reduced.*

Reverse Passive Anaphylaxis. Passive anaphylaxis can also be evoked by reversing the order of injections if the Ag is itself a foreign Ig of the type that is readily bound to mast cells. In that event if the Ag is injected first, and a latent period is allowed to elapse, the intravenous injection of specific antiserum, (anti-Ig) can cause anaphylaxis. This procedure, reverse passive anaphylaxis, is not effective with other Ags because they do not bind to mast cells. **Reverse passive cutaneous anaphylaxis** is used occasionally to evaluate an Ig's ability to bind to mast cells: e.g., the Ig under test is injected into a normal guinea pig's skin, and then antiserum to the Ig (plus blue dye) is injected intravenously.

Quantities of Antibody and Antigen Required for Anaphylaxis. The levels of Ag required are substantially greater than those necessary for precipitation in vitro: guinea pigs sensitized with 180μg of anti-egg albumin (anti-EA) require for a fatal response over 500μg of EA, about 25-fold more than is usually needed for maximal precipitation of this amount of Ab in the EA/anti-EA

precipitin reaction. Much of the injected Ag probably never has a chance to react with Abs in vivo, because it is taken up by phagocytic cells or excreted. Ags that form large complexes with circulating, soluble Abs tend to be rapidly phagocytized, and are also not efficient in provoking anaphylaxis. In fact, as suggested above, *high levels of circulating Abs may protect against anaphylaxis* because they compete with mast-cell-bound Abs for the Ag. Thus when an animal is passively sensitized with a small amount of antiserum and then given a large dose of the same antiserum immediately before the Ag, fatal shock can be replaced by mild signs (see Desensitization).

IgE RECEPTORS (Fc$_\epsilon$) ON MAST CELLS AND BASOPHILS

Immunofluorescence and electron micrographs (Fig. 21-3) show that IgE binds to the surface of mast cells and basophils. Mast cells are nonmotile connective tissue cells found next to capillaries throughout the body, with especially high concentrations in lungs, skin, gastrointestinal and genitourinary tracts; basophilic polymorphonuclear leukocytes (basophils) are motile white blood

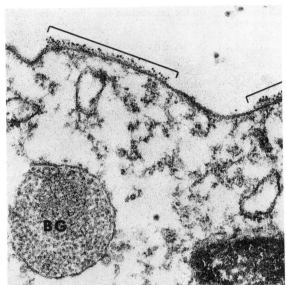

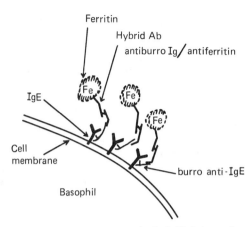

FIG. 21-3. Human IgE on the surface of a human basophil. Washed white blood cells were incubated successively (with intervening washes) with human IgE (a myeloma protein), burro antihuman IgE, hybrid Abs in which one combining site was specific for burro Ig and the other for ferritin, and finally with ferritin–the iron-rich, electron-dense particles seen under **brackets** on the cell surface (see Ferritin and Hybrid Abs; Appendix, Ch. 16). Basophil granules (**BG** is a typical one) resemble the histamine-containing granules of tissue mast cells (Fig. 21-7). **Diagram** shows enlargement of the bracketed region. (Sullivan AL et al: J Exp Med 134:1403, 1971. Electron micrograph, ×77,500)

cells derived from hematopoietic stem cells: they account for about 1% of blood leukocytes, and are the only ones that bind IgE and contain mediators of anaphylaxis.

The mediators in mast cells and basophils are stored in granules that also contain acidic proteoglycans: heparin in mast cells and chondroitin sulfate in basophils. The proteoglycans bind acid dyes and account for the characteristic metachromatic staining properties of these cells.

Human skin sites can be sensitized for P-K reactions with serum having an IgE Ab level as low as $0.2\mu g/ml$ (10^{-12} moles/liter), implying that receptors on mast cells have a high affinity for IgE (Fc domain, see below).

The affinity of IgE for cell receptors has been measured with basophil leukemia cells; the values observed are probably close to those for normal mast cells. Basophils have about 500,000 receptors (R) per cell for IgE, and for the reversible binding reaction

$$IgE + R \rightleftharpoons IgE \cdot R$$

the equilibrium constant (a measure of affinity, Ch. 16) is extremely high, about 10^9–10^{10} liters/mole. This value is the ratio of the association and dissociation rate constants (Rates of Reaction, Ch. 16); and from the measured dissociation rate constant ($10^{-5}sec^{-1}$) it is estimated that about 20 h are required for one half of the bound IgE molecules to dissociate from a basophil or mast cell.* *This re-*

markable stability accounts for the long persistence of skin sensitization in passive cutaneous anaphylaxis (e.g., the P-K reaction). (The less persistent sensitization caused by cytotropic Abs of other classes, such as IgG1, probably means that these other Igs have a lower affinity than IgE for the mast cell and basophil receptor.)

The basophil receptor (130,000 daltons), when isolated from the cell surface membrane, forms 1:1 molar complexes with IgE. The receptors are evidently specific for the Fc_ϵ region, since the Fc_ϵ fragment of IgE molecules, but not Fab_ϵ fragments, can block P-K skin reactions (by competitively displacing intact IgE Abs from skin mast cells).

CROSSLINKING OF IgE RECEPTORS AND MAST CELL ACTIVATION

Anaphylaxis is normally initiated by Ags that, by crosslinking the Abs bound to the Fc_ϵ receptors on mast cells, cause the receptors to aggregate. Thus when a skin site is passively sensitized (with antihapten reagins) multivalent haptens can elicit a wheal-and-erythema response, but univalent haptens are specifically inhibitory, just as they competitively block lattice formation in the precipitin reaction in vitro (Ch. 16).† The following observations emphasize the key role of Fc_ϵ receptor aggregation in anaphylaxis.

---

* If the binding of IgE to the cell surface receptor is a simple bimolecular reaction ($A + B \rightleftharpoons A \cdot B$), the half-time ($t_{1/2}$) and the rate constant for dissociation (K_d) show the following relation:

$$t_{1/2} = \frac{0.693}{K_d}$$

† In the unusual circumstances where certain univalent ligands seem to elicit cutaneous anaphylaxis, they probably aggregate in tissues and become functionally multivalent.

1) The experimental addition of dimers or higher multimers of IgE (previously crosslinked chemically), but not monomeric IgE, causes basophils, in the absence of Ag, to release histamine.

2) Abs produced against the isolated receptors (antireceptor Abs), but not the Fab fragments of these Abs can elicit both cutaneous anaphylaxis in vivo and histamine release in vitro, though the antireceptor Abs are not cytotropic.

3) Wheal-and-erythema skin reactions can be elicited by injecting anti-IgE Abs or their bivalent $F(ab')_2$ fragments, but not their monovalent Fab' fragments. A similar effect is seen with aggregated Fc_ϵ fragments of IgE myeloma protein, but not with the monomeric Fc_ϵ fragment (Fig. 21-4).

Receptors for IgE diffuse independently of each other in the basophil's surface membrane. Thus, when fluorescein-labeled IgE (yellow-green fluorescence) and rhodamine-labeled IgE (red-orange fluorescence) are bound to the same basophil, antifluorescein Abs cause capping of the yellow-green fluorescent molecules, leaving the red-orange fluorescent molecules spread diffusely

on the cell surface (for the capping reaction see Fig. 18-15, Ch. 18).

Serum IgE. Concentrations of IgE in serum are measured by radioimmunoassay: anti-IgE Abs are attached covalently to particles of an inert adsorbent and mixed with a standard amount of ^{125}I-labeled IgE and variable amounts of the human serum to be tested. Unlabeled IgE in the test serum competitively reduces the specific binding of ^{125}I-IgE; hence radioactivity associated with the washed particles decreases in proportion to the serum IgE concentration (Fig. 21-5).

Another assay measures IgE Abs of a particular specificity, e.g., to dog dander (epithelial scales). Protein extracts of the dander are coupled to agarose particles which are then trapped in small cellulose discs. The discs are incubated with about 0.05 ml of a patient's serum, washed, treated with radioactive anti-IgE (labeled with ^{125}I), washed again, and counted. A positive test (adherent radioactivity) can detect a few nanograms per milliliter of IgE Abs of a particular specificity (Fig. 21-5). In the serum of persons with se-

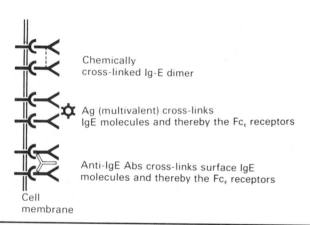

FIG. 21-4. Degranulation of mast cells and basophils and secretion of vasoactive amines induced by crosslinking the cell surface receptors that bind the Fc_ϵ domain of IgE.

Chemically cross-linked Ig-E dimer

Ag (multivalent) cross-links IgE molecules and thereby the Fc_ϵ receptors

Anti-IgE Abs cross-links surface IgE molecules and thereby the Fc_ϵ receptors

Cell membrane

Substance		+ = Degranulation and secretion of vasoactive amine − = No effect (or inhibition)
Multivalent ligand	✿	+
Univalent ligand	⬤	− ⎰ In excess can probably inhibit ⎱ activity of multivalent ligand
Anti-IgE (bivalent)	⅄	+
Anti-IgE F(ab')₂ (bivalent)	⋖	+
Anti-IgE F ab' (univalent)	⅂	− ⎰ In excess can probably inhibit ⎱ activity of bivalent Ab or F (ab')₂
IgE multimers (dimers, etc)		+
IgE monomer		−
Anti-Fc_ϵ receptor Abs		+
Aggregated Fc_ϵ of IgE		+
Monomeric Fc_ϵ of IgE		−

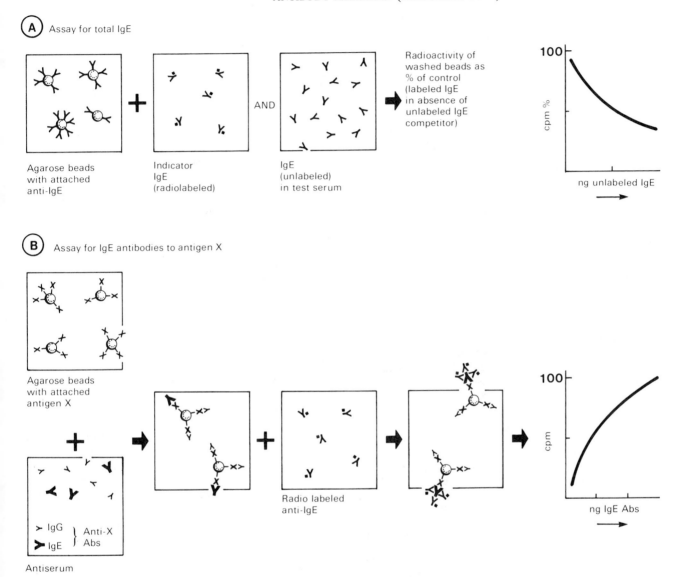

FIG. 21-5. Assays for total IgE concentration in serum **(A)** and for IgE Abs of a particular specificity (anti-X) **(B).** IgE molecules (in **heavy type**) function as Ag in the assay at top, and as both Ag and Ab in the one below.

vere atopic allergies the IgE levels are, on average, increased about two- to three-fold above the normal.

IgE, Intestinal Parasitism, and Eosinophils. While IgEs have obvious pathologic effects, it is likely that they also have beneficial ones, contributing to their evolutionary development. A hint of benefits is suggested by the exceedingly high serum levels of IgE in persons with chronic parasitic infections: values up to 10,000 ng/ml (about 30 times average normal levels) are found in persons with chronic intestinal roundworm infestations, but much of this IgE is not specific for the parasites. As is noted below, the mediators released by mast cells include chemotactic attractants for eosinophilic leukocytes. Since eosinophils can damage helminthic parasites, IgE-

producing plasma cells in the normal intestine could have evolutionary benefits. However, IgE-producers in the respiratory tract (derived embryologically from the fetal gut) contribute to the reactions underlying asthma and hay fever.

REGULATION OF IgE PRODUCTION

IgE-producing plasma cells are not rare. They are detected by immunofluorescence (see Fig. 18-10, Ch. 18) in the mucosa of the intestinal and respiratory tracts and particularly in the mesenteric lymph nodes, where there are at least as many B lymphocytes with surface IgE as

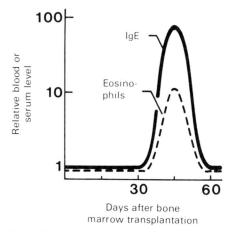

FIG. 21-6. Parallel increases in serum IgE and in blood eosinophils following human bone marrow allografts (Ch. 23). The changes in IgE (and eosinophils) accompany the transient depression of suppressor T cells. (Based on Reinherz EL, et al: N Engl J Med 300:1061, 1979 and Geha R, personal communication)

with surface IgG.* Hence the very low IgE levels in serum (about 1/50,000 of the IgG level) suggest that IgE production is tightly regulated, as expected from the hazards of anaphylaxis. A coherent picture of this regulation is not yet available, but many observations suggest that suppressor T cells are important:

1) IgE production requires T cell help: IgE is not detected in nude (athymic) mice, and T-independent Ags (Ch. 19) do not elicit IgE Abs.

2) It is probable that different T helper (T_h) cells cooperate with IgE-producing (IgE^+) and with IgG-producing (IgG^+) B cells. Thus mice that are primed with Ag adsorbed on alum develop T helper activity for both IgE and IgG Ab production, but in mice given the same Ag in complete Freund's adjuvant the resulting T_h cells help only the production of IgG Abs.

3) IgE Ab production is more susceptible than IgG Ab production to T cell suppression. In mice, for example, an overall reduction of suppressor T cell (T_s) activity (by low doses of x-ray or cyclophosphamide or by thymectomy) results in a greater enhancement of IgE than of IgG Ab responses. Similarly, in humans with acute graft-versus-host reactions (following bone marrow transplants) the transient depression of T_s cell activity is accompanied by an increase in serum IgE (and in blood eosinophils also; Fig. 21-6).

4) Immunization in mice, using tubercle bacilli labeled with 2,-4-dinitrophenyl (Dnp), elicits T cells that block production of anti-Dnp Abs of the IgE class, but not of the IgG class. (Incubation of cultured spleen cells, taken from these mice, with Dnp–protein also yields a soluble factor that blocks the production, by other cultured cells, of anti-Dnp Abs of the IgE class, but not of the IgG class.) Hence there may be different T_s cells for IgE and IgG pro-

---

* IgE-producing plasma cells are also conspicuous in human surgical specimens of tonsils, adenoids, and bronchial and intestinal mucosa; they are rare in spleen and in peripheral lymph nodes. In the respiratory and intestinal tracts IgE-producing cells are greatly outnumbered by IgA producers.

duction, or IgE^+ cells may be more susceptible than the IgG^+ cells to suppression by the same T cells.

5) Low Ag doses (e.g., 0.01μg) in mice can elicit persistent production of IgE Abs; intermediate doses (about 0.1μg) elicit only a short-lived IgE response; and after higher doses (> 1μg) none of the Abs produced are IgE. Since T_s cell activity (usually measured with isolated spleen cells) increases with Ag dose, these observations suggest an unusual sensitivity of IgE production to suppressor T cells.

IgE⁺ B Cells. Though much of IgG is produced by B cells that shift from production of μ to γ chains (Ch. 19), it appears that a special set of IgE^+ precursor B cells is the ultimate source of most serum IgE. Thus mesenteric lymph node cells produce IgE upon polyclonal activation of B lymphocytes with pokeweed mitogen (see Mitogens, Ch. 18), but not if the lymphocytes with surface IgE are first removed (by binding to a solid support with rabbit anti-IgE). However, in humans most IgE^+ B cells also have surface IgM. Whether, after antigenic stimulation, these cells first make IgM Abs and then IgE Abs of the same specificity (as in the IgM–IgG switch [Ch. 19]) is not known.

Correlations Between Experimental and Clinical Observations. Ags that are clinically associated with high IgE Ab responses (and with eosinophilia) are also especially effective elicitors of high IgE Ab responses in mice and rats: e.g., purified protein from ragweed pollen, extracts of *Ascaris* worm, hen's egg albumin. All of these proteins are also particularly effective carriers for haptens, eliciting anti-hapten IgE Abs.

These correlations, and experimental observations on T cell regulation of IgE production, suggest that allergic responses of the anaphylactic type involve not only a set of particular Ags, but also immunization conditions that have a minimal stimulatory effect on T_s cells: i.e., small doses of Ag and an absence of adjuvants, as in inhalation of ragweed pollen (where the Ag dose is estimated at about 1μg per person per year!), in bee stings, etc. By contrast, in the injection schedules used routinely to elicit immunity to tetanus and diphtheria toxins, etc., the use of much higher Ag doses and appropriate adjuvants results primarily in the production of IgG Abs, with little or no IgE and with minimal (though not negligible) danger of anaphylaxis following repeated ("booster") Ag injections.

Genetic Control of IgE Production. As noted before, the tendency to produce IgE Abs to trace amounts of environmental Ags is heritable (see Atopy, above). The production of IgE is controlled genetically at two levels: production of total IgE (an expression of the gene for the constant region of ε chains) and of IgE Abs of particular specificities (expression of V_H genes). Population studies in man suggest that total IgE is regulated by a single locus, not linked to the major histocompatibility (HLA) complex. Inbred mouse strains also differ in their ability to produce IgE Abs (especially in response to small Ag doses, e.g., 0.1μg per mouse), and one strain (SJL) seems unable to form any IgE Abs, while perfectly able to produce Abs of other classes.

Human IgE responses to some purified pollen Ags seem to be

TABLE 21-3. Anaphylaxis in Different Species

Species	Principal site of reaction ("shock organ")	Pharmacologically active agents implicated	Principal manifestations
Guinea pig	Lungs (bronchioles)	Histamine Kinins SRS-A	Respiratory distress; bronchiolar constriction, emphysema
Rabbit	Heart Pulmonary blood vessels	Histamine Serotonin Kinins SRS-A	Obstruction of pulmonary capillaries with leukocyte–platelet thrombi; right-sided heart failure; vascular engorgement of liver and intestines
Rat	Intestines	Serotonin Kinins	Circulatory collapse; increased peristalsis; hemorrhages in intestine and lung
Mouse	?	Serotonin Kinins	Respiratory distress; emphysema; right-sided heart failure; hyperemia of intestine
Dog	Hepatic veins	Histamine Kinins ? Serotonin	Hepatic engorgement; hemorrhages in abdominal and thoracic viscera
Man	Lungs (bronchioles) Larynx	Histamine SRS-A ? Kinins	Dyspnea; hypotension; flushing and itching; circulatory collapse; acute emphysema; laryngeal edema; urticaria on recovery

(Based mostly on Austen KF, Humphrey JH: Adv Immunol 3:1, 1963)

linked to particular major HLA haplotypes, suggesting that immune response genes, analogous to H-2–linked Ir genes in the mouse, may control the ability in man to make IgE Abs of particular specificities.

Species Variations. Guinea pigs are preferred for the study of anaphylaxis because they react uniformly and intensely. However, anaphylaxis can also be elicited in other vertebrates. The manifestations differ in different species (Table 21-3) and even, as noted above, when the Ag is injected by different routes. In a sensitized guinea pig, for example, intravenous injection leads to respiratory distress due to bronchial constriction and at autopsy the lungs appear bloodless and are greatly distended with air, whereas subcutaneous or intraperitoneal administration primarily produces hypotension and hypothermia, and death occurs after many hours, with engorged blood vessels in abdominal viscera as the main pathologic finding. These differences and the different manifestations of anaphylaxis among various species are probably due mostly to differences in the distribution or the reactivity of released pharmacologically active mediators (Table 21-3).

PHARMACOLOGICALLY ACTIVE MEDIATORS

Following Dale's observation, in 1911, that injections of histamine duplicate the manifestations of anaphylaxis, a number of vasoactive substances were found to be released from mast cells by Ab–Ag complexes. The direct action of these substances on blood vessels and smooth muscle accounts for nearly all manifestations of anaphylaxis.

Some of the active substances are **primary mediators:** stored in granules of mast cells (and basophils), they are promptly released when Ag crosslinks the Abs (IgE) and thereby the Fc$_\epsilon$ receptors on these cells (Fig. 21-7). Others are **secondary mediators:** they are synthesized and then released after the surface crosslinking occurs. The main properties of the mediators are summarized below and in Table 21-4.

Primary Mediators. 1) **Histamine,** formed by decarboxylation of L-histidine (Fig. 21-8A), is found in granules of mast cells (about 10^{13} molecules/cell) and blood basophils (and, to a lesser extent, in platelets in some species). Histamine levels decline in tissues and rise in plasma during anaphylaxis. In addition, sensitized animals that are temporarily depleted of histamine by certain drugs ("histamine liberators") are not susceptible to fatal shock until their histamine levels are restored.

Histamine has both inflammatory (or phlogistic) effects (it increases capillary permeability and smooth muscle contraction in small bronchi and intestines, and is a chemotactic attractant for eosinophils) and noninflammatory effects (e.g., it increases gastric secretion of HCl).

These actions are exerted via two surface receptors on many mammalian cells: **H$_1$ receptors** are blocked by the "classic" antihistamines (Fig. 21-8B) and **H$_2$ receptors** are blocked by thiourea derivatives (burimamide and others). Because the inflammatory ef-

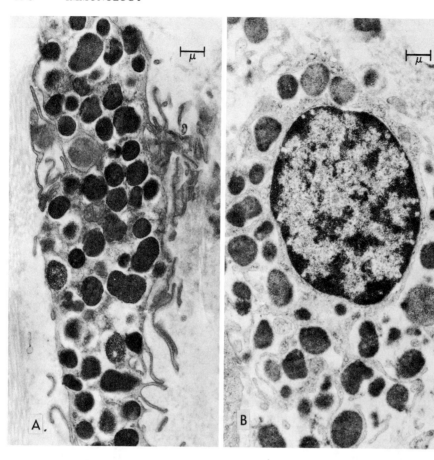

FIG. 21-7. Electron micrographs of mast cells from rat dermis. The intact cell **(A)** contains small, dense granules, each about the size of a mitochondrion. Mitochondria, which are generally scarce in mast cells, are not visible. The nucleus also is not visible in this section. The degranulating cell **(B)** contains larger, paler granules. In the release of granules (associated with secretion of histamine) the membrane surrounding each granule fuses with the cell membrane, releasing swollen granules into the extracellular space (Courtesy of S. L. Clark, Jr.; based on Singleton EM, Clark SL Jr: Lab Invest 14:1744, 1965. ×7000)

fects are exerted via H_1 receptors, the classic antihistamines are useful in the treatment of hives, hay fever, etc. The noninflammatory effects of histamine are exerted via H_2 receptors. Though anti-H_1 drugs block the anaphylaxislike effects of injected histamine, these drugs are less effective against Ag-elicited anaphylaxis because they do not antagonize the other mediators of anaphylaxis. Species vary widely in their susceptibility to histamine; humans and guinea pigs are exquisitely sensitive, whereas mice and rats are insensitive (Table 21-5).

2) **Serotonin (5-hydroxytryptamine, 5HT)** is formed by the hydroxylation and decarboxylation of L-tryptophan (Fig. 21-9). Its distribution in mast cells, basophils, and platelets, and its activity, differ in different species. In mice and rats 5HT is stored in mast cell and basophil granules, from which it is released by the standard IgE mechanism. In humans, however, 5HT is present in platelets, rather than in mast cells; and one of the primary mediators released from basophils, and perhaps mast cells, by the IgE mechanism is a platelet-activating factor (below) that causes human platelets to release their stored 5HT. These species also differ in their sensitivity to 5HT: mice and rats are extremely sensitive and humans are resistant (the reverse of histamine sensitivity; Table 21-5). Thus 5HT makes little if any contribution to the development of anaphylaxis in man; and 5HT antagonists (e.g., reserpine, lysergic acid diethylamide) have no value in its therapy. 5HT is abundant in brain and in intestinal chromaffin cells, but it is not released from these sources during anaphylaxis.

3) **Eosinophil chemotactic factor of anaphylaxis** (ECF-A). Eosinophilic leukocytes are often conspicuous at tissue sites of recurrent IgE-mediated reactions because granules of mast cells, and perhaps basophils, contain polypeptides that, when secreted, are preferentially chemotactic attractants for eosinophils (Fig. 21-10). The best characterized attractants are the two tetrapeptides called **ECF-A:**

Val-Gly-Ser-Glu
Ala-Gly-Ser-Glu

In addition to stimulating chemotaxis, the ECF-A tetrapeptides enhance the expression of complement receptors for C3b on eosinophils; they also augment the Ab- and C- dependent damage to some parasites (schistosomula) by eosinophils. Larger eosinophil-attracting polypeptides (1500–3000 daltons) are also released by activated mast cells. (A protein of much higher mol wt [750,000] from mast cells attracts neutrophilic leukocytes.)

4) **Heparin.** This intracellular proteoglycan is located in granules of mast cells and perhaps basophils. A polyanion, it is responsible for the characteristic metachromatic staining of granules with some cationic dyes (e.g., toluidine blue). Heparin can block the complement cascade (Ch. 20) at various levels (e.g., the binding of C1q to Ab-Ag complexes) and it inhibits both coagulation and fibrinolysis.

TABLE 21-4. Mediators of Anaphylactic Reactions

Category	Mediator	Structural properties	Assays	Other activities
I. Primary (stored in and released with granules of mast cells and basophils)	Histamine	β-imidazoleth-anolamine	Contraction of guinea pig ilieum; isotopic labeling with histamine N-methyl transferase	Increases vascular permeability; chemokinesis and chemotactic attractant for eosinophils
	Serotonin	5-hydroxytrypt-amine (5HT)	Contraction of guinea pig ileum and rat uterus; inhibition by lysergic acid	Increases vascular permeability
	Eosinophil chemotactic factor of anaphylaxis (ECF-A)	Val-Gly-Ser-Glu; Ala-Gly-Ser-Glu	Chemotactic attraction of eosinophils (and neutrophils)	Attractant and deactivator of eosinophils
	Intermediate-mol-wt eosinophil chemotactic factor	Peptides; mol wt 1500–3000	Chemotactic attracion of eosinophils (and neutrophils)	Attractant and deactivator of eosinophils
	Neutrophil chemotactic factor (NCF, high mol wt)	mol wt 750,000	Chemotactic attraction of neutrophils	Neutrophil deactivation
	Heparin	Acidic proteoglycan	Activation of anti-thrombin	Anticoagulant
	Chymase	Protein; mol wt 25,000	Hydrolysis of amino acid esters	Proteolysis
	N-acetyl-β-glucosaminase	Protein; mol wt 150,000	Substrate hydrolysis	?
	Arylsulfatase A	Protein; mol wt 115,000	Hydrolysis of p-nitrocatechol sulfate	Inactivates SRS-A
II. Secondary (not stored; formed and secreted after mast cells and basophils are activated)	SRS-A	Leukotriene C (see text)	Contraction of guinea pig ileum, not blocked by anti-histamines	Contracts bron-chioles, increases capillary permeability
	Platelet-activating factors (PAFs)	Neutral and polar lipids	Release of ^{14}C-serotonin from platelets	Aggregates and lyses platelets
	Lipidlike products of arachidonate metabolism	?	Chemokinetic and chemotactic activity on neutrophils	Attracts neutrophils
	Bradykinin	Peptide	Radioimmunoassay	Contracts bron-chioles; vasodilation; hypotension

5) Several **enzymes** are also released with mast cell granules. **Chymase,** a chymotrypsinlike proteolytic enzyme, has many basic amino acids. It is released slowly from the stable electrostatic complexes it forms with heparin (acidic) in the granules. **N-acetyl-β-glucosaminidase** is released rapidly along with histamine. It cleaves glycosaminoglycans of connective tissues and also of some cell surface glycoproteins. Its role in allergy is not known.

Arylsulfatases inactivate slow reactive substance of anaphylaxis (SRS-A), a potent lipid mediator of anaphylaxis (see Secondary Mediators, below). There are several types of arylsulfatase, differing in some physical properties and in substrate specificity: sulfate is cleaved from p-nitrocatecholsulfate (I) by types A and B, and from p-nitrophenylsulfate (II) by type C.

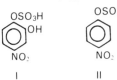

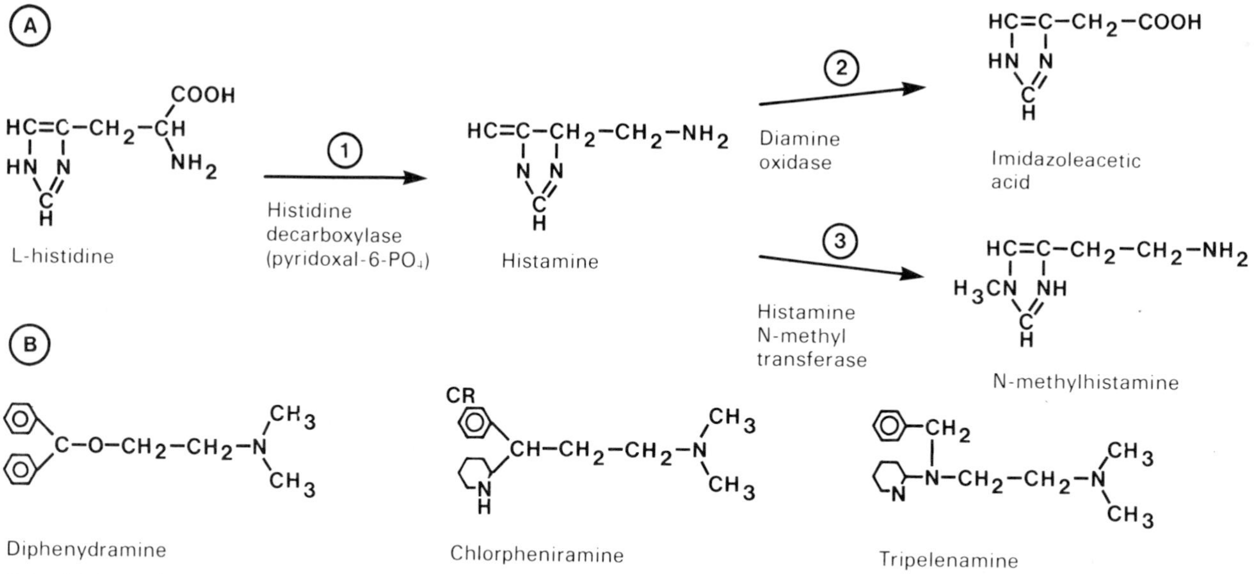

FIG. 21-8. A. Formation of histamine from L-histidine by histidine decarboxylase (with pyridoxal-6-phosphate as cofactor), an enzyme in mast cells and basophils. Once secreted, histamine is rapidly degraded to inactive derivatives by histaminase (diamine oxidase) and histamine N-methyl transferase. **B.** Some antihistamines that block H_1 receptors for histamine and are useful in the treatment of allergic reactions mediated by IgE Abs.

The type A enzyme is released rapidly along with histamine and probably is derived from mast cell granules. Type B is released by eosinophils, and is also present in basophils.

Secondary Mediators. 1) **Slow Reacting Substance of Anaphylaxis (SRS-A)** is a powerful constrictor of human bronchioles and is probably the most important mediator of asthma. It was first found after the addition of Ag to lung fragments from sensitized guinea pigs and humans: the liberation of histamine is complete within 5 min, but SRS-A is not detected for several minutes and its concentration then rises slowly. SRS-A is thus a secondary mediator, rather than one of the stored, preformed, primary mediators. It is identified by its ability to cause slow constriction of isolated human bronchioles and guinea pig ileum (Fig. 21-11) and by the failure of antihistamines and serotonin antagonists to block this activity.

An acidic lipid, SRS-A has a novel structure. Recently called **Leukotriene C**, it is a noncyclized C_{20} carboxylic acid with an S-ether-linked cysteine group, an OH substituent, and three conjugated double bonds. It is one of a group of **leukotrienes**, which are derived metabolically (by the lipoxygenase pathway) from **arachidonic acid.**, a S-glutathione derivative of which is shown on p. 481 in the form of leukotriene C (SRS-A) (based on Samuelsson, et al.).

2) **Platelet-Activating Factors (PAFs).** Platelets and their fragments are prominent at tissue sites of immediate-type allergic reactions. Their involvement in anaphylaxis has been clarified by the finding that PAFs aggregate platelets and stimulate them to release preformed serotonin (synthesized and stored in platelets).

The predominant PAFs are unique phospholipids, generated by, and released from basophils and mast cells. Besides aggregating platelets and stimulating their release of serotonin, PAFs also stimulate an increased level of cyclic GMP in platelets. Another PAF, a neutral lipid derived by immunologic challenge of mast-cell rich rat peritoneal cells, releases serotonin by lysing platelets.

3) **Other unstored lipid mediators** released from mast cells affect the motility of polymorphonuclear leukocytes (PMNs). One, a neutral lipid, has a chemokinetic effect (see Eosinophils, below) on the PMNs. The other, a polar lipid, has chemotactic activity; its formation is inhibited by indomethacin, suggesting that it is derived from arachidonic acid or a related unsaturated fatty acid (via the cyclooxygenase pathway).

4) **Kinins. Bradykinin,** a basic nonapeptide

Arg-Pro-Pro-Gly-Phe-Ser-Pro-Phe-Arg

becomes evident in the blood during anaphylaxis. Injected into

TABLE 21-5. Species Variation in Tissue Levels and Susceptibility to Histamine and Serotonin

Species	Lung content (μg/gm)		Bronchiolar sensitivity (minimal effective dose in μg)	
	Serotonin	Histamine	Serotonin	Histamine
Cat	>0.2	34	0.01	2
Rat	2.3	5	0.01	>5
Dog	<0.1	25	0.05	0.3
Guinea pig	<0.2	5–25	0.4	0.4
Rabbit	2.1	4	>8	0.5
Man	<0.3	2–20	>20	0.2

(From various sources summarized in Austen KF, Humphrey JH: Adv Immunol 3:1, 1963)

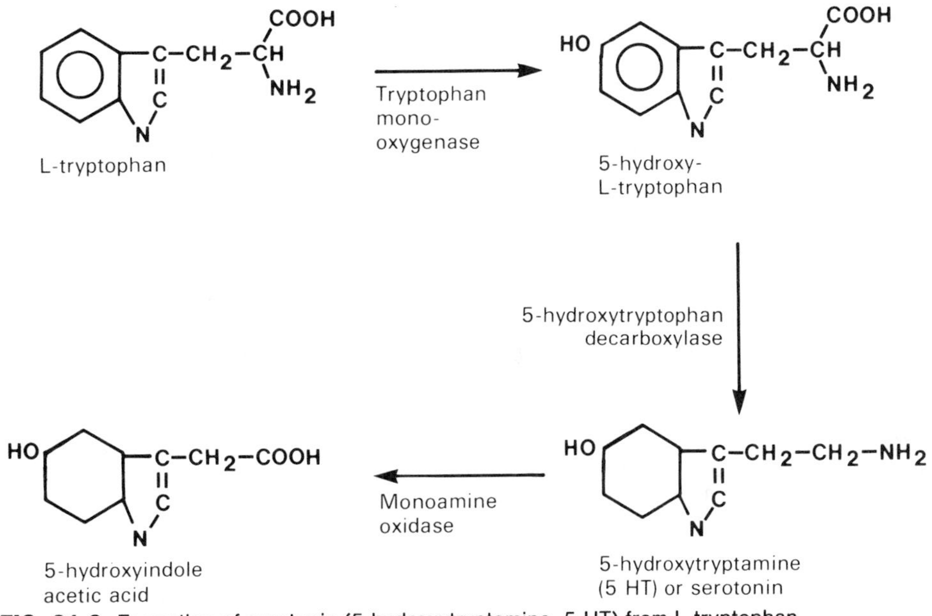

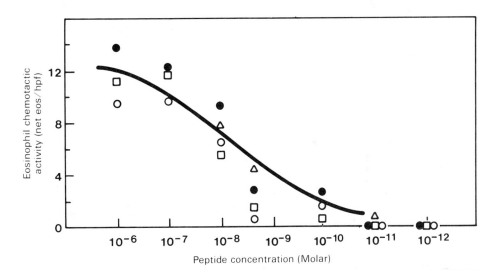

FIG. 21-9. Formation of serotonin (5-hydroxytryptamine, 5-HT) from L-tryptophan.

FIG. 21-10. Dose–response curve of native eosinophil chemotactic factor of anaphylaxis (ECF-A) isolated from human lung **(squares)** and released from human lung by anti-IgE Abs **(triangles),** compared to the synthetic tetrapeptides Ala-Gly-Ser-Glu **(open circles)** and Val-Gly-Ser-Glu **(solid circles).** ECF-A has considerable activity at 10^{-8} M; it is more chemotactic for eosinophils than for neutrophils and has essentially no attractive effect on macrophages. (Based on Goetzl EJ: Am J Path 85:42, 1976)

Leukotriene C
(SRS-A)

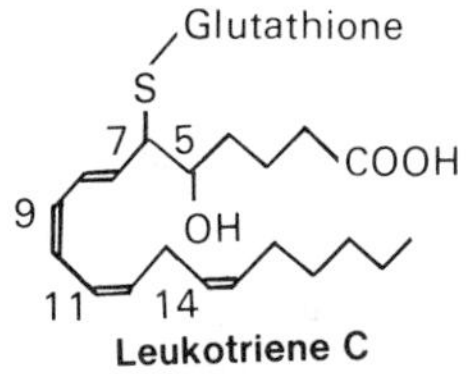

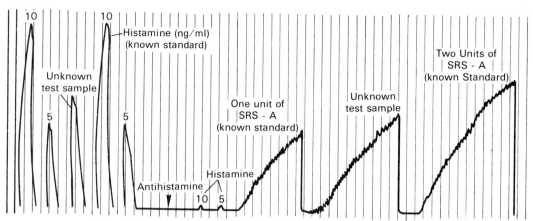

FIG. 21-11. Assays for histamine and SRS-A (slow reactive substance of anaphylaxis). Standards and test samples were added to an isolated strip of guinea pig ileum, whose contractile response was recorded by kymograph. Antihistamines block response to histamine. Methysergide can be added to block serotonin. Note the slow response to SRS-A and the faster response to histamine. Time scale is about 30 sec (vertical markers). (Based on Orange RP, Austen KF: In Good RA, Fisher DW (eds) Immunobiology. Stamford CI, Sinauer, 1971)

normal animals it causes vasodilation, contraction of smooth muscles, and increased capillary permeability. It is formed as the end-product of the sequential action of several plasma proteins that are involved in blood clotting and that can be activated by some Ab–Ag complexes. One component of the sequence, plasmin (a trypsinlike protease), cleaves from another component a fragment that converts a plasma proenzyme, **prekallikrein,** into the proteolytic enzyme kallikrein,* which cleaves bradykinin from a plasma globulin.

Tissue kallikreins form the same peptide with a lysine or a methionyllysine at the N terminus; aminopeptidases in blood remove the additional N-terminal residues, converting these kinins into bradykinin. Kallikrein also splits another bradykininlike peptide from complement; called **C-kinin,** the fragment has activities somewhat similar to those of bradykinin but differs in being susceptible to destruction by trypsin. Several mechanisms prevent the accumulation of kinins, which are potent hypotensive agents as well as inducers of bronchiolar constriction. Their activities are abolished, for instance, by 1) a plasma carboxypeptidase that removes the C-terminal arginine, and 2) an inhibitor of activated first component of complement (Ch. 20).

Release of Mediators. The aggregation of Fc_ϵ receptors on mast cells (and basophils) initiates an extensive set of reactions, culminating in secretion of the mediators. Only a few steps have been identified: 1) Initially, there is a rapid influx of Ca^{2+} ions; 2) from the inhibitory effect of isofluorphate (diisopropylphosphorofluoridate, DFP) it is inferred that in one of the early steps a critical serine esterase–type protease is activated; 3) intracellular cyclic AMP (cAMP) levels drop; 4) mediator-laden granules migrate to the cell's surface, fuse with the surface membrane and are discharged to the cell's exterior **(exocytosis).**

The discharge is due to secretion, not to cell lysis. Thus the releasing cells remain impermeable to ionic dyes (which penetrate

* Kallikrein was discovered by mixing tissue extracts, a source of the enzyme, with plasma which contains the substrate. The enzyme was named for its abundance in pancreas (Gr. *kallikreas,* pancreas); the product, bradykinin, was named for the slowness of the contraction it induces in isolated guinea pig ileum.

only into cells with damaged surface membranes); the cells are thought to survive and to slowly form new granules. The discharged granules promptly leak their stored histamine, serotonin (if present), ECF-A, and arylsulfatase A; heparin and chymase, combined in electrostatic complexes in the granules, first dissociate, and then diffuse away more slowly.

The level of cAMP in mast cells is crucial: as with many other secretory processes, *the rate and extent of granule release are enhanced when intracellular cAMP levels are low and are reduced when the levels are high.*

Drugs that either stimulate adenylcyclase, the enzyme that synthesizes cAMP, or inhibit the phosphodiesterase that degrades it, result in increased cAMP levels and block the relase of histamine and SRS-A (isoproterenol, epinephrine, aminophylline; Fig. 21-12). These drugs were used clinically for the control of anaphylaxis and allergic bronchospasm for many years before their mechanism of action was suspected. Another drug (cromolyn) is thought to prevent asthma by directly inhibiting granule secretion, possibly by enhancing phosphorylation of a particular protein in most cells.

Degradation of Mediators. Because the pharmacologically active mediators are rapidly degraded (Table 21-6) or excreted, they act only transiently. Their failure to accumulate, and their slow synthesis, account for the efficacy of repeated, closely spaced injections of small doses of Ag in depleting tissues of stored mediators (see Desensitization).

EOSINOPHILS

As noted above, eosinophils accumulate at tissue sites of recurrent IgE-mediated reactions (nasal and bronchial mucosa in respiratory allergies, intestinal mucosa in cer-

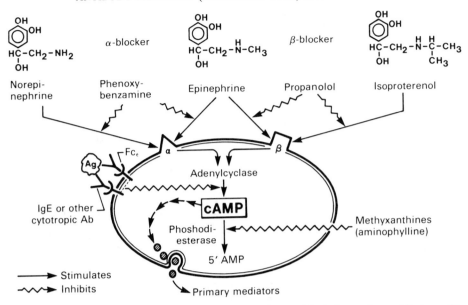

FIG. 21-12. Regulatory effect of the cAMP system on the secretion of granules and their mediators (histamine, serotonin, etc.) from mast cells and basophils. The therapeutic benefits of epinephrine, norepinephrine, and isoproterenol, and the worsening effects of propanolol on asthma and other acute allergic reactions are due also to effects on the cAMP system of other cells (e.g., mucin-secreting cells of bronchi), not just on mast cells.

tain helminthic infestations, etc.). Enzymes released by eosinophils degrade histamine and SRS-A (Table 21-6); hence these cells help to control the pathologic effects of these allergic reactions.

The factors that affect the migration of eosinophils and other leukocytes are detected with simple **Boyden chambers.** Each chamber is divided into two compartments by a filter whose pores allow the passage of macromolecules but not cells: by placing the cells in one compartment and the factor under test in the other, or in both compartments, and then counting the number of cells that migrate into and become trapped in the filter it is possible to

recognize both **chemotaxis** (stimulated cell movement along a concentration gradient) and **chemokinesis** (stimulated cell movement without regard to a concentration gradient).

Incubation of motile cells with some factors causes the cells to become **deactivated,** i.e., rendered unresponsive to subsequent exposure to the same or other chemotactic factors. ECF-A from mast cell granules (Fig. 21-10) can also deactivate eosinophils, suggesting that it attracts eosinophils to its site of release from mast cells and then deactivates the immigrant cells, causing their local accumulation.

ANAPHYLACTIC RESPONSES IN ISOLATED TISSUES

Many organs from sensitized animals respond to Ag in vitro. In the **Schultz–Dale reaction** the isolated uterus from a sensitized guinea pig contracts promptly when incubated with Ag, which evidently reacts with cytotropic Abs on tissue mast cells and causes the release of mediators (Fig. 21-13). Similar reactions are obtained with isolated segments of ileum, gallbladder wall, and arterial wall. These responses can also be elicited with tissues from passively sensitized animals, and with isolated normal tissues that are sensitized simply by incubation with antiserum. Because of the high affinity of IgE Abs for the mast cell surface, the isolated tissues retain their reactivity after extensive washing.

In one of the simplest in vitro reactions Ag elicits the release of histamine from washed leukocytes and the degranulation of basophils (demonstrated by staining smeared cells); extremely small amounts of Ag suffice (e.g., for purified ragweed Ag, 10^{-13} mg/ml). The degranulation has been used as a diagnostic assay for penicillin

TABLE 21-6. Degradation of Mediators

Mediators	Degrading enzymes*	Inactive derivatives
Histamine	Histamine (diamine oxidase)	Imidazoleacetic acid (Fig. 21-8A)
	Histamine *N*-methyl transferase	*N*-methylhistamine (Fig. 21-8A)
Serotonin	Monoamine oxidase	5-Hydroxyindole acetic acid (Fig. 21-9)
SRS-A	Arylsulfatase, types A and B	?
	Phospholipase, type D	?
ECF-A	Peptidases	Amino acids; smaller peptides
Kinins	Peptidases	

*Some are in eosinophils; others are in mast cells or basophils.

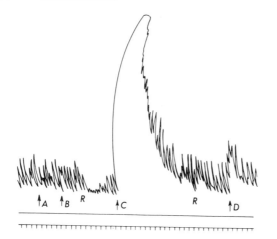

FIG. 21-13. Smooth muscle contraction in vitro in response to Ag (Schultz–Dale reaction). A uterine horn was excised from a guinea pig 13 days after a sensitizing injection of a horse serum euglobulin, and was suspended in Ringer's solution. Various protein fractions from horse serum were added **(arrows):** at **A,** 1 mg pseudoglobulin; at **B,** 10 mg pseudoglobulin; at **C** and at **D,** 10 mg euglobulin (the immunogen). Following the specific response at **C** the muscle was almost totally desensitized, either because the tissue-bound Abs were saturated with Ag or because the mediator content was depleted. Time scale markers at 30-sec intervals. **R** = changes of Ringer's solution. (Dale HH, Hartley P: Biochem J 10:408, 1916)

allergy (see Allergy to Drugs, Ch. 22). Human basophils can also be passively sensitized with atopic sera and with purified human IgE.

AGGREGATE AND CYTOTOXIC ANAPHYLAXIS

Some experimental forms of anaphylaxis can be elicited with Abs that do not bind to mast cells or basophils. With these Abs (unlike IgE and cytotropic IgG classes) no latent period for sensitization is required following their passive transfer. Instead, the transferred Abs cause **aggregate anaphylaxis,** in which relatively large amounts of soluble Ab–Ag complexes react with complement (C), and the resulting fragments C3a and C5a (Ch. 20) serve as **anaphylaxtoxins** that cause mast cells to release their granules and mediators. Thus in a normal guinea pig a single intradermal injection of such complexes, prepared by dissolving a specific precipitate in a concentrated solution of Ag, can evoke passive cutaneous anaphylaxis; and fatal shock can follow intravenous injection of antiserum that has been incubated for a few minutes with soluble Ag.[*]

Even some heat-aggregated Igs, without any Ag, can elicit cutaneous anaphylaxis, showing that the essential role of the Ag is simply to cross link certain classes of Ab molecules and trigger the C

[*] Normal serum becomes similarly toxic after incubation with suspensions of various particles (kaolin, talc, barium sulfate, inulin, agar), which apparently activate the alternate C pathway, again with formation of anaphylatoxin (Ch. 20). The response to these incubated sera (without Ag) is sometimes called **anaphylactoid** shock.

cascade. The only effective soluble immune complexes are those that fix C (e.g., with average molar composition of about Ag_3Ab_2; Ch. 20).

Another form of acute allergic reaction, called **cytotoxic anaphylaxis,** sometimes follows the injection of Abs to natural constituents of cell surfaces (prototype III or English type II, Table 21-1). For example, guinea pigs injected with rabbit Abs to the Forssman Ag, a constituent of all guinea pig cells, undergo acute shock. Acute hemolytic **transfusion reactions** in man (Ch. 23) are also sometimes associated with shock and could be considered a form of cytotoxic anaphylaxis.

COMPLEMENT AND ANAPHYLAXIS

Complement activation is not necessary for cytotropic responses: e.g., cutaneous anaphylaxis can be elicited in animals depleted of C3 by cobra venom factor (see Ch. 20). However, C fixation is essential for aggregate anaphylaxis.

DESENSITIZATION

Various strategies are used to prevent or treat allergic reactions in humans due to IgE Abs.

Prolonged Desensitization. Atopic individuals form blocking (i.e., nonhomocytotropic) Abs, as well as IgE Abs, especially after repeated injections of the Ag. Accordingly, such persons are commonly treated **(desensitized)** by repeated injections of small, increasing amounts of allergen, in doses and at intervals (e.g., weekly) that avoid systemic anaphylactic reactions. The level of blocking Abs, but not of IgE Abs, often rises considerably. However, therapeutic benefits are not always evident or regularly correlated with the titers of blocking Abs. (Blocking Abs can be measured by heating atopic sera [56 °C, 4 h] to inactivate IgE Abs.)

Acute Desensitization. Because the mediators are rapidly degraded and do not accumulate, the speed at which they are formed (determined by the rate of reaction between Ag and mast-cell–bound Abs) determines whether anaphylaxis will occur. Shock can therefore be prevented by administering Ag slowly; e.g., if $100\mu g$ of a particular Ag (injected intravenously) is required to provoke fatal shock experimentally, the same quantity given in ten divided doses at 15-min intervals would not elicit shock. Moreover, if the full dose were then given all at once, shortly after the last small injection, shock would probably still not be elicited, because the level of reactive Abs and mast cells would have been depleted.

Desensitization by repeated, closely spaced injections of small doses of Ag is thus often resorted to clinically when it becomes necessary to administer a substance, such as penicillin or horse antiserum, to a person known or suspected to be intensely allergic to it. The procedure is effective but requires great care to avoid anaphylaxis, and it has only temporary value. Several weeks afterward hypersensitivity is likely to be fully restored (in contrast to the desensitization based on the formation of protective [blocking] Abs).

T Suppressor Cells. The susceptibility of IgE production to suppression by T_s cells suggests that the selective stimulation of these cells may eventually offer a feasible approach to treatment and prevention of IgE-mediated hypersensitivity.

IMMUNE COMPLEXES AND SUBACUTE HYPERSENSITIVITY

ARTHUS REACTION

Shortly after the discovery of anaphylaxis, Arthus, a French physiologist, described a different kind of Ab-dependent allergic reaction. When rabbits were inoculated subcutaneously each week with horse serum there was at first no noticeable response, but after several weeks each injection evoked a localized inflammatory reaction. Similar responses were soon described in man and were called **Arthus reactions.** These reactions are not limited to the skin: they can take place when Ags are injected almost anywhere, e.g., into the pericardial sac or synovial joint spaces. *The principal requirement is the formation in tissues of immune aggregates that fix C; the resulting C fragments (C5a, etc.) attract polymorphonuclear leukocytes (PMNs, i.e. neutrophils). Their released lysosomal enzymes cause tissue damage, characteristically with destructive inflammation of small blood vessels (vasculitis).* Providing they can react with C in situ, Abs of any Ig class (and from any species) can mediate the reaction. Even the injection of Ab–Ag complexes formed in the test tube can evoke the response, though with less intensity than when the aggregates form in situ.

Patients with serum sickness or with certain forms of glomerulonephritis (below) develop similar lesions in small blood vessels and in kidney glomeruli, respectively. Similarly, persons with high serum levels of Abs to the thermophilic *Aspergillus* that thrives in decaying vegetation, or to molds used to produce cheese, develop severe localized Arthus-type lung lesions when they inhale these fungi or fungal spores (farmer's lung, cheese-maker's lung, and other examples, Table 21-7).

The main features are illustrated by the **passive cutaneous form of the Arthus reaction,** in which an antiserum is first injected intravenously into a nonsensitive recipient and the corresponding Ag is then injected into the skin. (Alternatively, to conserve antiserum, in the **reverse passive Arthus reaction** the antiserum is injected in the recipient's skin and the Ag is then injected into the same dermal site or intravenously.)

TABLE 21-7. Subacute Hypersensitivity in Lungs (Allergic Pneumonitis) Due to Inhaled Antigen

Occupationally related disease	Antigen
Farmer's lung	Moldy hay (*Thermoactinomyces vulgaris*)
Mushroom worker's lung	Compost (thermophilic actinomycetes)
Pigeon breeder's lung	Pigeon dander and droppings (avian proteins)
Bagassosis	Moldy bagasse (thermophilic actinomycetes)
Malt worker's lung	Moldy barley (*Aspergillus clavatus*)
Wheat weevil disease	Infested flour (*Sitophilus granarius*)
Sequoiosis	Moldy sawdust (*Coraphium* and *Pullularia* species)
Cheese washer's lung	Cheese casings (*Penicillin caseii* spores)

(Based on Barrett JT: Textbook of Immunology, 3rd ed. St. Louis, Mosby, 1978)

The passive Arthus reaction requires a large amount of Ab, about 10 mg when injected in a rabbit intravenously and about 100μg when injected into the skin. In contrast, about 100,000-fold less (i.e., 1 ng or less) of IgE Ab is sufficient for passive cutaneous anaphylaxis in humans (Prausnitz–Küstner reaction; see Passive Transfer, above).

Time Course. After intradermal injection of Ag the Arthus response becomes evident more slowly than cutaneous anaphylaxis and is more persistent. Local swelling and erythema appear after 1–2 h, followed by punctate hemorrhages. The changes are maximal in 3–4 h and are usually gone in 10–12 h; but severe reactions, with necrosis at the test site, subside more slowly (Table 21-8).

Histopathology. In anaphylaxis the inflammatory changes are largely limited to vasodilatation and exudation of plasma proteins; inflammatory cells are not conspicuous. The Arthus response, however, is characterized by classic inflammation: blood flow through small vessels is markedly retarded; thrombi rich in platelets and leukocytes form within small blood vessels, erythrocytes escape into the surrounding connective tissue, and after several hours the skin site becomes edematous and heavily infiltrated with PMNs (Fig. 21-14). Finally, localized patches of necrosis appear in the walls of the affected small blood vessels. As the lesion begins to subside, after 4–12 h, neutrophils become necrotic and are replaced by macrophages and eosinophils. Within a few days, the phagocytized immune complexes are degraded and inflammation disappears.

The response in the cornea emphasizes the role of blood vessels. The injection of Ag into an immunized rabbit's normal cornea, which is devoid of functional blood vessels, can result in concentric

TABLE 21-8. Comparison of Immediate and Subacute Allergic Skin Reactions

Properties	Immediate (anaphylactic) reactions	Subacute (immune-complex) reactions
Time course		
Onset	2–3 min	1–2 h
Peak	10 min	3–4 h
Disappearance	30 min	10–12 h
Class of mediating Abs	IgE (or other cytotropic Igs)	IgG or any other complement-fixing Ig
Amount of Ab necessary for passive sensitization of skin site	$<0.001 \mu g$	$>10 \mu g$
Latent period required between Ab and Ag injections	Yes	No
Complement fixation required	No	Yes
Major mediators	Histamine, serotonin, etc.	Lysosomal enzymes
Major cells involved	Mast cells, basophils, platelets (eosinophils)	Neutrophils (minor role for basophils, platelets)

opaque rings of Ab–Ag precipitates, like bands in gel precipitin reactions in vitro (Ch. 16), but little or no inflammation is observed. If, however, functional blood vessels are present (e.g., as a sequel of some earlier trauma to the cornea), then the Ag can elicit an Arthus response in the cornea, as in any other tissue.

Role of Complement and Granulocytes. At the site of the local reaction immunofluorescence reveals Ab–Ag aggregates with C components (C3) localized in blood vessel walls, between endothelial cells and the internal elastic membrane (Fig. 21-14). The aggregates are also evident within granulocytes in perivascular connective tissue. If an animal's C activity has been greatly reduced (e.g., by depleting C3 with cobra venom factor, Ch. 20), or if its level of circulating PMNs has been depressed (e.g., by an anti-PMN serum) no inflammatory reaction appears, even though the immune complexes form in blood vessel walls.

It appears therefore that the Arthus reaction depends on the following sequence: 1) Ag and Ab diffuse into blood vessel walls, where they combine, forming complexes, and fix C; 2) some C fragments and complexes (C5a, C5b,6,7) are chemotactic attractants for neutrophils, and other fragments (the anaphylatoxins C3a and C5a) probably cause the release of other leukocyte attractants from mast cells and basophils; 3) the accumulated neutrophils ingest the Ab–Ag complexes and release lysosomal enzymes; 4) the enzymes cause focal necrosis of the blood vessel wall and the other inflammatory changes. Since small peptides are released from radiolabeled Ags it is evident that lysosomal enzymes also degrade the immune complexes, leading to subsidence of the inflammation.

Increased permeability of the blood vessel endothelium (due to histamine and serotonin) probably aids the penetration of Ab–Ag complexes into blood vessel walls

(see Serum Sickness Syndrome, below), but antihistamines do not block the development of Arthus lesions.

SERUM SICKNESS SYNDROME

From about 1920 to 1940 various human bacterial infections were treated routinely by injecting large volumes of antiserum prepared in horses or rabbits. The recipients often developed, after a week or so, a characteristic syndrome called **serum sickness.** Heterologous antisera are now used only infrequently in medicine (e.g., duck antisera to rabies virus, to prevent or minimize the disease, antilymphocyte serum: horse antiserum to human lymphocytes, used to avoid rejection of kidney allografts; Ch. 23), but the same syndrome is commonly encountered as an allergic reaction to penicillin and other drugs (Ch. 22).

The syndrome includes 1) fever, 2) enlarged lymph nodes and spleen, 3) erythematous and urticarial rashes, and 4) arthritis. The disease usually subsides within a few days. In the few patients who have died at the height of the illness, autopsy has disclosed disseminated vascular and perivascular inflammatory lesions like those of the Arthus reaction.

The mechanisms have been analyzed in rabbits injected with large amounts of purified foreign protein. In addition, an opportunity to make detailed observations in humans arose in connection with attempts, during World War II, to use bovine serum albumin (BSA) as a plasma expander in the treatment of traumatic shock (Fig. 21-15).

Mechanisms. The illness usually becomes evident 7–14 days after the initial injection of Ag. During this interval the Ag level declines, but it is still high enough, after Ab production starts, to form the soluble Ab–Ag complexes

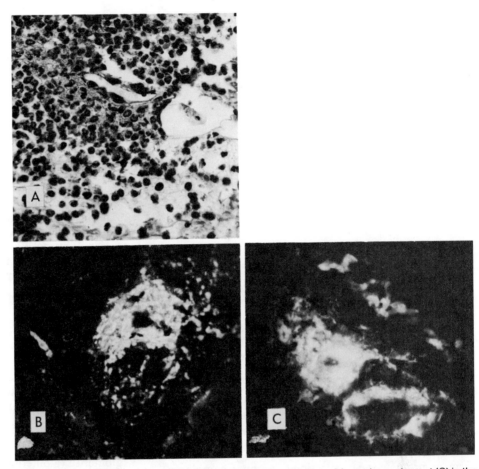

FIG. 21-14. Passive Arthus reaction in a rat, showing localization of Ag and complement (C) in the wall of an affected blood vessel. The skin site was excised 2–3 h after an intradermal injection of 300μg of rabbit Abs to bovine serum albumin (anti-BSA) and an intravenous injection of 6 mg BSA. (The Ab was injected intradermally, and the Ag intravenously, to conserve Abs.) **A.** Note intense neutrophil infiltration in and around the wall of a small blood vessel adjacent to skeletal muscle. **B.** Section was stained with fluorescent rabbit Ab to a purified component of rat C (C3; see Ch. 20). **C.** Section was stained with fluorescent anti-BSA to localize the aggregated Ag in the blood vessel wall and in the adjacent perivascular connective tissue. The same result would be obtained by staining the aggregated Ab (rabbit anti-BSA) with fluorescent anti–rabbit Ig. (Ward PA, Cochrane CG: J Exp Med 121:215 1965)

(in Ag excess) that initiate focal vascular lesions (in coronary arteries, glomeruli, etc.; Fig. 21-16). Serum sickness is thus usually observed only after exceptionally large amounts of foreign protein are injected, e.g., 25 g BSA in man, or 1 g in a rabbit. However, in a previously sensitized person, with an accelerated (anamnestic) Ab response (see Primary and Secondary Responses, Ch. 19), the reaction appears earlier and therefore requires much less Ag: e.g., 3 or 4 days after 1 ml of horse serum.

As the manifestations of serum sickness appear, soluble Ab–Ag complexes can be detected in the serum, and the decline in the level of free Ag is markedly accelerated (Figs. 21-15 and 21-17). Moreover, the Ab–Ag complexes fix C, and the serum C level is depressed at the height of the illness (Fig. 21-17); as in the Arthus reaction the most abundant C component (C3) can be detected by immunofluorescence in immune aggregates within the focal blood vessel lesions. (Fig. 21-14). As the complexes disappear, free Abs become detectable and inflammatory lesions regress.

The injection of some preformed complexes, prepared in vitro in moderate Ag excess (Ag_3Ab_2), also elicits the characteristic lesions in rabbits. Other complexes are ineffective: those formed at equivalence are generally par-

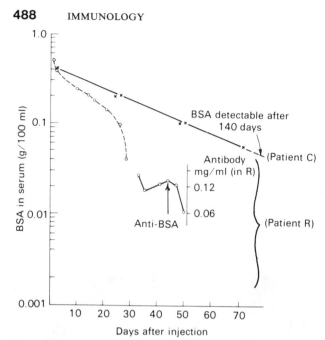

FIG. 21-15. Serum sickness syndrome in man following the injection of 25 g bovine serum albumin (BSA) at zero time. In patient R BSA levels declined abruptly (o - - - o), anti-BSA Ab levels rose (o——o), and serum sickness became evident, days 24 to 31; patient C did not form Abs to BSA or develop serum sickness. (Data of FE Kendall; modified from Seegal B: Am J Med 13:356, 1962)

ticulate and removed by macrophages, and those formed in extreme Ag excess (Ag_2Ab) fail to fix C (see Appendix, Ch. 20).

Vasoactive Amines. Increased vascular permeability seems to promote the penetration of immune complexes from plasma through endothelium into the blood vessel wall, and vasoactive amines probably aid at this step.

Thus vascular deposition of injected preformed complexes is diminished in rabbits that are treated with antihistamines and serotonin antagonists, or are depleted of platelets (the major circulating source of serotonin in rabbits). The immune complexes may contain enough cytotropic Abs to activate blood basophils, which produce the platelet-activating factors that cause platelets to clump and release vasoactive amines. These amines could also be released from basophils by C fragments (C3a, C5a; Ch. 20).

The pathogenic steps are summarized as follows:

1. Ab–Ag complexes fix C, resulting in C fragments that cause mast cells, basophils, and platelets to release histamine and serotonin.
2. Permeability of vascular endothelium increases.
3. Ab–Ag complexes penetrate into blood vessel walls or <u>form</u> within the walls; C chemotactic factors (C5a, C5b,6,7) attract PMNs (neutrophils).
4. Neutrophils penetrate into blood vessel walls, ingesting immune complexes and releasing lysosomal enzymes.
5. Lysosomal enzymes damage neighboring cells and connective tissue elements, leading to more inflammation.
6. If immune complexes are formed in an acute episode ("one-shot" serum sickness) the lesions abate as complexes are degraded.
7. If immune complexes are formed repeatedly (Ag is continuously present, as in persistent viremia, malaria, some forms of human glomerulonephritis) widespread chronic inflammatory disease can develop in small blood vessels (vasculitis) and kidney glomeruli (glomerulonephritis).

Relation to Other Allergic Reactions. Serum sickness has both Arthus and anaphylactic aspects. The focal vas-

FIG. 21-16. Representative cardiovascular and renal lesions in experimental serum sickness in the rabbit. The Ag was BSA (see Fig. 21-17). **A.** Medium-sized coronary artery: endothelial cell proliferation, necrosis of media, polymorphonuclear leukocyte infiltration through all layers, and mononuclear cells in the media and adventitia are evident. **C.** An affected glomerulus showing increase in size, proliferation of endothelial and epithelial cells, and obliteration of capillary spaces. **B.** Section through a normal glomerulus of a control rabbit for comparison with **C:** note the much lower density of glomerular cells and patency of capillaries. (Dixon FJ: In Samter M (ed): Immunological Diseases. Boston, Little, Brown, 1965)

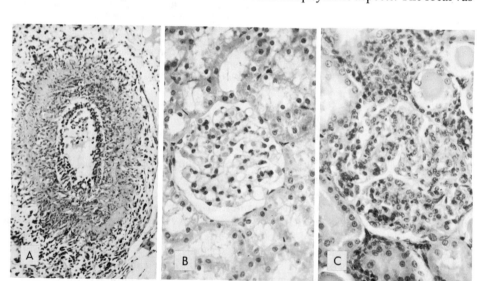

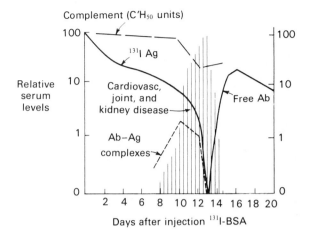

FIG. 21-17. Serum sickness in the rabbit. Changes in serum levels of free Ag (^{131}I-labeled BSA), free Ab (anti-BSA), Ag–Ab complexes, and complement (C) activity (CH$_{50}$ units; see Ch. 20) following the injection of rabbits at zero time with 250 mg ^{131}I-BSA/kg body wt. Ammonium sulfate at 50% saturation precipitates the ^{131}I-BSA–anti-BSA complexes, but not free ^{131}I-BSA. **Ordinate** (log scale) refers to free ^{131}I-BSA in total blood volume, as % of amount injected; anti-BSA, measured as μg of Ag bound per ml of serum; and C activity, as % of normal serum. All animals had cardiovascular, joint, and kidney lesions (Fig. 21-16), shown by shaded area, on day 13. (Dixon FJ: In Samter M (ed): Immunological Diseases. Boston, Little, Brown, 1965)

cular lesions and the requirement for immune complexes and for C suggest that the syndrome is essentially a disseminated form of the Arthus reaction, with the same injected substance, given in a large amount, serving first an immunogen and then as reacting Ag. However, the role of cytotropic Abs seems larger than in the Arthus reaction, since urticarial skin lesions are also prominent in serum sickness. Moreover, a person who has recovered from the disease will generally show a wheal-and-erythema response to intradermal injection of the responsible Ag. The amines probably contribute also to the development of the focal vasculitis, as in the Arthus reaction.

IMMUNE-COMPLEX DISEASES

Glomerulonephritis. The pathogenesis of experimental Arthus lesions and serum sickness probably accounts for some forms of glomerulonephritis, a kidney disease in which obstructive inflammatory lesions of glomerular blood vessels can lead to renal failure. Immunofluorescence of biopsies usually reveal lumpy deposits of Ig and C3 (probably complexed with some unknown Ag) beneath the glomerular endothelium (Fig. 21-18). The deposits resemble those of serum sickness, especially in the chronic experimental model of Dixon et al., in which Ag is administered almost daily for many weeks at a rate that approximates Ab synthesis and provides continuous production of immune complexes.

Ags have been identified in human glomerular lesions in special circumstances. *Plasmodium malariae* Ags have been recognized by immunofluorescence in kidneys of patients with the chronic nephritis associated with malaria, and Abs to malarial Ags have been eluted from kidney biopsies (at pH 2–3 to dissociate Ab–Ag complexes; see Appendix, Ch. 16). Similarly, Abs to single- and double-stranded DNA have been eluted from kidney tissue of patients with systemic lupus erythematosus; these patients often have high serum levels of Abs to nucleic acids and develp progressive glomerulonephritis with lumpy glomerular deposits containing Ig, C, and DNA (see Autoimmune Diseases Ch. 22).

Most cases of human glomerulonephritis occur as a sequel to infection with β-hemolytic streptococci (especially type XII, the "nephritogenic" strain); but streptococcal Ags have not been consistently detected in the associated glomerular deposits of Ig and C, perhaps because reactive sites may be covered by antistreptococcal Abs.

Viral complexes. Chronic viral infection appears to be a source of immune-complex disease in mice. Animals infected at birth with lymphocytic choriomeningitis (LCM) virus become chronic carriers of the virus, producing large amounts of antiviral Abs that do not neutralize infectivity. Virus–Ab–C complexes in serum can be precipitated if antiserum to mouse Igs or to mouse C3 is added. The mice develop progressive renal disease associated with inflammatory vascular lesions and with lumpy glomerular deposits containing virus, antiviral Abs, and C.

Immune-complex disease with glomerulonephritis also occurs in mice as a result of neonatal infection with murine leukemia viruses, coxsackie B virus, and polyomavirus; and similar lesions seem to be responsible for the high mortality rate in an economically important disease of mink that probably also derives from a neonatal viral infection (Aleutian mink disease).

Antikidney Antibodies. In a rare form of human glomerulonephritis (Goodpasture's disease) immunofluorescence reveals not lumpy but linear glomerular deposits of Ig that follow the basement membrane continuously (Fig. 21-18). The pattern resembles that seen in the experimental nephritis produced with heteroantiserum to basement membrane (e.g., the so-called Masugi nephritis produced in rabbits with duck antiserum to rabbit kidney). Some monkeys have developed glomerulonephritis when inoculated with Igs eluted from human kidneys with linear deposits suggesting that the human lesion could be due to autoantibodies to glomerular basement membranes in lung (see Characteristics of Autoimmune Diseases, Ch. 22).

Rheumatoid Arthritis. In this common chronic inflammatory disease of joints the synovial fluid in joint cavities contains high levels of Ig (much of it synthesized locally in the synovial membrane) and C-fixing aggregates of Igs, as well as granulocytes and C components that attract granulocytes (C5a, C5b,6,7). Hence the joint fluids con-

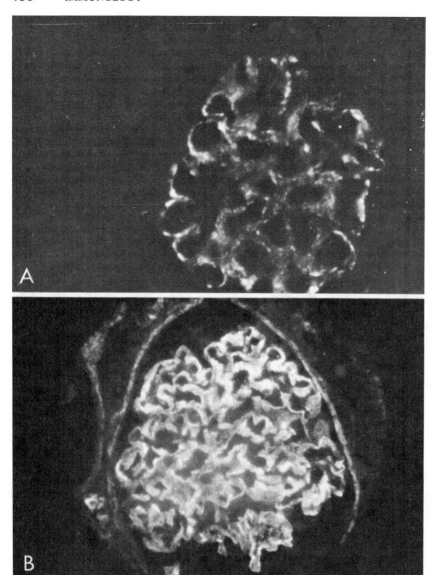

FIG. 21-18. Immune complexes in glomeruli revealed by immunofluorescence. Kidney biopsies from patients with glomerulonephritis were stained with fluorescein-labeled Abs to human Igs. **A.** Lumpy deposits due to Abs–Ag–C in glomerulus from a patient with systemic lupus erythematosus. **B.** Linear deposits due to Igs attached specifically to the glomerular basement membrane (Goodpasture's syndrome; see Characteristics of Autoimmune Disease, Ch. 22). (Courtesy of Dr. C. Kirk Osterland)

tain all the ingredients for an Arthus reaction. However, as in most human diseases that arise from immune complexes, the actual Ags remain unknown.

Some of the IgG in joint fluid probably functions as Ag for the characteristic **rheumatoid factors** of rheumatoid arthritis, which are IgM and IgG molecules that react specifically with antigenic determinants on Fc domains of various IgGs (see Human Allotypes, Ch. 17). The IgG–IgM and the IgG–IgG complexes could then initiate the C–granulocyte–lysosomal enzyme sequence that results in Arthus inflammation. Certain IgG–IgG dimers stimulate the release of serotonin from platelets. Some Igs might also function as Abs that bind special Ags in affected joints, such as DNA (and probably other, unidentified Ags): in this case the rheumatoid factors would not be essential participants but would arise secondarily as Abs to new antigenic sites on conformationally altered IgG molecules in immune complexes. Indeed, *IgM molecules that behave like rheumatoid factors are found in diverse situations where Abs and immune complexes are present at high levels for protracted periods* (e.g., experimental chronic serum sickness).

Evidence for Persistent Soluble Immune Complexes. In rheumatoid arthritis, glomerulonephritis, systemic lupus erythematosus, and other chronic diseases in which immune complexes are probably pathogenic their presence is often revealed by the formation of precipitates when serum (or joint fluid in rheumatoid arthritis) is simply stored at 4 °C. These **cyroprecipitates** contain IgM (probably anti-Abs), IgG, and sometimes additional components that could represent Ags (e.g., single-stranded DNA in patients who form anti-DNA, as in those with systemic lupus or with rheumatoid arthritis).

The presence of soluble immune complexes can also be revealed by: 1) ultracentrifugation (Fig. 21-19); 2) the appearance of breakdown products of C3 (recognized by immunoelectrophoresis with specific antisera); 3) increased levels of **immune conglutinins,** which are Abs (of IgM class) to antigenic sites that appear on activated C3 and C4 components of C (see Immune Conglutination, Ch. 20); 4) precipitation with C1q, the first C component, which reacts specifically with soluble immune complexes if the Ab moiety of the complex belongs to certain Ig classes (e.g., IgG1 and IgM in man; Ch. 20); 5) polyethylene glycol, which precipitates soluble immune complexes.

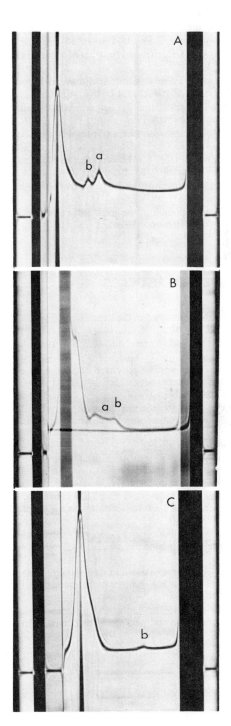

FIG. 21-19. Immune complexes in human serum. Sera were diluted with buffered saline and subjected to velocity sedimentation at about 50,000 RPM in the analytic ultracentrifuge. **A** Serum from a patient with rheumatoid arthritis showing (at **a**) specific complexes (22S) of "rheumatoid factor" of the IgM class with IgG (ligand); the 19S at **b** represents unbound IgM. **B** Serum from a patient with rheumatoid arthritis showing a less common pattern with polydisperse immune complexes (at **a**) involving rheumatoid factor of the IgG class with normal IgG (ligand **(a)**; the normal (unbound) 19S IgM peak is at **b. C** Control: serum from a normal person; **b** represents the normal IgM peak. No complexes are evident. (Courtesy of Dr. C. Kirk Osterland)

SELECTED READING

BOOKS AND REVIEW ARTICLES

AUSTEN KR: Biologic implications of the structural and functional characteristics of the chemical mediators of immediate type hypersensitivity. Harvey Lect 73:93, 1979

BECKER EL (ed): Molecular studies in the chemotaxis of leukocytes: A symposium. Mol Immunol 17:149, 1980

BENNICH H, JOHANSSON SGO: Structure and function of human IgE. Adv Immunol 13:1, 1971

COCHRANE CG, KOFFLER D: Immune complex disease in experimental animals and man. Adv Immunol 16:186, 1973

KATZ DH: Lymphocyte Differentiation, Recognition, and Regulation. New York, Academic Press, 1977

MCCLUSKEY RT, HALL CL, COLVIN RB: Immune complex mediated diseases. Hum Pathol 9:71, 1978

OSLER AG, LICHTENSTEIN LM, LEVY DA: In vitro studies of human reaginic allergy. Adv Immunol 8:183, 1968

WELLER PF, GOETZL EJ: The regulatory and effector roles of eosinophiles. Adv Immunol 27:339, 1979

ZVAIFLER NJ: The immunopathology of joint inflammation in rheumatoid arthritis. Adv Immunol 16:265, 1973

SPECIFIC ARTICLES

COCHRANE CG: Mechanisms involved in the deposition of immune complexes in tissue. J Exp Med 134:75S, 1971

DAWSON W: SRS-A and the leukotrines. Nature 285:68, 1980

KATZ D: Regulation of IgE antibody production of serum molecules. III. Induction of suppressive activity by allogeneic lymphoid cell interaction and suppression of IgE synthesis by the allogeneic effect. J Exp Med 149:539, 1979

GOETZL EJ, GORMAN RR: Chemotactic and chemokinetic stimulation of human eosinophil and neutrophil polymorphonuclear leukocytes by 12-L-hydroxy-5,8,10-heptadecatrienoic acid (HAT). J Immunol 120:526, 1978

KOFFLER D, AGNELLO V, THOBURN R, KUNKEL HG: Systemic lupus erythematosus: prototype of immune complex nephritis in man. J Exp Med 174:169S, 1971

KULCZYCKI A JR, MCNEARNEY TA, PARKER CW: The rat basophilic leukemia cell receptor for IgE I. Characterization as a glycoprotein. J Immunol 117:661, 1976

MENDOZA G, METZGER H: Distribution and valency of receptors for IgE on rodent mast cells and related tumor cells. Nature 264:548, 1976

ROSSI G, NEWMAN SA, METZGER H: Assay and partial characterization of the solubilized cell surface receptor for immunoglobulin E. J Biol Chem 252:704, 1977

SCHLESSINGER J, WEBB WW, ELSON EL, METZGER H: Lateral motion and valence of Fc receptors on rat peritoneal mast cells. Nature 264:550, 1976

SEGAL DM, TAUROG JD, METZGER H: Dimer immunoglobulin F serves as a unit signal for mast cell degranulation. Proc Natl Acad Sci USA 74:2993, 1977

WINCHESTER RJ, KUNKEL HG, AGNELLO V: Occurrence of γ-globulin complexes in serum and joint fluid of rheumatoid arthritic patients: use of monoclonal rheumatoid factors as reagents for their demonstration. J Exp Med 134:286S, 1971

BENNICH H, ISHIZAKA K, ISHIZAKA T, JOHANSSON SGO: Immunoglobulin E. A comparative study of γE globulin and myeloma IgND. J Immunol 102:826, 1969

CONRAD DH, FROESE A, ISHIZAKA T, ISHIZAKA K: Evidence for antibody activity against the receptor for IgE in a rabbit antiserum prepared against IgE-receptor complexes. J Immunol 120:507, 1978

CONROY MC, ADKINSON NF JR, LICHTENSTEIN LM: Measurement of IgE on human basophils: relation to serum IgE and anti-IgE induced histamine release. J Immunol 118:1317, 1977

ISHIZAKA K, ADACHI T: Generation of specific helper cells and suppressor cells in vitro for the IgE and IgG antibody response. J Immunol 117:40, 1976

ISHIZAKA K, ISHIZAKA T, OKUDAIRA H, BAZIN H: Ontogeny of IgE-Bearing lymphocytes in the rat. J Immunol 120:655, 1978

KIMOTO M, KISHIMOTO T, NOGUCHI S, WATANABE T, YAMAMURA Y: Regulation of antibody response in different immunoglobulin classes. II. Induction of in vitro IgE antibody response in murine spleen cells and demonstration of a possible involvement of distinct T helper cells in IgE and IgG antibody responses. J Immunol 118:840, 1977

KISHIMOTO T, HIRAI Y, SUEMURA M, YAMAMURA Y: Regulation of antibody response in different immunoglobulin classes. I. Selective suppression of anti-DNP IgE antibody response by preadministration of DNP-coupled mycobacterium. J Immunol 117:396, 1976

KULCZYCKI A JR, METZGER H: The interaction of IgE with rat basophilic leukemia cells. II. Quantitative aspects of the binding reaction. J Exp Med 140:1676, 1974

ÖRNING L, HAMMARSTRÖM S, SAMUELSSON B: Leukotrine D, a slow reacting substance from rat basophilic leukemia cells. Proc Natl Acad Sci USA 77:2014, 1980

SAMUELSSON B, BORGEAT P, HAMMARSTRÖM S, MURPHY RC: Introduction of a nomenclature: Leukotrienes. Prostaglandins 17:785, 1979

SIRAGANIAN RP, HOOK WA, LEVINE BB: Specific in vitro histamine release from basophils by bivalent haptens: evidence for activation by simple bridging of membrane-bound antibody. Immunochemistry 12:149, 1975

TAKATSU K, ISHIZAKA K: Reaginic antibody formation in the mouse. VII. Induction of suppressor T cells for IgE and IgG antibody responses. J Immunol 116:1257, 1976

TAKATSU K, ISHIZAKA K: Reaginic antibody formation in the mouse. VIII. Depression of the on-going IgE antibody formation by suppressor T cells. J Immunol 117:1211, 1976

TAMURA S, ISHIZAKA K: Reaginic antibody formation in the mouse. X. Possible role of suppressor T cells in transient IgE antibody response. J Immunol 120:837, 1978

chapter 22

CELL-MEDIATED HYPERSENSITIVITY AND IMMUNITY

GENERAL PROPERTIES OF CELL-MEDIATED IMMUNE (CMI) RESPONSES

Long before T and B cells were distinguished it was recognized that slowly developing tissue inflammatory reactions to antigen (Ag), known as **delayed-type hypersensitivity (DTH),** could be transferred, from immunized donors to "naive" recipients, with living leukocytes but not with serum. Since then other forms of cell-mediated immunity (CMI) have been recognized: increased phagocytic and bactericidal activity of macrophages, lysis of virus-infected cells and tumor cells, and rejection of allografts (Table 22-1). *In all these reactions the specific recognition of Ag is due to receptors on T lymphocytes, not to conventional antibody (Ab) molecules.*

There are two fundamentally different forms of CMI, mediated by different T cell subsets. One type of T cell is active in DTH: upon reacting with Ag, it secretes **lymphokines,** substances that attract and activate macrophages or sometimes other leukocytes, such as basophils; the attracted cells lack specificity for Ag and serve as nonspecific effectors of tissue inflammation. The second type of T cell recognizes and reacts specifically with Ag on target cells, causing their lysis. The latter T cells are referred to as **T killer,** or **T_k, cells** or, more often, as **cytotoxic T lymphocytes (CTLs).**

Though the two forms of CMI are readily distinguished experimentally, both are probably involved, to varying degrees, in many natural CMI reactions.

DELAYED-TYPE HYPERSENSITIVITY (DTH)

The response to proteins of the tubercle bacillus, studied for almost a century, serves as a general model for DTH reactions to soluble proteins and to microbial Ags.

Koch observed in 1890 that filtrates of cultures* of *Mycobacterium tuberculosis* elicited an inflammatory reaction many hours after injection into tuberculous animals but not into normal ones. Similar preparations from other bacterial and fungal cultures elicit similar delayed-type responses in those infected with the corresponding organisms. Such responses have long been used to diagnose infectious diseases and to screen populations for individuals with previous or current infections (Table 22-2).

Cutaneous Reaction. After 0.1μg of tuberculin* is injected intradermally into a sensitized individual no change is observed at the inoculated site for at least 10 h. Erythema and swelling then gradually appear and increase; maximal intensity and size (up to about 7 cm diameter) are reached in 24–72 h and the response then subsides over several days.

Species vary widely in responsiveness. Thus in highly sensitive humans 0.02μg of tuberculin can cause necrosis, ulceration, and scarring at the inoculated site. In contrast, more than 0.5μg is required to elicit a faint response in highly sensitized guinea pigs; even more is necessary in cattle and rabbits, and in sensitized rats and mice erythema is not discernible and histologic examination is required.

Histology. The response is characterized histologically by the accumulation of inflammatory cells around postcapillary venules, first detected after 6 h and maximal at 24–48 h (Fig. 22-1). Some of the inflammatory cells are lymphocytes, but most are monocytes that are presumed to differentiate later into macrophages. Basophils can also be prominent (see Cutaneous Basophil Hypersensitivity, below). Fibrin accumulates and probably accounts for the firmness of the inflammatory lesion.

Comparison Between Arthus and Delayed-Type Skin Reactions. Arthus reactions (Ch. 21) sometimes simulate delayed-type responses, but they usually appear about 2 h after Ag is injected into the skin, are maximal at 4 or 5 h, and subside by 24 h. They are also more boggy than indurated, reflecting a large amount of edema fluid and only a modest accumulation of inflammatory cells, mainly polymorphonuclear leukocytes. DTH reactions are not only later in onset and more persistent but they are indurated because of the intense accumulation of mononuclear cells and fibrin.

A **bimodal reaction** sometimes follows a single test injection; one phase is maximal at about 4 h (Arthus) and the other at about 24 h

---

* After its concentration by boiling, and the removal of debris, the culture filtrate was called **tuberculin** (now called **old tuberculin,** or **OT).** The active proteins are now concentrated from autoclaved cultures by precipitation with ammonium sulfate; the somewhat purified product is **purified protein derivative,** or **PPD.** A standard batch, PPD-S, has been designated as the basis for the international unit: **1 tuberculin unit (TU)** equals 0.02μg of total protein.

TABLE 22-1. Cell-Mediated Immune Responses

Delayed-type hypersensitivity
Resistance to many infectious agents (especially intracellular
  pathogens)
Resistance to many tumors
Rejection of allografts
Graft-vs.-host reactions
Some drug allergies
Some autoimmune diseases

(delayed-type). The situation is, however, often more complex: a severe Arthus reaction can remain conspicuous for 24 h or more; and its distinction from the DTH reaction then depends on whether transfer to a naive recipient is accomplished with serum or with lymphoctyes.

Noncutaneous and Systemic Reactions. Delayed responses also occur in tissues other than skin; e.g., tuberculin can cause severe inflammation and necrosis in the cornea of highly sensitized guinea pigs. A systemic response, **tuberculin shock,** ensues when a sensitized guinea pig is injected intraperitoneally with a relatively large amount (e.g., 5 mg) of tuberculin. Prostration develops after 3–4 h, body temperature falls about 4 or 5 degrees, and death may follow in 5–30 h. A systemic reaction (headache, malaise, prostration, but only rarely fatal) also occurs in highly sensitized persons who are injected with an excessive amount of tuberculin or who inhale aerosols of the material in the laboratory.

Tuberculous individuals exposed to large amounts of tuberculin also develop **focal reactions,** with increased inflammation in tuberculous lesions in the lungs and elsewhere, resembling histologically the responses elicited in the skin with tuberculin. The reactions are probably due to interaction of the Ag with specifically reactive lymphocytes in the infected lesions.

Systemic responses have also been evoked with Ags of histoplasma, brucella, vaccinia, and pneumococci in persons with delayed-type sensitivity to these organisms.

Sensitization. The time required for induction of DTH is about the same as for induction of Ab formation, i.e., roughly 4–14 days after Ag is first administered.

Small quantities of Ag, especially when on the surface of living cells, are the most effective **immunogens.** Sensitization to tuberculin is most effectively induced by infection with virulent tubercle bacilli (or with the attenuated BCG strain). Killed bacilli are less effective. Purified tuberculin is ineffectual (unless given in appropriate adjuvants, as described in the next paragraph), but it can apparently stimulate a secondary response (i.e., enhance a declining DTH).

Through efforts to improve the immunogenicity of soluble proteins, Dienes found that egg albumin injected into an animal's tuberculous lesions established intense generalized DTH to this protein. This observation led eventually to Freund's development of the adjuvant that bears his name (water-in-oil emulsions with suspended tubercle bacilli). The active component of the tubercle bacillus appears to be a muramyl dipeptide (contained in a lipid called wax D), and a water-in-oil emulsion containing the active component and purified tuberculin can establish DTH to tuberculin.

Complete Freund's adjuvant also strongly enhances Ab formation (see Adjuvants, Ch. 19), but the dose of incorporated protein is important: small doses (e.g., 1–50 μg in a guinea pig) elicit intense DTH, while larger amounts tend to induce vigorous Ab formation, not DTH. Other adjuvants such as alumina aid only Ab formation and even appear to inhibit the induction of DTH (see Immune Deviation, below).

T_{DTH} **Lymphocytes.** Ab molecules have been excluded from a role in DTH reactions by the failure of countless

TABLE 22-2. Some Delayed-Type Skin Reactions Used for Diagnosis and for Epidemiologic Surveys

Disease	Type of etiologic agent	Antigenic preparation used in skin test
Tuberculosis	Bacterium	Tuberculin
Leprosy	Bacterium	Lepromin
Brucellosis	Bacterium	Brucellin
Psittacosis	Bacterium	Heat-killed organisms
Lymphogranuloma venereum	Bacterium	Extract of chorioallantoic membrane of infected chick embryo
Mumps	Virus	Noninfectious virus from yolk sac of infected chick embryo
Coccidioidomycosis	Fungus	Concentrated culture filtrate
Histoplasmosis	Fungus	Concentrated culture filtrate
Blastomycosis	Fungus	Concentrated culture filtrate
Leishmaniasis	Protozoan	Extract of cultured *Leishmania*
Echinococcosis	Helminth	Fluid from hydatid cyst
Contact dermatitis	Simple chemical	Patch tests with simple chemicals

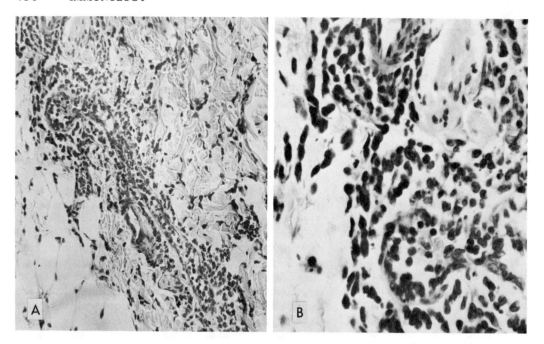

FIG 22-1. Delayed-type allergic reaction in guinea pig skin. The animal was sensitized by injecting 5 μg of hen's egg albumin (HEA) in complete Freund's adjuvant into its toepads. Six days later it was injected intradermally with 5 μg of HEA in saline. The skin site was excised 24 h later, when induration and redness (which had probably first become evident at about 12 h) were maximal. The infiltrating mononuclear lymphoid cells appear, by electron microscopy and with special stains, to be lymphocytes and macrophages. (Coe JE, Salvin SB: J Immunol 93:495, 1964) (**A,** ×64; **B,** ×355)

attempts to transfer the capacity for delayed-type skin reactions with serum. Moreover, DTH and other CMI reactions can be elicited in individuals who are almost totally devoid of serum Igs (e.g., children with X-linked agammaglobulinemia [Ch. 19], or mice treated at birth with Abs to μ chains).

Cells were shown in 1942 to mediate delayed-type allergy, when Chase transferred delayed sensitivity to tuberculin from sensitive to normal guinea pigs, using viable lymphoid cells: crude mixtures of lymphocytes, macrophages, and other cells (RBCs, granulocytes). Fractionation of crude cell mixtures and analysis of cell surface Ags on mouse lymphocytes showed, eventually, that the cells responsible for DTH are T cells of the Ly-1 type (see Ly Antigens, Ch. 18): i.e., they are killed (in the presence of complement) by antiserum to Lyt-1, and not by antiserum to Lyt-2 or Lyt-3.

T helper cells (which aid B cells in Ab formation; Chs. 18 and 19) are also Ly-1 cells; and both T helpers and T cells of DTH are alike in the restriction of their specificity by products of the I region of the major histocompatibility complex (see H-2 Restriction, below). However, the T helpers of Ab formation and the T cells of DTH are probably different cells, and so we shall refer to the mediators of DTH as T_{DTH} cells.

MACROPHAGES

Macrophages (Mφs) are also required for DTH reactions. Thus lethally irradiated animals (900 R) develop DTH responses only when given both lymphocytes and bone marrow cells (a source of stem cells that give rise to monocytes and macrophages; see Ch. 18): blood-borne monocytes (immature macrophages) infiltrate developing DTH skin lesions, and differentiate there into mature macrophages. Unlike the lymphocytes, the **macrophages act nonspecifically,** for they are equally effective from nonsensitized or sensitized donors. Relatively few specific T_{DTH} cells are needed to initiate the allergic response, and Mφs are much more abundant at the reaction site.

Thus after a normal recipient was sensitized with labeled lymphocytes from an actively sensitized donor (who had been treated with ³H-thymidine) fewer than 5% of the mononuclear cells in the recipient's DTH skin lesion (elicited later with Ag) were labeled; but when normal recipients were given ³H-thymidine and sensitized lymphocytes were then transferred from an unlabeled donor over 90% were labeled. Studies with cultured cells show that on reacting with Ag the sensitized lymphocytes release substances (lymphokines; see below) that cause the chemotactic accumulation and differentiation of macrophages (Fig. 22-2).

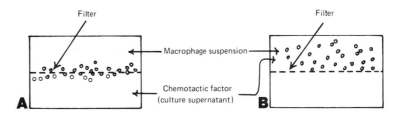

FIG. 22-2. A two-compartment chamber demonstrates **chemotaxis**: the movement of cells along a concentration gradient of an attracting substance (chemotactic factor). **A.** Movement of macrophages to and through the pores of a Millipore filter reveals chemotactic activity in culture supernatants of Ag-activated T lymphocytes. **B.** Uniform mixing of the culture supernatant with macrophages immobilizes the cells, as in the macrophage inhibition assay of Fig. 22-4. Culture supernatants of Ag-activated lymphocytes are also chemotactic for normal lymphocytes (probably owing to another attractant).

Mϕs have two functions in delayed-type hypersensitivity. 1) In the initial ("afferent") step, they activate specific T cells by presenting Ag to the lymphocyte and perhaps by releasing soluble factors that enhance the development of precursor cells into effector (T_{DTH}) cells. 2) In the final ("efferent") steps, under the influence of lymphokines released by Ag-stimulated T_{DTH} cells, the Mϕs themselves become activated and contribute to the local inflammatory reaction and to increased resistance against many microbial pathogens (see Activation of Macrophages, p. 498).

Ag Presentation and H-2 Restriction. In Chapter 19 we noted that T helper cells, when aiding B cells in Ab formation, recognize not soluble Ag molecules, but Ag on the Mϕ surface. The recognition is **H-2 restricted** (Ch. 18): i.e., the bound Ag is reactive only if the macrophage also has the appropriate alleles of the major histocompatibility complex (I region in mice). The same situation applies to T_{DTH} cells, which seem also to recognize Ag on Mϕs only in conjunction with the appropriate products of the H-2 I region (Fig. 22-3; see also Fig. 18-26, Ch. 18).

FIG. 22-3. H-2 restriction of T_{DTH} cells in delayed-type hypersensitivity reactions. T cells from F1 hybrid mice (H-2^a × H-2^b, or a × b) were injected into athymic (nude) mice of the two parental types: H-2^a (a) or H-2^b (b). An injected Ag X (fowl γ-globulin) was presumably bound by the athymic mouse's own macrophages (a or b) and stimulated the F1 (a × b) T cells. The T cells (which had settled in the spleen) were then transferred to normal mice of parental type (a or b). Delayed-type skin reactions were elicited by the Ag only in recipients with the same H-2 type as that of the athymic mouse whose macrophages originally presented the Ag to the (a × b) cells. Judging from studies of mice with various H-2 complexes, the critical difference between a and b in this study was in the I-A locus. (Based on Miller JFAP, Vadas MA: Cold Spring Harbor Symp Quant Biol 41: 579, 1976; see also Fig. 18-26, Ch. 18)

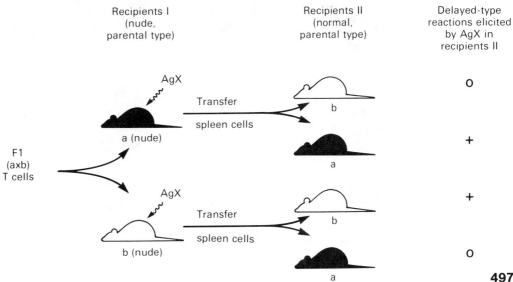

TABLE 22-3. Cells Affected by Lymphokines Released from Activated Lymphocytes

Cells affected	Lymphokine
Macrophages	Migration inhibitory factor (MIF)
	Macrophage activating factor (MAF)
	Chemotactic factor
Neutrophilic leukocytes	Leukocyte inhibition factor (LIF)
	Chemotactic factor
Lymphocytes	Mitogenic factor
	T cell replacing factor
	Suppressor factors
	Chemotactic factor
Eosinophils	Chemotactic factor
	Migration stimulatory factor
Basophils	Chemotactic factor
Other cells	Cytotoxic and cytostatic factors (lymphotoxin)
	Interferons (NK cells)
	Osteoclast activating factor
	Colony stimulating factor (CSF)

(Bloom BR: Fed Proc 37:2741, 1978)

Lymphokines. Specific reactions with Ag cause T_{DTH} cells to release soluble factors, known as lymphokines, that act on a variety of other cells (Table 22-3). About 50 such factors have been described, but how many distinct molecular entities they represent is not known. The **migration inhibitory factor (MIF)** blocks the migration of Mφs in culture; this effect has been used in a general assay for Ag reactions with T cells (Fig. 22-4). The **macrophage activating factor (MAF)** stimulates Mφs to change in appearance and in metabolic activities, as dis-

cussed below, and to become more effective killers of various microbial cells; hence these **activated macrophages** are thought to play an important role in the local resistance to many pathogens that accompanies DTH reactions.

Activated Macrophages. Tubercle bacilli, leprosy bacilli, and many other bacteria survive after having been phagocytized and even multiply within Mφs taken from normal (i.e., uninfected) individuals. However, Mφs from infected individuals generally have an augmented ability to kill the infecting bacteria and also many other antigenically unrelated ones. Though the increased antimicrobial activity of these activated Mφs (Fig. 22-5) is nonspecific, their conversion into "killers" is due to a specific Ag-triggered step: the release of a lymphokine from Ag-stimulated T_{DTH} cells. For instance, when an animal is primed by infection with a bacterial organism, then cured and later reinfected (challenged) with a small number of bacteria of the same type, its Mφs promptly become nonspecific killers, but a similar number of antigenically different bacteria as a challenge would not have triggered this conversion. Like other CMI reactions the specific ability of an Ag to activate Mφs can be transferred from an immunized to a normal animal with Ag-specific T cells, not with antiserum. Mφs can also be activated in culture by a soluble factor from Ag-stimulated spleen cells (undoubtedly T cells).

Mouse peritoneal Mφs infected with *Trypanosoma cruzi* (a protozoan that causes Chagas' disease) provide

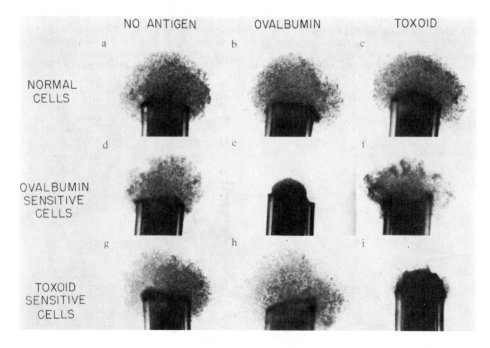

FIG. 22-4. Inhibition of macrophage migration: a response in vitro that parallels delayed-type hypersensitivity reactions in vivo. Peritoneal exudate cells (predominantly macrophages and lymphocytes) are cultured in capillary tubes in the presence or absence of Ags (ovalbumin, diphtheria toxoid). An Ag blocks outward migration of macrophages from the cell population derived from guinea pigs with delayed-type hypersensitivity to that Ag. The response is also carrier-specific: for example, 2,4-dinitrophenyl-guinea pig albumin (Dnp–GPA) blocks migration of cells from animals sensitized with Dnp–GPA, but not cells from those sensitized with Dnp on another protein, say Dnp–BγG (bovine γ-globulin). (David JR et al: J Immunol 93:264, 1964)

FIG. 22-5. Increased bactericidal activity of activated macrophages. *Salmonella typhimurium* coated (opsonized) with anti-*Salmonella* Abs are ingested and killed more rapidly and in greater numbers (decreased "percent survival") by activated than by normal macrophages. (Mackaness GB: Hosp Practice 73: 1970)

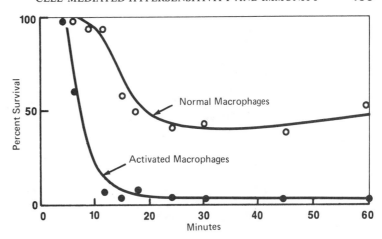

an illuminating example. When trypanosomes are phagocytized by "resident" macrophages (i.e., those found in the normal peritoneal cavity) they escape from phagocytic vacuoles, multiply in the cytoplasm, and eventually destroy the Mφs. However, Mφs from the peritoneal cavity of a mouse that has been immunized with *T. cruzi,* or with antigenically unrelated organisms such as tubercle bacilli (BCG), and then boosted with the homologous organism, will kill the ingested trypanosomes (Fig. 22-6).

In comparison with resting Mφs, the activated ones are larger, have more lysosomes and lysosomal enzymes per cell, and secrete enzymes (e.g., collagenase, plasminogen activators) that contribute to inflammation (Fig. 22-7). The activated Mφs also spread more actively on glass and other surfaces, become more sticky, and have a number of metabolic differences from resting cells, e.g., increased

FIG. 22-6. Correlation in activated macrophages between release of H_2O_2 and destruction of phagocytized microbes (trypanosomes). Freshly explanted peritoneal macrophages were derived from four groups of mice: 1) **IB**—immunized with viable *Trypanosoma cruzi* and later boosted with heat-killed *T. cruzi*; 2) **CB**—controls given the booster injection only; 3) **PP**—controls injected with proteose peptone, which elicits inflammation and altered macrophages that lack microbicidal activity but look like lymphokine-activated cells: 4) **Res**—untreated controls ("resident" macrophages from the normal peritoneal cavity). **A.** H_2O_2 release by the cultured macrophages. **B.** Destruction of a test dose of trypanosomes added to the macrophages in culture. (Nathan C et al: J Exp Med 149:1056, 1979)

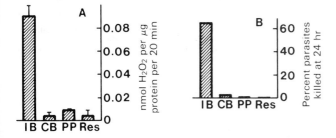

oxidation of glucose to CO_2, increased glucose transport, decreased oxygen consumption, increased lactate production, increased formation of hydrolytic enzymes (in lysosomes), and a decrease in membrane 5′ nucleotidase.

Lymphokine-activated Mφs also form the superoxide anion, O_2^-, and release relatively large amounts of H_2O_2:

$$2\ O_2 + NADPH \rightarrow 2\ O_2^- + NADP^+ + H^+$$
$$2\ O_2^- + 2\ H^+ \rightarrow O_2 + H_2O_2$$
$$\text{Superoxide}$$
$$\text{dismutase}$$

The superoxide anion and H_2O_2 probably account for much of the destructive effects of activated Mφs on bacteria, fungi, and protozoa, and perhaps also on tumor cells (Fig. 22-6; see also Tumor Immunology, Ch. 23). The importance of activated Mφs in resistance to infection is also evident from the much greater severity of clinical infections in those with diminished CMI due to heritable or acquired defects (such as those discussed later in this chapter) or to immunosuppressive agents (see Nonspecific Suppression, section).

Cutaneous Basophil Hypersensitivity. Histamine levels are increased at the sites of DTH reactions to tuberculin. This previously puzzling observation became understandable when it was found that the infiltrating leukocytes include many basophils, which have high levels of histamine (Ch. 21). In the guinea pig, DTH is usually elicited by injecting the Ag in complete Freund's adjuvant; if the same Ag is given in saline, the allergic skin reaction evoked shortly afterward (formerly called the Jones–Mote reaction) is also delayed-type, but distinctive: it follows the same time course as the classic DTH reaction, but it lacks fibrin and is less indurated. *Improved histologic technics showed that about 20%–60% of the infiltrating inflammatory cells are basophils* (which normally account for about 1% of the leukocytes in blood).

The cutaneous basophil hypersensitivity reaction is transferable from a sensitized to a normal animal with T cells. As mentioned above, the reaction of Ag with T_{DTH} cells yields lymphokines; one of these is chemotactic for basophils (Table 22-3). However, the cutaneous basophils that accumulate at the site of the reaction seem to

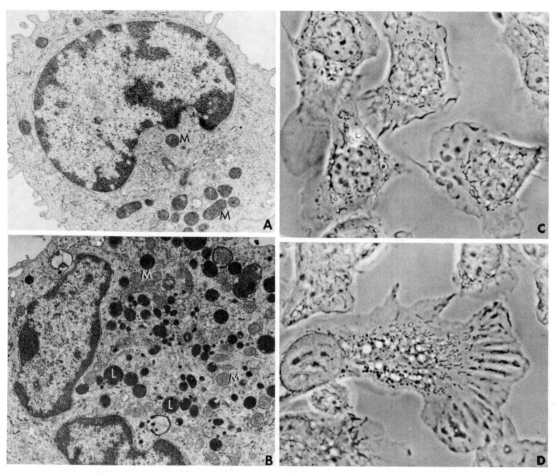

FIG. 22-7. Comparison of resting **(A,C)** and activated **(B,D)** macrophages. All are peritoneal cells from normal **(A,C)** or infected **(B,D)** mice 14 days after injection of tubercle bacilli (BCG, bacillus Calmette–Guérin). In the phase-contrast photographs **(C,D,** × 1900) the activated macrophages **(D)** are larger (the field is almost filled by half of a cell), are more spread out, and contain more organelles, especially lysosomes (dense, spherical bodies); the translucent spherical bodies represent ingested culture medium (pinocytotic vesicles). In the electron micrographs **(A,B;** magnification uncertain) the dense spherical lysosomes **(L)** are abundant in activated cells and rare in resting cells, most of whose organelles are mitochondria **(M)**. **(AB**: Blanden RV et al: J Exp Med 129:1079, 1969; **C,D**: courtesy of Dr. G. B. Mackaness)

carry on their surface few receptors for the Fc domain of IgE molecules: much less histamine is released than in anaphylaxis.

Transfer Factor. Though cell-mediated responses are readily transferred with living lymphocytes, disrupted cells are ineffective in all animal species tested except man, in whom Lawrence found that extracts of blood leukocytes can apparently transfer the cell-mediated responses of the leukocyte donor. Recipients become reactive to Ag 1–7 days after receiving the extract, and sensitivity can persist for years, even when the extract is prepared from as little as 5 ml of blood (corresponding to about 10^7 lymphocytes).

The active component, **transfer factor,** has been difficult to characterize since it can be assayed only in humans. It is dialyzable, is stable on prolonged storage, and has a mol wt less than 10,000; it is inactivated by heat (56° C, 30 min) but not by trypsin, RNase, or DNase.

Transfer factor seems to lack antigenicity; it is not inactivated by anti-Igs; and it seems too small to be an informational molecule, such as mRNA. Possibly it is a potent lymphokine or adjuvant, enhancing the induction of cell-mediated immunity to prevalent Ags. Clarificaton of the nature and the mode of action of transfer factor may illuminate fundamental mechanisms of cellular immunity. Clinical trials indicate that transfer factor may be effective therapeutically even in those with deficient CMI. For instance, patients with disseminated *Candida* infections (chronic mucocutaneous candidiasis) lack delayed skin responses to *Candida* proteins, but after receiving transfer factor from *Candida*-sensitive donors their previously intractable infections sometimes subside, and some patients manifest delayed-type skin reactions to *Candida*.

Similar events have occurred in patients with disseminated vaccinia, as well as in children with severe congenital deficiencies in CMI. However, nonspecific effects are sometimes also evident: e.g., the transfer factor recipient may have DTH skin reactions to Ags that the donor did not react against.

CYTOTOXIC T LYMPHOCYTES (CTLs)

Cytotoxic T lymphocytes (CTLs, or T killer [T_k] cells) lyse target cells. CTLs are elicited in vivo by "stimulator" cells, i.e. any cells that carry new surface Ags (neoantigens), such as those introduced by virus infection, or by tissue allografts or injections of allogeneic cells. In animals thus immunized CTLs are generally found among spleen cells and in the peritoneal fluid (**peritoneal exudate cells**).

CTLs can also be elicted readily in vitro by culturing spleen cells for about 5 days with stimulator cells, which are usually irradiated to prevent their proliferating. Even better than normal spleen cells are primed spleen cells, obtained from mice injected a few weeks earlier with the Ag or with Ag-bearing stimulator cells. Regardless of the procedure, CTLs are recognized, and their activity is measured, by test tube assays in which the target cells are labeled with a radioisotope (usually ^{51}Cr); lysis is then revealed by release of the label into the culture medium.

In studying the CTLs formed in virus-infected mice, Zinkernagel and Doherty discovered that antivirus CTLs are specific for viral Ag in conjunction with normal products of the major histocompatibility complex (MHC) on the infected cell's surface (see H-2 Restrictions, Ch. 18). For instance, when a mouse of $H\text{-}2^a$ haplotype (Table 18-6, Ch. 18) is inoculated with virus X, the resulting CTLs, elicited by X-infected $H\text{-}2^a$ cells, lyse any target cells that are infected with the same virus (or a crossreacting one) and that have the same H-2 haplotype; they lyse poorly or not at all X-infected cells of different H-2.

As was noted before (H-2 Restrictions, Ch. 18), the use of target cells from mice with recombinant H-2 complexes showed that some CTLs are specific for Ag plus products of the H-2K locus and others for Ag plus products of the H-2D locus. In a hybrid mouse ($H\text{-}2^a \times H\text{-}2^b$) infected with virus X, there are four sets of virus-specific CTLs, specific for $X + H\text{-}2K^a$, $X + H\text{-}2D^a$, $X + H\text{-}2K^b$, and $X + H\text{-}2D^b$.

Parallel observations have been made with stimulator and target cells whose new surface Ags are formed by coupling haptenic groups to cell surface proteins (e.g., 2,4,6-trinitrophenyl, or Tnp). In addition, minor histocompatibility Ags ("minors"; see Non–H-2 Loci, Ch. 23) can serve as the foreign Ag, and these Ags also are recognized only in conjunction with cell surface major histocompatibility Ags. Mouse cells that possess the necessary foreign Ag but lack surface H-2 Ags (or that have their surface H-2 molecules obscured by other cell surface glycoproteins) can neither elicit nor be destroyed by CTLs.

The models proposed to account for the double requirement, i.e., for foreign Ag plus cell surface H-2K or H-2D molecules, are described in Chapter 18, under Dual Receptor and Altered-Self Hypotheses.

MECHANISMS OF LYSIS

CTLs have surface Lyt-2 and Lyt-3 glycoproteins, but little or no Lyt-1 (see Ly Antigens, Ch. 18); hence (because Lyt-2 and Lyt-3 are always coexpressed) they are called Ly-23 cells. Contact between CTL and target cell is required for lysis, but accessory cells, such as macrophages, are not. Hence lysis is strictly confined to the target cells: *"innocent bystander" cells are not damaged.* (Recall that in DTH, in contrast, the reaction of T_{DTH} cells with Ag produces a local inflammatory lesion, which often results in the destruction of neighboring normal cells).

Several stages in the lytic process are distinguished (Fig. 22-8):

1) The CTL binds the target cell in a **conjugate** that persists for several minutes.

2) During conjugate formation the target cell becomes irreversibly committed to undergo subsequent lysis (i.e., it is **"programmed for lysis"**).

3) After binding and inflicting damage (Fig. 22-9), the CTL leaves the mortally wounded target cell, moves to another target cell, and repeats the cycle; one CTL can probably kill a dozen target cells over the course of a few hours.

The chemical changes underlying the lytic reaction are obscure. Mg^{2+} is required for the formation of conjugates, and Ca^{2+} is required for the target cell to become programmed to undergo lysis. Conjugate formation is prevented by cytochalasin, which disrupts cytoplasmic microfilaments; and programming the target cell for lysis can be blocked either by colchicine, which disrupts cytoplasmic microtubules, or by cholera toxin, which increases intracellular levels of cAMP.

Inhibiting CTL activity at particular stages provides a way to estimate the number of CTLs in a mixed cell population. The addition of cytochalasin after a brief incubation of the mixed population with target cells limits the lysis to a *single round of killing* (because further pairing of CTLs and target cells is prevented); the number of lysed target cells then equals the number of active CTLs. By this approach, approximately 1%–2% of stimulated spleen cells appear to be CTLs.

Killing is Unidirectional. Though not damaged when they kill their target cells, CTLs can themselves serve as

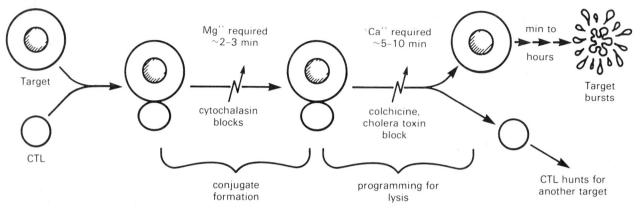

FIG. 22-8. The lytic cycle in the attack of a cytotoxic T lymphocyte (CTL) on a target cell.

targets for other CTLs. For instance, when CTLs produced by H-2^a mice (CTLs no. 1) are incubated with anti–H-2^a CTLs (no. 2), elicited in mice of a different strain, say H-2^b, the first CTLs (no. 1) are lysed but the second ones (no. 2) remain viable and active. It appears, therefore, that when two CTLs bind together to form a conjugate there is a distinction between the aggressor (no. 2) and the victim (no. 1): *delivery of the lethal hit is one-way.*

Difference Between CTL Lysis and Complement Lysis. The nature of the CLT's killing apparatus and of the fatal lesion it produces in the target cell are not known. There is, however, a clear difference between the lysis caused by CTL and that caused by activated complement (C; Ch. 20). Cells lysed by Abs and C develop a leaky membrane and release first their low-, then their high-molecular-weight cytoplasmic contents; but the cell's shape and nuclear contents are retained, and the cell is easily recognized in the microscope (by its uptake of ionic dyes such as eosin or trypan blue). In contrast, *cells lysed by CTLs explode* (Fig. 22-9), leaving debris rather than a recognizable cell.

Specificity. CTLs discriminate between similar target cells about as well as Abs distinguish between similar li-

gands (see Specificity and Affinity, Ch. 16). For instance, CTLs elicited against cells infected with a particular influenza virus will lyse target cells infected with that virus more extensively than targets infected with a variant influenza virus having a slightly different hemagglutinin. And CTLs elicited by stimulator cells with a surface haptenic group will lyse target cells with the same hapten better than targets with a crossreacting hapten (see NIP and NP, Fig. 22-10).

Identification of Antigens on Target Cells. When an Ab response is elicited by a complex immunogen, such as a bacterium, the particular Ag molecules recognized by the Abs can usually be identified by their specific inhibition of the Ab's activity on the bacterium; but with CTLs, soluble components of the stimulators compete poorly, if at all. However, some specific Ags can be identified through their ability to stimulate a vigorous CTL response (secondary response) in spleen cells from a previously immunized (primed) animal.

Induction of CTLs. As with other immune lymphocytes, the induction of CTLs is less well understood than are their effector activities. The use of purified Ags and H-2 molecules in synthetic phospholipid vesicles (liposomes) indicates that Ag and H-2 molecules in close proximity (<60–100 nm apart), in the same membrane, are required to trigger the development of precursor cells into mature CTLs. It is suspected that T helper (Ly-1) cells and macrophages are also necessary, as they are for the development of B cells and other T effector cells.

FIG. 22-9. Lysis of a fibroblast (target cell) by a CTL that is specific for H-2K or H-2D molecules on the fibroblast surface. Note the irregular shape of the crawling CTL **(arrow)**, and that the lysed target cell virtually explodes **(right)**, releasing its intracellular contents to the medium. (Chang TW, et al: Proc Natl Acad Sci 76:2917, 1979).

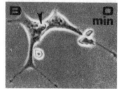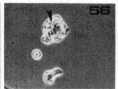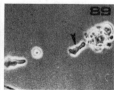

FIG. 22-10. Specificity of (CTLs) against target cells modified by attachment of small haptenic groups (**NP** [3-nitro-4-hydroxy-phenyl-acetate] or **NIP** [its 5-iodo analogue]). CTLs elicited by NP–cells lysed NP–target cells better than they did NIP–targets, whereas those elicited by NIP–cells lysed NIP–target cells better than NP–targets. CTLs elicited by NP–syngeneic cells do not lyse NP–target cells with different H-2K and H-2D alleles (**H-2 restriction**). (From Wall, K.A.)

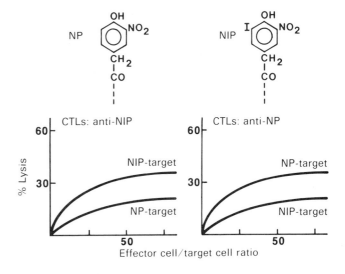

CELL-MEDIATED RESISTANCE TO INFECTION: COMPARISON OF T_{DTH} CELLS AND CTLS

The two forms of CMI discussed above, one due to T_{DTH} cells and the other to CTLs, represent fundamentally different mechanisms for controlling microbial infections. In virus infections, virus-specified Ags on the infected cell's surface elicit CTLs that later lyse cells infected with the same virus. If lysis occurs before mature virions are formed, the spread of virus to additional cells is prevented: *CTLs thus resemble a strategic defensive weapon.*

In contrast, CTLs seem to offer little protection against microbial cells that establish infection within host cells (usually macrophages).* Even if the infected host cells were to have surface neoantigens and were to be lysed by CTLs, the liberated pathogens, being microbial cells, would probably be free to reinfect other host cells. On the other hand, when the T_{DTH} cells react with Ags of these microbial cells the resulting lymphokines cause macrophages to differentiate into nonspecific "killers" with enhanced potency for ingesting and destroying the pathogens: in this situation T_{DTH} cells resemble tactical defense weapons.

Though CTL lysis and the T cell–macrophage system of DTH seem well suited to cope with viral infection and intracellular microbial infections, respectively, in many infections both types of mechanisms are probably operative. For instance, in many viral infections, such as mumps, DTH skin reactions can be elicited by viral Ags,

---

* Some better-known intracellular pathogens in humans are 1) **bacteria:** *Mycobacterium tuberculosis, Salmonella typhosa, Brucella abortus;* 2) **fungi:** *Cryptococcus neoformans,* yeast forms of *Histoplasma capsulatum* and *Candida albicans;* 3) **Protozoans:** *Trypanosoma cruzi, Toxoplasma gondii, Leishmania donovani,* malarial plasmodia.

though the CTLs are very likely more important in controlling the infection. The roles of the two systems in resistance to allografts and to tumors are considered in Chapter 23.

CONTACT SKIN SENSITIVITY

Contact skin sensitivity (**allergic contact dermatitis**) is induced, and its expression is elicited, simply by contact of low-molecular-weight chemicals (<1000 daltons) with intact skin. It is mediated by T cells, most likely T_{DTH}, but perhaps also CTLs. Reactions of this type are responsible for common allergic skin diseases in man, caused by contact with such simple substances as catechols of poison ivy plants and a great variety of synthetic chemicals, drugs, and cosmetics.

Induction. Sensitivity is normally induced by the percutaneous application of the sensitizer, but the skin can be bypassed by injecting the sensitizer in complete Freund's adjuvant. It has been suggested that skin lipids may exert an adjuvant effect comparable to the lipids of *M. tuberculosis* (see Adjuvants, Ch. 19).

Elicitation. Beginning 4 or 5 days after the initial exposure an application of the sensitizer almost anywhere on the skin surface elicits delayed inflammation.†

Guinea pigs are tested with a drop of dilute solution on the skin (after hair has been removed), followed by light massage. Humans are tested with a **patch test:** a piece of filter paper soaked with a

---

† Allergic contact dermatitis can be elicited on the ears of mice, but not elsewhere on the skin in this species, or in many others, perhaps because of anatomic differences in small cutaneous blood vessels.

dilute solution of sensitizer is placed on the skin, covered with tape, and left for 24 h. The degree of response in the area of contact is scored a few hours after the patch is removed, and again the following day.

Since sensitizers are potential irritants (because they react indiscriminately with most proteins; see Reactions with Proteins, below), they must be used at concentrations that avoid nonspecific inflammation; these are determined by testing nonsensitized individuals. With many potent sensitizers 0.01 M solutions are suitable.

Most simple sensitizers are relatively hydrophobic and readily penetrate the intact skin. They are usually administered in solvents that are partly nonvolatile, such as 1:1 acetone–corn oil, to prevent excessive concentration by evaporation. Hydrophilic sensitizers penetrate the skin less well, and relatively high concentrations are required (e.g., 0.5 M penicillin); these substances probably penetrate human skin via sweat ducts, and their penetration of guinea pig skin, which lacks sweat glands, can be augmented by the addition of nonionic detergents to the solvent.

The **time course** is the same as that of the tuberculin reaction: erythema and swelling appear at about 10–12 h and increase to a maximum at 24–48 h. Unusually intense reactions produce necrosis, and even without necrosis complete recovery can take several weeks.

Histologically (cf. Figs. 22-1 and 22-11), the dermis at the site of contact is invaded by monocytes (macrophages), lymphocytes, and basophils in small numbers, as in delayed-type responses to tuberculin. The more superficial epidermis, however, looks different: it is hyperplastic, invaded by monocytes and lymphocytes and, in addition, contains (in man but not in the guinea pig) intraepidermal vesicles, or blisters, filled with serous fluid, granulocytes, and mononuclear cells (Fig. 22-11). Lesions evolve slowly; they reach their peak histologically at 3–6 days.

Reactions with Proteins in Vivo. In accord with the general requirements of immunogenicity (Ch. 19), the actual immunogens are not the simple substances themselves, but the covalent derivatives they form with tissue (skin) proteins in vivo. For instance, among a group of 2, 4-dinitrobenzenes those that can form stable, covalent derivatives of protein SH and NH$_2$ groups in vivo are sensitizers, while those that cannot form such derivatives are not (Fig. 22-12).*

Persistence. Once established, in guinea pig or human, contact skin sensitivity persists for years, though it tends to wane. A patch test to evaluate its persistence can boost the level of sensitivity.

* Some macromolecules, such as denatured nucleic acids, become immunogenic on forming noncovalent complexes with proteins; the complexes are extremely stable because they involve many noncovalent bonds per interacting molecule. However, the inducers of contact sensitivity are small and can form only few noncovalent bonds per molecule; hence they must form covalent derivatives.

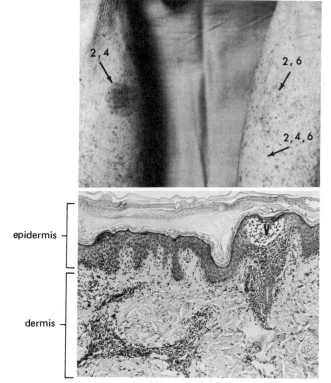

FIG. 22-11. Allergic contact dermatitis in man. The subject was sensitized to 2,4-dinitrofluorobenzene (DNFB) and then tested with 2,4-dinitrochlorobenzene **(2,4)**, 2,6-dinitrochlorobenzene **(2,6)**, and 2,4,6-trinitrochlorobenzene **(2,4,6**-TNCB). A positive response was evident at 24 h and photographed **(Top)** at 72 h. Specificity is shown by the strong reaction to 2,4-dinitrochlorobenzene, and the absence of reactions to the 2,6 and 2,4,6 analogs. DNFB and all the analogs tested form dinitrophenyl (or trinitrophenyl) derivatives of skin proteins in vivo (see Fig. 22-12). **Bottom.** Histology of the skin reaction. Note the characteristic intraepidermal vesicle **(V)** and the dense infiltration of dermis and epidermis by lymphoid cells. Epidermal cells around vesicles generally have a foamy cytoplasm (spongiosis).

The persistence of sensitizers in tissues is evident in the **flare reaction** to a small amount of sensitizer applied percutaneously as a diagnostic test. With an insensitive person this causes no reaction, but 10–20 days later the test site may flare up with a typical contact reaction. At that time repetition of the test anywhere on the skin elicits the characteristic response. Evidently enough sensitizer remains at the first site to provide an effective test dose when the subject becomes sensitized a few weeks later. The flare reaction thus resembles serum sickness (Ch. 21), though more extended in time and involving cell-mediated, rather than humoral, allergy.

Mechanisms. *T cells transfer contact dermatitis. The reaction is restricted by the major histocompatibility complex:* i.e., an inbred mouse recipient of T cells from a sensitized donor will develop lesions in response to a skin challenge

Substituents (X)
on C-1 of

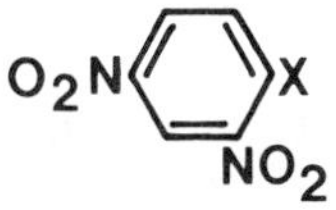

O_2N—⟨ring⟩—X (NO_2)	combine with protein in vivo		Ability to induce and elicit contact skin sensitivity
	with ε-NH$_2$ of lysine residues	with SH of cysteine residues	
X = -F	+	+	+
-Cl	+	+	+
-Br	+	+	+
-SO$_3$	-	+	+
-SCN	-	+	+
-SCl	-	+	+
-H	-	-	-
-CH$_3$	-	-	-
-NH$_2$	-	-	-

FIG. 22-12. Correlation among the C-1 substituted 2,4-dinitrobenzenes between the ability to form 2,4-dinitrophenylated proteins in vivo and the ability to induce and to elicit contact skin sensitivity. (Based on Eisen HN: In HS Lawrence (ed): Cellular and Humoral Aspects of the Hypersensitive States. New York, Hoeber, 1959)

only if it and the T cell donor have the same H-2 haplotype. It is probable that T$_{DTH}$ cells are specific for hapten–tissue protein conjugates that form in tissues. And it is possible that CTLs also contribute by lysing cells that have the sensitizing chemical attached to cell surface proteins.

Specificity. Contact skin reactions can distinguish between structures as similar as 2,4-dinitrophenyl, 2,6-dinitrophenyl, and 2,4,6-trinitrophenyl (Fig. 22-11).

However, the determinants of specificity appear to be larger and more complex than the smallest ligands involved in Ab-mediated reactions. For instance, guinea pigs sensitized with 2,4,6-trinitrophenyl–bovine γ-globulin (Tnp–BγG) respond more intensely to the whole immunogen than to the unsubstituted protein (BγG), and not at all to Tnp conjugated onto unrelated carriers, such as ovalbumin (Tnp–EA). Hence the response could be specific for the entire immunogen (Tnp plus carrier protein) or a large part of it: *delayed-type responses are usually carrier-specific.* In Ab-mediated reactions, in contrast, anti-Tnp Abs produced in response to Tnp–BγG form precipitates with Tnp–EA, bind Tnp–lysine, and mediate anaphylactic and Arthus responses with Tnp conjugated onto almost any protein.

The requirement for large antigenic units helps to clarify the specificity of contact skin allergy: for instance, a guinea pig that has been sensitized with Tnp–BγG, and that gives delayed-type responses to an injection of this conjugate, will not respond to a contact test with Tnp–chloride (TNCB, picryl chloride), though this compound readily forms Tnp conjugates with skin proteins in

vivo. Conversely, guinea pigs sensitized with TNCB give a delayed-type skin reaction to this substance, but not to conjugates of Tnp on BγG or other common proteins. It appears, therefore, that *self-proteins of skin, specifically modifed by covalent attachment of particular sensitizer groups (such as Tnp), are the actual antigenic units in contact skin responses.*

The simple group, however, is also required: for example, guinea pigs with contact skin sensitivity to TNCB do not give contact skin reactions to DNCB (see Fig. 22-11 for discrimination between TNCB and DNCB in man). Yet DNCB readily forms Dnp conjugates in vivo, probably by reacting with the same NH$_2$ and SH groups of the same skin proteins as TNCB (Fig. 22-12). The apparent requirement for a larger antigenic unit may be due to differences in the Ag recognition sites of T cells and of Ab molecules.

INTERFERENCE WITH CELL-MEDIATED IMMUNITY

TOLERANCE

Like Ab formation, CMI responses are normally elicited only by foreign Ags. The mechanisms responsible for unresponsiveness to self-Ags **(tolerance)** have been analyzed largely with model foreign Ags. Chapter 19 described tolerance with respect to Ab formation. In general, the same properties apply to tolerance to CMI.

1) Newborns. Early evidence that tolerance is more readily established in the fetus and the newborn than in the adult was provided by Billingham, Brent, and Medawar in their classic demonstration of tolerance to allografts, which are normally destroyed by CMI responses (see Tolerant Chimeras, Ch. 23). The injection of living cells from mice of one inbred strain into the newborn of a second strain established tolerance in the recipients for allografts from the first strain (but not for allografts from other strains). The tolerance persist as long as the inoculated cells continue to proliferate (sometimes for the recipient's lifetime), indicating that *persistence of Ag is necessary for the maintenance of tolerance.*

2) Specificity. Tolerance in CMI is as specific as in any other active immune response: for instance, mice rendered tolerant to 2,4-dinitrochlorobenzene (evaluated by contact skin sensitivity) can be sensitized to 2,4,6-trinitrochlorobenzene, and vice versa (see Fig. 22-11 for discrimination between these molecules in human contact skin reactions).

3) Tolerogens are immunogens. Just as in the induction of contact skin sensitivity (Fig. 22-12), tolerance can be elicited by simple chemicals that combine in vivo with proteins (such as DNFB), but not by those unable to combine (such as Dnp–lysine).

4) Route. As in Ab formation, whether an Ag functions in CMI as an immunogen or as a tolerogen depends upon the circumstances under which the lymphatic system is exposed to it. Thus, as was noted before, DNFB applied on mouse skin elicits contact skin sensitivity, but injected intravenously it elicits tolerance, i.e., contact skin sensitivity is not evoked when this sensitizer is subsequently applied to the skin. Similarly, guinea pigs fed 2,4,6-trinitrochlorobenzene (with care taken to avoid contact with skin) become specifically unable, later on, to develop contact skin sensitivity to this potent sensitizer (or to form anti-Tnp Abs against Tnp–proteins), while remaining normally responsive to other Ags.

5) Suppressor T cells and other mechanisms. Experimental and clinical evidence suggests that *tolerance is often due to T suppressor cells.* For instance, T cells from mice rendered tolerant to Dnp sensitizers can transfer specific Dnp unresponsiveness to normal syngeneic recipients. The cells responsible are probably T suppressors (Ly-23), which block the development of anti-Dnp T_{DTH} (Ly-1) cells. The intravenous route may be especially tolerogenic because 1) it permits the Ag to interact directly with T cells, rather than with macrophages, and 2) macrophage binding and presentation of Ag (in conjunction with products of the major histocompatibility complex) are probably necessary for inducing T_{DTH} cells but not for eliciting T suppressor cells.

Long after an intravenous injection of Ag, T cells from a still-tolerant mouse sometimes no longer transfer the unresponsive state to normal recipients, suggesting that mechanisms other than suppressor T cells can also account for tolerance. Two possibilities are 1) the deletion of clones of effector T cells, or 2) the production of Abs

that bind the Ag, preventing its stimulation of effector T cells (see Reaction with Antigen, below, and Enhancement in Ch. 23).

Desensitization. In contrast to the establishment of tolerance, where Ag acts as tolerogen *before it* has a chance to function as immunogen, densensitization is established by administering an Ag *after* an individual has already been sensitized to it. Partial densensitization probably occurs whenever an animal with DTH is tested simultaneously at many skin sites with the same Ag: the reaction at each site is usually less intense than when a single test is applied, probably because the number of sensitized T cells is limited. Similarly, in overwhelming infections (such as miliary tuberculosis or disseminated fungal or protozoan infections) DTH skin responses can no longer be elicited, perhaps because the large amount of released Ag saturates sensitized lymphocytes.

Comparable events occur with industrial workers, who sometimes develop and then lose skin sensitivity after prolonged and intense exposure to the sensitizer. However, deliberate desensitization in CMI by repeated administration of Ag or simple sensitizers has been difficult to achieve regularly in patients without frequently precipitating severe allergic reactions. Moreover, any desensitization that is achieved is short-lived (e.g. 5–10 days).

It is reported that American Indians formerly chewed poison ivy leaves as prophylaxis or treatment of poison ivy dermatitis. Recently a similar approach, refined by feeding the purified catechol responsible for poison ivy sensitization, was found to be of dubious value: the presence of the catechol in the feces occasionally produces in sensitized individuals an unusually severe perianal contact dermatitis (once referred to as the "emperor of pruritus ani").

Immune Deviation. Administration of Ag under some conditions (e.g., adsorbed on alumina gel) seems to induce Ab formation without DTH. When this happens CMI is not likely to be induced by a subsequent injection of the same Ag under what would ordinarily be effective conditions (e.g., incorporation in complete Freund's adjuvant). This effect borders on selective tolerance with respect to CMI. The mechanism is obscure. It may not be due entirely to the interception of Ag by the Abs formed as a result of the first inoculum, because the phenomenon has not been regularly transferred with antisera. If it could be consistently achieved, this **immune deviation** could have clinical value in enhancing the survival of allografts.

INTERFERENCE BY ANTIBODIES

Reaction with Antigen. Antisera can sometimes also block cell-mediated rejection of tumor grafts, resulting in **immune enhancement** of tumor growth (Ch. 23). This effect could arise in two ways: Ab molecules covering tumor-specific antigenic sites on the surface of tumor cells could block the induction of CMI (**afferent enhancement**) or they could block its expression (i.e., attack by T cells, or **efferent enhancement**). Similarly, Abs to histocompatibility Ags may prolong the survival of allografts of normal cells (see Enhancement Ch. 23), and lysis of some virus-infected cells by CTLs can be specifically blocked by Abs to viral Ags on the infected cells.

The effects of Abs on the induction of CMI are variable: some Ab–Ag complexes reduce the Ag's immunogenicity and others enhance it. The differences may derive from the variable effects of Abs with different heavy chains, and of Ab–Ag complexes with different mole ratios.

Antilymphocyte Serum (ALS). Studies with ^{51}Cr-labeled cells show that the antiserum made in one species against the thymus cells of another destroys particularly the latter's recirculating pool of lymphocytes, which are mostly T cells (Ch. 18); lymphocytes in lymph nodes and in other secondary lymphatic organs are much less affected. The differences could account for the selective depression of CMI reactions—i.e., the greatly prolonged survival of allografts by antilymphocyte serum (ALS: horse antiserum to human lymphocytes; see Ch. 23). Evidently ALS eliminates circulating T cell effectors of cell-mediated responses but leaves enough T cells in lymphatic organs to support Ab production, including Abs against the serum proteins of ALS. (The resulting risk of anaphylaxis and the serum sickness syndrome limits the clinical usefulness of ALS in preventing or treating rejection of allografts [Ch. 23].)

NONSPECIFIC SUPPRESSION

The cytotoxic agents that block Ab formation by preventing mitotic division (of cells in general), or by causing massive destruction of lymphocytes, were considered in Chapter 19. No new principles are involved in their application to CMI. The agents that block the proliferation of lymphocytes (γ- or x-rays; alkylating drugs, purine and folate analogs—see Fig. 19-18, Ch. 19), block the induction of CMI but not its expression: Ag-sensitive T cells and macrophages can evidently generate inflammatory responses without themselves undergoing cell division.

The agents most widely used clinically to inhibit cell-mediated allergic responses are the **11-oxycorticosteroids.** In therapeutic doses they act as nonspecific suppressors of inflammation, but they probably do not block induction of the sensitive state. Indeed, it is possible that mature T cells are particularly steroid-resistant, for the small proportion of mouse thymus cells (about 5%) that resist steroid action are both more mature than the others (they have less

Thy-1 and more histocompatibility Ags [Ch. 18]) and more effective in carrying out the graft-versus-host reactions (Ch. 23).

Immunosuppressants, such as ALS, prednisone (or another corticosteroid), and azathioprine (Imuran, a purine analog; see Fig. 19-18, Ch. 19) are often used clinically to control autoimmune disorders (such as those discussed below) or to prevent cell-mediated rejection of human allografts. Though they are relatively effective, the resulting suppression of all cell-mediated immune responses increases susceptibility to overwhelming infection by prevalent opportunistic bacteria and fungi, and also seems to increase the risk of lymphatic cancer.

GENETIC REGULATION OF CELL-MEDIATED IMMUNITY

Immune response (Ir) genes, associated with the major histocompatibility complex (Chs. 18 and 19), regulate CMI as well as Ab responses. Thus an animal that is unable, because of its Ir genes, to produce Abs to a particular Ag is also unable to develop DTH to that Ag. (For example, guinea pigs unable to make anti-Dnp Abs in response to Dnp–poly-L-lysine [Dnp–PLL] are unable to develop DTH to Dnp–PLL.) Because there are similar requirements for T cell activation in both Ab and DTH responses, the models proposed to account for Ir gene control of Ab production (see Genetic Controls, Ch. 19) can also be applied to DTH.

Thus both types of response require the activation of specific Ly-1 T cells by Ag bound to macrophages (Mφs). In Ab formation the activated T cell becomes a specific T helper; in DTH it becomes a specific T_{DTH} effector cell. T cell activation by the Mφ-T cell interaction depends on Ir alleles, which evidently determine whether macrophages function as **responders** for particular Ags: in responders, the macrophage's surface glycoproteins (I-region encoded Ia [I-associated] molecules in the mouse; see Ch. 18) somehow allow the Ag bound on the same cell to be presented effectively to T cells (see H-2 Restrictions, Ch. 18). It is also likely that Ir alleles regulate DTH by determining the extent to which T suppressor cells are formed.

SOME IMMUNE DISORDERS

HLA AND DISEASE

The high frequency of certain alleles of the human major histocompatibility complex (HLA) in certain diseases is probably related to the role of the major histocompatibility complex in immune reactions. A few of the associa-

tions between HLA Ags and disease, listed in Table 22-4, are already useful clinically: e.g., determining whether a person suspected of having ankylosing spondylitis (a form of arthritis) has the HLA-B27 Ag.

Some of the diseases listed probably have no immunologic basis (e.g., the iron-storage disease hemochroma-

TABLE 22-4. Diseases Associated with HLA Antigens

Disease*	Antigen	Frequency (%)§		Relative risk†
		Patients	Controls	
Celiac disease	Dw3	96	27	64.5
	B8	67	20	8.1
Chronic active hepatitis	DRw3	41	17	3.4
	B8	52	15	6.1
Myasthenia gravis	DRw3	32	17	2.3
	B8	39	17	3.1
Graves' disease	Dw3	53	18	5.1
	B8	44	18	3.6
Juvenile-onset diabetes	DRw3	27	17	1.8
	B8	32	16	2.5
	DRw4	39	15	3.6
Rheumatoid arthritis	DRw4	56	15	7.2
Myasthenia gravis: Japanese	DRw4	59	35	2.7
Juvenile-onset diabetes: Japanese	DRw4	65	35	3.4
Multiple sclerosis	DRw2	41	22	2.5
Ankylosing spondylitis	B27	90	8	103.5
Reiter's disease	B27	80	9	40.4
Hemochromatosis	A3	72	21	9.7
Psoriasis	Cw6	50	23	3.3
	B13	23	5	5.7
	B17	19	9	2.4
	B37	5	2	2.6
Psoriasis: Japanese	Cw6	53	7	15
	B13	18	1	22
	B37	35	2	26
	A1	30	2	21

*Subjects were Caucasoids, except as indicated.

†Relative risk = $P_d (1- P_c)/P_c (1 - P_d)$, where P_d = frequency in patients and P_c = frequency in controls.

§Frequency = proportion (%) of patients and of age- and sex-matched controls with the indicated Ag.

(Bodmer WF, Bodmer JG: Br Med Bull 34:314 1978)

tosis). However, many of the others, especially the inflammatory diseases (e.g., arthritis), could be due to immune responses that are abnormally active (as in the autoimmune reactions described below) or inadequate (as in susceptiblity to particular pathogenic microbes).

A disease associated with the D locus of the HLA complex (homologous to the I region of the mouse H-2 complex; see The Major Histocompatibility Complex, Ch. 18) could be due to immune response (Ir) or immune suppressor (Is) genes (see Genetic Controls, Ch. 19); and diseases associated with the A, B, or C loci (homologous to the K and D loci of the H-2 complex) could arise from anomalous H-2 restricted reactions of CTLs (see H-2 Restriction, above, and Ch. 18). It is necessary, however, to consider other possibilities also:

1) The associations shown in Table 22-4 are not absolute. Even in the most striking case, the presence of the B27 Ag in nearly all persons with ankylosing spondylitis, the opposite does not hold: only a small proportion of all persons with this Ag have the disease. There are several possible reasons: environmental factors, such as infection, may be necessary to precipitate the disease in susceptible individuals, or other genetic loci, linked or not linked to HLA, may also be required.

2) Because alleles at different loci can be inherited together (linkage disequilibrium; Ch. 18), a disease associated with the serologically detected product of a **marker allele** at one locus (such as the

B27 Ag) could be due to the effects of an unknown allele at a linked locus.

3) Strong positive associations with some alleles must mean that there are negative associations with other alleles at the same locus. The negative associations could be more significant than the positive ones (perhaps, as an example, because of the absence of a critical cell surface receptor); but the negative ones are also more difficult to recognize, because the frequency of most HLA Ags is low to start with and a further decrease is not likely to be regarded as significant.

CLINICAL T CELL DEFICIENCIES

In Chapter 19 we considered clinical procedures for evaluating suspected immunodeficient states (see especially Tables 19-6 and 19-7), and we described defects in B cells leading to deficiencies in immunoglobulin (Ig) and Ab formation. Some T cell deficiencies are considered here. The *fundamental distinctions between cell-mediated and Ab-mediated immunity came to be appreciated partly from clinical experience with such disorders.*

Thymic Aplasia. The principal form of this rare disorder, the DiGeorge syndrome, arises from defective embryonic development of the third and fourth pharyngeal outpouchings that give rise to the thymus, parathyroid, and thyroid glands (Ch. 18); hence hormonal deficiencies (hypoparathyroidism, rarely also hypothyroidism) are present in addition to profound immune defects. Another form of thymic aplasia, inherited as an autosomal recessive trait, occurs without associated hormonal deficiencies (sometimes called the Nezelof syndrome). Infants with either form suffer from severe recurrent infections; they also lack DTH to Ags of prevalent microbes, and fail to develop contact skin sensitivity to potent sensitizers, such as DNCB (Fig. 22-12). Their blood lymphocytes, being nearly all B cells (with normal distribution of surface Igs; Ch. 18), respond poorly to stimulation with phytohemagglutinin (PHA) or to allogenic cells in mixed lymphocyte cultures (see Table 19-7, Ch. 19; also see Mitogens, and Mixed Lymphocyte Reaction, Ch. 18).

Surprisingly, these infants can form Abs to many Ags, including proteins that are expected (from responses in mice) to be T-dependent; and their serum Abs are usually in the normal range. The probable explanation is that the thymic aplasia is incomplete in those infants who survive beyond 1 or 2 years after birth; indeed, in some who survive longer some T cell functions (such as stimulation of thymidine uptake by PHA) seem to recover spontaneously.

Severe Combined Immunodeficiency Disease. Infants with this disease suffer from crippling absence of both T and B cell functions, probably because they have defective hematopoietic stem cells (Ch. 18). Attempts to reconstitute the defective system with transplants of allogenic bone marrow cells from normal unrelated donors result in fatal graft-versus-host reactions (Ch. 23). However, when the donor is a sibling with well-matched histocompatibility Ags the transplanted cells and the host often survive, with prolonged recovery of all immune functions (see Bone Marrow Allografts, Ch. 23). With a female donor and a male recipient the female sex chromatin (Barr body) demonstrates the long-term persistence of donor cells in the recipient (chimerism).

Deficiencies of Adenosine Deaminase and Nucleoside Phosphorylase. Severe combined immunodeficiency disease is sometimes inherited as an autosomal (i.e., not sex-linked) recessive trait. About half of these children produce a defective form of the enzyme **adenosine deaminase (ADA)**, with greatly diminished activity (about 15% of the normal enzyme). The mutant enzyme is detectable in fibroblasts and lymphocytes, permitting prenatal diagnosis in amnion fibroblasts. Cells of affected individuals accumulate adenosine and deoxyadenosine triphosphate (dATP), and it has been suggested that the accumulated purine might block methyl transfer from S-adenosylhomocysteine, obstructing DNA synthesis and methylation: hence the defect has its greatest effect in rapidly dividing cells. Therapeutic trials indicate that transfused human RBCs, with normal ADA levels, can lower dATP levels and restore immunologic function in lymphocytes of ADA-deficient children. (These findings have been turned to possible advantage through the attempt to deliberately destroy lymphocytes in patients with

lymphocytic leukemia by using potent inhibitors of ADA [e.g., deoxycoformycin]; ADA inhibitors also offer possibilities as immunosuppressive agents.) Inhibition of methylation reactions (of DNA and RNA) should affect many cells, especially rapidly dividing ones. Why normal and neoplastic lymphocytes are especially vulnerable is not clear.

Another enzyme involved in purine degradation, **nucleoside phosphorylase,** is defective in some children with severe T cell deficiency; this defect is also inherited as an autosomal recessive trait. The patients have a severe T cell deficiency, but increased Ab responses and auto-Abs, perhaps because T suppressors are more affected than T helper cells (see Autoimmune Responses, below); lymphocytes in these children respond poorly to mitogens in vitro. Whether exogenously supplied enzyme can correct the defect is not yet known.

Acquired T Cell Deficiencies. In certain diseases of adults widespread involvement of the reticuloendothelial system with neoplasia (Hodgkin's disease), with intracellular parasitism (lepromatous leprosy or disseminated leishmaniasis), or with chronic granulomas (Boeck's sarcoid) is associated with striking deficiencies in cell-mediated immunity but preservation of Ab-forming ability. Affected individuals do not have delayed-type skin reactions to common Ags and their blood lymphocytes do not respond to PHA. However, their serum Igs are normal (or even markedly elevated) and they form Abs normally in response to immunization. The reasons for their selective deficiencies will probably become clear when lymphocytes from these patients are analyzed with the recently developed antisera for identifying subsets of human T cells (suppressors, helpers, etc.).

AUTOIMMUNE RESPONSES

Clinical and experimental observations show that individuals sometimes respond immunologically to certain of their self-Ags (auto-Ags). These important exceptions to the principle of self-tolerance help to clarify its fundamental mechanisms. Although they are frequently associated with disease it is often not clear whether these anomalous responses cause the disease or are its result, as we shall see below; hence it is necessary to emphasize the distinction between an **autoimmune response**, in which an individual makes Abs or produces T cells to a self-Ag, and an **autoimmune disease** (Table 22-5), which is a pathologic condition arising from an autoimmune response.

MECHANISMS

Autoimmunity could arise, in principle, from the following mechanisms:

1) A change in the distribution of a self-component, allowing its access to lymphoid cells that never encountered it previously.

2) A change in the structure of a self-component, or introduction of a crossreacting Ag. Either of these could lead to the formation of Abs and the appearance of

TABLE 22-5. Some Autoimmune Disorders in Man

Organ or tissue	Disease	Antigen	Detection of antibody*
Thyroid	Hashimoto's thyroiditis (hypothyroidism)	Thyroglobulin	Precipitin; passive hemagglutination; IF on thyroid tissue
		Thyroid cell surface and cytoplasm	IF on thyroid tissue
	Thyrotoxicosis (hyperthyroidism)	Thyroid cell surface	Stimulates mouse thyroid (bioassay)
Gastric mucosa	Pernicious anemia (vitamin B_{12} deficiency)	Intrinsic factor (I)	Blocks I binding of B_{12} or binds to I:B_{12} complex
		Parietal cells	IF on unfixed gastric mucosa; CF with mucosal homogenate
Adrenals	Addison's disease (adrenal insufficiency)	Adrenal cell	IF on unfixed adrenals; CF
Skin	Pemphigus vulgaris	Epidermal cells	IF on skin sections
	Pemphigoid	Basement membrane between epidermis and dermis	IF on skin sections
Eye	Sympathetic ophthalmia	Uvea	Delayed-type hypersensitive skin reaction to uveal extract
Kidney glomeruli plus lung	Goodpasture's syndrome	Basement membrane	IF on kidney tissue; linear staining of glomeruli (see Fig. 21-18B, Ch. 21)
Red cells	Autoimmune hemolytic anemia	Red cell surface	Coombs' antiglobulin test
Platelets	Idiopathic thrombocytopenic purpura	Platelet surface	Platelet survival
Skeletal and heart muscle	Myasthenia gravis	Muscle cells and thymus "myoid" cells	IF on muscle biopsies
Brain	Allergic encephalitis	Brain tissue	Cytotoxicity on cultured cerebellar cells
Spermatozoa	Male infertility (rarely)	Sperm	Agglutination of sperm
Liver (biliary tract)	Primary biliary cirrhosis	Mitochondria (mainly)	IF on diverse cells with abundant mitochondria (e.g., distal tubules of kidney)
Salivary and lacrimal glands	Sjögren's disease	Many: secretory ducts, mitochondria, nuclei, IgG	IF on tissue
Synovial membranes, etc.	Rheumatoid arthritis	Fc domain of IgG	Antiglobulin tests: agglutination of latex particles coated with IgGs, etc.
	Systemic lupus erythematosus (SLE)	Many: DNA, DNA–protein, cardiolipin, IgG, microsomes, etc.	Precipitins, IF, CF, LE cells

*IF = immunofluorescence staining, usually with fluorescent anti–human Igs (see Immunofluorescence, Ch. 16); CF = complement fixation (see Appendix, Ch. 20).

(Based on Roitt I: Essential Immunology, Oxford, Blackwell, 1971)

lymphocytes that react (or crossreact) with the native self-component.

3) Defective production of T suppressor cells, allowing the development of "forbidden" clones of B and T cells and production of anti-self Abs and T effector cells.

4) Immune responses to persistent and unrecognized extrinsic Ags, as in certain chronic viral infections. These can generate chronic allergic disorders that are difficult to distinguish from anti-self immune reactions.

1) **Altered distribution of self-antigens.** Many auto-Ags (e.g., from eye lens, spermatozoa, brain tissue) have little or no opportunity to establish tolerance because they are normally confined anatomically to sites that prevent their access to lymphocytes. Immune responses to such an Ag are readily produced experimentally by removing the Ag and injecting it back into the same animal. Clinical incidents that release such Ags have the same effect: Abs to heart muscle appear after myocardial infarction, and Abs to eye Ags after the trauma of eye injury. In these instances *disease gives rise to the autoimmune response,* rather than the reverse. Nonetheless, *such responses can be self-perpetuating:* once the response is initiated the resulting allergic inflammation in the target organ probably leads to contact of Ag with infiltrating lymphocytes, resulting in further sensitization.

2) **Altered forms of self-antigens.** Few self-Ags are totally isolated from lymphocytes: for example, thyroglobulin (Tg), once thought to be completely confined to thyroid nodules, evidently reaches the lymphatics, for sensitive assays reveal its presence in normal serum at concentrations (about 0.01 $\mu g/ml$) that could establish T cell tolerance. Hence B cells that make Igs to self-Tg are probably present but unable to respond to self-Tg. However, if a foreign determinant appears on Tg it will stimulate T helper cells that are specific for the new determinant. These helpers can then cooperate with the existing (quiescent) anti-Tg B cells, which eventually secrete anti-Tg Abs (see Fig. 19-14).

Thus rabbits immunized with chemically modified rabbit Tg, or with crossreacting hog Tg, develop thyroiditis, associated with Abs that react with native rabbit Tg. Similar lesions and Abs can even appear when rabbit Tg itself, incorporated into Freund's adjuvant, is injected back into rabbits: denaturation during preparation could expose new carrierlike groups. Similarly, encephalitis occasionally develops in people injected with rabies vaccine made from infected rabbit brain (Allergic Encephalomyelitis; see below; also T and B Cells in Immunologic Tolerance, Ch. 19).

Some bacteria carry antigenic determinants that resemble those of the host; e.g., in rheumatic fever some of the Abs produced against group A streptococci seem to crossreact with human heart tissue. In addition, in some bacterial infections the adjuvant effects of certain products (e.g., mycosides, endotoxins) might facilitate an autoimmune response.

3) **Defective production of suppressor T cells and emergence of anti-self clones.** In developing the clonal selection hypothesis Burnet assumed that lymphocyte clones are somehow eliminated as they arise if they are specific for auto-Ags. A breakdown in the hypothetic elimination mechanism could lead to the emergence of autoreactive **forbidden clones.** Clinical findings support this suggestion: when an individual who suffers from one autoimmune disorder is examined closely, signs of other autoimmune responses are likely to be found (see Multiplicity of Responses, also p. 512).

Defective production of **T suppressor (T_s) cells** could account for this multiplicity of autoimmune responses because, as the following observations suggest, these cells normally appear to be responsible for preventing the development of anti-self B and T cell clones.

Bone marrow transplants from one human monozygotic (genetically identical) twin to another have occasionally resulted in graft-versus-host (GvH) reactions in the recipient. The transplants, performed to save a mortally ill leukemic twin, were carried out after the recipient had been given a large dose of total-body x-irradiation and had his thymus removed (to eliminate all leukemic cells). Hence, T_s cells (Ch. 18), which are continuously produced in the thymus and circulate as short-lived cells, were virtually eliminated. Without them, the donor's lymphocytes were free to produce immune responses against what were, in effect, auto-Ags in the recipient (anti-RBC Abs, etc.). The resulting rash, diarrhea, hepatitis, etc. was indistinguishable from the ordinary graft-versus-host (GvH) syndrome, which invariably occurs when allogenic lymphocytes are transferred to immunologically unresponsive recipients (see Graft-versus-Host Reactions, and Bone Marrow Allografts, Ch. 23).

An experimental parallel to this tragic clinical situation is seen when lethally irradiated, thymectomized mice are given, from a syngeneic donor, a transplant of both bone marrow cells (a source of B cells) and T cells from which Ly-23 cells, including T suppressors, have been eliminated (by anti–Lyt-2 Abs and complement). Because the remaining T cells (Ly-1 helpers) and the B cells are unchecked by T suppressors, they can produce Abs to auto-Ags; e.g., to RBCs (resulting in autoimmune hemolytic anemia).

4) **Chronic viral infection.** Mice infected at birth with certain temperate viruses (e.g., lymphocytic choriomeningitis virus, lactic dehydrogenase virus, and others that also seem not to injure infected cells) become lifelong carriers, producing Abs that form virus•Ab complexes without neutralizing the infectivity of the virus (see Viral Complexes, under Immune Complex Diseases, Ch. 21). Budding virions on infected cells, or viral Ags adsorbed on RBCs or other cells, simulate self-Ags: their combination with antiviral Abs (or perhaps with specific T cells) can give rise to chronic disorders, such as hemolytic anemia, that resemble autoimmune diseases. Some of the ostensibly autoimmune human diseases might similarly

be due to unrecognized chronic viral infections (an analogous situation, hemolytic anemia due to penicillin, is discussed below).

CHARACTERISTICS OF AUTOIMMUNE DISEASES

Antibody- versus Cell-Mediated Reactions. All the mechanisms that cause allergic reactions to foreign Ags (see Table 21-1, Ch. 21) can participate in autoimmune diseases. Various **auto-Abs** (with or without C) can **act directly on cells,** e.g., lysing RBCs or platelets, or injuring cells of the thyroid gland (thyroiditis). These Abs can also form **autoimmune Ab•Ag aggregates,** illustrated by systemic lupus erythematosus (SLE), in which some complexes formed from DNA, anti-DNA, and C become lodged in kidney glomeruli and lead eventually to progressive glomerulonephritis and renal failure (see Immune Complex Diseases, Ch. 21). Similarly, IgGs that function as auto-Ags combine with certain IgM (or IgG) molecules that serve as auto-Abs **(rheumatoid factors),** forming aggregates that localize around synovial membranes and probably contribute to the joint inflammation of rheumatoid arthritis (Ch. 21)

Autosensitized T lymphocytes have not been shown to cause human disease, but hints are provided by the striking mononuclear infiltrates in many autoimmune disorders (atrophic gastritis in pernicious anemia, thyroiditis, allergic encephalomyelitis, etc.). Stronger support for cell-mediated mechanisms is provided by much experimental work in which bits of tissue (from thyroid, brain, adrenal, etc.) are removed surgically, emulsified in Freund's adjuvant, and injected back into the same animal. After 1–2 weeks the recipient develops lesions infiltrated with mononuclear cells in the corresponding organ (as well as specific serum Abs). Moreover, viable lymphocytes from the affected animal, but not serum, will usually transfer the disease to a normal recipient.

Failure to transfer the disorder with Ab-containing serum is not decisive evidence, however. Auto-Abs, like other Abs, are likely to be heterogeneous; and with the continuous presence of the target self-Ag those Abs with highest affinity are probably bound preferentially, leaving as free auto-Abs in serum the relatively ineffectual low-affinity molecules.

Localized versus Disseminated Disease. Autoimmunity can affect almost every part of the body (Table 22-6). Some responses are directed to **organ-specific** Ags and may even be limited to a particular cell type (e.g., parietal cells of the gastric mucosa in pernicious anemia). Other responses are directed to widely distributed Ags and are associated with **disseminated** disease (e.g., antinuclear Abs in systemic lupus erythematosus). In still other diseases the responses are **intermediate** between these extremes: for instance, in Goodpasture's disease, characterized by chronic glomerulonephritis and pulmonary hemorrhages, Abs are deposited on the basement membrane of both kidney glomeruli and lung parenchyma. (The strong crossreaction between these two basement membranes can be shown by localization in vivo of ^{125}I-labeled heteroantiserum to either.)

Multiplicity of Responses. As mentioned above, *an individual who makes one autoimmune response is likely to make others.*

For instance, 10% of persons with autoimmune thyroiditis have pernicious anemia, which is present in only 0.2% of the population at large (of the same age and sex distribution). Similarly, thyroid disease is found with excessive frequency in those who suffer from pernicious anemia. Serologic evidence for multiple reactions is even more frequent: 30% of those with autoimmune thyroiditis have Abs to gastric parietal cells, and 50% of those with pernicious anemia have Abs to thyroid, though the two kinds of Abs are entirely noncrossreacting. Similar associations are found among the disseminated group of autoimmune diseases: persons with systemic lupus erythematosus often have evidence of rheumatoid arthritis, autoimmune hemolytic anemia, or thrombocytopenia.

Genetic Factors. Relatives of patients with autoimmune thyroiditis commonly have antithyroid Abs, and relatives of those with pernicious anemia have a high incidence of Abs to gastric parietal cells. Moreover, some inbred animal strains suffer from a high frequency of autoimmune disorders.

All mice of the NZB strain (New Zealand Black) eventually develop autoimmune hemolytic anemia, and most of the hybrids (B × W) made by crossing NZB with another inbred strain (New Zealand White, NZW) develop a syndrome that is strikingly similar to systemic lupus erythematosus in man.

In accord with their diverse autoimmune responses, these mice are made tolerant to foreign Ags only with great difficulty. Their anomalies can be accounted for by several genetic defects: 1) they produce excessive amounts of IgM, and 2) they produce few T_s cells, especially as they age. Repeated *injection of aging NZB mice with thymocytes (a rich source of T_s cell precursors) from young NZB mice seems to postpone the development of autoimmune Abs and autoimmune disease.*

Age and Sex. The frequency of autoimmune responses increases with age. For unknown reasons, they are also more frequent in women than in men.

SOME REPRESENTATIVE AUTOIMMUNE DISEASES

Acquired Hemolytic Anemia. In this disorder a person's Abs react with his own red blood cells (RBCs). Such Abs are sometimes found in persons with other diseases that might alter self-Ags or introduce crossreacting ones, e.g., malignant neoplasms of lymphoid tissue, syphilis, mycoplasma pneumonia, infectious mononucleosis. More often, however, this disorder is idiopathic (i.e., it is unassociated with other diseases or with known exposure to agents toxic to RBCs).

Abs to the patient's red cells are demonstrated by rabbit antiserum to human Igs in two ways. 1) In the **direct antiglobulin (Coombs) test** the patient's washed RBCs are agglutinated by the antiserum. 2) In the **indirect test,** RBCs from another person are incubated with the patient's serum, washed, and examined for clumping by the rabbit antiserum.

The anti–red cell Abs have no hemolytic activity in vitro, but they accelerate destruction of RBCs in vivo. Thus, transfused normal RBCs have a shortened survival in affected patients only if the injected cells have the particular surface Ag that binds the patient's auto-Ab, indicating that red cell breakdown is due to the bound Abs. The Ab-coated (opsonized) cells are phagocytized by macrophages and their breakdown occurs especially in the spleen. Therapy includes nonspecific immunosuppressants (Ch. 19) and splenectomy.

Mice of the inbred NZB strain almost invariably develop hemolytic anemia as they age, and viable lymphoid cells (including Ab-forming clones) from older (affected) animals transfer the disease efficiently to young, unaffected mice of the same strain. These mice are chronically infected from birth with type C leukoviruses (see murine leukemia virus; see Ch. 66), and it is possible that what looks like autoimmune disease is really due to a conventional immune reaction to viral Ags adsorbed on RBCs or present on budding virions of infected cells.

Thrombocytopenic Purpura. In this disease platelets can decline to about 1/10 the normal level and bleeding occurs in many organs, including the skin, causing petechial rash and purpura. In the dramatic experiment that first provided evidence of the immune nature of this disease, a human volunteer injected with a patient's plasma suffered a precipitous fall in platelets (Fig. 22-13) and extensive bleeding into internal organs and skin. A patient's serum can also cause clumping of normal platelets, and lysis if complement (C) is present. Infants born to mothers with the disease may

have transient thrombocytopenia and bleeding, due to placental transmission of the maternal Abs.

Allergic Encephalomyelitis. When laboratory animals (e.g., rats, guinea pigs, monkeys) are injected with suspensions of central nervous system tissue from individuals of the same or other species they develop patchy areas of vasculitis and demyelination in the brain and spinal cord (Fig. 22-14). This response is readily evoked with a single injection of a small amount of brain or spinal cord tissue, even from the same animal, in Freund's adjuvant.

The immune nature of the disease is indicated by the following. 1) Lesions appear 9 days or more after the primary injection, but the onset is more rapid in animals that have recovered and are then reinjected (secondary response). 2) The lesions are specific: they follow inoculation only of myelin-containing tissues, and appear only in myelinated tissue, especially white matter of the brain. 3) Lymphocytes from inoculated animals produce cytopathic effects on myelinated brain tissue and glial cells in culture. 4) The lymphocytes can also cause specific neural lesions in nonsensitized syngeneic recipients. 5) Intradermal injection of myelinated tissue evokes a delayed-type skin response in affected animals. 6) Serum Abs that react with brain tissue can be demonstrated. However, they seem not to play a major role in the disease: the intensity of the lesions does not parallel their levels, and serum fails to transfer lesions to recipient animals.

The responsible self-Ag is a small protein (18,000 daltons) that 1) is heterogenetic (i.e., present in many species), 2) is organ-specific (found only in the central nervous system), and 3) constitutes at least 30% of the protein in myelin. Only 0.1μg (in complete Freund's adjuvant) is sufficient to elicit severe allergic encephalomyelitis in a guinea pig.

The protein's amino acid sequence is fully established. Of its many proteolytic fragments, the smallest with encephalitogenic activity is a nine-residue peptide (**Phe-Ser-Trp-Gly-Ala-Glu-Gly-Gln-Lys**) that contains the protein's only tryptophan residue. 2-Hydroxy-5-nitrobenzyl bromide

$$HO-\text{\raisebox{0pt}{⬡}}-NO_2$$
$$CH_2 \cdot Br$$

attaches only to this tryptophan, and eliminates the protein's encephalitogenic activity.

In view of the experimental background it is understandable that encephalitis often occurs in humans after vaccination with standard rabies virus, which is a suspension of infected rabbit brain. Rabies vaccine from virus grown in human cultured cell lines (not nerve cells) or in duck embryo tissues that lack myelin is now preferred (Ch. 61). The demyelinating encephalitides that occasionally follow measles and vaccinia could arise because infection of the nervous system brings the encephalitogenic Ag into contact with inflammatory (lymphoid) cells. A similar mechanism is suspected for **multiple sclerosis,** a common disabling neurologic disorder of unknown etiology with characteristic focal demyelinating lesions.

Thyroiditis. Patients with chronic inflammation of the thyroid (Hashimoto's disease) suffer destruction of secretory cells and loss of thyroid function. Their serum contains Abs that react in high titer with thyroglobulin (Tg), and others that react with Tg-free particulate fractions of thyroid (Fig. 22-15). Moreover, in an inbred strain of chickens that spontaneously develops antithyroid Abs and

FIG. 22-13. Passive transfer of thrombocytopenic purpura. The platelet count of a normal human volunteer drops following transfusion (**arrows**) of blood from a patient with idiopathic thrombocytopenic purpura. Ig fractions from other donors with the disease produce similar results. Some of the recipients suffer bleeding in internal organs simulating the natural course of the disease. (Harrington WJ et al: J Lab Clin Med 38:1, 1951)

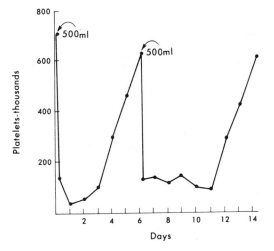

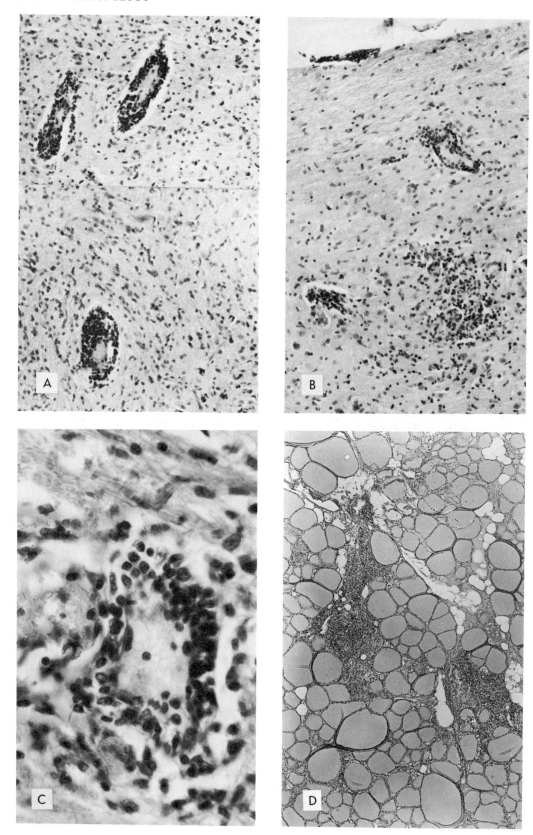

thyroiditis (and finally hypothyroidism and obesity) the Abs appear to be causal, since neonatal bursectomy, but not thymectomy, prevents the anomalies. However, experimental thyroiditis induced by injection of Tg or thyroid tissue, in other animals, is very similar in pattern to experimental allergic encephalomyelitis. The suspicion therefore remains that cell-mediated rather than (or in addition to) humoral autoimmunity is involved in the lesions of the human disease.

The interaction of Abs with Ags on cells sometimes causes **cell proliferation and differentiation rather than destruction**—e.g., when anti-Igs react with B lymphocytes (Blast Transformation, Ch. 18), or when appropriate Abs react with sea urchin eggs. Similar mechanisms may be involved in some patients with hyperplastic and hyperfunctional thyroid glands, who have auto-Abs to thyroid (often called **long-acting thyroid stimulators**, or LATS). If present during pregnancy, these autoimmune Abs cross the placenta and cause neonatal hyperthyroidism, which subsides within a few weeks of birth (as the maternal IgG is degraded; Ch. 19). The auto-Ab might bind specifically to the same cell surface receptors as thyroid-stimulating hormone, for the actions of both are potentiated by theophylline, indicating that they act through the adenyl cyclase system (see Fig. 21-12, Ch. 21).

Pernicious Anemia. Defective red cell maturation in this disease is due to lack of vitamin B_{12}. Affected persons have atrophic gastritis; their poor absorption of ingested vitamin B_{12} is due to lack of intrinsic factor (IF), a protein that is necessary for this absorption and is secreted into the stomach by special (parietal) cells of the gastric mucosa. Most patients with the disease have Abs to parietal cells (revealed by immunofluorescence of gastric biopsies) and CMI to these cells could also be present. In addition, auto-Abs to IF itself are also involved, for large amounts of oral B_{12} (which ordinarily cure the anemia) are ineffective if given together with serum from many patients, *and the active serum contains Abs to IF.* Evidently the inhibitory effect of anti-IF is exercised within the stomach, for absorption of the fed vitamin remains normal in immunized human volunteers even when they develop a high serum titer of these Abs, or intense delayed-type hypersensitivity to IF. The gastric Abs, produced by plasma cells in the gastric lesions, are able to function within the stomach (normally extremely acidic) because the atrophy of the affected mucosa reduces the secretion of HCl. Thus, *by decreasing production of both IF and HCl, and by secreting Abs to IF, the autoimmune atrophic gastritis contributes to (and may cause) B_{12} deficiency and pernicious anemia.*

Systemic Lupus Erythematosus. At various stages in this complex disease patients produce several of an immense variety of auto-Abs to various blood cells, clotting factors, and intracellular components (e.g., mitochondria, nuclear components; Fig. 22-15). The resulting difficulties include hemolytic anemia, leukopenia, thrombocytopenic purpura, and bleeding tendencies. Some Abs (e.g., to DNA) seem to cause difficulty mainly through deposition of Ab•Ag•C complexes in the walls of small blood vessels, leading to widespread vasculitis (Arthus lesions); deposits in the capillaries of renal glomeruli are especially threatening, for they can lead to glomerulonephritis and renal failure (see Immune Complex Diseases, Ch. 21). The production of so many auto-Abs is accompanied by elevated serum Ig levels; during the active phases of the disease serum C falls to low levels as C is bound by Ab•Ag aggregates.

The multiplicity of Abs to self-Ags and the appearance of **lupus erythematosus (LE) cells** are diagnostic. These cells appear simply on incubation of blood or bone marrow from patients: the breakdown of some cells evidently releases DNA, which combines with serum anti-DNA; the resulting Ab•Ag aggregates are taken up by granulocytes, which acquire a characteristic appearance (a large amorphous mass in the center of the cell, with the surrounding multilobed nucleus pushed to the periphery).

PATHOLOGIC CONSEQUENCES OF AUTOIMMUNE REACTIONS

When auto-Abs were first discussed, at the turn of the century, they were imagined to occur either not at all or with disastrous consequences. Now, with the development of increasingly sensitive and reliable assays for Abs and for hypersensitivity, it is clear that autoimmune responses are not so uncommon. However, their pathologic effects are often conjectural, with three possibilities to be considered: 1) The response can be innocuous. For example, though Abs to cardiolipin in a person with syphilis can react with cardiolipin extracted from his own tissues, these Abs do not appear to be pathogenetic: they may be present at high titer in apparently healthy persons. Cardiolipin is buried within cell membranes and is probably inaccessible. 2) The response can be secondary to another disease (e.g., antithyroglobulin Abs following the trauma of partial thyroidectomy); but it may then be responsible for continuing disease (i.e., inflammation in the affected organ introduces new lymphocytes that increase autosensitivity). 3) The response can be the causal factor in disease, e.g., in autoimmune hemolytic anemia or thrombocytopenic purpura.

Witebsky's Criteria. Because the spectrum of allergic effects is so wide—from hemolysis to demyelinating inflammatory lesions—autoimmune processes are com-

---

Fig. 22-14. Autoimmune allergic reactions. **A–C.** Allergic encephalomyelitis in a rat sensitized adoptively with viable lymph node cells from a donor rat that had been immunized with rat spinal cord (injected in complete Freund's adjuvant). The recipient, which had severe ataxia and hind-leg paralysis 5 days after receiving the donor's lymphocytes, was sacrificed at 7 days. Sections of brain show a focal inflammation with perivascular infiltration by mononuclear cells. H & E stain. **D.** Thyroiditis in a rabbit that had been immunized repeatedly with hog thyroglobulin. A small proportion of the rabbit's Abs (to the hog protein) reacted also with rabbit thyroglobulin. Note intense focal infiltration of the immunized rabbit's thyroid with mononuclear cells. Similar lesions appear in animals injected with their own thyroglobulin or with homogenates of a bit of their own thyroid tissue. (**A–C**: Paterson PY: J Exp Med 111:119, 1960; **D**: Witebsky E, Rose NR: J Immunol 83:41, 1959) **A** and **B**, ×130 reduced; **C**, ×500 reduced; **D**, ×60 reduced.)

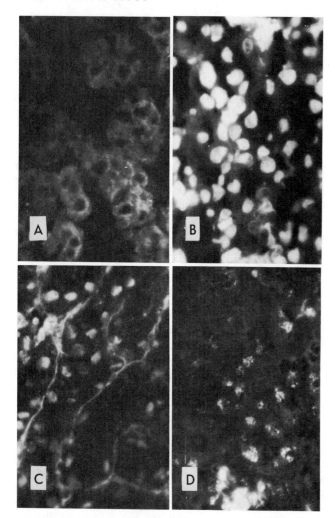

FIG. 22-15. Variety of antitissue Abs in sera of different patients with systemic lupus erythematosus. Mouse kidney sections were incubated with the sera and, after washing, were stained with fluorescein-labeled rabbit antiserum to human IgG. Similar reactions occur with human tissues. **A.** Normal human serum control. **B.** Abs in a lupus serum react uniformly with all nuclei, giving homogeneous nuclear staining. **C.** Abs react with nuclei and with basement membrane of renal tubules. **D.** Abs react with selected parts of nuclei, giving speckled staining. (Courtesy of Drs. E. Tan and H. Kunkel) (×250)

monly invoked to explain many diseases of unknown etiology; and highly sensitive assays often reveal auto-Abs. In an effort to provide guidelines for interpretation Witebsky suggested the following criteria (reminiscent of Koch's postulates for bacterial etiology): an autoimmune response should be considered the cause of a human disease if 1) it is regularly associated with that disease, 2) immunization of an experimental animal with Ag from

the appropriate tissue causes it to make an immune response, 3) the responding animal develops pathologic changes similar to those of the human disease, and 4) the experimental disease can be transferred to a nonimmunized animal by serum or by T cells.

ALLERGY TO DRUGS

Hypersensitivity reactions to drugs are among the most common allergic disorders in man. Most of the principles involved were noted in early sections of this chapter and in Chapter 21. *Sensitizing drugs, like inducers of contact skin sensitivity, must form stable covalent derivatives of proteins in vivo;* the derivatives can induce Ab formation and virtually any type of hypersensitivity. In the following section both the immediate- and delayed-type responses are discussed.

Most drugs would be too toxic to be useful if they reacted rapidly with cell proteins in vivo. If, however, the reaction is sufficiently slow a drug may be tolerably nontoxic and yet may form effective hapten–protein immunogens. For most drugs that cause allergic reactions, however, it is likely that reactive contaminants, or reactive metabolic derivatives are solely (or additionally) responsible for the formation of immunogenic conjugates. The mechanisms involved are well illustrated with penicillins, which are models for drug allergies in general.

PENICILLIN ALLERGY

Penicillin is probably the least toxic drug in use: 20 g/day can be given safely for prolonged periods. Nevertheless, some 5%–10% of persons given penicillin repeatedly can become sensitized; about 1% of these can undergo anaphylaxis.* (It is estimated that this reaction causes about 300 deaths/year in the United States.)

Immediate-Type Hypersensitivity (Table 22-6). Penicillin is unstable, and most of its solutions contain at least small amounts of **penicillenate,** a highly reactive derivative that forms **penicilloyl** and other substituents of amino and sulfhydryl groups of proteins (Fig. 22-16). Through its strained β-lactam ring penicillin itself can be directly attacked by nucleophilic groups (e.g., ε-NH_2 of lysyl residues), leading also to the formation of **penicilloyl–protein conjugates.** Penicilloic acid also forms a variety of minor determinant groups (Fig. 22-16). Conjugates with different haptenic moieties stimulate the formation of specific noncrossreacting Abs.

---

* Oral administration is less hazardous because the drug is absorbed more slowly, but it can also lead to severe reactions, such as the serum sickness syndrome (Ch. 21).

TABLE 22-6. Antibody-Mediated Reaction to Penicillin

Time of onset after penicillin administration	Clinical findings	Antibodies		Comments
		Class	Specificity	
2–30 min	Diffuse urticaria, hypotension, shock, respiratory obstruction	IgE	Penicilloyl and minor determinants	Anaphylactic Shock
3–72 h	Diffuse urticaria, pruritus; rarely respiratory symptoms	IgE IgM	Penicilloyl Penicilloyl	IgG antipenicilloyl eventually rises to high titer and probably acts as blocking Ab, leading to spontaneous termination of reaction
3 days to several weeks	Urticaria (sometimes recurrent), arthralgias, erythematous eruptions	IgE IgM	Penicilloyl and minor determinants Penicilloyl	Serum sickness syndrome

(Derived from Levine BB: In Textbook of Immunopathology, Miescher PA, Mueller-Eberhard HJ (eds): Vol. 1, NY, Grune and Stratton, 1968)

Intradermal injection of small amounts of the protein conjugates (e.g., 0.01μg) could provide a diagnostic test for allergy to penicillin.* However, even small amounts of the conjugates are potentially immunogenic. A safer reagent has been prepared by introducing penicilloyl substituents into poly-D-lysine, with an average of 20 residues per molecule, to form a ligand that is multivalent but not detectably immunogenic. This reagent elicits specific wheal-and-erythema responses as effectively as penicilloyl–proteins, and its specificity can be demonstrated by inhibition with univalent ligands, e.g., ε-penicilloyl-aminocaproate:

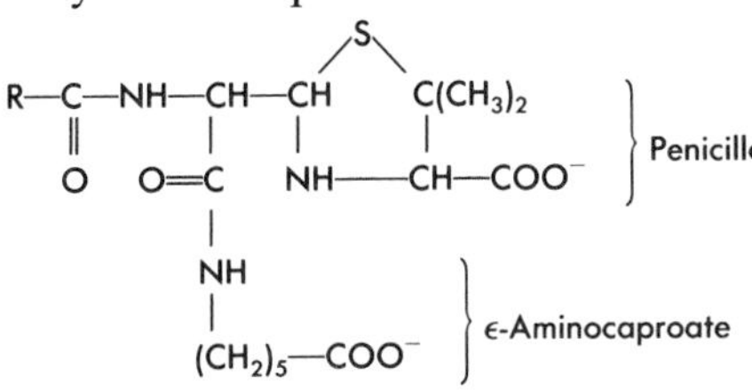

High-molecular-weight contaminants in solutions of penicillin can sometimes also cause anaphylactic reactions. The most significant are **penicilloyl–proteins** in which the proteins are probably acquired, during manufacture, from the fungus (Penicillium) or from the E. coli enzymes that are sometimes added to cleave R–CO side chains in the preparation of 6-aminopenicillanic acid (6-APA) for the production of semisynthetic penicillins (Fig. 22-16; see also Ch. 7). Removal of the protein (e.g., by proteolysis with enzymes attached covalently to solid Sepharose beads, to avoid introducing another immunogenic impurity) reduces the frequency of reactions to penicillin. However, reactions can still be elicited with other high-molecular-weight contaminants (polymers of 6-APA or of penicillins).

* Penicillin itself only infrequently yields wheal-and-erythema responses, even in highly sensitive persons. Positive skin reactions, when they occur, are probably due to high-molecular-weight contaminants. The unconjugated drug is univalent: if bound by Abs it should inhibit rather than elicit allergic reactions.

Ligand and Antibody Competition. Many persons with reaginic (IgE) Abs to penicilloyl do not suffer anaphylactic reactions after injections of penicillin, even when the solutions contain penicillenate, a major source of penicilloyl conjugates. One probable reason is that the penicillin molecule itself, and the penicilloic acid formed as its major degradation product (Fig. 22-16), are **univalent ligands for antipenicilloyl Abs:** these ligands are probably bound only weakly by the Abs, but they are present initially in great excess over the penicilloyl proteins that form in vivo. They are thus likely to act as specific inhibitors of anaphylaxis. However, the univalent small molecules are excreted fairly rapidly, and so they are unlikely to influence the later, serum-sickness–like reactions (e.g., rash, fever, joint pains) that often occur 3–14 days or more after penicillin is injected into persons with high antipenicilloyl Ab titers. In addition, these persons also usually have nonreaginic Abs (IgGs, IgM) of the same specificity, which probably act as blocking Abs (Ch. 21) (Table 22-6).

Hemolytic anemia due to penicillin allergy sometimes occurs in those receiving high doses of penicillin for several weeks: antipenicillin Abs react with penicillin adsorbed on RBCs, causing excessive cell destruction and anemia. (Similar events with other, unrecognized exogenous Ags could simulate autoimmune hemolytic anemia; see Chronic Viral Infection, above.)

Delayed-Type Hypersensitivity. This is also common, as allergic contact dermatitis, particularly among handlers of penicillin in bulk (e.g., nurses, pharmaceutical workers). One of the determinants that specify DTH reactions involves D-penicillamine, which elicits patch test reactions as effectively as penicillin in some persons with contact skin sensitivity to the drug. This compound forms stable mixed disulfides with cysteine in vitro, and probably acts in vivo by combining with protein sulfhydryl groups (X in Fig. 22-16).

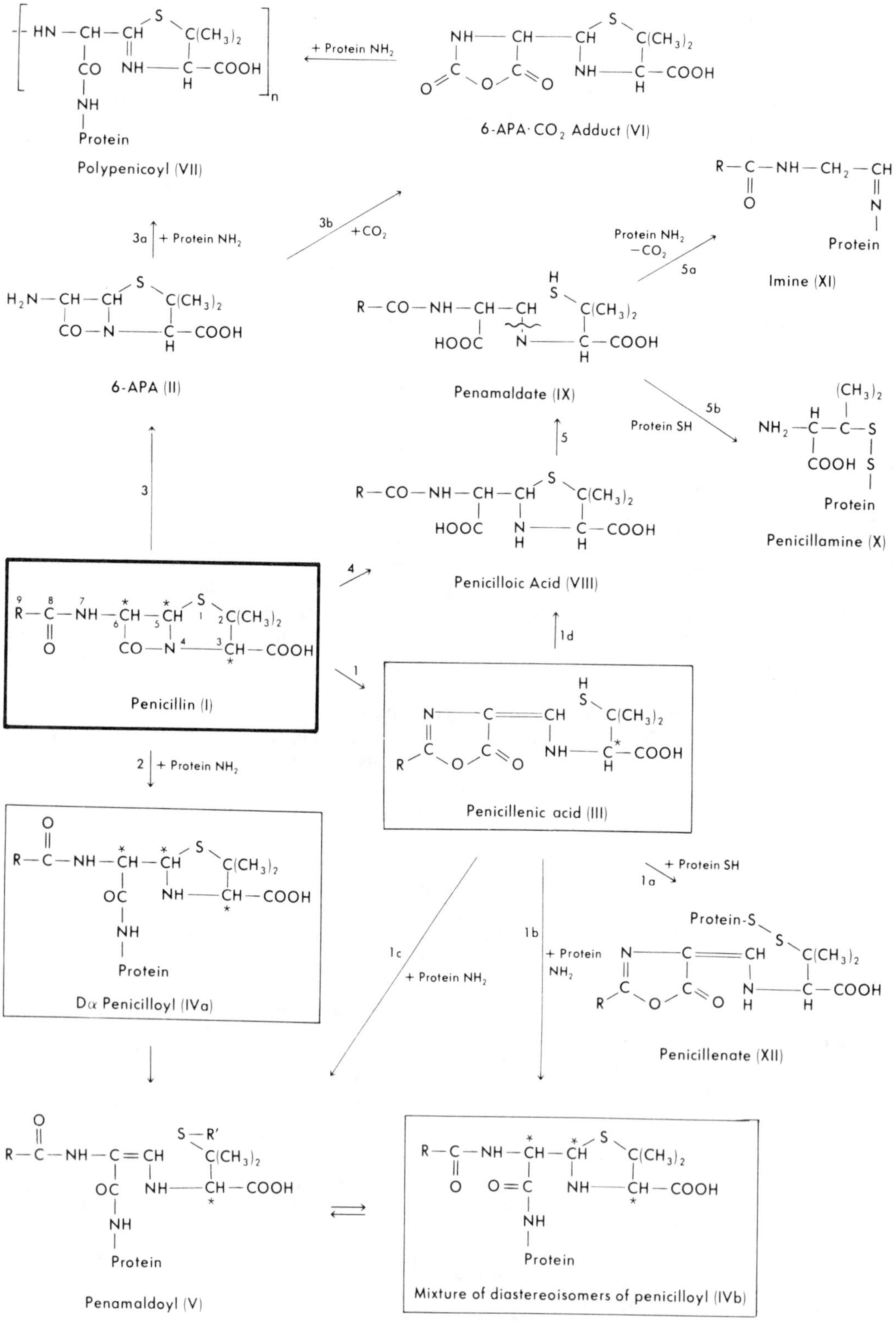

Polypenicoyl (VII)

6-APA·CO_2 Adduct (VI)

3a ↑ + Protein NH_2

3b $+CO_2$

Protein NH_2 $-CO_2$ 5a

Imine (XI)

6-APA (II)

Penamaldate (IX)

5b Protein SH

Penicillamine (X)

3

Protein SH 5

Penicilloic Acid (VIII)

4

1d

Penicillin (I)

1

Penicillenic acid (III)

2 + Protein NH_2

+ Protein SH 1a

Protein-S

Penicillenate (XII)

1b + Protein NH_2

1c + Protein NH_2

Dα Penicilloyl (IVa)

Penamaldoyl (V)

Mixture of diastereoisomers of penicilloyl (IVb)

Diagnosis of Immediate- and Delayed-Type Penicillin Allergy. An important difference between immediate-type and delayed-type drug hypersensitivity is brought out by the diagnostic procedures used to detect penicillin allergy. The intradermal injection of penicillin itself is often not reliable for detecting immediate-type sensitivity; and even the available synthetic multivalent ligands are not effective in all sensitive individuals. In contrast, patch tests with penicillin itself are almost infallible for identifying DTH reactions to the drug. One reason is that intradermally injected penicillin diffuses away too rapidly to form many local protein conjugates, whereas penicillin from a patch test percolates into the skin slowly and has more time to form various derivatives that become conjugated with skin proteins in situ. In addition, because T cells are specific for Ag plus products of the major histocompatibility complex (H-2 restriction; see above), soluble Ags inhibit T-cell–mediated reactions poorly, if at all.

Other Drugs. The diversity of haptenic groups introduced into proteins by penicillin is unusual, but the principles are undoubtedly the same for all allergenic drugs. Perhaps the most important are 1) *an allergic reaction to a drug is specific for the haptenic group(s) introduced by the drug, or by its derivatives, into proteins in vivo, rather than for the drug itself;* and 2) *the haptenic derivatives may differ greatly, in structure and in chemical properties, from the drug itself.*

Variations among Individuals. A striking feature of drug allergies is the extreme range in susceptibility of different individuals. Though the explanation is not certain, several possibilities are apparent. 1) Genetic differences could be important if enzymatic reactions are involved in converting a drug into metabolic derivatives that introduce haptenic groups into proteins; 2) the levels of Abs formed to any immunogen vary widely among different individuals, sometimes because of genetic differences (Ch. 19); 3) the tendency to form reaginic (IgE) Abs is also variable, and is especially pronounced in atopic individuals (Ch. 21). In addition, certain infections could influence the response to concomitantly administered drugs; for example, killed *Bordetella pertussis* (a commonly used adjuvant in mice and rats) seems to selectively favor the formation of reaginic Abs, and mycobacterial mycosides and endotoxins of gram-negative bacteria enhance the immunogenicity of most Ags.

Special Tissues. Most simple sensitizers react with proteins indiscriminately. Allergy to certain drugs, however, involves special tissue elements, which implies selective reactions with certain proteins. An example is the thrombocytopenia caused by quinidine or apronalide (Sedormid: 2-isopropyl-4-pentenoyl urea), resulting in bleeding in various tissues, including the skin (purpura). In vitro, in the presence of either drug, the serum of a sensitive person can agglutinate platelets (his own or a normal person's), and if C is added the platelets are lysed; but platelet lysis occurs at much lower concentrations of free quinidine in vivo than in vitro. It is possible that the drug, or one of its metabolic derivatives, reacts selectively under physiologic conditions with proteins on the platelet surface to form effective haptenic groups.

ALLERGY AND IMMUNITY

Many of the allergic reactions described in this and in the preceding chapter can be elicited with microbial Ags in persons who have, or have recovered from, the associated infectious disease. Two questions arise: 1) Do allergic responses occur naturally during the infectious disease? 2) If they occur, do they enhance or diminish host defenses?* Broadly speaking, the answers seem to depend on 1) the mass and distribution of microbial Ags, 2) the concentration and types of Abs present, and 3) the intensity and localization of CMI responses. The range of possibilities is large, as is illustrated by the following examples.

1) **No detectable allergic reactions.** In some infectious diseases allergic reactions are not likely to occur, or to be significant if they do. For example, in diseases caused primarily by exotoxins, such as diphtheria and tetanus, the quantities of toxin involved are too small to elicit an obvious allergic response even if the host were hypersensitive. Moreover, only a minority of persons successfully immunized with the corresponding toxoids exhibit allergic responses to intradermal tests with these proteins.

2) **Coincidental allergic reactions.** In the course of

* These questions are also considered, from different viewpoints, in Ch. 24 and in chapters dealing with particular microorganisms.

FIG. 22-16. Haptenic substituents derived from reactions of penicillin and its degradation products. **R** differs in different penicillins. (In penicillin G, the most widely used, it is benzyl ⬡—CH₂—....) **Asterisks** designate asymmetric carbons.

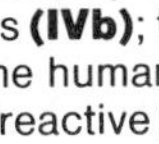

The Abs most often identified in penicillin-treated persons are specific for penicilloyl **(IVa,b)**. Penicilloyl groups formed from penicillenic acid **(III)** are a mixture of disatereoisomers **(IVb)**; those formed by direct reaction in vitro with penicillin at high pH largely retain the D-α configuration of penicillin itself **(IVa)**. Some human antipenicilloyl Abs react better with D-α-penicilloyl than with the diastereoisomeric mixture. Penicillenate **(III)** is much more reactive than penicillin at physiologic pH values, and it is probably the usual source of the penicilloyl groups formed in vivo. The immune responses induced by penicillin are sometimes specific for other, **minor-determinant** groups derived from the drug. 6-APA **(II)** is 6-aminopenicillanic acid, the intermediate to which many different acyl **(R)** groups may be attached synthetically to form a variety of penicillins (see Ch. 7). The structures of particular importance **(I, III, IVa** and **IVb)** are enclosed. (Parker CW: In Samter, M ed: Immunological Diseases. Boston, Little, Brown, 1965).

many infections various sterile skin eruptions appear, probably due to allergic reactions to microbial Ags. For example, transient erythematous rashes and urticaria, resembling cutaneous anaphylaxis, often occur in group A streptococcal infections; erythematous skin nodules (erythema nodosum), resembling the Arthus reaction, occur sometimes in the course of tuberculosis, coccidioidomycosis, and many other infectious diseases; and vesicular lesions ("ids"), resembling DTH reactions, occur often in fungal infections (dermatophytosis and candidiasis). Though these reactions are useful for clinical diagnosis there is no evidence that they affect the host's resistance.

3) **Significant allergic reactions.** Mycobacterial diseases regularly produce delayed-type allergy to the bacillary proteins, and the mass of bacteria in tissues is relatively large. Hence it is probable that intense allergic reactions occur regularly at infected foci in these diseases and account for much of the observed inflammatory reaction and tissue necrosis. As noted elsewhere (Chs. 24 and 37), the consequences of these severe reactions depend on their localization: a skin reaction that causes ulceration can lead to the drainage of virulent microbes from the body, but the same response on the meninges or in the bronchi can lead to fatal meningitis or to a spreading pulmonary infection.

The sterile cardiac lesions that sometimes develop after streptococcal infections (Ch. 28) can also cause severe disease (e.g., rheumatic fever), and seem to be allergic in origin. Thus the Abs to a cell wall Ag of β-hemolytic streptococci crossreact with an Ag from human cardiac muscle, suggesting that rheumatic carditis may be an autoimmune disease. A similar mechanism may contribute to the glomerulonephritis that frequently follows recovery from infection with the type 12 group A streptococcus.

The following considerations suggest, however, that *allergic responses to microbial Ags generally favor host defenses.* 1) A given microbe contains many Ags (probably hundreds with bacteria and fungi), of which only a few are directly concerned with virulence. Thus very few immune reactions that occur during an infectious disease are likely to result in specific neutralization of substances that cause toxicity and virulence. Nevertheless, many of the other reactions could well contribute indirectly to the host's resistance by provoking localized allergic responses around the microbial agents, leading to the accumulation of granulocytes, lymphocytes, macrophages, and serum proteins, including Abs and C. 2) The "activated" macrophages in delayed-type hypersensitivity can destroy bacteria (such as *M. tuberculosis*) that normally proliferate within phagocytic cells. 3) The major role of T cells in overcoming viral infections has been noted above (see Cell-Mediated Resistance to Infection, above). On balance, allergic responses are likely to benefit the host. Perhaps it is for this reason that allergic reactivity has persisted through evolution, being represented in all species of vertebrates, along with Ab formation, from primitive cyclostomes up.

SELECTED READING

SPECIFIC ARTICLES

BEVAN MJ: Alloimmune cytotoxic T cells: evidence that they recognize serologically defined antigens and bear clonally restricted receptors. J Immunol 114:316, 1975

BORELL Y, KILHAM L, KURTZ SE, REINISCH CL: Dichotomy between the induction of suppressor cells and immunologic intolerance by adult thymectomy. J Exp Med 151:743, 1980

CHU G: The kinetics of target cell lysis by cytotoxic T lymphocytes: a description by Poisson statistics. J Immunol 120:1261, 1978

CONLON PJ, MOORHEAD JW, MILLER SD, CLAMAN HN: Induction of tolerance to l-fluoro-34-dinitrobenzene (DNFB) contact sensitivity with hapten-modified lymphoid cells. II Selective tolerance in F_1 mice of T cell subsets recognizing DNFB associated with parental histocompatibility complex antigens. J Exp Med 151:959, 1980.

EYLAR EH, CACCAM J, JACKSON JJ, WESTFALL FC, ROBINSON AB: Experimental allergic encephalomyelitis: synthesis of disease-inducing site of the basic protein. Nature 1974

FEIGHERTY C, STASTNY P: HLA-D region-associated determinants serve as targets for human cell-mediated lysis. J Exp Med 149:485, 1979

GORDON S: Regulation of enzyme secretion by mononuclear phagocytes: studies with macrophage plasminogen activator and lysozyme. Fed Proc 37:2754, 1978

GOTTLIEB AR, MAZIARZ GA, TAMAKI N, SUTCLIFFE SB: The effect of dialyzable products from human leukocytes extracts in cutaneous delayed hypersensitivity responses. J Immunol 124:885, 1980

HUNG TC, RAUCH HC: Antibody response to synthetic encephalitogenic peptide. Molecular Immunol 17:527, 1980

JOHNSTON RB: Oxygen metabolism and the microbicidal activity of macrophages. Fed Proc 37:2759, 1978

KARNOVSKY ML, LAZDINS JK: Biochemical criteria for activated macrophages. J Immunol 121:809, 1978

KUPPERS RC, HENNEY CS: Evidence for direct linkage between antigen recognition and lytic expression in effector T cells. J Exp Med 143:684, 1976

MILLER SD, WETZIG RP, CLAMAN HN: The induction of cell-mediated immunity and tolerance with protein antigen coupled to syngeneic lymphoid cells. J Exp Med 149:758, 1979

MITWUOKA A, MORIKAWA S, BABA M, HARADA T: Cyclophosphamide eliminates suppressor T cells in age-associated central regulation of delayed sensitivity in mice. J Exp Med 149:1018, 1979

NATHAN C, NOGUEIRA N, JUANGBHANICH C, ELLIS J, COHN Z: Activation of macrophages in vivo and in vitro. Correlation between hydrogen peroxide release and killing of *Trypanosoma cruzi*. J Exp Med 149:1056, 1979

PTAK W, ROZYCKA D, ASKENASE PW, GERSHON RK: Role of antigen-presenting cells in the development and persistence of contact hypersensitivity. J Exp Med 151:362, 1980

SHAW S, BIDDISON WE: HLA-linked genetic control of the specificity of human cytotoxic T-cell responses to influenza virus. J Exp Med 149:565, 1979

STRELKAUSKAS AJ, CALLERY RT, MCDOWELL J, BOREL Y, SCHLOSSMAN SF: Direct evidence for loss of human suppressor cells during active autoimmune disease. Proc Natl Acad Sci USA 75:515D, 1978

WEINBERGER JZ, GERMAIN RN, JU S-T, GREENE MI, BENACERRAS B, DORF ME; Hapten-specific T cell responses to 4-hydroxy-3-nitrophenylacetyl. Demonstration of idiotypic determinants on suppressor T cells. J Exp Med 151:761, 1980

ZINKERNAGEL RM, DOHERTY PC: H-2 compatibility requirement for T-cell-mediated lysis of target cells infected with lymphocytic choriomenigitis virus. Different cytotoxic T cells specificities are associated with structures coded for in H-2K or H-2D loci. J Exp Med 141:1427, 1975

DOHERTY PC, BLANDEN RV, ZINKERNAGEL RM: Specificity of virus-immune effector T cells to H-2K and H-2D compatible interactions: implications for H-antigen diversity. Transplant Rev 29:89, 1976

DVORAK HF: Cutaneous basophil hypersensitivity. J Allergy Clin Immunol 58:229, 1976

HENNY C: T cell mediated cytolysis. Contemp Topics Immunobiol 7:245, 1978

Immunodeficiency. Report of a WHO Scientific Group. World Health Organizaiton Technical Report, Series 630, Geneva, 1978

LAWRENCE HS: Transfer factor in cellular immunity. Harvey Lect 68:239, 1974–75

LURIE MB: Resistance to Tuberculosis. Experimental studies in native and acquired defense mechanisms. Cambridge, Harvard University Press, 1964

MARTZ E: Multiple target cell killing by the cytolytic T lymphocyte and the mechanism of cytotoxicity. Transplantation 21:5, 1976

SHEARER GM, SCHMITT-VERHULST A-M: Major histocompatibility Complex-restricted cell-mediated immunity. Adv Immunol 25:55, 1977

WAKSMAN BH: Cellular hypersensitivity and immunity: inflammation and cytotoxicity. In Parker CW (ed): Clinical Immunology. Philadelphia, Saunders (in press)

BOOKS AND REVIEW ARTICLES

BLOOM BR, GLADE P (eds): In Vitro Methods in Cell-Mediated Immunity. New York, Academic Press, 1971

chapter

23

CELL SURFACE ANTIGENS: TRANSFUSION, TRANSPLANTATION, AND TUMOR IMMUNITY

This chapter deals with the host's immune responses to 1) transfused red blood cells (RBCs), 2) transplanted tissue grafts, and 3) tumor cells. All of these responses are directed to **cell surface Ags** that are recognized as **not self** because they derive from another individual (the blood or graft donor) or from tumor cells. In the analysis of these responses we will consider, from a different vantage point, many of the issues considered in earlier chapters.

The key Ags on RBCs and transplanted tissue cells are **alloantigens (alloAgs):** as noted before, these are Ags encoded by polymorphic genes, so that various alleles are present in different individuals from the same species.

The RBC alloAgs are often called **isoantigens** (Gr. *iso*, same) to emphasize that their variations occur among individuals of the same species, whereas **alloantigen** is the term applied almost universally to the trans-plantation Ags. We shall use alloAg for both, because the differences among these Ags are more pertinent than the sameness of the species.

AlloAgs were first recognized on RBCs some 80 years ago. Those on nucleated cells were later shown to account for the general observation that tissues transplanted from one individual to another are ultimately rejected, unless donor and recipient are genetically identical (**syngeneic**). Though Ags of this type were thought for many years to be limited to cells, many soluble proteins have recently also been recognized as alloAgs (e.g., haptoglobins, transferrins, and some β-lipoproteins). The alloantigenic forms of immuno-globulins (Igs), known as **allotypes**, have been exploited to great advantage in clarifying the genetics and structure of antibody (Ab) molecules (Ch. 17).

BLOOD TRANSFUSION

THE ABO SYSTEM

Harvey's discovery of the circulation led to repeated at-tempts to transfuse blood from one individual to another, usually with disastrous reactions. These were not under-stood until Landsteiner discovered the alloAgs of human RBCs in 1900.

Stimulated by observations in the 1890s that animal species could be distinguished by the reactions of their RBCs and serum proteins with specific antisera, Land-steiner sought to determine whether individuals of the **same species** could be distinguished in the same way. When individual samples of serum and RBCs from 22 human subjects were mixed in all possible combinations the red cells of some persons were found to be clumped by the serum of certain other individuals, leading to a classification of the subjects into four groups (Table 23-1).

From the results Landsteiner concluded that 1) the RBCs of any human carry one, both, or neither of two different Ags (or blood group substances), A and B, and 2) Abs for these alloAgs (called **alloAbs** or **isoAbs**) are regularly present in the serum of those individuals who lack the corresponding alloAg and are never in the serum of those who possess it.

The **blood groups** or **types** in human populations are named for the red cell Ags: group A has the A alloAg, group B has the B alloAg, group O has neither, and in group AB **each** red cell has **both** A and B. The corre-sponding serum alloAbs are anti-A in group B persons and anti-B in group A; group O has both anti-A and anti-B; group AB has neither.

GENETIC DETERMINATION OF THE ALLOANTIGENS

Family studies established the heritability of blood cell alloAgs. Analysis of large populations led Bernstein to propose that the *ABO* gene has three alleles—*A, B,* and *O*—with *A* and *B* dominant over *O*. Since man is diploid, the two alleles per individual provide the six genotypes and four phenotypes shown in Table 23-2.

Moreover, the Bernstein theory accounts for the dis-tribution of blood groups within families. For example, children of an O and an AB parent (OO × AB) average

TABLE 23-1. Division of Human Populations into Four Blood Groups on the Basis of Red Cell Agglutination by Normal Human Sera

Serum from group	Red cells from group			
	A	**B**	**O**	**AB**
A	0	+	0	+
B	+	0	0	+
O	+	+	0	+
AB	0	0	0	0

+ = clumping; 0 = no clumping.

TABLE 23-2. Genotypes and Phenotypes in the ABO Blood Group System

Genotype	Phenotype
AA ⎫ AO ⎭	A
BB ⎫ BO ⎭	B
OO	O
AB	AB

50% A (genotype AO) and 50% B (genotype BO); none are AB or O.

H Antigen. Though the A gene controls the formation of the A substance, and the B gene the B substance, the O gene, an **amorph**, does not specify a particular red cell alloAg. Nevertheless, O cells have a distinctive Ag, recognized from their agglutination by some normal animal sera (from eels and cattle), antisera (goat anti-*Shigella*), or plant proteins (extracted from the seeds of *Ulex europeus*). This Ag (as well as the A and B) Ags is lacking in the red cells of certain rare humans, first recognized in Bombay; and such **Bombay-type individuals** can form Abs in high titer to the characteristic Ag of O cells, either spontaneously or in response to injections of O cells. The O cell Ag is called **H substance** because it is **heterogenetic**,*† i.e., present not only in man, but also in other species. This Ag however, cannot be considered the product of the O gene, since it is also detectable, with anti-H

---

* The best known examples of heterogenetic Ags are the Forssman Ags, found in some human gastrointestinal cancers (see Tumors Induced by Chemical Carcinogens, below) and of clinical importance in the diagnosis of infectious mononucleosis (Ch. 55). These Ags, defined by their ability to induce rabbits to form hemolytic Abs to sheep RBCs (sheep hemolysins), are found in some bacterial and many animal species, e.g., in tissues of guinea pig, horse, cat, and chicken, but not rat or rabbit. The anti-Forssman Abs, sometimes called **heterophil** Abs, are of two types, differing in their ability to distinguish human A and AB from O and B cells. A and B substances are probably also heterogenetic; similar substances are found in a variety of animal and bacterial species.

† This H Ag is completely unrelated to histocompatibility Ags (Ch. 18 and this Ch.), which are sometimes also abbreviated as H Ags.

---

serum, on the RBCs of persons who **lack** the O gene, such as homozygous B individuals.

As we shall see below (section on Chemistry of the A, B, and H Alloantigens), the *H substance provides an obligatory precursor to which the A and B groups are attached. In O individuals the H substance is exposed and fully expressed.*

DISTRIBUTION OF ALLOANTIGENS: SECRETORS AND NONSECRETORS

The A, B, and H substances are surface components not only of RBCs but also of many epithelial and virtually all endothelial cells (Fig. 23-1), though not of connective tissue and muscle cells. They are also abundant (in some persons) in many mucinous secretions: saliva, gastric juice, pancreatic secretions, sweat, meconium, ovarian cyst mucin, etc.

The presence of A, B, and H in secretions is controlled by a separate genetic locus. About 80% of all persons are **secretors** (*Se/Se* or *Se/se*); depending on their blood type, their secretions contain water-soluble glycopeptides with A, B, or H activity. The other 20% are **nonsecretors** (*se/se*); their secretions lack A, B, and H.

Because of their abundance and solubility, blood group substances in secretions are valuable for chemical and antigenic analyses. They are serologically indistinguishable from the blood group substances that are isolable in trace amounts, and with great difficulty, from RBC membranes. The purified substances from secretions are water-soluble glycopeptides, whereas those from membranes of RBCs and other cells are alcohol-soluble and probably glycolipids; but the oligosaccharides appear to be the same on both.

A, B, and H Ags when present in saliva are all determinants on the same macromolecule, for all are coprecipitated by a monospecific antiserum to any one of them.

CHEMISTRY OF THE A, B, AND H ALLOANTIGENS

Regardless of their A, B, or H activity, the purified water-soluble glycopeptides have a similar overall structure: they are polydisperse macromolecules (mol wt 200,000–1,000,000) of similar composition (75%–80% carbohydrate, 15%–20% protein); they all consist of multiple heterosaccharide branches attached by glycosidic linkage at their internal, reducing ends to serine or threonine of the polypeptide backbone (Fig. 23-2). Stepwise enzymatic removal of sugars from the free, nonreducing ends of branches eliminates some determinant groups and exposes others that were previously detected only feebly or not at all (Fig. 23-3). These findings led Watkins and Morgan, and Ceppellini, to postulate that *the blood group macromolecule of the ABO system is built from a single large glycopeptide, on which the various*

FIG. 23-1. Localization of A and B alloAgs in human tissue by immunofluorescence. The A substance is shown in lymph node **(A)**, in epidermis (concentrated in the stratum corneum) **(B)**, in Hassall's corpuscle of the thymus **(D)**, and in goblet cells of a villus in the small intestine **(F)**. The B substance is shown in squamous epithelium of the tongue **(C)** and in transitional epithelium of renal calyces **(E)**. (Szulman AE: J Exp Med 111:789, 1960)

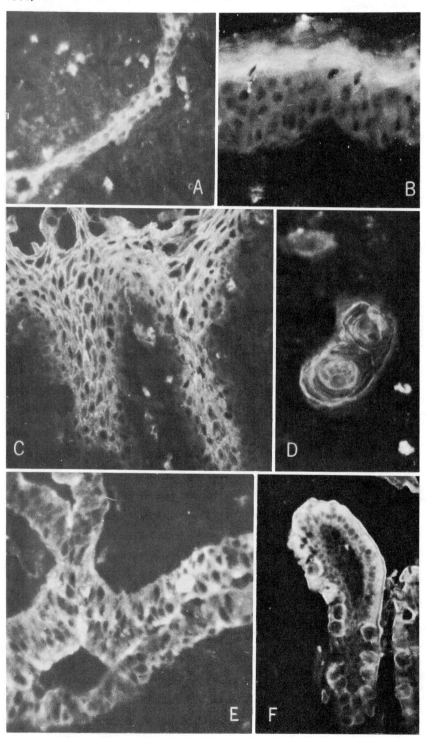

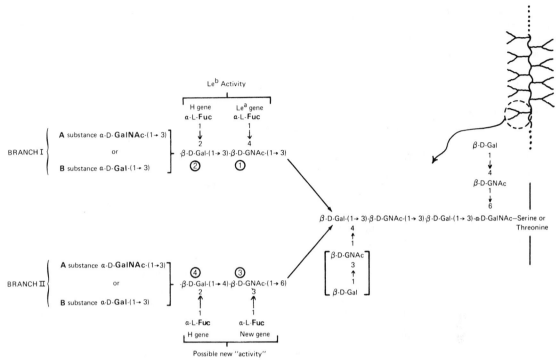

FIG. 23-2. Proposed structure of the ABO megalosaccharide. Branched heterosaccharides project from a polypeptide backbone in secreted water-soluble substances, or are associated with lipid in cell membrane glycolipids. A representative branched structure (enclosed in **dotted line**) is shown in detail. The residues responsible for A, B, H, Le^a, and Le^b activities are in **heavy type;** they are added by glycosyltransferases to the "core" structure, whose residues are in **light type.** The core, devoid of the heavy-type substituents, reacts with antisera to type 14 pneumococcal capsular polysaccharide; a fragment of the core (branch II) reacts with certain monoclonal human Abs (cold agglutinins called anti-I). (Modified from Lloyd KO, Kabat EA: Proc Natl Acad Sci USA 61:1470, 1968, and Lloyd KO et al: Biochemistry 9:3414, 1970)

sugars that correspond to the H, A, and B specificities are added sequentially by enzymes that are specified by the corresponding genes (H, A, B). As each sugar is added it introduces a new antigenic specificity, masking the previous one.

FIG. 23-3. Antigenic determinants of blood group mucopeptides revealed by sequential enzymatic removal of terminal sugar residues from A and B substances. (Watkins WM: Science 152:172, 1966)

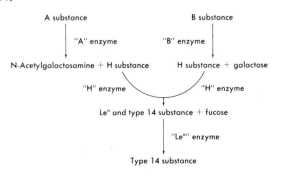

Thus, when purified A substance is digested with a crude enzyme preparation from clostridial "A" enzyme the A activity is lost, and H activity increases. Similarly, when B substance is digested with another enzyme ("B," an α-glycosidase from coffee) B activity is lost and H activity appears. Furthermore, when the H-active glycopeptides derived from these procedures (or isolated directly from O individuals) are treated with still another enzyme preparation, the H activity disappears and the capacity to react with Abs to another determinant (called Le^a, see Lewis Blood Groups, p. 528) appears. Finally, with further degradation Le^a activity is lost; the remaining glycopeptide crossreacts with type 14 pneumococcal polysaccharide (Fig. 23-3).

The Watkins-Morgan hypothesis has also been substantiated by chemical fragmentation of purified blood group glycopeptides (by periodate oxidation, alkaline borohydride reduction, and partial acid hydrolysis). From the resulting variety of small oligosaccharides the structure shown in Figure 23-2 was deduced.

Chemical Groups Responsible for A, B, and H Serologic Activities. Since many of the fragments specifically inhibit various alloAbs (e.g., in hemagglutination or precipitin assays) specificities can be assigned to particular saccharides. For instance, with fucose residues attached ($\alpha 1 \rightarrow 2$) at residues ② and ④ of Figure 23-2 the heterosaccharide has H activity; but if N-acetyl galactosamine is also present ($\alpha 1 \rightarrow 3$) on ② and ④, A activity is established and H activity is negligible, even though fucose is still present on ② and on ④. If galactose ($\alpha 1 \rightarrow 3$) is the additional residue at ② and at ④ the substance has B activity, rather than A, and H activity is also masked.

Thus the antigenic difference between A and B determinants, of great clinical importance for blood transfusions, is determined by the presence or absence of acetylated amino groups on the terminal sugars of complex, branched glycopeptides. While these few atoms are the crucial elements, the actual A and B determinants are probably as large as the terminal tetra- or pentasaccharides shown in Figure 23-2.

Various forms of A. When anti-A serum is adsorbed with cells from certain A persons (A_2), making up about 20% of the A population, it loses its ability to clump these cells but retains its ability to clump A cells from the remaining 80% (A_1). On the other hand, adsorption of the serum with A_1 cells abolishes its capacity to agglutinate all A cells. Thus there are two kinds of A: A_1 adsorbs anti-A completely, and A_2 adsorbs only some of the anti-A. Correspondingly, there are two types of AB cells: A_1B and A_2B.

The difference between A_1 and A_2 is evident in Fig. 23-2: In A_2, *N*-acetyl galactosamine is present only on ④ (branch II), and branch I lacks a terminal substituent on ②; hence A_2 also has H activity. In A_1, however, *N*-acetyl galactosamine is the terminal residue on both branches I and II (and H activity is lacking).

Lewis Blood Groups. Another pair of blood group activities, called **Lewis a and b** (**Le^a** and **Le^b**), are associated with the A, B, H substances. The Le^a activity is specified by another gene, Le, that segregates from the ABO and H genes. The Le^b activity is not due to a separate allele, but represents instead a novel specificity formed by the combined presence of the individual structures that, by themselves, determine the Le^a and H activities (Fig. 23-3).

It was noted above that mucinous secretions of most individuals (secretors) have water-soluble blood group glycopeptides with A, B, and H activity. The secretions of nonsecretors contain instead a similar glycopeptide with Le^a activity: L-fucose is attached at position ①, but not at ② or ④; hence H activity is lacking. As noted below, without fucose at ② and ④ the terminal residues characteristic of A and B cannot be added. In about 1% of all persons the secretions lack all these activities but contain a related glycopeptide that corresponds to the "core" of the blood group glycopeptides (crossreacting with pneumococcus type 14 polysaccharide; Figs. 23-2 and 23-3).

The A, B, H, and Le alleles code for enzymes. The Watkins-Morgan hypothesis postulates that the direct products of the A, B, H, and Le genes are enzymes (glycosyltransferases) that carry out the additions of particular sugars in the biosynthesis of the heterosaccharides, rather than the antigenic sugars themselves.

In support of this idea Ginsburg et al identified in human milk the four postulated enzymes: they transfer the appropriate sugar moieties from nucleotide-activated substrates to a milk tetrasaccharide that resembles the termini of branches I and II of the core heterosaccharide in the blood group glycopeptide (Fig. 23-2). The presence or absence of these glycosyltransferases in milk of different individuals coincides with their A, B, O, Le blood type (Fig. 23-4)

The A and B glycosyltransferases can function only if the core heterosaccharide has terminal fucose residues, corresponding to H activity. In the rare individuals with Bombay-type blood the absence of these fucose residues—owing to an absent or defective H gene—prevents addition of the terminal hexoses; the resulting glycopeptide in these persons lacks A and B as well as H determinants. The O allele evidently produces an inactive enzyme or no protein at all.

From the foregoing results it is clear that *both allelic (A, B) and nonallelic (H, Le) genes cooperate in sequential additions of different sugar determinants on a complex heteropolymer.* The process resembles other complex interactions of genes that determine the numerous different specificities of the cell wall antigens (O polysaccharides) in the salmonellae and other bacteria.

THE AB ALLOANTIBODIES

The anti-A and anti-B alloAbs are natural Abs (see Definitions, Ch. 15), probably formed in response to ubiquitous A- and B-like polysaccharides of many intestinal bacteria and of some foods. Additional inconspicuous immunogenic stimulation probably comes from inapparent infections. Because of self-tolerance (Ch. 19) an individual forms only those natural Abs that are specific for the alloAgs he lacks; for example, a person of type A forms anti-B, not anti-A. (Chickens have alloAgs and alloAbs that resemble the human AB system; but when raised under germ-free conditions they do not form these alloAbs.)

Classes of Immunoglobulin. Natural anti-A and anti-B are predominantly IgM molecules in persons of groups B and A, but they are both IgM and IgG in those of group O. The Abs formed in response to intensive antigenic stimulation by injections of A and B substance are largely of the IgG class.*

OTHER RED CELL ALLOANTIGENS

The discovery of the ABO system was facilitated by the presence of anti-A and anti-B Abs in most normal human sera. Since then many other red cell alloAgs have been recognized, but different means were required for

---

* Purified A and B substances are sometimes injected deliberately to raise the Ab titer, as in the preparation of typing sera.

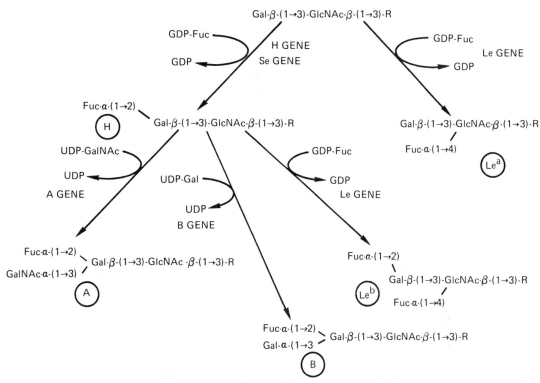

FIG. 23-4. Glycosyltransferases in the synthesis of A, B, H, and Le antigenic specificities. Enzymes specified by A, B, H, and Le genes transfer hexoses from activated precursors (uridine or guanosine diphosphosugars) to a tetrasaccharide (from human milk) that corresponds to branch I of the core heterosaccharide of Fig. 23-2. **R** is lactose. (Kobata A et al: Biochem Biophys Res Commun 32:272, 1968)

their detection, because the corresponding natural Abs are not found in normal human sera (perhaps because these Ags are not supplied in food or by infections).

Several approaches are illustrated with the blood group systems described below. One uses skillfully adsorbed animal antisera to human RBCs. Another depends on the appearance of alloAbs in persons who have received multiple blood transfusions or have had multiple pregnancies; for a fetal alloAg, inherited from the father and absent from the mother, can stimulate maternal Ab formation if it enters the maternal circulation.

MNSs BLOOD GROUPS

MN Antigens. About 25 years after the ABO system was recognized Landsteiner and Levine found that occasional rabbit antisera to human blood would, after adsorption with red cells from certain persons, agglutinate many but not all human RBCs, regardless of their ABO type. The adsorbed antisera were thus specific for an additional RBC alloAg, which they called M. Rabbit antisera subsequently elicited with certain samples of M⁻ cells revealed still another RBC alloAg, which was called N. All human RBCs react with either anti-M or anti-N

sera, or with both. From the distribution of the M and N Ags in many human families, it has become clear that these Ags are governed by a pair of allelic genes that segregate independently from the ABO locus: irrespective of an individual's ABO type, his red cells also have M or N, or both. The MN alleles are codominant; hence genotypes MM, NN, and MN correspond to phenotypes M, N, and MN.

Chemistry of MN. Glycophorin—a small protein, of about 150 amino acid residues, that spans the RBC membrane—has M or N blood group activity: it inhibits the agglutination of RBCs by anti-M or anti-N Abs. Approximately 16 oligosaccharides, most of them ending in sialic acid, are attached to various serine, threonine, and asparagine residues in the 80 amino-terminal residues of glycophorin that project from the outer face of the RBC membrane.

Glycophorin molecules with M activity differ from those with N activity in amino acid sequence at positions 1 and 5 from the NH_2 end (M has leucine and glutamic acid, whereas N has serine and glycine). However, inhibition of anti-M and anti-N sera with fragments of glycophorin suggests that the serologic activity resides in oligosaccharides, not in the protein. Perhaps amino acid differences in the peptide backbone result in different patterns of glycosylation. Glycophorin also acts, probably via its sialic acid tips, as receptor for the hemagglutinin of influenza virus.

Ss Antigens. Human serum (e.g., from a mother with Abs to her fetus's red cells) later led to the identification of another pair of alloAgs, called S and s, that are associated genetically with M and N. S and s are distributed in families as though governed by a pair of codominant allelic genes: i.e., every person has one or the other or both.

S is much more frequently found on M or MN than on N cells. Perhaps the S specificity is due to a mutation of the M gene (and less frequently of N) that adds a determinant group (S) without modifying the original determinants: a similar basis for the s specificity would mean that there are four alleles for the MN locus (MS, Ms, NS, Ns). However, family studies are also consistent with the alternative possibility that these alloAgs represent two separate but tightly linked genes (M,N and S,s): for example, the child of homozygous MS and Ns parents (MMSS × NNss) would be MSNs but would contribute to each of his offspring **either MS or Ns,** as though each pair were inseparable. If S and M are separate genes, it is not clear why the association of S with M is nonrandom. There may be selective advantages to the M–S combination, or contemporary populations may have resulted from relatively recent interbreeding of different racial groups that originally had considerably different frequencies of these genes (see Linkage Disequilibrium in the HLA Complex, Ch. 18).

Rh ANTIGENS

Like ABO, the Rh system was discovered as a byproduct of curiosity about alloAgs and was then found to solve an important clinical problem. In the 1930s, Landsteiner and Wiener, investigating the phylogeny of M Ag, found that rabbit antiserum to rhesus monkey RBCs (after removal of anti-M) clumped the red cells of about 85% of all persons (Rh^+) but not those of the remaining 15% (Rh^-). Wiener then showed that some severe reactions to blood transfusions could be explained by Rh incompatibility: donors were Rh^+ while recipients were Rh^- but had anti-Rh Abs.

At about the same time Levine and Stetson described the case of a mother who, after giving birth to a baby with the hemolytic disease erythroblastosis fetalis, suffered a severe reaction upon receiving blood from her husband, though both parents had the same ABO type. The newly discovered anti-Rhesus serum was subsequently found to clump the father's cells (Rh^+) but not the mother's (Rh^-); and in similar cases erythroblastotic babies were found to be Rh^+ with an Rh^- mother and an Rh^+ father. Evidently *fetal RBCs, carrying the paternal Ag, can cross the placenta and immunize the mother; maternal Abs can then enter the fetal circulation and react with the fetus's RBCs, causing massive hemolysis.*

Nonagglutinating Rh Antibodies. The foregoing explanation presented one difficulty: serum from most affected Rh^- mothers did not clump RBCs from the erythroblastotic Rh^+ babies or from other Rh^+ individuals. It was later discovered, however, that human anti-Rh Abs are often of the **incomplete** type (Ch. 16): they combine specifically with Rh^+ cells without causing agglutination, and when they cross the placenta (being IgG), they cause destruction of the fetal red cells, probably by acting as opsonins. As an additional complication, some mothers have anti-Rh Abs of the IgM class, which do agglutinate Rh^+ cells but do not cross the placenta and therefore do not affect the Rh^+ fetus.

The search for anti-Rh Abs in hemolytic disease provided the main impetus for the development of a number of ingenious methods for detecting incomplete Abs (see, for example, the antiglobulin, or Coombs, test, Ch. 16). Though these Abs were originally considered to be univalent they almost certainly are bivalent: they can clump RBCs under special circumstances, e.g., when 1) reactions are carried out in the presence of a high protein concentration, such as 30% serum albumin, or 2) RBCs are first treated with proteolytic enzymes, such as trypsin. (See Monogamous Bivalency, Ch. 16, for other bivalent Abs that do not cross link Ag particles.) Nonagglutinating anti-Rh Abs also played a key role in the discovery of diverse allotypes of human Igs (see Fig. 17-19, Ch. 17).

Frequency of Hemolytic Disease in the Newborn Due to Blood Group Incompatibility. Fetal red cells probably enter the maternal circulation in small numbers in most pregnancies, and yet only about 1 in 100 Rh^+ babies with Rh^- mothers have significant hemolytic disease. One reason is that few fetal red cells enter the maternal circulation until expulsion of the placenta at parturition. Hence it is the immunogenic stimulus experienced at the delivery of her first Rh^+ baby that causes the Rh^- mother to make a substantial anamnestic response to the small numbers of Rh^+ fetal cells that cross the placenta during subsequent pregnancies. Thus the first Rh^+ baby is usually unaffected and the risk of hemolytic disease increases in successive pregnancies.

The Rh factor is more immunogenic than most other alloAgs and it accounts for the vast majority of fetal deaths due to red cell incompatibilities. On rare occasions incompatibilities in the ABO and the Duffy and Kell blood group systems (see below) also cause severe hemolytic disease of the newborn.

Prevention of Rh Disease. The incidence of Rh disease is less when Rh^- mother and Rh^+ baby are also incompatible in ABO type (e.g., O mother, B baby): rapid removal of fetal cells from the maternal circulation by anti-B (or anti-A) evidently reduces the antigenic Rh stimulus (through the suppression of Ab formation by other Abs acting as antiimmunogens, Ch. 19). This observation led eventually to the development of a simple and effective means for reducing the incidence of erythroblastosis in the successive children born to high-risk parental combinations (Rh^+ father, Rh^- mother). At each delivery the mother is injected with anti-Rh Abs (Ig fraction of human anti-Rh serum), causing rapid clearance of Rh^+

fetal cells from the maternal blood: this minimizes immunogenic priming of the mother.

Multiplicity of Rh Antigens.

Analyses of many fetal–maternal incompatibilities have yielded a large variety of anti-Rh sera, of which there now seem to be about 30 distinctive types, each corresponding to a particular alloantigenic determinant or specificity. Family studies have established that these serologically distinctive specificities are all genetically linked. However, since structural information about Rh Ags is lacking the genetic basis for their extreme polymorphism remains controversial.

An early view was that three allelic pairs of closely linked genes (Cc, Dd, Ee) were responsible. The multiplicity of specificities now known cannot be reconciled with this scheme, but they can be accommodated by the view that all Rh specificities are due to a single Rh locus with many alleles. Various terminologies for the Rh Ags are still widely used (Table 23-3).

Antigen D.

Despite the large number and complexity of the genetically linked Rh Ags their clinical utility is simplified by the outstanding importance of one of them, the factor called D. This alloAg accounts for over 90% of all cases of erythroblastosis fetalis. Hence anti-D serum, from transplacentally immunized women, is used routinely to establish Rh type. Persons who are D^- are commonly referred to as being Rh^-, they will, however, usually possess some of the other antigenic determinants of the Rh locus.

THE KELL AND DUFFY BLOOD GROUPS

Besides ABO and Rh, several other Ag systems are important causes of severe transfusion reactions (and, rarely, of hemolytic disease in the newborn). The most common Ags of the **Kell system** are called K and k. K, present in about 10% of the population (Table 23-4), is a potent immunogen. Abs to K, formed as a result of pregnancy or transfusion, are usually incomplete and are detected by the Coombs (antiglobulin) test (see Blocking or "Incomplete" Abs, Ch. 16). Unlike the RBCs coated with Abs to most of the other blood group Ags, cells coated with anti-K are lysed by complement.

TABLE 23-4. Incidence of Some Red Blood Cell Phenotypes in the United States

Blood group system	Phenotype	Frequency (%)	
ABO	O	44	
	A ($A_1 + A_2$)	42	
	B	10	
	AB ($A_1B + A_2B$)	4	
MN	M	27	
	N	24	
	MN	50	
Ss*	S	11	
	s	45	
	Ss	44	
P	P_1	80	
	P_2	14	
	p	rare	
Rh†	DCe (Rh_1, R_1)	54	⎫
	DCE (Rh_z, R_z)	15	85% react with anti-D
	DcE (Rh_2, R_2)	14	(= "Rh-positive")
	Dce (Rh_0, R_0)	2	⎭
	dce (rh, r)	13	⎫
	dCe (rh', R')	1.5	15% do not react with
	dcE (rh", R")	0.5	anti-D (= "Rh-
	dCE (rh_y, R_y)	rare	negative") ⎭
Lutheran	Lu^a	6	
	Lu^b	94	
Kell	K+	6	
	K−	94	
Lewis	Le^a	22	
	Le^b	78	
Duffy	Fy^a	38	
	Fy^b	28	
Kidd	Jk^a	83	
	Jk^b	17	

TABLE 23-3. Notations for Rh Alloantigens and Genes*

Some Rh alloantigens		Some genes or gene combinations		
Fisher–Race	Wiener	Fisher–Race	Wiener	Commonly used
D	Rh_0	cDe	Rh_0	R_0
C	rh'	CDe	Rh_1	R_1
E	rh"	cDE	Rh_2	R_2
c	hr'	CDE	Rh_z	R_z
e	hr"	cde	rh	r
		Cde	rh'	R'
		cdE	rh"	R"
		CdE	rh_y	R_y

If a person's RBCs react only with anti-C, anti-D, and anti-e sera, the cells are called CDe in the Fisher–Race terminology, and the genotype is assumed to be CDe/CDe. In the Wiener terminology the same antisera are called anti-rh', anti-RH₀, and anti-hr", and the RBCs that react with all three are called Rh_1, as though a large antigen, Rh_1, is composed of three distinguishable antigenic determinants (rh¹, Rh₀, hr"). The terminology of Rosenfield, Allen, Swisher, and Kochwa (Transfusion 2:287, 1962) is purely descriptive: the antisera are numbered and the RBC's antigenic structure is given by listing the sera with which they have been tested, with a minus sign before the number meaning that no reaction was observed. Thus, anti-Rh1 = anti-D; anti-Rh2 = anti-C; anti-Rh3 = anti-E; anti-Rh4 = anti-c; anti-Rh5= anti-e; anti-Rh6 = anti-ce or anti-f, etc. The CDe (Rh_1) cells receive the following designation: Rh: 1, 2,−3, −4, 5, −6.

*The incidence of S, s, and Ss is from Race R, Sanger R: Blood Groups in Man, 2nd ed., Oxford, Blackwell, 1954. Anti-S sera agglutinate 73% of M, 54% of MN, and 32% of N cells.

†From Levine P. et al: In Dubos, RJ (ed.) Bacterial and Mycotic Infections of Man, 3rd ed. Philadelphia, Lippincott, 1958.

(Based on Smith DT et al [eds]: Zinsser's Microbiology, 13 ed. New York, Appleton, 1964)

TABLE 23-5. Frequency of ABO Blood Types in Various Ethnic Groups.

Population	Phenotypes (%)			
	O	A	B	AB
Scotland (Stornoway)	50	32	15	3
Sweden (Uppsala)	37	48	10	6
Switzerland (Berne)	40	47	9	4
Pakistan (NW frontier province)	25	33	36	5
India (Hindus in Bombay)	32	29	28	11
United States (Chippewa Indians)	88	12	0	0
Eskimos (Hudson Bay)	54	43	1.5	1.5

(Based on Mourant AE: Distribution of the Human Blood Groups. Springfield, Ill., Thomas, 1954)

The most common Ags of the **Duffy system** are termed Fy^a and Fy^b. Abs to Fy^b, formed as a result of transfusion or pregnancy, are being detected with increasing frequency in transfusion reactions and usually also require the Coombs test for their detection.

PATHOGENS AND EVOLUTION OF BLOOD GROUPS

It has been suggested that some severe infectious diseases could have served as selective agents in the evolution of blood group substances and accounted for the different frequencies of various blood types in different ethnic groups (Table 23-5). If, for instance, a particular blood group substance were the same as a surface Ag of a microbial pathogen, persons with that alloAg would be immunologically unresponsive to it and might be more susceptible to infection (see Tolerance, Ch. 19), assuming that the corresponding Abs (if formed) would be protective. So far the lack of correlation between blood type and incidence or severity of certain infectious diseases (e.g., smallpox, whose virion has been thought to contain an A-like substance) does not support this idea, but it nevertheless remains an attractive possibility.

That pathogens might operate as selective agents in another way is suggested by evidence that a blood group Ag of chicken RBCs serves as the specific receptor for an avian leukosis–sarcoma virus. Similarly, the MN blood group substances probably act as cell surface receptors for influenza virus on human RBCs. It is particularly striking that Duffy-negative Nigerian children are resistant to vivax malaria infection, evidently because their RBCs lack an appropriate receptor.

BLOOD TYPING

The identification of RBC alloAgs, called blood typing or grouping, is required for **blood transfusions.** Typing is also commonly used for genetic analysis in cases of disputed paternity, and for anthropologic surveys of human populations. It has even been used in archeologic work, since the red cell Ags are extraordinarily stable: the mummy of King Tutankhamen, for instance, has been typed A_2MN.*

* Blood typing and other serologic procedures are widely used in forensic medicine to distinguish human from animal blood; to

Blood typing is performed by agglutination reactions: unknown RBCs are typed with known antisera and unknown alloAbs with RBCs of known type. The C fixation test is not used since RBCs coated by Abs to most alloAgs are not lysed by C. This might be due to a peculiar distribution of most alloAgs on the RBC surface, or blood group Abs might often belong to classes of Igs that do not react effectively with C (see The Classic Pathway, Ch. 20).

Transfusion Reactions. Blood transfusion can lead to serious reactions if massive intravscular clumping and lysis of red cells take place.† Such reactions are usually the result of an attack on the donor's RBCs by the recipient's alloAbs. Rarely is the reverse reaction serious, between the donor's Abs and the recipient's cells, since 1) the alloAb titer in normal donors is usually low to begin with, and 2) the dilution of donor blood in the recipient (usually about tenfold) is generally sufficient to lower the titer to an innocuous level. Nevertheless, each lot of blood drawn from a prospective donor is screened against standard samples of RBCs with 8–10 of the Ags that are most often responsible for transfusion reactions; blood with a high titer of either agglutinating or incomplete Abs is used for other purposes (for example, the washed RBCs might be injected).

The blood considered for a particular transfusion must, of course, have the same major red cell Ags as those in the prospective recipient's blood. In addition, "major" crossmatching must be performed routinely, with the prospective donor's RBCs and the recipient's serum; if no agglutination is observed the RBCs must be washed and tested for adsorbed incomplete Abs by the addition of rabbit antiserum to human Igs (Coombs test, Ch. 16). As a further precaution "minor" crossmatching is similarly carried out, i.e., with the prospective donor's serum and the recipient's RBCs.

Universal Donors and Recipients. Under emergency conditions certain persons can donate blood to any recipient, and certain others can accept blood from any donor. The "universal donors" are type O; their red cells cannot react with either anti-A or anti-B in the recipient. However, type O donors are avoided if they have unusually

---

identify human blood types in blood stains, semen, or saliva; to distinguish horse meat from beef; etc.

† After multiple transfusions or multiple pregnancies alloAbs are also sometimes formed to HLA alloAgs on white blood cells and on platelets (see Histocompatibility Genes, p. 538 and Ch. 18). These Abs sometimes also cause serious transfusion reactions, characterized by high fever when leukoctyes are involved, or by thrombocytopenic purpura when platelets are involved (see Fig. 22-14, Ch. 22). AlloAbs to allotypes of IgA_2 Igs (see Am Allotypes, Ch. 17) can also cause transfusion reactions.

high titers of anti-A and anti-B. The "universal recipients" are type AB; they lack both anti-A and anti-B. As an additional precaution under such circumstances, soluble A and B substances are often added to the transfused blood as specific inhibitors to minimize the possibility of Ab alloimmune reactions.

Except for life-saving circumstances, A and B substances are not added to blood that is to be transfused into women of reproductive age: any enhancement of their immune response to A or B would increase the possibility that any babies they subsequently bear might have hemolytic disease due to A or B incompatibilities.

Uniqueness of the Individual's Red Cell Alloantigen Pattern. The foregoing precautions are not always reliable, however, since there are other red cell alloAgs that can cause serious incompatibilities (Table 23-4). In fact, so many different blood group alloAgs have now been identified (at least 60) that no two individuals are likely to be found with identical combinations, except for monozygotic twins.

In spite of this extreme diversity blood transfusions are extraordinarily successful, even when recipients undergo multiple transfusions from many different donors. Aside from the care exercised in the selection of prospective donors, through scrupulous typing procedures, the infrequency of transfusion reactions appears to be due to the fortunate fact that most red cell alloAgs are only feebly immunogenic. Moreover, transfused RBCs survive for only a limited time (an average of perhaps 3–4 weeks), so that even if Ab formation should be stimulated in the recipient the transfused cells are likely to be few in number by the time his alloAbs reach an effective level; and the destruction of a small number of cells is not likely to be serious. As we shall see below, however, nature is less benevolent in the immune response to the alloAgs of transplanted tissues.

TRANSPLANTATION IMMUNITY: THE ALLOGRAFT REACTION

Some properties of the major histocompatibility complex and its effects on lymphocyte interactions were considered in earlier chapters. This complex, and other genetic loci that encode cell surface transplantation Ags, will now be considered from the viewpoint of tissue and organ grafts.

DEFINITIONS

Four terms are used to describe tissue grafts.

1) **Autografts** are transplants from one region to another of the same individual.

2) **Isografts** are transplants from one individual to a genetically identical individual. These are possible only between monozygotic twins or between members of lines of mice and other rodents that have been so highly inbred as to be **syngeneic** or **isogenic**, i.e., genetically identical.*

3) **Allografts**, or **homografts**, are transplants from one individual to a genetically nonidentical (i.e., **allogenic**) individual of the same species.

4) **Heterografts**, or **xenografts** (Gr. *xenos,* foreign), are transplants from one species to another.

In these four types of grafts the donors are designated respectively as **autologous, isologous, homologous,** or **heterologous** with respect to the recipient.

---

* As is noted in Chapter 18, **syngeneic** refers to the absence of any discernible tissue incompatibility, i.e., to identity of the genes controlling histocompatibility Ags. It approximates the situation in **isogenic** individuals, who are identical with respect to all their genes.

THE ALLOGRAFT REACTION AS AN IMMUNE RESPONSE

Studies of tissue transplantation grew largely out of efforts to understand cancer cells. Beginning with transplants of mouse tumors in the early 1900s, these studies led to the development of inbred mouse strains and to the eventual realization that the results illuminated the general rules of transplantation—for normal as well as cancer tissue—rather than the nature of the cancer cell.

Successful transfers of tumors between mice of an inbred strain, but not of different strains, originally suggested that rejection of a graft is due to genetic differences between host and donor. More specifically, **autografts and isografts endure**, whereas **allografts and heterografts are rejected.**† Especially illuminating are observations on the hybrid offspring mice (*a/b*) of a cross between inbred parental strains (*a/a* × *b/b*). The *a/b* hybrids are tolerant of grafts from either parental strain, but mice of either parental strain reject grafts from the hybrids. Thus *a graft is permanently accepted only when essentially all of its histocompatibility Ags are present in the recipient.* These and further studies showed that grafted cells with genetically determined surface histocompatibility (transplantation) Ags elicit, in a recipient that lacks these Ags, an immune response that leads to destruction (rejection) of the graft.

In investigations into the mechanisms of the allograft

---

† Allografts are accepted by most invertebrates, which display few if any adaptive immune responses.

reaction skin grafts have been used most extensively, because of technical advantages: they are easy to prepare, their rejection is readily detected, and their median survival times provide an estimate of the intensity of the host's immune response. However, the same principles are involved in grafts of other tissues and cells.

THE SECOND-SET REACTION

When an allograft of skin is placed on a recipient animal, in a bed created by excising a slightly larger piece of skin, the graft at first becomes vascularized and its cells proliferate; but after about 10 days it quite abruptly becomes the seat of intense inflammation, withers, and is sloughed (**first-set reaction**). If a second graft is then made to the same recipient, with another piece of skin from the same donor, it is rejected much more rapidly than the first graft, perhaps in 5–6 days. This accelerated rejection, the **second-set reaction,** is specific for a particular donor: if after the accelerated rejection another donor, antigenically different from the first, provides skin grafts to the same recipient, first- and second-set reactions to successive grafts are again seen. Thus, *the capacity to reject an allograft (transplantation immunity) is acquired by virtue of exposure to the donor's cells; and it is specific for transplantation Ags of that donor.* The shorter survival of the second graft results from persistence of the immunity acquired from the first graft or from an anamnestic response (see Primary and Secondary Responses, Ch. 19).

The second-set reaction to a skin graft can be induced not only by a prior skin graft, but by prior inoculation of various other cells (e.g., spleen cells) from the same donor. In fact, virtually all cells induce transplantation immunity except RBCs; some histocompatibility Ags are present on red cells, but apparently not in immunogenic form (see H-2 Complex in the Mouse, Ch. 18).

HUMORAL VERSUS CELL-MEDIATED IMMUNITY

The second-set reaction can be **transferred "adoptively"** (see T Helper Cells, Ch. 18) from an immunized donor to a nonimmune recipient by T lymphocytes, but not by serum. (When the donor of sensitized lymphocytes is syngeneic with the recipient the adoptive immunity can be enduring; when the donor is allogenic the recipient's adoptive immunity is short-lived [1–2 weeks] owing to an allograft reaction against the donor's lymphocytes.)

Rejection of an allograft is probably due, in part, to a local delayed-type hypersensitivity (DTH) reaction (Ch. 22): thus lymphocytes from a recipient who has rejected a graft can elicit a delayed cutaneous reaction if injected into the skin of the donor. As in other DTH reactions the site of an allograft undergoing rejection is intensely in-

filtrated with macrophages and lymphocytes; granulocytes and plasma cells are much less conspicuous.

Cytotoxic T lymphocytes (CTLs; Ch. 22) probably also help to eliminate allografts: after a graft is rejected CTLs are found in the spleen and among peritoneal exudate cells, and are able to lyse any target cell that has the transplantation Ags of the donor.

A mouse that has rejected an allograft will nevertheless often have serum Abs for the donor's histocompatibility Ags. Some of these Ags are also present on RBCs (in the mouse); **hemagglutination reactions** have therefore provided a simple but powerful means of characterizing these Ags and mapping their genes (see H-2 Complex, and Fig 18-16, Ch. 18). However, these serum Abs seem not to be responsible for rejection of skin grafts: they generally appear late in the course of the allograft response, their levels bear no consistent relation to the intensity or rapidity of the rejection reaction, and as noted above they are incapable of transferring the ability to mount a typical second-set reaction.

However, some antisera from intensively immunized individuals can interfere specifically with the healing-in of a fresh graft, causing unusually rapid rejection (the **white-graft reaction**). Moreover, allogenic lymphocytes, unlike "solid" tissues, are readily destroyed by Abs to histocompatibility Ags, and complement (C) is required. It is possible that rejection of some grafts is Ab-mediated (see Human ABO Blood Group Substances, p. 538).

ACCEPTANCE OF ALLOGRAFTS

Under a number of exceptional circumstances allografts are not rejected.

Privileged Sites. In a few special ("privileged") sites allografts flourish for prolonged periods without inducing immunity because lymphatic drainage is lacking and so stimulation of the host's lymphocytes is minimal. These sites include the meninges of the brain and the anterior chamber of the eye.

Pregnancy. Histocompatibility Ags are formed early in embryonic life (by day 7 after fertilization in the mouse), and many of the Ags inherited from the father are alien to the mother. Hence in man and other mammals *the intrauterine fetus is actually an allograft.* Its failure to evoke an allograft rejection response, even when the mother has been previously immunized against the father's histocompatibility Ags, is not understood. One explanation offered is that mucinous secretions mask the histocompatibility Ags on the special fetal cell layer (trophoblast) at the placental interface between the fetus and its maternal host.

Tolerant Chimeras. Modern interest in immunologic tolerance (Chs. 19 and 22) grew out of a crucial observation on RBC alloAgs in nonidentical cattle twins, which frequently have in utero anastomoses of their placental blood vessels. Owen observed in 1945 that each twin often had two antigenically different kinds of RBCs, which persisted as the twins grew to maturity. It was inferred that hematopoietic stem cells from each twin, transferred through the common circulation in utero, had settled in the marrow of the other and then survived in the genetically foreign soil through the animal's lifetime. Each twin thus produced two kinds of RBCs, with its own and with its twin's characteristic alloAgs. Such individuals with mixtures of genetically different cells are called **chimeras,** after the monster in Greek mythology with a lion's head, a goat's body, and a serpent's tail.* Rare blood group chimeras among nonidentical human twins have also been found.

Surmising that such chimeras might also be generally tolerant of each other's tissues, Billingham, Brent, and Medawar demonstrated that they accept skin grafts from each other without an allograft reaction, and subsequently showed that mice of one inbred strain (A) could be rendered permanently tolerant of skin allografts from another strain (CBA) if embryonic or newborn A mice were inoculated with viable cells (e.g., of spleen) from CBA animals. When the inoculated animals matured they accepted allografts permanently from CBA donors, though they rejected grafts from any other strain in a normal manner.

The newborn mouse appears to develop allograft tolerance with ease because its immune apparatus is relatively immature, and it therefore cannot reject the foreign cells by an allograft reaction. Once established, the tolerance persists, allowing the foreign cells to remain and to proliferate. An individual rendered tolerant in this manner is a chimera; and its speen cells or white blood cells can induce, in a third strain, allograft sensitivity to both A and CBA.

Several mechanisms probably contribute to the enduring chimerism and tolerance. Specific T suppressor (T_s) cells are present but special procedures are required to demonstrate them. Consider, for instance, a chimeric H-2^d mouse that had been inoculated as a newborn with H-2^b lymphocytes: the transfer of its separated H-2^a lymphocytes to a heavily irradiated H-2^a mouse will prevent regeneration of the recipient's ability to reject an H-2^b graft but not the ability to reject an altogether H-2–different graft, say H-2^d.

Another suggested mechanism contends that blocking Abs (see Enhancement, below) protect the tolerated allogenic cells: thus spleen cells from the chimera are reported, in cytotoxicity assays in vitro, to destroy donor cells that would be tolerated in vivo, and this effect appears to be blocked specifically by the chimera's serum (presumably by noncytotoxic Abs).

The immunologic balance in the chimera can be abrogated by introducing immunologically competent cells that recognize the tolerated tissue as genetically alien. Thus in a mouse of strain X, rendered tolerant as a newborn to tissues of strain Y, a successful Y graft will undergo prompt rejection if the mouse is inoculated with viable lymphoid cells from a strain X animal that was previously immunized against Y (Fig. 23-5). With lymphoid cells from a normal, nonimmune animal of strain X, rejection requires many more cells and takes longer to develop.

* Other chimeras, discussed in Ch. 18, have been produced by rescuing lethally irradiated recipients with transplants of semisyngeneic bone marrow stem cells (e.g., A cells injected into [A × B] F_1 hybrids).

Total Lymphoid Irradiation. A special procedure for whole-body irradiation, originally developed to eliminate disseminated lymphatic cancer in humans can also promote prolonged survival of allografts. Recipient animals are prepared for the allograft by whole-body irradiation that is delivered while vital bone marrow regions are shielded with lead to protect the stem cells responsible for the production of erythrocytes, leukocytes, lymphocytes, etc.; peripheral lymphocytes are thus selectively destroyed. Transplants of allogenic bone marrow cells can then survive, proliferate, and differentiate into diverse blood cells, including lymphocytes. Meanwhile the recipient's own stem cells, having been protected, also continue to differentiate and produce blood cells, including lymphocytes. Mutual tolerance ensues, leading to peaceful coexistence, within these stable chimeras, of the allogenic donor and recipient lymphocytes. In such chimeric dogs and rats, transplanted organs (including the heart), taken from the bone marrow donor, have survived and functioned normally for prolonged periods, even without immunosuppression.

Kidney Allografts. These grafts in mice tend to survive much longer than skin grafts between the same donor–recipient pair. For instance, with a pair that differs at just the K end of the H-2 complex (see Fig. 18-16) a skin graft is promptly rejected (at about 13 days), whereas kidney grafts, introduced in place of the recipient's own kidneys, survive and function normally, supporting the recipient mouse's life for many weeks, even without immunosuppressive therapy.

These long-term survivors appear to become tolerant of the donor strain, since they tolerate subsequent skin grafts from the donor strain for a long period, whereas such grafts from an unrelated third strain (**third-party grafts**) are promptly rejected. Moreover, spleen cells from the long-term survivors lack CTLs against target cells prepared from the donor strain, and CTLs are not even elicited in vitro by incubating spleen cells from these long-term survivors with lymphocytes derived from the donor strain. However, the tolerance is precarious and easily reversed: the long-tolerated kidney grafts are rejected after nonspecific stimulation of the recipient's immune system by injections of attenuated tubercle bacilli (BCG), especially when it is administered with additional donor strain lymphocytes (as a source of more immunogen) and with cyclophosphamide at a dose that preferentially inactivates T suppressor cells (Ch. 22).

Enhancement. Paradoxically, the survival of many tumor allografts can be prolonged by previously immunizing recipients with special regimens: either actively (with the corresponding allogenic cells), passively (with antisera to the histocompatibility Ags of these cells), or with a combination of the two. The same kind of **enhancement** of allograft survival can be demonstrated, though not as readily, with allografts of normal tissues.

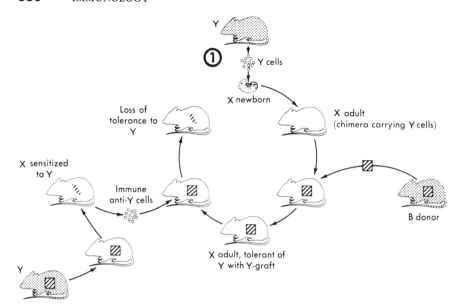

FIG. 23-5. Establishment and loss of tolerance to histocompatibility Ags. Newborn X was injected with Y cells. The subsequent injection of the Y-tolerant X animal (shown carrying a Y skin graft) with lymphocytes from a Y-sensitive X donor leads to adoptive immunity, with rejection of the previously tolerated graft. The risk of a graft-vs.-host reaction can be eliminated in step **1 (top)** if the donor of cells to the newborn is an X × Y hybrid, rather than a purebred Y-strain mouse, as shown.

The mechanisms are not completely understood. T_s cells, elicited by the immunization regimen, probably block the development of the effector T cells (T_{DTH} cells and CTLs) that normally cause allograft rejection. Thus, with an immunized, "enhanced" rat (i.e., carrying a long-surviving allograft) as donor, thymoctyes are far more effective than serum in causing prolonged survival, in recipient rats, of the corresponding allograft.

However, Abs can also contribute to enhancement. When Abs are bound to histocompatibility Ags on a cell's surface they can specifically block the destruction of that cell by CTLs (**efferent enhancement**). It is also possible that these Abs block the immunogenicity of a graft's histocompatibility Ags, preventing the induction of cell-mediated immunity (**afferent enhancement**).

GRAFT-VERSUS-HOST REACTIONS

Because the newborn mouse is immunologically unresponsive it does not reject lymphocytes from a mature, immunologically competent allogenic donor. However, the donor's lymphocytes can react against the recipient's alloAgs. In the ensuing graft-versus-host (GvH) reaction the inoculated newborn fails to gain weight normally, develops skin lesions and diarrhea, and usually dies after a few weeks (**runting syndrome**).* This syndrome is also seen when immunologically competent lymphocytes from an allogenic donor are injected into an adult that cannot reject the transplant, e.g., a recipient depleted of

* Surprisingly, the GvH reaction was only occasionally observed in the initial demonstration of newborn tolerance (described under Tolerant Chimeras, p. 534), probably because the allogenic mouse strains that were used differed at only part of the H-2 complex. The GvH reaction occurs regularly when the same procedure (Fig. 23-5) is carried out with most other pairs of histoincompatible strains.

his own lymphocytes by x-irradiation or cytotoxic drugs, as in human bone marrow transplants (see p. 538).

HISTOCOMPATIBILITY GENES

Histocompatibility Ags are specified by histocompatibility (H) genes.† Because a host's immune response to any of a donor's transplantation Ags can lead to rejection of a graft, permanent acceptance (in the absence of chronic immunosuppressive therapy) requires that essentially all of the donor's histocompatibility alleles be present in the recipient. Hence the probability that an allograft will be accepted when donor and recipient are drawn at random from an outbred population, such as humans, depends upon 1) the number of H genes or loci in the species, and 2) the number of alleles at each locus and their frequencies in the population.

The H loci and their alleles have been characterized most extensively in the mouse, because its many inbred strains permit detailed genetic analysis with skin grafts.

HISTOCOMPATIBILITY GENES IN THE MOUSE

The number of independent H loci at which two inbred strains differ can be estimated by mating them (AA × BB), crossing the F1 progeny (AB × AB), and using animals of the F2 generation as recipients for skin grafts from the purebred parental strains. As is illustrated in Figure 23-6, the number, *n*, of H loci, is provided by the

† This H must obviously not be confused with the completely unrelated H (heterogenetic) Ag of blood group substances mentioned earlier in this chapter.

FIG. 23-6. Estimate of the number of independent histocompatibility **(H)** genes at which two inbred strains of mice differ. **Shaded squares** correspond to progeny in the F_2 generation that accept a graft from parent A. (Courtesy of Dr. R. J. Graff)

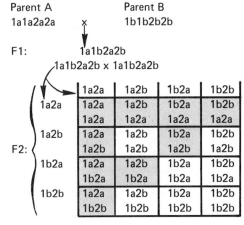

Parental Strains Differ by One H Gene

Parent A		Parent B
1a1a	x	1b1b

F1: 1a1b

1a1b x 1a1b

F2:

	1a	1b
1a	1a1a	1a1b
1b	1a1b	1b1b

$(\frac{3}{4})^1$ or 75% of F2 progeny accept grafts from parent A or parent B

Parental Strains Differ by Two Independent H Genes

Parent A		Parent B
1a1a2a2a	x	1b1b2b2b

F1: 1a1b2a2b

1a1b2a2b x 1a1b2a2b

F2:

	1a2a	1a2b	1b2a	1b2b
1a2a	1a2a	1a2b	1b2a	1b2b
	1a2a	1a2a	1a2a	1a2a
1a2b	1a2a	1a2b	1b2a	1b2b
	1a2b	1a2b	1a2b	1a2b
1b2a	1a2a	1a2b	1b2a	1b2b
	1b2a	1b2a	1b2a	1b2a
1b2b	1a2a	1a2b	1b2a	1b2b
	1b2b	1b2b	1b2b	1b2b

$(\frac{3}{4})^2$ or 56% of F2 progeny accept grafts from parent A or parent B

proportion (x) of F2 animals that accept grafts from one of the purebred parental strains: $x = (\frac{3}{4})^n$. This method yields minimal values because some H genes are linked to one another and behave as single rather than as multiple genes, and some inbred strains have certain H alleles in common.

A better estimate of the number of loci has been obtained through the use of **congenic mouse strains,** originally developed by Snell. As was shown in Figure 18-9, a congenic strain A.B is identical genetically with its inbred partner A except for a foreign chromosomal segment, derived by appropriate matings, from a second strain, B. If the foreign segment carries an H locus, the congenic strain is called **congenic resistant (CR),** because grafts exchanged between it and the A partner will be rejected (or **resisted,** a term that persists because in the original studies the exchanged grafts were tumors). Many CR lines can be derived from the progeny of a single A × B mating, and grafts exchanged between different pairs of CR lines [A.B(1), A.B(2), etc.) reveal whether the lines do or do not have the same H loci: the exchanged grafts are accepted if the loci are the same and rejected if different. This and other less laborious approaches have revealed over 40 H loci; they are designated H-1, H-2, etc.

Other approaches demonstrate **multiple alleles at each locus,** e.g., 12 at H-1, 5 at H-3, 6 at H-4, 4 at H-7, 7 at H-8. At H-2, as noted in Chapter 18, polymorphism is extreme, with at least 50 alleles at H-2K and 50 at H-2D.

The **number of alleles** at each H locus can be estimated by the **F-1 test,** which involves three strains: a congenic pair A and A.B, and an entirely different inbred strain C. Suppose A and A.B differ at H-1, with A having the H-1^a allele, A.B the H-1^b allele, and C an unknown H-1 allele. A graft of A skin to an (A.B × C) F1 hybrid will be accepted if C has the H-1^a allele but rejected if C's H-1 allele is not a. On the other hand, a graft of A.B skin to an (A × C) hybrid will be accepted if C has the H-1^b allele but rejected if C's H-1 allele is not b. If the H-1 allele of C turns out to be not a and not b, then there must be *at least three alleles* at H-1. When many inbred strains were typed in this way it was found that only a few have H-1^a or H-1^b alleles, and it could be estimated (from the number of non-a and non-b strains) that there are about 12 alleles at H-1.

H-2 Locus. As is noted in Ch. 18, the transplantation Ags specified by H-2 elicit the most intense rejection reactions: allografts exchanged between mice that differ only at this locus are usually rejected in about 11 days (first-set rejection), whereas those exchanged between mice that have the same H-2 alleles but differ at any one of the other H loci have median survival times that range from 20 to upward of 200 days. Indeed, skin allografts that differ only at certain H loci are not usually rejected unless the recipient is first immunized by several injections of cells from the prospective donor. The H-2 locus is thus the **major histocompatibility locus,** and the non–H-2 loci are the **minor histocompatibility loci.**

Non–H-2 Loci. Not only are survival times much longer when grafts are exchanged between congenic lines that differ only at one non–H-2 locus, but many more inoculated cells are required to induce transplantation immunity, and many fewer are needed to establish tolerance, than with combinations that differ only at H-2. Nevertheless, the cumulative effect of many non–H-2 differences can approximate that of an H-2 incompatibility and can lead to virtually as rapid allograft rejection.

H-Y Locus. Diverse H loci are associated with different genetic markers on various linkage groups (i.e., chromosomes): for example, the H-1 locus is linked with albinism, and H-2 with fused tail vertebrae. One of the non–H-2 loci appears to be associated with the Y sex chromosome. Thus within an inbred strain, grafts (of skin, spleen, thyroid, etc.) from female to male are accepted permanently, but those from male to female are rejected (after long survival times). Male tolerance of female cells is due to the presence of X chromosomes in both sexes, but some product of the Y chromosome on cell surfaces of the male can apparently function as a weak histocompatibility Ag. Because this Ag is weak, tolerance of it is readily established; e.g., by transplanting an extra large piece of skin from male to female.

Tissue-Specific Alloantigens. Some transplantation Ags (H-2 in mice, HLA in humans) are present on virtually all cells; others are present only on particular tissues. Thus, in the stable chimeras produced by inoculating lymphocytes of strain Y mice into newborn mice of strain X, the X and Y lymphocytes are mutually tolerant (see Tolerant Chimeras, p. 534, and Fig. 23-5), but with certain strain combinations the chimeric mice will nevertheless reject a skin graft from the donor (Y) strain. The reason is that some special alloAgs (SK) are present on skin epithelial cells but absent on lymphocytes and other hematopoietic cells.

HISTOCOMPATIBILITY GENES AND HUMAN ALLOGRAFTS

The major histocompatibility complex in man, HLA, was described in Ch. 18. To match donors of human allografts with recipients, some of the Ags encoded by this complex and by other H loci are determined by several assays:

1) **HLA typing.** In "microcytotoxicity" assays minute volumes ($1\mu l$–$2\mu l$) of anti-HLA serum, a source of complement, and about 2000 lymphocytes are incubated (beneath mineral oil to prevent evaporation). The killed cells become permeable to ionic dyes (such as eosin or trypan blue) and hence are selectively stained. The sera used for typing are generally obtained from multiparous women, of whom about 30% have anti-HLA Abs (produced against their offspring's alien HLA Ags, inherited from the father).

2) **Mixed lymphocyte reactions (MLRs).** As noted before (Ch. 18) this reaction is indicated by a stimulatory response. After incubating peripheral blood lymphocytes from the recipient with those from a prospective donor for 5 days ^{3}H-thymidine is added, and the next day the ^{3}H-labeled DNA is counted to measure the extent of cell proliferation. The intensity of the response depends upon differences at the HLA-D locus (the region homologous to the I region of the H-2 complex, which is chiefly responsible for the proliferative MLR responses of mouse lymphocytes; see Ch. 18). Blocking the proliferation of one of the cell populations (by treating it with mitomycin C or exposing it to x- or γ-irradiation) causes the reaction to become one-way: it distinguishes between the recipient's response to the donor and the reverse. The cells that respond in the one-way reaction are sometimes tested for their ability to lyse ^{51}Cr-labeled lymphocytes from the other ("stimulator") partner.

3) **Crossmatching.** As in blood transfusions, in an additional test the recipient's serum, together with a fresh source of complement, is tested for cytotoxic Ab activity against the donor's cells (e.g., detected by dye uptake).

Influence of HLA on Allograft Survival. Skin grafts exchanged between members of the same family clearly indicate the importance of HLA typing: grafts between sibs who are identical in HLA last longer (20–25 days) than those between sibs who differ (11–12 days). However, the rejection of HLA-identical grafts shows that non-HLA Ags, probably homologous to the many non–H-2 Ags of the mouse, can also elicit human rejection reactions.

Human Kidney Allografts. Kidney allografts are now used extensively to treat patients whose own kidneys have failed: over 20,000 of these grafts have been performed. Their survival is influenced by several factors, in addition to HLA incompatibility. For instance, 1) grafts can fail because they acquire the same disease that originally destroyed the recipient's kidneys, and 2) skillful use of immunosuppression therapy* can greatly prolong the survival of grafts that would otherwise be promptly rejected. Nevertheless, a clear effect of HLA on graft survival is evident when observations are focused on sibs as donor–recipient pairs: when the sibs are HLA-identical, failure of the kidney graft is rare (Fig. 23-7).

Human ABO Blood Group Substances and Transplantation Immunity. ABO blood group substances are present on many cells in addition to RBCs, and these substances can also function as transplantation Ags: when donor and recipient belong to different ABO groups, kidney grafts seem to be subject to especially rapid and violent rejection. Group O human volunteers injected with AB red cells (or with purified A and B substances) reject subsequent skin grafts from any AB donor, but not those from an O donor, in an accelerated manner (second-set rejection). It is not clear if this accelerated rejection is due to conventional anti-A and anti-B alloAbs or to cell-mediated reactions.

If donor and recipient belong to different ABO groups, injected donors RBCs do not induce a second-set rejection of a subsequent graft from the same donor: hence RBCs lack HLA Ags, at least in immunogenic form (or amount).

Bone Marrow Allografts. These grafts are used increasingly to replace bone marrow destroyed by intensive whole-body x-irradiation (therapy for individuals with leukemia) or by other, unknown causes (? unidentified toxic chemicals). Until recently these grafts nearly always resulted in severe graft-vs-host (GvH) reactions and ended fatally. However, with HLA-matched sibs the grafts usually take and persist, eliciting only mild, transient GvH reactions. About 300 successful bone marrow

---

* Human recipients of allografts are usually given antilymphocytic serum (ALS) immediately after receiving the graft, and they are maintained indefinitely thereafter on immunosuppressive drugs, e.g., 11-oxycorticosteroids and azathioprine (Imuran) or 6-mercaptopurine (see Nonspecific Immunosuppression, Chs. 19, 22).

FIG. 23-7. Correlation between HLA compatibility and survival of human kidney grafts. Number of incompatibilities, based on leukocyte typing, is given in parentheses. Every parent–child combination and half the sib–sib combinations have at least one haplotype in common. (Dausset J, Hors J: Transplant Proc 3:1004, 1971)

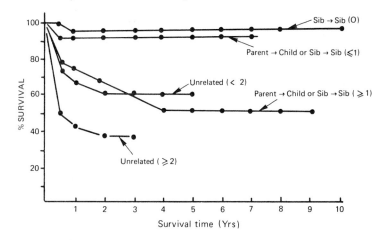

grafts, establishing stable chimeras, have now been performed in the United States.

With genetically identical (monozygotic) twins bone marrow grafts are usually successful, as expected. However, as noted before (see Autoimmune Responses, Ch. 22), when the recipient twin has been heavily irradiated and thymectomized (to eliminate leukemic cells) GvH reactions have occurred even with these grafts. The immunosuppressive procedures (X-irradiation, thymec-tomy) seem generally to depress or eliminate T_s cells more than other T cells, and the loss of T_s cells can evidently allow the donor's lymphocytes to make what amounts to an autoimmune response to RBC and other Ags. The T_s cell loss can also explain why recipients of bone marrow allografts commonly exhibit transient elevations of serum IgE, and accompanying eosinophilia (See Fig. 21-6, Ch. 21).

TUMOR IMMUNOLOGY

COMPARISON OF TUMOR CELLS AND NORMAL CELLS

Before considering immune responses to tumor cells some properties of these cells will be reviewed.

1) Tumors are characterized by **unrestrained growth** in the animal and in culture: tumor cells usually respond poorly, if at all, to the signals that regulate the growth of normal cells.

2) Tumor cells of the more malignant variety are **invasive:** as they grow in vivo they penetrate normal membranes and cell layers and spread by direct extension into neighboring tissues.

3) Tumors also spread by **metastasis:** some cells separate from the main tumor mass, are carried in the blood and lymph to distant organs or lymph nodes, and there invade normal tissues and establish satellite or secondary tumor masses (metastases). Death from cancer is usually due more to widespread metastases than to the primary tumor mass, which can often be excised.

4) Rapid growth and metastatic spread are not unique to tumor cells: stem cells in the normal epidermis, intestine, and bone marrow divide more rapidly than most tumor cells; and the migration of cells from one part of the body to another is characteristic of normal fetal development (as of the thymus; Ch. 18). What sets tumor cells apart is **a failure to respond to the regulatory signals** that are responsible for the orderly and balanced growth patterns of normal cells. The identification of these signals, and the basis for their inability to control the growth of tumor cells, are key problems in tumor biology.

5) The individual tumor is usually **a single clone** (i.e., it is derived from a single progenitor cell). Products of genes on the X chromosome demonstrate this principle. Thus in female mammals one of the two X chromosomes becomes inactive, at random, in each embryonic cell, and remains inactive thenceforth in all progeny of that cell **(Lyon effect).** Hence heterozygosity at any recognizable gene on the X chromosome normally leads to **mosaicism:** some cells express one of the alleles of that gene and other cells the other allele. In such an individual all cells of a tumor usually express only one of the two X-carried alleles.

6) Tumors, nevertheless, **are usually heterogeneous:** within a single tumor the cells often differ in shape, surface Ags, and tendency to spread by metastasis; and some tumors also have both self-renewing tumor "stem" cells and more differentiated tumor cells, derived from the stem cells.

7) Normal cells can be converted into **"transformed"** or tumor cells by various agents **(carcinogens).** Physical and chemical carcinogens are **mutagens** and probably cause mutations in critical but unidentified cellular genes, whereas in viral oncogenesis (Ch. 66) the transforming gene is part of the viral genome (which in a small virus, such as polyoma or SV40, can have as few as five genes).

8) Transformed and untransformed cells from the same clone of cultured cells show major differences. These include differences in cell surface glycoproteins, in surface Ags (see below), and in growth behavior, seen both in culture and when the cells are transplanted into animals.

In culture the division of normal fibroblasts is inhibited by contact with neighboring cells; hence proliferation ceases when the bottom of a culture dish is covered by a layer one cell thick. Transformed fibroblasts, however, are not subject to such **contact inhibition:** their continued proliferation results in heaped up multilayered cell masses, with growth ceasing only when nutrients are exhausted.

When suspended in a gel, normal fibroblasts do not proliferate, because their division requires attachment to a solid surface. Transformed fibroblasts, however, are **anchorage-independent** and readily form colonies when suspended in a gel.

To determine whether or not a clone of cells is cancerous, the ultimate test is its unrestrained growth in an animal. In T-cell–deficient athymic nude mice *(nu/nu)* transplanted cells that had been transformed in culture or removed from a tumor that originated in a person or in an experimental animal usually grow progressively and cause the animal's death, whereas transplanted normal cells grow to only a limited extent, if at all.

9) Tumor cells often **express genes that are silent in the corresponding normal cells.** For intance, tumors arising in an adult can produce **fetal Ags** (which are normally expressed only by some fetal cells); and nonendocrine tumors (e.g., of the bronchus) can produce **hormones** (such as insulin or parathyroid hormone) that ordinarily are made only by special endocrine cells. These aberrations reflect abnormal regulatory processes in tumor cells, and they are useful for the early detection of some forms of cancer.

IMMUNE RESPONSES TO TUMOR CELLS

Tumor immunology is based on evidence that 1) certain surface Ags of tumor cells **(tumor-associated Ags, or TAAs)** distinguish these cells from normal cells of the same type, 2) some TAAs can elicit immune responses in the tumor-bearing host, and 3) some host immune responses to TAAs can eliminate the tumor cells. Nevertheless, many tumors grow progressively and kill the host.

The following sections consider general properties of TAAs, the diversity of responses they elicit, the role of immunity in preventing the development of tumors, and some mechanisms that enable tumor cells to escape immune destruction.

TUMOR-ASSOCIATED ANTIGENS

Convincing evidence that tumors have TAAs required the development of **inbred strains of mice and rats,** which permit a tumor arising in one animal (the **autochthonous host)** to be continuously propagated by serial transplantation in recipients that are the same genetically as the autochthonous host (see Allograft Reaction as an Immune Response, p. 533). In an early classic study (Fig. 23-8) a sarcoma induced by methylcholanthrene was excised and was transplanted into syngeneic mice, while the surgically cured host was immunized with x-irradiated cells of its original tumor (x-irradiation blocks cell division by damaging chromosomes but leaves the cell

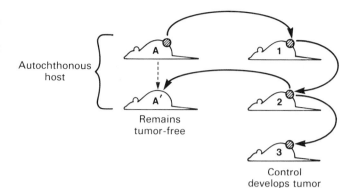

FIG. 23-8. Immunity to a transplantable tumor. A methylcholanthrene-induced tumor is completely excised from original (autochthonous) host **(A),** "passaged" by serial transplantation through syngeneic mice **(1,2),** and then injected back into the surgically cured original host **(A′)** and into normal control mice **(3).** The original host, immunized ("primed") by its growing tumor cells, rejects the transplant (remains tumor-free), whereas the controls develop tumors that grow progressively and fatally. In the original study, the autochthonous host, after excision of the tumor, was injected repeatedly with irradiated tumor cells (to increase immunity), but this is not usually necessary: the excision of a small, growing tumor often leaves sufficient immunity to reject a subsequent inoculum of the same tumor. (Based on Klein G et al: Cancer Research 20:1561, 1960)

Ags intact). The immunized animal was then able to reject a graft of its own tumor. Resistance induced in this way can be transferred to normal syngeneic recipients with T lymphocytes from the immunized animal. TAAs that elicit such transplantation-rejection responses are often referred to as **tumor-specific transplantation Ags (TSTAs).**

As we shall see below, the tumors induced by a chemical carcinogen differ greatly in their TSTAs, whereas the tumors induced by an oncogenic virus share a TSTA that is associated with that virus, though not necessarily a part of the virion.

TUMORS INDUCED BY CHEMICAL CARCINOGENS

Many of these tumors have individually distinctive ("private") TAAs. For instance, mice immunized against one methylcholanthrene-induced tumor are resistant to transplanted cells from that tumor but not, as a rule, to cells from other tumors, including those of the same histologic type elicited by the same carcinogen in the same mouse strain. Indeed, even in the same autochthonous host methylcholanthrene, painted on different skin areas, elicits multiple fibrosarcomas which exhibit little or no crossimmunity in transplantation-resistance tests. These private TAAs have not been characterized structurally or genetically.

Tumors induced by different chemical carcinogens or by different doses of the same carcinogen can also differ widely in immunogenicity and in **latent period** (the interval between administering the carcinogen and appearance of the tumor). *The least immunogenic tumors generally appear after the longest latent periods,* as though host immune responses to a slowly growing mass of nonidentical tumor cells results in the selective elimination of the most immunogenic variants. Such **immune selection** probably can also account for the gradual decrease in the immunogenicity of some tumors as they are "passaged" by transplantation through successive hosts.

However, repeated injections of irradiated cells from chemically induced tumors sometimes elicit Abs to **"public" TAAs,** those shared by many of the tumors elicited by a chemical carcinogen. Some of these TAAs appear to be fetal Ags (discussed below), or products of activated endogenous viral genes in the mouse genome (see below), or **anomalous expression of what are normally silent genes.** The following are two examples.

1) **Forssman Ag of human gastrointestinal cancer.** This ceramide-pentasaccharide

$$GalNAc\alpha 1 \rightarrow 3GalNAc\beta 1 \rightarrow 3Gal\alpha 1 \rightarrow$$
$$4Gal\beta 1 \rightarrow 4Glu\text{-ceramide}$$

is synthesized by many Forssman-positive (F^+) species. Man has been considered to be Forssman-negative (F^-), but the gastrointestinal mucosa in some persons is F^+; hence the Forssman Ag is an alloAg in man. F^- individuals make a precursor that lacks the terminal *N*-acetyl galactose (GalNAc) of the complete Ag. However, gastrointestinal cancers in F^- individuals produce the complete Ag; the terminal disaccharide (GalNAc$\alpha 1 \rightarrow$3GalNAc) appears to be the main serologic determinant.

2) **Thymus leukemia (TL) Ag of lymphocytic leukemia in mice.** This surface glycoprotein is a normal alloAg in mice, present on thymocytes of certain inbred strains (Ch. 18). However, when mice of other (TL^-) strains develop lymphocytic leukemia (induced by chemical carcinogens, x-irradiation, or murine leukemia viruses), the transformed lymphocytes can also express TL as a TAA. Hence, in TL^- strains the structural gene for TL must be present but suppressed; in transformed lymphocytes a perturbation of regulation evidently results in its expression.

TUMORS INDUCED BY VIRUSES

Injection of oncogenic viruses (Ch. 66) into immunologically unresponsive individuals (e.g., polyomavirus into the newborn mouse) usually results in progressively growing, fatal tumors. But in immunologically competent individuals (e.g., adult mice) most oncogenic viruses elicit either no tumors or tumors that grow for a short time and then regress, leaving a high level of specific immunity. Evidently TAAs of cells transformed in vivo elicit responses that cause destruction of the developing tumor cell clones. These self-cured animals also reject transplants of any other tumor induced by the same virus; the immunity is specific, for tumors induced (in syngeneic animals) by other viruses are not rejected.

Unlike the tumors elicited by a chemical carcinogen, the tumors elicited by a given oncogenic virus have shared TAAs, which are identical in different individuals and even in tumors of different histologic types (e.g., parotid gland adenomas and fibrosarcomas elicited by polyomavirus).

With the tumors induced by some viruses the shared TAAs are also found in the virions, while with others the shared TAAs are not present in virions, though they are specified by the viral genome.

For example a 70,000-dalton envelope glycoprotein (gp70) and a 30,000-dalton internal capsid protein (p30) of murine leukemia virus (MuLV) are TAAs of lymphoma cells that are transformed by and continue to produce MuLV virions. In contrast, with the oncogenic complex of feline sarcoma virus (FSV), a defective transforming oncornavirus, and feline leukemia virus (FLV), a helper oncornavirus that is necessary for FSV replication, the TAA on transformed cells is not present in the virions, though it is specified by the FSV genome.

Thus, adult pet cats infected with FLV–FSV often develop lymphomas that regress, and the cats concomitantly produce serum Abs that react with a surface TAA, called **FOCMA (feline oncornavirus cell membrane Ag),** on all FSV-transformed cat cells. However, these Abs are not absorbed by FLV-FSV virions, even when the virions are disrupted to expose internal proteins. FOCMA nevertheless appears to be specified by a viral rather than a cellular gene, since the same shared TAA is also present on FSV-transformed canine, primate, and mink cells. Moreover, some unusual FSV virions (rescued from FSV-transformed cells by superinfection with helper viruses) have in their envelope a protein with FOCMA antigenic activity.

In another pattern, the viral gene for a TAA appears to be derived from the host. The Abelson murine leukemia virus was originally isolated from a BALB/c mouse that had been infected with another mouse leukemia virus (the Moloney virus), and its characteristic surface Ag, the Abelson Ag, is found not only on Abelson-transformed lymphoma cells, but also on some normal lymphocytes of mice of the BALB/c strain, but not of other strains. It thus seems possible that the Abelson Ag is the expression of a host (BALB/c) gene that was fortuitously incorporated into the Abelson virus genome as this virus arose from the Moloney virus genome. The gene for the FOCMA Ag of FSV might have been derived from a similar ancient event in a cat, or it might be an intrinsic component of the FSV genome. The question is not likely to be clarified, because inbred strains of pet cats are not available, and the history of the FSV Ag is not known.

Private TAAs on Virus-Induced Tumors. Some virus-induced tumors seem to express individually distinctive ("private") TAAs, as well as shared ones. For instance, mice immunized with cells from a tumor induced by the mouse mammary tumor virus become more resistant to that particular tumor than to others induced by the same virus.

Spontaneous Tumors in Laboratory Rodents. As noted before, the immunogenicity of tumors induced by chemical carcinogens varies inversely with the latent period. Since the tumors that arise sporadically ("spontaneously") in old mice often seem to lack TAAs, they may arise after a very long latent period, during which the tumor cell population is exposed to immune selection against cells with potent surface TAAs.

FETAL ANTIGENS

These Ags, expressed by various cells at certain stages of normal fetal development, are produced in normal adults only at trace levels, if at all. However, some of these Ags are also produced by certain tumors (as the shared "public" TAAs induced by chemical carcinogens). Thus, sera from repeatedly pregnant rats contain Abs to fetal Ags which react with surface Ags on many rat tumors. Moreover, some fetal Ags seem to elicit resistance against tumors: in adult hamsters that had been injected repeatedly with fetal hamster cells the oncogenic virus SV40 elicited fewer tumors than in control (uninjected) animals.

Several fetal Ags are present at high concentration in the serum of individuals with certain forms of cancer. Two of these **oncofetal Ags,** α-fetoprotein and carcinoembryonic Ag, are currently used to monitor the clinical course of some cancer patients.

α-Fetoprotein (AFP). This Ag, synthesized in the fetal liver, yolk sac, and gastrointestinal tract, is the main fetal serum protein in vertebrates. At its peak level (2–3 mg/ml) in the human fetus (third to sixth month of gestation) it amounts to about 1/3 of all serum proteins. At birth the level drops about one million-fold, and only traces are present in normal human serum (about 5–25 ng/ml). However, it is greatly elevated in the serum of mice carrying transplanted hepatomas (as was first discovered by Abelev in the USSR), and in about 2/3 of patients with primary liver cancer (which is frequent in Asia and Africa, but rare in the United States). High levels are also found in patients with teratomas (embryonal cell tumors), and in some with cancer of the stomach and pancreas. The levels may also be high, but fluctuate more widely, in individuals with acute or chronic hepatitis or cirrhosis of the liver.

After surgical excision of a tumor, the persistence of an elevated AFP level, or its later increase, suggests that tumor cells remain or that the tumor has recurred. AFP is not immunogenic in the species of origin, but rabbits immunized with human AFP develop Abs that crossreact with rabbit AFP; the crossreacting Abs seem not to influence the progress of AFP-producing tumors.

AFP may represent a fetal form of serum albumin: it is a single polypeptide chain of 70,000 daltons, showing some amino acid sequence homology with albumin; like serum albumin it also binds estrogens and a variety of anionic dyes. As with some other pairs of proteins showing partial amino acid sequence homology, native human AFP does not crossreact with Abs to native human serum albumin (HSA), but when the AFP is denatured it reacts extensively with Abs to denatured HSA.

Carcinoembryonic Antigen (CEA). Gold and Freedman immunized rabbits with surgically excised specimens of human colon cancer, and then absorbed antisera from these rabbits with normal tissue from the same surgical samples. The residual Abs were found to react specifically with 1) colon cancer cells from many patients, 2) some other human cancers of gastrointestinal origin, and 3) endodermal tissues from the normal human fetus but not from normal adults. For use as a radioactive ligand in radioimmunoassays (Ch. 16), the corresponding Ag, called CEA (carcinoembryonic Ag), is currently isolated from liver metastases of colon cancers.

CEA is generally present at trace levels ($<5\mu g/ml$) in normal human serum. Elevated levels are found in serum from most patients with cancer of the lower gastrointestinal tract and pancreas, and less frequently from patients with cancer of other organs (breast, lung, etc.) or with certain chronic or recurrent inflammatory conditions (e.g., patients with colitis or chronic bronchitis, heavy cigarette smokers). Many of these conditions have in common excess mucin production. The significance of elevated levels of CEA is complicated by the heterogeneity of the Ag and by difficulty in distinguishing it from substances that crossreact with anti-CEA Abs.

CEA is a glycoprotein (180,000 daltons) with more carbohydrate (65%) than protein. Immunofluorescence with antisera to CEA suggests that it is present in a superficial mucinous layer (glycocalyx) that surrounds certain cells, from which it is probably easily shed. Most Abs to CEA are probably directed to the protein moiety, because the association constant is high (10^{11} liters/mole), and serologic reactivity persists after removal of many of the sugar residues. However, some Abs to CEA probably react with oligosaccharide groups and account for low-level crossreactivity with blood group substances.

Screening Human Populations for Cancer. A major medical goal is the systematic detection of cancer at an early stage (when treatment is more effective), by testing large human populations with sensitive, accurate, and inexpensive procedures. One approach is through serologic detection, in serum, of TAAs that are shed by tumor cells. Large-scale tests of serum CEA levels reveal some of the difficulties associated with this approach. Though the CEA levels in persons with colon cancer generally differ substantially from those in normal persons, the affected and normal populations overlap significantly. The test for CEA has accordingly been abandoned as a screening procedure. Instead, CEA levels are used to follow the course of individual patients, especially after therapy. (If, for instance, the CEA level drops after surgical excision of a colon cancer, the subsequent reappearance of an elevated level strongly suggests recurrence.)

HOST IMMUNE RESPONSES TO TUMORS

In the tumor-bearing host, TAAs elicit diverse immune responses, involving almost all classes of Abs and types of T cells. Under various experimental conditions the following effectors can destroy tumor cells:

1. Antibodies plus complement (C)
2. Antibody-dependent, complement-independent, cell-mediated cytotoxicity (ADCC)
3. Cytotoxic T lymphocytes (CTLs)
4. Macrophages, activated by lymphokines from Ag-stimulated T_{DTH} cells
5. Natural killer (NK) cells

1) **Antibodies and complement.** Tumor cells vary considerably in their susceptibility to lysis by Abs (to cell surface Ags) and C. Transformed (and normal) lymphocytes are generally more susceptible than other cells.

As was noted in Chapter 20, the specific binding of a single IgM Ab molecule to a cell is sufficient to generate a C-mediated lytic lesion, but at least two IgG molecules, bound in close proximity by Ags on the cell, are necessary. Hence if cytotoxic Abs are of the IgG class, the susceptibility of a cell depends on both the density and the distribution of its surface Ags, not merely on their presence. Susceptibility is probably also influenced by variations in the membrane's ability to bind C3 fragments and perhaps also in its repair mechanisms.

Mice of many inbred strains (e.g., DBA/2) do not synthesize the fifth component of complement (C5). These mice are, nevertheless, not unusually susceptible to spontaneous or transplanted tumors. Hence the complement system is probably not a major antitumor defense.

2) **Antibody-dependent, complement-independent, cell-mediated cytotoxicity (ADCC).** In this process Abs to tumor cell Ags provide specificity, while various effector cells provide the cytotoxic mechanisms: as is shown schematically in Fig. 23-9, Ab molecules specifically link target cells to surface receptors for Fc domains of Igs on the effector cells. Target cell lysis is not accelerated or augmented by the addition of C, nor is it inhibited by antisera to C components. The effective Abs can belong to any IgG class. Aggregated IgG is inhibitory, because it blocks Fc receptors on the effector cells.

Several types of normal leukocytes are effectors: granulocytes, macrophages (see below), and K (for killer) cells. **K cells,** which have the required Fc receptors, look like lymphocytes but they are not lysed by antisera to Igs or to Thy-1 (Ch. 18), and they do not adhere to glass, differentiating them, respectively, from B cells, T cells, and macrophages (see Null Cells, Ch. 18).

ADCC has been studied extensively in the test tube, but its role in the host's defenses against autochthonous tumors is not clear. However, some evidence suggests that this process can provide resistance in vivo: passively transferred Abs can provide specific protection against a transplanted tumor in C-deficient mice. Because C is not involved and resistance depends upon the transferred anti-tumor Abs, it is possible that the tumor cells are destroyed by ADCC.

3) **Cytotoxic T lymphocytes (CTLs).** Because these cells probably contribute more than Abs, macrophages, and T_{DTH} cells (responsible for delayed-type hypersensitivity; Ch. 22) to the rejection of allografts, they probably also have the greatest activity against autochthonous tumors and against transplanted syngeneic tumors. In inbred mice CTLs can be readily elicited against a transplantable syngeneic tumor by a few injections of irradiated tumor cells, or by allowing the transplanted tumor to grow for a few weeks and then excising it. The mechanism of action of CTLs against tumor cells are the same as against normal target cells, as described in Chapters 18 and 22 (Figs. 22-8 and 22-9).

As noted earlier (Ch. 18 and 22), CTLs recognize Ags (including TAAs) only in conjunction with products of the major histocompatibility complex (MHC) on the same tumor cells. Accordingly, CTLs elicited by a lymphoma would be expected to lyse the eliciting tumor cells but not the cells of another lymphoma with different MHC alleles, even if both tumors had the same TAAs. Moreover, soluble TAAs (secreted or shed by tumor cells) should not block the lytic activity of CTLs unless the soluble TAA molecules are subsequently adsorbed onto macrophages or other cells, whose MHC molecules could help meet the dual requirement of CTLs.

MHC restriction also hinders the search for CTLs in humans with cancer, for unrelated persons nearly always differ in their MHC alleles. Accordingly, CTLs are expected to lyse only autologous tumor cells (i.e., derived from the same individual), not similar tumors, even with the same TAAs, from other individuals.

4) **Macrophages.** It has long been suspected that macrophages destroy tumor cells: they are prominent in tumors undergoing rejection, and tumor cells transplanted in a mixed cell inoculum along with various leukocytes grow especially slowly when macrophages are

FIG. 23-9. Ab-dependent, complement-independent, cell-mediated cytotoxicity (ADCC). Ab molecules, specific for tumor-associated Ags (TAAs), link target cells to K (killer) cells via the latter's Fc receptor.

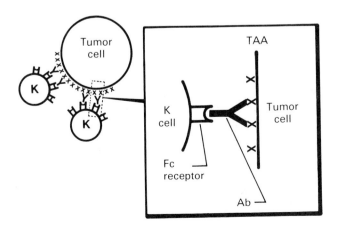

included. It is unlikely that the tumoricidal activity of macrophages arises from the enhanced phagocytosis of target cells coated with opsonins (Abs or C3b fragments or both), because Abs to the TAAs of autochthonous tumors are seldom demonstrable.

However, tumoricidal macrophages are activated cells (see Activated Macrophages, Ch. 22), and it is possible that they kill tumor cells by the same mechanisms that enable them to destroy diverse microbes (Figs 22-5 and 22-6). These macrophages do not discriminate among different tumors, but they do seem to be selective in destroying tumor cells more actively than normal ones. Though it is only poorly understood, this property is currently exploited for cancer therapy: activated macrophages are generated either systemically (by repeated injections of bacterial vaccines, such as BCG or *Corynebacterium parvum*) or locally, in accessible skin cancers (e.g., by injecting BCG or applying dinitrochlorobenzene to elicit a local delayed-type hypersensitivity reaction; see Contact Skin Sensitivity, Ch. 22).

With some forms of human cancer repeated injections of BCG seem to prolong life and increase the disease-free interval after excision of the primary tumor mass. With lung cancer, the benefits are more pronounced when BCG is administered to patients with early tumors (low "load tumor") and to those with evidence of generally diminished T cell activity (e.g., diminished activity in mixed lymphocyte reactions with allogeneic cells).

5) **Natural Killer (NK) Cells.** Some cells in the spleen or peritoneal cavity of normal mice can lyse a variety of tumors, especially when the tumor cells are persistently infected with certain enveloped viruses (e.g., murine leukemia viruses). The cytotoxic cell populations are predominantly made up of lymphocytes, but the tumoricidal cells lack the properties of B or T lymphocytes, of K cells (see ADCC, above), or of macrophages. Their specificity and mechanism of action are obscure. NK activity is enhanced by **interferon (IF)**, especially "immune IF" from Ag-stimulated lymphocytes. Stimulation of NK cells may be largely responsible for the anti-tumor effects of IF.

NK cells are abundant in nude mice and may be responsible for the resistance of these T-cell–deficient animals against some tumors. Thus though most transplanted tumors (of xenogeneic as well as of allogenic origin) grow readily in these mice metastases are rare; moreover, these mice do not suffer from a high incidence of spontaneous tumors or of tumors following the administration of chemical carcinogens.

IMMUNOLOGIC SURVEILLANCE

Since tumors frequently elicit reactive T cells and Abs in the autochthonous host or in syngeneic hosts, it has seemed possible that transformation of normal cells into tumor cells occurs much more frequently

than the clinical appearance of tumors would indicate, the bulk of the neoplastic clones being eradicated early in their development by immune responses (**immunologic surveillance**). However, this concept predicts an increased incidence of cancer in chronically immunodeficient individuals; and, as the following paragraphs show, except for lymphoreticular tumors in certain groups, this prediction has not been borne out.

1) Nude mice, with a virtual absence of T cells, have no increased incidence of spontaneous tumors, or even of tumors induced by chemical carcinogens. However, as noted above, these mice have NK cells and may not be entirely devoid of antitumor immune responses.

2) Individuals with multiple forms of cancer, arising in different organs and tissues, are rare; but they should be commonplace if a defective immune system was responsible for the emergence of clinically apparent tumors.

3) Persons with lepromatous leprosy are severely and chronically deficient in cell-mediated immune responses, but they have no increased incidence of cancer.

4) Children with primary immunologic deficiencies show no increased incidence of the common tumors of children (retinoblastoma, neuroblastoma, Wilms' tumor) or of adults (see below), although they do have an approximately tenfold greater frequency of lymphosarcoma and reticulum cell sarcoma than age- and sex-matched controls.

5) Individuals who receive long-term immunosuppressive drug therapy (to prevent kidney allograft rejection) have no increased frequency of the common human cancers (of lung, breast, gastrointestinal tract, prostate). However, they do have an approximately 300-fold increase in the incidence of reticulum cell sarcomas. (They are also reported to have a higher than normal frequency of skin cancer, perhaps because they are subjected to unusually meticulous scrutiny.)

In contrast to these negative findings with common tumors, immunologic surveillance seems to be effective against certain tumors induced in animals by oncogenic viruses that spread "horizontally", i.e., from one infected individual to another in the same population. For instance, pet cats infected with FLV-FSV oncornaviruses (see Tumors Induced by Viruses, above) do not develop leukemia if they have appreciable titers of Abs against the characteristic membrane Ag of FSV-transformed cells (FOCMA; Fig. 23-10). In addition, viral lymphomas of chickens (Marek's disease) can be prevented by immunization with an attenuated form of the virus (herpesvirus of turkeys), and mice deprived of the T cell immune system (through neonatal thymectomy or administration of antilymphocytic serum) are much more susceptible than littermate controls to tumor induction by polyoma and murine leukemia viruses (Table 23-6).

Taken together, *all of this evidence argues against im-*

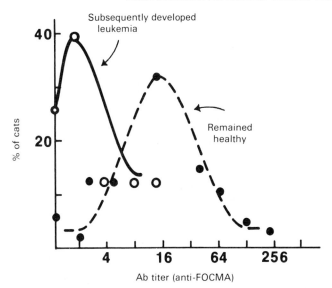

FIG. 23-10. Correlation between the development of leukemia and the titer of Abs to the distinctive cell membrane Ag (FOCMA: feline oncornavirus-associated cell membrane Ag) of tumors induced by the feline leukemia–feline sarcoma virus complex (FLV–FSV). Data are from a prospective survey of about 50 cats from one household with a high incidence of cat leukemia. Immune anti-FOCMA serum titers at the time of the first sampling were about eightfold higher in cats that remained healthy than in those that later succumbed to leukemia. (Based on Essex M et al: Science 190:790, 1975)

TABLE 23-6. Increased Incidence of Polyomavirus-Induced Tumors in T-Cell-Deficient Mice‡

Preliminary treatment of adult mice*	Subsequent treatment†	No. of mice	% That developed tumors
Normal rabbit γ-globulin (control)	None	24	0
Thymectomy + ALG	None	14	100
Thymectomy + ALG	Inject lymph-node cells from normal syngeneic mice	10	90
Thymectomy + ALG	Inject lymph node cells from syngeneic mice immunized against poly-omavirus	11	0

*Polyomavirus was injected after thymectomy and the first of three biweekly injections of ALG, which is more destructive for T than for B lymphocytes (see Ch. 22, Interference with Cell-Mediated Immunity). ALG: globulin fraction of rabbit antiserum to mouse lymphocytes.

†Given 7 weeks after rabbit γ-globulin, or 10 days after the final injection of ALG.

‡(Based on Allison AC: Proc Royal Soc Lond [Biol] 177:23, 1971)

munologic surveillance as a potent mechanism for controlling the more frequent of major forms of human cancer. However, it seems to be effective when oncogenic viruses elicit tumors with unusually immunogenic TAAs, such as FOCMA in cat leukemia.*

TUMOR ESCAPE MECHANISMS

Why is immunologic surveillance against tumors not more effective? The answer is not known, but various factors, described in the following paragraphs, enable tumors to grow progressively in spite of a competent host immune system.

Suppressor Cells. Preliminary evidence suggests that many TAAs readily elicit T suppressor (T_s) cells that block the development of immune responses against those TAAs.

Thus, in culture, some tumor cells stimulated syngeneic spleen cells from thymectomized mice (which are deficient in T_s cells), but

* The human lymphomas that are frequent in chronically immunodeficient individuals (lymphosarcoma, reticulum cell sarcoma) might have unusually potent TAAs, possibly because they might also have a viral etiology: hence their low frequency in individuals with normal immune defenses.

not from intact mice, to differentiate into antitumor CTLs. In a mixture of spleen cells from both the thymectomized and the intact mice, the T_s cells from the intact mice prevented the development of the CTLs. Another indication of the dominant effect of T_s cells is provided by the effects of anti-IJ serum, which is directed to a differentiation Ag on T_s cells (specified by the IJ locus of the H-2 complex; Chs. 18 and 19) and which probably destroys T_s cells selectively, thereby enhancing other T cell responses. Some tumors growing in normal mice were rejected following injections of this antiserum.

Weakly Immunogenic TAAs. It is possible that immunologic surveillance, acting on a heterogeneous tumor cell population, eliminates those variant tumor cells with the most immunogenic TAAs, leaving those with weak TAAs to proliferate. Why some TAAs are poorly immunogenic is not known: some may be weak just because they elicit more T suppressor than T helper activity.

Escape Mechanisms and Concomitant Immunity. Tumors often grow progressively in hosts that are making anti-TAA responses, demonstrable in vitro with lymphocytes and macrophages from the host. Yet when some of the tumor cells are reinjected into the same individual at a distant site they are usually eliminated, while the main tumor continues to grow. This **concomitant immunity** may be responsible for a commonplace event in the natural history of most tumors: the small numbers of tumor cells that frequently separate from a primary tumor mass and spead through blood and lymph to distant sites only rarely succeed in establishing a metastatic tumor.

Several explanations for the paradox have been suggested: 1) In the main tumor mass there are probably high local concentrations of TAAs, released from tumor cells; these TAAs could block antitumor Abs and possibly anti-TAA T lymphocytes. 2) Some tumor masses seem to produce antiinflammatory factors that, acting locally, block the penetration of macrophages and other leukocytes into the tumor and inhibit the cytotoxic activity of macrophages. 3) Some tumor cells become resistant to lysis by CTLs, in some instances because the tumor cells produce a surface glycoprotein that covers products of the major histocompatibility complex and perhaps other cell surface Ags.

Blocking Antibodies and Enhancement. Some Abs to TAAs are not cytotoxic, e.g., because their C_H domains (see Fig. 17-14, Ch. 17) are unable to fix C or to participate in ADCC (see above). By binding to TAAs on tumor cells these Abs can block cytotoxic reactions (**efferent blockade**) or can interfere with the induction of other cytotoxic responses (**afferent blockade**). *These blocking Abs might thereby enhance a tumor's growth in a host who can make an anti-TAA response.*

Enhancement was initially observed with tumors growing in allogeneic hosts that had been subjected, before receiving the tumor graft, to a special immunization program against the graft's major histocompatibility Ags (see Enhancement, above, under Acceptance of Allografts). Though blocking Abs may contribute to the graft-enhanced tumor growth, it appears that specific T_s cells, elicited by the enhancing immunization, prevent the usual cell-mediated allograft rejection response (of T_{DTH} cells and CTLs).

Serum from tumor-bearing individuals has also been frequently reported to block, in in vitro assays, the cytotoxic activity of the same individual's leukocytes on his own tumor cells. The blocking serum activity has not been well characterized but appears in various instances to be due to Abs, or to **Ab–TAA complexes.**

Immunostimulation. At concentrations far below those required for cytoxicity or blocking effects, some Abs (together with C) actually stimulate the proliferation, rather than the destruction, of tumor cells.* In addition, intravenous injections of syngeneic tumor cells clumped by tumor-specific T lymphocytes have elicited more growing lung tumor nodules than an injection of the same cells in dispersed form, without the T cells. Whether the clumped cells tend more readily than the dispersed cells to lodge in the pulmonary capillaries, or whether the specific T cells exert a more subtle influence in promoting the growth of metastatic tumor nodules, is not known.

SUMMARY

The three groups of cell surface alloAgs considered in this chapter differ in structure and in the immune responses they elicit. The blood group substances on RBCs are polysaccharides. Molecules of this type are generally bound weakly by Abs (Ch. 16) and rarely, if ever, elicit cell-mediated immune responses, perhaps because they bind too weakly to T cells. Hence Abs, not CMI, are responsible for immune reactions against RBCs in mismatched blood transfusions.

In contrast, the cell surface alloAgs on normal tissue and organ allografts are glycoproteins, whose antigenic activity resides in the protein moiety. These Ags can bind to Abs (and probably also to T cells) with high avidity and they can elicit the entire spectrum of humoral and cellular immune responses. However, T cells (T_{DTH} and CTLs) are probably largely responsible for the rejection reactions.

The tumor-associated Ags of tumor cells (TAAs) are least understood. They include a wide range of chemically diverse structures (Forssman Ag, glycolipids, polypeptides, etc.), but some are likely to be glycoproteins for, like transplantation Ags on normal cells, the TAAs elicit (in addition to Abs) various T cells that destroy the tumor cells. Unlike the cells of normal tissue allografts, however, tumor cells proliferate without restraint and generate variant cells with altered TAAs. It is probably because the resulting host immune response eliminates the cells with the most potent TAAs that well-established tumors usually have weakly immunogenic TAAs. These Ags and the responses they elicit are thus particularly difficult to analyze.

* The proliferation response is reminiscent of the stimulation of thyroid cells by some antithyroid Abs (see Thyroiditis, Ch. 22).

SELECTED READING

RED CELL ALLOANTIGENS

KABAT EA: In Aminoff D (ed): Blood and Tissue Antigens. New York, Academic Press, 1970

KOBATA A, GINSBURG V: Uridine diphosphate N-acetyl-D-galactosamine: D-galactose α-3-N-acetylgalactosaminyltransferase, a product of the gene that determines blood type A in man. J Biol Chem 245:1484, 1970

MARCHESI VT, ANDREWS EP: Glycoproteins. Isolation from cell membranes with lithium diiodosalicylate. Science 174:1247, 1971

MARCUS DM: The ABO and Lewis blood group system. Immunochemistry, genetics, and relation to human disease. N Engl J Med 280:994, 1969

MILLER LH, MASON SJ, DVORAK JA, MCGINNISS MH, ROTHMAN IK: Erythrocyte receptors for (Plasmodium knowlesi) malaria: Duffy blood group determinants. Science 189:561, 1975

MORGAN WT: Croonian lecture: a contribution to human biochemical genetics: the chemical basis of blood group specificity. Proc R Soc Lond [Biol] 151:308, 1960

RACE RR, SANGER R: Blood Groups in Man, 5th ed. Springfield, IL, Thomas, 1968

SPRINGER GF, CODINGTON JG, JEANLOZ RW: Surface glycoprotein from a mouse tumor cell as specific inhibitor of antihuman blood group N antigen. J Natl Cancer Inst 49:1469, 1972

SZULMAN AE: The ABH and Lewis antigens of human tissues during prenatal and postnatal life. In Human Blood Groups. Basel, Karger, 1977

WATKINS WM: Blood-group substances. Science 152:172, 1966

TRANSPLANTATION IMMUNITY

BACH F, VAN ROOD JJ: The major histocompatibility complex. Genetics and biology. N Engl J Med 295:806, 872, 927, 1976

BILLINGHAM RE, BRENT L, MEDAWAR PB: Actively acquired tolerance of foreign cells. Nature 172:603, 1953

BILLINGHAM RE, SILVERS W: The Immunobiology of Transplantation. Englewood Cliffs, Prentice-Hall, 1970

BODMER WF (ed): The HLA system. Br Med Bull 34:213, 1978 (Includes extensive review of HLA and disease)

HENDRY WS, TILNEY NL, BALDWIN WM, GRAVES MJ, MILFORD E, STROM TB, CARPENTER CB: Transfer of specific unresponsiveness to organ allografts by thymocytes. J Exp Med 149:1042, 1979

MEDAWAR PB: The immunology of transplantation. Harvey Lect 52:144, 1956–57

MEDAWAR PB: The Uniqueness of the Individual. New York, Basic Books, 1958

RUSSELL PS, CHASE CM, COLVIN RB, PLATE JMD: Kidney transplants in mice. An analysis of the immune status of mice bearing long-term H-2 incompatible transplants. J Exp Med 147:1449, 1469, 1978

SNELL GD, DAUSSET J, NATHENSON S: Histocompatibility. New York, Academic Press, 1976

SNELL GD, STIMPFLING JH: Genetics of tissue transplantation. In Green EL (ed): The Laboratory Mouse. New York, Dover, 1966

TUMOR IMMUNITY

BERENDT MJ, NORTH RJ: T-cell mediated suppression of anti-tumor immunity. An explanation of progressive growth of an immunogenic tumor. J Exp Med 151:69, 1980

CAIRNS J: Cancer, Science, and Society, San Francisco, WH Freeman, 1978

CARLSON GA, TERRES G: Antibody-induced killing in vivo of L1210/MTX-R cells quantitated in passively immunized mice with ^{131}I-iododeoxyuridine-labeled cells. J Immunol 117:822, 1976

CHAPDELAINE JM, PLATA F, LILLY F: Tumors induced by murine sarcoma virus contain precursor cells capable of generating tumor-specific cytolytic T lymphocytes. J Exp Med 149:1531, 1979

DJEU JY, HUANG K-Y, HERBERMAN RB: Augmentation of mouse natural killer activity and induction of interferon by tumor cells in vivo. J Exp Med 151:781, 1980

GIPSON TG, IMMAMURA M, CONLIFFE MA, MARTIN J: Lung tumor associated depressed alloantigen coded for by the K region of the H-2 major histocompatibility complex. J Exp Med 147:1363, 1978

GOLD P, FREEDMAN SO: Specific carcinoembryonic antigens of the human digestive system. J Exp Med 122:467, 1965

HAKAMORI S, WANG S-M, YOUNG WW: Isoantigenic expression of Forssman glycolipid in human gastric and colonic mucosa: its possible identity with "A-like antigen" in human cancer. Proc Natl Acad Sci USA 74:3023, 1977

HENNEY CS: Mechanisms of tumor cell destruction. In Green I, Cohen S, McCluskey RT (ed): Mechanisms of Tumor Immunity. New York, John Wiley & Son, 1977

KAMO I, FRIEDMAN H: Immunosuppression and the role of suppressive factors in cancer. Adv Cancer Res 25:271, 1977

MELIEF CJM, SCHWARTZ RS: Immunocompetence and malignancy. In Becker FF (ed): Cancer: A Comprehensive Treatise, Vol. 1. New York, Plenum, 1975

NORTH RJ, KIRSTEN DP, TUTTLE RL: Subversion of host defense mechanisms by murine tumors. I. A circulating factor that suppresses macrophage-mediated resistance to infection. J Exp Med. 143:559, 574, 1976

RISSER R, STOCKER TE, OLD LJ: Abelson antigen. A viral tumor antigen that is also a differentiation antigen of BALB/c mice. Proc Natl Acad Sci USA 75:3918, 1978

RUCO L, MELTZER MS: Macrophage activation for tumor cytotoxicity: induction of tumoricidal macrophages by PPD in BCG-immune mice. Cell Immunol 32:203, 1977

RUSSELL JH, GINNS LC, TERRES G, EISEN HN: Tumor antigens as inappropriately expressed normal alloantigens. J Immunol 122:912, 1979

STUTMAN O, DIEN P, WISUN RE, LATTIME EC: Natural cytotoxic cells against solid tumors in mice. Blocking of cytotoxicity by D-mannose. Proc Natl Acad Sci USA 77:2895, 1980

WEISSMAN IL: Tumor immunity in vivo: evidence that immune destruction of tumor leaves "bystander" cells intact. J Natl Cancer Inst 51:443, 1973

part **IV**

BACTERIAL AND MYCOTIC INFECTIONS

chapter

24

HOST–PARASITE RELATIONS IN BACTERIAL INFECTIONS*

W. BARRY WOOD, JR.†
BERNARD D. DAVIS

* Revised with the generous help of Zanvil Cohn, M.D.
† Deceased

After Koch defined the criteria for establishing a microorganism as the etiologic agent of a disease (Koch's postulates, Ch. 1), the Russian zoologist Elie Metchnikoff provided the first insight into the reaction of the host to such invaders. In 1882 he noted that the introduction of a thorn into the body of a transparent starfish larva caused mobile cells to surround the foreign body, and he then showed that these cells could also ingest and destroy small foreign particles, including bacteria. Cohnheim had earlier seen leukocytes migrating from tissue capillaries (in mammals) to form pus at sites of injury, and in patients who had died of bacterial infections bacteria were observed in such cells; but the cells were thought merely to transport the microbes. Metchnikoff, however, concluded that they served as an important defense reaction of the body.

Metchnikoff's theory was not readily accepted, especially since the discovery of bactericidal substances and antitoxins in serum in the 1890s provided an alternative explanation for resistance. As has happened so often in science, an artificial dichotomy was created: many pathologists and clinicians adhered to the doctrine of humoral immunity, while the followers of Metchnikoff favored cellular immunity. The controversy was resolved when Neufeld and Rempau in 1904 established a joint role for the two kinds of factors: animals injected with killed pneumococci were found to contain Abs in their serum, called opsonins (Gr. *opsonein*, to prepare food for), that greatly accelerated the in vitro phagocytosis of the same (but living) organisms.

With the subsequent development of the field of immunology the concept of immunity ceased being synonymous with resistance. As discussed in the preceding chapters, immunity is now restricted to the aspects of resistance due to specific antibodies (Abs) and specific lymphocytes that react with particular foreign substances (antigens = Ags). This chapter will review the additional role of various nonspecific factors in resistance, as well as mechanisms of bacterial pathogenicity.

PATHOGENICITY: INFECTION VERSUS DISEASE

Bacteria vary enormously in their pathogenicity, i.e., their capacity to cause disease. Of those that infect humans soon after birth most are nonpathogenic commensals (see Ch. 44), which cannot penetrate the natural defenses of the host unless these are seriously impaired. Others (pneumococci, staphylococci, tubercle bacilli) may coexist with the host in a truce that is only occasionally broken. Highly virulent pathogens, in contrast, usually or always produce disease when they infect. Some (e.g., typhoid bacilli, diphtheria bacilli, rickettsiae) are usually eradicated after causing an acute disease but may also persist, after recovery, in healthy carriers; while others (e.g., treponemes of syphilis, brucellae) can lead to a dormant infection, which may relapse. Still others (e.g., the bacilli of plague, tularemia, or anthrax) are present only during active disease. Accordingly, though the two terms are often used synonymously, even in this text, the distinction between infection and disease is often fundamental in interpreting the significance of a positive culture.

For example, viridans streptococci can be recovered from over 90% of human throats, where they are not a cause of disease. However, when the same organisms are repeatedly cultured from the blood, which is normally sterile, subacute bacterial endocarditis can be inferred. Moreover, in many lesions mixed infections are found, and it is often possible to distinguish primary and secondary invaders. This distinction is based on the kinetics of the disease process and on the characteristic pathogenicity of the different organisms.

Modest impairment of host resistance shifts the precarious balance existing with mild pathogens, while severe impairment permits even classic nonpathogens to become invaders (see Depression of Host Defense, below, and Chs. 31 and 32). Such opportunistic infections have become more common, as a byproduct of therapeutic advances: cytotoxic drugs and radiation depress the bone marrow and lymphopoiesis, and cortisone eliminates certain thymus-associated lymphocytes (see Ch. 18) and depresses the inflammatory response. Antibiotics may also, paradoxically, lower host resistance (to antibiotic-resistant organisms) by inhibiting the competing normal flora that are present on body surfaces and mucosa (see Microbial Antagonism, below).

VIRULENCE

The term virulence (degree of pathogenicity) is used to encompass two features of a pathogenic organism: its infectivity (the ability to colonize a host), and the severity of the disease produced. Virulence varies not only among bacterial species but also among strains. **Virulence factors** are those components of an organism whose loss (usually by mutation) specifically impairs virulence but not viability. The two main classes of virulence factors are toxins and surface molecules: some of the latter act by promoting resistance to phagocytosis (antiphagocytic capsules), and others by promoting adherence to susceptible cells, as will be discussed below.

It should be emphasized that virulence is a highly **polygenic** property of bacteria: it can be affected by variations not only in major virulence factors but in virtually any aspect of the physiology of an organism (e.g., nutrition, growth rate). Moreover, as in a higher organism, the adaptation of a microorganism to its ecologic niche depends on the evolution of a balanced genome. Hence the insertion of the gene for a virulence factor into a distantly related nonpathogenic microbe cannot be expected to convert it into a well-adapted, effective pathogen.

Measurement of Virulence. For both epidemiologic and pathogenetic studies it is often important to quantitate the virulence of a bacterial strain. Important variables include the route of inoculation, the strain of the test animals and their physiologic state (age, nutrition, freedom from other disease, immunologic experience), and the culture medium and growth stage of the inoculated culture.

The virulence of an organism (or of a toxin) is usually expressed as the **LD_{50}**: the dose that will kill 50% of the inoculated animals within a given time (see Appendix, Ch. 46). The reason for choosing this endpoint is the shape of the dose–response curves (Fig. 24-1), which shows that the rate of change in mortality with dose (C and D) is greatest around the point of 50% survival. Moreover, with a given number of animals the LD_{50} can be measured more precisely when the curve is steep, as in Figure 24-1B, rather than shallow, as in A.

The sigmoid shape of the dose–response curve is due to the heterogeneity of the animal population and also the distribution of chance events in the infection. When the heterogeneity is minimized by using animals of the same age, weight, sex, and inbred strain the dose–response curves are steeper but still sigmoid.

FIG. 24-1. Examples of dose–response curves used to measure bacterial virulence. The infecting dose (abscissa) is plotted in logarithmic units. Measurements of the lethal effects of toxins result in much the same kind of dose–response curves. (Wilson GS, Miles AA, eds: Topley and Wilson's Principles of Bacteriology and Immunology. Baltimore, Williams & Wilkins, 1964)

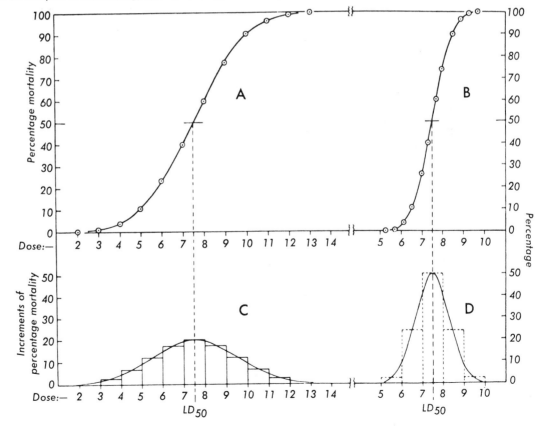

The **transmissibility** of a bacterial strain refers to its ability to cause a demonstrable infection in a given animal host, usually detected by cultural methods. The dose required is known as the **ID₅₀** (infectious dose in 50% of the animals). **Communicability,** an epidemiologic concept, will be discussed briefly at the end of this chapter.

DORMANT AND LATENT INFECTIONS

In **covert (inapparent)** bacterial infections pathogenic bacteria are present but they do not cause disease. Two types are distinguished, somewhat arbitrarily: **dormant infections,** from which the organisms can be recovered, and **latent infections,** in which they cannot be recovered by available methods but are inferred on indirect grounds.

With **dormant** infections the term **carrier** is used for those persons who may continue to spread the organism after recovery. Under epidemic conditions the carrier rate may approach 100%, presumably because of continual reinfection. The carrier state is often unstable, persisting for a number of weeks (as after β-streptococcus infection) and then disappearing. However, following some diseases, such as typhoid or louse-borne typhus, a few individuals may remain carriers indefinitely.

Latent infections are recognized in two ways: retrospectively through the later emergence of overt illness, or concurrently through the presence of immunologic reactivity. In syphilis a positive serologic test persists between the early and the late stages. In untreated infection with the tubercle bacillus (Ch. 37) the tuberculin test almost always remains positive, and the disease is occasionally activated many years later. Immunologic reactivity persisting for many years probably always reflects the persistence of Ag, but reactivity may be absent if the antigenic mass is too small.

The distinction between a dormant and a latent infection depends on the sensitivity of the detection methods. Detection of organisms by visualization, in stained sections or in smears, requires several thousand bacteria per milliliter. While cultural methods are in principle much more sensitive, they are limited by the ability of some bacteria to be sequestered in tissues in an ill-defined, sluggish state, possibly as L forms (Ch. 43). Thus tubercle bacilli from inactive lesions will sometimes grow out in cultures only after months, rather than after weeks.

The principles involved in inapparent bacterial infections are illustrated by studies on experimental tuberculosis in mice treated with pyrazinamide and isoniazid (Fig. 24-2). With either treatment alone the population of cultivable tubercle bacilli in the tissues fell steadily for some weeks and then stabilized at a low detectable level (i.e., as a dormant infection). Combined therapy for 60 days, in contrast, resulted in the complete elimination of cultivable tubercle bacilli. However, a latent infection was present: after 90 days without chemotherapy, over a third of the animals yielded tubercle bacilli, and virtually all did so if given cortisone. To take another illustration, treatment with cortisone in man (and in experimental animals) occasionally activates various unsuspected latent infections.

FIG. 24-2. Influence of pyrazinamide and isoniazid, used singly and together, on mice infected with tubercle bacilli. The inoculum was 2.0×10^6 culturable units of tubercle bacilli. Only numbers of bacilli in the spleen are charted, since the bacterial counts were consistently highest in this organ. (McDermott W: Public Health Rep 74:485, 1959)

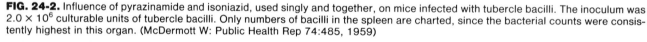

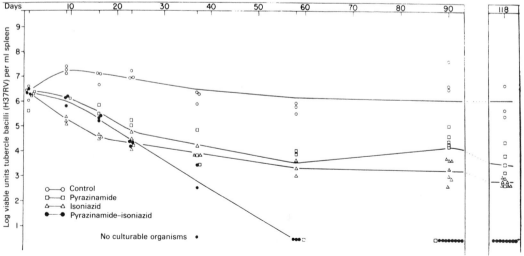

PATHOGENIC PROPERTIES OF BACTERIA

Invasiveness and Toxigenicity. Bacteria cause disease mainly by three basic mechanisms: **invasion** of tissues or surfaces, which leads to damage of host cells in the immediate vicinity; production of potent **toxins,** which may act at remote sites; and **hypersensitivity** reactions (discussed under Immunologic Factors, below).

The concept of **invasiveness** is broad, and includes colonization of epithelial surfaces (e.g., *Corynebacterium diphtheriae, Bordetella pertussis, Vibrio cholerae*), penetration of epithelial cells (*Shigella dysenteriae*), or penetration into deeper tissues (*Staphylococcus aureus; Streptococcus pneumoniae; Mycoplasma pneumoniae*). Organisms considered primarily toxinogenic, or toxigenic (e.g., *C. diphtheriae*), must ordinarily be invasive enough to achieve surface colonization (which is often promoted by damage to underlying tissue). Conversely, organisms that are strongly invasive usually damage tissues through the short-range toxicity of their metabolic products, though some (e.g., rickettsiae) may owe much of their pathogenicity to mechanical blocking of capillaries by hyperplasia of infected endothelial cells.

Extracellular versus Intracellular Parasitism. Pathogenic bacteria may be divided further on the basis of their fate after being phagocytized. Organisms that are promptly destroyed when phagocytized (e.g., *S. pneumoniae, Streptococcus pyogenes*) behave as **extracellular** parasites, damaging tissues only so long as they remain outside phagocytic cells. Since their presence usually stimulates the production of opsonizing Abs, which render them susceptible to phagocytosis, they usually produce acute diseases of relatively short duration (e.g., pneumococcal pneumonia). In contrast, **intracellular** parasites, which can multiply within the phagocytic cells, frequently give rise to chronic diseases, such as tuberculosis or brucellosis. When the poorly understood interactions between the ingested bacterium and the phagocyte in such infections are in balance a state of intracellular parasitism persists, and when they are out of balance either the parasites or the host cells are destroyed.

A third class of pathogens are the **obligate intracellular parasites** (rickettsiae, chlamydiae). For these organisms, as for viruses, subtle differences in the cell surface receptors essential for their uptake may be important in resistance.

ANTIPHAGOCYTIC FACTORS

Bacteria that function as extracellular parasites owe their virulence to antiphagocytic surface components, often demonstrable as definite capsules (Fig. 24-3). Most capsules are high-molecular-weight polysaccharides. In *S.*

pyogenes the major antiphagocytic factor is the M protein, which is associated with a dense pile of surface filaments rather than with a capsule.

The relation between capsules, phagocytosis, and virulence is clearly exemplified by *S. pneumoniae.* A fully encapsulated S (smooth) strain is found to resist phagocytosis (in the absence of Ab) and is highly virulent for mice, whereas its nonencapsulated R (rough) mutant is readily phagocytized and is virtually avirulent. Moreover, enzymatic removal of the capsular polysaccharide, or combination with Ab, renders the S organisms both nonpathogenic and susceptible to phagocytosis. Finally, the organisms are more virulent when harvested under conditions in which they have larger capsules: when in the exponential rather than the stationary phase of growth, or when grown in media rich in carbohydrate.

ORGANOTROPISM

Many bacteria are **organotropic:** i.e., highly selective in regard to the surfaces or tissues that they infect or invade. For example, in healthy individuals staphylococci tend to inhabit the skin and the vestibule of the nose and may invade the underlying tissues, whereas pneumococci are usually confined to the throat and may invade the lower respiratory tract to cause pneumonia. *Neisseria meningitidis* is also found in the nasopharynx, but it may invade the meninges; and the closely related species *N. gonorrhoeae* characteristically infects the genitourinary tract.

When the chorioallantoic fluid of the embryonated egg is inoculated with *B. pertussis,* the agent of whooping cough, the microorganism tends to localize on the ciliated bronchial epithelium in 15-day-old embryos (just as in diseased children—see Fig. 34-7), but not in younger embryos, which lack ciliated cells. Likewise, in the chick embryo *N. meningitidis,* which causes acute meningitis in man, localizes in the meningeal tissues, while *Streptobacillus moniliformis,* which causes severe polyarthritis, invades the joints.

One kind of determinant of tissue tropism is **nutritional.** The best-defined example is the production in the

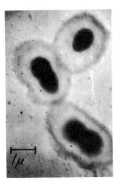

FIG. 24-3. Electron micrograph of *Streptococcus pneumoniae,* type 1. The capsule has been reacted with type 1 Ab to accentuate its visibility (quellung reaction). (Mudd S et al: J Exp Med 78:327, 1943)

FIG. 24-4. Scanning electron micrograph of bacteria adherent to the gastrointestinal epithelium. **A.** Small intestine of monkey. Note the virtual carpet of adherent spirochetes. (×6000) **B.** Esophagus of monkey. Note the multiple layers of adherent bacteria. (×2,000) (Courtesy of Dr. Z. Skobe)

bovine uterus of erythritol, a growth factor for those brucellae that cause infectious abortion in cattle. A more general determinant, however, is specific adherence.

ADHERENCE (COLONIZATION)

Many bacteria have surface macromolecules that promote adherence (Fig. 24-4) to specific receptors that are present on some animal cells but not on others: these adhesive elements on bacteria may reside on pili, on a fringe of finer filaments, or on a feltwork (glycocalyx). The role of specific bacterial adherence is now increasingly recognized, especially for the selective **colonization** of body surfaces by the ubiquitous normal flora, and in locations (e.g., the upper intestinal tract) where free organisms would be washed out faster than they could multiply. Presumably in such in vivo circumstances one end of a bacterium serves as a hold-fast while the other end elongates and is released by cell division, much like the case with bacteria that have been wedged in the pores of filters and then serve as a source of synchronized cells (Ch. 5).

In the development of this field, studies in oral microbiology (Ch. 44) have played a key role, both because they deal with readily accessible surfaces and because a

key problem has been the persistence of organisms despite the constant flow of sterile saliva. The answer to this problem has been the presence of a highly specific, tightly adherent population: one species of *Streptococcus* adheres to the enamel surface, another to the epithelial surfaces, and still others to underlying bacteria.

The concept of specific adherence clearly extends to other regions of the body, but only limited evidence is yet available, mostly on the basis of studies with tissue cultures. Thus *Escherichia coli* adheres better to intestinal lining cells (or to those of the urinary bladder) than to those of the oral mucosa, while the opposite is true of streptococci. Adherence to the intestinal epithelium has also been demonstrated for the important intestinal pathogen *Shigella;* and the enteropathogenicity of *E. coli* strains for pigs has been correlated with filamentous surface Ags that increase affinity for the intestinal epithelium. In similar studies the pathogenicity of *M. pneumoniae* has been correlated with adherence to cultured tracheal epithelium, and that of gonococci with the presence of pili that adhere to urethral epithelium.

The study of animal cell receptors for bacteria has lagged far behind that of receptors for viruses (Ch. 53). However, the recent emergence of this field offers promise of providing means for intercepting bacterial spread by blocking the receptor, in addition to using Abs to cover the bacterial Ags. For example, some strains of *E. coli* appear to attach to mannose-containing determinants in a surface receptor on the intestinal epithelium: adherence can be inhibited by mannose (competing with the animal cells for attachment to the bacteria), and also by the mannose-binding plant protein concanavalin A (competing with the bacteria for attachment to host cells).

Antibodies to bacterial surface constituents probably promote immunity not only by opsonization but also by covering Ags involved in adherence. Moreover, in body secretions **secreted glycoproteins** closely related to cell surface constituents (e.g., blood group substances in saliva) may play a role in host defense by competing with fixed receptors and thus **preventing adsorption** or causing **desorption** of adherent bacteria. **Desquamation** of surface epithelial cells is another defense mechanism, releasing adherent bacteria along with the cells.

Specificity of adherence seems likely to be the main cause of the striking host specificity of most pathogens. Ironically, this specificity has delayed the study of adherence: precisely because it has to be bypassed in the use of animal models, experimental studies on pathogenicity have generally employed methods (intravenous or intraperitoneal inoculation, overwhelming doses) that ignore this aspect of virulence.

The kinetics of adherence undoubtedly contribute to the usual requirement for a **large infecting dose** in infections on a surface bathed by secretions (e.g., the intestinal tract): in experimental studies only one bacterial cell in thousands becomes firmly attached and initiates a persistent colonization.

Recent studies have revealed a **dynamic state** of the **bacterial cell surface** in vitro, with shedding of Ags and with changes in their concentrations under various circumstances. The significance for pathogenesis has only begun to be explored.

ENZYMES THAT ACT EXTRACELLULARLY

The invasive properties of some bacteria are attributed to their elaboration of enzymes that act extracellularly.* Many gram-positive bacteria, for example, produce **hyaluronidase**. Originally referred to as "spreading factor," this enzyme promotes diffusion through connective tissue by depolymerizing hyaluronic acid in the ground substance. This action has been assumed to facilitate bacterial invasion, but there is no conclusive evidence that it does so. In fact, in experimental *Clostridium perfringens* infections the administration of antihyaluronidase Abs fails to influence the spreading of the lesion.

Equally unconvincing is the evidence concerning the pathogenetic role of staphylococcal **coagulases,** which

---

* Bacterial extracellular enzymes known to be cytotoxic are usually classified as toxins; these are discussed in the next section.

clot plasma by a thrombokinaselike action. These enzymes, produced by most virulent strains, have been thought to be responsible for the fibrin barrier that characteristically surrounds and appears to localize acute staphylococcal lesions, including antiphagocytic fibrin envelopes about the cocci themselves. However, some non–coagulase-producing strains are pathogenic, produce identical lesions, and are no more readily phagocytized. It therefore seems unlikely that coagulases of staphylococci per se account either for virulence or for the tendency of the lesions to remain circumscribed.

Streptococci and staphylococci also elaborate enzymes (**streptokinase** and **staphylokinase**) that catalyze the lysis of fibrin. Whether fibrinolysis actually influences the invasiveness of the organisms is also uncertain, although it has long been assumed that the relative absence of fibrin in spreading streptococcal lesions may be due to the action of streptokinase. On the other hand, the **collagenases** of *Clostridium histolyticum* and *C. perfringens,* which break down the collagen framework of muscles, clearly facilitate the extension of gas gangrene due to these organisms.

EXOTOXINS

The exotoxins of *Corynebacterium diphtheriae, Clostridium tetani, Clostridium botulinum,* and *S. dysenteriae* (Shiga's bacillus) are the most powerful poisons known. One nanogram (1 ng; 10^{-9}g) of tetanus or botulinum toxin is enough to kill a guinea pig (Table 24-1). These two toxins act by selectively blocking certain types of presynaptic terminals in the central nervous system. The somewhat lower potency of diphtheria toxin may be due to its general affinity for host cells, causing a broader distribution. The pharmacologic actions of the known bacterial exotoxins are comparatively slow, some requiring several days. Most exotoxins are inactivated by heat, acid, or proteolytic enzymes (e.g., in the gastrointestinal tract). However, with several types of botulinum toxin limited proteolysis enhances potency, by cleaving a precursor. Chemical treatment (e.g., by formalin) can convert exotoxins to toxoids (nontoxic but immunogenic derivatives), often used for prophylactic immunization. Certain mutations in *tox* genes, which cause production of inactive but serologically cross-reacting toxins, are also under investigation for this purpose, and may lead to live vaccines.

The potencies of bacterial exotoxins vary widely in different host species. For example, humans, horses, and guinea pigs are 1000 times more sensitive (per unit of body weight) than mice or rats to diphtheria toxin, because of its much readier uptake by their cells.

Koch's emphasis on the lethal action of toxins long delayed the recognition of those toxins (e.g., enterotoxins, the short-range toxins of *Bordetella pertussis*) that

TABLE 24.1 Exotoxins Produced by Principal Toxigenic Bacteria Pathogenic for Man

Bacterial species	Disease	Toxin	Action	Toxicity per mg. expressed as LD_{50}^{kg}*
Clostridium botulinum	Botulism	Six type-specific neurotoxins	Paralytic	1,200,000 (G)
Clostridium tetani	Tetanus	Tetanospasmin	Spastic	1,200.000 (G)
		Tetanolysin	Hemolytic cardiotoxin	
Clostridium perfringens	Gas gangrene	α-Toxin†	Lecithinase: necrotizing, hemolytic	200 (M)
		β-Toxin		
		γ-Toxin		
		δ-Toxin		
		ε-Toxin	Necrotizing	
		η-Toxin		
		θ-Toxin	Hemolytic cardiotoxin	
		ι-Toxin	Necrotizing	
		κ-Toxin	Collagenase	
		λ-Toxin	Proteolytic	
Clostridium septicum	Gas gangrene	α-Toxin	Hemolytic	
Clostridium novyi	Gas gangrene	α-Toxin	Necrotizing	50,000 (M)
		β-Toxin	Lecithinase: necrotizing, hemolytic	
		γ-Toxin	Lecithinase: necrotizing, hemolytic	
		δ-Toxin	Hemolytic	
		ε-Toxin	Lipase: hemolytic	
		ζ-Toxin	Hemolytic	
Corynebacterium diphtheriae	Diphtheria	Diphtheritic toxin	Enzyme altering transferase II	3,500 (G)
Staphylococcus aureus	Pyogenic infections	α-Toxin	Necrotizing, hemolytic, leukocidic	50 (M)
		Enterotoxin	Emetic	
		Leukocidin	Leukocidic	
		β-Toxin	Hemolytic	
		γ-Toxin	Necrotizing, hemolytic	
		δ-Toxin	Hemolytic, leukolytic	
Streptococcus pyogenes	Pyogenic infections and scarlet fever	Streptolysin O	Hemolytic	5.0 (M)
		Streptolysin S	Hemolytic	
		Erythrogenic toxin	Causes scarlet fever rash	
		Streptococcal DPNase	Cardiotoxic	
Yersinia pestis	Plague	Plague toxin	Necrotizing (?)	25 (M)
Bordetella pertussis	Whooping cough	Whooping cough toxin	Necrotizing	
Shigella dysenteriae	Dysentery	Neurotoxin	Hemorrhagic, paralytic	1,200,000 (R)
Vibrio cholerae	Cholera	Enterotoxin	Diarrhea	
Escherichia coli strains	Gastroenteritis	Enterotoxin	Diarrhea	

*LD_{50}^{kg} denotes median lethal dose (LD_{50},) per kilogram of body weight for guinea pig (G), mouse (M), or rabbit (R).

†The designation of toxins by Greek letters is purely arbitrary and is based on the order in which they were identified.

(Modified from Van Heyningen WE: in Florey HW, ed: General Pathology, p. 754. Philadelphia, Saunders, 1962)

were revealed by their action on tissue preparations (e.g., an isolated loop of intestine). At the molecular level, enzymatic action was recognized early for the α-toxin of *Clostridium perfringens,* which damages cell membranes by its action as a lecithinase. More recently another, more subtle enzymatic reaction, transfer of ADP-ribose from NAD to a specific host protein, has been found to explain two very different kinds of action by other toxins: 1) diphtheria toxin and pseudomonas toxin use this reaction to inactivate an elongation factor (EF2) in host cell protein synthesis; and 2) the enterotoxins of *Vibrio cholerae* and some *E. coli* strains use the reaction to activate host adenyl cyclase, thereby causing increased fluid secretion in the intestine. The action of diphtheria toxin will be further described in Chapter 26, and that of cholera toxin in Chapter 31 (Fig. 31-6).

The toxins that act intracellularly have two moieties: one enzymatic and the other concerned with binding to a specific receptor in the host cell membrane and transferring the enzyme across the membrane. The enzymatic moiety is not active until released from the native toxin. The molecular basis of these steps in toxin action will be further considered in Chapter 26.

There is conclusive evidence for the pathogenic role of diphtheria, tetanus, and botulinum toxins, various enterotoxins, and the erythrogenic toxin of *Streptococcus pyogenes.* The roles also seem clear for the short-range exotoxin of *Bordetella pertussis,* the α-toxin of *Staphylococcus aureus,* the neurotoxin of *Shigella dysenteriae,* the lethal toxin of *Bacillus anthracis,* the plague and cholera exotoxins, and the necrotoxic proteases elaborated by gas gangrene organisms.

The genes coding for exotoxin formation are generally located in **temperate bacteriophages** (diphtheria, botulinum, streptococcus erythrogenic toxin) or in **plasmids** (*E. coli* or staphylococcal enterotoxin, staphylococcal exfoliatin), rather than in the bacterial chromosome. This property is consistent with a role of toxins in adapting a bacterium to a specific parasitic niche without being essential for viability.

Many bacteria form **hemolysins,** which attack cell membranes by mechanisms not clearly defined. Hemolysis is not ordinarily important in pathogenesis but the same products may act as **leukocidins,** lysing white blood cells and hence decreasing local host resistance.

ENDOTOXINS

Unlike the heat-labile protein exotoxins, endotoxins are heat-stable lipopolysaccharides, associated with the outer membranes of gram-negative bacteria (see Ch. 6 and Endotoxins in Ch. 31); their toxicity resides in the lipid moiety (lipid A). They are less potent and less specific than most exotoxins in their cytotoxic activities, and they do not yield toxoids.

Biologic Effects. When injected in sufficient amounts an endotoxin causes shock within an hour or two, usually accompanied by severe diarrhea; experimental findings indicate that its absorption from the bowel is a major cause of terminal irreversibility in hemorrhagic shock. The significance of endotoxin is thus not limited to recognized infections.

After lethal injection of an endotoxin often the only discernible lesions are a few hemorrhages in the wall of the gastrointestinal tract. In smaller doses endotoxins cause fever, transient leukopenia followed by leukocytosis, hyperglycemia, hemorrhagic necrosis of certain tumors, abortion, altered resistance to bacterial infections (see next paragraph), various circulatory disturbances, and vascular hyperreactivity to adrenergic drugs. They are also capable of eliciting the Shwartzman phenomenon.*

Endotoxin appears to cause fever through an **endogenous pyrogen** released from polymorphonuclear leukocytes. The nonspecific effects of endotoxin on resistance to bacterial infection are complex: large doses depress resistance (including phagocytosis), while small doses enhance Ab synthesis (Ch. 19) and phagocytosis. Indeed, an appropriate dose of endotoxin yields a biphasic response, with depression followed by enhancement (as its concentration decreases).

Animals given repeated injections of small amounts of an endotoxin become relatively unresponsive to its effects. This **tolerance** is nonspecific, i.e., it extends to other endotoxins, and it is associated with increased resistance to stresses that increase the transfer of bacteria from the gut to the blood stream (e.g., traumatic shock, radiation damage). Tolerance subsides within a week or two after the last injection, in contrast to the much longer persistence of active immunity. Endotoxin tolerance results primarily from an **enhanced clearance of endotoxin by cells of the reticuloendothelial system:** it can be nullified by "reticuloendothelial blockade" with such agents as thorium oxide (Thorotrast) or India ink.

---

* When endotoxins are injected subcutaneously into rabbits, in doses of a few micrograms, a mild inflammatory reaction occurs in the skin. If, 24 h later, an intravenous injection of the same, or another, endotoxin is given in the same amount, the **originally injected** skin site becomes hemorrhagic within a few hours. The lesion is characterized by the presence of leukocyte–platelet thrombi, particularly in venules. Indeed, the reaction, first described by Shwartzman in 1928, will not occur unless a sufficient number of leukocytes are present in the circulation. If both injections are given by vein the animal usually dies within 24 h after the second, and at autopsy bilateral cortical necrosis of the kidneys is regularly found. Neither this **generalized Shwartzman reaction** (which is accentuated by cortisone), nor the localized form, has been satisfactorily explained in immunologic terms.

ANTIBACTERIAL DEFENSES OF THE HOST

The host must constantly defend itself against bacterial invasion. Differences in both the genetic makeup and physiologic state of the host influence the many factors that contribute to this resistance.

SKIN AND MUCOUS MEMBRANE BARRIERS

Mechanical Factors. The intact surface of a healthy epidermis seems to be rarely, if ever, penetrated by bacteria. Even the widely held opinion that *Francisella tularensis* and brucellae may do so is unsubstantiated. When the integrity of the epithelial surface is broken, however, subcutaneous infection may develop, often by staphylococci that normally inhabit the hair follicles and sweat glands. Dermal infections are particularly common when the skin is moist, as in hot, humid climates.

Many pathogenic bacteria, on the other hand, are able to penetrate **mucous membranes,** on whose moist surfaces they often thrive. These barriers include the conjunctiva and the oral, respiratory, gastrointestinal, and genitourinary surfaces.

Though efforts to demonstrate microscopic lesions at the sites of mucosal penetration have not been successful, it seems likely that damage to epithelial cells by toxic products of the bacteria plays a role. In addition, there is much indirect evidence that in the respiratory tract entry is facilitated by agents that injure mucosal cells: viruses, noxious gases (including certain inhalation anesthetics), cold, or low humidity.

Mechanical factors that help to protect the epithelial surfaces include the lavaging action of physiologic fluids (tears, saliva, urine), trapping on mucous layers, removal by cilia,* and elimination by coughing, sneezing, or desquamation. Nevertheless, the protection afforded by mucous membranes against penetration is very incomplete, especially when the surface is placed under mechanical tension or the epithelium is injured. Thus frequent transient **bacteremia** (i.e., presence of viable bacteria in blood), involving normal surface flora, has been shown to follow dental extraction (or even chewing), straining at stool, or childbirth; and the organisms sometimes initiate infection in distant vulnerable sites (e.g., damaged heart valves; prostate). It thus appears that *the tissue defenses are engaged continually in eliminating organisms* of low pathogenicity, as well as occasionally in combatting more virulent ones.

Chemical Factors. Mechanical factors alone do not account for the remarkable resistance of the skin and mucous membranes to bacterial invasion. The **acidity** of the gastric secretions preserves the usual sterility of the stomach; absence of acid (achlorhydria) is accompanied by heavy colonization of the gastric mucosa. The low pH of the skin (3–5) and of the adult vagina (4.0–4.5), due in part to products of bacterial metabolism, undoubtedly discourages the growth of many microorganisms. **Unsaturated fatty acids** on the skin kill certain pathogenic bacterial species (e.g., *S. pyogenes*) but stimulate the growth of others (diphtheroids). The selectivity of these processes may help to account for the fact that so few kinds of bacteria ordinarily inhabit the skin. The principal self-sterilizing factor in tears, nasal secretions, and saliva is **lysozyme,** and that in semen appears to be **spermine,** which is bactericidal to many pathogenic bacteria.

Mucous secretions (e.g., of the respiratory tract; saliva; tears) also contain **antibodies,** especially of the IgA class. The protective effect of these secreted Abs is suggested by the finding that immunization with a pneumococcal polysaccharide lowers not only the incidence of disease but also the carrier rate. IgA Abs to various enteric bacteria are similarly found in the feces (**coproantibodies**), but their protective function has not been demonstrated.

Microbial antagonism by the normal flora suppresses the growth of many potentially pathogenic bacteria and fungi at superficial sites, both through competition for essential nutrients and through production of inhibitory substances (e.g., colicins; Chs. 12 and 31). The importance of the microbial balance in the mouth and gastrointestinal tract is forcefully illustrated by the *Candida* and *Staphylococcus* infections that often occur during prolonged treatment with broad-spectrum antibiotics. Moreover, elimination of the normal intestinal flora (in germ-free animals; Ch. 44), or its reduction by antimicrobial agents, strikingly decreases the minimal infecting dose of an intestinal pathogen.

PHAGOCYTIC DEFENSE

PHAGOCYTIC CELLS

When bacteria or other invading parasites penetrate the skin or mucous membranes cellular defense mechanisms are immediately brought into play: local and blood-borne macrophages and polymorphonuclear leukocytes (PMNs) accumulate around the invaders and initiate the phagocytic process. We shall briefly review the properties of these cells.

Mononuclear phagocytes arise when a rapidly dividing precursor in the bone marrow gives rise to the immature **promonocyte.** This slightly phagocytic cell can divide, and it develops into a **monocyte,** which is released into the blood stream. Monocytes

---

* In man the ciliary "escalator" of the bronchi and trachea keeps the mucous "blanket" moving upward toward the pharynx at 1–3 cm/h.

normally make up 3%–7% of the circulating white cells. The monocyte is actively phagocytic and bactericidal, but it has only a few lysosomal granules, similar to the azurophil granules of the PMN (see below). Within 2 days in the blood stream monocytes emigrate into tissues, where they enlarge and become **macrophages (tissue histiocytes).** These are more active in phagocytosis (and pinocytosis), and they develop many more granules by packaging hydrolytic enzymes in Golgi vesicles. In addition, phagocytosis or pinocytosis stimulates further granule formation. Macrophages may also divide under certain inflammatory stimuli, but the vast majority of tissue and inflammatory macrophages are matured blood monocytes.

The total body pool of macrophages constitutes the **reticuloendothelial system.** In contrast to the relatively small number of circulating monocytes, the macrophages are numerous. They are widely scattered throughout the connective tissue and just outside the basement membrane of most small blood vessels, and they are present in larger numbers in liver sinusoids (Kupffer cells), spleen, lung (alveolar macrophages), bone marrow, and lymph nodes. In the latter sites the cells are continually exposed to foreign material and function as scavengers. Tissue macrophages are thought to be long-lived cells.

The **neutrophil granulocytes (polymorphonuclear leukocytes, PMNs:** Fig. 24-5) of man (and the equivalent heterophils of laboratory rodents) are critical for host defense. Through a complex series of divisions and maturational steps a multipotential stem cell in the

bone marrow gives rise to the adult PMN. In contrast to mononuclear phagocytes, granulocytes are matured and contain their full complement of bactericidal agents before their release. These short-lived cells, which constitute 30%–70% of the circulating white cells, emigrate into the tissues within a day and soon disintegrate.

During differentiation in the marrow the nucleus of the PMN becomes multilobed, cell division ceases, and mitochondria and endoplasmic reticulum are lost from the cytoplasm. At the same time the cell becomes motile and actively phagocytic. Cytoplasmic granules of two kinds (Fig. 24-5) are formed from the Golgi apparatus, where newly synthesized proteins are packaged within membranes. The earliest visible granule, termed the **azurophil** or **primary granule,** contains typical lysosomal digestive enzymes. Later the **secondary granule** becomes predominant: it is smaller, less dense, and contains alkaline phosphatase, lactoferrin, and lysozyme. The neutrophil also contains large stores of glycogen; since most of its metabolic energy is derived from glycolysis, it can function efficiently in anaerobic environments.

Radioactive tracer experiments have established the following additional features of granulocyte physiology: 1) Maturation from stem cells in the marrow takes 2–3 days. 2) Only half of the granulocytes (25×10^9) in the human circulation are detectable in peripheral blood; the rest adhere to vessel walls. 3) For every circulating granulocyte approximately 100 mature cells are held in the marrow reserve pool. 4) Once a granulocyte enters the tissues, bowel, or respiratory tract it never returns to the circulation.

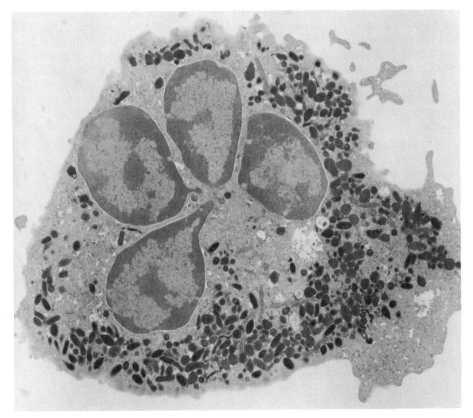

FIG. 24-5. Mature human polymorphonuclear leukocyte (PMN). The four large bodies are sections of the multilobulated nucleus. The cytoplasm contains two kinds of lysosomal granules: the generally elongated, dense primary (azurophil) granule, and the rounder, less dense secondary granule. (Briggs RT et al: J Cell Biol 67:566, 1975)

Eosinophilic and **basophilic granulocytes,** present in the circulation in much smaller numbers, are involved in allergic reactions rather than in antibacterial defense (see Ch. 21). Eosinophils also appear to play a particular role in the reaction to helminthic parasites.

THE SEQUENCE OF CELLULAR DEFENSE

The lodgment and multiplication of microbes or the presence of various foreign substances results in an **inflammatory response** characterized by 1) dilation of surrounding blood vessels; 2) an increase in vascular permeability, permitting the escape of plasma and a few RBCs; and 3) **diapedesis** (see below) of monocytes and PMNs. Early in the response granulocytes predominate in the inflammatory exudate as a result of their greater numbers in the blood stream. In the later phases, after many of the granulocytes have died, macrophages become the predominant cell type.

Organisms that escape phagocytosis in the local lesion may be transported by the lymphatics to regional lymph nodes. Here they initially encounter the resident macrophages of the nodal sinusoids, which filter particulates from the afferent lymph. If massive seeding of bacteria occurs there follows acute inflammation of the node **(lymphadenitis),** with an influx of blood granulocytes and monocytes as in the local lesion. If the defenses of the regional node are breached microbes may then be transported via the efferent lymph into the blood stream.

The presence of **bacteremia** (except for the transient invasion by normal flora noted above) usually indicates an invasive infection with unfavorable prognosis. Microbes are removed from the circulation by **clearance** mechanisms that depend largely on fixed macrophages lining the sinusoids of a number of parenchymatous organs, especially the liver and spleen. On occasion, en-

TABLE 24-2. Comparative Rates of Clearance from Blood Stream (in Normal Rabbits) of Intravenously injected Pneumococci of Varying Degrees of Virulence

Time after injection (h)	No. of viable pneumococci per ml of blood		
	Avirulent	Slightly virulent	Highly virulent
0	8,900,000	1,030,000	1,070,000
2	206	20,800	137,000
5	20	340	25,000
24	0	1,300	1,500,000*
48	—	134	Death
96	—	0	—

*The recrudescence of the bacteremia, leading to death, results from failure of the cellular defenses to destroy all the bacteria sequestered in the vascular beds of such organs as liver and spleen, where they eventually establish a firm foothold and reinvade the circulation.

(Wilson GS, Miles AA, eds: Topley and Wilson's Principles of Bacteriology and Immunology, p. 1180. Baltimore, Williams & Wilkins, 1964)

dothelial cells of capillaries are capable of engulfing certain microbial agents. The kinetics of clearance are illustrated in Table 24-2.

The Delivery of Phagocytic Cells. In the 1850s both Addison and Waller noted the adherence of white blood corpuscles to the inner lining of blood vessels in areas of injury (Fig. 24-6), and they inferred, correctly, that these cells somehow emigrate into the tissues to form pus. Through electron microscopy, and through use of the rabbit-ear chamber (which permits direct microscopy in vivo at oil-immersion magnifications), the process of **diapedesis** (the migration of cells across vascular walls) has been elucidated. For a description of this process the reader is referred to pathology texts.

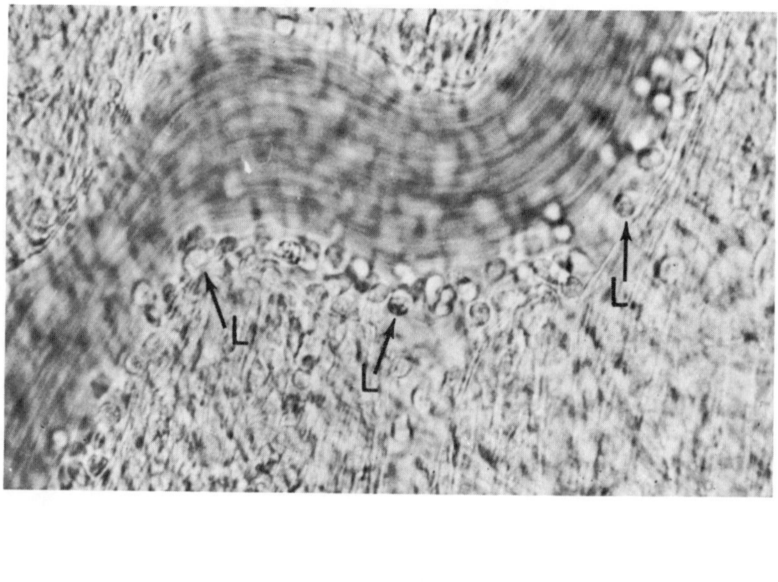

FIG. 24-6. Unilateral sticking of leukocytes to the endothelium of the venule in the rabbit-ear chamber 30 min after injury. The leukocytes **(L)** are adhering to the endothelial surface nearest to the zone of injury, which is located below the vessel. (Allison F et al: J Exp Med 102:655, 1955) (×200)

The **chemotaxis** that leads to the eventual concentration of migrating cells at a site of injury resembles bacterial chemotaxis (see Ch. 6): it involves a biased random walk toward the site, rather than unidirectional movement; and it is blocked by interference with methylation.

Studies in vitro—for example, in vessels with a filter separating the cells from a chemotactic stimulus—have identified a number of **chemotactic factors,** both for leukocytes and for monocytes. These include various bacterial products, fractions of leukocytes and other tissue cells, and peptides derived from some complement components. But because of the large number of these possible factors their roles, and the molecular basis of inflammation, are still obscure.

Macrophage Secretions; Role of Antigens. Macrophages not only ingest foreign particles: they serve other functions as well. Their key role in presenting Ags in active form to specific lymphocytes is described in Chapter 19. They also secrete many enzymes and mediators, when activated either by phagocytosis or by interaction with lymphocytes. These products include proteinases (collagenase, elastase, and plasminogen activator), prostaglandins, complement components, and unknown factors that stimulate the growth of fibroblasts. In chronic inflammation these products may be involved in tissue repair and matrix remodeling, or, conversely, in the destruction of tissues. The interplay between their actions and those of inhibitors, such as α_2 macroglobulin and α_1 trypsin inhibitors, may govern the extent of a lesion.

Fever. Most forms of fever are related to inflammation. A "messenger molecule" that acts upon the thermoregulatory center of the hypothalamus is detectable in the blood during fever. The role of leukocytes, macrophages, and lymphokines (Ch. 22) in releasing these endogenous pyrogens, in response to various foreign substances (exogenous pyrogens), is described in pathology texts. Despite much clinical speculation about the possible **beneficial effects of fever** in infections, definitive evidence is sparse (see Fig. 24-7).

THE PHAGOCYTIC PROCESS

Adherence of a particle to the surface of the plasma membrane of the phagocyte initiates phagocytosis. This step has little temperature dependence, and it involves various surface receptors.

In the opsonization process macrophages in mice have shown two major receptors that recognize the Fc portion of immunoglobulin (Ig) molecules (Ch. 17), thereby initiating phagocytosis; human macrophages have not been available for such detailed studies. One receptor is specific for monomeric Igs (of the IgG_{2a} class) and the other for Ag-crosslinked Igs. Another type of receptor binds a complement fragment, C3b. Activated macrophages

FIG. 24-7. Effect of temperature on the progression of syphilitic lesions as illustrated by their localization in shaved areas of rabbits after intravenous inoculation of *Treponema pallidum.* The anterior half of the white rabbit's body **(A)** and the posterior half of the black rabbit's **(B)** were shaved lightly (to avoid trauma) on the day of inoculation. Lesions appreared in the shaved (cooler) skin areas of both rabbits on the 22nd day; 5 days later the rest of each rabbit's body was shaved and no further lesions were found. (Hollander DH, Turner TB: Am J Syph 38:489, 1954)

(see below) ingest particles opsonized with IgM and complement, as well as those opsonized with IgGs. In general, opsonized particles (coated with Ab molecules) are more rapidly ingested, both by PMNs and by macrophages.

Adherence of a particle may somehow induce the next phase, **ingestion** (Fig. 24-8): an invagination of the plasma membrane envelops the particle, is pinched off, and enters the cytoplasm as a **phagocytic vacuole** (Fig. 24-9A and B).

Both the ameboid movement of the phagocytic cells and their engulfment of particles involve the formation of pseudopods (Fig. 24-8), which evidently requires both mechanical thrust and contraction. Actin microfilaments associated with myosin are undoubtedly involved in the contraction, and the thrust may involve an actin-binding protein (filamin), which forms reversible crosslinks between the filaments and thus causes their gelation into a relatively firm cytoskeleton. Immunofluorescent studies show that the contractile proteins are concentrated in the periphery of the

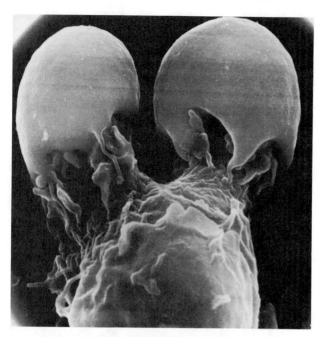

FIG. 24-8. Beginning of phagocytosis of antibody-coated erythrocytes by a macrophage. The active role of pseudopods in engulfment is similar, but less strikingly seen (because of size), with Ab-coated bacteria. (Scanning electron micrograph courtesy of M.E. Bayer).

cell. The changes in cell shape may involve cAMP, since filamin can be phosphorylated by a cAMP-dependent kinase.

In PMNs the **energy** required for phagocytosis is evidently provided by the glycolytic pathway, since iodoacetate and fluoride inhibit phagocytosis. (These cells also contain a large store of glycogen, which provides energy in poorly nourished and anaerobic tissues.) In macrophages either glycolysis or respiration is effective. In both cells phagocytosis is accompanied by a burst of oxygen consumption, increased glucose utilization and lactate production, and a tenfold increase in activity of the hexose monophosphate shunt (which provides NADPH for lipid and other biosyntheses). Increased turnover of phospholipids has been reported. Net synthesis of lipid is not increased in the leukocyte, but in the macrophage (which can form new lysosomes) it is increased in proportion to the amount of membrane interiorized.

POSTPHAGOCYTIC EVENTS

Formation of the Phagolysosome. The phagocytic vacuole formed at the surface migrates into the cytoplasm and collides with lysosomal granules, which explosively discharge their contents into it. Electron microscopy has revealed that the membranes of both granule and vacuole actually fuse (Fig. 24-9F), resulting in a **phagolysosome (digestive vacuole).** The number of granules lost by the cell through this **degranulation** (Fig. 24-9C and D) is directly related to the number of particles ingested. With very large particles lysosomal granules may discharge into an incompletely closed surface vacuole, thereby liberating their contents into the extracellular medium.

Intracellular Killing of Organisms. By 10–30 min after ingestion many pathogenic and nonpathogenic bacteria are killed: they fail to grow on artificial media, and structural degradation is often visible (Fig. 24-9E). Metabolic products as well as lysosomal constituents contribute to this bactericidal activity, in both the leukocyte and the macrophage. The production of **lactic acid** lowers the pH within the phagocytic vacuole to 4.0, which is low enough to kill some pathogens.

A number of components of the lysosome granule are potent antibacterial agents. **Lysozyme** (muramidase) hydrolyzes the peptidoglycan of gram-positive bacteria, but not that of gram-negative bacteria unless they are rendered susceptible by Ab plus complement. Klebanoff has described an important **bactericidal halogenating system,** which utilizes myeloperoxidase (a granule constituent), hydrogen peroxide, and chloride ion to cause lethal halogenation of bacteria and viruses. The killing, however, may also be due to the formation of reactive oxygen intermediates, such as superoxide anion (O_2^-), hydroxyl radicals (OH•), or singlet oxygen ($^1O_2$).

Granules of PMNs (but not of macrophages) also contain **cationic proteins,** which damage the permeability barrier of both gram-positive and gram-negative bacteria, and **lactoferrin,** an iron-binding protein that inhibits bacterial growth by depriving the organisms of Fe^{3+}.

Many species of bacteria, particularly those that produce chronic diseases, are not killed within phagocytes: the tubercle bacillus, for example, may multiply within the phagolysosome and may eventually destroy the cell and spread. This organism, and several others (e.g., the agents of tularemia and brucellosis), sometimes multiply only enough to maintain viable foci for months or years, thus giving rise to prolonged immunity and/or serologic reactivity.

Intracellular Digestion. Studies with radioactively labeled bacteria show that dead microbes are rapidly degraded in phagolysosomes to low-molecular-weight components. The various hydrolytic enzymes involved are all most active at an acid pH, which prevails in the digestive vacuole. Certain bacterial polysaccharides and lipids, however, are only slowly digested and may be retained within macrophages for weeks or months.

Stimulation of phagocytosis, as a means of enhancing nonspecific immunity, would obviously be of great value. However, efforts to achieve this goal have had only limited success. Macrophage activation (see below) by certain intracellular parasites, which increase the number of lysosomes, is promising, and repeated injections of a mycobacterium (BCG; see Genetic Variation, Ch. 37) or

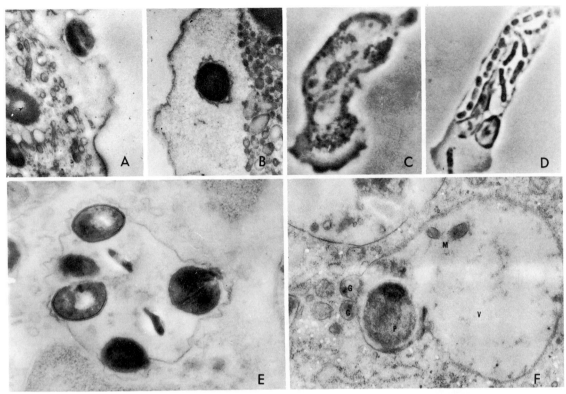

FIG. 24-9. A,B. Invagination of the plasma membrane of a leukocyte during phagocytosis. (×20,000, reduced) **C,D.** Degranulation of PMNs, resulting from phagocytosis of gram-poisitve bacilli. (×2000, reduced) **E.** Ingested cocci undergoing destruction in a phagocytic vacuole of a PMN. The cell walls of three of the organisms are partially or completely eroded. (×30,000, reduced) **F.** Fusion of the membranes of leukocytic granules **(G)** with the membrane of a phagocytic vacuole **(V),** resulting in discharge of the granular contents into the valuole containing a phagocytized pneumococcus **(P).** (×20,000, reduced) (**A,B,** Goodman JR et al: J Bacteriol 27:736, 1956. **C,D** courtesy of Hirsch JG. **E,** Florey HW: General Pathology. Philadelphia, Saunders, 1962. **F,** Lockwood WR, Allison F: Br J Exp Pathol 44:593, 1963)

Corynebacterium parvum have been reported to increase macrophage activity against certain kinds of cancer cells in patients. Endotoxins can also enhance the phagocytic activity of leukocytes, but they cause unpleasant side reactions; moreover, the heightened resistance that they induce is relatively short-lived. Prior **mobilization** of an inflammatory exudate—as by a chemical irritant in a pleural or peritoneal cavity—increases **local nonspecific resistance** to subsequent infection by its effect on cell numbers, but this procedure is obviously impractical.

TISSUE FACTORS IN ANTIBACTERIAL DEFENSE

Surface Phagocytosis. With agents of acute bacterial disease that possess antiphagocytic capsules there is some balance between invasion and host resistance, even though no corresponding Ab can usually be detected during the first 4 or 5 days of illness. This balance depends on a less efficient but essential mechanism discovered by Wood: phagocytosis, by granulocytes or monocytes, of organisms trapped against a surface.

This preantibody defense mechanism was detected in experimental subcutaneous and pulmonary infections with pneumococci: phagocytosis could be seen within a few hours. Subsequent in vitro investigations revealed a trapping of the encapsulated bacteria by the leukocytes, either **against tissue surfaces** (Fig. 24-10), or **between adjacent leukocytes** in concentrated exudates (intercellular surface phagocytosis), or in the interstices of **fibrin clots.** Studies with the rabbit-ear chamber have further shown that these phagocytic processes also occur in the blood stream during experimental bacteremia. This early defense mechanism often determines whether or not infection will become disease, and how severe the latter will become.

Architecture of Tissues. The efficiency of the preantibody phagocytic defense is greatly influenced by the general architecture of the tissues involved. In potentially "open" cavities (pleural, pericardial, peritoneal, joint, and meningeal), where the leukocytic exudate is initially

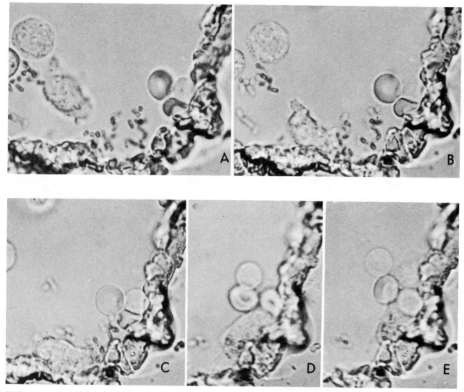

FIG. 24-10 A–E. Surface phagocytosis of encapsulated Friedlander's bacilli suspended with leukocytes in an antibody-free medium and applied to a cut section of normal rat lung. The leukocyte can be seen to trap the organisms against the alveolar wall in the process of ingesting them. Photographed action took place in a period of 10 min. (Smith MR, Wood WB, Jr: J Exp Med 86:257, 1947) (×1250, enlarged)

diluted with large amounts of fluid, surface phagocytosis is much less efficient than in tissues of tightly knit structure, where the exudate tends to be more concentrated and where tissue surfaces are more accessible to the leukocytes. The case fatality rate in untreated pneumococcal infections illustrates the importance of these factors: in pneumonia (in a tightly knit tissue) it is about 30%, whereas in meningitis (open surface) it is over 99%.

Necrosis and Abscess Formation. Necrosis also profoundly affects antibacterial defense. Whether caused by bacterial products, trauma, or vascular occlusion, necrosis of infected tissue eventually leads to suppuration and formation of walled-off abscesses. *In acute abscesses (e.g., staphylococcal) the exudate is predominantly granulocytic; in chronic abscesses (e.g., tuberculous) it is largely monocytic.* In both kinds the circulation cannot supply nutrients and Abs, and it cannot remove waste products. As a result the phagocytic cells are mostly inactive and in various stages of disintegration, and instead of protecting they provide a medium in which the infecting microorganisms multiply very little but survive for long periods. Al-

though abscesses sometimes heal spontaneously, they usually require drainage to remove the exudate and "resting" bacteria.

Chemical Factors. Many substances with antibacterial properties in vitro can be extracted from animal tissues and isolated from body fluids. However, their significance for antibacterial resistance is uncertain, except for lysozyme in various body fluids and lysosomal enzymes in the phagocytes. Part of the antibacterial action of serum appears to be due to **transferrin** (which binds Fe^{3+} very tightly), since the effect can be reversed by added Fe^{3+}.

IMMUNOLOGIC FACTORS IN ANTIBACTERIAL DEFENSE

ANTIBODIES

The roles of **toxin neutralization** and **opsonization** by Abs have been described above. In addition, **agglutina-**

tion, by Abs, of bacteria, such as pneumococci in edema-filled alveoli, impedes their spread. These responses make certain diseases self-limiting: if the host can survive long enough to mobilize a sufficient amount of protective Ab the illness will end within a few days.

In another antibacterial mechanism, Abs to surface Ags of gram-negative bacilli cause **bacteriolysis** in the presence of complement (Ch. 20). While immune bacteriolysis probably plays a role in host defense this reaction may not be sufficient to eradicate an extensive infection. In typhoid fever, for example, such bacteriolytic anti-O Abs are regularly found during the second week of illness, when the disease is usually still in full progress.

Immunologic mechanisms may also be harmful, by producing hypersensitivity reactions (see below), or by leading to the precipitation of immune complexes (of Ag, Ab, and complement) in tissues, especially the glomeruli (see Ch. 21).

Natural Antibodies. Specific Abs are sometimes found in a host not known to have been exposed to the corresponding Ag. These are termed natural Abs. Despite their low level, *natural Abs can markedly promote resistance, for any crosslinking of Ags by Abs can enhance immunogenicity.*

Natural Abs, as well as differences in sites of bacterial adherence, may account for some of the great variation among animal species in susceptibility to specific organisms. Thus Abs to pneumococci are present in the serum of resistant species (pigeons, chickens, dogs, cats) and are absent in susceptible ones (mice, guinea pigs, rabbits).

The distinction between natural and acquired Abs may be artificial, because of the possibility of an unrecognized earlier contact with the corresponding Ag or a closely related one. For example, more than 80% of humans over 1 year of age have Ab to type 7 pneumococcus; and though the carrier rate for this organism is less than 1%, a crossreacting Ag is found in streptococci of the universally carried viridans group. Similarly, natural Abs may result from continuous exposure of the host to its own gut flora, and from contact with foods, plants, and inhalants.

Cross-Tolerance. The presence of a **host Ag that resembles a bacterial Ag immunologically** may produce the opposite effect, for the tolerance to the former would **impair the formation of protective Ags** to the latter. There has been much speculation about the possible contribution of such cross-tolerance to the variation in susceptibility observed between species, and also observed between individuals with different blood group and transplantation Ags (Ch. 23). Thus the great susceptibility of the mouse to *Salmonella typhimurium* may be explained on this basis. However, firm evidence in man is lacking.

HYPERSENSITIVITY

A **delayed type** of hypersensitivity, originally referred to as "bacterial allergy," may develop during the course of bacterial disease. The resulting accelerated and intensified tissue response to the infecting organism was first demonstrated in Koch's classic observations on the response to two successive injections of tubercle bacilli (see The Koch Phenomenon, Ch. 37). Delayed hypersensitivity can be transferred by specific thymus-derived lymphocytes (Chs. 18 and 22): on interaction with the provoking Ag (probably on macrophages) these T cells release various substances that produce inflammation and necrosis.

Discovery of the Koch phenomenon engendered a prolonged argument about whether or not the hypersensitivity reaction benefits the host. The resulting inhibition of lymphatic spread of infection is clearly beneficial, but if spread does occur the localized necrosis promoted by the hypersensitivity is detrimental. Tuberculin sensitivity thus can either help or harm the host, depending upon the circumstances.

Delayed hypersensitivity is also encountered in brucellosis, leprosy, and mycoplasmal infection, as well as in fungal, viral, and parasitic* diseases. It is also frequently demonstrable by skin tests during (or following) pneumococcal, streptococcal, and mycoplasmal infections, but there is no convincing evidence for its role in the pathogenesis of these diseases (except for the rash in scarlet fever). In experimental staphylococcal infections in animals a hyperreactivity to Ags appears to be correlated with the amount of tissue necrosis.

MACROPHAGE ACTIVATION

During chronic infections with certain "intracellular" parasites, such as tubercle bacilli, the development of delayed hypersensitivity is accompanied by the development of enhanced bactericidal activity, of varying degree, in the macrophages. The **induction** of this macrophage activation is **immunologically specific:** only the infecting organism (or antigenically related ones) can evoke an accelerated immunity to reinfection. However, the **state of activation** is **nonspecific:** as Mackaness showed in mice, macrophages activated by *Listeria* infection also exhibit augmented bactericidal properties with antigenically unrelated organisms (*Salmonella, Brucella, Mycobacterium tuberculosis*).

Activation begins with an Ag-triggered release of **lymphokines** from T lymphocytes (see Ch. 18). These products cause the local macrophage population to develop an **increased number of lysosomes** (see Fig. 22-5) and also **increased secretion** of microbicidal products.

---

* According to accepted medical terminology, parasitic diseases are those caused by parasitic animals.

Macrophages probably vary over a continuous range in their degree of activation.

A significant degree of macrophage activation may occur even in the absence of T cells, in response to nonspecific mediators generated during inflammation. These include components of both the clotting and the complement cascades.

Activated ("angry") macrophages may play an important role in humans in the recovery from chronic bacterial infections, and also in resistance to some tumors. With many viral infections, in contrast, viral Ags appear in the host cell membrane before viable virions have assembled in the cytoplasm, and so a T lymphocyte attack on the host cell can eliminate it as a source of infectious units without requiring intracellular destruction of the virus.

DEPRESSION OF HOST DEFENSE

A wide variety of conditions may impair antibacterial resistance. **Circulatory disturbances,** both local (causing ischemia or congestion and edema) and general (as in shock), often interfere with the mobilization and functioning of phagocytic cells. **Mechanical obstruction** of drainage systems, such as the biliary or urinary tracts, interferes with circulation in the tissues under pressure and also prevents the washing out of infected contents of the organ. Excessive intake of **ethyl alcohol** has also been shown experimentally to depress the inflammatory response to bacterial infection. **Nutritional deficiency,** as illustrated by the frequent historical association of famine and pestilence, may profoundly influence the resistance of large populations. Thus some infectious diarrheas (cholera—Ch. 31; rotavirus—Ch. 64) are frequently fatal to natives in underdeveloped countries but rarely to visiting Westerners. Diet may also affect resistance (including immunity) through its action on the gastrointestinal flora.

Chronic debilitating diseases, as well as certain acute viral illnesses (e.g., influenza), may likewise depress antibacterial immunity. **Hormonal factors** are illustrated by the frequency of both acute and chronic bacterial diseases in diabetics, and in patients receiving adrenal steroids. Moreover, both diabetic acidosis and large doses of cortisone are known to depress the inflammatory response, and high glucose levels in acute exudates impair phagocytosis. **Physical and emotional fatigue** and overexposure to **cold** apparently increase susceptibility to bacterial illnesses, particularly of the respiratory tract; the mechanisms probably include impairment of cleansing reflexes and of ciliary action. These effects of cold may contribute, along with increased exposure to infectious aerosols in closed rooms, to seasonal variations in incidence. **Smoking** strongly depresses ciliary action.

Individuals with **agammaglobulinemia** are prone to bacterial but not to viral infections; **hereditary defects in thymic lymphocytes** impair resistance to both classes. Moreover, specific **immune response genes** have been shown to govern the formation of specific Abs in experimental animals (Ch. 18), and genetic diversity in these genes can be expected in man.

Agranulocytosis of various origins (including genetic and drug-induced) is associated with a marked impairment of resistance to bacterial disease. Genetic defects in phagocyte function are described below. The key role of leukocytes in resistance to bacterial invasion is illustrated by acute radiation injury, in which leukocytes decline sharply (since their half-life in the circulation is only 6–12 h), while Ab levels and the phagocytic capacity and number of the monocytic cells are not seriously disturbed.

Thus when mice were exposed to enough x-irradiation to cause bone marrow aplasia the peripheral white cell count fell precipitously (Fig. 24-11). When the tissue defense was challenged with

FIG. 24-11. The inverse relation between postirradiation blood leukocyte counts (upper curve) and the number of viable pneumococci in 24-h experimental myositis lesions (bar graph, below) during the recovery of mice from an acute radiation injury (650 r). The points within the squares represent average preinfection leukocyte counts done on animals in which myositis was produced. Other intermediate points indicate average postirradiation counts made on uninfected mice. The numeral above each point refers to the number of mice on which blood counts were made. (Smith MR et al: J Immunol 90:914, 1963. Copyright 1963 by Williams and Wilkins, Baltimore.)

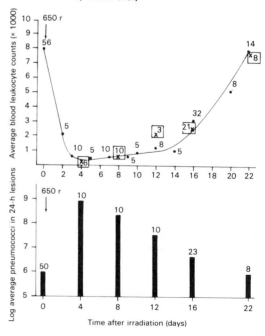

pneumococci, injected intramuscularly, the acute inflammatory response was absent, and the bacteria in the lesions reached a density approximately 1000 times greater than that in unirradiated controls. In later challenges, during recovery from irradiation, the *bacterial growth was inversely related to the number of leukocytes in the circulation* (Fig. 24-11).

GENETIC DEFECTS OF PHAGOCYTE FUNCTION

Some individuals with low resistance to infection have a hereditary defect in their phagocytes. One such defect is found in **chronic granulomatous disease,** a fatal disease of childhood associated with repeated, widespread in-fections with *Staphylococcus aureus* and gram-negative organisms. Leukocytes from such patients phagocytize these bacteria normally but fail to kill them; yet they kill pneumococci and streptococci (which produce peroxide) quite readily. This selectivity is now understood: the defective granulocytes fail to produce significant quantities of hydrogen peroxide, which is required by the myeloperoxidase–halogenation system described above. In another defect the phagocytes cannot migrate because their actin cannot be gelled by the actin-binding protein. Additional heritable defects of phagocytes will no doubt be uncovered in the future.

COMMUNICABILITY OF INFECTIONS

Epidemiology is the study of the factors that affect the distribution of diseases in populations, including the infectious diseases that are spread by pathogenic organisms. This area may be viewed in large part as an applied branch of evolutionary biology, involving primarily environmental (ecologic) factors but also occasionally genetic ones (e.g., epidemics of new genotypes of influenza virus, or variations in the resistance of different host populations to malaria or to tuberculosis). Some basic principles, which apply to viruses as well as to bacteria, will be briefly reviewed here; the epidemiologic aspects of specific infections will be discussed in later chapters. The reader is urged to consult a textbook of epidemiology or of preventive medicine for a more adequate background in this important subject.

The **incidence** of an infectious disease is the number of new cases per unit population and per unit time, while its **prevalence** is the fraction of the population that is ill at a given time. Incidence depends on virulence (see Pathogenicity, above), communicability, frequency of contacts, and host susceptibility. An **endemic** disease has a more or less steady incidence. An **epidemic** is a peak in the varying incidence of a disease: the size of the peak required for this designation is arbitrary and is related to the background endemic rate, the morbidity (frequency of the illness), and the anxiety that the disease arouses (e.g., a few cases of poliomyelitis, but not of the common cold, constitute an epidemic). **Pandemics** are unusually widespread epidemics.

To be **naturally pathogenic** a microbe must be **communicable** to a susceptible individual; otherwise it can cause disease only if artificially inoculated. For example, mice are highly susceptible to inoculated pneumococci, but even when dying of infection of the respiratory tract they rarely transmit pneumonia to normal cage-mates. Conversely, avirulent bacteria may be highly communicable.

Mechanisms of Transmission. Many bacteria are transmitted from one person to another primarily by **direct contact,** mostly via the hands, mouth, or sexual organs. A few bacterial diseases (e.g., syphilis) and viral diseases (e.g., rubella) can be transmitted **transplacentally** from mother to child. **Indirect transmission** occurs via several kinds of intermediate agents. **Enteric infections** are characteristically **waterborne** or **foodborne,** via fecal discharges, but some viruses exhibit both enteric and respiratory transmission. **Respiratory infections** are transmitted both by contact and by **droplets (aerosols)** emitted during coughing, sneezing, or talking (and in laboratories by any procedure involving splashing, such as blowing the last drop from a pipette). Particularly important in the transfer of staphylococci is dust, especially from contaminated clothing or bedding (which, with books, combs, and other objects that can harbor pathogens, are collectively referred to as **fomites**). Some systemic infections may be transmitted from **infected animals,** which serve as a **reservoir,** via milk (brucellosis, tuberculosis), carcasses (anthrax, tularemia), contaminated swimming water (leptospirosis), or wounds. **Arthropod vectors** may transfer rickettsiae or certain viruses, largely from an animal reservoir, by a cycle of infection (in lice, fleas, or ticks); flies may also transfer enteric bacteria mechanically.

Contact infections can be largely **controlled** by personal hygiene (with especial attention to washing hands after contact); and water- and foodborne infections, by sanitation. Airborne infections are the least successfully controlled, and they are even occasionally spread by central air-handling units in public buildings. The relative importance of contact and of airborne transmission in respiratory infections has been a matter of much dispute, and probably varies widely.

In airborne infection patients produce both **large droplets,** which settle before evaporating, and **microdroplets,** which evaporate immediately and yield dry **droplet nuclei,** of $<5\mu m$ diameter (often a single bacterium of $1\mu m$–$2\mu m$). The distinction is important in two ways. Droplets travel only a few feet before settling whereas droplet nuclei remain suspended for a long period and so can travel far. In addition, inhaled droplets are trapped in the mucous layer covering the upper respiratory tract and the bronchi, and are largely eliminated by ciliary action, followed by swallowing or expectoration. Hence infection with bacteria that invade the upper respiratory tract (e.g., streptococci) requires a large inoculum. Droplet nuclei, in contrast, penetrate with high frequency beyond the mucociliary blanket into the terminal bronchioles and the alveoli, and so the infecting dose is much smaller.

In fact, in experiments with highly susceptible guinea pigs exposed to a suspension of tubercle bacilli almost every inhaled organism yielded a pulmonary focus. Because measures for the control of airborne infection (ventilation, filtration of air, ultraviolet irradiation) are concerned with droplet nuclei, while control of short-range transmission by larger droplets depends on personal hygiene, the latter mode of transmission is often included in the concept of contact infection.

Efficient transmission depends upon several factors:

1) The **source** of the infecting agent. When it is restricted to individuals who are actively ill, or recently convalescent (e.g., smallpox), **isolation** measures are effective, and total **eradication** is possible. With many bacterial diseases (e.g., meningococcal meningitis) healthy **carriers** provide a **reservoir** and serve as a focus more than ill patients do; in addition, with carrier-borne agents it is suspected that epidemics are sometimes initiated by novel genetic recombinations between strains from different carriers. When the natural reservoir is an **animal host** or the **soil**, or when the organism is a widespread **human commensal** that only occasionally produces disease (e.g., most bacteria that cause respiratory disease), eradication is not possible.

2) The **number of organisms released.** "Open" and "closed" lesions differ greatly in various cases or stages of the same disease. Thus the bubonic form of plague, characterized by closed infections of lymph nodes and the blood stream, is transmitted from rat to man (by fleas) but not from man to man. However, if the lungs become involved the open pulmonary lesions make the disease (pneumonic plague) highly contagious. Similarly, carriers with hemolytic streptococci in their noses are more infectious than those with the organism confined to the pharynx; nursery epidemics of staphylococcal infections have been traced to "cloud babies," who emit veritable clouds of staphylococci into their surroundings; and lesions of pulmonary tuberculosis vary enormously in their release of bacilli. For most enteric infections the required infecting dose (e.g., about 10^8 cells for the typhoid bacillus) is much larger than in respiratory infections, because the organisms are diluted in a much larger volume of fluid in the recipient, and they have to compete with a large normal flora for adherence.

3) **Retention of virulence** in transit to a new host is required. Mere survival does not guarantee infectiousness: **phenotypic changes** in an unfavorable environment may depress pathogenicity without destroying viability. For example, viable group A streptococci in floor dust did not infect human volunteers even when large inocula were instilled into the nasopharynx; transmission by moist secretions from an infected subject appears to be necessary for this organism (unlike staphylococci). Similarly, with suspensions of *Francisella tularensis* disseminated in aerosols infectiousness decays more rapidly than viability.

4) The **frequency of effective contacts** between infected and susceptible individuals is an important factor in determining how fast infection spreads. Crowding and limited air flow may thus constitute critical terms in the crossinfection "equation." These factors undoubtedly contribute to the high rates of respiratory disease during the colder months, when persons are crowded together indoors.

5) For rapid spread in man-to-man transmission a relatively **high proportion** of the population must be **susceptible.** Hence to curtail the spread of such a communicable disease it is not necessary to immunize every individual at risk. The statistical principles underlying this important concept of **herd immunity** are discussed in textbooks of epidemiology.

SELECTED READING

BOOKS AND REVIEW ARTICLES

BERNHEIMER AW (ed): Mechanisms of Bacterial Toxinology. New York, Academic Press, 1976

BERNHEIMER AW (ed): Perspectives in Toxinology. New York, Wiley, 1977

BRADLEY SG: Cellular and molecular mechanisms of action of bacterial endotoxins. Annu Rev Microbiol 33:67, 1979

BULLEN JJ, ROGERS HJ, GRIFFITHS E: Role of iron in bacterial infection. Curr Top Microbiol Immunol 80:1, 1978

BURNET FM, WHITE DO: Natural History of Infectious Disease, 4th ed. Cambridge University Press, 1972

CAMPBELL PA: Immunocompetent cells in resistance to bacterial infections. Bacteriol Rev 40:284, 1976

COLLIER RJ, MEKALANOS JJ: ADP-ribosylating exotoxins. In Bisswanger H, Schmincke-Ott E (eds): Multifunctional Proteins. New York, Wiley-Interscience, 1980

GIBBONS RJ, VAN HOUTE J: Bacterial adherence in oral microbial ecology. Annu Rev Microbiol 29:19, 1975

GORDON S, COHN ZA: The macrophage. Int Rev Cytol 36:171, 1973

GOREN MB: Phagocyte lysosomes. Annu Rev Microbiol 31:507, 1977

JONES GW: The attachment of bacteria to the surfaces of animal cells. In Reissig JL (ed): Receptors and Recognition, Sec B, Vol 3, Microbial Interactions. New York, Wiley, 1978

KEUSCH GT: Specific membrane receptors: pathogenetic and therapeutic implications in infectious disease. Revs Infec Dis 1:547, 1979

KLEBANOFF SJ: Intraleukocytic microbicidal defects. Annu Rev Med 22:39–62, 1971

KLEBANOFF SJ, CLARK RA: The Neutrophil. Function and Clinical Disorders. New York, Elsevier, 1978

MACKANESS GB: Resistance to intracellular infection. J Inf Dis 123:439, 1971

MURPHY P: The Neutrophil. New York, Plenum, 1976

SCHLESSINGER D (ed): Microbiology—1979. Washington, Amer Soc Microbiol, 1979. Symposia on "Mechanisms of Microbial Virulence" and "Biochemical Genetics of Pathogenicity."

SILVERSTEIN S, STEINMAN RM, COHN ZA: Endocytosis. Annu Rev Biochem 46:669, 1977

SMITH H: Microbial surfaces in relation to pathogenicity. Bacteriol Rev 41:474, 1977

SMITH H, PEARCE JH (eds): Microbial Pathogenicity in Man and Animals. 22nd Symposium, Society of General Microbiology. Cambridge University Press, 1972

SMOLEN JE, WEISSMAN G: Polymorphonuclear leukocytes. In McCarty DJ (ed): Arthritis. Philadelphia, WB Saunders, 1977

SPECTOR WG, WILLOUGHBY DA: The inflammatory response. Bacteriol Rev 27:117, 1963

STOSSEL TP: Endocytosis. In Cuatrecasas P, Greaves MF (eds): Receptors and Recognition. London, Chapman and Hall, 1977, p 104

SYMPOSIUM: mechanisms of microbial virulence; biochemical genetics of pathogenicity. In Schlessinger D (ed): Microbiology—1979, Amer Soc for Microbiol, Washington DC, 1979

VAN FURTH R (ed): Mononuclear Phagocytes in Immunity, Infection, and Pathology. Oxford, Blackwell Scientific, 1975

WARD PA, BECKER EL: Biology of leukotaxis. Rev Physiol Biochem Pharmacol 77:125, 1977

WEINBERG ED: Iron and susceptibility to infectious disease. Science 184:952, 1974

WOOD WB JR: Studies on the cellular immunology of acute bacterial infections. Harvey Lect 47:72, 1951–1952

ZWEIFACH BW, GRANT L, MCCLUSKEY RT (eds): The Inflammatory Process, 2nd ed, New York, Academic Press, 1974

chapter **25**

CHEMOTHERAPY OF BACTERIAL DISEASES

ROBERT AUSTRIAN

The results of antibacterial chemotherapy are determined by a complex set of interactions between drug, host, and offending microbe. These may be diagrammed as follows:

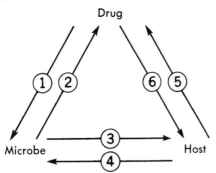

The action of the drug on the microbe (1) has been discussed in Chapter 7, as has its converse (2), illustrated by the inactivation of penicillin by microbial penicillinase. The fate of the drug in the host (5) and its side effects on the host (6) belong to pharmacology and will not be dealt with in detail in this text. The effects of the microbe on the host (3) and the host's defenses against the microbe (4) are extensively treated in several chapters (e.g., 15, 24, and 27). The present one discusses the host factors that influence the curative effects of antibacterial drugs and defines their uses and limitations as chemotherapeutic agents.* The distinction made in Chapter 7 between **bacteriostatic** and **bactericidal** drugs will be seen to be of paramount importance.

EXTRACELLULAR AND INTRACELLULAR PARASITISM

The preceding chapter emphasized the distinction between **extracellular** and **intracellular** pathogenic bacteria. The former survive and are harmful to the host only as long as they remain outside phagocytic cells, while the latter are capable of surviving and multiplying within phagocytes. Extracellular bacterial parasites tend to produce acute, short-lived diseases (e.g., streptococcal pharyngitis), which often terminate spontaneously as soon as the host has generated enough specific opsonins to bring about phagocytosis. In contrast, intracellular bacterial parasites ordinarily cause chronic diseases (e.g., tuberculosis), in which neither opsonization nor phagocytosis necessarily constitutes a critical event in the progress of the infection. Since host defenses, as well as the metabolic state of the microbe, influence the effectiveness of antimicrobial drugs, the distinction between extracellular and intracellular parasitism is relevant to chemotherapy.

PREDOMINANTLY EXTRACELLULAR BACTERIA

BACTERIOSTATIC DRUGS

No drug that merely slows down or prevents bacterial multiplication can be curative per se. For a cure most, if not all, of the bacteria invading the tissues must ultimately be destroyed. The role of phagocytic defenses is readily demonstrable in animal models. For example, in experimental pneumococcal pneumonia in rats (Fig. 25-1) sulfonamides suppress the multiplication of the pneumococci in the advancing edema zone of the lesion (see Pathogenesis of Pneumococcal Pneumonia, Ch. 27) and thus stop its spread, but the ultimate destruction of the pneumococci is brought about by the phagocytic cells of the inflammatory exudate (Fig. 25-1C). Thus the cure of the disease results from the **combined actions** of the **drug** and the **phagocytes** of the host. Such a dual mechanism appears to operate generally in acute bacterial diseases that are responsive to bacteriostatic drugs.

In **suppurative lesions,** on the other hand, assistance by cells is virtually nonoperative, since the phagocytes have become necrotic and nonfunctional (see Necrosis and Abscess Formation, Ch. 24). Hence such lesions respond poorly to bacteriostatic drugs, even to those that are not inhibited by products of tissue destruction (see Sulfonamides, Ch. 7).

* Antifungal drugs are discussed in Chapter 45 and antiviral agents in Chapter 51.

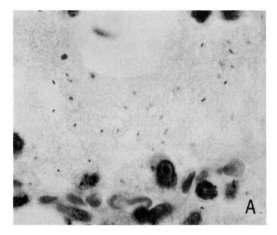

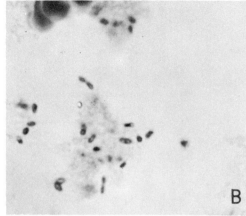

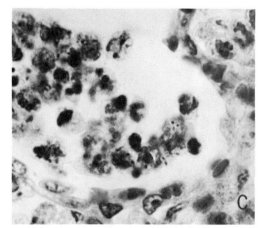

FIG. 25-1. Effects of sulfonamide therapy on experimental pneumococcal pneumonia in rats. **A.** Pneumococci in the edema zone of the pneumonic lesion shortly after onset of treatment. (×800) **B.** Pneumococci in edema zone 12 h later showing morphologic signs of bacteriostasis (pleomorphism and irregular staining). (×1500) **C.** Phagocytosis of pneumococci by leukocytes in alveolar exudate during period of treatment. (×800) (Wood WB Jr, Irons EN: J Exp Med 84:365, 1946)

BACTERICIDAL DRUGS

The relations that obtain with bactericidal drugs are more complicated. These drugs (with the exception of the polymyxins) *are bactericidal only to growing organisms.* When penicillin, for example, is added to a culture of pneumococci during the exponential phase of growth the organisms are rapidly killed, and the culture is eventually sterilized (Fig. 25-2, curve B). If the drug, on the other hand, is added to the culture in the stationary phase, when multiplication has greatly slowed, no killing occurs (Fig. 25-2, curve C). The same relation is demonstrable in vivo in rat pneumococcal pneumonia. In the outer edema zone of the lesion, where the organisms multiply rapidly in the untreated animal (Fig. 25-3A), penicillin causes lysis of pneumococci (Fig. 25-3B); in the more central portions of the lesion, where multiplication has slowed, phagocytosis plays the major role in destroying the organisms (Fig. 25-3C and D).

The part played by phagocytes can be roughly measured by comparing normal with leukopenic animals. The results (Fig. 25-4) indicate that the destruction of the invading bacteria is hastened by the combined bactericidal actions of the drug and the phagocytic cells.

PREDOMINANTLY INTRACELLULAR BACTERIA

With bacteria that can survive for long periods in phagocytic cells, such as tubercle bacilli, brucellae, and typhoid bacilli, the curative effect of chemotherapy is not nearly as rapid or reliable as with extracellular parasites, even when the organisms are highly drug-sensitive in vitro. One factor may be failure of the drug to reach the intracellular bacteria; another may be its relative ineffectiveness in the environment of the cell's cytoplasm (streptomycin; see Aminoglycosides, Ch. 7). Some drugs, however, penetrate animal cells and act upon susceptible microorganisms growing within them. Thus isoniazid can act upon tubercle bacilli within monocytes, and

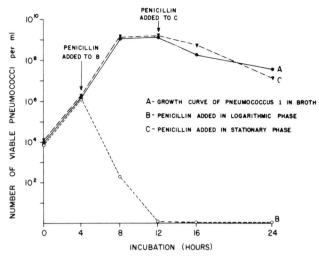

FIG. 25-2. Action of penicillin on type 1 pneumococci in broth culture. (Smith MR, Wood WB Jr: J Exp Med 103:487, 1956)

chloramphenicol can slow the growth of *Salmonella typhi* in cultured mouse fibroblasts. Nevertheless, the intracellular location of the parasites promotes sluggish metabolism and dormancy, as evidenced by the chronic nature of the diseases themselves and the frequent appearance of the carrier state; and dormancy decreases bactericidal drug action. Hence the actual destruction of the organisms must depend primarily upon the cellular defenses of the host. Since these defenses are slow to cure in the absence of treatment, it is hardly surprising that cure is also slow in the presence of drug-induced bacteriostasis.

The immune bactericidal processes in monocytic phagocytes, it will be recalled, involve functional as well as anatomic changes in the cytoplasm, which coincide with the onset of delayed hypersensitivity (see Phagocytic Defense in Ch. 24).

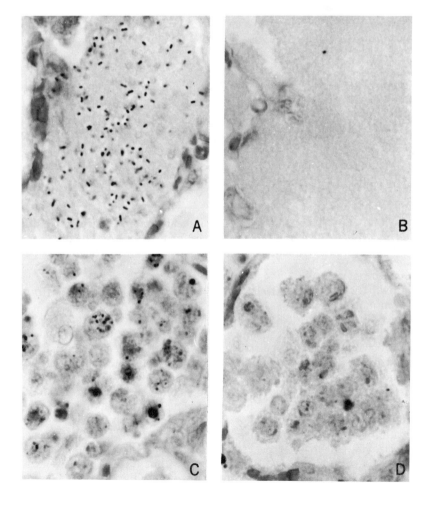

FIG. 25-3. Effect of penicillin therapy on experimental pneumococcal pneumonia. **A.** Pneumococci in edema fluid contained in alveolus at advancing margin of 18-h lesion (see Ch. 27). **B.** Similar alveolus 12 h after the start of penicillin treatment. Note that all but a few pneumococci have already been destroyed by the direct lytic action of the drug. **C.** Alveolus in zone of advanced consolidation (Ch. 27); same lesion as in **B.** Note that in this portion of the lesion, where bacterial growth has slowed to a standstill, most of the organisms are being destroyed by phagocytosis rather than by lysis. **D.** Similar alveolus to that pictured in **C** 18 h later. Most of the phagocytized pneumococci have been destroyed by the cells. (Smith MR, Wood WB Jr: J Exp Med 103:487, 1956) (×900)

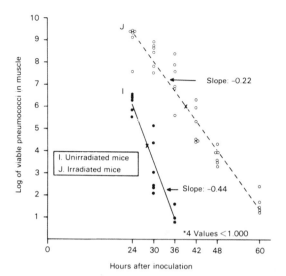

FIG. 25-4. Comparative rates at which pneumococci are killed by penicillin in experimental pneumococcal myositis lesions in unirradiated mice and in mice made severely leukopenic by exposure to 650 R of x-ray. Penicillin was administered 24 h after inoculation. Killing of organisms in tissues occurred less rapidly in the leukopenic animals. (Smith MR, Wood WB Jr: J Exp Med 103:499, 1956)

LIMITATIONS TO EFFECTIVE CHEMOTHERAPY

LOCAL TISSUE FACTORS

Abscess Formation and Necrosis. When bacteria reach a high enough extracellular population density in a localized tissue site the surrounding host cells die and **suppuration** results, forming an abscess. Antibacterial chemotherapy, even with potent bactericidal drugs, then becomes relatively ineffective. The reason is not failure of the drug to reach the organisms, as is often assumed: subcutaneous pneumococcal abscesses (Fig. 25-5) do not respond to penicillin even when it is injected directly into the abscesses at frequent intervals, and streptomycin has been shown to penetrate caseous tuberculous lesions that it cannot sterilize.

Several factors contribute to the persistence of infection under these conditions. The impairment of circulation in the lesion reduces the flow of Abs and leukocytes, of oxygen, and of nutrients; it also impedes the removal of waste products, including acids (produced by both host cells and bacteria) and specific bacterial toxins. As a result of these deficiencies and accumulations the leukocytes cease to function and eventually die and disintegrate (Fig. 25-5). Moreover, since acidity, lack of oxygen, and lack of nutrients also impair their multiplication the bacteria are not killed by penicillin or aminoglycosides. Such abscesses are cured only after **drainage,** which permits the bacteriostatic pus to be replaced by fresh serous exudate, providing nutrients as well as a wave of new leukocytes.

The action of some chemotherapeutic agents is also affected directly by tissue destruction. As noted earlier (Ch. 7), anaerobiosis and acidity markedly decrease the potency of aminoglycosides, while certain metabolites released by tissue autolysis antagonize the bacteriostatic action of sulfonamides: hence these drugs are ineffective in necrotic lesions.

Foreign Bodies and Obstruction of Excretory Organs. Chemotherapy is also relatively ineffective when there is a foreign body in the lesion, such as splinters of wood, spicules of dead bone (in osteomyelitis), a prosthesis, or sutures. These provide foci protected from leukocytes, where bacteria may accumulate their products or persist in a quiescent state. Similarly, bacterial infections associated with obstructions of the urinary, biliary, or respiratory tracts (bronchi or paranasal sinuses) tend to persist despite intensive chemotherapy: the increased fluid pressure impairs circulation and lymphatic drainage in the tissues surrounding the obstructed cavity, and so toxic products and eventually quiescent bacteria accumulate within the cavity.

Both these types of lesions are rarely cured by chemotherapy unless the foreign body or obstruction is removed. Surgical intervention alone may permit the natural defenses to eradicate the infection, but chemotherapy is ordinarily continued until the systemic manifestations (fever, leukocytosis, etc.) have subsided.

SYSTEMIC HOST FACTORS

Many disease states that depress natural resistance to bacterial infections also impair response to antimicrobial therapy. Some reduce the number of effective **phagocytic cells** available for mobilization at the infected site (e.g., agranulocytosis, leukemia, radiation injury, antitumor chemotherapy, corticosteroid therapy, general anaesthetics or ethanol), while others interfere with **Ab formation** (agammaglobulinemia, multiple myeloma, chronic lymphatic leukemia). **Uremia** and uncontrolled **diabetes mellitus** also depress resistance of the host and thus affect the response to chemotherapy, although the precise mechanisms are not known. Experiments with alloxan-diabetic mice indicate that hyperglycemia (even without acidosis) impairs the phagocytic activity of polymorphonuclear leukocytes.

SUPERINFECTIONS AND DRUG RESISTANCE

One of the most serious and frequent complications of antimicrobial therapy is **superinfection** with organisms resistant to the drug being used and unfortunately often resistant to many other drugs as well. This development

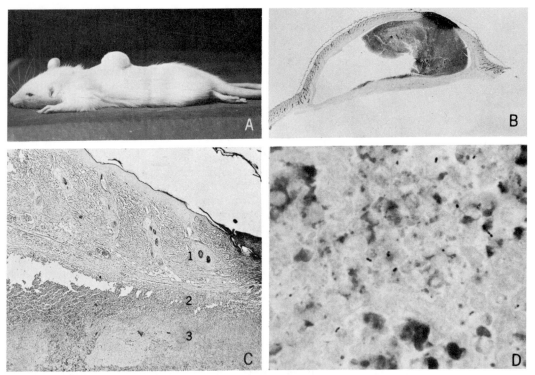

FIG. 25-5. Penicillin treatment of experimental pneumococcal abscesses in rats. **A.** Subcutaneous abscess on back of rat 23 days after inoculation. (×⅓, reduced) **B.** Cross section of abscess cavity. (×3, reduced) **C.** Wall of abscess showing small vessel **(1)** in subcutaneous tissue; periphery of abscess cavity **(2)** where leukocytes, recently arrived from **1** and from other vessels, are still functional; and necrotic pus **(3)** within core of lesion. (×40, reduced) **D.** Higher magnification of zone **3** in **C,** showing leukocytic debris and unphagocytized pneumococci. (×1200, reduced) (Smith MR, Wood WB Jr: J Exp Med 103:509, 1956)

may arise from the outgrowth of **indigenous** bacteria or fungi that are normally held in check by the growth of other organisms. For example, bacterial **enteritis** or candidal **stomatitis** occasionally develops during prolonged treatment with tetracycline. In patients with seriously impaired resistance (e.g., those with cachexia or with severe burns) penicillin may prevent infection of the local lesion and may decrease the frequency of terminal bronchopneumonia, but gram-negative bacillary septicemias then frequently occur after 2–3 weeks of therapy.

The enteritis induced by antibiotics, called **pseudomembranous colitis,** was long ascribed to outgrowth of staphylococci. However, anaerobic cultures showed that the disease, now most frequently induced by clindamycin, is more frequently correlated with *Clostridium difficile* (see Ch. 36). Moreover, the specific enterotoxin of that organism is frequently demonstrable in filtrates of the diarrheal feces.

Drug-resistant bacteria may also arise by mutation and selection or be acquired **exogenously** during therapy, usually from other patients or hospital personnel. Superinfections of either origin may be more serious than the original infection.

CLINICAL IMPLICATIONS

The foregoing considerations suggest the following general principles for the chemotherapy of bacterial diseases.

1) The earlier treatment is begun the more likely it is to be effective. Several reasons may be seen: a) organisms that are allowed to reach the stationary growth phase in tissues become refractory to bactericidal drugs; b) some species (e.g., *Staphylococcus aureus*) tend eventually to produce suppurative lesions which may require surgical drainage and may cause irreversible tissue damage in vital organs (e.g., the brain); c) drug-resistant mutants are more likely to arise in advanced bacterial lesions, owing simply to the greater numbers of bacteria involved; and d) late lesions are more likely to become superinfected with drug-resistant organisms.

2) In order to eliminate the last infecting bacterial cell, and thus prevent relapse, therapy must ordinarily be continued for some time after symptoms have subsided. This period is longer when a bacteriostatic agent is used, or when treatment is begun late with a bactericidal agent, since most of the invading organisms will have to be

eliminated by the relatively slow processes of cellular and humoral defense.

3) In infections of the endocardium, where the cellular defenses of the host are comparatively ineffective, cure almost always requires the use of a bactericidal drug.

4) Chemotherapy can rarely cure established abscesses, lesions containing foreign bodies, or infections associated with excretory duct obstruction, unless they drain spontaneously or are drained surgically.

5) In the presence of complications that impair natural defenses, larger doses of antimicrobial drugs are required.

6) Short-term chemotherapy is generally ineffective in chronic bacterial diseases caused by intracellular parasites. Only if treatment is prolonged, to the point where the relatively ponderous mechanisms of intracellular immunity can effectively intervene, is subsequent relapse likely to be avoided.

ANTIBACTERIAL DRUGS IN GENERAL USE

The presently accepted applications and limitations of the useful antibacterial drugs may be briefly summarized as follows.

THE SULFONAMIDES AND RELATED DRUGS

The sulfonamides have been largely replaced by more effective and less toxic antimicrobial agents, although they are still used widely in the management of urinary tract infections and trachoma. They penetrate the blood–brain barrier rapidly and were employed extensively to treat bacterial meningitis prior to the advent of penicillin and other antibiotics. Prophylactic doses are effective in preventing reinfection with group A streptococci in patients with rheumatic fever but may result in selection of resistant mutants, a phenomenon not observed with penicillin prophylaxis. Poorly absorbed sulfonamides (succinylsulfathiazole, sulfaguanidine, phthalylsulfathiazole) have been used preoperatively to lower the bacterial content of the bowel in preparation for abdominal surgery but have been replaced largely by more effective agents.

Sulfonamides act synergistically with trimethoprim (see Biosynthesis, Ch. 4) in inhibiting bacterial dihydrofolate synthesis. A combination of sulfamethoxazole and trimethoprim in a 5:1 ratio is available to treat some infections of the urinary tract and as an alternative to chloramphenicol in the therapy of typhoid fever. It may also be useful to treat infections caused by *Pneumocystis carinii.*

The principal disadvantages of the sulfonamides are 1) their limited antimicrobial spectrum at attainable drug levels, 2) their purely bacteriostatic action, 3) their delayed effect on bacterial growth, and 4) their tendency to cause such serious side effects as granulocytopenia, thrombocytopenia, aplastic anemia, dermatitides, and periarteritis nodosa.

Two compounds related to the sulfonamides, described in Chapter 7, have a very restricted antimicrobial spectrum: **dapsone** (diaminodiphenyl sulfone) is useful only in the treatment of leprosy, and *p*-**aminosalicylic acid (PAS)** in the treatment of tuberculosis. The choice of drugs for the latter disease will be discussed in Chapter 37.

PENICILLINS AND OTHER DRUGS THAT INHIBIT BACTERIAL CELL WALL SYNTHESIS

Penicillin G (benzylpenicillin) is effective, at readily attainable blood levels, against virtually all **gram-positive cocci** (pneumococci, streptococci, staphylococci); exceptions are penicillinase-producing strains of staphylococci and some enterococci (e.g., *Streptococcus faecalis*). It is active also against certain **other gram-positive bacteria,** such as *Corynebacterium diphtheriae* and some clostridia; the **gram-negative** neisseriae (menigococci and penicillin sensitive gonococci), some strains of bacteroides, and *Hemophilus ducreyi* (the cause of chancroid); various **spirochetes** (notably the treponeme of syphilis); and some **actinomycetes.**

In the usual doses *penicillin G is not effective in vivo against mycobacteria, many gram-negative bacilli of the enteric tract, and Hemophilus influenzae;* but their insusceptibility is only relative. Indeed, the mechanism of action of penicillin has been demonstrated in part through its effect, in high concentrations, in interfering with the normal synthesis of the cell wall of *Escherichia coli.* In contrast, fungi, protozoa, and viruses totally lack the penicillin-sensitive reaction (Ch. 6).

Because penicillin can cause a loss of viability by several logarithmic units within a fraction of a bacterial generation time its effect is often more dramatic than that of bacteriostatic agents. Furthermore, its action is not antagonized by products of tissue breakdown; indeed, enrichment of the medium accelerates bactericidal action in vitro by accelerating bacterial growth. Finally, except for its allergenicity (which may be very serious), it has virtually no toxicity for mammals except in huge doses.

Penicillin G can be administered orally, although a substantial fraction is not absorbed; penicillin V (phenoxymethylpenicillin) is better absorbed. Moreover, with penicillin-sensitive organisms penicillin G is considerably more potent than the congeners resistant to penicillinase (e.g., **methicillin, oxacillin);** hence the latter are valuable primarily against the frequent penicillinase-producing staphylococci, which fortunately are responsive even to penicillins of lower potency. Thus *penicillin G is almost always the drug of choice for the treatment of serious infections caused by susceptible bacteria.* In mixed infections, however, its action on naturally susceptible species may, rarely, be blocked by the presence of penicillinase-producing staphylococci.

Ampicillin and **carbenicillin** have extended the advantages of penicillin therapy to certain gram-negative bacilli. Ampicillin has proved useful against sensitive strains of *H. influenzae,* of certain gram-negative enteric bacilli, and of the gram-positive bacillus *Listeria monocytogenes.* Carbenicillin finds its chief application against sensitive strains of *Pseudomonas aeruginosa, Enterobacter,* and *Serratia.* Both these penicillins are inactivated by penicillinase. Ampicillin can be given orally or parenterally, but carbenicillin must be given intravenously.

The **cephalosporins,** which are closely related to penicillins in structure and action, are described in Chapter 7. Because of their relatively broad antibacterial spectrum, their resistance to β-lactamases, and their infrequent allergic crossreactivity with penicillins, they are widely used. They are significantly less potent than penicillin G, however, against the common penicillin-sensitive organisms, and less effective than ampicillin against most strains of *H. influenzae;* but they are the drugs of choice in infections due to certain enterobacteria, usually in combination with an aminoglycoside. Cephalosporins do not cross the blood–brain barrier readily and should not be used to treat meningitis. Most cephalosporins must be administered parenterally, although one derivative, cephalexin, can be given orally.

Vancomycin is useful in treating serious infections caused by susceptible gram-positive bacteria in patients allergic to penicillins and cephalosporins, and by some strains of *Bacteroides* and of *Flavobacterium.* It must be given intravenously and frequently causes chemical thrombophlebitis. It is moderately ototoxic and nephrotoxic.

THE MACROLIDES AND LINCOMYCINS

The macrolides (erythromycin, trioleandomycin) and lincomycins (lincomycin, clindamycin) are chemically dissimilar groups of antibiotics that inhibit protein synthesis (Ch. 13). They are effective against gram-positive bacteria; macrolides also inhibit some gram-negative bacteria (neisseriae, *H. influenzae*), mycoplasmas, treponemes, and chlamydiae. Clindamycin is active against many gram-negative anaerobes and is widely used when their presence is suspected. These drugs may be given orally or parenterally, and they are useful when allergy precludes the use of penicillins or cephalosporins. Resistant mutants may arise during treatment.

AMINOGLYCOSIDES

The aminoglycosides (streptomycin, gentamicin, kanamycin, tobramycin, amikacin, and neomycin) have a broad antibacterial spectrum, including many gram-positive as well as most gram-negative organisms. They are used alone, or, more often, in combination with another appropriate antimicrobial drug, to treat tuberculosis, bacterial endocarditis caused by sensitive gram-positive or gram-negative bacteria, and infections with various gram-negative bacteria including *Klebsiella* pneumonia, systemic or urinary tract infections caused by enteric bacteria, *Shigella* dysentery, plague, and tularemia.

Because of intracellular residence, salmonellae and brucellae, despite their sensitivity in vitro, can rarely be eradicated by these drugs (see Aminoglycosides, Ch. 7, and Pathogenicity, Ch. 33). The antagonistic effects of salts, low pH, and anaerobiosis limit the action of aminoglycosides in septic foci and in the urinary tract.

Many bacterial species, by a mutation affecting the ribosome, readily develop high-level resistance to streptomycin that usually does not extend to other aminoglycosides. In addition, in gram-negative bacilli, resistance to various aminoglycosides may be mediated by R factor–determined enzymes, which inactivate specific compounds by acetylation, adenylation, or phosphorylation (see Table 7-1, Ch. 7). Streptomycin dependence (Ch. 13) may also occasionally have clinical significance, particularly in the treatment of tuberculosis, but the presence of such dependent tubercle bacilli in treated patients does not necessarily contraindicate further use of the drug, because they may constitute only a minority of the infecting population.

Kanamycin is less effective than streptomycin against *Mycobacterium tuberculosis* but is more active against gram-positive cocci; moreover, resistant mutants are encountered less frequently. **Gentamicin** is more toxic than kanamycin or streptomycin and is given in smaller doses. It is effective against many gram-positive cocci (but not pneumococci) and many strains of *Pseudomonas aeruginosa,* as are **tobramycin** and **amikacin,** a semisynthetic aminoglycoside. **Neomycin,** because of its marked ototoxicity, should not be given parenterally or be used locally to irrigate well-vascularized lesions. It may be applied topically in various forms of infectious dermatitis, and it is still employed occasionally as an intestinal antiseptic.

RIFAMPIN

Rifampin (rifampicin, rifamycin B) selectively inhibits bacterial RNA polymerase (Ch. 11) but not that of mammals. Its principal use appears to be in the combined treatment of tuberculosis. It also inhibits many gram-positive and some gram-negative bacteria, but resistant mutants arise frequently.

BROAD-SPECTRUM ANTIBIOTICS: TETRACYCLINES AND CHLORAMPHENICOL

Tetracyclines are bacteriostatic inhibitors of protein synthesis (Ch. 13). They are effective against many gram-positive and gram-negative bacteria (though not against *Salmonella, Proteus,* or *Pseudomonas*), and they have largely displaced the sulfonamides. They are the agents of choice for rickettsiae, chlamydiae, and mycoplasmas, and they may indirectly suppress protozoal infections of mucous surfaces (amebic dysentery) by depriving the parasites of their bacterial "fodder." Tetracyclines are often administered to nonhospitalized patients, but they are less effective than the bactericidal penicillins and cephalosporins in the treatment of life-threatening infections.

The tetracyclines are well absorbed when administered orally. They evidently penetrate mammalian cells, since they are effective against intracellular parasites (e.g., rickettsiae). Serious toxic reactions are infrequent (except for hepatic damage following large doses), but diarrhea, pruritus ani, and oral inflammation are common and are believed to be due largely to replacement of the normal flora by yeasts (especially *Candida*) and by resistant staphylococci. Concomitant antifungal therapy, however, has been of uncertain value. Other serious fungal infections (vaginitis, cystitis, pneumonitis, septicemia) may arise in debilitated patients given tetracyclines; and, in treated children, *Candida* may cause oral infections (thrush). Tetracyclines should not be given to pregnant women and young children, save in exceptional circumstances, because they stain developing teeth. Partially degraded tetracyclines may cause a Fanconi-like syndrome.

Chloramphenicol is also a bacteriostatic inhibitor of protein synthesis (Ch. 13) with essentially the same antibacterial spectrum as the tetracyclines; in addition, it is moderately effective against salmonellae. Chloramphenicol causes less gastrointestinal toxicity than tetracycline, probably because it is more completely absorbed from the gut. However, it is inhibitory to the bone marrow and often gives rise to anemia, leukopenia, or thrombocytopenia; and though these effects are reversible if administration of the drug is stopped early enough, continued or repeated administration may be followed by **irreversible fatal aplastic anemia.** Chloramphenicol should therefore not be administered when an alternative form of therapy is available. Its use should be restricted to typhoid fever; other salmonelloses; selected cases of meningitis caused by *H. influenzae*, meningococci, or pneumococci (especially in patients allergic to penicillins); brain abscess caused by anaerobic bacteria; and infections caused by organisms resistant to other agents. Periodic hematologic examination during therapy is imperative.

Because of their wide range of action broad-spectrum antibiotics may be valuable in the treatment of mixed bacterial infections (chronic bronchitis, bronchiectasis, abscesses, septic wounds). Unfortunately, physicians often rely on such drugs for "shotgun" therapy in acute febrile illnesses when correct diagnosis (bacteriologic and clinical) would allow the use of more effective and less toxic agents.

POLYMYXINS

Polymyxins are antibiotics with a detergent-like action, which destroys the integrity of the membranes of gram-negative but not of gram-positive bacteria. Because of their nephro- and neurotoxicity their use is limited to serious infections caused by susceptible organisms. They are among the few drugs effective against *Pseudomonas aeruginosa*, which is a frequent and persistent secondary invader in patients undergoing prolonged antimicrobial therapy. Polymyxins are also used topically. They are ineffective against most strains of *Proteus*.

SPECIAL ASPECTS OF ANTIMICROBIAL THERAPY

TOPICAL THERAPY

Chemotherapeutic agents are often effective when applied directly to superficial infections, including those of skin, wounds, and eyes. (No benefit can be expected, however, from application of ointments to the superficial drainage of a deep focus.) For such topical application some of the more toxic antibiotics, including neomycin, polymyxin B, and bacitracin are used. A number of other bactericidal compounds (e.g., oxyquinolines, cationic detergents) are also employed for the treatment of skin infections.

Penicillins and aminoglycosides (other than neomycin) *should not be used on the skin* because application by this route frequently induces allergy, which may then hinder systemic use of the drug in a life-threatening illness. To provide high local concentrations at deeper sites of local infection, however, these drugs may be injected under special circumstances into the pleural, pericardial, or joint cavities or into the subarachnoid space.

DRUG RESISTANCE: CLINICAL AND EPIDEMIOLOGIC SIGNIFICANCE

Resistance to an antimicrobial agent may be an inherent property of the infecting organism, or it may result from mutation or from transfer of an extrachromosomal genetic determinant followed by selection of resistant organisms during therapy. With streptomycin and with nalidixic acid one-step mutations to extremely high levels of insensitivity are encountered. Mutants with significant degrees of resistance are encountered with varying frequency during treatment with other aminoglycosides, tetracyclines, macrolides, lincomycins, and sulfonamides but less commonly with penicillins, cephalosporins, and chloramphenicol.

In recent years there has been increasing recognition of the role of extrachromosomal determinants of heredity **(plasmids)** in the ecology of drug resistance among organisms pathogenic for man. These factors, designated as **resistance factors (R factors, RFs;** Ch. 12), can endow a pathogenic organism with simultaneous resistance to multiple antimicrobial agents. Their properties are discussed in Chapters 9 and 46 and the mechanisms of resistance in Chapter 7. RFs appear to spread, as extrachromosomal segments of DNA, between members of the Enterobacteriaceae and Pseudomonadaceae and such diverse genera as *Yersinia, Vibrio,* and *Hemophilus.* The inclusion of antibiotics in the feed of farm animals appears to favor the selection of resistant organisms in animals' intestinal flora, and legislation in Great Britain bans this practice.

Bacterial species differ in their likelihood of yielding variants resistant to a given drug. Some examples follow.

No group A β-hemolytic **streptococcus** resistant to penicillin has been isolated from man, but strains resistant to sulfonamides, tetracyclines, and macrolides have been recovered. Pneumococci resistant to penicillins, cephalosporins, tetracyclines, chloramphenicol, macrolides, and/or sulfonamides have all been identified as causes of human infection. **Staphylococci** are prone to acquire plasmids that determine resistance to multiple antibiotics, including penicillins, macrolides, tetracyclines, and chloramphenicol. In the treatment of **shigella dysentery** and of **typhoid fever** the spread of resistance transfer factor (RTF), carrying genes for resistance to several drugs (sulfonamides, chloramphenicol, ampicillin, tetracycline, streptomycin), has created serious problems.

In **tuberculous** patients with large caseating lesions, in whom streptomycin is not likely to be dramatically effective, resistant mutants emerge with some regularity after 1–3 months of treatment. Although this time may seem long, it should be recalled that the tubercle bacillus has a mean generation time 30 times longer than that of most bacteria. Recently **gonococci** that carry a plasmid with a gene for the production of penicillinase have been isolated. The appearance of such strains poses a potentially serious problem in controlling this most prevalent of venereal diseases.

In hospitals penicillin-resistant organisms are frequently found not only among patients but also in the rest of the hospital population. Their rapid dissemination is favored by several factors: 1)

staphylococci not only are pathogens but also are part of the normal body flora and are transferred readily from one person to another; 2) elimination of sensitive organisms from treated patients creates an ecologic niche into which resistant strains can move; 3) extrachromosomal determinants of drug resistance in nonpathogenic components of the flora provide a source for the transfer of drug resistance to sensitive pathogens; and 4) person-to-person transmission of bacteria is facilitated in many hospitals by relaxation of the standards of aseptic medical practice and environmental sanitation, an unfortunate byproduct of the era of antibiotics.

The widespread use of a drug may thus lead not only to the emergence of resistant mutants within individual patients but also to their spread in response to the ecologic alteration produced by the treatment. Such strains are often as virulent as their sensitive parents, and they may persist, therefore, in the population. However, when the wild type of a species is regularly sensitive to a drug the resistant mutants presumably have less evolutionary survival value than the parent; otherwise the resistant strain would be the wild type of that species. Hence if streptomycin were no longer used, streptomycin-resistant tubercle bacilli should disappear in time.

It should be emphasized that much of the present hard core of "natural" drug resistance, as exhibited by many strains of *Staphylococcus, Pseudomonas, Proteus,* and various other gram-negative bacterial pathogens, does not involve the genetic problem of new mutation to drug resistance; it simply reflects the fact that each drug has limits to its useful antibacterial spectrum, and that certain wild-type strains lie outside these limits.

COMBINED THERAPY

Though various combinations of chemotherapeutic agents have been promoted commercially, on the basis of alleged or real synergistic or broad-spectrum activity, they should be used rarely, if ever, in preference to treatment designed for the individual patient. Combinations of antimicrobial drugs may be employed rationally on the following grounds: 1) to prevent or to minimize the emergence of resistant mutants; 2) to take advantage of the synergistic action (Ch. 7) of certain drugs (penicillins and aminoglycosides in the treatment of certain types of bacterial endocarditis; sulfonamides and trimethoprim in the treatment of selected infections with gram-negative bacilli); 3) to provide the optimal potentially effective therapy in life-threatening emergencies before an etiologic diagnosis can be established with certainty; and 4) to lessen the toxicity of individual drugs by reducing the dose of each in a combination (triple sulfonamides).

Various combinations of isoniazid, rifampin, ethambutol, and streptomycin have been especially successful in the treatment of tuberculosis. However, multiply-resistant tubercle bacilli do emerge occasionally nonetheless, presumably because of the unequal distribution of the drugs in the body and the difficulty in maintaining adequate levels at all times. Combinations of drugs, therefore, have increased the usefulness of drugs that, used alone, result in selection of resistant mutants (especially in the treatment of tuberculosis), but they have not solved the problem completely.

Combinations of drugs that are antagonistic rather than synergistic (Ch. 7) should be avoided. Bacteriostatic drugs such as tetracyclines, chloramphenicol, and sulfonamides may interfere with the bactericidal action of drugs that block the synthesis of the bacterial cell wall (e.g., penicillins, cephalosporins, and vancomycin), and chloramphenicol blocks the irreversible action of aminoglycosides on the ribosome.

CHEMOPROPHYLAXIS

Antimicrobial prophylaxis is likely to be successful only when directed against a sensitive organism that rarely gives rise to mutants which are resistant to the drug employed. It has been used most effectively 1) in preventing a) recurrent streptococcal infections in patients with rheumatic fever, b) subacute bacterial endocarditis in patients with valvular heart disease undergoing dental or other surgical procedures, and c) certain venereal diseases, and also 2) in terminating epidemics of meningococcal infection and of shigellosis in closed populations (e.g., schools and military installations). 3) Chemoprophylaxis has also been successful in reducing infection with staphylococci following certain surgical procedures, such as the insertion of a prosthetic heart valve or hip;

but it has not been uniformly effective, and it is not a substitute for good aseptic technic. 4) Isoniazid appears useful in preventing tuberculosis in certain high-risk settings, and even more in preventing overt disease in those who have become infected (as shown by a positive tuberculin test).

Although chemoprophylaxis is often successful in preventing infection by specific organisms, one cannot expect it to be effective when susceptible bodily sites are exposed to a wide variety of bacterial species. It has proved unavailing in the prevention of bacterial pneumonia in patients with stroke or with viral infection of the lung, and it cannot be expected to keep an especially vulnerable organ, such as the urinary bladder containing an indwelling catheter, free of bacteria. Indeed, its indiscriminate use may do more harm than good by fostering the development of infection with drug-resistant organisms. For this reason, and because of the hazard of toxic and allergic drug reactions, chemotherapy for upper respiratory infections of viral origin is not justified. The practice of so treating such minor respiratory illnesses, though unfortunately common, should be condemned.

The chemotherapy of specific diseases is discussed further in the chapters that follow.

TESTS FOR MICROBIAL SENSITIVITY TO ANTIMICROBIAL DRUGS

Microbial sensitivity to various antimicrobial drugs may be determined by tests based on tube dilution, agar dilution, disc diffusion, and automated measurements of growth. The last two types are most widely employed because of their speed and simplicity. Microbial inhibition tests are used also to assay the levels of antimicrobial drugs in serum and in other body fluids. This information is useful in selected clinical situations, such as bacterial endocarditis, to determine whether or not inhibitory (or potentially toxic) levels of the drug are present.

To be meaningful, tests of microbial sensitivity must be rigidly standardized and carried out with pure cultures. Among the factors to be considered are the source of the inoculum (ideally several clones should be used to avoid selection of an atypical mutant), the size of the inoculum, the growth medium, and the period of incubation. Only standardized preparations of antimicrobial drugs should be used.

Tube dilution tests are usually performed with twofold dilutions of the antimicrobial agent in liquid medium. After a standard inoculum of organisms is added and the culture is incubated for 16–20 h the tube containing the highest dilution of drug and showing no turbidity defines the **minimal inhibitory concentration (MIC)** of that agent. An alternative procedure is the **agar dilution test,** which is performed with agar plates of medium each containing a serial dilution of the antimicrobial drug. The surface of the medium is inoculated with a standard loop in one spot without spreading, or

with a replicating device for simultaneously inoculating multiple cultures onto a single plate. Modifications of the tube dilution test may be employed when the simultaneous use of two antimicrobial drugs is contemplated, to determine if the effect of the combination will by synergistic, additive, or antagonistic.

Diffusion tests are employed in many laboratories because of their simplicity, speed, and economy. These tests employ standardized filter paper discs impregnated with fixed amounts of antimicrobial drugs and then dried. A "lawn" of bacteria is seeded on the surface of an agar plate with a swab moistened with a standardized liquid culture, and several different drug-impregnated discs are then placed on the surface of the agar. The plates are incubated for 18–24 h and the diameter of the zone showing no growth about a disc is then measured. The diameter of this zone of inhibition will indicate whether the organism under scrutiny has high, intermediate, or inadequate sensitivity to the drug in question. It has been shown experimentally that *the diameter of the inhibitory zone about the disc is related linearly to the logarithm of the MIC of the drug measured in tube dilution tests.*

Recently, **automated tests** of microbial sensitivity have been introduced that measure the inhibitory effect of drugs in liquid medium by using **light scattering** to determine bacterial growth at a standard time. This technic yields a result within a few hours, and it is being used increasingly by laboratories called upon to perform large numbers of tests.

For the details of microbial sensitivity tests, the reader is referred to the *Manual of Clinical Microbiology* edited by Lennette et al (see Selected References).

Tests of microbial sensitivity distinguish those drugs that *may* be useful from those that *cannot* be useful in effecting a cure. Among those drugs that may be useful (i.e., that inhibit the organism in vitro) the choice is based upon the pharmacologic and toxic properties of the drugs, their mode of action (i.e., bacteriostatic or bactericidal), the anatomic site of the infection, and the presence of underlying illnesses which, by their effects on renal, hepatic, or cardiac function, may modify the pharmacology of a drug. In particular, *the disc with the largest zone of inhibition does not necessarily identify the most suitable drug.*

SELECTED READING

BOOKS AND REVIEW ARTICLES

BARTLETT JG: Antibiotic-associated pseudomembranous colitis. Revs Infe Dis 1:530, 1979

FALKOW S (ed): Symposium on Infectious Multiple Drug Resistance. Washington DC, US Government Printing Office, 1976

The Medical Letter on Drugs and Therapeutics. Revised Edition. The Medical Letter, Inc., New Rochelle, NY, 1979

LENNETTE EH, SPAULDING EH, TRUANT JP (eds): Manual of Clinical Microbiology, 2nd ed. Washington DC, American Society for Microbiology, 1974, pp 407–442

SYMPOSIUM: Antibiotic-associated colitis. In Schlessinger D (ed): Microbiology-1979, Amer Soc for Microbiol, Washington, DC, 1979

WEINSTEIN L: Chemotherapy of microbial disease. In Goodman LS, Gilman A (eds): The Pharmacological Basis of Therapeutics, 5th ed. New York, Macmillan, 1975

SPECIFIC ARTICLES

LEPPER MH, DOWLING HF: Treatment of pneumococcic meningitis with penicillin compared with penicillin plus aureomycin: studies including observations on an apparent antagonism between penicillin and aureomycin. Arch Intern Med 88:489, 1951

SHOWACRE JL, HOPPS HE, DUBUY HS, SMADEL JE: Effect of antibiotics on intracellular Salmonella typhosa. I. Demonstration by phase microscopy of prompt inhibition of intracellular multiplication. II. Elimination of infection by prolonged treatment. J Immunol 87:153, 162, 1961

SMITH MR, WOOD WB JR: An experimental analysis of the curative action of penicillin in acute bacterial infections. I. The relationship of the antimicrobial effect of penicillin. II. The role of phagocytic cells in the process of recovery. III. The effect of suppuration upon the antibacterial action of the drug. J Exp Med 103:487, 499, 509, 1956

THORNSBERRY C, GAVAN TL, SHERRIS JC et al: Laboratory evaluation of a rapid, automated susceptibility testing system: report of a collaborative study. Antimicrob Agents Chemother 7:466, 1975

chapter

26

CORYNEBACTERIA

BERNARD D. DAVIS

The corynebacteria are gram-positive, rodlike organisms, which often arrange themselves in palisades, possess club-shaped swellings at their poles, and stain irregularly. Taxonomically they appear to be related to the mycobacteria (Ch. 37) and nocardiae (Ch. 38), since the principal cell wall antigens of all three genera are closely related chemically and serologically. In the species that can cause diphtheria, *Corynebacterium diphtheriae,* lysogenization by a bacteriophage causes synthesis of a potent, heat-labile protein toxin.

CORYNEBACTERIUM DIPHTHERIAE

No other bacterial disease of man has been as successfully studied as diphtheria. Its etiology and mode of transmission were established early, and a highly effective method of prevention—immunization with diphtheria toxoid—was subsequently developed. As a result of mass immunization diphtheria has become a rare disease in many countries, including the United States, and toxigenic strains of *C. diphtheriae* have virtually disappeared. Nevertheless, we shall discuss diphtheria in some detail as a prototype of toxigenic disease, and as an example of successful control.

HISTORY

Records of the existence of diphtheria or croup date back to Hippocrates in the fourth century B.C. Aetius, who lived in the sixth century A.D., wrote that "pestilential lesions of the tonsils . . . occur most frequently in children, but also in adults. . . . Usually in children the evils known as aphthae develop. These are white, like blotches; some are ashen in color or like eschars from the cautery. The patient suffers from a dryness of the gullet and frequently attacks of choking. . . . A spreading sore supervenes in the region afflicted. . . . In some cases the uvula is eaten up and when the sores have prevailed a long time and deepened, a cicatrix forms over them and the patient's speech becomes rather husky and, in drinking, liquid is diverted upward to the nostrils."

Despite its prevalence diphtheria was not recognized as a specific disease until 1821, when Pierre Bretonneau proposed that the throat "distemper" could be differentiated from other afflictions of the throat by the formation in the respiratory tract of a false membrane (composed largely of fibrin, bacteria, and trapped leukocytes,

rather than true epithelium). Because of this distinctive feature Bretonneau called the malady **diphtheritis** (Gr. *diphthera,* skin or membrane). He was convinced that diphtheria was a communicable disease, transmissible from person to person, but it was not until 1883 that the causative organism was described by Klebs in stained smears from diphtheritic membranes. Its etiologic role was proved a year later by Loeffler, who grew the organism on artificial media and produced fatal infections in guinea pigs, rabbits, and pigeons with lesions similar to those seen in the human disease. Loeffler noted that certain species, such as rats and mice, were resistant.

When the tissues of the guinea pigs were studied histologically Loeffler was surprised to find bacilli only in the local lesions at the site of inoculation, though damage was visible in the heart, liver, kidneys, adrenal glands, and other tissues. He concluded that the bacilli growing at the primary site must have produced a soluble poison, which was transported to remote tissues of the body by the blood stream. In 1888 Roux and Yersin demonstrated a soluble, heat-labile toxin, in the fluid phase of diphtheria bacillus cultures, whose injection into appropriate animals caused all the systemic manifestations of diphtheria. Two years later von Behring and Kitasato succeeded in immunizing animals with toxin modified with iodine trichloride; and they demonstrated that the serum of such immunized animals protected susceptible animals against the disease. On Christmas night 1891, in Berlin, the antitoxin was first given to a diphtheritic child.

Although the modified toxin was suitable for immunizing animals to obtain antitoxin, it still caused severe local reactions when injected and therefore could not be used as an immunizing agent in man. In 1909 Theobald Smith suggested the use of toxin neutralized by an equivalent amount of antitoxin as an immunizing agent, but it was not until 1922 that **toxin-antitoxin** was used on a large scale by W. H. Park for protection of children in New York City. Its use was greatly facilitated by Schick's discovery in 1913 of a practical test (see below) for distinguishing immune from nonimmune individuals by their reactions to an intradermal injection of a small dose of toxin. Finally, in 1923, G. Ramon introduced the use of formalin-treated toxin, or **toxoid,** as an immunizing agent. Toxoid is noninjurious to tissues but is fully immunogenic; it is now used universally for active immunization against diphtheria.

In 1951 Freeman made the remarkable discovery that all toxigenic strains of *C. diphtheriae* are lysogenic, i.e., are infected with a temperate bacteriophage (Ch. 48). If such strains lose their specific phage they cease to produce toxin and thus become relatively avirulent.

MORPHOLOGY

Corynebacteria (Gr. *coryne,* club) are gram-positive, non–spore-bearing rods, tapered from their septal ends, without flagella or demonstrable capsules. They vary from $2\mu m$ to $6\mu m$ in length and from $0.5\mu m$ to $1\mu m$ in diameter. Because of the way the individual bacilli divide they tend, in stained smears, to form sharp angles with one another, making characteristic figures resembling Chinese letters. When grown on coagulated serum slants rich in phosphate (Loeffler's medium) they contain polymerized polyphosphate granules, called Babès–Ernst bodies, which stain metachromatically with methylene blue or toluidine blue. In some strains well-defined polar bodies are discernible at the ends of each bacillus. These bodies are most frequently seen in organisms growing slowly on suboptimal media; they are less prominent during rapid growth, especially on media containing potassium tellurite (K_2TeO_3). On Loeffler's medium, and on blood agar, surface colonies are cream-colored or grayish white; on tellurite agar they are dark gray or black because of intracellular reduction of the tellurite to tellurium.

Three morphologically distinct types of *C. diphtheriae* have been described. On tellurite agar the *gravis* strains form large, flat, gray to black "daisy-head" colonies with dull surfaces. The colonies of the *mitis* strains are smaller, blacker, and more convex, with glossy surfaces. The colonies of the *intermedius* strains are still smaller and either smooth or rough. In broth the rough strains form a pellicle, whereas smooth strains grow diffusely. An initially postulated relation between virulence and colony morphology is no longer considered valid. Both smooth and rough strains (Fig. 26-1) may be either toxigenic or nontoxigenic.

Corynebacteria contain in their cell walls mycolic acids (very long-chain branched fatty acids), somewhat smaller (C_{28}–C_{40}) than those of mycobacteria (ca. C_{80}: see Ch. 37).

CULTIVATION

C. diphtheriae is an obligate aerobe. Most strains grow as a waxy pellicle on the surface of liquid media. For primary isolation Loeffler's coagulated serum medium is still useful since it permits growth of diphtheria bacilli with characteristic morphology but fails to support growth of streptococci and pneumococci, commonly present in the throat. Blood or chocolate agar with po-

FIG. 26-1. Smooth (small) and rough (large) colonies of *C. diphtheriae* on chocolate agar. (Barksdale WL: J Bacteriol 81:531, 1961)

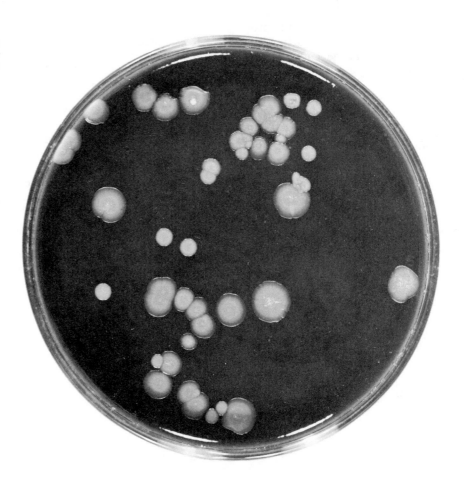

tassium tellurite to inhibit the growth of most other bacteria serves as an even better selective medium for *C. diphtheriae.*

Although the black colonies formed by the diphtheria bacilli are characteristic, other organisms found in the respiratory tract, particularly staphylococci and nonpathogenic corynebacteria (e.g., *C. pseudodiphtheriticum* [*C. hofmannii*]), may also form black colonies.

C. diphtheriae grows well on relatively simple media containing essential amino acids and an energy source such as glucose or maltose. Most strains are unable to synthesize nicotinic acid, pantothenic acid, or biotin. Under optimal conditions of pH, oxygen supply, and nutrients growth of 20–30 g dry weight/liter may be obtained.

Diphtheria bacilli typically ferment glucose rapidly and maltose slowly, producing acid but no gas. A comparison of the fermentation reactions of the common corynebacteria is shown in Table 26-1.

LYSOGENY AND TOXIN PRODUCTION

Only those strains of *C. diphtheriae* that are lysogenic for β-prophage or a closely related phage produce diphtheria toxin. The phage clearly contains the structural gene (*tox*) for the toxin molecule, since lysogenization by various mutated β-phages leads to production of a nontoxic, immunologically crossreacting material, CRM, instead of the toxin. Phage multiplication is not required for toxin production.

Regulation of Toxin Formation. The expression of the phage *tox* gene is controlled by the metabolism and physiologic state of the host bacterium. With an adequate supply of essential nutrients, including O_2, the most important factor is the inorganic iron (Fe^{2+}) concentration: toxin is synthesized in high yield only after the exogenous Fe supply has become exhausted. This regulation by Fe^{2+} evidently also involves a bacterial gene product (repressor protein): bacterial mutants have been isolated in which Fe does not repress toxin formation; and in an in vitro system from *Escherichia coli,* forming toxin by coupled transcription–translation of *tox* gene DNA, an iron-containing protein fraction from *C. diphtheriae* (even from a nontoxigenic, nonlysogenic strain) inhibits transcription, while Fe^{2+} is without effect unless *C. diphtheriae* protein is present.

The classic Park–Williams strain (PW8) of *C. diphtheriae,* isolated in 1898, is still used for commercial production of toxin. It possesses the unusual capacity to increase its bacterial mass 5- to 6-fold after depletion of the exogenous iron supply. During this period toxin synthesis may reach 5% of the total protein synthesis.

STRUCTURE AND MODE OF ACTION OF TOXIN

Diphtheria toxin may be isolated from culture filtrates of *C. diphtheriae* as an iron-free, crystalline, heat-labile protein. It is secreted as a single polypeptide chain of about 60,000 daltons, derived from a slightly larger intracellular precursor.

In highly susceptible species (rabbit, guinea pig) the pure protein is lethal at 0.1 µg/kg or less, and in man intradermal injection of 10^{-4} µg will produce a visible skin reaction. The key to its action was the finding of early inhibition of protein synthesis in cultured mammalian cells. However, inhibition in vitro depends on cleavage at a bond 194 residues from the N-terminus, the resultant conformational change activating a latent enzyme; pure, undamaged toxin is inactive in extracts. Fortunately, the early preparations had undergone slight proteolysis, which permitted the inhibition to be reproduced and analyzed in solution.

The activating cleavage occurs in an exposed loop that is subtended by a cystine S-S near the N-terminus (Fig. 26-2). On reduction of the S-S bond, plus exposure to denaturing reagents, the activated toxin can be separated into two fragments, the N-terminal **fragment A** (mol wt 21,150) and the larger **fragment B** (mol wt around 39,-000). **A** carries the **enzyme activity,** and **B** must be complexed with A if the latter is to be **bound to** and enter the cell. Fragment A is very stable, and after denaturation it quickly renatures to full enzymic activity.

The enzymatic reaction catalyzed by fragment A transfers the ADP-ribosyl group from NAD to a modified histidine in EF2, the elongation factor required for translocation of polypeptidyl-tRNA from acceptor to donor site on the eukaryotic ribosome:

$$NAD^+ + EF2 \rightleftharpoons ADP\text{-ribosyl}-EF2 + nicotinamide + H^+$$

The factor is thereby inactivated.

In cells with a high ratio of NAD to nicotinamide the equilibrium lies far to the right, and the reaction is in effect irreversible. However, after cells have been intoxicated, yielding extracts that can no longer incorporate amino acids, this activity can be restored by measures that reverse the above reaction (dialysis of the extract to remove NAD, lowering of the pH, and addition of excess nicotinamide); fragment A must be added to catalyze the reverse reaction. This finding proves that the enzyme reaction observed in extracts is the basis of the intoxication of cells.

Fragment A is a potent enzyme. Studies with cell fusion technics have shown that *a single molecule within a cell is lethal,* halting protein synthesis within a few hours,

TABLE 26-1. Fermentation Reactions of Corynebacteria Commonly Cultured from Man

Corynebacteria	Glucose	Maltose	Sucrose
C. diphtheriae	+	+	– (+)
C. xerosis	+	–	+
C. hofmannii	–	–	–

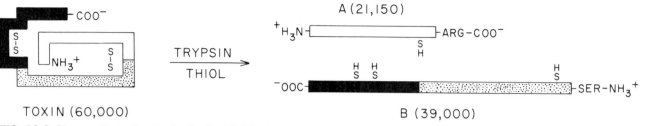

FIG. 26-2. Diagram illustrating "activation" of diphtheria toxin by trypsin in the presence of thiol. All of the ADP-ribosylating activity is present on the N-terminal fragment A, which is masked in intact toxin. Fragment B is required to enable fragment A to reach the cytoplasm of susceptible animal cells. The C-terminal amino acid sequence of B **(solid bar),** which contains determinants that interact with specific receptors on the sensitive cell membrane, is joined to A through a hydrophobic region **(stippled bar),** which like fragment A is masked in intact toxin.

even in cells from a species normally resistant to diphtheria toxin.

Specificity. EF2 is the only protein, in crude tissue extracts, that is significantly ADP-ribosylated by fragment A at low concentrations. The reaction is observed in extracts of cells of all eukaryotic species tested, ranging from yeasts and protozoa to higher plants and man. The prokaryotic translocation factor (EFG) does not respond.

INTERACTION OF TOXIN WITH MAMMALIAN CELLS

Fragment B has two important functions in the entry of fragment A into a cell: it recognizes and interacts with a specific **surface receptor** present on sensitive cells, and it facilitates the transport of A across the membrane to reach the cytoplasm. In studies on this mechanism mutant proteins (CRMs), serologically related to toxin but with markedly reduced or no toxicity, are proving very useful. The properties of some of these CRMs are compared with those of toxin and its fragments in Table 26-2.

Of particular interest are CRMs 45 and 197. CRM 45 yields an active A fragment but lacks the C-terminal, positively charged, 15,000-dalton sequence of B, which reacts with surface receptors.

Moreover, the loss of this hydrophilic "tail" (depicted as a solid bar in Fig. 26-2) evidently exposes a hydrophobic region (stippled in Fig. 26-2), since this CRM, unlike toxin, can bind much nonionic detergent and can enter membrane vesicles and detergent micelles. CRM 197, on the other hand, has a normal B region and can compete with toxin for specific membrane receptors, but its A fragment lacks enzyme activity because of a missense mutation. If CRM 45 and CRM 197 are cleaved, reduced, mixed, and allowed to reoxidize, fully toxic hybrid molecules are reconstituted.

We still know very little about the specific membrane receptors for diphtheria toxin, and about how fragment A crosses the lipid bilayer. The first step appears to be reversible binding of the C-terminal hydrophilic portion of fragment B to the receptor. It has been proposed that the resulting conformational change initiates the following sequence: the strongly hydrophobic region of B is irreversibly inserted into the lipid bilayer; its interaction with membrane components forms a channel across the membrane; a specific membrane protease (perhaps the receptor itself) nicks the loop that links A to B; fragment A (after reduction of the disulfide) passes through the channel as an extended chain; and in the aqueous cytoplasm this readily renatured molecule spontaneously recovers its active conformation.

The mechanism by which toxins are transported across the plasma membrane of animal cells is of much current interest as a possible model for the transport of other proteins, such as peptide hormones and growth factors.

TABLE 26-2. Some Properties of Diphtheria Toxin and Related Proteins

Protein	Approximate mol wt	Toxicity (MLD/µg)	Enzyme activity (%)*	Binding	Number of half-cystines
Toxin	60,000	25–30	100	(+)	4
Toxoid	60,000	0	0	–	4
Fragment A	21,150	0	100	–	1
Fragment B	39,000	0	0	+	3
CRM 45	45,000	0	100	–	2
CRM 176	60,000	ca. 0.1	8–10	+	4
CRM 197	60,000	0	0	+	4
$A_{45}B_{197}$‡	60,000	25–30	100	(+)	4

CRM = Serologically crossreacting material.

*ADP-ribosylating activity after "nicking" and reduction, relative to fragment A as 100%.

+Ability to compete with toxin for specific membrane receptors.

‡Toxic hybrid molecule formed by allowing a mixture of reduced, "nicked" CRM 45 and CRM 197 to reoxidize.

PATHOGENICITY

Experimental Models. The only known natural reservoirs for toxigenic as well as nontoxigenic *C. diphtheriae* are the upper respiratory tracts of men and horses, and indolent cutaneous lesions of man that are seen in tropical climates. Natural infections have been observed only in man.

The closely related *C. ulcerans* and *C. pseudotuberculosis* (*C. ovis*), frequently found in the nasopharynges of horses and sheep, can be converted to toxigenicity by a *tox*+ phage, and can be pathogenic for man, but neither of these strains has been isolated from typical cases of diphtheria.

Although protein synthesis is blocked by activated toxin in extracts from all eukaryotes, certain animal species are resistant because their cells do not bind toxin and so the enzymatically active fragment A cannot gain access to the cell cytoplasm. For this reason rats and mice are more than 1000 times as resistant per unit body weight as susceptible species, such as humans, monkeys, rabbits, guinea pigs, pigeons, and chickens.

Subcutaneous injection of diphtheria toxin into a guinea pig is followed by death within 12 h to several days, depending upon the dose. Local swelling and apparent tenderness develop within a few hours, and at autopsy edema, hemorrhage, and necrosis are noted at the site of injection. There is usually marked congestion of the adrenal cortices, and degenerative changes can always be demonstrated in the heart, liver, and kidneys. Sublethal doses of toxin may cause late paralyses similar to those observed in man.

As was first recognized by Loeffler, when virulent bacilli are injected instead of the toxin the end result is much the same. The systemic lesions that develop are indistinguishable from those caused by the toxin itself, and the injected bacilli remain localized at the original site of inoculation.

Human Disease. Diphtheria in man usually begins in the upper respiratory tract. When virulent diphtheria bacilli become lodged in the throat of a susceptible individual they first multiply in the superficial layers of the mucous membrane. There they elaborate toxin, which causes necrosis of neighboring tissue cells and establishes a nidus for further multiplication of the bacteria. The inflammatory response results in the accumulation of a grayish exudate, which eventually forms the characteristic **diphtheritic pseudomembrane.** It usually appears first on the tonsils or posterior pharynx and may then spread either upward into the nasal passages (**nasopharyngeal diphtheria**) or downward into the larynx and trachea (**laryngeal diphtheria**). The nasopharyngeal form of the disease is frequently accompanied by marked prostration and severe toxemia. If recovery ensues, late neurologic and cardiac complications are relatively common. Laryngeal diphtheria is particularly hazardous because

mechanical obstruction may cause suffocation unless the airway is restored by intubation or tracheotomy.

In contrast to streptococcal pharyngitis (with which it may be easily confused), diphtheria of the upper respiratory tract tends to remain localized. Enlargement of regional lymph nodes in the neck is common, but invasion of other tissues seldom occurs and bacteremia is not seen. Even the redness and swelling of the mucous membranes in the pharynx are much less pronounced than in streptococcal infections. Fever is usually only moderate.

Although typical clinically severe diphtheria is always caused by a toxin-producing organism, mild cases of sore throat with fever and an atypical membrane can be produced by certain *tox*⁻ strains. Epidemics of similar mild diphtherial throat infections caused by *tox*+ strains have been described among immune populations in persons whose serum already contains circulating antitoxin. In the tropics indolent, ulcerative cutaneous lesions with a diphtheritic membrane and *tox*+ *C. diphtheriae* are common.

IMMUNITY AND EPIDEMIOLOGY

Acquired immunity to diphtheria is primarily antitoxic. Newborn infants whose mothers are resistant acquire temporary immunity from transplacental antitoxin. Such passive immunity lasts at most only a year or two. It is probable that active immunity can be produced by a mild or inapparent infection in infants who still retain some circulating maternal antitoxin, and in susceptible adults infected with a strain of low toxigenicity. However, in areas where diphtheria is *not* endemic most children, unless artificially immunized, become highly susceptible within a few months after birth. Artificial immunization at an early age is, therefore, universally advocated.

Persons who recover from diphtheria may continue to harbor the organism in the nose or throat for weeks or even months. In the past, it was mainly through such healthy carriers that the disease was spread by droplet infection and toxigenic bacteria were maintained among the population. Before mass immunization the carrier rate for toxigenic *C. diphtheriae* in large cities was often 5% or higher. The advent of universal immunization resulted in a dramatic fall in the carrier rate; hence the *tox* gene must have survival value both for *tox*+ phage and for its bacterial host under natural conditions.

Schick Test. Whether or not a given individual is in need of active immunization can be readily determined by means of the Schick test: the intradermal injection of 1/50 MLD* of diphtheria toxin. In the absence of circulating antitoxin this small amount of toxin will cause a local necrotic reaction characterized by a pigmented area

---

* One minimum lethal dose (MLD) was originally defined by Ehrlich as that amount of diphtheria toxin which, when injected subcutaneously into a 250-g guinea pig, causes death on the 4th or 5th day.

of swelling and tenderness, reaching a maximum after 4–5 days, and gradually turning from red to brown. If, on the other hand, the blood stream contains a sufficient level of antitoxin the toxin will be neutralized and no reaction will occur. Thus, a positive Schick test indicates that little or no antitoxin is present in the serum (<0.01 U/ml).*

In practice the reading of the Schick test is not so simple, especially among adults and older children in areas where diphtheria is endemic, because many persons will have become hypersensitive to the toxin itself or to other antigens in the toxin preparation, either as a result of naturally acquired infections or from artificial immunization with toxoid. Such individuals will react allergically to the intradermally injected toxin. In order to distinguish these reactions from the primary action of the toxin a control injection of toxoid (about 0.005 Lf) is administered intradermally in the opposite arm. If the individual is immune but is sensitive to one or more antigens in the toxin preparation he will react to both the toxin and the toxoid. The allergic (delayed-type) reactions, however, usually reach a maximum within 48–72 h and then fade, whereas the true positive Schick reaction persists for many days. Persons showing delayed hypersensitivity **pseudoreactions** to Schick test materials respond, almost without fail, with a brisk secondary antitoxin response to the test itself.

Occasionally subjects who exhibit reactions on both arms will have insufficient circulating antitoxin for complete neutralization, so that the reaction at the toxin site may persist after that at the control site has subsided. Individuals who show such **combined reactions** to purified Schick test materials usually have experienced previous exposure to toxin and generate sufficient antitoxin to be Schick-negative upon retest.

Reimmunization of adults and older children should be approached with caution in order to avoid serious local and systemic reactions in persons sensitive to toxoid. For anyone who shows a pseudoreaction in the Schick test it is inadvisable to inject a full immunizing dose of toxoid, and the test itself will usually have served as a booster.

LABORATORY DIAGNOSIS

A definitive diagnosis of diphtheria can ordinarily be made only by isolating toxigenic diphtheria bacilli from the primary lesion. Exudate from the lesion, taken preferably from the membrane if present, should be immediately transferred to a Loeffler slant, a blood agar plate,

and tellurite agar. After 24 h incubation each culture should be carefully examined, and smears should be made from each type of colony that has grown out on any one of the media. If growth appears on the blood agar plate, but not on the tellurite plate, it may be tentatively concluded that no diphtheria bacilli are present. As a precaution, however, the tellurite plate should be reincubated for an additional 24 h before being discarded as negative. Smears should immediately be made of any colonies appearing on the tellurite agar and should be stained with methylene blue. The presence of corynebacteria should be readily detected in such smears. Once identified, they should be promptly subcultured on a Loeffler's slant and tested for toxigenicity, either by the guinea pig virulence test or by the in vitro gel diffusion method.

The **guinea pig test** is performed by injecting intradermally, into the shaved side of a guinea pig, 0.1 ml of a heavy suspension of the bacilli washed from the Loeffler's slant. After 4 h the guinea pig is injected intraperitoneally with 500 U antitoxin, and 30 min later a second sample of the test suspension is injected intradermally on the opposite side. Nonspecific inflammatory reactions may occur at both sites within 24–48 h, but if toxigenic bacilli are present only the site injected before the antitoxin was administered will progress to form a characteristic necrotic lesion at 48–72 h. Rabbits may be used instead of guinea pigs.

The **gel diffusion test** is performed by pouring a Petri plate of peptone maltose agar containing antitoxin-free calf serum, onto which has been placed a sterile strip of filter paper impregnated with diphtheria antitoxin. After the agar has solidified a heavy inoculum of the test culture is streaked at right angles across the paper strip and the plate is incubated for 24 h. If the inoculated organisms are toxigenic a visible line of Ag–Ab precipitate will form, as shown in Figure 26-3. The antitoxin used in this test must contain no significant amount of Ab against diphtherial proteins other than toxin.

TREATMENT

Once circulating diphtheria toxin has gained entrance to susceptible tissue cells it can no longer be neutralized by antitoxin. Accordingly, in suspected cases of diphtheria, **antitoxin therapy** must be administered without delay. Indeed, the time factor is so critical (Table 26-3) that the clinician is amply justified in giving antiserum without waiting for the results of the bacteriologic tests, provided the circumstances of the infection and the clinical signs of the disease suggest diphtheria. To ensure the maximum immediate therapeutic effect the antitoxin should be injected intramuscularly in a single large dose. The amount given should be 100–500 U/lb of body weight, depending upon the severity of the disease.† Since the

---

* Antitoxin titers in human serum are measured by comparing its ability to neutralize toxin with that of a standard antitoxin in the rabbit intradermal test. An international standard, kept at the State Serum Institute in Copenhagen, is available for reference; 1 U will neutralize about 75 MLD of freshly prepared toxic protein. Toxin itself cannot be used as a standard because it loses toxicity upon storage, without a parallel loss of antitoxin-binding capacity.

† Doses greatly in excess of those theoretically needed to neutralize all the toxin in the patient's tissues are indicated in order to achieve as rapid inactivation of the toxin as possible (including molecules **on** but not yet **in** cells).

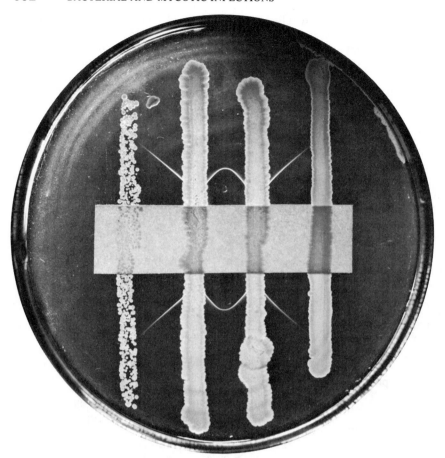

FIG. 26-3. Gel diffusion test of toxigenicity. Outer strains are nontoxinogenic, whereas those streaked in center of plate produce toxin. Note that lines of precipitate generated by the two toxinogenic strains merge to form arcs, indicating that the toxins elaborated are immunologically identical. The antitoxic serum used had been previously absorbed with nontoxic proteins, obtained from a nontoxigenic culture, and thus behaved as a **monospecific** antiserum. The intersection of the bands with the bacterial growth is removed some distance from the piece of filter paper impregnated with antiserum because horse antitoxin was employed, which inhibits the precipitin reaction in the region of Ab excess (King EO et al: Am J Public Health 39:1314, 1949)

antiserum used is derived from hyperimmunized horses, the usual **skin test for sensitivity to horse serum proteins** should be performed first, and a syringe containing epinephrine should be available for immediate use in the event of an anaphylactic reaction. If the results of the preliminary skin test indicate that the patient is sensitive to horse serum the antitoxin should be given with the greatest possible caution, beginning with a small, highly diluted (e.g., 1:10,000) subcutaneous dose, followed by gradually increasing doses administered intramuscularly, until the full dose has been injected.

TABLE 26-3. Relation of Case Fatality Rate to Time of Treatment with Antitoxin

Day of disease antitoxin administered	Cases treated	Case fatality rate (%)
1	225	0
2	1,441	4.2
3	1,600	11.1
4	1,276	17.3
5 or > 5	1,645	18.7

Although diphtheria bacilli are susceptible to the antimicrobial action of penicillin, the tetracyclines, and erythromycin, antibiotic therapy alone should never be relied upon in the treatment of diphtheria. These drugs, however, in conjunction with antitoxin, may hasten the elimination of the causative organisms from the primary lesion.

PREVENTION

The prevention of diphtheria is simpler than that of many infectious diseases, because only man appears to be an important reservoir for *C. diphtheriae* and because all strains elaborate the same antigenic type of toxin. For these reasons it should be possible to eradicate the disease by immunization.

In certain areas of the world, including the United States, where children are immunized at an early age and are given booster injections after 1 year and when they enter school, diphtheria has become a rarity. In other countries, however, where diphtheria immunization is rarely practiced, the disease is still prevalent. Its continued spread by droplet infection results from the contact of susceptible individuals not only with active cases but also with asymptomatic carriers. Persons with cutaneous diphtheria or diphtheritic ulcers (who are invariably Schick-negative) also may transmit toxigenic *C. diphtheriae* by contact.

As we have already discussed, carrier rates may be high among urban populations in areas where the disease is endemic. The treatment of carriers presents a troublesome problem. Daily treatment for 1 or 2 weeks with antibiotics such as penicillin, to which diphtheria bacilli are highly sensitive, causes apparent elimination of the organisms from the nose and throat. When treatment is stopped, however, the bacilli often reappear. Fortunately, with the almost complete elimination of the disease itself very few healthy carriers of virulent diphtheria bacilli remain.

When diphtheria toxin is treated with dilute formaldehyde under suitable conditions, it is converted to toxoid. **Formol toxoid** is devoid of toxicity, but is virtually indistinguishable antigenically from toxin. It is usually injected either as an alum precipitate or adsorbed on aluminum phosphate gel. The appropriate immunizing dose for a child is about 10 Lf.* Two doses given 1 month apart are usually adequate for primary immunization. A booster injection is given about a year later.

It is common practice in the United States to immunize infants with a combined vaccine containing diphtheria

* L stands for the Latin word *limes* (limit), and 1 Lf is that quantity of toxin or toxoid that flocculates most rapidly when mixed with 1 U of antitoxin.

toxoid, tetanus toxoid, and pertussis vaccine: the vaccine exerts an adjuvant effect. The primary course should be initiated at 3 or 4 months of age. The immunity ordinarily lasts for several years, although in some cases it may persist for only a few months. In any event, the antitoxin levels reached after primary immunization are usually rather low, and in order to ensure continued protection it is important to administer several booster injections during childhood.

OTHER CORYNEBACTERIA (DIPHTHEROIDS)

Corynebacteria are widely distributed in nature. Many species inhabit the soil, and a number cause disease in animals. *C. pseudotuberculosis* (*C. ovis*), for example, is responsible for pseudotuberculosis in sheep, horses, and cattle. It resembles the diphtheria bacillus morphologically and produces a potent toxin, which is immunologically unrelated to diphtheria toxin. *C. kutscheri* causes a similar disease in mice and commonly gives rise to latent infections (see Dormant and Latent Infections, Ch. 24).

The two nonpathogenic species of corynebacteria most often found in man are *C. pseudodiphtheriticum* and *C. xerosis*. They grow on tellurite agar and may be easily confused with diphtheria bacilli. *C. pseudodiphtheriticum* is commonly encountered in throat cultures, whereas *C. xerosis* normally inhabits the conjunctival sac. Both are nontoxigenic and may be readily differentiated from *C. diphtheriae* by their fermentation reactions (Table 26-1).

SELECTED READING

BOOKS AND REVIEW ARTICLES

ANDREWES FW, BULLOCH W, DOUGLASS SR, FREYER G, GARDNER AD, FILDES P, LEDINGHAM JCG, WOLF CGL: Diphtheria, Its Bacteriology Pathology and Immunity. London, HMSO, 1923

COLLIER RJ: Diphtheria toxin: mode of action and structure. Bacteriol Rev 39:54, 1975

PAPPENHEIMER AM, JR: Diphtheria toxin. Annu Rev Biochem 46:69, 1977

PAPPENHEIMER AM, JR, GILL DM: Diphtheria. Science 182:353, 1973

SINGER RA: Lysogeny and toxinogeny in *Corynebacterium diphtheriae*. In Bernheimer AW (ed): Mechanisms in Bacterial Toxinology. New York, Wiley, 1976, pp 31–51

SPECIFIC ARTICLES

BOQUET P, SILVERMAN MS, PAPPENHEIMER AM, JR, VERNON WB: Binding of Triton X-100 to diphtheria toxin, CRM45 and their fragments. Proc Natl Acad Sci USA 73:4449, 1976

COLLIER RJ: Effect of diphtheria toxin on protein synthesis. J Mol Biol 25:83, 1967

COLLIER RJ, KANDEL J: Structure and activity of diphtheria toxin. J Biol Chem 246:1492, 1971

FREEMAN VJ: Studies on the virulence of bacteriophage infected strains of *Corynebacterium diphtheriae*. J Bacteriol 61:675, 1951

GILL DM, PAPPENHEIMER AM JR: Structure-activity relationships in diphtheria toxin. J Biol Chem 246:1492, 1971

HONJO T, NISHIZUKA Y, KATO I, HAYAISHI O: Adenosine diphosphate ribosylation of aminoacyl transferase II by diphtheria toxin. J Biol Chem 246:4251, 1971

MUELLER JH: Nutrition of the diphtheria bacillus. Bacteriol Rev 4:97, 1940

MURPHY JR, PAPPENHEIMER AM JR, TAYARD DE BORMS S: Synthesis of diphtheria *tox* gene products in *E. coli* extracts. Proc Natl Acad Sci USA 71:11, 1974

UCHIDA T, GILL DM, PAPPENHEIMER AM JR: Mutation in the structural gene for diphtheria toxin carried by beta phage. Nature New Biol 233:8, 1971

YAMAIZUMI M, MEKADA E, UCHIDA T, OKADA Y: One molecule of diphtheria toxin fragment A introduced into a cell can kill the cell. Cell 15:245, 1978

chapter 27

PNEUMOCOCCI

ROBERT AUSTRIAN

The pathogenic bacteria to be discussed in the next chapters are often classed together as the **pyogenic cocci.** They include the pneumococci, streptococci, staphylococci, and neisseriae; all but the neisseriae are gram-positive. They are predominantly **invasive** pathogens, which tend to produce acute purulent lesions. Behaving as **extracellular parasites,** they cause tissue damage only as long as they remain outside phagocytic cells; once ingested, they are promptly destroyed. The diseases they cause are **acute,** except when they form abscesses or become lodged on heart valves. They are generally susceptible to antimicrobial drugs, including penicillin, but differ greatly in tendencies to yield drug-resistant mutants. Whereas penicillin-resistant variants are uncommon among pneumococci and virtually nonexistent among group A streptococci, they are common among staphylococci and group D streptococci and fairly common among α- and nonhemolytic streptococci and neisseriae.

HISTORY

The systematic elucidation of the properties of *Streptococcus pneumoniae* (pneumococcus, formerly *Diplococcus pneumoniae*) as an agent of disease has resulted in some of the most important discoveries of biomedical science. First isolated from human saliva in 1880 by Pasteur in France and by Sternberg in the United States, its relation to lobar pneumonia was established a few years later. Recognition of serologically different types in 1910 led to specific antisera and thus to the first effective treatment for pneumococcal pneumonia. There followed the fundamental observations of Avery, Heidelberger, and Goebel on the chemical structure of capsular antigens and their role in bacterial virulence. The discovery of genetic transformation of pneumococci by DNA, announced by Avery, MacLeod, and McCarty in 1944, opened the door to molecular genetics and the consequent revolution in biology.

Formerly a leading cause of death, pneumococcal pneumonia can now be effectively treated with penicillin and other antimicrobial drugs. Although it is still a common and serious illness, recovery is usual unless therapy is delayed or the patient is an infant, is over age 55, or is debilitated by a complicating illness. Bacteremic illness treated with penicillin has a mortality of 17%.

MORPHOLOGY

Pneumococci in their most typical form are encapsulated, gram-positive, lancet-shaped diplococci. In sputum, pus, serous fluid, and body tissues they may be found in short chains and occasionally as individual cocci. Their tendency to form chains is exaggerated when they are grown in relatively unfavorable media (particularly with a low Mg^{2+} concentration) or in the presence of type-specific antibody. Though gram-positive during the exponential phase of growth in artificial media, more and more cells become gram-negative as the culture ages. If incubation is continued the viable count falls and the culture tends to clear. These changes are due to **autolytic enzymes,** which first render the cell gram-negative and later bring about lysis. Autolysis is stimulated by surface-active agents, such as bile salts or sodium deoxycholate, and tests for "bile solubility" are useful in identifying pneumococci.

Deoxycholate activates an amidase that splits the tetrapeptide from the muramic acid in peptidoglycan. This reaction, properly regulated, normally functions in wall morphogenesis and cell division.

In liquid cultures most strains of encapsulated pneumococci grow diffusely, tending to sediment only when the medium becomes acid; unencapsulated strains, particularly those that tend to grow in chains, exhibit a granular growth, which results in relatively rapid sedimentation. On the surface of solid media (e.g., blood agar plates) the encapsulated organisms form round, glistening, unpigmented colonies with a diameter of 0.5–1.5 mm after 24–36 h. In general, the larger the capsule the bigger and more mucoid are the colonies; those of type 3, for example, may reach a diameter of 3 mm. As the mucoid colonies age on blood agar their centers often collapse from autolysis. Both surface and deep colonies become surrounded by a zone of incomplete α-hemolysis (see History and Classification, Ch. 28) unless grown anaerobically, in which case hemolysis may be absent or β-hemolysis may occur.

Pneumococcal capsules are demonstrable by suspending the encapsulated organisms in India ink. They can be seen most easily, and the cells can be typed, by treatment with homologous type-specific Ab, which combines with the capsular polysaccharide and renders it refractile **(quellung reaction;** Fig. 27-1).

FIG. 27-1. A. Quellung reaction of type 3 pneumococcus as seen by light microscopy. (×2000) **B.** Electron micrograph showing type 1 pneumococci reacted with ferritin-labeled Ab; it reveals that the Ab interacts primarily with the surfaces of the capsules. (×21,000) **C.** Electron micrograph of pneumococcus type 1 suspended in India ink. Note how closely ink particles abut on cell body **(CB)** of organism **(arrow).** (×32,-000) **D.** Type 1 Ab has been added to suspension, causing capsule **(C)** to swell and separate India ink particles from body **(CB)** of cell **(arrow).** **(B–D,** Baker RF, Loosli CG: Lab Invest 15:716, 1966. Copyright 1966 by U.S.–Canadian Division of the International Academy of Pathology)

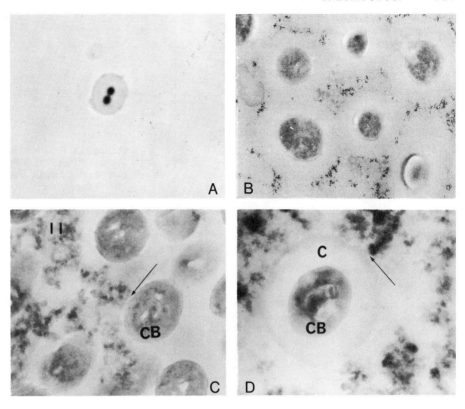

The word *Quellung* means "swelling" in German. Although there was once much dispute over whether the capsule actually swells when combined with Ab, electron microscopic observations have clearly shown that it does (Fig. 27-1). The quellung reaction, like other polysaccharide precipitin reactions, is inhibited by excess polysaccharide, i.e., in the zone of Ag excess (see Precipitin Reaction, Ch. 16). Indeed, the reaction may be reversed by the addition of enough homologous polysaccharide or by raising the ionic strength of the preparation to the point where the Ag–Ab complexes dissociate (see Ch. 16). A **nonspecific quellung reaction** may also be produced with nonantibody proteins that have the capacity to form strong ionic bonds with the capsular polysaccharide at an appropriate pH.

In electron photomicrographs pneumococcal capsules can be made out only in hazy outline (Fig. 27-2) unless the cells have been pretreated with homologous Ab, preferably labeled with ferritin (Fig. 27-1). Capsules tend to be largest during the exponential phase of growth, and to become smaller in later phases, owing to diffusion of the polysaccharide into the medium (Fig. 27-2).

The large capsules of pneumococcus type 3, like those of certain group A hemolytic streptococci, stain metachromatically (red) with methylene blue.

Pneumococcal L forms (Ch. 43), deficient in their cell walls, have been cultivated on hypertonic agar media containing penicillin. DNA bacteriophages have been isolated from pneumococci; they do not adsorb to encapsulated cells.

METABOLISM

Pneumococci need a complex medium for growth. Their energy requirements are met primarily by a lactic fermentation, and they are classified as **lactic acid bacteria.** Their ability to metabolize inulin is useful in differentiating them from most α-hemolytic streptococci.

The most satisfactory liquid medium is fresh **beef infusion broth,** containing 10% serum or blood, and at a pH of 7.4–7.8. Infusion broth tends to become inhibitory when exposed to air but is protected by reducing agents, such as cysteine or thioglycollate, which also permit a smaller inoculum to initiate growth even in fresh medium.

A synthetic culture medium for pneumococcus has been described by Tomasz. Although not satisfactory for routine use, this medium is useful for recovering macromolecular fractions from pneumococcal cells, since its constituents are all dialyzable.

Pneumococci are facultative anaerobes. Since they produce neither catalase nor peroxidase the cultures accumulate hydrogen peroxide, which decreases viability. In the presence of a source of catalase, such as red blood cells, the organisms survive in broth cultures at 0°–4° C for long periods.

Ethanolamine ($HO-CH_2-CH_2-NH_2$), substituted for choline

$$HO-CH_2-CH_2-\overset{\overset{\displaystyle OH}{\displaystyle |}}{N} \equiv [CH_3]_3)$$ in the synthetic culture medium, appears in the teichoic acid component of the cell wall. The following

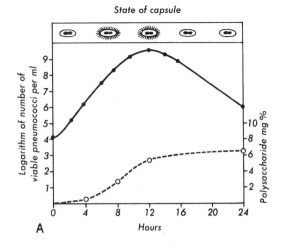

A

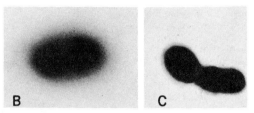

FIG. 27-2. Relation of the growth of type 3 pneumococcus in broth culture to the state of the capsule and cumulative synthesis of capsular polysaccharide. **(A)** Electron photomicrographs of type 3 pneumococci from 4-h **(B)** and 24-h **(C)** cultures demonstrate loss of outer portion of capsule (slime layer) as culture ages and synthesis of capsular polysaccharide slows (note change in slope of broken curve in chart) (Wood WB Jr, Smith MR: J Exp Med 90:85, 1949) (×6700)

abnormalities result: 1) the cells fail to divide normally and form long chains, 2) they are resistant to autolysis even when grown in the presence of penicillin, 3) they lose their ability to undergo genetic transformation, and 4) they do not adsorb phages. The $-N(CH_3)_3$ moiety of choline evidently plays a critical role in the pneumococcal cell wall.

ANTIGENIC STRUCTURE

More than 80 serologic types of pneumococci have been differentiated by the immunologically distinct polysaccharides of their capsules. Three different kinds of somatic Ags* have also been described: a poorly defined **R antigen** and a carbohydrate **C substance** are species-specific, and a type-specific protein (**M antigen**) is genetically and immunologically independent of the type-specific capsular polysaccharide.

CAPSULAR ANTIGENS

Structure. Pneumococcal capsules are composed of large polysaccharide polymers which form hydrophilic gels on the surface of the organisms. Although the composition of many of these capsular carbohydrates is known, the structure of only some (e.g., types 3, 6, and 8) has been definitely determined. Type 3, for example, consists of repeating cellobiuronic acid units (D-glucuronic acid $\beta 1 \rightarrow 4$ linked to D-glucose) joined by $\beta 1 \rightarrow 3$ glucosidic bonds (see Fig. 16-19).

Biosynthesis. The pathway of biosynthesis of type 3 polysaccharide has been elucidated (Fig. 27-3).

* The term **somatic antigen** is used to designate antigenic components of the body (*soma,* Gr.) of a bacterial cell, exclusive of its capsule.

Crossreactions. Although pneumococcal capsular polysaccharides exhibit an extraordinary degree of type specificity, some of them crossreact with capsular Ags of other pneumococcal types and of certain other bacterial species (e.g., α-hemolytic and nonhemolytic streptococci, klebsiellas, and salmonellae).

For example, pneumococcal type 8 capsular Ag contains D-glucose, D-galactose, and D-glucuronic acid in a sequence that includes cellobiuronic acid; hence it is not surprising that this polymer should crossreact with the type 3 capsular Ag.

Pneumococci that contain the polysaccharide Ags of two capsular serotypes have been produced by transformation. The identification of such binary encapsulated strains is illustrated in Figure 27-4.

Pathogenetic Role. The importance of the pneumococcal capsular Ags for pathogenicity can be demonstrated in a number of ways. Thus only encapsulated strains are pathogenic for man and most laboratory animals, and active† or passive immunization against a specific polysaccharide produces a high level of resistance to infection with pneumococci of the homologous type. **Intermediate variants** of type 3, which produce small capsules, are less virulent than fully encapsulated strains but more virulent than rough variants. However, pneumococci of different types with capsules of the same size may vary widely in virulence (e.g., types 3 and 37).

SOMATIC ANTIGENS

In 1930 Tillett, Goebel, and Avery isolated from pneumococcal cells a carbohydrate that is specific for species

† The purified polysaccharides, unlike whole pneumococcal cells, are weak immunogens for rabbits. In mice, however, those of many types are highly immunogenic; but when given in doses of more than 1μg they produce immunologic paralysis.

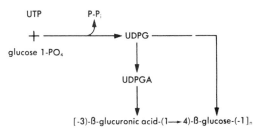

$$UTP + glucose\ 1\text{-}PO_4 \xrightarrow{\quad P\text{-}P_i \quad} UDPG$$

$$UDPG \longrightarrow UDPGA$$

$$UDPGA \longrightarrow [\text{-3})\text{-}\beta\text{-glucuronic acid-}(1 \longrightarrow 4)\text{-}\beta\text{-glucose-}(\text{-}1]_n$$

FIG. 27-3. Metabolic pathway of type 3 capsular polysaccharide synthesis. **UTP** = Uridine triphosphate; **UDPG** = uridine diphosphoglucose; **UDPGA** = uridine diphosphoglucuronic acid; **P-P$_i$** = inorganic pyrophosphate. (Austrian R et al: J Exp Med 110:585, 1959)

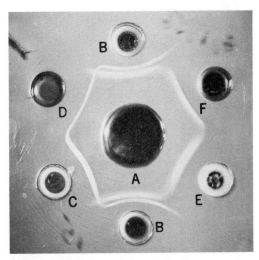

FIG. 27-4. Gel diffusion reactions (Ouchterlony technic) of antisera to type 1 and type 3 pneumococci (in center well, **A**) with type 1,3 pneumococci (**B**), type 1 pneumococci (**C**), type 1 capsular polysaccharide (**D**), type 3 cells (**E**), and type 3 polysaccharide (**F**). (Austrian R, Bernheimer HP: J Exp Med 110:571, 1959)

rather than for type. This cell wall Ag, referred to as pneumococcal **C substance**, appears to be analogous to (though antigenically different from) the group-specific C Ags of hemolytic streptococci (see Cellular Antigens, Ch. 28).

As purified from autolysates or deoxycholate lysates of pneumococci, C substance contains galactosamine-6-phosphate as a major constituent in addition to phosphorylcholine, a diaminotrideoxyhexose, ribitol, glucose, and elements of the cell wall peptidoglycan. Similar material obtained by extraction with trichloroacetic acid lacks the elements of the peptidoglycan, suggesting a teichoic acid.

The C substance, which also forms a portion of the **Forssman antigen** (see Blood Group Substances, Ch. 23)

of the pneumococcus, has the interesting property of being precipitated in the presence of Ca^{2+} by a β-globulin of the serum. This **C-reactive protein** (CRP) is not an Ab, (although it is composed of five to six noncovalently bound subunits having an amino acid composition similar to that of immunoglobulins): it combines also with a variety of phosphorylated compounds, including some bacterial capsular polysaccharides and sphingomyelin. Such complexes can activate the complement pathway. CRP is detectable in the blood only during the active phase of certain acute illnesses, not necessarily pneumococcal in origin. A precipitin test designed to determine its concentration in serum is used as a measure of "activity" in inflammatory diseases such as rheumatic fever.

The **R antigen,** so named because it was extracted originally from unencapsulated (rough, R) pneumococci, is believed to be a protein on or near the surface of the cell.

Sera prepared against unencapsulated pneumococci derived from a given encapsulated (smooth, S) type often agglutinate to a higher titer R pneumococci derived from homologous S types than R strains derived from heterologous S types. This difference is due to **type-specific somatic M antigens,** which are proteins similar to the type-specific M Ags of group A hemolytic streptococci (see Cellular Antigens, Ch. 28). Pneumococcal M Ags, however, do not exert a significant antiphagocytic effect; Abs against them, therefore, are not protective, as are the Abs to streptococcal M Ags.

Immunization of rabbits with heat-killed R pneumococci causes a slight increase in resistance to infections with pneumococci of any type. This broad immunity is believed to be due to the formation of Abs to the species-specific C polysaccharide and R Ags.[*] The degree of immunity thus induced is negligible, however, compared with that mediated by type-specific Ab.

GENOTYPIC VARIATIONS

Cultures of encapsulated pneumococci generate unencapsulated mutants at a low but definite rate. When such cultures are grown in the presence of homologous type-specific antiserum the encapsulated S cells become agglutinated and the R mutants are selected.[†] After a few serial passages in such a medium most of the cells recovered will be devoid of capsules and will have lost concomitantly both their type specificity and their virulence. R mutants are also selected, but less rapidly, when S cells are subcultured repeatedly without antiserum.

Conversely, in cultures of R pneumococci, occasional

---

[*] Although C polysaccharide is ordinarily confined to the cell wall a mutant has been isolated that makes a capsule of "C-like" polysaccharide.

[†] The crowding together of the agglutinated S cells probably places them at a metabolic disadvantage.

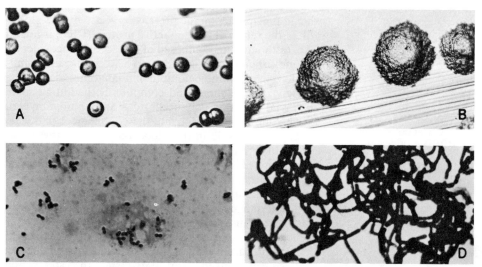

FIG. 27-5. A. Colonies of a nonfilamentous (fil⁻), unencapsulated (S⁻) variant (usual rough form) of pneumococcus type 2 on blood agar after 24 h at 36˚ C. (×18) **B.** Colonies of a filamentous (fil⁺), unencapsulated (S⁻) variant (very rough form) grown under same conditions. (×18) **C.** Cells from colony of nonfilamentous variant. (Gram stain; ×900) **D.** Cells of filamentous variant. (Gram stain; ×1100) (Austrian R: J Exp Med 98:21, 1953)

cells revert (back-mutate) to S. If anti-R serum is present in the medium the S cells are favored, and on repeated subculture, they will eventually replace the R cells. When R pneumococci are injected into a mouse the S mutants are given an even greater selective advantage, since the animal's phagocytes rapidly destroy the unencapsulated cells while the encapsulated ones continue to multiply and eventually kill the host. This principle underlies the conventional method of maintaining the maximum virulence of pneumococcal cultures by passing them through mice at frequent intervals.

Two kinds of surface colonies may be formed by R pneumococci: the usual small, moderately granular, rough colony and a large colony containing long chains of cocci which form intricate filaments responsible for the excessive roughness of its surface (Fig. 27-5). Austrian has also described filamentous smooth forms and has demonstrated that the chaining trait is a heritable characteristic.

Transformations of pneumococcal types by exogenous DNA have been discussed (see Ch. 9). How frequently such transformations occur in nature is not known. However, growing populations of pneumococci release transforming DNA into the culture medium, and this release is maximal during the phase of growth when the culture is most responsive to added DNA (Fig. 27-6). Moreover, *transformations have been shown to occur in experimental pneumococcal lesions in mice.*

Genetic markers that have been successfully transformed in vitro include those relating to 1) the type specificity of capsular Ags; 2)

FIG. 27-6. Development of transformability and release of transforming material by pneumococci growing in broth culture. The strain in the culture was sulfonamide-resistant but streptomycin-sensitive. Filtrates prepared from the culture at various intervals were used as donor material to transform a sulfonamide-sensitive strain **(transforming activity). Transformability** of the cultured strain was tested at various stages of the growth phase by adding purified DNA from a streptomycin-resistant culture. The cells from the original culture were exposed to the DNA for 30 min and the reaction was terminated by DNase. (Ottolenghi E, Hotchkiss RD: J Exp Med 116:491, 1962)

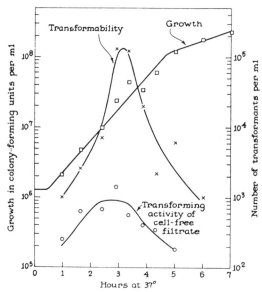

the amount of capsular polysaccharide produced (intermediate strains); 3) resistance to drugs, such as penicillin, streptomycin, sulfonamides, and ethylhydrocupreine (optochin); 4) the type specificity of somatic M proteins; 5) the capacity to produce certain inducible enzymes, e.g., mannitol–phosphate dehydrogenase; and 6) chain formation affecting the colonial morphology of R and S mutants.

Pneumococci have also been transformed with DNA prepared from streptococci, with which pneumococci are now classified.

PATHOGENICITY

Host Range. In addition to being pathogenic for man, most encapsulated strains of pneumococci will produce experimental disease in mice, rats, rabbits, monkeys, and dogs. Cats and birds are relatively resistant. The reasons for these wide variations in susceptibility are not clear. It is obvious, however, that the results of virulence studies performed in one animal species can be applied to another only with caution. In fact, strains within a given animal species may vary widely in susceptibility. Severe epizootics of pneumococcal infection occur occasionally in monkeys, rats, and guinea pigs.

Incidence and Significance of Types. The antigenic types of pneumococci that most commonly caused pneumonia in the 1930s are listed in Table 27-1. More than half the cases in adults were due to types 1, 2, and 3, while in children type 14 was especially common. Perhaps because of the crossreaction of its capsular polysaccharide with the ABO blood group substances, hemolytic anemia was sometimes observed in patients treated with type 14 antipneumococcal horse serum.

In a more recent study of bacteremic pneumococcal pneumonia in selected hospitals throughout the United States the most common types in adults were 8, 4, 1, 14, 3, 7, and 12, in that order. Type 2, formerly a major cause of infection, has rarely been encountered in this country and in northern Europe in recent years, though it remains important in South Africa. In children types 19, 6, 23, 14, 18, and 3 are mainly responsible for otitis media and for bacteremic pneumonia.

Type 3 pneumococcus causes the highest fatality rate in man. Other highly virulent types are 1, 2, 4, 5, 7, 8, and 12.

Toxin Production. Pneumococci produce a hemolytic toxin, called **pneumolysin,** which is related immunologically to the oxygen-labile O hemolysins of hemolytic streptococci, *Clostridium tetani,* and *Clostridium welchii* (*C. perfringens*). They also produce a toxic neuraminidase and release during autolysis a **purpura-producing principle** which causes dermal and internal hemorrhages when injected into rabbits. Although there is no conclusive evidence that any of these products play a role in the pathogenesis of pneumococcal infections, they may exert short-range toxic effects that have remained undetected.

The type-specific capsular polysaccharides released into the fluid phase by multiplying pneumococci are relatively nontoxic, but they are significant in pathogenesis because they neutralize Ab before the latter has a chance to become bound to the pneumococci invading the tissues.

Since there is no pneumococcal toxin known to be related to pathogenicity, and the virulence of the organism appears to be due primarily to its invasiveness, the cause of death in pneumococcal infection remains a mystery. It has been conjectured that in the terminal stages a lethal toxin analogous to that described in anthrax (Ch. 35) might be generated, but attempts to demonstrate such a toxin have been unsuccessful.

TABLE 27-1. Comparative Distributions of Pneumococcal Types in Adults and Children (Under Age 12) with Lobar Pneumonia

Adults		Children	
Type	% of Cases	Type	% of Cases
1	28.6	14	12.0
3	13.5	1	11.2
2	11.4	6	7.7
5	8.0	19	3.9
8	7.7	5	3.6
7	6.5	4	3.1
4	3.5	3	2.7
6	1.8	7	2.1
All other	19.0	All other	53.7
Total	100.0		100.0

(Data from Heffron, R: Pneumonia, with Special Reference to Pneumococcus Lobar Pneumonia, pp. 53, 76. New York, Commonwealth Fund, 1939)

PATHOGENESIS OF PNEUMOCOCCAL PNEUMONIA

Defense Barriers of the Respiratory Tract. Between 40% and 70% of normal human adults carry one or more serologic types of pneumococci in their throats; yet epidemics of pneumococcal pneumonia are rare and morbidity is low. Bacterial antagonism (see Microbial Antagonism, Ch. 24), involving primarily α-hemolytic streptococci, tends to limit the growth of pneumococci in the pharynx. The extraordinarily efficient defense barriers of the lower respiratory tract include 1) the epiglottal reflex, which prevents gross aspiration of infected secretions; 2) the sticky mucus to which airborne organisms adhere on the epithelial lining of the bronchial tree; 3) the cilia of the respiratory epithelium, which keep the infected mucus moving upward into the pharynx (see Mechanical Factors, Ch. 24); 4) the cough reflex, which aids the cilia in propelling accumulated mucus from the lower tract; 5) the lymphatics draining the terminal bronchi and bronchioles; and 6) the mononuclear "dust cells" (macrophages), which patrol the normal alveoli.

Predisposing Factors. Pneumococcal pneumonia develops most often during the course of viral infections of the upper respiratory tract, when the resulting flood of mucous secretion in the nose and pharynx enhances the likelihood of aspiration. Once past the epiglottal barrier, the thin mucus, laden with bacteria (including encapsulated pneumococci), is carried by gravity to the farthest reaches of the bronchial tree, where it establishes the initial focus of pulmonary infection. Aspiration is promoted by factors that slow the epiglottal reflex, including chilling of the body, anesthesia, morphine, and alcoholic intoxication.

Aerosol experiments with mice have also revealed that an edematous lung is far more susceptible to pneumococcal infection than the normally dry lung. Pulmonary edema fluid provides a suitable culture medium for aspirated pneumococci and interferes with the phagocytic activities of the tissue macrophages, which constitute the first line of cellular defense in the alveoli. It is not surprising, therefore, that factors which produce either local or generalized pulmonary edema should also predispose to pneumococcal pneumonia. These include inhalation of irritating anesthetics, trauma to the thorax, cardiac failure, influenza virus infections involving the lungs, and pulmonary stasis resulting from prolonged bed rest.

Evolution of the Lesion. Once infection has become established in a bronchial segment, the pneumonic lesion spreads centrifugally, as depicted in Figure 27-7. In the **edema zone,** at the outer margin of the spreading lesion, the alveoli are filled with acellular serous fluid, which serves as a suitable culture medium for the organisms.

Following the outpouring of edema fluid, polymorphonuclear leukocytes (PMNs) and a few RBCs begin to accumulate in the infected alveoli and eventually fill them with a densely packed leukocytic exudate (**consolidation**). As they accumulate in sufficient numbers, the leukocytes phagocytize and destroy the infecting pneumococci. When the bacteria have been disposed of, macrophages replace the granulocytes in the exudate, and **resolution** of the lesion ensues. Thus all stages of the inflammatory process are demonstrable simultaneously in the spreading pneumonic lesion. The earliest stage, characterized by increased capillary permeability and the accumulation of serous fluid, is represented by the peripheral edema zone. The late **macrophage reaction,** characteristic of subsiding inflammation, is prominent in the central "burned-out" portion of the lesion.

When alveoli underlying the pleura become involved, **pleurisy** develops and the pleural cavity frequently becomes infected. If unchecked, pleural infections may develop into extensive intrapleural abscesses (**empyema**). The adjacent pericardium may also be affected (**pericarditis**).

Pneumococcal pneumonia is often multilobar. Spread of the infection from one lobe to another results from the flow of infected edema fluid, propelled by the combined effects of coughing and the force of gravity.

Although the majority of pneumococci in the pulmonary lesion are destroyed by phagocytosis, some may be carried by the lymphatics to the regional lymph nodes at the hilus of the lung and thence, via the thoracic duct, to the blood stream. Once bacteremia has developed organisms may settle on the heart valves or in the meninges or joints, and occasionally can be cultured from the urine. In less serious cases the primary pneumonic lesion resolves spontaneously without complications.

The phagocytic mechanisms that operate at the various lines of cellular defense have been described in Chapter 24. As pointed out, these mechanisms are relatively inefficient in fluid-filled cavities (subarachnoid space, pleura, pericardium, joints) and are incapable of destroying all pneumococci after an abscess has formed. Furthermore, many strains of type 3 pneumococci, which usually produce large capsular envelopes, are highly resistant to surface phagocytosis in the absence of homologous type-specific Ab, and so they tend to reach considerably higher population densities than other types of pneumococci, resulting in irreversible tissue damage and abscess formation. Type 3 is virtually the only type that causes lung abscesses in man.

Factors affecting the PMNs may modify the course of pneumococcal pneumonia. The disease is more severe in those with granulocytopenia of any cause; glucocorticosteroid hormones, alcohol, and general anesthetics interfere with the migration of leukocytes from capillaries; and high levels of glucose, as in diabetes, impede their phagocytic activity.

OTHER PNEUMOCOCCAL DISEASES

Primary pneumococcal diseases of the upper respiratory tract include sinusitis and otitis media. The latter occurs most commonly in children and may spread to involve

FIG. 27-7. Schematic diagram of spreading pneumonic lesion, showing a characteristic microscopic field in each of its four histologically distinguishable zones. (Wood WB Jr: Harvey Lect 48:72, 1951–1952)

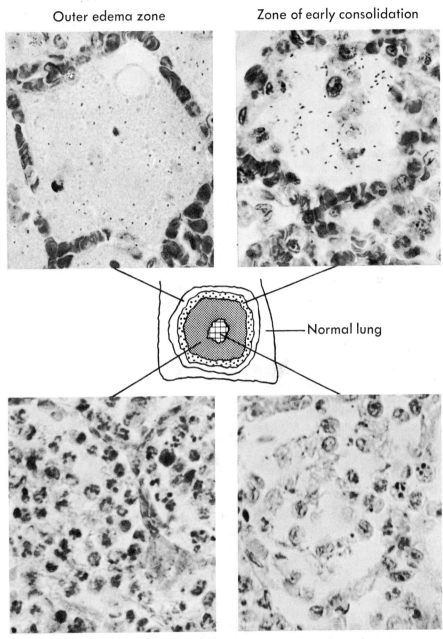

Outer edema zone

Zone of early consolidation

Normal lung

Zone of advanced consolidation

Zone of resolution

the mastoid. Progressive infections of the mastoid or respiratory sinuses sometimes extend directly to the subarachnoid space to cause pneumococcal meningitis. Secondary pneumococcal peritonitis, resulting from transient bacteremia following a primary respiratory infection, occurs most commonly in children with ascites due to nephrosis and in adults with cirrhosis or carcinoma of the liver.

IMMUNITY

An increase in type-specific anticapsular Ab is not usually demonstrable in the serum of patients with pneumococcal pneumonia before the fifth or sixth day of the disease, at which time a spontaneous crisis* may occur (unless, of course, the patient has already responded to specific treatment). Recovery occurs in approximately 70% of untreated cases. The spontaneous recovery may take place even in the "preantibody" stage of illness, as a result of the primary cellular defenses that operate in the lung (see Surface Phagocytosis, Ch. 24) in conjunction with the heat-labile opsonins activated via the alternate complement pathway. Human Abs to pneumococcal capsular polysaccharides do not activate the classic complement pathway. Following recovery, anticapsular Ab usually remains detectable in the patient's serum for months.

Patients recovering from pneumococcal pneumonia may continue to carry the infecting pneumococcus in the upper respiratory tract for many days or weeks despite the presence of circulating homologous Ab. A more or less permanent carrier state may result from chronic pneumococcal sinusitis. Recurrent attacks of pneumococcal disease are almost invariably due to different serologic types of the organism, except in cases involving a persistent focus of infection (e.g., bronchiectasis or chronic sinusitis) or in patients with immunologic defects, notably dysgammaglobulinemia.

The presence of type-specific anticapsular Ab may be detected by precipitin tests with the homologous polysaccharide or by agglutination, phagocytic, or capsular precipitin (quellung) tests performed with homologous pneumococcal cells. Antibody to those types virulent for the mouse may be assayed by protection tests in this species. Circulating Ab may also be detected by injecting homologous capsular polysaccharide into the patient's skin (**Francis' skin test**). The interaction of the Ab with the injected Ag causes an immediate wheal and erythema at the site of injection.

Pneumococcal capsular polysaccharides may reach amounts in the lung exceeding 1 g during the course of

---

* The term **crisis** refers to the dramatic defervescence and subsidence of symptoms which terminate some cases of uncomplicated pneumococcal pneumonia.

pneumonia and may remain in the tissues for relatively long periods. In experimental pneumonia they have been demonstrated in alveolar macrophages many weeks after recovery from the acute illness. The polysaccharide retained in the lung is released gradually and if the amount is large some may be excreted in the urine and detected by precipitin tests weeks or even months after recovery. Polysaccharide may be detected also, when present in the serum or infected body fluids, by counterimmunoelectrophoresis.

LABORATORY DIAGNOSIS

A tentative diagnosis of pneumococcal pneumonia can be made most rapidly by examining the patient's sputum. Smears made from a fresh sample, raised directly from the bronchial tree in the presence of the physician, should be stained by the Gram method to distinguish *S. pneumoniae* from *Klebsiella pneumoniae* (Ch. 31) and staphylococci (Ch. 29), which also produce acute bacterial pneumonia. If typical lancet-shaped diplococci in significant numbers are seen in the smear, together with PMNs and alveolar macrophages, a presumptive diagnosis of pneumococcal pneumonia may be made and treatment begun. At the same time, the sputum should be cultured on blood agar for final identification (see below). When sputum cannot be obtained (e.g., from a small child or comatose patient) a pharyngeal culture is utilized in the same fashion, or material may be obtained by transtracheal or lung puncture.

With typing sera available, pneumococci in the sputum may be typed immediately by the quellung reaction. (Although this procedure is now performed infequently in diagnostic laboratories, it was formerly done routinely to permit prompt treatment of the patient with the correct type of antiserum.) If typical gram-positive diplococci are not seen in the sputum of a patient strongly suspected of having acute bacterial pneumonia, a sample of the sputum should be emulsified with a small amount of broth in a 1- or 2-ml syringe and injected intraperitoneally into a mouse. When virulent pneumococci are present, the mouse will usually die within 4 days and the offending pneumococcus can then be recovered from cardiac blood in pure culture and typed directly from the peritoneal washings. The quellung reaction is especially useful for the prompt identification of pneumococci in body fluids such as spinal fluid.

Because many healthy humans carry pneumococci in their throats, demonstration of the organism in sputum or a throat culture does not provide conclusive evidence of pneumococcal infection. Culture of pneumococci from the patient's blood, on the other hand, is diagnostic. For this reason, blood should be obtained by venepuncture prior to the administration of antimicrobial drugs for immediate culture in both beef infusion broth and thioglycollate medium (the latter promotes the growth of small numbers of pneumococci). Measured samples of blood should be used also to make one or more pour plates for bacterial counts. Serous fluids obtained from pleural, pericardial, peritoneal, or synovial cavities should be cultured in beef infusion broth, in thioglycollate medium, and on blood agar in an atmosphere containing 5% CO_2 (candle jar). At the same time, direct smears of the fluid should be examined by the Gram method. If organisms are present they can be identified immediately as pneumococci by the quellung reaction. Cultures of spinal fluid are performed in the same manner. In pneumococcal meningitis the spinal fluid frequently contains sufficient organisms to permit presumptive diagnosis by microscopic examination. Capsular polysaccharide can sometimes be identified in the spinal fluid of a treated patient by counterimmunoelectrophoresis.

Because of their morphologic similarity, pneumococci are easily confused with other streptococci of the viridans group (see α-Hemolytic Streptococci, Ch. 28), but they usually differ in being bile-soluble, virulent for mice, and sensitive to optochin (ethylhydrocupreine), an antibacterial compound relatively specific for pneumococci.

The bile-solubility test is best done with deoxycholate and should be performed with saline suspensions of living bacterial cells, because extraneous proteins in liquid media inhibit the reaction, presumably by binding deoxycholate. A paper disc impregnated with optochin may be used to test for inhibition of bacterial growth on the surface of a blood agar plate.

TREATMENT

Type-specific antiserum is no longer available and sulfonamides are not recommended. The drug generally used is penicillin, although strains manifesting increased resistance to this drug and to other antibiotics are being isolated with increasing frequency. Penicillin for therapy of uncomplicated pneumococcal pneumonia caused by a sensitive strain is usually effective unless started too late. Its limitations, particularly in the management of established extrapulmonary abscesses, such as empyema, have been discussed in Chapter 25. Erythromycin or clinda-

mycin may be used in patients with known hypersensitivity to penicillin, but the pneumococcal isolate should be tested for sensitivity to the drug employed. Broad-spectrum antibiotics (e.g., the tetracyclines) may also be effective, although tetracycline-resistant strains are not rare. Chloramphenicol should be used only to treat pneumococcal meningitis in patients allergic to penicillin. Pneumococci resistant to all potentially useful antimicrobial drugs except vancomycin have been isolated from patients in Africa.

PREVENTION

Although pneumococcal diseases usually respond to early antimicrobial therapy, pneumococcal pneumonia afflicts approximately 1 in 500 persons per year, and renewed interest in prophylaxis has arisen following recognition of the significant mortality from pneumococcal infection among those over 50 years of age and in individuals with underlying systemic illness. Occasionally, in closed communities such as military or industrial installations and custodial institutions, the pneumococcal carrier rate will become unusually high and an epidemic will result. Vaccines containing 12–14 pneumococcal capsular polysaccharides have been shown to be safe and effective in preventing pneumococcal pneumonia and bacteremia, and their administration to individuals at high risk of fatal infection and in epidemic situations is potentially useful. Indiscriminate treatment of acute respiratory infections with antimicrobial drugs to prevent pneumonia should be discouraged, not only because of its relative inutility but also because of the hazard of drug reactions and of promoting the selection of drug-resistant bacteria being carried by the host.

Ideally, every patient with pneumococcal pneumonia should be isolated. Although isolation rules are often disregarded because of the relatively low crossinfection rates, such patients should not be placed in crowded hospital wards containing persons with congestive heart failure and other debilitating diseases.

SELECTED READING

BOOKS AND REVIEW ARTICLES

HEFFRON R: Pneumonia, with Special Reference to Pneumococcus Lobar Pneumonia. New York, Commonwealth Fund, 1939*

LARM O, LINDBERG B: The pneumococcal polysaccharides: a reexamination. Adv Carbohydr Chem Biochem 33:295, 1976

WHITE B: The Biology of Pneumococcus: The Bacteriological, Bio-

chemical and Immunological Characters and Activities of *Diplococcus pneumoniae*. New York, Commonwealth Fund, 1938*

WOOD WB JR: Pneumococcal pneumonia. In Beeson PB, McDermott W (eds): Cecil-Loeb Textbook of Medicine. Philadelphia, Saunders, 1971

* Although unrevised, these two monographs remain classics in the field. They have been reprinted recently by the Harvard University Press.

SPECIFIC ARTICLES

AUSTRIAN R: Morphologic variation in pneumococcus. J Exp Med 98:21, 1953

AUSTRIAN R: *Streptococcus pneumoniae* (Pneumococcus). In Lennette EH, Spaulding EH, Truant JP (eds): Manual of Clinical Microbiology, 2nd ed. Washington DC, American Society for Microbiology, 1974

AUSTRIAN R, GOLD J: Pneumococcal bacteremia with especial reference to bacteremic pneumococcal pneumonia. Ann Intern Med 60:759, 1964

BORNSTEIN DL, SCHIFFMAN G, BERNHEIMER HP, AUSTRIAN R: Capsulation of pneumococcus with soluble C-like (C_s) polysaccharide. J Exp Med 128:1385, 1968

KNECHT JF, SCHIFFMAN G, AUSTRIAN R: Some biological properties of pneumococcus type 37 and the chemistry of its capsular polysaccharide. J Exp Med 132:475, 1970

MOSSER JF, TOMASZ A: Choline-containing teichoic acid as a structural component of pneumococcal cell wall and its role in sensitivity to lysis by an autolytic enzyme. J Biol Chem 245:287, 1970

OTTOLENGHI E, HOTCHKISS RD: Release of genetic transforming agent from pneumococcal cultures during growth and disintegration. J Exp Med 116:491, 1962

OTTOLENGHI E, MACLEOD CM: Genetic transformation among living pneumococci in the mouse. Proc Natl Acad Sci USA 50:417, 1963

ROBINS-BROWN RM, GASPAR MN, WARD JI, WACHSMUTH IK, KOORNHOF HJ, JACOBS MR, THORNSBERRY C: Resistance mechanisms of multiply resistant pneumococci; antibiotic degradation studies. Antimicrob Agents and Chemother 15:470, 1979

TOMASZ A, JAMIESON JD, OTTOLENGHI E: The fine structure of *Diplococcus pneumoniae*. J Cell Biol 22:453, 1964

WINKELSTEIN JA, SMITH MR, SHIN HS: The role of C3 as an opsonin in the early stages of infection. Proc Soc Exp Biol Med 149:397, 1975

chapter 28

STREPTOCOCCI

MACLYN McCARTY

HISTORY AND CLASSIFICATION

Globular microorganisms growing in chains were first described by Billroth in 1874 in purulent exudates from erysipelas lesions and infected wounds. Similar organisms, eventually named streptococci (Gr. *streptos,* winding, twisted), were isolated from the blood in puerperal fever and from the throat in scarlet fever. It is now known that *a single streptococcal species may be responsible for a variety of diseases.* However, a number of different kinds of streptococci may be cultured from human patients and animals, and the first classifications were based on their capacities to hemolyze RBCs. In 1919 Brown introduced the terms **alpha, beta,** and **gamma** to describe the three types of **hemolytic** reactions observed on blood agar plates.

Primarily through the efforts of Lancefield in the early 1930s, the β-hemolytic streptococci were further differentiated into a number of **immunologic groups** designated by the letters A through O. Most strains causing human infections were found to belong to **group A.** That group in turn contains a variety of **antigenic types,** later demonstrated by precipitin tests (Lancefield) and by agglutination reactions (Griffith). The group-specific Ags were identified as carbohydrates and the type-specific Ags as proteins.

More than 55 types of group A β-hemolytic streptococci have been identified.

Hemolytic Classes. β-**Hemolytic streptococci** produce a wide clear zone of complete hemolysis (Fig. 28-1A) in which no red cells are visible on microscopic examination. Two types of β-hemolysin are released. **Streptolysin O** is destroyed by atmospheric oxygen and is therefore demonstrable only in deep colonies. **Streptolysin S** is oxygen-stable and is responsible for surface colony hemolysis. Since most strains produce both S and O hemolysins, they can usually be recognized as β-hemolytic by their surface colonies. To be certain of the hemolytic characteristics of a given strain, however, it may be nec-essary to examine colonies located beneath the surface of a pour plate. The most common species causing disease in man is *Streptococcus pyogenes,* which constitutes antigenic group A.

α-**Hemolytic** streptococcus colonies are surrounded by a narrower zone of hemolysis, with unhemolyzed RBCs persistent in an inner zone and complete hemolysis in an outer zone (Fig. 28-1B). (The mechanism responsible for the sparing of some of the red cells is not known.) Green discoloration of the colonies (due to formation of an unidentified reductant of hemoglobin) frequently occurs, depending upon the type of blood in the medium and the duration of incubation. This feature has given rise to the synonym term **viridans group.** *Streptococcus salivarius* is the most commonly encountered species of the α-hemolytic category.

γ-**Streptococci** produce no hemolysis, either on the surface or within the agar. *Streptococcus faecalis* is a typical nonhemolytic species.

This classification of streptococci based on hemolysis is far from satisfactory, for the following reasons:

1) Many species classified in the β-hemolytic category are, in fact, nonhemolytic, e.g., certain members of antigenic groups B, C, D, H, K, and O, as well as all of group N (Table 28-1).

2) Most streptococci found in the gastrointestinal tract (enterococci) are nonhemolytic, but some strains produce β-hemolysis (group D).

3) Certain strains of *S. faecalis,* classified as nonhemolytic (γ) after 24 h incubation, may show α-hemolysis if incubated for an additional 24–48 h.

4) Anaerobic streptococci, though generally nonhemolytic, are conventionally not classified with the aerobic γ-streptococci, but are considered in a separate category.

Despite its deficiencies, this system of classification has become firmly established.

Since streptococcal disease in man is due primarily to group A organisms, which produce β-hemolysis, most of the present chapter will be devoted to β-hemolytic streptococci.

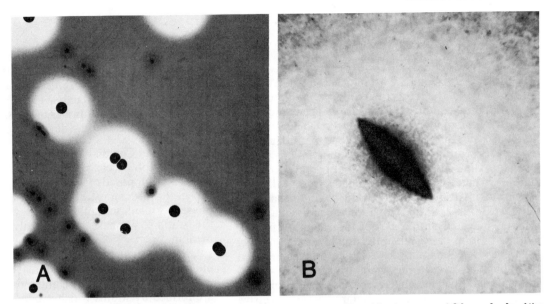

FIG. 28-1. A. Surface colonies of *S. pyogenes* and viridans streptococci on blood agar plate. The wide clear zone of *β*-**hemolysis** of the former is much more prominent than the latter's narrower dark zone of incomplete *α*-**hemolysis.** A few small viridans colonies can be seen to have grown within some of the zones of *β*-**hemolysis.** (×4) **B.** Higher-power (×50) view of a deep *α*-hemolytic colony on blood agar (note oblong shape) showing border of incomplete hemolysis about the colony. No such band of intact red cells is present in the much wider zones of *β*-hemolysis that surround comparable colonies of *β*-hemolytic streptococci. (Preparation by Dr. E. D. Updyke. Photograph courtesy of Center for Disease Control, Atlanta, Ga.)

TABLE 28-1. Group Classification of *β*-Hemolytic and Other Immunologically Related Streptococci

Group	Hemolysis	Usual habitat	Pathogenicity
A	+	Man	Many human diseases
B	±	Man	Neonatal meningitis
C	+	Cattle	Mastitis
	±	Many animals	Many animal diseases
		Man	Mild respiratory infections
D	±	Dairy products, intestinal tract of man and animals (enterococci)	Urinary tract and wound infections; endocarditis
E	+	Milk	Unknown
		Swine	Pharyngeal abscesses
F	+	Man	?; Found in respiratory tract
G	+	Man	Mild respiratory infections
		Dogs	Rare genital tract infections
H	±	Man	?; Found in respiratory tract
K	±	Man	?; Found in respiratory tract
L	+	Dogs	Genital tract infections
M	+	Dogs	Genital tract infections
N	–	Dairy products	None
O	±	Man	Carried in upper respiratory tract; endocarditis

+ = All strains hemolytic; ± = some strains hemolytic, others nonhemolytic; − = all strains nonhemolytic.

(Modified from McCarty M: Hemolytic streptococci. In Dubos RJ, Hirsch JG, eds: Bacterial and Mycotic Infections of Man, Philadelphia Lippincott, 1965)

β-HEMOLYTIC STREPTOCOCCI

MORPHOLOGY

Cell Division. β-Hemolytic streptococci (often simply called hemolytic streptococci), like all other streptococci, are gram-positive and characteristically grow in chains; in vivo they commonly occur as diplococci. The length of the chains tends to be inversely related to the adequacy of the culture medium; in actively spreading lesions within the tissues diplococcal and individual coccal forms are common, whereas in purulent exudates from walled-off lesions and in artificial culture media chain formation is the rule (Fig. 28-2). Prior to division the individual cocci become elongated on the axis of the chain, eventually dividing to form pairs. When the dividing pairs of cocci do not separate, chaining results. The bridges between the individual cocci in the chain are composed of cell wall material which has not cleaved. Uncleaved cell walls are particularly striking in mutants with excessive chaining, which produce opaque colonies on clear agar (Fig. 28-3).

Factors tending to promote preservation of the intercoccal junctions and thus to exaggerate chaining include not only conditions which impair growth (unfavorable medium, cold, antimicrobial agents, etc.), but also the presence of Abs that react with cell wall Ags. For example, the presence of Ab that reacts with the M protein surface Ag (see below) causes the organisms to grow in long chains in broth culture. Evidently such Abs may "cover up" substrates that are normally attacked enzymatically in the cleavage process. Formation of long chains may be used to test for the presence of anti-M Ab (see Laboratory Diagnosis, below).

Because streptococcal chains are difficult to disrupt

without killing the organisms, individual cocci cannot be readily counted by conventional plating methods. It is customary, therefore, to record streptococcal colony counts in **streptococcal units.** These values provide only a rough index of the number of cells, since they are obviously influenced by the degree of chaining.

Capsules and Colonial Morphology. Many strains of hemolytic streptococci produce capsules, which in groups A and C are composed of hyaluronic acid (Fig. 28-4). These capsules are demonstrable throughout the logarithmic phase of growth in liquid cultures, but after the onset of the stationary phase they dissolve rapidly into the medium.

On blood agar plates group A hemolytic streptococci may form any one of three colony types, designated **mucoid, matt** (Ger. *matt,* dull), and **glossy.** Mucoid colonies are formed by strains which produce large capsules: the abundance of hyaluronic acid gel gives the colony a glistening, watery appearance. The flatter, rougher, matt colonies were originally thought to reflect the production of M protein (the designation M was based on this apparent relation), but they are simply dried out mucoid colonies: as Figure 28-5 shows, as the gel becomes dehydrated the surface of the colony shrinks and becomes roughened.

Glossy colonies are smaller; they are formed by cells that do not generate hyaluronate or do not retain it as a capsular gel. Groups F and G include **minute streptococci,** which produce not only smaller cells than other streptococci but also smaller colonies.

L Forms and Protoplasts. L forms (Ch. 43) of group A streptococci, which lack many cell wall constituents, may be isolated from

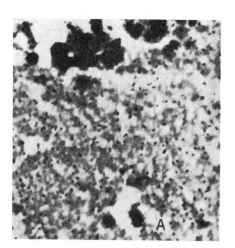

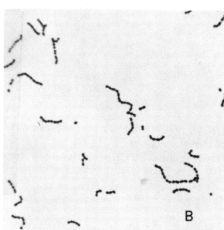

FIG. 28-2. A. Group A β-hemolytic streptococci in edema zone of experimental pneumonic lesion (rat). (See Evolution of Lesion, Ch. 27). Note diplococcal morphology. (×600). **B.** Chain formation characteristic of usual growth of streptococci in artificial media. Smear made from 24-h culture in serum broth (×1000) (**A,** Glaser RJ, Wood WB Jr: Arch Pathol 52:244, 1951. Copyright 1951, American Medical Association)

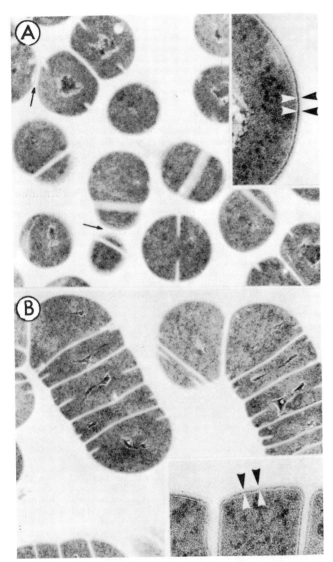

FIG. 28-3. Thin section electron micrographs of parent M⁻ strain of *S. pyogenes* **(A)** and a mutant **(B)** that produces opaque colonies. Whereas intercellular septa are completely cleaved in the parent strain to form individual cocci and diplococci, most of the mutant's septa remain uncleaved so that the organism grows in long chains. The fine structure of the cell walls in the two strains, however, appears to be identical (see **insets,** between arrows). (Swanson J, McCarty M: J Bacteriol 100:505, 1969)

anaerobic cultures on hypertonic media containing penicillin. Complete removal of the wall to form protoplasts may be achieved by treating the cells with a phage-associated mucopeptidase (see Group-specific C Antigens, below). Unlike the protoplasts isolated from most other bacterial species these will multiply and produce typical L form colonies in hypertonic agar medium. During growth they release cell wall M Ag into the medium, as well as hemolysin and deoxyribonuclease.

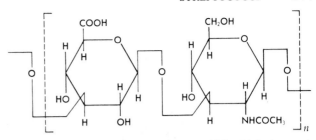

FIG. 28-4. Repeating unit of hyaluronic acid, in which glucuronic acid **(left)** is linked to *N*-acetyl glucosamine **(right)** by a β,1,3-glycosidic bond.

METABOLISM

Hemolytic streptococci are routinely grown in beef infusion media containing blood or serum. For the isolation of specific Ags and extracellular enzymes a medium containing only the dialyzable components of the complex meat infusion peptone medium may be used. The growth requirements are very similar to those of pneumococci (see Metabolism, Ch. 27). A group of peptides supplied by the complex medium have been shown to be essential for optimal growth, but there is no specific peptide requirement.

All streptococci are lactic acid bacteria, which derive their energy primarily from the fermentation of sugars, regardless of whether they are growing aerobically or anaerobically. Accumulation of lactic acid in media of high glucose content limits growth unless the pH is corrected.

CELLULAR ANTIGENS

Capsular Hyaluronic Acid. One potential surface Ag of group A hemolytic streptococcal cells, the hyaluronate of the capsule (see Figs. 28-4 and 28-8), is **not immunogenic,** presumably because it is chemically indistinguishable from the hyaluronate in the ground substance of connective tissue. The hyaluronate capsule is retained only by actively growing streptococci or by cells that have been rapidly chilled while in the logarithmic phase. The process responsible for release of the capsule in the stationary phase is not known, but there is no evidence that production of hyaluronidase by the cell is involved.

Group-Specific C Antigens. As already indicated, the separation of β-hemolytic streptococci into immunologically specific groups (A to O) depends upon the presence of group-specific carbohydrate Ags in their cell walls. These **C carbohydrates** may be extracted by a number of technics.

In the one used routinely in grouping streptococci the cells are suspended in dilute hydrochloric acid (pH 2) at 100° C for 10 min, neutralized, and centrifuged; the clear

Colony forms:
group A

Mucoid

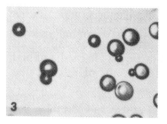

Matt

Glossy

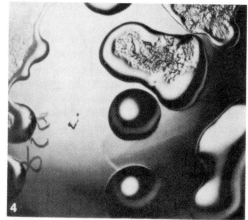

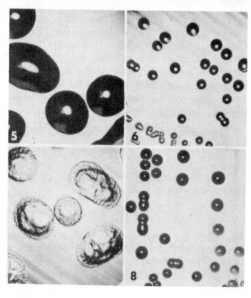

FIG. 28-5. Interrelations among mucoid, matt, and glossy colony forms of group A hemolytic streptococci. **1–3.** The three types of colonies that form on the surfaces of blood agar plates. **4.** Conversion of mucoid to matt form as a result of aging (20 h) and drying out of colonies. (×6.8) **5.** 19-Hour mucoid colonies of type 17 strain grown on Todd–Hewitt sheep blood agar. **6.** 19-Hour glossy colonies of same strain on same medium containing hyaluronidase. **7.** 24-Hour matt colonies of type 14 strain grown on Todd–Hewitt blood agar. **8.** Glossy colonies of same type 14 strain grown on same medium containing hyaluronidase. (**5–8,** ×5.5) Note that presence of hyaluronidase in agar prevents capsule formation and results in formation of glossy rather than mucoid (or matt) colonies. (**1–3,** Lancefield RC: Harvey Lect 36:251, 1942; **4–8,** Wilson AT: J Exp Med 109:257, 1959)

supernate contains the Ag. It may also be extracted by treatment with formamide at 150° C, by autoclaving, or by treatment with an enzyme from *Streptomyces albus* (or from bacteriophage lysates of group C streptococci*) that dissolves the peptidoglycan.

The group-specific Ag reacts with antiserum produced by immunizing rabbits with hemolytic streptococci of the same group. Precipitin reactions performed with appropriate sera permit the grouping of unknown strains. Although most hemolytic streptococci that cause disease in man fall into group A, human disease is occasionally due to members of other groups.

The carbohydrate Ag of the cell wall makes up approximately 10% of the dry weight of the organism and, in the case of groups A and C, is composed of rhamnose and

hexosamine (Fig. 28-6). Its specific antigenicity depends largely upon the nature of the terminal sugar residue on its oligosaccharide rhamnose side chains. This determinant is *N*-acetyl glucosamine in the group A Ag and *N*-acetyl galactosamine in the group C Ag.† Closely related strains have also been described (A-variant and C-variant) whose group-specific Ag lacks a terminal hexosamine; the antigenic specificity then appears to reside in the oligosaccharide side chains of rhamnose. Intermediate mutants (A-intermediate and C-intermediate) have been isolated which possess both antigenic determinants, some side chains terminating in a hexosamine and others in rhamnose. These relations are summarized in Figure 28-6.

---

* The phage-associated lysin acts on group A and E strains as well as on group C.

† Recent evidence indicates that a terminal disaccharide of *N*-acetyl galactosamine is the determinant in group C.

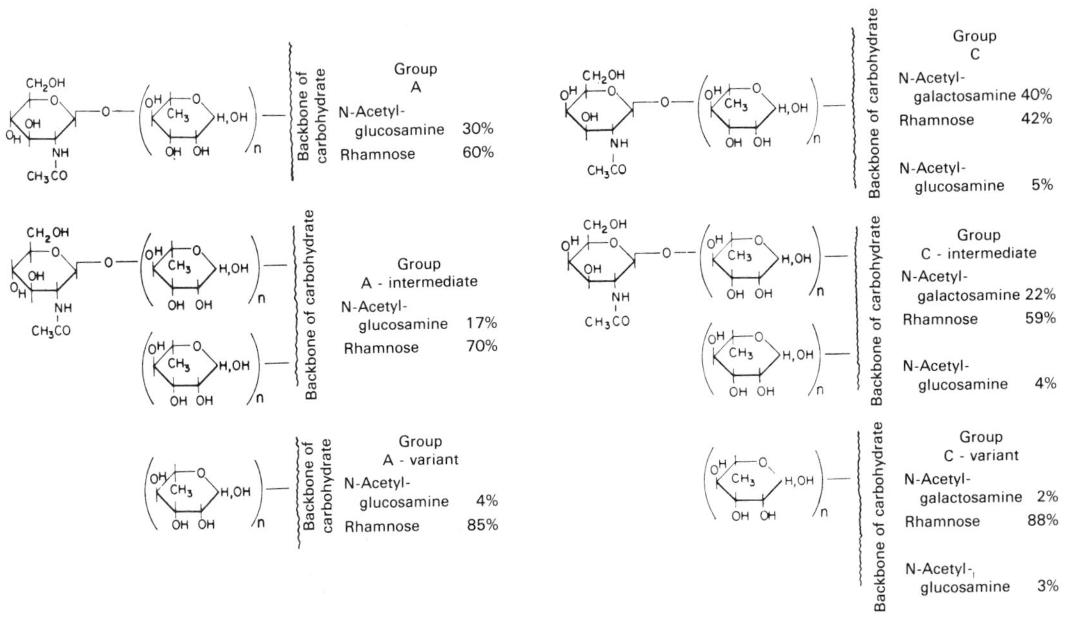

FIG. 28-6. Composition and antigenic determinants of groups A, A-intermediate, and A-variant carbohydrates (**left**) and groups C, C-intermediate, and C-variant carbohydrates (**right**) of β-hemolytic streptococci. (Recent evidence indicates that a terminal disaccharide of N-acetyl galactosamine is the determinant in group C.) (Krause RM: Bacterial Rev 27:369, 1963)

Type-specific M Antigens. Group A hemolytic streptococci can be further broken down into more than 55 immunologic types which differ in their cell wall M Ags. M protein is distributed on the surface of the cell attached to fimbriae (Fig. 28-7), which are lacking on most M⁻ strains (Fig. 28-3). The M proteins may be extracted either by the relatively drastic acid treatment (boiling at pH 2) used to solubilize the C Ag, or by enzymatic lysis of the cell (in the absence of proteolysis) with the phage-associated lysin (see Group-Specific C Antigens, above). More recently it has been solubilized with nonionic detergents, with guanidine, and with pepsin at pH 5.8.

Typing is usually done by precipitin tests with the extracted M protein and specific rabbit antisera, preabsorbed with cells of heterologous types to prevent cross reactions. Although streptococci may also be typed by agglutination tests the precipitin method is preferred because of the effect of the T Ags (see below) on the agglutination reactions.

The M protein, being on fimbriae, is readily accessible to anti-M Ab even when a hyaluronate capsule is present. Furthermore, anti-M Ab is protective, showing that the M Ag is directly involved in streptococcal virulence; indeed, both the hyaluronate capsule and the M protein are antiphagocytic (see Pathogenicity, below). Finally, the M antigenicity of intact cells may be destroyed by trypsin without affecting their viability or removing the group Ag.

Occasional strains of group A streptococci have been found to contain more than one M Ag, but there is as yet little information on the structure of the individual M proteins.

Other Streptococcal Antigens. Two other kinds of cell wall proteins, which appear to act as surface Ags of group A streptococci, have been identified. Neither seems to influence virulence. The T antigens include a number of immunologically distinct proteins, from different strains, which resist digestion by proteolytic enzymes but are readily destroyed by heat at an acid pH and hence are not present in the usual M-containing acid extract. Their distribution is not related to that of the M Ags.

Two immunologically distinct **R proteins** have thus far been identified: one (designated 3R) is destroyed by either trypsin or pepsin, the other (28R) only by pepsin. A **nucleoprotein fraction (P antigen)** is antigenically similar in hemolytic and nonhemolytic streptococci and in pneumococci; it also crossreacts with staphylococcal nucleoproteins. The **glycerol teichoic acids,** which make up approximately 1% of the dry weight of the cells, act as the group-specific Ag in groups D and N, but not in other groups. In group A strains the glycerophosphate polymer has no substituents other than ester-linked D-alanine. The Abs to this Ag reflect the glycerophosphate specificity, and occasionally also D-alanine specificity. When isolated with attached lipid (lipoteichoic acid) the Ag adsorbs readily to RBCs and other mammalian cells. There is evidence that surface lipoteichoic acid may be involved in the attachment and colonization of streptococci on mucosal surfaces in vivo.

The **peptidoglycan** which crossreacts with those of many other bacteria, produces many of the same biologic reactions as the endotoxins of gram-negative bacteria (see Endotoxin, Ch. 24), e.g.,

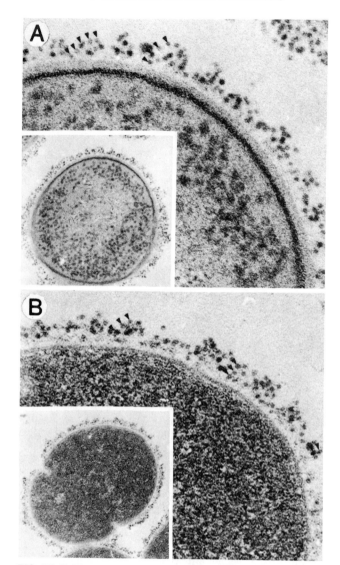

FIG. 28-7. Electron micrographs of fimbriae on surfaces of group A β-hemolytic streptococci. The fimbriae have taken up homologous ferritin-conjugated anti-M Ab. The section of an intact cell shown in **A** is compared in **B** with a nitrous acid–extracted cell. This treatment removes most of the C polysaccharide and teichoic acid of the wall, leaving the M protein intact. The ferritin particles can be seen to have assumed a linear distribution **(arrows)** along the surfaces of the fimbriae. (Swanson J et al: Exp Med 130:1063, 1969) (×250,000; insets ×60,000)

fever, dermal and cardiac necrosis, lysis of RBCs and platelets, enhancement of nonspecific resistance. The **cytoplasmic membranes,** prepared from osmotically shocked protoplasts, exhibit distinctive antigenic differences when derived from streptococci of different immunologic groups.

Apparent Spatial Relations. In summary, virulent group A hemolytic streptococci may possess, in addition to their nonimmunogenic antiphagocytic capsules, at least three distinct protein Ags on their cell walls (Fig 28-8). One of these, the type-specific M protein, is of major importance since it too is antiphagocytic and is therefore directly involved in virulence; the other two (T and R Ags) are unrelated to virulence. The group-specific C Ag is covalently linked to the wall's rigid mucopeptide matrix, which is itself immunogenic. The glycerol teichoic acids appear to reside in the wall but are not bound to this essentially insoluble structure. Other streptococcal antigens, ordinarily not exposed to the surface of the intact cell, include the lipoprotein Ags of the cell membranes and a nucleoprotein (P) Ag, which is presumably located within the cell.

EXTRACELLULAR PRODUCTS

The exceptionally wide variety of diseases in man caused by group A hemolytic streptococci may well be related to the large number of extracellular products they are known to produce. Since new streptococcal toxins and enzymes continue to be discovered, the list in Table 28-2 is surely incomplete. By means of immunoelectrophoresis, for example, there have been found in pooled human γ-globulin 20 different Abs that react with the extracellular Ags elaborated by the C-203 strain of group A hemolytic streptococcus (Fig. 28-9). This finding suggests that group A streptococci may release as many as 20 extracellular Ags when growing in human tissues. Of those identified thus far, the following appear to be of greatest clinical significance.

Erythrogenic Toxin. The erythrogenic toxin is known to be responsible for the rash in scarlet fever. Strains of group A streptococci that produce this toxin are lysogenic,* like toxigenic strains of *Corynebacterium diphtheriae* (see Lysogeny and Toxin Production, Ch. 26). The amount of toxin produced by different lysogenic strains, however, varies widely.

The mode of action of erythrogenic toxin is not clear. When injected into the skin of susceptible children, it causes localized erythematous reactions, which reach a maximum at about 24 h **(Dick test)**. This erythrogenic effect is neutralized by Ab: during covalescence, when the patient's serum contains demonstrable antitoxin, the skin text becomes negative; and an injection of homologous antitoxin intradermally at the height of scarlet fever causes a local blanching of the rash **(Schultz–Charlton test)**. Accordingly, a positive Dick test is interpreted as

* The finding that toxigenicity was transmissible from one streptococcal strain to another was first reported by Frobisher and Brown in 1927 (Bull. Johns Hopkins Hosp 41:167); at that time, of course, the mechanism involved was not understood.

FIG. 28-8. Schematic diagram of capsule, cell wall, and cytoplasmic membrane of group A hemolytic streptococcal cell. (Modified from Krause RM: Bacterial Rev 27:369, 1963)

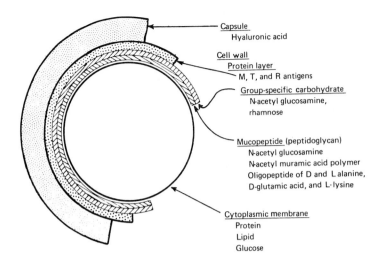

Capsule
Hyaluronic acid

Cell wall
Protein layer
M, T, and R antigens

Group-specific carbohydrate
N-acetyl glucosamine,
rhamnose

Mucopeptide (peptidoglycan)
N-acetyl glucosamine
N-acetyl muramic acid polymer
Oligopeptide of D and L alanine,
D-glutamic acid, and L-lysine

Cytoplasmic membrane
Protein
Lipid
Glucose

indicating an absence of circulating antitoxin and thus a state of susceptibility to scarlet fever.

There are at least three immunologically distinct forms of erythrogenic toxin (types A, B, and C), produced by different streptococcal strains. They are presumably proteins. Certain strains of group C and group G hemolytic streptococci, as well as staphylococci, produce erythrogenic toxins closely related to those of group A streptococci.

Streptolysins S and O. The two hemolysins responsible for the zones of hemolysis around streptococcal colonies have already been mentioned. **Streptolysin S** is stable in air and is largely cell-bound; its name derives from the fact that it can be extracted from intact streptococcal cells with serum. This extraction is dependent upon association with serum albumin as a macromolecular carrier, and other carriers (e.g., RNA) will similarly form a complex with the hemolysin. No Ab capable of neutralizing the hemolytic action of streptolysin S has been

described, but this action is inhibited by serum lipoproteins. Recent evidence indicates that this cell-bound hemolysin is responsible for the leukotoxic action of group A streptococci, manifested by the killing of a proportion of the leukocytes that phagocytize them.

Streptolysin O has been so named because it is reversibly inactivated by atmospheric oxygen. It gives rise to Abs which neutralize its hemolytic action. Since most strains of group A streptococci produce streptolysin O, patients recovering from streptococcal disease usually have antistreptolysin O Abs in their serum (see Tests for other Antibodies, below). As with other oxygen-labile bacterial exotoxins, the antigenicity of streptolysin O survives its detoxification by oxidation. Although its hemolytic activity is inhibited by cholesterol, the protein-bound cholesterol in normal serum does not have this effect and therefore does not interfere with the measurement of antistreptolysin O Abs.

The S and O hemolysins can injure the membranes of cells other than RBCs. The mechanism of action of streptolysin O is illustrated in Figure 28-10. When added to suspensions of leukocytes in vitro, it causes lysis of the cell's cytoplasmic granules, having first affected and penetrated its outer (plasma) membrane. As a result, the destructive hydrolytic enzymes contained in the granules (see Phagocytic Cells, Ch. 24) are released into the cytoplasm and irreversibly damage the cell. Macrophages are similarly injured, and streptolysin S, though less active, has much the same effect. Hydrolytic enzymes released from the leukocytic lysosomes may well damage other cell structures and thus intensify streptococcal lesions.

When injected intravenously into laboratory animals in sufficient quantities, streptolysin O causes fatal cardiac standstill. A possible relation to the pathogenesis of rheumatic fever has been suggested (see Nonsuppurative Sequelae, p. 618).

TABLE 28-2. Some Extracellular Products of Group A β-Hemolytic Streptococci

	Stimulates production of inhibitory antibody
Erythrogenic toxins (A, B, and C)	+
Streptolysin O	+
Streptolysin S	−
Diphosphopyridine nucleotidase	+
Streptokinases (A and B)	+
Deoxyribonucleases (A, B, C, and D)	+
Hyaluronidase	+
Proteinase	+
Amylase	?
Esterase	−

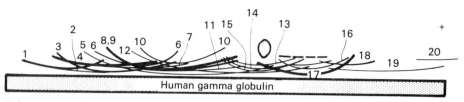

FIG. 28-9. Diagrammatic representation of group A streptococcal extracellular Ags detectable by immunoelectrophoretic analysis performed with crude streptococcal concentrate and 16% solution of pooled human γ-globulin. (Halbert SP, Keatings SL: J Exp Med 113:1015, 1961).

DPNase. Streptococcal cultures contain a diphosphopyridine nucleotidase (DPNase, also called nicotinamide adenine dinucleotidase or NADase) that liberates nicotinamide from DPN. Nephritogenic (type 12) strains are particularly prone to produce this enzyme, but there is no evidence that it plays a role in the pathogenesis of glomerulonephritis. Antibodies that inhibit its action are frequently found in the serum of patients convalescing from streptococcal disease.

Streptokinases. In 1939 Tillett and Garner described a substance in streptococcal culture filtrates that promotes the lysis of human blood clots. First termed streptococcal fibrinolysin, it was later shown to catalyze the conversion of plasminogen to plasmin, and so it was renamed

streptokinase. Two molecular species of streptokinase (A and B), differing in antigenicity and electrophoretic mobility, have been isolated from group A strains. They are immunogenic and induce antistreptokinase Abs in the course of most diseases caused by group A streptococci. Although their action has often been assumed to prevent the formation of effective fibrin barriers at the periphery of streptococcal lesions, thus permitting the organisms to spread with unusual rapidity, there is no conclusive evidence to support this attractive hypothesis. In fact, the invasiveness of streptococcal lesions appears to be uninfluenced by antistreptokinase Abs.

Deoxyribonucleases. Group A streptococci also elaborate enzymes that degrade DNA (DNases). Four immunologically and electrophoretically different types,

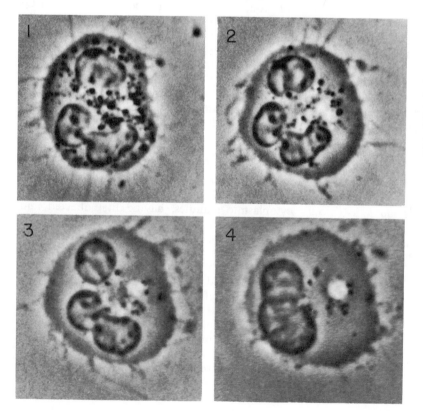

Fig. 28-10. Dissolution of leukocytic granules (lysosomes) and eventual destruction of cell resulting from action of streptolysin O. Besides obvious degranulation and formation of cytoplasmic vacuole, cell shows profound nuclear changes, which eventually result in fusion of individual nuclear lobes. First photograph **(1)** was taken after cell was already partially degranulated and had developed hairlike processes on its membrane. Thereafter pictures were taken at intervals of approximately 1 min **(2, 3, 4)**. (×2000) (Hirsch JG et al: J Exp Med 118:223, 1963)

A,B,C, and D, have been found in streptococcal filtrates. Since these enzymes do not penetrate the plasma membranes of living mammalian cells, they are not cytotoxic. They are capable however, of depolymerizing the highly viscous DNA which accumulates in thick pus as a result of the disintegration of polymorphonuclear leukocytes. Enzyme preparations containing both streptokinase and streptococcal deoxyribonuclease (streptodornase) were introduced by Tillett to liquefy purulent exudates (enzymatic debridement) in such diseases as pneumococcal empyema.

Hyaluronidase is of particular interest. Its substrate occurs in the streptococcal capsule. However, only certain group A strains (notably those of types 4 and 22) produce the enzyme in vitro, and they never form capsules; others fail to produce measurable hyaluronidase, even after prolonged growth and complete loss of capsules. Nevertheless, most patients recovering from streptococcal disease have Abs to the type 4 or 22 hyaluronidase; hence all streptococci may well produce a related protein, at least in vivo.

Originally called the "spreading factor" because of its striking lytic effect on the ground substance of connective tissue, hyaluronidase has long been thought to play a role in the characteristic tendency of streptococci to spread rapidly through mammalian tissues. How important its action really is in this regard has never been determined.

Streptococcal proteinase is capable of destroying another cell factor involved in pathogenesis, the M protein. Since this enzyme exhibits a relatively broad specificity it may also affect other extracellular proteins such as the streptolysins and streptokinase. It is released from streptococcal cells only when the pH of the medium is between 5.5 and 6.5. Under these conditions large amounts of the enzyme may appear in culture filtrates, from which it has been obtained in crystalline form. Like many other proteases, it is activated by sulfhydryl compounds and causes necrotic myocardial lesions in laboratory animals when injected intravenously. Whether or not it plays a role in the pathogenesis of poststreptococcal diseases, such as rheumatic fever, is unknown.

GENOTYPIC VARIATIONS

Genetic variations in hemolytic streptococci have been demonstrated to affect numerous traits, including elaboration of hemolysins, colonial morphology (capsule formation), production of M protein, synthesis of other cell wall antigens (including alterations in C polysaccharides), and resistance to antimicrobial drugs. The most thoroughly studied variants are those related to colony formation. When repeatedly subcultured on artificial media mucoid strains tend to become glossy, and on passage through mice glossy mutants revert to the mucoid form. Similarly, strains producing the M antigen (M$^+$ strains), which are carried in the throat after an attack of streptococcal pharyngitis, may eventually lose

their M antigens. When such M$^-$ strains are passed through mice, they frequently revert to M$^+$. Both these variations affect virulence.

Drug-resistant mutants of group A streptococci have been encountered to sulfonamides and tetracyclines. Following mass prophylaxis with sulfonamides in military personnel during World War II, sulfonamide-resistant mutants became prevalent, but the introduction of penicillin promptly suppressed their spread. No significant change in the sensitivity of naturally occurring hemolytic streptococci to penicillin has occurred. Even in the laboratory it is extremely difficult to obtain penicillin-resistant streptococci, and those resistant mutants which have been isolated have been found to lack virulence.

PATHOGENICITY

S. pyogenes causes both suppurative diseases and nonsuppurative sequelae. The first group includes acute **streptococcal pharyngitis** (with or without scarlet fever) and all its suppurative complications, including cervical adenitis, otitis media, mastoiditis, peritonsillar abscesses, meningitis, peritonitis, and pneumonia. It also includes streptococcal postpartum infections of the uterus (**puerperal sepsis), cellulitis** of the skin, **impetigo, lymphangitis,** and **erysipelas.** The principal diseases of the nonsuppurative category are **acute glomerulonephritis,** and **rheumatic fever.**

Suppurative Disease. The pathogenesis of suppurative streptococcal disease is fairly well understood. The factors that determine invasiveness are particularly important. Since hemolytic streptococci ingested by phagocytic cells are almost all killed within minutes, their antiphagocytic properties play a critical role in invasiveness. These in turn depend upon the hyaluronic acid capsule and the M protein. The combined antiphagocytic action of these two factors has been demonstrated both in vivo and in vitro.

The in vivo evidence was obtained in mice and rats infected intraperitoneally with strains of type 14 streptococci that differ in their capacities to produce capsules and M protein. Those strains producing large capsules and generous amounts of M protein were most virulent; those deficient in capsule formation, but producing a full complement of M protein, or vice versa, were of intermediate virulence; while strains deficient in both capsules and M protein were least virulent. These differences were directly correlated with the amount of phagocytosis occuring during the early hours of the infection. Confirmatory evidence has been provided in vitro, as is illustrated in Table 28-3, by using enzymes to deprive cells of these surface factors.

Although nearly all phagocytized streptococci are promptly killed, an occasional organism will escape unharmed. One mechanism is egestion, in which the engulfed organism is ejected from the cell. This event occurs only rarely in vitro, and there is no evidence that it is frequent enough in vivo to influence the course of the in-

TABLE 28-3. Relative Antiphagocytic Effect of the Hyaluronate Capsule and the M Protein of a Fully Virulent Strain of Group A β-Hemolytic Streptococcus*

Treatment of organism	State of capsule†	Amount of M protein†	% Phago-cytosis†
None	+++	+++	3 (± 1.8)
Trypsin	+++	0	49 (± 5.4)
Hyaluronidase	±	+++	41 (± 2.8)
Trypsin and hyaluronidase	±	0	64 (± 0.85)

*Type 14.

†Number of plus signs indicates approximate size of envelope, as visualized in India ink preparations; or amount of M protein demonstrable by quantitative precipitin tests. Figures in parenthesis are standard deviations. The phagocytic tests were performed in the absence of serum (i.e., surface phagocytosis; see in Ch. 24).

Foley MJ, Wood WB Jr: J Exp Med 110:617, 1959)

fection. A second escape mechanism results from elaboration of a leukotoxic factor, identified as the cell-bound streptolysin S. The possible effect of streptococcal leukocidins (including streptolysin O) on leukocytes and other cells prior to ingestion has already been discussed.

Scarlet fever occurs as a complication of pyogenic streptococcal disease when the infecting strain produces erythrogenic toxin and the patient (usually a child) is susceptible to the toxin. It has long been a matter of controversy, however, whether the rash is due to a direct action of the circulating toxin or to a generalized cutaneous hypersensitivity reaction. In favor of the latter possibility is the observation that infants under the age of 2 rarely have the disease and do not show positive Dick reactions, regardless of the immune state of the mother.

Nonsuppurative Sequelae. Acute glomerulonephritis results from infections caused by a limited number of types of group A streptococci. The majority of **nephritogenic strains** belong to type 12; a few have been of types 4, 18, 25, 49 (formerly designated the Red Lake strain), 52, and 55. The manner in which these strains cause acute glomerulonephritis is not fully understood, but several findings are consistent with a reaction to an **immune complex.** Thus the characteristic symptoms of hematuria, edema, and hypertension do not appear until about a week after the onset of the acute pyogenic infection, usually of the pharynx or skin; the serum titer of complement often falls during an attack; and immunoglobulin, the third component of complement (C3), and streptococcal Ags have been demonstrated by immunofluorescence in glomerular lesions. In addition, an Ag in the membrane of group A streptococci crossreacts with glomerular basement membrane, suggesting that the disease may have an **autoimmune** component.

The pathogenesis of **rheumatic fever** is even more ob-

scure, since it may follow pharyngeal infection with practically any type of group A streptococcus. The latent period between the onset of the acute streptococcal pharyngitis and the symptoms and signs of rheumatic fever is usually 2 or 3 weeks. The consistent finding of antistreptococcal Abs in the serum of patients with acute rheumatic fever strongly suggests that the disease is due to contact with *S. pyogenes*. Moreover, following a streptococcal epidemic most patients who develop rheumatic fever (roughly 3%) have higher titers of antistreptococcal Abs in their serum than do those who escape the disease. Nevertheless, it is far from clear how **immunologic hyperreactivity** to streptococcal products could cause the recurring cardiac, joint, and skin lesions that characterize rheumatic fever. Nor is it clear what particular streptococcal products may be involved. M protein, for example, when injected intravenously, becomes deposited beneath the endocardium as well as in the glomeruli; streptococcal protease injected intravenously causes subendocardial lesions (as do proteases from other sources). Streptolysin O is known to be cardiotoxic, and Halbert has postulated that its slow release from combination with Ab in the plasma is responsible for the characteristic recrudescences of acute rheumatic fever. Moreover, immunologic cross reactions have been described between a streptococcal Ag and tissue Ags of cardiac muscle, and between the group A carbohydrate and structural glycoprotein of heart valves. Suggestive as these various findings are, there is still no convincing evidence to support the implication of any one streptococcal product in the pathogenesis of rheumatic fever.

Although **erythema nodosum** is known to occur in association with a variety of diseases (tuberculosis, coccidioidomycosis, sarcoidosis), there is both clinical and experimental evidence that it may also be a poststreptococcal illness. By intradermally injecting suspensions of sonically disintegrated group A streptococcal cell walls, Schwab et al have produced in rabbits chronic remittent skin lesions which resemble erythema nodosum in man. There is evidence that the peptidoglycan portion exerts the primary toxic effect. The relation of this model to the human disease remains to be determined.

The possibility that **streptococcal L forms** may be involved in the pathogenesis of rheumatic fever, and perhaps even of glomerulonephritis, is currently under investigation. Streptococci in this form might well be difficult to culture and also to visualize in tissue lesions, even by electron microscopy, because of their lack of cell walls. Furthermore, they could survive intensive treatment with penicillin. The cultivation of group A streptococci from the cardiac tissues of an occasional patient who has died of acute rheumatic fever, and the experimental production of latent streptococcal infections in rabbits, has added inconclusive support to this line of reasoning.

DISEASE CAUSED BY STREPTOCOCCI OF OTHER GROUPS

Disease in man is only occasionally caused by streptococci of groups B to O (Table 28-1). Though some members of these immunologic groups produce no hemolysin they are all usually classed as β-hemolytic streptococci, as explained earlier in the chapter. Many of them are found primarily in animals, although strains of group C, E, G, H, K, and O are often isolated from the respiratory tract of man, and nonhemolytic strains of group D (*S. faecalis*) are common inhabitants of the human gastrointestinal tract (see Nonhemolytic Streptococci, below). Group B strains are a common cause of circumscribed epidemics of meningitis in newborn nurseries.

IMMUNITY

Of the many varieties of Abs that are generated in response to acute hemolytic streptococcal disease, only anti-M is known to protect the host against the invasiveness of the organisms. In acute streptococcal disease this Ab ordinarily becomes detectable in the serum within a few weeks to several months, and it usually persists for 1–2 years; in some individuals it may still be present after 10–30 years. Inasmuch as there are more than 55 serologic types of group A streptococci, no individual is likely to become immune to group A streptococcal infections in general.

Only a relatively few types of group A steptococci, on the other hand, are nephritogenic; therefore, persistence of anti-M Abs to one of these types may account for the observation that an initial attack of acute glomerulonephritis greatly decreases the probability of a subsequent attack. No such protective effect, of course, occurs in rheumatic fever because of the wide variety of streptococcal types that may cause the disease.

Immunity to scarlet fever is associated with the presence of erythrogenic antitoxin in the serum. Since there are at least three immunologic types of erythrogenic toxin occasional second attacks of scarlet fever may be expected.

LABORATORY DIAGNOSIS

Identification of Group A Organisms. The technics used to culture, group, and type hemolytic streptococci have already been described, and the need to employ pour plates to be certain of detecting β-hemolysis has been emphasized. A simple method of recognizing *S. pyogenes* (group A) depends on the fact that most group A strains are significantly more sensitive to **bacitracin** than are strains of other groups. An agar plate test using paper discs impregnated with bacitracin may be useful if facilities for serologic grouping are not available.

Tests for Anti-M Antibodies. Tests for type-specific streptococcal Abs in patients' sera based on precipitation, agglutination, or complement-fixation technics are often misleading because of crossreacting Ags. The most widely used technic, based on the bactericidal properties of whole blood, is more specific: it measures indirectly the opsonizing action of the homologous anti-M Ab. The mouse protection test depends on the same principle.

A microscopic method, based on the tendency of group A streptococci to grow in **long chains** when cultured in the presence of homologous type-specific Ab, appears to be almost as sensitive. However, it requires the use of strains that produce the right amounts of M Ag and that grow in short chains in the absence of Ab.

Tests for Other Antibodies. Serologic tests for Abs to extracellular products are much easier to perform and are therefore more widely used. The **antistreptolysin test** measures Abs against streptolysin O. Technics designed to measure Abs against streptokinase, hyaluronidase, or DNase may be similarly employed.* The antistreptolysin test, however, is routinely used in most diagnostic laboratories.

TREATMENT

β-Hemolytic streptococci are among the most susceptible of all pathogenic bacteria to the action of antimicrobial drugs. The sulfonamides readily suppress growth both in vitro and in vivo but since they are only bacteriostatic their use will not eliminate the organisms from the upper respiratory tract, nor will it significantly modify the Ab response of the host. Penicillin is bactericidal and hence is far more effective. When used in adequate dosage for a sufficient length of time, it will often rid the pharynx of hemolytic streptococci. Persistence usually indicates a suppurative complication, such as an intratonsillar abscess or purulent sinusitis. When given early in the course of acute streptococcal pharyngitis, penicillin will also depress the patient's Ab response, and such treatment of rheumatic individuals greatly reduces the rate of rheumatic attacks.

Other antibiotics, such as erythromycin, may also be used in the treatment of group A hemolytic streptococcal infections, particularly in patients who have a history of hypersensitivity to penicillin. Many strains are now resistant to the tetracyclines, which are therefore no longer generally recommended.

PREVENTION

Penicillin is often given continually in small doses to rheumatic patients to prevent streptococcal infections and thus reduce the likelihood of recurrences of rheumatic fever. Prophylactic therapy of this kind is possible only because group A hemolytic streptococci do not generate mutants that are significantly resistant to penicillin. Prevention of rheumatic fever with continuous sulfonamide treatment has also been reasonably successful, but because these drugs are not bactericidal and

---

* In streptococcal skin infections the test for Ab to streptococcal DNase B appears to be more reliable than the antistreptolysin test.

occasionally cause severe reactions (e.g., periarteritis nodosa), penicillin is usually preferred. The responsibility of the physician to eradicate (with penicillin) nephritogenic streptococci from the families of patients with acute glomerulonephritis should be evident.

The custom in the past of isolating scarlet fever patients but not patients with acute streptococcal pharyngitis was based on the erroneous view that the two diseases are basically dissimilar. From an epidemiologic standpoint there is no justification for such a distinction.

As already intimated, immunization of the general population against group A streptococcal infections is not practical because of the very large number of types. Efforts to prepare suitable M protein vaccines, however, are being continued in the hope that they will be useful, particularly in military installations, in controlling the spread of individual epidemic strains of streptococci. Active immunization against scarlet fever has been practiced in the past, but is no longer advocated because of the effectiveness of penicillin therapy. Indeed, scarlet fever has become relatively uncommon in the United States, as have streptococcal mastoiditis and puerperal infections, presumably due to the early treatment of suspected streptococcal illnesses with effective antimicrobial drugs.

EPIDEMIOLOGY

The incidence of streptococcal infections varies widely in different geographic areas and appears to be related to climate. Streptococcal diseases are most common in cold, relatively dry areas and occur most often in the winter and spring. Endemic rates are particularly high in the Rocky Mountain states, such as Colorado and Wyoming. Although overt streptococcal **disease** is less prevalent in the southern United States, culture surveys and Ab studies reveal that streptococcal **infections** are not uncommon.

Group A streptococcal **carrier rates** ordinarily run well below 10%. Just before an epidemic, however, they become much higher. Infection is transmitted from the respiratory tract of one person to that of another by relatively intimate contact. The epidemiologic studies of Rammelkamp et al have shown that group A streptococci in the air, in dust, on blankets, etc., are far less infectious than the moist secretions ejected from the respiratory tract during speech, coughing, and sneezing. Nasal carriers are known to disseminate many more streptococci into their environments than pharyngeal carriers do. The source of puerperal sepsis has often been traced to the upper respiratory tract of the obstetrician or of someone else in the delivery room.* Milk-borne epidemics of streptococcal disease are now uncommon because of the effectiveness of pasteurization. Isolated outbreaks due to infected food still occur on rare occasions.

OTHER STREPTOCOCCI

α-HEMOLYTIC STREPTOCOCCI

The α-hemolytic streptococci are often referred to collectively as the **viridans group;** they have never been satisfactorily classified. Antigenic analysis of many strains has led to the recognition of a number of immunologic varieties, only a few of which seem to fall into the Lancefield groups (A to O).* One of the most common species is *S. salivarius.*

Viridans streptococci colonize the human upper respiratory tract within the first few hours after birth; rarely does a carefully performed throat culture fail to reveal their presence. They have a very low degree of pathogenicity compared with pneumococci, which also are α-hemolytic and are often cultured from the throat. Unlike pneumococci, however, viridans streptococci are neither bile-soluble nor sensitive to ethylhydrocupreine (optochin).

The principal significance of α-hemolytic streptococci in clinical medicine relates to **subacute bacterial endocarditis.** This serious illness results from infection of an endocardial surface already damaged by either rheumatic fever or congenital heart disease. Since α-hemolytic streptococci are continually present in the throat and about the teeth, even minor trauma such as that due to vigorous chewing may result in their entry into the blood stream. The transient bacteremias that follow dental extraction or tonsillectomy may initiate subacute bacterial endocarditis in patients with abnormal valves. The seriousness of this type of infection is greatly increased by the frequency of drug-resistant strains of α-hemolytic streptococci.

In order for viridans streptococci to gain a permanent foothold on the endocardium the organisms must be trapped in a suitable

---

* One set of group F strains, designated "streptococcus MG," produce relatively small colonies and are of clinical interest because they are agglutinated by the serum of patients with mycoplasmal pneumonia.

* In the mid-nineteenth century Ignaz Semmelweis in Vienna and Oliver Wendell Holmes in Boston independently concluded that childbed fever was transmitted to patients by the unclean hands and contaminated clothing of the attending obstetrician. Their conclusions were vehemently rejected by their contemporaries (to whom the germ theory of disease was anathema) but have been fully substantiated by modern epidemiologic studies, made possible by the Lancefield typing technic.

nidus on the endocardial surface, usually provided by a tiny fibrin clot over an area of endocardial trauma. Dogs with artificially induced arteriovenous aneurysms in the peripheral circulation regularly develop microscopic clots of this kind on the endocardium. The clot formation results from the severe circulatory disturbance (widened pulse pressure, increased blood flow, etc.) created by the arteriovenous shunt. Similar endocardial lesions have been produced in rats by exposing them to prolonged anoxia, which also causes a hyperkinetic circulatory response. The remarkable frequency of bacterial endocarditis in such animals illustrates the importance of preexisting endocardial damage in the pathogenesis of the disease.

Histologic examination of the vegetations on the heart valves in both naturally acquired and experimentally induced viridans endocarditis reveals that the organisms grow in large colonies embedded in a fibrinous, relatively acellular exudate. The avascularity of the heart valves accounts for the paucity of phagocytic cells in the lesions and explains why an organism so easily phagocytized in vitro can survive and multiply in the valvular vegetations. As the organisms continue to multiply at the periphery of the lesion, they break off and are carried away in the blood stream. Quantitative blood cultures, taken at repeated intervals over long periods in patients with subacute bacterial endocarditis, show that the organisms are shed from the vegetation at a surprisingly constant rate. Thrombotic lesions, resulting in the formation of petechiae, commonly develop and constitute a hallmark of the disease.

NONHEMOLYTIC STREPTOCOCCI

The term **nonhemolytic streptococcus** is confusing because it is often used to include any streptococcus that is not β-hemolytic, and because many nonhemolytic species (including a particularly common one, *S. faecalis*) possess the same group-specific cell wall Ags as certain hemolytic streptococci.

The organisms that fall in the nonhemolytic group are generally of low pathogenicity for man and, like α-hemolytic streptococci, are of concern to physicians primarily as causative agents of subacute bacterial endocarditis. *S. faecalis,* often referred to as **enterococcus** because of its frequent presence in the human gastrointestinal tract,* is an exceptionally hardy microorganism.

Its ability to grow in the presence of 0.05% sodium azide is often utilized in the laboratory to separate it from other streptococci. Enterococci also tend to be relatively resistant to heat (62 ° C for 30 min) and will multiply in media containing 6.5% sodium chloride. Many strains encountered in clinical practice are highly resistant to antimicrobial drugs, making the treatment of enterococcal endocarditis especially difficult.

ANAEROBIC STREPTOCOCCI

All the varieties of streptococci thus far considered are facultative anaerobes. Obligate anaerobic (or microaerophilic) streptococci do exist, however, and may also cause human disease. They are usually nonhemolytic and are smaller than other streptococci. Although a number of different species have been described, they have not been systematically classified.

Because anaerobic streptococci are normal inhabitants of the female genital tract, they occasionally give rise to intrauterine infections. Their virulence for humans, as well as for other animals, is low and they tend to multiply only in necrotic or frankly gangrenous lesions. When growing in purulent exudates, they produce a fetid odor; hence their presence in lung abscesses is often suggested by the foul odor of the sputum. Although most anaerobic streptococci are susceptible to the action of antimicrobial drugs, the lesions in which they are found often require surgical drainage as well.

* Enterococci may also be cultured from the oropharynx.

SELECTED READING

BOOKS AND REVIEW ARTICLES

FOX EN: M proteins of group A streptococci. Bacteriol Rev 38:57, 1974

KUTTNER AG, LANCEFIELD RC: Unsolved problems of the non-suppurative complications of group A streptococcal infections. In Mudd S (ed): Infectious Agents and Host Reactions. Philadelphia, Saunders, 1970

MCCARTY M (ed): Streptococcal Infections. New York, Columbia University Press, 1954

MCCARTY M: The streptococcal cell wall. Harvey Lect 65:73, 1971

PATTERSON MJ, HAFEEZ AEB: Group B streptococci in human disease. Bacteriol Rev 40:774, 1976

RAMMELKAMP CH JR: Epidemiology of streptococcal infections. Harvey Lect 51:113, 1957

UHR JW (ed): The Streptococcus, Rheumatic Fever, and Glomerulonephritis. Baltimore, Williams & Wilkins, 1964

SPECIFIC ARTICLES

BEACHEY EH, OFEK I: Epithelial cell binding of group A streptococci by lipoteichoic acid on fimbriae denuded of M protein. J Exp Med 143:759, 1976

COBURN AF, FRANK PF, NOLAN J: Studies on the pathogenicity of Streptococcus pyogenes. IV. The relation between the capacity to induce fatal respiratory infections in mice and epidemic respiratory diseases in man. Br J Exp Pathol 38:256, 1957

FREIMER EH, MCCARTY M: Rheumatic fever. Sci Am 213:6, 1965

GOLDSTEIN I, HALPERN B, ROBERT L: Immunological relationship between streptococcus A polysaccharide and the structural glycoprotein of heart valve. Nature 213:44, 1967

HALBERT SP, KEATINGE SL: The analysis of streptococcal infections. VI. Immunoelectrophoretic observations on extracellular antigens detectable with human antibodies. J Exp Med 113:1013, 1961

KAPLAN MH, SVEC KH: Immunologic relation of streptococcal and tissue antigens. III. Presence in human sera of streptococcal antibody cross-reactive with heart tissue. Association with streptococcal infection, rheumatic fever, and glomerulonephritis. J Exp Med 119:651, 1964

KRAUSE RM: Antigenic and biochemical composition of hemolytic streptococcal cell walls. Bacteriol Rev 27:369, 1963

LANCEFIELD RC: Current knowledge of type-specific M antigens of group A streptococci. J Immunol 89:307, 1962

LILLEHEI CW, VARGO JD, HAMMERSTROM RN: Experimental bacterial endocarditis and proliferative glomerulonephritis. Dis Chest 24:421, 1953

SCHWAB JH: Analysis of the experimental lesion in connective tissue produced by a complex of C polysaccharide from group A streptococci. I. In vivo reaction between tissue and toxin. II. Influence of age and hypersensitivity. J Exp Med 116:16, 1962; 119:401, 1964

STOLLERMAN GH, SIEGEL AC, JOHNSON EE: Evaluation of the "long chain reaction" as a means of detecting type-specific antibody to group A streptococci in human sera. J Exp Med 110:887, 1959

SWANSON J, HSU KC, GOTSCHLICH EC: Electronmicroscopic studies on streptococci. I. M antigen. J Exp Med 130:1063, 1969

TILLETT WS: Studies on the enzymatic lysis of fibrin and inflammatory exudates by products of hemolytic streptococci. Harvey Lect 45:149, 1952

WANNAMAKER LW: Differences between streptococcal infections of the throat and of the skin. N Engl J Med 282:23, 1970

ZABRISKIE JB: The role of temperate bacteriophage in the production of erythrogenic toxin by group A streptococci. J Exp Med 119:761, 1964

ZABRISKIE JB: The relationship of streptococcal cross-reactive antigens to rheumatic fever. Transplant Proc 1:968, 1969

chapter 29

STAPHYLOCOCCI

STEPHEN I. MORSE[†]

[†] Deceased

In the family of gram-positive cocci constituting the Micrococcaceae the most important genus is *Staphylococcus*. These organisms are facultatively anaerobic and nonmotile, and the earliest work on the chemistry of bacterial peptidoglycans (see Ch. 6) revealed their characteristic pentaglycine cross-bridge. A second, largely nonpathogenic genus, *Micrococcus*, is widely separated in evolution from staphylococci, as indicated by the marked difference in DNA composition, the absence of pentaglycine, and the obligate aerobiosis (Table 29-1); micrococci are common in the environment.

The term staphylococcus (Gr. *staphyle*, bunch of grapes) reflects the grouping of organisms in irregular clusters (Figs. 29-1 and 29-2). There are three species. *S. aureus* causes a wide variety of suppurative diseases in man, including superficial and deep abscesses, wound infections, osteomyelitis, pneumonia and empyema, meningitis, purulent arthritis, and septicemia and endocarditis (often with metastatic abscesses). In addition, it causes two toxinoses: food poisoning and exfoliative skin disease. *S. aureus* is also an important pathogen for domestic animals. *S. epidermidis* usually does not cause serious disease except in the compromised host. *S. saprophyticus* is a newly recognized species which can cause urinary tract infections. Differential characteristics are shown in Table 29-2.

STAPHYLOCOCCUS AUREUS

HISTORY

The gram-positive cocci were first differentiated by Koch (1878), who recognized that distinct diseases were produced by organisms with different patterns of growth: in pairs, chains, or clusters. Rosenbach (1884) differentiated species of staphylococci on the basis of colonial **pigmentation;** the most pathogenic species, *S. aureus,* typically forms colonies with a golden yellow pigment, whereas colonies of the usually nonpathogenic *S. epidermidis* (formerly *S. albus*) are characteristically white. However, colonial pigmentation is highly variable and does not always provide species identification. The **ability to clot plasma** correlates much better with pathoge-

nicity, and all coagulase-positive staphylococci are now identified as *S. aureus* irrespective of pigmentation.

MORPHOLOGY AND STRUCTURE

The diameter of individual cocci is $0.7\mu m$–$1.2\mu m$. Individual cells in old cultures, or cells ingested by phagocytes, may be gram-negative. The characteristic **cell clustering** is most striking on solid media; it arises because staphylococci divide in three successive perpendicular planes, and the daughter cells do not separate completely (Fig. 29-2). The formation of irregular aggregates is due to attachments eccentric to the plane of division, as well as to actual movement of cells from the centric position. In liquid media short chains appear often, but unlike streptococci, staphylococci rarely form chains containing more than four members.

Wall-deficient L forms (see L-Phase Variants, Ch. 43) induced by penicillin or lysostaphin can be propagated in hyperosmolar media. **G (gonidial) colonial variants,** which form dwarf, pinpoint colonies, are selected when growth of normal forms is inhibited by various agents (lithium or barium salts, gentian violet, or antimicrobials). G forms are resistant to antibiotics and display reduced or absent production of pigment, coagulase, and hemolysins. Although G forms do not require hyperosmolar conditions their walls are probably partially defective. It has been proposed that recrudescent staphylococcal disease results from reversion of cryptic L or G forms.

GROWTH AND METABOLISM

Growth is usually more luxuriant under aerobic than under anaerobic conditions; multiplication of some strains is enhanced by increased CO_2 tension. Individual colonies on nutrient agar are sharply defined, round, convex, and 1–4 mm in diameter. The classic golden yellow color is due to carotenoids. Pigmentation is usually apparent after 18–24 h of growth at $37°C$ but is more pronounced when cultures are held at room temperature for a further 24–48 h; it is also enhanced if the medium is enriched in glycerol monophosphate or monoacetate. Pigment is not produced during anaerobic growth or in liquid culture. There is marked variation,

TABLE 29-1. Characteristics of the Genera *Staphylococcus* and *Micrococcus*

	Staphylococcus	*Micrococcus*
Cells		
Spherical, gram-positive	+	+
Irregular clusters	+	+
Regular clusters	0	tetrads
Anaerobic growth	+*	0
Glucose fermentation	+*	0
Cell wall teichoic acid	+	0
Pentaglycine cross-bridge		
in peptidoglycan	+	0
Sensitivity to lysostaphin	+	0
DNA base composition		
(% G C)	30–40	66–75

*S. saprophyticus grows poorly under anaerobic conditions and ferments glucose only weakly.

(Adapted from Baird-Parker AC: In Buchanan RE, Gibbons NE, eds: Bergey's Manual of Determinative Bacteriology, 8th ed, Baltimore, Williams and Wilkins, 1974; and Ann NY Acad Sci 236:7, 1974)

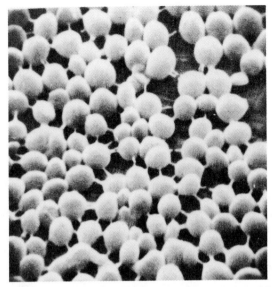

Fig. 29-2. Scanning electron micrograph of *S. aureus*. (Klainer AS, Betsch CJ: J Infect Dis 121:339, 1970) (×7500)

Fig. 29-1. Gram stain of exudate containing intracellular and extracellular staphylococci. (White A, Brooks GF. In Hoeprich PD ed: Infectious Diseases, 2nd ed. Hagerstown, Harper and Row, 1977) (×650)

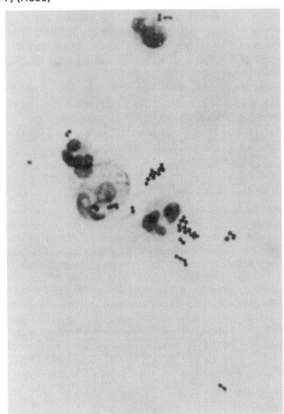

from white to deep orange, among strains and even among individual colonies of a single strain.

On blood agar a clear zone of hemolysis (*β*-hemolysis) is frequently seen around colonies. Since *S. aureus* can produce four distinct hemolysins, with different lytic spectrums, the extent of hemolysis depends on both the strain and the type of blood. Human blood may inhibit growth or alter the appearance of colonies and should not be used.

Most strains will grow readily in a chemically defined medium containing glucose, salts, 14 amino acids, thiamine, and nicotinic acid. On meat digest media devoid of blood or serum they grow well over a wide range of pH (4.8–9.4). Under aerobic conditions catalase is produced, and acid is formed from glucose, mannitol, xylose, lactose, sucrose, maltose, and glycerol. Both glucose and mannitol are fermented; neither *S. epidermidis* nor *S. saprophyticus* ferments mannitol.

Staphylococci are among the **hardiest** of all non–spore-forming bacteria. They remain alive for months on the surface of sealed agar plates stored at 4° C and may be cultured from samples of dried pus many weeks old. Some strains are relatively heat-resistant, withstanding temperatures as high as 60° C for 30 min. Though highly susceptible to the bactericidal action of certain basic dyes (e.g., gentian violet) they are more resistant than most bacteria to disinfectants such as mercuric chloride and phenol. *S. aureus* also has a high **salt tolerance,** and media containing 7.5%–10% NaCl permit its growth but inhibit other organisms.

TABLE 29-2. Differential Characteristics of Staphylococcal Species

	S. aureus	S. epidermidis	S. saprophyticus
Coagulase	+	0	0
Anaerobic growth	+	+	0*
Glucose fermentation	+	+	0*
Mannitol fermentation	+	0	0
Phosphatase	+	+	0
Heat-resistant endonucleases	+	0	0
Novobiocin†	S	S	R
Cell wall teichoic acids			
Ribitol-N-acetylglucosamine	+	0	+
Glycerol-glucose	0	+	0
Glycerol-N-acetyl-glucosamine	0	0	+

* *S. saprophyticus* strains usually do not grow anaerobically, or grow only slowly, and ferment glucose weakly.

† S = Sensitive (MIC $<0.6\mu g/ml$); R = resistant (MIC $>2.0\mu g/ml$)

(Adapted from Baird-Parker AC: Ann NY Acad Sci 236:7, 1974)

GENETICS

In *S. aureus* extrachromosomal DNA has particular importance in the major clinical problem of **antibiotic resistance.** When Fleming (1929) first noted the bactericidal action of the mold *Penicillium notatum* on this organism most strains isolated from patients were sensitive to the antibiotic. Now 60%–90% of strains are resistant. This resistance is caused by **penicillinase (β-lactamase),** a plasmid-coded enzyme which splits the β-lactam ring of the penicillin nucleus. Some penicillinase plasmids also encode for resistance to heavy metals and erythromycin, and other plasmids carry genes determining resistance to chloramphenicol, tetracyclines, neomycin, and kanamycin.

Unlike enterobacteria, *S. aureus cannot conjugate, and its plasmids are transferred by transduction.* Virtually all strains are lysogenic, and one group of phages is capable of generalized transduction (Ch. 48) . Transduction of penicillinase plasmids has been observed in mixed infections in mouse kidneys and on human skin. There is also considerable epidemiologic evidence of naturally occurring *S. aureus* transduction in man.

Plasmid genes are also responsible for the production of a bacteriocin termed **staphylococcin:** a heat-stable protein that inhibits the growth of many gram-positive organisms. Plasmids are also implicated in the production of other biologically active components, e.g., exfoliatin and enterotoxin. The lysogenic state itself influences susceptibility to the lytic activity of other staphylococcal bacteriophages, and also the production of hemolysins and enzymes (lysogenic conversion).

In the presence of "helper" bacteriophages and high concentrations of Ca^{2+} a number of *S. aureus* strains become competent recipients for transformation and transfection.

CLASSIFICATION

Strains are routinely identified by **bacteriophage typing.** For this purpose a selected set of lytic bacteriophages is divided into four host-range groups (Table 29-3); within each group are phages of different serologic types and morphology. One drop of each standardized suspension (routine test dilution, or RTD) is placed on separate areas of an agar surface previously seeded with a heavy culture of the staphylococcal strain to be tested. The pattern of the zones of lysis is recorded after incubation overnight at 30° C (Fig. 29-3).

The phage receptor, contained in the peptidoglycan–teichoic acid complex, is not specific for any phage group: the differences in phage susceptibility are related instead to the lysogenic immunity (see Ch. 48) of the test strain. Approximately 30% of strains are nontypable at the RTD, often because surface components mask the receptor site; more concentrated suspensions (RTD × 100) may then be useful.

S. aureus strains from different animal sources tend to have different physiologic properties, which are useful in determining whether a strain isolated from humans originated in animals. Human strains can be differentiated by agglutination and agglutination-inhibition reactions, but the typing systems used are complex and have not been codified, and specific antisera are not available for routine use. The Ags involved have not been identified. Identification of individual strains by their pattern of suscepti-

TABLE 29-3. Lytic Groups of Staphylococcal Typing Phages*

Group	Phage no.				
I	29	52	52A	79	80
II	3A	3C	55	71	
III	6	42E	47	53	54
	75	17	83A	84	85
Miscellaneous	81	94	95	96	

*Recommended for the typing of *S. aureus* of human origin by the International Subcommittee on Phage Typing of Staphylococci.

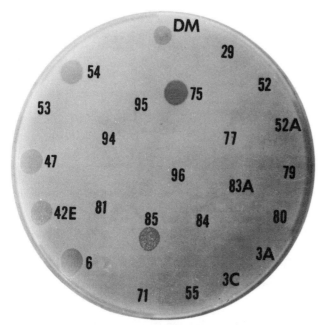

Fig. 29-3. Phage typing of *S. aureus*. The test strain is lysed by group III phages 6, 42E, 47, 54, 75, and 85 (see Table 29-3). Note colonies resistant to phages 42E and 85. The strain is correspondingly referred to as a group III *S. aureus*, phage type 6/42E/47/54/75/85. **DM,** dye marker to orient the petri dish.

bility or resistance to antimicrobials (**resistogram**) is useful for epidemiologic purposes but is relatively imprecise.

CELLULAR ANTIGENS

Capsules. A few strains of *S. aureus* are encapsulated. They tend to be more virulent in animals, and anticapsular Abs protect against experimental disease. It appears that capsules may be formed in vivo more often than is appreciated but are lost on cultivation. Most encapsulated strains are not phage-typable and lack bound coagulase (clumping factor).

The capsule of the Smith strain and related strains is composed of glucosaminuronic acid; another contains mannosaminuronic acid. The capsule of some strains has been reported to consist of peptidoglycan components (lysine, glutamic acid or isoglutamine, glycine, alanine, and glucosamine).

Polysaccharide A. Species-specific carbohydrate Ags of *S. aureus* and *S. epidermidis* (polysaccharides A and B, respectively) were recognized by Julianelle in 1935 and have now been identified as teichoic acids (see Ch. 6). Polysaccharide A of *S. aureus* is a linear ribitol teichoic acid with *N*-acetylglucosamine attached at the C-4 position and D-alanine at approximately 50% of the C-2 positions of ribitol. The antigenic determinant is the glucosamine residue, which may be in either α or β glycosidic linkage; most strains have teichoic acids with both anomers, but some have only one. Therefore antisera and teichoic acids with both specificities are required to test for species identification and for Ab.

Protein A. Strains of *S. aureus* frequently possess a surface component known as protein (or agglutinogen) A, the bulk of which is covalently linked to the peptidoglycan; some is also released extracellularly. Depending upon the method of isolation, the mol wt is 13,000 or 42,000. Protein A can induce specific Abs and will react with their Fab portion. In addition, *protein A interacts nonspecifically with the Fc portion of immunoglobulins of virtually all mammalian species, forming precipitates or soluble complexes.* All subclasses of human IgG except IgG3 are precipitated; some IgM and IgA2 samples also interact.

Protein A–Fc interactions cause a variety of biologic effects: local and systemic anaphylaxis in animals; wheal-and-flare reactions in humans; Arthus reactions; activation of complement by both the alternate and classic pathways with generation of chemotactic factors; inhibition of opsonic antibody activity because of competition with the Fc receptors of phagocytes; induction of histamine release from leukocytes; and proliferation of human B, but not T, lymphocytes.

Clumping Factor (Bound Coagulase). The majority of nonencapsulated strains of *S. aureus* will clump when suspended in plasma or in fibrinogen solutions. It is believed that clumping factor is a surface component that reacts with both the α and β chains of fibrinogen and causes crosslinking.

Other Antigens. A glycerol lipoteichoic acid is present. Several undefined Ags are involved in agglutination.

EXTRACELLULAR ENZYMES

Coagulase. The plasma of many animal species is readily clotted by the extracellular or free coagulase of *S. aureus.* The production of coagulase is the marker for the species, although a very few wild-type strains of *S. aureus*, as well as mutants, may be coagulase-negative. Antigenically distinct coagulases occur but all seem to have the same mechanism of action.

Clotting requires interaction with a **coagulase-reacting factor (CRF)** in plasma, which is either **prothrombin** or a prothrombin derivative. The interaction yields a substance that cannot be distinguished from thrombin: it converts fibrinogen to fibrin and the same fibrinopeptides as those formed by thrombin, and both activities are blocked by diisopropylfluorophosphate, an esterase inhibitor.

Other Enzymes. Staphylokinase, like streptokinase and urokinase, causes clot dissolution by activating conversion of the proenzyme plasminogen to the fibrinolytic enzyme plasmin.

The **nuclease** of *S. aureus* is a phosphodiesterase with both endo- and exonuclease activity that cleaves DNA and RNA to produce 3'-phosphomononucleotides. The purified enzyme has been extensively studied and found to be composed of a single polypeptide chain of mol wt 16,800; Ca^{2+} is required for maximal activity. Several **lipases** are produced; they are assayed by testing strains for the ability to produce opacity on egg yolk agar or to split Tween detergents. Lipase production is inhibited by lysogeny with certain prophages. **Hyaluronidase** is formed by most *S. aureus* strains, and several isoenzymes exist.

HEMOLYSINS AND TOXINS

In 1928 in the Australian city of Bundaberg a vial of diphtheria toxin–antitoxin, held at room temperature in the summer heat without a preservative, was inadvertently used to inoculate 21 children: 16 became acutely ill and 12 died within 2 days with symptoms of septicemia and toxemia. Hemolytic *S. aureus* was isolated from both the vial and the patients. This tragic event provoked intensive research on the pathogenicity of staphylococci, in which Burnet (1929) found a toxic hemolysin, termed α-hemolysin (α-toxin).

Hemolysins. Four chemically and serologically distinct hemolysins of *S. aureus* are now recognized, α, β, γ, and δ; a single strain may produce more than one. All are proteins that produce clear β-hemolysis, but they differ in RBC species specificity and in mechanism of action. The hemolysins are immunogenic, and Ab neutralizes their activity. Cells other than RBCs may also be damaged: some of the hemolysins produce local tissue necrosis and are lethal for experimental animals.

α-**Hemolysin** (α-toxin) is the principal hemolysin of human strains of *S. aureus*. It has a mol wt of approximately 33,000 but large, 12S aggregates with minimal hemolytic activity occur. Rabbit RBCs are the most sensitive; human RBCs are insusceptible, but human platelets and tissue culture cells are affected.

α-Hemolysin also causes leakage of ions from artificial liposomes. In experimental animals it causes dermal necrosis after local injection and is lethal when given systemically; the main effect appears to be spasm of vascular smooth muscle. Its mode and site of action on cell membranes are unknown.

β-**Hemolysin** is produced commonly by animal strains but by only 10%–20% of strains isolated from humans. It is a "hot–cold"

hemolysin: its lytic effects are not fully developed unless mixtures of blood and toxin, or blood agar cultures, are placed at low temperatures following incubation at 37 ° C. β-Hemolysin is a sphingomyelinase C, of mol wt 30,000, that is activated by Mg^{2+} but not by Ca^{2+}; the reaction is

$$\text{Sphingomyelin} \xrightarrow[\text{Mg}^{2+}]{\beta\text{-Hemolysin}} \text{N-Acylsphingosine} + \text{Phosphorylcholine}.$$

Sheep, human, and guinea pig RBCs, in that order, contain decreasing amounts of sphingomyelin and are decreasingly sensitive to β-hemolysin. The hemolysin is cytotoxic for a variety of tissue culture cells, and in relatively large doses it is toxic for experimental animals.

γ-**Hemolysin** consists of two basic proteins acting in concert. Rabbit, human, and sheep RBCs are susceptible, while horse and fowl RBCs are not. Agar and other sulfated polymers inhibit γ-hemolysin, and so it is not active on blood agar plates. Cholesterol and many other lipids are also inhibitory. Because of these interactions the existence of γ-hemolysin was long debated.

δ-**Hemolysin** occurs as heterogeneous aggregates of subunits of mol wt 5000; it is produced by most human strains of *S. aureus*. In addition to RBCs of different species, various other cell types, such as leukocytes, cultured mammalian cells, and bacterial protoplasts, are injured by δ-hemolysin, possibly by direct surfactant properties. Activity is blocked by phospholipids present in normal serum, but specific Abs are also produced. Toxicity for animals is minimal.

Toxins. Panton–Valentine (P-V) leukocidin is produced by most *S. aureus* strains. It is composed of two electrophoretically separable proteins: F (fast) and S (slow). Both components are antigenic and can be neutralized by Ab. They act synergistically, in the presence of Ca^{2+}, to cause degranulation of human and rabbit polymorphonuclear leukocytes (PMNs) and macrophages, which are the only susceptible cells known.

Exfoliatin (epidermolytic toxin) causes a variety of dermatologic lesions (see Toxinoses, below). This relatively heat-stable and acid-labile protein, of mol wt 24,000, is produced by approximately 5% of random isolates of *S. aureus* (mostly strains of phage group II). Plasmid genes play a role in its formation in most strains. More than one antigenic type may exist. The toxin acts by cleaving the stratum granulosum of the epidermis, probably by splitting desmosomes which link the cells of this layer.

Enterotoxin. Approximately 50% of *S. aureus* strains produce enterotoxin. There are five serologically distinct enterotoxins (A–E). They are a major cause (especially types A and D) of **food poisoning,** which is due to ingestion of preformed toxin in foods contaminated with *S. aureus*. In addition, enterotoxin B is often associated with staphylococcal **pseudomembranous enterocolitis.**

Staphylococcal enterotoxins are markedly heat-resistant and relatively trypsin-resistant proteins of mol wt

35,000. Their site and mechanism of action are uncertain, but both local and central neural effects appear to occur.

PATHOGENESIS OF INVASIVE DISEASE

Ecology. The skin and nares of infants are colonized by *S. aureus* within a few days after birth; the carrier rate then drops, only to increase during childhood to the adult rate of approximately 30%. The organisms are most commonly found in the anterior nares and on skin and mucous membranes; the nasal epithelium of carriers may have a special affinity for *S. aureus*. Colonization of the nares leads to dissemination to body surfaces.

Disease. Suppuration is the hallmark of staphylococcal disease, and the most frequent lesion is a cutaneous abscess or **boil** which begins as an infection of sebaceous glands or hair shafts. At full development the center of the abscess shows liquefaction necrosis (pus) consisting of dead bacteria, phagocytes, and fluid, surrounded by a firm wall of fibrin, inflammatory cells, and viable bacteria. Sometimes a series of interconnected abscesses occurs (**carbuncle,** Fig. 29-4). In addition *S. aureus* may cause acute hematogenous osteomyelitis ("osteomyelitis is a furuncle of the osseous medulla"—Pasteur); endocarditis, usually with metastatic abscesses and a fulminating course; and pneumonia (which is almost invariably preceded by viral influenza).

Although minor, superficial disease is very common, serious disease rarely occurs in normal persons. However, deep-seated invasive disease, sometimes accompanied by bacteremia, occurs in various conditions: in newborns; as a complication of traumatic or operative wounds, burns, or other serious skin lesions; and during chronic debilitating disorders such as diabetes mellitus, cancer, and cystic fibrosis. It is therefore not surprising that serious staphylococcal disease is most often the result of hospital-acquired (nosocomial) infections.

In the 1950s many hospital epidemics throughout the world were caused by antibiotic-resistant organisms of the 80/81 phage type complex. The prevalence of these epidemiologically virulent organisms was in part related to the selective effect of the widely used antibiotics, but the reason for the uncommonly high incidence of serious disease is unknown.

The relative avirulence of *S. aureus* for normal persons has been demonstrated in human volunteer experiments. Large numbers of organisms injected under the skin may cause no more than a barely discernible lesion. If, on the other hand, a silk suture contaminated with a relatively small number (<100) of cells of the same strain is inserted, suppurative lesions invariably result. The foreign body permits the organisms to gain a foothold, perhaps by providing sites of bacterial multiplication inaccessible to phagocytes (see Ch. 24).

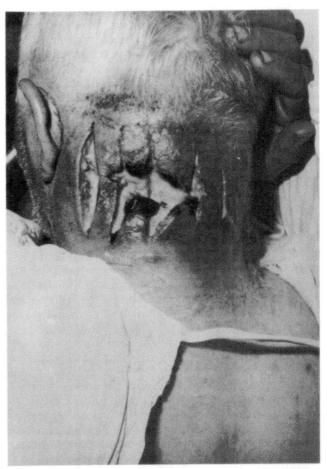

Fig. 29-4. Carbuncle caused by *S. aureus* in a patient with diabetes mellitus. (Smith IM, Rabinovitch S. In Top FH Sr., Wehrle PF, eds: Communicable and Infectious Diseases. St. Louis, Mosby, 1972)

Mice are also resistant to subcutaneous injections of staphylococci but susceptible to staphylococcal stitch abscesses. In rabbits other factors that cause tissue damage, such as thermal burns, chemical irritants, or the injection of other bacteria, have been shown to increase the local infectivity of pathogenic staphylococci. The most striking effects result from tissue necrosis, but even when the inciting procedure causes only acute inflammation, some loss of local resistance is demonstrable during the early, acellular phase of the inflammatory response. If, however, the staphylococci are injected into the inflammatory lesions after two days (when phagocytic cells have had time to mobilize), local resistance is found to be enhanced.

Special mechanisms of pathogenesis operate in staphylococcal diseases of the **gastrointestinal tract.** Because of their hardiness and their prevalence on the skin and mucous membranes staphylococci are frequently found in the feces. Their proliferation in the gastrointestinal tract, however, is controlled by the antagonistic action of the many other bacterial species present. When this balance is upset by a broad-spectrum antibiotic (e.g., a tetracycline) resistant staphylococci may achieve a high population density in the feces. Under such circumstances they sometimes invade the bowel wall and produce **acute staphylococcal enterocolitis,** which may be fatal. Many of the strains produce enterotoxin B, which probably also plays a causative role.

Determinants of Pathogenicity. Neither the host factors that prevent serious disease in normals nor the bacterial factors responsible for serious disease in susceptibles are known. Virtually all of the toxic and enzymatic products have been implicated: α-toxin is lethal for animals and seems likely to play a role in overwhelming septicemic disease; δ-hemolysin and P-V leukocidin both destroy human PMNs; lipase may be important in the development of boils; coagulase could wall off the lesions from phagocytic cells (although a fibrin coating on individual bacteria does not impede their phagocytosis); hyaluronidase breaks down interstitial hyaluronic acid and staphylokinase causes clot lysis, both of which could enhance the spread of lesions. On the other hand, strains that lack one or more of these substances have been isolated from lesions; and when such strains are produced by mutation they appear to be as virulent as the parental strain in experimental infections. Moreover, disease often develops despite the presence of Abs against these substances in the serum.

The role of cellular Ags is also unclear. *S. aureus* is taken up efficiently by human PMNs in normal human serum; all human sera apparently contain sufficient amounts of heat-stable and heat-labile opsonins for the organisms. Encapsulated strains tend to be taken up at a slightly lower rate and are more virulent for animals; and it also may be that capsule formation occurs in vivo but not in vitro. If so, these strains may also be more virulent for man, but proof is required. PMNs kill *S. aureus* extensively, but not as completely as coagulase-negative species. The importance of the few survivors is unknown. Protein A, despite its wide variety of biologic effects, particularly inhibition of phagocytosis, has not been shown to be a primary virulence factor.

It must be emphasized that virulent and avirulent strains of *S. aureus* cannot be defined in the same fashion as rough and smooth pneumococci, or *tox*$^+$ and *tox*$^-$ strains of *Corynebacterium diphtheriae*. Even epidemic strains (e.g., 80/81) demonstrate no qualitative or quantitative difference in presumed virulence factors, and indeed may be less virulent for animals than conventional strains. It is therefore likely that a number of bacterial factors interact with human fluids and cells in a complex, semistable equilibrium which is shifted in favor of the organism by a variety of local or systemic factors.

Humoral **antibodies** must play some role in staphylococcal disease, since agammaglobulinemic patients are at high risk. In addition, **delayed hypersensitivity** to *S. aureus* can result from repeated skin infections; and in experimental animals it is associated with more severe though more localized disease. Accordingly, delayed hypersensitivity may conceivably participate in the pathogenesis of staphylococcal disease much as it does in tuberculosis (Ch. 37).

TOXINOSES

Staphylococcal Food Poisoning. There is an abrupt onset 1–6 h after ingestion of contaminated food, with nausea, cramps, vomiting, and usually diarrhea. Recovery generally occurs within 24 h; rarely, deaths occur in infants and the elderly. Foods commonly implicated are pastries, custards, salad dressing, sliced meats, and meat products, contaminated by food handlers who have open lesions due to enterotoxin-producing *S. aureus* or who are asymptomatic carriers. Staphylococcal enterotoxins are heat-stable (100° C for 30 min); it is therefore essential to prevent multiplication of enterotoxin-producing organisms, and so *food must be refrigerated before as well as after cooking.*

Exfoliative Skin Disease. Exfoliatin causes a variety of syndromes: **generalized exfoliative dermatitis** in newborns (Ritter's disease) (Fig. 29-5); a form of **toxic epidermal necrolysis** (TEN) in children and occasionally adults; bullous **impetigo;** and **staphylococcal scarlatina,** which, unlike streptococcal scarlatina, spares the tongue and palate. The group of entities is collectively termed the **staphylococcal scalded skin syndrome.** Recovery is usual because the lesion is in the stratum granulosum and a sufficient layer of epidermis remains to avoid massive fluid loss and to protect against secondary deep skin infection.

Exfoliatin is produced by phage group II organisms, particularly those lysed by phage 71. Sometimes the organisms are found at the site of the skin lesions, but often they are at distant sites, with or without evidence of a primary staphylococcal lesion. In both cases it is clear that the skin lesions are mediated by a soluble product, which can be disseminated, and are distinct from those due to the organisms alone. The most severe form of the human disease is mimicked in newborn mice following injection of purified exfoliatin or of organisms from patients with the scalded skin syndrome. Exfoliatin is antigenic; the variation in the clinical picture and the infrequent occurrence of the severe syndrome in adults may reflect varying degrees of antitoxic immunity.

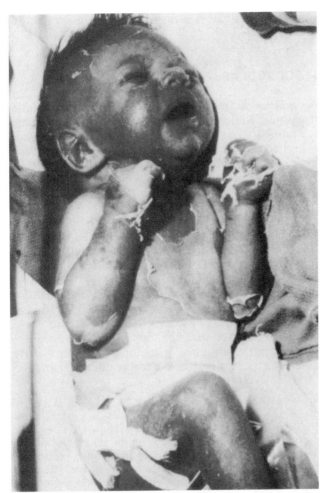

FIG. 29-5. Staphylococcal scalded skin syndrome in an infant. (Melish ME, Glasgow LA: N Eng J Med 282:1114, 1970. Reprinted by permission from the New England Journal of Medicine)

LABORATORY DIAGNOSIS

The finding of gram-positive cocci in stained smears of purulent exudates provides only suggestive information, since staphylococci cannot be differentiated from other gram-positive cocci on purely morphologic grounds. Specimens should be inoculated directly onto sheep blood, or preferably rabbit blood, agar plates and into thioglycollate broth. If contamination with other organisms is likely, mannitol–salt agar or phenylethyl agar should also be used. Blood cultures should be inoculated into blood agar pour plates as well as into broth.

Identification of *S. aureus* is suggested by pigment production, hemolysis, mannitol fermentation, and growth of the organism at a high salt concentration, but these properties are variable, whereas a positive coagulase test certifies the diagnosis. Although the presence of clump-

ing factor (bound coagulase) often correlates with free coagulase production, testing for free coagulase is preferred.

To perform the **coagulase test** a loopful of a colony on an agar plate, or 0.1 ml of a broth culture, is added to 0.5 ml of undiluted or diluted citrated rabbit plasma and the mixture is incubated at 37° C. Most positive strains cause clot formation within 4 h and many within 1 h, but tubes are observed for 24 h before being considered negative. (False-positive tests occur in mixed cultures containing citrate-utilizing gram-negative rods, e.g., *Pseudomonas.*) Specific strains are identified by bacteriophage typing.

Enterotoxins in suspected food are demonstrated by immunologic technics, such as gel diffusion, with specific antiserum. Diagnosis of the staphylococcal scalded skin syndrome is usually made on clinical grounds and by the isolation of *S. aureus* of phage group II. Antiserum for immunologic identification of exfoliatin is thus far not generally available.

TREATMENT

The therapeutic problem posed by drug-resistant staphylococci is self-evident. Because of the prevalence, particularly in hospital populations, of staphylococci resistant to one or more antimicrobials, it is essential to determine the drug sensitivity of the infecting organism. Penicillinase-producing strains are usually susceptible to semisynthetic penicillin derivatives that are not hydrolyzed by the enzyme; hence until the drug sensitivity is known the patient should be treated with such agents (e.g., methicillin or oxacillin) or a cephalosporin. Chemotherapy of severe staphylococcal disease should be intensive. Frankly suppurative lesions must be drained because established abscesses do not respond to antimicrobial therapy (see Abscess Formation, Ch. 25). Minor lesions do not generally require antimicrobial treatment.

PREVENTION

The principal reservoir of staphylococci in nature is man. Direct crossinfection from one human to another is extremely common and may either be airborne or result from direct contact.

Spread of Disease. How staphylococcal disease is acquired is not always clear. It seems likely that airborne staphylococci may infect exposed burns, and open wounds in operating rooms. Under other conditions, where trauma is not involved, more intimate contact may be necessary. In hospital nurseries the site of primary colonization in the newborn is the umbilicus, and spread of the infection can usually be shown to result from direct contact with infected attendants. To trace the mode of transmission in any given outbreak, however, is often difficult.

Hospital Epidemics. During recent years an increasing number of hospital epidemics of staphylococcal disease in the United States have been due to group III phage types, especially among surgical patients and newborn infants. The strains isolated have tended to be resistant to the antimicrobial drugs being used in the hospital. Although most of the surgical infections appear to have been acquired in the operating room some may be acquired on hospital wards. Attendants known to be nasal carriers of drug-resistant strains should be excluded from operating rooms, infant wards, and rooms of patients with known impairment of the immune system. Patients with open wounds infected with staphylococci should be separated, when feasible, from other susceptible patients. Although none of these methods will eliminate cross-infection their combined effects, together with strict adherence to accepted hygienic technics of patient care (including frequent washing of hands), may significantly lower the incidence of staphylococcal disease.

Although routine use of hexachlorophene for infants is banned because of its toxicity, in nursery outbreaks a 3% solution may be used for bathing infants.

Bacterial Interference. Deliberate colonization of newborns with an epidemiologically avirulent *S. aureus* (strain 502A) effectively prevents superinfection with epidemic strains and will abort nursery outbreaks. Similarly, this procedure is useful in managing patients with recurrent furunculosis: local inoculation of strain 502A, following antimicrobial therapy, may prevent reinfection. However, since strain 502A may cause disease this technic is not to be used indiscriminately. The mechanism of interference is uncertain but may involve bacteriocin production.

There is no convincing evidence that whole-cell staphylococcal vaccines or toxoids have a specific immunizing effect.

COAGULASE-NEGATIVE STAPHYLOCOCCI

S. epidermidis (S. albus). Colonies of this organism usually are white; occasionally, yellow or orange colonies are formed. The species-specific Ag is a cell wall teichoic acid (polysaccharide B) with a glycerol phosphate backbone; glucose residues are attached in either α- or β-glycosidic linkage to some of the glycerols, and D-alanine to others. Glucose is the antigenic determinant.

Strains of *S. epidermidis* are classified on the basis of different biochemical activities into four **biotypes** but recently a phage-typing system has been developed. Coagulase and the wide variety of extracellular products formed by *S. aureus* are not produced. Some strains are hemolytic, due to ϵ-hemolysin, which is very similar to or identical with δ-hemolysin of *S. aureus*. *S. epidermidis* is a frequent inhabitant of normal skin, where it is often responsible for minor abscesses. However, it also causes systemic disease on occasion, postoperative endocarditis in particular. Resistance to multiple antibiotics is frequent and therapy is often even more difficult than for *S. aureus* infections.

S. saprophyticus. This organism, previously classified as a micrococcus, is a new addition to the staphylococcal genus. It grows poorly under anaerobic circumstances, and unlike the other staphylococcal species it does not ferment glucose. However, it does have the characteristic lysostaphin-sensitive pentaglycine bridges in the peptidoglycan, and the cell walls contain ribitol teichoic acid of two types linked to glucosamine or to glucose; both may be present in the same strain. *S. saprophyticus* was at first regarded as avirulent, but there is evidence that it can be responsible for urinary tract infections. Unlike other staphylococci, *S. saprophyticus* is resistant to novobiocin.

PEPTOCOCCI AND PEPTOSTREPTOCOCCI

Peptococcus and *Peptostreptococcus* are strictly anaerobic gram-positive cocci that do not require fermentable carbohydrates for growth. The organisms may occur singly, in pairs, in chains, or, with the peptococci, in irregular clusters. They are part of the indigenous flora of the oral cavity, the skin, and the gastrointestinal and genitourinary tracts. They have been described as causative agents in a variety of infections, mostly in synergy with other organisms; mixed infection with *Bacteroides* or with *S. aureus* often produces severe necrosis. They are considered to be opportunistic organisms and are most frequently isolated from abdominal or pelvic surgical wound infections, although infection unassociated with trauma is seen. Bacteremia may occur.

On anaerobic blood agar plates colonies are small (0.5–1 mm), with color varying from grey to white. Occasional strains may produce α- or β-hemolysis. Species identification is based on morphology, biochemical reactions, and identification of metabolic products by gas–liquid chromatography. Immunologic technics are being developed but specific reagents are not yet widely available. Most strains are sensitive to penicillin.

SELECTED READING

BOOKS AND REVIEW ARTICLES

ARBUTHNOT JP: Staphylococcal α-toxin. In Montie TC, Kadis S, Ajl SJ (eds): Microbial Toxins, Vol III. New York, Academic Press, 1970

BAIRD-PARKER AC: Micrococcaceae. In Buchanan RE, Gibbons NE (eds): Bergey's Manual of Determinative Bacteriology, 8th ed. Baltimore, Williams & Wilkins, 1974

BERNHEIMER AW: Interactions between membranes and cytolytic bacterial toxins. Biochim Biophys Acta 344:27, 1974

COHEN JO (ed): The Staphylococci. New York, Wiley, 1970

JELJASZEWICZ J (ed): Staphylococci and Staphylococcal Disease. Proceedings of III International Symposium on Staphylococci and Staphylococcal Infections. New York, Gustav Fischer-Verlag, 1976

LACEY RW: Antibiotic resistance plasmids of *Staphylococcus aureus* and their clinical importance. Bacteriol Rev 39:1, 1975

MARTIN WJ: Anaerobic cocci. In Lennette EH, Spaulding EH, Truant JP (eds): Manual of Clinical Microbiology, 2nd ed. Washington DC, American Society for Microbiology, 1974

ROGOLSKY M: Nonenteric toxins of *Staphylococcus aureus*. Microbiol Rev 43:320, 1979

WISEMAN GM: The hemolysins of *Staphylococcus aureus*. Bacteriol Rev 39:317, 1975

YOTIS WW (ed): Recent advances in staphylococcal research. Ann NY Acad Sci 236:1, 1974

SPECIFIC ARTICLES

BERGDOLL MS, HUANG IY, SCHANTZ EJ: Chemistry of the staphylococcal enterotoxins. J Agric Food Chem 22:9, 1974

CHATTERJEE AN: Use of bacteriophage-resistant mutants to study the nature of the bacteriophage receptor site of *Staphylococcus aureus*. J Bactiol 98:519, 1969

FACKRELL HB, WISEMAN GM: Immunogenicity of the delta hemolysin of *Staphylococcus aureus*. J Med Microbiol 7:411, 1974

FORLANI L, BERNHEIMER AW, CHIANCONE E: Ultracentrifugal analysis of staphylococcal alpha toxin. J Bacteriol 106:138, 1971

FORSGREN A: Immunological aspects of protein A. In Schlessinger D (ed): Microbiology—1977. Washington DC, American Society for Microbiology, 1977

GROV A: Biological aspects of protein A. In Schlessinger D (ed): Microbiology—1977. Washington DC, American Society for Microbiology, 1977

KARAKAWA WW, KANE JA: Immunochemistry of an acidic antigen isolated from a *Staphylococcus aureus*. J Immunol 114:310, 1975

KLAINER AS, BETSCH CJ: Scanning-beam electron microscopy of selected microorganisms. J Infect Dis 121:339, 1970

KREGER AS, BERNHEIMER AW: Disruption of bacterial protoplasts and spheroplasts by staphylococcal delta hemolysin. Infect Immun 3:603, 1971

MARKOWITZ A, LERNER AM: Differentiation of several isolates of *Peptococcus magnus* by counterimmunoelectrophoresis. Infect Immun 16:152, 1977

MELISH ME, GLASGOW LA: The staphylococcal scalded-skin syndrome: development of an experimental model. N Engl J Med 282:1114, 1970

NAGEL JG, TUAZON CU, CARDELLA TA, SHEAGREN JN: Teichoic acid serological diagnosis of staphylococcal endocarditis. Ann Intern Med 82:13, 1975

NOVICK RP, MORSE SI: In vivo transmission of drug resistance factors between strains of *Staphylococcus aureus*. J Exp Med 125:45, 1967

NOVICK RP, WYMAN L, BOUANCHAUD D, MURRAY E: Plasmid life cycles in *Staphylococcus aureus*. In Schlessinger D (ed): Microbiology—1974. Washington DC, American Society for Microbiology

PETERSON PK, VERHOLF J, SABATH LD, QUIE PG: Effect of protein A on staphylococcal opsonization. Infect Immun 15:760, 1977

SMITH IM RABINOVITCH S: Staphylococcal infections. In Top FH Sr, Wehrle PF (eds): Communicable and Infectious Diseases. St. Louis, CV Mosby, 1972

TAYLOR AG, BERNHEIMER AW: Further characterization of staphylococcal gamma hemolysin. Infect Immun 10:54, 1974

THOMPSON NE, PATTEE PA: Transformation in *Staphylococcus aureus*: role of bacteriophage and incidence of competence among strains. J Bacteriol 129:778, 1977

TZAGALOFF H, NOVICK RP: Geometry of cell division in *Staphylococcus aureus*. J Bacteriol 129:343, 1977

WHITE A, BROOKS GF: Furunculosis, pyoderma, and impetigo. In Hoeprich PD (ed): Infectious Diseases, 2nd ed. Hagerstown, Harper & Row, 1977

YOSHIDA K, NAKAMURA A, OHTOMO T, IWAMI S: Detection of capsular antigen production in unencapsulated strains of *Staphylococcus aureus*. Infect Immun 9:620, 1974

chapter

THE NEISSERIAE

EMIL C. GOTSCHLICH

THE GENUS NEISSERIA

This genus includes two gram-negative species of pyogenic cocci that are pathogenic for man, and that have no other known reservoir: the **meningococcus** (*Neisseria meningitidis*) and the **gonococcus** (*Neisseria gonorrhoeae*). A number of nonpathogenic species also inhabit the upper respiratory tract and may be confused with meningococci.

Epidemic cerebrospinal meningitis, now usually referred to as **meningococcal meningitis,** was clearly recognized as a contagious disease early in the nineteenth century and was later found to be especially prevalent among military personnel. The causative organism was first isolated by Weichselbaum in 1887 from the spinal fluid of a patient with a purulent form of meningitis. However, the infection frequently does not extend to the meninges but remains a septicemia, referred to as **meningococcemia.** Meningococcal meningitis occurs endemically as well as in epidemics, and nasopharyngeal carrier rates for meningococci in the general population are relatively high and fairly constant.

Gonorrhea, the human ailment caused by the gonococcus, was known in the thirteenth century to be of venereal origin.

The term gonorrhea, meaning "flow of seed," was introduced by Galen (A.D. 130) under the mistaken impression that it was due to spermatorrhea. Gonorrhea was thought to be an early symptom of syphilis by Paracelsus, a teacher of great influence in the 1500s. To distinguish gonorrhea from syphilis the celebrated English physician John Hunter inoculated himself with the purulent exudate from a patient with gonorrhea, but unfortunately the donor had a double infection. Not until the mid-nineteenth century were syphilis and gonorrhea clearly differentiated.

N. gonorrhoeae, the causative agent of gonorrhea, was described by Neisser in 1879 and first cultivated by Leistikow and Loeffler in 1882. Treatment was unsatisfactory until the advent of the sulfonamide drugs, but sulfonamide-resistant strains of gonococcus became common within a few years; introduction of penicillin in 1943 restored control of the sulfonamide-resistant strains. Although penicillin has proved highly effective and has markedly decreased the serious consequences of the disease, its use has not lowered the **incidence** of gonococcal infections in the general population. Indeed, *gonorrhea is probably the most prevalent communicable bacterial disease in man today.*

MORPHOLOGY

The neisseriae are nonmotile, gram-negative cocci* which grow in pairs or occasionally in tetrads or clusters. The individual cocci are small (about $0.8\mu m \times 0.6\mu m$) and often assume bizarre shapes as a result of partial autolysis. The cytoplasm and the wall of the meningococcus and gonococcus are very similar in ultrastructure (Fig. 30-1). Many meningococci have a polysaccharide-containing capsule. Gonococci may also possess capsules, but their chemical composition is unknown. Both gonococci and meningococci may bear pili, and with the gonococcus these have been related to virulence.

METABOLISM

The principal species included in the genus *Neisseria* are listed in Table 30-1, together with the growth and fermentation characteristics that permit their differentiation. Only the nonpathogenic species grow at 22° C and on nutrient agar devoid of blood; also, each species exhibits a characteristic fermentation pattern when cultured in media containing glucose, maltose, or sucrose. The fermentation tests are done in semisolid cystine–typticase agar containing 0.5%–1% of the specific carbohydrate, and only small amounts of acid are generated in the positive reactions.

A group of neisseriae that ferment lactose but are otherwise metabolically indistinguishable from meningococci are designated as *Neisseria lactamica.* This species is commonly encountered in nasopharyngeal cultures of children below the age of 7 years, yet in only two cases has it been reported to cause disease.

All the species are essentially aerobic but will multiply under microaerophilic conditions. Moreover, members of the genus rapidly oxidize dimethyl- or tetramethyl-p-

---

* Although the neisseriae stain gram-negatively, contain endotoxins, are lysed by antibody and complement, and have an ultrastructure typical of gram-negative bacteria, they resemble pyogenic cocci in their pathogenicity and their sensitivity to penicillin.

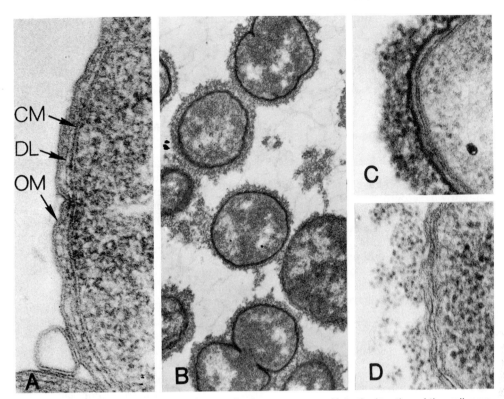

Fig. 30-1. A. Thin section electron micrograph of a gonococcus. Note the location of the cell membrane **(CM),** the dense layer **(DL),** which consists at least in part of the peptidoglycan, and the cell wall membrane **(OM)** containing the endotoxic lipopolysaccharide. (×160,000) **B–D.** Cells labeled with purified Ab against the group A meningococcal polysaccharide conjugated with horseradish peroxidase or ferritin. **B.** Group A meningococci stained with peroxidase-conjugated Ab. The polysaccharide forms a capsule around the organisms. (×25,000) **C.** Higher magnification (×120,000) showing an intensely opaque region immediately external to the cell wall and a less densely staining peripheral zone. **D.** Appearance of the capsule of the group A meningococcus stained with ferritin-conjugated Ab. (×120,000) (Electron-micrographs by Dr. John Swanson.)

phenylenediamine, causing surface colonies to turn first pink and then black when plates are flooded with a 1% solution of these reagents. The usefulness of this **oxidase test** is limited, since the nonpathogenic species of the genus also react.

Meningococci and gonococci are difficult to cultivate, primarily because of their sensitivity to the toxic fatty acids and trace metals contained in peptone and agar.

The inhibitory effect of these toxic components may be eliminated by adding to the medium blood, serum, starch, or charcoal, which bind the toxic substances. Blood agar, heated at $80°–90°C$ to form "chocolate agar," is even more suitable for the growth of gonococci. Growth of both species is stimulated by 5%–10% CO_2; hence cultures are usually incubated in a candle jar. Defined media for use with the gonococcus have been devel-

TABLE 30-1. Principal Differential Characteristics of the Common Species of the Genus *Neisseria*

			Fermentation		
Organism	Growth on agar devoid of blood	Growth at 22°C	Glucose	Maltose	Sucrose
N. meningitidis	–	–	+	+	–
N. gonorrhoeae	–	–	+	–	–
*N. catarrhalis**	+	+	–	–	–
N. sicca	+	+	+	+	+
N. flavescens	+	±	–	–	–

*Now *Branhamella catarrhalis.*

oped, and strain variations in nutritional requirements have proved useful for classification (**auxotyping**).

The resistance of meningococci and gonococci to unfavorable physical and chemical conditions is exceptionally low. They are particularly susceptible to desiccation, and they are killed when heated at 55° C for 30 min.

Both organisms also tend to undergo rapid autolysis, and in most laboratory cultures they die out in a few days. In preparing killed cell suspensions for serologic tests it is useful to inactivate the autolytic enzymes by heating the culture at 65° C for 30 min or adding cyanide or formalin.

NEISSERIA MENINGITIDIS

ANTIGENIC STRUCTURE

Most meningococci have a **polysaccharide capsule**. The ten known serogroups are described in Table 30-2. Group A strains are the classic epidemic strains, and have caused most of the major epidemics. Groups B and C are generally endemic strains, but C on occasion may cause an epidemic outbreak. The other serogroups infrequently cause disease but are quite commonly found as carrier strains. Occasionally nonencapsulated strains may be found in nasopharyngeal cultures, and these tend to autoagglutinate or to be agglutinated by several of the grouping sera.

Group C strains can be subclassified serologically into C1+ and C1−; the difference is based on whether the polysaccharide is O-acetylated. Both variants have been isolated from patients with disease.

All known group A strains have the same protein Ags of the outer membrane. Group B strains, however, have a dozen serotypes based on the principal outer membrane proteins.* These serotypes are shared by the other serogroups (C, Y, and W-135) that also contain sialic acid in their capsules. The endotoxin of meningococci has not been fully characterized.

GENETIC VARIATION

When smooth encapsulated strains of meningococci are repeatedly subcultured on artificial media they often yield rough (R) variants, which are unencapsulated and relatively avirulent. If mice are infected with a rough variant suspended in mucin, smooth (S) revertants may be recovered from the heart blood.

Various drug-resistant mutants have arisen. Strains highly resistant to sulfonamides are now very common in both military and civilian populations in many parts of the world. So far *no plasmids have been found in meningococci*.

Genetic **transformation** by DNA has been observed with the meningococcus: it is more efficient in piliated organisms. DNA accumulates in the slime of broth cultures, and is just as active in transformation as the DNA extracted from intact cells, suggesting that *transformation affecting virulence may occur in mixed populations in nature*. Such genetic recombination could contribute to the frequency of epidemics among army recruits brought together from diverse locations.

PATHOGENICITY

The meningococcus usually inhabits the human nasopharyngeal area without causing any symptoms. This **carrier state** may last for a few days to months; it provides the reservoir for the meningococcus, and it enhances the immunity of the host. The carrier rate tends to be appreciable at all times in the general population, and in military populations it may exceed 90%. When individuals without adequate levels of immunity acquire the meningococcus, usually through contact with a healthy carrier, the resulting nasopharyngeal infection may lead rapidly to bacteremia, which is ordinarily followed by acute purulent meningitis.

Meningococcal disease occurs with an **endemic** incidence of approximtely 2:100,000 per annum or less and shows a marked predilection for children below the age of 5 years, with the highest incidence in the first year of life. During the first 3 months, however, maternal antibodies still play a protective role. Rates in an **epidemic** may exceed the endemic rate by more than 100-fold, and the incidence above 2 years of age increases disproportionately.

The **mortality** of meningococcal meningitis is approximately 85% without treatment, but in young adults, such as military recruits, it can be less than 1% when early, vigorous antibiotic and supportive therapy is administered. It is approximately 10% in the general population. Mercifully, the rate of neurologic sequelae is low, in contrast to the rate following other prevalent meningitides.

Infrequently, meningococcemia fails to lead to meningitis, but metastatic lesions often develop in the skin, joints, lungs, ears, and adrenal glands. These lesions are usually thromboembolic and frequently contain cultivable meningococci. Acute adrenal insufficiency (Waterhouse–Friderichsen syndrome) occurs in fulminating cases and is commonly associated with hemorrhagic in-

---

* **Serogroup** refers to the classification on the basis of the capsular polysaccharides, while **serotypes** are based on the principal outer membrane protein Ags. Additional typing systems based on lipopolysaccharides are under development.

TABLE 30-2. Structures of the Polysaccharide Antigens of *N. meningitidis*

Serogroup	Repeating unit	Linkage	*O*-Acetyl content*	Location of *O*-acetyl groups
A	2-Acetamido-2-deoxy-D-mannopyranosyl phosphate	1→6-α	0.7	C-3 of mannosamine
B	D-N-Acetylneuraminic acid	2→8-α	None	—
C1+	D-N-Acetylneuraminic acid	2→9-α	1.3	C-7 and C-8 of sialic acid
C1-	D-N-Acetylneuraminic acid	—	None	—
D	Unknown	—	—	—
W-135	4-*O*-α-D-Galactopyranosyl-*N*-acetylneuraminic acid	2→6-α	None	—
X	2-Acetamido-2-deoxy-D-glucopyranosyl phosphate	1→4-α	None	—
Y	4-*O*-α-D-Glucopyranosyl-*N*-acetylneuraminic acid	2→6-α	1.1	Not specifically located
Z′	7-*O*-α-D-2-Acetamido-α-deoxy-galactopyranosyl-2-keto-3-deoxy-D-octulosonic acid	2→3-β	1.0	C-4 and C-5 of ketodeoxyoctonate
Z	Unknown	—	—	—

*Expressed in moles per mole of repeating unit

volvement of both adrenal glands.* Very rarely, meningococcemia becomes chronic and lasts for weeks as a fever of unknown etiology.

Meningococci behave in the animal host as **extracellular parasites**, i.e., they are unable to survive phagocytosis. Although they are often visible within leukocytes (Fig. 30-2), there is no evidence that they can multiply intracellularly. Virulence depends in part upon the antiphagocytic properties of the capsule, and in part on the principal outer membrane protein (serotype 2 is disproportionately common in isolates of patients with disease). As with pneumococci, the nature of the capsular polysaccharide is important: 90% or more of meningococcal disease is caused by strains of groups A, B, and C, while the other encapsulated strains of groups X, Y, Z, Z′, and W-135 are commonly found in carriers but rarely cause disease.

Endotoxins cause extensive vascular damage and are capable of producing both localized and generalized Shwartzman reactions (see Endotoxin, Ch. 24), with pathologic findings resembling those of the thromboembolic lesions of meningococcemia. It seems plausible, but has not been proved, that such reactions are involved in the pathogenesis of meningococcal disease.

Laboratory Models. Studies on pathogenesis have been limited, since meningococci are relatively nonpathogenic for most laboratory animals. Meningitis has been produced in various animals by direct injection into the subarachnoid space, but such large num-

bers of organisms are required that the relevance to pathogenesis in man is questionable. On intraperitoneal inoculation of mice with meningococci suspended in gastric mucin, which protects the organisms from prompt destruction by the peritoneal phagocytes, inocula of as few as ten meningococci may cause fatal infection.

Studies in chick embryos have been more interesting: regardless of the site of inoculation (amniotic sac, body wall, or vein), the meningococci eventually localized in the cranial sinuses, lungs, and meninges (see Organotropism, Ch. 24). Serial sections of the embryos revealed no evidence of direct extension of the infection from the nasopharynx to the brain, but bacteremia always preceded involvement of the meninges. The embryo can be protected by administration of Ab.

IMMUNITY

Antibodies are present in the serum of most adults, and they play a role in preventing meningococcal disease.

Fig. 30-2. Intracellular meningococci in smear of exudate from spinal fluid of patient with meningococcal meningitis. (×1000)

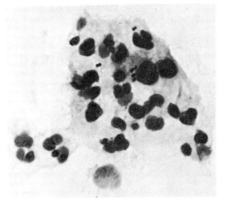

* Acute adrenal insufficiency of this sort may also be associated with overwhelming bacteremia due to other organisms, such as staphylococci, pneumococci, and streptococci. It may also be produced experimentally in rabbits given intravenous endotoxin within a few hours after stimulation of their adrenal cortices with ACTH, suggesting that the stimulated glands are hypersusceptible to the action of meningococcal endotoxin.

Thus, of 15,000 military recruits, 54 contracted group C meningococcal disease; serum bactericidal activity against this organism was initially absent in 51 of these individuals but was present in the majority of those who did not contract the disease. Moreover, the newborn very rarely contracts the disease, but with the loss of maternal Abs susceptibility appears; and children with agammaglobulinemia are protected by monthly administration of pooled concentrated human γ-globulin. The risk of disease is greatest between the ages of 6 and 24 months, and almost half the cases occur before age 5 years.

Protective Abs are evoked within a week by the meningococcal carrier state. These Abs are directed not only against the capsular group-specific polysaccharide but also against other species-specific Ags, including the endotoxin. Low-level immunity may explain why young children rarely contract disease on their first contact with a potentially virulent meningococcus: the immunity might result from prior contact with a nonencapsulated strain, or with an encapsulated strain of low virulence (such as *Neisseria lactamica* frequently found in nasopharyngeal cultures of children), or with a **crossreacting polysaccharide** from some other source. Thus, the majority of children and young adults in the United States have Abs to the group A capsular polysaccharide, although group A meningococci have rarely been encountered in the last 15 years; immunization by some crossreacting organism (such as some strains of *Bacillus pumilus* and of group D *Streptococcus*) must be suspected. Furthermore, the *Escherichia coli* capsular antigen K92 crossreacts with meningococcal group C polysaccharide, and *E. coli* antigen K1 is identical to the group B polysaccharide. (Curiously, 75% of the neonatal meningitis caused by *E. coli* is due to strains carrying the K1 antigen.)

Multiple attacks of meningococcal meningitis appear to be associated frequently with deficiencies of complement components C6, C7, or C8. The patients also show increased susceptibility to systemic infections with the gonococcus, but not with other pyogenic organisms.

LABORATORY DIAGNOSIS

In cases of suspected meningococcal disease specimens of blood, spinal fluid, and nasopharyngeal secretions should be examined for the presence of *N. meningitidis*. In addition, stained smears made from petechial lesions in the skin will occasionally reveal the organism.

The accepted **blood culture** procedure is to add 10 ml of blood to 100 ml of a suitable liquid medium (e.g., tryptose phosphate broth) and also to spread 0.1 ml of blood on the surface of a blood (or chocolate) agar plate. Both cultures are incubated at 37° C in a candle jar and are inspected daily for 7 days before being discarded. When heavy bacteremia is present a diligent search may reveal gram-negative diplococci in routine blood smears.

Spinal fluid may be permitted to drop directly from the lumbar puncture needle onto the surface of blood agar (or chocolate agar) plates, and a tube of broth should also be inoculated. In addition, a sample of the spinal fluid by itself should be incubated in a candle jar at 37° C and later subcultured; this procedure will occasionally reveal meningococci when the original drop cultures are negative. Smears of spinal fluid that are to be stained for bacteria must be promptly fixed, to prevent autolysis. Capsular antigen, which is generally present in the spinal fluid, may be demonstrated rapidly by countercurrent immunoelectrophoresis with appropriate antisera; these are becoming commercially available. Spinal fluid should also be analyzed for glucose and protein.

Nasopharyngeal cultures are best performed with a cotton swab on the end of a bent wire. Care must be taken to avoid touching the tongue: secretions must be obtained from the posterior nasopharyngeal wall, behind the soft palate.* The secretions should be swabbed directly and then streaked with a wire loop on a selective medium containing antibiotics to inhibit the contaminating flora. Prompt incubation in a candle jar is essential. If the cultures cannot be made immediately the swab should be placed in a transport medium in which meningococci remain viable for a number of hours. Although a positive nasopharyngeal culture does not prove meningococcal disease, meningococcal carriers can be detected only in this way.

Neisseria colonies may be tentatively recognized by the oxidase test. To identify meningococci, fermentation and serologic tests must be made. The latter are done with either monotypic or polytypic antisera and bacterial suspensions in saline containing 0.1% potassium cyanide to inhibit autolysis. Since nonspecific agglutination may be encountered, a control test with normal serum must be included.

In population surveys the identification of meningococci (or of crossreactive organisms) is markedly simplified by inoculating the cultures on agar supplemented with group-specific serum. The positive colonies can then be identified by a halo of immune precipitate.

TREATMENT

The drug of choice is penicillin, administered intravenously. Although it does not penetrate the normal blood–brain barrier, it does so readily when the meninges are acutely inflamed. If a history of anaphylactic re-

---

* Whereas meningococci usually inhabit only the nasopharyngeal mucosa behind the soft palate, staphylococci reside in the anterior nasopharynx and vestibule of the nose, and pneumococci, β-hemolytic streptococci, viridans streptococci, and *Hemophilus influenzae* are found most frequently on the mucous membranes of the tonsils and lower pharynx. (An alternative method of obtaining nasopharyngeal cultures, particularly in children, is described in *B. pertussis*, Laboratory Diagnosis, Chapter 34.)

activity to penicillin contraindicates its use, combined therapy with erythromycin and chloramphenicol is recommended. The success of antimicrobial treatment is much greater if therapy is started early. If signs of adrenal insufficiency develop therapy with hydrocrotisone, pressor amines, and parenteral fluids is required.

Except where facilities for intravenous therapy are unavailable, such as the rural areas of developing countries, sulfonamides are no longer used because of their greater toxicity and because resistant organisms have become prevalent.

PREVENTION

Two methods for preventing meningococcal disease exist: chemoprophylaxis and immunoprophylaxis. **Chemoprophylaxis** with sulfonamides was first used in World War II among military recruits. The meningococcal carrier rate and the incidence of disease were lowered, but by the early 1960s this measure was no longer effective because of the prevalence of resistant meningococci. At present rifampin and minocycline are used, but because they have several disadvantages they should be considered only for individuals in close contact with a case, such as those living in the same household or participating in mouth-to-mouth resuscitation. There is no evidence that classroom contact or the usual hospital personnel contact entails a significant added risk.

Vaccines consisting of purified group A and group C **capsular polysaccharide** are nontoxic, and wide use under epidemic conditions in several countries have shown them to be highly effective in preventing meningococcal disease. As with many Ags, the immune response of infants and young children is weak: with the group A vaccine children as young as 3 months old can be protected, but with the group C vaccine no significant protection is obtained below the age of 18 months. The Ab responses in those above the age of 7 years last for several years, but in younger children the persistence is shorter. Because of the low incidence of the disease in the United States at present (1:100,000 per annum) the vaccines are recommended only for epidemic control. Vaccines are under development for the prevention of disease due to other groups.

NEISSERIA GONORRHOEAE

ANTIGENIC STRUCTURE

There are at least 16 distinct serotypes of genococci, differing in their principal outer membrane proteins (mol wt 32,000–39,000). This serotyping system is closely analogous to that developed for group B meningococci, but these Ags of the two species are serologically unrelated.

Gonococci have also been classified serologically by means of an immunofluorescent technic patterned on the one used successfully for the classification of certain chlamydiae. In this classification, strains that have been isolated from patients with disseminated gonococcal disease fall primarily into two antigenic subclasses.

Gonococcal **pili** (Fig. 30-3) have been isolated and found to consist primarily, if not exclusively, of a single kind of protein subunit of about 20,000 mol wt. Several serologically distinct types are found.

Serologic evidence indicates appreciable heterogeneity in the determinants of the **lipopolysaccharide** (LPS) of the gonococcus. However, it has proved difficult to isolate LPS with polysaccharide attached to its core; the LPS consists of Lipid A, ketodeoxyoctonate (KDO), heptose phosphate, glucosamine, glucose, and galactose. The β-1, 4-linked galactose–glucose residues (lactose) constitute the immunodominant determinant.

GENETIC VARIATION

Gonococci grown on agar vary in colonial morphology, including colony size, sharpness of borders, and opacity. Piliated gonococci form small colonies, with sharply defined borders if heavily piliated; nonpiliated organisms give rise to large colonies, especially upon subculture. As observed with streptococci (see Fig. 28-3), bacteria that adhere in large aggregates form opaque colonies, while nonadherent strains form transparent colonies (Fig. 30-3). Opacity depends on additional outer membrane proteins; trypsin attacks these proteins, and it is generally lethal to opaque but not to transparent variants.

Piliated strains are capable of genetic **transformation,** and strains with an appropriate plasmid exhibit **conjugation.** Transformation may be involved in the high rate of colonial variation.

The chromosomal genes for resistance to several antibiotics (streptomycin, spectinomycin, tetracycline, chloramphenicol), and to rifampin, form a close cluster. The genes for resistance to penicillin and to erythromycin map elsewhere.

Recently penicillin-resistant gonococci that carry a **plasmid** coding for the production of β-lactamase have been isolated. Hybridization analysis indicates that the resistance gene may have been derived from the enterobacterial R factor pool, and a similar plasmid has also been found in *Hemophilus influenzae* in recent years.

The resistance plasmids that have become prevalent in gonococci in the Far East and in England have mol wts of 4.4×10^6 and 3.2×10^6, respectively, suggesting two independent transfers of the same resistance gene to a plasmid that was then selected for by the wide use of penicillin. Gonococci contain additional cryptic plasmids.

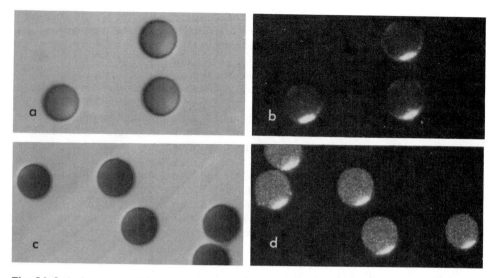

Fig. 30-3. A. Appearance of gonococci as seen through a colony scope. Two modes of substage lighting are used: light reflected from a ground glass diffuser **(a,c)** or transmitted from a polished mirror **(b,d)**. Panel a and b show heavily piliated transparent organisms, panel c and d heavily piliated opaque gonococci. Note the sharply defined borders indicative of profuse piliation (a,c) and the opacity to transmitted light (d). (x60) For an illustration of pili see Fig. 2-6.(Photomicrographs and electron-micrographs by Dr. John Swanson.)

PATHOGENICITY

As with meningococcal disease, study of the pathogenesis of gonorrhea has been greatly hampered by the absence of a satisfactory laboratory model. A progressive infection may be established in chick embryos, or in the anterior chamber of the rabbit's eye, and in mice a fatal disease may be produced by intraperitoneal injection of gonococci suspended in hog gastric mucin. Manifestly, none of these models simulates the human disease. Recently the disease has been established in chimpanzees by artificial inoculation, and natural transmission by coitus has been observed. Inoculation studies in human male volunteers have shown that piliated gonococci can cause genitourinary disease, whereas non-piliated strains are not virulent.

Studies of material from patients, and of short-term organ cultures of human fallopian tubes, showed that the *piliated gonococci stick to the epithelial cells* much more readily than nonpiliated organisms. They then enter these cells (presumably by phagocytosis), where they multiply freely, and finally burst the cells*, thereby gaining access to the subepithelial space. There the organisms evoke an acute inflammatory response, giving rise to the purulent urethral or cervical discharge characteristic of this disease.

* Gonococci are readily killed when phagocytized by "professional" phagocytic cells, such as the polymorphonuclear leukocyte.

The initial attachment of gonococci to polymorphonuclear leukocytes (PMNs) in vitro depends on a **leukocyte association factor** on the gonococcal surface. In some male urethral discharges the organisms appear to be mainly ingested by or attached to PMNs, while in others they are mainly free and unattached. Upon isolation the former strains have been found to have the leukocyte association factor, while the latter lack it. **Pili** have little bearing on the initial attachment to PMNs (in contrast to their affinity for epithelial cells), but they impair the subsequent ingestion.

In the male, infection may extend to the prostate and the epididymis. If untreated these invasions are often followed by fibrosis and stricture.

In the female the gonococcal population varies strikingly during the menstrual cycle: the endocervix harbors primarily transparent, trypsin-resistant variants at the time of menses, and opaque, trypsin-sensitive variants at midcycle. The former are more virulent in the chick embryo; and in women it is particularly at the time of their preponderance, during the first menses after infection, that the disease may spread to the fallopian tubes, causing acute or chronic pelvic inflammatory disease. In 1% or less of infected individuals bacteremia occurs, which then leads commonly to purulent arthritis and dermatitis and infrequently to endocarditis.

In general the strains isolated from disseminated gonococcal infection are resistant to the bactericidal action of fresh human serum; they also are frequently of the auxotype requiring arginine, uracil, and hypoxanthine for growth and are exquisitely sensitive to penicillin. However, these properties are genetically not closely linked.

Newborn infants may contract serious gonococcal infections of the eye (**ophthalmia neonatorum**) from passage through an infected birth canal. Unless promptly treated the disease may result in blindness.

The gonococcus may also establish itself in the **pharynx** or in the **rectal mucosa**. Generally, these infections are asymptomatic but occasionally severe proctitis may occur. In prepubertal girls, whose vaginal epithelium is thin and not cornified, **vulvovaginitis** may occur. This disease is ordinarily due to other species of pyogenic organisms; when caused by the gonococcus it has usually been acquired by sexual contact.

Repeated attacks of gonorrhea are common. Whether this occurs because local immunity cannot be developed or because of the large number of serotypes is not known.

LABORATORY DIAGNOSIS

In acute gonococcal disease **stained smears of fresh exudate** will often reveal the presence of **intracellular gram-negative diplococci**. This finding, together with a convincing clinical history, may permit the physician to make the provisional diagnosis of acute gonorrhea and to institute specific therapy. Whenever possible the exudate should be cultured for gonococci before treatment is begun. The diagnosis is definitely established by recovery of typical gram-negative diplococci that ferment glucose but not maltose or sucrose (Table 30-1). Rectal cultures are sometimes positive when urethral or cervical cultures are negative.

In chronic gonorrhea examination of gram-stained material is much less helpful and reliance must be placed on cultural or fluorescent Ab technics: particularly in chronic cervicitis commensal species of neisseriae may be present. Because gonococci are not hardy, specimens should be cultured immediately or placed in a special transport medium, e.g., charcoal-adsorbed semisolid thioglycollate agar. The medium of choice for the cultivation of gonococci contains antibiotics that inhibit any contaminating flora, and it is incubated in candle jars. After 12–16 h smears of the early growth may be stained with specific fluorescent Ab for rapid identification of *N. gonorrhoeae*. Immunologic tests for antigonococcal Abs in acute and convalescent sera are helpful when patients pose a diagnostic problem.

TREATMENT

The sulfonamide drugs provided the first effective therapy. During World War II, however, sulfonamide-resistant strains of gonococcus were increasingly encountered. Fortunately penicillin was developed at this time, and it gradually replaced the sulfonamides in the treatment of gonorrhea. As a rule, male patients require less intensive treatment than females. When walled-off abscesses are formed, as in chronic salpingitis, combined surgical and antibiotic therapy is necessary.

Recently, strains with plasmid-mediated penicillinase have appeared, as noted earlier. Spectinomycin is the favored alternative.

PREVENTION

Although present-day methods of treating gonococcal disease are extremely effective, the morbidity rates are of pandemic proportions. Since the incubation period of symptomatic gonorrhea is short, varying from 2 to 7 days, tracing of the recent sexual contacts, followed by treatment, lessens the spread of the disease. However, at least 10% of males and 50% of females with recently acquired gonorrhea have insufficient symptoms to seek treatment, and unfortunately immunologic tests are not reliable enough to be a cost-effective method of identifying the asymptomatically infected members of the population. Greater mobility, greater sexual freedom, and replacement of the condom by other birth control measures make the containment of the disease next to impossible. Research on vaccines is progressing, but none is presently available. For the individual the best preventive measures are discretion and the condom.

The time-honored method of preventing ophthalmia neonatorum (the **Credé procedure**) consists of dropping 1% silver nitrate solution into the infant's eyes immediately after birth. Many hospitals now use an ophthalmic ointment containing penicillin.

OTHER SPECIES OF *NEISSERIA* AND RELATED GENERA

Nonpathogenic species of neisseriae may be mistaken for meningococci or gonococci. For example, *N. catarrhalis* (*Branhamella catarrhalis*) and *N. sicca* (Table 30-1), and *N. lactamica* (as mentioned earlier), are frequently present in secretions of the normal pharynx, and other nonpathogenic species occasionally inhabit the female genital tract. Despite their lack of general pathogenicity, the first two occasionally cause bacterial endocarditis.

Several other species of gram-negative diplococci may be confused with the neisseriae. An **anaerobic** group, given the generic name *Veillonella*, inhabit the mouth and gastrointestinal tract of humans and certain animals and have been found in dental abscesses and in chronic urinary tract infections. An **aerobic** group, members of the genus *Acinetobacter* (formerly *Mima* and *Herellea*; see Ch. 32), including a species originally named *Bacterium ani-*

tratum, are tiny aerobic coccobacilli that may simulate gonococci in morphology and are occasionally cultured from the genitourinary tract. A third group includes the Morax–Axenfeld bacillus (genus *Moraxella;* see Ch. 32) which is sometimes cultured from patients with conjunctivitis.

SELECTED READING

BOOKS AND REVIEW ARTICLES

BROOKS GF JR et al (eds): Immunobiology of the Gonococcus. American Society for Microbiology, Washington, DC, 1978

SPECIFIC ARTICLES

APICELLA MA: Serogrouping of *Neisseria gonorrhoeae:* identification of four immunologically distinct acidic polysaccharides. J Infect Dis 134:377–383, 1976

BHATTACHARJEE AK, JENNINGS HJ, KENNY CP, MARTIN A, SMITH ICP: Structural determination of the sialic acid polysaccharide antigens of *Neisseria meningitidis* serogroups B and C with carbon 13 nuclear magnetic resonsance. J Biol Chem 250:1926–1932, 1975

BHATTACHARJEE AK, JENNINGS HJ, KENNY CP, MARTIN A, SMITH ICP: Structural determination of the polysaccharide antigens of *Neisseria meningitidis* serogroups Y, W-135, and BO. Canad J Biochem 54:1–8, 1976

BLAKE M, HOLMES KK, SWANSON J: Studies on gonococcus infection XVII. IgA$_1$-cleaving protease in vaginal washings from women with gonorrhea. J Infect Dis 139:89–92, 1979

BUMGARNER LR, FINKELSTEIN RA: Pathogenesis and immunology of experimental gonococcal infection: virulence of colony types of *Neisseria gonorrhoeae* for chicken embryos. Infect Immun 8:919–924, 1973

BUNDLE DR, SMITH ICP, JENNINGS HJ: Determination of the structure and conformation of bacterial polysaccharides by ^{13}C nuclear magnetic resonance studies on the group-specific antigens of *Neisseria meningitidis* serogroups A and X. J Biol Chem 249:2275–2281, 1974

DEHORMAECHE RD, THORNLEY MJ, GLAUERT AM: Demonstration by light and electron microscopy of capsules on gonococci recently grown *in vivo.* J Gen Microbiol 106:81–91, 1978

FRASCH CE, GOTSCHLICH EC: An outer membrane protein of *Neisseria meningitidis* Group B responsible for serotype specificity. J Exp Med 140:87–104, 1974

FRASCH CE, MCNELIS RM, GOTSCHLICH EC: Strain-specific variation in the protein and lipopolysaccharide composition of the Group B meningococcal outer membrane. J Bacteriol 127:973–981, 1976

FROHOLM LE, JYSSUM K, BØVRE K: Electronmicroscopical and cultural features of *Neisseria meningitidis* competence variants. Acta Pathol Microbiol Scand [B] 81:525–537, 1973

HERMODSON MA, CHEN KCS, BUCHANAN TM: *Neisseria pili* proteins: amino-terminal amino acids sequences and identification of an unusual amino acid. Biochem 17:442–445, 1978

JAMES JF, SWANSON J: Studies on gonococcus infection. XIII. Occurrence of color/opacity colonial variants in clinical cultures. Infect Immun 19:332–340, 1978

JAMES JF, SWANSON J: The capsule of the gonococcus. J Exp Med 145:1082–1086, 1977

JOHNSON AP, TAYLOR-ROBINSON D, MCGEE ZA: Species specificity of attachment and damage to oviduct mucosa by *Neisseria gonorrhoeae.* Infect Immunity 18:833–839, 1977

JOHNSTON KH, HOLMES KK, GOTSCHLICH EC: The serological classification of *Neisseria gonorrhoeae.* I. Isolation of the outer membrane complex responsible for serotypic specificity. J Exp Med 143:741–758, 1976

KING GJ, SWANSON J: Studies on gonococcus infection. XV. Identification of surface proteins of *Neisseria gonorrhoeae* correlated with leukocyte association. Infect Immun 21:575–584, 1978

LEPOW ML, GOLDSCHNEIDER I, GOLD R, RANDOLPH M, GOTSCHLICH EC: Persistence of antibody following immunization of children with Group A and Group C meningococcal polysaccharide vaccines. Pediatrics 60:673–680, 1977

MORSE SA, CACCIAPUOTI AF, LYSKO PG: Physiology of *Neisseria gonorrhoeae.* Adv Microb Physiol 20:251–320, 1979

MUNFORD RS, SUSSUARANA DE VASCONCELOS ZJ, PHILLIPS CJ, GELLI DS, GORMAN GW, RISI JB, FELDMAN RA: Eradication of carriage of *Neisseria meningitidis* in families: a study in Brazil. J Infect Dis 129:644–649, 1974

PLAUT AG, GILBERT JV, ARTENSTEIN MS, CAPRA JD: *Neisseria gonorrhoeae* and *Neisseria meningitidis:* Extracellular enzyme cleaves human immunoglobulin A. Science 190:1103–1105, 1975

ROBERTS M, FALKOW S: Plasmid-mediated chromosomal gene transfer in *Neisseria gonorrhoeae.* J Bacteriol 134:66–70, 1978

SALIT IE, GOTSCHLICH EC: Gonococcal color and opacity variants: virulence for chicken embryos. Infect Immunity 22:359–364, 1978

SCHOOLNIK GK, BUCHANAN TM, HOLMES KR: Gonococci causing disseminated gonococcal infection are resistant to the bactericidal action of normal human sera. J Clin Invest 58:1163–1173, 1976

SIPPEL JE, QUAN A: Homogeneity of protein serotype antigens in *Neisseria meningitidis* Group A. Infect Immun 16:623–627, 1977

SOX TE, MOHAMMED W, BLACKMAN E, BISWAS G, SPARLING PF: Conjugative plasmids in *Neisseria gonorrhoeae.* J Bacteriol 134:278–286, 1978

SWANSON J: Studies on gonococcus infection. XIV. Cell wall protein differences among color/opacity colony variants of *N. gonorrhoeae.* Infect Immun 21:292–302, 1978

chapter **31**

THE ENTERIC BACILLI AND BACTEROIDES

ALEX C. SONNENWIRTH

THE ENTERIC ORGANISMS

In stool cultures incubated **aerobically** the predominant organisms are members of the family Enterobacteriaceae: facultative gram-negative rods, with or without peritrichous flagella, that can grow in simple media. These organisms (Table 31-1) are found mostly in the vertebrate gut as normal flora or as pathogens, though some genera are saprobes or plant parasites. Enterobacteriaceae (enterobacteria, in the vernacular), together with the gram-negative vibrios, are also known unofficially as **enteric bacilli** or **the enterics,** though these terms have sometimes been extended to include other gram-negative rods found in the gut. The vibrios (Vibrionaceae) include the important agent of cholera.

Aerobic stool cultures often also include a small number of **nonfermenters,** which are negative in tests for fermentation of various sugars. These organisms are found primarily as nonpathogens in nature, but they have become increasingly important in infections outside the gut; they will be considered in the next chapter.

One genus of the Enterobacteriaceae, *Yersinia,* was formerly included with *Pasteurella;* because of its special distribution and pathogenic characteristics it is considered with that group in Chapter 33. Chapters 33 and 34 will also present other, more fastidious, gram-negative bacilli (*Francisella, Brucella, Hemophilus,* and *Bordetella*) which are not found in the gut.

Escherichia coli is the predominant facultative organism in stool cultures. However, to designate it and other "coliform" facultative bacteria as the enteric bacilli is a misnomer: **95%–99% of the gut flora** (amounting to some 10^{11} bacteria per gram of feces) are **obligate anaerobes** with complex growth requirements. Gram-negative rods (Bacteroidaceae) and gram-positive rods (*Bifidobacterium,* anaerobic lactobacilli) are major components. These give the stool most of its characteristic odor. Bifidobacteria have not been identified as a cause of disease; and while Bacteroidaceae also have not been demonstrably associated with enteric disease, they are being increasingly recognized in other infections, especially among the mixed flora often encountered in deep tissue sepsis and in septicemia. Bacteroidaceae will be considered in this chapter.

A number of other gram-positive organisms also appear in the gut flora, more variably and in smaller numbers. The facultative members include enterococci (chiefly *Streptococcus faecalis,* Ch. 28) and lactobacilli, and under some circumstances staphylococci (Ch. 29). The obligate anaerobes include anaerobic streptococci (Ch. 28) and clostridia (Ch. 36). The shifts in gut flora with age are discussed in Chapter 44.

The several groups of facultative and aerobic gram-negative rods are differentiated by a variety of criteria, listed in Table 31-2. The initial division between **fermenters** of sugars (Enterobacteriaceae, Vibrionaceae) and **nonfermenters** (i.e., either **oxidizers** or **nonutilizers** of sugars) is conveniently made by growth in two tubes of oxidation–fermentation (O–F) medium, one aerobic (open) and the other anaerobic (sealed with mineral oil). Fermenters produce acid in both tubes, oxidizers only in the open tube, and nonutilizers in neither. Also of particular value in differentiating major groups are 1) the **oxidase test** for the presence of an electron transport system (which oxidizes derivatives of *p*-phenylenediamine to a colored product), and 2) the type of **flagella:** peritrichous (when present) in Enterobacteriaceae, polar in pseudomonads and vibrios. The wide evolutionary distances between these groups are reflected in their DNA composition (Table 31-2).

Because the fecal flora is so mixed the use of selective media is indispensable, especially for isolating pathogens in the presence of a predominantly normal flora.

DNA hybridization patterns have been determined in most genera of Enterobacteriaceae. Most of the resulting groups (genospecies) correspond well to existing phenotypic species.

MEDICAL IMPORTANCE

Accurate identification of the organisms described here is of great value for choosing appropriate antimicrobial agents, for prognosis, for recognizing potential danger to contacts, and for epidemiologic investigation of the sources of infection.

For example, an epidemic of hospital-acquired septicemia (at least 400 cases and 40 deaths) was traced to the production of contaminated intravenous solutions; the key was a sharp rise in the isolation of two relatively unusual organisms, *Enterobacter cloacae* and *Enterobacter agglomerans,* in geographically dispersed institutions. Without identification to the species level the organisms

TABLE 31-1. Principal Genera of Enterobacteriaceae

Genera	Pathogenicity
Shigella	Bacillary dysentery
Escherichia*	Pathogenic only under special circumstances; certain types produce diarrheal or invasive (dysenteric) disease
Edwardsiella	Can cause gastroenteritis and other salmonellosis like diseases
Salmonella } Arizona }	Gastroenteritis, septicemia, enteric fever
Citrobacter† Klebsiella Enterobacter‡ Hafnia Serratia Proteus Providencia	Pathogenic only under special circumstances ("opportunistic," "secondary" pathogens)
Yersinia	Plague§; enterocolitis, mesenteric lymphadenitis¶
Erwinia } Pectobacterium }	Plant pathogens or saprophytes#

*E. coli; includes alkalescens–dispar organisms, formerly in Shigella.

†Formerly Escherichia freundii; includes Bethesda–Ballerup "paracolon" organisms.

‡Formerly Aerobacter.

§Y. pestis (Ch. 33).

¶Y pseudotuberculosis and Y. enterocolitica (Ch. 33).

#Not medically significant; one species isolated from animal and human hosts.

(Nomenclature based on Ewing WH: Differentiation of Enterobacteriaceae by biochemical reactions, revised. Center for Disease Control; Atlanta, 1973, Yersinia added according to Buchanan RE, Gibbons NE, eds: Bergey's Manual of Determinative Bacteriology, ed. 8. Baltimore, Williams & Wilkins, 1974)

might have easily been confused with common *Klebsiella* and *Enterobacter* species, and detection of the source of the epidemic might have been long delayed.

Typhoid fever (*Salmonella typhi*), bacillary dysentery (*Shigella*), and cholera (*Vibrio cholerae*) are now largely controlled in the Western world through public health measures, but in underdeveloped countries they still represent serious and periodically recurring problems. In addition, in developed countries milder forms of infectious diarrhea, due primarily to certain strains of *E. coli*, are second in incidence only to respiratory infections; they include sporadic cases, institutional outbreaks, and

"traveler's diarrhea" due to strains to which the local population had evidently acquired immunity. Viruses are also common diarrheal agents, while *Vibrio parahaemolyticus* and *Campylobacter* (*Vibrio*) *fetus* (this Ch.), and *Yersinia enterocolitica* (Ch. 33), play a smaller role. In underdeveloped countries, where malnutrition and intestinal infection form a vicious circle, acute gastroenteritis is the major cause of death among small children.

The basic pathogenesis of enteric disease is **enterotoxic** with some organisms (yielding a watery diarrhea) and **enteroinvasive** with others (yielding dysentery, usually with blood in the stools). The two mechanisms will be described under *Escherichia coli* and then further under the major enteropathogens; the main agents are listed in Table 31-3.

While the severe enteric infections have been declining in importance in the developed countries, various kinds of extraintestinal disease are increasingly associated with enteric organisms, nonfermenters, and anaerobes once thought to have little or no virulence (*E. coli, Klebsiella, Enterobacter, Serratia, Pseudomonas,* Bacteroidaceae: see Appendix to this chapter). The increase has been not only relative but absolute as these ordinarily harmless organisms have found new ecologic niches as a result of the elimination of other organisms by antibiotics, the use of immunosuppressive and cytotoxic agents, and the survival of patients with impaired immune responses. Enteric organisms have always been the most common agents of urinary tract infections, but they are now also the predominant etiologic agents in various **endogenous** infections (i.e., due to indigenous organisms) and **nosocomial** infections (i.e., hospital-acquired, such as surgical wound infections, pneumonia, and septicemia). These two classes now represent a large proportion of the serious bacterial diseases in Western countries, especially since staphylococcal, streptococcal, and pneumococcal infections have been declining in importance. Their treatment is often made difficult by resistance of the organisms to most antimicrobial agents and by the presence of underlying serious diseases in the patients.

To complete the introduction of this group of organisms we should note the prominence of Enterobacteriaceae, and their phages, in the development of bacterial and molecular genetics. This prominence derives partly from their ease of handling and partly from their familiarity to the early investigators.

ENTEROBACTERIACEAE

The gram-negative, non–spore-forming rods included in the family Enterobacteriaceae are relatively small (2μm–3μm × 0.4μm–0.6μm). *Shigella* and *Klebsiella* are

nonmotile, lacking flagella; the rest have peritrichous flagella, but nonmotile variants occur. Capsule production (e.g., by *Klebsiella pneumoniae* and *Enterobacter*

TABLE 31-2. Some Differential Characteristics of Enteric Bacilli and Nonfermenting Gram-Negative Bacteria

Family, genus, or species	Ox/F*	Oxidase	Growth on MacC	SS	Motility	Nitrate to Nitrite	Gas	Ly-sine†	Argi-nine‡	Orni-thine†	DNA base composition (% G+C)
Enterobacteriaceae	F	–	+	+	+or–	+	–	d§	d§	d§	39–58¶
Yersinia***	F	–	+	+,–	–,+#	+	–	–	–	+or–	46–47
Vibrio	F	+	+	–§	+	+	–	+	–	+	40–50
Aeromonas hydrophila	F	+	+	+	+	+	–	–	+	–	57–63
Pasteurella	F	+	–§	–	–	+	–	–	–	d	36–43
Pseudomonas	Ox	+**	+	+	+	–	+or–	– (NC	+ NC	– NC)	57–70
Achromobacter	Ox	–	d	d	+(–)	+or–	–	–	+,–	–	40–70
Acinetobacter anitratum (Herellea)	Ox	–	+	–§	–	–	–	NC	NC	NC	40–46
A. lwoffi (Mima)	In	–	+	–§	–	–	–	NC	NC	NC	40–46
Moraxella	In	+	–or+	–	–	–or+	–				41–43
Flavobacterium	Ox or In	+(–)	+(–)	–	+(–)	–	+or–	NC	NC	NC	30–42
Alcaligenes	In	+	+	(+)–	+	+	+or–	NC	NC	NC	58–70

F = Fermentative; Ox = oxidative; In = inactive (in O-F medium; see text Introduction); MacC = MacConkey agar; SS = Salmonella-Shigella agar.

+ = Positive reaction or growth; – = no reaction or growth; (+) = delayed positive reaction, or poor growth; d = different reactions: +, (+), or –; NC = no change; +(–) = majority positive, occasional strain negative.

*Based on reactions in Hugh and Leifson's O-F medium (0.2% peptone, 1.0% glucose, 0.5% NaCl, 0.03% K_2HPO_4, 0.2% agar, and a pH indicator). Glucose can ordinarily be used if any sugar can.

†Decarboxylases.

‡Dihydrolase.

§Rare strains grow.

#Motility, when present, demonstrable at 20°–25°C but not at 37°C.

**Except *Pseudomonas maltophilia.*

¶Range of DNA base composition of practically all Enterobacteriaceae is 50–58 moles % of G+C, except for *Proteus-Providencia* organisms, with a value of 39–53, and *Yersinia*, with 46–47.

*** *Yersinia* is now included in the family Enterobacteriaceae

---

TABLE 31-3. Infectious Gram-Negative Bacterial Agents That Commonly Cause Diarrhea and/or Dysentery

Diarrhea*
Salmonella[a,b]
Shigella[a,b]
Enterotoxigenic *Escherichia coli*
Vibrio cholerae
"Noncholera" vibrios
Vibrio parahaemolyticus
Yersinia enterocolitica
Campylobacter (Vibrio) fetus

Dysentery†
Shigella[a,b]
Salmonella typhi[a,c]
Salmonella, other[a,b]
Enteroinvasive *Escherichia coli*
Yersinia enterocolitica

*Profuse, watery diarrhea, no blood in stool.

†Abdominal cramps, tenesmus, usually blood in stool.

[a]Fecal leukocytes present.

[b]Predominant leukocytes: polymorphonuclear.

[c]Predominant leukocytes: mononuclear.

aerogenes) gives rise to large mucoid colonies that are easily distinguished from the usual smooth variety.

The classification of Enterobacteriaceae has been controversial and continues to change. We shall use the classification of Edwards and Ewing (1962), as revised and expanded by Ewing (1973): despite reservations of taxonomists over its "splitting" tendencies it has proved most useful for diagnostic and epidemiologic purposes. Table 31-4 shows the formal system of nomenclature and the differential characteristics.

The tribes and genera are distinguished by a variety of biochemical characteristics (Table 31-4) and are subgrouped (speciated) by means of further biochemical tests and by their antigenic structure.

METABOLIC CHARACTERISTICS

The Enterobacteriaceae grow readily on ordinary media under aerobic or anaerobic conditions. Most will grow in simple synthetic media, often with a single carbon source. They utilize glucose fermentatively with the for-

TABLE 31-4. Biochemical Reactions of Enterobacteriaceae (incl. *Yersinia*), *Aeromonas*, and *Vibrio*.

	Escherichieae		Edwardsielleae	Salmonelleae		Citrobacter		Klebsielleae	Enterobacter				Serratia			Proteeae — Proteus				Providencia		Yersinia			Aeromonas	Vibrio
	Escherichia	Shigella	Edwardsiella	Salmonella	Arizona	freundii	diversus	Klebsiella pneumoniae	cloacae	aerogenes	hafniae	agglomerans	marcescens	liquefaciens	rubidaea	vulgaris	mirabilis	morganii	rettgeri	alcalifaciens	stuartii	entero-colitica	pseudo-tuberculosis	pestis	hydrophila (other sp. vary)	cholerae
Oxidase test	−	−	−	−	−	−	−	−	−	−	−	−	−	−	−	−	−	−	−	−	−	−	−	−	+	+
Indole	+	−or+	+	−	−	−	+	−	−	−	−	−or+	−	−	−	+	−	+	+	+	+	−or+	−	−	+	+
Methyl red	+	+	+	+	+	+	+	−or+	−	−	−or+	−or+	−or+	−or−	−or+	+	+	+	+	+	+	+	+	+	+	±
Voges-proskauer	−	−	−	−	−	−	−	−or+	+	+	+	+	+	+	+	−	−or+	−	−	−	−	−	−	−	−or+	−or+
Simmons' citrate	−	−	−	d	+	+	+	+	+	+	d	d	+	+	+or−	d	+or(+)	−	+	+	+	−	−d	−d	+or−	(+)or+
Hydrogen sulfide (TSI)	−	−	+	+	+	+or−	−	−	−	−	−	d	−	−	−	+	+	−	−	−	−	−	−	−	−	−
Urease	−	−	−	−	−	d^w	d^w	+	+or−	−	−	d^w	d^w	d^w	d^w	+	+	+	+	−	+	+	+	−d	−	−
KCN	−	−	−	−	−	+	+	+	+	+	+	−or+	+	+	+	+	+	+	+	+	+	−	−	−	−or+	+
Motility	+or−	−	+	+	+	+	+	−	+	+	+	+or−	+	+	+or−	+or−	+	+or−	−or+	+	+	+or− (37°C − / 22°C +)	−or+ (37°C − / 22°C +)	− (37°C − / 22°C −)	+	+
Gelatin (22°C)	−	−	−	−	(+)	−	−	−	(+)or−	−or+	−	d	+or(+)	+	+or(+)	+	+	−	−	−	(+)or+	−	−	−	+	+
Lysine decarboxylase	d	−	+	+	+	−	−	+	−	+	+	−	+or(+)	+	+	−	−	+or−	−	−	−	−	−	−	−or+	+
Arginine dihydrolase	d	d	−	+or(+)	+or(+)	d	+or(+)	−	+	−	−	−	−	−	−	−	−	−	−	−	−	−	−	−	d	−
Ornithine decarboxylase	d	d(1)	+	+	+	d	+or(+)	−	+	+	+	d	+	+	−	−	+or(+)	+	−	−	−	+	−	−	+	+
Phenylalanine deaminase	−	−	−	−	−	−	−	−	−	−	−	−	−	−	−	+	+	+	+	+	+	−	−	−	−	−
Malonate	−	−	−	−	+	−or+	−or+	+	−or+	+or−	+or−	−or+	−or+	−or+[3]	+or(+)	+or−	+	+or−	−or+	+	d	−or+	−or+	−	+or−	−
Gas from glucose	+	−(1)	+	+	+	+	+	+	+	+	+or−	−or+	+or−[3]	d	+	+or−	+	+	−or+	d	(+)or+	−	−	−	−or+	−
Lactose	+	−(1)	−	−	d	d	−or+	+	(+)or+	+	d	d	−	d	+	+or−	−	−	−or+	−	d	−or+	−	−	+or−	−
Sucrose	d	−(1)	−	−	−	d	−or+	+	+	+	−	d	+	+	+	+	d	−	d	d	(+)or+	+	−	d	+	+
Mannitol	+	+or−	+	+	+	+	+	+	+	+	+	+	+	+	+	−	−	−	+or−	+	−	+	+	d	+	+
Dulcitol	d	d	−(2)	d(2)	−	d	−or+	d	−or+	−	−	−or+	−	−	+or(+)	−	−	−	d	d	−	−	−d	d	d	−
Salicin	d	−	−(2)	−	−	d	(+)or+	+	+or(+)	+	+	d	d	+	+or(+)	d	d	−	d	−	−	−or+	−d	d	+	−
Adonitol	−	−	−	−	−	−or+	+	+	−or+	+	−	−or+	d	d	d	−	−	−	d	+	−or+	−	−d	−	+or−	−
Inositol	−	−	−	d	−	−or+	d	+	d	+	−	d	d	+or(+)	d	−	+or−	−or+	+	+	+or−	d	−d	d	−or+	(+)
Sorbitol	d	d	−	+	+	d	−or+	+	+	+	−	d	d	+	−	−	d	−	d	d	d	d	−	d	d	+
Arabinose	+	−or+(2)	−	+(2)	+	+	+	+	+	+	+	+	−	+	+	+	−	−	−	+	−	+	+	+	−or+	−
Raffinose	d	d	−	−	−	d	d	+	+	+	−	d	−	+	+	−	−	−	−	−	−	−	+	d	d	−
Rhamnose	d	d	−	+	+	+	+	+	+	+	+	+or(+)	−	d	+	−	−	−	−or+	−	−	+	+	−	−	−

(1) Certain biotypes of *S. flexneri* produces gas; cultures of *S. sonnei* ferment lactose and sucrose slowly and decarboxylate ornithine.
(2) *S. typhi*, *S. cholerae-suis*, *S. enteritidis* bioser. Paratyphi-A and Pullorum, and a few others ordinarily do not ferment dulcitol promptly. *S. cholerae-suis* does not ferment arabinose.
(3) Gas volumes produced by cultures of *Serratia*, *Proteus*, and *Providencia* are small.

+ = 90% or more positive in 1 or 2 days; − = 90% or more negative; d = different biochemical types [+, (+), −]; (+) = delayed positive (decarboxylase reactions, 3 or 4 days); + or −
= majority of cultures positives; − or + = majority negative, w = weakly positive reaction.

(Compiled by A. C. Sonnenwirth, November, 1973. Data on Enterobacteriaceae from "Differentiation of Enterobacteriaceae by biochemical tests," USPHS Center for Disease Control, Atlanta, 1973)

mation of acid or of acid and gas, reduce nitrates to nitrites, and give a negative oxidase reaction. Genera and species can be tentatively identified by their capacity to ferment specific carbohydrates, to utilize certain other substrates (e.g., citrate) as a sole source of carbon, and to give rise to characteristic endproducts (e.g., indole from tryptophan; ammonia from urea; hydrogen sulfide). Many commonly employed tests are listed in Table 31-4.

Fermentation of lactose is a time-honored differential characteristic in the preliminary examination of suspected cultures. It early became a major criterion because the major enteric pathogens (*Salmonella* and *Shigella*) are lactose-negative, while prompt lactose fermentation, recognized by the formation of colored colonies on solid media containing lactose and an appropriate indicator (e.g., neutral red), serves well in delineating the **coliform** organisms (most *Escherichia, Citrobacter, Klebsiella,* and *Enterobacter*). However, a number of other Enterobacteriaceae (*Proteus, Providencia, Edwardsiella, Serratia, Yersinia,* and certain types of *Escherichia* and *Enterobacter*) are lactose-negative (or delayed), while many *Arizona* strains, resembling salmonellae in pathogenicity, are lactose-positive. Hence identification requires a combination of tests.

For many years the term **"paracolon"** was used to denote organisms that did not ferment lactose promptly and could not be unequivocally identified as members of recognized genera. Since all such organisms can now be identified with discrete groups this "wastebasket" designation is no longer used.

Selective Media. The enteric bacilli are resistant to inhibition by certain bacteriostatic dyes (e.g., brilliant green) and surface-active compounds (e.g., bile salts), compared with most gram-positive bacteria. Selective media containing such substances (e.g., MacConkey agar, deoxycholate agar) greatly facilitate the isolation of enteric bacilli from cultures of feces.

Shigellae and salmonellae are less sensitive than the coliform organisms to inhibition by citrate (which may depend on chelation of divalent metal ions). Agar containing both citrate and bile salts, e.g., **SS** (*Salmonella–Shigella*) **agar,** is therefore used for selective cultivation of pathogenic species. However, since some strains of shigellae are inhibited, one cannot rely on such media alone.

Screening Media. The preliminary screening of enteric bacilli employs various ingenious media that reveal a number of differential characteristics in a single culture tube. One example, **triple sugar iron (TSI) agar,** has the following formula:

Peptone	20g
Sodium chloride	5g
Lactose	10g
Sucrose	10g
Glucose	1g
Ferrous ammonium sulfate	0.2g
Sodium thiosulfate	0.2g
Phenol red	0.025g
Agar	13g
Distilled water	1000ml

The medium (pH 7.3) is poured in slants, and the inoculum (picked from a single colony on an original isolation plate)* is stabbed through the butt (i.e., into the depth) and also streaked onto the surface. The difference between the concentration of glucose and that of lactose and of sucrose in TSI agar allows one to distinguish their fermentations. In addition, production of H_2S causes formation of black ferrous sulfide, and gas production (H_2 and CO_2; see Ch. 3) forms bubbles in the agar.

The "decolorization"† of the phenol red indicator in the fermentation tests depends upon several factors: 1) a delicate balance of the concentrations of the several sugars and the nitrogenous consituents of the medium; 2) inclusion of ten times as much lactose and sucrose as glucose (see formula); 3) slow diffusion through the agar of the acid products of the fermentations and the alkaline products of nitrogen metabolism; and 4) greater production of acid in the anaerobic environment of the butt (fermentation) compared with the aerobic surface of the slant (respiration).

Thus an organism that ferments glucose only (e.g., *S. typhi*) will produce detectable amounts of acid in the butt of the TSI medium (anaerobic) but not on the surface of the slant (aerobic), where the nitrogenous bases formed will neutralize the less acid endproducts of respiration. Species that ferment the more concentrated lactose (e.g., *E. coli*) or sucrose (*Proteus vulgaris*), on the other hand, will generate in the butt enough acid to diffuse throughout the medium and thus acidify even the surface (Table 31-4). **Nonfermenters,** incapable of utilizing the sugars fermentatively, will give rise to an alkaline or neutral reaction throughout the medium (slant and butt pink to red).

Identification Methods. Although multiple sugar media like TSI are useful for preliminary identification of enteric bacilli, conventional methodology has required the confirmation of observed fermentation reactions with appropriate liquid media, each tube containing a single sugar as the sole source of carbohydrate; other tests (up to a total of 26) are added as needed (Table 31-4). To recognize **gas formation** an inverted small test tube is submerged in the medium to trap a portion of any gas generated. (During autoclaving the air initially present in the inverted tube will have been expelled and replaced by medium.) Recently, miniaturized multiple test systems

---

* Although it is permissible for diagnostic purposes to "fish" a colony from an original isolation plate to a screening medium, a second subculture on agar should always be made for future reference.

† Note that acid "decolorizes" phenol red to a pale yellow but makes neutral red colored.

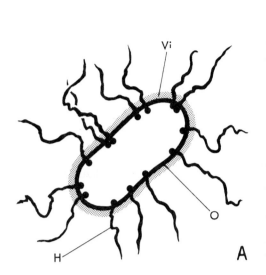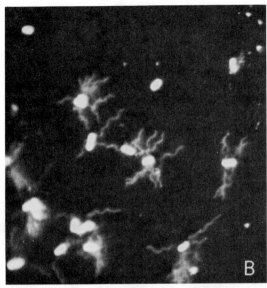

FIG. 31-1. A. Schematic diagram of cellular locations of H, O, and Vi antigens of enteric bacilli. **B.** *S. typhi* stained with fluorescein-labeled immune serum containing anti-H, anti-O, and anti-Vi Abs. Note that the flagella of different bacilli have in some cases become agglutinated with one another. Note also, however, that the presence of the flagella has prevented the bodies of the bacilli from coming close enough together to agglutinate, despite the presence of the antisomatic (Vi and O) Abs in the serum. The interflagellar agglutinate is, therefore, understandably less densely packed than if the flagella had been absent and intersomatic agglutination had occurred. (×3500, reduced) (Thomason BM et al: J Bacteriol 74:525, 1957)

have been developed which greatly simplify identification of enterobacteria. Compared to conventional tests, the "minikits" require far smaller amounts of material and yield results sooner.

The variety of kits that are available include a system of cupules (microtubes) or paper disks or strips containing dried substrates or reagents, and a multicompartmental tube containing various substrates incorporated in agar. Several of the systems utilize computer-based identification schemes. These systems have greatly improved identification, including recognition of unusual or aberrant strains.

ANTIGENIC STRUCTURE

Although the principal varieties of enteric bacilli can be identified by fermentation and other metabolic reactions in differential media, final identification of many individual species is based on antigenic structure. Strains with the same antigenic pattern may nevertheless exhibit different metabolic reactions (fermentative variants or biotypes).

As is shown diagrammatically in Figure 31-1A, three kinds of surface Ags **(H, O,** and **Vi)*** determine the orga-

* The designation **H** (Ger. *Hauch,* breath) was first used to describe the growth of proteus bacilli on the surfaces of moist agar plates: the film produced by the swarming of this highly motile organism resembles the light mist caused by breathing on glass. The designation **O** (Ger. *ohne,* without), first applied to nonswarming

nism's reactions with specific antisera. When motile species are treated with formalin the labile protein H Ags of their flagella are preserved, and the cells are agglutinated only by specific antiflagellar (anti-H) Abs, forming a light, fluffy precipitate. As Figure 31-1 suggests, the numerous peritrichous flagella explain the character of the precipitate and also the failure of anti-O Ab to agglutinate such cells. When, on the other hand, flagella are absent from a strain, or are denatured by heat (100° C for 20 min), acid, or alcohol, the somatic O Ag at the surfaces of adjacent bacilli can be linked by anti-O Abs, resulting in closely packed, granular clumps.

Certain species have an outermost polysaccharide layer, called a **Vi Ag,** that is usually too thin to be seen as a capsule. It does not cover the O Ag completely, since it allows adsorption of anti-O Abs. Nevertheless, Vi cells can be agglutinated only by anti-H or anti-Vi Abs, and O agglutination becomes possible only if the Vi and H Ags are destroyed (e.g., by boiling for 2 h). The Vi Ag of *S. typhi* is a homopolymer of *N*-acetyl galactosaminuronic

(i.e., nonflagellated) forms, is now used as a generic term for the lipopolysaccharide somatic Ags of all enteric bacilli and more specifically for their antigenically active polysaccharide components. The term **Vi** designates an additional surface Ag of *S. typhi* originally thought to be primarily responsible for virulence. Its precise relation to pathogenicity is still not clear, and it is a special example of a **K** (Ger. *Kapsel*), or capsular, Ag.

acid: because of its free carboxyl groups the cell surface is highly acidic. Other Vi Ags are found in other salmonellae.

The O Ags in smooth (S) strains cover the underlying **R Ags,** which become accessible to Ab in rough (R) mutants. The change from S to R may take place without the loss of flagellar or of Vi Ags. Rough strains tend to agglutinate spontaneously unless suspended in media of proper ionic strength.

In addition to the Ags that differentiate various Enterobacteriaceae from each other, many of these organisms (including *Yersinia*) share a **common enterobacterial antigen** (Kunin antigen), which can be extracted from the cells. Abs to this Ag do not precipitate it, and they do not agglutinate cells, but they can be detected by hemagglutination or hemolysis of Ag-coated RBCs.

When animals are immunized with formalin-killed motile strains of enteric bacilli the antisera contain both anti-H and anti-O Ags (as well as anti-R). The Abs to H usually have a higher titer than those to O.* If the immunizing strain also possesses a capsular (Vi) Ag the antiserum will contain, in addition, anti-Vi.

Specific antisera, reacting with individual surface Ags in agglutination tests, are prepared by selective adsorption. For example, anti-H Ab may be removed by suspensions of homologous flagella (mechanically removed), or by suspensions of mutants possessing the immunogenic cells' flagella but neither their Vi nor O Ags. Similarly, anti-O or anti-Vi Abs may be selectively removed by appropriate extracts of bacilli.

Kauffmann–White Classification of the Salmonellae.

The antigenic complexities of the enteric bacilli have been most thoroughly documented in the genus *Salmonella.* Largely as a result of the studies of Kauffmann in Denmark and White in England a system has been developed for distinguishing a wide variety of serotypes by the use of exhaustive crossadsorption and crossreaction tests.

Assume that two strains of *Salmonella,* **a** and **b,** have been isolated, each from a separate outbreak of salmonellosis. Antisera prepared against them (anti-a and anti-b) are first reacted with suspensions of each strain (standardized to the same density), and the intensity of the reactions is recorded in roughly quantitative terms (see below). Each serum is then adsorbed with the heterologous strain (i.e., anti-a with **b,** and vice versa). The adsorbed sera are similarly reacted with the standardized suspensions, and the results are measured in the same manner. If the observed agglutination reactions were as follows

Antisera	Organisms	
	a	**b**
Anti-a, unadsorbed	4+	2+
Anti-b, unadsorbed	2+	4+
Anti-a, adsorbed with b	2+	0
Anti-b, adsorbed with a	0	2+

* This difference may in effect be due to both the greater immunogenicity of the protein H Ag, as compared with the polysaccharide O Ag, and the greater sensitivity of the H agglutination test.

it would be concluded that **a** and **b** have unique, as well as shared, antigenic determinants. Using an arbitrary numbering system, **a** might be said to possess determinants (or factors) 1,2, and **b** determinants 2,3. If, on the other hand, strains **a** and **b** were equally effective in removing all the agglutinating Ab from the two antisera, they would be considered antigenically identical. Every new strain isolated would be similarly tested with the existing antisera (anti-a and anti-b, both adsorbed and unadsorbed) and with antiserum to the new strain itself, before and after adsorption. Whenever a new determinant was detected it would be given a new number, and according to the Kauffmann–White scheme the organism possessing it would be designated a new "species" (type).

Application of this technic to O, H, and Vi agglutination reactions has resulted in the identification of more than 1500 *Salmonella* "species" (serotypes). Only large *Salmonella* typing centers (Copenhagen, London, Atlanta, etc.) have the collections of specific antisera necessary for such work. In most diagnostic laboratories *Salmonella* strains are merely **grouped** by means of fermentation tests and by agglutination reactions performed with group-specific antisera.

O Antigens. Crossreactions show that each *Salmonella* O Ag has two or three distinguishable **determinants (factors),** each given a number and each shared with several other O Ags. Each Ag has a strongly reacting major determinant (heavy type in Table 31-5), and on this basis the salmonellae have been classified into O groups, the first 26 designated by letters (A–Z) and subsequent ones by numbers (50, 51, etc.). In group C (O Ag 6,7), for example, the major determinant is factor 6, and in group E (O Ag 3,10) it is factor 3 (Table 31-5). (About 90% of strains isolated from humans fall into groups A to E.) Additional divisions within a group are made on the basis of minor O determinants.

The members of each O group may be differentiated still further into serotypes ("species") on the basis of their flagellar (H) Ags. A given strain may form at different times either one of two kinds of H Ag. The first, **phase 1 flagellar Ags,** are shared with only a few other species of *Salmonella;* the second, **phase 2 Ags,** are less specific. In the Kauffmann–White classification the former are indicated by small letters and the latter by numbers (Table 31-5). Although the organisms in a given culture may be entirely in one phase (monophasic culture), they frequently can give rise to mutants in the other phase (diphasic culture), especially if the culture is incubated for more than 24 h. Such **phase variation,** which depends on a reversible DNA transposition (see Phase Variation, Ch. 8; 12), can be accentuated by growing the organisms in serum containing Abs to their flagellar Ags, thereby favoring the growth of mutants with the alternate (allelic) Ag which does not react with the antiserum.

TABLE 31-5. Serologic (Kauffmann–White) and Chemical Classification of Common *Salmonella* Species

Species	Group	O antigen*	H antigens Phase 1	H antigens Phase 2	O-specific sugars in determinants
S. paratyphi-A	A	(1), **2**, 12	a	—	Mannose, rhamnose, paratose
S. schottmülleri	B	(1), **4**, (5), 12	b	1, 2	Mannose, rhamnose, abequose
S. typhimurium	B	(1), **4**, (5), 12	i	1, 2	Mannose, rhamnose, abequose
S. paratyphi-C	C$_1$	**6**, 7, Vi	c	1, 5	Mannose
S. cholerae-suis	C$_1$	**6**, 7	c	1, 5	Mannose
S. montevideo	C$_1$	**6**, 7	g,m,s	—	Mannose
S. newport	C$_2$	**6**, 8	e,h	1, 2	Mannose, rhamnose, abequose
S. typhi	D	**9**, 12, Vi	d	—	Mannose, rhamnose, tyvelose
S. enteritidis	D	(1), **9**, 12	g,m	—	Mannose, rhamnose, tyvelose
S. gallinarum	D	1, **9**, 12	—	—	Mannose, rhamnose, tyvelose
S. anatum	E	**3**, 10	e,h	1, 6	Mannose, rhamnose

*Boldface number signifies major determinant (factor) of group. The determinants represent a subdivision of the O Ags. Parentheses indicate that Ag determinant may be difficult to detect.

(Data on O-specific sugars from Lüderitz O: Angew Chemic 9:649, 1970). Reprinted with permission of Verlag Chemie GMBH, Weinheim, Germany.

Elaboration of the polysaccharide Vi Ag by a species is indicated in the Kauffmann–White scheme by the letters Vi, placed by convention after the numbers indicating the individual O factors (e.g., *S. typhi* in Table 31-5). Since the surface coating of Vi prevents agglutination with homologous anti-O Ab (though it permits adsorption of the Ab), it must be removed to permit serologic identification.

Chemistry of the O Antigenic Determinants. As is described in Chapter 6, the O Ags and their associated endotoxin have the following overall structure: **O-specific chains,** consisting of **repeating** (oligosaccharide) **units,** are attached to a basal **core polysaccharide,** which in turn is attached to **lipid A;** the whole forms a **lipopolysaccharide** (LPS). The structure of the core (see Fig. 6-15) is identical or very similar in all salmonellae but differs in other genera. The serologic specificity of each LPS is determined by the O-specific chains, whose many variations form the main basis for classification of the Salmonellae.

A total of 18 different sugars (monosaccharides) have been identified in various *Salmonella* LPSs; some species have as many as 9. Among the 5 "basal" sugars of the core, L-glycero-D-mannoheptose and 2-keto-3-deoxy-D-mannooctonic acid (ketodeoxyoctonate, KDO) are unique to bacterial LPS (KDO linking the core to lipid A; see Fig. 6-15). The core is responsible for the serologic specificity of R (rough) mutants, which are blocked in synthesis of the complete O Ag: R$_a$ mutants contain the complete core, while R$_e$ have lost their basal sugars except KDO. This minimal (R$_e$) core (KDO plus lipid A) appears to be identical (except for slight differences in the lipid) in all Enterobacteriaceae studied.

Antigenic determinants of the LPS of R$_e$ mutants have been detected on the surface of S-phase organisms despite their additional layer of side chains. Moreover, Ab to such Ags protects animals against challenge by various gram-negative bacilli or by endotoxin, and reduces the severity of bacteremia in humans. These recent findings suggest the possibility of general immunoprophylaxis of enterobacterial infection.

Various **O-specific chains** contain a group of 3,6-dideoxyhexoses that so far are unique to enterobacteria and that are named for the organisms from which they were first derived: colitose, abequose, paratose, and tyvelose (Fig. 31-2).* Both ordinary sugars and dideoxyhexoses contribute to the specificity of major, group-defining determinants (see Immunologic Determinants, below). Figure 31-3 illustrates the repeating mannose–rhamnose–galactose sequence that constitutes the distinctive determinants of group E$_2$ salmonellae (Ag3,-15). The characteristic sequences of several *Salmonella* groups are listed in Table 31-6. Groups A, B, D, and E all contain a repeating mannose–rhamnose–galactose unit, but their linkages and substituents differ (see also Table 31-5).

The structural differences that have been observed in closely related Ags include 1) changes of position of glycosidic linkages, i.e., 1-4 versus 1-6; 2) altered anomeric configurations, i.e., α versus β

* Colitose is also found in O Ags of certain *E. coli* strains. The other 3,6-dideoxyhexoses have been found only in relatively virulent *Salmonella* species.

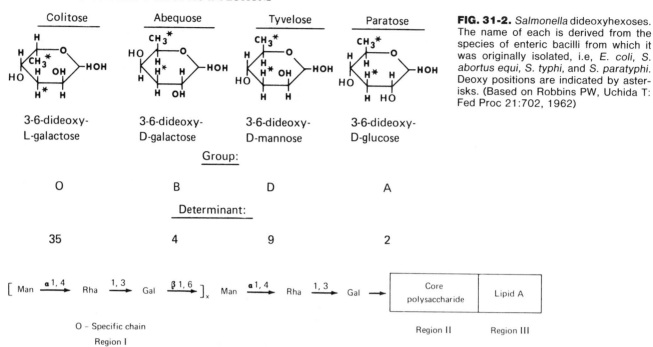

FIG. 31-2. *Salmonella* dideoxyhexoses. The name of each is derived from the species of enteric bacilli from which it was originally isolated, i.e, *E. coli, S. abortus equi, S. typhi,* and *S. paratyphi*. Deoxy positions are indicated by asterisks. (Based on Robbins PW, Uchida T: Fed Proc 21:702, 1962)

FIG. 31-3. Structure of the O-specific chains in the lipopolysaccharide of *S. newington* (group E₂). (Lüderitz O: Angew Chem 9:649, 1970. Reprinted with permission of Verlag Chemie GMBH, Weinheim, Germany)

linkage; 3) attachment of additional monosaccharides, such as glucose; 4) presence or absence of acetyl groups; and 5) deletion of, or substitution for, one of the monosaccharides in the basic trisaccharide units.

Immunologic Determinants. The determinants of an O Ag are identified from the capacity of fragments of the chain (e.g., di-, tri-, tetrasaccharides) to inhibit serologic reactions of monospecific antisera, containing Abs only for a particular determinant. *The length of a determinant is less than that of a repeating unit,* and so a given sugar can contribute to more than one determinant, i.e., *different determinants in a given sequence can overlap.* For example, factors 3 and 10 of Ag 3,10 may be represented as:

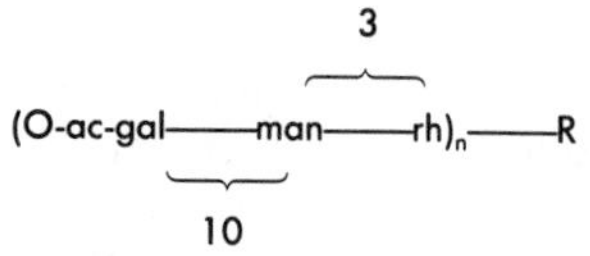

$$
\underset{10}{\underbrace{(\text{O-ac-gal}\underset{}{\overbrace{\text{---man---rh}}^{3}})_n}}\text{---R}
$$

The **immunodominant** sugar of a given determinant (e.g., tyvelose for group D factor 9, mannose for group E factor 3), as a monosaccharide at high concentration, can specifically inhibit the reactions to some extent, but strong inhibition is obtained only when the sugar is part of a larger fragment. Because of the low affinity of the Abs for monosaccharides, individual sugar residues common to different determinants are not usually sufficient for crossreactions.

The structure of Ag (3),(15),34 (Table 31-6) may be written as:

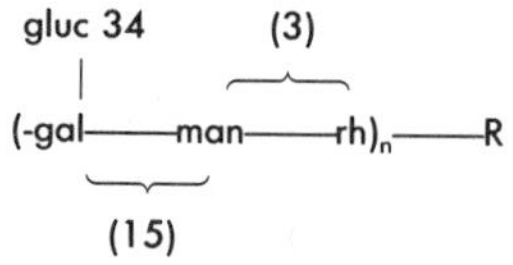

$$
\underset{(15)}{\underbrace{(\text{-gal}\underset{}{\overbrace{\text{---man---rh}}^{(3)}})_n}}\overset{\text{gluc 34}}{\overset{|}{}}\text{---R}
$$

The parentheses around 3 and 15 indicate that the addition of glucose (immunodominant in factor 34) has weakened antigenic determinants 3 and 15 to the point where they are difficult to detect.

ENDOTOXINS AND EXOTOXINS

The biologic effects of the endotoxins of gram-negative bacilli, present in the cell wall LPS, have been discussed (see Endotoxin, Ch. 24). The extracted LPSs, recovered as a colloidal suspension, may be split by mild acid hydrolysis into lipid A and degraded polysaccharides (see Ch. 6). Lipid A from salmonellae consists of β-1,6-glucosamine disaccharides (see Fig. 6-14) with the amino groups substituted by β-hydroxymyristic acid and the other OH groups by other long-chain fatty acids. Lipid A from other Enterobacteriaceae has a similar composition.

By itself, lipid A, unlike LPS, is insoluble in water and

TABLE 31-6. Simplified Structures of O Repeat-units of Some Salmonella Serogroups*

Serogroup	Species	O antigen	Structure
A	*S. paratyphi A*	1,2,12	Par OAc Glc ↓ ↓ ↓ Man→Rha→Gal
B	*S. typhimurium*	1,4,5,12	OAc-Abe Glc ↓ ↓ Man————————→Rha→Gal→
C$_1$	*S. cholerae-suis*	6,7	Glc ↙ ↘ Man→Man→Man→Man→GlcNac
C$_2$	*S. newport*	6,8	Abe OAc-Glc ↓ ↓ Rha → Man→Man→Gal →
D$_1$	*S. typhi*	9,12	Tyv OAc-Glc ↓ ↓ 2α-Man————→Rha →Gal →
E$_1$	*S. anatum*	3,10	OAc ↓ 6α-Man →Rha— Gal→
E$_2$	*S. newington*	3,15	6β-Man→Rha→Gal
E$_2$	*S. minneapolis*	(3),(15),34	Glc ↓ 6β-Man→Rha→Gal

Par = Paratose; OAc = acetyl; Glc = glucose; Abe = abequose; Tyb = tyvelose; Man = mannose; Rha = rhamnose; Gal = galactose.

*Only those type-linkages and anomeric positions of sugars are included which may explain differences between groups.

(Modified from Roantree RJ: In Microbial Toxins, vol 5 p 18. Kadis S, Weinbaum G, Ajl SJ, eds: New York, Academic Press, 1971. Reprinted with permission of Verlag Chemie GMBH. Weinheim, Germany)

lacks biologic activity. However, the polysaccharide in the intact LPS seems to contribute to the toxicity only by increasing solubility in water. Thus lipid A also becomes a highly active endotoxin when solubilized by the addition of serum albumin; and *Salmonella* R$_e$ mutants, whose "LPS" contains only lipid A plus KDO, yield "LPS" with full endotoxic activity.

Endotoxin is generally assumed to play a large role in the vascular, metabolic, pyrogenic, and hematologic alterations in severe gram-negative infections (see Endotoxins, Ch. 24), but the evidence is indirect: unlike exotoxins no protective specific Ab is available. Moreover, the correlation of symptoms with blood levels of endotoxin has been hampered because the assays, in whole animals, were cumbersome and imprecise. However, a recently introduced **in vitro assay,** based on the gelation of extracts of blood cells of the horseshoe crab *Limulus polyphemus,* can detect picogram quantities of LPS and is proving useful in such studies. The test is also useful for detection of gram-negative bacterial meningitis or bacteriuria, for pyrogen screening of parenteral solutions, and for evaluating the role of endotoxemia in shock.

Certain enteric bacteria cause disease by the secretion of locally acting protein exotoxins: see below *Escherichia coli* (Pathogenesis: Intestinal Diseases) and Cholera Vibrios (Cholera Enterotoxin).

PHAGE AND COLICIN TYPING

Some enterobacterial species can be subdivided into strains on the basis of their susceptibility to lysis by specific **bacteriophages,** or on the basis of their production of, or their susceptibility to, specific **colicins** (Ch. 12). This typing is useful in tracing the source and the progress of an epidemic.

For example, 72 subtypes of *S. typhi* have been distinguished through the use of various phage strains. These phages are descended from a single strain, some being mutants and others host-modified variants (see Role of DNA Modifications: Host-Induced Restriction and Modification, Ch. 47). Similarly, *Shigella sonnei* (discussed below), which is homogeneous according to phage typing and antigenic analysis, has been divided into 15 types by colicin typing.

Colicins are members of the broader class of bacteriocins (Ch. 12). These proteins, coded for by plasmids (colicinogens), attach to specific receptors on susceptible bacteria and kill the cells by various mechanisms. Responsible for a special form of **bacterial antagonism** (see Beneficial Actions, Ch. 44), colicinogeny may help to stabilize the microbial population in the intestinal tract.

GENETIC RELATIONS

The gastrointestinal tract provides an ideal environment for developing a wide variety of strains of bacteria and bacterial viruses: the microbial population has a high density; it grows as a continuous culture, as in a chemostat; nutritional conditions fluctuate; and genes may be transferred from one strain to another by viral transduction and lysogenization, and by the transfer of plasmids and of bacterial genes by conjugation (Ch. 9). One result is the accumulation of a wide variety of recombinants with overlapping patterns of metabolism and antigenicity. Another result is that these organisms continue to exhibit genetic instability in nature.

For example, lactose-fermenting strains of the traditionally lactose-negative genus *Salmonella* have now been isolated from acute diarrheal epidemics in Brazil (1974–1975). Other biochemically atypical organisms are increasingly recognized among Enterobacteriaceae.

The alteration of somatic O Ags by lysogenic conversion (Ch. 48) is illustrated in Figure 31-4, in which lysogenization by phage ϵ^{15} provides a new enzyme that causes replacement of factors 3,10 by factors 3,15 in the O Ag, while double infection with phages ϵ^{15} and ϵ^{34} causes still another antigenic change: an enzyme specified by ϵ^{34} adds a terminal D-glucosyl residue to determinant 15, changing it to determinant 34.

A set of plasmids of particular medical importance are the **R factors,** which carry multiple genes for drug resistance. These plasmids, discovered in shigellae in Japan, are now known to be widespread in salmonellae, coliforms, and many other organisms. They are readily transferred, both in vitro and in the intestine, among various enterobacteria. Their genetic properties are discussed in Chapter 12 and their phenotypic effects in Chapter 7.

The medical implications of the R factors are obvious. If the factor is transferred from *E. coli* to a pathogen intraintestinally an individual can become the source of a particularly serious epidemic. This process is certain to be accentuated by the selection pressure of antimicrobial therapy. Indeed, R factors have spread remarkably in shigellae and salmonellae. A survey in London (1969) showed that over 50% of healthy individuals harbored enteric bacilli containing an R factor, and often these were the predominant strain of *E. coli.*

COLIFORM BACILLI

Coliform is a general term for facultative gram-negative rods that inhabit the intestinal tract of man and other animals without usually causing disease. Some limit the

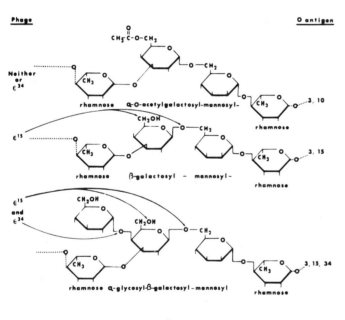

FIG. 31-4. Structures of determinant trisaccharide units of O polysaccharides of group E *Salmonella* (Table 31-6). The changes in structure brought about by lysogenic conversions of determinant 3,10 (*S. anatum*, E$_1$) to 3,15 (*S. newington*, E$_2$), and to 3,15,34 (*S. minneapolis*, E$_3$), by phage ϵ^{15} and phages ϵ^{15} plus ϵ^{34}, respectively (see bottom diagrams) are indicated by the **curved arrows.** (Modified from Robbins PW, Uchida T.: Fed Proc 21:702, 1962, and J Biol Chem 240:375, 1965)

term to lactose fermenters (*E. coli, Klebsiella, Enterobacter*), but we shall include several genera, previously classified as paracolon organisms (see Metabolic Characteristics, above), that are either lactose-negative (*Serratia, Edwardsiella*) or delayed lactose fermenters (*Citrobacter*). *Arizona* and *Providencia* are often also included among coliforms, but we shall discuss them along with *Salmonella* and with *Proteus,* respectively.

In the intestinal tract coliforms generally do not cause disease (except for *Edwardsiella* and some types of *E. coli;* see below). *E. coli* is normally a bowel organism, as are probably some of the *Klebsiella,* whereas the *Enterobacter, Serratia,* and *Citrobacter* groups occur infrequently in the normal intestine and are ordinarily free-living saprobes. The growing role of these organisms in endogenous and nosocomial infections has already been noted in the introduction to this chapter (see also the Appendix). The organisms illustrate the difficulty of defining pathogenicity in absolute terms.

ESCHERICHIA COLI

E. coli is referred to as the **colon bacillus** because it is the predominant facultative species in the large bowel. Its presence in a water supply usually indicates continuing fecal contamination; hence tests for its presence are widely used in public health laboratories (see Klebsiella-Enterobacter-Serratia, below).

Cultural and Antigenic Properties. When grown in broth *E. coli* produces a well-dispersed turbidity. Most strains (though not all) have flagella and are motile. Smooth (S) strains form shiny, convex, colorless colonies but when repeatedly subcultured they become rough (R) and form lusterless, granular colonies. Encapsulated variants produce mucoid colonies, particularly when incubated at low temperatures and when grown in media low in nitrogen and phosphorus but high in carbohydrate. Typical *E. coli* colonies are usually easy to recognize by their characteristic appearance on certain differential media: they are lactose fermenters, and on eosin–methylene blue and Endo agars they have a metallic sheen. However, some strains ferment lactose late, irregularly, or not at all. On blood agar some strains produce β-hemolysis.

E. coli produces acid and gas from a variety of carbohydrates, but some strains, such as the alkalescens–dispar organisms, do not form gas and hence were formerly classified as *Shigella*. Characteristically, *E. coli* produces indole in media containing tryptophan and is methyl red–positive (Table 31-4), but does not produce acetoin or utilize citrate as a sole source of carbon (IMViC reactions; see *Klebsiella-Enterobacter-Serratia,* below). As mentioned, coliform bacilli are more sensitive than salmonellae or shigellae to inhibition by high concentrations of citrate.

More than 160 different O Ags of *E. coli,* designated O55, O111, etc., have been identified by specific agglutination reactions. In addition, approximately 50 H Ags and over 90 capsular K Ags have been described. Like the Vi Ags of *Salmonella,* K Ags at the cell surface often mask the deeper O Ags.

Pathogenesis: Extraintestinal Diseases. *E. coli* most commonly causes **disease in the urinary tract.** The organisms may conceivably travel from the intestinal tract to the urinary passages and kidneys via hematogenous or lymphatic routes, but most often it follows the ascending route from the urethra through the bladder. The normal urinary tract is relatively free of bacteria, but asymptomatic bacilluria is common, particularly in women. When the bacterial count in the urine is greater than 100,000/ml bacterial disease of the urinary tract is usually present. It occurs most commonly in infants (during the diaper period), in pregnant women, and in patients with obstructive lesions or instrumentation of the urinary tract or with neurologic diseases affecting micturition.

When confined to the bladder (cystitis) the infection is usually well controlled by antimicrobial therapy. In the kidneys, on the other hand, once suppuration has occurred lesions may continue to progress despite treatment and may eventually lead to chronic pyelonephritis, with scarring, destruction of tubules and glomeruli, hypertension, and frequently terminal uremia.

E. coli is often found, along with other enteric bacteria, in **sepsis adjacent to the gut:** peritonitis, appendicitis, and infections of the gallbladder and biliary tract. It occurs on the skin of the perineum and genitalia and frequently infects **wounds** that become contaminated with urine or feces.

E. coli is now the most frequently encountered species in gram-negative **septicemia,** resulting in severe shock resembling that produced by intravenous injections of endotoxin in laboratory animals. This disease has been occurring with increasing frequency in the very young (often leading to meningitis), and in patients debilitated by corticosteroid therapy, immunosuppressive agents, or leukemia. Surgery or instrumentation of the intestinal, biliary, or genitourinary tract may also precipitate such sepsis.

Out of some 160 serogroups 9 are found in 40%–60% of *E. coli* meningeal, peritoneal, and urinary tract infections, but none of these belong to the enteropathogenic *E. coli* serotypes (see Pathogenesis, below). The presence of a particular capsular Ag (K1) is associated with increased invasiveness: such strains require far fewer cells to produce fatal septicemia in mice, and they are found more often among *E. coli* from neonatal meningitis or pyelonephritis than among strains from cystitis or from normal stools.

Some *E. coli* K Ags are immunochemically similar to capsular polysaccharides of other invasive organisms—*Neisseria meningitidis* (B and C), *Streptococcus pneumoniae* (several types), and *Hemophilus influenzae* (type b).

Pathogenesis: Intestinal Diseases. Certain strains of *E. coli* exhibit pathogenicity, either **enterotoxic** or **enteroinvasive** (Table 31-3). Some strains display both.

Enterotoxic strains (ETEC) produce a heat-labile enterotoxin, or a heat-stable one, or both. The **heat-labile toxin (LT)** closely resembles the toxin of *V. cholerae* (see p. 666) in mol wt (about 80,000), serologic specificity, and action **(activation of adenyl cyclase).** The much smaller **heat-stable toxin (ST),** mol wt about 8500, is poorly antigenic and is resistant to protease and to heating at 100° C; it **activates guanyl cyclase.** Tissue culture systems provide a sensitive assay for LT but are cumbersome and expensive; simpler immunologic tests are being developed. ST is assayed by injection into the stomach of the infant mouse or into a ligated ileal loop of a larger animal.

To cause disease, toxigenic strains must also produce a **surface factor** (pilus) for colonization of the small bowel, since the toxins are active there but not in the colon. All three factors are mediated by a group of plasmids that are absent in most strains of *E. coli.* These plasmids often also carry genes causing resistance to various drugs (see Ch. 12, Resistance Plasmids).

Enterotoxic strains are now recognized as worldwide agents of disease, ranging from a choleralike syndrome and severe infantile diarrhea to the milder "traveler's diarrhea." These strains apparently depend on poor sanitation for their continued prevalence in a population. The diarrhea is nonbloody and without pus cells.

The role of *E. coli* in **traveler's diarrhea** was in doubt until established by prospective studies, which demonstrated the appearance of ETEC strains (and occasionally of antibodies to them) in persons who apparently were free of these strains when tested before going abroad.

The similar designation "enteropathogenic *E. coli*" was applied earlier to 15 particular antigenic types that were found in epidemics of severe diarrhea in infants. However, serotyping in sporadic cases of infantile diarrhea is not useful; the association with toxin production is irregular, and the plasmids responsible are transmissible.

Invasiveness has recently been identified as a distinct, less common pathogenetic mechanism. **Enteroinvasive** *E. coli* strains (EIEC) can penetrate the intestinal epithelium, primarily in the large intestine (less frequently the small intestine), and without systemic disease. It causes a syndrome **(colitis)** much like *Shigella* (see Shigellae, below), in both infants and adults, with severe abdominal cramps and with pus and blood in the stool.

Invasiveness can be detected by a strain's ability to produce keratoconjunctivitis in the guinea pig's eye (Sereny test), or by its capacity to penetrate cells in tissue culture.

Treatment. Most *E. coli* strains are sensitive to sulfonamides, aminoglycosides, chloramphenicol, tetracyclines, ampicillin, carbenicillin, and cephalosporins. Resistance, often plasmid-mediated, is frequently encountered.

Treatment with a sulfonamide or ampicillin is usually adequate in acute, uncomplicated urinary tract infections. If there is no response an alternative drug is used, on the basis of sensitivity tests. In chronic or recurrent infections methenamine mandelate, nalidixic acid, trimethoprim–sulfamethoxazole, cephalexin, and carbenicillin are useful. More toxic drugs, such as aminoglycosides or chloramphenicol, should be used only if all other forms of treatment fail, or in life-threatening systemic infections (sepsis, etc.). In infants with **diarrhea** due to enteropathogenic *E. coli* oral neomycin or colistin are used. The efficacy of antimicrobials in diarrhea due to either toxigenic or invasive *E. coli* has not been established. The most important therapy in severe diarrheal disease is electrolyte and fluid replacement.

The efficacy of prophylaxis against traveler's diarrhea has not been convincingly proven (in Great Britain a combination of oral streptomycin with three sulfonamides is recommended). The efficacy of nonspecific antidiarrheal agents (e.g., kaolin, pectin), and the safety of diphenoxylate with atropine (Lomotil), have been questioned. Recently a bismuth–salicylate preparation (Pepto-Bismol) has been reported to possess demonstrable activity against *E. coli* enterotoxin, experimentally and in human clinical trials.

KLEBSIELLA–ENTEROBACTER–SERRATIA

The genera *Klebsiella, Enterobacter,* and *Serratia,* in the tribe Klebsiellae, can be identified by biochemical tests, including decarboxylase reactions (Table 31-4). Most strains of *Klebsiella* and *Serratia* can also be typed serologically. Members of the group are second to *E. coli* as causes of gram-negative bacteremia, which sometimes results from their ready proliferation in contaminated intravenous solutions containing glucose. The necessity for differentiating these three genera from each other, and from other enteric bacilli, is underlined by their wide differences in antibiotic sensitivity and in pathogenicity. *Klebsiella* predominates in clinical isolates and as a primary pathogen, and usually produces more severe illness; the other two are opportunistic pathogens.

In testing water supplies, it is important to distinguish *Escherichia coli,* which is an index of fecal contamination, from *Enterobacter aerogenes* (formerly *Aerobacter aerogenes*), which is found widely on plants. Accordingly, biochemical tests for this purpose were developed early. Some of these are based on the fact that *E. coli* carries out the **mixed acid fermentation** (Ch. 3), while *Klebsiella* and *Enterobacter* carry out the **butylene glycol fermentation** (Ch. 3), which forms large quantities of the neutral product 2,3-butylene glycol (CH_3-CHOH-CHOH-CH_3) and hence forms less acid.

Four metabolic tests (indole, methyl red, Voges–Proskauer, and citrate utilization), collectively referred to as the **IMViC tests,** were originally used to distinguish *Enterobacter aerogenes* from *Escherichia coli.* Only the latter produces **indole** in media containing tryptophan. The

methyl red test distinguishes heavy and light production of acid, because this indicator shifts from yellow to red below pH 4.5, and in glucose–peptone broth cultures incubated for 48 h only the mixed acid fermentation produces enough acid to turn the indicator red. The **Voges–Proskauer** reaction is a color test for acetoin, a product of the butylene glycol type of fermentation. **Citrate** can serve as the sole carbon source for *Enterobacter aerogenes* but not for *Escherichia coli*.*

The three genera are differentiated by a combination of tests (Table 31-4), primarily for motility, lactose utilization, decarboxylases, Voges–Proskauer reaction, fermentation of several carbohydrates, and urease production.

Klebsiella. *Klebsiella pneumoniae* (Friedländer's bacillus) is the most important human "pathogen" of the *Klebsiella* group. It forms a capsule and hence produces large, moist, often very mucoid colonies. A total of 5 O Ags and 72 capsular (K) polysaccharide Ags have been identified. Type-specific (K) antisera are useful in determining the epidemiology of hospital-acquired *Klebsiella* infections, which represent about two-thirds of all infections due to these organisms.

K. pneumoniae is found in the respiratory tract and the feces of 5%–10% of healthy subjects and is frequently present as a secondary invader in the lungs of patients with chronic pulmonary disease. It causes approximately 3% of all acute bacterial pneumonias; and it is the second most common urinary tract pathogen. Its invasive properties depend on the antiphagocytic effect of its capsule: unencapsulated (R) strains are avirulent.

Pneumonia caused by *K. pneumoniae* is characterized by the production of thick gelatinous sputum and a high bacterial population density in the edema zones of the active lesions, and by destructive action of the unphagocytized organisms on the pulmonary tissue. These features interfere with antimicrobial therapy (see Abscess Formation and Necrosis, Ch. 25), often resulting in lung abscesses requiring surgical resection.

Although *Klebsiella* has been found in acute diarrhea in children its enteric pathogenicity has been questioned. Recently, however, strains isolated from patients with **tropical sprue** were shown to elaborate an **enterotoxin** that resembles *E. coli* ST enterotoxin. *Klebsiella* species have been implicated in chronic inflammatory diseases of the upper respiratory tract: *K. ozaenae* in **ozena,** a progressive fetid atrophy of the nasal mucosa; and *K. rhinoscleromatis* in **rhinoscleroma,** a destructive granuloma of the nose and pharynx.

In acute, uncomplicated urinary tract infections oral sulfonamides, nalidixic acid, ampicillin, and tetracyclines are usually ef-

fective; in recurrent or chronic infection sensitivity tests are needed for selection of effective agents. In life-threatening infections aminoglycosides, cephalosporins, and chloramphenicol are the drugs of choice. Most strains of *Klebsiella,* unlike *Enterobacter* and *Serratia,* are sensitive to cephalothin.

Enterobacter. Organisms in the *Enterobacter* group occur in soil, dairy products, water, and sewage, as well as in the intestinal tract of man and animals. Enterobacters are frequently isolated from sputum (often after antibiotic therapy), urinary tract infections (often hospital-acquired), blood, and wound infections. They are usually considered **secondary pathogens** (i.e., superinfecting an underlying primary infection), or **opportunistic** (e.g., in urinary tract infections following catheterization), or **commensals** (i.e., not causally associated with disease).

In this genus (Table 31-4) the most common species in man is *E. cloacae,* followed by *E. aerogenes,* and *E. hafniae* (now renamed *Hafnia alvei*). As was discussed, *E. agglomerans* (formerly *Erwinia*) achieved notoriety, together with *E. cloacae,* in a nationwide epidemic of septicemia caused by contaminated intravenous products (1971), and it is now increasingly recognized as an opportunistic pathogen in a wide variety of settings. *E. agglomerans* includes both aerogenic (gas-producing) and anaerogenic organisms; over half of both groups produce yellow pigment, useful in identification.

Enterobacters are susceptible to aminoglycosides, chloramphenicol, the tetracyclines, trimethoprim–sulfamethoxazole, nalidixic acid, and nitrofurantoin (the latter three being used in urinary tract infections). Many strains are sensitive to colistin, but most are resistant to cephalothin and ampicillin.

Serratia. Cultures of *Serratia marcescens* have been used for many years by bacteriologists for demonstration purposes (e.g., to demonstrate bacteremia after dental extraction) because the bright red pigment of some strains is so easily observable.† The organism was long considered a harmless saprobe, but since 1960 it has been isolated with increasing frequency in humans, mostly in nosocomial infections (often severe). The earlier incidence is not known, since many nonpigmented strains were probably included with other slow lactose fermenters as paracolon organisms. Subdivision into three species is achieved by biochemical tests (Table 31-4); *S. marcescens* is the most common in clinical practice.

Serratiae are motile. Unlike most other coliforms, most strains do not ferment lactose or do so slowly. Although the organisms may produce a red pigment **(prodigiosin),** especially at room temperature, at least three-fourths of strains isolated at present are nonpigmented. Fifteen O Ags and sixteen H Ags of *S. marcescens* have been identified; serotyping and bacteriocin typing are useful for epidemiologic studies.

Serratiae are mostly involved in patients with debilitating disor-

---

* These tests are standard procedures in the study of Enterobacteriaceae. The IMViC formula for *Escherichia coli* is ++−−, for *Enterobacter aerogenes* and *Enterobacter cloacae,* −−++. All 16 possible combinations of the IMViC test have been found.

† Instances of the "miraculous" appearance of "blood" on communion wafers and other foods were most likely due to strains of *S. marcescens.*

ders, or under treatment with broad-spectrum antibiotics, or subjected to instrumentation such as tracheostomy tubes or indwelling catheters. Outbreaks have occurred in nurseries, intensive-care units, and renal dialysis units. Serratiae can cause endocarditis, osteomyelitis, septicemia, and wound, urinary tract, and respiratory tract infections.

Most strains are resistant to several antibiotics, especially to cephalosporins and ampicillin. Aminoglycosides (especially amikacin), chloramphenicol, and carbenicillin are useful. In urinary tract infections nalidixic acid and trimethoprim–sulfamethoxazole are possible alternatives; sensitivity testing is required. Drug-resistance plasmids have been demonstrated.

EDWARDSIELLA

The genus *Edwardsiella* includes a group of motile, H_2S-producing, lactose-negative organisms that resemble salmonellae in some biochemical features (Table 31-4) and sometimes in pathogenicity. *E. tarda* has been isolated from a variety of mammals and reptiles. It is occasionally also found in the human intestinal tract, especially in acute gastroenteritis, and it has been associated with meningitis, septicemia, and wound infections. A total of 49 O Ags and 37 H Ags have been identified. Kanamycin, ampicillin, cephalothin, and chloramphenicol are the drugs of choice.

CITROBACTER

The *Citrobacter* group (tribe Salmonelleae) is composed of Enterobacteriaceae previously designated as *Escherichia freundii* and the Bethesda–Ballerup group of paracolon organisms. Many strains ferment lactose, but this is frequently delayed. They are differentiated from salmonellae by their possession of β-galactosidase and lack of lysine decarboxylase. *C. freundii* and *C. diversus* can be differentiated by biochemical tests (Table 31-4).

Citrobacter strains are found infrequently in normal feces; they have been associated with diarrhea, with secondary invasion in compromised patients, and occasionally with severe primary septic processes. They are susceptible to chloramphenicol, gentamicin, kanamycin, and colistin, but strains vary in sensitivity to ampicillin, carbenicillin, and cephalothin.

PROTEUS–PROVIDENCIA GROUP

These lactose-negative, motile bacilli are unusual among Enterobacteriaceae in being able to deaminate phenylalanine* (Table 31-4) and lysine. **Rapid** and **abundant urease production** distinguishes *Proteus* from *Providencia*.

Proteus. These organisms are commonly found in soil, sewage, and manure. They are found with some frequency in normal human feces, but often in much increased numbers in individuals receiving antibiotic therapy or during diarrheal diseases due to other organisms.

* The product, phenylpyruvic acid, develops an intense green color on addition of $FeCl_3$.

Proteus organisms are frequent causes of urinary tract infection and are also involved in other, often serious infections.

The antigenic structure of *P. vulgaris* is of particular medical interest because strains possessing certain O Ags (OX2, OX19, and OXK) are agglutinated by the serum of patients with various rickettsial diseases (Weil–Felix reaction; see Laboratory Diagnosis, Ch. 40). These particular O Ags seem to be fortuitously related to antigenic determinants of rickettsiae.

P. vulgaris and *P. mirabilis* form a thin spreading growth (swarm) on the surface of moist agar media. To obtain isolated colonies the swarming is prevented by cultivation on the relatively dry surface of 5% agar or on ordinary 1%–2% agar containing 0.1% chloral hydrate. They also produce abundant H_2S and liquefy gelatin; *P. rettgeri* and *P. morganii* do not possess these characteristics. Fermentation of most *Proteus* strains is of the mixed acid type.

Gentamicin is active against all four species, but only the most commonly occurring, indole-negative *P. mirabilis* is sensitive to ampicillin, several aminoglycosides, cephalothin, trimethoprim–sulfamethoxazole, and often penicillin G. The three indole-positive species are generally highly resistant to a variety of antibiotics.

Providencia. Organisms in the genus *Providencia* are closely related to *Proteus morganii* and *Proteus rettgeri*. Members of the group have been isolated from human feces during outbreaks of diarrhea but also occur in normal individuals. They are primarily associated with urinary tract infections; *Providencia stuartii* has recently emerged as a major agent in burn infections, displaying marked resistance to many antibiotics, while *P. alcalifaciens* is less frequently encountered and is more susceptible.

SALMONELLAE†

The genus *Salmonella* contains a wide variety of "species" pathogenic for man or animals, and usually for both. They ferment neither lactose‡ nor sucrose, and with few exceptions they produce abundant H_2S (Table 31-4). Essentially all are motile and decarboxylate lysine and ornithine.¶ Their complex antigenic structures and genetic relations have already been discussed. Selective media for their isolation from feces contain brilliant green, cholate, selenite, tetrathionate, or citrate to suppress the growth of coliforms. Some of the salmonellae more often encountered in medical practice are listed in Table 31-7.

Originally salmonella species were named according to the disease they caused or the animals from which they were first isolated. Later each new antigenically distinguishable type was accorded

† This genus was discovered by the great bacteriologist Theobald Smith, working as a young man in the US Department of Agriculture; it was named for a coauthor, D. E. Salmon, his laboratory chief.

‡ For biochemically atypical strains see Genetic Relations, above.

¶ *S. typhi* fails to produce gas from carbohydrates and to decarboxylate ornithine. It also produces little or no H_2S in TSI agar.

TABLE 31-7. The Ten Most Frequently Isolated Serotypes of *Salmonella*, United States, 1978

Serotype	Number	Percent
S. typhimurium	10,015	34.8
S. heidelberg	2,078	7.2
S. enteritidis	1,934	6.7
S. newport	1,879	6.5
S. infantis	1,225	4.3
S. agona	1,186	4.1
S. montevideo	703	2.4
S. typhi	604	2.1
S. saint-paul	602	2.1
S. javiana	528	1.8
Subtotal	20,754	72.2
Total (all serotypes)	28,748	100.0

(Salmonella Surveillance Annual Summary 1978, USPHS Center for Disease Control, Atlanta, 1979)

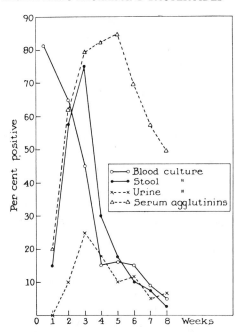

FIG. 31-5. Relative frequencies of positive blood, urine, and stool cultures and serum agglutination tests during the course of typhoid fever. (Morgan HR: In Dubos RJ, Hirsch JG [eds]: Bacterial and Mycotic Infections of Man, 4th ed, p 381. Philadelphia, Lippincott, 1965)

specific rank and was named after the geographic place at which it was first isolated. Recent United States usage, in contrast, recognizes only three species (*S. cholerae-suis, S. typhi,* and *S. enteritidis*): the first two do not contain subtypes (serotypes) but *S. enteritidis* contains over 1500 serotypes, each written in a nonitalicized form (e.g., *S. enteriditis* serotype paratyphi A, instead of *S. paratyphi* A). However, the classic names will be employed here because of their prevalence in the current clinical literature. About 95% of the strains causing disease in man belong to five O Ag groups (A–E) and comprise less than 40 serotypes. Table 31-7 lists the ten serotypes most commonly isolated from human sources in the United States in 1975.

Pathogenesis. Three clinically distinguishable forms of salmonellosis occur in man: enteric fever, septicemia, and acute gastroenteritis.

Of the **enteric fevers,** the prototye is **typhoid fever,** caused by *S. typhi.* The disease usually begins insidiously after an incubation period of 7–14 days, with malaise, anorexia, and headache, followed by the onset of fever. The ingested organisms multiply in the gastrointestinal tract, and some enter the intestinal lymphatics, from which they are disseminated throughout the body by the blood stream and are excreted in the urine. The bile is a good culture medium for *S. typhi,* and so luxuriant growth occurs in the biliary tract and provides a continuous flow of organisms into the small bowel, where they tend to localize in the Peyer's patches. Their ability to persist in the biliary tract may result in a chronic carrier state (see Carriers, below), with continued excretion in the feces. Figure 31-5 depicts the kinetics of the bacteremia, shedding of organisms in the feces and urine, and formation of Ab.

Fever often increases in a steplike manner and is accompanied by relative bradycardia. Prostration may be marked, particularly during the first week; and though diarrhea is usually absent, abdominal tenderness and distention are common. Cough and signs of bronchitis may be present; "rose spots," which last for only a few days, may appear on the trunk; and splenomegaly and leukopenia are common. In more severe cases the sensorium is dulled and the patient may become delirious. After the third week the fever usually subsides by gradual lysis. In fatal cases the most prominent lesion found at autopsy is lymphoid hyperplasia; ulcerations in the Peyer's patches may lead to intestinal hemorrhages or perforations of the bowel.

In active lesions bacilli are often detectable in the phagocytic mononuclear cells, where they can multiply (as is readily demonstrated in vitro). Intracellular bacilli seem to be protected from the bacteriolytic action of specific Ab, which appears in the blood long before the disease subsides (Fig. 31-5).

The pathogenesis of typhoid fever is difficult to study because most laboratory animals are resistant. Mice, for example, can be readily infected only by intracerebral injection, or by intraperitoneal inoculation with organisms suspended in mucin. Chimpanzees, however, can be infected by the oral route. In human volunteers 10^5–10^7 ingested bacilli have been required to induce the disease.

Enteric fevers caused by other salmonellae (**paratyphoid** fevers) are usually milder and have a shorter incubation period (1–10 days). Bacteremia occurs early, fever usually lasts for 1–3 weeks, and rose spots are rare. Almost any *Salmonella* may cause enteric fever, but in the United States only *S. paratyphi* B (*S. schottmülleri*) and *S. typhimurium* are common.

Salmonella **septicemias** are characterized by high, remittent fever and bacteremia, ordinarily without apparent involvement of the gastrointestinal tract. Focal suppurative lesions may develop almost anywhere in the body, including the biliary tract, kidneys, heart, spleen, meninges, joints, and lungs. Prolonged septicemia of this type is most commonly caused by *S. cholerae-suis*. Extraintestinal infections with *Salmonella* often occur in association with underlying chronic disease of defects in host defense, e.g., metastatic cancer or sickle cell disease.

Gastroenteritis, a form of the disease confined primarily to the gastrointestinal tract, is the most common kind of *Salmonella* infection. The most frequent cause in the United States is *S. typhimurium.* Symptoms begin 8–48 h after the consumption of contaminated food, with diarrhea ranging from mild to a fulminant form with sudden and violent onset (**"food poisoning"**). Headache, chills, and abdominal pain are followed by nausea, vomiting, and diarrhea, accompanied by fever lasting from 1–4 days. Blood cultures are rarely positive, but the organisms can usually be cultured from the feces.

In contrast to cholera (see Cholera Vibrios, below), the diarrhea of *Salmonella* enteritis involves **both small and large intestine,** with **mucosal invasion.** The mechanism of fluid secretion is not known: an "enterotoxic principle" has recently been isolated from a group D *S. enteritidis,* but its role has not been elucidated.

Salmonellae can be divided into three groups with respect to their host preferences. 1) *S. typhi, S. paratyphi* A and C, and *S. sendai* are more or less strictly adapted to humans. 2) Others are adapted to particular nonhuman animal hosts: e.g., the primary host of *S. cholerae-suis* is swine, but it occurs in other animals as well as in humans. 3. A far larger number of types produce disease equally in man and in other animals. The reasons for these varying interactions are unknown. There is evidence that virulent species multiply intracellularly, whereas avirulent ones do not.

Carriers. Following active salmonellosis the organisms occasionally become established in the host, who then continues indefinitely to excrete as many as 10^6–10^9 *S. typhi* per gram of feces. The source is usually a chronic suppurative focus in the biliary tract. When treated with broad-spectrum antibiotics carriers may come down with active *Salmonella* disease, just as mice are made more susceptible to experimental salmonellosis by treatments that upset the normal balance of their intestinal flora (see Microbial Antagonism, under Skin and Mucous Membrane Barriers, in Ch. 24).

Immunity. Both anti-O and anti-Vi Abs play a role in immunity to salmonellosis: maximal protection is obtained in chimpanzees only by immunization with both Ags. Anti-H Abs appear to have no protective effect. In experimental salmonellosis in the mouse two mechanisms have been demonstrated: 1) specific O Ab opsonizes *Salmonella* cells, with subsequent engulfing and destruction of a large proportion of these by macrophages; 2) in cellular immunity (appearing in the course of the infection or induced by living vaccines) macrophages become more efficient in killing intracellular bacteria.* In addition, it appears that in man an **orally** administered attenuated *S. typhi* strain can enhance **local immunity** of the intestinal tract and effectively prevent initiation of foci by typhoid bacilli.

Laboratory Diagnosis. A diagnosis of *Salmonella* infection is made by isolation of the organism. Isolation from blood or urine establishes the diagnosis, but a salmonella organism isolated from the feces is not necessarily the cause of the individual's illness (see Carriers, above).

Blood cultures are particularly important during the first week of illness (Fig. 31-5); occasionally bone marrow cultures are positive when blood cultures are negative. Urine and feces should be cultured repeatedly on differential and selective media. Preliminary incubation of fecal specimens in selenite or tetrathionate broth, which inhibit coliforms but not salmonellae, may facilitate the subsequent isolation of salmonellae on differential solid media. Final identification is based on biochemical reactions (Table 31-4) and agglutination tests with monospecific antisera.

Serologic tests for specific agglutinins (**Widal test**) should be performed on at least two serum specimens, the first obtained as early as possible and the second 7–10 days later. Dilutions should be tested with the infecting organism (if available) as well as with standard *Salmonella* O and H Ags. A significant rise in Ab titer to the organism isolated from the patient helps confirm the diagnosis.

O agglutinins are ordinarily of greater diagnostic significance than H agglutinins, which tend to persist longer following vaccination and occasionally are not produced at all in active infections. Most individuals have Abs to several *Salmonella* serogroups, owing to inapparent infection or previous immunization, and many Ags are shared by different salmonellae. Hence Widal tests with group Ags, as performed in most clinical laboratories, are often not helpful and occasionally are misleading.

Treatment. For typhoid fever and *Salmonella* septicemias chloramphenicol has long been the drug of choice, but chloramphenicol-resistant strains have appeared. Accordingly, drug sensitivity testing is required. Paren-

---

* This immunity is nonspecific, i.e., it may be induced by various intracellular bacterial parasites (see Macrophage Activation, Ch. 24).

teral ampicillin and trimethoprim–sulfamethoxazole are usually effective alternatives. Plasmid-mediated resistance to many antimicrobials has been observed. Response to therapy is usually rapid, but because the organisms tend to survive within phagocytic cells relapses frequently occur unless the patient is treated for at least 2 weeks.

⚱ Antibiotics are **not** indicated in *Salmonella* **gastroenteritis** (except in the very young and those over 60) since the disease is brief and limited to the gastrointestinal tract. In addition, the unnecessary use of antibiotics prolongs *Salmonella* excretion, promotes the incidence of the carrier state, and favors the acquisition of resistance by the infecting strain.

In eliminating the **carrier state** the bactericidal ampicillin is much more effective than the bacteriostatic chloramphenicol. Prolonged treatment with large doses of ampicillin is effective in 60%–80% of cases. In carriers who relapse after one or more courses of therapy cholecystectomy terminates the carrier state in 9 out of 10 cases.

Prevention. Most important in preventing the man-to-man transmission of typhoid fever have been 1) proper sewage disposal, 2) pasteurization of milk, 3) maintenance of unpolluted water supplies, and 4) scrupulous exclusion of chronic carriers as food handlers. Since *S. typhi* infects only humans its control is relatively feasible, and the incidence of typhoid fever in the United States has been declining steadily (e.g., from 5593 cases in 1942 to 398 in 1977).

In contrast, laboratory-confirmed nontyphoid salmonellosis has increased 20-fold in the same period. Since the milder food- and waterborne enteritis cases are rarely reported, it is estimated that the 27,850 cases of salmonellosis reported in 1977 represent more than 2 million cases per year.

Eradication of human disease due to salmonellae that infect animals as well as man is difficult, for elimination of the animal reservoir is impossible. Domestic fowl probably constitute the largest reservoir, salmonellae having been isolated from 40% of apparently healthy turkeys, and from many other domestic and wild animals including pet turtles. Meats and pooled preparations of dried eggs are notorious sources. Practical measures for controlling salmonellosis in animals (only recently begun) and improved technics of food processing should eventually lower the incidence of human infections.

Although **typhoid vaccines** have been employed for many decades, definitive evidence of their effectiveness has become available only recently. In controlled field studies in the 1960s in Eastern Europe and Guyana an acetone-inactivated dried typhoid vaccine (with the Vi Ag retained) conferred protection of up to 90% for 3 years. Conventional heat-killed, formalin-preserved ty-

phoid vaccine (with most of the Vi Ag destroyed) was less effective, and cell-free extracts containing either O or Vi Ags were virtually ineffective. An orally administered attenuated *S. typhi* strain appears to afford good protection, but additional studies are needed.

In human volunteers one vaccine protected 67% against a challenging dose of 10^5 organisms, but not at all against 10^7 organisms. Since immunity is thus obviously relative, and since contaminated food may contain large numbers of organisms, it is not realistic to expect complete protection by any vaccine. Routine typhoid immunization is not recommended in the United States but is widely used for military personnel and for travelers to endemic areas. It is also used in household contacts of known carriers and in community and institutional outbreaks. The endotoxin in the vaccine generally causes a mild toxic reaction.

The use of triple typhoid vaccine (TAB, containing *S. typhi* and *S. paratyphi* A and B) is no longer recommended, for the effectiveness of the A vaccine has never been experimentally established and recent field trials of the B vaccine failed to show any beneficial effect.

ARIZONA GROUP

Arizona, recognized as a separate genus in the United States but considered a subgenus of *Salmonella* elsewhere, contains a single species, *A. hinshawii*. It is antigenically closely related to salmonellae but differs in several metabolic tests (Table 31-4). Cold-blooded animals, especially snakes, seem to be the natural reservoir, but *Arizona* organisms have also been isolated from fowl and domestic animals and occur in dried egg powder and other foods. In man they are associated with salmonellosislike diseases, i.e., gastroenteritis, enteric fevers, bacteremia, or localized infections. However, the apparent high degree of invasiveness may reflect selective recognition of such cases: *many strains ferment lactose rapidly, and so they are likely to be recognized only when cultured from organs and not in cases of gastroenteritis.* Ampicillin, chloramphenicol, and cephalothin are useful in treatment; sensitivity tests are required.

SHIGELLAE

The shigellae are much less invasive than the salmonellae, rarely causing bacteremia. They also have a much narrower distribution in nature, inhabiting only the intestinal tracts of primates (or rarely of dogs). They cause in man a disabling disease known as **bacillary dysentery** (Gr. *dys* + *enteron,* sick gut). Described in the fourth century B.C., this common illness has been of great military importance through the ages, often rendering whole armies temporarily (but ignominiously) unfit for combat. It spreads rapidly under conditions of overcrowding and poor sanitation (e.g., disaster areas, prisoner-of-war camps, and overtaxed mental hospitals). In 1896 the Jap-

TABLE 31-8. Simplified Outline of Shigella Classification

Species	Serologic subgroup	Serologic type(s)*	Mannitol	Ornithine decarboxylase
S. dysenteriae	A	1-10	–	–
S. flexneri	B	1-6	+	–
S. boydii	C	1-15	+	–†
S. sonnei	D	1	+	+

+ = 90% or more positive in 1 or 2 days; – = 90% or more negative.

*Serotypes are separately numbered for each species; crossreactions between those with corresponding numbers of different species are absent or minimal.

†Cultures of *S. boydii* 13 are positive.

anese bacteriologist Shiga isolated the first species, now known as *Shigella dysenteriae.*

Properties that distinguish *Shigella* from most salmonellae (Table 31-4) include lack of motility, failure to produce gas during fermentation (except for some biotypes of *S. flexneri* 6 and rare strains of *S. boydii* 13), and lack of lysine decarboxylase. Shigellae possess specific polysaccharide O Ags, but since they are nonmotile they have no H Ags. Certain smooth (S) strains have heat-labile K (envelope) Ags and hence agglutinate more readily in homologous anti-O antiserum after they are heated.

The four species of *Shigella* (Table 31-8) are differentiated from other Enterobacteriaceae and from each other by biochemical and antigenic characteristics. Unlike the others, *S. sonnei* ferments lactose, but slowly. Shigella Ags, like those of salmonellae, may be altered by lysogenic conversion.

Pathogenicity. All known species of the genus *Shigella* are pathogenic for man.* The organisms invade intestinal epithelial cells and the lesions produced are mostly confined to the terminal ileum and the colon; they consist primarily of mucosal ulcerations, which are covered by a pseudomembrane composed of polymorphonuclear leukocytes, cell debris, and bacteria, all enmeshed in fibrin. It is believed that these lesions arise when the organisms cross the epithelial barrier and enter the lamina propria: local accumulation of metabolic products and release of endotoxin then cause death of the epithelial cells. *Penetration beyond the submucosa is rare.*

S. dysenteriae type 1 (*S. shigae,* Shiga bacillus) produces a **neurotoxin,** distinct from the endotoxin common to all shigellae; it causes peripheral paralysis and death when injected into rabbits and mice. Recently it was found that a purified preparation is also **cytotoxic** to various mammalian cells and exhibits **enterotoxin** activity. The toxin acts by **inhibiting protein synthesis,** as shown in extracts of mammalian cells.

* The organisms formerly known as *Shigella alkalescens* and *S. dispar,* of doubtful pathogenicity in man, are now classified as *E. coli.*

The role of the toxin remains to be defined, but it is clear that toxin production alone is not sufficient to confer virulence on shigellae: noninvasive toxigenic mutant strains do not cause shigellosis in volunteers or in monkeys, while invasive but nontoxigenic strains do produce disease. Hence epithelial penetration is apparently the major determinant of virulence.

Shigella dysentery is characterized by sudden onset of abdominal cramps, diarrhea, and fever, following an incubation period of 1–4 days. Both mucus and blood in the feces are common. The loss of water and salts may cause dehydration and electrolyte imbalance, particularly in infants and young children. Specific agglutinins appear in the blood during convalescence, and Abs can often be demonstrated in the feces (**coproantibodies**). Patients also develop neutralizing Abs to the *S. dysenteriae* toxin.

Stool cultures usually become negative within a week or two of convalescence but may remain positive for a month or more. *Shigella* dysentery varies in severity according to species (in the ascending order *S. sonnei, S. boydii, S. flexneri,* and *S. dysenteriae*), from a mild intestinal upset with little systemic disturbance to profuse bloody diarrhea with fever and severe prostration. A *S. dysenteriae* type 1 epidemic in Central America was characterized by fatality rates up to 20% among children and by multiple drug resistance of the organism.

Bacterial antagonism appears to play a major role in natural resistance of the gastrointestinal tract to invasion by *Shigella* organisms. Thus germ-free animals are susceptible to orally induced shigellosis; and though they are not protected by immunization with *Shigella* vaccine, they become highly resistant if their intestinal tracts are infected with *E. coli* before challenge.

Immunity. Persons residing in regions where bacillary dysentery of a given type is endemic seem to acquire immunity to the disease, presumably from inapparent infection. Patients with circulating Abs, however, often suffer second attacks of shigellosis when exposed to a large enough reinfecting dose. The ineffectiveness of Abs has been attributed to the superficial nature of the lesions in the intestinal tract. Intestinal immune factors (**co-**

proantibodies), in contrast, may play an important role, for encouraging results were obtained in recent field trials with live oral vaccines made from strains (a *Shigella*-*E. coli* hybrid and a streptomycin-dependent *Shigella* strain) that had lost the ability to penetrate the intestinal epithelial lining.

Laboratory Diagnosis. Although dysentery bacilli are usually excreted in large numbers during the active disease they often remain viable in the feces for only a short time and therefore must be cultured fairly promptly. Indeed, the best method of obtaining specimens for culture is by means of a rectal swab. Specimens from ulcerative intestinal lesions, obtained under direct vision through a sigmoidoscope, are most likely to contain the organisms. Cultures are best made on MacConkey, xylose–lysine–deoxycholate, or eosin–methylene blue agar,* with final identification of the organism based on biochemical (Table 31-4) and specific agglutination tests.† Identification of *S. dysenteriae* is important, since it causes a particularly severe dysentery.

Bacillary dysentery can often be differentiated from amebic dysentery (due to the protozoon *Entamoeba histolytica*) by microscopic examination of the feces: the bacillary lesions are frankly purulent, whereas those caused by amebae are remarkably free of white cells.

The demonstration of a rising titer of serum agglutinins may permit retrospective recognition of the disease, but it has little immediate diagnostic value because of the brevity of the illness. Tests for hemagglutinating and toxin-neutralizing Abs are useful in epidemiologic surveillance programs.

Treatment. Since the infection, especially that due to *S. sonnei*, is essentially self-limiting, opinion is divided on the advisability of antimicrobial treatment in mild cases, especially since drug therapy often results in the emergence of multiply drug-resistant strains. For severely ill patients, and for young children and debilitated adults, antimicrobial therapy as well as supportive intravenous fluid and electrolyte therapy is needed. At present ampicillin is considered the drug of choice, but resistance is common in some parts of the United states; trimethoprim–sulfamethoxazole and chloramphenicol are the favored alternatives. Shigellae are usually also sensitive to aminoglycosides, tetracyclines, nalidixic acid, and colistin but it is necessary to determine the sensitivity pattern of the patient's (or of the epidemic) strain. Resistance plasmids are widespread.

Prevention. Since the only important source of bacillary dysentery is man, and the disease is spread by "food, feces, fingers, and flies," sanitary measures are of the greatest importance. Control is complicated by the high incidence of **inapparent infections.** Moreover, in contrast to the findings with typhoid in volunteers (see above), *the ingestion of as few as 180* Shigella *cells can cause dysentery in man.* This property may explain the person-to-person mode of transmission and the recurrence in institutionalized populations.

All patients with clinically recognizable dysentery should be isolated, if possible, until their stool cultures have become negative. Proper chemotherapy significantly shortens the period of infectiousness. Sanitary measures should be directed against infected food handlers, contaminated water and milk supplies, and improper methods of sewage disposal. In widespread epidemics mass chemoprophylaxis may be indicated. Effective control may eventually be achieved through use of oral live vaccines: Streptomycin-dependent strains of *S. sonnei* and *S. flexneri* confer a high degree of protection and have seemed safe in small trials.

VIBRIONACEAE

Vibrios are curved, motile, gram-negative, facultative, oxidase-positive bacilli with a single polar flagellum. Other differentiating characteristics are given in Table 31-4. Of the five species described, *V. cholerae* is the major pathogen, while *V. parahaemolyticus* has been increasing in medical importance. Some additional species cause disease in humans and animals, while others are commonly found in the normal human flora. The genus *Aeromonas* is also classified in the family of Vibrionaceae.

---

* Deoxycholate citrate and SS (*Salmonella–Shigella*) agars are often inhibitory.

† Subgroups can be identified in most clinical laboratories, but serotypes can often be determined only in larger laboratories, where specific sera are available.

CHOLERA VIBRIOS

Cholera, endemic in India for centuries, has from time to time caused devastating epidemics in other parts of the world. Since 1817 seven worldwide pandemics have occurred. The latest one spread in the 1960s from Indonesia to the Far East, India, the Middle East, and Africa. Sporadic outbreaks occurred later in parts of Europe, and the first confirmed case since 1911 in the United States (except for laboratory-acquired illness) was reported in 1973. By 1975 the pandemic gave signs of waning.

The role of drinking water as a source of cholera was established in the London outbreak of 1854, when an anesthetist, John Snow, "with a notebook, a map, and his five senses," proved epidemiologically that "spoiled" water from the Broad Street pump spread the disease. In 1883 the causative agent, the cholera vibrio, was discovered by Robert Koch, who demonstrated the organism regularly in patients in epidemics in Egypt and India. Since vibrios were also seen in the absence of disease the pathogenicity and specificity of the cholera vibrio were not established until 10 years later, when the cells were found to be rapidly lysed in the peritoneal cavities of previously immunized guinea pigs (**Pfeiffer phenomenon**). Shortly thereafter the classic studies of Bordet showed that immune bacteriolysis involved both Abs and a heat-labile substance, now known as the complement system.

V. cholerae differs from Enterobacteriaceae in several important respects. 1) In fresh isolates it is shaped like a **comma** rather than a straight rod (hence it was earlier called *V. comma*), though on repeated subculture it assumes the shape of the enteric bacilli. 2) It is **oxidase-positive** (Table 31-2). 3) It grows luxuriantly in media that are too **alkaline** (pH 9.0–9.6) for the growth of most other bacteria, and it is sensitive to acid; it is rapidly killed in cultures containing fermentable carbohydrate. (This property is utilized in designing selective media for its primary isolation.) 4) The organism produces a **neuraminidase** that hydrolyzes *N*-acetylneuraminic acid, a sialic acid found in human plasma, in various mucins, and in surface structures of many mammalian cells (see Hemagglutination, Ch. 46).

In many respects, however, *V. cholerae* resembles Enterobacteriaceae (Table 31-4). It ferments lactose slowly (after 3 days or more) but promptly attacks mannose and sucrose, producing acid but not gas; it possesses lysine and ornithine decarboxylases; it can grow in selective media containing bile salts, bismuth sulfite, or tellurite; it produces a variety of O and H Ags, as well as endotoxin. *V. cholerae* is subdivided into three antigenic types (AB, or Ogawa; AC, Inaba; ABC, Hikojima); they share a group-specific O Ag (A), while the secondary Ags (B and C) are type-specific.

El Tor Biotype. Unlike most cholera vibrios, some strains form a soluble **hemolysin;** the first was isolated in 1906 at the El Tor quarantine station on the Gulf of Suez. Such strains are also unusual in their phage susceptibility, but their pathogenic properties are typical of cholera vibrios. The El Tor vibrios are more resistant to chemical agents, persist longer in man, and survive longer in nature than the classic strain. They are responsible for the cholera pandemic that started in 1961.

Pathogenesis. Naturally acquired cholera has been described only in humans. Suckling rabbits are susceptible to oral infection, as are guinea pigs if pretreated by starvation and administration of streptomycin and calcium carbonate. The resulting infection remains localized in the intestine and superficially resembles human cholera. In human volunteers challenges with varying doses resulted in a wide gradation of responses. *Gastric acidity seems an important defense against cholera:* its neutralization lowered the infecting dose from 10^8 to 10^4 organisms, and the first United States cholera patient in modern times had had a subtotal gastrectomy.

The disease starts, 2–5 days after infection, with a sudden onset of nausea, vomiting, diarrhea, and abdominal cramps. In severe cases the voluminous liquid feces ("rice-water stools") contain mucus, epithelial cells, and large numbers of vibrios (10^6 or more/ml). The loss of fluid from the gut may reach 10–15 liters/day, leading to extreme dehydration. The liquid feces are virtually free of protein and contain about the same concentration of sodium, about twice the concentration of bicarbonate, and nearly five times the concentration of potassium as normal plasma. The patient thus develops an extracellular fluid deficit, metabolic acidosis, and hypokalemia. Shock frequently intervenes, and death may occur in a few hours. The mortality rate in hospitalized cases without adequate therapy may be as high as 60% but is reduced to less than 1% by replacement of fluid and electrolytes (see Treatment). In milder cases symptoms may be minimal. Regardless of severity, the active disease rarely lasts for more than a few days.

Biopsies taken from both the small and the large intestine reveal only hyperemia and signs of superficial inflammation. In postmortem examination the organisms are found in large numbers "adsorbed" to the surface of the mucosa and in the luminal fluid, but in contrast to shigellae (see above) *they do not invade the epithelium:* the mucosa remains intact. In experimental animals and human volunteers cell-free filtrates of *V. cholerae,* containing cholera enterotoxin, cause intestinal secretory changes characteristic of the disease.

Cholera Enterotoxin. The extremely potent enterotoxin **(choleragen)** elaborated by *V. cholerae* acts upon the luminal surface of small bowel mucosal cells, causing hypersecretion of chloride and bicarbonate and hence accumulation of fluid within the intestinal lumen. Like *E. coli* LT enterotoxin, *cholera toxin activates cell membrane adenyl cyclase,* resulting in a large increase of cAMP. Moreover, addition of cAMP to the mucosal surface of isolated ileum has the same effect as the toxin on electrolyte excretion. The activation of the cyclase involves GTP and NAD, by a mechanism described in Figure 31-6.

Because most eukaryotic cells tested respond to its cyclase-activating effect, the cholera enterotoxin is widely used as a cyclase/cAMP probe. With the pathogenesis of cholera so well understood, cholera has become the prototype of the enterotoxic enteropathies (secretory diarrheas), including diarrheas due to enterotoxigenic *E. coli* and noncholera vibrios. Enterotoxins elaborated by these organisms are related to cholera enterotoxin, both immunologically and in their mode of action. Though cholera antitoxin neutralizes *E. coli* LT enterotoxin, *E. coli* antitoxin does not neutralize cholera toxin (example of a "one-way cross").

Toxigenicity of *V. cholerae* is determined by a gene (*tox*) located

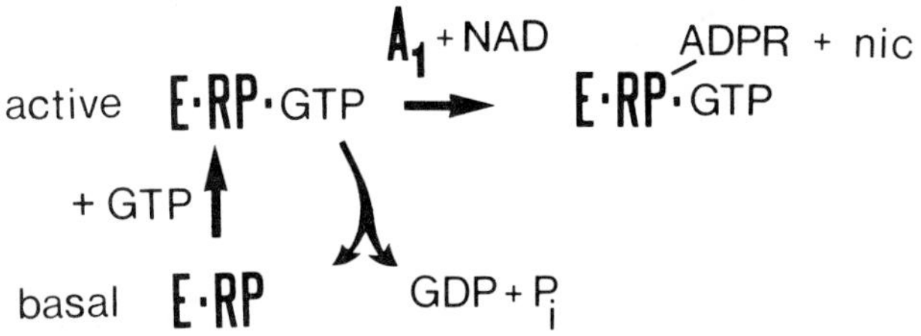

$$\text{active} \quad \underset{\text{basal}}{\overset{+\text{GTP}\uparrow}{E \cdot RP \cdot GTP}} \xrightarrow{\overset{A_1 + NAD}{}} \overset{ADPR + nic}{E \cdot RP \cdot GTP}$$

$$E \cdot RP \searrow GDP + P_i$$

FIG. 31-6. Activation of adenylcyclase by cholera toxin. The toxin, a complex of an A polypeptide (mol wt 29,000) and 5 copies of B polypeptide (each 11,500), binds to **ganglioside GM₁** receptors on host cells; the binding is very tight because of the cooperative action of the 5 B chains. Cleavage of the A chain and SS reduction separates it into polypeptides A_1 and A_2, and A_1 enters the cell. There it transfers adenosine diphosphate ribose (ADPR) from NAD to covalent linkage with regulatory protein RP, which is complexed with the cellular enzyme adenylcyclase (E). This complex has little enzymatic activity unless it binds GTP; and the steady-state of this binding is normally low because the complex also hydrolyzes GTP. However, this reaction is blocked by the attachment of ADPR to RP. The adenyl cyclase is thereby stabilized in an active conformation, and the resulting increase in the level of cAMP, in intestinal cells, stimulates secretion of ions (and fluid) into the intestinal lumen.

on the bacterial chromosome, whereas in *E. coli* it is determined by transmissible plasmids.

Immunity. Cholera occurs more often when the natural resistance of the host has been impaired; malnutrition seems to be a factor in epidemics. Following recovery specific Abs are demonstrable in both the serum (agglutinating, bactericidal, and toxin-neutralizing) and the feces (coproantibodies). Human volunteers after recovery from cholera had Ab to the homologous organism and a high degree of resistance to it but not to heterologous strains. The immunity induced by vaccination with killed organisms persists for only a few months and is therefore of limited value.

Since cholera enterotoxin is highly antigenic, and there is only one immunologic type, an effective antitoxic immunity (not yet developed) should protect against the various serotypes and biotypes of *V. cholerae*.

Laboratory Diagnosis. Provisional, rapid diagnosis of *V. cholerae* can be made by darkfield microscopic observation of a fecal sample: the organisms are comma-shaped, have a characteristic motility, and are clumped and immobilized by homologous (or polyvalent) antiserum. In cultures alkaline peptone water (pH 8.4), inoculated with a few drops of feces, allows much more rapid growth of vibrios than of Enterobacteriaceae; after 6–8 h smears can be stained with fluorescein-labeled specific antisera for quick presumptive identification. The feces should also be streaked on thiosulfate–citrate–bile-salt–sucrose

agar (TCBS) and taurocholate gelatin agar (TGA): the usual enteric media are suboptimal and often inhibitory for growth of *V. cholerae.** For final identification the organisms in typical colonies are subjected to biochemical (Table 31-4) as well as agglutination tests.

A gram-negative rod (isolated from diarrheal feces) that is oxidase-positive, gives rise to an acid butt–acid slant reaction in TSI agar (see Screening Media, above), and has lysine and ornithine decarboxylase, should be suspected of being a cholera vibrio. Regardless of the outcome, or availability, of agglutination tests, a subculture of such an organism should be submitted promptly to the state public health laboratory for verification. The so-called noncholera vibrios, and organisms in the genus *Aeromonas* (see below), may be misidentified by biochemical tests as *V. cholerae.* A rise in serum agglutinins may be of some value in establishing retrospective diagnosis.

Treatment. Replacement of the lost fluid and electrolytes is mandatory. The response to intravenous isotonic sodium chloride, supplemented with appropriate amounts of sodium bicarbonate (or sodium acetate) and potassium chloride, is dramatic; indeed, **prompt** and **adequate intravenous fluid** therapy is life-saving in virtually all cholera patients. Tetracycline, chloramphenicol, or furazolidone therapy significantly reduces gastrointestinal fluid loss, and also eliminates *V. cholerae* from the feces of a large majority of patients, but antimicrobial therapy is **not** a substitute for adequate fluid replacement.

Since intraluminal glucose promotes reabsorption of sodium, maintenance therapy with an **orally** administered solution of glucose and electrolytes,† after initial intravenous rehydration in severely dehydrated patients, should be used where the supply of intravenous fluids is limited. Oral rehydration alone is adequate in treating moderate dehydration.

Prevention. Cholera is spread primarily by contaminated water and food, and by contact with infected patients and carriers. Although explosive waterborne outbreaks occasionally occur the disease is most commonly disseminated by convalescent carriers or individuals with inapparent infection, whose fecal matter contaminates water supplies; it is persistently endemic in Calcutta. Mild forms of the disease often go unrecognized, making prevention of new infections difficult. Unlike shigellosis, cholera is not easily spread by person-to-person contact under reasonable conditions of personal hygiene, for a huge number of organisms (10^8–10^{10}) are needed to cause illness, as shown in volunteers.

---

* Imported cases of cholera would probably not be recognized in many United States clinical laboratories because of the inadequacy of the methods employed and the low index of suspicion.

† Sodium chloride, 3.5 g; sodium bicarbonate, 2.5 g; potassium chloride, 1.5 g; glucose, 20 g/liter of potable water.

Sanitary control of water and food, and adequate sewage disposal, are the most effective means of stopping the spread of the disease. Patients should be isolated and adequately treated, and their excreta should be disinfected. Contacts should have stool cultures done but need **not** be quarantined. Contacts who are culture-positive or are symptomatic should be treated with 1 g tetracycline daily for 5 days. Vaccination with killed cholera vibrios reduces the risk to the individual for a few months and is generally used where the disease is endemic, but it plays only a minor role in preventing the spread of cholera across borders.

OTHER VIBRIOS

Noncholera vibrios (NCVs). These organisms are very similar biochemically and morphologically to the cholera vibrio, but are not agglutinated by cholera polyvalent O antiserum; they are agglutinated by their homologous antisera. These vibrios are found in surface waters and in shellfish throughout the world, and they may cause a mild diarrhea or a frank choleralike disease; they have been responsible for recent epidemics in Czechoslovakia, Malaysia, and Sudan, and for rare cases in the United States. The relation between classic *V. cholerae* and NCVs is not well understood.

Vibrio parahaemolyticus. This **halophilic** organism (which requires 2%–7% sodium chloride for growth) is found in marine water and marine fauna throughout the world. It has been recovered from seafoods, especially shellfish and crustaceans, and is found with increasing frequency in cases of **food poisoning** in various countries (including the United States). In Japan it accounts for about half of the cases of bacterial food poisoning. The gastroenteritis ranges from the usual moderate, short-term illness to severe cases requiring hospitalization.

Antibiotics are used only in protracted or dysenterylike cases. The organism is sensitive in vitro to tetracycline, chloramphenicol, penicillin, ampicillin, aminoglycosides, and colistin. *V. parahaemolyticus* infections of the extremities, eyes, and ears, as well as the blood stream, have also recently been recognized in the United States, usually in persons scratched by the sharp edges of clams or oysters.

The organism does not grow on most routine enteric media. It is positive for oxidase, lysine and ornithine decarboxylase, and indole

formation; it ferments glucose, mannitol, and mannose, and utilizes citrate as the sole source of carbon. Most strains isolated from diarrheal stools are β-hemolytic on a special blood agar (**Kanagawa phenomenon**), while most marine isolates are not. A total of 12 O and 50 K (envelope) Ags are currently recognized.

Vibrio (recently renamed *Campylobacter*) *fetus* infections occur in cattle, sheep, and goats, resulting in abortions and sterility. Human infections are recognized with increasing frequency in infants and in the aged, especially in individuals suffering from debilitating disorders. The organism has been isolated most often from the blood stream, but also from the placenta, synovial fluid, and spinal fluid. Several subtypes are recognized: *C. fetus* var. *intestinalis* seems to be the main human pathogen. The organism grows in an atmosphere with 10% CO_2, or in an anaerobic environment, but not in ordinary air: it ferments components of peptone but not carbohydrates. *C. fetus* is usually sensitive to chloramphenicol, erythromycin, cephalothin, kanamycin, and streptomycin. *C. fetus* var. *jejuni* is a cause of human enteritis.

AEROMONAS

Organisms of the *Aeromonas* group (aeromonads) are found in natural water sources and soil and are frequent pathogens for cold-blooded marine and freshwater animals. They have been associated in man with septicemia, pneumonia, and moderate to severe gastroenteritis. Their recognized incidence in serious human disease has been steadily increasing, and many isolates are probably misdiagnosed as coliforms. The organisms have also been isolated from urine, sputum, feces, and bile without evident pathogenic significance.

A. hydrophila and *A.* (*Plesiomonas*) *shigelloides* may easily be mistaken for *E. coli* because of their similar reactions (Table 31-4); some strains ferment lactose. However, aeromonads are **oxidase-positive** and possess a **single polar flagellum.**

Enterotoxic, cytotoxic, and hemolytic exotoxins have recently been detected in *A. hydrophila.* Aeromonads are resistant to penicillin, ampicillin, and carbenicillin; most strains are sensitive to gentamicin, the tetracyclines, colistin, and trimethoprim-sulfamethoxazole.

THE OBLIGATE ANAEROBES: BACTEROIDACEAE

The family Bacteroidaceae encompasses a large, heterogeneous group of non–spore-forming, obligately anaerobic, mostly nonmotile, gram-negative bacilli, which are numerically preponderant in the gastrointestinal tract of humans and animals. *In the human bowel the gram-negative anaerobes outnumber coliforms at least 100:1.* They are also present in the mouth, nasopharynx, oropharynx, vagina, and urethra, and on the external genitalia.

Two genera, *Bacteroides* and *Fusobacterium* (both commonly known as bacteroides), are of medical impor-

tance, while a third, *Leptotrichia*, is not known to be pathogenic. Originally identified on morphologic grounds, they are now differentiated on the basis of the acids they produce.

Fusobacterium species yield butyric acid as a major product from glucose, often with acetic, formic, propionic, or lactic acid, while most *Bacteroides* instead produce combinations of succinic, lactic, acetic, formic, propionic, and isovaleric acid; some produce a mixture of butyric and acetic acid. Patterns of sensitivity to antibacterial agents are also used as additional criteria for classification (see Laboratory Diagnosis, p. 670).

Little is known about the physiologic role of bacteroides. *B. fragilis* is capable of synthesizing vitamin K (as is *E. coli*). Bacteroides can deconjugate bile acids, which might alter fat absorption in certain pathologic conditions. The organisms may have an important role in our defense against infection: mice become much more susceptible to experimental *Salmonella* infection after their normal colonic *Bacteroides* count is reduced by antibiotic treatment.

Infections due to bacteroides, once thought to be rare, are now identified much more frequently, at least partly because of improvements in anaerobic methodology. Bacteroides are commensal organisms, but they can cause severe, often life-threatening infections, mostly in proximity to the mucosal surfaces where they exist normally (as "potential pathogens"). These infections are at least ten times as common as those due to the classic (spore-forming) anaerobes (Ch. 36). Bacteroides are responsible for about 10% of the gram-negative bacteremias in the United States; they can form abscesses in all regions of the body, and they cause such diverse infections as peritonitis, endocarditis, septic arthritis, uterine sepsis after abortion, and wound infections after bowel surgery or human or animal bites. Some of these infections respond to surgical drainage, alone or together with antibacterial therapy, but many result in prolonged hospitalization and even death. Bacteroides are often present together with various other anaerobes (particularly cocci) and/or aerobes, but in about one-third of infections they yield pure cultures.

The presence of **foul-smelling discharge** or gas in tissue strongly suggests (but is not required for) infection by bacteroides or other anaerobes.* Vincent's angina (fusospirochetal disease), believed by many to be due chiefly to "fusiform" bacilli and spirilla, may be primarily caused by *B. melaninogenicus* (see below). Bacteroides infections are not communicable.

Major predisposing factors are surgical or accidental trauma, edema, anoxia, and tissue destruction (resulting from malignancies or from infection with aerobic organisms). Other predisposing factors are prior antibiotic treatment, cytotoxic or immunosuppressive therapy, and diabetes. In particular, preoperative bowel "sterilization" with oral nonabsorbable drugs such as kanamycin or neomycin temporarily eliminates practically all aerobes and thus may foster invasiveness of the more resistant bacteroides.

Some strains produce collagenase, heparinase, neuraminidase, hemolysin, deoxyribonuclease, and fibrinolysin. The lipopolysaccharides (LPSs) are quite different from those of Enterobacteriaceae: *B. fragilis* LPSs lack heptose and 2-keto-3-deoxyoctonate, apparently have no typical lipid A, and have low endotoxic activity. The O Ags are serologically divisible into a number of types, and Abs are present in healthy individuals and rise in titer in infections. However, no diagnostically useful serologic procedures are yet in general use.

* Neither *E. coli* nor *S. faecalis* produces putrid or fecal odor.

The *B. fragilis* group consists of 5 species (previously considered subspecies), differentiated by biochemical tests. They are resistant to penicillin and kanamycin, and are not inhibited by 20% bile.

B. fragilis is by far the most prevalent in infections, but it comprises only about 0.5% of the normal intestinal bacteria; *B. vulgatus* and *B. thetaiotaomicron* are the numerically dominant fecal isolates. The other two (*B. distasonis* and *B. ovatus*) are less frequently isolated from clinical specimens. A capsular polysaccharide may represent a virulence factor in *B. fragilis*, for encapsulated strains can alone cause abscess formation in animals, while uncapsulated strains in abscesses are generally mixed with another organism.

The coccoid *B. melaninogenicus* (3 subspecies: ss. *melaninogenicus*, ss. *intermedius*, ss. *asaccharolyticus*) forms a brown to black pigment (hematin) on blood agar. Under long-wave ultraviolet (Wood's) light colonies fluoresce and appear brilliant red; the fluorescence can also be noted in purulent drainage and ulcers containing *B. melaninogenicus*. The organism is inhibited by 20% bile, is sensitive to penicillin, and requires vitamin K and hemin for growth. It is frequently associated, often in mixed infections, with brain and lung abscesses.

B. oralis, present in the normal flora of the mouth and vagina, is rarely isolated from infections. *B. corrodens* is generally isolated from the oral cavity and from related infections; it forms small colonies that show erosion or pitting of agar (hence the name).

The most often encountered species of *Fusobacterium* in clinical infections are *F. nucleatum*, *F. necrophorum*, *F. varium*, and *F. mortiferum*. Of these, *F. nucleatum* is predominant in infections involving the brain, lungs, and mouth. The cells are slender, with tapered ends. *F. necrophorum* cells are often pleomorphic; they may be "beaded," and some have large swellings. *F. nucleatum* and *F. necrophorum* are inhibited by 20% bile; most fusobacteria are sensitive to penicillin.

ANAEROBIC METHODS

Laboratory diagnosis of anaerobic infections depends on careful collection of specimens: exposure to air for even a few minutes may drastically reduce viability. Swabs must be transported to the laboratory in tubes filled with oxygen-free CO_2, and inoculation should take place as promptly as possible. To avoid contamination with anaerobes present as normal flora on mucous membranes, special procedures are required (transtracheal needle aspiration or direct lung puncture in pneumonia or a lung abscess; urine for anaerobic culture, rarely needed, should be obtained by suprapubic bladder puncture).

Several methods are used for culture of anaerobes: 1) The **anaerobic jar,** employing a catalyst and hydrogen to eliminate oxygen, is most commonly used for incubation of both plates and tubes. 2) The **anaerobic chamber** ("glove box") is filled with oxygen-free gas, and specimens are introduced through an entry lock. 3) In the anaerobic tube or **"roll tube"** system each culture tube, containing prereduced medium, functions as its own anaerobic incubator. Oxygen-free gas is passed through the tubes whenever they are opened.

Anaerobic cultivation should be prompt, on prereduced (oxygen-free) or fresh media. For culture in the anaerobic jar or chamber it is convenient to use either blood agar (often with kanamycin and vancomycin or other appropriate antibiotics for inhibiting aerobes), hemolyzed blood agar with menadione, or chopped meat medium. Cultures in anaerobic jars are usually incubated for 48 h, while those in the anaerobic chamber or roll tubes can be examined after 24 h.

Laboratory Diagnosis. A **direct Gram stain** of the specimen is of great value because of the distinctive morphology of some bacteriodes; moreover, organisms that are seen but fail to grow aerobically strongly suggest anaerobic infection.

Simple screening tests (tolerance to bile, hydrolysis of esculin, resistance to various antibiotics) are available for the **presumptive identification** of various bacteroides, which helps to guide therapy. **Species determination** requires a relatively large array of fermentation tests and/or use of gas–liquid chromatography for the identification of the fatty acids produced. Antimicrobial sensitivity testing of anaerobes has not yet been standardized; the usual disc methods used for aerobes and facultative anaerobes are not applicable. Since the majority of bacteroides have predictable antimicrobial susceptibilities, presumptive identification may be sufficient.

Treatment. Penicillin G is useful in many infections with bacteroides, except *B. fragilis* and some strains of *F. varium*. In life-threatening infections where the agents are still not defined chloramphenicol is the drug of choice. Clindamycin has significant activity against many bacteroides, but it may cause enterocolitis; it is often used when *B. fragilis* is identified or suspected. Tetracycline was formerly regarded as the antibiotic of choice but about half the bacteroides strains are now resistant. Aminoglycosides are not useful.

Bacteroides (and other anaerobes) often appear with aerobes or facultative anaerobes in mixed infections; antimicrobial therapy should be directed against each type present. In addition to antimicrobial therapy, surgical resection and drainage are often imperative.

APPENDIX

FREQUENCY OF GRAM-NEGATIVE RODS ISOLATED FROM HUMAN INFECTIONS

The following tabulation lists gram-negative rods isolated from **extraintestinal** specimens during 1978 at the Jewish Hospital of Saint Louis, Missouri (a 630-bed, general, university teaching hospital; no pediatrics). Chapter numbers indicate the location of discussions in this book.

Ch.31	
E. coli	2122*
Klebsiella spp.	944
Proteus mirabilis	646
Enterobacter spp.	304
Proteus—indole-positive	168
Bacteroidaceae	298
Citrobacter	174
Serratia	133
Providencia	111
Salmonella	3(20)†
Shigella	0(6)
Aeromonas	10(4)
Edwardsiella	0
Arizona	1(1)
Vibrio, "related"	1
Ch.32	
Pseudomonas aeruginosa	634
Pseudomonas other than *aeruginosa*	51
Acinetobacter spp.	57
Achromobacter	4
Moraxella	11
Flavobacterium	5
Ch. 33	
Pasteurella multocida	4
Yersinia enterocolitica	0(6)
Brucella	0
Ch. 34	
Hemophilus	54
Bordetella	1

* Number of isolates.
† Numbers in parentheses indicate isolations from feces.

SELECTED READING

BOOKS AND REVIEW ARTICLES

BARTLETT JC: Antibiotic-associated pseudomembranous colitis. Rev Infect Dis 1:530, 1979

BAUER H: Growing problem of Salmonellosis in modern society. Medicine 52:323, 1973

CARPENTER CCJ: Cholera and other enterotoxin-related diarrheal diseases. J Infect Dis 126:551, 1972

CRAIG JP (ed): The structure and functions of enterotoxins. J Infect Dis 133:S1–S156, 1976

ELIN RJ, WOLFF SM: Biology of endotoxin. Annu Rev Med 27:127, 1976

FEINGOLD DC: Hospital-acquired infections. N Engl J Med 283:1384, 1970

FINEGOLD SM: Anaerobic Bacteria in Human Disease. New York, Academic Press, 1977

FINKELSTEIN RA: Cholera. CRC Crit Rev Microbiol 2:553, 1973

FINLAND M: The changing ecology of bacterial infections as related to antibacterial therapy. J Infect Dis 122:419, 1970

GORBACH SL: Intestinal microflora. Gastroenterology 60:1100, 1971

GORBACH SL, KEAN BH, EVANS DG JR, EVANS DJ JR, BESSUDO D: Travelers' diarrhea and toxigenic Escherichia coli. N Engl J Med 292:933, 1975

HOFSTAD T: Serological responses to antigens of Bacteroidaceae. Microbiol Rev 43:103, 1979

JAHN K, WESTPHAL O: Microbial polysaccharides. In Sela M (ed): The Antigens, Vol 3. New York, Academic Press, 1975, p 1

KANTOR HS: Enterotoxins of Escherichia coli and Vibrio cholerae: tools for the molecular biologist. J Infect Dis 131:S22, 1975

KASS EH, WOLFF SM (eds): Bacterial Lipopolysaccharides. University of Chicago Press, 1973

KEUSCH GT, DONTA ST: Classification of enterotoxins on the basis of activity in cell culture. J Infect Dis 131:58, 1975

LÜDERITZ O, WESTPHAL O, STAUB AM, NIKAIDO H: Isolation and chemical and immunological characterization of bacterial lipopolysaccharides. In Weinbaum G, Kadis S, Ajl SJ (eds): Microbial Toxins, Vol 4. New York, Academic Press, 1971, p 145

MACY JM, PROBST I: The biology of gastrointestinal bacteroides. Annu Rev Microbiol 33:561, 1979

MAKELA PH, MAYER H: Enterobacterial common antigen. Bacteriol Rev 40:591, 1976

MCCABE WR: Immunoprophylaxis of gram-negative bacillary infections. Annu Rev Med 27:335, 1976

MOSS J, VAUGHAN M: Activation of adenylate cyclase by choleragen. Annu Rev Biochem 48:581, 1979

ORSKOV I, ORSKOV F, JANN B, JANN K: Serology, chemistry, and genetics of O and K antigens of Escherichia coli. Bacteriol Rev 41:667, 1977

ROANTREE RJ: The relationship of lipopolysaccharide structure to bacterial virulence. In Kadis S, Weinbaum G, Ajl SJ (eds): Microbial Toxins, Vol 5. New York, Academic Press, 1971, p 18

SACK RB: Human diarrheal disease caused by enterotoxigenic Escherichia coli. Annu Rev Microbiol 29:333, 1975

SANDERSON KE: Genetic relatedness in the family Enterobacteriaceae. Annu Rev Microbiol 30:327, 1976

SONNENWIRTH AC: Antibody response to anaerobic bacteria. Rev Inf Dis 1:337, 1979

SONNENWIRTH AC (ed): Bacteremia: Laboratory and Clinical Aspects. Springfield, IL, Charles C Thomas, 1973

WORLD HEALTH ORGANIZATION SCIENTIFIC GROUP: Oral Enteric Bacterial Vaccines. WHO Tech Rep Ser No. 500, Geneva, 1972

SPECIFIC ARTICLES

ARROYO JC, SONNENWIRTH AC, LIEBHABER H: Proteus rettgeri infections: a review. J Urol 117:115, 1977

BARKER WH, GANGAROSA EJ: Food poisoning due to Vibrio parahaemolyticus. Annu Rev Med 25:7096, 1974

BOKKENHEUSER V, DUNSTON T: Vibrio fetus infection in man: occurrence, clinical picture, serology, and source of infection. In von Gravenitz A, Sall T (eds): Pathogenic Microorganisms From Atypical Clinical Sources, Vol 1. New York, Marcel Dekker, 1975, p 25

CROSA JH, BRENNER DJ, EWING WH, FALKOW S: Molecular relationships among the Salmonelleae. J Bacteriol 115:307, 1973

DORNER F: Escherichia coli enterotoxin: purification and partial characterization. In Schlessinger D (ed): Microbiology—1975. Washington DC, American Society for Microbiology, 1975, p 242

DUPONT HL, OLARTE J, EVANS DG, PICKERING LK, GALINDO E, EVANS DJ: Comparative susceptibility of Latin American and United States students to enteric pathogens. N Engl J Med 295:1520, 1976

ECHEVERRIA PD, CHANG CP, SMITH D: Enterotoxigenicity and invasive capacity of "enteropathogenic" serotypes of Escherichia coli. J Pediatr 89:8, 1976

FINKELSTEIN RA: Cholera enterotoxin. In Schlessinger D (ed): Microbiology—1975. Washington DC, American Society for Microbiology, 1975, p 236

FINKELSTEIN RA, LARUE MK, JOHNSTON DW, VASIL ML, CHO GJ, JONES JR: Isolation and properties of heat-labile enterotoxin(s) from enterotoxigenic Escherichia coli. J Infect Dis 133:S120, 1976

GIANNELLA RA: Pathogenesis of Salmonella enteritis and diarrhea. In Schlessinger D (ed): Microbiology—1975. Washington DC, American Society for Microbiology, 1975, p 170

GILL DM: Involvement of NAD in the action of cholera toxin in vitro. Proc Natl Acad Sci USA 72:2064, 1975

GILL DM, MEREN R: ADP-ribosylation of membrane proteins catalyzed by cholera toxin: basis of the activation of adenyl cyclase. Proc Natl Acad Sci USA 75:3050, 1978

GOLDSCHMIDT MC, DUPONT HL: Enteropathogenic Escherichia coli: lack of correlation of serotype with pathogenicity. J Infect Dis 133:153, 1976

GYLES C, SO M, FALKOW S: The enterotoxic plasmids of E. coli. J Infect Dis 130:40, 1977

HOLMGREN J, LÖNNROTH I: Cholera toxin and the adenylate cyclase-activating signal. J Infect Dis 133:S64, 1976

HORNICK RB, GREISMAN SE, WOODWARD TE, DUPONT HL, DAWKINS AT, SNYDER MJ: Typhoid fever: pathogenesis and immunologic control. N Engl J Med 283:686, 739, 1970

JONES SR, RAGSDALE AR, KUTSCHER E, SANFORD JP: Clinical and bacteriologic observations on a recently recognized species of Enterobacteriaceae, Citrobacter diversus. J Infect Dis 128:563, 1973

KASPER DL: Chemical and biological characterization of the lipopolysaccharide of *Bacteroides fragilis* subspecies *fragilis*. J Infect Dis 134:59, 1976

KEUSCH GT, JACEWICZ M: The pathogenesis of Shigella diarrhea. V. Relationship of Shiga enterotoxin, neurotoxin, and cytotoxin. J Infect Dis 131:S33, 1975

KEUSCH GT, JACEWICZ M, LEVINE MM, HORNICK RB, KOCHWA S: Pathogenesis of Shigella diarrhea. J Clin Invest 57:194, 1976

LEVINE MM, DUPONT HL, FORMAL SB, HORNICK RB, TAKEUCHI A, GANGAROSA EJ, SNYDER MJ, LIBONATI JP: Pathogenesis of *Shigella dysenteriae* 1 (Shiga) dysentery. J Infect Dis 127:261, 1973

MCIVER J, GRADY GF, KEUSCH GT: Production and characterization of exotoxin(s) of *Shigella dysenteriae* Type 1. J Infect Dis 131:559, 1975

MIWATANI T, TAKEDA Y: *Vibrio parahaemolyticus:* A Causative Bacterium of Food Poisoning. Tokyo, Saikon, 1976

MOORE WEC, CATO EP, HOLDEMAN LV: Anaerobic bacteria of the gastrointestinal flora and their occurrence in clinical infections. J Infect Dis 119:641, 1969

NETER E: *Escherichia coli* as a pathogen. J Pediatr 89:166, 1976

ONDERDONK AB, BARTLETT JG, LOUIE T, SULLIVAN-SEIGLER N, GORBACH SL: Microbial synergy in experimental intra-abdominal abscess. Infect Immun 13:22, 1976

QUICK JD, GOLDBERG HS, SONNENWIRTH AC: Human antibody to Bacteroidaceae. Am J Clin Nutr 25:1351, 1972

RYDER RW, WACHSMUTH IK, BUXTON AE, EVANS DG, DUPONT HL, MASON E, BARRETT FF: Infantile diarrhea produced by heat-stable enterotoxigenic *Escherichia coli*. N Engl J Med 295:849, 1976

SO M, HEFFRON F, MCCARTHY BJ: The *E.coli* gene encoding heat stable toxin is a bacterial transposon flanked by inverted repeats of IS1. Nature 277:453, 1979

SONNENWIRTH AC: Incidence of intestinal anaerobes in blood cultures. In Balows A, DeHaan RM, Dowell VR, Guze LB (eds): Anaerobic Bacteria: Role in Disease. Springfield, IL, Charles C Thomas, 1974, p 157

VAN HEYNINGEN S: The subunits of cholera toxin: structure, stoichiometry, and function. J Infect Dis 133:S5, 1976

YOUNG LS, MARTIN WJ, MEYER RD, WEINSTEIN RJ, ANDERSON ET: Gram-negative rod bacteremia: microbiologic, immunologic, and therapeutic considerations. Ann Intern Med 86:456, 1977

chapter

PSEUDOMONADS AND OTHER NONFERMENTING BACILLI

ALEX C. SONNENWIRTH

This chapter will include several genera that are identified, in the initial division of the gram-negative rods, as "nonfermenters," i.e., organisms that do not ferment glucose in the oxidation-fermentation (O-F) medium (see footnote to Table 31-2). These organisms comprise a heterogeneous group, in which the pseudomonads (family Pseudomonadaceae) are particularly prominent. Though pseudomonads superficially resemble the enteric bacilli they differ in several fundamental respects, and their habitat is primarily extraenteric. The pseudomonads, and some other nonfermenters (e.g., *Flavobacterium*), can oxidize sugars, but cannot ferment them, while other members of this group (*Alcaligenes, Moraxella,* some strains of *Acinetobacter*) are in fact fermenters, but of substrates other than carbohydrates.

Pseudomonads are found primarily in the soil or waters, or on plants, and as a group they are able to attack a great variety of organic compounds. Psychrophilic pseudomonads, i.e., those that grow in the cold, are prominent in the spoilage of refrigerated foods.

Some pseudomonads and other nonfermenters are found on the skin or other body surfaces, and often in small numbers also in the intestine; several occasionally cause human disease, largely extraintestinal. These opportunistic infections have become much more frequent with the use of antibiotics, immunosuppressive agents, cytotoxic agents, and indwelling instruments. The ability of many pseudomonads to grow in water with minimal nutrient contamination makes them frequent agents of infection from intravenous solutions or from respiratory ventilators.

Nonfermenters comprise about one-fifth of all gram-negative bacilli recovered in aerobic cultures from extraintestinal lesions. *Pseudomonas aeruginosa* is by far the most prevalent, followed by *Acinetobacter anitratum* (formerly *Herellea*) and *Pseudomonas maltophilia*.

Pseudomonas mallei and *P. pseudomallei* are serious obligate pathogens rarely encountered in the United States. An important pathogenic group of nonfermenting gram-negative rods, *Brucella*, will be considered in Chapter 33; they have fastidious growth requirements and do not turn up on enteric media.

PSEUDOMONAS

Pseudomonads resemble the Enterobacteriaceae morphologically, and most grow well on differential enteric media. However, they have polar flagella, are strongly oxidase-positive (except for *P. maltophilia* and some strains of *P. cepacia*), and they are strict aerobes (except that some species can use nitrate for anaerobic respiration). Pseudomonads have a considerably higher G + C content than Enterobacteriaceae in their DNA (see Table 31-2), and most metabolize sugars via the 2-keto-3-deoxygluconate (Entner–Doudoroff) pathway (see Ch. 3), rather than via glycolysis. Some strains produce water-soluble pigments. The species described below vary markedly in sensitivity to antibiotics, often exhibiting broad resistance. Sensitivity testing is therefore particularly important.

The study of the genetics of pseudomonads began much later than that of the Enterobacteriaceae, but *P. aeruginosa* is now known to be at least as richly endowed as *Escherichia coli* with plasmids, of at least ten compatibility groups. Some of these serve

as R factors, and others as sources of degradative enzymes that account for the broad range of carbon sources.

The outer membrane, which is responsible for some of the drug resistance, is usually dependent on Mg ions for its integrity: pseudomonads are highly sensitive to killing by a chelating agent, edetic acid (EDTA), which disrupts their outer membrane. The core and lipid A of the lipopolysaccharide closely resemble those of Enterobacteriaciae.

FLUORESCENT GROUP: *PSEUDOMONAS AERUGINOSA* AND RELATED FORMS

Distribution. *P. aeruginosa* (long known as *Bacillus pyocyaneus*), a common soil organism, has frequently been isolated from lesions in man, but its presence was long considered insignificant. However, because of its resistance to many antibiotics it often becomes dominant, both as a commensal and in lesions, when the more susceptible bacteria are suppressed. Hence after the intro-

TABLE 32-1. Some Differential Characteristics of Nonfermentative Gram-negative Bacilli

	Pseudomonas aeruginosa	P. fluorescens	P. pseudomallei	P. mallei	P. cepacia	P. maltophilia	P. stutzeri	P. acidovorans	Acinetobacter anitratum (Herellea)*	A. lwoffi (Mima)†	Flavobacterium spp.	Alcaligenes faecalis
Fluorescein	+	+	–	–	–	–	–	–	–	–	–	–
O-F medium	O	O	O	O	O	O	O	O	O	N	O/F	N
MacConkey agar	+	+	+	–	+	+	+	+	+	+	+/–	+
Motility	+	+	+	–	+	+	+	+	–	–	–	+/–
Oxidase	+	+	w+	w+	+	–	+	+	+	–	+	+
Growth: 42°C	+	–	+	+	+/–	+/–	+/–	–	+	+/–	–	+/–
Growth: 4°C	–	+	–	–	–	–	–	–	–	–	–	?
Gluconate oxidation	+	+/–	–	–	–	–	–	–	–	–	–	–
Glucose (O-F)	+	+	+	+	+	–	+	+	+	–	+/–	–
Gelatinase	+	+	+	–	+/–	+	–	–	–	+/–	+	–
Arginine dihydrolase	+	+	+	+	–	–	–	–	–	–	–	–

†No acid from 10% lactose medium.

O = oxidative; N = nonoxidative; F = fermentative; ? = not determined; +/– = variable; + = positive; – = negative; w = weak.

*Acid from 10% lactose medium.

duction of broad-spectrum antibiotics *P. aeruginosa* became a major agent of nosocomial (hospital-acquired) infections. Airborne spread appears to be rare, since the organism is very sensitive to drying.

P. aeruginosa is a resident of the intestinal tract in only about 10% of healthy individuals, and it is found sporadically in moist areas of the human skin (axilla, groin) and in the saliva. Its nutritional requirements are simple (including use of NH_3 as a source of nitrogen) and it can metabolize a large variety of carbon sources. It can thus multiply in almost any moist environment containing even trace amounts of organic compounds, e.g., eyedrops, weak antiseptic solution, soaps, anesthesia and resuscitation equipment, sinks, fuels, humidifiers, and even stored distilled water.

Identification. *P. aeruginosa* grows readily on standard laboratory media. In triple sugar iron (TSI) medium no change occurs in the slant or butt (see Screening Media, Ch. 31). Strains from clinical material are usually β-hemolytic. Most strains produce a bluish green phenazine pigment, pyocyanin (Gr., blue pus), as well as fluorescein, a greenish yellow pteridine that fluoresces;* the pigments color the medium surrounding the colonies. About 10% of strains do not form pigment. The useful biochemical reactions are shown in Table 32-1.

Individual strains of *P. aeruginosa* can be identified in epidemiologic studies by reactions with specific Abs (16 antigenic groups are recognized), phages, and bacteriocins. Serologic typing may become important for im-

* Wood's ultraviolet light can be used for early detection of *P. aeruginosa* infection of burns and wounds.

munotherapy. Strains vary widely in virulence for mice. Encapsulated, mucoid strains (resembling *Klebsiella*) are most commonly isolated from the sputum of patients with cystic fibrosis.

Pathogenicity. *P. aeruginosa* elaborates three potent, serologically distinct exotoxins, which are more important in its pathogenicity than the endotoxin. Exotoxin A, thought to be a major virulence factor, is a protein with mol wt of 50,000–55,000 and an LD_{50} in mice of about 0.1μg. It inhibits protein synthesis, by the same mechanism as diphtheria toxin (transfer of ADP–ribose to elongation factor EF2). Antiserum neutralizes the toxin and protects mice against infections with homologous strains. The organism also elaborates collagenase, lipase, lecithinase, and hemolysins.

P. aeruginosa infection is rare in normal individuals, except for minor infections such as chronic external otitis. However, it often occurs in those with compromised host defenses, including natural immunologic deficiency, immunosuppressive or corticosteroid therapy, immunologic immaturity, extensive burns, chronic pulmonary disease (e.g., cystic fibrosis), or intravenous narcotic use. The most common sites are the urinary tract and burns. *P. aeruginosa* can also cause septicemia, abscesses, corneal infections, meningitis, bronchopneumonia, and subacute bacterial endocarditis. Treatment often fails, and the mortality rate in *Pseudomonas* septicemia may exceed 80%.

Therapy. Gentamicin, tobramycin, amikacin, and carbenicillin have proved useful. Resistance to these agents has appeared. **Immunotherapy,** both active and passive, has recently been used with encouraging results in burn patients, in whom *P. aeruginosa* infection is often fatal.

Pseudomonas fluorescens and *P. putida*, which produce a yellow fluorescent pteridine pigment, have been isolated from hospital water sources, nebulizers, and intravenous catheters. Rarely, they have been associated with emphysema, wound infections, septicemia, and abscesses. *P. fluorescens* is sometimes the source of endotoxic reactions following infusion of contaminated material (it can grow at 4°C and has been isolated from refrigerated blood and medication). Unlike *P. aeruginosa*, both organisms are sensitive to kanamycin, tetracycline, and gentamicin, but are resistant to carbenicillin.

PSEUDOMALLEI GROUP

This group consists of three species.

Pseudomonas mallei (formerly *Actinobacillus*) is the agent of **glanders,** a severe infectious disease of horses that can be transmitted to man. This disease, now extremely rare in the West, still occurs in Asia, Africa, and parts of the Middle East. It is characterized by nodular, eventually necrotic, involvement of the nasal mucous membranes, lymphatics, lymph nodes, and skin, or by an acute or chronic pneumonitis.

P. pseudomallei is the agent of **melioidosis** long known as a disease of man and other animals in Southeast Asia; it has been observed in Americans who have returned from combat in Vietnam. The clinical manifestations range from a relatively benign pulmonary infection, often mimicking tuberculosis or mycotic infection, to a rapidly fatal septicemia characterized by abscesses in many organ systems. Latency and recrudescence are frequent occurrences. The disease can appear years after exposure, especially when host resistance is reduced.

Appropriate antibiotics are determined by sensitivity testing. Tetracycline is considered the drug of choice, sometimes combined with chloramphenicol. Patients require treatment over many months.

P. cepacia (formerly *P. multivorans, P. kingii,* or EO-1 group), genetically related to *P. pseudomallei* and *P. mallei,* is a plant pathogen that is also frequently recovered in the hospital environment and, as an opportunistic pathogen, in a variety of infections. It responds to chloramphenicol and to trimethoprim–sulfamethoxazole.

OTHER PSEUDOMONADS

Pseudomonas maltophilia, the only consistently oxidase-negative pseudomonad (Table 32-1), is an occasional opportunistic pathogen, associated with a variety of serious infections. It is resistant to most antimicrobials, except chloramphenicol, colistin, and nalidixic acid. Other occasional opportunistic pathogens are *P. alcaligenes, P. stutzeri, P. mendocina,* and *P. putrefaciens.* Others, in the *P. acidovorans* and *P. diminuta* groups, have been isolated from a variety of human sources but appear to have no clinical significance.

OTHER NONFERMENTERS

Acinetobacter. Organisms in the genus *Acinetobacter* are widely distributed in nature (water, soil, milk). They are also carried on the skin and in the gastrointestinal, genital, and respiratory tracts of up to 25% of healthy individuals. Virulence of these organisms appears to be low, and they are mainly opportunistic pathogens, but severe primary infections (meningitis, septicemia) do occasionally occur.

The organisms are nonmotile (hence the name: Gr. *akinetos,* immovable), plump, paired gram-negative rods (diplobacilli). In the stationary phase they often appear as diplococci, easily mistaken for neisseriae (see Ch. 30). They are obligately aerobic but oxidase-negative.

Acinetobacter lwoffi (previously *Mima polymorpha* and *Achromobacter lwoffi*) cannot metabolize carbohydrates, but *Acinetobacter anitratum* (previously *Herellea vaginicola* and *Achromobacter anitratum*) can (Table 32-1).

Kanamycin, gentamicin, colistin, trimethoprim–sulfamethoxazole, and carbenicillin are useful.

Moraxella. Moraxellas are similar to *Acinetobacter* (Table 31-2) but are oxidase-positive and highly sensitive to penicillin. Most of these organisms are nutritionally exacting and do not utilize carbohydrates. Because of their microscopic appearance and positive oxidase reaction they are easily confused with *Neisseria.*

Moraxella lacunata (the Morax-Axenfeld bacillus) is a rare cause of conjunctivitis and corneal infections. *M. osloensis, M. nonliquefaciens* (previously *Mima polymorpha* var. *oxidans*), and *M. phenylpyruvica,* are members of the normal flora of man and also are opportunistic pathogens, involved in often serious infections.

Alcaligenes. *Alcaligenes faecalis* (Table 32-1) fails to metabolize any of the usual carbohydrates but oxidizes organic acids or amino acids, making the medium more alkaline (hence the name). It may be encountered in feces or in sputum as a harmless saprophyte, but it has also been associated with serious infections. The organism has been involved in contamination of irrigating fluids and intravenous solutions, causing epidemics of urinary tract infections and postoperative septicemia. *A. faecalis* is oxidase-positive and usually motile.

Achromobacter. *Achromobacter xylosoxidans,* first described in 1971, occurs in lower animals and free in nature. It has been associated with a variety of illnesses, including meningitis, septicemia, and otitis media. The organism is motile and oxidase-positive.

Flavobacterium. Although flavobacteria are fermentative, they are often so slow that they are mistaken for oxidizers. Members of the group are oxidase-positive, form a yellow pigment (hence the name: L. *flavos*, yellow), and are usually nonmotile. Flagella, when present, are peritrichous.

Flavobacteria are widely distributed in nature. *Flavobacterium meningosepticum* has high virulence for the newborn, especially the premature, in whom it causes epidemics of septicemia and meningitis with a high fatality rate. Though these infections are usually attributed to contaminated hospital equipment and solutions, the organism has recently been isolated from the female genitalia. *F. meningosepticum* also occurs in postoperative bacteremia of adults, in whom the illness is much milder.

Infants with flavobacterial meningitis have sometimes had temperatures below normal, whereas adults with septicemia usually have high temperatures and recover rapidly. Since many *F. meningosepticum* strains cannot grow at 38°C, it has been suggested that body temperature is an important factor in the adult's and the infant's response.

Six separate serotypes (A–F) of *F. meningosepticum* are known. The organism has an unusual antibiotic sensitivity pattern for a gram-negative bacillus: it is susceptible in vitro to erythromycin, novobiocin, rifampin, and trimethoprim–sulfamethoxazole, and to a lesser degree to chloramphenicol and streptomycin; it is resistant to gentamicin and colistin.

Other groups of flavobacteria have been isolated from outbreaks of hospital-acquired bacteremia and from a variety of clinical specimens.

Campylobacter. See *Vibrio*, Ch. 31.

SELECTED READING

BOOKS AND REVIEW ARTICLES

CLARKE PH, RICHMOND MH (eds): Genetics and Biochemistry of Pseudomonas. New York, John Wiley, 1975

DOGGETT RG (ed): *Pseudomonas aeruginosa:* Clinical manifestations of infection and current therapy. New York, Academic Press, 1979

HENRIKSEN SD: Moraxella, Neisseria, Branhamella, and Acinetobacter. Annu Rev Microbiol 30:63, 1976

HOLLOWAY BS, KRISHNAPILLAI V, MORGAN AF: Chromosomal genetics of Pseudomonas. Microbiol Rev 43:73, 1979

HOWE C, SAMPATH A, SPOTNITZ M: The pseudomallei group: a review. J Infect Dis 124:598, 1971

JACOBY GA: Classification of plasmids in *Pseudomonas aeruginosa.* In Schlessinger D (ed): Microbiology—1977. Washington DC, American Society of Microbiology, 1977

PEDERSEN BA, MARSO E, PICKETT MJ: Nonfermentative bacilli associated with man. III. Pathogenicity and antibiotic susceptibility. Am J Clin Pathol 54:178, 1970

PICKETT MJ, MANCLARK CR: Nonfermentative bacilli associated with man. I. Nomenclature. II. Detection and identification. Am J Clin Pathol 54:155, 164, 1970

STANIER RY, PALLERONI NJ, DOUDOROFF M: The aerobic pseudomonads: a taxonomic study. J Gen Microbiol 43:159, 1966

SPECIFIC ARTICLES

FISHER MW: Development of immunotherapy for infections due to *Pseudomonas aeruginosa.* J Infect Dis 130:S149, 1974

GILARDI GL: Pseudomonas species in clinical microbiology. Mt Sinai J Med NY 43:710, 1976

LIU PV: Extracellular toxins of *Pseudomonas aeruginosa.* J Infect Dis 130:S94, 1974

STANISICH VA: Isolation and characterisation of plasmids in *Pseudomonas aeruginosa.* Bull Inst Pasteur 74:285, 1976

chapter **33**

YERSINIA, FRANCISELLA, PASTEURELLA, AND BRUCELLA

ALEX C. SONNENWIRTH,
MORTON N. SWARTZ

ZOONOSES

Zoonoses are infections that are naturally transmitted between lower vertebrates and man. This chapter will describe the agents of several important members of this group: plague and other yersinioses, tularemia, brucellosis, and less common infections produced by various pasteurellae. Several other groups of bacteria also cause zoonoses. The gram-negative members include salmonellae and vibrios (Ch. 31), pseudomonads of glanders and melioidosis (Ch. 32), and *Streptobacillus moniliformis* (Ch. 43). The gram-positive members include the bacilli of anthrax (Ch. 35), listeriosis, and erysipeloid (Ch. 43). Others are the spirochetes of leptospirosis, relapsing fever, and one form of rat-bite fever (Ch. 39); the rickettsiae (Ch. 40); some chlamydiae (Ch. 41); and the agents of bovine and avian tuberculosis (Ch. 37).

The organisms described in this chapter are all gram-negative, facultative or aerobic rods. Except for *Brucella,* which is quite distinctive in its chemical properties and pathogenicity, all were assigned until recently to the genus *Pasteurella.*

Yersin named this genus after his teacher when he (and Kitasato independently) discovered the agent of plague, in Hong Kong in 1894. However, this organism (*Pasteurella pestis*) and *P. pseudotuberculosis* have recently been shifted to the new genus *Yersinia,* and *P. tularensis* to the new genus *Francisella.* Yersiniae, in contrast to the other organisms described here, have simple growth requirements, and this genus is now included in the family Enterobacteriaceae (Ch. 31); moreover, a recently recognized additional member, *Yersinia enterocolitica,* has a pathogenicity similar to that of several other Enterobacteriaceae. However, because of the prominence of animal hosts *Yersinia* is retained in this chapter.

YERSINIA PESTIS

No infectious disease has created greater havoc in the world than plague.* The first adequately described pandemic, in the sixth century A.D., is believed to have killed more than 100 million people in its 50-year rampage. In the fourteenth century plague again assumed catastrophic proportions, presumably because of a vicious cycle of deteriorating social conditions that drove rats closer to humans. This pandemic, known as the **black death** because of the severe cyanosis of the terminally ill, destroyed approximately a quarter of the population of Europe and spread into the Middle and Far East. With improvement in living conditions the disease receded in Europe, but serious epidemics recurred elsewhere. The last pandemic developed in China at the close of the nineteenth century.

Plague declined worldwide in the first half of the twentieth century, with occasional outbreaks in Asia and Africa and sporadic cases in South America and the southwestern United States. However, since 1960 there has been a gradual increase, with a recent resurgence (1972–1975) on various continents. Worldwide there

were a total of 2737 cases reported in 1974 (including eight in the United States), with 164 deaths; in 1977 there were 18 reported cases in the U.S.

Urban (domestic) plague, the **epidemic** form of the disease, is transmitted by fleas from domestic rats to man (and from rats to rats), as was established in 1906 by the British Plague Research Commission in Bombay. Meyer later discovered **sylvatic (wild) plague,** epizootic among squirrels, prairie dogs, rabbits, and pack rats. This reservoir leads to sporadic cases in man and also constitutes a potential source of future epidemics. In the United States sylvatic plague exists in some 15 western states.

THE ORGANISM

Morphology and Cultivation. The plague bacillus, *Yersinia pestis,* is a nonmotile, short, ovoid bacillus. When grown under suboptimal conditions it produces involutional forms (filamentous, ring-shaped, or yeastlike organisms). Its tendency to stain in a **bipolar** "safety-pin" fashion in preparations from tissues, from buboes, and to a lesser extent from cultures is best demonstrated by Wayson's stain (methylene blue and carbolfuchsin) but can also be seen with Giemsa's or Gram's stain. Freshly

---

* The term as used here refers to the specific disease caused by *Y. pestis;* it is also often applied generically to any epidemic disease with a high mortality rate.

isolated virulent strains produce a generous capsule (Fig. 33-1). The organism grows well on initial isolation, producing brown, nonhemolytic colonies on blood agar after 2 days.

Unlike most pathogenic species, *Y. pestis* multiplies rapidly at 28°C (temperature optimum). Glucose and mannitol are fermented, without gas production; lactose, sucrose, rhamnose, and adonitol are not attacked (Table 31-4). *Y. pestis* may remain viable for weeks in dry sputum or in flea feces at room temperature.

Antigens. *Y. pestis* strains produce a number of antigenic components. The capsular or envelope Ag, a heat-labile protein, is known as **Fraction 1 (F1).** The VW Ag system is made up of a protein V (mol wt 90,000) and a lipoprotein W (mol wt 90,000), which act in combination to render the organism resistant to phagocytosis in the absence of a demonstrable capsule. The V Ag appears to be cell-bound and the W Ag is excreted into the medium. The F1 and the VW Ags are produced during growth at 37°C but not at 28°. The so-called **murine toxin** is intracellular but appears to be independent of endotoxin; it is found as two active proteins: toxin A (mol wt 240,000) and toxin B (mol wt 120,000). The former may be a dimer of the latter. The toxin is lethal for the mouse (IV LD_{50} <1μg) and rat, but not for rabbit, dog, or monkey. A lipopolysaccharide **endotoxin** is also produced, similar in pharmacologic action to those produced by the enteric bacilli, but its LD_{50} for mice is 50–100 times greater. The organism also shares an extractable Ag with all other Enterobacteriaceae, designated as **common enterobacterial antigen** or Kunin antigen (see Antigenic Structure, Ch. 31).

FIG. 33-1. Capsules of smooth phase *Yersinia pestis* stained by the indirect fluorescent Ab technic. (Cavanagh DC, Randall R: J Immunol 83:348, 1959 Copyright 1959 by The Williams & Wilkins Co., Baltimore)(×1200).

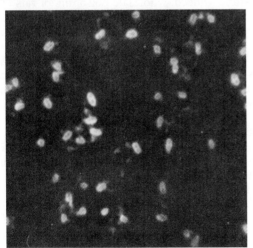

Genetic Variation. The genetics of virulence of *Y. pestis* have been admirably reviewed by Burrows. Mutants that have lost virulence, for a variety of reasons (see under Pathogenicity), are commonly encountered, especially when the organisms are grown under suboptimal conditions. Drug-resistant mutants have been produced in the laboratory, both in vitro and in vivo, but have not created a serious therapeutic problem in natural infections. Nonvirulent variants have been extensively used in the production of living vaccine.

PATHOGENICITY

Plague is a natural disease of both domestic and wild rodents. **Rats** are the primary reservoir: they usually die acutely, with a high-grade bacteremia, but they occasionally develop a more chronic form of infection. The disease is transmitted by the bites of **fleas** (e.g., *Xenopsylla cheopis,* the rat flea) which have previously sucked blood from an infected animal. The ingested bacilli proliferate in the intestinal tract of the flea and eventually block the lumen of the proventriculus. The hungry flea, upon biting another rodent, regurgitates into the wound a mixture of plague bacilli and aspirated blood. If its host dies the flea promptly seeks a replacement. If no rodent is available it will accept a human host, an accidental intruder in the rat → flea → rat transmission cycle.

A small pustule may be present at the portal of entry in the skin, but more often there is no discernible lesion. The bacilli enter the dermal lymphatics and are transported to the regional lymph nodes, usually in the groin, where they cause the formation of enlarged, tender **buboes.** In severe **bubonic plague** the regional lymph nodes fail to filter out all the multiplying bacilli; organisms that gain entrance to the efferent lymphatics disseminate via the circulation (septicemic plague) to the spleen, liver, lungs, and sometimes the meninges. The parenchymatous lesions produced are hemorrhagic; disseminated intravascular coagulation may occur. In the terminal stages bacteremia is often intense.

When metastatic pneumonia develops the sputum may become heavily contaminated, and infection may then be transmitted by way of respiratory droplets. **Pneumonic plague,** particularly under conditions of crowding in cold climates, is relatively contagious, and since the inoculum of virulent bacilli in the infected droplets tends to be large, this form of the disease is extraordinarily malignant. Both the bubonic and the pneumonic forms of the disease can be produced experimentally in rodents and monkeys.

The incubation period of bubonic plague in man varies from 1 to 6 days, depending upon the infecting dose. Onset is usually abrupt, with high fever, tachycardia, malaise, and aching of the extremities and back. If the disease progresses to the fulminant bacteremic stage it causes prostration, shock, and delirium; death usually occurs within 3–5 days of the first symptoms. The course of plague pneumonia is even more fulminant; untreated patients rarely survive longer than 3 days. Pulmonary signs may be totally lacking until the final day of illness, making early diagnosis particularly difficult. Late in the disease copious bloody, frothy sputum is produced.

The occurrence of asymptomatic cases is suggested by serologic studies in areas where the disease is endemic, and by the finding in Vietnam of a pharyngeal carrier rate of about 10% in family members of plague patients.

An injection of less than ten cells of a fully virulent strain of *Y. pestis* in a mouse is lethal. The factors responsible for virulence are complex and only partially understood. Infectivity appears to depend on the presence of the combined VW proteins (VW). Virulence, on the other hand, seems to depend on other factors, including the envelope protein (F1), coagulase, fibrinolytic activity, and the murine toxin. Endotoxin can be detected in patients with plague septicemia and meningitis (by *Limulus* assay; see Endotoxins, Ch. 31); its possible pathogenetic role is suggested by the occurrence of disseminated intravascular coagulation in plague patients.

The bacilli contained in the gut of the rat flea possess neither capsular nor VW Ag; consequently they are promptly ingested and destroyed by polymorphonuclear leukocytes. How then does the flea serve as an effective vector? This riddle was solved by Cavanaugh and Randall. *The virulence of the bacilli in the flea is masked by the low temperature (about 25° C) at which they have proliferated.* When such bacilli are phagocytized at 37° by monocytes (in contradistinction to granulocytes) they survive and multiply intracellularly, and they then emerge as fully virulent organisms possessing both the F1 and VW antiphagocytic factors.

IMMUNITY

Recovery from plague appears to confer relatively solid immunity, but rare reinfections have occurred. The Abs primarily involved are those to the antiphagocytic Ags, F1 and VW complex; antitoxic sera are not protective. Effective vaccines therefore must contain both antiphagocytic Ags.

LABORATORY DIAGNOSIS

Rapid preliminary diagnosis is of paramount importance, in view of the swift progression of the untreated disease. Because of the danger of serious laboratory infections great care must be exercised in handling specimens suspected of containing *Y. pestis.* Smears of sputum or of fluid aspirated from lymph nodes should be stained by the Gram method and also with either methylene blue or Wayson's reagent to identify bipolar staining. In epidemics of plague pneumonia **fluorescent Ab** is of great value for rapid identification of *Y. pestis* in the sputum.

Aspirates from buboes, sputum, throat swabs, and autopsy materials should be cultured on blood agar and in infusion broth con-
taining blood or 0.025% sodium sulfite. Blood samples should be placed in infusion broth and spread on plain extract agar. Cultured organisms are identified by colony characteristics, stained appearance, fluorescent Ab staining, lysis with specific bacteriophage, agglutination by specific antiserum, biochemical characteristics (for differentiation from other yersiniae), and animal inoculation (usually lethal in mice or guinea pigs, with typical lesions). Animals so inoculated must be free of ectoparasites and should be kept strictly isolated. Solid media containing Ab to F1 Ag have been used to identify *Y. pestis* colonies in mixed culture, since on such plates treatment with chloroform vapor releases the Ag and a precipitin ring forms around each colony.

Diagnostic serologic tests are of only retrospective value.

TREATMENT

The case fatality rate in untreated bubonic plague is 50%–75%, and in plague pneumonia approaches 100%. Fortunately, *Y. pestis* is responsive to streptomycin, chloramphenicol, and tetracyclines. A few strains resistant to streptomycin have been noted. Penicillin is totally ineffective, though the organisms are often sensitive in vitro.

If instituted early enough antimicrobial therapy dramatically alters the course of pneumonic plague and reduces the mortality of bubonic plague to 1%–5%. *Time is of the essence,* particularly in pneumonic plague, which can rarely be controlled after 12–15 h of fever.

PREVENTION

While it is clear that plague is initially transmitted to man primarily by the rat fleas, epidemiologic studies have revealed that in epidemics of bubonic plague man-to-man transmission also occurs, the principal vector being the human flea, *Pulex irritans.*

Prevention of the disease is difficult, since elimination of the animal reservoir through rodent control is virtually impossible. Indeed, wholesale poisoning of rats may accentuate an epidemic by forcing infected fleas to leave the dying rats and seek human hosts. However, insecticides (e.g., DDT) properly directed against human fleas may lower the transmission rate in epidemics. All patients with pneumonic plague should be strictly isolated.

Immunization with killed or attenuated vaccines, or with antigenic fractions of the bacilli, appears to provide short-term relative immunity. Formalin-treated vaccine is recommended for persons entering an area where plague is endemic and for laboratory personnel working with *Y. pestis.* Close contacts of patients with pneumonic plague should be treated prophylactically with tetracycline.

To prevent plague from entering uninfected areas ships from ports infected with the disease are quarantined, their cargoes are fumigated, and metal shields are placed around each hawser to prevent rats from leaving the ship.

OTHER YERSINIAE

Besides *Y. pestis* the genus *Yersinia* contains two other species that can produce disease (yersiniosis) in man: *Y. pseudotuberculosis* and *Y. enterocolitica*. Both are relatively large gram-negative coccobacilli. Both organisms have an extensive animal reservoir, including domestic and wild mammals and birds. *Y. enterocolitica* has also been isolated from frogs, snails, fleas, and oysters, and from rivers and lakes. The organisms may cause extensive epizootic outbreaks with various patterns of disease. They may also persist in healthy animal carriers.

Transmission of *Y. pseudotuberculosis* to man is probably through foods contaminated by animal excreta; other possibilities include direct contact with infected animals and consumption of infected meat. For *Y. enterocolitica* food and water appear to be the main vehicles, but interfamilial and institutional outbreaks and hospital-acquired infections suggest person-to-person transmission as well.

Y. enterocolitica infections were first identified in Europe in animals and have been increasing since the 1960s all around the world. The same organism, then known as *Bacterium enterocoliticum,* had been identified in 1933 in the United States as a rare cause of human illness. By 1977 over 300 cases had been recognized in the United States and several thousand in Europe. Whether this increase represents improved identification or a true increase in incidence is not known.

The most common clinical infection due to *Y. enterocolitica* is an acute self-limited **gastroenteritis** or **enterocolitis,** predominantly affecting young children. Clinically it is indistinguishable from gastroenteritis due to *Salmonella, Shigella,* or toxigenic *E. coli* (Ch. 31). *Y. enterocolitica* often also causes acute **mesenteric lymphadenitis,** and less often **terminal ileitis,** occurring mostly in older children and young adults; the symptoms resemble those of appendicitis. *Y. pseudotuberculosis,* in contrast, is not associated with acute gastroenteritis; mesenteric lymphadenitis or terminal ileitis are its commonest manifestations.

In a recent *Y. enterocolitica* (serotype 8) outbreak involving 218 children 13 underwent appendectomies because of abdominal pain and fever; none had appendicitis.

Septicemia due to either species is uncommon, occurring mainly in patients debilitated by underlying disease. It may present in an acute (typhoidlike) form, or subacutely with hepatic or splenic abscesses. The prognosis is favorable in adequately treated acute septicemic cases but is poor for the subacute localizing form.

Additional manifestations attributed to *Y. enterocolitica* (primarily on serologic grounds), mostly in adults, include erythema nodosum, acute nonsuppurative polyarthritis (often preceded by fever and diarrhea), and Reiter's syndrome (arthritis with urethritis and conjunctivitis). *Y. enterocolitica* has also been isolated in meningitis, abscesses, suppurative arthritis, cholecystitis, and eye infections.

Both organisms grow on ordinary media, in air or anaerobically, somewhat slower than other enteric bacilli. Both are motile at 22°–28°C but not at 37°C, which helps to distinguish them from the nonmotile *Y. pestis* and from other Enterobacteriaceae. Their differential metabolic characteristics are summarized in Table 31-4. On initial isolation yersiniae (especially *Y. enterocolitica*) are easily mistaken for certain other Enterobacteriaceae (Ch. 31), especially the anaerogenic "atypical coliforms": *Proteus, Providencia,* and *Shigella.*

Y. enterocolitica and *Y. pseudotuberculosis* can be distinguished from each other by biochemical tests (Table 31-4), by susceptibility of *Y. pseudotuberculosis* to specific bacteriophages, and by agglutination with specific antisera. Differences in pathogenicity for laboratory animals are not reliable.

DNA hybridization shows DNAs from all *Y. pseudotuberculosis* strains to be highly related, but only 40%–60% related to *Y. enterocolitica.* Some degree of crossimmunity exists between *Y. pestis* and *Y. pseudotuberculosis* in experimental animals; the antigenic basis is not known.

YERSINIA ENTEROCOLITICA

Metabolism and Identification. *Y. enterocolitica* is easily isolated from usually sterile body areas on ordinary media, but isolation from feces requires additional technics. The organism grows slower at 37°C than other enteric organisms, and so it is preferable to use MacConkey and *Salmonella–Shigella* agar plates (see Metabolic Characteristics, Ch. 31) incubated at room temperature. **Cold enrichment** of the stool (refrigeration in isotonic saline at 4°C for 1–3 weeks, with weekly culture) markedly enhances isolation. In contrast to *Y. pestis, Y. enterocolitica* shows little or no bipolarity ("safety-pin" appearance). On triple sugar iron (TSI) agar (see Screening Media, Ch. 31) its reactions are those of anaerogenic *E. coli.* The tests used to identify the organism are listed in Table 31-4.

Many of the tests show a marked temperature dependence: motility, β-galactosidase production, maltose fermentation, and acetoin production are usually positive at 22°–25°C but negative at 37°C.

Atypical strains, which have been isolated mostly from nonenteric syndromes in man and from water, ferment lactose, rhamnose, and melibiose.

Antigens and Antibody Response. On the basis of O Ags, 34 serotypes have been established. Serotype 8 predominates in isolates from humans in the United States and types 3 and 9 in Europe, Africa, Japan, and Canada. In

Europe type 3 is also frequent among strains from pigs, which may be a major reservoir for human infection.

Antibodies, demonstrable by agglutination or hemagglutination, are usually absent at the onset and peak during convalescence. Marked crossagglutination reactions have been observed between *Y. enterocolitica* serotype 9 and *Brucella abortus.*

Pathogenicity. Until recently the disease could not be reproduced in laboratory animals. However, certain strains are virulent in mice. As with *Y. pestis,* virulence is temperature-dependent: organisms grown at 37°C are cleared from the animal much more rapidly than those grown at 25°C. No exotoxin has been detected.

Diagnosis depends on isolation and identification of the organism. Serodiagnosis is limited by the large number of serotypes.

Treatment. *Y. enterocolitica* is usually sensitive in vitro to aminoglycosides, colistin, chloramphenicol, tetracycline, and trimethoprim–sulfamethoxazole.

YERSINIA PSEUDOTUBERCULOSIS

Six main serotypes and eight subtypes, based on combinations of 15 somatic (O) Ags, have been established. Type I is found in about 90% of human cases. Several of the Ags crossreact with some salmonellae. The lipopolysaccharide contains 6-deoxy-D-mannoheptose, a sugar not yet found elsewhere in nature.

Antibodies are normally demonstrable, by agglutination or indirect hemagglutination, at the onset of illness; they wane or disappear in 1–4 months.

Y. pseudotuberculosis is generally sensitive in vitro to kanamycin, tetracycline, and chloramphenicol.

FRANCISELLA TULARENSIS

The history of **tularemia** is less dramatic and more recent than that of plague. While attempting to culture plague bacilli from ground squirrels in Tulare County, California, in 1912, McCoy and Chapin isolated a new bacterial species which became known as *Bacterium tularense.* The human illness caused by this organism was described 2 years later. In an extraordinary series of field, laboratory, and clinical investigations in Utah in 1919 Francis proved that jack rabbits are an important source of human tularemia, and that the disease may be transmitted to man by the bite of a **deer fly. Ticks** were subsequently shown to be important also, not only as vectors but also as reservoirs, since they can transmit the organism transovarially.

While bites of flies and ticks are a significant mode of transmission to man, **direct contact** with the tissues of infected **rabbits** is more common. A major outbreak in 1968 resulted from handling infected muskrats. In some parts of the world water polluted by carcasses of infected rodents is an important source. The organism may gain entrance through an abrasion in the skin (including an animal bite) or through the conjunctivae, or by ingestion of improperly cooked meat or inhalation of aerosols. A very small dose is infectious. These features make the organism exceptionally dangerous in the laboratory.

Tularemia has been reported throughout North America, in many parts of Europe, in the USSR, and in Japan; it seems to be a disease of the northern hemisphere. In the United States, from 1966 to 1975 there were 1567 cases reported.

THE ORGANISM

Morphology and Cultivation. *F. tularensis* is a short, nonmotile, unencapsulated bacillus. In young cultures its morphology is relatively uniform but in older cultures it is markedly pleomorphic, exhibiting bean-shaped, coccoid, bacillary, and filamentous forms. The organism is very small ($0.2\mu m \times 0.2$–$0.7\mu m$), stains poorly, and does not grow on ordinary media. Minute coccoid forms are visible in hepatic cells of experimental animals, but identifiable organisms are only rarely observed in tissues of humans dead of the disease.

Colonies are slow to grow on primary inoculation, taking 2–10 days even on appropriate media. The colonies are minute, transparent, and easily emulsified (smooth phase) even though the cells lack a capsule. Rough (R) mutants are readily recognized by their granular colonial morphology. They are generally less virulent and less immunogenic than smooth (S) strains.

The outstanding growth characteristic of *F. tularensis* is its **requirement for cysteine** (or other sulfhydryl compounds) in amounts exceeding those usually present in nutrient media. It grows best on cysteine–glucose–blood agar and on coagulated egg yolk medium, and less well in thioglycollate broth. Multiplication is most rapid at 37°C. Factors required for growth in synthetic media include 13 amino acids, thiamine, and spermidine. A blood-free tryptose broth medium containing added thiamine, cysteine, glucose, and iron also supports rapid growth. Though a facultative anaerobe, the organism grows best under aerobic conditions.

Antigens. Only a **single immunologic type** of *F. tularensis* has been identified. The immunizing Ags appear to reside in the cell wall. Several kinds of Ags have been extracted: 1) a polysaccharide that causes an immediate wheal-and-erythema reaction when injected into the skin of patients convalescing from tularemia, 2) a protein Ag that crossreacts with agglutinating Ags of the genus *Brucella,* and 3) an endotoxin whose role in pathogenesis appears to be similar to that of *S. typhi* endotoxin. No exotoxin has been identified.

PATHOGENICITY

The factors responsible for the pathogenicity of *F. tularensis* are poorly defined. The general correlation of virulence with colonial morphology suggests that surface components of the bacterial cell may be involved. In general, strains of high virulence for man tend to ferment glycerol and have usually been isolated from tick-borne tularemia in rabbits; isolates of low virulence do not ferment glycerol and have commonly been isolated from waterborne disease of rodents.

The organism behaves primarily as an **intracellular parasite,** surviving for long periods in monocytes and other body cells; no antiphagocytic properties of virulent strains have been demonstrated. The prolonged intracellular survival helps to explain the persistent immune response (see below) and the occasional tendency of the disease to relapse and to remain chronic.

At the site of primary lodgment in the skin or mucous membrane an ulcerating papule often develops. The organisms are carried by the lymphatics to regional lymph nodes, which become enlarged and tender and may suppurate. Further penetration to the blood stream causes transitory bacteremia in the acute phase of the illness, and results in spread to parenchymatous organs, particularly the lungs, liver, and spleen. The characteristic lesions are granulomatous nodules in the reticuloendothelial system which may caseate or form small abscesses.

The **ulceroglandular** and the **oculoglandular** forms of the disease result from primary infection of the skin or the conjunctivae, respectively. **Pneumonic** tularemia, produced by inhalation of infected droplets, is apt to occur in laboratory workers, but as in plague it may also result from hematogenous dissemination from local infection elsewhere. **Typhoidal** tularemia follows ingestion of the organism; it resembles typhoid fever, with gastrointestinal manifestations, fever and toxemia.

The incubation period in tularemia ranges from 3 to 10 days and is followed by headache, fever, and general malaise. If specific treatment is not instituted the course of the disease is usually protracted; delirium and coma may develop, and the outcome may be fatal. The case fatality rate in untreated ulceroglandular tularemia is about 5%, and in the typhoidal and pulmonary forms it approaches 30%.

Immunity. Although naturally acquired immunity to tularemia is usually permanent, second attacks have been described. Agglutinins are usually demonstrable in the serum by the second or third week of illness and persist for many years after recovery. The tendency of the disease to progress and even to relapse, despite high titers of serum Abs, is undoubtedly due to the ability of the organism to survive within cells of the host.

Altered cellular reactivity probably plays the major role in immunity. Resistance has been passively transferred by spleen cells from mice that have recovered from infection with an attenuated strain.

LABORATORY DIAGNOSIS

A definitive diagnosis of tularemia from exudate smears requires specific **fluorescent Ab.** Cultures must be made with **special media** (e.g., cysteine–glucose–blood agar), and they should be incubated for 3 weeks before being discarded as negative. If an organism grows in the special medium, and not in ordinary media, it may well be *F. tularensis.* Its identity should be established by staining with fluorescent Ab or by an **agglutination test.** *F. tularensis* has been isolated from gastric washings, sputum, and tissue specimens by guinea pig inoculation. In all laboratory work with *F. tularensis* great care must be taken to avoid infection.

Serologic tests are of diagnostic value. **Agglutinins** (and hemagglutinins) appear within 8–10 days of the onset of illness, continue to rise for up to 8 weeks, and may be detectable for years after the disease. The demonstration of a rising titer is confirmatory evidence of recent infection. *Brucella* Abs crossreact, but they may be distinguished by comparative titers with Ags of both species.

A delayed hypersensitivity reaction to the tularemia **skin test Ag** appears to be a sensitive indicator of present or past infection. It becomes positive earlier than the agglutination reaction and remains positive for years.

TREATMENT AND PREVENTION

Because of its bactericidal properties, **streptomycin** is the drug of choice in the treatment of tularemia. The bacteriostatic tetracyclines and chloramphenicol are also effective, but relapses tend to occur when treatment is discontinued prematurely. Even when streptomycin is used relapses occasionally occur, probably because of the failure of the drug to affect many of the intracellularly located organisms. Appropriate antibiotic therapy reduces the overall case fatality rate to approximately 1%, but it is higher with tularemic pneumonia.

Precautions include wearing of gloves while skinning and dressing rabbits. An attenuated live **vaccine** is available, and its use is indicated in laboratory workers and other individuals who are likely to be exposed to *F. tularensis*. It causes a local reaction on intradermal administration, but it affords significant protection against respiratory (though not against cutaneous) challenge.

PASTEURELLAE*

Organisms in the genus *Pasteurella* are primarily animal pathogens, but they are also responsible for a variety of syndromes in man ranging from localized abscesses to septicemias. They are nonmotile, ovoid or rod-shaped bacilli, frequently with bipolar staining. They grow best on media containing blood, and in contrast to yersiniae and francisellas they are oxidase-positive.

Pasteurella multocida is the species most often encountered in human infections. It was described by Pasteur as the cause of fowl cholera and it is the cause of hemorrhagic septicemia in a variety of animals. The organism is frequently carried in the respiratory tract of healthy domestic animals and rats. Local soft tissue infection, or even osteomyelitis, sometimes follows an animal bite (most commonly that of a cat). The organism is also found, without any antecedent animal exposure, in **systemic** infection, in **respiratory tract** infection, and in **localized** purulent lesions.

When resistance in animals is lowered, as when herds of cattle are shipped, the organisms may become invasive, producing fulminating septicemia or pneumonia (**shipping fever**) and spreading to other animals. Killed vaccines are used to protect cattle from shipping fever, and to control fowl cholera in areas where the disease is endemic.

Diagnosis depends on the isolation and identification of *P. multocida,* a small, nonmotile coccobacillus that forms small, nonhemolytic, gray colonies on blood agar. Smooth variants (iridescent or blue by oblique transmitted light on dextrose starch agar) and mucoid variants exhibit marked pathogenicity in mice, in contrast to rough variants. The organism cannot grow on MacConkey agar. It produces acid, but no gas, from glucose, sucrose, mannitol, and fructose, but not from lactose, maltose, inositol, or salicin. It also produces H_2S, catalase, and usually indole, but not urease. Based on O antigens, at least 16 serotypes are known.

Most strains of *P. multocida* are sensitive to penicillin; this serves in its rapid laboratory differentiation from other gram-negative bacilli. Tetracycline is also effective.

Pasteurella pneumotropica causes respiratory infections and abscesses in various animals, and it also occurs in the mouth of healthy dogs: following dog or cat bites it occasionally is involved in human disease, with a few fatal cases recorded. *P. ureae* is occasionally found in the sputum of patients with chronic bronchitis and bronchiectasis, but its pathogenicity has not been demonstrated.

BRUCELLAE

The brucellae are gram-negative coccobacilli. The genus contains three principal species pathogenic for man, originally differentiated on the basis of their major animal sources: goats and sheep for *B. melitensis,* cattle for *B. abortus,* and swine for *B. suis.* This speciation has since been supported by metabolic and antigenic differences. Three other species (*B. ovis, B. neotomae,* and *B. canis*) have been described more recently, of which only the last appears to have any role in human disease.

All brucellae are obligate parasites capable of causing acute or chronic illness or inapparent infection. The chronicity depends on the marked capacity for multiplication in phagocytic cells, which is opposed by the development of cellular immunity. In their natural animal reservoirs brucellae show a striking propensity to localize in the pregnant uterus (frequently causing abortion) and in the mammary glands; *apparently healthy animals may shed brucellae in their milk for years.*

Man becomes infected through the ingestion of unpasteurized milk or cheese, or through contact with the tissues of infected animals. Human brucellosis may be an acute or relapsing febrile illness, a chronic illness, or a subclinical infection. Unlike its counterpart in animals it does not tend to localize in the genital tract but rather involves the reticuloendothelial system.

Brucellae were first isolated in 1887 by Bruce from the spleens of British soldiers dying on the island of Malta from a disease known as **Malta fever.** The source of the organism was not discovered until 1904, when it was cultured from milk and urine of apparently healthy goats. When the consumption of raw goat's milk was stopped the incidence of the disease declined sharply. The second organism of the group was isolated in Denmark by Bang in 1897 from cattle suffering from infectious abortion (**Bang's disease**), and the third was cultured in the United States in 1914 from the fetus of a prematurely delivered sow. In the 1920s Evans recognized that all three organisms were closely related, and they were placed in a separate genus, *Brucella.*

---

* A closely related genus, **Actinobacillus,** includes gram-negative bacilli that cause septicemia, or granulomatous or suppurative abscesses, in various domestic animals and occasionally in man. *A. actinomycetem-comitans* has been implicated in endocarditis in man.

THE ORGANISMS

Brucella organisms are nonmotile, pleomorphic, short, slender bacilli. Bipolar staining is sometimes present. Colonies are small, convex, smooth, moist-appearing, and translucent. Growth is slow, particularly on initial cultivation, and colonies are not usually visible for 2 days or more. As a rule, *Brucella* isolates recovered from tissues are smooth (S) colonies, and in freshly isolated strains small capsules can be demonstrated with appropriate staining.

The brucellae are aerobes but can use nitrate anaerobically as an electron acceptor. Their nutritional requirements are relatively complex. They may be cultivated on trypticase soy plain (or blood) agar, in trypticase soy broth, or in synthetic media containing a variety of amino acids and vitamins. *B. abortus* differs from the other species infecting man in requiring, on primary isolation, an atmosphere containing 5%–10% CO_2. All produce catalase and decompose urea; sugars are not fermented. Various differences, presented in Table 33-1, form the basis for the metabolic differentiation of the six species.

Despite these several differences, *Brucella* species are often difficult to distinguish. Moreover, stable variant strains (biotypes) often appear as the major strain in a given geographic area. The genus appears to be quite homogeneous as judged by DNA–DNA homology.

Brucellae are killed by pasteurization but they may survive for many weeks in discarded infected fetal tissues. In cheese they may be killed within a few days by the accumulated lactic acid, but in butter they remain viable for more than a week.

Genetic Variation. When serially grown on laboratory media the smooth (S) form of *Brucella* isolated from infected tissues tends to be replaced by rough (R) forms, which are less virulent and also exhibit less specific and more nonspecific agglutination. The selection responsible for this "dissociation" is due to the greater resistance of R cells to alterations of the medium produced by the S cells, including accumulation of D-alanine and lowering of pO_2 (see Ch. 8). Intermediate (I) and mucoid (M) forms, exhibiting reduced virulence, may also emerge. In 1:1000 acriflavine S-type cells remain evenly suspended, R-type cells clump, and M-type cells form a slimy, threadlike precipitate. Non–S-type colonies do not readily revert either in vitro or in vivo to the fully virulent S type.

Antigenic Structure. Antisera from animals immunized with a smooth strain agglutinate the three principal *Brucella* species. Two shared determinants (A and M) have been proposed to account for these crossreactions. Abortus (A) Ag is the major surface determinant in both *B. abortus* and *B. suis,* and is a minor determinant (a) in *B. melitensis,* while the M Ag predominates in the latter and is a minor Ag (m) in the others. By adsorbing anti-Am (*B. abortus* or *B. suis*) Abs with *B. melitensis* organisms (aM), in an amount that will remove all the minor (m) agglutinin but only a small fraction of the major (A) agglutinin, it is possible to prepare monospecific serum to A Ag. Monospecific anti-M serum may be prepared similarly. These sera are useful in diagnosis (Table 33-1).

The failure of monospecific serum to agglutinate the species with the corresponding minor Ag may be due to a difference between the minor and the major determinants, or to a sparse distribution of the former on the cell surface.

The surface Ags (A, M) resemble, in composition and properties, the protein–lipopolysaccharide complex of the Enterobacteriaceae.

PATHOGENESIS

Although each of the three principal *Brucella* species has preferred hosts all are pathogenic in a wide range of mammals; experimental infections are readily produced in guinea pigs, rabbits, mice, and monkeys.

In naturally acquired brucellosis the organisms gain entrance via the broken skin, the conjunctivae, the alimentary tract, or possibly the aerosol route. At the site of lodgment in the skin or mucous membranes the organisms are ingested by polymorphonuclear cells, multiply within them, and are carried (mostly in these cells) via the lymphatics to the regional lymph nodes. There the bacteria enter and multiply within mononuclear cells; some of these cells die, and the released bacteria and cell contents stimulate local mononuclear cell activation and proliferation. The outcome of this confrontation determines whether or not the invasive infection is contained. If not, polymorphonuclear and mononuclear cells carrying the bacteria reach the blood and soon accumulate in the sinusoids of the liver. These focal aggregations of Kupffer cells containing large numbers of organisms develop and after another few days form typical small granulomas. Similar lesions appear in spleen, bone marrow, and kidney (Fig. 33-2). In certain mammals other than man (cattle, swine, sheep, goats, etc.) brucellae also accumulate in the mammary glands (causing infection of the milk), in the genital organs and in the pregnant uterus (often resulting in abortion).

In bovine infectious abortion the organisms are found mostly in the fetal portion of the placenta, the birth fluids, and the chorion. This remarkable **viscerotropism** (see Organotropism, Ch. 24) depends upon the presence of **erythritol,** which may be an important factor in the pathogenesis of infectious abortion. This four-carbon polyhydric alcohol ($HOCH_2–CHOH–CHOH–CH_2OH$) was found to be the factor in bovine allantoic and amniotic fluids that stimulated the growth of *B. abortus;* it also enhances experimental infections with *B. melitensis.*

TABLE 33-1. Differential Characteristics of *Brucella* Species*

Species	CO₂ requirement	H₂S production	Hydrolysis of urea	Growth on dye† media		Agglutination in			Lysis by phage		Oxidative metabolic tests			
						Mono-specific antisera to		Anti-rough sera‡						
				Thionin	Basic fuchsin	A	M		RTD§	10⁴ × RTD	L-glutamic acid	DL-ornithine	L-lysine	D-ribose
B. melitensis	−	−	Slow	−	+	−	+	−	−	−	+	−	−	−
B. abortus	+¶	+	Slow	−	+	+	−	−	+	+	+	−	−	+
B. suis	−	+	Rapid	+	−	+	−	−	−	+	±	+	+	+
B. canis	−	−	Rapid	±	±	−	−	+‡	−	−	±	+	+	±
B. ovis	+	−		+	+	−	−	+‡	−	+	+	−	+	−
B. neotomae	−	+		−	−	+	−	−	−	−	+	−	+	±

*To accommodate significant intraspecies heterogeneity the classification scheme has been expanded to include species variants as biotypes (not shown here): 4 biotypes of *B. melitensis*; 9 of *B. abortus*; 4 of *B. suis*. The properties listed in this table are those of reference strains and are generally characteristic of the majority (but not all) of the biotypes of each species.

†Species differentiation is obtained on Albimi or tryptose agar with thionin at 1:25,000, basic fuchsin at 1:100,000.

‡*B. canis* (and also *B. ovis*) grow as "rough" colonies on primary isolation and appear to be deficient in somatic O antigen of the cell wall. Only antisera to *B. canis* (or *B. ovis*) will produce agglutination of the homologous cells; the same antisera will not produce agglutination when tested with the usual type of brucella antigen prepared from "smooth" cells of *B. abortus* (employed in the diagnostic laboratory).

§Lysis by reference brucella phage, *Tbilisi* (Tb) phage, at routine test dilution (RTD) and at 10⁴ x that concentration. (Other brucella phages have recently been described which are lytic for *B. suis* and *B. neotomae*. None of the phages are lytic for *B. meliensis*, *B. ovis*, or *B. canis*.)

¶*B. abortus* has a need for increased CO₂, especially on primary isolation. An attenuated strain (*B. abortus* strain 19), widely employed as a living vaccine in cattle, does not require added CO₂ for growth.

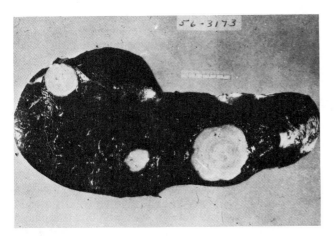

FIG. 33-2. Longitudinal section of the spleen of a patient with chronic *B. suis* infection, showing splenomegaly and scattered, large, partially calcified areas of caseation.

Erythritol is present in appreciable quantities only in the chorion, cotyledons, and fetal fluids of animals prone to infectious abortion (cows, sheep, pigs, goats, and dogs). It is not detected in the human placenta, which is not a site of localization of the infection.

B. abortus US-19, the attenuated strain used for immunization of cattle in the United States, is unique among brucella strains in that its growth is inhibited, rather than stimulated, by the presence of erythritol. This strain appears to lack the enzyme D-erythrulose-1-phosphate dehydrogenase on the erythritol pathway. The inability to utilize erythritol may be a significant (but not the sole) factor in the decreased virulence of this strain for cattle. While erythritol catabolism undoubtedly is important in specifying tissue localization in pregnant cattle, etc., it is not the only determinant of *Brucella* infection. There is no erythritol within macrophages, yet the reticuloendothelial system is commonly involved in chronic brucellosis.

No exotoxins or antiphagocytic capsular or cell wall constituents have been detected among *Brucella* species. Instead, **intracellular** events largely determine the course of the disease: smooth *Brucella* organisms multiply abundantly in nonimmune monocytes, while rough, avirulent organisms do not. A possible **virulence factor,** made only in vivo, appears to enhance the intracellular survival. Thus virulent *B. abortus* from cultures of monocytes or from infected bovine placenta survives in mononuclear cells better than the same strain grown on artificial media. Moreover, cell walls from virulent *Brucella* obtained from the bovine placenta, but not from the same organisms grown on artificial media, inhibit the normal intracellular destruction of an avirulent (R) strain by mononuclear cells.

Clinical Features. The incubation period in human brucellosis is long, often several weeks or even months. The onset of symptoms is usually insidious, with malaise, chills, fever, sweats, weakness, myalgia, and headache. Fever may be remittent, particularly with *B. melitensis* **(undulant fever).** Vague gastrointestinal and nervous symptoms are common. The acute illness may be associated with enlarged lymph nodes, spleen, and liver, and with localized vertebral spondylitis. Bacteremia is present in about 20% of cases. Meningoencephalitis, osteomyelitis, endocarditis, and interstitial nephritis with focal glomerular lesions sometimes occur in the course of the acute disease. Epididymoorchitis occasionally occurs with *B. melitensis* infection. In the later stages of the disease the persistence of vague complaints often suggests a psychoneurosis. The establishment of a definitive diagnosis at this "chronic" stage is especially difficult.

Hepatic involvement is common in human brucellosis, with small noncaseating granulomas observed in liver biopsies. In acute brucellosis due to *B. melitensis* the pathologic picture may instead be that of diffuse hepatitis with focal necrosis. *B. suis* infection may cause chronic suppurative abscesses and large areas of caseation in the liver and spleen of many years' duration. These lesions usually calcify after some years (Fig. 33-2).

IMMUNITY

Circulating Abs, measured by agglutinin, precipitin, opsonin, or bactericidal tests, are usually detectable from the time the first signs and symptoms of the disease develop (see Laboratory Diagnosis, below), but their presence does not prevent bacteremia or reinfection (which is common). However, relative immunity to brucellosis may be acquired: asymptomatic but seropositive abattoir workers are less likely to develop clinical brucellosis than previously noninfected workers.

Cellular immunity plays a critical role in this disease: the survival of brucellae is shorter in cultured macrophages from actively immunized animals. The mechanism of this nonspecific activation of macrophages, and its relation to hypersensitivity, are discussed in Chapter 24.

A hypersensitivity reaction to suddenly released brucella Ags is presumed to be responsible for the Herxheimer-like reaction (see Treatment, Ch. 39) that sometimes follows vigorous chemotherapy. Similarly, severe systemic as well as local reactions are observed after accidental inoculation of previously infected veterinarians with a vaccine strain.

LABORATORY DIAGNOSIS

The patient with brucellosis commonly presents with an unexplained fever. The diagnosis is usually suggested by the combination of clinical and epidemiologic considerations. It may also be suspected when liver biopsy reveals noncaseating granulomas, but these also occur in other infectious and noninfectious processes.

The establishment of a definitive diagnosis requires **cultivation** of the organism from the blood or from a biopsy of bone marrow, liver, or lymph node; blood should be cultured repeatedly in all suspected cases. Specimens should be incubated in trypticase soy broth under 10% CO_2. (The Casteñada bottle, containing both solid and liquid media, is often used.) At 4-to 5-day intervals, or when visible growth is first noted, a sample is removed for staining and for subculture on blood trypticase soy agar under 10% CO_2. The primary cultures should be incubated for at least 4 weeks before being discarded. Since the number of viable bacilli present in specimens is usually small, cultivation is often difficult. Species are identified on the basis of the tests summarized in Table 33-1.

Owing to the increased use of antibiotics prior to obtaining cultures the majority of cases of brucellosis are now diagnosed serologically. Of 1644 abattoir-associated cases reported from 1960 to 1971 in this country 85% were diagnosed serologically; in only 17% was the diagnosis confirmed by culture.

Because of the prolonged incubation period in human brucellosis **Abs** are frequently demonstrable in the serum by the time the disease is first recognized. The initial Ab response is the appearance of IgM, followed shortly by IgG. **Agglutination** tests are performed with phenolized suspensions of heat-killed smooth (S) bacilli. A standard strain of *B. abortus*, no. 456, which is avirulent for man, is generally used for this purpose; it detects Abs to all three principal species equally well. Titers above 1:80 are usually considered indicative of either past or present infection, and a fourfold or greater rise in titer during the course of the illness is strong evidence for the diagnosis.

The titer remains elevated (as high as 1:640 to 1:2560) during the active phases of the disease and usually declines as the patient improves. High titers of agglutinins (IgM but not IgG) can persist following recovery from infection and in the absence of overt disease; they may even be found with no history of clinical disease, usually in individuals who have been exposed to farm animals or unpasteurized dairy products. However, the persistence or recrudescence of chronic active brucellosis is usually associated with the presence of IgG, even in sera with low agglutinating titers (less than 1:100). Thus the finding of 2-mercaptoethanol–resistant (IgG) agglutinins in a **symptomatic patient** is suggestive of active disease; their absence is also significant and may be helpful in evaluating possible chronic brucellosis.

As in all bacterial agglutination tests, care must be taken to avoid misinterpretation of the prozone phenomenon (see Prozone, Ch. 16). Occasionally, in chronic brucellosis particularly, the Abs are of the incomplete, or blocking, type (see Blocking Antibody, Ch. 16). They may be detected by diluting sera to at least 1:1280 (in 5% sodium chloride or albumin solution) in the agglutination test, or by adding the Coombs reagent (anti-human-immunoglobulin serum) to the test for agglutinating Abs.

Serum containing Abs to *F. tularensis, Yersinia enterocolitica, Vibrio cholerae* (due to infection or to immunization within the year with cholera vaccine), or *V. fetus* can crossreact in the *Brucella* agglutination test. The recent performance of a *Brucella* skin test may cause a low titer of agglutinins to appear.

An intradermal skin test employing suspensions of killed *Brucella* cells or a nucleoprotein fraction (Brucellergen) has been widely used to demonstrate specific delayed hypersensitivity. Like the tuberculin test, it reveals only prior exposure to the organism, and is of no value in the diagnosis of clinical brucellosis.

TREATMENT

Most strains of *Brucella* are sensitive in vitro to the tetracyclines and streptomycin, but streptomycin does not reach organisms sequestered within mononuclear phagocytes. Tetracycline alone, or in combination with streptomycin in severe infections, is usually effective, often within a few days. However, relapse commonly occurs unless therapy is prolonged for at least 3–4 weeks. The difficulty in eradicating the organisms undoubtedly arises because an intracellular location protects them from the action of both drugs and Abs.

EPIDEMIOLOGY AND PREVENTION

B. abortus infects cattle almost worldwide, including the United States. *B. suis* infects cattle as well as swine and is presently the most frequently isolated species from human cases in the United States. Goats are important as sources of *B. melitensis* in Mexico and in Mediterranean countries, but this organism is rarely isolated from cases in the United States.

An attenuated live vaccine (*B. abortus*, strain 19) has been used in calves to decrease the incidence of brucellosis in cattle; it produces a limited infection, followed by reasonable immunity. The identification of infected cattle by an agglutination test (on serum or milk) or by direct fluorescent Ab testing of bovine abortion material, followed by segregation or slaughter, has also been utilized.

Through the use of public health measures (testing of cows, vaccination of calves, pasteurization of milk products) the incidence of human brucellosis in the United States has decreased in the past 25 years from 6300 to 230 cases annually. Most cases now occur in workers in meat-packing plants (90% of cases related to slaughtering of hogs), in livestock raisers, and in veterinarians. About 10% of cases are attributable to ingestion of raw milk or imported cheeses.

The identification of epidemic disease due to *B. canis* in dogs has revealed yet another reservoir for brucellosis. The infection may produce little evidence of illness, even though bacteremia may be demonstrable for months or years. A few cases have occurred in man.

Although active immunization of humans at high risk is practiced in Russia, public health authorities in the United States have been reluctant to use the currently available vaccines because of their potential pathogenicity for man.

SELECTED READING

BOOKS AND REVIEW ARTICLES

BOTTONE EJ: *Yersinia enterocolitica:* a panoramic view of a charismatic organism. CRC Crit Rev Microbiol 5:211, 1977

BUCHANAN TM, FABER LC, FELDMAN RA: Brucellosis in the United States, 1960–1972. Part I. Clinical features and therapy. Medicine 53:403, 1974

BUCHANAN TM, HENDRICKS SL, PATTON CM, FELDMAN RA: Brucellosis in the United States, 1960–1972. Part III. Epidemiology and evidence for acquired immunity. Medicine 53:427, 1974

BUCHANAN TM, SULZER CR, FRIX MK, FELDMAN RA: Brucellosis in the United States, 1960–1972. Part II. Diagnostic aspects. Medicine 53:415, 1974

EIGELSBACH HT: *Francisella tularensis.* In Lennette EH, Spaulding EH, Truant JP (eds): Manual of Clinical Microbiology, 2nd ed. Washington DC, American Society for Microbiology, 1974, p 316

EISENBERG HG, CAVANAUGH DC: *Pasteurella.* In Lennette EH, Spaulding EH, Truant JP (eds): Manual of Clinical Microbiology, 2nd ed. Washington DC, American Society for Microbiology, 1974, p 246

ELBERG SS: Immunity to brucella infection. Medicine 52:339, 1973

HUBBERT WT, ROSEN MN: I. *Pasteurella multocida* infection due to animal bite. Am J Public Health 60:1103, 1970

HUBBERT WT, ROSEN MN: II. *Pasteurella multocida* infection in man unrelated to animal bite. Am J Public Health 60:1109, 1970

MCCULLOUGH NB: Microbial and host factors in the pathogenesis of brucellosis. In Mudd S (ed): Infectious Agents and Host Reactions. Philadelphia, Saunders, 1970

MONTIE TC, AJL SJ: Nature and synthesis of murine toxins of *Pasteurella pestis.* In Montie TC, Kadis S, Ajl SJ (eds): Microbial Toxins, Vol III. New York, Academic Press, 1970

REED WP, PALMER DL, WILLIAMS RC JR, KISCH AL: Bubonic plague in the southwestern United States: a review of recent experience. Medicine 49:645, 1970

SONNENWIRTH AC: *Yersinia.* In Lennette EH, Spaulding EH, Truant JP (eds): Manual of Clinical Microbiology, 2nd ed. Washington DC, American Society for Microbiology, 1974, p 222

SONNENWIRTH AC: Isolation and characterization of *Yersinia enterocolitica.* Mt Sinai J Med 43:736, 1976

SPINK WW: The Nature of Brucellosis. Minneapolis, University of Minnesota Press, 1956

WHO Expert Committee on Plague: Fourth Report. WHO Tech Rep Ser No. 447, Geneva, 1970

WINBLAD S (ed): *Yersinia, Pasteurella* and *Francisella.* In Contributions to Microbiology and Immunology, Vol 2. Basel, Karger, 1973

SPECIFIC ARTICLES

BURKE S: Immunization against tularemia: analysis of the effectiveness of live *Francisella tularensis* vaccine in prevention of laboratory-acquired tularemia. J Infect Dis 135:55, 1977

BURROWS TW: Genetics of virulence in bacteria. Br Med Bull 18:69, 1962

BUTLER T: A clinical study of bubonic plague. Observations of the 1970 Vietnam epidemic. Am J Med 53:268, 1972

CAVANAUGH DC, STEELE JH (eds): Trends in research on plague immunization. J Inf Dis [Suppl] 129:S1–S120, 1974

FITZGEORGE RB, SMITH H: The chemical basis of the virulence of *Brucella abortus.* VII. The production in vitro of organisms with an enhanced capacity to survive intracellularly. Br J Exp Pathol 47:558, 1966

GUTMAN LT, OTTESEN EA, QUAN TJ, NOCE PS, KATZ SL: An inter-familial outbreak of *Yersinia enterocolitica* enteritis. N Engl J Med 288:1372, 1973

HARTLEY JL, ADAMS GA, TORNABENE TG: Chemical and physical properties of lipopolysaccharide of *Yersinia pestis.* J Bacteriol 118:848, 1974

HEDDLESTON KL, WESSMAN G: Characteristics of *Pasteurella multocida* of human origin. J Clin Microbiol 1:377, 1975

HOLLAND JJ, PICKETT MJ: A cellular basis of immunity in experimental brucella infection. J Exp Med 108:343, 1958

KANAMORI M: Biological activities of endotoxins from *Yersinia enterocolitica.* Jpn J Microbiol 20(4):273, 1976

KEPPIE J, WILLIAMS AE, WITT K, SMITH H: The role of erythritol in the tissue localization of the brucellae. Br J Exp Pathol 46:104, 1965

KEPPIE J, WITT K, SMITH H: The chemical basis of virulence of *Brucella abortus.* IX. The increased immunogenicity of *Brucella abortus* grown in media which enhance its ability for survival within bovine phagocytes. Br J Exp Pathol 50:219, 1969

LAWTON WD, ERDMAN RL, SURGALLA MJ: Biosynthesis and purification of V and W antigen in *Pasteurella pestis.* J Immunol 91:179, 1963

LEINO R, KALLIOMAKI JL: Yersiniosis as an internal disease. Ann Intern Med 81:458, 1974

MEYER KF: The natural history of plague and psittacosis. Public Health Rep 72:705, 1957

REDDIN L, ANDERSON RK, JENESS R, SPINK WW: Significance of 7S and macroglobulin brucella agglutinins in human brucellosis. N Engl J Med 272:1263, 1965

SMITH H, FITZGEORGE RB: The chemical basis of the virulence of *Brucella abortus.* V. The basis of intracellular survival and growth in bovine phagocytes. Br J Exp Pathol 45:174, 1964

SMITH H, KEPPIE J, PEARCE J, WITT K: The chemical basis of the virulence of *Brucella abortus.* Br J Exp Pathol 42:631, 1961; 43:31, 530, 538, 1962

SPERRY JF, ROBERTSON DC: Inhibition of growth by erythritol catabolism in *Brucella abortus.* J Bacteriol 124:391, 1975

WEBER J, FINLAYSON NB, MARK NBD: Mesenteric lymphadenitis and terminal ileitis due to *Yersinia pseudotuberculosis.* N Engl J Med 283:172, 1970

chapter

34

THE HEMOPHILUS–BORDETELLA GROUP

STEPHEN I. MORSE[†]

† Deceased

The organisms of the hemophilus–bordetella group are small, gram-negative, nonmotile, non–spore-forming bacilli with complex growth requirements. Species of the genus *Hemophilus,* which require either one or two growth factors (X, V) provided by blood (Table 34-1), include *H. influenzae* (a major cause of bacterial meningitis in children), *H. aegyptius* (conjunctivitis), and *H. ducreyi* (chancroid). In the closely related genus *Bordetella* the important species is *B. pertussis,* which causes whooping cough.

HEMOPHILUS INFLUENZAE

This prevalent pathogenic species, first isolated by Pfeiffer during the influenza pandemic of 1890, was erroneously thought to be the cause of the disease and was named *H. influenzae.* The etiologic agent of epidemic influenza is, of course, now known to be a virus. In the pandemics of 1890 and 1918 *H. influenzae* could have been an important secondary invader, since at autopsy it was often the predominant or even the only bacterial species that could be cultured from the lungs. Moreover, Shope discovered in 1931 that swine influenza is caused by an influenza virus plus *H. suis* (a bacterial species closely related to *H. influenzae*), and a similar synergistic effect has been demonstrated in chick embryos and in rats infected with a human influenza virus strain and *H. influenzae.*

H. influenzae is the most common cause of bacterial meningitis in children under 4 years of age (Fig. 34-1). *H. influenzae* may also cause a fatal epiglottitis (see Pathogenesis, below). The organism is sometimes isolated from children with obstructive bronchiolitis, and in up to 25% of children with otitis media, but its etiologic role in those diseases is not certain.

Although gram-negative, *H. influenzae* shares many properties with *Streptococcus pneumoniae:* both organisms are primarily invasive and gain entrance to the body through the respiratory tract; both have polysaccharide antiphagocytic capsules, some of which cross-react between the two species; when grown in artificial media both species tend to undergo autolysis and are bile-soluble; and genetic transformation by DNA occurs with both.

MORPHOLOGY

Organisms seen in pathologic samples such as cerebrospinal fluid are generally small, gram-negative, encapsulated coccobacilli ($1\mu m$–$1.5\mu m \times 0.3\mu m$) (Fig. 34-2), but short chains and rods are often found. Unless the Gram stain is properly decolorized bipolar staining may lead to the mistaken identification of gram-positive organisms, especially streptococci or pneumococci. Unencapsulated strains are markedly pleomorphic and filamentous.

On solid medium encapsulated virulent strains appear as small "dewdrop" colonies, and on transparent medium, such as Levinthal agar, these are characteristically iridescent in obliquely transmitted light. After 24–48 h the capsules and iridescence disappear, and autolysis and a variable gram-stain reaction occur. Both the autolysis and capsular destruction are apparently caused by activation of endogenous enzymes. Similarly, in liquid medium the capsules disappear early in culture.

Even in early cultures a few noniridescent colonies containing unencapsulated variants of *H. influenzae* are always demonstrable. It is evident, therefore, that the rate of spontaneous smooth to rough (S → R) mutation is relatively high. When cultural conditions are suboptimal R variants will often predominate. Such S → R shifts have been observed in the upper respiratory tracts of patients recovering from influenzal meningitis.

Transformation mediated by DNA from *H. influenzae,* extensively studied by Alexander, has included drug resistance and specific capsular antigen synthesis; interspecific transformation occurs. The demonstration of transformation between pneumococci in host tissues suggests that it may also take place with *Hemophilus.* Restriction endonucleases from *Hemophilus* species are widely used in the analysis and cloning of DNA (see Ch. 12).

GROWTH AND METABOLISM

H. influenzae is a facultative anaerobe which requires two growth factors present in blood, the heat-stable X and the labile V (Table 34-1). X factor can be replaced by hemin

TABLE 34-1. Differential Characteristics of Species of Hemophilus

Species	Growth factor X	Growth factor V	Hemo-lysis	Increased CO$_2$ requirement
H. influenzae	+	+	−	−
H. aegyptius	+	+	−	−
H. suis	+	+	−	−
H. haemolyticus	+	+	+	−
H. ducreyi	+	−	Slight	+
H. aphrophilus	+	−	−	+
H. parainfluenzae	−	+	−	−
H. parahaemolyticus	−	+	+	−

(hematin), which is a precursor for the prosthetic groups of the respiratory enzymes. (Under anaerobic conditions the X factor requirement is usually reduced.) V factor can be replaced by NAD or NADP, both of which are destroyed by excessive heat.

H. influenzae lacks the enzymes required to synthesize protoporphyrin from γ-aminolevulinic acid. Protoporphyrin can substitute for hemin in H. influenzae but not in other species, such as H. aegyptius, which lack the iron-adding enzyme. V factor can be supplied by other microorganisms, such as staphylococci, growing in the immediate vicinity; this **satellite phenomenon** is illustrated in Figure 34-3.

H. influenzae grows slowly in blood broth or on blood agar containing rabbit, horse, or guinea pig blood. Fresh human and sheep blood inhibit growth but can be used in the preparation of **chocolate agar,** on which H. influenzae

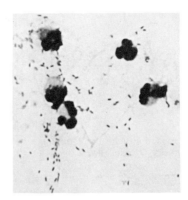

FIG. 34-2. Gram-stained smear of exudate from spinal fluid of patient with H. influenzae meningitis. (Courtesy of R. Drachman) (×1000)

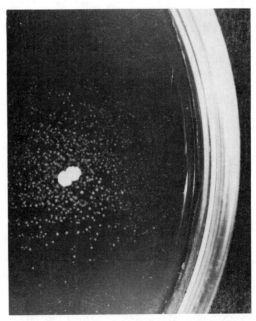

FIG. 34-3. Satellite phenomenon: heavily seeded colonies of H. influenzae growing only in vicinity of staphylococcal colonies on autoclaved blood agar. Autoclaving destroys V factor but not X factor. (Courtesy of PH Hardy and EE Nell)

grows profusely. Chocolate agar is prepared by adding blood to an agar base at 80°C and maintaining the temperature at 50° for approximately 15 min or until a brown color appears. The mild heat releases X and V factors from RBCs and also destroys inhibitors of V factor without inactivating V factor itself. Other useful media in which RBCs are also lysed include the transparent Levinthal agar, Fildes agar, and Levinthal broth.

Growth is optimal at pH 7.6, and in liquid media it may be stimulated by aeration. On primary isolation

FIG. 34-1. Age-specific incidence of meningitis caused by H. influenzae **(solid line),** N. meningitidis **(broken line),** and S. pneumoniae **(dotted line)** in children under 30 months of age. (Fraser DW et al: J Infect Dis 127:271, 1973)

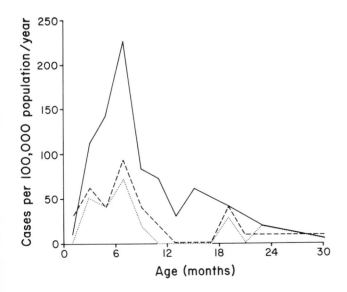

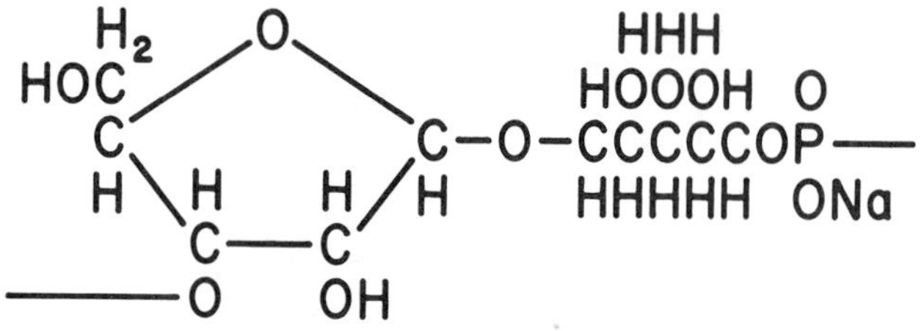

FIG. 34-4. Proposed structure of repeating unit of type b capsular polysaccharide. (Crisel RM et al: J Biol Chem 250:4926, 1975)

some strains grow best in the presence of 5%–10% CO_2 (i.e., in a candle jar). Most strains are capable of converting tryptophan to indole and of utilizing nitrate as an electron acceptor in the absence of oxygen; these properties are of diagnostic value. Fermentation reactions of *Hemophilus* organisms are variable and are of no differential value.

ANTIGENIC STRUCTURE

As with the pneumococcus, the capsular carbohydrates of *H. influenzae* evoke protective Abs. **Six types,** designated types a to f, have been described; these are serologically identified by agglutination, precipitation, or quellung tests performed with specific antisera.

In type b, the most important pathogen for humans, the polysaccharide (mol wt 550,000) contains ribose, ribitol, and phosphate, probably linked as shown in Figure 34-4. The specific carbohydrates of types a and c are polyhexosephosphates. Hexose is found in type d, an unknown hexosamine in type 3, and galactosamine in type f.

Crossreacting polysaccharides from other organisms are common. Ab to type b polysaccharide crossreacts with Ags from a number of bacteria, including strains of pneumococci and streptococci, *Bacillus subtilis, Staphylococcus aureus,* and *Escherichia coli.* The gram-positive organisms have ribitol phosphate in their teichoic acids, while both ribitol and ribose are found in Ags of the most crossreactive strains of *E. coli.*

Much less is known about the somatic Ags of *H. influenzae.* A protein (M) Ag common to all types has been described, as well as an endotoxin that resembles those from other gram-negative bacilli.

PATHOGENESIS

Naturally acquired disease due to *H. influenzae* seems to occur only in man. Experimental bronchopneumonia and meningitis have been produced in monkeys, and fatal peritonitis in mice, but the most useful experimental model is the infant rat, in which bacteremia and meningitis follow intranasal or intraperitoneal inoculation with type b *H. influenzae.* Diseases due to *H. influenzae* are common in young children but rare in neonates and adults. The reason is explained under Immunity, below.

Virulence for man is directly related to capsule formation; the organisms produce no demonstrable exotoxin. Moreover, endotoxin does not appear to play a significant role in pathogenicity, since homologous anticapsular Abs are protective but antiendotoxic Abs are not. Virtually all severe infections are caused by type b.

The organism gains entrance to the tissue via the respiratory tract, where it frequently resides without causing trouble. Carrier rates in children may be as high as 30%–50%, but the organisms are usually unencapsulated. Of the encapsulated strains found, most are type b.

Disease due to *H. influenzae* usually begins as a nasopharyngitis, probably precipitated by a viral infection of the upper respiratory tract. The resulting coryza may be followed by sinusitis or otitis media and may lead to pneumonia; the latter is often complicated by empyema. Bacteremia occurs early in severe cases and frequently results in metastatic involvement of one or more joints or in the development of acute bacterial meningitis—the most important clinical entity caused by *H. influenzae.* Indeed, in children *H. influenzae* is the commonest cause of bacterial meningitis (Fig. 34-1), except during epidemics of meningococcal meningitis. The clinical signs are like those of other forms of acute bacterial meningitis. Unless vigorously treated the patient rarely recovers.

A less common but even more serious disease caused by *H. influenzae* type b is epiglottitis and obstructive laryngitis. The onset is sudden and the course fulminating, often ending fatally within 24 h. Infection starts in the pharynx and spreads to the epiglottis, which becomes cherry red and grossly edematous. Laryngeal obstruction ensues. The patient should be immediately hospitalized when the clinical diagnosis is made, for survival may require prompt tracheotomy. Bacteremia is usually a feature of the disease.

IMMUNITY

The incidence of *H. influenzae* meningitis as a function of age is inversely related to the titer of Ab in the blood (Fig. 34-5), whether passively acquired from the mother or actively formed. In children aged 2 months to 3 years Ab levels are minimal; thereafter, Ab increases and the disease becomes much less common.

Like other gram-negative bacilli, *H. influenzae* is susceptible to lysis by Ab and complement. But though the immunologic test used in the studies in Figure 34-5 primarily measures bacteriolysis, immunity is not necessarily due to this action. Thus anticapsular Ab, which pro-

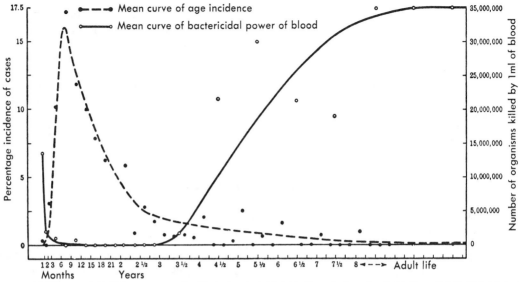

FIG. 34-5. Relation of the age incidence of *H. influenzae* meningitis to bactericidal Ab titers in the blood. (Fothergill LD, Wright J: J Immunol 24:281, 1933. Copyright 1933 by The Williams & Wilkins Co, Baltimore)

motes phagocytosis, is known to be the factor in serum correlated with protection; phagocytosis is prominent in the meningeal lesions; complement is rarely detectable in the spinal fluid of patients with bacterial meningitis; and not all fresh isolates of *H. influenzae* are susceptible to lysis by Ab and complement. Phagocytosis is thus evidently an important defense mechanism in influenza bacillus meningitis, whereas the role of complement remains undefined.

LABORATORY DIAGNOSIS

In all cases of suspected bacterial meningitis a sample of blood, as well as of spinal fluid, should be cultured. A smear of the spinal fluid, or preferably a smear made with the sediment from a centrifuged specimen, should be stained by the Gram method. The detection of small, pleomorphic, gram-negative bacilli warrants a provisional diagnosis of influenza bacillus meningitis. If exposure of the organisms in the spinal fluid to specific antiserum results in a positive quellung reaction (see Fig. 27-1), the diagnosis is established. The specimen of spinal fluid should be streaked on chocolate agar and incubated in a candle jar. Organisms suspected of being *H. influenzae* should be subjected to the quellung test. If the results are negative, the requirement of the organism for X and V factors (Table 34-1) is determined by adding commercially available filter paper strips impregnated with each factor onto a culture seeded on nutrient agar. Detection of free capsular Ag in the spinal fluid by countercurrent immunoelectrophoresis with type b an-

tiserum is exceedingly sensitive and provides rapid, accurate diagnosis.

TREATMENT AND PREVENTION

Virtually all patients treated early in the course of influenza bacillus meningitis can now be cured. Ampicillin and chloramphenicol are effective, and ampicillin has been the drug of choice. However, approximately 10% of strains are now resistant to ampicillin because of a plasmid-mediated β-lactamase. At present, therefore, initial therapy consists of both ampicillin and chloramphenicol; ampicillin alone is continued if the organism is sensitive. Ampicillin resistance is conveniently and rapidly assessed by testing for β-lactamase formation. A few chloramphenicol-resistant strains of *H. influenzae* have been isolated, but combined ampicillin–chloramphenicol resistance has not been found.

The mortality rate of treated *H. influenzae* is less than 10% and most children acquire effective immunity by age 10. Nevertheless, *H. influenzae* meningitis causes 1500–2000 deaths/year in the United States, mostly in young children. Moreover, a significant proportion, perhaps 30%, of those who recover have residual neurologic defects; about 5% must be institutionalized. For these reasons, the immunizing and protective effects of the type b Ag are now being studied. Unfortunately, young children, who are at greatest risk, mount a poor immunologic response. Immunization by colonization with a crossreacting strain of *E. coli* remains a possibility. Widespread outbreaks of *H. influenzae* meningitis are rare, and although secondary cases may occur in families, prophylactic chemotherapy is not routine.

OTHER *HEMOPHILUS* BACILLI PATHOGENIC FOR MAN

H. ducreyi causes **chancroid,** or soft chancre. The paragenital ulcerative lesions in this venereal disease lack the firm indurated margins of syphilitic ulcers. Response to sulfonamides and various antibiotics is usually prompt. *H. aegyptius* (the Koch–Weeks bacillus) produced purulent conjunctivitis, and *H. parainfluenzae* is occasionally the cause of bacterial endocarditis. *H. haemolyticus* is nonpathogenic but its β-hemolytic colonies may be easily mistaken for those of *Streptococcus pyogenes* on blood agar, particularly if rabbit blood is employed.

Some differential properties of these organisms are summarized in Table 34-1.

BORDETELLA PERTUSSIS

B. pertussis, the main causative agent of whooping cough, was first isolated in 1906 by Bordet and Gengou. It seldom penetrates the mucous membranes of the respiratory tract and consequently is not found in the blood. It is a strict aerobe and does not require either X or V factor for growth.

MORPHOLOGY, GROWTH, AND VARIATION

Morphologically *B. pertussis* is very similar to *H. influenzae:* a small, nonmotile, gram-negative bacillus which forms a capsule when virulent and tends to be pleomorphic, particularly in older children. It is an extremely delicate organism and survives in vitro for only a few hours in respiratory secretions. Although nutritional requirements are simple, to ensure maximum growth of virulent (phase 1) organisms it is usually necessary to add blood or albumin, charcoal, starch, or anion exchange resins to media. These substances bind fatty acids, to which the organism is very susceptible. Bordet–Gengou agar, which contains fresh sheep blood, is the medium most utilized for primary isolation of *B. pertussis* from patients. Narrow zones of hemolysis often surround the colonies. Carbohydrates are not fermented.

When first isolated (phase 1) the organism is a small, encapsulated coccobacillus, fully virulent, and possessing pili; it will not grow on plain nutrient agar. Prolonged laboratory passage leads to selection of phase 4 organisms: pleomorphism becomes striking, growth occurs on nutrient agar, and the capsule, virulence, and pili are lost. The intermediate phases 2 and 3 are not precisely defined. Clearly, this change is equivalent to the classic S → R variation.

CELLULAR AND EXTRACELLULAR PRODUCTS

Phase 1 *B. pertussis* is encapsulated; the nature of the capsule is unknown and anticapsular Abs have not been shown. The pili of *B. pertussis* are 2.5 nm in diameter and are hemagglutinating; hemagglutination-inhibition Abs are formed. At least seven surface Ags (factors) play a role in the agglutination of *B. pertussis* by Ab; some are genus- or species-specific and others are strain-specific (Table 34-2). The factors are detected by agglutination of organisms with appropriately absorbed antiserum and by the capacity of the organisms to absorb out specific factor agglutinins from immune sera. None of the factors has been isolated or characterized. A heat-stable endotoxin forms part of the cell wall. Intact cells also have an adjuvant effect separate from the endotoxin.

The heat-labile **exotoxin,** which is dermonecrotic in rabbits and lethal for mice, is a soluble protein but is not excreted freely; substantial amounts are released only when the integrity of the cell is broken. The supernatant fluids of phase 1 cultures produce a variety of unusual biologic effects including **lymphocytosis;** heightened sensitivity to histamine, serotonin, and anaphylaxis; and hypoglycemia and unresponsiveness to the hyperglycemic effect of epinephrine. A single product, **lymphocytosis-promoting factor (LPF),** is probably responsible. LPF is also an in vitro mitogen for T lymphocytes.

PATHOGENESIS

B. pertussis has been found only in man. Experimental infection has been produced in chimpanzees, monkeys,

TABLE 34-2. Characteristics of *Bordetella* Species

| Species | Hemolysis | Motility | Agglutinogen (factor) designation | | |
			Genus-specific	Species-specific	Others
B. pertussis	+	0	7	1	2,3,4,5,6
B. parapertussis	±	0	7	14	8,9,10
B. bronchiseptica	±	+	7	12	8,9,10,11

dogs, rabbits, rats, mice, ferrets, and chick embryos.

The remarkable **viscerotropic** properties that cause *B. pertussis* to attach preferentially to ciliated bronchial epithelial cells has already been noted (see Organotropism, Ch. 24) and is depicted in Fig. 34-6. A similar localization is found in whooping cough (Fig. 34-7). In addition, mice become 1000 times more susceptible to *B. pertussis* infection when inoculated intranasally rather than intraperitoneally. In intracerebral infection of mice the organisms accumulate on the ciliated ependymal cells lining the ventricles. The specificity of ciliary attachment of the organism may be related to its pili.

Whooping cough is a **surface infection,** the organism rarely penetrating the mucosa. However, **subepithelial necrosis and inflammation** are characteristic features. These may be due to local effects of the heat-labile exotoxin; systemic toxic effects do not occur.

The disease ordinarily begins after an average incubation period of 10 days, with benign nasopharyngeal symptoms, including coryza, sneezing, and a mild cough. After 10–14 days the lower respiratory tract becomes involved and paroxysms of severe coughing develop, usually terminated by a characteristic inspiratory "whoop" and often followed by vomiting. Secondary bronchopneumonia is common, and obstruction of the bronchi by mucous plugs may cause severe anoxia, resulting in convulsions. A marked lymphocytosis is often demonstrable during the paroxysmal stage. This stage ordinarily lasts approximately 2 weeks and is followed by continued, milder cough for another 2–3 weeks. Encephalitis occasionally occurs as a late complication. The mortality rate is highest in infants.

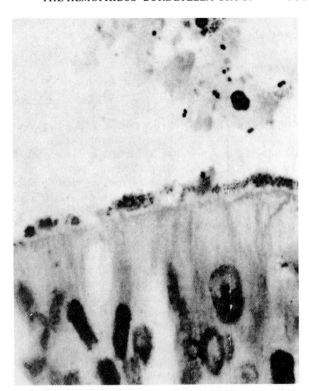

FIG. 34-7. *B. pertussis* lodged among cilia on bronchial epithelium of child who died of whooping cough. (Brown JH, Brenn L: Bull. Johns Hopkins Hosp. 48:69, 1931) (×1000)

FIG. 34-6. *B. pertussis* adhering only to ciliated cells of hamster tracheal rings in culture. Note cluster of six bacilli on tips of cilia in center of photograph. (Muse KE et al: In O Jahari, RP Becker [eds]: Scanning Electron Microscopy, Vol. II, Chicago, IIT Research Institute, 1977, p 263) (5000)

Communicability is greatest in the catarrhal stage and the organism is only rarely recovered after the fourth week of illness. A carrier state has not been identified.

IMMUNITY

Antibodies against *B. pertussis* are not usually found until the third week of the disease and are detected by complement fixation, complement-dependent bacteriolysis, agglutination, hemagglutination inhibition, or mouse protection. The last is assayed by the ability of the serum to protect mice from lethal intracerebral infection, and correlates best with clinical immunity; the nature and mechanism of the protection are unknown. Abs against LPF also occur. All decrease in titer after several years.

Naturally acquired immunity is impermanent, but second attacks are usually mild and pass unrecognized; in older persons full-blown whooping cough causes severe distress. The lack of a carrier state, and hence of a repetitive or continuing antigenic stimulus, is probably important in the loss of immunity. The newborn is highly susceptible to whooping cough, in contrast to *H. influen-*

zae infection, since maternal serum has little or no protective Ab.

LABORATORY DIAGNOSIS

The time-honored "cough plate" procedure has been supplanted by a nasal swab technic in which a small cotton swab, wrapped on the end of a flexible copper wire, is passed through the nose to the posterior pharyngeal wall, where it is allowed to remain in place while the patient coughs. Upon withdrawal the swab is passed through a drop of penicillin solution previously placed on the surface of a Bordet–Gengou agar plate (to inhibit contaminants), and the drop is then streaked on the plate. Two to three days' incubation at 37 ° C is required for the characteristic small, pearl-white, glistening colonies of *B. pertussis* to form. Final identification depends upon either agglutination by antiserum or immunofluorescence. A definitive bacteriologic diagnosis can be made rapidly by direct staining of nasopharyngeal smears with fluorescein-labeled Ab. Serologic tests are of little diagnostic value, because Abs appear late.

B. parapertussis is a less common cause of whooping cough. Although it shares common Ags with *B. pertussis,* it may be differentiated by fluorescent Ab (or agglutination) tests performed with properly absorbed antiserum. *B. parapertussis* differs from *B. pertussis* in producing larger colonies, splitting urea, and utilizing citrate as a sole source of carbon. *B. bronchiseptica,* the only motile species, is associated with bronchopneumonia in rodents and dogs, and rarely causes whooping cough in man. Some characteristics of these species are shown in Table 34-2.

Adenoviruses have been isolated from some patients with clinical whooping cough who have no evidence of bordetella infection, and who develop a rise in Ab titer against adenoviruses. The significance of these observations is not clear.

TREATMENT

Mild whooping cough requires no specific treatment. The severe form of the disease, particularly in infants, is usually treated with erythromycin. Tetracyclines and chloramphenicol are also effective but ampicillin is not. The clinical response may not be dramatic, although the drugs rapidly render the patient noninfectious. Secondary pneumonia due to other bacteria may occur and must be treated appropriately. Hyperimmune human pertussis γ-globulin is still used occasionally but there are no reliable data on its efficacy. Nonspecific measures such as maintaining adequate oxygenation, hydration, and electrolyte balance are of great importance.

PREVENTION

Whooping cough occurs throughout the world, and in susceptible children the attack rate may exceed 90%. Although mortality in the United States fell markedly in the period 1920–1950, the incidence of the disease began to decrease substantially only after the introduction of an effective vaccine in the late 1940s (Fig. 34-8). There were only 2400 cases in 1974 compared to 120,000 in 1950. Most of the 14 deaths occurred in infants less than 1 year old.

Active immunization of all infants, beginning at the age of 2 months, is strongly advocated. The vaccine consists of either killed phase 1 organisms or a crude extract prepared from them, and it is usually combined with diphtheria and tetanus toxoids, since *B. pertussis* has a marked adjuvant effect on their immunogenicity. After the primary course of three injections at monthly intervals booster injections are given 1 and 5 years later. The exact incidence of serious pertussis vaccine reactions

FIG 34-8. Morbidity and mortality rates for pertussis in the United States, 1922–1968. (Brooks GF, Buchanan TM: J Inf Dis 122:123, 1970)

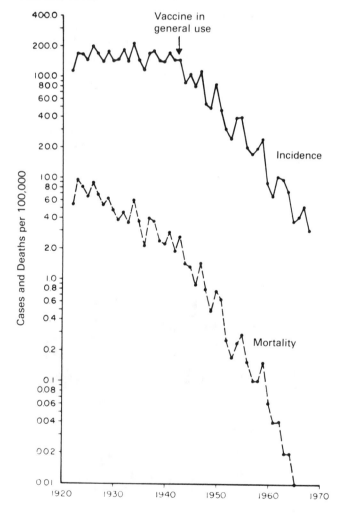

such as encephalopathy is unknown, but the risks are far outweighed by the risk from whooping cough in infants.

Vaccine potency is assayed by the capacity to immunize mice against intracerebral infection. The protective component of the vaccine is unknown. Vaccine failures are usually due to the use of impotent preparations; antigenic differences between vaccine strains and the infecting strain may sometimes be responsible. Like natural disease, pertussis vaccine does not produce permanent immunity nor does it provide protection against *B. parapertussis.*

Prevention of disease in susceptible **contacts** is not likely to be accomplished by isolation of the patient with overt whooping cough, since the period of maximum communicability will already have passed. Nevertheless, isolating the patient from infants under 2 years of age is warranted. **Chemoprophylaxis** with erythromycin is recommended for unimmunized susceptible contacts, and contacts under 4 years of age who have been immunized should receive a booster dose of vaccine as well as erythromycin. Hyperimmune pertussis γ-globulin is not an effective preventive. When it is necessary to protect older individuals, as in outbreaks among pediatric hospital personnel, low doses of vaccine may be effective. However, side effects are relatively frequent and erythromycin chemoprophylaxis may be preferable.

SELECTED READING

BOOKS AND REVIEW ARTICLES

MORSE SI: Biologically active components and properties of *Bordetella pertussis.* Adv Appl Microbiol 20:9, 1976

OLSON LC: Pertussis. Medicine 54:427, 1975

SELL SHW, KARZON DT (eds): Hemophilus Influenzae. Nashville, Vanderbilt University, 1973

TURK DC, MAY JR: Haemophilus Influenzae: Its Clinical Importance. London, English Universities Press, 1967

SPECIFIC ARTICLES

BROOKS GF, BUCHANAN TM: Pertussis in the United States. J Infect Dis 122:123, 1970

CRISEL RM, BAKER RS, DORMAN DE: Capsular polymer of *Haemophilus influenzae,* Type b. I. Structural characterization of the capsular polymer of strain Eagan. J Biol Chem 250:4926, 1975

FRASER DW, DARBY CP, KOEHLER RE, JACOBS CF, FELDMAN RA: Risk factors in bacterial meningitis: Charleston County, South Carolina. J Infect Dis 127:271, 1973

LINNEMAN CC JR, RAMUNDO N, PERLSTEIN PH, MINTON SD, ENGLENDER GS, MCCORMICK JB, HAYES PS: Use of pertussis vaccine in an epidemic involving hospital staff. Lancet 2:540, 1975

MORSE SI, MORSE JH: Isolation and properties of the leucocytosis- and lymphocytosis-promoting factor of *Bordetella pertussis.* J Exp Med 143:1483, 1976

MYEROWITZ RL, NORDEN CW: Immunology of the infant rat experimental model of *Haemophilus influenzae* Type b meningitis. Infect Immun 16:218, 1977

SCHNEERSON R, ROBBINS JB: Induction of serum *Haemophilus influenzae* Type b capsular antibodies in adult volunteers fed cross-reacting *Escherichia coli* 075:K100:H5. N Engl J Med 292:1093, 1975

SELL SHW, MERRILL RE, DOYNE ED, ZEMSKY EP JR: Long-term sequelae of *Haemophilus influenzae* meningitis. Pediatrics 49:206, 1972

SHAW S, SMITH AL, ANDERSON P, SMITH DH: The paradox of *Haemophilus influenzae* type b bacteremia in the presence of serum bactericidal activity. J Clin Invest 58:1019, 1976

SMITH AL: Antibiotics and invasive *Haemophilus influenzae.* N Engl J Med 294:1329, 1977

SMITH DH, PETER G, INGRAM DL, HARDING AL, ANDERSON P: Responses of children immunized with the capsular polysaccharide of *Haemophilus influenzae,* Type b. Pediatrics 52:637, 1973

WHITE DC, GRANICK S: Hemin biosynthesis in *Haemophilus.* J Bacteriol 85:542, 1963

chapter

AEROBIC SPORE-FORMING BACILLI

MORTON N. SWARTZ

The genus *Bacillus* is composed of large gram-positive rods that form spores and grow best under aerobic conditions. Most species are saprobic and are found on vegetation and in soil, water, and air. The only species that is highly pathogenic for man is *B. anthracis,* which causes **anthrax,** a disease primarily of domestic livestock.

Human anthrax is rare in the United States: It has occasionally been contracted from infected livestock by farmers, veterinarians, and slaughterhouse workers (**agricultural** anthrax), but it now occurs almost exclusively in workers at plants processing imported goat hair, wool, or hides (**industrial** anthrax).

BACILLUS ANTHRACIS

The anthrax bacillus is unusually large and was the first bacterium shown to cause a disease. As early as 1850 it was seen in the blood of sheep dying of anthrax, and in 1877 Robert Koch grew it in pure culture, demonstrated its ability to form spores, and produced experimental anthrax by injecting it into animals.

In 1881, at the celebrated field trial at Pouilly-le-Fort, Pasteur vaccinated 24 sheep, 1 goat, and 6 cows with two injections of a culture of the bacillus attenuated by growth at 42°–43°C. Two weeks later all the vaccinated animals were injected with a highly virulent culture, as were a similar number of unvaccinated controls. Two days later all the vaccinated animals were well, whereas the unvaccinated sheep and goat died, and the unvaccinated cows were obviously ill. This dramatic demonstration provided a potent stimulus to the development of immunology.

MORPHOLOGY

B. anthracis is a large, gram-positive, nonmotile, spore-forming rod, $1\mu m$–$1.5\mu m$ in width and $4\mu m$–$10\mu m$ in length. In smears from infected tissues it appears singly or in short chains, and its capsule is readily demonstrable by Giemsa's stain. It does not form spores in the living animal. Spores are formed under conditions unfavorable for continued multiplication of the vegetative form.

The organism grows well on blood agar, where it rarely causes hemolysis (in contrast to its saprobic relatives). Surface colonies of virulent strains are large, gray-white, and rough, with comma-shaped outgrowths; when viewed under a hand lens or colony scope they usually exhibit the "medusa head" or "curled hair-lock" appearance illustrated in Figure 35-1. If cultivated in the presence of high CO_2 the organisms form capsules and the colonies are smooth and mucoid. Spores begin to appear at the end of the logarithmic phase of growth and are numerous after 48 h of incubation. The oval spores are clearly visible in the centers of the bacilli in specially stained smears (Fig. 35-2).

METABOLISM

The anthrax bacillus is readily cultivated on ordinary nutrient media, and although it grows best aerobically, it will also multiply under anaerobic conditions. Aerobic conditions are required for sporulation, but not for germination.

The nutritional requirements include thiamine and certain amino acids. Uracil, adenine, guanine, and manganese stimulate the growth of many strains. Glucose, sucrose, maltose, and trehalose are fermented in most cultures. Weak proteolytic and lecithinase activities are produced.

Anthrax spores are relatively resistant to heat and to chemical disinfectants. They are usually destroyed by boiling for 10 min and by dry heat at 140°C for 3 h, but they may survive for 70 h in 0.1% mercuric chloride. They persist for years in dry earth and may remain viable for months in animal hides.

ANTIGENIC STRUCTURE

Two major groups of Ags are known to be associated with *B. anthracis:* 1) **cellular (somatic) Ags** and 2) **components of the complex exotoxin** elaborated by the organism both in vivo and in vitro.

Cellular Antigens. In addition to a cell wall polysaccharide the organism forms a single antigenic type of capsular α-polypeptide of D-glutamic acid. Capsule formation by virulent strains occurs in infected tissues; in culture, capsule production is enhanced by incubation in bicarbonate- and serum-containing medium under increased CO_2, and the resulting colonies are mucoid. (In

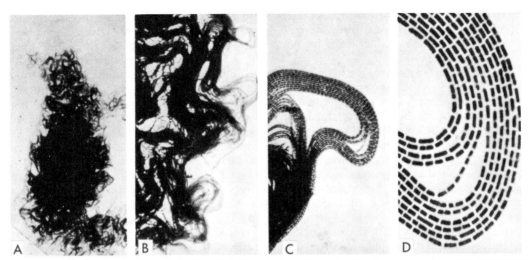

FIG. 35-1. Basis of "medusa head" appearance of surface colonies of virulent strains of *B anthracis.* Organism grown on nutrient agar and stained with methylene blue. **A.** Whole colony. (×45, reduced) **B–D.** Border of colony. (×145, 400, and 1600, all reduced) (Stein CD: Ann NY Acad Sci 48:507, 1947)

the absence of CO_2 enrichment the same organisms produce rough-appearing colonies.) Smooth (S) and rough (R) variants of *B. anthracis* occur and the S → R transition is associated with the selection of mutants that have lost the ability to synthesize capsular polypeptide. Encapsulated (S) strains may be selected for in cultures of unencapsulated (R) strains by addition to the medium of

FIG. 35-2. Anthrax bacilli, from 48-h plate culture, stained with crystal violet. Spores are unstained and clearly visible. (1200, reduced) (Burrows W: Textbook of Microbiology. Philadelphia, Saunders, 1963)

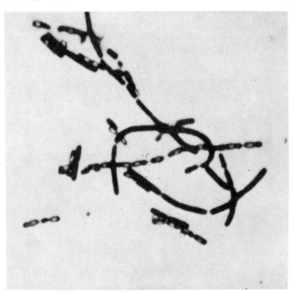

W_α phage, which attacks only the unencapsulated cells. The virulence-enhancing effect of adding egg yolk to a relatively hypovirulent inoculum of *B. anthracis* is associated with the development of capsules in the inoculated strain.

The capsule is antiphagocytic and appears to play an important role in pathogenicity, since nonencapsulated (even when grown under increased CO_2) mutants are avirulent. However, this role appears to be limited to the establishment of the infection and is not apparent in the terminal phase of the disease, which is much more closely linked to in vivo toxin production and toxemia.

The somatic polysaccharide Ag contains equimolar quantities of *N*-acetyl glucosamine and D-galactose, and cross reacts with type 14 pneumococcus polysaccharide and with human blood group A substance. In the cell wall it seems to be attached to a peptide containing diaminopimelic acid. The polysaccharide appears to play no role in virulence.

Exotoxin Components. Smith and Keppie first demonstrated an exotoxin in *B. anthracis* in 1954: in guinea pigs dying of experimental anthrax a toxic material was found in all infected tissues and exudates but it was most concentrated in edema fluid and plasma. This crude toxin produces extensive edema when injected subcutaneously in guinea pigs or rabbits and is lethal when injected intravenously in mice; it is also immunogenic. The same toxin is produced in vitro, but it is present in the culture medium only for a short time, when the cell density is about 1×10^8 chains/ml.* Bicarbonate ion is required

* The discovery of the anthrax toxin illustrates how the existence of a toxin may escape detection because of unfavorable conditions of artificial cultivation.

early in the growth cycle for production of toxin and possibly later for its release from the cell. Almost all strains of *B. anthracis* produce toxin, including avirulent (unencapsulated) ones.

Purification has separated the toxin into three distinct antigenically active components. All are thermolabile and appear to be proteins or lipoproteins. One component (necessary for the edema-producing activity of the toxin) is designated the **edema factor (EF),** or factor I. A second component, known as the **protective antigen (PA),** or factor II, induces protective Abs in guinea pigs. It can be assayed by its immunogenic activity, by complement fixation with appropriate antisera, or by immunodiffusion in agar. The third component (essential for the lethal effect) is known as the **lethal factor (LF)** or factor III. Upon extensive purification neither EF nor LF retains biologic activity individually. However, EF in combination with PA produces edema in guinea pigs; LF in combination with PA is lethal in rats.

GENETIC VARIATION

Mutants derived from wild-type strains of *B. anthracis* exhibit variations in virulence, in nutritional requirements, and in sensitivity to antimicrobial drugs, bacteriophages, and lysozyme. When repeatedly subcultured on laboratory media at elevated temperatures (42.5° C) wild-type strains gradually become avirulent. Indeed, it was by this method that Pasteur prepared his famous attenuated vaccine.

For some time the relation of **capsule formation** to virulence was confusing, for in growth under ordinary laboratory conditions some attenuated strains form capsules and many virulent strains do not, suggesting a negative correlation between virulence and encapsulation. Further studies, however, revealed that virulent strains which are unencapsulated when cultivated in air do indeed generate capsular envelopes both in vivo and when grown on appropriate media with added CO_2. In contrast, avirulent mutants, though encapsulated when cultured in the absence of CO_2, are incapable of forming capsules in vivo. Inasmuch as virulence is apparently due to **both capsule formation and toxin production** (see below), only those strains that produce both capsules and toxin in the infected host are highly pathogenic.

Sporulation and virulence are unrelated, for many nonsporulating mutants are still virulent. Nonproteolytic mutants likewise retain virulence.

PATHOGENICITY

Anthrax is primarily a disease of domesticated and wild animals. Humans become infected only incidentally, when brought into contact with diseased animals, their hides or hair, or their excreta. Many species of mammals

and birds acquire the natural disease. Anthrax epizootics still occur in wildlife sanctuaries in Africa. Among laboratory animals mice, guinea pigs, rabbits, goats, sheep, and monkeys are all highly susceptible. The LD_{50} for mice, for example, is about five spores. Rats, cats, dogs, and swine, on the other hand, will usually survive subcutaneous injections of at least a million spores. Susceptibility also varies widely with the site of inoculation.

The most common form of anthrax in humans is the cutaneous variety known as **malignant pustule.** The primary lesion usually develops at the site of a minor scratch or abrasion on an exposed area of the face, neck, or upper extremities into which anthrax spores have been accidentally inoculated. The spores germinate, and after an incubation period of 2–5 days an inflamed papule develops, later becoming a vesicle. Eventually the vesicle breaks down and is replaced by a black eschar (Fig. 35-3). A striking "gelatinous" nonpitting edema surrounds the eschar for a considerable distance. At no stage is the lesion particularly painful. In severe cases of cutaneous anthrax the regional lymph nodes become enlarged and tender, and the blood stream is eventually invaded. The systemic form of the disease is frequently fatal.

Another form is inhalation anthrax (**woolsorters' disease),** which results most commonly from exposure to spore-bearing dust in industrial plants where animal hair or hides are being handled. It not uncommonly leads to hemorrhagic mediastinitis and to hemorrhagic meningitis. The disease begins abruptly with high fever, dyspnea, and chest pain; it progresses rapidly and is often fatal before treatment can halt the invasive aspect of the infection.

Studies on the pathogenesis of pulmonary anthrax in

FIG. 35-3. Cutaneous anthrax, 3- to 4-day-old lesion. Note black eschar in center surrounded by rim of edema. (Gold H: Arch Intern Med 96:387, 1955. Copyright 1955, American Medical Association)

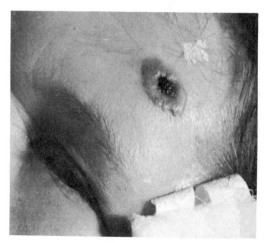

susceptible laboratory animals have revealed that spores inhaled in aerosols are phagocytized by alveolar macrophages, which in turn are carried via the pulmonary lymphatics to the regional tracheobronchial lymph nodes. There the spores germinate, multiply rapidly, and cause an active bacterial infection of the nodes. Although many of the vegetative bacilli are destroyed by the cellular defenses of the lymph nodes, some escape and are carried by efferent lymphatics to the blood stream. Subsequently, they are rapidly cleared by the reticuloendothelial system (particularly the spleen), but soon overgrow this defense system and establish a massive, fatal bacteremia.

To produce a fatal pulmonary infection, however, it is necessary to introduce a relatively large number of spores into the respiratory tract. The LD_{50} for the guinea pig, for example, is approximately 20,000 spores. Moreover, since only particles (droplets) less than $5\mu m$ in diameter are likely to penetrate to the alveoli, the average particle size in the aerosol is critical: the LD_{50} varies directly with their median size.

The pathologic changes are similar in human cases of inhalation anthrax and in the aerosol-induced disease in laboratory primates. Tracheobronchial and mediastinal lymph nodes are markedly enlarged and hemorrhagic. The striking finding in all cases is an extensive acute hemorrhagic mediastinitis characterized by marked gelatinous edema. Focal hemorrhagic pulmonary edema and pleural effusion may be present but there is no pneumonia except in a rare patient, usually with preexisting pulmonary pathology. Thus, the term "anthrax pneumonia" is less appropriate than "inhalation anthrax."

The intestinal tract, commonly the portal of entry in cattle, is rarely so in man. A few human cases of intestinal anthrax have resulted from the ingestion of poorly cooked meat from infected animals. Abdominal pain, fever, vomiting, bloody diarrhea, and shock are the principal manifestations. The mortality is extremely high and autopsy reveals hemorrhagic inflammation of the small intestine with lymphadenopathy.

Until recent years there was great confusion concerning the pathogenesis of anthrax infections, due primarily to the failure to recognize the toxigenic properties of *B. anthracis*. Discovery of the lethal toxin resulted from a systematic study of the mechanism of death in experimental anthrax. Because of the tremendous number of organisms demonstrable in the blood of animals dying of the infection, it was long assumed that death was due to blockage of the capillaries ("log-jam" theory). However, this concept was rendered untenable by the observation that, although streptomycin administered a few hours before death promptly controlled the bacteremia, death inevitably occurred if the number of bacteria in the blood had exceeded 3×10^6/ml. Inasmuch as untreated animals usually had 300 times this number of organisms in their blood at the time of death it was evident that factors other than the physical presence of the organisms in the circulation were responsible for the fatal outcome. Sterile heparinized plasma obtained from streptomycin-treated guinea pigs dying of the infection contains a toxin that causes extensive edema when injected into the skin of normal guinea pigs and is lethal when injected intravenously or intraperitoneally. There is now little doubt that the exotoxin plays a major role in the pathogenesis of the disease.

Anthrax toxin also may play a role in early stages of infection through a direct harmful effect on phagocytes. When virulent anthrax bacilli are injected subcutaneously into the susceptible host the encapsulated organisms proliferate freely and appear to resist phagocytosis by polymorphonuclear leukocytes, which accumulate in the lesion. Moreover, animal species with leukocytes more sensitive to toxin tend to be more susceptible to anthrax than species with highly resistant leukocytes. Finally, protective antitoxic immunity is induced by injections of the protective Ag of the lethal toxin.

The level of lethal toxin in the circulation increases rapidly quite late in the disease, and it closely parallels the concentration of organisms in the blood. The primary site of action of the toxin is still unknown. Death from anthrax in humans or experimental animals frequently occurs suddenly and unexpectedly. Cardiac failure, increased vascular permeability, shock, hypoxia, and respiratory failure have all been implicated as the cause. Respiratory failure is regularly seen and may be of cardiopulmonary origin (pulmonary capillary thrombosis) or due to central nervous system depression.

Considerable variation in innate susceptibility to anthrax exists among animal species. Resistant animals appear to fall into two groups: 1) those resistant to establishment of anthrax but, once infection is established, sensitive to the toxin; and 2) those susceptible to establishment of the disease but resistant to the toxin (Table 35-1).

IMMUNITY

Animals surviving naturally acquired anthrax are resistant to reinfection; second attacks in humans are likewise extremely rare. Nevertheless, vaccines composed of killed bacilli produce no significant immunity, and anticapsular Ab is protective only in certain species, e.g., mice. Hence the attenuated bacilli of Pasteur's original vaccine were probably capable of producing enough toxin to induce an antitoxic immunity. The Pasteur strain, however, was later found to be sufficiently virulent to cause serious disease in vaccinated animals. Although it was later replaced by a safer nonencapsulated toxigenic strain for use in livestock, in the United States no attenuated spore vaccine has ever been considered to be satisfactory for human use. The best vaccine for humans now appears to be an alum-precipitated preparation of the protective antigen of the lethal toxin recovered from

TABLE 35-1. Relation Between Dose to Establish Anthrax Infection, Number of Organisms per Milliliter of Blood at Death, and Susceptibility to Toxin Challenge

Species	Relative resistance to parenteral challenge of spores	Parenteral spore dose to establish anthrax	IV toxin dose causing death (units/kg)	Quantitation of blood at death	
				Bacilli/ml	Toxin units/ml
Mouse	Very susceptible	5	1000	$10^{6.9}$	
Guinea pig	Susceptible	50	1125	$10^{8.3}$	50
Rhesus monkey	Susceptible	3000	2500	$10^{6.8}$	35
Chimpanzee	Susceptible		4000	$10^{8.9}$	110
Rat	Resistant	1×10^6	15	10^4–10^6	15
Dog	Very resistant	50×10^6	60		

(Modified from Lincoln RE et al: Fed Proc 26:1558, 1967)

culture filtrates. Frequent boosters are necessary to maintain resistance to anthrax challenge.

Acquired immunity to anthrax seems, then, to be due to Abs to both the thermolabile toxin and the capsular polypeptide. The relative importance of these two kinds of Abs appears to vary widely in different species of hosts.

LABORATORY DIAGNOSIS

Anthrax bacilli are readily cultured from skin lesions in the vesicular stage. The organisms may even be seen in stained smears of the exudate and may sometimes be tentatively identified by their characteristic morphology. As the lesion ages demonstration of bacilli becomes increasingly difficult. A specimen should always be obtained for culture prior to the administration of antimicrobial therapy. The organism is encapsulated when grown under CO_2 on a rich bicarbonate-containing medium, and the cells form long chains giving a bamboo-like appearance with Gram's stain. Unlike the numerous other saprobic members of the genus *Bacillus, B. anthracis* is nonhemolytic on sheep blood agar. It is also nonmotile, unlike *B. subtilis* and many strains of *B. cereus.*

In pulmonary anthrax sputum cultures are rarely positive, because the inhaled spores usually do not germinate and multiply until they have reached the mediastinal lymph nodes. Whenever anthrax is suspected the blood should be cultured. Since other aerobic spore-forming bacilli, such as *B. subtilis,* may sometimes be cultured from the skin, respiratory tract, and even the blood, organisms thought to be anthrax bacilli should be tested for pathogenicity by intraperitoneal inoculation of a guinea pig or mouse with a washed culture. The animals succumb in 48–96 h with respiratory failure, and enormous numbers of bacilli can be found in the blood and in smears of the cut surface of the spleen.

Susceptibility to a specific gamma bacteriophage is helpful confirmatory evidence that an isolate is a strain of

B. anthracis. The fluorescent Ab technic may be helpful in evaluation of smears obtained from lesions of patients after initiation of antibiotic therapy, or in examination of isolates grown in the appropriate medium for capsule production.

TREATMENT

Most strains of *B. anthracis* are sensitive to penicillin, tetracycline, erythromycin, and chloramphenicol. Penicillin G is the drug of choice and tetracycline is an alternative for the patient who is allergic to penicillin. These drugs are usually effective in cutaneous anthrax, although the toxin in the primary lesion often causes the formation of an eschar despite early treatment. In respiratory anthrax chemotherapy is usually ineffective, because the disease is rarely recognized before bacteremia has developed; hence the mortality rate is extremely high. Although antiserum, presumably containing antitoxin, was formerly used in the treatment of human anthrax, it has been superseded by chemotherapy.

PREVENTION

Anthrax still causes heavy loss of livestock, particularly in the Middle East, Africa, and Asia. In the United States the disease is endemic in cattle in Louisiana, Texas, South Dakota, Nebraska, and California. It is transmitted primarily through the ingestion of contaminated forage or from the carcasses of infected animals. Spores may remain viable in the soil for many years. Epidemiologic evidence indicates that in suitable soils anthrax bacilli can survive ecologic competition with other organisms and can maintain an organism → spore → organism cycle for years without infecting livestock. Whether the organism also grows saprobically is not known.

Animal anthrax can be at least partially controlled by immunization with attenuated vaccines (see Immunity, p. 707). The disease has been virtually eliminated in

South Africa by this means. The carcasses of animals dying of anthrax should be disposed of either by cremation or by deep burial.

No more than five cases of human anthrax have been reported annually over the past decade in the United States, but the disease is much more prevalent in some other countries.

In the United States, industrial anthrax was formerly most common among textile workers employed in plants that processed goat hair imported from the Middle East.

Its control poses many practical problems. Economically feasible methods of disinfecting hides without damaging them are extremely difficult to devise. Industrial workers at high risk should probably be immunized with the protective Ag, as should veterinarians practicing in "anthrax districts" and laboratory workers who have contact with *B. anthracis.* It is important to continue to enforce hygienic measures at plants (clothing changes, etc.), to keep the incidence of disease at its current low level.

OTHER AEROBIC SPORE-FORMING BACILLI

Many saprobic species of aerobic spore-forming bacilli are hard to distinguish from *B. anthracis,* except on the basis of pathogenicity. The most commonly encountered are *B. cereus, B. subtilis,* and *B. licheniformis.* The DNA base composition indicates that *B. cereus* is most closely related to *B. anthracis. Bacillus* species other than *B. anthracis* have occasionally been responsible for infections in man: pulmonary and disseminated infections in immunologically compromised hosts; localized infections in a closed space (e.g., ophthalmitis, meningitis); wound infections following trauma, surgery, or the introduction of foreign prosthetic material. *B. cereus* is involved as such an opportunistic invader more often than other *Bacillus* species. Several properties serve to differentiate it from *B. anthracis* (Table 35-2).

FOOD POISONING

B. cereus may cause outbreaks of food poisoning characterized by an incubation period of less than 12 hours and a short duration. The vehicles have usually been fried rice or dairy products. Pathogenesis appears to be due to the production of an enterotoxin, rather than to invasion, since culture filtrates induce fluid accumulation in ligated loops of rabbit ileum. Enterotoxin is secreted by actively growing cells, without cell lysis.

Unlike *B. anthracis, B. cereus* is resistant to penicillin and the cephalosporins by virtue of production of extracellular penicillinase and cephalosporinase.

TABLE 35-2. Properties Differentiating *Bacillus anthracis* from *B. cereus*

	B. anthracis	B. cereus
Colonies on blood agar (Comma-shaped outgrowths)	Many	Few or none
Colonies on bicarbonate medium (CO$_2$)	Raised, mucoid	Flat, dull
Hemolysis (sheep RBCs)	None or weak	Usually β-hemolytic
Motility	–	+
Gelatin liquefaction	– or slow	+
Lecithinase reaction	Weakly +	Strongly +
Fluorescent Ab vs. *B. anthracis*	+	–
Lysis by gamma phage	+	–
Animal pathogenicity (mouse, guinea pig)	+	–

SELECTED READING

BOOKS AND REVIEW ARTICLES

LINCOLN RE, FISH DC: Anthrax toxin. In Montie TC, Kadis S, Ajl SJ (eds): Microbial Toxins, Vol III. New York, Academic Press, 1970

NUNGESTER WJ: Proceedings of the conference on progress in the understanding of anthrax. Fed Proc 26:1491, 1967

SMITH H: The use of bacteria grown in vivo for studies on the basis of their pathogenicity. Annu Rev Microbiol 12:77, 1958

SPECIFIC ARTICLES

ALBRINK WS: Pathogenesis of inhalation anthrax. Bacteriol Rev 25:268, 1961

Bacilli: biochemical genetics, physiology, and industrial applications. In Microbiology—1976. Washington DC, American Society for Microbiology, 1976, pp 5–444

BERDJIS CC, GLEISER CA, HARTMAN HA: Experimental parenteral anthrax in Macaca mulatta. Br J Exp Pathol 44:101, 1963

BRACHMAN PS: Anthrax. Ann NY Acad Sci 174:577, 1970

DALLDORF FG, KAUFMANN AF, BRACHMAN PS: Woolsorters' disease. Arch Pathol 92:418, 1971

IHDE DC, ARMSTRONG D: Clinical spectrum of infection due to *Bacillus species*. Am J Med 55:839, 1973

MEYNELL E, MEYNELL GG: The roles of serum and carbon dioxide in capsule formation of *Bacillus anthracis*. J Gen Microbiol 34:153, 1964

SMITH H, STANLEY JL: The three factors of anthrax toxin: their immunogenicity and lack of enzymic activity. J Gen Microbiol 31:329, 1963

SPIRA WM, GOEPFERT JM: Biological characteristics of an enterotoxin produced by Bacillus cereus. Can J Microbiol 21:1236, 1975

THORNE CB: Capsule formation and glutamyl polypeptide synthesis by *Bacillus anthracis* and *Bacillus subtilis*. Symp Soc Gen Microbiol 6:68, 1956

chapter

36

ANAEROBIC SPORE-FORMING BACILLI: THE CLOSTRIDIA

MORTON N. SWARTZ

GENERAL PROPERTIES

The anaerobic spore-forming bacilli belong to the genus *Clostridium.* (L. *clostridium,* spindle). Their natural habitat is the soil and the intestinal tracts of animals and man. A few of these saprobes are also human pathogens under appropriate circumstances, causing, three very different diseases: **botulism, tetanus,** and **gas gangrene.** The pathogenicity of these organisms depends on the release of powerful exotoxins or highly destructive enzymes.

MORPHOLOGY AND GROWTH

The clostridia are relatively large pleomorphic, gram-positive, rod-shaped organisms. Filamentous forms are common. In 48-h cultures many of the bacilli may be gram-negative. All species form spores, but with considerable variation in the required conditions. The highly refractile spores (Fig. 36-1) are oval or spherical and usually wider than the parental cell; they may be central, terminal, or subterminal in the cell. Most species of clostridia possess peritrichous flagella and are motile (Fig. 36-2). A few (e.g. *C. perfringens*) are encapsulated (Fig. 36-3). Colony forms are variable; hemolysis on blood agar is frequent.

Clostridia lack the cytochromes required for electron transport to O_2. They contain flavoprotein enzymes that reduce O_2 to H_2O_2 and to superoxide (O_2^-; Ch. 3), and they lack the catalase, peroxidases, and superoxide dismutase that would destroy these toxic products; hence their obligate anaerobiosis. Not all clostridia are equally oxygen-sensitive: *C. tetani* requires strict anaerobic conditions; *C. perfringens* is much less fastidious; and *C. histolyticum* produces small but visible colonies on aerobic blood agar plates. Clostridial spores not only are re-sistant to heat and disinfectants but survive long periods of exposure to air; they germinate only under strongly reducing conditions. Specimens for anaerobic culture should be collected carefully, transported in "gassed-out" (evacuated) tubes, inoculated promptly on prere-duced media, and incubated anaerobically by one of the methods previously described (Chap. 31).

METABOLISM AND CLASSIFICATION

Most clostridia produce large amounts of gas (mainly CO_2 and H_2) in butyric fermentation (Ch. 3). The fermentation of various sugars is of value in differentiating species. Other biochemical tests include reactions in milk, the liquefaction of gelatin, and the production of indole from tryptophan (Table 36-1). The characteristic "stormy fermentation" of milk by *C. perfringens* is due to the formation of a clot which becomes torn by the accumulating gas. Some clostridia are predominantly proteolytic and others saccharolytic.

A wide variety of **enzymes** have been identified in the filtrates of clostridial cultures, including collagenase, other proteinases, hyaluronidase, deoxyribonuclease, lecithinase, and neuraminidase. Some of these are known to act as toxins in the animal host; other potent protein exotoxins are **botulinum toxins** and **tetanus toxin.** Many species produce hemolysins. Antibodies to some of these products may be of use in identifying individual clostridial species. Nontoxigenic mutants of various species have been described.

The somatic and flagellar Ags of clostridia have not provided a practical system for identification of species or for division of species into serotypes.

BOTULISM

Botulism is usually **not an infectious disease** in adults but an **intoxication:** it results from the ingestion of food contaminated by preformed botulinum toxin, rather than from multiplication (and toxin production) of *C. botulinum* in the gastrointestinal tract. In this respect it resembles staphylococcal food poisoning. Uncooked meat or sausage (L. *botulus*) was formerly one of the commonest sources of the disease; currently in the United States the principal vehicles are improperly canned fruits and vegetables (usually home-processed), condiments, and fish products. From 1970 to 1975 there were 68 outbreaks (152 cases) of food-borne botulism in the United States. Rarely, wound contamination with **C. botulinum** has been associated with noninvasive infection and in

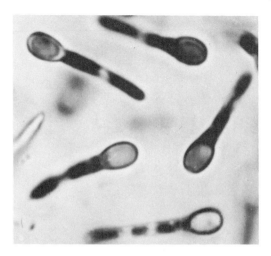

FIG. 36-1. Terminal clostridial endospores in stained wet-mount preparation. (Courtesy of C. F. Robinow) (3600)

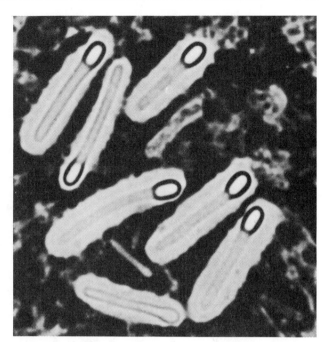

FIG. 36-3. Nigrosin preparation of *C. pectinovorans* showing both capsules and endospores. (Courtesy of C. F. Robinow) (3600)

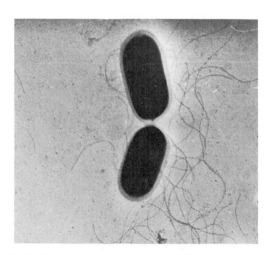

FIG. 36-2. Electron photomicrograph of *C. tetani* showing cell wall and peritrichous flagella. (Courtesy of Stuart Mudd) (11,000, reduced)

vivo toxin formation, producing clinical **wound botulism.** Recent evidence suggests that, infrequently, ingested *C. botulinum* can colonize the gastrointestinal tracts **of infants** and elaborate toxin there.

Botulinum toxin acts on the nervous system, producing a life-threatening illness with cranial nerve impairment and symmetric weakness or paralysis.

THE ORGANISM AND ITS TOXIN

C. botulinum is widely distributed in soil, in the silt of lake and pond bottoms, and on vegetation; hence the intestinal contents of mammals, birds, and fish may occa-

sionally contain these organisms. Seven types (A–G) have been recognized; each elaborates an **immunologically distinct form of neurotoxin.** In the United States only types A (62% of cases), B (28%), and E (10%), have been significant causes of human botulism; and the geographic distribution of cases is in keeping with the isolation of various toxin types from regional soil samples: type A, from the West; type B, from the Northeast and Central states; type E, from the Pacific Northwest, Alaska, and the Great Lakes area (particularly in marine sediments). Type F is rare; C and D are usually associated with botulism in fowl and mammals; G has been isolated only from soil thus far.

These are the most powerful biologic toxins known; $1\mu g$ contains 200,000 minimal lethal doses (MLD) for a mouse, and is nearly a lethal dose for man.

Toxin production is associated both with germination of spores and with growth of vegetative cells, but little toxin appears in the medium until late in growth, when it is *released by cell lysis.* With some types a slightly active "progenitor toxin" is produced initially and is *activated subsequently by a protease* appearing in the culture filtrate; hence the toxicity of early culture filtrates of types A, B, and E organisms can be enhanced by treatment with trypsin.

Toxin production in some strains is dependent on the continued presence of specific **prophages:** type C strains produce primarily C_1

TABLE 36-1. Properties of Common Species of Pathogenic Clostridia

Disease and causative species	Aerobic growth	Spores	Motility	Lecithinase	Toxicity (mice)	Milk	Gelatinase	Lactose	Glucose	Maltose	Sucrose	Indole	Toxin neutralization for specific identification	Organic acids detected by GLC*
Botulism														
C. botulinum	−	ST	+	−	+	V	V	−	A	V	V	−	+	
Tetanus														
C. tetani	−	T	+	−	+	NC	+	−	−	−	−	V	+	A,P,B
Gas gangrene														
C. perfringens	−	ST†	−	+	V	CG	+	A	A	A	A	−		A,B
C. novyi	−	ST	+	+	V	V	+	−	A	V	−	−	+	A,P,B
C. septicum	−	ST	+	−	+	CG	+	A	A	A	−	−	+	A,B
C. histolyticum	+	ST	+	−	V	CD	+	−	−	−	−	−		A,L
C. tertium	+	T	+	−	−	V	−	A	A	A	A	−		A,B,L
C. bifermentans	−	ST	+	+	−	D	+	−	A	A	−	+		A,P,IB,IV,IC
C. sporogenes	−	ST	+	−	−	CD	+	−	A	V	V	−		A,IV
Pseudomembranous enterocolitis														
C. difficile	−	ST	+	−	V	NC	V	−	A	−	−	−	+	A,P,IB,B,IV,V,IC

A = acid; C = coagulated; D = digested; G = gas; NC = no coagulation; ST = subterminal; T = terminal; V = variable.

*Gas liquid chromatography: A = acetic acid; B = butyric acid; IB = isobutyric acid; IV = isovaleric acid; IC = isocaproic acid; L = lactic acid; P = propionic acid; V = valeric acid.

†Spores are only rarely seen in direct smears from wounds or cultures but can be demonstrated upon growth in special media.

toxin but also small amounts of a second type, C_2; type D strains produce mainly type D toxin but also small amounts of C_2.

Cultures of types C and D cured of their prophages stop producing their dominant toxins (but still produce low levels of C_2) and become indistinguishable. Such cured strains subsequently can be converted to either type C or D toxin production by infection with the appropriate phage. The relation between the host bacteria and the phage is unstable (pseudolysogeny) and may account for the occasional isolation of nontoxigenic C. botulinum in areas where types C and D are prevalent.

Bacteriophages may be responsible even for interspecies conversion: infection of a cured strain of C. botulinum type C by phage NA_1 (from C. novyi type A) can induce production of the α-toxin of C. novyi. The resultant organism would then be identified as C. novyi type A. Evidently a prototype clostridial organism may acquire the characteristics of various species, depending on the presence of one or another resident phage.

Crystalline type A toxin is a protein with a mol wt of 900,000. It is made up of three constituents: a neurotoxin (mol wt 150,000), a hemagglutinin, and a nontoxic component lacking hemagglutinin activity. The hemagglutinin is not present in the toxin first elaborated in culture filtrates but appears to become bound to it later, and its production can be transmitted by bacteriophages either with toxigenicity or separately.

The **spores** of C. botulinum are relatively heat-resistant, and pressure sterilization is necessary to ensure their destruction. Effective sterilization is routine in the canning industry; home-canned foods, especially "low-acid" vegetables (pH above 4.5), have been the source of most

outbreaks of botulism in this country. In the past tomatoes have been sufficiently acid to inhibit production of C. botulinum toxin, but "low-acid" tomatoes are now available and are a potential source of botulism in home-canned products.

In contrast to the spores, botulinum toxin is relatively heat-labile, being completely inactivated in 10 min at 100°C; hence *boiling just prior to ingestion renders home-canned vegetables or processed fish safe.* Care should be exercised in handling suspect food samples in the laboratory since the toxin may be absorbed from fresh wounds and mucosal surfaces. Most outbreaks in the United States now involve types A or B in home-canned vegetables. During the last decade type E and occasionally A in commercial smoked fish became more frequent.

PATHOGENESIS

Botulism results from the ingestion of uncooked foods in which contaminating spores have germinated and elaborated toxin. *Food that is not visibly fermented or spoiled to taste may still contain botulinum toxin.*

The toxin is not inactivated by gastric acid or by the proteolytic enzymes of the stomach and upper bowel. In fact, crude toxin of C. botulinum type E can be potentiated 10- to 1000-fold by partial proteolysis (e.g., limited trypsin treatment). Since activation cannot be shown with toxin-containing serum from patients with botulism, the toxin may have already undergone proteolysis in the gastrointestinal tract.

After absorption from the gastrointestinal tract the toxin reaches susceptible neurons, at neuromuscular junctions and peripheral autonomic synapses, by way of the bloodstream. There it becomes bound to **presynaptic terminals,** where electrophysiologic and electron microscopic studies have shown that it *blocks the exocytosis of acetylcholine-containing vesicles.* The nature of the toxin receptor is suggested by the finding that certain gangliosides bind botulinum toxin as a function of their sialic acid content.

Although the various types of botulinum toxin appear to have the same pharmacologic action, their potency varies with the test species. The ratio of the lethal doses for mice and fowl, for example, are 1:15 for type A, 1:2000 for type C, 1:100,000 for type D, and 1:25 for type E. The differences may well be of epidemiologic significance.

Clinical Features. Symptoms in man usually begin 18–36 h after ingestion and are due to cholinergic blockade. Weakness, dizziness, and severe dryness of the mouth and pharynx are early manifestations; nausea and vomiting are common with type E disease. Fever is absent. Neurologic features soon develop: blurring of vision, dilatation of pupils, inability to swallow, difficulty in speech, urinary retention, generalized descending weakness of skeletal muscles, and respiratory paralysis. Sensory abnormalities are absent.

A curious neurophysiologic response, a facilitated muscle action potential after repetitive nerve stimulation, is typical of patients with botulism.

The fatality rate has been extremely high but has declined to 21% in this country during 1970–1975. Fortunately, the disease is much less common in man than in animals.

Wound botulism should be considered in an isolated case when characteristic neurologic abnormalities develop but no food source is implicated epidemiologically. However, *C. botulinum* spores do not germinate readily in tissue: approximately 2×10^7 spores are needed to produce symptoms of botulism in experimental animals. The incubation period is 4–14 days after injury. Only type A has been incriminated in wound botulism.

Infant botulism, due to **infection,** has been described only since 1976, and 60 cases have been identified in the United States. Constipation, weak sucking ability, cranial nerve deficits, generalized weakness, and sudden apneic episodes have been the principal features. The infants, 5–20 weeks of age, had all been exposed to some solid foods, presumably the source of infection. This "infection–intoxication" is restricted to infants, probably because of their less well established competing intestinal flora. *C. botulinum* organisms as well as toxin have been found in the stools for as long as 160 days after onset of symptoms, but not in the stools of healthy infants or adults. All reported cases have recovered, usually without administration of specific antiserum. The possible role of infant botulism in a few cases of the "sudden infant death syndrome" has

been suggested by the finding of *C. botulinum,* its toxin, or both in the bowel contents of several infants who died suddenly and unexpectedly.

LABORATORY DIAGNOSIS

In cases of suspected botulism mice should be injected intraperitoneally with the patient's serum (up to 1.0 ml) and with aqueous cell-free extracts of the implicated food. Heat-treated samples should also be inoculated to serve as controls for the presence of nonspecific toxic substances. Trypsin treatment of a portion of food samples may enhance toxin activity (particularly type E) and make detection easier. If significant amounts of toxin are present the mice will develop paralysis and will succumb within 1–5 days; neutralization of toxin by type-specific antiserum provides protection.

Samples of the suspected food should also be cultured anaerobically, with heat (or alcohol) treatment of part of the sample to select for spores. Since only vegetative cells or relatively heat-sensitive spores (type E) may be present, part of the sample should be incubated without heating. If an anaerobic gram-positive bacillus is recovered from the food it should be tested for toxin production. The cultural approach is generally less useful than direct tests for the toxin.

In occasional cases of typical botulism the toxin has been demonstrable in the feces but not in the serum; in infant botulism both *C. botulinum* and toxin have been found in the stool.

TREATMENT AND PREVENTION

The toxins that cause botulism in man are each specifically neutralized by its antitoxin. Botulinum toxin, as toxoid, is a good Ag; as little as 1 μg will induce high levels of protective Ab in the mouse. Clinical botulism, however, does not induce demonstrable Ab because an amount of toxin sufficient to induce an immune response would be lethal.

Once toxin has become "fixed" at susceptible nerve endings its harmful action is unaffected by antitoxin (cf. diphtheria; see Treatment Ch. 26). With type A botulism, antitoxin does not appear to alter the clinical course once neurologic symptoms have occurred, but beneficial results have followed the use of antitoxin in type E botulism. To neutralize any circulating or "unfixed" toxin, equine polyvalent *antiserum (types A, B, and E) should be given intravenously in suspected cases at once,* without awaiting the results of laboratory tests. The usual precautions should be taken against hypersensitivity reactions to the foreign serum.

Other individuals known to have ingested the same food as the patient should also be treated, even if asymptomatic.

As already mentioned, boiling of any improperly canned or processed food for at least 10 min will destroy botulinum toxin. A pentavalent toxoid evokes a good Ab response but its use in man is justified only in frequently exposed laboratory workers. Toxoids have been used successfully to immunize cattle in areas where the disease is endemic.

TETANUS

Tetanus is an acute, often fatal bacterial disease in which the clinical manifestations stem not from invasive infection but from a potent neurotoxin **(tetanospasmin)**, elaborated when spores of *C. tetani* germinate after gaining access to wounds. The disease develops in the setting of penetrating trauma, chronic skin ulcers, infections about the umbilical stump in the newborn **(neonatal tetanus)**, obstetrical procedures **(postabortal tetanus)**, and infected injection sites in narcotics addicts. The disease is of particular importance in military medicine. Because only **one antigenic type of toxin** is involved an effective **monotypic toxoid** could be developed. Because of its widespread use for prophylactic immunization fewer than 150 cases of tetanus occur annually in the United States. However, the disease remains a major problem worldwide (approximately 300,000 cases a year).

THE ORGANISM AND ITS TOXIN

C. tetani is a strict anaerobe without a capsule but with spherical terminal spores that give a characteristic "drumstick" appearance (Fig. 36-1). It is found in the soil and in feces of various animals. In humans fecal carrier rates are very variable (0%–25%), suggesting that the organism is a transient whose presence depends on its ingestion.

Tetanus toxin is only slightly less potent than type A botulinum toxin. It is also produced by growing cells and released only on cell lysis: hence the disease appears only when spores of *C. tetani* germinate after gaining access to wounds. Tetanus toxin ("intracellular" or "progenitor" toxin) is produced initially as a single polypeptide chain of mol wt 160,000. As with progenitor toxin of *C. botulinum,* the toxicity of autolysates can be **increased** by autologous or trypsin-induced **proteolysis.** Such "nicked" ("extracellular") toxin molecules can be dissociated by thiol reducing agents into two polypeptide chains: a heavy (β) chain (mol wt 105,000) and a light (α) chain (mol wt 55,000). Tetanus toxin is produced in vitro in amounts up to 5%–10% of the bacterial weight, but it serves no known useful function for the bacillus.

Animals vary widely in their susceptibility to tetanus toxin: mammals are most sensitive, and birds are relatively resistant (e.g., the pigeon is 24,000 times as resistant as the guinea pig).

In addition to its neurogenic toxin, *C. tetani* produces an oxygen-labile hemolysin, antigenically similar to streptolysin O (Ch. 28), known as **tetanolysin.** There is no evidence that it plays a significant role in pathogenesis.

Ten serologic types of *C. tetani* can be distinguished by their flagellar Ags. All, however, share a common O Ag and produce the same exotoxin. Immunity to tetanus involves Ab to toxin: Abs to somatic Ags are not protective. Tetanus toxoid is an excellent immunogen in man, but *clinical tetanus does not induce immunity* because so little toxin is released. Actively immunizing all patients on recovery from tetanus is therefore extremely important.

PATHOGENESIS

Most cases result from small puncture wounds or lacerations; the infection remains localized to the traumatized tissue at the site of entry, usually with only a minimal inflammatory response. Mixed infection with other organisms or the presence of a foreign body may induce more marked inflammation (and lower the oxidation-reduction potential locally), promoting the growth of *C. tetani.* However, in 5%–10% of cases the initial injury is so trivial as to have been forgotten by the patient and to have left no residue.

Access of tetanus spores to open wounds does not necessarily result in disease: *C. tetani* sometimes can be cultured from wounds of patients without tetanus; for in clean wounds, where the blood supply is good and the oxygen tension remains high, germination will rarely occur. In necrotic and infected wounds, on the other hand, the anaerobic conditions will permit germination, which may occur rapidly. In mice challenged intramuscularly with 10^3–10^5 spores (mixed with $CaCl_2$) the first signs of tetanus appear in 20 h, only 7 h later than in mice given 2 MLD of tetanus toxin. On the other hand, spores may occasionally remain dormant in healed human wounds for months (a latent period as long as 10 years has been recorded); trauma to the area may then cause germination and disease.

The importance of tissue necrosis is readily demonstrable in laboratory experiments. Mice inoculated intramuscularly with heavy suspensions of tetanus spores fail to develop the disease unless necrotizing chemicals, such as $CaCl_2$ or lactic acid, are injected with the spores.

Toxin Action. Toxin elaborated in a local wound migrates to its sites of action in the central nervous system.

Whether it is carried via retrograde axonal transport in peripheral nerves or via the bloodstream remains a matter of controversy. It is possible that the toxin can spread to the spinal cord by either path, but the circulatory route appears to be the predominant one. Studies of the movement of other labeled proteins, as well as cord transection experiments, strongly suggest that once toxin has reached the spinal cord it ascends within it.

When tetanus toxin reaches the central nervous system it becomes rapidly fixed. Indeed, when mixed with brain emulsion, or with subcellular fractions rich in synaptosome membranes, a comparatively large dose can be injected into a susceptible animal without causing damage **(Wassermann–Takaki phenomenon)**. The substance responsible for this fixation is a ganglioside containing stearic acid, sphingosine, glucose, galactose, *N*-acetyl galactosamine, and *N*-acetyl neuraminic acid (sialic acid). Fixation depends on the number and positions of the sialic acid residues. Conversion to toxoid destroys reactivity with the ganglioside. Tetanus toxin appears to bind to ganglioside-containing membranes without modifying the ganglioside or causing disruption of the bilayer structure, observations consistent with the absence of pathologic lesions in tetanus.

The ganglioside-binding site of the intact toxin molecule resides in its β chain: isolated β chains, but not α chains, bind to ganglioside. Analogy with diphtheria or with cholera toxin suggests that the α chain may have some toxic action within the cell, after the β chain secures its entry, but such an action has not been identified for tetanus toxin. (Unlike diphtheria and cholera toxins, tetanus toxin, or its α chain, has no known effect on specific enzymatic reactions.)

The spasmogenic effect of the toxin has been shown by Eccles to be due to its action on polysynaptic reflexes involving interneurons in the spinal cord. It *blocks the normal postsynaptic inhibition of spinal motor neurons* following afferent impulses, probably by *preventing the release of inhibitory transmitter* (glycine). The resulting sensitivity to excitatory impulses, unchecked by inhibitory mechanisms, produces the generalized muscular spasms characteristic of tetanus.

Clinical Patterns. The incubation period of tetanus may range from several days to many weeks; an incubation period of less than 4 days is associated with a very high mortality. Although the portal of entry is most commonly a puncture wound or laceration, other foci of infection include burns, skin ulcers, compound fractures, operative wounds, and sites of subcutaneous injection of adulterated narcotics by addicts.

Generalized tetanus, the usual form of the disease, is characterized by severe, painful spasms and rigidity of voluntary muscles. The usual sequence is local injury followed after some days by mild intermittent muscular contractions near the wound, then trismus ("lockjaw," spasm of the masseter muscles), generalized rigidity, and violent spasms of the trunk and limb muscles. Spasm of the pharyngeal muscles causes difficulty in swallowing. Death ordinarily results from interference with the mechanics of respiration. The patient's sensorium remains clear. Fever is not seen except with increased metabolism (in patients with severe spasms), complicating infection, or autonomic nervous system complications. Occasional patients with severe tetanus may show manifestations of sympathetic dysfunction (hypertension, tachycardia, fever, hypotension).

Local tetanus is a much rarer form of the disease, usually occurring in individuals with partial immunity or as the result of minor wounds containing only a few organisms. It is characterized by localized twitching and spasm in muscles near the wound. It may persist for weeks or months and then subside, or mild trismus may follow.

Generalized tetanus can readily be reproduced in experimental animals by intravenous injection of toxin, and local tetanus by intramuscular injection. Local tetanus is apparently due not to the action of the toxin directly on the muscle but to early involvement of anterior horn cells at the level of initial entrance of the toxin. However, when excessive amounts are elaborated in the tissues enough may reach the circulation to cause generalized involvement before the signs of local tetanus have developed.

LABORATORY DIAGNOSIS

The diagnosis of tetanus is usually based on clinical findings alone. Although attempts should be made to culture *C. tetani* from all suspicious lesions, antitoxin should not be withheld for the 2–3 days needed to identify the organism.

The organism is cultured from an infected focus in only about 30% of cases; moreover, its isolation does not necessarily establish a clinical diagnosis of tetanus. The tetanus bacillus may be provisionally recognized by its round terminal spores, the narrow zone of hemolysis on blood agar incubated anaerobically, and the granular gray surface colonies surrounded by an area of swarming (see Proteus–Providencia Group, Ch. 31).* Since, however, other gram-positive anaerobic spore-forming bacilli may be cultured from infected wounds,† positive identification requires the demonstration of production of toxin and its neutralization by specific antitoxin. A mouse protection test is commonly used for this purpose.

---

* Staining with fluorescent Ab to the somatic O Ag will probably prove the best method for rapid identification.

† About 30% of war wounds, for example, become contaminated with various *Clostridia*.

TREATMENT AND PREVENTION

Antitoxin should be promptly administered in all cases of suspected tetanus to neutralize accessible toxin; it is ineffective against toxin already fixed in the central nervous system. Tetanus immune globulin (TIG, prepared from the plasma of persons who have been hyperimmunized with tetanus toxoid) has recently become available in this country and has supplanted conventional equine antitoxin. The risk of sensitivity reactions with human material is minimal, and Abs persist longer (half-life of 3–4 weeks) than with equine antitoxin. A dose of 3000–6000 U* should be administered intramuscularly.

If human TIG is not available equine antitoxin should be administered in a dose of 100,000 U, half given intravenously and half intramuscularly. This amount is far in excess of that needed to neutralize all the performed toxin present in the body of an adult, as well as that which may be subsequently elaborated. It is essential to test for immediate-type hypersensitivity to horse serum before injecting antitoxin, and to take appropriate precautions if the result is positive (see Serum Sickness Syndrome, Ch. 21). Local injections of antitoxin, totaling 10,000–20,000 U, may also be made around the suspected primary lesion.

Careful surgical debridement of the lesion, and removal of any foreign bodies present, should be carried out only after the TIG or antitoxin has been given. Penicillin should also be administered to prevent the germination of spores and further bacterial multiplication (and toxin production) in infected tissue; but chemotherapy alone should never be relied upon.

Supportive measures to minimize spasm and aid respiration are of the utmost importance. These measures may be needed for one or more weeks since fixed toxin in the central nervous system decays slowly. About half of the patients with severe generalized tetanus die despite intensive treatment.

Because the spores of *C. tetani* are so widely disseminated, the only effective way to control tetanus is by **prophylactic immunization. Tetanus toxoid,** in combination with diphtheria toxoid and pertussis vaccine (DPT), is usually given during the first year of life. Three injections of either fluid or alum-precipitated toxoid should be administered in the initial course of immunization. A single booster injection (of DPT) should be given about a year later (see Secondary Response, Ch. 19) and again on entrance to elementary school. Subsequent booster immunization with a single dose of adult-type tetanus and diphtheria toxoid is recommended at 10-year intervals for persons at relatively high risk.

Whenever a previously immunized individual sustains a potentially dangerous wound, a booster of toxoid should be injected. (Booster injections may be effective even after 10–20 years, inducing protective levels of Ab within 1–2 weeks.) Seriously wounded subjects who have not previously been immunized, on the other hand, should be given prophylactic human TIG (250 U), since the Ab response to an initial toxoid injection is too slow to be useful. Prompt, thorough surgical debridement of wounds is also essential in tetanus prophylaxis.

Intensive programs of prophylactic immunization with toxoid in both civilian and military populations† have led to a striking reduction in the incidence of tetanus.

HISTOTOXIC CLOSTRIDIAL INFECTIONS

A variety of toxigenic species of the genus *Clostridium* are associated with invasive infection in man (listed under gas gangrene in Table 36-1). They are not highly pathogenic when introduced into healthy tissues; but in the presence of preexisting tissue injury, particularly damaged muscle, they can be responsible for a rapidly progressive, devastating infection characterized by the accumulation of gas and the extensive destruction of muscle and connective tissue (**clostridial myonecrosis** or **gas gangrene**). *C. perfringens* (formerly known as *C. welchii*), the species most commonly involved in human disease, is found in 1) wound infection, 2) uterine infection, 3) bacteremia, 4) unusual localized infections (pneumonia, meningitis), and 5) enteric infections. *C. perfringens* is widely distributed in soil, is normally a commensal in the lower gastrointestinal tract of animals and man, and is frequently isolated from contaminated skin surfaces or clothing.

THE TOXINS

Invasive strains of clostridia produce, during active multiplication, exotoxins with necrotizing (histolytic), hemolytic, and/or lethal properties. In addition, enzymes such as collagenase,‡ proteinase, deoxyribonuclease, and

---

* An American unit of antitoxin is defined as 10 times the smallest amount needed to protect, for 96 h, a 350-g guinea pig inoculated with a standard dose of toxin furnished by the National Institutes of Health.

† Of 855 cases of tetanus reported in the United States from 1967 to 1971, only one was in a person who had previously had three or more tetanus toxoid injections. During the same period, 23 cases of tetanus occurred in unimmunized individuals who received either equine antitoxin or human TIG prophylactically at the time of wounding. The greater reliability of active immunization is clear.

‡ Collagen, unless denatured to gelatin, is not digested by the usual proteases.

TABLE 36-2. Toxins and Toxigenic Types of *C. perfringens*

Toxins†	Bacterial types*				
	A‡	B	C	D	E
α (Lethal, lecithinase, necrotizing)	+	+	+	+	+
β (Lethal, necrotizing)	−	+	+	−	−
ε (Lethal, necrotizing)	−	+	−	+	−
ι (Lethal, necrotizing)	−	−	−	−	+
θ (Hemolytic)	±§	+	+	+	+

*C. perfringens is classified into 5 types (A–E) by the ability of antisera to neutralize the lethal toxins. For example, antiserum to type A (containing antitoxin to α) will protect mice against toxin-containing filtrates from type A organisms but not against those of other types (e.g., type B). Antiserum to type B (containing antitoxin to α, β, and ε) will protect mice against toxin-containing filtrates of types B,A,C, and D, but not type E.

†Other toxins (γ, δ, η, λ, [proteinase], κ [collagenase], μ [hyaluronidase], ν [deoxyribonuclease]) are elaborated as well to varying degrees by the different types of *C. perfringens*.

‡Type A is the principal type involved in human disease; type C is implicated in enteritis necroticans in man; types B,D, and E produce enteritis and enterotoxemia in sheep and cattle.

§θ toxin is produced by only some strains of *C. perfringens* type A (rarely by strains implicated in food poisoning outbreaks).

hyaluronidase, elaborated by the growing organisms, cause accumulation of toxic degradation products in the tissues. The toxins elaborated by *C. perfringens* have been most thoroughly studied (Table 36-2).* One of these, the **α-toxin,** is a calcium-dependent **lecithinase (phospholipase-C);** it is lethal and necrotizing. These actions, which can be neutralized by antilecithinase serum, are due primarily to splitting of **lecithin in cell membranes** (Fig. 36-4). (This enzyme also hydrolyzes cephalin and sphingomyelin.) Since lecithin is present in the membranes of many different kinds of cells the toxin can cause extensive damage in many tissues. The paucity of leukocytes in the exudate of gas gangrene may be due in part to this effect.

The **θ-toxin** of *C. perfringens* is hemolytic but has no phospholipase-C activity. It is oxygen-labile, is inactive in the presence of cholesterol, and crossreacts immunologically with both streptolysin O and tetanolysin. Other species of clostridia elaborate similar exotoxins, each of which is immunologically specific, but their roles in pathogenesis remain obscure.

Studies with inhibitors (EDTA, blocking phospholipase-C; heat treatment, blocking hemolytic activity), and with purified preparations, indicate that the initial action of the θ-toxin on RBC membranes exposes buried phospholipid groups to subsequent phospholipase-C action. This combined action may account for the

* For detailed review of others see MacLennan JD: Bacteriol Rev 26:177, 1962.

intravascular hemolysis observed with intense bacteremia due to *C. perfringens* (accompanying uterine infection or occasionally with gas gangrene).

An **enterotoxin,** quite distinct from the histotoxic and hemolytic toxins, is produced by certain strains of *C. perfringens* type A, which cause **acute food poisoning** in man. This toxin is produced only during sporulation and, unlike other *C. perfringens* toxins, not during vegetative growth.

The enterotoxin is produced intracellularly within 3 h after the inoculation of vegetative cells into sporulation medium, and it appears in the culture filtrate in parallel with the release of mature spores. Large quantities of the toxin (over 11% of the total cellular protein) are present intracellularly just prior to spore release. What, if any, role it has in the cell's economy is unknown.

PATHOGENESIS: HISTOTOXIC

Wound Infections. There are three types of clostridial wound infection: wound contamination, anaerobic cellulitis, and true myonecrosis (gas gangrene). From 80% to 90% of isolations of *C. perfringens* from hospitalized patients represent simple saprobic **wound contamination,** especially of operative wounds, skin ulcers, etc., and usually in association with other organisms. The presence of the organism in this setting does not herald invasive infection.

Anaerobic cellulitis is a clostridial infection that does not involve the muscles and is much less aggressive than gas gangrene. It begins with the introduction of spores into an open wound. Severe wounds, such as those acquired in automobile accidents and military combat, grossly contaminated with soil or fecal matter and harboring a mixture of organisms, are particularly likely to contain clostridial spores. Germination occurs in devitalized tissue where damage to the blood supply and the presence of foreign material has caused a lowering of the redox potential. The vegetative bacilli multiply, and anaerobic cellulitis usually develops after several days and extends widely, spreading in the fascial planes; the marked gas formation is detected by the resulting crepitus. The infection rarely produces intense local pain, toxemia, or invasion of healthy muscle.

Gas gangrene is an intensely aggressive, highly lethal infection, primarily of muscle. After the germination of clostridial spores in injured muscle, bacterial multiplication and toxin production occur. A self-perpetuating cycle of progressive tissue injury ensues. The onset of the disease is usually sudden, following an incubation period of 6–72 h after injury or abdominal surgery. The involved area is edematous and the skin has a bronze discoloration. Thin dark fluid is exuded, often in tense blebs; aspiration reveals many clostridia and few polymorphonuclear leukocytes. Gas is present in the subcuta-

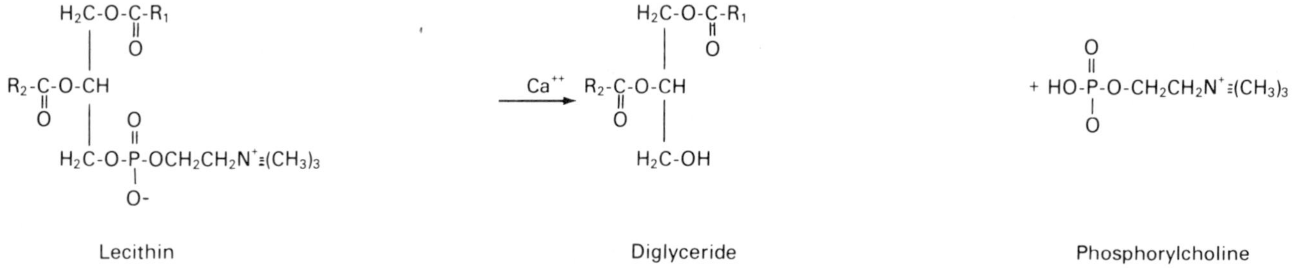

$$H_2C\text{-}O\text{-}\underset{\underset{O}{\|}}{C}\text{-}R_1$$

$$R_2\text{-}\underset{\underset{O}{\|}}{C}\text{-}O\text{-}CH$$

$$H_2C\text{-}O\text{-}\underset{\underset{O^-}{|}}{\overset{\overset{O}{\|}}{P}}\text{-}OCH_2CH_2N^+\!\equiv\!(CH_3)_3$$

$$\xrightarrow{Ca^{++}}$$

$$H_2C\text{-}O\text{-}\underset{\underset{O}{\|}}{C}\text{-}R_1$$

$$R_2\text{-}\underset{\underset{O}{\|}}{C}\text{-}O\text{-}CH$$

$$H_2C\text{-}OH$$

$$+\ HO\text{-}\underset{\underset{O}{\|}}{\overset{\overset{O}{\|}}{P}}\text{-}O\text{-}CH_2CH_2N^+\!\equiv\!(CH_3)_3$$

| Lecithin | Diglyceride | Phosphorylcholine |

FIG. 36-4. Action of *C. perfringens* α-toxin (lecithinase; phospholipase C) on lecithin (choline phosphoglyceride).

neous tissue and muscles.* Exploration of the wound reveals the muscle involvement: swollen, grayish or purple in color, and ischemic. Gas and fluid accumulation in tissue spaces confined by fascial compartments produces additional pressure which in turn causes further muscle necrosis and ischemia, allowing rapid spread of the anaerobic causative organism from one group of muscles to the next; this progressive process rapidly leads to irreversible shock.

Only six clostridial species (all those, except *C. tertium*, listed under gas gangrene in Table 36-1) are capable on their own of producing gas gangrene in man, although other toxigenic and nontoxigenic species can sometimes be isolated from the lesions of gas gangrene (particularly in war wounds) due to one or more of the six principal species. *C. perfringens* is the cause of over 90% of gas gangrene in the United States. It is widely distributed in soil and excreta and is often present on skin and clothing, and it frequently (4%–40%) contaminates major traumatic wounds. However, the incidence of gas gangrene in such wounds is less than 2%, indicating the importance of the type of wound and circumstances of initial treatment in the genesis of clostridial myonecrosis.

Uterine Infection. *C. perfringens* is present in the genital tract of about 5% of women. Following septic abortion, or uncommonly after prolonged labor, this organism may invade the uterine wall, producing extensive necrosis, high fever, and circulatory collapse. **Severe bacteremia** is characteristic of this process.

Bacteremia. Bacteremia due to *C. perfringens*, or to nonhistotoxic clostridia (e.g., *C. sporogenes, C. difficile, C. tertium*), occasionally occurs also in patients with leukemia, malignant tumors, gastrointestinal bleeding, bowel necrosis, gangrenous extremities, or decubitus ulcers. In patients with ulcerative lesions of the bowel prior treatment with aminoglycoside antibiotics may favor the overgrowth of clostridial species and predispose to enterogenous bacteremia. In occasional patients with

advanced leukemia the bowel becomes gangrenous and intense clostridial bacteremia may develop, accompanied by massive intravascular hemolysis. *C. septicum* infections (especially bacteremia) occur in patients with intestinal malignancy (in the absence of traumatic wounds). Clinical features may resemble those of a gram-negative bacillus bacteremia.

Bacteremia occurs in 15% of patients with gas gangrene, and in most uterine clostridial infections. Occasionally the bacteremia is intense and intravascular hemolysis develops rapidly, leading to acute renal shutdown.

Unusual localized infections. Necrotizing pneumonia or meningitis due to *C. perfringens* usually occurs after a penetrating injury, is not associated with toxemia, and follows the course of infections due to the more common bacteria.

PATHOGENESIS: ENTEROTOXIC

Enteric Infections. *C. perfringens* is a common cause of **food poisoning,** ranking second only to *Staphylococcus aureus* in the frequency of outbreaks in the United States. It produces a self-limited gastroenteritis, lacking constitutional symptoms, about 12 h following the ingestion of heavily contaminated food. The "food poisoning strains" belong to type A (based on the toxin produced—Table 36-2). The incriminated food is usually a meat product that had been stewed or boiled and then held before reheating and serving: heat-resistant spores would survive the cooking and later germinate during cooling in a medium from which oxygen had been driven out. An **enterotoxin** is implicated in the production of the diarrheal symptoms of *C. perfringens* food poisoning. A large number of viable organisms must be ingested to cause the disease (10^5/g are found in foods incriminated in outbreaks), and large numbers of *C. perfringens* (10^5/g) are subsequently found in the feces.

A serotyping system for *C. perfringens* type A (based on 57 capsular Ags) has been of value in the epidemiologic investigation of food-borne outbreaks. In addition to strains that form heat-resistant spores, heat-sensitive strains are also sometimes involved.

* The gas that accumulates is predominantly hydrogen, which is less soluble than CO_2 (see Butyric Fermentation, Ch. 3).

A much more serious form of clostridial enteric infection, **enteritis necroticans,** occurred in epidemic form in Germany at the close of World War II and is currently seen among the natives of New Guinea. This acute, severe inflammatory process in the small bowel is associated with a high mortality. *C. perfringens* type C has been isolated from the lesions. This organism produces large amounts of β-toxin, and during the disease β-antitoxin levels rise. A similar disease, **necrotizing enterocolitis** of formula-fed premature infants, is a fulminant, often fatal illness characterized by ileocecal ischemic necrosis, ileus, and shock. In a few infants *C. perfringens* or *C. butyricum* has been implicated, but the etiology is unknown in most cases.

A clostridial etiology has been ascribed recently to **antibiotic-associated pseudomembranous enterocolitis,** a severe and occasionally fatal illness of humans that is characterized by diarrhea, multiple small colonic mucosal plaques, and toxic megacolon. Clindamycin and lincomycin are the antibiotics most frequently associated with this syndrome. The selection pressure of the antibiotic, leading to the colonic overgrowth of *C. difficile* strains that produce a heat-labile enterotoxin, appears to be responsible for this disease in humans and in a hamster model. *C. difficile* is found in the stools of only a small percentage of normal individuals, and the strains are often resistant to clindamycin.

Fecal filtrates from patients with enterocolitis (and cell-free supernates of cultures of *C. difficile*) produce cytotoxicity in tissue culture and cause enterocolitis when injected intracecally into hamsters. Each effect can be neutralized by polyvalent gas gangrene antitoxin and *C. sordelli* antitoxin (both of which appear to contain crossreacting Abs to *C. difficile* toxin).

LABORATORY DIAGNOSIS

The bacteriology of gas gangrene is complicated, because most of the infections are mixed (aerobes and anaerobes). Furthermore, the presence of gas in the tissues does not necessarily incriminate clostridia, for klebsiellae, bacteroides, *E. coli,* and anaerobic streptococci may also cause the accumulation of enough gas to be detectable by palpation or roentgenography. The finding of many plump, blunt-ended, gram-positive bacilli but few granulocytes in smears of exudate from crepitant wounds (or from adjacent bullae), or from the uterine cervix, suggests clostridial infection; *C. perfringens* rarely forms spores in infected wounds. In wound smears *C. septicum,* the cause of about 10% of wartime cases of gas gangrene, sometimes appears as chains made up of elongated cells or as gram-positive rods with large subterminal spores. Smears of the peripheral blood (or of the buffy coat) from patients with intense bacteremia occasionally show the bacilli, providing a rapid diagnosis.

Exudate for anaerobic culture should be obtained if possible (using a needle and syringe from which all air has been expelled) and processed promptly. Solid media (enriched blood agar; or selective media such as kanamycin–vancomycin–laked blood agar or phenethyl alcohol blood agar if mixed bacterial populations are likely) should be incubated aerobically and anaerobically, and liquid media (chopped meat–glucose or supplemented brain heart infusion broth available as prereduced medium; thioglycollate medium) should also be employed. If gram-positive spore-bearing bacilli are found in the liquid media, and if colonies of such bacilli appear on the blood agar incubated anaerobically but not on that incubated aerobically, the genus *Clostridium* should be suspected.

The identification of individual clostridial species is based primarily on the biochemical tests listed in Table 36-1. *C. perfringens* can be recognized by its failure to show spores in direct smear or culture, by double-zoned hemolysis around colonies on blood agar, and by the Nagler reaction (formation of a visible opacity due to lecithinase action around colonies grown on egg yolk agar—Fig. 36-5); type A antitoxin in the agar prevents this opacification without inhibiting growth. Other helpful properties in distinguishing *C. perfringens* are capsule formation, lactose fermentation, and lack of motility. With other clostridia found in wounds the demonstration of histotoxin production in animals, and its neutralization with specific antitoxin, is useful.

PREVENTION AND TREATMENT

Prompt surgical debridement of dirty wounds is the most effective means of preventing anaerobic cellulitis and gas

FIG. 36-5. Colonies of *C. perfringens* grown on egg yolk medium. The halo of precipitate about the colonies is due to splitting of lecithin by lecithinase released by the organisms; the diglyceride released is insoluble. (McClung LS, Toabe R: J Bacteriol 53:139, 1947) (×5, enlarged)

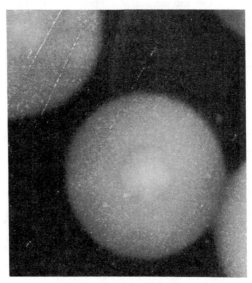

gangrene; antibiotic therapy alone does not suffice. Established clostridial wound infection requires surgical debridement, as promptly as possible. It is important to distinguish between anaerobic cellulitis and gas gangrene, since the former requires wide excision and debridement while the latter requires complete extirpation of involved muscle (usually amputation of a limb). Penicillin is administered in large doses to prevent or treat bacteremia and to inhibit further bacterial multiplication in the wound. Clindamycin (occasional strains are resistant) and chloramphenicol are other antibiotics with good activity against *C. perfringens*. (However, the recent observation of plasmid-mediated resistance to tetracycline and chloramphenicol, and to erythromycin and clindamycin, in *C. perfringens* portends possible future problems in the use of these antimicrobial agents.) Hyperbaric oxygen at 3 atm in a compression chamber, to raise the oxygen tension in healthy tissue, may have a place in therapy, especially in gas gangrene of the trunk, where complete debridement may be precluded.

Polyvalent equine antitoxin prepared against toxic filtrates of *C. perfringens*, *C. novyi*, *C. septicum*, and *C. histolyticum* has been used in the past in the prophylaxis and treatment of gas gangrene. Its efficacy in human gas gangrene has never been established (a better rationale could be offered for its use in severe *C. perfringens* bacteremia with hemolysis), and it is no longer available for clinical use.

Oral vancomycin, to which *C. difficile* is susceptible, appears to be effective in the treatment of clindamycin-induced enterocolitis.

SELECTED READING

BOOKS AND REVIEW ARTICLES

BALOWS A, DEHAAN RM, GUZE LB, DOWELL VR (eds): Anaerobic Bacteria: Role in Disease. Springfield, IL, CC Thomas, 1974

BOROFF DA, DAS GUPTA BR: Botulinum toxin. In Kadis S, Montie TC, Ajl SJ (eds): Microbial Toxins, Vol IIA. New York, Academic Press, 1971, p 1

DOWELL VR, HAWKINS TM: Laboratory Methods in Anaerobic Bacteriology. CDC Laboratory Manual. Washington DC, DHEW Publ No. (CDC) 74–8272, 1974

FINEGOLD SM: Anaerobic Bacteria in Human Disease. New York, Academic Press, 1977

MACLENNAN JD: The histotoxic clostridial infections of man. Bacteriol Rev 26:177, 1962

NAKAMURA M, SCHULZE JA: *Clostridium perfringens* food poisoning. Annu Rev Microbiol 24:359, 1970

SUTTER VL, VARGO VL, FINEGOLD SM: Wadsworth Anaerobic Bacteriology Manual, 2nd ed. Department of Continuing Education in Health Sciences and the School of Medicine, UCLA, 1975

VAN HEYNINGEN WE, MELLANBY J: Tetanus toxin. In Kadis S, Montie TC, Ajl SJ (eds): Microbial Toxins, Vol IIA. New York, Academic Press, 1971, p 69

WRIGHT GP: The neurotoxins of *Clostridium botulinum* and *Clostridium tetani*. Pharmacol Rev 7:413, 1955

SPECIFIC ARTICLES

ARNON SS, MIDURA TF, CLAY SA, WOOD RM, CHIN J: Infant botulism. Epidemiological, clinical and laboratory aspects. JAMA 237:1946, 1977

BARTLETT JG, CHANG TW, GURWITH M, GORBACH SL, ONDERDONK AB: Antibiotic-associated pseudomembranous colitis due to toxin-producing clostridia. N Engl J Med 298:531, 1978

BROOKS VB, CURTIS DR, ECCLES JC: The action of tetanus toxin on the inhibition of motoneurones. J Physiol (Lond) 135:655, 1957

Center for Disease Control, US Department of Health, Education, and Welfare Report. Tetanus Surveillance, Report No. 4, 1974

DUNCAN CL: Time of enterotoxin formation and release during sporulation of *Clostridium perfringens* type A. J Bacteriol 113:932, 1973

EKLUND MW, POYSKY FT: Interconversion of type C and D strains of *Clostridium botulinum* by specific bacteriophages. Appl Microbiol 27:251, 1974

EKLUND MW, POYSKY FT, MEYERS JA, PELROY GA: Interspecies conversion of *Clostridium botulinum* type C to *Clostridium novyi* type A by bacteriophage. Science 186:456, 1974

GEORGE WL, SUTTER VL, FINEGOLD SM: Antimicrobial agent-induced diarrhea—a bacterial disease. J Infect Dis 136:822, 1977

HELTING TB, ZWISLER O, WIEGANDT H: Structure of tetanus toxin. II. Toxin binding to ganglioside. J Biol Chem 252:194, 1977

HOWARD FM, BRADLEY JM, FLYNN DM, NOONE P, SZAWATKOWSKI M: Outbreak of necrotizing enterocolitis caused by *Clostridium butyricum*. Lancet 2:1099, 1977

KOENIG MG, SPICKARD A, CARDELLA MA, ROGERS DE: Clinical and laboratory observations on type E botulism in man. Medicine 43:517, 1964

MILLER PA, EATON MD, GRAY CT: Formation of tetanus toxin within the bacterial cell. J Bacteriol 77:733, 1959

National Communicable Disease Center, US Department of Health, Education and Welfare report. Botulism in the United States: Review of Cases 1899–1973

SMITH JWG, MACIVER AG: Growth and toxin production of tetanus bacilli in vivo. J Med Microbiol 7:497, 1974

SMYTH CJ, FREER JH, ARBUTHNOTT JP: Interaction of *Clostridium perfringens* theta-haemolysin, a contaminant of commercial phospholipase-C, with erythrocyte ghost membranes and lipid dispersions. Biochim Biophys Acta 382:479, 1975

SUTTON RGA, HOBBS BC: Food poisoning caused by heat-sensitive *Clostridium welchii*. A report of five recent outbreaks. J Hyg (Camb) 66:135, 1968

VAN HEYNINGEN S: Binding of ganglioside by the chains of tetanus toxin. FEBS Lett 68:5, 1976

VAN HEYNINGEN WE, MILLER PA: The fixation of tetanus toxin by ganglioside. J Gen Microbiol 24:107, 1961

chapter 37

MYCOBACTERIA

EMANUEL WOLINSKY

The mycobacteria are a group of nonmotile rods that are defined on the basis of a distinctive staining property: they are relatively impermeable to various basic dyes, but once stained they retain dyes with tenacity. Specifically, they resist decolorization with acidified organic solvents and are therefore called **acid-fast.** This property, and their relatively slow growth, are due to the presence of a lipid-rich cell wall. Mycobacteria range from widespread innocuous saprobic inhabitants of soil and water to organisms that are responsible for **tuberculosis** and **leprosy.**

In both these chronic diseases intracellular infection, delayed hypersensitivity, and cellular resistance play important roles, and slowly evolving granulomatous lesions result in extreme tissue destruction. Leprosy largely involves the skin, sometimes with shocking disfigurement, whereas tuberculosis is usually confined to internal organs. Moreover, the contagious nature of leprosy was recognized in Biblical times, whereas tuberculosis, though even more contagious, was not recognized as such until the last century. Hence the leper became a social outcast, while in Europe a century or two ago tuberculosis was regarded romantically as a source of enhanced esthetic sensitivities.

The leprosy bacillus, discovered by Hansen in 1879, was the first bacterium shown to be associated with human disease. It is strikingly adapted to man: it has been difficult to transfer to any other host, and it has not been grown on artificial culture media. In contrast, the agents of human tuberculosis (*Mycobacterium tuberculosis* and the closely related *M. bovis*) are readily cultivated on simple media and are pathogenic for various lower animals, especially guinea pigs and mice. With the decreasing prevalence of tuberculosis it has been recognized that similar but generally milder infections are caused by a variety of previously ignored mycobacterial species that do not infect these test animals. These organisms have reservoirs in the environment or in lower animals, rather than in man.

TUBERCULOSIS

Pulmonary tuberculosis **(consumption, phthisis)** has been recognized as a widespread and grave clinical entity for many centuries, and its incidence was probably increased by the social consequences of the Industrial Revolution. However, its communicable nature was not recognized until Villemin produced a similar disease in rabbits, in 1868, by injecting material from tuberculous lesions of man; and in 1882 Koch discovered the tubercle bacillus, impressively fulfilling the criteria that he had developed for identifying the etiologic agent of an infectious disease (see Koch's postulates, Ch. 1). Since then tuberculosis has been one of the most intensively studied infectious diseases. Not only has it been a major cause of death and prolonged disability, but until recent times it usually struck people at the age of greatest vigor and promise.

PROPERTIES OF *MYCOBACTERIUM TUBERCULOSIS*

The mycobacteria are considered transitional forms between eubacteria and actinomycetes (Ch. 38): some of the latter, of the genus *Nocardia,* are weakly acid-fast, while some mycobacteria may exhibit branching. Accordingly, the mycobacteria have been classified in the order Actinomycetales.

MORPHOLOGY

Acid-Fastness. To stain the tubercle bacillus, in smears or in tissues, methods must be used that promote penetration of the dye. In the widely used Ziehl–Neelsen method the smeared specimen is heated to steaming for 2–3 min in carbolfuchsin (a mixture of the triphenylmethane dyes rosaniline and pararosaniline in aqueous 5% phenol). Subsequent washing in 95% ethanol–3% HCl (acid alcohol) decolorizes most bacteria in a few seconds, whereas acid-fast organisms retain the stain much longer. The mechanism of acid-fastness will be discussed below, under Lipids.

Mycobacteria may appear to be gram-positive, but they take up the stain weakly and irregularly and without requiring iodine treatment to retain it.

Other acid-fast objects include some bacterial and fungal spores, and a substance known as ceroid (found in the liver in certain states of nutritional deficiency). These materials, however, do not require the presence of phenol (or aniline) in the staining solution. The acid-fast nocardiae are more easily stained and more easily decolorized than are mycobacteria.

Structure. Tubercle bacilli in the animal are typically slightly bent or curved slender rods, about $2\mu m$–$4\mu m$ long and $0.2\mu m$–$0.5\mu m$ wide. The rods may be of uniform width but more often appear beaded, with irregularly spaced, unstained vacuoles, or heavily stained knobs. In culture media the cells may vary from coccoid to filamentous. Strains differ in their tendency to grow as discrete rods or as aggregated long strands, called **serpentine cords** (Fig. 37-1).

The walls of mycobacteria contain a peptidoglycan with diaminopimelate, and the cells can be converted to spheroplasts by lysozyme. The walls have a remarkably high lipid content (up to 60%), much of which is attached to polysaccharide; the polysaccharides, which include glucan, mannan, arabinogalactan, and arabinomannan, are also found in culture filtrates. The glycolipids and protein are located in a firmly attached outer layer of the wall (Fig. 37-2), and the external location of the lipid accounts for the hydrophobic character of the cells.

Electron microscopy reveals a rather thick wall, and large lamellar mesosomes are common. Glycogen granules and polymetaphosphate (volutin) bodies are also seen: the latter stain metachromatically with cationic dyes. These inclusion bodies contribute to the frequently beaded, irregular staining of tubercle bacilli.

GROWTH

Unlike most other pathogenic bacteria, which are facultative aerobes or anaerobes, the tubercle bacillus is an **obligate aerobe.** It can grow in **simple synthetic media,** with glycerol or other compounds as sole carbon source

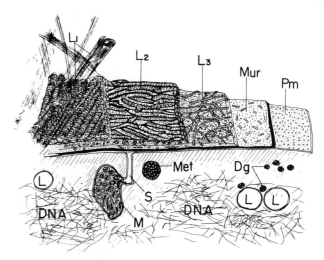

FIG. 37-2. Mycobacterial cell wall as diagrammed from electron micrographic data. **Dg** = Dense granules; **L** = lipoidal bodies; **L1** = surface glycolipid; **L2** and **L3** = fibrous ropelike structures; **M** = mesosome; **Met** = metachromatic granule; **Mur** = murein or peptidoglycan; **Pm** = plasma membrane; **S** = septum. (Barksdale L, Kim K-S: Bacteriol Rev 41:217, 1977)

and ammonium salts as nitrogen source; asparagine or amino acid mixtures are usually added to promote the initiation and improve the rate of growth.

In ordinary synthetic liquid media the bacilli grow in adherent clumps which form a **surface pellicle.** (This "moldlike" property is responsible for the name *Mycobacterium.*) Nonionic detergents, such as Tweens,* which cause the bacilli to grow in more dispersed form, though still not as single cells.

Mycobacteria generally show a marked nutritional preference for lipids; egg yolk has been a prominent constituent of the rich media used for diagnostic cultures. Thus, though the tubercle bacillus is very sensitive to inhibition by long-chain **fatty acids,** it is stimulated by them at very low concentrations. A satisfactory concentration is maintained by adding to the medium **serum albumin,** which binds the fatty acids with sufficient affinity to maintain a low free concentration.

Mycobacteria produce iron chelators **(mycobactins),** whose competition with host chelators (transferrins) may play a role in pathogenesis and resistance.

Growth of tubercle bacilli in culture media (and in animals) is characteristically slow: *the shortest doubling time observed, in rich media, is about 12 h.* Saprobic mycobacteria grow more rapidly, but not so rapidly as most other bacteria (which have doubling times as short as 20

FIG. 37-1. Tubercle bacilli in stained smear from colony. Note parallel growth in ''cords.'' From original drawing of Koch. (Courtesy of Robert Koch-Institut, Berlin)

* Polyoxyethylene ethers of sorbitol, or of other polyhydric alcohols, esterified to long-chain fatty acids.

min). In harmony with the slow growth, the ribosomes of *M. tuberculosis* translate in vitro only 1/10 as rapidly as those of *E. coli.*

GENETIC VARIATION AND VIRULENCE TESTS

Human tubercle bacilli freshly isolated from pulmonary lesions produce progressive disease in guinea pigs, which die within 1–6 months after infection, depending in part on the size of the inoculum. Tuberculous skin lesions, called **lupus vulgaris,** often yield less virulent organisms, as do lung lesions from some patients in India and Africa. As with other pathogenic bacteria, virulence varies quantitatively: serial passage through artificial culture media selects indirectly for less virulent mutants, while animal passage of large inocula of such strains selects directly for mutants with restored virulence. One attenuated bovine strain,* carried through several hundred serial cultures on unfavorable (bile-containing) media, is known as **bacille Calmette-Guérin (BCG).** This strain, which is used to immunize humans against tuberculosis (see Immunization), has remained avirulent for over 40 years.

Most virulent strains produce rough colonies, while many avirulent laboratory strains produce less rough colonies (called smooth, though the cells lack capsules). Thus the correlation between colonial morphology and virulence, though not highly consistent, is the reverse of that seen with many other bacteria.

Two widely used rough laboratory strains, H37Rv (virulent) and H37Ra (avirulent), were derived from the H37 strain of human tubercle bacilli.

Virulence Tests. In contrast to pneumococci, virulent strains of tubercle bacilli cannot be reliably distinguished from avirulent strains by cellular or colonial morphology, or by serologic tests; and the direct test for virulence in animals is slow and inconvenient. Properties correlated with virulence have therefore been intensively sought. The most significant finding has been that virulent strains grow, on the surface of liquid or on solid media, as intertwining **serpentine cords,** in which the bacilli aggregate with their long axes parallel (Fig. 37-1), while most avirulent strains grow in a more disordered manner. The correlation, however, is imperfect. Growth in cords can be correlated with the content of a surface lipid (cord factor, described below). Nonionic detergents (e.g., Tween 80) reduce cord formation (but not virulence), presumably by coating the cell surface. Another property that is lost in many avirulent mutants is binding of the dye **neutral red.**

* A strain is considered **attenuated** for a given host if it multiplies in that host to only a limited extent, causing at most minor, transient lesions.

The failure to find a reliable virulence test is not surprising: since virulence is a multifactorial property a given bacterial component may be necessary but it cannot be sufficient to make an organism virulent.

LIPIDS

The most striking chemical feature of the mycobacteria is their extraordinarily high lipid content. Various lipid fractions are defined by the conditions used for their extraction from dried organisms. The striking abundance of lipids in the cell wall (up to 60% of its dry weight) accounts for the hydrophobic character of the organisms. The lipid-rich wall probably also accounts for some of the other unusual properties of mycobacteria, e.g., relative impermeability to stains, acid-fastness, unusual resistance to killing by acid and alkali, and resistance to the bactericidal action of Abs plus complement.

Among the lipids extracted with neutral organic solvents are **true waxes** (esters of fatty acids with fatty alcohols) and **glycolipids** (**mycosides:** lipid-soluble compounds with covalently linked lipid and carbohydrate moieties).

Mycolic Acids. While many different fatty acids are found in mycobacteria, the **mycolic acids** appear to be unique to the cell walls of these organisms, nocardiae, and corynebacteria. These very large, saturated, α-alkyl, β-hydroxyl fatty acids (Fig. 37-3) are found in both waxes and glycolipids. A large arabinogalactan (mol wt 31,000), covalently attached to the peptidoglycan and to about 30 mycolic acid residues, forms a bridge between the rigid layer and the outer, lipophilic layers of the cell wall.

Three classes of mycolic acids have been found in *M. tuberculosis,* with cyclopropane, methoxy, or keto groups, and with slight variation in chain length (mostly C_{78} and C_{80}). Mycolyl acetyl trehalose is a precursor of wall mycolic acid (and probably of "cord factor," described below); the disaccharide solubilizes the very hydrophobic, unusually long-chain fatty acid.

Cord Factor. This factor, which may be essential for both virulence and serpentine growth, was extracted by Bloch from virulent cells by petroleum ether. The extracted cells disperse in aqueous media. Cord factor has been identified as a mycoside, 6,6′-dimycolyltrehalose (Fig. 37-3). Several lines of evidence relate it to virulence. 1) Its extraction renders cells phenotypically nonvirulent (though they remain viable). 2) It is toxic: it inhibits migration of normal polymorphonuclear leukocytes in vitro (as do virulent tubercle bacilli), and 10μg given subcutaneously will kill a mouse. 3) It is more abundant in virulent strains. 4) Tubercle bacilli recovered from animals or from young cultures are more virulent, and they have a higher content of cord factor, than cells of the

FIG. 37-3. Structures of some distinctive myco-bacterial lipids. Trehalose is D-glucose-α-1, 1′-D-glu-coside

Mycolic acids

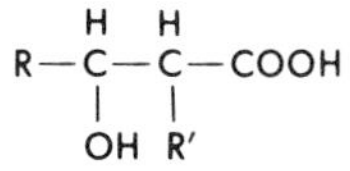

$$R - \overset{\displaystyle \underset{|}{\overset{|}{H}}}{\underset{OH}{C}} - \overset{\displaystyle \underset{|}{\overset{|}{H}}}{\underset{R'}{C}} - COOH$$

Human and bovine tubercle bacilli: $R' = C_{24}H_{49}$

Avian and saprophytic bacilli: $R' = C_{22}H_{45}$

R has about 60 C atoms, 1 or 2 O atoms

6, 6′-Dimycolyltrehalose (cord factor)

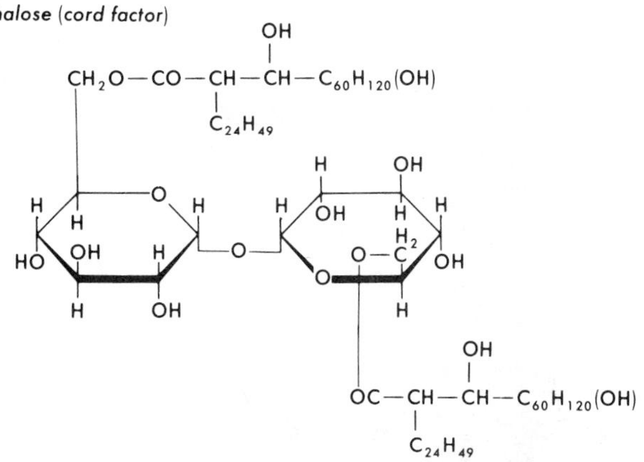

same strain from older cultures. 5) Mice have been pro-tected against tuberculous infection by active immuniza-tion with a complex of cord factor and methylated bovine serum albumin, or by passive transfer of rabbit anti–cord-factor serum. The toxicity of cord factor has been related to profound disturbances of microsomal enzymes, mitochondria, and lipid metabolism in the livers of mice.

A **sulfolipid** (trehalose-2′-sulfate esterified with long-chain fatty acids) also shows a correlation, but not with complete consistency, between its concentration and virulence.

Wax D and the Immune Response. Another mycoside of interest is the high-molecular-weight **wax D,** which is not a true wax but contains mycolic acids and a glycopeptide. It is apparently extracted from the basal wall layer, since the peptide contains the characteristic amino acids of that layer, linked to a polymer of hexoses and hexosa-mines. In a water-in-oil emulsion (Freund's adjuvant) this fraction, like whole tubercle bacilli, **enhances the immunogenicity** of a variety of added antigens (see Ch. 19). Moreover, a mixture of wax D and proteins of the tubercle bacillus **induces delayed-type hypersensitiv-ity** to tuberculin, whereas the protein alone is poorly immunogenic.

The **crude phosphatide** fraction has the interesting property of evoking a **cellular response resembling tu-bercle formation,** including caseation necrosis (see

Pathogenesis, p. 729). Rather large amounts must be in-jected, and comparable fractions from saprobic myco-bacteria are also effective.

Acid-Fastness. After exhaustive extraction of the above-described lipids by organic solvents the residual ghosts remain acid-fast, and they retain **firmly bound lipids** largely in the wall fraction. These lipids are removed by hot acid, which also destroys the acid-fastness. Since this property is also lost on sonic disruption of normal cells it appears to depend on the integrity of the cell wall, in-cluding certain lipids.

Presumably because of their lipid layer, tubercle bacilli are un-usually resistant to acid and alkali, and this property is exploited in their diagnostic isolation (see Laboratory Diagnosis, below). They are also relatively insensitive to cationic detergents, but no more resistant than other nonsporulating bacteria to heat, ultraviolet ir-radiation, or phenol.

ANTIGENIC STRUCTURE

Chromatographic fractionation of extracts, followed by immunodiffusion, has revealed at least 20 Ags in *M. tu-berculosis.* Figure 37-4 illustrates 11 Ags produced by electrophoresis of a reference culture filtrate. Some are species-specific; others are common to many mycobac-teria and even nocardiae and corynebacteria. Strain- or type-specific Ags are also identified by agglutination re-

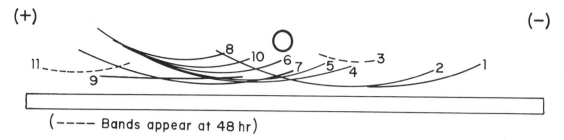

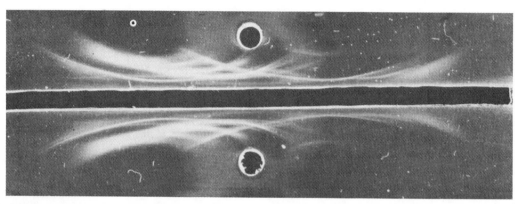

FIG. 37-4. Multiplicity of mycobacterial antigens. A reference culture filtrate of *M. tuberculosis*, strain H37Rv, was subjected to electrophoresis and analyzed with a goat antiserum. (Janicki BW et al: Am Rev Respir Dis 104:602, 1971)

actions and by skin testing guinea pigs with partially purified proteins (PPD; see Tuberculin Test, below) from diverse mycobacteria.

IMMUNE RESPONSE TO INFECTION

ACQUIRED IMMUNITY

The Koch Phenomenon. Shortly after isolating the tubercle bacillus Koch demonstrated that it induces an altered response to superinfection. In the original study bacilli were injected subcutaneously into a guinea pig, and 10–14 days later the inoculation site developed a nodule which then became a persistent ulcer. In addition, the bacilli spread to the regional lymph nodes, causing them to enlarge and then become necrotic. When a similar injection was then made at another site the response was faster and more violent, but also more circumscribed: a dusky, indurated lesion appeared in 2–3 days and soon ulcerated; but the ulcer healed promptly and the regional lymph nodes remained virtually free of infection. Koch later showed that a similar second response could be obtained with culture filtrates of the organism (see Tuberculin Test).

The superinfecting bacilli are evidently better localized than the initial inoculum, and they multiply more slowly in the tissues. The infected animal has thus acquired increased resistance (i.e., partial immunity). But though this immunity may lead to elimination of the bacilli of superinfection, it cannot accomplish the same for the primary lesion, which has developed a denser bacterial population and a different histologic pattern. Evidently the level of immunity acquired in the guinea pig is limited, and it can be overcome by local factors in the lesion.

Nonhumoral Immunity. Serum Abs do not account for this immunity. Thus, though the tubercle bacillus enhances formation of Abs to a wide variety of immunogens, Abs to proteins and polysaccharides of the tubercle bacillus itself are generally found only in low titers in tuberculous individuals, and the levels observed have no prognostic value. Moreover, these Abs are not bactericidal in vitro, even in the presence of complement; and though they promote phagocytosis of tubercle bacilli in vitro, the organisms multiply within the phagocytes. Finally, increased resistance to infection cannot be transferred passively with serum, but it can with viable lymphoid cells, which transfer delayed-type hypersensitivity to tuberculin.

ACTIVATED MACROPHAGES AND HYPERSENSITIVITY

Though macrophages of normal animals permit ingested tubercle bacilli to flourish, macrophages from tuberculous animals destroy them more effectively. Some kind of specific cellular immunity in macrophages was therefore long suspected, analogous to the altered mononuclear cells associated with delayed hypersensitivity. However, similar activated macrophages can be obtained from animals infected with various other bacteria (such as *Listeria*) that also survive in normal macrophages. This increased bactericidal capacity is nonspecific—i.e., it extends to a wide variety of microbes—and it is associated with an **increased number of lysosomes** in the activated macrophage.

While the enhanced activity of altered macrophages is nonspecific, its induction is immunologically specific: Mackaness has shown that lymphocytes modified by specific interaction with the infecting organism can cause this activation in the macrophage population (see Macrophage Activation, Ch. 24). The effect lasts only a few weeks in experimental animals unless antigenic stimulation is repeated or sustained.

Though tuberculin sensitivity is thus associated with an increase in host defenses, it can also lead to harmful reactions. For example, in sensitive individuals the administration of a large amount of tuberculin can provoke severe systemic illness, including chills, fever, and increased inflammation around existing lesions.*

Attempts to determine whether immunity to tuberculosis is due to the allergic state or develops in spite of it have been indecisive. A residual immunity was observed in animals that had been desensitized by the gradual administration of large doses of tuberculin, and a labile immunizing factor, in a fraction consisting primarily of ribosomes, was reported to engender specific immunity without apparent tuberculin hypersensitivity. Delayed hypersensitivity and immunity may be mediated by different classes of T cells, in response to different components of the bacillus (see Ch. 18).

Tuberculin sensitivity is clearly a double-edged sword; it is associated with increased resistance, but under some circumstances the response can also exacerbate symptoms and increase spread of the bacteria.

PATHOGENESIS

The consequences of inhaling or ingesting tubercle bacilli depend upon both the **virulence** of the organism and the **resistance** of the host (as well as the size and the location of the inoculum). At one extreme, organisms with little virulence for the particular host disappear completely, leaving no anatomic trace behind. At the opposite extreme (e.g., in guinea pigs inoculated with human tubercle bacilli) the bacilli flourish in macrophages as well as extracellularly, are disseminated widely, and cause death within a few months.

Humans exhibit a range of responses. The initial infection in a tuberculin-negative individual most often produces a self-limited lesion, but sometimes the disease progresses, presumably because of low resistance or a large inoculum. Because of the delicate balance of resistance, involving local as well as systemic factors, healing and progressing lesions may coexist in the same individual, and the disease often has a chronic, cyclic course (especially without chemotherapy). This pattern stands in marked contrast to the acute course of those infections that give rise to well-defined humoral immunity.

Histologically the tubercle bacillus evokes two types of reaction. 1) **Exudative** lesions are seen in the initial infection, or in the individual in whom the organism proliferates rapidly without encountering much host resistance. Acute or subacute inflammation occurs, with exudation of fluid and accumulation of polymorphonuclear leukocytes around the bacteria. 2) **"Productive"** (granulomatous) lesions form when the individual becomes hypersensitive to tuberculoprotein. The macrophages then undergo a dramatic modification on contact with tubercle bacilli or their products, becoming concentrically arranged in the form of elongated **epithelioid cells,** to form the **tubercles** characteristic of this disease. In the center of the tubercles some of these cells may fuse to form one or more **giant cells,** with dozens of nuclei arranged at their periphery and viable bacilli often visible in their cytoplasm. Outside the multiple layers of epithelioid cells is a mantle of lymphocytes and proliferating fibroblasts (Fig. 37-5), leading eventually to extensive fibrosis. The subsequent development of caseation necrosis, and its various fates, are noted below.

In early lesions tubercle bacilli are localized primarily within macrophages, in which they multiply (at least for a time). In more advanced lesions only extracellular bacilli are prominent, probably because the activated macrophages destroy the intracellular bacilli more efficiently.

PRIMARY INFECTION

After inhalation of tubercle bacilli the initial lesion appears as an area of nonspecific pneumonitis, usually located in a well-aerated peripheral zone. It is only after delayed hypersensitivity develops, in 2–4 weeks, that granulomatous inflammation supervenes and the characteristic **tubercles** (L. *tuberculum*, small lump) are formed. In the meantime bacilli are carried to the draining lymph nodes and then, by way of the lymph and blood, throughout the body of the host. The pulmonary focus and the granulomatous lesion in the hilar lymph node to-

---

* This effect was reluctantly recognized by Koch, who enthusiastically injected tuberculin in tuberculous patients in the hope of enhancing immunity.

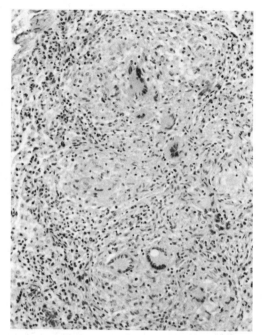

FIG. 37-5. Section of a tubercle, showing several giant cells containing a peripheral ring of nuclei, epithelioid cells, and mononuclear cells toward the periphery of the lesion. (Courtesy of B. W. Castleman) (×180, reduced)

gether are known as the **primary complex.** The next stage in the inflammatory response consists of **caseation necrosis** (L. *caseus,* cheese), in which the necrotic centers of tubercles remain semisolid rather than softening to form pus, presumably because the enzymes that usually liquefy dead cells and tissue are inhibited.

Primary tuberculosis in children usually stabilizes and heals. Caseous lesions heal by **fibrosis** and **calcification,** which may result in extensive scar formation and shrinkage. The healed and frequently calcified primary complex lesions are referred to as the **Ghon complex,** which may be recognized in chest radiographs for the remainder of the person's life. In a small proportion of individuals, however, the infection is not brought under control and the primary lesions become progressively larger, coalesce, and **liquefy.** When this material is released, a **cavity** is formed in the lung and there may be extension of the disease via the bronchi.

REACTIVATION DISEASE

Most tuberculosis in adults is due to reactivation of long-dormant foci remaining from the primary infection. The foci are located mostly in the posterior and the apical or subapical portions of the lung, whose persistent infection, after hematogenous spread, may be due to the high P_{O_2} resulting from their favorable ventilation–perfusion ratio. Viable tubercle bacilli are rarely found in the healed lesions of the primary complex.

By the time disease is recognized liquefaction of the caseous lesion usually has occurred and a cavity has provided a favorable site for the rapid proliferation of the bacilli. These may then be transmitted to other individuals via droplet nuclei produced by aerosols of infected sputum, and to other parts of the lungs by bronchogenic spread. Almost every organ of the body may be the site of **extrapulmonary** tuberculosis. The most common locations are the genitourinary system, bones and joints, lymph nodes, pleura, and peritoneum. Extrapulmonary disease commonly develops as a result of reactivation of dormant lesions seeded during the primary infection. **Disseminated,** or **miliary,** tuberculosis may be seen following the rupture of a caseous lesion into a pulmonary vein.

The concept of reactivation disease is based on the following evidence: 1) Phage types of bacilli isolated from multiple sites are usually identical. 2) In contrast to healed primary lesions, apical residuals are apt to contain viable tubercle bacilli. 3) It is unusual to find a source case to explain apparently new disease of the reactivation type, whereas in adults with primary type infection an index case often can be identified. 4) Tuberculin sensitivity persists for a lifetime; and this sensitivity is occasionally eliminated by chemotherapy. The most suggestive initial evidence was the striking upward shift in the age of peak mortality ever since the total incidence of the disease began declining around 1870 (Fig. 37-6).

Since the present group of elderly tuberculin reactors had a high risk of acquiring initial infection at an earlier age, and would then have remained tuberculin-positive, their frequent development of active tuberculosis is best explained as a reactivation of dormant foci acquired much earlier in life. It is thus no longer advantageous to be tuberculin-positive as a result of an earlier inapparent infection: though the hypersensitive state partially protects against the production of disease by a fresh infection, the level of exposure to such an infection is now so low that this advantage is outweighed by the possibility of progression or activation of the previous infection.

Although most cases of adult disease can be attributed to reactivation, exogenous reinfection has also been demonstrated, by the recognition of more than one phage type from the same patient, and by the isolation of bacilli resistant to drugs which the patient could not have taken during his primary infection.

VARIATIONS IN HOST RESISTANCE

Among persons infected with the tubercle bacillus, as detected by a positive tuberculin test (see below), only a small proportion develop overt disease, and even before the advent of chemotherapy only about 10% of these progressed to fatal disease. The transition from **infection**

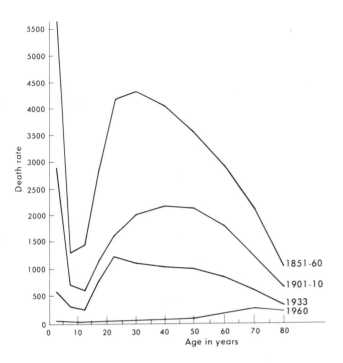

FIG. 37-6. Death rates per million from tuberculosis (all forms) at various ages, in England and Wales. (Wilson GS, Miles AA: Principles of Bacteriology and Immunity, 5th ed. Baltimore, Williams & Wilkins, 1964. Copyright 1964, Edward Arnold Publishers, Ltd, London)

to mild or to severe **disease** depends strongly on various factors besides the presence of the bacilli; hence in an earlier era, when nearly everyone eventually became tuberculin-positive, tuberculosis could almost be regarded as one of the "endogenous diseases" (Ch. 44).

While many of the observed variations in the response to infection may involve differences in the size, the site, or the virulence of the inoculum, there is also no doubt that resistance in man varies more strikingly with tuberculosis than with most infectious diseases. Thus in a tragic accident in Lübeck, Germany, in 1930, in which 251 children received identical inocula of a virulent strain instead of an attenuated vaccine, 77 developed fatal disease, 127 developed radiologically detectable lesions that healed, and 47 showed no signs of disease. Such variations in resistance involve both genetic and nongenetic (physiologic) factors.

Genetic Differences. The importance of genetic factors in host resistance has been unequivocally established by Lurie's development of inbred lines of rabbits with a high or a low tendency to acquire progressive tuberculosis from experimental infection. Multiple factors are evidently involved: some lines are more resistant to **initia-**

tion of disease by small inocula, while in others the resistance primarily influences the **rate of progress** of the disease.

In humans it seems likely that there are similar **racial differences** in resistance, associated with different lengths of exposure of the race to the selective pressures of an environment with widespread tuberculosis. Thus in the United States the incidence of tuberculosis and the ratio of deaths to cases are especially high among American Indians, Eskimos, and Negroes. These races may have had little or no exposure to tuberculosis until the last two or three centuries, contrasted with descendants of a European population. However, this evidence is complicated by the concomitant differences in environmental and socioeconomic factors. Environmental factors, however, were relatively equalized in two kinds of studies. In the US Army (from 1922 to 1936) the case fatality rate was four times higher among black soldiers than among whites. In a study of twins the frequency of the disease in both members of the pair, compared with that in one member, was three times higher among monozygotic than among dizygotic pairs.

Resistance to tuberculosis has not been correlated with genetically determined variations in any component of the immune response. The hereditary component of this resistance is probably quite polygenic. Accordingly, it cannot be as precisely defined, in its nature and its distribution, as an independently measurable monogenic trait, such as the production of sickle cell hemoglobin or a deficiency in glucose-6-phosphate dehydrogenase (G6PD)—both selected for in falciparum malaria. Nevertheless, there is no doubt that tuberculosis has exerted a similar selective influence on the gene pool of populations where it has been endemic. Indeed, a few generations of selection were sufficient with rabbits (Lurie's experiments, above) to yield marked variations in heritable resistance.

Physiologic Factors. Epidemiologic evidence on the frequency of tuberculosis in various populations has long suggested that **malnutrition, overcrowding,** and **stress** decrease resistance to the disease; but this kind of evidence does not firmly establish a causal relation because these conditions are generally also associated with a high rate of infection. However, the ability of one or more of these factors to overcome innate resistance is well illustrated by the experience of inmates of Nazi concentration camps during World War II. Tuberculosis was exceedingly prevalent in this population, which was subjected to extreme stress and prolonged starvation; yet many seemingly moribund persons with far advanced disease exhibited a remarkable recovery when liberated from the camps and renourished. In contrast, before the era of chemotherapy many persons who entered a sanatorium with minimal tuberculosis progressed inexorably to a fatal outcome despite good nutrition and general care.

The most specific and direct evidence on the effect of starvation has been obtained in experimental infections in mice: a reduced **protein** consumption increased susceptibility, while variations in vitamin and total caloric intake had no demonstrable effect.

The significance of **hormonal** factors is suggested by the striking variations in resistance with **age** and, to a smaller extent, with **sex.** Tuberculosis is apt to be very severe in infants, perhaps because of the immaturity of the immune mechanisms. It then decreases rapidly in both incidence and severity with increasing age, and between 3 and 12 years of age progressive disease is almost unknown, even in children heavily infected by exposure to a tuberculous parent. Susceptibility to the disease increases rapidly at adolescence, and among young adults tuberculosis is more frequent in females. Curiously, in experimental animals no marked influence of age on susceptibility has been noted.

Administration of **cortisone,** in humans and in experimental animals, decreases resistance to tuberculosis and tends to mask its symptoms and diminish reactivity to tuberculin. Hence in the therapy of other diseases with this hormone the arousal of dormant tuberculosis is occasionally a serious and readily overlooked complication. Increased secretion of cortisone may also well be involved in the presumptive role of stress, noted above.

Silicosis. Tuberculosis is notoriously frequent among miners and others exposed to dust containing silica. Moreover, in guinea pigs that have inhaled silica suspensions even the avirulent BCG strain produces fatal tuberculosis. The decrease in resistance appears to be due to damage to phagocytic cells: phagocytized silica particles cause rapid disruption of lysosomes, followed by cell lysis; and since the particles are never digested, they can attack cell after cell. Silica is exceedingly insoluble, and it is believed to act by direct contact with lysosomal membranes.

LABORATORY DIAGNOSIS

A provisional diagnosis of tuberculosis is usually made by demonstrating acid-fast bacilli in stained smears of sputum or of gastric washings (containing swallowed sputum). For rapid screening of smears some laboratories employ fluorescence microscopy following staining with the fluorescent dye mixture rhodamine–auramine.

Whether or not the smear is positive the material should be cultured, for several reasons: 1) cultivation can detect fewer organisms; 2) cultural characteristics distinguish human tubercle bacilli from other acid-fast bacilli, including avirulent contaminants, other pathogenic mycobacteria (see below), and *Nocardia* (Ch. 38); 3) tests for drug sensitivities should be initiated. During chemotherapy frequent sputum smears and cultures provide a rapid, objective index of the response.

Sputum is prepared for culture by exploiting the un-

usual resistance of tubercle bacilli to strong alkalis and acids. For example, the material may be shaken at 37° C in 0.1 N NaOH. The resulting liquefaction permits concentration of bacteria by centrifugation; and more rapidly growing bacterial contaminants are mostly destroyed by this treatment, whereas a fraction of the mycobacterial cells remain viable (and virulent). The centrifuged sediment should be smeared and stained, as well as cultured.

Solid culture media are preferred for primary isolation; they often contain egg yolk to promote growth of macroscopic colonies from small inocula. However, oleic acid–albumin agar medium, which is transparent, allows detection of smaller colonies than those required on an opaque egg medium. These media contain malachite green or an antibiotic at levels that selectively suppress growth of pyogenic contaminants. A positive culture usually grows out in 2–4 weeks. The same methods are used to process body fluids and exudates not likely to be free of pyogenic organisms, but normally sterile fluids such as cerebrospinal fluid may be planted directly on culture media.

Part of the sediment from sputum is sometimes inoculated subcutaneously into the groin of a guinea pig. The animal is skin-tested with tuberculin biweekly, and when hypersensitivity becomes evident the lesions are examined for bacilli by smear and culture. This test can detect small inocula, but it is slow and expensive. It was long employed to distinguish virulent mycobacteria from contaminants, but it is no longer considered reliable for this purpose (see Other Pathogenic Mycobacteria, below). Modern cultural technics have made guinea pig inoculation unnecessary, except possibly for specimens heavily contaminated with organisms that are difficult to suppress, such as *Pseudomonas.*

TUBERCULIN TEST

Significance. Delayed-type hypersensitivity to tuberculin is highly specific for the tubercle bacillus and closely related mycobacteria. Reactivity appears about 1 month after infection in man and persists for many years, often for life; hence the frequency of reactors in the population increases cumulatively with age. *A positive test thus reveals previous mycobacterial infection; it does not establish the presence of active disease.*

The persistence of hypersensitivity probably depends on persistence of bacilli in dormant foci, and reactivity may disappear following chemotherapy of recent infections.

Koch's discovery of the tuberculin reaction in 1890 provided a powerful diagnostic and epidemiologic tool, which revealed much more widespread infection than had previously been suspected. Moreover, the importance of the tuberculin test has grown as control measures have aimed increasingly at persons with inapparent infection.

The capacity of the skin to react may be suppressed by advanced age or by terminal or severe acute illness, perhaps because of circulatory changes. Reactivity can also be suppressed by rapidly progressive tuberculosis, perhaps through massive release of Ag; by measles and certain other viral infections or live vaccines, through unknown mechanisms; by cortisone; and by diseases such as lymphoma which are associated with depressed cell-mediated immune reactions.

Preparation of Tuberculin. Tuberculin as originally described by Koch, known as **old tuberculin (OT)**, is prepared by autoclaving or boiling a culture of tubercle bacilli, concentrating it tenfold on a steam bath, filtering off the debris, and adding glycerol as a preservative. In this impure product the active constituent is a protein remarkable for its heat stability: after being autoclaved it remains soluble and retains specific determinants of the protein in the infecting bacilli. Stock solutions retain full potency for years when stored at 5° C.

A slightly more refined tuberculin, called **purified protein derivative (PPD)**, is prepared by precipitation several times with 50% saturated ammonium sulfate. The product is mostly a mixture of small proteins (average mol wt 10,000).

Testing Procedures. Tuberculin hypersensitivity is generally tested for in the United States by **intradermal** injection of 0.1 ml of an appropriate dilution of standardized PPD into the most superficial layers of the skin of the forearm (**Mantoux** test). The average diameter of **induration** (and not simply erythema) at the injected site is measured at 48 h, and reactions less than 10 mm in diameter are recorded as doubtful. Intense reactions can cause necrosis and subsequent scarring. Persons infected with crossreacting "atypical" mycobacteria (see Other Pathogenic Mycobacteria, below) frequently exhibit weak reactions (4–9 mm in diameter). Ordinarily, however, the population is quite sharply divided into tuberculin-positive or -negative.

In epidemiologic work the standard test dose generally used is 5 tuberculin units (TU) of PPD (0.1μg) in 0.1 ml, corresponding to OT 1:2000. In children suspected of having tuberculosis a fivefold lower concentration is first used (**first strength**), to avoid a severe reaction, and if there is no response within 2–3 days the standard strength is used. **Second strength** PPD, containing 20–50 times the standard dose, may be used in those patients with negative or doubtful standard tests but still suspected of infection with tubercle bacilli or with crossreacting mycobacteria.

Other commonly used methods of skin testing involve multiple punctures of the skin by means of various mechanical devices. These methods are less accurate than the Mantoux test and positive or doubtful results should be rechecked by the intradermal technic.

THERAPY

Until specific therapy became available, with the discovery of streptomycin in 1945, the treatment of tuberculosis was limited to rest, good nutrition, and artificial collapse of the lung. Unfortunately, the value of streptomycin was restricted by its toxicity to the eighth cranial nerve on prolonged administration, as well as by the frequent emergence of resistant tubercle bacilli during therapy. The real revolution in tuberculosis therapy came with the introduction of isoniazid (INH), which is more effective, because it is bactericidal and acts on intracellular as well as on extracellular tubercle bacilli. Moreover it is remarkably free from toxicity, inexpensive, and easy to take (i.e., small doses, given orally). The rapid response to INH and the low toxicity have made ambulatory treatment feasible.

Rifampin (RMP), introduced later, is also very effective, relatively nontoxic, and absorbed very well when given by mouth. The combination of rifampin and INH may prove to be the most effective regimen. *p*-Aminosalicylate (PAS), used as a companion to INH, has been largely replaced by ethambutol (EMB). Other useful drugs include kanamycin and capreomycin (injected), and cycloserine, ethionamide, and pyrazinamide (given orally); they are generally used only in the presence of allergy or resistance to the "primary" drugs.

Streptomycin is bactericidal (see Ch. 7) but is not effective against organisms in macrophages. **PAS** is only bacteriostatic: it interferes with the conversion of *p*-aminobenzoate to folic acid (see Ch. 7). It is also relatively ineffective against bacteria localized within macrophages, presumably because of the presence of metabolites that reverse its action. **RMP** is bactericidal for tubercle bacilli, including those in macrophages. **INH** acts by interfering with mycolic acid synthesis, which explains why it is effective only against mycobacteria (see Ch. 7).

The peculiar lesions in tuberculosis, which protect the organisms from host defenses, also promote a metabolically inactive state which protects them from the action of bactericidal drugs. Hence even in minimal tuberculosis *chemotherapy must be prolonged* to bring about a cure. With the usual two-drug regimen of INH with EMB or PAS it was customary to treat for 18–24 months, but recent experience has provided evidence that 6–9 months of intensive treatment with various combinations of INH, RMP, and pyrazinamide, EMB, or streptomycin may be sufficient. More than 95% of patients now have their disease arrested by appropriate drug therapy.

Drug Resistance and Combination Therapy. The selection of drug-resistant mutants in patients with tuberculosis is favored by several features of the disease: the numerous bacteria in the lesions, especially in the walls of cavities; their further multiplication during the required long periods of therapy; and the limited degree of host resistance, which provides the rare mutant cell with a good opportunity to proliferate before being eliminated by host defenses. Hence specifically resistant strains often appear in patients treated with any one of the drugs used singly. The rationale for preventing the emergence of resistant strains by **combined therapy** with two or more drugs, has been presented in Chapter 7.

The INH-resistant mutants isolated from treated patients differ from wild-type tubercle bacilli in several additional respects: they lack catalase and peroxidase activity, grow more slowly, and are much less virulent for guinea pigs. However, they retain virulence for man: they may continue to be shed in sputum over months and even years, and they are capable of producing primary tuberculosis in children. Indeed, the frequency of childhood tuberculosis caused by drug-resistant bacilli has been reported to be in the range of 10%–20%.

While drug-resistant tubercle bacilli have been transmitted from patients to others their frequency among freshly discovered cases has remained low.

EPIDEMIOLOGY

The human tubercle bacillus is spread principally by droplets and sputum from individuals with open pulmonary lesions. Small droplets produced by coughing are probably the most effective vehicles, since they rapidly dry in the air to yield droplet nuclei, of less than $5\mu m$ diameter, which can reach the alveoli. The organism can survive in moist or dried sputum for up to 6 weeks, but it is killed by a few hours' exposure to direct sunlight.

The incidence of tuberculosis may be estimated by tuberculin test surveys as well as from reports of active cases or of deaths. The frequency is higher in impoverished social groups that live under crowded conditions, compared with the more affluent and with those that live in sparsely populated regions. The incidence is also high, however, in certain nonurban groups (American Indians and Eskimos) that probably have a high genetic susceptibility. Socioeconomic conditions undoubtedly affect both host resistance and frequency of infection.

With the progressive decline in prevalence over the past century the peak incidence has shifted from young adulthood, with a preponderance of females, to the age group over 65, with a severalfold preponderance of males. The highest incidence of active tuberculosis is now found among elderly, impoverished men, living alone in city slums and often alcoholic and malnourished.

The worldwide decline of tuberculosis over the past century probably reflects general social and economic improvement. In addition, since about 1950 death rates in the United States have declined especially sharply as a result of effective chemotherapy (Fig. 37-7), and the incidence of new cases has dropped from 52 per 100,000 in 1953 to 14 per 100,000 in 1977.

While tuberculosis is no longer the "white plague" of previous centuries, it is still an immense problem. In the world at large it is estimated that some 15 million persons have active tuberculosis, and annually 3 million die of the disease. Currently in the United States tuberculosis causes about 3000 deaths each year, and about 30,000 new active cases are recognized. Nevertheless, the tuberculosis problem in this country today is largely one of reactivation, for over 95% of the population now reach adulthood without infection.*

PREVENTION

Surveys. A major weapon in tuberculosis control programs in the recent past has been case-finding by mass surveys of apparently well persons (by means of tuberculin tests and chest x-rays), followed by isolation and treatment. However, the yield of new cases in the United States is now too low to justify this program. Instead, x-ray surveys are reserved for high-risk groups such as jail inmates and male derelicts; and additional cases are uncovered by examination of the home and school contacts of recent tuberculin converters and newly recognized tuberculosis patients.

Chemoprophylaxis. The effectiveness and the low toxicity of INH have made possible prophylactic chemotherapy of tuberculosis. This approach was first applied, with striking success, to high-risk groups, such as tuberculin-positive household contacts of cases. Subsequently, in carefully studied populations a 50% decrease in the incidence of active disease was achieved by giving a course of treatment to all persons who had recently become tuberculin-positive (though only a small fraction of the treated population lost their tuberculin hypersensitivity).

Although prophylactic chemotherapy of all tuberculin reactors would seem to offer the best possibility of interrupting the chain of transmission, the chance of INH hepatotoxicity, especially in older persons, should restrict its use to those groups at highest risk. Also limiting the usefulness of INH prophylaxis is the difficulty in obtaining sufficient cooperation from the patient to allow completion of the one year of treatment.

Immunization. Live vaccines can produce an increase in resistance to tuberculosis, but not complete protection; in experimental animals the effect is judged by the more prolonged course of the disease, and in man by its reduced incidence and case fatality rate. The most widely used preparation has been **BCG** (see Genetic Variation, above), which was introduced as a vaccine in France in 1923. In the 1940s the **vole bacillus** (or murine tubercle bacillus; see Mycobacteria Not Associated with Man, below) was similarly introduced in England: as isolated from field mice, without any attenuation, it is quite avirulent for man, as well as for guinea pigs and rabbits.

---

* This figure is arrived at by correcting the 10% tuberculin reactors for the estimated frequency of crossreactions (see Fig. 37-8).

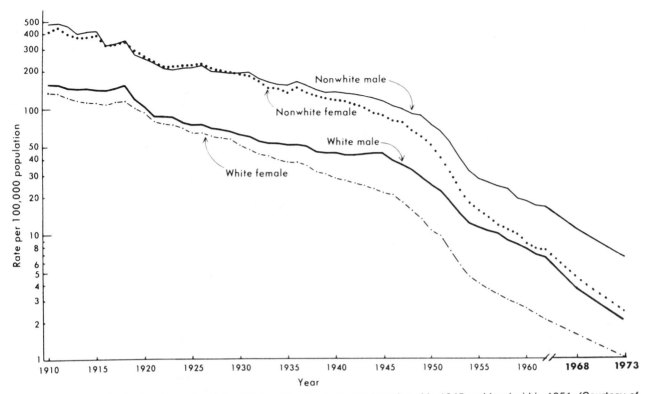

FIG. 37-7. Tuberculosis death rates in the United States. Streptomycin was introduced in 1945 and isoniazid in 1951. (Courtesy of Tuberculosis Program, US Public Health Service, Center for Disease Control)

Killed vaccines, and various fractions of tubercle bacilli, have given at best only weak and transient protection.

The safety and efficacy of BCG were the subject of bitter dispute over a 40-year period. Fear that the bacilli might revert to virulence has by now been allayed: over the course of the past several decades more than 150 million persons, in various countries, have been injected by the WHO without severe reactions. Controversy with respect to its efficacy arose from several sources: some questionable experimental statistics; the probable loss of viability of many lots of vaccine (before the introduction of a stable lyophilized preparation); and its trial in populations already naturally immunized by widespread infection with other mycobacteria. The most definitive study is that of the British Medical Research Council. Over 50,000 tuberculin-negative children, at about age 15, were randomly assorted into three groups which were, respectively, immunized with BCG, immunized with vole bacillus, or left unvaccinated. The incidence of tuberculosis in either vaccinated group over the course of the next 9 years was only 20% of that observed in the unvaccinated group. Similar favorable results had been obtained earlier in a smaller controlled study in American Indians, an especially high-risk group.* However, BCG was not accepted by public health authorities in the United States

* Much less favorable results were obtained in a carefully controlled study by the US Public Health Service in Alabama and Georgia.

when it might have been useful, and by now the prevalence of the disease here has become too low to justify its widespread use.

Vaccination has a major disadvantage: the resulting development of tuberculin hypersensitivity *destroys the usefulness of the tuberculin test.* Because recognition and prophylactic treatment of the recent tuberculin converter have proved to be safe and effective, vaccination of medical personnel is no longer recommended. Young household contacts of infectious cases also should be given preventive treatment even before tuberculin conversion, but they probably do not require a full year of treatment.

MYCOBACTERIUM BOVIS

M. bovis causes tuberculosis in cattle and is also highly virulent for man; unpasteurized milk (and occasionally other dairy products) from tuberculous cows have been responsible for much human tuberculosis. The ingested organisms presumably penetrate the mucosa of the oropharynx and intestine (though without apparent damage), giving rise to early lesions in the **cervical lymph nodes (scrofula)** or in the **mesenteric nodes.** Subsequent dissemination from these sites principally infects **bones and joints;** such infection of vertebrae was largely responsible for the hunchbacks of previous generations. When inhaled (e.g., by dairy farmers) the organism can also cause pulmonary tuberculosis indistinguishable from that caused by *M. tuberculosis.*

Tuberculosis due to *M. bovis* has now become very rare in many countries as a result of the widespread pasteurization of milk and the virtual elimination of tuberculosis in cattle. The conquest of bovine tuberculosis vividly illustrates the effectiveness of public health legislation when the reservoir of a disease can be controlled.

Eradication of bovine tuberculosis was first aimed at in Denmark, through the isolation of tuberculin-positive cattle in separate herds and the separation of their calves at birth; but while this program was successful on individual farms, it required too much cooperation and vigilance to succeed on a national scale. In 1917 the US Department of Agriculture, with widespread support from veterinarians, undertook an audacious program: tuberculin testing of all cattle and **the slaughter of all positive reactors.** As a result the proportion of tuberculin reactors in American cattle has been reduced from 5% to 0.5% (most of which now have no visible lesion at autopsy and may be crossreactors).

M. bovis is a bit shorter and plumper than *M. tuberculosis.* For 20 years after discovering the tubercle bacillus Koch maintained that all mammalian tuberculosis was due to the same organism. By about 1900, however, largely through Theobald Smith's work, *M. bovis* and *M. tuberculosis* were clearly distinguished. For example, *M. bovis* is highly pathogenic in rabbits, whereas *M. tuberculosis* is much less so; and in cultures *M. bovis* tends to grow more slowly (so-called dysgonic growth), and cannot tolerate as high a concentration of glycerol. *M. bovis* also differs from *M. tuberculosis* in being niacin-test–negative, in not reducing nitrate, and in being resistant to pyrazinamide and susceptible to thiophen-2-carbonic acid hydrazide. Serologic tests and skin tests, however, do not differentiate between the two organisms.

M. africanum, an organism isolated from tuberculosis patients in Africa, resembles *M. bovis* in its slow growth and in several biochemical reactions. The question of species status has not been decided.

OTHER PATHOGENIC MYCOBACTERIA

In addition to the mammalian tubercle bacilli described above, various other acid-fast bacilli possessing little or no virulence for guinea pigs or rabbits occasionally have been cultured from respiratory secretions of apparently tuberculous individuals. Because virulence for guinea pigs was a sacred criterion since Koch's time in differentiating tubercle bacilli from contaminating saprobes, there was enormous delay in recognizing that the mycobacteria, like the gram-positive cocci, include a variety of distinct, medically important organisms. Recognition of the role of the "atypical" mycobacteria in human disease was crystallized in the 1950s by repeated recovery of such organisms from the sputum and **directly from diseased tissue,** in the absence of classic tubercle bacilli.

As the incidence of tuberculosis decreases the relative importance of pulmonary disease attributable to nontuberculous mycobacteria should increase. This proportion has been reported to be as high as 15% in certain areas of the world; their frequency in skin and other soft tissue infections is probably much higher. Chronic destructive pulmonary disease, especially silicosis, predisposes to infection with these mycobacteria, and in the immunodeficient host they may become opportunistic pathogens and cause disseminated disease.

The initial grouping was based primarily on pigment production, i.e., **photochromogens,** whose carotenoid pigment was photoactivated; **scotochromogens,** which developed pigment in the dark; and **nonphotochromogens,** which resembled the classic avian tubercle bacilli and had variable pigmentation not related to light exposure. Subsequent studies have allowed a better classification. A few simple tests (Table 37-1) will serve to differentiate the major species; a more detailed description and illustrations may be found in the Center for Disease Control publication by Vestal.

Mycobacterium avium Complex. This complex includes **M. avium,** which was recognized as a distinct species prior to 1900, and the more recently described **M. intracellulare.** The avian tubercle bacillus causes tuberculosis in chickens, pigeons, and other birds, and in swine. There are many well-documented cases of human tuberculosis caused by this organism, mostly in farmers or their children, and in men with silicosis.

The avian bacillus is readily distinguishable from *M. tuberculosis* and *M. bovis.* The individual cells are smaller; the colonies are smooth; it grows optimally at about 41 ° C (at which human and bovine bacilli will not grow); it is resistant to most of the antituberculosis drugs; and it is pathogenic for chickens and rabbits but not for guinea pigs. Antigenic differences are also readily demonstrated.

M. intracellulare (formerly known as the Battey bacillus) is generally less thermophilic than *M. avium,* and most strains are not pathogenic for birds or animals. Some, however, have intermediate virulence (as is also true of some *M. avium* strains). In seroagglutination tests the avian strains are included in *M. avium* complex serotypes 1, 2, and 3, and the other presently recognized 25 serotypes are *"intracellulare"* types. The two members of the complex are not readily distinguishable by other laboratory procedures.

Mycobacterium scrofulaceum. This species is a common cause of lymphadenitis in children. The few cases of pulmonary disease associated with it have been in adults with previously damaged lungs. Seroagglutination has revealed three or four types, but some Ags are shared with the *M. avium* complex. In contrast to the nonpigmented growth of *M. avium* complex strains, *M. scrofulaceum* colonies are usually yellow-orange even when grown in the dark (scotochromogenic). Some strains derived from human material and from the environment have intermediate biochemical and pigment characteristics; Australian workers have lumped these strains with those of the avian complex into the MAIS (*M. avium-intracellulare-scrofulaceum*) complex. They are usually resistant to antituberculosis drugs in vitro.

Mycobacterium kansasii. M. kansasii and M. avium-intracellulare together account for most of the human mycobacterial disease at-

TABLE 37-1. Some Characteristics of Nontuberculous Mycobacteria That May Be Encountered in Human Material

Species	Clinical signifi-cance	Growth at 25°	Growth at 37°	Growth at 42°	Pigment	Catalase 25–37°	Catalase 68°	Nitrate	Tween hydrolysis	Urease	Serotype by agglutination Helpful	Serotype by agglutination No. of types recorded
M. kansasii	+	+	+	–	+*	Strong	+	+	+	+	+	1
M. simiae	+	+	+	±	+*	Strong	+	–	–	+	+	2
M. marinum	+	+	±	–	+*	Weak	+	–	+	+	+	2
M. scrofulaceum	+	+	+	–	+	Strong	+	–	–	+	+	3
M. szulgai	+	+	+	±	+†	Strong	+	+	±	+	+	1
M. gordonae	–	+	+	–	+	Strong	+	–	+	–	?	7
M. flavescens	–	+	+	–	+	Strong	+	+	+	+	?	?
M. avium-intracellulare	+	±	+	±	±‡	Weak	+	–	–	–	+	28
M. xenopi	+	–	+	+	+	Weak	+	–	–	–	?	?
M. gastri	–	+	+	–	–	Weak	–	–	+	+	+	1
M. terrae-triviale	–	+	+	–	–	Strong	+	+	+	–	–	
M. fortuitum	+	+	+	–	–	Strong	+	+	±	+	+	2
M. chelonei	+	+	+	–	–	Strong	+	–	–	+	+	1
M. smegmatis	–	+	+	+	–	Strong	+	+	+	+	–	

*Photochromogenic (some *M. simiae* strains are only weakly photochromogenic).

†Scotochromogenic when grown at 37°C but variably photochromogenic at 25°C.

‡Some strains demonstrate light yellow pigment that intensifies with age.

tributable to acid-fast organisms other than mammalian tubercle bacilli. *M. kansasii* is photochromogenic; the overnight change of the colonies to yellow is followed by the formation of red crystals of β-carotene on exposure to light for several more days. Most strains are sensitive to rifampin and to several other drugs, and the disease responds well to treatment.

Mycobacterium ulcerans. This organism, found mainly in Africa and Australia, **will grow only below 33 °C.** It causes chronic, deep **cutaneous ulcers** in man, and in mice and rats it produces lesions in the cooler parts of the body. Its usual drug sensitivity pattern—resistance to INH and ethambutol, and susceptibility to streptomycin and rifampin—is unique. Human disease responds poorly to drug treatment, and extensive excision followed by skin grafting is often necessary. Although rare, the disease is of considerable theoretic interest, since the low temperature range of the organism provides a simple explanation for its unusual pathogenetic properties.

Mycobacterium marinum. This organism (synonyms: *M. balnei, M. platypoecilus*) also grows best at 30°–33 ° C, and relatively poorly at 37°. It causes a tuberculosislike disease in fish and a chronic skin lesion, known as **"swimming pool granuloma,"** in humans. Infection acquired by injury of a limb around a home aquarium or a marine environment also can lead to a series of ascending subcutaneous abscesses not unlike the lesions of sporotrichosis. *M. marinum* resembles *M. kansasii* in being photochromogenic, but it can be differentiated by the optimum growth temperature, negative nitrate test, agglutination with specific antisera, and distinctive drug sensitivity pattern (resistance to INH, sensitivity to the other major antituberculosis drugs).

Mycobacterium fortuitum Complex. Most of the fast-growing disease-associated mycobacteria are members of this complex, which may be divided into two accepted species, *M. fortuitum* and *M.*

chelonei. They abound as saprobes in soil and water. Human disease consists mainly of soft tissue abscesses, but disseminated disease, osteomyelitis, and infections of prosthetic heart valves have been described. These strains are resistant to most of the presently available antimicrobial drugs.

Mycobacterium xenopi, M. szulgai, and M. simiae. *M. xenopi,* first isolated in 1959 from a South African toad, has been reported to cause chronic human pulmonary disease in England, Europe, Australia, and the United States. The organisms are thermophilic, scotochromogenic, and relatively sensitive to INH, rifampin, and streptomycin. Stained cells are long, fusiform, and palisaded. Hot water tanks represent an important reservoir of infection.

M. szulgai recently was recognized as a distinct species with the potential ability to produce chronic lung disease, as well as infection of lymph nodes and bursae. Almost all strains are scotochromogenic at 37 ° C, but many apparently are photochromogenic at 25°.

M. simiae was first recovered from monkeys in Budapest. Only a few documented cases of human lung disease have been reported. The bacilli have distinctive properties of variable photochromogenicity and niacin positivity.

Information on the pathogenicity for experimental animals of mycobacterial species is presented in Table 37-2, and a summary of the nontuberculous mycobacterial diseases of humans in Table 37-3.

Epidemiology. Many of these species (*M. scrofulaceum, M. avium-intracellulare,* and *M. fortuitum-chelonei*), including serotypes causing human disease, are found in dust, soil, water, and the tissues of domestic animals. *M. kansasii* and *M. xenopi* have been isolated from water

TABLE 37-2. Pathogenicity of Mycobacteria for Experimental Animals*

Species	Guinea pig	Rabbit	Mouse	Chick
M. tuberculosis	+	–	+	–
M. bovis	+	+	+	–
M. avium	–	+	±	+
M. kansasii	–	–	±	–
M. marinum	–	–	Cool areas	–
M. scrofulaceum	–	–	–	–
M. intracellulare	–	–	–	±
M. xenopi	–	–	±	±
M. fortuitum-	–	–	+	–
M. chelonei				

* *M. gordonae, M. terrae* complex, *M. gastri, M. smegmatis,* and *M. phlei* are essentially nonpathogenic. Adequate information concerning *M. simiae* and *M. szulgai* is lacking.

taps and water storage tanks. Local soft tissue infections (*M. marinum, M. scrofulaceum, M. fortuitum-chelonei*) may result from skin or mucous membrane contamination originating from the environment. However, in pulmonary and disseminated disease it is often not clear how the bacilli enter the human host, or whether the disease in adults represents reactivation of dormant lesions or a new infection. Person-to-person transmission rarely, if ever, occurs.

An interesting geographic variation in the frequency of specific disease-associated strains has been documented: *M. kansasii* predominates in Great Britain and Europe, and in certain cities in the United States. In Australia, Japan, and the states of Florida and Georgia, on the other hand, *M. avium-intracellulare* is more common.

The skin test Ags (**sensitins**) prepared from various mycobacteria show both specificity and crossreactivity; hence patients with disease due to nontuberculous mycobacteria may or may not respond to a standard tuber-

culin test. A large-scale epidemiologic study by Palmer, of the US Public Health Service, revealed that in general humans infected with various mycobacteria react most intensely to the PPD from the homologous organism (Fig. 37-8).

This approach revealed a high level of **inapparent infection** with nontuberculous mycobacteria in certain areas. In a study of over 200,000 US Navy recruits only 8.6% reacted to the tuberculin of *M. tuberculosis,* while 35% reacted to the sensitin of *M. intracellulare.* The PPDs of several other mycobacteria, including some from soil, also gave reactions with striking frequency, and crossreactions among various strains were frequent. Moreover, comparative studies with several Ags have provided strong evidence that *in some regions the bulk of the weak reactions to tuberculin PPD are crossreactions.* Hypersensitivity to the Ags of *M. scrofulaceum* and *M. avium* complex organisms is especially frequent in the southeastern part of the United States. Because of the need for testing repeated cultures during chemotherapy the isolated organisms are now being studied more intensively; hence strains with unusual characteristics are more likely to be detected, and are no longer automatically discarded as nonpathogens. The nontuberculous acid-fast bacilli, with their ineradicable reservoir, may eventually become the major residue of the problem of mycobacterial infection.

Guinea pigs inoculated with various nontuberculous mycobacteria isolated from humans show some degree of protection against subsequent challenge with *M. tuberculosis.* Hence it seems quite possible that *inapparent infection of man with these organisms may serve as a sort of natural vaccination against tuberculosis,* much as exposure to cowpox was noted, in Jenner's day, to protect against smallpox.

The delayed recognition of the other pathogenic mycobacteria illustrates the complexity of the problem of virulence, and the falli-

TABLE 37-3. Nontuberculous Mycobacterial Diseases of Man

Disease	Common associated species	Other associated species
Chronic cavitary lung disease in adults	*M. avium-intracellulare* *M. kansasii*	*M. xenopi, szulgai, simiae, scrofulaceum*
Local lymphadenitis in children	*M. scrofulaceum*	*M. avium-intracellulare* *M. kansasii*
Bone and joint infection	*M. kansasii* *M. avium-intracellulare*	
Disseminated disease	*M. kansasii* *M. avium-intracellulare*	*M. scrofulaceum* *M. fortuitum-chelonei*
Skin and soft tissue diseases		
Swimming pool granuloma	*M. marinum*	
Sporotrichoid	*M. marinum*	*M. kansasii*
Cutaneous "tuberculosis"	*M. kansasii*	*M. avium-intracellulare*
Buruli or Bairnsdale ulcer	*M. ulcerans*	
Inoculation-site abscess	*M. fortuitum-chelonei*	
Deep hand infection	*M. kansasii*	
Bursitis	*M. kansasii*	*M. szulgai*
Leprosy	*M. leprae*	

FIG. 37-8. Correlation of sizes of reactions of sanatorium patients, with proved typical or atypical mycobacterial infections, to standard tuberculin (PPD-S) and to comparable preparation (PPD-B) from a strain of *M. intracellulare* isolated from a patient in the same sanatorium. (Edwards LB et al: Acta Tuberc Scand Suppl 47:77, 1959. Copyright 1959, Munksgaard International Publishers Ltd, Copenhagen, Denmark)

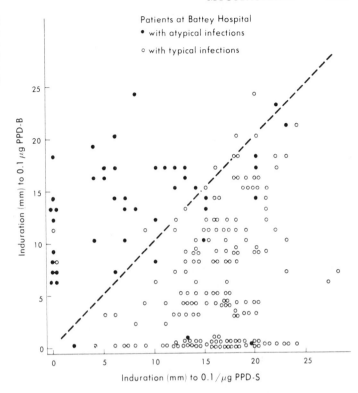

Patients at Battey Hospital
- with atypical infections
- with typical infections

bility of laboratory tests for this property. The distinction between pathogens and nonpathogens is evidently not sharp among the mycobacteria. On the one hand, most of the nontuberculous pathogens are probably saprobes that can become pathogenic under appropriate conditions; and though they are much less virulent than *M. tuberculosis,* they occasionally cause chronic and even life-threatening disease (just as do many saprobic fungi; see Ch. 45). On the other hand, though the tubercle bacillus is identified with a grave illness, its most frequent fate, after infecting an individual, is to persist for a lifetime without causing disease.

MYCOBACTERIA NOT ASSOCIATED WITH MAN

In spite of their sluggish growth the mycobacteria are extremely widespread in nature, not only as saprobes but as parasites throughout the animal kingdom.

The **vole bacillus** (*Mycobacterium microti*) produces an epizootic chronic disease resembling tuberculosis in the field mouse. The bacillus is indistinguishable from *M. tuberculosis* in morphology, staining, and cultural behavior. In guinea pigs it produces an infection that spontaneously regresses and establishes increased resistance to subsequent infection with virulent tubercle bacilli; it has therefore been used as a vaccine in man (see Immunization, above).

Mycobacterium paratuberculosis causes an often fatal enteritis of cattle and sheep, called **Johne's disease,** which is characterized by chronic infiltration of the intestinal mucosa and the mesenteric lymph nodes. The lesions resemble those caused by tubercle bacilli except that they are less localized and lack caseation. Infected cattle give delayed-type reactions to intradermal injections of culture

filtrates (called johnin). The organism does not grow on the media used to cultivate other mycobacteria; it was initially cultivated by the ingenious expedient of enriching the medium with killed tubercle bacilli, and this special requirement can now be satisfied by a novel lipoidal growth factor, mycobactin (an iron-complexing sideramine), obtained from a mycobacterium.

Mycobacterium lepraemurium causes an epizootic chronic disease of wild rats known as rat leprosy; the organism is observed within macrophages in lesions. The disease may be transmitted to mice, guinea pigs, and white rats by inoculation of infected tissue, but the organism, like that of human leprosy, has not been cultivated. Histologically the lesions resemble those of human leprosy, but rats are not susceptible to infection with the human leprosy bacillus.

Mycobacteria of Poikilothermic Vertebrates. Acid-fast bacilli have been isolated from a variety of lesions, some of which resemble tuberculosis, in a number of cold-blooded vertebrates. These bacilli were named according to the species from which they have been isolated but later developments have necessitated a reevaluation of some of these names. Thus *M. ranae* (from frogs) is the same as *M. fortuitum; M. chelonei* (from turtles) is a member of the *M. fortuitum* complex; "*M.*" *thamnopheos* (from snakes) is more like a nocardia species than a mycobacterium; *M. marinum* (from saltwater fish), as was discussed, is capable of producing superficial lesions in man.

Saprobic Mycobacteria. A number of acid-fast bacilli found chiefly on plants and in soil and water grow more rapidly than tubercle bacilli in culture; their role in nature is primarily concerned with the degradation of lipids. These organisms have not been established as pathogens. They include *M. phlei* (the timothy hay bacillus), and *M. smegmatis,* found in smegma and butter; both are also found in dust, soil, and water.

A group of acid-fast bacilli with little or no pathogenicity for man or animals may be found in human specimens and mistaken for potential pathogens. *M. gordonae* is a scotochromogen resembling *M. scrofulaceum; M. gastri* and *M. terrae-triviale* are nonchromogenic; and *M. flavescens* is a scotochromogen with a moderately rapid growth rate.

LEPROSY

MYCOBACTERIUM LEPRAE

Though *M. leprae* **(Hansen's bacillus)** grows profusely in lesions it has never been cultivated in the test tube, and until 1959 it had never been convincingly transmitted to laboratory animals. Despite these limitations, it has long been recognized as the etiologic agent of human leprosy, for the organism is readily demonstrated, often in great numbers, in stained smears of exudates of persons with leprosy, and in tissue sections from lesions, whereas no other organism has been consistently identified in these preparations. Moreover, *M. leprae* is virtually indistinguishable in morphology and staining properties from *M. tuberculosis,* and leprosy has many clinical features in common with tuberculosis.

The failure to cultivate *M. leprae* has hampered its investigation and has made it difficult to test strains for drug sensitivity. The absence of an animal host presented another obstacle. However, in 1960 Shepard discovered that the bacilli could be propagated, though slowly, in the relatively cool environment of the foot pads of mice. This experimental model has made it possible to study the effects of various drugs and also the properties of the bacilli obtained from patients with various forms of leprosy. More recently the nine-banded armadillo was found to develop disseminated disease, and it has yielded enough material for the production of a skin-test Ag.

Lepromin. Though *M. leprae* cannot be cultivated, an antigenic bacillary preparation for skin testing was developed by Mitsuda in 1919 by boiling human lepromatous tissue rich in bacilli. This material, called lepromin, is now standardized to contain 160×10^6 acid-fast bacilli/ml. Preparations from infected armadillo tissues have given comparable skin reactions.

A positive **Mitsuda reaction** consists of the slow development of a papule at the intradermal injection site which, when biopsied at 2–4 weeks, reveals a hypersensitivity granuloma (not merely a granuloma of the foreign-body type). The test is not very helpful in diagnosis, because of false-positive reactions in tuberculosis and in apparently healthy individuals, but it is useful in determining the prognosis and the phase of the disease (see Pathogenesis, below). Since lepromin also contains some soluble material an earlier reaction (24–48 h), typical of delayed hypersensitivity, is often also seen (the **Fernandez reaction).**

PATHOGENESIS

M. leprae causes chronic granulomatous lesions closely resembling those of tuberculosis, with epithelioid and giant cells, but without caseation. The organisms in the lesions are predominantly intracellular and can evidently proliferate within macrophages, like tubercle bacilli.

Leprosy is distinguished by its chronic, slow progress and by its mutilating and disfiguring lesions. These may be so distinctive that the diagnosis is apparent at a glance; or the clinical manifestations may be so subtle as to escape detection by any except the most experienced observers armed with a high index of suspicion. The organism has a predilection for skin and for nerve. In the **cutaneous** form of the disease large, firm nodules (lepromas) are distributed widely, and on the face they create a characteristic leonine appearance. In the **neural** form segments of peripheral nerves are involved, more or less at random, leading to localized patches of anesthesia. The loss of sensation in fingers and toes increases the frequency of minor trauma, leading to secondary infections and mutilating injuries. Both forms may be present in the same patient.

Phases of the Disease. In either form of leprosy three phases may be distinguished. 1) In the **lepromatous** or **progressive** type the lesions contain many **lepra cells:** macrophages with a characteristically foamy cytoplasm, in which acid-fast bacilli are abundant (Fig. 37-9). When these lesions are prominent the lepromin test is usually negative, presumably owing to desensitization by massive amounts of endogenous lepromin, and the cell-mediated immune reactions to specific and nonspecific stimuli are markedly diminished. The disease is then in a progressive phase and the prognosis is poor. 2) In the **tuberculoid** or **healing** phase of the disease, in contrast, the lesions contain few lepra cells and bacilli, fibrosis is prominent, and the lepromin test is usually positive. 3) In the **intermediate** type of disease bacilli are seen in areas of necrosis but are rare elsewhere, the skin test is positive, and the long-range outlook is fair. Shifts from one phase to another, with exacerbation and remission of the disease, are common.

Hansen's bacillus may be widely distributed in the tissues of persons with leprosy, including the liver and spleen. Nevertheless,

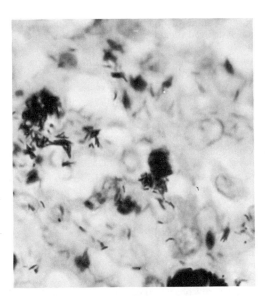

FIG. 37-9. Leprosy bacilli in cells in section of human tissue. Acid-fast stain, methylene blue counterstain. (Courtesy of C. H. Binford, Leonard Wood Memorial) (×1200)

no destructive lesions or disturbance of function are observed in these organs. Most deaths in leprous patients are due not to leprosy per se but to intercurrent infections with other microorganisms—often tuberculosis. Leprosy itself often causes death through the complication of **amyloidosis,** which is characterized by massive waxy deposits in kidneys, liver, spleen, and other organs. This curious disorder also occurs as a sequel to a variety of other chronic diseases with extensive necrosis and suppuration, or to prolonged, intensive immunization of animals with diverse antigens, especially some bacterial toxins. The waxy deposits contain abundant precipitates of fragments of immunoglobulin light chains.

DIAGNOSIS AND TREATMENT

Bacteriologic diagnosis is accomplished by demonstrating acid-fast bacilli in scrapings from ulcerated lesions, or in fluid expressed from superficial incisions over non-ulcerated lesions. Also useful is the skin test with lepromin in the tuberculoid phase. No useful serologic test is available, but patients with leprosy frequently have a false-positive serologic test for syphilis.

Therapy with **dapsone (diaminodiphenylsulfone)** or related compounds usually produces a gradual improvement over several years, and is continued for a prolonged period after apparent clinical remission. Resistance to sulfone drugs may be noted after years of treatment. Rifampin and clofazimine (B663, a phenazine derivative) are promising agents now under investigation, and transfer factor (Ch. 22) has been reported to be beneficial in a few patients with lepromatous disease. Treatment results may be evaluated by counting the acid-fast bacilli in serial biopsies and skin scrapings.

EPIDEMIOLOGY

Leprosy is apparently transmitted when exudates of mucous membrane lesions and skin ulcers reach skin abrasions; it is not highly contagious and patients need not be isolated. Young children appear to acquire the disease on briefer contact than adults. The incubation period is estimated to range from a few months to 30 years or more. Apparently *M. leprae* can lie dormant in tissues for prolonged periods. The prophylactic use of BCG vaccine or dapsone for contacts has not been successful, although immunologic tests with lepromin (lymphocyte transformation, leukocyte migration inhibition) suggest that inapparent infection is common among intimate contacts.

In ancient times leprosy was rampant throughout most of the world, but for unknown reasons it died out in Euorpe in the sixteenth century and now occurs there only in a few isolated pockets. In the United States leprosy occurs particularly in Texas, California, Louisiana, Florida, New York City, and Hawaii, usually in persons from Puerto Rico, the Philippines, Mexico, Cuba, or Samoa. Several hundred patients are cared for at the national leprosarium in Carville, Louisiana. Recent estimates place the number of lepers in the world at large at about 3 million, with the greatest density in central Africa and in parts of the Orient.

SELECTED READING

BOOKS AND REVIEW ARTICLES

American Thoracic Society and Center for Disease Control. BCG vaccines for tuberculosis. Am Rev Respir Dis 112:478, 1975

ARNASON BG, WAKSMAN BH: Tuberculin sensitivity. Adv Tuberc Res 13:1, 1964

ARONSON JD: Tuberculosis of cold blooded animals. A review. Leprosy Briefs 8:21, 29, 1957

BARKSDALE L, KIM KS: Mycobacterium. Bacteriol Rev 41:217, 1977

BCG: A discussion of its use and application. Adv Tuberc Res 8:1, 1957

CHAPMAN JS: The Atypical Mycobacteria and Human Mycobacteriosis. New York, Plenum Medical, 1977

Ciba Foundation Study Group No. 15: The Pathogenesis of Leprosy. Boston, Little, Brown, 1963

DUBOS RJ, DUBOS J: The White Plague. Boston, Little, Brown, 1952

FOGAN L: Atypical mycobacteria: their clinical, laboratory, and epidemiologic significance. Medicine 49:243, 1970

FOX W, MITCHISON DA: Short course chemotherapy for pulmonary tuberculosis. Am Rev Respir Dis 111:325, 1975

GODAL T, MYRVANG B, STANFORD JL, SAMUEL DR: Recent advances in the immunology of leprosy with special reference to new approaches in immunoprophylaxis. Bull Inst Pasteur 72:273, 1974

GOREN MB: Mycobacterial lipids: selected topics. Bacteriol Rev 36:33, 1972

Immunization in Tuberculosis. Fogarty International Center Proceedings No. 14. Washington DC, DHEW Publ No. (NIH) 72–68, 1971

International Conference on Mycobacterial and Fungal Antigens. Am Rev Respir Dis 92:Suppl, 1965

JOHNSTON RF, WILDRICK KH: The impact of chemotherapy on the care of patients with tuberculosis. Am Rev Respir Dis 109:636, 1974

KUBICA GP, GROSS WM, HAWKINS JE et al: Laboratory services for mycobacterial diseases. Am Rev Respir Dis 112:773, 1975

LINCOLN EM: Disease in children due to mycobacteria other than *Mycobacterium tuberculosis*. Am Rev Respir Dis 105:683, 1972

LURIE MB: Resistance To Tuberculosis. Cambridge, Harvard University Press, 1964

NOLL H: The chemistry of cord factor, a toxic glycolipid of M. tuberculosis. Adv Tuberc Res 7:149, 1956

RATLEDGE C: The physiology of mycobacteria. Adv Mircrob Physiol 13:115, 1976

RICH AR: The Pathogenesis of Tuberculosis, 2nd ed. Springfield, IL, Thomas, 1951

RUNYON EH: Anonymous mycobacteria in pulmonary disease. Med Clin North Am 43:273, 1959

RUNYON EH: Pathogenic mycobacteria. Adv Tuberc Res 14:235, 1965

SHEPARD CC: The first decade in experimental leprosy. Bull WHO 44:821, 1971

VESTAL A: Procedures for the Isolation and Identification of Mycobacteria. Washington DC, DHEW Publ No. (CDC) 76–8230, 1976

WOLINSKY E: Nontuberculous mycobacteria and associated diseases. Am Rev Respir Dis 119:107, 1979

SPECIFIC ARTICLES

ALBERT RK, SBARBORO JA, HUDSON LD, ISEMAN M: High dose ethambutol: its role in intermittent chemotherapy. Am Rev Respir Dis 114:699, 1976

BIRNBAUM SE, AFFRONTI LF: Mycobacterial polysaccharides. II. Comparison of polysaccharides from strains of four species of mycobacteria. J Bacteriol 100:58, 1969

BRITISH THORACIC AND TUBERCULOSIS ASSOCIATION: Short course chemotherapy in pulmonary tuberculosis. Lancet 2:1102, 1976

BULLOCK WE: Studies of immune mechanisms in leprosy. I. Depression of delayed allergic response to skin test antigens. N Engl J Med 278:298, 1968

CHAPMAN JS: The ecology of the atypical mycobacteria. Arch Environ Health 22:41, 1971

COMSTOCK GW, LIVESAY VT, WOOLPERT SF: Evaluation of BCG vaccination among Puerto Rican children. Am J Public Health 64:283, 1974

FELDMAN RA, STURDIVANT M: Leprosy in the United States, 1950–1969: an epidemiological review. South Med J 69:970, 1976

FOX W: Changing concepts in the chemotherapy of pulmonary tuberculosis. Am Rev Respir Dis 97:767, 1976

GROVE DI, WARREN KS, MAHMOUD AAF: Algorithms in the diagnosis and management of exotic diseases. IV. Leprosy. J Infect Dis 134:205, 1976

KANETSUNA F: Chemical analysis of mycobacterial cell walls. Biochim Biophys Acta 158:130, 1968

KATO M, MIKI K, MATSUNAGA K, YAMAMURA Y: Biologic and biochemical activities of "cord factor" with special reference to its role in the virulence of tubercle bacilli. Am Rev Respir Dis 77:482, 1958

LEDERER E: The mycobacterial cell wall. Pure Appl Chem 25:135, 1971

LEFFORD MJ: Delayed hypersensitivity and immunity in tuberculosis. Am Rev Respir Dis 111:243, 1975

LURIE MB: The fate of tubercle bacilli ingested by phagocytes derived from normal and immunized animals. J Exp Med 75:247, 1942

MACKANESS GB: The immunologicial basis of acquired cellular resistance. J Exp Med 120:105, 1964

MCCONVILLE JH, RAPOPORT MI: Tuberculosis management in the mid-1970s. JAMA 235:172, 1976

MEDICAL RESEARCH COUNCIL: BCG and vole bacillus vaccines in the prevention of tuberculosis in adolescence and early life. Third report. Br Med J 1:973, 1963

RUNYON EH: Identification of mycobacterial pathogens utilizing colony characteristics. Am J Clin Pathol 54:578, 1970

SALVIN SB, NETA RA: A possible relationship between delayed hypersensitivity and cell-mediated immunity. Am Rev Respir Dis 111:373, 1975

TOIDA I: Effects of cord factor on microsomal enzymes. Am Rev Respir Dis 110:641, 1974

YOUMANS GP: Relation between delayed hypersensitivity and immunity in tuberculosis. Am Rev Respir Dis 111:109, 1975

chapter

ACTINOMYCETES: THE FUNGUS-LIKE BACTERIA

GEORGE S. KOBAYASHI

GENERAL CHARACTERISTICS

The actinomycetes are gram-positive organisms that tend to grow slowly as branching filaments. In some genera the filaments readily segment during growth and yield pleomorphic, club-shaped cells that resemble corynebacteria and mycobacteria; some are acid-fast.

Since the filamentous growth leads to mycelial colonies, and some actinomycetes cause chronic subcutaneous granulomatous abscesses much like those caused by fungi (Ch. 45), the actinomycetes (Gr. *aktino,* ray, and *mykes,* mushroom or fungus) were long regarded as fungi. (The term ray refers to the characteristic radial arrangement of club-shaped elements seen in microcolonies in infected tissues.)

Nevertheless, the following properties establish the actinomycetes as bacteria.

1) They are **prokaryotic** (i.e., lack a nuclear membrane).

2) The diameter of their filaments is 1μm or less, i.e., much narrower than the characteristic tubular structures (hyphae) of molds (see Molds, Ch. 45). The filaments of actinomycetes readily segment into bacillary and twig-like forms, with dimensions typical of bacteria.

3) Their cell walls contain muramic acid and diaminopimelic acid (lysine in some species), characteristic of bacterial walls (see Ch. 6), and they lack the chitin and glucans that are characteristic of cell walls of molds and yeasts (see Cell Wall, Ch. 45).

4) Their growth is inhibited by penicillins, tetracyclines, sulfonamides, and other antibacterial drugs that are innocuous for fungi.

5) Their cell membranes do not contain sterols and are therefore insensitive to the fungicidal polyenes (see Chemotherapy, Ch. 45).

6) Genetic recombination, observed in one group, *Streptomyces,* involves a transmissible fertility factor and the formation of merozygotes, rather than the zygote formation seen with fungi.

7) Motile forms possess simple flagella of bacterial type.

8) Anaerobic and chemoautotrophic forms are known. Together with the mycobacteria, the actinomycetes make up the order **Actinomycetales*** (Table 38-1). *Actinomyces* organisms grow as branching filaments when freshly cultivated but in older cultures and in lesions an increased septation leads to spontaneous fragmentation into bacillary and even coccoid elements indistinguishable from diphtheroids. Many *Nocardia* organisms are acid-fast and contain mycolic acids, like mycobacteria (see Fig. 37-3). *Streptomyces* organisms are more funguslike, forming aerial mycelia and chains of asexual spores.

On primary isolation actinomycetes can usually be identified to the genus level on the basis of morphologic features alone, but organisms that have been repeatedly transferred in culture may not retain the typical morphologic features and therefore require biochemical methods for classification (Table 38-2). Serologic methods are also used, but crossreactions among actinomycetes are extensive.

Actinomycosis was long assumed to arise from exogenous infection by soil organisms. However, it is now clear that the bulk of the infections are endogenous, arising from normal inhabitants of the oral cavity. Whether the infections due to inhalation of soil organisms should be referred to as nocardiosis, or should still be included under actinomycosis, is not yet settled. With both groups inexorable extension of the well-established lesions reflects a poor immune response.

Another disease, called **farmer's lung,** occurs among agricultural workers who have inhaled dust from moldy plant material; it has been traced to allergy to at least three thermophilic actinomycetes, *Thermopolyspora polyspora, Micromonospora vulgaris,* and *Micropolyspora faeni.* A similar allergic pneumonitis is caused by allergens produced by various species of fungi, particularly those in the genus *Aspergillus* (Ch. 45).

ACTINOMYCES

Several species of *Actinomyces,* including *A. israelii, A. bovis, A. naeslundii, A. odontolyticus,* and *A. viscosus,* and

---

* In this chapter we follow the practice of referring to all members of the order Actinomycetales, except for the mycobacteria, as actinomycetes (see Table 38-1).

TABLE 38-1. Members of the Order Actinomycetales*

Actinomycetes	Family: Mycobacteriaceae
	Genus: *Mycobacterium*
	Family: Actinomycetaceae
	Genus: *Actinomyces*
	Genus: *Arachnia*
	Family: Dermatophilaceae
	Genus: *Dermatophilus*
	Family: Actinoplanaceae
	Genus: *Actinoplanes*
	Genus: *Spirillospora*
	Family: Micromonosporaceae
	Genus: *Micromonospora*
	Genus: *Thermoactinomyces*
	Family: Streptomycetaceae
	Genus: *Streptomyces*
	Family: Nocardiaceae
	Genus: *Nocardia*

*Based on the classification of HA and MP Lechevalier, and of L Pine and LK Georg.

the related organism *Arachnia propionica,* have been implicated as the cause of actinomycosis in humans and animals. *A. israelii* is usually responsible for the disease in man and *A. bovis* in cattle. Differences in cell wall composition, tolerance to oxygen, and metabolic products distinguish the various species. These organisms are anaerobic or microaerophilic and require rich media (e.g., containing blood or brain–heart infusion); growth is stimulated by CO_2 and is poor at temperatures below 37° C.

Most strains of *A. israelii* form rough colonies on agar and grow at the bottom of broth tubes as discrete aggregated clumps ("bread crumbs"; Fig. 38-1). A few strains, however, and most strains of *A. bovis,* form smooth colonies and tend to grow more diffusely in broth, and some strains of both species form intermediate colonies. In anaerobic cultures on brain–heart agar macroscopic colonies commonly mature in about 7 days. Gram stains of cultures reveal highly pleomorphic, irregular, club-shaped, gram-positive rods. Occasional cells appear as branched twigs, but long branching filaments are not observed (Fig. 38-1).

PATHOGENESIS

Because most actinomycetes are found in soil, and because human and bovine actinomycosis often affects the jaw and face, this disease was long held to arise from chewing straw and grass. More recent studies, however, have shown that *A. israelii* and *A. bovis* are apparently not present in soil or on vegetation, but are quite regularly isolable from the normal mouth of man and of cattle, respectively. Like many respiratory pathogens, they are much more frequently commensal than pathogenic. *A. israelii,* for example, can be cultured from the majority of human tonsils and is nearly always found in scrapings of gums and teeth. Its fastidious growth requirements, already noted, are in accord with this parasitic (or saprobic) existence.

The conditions that lead the organism to become invasive are not definitely known but may involve trauma (including dental surgery), aspiration of a heavily contaminated tooth or detached bits of dental tartar, or, rarely, human bites.

Actinomycosis is characterized by chronic, destructive

TABLE 38-2. Major Constituents in Cell Wall Preparations of the Most Important Actinomycetes*

Cell wall type	L-DAP†	Meso-DAP	Lysine	Orni-thine	Aspartic acid	Gly-cine	Arabi-nose	Galactose	Distribution and growth requirements
I. *Streptomyces*	+					+			
II. *Micromonospora* *Actinoplanes*		+				+			
III. *Streptosporangium* *Dermatophilus* *Thermoactinomyces* *Microbispora* *Actinomadura*‡		+							Aerobic; simple media; usually found in soil
IV. *Nocardia* *Mycobacterium*§		+					+	+	
V. *Actinomyces* (*israelii*-type)			+	+					Anaerobic; rich media; usually associated with animals
VI. *Actinomyces* (*bovis*-type)			+		+				

*All preparations contain major amounts of glucosamine, muramic acid, alanine, and glutamic acid.

†DAP = 2, 6-diaminopimelic acid.

‡Formerly *Nocardia* (*madurae*-type).

§See Chapter 37.

(Modified according to suggestions of HA Lechevalier, from Lechevalier HA, Lechevalier MP: Annu Rev. Microbiol 21:71, 1967). Reproduced with permission from the Annual Review of Microbiology, Vol. 21. © 1967 by Annual Reviews Inc)

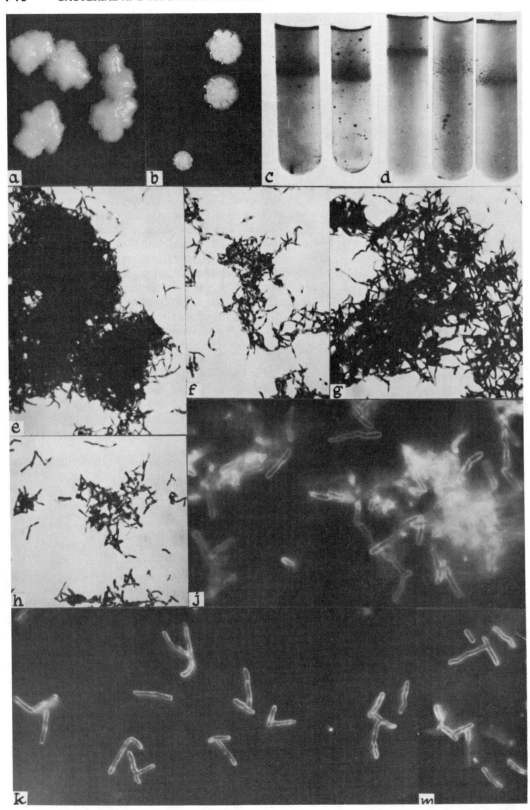

abscesses of connective tissues. In one series of patients in the United States about 50% of the infections involved the abdomen, particularly the cecum and appendix; 20% the lungs and chest wall; and 30% the face and neck. In a larger series in England 60% of all cases involved the cervicofacial area.

Wherever the lesions occur they are basically the same. Abscesses expand into contiguous tissues and eventually form burrowing, tortuous sinuses to the skin surface, where they discharge purulent material. Penetration into mucous membranes is much less frequent. Connective and granulation tissues tend to form a wall around the abscess. Histologically, the lesions are not distinctive except for the presence of the organisms as small colonies, described under Diagnosis, below ("sulfur granules"; Fig. 38-2).

Actinomycotic lesions of cattle are characteristically large, bone-destroying abscesses of the lower jaw, often referred to as "lumpy jaw." In humans, however, lesions of bone are infrequent.

Like most diseases caused by organisms that ordinarily are saprobes actinomycosis is not transmissible from man to man or from animals to man. It is, in fact, even difficult to establish the infection experimentally in laboratory animals. Repeated inoculation of guinea pigs with *A. israelii* does, however, occasionally establish progressive and fatal infection, while the hamster develops localized abscesses when injected intraperitoneally. The actinomycetes isolated from abscesses are no more effective in establishing experimental infections than those isolated from the normal mouth.*

DIAGNOSIS

When pus from an abscess or infected sputum is examined carefully yellow **sulfur granules,** named for their color, are occasionally seen. These small clusters of colonies of actinomycetes range from barely visible to several millimeters in diameter. Detection of granules is not required to establish a diagnosis of actinomycosis, but their presence facilitates identification of the organism. The granules, made up of one or more colonies embedded in a matrix of calcium phosphate, consist of a central filamentous mycelium surrounded by club-shaped structures in a characteristic radial arrangement

* *Streptobacillus moniliformis* (Ch. 43), one of the causes of rat-bite fever, may be mentioned here because it is a normal inhabitant of the mouth of rats, is also pleomorphic, and often grows in filamentous form, and so it has been called *Actinomyces muris.* The disease that it causes in man, however, is an acute, self-limiting septicemia, with rash and arthritis.

(Fig. 38-2). When granules are crushed and stained the organisms appear as gram-positive bacillary and diphtheroid forms. Branched filaments are not readily discerned, and the club-shaped elements are gram-negative.

Actinomycotic abscesses are almost invariably mixed infections, like many other abscesses; even the washed sulfur granules may contain colonies of fusiform bacilli, anaerobic streptococci, and a tiny anaerobic gram-negative bacillus that bears the formidable name *Bacterium actinomycetemcomitans.* The accompanying bacilli may secrete collagenase and hyaluronidase and thus facilitate extension of the lesion.

Anaerobic diphtheroids and a facultative anaerobic actinomycete, *A. naeslundii,* look very much like *A. israelii* and are often found in anaerobic cultures of the oral cavity. *A. naeslundii* has been isolated from a variety of human lesions, and it can cause lesions in mice.

EPIDEMIOLOGY

Actinomycosis is worldwide in distribution but is relatively rare. Its incidence is higher in men than in women, and in persons over 20; in older reports it was much more frequent in rural than in urban areas. The allegedly greater incidence among farmers is not readily reconcilable with the prevailing view that the disease arises from invasion by an indigenous organism.

Actinomycosis occurs in a variety of wild and domesticated animals besides man and cattle.

TREATMENT AND PROGNOSIS

Actinomyces, like the fungi that cause subcutaneous mycoses (Ch. 45), does not give rise to an effective immune response, and in the absence of therapy these chronic lesions tend to spread slowly but inexorably. Prior to the availability of chemotherapy the prognosis was poor.

A. israelii is sensitive to penicillin, tetracyclines, chloramphenicol, and streptomycin. Penicillin is reported to be most effective clinically. Surgical drainage of abscesses and resection of damaged tissue are important adjuncts to chemotherapy. With combined chemotherapy and surgery the cure rate is now about 90% for cervicofacial actinomycosis, but somewhat less for abdominal and thoracic actinomycosis.

FIG. 38-1. *Actinomyces israelii.* **a,b.** Rough colonies grown anaerobically on brain–heart agar. **c,d.** Shake cultures incubated anaerobically; note growth in a layer below the surface. **e–h.** Gram-stained smears, showing masses of filaments. (×750, reduced). **j–m.** Unstained wet films under darkfield illumination; note distinct branching and twiglike forms. (×1200, reduced) (Rosebury T et al: J Infect Dis 74:131, 1944)

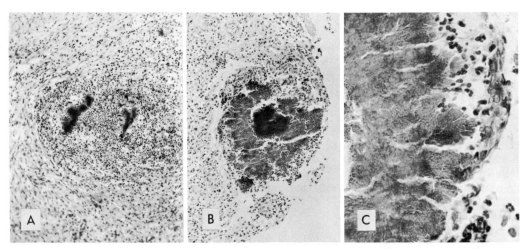

FIG. 38-2. A. Colonies of *A. israelii* (sulfur granules) in a lung abscess. A higher magnification of the periphery of the granule in **B** is reproduced in **C**, showing characteristic radial clublike structures. (Hematoxylin–eosin; **A** and **B**, ×130, reduced; **C**, ×600, reduced) (**A**, Emmons CW et al: Medical Mycology. Philadelphia, Lea & Febiger, 1963; **B** and **C**, courtesy of Dr. A. C. Sonnenwirth)

NOCARDIA

In contrast to *Actinomyces,* species of *Nocardia* are inhabitants of soil rather than commensals in animals. They are aerobic, grow readily over a wide temperature range, and grow on relatively simple media (e.g., Sabouraud's glucose agar*).

When grown on agar the colonies of *Nocardia* species may be smooth and moist, or rough with a velvety surface due to a rudimentary aerial mycelium. However, when smeared and stained the filaments fragment into bacillary and coccoid bodies. Examination of liquid cultures, especially slide cultures, may be helpful in visualizing branching filaments. Growth in liquid media usually produces a dry, waxy surface pellicle, as with mycobacteria.

Nocardia species are all gram-positive, and the two species most often pathogenic in man, *N. asteroides* and *N. brasiliensis,* are also somewhat acid-fast: they are more easily stained with fuchsin and retain the stain less tenaciously than mycobacteria.† Young cultures tend to be more acid-fast than older cultures.

N. asteroides also resembles mycobacteria in being resistant to dilute alkali and to some of the dyes (such as brilliant green) used to inhibit the growth of rapidly growing gram-positive bacteria; hence they may grow on

---

* This medium is a particular favorite for the cultivation of fungi, and its occasional use with *Nocardia* is another vestige of the former view that these organisms are fungi.

† The acid-fast nocardiae are readily stained with basic fuchsin without heating, and to test for acid-fastness 1% H_2SO_4 without ethanol is used, since acid plus alcohol is too effective as a decolorizing solvent (cf. Morphology, Ch. 37).

the same medium and after the same manipulations as used for routine selective isolation of tubercle bacilli from sputum and exudates (see Laboratory Diagnosis, Ch. 37). Moreover, the colonies formed by nocardiae resemble those of saprobic mycobacteria and the unclassified mycobacteria (Ch. 37 and Fig. 38-3). Finally, extensive serologic crossreactions (in agglutination and complement fixation) are observed with antisera to mycobacteria and to nocardiae. Despite these similarities, however, it is not difficult to distinguish between the two groups: nocardiae grow more rapidly, are less acid-fast, and tend to branch.

PATHOGENESIS

As was noted, *Nocardia* species are regularly isolable from soil, and there are two common modes of infection. Pulmonary nocardiosis arises from inhalation of the organisms, while chronic subcutaneous abscesses (**mycetomas**) arise from contamination of skin wounds, usually on the feet and hands of laborers.

Infections with *Nocardia* are hard to establish in laboratory animals. Suspension of the organisms in gastric mucin enhances their pathogenicity, and guinea pigs inoculated with such suspensions regularly develop abscesses and occasionally succumb.

Pulmonary Nocardiosis. In pulmonary nocardiosis, caused by *N. asteroides,* the lesions may be scattered through the lung parenchyma, simulating miliary tuberculosis or histoplasmosis. Alternatively, they may take the form of larger, confluent, partially excavated abscesses, which superficially resemble the cavities of chronic pulmonary tuberculosis. The lesions are charac-

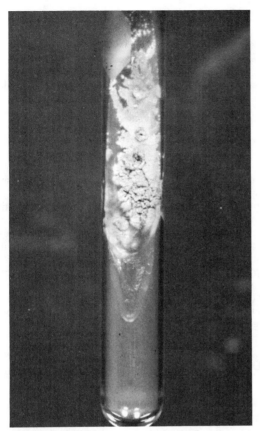

FIG. 38-3. Colonies of *Nocardia asteroides* after 4 weeks on Sabouraud's glucose agar. Note the typical heaped-up irregularly folded appearance. (Courtesy of Dr. A. C. Sonnenwirth)

terized histologically by suppuration, with granulation and fibrous tissue surrounding the areas of necrosis. Neither the characteristic granulomas of tuberculosis nor the burrowing sinuses of *Actinomyces* abscesses are seen. *N. asteroides* lies scattered through the abscesses in the form of tangled, fine, branching filaments: aggregation into granules does not occur, in contrast to the actinomycotic lesions due to *A. israelii* or *A. bovis*.

In histologic sections the bacterial filaments are not seen with hematoxylin–eosin stains. When visualized with bacterial stains the organisms (which tend to fragment during staining) appear as gram-positive, weakly acid-fast diphtheroid and bacillary forms, which may easily be mistaken for tubercle bacilli.

N. asteroides often spreads from pulmonary lesions by way of the blood stream and establishes metastatic abscesses, usually in subcutaneous tissues and in the central nervous system. Lesions in the brain and the meninges are usually fatal.

N. asteroides is occasionally identified in sputum from patients with chronic pulmonary diseases of unknown etiology. When it is isolated consistently from an individual a presumptive diagnosis of pulmonary nocardiosis is warranted, but isolation in a solitary

specimen raises the possibility of its occasional presence as a saprobe in the upper respiratory tract.

Mycetoma Due to *Nocardia*. Different species of *Nocardia* are associated with mycetomas in different parts of the world, e.g., *N. brasiliensis* in Mexico. These chronic subcutaneous abscesses are clinically very similar to those due to *Streptomyces* and to various fungi. The abscesses spread locally by direct extension, destroy soft tissue and bone, and form burrowing sinus tracts; colonial aggregates (granules) of the causative microorganism are often present in the abscesses and in their purulent discharge. The detection of granules helps in the isolation and identification of the etiolotic agent.

A variety of properties distinguish among *N. asteroides*, *N. brasiliensis*, and some species of *Streptomyces* that also cause mycetomas. These are listed in Table 38-3.

EPIDEMIOLOGY

Though nocardiae are worldwide in distribution, nocardiosis is a rare disease. It seems likely, therefore, that man has a high degree of innate resistance. Indeed, untreated pulmonary nocardiosis is sometimes a self-limited disease, in contrast to actinomycosis due to *Actinomyces*. Generally, however, once nocardiosis becomes evident the disease tends to be progressive and fatal; even with intensive therapy about 50% of patients succumb.

TREATMENT

Various antibacterial drugs are used in the treatment of experimental nocardiosis, but sulfonamides are reported to be most effective. Draining of abscesses is an important adjunct. The distinction from fungal mycetomas is essential because entirely different chemotherapeutic agents are indicated for the latter (see Chemotherapy, Ch. 45).

STREPTOMYCES

Streptomyces species are characterized by the stability of their filaments and by the formation of spores on the aerial mycelia that project above the surface of the culture medium. Mutants that lose the ability to form aerial mycelia and spores are difficult to distinguish from nocardiae.

With increasing appreciation of the distinction between nocardiae and streptomycetes (Table 38-3) it has been realized that both cause actinomycotic abscesses. Since streptomycetes are ubiquitous in soil, infection is attributed to contamination of scratches and penetrating

TABLE 38-3. Some Aerobic Organisms that Cause Actinomycotic Mycetomas in Man

Species	Microscopic appearance	Hydrolysis of casein	Hydrolysis of starch	Pathogenicity in mouse and guinea pig
Nocardia asteroides	Partially acid-fast	0	0	Abscesses without granules; frequently causes death
Nocardia brasiliensis	Partially acid-fast	+	0	Abscesses with granules caused by some strains
*Streptomyces madurae**	Not acid-fast	+	+	Not pathogenic
*Streptomyces pelletierii**	Not acid-fast	+	0	Not pathogenic

*Sometimes placed in genus *Nocardia* rather than *Streptomyces*. In complement-fixation tests they react with antisera to various mycobacteria, nocardiae, and streptomycetes.

wounds. Mycetomas due to streptomycetes are indistinguishable clinically from those due to other actinomycetes. Identification of these organisms is important since they are not generally susceptible to antimicrobial agents and surgical removal by excision or amputation of the affected area must be considered.

Since the isolation by Waksman of actinomycin in 1940, and streptomycin in 1943, the streptomycetes have received a phenomenal amount of attention. Innumerable isolates from soil samples, taken from all parts of the world, have been systematically scrutinized and have yielded over 500 different compounds, including

most antibiotics of practical value except penicillin and griseofulvin (see Ch. 7).

Streptomycetes produce more than 90% of the therapeutically useful antibiotics. Problems exist in the identification and taxonomic grouping of these bacteria, partly because many strains have great economic value and hence are not generally available to the scientific community for taxonomic studies.

Streptomycetes and other actinomycetes are widespread scavengers in soil, breaking down proteins, cellulose, and other organic matter (including paraffin). They probably contribute as much as all other bacteria and fungi combined to the fertility of the soil and to the geochemical stability of the biosphere.

SELECTED READING

BOOKS AND REVIEW ARTICLES

BRADLEY SG: Significance of nucleic acid hybridization to systematics of actinomycetes. Adv Applied Microbio 19:59, 1975

BRONNER M, BRONNER M: Actinomycosis. Bristol, John Wright & Sons, 1971

BROWNWELL G, GOODFELLOW M, SERRANO J (eds): The Biology of the Nocardiae. New York, Academic Press, 1976

EMMONS CW, BINFORD CH, UTZ JP, KWON-CHUNG KJ: Medical Mycology, 3rd ed. Philadelphia, Lea & Febiger, 1977

GEORG LK: Diagnostic procedures for the isolation and identification of the etiologic agents of actinomycosis. In Proceedings, International Symposium on Mycoses. Scientific Publ No. 205, p 71. Washington DC, Pan American Health Organization, 1970

LECHEVALIER HA: Production of the same antibiotic by members of different genera of microorganisms. Adv Applied Microbiol 19:25, 1975

ROSEBURY T: Microorganisms Indigenous to Man. New York, McGraw-Hill, 1962

WAKSMAN S: The Actinomycetes, Vols 1, 2, 3 (with Lechevalier HA). Baltimore, Williams & Wilkins, 1962

SPECIFIC ARTICLES

BEAMAN BL: Structural and biochemical alterations of *Nocardia asteroides* cell walls during its growth cycle. J Bacteriol 123:1235, 1975

BERD D: Laboratory identification of clinically important aerobic actinomycetes. Appl Environ Microbiol 25:665, 1973

BERD D: *Nocardia asteroides:* a taxonomic study with clinical correlations. Am Rev Respir Dis 108:909, 1973

CUMMINS CS, HARRIS H: Studies on the cell-wall composition and taxonomy of Actinomycetales and related groups. J Gen Microbiol 18:173, 1958

EDWARDS JH: The isolation of antigens associated with farmer's lung. Clin Exp Immunol 11:341, 1972

HARDISSON C, MANZANAL MB: Ultrastructural studies of sporulation in Streptomyces. J Bacteriol 127:1443, 1976

LECHEVALIER MP, HORAN AC, LECHEVALIER HA: Lipid composition in the classification of nocardiae and mycobacteria. J Bacteriol 105:313, 1971

REDSHAW PA, MCCANN PA, SANKARAN L, POGELL BM: Control of differentiation in streptomycetes: involvement of extrachromosomal deoxyribonucleic acid and glucose repression in aerial mycelial development. J Bacteriol 125:698, 1976

chapter 39

THE SPIROCHETES

P. FREDERICK SPARLING
JOEL B. BASEMAN

GENERAL CHARACTERISTICS

Spirochetes are motile, unicellular, spiral-shaped organisms, morphologically quite different from other bacteria. Certain structural features are shared by all spirochetes (Fig. 39-1), but in physiology they range from obligate anaerobes to aerobes, and from free-living forms to obligate parasites. Many are not yet cultivable.

Three genera are pathogenic for man. The genus *Treponema* includes the pathogens that cause syphilis (*T. pallidum*), yaws (*T. pertenue*), and pinta (*T. carateum*). The genus *Borrelia* includes the causative agents of epidemic and endemic relapsing fever. The genus *Leptospira* includes a wide variety of small spirochetes that cause mild to serious systemic human illness (Fig. 39-2). In the United States syphilis remains a disease of high incidence, ranking third (behind gonorrhea and varicella) in reported infectious diseases. Over 23,000 cases of early infectious syphilis were reported in 1976, which may be less than one-third of the actual cases. Only approximately 100 new cases of leptospirosis are reported in the United States yearly, and fewer of borreliosis.

Spirochetes (Gr. *spira,* coil, and *chaete,* hair) are relatively long, slender, flexible organisms, and most appear as helical coils, resembling a spring, or as undulating waves of varying amplitude. Many are too slender to be seen by ordinary light microscopy. They can be visualized by darkfield microscopy* (Fig. 39-3), or by staining with special reagents such as silver salts. When visualized by darkfield illumination spirochetes have a characteristic motility, including apparent rotation around their long axis, flexion, and a boring corkscrew motion. Species vary somewhat in motility, which may be useful in diagnosis. **Spirochetal motion persists at high viscosities,** which block the flagellar motion of ordinary bacteria; this finding suggests a possible basis for the evolution of the complex structure of spirochetes.

All spirochetes contain a central protoplasmic cylinder, consisting of cytoplasmic and nuclear regions surrounded by a cytoplasmic membrane and a closely adherent cell wall. The cell wall is a peptidoglycan similar to that of other bacteria, containing either lysine, diaminopimelic acid, or ornithine. The peptidoglycan, when isolated, retains a helical configuration, indicating that it determines the shape of the cell.

Structures resembling microtubules can be seen in the cytoplasm of most treponemes. Their function and relationship to microtubules of eukaryotic cells are unknown.

Between the helical cylinder and the outer envelope are one or more filamentous structures, termed **axial fibrils.** These are similar to flagella of other bacteria, both in appearance under the electron microscope (Fig. 39-4 and 39-5) and in chemical content. The axial fibrils are attached subterminally by characteristic organelles to either end of the protoplasmic cylinder and extend along the body of the organism, overlapping in the center. They probably play a role in motility, but the mechanism is unknown.

The outer envelope is a membrane similar in structure to the outer membrane of gram-negative bacteria. Large spirochetes stain gram-negative, but there is no clear evidence for endotoxin in any spirochetes. Previously reported treponemal encystment is probably an artefact of cell death.

SYPHILIS

HISTORY

During the last decade of the fifteenth century syphilis became rampant in Western Europe. The story that it was brought from the West Indies by members of Columbus' crew is probably apocryphal. Records indicate, however, that the disease in Europe, in its acute phase, was extraordinarily malignant for about 60 years, beginning in the 1490s; thereafter, it became milder, though its late consequences remained serious. It now exists endemically in nearly all parts of the world.

Starting with the poem in which the early epidemiologist Fracastorius (1530) named the disease after a mythical shepherd, syphilis has been the subject of an extensive and colorful literature. The causative agent was discovered by Schaudinn and Hoffmann in

* Darkfield microscopy requires the use of a special substage condenser which throws the light across rather than directly through the field. Only light that is reflected, refracted, or diffracted by an object in the field reaches the eye. Hence observed objects are light and the background is dark (Fig. 39-3).

FIG. 39-1. Schematic representation of a spirochete. **Broken line** represents the outer envelope. **Solid heavy line,** adjacent to the broken line, delimits the protoplasmic cylinder. **Thin solid lines** represent axial fibrils, one of which is inserted at each end of the protoplasmic cylinder. (Canale-Parola E: Bacteriol Rev 41:181, 1977)

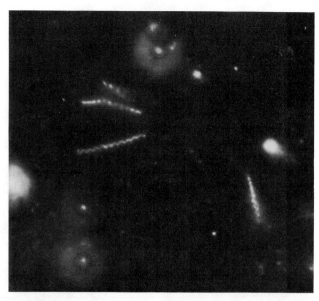

FIG. 39-3. *T. pallidum* in a darkfield preparation. (From Microbiology by Pelczar MJ, Reid RD. Copyright 1958, McGraw-Hill. Used with permission of McGraw-Hill Book Company) (× about 2000)

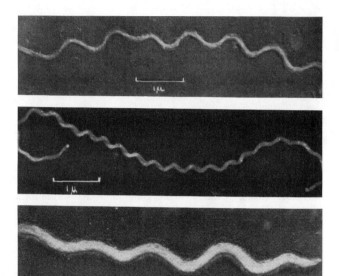

FIG. 39-2. Electron microscopic study of the morphology of pathogenic spirochetes. **Top:** *Treponema pallidum* from exudate of infected rabbit. (×16,500) **Middle:** *Leptospira interrogans* serotype *icterohaemorrhagiae* from broth culture. (×17,000) **Bottom:** *Borrelia duttonii* from the blood of an infected mouse. (×8,500) (Swain RHA: J Path Bact 69:117, 1955)

1905, and the "Wassermann reaction" for detecting antisyphilitic antibodies was described by Wassermann, Neisser, and Bruck in 1906. Widespread treatment of syphilis in the United States with penicillin, beginning in the mid-1940s, led to a decline of more than 85% in the incidence of early cases, but after 1955 newly acquired infections increased again. This resurgence has been attributed to relaxation of control measures and to increased sexual promiscuity, but a decrease in the indiscriminate use of penicillin for minor ailments is probably also a significant factor.

MORPHOLOGY

T. pallidum, which is morphologically indistinguishable from other pathogenic treponemes, is an extremely slender spiral organism measuring 5μm–20μm in length and less than 0.2μm in thickness. It has 4–14 spirals that ap-

pear uniform near the center of the cell but frequently increase in periodicity and decrease in amplitude toward the ends, thus giving the cell a tapered appearance. Darkfield illumination is best for recognizing the organism, for it permits observation of the characteristic motility of live spirochetes.

In aqueous media young treponemes demonstrate seemingly random vigorous movements. However, in a more viscous environment or on surfaces they achieve sufficient traction to propel themselves in a snakelike fashion; in tissues they exhibit remarkable flexibility as they adapt themselves to the intercellular spaces.

T. pallidum and other noncultivable pathogenic treponemes all contain three axial fibrils inserted at each end of the protoplasmic cylinder, overlapping in the midregion of the organism. The pathogenic treponemes are pointed at each end, unlike nonpathogens, which have blunt rounded ends; the pointed ends may be used for attachment to host cells (Fig. 39-6). There is suggestive evidence for an external "slime" layer outside the outer envelope of *T. pallidum,* which could explain the serologic nonreactivity of freshly isolated treponemes and may contribute to their virulence.

METABOLISM

The pathogenic treponemes can be propagated in vivo in laboratory animals (see Pathogenicity, p. 755), but not in vitro, despite numerous reports to the contrary. Results

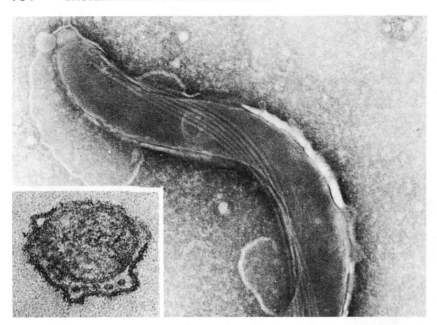

FIG. 39-4. Electron photomicrograph of the cultivable Reiter treponeme (negative staining technic) showing axial fibrils arising in end bulb at tip of organisms. Outpouchings of periplastic membrane are probably an artifact. (×50,000) Cross-section **inset** (thin section) reveals spatial relations between outer envelope, axial fibrils, and central protoplasmic cylinder. (×120,000) (Ryter A, Pillott J: Ann Inst Pasteur 104:496, 1963)

in tissue culture have been marginal and inconsistent, and attempts in chick embryos have also failed. Many nonpathogenic treponemes (e.g., some oral species, and the cultivable strains such as Reiter, Noguchi, and Kazan), have been grown in a variety of artificial media; they divide by binary fission. All cultivable strains are obligate anaerobes and grow slowly, with division times ranging from 4 to 18 h.

Suspensions of *T. pallidum* have remained viable and motile for periods up to 15 days when kept at 35 ° C under supposedly anaerobic conditions in a medium containing serum albumin, glucose,

carbon dioxide, pyruvate, cysteine, glutathione, and serum ultra-filtrate. Although growth as measured by an increase in cell numbers is not evident, such organisms are killed by penicillin after 5 h or more, suggesting some biosynthetic activity.

Recent studies have challenged the concept that *T. pallidum* is an obligate anaerobe. The presence of 10%–20% O_2 considerably enhances in vitro protein and RNA synthesis and glucose degradation during short-term experiments. Incomplete oxidation of glucose

FIG. 39-5. Axial fibril released from Nichols strain of *T. pallidum,* negatively stained with 1 % ammonium molybdate. **Bar** represents 100 nm. Shaft **(S)** is sometimes seen to be surrounded by a sheath, not easily visualized here. The fibril terminates in a hook **(H),** collar **(C),** and basal knob **(B),** which are used for attachment to the protoplasmic cylinder. Remarkable structural similarity exists between the axial fibril and bacterial flagella. (Hovind-Hougen K: Acta Path Microbiol Scand B80:297, 1972) (×160,000)

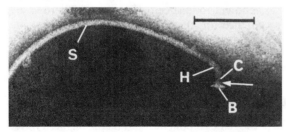

FIG. 39-6. Scanning electron micrograph demonstrating the specific attachment of *T. pallidum* to rabbit testicular cells in monolayer. Note the orientation of the treponemes as mediated by their tapered ends and apparent terminal disclike organelle. (Hayes NS et al: Infect Immun 17:174, 1977) (×28,000)

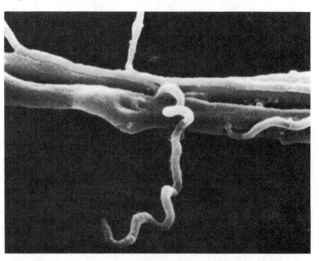

proceeds via the Embden–Meyerhof and the hexose monophosphate shunt pathways, with terminal electron transport to O_2 mediated by flavoproteins and specific cytochromes, but without a complete Krebs cycle. In vitro survival, and retention of motility and virulence, are improved by coincubation with certain tissue cell lines under low O_2 concentrations.

The cultivable treponemes utilize glucose or other fermentable carbohydrates as the primary energy source, and they have complex nutritional requirements, including multiple amino acids, purines, and pyrimidines. Many strains also need bicarbonate and one or more coenzymes, and all require exogenous fatty acid, usually supplied as protein-bound lipid in a serum supplement to the growth medium.

Suspensions of treponemes can be kept viable for years when frozen at temperatures below $-70°$ C in the presence of glycerol or other cryoprotective agents. Of practical importance for the problem of transfusion-associated syphilis is the fact that in blood, serum, or plasma stored at refrigerator temperatures viable *T. pallidum* organisms have not been recovered after 48 h.

ANTIGENS

Wassermann Antigen. The first serologic test for syphilis was a complement-fixation (CF) reaction—the so-called Wassermann test. In its original form this test used as Ag an extract of liver, containing many treponemes, from human fetuses with congenital syphilis. However, the specific ligand involved was later found to be present in alcoholic extracts of normal liver and other mammalian tissues as well. It was subsequently isolated from cardiac muscle and identified as a phospholipid, diphosphatidylglycerol (designated earlier as **cardiolipin**).

Many modifications and variations of the Wassermann test have been devised in hopes of increasing specificity. In contrast to the original CF reaction, most of the later tests are flocculation reactions. These are analogous to precipitin reactions but instead of a soluble Ag they use cardiolipin in an aqueous suspension finely dispersed with the aid of cholesterol and lecithin. The anticardiolipin tests are designated **serologic tests for syphilis (STS)**, to distinguish them from the antitreponemal tests developed in recent years.

Since cardiolipin is a normal constituent of host tissue, there arose a controversy about whether the primary antigenic stimulus for Wassermann Ab (anticardiolipin) comes from the invading organism or from the host, in an autoimmune response. The appearance of Wassermann Ab in other disorders, notably in lupus erythematosus, supports the autoimmune theory.

To serve as an immunogenic stimulus for Wassermann Ab, free cardiolipin, a hapten, must be attached to a suitable carrier. The lipid composition of cultivable treponemes depends to a large extent on the lipids in the growth medium; and since pathogenic treponemes growing in vivo have access to a great deal of cardiolipin they might incorporate it. With the microbial cell as a foreign carrier the bound cardiolipin could serve as an effective immunogenic determinant.

Treponemal Antigens. Two classes of Ags have been recognized in treponemes: those shared by many different serotypes and those restricted to one or a few species.

Progress in characterizing treponemal Ags has been impeded greatly by the inability to grow pathogenic treponemes free of animal cells or tissue. The Reiter treponeme, a spirochete once reputed to be a cultivable, nonvirulent variant of *T. pallidum,* was formerly used in a diagnostic test, but DNA hybridization has now shown the organisms to be only weakly related.

Specific treponemal Ags have been detected by a variety of serologic procedures. In the *T. pallidum* immobilization (TPI) test, a complement-dependent bactericidal reaction, treponemes obtained from syphilitic lesions in rabbits are mixed with a patient's serum and fresh guinea pig serum, and after anaerobic incubation for 18 h at $35°$ C the mixture is examined microscopically. In a positive reaction more than half the organisms are immobilized (i.e., killed). In a parallel reaction with heat-inactivated complement, to control for nonspecific effects, the treponemes should remain motile.

The TPI test is highly specific for the detection of syphilis, yaws, and other treponematoses, but it cannot distinguish among these diseases. Evidently the responsible Ag is shared by the various pathogenic species but is not present in indigenous organisms. The TPI test is expensive and difficult, and it has been replaced in the United States by other treponemal tests (see p. 757, Laboratory Diagnosis).

GENETIC VARIATION

Since only in vivo methods are available for studying variations of *T. pallidum,* little is known about its mutations. Tests of cross-immunity and pathogenicity are crude and have yielded only fragmentary information. Rapid passage in rabbits usually results in a detectable increase in virulence for rabbits. Some treponemal strains obtained by primary isolation from patients with yaws produce the typically mild lesions of experimental yaws, but after repeated passage in rabbits they cause the malignant lesions of experimental syphilis. Such changes may involve the in vivo selection of mutants.

No penicillin-resistant *T. pallidum* strains have been reported.

PATHOGENICITY

Syphilis in Man. Human syphilis is ordinarily transmitted by sexual contact. In heterosexual males the offending organisms either are present in lesions on the penis or are discharged from deeper genitourinary sites along with the seminal fluid. In women the infectious lesions are most commonly located in the perineal region or on the labia, vaginal wall, or cervix. Homosexual men account for a large proportion of early syphilis cases in many cities; the infectious lesions in these individuals commonly occur in or about the rectum. Primary infection may also occur in the mouth or other areas.

The organism penetrates mucous membranes but seems to enter the skin only through small breaks. Multiplication at the site of entrance results, within 10–60 days, in the formation of a characteristic **primary** inflammatory lesion known as a **chancre,** which begins as a papule and breaks down to form a superficial ulcer with a clean firm base. The predominant inflammatory cells in the lesion are lymphocytes and plasma cells. Although the chancre heals spontaneously, organisms escaping from it at an early stage invade the regional lymph nodes, forming **"satellite buboes,"** and eventually reach the blood stream, where they establish a systemic infection.

Two to twelve weeks after the appearance of the primary lesion a generalized skin rash (Fig. 39-7) usually

FIG. 39-7. Typical papulosquamous lesions on palms and other skin surfaces during secondary syphilis.

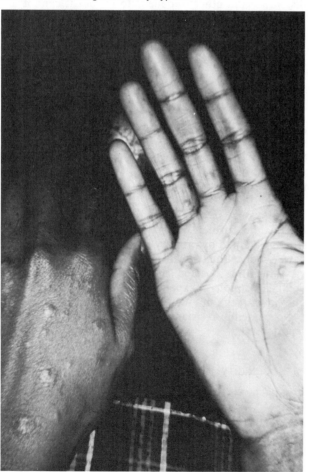

appears, which also often involves the mucous membranes. During this systemic **secondary** stage of the disease patients usually experience low-grade fever and a generalized enlargement of lymph nodes. Lesions may develop in bones, liver, kidneys, central nervous system, or other organs. The pathogenesis of the renal lesion in some instances is deposition of immune complexes. Tissue damage in other organs may be the result of local proliferation of treponemes. No treponemal toxins are known.

Secondary lesions often contain large numbers of spirochetes, and so they are highly contagious. In time they subside spontaneously, and once they have healed the patient is no longer dangerous to others, except for transplacental transmission.

Little is known concerning the mechanisms that bring about the destruction of the myriads of treponemes contained in the primary and secondary lesions. It is likely that the host defense in syphilis includes cellular immune mechanisms, as well as immobilizing (treponemicidal) Abs: the late primary and secondary stages of illness are frequently accompanied by temporary depression of T-lymphocyte responsiveness to certain treponemal Ags and common mitogens. Electron microscopy reveals *T. pallidum* organisms within a variety of host tissue cells, but there is no evidence that they survive there.

In approximately one-third of untreated patients enough treponemes persist in the tissues to give rise to the **late,** or **tertiary, lesions** of the disease several years after the primary infection. (Rarely, the primary and secondary stages go unrecognized, only to be followed by the late tertiary lesions.) In about another third of untreated cases the disease remains asymptomatic **(latent)** and is recognized only by the detection of Abs in the serum; in the remaining third the primary and systemic lesions heal so completely that even the STS become negative.

Tertiary lesions often contain very few organisms but frequently result in necrosis, scar formation, and extensive tissue damage, probably involving a **delayed hypersensitivity response** to products of the small number of persisting organisms. Among tertiary lesions **gummas** of the skin or bones may cause relatively little trouble (gummas are now rare). Serious manifestations, however, usually result from lesions in the central nervous system, causing **general paresis** or **tabes dorsalis;** in the cardiovascular system, where they affect the aortic valves or cause aortic aneurysms; and in the eyes, where they may produce permanent blindness.

Since *T. pallidum* readily passes the placental barrier a syphilitic mother may transmit the disease to her child. The lesions of **congenital syphilis** resemble, in general, those of acquired syphilis of comparable duration. In infants who fail to survive for more than a few weeks, or are stillborn, the syphilitic process is usually acute and is characterized by massive invasion of nearly all body tissues.

Experimental Syphilis. Although humans are the only known natural host of *T. pallidum*, experimental infections may be produced in a variety of laboratory animals. The experimental disease in rabbits, monkeys, and chimpanzees simulates the early course of the natural disease in humans, but late lesions are rarely encountered. However, in infected rabbits viable organisms persist for life and can be readily isolated from the lymph nodes.

Rabbits may be experimentally infected by inoculation of organisms into the eye, skin, testis, or scrotum. A single viable organism injected intratesticularly is said to be infectious. From the lag with different infecting doses it can be estimated that the time required for the organisms to divide in vivo is about 24–36 h.

In both experimental and human infections *T. pallidum* tends to proliferate at sites of trauma. In the rabbit, treponemes multiply best at peripheral tissue sites in which the temperature is several degrees below that of the internal tissues (see Fig. 24-7, in Ch. 24). Artificial cooling of the skin, for example, may strikingly increase the number of skin lesions that develop in the systemic phase of the infection. (Induced fever was formerly used successfully as treatment of human syphilis.) Pretreatment of rabbits with the antiinflammatory agent cortisone causes the formation of enlarged cutaneous syphilomas, which contain excessive amounts of hyaluronic acid and increased numbers of motile treponemes.

IMMUNITY

Resistance to reinfection, as determined both in human volunteers and in experimentally infected rabbits, usually begins about 3 weeks after appearance of the primary lesion. The persistence of a latent infection then maintains resistance, but if the disease is eradicated in the early stages by adequate treatment the host may again become fully susceptible to infection. However, treatment late in the course of the disease does not result in loss of resistance to reinfection. Immunity to *T. pallidum* in rabbits shows considerable strain specificity.

It is clear that good stimulation of humoral immunity accompanies infection, yet treponemes are not completely eliminated and the disease can progress through its characteristic stages. Passive transfer of large amounts of antitreponemal Ab to uninfected rabbits prior to challenge with virulent *T. pallidum* attenuates the disease, but it does not provide complete protection.

Immunity can be produced in rabbits by a series of injections of freshly harvested treponeme suspensions that have been appropriately inactivated. However, the protective immunogens have not been identified, nor have the roles of humoral Ab and immune lymphocytes. Prospects for an effective human vaccine depend on the successful cultivation of *T. pallidum* and the identification of virulence determinants.

LABORATORY DIAGNOSIS

Detection of *T. pallidum* in Lesions. In its primary and secondary stages syphilis can often be diagnosed by darkfield examination of fresh exudate fluid obtained from open or abraded lesions. Since occasionally the exudates from nonsyphilitic lesions may also contain spiral organisms, particularly near the gingival crevices and rectum, darkfield observations must be interpreted with care. An experienced observer can often differentiate *T. pallidum* from other spiral organisms on the basis of its characteristic morphology and motility. Negative darkfield examinations, on the other hand, do not necessarily exclude the diagnosis of syphilis, since lesions in the later stages of the disease may contain relatively few organisms.*

T. pallidum may also be identified in fluid from active lesions by immunofluorescent staining (Ch. 16) with *T. pallidum* Abs, freed of crossreacting treponemal Abs by absorption with cultivable treponemes.

Serologic Tests. Flocculation tests for Wassermann Abs are of great value in the diagnosis of syphilis. The standard is the Venereal Disease Research Laboratory (VDRL) test. A wide variety of other diseases, including malaria, lupus erythematosus, leprosy, and other infections, yield a "biologic false positive" (BFP) reaction, but in general Wassermann Ab develops a much higher titer in syphilis than in nontreponemal disorders. It is usually first detected 1–3 weeks after the primary lesion appears and reaches a maximum during the secondary stage of infection. Subsequently this Ab may remain at an elevated level, or it may disappear from the serum. It is positive in about 75% of patients with late (tertiary) stages of the disease; thus a *negative VDRL test does not exclude late syphilis of the heart or central nervous system.*

Tests for treponemal Abs are more specific. The TPI test (see Treponemal Antigens, p. 755) is the most specific but is not generally available. In the widely performed FTA (fluorescent treponemal antibody) test (also refer to Fluorescence and Immune Reactions, Ch. 16), *T. pallidum* cells extracted from animal lesions are exposed first to the patient's serum being tested and then to fluorescein-labeled Abs to human immunoglobulins. In its simplest form the FTA test also detects

---

* It must be remembered that each oil-immersion field in the usual coverslip preparation contains only about 10^{-6} ml of fluid. Accordingly there must be approximately 10^6 organisms/ml of fluid under the coverslip for the observer to find an average of one per field. Despite the obvious insensitivity of the method, the spirochetes in early lesions are usually present in sufficient numbers to be detected.

crossreactive and *T. pallidum*-specific Abs. Several modifications have therefore been devised in an attempt to increase specificity.

One of these, the fluorescent treponemal antibody absorption (FTA-ABS) test, uses an extract of the Reiter treponeme to absorb or competitively inhibit crossreacting Abs, but this procedure does not achieve complete immunologic specificity. This test is positive in only about 80% of patients with primary syphilis, but in almost all patients with secondary or tertiary syphilis. It should not be used as a screening test. A hemagglutination test for treponemal Ag (TPHA) is similar to the FTA-ABS in its sensitivity and specificity.

Antitreponemal Abs decline more slowly than Wassermann Ab following treatment, and positive reactions may persist for years. Positive serologic tests after therapy are not necessarily evidence of continuing disease. However, Wassermann Abs in the spinal fluid ordinarily reflect active central nervous system syphilis, since serum Abs do not penetrate the normal blood–brain barrier.

TREATMENT

When penicillin superseded arsenicals, bismuth, and mercury the treatment of syphilis was greatly simplified and improved. Patients in the early stages of the disease can be treated adequately by a single injection of long-acting benzathine penicillin G, which provides treponemicidal serum levels of penicillin for several weeks. Alternatively, procaine penicillin (which is more rapidly excreted than benzathine penicillin) may be given by daily injection for at least 10 days.* Only rarely is a second course of treatment necessary.

In the late stages of the disease penicillin is less effective, owing to irreparable damage to tissues. In addition, it has been suggested that penicillin occasionally fails to eradicate treponemes from the central nervous system and other sites in late syphilis.

PREVENTION

Many patients with early syphilis are unaware that they have the disease; in women especially the early lesions may be symptomless and unnoticed. In addition, a single promiscuous person may transmit the organism to many sexual partners. Hence control of the spread of syphilis is extremely difficult. When the diagnosis of early syphilis is established in an individual all known sexual contacts should be examined. Most authorities recommend antibiotic treatment of all recently exposed contacts of patients with infectious (primary or secondary) syphilis.

Prophylaxis by thorough cleansing of the genitalia and adjacent areas with soap and water is often ineffective, unless applied early and with great diligence. The prophylactic use of a single dose of medium- to long-acting penicillin is probably highly effective, since such therapy is known to cure early incubating syphilis. Prophylaxis is best achieved by the use of condoms, or by avoidance of sexual contact with persons likely to be infectious.

RELATED DISEASES

YAWS (FRAMBESIA)

T. pertenue, which causes the **nonvenereal tropical disease** known as yaws, is virtually indistinguishable from *T. pallidum,* except for the character of the lesions it produces. Both organisms induce the formation of Wassermann Abs and serologically identical antitreponemal Abs, and a striking degree of cross-resistance has been demonstrated in both man and laboratory animals. A similar disease has recently been found among apes in central Africa, and the responsible spirochete is indistinguishable from *T. pertenue.*

The **mother yaw,** the primary lesion in the human disease, usually appears 3–4 weeks after exposure. It begins as a painless red papule surrounded by a zone of erythema, often referred to as a **framboise** (Fr., raspberry). Eventually it ulcerates, becomes covered with a dry crust, and heals. Generalized secondary lesions of a similar character make their appearance 6 weeks to 3 months later and commonly occur in successive crops over a period of months or even several years; on the soles of the feet tender, hyperkeratotic lesions, known as **crab yaws,** often appear. The late, tertiary lesions are generally restricted to the skin and bones; gummatous nodules and deep chronic ulcerations may disfigure the nose and face and are often disabling. The disease, however, is not as grave as syphilis, since it **rarely involves the viscera,** and congenital yaws is very uncommon.

Experimental yaws also differs from experimental syphilis. Intracutaneous inoculation of *T. pertenue* in hamsters ordinarily causes a local lesion, whereas *T. pallidum* usually does not.

Epidemiology. Yaws occurs in the tropics, where the combination of high temperatures, high humidity, and poor hygiene promotes the persistence of open skin lesions and thus facilitates nonvenereal transmission by direct contact. Children often become afflicted at an early age. In areas of high endemicity more than 75% of the population may acquire the disease before the age of 20. Flies that feed on open lesions are thought by some investigators to act as vectors.

---

* Within 6–10 h after the start of treatment in early syphilis transient fever (2–4 h) and a brief exacerbation of visible lesions **(Herxheimer reaction),** often occur, apparently caused by an immunologic response to Ags released by the organisms being killed in the tissues. In addition, some of the products released may have the properties of endotoxins.

Treatment. Like syphilis, yaws responds dramatically to treatment with penicillin, often requiring only a single long-acting injection. However, yaws occurs primarily in areas where medical services are too limited for general application of this effective control measure.

PINTA

Pinta also occurs primarily in the tropics, especially Central and South America. The causative organism, *T. carateum,* is morphologically indistinguishable from *T. pallidum* but differs in being difficult to propagate in laboratory animals. An infection similar to the human disease, though less severe, has been produced in chimpanzees.

The human disease, which may be contracted at any age, is usually transmitted by nonvenereal person-to-person contact, although flies have been implicated as possible vectors. The primary lesion is nonulcerating and is followed within 5–18 months by successive crops of flat, erythematous skin lesions which first become hyperpigmented and then, after several years, depigmented and hyperkeratotic. They are most often seen on the hands, feet, and scalp. Late visceral manifestations are extremely rare.

Experimental pinta has been produced in syphilitic patients, and though subjects with pinta regularly develop Wassermann and treponemal Abs, they occasionally contract syphilis. These results suggest significant differences in host responses and in antigenic properties of *T. carateum* compared to other pathogenic treponemes.

Penicillin therapy is very effective.

OTHER TREPONEMATOSES

Endemic treponemal diseases occur in many areas of the world where people live under relatively unhygienic conditions. Transmitted by direct contact, these afflictions are often given local names, such as **bejel** in Syria and **siti** in British West Africa. All resemble yaws and are not readily distinguishable from one another.

A venereal form of treponematosis also occurs naturally in **rabbits.** The causative organism, *Treponema paraluis-cuniculi,* is morphologically indistinguishable from *T. pallidum.* Its natural occurrence in rabbits may complicate experimental studies on treponemes pathogenic for man.

Treponema hyodysenteriae, an anaerobic spirochete implicated in swine dysentery, is currently the only cultivable potentially pathogenic treponeme.

RELAPSING FEVER

Spirochetes of the genus *Borrelia* are usually longer than the treponemes, **their spirals are more loosely wound** and more flexible, and they are thick enough to be readily visible when stained with ordinary aniline dyes. They usually contain 12–15 axial fibrils inserted at each end. They can be propagated in young mice and rats, but host susceptibility varies from one strain to another. Some *Borrelia* strains may also be grown in chick embryos, and some can be grown under microaerophilic conditions in a complex medium. The organisms have an average division time of 18 h and retain animal pathogenicity through numerous subcultures.

In man spirochetes of this genus cause **relapsing fever,** in an epidemic and an endemic form. The **epidemic** form, caused by *B. recurrentis,* is transmitted only from man to man, by the human body louse. There is no animal reservoir, and lice do not transmit the organism transovarially. The **endemic** form, caused by various species of *Borrelia,* is spread from various **animal reservoirs** (often rodents) to man; each species is named for the species of *Ornithodorus* tick that transmits it to man.

After an incubation period of 3–10 days there is a sudden onset of fever, which ordinarily lasts for about 4 days: during this time large numbers of organisms may be demonstrated in the blood and urine, and rarely in the spinal fluid. The fever then declines and the borreliae disappear from the blood. As they decrease in number they become less motile, assume pleomorphic forms, and agglutinate, often in **rosettes.** During the ensuing afebrile period the blood is not infectious, but after 3–10 days it teems again with organisms and the fever returns. The ensuing febrile attacks usually number from 3 to 10 and become progressively less severe, until they subside altogether.

Although the case fatality rate is less than 5% for the endemic disease it may exceed 50% in severe louse-borne epidemics, presumably owing to increased adaptation to the human host in man-to-man transmission. There is evidence that circulating endotoxin is present during the height of symptomatic relapsing fever. Autopsy usually reveals miliary necrotic lesions containing large numbers of organisms, particularly in the spleen and liver. Gross hemorrhagic lesions may also be prominent in the gastrointestinal tract and kidneys. Transplacental transmission has been reported.

Experimental infections may be produced in monkeys, mice, rats, and guinea pigs. The experimental disease in monkeys follows the characteristic relapsing course of the illness in man.

Mechanism of Relapse. The most remarkable feature of this disease is its tendency to relapse at regular intervals. The pathogenesis of the sequential relapses is unique: the *organisms in each successive attack show antigenic differences,* and circulating Abs specific for the organisms of each onset appear in the blood. These Abs are responsible for agglutination and presumably for the subsequent disappearance of the spirochetes; the next relapse depends on the outgrowth of antigenically distinct mutants, for which the host then elaborates new Abs. The rapidity and the regularity of the successive relapses imply that the mutation rates involved are high, and that a reproducible and limiting sequence of antigenic variation exists.

Diagnosis. The diagnosis is usually made by microscopic examination of blood samples obtained during a febrile attack. The organisms can be identified by darkfield microscopy or in stained smears. Because of the numerous antigenic variants encountered serologic tests are of little diagnostic value. Animal inoculation may help when direct microscopic examinations are negative: infected blood injected into young white rats will usually result in a readily demonstrable spirochetemia within 24–72 h.

Treatment. Relapsing fever responds well to penicillin. Tetracyclines and chloramphenicol are also effective.

Prevention. In the United States sporadic cases are encountered in the South and West. Infected ticks* are plentiful in endemic areas and are thought to be the principal vector, both from animal to animal and from animal to man. Relapsing fever may also be louse-borne, and is not uncommonly associated with typhus epidemics. Tick- and louse-control measures constitute the most effective means of prevention.

Other Borrelia Infections of Man. Ulcerative lesions of the skin, mucous membranes, and lungs, occurring most commonly in debilitated individuals, often contain *Borrelia* organisms; their etiologic significance is uncertain.

LEPTOSPIROSIS

The first human leptospiral disease to be described was a severe febrile illness characterized by jaundice, hemorrhagic tendencies, and involvement of the kidney. Known as **Weil's disease,** it was shown in 1915 by Inada to be caused by *Leptospira interrogans* serotype *icterohaemorrhagiae* (Gr. *leptos,* thin), transmitted to man from infected rats.

It has since been learned that many immunologically different strains of leptospiral organisms cause a variety of human illnesses, most of which are not associated with jaundice. Each also seems to have a different natural host. Thus *L. interrogans* serotype *icterohaemorrhagiae* is most commonly found in rats, serotype *canicola* in dogs, and serotype *pomona* in swine. All the common species are pathogenic for man (except for the saprobic *L. biflexa,* found in small streams, lakes, and stagnant water). Each species can be identified by specific immunologic reactions, but there is considerable antigenic overlap and so absorbed sera must be used for species identification.

Current classifications consider all pathogenic leptospires to belong to a single complex, which can be subdivided into 19 serogroups and over 250 serotypes. On the basis of G + C content of DNA, and DNA hybridization, the leptospires have also been divided into four distinct genetic groups, which bear little relationship to groupings based on antigenic structure.

MORPHOLOGY

All the leptospires are characterized by **extraordinarily fine spirals,** wound so tightly as to be barely distinguishable under the darkfield microscope.

The cells vary in length from $4\mu m–20\mu m$. The fine structure is basically similar to that of other spirochetes. Motility results from rotational, flexing, and translational movements of the organism, one or both ends of which are often bent into a hook. Because the tightly coiled spirals are so difficult to recognize, diagnoses made solely on the basis of direct microscopic examination of the blood are unreliable. Tiny strands extending from the surface of RBCs in blood preparations are often mistaken for leptospires.

CULTIVATION

Leptospires are obligate aerobes, and can be readily grown in a variety of artificial media supplemented with 10% heat-inactivated (56° C, 30 min) rabbit serum. They can also be grown in a synthetic medium containing inorganic salts, fatty acids, and vitamin B_{12}. Growth is stimulated by thiamine, and is best at pH 7.2, at 25°–30° C, and at a slightly increased pCO_2. Pathogenic leptospires cannot synthesize long-chain fatty acids de novo. These compounds (at least 15 carbon atoms long, and both saturated and unsaturated) are essential for growth, being a major energy source, and they must be supplied in the medium in a bound form. In contrast, certain aquatic leptospires can grow in a defined medium with acetate as the sole source of carbon and energy. The lipid content of leptospires as well as of the other spirochetes is high (about 20% of the cell dry weight), and it

* Transovarial transmission is known to occur in ticks.

largely reflects the fatty acid composition of the environment. On solid media leptospires produce a variety of subsurface colonies, some of which may be difficult to visualize unless stained with oxidase reagent (see Metabolism, Ch. 30).

Saprobic leptospires can be differentiated from the pathogenic types by several cultural differences: they produce more oxidase, and their requirements for environmental CO_2 are less. Neither type of leptospire incorporates exogenous pyrimidine; hence cultivation in the presence of fluorouracil permits selective growth of leptospires from material contaminated with other microorganisms.

ANTIGENICITY AND GENETIC VARIATION

All leptospires possess a common somatic Ag (lipopolysaccharide) but antigenic types vary in their surface (agglutinating) Ag. When a given antigenic type is grown in the presence of homologous immune serum antigenic variants frequently appear in the culture; these are analogous to the *Borrelia* variants that appear in patients with relapsing fever (see above). Differences in surface Ags are most readily demonstrated by agglutination of suspensions of either living or formalinized organisms; CF and precipitin procedures are also sometimes employed. Mutations affecting virulence, morphology (nonhooked forms), motility, hemolysin production, and colony formation have also been noted. No leptospiral toxins other than a hemolysin and lipase have yet been described.

PATHOGENICITY

Lower Animals. Leptospirosis is primarily a zoonosis, wild rodents and domestic animals providing the principal reservoirs. The disease can be produced experimentally in many species of rodents; the young of the guinea pig and the Syrian hamster are particularly susceptible. When injected with *L. interrogans* serotype *icterohaemorrhagiae* they develop fever within 3–5 days, followed by jaundice and hemorrhages into the skin, subcutaneous tissues, and muscles. Death usually occurs in less than 2 weeks. At necropsy the tissues are jaundiced and hemorrhagic, the liver and spleen are enlarged, and leptospires are recoverable from the blood, cerebrospinal fluid, and urine.

In contrast, in species that are naturally infected with leptospires (e.g., wild rats) the infection is usually mild, often lifelong, and characterized by chronic involvement of the kidneys. The more or less continuous shedding of organisms in the urine is often responsible for transmission of leptospirosis to man. Because the organisms will not survive long in an acid medium, the urine is usually infectious only when alkaline or excreted into alkaline water.

Humans. Different serotypes cause a similar pattern of illness, after an incubation period of 8 to 12 days. Onset is abrupt, often with chills followed by high fever. Headache, photophobia, and severe muscular pains, particularly in the back and legs, are prominent symptoms. The most constant physical sign is conjunctivitis. The classic icterohemor-rhagic picture is rarely seen: albuminuria is common but jaundice occurs in fewer than 10% of clinically recognizable cases. Lymphocytic meningitis is often present. In some patients meningitis occurs as the second phase of a biphasic illness, suggesting that immune mechanisms may play a pathogenetic role. The acute illness ordinarily lasts for 3–10 days. The mortality rate in clinically recognized cases is approximately 5%–10%, but serologic surveys in heavily endemic areas indicate that **subclinical cases are extremely common.**

During World War II a novel disease called **pretibial fever,** characterized by an erythematous rash most frequently over the shins, was encountered among soldiers stationed at Fort Bragg, North Carolina. The agent, obtained from the blood of a patient, was maintained for years by passage in guinea pigs and hamsters on the assumption that it was a virus, but nearly a decade later it was identified as *L. interrogans* serotype *autumnalis*, a leptospiral species previously isolated in Japan. Moreover, subsequently paired acute and convalescent sera, saved from the Fort Bragg epidemic, uniformly showed a rise in titer of specific agglutinins for the same organism.

Human convalescent serum protects guinea pigs against otherwise fatal inoculations of homologous leptospires. These protective (and agglutinating) Abs persist in the patient's serum for many years.

LABORATORY DIAGNOSIS

A diagnosis based on direct microscopic examination of blood should be made only by experienced observers. Leptospiremia may be detected by culturing the blood, preferably at 30° C, in broth or agar media enriched with 10% serum, or by inoculating it intraperitoneally into young guinea pigs or hamsters.

Serum Abs usually appear during the second week of illness and reach a maximum titer during the third or fourth week, but they are not easily identified because of the numerous antigenic types encountered. Agglutination tests, however, may be done with suspensions of killed leptospires, pooled to contain the most common antigenic strains. A broader serologic test, for hemolysis, involves incubation of serum and complement with RBCs previously sensitized by **genus-specific antigen,** extracted from the nonpathogenic *L. biflexa.*

TREATMENT

Penicillin, the tetracyclines, and chloramphenicol all are effective against leptospires in vitro, and also, if given early, against experimental infections in animals. Nevertheless, the treatment of leptospirosis is relatively unsatisfactory, in part because the disease is generally recognized late. Penicillin appears to be the drug of choice.

PREVENTION

Man usually acquires leptospirosis from infected animals through contact with water contaminated with their urine, or through direct contact with their tissues. Workers in rat-infested slaughterhouses and fish-cleaning establishments, miners, farmers, sewer workers, and swimmers in stagnant ponds and canals run the greatest risk. Dogs vaccinated for leptospirosis may carry leptospires in their urine; they have been the source of urban outbreaks. Because the natural reservoir in wild animals is far too vast to be attacked directly, and because of the many serotypes, preventive measures must be directed at diminishing the chances of contact with contaminated water. The organisms are believed to gain entrance through abrasions in the skin and through the mucous membranes of the conjunctiva, nose, and mouth. Because of the acidity of the gastric juice, infection via the intestinal tract is probably rare.

RAT-BITE FEVER (SODOKU)

Although ordinarily not classified as a spirochete, *Spirillum minor* is a short, rigid, spiral organism possessing polar flagella. It is commonly carried by rats and causes one of the two forms of rat-bite fever in man. (The other is due to *Streptobacillus moniliformis,* Ch. 43). From the primary rat-bite lesion the organism invades the regional lymph nodes and eventually the blood stream, causing lymphadenitis, rash, and relapsing fever.

S. minor has never been cultivated on artificial media; its isolation from human patients depends upon animal inoculation. Guinea pigs and mice are susceptible; when infected, they harbor in their blood large numbers of organisms, which are often visible in Wright-stained smears. The disease occasionally causes a false-positive serologic test for syphilis. Penicillin, streptomycin, and the tetracyclines are all effective.

SELECTED READING

BOOKS AND REVIEW ARTICLES

CANALE-PAROLA E: Physiology and evolution of spirochetes. Bacteriol Rev 41:181, 1977

HOLT SC: Anatomy and chemistry of spirochetes. Microbiol Rev 42:114, 1978

JOHNSON RC (ed): The Biology of Parasitic Spirochetes. New York, Academic Press, 1976

SMIBERT RM: Spirochaetales, a review. CRC Crit Rev Microbiol 2:491, 1973

SOUTHERN PM JR, SANFORD JP: Relapsing fever: a clinical and microbiological review. Medicine 48:129, 1969

SPARLING PF: Diagnosis and treatment of syphilis. N Engl J Med 284:642, 1971

Syphilotherapy 1976: Position papers for the current United States Public Health Service recommendations. J Am Vener Dis Assoc 3:98, 1976

SPECIFIC ARTICLES

BASEMAN JB, NICHOLS JC, HAYES NS: Virulent *Treponema pallidum:* aerobe or anaerobe. Infect Immun 13:704, 1976

BERMAN SJ, TSAI C-C, HOLMES K et al: Sporadic anicteric leptospirosis in South Vietnam: a study in 150 patients. Ann Intern Med 79:167, 1973

BISHOP NH, MILLER JN: Humoral immunity in experimental syphilis. II. The relationship of neutralizing factors in immune serum to acquired resistance. J Immunol 117:197, 1976

BRAUNSTEIN GD, LEWIS EJ, GALVANEK EG et al: The nephrotic syndrome associated with secondary syphilis: an immune deposit disease. Am J Med 48:643, 1970

FEIGIN RD, LOBES LA JR, ANDERSON D, PICKERING L: Human leptospirosis from immunized dogs. Ann Intern Med 79:777, 1973

FULFORD KWM, JOHNSON N, LOVEDAY C et al: Changes in intra-vascular complement and anti-treponemal antibody titres preceding the Jarisch-Herxheimer reaction in secondary syphilis. Clin Exp Immunol 24:483, 1976

JOSEPH R, CANALE-PAROLA E: Axial fibrils of anaerobic spirochetes: ultrastructure and chemical characteristics. Arch Microbiol 81:146, 1972

MAGNUSON HJ, THOMAS EW, OLANSKY S et al: Inoculation syphilis in human volunteers. Medicine 35:33, 1956

MUSHER DM, SCHELL RF, JONES RH, JONES AM: Lymphocyte transformation in syphilis: an in vitro correlate of immune suppression in vivo? Infect Immun 11:1261, 1975

SCHILLER NL, COX CD: Catabolism of glucose and fatty acids by virulent *Treponema pallidum.* Infect Immun 16:60, 1977

chapter

40

RICKETTSIAE

EDWARD S. MURRAY†

† Deceased

INTRODUCTION TO THE INTRACELLULAR BACTERIA

The **rickettsiae** and the **chlamydiae,** described in this and the following chapter, are obligate intracellular parasites. This feature reflects fundamental differences from ordinary bacteria: 1) a smaller size, with a smaller genome providing fewer enzymes; 2) a longer generation time; 3) a requirement for an exogenous energy supply for growth. In addition, unlike optional intracellular parasites (e.g., *Mycobacterium tuberculosis, Brucella*), they infect cells that are ordinarily nonphagocytic, and they are protected from lysosomal degradation by envelopment in a vesicle of the host cell membrane. Chlamydiae have a complex life cycle, with different forms for intracellular multiplication and for extracellular transmission. Rickettsiae, in contrast, have only a single, fragile form, whose transmission depends on arthropod vectors. These two sets of pathogens cause profoundly different diseases, some of worldwide significance.

HISTORY OF RICKETTSIAE

Fracastorius in 1546 first described an epidemic with enough precision to permit its later definitive identification as typhus fever.

This disease has played a large role in history ever since people began to congregate densely under conditions where body lice flourished. The effect of typhus on military campaigns has often been more decisive than the battles: as Zinsser remarked in *Rats, Lice, and History,* Napoleon's retreat from Moscow "was started by a louse."

Typhus has waxed and waned with economic conditions. Between 1918 and 1922 in Eastern Europe and Russia it is estimated to have infected 30 million people and to have caused at least 3 million deaths. During World War II millions of cases again occurred in prison camps, in the Eastern European combat zone, among Yugoslav partisan forces, and in North Africa; in addition, **scrub typhus** was a continuing problem in the Pacific Theater.

Since World War II rickettsial infections have come under control in Europe and Asia, but tens of thousands of cases of typhus fever have been reported in Central Africa. Moreover, during the same period in the United States **Rocky Mountain spotted fever,** one of the most severe rickettsial infections, has more than quadrupled in incidence and now exceeds 1000 cases per year.

The microorganisms causing this group of diseases were named rickettsiae to honor H. T. Ricketts, who discovered the agent of Rocky Mountain spotted fever early in the 1900s and later died of typhus fever while investigating its etiology.

GENERAL CHARACTERISTICS OF RICKETTSIAE

Rickettsiae (other than that of Q fever) have a number of common characteristics: 1) they are similar in size and shape, and they stain red with the Gimenez stain; 2) they multiply only within certain cells of susceptible hosts; 3) the characteristic pathologic lesion is a widespread peripheral vasculitis; 4) acute infections are characterized clinically by fever, headache, and a rash; 5) the early stages of the disease respond well to broad-spectrum antibiotics; 6) the organisms occur under natural conditions not only in mammals but also in arthropods—either insects (lice and fleas) or arachnids (ticks and mites); 7) these arthropods are the primary means of transmission to man.

In general the following modes of transmission (see Table 40-3, below) are employed: 1) from arthropod (louse, flea, mite, or tick) to mammal by direct inoculation into the skin, in association with feeding; 2) from

mammal to arthropod by ingestion of blood; or 3) from arthropod to arthropod by transovarial passage. The one exception, the agent of Q fever (*Coxiella burnetii*), is ordinarily transmitted to man by infected dust or droplets. (Other rickettsiae may also infect by the airborne route under some circumstances, e.g., laboratory exposure to dust from dried louse feces.)

With the important exception of louse-borne typhus, rickettsial infections in humans are only incidental to the survival of the organisms. The diseases vary in severity from benign and self-limited to some of the most fulminating illnesses known.

MORPHOLOGY

The rickettsiae are small coccobacillary microorganisms measuring 0.3–0.5μm in diameter and 0.3–0.4μm in

FIG. 40-1. Electron micrograph of a thin section through midgut cells (between basement membrane and lumen) of a human body louse, *Pediculus humanus*, heavily infected with *R. mooseri*. Although the rat flea is the usual arthropod vector for this rickettsia, proliferation in infected lice is also luxuriant. Two cells are swollen by their densely packed rickettsiae, and a third cell, sectioned through a nucleus, is more lightly infected. At **bottom center** is a circular collection of dark, degenerating rickettsiae, possibly in a lysosome. (Courtesy of S. Ito, Harvard Medical School) (×6000)

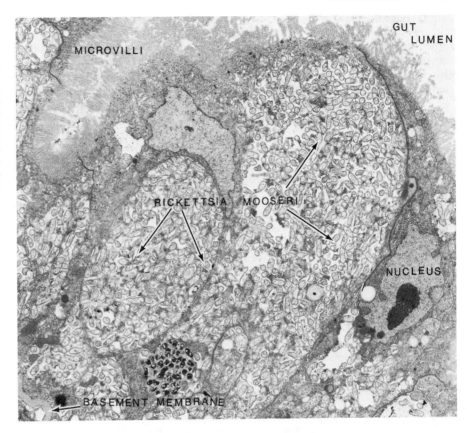

FIG. 40-2. Electron micrograph of a small portion of a cell in culture infected with *R. tsutsugamushi*. The organisms are directly in the host cell cytoplasmic matrix and intermingle with mitochondria and smooth-surfaced endoplasmic reticulum; they exhibit a cell wall and underlying plasma membrane. (Courtesy of S. Ito, Harvard Medical School) (×70,000)

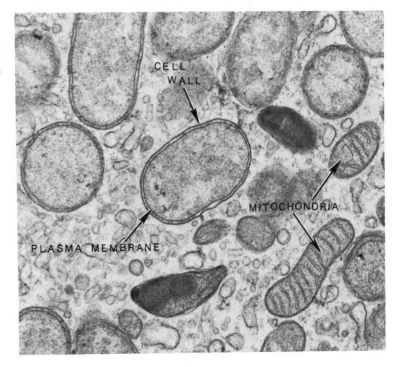

length (Figs. 40-1 and 40-2). They occur singly, in pairs, and occasionally in strands. The most typical stained form is a diplobacillus with slightly pointed ends and a transparent band between the two bacilli. Electron microscopy reveals a cell envelope, ribosomes, and a nucleoid. The obligate intracellular parasitism of the rickettsiae is thus not reflected in unusual morphologic characteristics. Most species are found only in the cytoplasm of host cells, but those of the spotted fever group occasionally multiply in nuclei as well.

GROWTH AND METABOLISM

Rickettsiae are most easily grown in large numbers in the yolk sac of the chick embryo, but growth requirements are best studied in cell cultures.

When scrub typhus rickettsiae are added to cultured mouse lymphoblasts almost all the cells become infected within 1–2 h. Penetration seems to involve an expenditure of energy by the microorganisms (an uncommon finding among pathogenic bacteria); it is enhanced by L-glutamate and blocked by 2,4-dinitrophenol, and rickettsiae killed by heat or formalin are not taken up.

Rickettsiae multiply, by transverse binary fission, within vesicles in the cytoplasm of the host cell, which protect them from lysosomal degradation. Optimal growth is obtained in well-nourished cells, but the cells need not multiply. The generation time is much longer than that of most bacteria. Rickettsiae are released by host cell lysis or via a vesicle in the host cell membrane.

The rickettsial cell wall contains peptidoglycan and lipopolysaccharide, typical of gram-negative bacteria. Analyses yield a relatively high content of DNA per unit weight—perhaps because RNA and protein leak out of rickettsial cells during purification.

In dilute buffered salt solutions isolated typhus and spotted fever rickettsiae are extremely unstable, rapidly losing both metabolic activity and infectivity. However, in media enriched with potassium, glutamate, and sucrose the isolated organisms can survive for many hours at 35° C.

Various rickettsiae fail to attack glucose but do oxidize pyruvate and several components of the Krebs cycle. Glutamate oxidation, accompanied by oxidative phosphorylation, is probably the principal energy-yielding process in typhus rickettsiae. Isolated typhus rickettsiae incorporate amino acids into protein, but are unusual for intact cells in requiring all amino acids plus ADP or ATP.

The instability of isolated typhus rickettsiae probably arises from their *tendency to leak essential metabolites.* Thus infectivity is lost when the organisms are maintained at 0° C in balanced salt solution, but it is restored by addition of NAD or CoA (which cannot be taken up by most bacteria). The organisms evidently have most of the metabolic capabilities of bacteria but require an exogenous supply of nucleotide cofactors.

The structural basis for this permeability has not been characterized, but the property explains the peculiar ecology and epidemiology of these organisms. It seems likely that most rickettsiae may borrow ATP, NAD, and CoA from the host cell; at the same time they have lost the ability to be self-sufficient outside host cells. Their survival in nature therefore requires that their means of transmission minimize their exposure to the extracellular world. The exception, *C. burnetii,* is much more stable.

PATHOGENESIS

Rickettsial infections in man usually begin in the vascular system following the bite of an infected arthropod. The microorganisms proliferate mainly in endothelial cells, and they become widely disseminated by the blood. They establish **focal areas of obstruction in the small blood vessels** due to hyperplasia of infected endothelial cells and to small thrombi; RBCs then escape, and inflammatory cells migrate into the surrounding tissue. These widespread small lesions appear to account for the more prominent clinical manifestations, such as rash, headache, and stupor, as well as the terminal heart failure and shock; gangrene is a rare complication.

Rickettsial infections can also be acquired by inhalation: of dried louse feces or infected dust in Q fever, and of aerosols in laboratory infections. Pneumonitis is a regular and prominent feature of Q fever but is infrequent in other rickettsial infections.

Rickettsial Toxin. When large numbers of rickettsiae are injected intravenously into mice or rats the animals die within 2–8 h. Death results from damage to endothelial cells, with consequent leakage of plasma, decrease in blood volume, and shock. A direct toxic effect is inferred because 1) death occurs so rapidly; 2) ultraviolet irradiation, in doses that greatly diminish infectivity, does not reduce this effect; 3) it is not prevented by antirickettsial drugs; and 4) it is prevented by antisera specific for Ags of the rickettsial cell wall. The toxin has never been isolated; its physiologic effects differ from those of endotoxins of Enterobacteriaceae.

LABORATORY DIAGNOSIS

The organism responsible for a particular rickettsial disease is most often identified serologically, by comparing an early serum with one taken during convalescence. Abs appear during the second week of illness and reach maximal titer during, or after, recovery. When Ags from several related organisms are used the one that reacts most intensely is assumed to represent the etiologic agent. This useful rule suffers from the limitation that still other Ags, not tested, might have given even higher titers.

TABLE 40-1. Complement-Fixation and Weil-Felix Reactions in Rickettsioses*

Group	Disease	Complement fixation Group antigen type			Weil-Felix agglutination Proteus strain		
		Typhus	RMSF	Q Fever	OX19	OX2	OXK
I	Primary louse-borne typhus	+++	±	0	+++	+	0
	Brill–Zinsser disease	+++	±	0	0 or +	0	0
	Murine typhus	+++	±	0	+++	+	0
II	RMSF	±	+++	0	+++†	+++†	0
	Tick typhus	±	+++	0	+++†	+++†	0
	Rickettsialpox	±	+++	0	0	0	0
III	Scrub typhus	0	0	0	0	0	+++
IV	Q Fever	0	0	+++	0	0	0

RMSF = Rocky Mountain spotted fever.

+++ = Strong reactions (1:80 to 1:5120 in CF; 1:320 to 1:2560 in WF); + = relatively weaker reactions; 0 = negative at 1:5 dilution in CF; negative at 1:80 dilution in WF; ± = crossreactions between the typhus and RMSF groups occur frequently.

*Immunofluorescence, a sensitive Ab assay, is available chiefly in research laboratories.

†In RMSF or tick typhus, agglutinins to OX19, OX2, or both can be present in either high or low titer.

Cultivation in yolk sacs has made rickettsiae available in sufficient quantity and purity to provide Ags for detection of Abs. After ether extraction of lipids from crude suspensions of infected yolk sac, two kinds of Ags are obtained: 1) **soluble, group** Ags, which are released into the suspending medium, show a broad specificity characteristic for each of the major rickettsial groups in Table 40-1; and 2) **insoluble, type-specific** Ags, which are presumably associated with the cell envelope, help to differentiate species and strains.

Abs to both kinds of Ags can be detected by complement fixation (CF). However, in most laboratories only soluble Ags are used, leaving the clinician to make a more specific diagnosis through clinical and epidemiologic considerations. Other useful immunologic tests include **agglutination** of rickettsial suspensions, **immunofluorescence,** and ability to **neutralize** the toxicity of rickettsial suspensions in mice (see Rickettsial Toxins, above), or to **protect** laboratory animals from infection.

One widely used diagnostic test, the **Weil–Felix reaction,** is based on the curious fact that Abs formed in certain rickettsial diseases crossreact with the polysaccharide O Ags of certain strains of the *Proteus* bacillus, called OX, whose O Ags are accessible (i.e., are not covered by H Ags). The Weil–Felix test is helpful under circumstances that preclude the use of more specific serologic procedures. Three strains are used: *Proteus* OX2, OX19, and OXK. The agglutination pattern helps to differentiate various rickettsial diseases (Table 40-1). However, since *Proteus* infections in man are fairly common (particularly in the urinary tract) low-titer Abs are often found in normal human sera.

Transmission to Laboratory Animals. Rickettsial diseases are characterized by rickettsemia during the early febrile phase of illness, and intraperitoneal inoculation of the patient's blood into guinea pigs or gerbils (or, with scrub typhus, into mice) will often provide a means of identifying the organism. (After the first week of illness the inoculations should be made with crushed blood clots from which as much serum as possible has been removed, in order to minimize inhibitory Abs.) In male guinea pigs characteristic scrotal lesions may develop with some rickettsial species. Although adapted strains of rickettsiae can be propagated readily in yolk sacs of chick embryos, primary isolation by this means is difficult.

Isolation is rarely indicated. Great care must be taken: laboratory infections have occurred frequently in rickettsial research laboratories, with a number of fatalities.

CHEMOTHERAPY

The tetracyclines and chloramphenicol are highly effective in all rickettsial diseases. Both these antibiotics are known to be effective against intracellular bacteria.

SPECIFIC DISEASES DUE TO RICKETTSIAE

The various rickettsial diseases differ in 1) arthropod vector; 2) specificities of the Abs formed; 3) the localization of the organism (in the cytoplasm or also in the nucleus); 4) clinical manifestations, especially the distribution and appearance of the rash; and 5) behavior of the organism in experimental animals.

TABLE 40-2. Important Epidemiologic Characteristics of Rickettsial Diseases

| Disease | Agent | Epidemiologic features | | |
		Geographic occurrence	Usual mode of transmission in man	Reservoir
Typhus group				
Primary louse-borne typhus	*Rickettsia prowazekii*	Worldwide	Infected louse feces rubbed into broken skin or as aerosol to mucous membranes	Man
Brill–Zinsser disease	*R. prowazekii*	Worldwide	Recrudescence months or years after primary attack of louse-borne typhus	
Murine typhus	*R. mooseri (R. typhi)*	Scattered pockets world wide	Infected louse as in primary louse-borne typhus	Rodents
Spotted fever group				
Rocky Mountain spotted fever	*R. rickettsii*	Western hemisphere	Tick bite	Ticks/rodents
Tick typhuses (boutonneuse)	*R. conorii**	Mediterranean, Caspian and Black Sea littoral, Africa, SE Asia	Tick bite	Ticks/rodents; dogs
Rickettsialpox	*R. akari*	USA, Russia, Korea	Mite bite	Mites/mice
Scrub typhus	*R. tsutsugamushi*	Japan, SE Asia, W & SW Pacific	Mite bite	Mites/rodents
Q Fever	*Coxiella burnetii*	Worldwide	Inhalation of infected particles from environment of infected animals	Ticks/mammals

*In addition, in Australia *R. australis* (Queensland tick typhus) and in North Asia *R. sibirica* (Siberian tick typhus) are antigenically and geographically distinct entities.

On the basis of these properties rickettsial diseases fall into several groups, which are compared in Table 40-2; their means of transmission are outlined in Table 40-3.

TYPHUS GROUP

The typhus group* consists of primary louse-borne typhus, caused by *R. prowazekii;* a clinical variant, Brill–Zinsser disease (or recrudescent typhus); and murine or endemic typhus, caused by *R. mooseri.* These three diseases are quite similar clinically and pathologically but differ markedly in their epidemiologic features.

LOUSE-BORNE TYPHUS

Transmission. Under natural conditions *R. prowazekii* infects only humans and the human body louse, and the transmission cycle is man → louse → man, with humans being the reservoir. The louse, *Pediculus humanus* var. *vestimentorum,* nests in clothing and emerges several times a day to take a blood meal from its host's skin. When the ingested blood contains the organism, the cells of the louse's intestinal tract become infected and eventually rupture, discharging the organisms into the insect's

* Gr. *typhos,* stupor. Typhus and typhoid fever (due to *Salmonella typhi;* Ch. 31) were not clearly distinguished until their agents were discovered. In German *Typhus* refers to typhoid fever, and *Flecktyphus* to typhus.

feces. While feeding on the skin the louse defecates, and because its bite causes intense itching the victim often scratches and rubs the infected feces into the bite wound. Infection can also be transmitted when dried infected louse feces gain access to the mucous membranes of the eye or respiratory tract.

Lice seek an environment of approximately 29°C, found in the folds of garments on healthy humans; they tend to leave the bodies of those who have died or who have high fevers. Since lice can neither fly nor jump, and can crawl but a few yards, louse-borne typhus flourishes only under conditions where crowding permits the infected lice to find new hosts.

In lice *R. prowazekii* invades only cells of the intestinal tract. The organism is not passed transovarially, and the infected lice die within 2–3 weeks. Lice therefore do not serve as reservoirs of typhus, and no animal reservoir has been demonstrated. *Human carriers represent the only proven reservoir* in the louse-borne typhus cycle (see Brill–Zinsser Disease, below).

The head (hair) louse is also capable, and the pubic (crab) louse is probably capable, of transmitting louse-borne typhus fever; but neither of these species has been proven to be an important vector. Virulent *R. prowazekii* organisms have been isolated from flying squirrels in the southeastern United States, but their relation to human disease is as yet unknown.

Pathogenesis. The clinical and pathologic manifestations of louse-borne typhus can be largely accounted for by the vascular lesions. The rickettsiae multiply in the

TABLE 40-3. Modes of Transmission of the Principal Rickettsial Diseases

Disease in man	Etiologic agent	Chain of transmission
Louse-borne typhus	*R. prowazekii*	Man→Louse→Man→Louse . . . ↓ Recrudescent Same man→(Brill–Zinsser Disease)→Louse→Man→Louse . . . (carrier)
Flea-borne typhus	*R. mooseri (R. typhi)*	Rat→Rat flea→Rat→Rat flea→Flea . . . ↓ Man
Rocky Mountain spotted fever (boutonneuse fever, other spotted fevers)	*R. rickettsii* (prototype)	Tick→Tick→Tick→Tick . . . ↓ ↓ Dog Man ↓ Tick→Man
Scrub typhus (tsu-tsugamushi fever)	*R. tsutsugamushi*	Mite→Field mouse→Mite→Field mouse . . . ↓ Man
Rickettsialpox	*R. akari*	Mite→House mouse→Mite→House mouse . . . ↓ Man
Q fever	*C. burnetii*	Tick→Small mammal→Tick→Cattle, Sheep, other domestic animals . . . airborne (airborne) ↓ Man

(Modified from Moulder JW: The Biochemistry of Intracellular Parasitism Ch. 3. Chicago, University of Chicago Press, 1962)

endothelial cells lining the small blood vessels, causing endothelial proliferation, thrombosis, perivascular hemorrhages, extravasation of plasma, and finally hemoconcentration and shock. Vascular lesions occurring in the skin produce the rash; those in the meninges probably account for the highly characteristic rickettsial headache; and those in the heart and kidneys may lead to myocardial or renal failure.

From 1 to 2 weeks after a bite from an infected louse illness usually begins abruptly, with chills, fever, headache, generalized aches and pains, and exhaustion. Body temperature reaches 40° C or more within 2–3 days, and it tends to remain high in untreated patients until death or recovery ensues in 2–3 weeks. A rash usually appears on the trunk by 4–7 days and spreads peripherally to the extremities, usually sparing the face, palms, and soles. At first macular and fading on pressure, the rash soon becomes fixed and maculopapular, and, later, if the patient remains untreated, petechial and hemorrhagic. Severe cases progress to prostration, stupor, and delirium, with myocardial and renal failure. The fatality rate in untreated cases is correlated with age: low in children, about 10% in young adults, and as high as 60%–70% in those over 50. Recovery from an attack gives rise to an enduring immunity.

Diagnosis by laboratory means is based primarily on three types of serologic tests (see Table 40-1):

1) The Weil–Felix agglutination reaction is of great value: it is positive in over 90% of primary louse-borne typhus cases, and in a rapid slide method it can be performed at the bedside in 3–5 min with a drop of the patient's blood.

2) CF Abs to both group and type-specific Ags appear regularly.

3) Specific Abs are also measured by immunofluorescence, the most sensitive of all the tests, especially for detecting residual IgG Abs. These persist for decades following an acute attack of typhus. Unfortunately, this test is not widely available at present outside research laboratories. Crossreactions occur frequently between the typhus and the spotted fever groups of organisms. Abs cannot be detected by any of the tests before the eighth or ninth day after onset of illness, and by this time untreated patients may be moribund or dead. Accordingly the diagnosis must be suspected clinically, and serologic tests are mainly useful as confirmatory evidence.

Prevention and Control. Two powerful means of controlling louse-borne typhus have been developed: **vaccination** and **louse control**. A number of effective vaccines have been developed, such as the Cox vaccine (*R. prowazekii* cultivated in the yolk sac and treated with formalin). Immunization does not protect fully against infection, but it makes the illness shorter and milder.

Other killed vaccines were administered successfully on a large scale to military and civilian populations during World War II. A small percentage of vaccinees are sensitive to egg contaminants in the vaccines. An attenuated live vaccine is also effective.

DDT offers an effective weapon against typhus epidemics: simply blowing it under clothing eliminates lice, and it can be used in large populations very quickly; it remains effective for 2–4 weeks. In some regions lice have become resistant to DDT. Other insecticides, such as lindane or malathione, are then used.

BRILL–ZINSSER DISEASE AND THE LATENT STATE

Brill–Zinsser disease was first described by Brill in 1898 in New York City among immigrants from Eastern Europe; it represents the reservoir (or latent state) of louse-borne typhus. After the primary attack the rickettsiae remain dormant, probably in cells of the reticuloendothelial system. Years later they may multiply and cause a second acute infection when stress, lowered immunity, or some unknown factor reduces the host's resistance (as in a relapse of malaria).

Brill–Zinsser disease differs from primary louse-borne typhus in several respects: 1) no louse vector or exogenous source of the infection is apparent; 2) the disease is milder, shorter, and frequently without a rash, like primary typhus in a vaccinated and partly immune individual; 3) the immune response is of the secondary type: Abs appear within 4–5 days, they are IgG, and they reach higher titers than in the primary response; 4) Weil–Felix reactions, which develop regularly following primary typhus, are almost uniformly negative in relapse attacks within 5 years of the primary attack (i.e., when immunity from the original illness is presumably relatively high). As the interval between the original and the recrudescent attack lengthens, positive Weil–Felix reactions occur more frequently.

Weil–Felix reactions usually also fail to develop in those who contract primary louse-borne typhus despite prophylactic vaccination, as well as in typhus patients who receive early antimicrobial therapy. Evidently *R. prowazekii* contains only a small amount of the Ag that crossreacts with *Proteus,* and so only the large mass of organisms appearing in full-blown primary typhus produces sufficient antigenic stimulus.

Brill–Zinsser disease was common in Eastern Europe after World War II, and the flow of *R. prowazekii* through the Bosnian community in Yugoslavia has been studied carefully. Up to 1970 some 50–100 cases were reported annually, sometimes followed after several weeks by primary louse-borne typhus infections in the children. In the past these family outbreaks would undoubtedly have spread to become community epidemics, but louse infestation has come under control. The incidence of Brill–Zinsser disease has steadily decreased, probably through the dying off of the population infected with typhus during World War II. The last recorded case of primary louse-borne typhus in Bosnia occurred in 1967.

MURINE (ENDEMIC) TYPHUS

Murine typhus (L. *mus, muris,* mouse), a natural infection of rats, is occasionally transmitted to man by the rat flea. Infected rats are scattered in circumscribed areas throughout the world, particularly in coastal and port areas: in the United States such reservoirs are found along the southern Atlantic seaboard and in large areas fanning out from the Gulf of Mexico seaports. The disease closely resembles louse-borne typhus clinically but is milder.

The etiologic agent, *Rickettsia mooseri (R. typhi* in Bergey's Manual), is similar to *Rickettsia prowazekii* but is slightly smaller. The two organisms possess a common soluble Ag, and their surface Ags crossreact. Accordingly, recovery from either infection leads to enduring immunity to both, but immunization with either killed organism induces only homologous immunity.

This difference fits experience with other organisms: natural infection, or vaccination with viable microbes, usually establishes more solid and more enduring immunity than killed microbes, presumably because of more prolonged persistence of the immunogens (Ch. 22); and an immunogen produces a lower titer of Ab to crossreacting Ags than to itself.

Laboratory diagnosis of murine typhus (Table 40-1) is similar to that for louse-borne typhus, using the Weil–Felix, CF, and immunofluorescence tests: the predominant Abs are to *R. mooseri* rather than to *R. prowazekii.*

Epidemiology and Control. Murine typhus is worldwide in distribution. It is called Toulon typhus, Manchurian typhus, Moscow typhus, shop typhus (Malaya), tarbadillo (Mexico), and red fever (Congo). In the United States from 1930 to 1950 there were about 3000–5000 cases per year, mostly in the southeastern states. Intensive rat control measures have now reduced the incidence to a few dozen cases a year.

R. mooseri is transmitted from rat to rat, and occasionally from rat to man, by the rat flea (*Xenopsylla cheopis*).* The flea becomes infected by feeding on a rat that is acutely ill (i.e., has rickettsemia), and when rats are not available, it will bite a human, infecting via its feces (as in louse-born infections). However, while *R. prowazekii* causes severe disease in its natural hosts (man and the body louse), *R. mooseri* causes only latent infection in its natural hosts (rat and the rat flea). These provide the reservoir of *R. mooseri;* no reservoir in man has been demonstrated.

The prevention and control of murine typhus depend on limiting rat and flea populations. The first step is to kill off fleas by dusting rat runs with appropriate insecti-

---

* It is also transmitted among rats by the rat louse, *Polypax spinulosa.*

cides; the rat population can then be reduced by poisoning and by rat-proofing buildings. (If the insecticides are not applied before the rat poisons the fleas will depart from the dying rats and will infest man.)

SPOTTED FEVER GROUP

The spotted fevers are a group of infectious diseases caused by *Rickettsia rickettsii* and its antigenic variants. They are also called **tick typhuses** because all are transmitted by ticks. Rocky Mountain spotted fever (RMSF) occurs primarily throughout the temperate zones in the tick areas of North and South America. Three other clinical forms of tick typhus are distinguished by geographic location and by antigenic differences in the causative organisms.

The rickettsiae of tick typhus multiply in the cytoplasm and occasionally in the nucleus of susceptible animal cells, whereas other rickettsiae grow exclusively in the cytoplasm.

ROCKY MOUNTAIN SPOTTED FEVER

This disease, caused by *R. rickettsii*, was first recognized in Idaho and Montana at the turn of the century; it has now been observed in almost every state in the United States, and is especially prevalent in the zone of coastal states from Georgia to Maryland. A rapid recent shift in distribution in the United States makes the name "Rocky Mountain" archaic (Fig. 40-3).

Pathogenesis. The disease usually begins 2–8 days after the bite of an infected tick. RMSF resembles typhus in causing fever, headache, and a rash. However, the rash begins peripherally on the ankles and wrists and spreads centripetally to the trunk, while the reverse is characteristic of the louse- and flea-borne typhus fevers. A highly diagnostic feature of the RMSF rash is its regular occurrence on the palms and soles. As in typhus and all other rickettsial diseases (except Q fever), the pathologic lesions are primarily those of angiitis and intravascular thrombosis. In RMSF the focal lesions are particularly severe, with the frequently grave complication of disseminated intravascular coagulation.

In preantibiotic days fatality rates were 25% and upwards. Though the tetracyclines and chloramphenicol are almost uniformly curative if given before the sixth day of illness, RMSF still runs a 5%–7% mortality in the United States, largely because of incorrect or late diagnosis.

Diagnosis and Prevention. The same diagnostic tests (Weil–Felix, CF, and immunofluorescence) are used as for louse-borne typhus (Table 40-1). Unfortunately, the late development of Abs (approximately the eighth to tenth day of the disease) again limits the value of serologic diagnosis.

To be life-saving specific treatment must be started early, when the clinician has little else than symptoms, signs, and epidemiologic considerations to aid him in a provisional diagnosis. During the first week of illness a low WBC count (uncommon in severe bacterial infections), and/or thrombocytopenia (which is also present in other infectious diseases), are helpful signs. Recently, success has been reported in immunofluorescent examination of skin lesion biopsies for RMSF organisms as early as the fourth or fifth day of illness.

Vaccines from eggs have not been satisfactory, but a promising killed vaccine has been prepared from infected, cultured chick embryo cells. Cultivation requires scrupulous care, because penicillin cannot be used to prevent contamination. The vaccine could be useful for people living in the "tick belt."

Epidemiology and Transmission. Spotted fever rickettsiae are primarily parasites of ticks. They do not kill their arthropod hosts and are passed transovarially, via the eggs. These undergo a series of metamorphoses to larvae, nymphs, and finally adults, over a 2-year period; and the adults can survive for as long as 4 years without feeding. Congenitally acquired rickettsiae usually persist through all these stages.

In the infected tick, unlike lice infected with *R. prowazekii*, the rickettsiae are not limited to the cells of the intestinal lining but are widely distributed. Those in the ovaries give rise to infection in eggs and in progeny, and those in the saliva are transmitted to man during the

FIG. 40-3. Changing annual incidence of Rocky Mountain spotted fever in the eastern and the western regions of the United States.

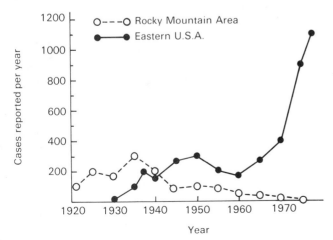

tick bite. Ticks are thus both **primary reservoirs** and **vectors** of the organism.

The wood tick, *Dermacentor andersoni,* is mainly responsible for the transmission of RMSF in the western United States, and the dog tick, *Dermacentor variabilis,* in the eastern United States. Only a very small percentage of ticks are infected with RMSF organisms, but infected ticks may be concentrated in focal areas. For example, Long Island reports about 80% of all the RMSF cases for New York State; and Cape Cod and the offshore islands account for almost all the cases in Massachusetts.

Many small wild animals possess Abs to RMSF and hence may be involved in the tick → mammal → tick cycle. Since ticks themselves are reservoirs the mammals may be important mainly as a source of blood meals, which are indispensable for tick survival and maturation.

During 1977 on Cape Cod classic RMSF strains of rickettsiae were isolated from at least five acutely ill dogs (one of which died with a fulminating infection). In addition, rising titers of RMSF Abs have been observed in sick dogs, and significant titers in dogs without a history of a prior severe illness. Nevertheless, the role of dogs in the RMSF cycle remains obscure: they may be accidental hosts of the rickettsiae, like man; or they may transport ticks to the proximity of man.

TICK TYPHUS FEVERS

Varieties of tick-borne rickettsial diseases that are similar to RMSF are widely distributed geographically: in different localities they are known by different names and vary in severity. The etiologic agents share the same group Ag as *R. rickettsii* but have distinguishing type-specific Ags, as revealed in CF and neutralization tests. Three general antigenic types (in addition to RMSF), have been demonstrated: **boutonneuse (Marseilles) fever, Siberian tick typhus,** and **Queensland tick typhus.** They have similar clinical, pathologic, and epidemiologic patterns.

Boutonneuse fever is the most widely distributed, occurring along the Mediterranean and the Black Sea and Caspian Sea littorals. In the African interior it is known as Kenya tick typhus and as South African tick-bite fever, and in Southwest Asia as Indian tick typhus.

The principal mammalian reservoir of the organism is the dog, and infection is also transmitted transovarially in ticks. The disease in humans is much milder than RMSF, the mortality rate being only 1% in untreated patients. A distinctive feature is the **small indurated lesion** that develops about one week before the onset of overt disease at the site of the tick bite. This buttonlike lesion, referred to as **tache noire,** becomes necrotic at its center, develops an eschar, and gives rise to enlargement of the regional lymph nodes.

RICKETTSIALPOX

This disease, first recognized in New York City in 1946, has been observed in a number of other cities in the northeastern United States, and what is probably the same disease has been described in urban centers within the Soviet Union. The causative agent is *R. akari.* Like the spotted fever rickettsiae, it multiplies in both the nucleus and the cytoplasm of host cells; and its particulate type-specific Ag crossreacts extensively in CF assays with spotted fever organisms. However, it is transmitted by a **mite** rather than a tick, and it only occasionally elicits Abs to *Proteus* OX strains.

The **house mouse** constitutes the natural mammalian reservoir for *R. akari.* **A blood-sucking mite** transmits the organism among mice and occasionally to humans. Since house mice are particularly common near garbage and refuse piles in apartment houses the disease is characteristically seen among apartment dwellers in urban centers. At the skin site where rickettsiae are introduced by the mite a papule occurs that subsequently develops a vesicle and then an eschar (**tache noire,** as in boutonneuse fever); it resembles the site of an inoculation with vaccinia.

As with other rickettsial diseases, the systemic illness is characterized by abrupt onset, fever, headache, and rash. The rash is unusual for a rickettsial disease in that it is vesicular and simulates chickenpox. The disease usually lasts 1–2 weeks, and no fatalities have been reported. **Diagnosis** can be made by CF or immunofluorescence, using either RMSF or rickettsialpox organisms as Ags. Abs to *Proteus* strains are not frequent enough to be useful.

TSUTSUGAMUSHI DISEASE (SCRUB TYPHUS)

Scrub typhus is an acute infectious disease transmitted to man by the bite of **trombiculid mites.** The disease occurs almost exclusively in a roughly triangular area including Japan, Southeast Asia, and the islands of the South Pacific. The organism responsible, *R. tsutsugamushi,* is common in mites and in wild rodents. Mites act as both reservoir and vector, transmitting the rickettsiae among rodents and also, via infected ova, to their own progeny. Man becomes an incidental host when infected larvae of the mites attach to the skin and suck blood.

Clinical Manifestations. One to two weeks after being bitten the human host develops the characteristic rickettsial manifestations: fever, intense headache, and a rash. At the time of onset the site of the infected mite's bite becomes an ulcerated papulovesicle, which forms a black scab (eschar) and is associated with enlarged, tender regional lymph nodes. Later generalized lymphadenopathy occurs—a manifestation rare or absent in other rickettsial diseases. The macular rash, commonly evanescent, appears on the trunk between the fifth and eighth day of illness. Case fatality rates in different localities vary widely without treatment (from 1% to 35%), probably reflecting strain differences.

Patients who recover from scrub typhus are immune for a time, but they may suffer second attacks. Studies on recovered human volunteers have shown that immunity to the homologous strain of *R. tsutsugamushi* persists for several years, but immunity to other strains lasts for only a few months. The diversity of strains greatly complicates the problem of developing an effective vaccine.

Scrub typhus may exist in a **latent** form, like the carrier state of louse-borne typhus: viable *R. tsutsugamushi* organisms have been found in lymph node biopsies taken as long as 1 year after recovery.

Diagnosis and Prevention. A Weil–Felix text with the OXK *Proteus* strain is positive in over half of the cases. Complement fixation and immunofluorescence tests require the use of three to eight different strains to be useful. Isolation of *R. tsutsugamushi* from the blood by inoculation of mice is frequently employed to confirm the diagnosis. As in spotted fevers, the clinician must usually rely on clinical and epidemiologic considerations in order to institute life-saving therapy during the acute phase of the illness.

During World War II the incidence of scrub typhus in military personnel was greatly lessened by clearing away brush, spraying insecticides, and treating clothes and exposed skin with mite repellents such as benzyl benzoate and phthalate esters. Vaccines are not feasible.

Q FEVER

Q fever, an acute infectious disease of worldwide occurrence, is primarily a disease of animals **transmitted to humans through inhalation.** The rickettsial agent, discovered by Burnet and Freeman in Queensland, Australia, and by Cox in the United States, is therefore referred to as *Coxiella burnetii.* It is placed in a different genus from the other rickettsiae because it differs in several major respects: 1) it is unusually stable outside host cells; 2) infection in man is acquired not by the bite of an infected arthropod but by inhalation of infected particles, though in nature the organism is arthropod-borne; 3) disease in man is usually characterized by pneumonitis, without a rash; and 4) *C. burnetii* does not elicit Abs for OX strains of *Proteus,* nor does it crossreact with other rickettsiae in

CF tests and in protection against infection in experimental animals.

Pathogenesis. The infection is usually acquired by inhaling contaminated dusts or aerosols. Onset of the disease in man is abrupt, with fever and a severe, intractable headache, but no rash. On x-ray well over 50% of patients reveal a patchy pneumonic process which resembles a viral or atypical pneumonia, although physical findings in the chest are few. In addition, Q fever is, like the other rickettsioses, a widely systemic disease. Complications and death are uncommon, but severe chronic Q fever may occur in the form of myocarditis, pericarditis, or endocarditis.

Diagnosis by serologic laboratory methods includes CF and immunofluorescence tests. Isolation procedures are dangerous to laboratory personnel. While clinical features of Q fever are not specific, epidemiologic data can be highly valuable.

Epidemiology. Q fever has been reported from almost every region in the world. In nature it is arthropod-borne, like other rickettsial diseases; the organism has been demonstrated in at least 20 different species of ticks. However, transovarial passage is apparently rare, so that a tick → mammal → tick cycle is important in maintaining the disease; natural infection has been found in many mammalian species.

With man, in contrast, transmission from ticks is uncommon (and man-to-man transmission is rare if it occurs at all). *C. burnetii* is spread to man primarily by **inhalation of aerosols from domestic livestock.** In these animals the infection ordinarily remains latent until the stress associated with parturition. The organisms then multiply and are found in the birth tissue, and also in urine and the stool; after these contaminate the soil dried dust particles containing the hardy organisms can remain a source of infection for many months.

Coxiella also appears in the milk of cows and sheep, but man is not very susceptible to Q fever infection via the gastrointestinal tract. Of most importance is the **massive infection of placentas** of cattle, sheep, and goats; and at slaughter infected pregnant animals may produce highly infectious aerosols and thus cause epidemics in abattoirs. In addition, **contaminated wool** can produce epidemics in textile plants far removed from the original source of infection. Most cases of Q fever have been recognized in populations of slaughterhouse workers, wool handlers, livestock tenders, and laboratory workers.

Prevention and Control. Experimental vaccines have proven impractical because of adverse reactions. After treatment with tetracyclines relapses are rarely observed.

SELECTED READING

BOOKS AND REVIEW ARTICLES

HOEPRICH P (ed): Infectious Diseases, Chapters 29, Q fever; 92, Spotted Fevers; 93, Rickettsialpox; 94, Typhus Fevers; 95, Scrub Typhus; 2nd ed. Hagerstown, Harper & Row, 1977

LENNETTE EH, SCHMIDT NF (eds): Diagnostic Procedures for Viral and Rickettsial Diseases, 4th ed. New York, American Public Health Association, 1969

Proceedings of the International Symposium on the Control of Lice and Louse-Borne Diseases, Scientific Publ No. 263. Washington, DC, Pan American Health Organization, 1972

ROUECHE B: The alerting of Mr. Pomerantz. In Eleven Blue Men and Other Narratives of Medical Detection. Boston, Little, Brown, 1954

WEISS E: Growth and physiology of rickettsiae. Bacteriol Rev 37:259, 1973

ZINSSER H: Rats, Lice and History. Boston, Little, Brown, 1955

SPECIFIC ARTICLES

BOZEMAN FM, ELISBERG BL: Serologic diagnosis of scrub typhus by indirect immunofluorescence. Proc Soc Exp Biol Med 112:568, 1963

BREZINA R, MURRAY ES, TARRIZO ML, BÖGEL K: Rickettsiae and rickettsial diseases. Bull WHO 49:433, 1973

HATTWICK MA, O'BRIEN RJ, HANSON BF: Rocky Mountain Spotted Fever: epidemiology of an increasing problem. Ann Intern Med 84:732, 1976

KENYON RH, CANONICO PG, SAMMONS LS: Antibody response to Rocky Mountain Spotted Fever. J Clin Microbiol 3:513, 1976

MURRAY ES: On the alert for Rocky Mountain spotted fever. N Engl J Med 292:1127, 1975

OSTEN CN, BURKE DS, KENYON RH et al: Laboratory acquired Rocky Mountain spotted fever. N Engl J Med 297:859, 1977

OSTEN CN, KENYON RH, ASCHER MS: Initial clinical evaluation in man of a new Rocky Mountain spotted fever vaccine of tissue culture origin. Clin Res 25:382, 1977

WOODWARD TE, PEDERSEN CE JR, OSTEN CN et al: Prompt confirmation of Rocky Mountain spotted fever: identification of rickettsiae in skin tissues. J Infect Dis 134:297, 1976

chapter 41

CHLAMYDIAE

ROGER L. NICHOLS,
G. PHILIP MANIRE

GENERAL CHARACTERISTICS

The genus *Chlamydia* comprises two species, *C. trachomatis* and *C. psittaci*. These organisms are differentiated from other bacteria by their morphology, by a common group antigen, and by a unique developmental cycle involving two morphologic forms—one adapted to extracellular survival and the other to intracellular multiplication within cytoplasmic vesicles, commonly termed inclusions.

Because chlamydiae are small and multiply only within susceptible cells they were long thought to be viruses. However, they have many characteristics in common with other bacteria (Table 41-1): (1) they contain both DNA and RNA, (2) they divide by binary fission, (3) their cell envelopes resemble those of other gram-negative bacteria, (4) they contain ribosomes similar to those of other bacteria, and (5) they are susceptible to various antibiotics. Chlamydiae can be seen in the light microscope, and the genome is about one-third the size of the *Escherichia coli* genome.

Many different strains of chlamydiae have been isolated from birds, man, and other mammals, and these strains can be distinguished on the basis of host range, virulence, pathogenesis, and antigenic composition. They have been classified in two species on the basis of sulfonamide susceptibility, presence of glycogen in the inclusion, type of inclusion, and DNA homology and content (Table 41-2). There is strong homology of DNA within each species, but surprisingly little between the two, suggesting long-standing evolutionary separation.

C. trachomatis has a high degree of host specificity, being almost completely limited to man; it causes ocular and genitourinary infections of widely varying severity. Burnet lists trachoma among the three most important infectious diseases of man, and its sequelae hamper socioeconomic progress in some countries. In contrast, *C. psittaci* strains are rare in man but are found in a wide range of birds and also in wild, domestic, and laboratory mammals, where they multiply in cells of many organs. They also cause economically important diseases of livestock and are a major problem in the poultry industry.

In general chlamydiae exhibit low pathogenicity except when the host is stressed. Apparently healthy birds develop the disease under crowded conditions or when deprived of essential nutrition, and in man trachoma increases in severity as well as in frequency as hygiene and public sanitation deteriorate.

MORPHOLOGY AND DEVELOPMENTAL CYCLE

Shortly after the discovery of chlamydiae, Bedson showed that their morphology varies during the developmental cycle. This observation has been extended in recent years by electron microscopic examination of thin sections of infected mammalian cells in culture.

The developmental cycle alternates between two forms: the **elementary body (EB),** 300 nm in diameter, is specialized for survival when released from the cell, and the **reticulate body (RB),** up to 1000 nm, is engaged in intracellular multiplication (Fig. 41-1). These and the many intermediate forms all stain readily with Giemsa, Macchiavello, Castaneda, and Gimenez stains.

The chlamydiae exhibit affinity for the epithelial cells of mucous membranes. With such cultured cells, or with mouse fibroblasts (L cells), attachment and penetration seem to involve a heat-labile surface component on the chlamydiae and a trypsin-sensitive receptor on the host cells. On HeLa cells sialic acid residues appear also to serve as receptors. The EBs enter by a phagocytic process, creating a phagosome or inclusion in which the microcolony develops. Uptake of chlamydiae by macrophages or granulocytes apparently does not depend on such specific attachment mechanisms.

Within a few hours after entry into the host cell the EBs begin to undergo profound changes in their cell envelopes and the characteristic central condensate begins to disperse, to form a more homogeneous cytoplasm in which strands of nucleic acid and ribosomes are seen. The resulting RBs continue to grow in size, and 10–15 h after infection binary fission begins. At 20–30 h after infection some of the RBs develop a central condensation of cytoplasmic contents, decrease in size, and become typical EBs. Most of the RBs, however, continue to multiply until the host cell cytoplasm is almost filled by the

TABLE 41-1. Comparison of Chlamydiae with Other Pathogenic Agents

	Chlamydiae	Bacteria	Rickettsiae	Mycoplasmas	Viruses
Growth outside host cell	0	+	0*	+	0
Independent protein synthesis	+	+	+	+	0
Generation of metabolic energy	0	+	+	+	0
Rigid cell envelope	Variable	+	+	0	Variable
Antibiotic susceptibility	+	+	+	+	0
Mode of reproduction	Fission	Fission	Fission	Fission	Host cell synthesis of sub-units; then assembly of virion
Nucleic acids	DNA & RNA	DNA & RNA	DNA & RNA	DNA & RNA	DNA or RNA, not both

*Except *Rochalimaea quintana.*

colony. Little is known about the mechanism of release of organisms, but in cell culture systems the host cells die and autolyze 40–60 h after infection.

The evolution of the EB into the RB involves major changes in structure and composition of the cell envelope. The EB envelope is rigid and resistant to sonic treatment, is relatively impermeable to macromolecules, contains hemagglutinin, and has an inner layer composed of hexagonally arranged subunits. The subunit layer probably accounts for the rigidity of the Eb. Soon after the EB enters into the phagosome its envelope loses its rigidity and the subunit layer is disrupted and disappears. The RB cell envelope is fragile, easily disrupted, and highly permeable to macromolecules; it contains no hemagglutinin; and the subunit layer is missing (Fig. 41-2). *RBs are not infectious.*

In cells infected with *C. trachomatis* a glycogen-containing substance appears within the inclusion concurrently with EB production. Its origin and function are unknown.

CHEMICAL AND METABOLIC CHARACTERISTICS

Chlamydiae are exquisitely adapted for extracellular survival and intracellular growth. Although the mechanisms are not known, they prevent their phagosomes from fusing with lysosomes, and they inhibit host synthesis of proteins and DNA. They make all of their own macromolecules, but they parasitize their host cells for metabolic energy.

The dependence on host cells is even greater than that of rickettsiae (Ch. 40), which make at least some of their ATP by oxidative phosphorylation and also possess more extensive biosynthetic capabilities. In contrast, the chlamydiae do not oxidize pyruvate by way of the tricarboxylic acid cycle and do not appear to generate ATP. However, the purified suspensions of chlamydiae used in studies of these processes may give an incomplete picture, since the more active but fragile RB form has been destroyed.

TABLE 41-2. Division of Chlamydiae into Species

Property	*C. trachomatis*	*C. psittaci*
Susceptibility to sulfonamides	Sensitive	Resistant
Type of inclusion produced	More compact	More diffuse
Presence of glycogen in inclusion	Yes	No
% G + C	~44%	~41%
Principal hosts	Man	Birds and mammals
Diseases induced	Trachoma	Ornithosis, psittacosis
	Inclusion conjunctivitis	Many clinical and subclinical
	Lymphogranuloma venereum	diseases of mammals and birds
	Urethritis	
	Mouse pneumonitis	

(Adapted from Gordon FB, Quan AL: J Infect Dis 115:186, 1965)

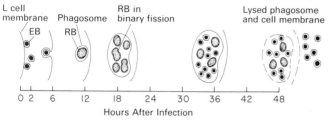

Elementary body	Reticulate body
Size about 0.3 μm	Size 0.5 - 1.0 μm
Rigid cell wall	Fragile cell wall
Relatively resistant to sonication	Sensitive to sonication
Resistant to trypsin	Lysed by trypsin
Subunit in cell envelope	No subunit in envelope
RNA:DNA content = 1:1	RNA:DNA content = 3:1
Toxic for mice	Nontoxic for mice
Isolated organisms infectious	Isolated organisms not infectious
Adapted for extracellular survival	Adapted for intracellular growth

FIG. 41-1. Schematic representation of the infectious cycle of chlamydiae, with characteristics of the two developmental forms of the organism.

The genome of *C. trachomatis*, about 660×10^6 daltons, is sufficient to code for about 1000 protein molecules. Ribosomes of chlamydiae are typical of those of other bacteria.

ANTIGENS

Both **group-specific** and **type-specific** Ags can be identified by Abs elicited by infection or by vaccination. The group Ag can be demonstrated only in lysates (or heated suspensions) of the organisms, whereas the type-specific Ag is associated with the outer layers of envelopes in intact cells. As with many other pathogenic organisms, the internal Ags are relatively common to a generic group of organisms, whereas the surface Ags concerned with infectivity and virulence tend to be diverse.

The group Ag formed in all chlamydiae is heat-stable and resistant to trypsin, and it is inactivated by ether or periodate, suggesting the presence of glycolipids. Abs to this Ag can be detected by complement fixation (CF) or immunofluorescence. The type-specific Ags have not been well defined but they appear to be proteins, for they are resistant to periodate and highly sensitive to heat. Abs to these specific Ags appear in infected patients and infected animals.

TOXICITY

When approximately 10^9 infectious EBs of most chlamydiae (about 10μg–20μg dry weight) are injected intravenously into young mice death occurs in 12–24 h, and histologic findings are typical of shock rather than of overwhelming infection.

The toxin has not been isolated; it can be neutralized by antisera to homologous strains, and the Abs can be absorbed by EBs but not by RBs. Recent evidence indicates that when relatively large numbers of EBs are phagocytized by macrophages or L cells early death and lysis of the cells occur, resulting in release of lysosomal enzymes. The toxicity associated with this release probably plays a role in pathogenesis.

IMMUNITY

Diseases caused by chlamydiae tend to run chronic and relapsing courses: without treatment the same serotype of *C. trachomatis* can be found in the eye or the cervix of an individual patient for 2–5 years or longer, which implies continued infection rather than reinfection. Chronicity of infection in turn implies that chlamydiae do not usually evoke a thoroughly effective immune response, but ordinarily the host does restrain and localize the disease without serious sequelae.

Neutralizing Ab that combines with cell-envelope Ag of the parasite arrests the spread of infection to susceptible host cells. However, these Abs do not appear to inactivate chlamydiae already localized within the cells. Individuals free of apparent disease and relatively immune to reinfection may continue to shed virulent organisms, thus serving as healthy **carriers.** The cellular mechanisms underlying the carrier state are unknown.

In primates infection of the eye with *C. trachomatis* can induce quite solid immunity of a few months' duration, but in birds and mammals with *C. psittaci* pneumonic and other infections only partial protection has been observed. An immune response occurs, but it can be overwhelmed by large challenge doses or by adverse conditions.

The roles of circulating Ab, secretory Ab, and the cellular immune reaction in naturally occurring disease are not clear. In passive transfer experiments circulating Ab to inclusion conjunctivitis failed to protect guinea pigs from subsequent challenge, whereas animals with Ab in eye secretions (following eye infections) were protected. However, cell-mediated immunity may also have played a role in this protection.

FIG. 41-2. A. Shadowcast preparation of purified elementary bodies of *C. psittaci*. **B.** Shadowcast preparation of purified cell envelopes of elementary bodies of *C. psittaci* showing hexagonally assayed subunit structures on the inner aspect of the envelope. **C.** Shadowcast preparation of purified reticulate bodies of *C. psittaci*. **D.** Thin section electron micrograph of inclusion of *C. psittaci* in L cells 36 h after infection. Elementary bodies, reticulate bodies, cells undergoing binary fission, and transitional forms may be seen. (**A,D,** Courtesy of Dr. A. Matsumoto, Kyoto University; **B,** Matsumoto A, Manire GP: J Bacteriol 104:1332, 1970; **C,** Courtesy of Drs. A. Matsumoto and A. Tamura, Kyoto University)

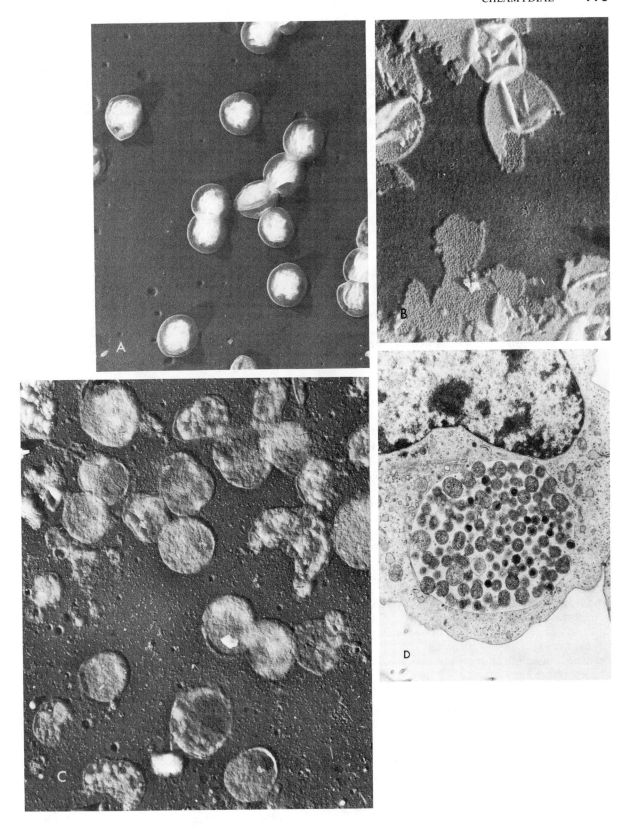

CHEMOTHERAPY

The sensitivity of chlamydiae to sulfonamides and penicillin provided the first evidence that they are bacteria and not viruses. Penicillin affects a specific stage in the cell cycle: it does not affect the penetration of host cells by EBs, or reorganization to form RBs, but it inhibits subsequent binary fission, synthesis of envelope subunits, and maturation to form EBs. Sulfonamides inhibit chlamydiae that synthesize folic acid but not those that

require it; in general *C. trachomatis* strains are susceptible whereas *C. psittaci* strains are resistant. Sulfonamides have been widely used in the treatment of trachoma and lymphogranuloma venereum.

Both *C. psittaci* and *C. trachomatis* are highly susceptible to tetracycline and chloramphenicol. Rifampin and related compounds can also be used. Drugs that penetrate host cells with difficulty (e.g., aminoglycosides) are relatively ineffective.

DISEASES DUE TO CHLAMYDIAE

TRACHOMA

HISTORY AND EPIDEMIOLOGY

The agent of trachoma is one of the most successful parasites of man. The Ebers Papyrus, the oldest known medical writing in the Western world, describes the clinical features of trachoma as well as several forms of treatment, at least one of which (copper sulfate) is still in use in certain parts of the world.

In nature *the disease is limited to man,* infecting only epithelial cells of the eye and possibly the nasopharynx; no systemic involvement has been described. Characteristic cytoplasmic inclusions in conjunctival cells were seen early in this century, but the agent was first grown in the yolk sac of chick embryos in 1955. Progress in understanding the organism has since been rapid.

Trachoma is found in nearly every country and *is the greatest single cause of blindness.* It is estimated that over 400 million people have the disease, of whom 6 million are totally blind. Although frequently associated with dry or sandy regions of the world, trachoma is also common in areas of high rainfall such as the equatorial regions of Africa, and even in populations living above the Arctic Circle. The disease flourishes in communities where public sanitation and personal hygiene are poor. It is thought to be transmitted primarily within the family by eye-to-finger-to-eye contact or by towels and clothing. Transmission by flies is suspected but has not been established. Subclinical infections are common and serve as important reservoirs.

Many enigmas remain in the epidemiology of trachoma. Why was the disease prevalent among Indians in the reservations of the United States and among whites in the border states (Kentucky, Tennessee, West Virginia), while Negroes in these regions remained relatively unaffected, though Negroes in widely distributed areas of Africa are quite susceptible? Why is the disease relatively infrequent in the southern portion of India but hyperendemic in the northern Punjab regions? Why does onset of the disease occur usually in infants in some countries, in children in others, and in adults in yet others?

PATHOGENESIS

In many communities almost all children are infected in their first few years. Under these holoendemic conditions the infection becomes chronic in a small percentage of patients; blindness ensues primarily in this group. However, the disease may begin at any age and adult volunteers have developed severe infections after administration of a few infectious doses of trachoma elementary bodies.

Onset of disease is usually abrupt, with inflamed palpebral and bulbar conjunctivae; within a few weeks accumulations of lymphocytes, polymorphonuclear neutrophils, and macrophages coalesce to form characteristic **follicles** beneath the conjunctival surface. Still later **vascularization** of the cornea begins, usually at the upper limbus, followed by an **infiltration** of the cornea (termed **pannus**), which may produce partial or complete blindness. Scarring of the conjunctiva may cause the eyelids to turn inward so that the lashes scratch the cornea. Distortion of the structures of the external eye also interferes with normal lacrimal flow, growth of lashes, and function of glands; as a result, **bacterial infections** of trachomatous eyes are common. These factors act in concert to destroy vision.

Study of the pathogenesis of trachoma is hampered by the lack of satisfactory animal models. Although Old and New World monkeys can be infected, the disease does not closely resemble that in man.

ANTIGENIC COMPOSITION

C. trachomatis strains isolated from trachomatous patients around the world are relatively homogeneous, with three serotypes (A, B, and C), determined by mouse toxicity and immunofluorescence tests, accounting for the vast majority of strains. Sporadic eye infections with most of the remaining nine serotypes (Ba, D–K) usually result from genital-to-finger-to-eye transmission. The biologic significance of these serotypes is not clear.

IMMUNITY

In populations where trachoma is holoendemic and not effectively treated over 90% of patients recover spontaneously without serious sequelae. As was noted, trachoma generates partial immunity, which may diminish the duration of infection or increase the dose required for reinfection. Vaccination protects mice against the toxic reaction described under Toxicity, above.

Abs appear in the serum and in the eye secretions within 2 or 3 weeks after infection. Eye secretions from human patients with active clinical trachoma contain IgG and secretory IgA Abs, which are type-specific for the infecting strain. When mixed with homologous EBs these Abs neutralized or delayed their capacity to infect the eyes of owl monkeys, but systemic administration failed to protect the eyes from challenge. Infection also induces a cellular immune response, which may be implicated in the development of pannus as well as in the immune reactions.

In animal models infection followed by spontaneous cure renders the animal almost totally immune to an ocular challenge, but only for a few months. Killed **vaccines** have temporarily heightened resistance to subsequent challenge in animals. In field trials clinical attack rates in vaccinated humans are significantly diminished, and those patients who become infected have fewer inclusions in conjunctival scrapings; the transmission rates may thereby be reduced and the epidemic curtailed. However, the short duration of immunity severely limits the usefulness of these vaccines, and none are recommended at present.

DIAGNOSIS

Clinical diagnosis of trachoma rests on the finding of characteristic follicles and scars in the conjunctiva, and vascularization and infiltration of the cornea. In the typical case diagnosis is easy; in mild cases, or after distortion of the anatomy of the external eye, the diagnosis of activity may be difficult.

Currently three technics are used for laboratory diagnosis. 1) The **agent is readily isolated** from conjunctival scrapings in active cases by inoculation in irradiated McCoy cells, bacterial contamination being suppressed by use of appropriate antibiotics. 2) Typical **cytoplasmic inclusion bodies** are found within conjunctival cells stained by fluorescent Ab, Giemsa's stain, or iodine. 3) A **serologic test of eye secretions** for trachoma-specific Abs correlates well with active trachoma as defined by the microbiologic tests.

TREATMENT

In the patient with early trachoma a sulfonamide or tetracycline given for 3 weeks is usually curative. Frequent regular application of ophthalmic ointments containing tetracycline has also been recommended, but the infection may persist under this regimen. In chronic trachoma anatomic changes in the lacrimal drainage system and scarring in the eyelids predispose to bacterial infections and render clinical cures uncertain.

When trachoma is holoendemic throughout an entire nation its management poses entirely different problems, in which improvements in standards of living and hygiene are more important than treatment.

INCLUSION CONJUNCTIVITIS

In developed countries the characteristic chlamydial infection of the conjunctiva is inclusion conjunctivitis. Its two synonyms—inclusion blenorrhea of the newborn and swimming pool conjunctivitis—arise from the different epidemiology of the disease in infants and adults, but in both the reservoir of infection is the human genitourinary tract, where the infection is often asymptomatic.

Clinical Description. In the newborn clinical signs appear 5–12 days after birth. The conjunctiva becomes inflamed and thickened, often with copious discharge of pus; follicles can develop on the conjunctiva. The disease is usually self-limiting, resolving spontaneously within a few months even without treatment.

In the adult, too, the conjunctivitis resembles the early stages of trachoma, but it usually does not progress to a chronic disease; blindness is not a threat. Indeed, in classic descriptions inclusion conjunctivitis, by definition, did not involve the cornea or cause scars on the conjunctiva. Most of these cases involve the serotypes characteristic of genital tract infection (see below). However, in epidemics of inclusion conjunctivitis, in communities without known trachoma, cases clinically indistinguishable from trachoma occasionally occur. Moreover, trachoma is often relatively mild in some communities where it is endemic, leading some authorities to postulate a continuous spectrum between the two classic diseases.

Pathogenesis and Transmission. In infections in humans **inclusion bodies** identical to those of trachoma are found in conjunctival cells, and also in epithelial cells of the genitourinary tract (see Genital Tract Infections). Transmission to the newborn is from the mother's genital tract, and occasionally adults contract the disease from affected babies. Before chlorination became widespread swimming pools were a major source of adult infection, the water being contaminated by genital or ocular secretions. The disease in adults is now more likely to result from contamination of the eyes by genitourinary exudate borne on fingers or towels. The reservoir of infection is maintained by sexual intercourse. Inclusion conjunctivitis is not believed to generate immunity to reinfection.

A distinctive pneumonia syndrome with respiratory tract colonization may follow inclusion conjunctivitis in infants, together with infection in multiple anatomic sites, including the vagina and rectum.

Diagnosis. Diagnosis of inclusion conjunctivitis, like that of trachoma, is not always easy on purely clinical grounds and should be confirmed by finding inclusions or isolating the agent. The serology has been studied very little compared with that of trachoma.

Prevention and Treatment. Insofar as inclusion conjunctivitis is venereal its prevention is extremely difficult, particularly since genitourinary infection by chlamydiae is often asymptomatic. Adequate chlorination of swimming pools and good personal hygiene make direct eye-to-eye spread of little importance.

In the newborn and in the adult the eye infection is quickly cured by erythromycin and by tetracyclines, respectively, given in full doses systemically for at least one week.

GENITAL TRACT INFECTIONS

The chlamydiae have recently been recognized as a major cause of sexually transmitted diseases, with frequencies exceeding those of *Neisseria gonorrhoeae*. Specific infections include urethritis, epididymitis, prostatitis, cervicitis, endometritis, salpingitis, proctitis, and lymphadenitis. Serotypes D through K are most frequently isolated, but some overlap with serotypes causing trachomatous infections (A, B, C) is found.

Since *C. trachomatis* causes epididymitis in men and pelvic inflammatory disease in women, it may be a cause of human **infertility.** In one study approximately one-third of men with sterility associated with poor semen samples were infected with chlamydiae; treatment of couples with chlamydial infections led to higher rates of conception. In women an association with abortion has not been demonstrated, although *C. psittaci* is an economically important cause of abortion in veterinary medicine.

NONGONOCOCCAL URETHRITIS

Chlamydiae are recovered from 40% or more of men with nongonococcal urethritis (NGU) when samples are taken with swabs placed deeply in the urethra. The etiologic role of chlamydiae has been demonstrated by many studies based on isolation rates, differential chemotherapeutic trials, and serologic evidence of recent infection. Repeated attacks are common, but whether each episode involves a new serotype is unknown.

Diagnosis by isolation in cell cultures is not routinely available to the clinician. Scrapings of Giemsa-stained urethral cells may demonstrate classic inclusions, but this is an unreliable method for routine identification. The diagnosis in practice is one of exclusion.

Treatment with 2 g tetracycline daily for 7 days results in cure. Contact tracing and treatment of sexual partners diminishes the opportunity for reinfection.

Nongonococcal urethritis may also be due to ureaplasmas (Ch. 42). The failure of therapy in a small percentage of patients suggests a role of yet unknown organisms.

INFECTION OF THE FEMALE GENITAL TRACT

Chlamydiae are found in the cervix in a relatively high percentage of women in the United States and elsewhere, often not in association with signs of disease: rates as high as 12% of pregnant women having routine examinations are reported. Infections may be long-lasting, even in the absence of opportunity for reinfection. Current studies suggest that *C. trachomatis* is responsible for a significant proportion of pelvic inflammatory disease and endometritis.

LYMPHOGRANULOMA VENEREUM (LGV)

This disease was first recognized late in the eighteenth century. Its epidemiology is poorly understood. The disease is not common except in sexually promiscuous individuals, and then only in certain groups.

Clinical Description. After a latent period of 7–12 days the first manifestation appears at the site of infection, usually as a herpetiform vesicle on the genitals. The vesicle ruptures and then heals without scarring. Since it is painless it may go unnoticed. From 1 week to 2 months later the regional lymph nodes become enlarged and tender and may suppurate. (The enlarged lymph nodes are sometimes referred to as buboes; thus one of the common synonyms of this disease is **venereal bubo**). The granulomatous inflammation in the lymph nodes heals with scarring, occasionally obstructing lymphatic channels and thus leading to edema of genital skin. Because lymphatic drainage of the perineum in women is directed to perirectal lymph nodes, perirectal scarring sometimes develops and severe rectal obstruction may be a late manifestation.

Pathogenesis and Transmission. The special characteristics of the infection arise from the severe damage to **regional lymph nodes,** a site not especially attacked by other chlamydial infections. Most strains that cause LGV (serotypes L-1, L-2, and L-3) differ from the spectrum of strains with varying degrees of predilection for the eye or the genitourinary tract. However, LGV strains may merely be at one end of this spectrum.

Transmission is virtually always by sexual intercourse. Since notification of cases is not required there are no accurate figures on prevalence; indeed, an overall figure has little epidemiologic meaning, for, like other venereal diseases, LGV varies in incidence with the sexual mores of the group.

Diagnosis. Clinical diagnosis is often confirmed by a delayed-type skin reaction to killed organisms from yolk sac (the **Frei test**). Affected individuals often have CF Abs in their serum. Both these tests, however, give crossreactions with psittacosis. The skin reactivity can last for many years, probably because of a continuing focus of infection (and possibly a state of "infective immunity"), since it can be eliminated by thorough treatment. It is not known whether the disease confers lasting immunity.

Prevention and Treatment. No effective vaccination exists. Treatment with tetracyclines cures the acute clinical signs but has no effect on scarring already present.

ORNITHOSIS

Chlamydial infection of birds was at first thought to be restricted to the parrot (psittacine) family, hence the name **psittacosis** for the disease. With the realization that a wide variety of birds harbor the organisms (*C. psittaci*), with or without disease, the term **ornithosis** became generally accepted. However, the term psittacosis is still often applied to the human infection contracted from birds. In birds latent infection is common, but with increased stress, such as overcrowding, epidemics of varying severity occur.

Pathogenesis and Transmission in Birds. Affected birds often show general debility, have discharges from the eyes and nostrils, and suffer diarrhea, but these signs are not specific to ornithosis. Death rates vary widely, sometimes reaching 30% of an affected flock. Chlamydiae can infect practically all the organs and are shed in the feces and discharges. Since the organisms remain infective in dried feces, transmission by inhalation of dust is common. Vertical transmission through the egg occurs in some species, but in domestic fowl artificial incubation and brooding usually result in uninfected offspring.

Human Infection. In man, too, ornithosis is frequently asymptomatic or causes only trivial respiratory symptoms. Severe and fatal pneumonia can develop, however, and is a recognized hazard of contact with birds as pets, on farms, or in poultry-dressing plants. After an incubation period of 1–3 weeks, the disease begins abruptly or insidiously with chills, fever, headache, and an atypical pneumonia. Physical findings may be surprisingly few, while x-ray examination may change from day to day. Infected individuals may harbor the organisms (as shown by isolation from sputum) for some years. Transmission from human patients is uncommon but has been observed in hospital personnel. The alleged existence of strains of chlamydiae especially well adapted to man and nonpathogenic for birds has not been confirmed.

Diagnosis. The clinical signs of ornithosis are not sufficiently characteristic to establish the diagnosis in either birds or man, and **isolation** of chlamydiae from blood or sputum, in eggs or cell culture, provides the best evidence of infection. Isolates are identified by the use of standard antisera in CF, by the fluorescent Ab technic, or by neutralization of infectivity in mice. As noted in Table 41-2, the organism differs in several respects from *C. trachomatis*.

Infection can also be diagnosed serologically, even in the absence of overt disease, from a rising titer of Ab against the *Chlamydia* group Ag; CF and fluorescent Ab staining are most useful, with Ags grown in yolk sac.

Ornithosis is associated with development of **delayed hypersensitivity** as well as of serum Ab. Both reactions may wane if the infection is eradicated, but they persist in the carrier state. The protection afforded is characteristically not absolute and can be overcome by severe challenge.

Prevention and Treatment. Ornithosis cannot be eradicated, since avian infection is widespread and does not cause high mortality. Infection can be avoided in flocks of domestic fowl, particularly where birds are kept in enclosed houses throughout their lives. Paradoxically, if infection is then introduced the results are likely to be catastrophic. The use of tetracyclines in the food of turkeys is effective both prophylactically and therapeutically. However, such treatment is likely to produce bacteria resistant to antibiotics useful in man. Vaccines against ornithosis have so far proved of little use.

In humans the pneumonia of ornithosis is cured by therapy with tetracyclines; the greatest problem is early diagnosis.

SELECTED READING

BOOKS AND REVIEW ARTICLES

BECKER Y: The chlamydia: molecular biology of procaryotic obligate parasites of eucaryocytes. Microbiol Rev 42:274, 1978

JONES BR: Prevention of blindness from trachoma. Trans Ophthalmol Soc UK 95:16, 1975

MOULDER JW: The Psittacosis Group as Bacteria. New York, Wiley, 1964

NICHOLS RL (ed): Trachoma and Related Disorders Caused by Chlamydial Agents. Int Congr Ser No. 223. Princeton, Excerpta Medica, 1971

SCHACHTER J: Chlamydial infections. N Engl J Med 298:428, 490, 540, 1978

SCHACHTER J, DAWSON CR: Human Chlamydial Infections. Littleton, MA, Publishing Sciences Group, 1978

THYGESON P: Historical review of oculogenital disease. Am J Ophthalmol 71:975, 1971

SPECIFIC ARTICLES

ABRAMS AJ: Lymphogranuloma venereum. JAMA 205:199, 1968

BECKER Y: Recent studies on the agent of trachoma: biology, biochemistry and immunology of a prokaryotic obligate parasite of eukaryocytes. Prog Med Virol 16:1, 1974

BEEM MO, SAXON EM: Respiratory-tract colonization and a distinctive pneumonia syndrome in infants infected with *Chlamydia trachomatis.* N Engl J Med 296:306, 1977

DAWSON CR, DAGHFOUS T, MESSADI M et al: Severe endemic trachoma in Tunisia. Br J Ophthalmol 60:245, 1976

DUNLOP EMC, JONES BR, DAROUGAR S: Chlamydia and nonspecific urethritis. Br Med J 2:575, 1972

FRIIS RR: Interaction of L cells and *Chlamydia psittaci:* entry of the parasite and host responses to its development. J Bacteriol 110:706, 1972

HOLMES KK, HANDSFIELD HH, WANG SP et al: Etiology of non-gonococcal urethritis. N Engl J Med 292:1199, 1975

MANIRE GP, TAMURA A: Preparation and chemical composition of the cell walls of mature infectious dense forms of meningopneumonitis organisms. J Bacteriol 94:1178, 1967

MCCOMB DE, NICHOLS RL: Antibodies to trachoma in eye secretions of Saudi Arab children. Am J Epidemiol 90:278, 1969

MORDHURST CH, DAWSON C: Sequelae of neonatal inclusion conjunctivitis and associated disease in parents. Am J Ophthalmol 71:861, 1971

ORIEL JD, REEVE P, THOMAS BJ et al: Infection with *Chlamydia* Group A in men with urethritis due to *Neisseria gonorrhoeae.* J Infect Dis 131:376, 1975

RICHMOND SJ, SPARLING PF: Genital chlamydial infections. Am J Epidemiol 103:428, 1976

chapter 42

MYCOPLASMAS

ROBERT M. CHANOCK
JOSEPH G. TULLY

GENERAL CHARACTERISTICS

Contagious bovine **pleuropneumonia,** a fatal disease of cattle that created considerable economic loss, was first recognized in Germany in 1693 and reached the United States in 1843. Pasteur suggested that the cause was a specific ultramicroscopic agent, since bacteria could not be seen in serous exudate capable of producing the disease in cattle. The causative agent was cultivated by Nocard and Roux, in 1898, in a collodion sac filled with serum-broth in the peritoneal cavity of a rabbit, and then on a serum-enriched medium. The organisms were difficult to detect—the colonies were very small, the individual organisms did not take up most stains—and they were too small to be defined morphologically at that time.

A number of similar microorganisms were isolated from animals and sewage between 1920 and 1940 and were termed **pleuropneumonialike organisms (PPLO).** These organisms, now termed mycoplasmas, are the smallest free-living cells known. They lack the rigid peptidoglycan cell wall of eubacteria, yet they contain the minimal macromolecular constituents for replication in a cell-free environment. For this reason, and because their membrane is readily accessible, they are of considerable interest to molecular biologists.

Over 70 organisms belonging to this group have now been recognized, including 11 species that infect man. They include important pathogens of animals, plants, and insects; they are also a significant part of the normal microbial flora of most animals; and they include saprobes. *Mycoplasma pneumoniae* has been established as the etiologic agent of primary atypical pneumonia, and mycoplasmas also appear to be involved in some cases of nongonococcal urethritis and pelvic inflammatory disease.

Mycoplasmas resemble chlamydiae, rickettsiae, and viruses in passing through 450-nm filters (Table 42-1), but they differ in being able to grow (though slowly) on artificial media. They often appear as contaminants during the passage of viruses (in animals, chick embryos, or tissue cultures), and many mycoplasmas were initially mistaken for viruses. Their filterability is due not only to their small size but also to the flexibility of their cell envelope.

Because mycoplasmas are bounded only by a membrane they have been designated as the class Mollicutes (L. *mollis* and *cutis,* soft skin). We shall follow the traditional practice of referring to all members of the class as mycoplasmas, one genus being *Mycoplasma.* Taxonomic divisions (Table 42-2) are based upon nutritional requirements (particularly involving sterols), biochemical activities, genome size, morphology, and serologic characteristics.

Among the genera of medical interest *Mycoplasma* and *Ureaplasma* species infect only animals; *Acholeplasmas* are found in plants as well. *Spiroplasmas* are helical, wall-free pathogens primarily of plants and insects. *Anaeroplasmas* are obligate anaerobes, recovered from the rumen of cattle and sheep.

L Forms of Bacteria. Mycoplasmas were long confused with the L-phase variants of various eubacteria, which also lack a rigid cell wall and hence resemble them in cellular and colonial morphology, filterability, and slow growth. However, the loss of wall in L forms is often reversible. These variants are discussed in Chapter 43.

MORPHOLOGY

Electron microscopy has shown that in cellular morphology the mycoplasmas are extremely variable and pleomorphic, even within a pure culture. (The name means plastic mold.) Some organisms assume a predominantly spherical appearance (300–800 nm in diameter) while others may form **filaments** of uniform diameter (100–300 nm) that vary in length from $3\mu m$ to over $150\mu m$. Most mycoplasmas stain gram-negative.

The spiroplasmas vary over a range that includes pleomorphic spherical cells (200–300 nm in diameter), helical filaments (Fig. 42-1), and branched, nonhelical filaments.

Variations in the size and shape of mycoplasmas, as well as their poor staining, can be ascribed to the lack of a cell wall. Also, preparative technics used in electron microscopy may distort the organisms. Thin sections reveal a simple ultrastructure consisting of cell membrane and cytoplasm, including ribosomes and the characteristic prokaryotic nucleoid. There are no intracellular membranous structures.

Some mycoplasmas have a specialized **terminal structure** that appears to play a role in attachment. These include *M. gallisepticum* (a respiratory pathogen of birds; Fig. 42-2) and *M. pneumoniae* (Fig. 42-3). These species exhibit a type of **gliding motility,** which may also involve the specialized terminal organelles.

TABLE 42-1. Characteristics of Mycoplasmas and Some Other Prokaryotic Organisms

Property	Mycoplasmas	Schizomycetes	Chlamydiae	Rickettsiae	Viruses
Grows on cell-free medium	+	+*	–	–*	–
Cell wall or cell wall peptidoglycan absent	+	–	–	–	+
Generates metabolic energy	+	+	–	+	–
Depends on host cell nucleic acid for multiplication	–	–	–	–	+
Can synthesize proteins by own enzymes	+	+	+	+	–
Requires sterols	+†	–	–	–	–
Visible in optical microscope (× 1500)	+	+	+	+	–*
Filterable through 450-nm pore-size filters	+	–*	+	+	+
Contains both RNA and DNA	+	+	+	+	–
Growth inhibited by antibody alone	+	–	+	+	+
Growth inhibited by antibiotics acting on protein synthesis	+	+	+	+	–

*With few exceptions.

†Except *Acholeplasma* species.

(Tully JG, Razin S: The Mollicutes "Mycoplasmas." In Handbook of Microbiology, Vol 1, 2nd ed. Cleveland, CRC Press, 1977)

Spiroplasmas also exhibit motility but of a different type, characterized by a rotary or "screwing" motion of the helix accompanied by flexional movements. Since they lack flagella and axial filaments, their movement may depend on intracellular contractile elements.

Colonial Morphology. On solid medium most mycoplasmas form very small colonies (50μm–600μm in diameter), which can be seen only with a hand lens or under the low power of a microscope. Some of the acholeplasmas, however, may yield visible colonies (up to 5 mm). The classic **"fried egg"** appearance of the usual colony is due to an opaque, granular central zone of growth down into the agar (whose fibers may help cell division),

TABLE 42-2. Taxonomy of the Mycoplasmas

Class: Mollicutes
  Order: Mycoplasmatales
    Family I: Mycoplasmataceae
      Sterol required for growth
      Genome size about 5×10^8 daltons
      NADH oxidase localized in cytoplasm
      Genus I: *Mycoplasma* (over 60 species)
        Do not hydrolyze urea
      Genus II: *Ureaplasma* (single species with serotypes)
        Hydrolyzes urea
    Family II: Acholeplasmataceae
      Sterol not required for growth
      Genome size about 10^9 daltons
      NADH oxidase localized in the membrane
      Genus I: *Acholeplasma* (8 species)
    Family III: Spiroplasmataceae
      Helical organisms during some phase of growth
      Sterol required for growth
      Genome size about 10^9 daltons
      NADH oxidase localized in cytoplasm
      Genus I: *Spiroplasma* (2 species)
  Genus of uncertain taxonomic position
    Anaeroplasma (two species)

and a flat, translucent peripheral zone of growth on the surface (Fig. 42-4A). Not all mycoplasmas produce "fried egg" colonies (Fig. 42-4B), and variations in colonial morphology, such as size and extent of the central zone, are frequently dependent on the constituents and the hydration of the medium and on aerobiosis. Ureaplasmas were designated earlier as tiny or **T-strains** because they form only 15-μm to 30-μm colonies on unbuffered medium.

MODE OF REPRODUCTION

Mycoplasmas usually divide, like other prokaryotes, by binary fission, but genomic replication and cytoplasmic division are not precisely synchronized and are easily dissociated. A lag in cytoplasmic division yields multinucleated filaments, which subsequently form chains of coccoid cells and then fragment into individual cells. Cells less than about 300 nm, like the "minicells" of certain mutant eubacteria (see Ch. 6), probably do not receive a complete genome. Fission by budding is observed in some species.

Mycoplasmas have a circular genome of double-stranded DNA, one-fifth to one-half as large as that in most bacteria. This is evidently the smallest genome that can code for all the products needed for self-reproduction at the expense of the foodstuffs in an artificial medium. DNA hybridization has demonstrated relatedness among different mycoplasmas of the same serotype.

Mycoplasma genomes have a low G + C content (23–45 moles %). The G + C of their rRNA (43%–48%) is outside the range of eubacteria (50%–54%), indicating much evolutionary separation. Their lipids are also unusual: more than 50% contain glucose. Both lytic and temperate **bacteriophages** have been recovered from mycoplasmas.

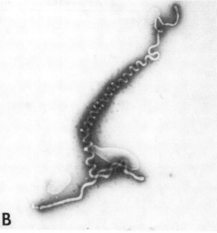

FIG. 42-1. A. Helical morphology of a spiroplasma (suckling mouse cataract agent) observed by dark-field microscopy. Photograph of the chorioallantoic fluid of embryonated hen's egg 4 days following yolk sac inoculation with spiroplasmas. (Tully JG et al: Nature 259:117, 1976) (×2800) **B.** Electron photomicrograph of a helical filament of *Spiroplasma citri* negatively stained with ammonium molybdate. Note saclike blebs emerging from main helical body. **Bar** represents 1.0µm. (Courtesy of R. M. Cole)

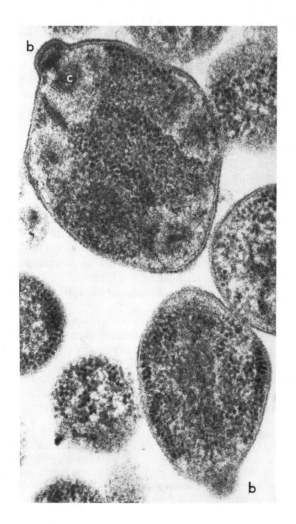

Mycoplasmas generally grow more slowly than bacteria, with a mean generation time of 1–3 h; for a few species it is so prolonged (6–9 h) that visible growth may require 1–2 weeks. *Acholeplasma laidlawii* is unusual in producing visible turbidity in 18–24 h in broth cultures enriched with serum. The optimum temperature for growth of all *Mycoplasma* and *Ureaplasma* species is 37° C, but spiroplasmas and acholeplasmas (found in plants or as saprobes) have a wide temperature range (22°–37° C).

Many mycoplasmas utilize either glucose (fermentative) or arginine (nonfermentative) as a major source of energy; those that utilize arginine convert it to ornithine via a three-enzyme system. Ureaplasmas require urea and convert it to ammonia by the action of a urease. These properties are useful in classification. Most mycoplasmas are facultatively anaerobic, but growth of some is enhanced by incubation in air with 5% CO_2. Mycoplasmas of the human oral cavity generally prefer an anaerobic environment (95% N_2 + 5% CO_2).

Nutrition. Mycoplasmas have exacting nutritional requirements, especially for lipids needed for synthesis of the plasma membrane. The most frequently employed medium is heart infusion broth (with peptone), supplemented with 0.5% glucose, 0.2% arginine, 10% fresh yeast extract, and 20% horse serum. Additional substances are used to demonstrate pH change (phenol red), to inhibit bacterial growth (penicillin and/or thallium acetate), or

FIG. 42-2. Electron photomicrograph of thin section of *Mycoplasma gallisepticum* showing specialized terminal structures. The densely stained structures consist of a "bleb" **(b)** and its infrableb core **(c).** (Maniloff J, Quinlan DC: Ann NY Acad Sci 225:181, 1973) (×70,000)

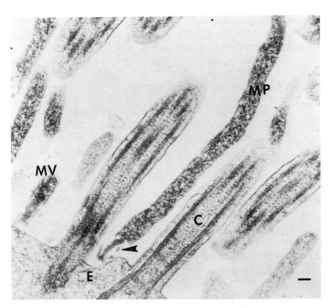

FIG. 42-3. Electron photomicrograph of *Mycoplasma pneumoniae* infection of a hamster tracheal organ culture. Note filamentous structure of the organism **(MP)** with its specialized tip **(arrow)** in close apposition to the epithelial surface **(E).** Also note proximity of several cilia **(C)** and a microvillus **(MV).** The tip has a dense central core and is separated by a lucent space from the limiting unit membrane of the organism. **Bar** represents 0.1 μm. (Courtesy of A. M. Collier)

to provide a solid medium (0.6%–0.8% washed or purified agar).

The nutritional requirement of many mycoplasmas for a **sterol** is unique among prokaryotes. It is usually met by animal serum, which contains cholesterol bound to a lipoprotein. **Acholeplasmas** do not need exogenous cholesterol but synthesize lipids that substitute for it in maintaining fluidity of the membrane lipid bilayer. However, some acholeplasmas, such as *A. laidlawii,* will incorporate cholesterol (up to 8% of the total membrane lipids) when grown in its presence. Sterol-containing mycoplasmas are lysed by digitonin and other substances that form complexes with cholesterol.

Selective media have been developed for isolation of *M. pneumoniae* and the ureaplasmas from clinical materials. Media for the latter must be supplemented with urea; liberation of NH_3 from urea during growth results in a change in color of the phenol red pH indicator.

Inhibitors. Mycoplasmas are generally not sensitive to inhibitors of cell wall synthesis (penicillin, cycloserine, bacitracin). Polyenes have variable effects depending upon the sterol content of the cell membrane. Mycoplasmas are generally sensitive to tetracycline, aminoglycosides, and erythromycin, but individual strains and species vary.

Strains of several *Mycoplasma* species (*M. hyopneumoniae* and *M. flocculare*) are inhibited by benzylpenicillin, and ampicillin is used instead for their isolation. Thallium acetate at 1:2000 is selec-

FIG. 42-4. A. Mycoplasma colonies (*Acholeplasma axanthum*) on the surface of solid medium. Note the classic "fried egg" appearance produced by central penetration of growth into the agar and peripheral growth on the surface. **Bar** represents 100μm. **B.** Morphology of *Mycoplasma pneumoniae* colonies on agar medium. Note the "mulberrylike" appearance of colonies and absence of a peripheral halo. **Bar** represents 100μm. (Chanock RM et al: Proc Nat Acad Sci USA 48:41, 1962)

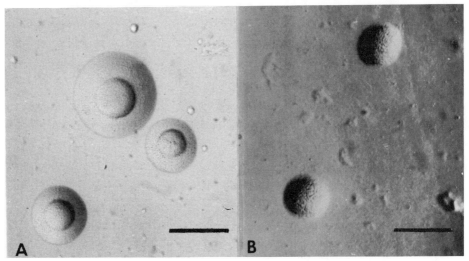

tively inhibitory for gram-negative bacteria and for aerobic spore-formers, but it also inhibits ureaplasmas and should be omitted from the medium used for their recovery.

IMMUNOLOGIC CHARACTERISTICS

The major antigenic determinants of mycoplasmas are membrane proteins and glycolipids. Capsulelike Ags, including galactan and a hexosamine polymer, have been reported in a few mycoplasmas. The protein Ags have not been well characterized.

The **glycolipids** of *M. pneumoniae,* which contain glucose and galactose, have received special attention since they are suspected of playing a role in autoimmunity in humans. They are haptens, losing immunogenicity when separated from membrane protein. Their surface location is confirmed by agglutination of the organism, and by complement-dependent lysis by rabbit Abs to the homologous glycolipid. These Abs crossreact with glycolipids of similar structure found in certain other mycoplasmas, and also in human brain and in many plants.

Species classification of mycoplasmas is based primarily on **serologic** reactions. Tests that are carried out with whole cell extracts (complement fixation [CF], immunodiffusion) tend to show group relations. Tests that detect membrane Ags (growth inhibition, metabolism inhibition, and immunofluorescence) are more specific. The direct identification of mycoplasma colonies on agar by **immunofluorescence** (see Fig. 42-5, below) is rapid and specific, and it has the additional advantage of detecting a mixture of mycoplasmas of different serotypes in the same culture—an important feature, since clinical material frequently contains more than one species.

In the **growth inhibition** (GI) test growth of the organism on agar

is inhibited in a zone around a filter paper disc saturated with specific antiserum. The **metabolism inhibition** test is based upon inhibition of growth by specific Ab, as indicated by the inhibition of a metabolic activity (glucose fermentation, hydrolysis of arginine or urea, or reduction of tetrazolium). This technic has also been applied to the measurement of specific Abs.

The ureaplasmas of human origin represent a special problem in classification. At least eight serotypes have been distinguished; they are temporarily designated as a single species, *Ureaplasma urealyticum,* with numbered serotypes.

MYCOPLASMAL FLORA OF MAN

At least 11 serologically and biologically distinct mycoplasmas have been found in man (Table 42-3). A majority of those recovered from the **oral cavity** are commensals, which do not appear to play a role in disease; *M. orale* and *M. salivarium* are found in almost every healthy adult. The reported frequency of *M. hominis* and ureaplasmas ranges from 0–9%, and others are found in less than 0.1% of various population groups.

In the genital tract *M. hominis* and the ureaplasmas are present in a large proportion of sexually active adults, while *M. fermentans* and *M. primatum* have been recovered infrequently. Genital mycoplasmas colonize 20%–30% of newborns at the time of birth, but usually only transiently. Following puberty colonization occurs primarily as a result of sexual activity: these organisms can be recovered from only 5%–10% of nuns and other sexual virgins, while in women attending venereal disease clinics ureaplasma recovery rates as high as 75%–80% have been observed. In the asymptomatic adult male

TABLE 42-3. Properties of Mycoplasmas that Infect Man

| Mycoplasma | Primary site of colonization | | Metabolic substrate | | | Aerobic growth (in vitro) | Hemadsorption of guinea pig RBCs |
	Respiratory tract	Genitourinary tract	Glucose	Arginine	Urea		
M. salivarium	+	—	—	+	—	—	—
M. orale	+	—	—	+	—	—	—
M. buccale	+*	—	—	+	—	—	—
M. faucium	+*	—	—	+	—	—	—
M. pneumoniae	+†	—	+	—	—	+‡	+
M. lipophilum	+*	—	—	+	—	—	—
Acholeplasma laidlawii	+*	—	+	—	—	+‡	—
M. hominis	+*	+	—	+	—	+‡	—
M. primatum	—	+*	—	+	—	—	—
M. fermentans	—	+*	+	+	—	—	—
Ureaplasma urealyticum	+*	+	—	—	+	—	—§

*Rare.

†Does not colonize, rather produces acute or subacute infection.

‡Also grow anaerobically.

§Serotype 3 hemadsorbs guinea pig red cells.

mycoplasmas may be present in the urine, semen, and distal urethra. The asymptomatic female may have organisms present over the entire genital mucosa, but not the bladder, uterus, or fallopian tubes.

MYCOPLASMAL PNEUMONIA

IDENTIFICATION OF THE ORGANISM

In the late 1930s a group of nonbacterial pneumonias was first recognized and was given the name **primary atypical pneumonia,** to distinguish them from typical (pneumococcal) lobar pneumonia. This syndrome was found to have a multiple etiology; viruses could be demonstrated in some patients, but not in an additional large group, who developed cold agglutinins (which agglutinate certain RBCs at low temperature). Attempts to isolate the etiologic agent were unsuccessful until Eaton reported that a filterable organism could be passaged serially in embryonated eggs: there were no discernible changes (**"blind passage"),** but tissue extracts and extraembryonic fluids from these eggs (like the original filtrate) produced pneumonia in cotton rats and hamsters. Confirmation was delayed for years, but the subsequent development of an immunofluorescence test, with sections of infected chick embryo lung as Ag, facilitated detection of the organism and measurement of the corresponding Abs. The tiny organisms were later seen as coccobacillary bodies, after Giemsa staining, on the surface of the bronchial epithelium of infected chick embryos.

The Eaton agent was considered to be a virus, but a problem arose when it was shown to be inhibited by tetracycline or streptomycin. (This feature of bacteria might have been recognized earlier in patients, but the results were inconsistent because the syndrome included viral infections.) The organism was finally cultivated on a cell-free medium in 1962 by Chanock, Hayflick, and Barile, identified by immunofluorescence with convalescent serum (Fig. 42-5), confirmed as an etiologic agent in human volunteers, and designated *Mycoplasma pneumoniae.* It forms circular colonies partially embedded in the agar, with a granular, "mulberry" surface (Fig. 42-4B) but without the light peripheral zone observed with many other mycoplasmas.

The colonies cause *β*-**hemolysis,** large enough to be seen without magnification, when overlaid with guinea pig RBCs (Fig. 42-6). This reaction is due to the organism's production of hydrogen peroxide in larger amounts than other human mycoplasmas (which produce a slow, *α*-type hemolysis).

PATHOGENESIS

Unlike other mycoplasmas that inhabit the respiratory tract, *M. pneumoniae* can attach to the surface of the respiratory epithelium, its tip binding to neuraminic acid receptors (Fig. 42-3). The organism does not enter host cells, nor does it penetrate beneath the epithelial surface. In tracheal organ cultures its attachment leads to direct damage of the epithelium, with ciliostasis, loss of cilia, and finally cell death. This cell damage may be produced by the hydrogen peroxide released by the organisms; there is no evidence for the production of a protein exotoxin.

Since children 2–5 years of age often possess mycoplasmacidal Abs, while the disease appears most often at age 5–15, it has been suggested that the pathogenic effects of *M. pneumoniae* may include an **immunopathologic reaction** in a host sensitized by prior infection. Reinfection occurs with appreciable frequency, and second attacks of *M. pneumoniae* pneumonia have been documented. Additional support for the sensitization hypothesis is provided by the observation that thymic ablation, preceding experimental infection of hamsters, prevents the development of characteristic pulmonary infiltration by mononuclear cells. However, it is not certain that this type of infiltrate is responsible for the symptoms experienced by patients with *M. pneumoniae* disease.

FIG. 42-5. Colonies of *Mycoplasma pneumoniae* stained by the indirect immunofluorescence technique using acute **(A)** and convalescent **(B)** phase sera from a patient with mycoplasmal pneumonia. (Chanock RM et al: Proc Nat Acad Sci USA 48:41, 1962)

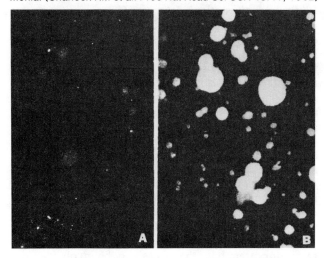

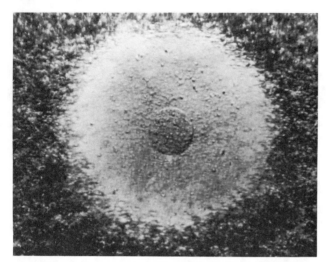

FIG. 42-6. β-Hemolytic plaque produced by a 5-day colony of *Mycoplasma pneumoniae* overlaid with guinea pig RBCs in agar and reincubated for 48 h. Mycoplasma colony is the small circular structure in the center of the plaque. (Chanock RM et al: Am Rev Resp Dis (Suppl): 223, 1963

An immunopathologic basis of *M. pneumoniae* disease represents an intriguing possibility. Suggestive evidence is the prolonged asthenia, which is out of proportion to the limited lung consolidation. In addition, some patients develop neurologic symptoms, without demonstrable *M. pneumoniae* in the cerebrospinal fluid but usually with serum Abs to brain tissue; and these Abs can be adsorbed by *M. pneumoniae* or its surface glycolipids. (However, the pathogenetic role of Abs reacting with brain remains unsettled, since these Abs also develop in most patients with mycoplasmal pneumonia who do not exhibit central nervous system symptoms.) Finally, *M. pneumoniae* produces other distant effects, whose mechanism is also obscure and could be immunologic.

Clinical Manifestations. In man *M. pneumoniae* causes inapparent infection, or mild upper respiratory tract disease, more often than it causes pneumonia. The onset of pneumonia is generally gradual and the symptoms are mild or moderately severe, with early complaints referable to the lower respiratory passages. Because the areas of consolidation are small, x-ray examination frequently reveals pneumonia before physical signs are apparent. Involvement is usually interstitial or bronchopneumonic, and limited to one of the lower lobes. The organism can also cause bronchitis, tracheobronchitis, laryngitis, otitis media, and bullous myringitis.

The course of the pneumonia is variable, with remittent fever, cough, and headache lasting for several weeks. Convalescence is prolonged, often extending for 4–6 weeks even in the absence of complications. Few fatal cases have been reported. Illness is usually limited to the respiratory tract but other areas may be involved.

Complications most often affect the central nervous system (Guillain–Barré syndrome, polyradiculitis, encephalitis, aseptic meningitis, or acute psychosis may occur); occasionally pericarditis or pancreatitis is seen. These sequelae may be related to immunopathology, as noted above. An infrequent complication is hemolytic anemia with crisis, brought about by the cold agglutinins in serum.

Immunologic Response. Infection with *M. pneumoniae* induces the development of IgM Abs and later IgG Abs that are detectable by CF, growth inhibition (GI), or cell lysis in the presence of complement. These serum Abs do not appear to provide effective protection against infection or disease: in one longitudinal study more than two-thirds of the individuals possessed a moderate to high titer of mycoplasmacidal Ab prior to becoming infected and developing pneumonia. IgA Ab also is found, in respiratory secretions, and it appears to play a role in host resistance. Finally, cell-mediated immunity to *M. pneumoniae* has been detected by skin test reaction, lymphocyte transformation, and inhibition of macrophage migration.

Glycolipids are the predominant antigenic determinants inducing circulating Ab, while a polysaccharide–protein fraction seems to be the component involved in cell-mediated immunity. As was noted, circulating Ab often develops within the first few years of life, but whether it has been induced by *M. pneumoniae,* or by related glycolipids found in many plants and eubacteria, is not clear.

Epidemiology. An unusual feature of mycoplasmal pneumonia, as we have noted, is its peak incidence in individuals 5–15 years of age. *M. pneumoniae* accounts for as much as 15%–50% of all pneumonias observed among school children and young adults. The incubation period is relatively long (2–3 weeks). *M. pneumoniae* infection is endemic in most areas. Infections occur throughout the year, with a predilection for late summer and early fall. Intensive exposure to infected persons appears to be required for transmission: spread of the organism from person to person is quite slow and generally occurs within a closely associated group rather than by casual contact. The organism is usually introduced into a household by a school-aged child.

TREATMENT AND PREVENTION

Tetracycline and erythromycin are effective in mycoplasmal pneumonia. Elimination of symptoms, however, is not accompanied by eradication of the organism.

Although these antibiotics are effective in reducing the morbidity of *M. pneumoniae* pneumonia, prevention through **vaccination** is a more logical means for control, particularly in populations where the risk of disease is high. In field trials the efficacy of experimental inactivated *Mycoplasma pneumoniae* vaccines in preventing pneumonia has ranged from 45% to 67%. In the hamster prior infection with *M. pneumoniae* induces greater resistance to disease than does parenteral inoculation of inactivated organisms: hence a live attenuated mutant might provide more effective protection than an inactivated vaccine. However, efforts in this direction must proceed with caution because of the possible role of immunopathology in *M. pneumoniae* disease.

LABORATORY DIAGNOSIS

M. pneumoniae infection is diagnosed by isolating the organism or by demonstrating a rise in serum Ab titer. Nasopharyngeal secretions are inoculated into a selective diphasic medium that contains mycoplasma broth on top of agar, supplemented with glucose and phenol red. When *M. pneumoniae* grows in this medium it produces acid, thereby changing the color of the medium from red to yellow. Broth from the diphasic medium should be subcultured to mycoplasma agar when a color change occurs, or at weekly intervals for a minimum of 8 weeks. Isolates are identified as *M. pneumoniae* by staining colonies directly with fluorescein-conjugated Ab or by demonstrating that specific Ab inhibits growth on agar.

The Ab response in mycoplasmal pneumonia is most easily demonstrated by CF. Since a relatively high Ab level may persist for a year or more after infection only a rise during convalescence (fourfold or more) is considered indicative of recent infection. The test for cold agglutinins is less useful: these Abs develop in only one-half of patients with mycoplasmal pneumonia, and they are also induced by several other diseases.

M. pneumoniae can occasionally be recovered from the oropharynx for 1–2 months after subsidence of fever, and for a shorter period in silent infections.

MYCOPLASMAS IN GENITAL TRACT DISEASE

M. hominis and *U. urealyticum* are common inhabitants of the genital tract, acquired primarily through sexual contact. They usually behave as commensals. However, on rare occasion either appears to be capable of causing pelvic inflammatory disease (tuboovarian abscess, salpingitis).

M. hominis has been recovered from the blood of a small proportion of women who develop fever post partum or after abortion, but its etiologic significance is not clear since 3%–5% of women have a transient invasion of the blood by mycoplasmas at the time of parturition.

There has been considerable dispute concerning the role of genital mycoplasmas in **nongonococcal urethritis**

(NGU). Although most *U. urealyticum* infections are silent, this organism appears to be one of the several causes, since the disease could be produced in volunteers by inoculating ureaplasmas intraurethrally. Moreover, ureaplasma-positive NGU patients who were free of chlamydiae (another etiologic agent of NGU) shed more ureaplasmas than did chlamydia-positive patients or healthy carriers of *U. urealyticum*. With *M. hominis*, in contrast, there is no evidence for an etiologic role.

OTHER INTERACTIONS OF MYCOPLASMAS

ANIMAL DISEASES

Contagious bovine pleuropneumonia (*M. mycoides* subsp. *mycoides*), the first mycoplasma infection recognized, has been eradicated from most countries by the application of strict control and/or vaccination programs, although the disease remains endemic in some tropical areas. However, other mycoplasmas in cattle, sheep, goats, poultry, and swine cause serious economic loss in most parts of the world. Mycoplasmas are also indigenous to common pets and laboratory animals (mice, rats, dogs, and cats), commonly causing respiratory or neurologic disease or arthritis. Chronic respiratory disease produced by *M. pulmonis* in laboratory mice or rats often complicates long-term studies with these animals.

The clinical and histopathologic features of arthritis produced by mycoplasmas in many animal species closely resemble those of human rheumatoid arthritis, but there is no evidence to support this etiology of the human disease.

Spiroplasmas as Pathogens. Recently, helical, wall-free prokaryotes were recovered from diseased plants and suspected insect vectors, and were termed spiroplasmas (Table 42-2). A filterable "suckling mouse cataract agent," initially recovered from a rabbit tick, was later shown to possess similar characteristics (Fig. 42-1A); it is pathogenic for several vertebrate hosts. In rodents various strains produce cataracts, central nervous system disease, or stunted growth.

TISSUE CULTURE CONTAMINATION

Mycoplasmas frequently contaminate animal cell cultures, producing alterations in cell metabolism, chromosomal aberrations, and inhibition of lymphocyte transformation. Since these effects may appear without morphologic alterations or loss of viability of the cells, and since tests for bacterial contamination fail to detect mycoplasmas, their presence has often not been recognized and their effects have been misinterpreted as spontaneous genetic or epigenetic changes in the cells.

Many mycoplasmas adsorb readily to the surface of tissue cells (Fig. 42-7) and derive part of their nutritional needs from the cell. They do not appear to penetrate cells, except for phagocytosis under appropriate conditions. However, the association of myco-

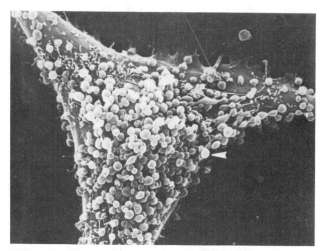

FIG. 42-7. Scanning electron photomicrograph of a mouse tissue culture cell line (A9) infected with *Mycoplasma hyorhinis*. Note numerous mycoplasma cells **(arrow)** adsorbed to surface of tissue cell. (Courtesy of D. M. Phillips)

plasma and tissue cell may be quite intimate. Recently it was shown that T lymphocyte and H-2 histocompatibility Ags of a lymphoblastoid cell line were transferred to *M. hyorhinis* which had established a persistent infection in a culture of these cells. Abs specific for *M. hyorhinis* were cytotoxic for the lymphoblastoid cells, providing additional evidence for a close relationship between host cell and mycoplasma membranes. These types of interactions may have relevance to tissue damage and autoimmune phenomena seen in diseases such as *M. pneumoniae* pneumonia.

Certain mycoplasmas that grow rapidly in cell culture can destroy cells by depleting the medium of arginine or by producing excess acid. Mycoplasmas can also drastically decrease the growth of DNA viruses that require arginine by depleting this amino acid, and they can enhance viral replication by suppressing interferon production.

Mycoplasmal contamination occurs most often in continuous-passage cell lines. Commercial bovine serum is the major source, but in laboratories not utilizing strict aseptic technics human oral mycoplasmas are also frequent contaminants. Contamination is usually detected by direct culture. However, *M. hyorhinis,* a common contaminant, often cannot grow on artificial media and is identified by immunofluorescence.

SELECTED READING

BOOKS AND REVIEW ARTICLES

BARILE MF: Mycoplasma contamination of cell cultures. In Fogh J (ed): Contamination in Tissue Cultures. New York, Academic Press, 1973, pp 131–172

BARILE MF, RAZIN S (eds): The Mycoplasmas, Vol 1, Cell Biology. New York, Academic Press, 1979

Cellular and molecular biology of mycoplasma and bacterial L-forms. In Schlessinger D (ed): Microbiology—1978. Washington DC, American Society for Microbiology, 1978

Ciba Foundation Symposium: Pathogenic Mycoplasmas. Amsterdam, Elsevier, 1972

CLYDE WA JR: *Mycoplasma pneumoniae* infections of man. In Tully JG, Whitcomb RF (eds): The Mycoplasmas, Vol 2. New York, Academic Press, 1979

COLE RM: Mycoplasma viruses. In Laskin AI, Lechevalier H (eds): Handbook of Microbiology, Vol 2, 2nd ed. Cleveland, Chemical Rubber Co, 1978, pp 683–690

GRAYSTON JT, FOY HM, KENNY GE: The epidemiology of mycoplasma infections of the human respiratory tract. In Hayflick L (ed): The Mycoplasmatales and the L-Phase of Bacteria. New York, Appleton-Century-Crofts, 1969, pp 651–682

KLEIN JO: Mycoplasma infections. In Remington JS, Klein JO (eds): Infectious Diseases of the Fetus and Newborn Infant. Philadelphia, Saunders, 1976, pp 587–615

MANILOFF J, DAS J, CHRISTENSEN JR: Viruses of mycoplasmas and spiroplasmas. Adv Virus Res 21:343, 1977

MCCORMACK WM, BRAUN P, LEE Y-H, et al: The genital mycoplasmas. N Engl J Med 288:78, 1973

MCGARRITY GJ, MURPHY DG, NICHOLS WW (eds): Mycoplasma Infection of Cell Cultures. New York, Plenum Press, 1978

RAZIN S: The physiology of mycoplasmas. Adv Microbiol Physiol 10:1, 1973

RAZIN S: The mycoplasmas. Microbiol Rev 42:414, 1978

TAYLOR-ROBINSON D, MCCORMACK WM: The genital mycoplasmas. N Engl J Med 302:1003, 1063, 1980

TULLY JG, RAZIN S: The mollicutes. In Laskin AI, Lechevalier H (eds): Handbook of Microbiology, Vol 1, 2nd ed. Cleveland, Chemical Rubber Co, 1977, pp 405–459

TULLY JG, WHITCOMB RF (eds): The Mycoplasmas, Vol 2, Human and Animal Mycoplasmas. New York, Academic Press, 1979

WHITCOMB RF, TULLY JG (eds): The Mycoplasmas, Vol 3, Plant and Insect Mycoplasmas. New York, Academic Press, 1979

WHITTLESTONE P: Immunity to mycoplasmas causing respiratory diseases in man and animals. Adv Vet Sci Comp Med 20:277, 1976

SPECIFIC ARTICLES

BOWIE WR, ALEXANDER ER, FLOYD JF, et al: Differential response of chlamydial and ureaplasma-associated urethritis to sulfafurazole (sulfisoxazole) and aminocyclitols. Lancet 2:1276, 1976

BRUNNER H, PRESCOTT B, GREENBERG H, et al: Unexpectedly high frequency of antibody to *Mycoplasma pneumoniae* in human sera as measured by sensitive techniques. J Infect Dis 135:524, 1977

CHANOCK RM, HAYFLICK L, BARILE MF: Growth on artificial medium of an agent associated with atypical pneumonia and its identification as a PPLO. Proc Natl Acad Sci USA 48:41, 1962

FERNALD GW, COLLIER AM, CLYDE WA JR: Respiratory infections due to *Mycoplasma pneumoniae*. Pediatrics 55:327, 1975

FORD DK, HENDERSON E: Non-gonococcal urethritis due to T-mycoplasmas (*Ureaplasma urealyticum*) serotype 2 in a conjugal sexual partnership. Br J Venen Dis 52:341, 1976

FOY HM, KENNY GE, SEFI R, et al: Second attacks of pneumonia due to *Mycoplasma pneumoniae*. J Infect Dis 135:673, 1977

HOPPS HE, MEYER BC, BARILE MF, DEL GIUDICE RA: Problems concerning "non-cultivable" mycoplasma contaminants in tissue cultures. Ann NY Acad Sci 225:265, 1973

SHEPARD MC, LUNCEFORD CD: Differential agar medium (A7) for identification of *Ureaplasma urealyticum* (human T-mycoplasmas) in primary culture of clinical material. J Clin Microbiol 3:413, 1976

TAYLOR-ROBINSON D, CSONKA GW, PRENTICE MJ: Human intraurethral inoculation of ureaplasmas. Q J Med 46:309, 1977

WISE KS, CASSELL GH, ACTON RT: Selective association of murine T lymphoblastoid cell surface alloantigens with *Mycoplasma hyorhinis.* Proc Nat Acad Sci USA 75:4479, 1978

chapter **43**

OTHER PATHOGENIC MICROORGANISMS; L-PHASE VARIANTS

ALEX C. SONNENWIRTH
ZELL A. McGEE
BERNARD D. DAVIS

OTHER PATHOGENIC MICROORGANISMS

LEGIONNAIRES' DISEASE (*LEGIONELLA*)*

In the summer of 1976 an outbreak of pneumonia, with 182 cases and 29 deaths, occurred among about 5000 persons who had attended an American Legion convention in Philadelphia. The epidemic aroused enormous public attention. Epidemiologic investigations showed that the focus was the lobby of a famous hotel, but the cause remained unidentified for months: conventional histopathologic staining of pulmonary lesions, and cultures, did not reveal any bacteria

The organism was eventually isolated from lesions by blind passage through guinea pigs (which did not become severely ill), and then chick embryos (which died with bacilli visible in the yolk sac). These bacteria slowly formed small colonies on a rich medium (e.g., enriched Mueller-Hinton agar plus 1% hemoglobin). The cultured bacteria and Abs in the patients' sera could be related to each other by indirect immunofluorescent staining: 90% of the patients showed a high (1:64) or a rising titer of Abs to the organism. Bacteria were eventually also identified in alveolar debris and in phagocytes in lesions, first by silver impregnation and then by specific immunofluorescent staining.

The organism, called **Legionnaires' Disease bacillus** initially, and now *Legionella pneumophila,* is a member of a previously unknown bacterial group, present in nature but difficult to cultivate. It appears to be, like the fungus *Histoplasma* (Ch. 45), a soil organism that can cause either mild or severe disease when inhaled in sufficient quantity. *Legionella* is a thin, nonmotile, gram-negative rod ($0.6 \times 2\text{--}3 \ \mu m$), occasionally filamentous. It is distinct, by serologic tests, from any known pathogen, and it has a unique fatty acid composition (about 90% branched chains). Various isolates, identified by their common surface Ag, have been found also to carry different **type-specific surface Ags;** four have been distinguished.

Pathogenicity and Epidemiology. The organism has turned out to be a widespread source of disease, some-

times quite mild. Thus Abs to it were found in stored sera from three earlier undiagnosed, localized epidemics of febrile illness. Within a year after the Ag (bacterial suspension) and antisera became available for diagnosis, sporadic cases of pneumonia were found to be associated with this organism in nearly every state of the United States and in several foreign countries.

In the Philadelphia epidemic circumstantial evidence pointed to the air conditioner in the hotel lobby as the probable source, and in one of the retrospectively diagnosed epidemics an air conditioner clearly was identified by exposing test animals to its outflow. However, in another epidemic there was no air conditioner, and the source may have been a nearby excavation of earth: serologically and morphologically similar organisms have been recovered from soil, and from the recirculating spray water of an airconditioning cooling tower. There is no evidence of person-to-person transmission. The possible spread by a central air-handling system opens a new chapter in public health practice.

In contrast to the severity of the disease in some epidemics (case fatality 16% in Philadelphia, and 12 of 16 cases in another outbreak), the 144 patients in a serologically related Pontiac, Michigan, epidemic had headache and myalgia but no pneumonia, and there were no fatalities. *Legionella* thus appears to comprise a serologically related group of novel potential pathogens with a wide range of virulence. On the other hand, many employees of the hotel in Philadelphia had elevated Ab titers without a history of pneumonic illness, suggesting that the organism had been present in the air there at earlier times. Accordingly, the severity of the disease among the Legionnaires might be at least partly due to the special characteristics of that population (e.g., middle age, concomitant alcohol intake, possible first exposure to an overwhelming dose).

Diagnosis. Clinical diagnosis of the severe form of the disease is suggested by pneumonia (especially in the summer) with patchy interstitial infiltration and later nodular consolidation, associated with a high fever, severe malaise, headache, nonproductive cough, and little elevation of the leukocyte count. Among the Legion-

* Contributed by Bernard D. Davis

naires the incubation period was 2–10 days following first exposure.

Diagnosis has been established mostly by immunofluorescence, using the organism to detect Abs in the patient's serum, or, less often, using convalescent serum to identify the organism in sputum. In a few cases the organism has been cultured from pleural fluid. Cultures should be incubated, preferably with CO_2, for at least 10 days.

Because of its slow growth and fastidious nutritional requirements, the organism is difficult to isolate from the mixed populations found in nature. It is curious that an organism able to grow in so thin a medium as cooling water should require a rich medium in the laboratory. The requirements have not been entirely defined: they include cysteine and a high concentration of Fe^{2+}.

Treatment. The organism responds in cultures to a variety of antibiotics, but in intraperitoneally infected guinea pigs only erythromycin and rifampin proved effective, presumably because the organisms frequently reside within host cells (where these agents are known to be effective). In the Philadelphia epidemic mortality among the patients treated with erythromycin was about half the general rate.

LISTERIA MONOCYTOGENES

Listeria monocytogenes infection has been recognized as an important economic problem in domestic and feral animals since 1911, but listeriosis in man, first described in 1929, was considered a rare disease for the next quarter century. In the last decade it has been voluntarily reported in the United States with increasing frequency, now exceeding that of tularemia, trichiniasis, lepto-spirosis, or psittacosis. Moreover, unrecognized disease is probably much more common.

The organism elicits a striking monocytic blood reaction, which gave rise to its name (and to the earlier belief that it was the cause of infectious mononucleosis). Injection of a lipid extract of the organism can produce a similar monocytosis in mice and rabbits. *Listeria* behaves as an intracellular parasite; it has been used in experimental models to study macrophage activation (Ch. 24).

The Organism. *L. monocytogenes* is a small, gram-positive, aerobic to microaerophilic, non–spore-forming rod; it exhibits a peculiar end-over-end tumbling motility at 20°–25° C, but often not at 37°.

The organism grows well on blood agar and other general media, but storage of clinical material at 4 C for several days before inoculation enhances its isolation rate, perhaps by releasing the organisms from their intracellular location. On blood agar, the colonies usually are surrounded by a narrow band of β-hemolysis, and if not carefully examined in stained smears and tested for motility at 20°–25° the organism may be mistaken for a hemolytic streptococcus. Because it often shows palisade formation on microscopic examination and also grows well on potassium tellurite it is frequently mistaken for a diphtheroid (see Other Corynebacteria, Ch. 26) and discarded as a contaminant. It may also be confused with enterococci because of its growth on bile–esculin agar (a selective medium for enterococci).

L. monocytogenes ferments carbohydrates, producing principally lactic acid without gas (Table 43-1). It elaborates catalase, hydrolyzes esculin, produces acetoin (Voges–Proskauer test) but not indole, and does not reduce nitrate. Characteristically, instillation into the conjunctival sac of a rabbit results in a purulent conjunctivitis in 3–6 days, followed by keratitis **(Anton test)**. DNA hy-

TABLE 43-1. *Listeria* and *Erysipelothrix:* **Comparison with Some Common Gram-Positive Bacteria**

	Shape	β-Hemo-lysis	Catalase	Motility	Nitrate reduction	Esculin hydrolysis	Carbohydrate & acid production*			Kerato-conjunc-tivitis
							Glucose	Salicin	Mannitol	
Listeria monocytogenes	R	+	+	+	–	+	+	+	–	+
Erysipelothrix rhusiopathiae	R	–	–	–	–	–	+	–	–	–†
Streptococcus pyogenes	C	+	–	–	–	–	+	+	–	–
Streptococcus faecalis	C	–/+	–	–/+	–	+	+	+	+	–
Corynebacterium	R	–+	+	–	+/–	–	+/–	–	+/–	–
Lactobacillus	R	–	–	–	–	–	+	+–	+/–	–

R = rod; C = coccus.

*Acid only; no gas. Some *Lactobacillus* species produce gas.

†Conjunctivitis but no keratitis.

(Based, in part, on Buchner L, Schneirson SS: Am J Med 45:904, 1968)

bridization fails to reveal significant relatedness to members of the major gram-positive families.

On the basis of somatic (O) and flagellar (H) Ags, 4 main serologic groups and 11 serotypes are recognized. Types 4b and 1 predominate among isolates from man and animals. **Serologic diagnosis** is highly unreliable because *L. monocytogenes* crossreacts with a number of common gram-positive bacteria. This is due, at least in part, to its sharing the so-called Rantz Ag (an Ag of undetermined chemical composition) with various streptococci and with *Staphylococcus aureus.*

Pathogenesis. The astonishingly wide host range of *Listeria* includes a great variety of mammals and birds, as well as ticks, fish, and crustaceans. Disease in animals is characterized by monocytosis, septicemia, and the formation of multiple focal abscesses in the viscera; meningoencephalitis and involvement of the uterus and fetus have also been described in domestic animals. Apparently healthy **enteric carriers** are common in many species including man.

Listeriosis is generally regarded as a **zoonosis** (see Ch. 33), and transmission from animals to man definitely occurs, e.g., from handling of newborn calves, contact with infected dogs, and drinking of infected milk. However, in the United States most cases occur among urban residents with few or no known animal contacts; and extensive distribution of the organism on plants and in soil has recently been demonstrated. Hence some now regard it as a **sapronosis** (a disease of man and animals caused by organisms from the environment).

In man the disease takes various forms (Table 43-2), the most common being a purulent **meningoencephalitis** or **meningitis.** In such cases the organism can usually be cultured from the spinal fluid and often from the blood and the throat. The mortality rate among untreated patients is approximately 70%. At least half the cases are associated with underlying disorders (collagen disease, diabetes), or with the administration of agents (corticosteroids, immunosuppressive agents, or radiation therapy) that impair cellular immunity.

A less common but even more serious manifestation is **perinatal listeric septicemia** (previously known as **granulomatosis infantiseptica**), arising from a low-grade uterine infection of the mother often associated with bacteremia but with few or no symptoms. The disease in the newborn has a high mortality rate. It is characterized by numerous foci of necrosis throughout the body, especially in the liver and meninges; the meconium usually contains large numbers of *L. monocytogenes.* Intrauterine infection can also lead to **abortion,** to **stillbirth,** or to delivery of a healthy child who develops **meningitis** within a few days or weeks after birth.

Prompt therapy with ampicillin, penicillin, or tetracycline appreciably lowers the mortality; combined therapy with penicillin and gentamicin is now recommended, especially if there is an underlying defect in the host.

ERYSIPELOTHRIX RHUSIOPATHIAE

E. rhusiopathiae (*E. insidiosa*) is a gram-positive, nonsporulating, aerobic to microaerophilic bacillus. It is similar to *L. monocytogenes* except for differences listed in Table 43-1; also, the cell wall lacks diaminopimelic acid. It causes **swine erysipelas** (a disease of economic importance) and is a parasite in fish, crustaceans, rodents, and a number of domestic animals (swine, sheep, turkeys, ducks). It is also found in decomposing organic matter.

The organism enters the human skin through minor abrasions, following contact with fish, shellfish, meat, or poultry, and it produces a disease known as **erysipeloid.** This disease is therefore an occupational hazard of fish and meat handlers, veterinarians, and housewives; epidemics have been reported among crab handlers. The localized cutaneous form of the disease is characterized by a spreading, painful, erythematous skin eruption, usually on the fingers and hands; the lesions exhibit sharp margins, extend peripherally, and clear centrally. Less commonly the disease appears in a severe generalized cutaneous form or in a septicemic form (often followed by endocarditis). The organism is sensitive to penicillin, cephalosporin, clindamycin, and erythromycin.

STREPTOBACILLUS MONILIFORMIS

S. moniliformis is a facultatively anaerobic, gram-negative, pleomorphic bacillus, named because of its tendency, in older cultures and on solid media, to form filaments and chains of bacilli with prominent yeastlike swellings (Fig. 43-1). It grows only in media enriched with blood, serum, or ascitic fluid. In young broth cultures and in infected tissues the organisms grow as typical

---

TABLE 43-2 Clinical Manifestations of Human Listeriosis

Meningoencephalitis, most common in neonates, immunosuppressed patients, and those with underlying chronic disease
Low-grade septicemia in gravida (flu like); premature or nonviable termination of pregnancy
Septicemia in perinatal period
Infectious mononucleosis–like syndrome
Septicemia in adults
Pneumonia
Endocarditis
Aortic aneurysm (bacterial)
Localized abscesses, external or internal
Papular or pustular cutaneous lesions
Conjunctivitis
Urethritis
Habitual abortion (likely but needs further study)

(Revised from Gray ML, Killinger AH: Bacteriol Rev 30:309, 1969)

FIG. 43-1. Many irregular filaments of *Streptobacillus moniliformis* with yeastlike swellings growing on solid medium. (× about 2520) **Inset** shows unstained colonies of L-phase variant derived from bacillary form. (×80) (Courtesy of L. Dienes)

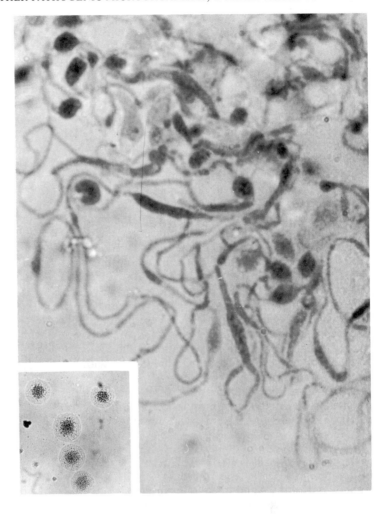

bacilli, and on solid media the L-phase variant (see below) is often demonstrable (Fig. 43-1 inset). Under the crowded conditions in surface colonies the organism has a tendency to form defective cell walls, thus leading to the growth of L forms. All strains appear to be antigenically the same.

S. moniliformis is a normal inhabitant of the upper respiratory tract of wild and laboratory rats (as well as squirrels and weasels). It is the causative agent of one form of human **rat-bite fever,** the other being caused by *Spirillum minor* (see Rat-Bite Fever, Ch. 39). Clinically the two diseases are quite similar, except that arthritis is more common in the *Streptobacillus* variety. Endocarditis, either acute or subacute, may also occur. The organism can usually be recovered, after 3–8 days' incubation, from blood cultures, where it grows on the surface of the sedimented blood cells in the form of "fluff balls." In untreated *Streptobacillus* rat-bite fever the mortality rate is approximately 10%. Penicillin is the agent of choice; streptomycin is also effective.

Streptobacillus disease may also be acquired through chance infection of skin abrasions or by ingestion of contaminated food. When unassociated with a rat bite the disease is often referred to as **Haverhill fever,** after a milk-borne epidemic that occurred in Haverhill, Massachusetts, in 1929.

CALYMMATOBACTERIUM (DONOVANIA) GRANULOMATIS

Granuloma inguinale is an indolent, granulomatous, ulcerative disease caused by a short, plump, gram-negative bacillus which is antigenically similar to, but not identical with, *Klebsiella pneumoniae* and *K. rhinosceleromatis* (Ch. 31). It is thought to be transmitted during coitus but is not highly communicable. The initial lesion commonly appears on or about the genital organs, beginning as a painless nodule. It soon breaks down, forming a sharply demarcated ulcer, which spreads by direct extension and often destroys large areas of skin in the groins and about the anus and genitalia. Occasionally, extragenital lesions occur on the face and neck, or in the mouth and throat; metastatic lesions in bones, joints, and viscera have also been reported. Histologic examination of the friable granulomatous tissue which forms the base of the coalescing ulcerative lesions reveals a heavy infiltration with polymorphonuclear leukocytes and monocytes.

In Wright-stained smears of scrapings from the lesions the encapsulated causative organism *C. granulomatis* can often be seen within the pathognomonic large mononuclear cells (Fig. 43-2); they are referred to as **Donovan bodies.** These bacilli were first cultivated in the yolk sacs of chick embryos. They cannot be cultured on artificial media when first isolated, but they can eventually be adapted to media containing egg yolk and even to beef heart infusion agar.

Antigenic extracts of *C. granulomatis* give positive skin and complement-fixation tests in patients with granuloma inguinale, but these procedures are not routinely used. Diagnosis rests with demonstraton of Donovan bodies in stained spreads of granulation tissue, or in the pathognomonic cells in biopsy material stained with hematoxylin and eosin.

Gentamicin and chloramphenicol are probably the most effective drugs, curing most lesions within 3 weeks. Tetracyclines or streptomycin are also useful.

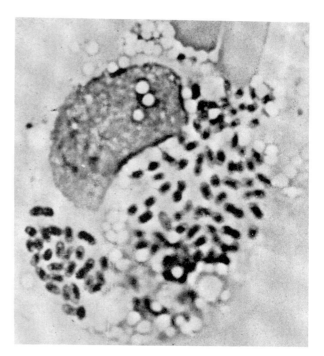

FIG. 43-2. Donovan bodies within large macrophage in film stained with pinacyanole technic. (Greenblatt RB, Dienst RB, West RA: Am J Syph 35:292–293, 1951) (× about 8000)

BARTONELLA BACILLIFORMIS

Inhabitants of the Andes mountains in Peru, Colombia, and Ecuador have long been known to suffer from two different clinical manifestations of the same bacterial infection, collectively designated as **Carrión's disease:** a severe, febrile, hemolytic anemia **(Oroya fever),** and a relatively benign eruption of hemangiomalike nodules **(verruga peruana).** The causative agent is *Bartonella bacilliformis,* a small, motile, gram-negative coccobacillus with flagella at one end. The organism is remarkable in its geographic restriction and its tropism for RBCs.

B. bacilliformis grows readily at either 28° or 37° C, with greater longevity at 28°, in semisolid nutrient agar containing 10% fresh rabbit serum and 0.5% hemoglobin. The growth first appears just below the surface of the medium after about 10 days. There is apparently only a single antigenic type.

Bartonellosis is transmitted from man to man by sandflies (*Phlebotomus*) indigenous to the region where the disease is endemic. No other reservoirs of the organism have been found. After an incubation period of 2–3 weeks the patients exhibit intermittent fever and severe constitutional symptoms (including myalgia, nausea,

vomiting, diarrhea, and headache), followed by increasingly severe signs and symptoms of **anemia.** Wright- or Giemsa-stained blood films usually reveal many organisms, either in or on the RBCs, and blood cultures are positive. This initial stage of infection is frequently complicated by *Salmonella* superinfection. Among untreated patients the mortality rate averages 40%, salmonellae being responsible for many of the deaths; in those who recover convalescence is slow and blood cultures may remain positive for many months.

The severe anemic stage of the disease may be followed in 2–8 weeks by a **cutaneous** stage, characterized by multiple eruptions of hemangiomalike nodules (verruga) which often ulcerate before healing. This form of the disease also occasionally develops without any obvious preceding anemia; it causes virtually no systemic manifestations and is not fatal. In this stage the organism is not seen in the blood, the basic pathologic process being endothelial proliferation with new vessel formation.

Chemotherapy with penicillin, streptomycin, or chloramphenicol is effective, even in the severe anemic phase. Control measures, including the use of DDT, are directed against the sandfly.

L-PHASE VARIANTS (L-FORMS)*

Some bacteria readily give rise to spontaneous variants that replicate as pleomorphic, filterable elements with defective or absent cell walls. Similar variants are formed by many other bacterial species when wall synthesis is impaired (e.g., by penicillin); hypertonic media also favor their development. Their passage through filters that retain bacteria depends more on flexibility than on small size. These **L-phase variants**, or **L forms** (named for the Lister Institute), probably arise by two mechanisms: **selection** of **wall-defective mutants,** and **induction** of persistent, nongenetic **phenotypic** changes (e.g., the absence of a preexisting wall to which building blocks can be added).

L-phase variants resemble protoplasts and spheroplasts (see Ch. 6), but they have developed compensatory changes that strengthen their cell membrane—for example, an increased ratio of saturated to unsaturated fatty acids. Accordingly, though they lack a wall, some L-phase variants (unlike protoplasts) can grow at the osmolality of serum. Growth is slow. Division appears to be assisted by pinching of the growing cell between the strands of an agar gel: hence these organisms grow as "fried-egg" surface colonies (Fig. 43-3), thickened in the center (where cells have burrowed into the agar) and consisting largely of dead cells at the periphery.

L-phase variants were long confused with mycoplasmas (Ch. 42), because both lack a cell wall and so they have similar cellular and colonial morphology, filterability, and slow growth. However, their immunologic and biochemical differences, and the absence of DNA homology, now make it clear that *L-phase variants and mycoplasmas are not closely related.* In addition, in the absence of inhibitors of wall synthesis *many L-phase variants can revert* to normal vegetative cells. Highly viscous media, which impair diffusion of wall precursors away from the cell, favor reversion in vitro, and the same effect might be expected from the constraining effect of surrounding tissue in the host. Reversion of such phenotypic variants is also favored by Mg^{2+} and antagonized by Ca^{2+}. In contrast genotypic variants, with a genetic defect in wall formation, are more stable and cannot be induced to revert.

RELATION TO DISEASE

The role of L-phase variants in human disease remains unclear. They have been cultivated from cases of pyelonephritis and endocarditis as well as from pulmonary, meningeal, and gastrointestinal infections. Such isolations alone fail to prove that the recovered variants were present as such in the patient, rather than emerging during cultivation. However, a few reports have provided reliable evidence.

Thus in patients with chronic pyelonephritis (including some who were not taking antibiotics) such variants have been recovered from urine passed through filters that retain vegetative cells. Similarly, in a case of endocarditis with negative blood cultures wall-defective variants were seen in blood cells, and they were recovered by special culture. It thus appears that L-phase variants may occasionally be produced and persist in patients.

Cultivation in hypertonic media (e.g., with 0.6 M sucrose) sometimes permits the outgrowth of typical bacteria from tissues while ordinary media yield no organisms: presumably organisms with damaged cell walls initiate growth in a hypertonic medium and then revert to the vegetative phase. Because L-phase variants are not detected by the usual technics for cultivation (or for staining in tissues), their presence may be suspected when no organism can be cultured from a patient exhibiting the general characteristics of a bacterial infection.

Although in culture L-phase variants produce the same exotoxins and endotoxins as the parent bacteria, in

FIG. 43-3. Variant phases. *Proteus mirabilis* is seen by phase contrast in identical media, and at the same magnification (×3400), as **(a)** vegetative cells, **(b)** spheroplasts, and **(c)** a clump of L-phase variants. Photo **(d)** shows a "fried egg" colony of L-phase variants. (×150) (After McGee ZA et al: J Infect Dis 123:433, 1971)

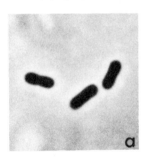
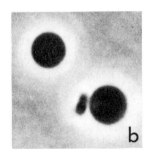
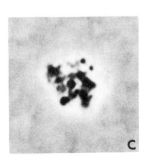
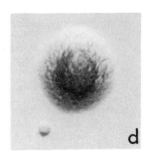

* Contributed by Zell A. McGee.

experimental animals they have not been shown to cause significant tissue damage, though they can persist for long periods. Because of this lack of a good animal model, and because Koch's postulates do not apply to an unstable form, their etiologic role in human disease is particularly difficult to determine: *they may be important largely as dormant forms,* resistant to antibiotics that act on wall synthesis, and able to revert and cause a relapse after therapy ceases.

However, L-phase variants need not depend on antibiotics for their production: they were first discovered as frequent **spontaneous** products in cultures of *Streptobacillus moniliformis,* and they have occurred spontaneously with *Neisseria gonorrhoeae,* and with a few other pathogens, in animals, tissue cultures, and embryonated eggs. Moreover, growth in the presence of certain fatty acids present in human tissues strengthens their membrane. Their development could well also be promoted by the attack of the Ab–complement system on the cell envelope. On the other hand, they are susceptible to phagocytosis and to killing by Ab and complement, they stimulate an immune response, and their persistence is enhanced by immunosuppression.

A search for L-phase variants may be indicated in selected patients, though the methods are tedious and are not standardized. A search for transient, revertible forms may also be useful: in one series the yield of blood cultures positive for vegetative bacteria was increased 20% by the addition of a hypertonic agent and an inhibitor of complement. However, this circumstantial evidence for frequent defective forms in vivo does not prove that the defect was in the wall, or that the defective forms multiplied in the host.

SELECTED READING

BACTERIA

BOOKS AND REVIEW ARTICLES

BALOWS A, FRASER DW (eds): International symposium on Legionnaires' disease. Ann Int Med, Vol 90, No 4, April 1979

GRAY ML, KILLINGER AH: *Listeria monocytogenes* and listeria infections. Bacteriol Rev 30:309, 1966

GRIECO MH, SHELDON C: *Erysipelothrix rhusiopathiae.* Ann NY Acad Sci 174:523, 1970

KALIS P, LEFROCK JL, SMITH W, KEEFE M: Listeriosis. Am J Med Sci 271:159, 1976

KILLINGER AH: *Listeria monocytogenes.* In Lennette EH, Spaulding EH, Truant JP (eds): Manual of Clinical Microbiology, 2nd ed. Washington DC, American Society for Microbiology, 1974, p 135

KING A, NICOL C: *Granuloma inguinale.* In Venereal Diseases, 3rd ed. Baltimore, Williams & Wilkins, 1975, p 252

LAVETTER A, LEEDOM JM, MATHIES AW JR et al: Meningitis due to *Listeria monocytogenes:* a review of 25 cases. N Engl J Med 285:598, 1971

LOURIA DB, BLEVINS A, ARMSTRONG D: *Listeria* infection. Ann NY Acad Sci 174:545, 1970

ROGOSA M: *Streptobacillus moniliformis* and *Spirillum minor.* In Lennette EH, Spaulding EH, Truant JP (eds): Manual of Clinical Microbiology, 2nd ed. Washington DC, American Society for Microbiology, 1974, p 326

SONNENWIRTH AC: In Sonnenwirth A, Jarett L (eds): Gradwohl's Clinical Laboratory Methods and Diagnosis, 8th ed. St. Louis, Mosby, 1980, pp 1593–1594, 1673–1679, 1841–1842

WEINMAN D: Bartonellosis. In Weinman D, Ristic M: Infectious Blood Diseases of Man and Animals, Vol 2, 2nd ed. New York, Academic Press, 1968, p 3

WOOBDINE M (ed): Problems of Listeriosis. Proceedings of the Sixth International Symposium. Leicester, Engl, Leicester University Press, 1975

SPECIFIC ARTICLES

BUSCH LA: Human listeriosis in the United States. J Infect Dis 123:328, 1971

FRASER DW, TSAI TR, ORENSTEIN W et al: Legionnaires' disease: Description of an epidemic of pneumonia. N Engl J Med 297:1189, 1977

MCDADE JE, SHEPARD CC, FRASER DW et al: Legionnaires' Disease: isolation of a bacterium and demonstration of its role in other respiratory disease. N Engl J Med 297:1197, 1977

MEDOFF G, KUNZ LJ, WEINBERG AN: Listeriosis in humans: an evaluation. J Infect Dis 123:247, 1971

SIMERKOFF MS, RAHAL JJ: Acute and subacute endocarditis due to *Erysipelothrix rhusiopathiae.* Am J Med Sci 266:53, 1973

L-PHASE VARIANTS

BOOKS AND REVIEW ARTICLES

CLASENER H: Pathogenicity of the L-phase of bacteria. Annu Rev Microbiol 26:55, 1972

DIENES L, WEINBERGER HJ: The L-forms of bacteria. Bacteriol Rev 15:245, 1951

GUZE LB (ed): Microbial Protoplasts, Spheroplasts and L-Forms. Baltimore, Williams & Wilkins, 1967

HIJMANS W, VAN BOVEN CPA, CLASENER H: Fundamental biology of the L-phase of bacteria. In Hayflick L (ed): The Mycoplasmatales and the L-Phase of Bacteria. New York, Appleton-Century-Crofts, 1969, p 67

SPECIFIC ARTICLES

GUTMAN LT, TURCK M, PETERSDORF RG, WEDGWOOD RJ: Significance of bacterial variants in urine of patients with chronic bacteriuria. J Clin Invest 44:1945, 1965

LEON O, PANOS C: Adaptation of an osmotically fragile L-form of *Streptococcus pyogenes* to physiological osmotic conditions and its ability to destroy human heart cells in tissue culture. Infect Immun 13:252, 1976

MCGEE ZA, WITTLER RG, GOODER H, CHARACHE P: Wall-defective microbial variants: terminology and experimental design. J Infect Dis 123:433, 1971

RYTER A, LANDMAN OE: Electron microscopic study of the relationship between mesosome loss and the stable L state (or protoplast state) of *Bacillus subtilis*. J Bacteriol 88:457, 1964

WITTLER RG, MALIZIA WF, KRAMER PE, TUCKETT JD, PRITCHARD HN, BAKER HJ: Isolation of a corynebacterium and its transitional forms from a case of subacute bacterial endocarditis treated with antibiotics. J Gen Microbiol 23:315, 1960

chapter 44

INDIGENOUS BACTERIA; ORAL MICROBIOLOGY

ALEX C. SONNENWIRTH
RONALD GIBBONS
SIGMUND SOCRANSKY

INDIGENOUS BACTERIA*

Born into an environment laden with microbes, the body of man becomes infected from the moment of birth. A variety of bacterial species (called **indigenous** or **autochthonous**) establish a more or less permanent residence on, and even in, the superficial tissues. Here they usually are symbiotic and benefit the host in a number of ways. Some of their actions, however, may not be beneficial, and when the general resistance of the host is sufficiently depressed they may even cause disease.

BENEFICIAL ACTIONS

The indigenous bacteria play a major role in preventing certain bacterial diseases. One mechanism is collectively referred to as **bacterial antagonism** (see Microbial Antagonism, under Chemical Factors, in Ch. 24). Its importance is illustrated by the adverse effects of antibiotics that suppress the normal bacterial flora. Thus when tetracycline or lincomycin is given in large doses for many days, resistant, potentially pathogenic organisms (staphylococci, *Clostridium difficile*), no longer held in check by the coliforms and other organisms, occasionally give rise to a serious enteritis. Moreover, in mice a single oral dose of streptomycin reduces by about 10^5 the oral dose of *Salmonella enteritidis* needed to induce infection.

Probable mechanisms in this antagonism include competition for nutrients, alterations in pH or in oxidation–reduction potential, and accumulation of organic acids. Elaboration of highly specific bacteriocins (Ch. 12) may also play a large role in maintaining a balanced flora, but this intriguing possibility has not been explored.

The indigenous flora also influences the **immune system.** Animals reared in a **germ-free** environment exhibit grossly underdeveloped and relatively undifferentiated lymphoid tissues, low levels of serum γ-globulins, and a primary response instead of the usual secondary response to certain immunogens. Such animals are unusually susceptible to challenge with pathogenic microbes, even at sites normally free of bacteria (e.g., the peritoneal cavity). Their **lowered resistance** may be due to defects both in specific immune mechanisms and in the nonspecific immunity that is conferred by exposure to endotoxin (Ch. 24).

The resident bacteria also **synthesize nutrients** essential to the welfare of the host. Thus vitamin K, which can be synthesized by coliforms, is required in the germ-free host but not in animals with normal bowel flora. Other vitamins supplied in part by intestinal bacteria include biotin, riboflavin, pantothenate, and pyridoxine.

DETRIMENTAL ACTIONS

The indigenous bacterial flora, however, does not always benefit the host. **Bacterial synergism** may enable pathogenic species to thrive in tissues which otherwise would not support their growth. For example, the crossfeeding of certain bacteria by specific metabolites excreted by others has been observed in vitro and may well also occur in the body. In another kind of synergism, gonococci in the urethra are protected against the action of penicillin by penicillinase-producing staphylococci, which do not themselves cause urethritis.

Endotoxins derived from bacteria that are normally present in the intestinal tract are thought to cause low-grade toxemia.† Furthermore, host susceptibility to the harmful effects of endotoxins may be due in part to hypersensitivity to endotoxin Ags derived from the intestinal flora, for germ-free animals are extraordinarily resistant to injections of endotoxins. Diets that suppress many normal organisms and low-grade pathogens in the bowel increase the growth of poultry and other animals raised for food, and various antibiotics are used for this purpose on a huge scale in animal feeds. The shifts in flora that are responsible, however, have not been clearly identified.

Under circumstances of lowered general or local resistance the bacteria of the resident flora may become **opportunistic pathogens.** For example, *Streptococcus mitis* and *S. salivarius,* the most common aerobic species inhabiting the oropharynx, often cause very transient bac-

* Contributed by Alex C. Sonnenwirth.

† Endotoxemia becomes more severe in the late stages of hemorrhagic shock and contributes to its irreversibility.

teremia following dental extraction, tonsillectomy, or even vigorous mastication. If the endocardium of the host has been previously injured by rheumatic fever the usually harmless streptococci may infect the sites of damage (see α-Hemolytic Streptococci, Ch. 28) and give rise to **subacute bacterial endocarditis.** Similarly, *Bacteroides* may contribute to serious infections and even produce bacteremia if introduced into the peritoneal cavity or into an open wound, *Escherichia coli* may infect the obstructed urinary tract, and *Pseudomonas* often infects open burns of the skin. The role of bacteria in the pathogenesis of dental caries and periodontal disease will be discussed later in this chapter.

DISTRIBUTION

Although many sites of the human body are usually sterile, the external and many internal surfaces are inhabited by various bacteria (listed in Table 44-1), and in some cases by yeasts and protozoa.

The **skin** has a different bacterial flora in different body regions. The flora of the facial or of the perineal area, for example, reflects that of the oropharynx or of the lower intestinal tract, respectively. Moreover, different local secretions select for quite different floras (which contribute to the characteristic odors) in, for example, the external auditory canal, the axillae, and the region between the toes. Although vigorous scrubbing with soap and water or other disinfectants, as in preparation for surgical operations, will temporarily rid the skin of most of its surface bacteria, organisms sequestered in hair follicles and sweat glands will soon reestablish the surface infection. Gram-positive aerobic cocci (*Staphylococcus epidermidis,* micrococci) and anaerobic corynebacteria (*Propionibacterium*) comprise the majority of the resident flora, with aerobic corynebacteria, *Staphylococcus aureus,* and nonhemolytic streptococci less prevalent, and gram-negative bacteria (Enterobacteriaceae, *Pseudomonas*) found sporadically.

The microbiology of the **oral cavity** will be discussed below. In the **pharynx** viridans streptococci, staphylococci, and *Branhamella* (*Neisseria*) *catarrhalis* are commonly found, while the **nose** is colonized chiefly by staphylococci and corynebacteria (diphtheroids). The trachea may contain a few bacteria, but the normal lower respiratory tract is usually sterile, owing primarily to the efficient cleansing action of the mucociliary "blanket" that lines the bronchi (see Mechanical Factors, Ch. 24).

In the **upper gastrointestinal tract** the esophagus contains only the bacteria swallowed with the saliva and food. Because of the high acidity of the gastric juice few organisms (mostly lactobacilli) can be cultured from the normal stomach, except immediately after meals. The proximal **small intestine** has a sparse, predominantly gram-positive microflora (10^5–10^7/ml of fluid), the major constituents being lactobacilli and enterococci. The flora gradually changes toward the lower small intestine, with the appearance of anaerobes and coliform bacteria; the flora of the lower ileum is qualitatively similar to that of feces. The normal **colon** flora is usually inferred from that of the feces, which contains 2–4×10^{11} bacteria/g of dry weight. At birth the entire intestinal tract is sterile but bacteria enter with the food. Enterobacteria (coliforms), micrococci, and streptococci appear first in the fecal flora, and within a few days anaerobic lactobacilli (bifidobacteria) predominate, especially in breast-fed infants. With liberalization of the diet gram-negative enterobacteria become more prominent and are joined by bacteroides, enterococci, and clostridia. *The bacteria in the feces of*

TABLE 44-1. Bacteria Most Commonly Found on Surfaces of the Human Body

Bacteria	Skin	Conjunctiva	Nose	Pharynx	Mouth	Lower intestine	External genitalia	Vagina
Staphylococci	+	±	+	±	±	±	±	+
Pneumococci		±	±	+	+			
Streptococci								
viridans	±	±	+	+	+	+	+	+
β-Hemolytic				±	±			
faecalis	±			±	+	+	+	±
Anaerobic					+	+	+	±
Neisseriae			±	+	+	±	+	±
Veillonellae				±	+	±	+	
Lactobacilli					+	+		+
Corynebacteria	+	+	+	+	+	+		+
Clostridia					±	+	±	
Hemophilic bacilli		+	+	+	+			
Enteric bacilli				±	+	+	+	
Bacteroides				+	+	+	+	
Mycobacteria	+		±	±		+	+	
Actinomycetes				+				
Spirochetes				+	+	+	+	
Mycoplasmas				+	+	+	±*	+

+ = Common; ± = rare.

*Prevalence related to sexual activity.

(Modified from Rosebury T: Microorganisms Indigenous to Man. New York, McGraw-Hill, 1962)

adults are predominantly bacteroides and bifidobacteria: these are obligate anaerobes, which outnumber *E. coli* (10^7–10^8/g) by 1,000:1 to 10,000:1. Studies in mice have revealed similar relations of the intestinal flora.

The **vagina** also becomes infected soon after birth, originally with lactobacilli and later with a variety of cocci and bacilli. During the childbearing years lactobacilli, *S. epidermidis,* and corynebacteria predominate, with a variety of anaerobes (*Bacteroides,* cocci) also present in about 70% of women. Other bacteria of lower incidence are viridans streptococci, enterococci, group B streptococci, nonpathogenic neisseriae, enteric bacilli, *Acinetobacter* spp., *Hemophilus vaginalis,* and *Mycoplasma* spp. The flora remains essentially the same in the postmenopausal years except for the greater incidence of gram-negative bacilli (other than *E. coli*).

The normal **urethra** usually contains a small number of contaminants from the external mucous membranes of the genital organs.

A variety of bacteria may be cultivated from the **conjunctivae,** but the number of organisms is usually small. *S. epidermidis* and *Propionibacterium acnes* (anaerobic diphtheroids) are predominant; nonhemolytic streptococci, *S. aureus,* anaerobic streptococci, lactobacilli, corynebacteria, and enteric bacilli occur less frequently.

ENDOGENOUS BACTERIAL DISEASES

A general knowledge of the normal bacterial flora of the human body is essential, not only for interpreting the results of bacteriologic findings but also because of the increasing incidence of **endogenous** bacterial diseases. Involving bacteria that are indigenous* yet potentially pathogenic, these diseases are "caused" by factors that lower the resistance of the host, either generally or at specific tissue sites. Such factors include radiation damage, prolonged use of corticosteroid hormones, severe malnutrition, shock, debilitation from other diseases (particularly those involving the bone marrow and/or lymphoid tissues, e.g., leukemia), superinfection and disturbances in bacterial antagonism resulting from antimicrobial therapy, localized obstruction of excretory organs, and predisposing lesions of unrelated etiology (see Ch. 24).

Unlike **exogenous** bacterial disease, those of **endogenous** origin 1) have no definable incubation period, 2) are not communicable in the usual sense, 3) do not result in clinically recognizable immunity and tend to recur or to progress slowly for years, and 4) involve bacterial agents found in the normal flora. These agents, characterized by low intrinsic pathogenicity, cause disease when they appear either in unusual body sites or in enormously increased concentration in or near their usual sites.

As the lives of more patients with major illnesses are prolonged by improved methods of treatment, and as more infections caused by virulent exogenous bacteria are controlled by effective antimicrobial drugs, *endogenous bacterial diseases have become a major proportion of the serious bacterial diseases encountered in clinical practice* (see Ch. 31).

ORAL MICROBIOLOGY†

ORAL ECOLOGY

In the oral cavity distinct collections of indigenous bacteria colonize the surfaces and crypts of several sites: the teeth, the tongue dorsum, the buccal mucosa, and the gingival crevice area. On the surfaces of teeth bacteria accumulate in large masses, referred to as **dental plaques;** these contain over 10^8 bacteria/mg. On mucosal surfaces desquamation limits their numbers: only 10–15 bacteria are usually present on a buccal epithelial cell and 100–150 on a cell of the tongue dorsum. Saliva averages over 10^8 bacteria/ml, which have been washed off oral surfaces (particularly the tongue and the teeth).

Nutrients available to bacteria within the oral cavity include constituents of the diet, gingival crevice fluid (a serous transudate), salivary secretions, and metabolites synthesized by associated organisms. This diversity permits growth of a wide spectrum of bacterial species. The differences in the types of organisms found in varous sites appear to be due largely to the selective **attachment** of organisms to different oral surfaces. Bacterial attachment is the first essential step for colonization, because of the flow of oral fluids and the creation of newly exposed epithelial surfaces by desquamation. Therefore the species of bacteria found initially colonizing tooth surfaces and various oral epithelial surfaces is mainly determined by their **adhesiveness** for a particular surface and by their numbers, in saliva, available for attachment. On teeth the first colonizing bacteria then provide new surfaces for attachment of additional species, and such distinct epiphytic formations are common (Fig. 44-1). Thus in the absence of desquamation population shifts occur, changing the composition and the complexity of dental plaques.

Tooth mineral consists primarily of a complex calcium phosphate salt, hydroxyapatite, which readily adsorbs glycoproteins. Adsorbed salivary components, including blood-group–reactive salivary mucins, form a thin film termed the **enamel pellicle** on teeth in vivo. The adsorbed mucins are thought to serve as bacterial receptors, while

---

* Dormant or latent infections (Ch. 24) with exogenous pathogens (e.g., *Mycobacterium tuberculosis*) may also give rise to disease that is in a sense endogenous, but is usually not classified as such.

† Contributed by Ronald Gibbons and Sigmund Socransky.

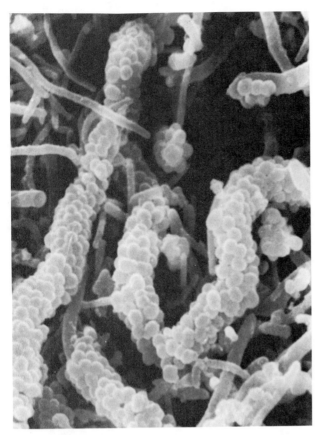

FIG. 44-1. Scanning electron photomicrograph of a sample of human dental plaque showing cocci colonizing the surface of filamentous forms. These epiphytic arrangements have been referred to as "corn cobs." (Courtesy of Dr. Z. Skobe, Forsyth Dental Center)

in the membranes of oral epithelial cells blood-group–reactive glycoproteins serve as similar receptors. Unadsorbed mucins in saliva mimic bacterial receptors and hence serve as potent inhibitors of bacterial attachment; they also promote the desorption of adherent organisms. This property probably explains how mucins contribute to the cleansing action of secretions.

Abs (the secretory IgA type predominating in saliva and the IgG type in crevicular fluid—i.e., in gingival crevices) may limit bacterial colonization by binding to surface Ags of the organisms and thus interfering with attachment. Abs to a wide range of indigenous species are commonly found in serum and saliva. However, antigenically shifting populations of *Streptococcus salivarius* colonize the dorsum of the tongue from birth to death in spite of an Ab response by the host.

Acquisition and Nature of the Oral Flora. The oral cavity is sterile at birth but bacteria are quickly introduced via food and other contacts. Buccal epithelial cells derived from infants immediately after birth (24–48 h) do not promote the attachment of indigenous organisms; consequently the levels of colonizing bacteria are low. By 3–5 days of age the cells permit bacterial attachment and so increased populations of organisms, such as *S. salivarius*, are found. The eruption of teeth at 6–9 months of age leads to colonization by organisms that require a nondesquamating surface, such as *Streptococcus sanguis* and *S. mutans*. In addition the creation of the gingival crevice area increases the variety of anaerobic species found. The complexity of the oral flora continues to increase with time, and *Bacteroides melaninogenicus* and spirochetes colonize around puberty. Conversely, loss of all teeth results in the elimination of *S. sanguis*, *S. mutans*, and many anaerobic species from the mouth. The distribution of organisms commonly found in various sites of the mouth is listed in Table 44-2. The mother is probably an important early source of colonizing organisms, since strains of *S. mutans* and *S. salivarius* isolated from infants frequently have the same bacteriocin sensitivity pattern as those possessed by the mother.

Effects of the Oral Flora. The indigenous oral flora contributes to host **nutrition** through the synthesis of vitamins. Though a much larger mass of indigenous bacteria is found in the lower bowel the greater efficiency of nutrient absorption in the small intestine increases the effective contribution of the oral flora. In addition, 1–2 g of bacterial cells are swallowed daily via saliva and these also contribute nutrients.

Indigenous organisms contribute to **immunity** by inducing low levels of circulating and secretory Abs that crossreact with pathogenic bacteria. For example strains of *Streptococcus mitior* and *S. sanguis* possess Ags that are crossreactive with some pneumococci. The oral flora also exerts microbial **antagonism** against certain pathogens. Thus the yeast *Candida albicans*, a frequent minor inhabitant of the mouth, may grow out and cause lesions when the indigenous bacterial flora is suppressed by antibiotics.

DISEASES INITIATED BY THE ORAL FLORA

Some members of the oral flora possess pathogenic potential. When dental plaque or saliva is injected subcutaneously into experimental animals transmissible purulent abscesses are formed. Oral organisms gaining entrance into tissues may cause abscesses of the alveolar bone or of the lung, brain, or extremities; they may also infect surgical wounds. Such infections usually contain mixtures of bacteria, with *B. melaninogenicus* often playing a dominant role. Other residents of the oral cavity associated with distant infections are *Eikenella corrodens* (abscesses), *Actinomyces* and *Arachnia* species (ac-

TABLE 44-2. Distribution of Frequently Encountered Oral Bacteria in Different Sites of the Month

	Supragingival plaque associated with			Subgingival plaque associated with				
	Pits & fissures of teeth	Smooth tooth surfaces	Gingivitis	Periodontosis	Periodontitis	Saliva	Cheek	Tongue
Gram-positive cocci								
Strep. sanguis	++	++	++	0	+	+	+	+
Strep. mutans	++	+	0	0	0	0	0	0
Strep. mitior-milleri	+	++	++	0	+	+	++	+
Strep. salivarius	0	0	0	0	0	++	+	++
Staph. epidermidis	0	+	+	0	0	+	+	+
Peptostreptococcus sp.	0	+	+	0	+	+	0	+
Gram-negative cocci								
Neisseria sp.	0	+	+	0	0	+	0	+
Veillonella sp.	+	+	+	+	+	++	0	++
Gram-positive rods								
Bacterionema matruchotii	+	+	+	0	0	0	0	+
Actinomyces viscosus	+	++	++	+	+	+	+	+
A. naeslundii	+	++	++	+	+	+	+	+
A. israelii	+	++	++	+	++	+	+	+
Other Actinomyces sp.	+	+	+	+	+	+	+	+
Lactobacillus sp.	+	0	0	0	0	0	0	0
Propionibacterium acnes		+	+	0	++	+	0	+
Rothia dentocariosa		+	+	0	0	+	0	+
Eubacterium alactolyticum	0	0	0	+				
Gram-negative rods								
Bacteroides melaninogenicus								
subsp. melaninogenicus	0	0	+	+	+	0	0	0
subsp. intermedius	0	0	+	++	++	0	0	0
subsp. asaccharolyticus	0	0	+	+	++	0	0	0
Bacteroides oralis	0	+	+	+	+	+	0	+
Eikenella corrodens		0	+	++	++	0	0	0
Fusobacterium nucleatum	0	+	++	++	++	+	0	+
Capnocytophaga sp.								
(Bacteroides ochraceus)		0	+	++	+	0	0	0
Hemophilus sp.	+	+	+	0	+	+	0	+
H. segnis; H. aphrophilus	0	+	+	0	0			
Campylobacter sp.	0	0	+	+	++	0	0	0
Anaerobic vibrios	0	0	+	+	++	0	0	0
Selenomonas sputigena	0	0	+	+	++	0	0	0
Small treponemes								
(Treponema macrodentium, T. orale, T. denticola)	0	0	+	+	+	0	0	0
Intermediate & large spirochetes	0	0	0	0	++	0	0	0

++ = Frequently encountered in high proportions; + = frequently encountered in low to moderate proportions; 0 = sometimes encountered in low proportions or not detectable.

tinomycosis), and *S. sanguis* and *S. mutans* (subacute endocarditis). In patients with subacute bacterial endocarditis it is important that *S. mutans* be distinguished from enterococci, since the choice of antibiotic is affected.

Dental plaque consists of mixed colonies of bacteria, which may reach a thickness of 300–500 cells on the more protected surfaces of teeth (Fig. 44-2). Bacterial cells make up 60%–70% of the volume of plaque and are embedded in a matrix composed of bacterial and salivary polymers and remnants of epithelial cells and leukocytes. These accumulations subject the teeth and gingival tissues to high concentrations of bacterial metabolites, which result in dental diseases.

Dental Caries. Dental caries is the destruction of the enamel, dentin, or cementum of teeth by bacterial action (Fig. 44-3). The mechanism appears to be a direct demineralization, caused by lactic and other organic acids produced from dietary carbohydrates by the overlying bacteria. Though many bacteria can produce high concentrations of acid in vitro, only a few species have been associated with dental caries in humans. Organisms of the *S. mutans* group are associated with the initial development of enamel lesions in humans; they are also highly cariogenic in monkeys, hamsters, and rats, as well as in gnotobiotic (germ-free) rats (Fig. 44-4). These organisms preferentially colonize occlusal fissures and the contact points between teeth; they are less frequently isolated

FIG. 44-2. Accumulation of bacterial dental plaque 3 days after cleaning. Stained with beta rose.

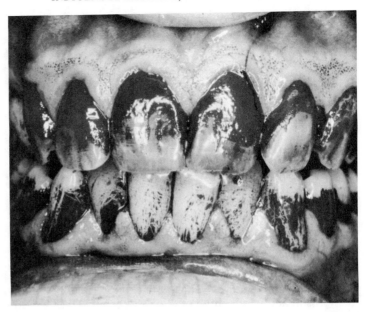

from buccal or lingual tooth surfaces. This pattern correlates with the incidence of decay on these surfaces.

The high cariogenic potential of *S. mutans* is partly due to its ability to accumulate on teeth through the formation of **extracellular glucans** (dextran) synthesized specifically from **sucrose** (see Polysaccharides, Ch. 4): predominantly α-1,6- and α-1,3-linked polymers, which are formed from one-half the sucrose molecule by glycosyl transferases present on the organism's surface. This effect of sucrose partially explains its well-known cariogenicity. Extracellular polymers present in colonies of *S. mutans* also reduce diffusion and promote accumulation of metabolic acids. Thus they attain higher acidity than

FIG. 44-3. Crown of a human tooth with dental caries. Note the colonies of adherent bacteria associated with the decay.

colonies of other oral organisms, which is another important determinant of pathogenicity.

S. mutans appears to be important in the initiation of enamel caries, but it is unlikely to be the only cause of dental decay. Lactobacilli are commonly found in occlusal fissures, and several strains have caused dental disease in experimental animals; some *Actinomyces* species also produce cemental and occasionally enamel carious lesions in experimental animals. Both lactobacilli and *Actinomyces* are commonly found in human carious dentin, which suggests they may also be common secondary invaders that contribute to the progression of lesions.

Periodontal Disease. Diseases that affect the supporting structures of teeth (i.e., the gingiva, cementum, periodontal membrane, and alveolar bone) are collectively referred to as periodontal disease. The most common form, **gingivitis,** is an inflammatory condition confined to the gingiva. It is essentially ubiquitous in adults and is associated with accumulations of bacterial plaque in the gingival crevice area. Its association with increased populations of *Actinomyces* suggests its etiology. Another form, **acute necrotizing ulcerative gingivitis (ANUG)** (also known as **trench mouth** or Vincent's infection), is an acute infection that causes necrosis of the interdental papillae and leads to the formation of ulcerated "craters" between the teeth. Unusual spirochetes observed at the forefront of such lesions are thought to be responsible for the disease, although they have not yet been cultivated. This periodontal disease is the only type in which bacterial invasion of the tissues is a predominant feature.

Diseases that are confined to the gingiva do not lead to loss of teeth. However, other types of periodontal disease commonly affect the periodontal membrane and alveolar

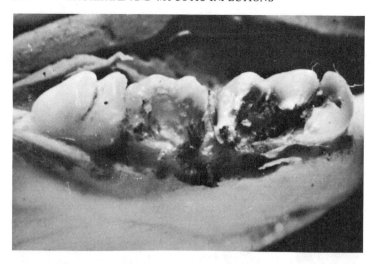

FIG. 44-4. Dental caries in a gnotobiotic rat monoinfected with *Streptococcus mutans* for 90 days.

bone, resulting in tooth loss. The most common cause of tooth loss in adults is a group of diseases known collectively as **chronic destructive periodontitis.** Microorganisms colonizing the gingival crevice area often lead to a swelling of the gingiva and an apical migration of the gingival epithelium attached to the tooth surface. The resulting space between the tooth and gingiva is termed a **periodontal pocket.** Bacteria accumulate in this protected niche, and its progress toward the apex of the root eventually destroys the alveolar bone and periodontal membrane, causing tooth loss.

The structure of subgingival plaque in periodontitis is complex. Microscopic observations reveal 1) a dense layer, consisting of *Actinomyces* and other gram-positive organisms, attached to the cemental and root surfaces, 2) a zone of gram-negative organisms between this layer and the pocket epithelium, and 3) spirochetes and other

motile forms at the base of the pocket. Large numbers of polymorphonuclear leukocytes (PMNs) are also present in the crevicular epithelium and pocket. The gram-positive organisms consist of *Actinomyces, Peptostreptococcus, Propionibacterium, Eubacterium,* and streptococci; the gram-negatives include spirochetes, *Selenomonas sputigena,* anaerobic *Vibrio, B. melaninogenicus, Fusobacterium nucleatum,* and *E. corrodens.*

Another periodontal disease, **periodontosis** (juvenile periodontitis), is a distinctive infectious process characterized by a rapid destruction of alveolar bone around the first molars and incisors of adolescents, often with minimal bacterial accumulations and gingival inflammation. The associated organisms include *Capnocytophaga* (*Bacteroides ochraceus*), *E. corrodens,* and fermentative *Bacteroides.* Patients frequently exhibit a defect in PMN chemotaxis. Periodontosis is relatively

FIG. 44-5. Periodontal destruction and root caries in gnotobiotic rat monoinfected with *Actinomyces naeslundii* for 90 days.

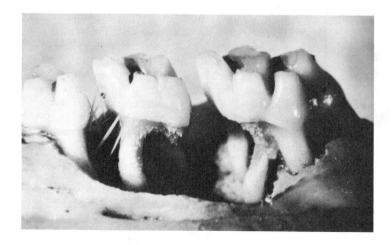

uncommon in North America but is prevalent in Asia and the Middle East.

The pathogenic potential of organisms associated with different forms of periodontal disease has been assessed in gnotobiotic animals. Not all organisms elicit disease. Gram-positive species, including *Actinomyces* and *S. mutans,* accumulate in large masses and lead to alveolar bone destruction and carious lesions on the root surfaces (Fig. 44-5). Gram-negative organisms, including *Capnocytophaga, E. corrodens, F. nucleatum, B. melaninogenicus,* and *S. sputigena,* do not form massive accumulations or carious root lesions, but they lead to extensive alveolar bone destruction characterized by large numbers of osteoclasts (Fig. 44-6).

The mechanisms of tissue destruction have not been clearly delineated in any form of human periodontal disease. Hydrolytic enzymes, endotoxins, and other bacterial toxic products have been suggested as the agents that lead to destruction of the intercellular matrix of periodontal connective tissues; they may also interfere with the metabolism of local tissue cells. Destruction may also result from the host's response to the organisms, involving inflammatory or hypersensitivity reactions. Numerous biochemical mediators of tissue destruction have been detected in diseased tissues.

Control of Dental Diseases. The problem of controlling dental infections is unique in that the responsible organisms are widespread in the population and are accompanied by benign indigenous organisms; in addition, the resultant tissue destruction has been incorrectly accepted by much of the population as inevitable. In fact, dental caries and periodontal diseases can be controlled by stringent mechanical removal of the bacterial accumulations on teeth. However, such procedures have not been widely successful because few individuals can be adequately motivated or possess the requisite skill.

The most effective practical measure for the control of dental caries is the administration of fluorides in drinking water or as dietary supplements during the period of tooth formation. Such procedures reduce the incidence of decay by more than half. Topically applied fluorides are less effective. It is thought that systemically administered fluoride results in the formation of fluoroapatite, which resists acid dissolution. Topically applied fluorides may affect bacterial attachment to teeth or may be

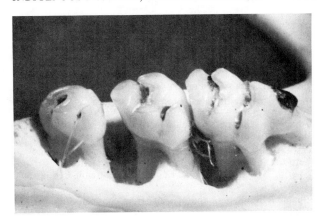

FIG. 44-6. Periodontal disease involving extensive loss of alveolar bone developing in gnotobiotic rat monoinfected with *Eikenella corrodens.* Note the absence of root surface carious lesions in contrast to those which develop in the presence of *Actinomyces* (FIG. 44-5).

directly antibacterial. Reducing sucrose ingestion can also reduce decay, but this is not generally practiced. **Vaccines,** consisting of *S. mutans* cells or glycosyl transferase preparations, have reduced *S. mutans* colonization and dental caries development in experimental animals. However, the effectiveness of such vaccines in human dental decay has not been assessed. Treatment with antibiotics and disinfectant agents is being explored, but to date the total elimination of the organisms has not been achieved, perhaps because of their inaccessibility, in fissures or in carious lesions, to topically or systemically administered agents, and because of their slow growth. In addition, such host defense mechanisms as phagocytosis, inflammatory responses, and desquamation are not operative in dental caries.

The frequent removal of gingival plaque by toothbrushing and using dental floss, or by professional dental personnel, will completely control gingivitis and destructive periodontitis, but these methods have not proven practical. The daily use of powerful antiseptic agents, such as chlorhexidine, can control gingivitis, but the teeth often become stained. Acute gingival infections such as ANUG are controlled by systemic treatment with penicillin and metronidazole, while rapidly destructive forms of periodontitis and periodontosis have responded favorably to tetracycline or spiramycin.

SELECTED READING

BOOKS AND REVIEW ARTICLES

CLARKE RTJ, BAUCHOP T (eds): Microbial Ecology of the Gut. New York, Academic Press, 1977

DRASAR BS, HILL MJ: Human Intestinal Flora. New York, Academic Press, 1974

GORBACH SL: Intestinal microflora. Gastroenterology 60:1110, 1971

Interactions at body surfaces. In Schlessinger D (ed): Microbiology—1975. Washington DC, American Society for Microbiology, 1975, pp 105

NOBLE WC, SOMERVILLE DA: Microbiology of Human Skin. Philadelphia, Saunders, 1974

ROSEBURY T: Microorganisms Indigenous to Man. New York, McGraw-Hill, 1962

SAVAGE DC: Microbial ecology of the gastrointestinal tract. Annu Rev Microbiol 31:107, 1977

SKINNER FA, CARR JG (eds): The Normal Microbial Flora of Man. New York, Academic Press, 1974

SONNENWIRTH AC: Incidence of intestinal anaerobes in blood cultures. In Balows A, DeHaan RM, Dowell VR Jr, Guze LB (eds): Anaerobic Bacteria: Role in Disease. Springfield, IL, CC Thomas, 1974, p 157

TABAQCHALI S: Ecology and metabolic activity of non-sporing anaerobes. In Phillips I, Sussman M (eds): Infection with Non-Sporing Anaerobic Bacteria. Edinburgh, Churchill-Livingstone, 1974

Williams REO: Benefit and mischief from commensal bacteria. J Clin Pathol 26:811, 1973

SPECIFIC ARTICLES

FINEGOLD SM, ATTEBERY HR, SUTTER VL: Effect of diet on human fecal flora: comparison of Japanese and American diets. Am J Clin Nutr 27:1456, 1974

FINLAND M: Changing ecology of bacterial infections as related to antibacterial therapy. J Infect Dis 122:419, 1970

LILJEMART WF, GIBBONS RJ: Suppression of *C. albicans* by human oral streptococci in gnotobiotic mice. Infect Immun 8:846, 1973

MILLER CP, BOHNHOFF M: Changes in the mouse's enteric microflora associated with enhanced susceptibility to salmonella infection following treatment with streptomycin. J Infect Dis 113:59, 1963

O'GRADY F, VINCE A: Clinical and nutritional significance of intestinal bacterial overgrowth. J Clin Pathol [Suppl] 24 (5):130 1971

PERKNIS RE, KUNDSIN RB, PRATT MV et al: Bacteriology of normal and infected conjunctiva. J Clin Microbiol 1:147, 1975

SELWYN S, ELLIS H: Skin bacteria and skin disinfection reconsidered. Br Med J 1:136, 1972

SPRUNT K, REDMAN W: Evidence suggesting importance of role of interbacterial inhibition in maintaining balance of normal flora. Ann Intern Med 86:579, 1968

TASHJIAN JH, COULAM CB, WASHINGTON JA: Vaginal flora in asymptomatic women. Mayo Clin Proc 51:557, 1976

ORAL MICROBIOLOGY

GIBBONS RJ: Adherence of bacteria to host tissue. Microbiology 1977, Amer Soc for Microbiol, Washington, DC, p 395

GIBBONS RJ, VAN HOUTE J: Dental caries. Annu Rev Med 26:121, 1975

GIBBONS RJ, VAN HOUTE J: Bacterial adherence in oral microbial ecology. Annu Rev Microbiol 29:19, 1975

LOESCHE WJ: Chemotherapy of dental plaque infections. Oral Sci Rev 9:65, 1976

SOCRANSKY SS, CRAWFORD ACF: Recent advances in the microbiology of periodontal disease. In Goldman, Gilmore, Irby, McDonald (eds): Current Therapy in Dentistry, CV Mosby, St. Louis, 1977, pp 3–13

SOCRANSKY SS, MANGANIELLO AD: The oral microbiota of man from birth to senility. J Periodontol 42:485, 1971

chapter

45

FUNGI

GEORGE S. KOBAYASHI

CHARACTERISTICS OF FUNGI

The fungi (L. *fungus,* mushroom) have traditionally been regarded as "plantlike" (see Plants, Animals, and Protists, Ch. 1). Most species grow by continuous extension and branching of twiglike structures. In addition, they are mostly immotile and their cell walls resemble those of plants in thickness and, to some extent, in chemical composition and ultramicroscopic structure.

Fungi grow either as single cells, the **yeasts,** or as multicellular filamentous colonies, the **molds** and **mushrooms.** The multicellular forms have no leaves, stems, or roots, and are thus much less differentiated than higher plants, but they are much more differentiated than bacteria. However, fungi do not possess photosynthetic pigments, and so they are restricted to a saprobic (Gr. *sapros,* rotten, *bios,* life) or a parasitic existence. A single uninucleated cell can yield filamentous multinuclear strands, yeast cells, fruiting bodies with diverse spores, and cells that are differentiated sexually (in many species). Moreover, a few species form remarkable traps and snares for capturing various microscopic creatures.

Fungi are abundant in soil, on vegetation, and in bodies of water, where they live largely on decaying leaves or wood. Their ubiquitous airborne spores are frequently troublesome contaminants of cultures of bacteria and mammalian cells. In fact, it was just such a contaminant in a culture of staphylococci that led to the discovery of penicillin (Ch. 7).

Though we shall be concerned mainly with those few fungi that cause diseases of man, other fungi have had an even more adverse effect on human welfare as causes of plant diseases: for example, the potato blight led to death from starvation in Ireland alone of over 1 million persons in the period 1845–1860. More recently, the turkey industry of England was threatened by a malady finally traced to metabolic byproducts of *Aspergillus flavus* which had contaminated the peanut meal used as feed. These **aflatoxins** (named from *A. flavus*) have been isolated and characterized as highly unsaturated molecules with a coumarin nucleus. They are of great interest because of their direct toxicity and long-term carcinogenic effects.

Nevertheless, the overall effects of fungi on man's condition are more benign than malignant. As scavengers in soil they make an essential contribution to the chemical stability of the biosphere (see Soil Microbiology: Geochemical Cycles, Ch. 1), and their biosynthetic capabilities are used in the industrial production of penicillins, corticosteroids, and numerous organic acids (e.g., citric, oxalic). Moreover, through their role in the production of certain cheeses, bread, and ethanolic beverages, the fungi help provide more than calories to man's food supply.

STRUCTURE AND GROWTH

Molds. The principal element of the growing or vegetative form of a mold is the **hypha** (Gr. *hyphe,* web), a branching tubular structure, about $2\mu m-10\mu m$ in diameter, i.e., much larger than bacteria. As a colony, or **thallus,** grows, its hyphae form a mass of intertwining strands, called the **mycelium** (Gr. *mykes,* mushroom). Hyphae grow by elongation at their tips (apical growth) and by producing side branches.

Those hyphae that penetrate into the medium, where they absorb nutrients, are known collectively as the **vegetative mycelium,** while those that project above the surface of the medium constitute the **aerial mycelium;** since the latter often bear reproductive cells or spores they are also referred to as the **reproductive mycelium.** Most colonies grow at the surface of liquid or solid media as irregular, dry, filamentous mats. Because of the intertwining of the filamentous hyphae the colonies are much more tenacious than those of bacteria. At the center of mycelial colonies the hyphal cells are often necrotic, owing to deprivation of nutrients and oxygen, and perhaps to accumulation of organic acids.

In most species the hyphae are divided by cross-walls, called **septa** (L. *septum,* hedge, partition; Fig. 45-1). However, the septa have fine, central, pores; hence even septate hyphae are **coenocytic,** i.e., their many nuclei are embedded in a continuous mass of cytoplasm.

Yeasts. Yeasts are unicellular oval or spherical cells, usually about $3\mu m-5\mu m$ in diameter. Sometimes yeast cells and their progeny adhere to each other and form chains or "pseudohyphae."

FIG. 45-1. The coenocytic nature of hyphae. Electron micrograph of a longitudinal section through two cells of *Neurospora crassa* partially separated by a septum **(s).** Note the streaming of mitochondria **(m)** through the septal pore **(p).** Other labeled structures are cell wall **(w),** outer frayed coat of the cell wall **(f),** cell membrane **(cm),** nucleus **(N),** nucleolus **(Nu),** nuclear membrane **(nm),** ribosomal particles **(p₁),** and endoplasmic reticulum **(er).** Fixed with OsO₄ and stained with uranyl nitrate. (Shatkin AJ, Tatum EL: J Biophys Biochem Cytol 6:423, 1959) (×47,000, reduced)

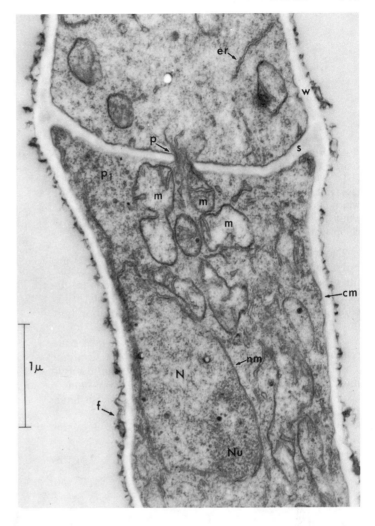

Cytology. Yeasts and molds resemble higher plants and animals in the anatomic complexity of their cells. They are eukaryotic, with several different chromosomes and a well-defined nuclear membrane, and they possess mitochondria and an endoplasmic reticulum (Fig. 45-2). Moreover, their membranes contain sterols, thus resembling higher forms rather than bacteria.

CELL WALL

The cell wall of a fungus, like that of a bacterium, lies immediately external to the limiting cytoplasmic membrane, and in some yeasts it is surrounded by an external capsular polysaccharide (e.g., *Cryptococcus,* Fig. 45-16). However, unlike bacteria, whose cell walls often contain bricklike structural units (see Ch. 2), fungus cell walls appear thatched (Fig. 45-3). Polymers of hexoses and

hexosamines provide the main structural wall elements of fungi. In many molds and yeasts the principal structural macromolecule of the wall is **chitin,** which is made up of *N*-acetyl glucosamine residues. These are linked together by β-1,4-glycosidic bonds, like the glucose residues in cellulose, the main cell wall material in higher plants.

Chitin also makes up the principal structural material of the exoskeleton in crustaceans. Hence this substance represents, at a molecular level, an interesting example of convergent evolution, in which distantly related organisms have independently evolved similar or identical structures to serve similar needs.

In yeasts extraction of the wall with hot alkali leaves an insoluble **glucan,** made up of β-1,6-linked D-glucose residues, with β-1,3-linked branches arising at frequent intervals. (Glucan resembles cellulose in its insolubility and rigidity.) An additional, soluble polysaccharide is **mannan,** an α-1,6-linked polymer of D-mannose with α-1,2 and α-1,3 branches.

CH₂OH ... O ... OH ... NH CO CH₃ — O — CH₃ CO NH ... OH ... CH₂OH Chitin — O — CH₂OH ... OH ... NH CO CH₃ — O — CH₃ CO NH ... OH ... CH₂OH — O ...

As with bacteria (see Polysaccharides, Ch. 4), the polysaccharides of fungi are synthesized from various sugar nucleotides (UDP-*N*-acetyl glucosamine, GDP-mannose, etc.). Chitin synthetase is attached to the cell membrane. It requires a primer in vitro, which must contain at least six or seven residues connected as in chitin. A wall digestion complex in fungi is a selective inhibitor of chitin synthetase.

The walls of a number of yeasts contain complexes of polysaccharide with proteins rich in cystine residues, and the reversible reduction of –S–S– bonds has been implicated in the formation of buds. In some yeasts lipids containing phosphorus and nitrogen are also abundant in the wall (up to 10% of dry weight).

Fungus cell walls can be digested by enzymes contained in the digestive juices of the snail *Helix pomatia.* These juices contain over 30 recognized enzymatic activities, including glucanase, chitinase, and mannanase. Bacteria that produce lytic enzymes for these walls have

also been isolated from soil samples, by application of the enrichment technic (see Genotypic Adaptation, Ch. 3) with purified cell walls as the carbon source.

As with bacteria, digestion of the walls of yeasts or molds in hypertonic solution yields viable **protoplasts.** Protoplasts are also produced by growth in media, or by mutations, that inhibit cell wall synthesis (Fig. 45-4). Yeast protoplasts are deficient in some of the hydrolases of the intact cells, such as invertase and β-fructosidase. These observations suggest that as in bacteria (see Ch. 6) secreted enzymes are located in the periplasmic space.

METABOLISM

All fungi are heterotrophic (see Ch. 3), requiring organic foodstuffs, and most are obligate aerobes. Some, however, are facultative; but none are obligate anaerobes.

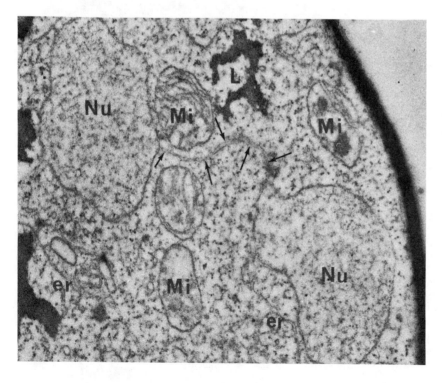

FIG. 45-2. Cytoplasmic organelles in a yeast cell of *Blastomyces dermatitidis,* showing two nuclei **(Nu),** each surrounded by a nuclear membrane. Also seen are mitochondria **(Mi),** vesicles and ribosomes of endoplasmic reticulum **(er),** and lipid droplets **(L). Arrows** point to membranous connections between nuclei. Osmium-fixed thin sections. (Edwards GA, Edwards ME: Am J Bot 47:622, 1960) (×35,000, reduced)

FIG. 45-3. Microfibrillar structure of the wall of a species of phycomycete. Chemical analysis and x-ray diffraction showed the thatched fibrils to be chitin. Electron micrograph shadowed with palladium-gold (Pd-Au) (Aronson JM, Preston RD: Proc R Soc Lond (Biol) 152:346, 1969) (×30,000, reduced)

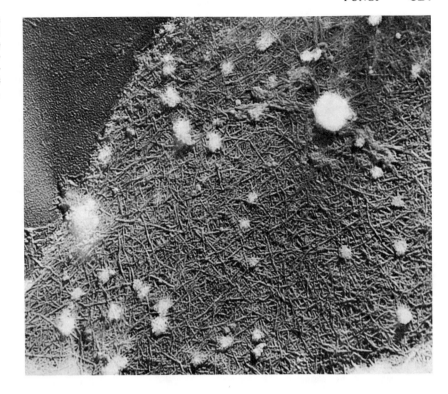

FIG. 45-4. Protoplast formation in hypertonic media by a mutant of *Neurospora crassa* with defective synthesis of cell wall. Cultures were grown in minimal medium with sorbose 0% **(A),** 5% **(B),** or 10% **(C** and **D).** Branching hyphae with septa are evident in **A** and **B.** Protoplasts with occasional cell wall fragments are present in **C** and **D.** Similar results were obtained with equimolar concentrations of glucose, sucrose, and fructose. (Hamilton JG, Calvet J: J Bacteriol 88:1084, 1964)

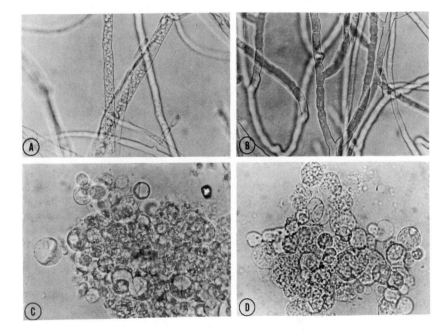

Except for the absence of autotrophs or obligate anaerobes, the fungi as a group exhibit almost as great a diversity of metabolic capabilities as the bacteria. Many species can grow in minimal media, given an organic carbon source and nitrogen as NH_4^+ or NO_3^-. Thermophilic species can grow at temperatures as high as $50°$ C and above; some can flourish in the high-salt media of cured meats, and others in highly acidic media. Some fungi can hydrolyze such complex organic substances as wood, bone, tanned leather, chitin, waxes, and even synthetic plastics.

Fungi can be induced by appropriate substrates and analogs to form the corresponding degradative enzymes, and their regulatory mechanisms for controlling enzyme synthesis and activity appear similar to those in bacteria, but differ in some interesting details.

Like bacteria, yeasts show induction, repression, and catabolite repression; in fact, the glucose effect on enzyme induction (Ch. 14) was first observed in yeast, at the turn of the century. In contrast to the case with bacteria, however, the structural genes for a given metabolic pathway (e.g., histidine synthesis) are scattered over the genome, and operator genes and operons have, accordingly, been difficult to identify. Since fungi are eukaryotic and yet can be handled with almost as much ease as bacteria, they are attractive for extending knowledge of molecular genetics and regulatory mechanisms to higher forms.

REPRODUCTION

In addition to growing by apical extension and branching, fungi reproduce by means of sexual and asexual cycles, and also by a parasexual process. We shall consider asexual reproduction and the parasexual process in particular detail, since the vast majority of fungi that are pathogenic for man lack sexuality.

ASEXUAL REPRODUCTION

The vegetative **growth** of a coenocytic mycelium involves nuclear division without cell division, the classic process of mitosis ensuring transmission of a full complement of chromosomes to each daughter nucleus. The further step of cell division leads to asexual (vegetative) **reproduction,** i.e., the formation of a new clone without involvement of gametes and without nuclear fusion. Three mechanisms are known: 1) sporulation, followed by germination of the spores; 2) budding; and 3) fragmentation of hyphae.

Asexual spores are classified in Table 45-1. They are sometimes all referred to as **conidia,** but more often this term is reserved, as in this chapter, for those spores that form by a process akin to budding at tips of specialized hyphae, called **conidiophores.**

Other asexual spores (**chlamydospores** and **arthrospores**) develop **within hyphae.** The spores germinate when planted in a congenial medium, and, if destined to become a mold, they send out one or more germ tubes (see Fig. 45-11) which elongate into hyphae. **Chlamydospores,** which can be formed by many fungi, are thick-walled and unusually resistant to heat and drying; they are probably formed, like bacterial spores (Ch. 6), by true endosporulation, and they similarly promote survival in unfavorable environments. In contrast, arthrospores and conidia are not unusually resistant; they probably function to promote aerial dissemination.

Spores vary greatly in color, size, and shape; they may contain more than one nucleus. Their morphology and mode of origin constitute the main basis for classifying fungi that lack sexuality. Some species produce only one kind of spore and others as many as four. Various common asexual spores are illustrated in Figures 45-5 and 45-6. They should not be confused with sexual spores (see Sexual Reproduction).

TABLE 45-1. Asexual Spores Formed by Certain Fungi

Conidia (Gr. *konis*, dust)	This term is sometimes used generically for all asexual spores, sometimes more specifically for spores borne singly or at tips of specialized hypal branches (**conidiophores**). Highly diversified in shapes, size, color, and septation.
Aleuriospores (Gr. *aleuron*, wheaten flour)	Spores that resemble conidia but develop on short lateral branches or directly on the hyphae, rather than on specialized condiophores.
Arthrospores (Gr. *arthron*, joint)	Cylindrical cells formed by double septation of hyphae, individual spores are released by fragmentation of hyphae, i.e., by disjunction.
Blastospores (Gr. *blastos*, bud, shoot)	Buds that arise from yeastlike cells.
Chlamydospores (Gr. *chlamy*, mantle)	Thick-walled, round spores formed from terminal or intercalated hyphal cells.
Sporangiospores (Gr. *angeion*, vessel)	Spores within saclike structures (**sporangia**) at ends of hyphae or of special hyphal branches (**sporangiophores**). Characteristically formed by species of *Phycomyces*.

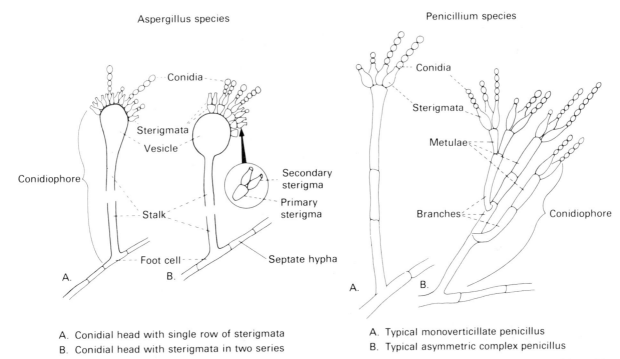

Aspergillus species

Penicillium species

Conidia

Sterigmata

Vesicle

Conidiophore

Secondary sterigma

Primary sterigma

Stalk

Foot cell

Septate hypha

A.

B.

Conidia

Sterigmata

Metulae

Branches

Conidiophore

A.

B.

A. Conidial head with single row of sterigmata
B. Conidial head with sterigmata in two series

A. Typical monoverticillate penicillus
B. Typical asymmetric complex penicillus

FIG. 45-5. Diagram of some representative filamentous fungi and their asexual spores (conidia). (Ajello L et al: Laboratory Manual for Medical Mycology, USPHS Publ. No. 994. Washington, D.C., GPO, 1963)

Budding is the prevailing asexual reproductive process in **yeasts,** though some species divide by fission **(fission yeasts).** Whereas in fission (the usual reproductive process of almost all bacteria) a parent cell divides into two daughter cells of essentially equal size, in budding the daughter cell is initially much smaller than the mother cell. As the bud bulges out from the mother cell the nucleus of the latter divides, and one nucleus passes into the bud; cell wall material is then laid down between bud and mother cell, and the bud eventually breaks away (Fig. 45-7). A **birth scar** on the daughter cell's wall, and a **budding scar** on the mother's wall, are visible in electron micrographs (Fig. 45-8). As a result of repeated budding old yeast cells bear many budding scars, but they have only a single birth scar.

Fragments of hyphae (e.g., formed by teasing a mycelium) are also capable of forming new colonies. This capacity is often exploited in the cultivation of fungi, but it is probably not important in nature.

SEXUAL REPRODUCTION

Fungi that carry out sexual reproduction go through the following steps: 1) A haploid nucleus of a donor (male) cell penetrates the cytoplasm of a recipient (female) cell. 2) The male and female nuclei fuse to form a diploid zygotic nucleus. 3) By meiosis the diploid nucleus gives rise

to four haploid nuclei, some of which may be genetic recombinants. In most species the haploid condition is the one associated with prolonged vegetative growth, and the diploid state is transient; but in other species, as in higher animals, the opposite is true.

In **homothallic** species the cells of a single colony (arising from a single nucleus) can engage in sexual reproduction. In some homothallic species (hermaphrodites) male and female cells are anatomically differentiated, but in others they are indistinguishable. In **heterothallic** species the cells that engage in sexual reproduction must arise from two different colonies, of opposite mating type. The reproductive cells of some heterothallic species are anatomically distinguishable as male and female, while those of others are only functionally differentiated into sexually compatible mating types. Among fungi with a sexual stage the anatomy of sex organs and the mating procedures are characteristic for any particular species; hence they are important for taxonomy.

Sexual Cycle in Neurospora. The sexual process in some fungi has played so vital a role in the development of biochemical genetics, as in the classic studies by Beadle and Tatum of *Neurospora crassa,* that a brief description of a representative cycle is warranted.

Neurospora contains in its nucleus seven different chromosomes, each a single copy (i.e., the vegetative organism is haploid). The haploid state is maintained during mycelial growth and during asexual reproduction (i.e., the formation of conidia). Sexual reproduction occurs when two cells (hyphae or conidia) of different mating type fuse to form a **dikaryotic** cell; the two kinds of haploid nuclei coexist in the same cytoplasm and for a time divide more or

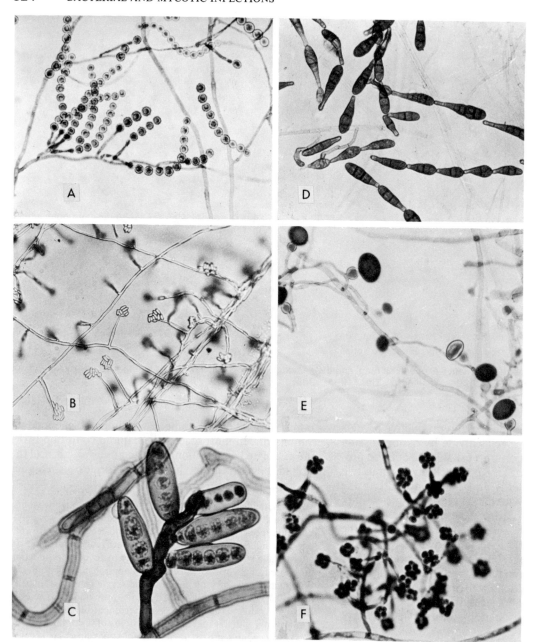

FIG. 45-6. Slide cultures of some representative molds showing diverse forms of asexual spores (conidia), which aid in identifying species, particularly with members of the class Fungi imperfecti (Deuteromycetes). (Ajello L et al: Laboratory Manual for Medical Mycology, USPHS Publ. No. 994. Washington, D.C., GPO, 1963)

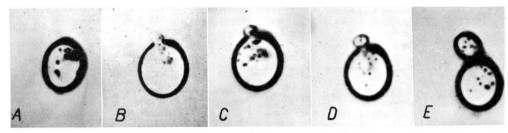

FIG. 45-7. Budding in a yeast cell (*Saccharomyces cerevisiae*). Wall-less bud in **B** was extruded in the 20-sec interval between the photos in **A** and **B**. Subsequent photos were taken at approximately 15-min intervals. Bud in **E** is nearly mature. (Nickerson WJ: Bacteriol Rev 27:305, 1963)

less in synchrony. If a cell initiates ascospore formation, however, two different haploid nuclei fuse and form a diploid nucleus, containing pairs of **homologous** chromosomes (Fig. 45-9).

The diploid cell then initiates the process of **meiosis**. The homologous chromosomes pair **(synapse)** with each other (i.e., assume parallel, closely adjacent positions), and each chromosome divides without duplicating its centromere (whose attachment to a spindle fiber subsequently guides its migration to one pole or the other during anaphase). Each chromosome thus becomes a pair of identical **chromatids** connected by a centromere (a bivalent), and each pair of homologous bivalents constitutes a **tetrad**. One or more exchanges of parts **(crossing over)** may then occur at random among the four chromatids of the tetrad, resulting in **genetic recombination**.

Two meiotic divisions follow. In the first there are no divided centromeres to separate (as in mitosis); instead, one member of each pair of homologous chromosomes (in the form of a bivalent) is drawn after its centromere to each daughter nucleus. In the second meiotic division the chromosome is already divided into two chromatids, and only the centromere divides; one product then migrates (as in mitosis) to each pole, drawing its chromosome with it. Thus the four chromatids of each tetrad are distributed to four different cells, each of which ends up with a haploid set of chromosomes. The individual members of each set are thus derived at random by segregation from either parent, and are further scrambled by the genetic recombination occurring at the tetrad stage (Fig. 45-9).

One further feature of this process is that the four haploid cells resulting from the meiosis remain together in the same sac **(ascus)**; and in *Neurospora* (though not in all Ascomycetes) before the ascus is fully matured each cell divides mitotically into two identical spores. In the ascus, which is shaped like a narrow pod, the eight resulting **ascospores** are held in a linear array, whose order reflects the meiotic segregation of their chromosomes.

Analysis of the genetic constitution of all four pairs of spores in the same ascus **(tetrad analysis)** allows the most complete possible description of the genetic events occuring during meiosis: the products of reciprocal recombination can be identified, rather than merely deduced (as with higher organisms) from the statistical distribution of genetic markers among the progeny. Moreover, the haploid nature of the vegetative phase eliminates the complicating effect of dominance on the relation between genotype and phenotype. These organisms have been especially valuable in studying the mechanism of genetic recombination (Ch. 11).

PARASEXUAL CYCLE (MITOTIC RECOMBINATION)

Some fungi go through a cycle that imparts some of the biologic advantages of sexuality (e.g., recombination of parental DNA) without involvement of specialized mating types or gametes. This process of parasexuality, first demonstrated by Pontecorvo with *Aspergillus*, involves the following steps (Fig. 45-10): 1) By **hyphal fusion** different haploid nuclei come to coexist in a common cytoplasm. The **heterokaryon** thus formed can be stable, the two sets of nuclei dividing at about the same rate. 2) Rare **nuclear fusion** will yield heterozygous diploid nuclei. These are usually greatly outnumbered by the haploid nuclei, but once formed they tend to divide at about the same rate as the latter; and stable diploid strains may be isolated. 3) Though homologous chromosomes are usually arranged independently on the equatorial plane in the mitosis of a diploid cell, rarely (about 10^{-4} per mitosis) sufficient **somatic pairing** will occur to permit crossing over, as in meiosis.

The result of such mitotic recombination between

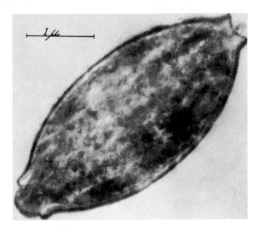

FIG. 45-8. Thin section of an OsO₄-fixed yeast cell (*Saccharomyces cerevisiae*), showing the concave birth scar at one pole and a convex bud scar at the other. (Agar HD, Douglas HC: J Bacteriol 70:247, 1955) (×29,000, reduced)

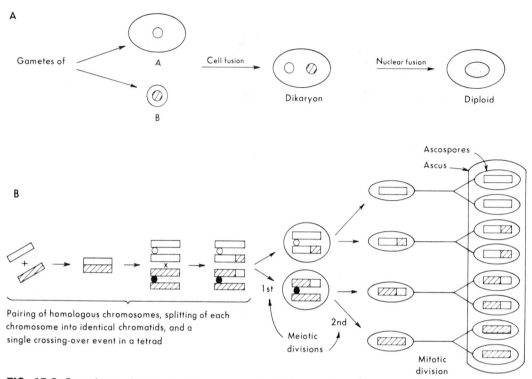

FIG. 45-9. Sexual reproduction in *Neurospora crassa*. **A.** Mating leads to formation of a dikaryon and then eventually, by fusion of nuclei, to a diploid cell. **B.** Meiotic divisions with recombination, followed by mitosis, produce eight haploid ascospores linearly arranged in a narrow pod (ascus).

FIG. 45-10. Somatic pairing and mitotic recombination in the parasexual cycle of fungi. After fusion of genetically different hyphae to form a heterokaryon, the unlike haploid nuclei occasionally fuse to yield the heterozygous diploid nucleus depicted in **A.** Rare somatic pairing **(B)** and recombination give rise, through the daughter nuclei depicted in **C,** to partially homozygous lines of diploid cells, as shown. These occasionally also segregate haploid strains, half of which are recombinant (compared with the original haploid parents) in respect to loci dD and eE. **p** = parental chromosome; **r** = recombinant chromosome.

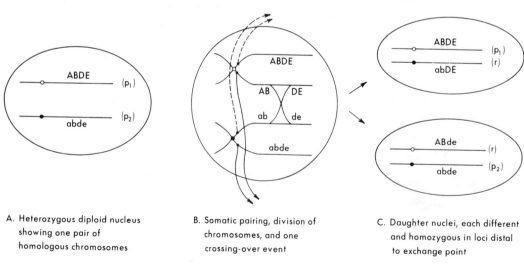

A. Heterozygous diploid nucleus showing one pair of homologous chromosomes

B. Somatic pairing, division of chromosomes, and one crossing-over event

C. Daughter nuclei, each different and homozygous in loci distal to exchange point

TABLE 45-2. Classes of Fungi

Class	Asexual spores	Sexual spores	Mycelia	Representative genera or groups
Phycomycetes	Endogenous (in sacs)	Anatomy variable	Nonseptate	*Rhizopus, Mucor,* watermolds (aquatic)
Ascomycetes	Exogenous (at ends or sides of hyphae)	Ascospores, within sacs or asci	Septate	*Neurospora, Penicillium, Aspergillus,* true yeasts
Basidiomycetes	Exogenous (at ends or sides of hyphae)	Basidiospores, on surface of basidium	Septate	Mushrooms, rusts, smuts
Deuteromycetes (Fungi imperfecti)	Exogenous (at ends or sides of hyphae)	Absent	Septate	Most human pathogens

heterozygous homologous chromosomes is to make the products **homozygous for genes distal to the exchange point.** Thus two diploid daughter cells with different properties result, each homozygous for some alleles for which the diploid parent cell is heterozygous (Figure 45-10). From this figure it can also be projected that when these new diploid strains yield haploid progeny half of these will be parental in genetic composition and half will be recombinant.

Mitotic recombination has provided unique opportunities for genetic analysis of asexual molds. Its possible application to the study of the genetics of somatic diploid cells, such as human cells in tissue culture, is of great interest. With the limited range of genetic markers available thus far in human cells the recombination rate has been low, but sufficient to be encouraging.

Though recombinants are infrequent in a parasexual cycle, compared with a sexual cycle, the process may be significant for the evolution of asexual fungi.

TAXONOMY

The four major classes of true fungi (division Eumycota) are summarized in Table 45-2.

The **Phycomycetes*** include all nonseptate as well as some septate filamentous fungi; their asexual spores are of various kinds, of which the sporangiospores, contained within sacs (sporangia) formed at the end of specialized stalks (sporangiophores), are unique to this class. Different species have different sexual cycles, and those that live in aquatic environments (Gr. *phyco,* seaweed) have **flagellated gametes.** The flagella resemble in structure the cilia of protozoa or higher animals, rather than bacterial flagella (Ch. 2).

The **Ascomycetes** are distinguished from other fungi by the

* The class Phycomycetes is not recognized by some taxonomists, who would separate these organisms into two subdivisions based upon the production of motile or nonmotile spores.

ascus, a saclike structure containing sexual spores (**ascospores).** The ascospores are the endproduct of mating, fusion of male and female nuclei, two meiotic divisions, and usually one final mitotic division, as described above for *Neurospora.* There are thus usually eight ascospores in an ascus. The yeasts are also ascomycetes, though they generally do not grow as molds (see Dimorphism).

The **Basidiomycetes** are distinguished by their sexual spores, called **basidiospores,** which form on the surface of a specialized structure, the **basidium.** They include the edible mushrooms. Basidiomycetes cause serious diseases of plants but no infectious diseases of man. However, some species synthesize toxic alkaloids, which may be lethal in man and are often of pharmacologic interest (e.g., ergotamine, muscarine).

The **Deuteromycetes (Fungi Imperfecti)** are particularly important for medicine, as they include the vast majority of human pathogens. Because no sexual phase has been observed they are often referred to as imperfect fungi. The hyphae are septate, and conidial forms are very similar to those of the ascomycetes; they have therefore long been suspected of being **special ascomycetes** whose sexual phase is either extremely infrequent or has disappeared in evolution. Indeed, typical ascomycetous sexual stages have recently been observed in several species of this class (Table 45-3). Superficially, the reversible change from sexual to asexual form resembles phase variations in bacteria (Ch. 8), suggesting that the imperfect fungi are mutants in genes that specify sexual development. Indeed, mutations in such a gene have revealed that the classic species *Microsporum gypseum* includes the imperfect stage of two different species of ascomycetes (Table 45-3).

DIMORPHISM

Some species of fungi grow only as molds, and others only as yeasts. Many species, however, can grow in either form, depending on the environment. This capacity is

TABLE 45-3. Some Pathogenic Imperfect Fungi Discovered to Have a Sexual (Perfect) Stage

Name of imperfect species	Perfect form
Microsporum gypseum	{ Nannizia incurvata { N. gypsea
M. fulvum	N. fulva
M. nanum	N. obtusa
M. cookei	N. cajetana
M. vanbreuseghemii	N. grubia
M. canis	N. otae
M. persicolor	N. persicolor
Trichophyton mentagrophytes	Arthroderma benhamiae
T. simii	A. simii
T. ajelloi	A. uncinatum
T. simii	A. simii

The species listed are dermatophytes: they infect the epidermis, nails, and hair of mammals. Only the asexual (imperfect) form is found in infected skin. Conversion to the sexual (perfect) form was facilitated by growth on sterilized soil enriched with keratin (e.g., hair, feathers). The perfect stages were then identified as Ascomycetes by observing production of fruiting bodies when compatible "imperfect" forms were mated (e.g., + and – strains). (A fruiting body contains many asci with their enclosed ascospores.)

known as **dimorphism**.* It is important clinically, since most of the more pathogenic fungi (in man) are dimorphic: they usually appear in infected tissues as yeastlike cells but in cultures as typical molds (Figs. 45-11 and 45-12).

Dimorphism can be experimentally controlled by modifying cultural conditions, a single factor sometimes being decisive. For example, the human pathogen *Blastomyces dermatitidis* grows as a mycelium at 25° C but as yeast cells at 37°. In general the mycelial forms, leading to aerial dissemination of asexual spores, are a response to unfavorable conditions, while yeast forms are favored by rich nutrition.

COMPARISON WITH BACTERIA

Fungi resemble bacteria in many respects: their role in maintaining the geochemical stability of the biosphere, the methods used for their isolation and cultivation, their capacity to cause infectious diseases, and the applications of their fermentations to industrial processes. Fungi differ greatly from bacteria, however (Table 45-4), in their reproductive processes, their growth characteristics (budding; branching filaments), their size, the greater complexity of their cytoplasmic architecture, their somewhat less diversified range of metabolic activities, the composition and ultrastructure of their cell walls, and the

* The alternatives are not limited to yeast and mold forms. One of the most virulent human pathogens, *Coccidioides immitis* (discussed later in this chapter), is dimorphic, growing in infected tissues as spherules and in conventional cultures as a mold.

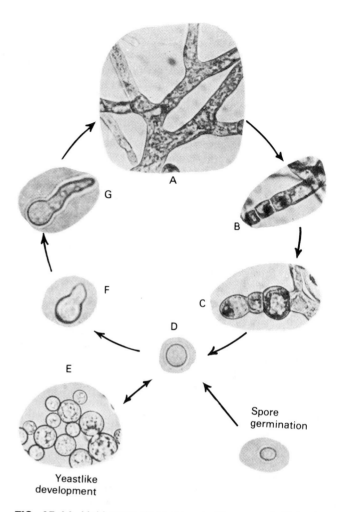

FIG. 45-11. Mold–yeast dimorphism in *Mucor rouxii*. Note absence of septa, typical of phycomycetes, in hyphae of the mold at **A.** Arthrospores are being formed in **B** and **C.** At **D** an isolated arthrospore is shown developing into yeastlike cells **(E),** or into a mold **(A)** by outgrowth of a filamentous tube **(F** and **G).** (Bartnicki-Garcia S: Bacteriol Rev 27:293, 1963)

chemotherapeutic agents to which they are susceptible. As we shall see, the human diseases caused by fungi are much less common and less varied than those caused by bacteria.

Genetic, biochemical, and antigenic differences have been studied much less among fungi than among bacteria. And although modern biochemical genetics had its inception in the classic studies of Beadle and Tatum with the mold *Neurospora crassa,* only recently have the mating types for three systemic pathogenic fungi for man been discovered, viz., *Blastomyces dermatitidis, Histoplasma capsulatum* and *Cryptococcus neoformans.* Hence the genetics of the medically important fungi remains virtually unexplored.

FIG. 45-12. Mold–yeast dimorphism in *Mucor rouxii*, growing in yeast extract–peptone–glucose medium. Gaseous phase for incubation is shown at left. **A.** Filamentous growth in submerged culture. **B.** Surface growth with active spore formation. **C.** Filamentous growth at low concentration of glucose (about 2%). **D.** Arthrospore formation stimulated by high concentration of glucose (10%). **E.** Yeastlike cells. **F.** Inhibition of yeastlike growth by a chelating agent (diethylene-triamino-pentaacetic acid). All the forms shown can develop under appropriate conditions from a single uninucleated cell (Fig. 45-11). (Bartnicki-Garcia S: Bacteriol Rev 27:293, 1963)

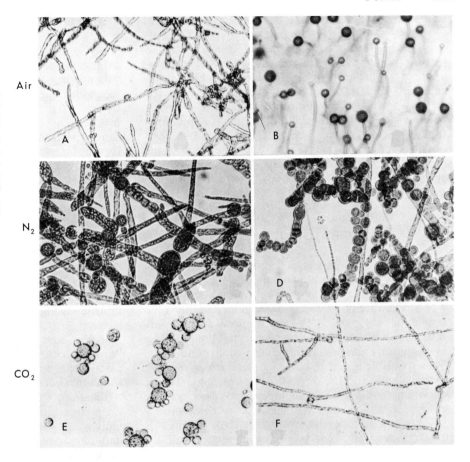

TABLE 45-4. Contrast Between Fungi and Bacteria

Property	Fungi	Bacteria
Cell volume (μ^3)	Yeast 20–50 Molds: Not definable because of indefinite size and shape and coenocytic form; but much greater than yeast	1–5
Nucleus	Eukaryotic (well-defined membrane)	Prokaryotic (no membrane)
Cytoplasm	Mitochondria, endoplasmic reticulum	No mitochondria or endoplasmic reticulum
Cytoplasmic membrane	Sterols present	Sterols absent (except for *Mycoplasma* grown on sterols)
Cell wall	Glucans; mannans; chitin, glucan- and mannan-protein complexes No muramic acid peptides, teichoic acids, or diaminopimelic acid	Muramic acid peptides; teichoic acids; some have diaminopimelic acids residues No chitin, glucans, or mannans
Metabolism	Heterotrophic, aerobic, facultative anaerobes; no known autotrophs or obligate anaerobes	Obligate and facultative aerobes and anaerobes; heterotrophic and autotrophic
Sensitivity to chemotherapeutic agents	Sensitive to polyenes and griseofulvin (dermatophytes); not sensitive to sulfonamides, penicillins, tetracyclines, chloramphenicol, streptomycin	Often sensitive to penicillins, tetracyclines, chloramphenicol, streptomycin; not sensitive to griseofulvin or polyenes
Dimorphism	A distinguishing feature of many species	Absent

GENERAL CHARACTERISTICS OF FUNGOUS DISEASES

Because fungi are larger than bacteria they were recognized earlier as agents of disease. However, of the estimated 50,000–200,000 species of fungi, only about 100 are known to cause infectious disease (**mycoses**) in man. A few of these, with a special predilection for epidermal structures, seem to depend on parasitic growth in animal tissue, some only in man. For all the other pathogenic species, however, such infection is only incidental to growth in soil or plants, and it is of no ecologic importance to the microbe.

TYPES OF MYCOSES

It is useful to divide the mycoses into four groups, differing in the level of the infected tissue. 1) The **systemic or deep mycoses** primarily involve internal organs and viscera. They are often widely disseminated and involve many different tissues. 2) The **subcutaneous mycoses** involve skin, subcutaneous tissue, fascia, and bone. 3) The **cutaneous mycoses** involve epidermis, hair, and nails. The responsible fungi are known as **dermatophytes** (Gr. *phyton,* plant), and the diseases as **dermatophytoses** or **dermatomycoses.** 4) The **superficial mycoses** involve only hair and the most superficial layer of epidermis.

The natural history of the diseases in these groups will now be summarized, and the following section will describe some representatives of each group.

The **systemic or deep mycoses** are caused by saprobic fungi in soil, through **inhalation of spores.** As will become evident, these infections resemble clinically the chronic bacterial infections due to mycobacteria (Ch. 37) and to actinomycetes (Ch. 38). The earliest lesions are usually pulmonary, and the initial acute, self-limited pneumonitis is easily overlooked or ascribed to bacteria or viruses. The subsequent chronic form (which is usually much less frequent) begins insidiously, progresses slowly, and is charcterized by suppurative or granulomatous lesions. These sometimes form pulmonary cavities and often spread by direct extension, e.g., into contiguous soft tissues such as the pleurae. These fungi are also prone to spread by way of the blood stream, yielding metastatic abscesses or granulomas in almost any organ including the skin. Prior to the development of effective antifungal chemotherapy these disseminated mycotic diseases were almost invariably fatal. They are not contagious.

The **subcutaneous mycoses** are also caused by saprobes in soil and on vegetation. Infection occurs by **direct implantation of spores or mycelial fragments,** commonly in scratches caused by thorns. Hence these diseases tend to be especially prevalent in rural and tropical regions, e.g., in jungle terrain. The diseases begin insidiously, progress slowly, and are characterized by localized subcutaneous abscesses and granulomas that spread by direct extension, often breaking through the skin surface to form chronic, draining, ulcerated and crusted lesions. Extension may also occur via lymphatics, leading to suppurative, granulomatous lesions in the regional chain of lymph nodes. These diseases are often extremely disfiguring and not infrequently fatal, though dissemination to the viscera is rare.

Those localized subcutaneous abscesses that are particularly invasive and destructive of soft tissues, fascia, and bone are known as **mycetomas.** They are characterized by burrowing, tortuous sinus tracts that open onto the skin surface. The purulent discharge and the abscesses frequently contain "granules," which are bits of colonies of the responsible microorganism. Clinically indistinguishable abscesses are also caused by certain bacteria, e.g., by various Actinomycetales (species of *Nocardia* and *Streptomyces,* Ch. 38). The latter are known as **actinomycotic mycetomas** to distinguish them from those due to fungi. (The fungal mycetomas are sometimes referred to as **maduromycoses.**)

The fungi that cause **cutaneous mycoses** have a striking predilection for growth in epidermis, hair, and nails. Only a few pathogenic species have been found in soil, and only one (*Microsporum gypseum*) with any frequency. The many other dermatophytes are found only in mammalian skin; they appear to be obligate parasites of man and other animals, transmitted by direct contact or by bits of infected hair or desquamated epidermal scales. The diseases they cause tend to be chronic, and the inflammatory response, mostly confined to the skin at the site of infection, is not especially destructive.

The dermatophytes acquired by man from contaminated soil (*M. gypseum*), or from animals (e.g., *M. canis* from dogs and cats), tend to produce relatively intense but transitory inflammatory lesions in human skin, while those that are indigenous in man, and apparently are obligatory parasites, usually evoke only trivial reactions. It would appear that man is immunologically more tolerant of his indigenous dermatophytes than of alien species.

The fungi that cause **superficial mycoses** are localized along hair shafts and in the more superficial, nonviable, cornified epidermal cells. The pathologic lesions are of minor importance.

Certain widespread fungal saprobes almost never establish infections in healthy humans but can cause serious illness in those with various conditions that lower their resistance. The diseases caused by these **opportunistic** fungi may be widely disseminated, or they may be localized in the respiratory tract or the mucous membranes and skin.

DIAGNOSIS

The fungal origin of a disease is usually first suspected on the basis of its clinical behavior and the appearance of the lesion. As with bacteria that are abundant in man's environment, serologic reactions with Ags of the suspected fungus, and delayed allergic skin reactions to intradermal injection of fungal Ags, may provide supporting evidence but it is not usually conclusive. The most convincing diagnostic evidence is usually provided by **detection of the fungus** in lesions and exudates by direct microscopic examination and by isolation and cultivation. With some fungous diseases transmission of the infection to experimental animals by inoculation of tissue suspensions or exudates facilitates isolation, and the lesions in the test animal may also aid in identification (e.g., *Histoplasma capsulatum* in mice, and *Coccidioides*

immitis in mice and guinea pigs; see below). In addition, animal inoculation serves to distinguish between pathogenic and nonpathogenic (saprobic) fungi, which can be identical in colonial and cellular morphology; the saprobes are usually innocuous in test animals.

Visualization of Fungi in Tissues. In one of the simplest procedures, scrapings of the lesion, or bits of exudate (e.g., sputum, pus), are warmed on a slide in 10% NaOH or KOH (see, for example, Fig. 45-27). Proteins, fats, and many polysaccharides are extensively solubilized (hydrolyzed) and the tissues become optically clear; but the cell walls of most fungi remain largely intact and visible, because of their alkali-resistant glucans. With some fungi, particularly small yeast cells, visualization of the alkali-resistant cell wall is aided by warming tissue in a mixture of NaOH or KOH and a suitable dye.

One of the most widely used staining procedures is based on the periodic acid–Schiff reaction (PAS stain). As is shown in Figure 45-13, periodate cleaves vicinal

FIG. 45-13. The periodic acid–Schiff (PAS) stain. Fuchsin leucosulfonate (Schiff's reagent) reacts with the aldehyde groups generated by periodate cleavage between vicinal hydroxyls of a polysaccharide, forming a red or magenta quinonoid dye. If the polysaccharide had the sugar residue linked 1,6, rather than 1,4 as shown, a further attack by periodate would have split out C-3 as formic acid, and left the two aldehyde groups on C-2 and C-4.

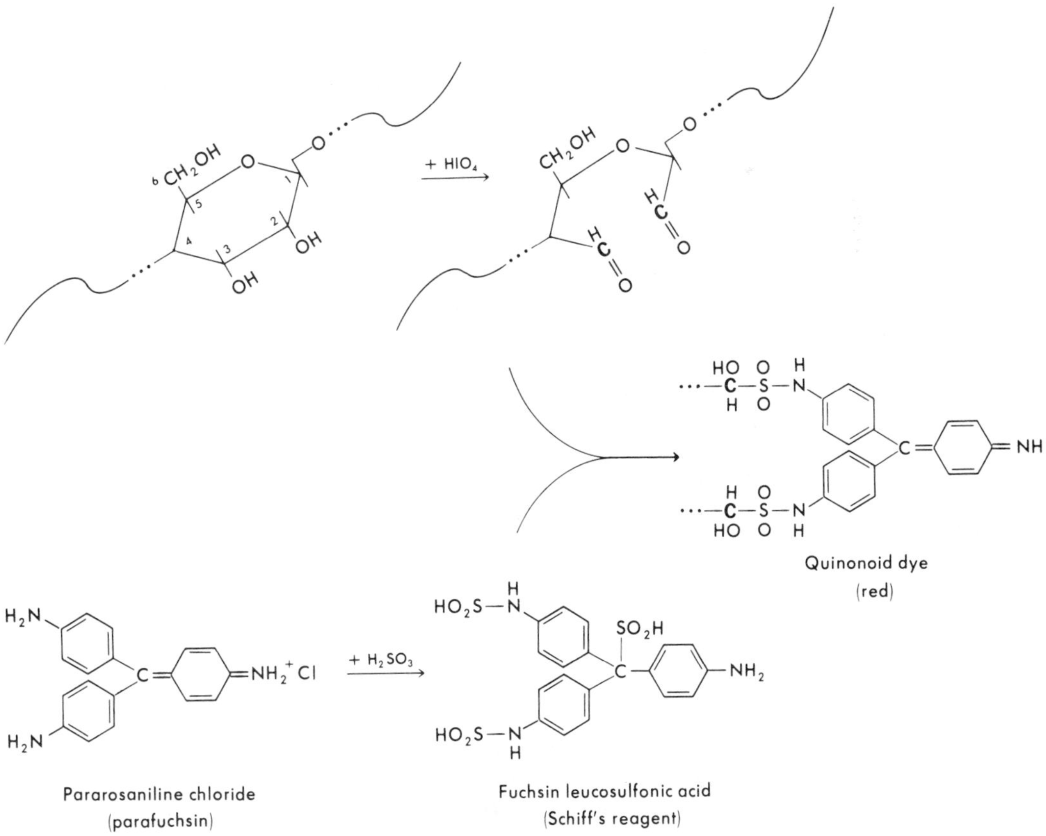

Pararosaniline chloride
(parafuchsin)

Fuchsin leucosulfonic acid
(Schiff's reagent)

Quinonoid dye
(red)

hydroxyl groups and forms dialdehydes; subsequent re-action of the aldehydes with fuchsin leukosulfonate (Schiff's reagent) forms colorful quinonoid dyes. Nearly all fungus walls are stained an intense red or magenta by this reaction, because a large number of aldehyde groups is produced on periodate oxidation of their insoluble glucans and mannans. Chitin, however, does not stain, owing to the absence of vicinal hydroxyls (see Cell Wall, above). The reaction is not, of course, specific for fungi, and some tissue polysaccharides (glycogen, hyaluronic acid) are also stained, though not as intensely.

Cultivation of Fungi. Under optimal conditions fungi grow much more slowly than bacteria. Cultures must therefore be maintained for prolonged periods, and it is essential to inhibit growth of bacterial contaminants, e.g., by drugs, or by maintaining low pH and low tempera-tures (25° C), which inhibit bacteria more than fungi. (For this reason contaminating fungi often overgrow bacterial cultures stored in the refrigerator.)

Sabouraud's agar is the most widely used medium for cultivating fungi. Devised by Sabouraud (a nineteenth century dermatologist) for dermatophytes, it is useful for virtually all fungi.

With glucose and peptone as the sole nutrients, it was originally devised with a pH of 5 to discourage bacterial growth. At present the medium is adjusted to neutral pH (which is more favorable for the growth of most fungi), and chloramphenicol (40µg/ml) and cycloheximide (500µg/ml) are added to reduce the growth of bacte-ria and saprobic fungi. Cycloheximide, however, inhibits *Crypto-coccus neoformans* and some species of *Candida*, and chloram-phenicol inhibits the yeast forms of some dimorphic fungi, so cultures are often prepared both with and without these drugs.

The anatomy of fungal spores and the manner in which they are produced are important determinative characteristics. With dimorphic forms, which generally grow in infected tissues or in rich media as yeasts, sporulation can be stimulated by cultivation under suboptimal nutritional conditions (e.g., rice medium, cornmeal agar). Mycelia are usually white and fluffy at first and then become colored as they develop pigmented spores.

Intact colonies may be examined directly under a dis-secting microscope, or fragments of colonies may first be teased gently in media containing a dye such as Poirrier's blue, lactic acid, and phenol (**lactophenol cotton blue**). Fungi are gram-positive.

CHEMOTHERAPY

In view of the substantial physiologic differences be-tween fungi and bacteria it is not surprising that they re-spond to different drugs. Effective antifungal chemother-

apy began late, with the development of griseofulvin and the polyene agents (Fig. 45-14).

Griseofulvin. Griseofulvin, synthesized by several species of *Penicillium*, produces hyphal distortions in cultures of dermatophytes and other fungi. However, its use in the treatment of dermatophytoses was delayed for many years because its direct application to infected skin le-sions had only a limited effect. When given orally over a period of time, however, it is highly effective: it does not sterilize the structures already infected at the start of therapy, but it accumulates in keratinous structures (cornified layer of epidermis, hair, and nails) as they are laid down and renders them resistant to infection. Hence as older, infected structures are desquamated by normal epidermal growth they are slowly replaced with struc-tures that contain a high level of griseofulvin. Treatment must therefore be prolonged (weeks or months), espe-cially with infected nails. Fortunately, griseofulvin has only minimal toxicity, even on prolonged administra-tion. As might be expected from its specialized location, griseofulvin is ineffective in the deep mycoses, but it may have value in one of the subcutaneous mycoses, sporotrichosis.

While little is known about the antifungal mechanism

FIG. 45-14. Structures of antifungal agents. Griseofulvin, first isolated from *Penicillium griseofulvum*, is produced by several species of *Penicillium*. Various polyenes are produced by differ-ent species of *Streptomyces* (e.g., nystatin by *S. noursei*, pimari-cin by *S. natalensis*). Nystatin, and pimaricin (shown here), are tetraenes, i.e., have four alternating double bonds); amphotericin B and candicidin are heptaenes.

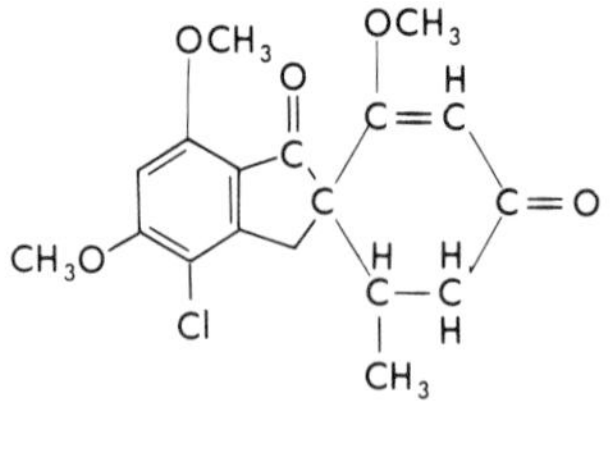

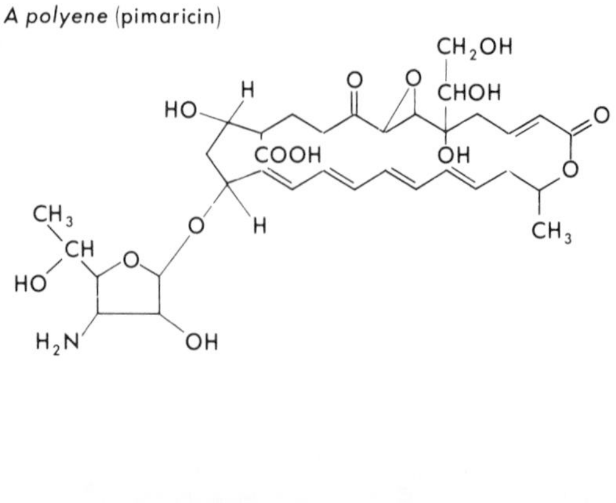

of griseofulvin, in animal cells it causes disorientation of the mitotic spindle and inhibits the movement of chromosomes at anaphase (much like colchicine and vinblastine, which bind to receptors on tubulin subunits). The microtubules in fungal cells therefore may be the site of action of griseofulvin.

Polyenes. The polyene antibiotics (Fig. 45-14) are highly effective in the treatment of many systemic mycoses and ineffective in the superficial and cutaneous mycoses. The most widely used member of the group, **amphotericin B,** must be given intravenously for weeks, and it frequently has toxic side effects. **Nystatin** (named for its discovery in a New York State laboratory) is used topically for *Candida* infections.

The polyenes are fungicidal at sufficiently high concentration in growing cultures.

Bacteria are not susceptible to the polyenes because their cell membranes lack sterols. However, some species of *Mycoplasma* are

a notable exception: when these wall-less bacteria are grown in sterol-containing media they incorporate sterols into their cytoplasmic membrane and become sensitive to polyenes, but when grown in sterol-free media they are resistant.

Red blood cells contain sterols in their membrane, and polyenes cause hemolysis in vitro. Correspondingly, in treatment with polyenes (especially amphotericin B) one of the frequent toxic effects is hemolytic anemia. That the anemia and other cytotoxic effects (e.g., kidney damage) are not more severe and frequent is probably due to the binding of polyenes by serum albumin, which reduces their free concentration in body fluids.

Flucytosine (5-Fluorocytosine). This synthetic compound is effective, taken orally, in the treatment of candidiasis, cryptococcosis, and chromomycosis. Fungi sensitive to this compound (but not man) deaminate it to yield fluorouracil, which is incorporated into RNA. Fungi rapidly develop resistance to flucytosine, but its combined use with amphotericin B appears to be better than the longer treatment required with amphotericin B alone.

SOME PATHOGENIC FUNGI AND FUNGOUS DISEASES

SYSTEMIC MYCOSES

CRYPTOCOCCUS NEOFORMANS (CRYPTOCOCCOSIS)

Fungi of the genus *Cryptococcus* appear as spherical cells that reproduce by budding; they look like true yeasts (ascomycetes), but the recent discovery of the sexual phase has established that this phase of *C. neoformans* is a Basidiomycete, now placed in the genus *Filobasidiella*. When two compatible isolates are crossed on sporulation medium the cells conjugate, hyphae sprout, and these develop the clamp connections and terminal basidiospores characteristic of Basidiomycetes (Fig. 45-15).

In rapidly growing cultures of the yeast phase, the buds separate from their mother cells precociously; hence suspensions often exhibit unusually large variations in cell diameter, from about 4μm to 20μm (Fig. 45-16). All strains produce capsules, some much thicker than the enclosed cells. About 12 species have been defined on the basis of antigenic and morphologic characteristics; only one, *C. neoformans,* is pathogenic in man. *C. neoformans* is readily differentiated from nonpathogenic species by its virulence for mice (LD_{50} about 1000 cells inoculated intracerebrally).

Since *C. neoformans* can produce a filamentous phase when reproducing sexually it is dimorphic. In infected tissues and ordinary cultures, however, it appears only as encapsulated yeast cells (Table 45-5). It grows readily on Sabouraud's agar at 37°C. Urease is produced by all species.

The capsule of *C. neoformans,* its most distinctive feature, is easily visualized in India ink suspensions of the cells (Fig. 45-16). It is composed of a polysaccharide containing xylose, mannose, and glucuronic acid, and it crossreacts with antisera to the capsular polysaccharides of types 2 and 14 pneumococci. Although *C. neoformans* evokes only a feeble immune response in infected humans, hyperimmunized rabbits yield capsule-specific antisera that differentiate four strains, called A, B, C, and D. When suspended in homologous antiserum the cells undergo a quellung reaction like encapsulated bacteria (Fig. 27-2).

Strains differ in the size of their capsules and in their virulence for mice, but there is no consistent relation between these variables. The capsule is highly acidic, and various cationic dyes, such as toluidine blue, stain it metachromatically.

Pathogenesis. The disease caused by *C. neoformans* is called **cryptococcosis.*** Inhalation of yeast cells is assumed to initiate pulmonary infection, with subsequent hematogenous spread to other viscera and the central nervous system, especially in immunosuppressed individuals. It seems likely that silent infections of the lung are common and that only a very small proportion become disseminated.

* In the European literature cryptococcosis is still referred to as **European blastomycosis,** the term blastomycosis formerly being used for any disease due to infection with yeastlike cells. In the United States *C. neoformans* was previously named *Torula histolytica* and the disease was called **torulosis.**

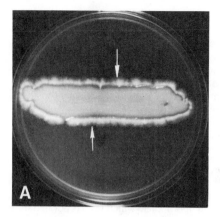

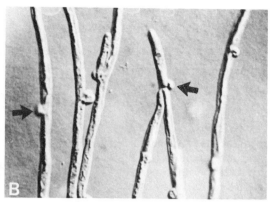

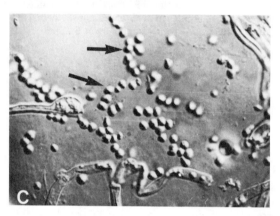

FIG. 45-15. *Filobasidiella neoformans,* the sexual state of *Cryptococcus neoformans.* **A.** Mated culture of yeast cells characterized by the production of dense white mycelium along the margin of yeast growth **(arrows).** (×10) **B.** Hyphae with clamp connections **(arrows).** (×1200) **C.** Terminal basidium from which subglobose to elliptical, finely roughened basidiospores, 1.8μm–2.5μm in diameter **(arrows),** are produced in chains by budding from four points on the apex of the basidium (×1200) (Courtesy of Dr. K. J. Kwon-Chung)

In the chronic, disseminated form of the disease the brain, lungs, other viscera, skin, and bones may be involved. Chronic meningitis is the most frequent pattern and mimics tuberculous meningitis; the lesions may also simulate brain abscess or brain tumor. Pulmonary lesions are usually inapparent clinically but are almost always found at autopsies. In a few individuals chronic pneumonitis is the most conspicuous clinical manifestation.

The microorganism appears in tissues as masses of budding encapsulated yeast cells. There is often little or no surrounding inflammatory reaction, but sometimes granulomatous lesions are formed, with multinuclear giant cells. The yeast cells may be observed within macrophages, particularly when the periodic acid–Schiff stain is used.

Diagnosis. The diagnosis is usually established either by visualizing *C. neoformans* in spinal fluid (the India ink technic being particularly useful, Fig. 45-16) or by cultivating it from spinal fluid, pus, or other exudates. Cycloheximide is inhibitory and should be omitted from the medium. The capsular antigen can be identified in spinal fluid, serum, or urine by means of the precipitin reaction with hyperimmune rabbit serum.

FIG. 45-16. *C. neoformans* suspended in India ink. Note budding, thick capsules, and variations in cell diameter. In fresh spinal fluid, without India ink, the yeast cells are easily mistaken for lymphocytes (×450, reduced)

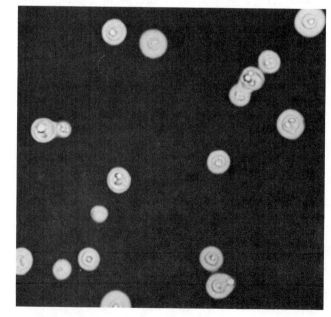

TABLE 45-5. Grouping of Most Frequently Encountered Pathogenic Fungi (for Man) in the United States, with Respect to Tissue Involved and Dimorphism

Type of mycotic disease	Representative fungus	Morphology in		Comment
		Infected tissue	Room temperature culture	
Systemic	*Cryptococcus neoformans*	Yeast (encapsulated)	Yeast (encapsulated)	No dimorphism*
	Coccidioides immitis	Spherules	Mycelia	Dimorphism
	Histoplasma capsulatum	Yeast	Mycelia	Dimorphism
	Blastomyces dermatitidis	Yeast	Mycelia	Dimorphism
Systemic, and particularly opportunistic	*Candida* (especially *C. albicans*)	Yeast and hyphae	Yeast and hyphae	Dimorphism
	Aspergillus (most often *A. fumigatus*)	Mycelia	Mycelia	No dimorphism
	Phycomycetes (*Mucor, Rhizopus* species)	Mycelia	Mycelia	No dimorphism
Subcutaneous	*Sporothrix schenckii*	Yeast	Mycelia	Dimorphism
Cutaneous	*Microsporum* species	Mycelia	Mycelia	No dimorphism†
	Trichophyton species	Mycelia	Mycelia	No dimorphism†
	Epidermophyton floccosum	Mycelia	Mycelia	No dimorphism†

*Except during sexual phase.

†The fungi that parasitize epidermis, nails, and hair (dermatophytes) all appear alike in infected skin, but in culture they develop a variety of specialized hyphae and spore structures that differentiate diverse genera and species; in a sense, they do exhibit a certain amount of dimorphism.

Therapy. Cryptococcosis involving the central nervous system, with or without disseminated visceral lesions, formerly was invariably fatal, but amphotericin B has cured the infection even when first administered in an extremely ill person.

Epidemiology. Cryptococcosis occurs sporadically throughout the world. The organism has been isolated from soil, particularly soil enriched with pigeon droppings. It is also found in pigeon roosts and nests far removed from the soil, e.g., on window ledges and towers of urban buildings. Small outbreaks of acute pulmonary infection have occurred among workers demolishing old buildings.

Since the fungus remains viable in dried material for many months, contaminated materials are a potent source of airborne infection. Pigeon droppings are often highly contaminated: 5×10^7 viable organisms per gram were found in one study. These organisms are evidently due not to intestinal infections of pigeons but to airborne contaminants that find a particularly fertile medium in the droppings.

Thus *C. neoformans* is unable to grow at the normal body temperature of birds (40°–42°C), and birds are highly resistant to cryptococcal infection; mice infected with *C. neoformans* have been observed to survive longer when maintained at 35°C than at 25°; and infected chick embryos also survive longer at 40° than at 37°.

BLASTOMYCES DERMATITIDIS (BLASTOMYCOSIS)*

B. dermatitidis grows as yeast cells in infected tissues or in cultures at 37°C and as a mold at 25°C on conventional media. The mycelia are white at first, darken with age, and have characteristic spherical aleuriospores, about 3μm–5μm in diameter, borne directly on the sides of hyphae or at the tips of short, slender lateral branches. The spore walls also darken with age and resemble chlamydospores. Conversion from yeast to mycelial growth, and vice versa, is readily accomplished in vitro simply by altering the temperature. In the sexual stage (called *Ajellomyces dermatitidis*) the fungus is a heterothallic ascomycete.

In sections of infected tissues (and in sputum or in pus expressed from skin lesions) *B. dermatitidis* appears as unusually thick-walled, multinucleated spherical cells, 8μm–15μm in diameter, without a capsule (Fig. 45-17). Buds, attached to the mother cell by a broad base, are characteristically **unipolar,** i.e., there is not more than one bud per mother cell at any time.

B. dermatitidis is readily isolated on Sabouraud's medium.

Pathogenesis. Infection apparently begins in the lungs and spreads hematogenously to establish focal destruc-

* Paracoccidioidomycosis, formerly referred to as South American blastomycosis, is due to a different fungus, *Paracoccidioides brasiliensis,* discussed later in this chapter.

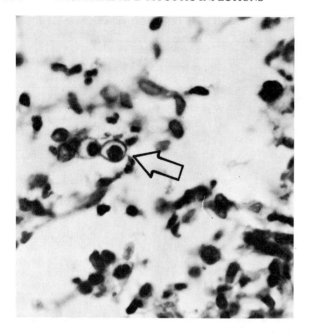

FIG. 45-17. *Blastomyces dermatitidis* in a tissue section; a thick-walled yeast cell **(arrow)** with bud. The broad connection between mother cell and bud (dumbbell shape), and the single bud per mother cell, are both characteristic of *B. dermatitidis* (compare with *Paracoccidioides brasiliensis*, Fig. 45-22). (Hematoxylin–eosin stain; ×450, reduced)

tive lesions in bones, skin, prostate, and other viscera; the gastrointestinal tract is spared. The skin lesions are often particularly conspicuous; they may arise as metastases, from the primary pulmonary lesion, that break through the skin surface and establish spreading, ulcerated, crusted lesions. Skin lesions seem also occasionally to be initiated by direct implantation of spores into broken skin.

The lesions are characterized by granulomatous inflammation, microabscesses, and extensive tissue destruction. The yeast cells are visible within the abscesses and granulomas (Fig. 45-17). Calcification is rare.

Immune Response. Serum from persons with blastomycosis often gives a complement-fixation (CF) reaction with intact *B. dermatitidis* yeast cells or soluble Ags prepared from them, and patients also often exhibit a delayed-type skin response to **blastomycin**, a crude filtrate of mycelial culture. However, crossreactions with *Histoplasma capsulatum* or *Coccidioides immitis* diminish the diagnostic value of either reaction.

These reactions may be negative in far-advanced illness. It is not clear whether the reason is 1) neutralization of Abs by excessive levels of Ags, 2) changes in the nature of the Abs formed with prolonged antigenic stimulation, or 3) progressive impairment of the host's capacity to respond to immunogens in general.

Diagnosis. The demonstration of nonencapsulated, thick-walled, multinucleate yeast cells in pus, sputum, or tissue sections, and their isolation by culture, establish the diagnosis of blastomycosis. The yeast form of *B. dermatitidis* is inhibited by chloramphenicol and cycloheximide. Inoculation of mice intraperitoneally with pus or sputum is occasionally helpful; the fungus is usually readily cultivated from the localized abscesses that appear in the peritoneal cavity in 3 or 4 weeks.

Therapy. Amphotericin B is effective but hydroxystilbamidine is occasionally preferred because of its lower toxicity.

Epidemiology. The disease caused by *B. dermatitidis*, **blastomycosis,** is largely confined to Canada and the United States. It is most often encountered in the Mississippi valley, but occasional infections have turned up in Mexico and in northern South America. Autochthonous cases have also been encountered in various parts of Africa. Most mycologists have therefore dropped the former descriptive term North American and the disease is now usually called just blastomycosis. Infections also occur sporadically in dogs, presumably from inhalation of conidia.

B. dermatitidis is thought to be a soil saprobe, and has been cultivated in the laboratory on sterilized soil. However, the many attempts to isolate it from soil have rarely been successful.

HISTOPLASMA CAPSULATUM (HISTOPLASMOSIS)

H. capsulatum appears in infected tissue as small, oval yeast cells (1μm–3μm in diameter), usually localized

FIG. 45-18. Yeast cells of *H. capsulatum* in a macrophage; an impression smear of liver from an infected mouse. (Wright's stain; ×1250)

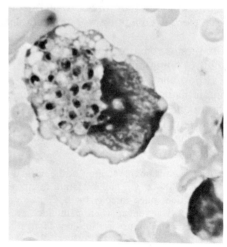

within macrophages and reticuloendothelial cells (Fig. 45-18). Budding is only rarely observed in tissues, since buds separate readily from mother cells. The organism is dimorphic; when cultivated on Sabouraud's or other media at room temperature it forms slowly growing mycelial colonies. These are white at first, become tan with aging, and have fine, silky aerial mycelia. The spores are small or large spherical aleuriospores, with characteristic regularly spaced, spiny projections (Fig. 45-19). The sexual stage of *H. capsulatum* has been designated as *Emmonsiella capsulata* and is classified as an ascomycete.

Hyphae are readily converted to yeast cells in vitro by enrichment of the medium with blood or yeast extracts, an increase in the temperature to 37° C, and the presence of adequate moisture. The conversion becomes apparent macroscopically at the edges of the mycelial mat, where creamy, round, smooth colonies develop. Microscopically, hyphae become constricted at the septa and then fragment into elongated cells, which become oval and multiply by budding. The conversion of mold to yeast also occurs when mycelial fragments are serially passed through animal cell cultures (e.g., HeLa cells) or inoculated into mice.

The reverse conversion (yeast to mold) is brought about simply by decreasing the incubation temperature of yeast cultures from 37° to 25° C; germinal hyphae grow out of the yeast cells, elongate, and develop septa.

Pathogenesis. *H. capsulatum* is present in soil, and inhalation of aleuriospores leads to pulmonary infection. Miliary lesions appear throughout the lung parenchyma, and hilar lymph nodes become enlarged. The initial infection is mild. It may pass unnoticed, or may appear as a self-limited respiratory infection. With healing the pulmonary lesions become fibrotic and calcified and give rise to the characteristic roentgenographic pattern of healed histoplasmosis, formerly confused with healed primary tuberculosis (see Epidemiology, p. 838).

In a very small number of infected individuals the infection becomes progressive and widely disseminated, with lesions in practically all tissues and organs. Fever, wasting, and enlargement of liver, spleen, and lymph nodes occur, and the disease may closely simulate miliary tuberculosis. In fact, this severe disseminated form of histoplasmosis often coexists in individuals who have tuberculosis or some other severe generalized disease, such as leukemia or Hodgkin's disease. Histoplasmosis is also occasionally seen as a chronic pulmonary disease with cavitation, simulating chronic pulmonary tuberculosis.

The tissue lesions are characterized by granulomatous inflammation (similar to that in tuberculosis), with epithelioid cells, occasional giant cells, and even caseation necrosis. The characteristic features, however, are swollen fixed and wandering macrophages containing many small, oval yeast cells (Fig. 45-18). These are readily visualized in tissue sections or cell smears treated with Wright's, Giemsa's, or the periodic acid–Schiff stain.

Immune Response. Antibodies to *H. capsulatum* are detected by tests with filtrates of mycelial or yeast-phase cultures or with heat- or formalin-killed yeast cells. Precipitation and agglutination reactions are observed in mild histoplasmosis. With more advanced disease CF Abs also appear and are regarded as having some prognostic significance. Crossreactions are seen with culture filtrates of *Blastomyces dermatitidis* and *Coccidioides immitis;* and unfortunately the clinical manifestations of these three mycoses may be indistinguishable, particularly in early infections confined to the lungs.

FIG. 45-19. Characteristic thick-walled spores with spiny projections (tuberculate chlamydospores) in a culture of *H. capsulatum* at room temperature (×360)

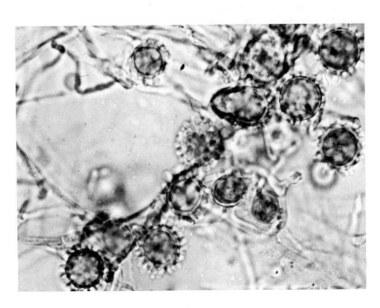

Persons previously infected with *H. capsulatum* regularly give delayed-type skin responses to intradermal injection of **histoplasmin,** a crude, sterile culture filtrate of mycelia grown in synthetic medium.* In both appearance and significance the response resembles the tuberculin reaction: it indicates past or present infection with *H. capsulatum.* Crossreactions also occur with coccidioidin (from *C. immitis*) and with blastomycin (from *B. dermatitidis*).

For both diagnostic and epidemiologic purposes it is often desirable to perform repeated skin tests with histoplasmin in the same persons. Fortunately, the response to one test is not invalidated by prior tests, since histoplasmin injected intradermally does not appear to induce delayed-type hypersensitivity. Such an injection, however, appears to be able to stimulate a secondary Ab response in some individuals who are hypersensitive.

Diagnosis. A provisional diagnosis of histoplasmosis is based upon clinical manifestations, serologic tests, and a positive skin response to histoplasmin. The latter has, of course, virtually no value in those localities where the fungus is so prevalent that most persons react to histoplasmin (see Epidemiology). For a definitive diagnosis it is necessary to identify *H. capsulatum* in tissues or exudates and to isolate it by culture. Staining of blood cells, sputum, and tissue biopsies with Wright's or Giemsa's stain often reveals monocytes or macrophages with characteristic intracellular small yeast cells; and the yeast has even been found in the urinary sediment of persons with progressive histoplasmosis. Cultivation is readily accomplished on Sabouraud's agar and a variety of other media.

Intraperitoneal inoculation of mice with sputum, to which penicillin, streptomycin, and chloramphenicol are added, can detect as few as 10 yeast cells. Autopsy of the mice 2–6 weeks later generally reveals the characteristic yeast within reticuloendothelial cells in smears of spleen cells; and culture of the tissues on Sabouraud's agar at room temperature yields characteristic mycelial growth.

Therapy. Disseminated histoplasmosis was formerly invariably fatal, but it can be treated effectively with amphotericin B.

Epidemiology. Histoplasmosis provides a striking example of the value of epidemiologic analysis in the discovery of previously unrecognized agents of disease—a value that is likely to increase since the more obvious agents have already been revealed by more classic methods.

Histoplasmosis first came to light in 1906 when the spherical yeast cells, observed in tissues taken postmortem in Panama, were

mistakenly classified (and named) as a protozoon. In the next 40 years a few additional cases were reported, all fatal, in scattered parts of the world. But in the 1940s an enormous background of mild or inapparent infection, underlying the rare, fatal disease, was discovered as an unexpected byproduct of a survey of tuberculosis in student nurses in different parts of the United States, undertaken by the US Public Health Service. At that time pulmonary calcification was considered to be almost invariably the endproduct of old tuberculosis, and tuberculin-negative persons with such lesions (revealed by x-ray) were assumed to have lost their previous tuberculin hypersensitivity (see Ch. 37). However, the nationwide survey revealed that in eastern United States cities few such persons were encountered, while in some midwestern communities they constituted the majority of those with pulmonary calcification.

The striking geographic distribution of the discrepancy between tuberculin sensitivity and pulmonary calcification, and the exclusion of a technical artefact by the employment of identical testing materials and personnel in the different localities, clearly implied that some disease other than tuberculosis was producing a large incidence of an indistinguishable pulmonary calcification. Skin testing with culture filtrates of several organisms revealed that 1) histoplasmin sensitivity was much more prevalent in the regions presenting the aberrant calcifications, and 2) the tuberculin-negative persons with these healed lesions were invariably histoplasmin-positive.

In subsequent studies *H. capsulatum* was isolated repeatedly from soil, particularly where contaminated heavily with droppings of birds and bats: barnyards, chicken coops, caves, and beneath trees in which birds nested. Bird droppings enrich the soil as a culture medium, but birds do not develop the disease or carry the organism, presumably because it does not thrive at their high body temperature.

Mass surveys, with chest roentgenograms and histoplasmin skin testing, have now revealed that infection is extremely widespread: perhaps 30 million persons in the United States have been infected. In some areas in the Ohio and Mississippi valleys about 80% of all adults, and over 97% of those with calcified pulmonary lesions, react positively to histoplasmin. Domesticated and wild animals are also naturally infected; in one area of Virginia 50% of the dogs in a large sample were found at necropsy to have histoplasmosis.

Infection occurs by inhalation of airborne aleuriospores. The yeast cells of *H. capsulatum* in sputum and other exudates are less stable than the aleuriospores, and are readily killed by drying, freezing, or heating. Transmission from man to man, and from animal to man, has not been established.

H. capsulatum var duboisii. Histoplasmosis in Africa is caused by a variant of *H. capsulatum,* distinguished by its large ovoid cells, $10\mu m$–$13\mu m$ in diameter, in infected tissues. In culture the mycelium is identical with that obtained from conventional strains.

COCCIDIOIDES IMMITIS (COCCIDIOIDOMYCOSIS)

The parallel between *H. capsulatum* and *C. immitis* is striking. *C. immitis* was also first observed in postmortem

---

* The same medium is used to cultivate other fungi and also *Mycobacterium tuberculosis* in the preparation of the corresponding culture filtrates as skin test reagents.

tissue and mistakenly identified, and named, as a protozoon. Though this error was soon corrected when the organism was grown in culture, almost 40 years elapsed before it was also recognized that *C. immitis* could cause an acute, benign respiratory infection as well as a fatal chronic systemic disease.

In infected tissues *C. immitis* appears as spherules, or sometimes as a mixture of spherules and hyphae. The **spherules** are thick-walled structures which may be as small as 5μm in diameter, but at maturity are usually 20μm–60μm (Fig. 45-20). They are filled with a few to several hundred globular or irregularly shaped **endospores,** from 2μm to 5μm in diameter. When the large spherules rupture the individual endospores are released and in turn develop into spherules: they enlarge, acquire thickened walls, and form multiple endospores.

Growth of *C. immitis* is not inhibited by chloramphenicol or by cycloheximide.

When the spherules are cultivated on Sabouraud's agar or other simple media, even at 37°C, they grow out in mycelial form. Growth is rapid, with fluffy white mycelia appearing within about 5 days. A characteristic feature of the hyphae are the cask-shaped **arthrospores,** which alternate with smaller, clear hyphal cells (Fig. 45-21). The hyphae of older cultures fragment easily and release huge numbers of arthrospores which are easily airborne and highly infectious. Unusual care is therefore required in handling the cultures.

The mycelial (saprobic) form is readily converted into the spherule (parasitic) form by inoculating mycelial fragments or arthrospores intraperitoneally into mice or guinea pigs. Several days after inoculation some arthrospores become enlarged, spherical, and thick-walled, and by 1 week, when they are about 40μm in diameter, radial

partitions appear and subdivide the nascent spherule into cells that subsequently become endospores.

In vitro enrichment of cultures with ascitic fluid, serum, or blood, and incubation at 37°C, favor the conversion of mycelia into spherules, but this is seldom extensive. Mixtures of spherules and hyphae are obtained in such enriched cultures, and some of the developing spherules remain linked together to form hyphal chains.

Pathogenesis. *C. immitis* grows as a saprobe in desert soils of the southwestern United States and northern Mexico. Infection is established by inhalation of airborne spores, and the disease is known as **coccidioidomycosis.**

During World War II the establishment of military bases in areas of the United States where the disease is endemic provided a unique opportunity to observe its evolution in large numbers of freshly exposed persons. In about 60% of infected persons infection is revealed only by acquisition of delayed-type hypersensitivity to Ags of *C. immitis*. In about 40%, however, acute pneumonitis develops, often with pleurisy. Various skin eruptions may also occur, such as erythema multiforme and erythema nodosum: sterile skin lesions that probably represent allergic responses to fungal Ags or perhaps fungus-modified tissue Ags. (Similar sterile skin eruptions appear in some other infections, e.g., with *Mycobacterium tuberculosis* and β-hemolytic streptococci.) About 5% of infected persons ultimately develop chronic pulmonary cavitary disease resembling pulmonary tuberculosis and occasionally leading to calcification. In less than 1% dissemination occurs, with granulomatous lesions in skin, bones, joints, and particularly the meninges.

The lesions of acute pneumonitis due to *C. immitis* are histologically like those caused by pyogenic bacteria,

FIG. 45-20. Spherules of *Coccidioides immitis* in tissue of an infected mouse. (Courtesy of Dr. D. Pappagianis)

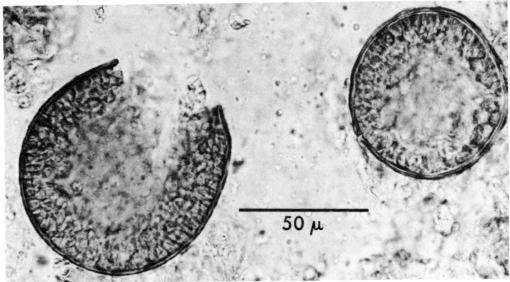

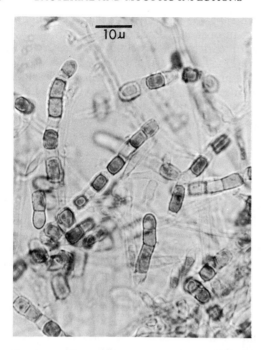

FIG. 45-21. Arthrospores in a culture of *C. immitis*. The chainlike arrangement of spores, separated by what appear to be empty or vacuolated cells, is characteristic. (Courtesy of Dr. D. Pappagianis)

while in the chronic pulmonary disease and in the disseminated lesions the inflammation is granulomatous and is characterized by abundant histiocytes, giant cells, and caseation necrosis. Small spherules are found within macrophages or giant cells, and the larger, more mature spherules lie freely in tissue spaces (Fig. 45-20). The spherules are readily visualized with a number of special stains, e.g., the periodic acid–Schiff stain. In the walls of pulmonary cavities **both** forms of this dimorphic fungus (spherules plus hyphae) may be seen, as in enriched cultures.

Immune Response. Coccidioidin is used in precipitin reactions, CF assays, and skin tests. It is a crude filtrate of a mycelial culture, grown in the same synthetic medium used for the preparation of tuberculin (Ch. 37), histoplasmin, and blastomycin. The skin reaction is of the delayed type, with erythema and induration; reactions greater than 5 mm in diameter are considered positive. The earliest immune manifestation of infection is a positive skin test; only with more protracted infection do precipitins and CF Abs become detectable.

During the first week with overt symptoms 80% of patients in one extensive study had positive coccidioidin skin tests, whereas only 50% and 10% gave positive precipitin and CF reactions, respectively. In those with self-limited disease (spontaneously cured), pre-cipitins in the serum gradually diminished and were not detectable 4–5 months later. The CF Abs tended to appear later than the precipitins and to persist longer, becoming negative only several years after apparent cure. With active, disseminated disease the titers of CF Abs in serum sometimes rose to relatively high levels (e.g., over 1:16) but decreased again as the disease progressed to its terminal stages. Crossreactions may occur with culture filtrates of *B. dermatitidis* and *H. capsulatum.*

A decrease in intensity of the skin response will often, but not invariably, occur in clinically well persons who move away from areas where *C. immitis* is endemic; their skin reaction may become entirely negative within 12 months. Hypersensitivity may, however, persist indefinitely in others without apparent disease. This persistence is probably due to survival of viable endospores within healed, walled-off scars: in experimental animals viable organisms have been found in fibrotic and calcified scars as long as 2–3 years after infection.

Coccidioidin is often injected repeatedly in the same individuals during clinical and epidemiologic studies. Fortunately, repeated injections in humans, in the amounts used in routine skin tests, have not induced delayed-type hypersensitivity, though this type of response to coccidioidin has been induced in guinea pigs by repeated intracutaneous injections. The discrepancy is curious, since delayed-type hypersensitivity to most Ags is induced at least as easily in man as in guinea pigs.

Coccidioidin, blastomycin, and histoplasmin at high concentrations elicit crossreacting skin responses in persons with any of the three corresponding diseases. At lower concentrations, however, coccidioidin seems to provide a relatively specific test, though its value is somewhat limited at present by variations in potency among different lots.

Diagnosis. A provisional diagnosis of coccidioidomycosis is usually based on epidemiologic considerations, clinical manifestations, the skin response to coccidioidin, and the detection of Abs. Definitive diagnosis, however, requires that the spherules be identified in sputum, exudates, or tissue sections.

Cultivation of *C. immitis* in vitro for this purpose is hazardous, because cultures release large numbers of airborne infectious arthrospores; it is best carried out by experienced personnel with access to ventilated hoods. Transmission of *C. immitis* to laboratory animals is, however, a relatively simple and safe procedure. Sputum or exudate is treated with penicillin, streptomycin, or chloramphenicol, and centrifuged; the sediment is injected either in the testes of guinea pigs or intraperitoneally in mice. If *C. immitis* is present mice develop disseminated disease, and the testes and the mouse tissue fluids should contain characteristic spherules; moreover, CF Abs may appear in the guinea pig.

Therapy. When confined to the lungs coccidioidomycosis usually heals with scarring, but the disseminated disease was invariably fatal until the polyene antibiotics became available. Treatment with amphotericin B is moderately successful.

Epidemiology. The fungus has a predilection for growth in desert soils, especially after winter and spring rains;

and windborne arthrospores readily infect man. Skin testing of large human populations has established that coccidioidomycosis is prevalent in the southwestern United States: central California (especially in the San Joaquin valley), Arizona, New Mexico, western Texas, and southern Utah. In some of these areas 50%–80% of the population reacts to coccidioidin. The organism has also been found in northern Mexico, parts of Argentina, and Paraguay.

Other mammals are also easily infected, e.g., wild rodents, dogs, and cattle. Histologic examination and culture of the lungs of trapped wild rodents, and skin testing of domestic cattle, have provided additional means for identifying regions of prevalence.

PARACOCCIDIOIDES BRASILIENSIS (PARACOCCIDIOIDOMYCOSIS)

The organism that causes South American blastomycosis, first recognized in Brazil, was thought to be very similar to *C. immitis:* hence its name, *Paracoccidioides brasiliensis.* It has also been suggested that it be classified in the genus *Blastomyces* and named *B. brasiliensis.* The fungus appears in infected tissues as large spherical or oval yeast cells, 10μm–30μm in diameter, and sometimes even 60μm. Characteristicially, multiple buds sprout from a single mother cell and remain attached to it by narrow constricted bands (Fig. 45-22). When the buds are about the same size and all quite small their distinctive arrangement around the mother cell is often referred to as a **pilot's wheel.** The buds may also be equal in size to

FIG. 45-22. Budding of yeast cells in a culture (37°C) of *Paracoccidioides brasiliensis.* Multiple buds, attached to their mother cell by constricted tubes, are characteristic. Compare with the broad-based, unipolar buds characteristic of *Blastomyces dermatitidis* (Fig. 45-17) (×450, reduced)

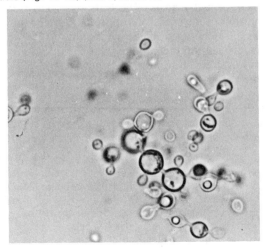

the mother cell and still remain attached, as **satellite cells.** Chains of budding cells are also seen.

P. brasiliensis is dimorphic. In culture at room temperature on Sabouraud's medium it grows as a mycelium with chlamydospores. Conversion to yeast cells is induced by enrichment of the medium (e.g., with brain–heart infusion), adequate moisture, and increase in incubation temperature to 37°C.

Pathogenesis. The disease produced by *P. brasiliensis* is called **paracoccidioidomycosis, paracoccidioidal granuloma,** or **South American blastomycosis.** The earliest lesions arise in the mucous membranes of the mouth or nose and spread by direct extension, e.g., over the mucocutaneous borders to involve the face. Dissemination also occurs, with frequent involvement of lymphoid tissue (including the spleen). In the intestinal tract lesions begin in submucosal lymphoid tissue and may lead to ulceration and even perforation. Subcutaneous abscesses can appear, and by extension to the skin surface they form large, unsightly crusted and ulcerated lesions.

Histologically, skin lesions appear as pyogenic abscesses and granulomatous inflammation with epithelioid cells, giant cells, and necrotic centers. Large spherical or oval yeast cells with multiple circumferential buds (pilot's wheel or satellite forms) may be observed in routine hematoxylin–eosin stains of tissue sections, but are more clearly brought out with the periodic acid–Schiff reaction. The yeast cells sometimes appear as chains and may be found within giant cells.

Diagnosis. Detection of *P. brasiliensis* in tissue sections or in exudates, and cultivation as yeast and mycelial forms, establish the diagnosis. Chloramphenicol and cycloheximide are added to Sabouraud's medium and the cultures are maintained as mycelia at room temperature, since these antibiotics inhibit the growth of the yeast cells but not the mycelia of this fungus. Transmission to laboratory animals is possible but is not often resorted to as a diagnostic procedure.

Therapy. The disseminated disease is slowly progressive and was formerly invariably fatal. However, amphotericin B arrests the spread of lesions.

Epidemiology. *P. brasiliensis* is probably a soil saprobe. The disease it causes has been reported mainly in Brazil, and also in most other South American and Central American countries. Experimental and autopsy data indicate that the lung is the primary focus of infection, a mild or asymptomatic respiratory form of the disease occurring as in histoplasmosis.

MYCOSES DUE TO OPPORTUNISTIC FUNGI

A number of fungi that are not pathogenic in healthy humans may behave as virulent pathogens in those suffering from a variety of disorders (e.g., malignant lymphomas, severe diabetes) or treated intensively with broad-spectrum antibacterial drugs or with immuno-suppressive measures. In addition, among the pathogenic fungi discussed above, *Cryptococcus neoformans, H. capsulatum, B. dermatitidis,* and possibly even *Coccidioides immitis* are also somewhat opportunistic, causing progressive infections more frequently under debilitating conditions. The frankly opportunistic fungi are mostly species of *Candida, Aspergillus, Rhizopus,* and *Mucor.*

CANDIDA ALBICANS (CANDIDIASIS)

C. albicans is dimorphic. At the surface of a rich agar medium it grows as oval budding yeast cells, but deeper in the medium hyphae are found. Both forms are characteristically seen in infected tissues and in most cultures. Some hypae, called *pseudohyphae,* have recurring constrictions, as though a chain of sausage-shaped cells were joined end to end (Fig. 45-23).

C. albicans is readily grown on conventional media at room temperature or at 37° C. In cultures on agar the early colonies are smooth, creamy, and bacterialike, but the older, larger colonies may appear furrowed and rough. Cultivation on cornmeal agar stimulates the formation of characteristic thick-walled chlamydospores, which distinguish *C. albicans* from other candidae (Fig. 45-23).

On the basis of colonial morphology and assimilation and fermentation reactions, several medically important species of *Candida* have been distinguished. *C. albicans* is by far the most frequent; rare species isolated from human lesions include *C. krusei, C. parakrusei,* and *C. parapsilosis.* Because of extensive crossreactions their serologic differentiation requires hyperimmune sera absorbed with various cells.

Pathogenesis. *C. albicans* and other species of *Candida* are frequently present on the normal mucous membranes of mouth, vagina, and intestinal tract. When they become invasive, under the special circumstances mentioned previously, they establish a variety of acute or chronic, localized or widely disseminated, lesions. The following are some examples.

1) **Thrush (oral candidiasis)** consists of discrete or confluent white patches, composed of hyphae and yeast cells, on the mucous membranes of the mouth and pharynx. They occur particularly during the first few days of life in the newborn (resulting from infection during birth), and in persons in the terminal stages of a wasting disease, e.g., carcinomatosis. 2) Vaginal mucous membranes are occasionally invaded (**vulvovaginal candidiasis**) during

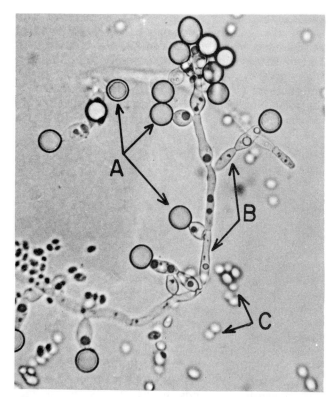

FIG. 45-23. Multiplicity of structural forms in a culture of *Candida albicans.* **A** = chlamydospores; **B** = pseudohyphae (elongated yeast cells, linked end to end); **C** = budding yeast cells (blastospores). (×450, reduced)

pregnancy and in diabetes. 3) Invasion of bronchial and pulmonary tissues (**bronchopulmonary candidiasis**) is usually secondary to chronic bronchial obstruction with impaired drainage of secretions (e.g., from bronchial carcinoma or bronchiectasis). 4) Infections in skin areas that are continuously wet and macerated (**intertriginous candidiasis**) are common in the perineum and inframammary folds, and on the hands of those whose occupation requires prolonged immersion in water.

Endocarditis due to *Candida* is rare but is occasionally seen in drug addicts and in patients with intravascular tubes or cardiac prostheses. The organisms are usually not *C. albicans,* as they are in the forms of candidiasis mentioned above, but other species, especially *C. parapsilosis.*

Some individuals with candidiasis develop sterile vesicular or papular skin lesions, called **monilids** (since the genus *Candida* was formerly called *Monilia* and candidiasis was called **moniliasis**). These lesions, presumably allergic, resemble the dermatophytids observed in dermatophyte infections (see below).

The clinical impression of the role of lowered resistance has been supported by experimental observations. For example, mice given cortisone are unusually susceptible to lethal infection with *C. albicans*. In one study of humans under treatment with a tetracycline over 60% had positive rectal cultures for *C. albicans,* whereas all had had negative cultures prior to therapy. In the latter case, it is not clear whether the mechanism is a decrease in competition for a nutrient by normal intestinal bacteria or a decreased production of inhibitory substances.

Diagnosis. It is not surprising that many normal human sera (15%–30% in one study) specifically agglutinate cells of *Candida* species, as these organisms are so commonly found on normal mucous membranes; and though agglutinin titers tend to be higher in those with frank candidiasis, serologic tests are of little diagnostic value. Similarly, skin tests with aqueous extracts of *C. albicans* **(oidiomycin)** are positive in most normal persons. A presumptive diagnosis of candidiasis is usually made by microscopic demonstration of abundant hyphae and yeast cells in scrapings of lesions, and the diagnosis is supported by isolation of the organism in cultures. However, *Candida* is so ubiquitous that it is often difficult to decide whether or not it is the causative agent. The response to therapy aids in arriving at a decision.

Therapy. Polyene antibiotics are effective: nystatin is generally applied locally to accessible lesions, and amphotericin B is administered in the treatment of severe visceral infections.

Epidemiology. Infection of the newborn, by *C. albicans* from the birth canal, is one of the few instances in which a fungus infection is clearly transmitted from one person to another. Usually, however, candidiasis is due to increased susceptibility to a normal commensal (see Ch. 44). The underlying mechanisms are obscure, but defects in cell-mediated immunity have been implicated in chronic mucocutaneous candidiasis.

ASPERGILLUS FUMIGATUS (ASPERGILLOSIS)

Many species of *Aspergillus* have been recognized in nature, and seven have been associated with human disease **(aspergillosis).** *A. fumigatus* accounts for over 90% of all infections. Most aspergilli, including *A. fumigatus,* are **not dimorphic:** they grow only in mycelial form.

Colonies grow well over a wide temperature range, and *A. fumigatus* can thrive up to 50° C. Growth is inhibited by cycloheximide. In culture the mycelia are powdery and have a dark bluish green cast, hence the name **fumigatus** (L., smoky). Conidiophores are up to 500μm long; each bears a dome-shaped vesicle (about 20μm–30μm in diameter) with a single row of sterigmata (see Fig. 45-5) arranged about the distal half. From these, green conidia (2μm–5μm in diameter) grow in parallel linear chains (Fig. 45-5), thus accounting for the generic name (L. *aspergillus,* brush).

As suggested earlier, most fungi that abound in the environment of birds do not grow well at their high body temperatures (e.g., *Cryptococcus neoformans, Histoplasma capsulatum*). Birds are, however, highly susceptible to fatal infection with various thermophilic species of *Aspergillus* (especially *A. fumigatus*). In cattle and sheep, abortions can be caused by aspergillosis but may be due to a toxin produced by the fungus rather than to overwhelming infection.

Airborne species of aspergilli are ubiquitous, and since these fungi thrive at elevated temperatures, they tend to be particularly abundant in damp, decaying vegetation heated by bacterial fermentations. In compost piles most microorganisms cease to grow as the temperature rises, and under these conditions aspergilli can become almost a pure culture.

Pathogenesis. Farmers and others who handle decaying vegetation are often heavily exposed to spores of *Aspergillus.* These spores are also disseminated from humidifiers and from air conditioner filters and ducts that have accumulated moisture. Hypersensitivity pneumonitis, asthma, and rhinitis due to allergy to Ags of these spores are common.

In addition, progressive infection can be established under predisposing conditions, including not only those previously mentioned but also chronic pulmonary diseases with impaired ciliary activity in the bronchi (e.g., bronchiectasis, pulmonary tuberculosis, and bronchial neoplasms).

Most initial infections are pulmonary, following inhalation of spores. These germinate, and hyphae grow and penetrate contiguous tissues by direct invasion (Fig. 45-24). They tend particularly to invade blood vessel walls, producing angiitis and thromboses; severe hemoptysis is

FIG. 45-24. Hyphae of *Aspergillus fumigatus* in the wall of a pulmonary cavity (Hematoxylin–eosin stain; ×600)

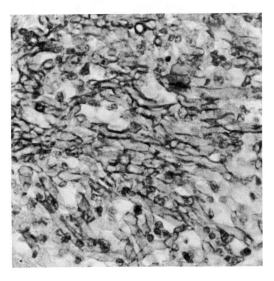

conspicuous and life-threatening in pulmonary aspergillosis. Infected emboli may also establish widespread metastatic granulomatous lesions in various organs.

Severe local infection can also arise by direct implantation of *Aspergillus* spores in the nasal sinuses, with resulting cellulitis of the sinuses and face; in the eye, especially during local treatment with corticosteroids; and in the external ear canal, in the presence of concomitant chronic inflammatory disease.

Experimental evidence supports the clinical impression that corticosteroid therapy increases susceptibility: untreated mice exposed to aerosolized spores developed only mild, transient pneumonitis, but cortisone-treated mice developed fatal pulmonary aspergillosis.

Diagnosis. Since *Aspergillus* frequently contaminates cultures the pathogenic significance of a single isolate is not clear. In general if a species of *Aspergillus* is consistently isolated from a particular patient, in repeated cultures of sputum, exudates, or scrapings of infected tissues, it is presumed to be significant. A definitive diagnosis is established by demonstrating hyphae, which are often abundant, in tissue sections. The hyphae are easily seen on hematoxylin–eosin and on Gram stains: they are 3μm–4μm in diameter, exhibit dichotomous branching, and are septate (Fig. 45-24). The tissue reaction takes the form of either suppurative or granulomatous inflammation.

Therapy. The prognosis in pulmonary aspergillosis is grave, but encouraging therapeutic results have been obtained with amphotericin B.

PHYCOMYCETES (PHYCOMYCOSES)

Several genera of the class Phycomycetes have been recognized as occasional causes of human disease. *Rhizopus* species are most often involved. The disease produced was formerly referred to as **mucormycosis,** but the generic term **phycomycosis** is more appropriate. Phycomyces, like *Aspergillus,* is **not dimorphic:** growth is mycelial in both infected tissues and cultures.

Pathogenesis. Infection occurs by inhalation of spores, and, rarely, by their traumatic implantation in broken skin or mucous membranes. Severe infections of the central nervous system occur in poorly controlled diabetes. Widely disseminated visceral lesions have been described as complications of severe malnutrition, uremia, and hepatic insufficiency, as well as in persons receiving corticosteroids or broad-spectrum antibacterial drugs.

Hyphae are abundant in infected tissues. They grow by direct extension through contiguous tissues and tend to invade blood vessel walls, producing angiitis, thrombi, and ischemic necrosis; in addition, emboli establish metastatic infections in many organs. Acute inflammation is characteristically present at sites of infection.

A useful experimental model for phycomycosis has been developed in rabbits made acutely diabetic with alloxan or treated with corticosteroids. In such animals *Rhizopus* spores introduced into the paranasal sinuses lead to severe nasal, pulmonary, and cerebral lesions, which simulate the disease in diabetic humans.

Diagnosis. Distinctive hyphae are seen in infected tissues stained with hematoxylin–eosin. In contrast to the hyphae of *Aspergillus,* those of *Phycomyces* are relatively huge (up to 15μm in diameter), lack septa, and branch haphazardly (Fig. 45-25; see also Fig. 45-11).

Because their walls are rich in chitin and probably correspondingly poor in glucans and mannans, phycomycetes stain poorly with the periodic acid–Schiff stain (Fig. 45-13). Identification of the individual species requires considerable experience. In the following paragraphs the main properties of the more commonly encountered genera are outlined.

Rhizopus. Species associated with phycomycosis grow rapidly on Sabouraud's medium and form coarse and woolly mycelia, which are initially white but subsequently become peppered with black

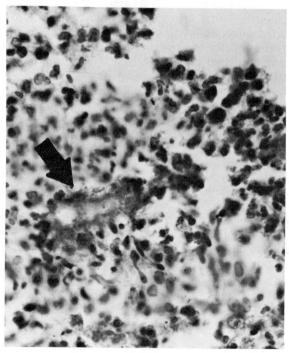

FIG. 45-25. Phycomycosis (due to *Basidiobolus haptosporus*). The hyphae in the tissue section are unusually wide (about 15μm in diameter; see **arrow**), lack septa, and are surrounded by an intense leukocytic infiltrate, including many eosinophils. (Hematoxylin–eosin stain; ×450, reduced)

and brown fruiting structures (sporangia). The hyphae are non-septate and colorless. Asexual reproductive elements consist of long, unbranched stalks (sporangiophores), which arise directly from a cluster of rhizoids (rootlike structures) and are each topped by a spherical sac (sporangium). The latter are dark, and when mature are filled with spores. The species isolated most often from human phycomycosis are *R. oryzae* (most frequently), *R. arrhizus,* and *R. nigricans.*

Mucor. Cultures of these species grow rapidly on Sabouraud's medium and form fluffy mycelia which are initially white, and subsequently are gray to brown. Hyphae are usually nonseptate, but in old cultures irregular cross-walls are occasionally seen. Sporangiophores, which do not originate from rhizoids, form thick upright tufts. Some are branched, and each terminates in a spherical sporangium. The species most often identified in human infections is *M. corymbifer* (*Absidia corymbifera*).

Phycomycosis is also occasionally due to species of *Absidia, Mortierella,* and *Basidiobolus.*

SUBCUTANEOUS MYCOSES

Subcutaneous mycotic infections are usually initiated by penetration of the skin with contaminated splinters, thorns, or soil. Once established, these infections tend to remain localized in subcutaneous tissues and to be extremely persistent. The diseases are classified as **sporotrichosis, chromomycosis,** and **maduromycosis.**

SPOROTHRIX SCHENCKII (SPOROTRICHOSIS)

Sporotrichosis is characterized by an ulcerated lesion at the site of inoculation, followed by multiple nodules and abscesses along the superficial draining lymphatics. Only rarely is there dissemination to the meninges. In infected tissues the organism appears as cigar-shaped, budding yeast cells (Fig. 45-26); these cells are usually scarce but may sometimes be recognized by the periodic acid–Schiff stain or with fluorescein-labeled Abs.

S. schenckii is dimorphic. When pus or curettings from skin lesions are cultured on Sabouraud's medium at room temperature (with chloramphenicol and cyclohex-imide) the fungus grows rapidly in mycelial form as a flat, moist colony. Hyphae are slender (about 2μm in diameter) and septate, and conidia are individually attached by sterigmata to a common conidiophore (Fig. 45-5). Pigmentation of the conidia accounts for the dark color of the mycelium, the intensity of pigmentation depending on the level of thiamine in the medium. The pigment is allegedly melanin, and tyrosinase has been identified in mycelia.

Identification of yeast cells is also of diagnostic value, since they are usually difficult to visualize in human lesions. Hyphae are converted to yeast forms by cultivation at 37°C, at slightly increased pCO_2, in media enriched

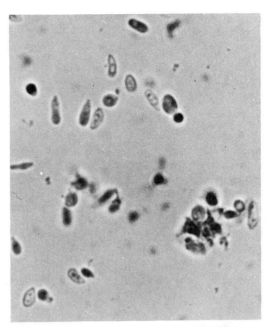

FIG. 45-26. Cigar-shaped budding yeast cells of *Sporothrix schenckii* in culture at 37°C. (×450)

with proteins, thiamine, and biotin. This change is also brought about by inoculating mycelial fragments into the testes of mice; pus withdrawn after 2–3 weeks contains abundant characteristic yeast cells. In addition, mouse inoculation differentiates *S. schenckii* from several saprobes that are morphologically similar but not pathogenic.

S. schenckii has been isolated from soil and plants, and can apparently grow in wood. There is little doubt that penetration of skin by contaminated splinters and thorns is the principal means of introducing infection. The disease is usually sporadic among farmers and gardeners, but a few industrial outbreaks have occurred among workers exposed to batches of heavily infected timbers and plants.

CHROMOMYCOSIS

A group of slowly growing, dimorphic fungi produce the subcutaneous mycoses known as chromomycosis. Though the first case was reported in Boston the disease occurs primarily in the tropics. The infection is seen mostly on the legs of bare-legged laborers, and lesions appear as warty, ulcerating, cauliflowerlike growths.

In draining lesions the fungus appear as thick-walled, round cells, about 6μm–10μm in diameter, whose dark brown color gives the disease its name. The yeast cells apparently multiply by fission rather than by budding. In cultures on Sabouraud's medium at room temperature darkly pigmented mycelial colonies are formed slowly.

Several species have been distinguished on the basis of conidial arrangements, but their classification is unsettled. According to one terminology, they are *Fonsecaea pedrosoi, F. compacta, F. dermatitidis, Cladosporium carrionii,* and *Phialophora verrucosa.*

MADUROMYCOSIS

Mycetoma is the generic term for localized destructive granulomatous and suppurative lesions that usually affect the foot or hand, and involve skin, subcutaneous tissues, bone, and fascia (see Ch. 38). Mycetomas usually originate with injuries. They are especially prevalent in the tropics.

Lesions are characterized by multiple burrowing sinus tracts that extend through soft tissues and penetrate to the skin surface. The draining pus contains granules (small pieces of colonies of the causative microorganism), which vary in texture, color, and shape, depending on the microorganism, and also vary in size (from about 0.1 to 2 mm).

The microorganisms that cause mycetomas are either fungi or actinomycetes (Ch. 38); the fungal mycetomas are often referred to as **maduromycosis.** Diagnosis is based on demonstration of fungus cells directly in KOH-treated granules and pus, and by culture. At least 13 species of fungi have been identified, including *Madurella mycetomi, M. grisea, Allescheria boydii, Phialophora jeanselmi,* and *Aspergillus nidulans.*

Surgical drainage is an important adjunct to therapy. Polyenes, which have not yet been tried extensively, may prove of value, since the causative agents are susceptible in vitro.

CUTANEOUS MYCOSES (DERMATOMYCOSES)

The dermatophytes are fungi that infect only epidermis and its appendages (hair and nails), i.e., structures in which keratin is abundant. The ensuing skin lesions are usually roughly circular, tend to expand equally in all directions, and have raised serpiginous borders. They were therefore thought in ancient times to be due to worms or lice, and they are still called **ringworm** or **tinea** (L., worm or insect larva). The names are usually qualified by the area of the skin involved: e.g., ringworm of the scalp **(tinea capitis),** of the body **(tinea corporis),** of the groin ("jock itch," **tinea cruris**), and of the feet (athlete's foot, **tinea pedis**).

Dermatophytes are not dimorphic. In infected skin lesions they all look alike, with septate hyphae and arthrospores (Figs. 45-27 and 45-28). However, additional differentiated structures appear in cultures and provide the basis for identification.

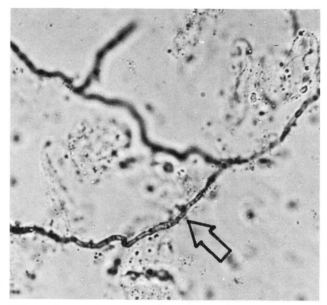

FIG. 45-27. Skin scraping treated with 10% KOH to show hyphae **(arrow)** of a dermatophyte among epidermal debris from a human skin lesion (×450, reduced)

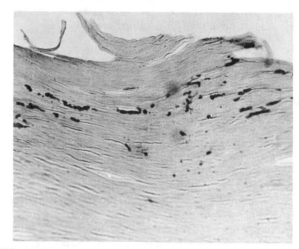

FIG. 45-28. Periodic acid–Schiff stain of skin section of a human lesion showing a dermatophyte in the stratum corneum. (×100, reduced)

ATTACK ON KERATIN

The predilection of dermatophytes for epidermis, firmly established by clinical observations, is also demonstrable experimentally. For example, when spores or mycelial fragments are injected intravenously into guinea pigs no lesions develop; but if an area of skin is abraded at the time of the injection dermatophyte infection appears in the scarified skin a few weeks later.

In view of the evident affinity of dermatophytes for keratin-rich tissues, one might expect these fungi to have an unusual capacity to degrade and utilize keratin. In fact, however, they do so at only a low rate in vitro. Evidently the pathogenicity of dermatophytes depends on more than their ability to attack keratin.

Keratin is a fibrous and very insoluble protein, stabilized by the disulfide groups of frequent cystine residues; and in its native state it is resistant to most proteolytic enzymes. An enzyme preparation of *Streptomyces* cleaves disulfide bonds effectively and digests keratin.* *Microsporum gypseum,* on the other hand, one of the few dermatophytes often isolated from soil and more keratinolytic than the others, does not rupture keratin disulfides and digests this protein to only a limited extent. Nevertheless, many dermatophytes can be cultured on sterile hair and they dissolve localized segments of hair fibers.

Though dermatophytes are epidermal parasites most species grow well in simple media, with ammonium salts and glucose as sole sources of nitrogen, carbon, and energy. Growth is more vigorous, however, in media enriched with proteins or amino acids.

Invasion of Hair. In infection of hairs hyphae grow first from the epidermis into hair follicles, and then into hair shafts. In the **endothrix** type of infection the hyphae then grow only **within** the hair shaft, where they form long, parallel rows of arthrospores (Fig. 45-29). In **ectothrix** infections they grow both **within and on the external surface** of the hair shaft (Fig. 45-30).

EPIDEMIOLOGY

About 15 species of dermatophytes are found primarily in human skin (**anthropophilic**). Many others are indigenous in domesticated and wild mammals (**zoophilic**); and a few may be free-living saprobes, since they are isolable from soil (**geophilic**), e.g., *M. gypseum, Trichophyton mentagrophytes,* and *T. ajelloi.* Infection is transmitted, though with difficulty, from man to man, or animal to man or vice versa, by direct contact or by contact with infected hairs and epidermal scales (e.g., from barber shop clippers, shower room floors, etc.). The reservoir of animal infection is huge: about 30% of dogs and cats in the United States are infected with *Microsporum canis,* a frequent cause of ringworm of the scalp in children.

The incidence of different dermatomycoses varies with age. For example, intertriginous infection of the feet (athlete's foot) is common in adults but rare in children, whereas the opposite is true for ringworm of the scalp.

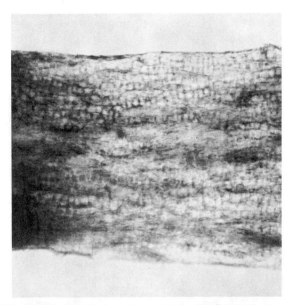

FIG. 45-29. Endothrix hair infection with *Trichophyton tonsurans.* Chains of arthrospores are localized within the hair shaft. (×450, reduced)

Resistance of adults to scalp infection has been linked to the increased secretory activity of sebaceous glands at puberty and the antifungal activity of the C-7 to C-11 saturated fatty acids in sebum.*

Most dermatophytes have a worldwide distribution, but a few species are restricted geographically. With increased travel in the past 25 years even the localized species are becoming more widely distributed. For example, *Trichophyton tonsurans,* endemic in Mexico, has become common in the United States.

IMMUNITY

Human resistance to some dermatophytes is emphasized by the low incidence of conjugal infections; for example, in one study of 60 *T. rubrum*-infected persons, followed up for 1–20 years, not one spouse acquired active infection. This natural human resistance is probably not due to conventional immune reactions: circulating Abs are not, as a rule, demonstrable in the serum of persons with dermatomycosis. However, a fungistatic factor, which seems not to be an immunoglobulin, is demonstrable even in normal serum. This factor may well be responsible for the limited penetration of dermatophyte infections. We do not understand the balance of forces that causes these infections to be so widespread yet difficult to

---

* This keratinase is used industrially to remove hair from hides in the preparation of leather.

* Undecylenic acid, an unsaturated C-11 fatty acid, is widely used for topical therapy of some dermatomycoses.

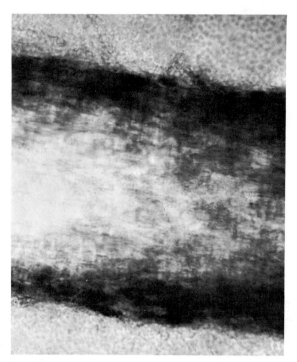

FIG. 45-30. Ectothrix infection of hair by *Microsporum audouini.* Spores are clustered on the surface of the hair shaft. (×450, reduced)

transmit deliberately, and so often self-limited yet difficult to cure.

Hypersensitivity. Persons with dermatophytosis sometimes have another kind of skin lesion, believed to represent an allergic reaction to fungal Ags that spread from the site of infection. These lesions, **dermatophytids,** are sterile and appear as vesicles symmetrically distributed on the hands.

Hypersensitivity to dermatophytes appears in the course of infection and is usually persistent. Trichophytin, a crude filtrate of broth in which a dermatophyte has been grown, elicits delayed cutaneous responses and sometimes also wheal-and-erythema responses. However, as with tuberculin, these responses do not distinguish between current and prior infection; they have little diagnostic or prognostic value.

TREATMENT

Many infections are eradicated by griseofulvin, but the response varies with the thickness of the keratin and the rate of its replacement. Infections of the scalp and smooth skin are usually cured after several weeks of therapy, but infections of the feet, especially of toenails, require many months of continuous treatment. Traditional

forms of local therapy with keratinolytic agents are therefore still widely used.

CLASSIFICATION

In identifying specific dermatophytes the form and arrangement of macroaleuriospores is especially important. Additional determinative characteristics are 1) the form and pigmentation of mycelia; 2) the quantity and disposition of microaleuriospores; 3) the development of special hyphal structures, e.g., racket-shaped ends of some hyphae (racket mycelia), helically coiled hyphae (spirals), and tightly coiled, twisted hyphae (nodular bodies); 4) the presence of arthrospores and chlamydospores; 5) some physiologic characteristics, e.g., growth in culture on sterile hair.

Identification of dermatophytes is hindered by their **pleomorphism**—a word used in mycology in a special sense, to describe the frequent loss of pigmentation and spore formation during laboratory cultivation. The resulting mycelia resist identification. However, transfer to a medium that stimulates sporulation (e.g., potato–glucose agar) reduces the frequency of this conversion. In order to maintain sporulating cultures frequent transfers are usually necessary (about once every 10 days), or else storage at −20° C.

Genetic crosses have shown that pleomorphic conversion is the result of one or more gene mutations. Thus pleomorphism in dermatophytes resembles phase variation in bacteria (Ch. 8), in which culture conditions select for a frequent, reversible mutation.

Dermatophytes fall into three genera: in general, *Microsporum* attacks hair and skin but not nails; *Trichophyton* attacks hair, skin, and nails; *Epidermophyton* infects skin and occasionally nails, but not hair. From 20 to 100 species, depending on the classification, cause human infection. A few representative species will be discussed briefly.

Microsporum. Hair infections are of the ectothrix type (Fig. 45-30), with spores packed closely on the external hair surface in a mosaic pattern. In culture the genus is characterized by the production of rough-walled multicellular aleuriospores (Fig. 45-31).

M. audouini is primarily a human pathogen and used to be the most frequent cause of epidemics of ringworm of the scalp in children in the United States. Adults are only rarely infected and animals are highly resistant. When the scalp is irradiated with ultraviolet light (366 nm) infected hairs emit a bright yellow-green fluorescence. This property makes possible the rapid diagnostic screening of large populations of children and also facilitates the selection of infected hairs for culture.

M. canis is primarily a parasite of domesticated and wild animals, and children commonly acquire infections from cats and dogs. An intense but localized inflammatory reaction (called a

FIG. 45-31. Macroaleuriospores in a culture of *Microsporum gypseum.* (×450, reduced)

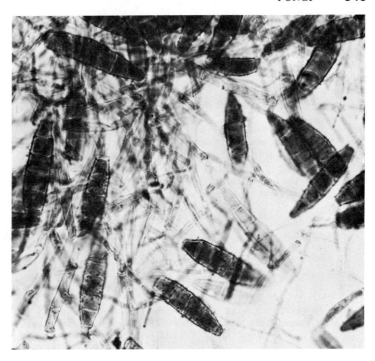

kerion) develops in the skin and subsides spontaneously after several weeks; it may represent an allergic response. Hairs infected with *M. canis* resemble those infected with *M. audouini* (Fig. 45-30) and also fluoresce in ultraviolet light.

M. gypseum is abundant in soil but is an infrequent cause of human infections. Infected hairs are not fluorescent. Arthrospores appear on the surface of infected hairs and are larger than those of other *Microsporum* species. Sexual forms, with typical asci, have been isolated from cultures (Table 45-3).

Trichophyton. Species of this genus usually produce smooth-walled macroaleuriospores in culture, but classification is difficult since spores are often sparse or lacking. A large number of so-called species have been described but are regarded increasingly as variants of 12 or 13 species.

T. schoenleinii is a major cause of **favus,** a severe form of chronic ringworm of the scalp, with destruction of hair follicles and permanent loss of hair. It was the first microbial pathogen of man to be identified (1843), isolated, and established as a pathogen through experimental infection of an animal host. In culture, hyphae are coarse and have knobby, broadened ends and many short lateral branches: these structures, reminiscent of reindeer horns, are called favic chandeliers.

T. violaceum causes endothrix infection in hair, and may also cause favus. It is a common cause of ringworm of the scalp and body in the Mediterranean area.

T. tonsurans causes endothrix infections of hair. With a large influx into the United States of persons from Latin America, ringworm of the scalp is now more frequently due, in large cosmopolitan centers, to *T. tonsurans* than to *M. audouini.* Early diagnosis of infection is difficult, however, because unlike *M. audouini*-infected hairs, those infected with *T. tonsurans* do not fluoresce.

T. mentagrophytes and *T. rubrum* are common causes of athlete's foot and of infections of smooth skin in adult humans. They also infect nails, and *T. mentagrophytes* can cause endothrix infections of hair (scalp and beard).

Epidermophyton. This genus is represented by a single species, *E. floccosum,* and is found only in man. It grows in epidermis (especially in intertriginous areas, as between toes) but does not invade hair.

SUPERFICIAL MYCOSES

Fungi that produce these infections are limited to invasion of the most superficial layers of skin and hair. Four superficial mycoses are fairly common.

Tinea Versicolor (Pityriasis Versicolor). The disease is common in all parts of the world, but is most frequently seen in warm climates. The infected lesions appear as white or tanned scaly areas on the trunk, and lesions are chronic but asymptomatic. The etiologic agent, *Pityrosporum orbiculare,* can be readily visualized in clinical specimens (treated with 10% KOH) as clusters of round budding cells ($3\mu m$–$8\mu m$ in diameter) and mycelial elements. The taxonomic status of this agent is no longer in question since it has been repeatedly isolated from clinical specimens. The yeastlike fungi that constitute this genus require long-chain fatty acids for their growth. *P. ovale,* which morphologically resembles *P. orbiculare,* is frequently isolated from sebaceous glands and hair follicles but is not pathogenic and is considered to be part of the normal flora of the skin.

A similar disease, **erythrasma,** is caused by a corynebacterium, *C. minutissimum.*

Tinea Nigra. These lesions are largely confined to the palms, where they appear as irregular, flat, darkly discolored areas. Infection is particularly prevalent in the tropics and is rare in the United States. The fungus that causes this disorder is *Cladosporium werneckii.* It appears in scrapings of skin lesions as branched hy-

phae, and it grows slowly on agar as greenish black colonies with numerous dark, budding cells.

White Piedra. The fungus that causes this disorder (*Trichosporon cutaneum*) grows on scalp or beard hair. Soft, pale nodules appear on the hair shafts; they consist of hyphae and oval arthrospores. On agar this fungus grows as soft, creamy colonies that become wrinkled and gray with age, and its septate hyphae fragment easily into arthrospores.

Black Piedra. In this tropical disorder an ascomycete, *Piedraia hortai,* forms hard, dark nodules on the shafts of infected scalp hairs. The nodules contain oval asci with two to eight ascospores. On agar the organism forms greenish black mycelia that bear chlamydospores.

SELECTED READING

GENERAL MYCOLOGY

AINSWORTH GC, SUSSMAN AS (eds): The Fungi: An Advanced Treatise, Vols I, II, III, IVA, IVB. New York, Academic Press, 1965 (Vol I), 1966 (Vol II), 1968 (Vol III), 1973 (Vols IVA and IVB)

BURNETT JH: Mycogenetics: An Introduction to the General Genetics of Fungi. New York, Wiley, 1975

CIEGLER A, KADIS S, AJL SJ (eds): Microbial Toxins: Fungal Toxins, Vol VI. New York, Academic Press, 1971

HAMILTON-MILLER JMT: Fungal sterols and the mode of action of the polyene antibiotics. Adv Appl Microbiol 17:109, 1974

KADIS S, CIEGLER A, AJL SJ (eds): Microbial Toxins: Algal and Fungal Toxins, Vol VII. New York, Academic Press, 1971

KADIS S, CIEGLER A, AJL SJ (eds): Microbial Toxins: Fungal Toxins, Vol VIII. New York, Academic Press, 1972

MOORE-LANDECKER E: Fundamentals of the Fungi. Englewood Cliffs, Prentice Hall, 1972

NORMAN AW, SPIELVOGEL AM, WONG RG: Polyene antibiotic–sterol interaction. Adv Lipid Res 14:127, 1967

PRESCOTT DM (ed): Methods in Cell Biology: Yeast Cells, Vol XI. New York, Academic Press, 1975

ROSE AH, HARRISON JS (eds): The Yeasts, Vols. I, II, III. New York, Academic Press, 1969 (Vol I), 1971 (Vol II), 1970 (Vol III)

MEDICAL MYCOLOGY

BENNETT JE: Chemotherapy of systemic mycoses. N Engl J Med 290:30, 1974

CAMPBELL CC: Serology in the respiratory mycoses. Sabouraudia 5:240, 1967

CHU FS: Mode of action of mycotoxins and related compounds. Adv Appl Microbiol 22:83, 1977

CONANT NF, SMITH DT, BAKER RD, CALLAWAY JL: Manual of Clinical Mycology, 3rd ed. Philadelphia, Saunders, 1971

EMMONS CW, BINFORD CH, UTZ JP, KWON-CHUNG KJ: Medical Mycology, 3rd ed. Philadelphia, Lea & Febiger, 1977

HILDICK-SMITH G, BLANK H, SARKANY I: Fungus Diseases and Their Treatment. Boston, Little, Brown, 1964

KAUFMAN L, HUPPERT M, FAVANETTO C et al: Manual of Standardized Serodiagnostic Procedures for Systemic Mycoses. Washington DC, Pan American Health Organization, 1974

LARONE DH: Medically Important Fungi: A Guide to Identification. Hagerstown, Harper & Row, 1976

REBELL G, TAPLIN D: Dermatophytes: Their Recognition and Identification, revised ed. University of Miami Press, 1970

RIPPON JW: Medical Mycology: The Pathogenic Fungi and the Pathogenic Actinomycetes. Philadelphia, Saunders, 1974

SCHOLER HJ: Grundlagen und Ergebnisse der antimykotischen Chemotherapie mit 5-Fluorocytosin. Chemotherapy 22:103, 1976

SERMONTI G: Genetics of Antibiotic-Producing Microorganisms. New York, Wiley, 1969

SMITH JE, BERRY DR: An Introduction to Biochemistry of Fungal Development. New York, Academic Press, 1971

VAN DEN ENDE H: Sexual Interactions in Plants: The Role of Specific Substances in Sexual Reproduction. New York, Academic Press, 1976

VILLANUEVA JR, GARCIA-ACHA I, GASCÓN S, URUBURU F (eds): Yeast, Mould and Plant Protoplasts. New York, Academic Press, 1973

part V

VIROLOGY

RENATO DULBECCO
HAROLD S. GINSBERG

THE NATURE
OF VIRUSES

DISTINCTIVE PROPERTIES

Viruses are a unique class of infectious agents. They were originally distinguished because they are **small** (Table 46-1, Fig. 46-1) (hence the original term "filterable viruses") and because they are **obligatory intracellular parasites.** These properties, however, are shared by some bacteria. Thus the larger viruses (e.g., vaccinia) are as large as certain small bacteria (e.g., mycoplasmas, rickettsiae, or chlamydiae); and chlamydiae are obligatory intracellular parasites. The truly distinctive features of viruses are now known to lie in their **simple organization** and their **mechanism of replication.** In fact a complete viral particle, or **virion,** may be regarded mainly as a bloc of **genetic material** (either DNA or RNA) capable of autonomous replication, surrounded by a protein coat, and sometimes by an additional membranous **envelope,** which protects it from the environment and serves as a vehicle for its transmission from one host cell to another.

Mode of Replication. Unlike cells, viruses do not grow in size and then divide, because they contain within their coats few or none of the biosynthetic enzymes and other machinery required for their replication. Rather, viruses multiply in cells by *synthesis of their separate components, followed by assembly.* Thus the viral nucleic acid, after shedding its coats, comes in contact with the appropriate cell machinery, where it specifies the synthesis of proteins required for viral reproduction. The viral nucleic acid is then itself replicated, through the use of both viral and cellular enzymes; the components of the viral coat are formed; and these components are finally assembled (see Chs. 47 and 50). With some viruses replication is initiated by **enzymes present in virions** (Table 46-2) (their functions will be discussed in the appropriate chapters of this book).

Host Range. Viruses are subdivided into three main classes: **animal viruses, bacterial viruses (bacteriophages),** and **plant viruses.** Within each class each virus is able to infect only certain species of cells. The host range is determined by the specificity of attachment to the cells, which depends on properties of both the virion's coat and specific receptors on the cell surface. These limitations disappear when **transfection** occurs, i.e., when infection is carried out by the naked viral nucleic acid, whose entry does not depend on virus-specific receptors. On the other hand, the host range, for nucleic acid as well as for virions, is affected by the availability of cellular factors required for viral replication.

Virus-Related Genomes. **Bacterial plasmids,** like viruses, are present in cells as naked genomes; however, they do not form a coat, and they are transmitted from one bacterium to another by conjugation rather than as free extracellular particles. **Viroids,** which produce diseases in plants, are small naked infectious molecules of circular single-stranded RNA with extensive internal base pairing. The tight folding of the RNA and the closed circle presumably protect them from extracellular nucleases.

VIRUSES AS INFECTIOUS AGENTS

Two properties of viruses explain why they are infectious, pathogenic agents: virions produced in one cell can invade other cells and thus cause a spreading infection, and viruses cause important functional alterations of the invaded cells, often resulting in their death.

The role of viruses as infectious agents was recognized long before their true nature was understood. In 1898 Loeffler and Frosch proved that the agent of foot-and-mouth disease in cattle can be transmitted by a cell-free filtrate; transmission for a plant virus was demonstrated even earlier (by Ivanovsky, in 1892). These findings paved the way for the recognition of many other viral agents of infectious disease. Soon the tumor-producing ability of viruses was also indicated by the discovery of the viral transmission of fowl leukosis (Ellerman and Bang, 1908) and of a chicken sarcoma (Rous, 1911). The discovery of bacteriophages, made independently by Twort in England and D'Herelle in France in 1917, was of great significance for the development of virology as a science, because it afforded an important model system for investigations of basic virology (Chs. 47 and 48).

ARE VIRUSES ALIVE?

Life can be viewed as a complex set of processes resulting from the actuation of the instructions encoded in nucleic acids. In the nucleic acids of living cells these are actuated all the time; in contrast, in a virus they are ac-

TABLE 46-1. Characteristics of Viruses

Morphologic class	Nucleic acid*	Example Virus family	Example Virus	Size of capsid (nm)	No. of capsomers	Size of virions (enveloped viruses)nm	Special features
Helical capsid Naked	DNA	Coliphage fd		5 × 800			Single-stranded DNA
	RNA	Many plant viruses	Tobacco mosaic	17.5 × 300			
			Beet yellow	10 × 1200			
Enveloped	RNA	Orthomyxoviruses	Influenza	9 diam.		90–100	Segmented RNA
		Paramyxoviruses	Newcastle disease	18 diam.		125–250	
		Rhabdoviruses	Vesicular stomatitis			68 × 175	Bullet shape
Icosahedral capsid Naked	DNA	Parvoviruses	Adeno-associated	20	12		Single-stranded linear DNA
			Coliphage φx 174	22	12		Single-stranded cyclic DNA
		Papovaviruses	Polyoma	45	72		Cyclic DNA
			Papilloma	55	72		Cyclic DNA
		Adenoviruses		60–90	252		
	RNA	Coliphage F2 and others		20–25			
		Picornaviruses	Polio	28	32		
		Many plant viruses	Turnip yellow	28	32		
		Reoviruses		70	92		Segmented double-stranded RNA
Enveloped	DNA	Herpesviruses	Herpes simplex	100	162	180–200	
Capsids of binal symmetry (i.e., some components icosahedral, others helical) Naked	DNA	Large bacteriophages	T2,T4,T6	Modified icosahedral head 95 × 65; helical tail 17 × 115			
Complex virions	DNA	Poxviruses	Variola } Vaccinia			250 × 300	Brick shape
			Contagious pustular dermatitis of sheep			160 × 260	

*DNA double-stranded, RNA single-stranded, unless specified in last column.

tuated only when the viral nucleic acid, upon entering a host cell, causes the synthesis of virus-specific proteins. Viruses are thus "alive" when they replicate in cells, while outside cells viral particles are metabolically inert and are no more alive than fragments of DNA (e.g., the DNA used in bacterial transformation).

When Stanley, in 1935, crystallized tobacco mosaic virus, there followed extensive debates on whether such a crystallizable substance was a living being or merely a nucleoprotein molecule. As Pirie pointed out, these discussions showed only that some scientists had a more te-

leologic than operational view of the meaning of the word "life." Just as physicists recognize light either as electromagnetic waves or as particulate photons, depending on the context, so biologists can profitably regard viruses both as exceptionally simple microbes and as exceptionally complex chemicals.

THE ANALYSIS OF VIRUSES

Physical and chemical determinations require fairly large amounts of highly purified virus, separated from infected host cell constitu-

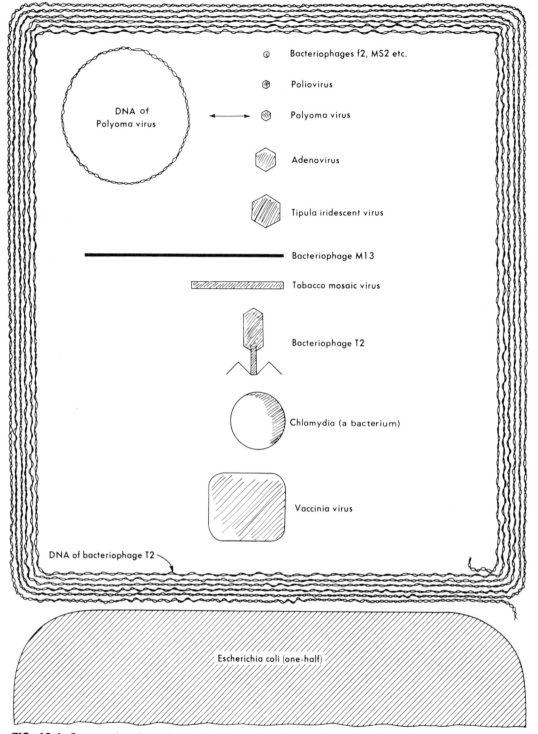

FIG. 46-1. Comparative sizes of virions, their nucleic acids, and bacteria. The profiles and the lengths of the DNA molecules are all reproduced on the same scale.

TABLE 46-2. Virion Enzymes

Enzyme	Virus	Product or function
Enzymes affecting interaction of virions with the host cell surface		
Neuraminidase	Orthomyxovirus, paramyxovirus	Splits off NANA from surface polysaccharides
Endoglycosidase	E. coli K bacteriophages	Breaks down surface polysaccharides
Fusion factor*	Paramyxovirus	Alters lipid bilayer
Enzymes transcribing the viral genome into messenger RNA		
DNA-dependent RNA polymerase	Poxvirus, polyhedrosis viruses of frogs, bacteriophages N4, SP02	Single-stranded mRNA
Double-stranded RNA transcriptase	Viruses with double-stranded RNA	Single-stranded mRNA
Single-stranded RNA transcriptase	Viruses with single-stranded RNA (negative strand)	Single-stranded mRNA (positive strand)
Enzymes adding specific terminal groups to viral mRNA made in virions		
Nucleotide phosphohydrolase	Viruses synthesizing mRNA in virions (e.g., poxviruses, reoviruses)	Converts terminal 5'-triphosphate to diphosphate as prelude to guanylylation
Guanylyl transferase	"	Adds guanylyl residue to 5'-end diphosphate in mRNA
RNA methylases	"	Methylate guanylyl residue at 5' end of mRNA and some riboses in 2' position
Poly(A) polymerase	"	Synthesizes poly(A) tail at 3' end of mRNA
Enzymes involved in copying virion RNA into DNA		
RNA-dependent DNA polymerase (reverse transcriptase)	Retroviruses	DNA–RNA hybrids; double-stranded DNA
RNase H (in association with reverse transcriptase)	Retroviruses	Breaks down RNA strand in RNA–DNA hybrids
Polynucleotide ligase	Retroviruses	Closes single-strand breaks in double-stranded DNA
Enzymes for nucleic acid replication or processing		
RNA-dependent DNA polymerase	Hepatitis B	Synthesizes double-stranded DNA
Deoxyribonucleases (exo- and endo-)	Poxviruses, retrovirus, adenovirus	Break DNA chains and crosslinks
Endoribonuclease	Viruses with single-stranded mRNA (e.g., poxvirus)	Processing of mRNA
Other enzymes		
Protein kinases	Retrovirus, orthomyxovirus, paramyxovirus, herpesvirus, adenovirus	Phosphorylate proteins
tRNA aminoacylases	Retrovirus	Aminoacylate tRNA

NANA = N-acetylneuraminic acid.

*No enzymatic activity known.

ents. The following methods of purification have proved especially valuable. 1) **Zonal sedimentation** in a density gradient separates particles on the basis of differences in sedimentation constants; 2) **Density gradient equilibrium (isopyknic) centrifugation** (usually in CsCl) separates particles according to buoyant density; 3) **Partition between two polymer solutions** (such as dextran and polyethylene glycol) separates particles according to surface properties; and 4) **Extraction with a solvent** (such as fluorocarbon) removes membrane fragments and other cellular constituents from the water phase, leaving most unenveloped viruses unaltered.

Danger of Contamination with Other Viruses. Since viruses are propagated in cells, which often harbor unrecognized viruses, a preparation may contain viral contaminants, and these may even be enriched during purification. This possibility can be minimized by monitoring the physical, immunologic, and biologic properties of the material during the various steps of viral propagation and purification.

Electron Microscopic Technics. The following technics reveal different structural details of viruses.

1) Shadowcasting (Fig. 46-2). A metal vapor, projected at an angle onto a membrane to which dried viral particles are attached, coats it with an electron-opaque layer of metal. Absence of metal deposition behind a particle forms a "shadow," whose features reveal the size and shape of the particle.

2) Negative Staining (Fig. 46-3). The viral particles are mixed with a solution of a salt highly opaque to electrons, usually sodium phosphotungstate. The mixture is then spread in a thin layer on a carbon membrane and dried. The parts of the particles that are not penetrated by the salt stand out as electron-lucent areas on an opaque background. The penetration between protruding parts of the salt reveals details of surface structure.

3) Positive Staining. Certain components of viruses can be stained by salts that become selectively absorbed. Uranyl acetate stains the viral nucleic acid and other components; Abs conjugated

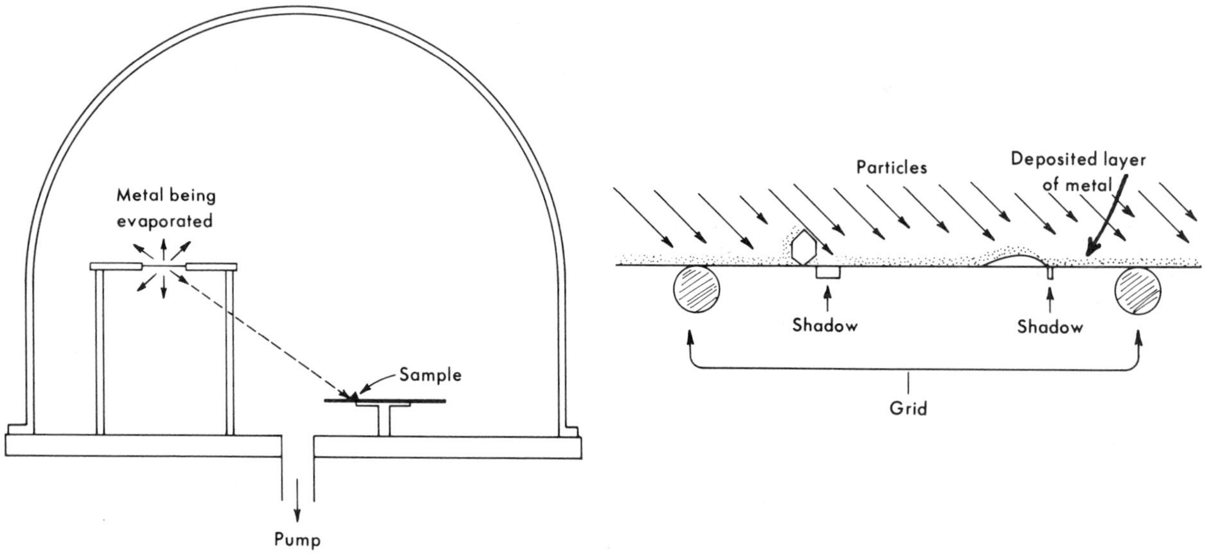

FIG. 46-2. Shadowcasting. A sample dried on a film supported by a grid is placed in an evaporation chamber, which is evacuated. Metal atoms projected from a glowing filament impinge at a predetermined angle on the film, where they form a relatively uniform layer of metal. The grid is then examined in the electron microscope. Particles present on the film cause the formation of "shadows," lacking deposited metal. The length and the shape of the shadows provide information on the three-dimensional shape of the particles.

to an electron-opaque molecule, such as ferritin, stain the proteins for which they have specificity. Positive staining can be combined with negative staining to improve the resolution.

4) Thin Sectioning. This is used to study viral particles in cells or in centrifugal pellets.

The size of a virion will be maximal in shadowed preparations, which enhance the contrast at the periphery of the particles; it will be smaller in negatively stained preparations, since the phosphotungstate penetrates the surface details; and it will be even smaller in sections, where the action of the knife tends to collapse the particles. Furthermore, the values obtained by all these methods are less than the size of the particles in water, since drying, as required for electron microscopic examination, causes shrinkage of as much as 30% in linear dimension.

THE VIRAL PARTICLES

GENERAL MORPHOLOGY

Electron microscopy shows that virions belong to several morphologic types (Fig. 46-4).

1) Some virions resemble small crystals. Extensive studies, especially by Klug and Caspar, have shown that these virions have an **icosahedral** protein shell (the **capsid**) surrounding a **core** of nucleic acid and proteins. The capsid and the core form the **nucleocapsid.** These virions are called **icosahedral virions.** (The icosahedron is a regular polyhedron with 20 triangular faces and 12 corners.) Examples are picornaviruses, adenoviruses, papovaviruses, and bacteriophage ϕX174 (Fig. 46-5A).

2) Some virions form long rods. The nucleic acid is surrounded by a **cylindrical capsid** in which a helical structure is revealed by high-resolution electron microscopy. They are called **helical virions.** Examples are tobacco mosaic virus (Fig. 46-5B) and bacteriophage M13.

3) Virions of more complicated morphology contain lipids. In most cases the nucleocapsid—in some viruses isosahedral, in others helical—is surrounded by a loose membranous **envelope.** Enveloped virions are roughly spherical but highly **pleomorphic** (i.e., of varying shapes) because the envelope is not rigid. Examples of **enveloped icosahedral** viruses are herpesviruses and togaviruses (Fig. 46-5C). **In enveloped helical** viruses, such as orthomyxoviruses (Fig 46-5D), the nucleocapsid is coiled within the envelope.

Virions of more complex structure belong to two groups. Those illustrated by poxviruses (Fig. 46-5E) do not possess clearly identifiable capsids, but have several coats around the nucleic acid; while certain bacteriophages (Fig. 46-5F) have a capsid to which additional structures are appended.

The morphologic types of the representative viruses are given in Table 46-1.

The morphology of virions is so characteristic that electron microscopy can establish the presence of a viral

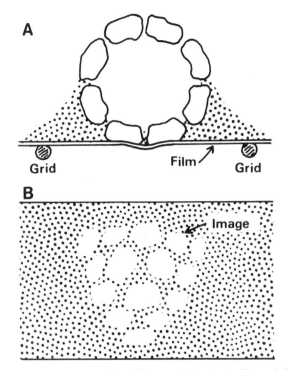

A

Grid Film Grid

B

Image

FIG. 46-3. Negative staining. The sample is mixed with a solution of a salt having high electron opacity (e.g., sodium phosphotungstate), and a thin layer is spread on the supporting film and dried. **A.** The salt solution surrounding the particles penetrates between particles and film and also into indentations in the surface of the particles. **B.** When the preparation is examined in the electron microscope, surface projections appear as transparent areas on an opaque background.

infection in a cell or organism, although (with some exceptions, such as rotaviruses) it cannot identify the virus. However, electron microscopy does allow identification by the binding of specific Abs to virions (see Immuno-Electron Microscopy, Ch. 52).

VIRAL NUCLEIC ACIDS

A given virus may contain either DNA or RNA, and this may be either single- or double-stranded. The proportion of nucleic acid in the virion varies from about 1% for influenza virus to about 50% for certain bacteriophages, and the amount of genetic information per virion varies from about 3 to 300 kilobases (Kb) per strand. If 1 Kb is taken as the size of an average gene, *small viruses contain perhaps three or four genes and large viruses contain several hundred.* The diversity of virus-specific proteins synthesized in the infected cell varies accordingly. With the exception of retroviruses (Ch. 66), virions contain only a single copy of the nucleic acid; i.e., they are **haploid.**

Double-Stranded Viral DNA. The molecular weights of viral DNAs, obtained by various physical methods, by radioautography (Fig. 46-6), or by electron microscopy, are given in Table 46-3. The base composition varies considerably in different viruses; some even contain **abnormal bases.** Thus, in the DNA of T-even coliphages (Ch. 47), 5-hydroxymethylcytosine (HMC) replaces the usual cytosine. This finding of Wyatt and Cohen made it possible to measure the replication of viral DNA in the presence of host cell DNA. The formation of this novel base also first revealed that phage-infected cells synthesize new enzymes, absent in uninfected cells. Moreover, the hydroxymethyl group of 5-hydroxymethylcytosine is glucosylated, with a pattern that depends on the glucosylating enzymes made in cells after infection by the corresponding phages (see Role of DNA Modifications: Host-Induced Restriction and Modification, Ch. 47).

Thymine is replaced by 5-hydroxymethyluracil or 5-dihydroxypentyluracil in certain *Bacillus subtilis* bacteriophages, and by putrescinyl-thymine in a *Pseudomonas* phage. These substitutions change certain physical properties of the DNA, such as the melting temperature or the buoyant density in CsCl.

Some viral DNAs have **special configurations.** Thus in some small viruses they are **cyclic** (Fig. 46-7). This property seemed surprising until it was recognized that it circumvents a replication problem of linear DNA, in which the strand that grows backwards via Okazaki segments cannot be completely replicated (Fig. 46-8A) because the last segment usually does not coincide with the end of the molecule. Special features eliminate this difficulty in linear DNAs (Fig. 46-8B–E). Thus **cohesive ends** (B) allow the DNA to cyclize upon entering the cells, and **terminal repetitions** (C) allow incomplete replicas to form complete tandem additions. In other cases (e.g., in adeno-associated viruses) a terminal **palindrome** (able to form a hairpin after denaturation and renaturation) can form a hairpin primer for the completion of the lagging strand (D); and a **protein covalently bound** to the 5′ end of each strand of the DNA of adenovirus and some phages apparently fulfills the same function (E). The free protein has a covalently bound nucleoside; during replication this protein becomes temporarily associated with the 3′ end of a strand, so that the bound nucleoside can act as primer for the synthesis of the complementary strand. The protein then remains covalently connected to the new strand.

Other special features of uncertain significance include **single-stranded breaks** at fixed positions in phage T5 DNA, which may identify parts with special functions during infection; and terminal covalent **crosslinks** between strands in poxvirus DNA, which cause the strands, when separated in alkali, to snap back after neutralization. During infection the crosslinks are cut by a virion endonuclease. In adenovirus DNA **inverted repetitions** of 100 bases allow a single strand to anneal into a "panhandle" structure. In herpes-

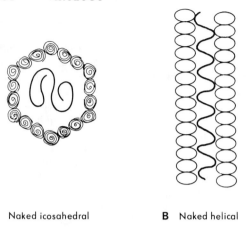

A Naked icosahedral

B Naked helical

C Enveloped icosahedral

nucleocapsid

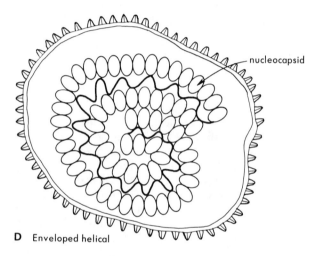

nucleocapsid

D Enveloped helical

protomers (protein)

capsomers (protein)

nucleic acid

spikes (glycoprotein)

envelope (protein and lipids)

FIG. 46-4. Simple forms of virions and of their components. The naked icosahedral virions (**A**) resemble small crystals; the naked helical virions (**B**) resemble rods with a fine regular helical pattern in their surface. The enveloped icosahedral virions (**C**) are made up of icosahedral nucleocapsids surrounded by the envelope; the enveloped helical virions (**D**) are helical nucleocapsids bent to form a coarse, often irregular, coil within the envelope.

virus DNA two sections of unique sequences are bracketed by inverted repeats, which undergo frequent recombination; hence the viral DNA contains four equimolar classes of molecules, with the two unique sections in all possible orientations.

Single-Stranded DNA. The DNA is single-stranded in very small bacteriophages (e.g., φX174, icosahedral; f1 and M13, helical) or animal viruses (parvoviruses). In the phages, the DNA is cyclic and is always of the same strand (called the **viral** strand), since molecules from different virions do not form helices on annealing. In contrast, in the adeno-associated viruses the two complementary strands exist in different virions.

Viral single-stranded DNAs also have **inverted repeat sequences** that form hairpins which are important in replication, or in the cyclization of progeny linear strands.

The great diversity of the structures involved in replication of single- and double-stranded viral DNAs suggests that when primordial viral DNAs were formed (probably from segments of cellular DNA that included an origin of replication), different preexisting features were incorporated into the replication scheme, and the other steps were adapted to them.

Double-Stranded RNA; Segmentation of the Genome. Double-stranded RNA, recognizable for its sharp melting, resistance to RNase, and complementary base ratios, is found in several unrelated icosahedral viruses of animals (reovirus of humans and other animals, bluetongue virus of sheep, cytoplasmic polyhedrosis virus of silkworm), and of plants (wound tumor virus and rice dwarf virus). A common feature of these viruses is the

FIG. 46-5. Electron micrographs of representative virions, with negative staining. **Markers** under each micrograph are 100 nm. **A.** Naked icosahedral: human wart virus (papovavirus, Ch. 66). **B.** Naked helical: segment of tobacco mosaic virus. **C.** Enveloped icosahedral: herpes simplex virus (herpesvirus, Ch. 55). **D.** Enveloped helical: influenza virus (orthomyxovirus, Ch. 58). **E.** Complex virus: vaccinia virus (poxvirus, Ch. 56). **F.** Coliphage λ (Ch. 48). (A, Noyes WF: Virology 23:65, 1964; B, Finch JT: J Mol Biol 8:872, 1964. Copyright by Academic Press, Inc. (London) Ltd.; C, courtesy of P. Wildy; D, Choppin PW, Stockenius W: Virology 22:482, 1964; E, courtesy of R. W. Horne; F, courtesy of F. A. Eiserling)

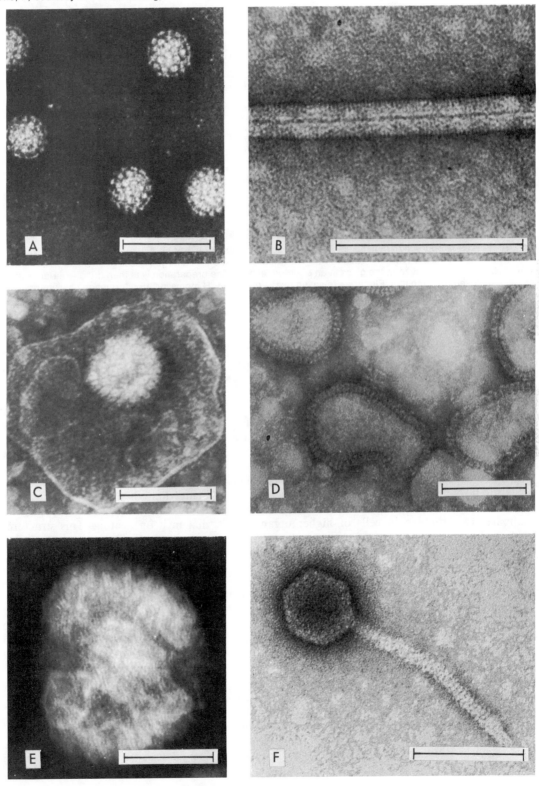

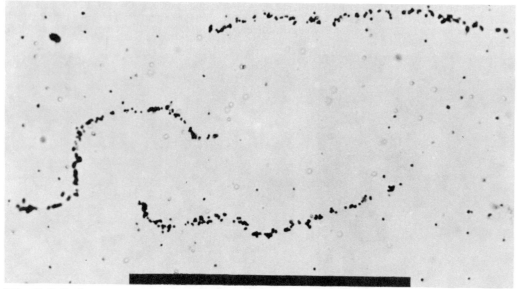

FIG. 46-6. Radioautograph of molecules of DNA of bacteriophage T2 (Ch. 47). A specially prepared glass slide was drawn through a solution containing T2 DNA heavily labeled with ^{3}H-thymidine, mixed with a large excess of unlabeled T2 DNA. This procedure caused some DNA molecules to stick to the glass in an oriented fashion. The preparation was then dried, overlaid with sensitive photographic film, and exposed for 2 months. Electrons released by the decaying ^{3}H atoms initiated formation of silver grains in the emulsion, which were made visible by photographic development. (Courtesy of J. Cairns)

segmentation of the genome; thus reovirus contains ten or more nonoverlapping segments totaling 21 Kb per strand (see Ch. 64). The segments are not produced by artifactual breakage, since they are present in intact virions, where their free 3′ ends can be labeled by suitable reagents. In virions the segments may be connected by noncovalent bonds, such as short stretches of base pairs. Each segment is transcribed into a separate mRNA and specifies a separate polypeptide chain.

As noted in Chapters 49 and 50, genome segmentation appears to be one of the devices for avoiding the internal initiation of translation in multicistronic mRNAs, which is apparently forbidden in cells of higher organisms. DNA can be multicistronic, because it can be transcribed into monocistronic mRNAs either directly or after processing of the primary transcript.

Single-Stranded RNA. Many viruses, either helical or icosahedral, contain single-stranded RNA of 1–17 Kb in total length. Some RNAs are in one piece, others are **segmented.** All have considerable **secondary structure,** which may have an important function in replication and translation (see RNA Phage, Ch. 47). Most of these viruses have a **positive RNA strand,** i.e., one functioning as mRNA in the cells. Animal viruses with a positive strand have structures found in mRNAs of the host cells: a **poly(A) sequence** at the 3′ end and a **cap** at the 5′ end (see Transcription of Chromatin, Ch. 49). An exception is picornavirus RNA, in which the cap is replaced by a special sequence, to which a small protein is covalently attached. This protein probably acts as a primer during synthesis of the viral RNA. In positive-strand plant viruses the RNAs have caps but no poly(A); instead their 3′ ends resemble those of tRNAs. The RNA of each of these viruses can be charged with a specific amino acid by the appropriate *E. coli* aminoacyl-tRNA synthetase. All these features suggest a relatedness of these RNA viruses to cellular polysomal components.

Retroviruses (Ch. 66) have the unusual feature of being diploid: they contain two equal positive-strand RNA molecules held together near the 5′ ends by bonds that melt on heating. This structure may have special meaning in replication.

Negative-strand viral RNAs are terminated at the 5′ end by a nucleoside triphosphate, not a cap. Each is transcribed into complementary mRNAs by a transcriptase present in the virions (see Table 46-2).

Defective Nucleic Acids. Stocks of either DNA or RNA viruses often have defective virions. Some contain only a fraction of their normal nucleic acid complement; others have a nucleic acid of nearly normal length, but made up of repetitions of a limited viral sequence (usually containing the origin of replication) or even of cellular nucleic acid. These virions are not infectious but in most cases their nucleic acids can replicate in cells coinfected by regular virus as **helper.** Often they replicate more rapidly than the nucleic acid of the helper, either because they are shorter or because they have several replication origins. Some defective virions retaining viral functions can cause cell transformation (Ch. 66); others in-

TABLE 46-3. Characteristics of Viral Nucleic Acids

Type of nucleic acid	Representative virus	Mol wt (in 10⁶ daltons)	Kilobases per strand*	Number of segments	Polarity
DNA, double-stranded					
Hepatitis B (cyclic)		1.6	2.5		
Papovavirus (cyclic)	Polyoma	3.5	5.0		
	Papilloma	6	9		
Pseudomonas phage PMS2 (cyclic)		6	9		
Adenovirus	Types 12,18	21	32		
	Types 2,5	23	35		
Coliphages T3,T7		25	38		
Coliphage Mu		26	39		
Coliphage λ		31	47		
Coliphate T5		77	117		
Herpesvirus	Herpes simplex	100	151		
Coliphages T2,T4,T6		110	167		
B. subtilis phage SP8		130	197		
Poxvirus	Vaccinia	160	242		
DNA, single-stranded					
Parvovirus	Adeno-associated†	1.5	4.5		
Coliphage φx174 (cyclic)		1.7	5.2		
Coliphage M13 (cyclic)		2.4	7.3		
RNA, double-stranded					
Reovirus		15‡	23	10	
Rice dwarf virus		15‡	23	10	
Cytoplasmic polyhedrosis of silkworms		15‡	23	10	
RNA, single-stranded					
Satellite necrosis virus†		0.4	1.2	1	
Coliphage R17		1.3	4	1	+
Tobacco mosaic virus		2	6	1	+
Turnip yellow mosaic virus		2	6	1	+
Picornavirus	Polio	2.5	7.5	1	+
Bunyavirus		3‡	9	3	−
Retrovirus§	Rous sarcoma virus	3.5	10.5	1	+
Alphavirus	Sindbis	4	13	1	+
Rhabdovirus	Vesicular stomatitis virus	4	13	1	−
Myxovirus	Influenza	6‡	18	8	−
Paramyxovirus	Newcastle disease	6	18	1	−

+ = Positive; − = negative.

*A **kilobase** (1000 bases) corresponds to a mol wt of about 700,000 for double-stranded and 350,000 for single-stranded nucleic acid; it can specify about 33,000 daltons of protein. The number of genes is approximately equal to the number of kilobases; φx 174 has less kilobases because *some genes overlap each other* (see Ch. 47).

†These viruses are defective and multiply only in cells infected by a helper virus (adenovirus or tobacco necrosis virus, respectively). They probably specify only their own capsid, perhaps with another small protein.

‡This value, as for other virions with segmented genomes, is the aggregate of all fragments.

§Retroviruses have diploid virions.

hibit the replication of regular virus infecting the same cell (see Defective Interfering Particles, Ch. 51).

THE CAPSID

The capsid accounts for most of the virion mass, especially in small viruses. In naked virions it protects the nucleic acid from nucleases in biologic fluids and promotes the attachment to susceptible cells.

As was pointed out by Crick and Watson, a virus cannot afford too many genes for specifying capsid proteins; hence the capsid must be formed by the association of many identical protein units (protomers). Thus, poliovirus RNA (7Kb) can specify at most 250,000 daltons of different proteins, and some of these must be used for replication; yet the poliovirus capsid weighs about 6×10^6 daltons.

The prediction has been borne out: after dissociation

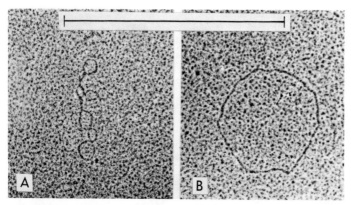

FIG. 46-7. Electron micrographs demonstrating the cyclic structure of polyoma (a papova-virus) DNA molecules (Kleinschmidt technic). The native molecules (**A**) are twisted (super-coiled); the twist is maintained as long as the two strands constituting the molecule are both intact. If one of them is broken the corresponding phosphodiester bond of the other strand acts as a swivel, allowing the DNA molecule to rotate around the helix axis until it forms an untwisted circle (**B**). **Marker**=1 μm. (Weil R, Vinograd J: Proc Natl Acad Sci USA 50:730, 1963)

into their component polypeptides, capsids were invariably found to contain a small number of different polypeptide chains separable by acrylamide gel electrophoresis and ranging in mol wt between 10,000 and 100,000 daltons. Helical capsids usually consist of a single type of protein unit (**protomer**); icosahedral capsids may have one or several types. The protein units of the capsid are **specified by viral genes,** because their amino acid sequence is altered by some viral mutations. The shape and dimensions of the capsid depend on characteristics of its protomers, and also, for helical capsids, on the length of the viral nucleic acid. Accordingly, the capsid is uniquely determined by the constitution of the viral genome.

The protomers forming the capsid must be connected by bonds between suitable chemical groups on their surfaces. With one or few kinds of protomer, only a small number of suitable chemical groups is available. Hence, the protomers must be arranged in a regular architecture that utilizes bonds between the same pairs of chemical groups. This goal is attained in different ways in the helical and the icosahedral capsids, as shown by extensive x-ray crystallographic studies.

HELICAL CAPSIDS

This roughly cylindrical capsid (Fig. 46-9) has the simplest organization. The identical protomers are bound end to end by identical bonds to form a ribbonlike structure, which is wound around the axis of the helix. The protomers in two successive turns of the ribbon form lateral bonds with each other, which are identical for all protomers; however, since these protomers are not aligned with each other laterally, each protomer forms bonds with two protomers of adjacent turns. This pattern confers great stability on the structure.

The straight line in the center of the cylinder is an **axis of rotational symmetry,** because rotation by a certain fixed angle around this axis leaves the capsid pattern unchanged.

The diameter of the helical capsid is determined by the characteristics of its protomers, while the length is determined by the length of the nucleic acid it encloses.

The capsids of naked helical viruses (e.g., tobacco mosaic virus) are very tight (Fig. 46-5B). In contrast, the capsids of enveloped helical viruses are very flexible, as they have to coil within the envelope; in some viruses they form rings, although the nucleic acid is not cyclic. Often the turns of the helices are visible in electron micrographs (Fig. 46-10), suggesting a loose structure; in these viruses the envelope rather than the capsid may provide the required barrier to nucleases.

ICOSAHEDRAL CAPSIDS

These capsids have two levels of organization. Short ribbons of protein monomers form structures known as **capsomers,** recognizable in electron micrographs with negative staining as rings with a central hole (Fig. 46-11) which usually does not extend through the thickness of the capsid. If the hole is too small, the capsomers appear as solid knobs (Fig. 46-12). Some capsomers, the **pentons,** are each connected to five other capsomers, whereas the **hexons** are each connected to six capsomers. Pentons are probably formed by five protomers, and hexons usually by six (Fig. 46-13); adenovirus hexons each contain only three protomers.

Protomers in a capsomer (Fig. 46-13), and capsomers

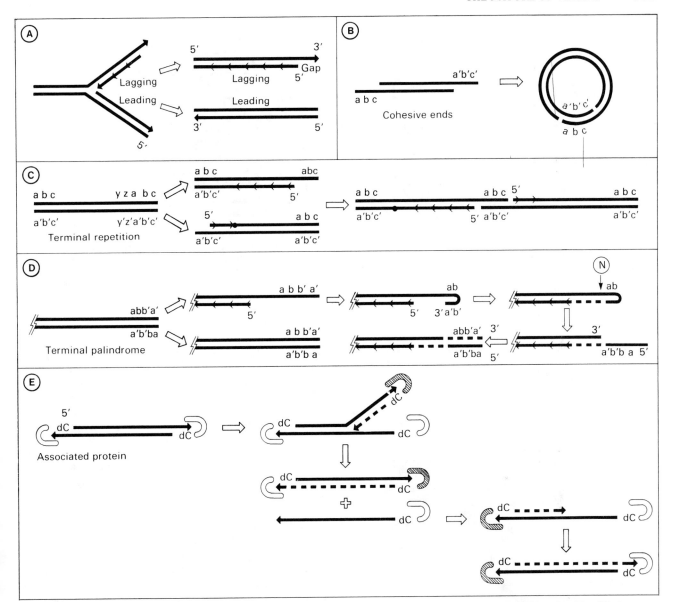

FIG. 46-8. Various kinds of structures ensure complete replication of the 5′ end of the lagging DNA strand (**A**), i.e., the strand that grows, by accretion of Okazaki segments, in a direction opposite to that of synthesis (which occurs at the 3′ end). **B.** Cohesive ends allow cyclization. **C.** Repetitious ends allow formation of tandem dimers from incompletely replicated progeny (see Viral DNA Replication, Ch. 47). **D.** A terminal palindrome allows complete replication through hairpin formation, extension of the 3′ end of the hairpin, nicking (N) of the parental strand by endonuclease, and its reconstitution by extension of the 3′ end. Single-stranded DNA can also use this method of replication. **E.** A protein bound to a nucleoside (here dC) recognizes the 3′ end of a strand and opens the helix; the dC serves as primer for the beginning of the new complementary strand to which it remains covalently bound; the other parental strand goes independently through the same process. The strands replicate one at a time, so there is no lagging strand.

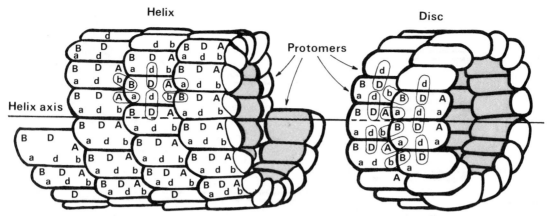

FIG. 46-9. Constitution of the helical capsid. All protomers are identical and establish regularly repeated bonds with their neighbors between chemical groups indicated by **letters**. Since each protomer is staggered in respect to its lateral neighbors, it forms bonds with the two of them on each side along the helix axes. This confers considerable stability on the capsid. The protomers assemble first to constitute flat nonhelical discs containing two rings of protomers, with the staggered arrangement found in the finished helix. Under physiologic conditions when the protomers of the discs associate with the RNA they shift slightly to produce a helix. The helix grows in length by the addition and assimilation of discs.

FIG. 46-10. The helical capsid of a paramyxovirus with negative staining. Two particles of the simian paramyxovirus SV5 are seen: from both particles segments of the helical capsid protrude (**arrows**), probably owing to rupture of the envelope. Note the loose arrangement of the protomers and the hole along the axis of the helix. The envelopes are covered by the characteristic spikes (**S**). (Choppin PW, Stockenius W: Virology 23:195, 1964)

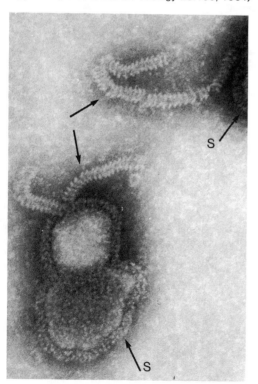

in a shell (Fig. 46-14), are held together by noncovalent bonds. The bonds between protomers are usually stronger than those between capsomers: empty capsids often disintegrate into intact capsomers during purification (Fig. 46-11).

Twelve pentons, without hexons, form the icosahedral capsids of the smallest viruses (Table 46-1). Hexons alone can only form flat sheets or cylinders (Fig. 46-15), but together with pentons they can form icosahedrons of many different sizes; hence most viral capsids have both types of capsomers.

Like the helical capsids, icosahedral capsids have axes of rotational symmetry. These are of three different kinds (Fig. 46-16). **Fivefold axes** connect the center of the capsid to the center of each penton at the 12 corners of the icosahedron. They are so named because an identically oriented pattern is reproduced five times when the capsid undergoes a complete rotation around one of these axes. **Threefold axes** go through the centers of each triangular face and **twofold axes** go through the middle of each ridge between the faces. The icosahedron is thus referred to as a solid with a 5:3:2 rotational symmetry.

Animal viruses usually have different protomers in pentons and in hexons, but plant viruses and RNA phages use the same protomer for both. The latter protomers are internally flexible in order to adapt to the different angles of the bonds with each other in pentons and in hexons.

Variations of the Icosahedral Capsid. In adenoviruses, hexons contain only three protomers, each able to bind to two neighbors. Large capsids, such as those of T-even phages, adenoviruses, or herpesviruses, contain several types of polypeptide chains; all of

FIG. 46-11. Preparation of rabbit papilloma virus (papovavirus) containing mostly empty capsids, i.e., devoid of nucleic acid. Some of the capsids have disintegrated during the preparation of the specimen for electron microscopy, each producing a small puddle of capsomers (some of the capsomers of the original capsid have been lost). The angular polygonal shape of the capsomers is evident, but it is not possible to differentiate hexamers from pentamers. (Breedis C et al: Virology 17:84, 1962)

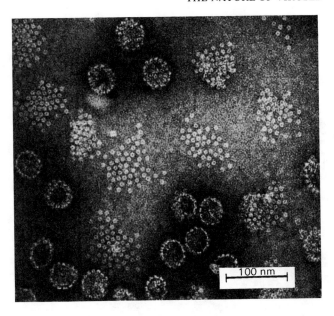

them conform to the symmetry of the icosahedron both for position and for bond arrangement. Similarly, in the capsid of reoviruses, which is made up of two layers of protomers (Fig. 46-17), each layer satisfies the requirements of an icosahedral capsid. Small fibers (presumably possessing fivefold rotational symmetry) protrude at the corners of the capsids in adenoviruses and in bacteriophage φX174.

FIG. 46-12. Electron micrograph of GAL virus (chicken adenovirus) by negative staining, showing capsomer structure. The **arrowed** capsomers are situated on the fivefold axes. (Wildly P, Watson JD: Cold Spring Harbor Symp Quant Biol 27:25, 1962)

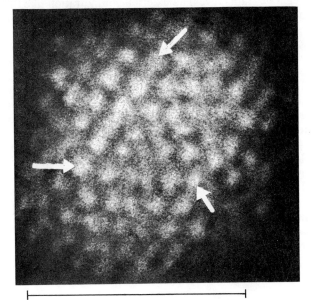

62.5 nm

Why are Capsids Icosahedral? The aggregation of protomers to form an icosahedral shell follows the laws of crystallography. That the capsid must be characterized by rotational symmetry can be seen by considering first the two-dimensional problem of making a ring on a sheet of paper with a number of identical **asymmetric** protomers. Figure 46-18 shows how this can be done with five protomers; the figure obtained has rotational symmetry with a fivefold axis passing through its center but the asymmetry of the protomers prevents bilateral symmetry. This is the basis of the helical capsid. The same reasoning extended to the three-dimensional

FIG. 46-13. Possible constitution of capsomers from protomers. In the icosahedral capsids, the protomers constitute oligomers of either five or six protomers, called capsomers, which are drawn as seen from the outside of a virion. Every protomer in a capsomer establishes bonds with two neighboring protomers, always through the same chemical groups (diagrammatically indicated as **A–a** and **B–b** in the hexon, and **c–C** and **d–D** in the penton). Each protomer also has other contacts with neighboring capsomers in the complete capsid.

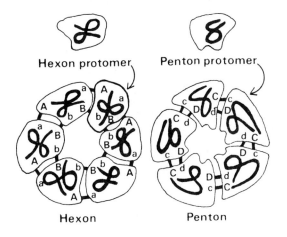

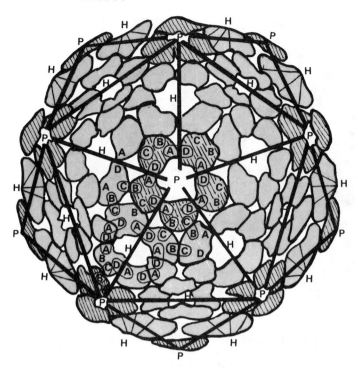

FIG. 46-14. Position of capsomers in an icosahedral capsid. The capsomers are shown with reference to the edges between triangular faces of the icosahedron. The icosahedron depicted here contains 42 capsomers, of which 12 are pentagonal (**P**) and 30 hexagonal (**H**). All are made up of identical protomers. The hypothetical chemical groups involved in the bonds between protomers are indicated by letters (**A, B, C, D**). Identical intracapsomeric A–D bonds connect protomers of the same capsomer, and identical intercapsomeric B–C bonds connect protomers of different capsomers.

case leads to the conclusion that here the capsid must also contain only rotational symmetry. Of several theoretically possible shapes, the icosahedron has the advantage of being more stable and of requiring smaller protomers, resulting in economy of genetic information. Many kinds of icosahedrons are employed in capsids of different viruses; they are discussed in the appendix to this chapter.

Allosteric Transitions of Capsids. The large number of individual bonds permits important conformational changes in capsid structure with little expenditure of energy. One example is the change produced by the association with the nucleic acid (see Location of the Nucleic Acid, below); in poliovirus a transition induced by lowering the pH or by interaction with Abs (see Ch. 52) markedly alters the exposure of ionized groups to the surface, with resulting changes in electrophoretic mobility. It is likely that conformational changes also determine the important functional effects that result from attachment of virions to cells or from heating (see Chs. 50 and 67).

LOCATION OF THE NUCLEIC ACID

The coiling of the nucleic acid in the very confined space within the capsid presents important topologic and thermodynamic problems, especially for the stiff double-stranded DNA. The arrangement of the nucleic acid is shown by electron microscopy, x-ray crystallography, and neutron diffraction.

In **icosahedral capsids** the nucleic acid, together with certain proteins, is tightly packed in a central **core,** forming a spool of parallel loops around a cylindrical hole. This folding is accompanied by some denaturation of the double helix. In icosahedral animal viruses the DNA, in conjunction with histones or histonelike pro-

teins, forms a **chromatinlike structure,** which can then fold tightly like cellular chromatin in condensed chromosomes (see Chromatin, Ch. 49).

Most preparations of icosahedral viruses, both with and without envelopes, include capsids without nucleic acid (**empty capsids,** Fig. 46-19). These can be separated from the nucleocapsids by equilibrium density gradient centrifugation, owing to their lower buoyant density. Negative stains reveal an external configuration like that in normal virions, and also a hollow center of the capsid.

The existence of empty capsids shows that the nucleic acid is not essential for capsid assembly (see Maturation and Release, Ch. 47, and Maturation and Release of Animal Viruses—Naked Icosahedral Viruses, Ch. 50). However, its presence increases the stability of the nucleocapsid: empty capsids are more subject to disintegration during preparation of specimens for electron microscopy (see Fig. 46-11).

In **helical capsids,** such as tobacco mosaic virus (TMV), the RNA is extended and is located in a helical groove between the protomers (Fig. 46-20) to which it is attached by multiple weak bonds. The axial location is confirmed by electron microscopy of virions with a partially digested capsid (Fig. 46-21). The interaction between nucleic acid and capsid often increases the stability of the capsid.

THE ENVELOPE

Envelopes, like cellular membranes (Ch. 49), contain lipid bilayers and proteins with special functions. There

FIG. 46-15. Very long filament of human wart virus (papovavirus) hexons, together with two regular virions. The filament, of a diameter close to that of the virions, is made up of hexagonal capsomers. (Noyes WF: Virology 23:65, 1964)

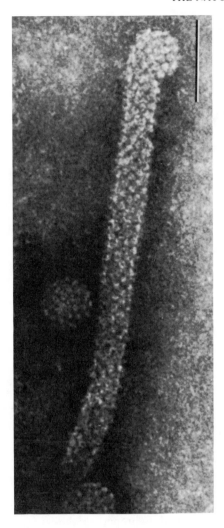

FIG. 46-16. Rotational axes in the icosahedron. The edges of the icosahedron, which limit the triangular faces, are drawn as **heavy lines**. The outlines of the capsomers are in **thin lines**. Pentons are **crosshatched. A.** The icosahedron of Figure 46-15 is viewed looking down the center of a pentagonal capsomer which corresponds to the corner of the polyhedron. Rotating the figures by 1/5 of a rotation reproduces the same figure. A fivefold rotational axis, therefore, passes through the center of the pentagonal capsomer. **B.** The same icosahedron viewed looking down the center of a triangular face. Through this point passes a threefold rotational axis which is situated between three hexagonal capsomers; rotating the figure by 1/3 of a rotation reproduces the same figure. **C.** The same icosahedron viewed looking down the middle of an edge between two triangular faces. This is an axis of twofold rotational symmetry and is situated in the center of a hexagonal capsomer.

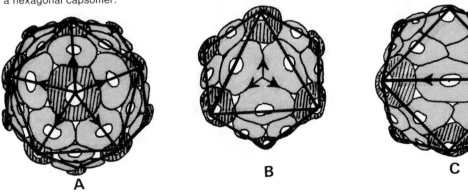

A   B   C

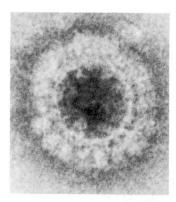

FIG. 46-17. The capsid of a reovirus particle examined with negative staining. The capsid is seen to be composed of two layers: the capsomers are especially evident in the outer layer. The capsid appears to be empty since it is penetrated by the stain. (Dales S, Gomatos PJ: Virology 25:193, 1965)

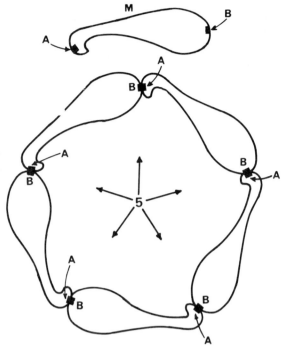

FIG. 46-18. Formation of a closed ring by five asymmetric protomers, **M**, in which group **B** can form a bond with group **A**. Since the distance between successive A–B bonds and the angle of the A–B bonds to the axis of the monomer are constant, a closed ring is formed which has fivefold rotational symmetry around an axis through its center.

are two kinds of **membrane proteins,** glycoproteins and matrix protein. **Glycoproteins** are exposed at the outer surface of virions; they can be labelled enzymatically with radioactive iodine in the intact virions, a procedure that does not label internal proteins. Glycoproteins have a glycosylated hydrophilic end protruding into the medium and a hydrophobic end held in the lipid bilayer. Some glycoproteins span the membrane and are connected to the underlying nucleocapsid or the matrix protein. Glycoproteins are often recognizable by electron microscopy as **spikes** on the outer surface of the virions (see Fig. 46-10).

In influenza virus the spikes, about 10 nm long and 7–8 nm apart, are of two kinds: some bind to RBCs, conferring on the virions hemagglutinating properties (**he-**

FIG. 46-19. Electron micrographs with negative staining of purified polyoma virus, a papovavirus. **A.** The full virions are not penetrated by the stain and show only the pattern on the surface of the capsid; **B.** The empty capsids are penetrated by the stain.

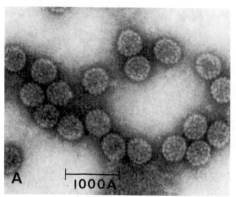

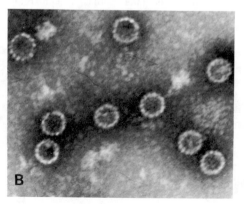

FIG. 46-20. Drawing of a segment of tobacco mosaic virus showing the helical nucleocapsid. In the upper part of the figure two rows of protein monomers have been removed to reveal the RNA. This drawing is based on results of x-ray diffraction studies. (Klug A, Caspar DLD: Adv. Virus Res 7:225, 1960)

magglutinins),* and others have **neuraminidase** activity (see Chemical and Physical Determinations, below). In paramyxoviruses a single kind of spike performs both these functions, while a second type causes **hemolysis** of RBCs and **fusion of tissue culture cells** to which the virus is adsorbed. Glycoproteins, although functionally essential, may not be required as structural elements of the envelope, because the virions of some paramyxovirus mutants, which lack a glycoprotein, have a normal morphology though they are noninfectious.

A second type of envelope protein is a nonglycosylated **matrix protein,** which forms a layer at the inner surface of the envelope in the virions of orthomyxo-, paramyxo-, and rhabdoviruses. This protein appears to establish the connection between the envelope and the capsid.

The proteins of the envelope, like those of the capsid, are specified by viral genes; they contribute the major antigenic specificities to the virions. However, the lipids and the carbohydrate moieties of glycoproteins depend on the host cell, and are often different in the same virus grown in different cells. Hence the *virion surface may contain polysaccharide-determined cellular Ags.*

In many enveloped viruses the liquid state of the lipids

---

* These hemagglutinins should not be confused with hemagglutinating Abs, which are often referred to by the same name.

and the absence of connections among the monomers of the envelope proteins prevent the formation of a rigid structure, leading to pronounced **pleomorphism** of the virions. Indeed, enveloped virions containing helical nucleocapsids may assume a bizarre tadpole-like shape when dried for electron microscopy, presumably because one end of the nucleocapsid unravels and pushes the envelope out. However, in other viruses a firm connection between envelope and nucleocapsid confers on the virions characteristic shapes. Thus the rhabdoviruses are notable for their **bullet-shaped virions,** with the helical nucleocapsids coiled under the outer layer (Fig. 46-22). In alphaviruses the lipid bilayer adheres closely to the icosahedral nucleocapsid, so that the virions appear nearly spherical.

The presence of lipids makes enveloped viruses sensitive to disinfection or damage by lipid solvents, such as ether.

OTHER VIRION COMPONENTS

In addition to the coat proteins, virions contain **internal proteins,** generally basic, which are bound to the nucleic acid in the core. In papovaviruses these proteins are regular cellular **histones,** while in adenoviruses they are histonelike but virus-specified. In both cases they form, with the viral DNA, complexes similar to the chromatin of animal cells. Small peptides and polyamines are present in bacteriophages. These polycationic compounds presumably help the folding of the nucleic acids by linking together different loops. Each reovirus virion contains about 2000 small oligonucleotides of unknown function.

COMPLEX VIRUSES

Poxviruses. The virions, brick-shaped or ovoid, hold the viral DNA, associated with protein, in a **nucleoid** shaped like a biconcave disc and surrounded by several lipoprotein layers. A layer of coarse fibrils near the outer surface gives the virions a characteristically striated appearance in negatively stained preparations (Fig. 46-5E; see also Ch. 56).

Large Bacteriophages. Some bacteriophages, such as the even-numbered coliphages T2, T4, and T6 (**T-even coliphages**), or phage λ, have very complex structures (Fig. 46-5F), including a head and a tail. They are said to have **binal symmetry** because they have some components with icosahedral and others with helical symmetry within the same virion.

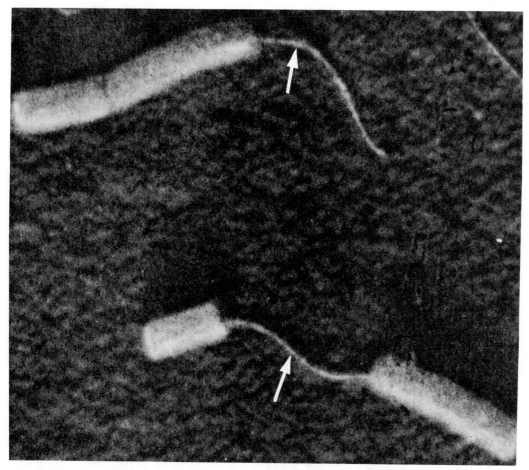

FIG. 46-21. Virions of tobacco mosaic virus in which segments of the capsid have been stripped away by detergent. Threads **(arrows)** protrude from the center of the remaining capsid segments; they can be destroyed by RNase, and represent the viral RNA centrally located in the nucleocapsid. (Corbett MK: Virology 22:539, 1964)

ASSAY OF VIRUSES

The methods used for the assay of viruses reflect their dual nature as both complex chemicals and living microorganisms. Viruses can be assayed by chemical and physical methods, by immunologic technics, or by the consequences of their interaction with living host cells, i.e., their **infectivity**. Assays carried out by different technics can differ vastly in their significance.

CHEMICAL AND PHYSICAL DETERMINATIONS

COUNTS OF PHYSICAL PARTICLES

Virions can be clearly recognized in the electron microscope; if a sample contains only virions of a single type their number can be determined unambiguously. Virions are counted by mixing the sample with a known number of polystyrene latex particles, viewing droplets of the mixture in the electron microscope, and counting the two types of particles present in the same droplet. Simple arithmetic then yields the number of virions in the total sample (Fig. 46-23). This technic does not distinguish between infectious and noninfectious particles.

HEMAGGLUTINATION

Many viruses, both small and large, can agglutinate RBCs. This important property, discovered independently for influenza virus by Hirst and by McClelland and Hare in 1941, affords a simple, rapid method for viral titration. Hemagglutination is usually caused by the virions themselves; in some cases, however, as with pox-

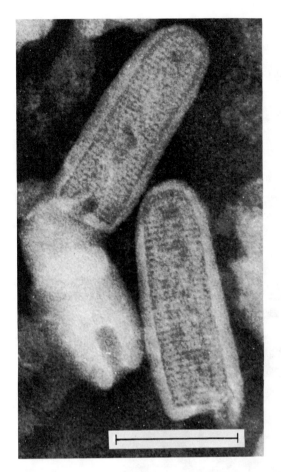

FIG. 46-22. Virions of vesicular stomatitis virus (a rhabdovirus) with negative staining. The helical filament, present in a deeper layer, is visible in two particles. (Howatson AF, Whitmore GF: Virology 16:466, 1962)

where they form a small, sharply outlined, round pellet. Aggregates, however, sediment to the bottom but do not roll; they form a thin film, which has a characteristic serrated edge. The proportion of aggregated cells can be counted more precisely by observing the sedimentation in a photoelectric colorimeter, since aggregated RBCs sediment faster than nonaggregated ones and can be separately measured.

The assay is usually carried out by an **endpoint procedure.** Serial twofold dilutions of the virus sample are each mixed with a standard suspension of RBCs (usually 10^7/ml). The last dilution showing complete hemagglutination is taken as the endpoint. The titer estimated by the pattern method has an inherent imprecision at least as large as the dilution step used (usually twofold); the colorimetric method reduces this imprecision. The titer obtained either way is expressed in **hemagglutinating units.**

Hemagglutination is inhibited by antiviral Abs; hence **hemagglutination inhibition** (Ch. 52), provides a convenient basis for measuring Abs to many viruses.

Mechanism of Hemagglutination. The phenomenon of hemagglutination throws considerable light on the interaction of orthomyxovirus (Ch. 58) and paramyxovirus (Ch. 59) virions with cell surfaces. It is caused by the attachment of the hemagglutinin spikes of the virion envelope to receptors on the RBC membrane. The receptors are N-acetylneuraminic acid (NANA) residues (Fig. 46-25) carried on the major glycoprotein of the RBC surface, the glycophorin. In fact, the receptors are inactivated by neuraminidase, which splits off the NANA. Moreover, hemagglutination is competitively inhibited by sialic acid–containing glycoproteins present in biologic fluids (e.g., serum, urine, submaxillary gland secretion), whose terminal NANA residues bind to virions. The removal of the terminal NANA by neuraminidase abolishes the activity of these inhibitors.

Neuraminidase is obtained from culture filtrates of the cholera vibrio. This enzyme is commonly referred to as **receptor-destroying enzyme** (RDE) because it destroys the receptor activity for orthomyxoviruses and paramyxoviruses on RBCs and susceptible host cells.

As noted above, neuraminidases are present on the surface of orthomyxovirus and paramyxovirus virions. At 37°C the viral neuraminidase ultimately dissociates the viruses from RBCs by splitting NANA from receptors; the virus then spontaneously elutes off the RBCs, which disaggregate. At 0°C, in contrast, the enzyme is much less active and the virion–RBC union is stable.

After the virus has eluted, cells cannot be agglutinated again by a new batch of virus, since they have lost the receptors; these cells are said to be stabilized. (The eluted virus, on the contrary, retains all its activities.) However, cells stabilized by a given orthomyxovirus or paramyxovirus can sometimes be agglutinated by another virus of these families. Indeed, it is possible to arrange the viruses in a series (called a **receptor gradient**) such that any virus will exhaust the receptors for itself and the viruses preceding it in the gradient,

viruses, it is caused instead by hemagglutinins produced during viral multiplication. Breakdown products of virions may also cause hemagglutination, e.g., the hemagglutinins released from influenza virus by ether treatment.

Although the spectrum of red cell species that are agglutinated and the required conditions differ for different viruses, the phenomenon is basically similar in all cases: a virion or a hemagglutinin attaches simultaneously to two RBCs and bridges them, and at sufficiently high viral concentrations multiple bridging yields large aggregates.

Hemagglutination Assay. The formation of aggregates can be detected in a number of ways. In the simplest, the **pattern method** (Fig. 46-24), the suspension of RBCs and virus are left undisturbed in small caps in a plastic plate for several hours. Nonaggregated cells sediment to the round bottom of the tube and then roll toward the center,

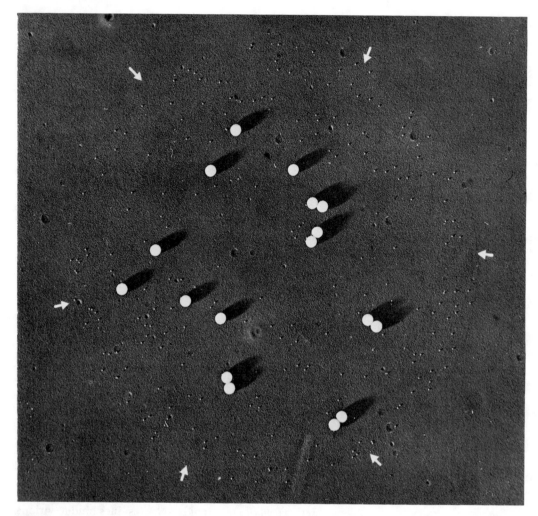

FIG. 46-23. Counting of particles of poliovirus (a picornavirus) mixed with polystyrene latex particles. The mixture was sprayed in droplets on the supporting membrane, dried, and shadowed. The micrograph shows a droplet, whose outline is partly visible (**arrows**). The small particles are virus, the large ones latex. There are 220 viral and 17 latex particles in the droplet. Since the latex concentration in the sample was 3.2×10^{10} particles/ml, the concentration of viral particles is $220/17 \times 3.2 \times 10^{10} = 4.1 \times 10^{11}$/ml. The precision of the assay based on this one droplet is only about ± 50% (see Appendix this chapter), owing to the small number of latex particles counted. To obtain a greater precision pooled counts from many similar drops would have to be used. (Courtesy of the Virus Laboratory, University of California, Berkeley)

but not for those that follow it. This result indicates that viruses differ quantitatively either in their neuraminidases or in their mode of attachment to the receptors.

Heating the virus inactivates its neuraminidase without destroying hemagglutinating activity. This **indicator** virus is useful for studying the union with receptors and mucoproteins without the complication of their enzymatic inactivation.

Receptors for other hemagglutinating viruses are probably also glycoproteins.

ASSAYS BASED ON ANTIGENIC PROPERTIES

Complement fixation (CF) or precipitation with antiserum can be used to measure amounts of virus. These methods have relatively low sensitivity and are used only for special purposes (Ch. 52).

ASSAYS OF INFECTIVITY

THE PLAQUE METHOD

This is the fundamental assay method in virologic research and it is also of great value in diagnosis: it combines simplicity, accuracy, and high reproducibility. First used with bacteriophages by D'Herelle and by Gratia, this method was a key factor in the spectacular

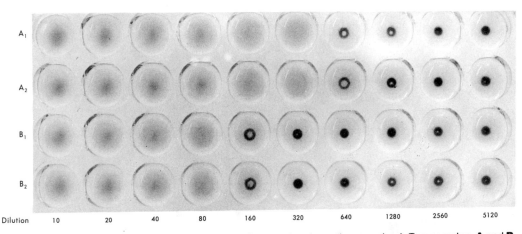

FIG. 46-24. Results of a hemagglutination assay by the pattern method with influenza virus (an orthomyxovirus). Two samples, **A** and **B**, were diluted serially by using twofold steps: 0.5 ml of each dilution was mixed with an equal volume of a red cell suspension, and each mixture was placed in a cup drilled in a clear plastic plate and left for 30 min at room temperature. Each assay was made in duplicate. Sample **A** causes complete hemagglutination until dilution 320; sample **B** until dilution 80; in either case the subsequent dilution still shows partial hemagglutination. The hemagglutinating titer of **A** is 320; that of **B,** 80.

advances of research on phage and later also on animal viruses.

Bacteriophages are assayed in the following way. A phage-containing sample is mixed with a drop of a dense liquid culture of suitable bacteria and a few milliliters of melted soft agar at 44° C; the mixture is then poured over the surface of a plate (Petri dish) containing a layer of hard nutrient agar. The soft agar spreads in a thin layer and sets, and the bacteriophages diffuse through it until each meets and infects a bacterium. After 20–30 min the bacterium lyses, releasing several hundred progeny virions. These, in turn, infect neighboring bacteria, which again lyse and release new virus. The uninfected bacteria, in the meantime, grow to form a dense, opaque lawn. After a day's incubation the lysed areas stand out as transparent **plaques** against the dense background (Fig. 46-26A). The soft agar permits diffusion of phage to nearby cells, but prevents convection to other regions of the plate; hence secondary centers of infection cannot form.

With **animal viruses** a similar method is possible, the bacteria being replaced by a suspension of cultured cells. More commonly, however, monolayers of cells growing on a solid support (Ch. 49) are used, and the nutrient medium is replaced by a solution containing the viral sample. Within an hour or so most of the virions attach to cells. Soft nutrient agar or some other gelling mixture is poured over the cell layer.* Plaques develop after 1 day to 3 weeks of incubation, depending on the virus (Fig. 46-26B).

Plaques are detected in a variety of ways.

---

* The sulfated polysaccharides in agar adsorb some viruses and prevent them from forming plaques. This complication is eliminated either by adding cationic substances that neutralize the negative charges of the polysaccharides, or by replacing the agar with methylcellulose.

FIG. 46-25. Action of neuraminidase on a serum inhibitor of influenza virus hemagglutination. This compound contains *N*-acetylneuraminic acid (NANA) linked to *N*-acetylgalactosamine as part of the mucoprotein; the NANA is released by the enzyme.

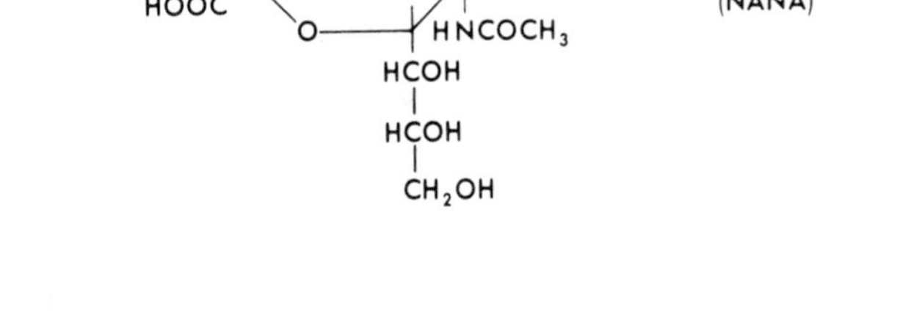

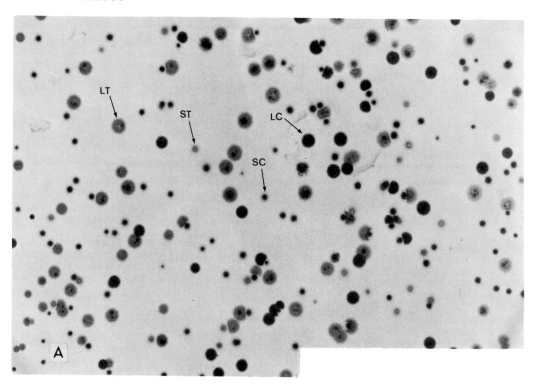

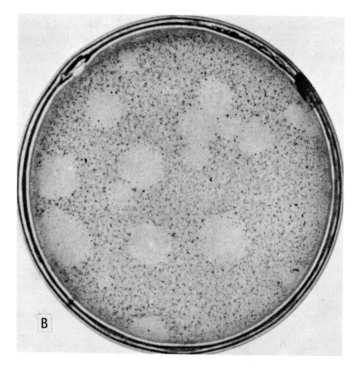

FIG. 46-26. Plaque assays.

A. Phage. The progeny of cells infected by two phage types was diluted by a factor of 10^7; 0.1 ml of the diluted virus was assayed. The plate was counted 18 h after plating. Four different plaque types, differing in plaque size and turbidity—large clear (**LC**), large turbid (**LT**), small clear (**SC**), and small turbid (**ST**)—can be distinguished, showing the great usefulness of plaque formation for genetic work with bacteriophages. Part of the plate is reproduced; a total of 407 plaques could be counted on the whole plate. The titer is 4.07×10^{10}/ml in the undiluted sample. The accuracy is $\pm$ 10%.

B. Poliovirus (picornavirus). A sample of poliovirus type 1 was diluted by a factor of 2×10^5 and 0.1 ml was assayed on a monolayer culture of rhesus monkey kidney cells, with an agar overlay containing neutral red. The culture was incubated for 3 days at 37° C in an atmosphere containing 7% CO_2, which constitutes a buffer with the bicarbonate present in the overlay. Some of the plaques show partial confluence, but they can still be identified as separate plaques; 17 plaques can be counted on the photograph. The corresponding titer is 3.4×10^7/ml in the undiluted sample. The accuracy is $\pm$ 50%.

FIG. 46-27. Dose–response curve of the plaque assay. The number of plaques produced by a sample of poliovirus type 1 at various concentrations was plotted versus the relative concentration of the virus; the accuracy of the assay ($\pm 2\sigma$) is indicated for each point. The data are in agreement with a linear dose–response curve which falls between lines 1 and 2, and, therefore, with the notion that a single particle is sufficient to give rise to a plaque. Curves 3 and 4 give the range of data that would be obtained if at least two viral particles were required to initiate a plaque; the deviation is such that the hypothesis is ruled out (see Appendix, this chapter).

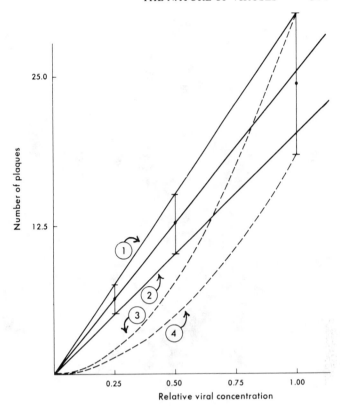

1) The virus often kills the infected cells, i.e., produces a **cytopathic effect**; the plaques are then detected by staining the cell layer either with a dye that stains only the live cells (e.g., neutral red) or only the dead cells (e.g., trypan blue).

2) With certain viruses the cells in the plaques are not killed, but acquire the ability to adsorb RBCs (see Maturation and Release of Animal Viruses—Enveloped Viruses, Ch. 50). The plaques are revealed by **hemadsorption,** i.e., by flooding the cell layer with a suspension of RBCs, then washing out those not attached to infected cells.

3) The infected cells may fuse with neighboring uninfected cells to form **polykaryocytes** (i.e., multinucleated cells), which are microscopically detectable (**syncytial plaques**).

4) Often the cells of the plaques contain large amounts of viral Ags, which can be detected by **immunofluorescence**.

If too many plaques develop on a plate some fuse with others and the titer is underestimated. The maximum density allowable varies with the size of the plaques and the sharpness of their margins.

The titer of the viral preparation is directly calculated from the number of plaques and the dilution of the sample, as shown in Figure 46-26. As discussed in the appendix to this chapter, the accuracy of the assay depends on the number of plaques counted. An assay estimated from n plaques will be within $(2/\sqrt{n})$ of the true value (e.g., with 100 plaques the range will be $\pm$ 20%).

The Dose-Response Curve of the Plaque Assay. The number of plaques in plates infected with different dilutions of the same viral sample is proportional to the concentration of the virus, i.e., the dose–response curve is linear (Fig. 46-27). *A single virion is therefore sufficient to infect a cell* (Fig. 46-27), a conclusion fundamental to virology. Moreover, it follows that the viral population contained in a plaque is the progeny of a single virion, i.e., a **clone,** provided cross-contamination from neighboring plaques is avoided.

Virus isolated from a single plaque thus represents a genetically pure line. Plaques also provide useful **genetic markers** through their visible characteristics such as size, shape, and turbidity.

POCK COUNTING

When the chorionic epithelium of the chorioallantoic membrane of a chick embryo (see Fig. 50-1, Ch. 50) is

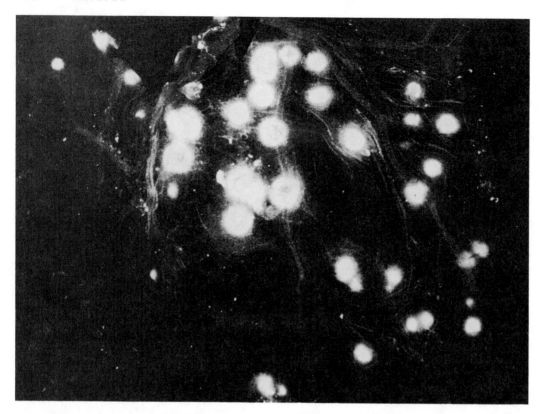

FIG. 46-28. Pocks formed by vaccinia virus (a poxvirus) on the chorioallantoic membrane of the chicken embryo. The membrane was removed, washed, and photographed against a dark background. The pocks appear as brilliant white foci, whereas the membrane, which is transparent, is barely visible. (Coriell LL et al: J Invest Dermatol 11:313, 1948, The Williams & Wilkins Co., Baltimore)

infected by certain viruses, characteristic lesions **(pocks)** appear. The counting of these pocks, introduced by Burnet for titrating poxviruses, has also been applied to other viruses (e.g., herpes simplex). The formation of a pock, like that of a plaque, begins with the infection of a single cell. Since the chorioallantoic membrane contains several different cell types and blood vessels, response to the local infection is complex, with both cell proliferation and cell death, accompanied by edema and hemorrhage. The pocks appear after 36–72 hours as opaque areas, usually white and often hemorrhagic, on the transparent membrane (Fig. 46-28). Viral mutants may be distinguishable by the appearance of the pocks.

Pock counting is satisfactory only for viruses that are released from the infected cells too slowly to give rise to secondary infectious centers. This method was important initially but is now largely superseded by the plaque method.

OTHER LOCAL LESIONS

Tumor-producing viruses, such as the Rous sarcoma virus (Ch. 66), can be assayed on monolayer tissue cul-

tures; they produce **proliferative foci,** each initiated by a single viral particle.

Many **plant viruses** can be titrated by counting the lesions produced on leaves rubbed with a mixture of virus and an abrasive material. The virus penetrates through ruptures of the cell walls caused by the abrasive, and the progeny spread to neighboring cells, probably through the protoplasmic bridges between cells. Recognizable spots, each started by a single virion, are thus produced.

ENDPOINT METHOD

This method, used for assaying animal viruses before the advent of the plaque method, is still employed for certain diagnostic assays and for quantitating virulence or host resistance. The virus is serially diluted and a constant volume of each dilution is inoculated into a number of similar **test units,** such as mice, chick embryos, or cell cultures. At each dilution the proportion of infected test units **(infectivity ratio)** is scored: for example, by 1) death or disease of an animal or embryo, 2) degeneration of a tissue culture, or 3) recognition of progeny virus in vitro (e.g., by hemagglutination).

The lower dilutions of the virus infect most of the test units and the highest dilutions infect none. A rough idea of the viral titer is given by those intermediate dilutions which produce signs of infection in only a fraction of the test units. The transition is not sharp, however, and only by combining the data from several dilutions is it possible to calculate the precise endpoint at which 50% of the test units are infected. At this dilution each sample contains on the average one **ID$_{50}$**, i.e., one **infectious dose for 50%** of the test units. One ID$_{50}$ can be shown mathematically to correspond to 0.7 plaque-forming units (see Appendix).

The interpolation to obtain the ID$_{50}$ can be carried out in a variety of ways. The method of Reed and Muench, though not mathematically derived, yields results in fair agreement with more rigorous methods. In the Reed and Muench method (see Appendix) the dilution containing 1 ID$_{50}$ is obtained by interpolation between the two dilutions that straddle the 50% value of the infectivity ratio. The interpolation assumes that in the proximity of the ID$_{50}$ the infectivity ratio varies linearly with the log dilution. Usually the accuracy of the method is low, since the number of test units used at each dilution is small.

When, for instance, 6 test units are employed, as is common in diagnostic titrations, the estimated titer has a **range of uncertainty** of at least 36-fold between the minimum and the maximum titers compatible with the result obtained (see Appendix). Under these conditions, the titration is useful only to ascertain large differences in viral titer: a 50-fold difference is considered significant and is adequate for many routine diagnostic procedures. The precision increases as the square root of the number of test units employed at each dilution.

Viral titers obtained by the endpoint method are expressed in various equivalents of the ID$_{50}$: LD$_{50}$ (lethal dose) if the criterion is death; PD$_{50}$ (paralysis dose) if the criterion is paralysis; TC$_{50}$ (tissue culture dose) if the criterion is degeneration of a culture.

COMPARISON OF DIFFERENT TYPES OF ASSAYS

The focal assay methods (plaques, foci, and pocks) are most satisfactory for their simplicity, reproducibility, and economy. For example, to match the precision obtained by counting 100 plaques on a single culture one would require more than 100 test units per decimal dilution in an endpoint titration. The precision of any type of assay is adversely affected by a variability in the response of the cells or organisms used in the assay; the variabilities can be quite large in the pock assay and even larger in endpoint assays using animals.

The various methods of assay have different sensitivities and measure different properties. Assays based on infectivity are as much as a millionfold more sensitive than those based on chemical and physical properties. Chemical and physical technics, moreover, titrate not only infectious but also noninfectious virions, such as empty capsids or particles with a damaged nucleic acid. These methods, therefore, can be useful for studies requiring measurement of the total number of viral particles. Hemagglutination or immunologic methods can also titrate soluble components, obtained from breakdown of the virions or produced during intracellular viral synthesis.

The **ratio of the number of viral particles** (determined by electron microscopy) **to the number of infectious units** measures the **efficiency of infection,** which varies widely among different viruses, and even for the same virus assayed in different hosts. As is shown in Table 46-4, for most viruses the ratio is larger than unity. This result is due in part to the presence of noninfectious particles, and in part to the failure of potentially infectious particles to reproduce. However, even with the highest ratio of particles to infectivity, infection is initiated by a single virion, since the dose–response curve remains linear.

The ratio of total viral particles to hemagglutinating units is related to the number of RBCs that must be aggregated to reveal hemagglutination. Since in the hemagglutination assay by the pattern method each diluted viral sample is mixed with an equal volume of a red cell suspension containing 10^7 cells per ml, at the endpoint the virus should theoretically contain about 10^7 hemagglutinating particles per ml.* Indeed, experimental determinations with influenza and polyoma viruses, under optimal conditions, come close to this value. But because the ratio of particles to infectious units exceeds 1.0, a hemagglutinating unit corresponds to

---

* If a single particle can form a stable bridge between two RBCs, the aggregates at the endpoint with n viral particles would consist of $n + 1$ cells.

TABLE 46-4. Ratio of Viral Particles to Infectious Units

Virus	Ratio
Animal viruses	
Picornaviruses	
Poliovirus	30–1000
Foot-and-mouth disease virus	33–1600
Papovaviruses	
Polyoma virus	38–50
SV40	100–200
Papilloma virus	~10^4
Reoviruses	10
Alphaviruses	
Semliki Forest virus	1
Orthomyxoviruses	
Influenza virus	7–10
Herpesviruses	
Herpes simplex virus	10
Poxviruses	1–100
Adenoviruses	10–20
Bacterial viruses	
Coliphage T4	1
Coliphage T7	1.5–4
Plant viruses	
Tobacco mosaic virus	5 × 10^4–10^6

only about $10^{6.3}$ infectious units of influenza virus and 10^5 of polyoma virus.

APPENDIX

QUANTITATIVE ASPECTS OF INFECTION

Distribution of Viral Particles
Per Cell: Poisson Distribution

In a cell suspension mixed with a viral sample individual cells are infected by different numbers of viral particles, and it is often important to know the distribution, i.e., the proportions of cells infected by zero, one, two, etc., viral particles.

These proportions depend on the **average number of viral particles per cell**, known as the **multiplicity of infection (m)**. The relevant viral particles are those that initiate infection of a cell; inactive particles or particles that, for whatever reason, never enter a cell are neglected. Hence m is related to the total number of viral particles (N) and of cells (C) by the relation $m = aN/C$, where a is the proportion of viral particles that initiates infection.

The proportion $P(k)$ of cells infected by k viral particles is given by the **Poisson distribution,** assuming that the cells are all identical in their ability to be infected. In fact, cells vary in size, surface properties, etc., but usually the deviations are small enough to be negligible, at least as a first approximation.

According to the Poisson distribution:

$$P(k) = \frac{e^{-m}m^k}{k!} \tag{1}$$

The value of m can be derived from the known values of N and C if a can be determined; otherwise m can be calculated from the experimentally determinable proportion of uninfected cells, $P(0)$. By making $k = 0$ in equation 1,

$$P(0) = e^{-m}, \text{ and} \tag{2}$$
$$m = -\ln P(0), \tag{3}$$

where ln stands for the natural logarithm.

The use of equations 1, 2, and 3 will now be illustrated with reference to two practical problems.

Problem 1. 10^7 cells are exposed to virus; at the end of the adsorption period there are 10^5 infected cells; what is the multiplicity of infection?

$$P(0) = 0.99, m = -\ln (0.99) = 0.01$$

This problem brings out the point that the multiplicity of infection can assume any value from 0 to ∞. Values smaller than unity indicate that a small fraction of the cells is infected, mostly by single viral particles.

Problem 2. What is the multiplicity of infection required for infecting 95% of the cells of a population?

$$P(0) = 5\% = 0.05, m = -\ln (0.05) = 3$$

The point of this problem is that even at very high multiplicities a certain proportion of the cells remains uninfected. The multiplicity of infection required to reduce the proportion of uninfected cells to a certain value can be calculated from equation 3.

Classes of Cells in an Infected Population

It is usually important to determine the proportion of only three classes of cells: **uninfected cells ($k = 0$); cells with a single infection ($k = 1$); and cells with a multiple infection ($k > 1$).** *The proportions are:*

Uninfected cells, $P(0) = e^{-m}$
Cells with single infection, $P(1) = m \, e^{-m}$
Cells with multiple infection, $P(>1) = 1 - e^{-m}(m + 1)$*

Problem 3. To determine the various classes of infected cells if the multiplicity of infection is 10.

$$P(0) = e^{-10} = 4.5 \times 10^{-5}$$
$$P(1) = 10 \times 4.5 \times 10^{-5} = 4.5 \times 10^{-4}$$
$$P(>1) = 1 - (4.5 \times 10^{-5})(10 + 1) =$$
$$1 - (4.95 \times 10^{-4}) = 99.95\%$$

With a population of 10^7 cells there are $4.5 \times 10^{-5} \times 10^7 = 450$ uninfected cells and 4500 cells with single infection; all the others have multiple infection.

Problem 4. To determine the composition of the population of infected cells if the multiplicity of infection is 10^{-3}, or 0.001.

$$P(0) = e^{-0.001} = 0.9990 = 9.99 \times 10^{-1} = 99.9\%$$
$$P(1) = 0.001 \times e^{-0.001} = 10^{-3} \times 9.99 \times 10^{-1} = 9.99 \times 10^{-4} = 0.0999\%$$
$$P(>1) = 1 - 0.9990 (0.001 + 1) = 0.000001 = 10^{-6}$$

With a population of 10^7 cells there are $9.99 \times 10^{-4} \times 10^7 = 9900$ cells with single infection, and $10^{-6} \times 10^7 = 10$ cells with multiple infection; most of the cells are uninfected.

MEASUREMENT OF THE INFECTIOUS TITER OF A VIRAL SAMPLE

To measure infectious titer, a viral sample containing an unknown number (N) of infectious viral particles is mixed with a known number (C) of cells. N is then calculated from the proportion of cells that remains uninfected according to equation 3: $m = -\ln P(0)$; since, as defined above, $m = a(N/C)$, $N = mC/a = -C \ln P(0)/a$, or

$$aN = -C \ln P(0) \tag{4}$$

Usually the factor a is not determinable and therefore the number (N) of infectious viral particles present in the sample to be assayed cannot be calculated. In its place one obtains the product aN, the number of infectious units.

---

* This value is obtained by subtracting from unity (the sum of all probabilities for any value of k) the probabilities $P(0)$ and $P(1)$.

This is the basis for all measurements of the infectious viral titer. Its **application** is different in the plaque method and in the endpoint method.

Plaque Method

In this method the number of plaques equals the number of infectious units. The actual number of cells employed in the assay is irrelevant, provided it is in large excess over the number of infectious viral particles, so that m is very small; uncertainties connected with the counting of the cells are therefore eliminated.

The Dose-Response Curve of the Plaque Assay.

As stated above the number of plaques that develop on a series of cell cultures infected with different dilutions of the same viral sample is proportional to the concentration of the virus. We shall now show that this linearity proves that a single infectious viral particle is sufficient to infect a cell (**single-hit kinetics**).

Let us assume that more than one particle, say two particles, is required. There would then be two types of uninfected cells: those with no infectious viral particles, and those with just one such particle. According to the Poisson distribution the proportions of cells in these two classes are e^{-m} and me^{-m}, respectively. Thus, under the foregoing assumption, $P(0) = e^{-m}(1 + m)$, which, for very small values of m, approximates to $P(0) = 1 - \frac{1}{2}m^2$. Therefore $P(i) = \frac{1}{2}m^2$, and the dose–response curve would be parabolic rather than linear (Fig. 46-27). If more than two particles were required to infect a cell, the curvature of the dose–response curve would be even more pronounced.

Endpoint Method

In this method the virus to be assayed is added to a number of test units (e.g., cultures or animals), each consisting of a large number of cells. A test unit is now equivalent to a single cell of the plaque assay. Therefore m is the multiplicity of infection of a test unit, rather than of a cell.

The virus titer can be calculated from the proportion of noninfected units, $P(0)$, at the endpoint dilution, according to equation 3: $m = -\ln P(0)$. If at the endpoint $m = 1$, i.e., there is one infectious unit per test unit, then on the average $P(0) = 0.37$.

Another approach, especially useful in quantitating virulence or host resistance, is the **method of Reed and Muench**, which is applicable to an assay involving a series of progressive dilutions of a virus. A constant volume of each dilution is inoculated in each animal of a group. An empirical pooling of the results obtained at all dilutions gives the dose at which 50% of the animals are infected (ID_{50}) or killed (LD_{50}).

In the example in Table 46-5, none of the dilutions gives a 50% endpoint; this lies between the third and the fourth dilution. The LD_{50} is calculated from the cumulated values, assuming that the proportion of the animals affected varies linearly with $\log_{10}$ dilution. The interpolated value is given by

$$h \frac{\text{\% animals affected at dilution next above 50\%} - 50\%}{\substack{\text{\% animals affected at dilution next above 50\%} - \\ \text{\% animals affected at dilution next below 50\%}}}$$

In this formula h is the log of a dilution step. The interpolated value is then added arithmetically (i.e., with the proper sign) to the log of the total dilution at the step just above 50% affected animals. In the example, interpolated value $= h \frac{71 - 50}{71 - 13} = h \frac{21}{58} = 0.36h$ (approximated to $0.4h$). If $h = 1/10$, $\log h = -1$; total dilution at the third step is $(1/10)^3 = 10^{-3}$, and log dilution $= -3$. Interpolated value then is $(-1) \times 0.4 = -0.4$. The $\log LD_{50} = -3 + (-0.4) = -3.4$. The LD_{50} titer is expressed as $= 10^{-3.4}$; i.e., the virus sample contains $10^{3.4}$ LD_{50} doses. If, instead, $h = \frac{1}{2}$, $\log h = -0.3$, total dilution at the third step $= (\frac{1}{2})^3 = \frac{1}{8}$, and log dilution $= -0.9$. Interpolated value $= (-0.3) \times 0.4 = -0.12$; $\log LD_{50} = -0.9 + (-0.12) = -1.02$, and LD_{50} titer $= 10^{-1.2}$.

The multiplicity corresponding to one LD_{50} (50% survival) is calculated from the relation $e^{-LD_{50}} = 0.50$. Therefore LD_{50}

TABLE 46-5. Example of Endpoint Titration

Dilution step	Mortality ratio	Died	Survived	Total dead*	Total survived*	Mortality Ratio	%
1	6/6	6	0	17	0	17/17	100
2	6/6	6	0	11	0	11/11	100
3	4/6	4	2	5	2	5/7	71
4	1/6	1	5	1	7	1/8	13
5	0/6	0	6	0	13	0/13	0

*Cumulated values for the total number of animals that died or survived are obtained by adding in the directions indicated by the arrows.

(Lennette EH: General Principles Underlying Laboratory Diagnosis of Virus and Rickettsial Infections. In Lennette EH, Schmidt NJ, eds: Diagnostic Procedures of Virus and Rickettsial Disease, p. 45. New York, American Public Health Association, 1964)

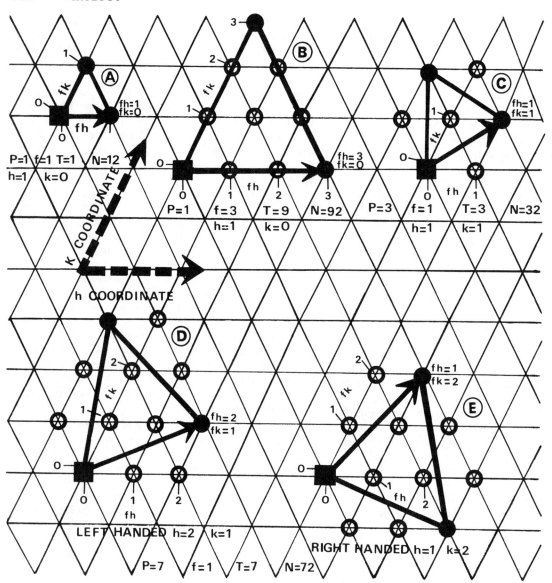

FIG. 46-29. Various forms of icosahedral capsids and their parameters. The triangulation number T is obtained by drawing the basic triangular face of the icosahedron (**heavy lines**) on a sheet of coordinates (**thin lines**) forming equilateral unit triangles. Capsomers are located at the intersections of the coordinates; the **black circles** are pentons, the **white ones** hexons. One of the sides of the triangular face of the icosahedron (**arrow**) is chosen as the vector going from an origin (**square**) to another penton. The coordinates of the end to the vector with respect to the origin (given near the vector) determine the parameters of the capsid. Capsids with equal values of h and k (hence of P) have equal general shapes; f measures how many times the basic vector crosses an intersection of the coordinates; hence different values of f change the number of capsomers but not the shape (compare **A** and **B**). T measures the number of unit triangles within the triangular face of the icosahedron.

If either h or k is zero (P = 1; **A, B**), or if h = k and both are different from zero (P = 3; **C**), the capsid is **symmetric** with respect to the coordinates; when h is different from k and both differ from zero (**D, E**) the capsid is **asymmetric** and can occur in either a left-handed (**D**) or right-handed (**E**) form. By convention it is left-handed when h > k.

On the basis of relations such as those exemplified in this figure, the parameters of a capsid can be determined in electron micrographs of negatively stained virions by identifying two adjacent pentons (as centers of fivefold rotational symmetry) and the neighboring hexons (see Fig. 46-12). Very often considerable technical difficulties are encountered in these determinations.

$= -\ln(0.50) = 0.70$. One LD_{50} corresponds to 0.70 infectious units.

PRECISION OF VARIOUS ASSAY PROCEDURES

Plaque Method. The statistical precision is measured by the **standard deviation** (σ) of the Poisson distribution, which is equal to the square root of the number of plaques counted. If the number of plaques counted is not too small, 95% of all observations made should fall within two standard deviations from the mean in either direction (i.e., $\pm 2\sigma$). Thus 4σ is the expected range of variability of the assay. If n replicate assays are made, $\sigma = \sqrt{\bar{x}/n}$, where $\bar{x}$ is the mean value of plaque numbers in the replicate assays. The standard deviation relative to the mean (**coefficient of variation**) serves as a relative measure of precision: $\sigma/\bar{x}$. This is $\sqrt{\bar{x}}/\bar{x} = 1/\sqrt{\bar{x}}$ for a single assay, and $1/\sqrt{n\bar{x}}$ for n replicate assays. The smaller the coefficient of variation, the higher the precision, which therefore increases as the square root of the number of plaques.

Example: If a total of 100 plaques is counted the standard deviation is 10. If the same assay is repeated many times its results will fall between 80 and 120 plaques in 95% of the cases; the coefficient of variation is 1/10. If 400 plaques are counted the coefficient of variation is 1/20.

Reed and Muench Method. An approximate value, empirically arrived at, of the standard deviation of the titer determined by this method is $\sigma = \sqrt{0.79\,hR/U}$, where h is the log of the dilution factor employed at each step of the serial dilution of the virus, U is the number of test units used at each dilution, and R is the interquartile range, namely, the difference between the log of the dilution at which P (i) is 0.25 and 0.75, respectively. In this calculation σ is expressed in logarithmic units. For the data of Table 46-5, with six assay units (animals) at each dilution, $h = 1.0$ and $R = 1.0$ (both in $\log_{10}$ units); $\sigma = \sqrt{0.79/6} = 0.36$ (in $\log_{10}$ units). The range of variation of the LD_{50} is therefore ± 0.72 in $\log_{10}$ units, and the highest expected value (within the 95% confidence limits) is 28 times (antilog of 1.44) the lowest value.

TABLE 46-6. Values of Capsid Parameters and Numbers of Capsomers found in Icosahedral Viruses

P^*	f^*	T^*	No. of capsomers	No. of hexons
1	1	1	12	0
	3	9	92	80
	4	16	162	150
	5	25	252	240
3	1	3	32	20
	7	147	1472	1460
7	1	7	72	60

*For explanation of P, f, and T, see text.

NUMBER OF CAPSOMERS IN ICOSAHEDRAL CAPSIDS

The laws of crystallography permit only certain numbers of capsomers in an icosahedral capsid, given by $10T + 2$, where T (the triangulation number) is Pf^2, with $f = 1$ or 2 or 3 . . . etc., and $P = h^2 + hk + k^2$ (h and k being any pairs of integers without common factors). Identically, $T = (fh)^2 + (fh)(fk) + (fk)^2$, the expression used in Figure 46-29. The minimal permissible number is 12 pentons, all located at the icosahedron corners. Higher permissible numbers of capsomers are 32, 42, 72, 92, etc., of which 12 are always corner pentons, the others hexons. Only some of the possible values of P and T are found in viruses (Table 46-6). The number of protomers in an icosahedral capsid is $5 \times 12 + 6(10T + 2 - 12) = 60T$, i.e., either 60 or a multiple thereof.

Examples of the arrangements of capsomers in a triangular face of the icosahedron in various types of capsids are given in Figure 46-29. These patterns allow identification of the capsid type by electron microscopy. For instance, it can be easily decided that the capsid in Figure 46-12 has $P = 1$ (several hexons in line between two pentons) and $f = 5$ (since the basic vector crosses the intersection of the coordinates 5 times); hence, $T = 25$ and the number of capsomers = 252.

SELECTED READING

BOOKS AND REVIEW ARTICLES

CASPAR DLD: Design principles in virus particle construction. In Horsfall F, Tamm I (eds): Viral and Rickettsial Infections in Man. Philadelphia, Lippincott, 1965, p 51

HALL R: Structure of tubular viruses. Adv Virus Res 20:1, 1976

RAGHOW R, KINSBURY DW: Endogenous viral enzymes involved in messenger RNA production. Annu Rev Microbiol 30:21, 1976

RUSSELL WC, WINTERS WD: Assembly of viruses. Prog Med Virol 19:1, 1975

SHATKIN AJ, BOTH GW: Reovirus mRNA: transcription and translation. Cell 7:305, 1976

TEMIN HM, BALTIMORE D: RNA-directed DNA synthesis and RNA tumor viruses. Adv Virus Res 17:129, 1972

WINNACKER EL: Adenovirus DNA: structure and function of a novel replicon. Cell 4:761, 1978

SPECIFIC ARTICLES

ABRAHAM G, RHODES DP, BANERJEE AK: Novel initiation of RNA synthesis in vitro by vesicular stomatitis virus. Nature 255:37, 1975

BALTIMORE D, HUANG AS, STAMPFER M: Ribonucleic acid synthesis of vesicular stomatitis virus. II. An RNA polymerase in the virion. Proc Natl Acad Sci USA 66:572, 1970

BRIAND JP, RICHARDS KE, BOULEY JP et al: Structure of the amino acid accepting 3'-end of high-molecular-weight eggplant mosaic virus RNA. Proc Natl Acad Sci USA 73:737, 1976

CHAMPNESS JN, BLOOMER AC, BRIGOGNE G et al: Structure of the protein disk of tobacco mosaic virus to 5 Å-resolution. Nature 259:20, 1976

CRICK FHC, WATSON JD: Structure of small viruses. Nature 177:473, 1956

DAVIDSON N, SZYBALSKI W: Physical and chemical characteristic of lambda DNA. In Hershey AD (ed): The Bacteriophage Lambda. Cold Spring Harbor, NY, Cold Spring Harbor Laboratory, 1971, p 45

DULBECCO R: Production of plaques in monolayer tissue cultures by single particles of an animal virus. Proc Natl Acad Sci USA 38:747, 1952

EARNSHAW WC, KING J, HARRISON SC, EISERLING FA: Structural organization of DNA packaged within the heads of T4 wild-type, isometric and giant bacteriophages. Cell 14:559, 1978

FEHMEL F, FEIGE U, NIEMANN H, STIRM S: *Escherichia coli* capsule bacteriophages. VII. Bacteriophage 29-host capsular polysaccharide interaction. J Virol 16:591, 1975

FURUICHI Y, SHATKIN AJ: 5'-termini of reovirus mRNA: ability of viral cores to form caps post-transcriptionally. Virology 77:566, 1977

GOTTSCHALK A: Influenza virus neuraminidase. Nature 181:377, 1958

GROSS HJ, DOMDEY H, LOSSOW C et al: Nucleotide sequence and secondary structure of potato spindle tuber viroid. Nature 273:203, 1978

HIRST GK: Agglutination of red cells by allantoic fluid of chick embryos infected with influenza virus. Science 94:22, 1941

KATES TR, MCAUSLAN BR: Poxvirus DNA-dependent RNA polymerase. Proc Natl Acad Sci USA 58:134, 1967

KEEGSTRA W, VAN WIELINK PS, SUSSENBACH JS: Visualization of a circular DNA–protein complex from adenovirions. Virology 76:444, 1977

KNIPE DM, RUYECHAN WT, ROIZMAN B, HALLIBURTON IW: Molecular genetics of herpes simplex virus: demonstration of regions of obligatory and nonobligatory identity within diploid regions of the genome by sequence replacement and insertion. Proc Natl Acad Sci USA 75:3896, 1978

LAI MMC: Phosphoproteins of Rous sarcoma viruses. Virology 74:287, 1976

MCGEOCH S, FELLNER P, NEWTON C: Influenza virus genome consists of eight distinct RNA species. Proc Natl Acad Sci USA 73:3045, 1976

MORRISON TG, MCQUAIN CO: Assembly of viral membranes: nature of the association of vesicular stomatitis virus proteins and membranes. J Virol 26:115, 1978

NOMOTO A, DETJEN B, POZZATTI R, WIMMER E: Location of the polio genome protein in viral RNAs and its implication for RNA synthesis. Nature 268:208, 1977

OWENS RA, ERBE E, HADIDI A et al: Separation and infectivity of circular and linear forms of potato spindle tuber viroid. Proc Natl Acad Sci USA 74:3859, 1977

RAVETCH JV, HORIUCHI K, ZINDER ND: Nucleotide sequences near the origin of replication of bacteriophage f1. Proc Natl Acad Sci USA 74:4219, 1977

ROBINSON AJ, BELLETT AJD: Circular DNA–protein complex from adenoviruses and its possible role in DNA replication. Cold Spring Harbor Symp Quant Biol 39-1:523, 1975

SALZMAN LA: Evidence for terminal S$_1$-nuclease-resistant regions on single-stranded linear DNA. Virology 76:454, 1977

SKEHEL JJ, WATERFIELD MD: Studies on the primary structure of the influenza virus hemagglutinin. Proc Natl Acad Sci USA 72:93, 1975

SPECTOR DH, VILLA-KOMAROFF L, BALTIMORE D: Studies on the function of polyadenylic acid on poliovirus RNA. Cell 6:41, 1975

STANLEY WM: Isolation of a crystalline protein possessing the properties of tobacco mosaic virus. Science 81:644, 1935

WINKLER FK, SCHUTT CE, HARRISON SC, BRICOGNE G: Tomato bushy stunt virus at 5.5 Å-resolution. Nature 265:509, 1977

WU M, ROBERTS FJ, DAVIDSON N: Structure of the inverted terminal repetition of adenovirus type 2 DNA. J Virol 21:766, 1977

YANAGIDA M: Molecular organization of the shell of T-even bacteriophage head. II. Arrangement of subunits in the head shells of giant phages. J Mol Biol 109:515, 1977

chapter 47

MULTIPLICATION AND GENETICS OF BACTERIOPHAGES

Model System. In spite of marked differences in structure and in the abundance of genetic information, all viruses are similar in the basic aspects of multiplication. Hence their general properties can best be studied by selecting a technically suitable model. An important cause of the rapid advance of virology has been the early focus on bacteriophages, first in the 1930s by Burnet in Australia and by the Hungarian chemist, Schlesinger; later by a large group of investigators working in close contact, including Delbrück, Luria, Hershey, and S. S. Cohen. These more recent workers concentrated especially on the **"T-even" phages** of *Escherichia coli* (T2, T4, and T6), which were thought to be the simplest possible organisms. They turned out, in fact, to be among the most complex of all viruses; but their complexity was instrumental for many discoveries, and the results could be extended, with some variations, to many other phages.

The special properties of phages containing single-stranded DNA or RNA, and of those with lipids in their coats, will be discussed at the end of this chapter. Phage λ, the model for lysogeny, will be discussed in Chapter 48.

MULTIPLICATION OF BACTERIOPHAGES

As pointed out by Lwoff, the same bacteriophage can exist in structurally different stages: **extracellular virions, vegetative phage,** and **prophage.** All bacteriophages exist at some stage as vegetative phages during their intracellular multiplication. Those able to become prophages are called **temperate;** those unable to do so are **virulent.** The properties of the extracellular virions were reviewed in the preceding chapter; those of vegetative phage will be studied in this chapter, mostly with the virulent T-even coliphages; and those of prophage will be discussed in the next chapter.

STRUCTURE OF BACTERIOPHAGES

The even-numbered T phages are made up of a head and a tail (Fig. 47-1). The **head** contains the DNA in association with polyamines, several internal proteins, and small peptides. It has the shape of two halves of an icosahedron connected by a short hexagonal prism. In other phages, such as *Salmonella* phages P1 and P2 and other coliphages, the head is strictly icosahedral (Fig. 46-5F). The **tails** of different phages vary greatly in dimensions and structure. The T-even phage tail consists of a central helical **tube** (through which the viral DNA passes during cell infection), surrounded by a helical **sheath** capable of contraction. The sheath is connected to a thin disc or **collar** at the head end and to a **base plate** at the tip end. The plate is hexagonal and of complex structure; it has a pin at every corner, and is connected to six long thin **tail fibers** which are the organs of attachment to the wall of the host cell.

Coliphages T1 and T5 have a sheathless tail; phages T3 and T7, as well as *Salmonella* phage P22, have a short stubby tail which terminates in a structure resembling a base plate with six short fibers. Some small icosahedral phages, such as φX174, have no tail.

DNA-less capsids (**ghosts**) of T-even phages are obtained by rapidly diluting the phage from a concentrated salt solution into water and thus osmotically bursting the head, with release of DNA.

INFECTION OF HOST CELLS

The first step in infection is a highly specific interaction of the phage's adsorption organelle, such as the tail, with **receptors** on the surface of the host cell, which leads to attachment of the phage to the cell (**adsorption**). Then the DNA is **released** from the capsid and enters the cell.

ADSORPTION

The initially reversible attachment of the phage to the receptors rapidly becomes irreversible (i.e., the phage cannot be washed away) at a suitable ionic strength and temperature. Adsorption can be abolished by bacterial **mutations to bacteriophage resistance;** thus B/2 of *E. coli* is resistant to T2. These mutations change the receptors and may also change the antigenic specificity of

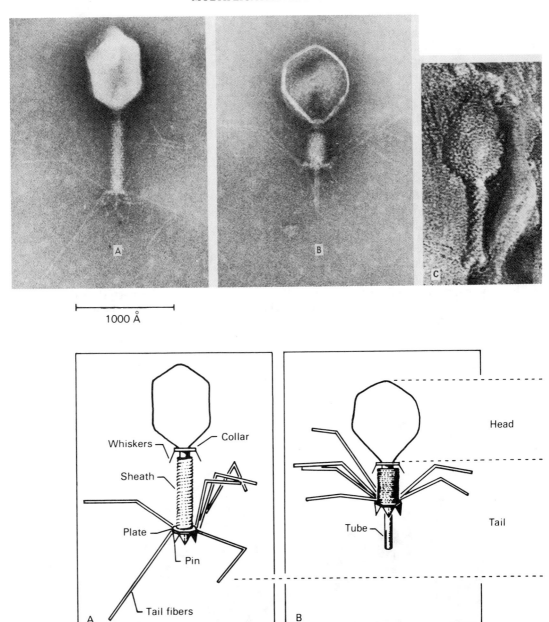

FIG. 47-1. Electron micrographs of bacteriophage T2. A. Phage before injection, with full head and extended sheath. B. Phage after injection, with empty head and contracted sheath. C. Helical structure of the sheath. A and B, negative staining. C, obtained by freeze-etching: the phage was embedded in ice which was then fractured; the place of fracture was covered with a thin layer of evaporated metal, which was then photographed. (A and B, courtesy of E. Boy de la Tour; C, courtesy of M. E. Bayer)

the cells. Conversely, if B/2 is exposed to a large concentration of T2, rare **host-range mutants** (T2h) can adsorb to the B/2 and initiate normal viral multiplication. Similar mutants occur with most phages. Host-range mutants played an important role in the early development of bacteriophage genetics. Their selection illustrates the connection between the evolution of viruses and that of the host cells.

Viral Sites for Adsorption. All virions have a specialized structure for adsorption. In the T-even coliphages it is a complex structure: the **tail fiber.** Thus isolated tail fibers adsorb to the same range of bacteria as the bacteriophage from which they were derived, and antiserum to the fibers inhibits phage adsorption. Electron microscopy shows that with the T-even coliphages the tips of the fibers attach first and reversibly to the cell surface and are followed by the **tail pins** which attach irreversibly. The adsorbed virion acquires a characteristic position with the head perpendicular to the cell wall (Fig. 47-2).

In RNA phages (discussed at the end of this chapter) the adsorption site is a simple protein molecule.

The host cell **receptors** are often complex polysaccharides with phage-binding and antigenic specificity. Thus phages used in *Salmonella* typing (Ch. 31) adsorb to various forms of the O Ag. Isolated receptors can bind

to the phage tail (Fig. 47-3), blocking adsorption of the phage to bacteria.

With capsulated hosts, an enzyme in the tip of the phage tail hydrolyzes specific capsular polysaccharides, digging a tunnel through which the phage reaches the cell wall. The receptor for phage λ is a mannose transport protein, suggesting that receptors usually perform an independent function in the cell. Some male-specific coliphages adsorb only to the sex pili of F$^+$ cells (see RNA Phage, below, and Physiology of Conjugation, Ch. 9). Phage f2 (RNA-containing) adsorbs laterally on the entire pilus, whereas fd (DNA-containing) adsorbs exclusively on its tip.

SEPARATION OF NUCLEIC ACID FROM COAT

In one of the most significant experiments of modern biology, Hershey and Chase demonstrated in 1952 the separation of the viral nucleic acid from the capsid (Fig. 47-4). They labeled the protein of T2 with ^{35}S or the DNA with ^{32}P, allowed the virus to infect bacteria, and exposed the bacteria to violent agitation in a Waring Blendor, which shears the tails of the adsorbed virions. The experiment yielded two results that, at the time, seemed astonishing. 1) With ^{35}S-labeled phage 80% of the label came off; but with ^{32}P-labeled phage essentially all the label remained with the cells and, since it was DNase-resistant, it was **within** the cells. 2) The blended

FIG. 47-2. Electron micrograph of particles of phage T5 adsorbed to an *Escherichia coli* cell. The virions attach by the tips of their tails. Note also that the heads of some particles are clear (electron-transparent), having injected their DNA into the cells; others are dark (electron-opaque) and still contain their DNA. (Anderson, T.F.: Cold Spring Harbor Symp Quant Biol 18:197, 1953)

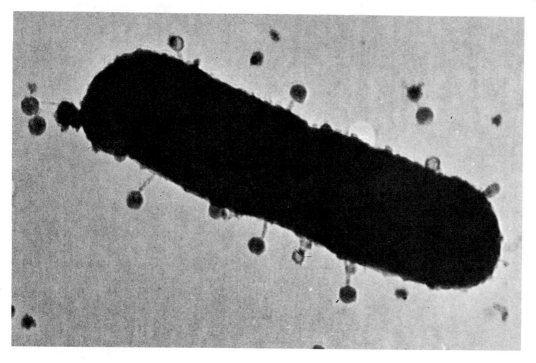

FIG. 47-3. Electron micrograph of bacteriophage T5 mixed with purified receptors. The receptors are in the form of spheres (**arrows**), about 30 nm in diameter, artificially produced upon cell disruption. Some receptors are free and others are adsorbed to the tip of the phage tails. Shadowed preparation. (Weidel W, Kellenberger E: Biochim Biophys Acta 17:1, 1955)

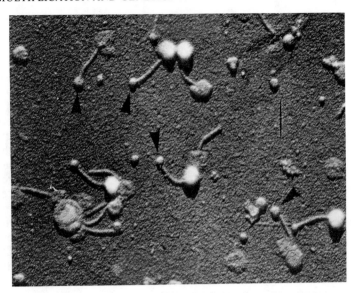

bacteria produced progeny phage as if they had not been blended. These results strongly suggested that *phage DNA carries the genetic information of the phage into the cell.*

Ejaculation of the nucleic acid from the coat could also be elicited with wall fragments instead of cells; the viral DNA was then released into the medium, where it was digestible by DNase. This result indicated that the injection of DNA into the cell does not require energy from the cell.

Eclipse. After the DNA is injected, the intact cells can produce plaques (see Assays of Infectivity, Ch. 46), but disrupted cells can not. However, infectivity reappears later when progeny virus is formed. The temporary disappearance of infectivity, called **eclipse**, is due to the in-

ability of the naked viral DNA to infect bacteria under ordinary conditions.

MECHANISM OF PENETRATION OF THE NUCLEIC ACID

The nucleic acid, generally associated with some internal proteins of the virion, is released from the capsid after the virions have become irreversibly adsorbed. It penetrates into the cells, apparently at sites at which the outer and inner membranes are in contact, and remains associated with the membrane. DNA released from virions is protected from membrane nucleases by the associated proteins and by DNA modifications (see Role of DNA Modifications, below).

The DNA of some phages (e.g., T5, T7) is always injected into the cells starting with the same end (i.e., with **constant polarity**),

FIG. 47-4. Hershey and Chase experiment, showing the separation of viral DNA and protein at infection. **A.** The phage protein was labeled by propagation in a medium containing $^{35}SO_4{}^{2-}$. **B.** Phage DNA was labeled by propagation in medium containing $^{32}PO_4{}^{2-}$. Phage was adsorbed to host bacteria, and after 10 min at 37° C, the culture was blended. Most of the ^{32}P remained associated with the cells, whereas most of the ^{35}S came off. Labeled components are **shaded.**

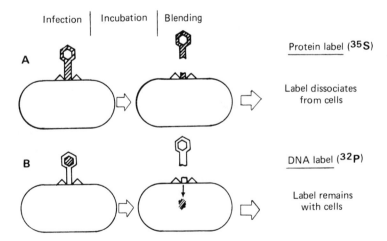

allowing some genes to be expressed first. Thus one of the first T7 genes entering the cells specifies a protein that protects the DNA from host nucleases. With T5, only 8% of the DNA penetrates in a first stage; invasion of the cell membrane by proteins specified by this segment then allows the injection of the rest.

The T-even phages have a highly specialized mechanism for releasing their DNA. Electron microscopy shows that after the tip of the tail has become anchored to the cell surface, the **contraction of the tail sheath** pulls the collar and the phage head toward the plate, pushing the tube through the cell wall (see Fig. 47-1). When the tube reaches the plasma membrane, the DNA is ejected. Because of this action, as well as its shape, the virion has been likened to a hypodermic syringe, and the release of the nucleic acid is called **injection.**

Contraction is the result of a **chain of conformational changes** initiated by the attachment of fibers and pins to the cell. The hexagonal base plate becomes starlike and separates from the tube; then the sheath shortens and thickens. Contraction is an irreversible reciprocal shift of the monomers, driven by the release of potential energy in the bonds between them.

Bacteria can also be infected by purified phage DNA after pretreatment with Ca^{2+} or conversion to spheroplasts; however, the efficiency of this process, called **transfection,** is very low, because the DNA is likely to be degraded by exonucleases. The efficiency is higher (about 10^{-4}) for DNAs that can rapidly cyclize in the cells (e.g., λ DNA—see Ch. 48), thus becoming resistant to the enzymes, and for the DNA of *Bacillus subtilis* phage $\phi29$, which has a protein covalently bound at its ends.

EFFECT OF PHAGE ATTACHMENT ON CELLULAR METABOLISM

The attachment of the T-even phage tail per se, without expression of viral genes (e.g., in the presence of chloramphenicol to prevent new protein synthesis), causes a **disorganization of the cell plasma membrane.** Its increased permeability then allows small molecules (e.g., nucleotides) to leak to the medium. Similar alterations are produced by the attachment of **phage ghosts** and of some bacteriocins.

These alterations cause profound **metabolic changes,** including an almost immediate cessation of cellular protein synthesis. In cells infected by ghosts these changes lead to cell lysis without viral multiplication (**lysis from without**). In cells infected by phage the membrane alterations and their metabolic effects are only transient, and are rapidly reversed by the incorporation of several phage-specified proteins in the membrane (see Fig. 47-29). These proteins cause other effects: they make the cell resistant ("immune") to superinfection by phage of the same type, and to lysis from without by superinfecting

ghosts, and they cause the phenomenon of lysis inhibition (see One-Step Multiplication Curve, below).

THE VIRAL MULTIPLICATION CYCLE

The process of viral multiplication, initiated by the penetration of the parental nucleic acid, involves many sequential steps, which end in the release of the newly synthesized **progeny virions.** This sequence is called the **multiplication cycle.** Kinetic analysis of the various steps requires large numbers of cells in which infection proceeds **synchronously,** i.e., the cells must be infected simultaneously, and secondary infection by progeny virus must be avoided (one-step conditions). These conditions can be achieved very simply, as in the classic studies of Delbrück, by allowing virus to infect cells for a brief time and then preventing further infection either by diluting the virus–cell mixture or by adding phage-specific antiserum. The infected cells are then freed of unadsorbed virus and antiserum by centrifugation.

The **multiplicity of infection** (i.e., the number of virions per cell) must be carefully selected according to the demands of the experiment, for if it is too high viral replication may be abnormal, and if it is too low many cells remain uninfected. Since not all virus particles adsorb during the brief initial incubation, the actual multiplicity of infection is usually monitored by determining the proportion of **productively infected cells** (see Appendix, Ch. 46). These cells are identified as **infectious centers,** i.e., cells able to produce a plaque in the regular assay used for the virus (see Assays of Infectivity, Ch. 46).

The viral growth cycle can also be studied in individual infected cells isolated in tubes or in small drops of medium under paraffin oil (**single-burst experiments**).

ONE-STEP MULTIPLICATION CURVE

A one-step multiplication curve describes the production of progeny phage as a function of the time after infection under one-step conditions. The **extracellular titer** is determined by assaying the medium after centrifuging down the cells, and the **total titer** (extracellular plus intracellular) by assaying the culture after disrupting the cells. These titers are usually expressed as infectious units per productive cell. Plotting their values versus the time of assay yields multiplication curves such as those of Figure 47-5, in which the following stages can be recognized.

1) The **eclipse period.** Total virus titer is almost nil (only residual from the inoculum). The end of the eclipse period is taken as the time at which an average of one infectious unit has been produced for each productive cell.

2) The **intracellular accumulation period.** Progeny phage accumulates intracellularly but is not yet released

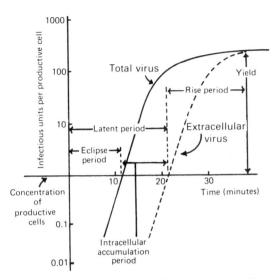

FIG. 47-5. Diagram of one-step multiplication curve of bacteriophage T2. Bacteria and phage were mixed and adsorption was allowed for 2 min; antiphage serum was then added. The bacteria were recovered by centrifugation and were resuspended in a large volume of medium at 37° C. A sample was immediately plated to determine the concentration of productive cells. Other samples were taken from time to time and divided into two aliquots: one was shaken with chloroform to disrupt the bacteria and was then assayed (**total virus**); the other was freed of bacteria by centrifugation and the supernatant was assayed (**extracellular virus**). The titers are compared with the concentration of productive cells as 1.0.

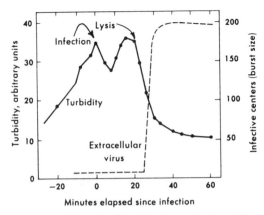

FIG. 47-6. Turbidity of an *E. coli* culture at various times after infection with phage T4r. Multiplicity of infection about 5. Onset of lysis (**arrow**) coincides with the end of the latent period. The temporary drop in turbidity immediately following infection (**arrow**) is determined by a transient increased permeability of the plasma membrane, which causes leakage of cellular components. (Modified from Doermann AH: J. Bacteriol 55:257, 1948)

into the medium. The end of this period is the time at which one viral infectious unit per productive cell, on the average, has appeared extracellularly. The end of this period also marks the end of the **latent period.**

3) The **rise period.** The extracellular phage titer increases until it reaches a constant titer, at the end of the multiplication cycle. The average number of infectious units of virus per productive cell at that time represents the **viral yield.**

Progeny phage is released into the medium by the **lysis (burst)** of each cell, which causes a **drop in turbidity** of the culture if most cells are infected (Fig. 47-6). With T-even phages the event can be seen in darkfield microscopy, where the virions are visible. With these phages lysis is delayed by more than an hour if a culture infected with a wild-type strain is heavily reinoculated, before the time of normal lysis, with the same strain (**lysis inhibition**). **Rapid-lysis (r) mutants** (which are defective in a membrane protein) do not delay lysis under these conditions. The different lysis properties of r+ (wild-type) and r phages are reflected in the **nature of their plaques:** r plaques are about 2 mm in diameter, clear, and with a sharp edge; r+ plaques are smaller and surrounded by a halo of increased turbidity. The halo is formed by bacteria infected more than once and therefore lysis-inhibited

(Fig. 47-7). This readily observed plaque difference was a valuable marker in the early development of phage genetics.

Lysis inhibition is very useful in phage research because *it extends the duration of the period of viral synthesis.* As a consequence, the cells make much more virus; moreover the time-dependent events of viral multiplication can be studied more easily.

SYNTHESIS OF VIRAL MACROMOLECULES

Infection of bacteria by T-even phages causes a *profound rearrangement of all macromolecular synthesis.* The first hint of this rearrangement was gained when Hershey, taking advantage of the unique presence of 5-hydroxymethylcytosine in viral DNA, showed that in the infected cells *the synthesis of cellular DNA stops and is replaced by viral DNA synthesis;* furthermore, the preexisting cellular DNA breaks down and its components are utilized as precursors of viral DNA. It is now clear that within a few minutes after infection the synthesis of all **cellular** DNA, RNA, and protein (i.e., that directed by the cellular genome) ceases. The effect is unrelated to the transient inhibition produced by membrane damage during adsorption; it requires the expression of viral genes. Soon the cellular syntheses are entirely replaced by viral syntheses. This shift represents the fundamental characteristic of viral parasitism: *the substitution of viral genes for cellular genes in directing the synthesizing machinery of the cell.*

This profound reshuffling is not reflected in big changes of **overall** synthetic rates. RNA synthesis is reduced because rRNAs and

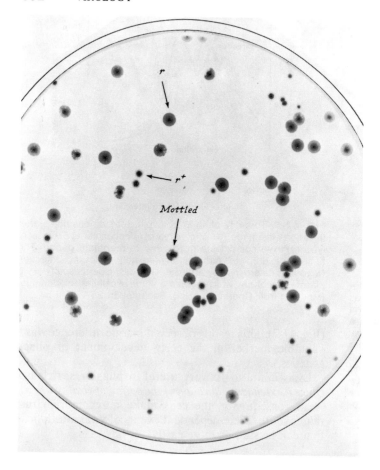

FIG. 47-7. Plaques produced by cells infected by a mixture of T2r and T2r⁺ phage. Plaques of **r** type are large and without a halo; those of **r⁺** type (wild-type) are small and surrounded by a halo. Cells infected by both r and r⁺ phage produce **mottled** plaques with a sectored halo (dark sectors, r phage; clear sectors, r⁺ phage). The halo is produced by infected cells with lysis inhibition caused by r⁺ phage. (Molecular Biology of Bacterial Viruses by Gunther S. Stent. W. H. Freeman and Company. Copyright © 1963)

most tRNAs are no longer synthesized. Protein synthesis is limited by the available number of ribosomes and thus remains at much the same level as before infection. DNA synthesis, after a brief halt, resumes at an increased rate, reflecting the rapid multiplication of the viral DNA.

The degree to which cellular functions are replaced by viral functions varies greatly for different phages. It is maximal with the largest and most virulent phages, and minimal for some small, filamentous phages (see Multiplication of Bacteriophages with Cyclic Single-Stranded DNA, below) whose replication does not impede the multiplication of the host cells. As the degree of replacement decreases, more and more functions of cellular genes are utilized for phage replication.

CESSATION OF SYNTHESIS OF HOST MACROMOLECULES

Following the discovery of deoxycytidylate hydroxymethylase in cells infected by T-even phages, an intense search revealed many enzymes that are synthesized only after infection; they must be **specified by viral genes**, because their properties are changed by phage mutations.

Some of these proteins contribute to the metabolic shift by carrying out viral syntheses, while others turn off cellular syntheses.

Turn-off proteins are of three kinds: 1) Some phage proteins add to the host transcriptase or cause phosphorylation and adenylylation of its subunits, making the enzyme unable to recognize host promoters (see Transcription of DNA, Ch. 11) and consequently **shutting off host RNA synthesis.** 2) Some cause the **bacterial nucleoid to unfold,** probably as a consequence of the inhibited synthesis of RNA which holds the host DNA together. The host DNA attaches to the cellular membrane where it is degraded (Fig. 47-8). 3) Some cause changes in tRNAs, inhibiting **host protein synthesis:** T7 induces a translational repressor.

Thus after infection with T-even phages a host leucine tRNA is inactivated by endonucleolytic cleavage and is replaced by a phage tRNA recognizing a different codon. Several new tRNAs are specified by these phages and by T5. These tRNA changes are *not essential for phage multiplication in the usual laboratory strains* of *E. coli,* since deletion of the phage tRNA genes has little effect on phage growth. The new tRNAs are needed for optimal growth in

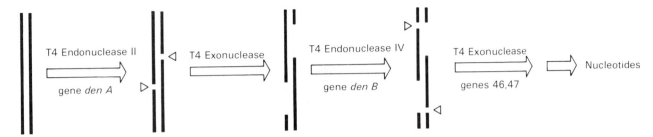

FIG. 47-8. Stepwise degradation of *E. coli* DNA after T4 infection by phage enzymes. Endonucleases II and IV are inactive on HMC-containing phage DNA. **Arrowheads** indicate endonuclease attack. (Reproduced, with permission, from the Annual Review of Biochemistry, Vol. 39. © 1970 by Annual Reviews Inc.)

some wild strains of *E. coli*, which may be closer to those in which the phage has evolved. Similar evolutionary needs may explain the existence of other "inessential" genes in the phage DNA, such as those for some aminoacylases and tRNA methylases.

REGULATION OF THE EXPRESSION OF PHAGE GENES

The genes introduced into the host cell by the phage DNA are expressed in an **orderly temporal sequence**, which allows the efficient coordination of many events participating in the production of progeny phage.

REGULATION OF TRANSCRIPTION

Regulation occurs mainly at the level of transcription. Of the following mechanisms some, but not necessarily all, are employed by a given phage: 1) injection of a **transcriptase present in the virions** (in coliphage N4), which immediately upon infection transcribes a section of the phage DNA; 2) **polar** (and, for T5, stepwise) **penetration** of the viral DNA, allowing the successive transcription of various genes as they enter the cell; 3) **sequential transcription** of genes in a transcriptional unit, those genes close to the promoter (see Transcription of DNA, Ch. 11) being transcribed first; 4) **modifications of the host transcriptase**, which alter its interaction with regulatory subunits (such as σ factor [Ch. 11]); 5) association of the host transcriptase with **phage factors** that modify its initiation or termination specificity: coliphage T7 and *Pseudomonas* phage PBS2 specify entirely **new transcriptase**; 6) introduction of **nicks and gaps** in the phage DNA which modify its template properties; 7) differences in the duration of **functional activity** (i.e., the ability to act as template for in vitro synthesis of functional proteins) of different mRNAs; 8) **processing (cleavage)** of polycistronic mRNAs into smaller messengers.

Characteristics of Phage mRNA. Viral messenger RNAs can be characterized 1) by hybridization to phage DNA in competition with RNA extracted at specific times after infection or from cells infected with mutants; 2) by translation in vitro and identification of the protein products from their size (in acrylamide gel electrophoresis) or their function (for enzymes).

Phage mRNA is usually **polycistronic.** Like host mRNA, it **turns over**; its degradation products return to the nucleotide pool and some of them are converted into deoxynucleotides and end up in viral DNA. However, phage transcription does not require a rapid change of gene action, and the rate of degradation of much phage mRNA is low.

Phage mRNA has considerable **secondary structure,** i.e., segments capable of forming double strands (**"hairpins"**), some of which are signals for processing by site-specific RNases.

Temporal Organization of Transcription. Certain groups of genes display different modes of transcription, which are characterized by 1) the time interval between infection and appearance of their messengers, 2) the effects of inhibitors (chloramphenicol to block protein synthesis, rifampin to prevent RNA initiation), 3) whether or not the messengers are synthesized in vitro by the regular host transcriptase, and 4) the polarity of the transcribed strand. The modes defined for large phages (coliphage T4, *B. subtilis* phage SPO1) are shown in Table 47-1 and in Figs. 47-9 and 47-10.

Immediate early and delayed early genes are transcribed during the first part of the infection cycle, forming products that shut off cellular macromolecule syntheses and participate in the replication of the phage DNA. These genes are usually clustered together. **Middle genes** are related to the early genes, since in T4 they are transcribed from the same ("left") DNA strand, and their products are required for DNA replication and recombination. In contrast to early genes they continue to be transcribed throughout infection. **Late genes** are a completely different class; they are transcribed from the "right" strand in T4 after both the transcriptase and the template DNA have been changed, and their products

TABLE 47-1. Definitions of the Transcription Modes for Bacteriophage T4

Classes of genes	Transcription in vivo: Time of initiation after infection (minutes) at 37° C	Transcription in vivo: Synthesis under rifampin added at 1 min	Transcription in vivo: Synthesis under chloramphenicol	Transcription in vitro	Inferred promoters	Template strand
Immediate early	0–1	+	+	+	Immediate early	Left
Delayed early	2	–	–	+	Immediate early or Delayed early	Left
Quasi-late (or middle)	<2	–	–	–	Quasi-late	Left
Late	5	–	–	–	Late	Right

are capsid proteins and enzymes for lysing the cells.

A gene may be transcribed in more than one mode. Thus the T4 $r_{II}B$ gene is represented in mRNAs of different sizes, since it is transcribed both in delayed early and in middle modes.

The immediate early genes of DNA phages are usually transcribed by the unaltered *E. coli* RNA polymerase. Some phages (like N4) use an enzyme present in the virions: they can multiply in the presence of rifampin, which inactivates the host polymerase. The N4 enzyme, which in vitro transcribes only single-stranded DNA, acts in vivo in concert with gyrase, which generates single-stranded segments by introducing negative supercoils in the DNA helix. With T4, immediate transcription is restricted in vivo to the promoter-proximal part of each transcription unit by the host terminator ρ factor (Ch. 11). Transcription of T4 delayed early genes in vitro requires proteins specified by immediate early genes: whether an antiterminator (as in λ infection; see Transcription, Ch. 48), or a factor causing recognition of new promoters, is not clear. Transcription of middle genes begins at middle promoters and requires a phage-specified protein (gene *mot* in T4, gene 28 in SPO1). The late genes of several phages are transcribed from new sets of promoters after the host transcriptase has undergone profound changes that alter its promoter specificity. In the case of T4, there is adenylylation of subunit α by phage enzymes and association with phage-specified proteins (the products of genes 33, 45, 55) that replace the host σ factor (Fig. 47-11). Proteins with similar functions are also specified by other phages (Fig. 47-12).

The interactions of the phage regulatory peptides with the host subunits are blocked by certain bacterial mutations to phage nonpermissiveness. In turn their effect is annulled by phage mutations. As with the host-range mutations discussed earlier (see Adsorption) these effects derive from the **interaction of host and viral macro-**

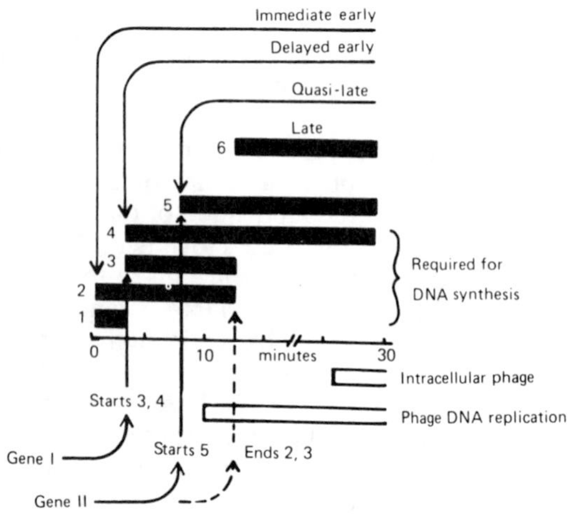

FIG. 47-9. Program of transcription of *Bacillus subtilis* phage SP01. **Black bars** (1,2,3 . . . 6) indicate the periods during which various classes are transcribed; **open bars** indicate the timing of other events. Phage genes I and II affect the beginning of synthesis of certain classes and the end of others, as indicated. Time in minutes from infection. (Modified from Gage P, Geiduschek EP: J Mol Biol 57:279, 1971. Copyright by Academic Press, Inc., (London) Ltd.)

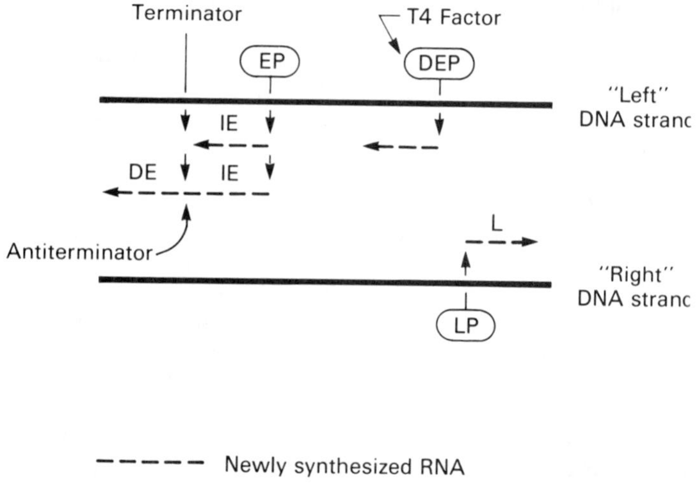

FIG. 47-10. Transcription of different sets of genes on T4 DNA. Dashed arrows indicate newly synthesized RNA. **EP** = early promoters; **DEP** = delayed early promoters; **LP** = late promoters; **IE** = immediate early RNA; **DE** = delayed early RNA; **L** = late RNA. Delayed early messengers are transcribed either by interfering with termination of immediate early transcription (antitermination) or by initiating at new promoters. Early and late messengers are transcribed on different strands.

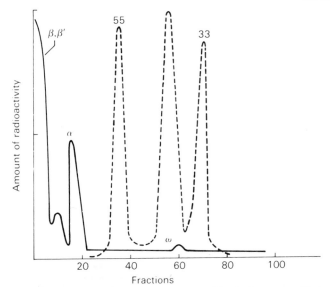

FIG. 47-11. Evidence for the presence of both phage-specified and host proteins in RNA polymerase extracted from late T4-infected cells. The cells were labeled with ^{35}S before infection to mark the bacterial units β, β', α and ω (**solid line**), and with ^{3}H after infection to mark the phage units (**dashed line**). Of the latter, two have been identified as the products of genes 55 and 33 (Modified from Horvitz HR: Nature New Biol 244:137, 1973)

molecules: the altered fit caused by a mutation is corrected by a second mutation in the other partner.

In T4 and some other phages mutations or inhibitors that block the replication of DNA inhibit late transcription. This shows that *transcription is coupled to structural DNA changes that accompany replication,* including single-strand breaks and gaps (see Viral DNA Replication, below).

Processing of Primary Transcripts. Polycistronic transcripts may be secondarily cleaved. Thus a transcript including all the T7 early genes is later cleaved by RNase III to yield separate monocistronic messengers (see Fig. 47-31). This step is not required for multiplication (since T7 grows in RNase-deficient cells) but favors it. Another example of processing is the cleavage of dimeric precursors into the mature T4-specific tRNAs.

FIG. 47-12. Model for the temporal control of *B.subtilis* phage SP01 by subunits that interact with the core host transcriptase. Phage peptides are represented as **squares**; host peptides as **circles**; σ is the host initiation factor; 28, 33, and 34 are peptides specified by phage genes; δ is a host factor required for late transcription. (Modified from Fox TD Nature, 262:748, 1976, with data from Tijan R, Pero J: Nature 262:753, 1976)

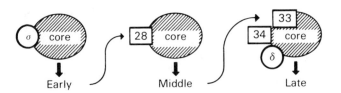

TRANSLATION OF VIRAL MESSENGERS

The synthesis of phage proteins reflects the temporal sequence of transcription but is also determined in part by additional **independent mechanisms directly affecting translation.** Some were noted above under Cessation of Synthesis of Host Macromolecules. In addition, with the T-even phages, viral genes determine the duration of translation of early messengers, as well as the rate of translation of specific messages. The result is accurate control of the synthesis of viral proteins.

The phage proteins are synthesized on preexisting host ribosomes. *Individual proteins are synthesized in widely different amounts,* probably owing to different efficiencies of initiation of transcription and translation. The most abundant T4 protein is the main component of the head capsid, a late protein.

VIRAL DNA REPLICATION

Phage DNA replication has been extensively investigated because its study is facilitated by the relatively small size of the molecules and their high rate of synthesis.

REPLICATIVE INTERMEDIATES

These intermediates contain newly synthesized DNA; hence they are labeled by exposing the cells to a short pulse of a radioactive precursor. Their structure is inferred from their shape in electron micrographs and from their sedimentation behavior. Different phages may have different kinds of replicative intermediates, which may also change in the course of infection. These differences in the method of replication will be illustrated using phages T4, T7, and λ. The **mature** DNA molecules (i.e., present in the virions) of these viruses are linear but differ in other features, which were reviewed in Chapter 46 (Fig. 46-8). Phage with mature cyclic DNA will be considered later in this chapter.

Phage DNA replicates in two phases. In the **first phase** some DNAs (e.g., T4, T7) remain linear, while others, especially those with cohesive ends (e.g., λ) become cyclic. Replication **begins at fixed locations,** and it proceeds **bidirectionally** (Fig. 47-13). Visualization of the origins of replication, in the form of growing loops like those of Fig. 47-13, shows that many phage DNAs have a unique origin, but the DNAs of T-even phages, which are very long, have several origins each.

After several rounds of replication the mode changes to a **second phase.** An increased sensitivity to certain nucleases reveals the appearance of single-strand nicks and gaps in the DNA. They are a prelude to the formation of **concatemers,** i.e., giant molecules which can be

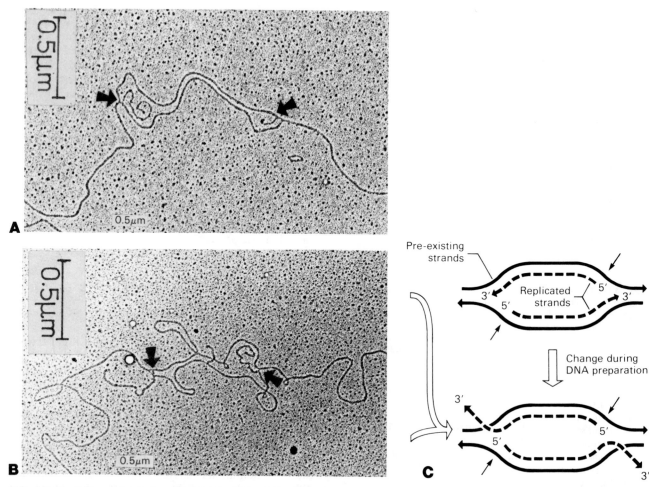

FIG. 47-13. Evidence for bidirectional replication of T4 DNA. The two electron micrographs of partially replicated DNA **(A,B)** show symmetric **growing loops**, each with a "whisker" of single-stranded DNA **(arrows)**. The sensitivity to specific nucleases shows that a whisker represents the 3′ end of each growing strand. In all likelihood the whisker is formed as shown at right **(C)**, because at each end of the loop the 3′-ended strand grows more rapidly than the 5′ end (which is synthesized backwards), and during the preparation of the DNA the two parental strands snap back to the point where replication is complete. (Modified from Delius et al: Proc Natl Acad Sci, USA 68:3049, 1971)

shown, from their electron microscopic appearance, to have the length of 20 or more mature molecules. Phage λ concatemers are generated by rolling-circle replication, but those of T4 and T7 are probably generated by **breakage and reunion** (i.e., **recombination**) between homologous ends, since they are absent with recombination-deficient mutants (Fig. 47-14). Recombination between molecular ends generates **linear concatemers;** recombination at intramolecular gaps leads to **branched molecules.** It is difficult to establish the precise configuration of the concatemers because in electron micrographs they appear as complex entanglements of filaments.

Formation of concatemers is required for the complete replication of linear molecules. In fact, in their bidirectional replication,

both strands remain incompletely replicated at opposite ends if the synthesis of the last Okazaki segment (which grows backwards) does not begin at the last nucleotide (Fig. 47-15A). (This complication does not arise in cyclic molecules). The incompletely replicated molecules can form complete dimers by pairing the single-stranded ends, which are complementary owing to the terminal repetitions of linear phage DNAs (Fig. 47-15B). After cohesion of the ends the gaps are closed by DNA polymerase and ligase.

At the end of the replication *concatemers generate mature molecules* in different ways (Fig. 47-16). T4 concatemers are cut by a mechanism that recognizes the length of the mature DNA but no specific sequences, yielding both permuted sequences and repetitious ends (see Association of DNA and Capsid, below). Phage λ concatemers are cut by nucleases that recognize the specific sequences

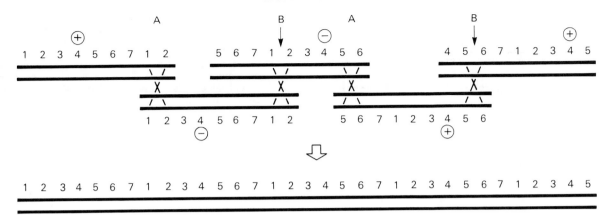

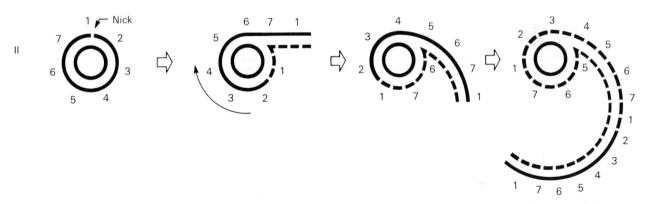

FIG. 47-14. Concatemer formation. I. By recombination. II. By rolling-circle replication. In I crossovers of type **B (arrows)**, between different parental molecules, cause genetic recombination, but those of type **A**, between identical molecules, do not. **Numbers** (arbitrary) indicate genes. ⊕ and ⊖ indicate molecules of different genetic constitution.

of the molecular ends, yielding nonpermuted molecules with cohesive ends.

Breakage and Reunion of Concatemers. Breakage and reunion continues to occur after concatemers are formed; when these processes occur at a high frequency (as with T-even phages) they lead to the dispersion of the parental DNA into many progeny molecules, each of which contains a small parental segment covalently linked to newly synthesized DNA.

This phenomenon was revealed by the abnormal results of experiments testing whether phage DNA replication is semiconservative (Fig. 47-17). Thus when cells were infected with ^{32}P-labeled phage T2 and then grown in a medium containing 5-bromouracil (5BU), which yields DNA of increased buoyant density, all the ^{32}P was expected to be in hybrid molecules of intermediate density having 5BU in one strand, with the bulk of the DNA having two heavy strands. However, on equilibrium density centrifugation the intact progeny phage DNA yielded a single band, at the density of 5BU–DNA, and it contained all the radioactivity. This result was due to the dispersion of segments of parental DNA into progeny molecules, as was shown by fragmenting the DNA by sonication. Banding in CsCl now produced the two bands predicted by the semiconservative model. It could be estimated that the dispersed parental segments were on the average 10% of a whole molecule and therefore increased only slightly the density of the molecules that contained them.

BIOCHEMISTRY OF REPLICATION

Replication of the phage DNA initiates **at the cell plasma membrane,** with which the DNA becomes associated after release from the virions. This association is stabilized by phage-specified proteins and seems to involve

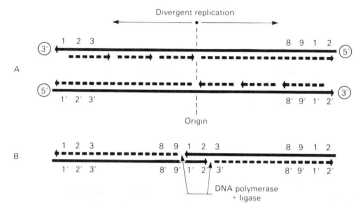

A

B

FIG. 47-15. Significance of concatemer formation. **Dashed lines** indicate new DNA strands; **numbers** (arbitrary) represent genes. The divergent replication of a linear molecule of T7 DNA (**A**) remains incomplete at the 3′ template ends (**circled**) if Okazaki segments cannot initiate at the terminal nucleotides. However, owing to terminal repetitions, the two unreplicated ends can pair (**B**), forming a completely replicated concatemer.

FIG. 47-16. Production of mature viral DNA molecules from concatemers. **A.** Mechanisms generating permuted molecules with repetitious ends, as with phage T4. Constant lengths are cut off the concatemer, irrespective of sequences. B, C. Mechanisms generating nonpermuted molecules with cohesive or repetitious ends. **B.** Nucleases make staggered cuts on the two strands, recognizing specific sequences, and generate λ-type molecules (with cohesive ends). **C.** If the short strand at each end is continued by DNA polymerase as a complement of the longer strand, T7-type molecules with repetitious ends are generated. Black **arrowheads** indicate endonucleolytic nicking. **Numbers** indicate sequences, **primed numbers** complementary sequences; **dashed segments** are replicated after cutting.

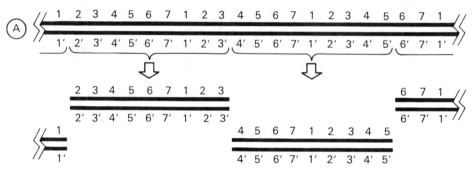

T4-type permuted molecules with repetitious ends

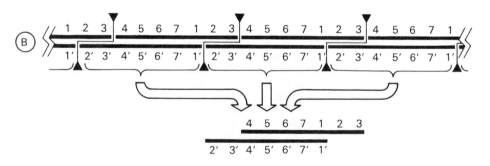

λ-type molecules with cohesive ends

T7-type molecules with repetitious ends

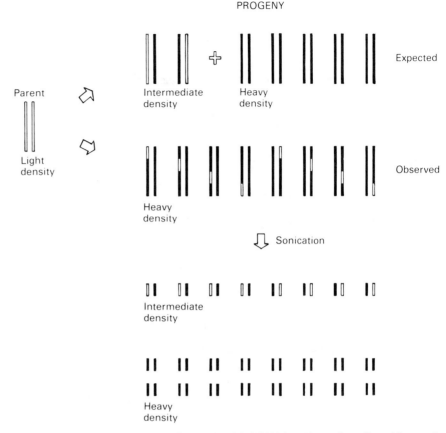

FIG. 47-17. Scheme of replication of T2 DNA with "light" DNA in a "heavy" medium. After equilibrium centrifugation in a CsCl density gradient the parental DNA (**open lines**) is expected entirely in a band of hybrid density, on the basis of semiconservative replication. The different result observed is explained if semiconservative replication is associated with breakage and reunion. If a progeny molecule has experienced, in its line of descent, many breakage and reunion events, the parental DNA may be present in small segments distributed over a large number of molecules, which therefore would have a density only slightly less than that of new DNA. Sonication, by breaking the molecules into small fragments, allows the identification of the hybrid segments, since their lower density will not be "diluted" by attachment to nonhybrid (all new) segments.

specific discrete sites, because a phage cannot replicate in a cell saturated by a related phage, whereas an unrelated phage can.

Labeling experiments show that precursors derive from the medium; most phages that destroy the host DNA also utilize as precursors its breakdown products. These precursors are elaborated and synthesized into phage DNA by host and phage enzymes. Large phages (such as the T-even) specify many such enzymes (Fig. 47-18), whereas the smallest ones depend almost entirely on those of the host. With all phages a specific protein (**primase**) is required for **initiating phage DNA replication;** it recognizes a specific initiation sequence on the phage DNA. As with bacterial DNA, phage primases participate in the synthesis of a short RNA segment, which is then elongated by DNA polymerase and is often found bound to nascent phage DNA. Some primases are phage-specified (gene 4 product of T7; gene 41 product, and others, of T4); other phages use host primases.

Many enzymes specified by large phages **duplicate host enzymes,** presumably to cope with the increased DNA synthesis but perhaps also to fulfill some more specific functions. Thus although T4 mutations that block the functions of many enzymes only slow down phage replication, those affecting phage DNA polymerase (gene 43) or ligase (gene 30) are lethal.

Phages with special bases in their DNAs—e.g., 5-hydroxymethylcytosine (HMC) instead of cytosine in T-even phages, or 5-hydroxymethyluracil instead of thymine in some *B. subtilis* phages—specify enzymes not only for synthesizing the special nucleotide but also for preventing the incorporation of the usual one. Thus T-

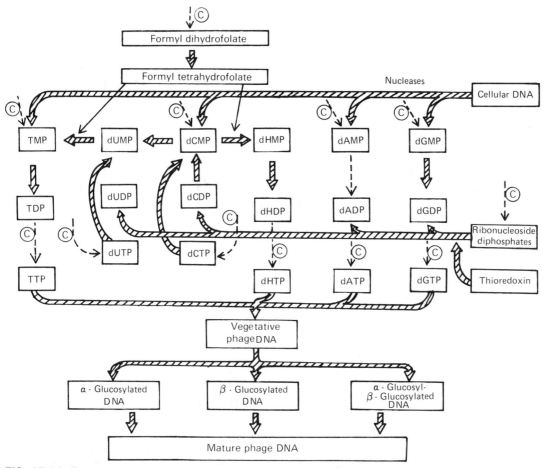

FIG. 47-18. Enzymatic reactions involved in the synthesis of T4 DNA. **Heavy arrows** indicate enzymes specified by phage genes (whether or not there is a cellular enzyme with the same function). **Dashed arrows** marked with a C indicate enzymes of cellular origin.

even phages specify enzymes that dephosphorylate dCTP, dCDP, and dCMP. This is an essential step because dCMP-containing phage DNA is unsuitable as a template for late transcription and is exposed to breakdown by the phage nucleases that degrade the host DNA.

ROLE OF DNA MODIFICATIONS: HOST-INDUCED RESTRICTION AND MODIFICATION

The effects of restriction endonucleases and of modifying enzymes were detected with bacteriophages before the mechanisms were known. The observations were puzzling because they showed that the host caused quasi-hereditary changes in the phage.

GLUCOSYLATION

The first observation of host-controlled restriction and modification involved glucosylation (Fig. 47-19A). With T-even phages the newly formed DNA is rapidly glucosylated by phage enzymes, which are responsible for the different patterns of glucosylation of the DNAs of the different phages. However, in bacteria lacking uridine diphosphate glucose (UDPG) the progeny viral DNA **remains unglucosylated.** The resulting virions, designated T*, are unable to grow, i.e., are **restricted,** in any E. coli B strains, because unglucosylated DNA is specifically broken down by restriction endonucleases present in the plasma membrane of B cells. These enzymes recognize insufficiently glucosylated 5-hydroxymethylcytosine residues in special sequences. T* phage can, however, multiply in *Shigella,* which lacks the restriction enzymes. Moreover, since *Shigella* makes UDPG, the phage DNA is glucosylated, and therefore the progeny is again unrestricted in E. coli B. In a parallel situation, a phage mutant that is unable to glucosylate (gt⁻) cannot grow in E. coli: it grows in *Shigella,* but it is not modified.

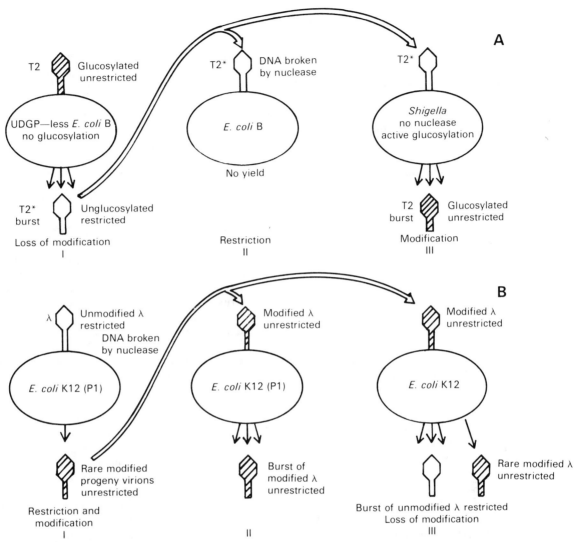

FIG. 47-19. Two examples of host-induced restriction and modification. **Shaded** outlines indicate modified phage which is unrestricted in the bacterial strains employed. **A.** Restriction and modification occur in different hosts. The DNA present in wild-type T2 is glucosylated, i.e., modified. Unglucosylated phage T* is produced after a growth cycle in a nonglucosylating host (I). T2* is restricted in *E. coli* B (II); it grows in *Shigella*, where it is modified (III) to yield regular T2 again. **B.** Restriction and modification in the same host. The usual phage is restricted and at the same time modified (by methylation) in *E. coli* K12(P1)(I); the modified phage then grows regularly in the restricting host (II). After one growth cycle in *E. coli* K12 the modification is lost, except for rare virions that inherit a DNA strand from the parent (III).

METHYLATION

Methylation-dependent restriction and modification (Fig. 47-19B) were discovered from observations with phage λ. In P1-lysogenic *E. coli* K12 cells, designated as K12(P1), the plating efficiency of λ is much lower than in the corresponding nonlysogenic cells, because the injected λ DNA is rapidly cleaved by a P1-specific restriction nuclease which recognizes a target sequence of several nucleotides. In rare DNA molecules methylation of a base in one strand of the target before it is cut protects it from the nuclease; and the progeny of the methylated strand are also protected because they are methylated in the parental strand. The new strand is then methylated shortly after synthesis. This DNA generates progeny phage that grows normally in P1-lysogenic cells. This rare successful infection *does not represent selection of mutant*, for when the progeny is grown through a single cycle in K12 not carrying P1 DNA the new progeny viral DNA is not P1-modified and is again restricted in K12(P1) (except for those few particles that contain a parental, and therefore modified, DNA strand).

Significance of Restriction and Modification. It is likely that restriction endonucleases developed in bacterial evolution as a **defense against phages;** at the same time modifying enzymes had to evolve, in order to protect the cellular DNA. Phages, in turn, have developed defenses against restriction. One of these is the replacement of base C with HMC in T-even phages, which renders them resistant to the usual *E. coli* restriction systems; this change, however, makes them susceptible to another system against which they have developed glucosylation as a defense.

MATURATION AND RELEASE

Maturation and release are the last two acts of the process of viral multiplication. In **maturation** the various components become assembled to form complete or **mature** infectious virions; in **release** these leave the infected cells. In the one-step multiplication curves of Figure 47-5 the **total** multiplication curve represents the maturation of the virions; the **extracellular** curve, their release.

Pools of Precursor Macromolecules. Hershey studied the relation between DNA synthesis and maturation by infecting cells with phage T2 labeled in the DNA. Samples of the mature progeny phage were collected by breaking open the cells at various times after infection. The label was recovered in all samples of the progeny phage, including those collected very late; hence the *replicating viral DNA in a cell forms a single pool from which molecules are withdrawn at random* for incorporation into virions. At the end of the eclipse period the viral DNA pool was found to be equivalent to about 50 virions; it remains constant afterward, when new synthesis balances withdrawal for virion formation.

Similar experiments with radioactive pulse-labeling of the viral proteins showed that maturation draws the capsid precursor proteins at random from a common pool. Indeed, mutations in any phage genes complement efficiently in the same cell, implying a common pool for their products.

ASSEMBLY OF THE CAPSID

Labeling the virion proteins at various times with radioactive amino acids shows that they belong to the **late** class, except three small **internal proteins,** which belong

to the immediate early group. Assembly begins with the aggregation of the "soluble" capsid proteins, which is detected by their increased sedimentation rate. At a very early stage components of tail fibers can be recognized **serologically,** but most partly assembled components can be identified only when they are large enough to have **distinguishing electron microscopic features.**

The assembly of large phages is an interesting model for the morphogenetic processes of higher organisms. Its elucidation has depended almost entirely on the use of **conditionally lethal mutations,** which under nonpermissive conditions block the morphogenetic process at a step dependent on the mutated gene. Upon lysis the cells yield partly and often erroneously assembled structures (often recognizable by electron microscopy; Fig. 47-20), just as an auxotrophic mutant accumulates the biosynthetic intermediate preceding the blocked reaction. In addition, certain pairs of defective lysates yield **complementation** in vitro, i.e., the accumulated incomplete structures can assemble spontaneously, when mixed, to form infectious virus. Such complementation studies have shown that the T4 capsid is assembled through three **independent subassembly lines,** which produce the phage head, the fiberless tails, and the tail fibers, respectively (Fig. 47-21). These structures then spontaneously assemble into capsids in vitro, and presumably in vivo also.

.

Method of Assembly. In contrast to the assemblies of isometric and helical capsids, discussed in Chapter 46, T4 proteins do not produce an organized structure by self-assembly alone. Thus the protein p23 (i.e., the product of gene 23), the main constituent of the head, by itself aggregates randomly into "lumps."

Each assembly occurs through a series of obligatory **sequential steps,** in which each protein (except the first) undergoes a conformational change as it is assembled, revealing the binding site that is recognized by the next protein. The first protein in the sequence may have a transient function as a **scaffolding protein,** not retained in the finished structure. Sometimes an assembly step is stablized by a **proteolytic cleavage.**

The head assembly of T4 illustrates these principles (Fig. 47-21). The assembly of the major capsid protein p23 around a **core** formed by the scaffolding protein p22 and the three internal proteins yields the rounded **prohead I.** Maturation to the cornered and slightly wider prohead III occurs after cleavage of p23 (from a mol wt of 55,000 to 45,000) and of all the core proteins. Of these p22 is split into small fragments, some of which remain in the head as **internal peptides.** (In *Salmonella* phage P22 the scaffolding protein is released intact from the maturing head and is reutilized for assembly of new proheads.)

The initiation of assembly may also rely on host components. Thus the assembly of the head and the base plate of phage T4 on

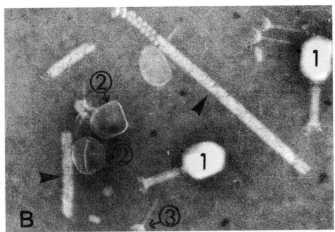

FIG. 47-20. Electron micrographs of polyheads (**A**) and polysheaths (**B, arrow**) present in lysates produced by nonsense mutants of phage T4. In B one also sees "empty" head membranes (**2**) and tubes attached to base plates (**3**). Both micrographs also contain some normal virions (**1**) because the cells were simultaneously infected with wild-type phage. The polyheads contain hexagonal capsomers which are no longer recognizable in regular phage, owing to further assembly steps; the polysheath has the diameter of a contracted regular sheath. (A courtesy of E. Boy de la Tour; B Boy de la Tour E: J Ultrastrut Res 11:545, 1964)

the inner layer of the cell's plasma membrane is prevented by bacterial mutations.

These mechanisms allow the regular assembly of a complex structure even if all components are synthesized at the same time and accumulate together in the cell. Such mechanisms must be the result of a long evolution, during which the ability of the main components to form an isometric capsid by self-assembly was lost in order to allow the addition of other structures.

ASSOCIATION OF DNA AND CAPSID

With phage, as with other viruses (see Mode of Replication, Ch. 46, and Maturation, Ch. 50), the *DNA enters preformed "empty" heads.* Thus when "empty" T4 proheads are accumulated at high temperature by a temperature-sensitive mutant (gene 49) they rapidly develop into full heads and infectious virions after the temperature is lowered. Moreover, when an extract of cells infected by T7 or λ mutants containing DNA but no heads is mixed with one containing heads but no DNA full heads are formed. The DNA becomes packed within the heads after the cleavage of the main capsid protein confers on them a cornered shape. The DNA probably binds to chemical groups revealed by the cleavage and the resulting protein rearrangement; this binding would afford the energy for "sucking in" the DNA and for compensating for its loss of entropy. Folding is stabilized by polyamines or basic peptides.

After the T4 head is filled with concatemeric DNA, the excess is cut off by an *endonuclease that works only when the head is complete.* Because this enzyme does not recognize DNA sequences a **"headful"** of DNA is packaged; its length is determined by the properties of the capsid. Thus smaller heads, the result of a petite mutation, contain less DNA; and giant heads, the result of other mutations, contain more DNA than normal. Assembly of a headful of DNA explains the encapsidation of random fragments of cellular DNA within phage capsids in generalized transduction (Ch. 48). With other phages (e.g., λ or T7) the nuclease recognizes special sequences, and their positions determine the length of the packaged DNA.

PHENOTYPIC MIXING

In bacteria mixedly infected by two related phages (e.g., T2h$^+$ and T2h) each causes the formation of a complete set of protein protomers. Cells infected by phage T2h$^+$ and T2h thus contain precursor tail fibers of both h$^+$ and h specificity. During assembly either fiber can be used in any virion; hence virions of a given genotype may contain tail fibers of that type, or of the other type, or a mixture of the two (Fig. 47-22). Virions in which the genotype and the phenotype are discrepant are called **phenotypically mixed.**

Phenotypic mixing may affect many capsid proteins, but the most common example is mixing of the host-range character, which is detected by the two-step procedure outlined in Figure 47-22.

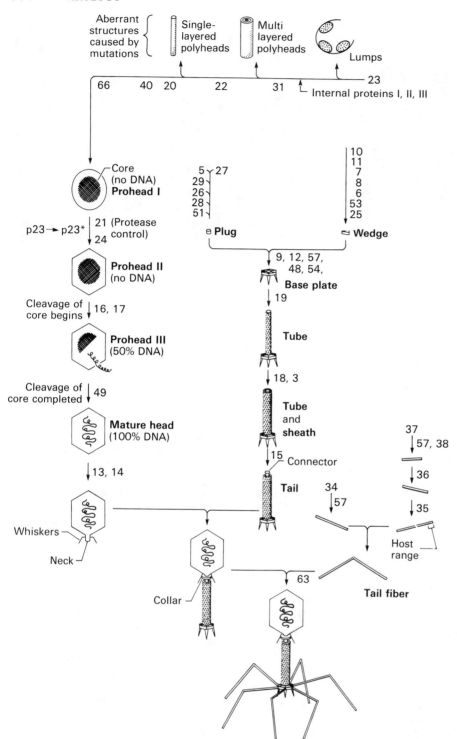

FIG. 47-21. Assembly of T4 virions. Assembly occurs in three major subassemblies: head, tail, and tail fibers. **Numbers** indicate the T4 genes participating in a given step. **p23** = Protein specified by gene 23; **p23*** = product of cleavage of protein p23. The aberrant structures in head assembly accumulate when a mutation prevents the function of the next gene in the assembly line. "Lumps" are disorganized masses of p23 at the plasma membrane. (Data from Laemmli VK et al: J Mol Biol 49: 99, 1970; Laemmli VK, Faure M: J Mol Biol 80:575, 1973; Dickson RC: J Mol Biol 79:633, 1973. Copyright by Academic Press, Inc., (London) Ltd.)

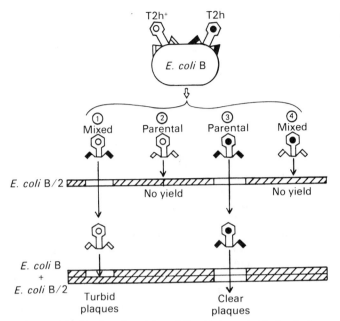

FIG. 47-22. Phenotypic mixing of the h character in bacteriophage T2. *E. coli* B infected by both T2h and T2h⁺ produces four kinds of progeny. Phage **1** is genotypically h⁺, phenotypically h; phage **4** is genotypically h, phenotypically h⁺; **2** and **3** are parental. After infecting *E. coli* B/2 only phenotypically h phage (**1** and **3**) produces progeny, which is phenotypically true to the genotype. The genotype is then determined by plating the B/2 yields on a mixture of *E. coli* B + *E. coli* B/2 (**mixed indicator strains**), where genotypically h⁺ phage (1, which does not infect B/2 cells) produces turbid plaques and genotypically h phage (**3**, which infects both types of cell) produces clear plaques. (see FIG. 47-23)

RELEASE

With T-even phages several gene products alter the plasma membrane, and then the **phage lysozyme** (from gene *e*) crosses the altered membrane and attacks the cell wall, causing lysis.

In the wild-type allele gene *t* promotes lysis by altering the plasma membrane of the cells and stopping cellular metabolism; the familiar rapid-lysis gene *r* and the gene *s* affect the membrane in ways that delay lysis (hence their mutations accelerate lysis). A block in energy metabolism (e.g., addition of cyanide or deprivation of oxygen) accelerate lysis.

With the very small phage φX lysis does not seem to involve lysozyme; some phages require the activity of a host autolysin. Filamentous phages are released by an entirely different mechanism, without lysis.

A PARASITE OF A PARASITE

In the same way that a phage parasitizes the machinery of the host cell, a phage can also *parasitize the machinery induced by another phage*. Thus the small **satellite** coliphage P4 (DNA length 10 Kb) replicates only in cells also infected by coliphage P2 (DNA length 30 Kb), turning some P2 genes to its own advantage. Two P4 early proteins switch on P2 late transcription, producing capsid proteins; another P2 protein activates two P4 late genes, which also specify capsid proteins. The combination of P2 and P4 proteins then allows the assembly of a capsid about one-third the size of a P2 capsid, which is suitable for accommodating the smaller P4 DNA. The DNAs of the two phages show no homology in DNA hybridization.

PHAGE GENETICS

Historically, the study of phage genetics provided the main incentive for undertaking the analysis of phage multiplication. The investigators who started the new wave of bacteriophage work in the late 1930s recognized that genetics had to turn to simpler organisms in order to determine the molecular basis of the previously established complex formalism. For this purpose phages offer outstanding advantages: about half their total mass is genetic material; even larger populations are conveniently available than with bacteria; and the genomes are smaller. The development of phage genetics in the last two decades has been an outstanding intellectual achievement and has contributed heavily to the foundations of molecular genetics.

A highly sophisticated mathematical analysis of phage genetics was built on the basis of technically simple observations, carried out with a few mutant types. In addition, fine-structure genetics was launched by Benzer through the study of mutants of a phage gene, even though its protein product was unknown. The subsequent extension of fine-structure genetics to bacteria led to a profound understanding of the relation between gene and protein. Further advances have resulted from the use of nucleic acid hybridization and restriction endonucleases. These technologies are especially suitable for phage because after the nucleic acid has been manipulated in vitro its genetic properties can be tested by transfection.

MUTATIONS

The genetics of phages, like that of all organisms, is based on the study of mutants. The types that can be recognized are limited because phages are haploid; hence mutations in many genes prevent propagation, i.e., are **lethal**. The useful mutations are of two main classes: plaque-type and conditionally lethal.

Mutants producing plaques with changed morphology (**plaque-type mutants**) were extensively used in the initial investigations of bacteriophage genetics, since they could be recognized easily on inspection of the assay plates. However, some of these mutants do not allow the use of selective technics; and they leave much of the chromosome unmapped. These difficulties were overcome by the use of **conditionally lethal mutants** (Ch. 8). These mutants, which can be isolated in most genes, have led to the establishment of a fairly complete **phage map** (see below). Moreover, they have been indispensable for **fine-structure** genetics, because in genetic crosses between two mutants even in the same gene, rare wild-type **recombinants** can be easily selected. Similarly, **complementation** is easily observed and establishes the limits of a gene: two mutants in different genes (but not those in the same gene), each unable to multiply alone under restrictive conditions, multiply together when they infect the same cell, in which each supplies the function missing in the other (see Ch. 11, Units of Information).

A number of mutant types employed in genetic work with T-even bacteriophages will be briefly examined.

PLAQUE-TYPE MUTANTS

Rapid-lysis (r) mutants (Fig. 14-7) have already been described above (see Release). These mutants have been used extensively in genetic work, since they incidentally turned out to be conditionally lethal.

Host-range (h) mutants were also described above (see Adsorption). It will be recalled that they can be identified on mixed indicator strains, where they produce clear plaques and h$^+$ (wild-type) phages produce turbid plaques (see Fig. 47-22).

Minute plaque and **turbid plaque mutants** produce plaques described by their names.

CONDITIONALLY LETHAL MUTANTS

The **r$_{II}$ mutants** multiply in *E. coli* B but not in *E. coli* K12 (λ) (see Effect of Prophage on Host Functions, Ch. 48), whereas r$_{II}^+$ phages multiply equally well in both hosts. In pioneering studies on fine-structure genetics Benzer used this property to detect minute proportions of r$_{II}^+$ mutants or recombinants in populations of r$_{II}$ phage.

Amber or ochre mutants are characterized by a **non-sense codon,** which causes premature termination of the synthesis of the polypeptide chain specified by the gene. These mutants can multiply only in bacterial strains carrying a suppressor mutation that specifies an altered tRNA able to place an amino acid at the site of the non-sense codon, allowing the completion of the polypeptide chain. These mutants, therefore, are **suppressor-sensitive.**

Temperature-sensitive mutants (ts) express the affected gene at 35°C but not at 43°C.

Restriction mutants can multiple more efficiently in a restricting host because they have lost a site for a restriction endonuclease.

GENETIC RECOMBINATION

Genetic recombination in T-even bacteriophages was discovered in 1946 by Delbrück and by Hershey, who performed genetic crosses by mixedly infecting bacteria with a host-range (h) and a rapid-lysis (r) mutant. The yields of these cultures contained four distinct types of particles (Fig. 47-23): the two parental types and two recombinant types with a marker from each parent.

The study of genetic recombination, especially with the larger bacteriophages, turned out to require complex statistics because *recombination takes place in the vegetative pool* (probably during formation of concatemers) involving many DNA molecules. Indeed, DNA molecules evidently continue to recombine throughout the period of multiplication, since the **proportion** of recombinants in a mixedly infected cell increases with the time after infection. Furthermore, in triparental crosses some of the progeny individually incorporate markers from all the parents, a result that requires at least two independent matings. (In molecular terms a **mating** corresponds to the breakage of a DNA molecule and its subsequent reunion to another.) On the average, T-even bacteriophages undergo four or five **rounds of mating** in an infection cycle; most other phages undergo a smaller number of matings. The consequences of a mating between two phage DNA molecules, therefore, cannot be inferred directly from the types and proportions of recombinants in the lysate of an infected culture, but require a statistical analysis of the effect of multiple matings. Some more direct insight into the results of a single mating can be obtained by studying the yield of single infected cells, especially with phages that undergo a small number of matings per replication cycle.

CONSTRUCTION OF A FORMAL GENETIC MAP

The formal methods for determining the sequence of phage markers are similar to those employed for bacteria or higher organisms and are based on the proportion of recombinants (**recombination frequencies**) in the lysate of a culture infected by two marked strains (see Ch. 11, Genetic Mapping). These recombination frequencies establish the **order** of markers but are not directly related to the **distances between** markers, even after correction for double crossovers. In fact, with a series of markers the corrected frequencies are not additive, as shown in Fig. 47-24; double crossovers seem to be too frequent (**nega-**

FIG. 47-23. Plaques formed by a mixture of T2 phages carrying mutations at the h and the r locus, placed on a mixture of *E. coli* B and *E. coli* B/2 (i.e., resistant to T2h$^+$ but sensitive to T2h). Phages with the h and those with h$^+$ allele produce, respectively, clear plaques (**dark** in the photograph) and turbid plaques (**gray** in the photograph); phages with the r allele are larger than those with the r$^+$ allele. Thus all four possible combinations can be distinguished: T2h$^+$r$^+$ (wild-type), T2hr, T2h$^+$r, and T2hr$^+$. (From Molecular Biology of Bacterial Viruses by Gunther S. Stent. W. H. Freeman and Company. Copyright © 1963)

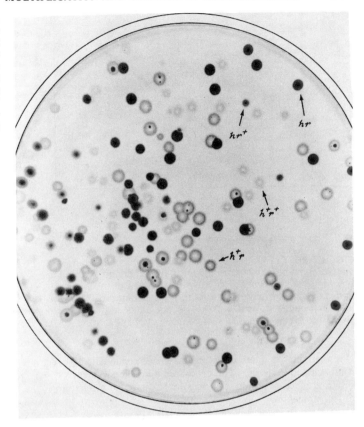

tive interference). This lack of additivity derives from the statistical characteristics of phage replication; i.e., the different progeny DNA molecules undergo an unequal number of rounds of matings before encapsidation. After they are multiplied by an empirical **mapping function** that restores additivity, the observed recombination frequencies establish the distances between markers in **map units** (where 1 unit corresponds to 1% corrected recombination frequencies). The length of the map of phage T4 determined in this way is 2500 map units, each corre-

sponding to about 100 nucleotide pairs. In λ, however, a map unit corresponds to about 2000 nucleotide pairs. The difference can be attributed to the much brisker breaking and rejoining activity of replicating T4 DNA.

Relative distances calculated from recombination frequencies are in good agreement with physical distances (Fig. 47-25), but deviations are observed in a few regions of the genome. For example, though the average probability of recombination between adjacent nucleotides in T4 is about 10^{-2} units, extremely low frequencies of recombination (10^{-4}–10^{-8} map units) are observed be-

FIG. 47-24. The order of genes in a segment of the T4 genome, determined by three-factor crosses. Each **vertical line** corresponds to a marker; the observed recombination frequency in each pairwise cross is recorded. Note the consistency of the results: the recombination frequencies between the closest pairs (top row) establish a sequence that yields increasing frequencies, as predicted, in crosses involving more distant pairs. The frequencies, however, are not additive. (Modified from Edgar RS, Lielausis I: Genetics 49:649, 1964)

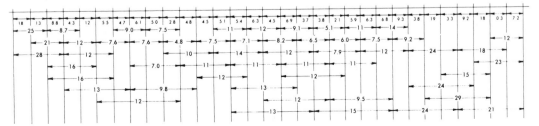

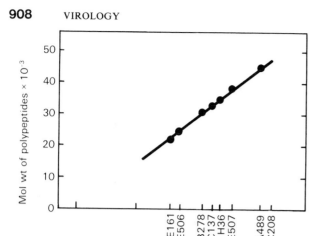

FIG. 47-25. Correlation between the map distances between amber (chain-terminating) mutations in p23 of phage T4 and the physical distances measured from the lengths of the polypeptiedes synthesized. There is a relatively constant ratio of 0.012 map units per amino acid, or 0.004 per nucleotide. (Modified from Cells, JE et al: Nature New Biol 241:130, 1973)

tween certain r_{II} mutants. Other regions have exceptionally high recombination frequencies. These localized differences evidently result from effects of certain sequences on the molecular mechanisms of recombination.

CONSTRUCTION OF A PHYSICAL MAP

Especially useful is the technology developed by Nathans for producing characteristic DNA fragments with restriction endonucleases (see Ch. 10, Restriction Endonucleases). The many available enzymes, which recognize different sequences, yield from a given DNA many well characterized fragments, in a wide range of sizes. The overlaps of the fragments with larger pieces obtained by incomplete digestion reveal the **sequential order** of the fragments, thus allowing a physical mapping of the restriction sites.

Genes can be located on these maps by determining which fragments overlap a given marker. One method is **marker rescue from fragments:** transfection of a purified wild-type DNA fragment into a cell infected by a mutant phage can generate wild-type progeny by recombination if the fragment overlaps the mutation. Another method involves **transfection of a partial heteroduplex** containing a mutant complete strand and a wild-type fragment strand overlapping the mutation (Fig. 47-26). Again, wild-type progeny can be obtained either by DNA synthesis completing the fragment strand or by error correction.

Physical distances between nonsense mutations in the same gene are precisely determined by the differences in length of the corresponding polypeptide chains, if they can be isolated (see Fig. 47-25).

OTHER CHARACTERISTICS OF BACTERIOPHAGE RECOMBINATION

Nonreciprocal Recombination. In the yield of a large population of infected bacteria a phage cross Ab × ab yields equal proportions of the two reciprocal recombinants, Ab and aB. However, *in the yields of individual bacteria one recombinant regularly predominates,* especially if the average number of rounds of matings is reduced by prematurely lysing the cells. In crosses involving phages with single-stranded cyclic DNA (see below) the yield of single cells is often one recombinant and one parental type when recombination precedes DNA replication. A possible model is shown in Fig. 47-27.

Formation of Heterozygous DNA. Mixed infection with phages carrying different alleles regularly yields some virions that are

FIG. 47-26. Marker rescue from synthetic partial heteroduplex. **A.** Wild-type (**wt**) viral DNA is cut by restriction endonucleases, yielding characteristic fragments. Mutant DNA (mutation at −) and purified fragment 5 (which contains the corresponding wt allele +) are denatured and hybridized together. One of the products is the partial heteroduplex (**B**), which is introduced into suitable cells by transfection. **C.** The heteroduplex is completed by DNA synthesis (**dashed lines**) and after replication yields a wt molecule (D). The mutant strand of the heteroduplex can also be converted to wt by error correction.

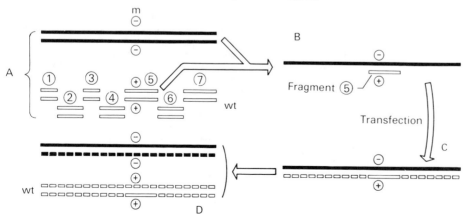

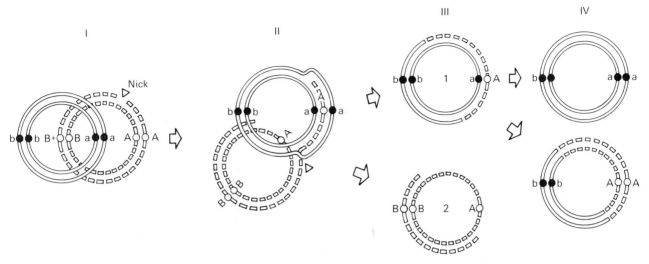

FIG. 47-27. Production of a recombinant and a parental type as a result of phage recombination. The model assumes the following steps: **I.** A strand of one of the parental molecules is nicked (**arrows**). **II.** The detached single strand is assimilated by another molecule; in cyclic DNAs this step is favored by supercoiling, with a deficiency of helical turns, in the recipient molecule, since stand assimilation eliminates the deficiency. The assimilated strand increases in length by displacing the regular strand. Then endonucleases (**arrows**) cut various single-stranded segments, yielding two molecules (**III**), of which 1 is repaired (by polymerase and ligase) whereas 2 is broken up by single-strand specific nucleases. Finally replication of molecule 1 (**IV**) yields one parental and one recombinant molecule. **Continuous** and **dashed lines** denote the sequences of the two parental molecules.

heterozygous for one of the markers: with a mixture of r and r$^+$ phage about 1% of the progeny virions is heterozygous and produces characteristic **mottled plaques,** i.e., with alternating r and r$^+$ sectors (see Fig. 47-7).

Extensive studies of this phenomenon demonstrated two classes of heterozygous particles: internal and terminal (Fig. 47-28). **Internal** heterozygotes contain heteroduplex DNA, which is regularly produced in all organisms during recombination; **terminal** heterozygotes are peculiar to DNA with **terminal duplications.** Both heterozygotes are formed by recombination; only terminal heterozygotes persist after replication.

High Negative Interference. In addition to the moderate interference of statistical origin discussed above (see Construction of a Formal Genetic Map), an apparently large excess of multiple crossovers is found in crosses between **close markers.** Transfection of cells with heteroduplex DNA made in vitro by annealing complementary strands from phages with different markers shows that the phenomenon is due to *correction of mispairings present at the mutated sites in the heteroduplex DNA* formed in recombination. A correction that eliminates a marker on a DNA strand, leaving two flanking markers unchanged, appears formally as a double crossover.

MOLECULAR AND ENZYMATIC MECHANISMS OF PHAGE RECOMBINATION

The effects of viral mutations on the frequency of recombination in T-even phages point to the importance of single-strand breaks in DNA and of unfolding of single strands. Thus the frequency of recombination is **increased** if many single-strand breaks are gen-

erated, as in mutants deficient in dCMP hydroxymethylase, or if the breaks remain unsealed, as in mutants deficient in ligase or DNA polymerase. Conversely, recombination frequency is **decreased** by mutations in gene 46 or 47, which together specify an exonuclease. Recombination is also reduced by mutations in gene 32, whose protein binds to single-stranded DNA, keeping it unfolded. Altogether T4 recombination requires proteins specified by at least nine phage genes.

Recombination occurs simultaneously with DNA replication, but it does not depend on it, since it can take place in thymine-starved cells. Hence the enzymatic mechanisms that produce recombination, although they may have been developed to favor DNA replication, continue to work in its absence.

ORGANIZATION OF THE GENETIC MATERIAL OF PHAGE T4

In bacteria the grouping of the genes in the chromosome has important regulatory aspects. The question whether the much simpler genetic system of viruses possesses a similar organization can best be approached with bacteriophages whose maps are known in considerable detail. We shall consider the T4 and T7 maps here (Figs. 47-29 through 47-31) and the λ map in the next chapter.

Circularity. If the total length of the T4 map (Fig. 47-29) is covered by a series of crosses between pairs of markers

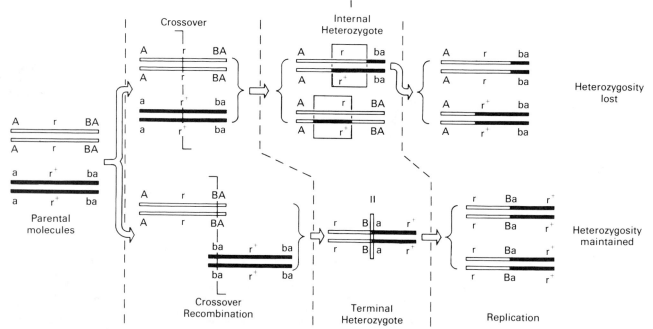

FIG. 47-28. Terminal and internal phage heterozygotes. Two types of internal heterozygotes are found: one with crossing over of the external markers A and B (**I**) and one without crossing over (**II**), depending on the type of recombination that generated them. The internal but not the terminal heterozygotes are dissolved at replication. Note that recombination generates a molecule whose ends differ from those of the parental molecule because it leads to formation of a concatemer, from which mature molecules are excised at random. Only molecules excised as indicated are terminally heterozygous. **Boxes** indicate heteroduplex DNA. In the lower crossover the heteroduplex DNA, in an area without markers, is undetectable.

at relatively close distances (20–30 units), the starting marker is always closely linked to the terminal marker. Thus the map is circular, even though the DNA of the virions is linear.

The apparent contradiction is resolved by the finding that the DNA molecules in virions are **circularly permuted** (see above, Viral DNA Replication): they are generated from a periodic concatemer by a cutting off of segments longer than the periodic unit, and so they can initiate at any gene and have long repetitious ends. Hence any two genes that are adjacent to each other in the concatemer are also adjacent in the mature DNA molecules, and therefore are genetically linked. Such molecules can be ideally derived by copying a circle beginning at various points (Fig. 47-30).

Other bacteriophages, such as λ and T7 (Fig. 47-31), have linear maps, because their DNA molecules are not permuted.

Number of Genes. About 140 genes have been identified in phage T4; the functions are known for almost all of them (Fig. 47-29). Still other genes without known functions remain to be discovered in the map intervals. About 80 of the genes have functions **essential for phage multiplication** and therefore are detectable by conditionally lethal mutations. **Inessential genes** make multiplication more efficient; some of them are especially useful in some wild *E. coli* strains, in which the phage probably evolved.

General Organization of the Map. The genetic maps of several phages (Figs. 47-29, 47-31, and 48-4) show that *genes are clustered according to their functions* (e.g., syn-

thesis of early enzymes, or of the head, etc.) though with a few exceptions. The clusters recall bacterial **operons,** which are groups of genes under common transcription and regulation. The reason for clustering may be related to morphogenesis.

PHAGE EVOLUTION

The three related T-even phages must have derived from a common precursor, since their DNAs show extensive homology. The localization of the limited nonhomologous parts, determined by heteroduplex formation between annealed single strands of different phages, illuminates how **evolutionary divergence** occurs. The major divergence occurred in the contiguous genes 37 and 38, which specify the part of the tail fibers that recognizes the bacterial receptors (see Fig. 47-29). It was probably promoted by the existence of bacterial strains that are resistant to some phage of this group (see Adsorption, above). In contrast, there is essentially *no divergence in the morphogenetic genes,* and in those for proteins that become incorporated in the host membrane. Obviously, changes in morphogenetic genes are only permissible if they affect all the interacting proteins at the same time so that their interactions are not disturbed, and this event is ex-

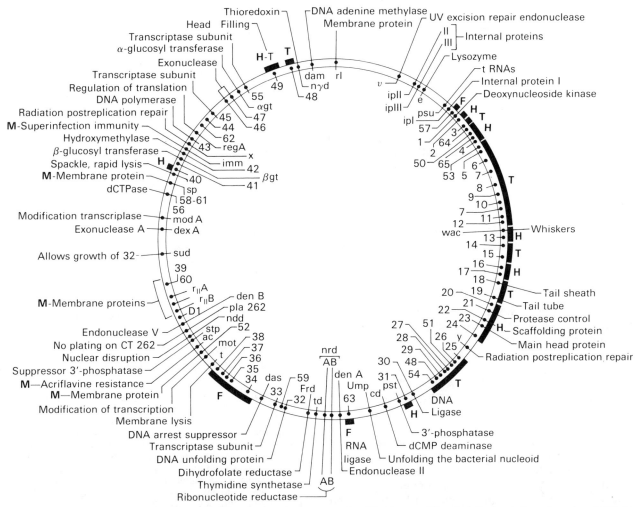

FIG. 47-29. Genetic map of bacteriophage T4. **Numbers** refer to genes identified by conditionally lethal mutations, in order of discovery. Genes identified by other means are indicated by **letters**. The known functions are indicated outside the circle. Late genes are indicated by a **black band**. Letters **H, T,** and **F** indicate that the corresponding genes are involved in head, tail, or fiber synthesis, respectively. **M** = membrane proteins. (Based on data from Wood WB, Revel HR: Bacter Rev 40:847 1976)

FIG. 47-30. Generation of the circular T4 map. Cutting the concatemer, precursor of mature virions, into equal segments longer than the period of the concatemer generates the indicated collection of permuted, terminally repetitious molecules. They can all be generated from the circle on the right, beginning at various points, as shown by these examples. The genetic map equals that circle.

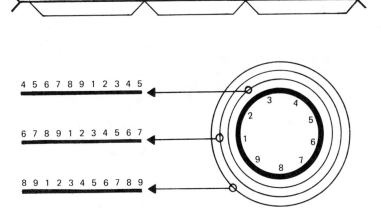

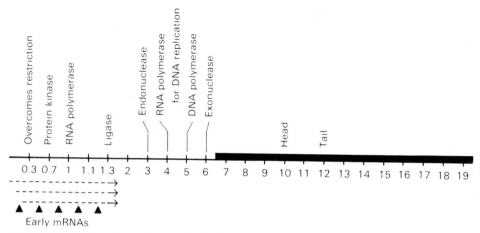

FIG. 47-31. The linear genetic map of phage T7. The **black bar** indicates the late genes. **Arrowheads** are processing points of the early mRNAs which initiate on several closely spaced but distinct promoters.

tremely rare. Similarly, since host components are constant, proteins interacting with these components are very unlikely to change.

GENETIC REACTIVATION OF ULTRAVIOLET-INACTIVATED PHAGE

Phages have contributed much to our understanding of the biology of radiations, because their genetic constitution is known in detail and the effect of radiation on individual genes can be measured accurately. Thus the **functional survival** of a phage gene can be determined directly in cells infected with ultraviolet-irradiated (UV'd) phage from the amount of a gene product synthesized, or indirectly from its ability to complement a mutant gene infecting the same cells (Fig. 47-32). The functional survival depends on both the size of the gene and its distance from the promoter (since a UV damage interrupts transcription).

Both viral and cellular genes are involved in repair of DNA damage (Ch. 11). Thus the effects of mutations show that in T4 there are two or three repair pathways: genes v (excision nuclease) and 30 (ligase) seem to be involved in **excision repair;** after the excision of dimers the DNA gaps are probably filled by the host polymerase I. More T4 genes are involved in **postreplication** repair: x and y (unknown functions), 43 (DNA polymerase), 30 (ligase), 32 (DNA helix-destabilizing protein), 46, 47, 59 (recombination genes). These pathways also repair damage produced by ionizing radiation (e.g., from ^{32}P decay) and by alkylating agents. As in bacteria, an inducible, error-prone type of postreplication repair causes mutations to occur.

The small phages (e.g., ϕX or λ) rely in large part or completely on repair mechanisms of the host, which is then said to carry out

host cell reactivation. The survival of these phages after UV irradiation depends on characteristics of the host cell.

In addition to the usual forms of repair of UV damage, bacteria infected by more than a single virion exhibit

FIG. 47-32. Kinetics of ultraviolet (UV) damage to a gene and to a genome. Functional survival of the $r_{II}B$ gene is compared with survival of the entire genome of T4 bacteriophage particles irradiated with UV light. Survival of the genome was measured from the fraction of $K12(\lambda)$ cells yielding phage after single infection by irradiated $T4r_{II}B^+$ particles (**dashed line**). Survival of the $r_{II}B$ gene was measured by simultaneously infecting the same cells with unirradiated $r_{II}B$ mutant phage, which cannot multiply in this host unless complemented with an undamaged $r_{II}B^+$ allele. (Data from Krieg D: Virology 8:80, 1959)

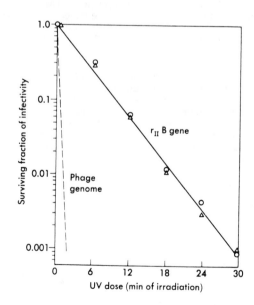

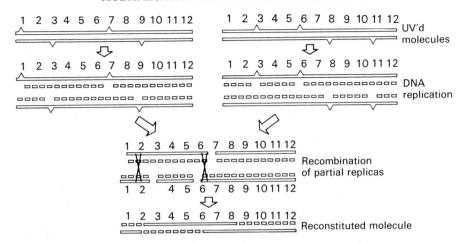

FIG. 47-33. Reconstitution of an intact molecule by recombination of partial replicas (between UV damages) of two UV-irradiated parental molecules. **Arrowheads** indicate UV damages.

mechanisms based on recombination between different genomes in the same cell.

Multiplicity Reactivation. When a cell is infected by irradiated T2 the probability of yielding infectious virus increases disproportionately with the multiplicity of infection. The explanation is that the DNA molecules will have their randomly distributed UV lesions in different locations, and replicas of the undamaged segments can recombine to form concatemers containing the complete information of an intact molecule (Fig. 47-33). Hence *multiplicity reactivation is a special form of postreplication repair.* The effect is especially striking in cells that have lost the other mechanisms for repairing UV damage. Conversely, if recombination is prevented by mutations, multiplicity reactivation does not occur. This form of repair is not mutagenic, i.e., it is error-free

For effective reactivation a cell must contain at least one undamaged copy of each of the phage genes involved in postreplication repair. This requirement determines the shape of the survival curves of multicomplexes as a function of the UV dose (Fig. 47-34). These curves are of the **multiple-hit type** (see Quantitative Aspects of Killing by Irradiation, Ch. 11, Appendix), i.e., they show an exponential decrease in the survival rate after a lag, because viral multiplication is absent only if all the representatives of a required gene are functionally inactivated. The curves have *final slopes smaller than those for phage inactivation*, because all the required genes present a smaller UV target than the whole genome.

In a genetic cross with phage capable of multiplicity reactivation, UV irradiation of the parents *increases the frequency of recombination* among the progeny, a useful tool in genetic studies. The explanation is that active progeny molecules arise by multiplicity reactivation, i.e., by many crossovers.

Marker Rescue. When a cell is simultaneously infected by a UV-irradiated virion carrying a marker allele of a particular gene and

FIG. 47-34. Multiplicity reactivation of UV-irradiated T2. In curve **A**, bacteria were each infected with an average of four T2 phages; the curve shows the fraction of the cells able to yield infectious phage, for different UV doses given to the phage. Curve **B** shows the results obtained when the cells were infected at low multiplicity (i.e., mostly single infection). (Modified from Dulbecco R: J. Bacteriol 63:199, 1952)

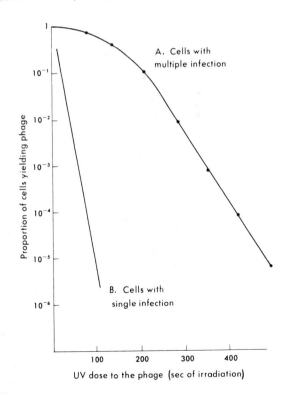

by one or more undamaged virions carrying another allele, recombination can incorporate the marker allele into the undamaged DNA and hence into the progeny. Effective recombination must involve a partial replica of the UV'd phage containing the marker allele. This replica will initiate at a replication origin and terminate at the nearest unrepaired UV damage on each side. Hence the probability of such marker rescue decreases as the number of lesions in the irradiated DNA increases, and is higher for genes near an origin of replication.

MULTIPLICATION OF BACTERIOPHAGES WITH CYCLIC SINGLE-STRANDED DNA

This class includes two groups of phages with very different virions: **icosahedral** in one, **filamentous** in the other. However, the methods of infection and replication are surprisingly similar for the two groups. The prototypes are ϕX174 and G4 for icosahedral phages; f1 and M13 for the filamentous phages. These phages have been extremely useful *probes for studying bacterial DNA replication* because they are completely dependent on cellular replication enzymes.

These phages have very small DNAs (5.4 Kb for ϕX174, which has been entirely sequenced; 6.4 Kb for filamentous phages). Phage ϕX174 has 9 genes, and M13 and f1 each have 8, several of them expressing similar functions in the two groups. The phages do not inhibit host macromolecule synthesis (except for DNA late in infection with icosahedral phages).

Infection. This process relies strongly on a **pilot protein**, located in ϕX174 in the spikes jutting out at the 12 corners of the capsid, and in M13 at one end of the filamentous virion. This protein performs several important and seemingly unrelated functions, as shown by the effects of mutations: 1) it causes the **adsorption** of the virions to the cell receptors; 2) after undergoing conformational changes it carries the viral **DNA into the cell**; 3) it **initiates the replication** of the viral DNA, probably by linking it to the replicating machinery of the host at the cell plasma membrane; 4) it is important for phage **morphogenesis**. The versatility of this protein is an example of the **genetic economy** of these phages. This economy finds another striking example in extensive gene overlaps (Fig. 47-35) in both ϕX174 and G4.

Most filamentous phages adsorb to the tip of pili specified by the F plasmid. As the DNA penetrates into the cell the capsid protein becomes incorporated into the cell plasma membrane and is later reutilized during virus release (see below).

DNA Replication. The replication of phage DNA takes place in three phases.

1) Synthesis of a **complementary (minus) strand** on the infecting viral (plus) strand to form the **parental replicative form (RF)** is carried out by host enzymes.

The virus initiates complementary strand synthesis by forming a short RNA segment at a constant origin, using host enzymes. After the rest of the strand is synthesized the RNA is removed, leaving a gap that is later filled with DNA. The covalently closed cyclic molecule is then made supercoiled by the enzyme **gyrase**. This parental RF is transcribed into **mRNAs with plus polarity**, which are templates for the viral proteins.

2) **Replication of the parental RF by the rolling-circle model** (see Fig. 47-14) generates 10–20 **progeny RFs** per cell for ϕX174, and 100–200 for f1.

Replication is initiated by a phage-specified endonuclease (the A protein), which nicks the viral strand of the supercoiled parental RF at a specific site and becomes covalently bound to its 5' end. The bound endonuclease moves along the negative strand as replication proceeds and cuts off the progeny molecules as soon as it is completed; at the same time the endonuclease, acting as ligase, causes the cyclization of the progeny viral strand. The energy of the protein–DNA bond is utilized to seal the nick: the energy of the broken phosphodiester bond is transferred to the protein–DNA bond and then back to the new phosphodiester bond.

FIG. 47-35. A gene within a gene. The sequence of the ϕX174 E gene (for a protein needed for cell lysis) is contained within that of the D gene (for a protein for virion assembly). Two entirely different proteins are generated from the same nucleotide sequence by a shift in the reading frame, an event that in most cases produces an inactive protein. The J gene (for a virion component) is also read on a frame shifted from that of D. The sequence indicated is in the viral DNA (plus strand), which has the same polarity as the mRNA; hence T replaces U, which is more familiar in codons. A similar situation exists for two other genes, A and B. (Modified from Barrell BG et al: Nature 264:34, 1976)

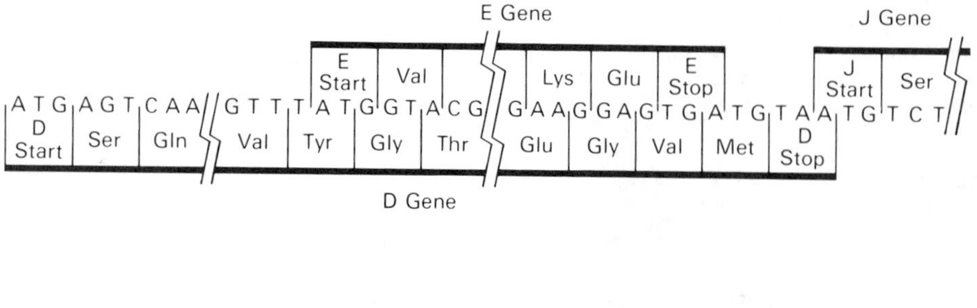

3) **Asymmetric synthesis of progeny viral strands** on progeny RFs occurs as in phase 2, except that association with viral proteins causes the elongating viral strand to remain single.

The minus strand of the parental RF, which remains unnicked, acts as the template for most or all progeny RFs, as is shown directly by examining the yields of cells transfected with f1 heteroduplex DNA made in vitro, which carry different markers on the two strands. Therefore replication of these phages follows a **stamping-machine model.**

In spite of the overall similarities, there are important differences among various phages. Thus whereas φX174 and filamentous phages use the host transcriptase, in association with other proteins, to synthesize the initial RNA segment, G4 uses the protein of the host *dnaG* gene; filamentous phages and G4 have a constant origin for synthesis of the complementary strands, whereas φX174 has multiple origins but requires a more complex enzymatic machinery; and in G4 this origin is different from that for RF replication, whereas it is the same in the other phages.

Morphogenesis. Major differences exist in the formation of mature virions and their release. With φX174 morphogenesis takes place together with the synthesis of the progeny viral strands, which become associated with virion proteins as they are synthesized. A mutation affecting any one of these proteins will block both single-stranded synthesis and virion formation. Presumably these proteins form with the DNA a kind of **core** by which the major capsid proteins of icosahedral viruses assemble. The virions are then released by cell lysis. In contrast, the progeny viral strand of filamentous phages first becomes associated with a phage-specified DNA-binding protein. The complex reaches the inner cell plasma membrane, which contains the newly synthesized capsid protein together with that imported by the infecting virions. As the viral DNA exits from the cells it picks up protein monomers of either origin and releases the DNA-binding protein, which is then reutilized for further single-strand synthesis.

The capsid protein of filamentous viruses has special structural features. It is elongated, has a hydrophobic center, a strongly basic COO^- end, and a somewhat acidic NH_3^+ end. In the nascent form it has an extra segment of 23 amino acids at the amino end, which are cleaved off after it has penetrated the inner membrane. This cleavage causes the permanent association of the protein with the membrane (see Cellular Membranes, Ch. 49); the COO^- end remains inside the cell while the NH_3^+ end reaches to the outside. During assembly and release of virions the basic COO^- end forms ionic bonds with the phosphates of the extruded DNA.

The interaction between the viral protein and the cell membrane is **very intimate and balanced.** Cell metabolism and growth are not impaired, although about 1000 virions are excreted in each cell generation. However, phage mutations affecting virion proteins kill the cells. This **steady-state virus–cell interaction** resembles that occurring with some animal viruses (see Ch. 50).

LIPID-CONTAINING PHAGES

Many bacteriophages contain lipids. They are interesting as tools for studying the interaction between proteins and lipids in biologic membranes.

The best studied is the small *Pseudomonas* phage PM2. It has icosahedral virions containing a double-stranded cyclic DNA about 9 Kb long. The capsid contains a lipid bilayer sandwiched between two icosahedral protein shells. A striking observation is that the phospholipid composition is very different from that of the host, although the lipids are synthesized by the host. Apparently the *lipids are selected from among the various host lipids by the protein of the outer shell* during assembly.

RNA PHAGE

The small, icosahedral RNA phages were isolated by Zinder as **male-specific;** they attach to F pili of male bacteria, apparently using them as channels for injecting their RNA into the cells. They are divided into several groups on the basis of RNA sequences and serology: e.g., f2, MS2, R17, in one group, Qβ in another. The phages contain single-stranded RNA of 3.5 Kb; that of MS2 has been completely sequenced, revealing **extensive self-complementarity** and therefore the ability to form a complex secondary and tertiary structure (Fig. 47-36).

The genetic organization of these phages is the simplest of all autonomous viruses. All their functions are expressed by only three genes, which form (in a $5' \rightarrow 3'$ sequence) an A protein, the protomer of the viral capsid, and a special replicase (RNA-dependent RNA polymerase). The A protein, with a single molecule per virion, is similar in function to the pilot protein of phages with single-stranded DNA. The RNA phages do not display recombination; the gene order was determined by replicating variable lengths of the phage RNA in vitro from the $5'$ end and identifying the products after translation in vitro.

The viral RNA acts as both genome and messenger in vivo; as messenger, it has been extremely valuable for studying protein synthesis in vitro. The knowledge of the sequence of the viral RNA provides important evidence

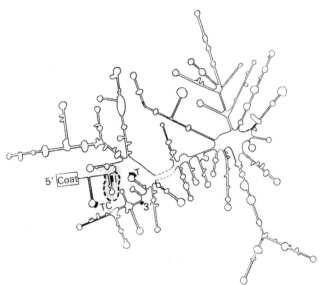

FIG. 47-36. Possible secondary structure of the replicase gene of phage MS2. The gene is at the 3′ end of the viral RNA (plus strand), following the coat protein gene. Translation begins at an AUG codon (**I**) and terminates at a UAG codon (**T**). **TC** is the terminating codon of the coat gene. **Parallel lines** indicate complementary sequences capable of forming helixes; **loops** are single-stranded segments. The gray area around the initiating codon identifies the ribosomal binding site; the **heavily dashed line** identifies the binding site for the coat protein (Data from Fiers W et al: Nature 260:500, 1976 [structure of the gene]; Gralla J et al: Nature 248:204, 1974 [coat protein binding site in R17]; and Steitz JA: Proc Natl Acad Sci USA 70:2605, 1973 [ribosome binding site in R17])

concerning intercistronic punctuation, ribosome-binding sites, and the significance of secondary structure in the regulation of translation.

RNA REPLICATION

RNA replication, which is similar to that of some animal viruses (see Ch. 50), involves special intermediates (Fig. 47-37). In cells infected with RNA-labeled virions the label is found in two forms: entirely double-stranded molecules, completely resistant to RNase, called **replicative form (RF)**; and molecules partially RNase-resistant, called **replicative intermediate (RI)**, which have a double-stranded backbone with one or two single-stranded tails. The pattern of labeling after brief pulses of a radioactive precursor shows that the RF is produced by building a **complementary** (minus) strand on the infecting **viral** (plus) strand; the subsequent synthesis of a third progeny plus strand on the double-stranded RF converts it into the RI, from which the progeny strands are then released. The synthesis of the plus strands can be **semi-conservative or conservative** in different RIs; the two RI

types are distinguishable by the different accessibility of labeled parental RNA to RNase degradation.

A special feature of the synthesis of phage RNA is the addition of an adenylic acid residue at the 3′ end of both strands after replication. This causes the end of the viral RNA to have a CCA sequence, as in tRNA, a feature that may have special functional significance (see Single-Stranded RNA, Ch. 46).

RNA Replicase. The replicases of various phages are **highly specific**, each recognizing only the RNAs of phages of the same group; they do not replicate cellular RNAs. *They recognize a proper steric arrangement of two CCC sequences,* one present at the 5′ end of all phage RNAs, the other at slightly different distances in various virus groups; the reciprocal position of these sequences is fixed by the secondary structure of the RNA. These replicases all accept C-containing synthetic polymers as templates: because these polymers are flexible, randomly located groups in them can interact with the enzyme.

The replicase of phage Qβ, first studied by Spiegelman, and now well characterized, is made up of **four subunits**, of which only one (subunit II) is phage-specified; *the others are cellular proteins involved in protein synthesis.* Subunit I is the ribosomal protein S1, which inhibits protein synthesis on other phage RNAs, such as those of R17 or MS2; subunits III and IV correspond to two elongation factors of protein synthesis, EF-Tu and EF-Ts. The complex of subunits II, III, and IV can replicate the Qβ **minus** strand; for replication of the **plus** (viral) strand subunit I and an additional host ribosomal protein, the host factor i, are required. The phage-specified subunit II is the true polymerase, for it can carry out chain elongation alone.

The RNA replicase makes mistakes at a much higher frequency than DNA polymerase, probably because it lacks the editing functions of the latter enzyme. As a result each viable phage differs at one or more bases from the average population.

The roles of the host factors in replication seem different from those in protein synthesis. Thus abolishing the translational function of the two elongation factors, by chemically crosslinking them or by means of specific inhibitors, does not interfere with their function in replication. The host factors may act by changing the conformation of the subunit II or the tridimensional structure of the RNA. The interaction of the phage subunit with the protein synthesis factors perhaps indicates an evolutionary relationship between phage RNA and cellular messengers, although it does not reveal which was the progenitor.

REGULATION OF TRANSLATION

In the infected cells the rates of synthesis of the three viral proteins are adjusted to the requirements of viral multiplication, since capsid protein is made in large excess over either replicase or A protein.

Regulation is based on changes of the **secondary structure of the RNA** or on its **interaction with proteins**. Thus the ribosome-binding region of the **A gene** is normally buried in the RNA folds; it may be accessible only once in the life of an RNA molecule, i.e., as it is synthesized, being permanently covered as soon as the RNA folds up. The translation of the **replicase gene** is favored early in

FIG. 47-37. Intermediates in RNA replication. In the semiconservative type of replication the parental strand (**straight heavy line**) is exposed to RNase attack during replication, whereas in the conservative model it is not. **Wavy lines** = progeny strands; **dashed arrows** = direction of replication.

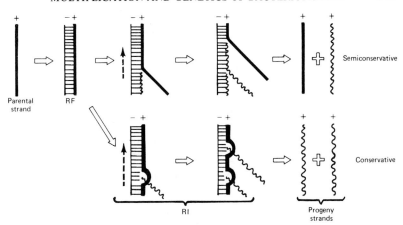

infection by the host factor i, which promotes ribosome binding to this gene; but this effect is itself regulated, because the replicase binds factor i, decreasing its availability. Later in infection the **coat protein** is formed and represses the translation of the replicase gene by binding strongly to its ribosome-binding site (which it protects from RNase digestion; see Fig. 47-36).

SELECTED READING

BOOKS AND REVIEW ARTICLES

BROKER TR, DOERMANN AH: Molecular and genetic recombination of bacteriophage T4. Annu Rev Genet 9:213, 1975

DOERMANN AH: T4 and the rolling circle model of replication. Annu Rev Genet 7:325, 1974

FRANKLIN RM: Structure and synthesis of bacteriophage PM2 with particular emphasis on the viral lipid bilayer. Curr Top Microbiol Immunol 68:107, 1974

MILLER RC: Replication and molecular recombination of T-phage. Annu Rev Microbiol 29:355, 1975

RABUSSAY D, GEIDUSCHEK EP: Regulation of gene action in the development of lytic bacteriophage. Comp Virol 8:1, 1977

WITMER HJ: Regulation of bacteriophage T4 gene expression. Prog Molec Subcell Biol 4:53, 1976

SPECIFIC ARTICLES

ALBERTS BH, FREY L: T4 bacteriophage gene 32: a structural protein in the replication and recombination of DNA. Nature 227:1313, 1970

BARRELL BG, AIR GM, HUTCHINSON CA III: Overlapping genes in bacteriophage φX174. Nature 264:34, 1976

CAILLET-FAUQUET P, DEFAIS M, RADMAN M: Molecular mechanisms of induced mutagenesis. Replication in vivo of bacteriophage φX-174 single-stranded, ultraviolet light-irradiated DNA in intact and irradiated host cells. J Mol Biol 117:95, 1977

CELIS JE, SMITH JD, BRENNER J: Correlation between genetic and translational maps of gene 23 in bacteriophage T4. Nature 241:130, 1973

CHAMBERLIN M, MCGRATH J, WASKELL L: New RNA polymerase from *E. coli* infected with bacteriophage T7. Nature 228:227, 1970

DELIUS H, HOWE C, KOZINSKI AW: Structure of the replicating DNA from bacteriophage T4. Proc Natl Acad Sci USA 68:3049, 1971

DUFFY JJ, GEIDUSCHEK EP: Virus-specified subunits of a modified *B. subtilis* RNA polymerase are determinants of DNA binding and RNA chain initiation. Cell 8:595, 1976

EGGEN K, NATHANS D: Regulation of protein synthesis directed by coliphage MS2 RNA. II. In vitro repression by phage coat protein. J Mol Biol 39:293, 1969

EISENBERG S, SCOTT JF, KORNBERG A: Enzymatic replication of viral and complementary strands of duplex DNA of phage φX174 proceeds by separate mechanisms. Proc Natl Acad Sci USA 73:3151, 1976

FALCO SC, ZIVIN R, ROTHMAN-DENES LB: Novel template requirement of N4 virion RNA polymerase. Proc Natl Acad Sci USA 75:3220, 1978

GAGE LP, GEIDUSCHEK EP: RNA synthesis during bacteriophage SPO1 development: six classes of SPO1 RNA. J Mol Biol 57:279, 1971

GUHA A, SZYBALSKI W, SALSER W et al: Controls and polarity of transcription during bacteriophage T4 development. J Mol Biol 59:329, 1971

HORVITZ HR: Polypeptide bound to the host RNA polymerase is specified by T4 control gene 33. Nature 244:137, 1973

HSIAO CL, BLACK LW: DNA packaging and the pathway of bacteriophage T4 head assembly. Proc Natl Acad Sci USA 74:3652, 1977

JACWINSKI SM, LINDBERG AA, KORNBERG A: Lipopolysaccharide receptor for bacteriophages φX174 and S13. Virology 66:268, 1975

KANO-SUEOKA T, SUEOKA N: Leucine tRNA and cessation of *E. coli* protein synthesis upon phage T2 injection. Proc Natl Acad Sci USA 62:268, 1975

KIKUCHI Y, KING J: Genetic control of bacteriophage T4 baseplate morphogenesis. I. Sequential assembly of the major precursor in vivo and in vitro. J Mol Biol 99:645, 1975

KIM JS, DAVIDSON N: Electron microscope heteroduplex study of sequence relations of T2, T4, and T6 bacteriophage DNA. Virology 57:93, 1974

KING J, CASJENS S: Catalytic head assembling protein in virus morphogenesis. Nature 251:112, 1974

LAEMMLI UK: Cleavage of structural proteins during the assembly of the head of bacteriophage T4. Nature 227:680, 1970

LAEMMLI UK, FAVRE M: Maturation of the head of bacteriophage T4. I. DNA packaging events. J Mol Biol 80:575, 1973

LANDERS TA, BLUMENTHAL T, WEBER K: Function and structure in ribonucleic acid phage Qβ ribonucleic acid replicase. J Biol Chem 249:5801, 1974

MATTSON R, RICHARDSON J, GOODIN D: Mutant of bacteriophage T4D affecting expression of many early genes. Nature 250:48, 1974

MAYNARD-SMITH S, SYMONDS N: Involvement of bacteriophage T4 genes in radiation repair. J Mol Biol 74:33, 1973

MAZUR BJ, ZINDER ND: Role of gene V protein in f1 single-strand synthesis. Virology 68:490, 1975

RIVA S, CASCINO A, GEIDUSCHEK EP: Uncoupling of late transcription from DNA replication in bacteriophage T4 development. J Mol Biol 54:103, 1970

SANGER F, AIR GM, BARRELL BG et al: Nucleotide sequence of bacteriophage φX174 DNA. Nature 265:687, 1977

SCHEKMAN R, WEINER JH, WEINER A, KORNBERG A: Ten proteins required for conversion of φX174 single-stranded DNA to duplex form in vitro. J Biol Chem 250:5859, 1975

SHOWE MK, ISOBE E, ONORATO L: Bacteriophage T4 prehead proteinase. II. Its cleavage from the product of gene 21 and regulation in phage-infected cells. J Mol Biol 107:55, 1976

STERNBERG N: Genetic analysis of bacteriophage λ head assembly. Virology 71:568, 1976

STREISINGER G, EMRICH J, STAHL MM: Chromosome structure in phage T4. III. Terminal redundancy and length determination. Proc Natl Acad Sci USA 57:292, 1967

chapter

LYSOGENY AND
TRANSDUCING
BACTERIOPHAGES

LYSOGENY

Most of the bacteriophages described in the preceding chapter are **virulent,** i.e., they multiply vegetatively and kill the cells at the end of the growth cycle. The **temperate** phages, in contrast, besides multiplying vegetatively can also produce the phenomenon of **lysogeny:** the indefinite persistence of the phage DNA in the host cells, without phage production. Occasionally, however, the viral DNA in a **lysogenic cell** will initiate vegetative multiplication, generating mature virions. Lysogeny is important as a method that favors the persistence and spreading of a virus in a more subtle way than virulence: as Burnet has pointed out in connection with viral diseases in higher organisms, the parasites best adapted to their environment are those that do not rapidly kill their hosts and thus deprive themselves of the opportunity to spread.

Temperate phages have provocative implications for many biologic problems: they throw light on the origin of viruses and the evolution of bacteria, they provide an important mechanism for gene transfer between bacteria **(transduction),** and they supply a model for viral oncogenesis (Ch. 66). Moreover, the switch between the lysogenic and the vegetative state is useful as a model in the study of differentiation.

NATURE OF LYSOGENY

Lysogeny characterizes many bacterial strains freshly isolated from their natural environment. Such lysogenic cultures contain a low concentration of bacteriophage, which can be recognized because it lyses certain other related bacterial strains, known as **sensitive** or **indicator** strains.*

When a sensitive bacterial strain is infected by a temperate bacteriophage two alternative responses are seen (Fig. 48-1): some cells are lysed by phage multiplication, and others are lysogenized. Lysogenic strains thus produced are designated by the name of the sensitive strain followed by that of the lysogenizing phage in parenthe-

ses, e.g., *Escherichia coli* K12(λ). Because temperate phages lyse only a fraction of the sensitive cells that they infect, they produce **turbid plaques.**

A bacterial strain can easily be recognized as lysogenic by streaking it on a solid medium across a strain sensitive to the phage released; a narrow zone of lysis is seen along the border of the lysogenic strain (Fig. 48-2). Since lysogeny cannot be recognized unless such a sensitive strain is available, many bacterial strains—perhaps most of those known—may be unrecognized as lysogens. Furthermore, many strains are lysogenic for several different phages.

The systems used most in experimental work on lysogeny are λ and related phages, active on *E. coli* K12; Mu, also active on *E. coli;* and P1 and P2 with *Shigella dysenteriae* or with several strains of *E. coli.*

Relation to Vegetative Growth Cycle. Lysogeny was recognized in the early 1920s; and it was soon realized that lysogenic strains are not simply phage-contaminated bacterial cultures, since the phage could not be eliminated by repeated cloning of the bacteria or by growing the cells in the presence of phage-specific antiserum to prevent cell infection by virions present in the medium. Bordet then recognized, in 1925, that the ability to produce phage was a hereditary property of the cells. Nevertheless, disruption of the lysogenic cells does not yield infectious phage; hence the phage must be present in the cells in a noninfectious form. However, the ability of lysogenic cultures to produce virus without obvious lysis remained puzzling until Lwoff, in 1950, patiently observing the behavior of single cells in microdroplets, showed that *phage is produced by a small proportion of the cells:* these lyse and release phage in a burst **(induction),** just like cells infected by phage T4. The other cells of the culture do not give rise to a productive infection and are said to be **immune**† to the released phage. The phage adsorbs to the immune cells and injects its DNA, but *the DNA does not multiply* and does not cause cell lysis. *Immunity, therefore, is different from resistance,* which, as

---

* Sensitive strains either are found by chance or, as noted later on, are obtained by "curing" lysogenic strains.

† This term is totally unrelated to immunity as studied in immunology.

FIG. 48-1. Development of a temperate bacteriophage.

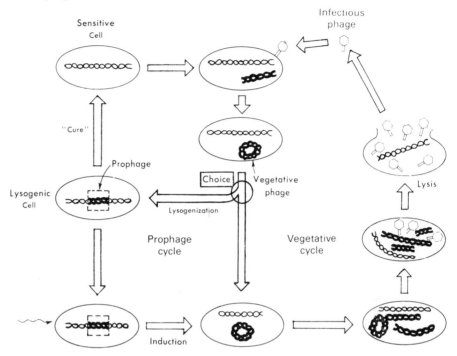

noted in the preceding chapter, prevents adsorption and injection (see Infection of Host Cells, Ch. 47).

These experiments made it clear that lysogeny involves a special, stably inherited, noninfectious form of the virus, called **prophage,** associated with immunity. The prophage occasionally shifts abruptly to the vegetative form and then reproduces just like a virulent phage. Lwoff further provided a powerful tool for studying lysogeny by showing that the shift from the prophage cycle to the lytic cycle, normally a rare event, could be induced in all the cells of a culture by certain environmental influ-

ences, such as moderate ultraviolet irradiation (Fig. 48-2).

THE VEGETATIVE CYCLE

The vegetative cycle of temperate phages is similar to that of virulent phages (Ch. 47) but with some modifications. Virions of λ contain a double-stranded linear DNA, 47 Kb long, with a complementary single-stranded segment 12 nucleotides long at each 5′ end (Fig.

FIG. 48-2. Cross-streaking of lysogenic and sensitive strains of *Escherichia coli* on nutrient agar. **A.** Untreated. **B.** Exposed to a small dose of ultraviolet (UV) light, after streaking, to induce the lysogenic cells. In **A** note the narrow bands of lysis of the sensitive strain (**vertical streak**) flanking the lysogenic strain (**horizontal streak**). In **B** note that the inducing treatment, by causing cell lysis, markedly reduces the colony density of the lysogenic streak, and the accompanying release of infectious phage causes pronounced lysis of the sensitive strain in the area of crossing.

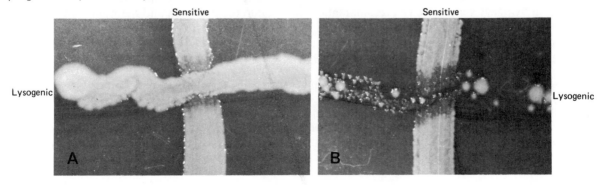

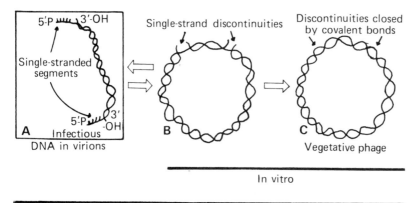

FIG. 48-3. Different forms of phage λ DNA. **A.** In the virions the DNA exists as a linear double-stranded molecule with complementary ends. **B.** Under conditions of nucleic acid hybridization, the linear molecule can reversibly close into a ring by base pairing of the single-stranded ends. **C.** Within the cell, the DNA forms completely covalently closed rings.

48-3). Under annealing conditions in vitro these **"cohesive" ends** pair, generating a cyclic molecule with two staggered nicks. Infection with labeled phage reveals a similar cyclization in vivo, with ligase closing the two nicks. The evolution of the cohesive ends can be explained by the requirement both for a linear DNA in the virions, to allow encapsidation, and for a cyclic form intracellularly, during replication (see Ch. 47) and lysogenization (see The Prophage Cycle, below).

THE GENETICS OF λ

As with other phages, the genes of λ are organized in blocks of related functions, as shown in Fig. 48-4. In this figure the genome is represented as circular because within the cells the viral DNA is circular while the genes are expressed, i.e., during transcription. The **vegetative genetic map** (i.e., based on recombination during vegetative growth) is linear, with the ends coinciding with those of the virion DNA (see Genetic Recombination, Ch. 47). This recombination is called **generalized** (to distinguish it from the specialized form, discussed below), and is promoted both by phage genes (*red*, i.e., recombination deficient, and *gam*) and by the bacterial *rec* system.

TRANSCRIPTION

The extraordinarily detailed study of λ transcription has led to a fairly complete understanding of the switching between vegetative and prophage cycle, which is an important model for the switching of genes in differentiation. Aspects important for the vegetative cycle will be reviewed now; those specifically related to lysogenization will be reviewed in a later section.

The general pattern of λ transcription (Fig. 48-4) is similar to that of T4 (Ch. 47). We distinguish **immediate early** messengers (which appear in the presence of chloramphenicol), **delayed early**, and **late** messengers, all synthesized by the host transcriptase. In vitro experiments suggest that, as with T4, the transition from immediate to delayed early is produced by **interference with termination**. Thus, in vitro, in the presence of the termination factor ρ, two immediate early RNAs are formed. One is transcribed from the "left" DNA strand, and extends leftward from promoter PL through gene N to the tL terminator; the other, transcribed from the "right" DNA strand, extends rightwards from promoter PR through gene *cro* to the tR1 terminator, and some of this transcription continues further, through genes O and P, to tR2. The products of these genes have important functions in viral development and lysogeny, which will be reviewed below. The product of gene N (indicated as pN) prevents the above terminations, apparently by interacting with the RNA polymerase at the promoters, and forming with it a ρ-resistant complex. Lack of termination then allows the leftward transcription to proceed into the b2 region and the rightward transcription to include gene Q. The product of Q in turn controls late transcription. Through this effect on transcription in both directions *gene N controls the expression of most viral functions.*

The gene Q product controls late functions by acting on transcription initiated on a promoter (PR') just to the right of Q, allowing this transcription to avoid terminator tR3 and thus to continue through the genes for the vegetative components all the way to gene J and into b2. Late transcription is strongly increased by DNA replication but also occurs in its absence.

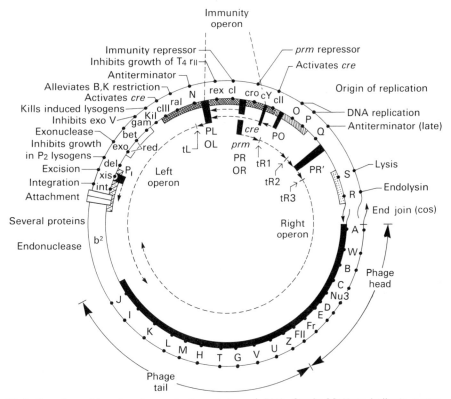

FIG. 48-4. Genetic and functional organization of phage λ DNA. **Capital letters** indicate genes identified by nonsense mutations, which include most of those required for the vegetative cycle. The genes are arranged in functional groups, identified by the following bands (from left to right): **dashed and dotted** = specialized recombination; **white** = generalized recombination; **cross-hatched** = regulation; **dashed** = DNA replication; **dotted** = lysis; **black** = capsid. The DNA is represented in the cyclic configuration in which transcription occurs. In the virions the DNA is linear, with ends at **end join.** The map is linear, i.e., markers close to the end join but at opposite sides have the highest recombination frequencies. Transcription is indicated by **dashed lines** and proceeds in the directions indicated by the **arrows.** There are three main transcription segments designed as operons: leftward, rightward, and immunity operons. **PL, OL** = leftward promoter and operator; **PR, OR** = rightward promoter and operator; **CRE** = controller for repressor establishment; **PRM** = promoter for repressor maintenance; **PO** = promoter for replication-related transcription and possibly for repressor establishment; **P$_I$** = promoter for *int* transcription; **tL, tRI, tR2** = terminators neutralized by the N gene product; **PR'** = promoter for late transcription; **tR3** = terminator neutralized by the Q gene product.

REPLICATION

The replication of λ DNA utilizes exclusively exogenous precursors, because the bacterial DNA is not destroyed. Replication occurs in two stages (Fig. 48-5). First the cyclic parental DNA associates with the cell membrane and replicates symmetrically, generating cyclic molecules. Initiation requires the functions of viral genes O and P, as well as host functions. Replication begins within gene O and, according to electron microscopic evidence similar to that of Fig. 47-13, then proceeds in opposite directions and terminates where the two forks meet.

In the second, **late,** stage the progeny DNA leaves the membrane and replicates according to the rolling-circle model, initiating at variable locations on the cyclic DNA and generating long concatemers. The control for the transition from early to late replication depends on phage genes under N dependence, because if N is defective the DNA replicates continuously in cyclic form (see Plasmidial Prophages, below). The function of gene *gam* is required in order to inhibit the host's *rec* B.C. exonuclease V, which would otherwise degrade the concatemers.

In the formation of **mature molecules with cohesive ends** the protein specified by phage gene A (pA) binds to concatemers at the sites corresponding to cohesive termini (*cos*) (Fig. 48-5) and produces staggered single-strand cuts, 12 nucleotides apart. Protein pA also carries

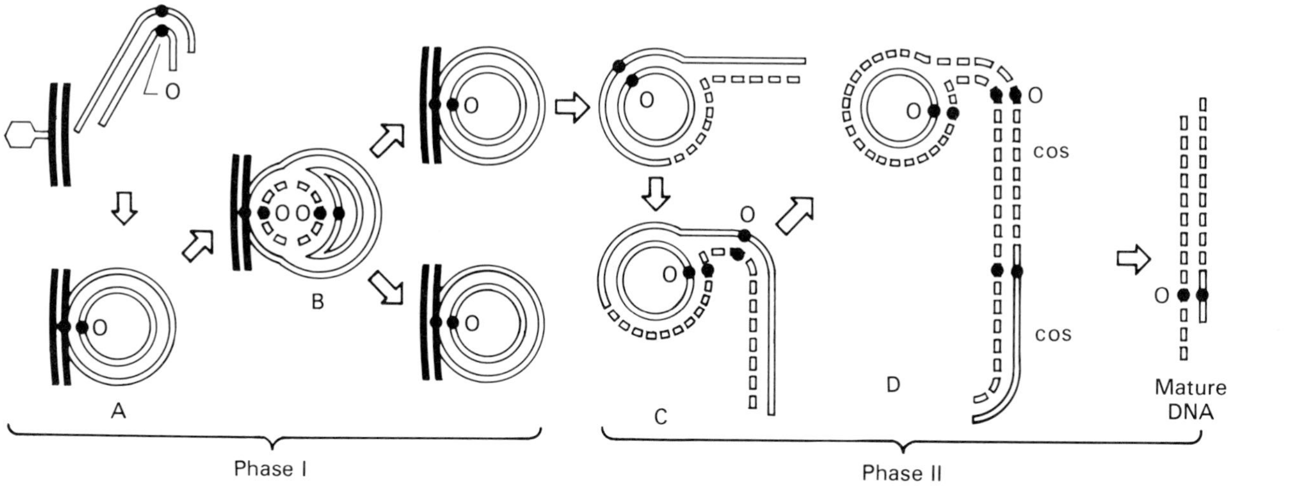

FIG. 48-5. Replication of λ DNA. In phase I the DNA injected by the infecting virion first cyclizes (**A**), then (**B**) it replicates symmetrically in association with the cell plasma membrane (**double line**), initiating at a fixed origin (**O**). In phase II the DNA, free of the membrane, replicates asymmetrically (**C**), by the rolling circle model, generating linear concatemers. These are then cut at *cos* sites (**D**) to generate mature molecules with cohesive ends.

a mature DNA molecule into a prohead. Further stages in head assembly (Fig. 48-6) and capsid maturation generally follow the T-even model. Virions are released through the action of gene S, whose product stops cellular metabolism and weakens the cellular membrane, allowing the lysozyme produced by gene R to lyse the cell wall.

THE LYSOGENIC STATE

When a sensitive bacterium is infected by a temperate phage the entering DNA has a **"choice"** between vegetative multiplication and lysogenization (Fig. 48-1). The proportion of infected cells that are lysogenized may vary from a few percent to nearly all, depending on both the system and the conditions.

The lysogenic state is determined by the activity of the **regulatory region** of the λ genome, which both generates immunity and causes integration of the phage genome in the cellular DNA.

IMMUNITY AND REPRESSION

Immunity originally appeared to be only a curious product of the lysogenic state. However, it emerged as the central feature of lysogeny when it was shown, by Jacob and Wollman, to be produced by repression of phage

FIG. 48-6. Model for λ head assembly showing the sequence in which the various gene products (**pX**) and the phage DNA must interact for normal assembly. The various genes appear on the genetic map of FIG. 48-4; *gro E* is a bacterial gene whose function is required for normal head assembly. (Sternberg N: Virology 71:568, 1976)

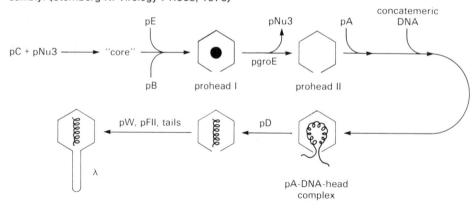

genes, much like the repression of bacterial operons (Ch. 14).

Repression is demonstrated in lysogenic cells by the **absence of vegetative mRNAs.** They are replaced by a small mRNA that transcribes part of the regulatory region, the **immunity operon** on the "left" DNA strand (see Fig. 48-4). Lysogenic cells contain the **immunity repressor** but no vegetative proteins, and in induced cells the opposite is true.

The repressor is specified by the cI gene, in the immunity region. Mutations that alter the repressor prevent lysogenization but allow vegetative growth, thus producing **clear** plaques.*

Temperature-sensitive cI mutants have been especially important in the elucidation of the regulation of repression; they produce a **thermosensitive repressor** which becomes ineffective at high temperature. In cells lysogenized at low temperature by a cI mutant the provirus cannot be maintained after the temperature is raised, and vegetative multiplication begins.

Immunity to exogenous infection is also the result of repression: it is not produced by cI$^-$ phage, and when normal lysogenic cells are induced it breaks down simultaneously with repression. In the presence of repression, infecting DNA cyclizes but is not replicated; it survives for many cell generations as an **abortive prophage,** which is transmitted unilinearly like the genes involved in abortive transduction (see under Generalized Transduction, below); it can be recognized by proper genetic tests.

The λ **immunity repressor,** isolated by Ptashne, is an acidic protein with a monomer mol wt of 26,000 daltons. In vitro this compound, in oligomeric form, binds strongly to DNA containing the immunity region (Fig. 48-7). Mutations identify the binding sites as two **operators** (OR, OL: Fig. 48-4) that regulate transcription starting at the **rightward and the leftward promoters** (PR, PL). By acting on these operators the repressor normally prevents the expression of most viral functions outside the immunity operon.

In **virulent mutants** (vir) both OR and OL are defective and fail to bind the repressor. Like the cI$^-$ mutants that fail to make repressor, vir mutants form **clear plaques,** because vegetative multiplication is constitutive. However, there is an important difference: immune cells, containing cI repressor, can be lysed by vir but not by cI$^-$ phages, because the latter are sensitive to the repressor.

Specificity of Immunity. Immunity is highly specific: even closely related phages form cI repressors of different specificities, and each phage recognizes exclusively its own. No point mutation in a repressor gene is known

* Clear plaques are also produced by mutations in genes cII, cIII, and cY (cre) which, as discussed below, participate in lysogenization but not in repressor function.

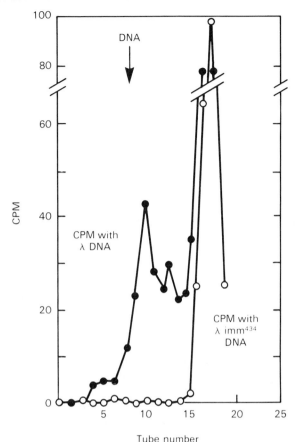

FIG. 48-7. Specific binding of λ repressor to λ DNA. Purified ^{14}C-labeled repressor was incubated with either λ DNA (**black circles**) or λimm[434] DNA (**white circles**). λimm[434] contains mostly λ sequences, except for the immunity region, which derives from the related phage 434 (see FIG. 48-8). The mixture was then sedimented in a sucrose gradient. The distribution of radioactivity in fractions collected from the bottom of the tube showed that part of the repressor cosediments with λ DNA (whose position is indicated by the **arrow**) but not with λimm[434] DNA (sedimenting at the same place as λ DNA), which does not contain the operator recognized by the λ repressor. (Modified from Ptashne M: Nature 214:232, 1967)

to change its specificity into that of a different phage, even a closely related one. **Heteroimmune** recombinants (Fig. 48-8), which have most genetic properties of one phage, but the immunity of another, have played an important part in unraveling the mechanisms of lysogeny.

REGULATION OF cI REPRESSOR FORMATION

During the late 1960s and the 1970s the problem of how lysogenization is produced, maintained, and reversed has attracted a large number of investigators. The problem

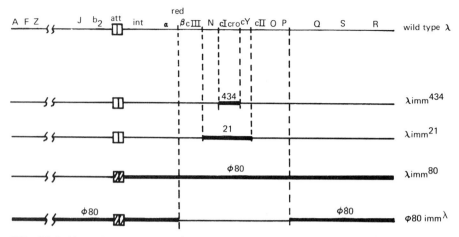

FIG. 48-8. Heteroimmune derivatives of phage λ. They arose from exchanges of λ with the related phages 434, 21, or φ80. Each was obtained by a series of back-crosses to one parent, always using recombinants with the immunity properties of the other parent. In this way, several different immunity regions, including the corresponding rightward and leftward operators, have been inserted on a more or less complete λ background, or, conversely, a λ immunity region was introduced in a φ80 background. **Thin lines** = λ DNA; **heavy lines** = substitutions; open squares = λ attachment sites; shaded squared = φ80 attachment sites.

has been solved almost completely and the basic facts, although difficult to unravel, are simple. The basic mechanism in the production and maintenance of the lysogenic state is the **antagonism of two repressors,** the immunity repressor, and the *cro* repressor, which prevents immunity.

These studies have centered on the **functional state of the immunity operon,** i.e., its rate of transcription (see Fig. 48-4 for the genes, control sites, and transcriptions involved). This rate can be determined directly, because the mRNA of the immunity operon can be isolated and quantitated, or indirectly, from the intracellular concentration of the immunity repressor (measured serologically). It can also be inferred from the degree of resistance of the cells to r_{II} mutants of phage T4 (see Conditionally Lethal Mutants, Ch. 47), which depends on the expression of the gene *rex* included in the operon.

Choice between Lysogenization and Vegetative Multiplication. In the early period after infection (about 20 min) the host transcriptase initiates transcription at the PL and PR promoters (Fig. 48-9A). Gene *cro* starts immediately to produce the *cro* repressor. At the same time the expression of gene N permits transcription of genes cII and cIII, whose combined products activate transcription of the immunity operon at the **controller for repressor establishment** (*cre*), initiating synthesis of the immunity repressor.

The role of cII, cIII, and *cre* (originally known as cY) was shown by the lack of lysogenization by phages with mutations at these genes, which therefore cause formation of clear plaques. Gene *cre*, like a promoter or an operator, was found to be *cis*-active, i.e., its

mutations affect the establishment of immunity only if *cre* and cI are on the same molecule. It is not clear whether *cre* is a promoter activated by the cII-cIII products, or a terminator neutralized by the cII–cIII products, regulating transcription initiated at promoter PO (see Fig. 48-4).

During this early period the cI and *cro* repressors accumulate; either inhibits initiation of transcription at PL and PR. Therefore once the two repressors have reached a sufficient concentration, all transcriptions, and soon also all syntheses of viral products, come to a halt. The role of *cre* is now ended.

In the following period (Fig. 48-9B) the choice between lysogenization and vegetative multiplication is determined by the **relative concentrations of the two repressors,** which act in opposite directions on another initiator of immunity transcription, *prm* (the promoter for cI repressor maintenance). This promoter requires cI repressor for its activity, and therefore is not active early after infection; in contrast the *cro* repressor inhibits initiation at *prm*. If the cI repressor (cIR) predominates (Fig. 48-9C), *prm* is activated, cI repressor synthesis restarts, and the lysogenic state becomes established; if on the contrary the *cro* repressor (cro R) predominates (Fig. 48-9D), *prm* is not activated, there is no cI repressor synthesis, and vegetative multiplication is established under the control of the *cro* repressor (see Plasmidial Prophages, below).

The choice is therefore determined during the first 20 min by the rates of synthesis of the two repressors. These rates are influenced by many factors, which in this way control the outcome of infection.

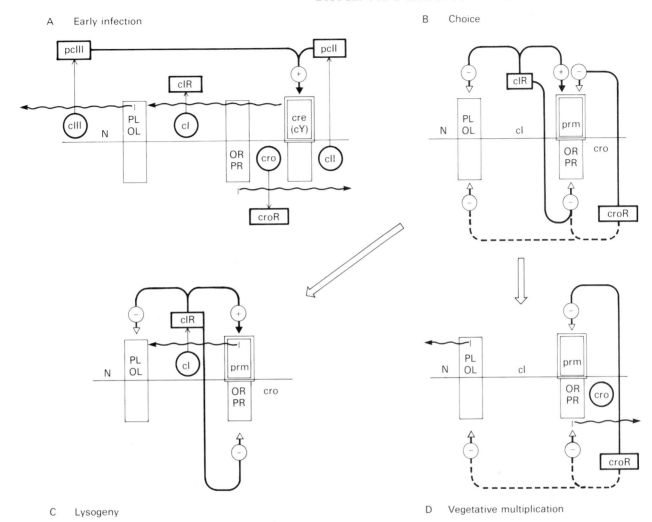

FIG. 48-9. Regulation of repression in phage λ and determination of choice between vegetative replication and lysogeny. In the early period of infection (**A**) the left and right promoters (**PL, PR**) become active, allowing transcription (**wavy lines**) which, in the presence of pN, expresses **cII** and **cIII**. The products of these two genes (**pcII** and **pcIII**) activate the **controller for cI repressor establishment (cre)**, allowing transcription of the immunity operon (positive control, indicated by **+** and **black arrowhead**) and synthesis of immunity repressor, **cIR.** Cro repressor, **croR,** is also made. After about 20 min the various products have accumulated, and new synthesis temporarily cease. **B.** The choice is then determined by the competition between the negative control (**—** , **white arrowheads**) of the cro repressor (**croR**) and the positive control of the immunity repressor (**cIR**) on prm. If **cIR** wins out (**C**), it activates prm, keeps it permanently activated by positive feedback control, and blocks **PL** and **PR**, resulting in the lysogenic state; if **croR** wins out (**D**), prm remains inactive and transcription initiated at **PL** and **PR** causes vegetative multiplication. Active genes are **circled. Heavy lines** indicate the main actions of the gene products (in rectangular **boxes**); **dashed lines** indicate the weaker repression of the cro repressor on **PL** and **PR**, which forms a negative feedback control of vegetative multiplication. **OL, OR,** leftward and rightward operators.

Thus the promoters PL and PR, having somewhat different base sequences, are not transcribed with equal efficiencies. Furthermore, their initiation rates are affected in different ways by many conditions such as temperature, ion concentrations, or rate of protein synthesis. Another interaction occurs between the rightward transcription initiated at OR, which is required for cro expression, and the transcription of the immunity operon from cre; these transcriptions cannot coexist because they overlap on the stretch of

DNA between OR and cre, where they proceed in opposite directions. Their head-on collision must create a **steric obstacle** to both, until finally the transcription that achieves the greater initiation frequency prevails.

The choice between lysogenization and vegetative multiplication also involves **cellular genes.** Thus E. coli gene hfl inhibits immunity repressor synthesis. This antagonism involves **cyclic AMP (cAMP):** lysogenization is less frequent in hosts lacking adenylate cyclase (which synthesizes cAMP) or cAMP-binding protein (which mediates its effect). This action of cAMP in promoting lysogenization over the more expensive lytic multiplication parallels its action in bacterial metabolism, i.e., mediating the adaptation to poorer food sources, in response to hard times.

Control of Viral Genes in the Lysogenic State (Fig. 48-9C).

Once established, the lysogenic state is maintained by a positive feedback: the cI repressor, acting on *prm*, promotes its own synthesis.

This can be shown with prophages having 1) a mutation in cI that makes the repressor **reversibly** thermosensitive, and 2) additional mutations in the N, O, and P genes that prevent vegetative multiplication after inactivation of the repressor. At 30°C these lysogens are fully repressed and manufacture only RNA of the immunity operon; after the temperature is raised to 42°C, the inactivation of the cI repressor causes loss of immunity, and, surprisingly, also **cessation of synthesis of the immunity mRNA.** This shows that active cI repressor is necessary for this synthesis. Moreover, if within 5 min the temperature is again lowered the renaturation of the cI repressor rapidly restores immunity and also a normal rate of synthesis of immunity mRNA. Genetic tests have located *prm* at the PR promoter.

The positive feedback of the **cI repressor on its own synthesis** ensures stability of cI repressor production in the presence of fluctuations in the rates of the reactions involved, and is responsible for the high stability of the lysogenic state.

Special Features of the Control of λ Immunity.

The preceding brief review of the control of the immunity operon shows that it has many unusual features. 1) The cI repressor **prevents transcription** of the rightward and leftward operons by acting on their operators, but **promotes transcription** of the immunity operon by acting on *prm*. 2) The immunity operon is independently controlled at two sites, *cre* and *prm,* in order to separate the **initial synthesis** of cI repressor, which takes place in all infected cells, from the **permanent synthesis,** which only occurs in lysogenic cells. 3) A single site, containing both OR and *prm,* controls two different operons, responding to the same repressor with opposite consequences. This site can be considered as resulting from the fusion of two promoters back to back (Fig. 48-10).

These features are probably related to the necessity of coordinating and finely tuning the action and synthesis of the cI repressor.

PLASMIDIAL PROPHAGES

The inhibition by the *cro* repressor of OR, which regulates *cro* transcription, generates a negative feedback with resulting **autoregulation of *cro* expression.** This in turn regulates, through the action of the *cro* repressor on PR and PL (and therefore on genes N and Q), all the genes for phage replication. Therefore *cro is the main regulator of vegetative λ growth.*

Autoregulation of phage DNA replication through *cro* is clearly demonstrated by the λdv plasmids, which are unintegrated prophages formed by greatly deleted cyclic λ DNAs; they persist indefinitely in cells, with a constant

FIG. 48-10. General features of the OR–*prm* promoter-operator site. The **dots** represent nucleotides; the **boxes**, the repressor-binding sites, separated by spacers (**S**). The beginning of the cI and *cro* genes are shown together with their mRNAs. The **heavy line** in the center indicates a rough rotational symmetry of the base sequences, suggesting that the regulatory complex results from the symmetric coupling of two distinct units, one for immunity transcription (*prm*, left), the other for rightward transcription (OR–PR, right). The transcriptase may at first recognize the center of the unit, close to the central binding site, and then bind either to the left or to the right. All repressor-binding sites have similar sequences in the direction of transcription. However, small sequence differences between the left and the right sites may explain the opposite effect of cI repressor on them, i.e., repression to the right, activation to the left. The sequences of binding sites have considerable internal symmetry and can form cloverleaf during interaction with proteins. (Data from Waltz A et al: Nature 262:665, 1976; Ptashne M et al: Science 194:156, 1976. Copyright 1976 by the American Association for the Advancement of Science)

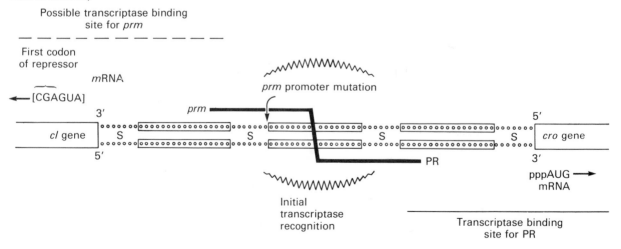

number per cell. Genetic studies show that these plasmids, which perform no other function but self-replication, must contain only PR, OR, *cro,* O, P, and the origin of DNA replication. Gene cI has no role in this form of lysogenization. The absence of N function keeps the DNA cyclic. If *cro* is inactive, the plasmid cannot be maintained, probably because overproduction of O and P products interferes with the replication apparatus of the host. Cells carrying λdv are immune to infection by wild-type λ, whose replication is inhibited by the *cro* repressor; they can be infected by mutants which have changes in the right operon, at the site of action of the *cro* repressor.

Similar mechanisms maintain the symbiotic association of cells with other plasmidial prophages. One is P1, which is present in lysogenic cells in about one copy per cell chromosome.

THE PROPHAGE CYCLE

When lysogenization occurs, the vegetative λ DNA becomes inserted into the cellular DNA as a **prophage.** Conversely, at induction the prophage becomes excised to generate cyclic vegetative DNA again. The prophage cycle is therefore **an alternative pathway** in the vegetative DNA cycle (see Fig. 48-1).

LOCATION AND STATE OF THE PROPHAGE IN LYSOGENIC CELLS

Most prophages are integrated at **fixed locations** on the bacterial chromosome (Fig. 48-11). They are transferred with the *Hfr* chromosome in bacterial crosses, and their sites can be mapped in relation to bacterial genes. In *E.*

FIG. 48-11. Map of the *E. coli* genome showing the regular attachment sites for some prophages (**circled**).

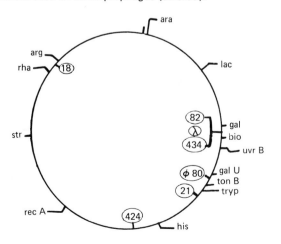

coli, λ and related phages usually settle in a unique site on the host chromosome; P2 can occupy at least nine distinct sites, but two preferentially, and Mu can integrate anywhere in the host genome. Some prophages, e.g., P1, are separate from the chromosome as plasmids (see Plasmidial Prophages, above). P1, however, must occasionally interact with the bacterial chromosome, because it gives rise to specialized transduction (see below).

In the lysogenic cells the prophage DNA is recognizable by hybridization with radioactively labeled phage DNA. Integrated prophages are connected to the bacterial chromosome by covalent bonds: after the DNA is denatured the viral DNA continues to sediment in an alkali sucrose gradient with the cellular DNA.

Multiplication of the Prophage. Whereas the vegetative multiplication of phage DNA is autonomous, i.e., is controlled by phage genes, the *replication of the prophage is regulated by the system that controls replication of the bacterial chromosome.* Thus at 40° C certain temperature-sensitive phage mutants persist as prophage, though they cannot initiate vegetative replication. Moreover, genetic evidence shows that a prophage doubles when a growing point of the replicating bacterial DNA reaches its integration site.

INSERTION

Genetic experiments demonstrate that prophage λ is **linearly inserted** in the bacterial chromosome (Fig. 48-12). Crosses of bacteria with genetically marked prophages yield a linear **prophage map;** the order of the genes, however, is permuted with respect to that of the vegetative map. Campbell explained this permutation by suggesting that in lysogenization the *viral DNA in cyclic form is inserted linearly into the bacterial chromosome by a single reciprocal crossover* which opens the ring at a point different from that where the ends of the mature DNA meet (Fig. 48-13). Insertion involves recombination between a specific **phage attachment site** in the vegetative λ DNA and a corresponding **bacterial attachment site,** creating two recombinant **prophage attachment sites** flanking the prophage.

Experiments with heavy isotopes have shown that *most prophages derive from replicated DNA;* however, the infecting phage DNA can be directly integrated if its replication is prevented. Since the chance of insertion of a DNA molecule is low, several cell generations may intervene between infection and the emergence of a lysogenic clone.

The **requirement for a special enzyme** in insertion is evident in experiments with a partially diploid strain of *E. coli* carrying two identical λ attachment sites. Initially λ integrates at either site with an equal and high probability. However, after the first prophage has established

| ara | gal | chlD | blu | int | exo | β | N | rex | cl | cll | O | P | Q | R | A | B | C | D | E | F | G | H | M | L | K | I | J | bio | uvr | chlA |

FIG. 48-12. Evidence for linear insertion of the λ prophage in the bacterial chromosome, and detailed prophage map. Prophage and bacterial genes were mapped by using deletions that entered the prophage from either side by taking advantage of the two *chl* (chlorate resistance) genes, A and D. To determine which prophage markers had also been deleted, the cells were induced and then superinfected with λ phage carrying distinguishable alleles of all the markers: the appearance (or nonappearance) of various prophage markers in the progeny phage indicated whether or not they were present in the partly deleted prophage. Phage DNA = **thin line**; bacterial DNA = **heavy lines**; *ara* = arabinose utilization; *gal* = galactose utilization; *blu* = stained blue by iodine; *bio* = biotin synthesis; *uvr* = ultraviolet light resistance. (Data from Adhya S et al: Proc Natl Acad Sci USA 61:956, 1968)

immunity a new λ infection only rarely causes integration at the unoccupied site. In contrast, if the infecting phage is heteroimmune but has the same attachment site (e.g., λimm[434], Fig. 48-8) it again integrates at the free site with high probability. Hence, insertion requires a product specified by a gene that is repressed in the lysogenic cells. The gene is *int* (see Fig. 48-4), because *int*⁻ mutants do not integrate.

Expression of int is regulated like that of cI. Thus early in infection it is stimulated by the cII-cIII products, which also start cI repressor synthesis; and after the lysogenic state is established *int* continues to be expressed at a low level because its transcription is initiated at a **private promoter**, P_I (see Fig. 48-4), which is independent of N product and is not repressed by the cI repressor. Accordingly, integration can occur in cells under cI repression. The similarity of *int* and cI regulation coordinates the two main events of lysogenization: repression and integration.

Another requirement for insertion is that the *attachment region of the phage DNA must be transcribed.* Thus a repressed phage DNA, such as an abortive prophage (see Immunity and Repression, above) does not integrate even if the *int* product is supplied by a coinfecting heteroimmune phage, such as λimm[434] (see Fig. 48-8), whose *int* gene is not repressed. Transcriptpion may be required for opening the DNA helix.

Unusual Insertions. The heteroimmune phage λimm[434] can integrate in a cell already lysogenic for λ, whose only attachment site is

already occupied; it uses for insertion one of the attachment sites flanking the existing prophage, and generates a **tandem double prophage**. On rare occasions λ can also integrate in a λ-lysogenic cell; but then integration does not occur at attachment sites, but **within the existing prophage,** using the bacterial *rec* recombination system, which recognizes long homologous DNA sequences. Normally the *rec* system plays little role in integration because the *int* system is much more efficient.

In hosts in which the attachment sites are deleted, λ can still integrate at **other chromosomal sites** whose sequences may be related to those of the regular site. These integrations still use the *int* product, but are infrequent. They are recognizable because the genes in which they occur are inactivated (see Phage Mu, below).

EXCISION

Excision of the prophage is explained in the Campbell model as a reversal of integration, i.e., a **reciprocal crossover between the attachment sites at the two ends** of a prophage, yielding a cyclic phage DNA molecule and an intact bacterial chromosome (Fig. 48-13). Two predictions of this model have been experimentally verified: excision of a λ prophage as an unreplicated, covalently closed DNA ring can be demonstrated if the phage genes for DNA replication are inactive; and the intactness of the bacterial DNA after excision is borne out by the "curing" experiments reviewed below.

However, detailed studies have shown that there are important **differences between insertion and excision.** Thus excision requires **another viral function in addition**

FIG. 48-13. Campbell model for prophage integration explaining the permutation of the vegetative and prophage maps of λ DNA. Both the vegetative and the prophage maps can be derived from the same circle by opening it at different points: **end join** for vegetative multiplication and maturation, and **att** for prophage insertion. The difference between the two maps is equivalent to shifting the block of markers (*int* to *mi*) from one end of the map to the other. For terminology, see FIG. 48-4. In addition, **h** = host range mutation; **mi** = minute plaque mutants; **heavy lines** = bacterial DNA; **thin lines** = phage DNA; **B⬛B'** = bacterial attachment site; **P☐P'** = phage attachment site; **B⬛P'** and **P⬛B'** = left and right prophage attachment sites respectively, resulting from recombination between bacterial and phage sites.

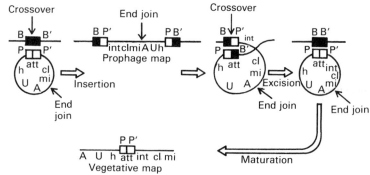

to *int,* since some λ mutants integrate but do not excise. These are altered in gene *xis,* adjacent to *int* (see Fig. 48-4). The *int* product may be the basic recombination enzyme for both insertion and excision, and the *xis* product would then make it conform to the different site configuration during excision (see Specificity of Attachment Sites, below).

Gene *xis,* in contrast to *int,* is under the control of the PL promoter (Fig. 48-4), and therefore is **not expressed in lysogenic cells.** The differential control of *int* and *xis* confers stability on the lysogenic state. In fact, *xis* expression, while repression persists, would cause loss of the prophage through "curing" (see below).

Specificity of Attachment Sites. Another important difference between insertion and excision is the **lack of equivalence** of bacterial, phage, and recombinant attachment sites. In fact, the probability of *int*- (or *int-xis*-) promoted recombination between a pair of sites depends on the **arrangements of half-sites** (on either side of the crossover point) in each of them; certain arrangements are more suitable for *int* alone (i.e., for insertion), others for *int-xis* (i.e., for excision).

The normal process of insertion and excision can then be represented by the equation:

$$\text{POP}' + \text{BOB}' \underset{\text{int-xis}}{\overset{\text{int}}{\rightleftharpoons}} \text{BOP}' + \text{POB}'$$

Phage site	Bacterial site	Left	Right
			prophage sites

where P and P′ represent phage half-sites, B and B′ bacterial half-sites, and 0 a common central sequence, 15 nucleotides long, where the crossover occurs.

These results are based on studies of recombination as well as lysogenization. In fact, in cells infected with two phages containing sites that are half viral, half bacterial, such as the transducing phages of Figure 48-18, *int* and *xis* functions can also cause **vegetative recombination, but limited to the attachment sites.** In the absence of generalized recombination (i.e., in *rec⁻* bacteria using *red⁻* phage strains) the recombination frequencies markedly depend both on the half-site combination and on the presence or absence of *xis* function (Fig. 48-14): whereas *int*-mediated recombination recognizes a broad spectrum of sites, the combined *int-xis* products recognize only the two sites normally flanking the prophage, as in excision.

In this recombination the sites are recognized by the enzymes, and then the short DNA homology provides the fine alignment. Either *rec*- or *red*-promoted recombination is rare at the attachment sites because *the DNA homology is very limited.* Another special feature of the *int*- or *int-xis*-promoted recombination is its **symmetry:** each event yields two symmetric recombinant types, in contrast to generalized phage recombination, which is asymmetric (see Genetic Recombination, Ch. 47).

DEFECTIVE PROPHAGES

Many mutations of the prophage in genes for vegetative functions permit its perpetuation but prevent synthesis of infectious phage. Such **defective prophages** can be recognized, when genetically marked, by inducing the cells (to break down immunity) and then superinfecting with a **helper phage** (usually the homologous wild type) which can complement the missing function of the defective prophage. The progeny then includes both wild-type and mutant virions.

Defective prophages are extremely useful tools for investigation, much like conditionally lethal mutants (see in Ch. 47): they can be propagated as prophages, and the effects of their defect on vegetative multiplication can be observed after induction.

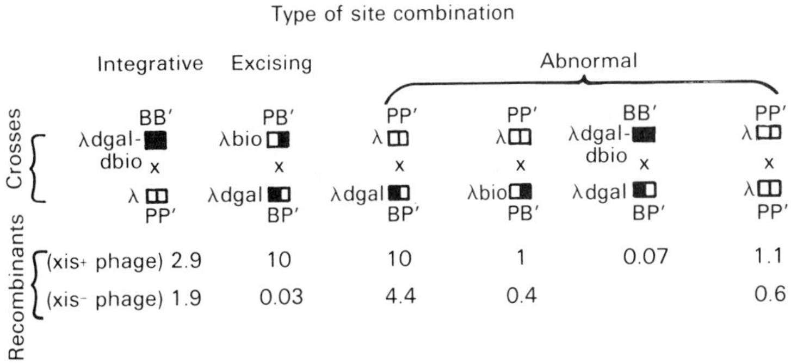

FIG. 48-14. Role of sites and *xis* product in *int*-mediated recombination. Vegetative crosses were made with the phages of FIG. 48-18, using *red⁻* phages and *rec⁻ E. coli* to abolish generalized recombination; the phages carried suitable markers on opposite sides of the attachment site. The values given are percent recombination between attachment sites. The first row gives the results obtained with both *int* and *xis* products present, and the second row shows the effect of eliminating the *xis* product, using a similar pair of *xis⁻* λ phages. The **integrative** sites (column 1) of recombination (**BB'** x **PP'**, corresponding to those of λ and of the bacterial chromosome) allow a high frequency of recombination with or without *xis.* The **excising** sites (column 2) (**BP'** x **PB'**, corresponding to the two hybrid sites flanking the prophage) require the *xis* product for a high frequency. The frequencies given by **abnormal** combinations vary not only with the kinds of half-sites used, but also with their arrangements (compare columns 3 and 4). (Data from Echols H: J Mol Biol 47:575, 1970. Copyright by Academic Press, Inc., (London) Ltd.)

Type of site combination

	Integrative	Excising	Abnormal			
Crosses	BB′ λdgal-■ dbio x λ □□ PP′	PB′ λbio □■ x λdgal ■□ BP′	PP′ λ □□ x λdgal ■□ BP′	PP′ λ□□ x λbio□■ PB′	BB′ λdgal-■ dbio x λdgal ■□ BP′	PP′ λ□□ x λ□□ PP′
Recombinants (xis⁺ phage)	2.9	10	10	1	0.07	1.1
(xis⁻ phage)	1.9	0.03	4.4	0.4		0.6

Whereas most known defective prophages contain the immunity system, some **extremely defective prophages** of λ have lost both the immunity genes and genes responsible for detachment; hence these prophages persist but the cells are not immune. These **cryptic prophages** are a useful model for understanding the consequence of infection of animal cells with certain tumorigenic DNA-containing viruses (Ch. 66).

CURING OF A LYSOGENIC BACTERIUM

Though excision of the prophage is usually followed by phage multiplication and cell lysis, this multiplication can sometimes be prevented; the cell then survives and gives rise to a normal, sensitive clone. *The excision is in perfect register with the insertion;* thus cells rendered proline-deficient by insertion of prophage within a *pro* gene return to prototrophy after curing.

An induced lysogen is cured if the excised prophage cannot replicate, either because the phage replicon is incomplete (e.g., owing to mutations in the O and P genes) or because immunity persists. Thus λimm[434], which does not recognize the λ cI repressor, can provide the *int-xis* products in the presence of λ immunity and can thus promote the cure of λ.

INDUCTION OF A LYSOGENIC CELL

The transition of prophage to vegetative phage represents a **failure of repression**: it can occur either spontaneously or in response to an inducing stimulus, as already indicated. The first sign is **loss of immunity and initiation of transcription of phage genes outside the immunity operon.** The phage begins to replicate in situ, from its own origin, and it is soon excised; capsid proteins are subsequently synthesized.

Spontaneous induction may occur in rare cells as a result of the accidental activation of the induction machinery. **Artificial induction** may occur in a large proportion of the cells if external conditions impair the activity of the cI repressor. Many prophages are induced by **ultraviolet (UV) light** in doses too small to inactivate the phage. Prophages that are poorly inducible by UV light can be induced by a variety of other means, such as thymine starvation, x-rays, alkylating agents, and some carcinogens. Some prophages cannot be induced at all; they do exhibit spontaneous induction but at a lower frequency than the inducible phages.

MECHANISM OF INDUCTION

Several mechanisms of induction are known.

cI Repressor Inactivation. Heating inactivates thermosensitive repressors specified by certain cI mutants. More generally, induction is produced by UV or other agents that produce *single-strand nicks or gaps* in DNA, either directly or indirectly (i.e., during repair). The altered DNA may be that of the lysogenic cell or even irradiated foreign DNA introduced into the cell, e.g., by the conjugation of an irradiated F^+ nonlysogen to an F^- lysogen **(indirect induction)**. In turn the altered DNA causes the activation of a cellular **protease** that cleaves the normal λ cI repressor. Induction can be prevented by protease inhibitors, or by *ind⁻* mutations, which yield an altered repressor.

This type of induction is part of a more general response of bacteria, whether or not lysogenic, to DNA damage. This response also includes activation of a new DNA repair pathway accompanied by enhanced mutagenesis (error-prone repair), and inhibition of bacterial division, with filamentous growth. These effects are designated as the **"SOS response"** because they rescue the cell from genetic damage, and they allow the phage genome to abandon an irreparably damaged host (see Repair of DNA Damage, Ch. 11).

The SOS response is controlled by several bacterial genes; the *rec A* gene may specify the protease. Some mutations in the *rec A* gene cause λ induction when the temperature is raised to 41°C; other mutations make induction **constitutive,** i.e., forbid the establishment of lysogeny, producing clear plaques. In the latter cells the cI repressor is cleaved as soon as it is produced.

Transfer of the Prophage to a Repressor-Free Environment. When a lysogenic *Hfr* cell conjugates with a nonlysogenic F^- cell, induction occurs as soon as the prophage is introduced into the nonrepressive F^- cytoplasm **(zygotic induction).** There is no induction if the recipient cell is lysogenic for the same phage.

Circumventing the Repressor. Since virulent λ is insensitive to the cI repressor, when it infects λ-lysogenic cells it produces all the gene products required for prophage excision and vegetative growth. The prophage is therefore induced despite the presence of its own repressor.

Applications. Induction of prophage λ has proved useful for the **identification of carcinogenic chemicals,** which are usually strong inducers. Special lysogens are used with mutations for high permeability (since many carcinogens do not enter normal cells) and for elimination of excision repair (to enhance induction).

EFFECT OF PROPHAGE ON HOST FUNCTIONS

Except for transducing phages (see below), which carry genes known to be derived from a recent bacterial host, most prophages exert no discernible effect on the bacterial phenotype other than immunity to superinfection. Certain prophages, however, change the cell's phenotype, either by expressing new functions (phage conversion) or by modifying quantitatively the expression of adjacent bacterial genes.

In **phage conversion** the new functions are viral, since they are abolished by phage mutations. Examples are changes of surface antigens in *Salmonella typhimurium* and the formation of toxins by *Corynebacterium diphtheriae* and other bacteria. These effects are expressed by both vegetative phage and prophage, but other converting functions are expressed only by the prophage.

For example, the resistance of *E. coli* K12 (λ) to T-even phages with an r_{II} mutation (see Conditionally Lethal Mutants, Ch. 47) is caused by the *rex* gene, which is located in the immunity operon and hence expressed only by prophage.

Prophages often interfere with the regulation of neighboring bacterial genes. The insertion of λ prophage, for instance, increases the repression of the adjacent *gal* operon. However, after the prophage is induced by UV irradiation, there is a burst of enzyme synthesis before the cells lyse. This effect is correlated with local DNA replication initiated within the prophage before excision, which continues into the *gal* operon (Fig. 48-12). The *gal* genes present in these replicas are transcribed as "read-through" from the derepressed phage promoter PL as soon as N product is available to prevent termination; hence they are not under control of the *gal* repressor **(escape synthesis)**.

More striking examples of the effect of prophage on host functions are shown by phage Mu (see below).

PHAGE λ AS DNA-CLONING VECTOR

Phage λ derivatives are very useful as vectors for molecular cloning (Ch. 12). The narrow host range of λ (which excludes all known native *E. coli* strains) provides a safety feature. Moreover, a widely used derivative, λgtWESλB (Fig. 48-15), has several mutations and deletions that strongly decrease its transmissibility (except in special suppressor host strains) by impairing its assembly, lysis, lysogenization, and generation of plasmids. Phage λ DNA has five sites for restriction endonuclease Eco R1, but in this derivative these have been reduced to

two; hence when it is cut by this endonuclease (which creates sticky ends) it yields three fragments. The two larger fragments together contain all the genes required for replication, and so they generate a hybrid suitable for cloning when they are annealed and then ligated with foreign DNA fragments (formed by the same enzyme), which replace fragment B. Moreover, elimination of this fragment removes the recombination genes.

The hybrid DNA can also be packaged in vitro by mixing it with lysates of two λ mutants that can complement each other, one producing empty heads, the other tails. The resulting λ particles containing the hybrid DNA can infect suitable *E. coli* strains efficiently.

PHAGE Mu

Special Features of the Virion's DNA. Phage Mu is a temperate coliphage of considerable interest both for its unique properties and for its important applications. It was discovered through its ability to **induce mutations** in bacterial genes (hence the name Mu).

The mature genome of Mu is linear double-stranded DNA, with 19 Kb per strand, and with **highly heterogeneous ends.** The heterogeneity is shown by the formation of heteroduplex molecules with frayed ends when a preparation of Mu DNA is melted and then reannealed. Moreover, *the heterogeneous ends are bacterial:* after separation from the rest of the molecules by restriction endonucleases they hybridize to *E. coli* DNA.

In a preparation of Mu DNA different molecules have different sequences in such ends, which together encompass essentially all host sequences. The mechanism that generates these ends will be-

FIG. 48-15. A λ vector for DNA cloning. A map of wild-type (wt) λ shows the five Eco R1 restriction sites and the fragments (**A** to **F**) produced by digesting the DNA with the enzyme. Elimination of the two crossed sites and deletion of **B** yields the vector, which contains the indicated amber mutations in genes **W, E,** and **S. nin 5** is a deletion in gene *cro* that reduces plasmid formation; **cI857** is a temperature-sensitive mutation in the repressor gene that at 37 °C causes induction and therefore prevents formation of stable lysogens. Lysogenization is also prevented by loss of the *int* function. The **heavy line** in the hybrid DNA denotes the inserted foreign DNA. (Modified from Leder P et al: Science 196:175, 1977. Copyright 1977 by the American Association for the Advancement of Science)

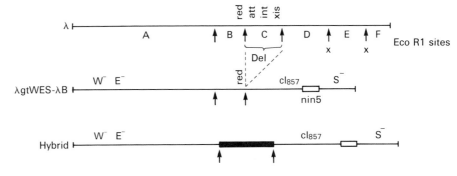

come apparent below. In another unusual feature, a sequence about 3 Kb long within the viral part can appear in either orientation in different molecules. It contains genes for adsorption functions. Hence a phage DNA preparation melted and then reannealed shows many molecules with an open loop (**G loop**) in electron micrographs. This invertible region is bounded by two identical inverted repeat sequences of almost 20 base pairs; recombination between these sequences is responsible for the frequent inversion. One of the orientations is lethal because the adsorption

functions are not expressed and so the phage is not adsorbed to the cells.

Lysogenization and Its Consequences (Fig. 48-16). The phage Mu DNA (excluding the terminal host sequences) can integrate **at any place** in the bacterial DNA. The ends of the prophage coincide with those of the phage components of the mature DNA molecules; therefore the genetic map of the phage and of the prophage are identical.

FIG. 48-16. Consequences of phage Mu lysogenization. **A.** Integration of Mu DNA in a site (**box**) of the bacterial chromosome (**a,b**) by recombination at the two ends (**m,n**), and excluding the bacterial DNA (**C,S**) present in the virions' DNA. **B.** Mobilization of Integrated F Plasmid by Mu. The model assumes integration of Mu DNA near **F**, replication of Mu DNA, and integration of two ends into a bacterial site near F (**dashed lines**), generating a hybrid cyclic DNA, partly Mu and partly bacterial (with **F**). **C.** Mu-mediated integration of **F**. Mu DNA integrates into **F** as shown in A; after replication of Mu DNA, two ends recombine with a site of the bacterial chromosome (**dashed lines**). The integrated **F** is flanked by two Mu DNA molecules in the same orientation. (Reproduced, with permission, from the Annual Review of Genetics, Vol. 10 © 1976 by Annual Reviews Inc.)

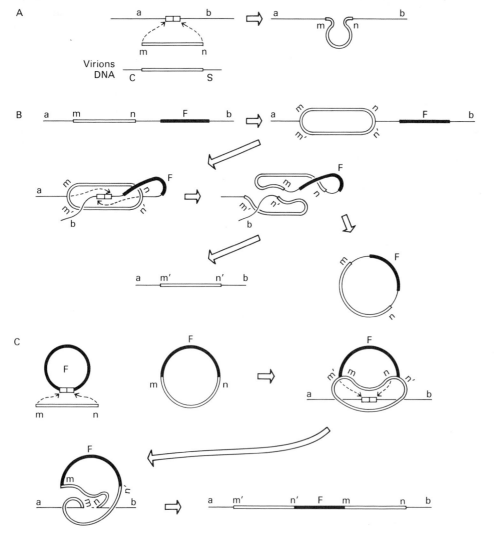

How integration (Fig. 48-16A) occurs is not clear. The prophage usually **inactivates** the baterial gene in which it is inserted; and by terminating transcription it also inactivates distal genes in the operons **(polarity)**. The prophage can be excised, but with wild-type phage excision is never precise; hence it does not restore the functions of bacterial genes inactivated by the integration. Apparently induction involves local replication of the prophage followed by recombination with an adjacent point on the bacterial chromosome, which generates a **hybrid DNA ring** that is partly phage, partly bacterial (Fig. 48-16B). This process causes **deletion in the host DNA.**

With some phage mutants restoration of function can rarely occur by an excision that is the exact reversal of integration.

Vegetative Integration. Similar hybrid rings appear in the middle of the lytic cycle, owing to vegetative integration followed by local DNA replication. In fact, phage Mu with a mutation that prevents DNA replication produces no rings. It seems that *the phage DNA can multiply only after integration.* Mature DNA molecules are probably derived from the hybrid rings by the cutting off of a **"headful,"** as with phage T4 (see Association of DNA and Capsid, Ch. 47). The derivation of the viral DNA from many different hybrid excisions of integrated genomes would explain the heterogeneity of the terminal host sequences (since each genome is integrated at a different place).

Mu as a Transposing Vector. The hybrid rings resulting from prophage excision may contain host markers or integrated episomes **(Mu-mediated mobilization of genes or episomes)**; a possible mechanism is shown in Fig. 48-16B. Conversely, integration of Mu into phage or plasmidial DNA, followed by Mu-directed replication and integration into the host chromosome, leads to the integration of the attached DNA into the chromosome, flanked by the two Mu genomes in identical orientation **(Mu-mediated integration of phage or episomes—Fig. 48-16C).**

Applications. The properties of phage Mu find many applications in molecular genetics. Thus their polar gene inactivation is useful for determining the composition of bacterial operons; Mu prophage inserted in a specific re-gion of DNA can be used as a recombination site for the introduction of other genes or episomes also carrying Mu DNA; during Mu induction genetic elements can be transposed from the bacterial chromosome to phages or plasmids.

SIGNIFICANCE OF LYSOGENY

Lysogeny indicates a close evolutionary relation between phages and bacteria. Lysogeny, in fact, depends on a remarkable congruence of attachment sites between the DNA of the phage and that of the host cells; and *most phages can lysogenize some bacterial strains.* The virulent T coliphages, though long viewed as the prototype phages, are thus exceptional: they represent extreme evolution in the direction of autonomous viral development.

Prophages give rise to **infectious heredity** in their host cells; i.e., they contribute new genetic characteristics to the lysogenic cells. **Conversion,** in which the prophage introduces new characters, raises the question of whether the converting genes have been viral components for a long time, or have, as in transducing phages, a recent cellular origin. In fact, converting genes are not essential for phage function, since phage mutations that abolish conversion do not impede either vegetative multiplication or lysogenization of the phage. However, they are probably advantageous in subtle ways. For instance, the conversion of surface antigens by *Salmonella* phage F_{15} prevents further adsorption of the phage, thus eliminating the waste of phage on immune cells. Such advantages may be especially evident in the natural environment in which the phage has evolved.

Infectious heredity is also seen with **highly defective prophages** that cannot be induced and do not confer immunity; one cannot distinguish these prophages from segments of cellular genetic material except by identifying their genes after they are recombined into a regular phage. Through evolutionary changes these prophages may acquire genetic significance for the cell and thereby become bona fide cellular genes.

PHAGES AS TRANSDUCING AGENTS

Transduction of bacterial genes from one cell to another by phages has been extensively used for mapping the bacterial chromosome (see Transduction, Ch. 9). We shall consider here the virologic aspects of the process.

Two types of transduction can be distinguished: generalized, which can transfer any bacterial genes, and restricted (specialized), which can transfer genes from only a very small region of the host chromosome adjacent to the prophage site.

GENERALIZED TRANSDUCTION

This type of transduction, due to the encapsidation of cellular DNA in a phage coat, was discovered by Zinder and Lederberg during a search for conjugation in *Salmonella.* Instead, they found that a culture filtrate caused the appearance of rare prototrophs in an auxotrophic culture; and they could show that the activity in the filtrate was due to phage particles.

Transduction can apparently occur for **any marker** of the donor bacterium. *Only closely linked markers can be cotransduced* by the same phage particle because the piece of bacterial DNA carried by a phage corresponds to about 1%–2% of the bacterial DNA. Transduction is usually carried out with a high-titer phage preparation obtained from the donor strain either by **lytic infection** or by **induction of lysogenic cells.**

Phage P1 is widely used in genetic studies of *Salmonella, E. coli,* and *Shigella.* Transduction has also been reported, with different phages, for genera as varied as *Pseudomonas, Staphylococcus, Bacillus,* and *Proteus.* With a recipient culture infected at multiplicity of about 5, the frequency of transduction of a given character ranges from 10^{-5} to 10^{-8} per cell.

Mechanism of Generalized Transduction. *Transducing particles contain cellular DNA instead of phage DNA.* Thus cells with DNA of heavy density (owing to incorporated 5-bromouracil) infected with P1 and grown in thymine plus $^{32}PO_4^{2-}$ yield transducing particles containing nonradioactive, heavy-density DNA (i.e., existing before infection); in contrast the infectious P1 parti-

cles in the progeny contain radioactive, light-density DNA. The incorporation of the host DNA in the phage capsid probably occurs by the same **"headful"** mechanism, without sequence recognition, that encapsidates T-even phage DNA (see Association of DNA and Capsid, Ch. 47).

Experiments with density-labeled phage show that in the recipient cells a **double-stranded** segment of the transducing DNA is incorporated into the host DNA **in exchange for bacterial genes.** Recombination is under the *rec* system of the host. The mechanism differs from that of transformation, in which only one strand of the entering DNA is integrated.

ABORTIVE TRANSDUCTION

The introduced fragment of bacterial DNA (the **exogenote**) does not always undergo either partial insertion (by recombination) or destruction. In a third choice, abortive transduction, the exogenote persists and expresses the function of its genes; but since it lacks an origin of replication it cannot multiply and is transmitted **unilinearly,** i.e., from a cell to only one of its two daugh-

FIG. 48-17. Complete and abortive transduction. In complete transduction part of the fragment of bacterial DNA (exogenote) introduced by the transducing phage (**thin line**) becomes inserted in the DNA of the recipient bacterium (**heavy line**) and continues to replicate with it. In abortive transduction, the exogenote is not integrated; it neither replicates nor is destroyed and so it is transmitted unilinearly (**heavy arrows**). A gene contained in the exogenote makes a cytoplasmic product (indicated by **x**) that is transmitted to both daughter cells. However, in those that do not receive the exogenote in abortive transduction it is progressively diluted out as cell multiplication proceeds.

Complete transduction

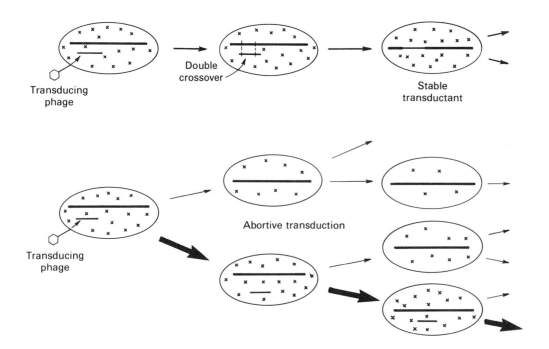

Abortive transduction

ters (Fig. 48-17). Abortive transduction is easily recognized when a gene of the exogenote codes for an enzyme required for growth on a selective medium, for though the amount is restricted by the unilinear inheritance, abortive transductants for a prototrophic gene can make **microcolonies** on minimal medium, which are detectable with magnification.

The proportion of microcolonies to large, prototrophic colonies reveals that *abortive transduction is several times as frequent as the corresponding complete transduction:* hence the probability of integration of the exogenote is rather small.

Abortive transduction is a useful tool in determining whether auxotrophic mutations affecting the same character are allelic. One mutant is used as donor and another as recipient in a transduction experiment, and the cells are plated on minimal medium. Large prototrophic colonies (indicating complete transduction) are formed by **recombination** between two mutations whether or not they are allelic; whereas the microcolonies of abortive transduction are formed by **complementation**, usually denoting that the mutations are not allelic, i.e., they are in different genes.

SPECIALIZED TRANSDUCTION

The Lederbergs discovered that phage λ can give rise to transduction in quite a different manner. It transfers only a **restricted group of genes** (the *gal* or *bio* regions) which are located near the prophage (see Fig. 48-12); and it is generated **only on induction of prophage,** and not (in contrast to generalized transduction) in lytic infection. The transducing genes are incorporated into the phage genome by abnormal excision of the prophage (Fig. 48-18).

The DNA of transducing particles has **deletions** that compensate in length for the insertion of phage genes (Fig. 48-18). Some types of transducing particles (e.g., λgal, λbio) are **infectious** because the missing phage genes are not essential for vegetative multiplication; more often transducing particles are **defective,** i.e., they cannot multiply by themselves (e.g., λdgal, λdbio, where d stands for defective). However, they can replicate in mixed infection with regular λ as a **helper** to complement the missing functions.

In a lysate from a normal lysogenic culture only a very small proportion of the virions are transducing; hence the

FIG. 48-18. Production of transducing λ derivatives by crossovers outside the attachment sites. Crossovers on the left of the prophage (**A**) generate molecules containing *gal* and other bacterial genes (λdgal) (where d stands for defective), which cannot replicate owing to loss of capsid genes. If the bacterium is deleted between *gal* and the prophage, *gal* genes can be incorporated in replacement of the inessential b₂ region, and the phage is not defective (λgal). Crossovers on the right (**B**) generate molecules containing *bio*, sometimes with other bacterial genes. Gene *bio* replaces recombination genes, and the phage can replicate (λbio); if the replacement is longer (including N) the phage is defective (λdbio). Phage λdgaldbio, which was useful for studying *int-xis*-promoted vegetative recombination, is obtained by recombination between λdgal and λdbio. The DNA between the two **broken lines** is not essential for replication. **Heavy lines** = bacterial DNA; **thin lines** = phage DNA.

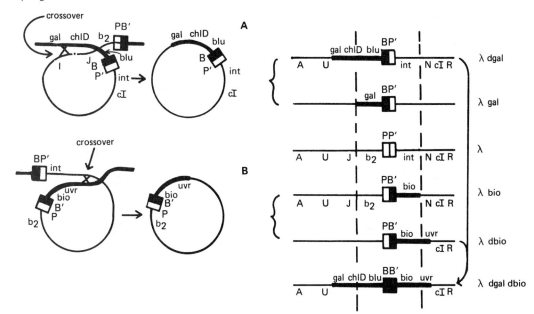

lysate causes **low-frequency transduction (LFT).** More-over, the transducing virions differ from each other in the precise DNA replacements, because they arose by independent recombination events.

High-frequency transduction (HFT) lysates, greatly enriched in transducing particles, can be produced by propagating phage recovered by induction of transduced cells (Fig. 48-19A). With nondefective transducing par-

ticles the phage from a single plaque is sufficient, but with defective phages the cells must be coinfected with wild-type λ as a helper; these **doubly lysogenic** cells then yield on induction a lysate containing nearly equal proportions of transducing and wild-type particles.

Thus a *lac⁻* cell infected by λ and λdgal⁺ (i.e., transducing the *gal⁺* allele) becomes lysogenic for both phages; it is a gal **hetero-genote** (indicated as *gal⁻/*λdgal⁺), containing both the preexisting

FIG. 48-19. A. Production of high-frequency transduction (HFT) lysate. Infection by a λdgal particle present in a low-frequency transduction (LFT) lysate causes the formation of a **heterogenote,** containing both the *gal⁻* gene preexisting in the bacterium and **gal⁺** introduced by λ**dgal,** plus regular λ as helper. When the cells of a clone deriving from such a heterogenote are induced, they lyse and produce the HFT lysate. **B.** Occasionally the heterogenote segregates, by loss of the λdgal prophage, a cell with only the original *gal⁻* gene (**I**); even more rarely it segregates, by a new exchange, a nonlysogenic cell with the *gal⁺* gene of the transducing prophage (**II**). The probable molecular mechanisms of these segregations are shown in **C.**

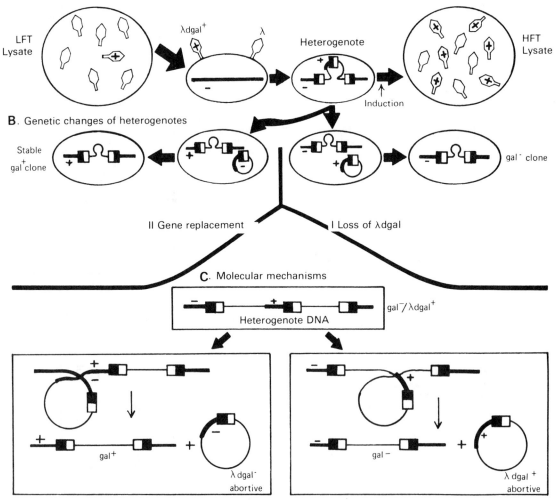

A. Formation of HFT lysate

B. Genetic changes of heterogenotes

II Gene replacement   I Loss of λdgal

C. Molecular mechanisms

rec - Promoted recombination between homologous bacterial sites

rec - Promoted recombination between homologous prophage sites

gal⁻ gene and the new *gal⁺* gene brought in by the phage. The heterogenote can be isolated by taking advantage of its *gal⁺* phenotype; after propagation it becomes the source of an HFT lysate containing λdgal⁺.

In contrast to the heterogeneity of the rare transducing particles in LFT lysates, an HFT lysate contains only one kind of transducing particle; and independently derived HFT lysates usually contain transducing particles with different properties since they were produced by different recombination events.

The properties of phages producing specialized or generalized transduction are given in Table 48-1.

Range of Specialized Transduction. The natural range of transducible genes is limited to those near the attachment sites of the available temperate phages. However, **the range can be enlarged** by obtaining the transducing phage from rare cells in which the desired gene is artificially integrated near a prophage.

Thus an F′ plasmid carrying the desired gene may be integrated near a phage attachment site. Alternatively, λ phage may be integrated at an unusual location near the desired gene in an *E. coli* strain that lacks the normal λ attachment site (see Unusual Insertions, above).

In addition to transduction by λ, φ80 is widely used to transduce *E. coli* genes for tryptophan synthetase, T1 resistance, and tyrosine tRNA; P1 (which is plasmidial but occasionally integrates) for *lac* or *pro* genes; and P22 in *Salmonella typhimurium* for genes for proline synthesis.

Control of Transduced Bacterial Genes. These genes often escape normal regulation. If they become disconnected from their operator they become regulated by the phage repressor. Moreover, those remaining connected to their operator may also lose the normal control if the number of copies of the operon increases markedly relative to the number of molecules of repressor; the genes then display **transduction escape synthesis** of their products (see Effect of Prophage on Host Functions, above).

Properties of heterogenotes. *Gal* heterogenotes obtained by λdgal transduction show marked **segregational instability**. About 1 in 10³

cells lose λdgal, yielding completely stable haploid cells with the *gal* gene that was present before their lysogenization. More rarely, heterogenotes segregate **stable** bacterial strains with the *gal* gene that was present in the λdgal (Fig. 48-19B); the resulting clone can no longer produce an HFT lysate. The mechanism is probably reciprocal crossover between homologous regions, as shown in Figure 48-19C.

USES AND SIGNIFICANCE OF SPECIALIZED TRANSDUCTION

Transducing phages have played an important role not only in mapping genes and in creating useful recombinants but also in studying gene regulation at a molecular level. Thus the transduced genes are enriched in an HFT lysate, from one per bacterial genome to one per phage genome (which is much smaller); and their products are enriched in cells containing multiple copies of the transducing DNA during the vegetative multiplication of a transducing phage. These materials have been of great value in purifying such components as the operator of the bacterial *lac* operon and its repressor, and in studying their interactions. Moreover, a variety of HFT phages, with different degrees of replacement of phage DNA by bacterial DNA, have been powerful tools for studying the organization of the genetic material, both viral and bacterial.

The evolutionary relationship between bacteria and phages is probably based on exchanges between their DNAs. One aspect of such a relationship is shown by the prophages, which, once inserted into the bacterial DNA, act like cellular genes in replication. Moreover, some prophages contain genes that affect cellular properties. The phenomenon of transduction shows the converse relationship, i.e., the incorporation of bacterial DNA into the phage DNA, where it acts like a viral gene in replication and regulation. In the cases we have examined, the viral and cellular genes inserted in the host DNA can be identified: but this is probably possible only because

TABLE 48-1. Comparison of Transducing Phages

Property	Specialized Example: *E. coli* K12 phages λgal and λdgal	Generalized Example: *Salmonella* phage P22
Genes transduced	Only genes closely linked to prophage	Any selectable marker
Source of transducing phage	Induction only	Induction or lytic infection
Capacity of transducing particles to produce infective viral progeny	Either infective (λgal), or defective (λdgal: multiply with helper)	No multiplication
Characteristics of the clones of transduced cells	Unstable heterogenotes, segregating stable haploids	Stable haploids
Efficiency of transduction per phage particle	LFT 10^{-6} (from haploid) to HFT 10^{-1} (from heterogenote)	10^{-5} to 10^{-6}

LFT = Low-frequency transduction; HFT = high-frequency transduction.

these insertions are the products of recent exchanges. It is conceivable that mutant genes exist whose function is inactivated, and which therefore are undetectable experimentally. After insertion into the host DNA such genes might remain there and then evolve until they developed a function useful for the host. The highly defective prophages may be such intermediaries in the evolution of viral to bacterial genes; and, conversely, transducing phages carrying defective bacterial genes might be intermediaries in the evolution of bacterial to viral genes. The heredity of bacteria and that of phages are therefore mixed together: much of the bacterial genome may be of viral origin; and phages may be fragments of bacterial DNA that have evolved in the direction of autonomy.

SELECTED READING

BOOKS AND REVIEW ARTICLES

BARKSDALE L, ARDEN SB: Persisting bacteriophage infection, lysogeny, and phage conversion. Annu Rev Microbiol 28:265, 1976

BUCHARI AI: Bacteriophage Mu as a transposition element. Annu Rev Genet 10:389, 1976

COUTURIER M: The integration and excision of the bacteriophage Mu-1. Cell 7:155, 1976

LOTZ W: Defective bacteriophages. Prog Molec Subcell Biol 4:53, 1976

PIRROTTA V: The λ repressor and its action. Curr Top Microbiol Immunol 74:21, 1976

SPECIFIC ARTICLES

ADHYA S, GOTTESMAN M, DE CROMBRUGGE B: Release of polarity in *Escherichia coli* by gene N of phage λ: termination and antitermination of transcription. Proc Natl Acad Sci USA 71:2534, 1974

CAMPBELL A: Segregants from lysogenic heterogenotes carrying recombinant λ prophages. Virology 20:344, 1963

EISEN H, BRACHET P, PEREIRA DASILVA L, JACOB F: Regulation of repressor expression in λ. Proc Natl Acad Sci USA 66:855, 1970

FAELEN M, TOUSSAINT A: Bacteriophage Mu-1: a tool to transpose to localize bacterial genes. J Mol Biol 104:525, 1976

HERSHEY AD, BURGI E, INGRAHAM L: Cohesion of DNA molecules isolated from phage lambda. Proc Natl Acad Sci USA 49:748, 1963

HONG JS, SMITH GR, AMES BN: Adenosine 3′5′-cyclic monophosphate concentration in the bacterial host regulates the viral decision between lysogeny and lysis. Proc Natl Acad Sci USA 68:2258, 1971

IMAE Y, FUKASAWA T: Regional replication of the bacterial chromosome induced by derepression of prophage lambda. J Mol Biol 54:585, 1970

JOHNSON A, MEYER BJ, PTASHNE M: Mechanism of action of the *vio* protein of bacteriophage λ. Proc Natl Acad Sci USA 75:1783, 1978

KAISER AD, MATSUDA T: Specificity in curing by heteroimmune superinfection. Virology 40:522, 1970

LANDY A, ROSS W: Viral integration and excision: structure of the lambda *att* sites. Science 197:1147, 1977

LWOFF A, SIMINOVICH L, KJELDGAARD N: Induction de la production de bacteriophages chez une bacterie lysogene. Ann Inst Pasteur (Lille) 79:815, 1950

MATSUBARA K: Genetic structure and regulation of a replicon of plasmid λdv. J Mol Biol 102:427, 1976

MOREAU P, BAILONE A, DEVORET R: Prophage λ induction in *Escherichia coli* K12 *envA uvrB*: a highly sensitive test for potential carcinogens. Proc Natl Acad Sci USA 73:3700, 1976

MORSE ML, LEDERBERG EM, LEDERBERG J: Transduction in *Escherichia coli* K$_{12}$. Genetics 41:142, 1956

MOUNT DW: Mutant of *Escherichia coli* showing constitutive expression of the lysogenic induction and error-prone DNA repair pathway. Proc Natl Acad Sci USA 74:300 1977

PARKER V, BUKHARI AI: Genetic analysis of heterogeneous DNA circles formed after prophage Mu induction. J Virol 19:756, 1976

PTASHNE M: Detachment and maturation of conserved lambda prophage DNA. J Mol Biol 11:90, 1965

PTASHNE M, BACKMAN K, HUMAYUN MZ et al: Autoregulation and function of a repressor in bacteriophage lambda. Science 194:156, 1976

REICHARDT L, KAISER AD: Control of λ repressor synthesis. Proc Natl Acad Sci USA 68:2185, 1971

ROBERTS JW, ROBERTS CW, MOUNT DW: Inactivation and proteolytic cleavage of phage λ repressor in vitro in an ATP-dependent reaction. Proc Natl Acad Sci USA 74:2283, 1977

SIGNER ER: Plasmid formation: a new mode of lysogeny in phage λ. Nature 223:158, 1969

SIGNER ER, WEIL J, KIMBALL PC: Recombination in bacteriophage λ. III. Studies on the nature of the prophage attachment regions. J Mol Biol 46:543, 1969

TAYLOR AL: Bacteriophage-induced mutation in *Escherichia coli.* Proc Natl Acad Sci USA 50:1043, 1963

WALZ A, PIRROTTA V, INEICHEN K: λ repressor regulates the switch between PR and Prm promoters. Nature 262:665, 1976

YARMOLINSKY MB, WIESMEYER H: Regulation by coliphage lambda of the expression of the capacity to synthesize a sequence of host enzymes. Proc Natl Acad Sci USA 46:1626, 1960

chapter **49**

ANIMAL CELL CULTURES

CHARACTERIZATION OF CULTURES OF ANIMAL CELLS

The development of improved methods for cultivating animal cells in vitro has been essential to the progress of animal virology and to the analysis of gene function, metabolic regulation, and other problems of cell physiology in animal cells. In particular, cell culture has provided quantitative technics for studying animal viruses comparable to those used for bacteriophages.

Cultures of animal cells are rather unstable. Thus **primary cultures** may die out on repeated secondary cultivation or they may give rise to altered but still diploid cell **strains**. These, in turn, will also eventually die out unless they give rise to more radically altered but permanent cell **lines**. These several types of cultures are all valuable for studies in virology as well as in cell physiology and genetics.

Primary and Secondary Cultures. To propagate separated animal cells, tissue fragments are first dispersed into the constituent cells, usually with the aid of trypsin. After removal of the trypsin the suspension is placed in a flat-bottomed **glass or plastic container** (a Petri dish, bottle, or test tube), together with a **liquid medium** (such as that devised by Eagle) containing required ions at isosmotic concentration, a number of amino acids and vitamins, and an animal serum in a proportion varying from a few percent to 50%. Bicarbonate is commonly used as a buffer, in equilibrium with CO_2 (about 5%) in the air above the medium. After a variable lag the cells attach and spread on the bottom of the container and then start dividing mitotically, giving rise to a **primary culture**. Attachment to a rigid support is essential for the growth of most normal cells (**anchorage dependence**).

In the primary culture the cells retain some of the characteristics of the tissue from which they were derived, and are mainly of two types: thin and elongated (**fibroblastlike),** or polygonal and tending to form sheets (**epitheliumlike).** In addition, certain cells have a roundish outline and resemble epithelial cells but do not form sheets (**epithelioid** cells). The cells multiply to cover the bottom of the container with a continuous but thin layer, often one cell thick (**monolayer);** if they are fibroblastic

they are **regularly oriented** parallel to each other (see Fig. 49-2A). Primary cell cultures obtained from **cancerous tissues** usually differ from those of normal cells and are like transformed cells (see Cell Transformation, below). The cultures are maintained by changing the fluid two or three times a week. When the cultures become too crowded the cells are detached from the vessel wall by either trypsin or the chelating agent EDTA, and a fraction is used to initiate new **secondary cultures (transfer).**

Cell Strains and Cell Lines (Fig. 49-1). Cells from primary cultures can often be transferred serially a number of times. This process usually causes a selection for some cell type, which becomes predominant. The cells may then continue to multiply at a constant rate over many successive transfers, and the primary culture is said to have originated a **cell strain** (often called a **diploid cell strain**) whose cells appear unaltered in morphologic and growth properties. These cells must be transferred at a relatively high cell density to initiate a new culture. Even so, the transferability of cell strains is limited: for instance, with cultures of human cells the growth rate declines after about 50 duplications and the life of the strain comes to an end.

During the multiplication of a cell strain some cells become **altered;** they acquire a different morphology, grow faster, and become able to start a culture from a small number of cells. The clone derived from one such cell, in contrast to the cell strain in which it originated, has **unlimited life;** it is designated a **cell line.** Cell lines derived from normal cells have a low saturation density under standard conditions (e.g., addition of 10% serum and changes of medium two or three times weekly); they grow rapidly to form a monolayer and then slow down considerably.

CELL TRANSFORMATION

Oncogenic viruses (Ch. 66), radiations, and certain chemicals can cause mutationlike changes in cultured cells that affect their growth and other properties. Such **transformation** (Fig. 49-2) can also occur during the serial growth of cell lines without evident exposure to any agent. Radiation and chemicals may cause transforma-

FIG. 49-1. Multiplication of cultures of cells derived from normal tissues. The primary culture gives rise to a cell **strain,** whose cells grow actively for many cell generations; then growth declines and finally the culture stops growing and dies. During multiplication of the cell strain altered cells may be produced, which continue to grow indefinitely and originate a cell **line.** The cumulative number of cells is calculated as if all cells derived from the original culture had been kept at every transfer. (Modified from Hayflick L: Analytic Cell Culture. National Cancer Institute Monograph, No. 7, 1962, p. 63)

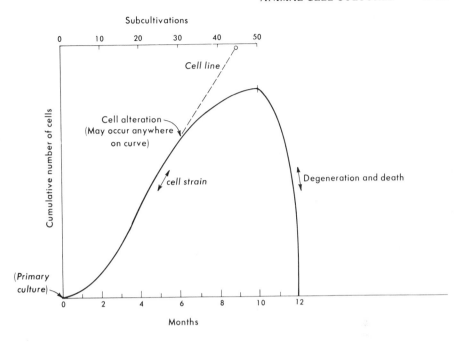

tion, as well as cancer in vivo, by inducing mutations (**somatic mutations,** because they are absent in the germ cells). In fact, Ames has shown that most oncogenic chemicals or their metabolic derivatives are also mutagenic.

Most **spontaneous cancers** in vivo appear to result from multiple mutations. Thus in individuals carrying a predisposing hereditary, or **germinal,** mutation, retinoblastoma (a cancer of children) appears early and is multiple and bilateral, whereas in individuals without the mutation it appears late and is unilateral. It seems that in the hereditary cases a single somatic mutation added to the germinal mutation is sufficient for starting a cancer (hence its early and multiple occurrence), whereas in the absence of the germinal mutation at least two somatic mutations are required. Support for the mutational hypothesis comes also from the high incidence of

FIG. 49-2. A. Uninfected crowded secondary culture of hamster embryo cells. Note that the cells are arranged in a thin (mostly single) layer with parallel orientation. **B.** Similar culture transformed by polyoma virus (Ch. 66). Note that cells lie on top of each other and are randomly oriented.

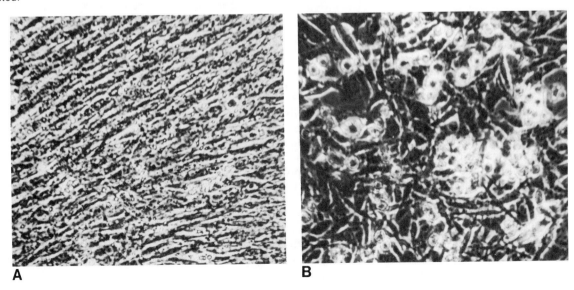

A   B

cancer in individuals with hereditary defects in DNA repair such as **xeroderma pigmentosum,** in which somatic mutations are induced by ultraviolet irradiation.

Transformed cells have many properties that are absent in resting untransformed cells:

A. Cultural behavior
   1. Increased culture thickness
   2. Random cell orientation
   3. Increased saturation density
   4. Decreased serum requirement
   5. Increased efficiency of clone formation
   6. Increased edge indentation and refractility
   7. Decreased anchorage dependence

B. Cell surface
   1. Decreased large external transformation-sensitive (LETS) protein
   2. Shorter gangliosides
   3. Reduced glycosyl transferase activity
   4. Reduced adhesion to plastic
   5. Increased receptor mobility (lectins)
   6. Increased agglutination by lectins
   7. Loss of receptors for hormones and toxins
   8. Increased and unregulated transport activity (e.g., glucose, phosphate, nucleosides, K^+)
   9. Increased glucose binding protein
   10. Presence of new antigens

C. Biochemical characteristics
   1. Increased protease production
   2. Disaggregated microfilaments and myosin
   3. Increased aerobic glycolysis
   4. Decreased intracellular concentration of cyclic AMP
   5. Increased, abnormal collagen synthesis
   6. Resurgent fetal functions

D. Other characteristics
   1. Production of tumor by 10^6 or fewer cells inoculated subcutaneously in immunologically accepting animals (syngeneic or nude mice)
   2. Formation of chromosomal abnormalities
   3. Absence of aging

Some of the changes seen in transformed cells are probably interrelated. Thus some are related to the persistent growth of transformed cells and are also found in growing untransformed cells. Other connections are functional: thus the reduction of LETS surface protein (see Constitution of Cellular Membranes, below) causes a disorganization of the cytoskeleton, perhaps by decreasing adhesion to the substrate. Not all transformed cells have all the characteristics listed; the spectrum varies widely among individual clones and depends on the selection imposed during isolation of the cells (e.g.,

whether by growth in agar, or in low serum, or in dense cultures) and on the inducing agent (e.g., whether chemical or viral). These conditions probably act by selecting for cells with different cellular mutations. Some kinds of transformed cells are malignant (i.e., they produce tumors in animals), others are not. The properties most closely associated with malignancy are loss of anchorage dependence, protease production, and reduction of LETS protein, but none is essential. The genetic and epigenetic events controlling the various characteristics are not known; this is understandable given the limited methodology available for studying the complex genome in animal cells.

Liquid Suspension Cultures. Transformed (and also some nontransformed) lines sometimes produce, by additional changes, cells with low adhesion to the container. These cells can be propagated in suspension by using a liquid medium poor in divalent ions and stirring constantly. Such suspension cultures (e.g., derivatives of L cells: see Table 50-2, Ch. 50), are very useful for experimental virology.

State of Differentiation of Cells in Culture. Animal cells in culture retain, at least in part, the state of differentiation they had in the animal, but it may not be easily recognizable, and so the cells are usually identified by descriptive names (such as fibroblastic or epithelioid) that bear little relation to the cells of origin. However, specific products can sometimes be formed, e.g., collagen from fibroblasts, casein from mammary gland cells, specific hormones from pituitary cells. A differentiation potential may also be expressed in vitro. Thus, skin keratinocytes differentiate into squamous cells, and 3T3 cells (skin "fibroblasts") into fat cells.

Most striking are cells of mouse **teratocarcinomas,** which are obtained by transplanting early embryos to an adult animal. Cultures of these cells, which are at first undifferentiated, undergo differentiation and develop morphologically recognizable tissues when they attain a high cell density. The differentiation potential of the teratocarcinoma cells after extensive growth in vitro is essentially unaltered, because after introduction into an early embryo they can contribute their genetic markers to the normal tissues of the resulting adult animal.

Storage. Because cell lines tend to change continually on repeated cultivation, it is useful to keep cells of early passages **in the frozen state.** Large batches of cells are mixed with glycerol or dimethylsulfoxide and subdivided in a number of ampules, which are then sealed and frozen. The additives allow the cells to survive the freezing. The frozen cells can be maintained in liquid nitrogen for years **with unchanged characteristics;** when the ampules are thawed most of the cells are viable and can initiate new cultures.

CLONING OF ANIMAL CELLS

Clones can be obtained from most cell strains or lines by transferring cells to a new culture at a very high dilution.

For cell lines the proportion of cells that survive and give rise to colonies (**efficiency of plating**) approaches 100%, but for primary cultures or cell strains it is very small. However, the efficiency of plating can be greatly increased if the cells are mixed with a **feeder layer** of similar cells made incapable of multiplication by x-irradiation or mitomycin; these cells are still metabolically active and supply factors that enable the unirradiated cells to survive and multiply. The efficiency of plating can also be increased by introducing individual cells into very small volumes of medium (as in sealed capillary tubes or in small drops of medium surrounded by paraffin oil), which allows the cell products to reach an adequate concentration.

AGING

Culture Senescence and the Establishment of a Cell Line. Cultures of cells obtained from an animal become able to grow indefinitely only after they have undergone an **"immortality event,"** whose occurrence is revealed by their ability to grow in very sparse cultures and with a high cloning efficiency. Unless this occurs, cultures undergo some kind of cellular senescence and finally die. Culture senescence is absent in transformed cells.

Senescence appears to depend directly on the **number of doublings** rather than on astronomical time. It does not depend on the special conditions of growth in vitro: *it also occurs in vivo,* as during serial transplantation from mouse to mouse of mouse mammary epithelium or of hemopoietic cells.

Aging in vitro is related to aging in vivo, since fibroblasts obtained from human donors of increasing ages show decreasing life spans in vitro. Furthermore, fibroblasts from patients with Werner's syndrome (premature aging), who have a shortened life span, can undergo fewer generations in vitro.

Cellular senescence is attributed to the **accumulation of unrepaired damage** in cellular constituents. Accumulation of mutations in DNA may lower the growth ability of the cells and finally become lethal. Accidental errors in translation, causing changes in proteins that can influence information transfer (such as the aminoacyl-tRNA synthetases, and DNA and RNA polymerases), may also cause progressive deterioration of cellular proteins, even independently of mutation. Indeed, in cultures of the mold *Neurospora* that are undergoing senescence, altered and partly inactive enzymes accumulate in the cells.

The key to senescence may be the **absence of efficient selection** against cumulative damage, because the culture of untransformed cells requires seeding at a relatively high cell density and the cells also have their growth limited soon by density-dependent growth inhibition. The immortality event, in contrast, allows the cells to initiate growth at low density, while transformation allows the cells to grow to a higher density; both changes increase selection against damaged cells.

Another factor determining the life span in vitro is **terminal differentiation.** Thus human keratinocytes on 3T3 feeder layers normally survive for about 50 cell generations (less if from old donors); at the end they differentiate into squamous cells and die. If the cultures are grown in the presence of epidermal growth factor (EGF) (see Growth Factors, below) their life span increases to about 150 generations, apparently because differentiation is delayed. Transformation may prevent aging because it prevents terminal differentiation.

RELATION OF CELLS TO A SOLID SUPPORT

Electron micrographs show that a cell attaches at a few points to the bottom of the vessel, from which it is elsewhere separated by a layer of medium (Fig. 49-3). The attachment points become evident after trypsin treatment: the cytoplasm retracts toward the nucleus but remains connected to the attachment points by thin **retraction fibers,** which are easily broken by mechanical action.

Cell Movement. Active locomotion is beautifully seen in slow-motion pictures of living cell cultures using phase contrast microscopy. The advancing part of a fibroblast is thin and rapidly moving (**ruffling**); after anchoring to the vessel wall it pulls the rest of the cell along. If the fibroblast meets another cell, both ruffling and locomotion stop (**contact inhibition of movement**). This effect is slight or absent with transformed cells, probably because they adhere less to other cells. The plasma membrane flows from the forward edge toward the nucleus, as shown by the movement of adhering particles.

Movements of parts of a cell are observed in **cytoplasmic streaming** and in **endocytosis.** The latter is an engulfment of either liquid droplets (**pinocytosis**) or solid particles after they are recognized by specific surface receptors (**phagocytosis;** Ch. 24). The engulfed material is contained in membrane-bound vesicles, which then fuse

FIG. 49-3. Schematic cross section of a culture on a solid support in liquid medium. Microfilaments are shown in black. Medium, both outside the cells and in pinocytotic vesicles, is gray. **Arrows** indicate the movement of pinocytotic vesicles. (Modified from Brunk U et al: Exp Cell Res 67:407, 1971)

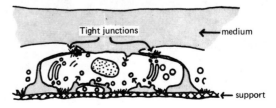

with **lysosomes,** where the content of the vesicles is attacked by lytic enzymes. The components of the internalized membranes are later reutilized for new membrane synthesis.

CYTOSKELETON

The acts of movement required for cell streaming and locomotion, and for endocytosis, secretion, and cell division, are produced by three types of filaments.

1) A contractile system is made up of **microfilaments,** 5–7 nm wide, composed of actin associated with myosin and other proteins, which act like primitive muscle fibers. They are abundant at the inner side of the plasma membrane, to which they are connected by specialized proteins, and connect the membrane to the nucleus. In normal fibroblasts the microfilaments occur in thick bundles **(stress fibers)** stretching across the cell (Figs. 49-4 and 49-6). In transformed cells (Fig. 49-4B) and some normal cells (e.g., lymphocytes) the microfilaments are dispersed and stress fibers are absent.

Microfilaments disappear under the action of **cytochalasin B.** Then the nucleus protrudes, and mild centrifugation causes the cell to break up into the **karyoplast** (containing the nucleus) and the **cytoplast** (containing most of the cytoplasm). These incomplete cells are useful in experimental work.

2) A rigid framework of 25-nm-wide **microtubules** (Fig. 49-5) is responsible for maintaining the reciprocal positions of cytoplasmic components. The microtubules present in interphase cells disappear in mitosis and are replaced by the microtubules of the spindle.

Microtubules are in a dynamic state; they grow at one

FIG. 49-4. Cytoskeleton of normal fibroblastic rat cells **(A)** and of their derivatives transformed by the Rous sarcoma virus **(B),** stained by antimyosin serum. Microfilaments are well organized in normal cells forming stress fibers, but are disrupted in transformed cells. (Ash JS, Vogt PK, Singer J: Proc Natl Acad Sci USA 73:3603–3607, 1976.)

end by the GTP-dependent aggregation of the two subunits of the protein **tubulin** (together with smaller amounts of other proteins), while they disaggregate at the other end. They disappear in the presence of **colchicine** and other alkaloids, which bind to tubulin and prevent its polymerization.

3) **Intermediate filaments,** 10 nm wide and heterogeneous, connect the nucleus to the surface of the cell adherent to the dish. Some intermediate filaments are stained by fluorescent Abs present as autoantibodies in the serum of some cancer patients. They are not affected by cytochalasin B; colchicine causes them to condense and coil around the nucleus. In epithelial cells these filaments are made up of keratin; in nerve cells they are known as **neurofilaments.**

CELLULAR MEMBRANES

Animal cell membranes have many important functions, including compartmentation of cells, regulatory interactions with external substances controlling growth or specific cellular functions, and transmission of signals resulting from these interactions to the sites of macromolecular synthesis.

The cell is surrounded by the **plasma membrane,** whose extreme plasticity allows changes of shape and movement. **Microvilli** are present on its external surface in actively growing or transformed fibroblasts (Fig. 49-7); they are always present on epithelial cells, whether or not transformed. Inside the cell the plasma membrane is continuous with intracellular membranes, such as the **endoplasmic reticulum,** which forms a system of double membranes enclosing narrow channels stretching from the inner side of the plasma membrane to the outer nuclear membrane. The **rough reticulum** is associated with polysomes and is the site of synthesis of proteins for "export," i.e., destined to leave the cell; the **smooth reticulum,** without polysomes, channels these proteins to the outside. The endoplasmic reticulum is in turn connected to **mitochondria,** which are the sites of oxidative phosphorylation, and to **lysosomes,** bags full of degradative enzymes whose function is to digest phagocytized material or, when the cell dies, the cellular constituents themselves.

The cell nucleus is enveloped by a **double nuclear membrane** provided with several thousand pores 40–80 nm in diameter, filled with an amorphous material connected to a lamina apposed on the outside of the nuclear membrane. It seems that the pores control communication between the nuclear and the cytoplasmic compartments. The inner nuclear membrane is connected to the chromosomes, probably at the pores, and it is therefore the counterpart of the plasma membrane of bacteria (Ch. 6). The outer membrane is studded with ribosomes. The nuclear membrane dissolves as the beginning of mitosis and is re-formed at the end of mitosis by the endoplasmic reticulum which surrounds the chromatin.

CONSTITUTION OF CELLULAR MEMBRANES

Membranes are "fluid mosaics" of proteins interspersed in a **lipid bilayer** made up of various phospholipids and cholesterol. **Integral** (or **transmembrane) proteins** are usually amphipathic, i.e., they contain both apolar groups interacting with the lipids in the bilayer and polar groups at either or both surfaces of the bilayer (Fig. 49-8). These proteins are extracted by agents that break hydrophobic bonds (e.g., nonionic detergents). **Peripheral proteins** are associated with integral proteins by polar bonds at either surface and are usually extracted with a concentrated salt solution. Proteins that protrude at the outer surface are **glycoproteins.** They have a vectorial orientation in the membrane: the part that protudes outside the cell is glycosylated, whereas that protruding in the cytoplasm (in those spanning the membrane) is not.

Both integral and peripheral membrane proteins, as well as proteins to be secreted, are synthesized on polysomes present on the cytoplasmic side of the membranes of the rough endoplasmic reticulum, in most cases as **preproteins.** The preproteins usually contain a **signal sequence,** rich in nonpolar amino acids, at the NH_2 end. During synthesis this end of the growing polypeptide chains (which is synthesized first) binds to proteins present on the inner membrane side and then penetrates vectorially through the membrane; as it emerges at the outer side the protein is glycosylated by membrane-bound enzymes. In most cases the signal sequence is cleaved away by a membrane protease. Both glycosylation and cleavage fix the protein in its relation to the lipid bilayer; glycosylation also stabilizes the conformation of the protein and protects it from enzymatic degradation. After synthesis the glycoproteins move from the rough endoplasmic reticulum to the plasma membrane.

Integral membrane proteins are responsible for the transport of solutes into the cells, for enzymatic activities, and for recognition of external ligands; they also form the **intramembranous particles** (7–8 nm in diameter, seen in freeze-fracture electron micrographs of plasma membranes) that are apparently involved in communication between cells (see below). Peripheral proteins contain many **receptors** recognizing hormones or growth factors.

Microfilaments at the inner face of the plasma membrane are especially abundant in areas of movement or endocytosis. In untransformed cells the outer face is partly covered by a thick glycoprotein layer, the **glycocalyx.** In fibroblasts an important constituent of this outer layer is the **LETS** (large external transformation-sensitive) protein (also called **fibronectin),** which is closely similar in structure to a plasma protein (cold-insensitive globulin).

The LETS protein in conjunction with collagen or precollagen forms a fibrillar network, especially abundant in resting cultures, on the underside of cells and at points of contact between cells. In this way it strengthens the adhesion of the cells to the dish and to each other, and affects cell morphology. In fact its addition to transformed cells temporarily confers on them normal morphology and adhesiveness. The LETS protein is almost absent from transformed cells, probably because it sticks only weakly to their surface.

Antigenic Structure of the Cell Surface. Both integral and peripheral proteins contain on their exposed surface antigenic determinants, which are responsible for **blood group** specificity in RBCs and **histocompatibility** specificities in tissue cells. In addition certain cells have differentiation Ags; and new Ags appear in cancer cells **(tumor-specific Ags)** and in cells infected by viruses, both oncogenic and nononcogenic. Both the blood group and the histocompatibility Ags have been shown to be glycoproteins. All these Ags can be recognized by various immunologic methods: **immunofluorescence** with live cells, which labels only surface Ags (membrane fluorescence); **cytotoxicity,** in which the cells are killed by Abs

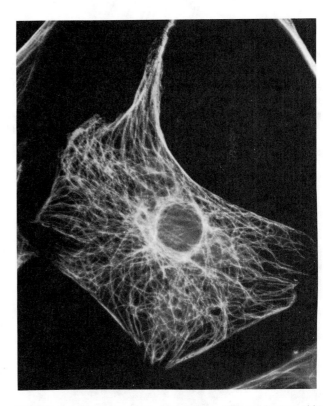

FIG. 49-5. The system of microtubules in a 3T3 cell (mouse skin fibroblast) stained by fluorescent antitubulin serum. (Cox SM, Rao PN, Brinkley BR: In Busch H, Crooke ST, Daskal I (eds): Effects of Drugs on the Cell Nucleus. New York, Academic Press, 1979)

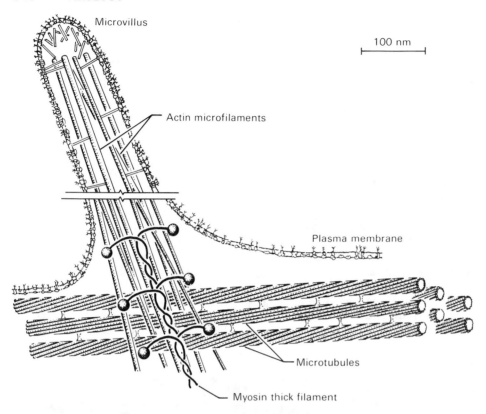

FIG. 49-6. Diagram of cell surface and cytoskeleton drawn to scale. Only the base and tip of a microvillus are shown; connections between plasma membrane and microfilaments are idealized. (Modified from Loor F: Nature 264:272, 1976)

FIG. 49-7. Cell surface structure in scanning electron micrograph, which records electrons bounced off the cell surface, faithfully reproducing its details. Chick embryo fibroblasts were transformed by a ts mutant of Rous sarcoma virus, which causes the phenotype, normal at 41°C **(A)**, to be transformed at 36° **(B)**. In the normal state **(A)** cells have a smooth surface with vey few microvilli; in the transformed state **(B)** both microvilli and ruffles are very prominent. (Ambros VR et al: Proc Natl Acad Sci USA 72:3144, 1975)

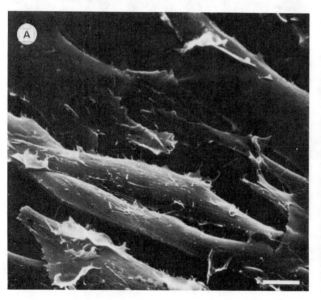

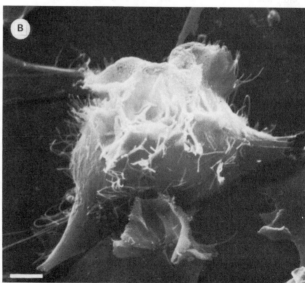

FIG. 49-8. A. Schematic cross section of a membrane formed by a lipid bilayer with incorporated integral proteins. The proteins have hydrophobic regions **(cross-hatched)** and polar (hydrophilic) regions (with + or − charges). **B.** The way that a protein can form a water channel within the hydrophobic bilayer. (Modified from Singer SJ: In Bolis L et al, eds: Comparative Biochemistry and Physiology of Transport. Amsterdam, North-Holland, 1974, p. 95)

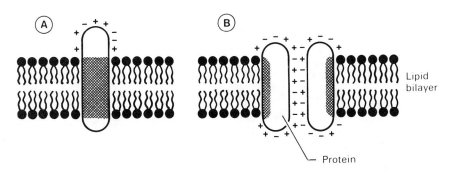

in the presence of complement; **colony reduction,** in which the colony-forming efficiency of the cell is reduced by immune lymphocytes; and binding of **Abs labeled by ferritin or enzyme,** which are recognizable by electron microscopy or by the enzymatic production of dye.

MEMBRANE DYNAMICS

All membrane components are capable of **lateral diffusion** in the plane of the membrane: phospholipids diffuse rapidly, proteins more slowly and with great individual differences (from 1 μm/sec to almost no displacement at all).

Protein mobility was first recognized when Edidin observed that the Ags of two cell types (revealed by fluorescent Abs) mingle after the cells are fused. Quantitative studies of mobility are made by labeling the membrane proteins in situ with fluorescein or by binding specific ligands (Abs, lectins) also labeled by fluorescein. After the fluorescence is bleached in a localized area by an intense laser beam, its return is due to diffusion of proteins from the unbleached areas; and the rate of reappearance measures the rate of protein diffusion.

The attachment of transmembrane proteins to microfilaments limits protein mobility, as shown by experiments with concanavalin A, a plant lectin. Each molecule of the lectin can bind to two or more receptors (mannose residues) on transmembrane proteins, crosslinking them. The result is that the receptors, which are normally distributed uniformly over the cell surface, bunch in **patches,** even in the absence of cell metabolism. In cells with disaggregated microfilaments (e.g., lymphocytes, transformed cells) the freely mobile proteins form large patches; if metabolism is active, the patches of a cell fuse in a mass that, after migrating toward the nucleus, is internalized **(capping).** In normal fibroblasts, in which the microfilaments form rigid fibers, only minute patches are formed, which do not cap (Fig. 49-9).

The connection of transmembrane proteins to microfilaments is shown by labeling the proteins protruding at the outer cell surface with ^{125}I (carried in an enzyme complex that does not enter the cell): the label follows actin (which is in microfilaments) during cell fractionation.

Capping is due to the ATP-dependent contraction of the microfilaments; hence it requires metabolic energy and is prevented by cytochalasin B, which disrupts microfilaments. The contraction of the microfilaments toward the nucleus, to which they are anchored, pulls the protein–lectin complexes inside the cell; in cytoplasts, which lack nuclei, capping does not take place.

The high receptor mobility of transformed cells makes them easily **agglutinable** by lectins: the formation of many lectin bridges, required for agglutination, is facilitated if the receptors can readily migrate to the points of cell contact.

COMMUNICATION BETWEEN CELLS

The cells of a culture, although individual in many respects, form a continuum in relation to the exchange of ions and small molecules through channels that form where the plasma membranes of two cells come in contact. The channels are revealed by a low electrical resistance between the cells, and by the passage of fluorescent substances, of mol wt up to 2000, to a cell adjacent to the one into which they were injected. Through the channels metabolic intermediates are exchanged between cells in sufficient quantities to allow cells with a metabolic defect to grow in mixed culture with normal cells **(metabolic cooperation).**

For instance, cells lacking the enzyme hypoxanthine: guanine phosphoribosyl transferase cannot utilize hypoxanthine as a purine source in a medium containing aminopterin (which blocks the normal purine pathway) and do not incorporate label derived from hypoxanthine into nucleic acids. However, if they are in contact with normal cells in mixed cultures the deficient cells incorporate label from hypoxanthine phosphorylated in the contiguous normal cells, as shown by radioautography. The channels cannot be crossed by nucleic acids or proteins.

The exchange of substances occurs through gap junctions (Fig. 49-10). Visible by freeze-fracture electron microscopy, these appear as areas of closely packed particles, 8–9 nm in diameter, spanning the membrane where the plasma membranes of two cells touch each other. These particles seem to have a central channel that admits water and solutes.

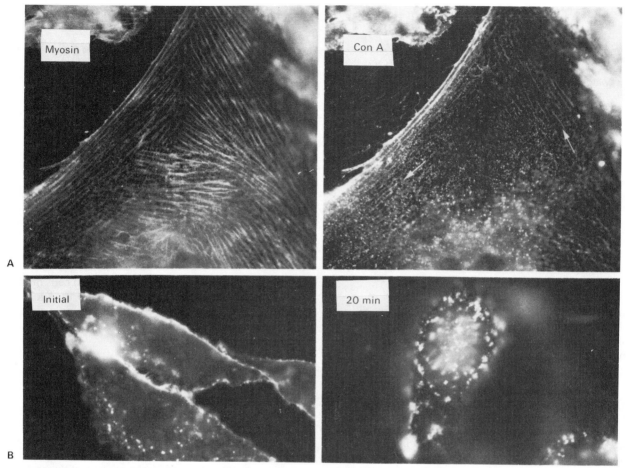

FIG. 49-9. Patching of concanavalin A (conA) receptors on the surface of normal and transformed cells. **A.** At 20 min after exposure to conA NRK (normal rat kidney) cells were fixed and treated with an anti-conA serum labeled with fluorescein (blue fluorescence, **right**), then treated with antimyosin serum labeled with rhodamine (red fluorescence, **left**). Using appropriate light filters the distributions of myosin **(left)** and of conA **(right)** were photographed in the same cell. The conA receptors, which were at first uniformly distributed over the cell surface, have formed very fine patches exactly aligned along the underlying myosin-containing stress fibers. **B. Right.** Large patches of conA receptors on Rous sarcoma virus–transformed NRK cells after 20 min of interaction with fluorescent conA. The initial uniform distribution is seen at **left.** (Ash JF, Singer SJ: Proc Natl Acad Sci USA 73:4575, 1976)

GROWTH REGULATION

GROWTH REQUIREMENTS

Animal cells have complex requirements, which are satisfied by a medium containing salts, amino acids, vitamins, and blood serum. The growth-promoting effect of serum is **stoichiometric,** i.e., if other requirements are fulfilled, the number of cells produced in a culture is proportional to the amount of serum added. Transformed cells require very little serum, while the serum requirement of several cell lines can be satisfied by defined serum-free mixtures of hormones, serum proteins, and growth factors.

Growth Factors. For fibroblastic cells an important growth factor is released from platelets during blood coagulation; hence plasma has a much lower growth-promoting activity than serum. Among other growth factors the most potent is **epidermal growth factor** (EGF) isolated by S. Cohen from the mouse submaxillary gland and also present in human urine. EGF stimulates the growth of both epithelial and fibroblastic cells.

Other factors with activities restricted to certain cell types include **fibroblastic growth factor** and **ovarian growth factor,** both extracted from the pituitary, **erythropoietin** (active on precursors of RBCs), and **colony-stimulating factor** (for progenitors of granulopoietic cells). Some of these factors have been purified and found to be small proteins. Many **hormones,** either alone or in combination with other hormones or with low concentrations of serum, have growth-promoting activity for certain cells; among these hormones

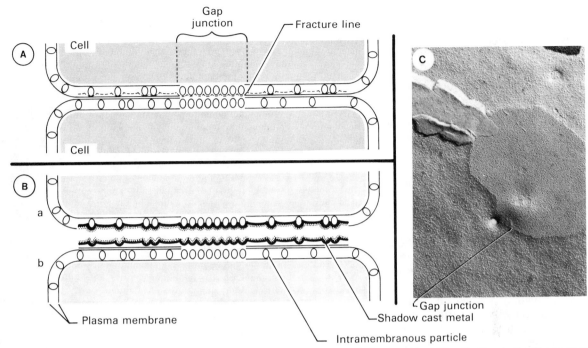

FIG. 49-10. Intramembranous particles and gap junctions. A frozen cell preparation is fractured by hitting it with a knife. The fracture frequently goes through membranes, separating them into two layers. Where two cells are in contact (**A**) the fracture line **(broken line)** sometimes goes through a junction. The products of fracture (**B**) are coated with evaporated metal, and a replica of the surface is examined in the electron microscope. **C.** Photograph of surface fracture **b** shows the gap junctions **(arrows)** and the intramembranous particles. (Courtesy of DA Goodenough)

are insulin and insulinlike serum peptides, glucocorticoids, prolactin, somatomedin, and prostaglandin $F_{2\alpha}$. Substances that potentiate the action of chemical carcinogens in the induction of cancer in animals, such as certain **phorbol esters,** are also potent mitogens. Some plant **lectins** (phytohemagglutinin, concanavallin A) are mitogenic for lymphocytes.

The cells themselves produce factors required for their growth. Thus the growth of sparse cultures is enhanced by the addition of **conditioned medium,** i.e., medium obtained from actively growing cultures.

Growth factors first interact with specific **receptors** at the cell surface; then some factors (e.g., EGF, insulin), together with their receptors, are internalized into the cells, where they are degraded by lysosomal enzymes or bind to internal receptors. Hence growth factors may act at the plasma membrane or internally; or their action might be due to degradation products of the factors or of their receptors. Transmission of signals from the plasma membrane, after receptor attachment, to the sites of macromolecular synthesis may occur through an internal messenger (as for certain peptide hormones, for which surface-synthesized cyclic AMP is such a messenger) or through the cytoskeleton, which connects all cell constituents.

THE CELL GROWTH CYCLE

As with bacteria (Ch. 5), the cells of a sparse culture in optimal medium multiply exponentially (i.e., with a fixed doubling time), although individual cells divide at random times. The cell growth cycle consists of four main phases (Fig. 49-11), each with a different biochemical and regulatory significance. DNA synthesis occupies only a fraction of the doubling time, the **synthetic (S) period.** Thus radioautographs of cultures show that a long exposure (e.g., 24 h) to ^{3}H-thymidine labels most nuclei, whereas an exposure of only 30 min leaves many still

FIG. 49-11. The cell growth cycle.

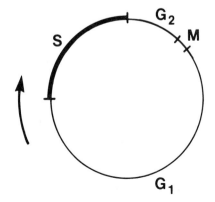

unlabeled. Between the S period and the **mitotic (M) period** there is a **G2 period** (G for gap) whose length is measured by determining, by radioautography, the time between a brief exposure to ³H-thymidine and the first appearance of label in metaphase chromosomes. After the mitotic phase and before the next S period there is the **G1 period,** which can vary enormously in length.

The distribution of the cells of a culture in the various phases can be ascertained by **flow cytofluorometry,** as portrayed in Figure 49-12.

From cells with a doubling time of 18 h typical lengths of the various periods are: G1, 10h; S, 6–7 h; G2, 1 h; M, about ½ h.

Growth Restriction in G1. Cultures of untransformed cells stop growing after the depletion of serum or growth factors, or, for certain cell types, of some amino acids or ions. The cultures then survive in a **quiescent state.** Growth resumes after addition of the depleted substance.

The growth of a quiescent culture can also be restored locally by removing a strip of cells (a **wound**) without replenishing the medium. Cells penetrating the wound from the edges initiate DNA synthesis within 12 h and then divide. This phenomenon indicates that in quiescent, dense cultures growth stoppage also depends on local conditions **(topoinhibition).** A major source of topoinhibition is a boundary layer of water in contact with a continuous cell layer, through which growth factors can penetrate only by diffusion. Elimination of the boundary layer at the edges of a wound allows the factors to reach the cells much more efficiently by convection.

Cytofluorometry shows that *in quiescent untransformed cultures the cells are arrested in the G1 phase.*

These cultures manifest a reduced uptake of glucose, phosphate, and other substances, and a reduced synthesis of RNA and proteins; they display faster protein degradation, and have most of the mRNAs free, rather than in polysomes. In contrast, with transformed cells medium depletion causes random arrest, each cell stopping at the phase it happened to be in at the time of depletion.

Under a variety of growth conditions the G1 phase is the one showing the greatest variation in length: its variation accounts for most of the variation in the total cycle time. Apparently, during the G1 phase the growth of untransformed cells must proceed through **restriction points** which are overcome by the availability of medium factors. The ease with which such restrictions can be overcome determines the length of the growth cycle.

At the end of the G1 phase the cell manufactures one or more intracellular substances that initiate DNA replication. In fact, the DNA of a quiescent cell starts replicating after fusion of that cell with one in S phase.

The G1 phase is probably the expression of growth regulation because it is not obligatory: rapidly growing cells of early embryos and certain hemopoietic cells lack a G1 phase. Mutant cells also lacking G1 (G1⁻) can be isolated from regular G1⁺ cells (i.e., hav-

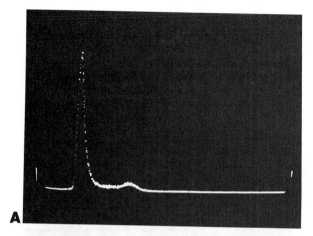

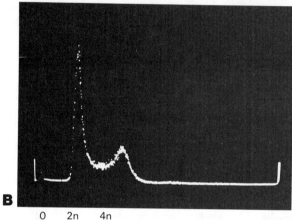

0   2n   4n

FIG. 49-12. Flow cytofluorometry of fibroblastic cells. A single-cell suspension is exposed to an acridine dye which, after intercalating in DNA, fluoresces green. The cells pass one at a time through a laser beam, and the excited fluorescence, which quantitates the amount of DNA, is measured. Each point in the graphs shows the number of cells that have the amount of DNA given in the abscissa. The tall peak at left measures the number of cells in G1 (with a diploid amount of DNA); the smaller peak at right measures the number of cells in G2 (with a tetraploid amount of DNA). Cells in S phase are between the two peaks. **A.** Quiescent population, with most cells in G1. **B.** Growing population, with a substantial proportion in S and G2. (Courtesy of RE Holley)

ing a G1 phase). The dominance of the G1⁻ state in G1⁻ × G1⁺ hybrid cells suggests that it is the wild-type condition. Various G1⁺ revertants from a G1⁻ line complement each other in cell fusions to yield a G1⁻ phenotype; they define five complementation groups. Therefore the G1 period seems to result from at least five different blocks to progression through the cycle.

The multiple blocks may correspond to the multiple requirements for restoring the growth of serum-depleted cells: thus the platelet factor must be supplemented by platelet factor–depleted plasma, which in turn may contain several other factors. Some of the blocks are abrogated in transformed cells. When cultures grow in the usual media only one of the factors seem to be limiting, causing the G1 lengths of different cells to follow a "single-event" distribution (see Appendix, Ch. 11).

Synchronization. The cell growth cycle is directly observable in synchronized cultures. The preferred method of synchronization is to start a culture with mitotic cells, which are weakly attached to the vessel and can be collected by shaking a randomly growing culture. In the synchronous culture all cells are initially at the M–G1 boundary. However, the first S phase is only partially synchronous, and in subsequent cycles synchronization is rapidly lost, because the G1 phases have different lengths in different cells.

Mechanism of Growth Control. Efforts directed at clarifying the biochemical nature of the blocks operating in G1 have concentrated on the events that occur in the 12 h or so before DNA synthesis resumes after the addition of a depleted growth factor to a culture. Many such events occur during this period **(pleiotypic response).** Some take place immediately: 1) increased transport of K^+, phosphate, glucose, and amino acids; 2) a decrease in the intracellular concentration of cAMP, and sometimes an increase in cGMP; 3) increased turnover of membrane lipids. Later the rate of protein synthesis increases, by reinitiation on mRNAs that had become dissociated from ribosomes in the period of quiescence. However, none of the immediate changes occurs uniformly with a variety of cell types and growth factors, suggesting that several changes act as a signal in different cases, perhaps depending on the specific blocks to be overcome in different cases.

ORGANIZATION AND EXPRESSION OF THE GENETIC INFORMATION

Genes in animal cells have several special features, which have been clarified by studying DNA fragments cloned in plasmids or phages.

1) Certain genes are highly **reiterated:** there are 24,000 genes for 5S ribosomal RNA (rRNA) in amphibians, and about 100 for histone in amphibians and *Drosophila.*

2) Genes specifying products required in equimolar amounts may form units in which they are separated by AT-rich **spacers.** The three rRNA genes and the five histone genes form such units, which are then repeated in tandem (i.e., with the genes in a constant order in the repeats).

3) Several genes (those for chicken ovalbumin, human or murine β-globin, and mouse immunoglobulin, and genes of animal viruses) contain several distinct **coding regions,** or **exons** (eight in the ovalbumin gene), separated by noncoding **intervening sequences,** or **introns.** The coding regions are in the same order as in mRNA. The complete sequence of each gene is transcribed into RNA; the exons are then removed, bringing the coding regions together in the mature mRNA (see Transcription of Chromatin, below). Therefore a gene can be considerably longer than its mRNA: thus the ovalbumin gene is 7 Kb long compared to 1.8 Kb of the coding fraction.

4) **Transposable genes** have been found in *Drosophila* by hybridizing labeled clonal DNA to polytene salivary chromosomes of closely related species; the hybridized sites show considerable variation. Like some bacterial transposons, they are flanked by two **direct repeat sequences.**

The functional significance of the reiteration of genes is that it allows the synthesis of large amounts of the corresponding product. In fact, genes normally present in single copies may be reiterated at stages of differentiation in which the specific products are in high demand, or in drug-resistant cells. Thus methotrexate-resistant cells (selected by growth in the presence of the drug) contain about 200 copies of the dihydrofolate reductase gene, whose increased product is responsible for drug resistance **(gene amplification).**

The regular arrangement of genes in a unit may allow equimolar synthesis of the mRNAs through coordinated regulation. Unclear is the significance of introns. The observation that in mice two β-globin genes with similar coding sequences have different introns suggests that introns may participate in the regulation of the expression of the genes. However, sequence heterogeneity (usually single-base differences) of the DNA of individuals of the same species is much higher in introns than in exons. The much higher divergence suggests that at least part of the intron sequences do not perform any required function.

Reiterated genes all specifying a product with the same function (e.g., the 5S rRNA genes) would be expected to undergo considerable divergence of sequences. In fact, different repeats of the 5S genes or of the histone genes show heterogeneity of length (in the spacers) and of sequences (in both the spacers and the genes, excluding the promoter regions); however, the heterogeneity is far less than expected from unrestricted evolutionary divergence. Extensive variation must be prevented by strong selection pressure or by a correction mechanism.

CHROMATIN

DNA in eukaryotic cells is associated with proteins, 60% basic **(histones)** and 40% acidic, to form chromatin, which has essentially the same structure in all eukaryotes. There are five histones. H3 and H4 **(arginine-rich)** are among the most conserved proteins in eukaryotes; H2A and H2B are **slightly lysine-rich,** and H1, which is heterogeneous, is **lysine-rich.** Octamers of two molecules each of histones H3, H4, H2A, and H2B, in conjunction with DNA, form **nucleosomes,** the basic units of chromatin. These are flat beads about 11 nm in diameter and 6nm thick. Nucleosomes assemble spontaneously in vitro after DNA is added to a mixture of histones H3 and H4, which are mainly responsible for nucleosome formation. Crystallographic evidence shows that the DNA forms a double loop (1.75 turns) around the core, in the process losing a helical turn and becoming supercoiled. The association does not depend on specific DNA sequences. In electron micrographs chromatin appears as chains of nucleosomes connected by short spacers of DNA, probably associated with histone H1 (a **"string of pearls,"** Fig. 49-13).

The arrangement of the DNA in the nucleosome is elucidated by the action of endonucleases. Mild digestion with micrococcal nuclease, which causes double-strand cuts, releases free nucleosomes, each containing 200 base pairs of DNA. Further digestion cuts off the **spacers,** releasing H1 and leaving a nucleosome **core** with 140 base pairs of DNA. This core is constant in chromatins. Intensive further digestion with DNAse 1 (which causes single-strand cuts) nicks the DNA at points spaced by integral numbers of groups of ten base pairs. Since ten base pairs is the periodicity of the supercoiled helix, the histones must leave only one narrow gap for access to the DNA at each helix turn.

Chromatin containing rRNA genes is smooth and more extended.

Replication of Chromatin. Each chromosome contains a continuous DNA helix, which is functionally subdivided into many independent **replicons.** Each of these replicates at a characteristic time in the S phase. Replication of DNA is **bidirectional,** i.e., it begins in the center of each replicon and spreads in both directions. Synthesis of DNA is initiated by **RNA primers,** which are later removed by RNase H (able to hydrolyze RNA in RNA–DNA hybrids).

Synthesis is probably carried out by DNA polymerases α and γ, which are abundant in growing cells, whereas a smaller polymerase, β, which is abundant in resting cells, may be devoted to repair synthesis. As with other linear DNAs, it is not clear how replication completes the DNA at the ends of a chromosome; a possible model involving a terminal palindrome is shown in Fig. 49-14. During DNA replication histones are synthesized and immediately move to the nucleus.

Nucleosomes persist during replication, but their structure loosens up, apparently owing to changes in the degree of acetylation and phosphorylation of histones, which alter the interaction of histones with DNA. Old nucleosomes retighten after replication, while additional nucleosomes are formed by newly synthesized histones associating with newly synthesized DNA.

Packing of DNA in Chromosomes during Mitosis. In acquiring the highly compact arrangement present in mitotic chromosomes the DNA coils to 1/10,000 of its extended length, through a hierarchical series of coilings. A sevenfold reduction in length is due to the coiling of DNA in chromatin; coiling of chromatin into 30-nm-wide filaments, probably due to interactions between molecules of histone H1, causes a further sixfold shortening; coiling of the fila-

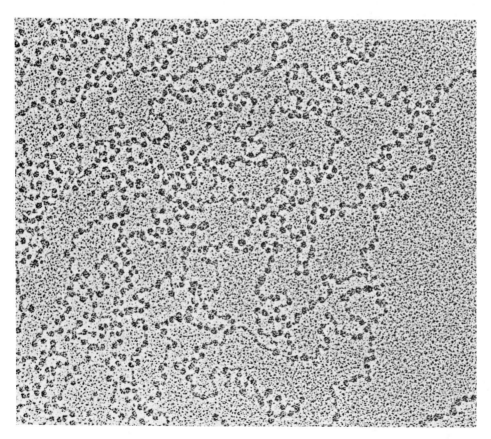

FIG. 49-13. Electron micrograph of *Drosophila melanogaster* chromatin, showing its "string of pearls" appearance. (Courtesy of SL McKnight and OL Miller)

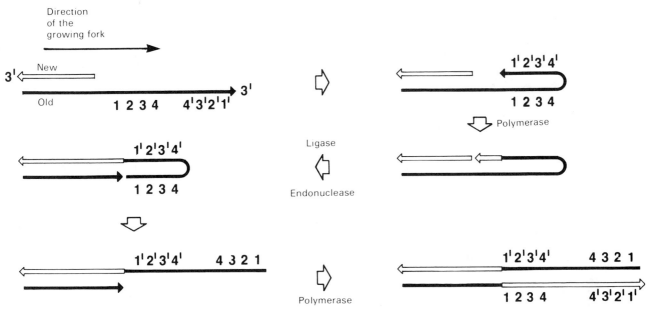

FIG. 49-14. Model for complete replication of the linear DNA in chromosomes. The chain that grows toward the 5' end **(new)** is synthesized backwards with respect to the direction of the growing fork and in general cannot be completed. Model assumes that a palindrome (1,2,3,4 . . . 4',3',2',1', where a primed number is a nucleotide complementary to the unprimed number) allows a hairpin folding of the unreplicated **old** strand, so that its terminal segment can be transferred to the **new** chain. Then the **old** chain can be completed in the regular 5' → 3' direction. (Modified from Cavalier-Smith T: Nature 250:467, 1974)

ments into thin-walled tubes, about 400 nm in diameter, contributes a factor of forty; and the final coiling of the tube into mitotic chromosomes adds a factor of five. Packing requires the cooperation of proteins, especially a group of **scaffolding proteins,** which retain the shape of metaphase chromosomes after removal of both DNA and histones. Packing is probably facilitated by the phosphorylation of histone H1, which occurs during the mitotic period.

TRANSCRIPTION OF CHROMATIN

Transcriptionally active and inactive chromatins have distinguishing cytologic properties, as evidenced, for instance, in female cells by the different behavior of the two X chromosomes, of which only one is active. In interphase nuclei one of the chromosomes, presumably the inactive one, is condensed and stainable (the so-called **Barr body**), whereas the other is dispersed. This correlation suggests that inactive chromatin corresponds to the cytologists' **heterochromatin** fraction of certain chromosomes, which remains condensed in interphase. Moreover, in labeling experiments heterochromatin shows *no RNA synthesis and late replication of its DNA,* compared with the active **euchromatin.**

The inactive X chromosome is designated **facultative** heterochromatin, since neither X chromosome is intrinsically heterochromatic, to distinguish it from **constitutive** heterochromatin, which contains highly repetitive sequences. Constitutive heterochromatin is found at various places but especially near the

centromeres, where it can be recognized cytologically by hybridizing the metaphase chromosome with radioactive RNA synthesized in vitro from highly repetitive DNA, using *E. coli* transcriptase. These repetitive sequences, by pairing with each other, can form a compact knob for anchoring the spindle fibers; they may also participate in chromosome pairing. Constitutive heterochromatin is also present at the nucleolus organizer, where it insulates the various rRNA genes from each other and from the rest of the genome. Heterochromatin interspersed in the chromosome may function in the regulation of gene expression (discussed under Genetic Mapping, below).

During transcription the basic nucleosome structure is retained but becomes more open, and therefore more susceptible to digestion by endonucleases. This effect may be due to loss of histone H1 and increased acetylation of H4. These changes are independent of the actual rate of transcription. Thus in erythroid cells transformed by the Friend leukosis virus the globin gene is in the open state even in the absence of the inducers required for starting transcription. The openness seems to depend on the state of differentiation of the cells, which determines the potential for transcription.

The Transcribing Enzymes. Transcription of chromatin is carried out by three large oligomeric DNA-dependent RNA polymerases, designated I, II, and III according to their order of elution in gel filtration. They have different specificities: I transcribes genes for rRNAs; II transcribes

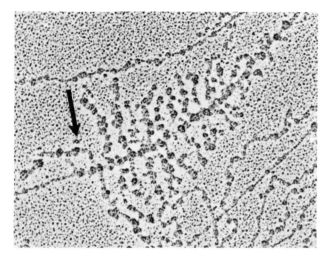

FIG. 49-15. Electron micrograph of transcribed *D. melanogaster* chromatin showing a gradient of RNA molecules at various degrees of completion. **Arrow** indicates the end of the transcription unit. (McKnight SL, Miller OL: Cell 8:305, 1976)

unique genes, and III transcribes genes for tRNAs and 5S RNA. Polymerase II is highly sensitive to the alkaloid α-amanitin. Initiation of transcription by this enzyme is also inhibited by rifamycin derivatives and some nucleoside analogs.

The number of transcription initiation points recognized by polymerase II in chromatin containing unique genes is measured by labeling in vitro the first nucleotide of each chain, which is a purine, using labeled purine nucleoside triphosphate as sole precursor. After further initiations are blocked with a rifamycin derivative, chain elongation is started by adding a complete, unlabeled nucleoside triphosphate mixture. The amount of label in RNA at the end of synthesis corresponds to about 15,000 initiation points per diploid nucleus. The transcripts initiated at one such point identify a **transcription unit** in DNA, which is recognizable in electron micrographs as a tree with branches of increasing lengths, corresponding to the growing RNA chains (Fig. 49-15).

The Primary Transcription Products. These products, revealed by brief exposure of a cell to a radioactive RNA precursor, include the **heterogeneous nuclear RNA (hnRNA),** 6–30 Kb long, and the **rRNA precursor** (45S). Hybridization to suitable labeled nucleic acids shows that the hnRNA contains both unique and repetitive sequences; it reproduces 10%–30% of the unique DNA sequences, depending on the organ examined. The hnRNA turns over rapidly, because when the radioactive "pulse" is followed by a "chase" with an unlabeled precursor 80%–90% of the label becomes acid-soluble with a half-life of ½ h. The remainder appears as mRNA in the cytoplasm. In contrast, most of the rRNA precursor is conserved and is precisely cleaved by RNase P (**processing;** Fig. 49-16). During transcription, the rRNA becomes associated with several proteins to form strings of 20–30 **nucleoprotein particles** reminiscent of nucleosomes.

Cytoplasmic mRNA. Depending on the growth state of the cells (see Growth Restriction in G1, above), mRNA is either in **polysomes** or free. The mRNA is isolated from polysomes by exposing cells to puromycin before extraction or by treating the extract with EDTA to chelate the divalent ions. Its length is broadly distributed around 2 Kb. Hybridization to cellular DNA shows that mRNA corresponds **almost exclusively to unique DNA sequences;** of these about 1% (corresponding to 10,000 to 20,000 genes) are represented in the mRNAs of a given

FIG. 49-16. Scheme of processing of the rRNA precursor in animal cells. **Heavy lines** indicate the sequences corresponding to mature ribosomal RNAs. **Thin lines** are spacers. **Parenthetic numbers** are molecular weights. (Modified from Weinberg RA, Penman S: J Mol Biol 47:169, 1970)

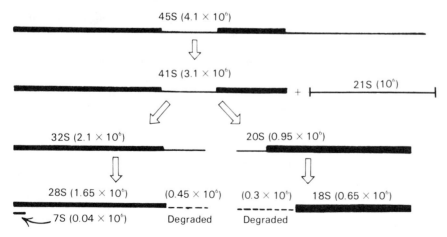

45S (4.1 × 10⁶)

⇓

41S (3.1 × 10⁶)   +   21S (10⁶)

32S (2.1 × 10⁶)   20S (0.95 × 10⁶)

28S (1.65 × 10⁶)   (0.45 × 10⁶)   (0.3 × 10⁶)   18S (0.65 × 10⁶)

7S (0.04 × 10⁶)   Degraded   Degraded

animal cell. The number of mRNA copies corresponding to a given gene varies from a few to 10^4 per cell, depending on the gene and its functional state. Eukaryotic mRNAs are rather stable, having a half-life up to several days.

Several mRNAs, for protein made abundantly in some cells (e.g., globin in reticulocytes), or produced by some viruses, have been purified. Their identity is established from the proteins synthesized in vitro after the RNA is added to a suitable extract of eukaryotic cells (e.g., from wheat germ) or is injected into frog oocytes, in which it is faithfully translated.

Sequence studies show that purified mRNAs contain five parts: a central coding segment, flanked by two untranslated segments, and terminated by two specialized structures (Fig. 49-17). Thus a **polyadenylate (polyA)** segment is present at the 3′ end of most mRNAs; how-

FIG. 49-17. Structure of mRNA. **A.** The central **coding segment** is flanked by two untranslated segments **(US)**, which are connected to **poly (A)** at the 3′ end and to a **cap** at the 5′ end. The cap is formed by one or two nucleotides **(Nm)** whose ribose is methylated at the 2′ position **(boxed)**, connected by a triphosphate bridge to 7-methylguanosine (m⁷G); each is connected through the 5′ position of its ribose **(circled)**. Therefore the 5′ end of the whole molecule is actually terminated by a 3′-OH **(arrow)**, just like its 3′ end. This makes the mRNA unattackable by 5′-exonucleases. **B.** Synthesis of the cap. (A modified from Adams JM, Cory S: Nature 255:28, 1975)

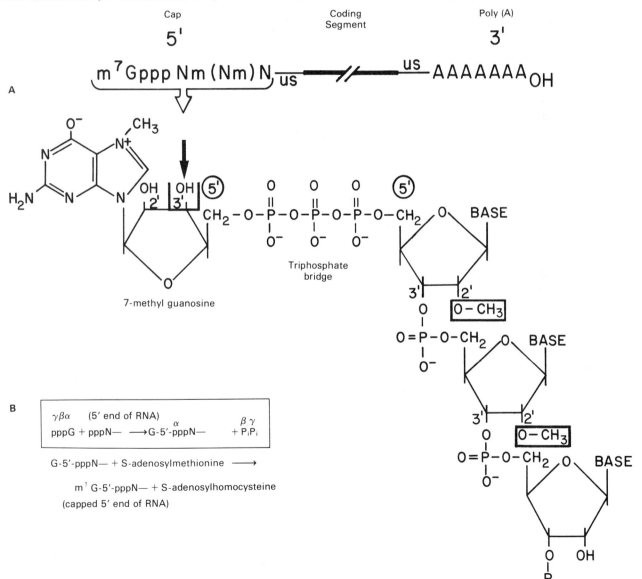

ever, it is absent in histone mRNAs. PolyA segments are 40–60 bases long in steady-state mRNA populations and up to 200 in newly produced mRNA. Through hybridization of the polyA to polyuridylate chains covalently bound to Sepharose, mRNAs can be separated from other RNAs. At the 5′ end mRNAs have **caps.** These short structures have the extraordinary feature of a terminal 7-methylguanosine connected backwards (i.e., in a 5′-triphosphate link) to the rest of the chain. Hence *mRNAs are terminated at both ends by a 3′-OH.*

The presence of a **cap** on mRNA increases the rate of translation in vitro, while analogs (7-methylguanosine mono- or diphosphate) inhibit translation. Presumably caps have this function in vivo. However, some viral RNAs that lack caps are translated very efficiently in infected cells (see Ch. 50). Poly(A) is not essential for translation in vitro. It evidently functions in the transport of mRNA from the nucleus (possibly after association with poly(A)-specific proteins), and it may slow down intracellular degradation.

Relation of hnRNA to mRNA (Fig. 49-18). We have already noted that hnRNA contains unique sequences; it has about ten times as many as mRNA, and also has

FIG. 49-18. Transcription of eukaryotic genes and processing of the primary transcript. **I:** A gene has alternating coding **(heavy lines)** and intervening **(thin lines)** sequences. The primary transcript, which reproduces both types of sequences **(II)** is capped at the 5′ end and polyadenylated at the 3′ end while in hnRNA **(III).** Removal of the intervening sequences by splicing generates the mature mRNA **(IV).** The differences between the mRNA and its precursor in hnRNA are revealed by examining the heteroduplexes they form with the DNA of the gene **(V).**

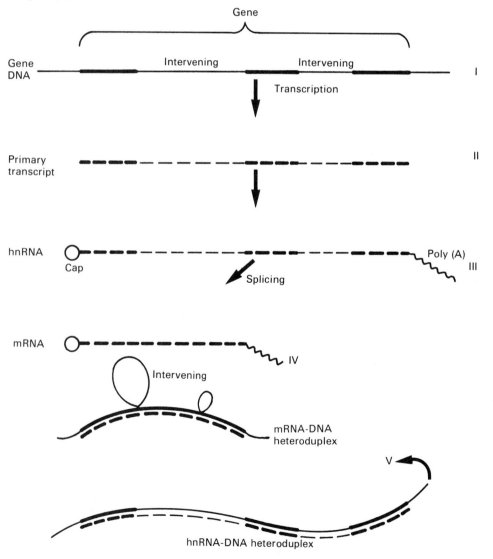

special features (such as palindromes) that are absent in mRNA. Hence, most of the sequences present in hnRNA are destroyed during its **maturation to mRNA.** Most of the lost unique sequences belong to intervening sequences (introns; see Organization and Expression of Genetic Information, above). Their removal occurs by a **splicing** together of separate segments of the hnRNA precursors of mRNAs. Splicing is carried out by special enzymes that apparently recognize short sequences present at the exon–intron junctions.

This method of mRNA maturation, which is costly in terms of metabolic energy, complements the genetic variability that occurs at the DNA level by removing blocks of sequences at the mRNA level; therefore it considerably increases the amount of variability that can be utilized for evolutionary purposes. In viruses it increases the informational content of the genome by allowing the same DNA segment to be used for expressing several functions (see Viral Multiplication in Tissue Culture, Ch. 66).

The mature mRNAs are transported to the cytoplasm in association with proteins as **ribonucleoprotein particles.** The mRNAs of some viruses are synthesized and mature in the cytoplasm.

Regulation of Transcription. The key regulatory role of transcription, discovered in prokaryotes, also extends to development and growth control in eukaryotes. With the increased synthesis of specific products, during development or in response to hormone action, new mRNAs appear in the cells and others become more abundant. For instance, steroid hormones freely enter cells and become associated with specific **cytoplasmic receptor** proteins (not to be confused with the surface receptors for polypeptide hormones), which then move to the nucleus, interacting with chromatin and changing its transcription. When the stimulus is withdrawn the new mRNAs rapidly disappear.

Transcription regulation is caused by the **acidic proteins** of chromatin, whereas histones have a predominantly structural role (except, perhaps, H1). Thus chromatin formed in vitro by adding histones to DNA is not transcribed by RNA polymerase; however, transcription is restored by the addition of acidic proteins extracted from chromatin. The action of the proteins is gene-specific; for example, acidic proteins from S-phase cells stimulate the synthesis of histone mRNAs by chromatin from G1-phase cells.

Further aspects of regulation emerge from experiments on cell hybridization, reviewed below.

PROTEIN SYNTHESIS AND BREAKDOWN

Monocistronic Messengers. One approach for determining whether messengers each transcribe a single gene or several genes is to grow cells in the presence of a radio-

active amino acid; polysomes are fractionated and the mRNA length is compared with the amount of label in the attached peptide (Fig. 49-19). The results reveal **monocistronic** (i.e., **single-gene**) messengers. Moreover, all animal cell mRNAs with a known function that have been studied so far (e.g., those for histones, hemoglobin, γ-globulin, myosin, lens protein) have a monocistronic length. However, extracts of animal cells regularly translate phage mRNA in vitro, although it is polycistronic. As discussed in Chapter 50, the monocistronic property of animal cells determines important features of the organization of genetic material in animal viruses.

Regulation of Translation. Regulation of gene expression in animal cells occurs at the level of translation as well as transcription. Regulation acts at the **initiation of protein synthesis,** which involves the formation of a complex between an initiation factor (eIF-2), methionyl-tRNA$_f$, GTP, and the 40S ribosomal subunit.

For example, in reticulocytes the main protein product is globin, which combines stoichiometrically with heme to form hemoglobin. In lysates heme is required for initiation, preventing excessive synthesis of globin. Heme acts by preventing the phosphorylation and consequent inactivation of eIF-2 by a specific protein kinase. Heme also enhances protein synthesis in extracts of some nonerythroid cells. Other cells contain similar systems of eIF-2 inactivation. The inactivation promoted by double-stranded RNA may be significant in the mode of action of interferon (see Mechanims of Interferon Action, Ch. 51).

The *binding of mRNA to ribosomes to form polysomes* is also regulated. Cells contain free (i.e., nonfunctional) mRNA, whose proportion varies, depending on the cells' growth stage, between 10% and 80%. The proportion increases when fibroblastic cultures become resting, and decreases when they resume growth. In liver cells about 50% of the mRNA for ferritin, the major intracellular iron storage protein, is normally free, and upon administration of iron this mRNA becomes polysome-associated, with a consequent increase of iron-associated protein. Differential regulation of the attachment of different messengers may be due to different sequences of the ribosome binding sites adjacent to the cap.

The fate of proteins synthesized in animal cells is related to the location of the polysomes, which are free for cytoplasmic proteins and bound to the rough endoplasmic reticulum for membrane proteins or proteins to be exported (e.g., immunoglobulins).

Protein Breakdown. The amino acid precursors for protein synthesis may be derived from the medium as well as from breakdown of cellular proteins. The average rate of protein breakdown is about 1% per hour but varies for individual proteins. Proteins subject to regulation (e.g., in the cell cycle, or in response to hormones) often have a very short half-life (1–2 h), so that after their synthesis stops their function is rapidly turned off. Examples are ribonucleotide reductase, which is induced at the beginning of the S phase and rapidly decays at its end; RNA polymerase of regenerating liver or of the estrogen-stimulated uterus; and certain catabolic enzymes, such as

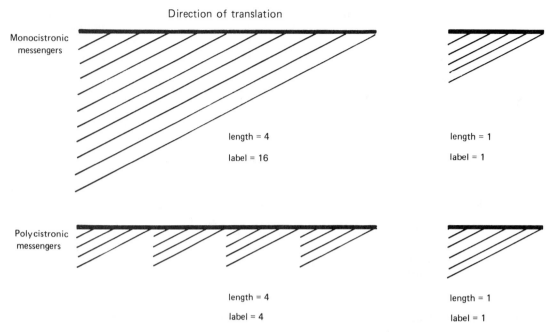

FIG. 49-19. Method for distinguishing mono- from polycistronic translation by comparing polysome label in nascent polypeptide chains with messenger length. The nascent chains, of different lengths, are indicated by parallel **thin lines.** For each messenger **(heavy lines)** the total amount of label is given by the area of the triangle covered by the polypeptide chains. It is clear that for polycistronic messengers the area is proportional to L (where L is the messenger length), whereas for monocistronic messengers it is approximately proportional to L^2 (beyond a certain length). The ratio of radioactive label to polysome length is expected to increase in proportion to L for monocistronic messengers, but to reach a constant value for polycistronic messengers. (Data from Kuff EL, Roberts NE: J Mol Biol 26:211, 1967)

tyrosine-α-ketoglutarate transaminase of liver cells. Protein breakdown is carried out largely by enzymes present in lysosomes.

As in bacteria (see Ch. 14), *breakdown is increased for abnormal proteins,* e.g., those that have incorporated amino acid analogs or are found in senescent cells.

GENETIC STUDIES WITH CULTURED CELLS

THE KARYOTYPE OF CULTURED CELLS

The analysis of the chromosomal constitution **(karyotype)** of tissue culture cells has acquired paramount importance in genetic studies since it became clear that karyotype anomalies of cultured cells are associated with certain human diseases. In addition, the karyotype gives an indication of the degree of abnormality that cells have attained during their cultivation in vitro.

Staining technics now allow not only the determination of the number of chromosomes in a cell, but also their precise cytologic identification. Thus characteristic bands are observed in mitotic chromosomes stained with the fluorescent dye quinacrine mustard and examined under ultraviolet light (Fig. 49-20), or stained after trypsin treatment.

In young cell strains most cells tend to maintain the **diploid** ($2n$) chromosome number characteristic of the animal. The types of chromosomes are also usually normal, and the cells are said to be **euploid.** In contrast, the cells of older strains and cell lines (especially of transformed cells) always contain deviations from the euploid chromosome number and distribution (**aneuploid** cells).

The number of chromosomes may be different from diploid (**heteroploid**), either higher (usually between $3n$ and $4n$, i.e., **hypertriploid**) or lower (**hypodiploid**). In *quasi-diploid* cells the number of chromosomes is $2n$ but their distribution is abnormal; for example, a chromosome of one pair may be missing and replaced by an extra chromosome of another pair. In addition, **chromosomal aberrations** (e.g., **translocations** and **deletions**) often involve highly characteristic morphologic abnormalities in individual chromosomes, which are useful as **markers** for cell identification.

Although in some cell lines most cells are diploid the majority of cell lines are constituted of heteroploid, especially hypertriploid, cells. *Individual lines are heterogeneous,* with cells containing different numbers of chromosomes. The most frequent (**modal**) number remains constant if the cells are grown under a constant set of conditions, but a change in the conditions often results in selection of a type with a different modal number. The

FIG. 49-20. Fluorescence photograph of the metaphase chromosomes of a human cell stained by quinacrine dihydrochloride. (Courtesy of Dr. WR Breg)

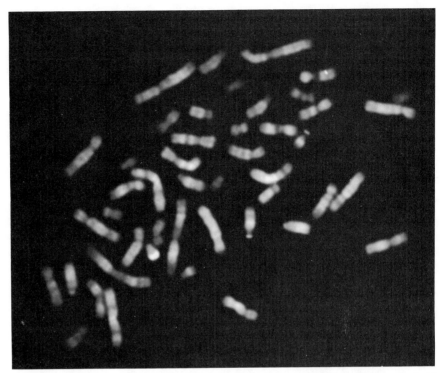

variation encountered in heteroploid cultures reflects frequent unequal segregation of chromosomes at mitosis.

Functional Chromosome Markers. Chromosomes carrying genes specifying recognizable proteins with different electrophoretic mobilities can be readily identified. In hybrid cells resulting from cell fusion, chromosomes with genes for oligomeric enzymes produce distinct hybrid oligomers (so-called **isozymes,** Fig. 49-21).

Somatic Mutations. Most mutations occurring in tissue culture cells consist of **structural changes in a protein** similar to those observed in prokaryotes. The mutations are revealed indirectly by physical changes of the protein

FIG. 49-21. Isozyme patterns of glucose phosphate isomerase (a dimeric enzyme) in homogenates of cells derived from two inbred mouse strains, BALB/c and C3H, and from hybrid cells obtained by fusing cells of the two strains. The enzyme of each strain contains a single type of monomer of different electrophoretic mobility; the hybrid cells have a third band, produced by enzyme molecules containing a monomer from each parent. The isozyme patterns were recognized by fractionating cell extracts by gel electrophoresis and then treating the gel with a substrate that releases a dye when acted upon by the enzyme. (Klebe RJ et al: J Cell Biol 45:74, 1970)

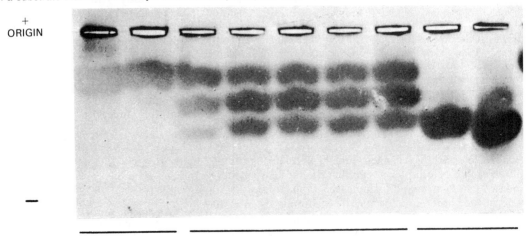

+
ORIGIN

–

BALB/C   HYBRIDS   C3H

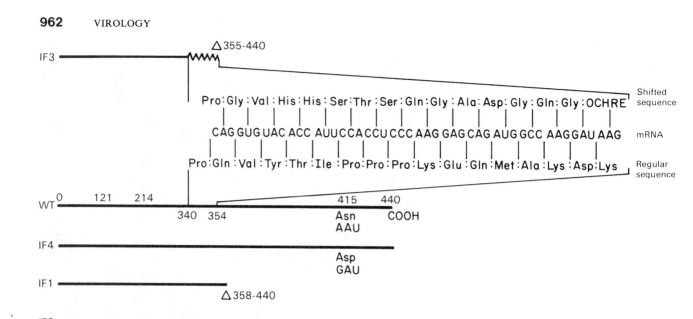

FIG. 49-22. Changes in protein sequences caused by various mutations in an immunoglobulin gene. IF3 is a frame shift followed by a deletion **(arrowhead)**; IF4 is a base substitution (transition); IF1 is a terminator mutation; IF2 is an internal deletion; WT is the wild-type protein. Numbers refer to positions in the amino acid sequence. (Modified from Adetugbo K et al: Nature 265:299, 1977)

(e.g., isoelectric point, electrophoretic mobility, heat stability), or directly by amino acid changes (Fig. 49-22). In aneuploid cell cultures a mutational phenotype can also be brought about by a **change in chromosome balance,** due to the loss or the gain of certain chromosomes by mitotic segregation, without structural protein changes. Variants of this type occur and revert at a relatively high frequency which is not enhanced by mutagenic agents.

It appears that in some cells of permanent lines diploid genes tend to become **functionally haploid;** and selective technics isolate mutations in these cells. Thus in quasi-diploid cell lines, such as one derived from the Chinese hamster ovary, many recessive mutations in diploid (autosomal) genes are phenotypically expressed almost as frequently as are mutations in haploid (X-linked) genes, though they would be expected to require two homologous mutations and hence to be much less frequent. Moreover, in α-amanitin–resistant mutant cells (see Transcribing Enzymes, above) essentially all the enzyme molecules are resistant to the drug, although it is unlikely that both the genes specifying the enzyme were mutated. Functional gene inactivation probably results from translocations that replace a promoter with heterochromatin, or that separate it from the rest of the transcription unit.

Sister Chromatid Exchanges (Fig. 49-23). During the S phase each DNA duplex (a **chromatid**) doubles, producing two sister chromatids, which at metaphase lie side by side in the same chromosome. The chromatids can be differentially labeled by incorporating 5-bromodeoxyuridine (BUdR) instead of thymidine during two successive S phases. After the first S phase the sister chromatids of a pair of daughter chromosomes are TB–BT (where T is a thymidine-containing strand and B a BUdR-substituted strand), and are TB–BB after a second S phase. The absence of BUdR in one of the strands makes a TB chromatid (or region) distinguishable cytologically from BB. Induction of chromatid exchange is a sensitive test for the mutagenic activity of chemicals.

GENETIC MAPPING

Mutations. Genetic mapping of mutations in cultured cells not only is much simpler than the classical breeding approach, but it also offers the opportunity to study the genetic organization in man; it is also important for locating integrated viral genomes (see Categories of Type C Oncoviruses, Ch. 66). Some **germinal mutations** present in the source of the cultured cells are recognizable in cultures; some **somatic mutations** induced in vitro by suitable mutagenic agents (e.g., nitrosoguanidine or alkylating agents) are also suitable (e.g., those altering

FIG. 49-23. Part of the chromosome complement of a mitotic human cell showing many sister chromatid exchanges. Darkly stained segments belong to thymidine-containing chromatids. The number of exchanges is high because the cell was treated with a mutagen, which increases exchanges, and was from a patient with xeroderma pigmentosum, in which repair of mutagen-induced DNA damages is impaired. (Wolff S et al: Nature 265:347, 1977)

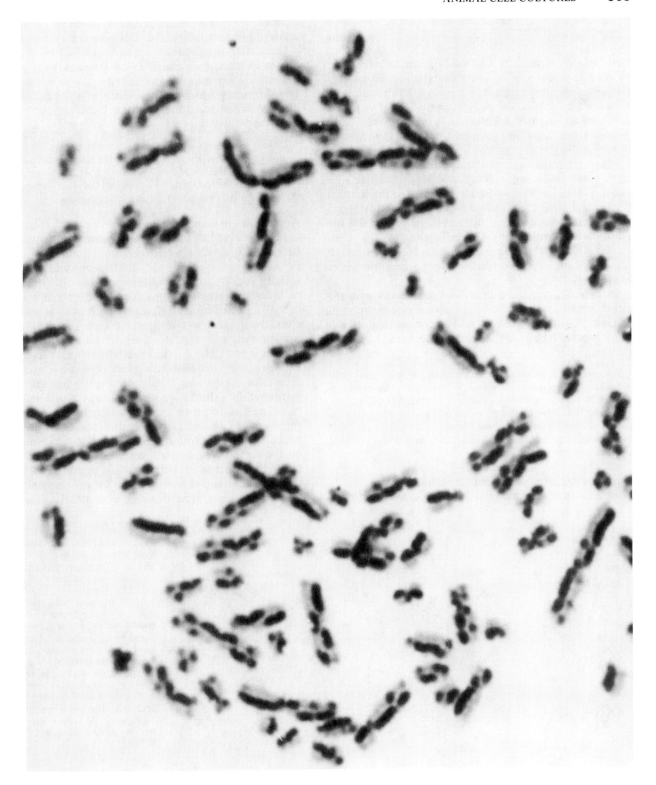

either isozymes, resistance to an inhibitor, or growth requirements).

Cell Hybridization. The location of genetic markers can be studied in cultures by using **somatic cell fusion.** The frequency of spontaneous fusion is low, but it is strongly enhanced by polyethylene glycol or by Sendai virus that has been inactivated by chemicals or ultraviolet light. Sendai virus (a paramyxovirus, Ch. 59) acts through a hemolysin in the viral envelope. Membrane fusion seems to result from the aggregation of intramembranous particles within each cell, making possible cell contacts between protein-free areas. Rearrangement of the lipids in the contact areas then leads to fusion (Fig. 49-24). Harris and Watkins showed that fusion results first in the formation of **heterokaryons,** i.e., cells with two (or more) nuclei; if the nuclei then enter mitosis at approximately the same time they also fuse, yielding a mononucleated **hybrid** cell.

In heterokaryons and in hybrid cells many genes of either parent continue to be expressed (see Regulation of Gene Expression, below), and so **complementation** of these genes can be used for isolating these cells from their parents.

For example, to grow in medium containing aminopterin (which blocks the endogenous purine and pyrimidine pathways) supplemented by thymidine and hypoxanthine (so-called HAT medium) cells must have both thymidine kinase and hypoxanthine:guanine phosphoribosyl transferase. With two parental strains that each lack one of these enzymes, HAT medium selects sharply for their hybrids.

Chromosome-Mediated Gene Transfer. Cells can incorporate small fragments of isolated metaphase chromosomes and express the genes they contain. This transfer is achieved by incubating the cells with the chromosomes, which, after being taken up, are attacked by lytic enzymes in lysosomes. Small DNA fragments can escape and survive, generating, at a frequency of about 10^{-8}, **transformants** which express some of the genes of the chromosomes (recognizable, for instance, on the basis of characteristic isozymes).

Usually a single gene is expressed, but two tightly linked genes can be expressed together. Some unstable transformants lose the transgenome at the rate of 1%–10% per cell generation. In the stable transformants the transgenome is integrated in the chromosomes of the recipient cell, but not at the homologous locus.

Approaches to Gene Mapping. One approach is based on the progressive loss of the chromosomes of one parent in heterospecific hybrids **(chromosome elimination).** Thus in human–mouse hybrids human chromosomes are usually lost; it is then possible to correlate the persistence of a certain human chromosome (recognized cytologically by the banding technique) with that of a certain character (such as an antigen or an enzyme). This approach assigns a gene to a chromosome. Another approach is to establish a linkage between characters that are lost simultaneously during chromosome elimination, thus establishing the chromosome as a linkage group **(synteny).** In a third approach a heavily x-irradiated parental cell is fused to a nonirradiated one; since only fragments of the irradiated chromosomes survive, the probability of the simultaneous transfer of two markers on the same chromosome decreases as the distance between them increases. This approach establishes the relative distances and therefore the order of the markers on the chromosome. The distance can also be estimated

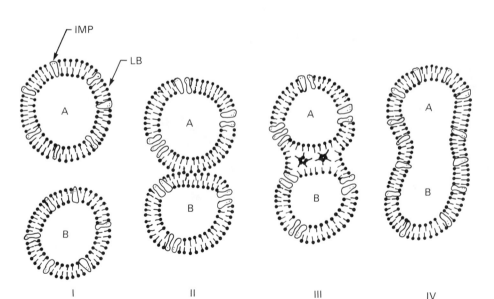

FIG. 49-24. Fusion of two cells to form a heterokaryon. **I.** Cells **A** and **B** are represented schematically as two membrane-bound vesicles. Membranes are formed by lipid bilayers **(LB)** containing integral membrane proteins **(IMP). II.** Fusion process is initiated by aggregation of IMPs, allowing the lipid bilayers to come into close contact. **III.** At the contact point the two lipid layers fuse, generating free lipid micelles. **IV.** A further rearrangement of the lipid bilayers produces the heterokaryon. (Modified from Cullis PR, Hope MJ: Nature 271:672, 1978)

from the rate of simultaneous recovery in chromosome transfer experiments.

Regulation of Gene Expression. With differentiated cells which express different genes, the study of hybrids reveals some basic properties of regulation of gene expression in animal cells. First, 80%–90% of the genes expressed in a given cell type are also expressed in all other cell types; these are **basic function genes.** The others are **differentiation genes,** specific for certain cell types. Moreover, if a gene is expressed in both parental cells it is also expressed in the hybrid, but if a differentiation gene is expressed only in one parent it is usually not expressed in the hybrid **(extinction).** Finally, an extinguished gene is sometimes **reexpressed** after the hybrid has lost chromosomes of the extinguishing parent. Extinction (e.g., of the globin gene) is accompanied by **absence of the corresponding mRNA** and therefore occurs at the level of transcription or of processing of the primary transcript.

These observations show that, like many bacterial genes, differentiation genes can be negatively regulated by products of other genes, i.e., repressed, at the level of transcription: they are expressed in a cell in which the repressor either is not made or is neutralized. There are also cases of **gene activation** in hybrids, suggesting the existence of endogenous inducers of **positive gene regulation.** In contrast, basic function genes are **expressed constitutively.**

Among known cases of extinction, the gene for tyrosine aminotransferase is expressed in hepatoma cells after **induction** by corticosteroids. The hormone probably counteracts a negative regulator, perhaps in the same way that inducers counteract repressors in bacterial systems. However, in hybrids of hepatoma cells and fibroblasts the gene is no longer inducible.

Control of Transformation and Malignancy (see Cell Transformation, above). As with gene regulation, the study of hybrids derived from the fusion of a malignant and a nonmalignant cell can determine the nature of the control of gene expression involved in malignancy. Two patterns are seen. With malignant cells derived from viral (SV40) transformation the hybrids are malignant, showing **dominance of the viral genome.** With nonviral malignancy, in contrast, the hybrids are not (or are less) malignant as long as they retain the complete chromosome complements of both parents. Loss of the chromosomes from the nonmalignant parent allows the reappearance of malignancy in the hybrids. This result suggests that nonviral malignancy results from the loss of a negative control.

SELECTED READING

BOOKS AND REVIEW ARTICLES

BASERGA R, STEIN G: Nuclear acidic proteins and cell proliferation. Fed Proc 30:1752, 1971

BASILICO G: Temperature-sensitive mutations in animal cells. Adv Cancer Res 24:223, 1977

DAVIDSON RL: Gene expression in somatic cell hybrids. Annu Rev Genet 8:195, 1974

DUTRILLAUX B, LEJEUNE J: New techniques in the study of human chromosomes: methods and applications. Adv Hum Genet 5:119, 1975

EDELMAN GM: Surface modulation in cell recognition and cell growth. Science 192:218, 1976

EDENBERG HJ, HUBERMAN JA: Eukaryotic chromosome replication. Annu Rev Genet 9:245, 1975

FELSENFELD G: Chromatin. Nature 271:115, 1978

GOSPODAROWICZ D, MORAN JS: Growth factors in mammalian cell cultures. Annu Rev Biochem 45:531, 1976

HARRIS H: Cell fusion and the analysis of malignancy: the Croonian Lecture. Proc R Soc Lond (Biol) 179:1, 1971

HOLLEY RW: Control of growth of mammalian cells in cell culture. Nature 258:487, 1975

HYNES RO: Role of surface alterations in cell transformation: the importance of proteases and surface proteins. Cell 1:147, 1974

MINTZ B: Gene control of mammalian differentiation. Annu Rev Genet 8:411, 1974

NICOLSON GL: Transmembrane control of the receptors on normal and tumor cells. Biochim Biophys Acta 457:57, 1976

RAFFERTY KA JR: Epithelial cells: growth in culture of normal and neoplastic forms. Adv. Cancer Res 21:249, 1975

ROBERTS K: Cytoplasmic microtubules and their functions. Prog Biophys Mol Biol 28:421, 1974

ROSS R, VOGEL A: Platelet-derived growth factor. Cell 14:203, 1978

SEGAL HL: Mechanisms of protein turnover in animal cells. Curr Top Cell Regul 11:183, 1976

SIMINOVITCH L: On the nature of hereditable variation in cultured somatic cells. Cell 7:1, 1976

SINGER SJ: Fluid mosaic model of membrane structure: some applications to ligand–receptor and cell–cell interactions. In Bradshaw RA, Frazier WA, Merrell RC et al (eds): Surface Membrane Receptors. p. 1, New York, Plenum, 1976

WESSELLS NK, SPOONER BS, ASH JF et al: Microfilaments in cellular and developmental processes. Science 171:135, 1971

YAMADA KM, OLDEN K: Fibronectins-adhesive glycoprotiens of cell surface and blood. Nature 275:179, 1978

SPECIFIC ARTICLES

ABERCROMBIE M: Contact inhibition in tissue cluture. In Vitro 6:128, 1970

ASH JF, SINGER SJ: Concanavalin A–induced transmembrane linkage of concanavalin A surface receptors to intracellular myosin-containing filaments. Proc Natl Acad Sci USA 73:4575, 1976

BAK AL, ZEUTHEN J, CRICK FHC: Higher-order structure of human mitotic chromosomes. Proc Natl Acad Sci USA 74:1595, 1977

BERGET SM, MOORE C, SHARP PA: Spliced segments at the 5′ terminus of adenovirus 2 late mRNA. Proc Natl Acad Sci USA 74:3171, 1977

BRACK C, TONEGAWA S: Variable and constant parts of the immunoglobulin light chain gene of a mouse myeloma cell are 1250 nontranslated bases apart. Proc Natl Acad Sci USA 74:5652, 1977

CATTERRALL JF, O'MALLEY BW, ROBERTSON MA et al: Nucleotide sequence homology at 12 intron–exon junctions in the chick ovalbumin gene. Nature 275:510, 1978

CORY S, ADAMS ME: Modified 5′-terminal sequences in messenger RNA of mouse myeloma cells. J Mol Biol 99:519, 1975

CUNNINGHAM DD, PARDEE AB: Transport changes rapidly initiated by serum addition to "contact inhibited" 3T3 cells. Proc Natl Acad Sci USA 64:1049, 1969

CURTIS PJ, WEISSMAN C: Purification of globin messenger RNA from dimethylsulfoxide-induced Friend cells and detection of a putative globin messenger RNA precursor. J Mol Biol 106:1061, 1976

DARNELL JE, PHILIPSON L, WALL R, ADESNIK M: Polyadenylic acid sequences. Role in conversion of nuclear RNA into messenger RNA. Science 174:507, 1971

DATTA A, DEHARO C, SIERRA JM, OCHOA S: Mechanism of translational control by hemin in reticulocyte lysate. Proc Natl Acad Sci USA 74:3326, 1977

DEISSEROTH A, BARKER J, ANDERSON WF, NIENHUS A: Hemoglobin synthesis in somatic cell hybrids: coexpression of mouse with human or Chinese hamster globin genes in interspecific somatic cell hybrids of mouse erythroleukemia cells. Proc Natl Acad Sci USA 72:2682, 1975

DERMAN E, GOLDBERG S, DARNELL JE: hnRNA in HeLa cells: distribution of transcript sizes estimated from nascent molecule profile. Cell 9:465, 1976

EFSTRATIADIS A, KAFATOS FC, MANIATIS T: Primary structure of rabbit β-globin mRNA as determined from cloned DNA. Cell 10:571, 1977

FLANAGAN J, KOCH GLE: Cross-linked surface Ig attaches to actin. Nature 273:278, 1978

FOE VE, WILKINSON LE, LAIRD CD: Comparative organization of active transcription units in *Oncopeltus fasciatus*. Cell 9:131, 1976

GAHMBERG CG, HAKOMORI S: Surface carbohydrates of hamster fibroblasts. J Biol Chem 250:2438, 1975

GEIGER B: A 130k protein from chicken gizzard: its localization at the termini of microfilament bundles in cultured chicken cells. Cell 18:193, 1979

GAREL A, ZOLAN M, AXEL R: Genes transcribed at diverse rates have a similar conformation in chromatin. Proc Natl Acad Sci USA 74:4867, 1977

GERMOND JE, HIRT B, OUDET P et al: Folding of the DNA double helix in chromatin-like structures from simian virus 40. Proc Natl Acad Sci USA 72:1843, 1975

GOSS SJ, HARRIS H: New method for mapping genes in human chromosomes. Nature 255:680, 1975

HYNES RO: Alteration of cell-surface proteins by viral transformation and by proteolysis. Proc Natl Acad Sci USA 70:3170, 1973

JI TH, NICOLSON GL: Lectin binding and perturbation of the outer surface of the cell membrane induces a transmembrane organizational alteration at the inner surface. Proc Natl Acad Sci USA 71:2212, 1974

KORNBERG RD: Chromatin structure: a repeating unit of histones and DNA. Science 184:868, 1974

LISKEY RM, PRESCOTT DM: Genetic analysis of the G1 period: isolation of mutants (or variants) with a G1 period from a Chinese hamster cell line lacking G1. Proc Natl Acad Sci USA 75:2873, 1978

MILLER JR, BROWNLEE GG: Is there a correction mechanism in the 5S multigene system? Nature 275:556, 1978

RHEINWALD JG, GREEN H: Epidermal growth factor and the multiplication of cultured human epidermal keratinocytes. Nature 265:421, 1977

SCHLESSINGER J, ELSON EL, WEBB WW et al: Receptor diffusion on cell surfaces modulated by locally bound concanavalin A. Proc Natl Acad Sci USA 74:1110, 1977

SCHRAMM M, ORLY J, EIMERL S, KORNER M: Coupling of hormone receptors to adenylate cyclase of different cells by cell fusion. Nature 268:310, 1977

SKLAR VEF, SCHWARTZ LB, ROEDER RG: Distinct molecular structures of nuclear class I, II, and III DNA dependent RNA polymerases. Proc Natl Acad Sci USA 72:348, 1975

STANBRIDGE EJ, WILKINSON J: Analysis of malignancy in human cells: malignant and transformed phenotypes are under separate genetic control. Proc Natl Acad Sci USA 75:1466, 1978

TILGHMAN SM, CURTIS PJ, TIEMEIER DC et al: Intervening sequence of a mouse β-globin gene is transcribed within the 15S β-globin mRNA precursor. Proc Natl Acad Sci USA 75:1309, 1978

WEINTRAUB H, GROUDINE M: Chromosomal subunits in active genes have an altered conformation. Science 193:848, 1976

MULTIPLICATION AND GENETICS OF ANIMAL VIRUSES

MULTIPLICATION

The multiplication of many animal viruses follows the pattern of bacteriophage multiplication described in preceding chapters, but the composition of the viruses and of their host cells also creates important differences. Some animal viruses, for instance, contain segmented RNA genomes; some DNA genomes have unusual termini (such as inverted, repeated nucleotide sequences and covalent crosslinking of linear strands); some contain a transcriptase and other enzymes in the virions; and some express their genetic information in a unique manner, by reverse flow from RNA to DNA. Moreover, animal viruses differ from bacteriophages in their interactions with the surface of the host cells (which do not have rigid walls) and in the mechanisms of release of their nucleic acid in the cell.

Like bacteriophages, animal viruses can be differentiated into virulent and moderate; the latter resemble temperate bacteriophages in their ability to establish stable relations with the host cells. The differentiation, however, is less sharp than with bacteriophages, and it is sometimes difficult to decide whether an animal virus is virulent or moderate.

The main characteristics of animal virus families are given in Table 50-1. Further details are found in later chapters of this book.

HOST CELLS FOR VIRAL MULTIPLICATION

The first hosts for experimental or diagnostic work with animal viruses were adult animals or embryos, but these have now been replaced almost completely by cultures of animal cells (Ch. 49) for detailed studies of viral replication.

CELL STRAINS AND CELL LINES

Every type of animal cell culture discussed in Chapter 49 has found application in virology. The choice of species, tissue of origin, and type of culture (primary, cell strain, or cell line) depends on the virus and the experimental objectives. The systems used for the individual viral families will be given in the appropriate chapters. The origin

and the characteristics of the most widely used cell lines are summarized in Table 50-2.

In the **production of viral vaccines** for human use, permanent cell lines are not employed, for they resemble malignant cells in their culture pattern and their aneuploidy, and so it is feared that viruses propagated in them may acquire genetic determinants, of malignancy. On the other hand, primary cultures or diploid cells strains, which are used often, contain a variety of latent viruses (Ch. 53); these may also constitute a health hazard.

Host Susceptibility. Each animal virus can replicate only in a certain range of cells. **Nonsusceptible cells** furnish a block at an early step, so that all expression of viral functions is prevented **(resistant cells);** others present a block at a later step, so that some viral activities are expressed **(nonpermissive cells).** In either case, a required cellular function is lacking (e.g., receptors for viral attachment or a factor required for expression of viral genes); a **heterokaryon,** formed by fusing a susceptible and a nonsusceptible cell, has the function and is usually susceptible. Infection with naked viral nucleic acid bypasses resistance that is due to lack of appropriate cell receptors.

Examples of resistance or nonpermissiveness will be frequently encountered in this chapter and in those dealing with specific viruses.

The Chick Embryo. Chick embryos have contributed in an important way to the development of virology by conveniently providing a variety of cell types susceptible to many viruses. Various cell types can be reached by inoculating the embryo by different routes (Fig. 50-1).

PRODUCTIVE INFECTION

INFECTION AS A FUNCTION OF THE NUCLEIC ACID

That cells of higher organisms can be infected by naked viral nucleic acid was first shown for tobacco mosaic virus RNA by Gierer and Schramm. This discovery was soon extended to show that the highly purified RNA or

TABLE 50-1. Main Characteristics of Animal Virus Families

Type	Nucleic acid strandedness	Symmetry of nucleo-capsid	Naked (N) or enveloped (E)	Diameter of virus (nm)	Family	Examples (specific viruses mentioned in this chapter)
RNA	Single-stranded	Icosahedral	N	21–30	Picornaviruses (Ch. 57)	Mengovirus, poliovirus
			E	45	Togaviruses (Ch. 62)	Semliki Forest virus, western equine encephalomyelitis virus
		Helical	E	80–120	Orthomyxoviruses (Ch. 58)	Influenza virus, fowl plaque virus
				125–300	Paramyxoviruses (Ch. 59)	Newcastle disease virus, Sendai virus, measles virus
				70–80 × 130–240	Rhabdoviruses (Ch. 61)	
				80–160	Coronaviruses (Ch. 60)	
		Unknown	E	10–130	Arenaviruses (Ch. 62)	
				100	Retroviruses (Ch. 66)	
	Double-stranded	Icosahedral	N	75–80	Reoviruses (Ch. 64)	Reovirus
DNA	Double-stranded	Icosahedral	N	70–90	Adenoviruses (Ch. 54)	Adenovirus
				43–53	Papovaviruses (Ch. 66)	Polyoma virus, SV40
			E	180–200	Herpesviruses (Ch. 55)	Herpes simplex virus, pseudorabies virus, equine abortion virus
		Complex	E*	200–250 × 250–350	Poxviruses (Ch. 56)	Vaccinia virus
	Single-stranded	Icosahedral	N	18–22	Parvoviruses (Ch. 54)	Adenoassociated virus

*Lipid in outer coat, but no distinct envelope

DNA of many animals viruses is infectious and gives rise to synthesis of normal virions. In conjunction with the Hershey and Chase experiment with bacteriophage (see Separation of Nucleic Acid from Coat, Ch. 47), this result established the exclusive genetic role of the viral nucleic acid.

There are several important differences between infections by nucleic acid and by virions.

1) *The efficiency of infection with nucleic acid is much lower,* by a factor of 10^{-6}–10^{-8} in ordinary media, showing the important role of the viral coat in infectivity. The infectivity of nucleic acid is increased by several orders of magnitude in hypertonic solutions (e.g., 1 M $MgCl_2$), in the presence of basic polymers (e.g., diethylaminoethyl-dextran), or by precipitation of the viral DNA onto cells with calcium phosphate. These additions appear to protect the nucleic acid against nucleases and to increase its uptake by the cells. Even under the most favorable conditions, however, the bare nucleic acid is no more than 1% as infectious as the corresponding virions. This limitation seems to arise from degradation of much of the nucleic acid within the cells.

2) *The host range is much wider with nucleic acids,* which can infect resistant cells. For instance, chicken cells, although resistant to poliovirus because they lack receptors for the virions, are susceptible to its RNA; but only a single cycle of viral multiplication takes place because the progeny are again virions.

3) *Infectious nucleic acid can be extracted even from heat-inactivated viruses,* in which the protein of the capsid has been denatured; the nucleic acid can withstand much higher temperatures than the protein.

4) Finally, *the infectivity of nucleic acid is unaffected by virus-specific Abs,* which suggest that this form of a virus could be an effective infectious agent even in the presence of immunity. However, nucleases in body fluids probably greatly limit its role, because a single complete break in a molecule abolishes its infectivity. Indeed, it is not clear whether naked viral nucleic acid plays any role in natural infection. Nevertheless, in the preparation of viral vaccines the ability of nucleic acid infectivity to survive damage to the viral coat must be considered (Ch. 67).

Of the animal viruses, only papovaviruses, adenoviruses, some herpesviruses, togaviruses, and picornaviruses yield infectious nucleic acids. With retrovirus (Ch. 66), infectious DNA can be extracted from infected cells. Failure in other cases can be ascribed either to the difficulty of extracting a large DNA molecule intact, or to **loss of virion enzymes** (e.g., a transcriptase), which are required for initiating the viral growth cycle.

INITIAL STEPS OF VIRAL INFECTION

Initially, as with bacteriophage infection, the viral nucleic acid must be made available for replication. Unlike the case with bacteriophage, however, the entire animal virus nucleocapsid enters the cell, and the nucleic acid is then released. These early events can be investigated by studying the changes of viral infectivity or the fate of ra-

TABLE 50-2. Cell Lines Commonly Used in Virology

Name of cell line	Species of origin	Tissue of origin	Morph-ology*	Ploidy†	Kariology Model no.	Markers
HeLa	Human	Carcinoma, cervix	Epi	Aneu	79	
Detroit-6	Human	Sternal bone marrow	Epi	Aneu	64	
Minnesota-EE	Human	Esophageal epithelium	Epi	Aneu	67	Yes
L-132	Human	Embryonic lung	Epi	Aneu	71	
Intestine 407	Human	Embryonic intestine	Epi	Aneu	76	
Chang liver	Human	Liver	Epi	Aneu	70	
KB	Human	Carcinoma, oral	Epi	Aneu	77	
Detroit-98	Human	Sternal bone marrow	Epi	Aneu	63	
AV$_3$	Human	Amnion	Epi	Aneu	74	
Hep-2	Human	Carcinoma, larynx	Epi	Aneu	76	Yes
J-111	Human	Peripheral blood‡	Epi	Aneu	111	Yes
WISH	Human	Amnion	Epi	Aneu	74, 75	
LLC-MK$_2$	Monkey§	Kidney	Epi	Aneu	70	
BS-C-1	Monkey‖	Kidney	Epi	Quasidip		
HaK	Syr. hamster#	Kidney	Epi	Aneu	57	
BHK	Syr. hamster					
B14-FAF-G3	Ch. hamster**	Kidney	Fib	Eu		
Don	Ch. hamster**	Peritoneal cells	Fib	Quasidip	22	
CHO	Ch. hamster	Lung	Fib	Eu	22	
		Ovary	Fib	Quasidip	22	
L	Mouse	Connective tissue	Fib	Aneu	78	Yes
NCTC, clone 929	Mouse	Connective tissue	Fib	Aneu	66	Yes
NCTC, clone 2472	Mouse	Connective tissue	Fib	Aneu	52	Yes
NCTC, clone 2555	Mouse	Connective tissue	Fib	Aneu	56	Yes
CCRF S-180 II	Mouse	Sarcoma 180	Fib	Aneu	86	Yes
3T3	Mouse	Connective tissue	Fib	Aneu	75	

Most of these cell lines are certified by the Cell Culture Collection Committee and are maintained frozen by the American Type Tissue Culture Collection

*Epi = epitheliumlike; Fib = fibroblastlike.

†Aneu = aneuploid; Eu = euploid; Quasidip = quasidiploid.

‡Monocytic leukemia.

§*Macaca mulatta.*

‖*Cercopithecus aethiops.*

#*Mesocricetus auratus.*

**Cricetulus griseus.*

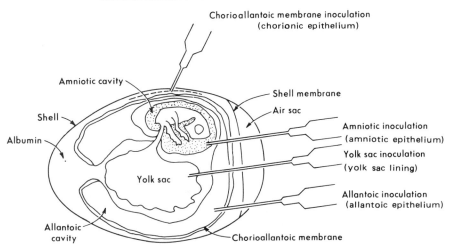

FIG. 50-1. Chicken embryo (10–12 days old) and routes of inoculation to reach the various cell types (as indicated). For chorioallantoic membrane inoculation a hole is first drilled through the eggshell and shell membrane; the shell over the air sac is then perforated, causing air to enter between the shell membrane and the chorioallantoic membrane, creating an artificial air sac, where the sample is deposited. The sample comes in contact with the chorionic epithelium. Yolk sac inoculation is usually carried out in younger (6-day-old) embryos, in which the yolk sac is larger.

dioactively labeled viral components, or by electron microscopy. The following steps can be identified.

1) **Attachment or adsorption.** The virus becomes attached to the cells, and at this stage it can be **recovered in infectious form** without cell lysis. Most animal viruses lack specialized attachment organs and probably have attachment sites distributed over the virion's surface. However, enveloped viruses such as orthomyxoviruses and paramyxoviruses attach through glycoprotein spikes, (see The Envelope, Ch. 46), and adenoviruses attach through the penton fibers (see Immunologic Characteristics, Ch. 54); these purified virion components also specifically adsorb to susceptible cells.

Adsorption occurs to **specific receptors:** cells are resistant if receptors are naturally lacking or have been artificially (e.g., enzymatically) destroyed. For orthomyxoviruses the receptors are mucoproteins, similar to the RBC receptors involved in hemagglutination (Ch. 46); for poliovirus they are lipoproteins. Whether or not receptors for a certain virus are present on a cell depends on the species and the tissue from which the cell derives, and on its **physiologic state.**

Adsorbed viruses can be recovered in infectious forms by various procedures that either destroy the receptors or weaken their bonds to the virions: poliovirus is removed by detergents, low pH, or high salt concentrations; influenza viruses by neuraminidase; and certain echoviruses by chymotrypsin.

2) **Penetration.** After this step, which rapidly follows adsorption, the attached virus can no longer be recovered from the intact cell. Electron micrographs show that with most **enveloped viruses** penetration occurs when the virion envelope fuses with the cellular membrane (Fig. 50-2). With **naked virions** the complete virion is most often taken into the cells by **phagocytosis** (Ch. 24), although rarely it penetrates through the cellular plasma membrane into the cytoplasmic matrix. Even enveloped viruses can also penetrate the cells by phagocytosis. Phagocytosis may enhance the rate of penetration by preventing the loss of virions through elution. However, it is possible that this phenomenon is no more than a preparatory step in penetration, since engulfed virions are still separated from the cytoplasmic matrix by the plasma membrane that surrounds the phagocytic vesicles, and the nucleocapsid must enter the cytoplasm either by envelope fusion or by direct penetration.

3) **Uncoating and eclipse.** Uncoating—removal of the capsid, freeing the nucleic acid—is detected by the accessibility of the viral nucleic acid to nucleases after disruption of the cells. Eclipse is recognized by failure to recover viral infectivity from the disrupted cells (although nucleic acid infectivity may be recoverable). For naked viruses the two events probably reflect the same process, i.e., alteration of the capsid and exposure of the viral genome. With enveloped viruses eclipse results from the loss of the envelope at penetration, whereas uncoating requires additional alteration of the capsid.

For some time after eclipse of **icosahedral viruses** the capsid does not appear morphologically altered (Fig. 50-2). However, invisible structural changes of the capsid probably occur: for example, poliovirus virions are al-

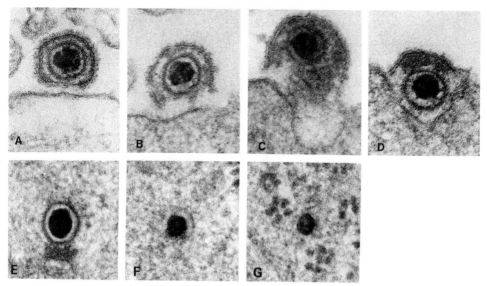

FIG. 50-2. Adsorption **(A)**, penetration **(B–D)**, and digestion of the capsid **(E–G)** of herpes simplex virus on HeLa cells, as deduced from electron micrographs of infected cell sections. The penetration involves local digestion of the viral and cellular membranes **(B, C)**, resulting in fusion of the two membranes and release of the nucleocapsid into the cytoplasmic matrix **(D).** The naked nucleocapsid is intact in **E,** is partially digested in **F,** and has disappeared in **G,** leaving a core containing DNA and protein. (Morgan C et al: J Virol 2:507, 1968)

tered while still attached outside the cells, since about half of those originally adsorbed elute and are unable to reattach (similar alterations are produced in vitro by exposure to fragments of membranes of susceptible cells). The poliovirus changes are accompanied by the loss of a small, superficially situated capsid protein (VP$_4$; see Maturation and Release of Animal Viruses, below), which is required for adsorption. Later the capsid breakdown becomes morphologically noticeable (Fig. 50-2F and G); the mechanism varies among different viruses. With **naked viruses** breakdown appears to be initiated by **interaction with cellular constituents** (such as the receptors).

Some viruses engage in more specialized steps. Poxvirions begin losing their outermost coat immediately after penetration, often in phagocytic vacuoles, releasing the complex cores into the cytoplasmic matrix. A transcriptase present in the core (see Viral Nucleic Acids, Ch. 46) then synthesizes some viral mRNAs and the newly made protein completes the uncoating. Reoviruses (Ch. 64) penetrate into lysosomes, where proteolytic enzymes strip the outer capsid, activating a transcriptase in the core.

ONE-STEP MULTIPLICATION CURVES

As with bacteriophages, multiplication curves for animal viruses are obtained under one-step conditions (Ch. 47).

The phage technics are applied to cells from suspension cultures or from monolayer cultures dispersed by trypsin. For viruses whose cell receptors can be easily destroyed (e.g., influenza viruses, polyoma virus) one-step conditions are possible even without dispersing the cell layers; the infected cultures are washed free of unadsorbed virus and then covered with a medium containing receptor-destroying enzyme, which prevents adsorption of released virus to the cells. Finally, for the viruses whose progeny is retained within the cells during the growth cycle (e.g., poxvirus, adenovirus, herpesvirus) the mere washing of the monolayers with viral Ab after infection creates approximate one-step conditions.

One-step conditions are also realized by infecting cultures with a large viral inoculum, which **infects all the cells;** the unadsorbed virus is then removed by washing the cells. Since all cells are infected at once, multiplication is necessarily confined to a single cycle. However, high multiplicity of infection may cause abnormal multiplication, and progeny virus may be lost by adsorption to cell debris.

One-step multiplication curves of animal viruses (Fig. 50-3) show the same stages observed with bacteriophages (Ch. 47). The length of the **intracellular accumulation period** varies, however, over a wide range with different viruses; it is very long with some whose progeny virions tend to remain within the cells, and is nonexistent with others, which mature and are released in the same act (by acquisition of an envelope at the cell surface). If the accumulation period is very short, the intracellular virus is

FIG. 50-3. One-step multiplication curves of two viruses with intracellular accumulation periods of different lengths. Extracellular virus is measured in the medium surrounding the intact cells; intracellular virus after removal of the medium and disruption of the cells. A. Western equine encephalitis virus multiplies in cultures of chick embryo cells with an extremely short accumulation period. B. Type 5 adenovirus multiplies in cultures of KB cells with a long accumulation period. Intracellular (i.e., cell-associated) virus is measured by disrupting cells after they have been washed free of extracellular virus (i.e., virus already released into the medium). PFU = plaque-forming unit. (A from data of Rubin H et al: J Exp Med 101:205, 1955; B from data of Ginsberg HS, Dixon M)

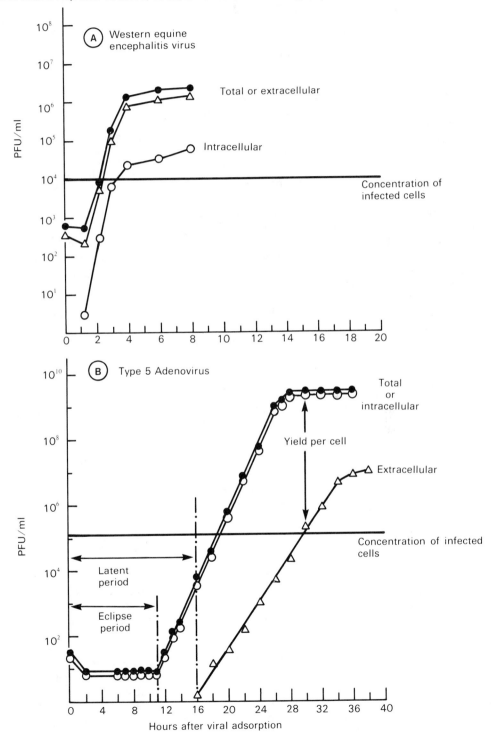

at any time a small fraction of the total virus, as in Figure 50-3A.

EFFECT OF VIRAL INFECTION ON HOST MACROMOLECULAR SYNTHESIS

In these studies viral and cellular nucleic acids are separated and identified by their size, buoyant density, and configuration (e.g., cyclic), or their hybridization to the nucleic acid present in purified virions. Viral and cellular proteins can be distinguished immunologically or by their different rates of migration in acrylamide gel electrophoresis.

As with bacteriophages, **virulent viruses,** either DNA-containing (e.g., adenovirus, vaccinia virus, herpesvirus) or RNA-containing (e.g., poliovirus, Newcastle disease virus, reovirus), turn off cellular macromolecular synthesis (Figs. 50-4 and 50-5) and disaggregate cellular polyribosomes (Fig. 50-6), favoring a **shift to viral synthesis** (Fig. 50-7). With most viruses the effect does not occur in the presence of inhibitors of protein synthesis (puromycin, cycloheximide), indicating that it is mediated by new proteins, probably virus-specified. The cellular DNA is not usually degraded, but **chromosome breaks** are often observed.

The mechanisms of inhibition of host synthesis are little understood. Cellular DNA synthesis seems to be blocked especially at initiation, since radioautography

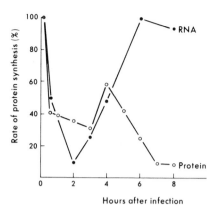

FIG. 50-5. Inhibition of cellular RNA and protein synthesis in L cells infected with mengovirus (a picornavirus). The decline in incorporation of radioactive precursors begins immediately after infection. The resumption of synthesis at about 3 h is due to synthesis of viral RNA and proteins. (Modified from Franklin RM, Baltimore D: Cold Spring Harbor Symp Quant Biol 27:175, 1962)

and biophysical analysis after ^{3}H-thymidine labeling show that the replicated segments are of normal length but are reduced in number. Inhibition of RNA synthesis often selectively affects the ribosomal RNA. Inhibition of protein synthesis also seems to occur at initiation: thus cell infection with poliovirus (whose RNA is not capped) inactivates a factor (eIF-4B) required in the initiation of translation of capped mRNA.

In contrast to virulent viruses, **moderate viruses** (e.g., polyoma virus) may **stimulate** the synthesis of host DNA, mRNA, and protein. This phenomenon is of considerable interest for viral carcinogenesis (see Viral Multiplication in Tissue Culture, Ch. 66).

SYNTHESIS OF DNA-CONTAINING VIRUSES

Biosynthesis of the DNA-containing animal viruses follows patterns similar to those described for bacteriophages, but the structural differences in the viral genomes appear to set unique regulatory strategies for their expression and replication. The variations in structure and complexity of the viral DNAs apparently dictate the modes of transcription, the methods of posttranscriptional processing, and the forms of DNA replication. There are in fact **three classes of DNA-containing viruses,** based on genome structure: class I, **double-stranded linear;** class II; **double-stranded circular;** and class III, **single-stranded linear.** All of the linear DNAs possess some form of inverted repetitions, either at the termini of the genomes (e.g., adenoviruses) or in their substructure (e.g., herpesviruses). In addition, within these classes distinctive structural variations also occur (Table 50-3). The general reactions will be described for

FIG. 50-4. Inhibition of cellular DNA synthesis in L cells infected by equine abortion virus (a herpesvirus) in the presence of ^{3}H-thymidine. Viral DNA was separated from cellular DNA because of its higher buoyant density in CsCl. (Modified from O'Callaghan DJ et al: Virology 36:104, 1968)

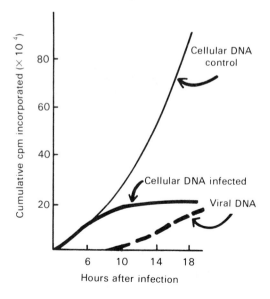

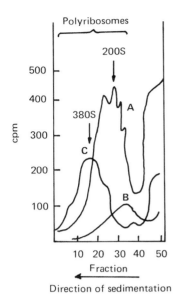

FIG. 50-6. Effect of poliovirus infection on polyribosomes of HeLa cells. Polyribosomes are studied by exposing the cells briefly to radioactive amino acids, then disrupting the cells and sedimenting the cytoplasmic extract through a sucrose gradient. The normal polyribosomes present in uninfected cells (most frequent sedimentation rate 200S; curve **A**) tend to disappear after infection (curve **B**), and after viral protein synthesis has begun, they are replaced by polyribosomes around 380S, which contain the viral RNA as messenger (curve **C**). (Modified from Penman S et al: Proc Natl Acad Sci USA 49:654, 1963)

synthesis of these classes of DNA-containing viruses; more specific details will be found in the chapters that follow.

TRANSCRIPTION

As in eukaryotic cells, and unlike the process in prokaryotes, transcription and translation are not coupled in the synthesis of DNA-containing animal viruses. Consequently, transcription and the posttranscriptional events of selection, processing, modification, and transport are prominent features of synthesis of the mRNAs. Generally the **primary transcripts** are larger than the mRNAs found on polyribosomes, and in some cases as much as 30% of the transcribed RNA remains untranslated in the nucleus (e.g., adenovirus transcripts). The viral messengers, however, like those of animal cells, are monocistronic (see Monocistronic Messengers, Ch. 49 and Synthesis of RNA Viruses, below).

As with DNA phages, transcription has a **temporal organization.** With most DNA-containing viruses (herpesviruses are the exception), before DNA replication only a fraction of the genome is transcribed into **early** messengers, and after DNA synthesis the remainder is transcribed into **late** messengers. Studies with DNA-defective, temperature-sensitive mutants, and with inhibitors of DNA synthesis (such as Cytarabine [arabinosyl cytosine]), show that the switch to late transcription requires prior DNA replication. With herpes simplex virus, although 40%–50% of the genome (i.e., equivalent to most of one entire strand) is transcribed early and late, only

FIG. 50-7. Shift in synthesis from cellular proteins to viral proteins after infection of BHK cells with influenza virus. The two superimposed acrylamide gel electropherograms were obtained from cells labeled with a radioactive amino acid either 1 h before infection (**circles**) or 4–7 h after infection (**crosses**). The synthesis of viral proteins has completely replaced that of cellular protein. (Modified from Holland JJ, Kiehn ED: Science 167:202, 1970)

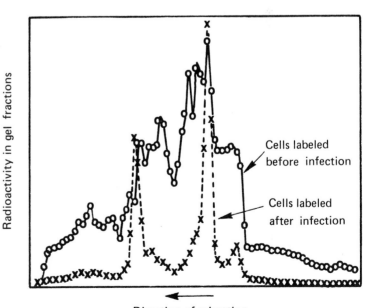

TABLE 50-3. Classes of Viral DNAs

Class	Viral family	Prototype virus	Unique features*
I. Double-stranded linear	Adenovirus	Type 2 adenovirus	Inverted terminal repetitions; 5′ terminal covalently linked protein
	Herpesvirus	Herpes simplex type I	Terminal and internal redundancies
	Poxvirus	Vaccinia	Terminal covalent crosslinking of strands; inverted terminal repetitions
II. Double-stranded circular	Papovavirus	SV40	Supercoiled circular
III. Single-stranded linear	Parvovirus	Adeno-associated virus	Inverted terminal repetitions and terminal repetitions

*Diagrams of unique features of viral genomes.

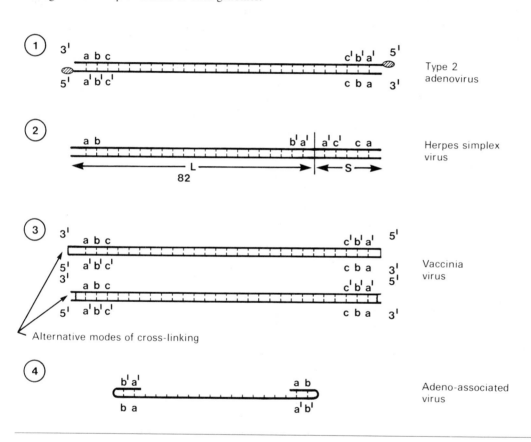

FIG. 50-8. Transcription map of the type 2 adenovirus genome and a model for processing a late mRNA, the hexon message. **A.** The encoded regions for the early and late mRNAs are indicated. The possible sites for cleavage and polyadenylation of the late transcript to produce the 3′ termini of the messages are designated by **Xs** or **arrowheads.** The cleavage and splice sites (see B) are each recognized by a unique nucleotide sequence. The enlarged segment between position 51.0 and 63.0 represents two messages processed from large transcripts; the 3′ ends are identical; the region of the message translated into the encoded protein is designated by the heavy segment. Not every separate late mRNA on the r strand is shown (see Ch. 54, Fig. 54-11). **White arrows** = Early transcripts; **black arrows** = late transcripts. The thickened lines on the early transcripts indicate the regions that are included in one of the major mRNAs processed from transcripts of each region. **B.** Primary transcript for late mRNAs on the r strand, and suggested model for processing this transcript to form the hexon mRNA. Only one mRNA can be derived from each transcript. Electron micrograph **(C)** and its diagrammatic representation **(D)**, showing an RNA–DNA hybrid **(heavy line)** of the adenovirus DNA and the hexon mRNA. The looped out areas (**a, b,** and **c** in part D) are the regions of the adenovirus DNA that are not homologous to the mRNA, and therefore represent the nucleotide sequences intervening between the encoded hexon message and the l_3 leader sequences as well as the spacer sequences between l_1 and l_2 and between l_2 and l_3.

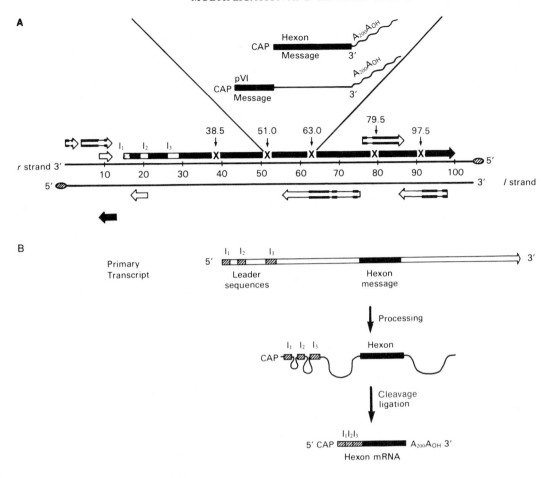

A

CAP Hexon Message 3' $A_{200}A_{OH}$

CAP pVI Message 3' $A_{200}A_{OH}$

38.5 51.0 63.0 79.5 97.5

I_1 I_2 I_3

r strand 3'

10 20 30 40 50 60 70 80 90 100

5'

5' 3' l strand

B

Primary Transcript 5' I_1 I_2 I_3 3'

Leader sequences Hexon message

Processing

I_1 I_2 I_3 Hexon

CAP

Cleavage ligation

$I_1I_2I_3$

5' CAP $A_{200}A_{OH}$ 3'

Hexon mRNA

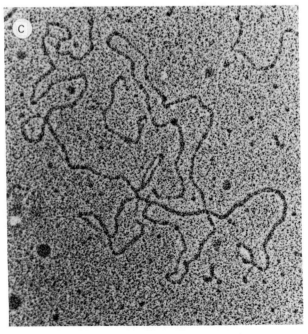

C

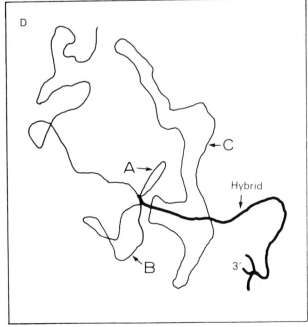

D

A B C Hybrid 3'

10%–12% of the early transcripts are processed into mRNAs; but coordinated translation and posttranscriptional processing leads to the sequential appearance of different classes of mRNAs: **immediate early (α); delayed early (β),** and **late (γ)** (see Viral Multiplication, Ch. 55).

Temporal regulation implies some topographical control, employing either specific regions or different strands of the genome to code for early and for late transcripts. Both the region and the strand can be determined by hybridizing mRNAs to separated strands of specific fragments of the viral DNA—obtained after cleavage with bacterial restriction endonucleases. Thus, in productive infections with SV40, a papovavirus, early and late mRNAs are transcribed from unique regions on opposite strands and therefore in opposite directions (see Productive Infection, Ch. 66).

The transcription of larger viruses may be even more complex. For example, in **adenovirus** (Fig. 50-8) **early transcription** occurs in five scattered regions of the genome, and it also proceeds, simultaneously, in opposite directions on the two strands. Synthesis of adenovirus **late messages** is also not confined to a single region, although most of the late mRNAs are encoded in the rightward reading **(r) strand,** and are derived from identical large primary transcripts, of about 28,000 nucleotides, initiated at a single promoter. Each large r-strand primary transcript may be cleaved at one of five possible sites, so that five groups of messages (with a total of at least 14) are produced. The mRNAs in each group have coterminal 3′ ends; and all of these messages have identical 5′ ends, each composed of a nontranslated cap (see Transcription of Chromatin, Ch. 49) and a leader which is derived from three short nucleotide sequences that are not contiguous to each other or to the message sequences in the genome. A similar arrangement pertains for the two late mRNAs of SV40 and polyoma viruses. This complex method of producing mRNAs by **posttranscriptional processing** of primary transcripts serves to remove **intervening sequences** not destined to be translated. It is also known for cellular mRNAs (see Transcription of Chromatin, Ch. 49) and appears to be essential for regulating the appearance of functional messages for eukaryotic cells as well as for viruses.

Viral transcripts undergo other modifications similar to eukaryotic cell mRNAs, such as the addition of a **poly(A) chain** (100–200 adenine residues long) to the 3′ end and a methylated cap at the 5′ end. Adenylation and capping may be accomplished either in the nucleus (e.g., adenoviruses) or in the cytoplasm (e.g., poxviruses). As with cellular mRNAs, the methylated capped 5′ terminus appears essential for the stable attachment of viral mRNAs to the 4OS ribosomal subunit and for effective translation of the message.

The early transcription and processing of most DNA-containing viruses are carried out by host enzymes, since these are not blocked when inhibitors of protein synthesis are added at the time of infection. With poxviruses, however, after the loss of the outer envelope a virus-specified transcriptase present in the virion's core transcribes a set of **immediate early genes** (about 26% of the genome), after which enzymes in the virion modify the transcripts into mRNAs. **Delayed early genes** are transcribed after poxvirus uncoating is completed.

Most viral DNAs are transcribed in the nucleus (in which they replicate), as shown by cell fractionation after a short pulse of ^{3}H-uridine. Within about 30 min after synthesis and processing the mRNAs are transported to the cytoplasm, where they form polysomes; the mechanisms appear to be similar to those used for cellular messengers. The DNAs of poxviruses, however, are transcribed and replicated in the cytoplasm (the immediate early RNAs, after modification, are extruded from the cores and form polysomes in the cytoplasm). A product of one poxvirus mRNA is thymidine kinase, whose synthesis is a useful indicator of immediate early transcription. By studying the synthesis of this enzyme it can be shown that immediate transcription ceases after uncoating: thus, if the delayed early transcription is inhibited by small doses of dactinomycin (actinomycin D), immediate transcription (apparently more resistant to the inhibitor) continues indefinitely. The effect is probably caused by a delayed early gene product, because a temporary halt of protein synthesis also prolongs the immediate transcription (Fig. 50-9).

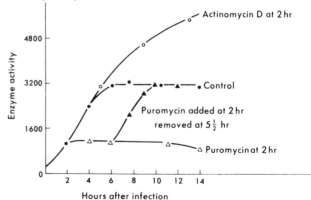

FIG. 50-9. Evidence that the expression of a gene (probably viral) controls the cessation of thymidine kinase synthesis in cells infected by vaccinia virus. The evidence derives from the effects of dactinomycin and puromycin. In the control cells, i.e., not exposed to any inhibitor, synthesis of the enzyme stops at about 5½ h after infection; but earlier addition of dactinomycin allows synthesis to continue, showing that the block in this specific synthesis requires new RNA synthesis. Puromycin at 2 h interrupts all enzyme synthesis, but after its removal at 5½ h synthesis of thymidine kinase resumes, although it has stopped in the control; this result shows that synthesis of the enzyme stops only after some required protein synthesis. (Modified from McAuslan B. Virology 21:383, 1963)

SYNTHESIS OF VIRAL PROTEINS

Viral proteins, identified by acrylamide gel electrophoresis of extracts of infected cells, are synthesized on cytoplasmic polysomes in a temporal sequence corresponding to that of mRNAs. With viruses that assemble in the cell nucleus, proteins then migrate to the nucleus: the early proteins participate in DNA replication and perhaps transcription, and the late gene products, which are predominantly structural proteins, form the virions.

The proteins can be identified as viral when they are made after host protein synthesis is shut off, or when they are absent in uninfected cells and have unique biochemical and immunologic characteristics. Thus, thymidine kinase is probably viral since viral mutants lacking the enzyme have been isolated, and thymidine kinase appears in thymidine kinase–free mutant cells after transfection with a specific piece of herpes simplex DNA.

As noted earlier, transcriptional controls appear to regulate the temporal organization of the biosynthesis of proteins of most DNA-containing viruses. However, with herpes simplex virus the temporal regulation of protein synthesis appears to be **posttranscriptional:** proteins are sequentially made in three sets, each depending upon the synthesis of the preceding one: $\alpha \rightarrow \beta \rightarrow \gamma$. This coordinate control—which is detected if protein synthesis is blocked by adding cycloheximide at increasing times after infection and then removing it, (Fig. 50-10)—demonstrates the sequential synthesis of sets of viral proteins required to process RNA transcripts. Thus, immediately after removal of cycloheximide α **proteins** appear; these modify primary transcripts, producing immediate early mRNAs which are translated to form β **proteins;** and these in turn process transcripts into early mRNAs which are used to synthesize γ **proteins.** Hence, although there is not a strict sequential transcription of early and late mRNAs that is dependent upon replication of viral DNA, there is a sequential synthesis of viral proteins, so that the virion structural proteins are produced last (after onset of DNA replication) and their synthesis is dependent upon temporally related posttranscription reactions.

DNA REPLICATION

DNA replication utilizes cellular pool precursors derived from the medium, since cellular DNA is not degraded. Synthesis begins toward the middle of the eclipse period (Fig. 50-3B), after biosynthesis of the early viral proteins. Indeed, replication of the DNA of animal viruses, like that of bacteriophages, depends upon the early viral proteins, since mutation in some early viral genes or inhibition of protein synthesis shortly after infection prevents viral DNA replication and production of infectious virus. The smaller DNA viruses, including adenoviruses, rely

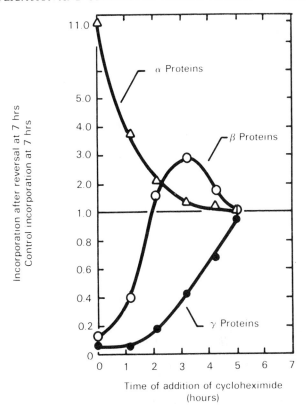

FIG. 50-10. Sequential appearance of three sets of type 1 herpes simplex virus proteins. These data show the rate of synthesis of representative α, β, and γ polypeptides following removal of cycloheximide 7 h after infection. Cycloheximide was added to infected cultures at different times after infection (noted on abscissa) and was removed from all cultures simultaneously. The relative rates of synthesis (plotted on the ordinate) were determined by measuring radioisotope incorporation into a specific polypeptide for 30 min after removal of cycloheximide as compared to the incorporation into the same viral polypeptide from infected cells to which inhibitor had not been added. Each point is based on analyses of autoradiograms of polypeptides separated by SDS–polyacrylamide gel electrophoresis. (Modified from Honess RW, and Roizman B: J Virol 14:8, 1974)

on the host cell DNA polymerase, but the more complex herpesviruses and poxviruses appear to require virus-encoded polymerases for DNA replication.

The mode of replication of the viral DNA is generally **semiconservative** (Fig. 50-11), but the nature of the replicating molecules (i.e., the **replicative intermediates**) depends upon the structure of the viral genome. Thus, since the DNA of each family of animal viruses has a unique structure (Table 50-3), the replicative intermediate of each family varies in form and mode of replication (Fig. 50-12).

1) **Adenoviruses** show **asymmetric** replication, which initiates at either 3′ end (but usually allows only one ini-

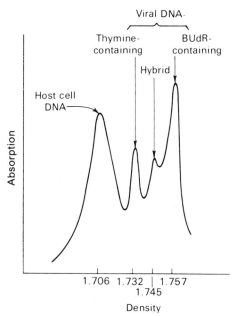

Viral DNA

Host cell DNA

Thymine-containing

Hybrid

BUdR-containing

Absorption

1.706 1.732 | 1.757
1.745

Density

FIG. 50-11. Evidence for semiconservative replication of the DNA of pseudorabies virus (a herpesvirus). The infected cells were transferred from a medium containing thymidine to one containing the heavy precursor, 5-bromodeoxyuridine (5-BUdR), where synthesis was allowed to continue. The DNA later extracted and analyzed by density gradient equilibrium centrifugation shows four density peaks, one of cellular DNA and three of viral DNA; one of the latter has a hybrid density, proving semiconservative replication. (Modified from Kaplan AS: Virology 24:19, 1964)

tiation per DNA molecule) and *displaces the unreplicated single strand.* The replicative intermediates sediment faster and have a greater buoyant density in CsCl than mature virion DNA, owing to the single-stranded component. Circular forms are not detected, and only hypotheses can be made as to the roles played by the 55,000-dalton protein covalently linked to the 5′ terminus of each strand and by the inverted terminal repetitions.

2) The modes of replication of **herpesvirus DNA** (Fig. 50-12), with its complex structure containing terminal and internal redundant sequences, is still obscure. Biochemical and electron microscopic examination of the replicating complexes reveals a variety of **unit length** and "head-to-tail" (i.e., 5′–3′) **concatemeric linear, complex linear–circular,** and **wholly circular** forms, but which of these forms is a precursor to virion DNA is unclear. Free single strands are not observed, so replication is probably symmetric.

3) **Poxvirus DNA,** in addition to its large size and inverted terminal repeat sequences, has a striking feature that must affect its mode of replication: a short polynucleotide that **covalently crosslinks** the strands of the linear double-stranded molecule at or near each terminus. The mechanism for reproducing this complex molecule is uncertain, but electron microscopic examination of replicating molecules reveals the forms shown in Fig. 50-12. These findings, in conjunction with biochemical studies, suggest that a) semiconservative DNA replication initiates at one end of the molecule, and proceeds bidirectionally, forming a lariatlike structure of increas-

FIG. 50-12. Proposed models for the replication of DNA based upon the structure of the virion DNA in each family. The intermediate stages noted are proposed from electron microscopic observations and physicochemical determinations made upon DNA replication forms sequentially extracted after infection.

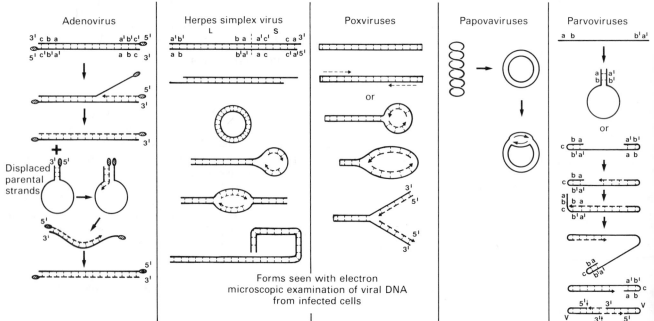

Adenovirus   Herpes simplex virus   Poxviruses   Papovaviruses   Parvoviruses

Displaced parental strands

Forms seen with electron microscopic examination of viral DNA from infected cells

ing diameter; and that b) an endonuclease cleaves the terminal polynucleotide crosslink at an appropriate stage in replication, permitting separation of the newly made molecules.

4) The **cyclic DNAs** of **papovavirus** (e.g., SV40) replicate via cyclic intermediates. Sedimentation analyses and electron microscopic studies show that they maintain continuous parental strands, that replication is **bidirectional** and **symmetric** (Fig. 50-12) and the **unwinding** is carried out by a **"swivel enzyme"** (see Enzymes Involved in Replication, Ch. 10, and Fig. 10-31).

5) Replication of **parvovirus DNA,** a single-stranded linear molecule, is directed by the **unusual structure of its ends,** which have both inverted terminal repetitions (single strands can form circles with panhandles) and terminal nucleotide repetitions (double-stranded molecules can form circles). The replicative intermediates of the defective **adeno-associated virus DNA** consist of concatemers, predominantly of double length, in which a plus and a minus strand are covalently linked in the same strand; larger concatemers are also present (Fig. 50-12). A specific endonuclease cleaves the larger molecules at sites identified by the inverted terminal repetition sequences.

As with bacteriophages, the newly synthesized viral DNA enters a pool from which it is subsequently removed to associate with virion structural proteins. Thus if the infected cells are exposed to a short pulse of a radio-active DNA precursor at any time during the eclipse period the label is distributed among virions finished at any subsequent time. In contrast to bacteriophage infection, viral DNA is made in excess and much remains unused in the infected cells at the end of the multiplication cycle, often as a constituent of inclusion bodies.

SYNTHESIS OF RNA VIRUSES

The RNA genomes of viruses infecting animal cells cannot initiate replication like RNA phages (Ch. 47), for in eukaryotic cells translation cannot be readily initiated internally (see Protein Synthesis and Breakdown, Ch. 49), and so virion RNAs cannot function as polycistronic messengers for the synthesis of viral proteins. This obstacle, however, is overcome in different ways: 1) With some viruses the virion RNA acts as a messenger, but since there are no internal initiation and termination signals it is **translated monocistronically** into a **giant peptide** which is then cleaved to generate distinct viral proteins. 2) In other viruses the virion RNA is **transcribed** to yield **monocistronic mRNAs.** 3) Occasionally the **genome** itself is a collection of **separate RNA fragments** which are **transcribed** into **monocistronic mRNAs.**

RNA-containing animal viruses can be placed in five different classes, according to the nature of the RNA in the virions and the relation of this RNA to the viral messenger (Fig. 50-13).

FIG. 50-13. The various classes of RNA viruses and their primary modes of expression. The numbers of multiple genome pieces, messengers, and gene products are only diagrammatic representations and are not indicative of precise numbers.

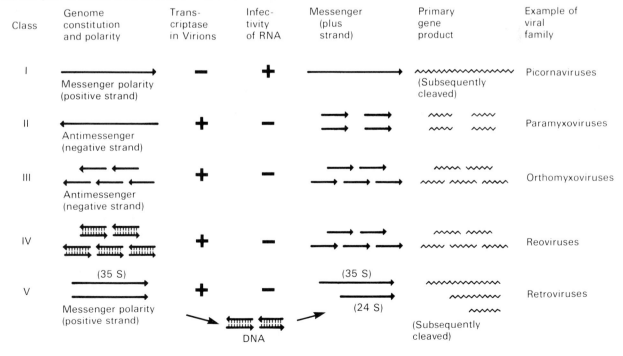

Class I viruses (e.g., picornaviruses) contain a molecule of single-stranded RNA which **acts as messenger** (**"positive"-strand** genomes), specifying information for the synthesis of nonstructural proteins (e.g., replicase) and structural proteins. In vivo the same RNA molecule must also initiate replication, because infection by a single viral particle occurs. Since RNA replication requires viral proteins (see below), the messenger function must be expressed first.

Class II viruses (e.g., paramyxoviruses) have a molecule of single-stranded RNA which cannot serve as messenger, having the opposite polarity (**antimessenger**). On this **"negative" viral strand,** a **virion transcriptase** initiates transcription at a single promoter to yield several (five to eight) complementary **messenger ("positive"-strand)** molecules. At present it is uncertain whether the transcriptase synthesizes a long poly-mRNA which is subsequently processed into monocistronic messages or whether the polymerase stops and restarts at each juncture between message sequences.

In **class III** viruses (e.g., orthomyxoviruses) the single-stranded RNA is in several distinct, nonoverlapping pieces, which, like Class II viruses, are of antimessenger polarity (**"negative"-strand genome**). Each RNA segment contains one gene and is separately transcribed into a complementary, monocistronic, messenger by a **virion transcriptase.**

Class IV viruses (e.g., reoviruses) contain distinct, nonoverlapping segments of **double-stranded RNA** (ten in reoviruses); each is transcribed into a monocistronic mRNA by a **virion transcriptase.**

The function of the virion transcriptases in classes II to IV is easily demonstrated in vitro. In vivo the complementary mRNAs are evidently made by the same mechanism, since inhibitors of protein production do not block their synthesis. In contrast to DNA viruses, there is no differentiation between early and late messengers in any of the preceding classes, with the possible exception of reoviruses (Class IV).

Class V viruses (e.g., retroviruses) contain two identical positive-strand (i.e., having **messenger polarity**) segments of single-stranded RNA, with a poly(A) tail at the 3'-OH terminus and a 7-methylguanine triphosphate cap at the 5' end. In productive infection each is transcribed into DNA by a **reverse transcriptase** present in the virion; the functional mRNAs are then transcribed from this DNA.

Two main points emerge from this classification: 1) viruses with an antimessenger, or negative (−) RNA strand (classes II and III) also have in the virions an RNA polymerase, i.e., a transcriptase, which synthesizes the complementary messenger; 2) antimessenger viral RNA is not infectious when purified and separated from its transcriptase. Hence viruses of class I, but not of classes II and III, yield infectious RNA. Viruses of class V behave aberrantly: the virion RNA is apparently of messenger, or positive (+), polarity but reverse transcription by a virion transcriptase, producing a DNA intermediate, is essential for infectivity; and mRNAs, different from the virion RNA, are made in the infected cells.

VIRAL PROTEINS

These proteins are synthesized in two different patterns which satisfy the monocistronic nature of the messengers.

1) When a virus has several messengers each appears to yield only one protein, since the number of viral proteins identifiable by acrylamide gel electrophoresis is very similar to the number of messengers.

2) When, as with picornaviruses, the genome serves as a single messenger, a giant **polyprotein** is made which is then cleaved to yield the several viral proteins. This cleavage, called **processing,** merits special description because it solves the paradox that the sum of the molecular weights of all the poliovirus proteins in infected cells corresponds to almost double the coding capacity of the RNA. Short-pulse labeling with radioactive amino acids followed by synthesis without an isotope ("chase") for periods of various lengths showed that the largest peptides are precursors of the smaller ones (Fig. 50-14), and therefore are counted two or more times in the total protein enumeration. Cleavage is carried out by a specific enzyme which recognizes the secondary and tertiary structures of the polyprotein and other precursor proteins. Inhibition of cleavage, by amino acid analogs incorporated in the precursor polyprotein or by high temperature, allows detection of the giant polypeptide, which is not normally recognized because it undergoes the first cleavage while being synthesized. A second cleavage produces three structural proteins. A third cleavage of a precursor capsid polypeptide (VP_0) occurs during virion maturation (see Association of the Capsid with the Nucleic Acid, below).

RNA Polymerase Activity. RNA-dependent RNA polymerase activity that is similar to that formed in bacteria infected with RNA phages (see RNA Replication, Ch. 47) makes its appearance in animal cells infected by viruses with single-stranded RNA. In crude extracts or after partial purification, the enzymes incorporate label from radioactive ribonucleoside triphosphates into RNAs with the characteristics of those formed in vivo (see Replication of Single-Stranded RNA, below). The incorporation requires all four triphosphates and suitable ionic conditions. The polymerase has two functions: transcription and replication.

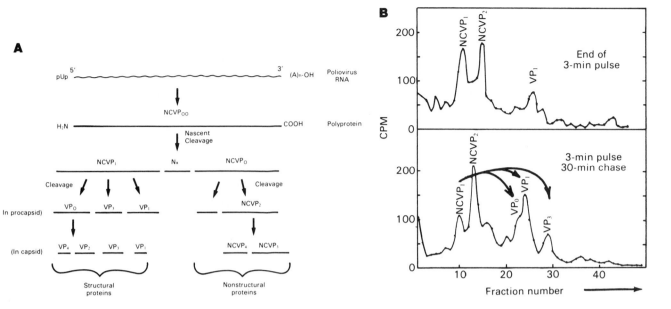

FIG. 50-14. Processing of the poliovirus precursor polypeptide. Similar processing of precursor polyproteins into functional proteins has been demonstrated for other viruses containing positive, single-stranded RNAs (picornaviruses and togaviruses). **A.** General scheme. $NCVP_{00}$ is a polypeptide of about 250,000 daltons; it represents a translation product of the entire viral RNA and is detected only when formed in the presence of amino acid analogs or at a high temperature which inhibits protein cleavage. The function of all the nonstructural proteins is still unclear, but at least one is the viral RNA polymerase (N_x is not utilized). **B.** Evidence for processing of the precursor polypeptide $NCVP_1$ during poliovirus synthesis to generate three new peptides, VP_0, VP_1, and VP_3. Infected cultures of HeLa cells were labeled for 3 min with a radioactive amino acid during the period of viral synthesis. The extract from a culture at the end of the labeling was fractionated by acrylamide gel electrophoresis. The electropherogram shows a prominent $NCVP_1$ peak, no VP_0 or VP_3, and a small VP_1 peak. After a 30-min chase with cold amino acids $NCVP_1$ is greatly reduced, having generated VP_0, VP_1, and VP_3. VP_0 is present in procapsids and is later processed into $VP_2 + VP_4$, which are found in the virions. NCVP = Noncapsid viral protein; VP = virion protein. (Modified from Jacobsen MF, Baltimore D: Proc Natl Acad Sci USA 61:77, 1968)

The **transcriptase** is found in virions from classes II to IV, and its activity is measured by synthesis of messenger strands. The **replicase** is extracted from infected cells and is active in synthesis of new viral RNA strands of both polarities. It is unclear whether different enzymes carry out these activities or whether both enzymes contain a common subunit whose specificity is modified by interaction with the virions' proteins, analogous to the way the core replicase of RNA phages changes its specificity after binding to a cellular subunit. In poliovirus-infected cells the *replicase is unstable,* since its level drops rapidly with inhibition of protein synthesis. This finding explains the *dependence of viral RNA synthesis on continuing protein synthesis:* RNA synthesis can be interrupted at any stage by adding puromycin or the amino acid analog *p-fluorophenylalanine.*

Regulation of Production of Viral Proteins. As would be predicted, the regulation of viral protein formation varies with the structure of the RNA genomes. Regulation of class I viruses (picornaviruses and togaviruses) is

governed by **posttranslation processing:** proteins resulting from the first cleavage are available earlier in infection than those from later cleavages. With class II viruses it is uncertain whether regulation is effected at transcription or by modification of primary transcripts. With reoviruses (class IV) there is evidence for both **transcriptional** and posttranscriptional regulation: early in infection the mRNAs from some segments appear in greater abundance, and late in infection certain polypeptides are synthesized at much greater rates than others.

REPLICATION OF SINGLE-STRANDED RNA (CLASS I, II, AND III VIRUSES)

The synthesis of RNA of many animal viruses can be studied by blocking the synthesis of cellular RNA by dactinomycin: viral RNA is still synthesized. The replication of single-stranded RNAs of class I, II, and III animal viruses is similar to that of phage RNA. The findings of poliovirus (class I) will be summarized here; this mode

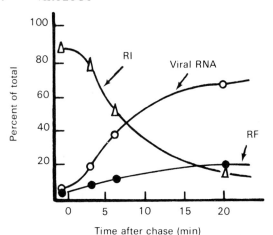

FIG. 50-15. Kinetic evidence that the replicative intermediate, RI is the intermediate that generates progeny viral RNA strands in a cell-free system from infected HeLa cells. Label was first incorporated from ^{3}H-uridine triphosphate into the RNAs; at time 0 the radioactive precursor was replaced with unlabeled uridine triphosphate, beginning the chase. Radioactivity disappearing from RI is found mostly in viral RNA, showing the precursor–product relation of the two molecular types. In contrast, radioactivity slowly accumulates in the replicative form (RF), showing that it, like the viral RNA, is an end product in the synthesis rather than an intermediate. (Modified from Girard M: J Virol 3:376, 1969)

Replicative Intermediate. When poliovirus-infected cells are labeled by a very short pulse of a radioactive precursor, extraction and fractionation of the RNA shows that most of the label is present in a partially double-stranded replicative intermediate with single-stranded tails; the label then chases into single-stranded viral strands (Fig. 50-15). The **replicative intermediate** (RI) contains a complete positive strand and several growing complementary (negative) strands (Fig. 50-16). The **messenger** or **virion** (positive), strands are replicated on **complementary** (negative) templates via a similar intermediate. Each growing strand is weakly held to the complementary strand, probably by a replicase molecule: if a homogenate of infected cells is treated with a protein denaturant, the strands separate. Both positive and negative strands have a protein (VP$_g$) that is attached covalently at the 5′ terminus concurrent with or soon after initiation of replication; it is subsequently removed from those plus strands that are destined to serve as messengers.

Isolated poliovirus minus strands are not infectious: they are antimessenger and cannot specify the viral replicase required for generating viral strands. *Minus strands are not found free;* presumably after they are formed on viral strands, they immediately generate RIs.

Site of Replication of Viral RNA (Classes I and II).
Viral RNA replicates in the cytoplasm, as shown by radioautographs of infected cells exposed to a brief pulse of tritiated uridine in the presence of dactinomycin to inhibit the synthesis of cellular RNA. Labeled RNA is exclusively in the cytoplasm, whereas in uninfected cells

of replication also applies to the other single-stranded RNAs (classes II and III; see Fig. 50-13). With viruses of class III (orthomyxoviruses) each genome segment has a nucleoside triphosphate at the 5′ end (where synthesis begins); hence each segment replicates independently.

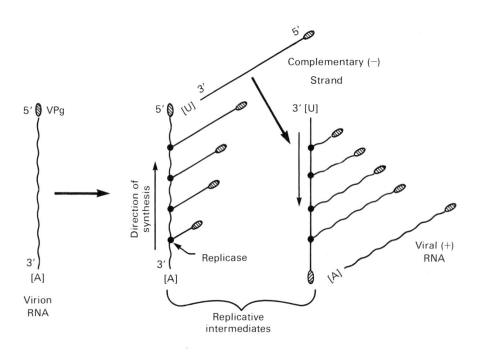

FIG. 50-16. Formation of the replicative intermediate (RI) and generation of viral RNA in poliovirus RNA replication. During replication progeny strands are held on the template strand mainly by the replicase; base pairing is probably prevented by some protein bound to the nascent RNA. (Extensive base pairing may occur as an artifact during phenol extraction.) [A] The polyadenine tail at the 3′ end of the virion and mRNAs; this is copied into a polyuridine ([U]) sequence at the 5′ end of the complementary RNA strand.

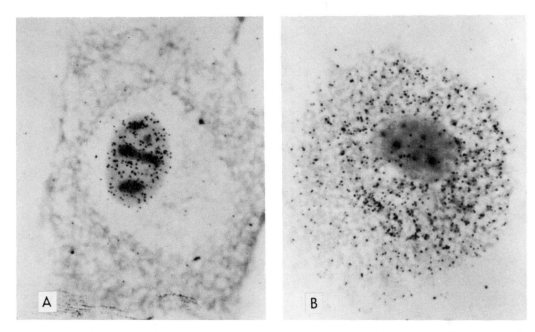

FIG. 50-17. Effect of infection with mengovirus (a picornavirus) on the distribution of RNA synthesis in the cell. An uninfected L cell **(A)** and one infected 6 h earlier **(B)** were radioautographed after exposure for 3 min to ³H-uridine. The infected cell was also exposed to dactinomycin. In the uninfected cell the silver grains produced by the β-radiation of decaying ³H atoms are concentrated over the nucleus; in the infected cells there are only a few grains over the nucleus, because the synthesis of cellular RNA is inhibited, and many grains over the cytoplasm, due to synthesis of viral RNA. (Franklin RM, Baltimore D: Cold Spring Harbor Symp Quant Biol 27:175, 1962)

(those without dactinomycin) it is only in the nucleus (Fig. 50-17). Cytoplasmic synthesis is also supported by the demonstration that complete replication of infectious poliovirus can occur in cells enucleated by cytochalasin B. Although orthomyxovirus (class III) RNA accumulates in infected nuclei its site of synthesis is still uncertain.

In homogenates of cells infected by poliovirus the RI is found to be associated with membrane proliferations resembling smooth endoplasmic reticulum (Fig. 50-18; see also Cellular Membranes, Ch. 49). The membrane component appears to play an important role, since in cells infected by Semliki Forest virus (a togavirus, class I) RI formation is inhibited after treatment with phospholipase C.

Requirement for a Cellular DNA Function (class III). Cellular DNA appears to have a role in the replication of the RNA of orthomyxoviruses (class III). If dactinomycin is added to the cells, or if cells are irradiated with ultraviolet light before or shortly after influenza virus (orthomyxovirus) infection, the synthesis of messengers by the virion's transcriptase is specifically inhibited, while these agents have much less effect toward the end of the eclipse period. This effect appears to result from interference with required cellular functions such as RNA polymerase II and reactions needed to prime

transcription and to cap the mRNAs (see Viral Multiplication, Ch. 58).

REPLICATION OF DOUBLE-STRANDED RNA (CLASS IV VIRUSES)

Each segment of reovirus RNA is replicated independently, since each has a nucleoside diphosphate at the 5′ end (derived from partial dephosphorylation of the original triphosphate). Replication is intimately tied to transcription of the genome by the **virions' transcriptase,** which generates mRNAs in the virion cores. Negative strands are then made from the nascent positive-strand templates by another enzyme, the **replicase.** These negative strands pair with newly made messenger strands to form double-stranded virion molecules (Fig. 50-19). Hence this replication, unlike that of DNA, is **asymmetric** and **conservative:** only the negative strand of the virion RNA serves as the initial template and the parental RNA is otherwise unused and does not end up in the progeny (Fig. 50-19). As with poliovirus RNA, minus strands are not found free. However, it differs from single-stranded RNA synthesis, because 1) it is conservative, 2) it does not proceed through a comparable RI form, and 3) only double-stranded RNA is packaged in virions.

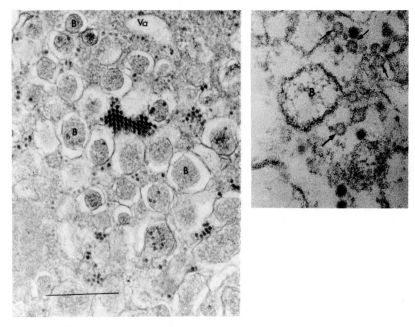

FIG. 50-18. Electron micrograph of part of the cytoplasm of HeLa cell infected by poliovirus (a picornavirus) showing a focus of viral reproduction. **Left.** Many viral particles are present in the cytoplasmic matrix (some in small crystals) around or within membrane-bound bodies **(B)** and vacuoles **(Va). Right.** Presence of empty capsids **(arrows).** (Dales S et al: Virology 26:379, 1965)

FIG. 50-19. Diagram comparing transcription and replication of virion double-stranded RNA and DNA molecules. In the replication of double-stranded RNA information flows from **only one strand** (asymmetric) and the molecule is replicated **conservatively** (i.e., both parental strands are conserved together). This is in contrast to the double-stranded viral DNA where information flows from **both strands.** New strands are shown as **broken lines.**

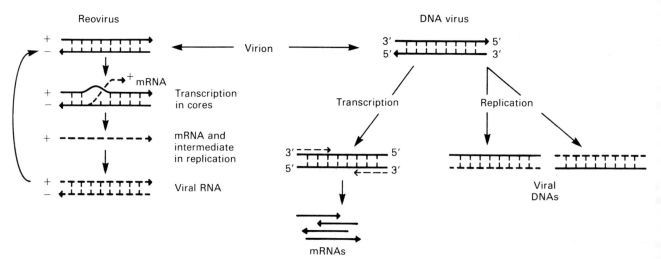

REPLICATION THROUGH A DNA-CONTAINING REPLICATIVE INTERMEDIATE (CLASS V VIRUSES)

The replication of retroviruses is blocked if dactinomycin or an inhibitor of DNA synthesis is added immediately after infection; later in infection viral replication becomes insensitive to inhibitors of DNA synthesis but remains susceptible to dactinomycin. To explain these findings Temin proposed that the viral replication proceeds through a double-stranded DNA RI (a **provirus**), and that progeny viral RNA is obtained from it by regular transcription. Strong support for this theory was afforded by the discovery that a **reverse transcriptase,** present in virions, synthesizes DNA using the single-stranded viral RNA as template, and that infectious DNA can be extracted from the infected cells. This method of replication can fulfill the requirement for monocistronic mRNAs since, as with DNA viruses, the viral DNA can act as template for separate mRNAs. However, retrovirus mRNAs are still translated into polyproteins which are then cleaved (Fig. 50-13).

MATURATION AND RELEASE OF ANIMAL VIRUSES

Maturation proceeds differently for naked, enveloped, and complex viruses; therefore, these several groups of viruses will be considered separately.

NAKED ICOSAHEDRAL VIRUSES

Maturation. As with bacteriophages, maturation of naked viruses consists of two main processes: the assembly of the capsid and its association with the nucleic acid. For **DNA viruses** the two steps are clearly separate, since DNA synthesis precedes the appearance of recognizable capsid components, sometimes by several hours. The association of the capsid with the nucleic acid proceeds almost concurrently with **naked icosahedral RNA viruses.** The synthesis of viral RNA, measured either chemically or by infectious titer, is followed shortly by synthesis of mature virions; for poliovirus the time difference is 30–60 min (For **enveloped viruses** the synthesis of virion RNA precedes the appearance of finished virions by a longer time interval—e.g., 2 h for western equine encephalomyelitis virus—the extra time probably being spent in adding the envelope since the RNA appears to be assembled into the nucleocapsid early.) The fairly close temporal connection between synthesis and assembly may have evolved because of the inherent instability of the naked viral nucleic acids in the cells.

Association of the Capsid with the Nucleic Acid. With icosahedral animal viruses, as with bacteriophages, **empty capsids** generated during replication appear to be precursors of virions **(procapsids):** labeled amino acids added to infected cells are incorporated first into nascent polypeptide chains which are rapidly assembled into capsomers, later into procapsids, and finally into complete virions.

With a number of animal viruses (e.g., picornaviruses, adenoviruses) the association of the nucleic acid with the procapsid is accompanied by polypeptide processing. For example, in poliovirus maturation the final step is accomplished when the precursor polypeptide VP_o is cleaved into $VP_2 + VP_4$ (Fig. 50-14). Presumably an RNA molecule penetrates between the capsomers from the outside, triggering the cleavage reaction. VP_4 polypeptides may be responsible for covering and locking in the RNA, because their loss at uncoating (see Initial Steps of Viral Infection, above), and during heat inactivation, is accompanied by separation of RNA from the capsid.

The morphogenesis of adenoviruses offers another instance in which the processing of precursor proteins effects the final phase of maturation. The procapsid contains not only properly folded proteins but two precursors, called pVI (28,000 daltons) and pVIII (26,000 daltons). Following entry of the immature nucleoprotein core (consisting of viral DNA covalently linked to its 55K protein, plus protein V and precursor pVII), protein VII is generated from pVII, and the final maturation is completed with the processing of pVI and pVIII to the capsid proteins VI and VIII.

After they are assembled, icosahedral virions may become concentrated in large numbers at the site of maturation, forming the intracellular crystals (Fig. 50-18) frequently observed in thin sections of infected cells.

With reoviruses, whose genome is segmented (class IV), the various progeny RNA pieces appear to assemble within an inner capsid in which they are held together, forming the virion core. The undefined final steps in morphogenesis lead to association of this core with the outer capsid (see Ch. 64).

Release. Naked icosahedral virions are released from infected cells in different ways, which depend on both the virus and the cell type. Poliovirus, for instance, is **rapidly released** from HeLa cells, in which it is highly virulent; the intracellular accumulation is then small. The study of single infected cells, contained in small drops of medium under paraffin oil (Fig. 50-20), shows a total yield of about 100 plaque-forming units released over a period of ½ h. During release the cells show rupture of surface vacuoles and surface bubbling with detachment of small cytoplasmic blebs. In contrast, virions of DNA viruses that mature in the nucleus do not reach the cell surface as rapidly; they are released when the cells undergo autolysis, but in some cases they are extruded without lysis.

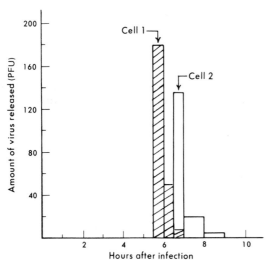

FIG. 50-20. Viral yield from two single monkey kidney cells infected by poliovirus. The cells were obtained by disrupting a monolayer of monkey kidney cells by trypsin; after being infected with poliovirus each was introduced into a separate small drop of medium immersed in paraffin oil. Every ½ h the medium of each drop was removed, replaced with fresh medium, and assayed for infectivity by plaque assay. Note that with either cell the release was rapid (most virus came out in ½ h), and note also the differences in the latent periods. (Data from Lwoff A et al: Virology 1:128, 1955)

In either case they tend to **accumulate** within the infected cells over a long period.

ENVELOPED VIRUSES

Maturation. Viral proteins must first be associated with the nucleic acid to form the nucleocapsid, which is then surrounded by the envelope.

In **nucleocapsid formation** the proteins are all synthesized on cytoplasmic polysomes and are rapidly assembled into capsid components (recognizable by immunofluorescence or electron microscopy). With paramyxoviruses these components accumulate in a perinuclear zone of the cytoplasm starting 3 h after infection (Fig. 50-21). In contrast, with orthomyxoviruses (e.g., influenza and fowl plague virus) the capsid protein is recognizable first in the nucleus (by 3 h—Fig. 50-22) and 1 h later in the cytoplasm.

The nuclear migration of the nucleocapsid protein in orthomyxovirus replication is probably required for protection and transport of the viral RNA processed in the nucleus. With measles virus (Ch. 59), the only paramyxovirus associated with intranuclear accumulation of nucleocapsids, the proportion of nucleocapsids found in the nucleus or in the cytoplasm varies in different cell types, and only cytoplasmic accumulation is accompanied by release of infectious virus.

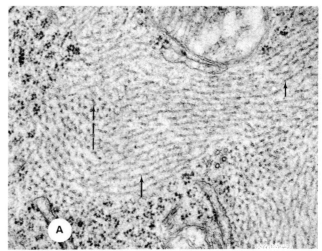

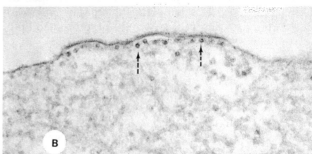

FIG. 50-21. Maturation of an enveloped virus with a helical nucleocapsid (Sendai virus, a paramyxovirus) as seen in electron micrographs of infected chick embryo cells. **A.** Accumulation of nucleocapsids in the cytoplasm, some cut transversely **(dashed arrow),** some longitudinally **(solid arrows). B.** Transversely cut nucleocapsids under thickened areas of the plasma membrane, covered by spikes, preliminary to budding and virion formation. (Darlington RW et al: J Gen Virol 9:169, 1970)

In **envelope assembly** virus-specified envelope proteins go directly to the appropriate cell membrane, replacing the host proteins. In contrast, *the lipids are those produced by the host cell.* Thus, the lipids are very similar to those of the cellular membranes in composition and also in relative specific activities if the virus-producing cells are labeled before infection with a radioactive precursor. Furthermore, when the plasma and the nuclear membranes of a cell differ in composition, the viral envelope has the lipid constitution of the membrane where its assembly takes place (e.g., the nuclear membrane for herpesvirus; the plasma membrane for orthomyxo- and paramyxoviruses).

The formation of the viral envelope is best understood for viruses that bud at the plasma membrane (e.g., orthomyxoviruses). The **viral glycoproteins** are synthesized on polysomes bound to the endoplasmic reticulum (see Cellular Membranes, Ch. 49), whereas all other viral

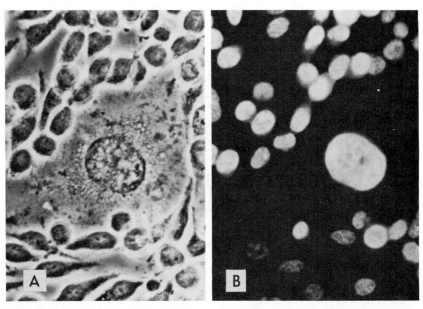

FIG. 50-22. Localization of the nucleocapsid antigen of fowl plaque virus (an ortho-myxovirus) in the nucleus 3 h after infection of a culture of L cells. The cells were fixed and treated with fluorescent Ab to the viral Ag. **A.** Phase contrast micrograph. **B.** Micrograph of the same field in ultraviolet light, where only the Ag bound to the viral Ab is visible. The absence of fluorescence in the cytoplasm is especially evident in the giant cell. (Franklin R, Breitenfeld P: Virology 8:293, 1959)

proteins are synthesized on free polysomes. As the polypeptide is synthesized its amino end is inserted into the membrane, and as it emerges on the other side it is glycosylated by host membrane enzymes. The carboxyl end may remain on the cytoplasmic side, so that the protein spans the lipid bilayer (transmembrane protein), or it may be embedded within the membrane. In contrast, when the viral envelope contains a **matrix protein,** it is made on polysomes and sticks to the cytoplasmic side of the plasma membrane, probably to the protruding ends of the viral glycoproteins.

The same virus will therefore differ in its lipids and carbohydrates when grown in different cells, with consequent differences in physical, biologic, and antigenic properties (see also The Envelope, Ch. 46). The viral proteins, however, to a certain extent also select the lipids with which they aggregate: thus two types of influenza virus (A and B) grown in the same cells may differ in the proportions of individual phospholipids.

Formation and Release of Virions. Envelopes are formed around the nucleocapsids by **budding of cellular membranes** (Fig. 50-23). This budding can be considered the result of an intimate adhesion of the nucleocapsid with the matrix (M) protein present at the cytoplasmic side of the cell membrane where the viral glycoproteins are embedded (Fig. 50-23); the adhesion causes the

membrane to curve into a protruding sphere surrounding the nucleocapsid. Virion maturation and budding differ slightly for a virus whose envelope does not contain a matrix protein (e.g., togaviruses): the nucleocapsid associates directly with the glycoprotein spikes which span the envelope and extend into the cytoplasm.

The bud detaches from the membrane by a process that can be considered the reverse of penetration. If budding is from the surface membrane (e.g., paramyxoviruses), release occurs at the same time. If budding occurs in a vacuole (e.g., togaviruses), release requires subsequent fusion of the vacuole with the cell membrane. Either method of release is compatible with cell survival and can be very efficient, allowing a cell to release thousands of viral particles per hour for many hours. Herpesviruses bud out of the nuclear membrane into the cytoplasm and reach the outside through cytoplasmic channels.

With some newly isolated influenza viruses (Ch. 58) a deviation from the normal pattern of maturation leads to the formation of infectious cylindrical **filaments** with a diameter similar to that of the spherical particles. Their formation depends on genetic properties of the virus, type of host cell, and environment (e.g., in cell cultures it is greatly enhanced in the presence of vitamin A, alcohol or surfactants).

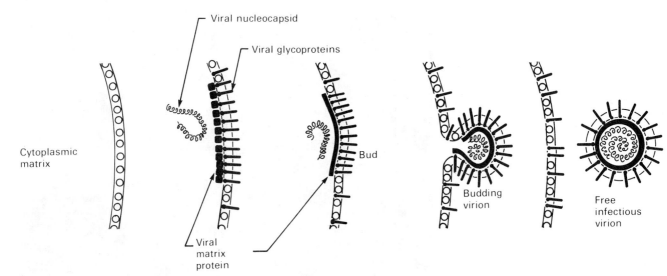

FIG. 50-23. Budding of an enveloped virus (orthomyxo- or paramyxovirus). **White circles** = host proteins of the plasma membrane (specified by cellular genes); **black spikes** = viral glycoproteins (the peplomers specified by viral genes), which become incorporated in the cell membrane, replacing host cell proteins, before budding of the viral particles begins. The viral **matrix (M) protein** attaches to the inner surface of the plasma membrane segment containing the viral glycoproteins and appears to serve as recognition site for the nucleocapsid as well as a stabilizing structure.

Cell Surface Alterations Produced by Viral Maturation. With orthomyxo- and paramyxoviruses the budding occurs in membrane regions containing the virus-specified building blocks of the envelope (e.g., the glycoprotein spikes [peplomers] and membrane protein). These viral constituents become incorporated in the membranes before budding begins (Fig. 50-23), conferring on the cell some properties of a giant virion. Thus the infected cells may bind RBCs (**hemadsorption,** Fig. 50-24, the equivalent of hemagglutination) or viral Ab, or they may fuse with uninfected cells to form multinucleated syncytia, called **polykaryocytes** (Fig. 50-25). This fusion is equivalent to adsorption of virions to uninfected cells.

Formation of **endogenous polykaryocytes** requires infectious virus, is slow, and is prevented by inhibitors of

FIG. 50-24. Hemadsorption of HeLa cells infected by Newcastle disease virus. Cells had been heavily irradiated with x-ray several days before infection; they stopped multiplying but increased in size and became giant cells, facilitating observations. The virus multiplies regularly in these cells. **A.** Cell 5 h after infection. The ability to adsorb chicken RBCs begins to appear at two opposite regions of the cell surface. At these regions new cell membrane appears to be laid down, allowing viral components to become incorporated together with the cellular components. **B.** Cells at lower magnification, 9 h after infection, showing that the entire cellular membrane has developed the capacity for hemadsorption. The RBCs are firmly attached to the cells and are not removed by repeated washing. (Marcus P: Cold Spring Harbor Symp Quant Biol 27:351, 1962)

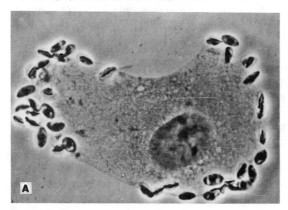
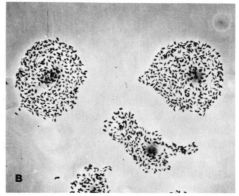

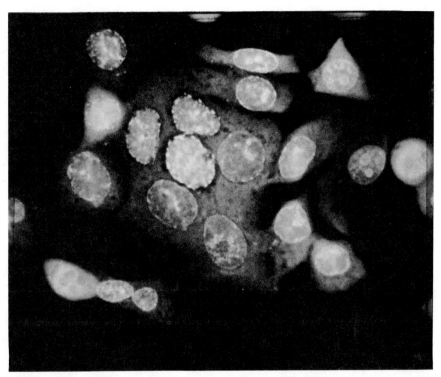

FIG. 50-25. Formation of multinucleated syncytia (polykaryocytes) by Hep-2 cells infected by herpes simplex virus. Five cells have fused completely, and several others partly, into a central mass. Cells were stained with the fluorescent dye acridine orange and photographed in a dark field. (Roizman B: Cold Spring Harbor Symp Quant Biol 27:327, 1962)

protein synthesis added to cells together with the virus. In contrast, formation of **exogenous polykaryocytes** by external viruses (especially for the purpose of cell hybridization; see in Ch. 49) can be caused by inactivated virus, is rapid, and is not affected by inhibitors. In both cases the polykaryocytes are formed by fusion of the altered membranes (cell–cell, or cell–virus).

COMPLEX VIRUSES

The process of maturation of the highly organized poxviruses can be inferred by a temporal ordering of electron microscopic images (Fig. 50-26). Synthesis of the viral constituents takes place in cytoplasmic foci called **"factories."** These foci of **viroplasm** contain granular and fibrillar material. At the beginning of particle formation fragments of bilaminar membranes appear, eventually forming trilaminar continuous membranes surrounding condensed viroplasm. Viral DNA appears to enter almost completely preformed membrane particles. A dense, fibrillar, immature nucleoid, which contains the viral DNA, develops within the completed membrane. The multilayered membranes thus differentiate into two membranes: the inner one envelops the maturing nu-

cleoid, while the external one acquires the characteristic pattern (recognizable with negative staining) of the surface of the virion (see Fig. 56-8, Ch. 56). As with bacteriophages and with other animal viruses, peptide cleavage during poxvirus maturation appears to perform an important morphogenetic role, since rifampin inhibits these late events (see Ch. 51 and Viral Multiplication, Ch. 56). In contrast to simple viruses, the poxvirus membrane contains newly synthesized lipids that differ in composition from cellular lipids.

The maturation of poxviruses has important differences from the maturation of simpler viruses: this differentiation after the precursors have been enclosed within the primitive membrane suggests that *poxviruses may be transitional forms toward a cellular organization,* where differentiation is all internal.

ASSEMBLY FROM SUBUNITS IN PRECURSOR POOLS

In a cell simultaneously infected by certain pairs of related viruses that differ in capsid Ags, such as poliovirus types 1 and 2, **phenotypic mixing** can occur because capsids made from building blocks of both types (or either

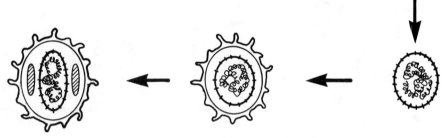

Filamentous matrix

Membrane
formation
in matrix

Completion of
nucleoid membrane
and entrance
of viral DNA

Differentiation of nucleoid and envelope

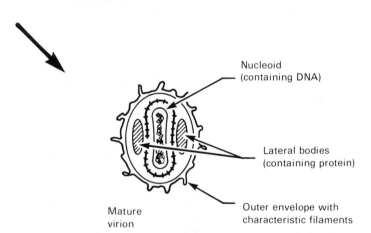

Nucleoid
(containing DNA)

Lateral bodies
(containing protein)

Outer envelope with
characteristic filaments

Mature
virion

FIG. 50-26. Scheme of the development of vaccinia virions, reconstructed from electron micrographs (see Fig. 56-8). The entire process proceeds in "cytoplasmic factories."

type) may enclose the same type of genome. Indeed antiserum to either Ag may neutralize particles with mixed capsids. Hence infection with a mixture of poliovirus of types 1 and 2 may yield six classes of virions (Fig. 50-27) with different combinations of genotype (RNA) and phenotype (protein) (Fig. 50-28). Similar mixing is seen with adenovirus, yielding capsids with fibers of different lengths and with random combinations of antigenic

types. Phenotypic mixing affecting envelope glycoproteins can occur even between unrelated viruses, such as rhabdoviruses and retroviruses.

Therefore, the virions of animal viruses and bacteriophages are assembled from building blocks more or less **randomly picked from pools.** Moreover, it can be shown that these building blocks are specified not solely by the nucleic acid of the infecting virions, but also by that of

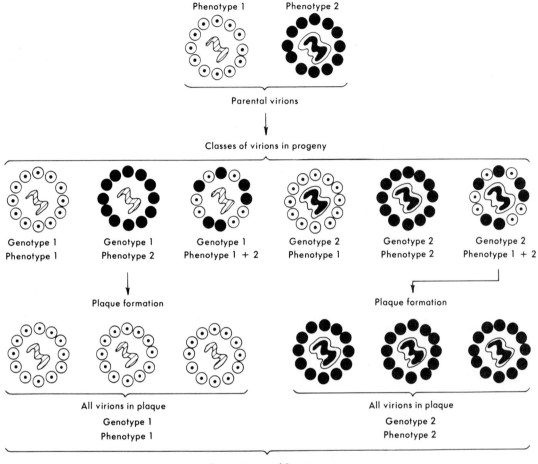

Genotype = the genetic information residing in the nucleic acid

Phenotype = the immunological characteristics residing in the capsid protein

FIG. 50-27. Mechanism underlying phenotypic mixing of the antigenic specificity of poliovirus. Cells mixedly infected by types 1 and 2 produce virions of six genotype and phenotype combinations. Cloning of these by plaque formation yields unmixed virions, with phenotype determined by the genotype of the initiating virion, irrespective of the latter's phenotype.

the progeny: preferential synthesis of one genotype in a mixedly infected cell leads to preferential synthesis of the corresponding proteins.

ABORTIVE INFECTION

Owing to a property either of the cells or of the virus, when some step in viral replication is defective the result is an **incomplete single cycle,** which does not yield infectious virus, although it may kill the cells.

Cell-Dependent Abortive Infection. Nonpermissiveness may be a property of the species of origin of the cells (e.g.,

with certain strains of influenza virus in HeLa [human] cells, or with some strains of herpes simplex virus in dog cells). In other cases, it depends on the tissue of origin (e.g., with measles virus in nerve cells).

The mechanisms differ in different cases. Thus in abortive influenza and measles virus infections nucleocapsids accumulate in the nucleus, suggesting that the defect is in a cellular component involved in the assembly in, or transport to, the cytoplasm. With herpesvirus the nuclear membrane, which normally provides the virion envelopes, fails to proliferate and the nucleocapsid remains in the nucleus unenveloped.

Monkey cells are not effective hosts for adenoviruses unless they are also infected with a **helper** SV40 virus, which allows complete

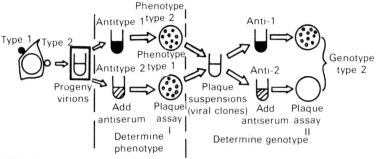

FIG. 50-28. Experimental basis for determining the phenotype and geno-type—in respect to capsid Ags—of virions produced in cells mixedly in-fected by polioviruses types 1 and 2. Only one genotype determination, on virus derived from a single plaque of assay 1, is shown here; it assigns a type 2 genotype to the progeny virion that generated that viral clone. Nor-mally a number of plaques from either plate of assay I would be tested.

transcription and increased translation of late adenovirus mRNAs in these cells.

These results show that some steps in viral multiplication are cell-dependent.

Virus-Dependent Abortive Infection. This results when defective viral mutants are formed during viral replica-tion. Thus infection at a high multiplicity with influenza virus, which has a segmented genome, produces defective viral particles lacking the largest RNA piece (**von Magnus phenomenon**). These defective particles will cause an abortive infection if infecting alone, but they replicate if the cells are also infected by fully infectious particles, which act as helper. In serial passages at high multiplicity the defective particles become enriched until they constitute almost the whole yield, probably because the smaller genome segments have a selective advantage by replicating faster. Production and enrichment of the

defective particles are partly cell-dependent. These de-fective viral particles (**defective interfering particles**), not only may interfere with multiplication of infectious virus but also may be responsible for inducing and maintain-ing persistent infections (Chs. 51 and 53).

In a parallel mechanism the small oncogenic DNA-containing viruses (polyoma and SV40) generate **dele-tion mutants** which express some gene functions and transform cells to the cancer state but do not produce progeny virions, except in cells coinfected by regular virus as helper.

An extreme case of viral defectiveness is observed with human **adeno-associated virus,** a parvovirus, which rep-licates completely only in cells coinfected by an adenovi-rus; this absolute dependence on helper may be permit-ted by the widespread occurrence of adenovirus infection in man.

GENETICS

The study of animal virus genetics has made impressive progress in recent years, owing to methodologic improve-ments and to the extensive use of molecular methods. Yet it is not known as well as the genetics of bacteriophages, owing to the greater technical difficulties and also to an absence of recombination with some animal viruses and a low frequency with most others.

MUTATIONS

Mutations of animal viruses occur spontaneously and can be induced by various mutagenic treatments, in-cluding nitrous acid, 5-bromodeoxyuridine (BUdR), hy-droxylamine, and nitrosoguanidine (fluorouridine is also

useful for RNA viruses). Small deletions can be readily isolated with some viruses (e.g., papovaviruses). The known mutant phenotypes are numerous and cover a larger range than bacteriophage mutations, because there are more ways for studying their properties, for in-stance, their various effects on animals. The frequency of mutation, either spontaneous or induced, is higher with RNA-containing than with DNA-containing viruses be-cause RNA polymerases are less accurate.

MUTANT TYPES

Only one class of **conditionally lethal mutations, tem-perature-sensitive (ts) mutations,** are useful in animal viruses. Suppressor-sensitive mutations occur and can be demonstrated by translation in vitro using yeast suppressor tRNAs; however, no suppressors have been recognized in animal cells. The ts mutations have been extremely valuable because they occur in many (possibly all) genes, both in DNA-containing and in RNA-containing viruses. In spite of some defects, such as leakiness and a high reversion rate, they form the basis for most of our present knowledge of animal virus genetics.

Other kinds of mutants affect special properties. 1) **Host-dependence (or host-range) mutants,** occurring in many viruses, fail to multiply in certain nonpermissive cell types, which differ for the various viruses. These mutants, like the suppressor-sensitive mutants of phage, can be propagated in permissive cells, are not leaky, and have a low reversion rate; however, they are limited to one or a few genes involved in overcoming the nonpermissiveness of the cells. 2) **Plaque size or type** (Fig. 50-29). These mutants are not as diverse as the corresponding phage mutants because animal virus plaques have less detail (they affect fewer and larger cells). Differences of plaque size may depend on differences either in features of the multiplication cycle or in the surface charges of the virions. Small charge differences affect plaque size because agar contains a sulfated polysaccharide which adsorbs the more highly charged virions, especially at certain pHs. 3) **Noncytopathic mutants** produce plaques that are not lytic but are recognizable by other criteria (e.g., hemadsorption). 4) **Pock type.** Some poxvirus types produce white smooth pocks on the chorioallantoic membrane of the chick embryo, while others produce red, ulcerated pocks (*u* character). 5) **Surface properties,** detected by physical methods. 6) **Antigenic properties.** 7) **Hemagglutination** (in orthomyxoviruses). 8) **Resistance to inactivation** by a variety of agents. 9) **Resistance toward or dependence upon inhibitory substances during multiplication.** 10) **Pathogenic effect** for organisms. 11) **Functions of certain viral genes in the infected cells,** such as production of thymidine kinase or induction of interferon.

PLEIOTROPISM

Mutants selected for a given phenotypic alteration are often found to be changed in other properties as well. For instance, poliovirus mutants with altered chromatographic properties often (though not always) have decreased neurovirulence for monkeys. This **pleiotropism** reflects the effect of a single viral protein on several properties of the virus. The connection between the properties is variable: for example, mutations that can modify different charged groups of the viral capsid may have similar effects on chromatographic adsorption but different effects on the more specific adsorption to cells.

Pleiotropism has useful applications. Thus by employing characters that are detectable in vitro (such as temperature sensitivity of various functions, or deficiency of cytopathic effect) it is possible to select **attenuated** (i.e., nonvirulent) strains for use in live virus vaccines, reserving the more cumbersome animal testing for final characterization. This approach has been used in the selection of poliovirus vaccines.

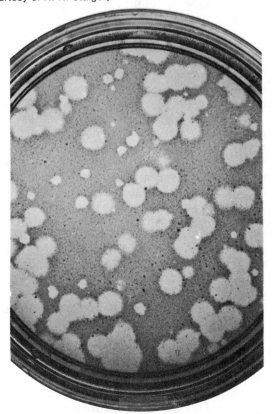

FIG. 50-29. Two plaque-type mutants of fowl plaque virus on a monolayer of chick embryo cells. The wild type produces large round plaques with fuzzy edges; a small-plaque mutant produces small plaques with irregularly indented outline and sharp edges. (Courtesy of H. R. Staiger)

COMPLEMENTATION

With many viruses complementation of ts mutants has been useful for determining the functional organization of the viral genome. When two ts mutants complement each other, the yield of a mixed infection, at the nonpermissive temperature, is greater than the sum of the separate yields at the same multiplicity of infection. Complementation is very efficient with the larger DNA viruses (adenoviruses, herpesviruses), for which the yield from mixed infection may approach 50% that of cells infected with wild-type virus. However, many animal viruses complement much less, especially those (e.g., picornaviruses) whose genome is translated into a large polyprotein, which is then cleaved. Interference with cleavage caused by mutations that alter the tertiary structure of the polyprotein prevents the formation of separate functional products, and therefore prevents complementation.

In either DNA- or RNA-containing viruses several complementation groups have been identified. Each corresponds to a different gene, because the mutations in different groups usually display different functional alterations or affect different proteins recognizable by gel electrophoresis; moreover, in viruses with a fragmented genome, in which each fragment is a gene, the number of the complementation groups is close to that of the fragments.

GENETIC RECOMBINATION

DNA VIRUSES

In the large DNA viruses with a linear DNA (adenoviruses, herpesviruses, poxviruses) recombination between pairs of ts mutations or between a ts and a plaque type mutation occurs at frequencies comparable to those observed in some phages or bacteria. The yield of mass cultures regularly contains the two reciprocal recombinant types, in comparable proportions. However, little is known about the individual recombination events. Recombination is also observed between different types of the same group, but at a lower frequency. The proportion of recombinants between markers tested pairwise are additive. By a series of two-factor crosses, or by three-factor crosses, the order of various mutations can be unambiguously established, in spite of the unusual structure or replication of these viral DNAs (see above). By this approach the herpesvirus map is 25–30 units long; hence the frequency of recombination per unit length is much less than for phage T4 but of the same order as that of bacteriophage λ (see Chs. 47 and 48). The maps are **linear,** i.e., distant markers are unlinked.

Maps have also been determined by molecular methods (Fig. 50-30) based on the use of characteristic fragments of the viral DNA generated by restriction endonucleases. The order of the fragments is established from the overlaps of fragments produced by different enzymes. In **intertypic crosses** differences in the fragments of the two parental DNAs reveal the crossover point in the DNA of a recombinant (Fig. 50-30A). In **marker rescue** a fragment from wild-type virus is introduced into the cells by transfection together with viral DNA from a ts mutant, if the fragment overlaps the mutational site, wild-type virus can be generated by recombination (Fig. 50-30B). With viruses having a cyclic DNA (e.g., polyoma virus) wild-type progeny virus can

be obtained without a need for recombination (Fig. 50-30C): a strand of the wild-type fragment is annealed to the complementary complete strand of the mutated DNA and the resulting hybrid is introduced into the cells. Extension of the fragment copying the complementary strand yields a wild-type strand if the fragment overlaps the mutation.

The maps obtained by molecular mapping are in good agreement with those obtained by recombination.

RNA VIRUSES

Burnet, studying influenza virus, first recognized recombination between animal viruses in 1951. The markers of this and other RNA viruses with **fragmented genomes** (classes III and IV) fall into groups that show high recombination frequencies, up to 50%, between groups but none within groups. The groups correspond to RNA pieces which assort nearly randomly at maturation. Thus in recombinants from an intratypic cross of two variant viruses having corresponding segments of different sizes, mutations or characteristic peptides can be assigned to fragments. This method of recombination is a very important cause of variation among viruses with fragmented genomes: thus much of the antigenic variation of influenza virus is probably caused by recombination between different species (of the same type) (see Evolution of Viral Antigens, Ch. 52). In contrast, viruses with **continuous genomes** show no or little recombination (less than 1% frequency for poliovirus). Absence or a low frequency of recombination derives in part from the short length of the genome, in part perhaps from the rare occurrence of RNA recombination. Most RNA virus recombinants seem to arise at the beginning of replication, probably between nonreplicated molecules.

An exceptionally high recombination frequency among RNA viruses with continuous genomes is found in retroviruses; this probably derives from the diploidy of the virions, because recombination appears to occur in cells infected by heterozygous virions, in which the two RNA molecules carry different markers.

Gene **maps** of viruses **with a positive-strand continuous genome** can be obtained biochemically (Fig. 50-30D). Thus the poliovirus gene order has been determined from the labeling of the various proteins when a radioactive amino acid is added to the cell culture together with the antibiotic pactamycin, which inhibits initiation of protein synthesis. Since the whole genome is translated into a single polypeptide chain there will be more label in proteins corresponding to the distal (3′ end) part of the RNA, which has the highest chance of being still untranslated when the label is added. The relative labeling of various proteins gives their location in the uncleaved polypeptide chain.

The gene order for several **negative-strand** viruses has

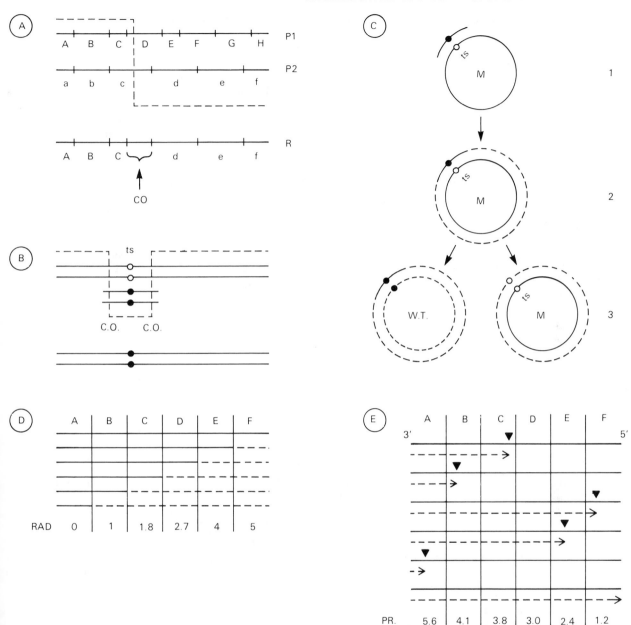

FIG. 50-30. Molecular mapping. **A.** In the intertypic cross of two viruses, **P1** and **P2,** the derivation of the DNA in the recombinant **(R)** can be ascertained from the restriction endonuclease fragments it contains; the position of the crossover point is delimited **(CO)**. **B.** Mixed infection of cells with a viral DNA carrying a temperature-sensitive (ts) mutation and a wild-type (wt) fragment overlapping the mutation **(marker rescue)** produces a wt molecule by two crossovers **(CO)**. **C.** 1) A strand of a wt DNA fragment is annealed to a complete strand carrying a ts mutation **(M);** 2) after transfection (i.e., infection with pure DNA) the fragment is elongated; 3) replication yields a wt DNA molecule. **D. Continuous lines** = a set of incomplete polyproteins at the time of addition of amino acid label + pacta-mycin. **Dashed lines** = radioactive parts of the polyproteins when they are completed. The gradient of radioactivity **(RAD)** incorporated, after processing, establishes the order of genes (**A to F**). **E. Continuous lines** = a set of viral genomes exposed to UV light; **arrowhead** = pyrimidine dimers. **Dashed line** = the transcribed segments. Numbers relative amounts of protein **(PR)** synthesized for each gene (**A to F**). The gradient of synthesis establishes the order of genes.

also been determined by measuring the effect of UV-ir-radiation of virions on the synthesis of the polypeptides specified by the various genes. The method is based on the blocking of the progress of transcription by the UV-induced pyrimidine dimers; therefore, a gene at the 5′ end on the template strand has the highest sensitivity to the radiation because its transcription is blocked by any dimer along the whole genome, and the synthesis of the peptide it specifies undergoes the greatest reduction (Fig. 50-30E).

Heterozygosity. True heterozygosity is observed in retroviruses, which have two molecules of single-stranded RNA in the virions. **Multiploid heterozygosity** occurs in enveloped viruses that generate a fraction of particles with two nucleocapsids. This event is frequent in orthomyxo- and paramyxoviruses because two nucleocapsids can be readily accommodated in the floppy envelope; rhabdoviruses form occasional diploid particles of double length.

REACTIVATION OF UV-IRRADIATED VIRUSES

Cross-Reactivation. With viruses that give rise to efficient recombination (large DNA viruses, influenza) markers from a strain inactivated with UV light can be rescued by active virus simultaneously infecting the cells. The survival of a given marker, as a function of the UV dose, is much greater than the survival of the whole virus, as already seen with phages (see Genetic Reactivation, Ch. 47).

Multiplicity Reactivation. Like bacteriophages, animal viruses capable of recombination also display **multiplicity reactivation** when UV-inactivated: as multiplicity of infection is raised the proportion of cells yielding infectious virus increases excessively.

Both cross- and multiplicity reactivations are much more pronounced for RNA viruses with a fragmented than with a continuous genome, in accord with the different recombination frequencies.

SELECTED READING

MULTIPLICATION

BOOKS AND REVIEW ARTICLES

BACHRACH HL: Comparative strategies of animal virus replication. Adv Virus Res 22:163, 1978

CHOPPIN PW, COMPANS RW: Reproduction of paramyxoviruses. Comp Virol 4:95, 1975

LEVINE AG, VAN DER VLEIT PC, SUSSENBACH JS: Replication of papovavirus and adenovirus DNA. Curr Top Microbiol Immunol 73:67, 1976

LEVINTOW L: Reproduction of picornaviruses. Comp Virol 2:109, 1974

LONBERG-HOLM K, PHILIPSON L: Early Interactions between Animal Viruses and Cells. Virology Monographs No. 9. Basel, Karger, 1974

LURIA SE, DARNELL JE JR, BALTIMORE D, CAMPBELL A: General Virology, 3rd ed. New York, Wiley, 1978

MOSS B: Reproduction of poxviruses. Comp Virol 3:405, 1974

RAGHOW R, KINGSBURY DW: Endogenous viral enzymes involved in messenger RNA production. Annu Rev Microbiol 30:21, 1976

ROIZMAN B, FURLONG D: Replication of herpesviruses. Comp Virol 3:229, 1974

SILVERSTEIN SC, CHRISTMAN JK, ACS G: Reovirus replication cycle. Annu Rev Biochem 45:375, 1976

SPECIFIC ARTICLES

ANDERSON CW, BAUM PR, GESTELAND RF: Processing of adenovirus 2–induced proteins. J Virol 12:241, 1973

BALL LA, WHITE CN: Order of transcription of genes of vesicular stomatitis virus. Proc Natl Acad Sci USA 73:442, 1976

CHOW LT, GELINAS RE, BROKER TR, ROBERTS RJ: An amazing sequence arrangement at the 5′ ends of adenovirus 2 mRNA. Cell 12:1, 1977

HAY AJ, LOMNICIZI B, BELLAMY AR, SKEHEL JJ: Transcription of the influenza virus genome. Virology 83:337, 1977

HONESS RW, ROIZMAN B: Regulation of herpesvirus macromolecular synthesis. I. Cascade regulation of the synthesis of three groups of viral proteins. J Virol 14:8, 1974

NOMOTO A, DETJEN B, POZZATTI R, WIMMER E: Location of the polio genome protein in viral RNAs and its implication for RNA synthesis. Nature 268:208, 1977

REKOSH DMK, RUSSELL WG, BELLET AJD, ROBINSON AJ: Identification of a protein linked to the ends of adenovirus DNA. Cell 11:283, 1977

TATTERSALL P, WARD DC: Rolling hairpin model for replication of parvovirus and linear chromosomal DNA. Nature 263:106, 1976

GENETICS

BOOKS AND REVIEW ARTICLES

FRAENKEL-CONRAT H, WAGNER RR (eds): Regulation and Genetics—Genetics of Animal Viruses. Comp Virol, Vol. 9, 1977

SPECIFIC ARTICLES

BALL LA, WHITE CN: Order of transcription of genes of vesicular stomatitis virus. Proc Natl Acad Sci USA 73:442, 1976

GESTELAND RF, WILLS N, LEWIS JB, GRODZICKER T: Identification of amber and ochre mutants of the human virus Ad2⁺ ND1. Proc Natl Acad Sci USA 74:4567, 1977

PALESE P, SCHULMAN JL: Mapping of the influenza virus genome: identification of the hemagglutinin and the neuraminidase genes. Proc Natl Acad Sci USA 73:2142, 1976

SAMBROOK J, WILLIAMS J, SHARP PA, GRODZICKER T: Physical mapping of temperature-sensitive mutations of adenovirus. J Mol Biol 97:369, 1975

SCHOLTISSEK C, KOENNECKI I, ROTT R: Host range recombinants of fowl plague (influenza A) virus. Virology 91:79, 1978

STOW ND, SUBAK-SHARPE JH, WILKIE NM: Physical mapping of herpes simplex virus type 1 mutations by marker rescue. J Virol 28:182, 1978

chapter **51**

INTERFERENCE WITH VIRAL MULTIPLICATION

Agents that interfere with viral multiplication are useful not only for therapy and prophylaxis but also for advancing our understanding of viral biology and of infection.

THE CONTROL OF VIRAL DISEASES BY INHIBITION OF REPLICATION

Viral diseases result from a series of growth cycles that kill or alter cells (see Ch. 53). The maximal goal of antiviral treatment—to restore function to the infected cells—is usually unattainable because cellular macromolecules are damaged early in viral infections. Accordingly, the realistic goal is to stop viral replication and thus prevent spread to additional cells. But even this more limited goal presents considerable difficulties.

A major one is the problem of inhibiting the viruses without harming the cells. This selectivity is possible in bacteria because of their many metabolic, structural, and molecular differences from animal cells. Thus sulfanilamide interferes with the function of p-aminobenzoic acid, which is a vitamin in bacterial but not in animal cells; penicillin interferes with the synthesis of the peptidoglycan, which is unique to bacteria; and streptomycin interacts with molecular features that are peculiar to bacterial ribosomes. The dependence of viral multiplication on cellular genes, in contrast, limits the points of differential attack (Fig. 51-1). Even the largest viruses (e.g., poxviruses) contain only about 7% as much genetic information as *E. coli* and should therefore have fewer biochemical reactions that are unique in relation to the cells of the host.

Another limitation is that diseases become evident only after extensive viral multiplication and cellular alteration have occurred. Therefore, the most general approach to control is prophylaxis. Therapy is chiefly limited to localized viral diseases, such as herpetic keratoconjunctivitis (Ch. 55), where the killing of some uninfected cells can be tolerated if the damage is subsequently repaired. In addition, the special properties of herpesviruses permit the use of certain drugs for treating systemic herpes infections (see Agents Interfering with DNA Synthesis, below).

As with bacterial chemotherapy, a third important limitation of antiviral therapy is the emergence of **resistant mutants.** In order to avoid their selection, the principles valid for bacteria are equally applicable to viruses: adequate dosage, multidrug treatment, and avoiding therapy unless clearly indicated. Fortunately, however, genetic resistance to two important antiviral agents—interferon and interferon inducers—does not seem to occur.

VIRAL INTERFERENCE

When viruses of more than one type infect the same cell each may multiply undisturbed by the presence of the others, except for possible recombination or phenotypic mixing. In certain combinations, however, the multiplication of one type of virus may be inhibited by the other. This inhibition is called **viral interference.**

The notion of interference developed first from observations with ring spot virus in tobacco plants. The initial lesions regress but the virus persists, and if the plant is reinoculated with the same virus no new lesions develop. Thus the first infection interferes with the expression of the second infection. Subsequently it was found that in monkeys infection with a mild strain of yellow fever virus (a togavirus) can prevent the usually lethal disease caused by a virulent strain, or even by an antigenically unrelated togavirus, showing that the protection is not due to Ab. Interference was later found also with viruses in bacteria, thus opening the way for quantitative studies.

The study of interference with animal viruses took an important turn when Isaacs and Lindenmann, in 1957, discovered that virus-infected cells produce a substance, which they called **interferon,** that accounts for many observed instances of viral interference.

DEMONSTRATION OF INTERFERENCE WITH ANIMAL VIRUSES

Interference was observed with many pairs of viruses in animals, but especially with influenza viruses in the allantoic epithelium of the chick embryo, and recently with a variety of viruses in cell cultures. A typical experiment

FIG. 51-1. Points in the growth cycle of animal viruses that are suitable for differential inhibition. At points **A** and **B** only certain viruses use a viral function: **A,** viruses with a virion's transcriptase use it for partial or complete transcription; **B,** release of enveloped virus depends on incorporation of the viral envelope in the cellular membrane. HBB = 2-(α-hydroxybenzyl)-benzimidazole.

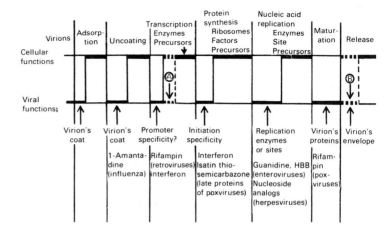

consists in inoculating the allantoic cavity with influenza A virus, as **interfering virus,** followed 24 h later by influenza B virus, as **challenge virus:** the multiplication of the second inoculum is partially or totally inhibited. The experiments can be simplified by using inactivated interfering virus; since it does not multiply interference can be determined by measuring the yield of the challenge virus without the need to distinguish it from the interfering virus.

Interference depends on timing and on viral concentrations. Thus, if influenza A and B are inoculated **simultaneously** and at equal multiplicity in the allantoic cavity they can multiply concurrently in the same cells, as shown by the production of phenotypically mixed progeny particles (see Assembly from Subunits in Precursor Pools, Ch. 50). Even viruses of different families can multiply in the same cells under proper circumstances.

Thus electron microscopy reveals adenovirus and simian virus 40 virions, which have a different size, in the same nucleus; immunofluorescence demonstrates the unrelated Ags of Newcastle disease virus and parainfluenza virus 3 (both paramyxoviruses) in the same cytoplasm; and the same cell may contain characteristic inclusions of herpes simplex in the nucleus and vaccinia (a poxvirus) in the cytoplasm.

We shall first consider in some detail the role of interferon in viral interference, and shall then consider a heterogeneous group of other mechanisms.

INTERFERON

Interferon was discovered in the course of studying the effect of influenza virus inactivated by ultraviolet (UV) light on fragments of the chick chorioallantoic membrane, maintained in an artificial medium. The supernatants, although devoid of viral particles, inhibited the multiplication of active influenza virus in fresh frag-

ments. Subsequently such "interferons" were shown to be produced by cells infected by almost any animal virus, either DNA- or RNA-containing, and in tissue culture or in the animal.

CHARACTERISTICS

Purified interferons from various sources consist of small glycoproteins that are usually stable at low pH (e.g., pH 2 in the cold) and fairly resistant to heat (mouse interferon loses half its activity after 1 h at 50° C, chick interferon at 70°). They bind strongly to certain polynucleotides [poly(I) or poly(U)].

Interferons are not virus-specific, but cell-specific, in both their production and their effects. Interferons induced by the same agent in different species, or even in different cells of the same species (such as human fibroblasts and leukocytes) differ in antigenicity, isoelectric point, and molecular weight. Thus a given species has more than one interferon-specifying gene. For example, two mouse interferons, of 38,000 and 22,000 daltons, differing in glycosylation, have been observed, and two human interferons, of 15,000 and 21,000 daltons.

A given interferon *inhibits viral multiplication most effectively in cells of the species in which it was produced.* For instance, purified interferon of chick origin is less than 0.1% as effective in mouse cells as in chick cells. However, interferon produced in monkey kidney cells is effective in human as well as in monkey cells, and human interferons are active in cells of several mammals, including nonprimates.

Interferon is usually **assayed** by determining its effect on plaque production by a test virus: usually vesicular stomatitis virus (VSV, a rhabdovirus), which is very sensitive to interferon and produces plaques on cells of many vertebrate species. Serial interferon dilutions are added to the agar overlay, and the endpoint (a **unit**) is a 50% reduction in the number of plaques. Interferons are among the most powerful drugs available: 10–20 molecules, and possibly even fewer, are sufficient to confer resistance on a cell.

Although interferons are produced only in small amounts they can be obtained in highly pure form because they are very stable and can be specifically adsorbed. Thus interferons containing more than 10^9 units/mg of protein are obtained by affinity chromatography through columns of poly(U) or of antiinterferon Abs attached to a solid support.

Relation to Viral Multiplication. Interferon is produced by cells infected with complete virions, either infectious or inactivated. Under one-step conditions of viral multiplication the **synthesis of interferon** begins after viral maturation is initiated; if not interrupted by an early block in synthesis of host macromolecules, it continues at the same rate for 20–50 h, then stops (Fig. 51-2). The interferon is mostly released extracellularly. If the cells survive for a longer time, as after infection by UV-inactivated virus, they cannot produce interferon again, in response to reinfection, until after a **refractory period** of at least two cell divisions.

The amounts of interferon produced in different systems differ widely. Good interferon inducers are usually viruses that multiply slowly and do not block the synthesis of host protein early or damage the cells markedly. For example, an attenuated mutant of poliovirus is a much better interferon inducer than the wild type, which multiplies better; and the paramyxovirus of Newcastle disease multiplies well in chick embryo cells but causes little interferon production, while in human cells it causes a defective infection but induces abundant interferon. However, many togaviruses induce large amounts of interferon even though they multiply rapidly and have a pronounced cytopathic effect. Possible variables are effects on host macromolecule synthesis and amounts of inducer formed (see Production of Interferon, below).

Viral strains capable of high interferon production give rise to **autointerference** in endpoint assays: the dilutions containing the most virus may produce less virus because of the presence of enough interferon to block further cycles of viral multiplication. Autointerference can also arise by a different mechanism (see Animal Viruses: Defective Interfering Particles, below).

Relation to Cells. Although all animal cells appear able to produce interferon, *cells of the bone marrow and spleen, and macrophages, appear to play a special role.* Thus lethally x-irradiated mice grafted with rat bone marrow cells produce only rat-specific interferon. Moreover, antilymphocytic serum can inhibit interferon production in the animal. In vitro studies have shown that interferon is produced by T lymphocytes stimulated either by mitogens or by Ag in the presence of macrophages. Since vaccination against viral diseases stimulates these cells, interferon production may play a role in the resulting protection.

In a confluent culture of mouse embryo cells added spleen cells from virus-injected mice protect the surrounding cells from VSV, providing a **plaque assay for interferon-producing cells.** The proportion of spleen cells releasing interferon increases, to a maximum of approximately 1%, with the dose of the inducing virus; the interferon concentration in the serum of the same animal rises in parallel.

CHEMICAL INDUCTION OF INTERFERON: DOUBLE-STRANDED RNA

The nature of the stimulus to interferon production has been clarified by the discovery, by the Hilleman group, that **double-stranded RNAs,** such as reovirus RNA and certain synthetic polynucleotides, can induce a large production of interferon in many animals and in tissue cultures. The activity resides in polyribonucleotides, of a high molecular weight and resistant to enzymatic degradation, in which the 2′ position of the ribose is unsubstituted. One of the best inducers is **poly(I:C),** the double-stranded synthetic polymer formed by one chain of polyriboinosinic and one of polyribocytidylic acid.

The poly(I) strand is the more important since discontinuities in it diminish the induction potency. Single-stranded polynucleotides able to form a stable secondary structure (e.g., polyguanylic, polyriboinosinic, and polyriboxanthylic acids) are also active, as are certain other synthetic anionic polymers. Polydeoxyribonucleotides and DNA–RNA hybrids are inactive. All the inducers mentioned so far cause a **new synthesis** of interferon. Others, such as bacteria, rickettsiae, bacterial endotoxin, and phytohemagglu-

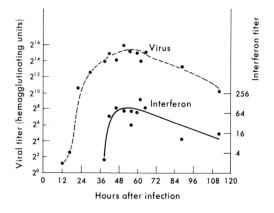

FIG. 51-2. Time course of viral multiplication and interferon synthesis in the allantoic membrane of the chick embryo infected with influenza virus. Owing to the low multiplicity of infection several cycles of viral growth were required before progeny virus could be detected; hence the lag observed is much longer than the regular eclipse period of the virus. Note the considerable delay in the synthesis of interferon. (Modified from Smart KM, Kilbourne ED: J Exp Med 123:309, 1966)

tinin, cause mostly **release** of preformed interferon, which occurs very rapidly and is insensitive to inhibition of protein synthesis.

The activity of synthetic polynucleotides in tissue cultures is greatly enhanced by the cationic polymer diethylaminoethyl-dextran (DEAE-dextran), (which also increases the infectivity of viral nucleic acids: see Productive Infection, Ch. 50). The DEAE-dextran and the polynucleotide form a compact electroneutral aggregate which is resistant to nucleases and binds to the cell surface, as shown by radioautographs using radioactive poly(I:C); some polynucleotide molecules may enter the cell and interact with cellular components.

Role of Double-Stranded RNA in Induction of Interferon by Viruses.

It is likely that, for most RNA viruses, double-stranded RNA segments produced during replication (Ch. 50) mediate the induction of interferon. Viruses containing double-stranded RNA in the virions may induce without replication. For DNA viruses the inducer may be double-stranded RNA resulting from symmetric transcription. Thus double-stranded viral RNA can be extracted from cells infected with vaccinia (a DNA virus), and it induces interferon in tissue cultures.

The role of the **partly double-stranded RNA replicative intermediates** (RIs) is based on observations (Fig. 51-3) with viruses whose virion transcriptase forms such an RI (see Replication of Single-Stranded RNA, Ch. 50) before the expression of viral genes. These viruses are poor inducers but become good inducers after mild UV irradiation; at high UV doses the interferon-inducing activity decays in parallel to the activity of the transcriptase itself (and hence parallels the ability to form double-stranded RNA).

By inactivating viral genes that are required for the release of progeny viral strands from the RI, low UV doses favor accumulation of the RIs; they do not affect the transcriptase, since proteins are much less UV-susceptible than nucleic acids. With viruses that do not use a virion transcriptase UV light does not enhance the inducing ability, apparently because it rapidly destroys the ability of the viral genome to code for the synthesis of new enzymes required for the formation of RIs.

The ability of some viruses to induce interferon is affected by temperature-sensitive mutations, but no clear pattern for the genetic control of this property has yet emerged.

PRODUCTION OF INTERFERON

In **virus-infected cells** the synthesis of interferon begins after viral maturation is initiated and then continues for many hours, unless the macromolecular syntheses of the host come to a halt. In **cells exposed to poly(I:C)** interferon production starts within about 2 h; the rate of synthesis increases for several hours and then rapidly declines **(shutoff).** After shutoff a new exposure of the cells to poly(I:C) does not restore interferon production for several hours **(hyporesponsiveness;** Fig. 51-4).

Interferon is synthesized on membrane-bound poly-

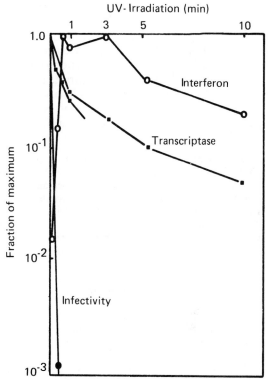

FIG. 51-3. Effect of ultraviolet (UV) irradiation on infectivity, interferon-inducing ability, and transcriptase activity of Newcastle disease virus (a paramyxovirus). There is a marked enhancement of inducing ability after low UV doses; then it decays with a slope similar to that of transcriptase activity but much smaller than that of infectivity. (Modified from Clavell LA, Bratt MA: J Virol 8:500, 1971)

FIG. 51-4. Refractory period following induction of interferon by double-stranded RNA injected into mice. After the first injection of RNA **(arrowhead)** there is a burst of interferon appearance in the serum, followed by the refractory period in which repeated injections **(arrows)** do not cause further production. During this period, however, the mice are fully resistant to infection with encephalomyocarditis virus (a picornavirus), because the induced antiviral resistance persists. (Modified from Sharpe TJ et al: J Gen Virol 12:331, 1971)

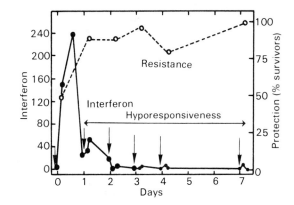

somes. As it is formed it is segregated into vesicles and is glycosylated; from the vesicles it is excreted outside the cells. Therefore until it is excreted interferon does not act on the cell that produces it. Large concentrations of extracellular interferon inhibit the induction of interferon by either a virus or poly(I:C); in contrast, low concentrations enhance induction by an RNA virus (**priming**). The latter effect may be due to the demonstrable interference, by interferon, with the regular evolution of the viral RI.

Regulation of Interferon Synthesis. The production of interferon begins with the transcription of its structural gene. After induction by poly(I:C) interferon mRNA makes its appearance, and this event—as well as interferon production—is prevented by actinomycin D (Fig. 51-5). The mRNA can be assayed after extraction of RNA from the cells because it acts as a template for the synthesis of interferon in cell extracts or in amphibian oocytes.

At the time of shutoff functional interferon mRNA disappears; however, if the cells are exposed at that time to actinomycin D the mRNA persists, shutoff does not occur, and interferon production is increased. This increase is even more pronounced if the cells are exposed for several hours to a reversible inhibitor of protein synthesis before actinomycin D (Table 51-1).

Such **superinduction,** characteristic of systems with **posttranscriptional regulation,** is interpreted according to the scheme of Fig. 51-6. The inducer simultaneously activates transcription of a gene for interferon and one for a posttranscriptional **repressor** that irreversibly inactivates the interferon mRNA (Fig. 51-6A); the repressor system (mRNA + protein) is labile (its measured half-life: 3–4 h), whereas the interferon mRNA is stable (half-life about 20 h); hence the accumulated repressor causes the early shutoff and the subsequent hyporesponsiveness (Fig. 51-6B). Addition of actinomycin D at the time of shutoff results in rapid disappearance of the labile repressor and its mRNA, while interferon continues to be

TABLE 51-1. Effect of Cycloheximide and Actinomycin D on Interferon Production*

Group	Cycloheximide (hours of treatment)	Actinomycin D at h 3–4	Interferon yield
1	None	None	500
2	0–18	None	1,200
3	0–4	Yes	56,000
4	0–18	Yes	2,000

*Data show the effect that inhibition of RNA and protein synthesis has on interferon production in rabbit kidney cell cultures after poly(I:C) induction. The combination of the two inhibitors at suitable times and concentrations brings about dramatic enhancement of interferon production (compare groups 3 and 1).

(Data from Tan Yh et al: Virology 3:503, 1971)

made on its stable mRNA (Fig. 51-6C). The inhibition of protein synthesis at the time of induction prevents the synthesis of both repressor and interferon, but not of their mRNAs; when the inhibitor is removed and actinomycin D is added there is a weak transient translation of the accumulated mRNA for repressor; and after the repressor system has decayed, interferon is synthesized on its persisting mRNA at a maximal rate (Fig. 51-6D).

The regulation of interferon production induced by viruses may differ somewhat from that described here for poly(I:C).

Because interferon is produced by cellular genes, it is clear why *viruses that block cellular mRNA or protein synthesis are poor inducers of interferon production.* Moreover, it is understandable that interferon synthesis fails in tissue culture cells simultaneously infected by a good inducer and a poor inducer: the poor inducer evidently inhibits the required cellular functions. Interactions of this type presumably also occur in animals, and may influence the pathogenesis of viral infections.

MECHANISM OF INTERFERON ACTION

Interferon causes antiviral resistance not directly but by **activating cellular genes for antiviral proteins:** it is ineffective in enucleated cytoplasts (see Cytoskeleton, Ch. 49) or in the presence of actinomycin D.

Interferon apparently induces the cell response by interacting with the cell surface. Thus interferon bound to polysaccharide particles (Sepharose) is as active as free interferon. (However, in this experiment it is difficult to exclude the penetration into the cells of interferon fragments generated by proteolytic cleavage.) At the cell surface interferon binds to receptors containing gangliosides (glycosylated phospholipids); transformed cells (Ch. 49), which are deficient in gangliosides, are less interferon-sensitive than normal cells. In human cells genes specifying the receptors are present on chromosome 21, and 21-trisomic (Down syndrome) cells are especially sensitive to interferon.

FIG. 51-5. Inhibition of interferon production by actinomycin D. Ehrlich ascites cells (a transplantable mouse cancer) were infected, at a multiplicity of 5 infectious units per cell, with Newcastle disease virus. Actinomycin D prevents interferon synthesis if added within 4 h; later additions have progressively less effect. (Modified from Wagner RR, Huang AS: Virology 28:1, 1966)

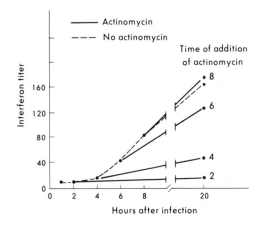

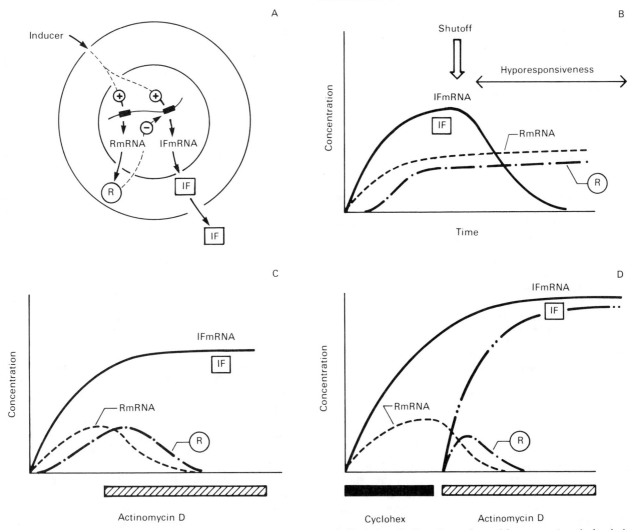

FIG. 51-6. Model of regulation of interferon synthesis induced by poly (I:C), the synthetic polymer formed from one strand of polyriboinosinic acid and one of polyribocytidylic acid, added at time 0. **A.** General features of the model. **B.** Hypothetic time dependence of transcription of the interferon and repressor mRNAs, showing the effect of repression. **C.** Superinduction caused by the later addition of actinomycin D. **D.** Superinduction produced by early inhibition of protein synthesis (using cycloheximide) followed by actinomycin D. IF = interferon; R = repressor; (+) = stimulation of transcription; (−) = inhibition of transcription.

After the removal of extracellular interferon the inhibition of viral multiplication persists for a considerable period, whose duration depends on the interferon concentration; inhibition may persist even after cell division. The inhibition can be overcome by virus at a large multiplicity of infection.

The **molecular mechanisms** of interferon-induced antiviral resistance are multiple, and probably differ in different cell–virus systems. Effects on uncoating of the virus, on the stability or methylation of the viral RNA, and on the assembly of virions have been reported. However, in vitro studies with extracts of interferon-treated cells show that the main target of interferon action is **translation,** which is blocked by two mechanisms, involving a protein kinase and a nuclease (Fig. 51-7). In both, the block requires the presence of minute amounts of **double-stranded RNA (dsRNA),** which seems to signal to the cells the presence of a viral infection. (As in the induction of interferon, the dsRNA may be that of a viral replicative intermediate, or it may result from symmetric transcription of the viral DNA.) Both translation blocks are therefore specific for virus-infected cells (containing dsRNA) although they do not distinguish cellular and viral messengers within such cells. But even if the blocks in translation kill the infected cells, they halt the progress of infection.

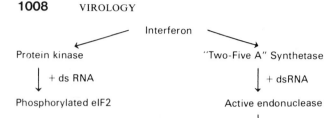

FIG. 51-7. Effects of interferon and double-stranded **(ds)** RNA on translation. "2-5A" = The adenine trinucleotide "two-fiveA"; eIF-2 = initiation factor.

One of the **translation blocks** is caused by the activation, by dsRNA, of a **protein kinase,** which apparently inactivates initiation factor eIF-2 by phosphorylating one of its subunits (a mechanism similar to that involved in the control of globin synthesis in reticulocyte lysates; see Protein Synthesis and Breakdown, Ch. 49). The other translation block is due to **nucleolytic destruction of mRNA:** Kerr found that interferon induces the production of an enzyme that, after activation by dsRNA, synthesizes an **adenine trinucleotide** with an unusual 2'-5' phosphodiester linkage (pppA2'p5'A2'p5'A—nicknamed "two-five A"). The trinucleotide in turn activates an endonuclease.

In mixed cultures of interferon-sensitive and interferon-resistant cells the antiviral resistance induced in the sensitive cells spreads to the resistant cells, presumably through channels of intercellular communication (see Communication Between Cells, Ch. 49). Such a spread may favor the establishment of resistance throughout an organism.

PROTECTIVE ROLE IN VIRAL INFECTIONS

The protective role of endogenous interferon in viral infection of **cell cultures** is demonstrated by the establishment of **carrier cultures,** in which interferon produced in the cultures makes most of the cells resistant but cannot wipe out the infection; hence only a small proportion are infected at any time (Ch. 53).

A protective role of interferon **in animals** is suggested by many observations. 1) In mice recovering from influenza virus infection the titer of interferon is maximal at the time when the virus titer begins to decrease and before a rise in Abs can be detected. At this stage the interferon titer in the animals is sufficient to protect them against the lethal action of a togavirus. 2) Suckling mice, which are susceptible to coxsackievirus B1 (a picornavirus), produce little interferon in response to this virus; whereas adult mice, which are resistant, produce large amounts. Cortisone, which suppresses interferon production, makes adult animals susceptible. 3) Administration of a potent antiserum to interferon markedly increases the lethality of mouse hepatitis virus infection.

These studies suggest that interferon has a major protective role in at least some viral infections. Much depends on the **dynamics of the disease,** i.e., the relation between virus titer and interferon titer at various times.

Interferon is most effective when present before infection, and when the dose of infecting virus is not too large (as at the beginning of most natural infections). The protection afforded appears to be especially useful because it develops more promptly than Ab production.

EFFECT OF INTERFERON ON THE FUNCTIONS OF UNINFECTED CELLS

In vitro, interferon **inhibits cell replication,** and also inhibits the activation of spleen lymphocytes by phytohemagglutinin. In vivo, high doses inhibit liver regeneration, as well as the production of platelets and leukocytes, and enhance the expression of histocompatibility Ags on the lymphocyte surface. Some of these effects determine a **pathogenetic role** of interferon.

Thus administration of high doses to newborn mice induces a lethal liver degeneration and an autoimmune glomerulonephritis. Moreover, production of endogenous interferon is responsible for the disease caused in mice by lymphocytic choriomeningitis virus (see Ch. 53), and administration of antiserum to interferon alleviates the disease even though it markedly increases the viral titer in the blood.

Interferon also profoundly affects the **immune response** in opposite ways: it reduces Ab production by inhibiting B cells, but it enhances T cell activity; it also increases the cytotoxic activity of natural killer cells against virus-infected cells (see Ch. 52). These effects denote a **shift from humoral to cell-mediated immunity,** which has a defensive role in many viral infections, but a pathogenetic role in some (e.g., the above-mentioned disease produced in mice by the lymphocytic choriomeningitis virus). The combination of cell growth inhibition and enhancement of cell-mediated immunity accounts for the **antitumor effect** of interferon in experimental animals.

Studies with highly purified interferon preparations show that all these effects are caused by the same molecule.

Possible Therapeutic Use. Interferon could theoretically be an ideal antiviral agent, since it acts on many different viruses, has high activity, and despite the adverse effects noted above at high doses, at moderate doses lacks serious toxic effects on the host. However, its therapeutic value is limited by various factors: interferon is effective only during relatively short periods, and it has no effect on viral synthesis that is already initiated in a cell. Moreover, interferon is difficult to produce in large amounts. However, the successful cloning of a human interferon gene in a bacterial plasmid may obviate this difficulty.

Attempts to use exogenous interferon for therapeutic purposes in human viral diseases have had some limited success. Thus interferon produced in cultures of human leukocytes had a prophylactic effect against influenza infection during epidemics; and local administration lessens the severity of respiratory diseases. Repeated administration reduces the extent of infection in carriers of hepatitis B virus, and it causes clearance of the hepatitis surface Ag from the blood in long-time chronic patients. Interferon also lessens the pain of herpes zoster, and local applications decrease the spread of herpes simplex keratitis. However, the lesions produced by these viruses are not cured.

Interferon treatment of human cancers is also in an experimental stage. Encouraging results were obtained in a small sample of osteosarcomas.

THERAPEUTIC POTENTIAL OF DOUBLE-STRANDED POLYNUCLEOTIDES

These substances appear promising because they are easily available and very effective—e.g., $10^{-3}\mu g$ of poly(I:C) is sufficient to induce resistance to virus infection in a culture. However, their use has been severely limited by their high toxicity and their rapid breakdown by enzymes in the blood and tissues. They appear only useful against **localized disease:** thus herpetic keratoconjunctivitis in rabbits could be controlled by local applications of poly(I:C) even 3 days after infection, when the disease was already moderate to severe. Also, intranasal administration of poly(I:C) can protect human volunteers from rhinoviruses (the agents of the common cold), and mice from an otherwise fatal subsequent infection with pneumonia virus of mice (a paramyxovirus).

The resistance to viral infection that is induced by polynucleotides appears to be due to interferon production: large doses of poly(I:C) induce the two effects in a parallel way, and resistance induced by poly(I:C) can be duplicated by exogenous interferon. Certain apparent exceptions can be readily explained. Thus development of resistance without the production of external interferon at low poly(I:C) concentrations can be attributed to rapid adsorption of the released interferon to the cells; and inhibition of resistance but not of interferon production by the addition of actinomycin D 2 h after poly(I:C) can be explained as an inhibition of the synthesis of the proteins required for the antiviral effect of the interferon produced.

INTRINSIC INTERFERENCE (I.E., NOT MEDIATED BY INTERFERON)

An inability to detect interferon does not exclude its participation in an instance of viral interference, because the detection methods are relatively insensitive. However, if interference is established early in the infectious cycle the participation of interferon can be considered unlikely because its production usually begins later.

Other mechanisms, grouped as intrinsic interference, have been studied with both bacteriophages and animal viruses.

BACTERIOPHAGE

Homologous or closely related phages (such as two mutants of the same strain, or T2 and T4) in the same cell must compete for the same precursors, cellular sites, and enzymes. The two strains can replicate more or less equally if they both infect the cells simultaneously and at low multiplicities. However, if they infect at different times or with different multiplicities, the phage with the advantage multiplies normally and interferes with (or even completely prevents) the multiplication of the other phage. In nonsimultaneous infection interference appears to result from a change in the bacterial plasma membrane that prevents penetration of the DNA of the second phage.

With **unrelated** phages one phage is excluded by a variety of mechanisms. Thus T-even bacteriophages probably exclude T1, T3, and λ by destroying the host cell DNA, whose functions are needed for the multiplication of the excluded phages. Phage T4 excludes the RNA phage F2 both by inactivating a translation initiation factor that it requires and by rapidly degrading its RNA. If infection is not simultaneous, the phage that injects its DNA after another phage has taken over control of the cell machinery is always excluded.

ANIMAL VIRUSES: DEFECTIVE INTERFERING (DI) PARTICLES

Homologous Interference. As with bacteriophages, interference involving homologous animal viruses is most pronounced if one virus has an advantage, either in time or in multiplicity. This type of interference takes place with Newcastle disease virus (a paramyxovirus) at **adsorption,** through destruction of cellular receptors; with retroviruses at **penetration;** and with many viruses at **replication.**

Interference at replication is usually generated by **defective interfering (DI) particles,** which accumulate after serial passages at a high multiplicity of infection. With some viruses (e.g., vesicular stomatitis virus [VSV]) the DI particles are smaller than regular particles and therefore can be obtained in pure form. They usually contain the normal virion proteins, but they always have a shorter genome than the regular particles. Most kinds of DI particles cannot multiply alone but require as **helper** a regular virus coinfecting the same cells. In the early serial passages DI particles rapidly increase in titer; then the yield of infectious virus, and finally the total particle yield, is progressively reduced (**autointerference**). In some cases these events lead to establishment of a **carrier state** (see Ch. 53). Through autointerference the DI particles appear to limit the spread of viral infections in humans or animals. Thus mice can be protected from a lethal infection of the central nervous system by the intranasal administration of DI particles.

With DI particles of some **negative-strand RNA viruses** (e.g., VSV, Sendai virus; see Ch. 50), the RNA is internally deleted, but both ends are conserved (Fig. 51-8). These ends contain short inverted repeat sequences recognized by the polymerase for initiation

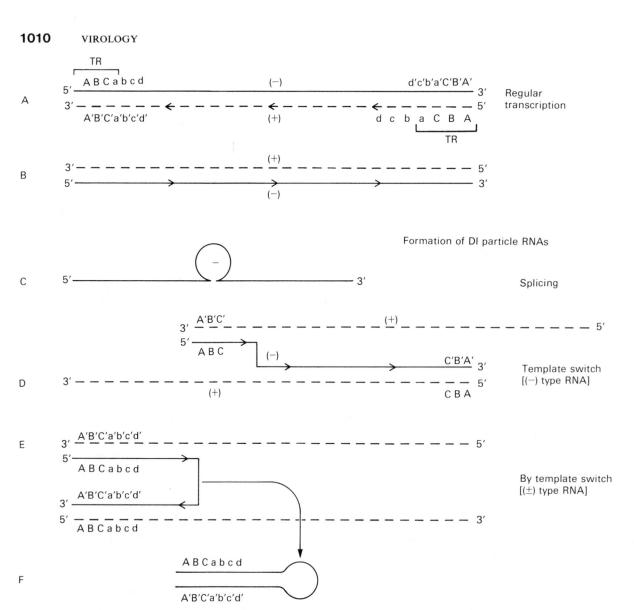

FIG. 51-8. Models of production of the RNAs of defective interfering (DI) particles of vesicular stomatitis virus. **A.** Regular transcription of virions RNA (−) to generate (+) RNA. ABC . . . C′B′A′ = short inverted terminal repetitions in regular virion RNA; ABC = recognition site for transcriptase (TR); a,b,c,d,e = nucleotide sequences of the 5′ half of the virion RNA. Primed letters = complementary sequences. **B.** Synthesis of a normal virion (−) strand on a (+) strand. **C–F.** Formation of defective RNAs by splicing (**C**) and by various forms of template switching (**D–F**). Those produced in **E** have long terminal repetitions that are absent in regular viral RNA and can self-anneal to form circles (**F**).

of replication of either strand; hence DI particle RNA can replicate. In the early serial passages the RNA of DI particles is heterogeneous and only slightly shorter than the normal viral RNA; with further passages particles with shorter RNA appear, and certain lengths become predominant. The spectrum of these RNAs varies with the viral clone at the beginning of a series of passages, and with the cell type in which the passages are performed. Because molecules of DI RNA are shorter than regular viral RNA, they can replicate more rapidly, thus causing their proportion to grow at every passage.

DI particles interfere because some or all of their genes are lacking, and they compete for the products of these genes that are being provided, in the same cell, by the nondefective genomes. The degree of interference depends on the type of **host cell:** thus DI particles of Sindbis virus that interfere strongly in hamster cells do not interfere in mosquito cells. Perhaps the amount of replicase made per genome is higher in the latter cells.

With VSV or Sendai virus the shortest molecules of DI particle RNA are the simplest possible replicons, consisting of the origin of replication together with the minimum amount of RNA compatible with efficient initiation and encapsidation. The deletions probably occur rarely, when RNA synthesis jumps from one template to another, skipping a segment (Fig. 51-8D); this process may even lead to the **incorporation of complementary sequences,** as in one kind of DI particle of VSV, whose RNA is capable of rapid self-annealing, forming a hairpin. Because such ± DI particles, in contrast to the majority of DI particles, contain (or can generate) dsRNA, they induce the synthesis of interferon in cells infected without helper. The jumping event causing deletion may be related to maturation of the viral messages, which for VSV are formed by splicing a constant leader sequence with various segments of the messenger strand. Occasional action of the splicing enzymes on replication would explain the host influence on the spectrum of DI particle RNA (since these enzymes are cellular), and also the high frequency of their formation with this virus.

DNA viruses also form DI particles. Owing to the needs of encapsidation these particles contain a DNA of normal length but with extensive substitutions, often including the presence of **multiple replication origins.** This feature confers a replicative advantage despite the normal length.

Heterologous Interference. The mechanisms vary in different systems, as shown by two examples. 1) Poliovirus arrests the multiplication of VSV, even when infecting later and at lower multiplicity. Apparently the poliovirus mechanism that blocks the translation of cellular messengers has the same effect on VSV messengers. The interference may be based on the recognition of specific sequences in mRNA, since it does not occur with many other viruses tested. 2) Newcastle disease virus (NDV) RNA fails to replicate in cells previously infected by certain other RNA viruses, e.g., Sindbis virus (a togavirus). NDV RNA may form an inactive complex with the replicase induced by the interfering virus, because

Sindbis mutants deficient in RNA replicase do not interfere.

This phenomenon is exploited for assaying noncytopathic viruses on the basis of their antagonism to the cytopathic effect of NDV.

SIGNIFICANCE OF VIRAL INTERFERENCE

Interference, both by interferon and by other mechanisms, is important in several aspects of viral infection. For instance, in human oral **vaccination** with attenuated poliomyelitis viruses (Ch. 57) the three strains must be administered in a precise sequence, or at specified concentration ratios, in order to avoid interference of one strain with another. Similarly, the presence of various enteroviruses in the normal intestinal flora may hinder the establishment of infection by the vaccine strains. Viruses already present may also influence the response to a **naturally infecting virus.** For example, dengue virus infection in man is milder in the presence of an attenuated strain of yellow fever virus (both togaviruses).

Both interferon and DI particles seem to play an important role in initiating recovery from some acute viral infections. DI particles may also maintain a persistent infection (see Ch. 53).

Finally, it seems likely that interferon and interferon inducers will play an important role in antiviral therapy, especially in local superficial infections, if the problems related to the large-scale production of the former and toxicity of the latter can be overcome.

CHEMICAL INHIBITION OF VIRAL MULTIPLICATION

SYNTHETIC AGENTS

The chemical structure of the more important of these agents is shown in Fig. 51-9.

SELECTIVE AGENTS

Amantadine (adamantanamine). This compound, of peculiar structure, inhibits the multiplication of several viruses, including influenza and rubella. It appears to interfere with an early step in virus–cell interaction after adsorption—perhaps the release of the viral nucleic acid in the cells. The compound appears to have some prophylactic value and even a marginal therapeutic effect if given early in influenza A infections.

Substituted Benzimidazole and Guanidine. 2-(α-Hydroxybenzyl)-benzimidazole (HBB), a substituted benzimidazole, inhibits multiplication of certain enteroviruses, while guanidine inhibits mul-

tiplication of the same viruses and a few more picornaviruses. These two compounds interfere with the synthesis of viral single-stranded RNA without affecting cellular RNA synthesis.

In spite of their marked effect in vitro, neither HBB nor guanidine has shown promise as a chemotherapeutic agent in animals, owing to the rapid emergence of **resistant viral mutants.** Viruses can also mutate to **dependence** on these drugs; and the mutations to resistance or dependence may affect the response to only one drug or to both (covariation).

With sensitive virus in the presence of guanidine, as with dependent virus in its absence, RNA synthesis is specifically impaired. Hence the drug evidently has the same target when inhibiting a susceptible system or activating a dependent one. A useful model is provided by streptomycin, for which bacterial sensitivity, dependence, and resistance involve alternate configurations of the ribosome (Ch. 13).

Isatin-β-Thiosemicarbazone. This drug and its N-methyl and N-ethyl derivatives inhibit the multiplication of several DNA and RNA viruses, especially poxviruses.

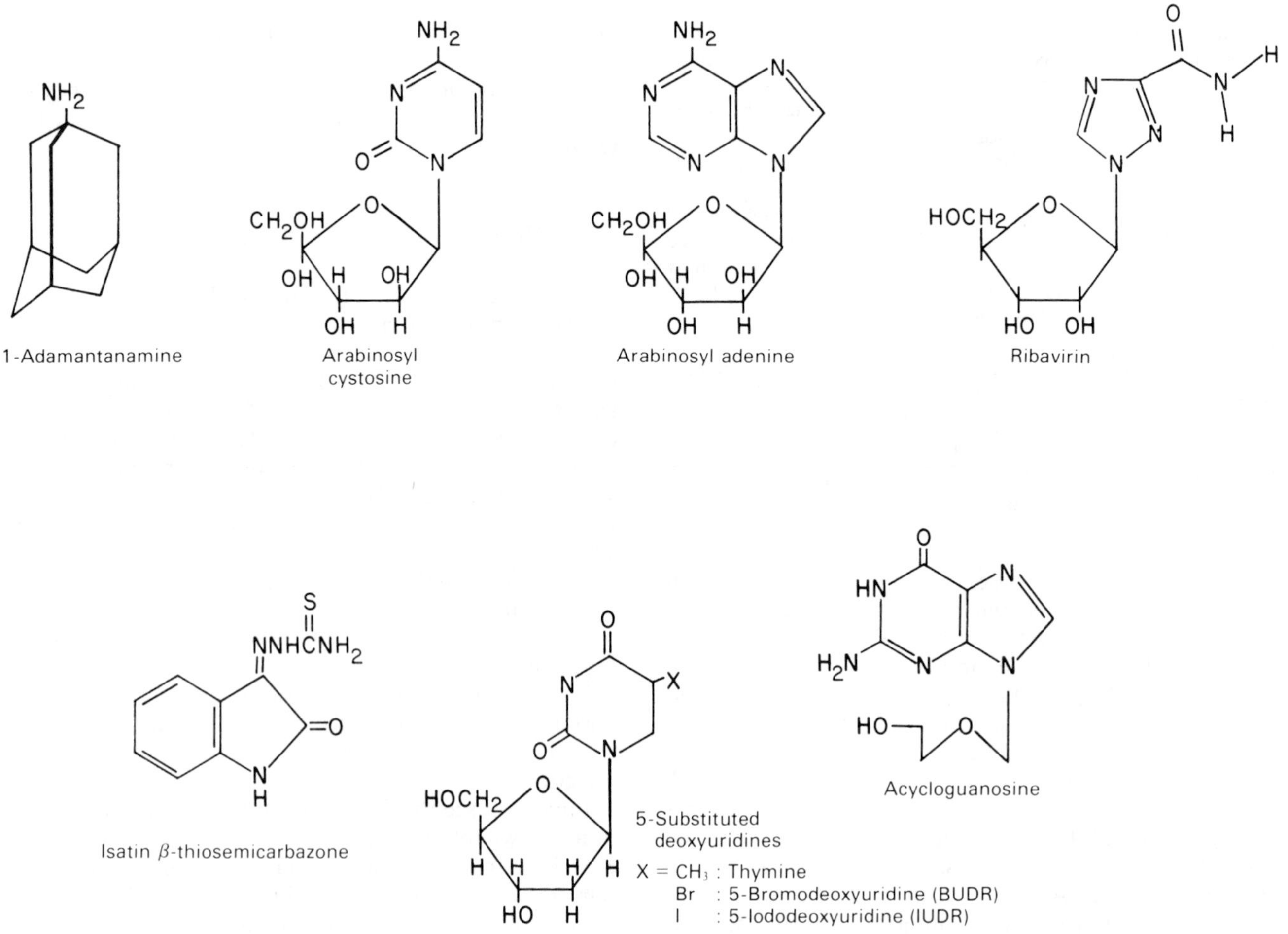

FIG. 51-9. Chemical constitution of some antiviral inhibitors.

They selectively inhibit the synthesis of late poxvirus structural proteins, apparently by inactivating the late viral mRNAs. The result is the production of nearly spherical defective particles.

N-methylisatin-β-thiosemicarbazone (methisazone) had an impressive **prophylactic** success in preventing the spread of smallpox to contacts in an epidemic in Madras, India, in 1963. Among 1101 contacts treated with methisazone and vaccinated 3 mild cases of smallpox occurred; in contrast, among 1126 vaccinated but not treated with the drug 78 contracted smallpox and 12 died. The much greater prophylactic effect of the drug can be attributed to its immediate effect on viral multiplication, contrasted to the delayed effect of the immune response to vaccination.

In contrast, the drug failed as a **therapeutic** agent in patients already suffering from smallpox. This failure is understandable, because the disease appears only after viral multiplication has reached a maximum (a principle that may adversely affect the therapeutic usefulness of all antiviral compounds).

AGENTS INTERFERING WITH DNA SYNTHESIS

A considerable advance in antiviral chemotherapy has been the synthesis of analogs of purines and pyrimidines that inhibit the replication of certain viral DNAs much more than that of the cellular DNA. At the doses required for effective antiviral therapy they show little toxicity for the organism. These compounds represent the closest approximation to a selective antiviral therapy without cellular toxicity.

These drugs are especially effective against herpesviruses (Ch. 55) and poxviruses (Ch. 56), which, like the T-even bacteriophages (Ch. 47), have a large genome and specify many enzymes for DNA synthesis. Viral enzymes that differ markedly from the corresponding cellular enzymes can be selectively inhibited. Viruses with smaller genomes, which depend much more on cellular enzymes, cannot be selectively inhibited by these drugs.

Among these compounds are **ara-T** (arabinosyl thymine) and **5-iodo-5'-amino-2'-5'-dideoxyuridine,** which effectively interfere with the replication of herpes simplex and varicella–zoster in vitro. The drugs are phosphorylated by the viral but not by the cellular thymidine kinase; and the phosphorylated products inhibit the viral DNA polymerase.

The most effective analog is **vidarabine (ara-A,** arabinosyl adenine); its high selectivity makes it the drug of choice for the clinical treatment of serious infections by herpesviruses in humans, including herpes simplex, herpes zoster, and cytomegalovirus. The drug is effective against vaccinia virus in mice and herpes simplex in hamsters and against both viruses in vitro. Vidarabine inhibits the DNA polymerases as well as the ribonucleotide reductase, which generates the deoxyribonucleotide precursors of DNA synthesis. The similar analog of cytidine, **Cytarabine (ara-C,** arabinosyl cytosine), has a similar action but is less selective. The guanosine analog **acycloguanosine** also has a high selectivity. It has proven effective against intracerebral infection of mice by type 1 herpesvirus, herpetic keratitis in rabbits (even after the development of ulcerations), and herpetic skin lesions in guinea pigs.

A purine analog, **ribavirin,** has a certain degree of selectivity, although it is phosphorylated in both virus-infected and uninfected cells. In both situations the drug interferes with the synthesis of guanosine monophosphate, with resultant inhibition of both RNA and DNA synthesis. A higher demand for nucleic acid precursors for the synthesis of viral nucleic acids accentuates the toxic effect of the drug in virus-infected cells. Given its mode of action, the drug has a broad antiviral spectrum, including both DNA viruses (herpes simplex, vaccinia) and RNA viruses (VSV, influenza, parainfluenza), both in vitro and in experimental animals. An adverse effect is its suppressive activity on the humoral immune response.

Phosphonoacetic acid, which inhibits the herpesvirus-specific DNA polymerase, markedly reduces the replication of this virus in cultures, as well as the severity of experimental infection in animals; however, it is too toxic for use in humans. Some mutants in the viral DNA polymerase are resistant to the drug.

Halogenated derivatives of deoxyuridine, such as 5-bromo-deoxyuridine (BUdR) and **idoxuridine** (5-iododeoxyuridine, IUdR) (Fig. 51-9), were among the first antiviral compounds to be synthesized; however, they are less selective and more toxic than the preceding compounds. These analogs are taken up by cells and phosphorylated by the viral and the cellular thymidine kinase; the phosphorylated derivatives are incorporated into DNA instead of thymidine, in both virus-infected and uninfected cells. The DNA continues to replicate, but causes the synthesis of altered proteins. Indeed, in cells infected by herpesvirus in the presence of idoxuridine viral maturation fails, owing to defects of the capsid protein. If the drug is later removed virions are formed; they contain DNA replicated in the presence of the drug, in which iodouracil replaces thymine.

In spite of its toxicity, idoxuridine is valuable for the treatment of topical surface lesions (e.g., keratitis) produced by herpes simplex or vaccinia virus. Some herpesvirus mutants are resistant.

Clinical Applications. Of the drugs interfering with DNA synthesis only ara-A is used for the treatment of **systemic** infections, even in patients with immunodeficiencies or under treatment with immunosuppressive drugs. The lower selectivity of most of the other compounds precludes their systemic administration because by affecting DNA synthesis they are toxic for essential rapidly dividing cells (e.g., hemopoietic cells). Several drugs (ara-A, ara-C, ara-T, acycloguanosine, and idoxuridine) are useful **topically,** as in the control of herpes simplex keratitis. Even after keratitis has appeared, the drugs, dropped directly into the eye every 1–2 h, drastically reduce the period of clinical disease and prevent the formation of corneal scars.

ANTIBIOTICS

Antiviral activity against poxviruses and retroviruses is displayed by derivatives of **rifamycin** (e.g., rifampin) and by the related antibiotics tolypomycin and streptovaricin. In bacteria rifampin is known to inhibit initiation by RNA polymerase (see Antibiotics Affecting Transcription, Ch. 11); and with retroviruses these antibiotics inhibit the activity of the reverse transcriptase (see Reverse Transcription, Ch. 11) and also cause the formation of defective, RNA-deficient virions. With poxviruses rifampin acts by **interfering with virus maturation:** viral membranes begin to form at the periphery of the filamentous matrix (see Complex Viruses, Ch. 50) but remain incomplete. If rifampin is then removed the membranes close and form virions (Fig. 51-10).

Since this maturation can occur even in the presence of actinomycin D and cycloheximide it can utilize proteins and DNA synthesized earlier, despite the presence of rifampin. Indeed, this antibiotic blocks the cleavage of several precursor peptides. In addition, though rifampin does not affect transcription by the poxvirus transcriptase in vitro (see Table 46-2, in Ch. 46), or overall transcription in infected cells, it prevents the appearance of a particle-associated transcriptase function and the appearance of other virions' enzymes. These defects might be the consequence of the maturation defect, if the enzymes require virions or structural proteins for activity or stability.

The mechanism of action of rifampin on poxvirus maturation is

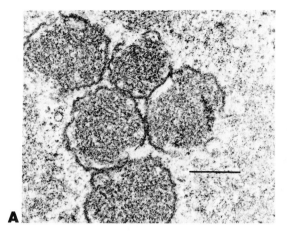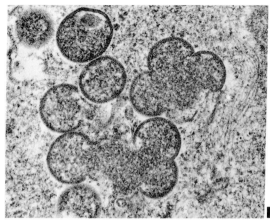

FIG. 51-10. Effect of rifampin on the maturation of vaccinia virus in HeLa cells. **A.** Electron micrograph from a thin section of infected cells treated for 8 h with rifampin (100 µg/ml), showing the incomplete and disorganized viral membranes, each surrounding a matrix. **Bar** = 300 nm. **B.** Thin section of similar cells 10 min after removal of rifampin, showing the rapid reorganization of the membranes to a morphology similar to that observed in normal maturation. (Courtesy of P. M. Grimley)

unknown. It may conceivably alter the structure of the uncleaved precursor peptides or inactivate the cleaving enzyme, or perhaps block transcription of a few viral genes important in maturation. The activities of the various rifamycin derivatives toward retrovirus, poxvirus, or *E. coli* transcriptase are uncorrelated, suggesting different modes of action. Poxvirus mutants resistant to rifampin and other antiviral rifamycin derivatives are readily isolated.

In vivo these compounds have some local effect: for instance, rifampin inhibits the development of vaccination lesions in man. They have little effect on systemic diseases in animals, perhaps because their high toxicity prevents the use of adequate doses.

SELECTED READING

BOOKS AND REVIEW ARTICLES

COHEN SS: Strategy for the chemotherapy of infectious diseases. Science 197:431, 1977

FINTER NB (ed): Interferon and Interferon Inducers. Amsterdam, North-Holland, 1973

HO M, ARMSTRONG JA: Interferon. Annu Rev Microbiol 29:131, 1975

MERIGAN TC (ed): Antivirals With Clinical Potential. Chicago, University of Chicago Press, 1976

METZ DH: Mechanism of action of interferon. Cell 6:429, 1975

WEHRLI W, STAEHLIN M: Actions of the rifamycins. Bacteriol Rev 35:290, 1971

SPECIFIC ARTICLES

BAUER DJ, ST. VINCENT L, KEMPE CH, DOWNIE AW: Prophylactic treatment of smallpox contacts with N-methylisatin-β-thiosemicarbazone. Lancet 2:494, 1963

CANTELL K: Towards the clinical use of interferon. Endeavour (New series) 2:27, 1978

COLE CN, BALTIMORE D: Defective interfering particles of poliovirus. J Mol Biol 76:345, 1973

DE CLERCQ E, MERIGAN TC: Requirement of a stable secondary structure for the antiviral activity of polynucleotides. Nature 222:1148, 1969

DE MAEYER-GUIGNARD M, TOVEY MG, GRESSER I, DE MAEYER E: Purification of mouse interferon by sequential affinity chromatography on poly (U)- and antibody-agarose columns. Nature 271:622, 1978

DIMMOCK NJ, KENNEDY SIT: Prevention of death in Semliki Forest virus-infected mice by administration of defective-interfering Semliki Forest virus. J Gen Virol 39:231, 1978

FARRELL PJ, SEN GC, DUBOIS MF et al: Interferon action: two distinct pathways for inhibition of protein synthesis by double-stranded RNA. Proc Natl Acad Sci USA 75:5893, 1978

GLASGOW LA: Transfer of interferon-producing macrophages: new approach to viral chemotherapy. Science 170:854, 1970

GRESSER I, MOREL-MAROGER L, VERROUST P et al: Anti-interferon globulin inhibits the development of glomerulonephritis in mice infected at birth with lymphocytic choriomeningitis virus. Proc Natl Acad Sci USA 75:3413, 1978

GRIMLY PM, ROSENBLUM EN, MIMS SJ, MOSS B: Interruption by rifampin of an early state in vaccinia virus morphogenesis: accumulation of membranes which are precursors of virus envelopes. J Virol 6:519, 1970

GUILD MG, STOLLAR V: Defective interfering particles of Sindbis virus. Virology 77:175, 1977

GURGO C, KAY RK, THIRY L, GREEN M: Inhibitors of the RNA and

DNA dependent polymerase activities of RNA tumor viruses. Nature 229:111, 1971

HOLLAND JJ, VILLAREAL LP: Persistent noncytocidal vesicular stomatitis virus infection mediated by defective T particles that suppress virion transcriptase. Proc Natl Acad Sci USA 71:2959, 1974

HUANG AS, WAGNER RR: Defective T particles of vesicular stomatitis virus: biological role in homologous interference. Virology 30:173, 1966

ISAACS A, LINDENMANN J: Virus interference. I. The interferon. Proc R Soc Lond [Biol] B147:1258, 1957

KANG CY, GLIMP T, CLEWLEY JP, BISHOP DHL: Studies on the generation of vesicular stomatitis virus (Indiana serotype) defective interfering particles. Virology 84:142, 1978

KERR IM, BROWN RE: pppA2′p5′A2′p5′A: an inhibitor of protein synthesis synthesized with an enzyme fraction from interferon-treated cells. Proc Natl Acad Sci USA 75:256, 1978

MERIGAN TC: Interferon stimulated by double-stranded RNA. Nature 228:219, 1970

OSBORN JE, WALKER DL: Role of individual spleen cells in the interferon response of the intact mouse. Proc Natl Acad Sci USA 62:1038, 1969

PERRAULT J, SEMLER BL: Internal gerome deletions in two distinct classes of defective interfering particles of vesicular stomatitis virus. Proc Natl Acad Sci USA 76:6191, 1979

SCHAEFFER HL, BEAUCHAMP L, DE MIRANDA P et al: 9-(2-Hydroxyethoxymethyl) guanine activity against viruses of the herpes group. Nature 272:583, 1978

SEHGAL PB, DOBBERSTEIN B, TAMM I: Interferon messenger RNA content of human fibroblasts during induction, shutoff, and superinduction of interferon production. Proc Natl Acad Sci USA 74:3409, 1977

STAMMINGER GM, LAZZARINI RA: RNA synthesis and autointerfered vesicular stomatitis virus infections. Virology 77:202, 1977

STARK C, KENNEDY SIT: Generation and propagation of defective-interfering particles of Semliki Forest virus in different cell types. Virology 89:285, 1978

WOODSON B, JOKLIK WK: Inhibition of vaccinia virus multiplication by isatin-β-thiosemicarbazone. Proc Natl Acad Sci USA 54:946, 1965

chapter

VIRAL IMMUNOLOGY

The immunologic properties of viruses are determined by both structural proteins of the virions and other, nonstructural, viral proteins made in the infected cells (Chs. 47 and 50). The Ags contained in these proteins elicit both a humoral and a cellular response.

HUMORAL FACTORS

Humoral Abs elicited by intact virions are specific for components of the virions' surface, whereas those induced by disrupted virions react with both surface and internal components. In virus-infected animals Abs are induced not only by complete infectious virions but also by viral products built into the surface of the infected cells or released by the cells (e.g., when they die). Thus, animals bearing a tumor induced by an oncogenic DNA virus develop Abs against the T Ag, a nonvirion protein present in the nuclei of the tumor cells (see Viral Multiplication in Tissue Culture, under Papova V, Ch. 66).

Antibodies are useful in the laboratory for identifying, quantitating, and isolating virions (and also some of the unassembled components), for classifying viruses, and for serologic diagnosis of viral diseases. In the host they provide the key to protection against many viral infections. However, Abs are sometimes pathogenic: deposits of Ab–Ag complexes in the kidneys of mice infected with lymphocytic choriomeningitis virus cause an immune complex disease, and cytotoxic Abs plus complement can lyse infected cells if their membranes contain viral proteins.

INTERACTIONS BETWEEN VIRIONS AND ANTIBODIES

Special features of virion–Ab interactions, especially in the quantitative aspects of viral neutralization, reflect the presence of different antigenic sites on the virions' surface and the multimeric structure of the coat (Ch. 46). Thus, influenza virions are covered by both hemagglutinins and neuraminidase molecules, and naked icosahedral virions often have different proteins in the two types of capsomers (see Icosahedral Capsids, Ch. 46). Each such component specifies different Abs whose interactions with the virions have different consequences. Moreover, owing to the multimeric nature of the capsid and the envelope, many identical antigenic sites, regularly repeated on viral surfaces, interact with Abs in ways that would be impossible with isolated sites.

Reactions of viral Ags with their corresponding antisera are studied by both the usual immunologic tests and some special ones, such as hemagglutination inhibition and neutralization. Because neutralization is particularly characteristic of viruses and has widespread application, we shall consider it in detail.

NEUTRALIZATION

Viral neutralization consists of a decrease in the infectious titer of a viral preparation following its exposure to Abs. As with other Ag–Ab complexes, the components can be dissociated and recovered in their original forms. However, the reversibility of the association often decreases with time.

Readily Reversible and Stable Virion–Antibody Complexes. These two classes of complexes are distinguished by an experiment of the following kind. Ab is added to a preparation of influenza virus of known titer at 0°C to form a neutralization mixture. If a sample of the mixture is added ½ h later **without dilution** to the viral assay system (e.g., the allantoic cavity of chick embryos or cell cultures) a decrease of the viral titer in comparison with untreated virus may be seen; but if the neutralization mixture is **diluted** by a large factor before assay the original titer is restored. Hence the combination between virions and Ab molecules under these conditions is freely reversible, i.e., there is an equilibrium between dissociation and re-formation of the Ab–virion complexes. Decreasing the concentration of the reactants diminishes the rate of complexes formation, but not the rate of dissociation.

If, however, the assays are made **several hours after the virus has been mixed** with the Abs, neutralization persists after dilution. Hence with time the virus–Ab complexes become more stable and are no longer reversed on dilution (at neutral pH and physiologic ionic strength).

Results of this type with several viruses show that in general virions unite with Ab molecules in at least two steps, the first forming freely reversible complexes, and the second converting these into practically irreversible complexes.

The equilibrium of the first step is reached very rapidly and as with many Ab-Ag combinations shows little temperature dependence; it merely reflects the attachment of Ab molecules to exposed antigenic sites of the virion. The second step, however, occurs very slowly at 0° C but at 37° C it occurs rapidly, so that most of the complexes become stable within minutes.

Reactivation of Stably Neutralized Virus. Stabilization of the virus–Ab complex is not associated with a permanent change of either reactant, since the unchanged components can be recovered. The neutralized virus can be reactivated by proteolytic cleavage of the bound Ab molecules into monovalent fragments; and intact Abs are recovered by dissociating the Ab–virus complexes at acid or alkaline pH (Fig. 52-1), by sonic vibration, or by extraction with a fluorocarbon.

FIG. 52-1. Reactivability of neutralized Newcastle disease virus at different pHs, showing the dissociability of the virus–Ab complexes at acid and alkaline pHs. Virus and Abs were incubated together until a relative infectivity of about 10^{-4} was obtained (assayed after dilution). Aliquots of the mixture were then diluted 1:100 in cold buffer at various pH values, and after 30 sec the pH was returned to 7 by dilution in a neutral buffer. The samples were then assayed. (Granoff A: Virology 25:38, 1965)

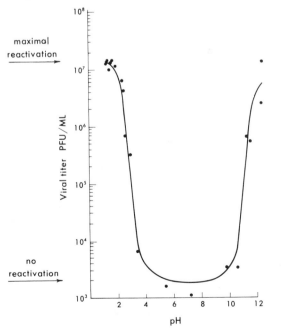

Physicochemical Basis of a Stable Antibody–Virion Association. The firm association of Abs with virions originates from the multimeric nature of the viral coat, which allows a single Ab molecule to establish specific bonds with two sites on a single virion: electron micrographs show that an Ab molecule can bridge two subunits of the virion's surface (Fig. 52-2). The stability of the complex is high, because whenever one of the bonds dissociates the other holds, and the dissociated one has time to become reestablished.

For example, the affinity of Ab for a chemical residue bound in many copies to the surface of a virion is about 10,000 times greater than for the free residue. Usually monovalent Ab fragments produced by proteolytic cleavage do not form stable complexes, even though they combine specifically with the virions, sometimes causing reversible neutralization. The high temperature coefficient for the stabilization of the virion–Ab complexes suggests that the interacting proteins undergo a conformational change.

MECHANISM OF NEUTRALIZATION; SENSITIZATION

Reversible neutralization probably occurs because virions that have bound many Ab molecules (or monovalent fragments) either do not adsorb to cells or else undergo aggregation, so that the number of foci of infection is decreased. In contrast, **stable neutralization** does not require saturation of the surface of the virion with Ab molecules. Thus phenotypically mixed virions with two types of capsomers, produced by double infection (Chs. 47 and 50) (Fig. 52-3), can be neutralized by antisera specific for **either** capsomer. Indeed, kinetic evidence, which will be discussed below, shows that *neutralization can be produced by a single Ab molecule*. In stable neutralization adsorption is not affected, because neutralized virions can still adsorb; and virions can still be neutralized after being adsorbed to cells.

Stable neutralization is probably due to a block in uncoating, created by two mechanisms. With poliovirus, neutralization changes the isoelectric point (i.e., the balance of all + and − charges on the virion's surface) from pH 7 to 5.5, and also changes the accessibility of chemical groups on the virion's proteins to external reagents. *The transition is all-or-none*, for there are no virions with intermediate isoelectric points. Apparently the binding of an Ab molecule to a capsomer **alters the capsomer's conformation;** the alteration then spreads to the whole capsid by a domino effect. With bacteriophage M13, whose DNA penetrates into the cells while the **protomers of the capsid enter the plasma membrane,** an Ab molecule blocks the latter process, presumably by holding two adjacent protomers together. Neutralization of enveloped animal viruses may follow the M13 model: perhaps an Ab molecule bridging two envelope glycoproteins prevents them from fusing with the cell plasma membrane and from releasing the nucleocapsid.

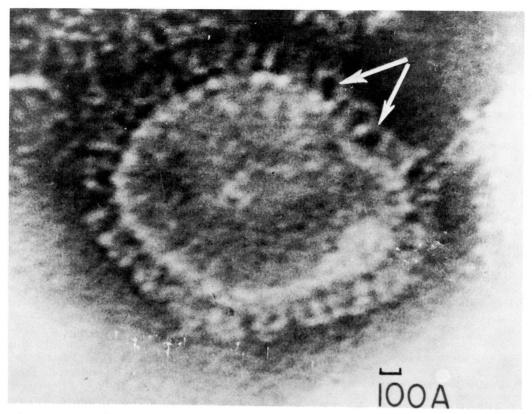

FIG. 52-2. Electron micrograph of an influenza virion with Ab molecules attached to the spikes, some of them forming bridges between two spikes (**arrows**). (Lafferty KJ: Virology 21:91, 1963)

Sensitizing Antibodies. The combination of Ab with virion elements does not necessarily cause neutralization. Thus Abs to adenovirus fibers at low concentrations do not neutralize the virus but do "**sensitize**" **it to neutralization by Abs to the bound immunoglobulin** (Ig). Presumably the anti-Ig Abs increases the stability of the initially bound Ab molecules by bridging them, with greater distortion of the virion's coat. Sensitization is also caused by Abs that bind weakly to the Ag, such as those formed early in immunization, or Ab fragments. The antigenic sites of sensitization are often different from those involved in primary neutralization, and include many group Ags. Some sensitized viruses can also be neutralized by complement, especially the polyvalent component C3 (Ch. 20).

Cooperation between Antibodies of Different Specificities. Exposure of virions to two Abs of different specificity (e.g., one to a type-specific Ag, the other to a group Ag) may result in far greater neutralization than would be expected from the effects of each Ab alone. Presumably the two Abs form two separate bridges between the same pair of subunits, producing a much greater distortion. In phenotypically mixed enveloped virions the co-

operation occurs even between Abs to the two different subunits.

QUANTITATIVE ASPECTS OF NEUTRALIZATION

Kinetics of Neutralization. The mechanism of neutralization is illuminated by studying the kinetics (i.e., the time course) of the reaction. A virus preparation and antiserum are mixed, and the mixture is then assayed for infectivity at regular intervals. The results are plotted as the logarithm of the fraction of the initial infectious titer that remains unneutralized (**relative infectivity**) versus the time of sampling.

1) **Relation of the kinetics of neutralization to the stability of the complexes.** The stability of the Ab–virion complexes affects the kinetics of neutralization, as revealed by an experiment of the following type: A virus–serum mixture with Ag excess is made and part is immediately diluted (e.g., fivefold); the kinetic curve is determined for the two mixtures by carrying out the infectivity assay without further dilution. (It is assumed that the concentration of Ab in the assay system is sufficiently low and will not interfere with the assay.) In the undiluted and the diluted mixtures the relative concentrations of virus and Ab are equal, but the absolute concentrations differ by a factor of five.

The curve obtained with **readily reversible Ab–virion complexes** begins without a shoulder, then decreases in slope and tends to a horizontal line (plateau), which is reached when the rate of disso-

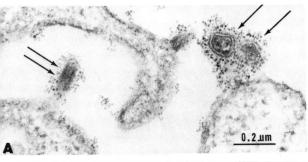

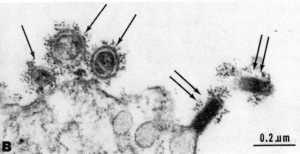

FIG. 52-3. Immuno-electron microscopy of virions with a mixed envelope (pseudotypes) obtained by superinfecting a murine leukemia virus (MuLV)-producing cells with vesicular stomatitis virus (VSV). The virions are stained with ferritin-labeled anti-MuLV Abs: the ferritin molecules are recognizable as black spots. **Single arrows** point to uniformly labeled MuLV virions (spherical); **double arrows** to VSV (bullet-shaped) virions, some uniformly labeled (**A**), i.e., completely coated by MuLV-specified glycoproteins, others (**B**) with some glycoproteins VSV-specified and unlabeled, and some MuLV-specified and labeled. (Chan JC et al: Virology 88:171, 1978)

ciation of virion–Ab complexes equals their rate of formation (equilibrium; Fig. 52-4), and which differs for the two mixtures. If the undiluted mixture is diluted fivefold after reaching equilibrium (time A in the figure), its equilibrium shifts to that of the originally diluted mixture.

The curves obtained with the **more stable complexes** also begin without a shoulder and then tend to a plateau. In contrast to the previous case, however, the final relative infectivity is equal in the two mixtures. Since every stable Ab–virion complex, once formed, persists, the plateau is reached when no more Abs combine; and the residual infectivity, which depends on the ratio of Ab concentration to virion concentration, is the same in the two mixtures. In the diluted mixture, however, the horizontal part of the curve is reached more slowly because collisions between reactants are fewer per unit time. If the mixtures are now further diluted their relative infectivity does not change, since the Ab molecules do not dissociate appreciably from the virions.

When the Ag–Ab complexes have a long life, as is usual in neutralization carried out at 37° C with hyperimmune or convalescent serum, the virus–serum mixture can be diluted for assay without altering the results, and so kinetic curves can be obtained for mixture in which **Ab is in large excess** over the virions. The curves obtained closely approach straight lines passing through the origin; the slopes are proportional to the concentration of the Ab and to its tendency to combine with the virions, and are independent of the virions' concentration (Fig. 52-5). These curves follow the equation:*

$$I = e^{-ktC} \tag{1}$$

and, taking the logarithm of both sides:

$$\ln I = -ktC$$
$$k = -\frac{\ln I}{tC} \tag{2}$$

where I is relative infectivity; t, time after mixing the virus and the Ab (in minutes); C, concentration of the serum; and k, a constant proportional to the concentration and the combining power of the Ab in the serum.

2) **The residual relative infectivity.** Kinetic curves of neutralization tend to show a plateau of residual infectivity as the time after mixing increases. This plateau is caused by attainment of equilibrium if the Ab–virion complexes are readily dissociable, and by exhaustion of free Ab if the complexes are more stable. In either case the residual relative infectivity corresponding to the plateau is a function of the quantities of virus and Ab in the mixture.

If the **complexes are readily reversible** (and are tested without dilution), the **percentage law** established by Andrews and Elford in 1933 applies: Under conditions of Ab excess the proportion (percentage) of virus neutralized by a given antiserum is constant, irrespective of the virus titer. This law can be deduced from the mass law.

$$V + A \underset{k_2}{\overset{k_1}{\rightleftharpoons}} \overline{VA}; \; \frac{k_1}{k_2} = \frac{(\overline{VA})}{(V)(A)}; \; k(A) = \frac{(\overline{VA})}{(V)}$$

where V indicates the free virus; A, the free Ab (assumed to have a much higher concentration than bound Ab); and $\overline{VA}$, the Ab–virion complexes.

If the **complexes are stable** the residual relative infectivity in Ag excess, determined after dilution of the virus–serum mixture (Fig. 52-6), theoretically obeys the relations

$$Res = e^{-c(A/V)}$$

where Res is residual infectivity, c is a constant, A denotes the total amount of Ab, and V the total amount of virus.† The problem is parallel to that of determining the proportion of uninfected cells at different multiplicities of infection (See Quantitative Aspects of Infection, Appendix, Ch. 46).

---

*This equation is equivalent to that used in Chapter 46 Appendix; ktc is the average number of neutralizing Ab molecules per virion after time t, since molecules of Ab (which is in large excess) continue to attach to the virions at a constant rate. Thus I is the fraction of virions that have **no neutralizing molecule.** As discussed in Chapter 11 Appendix, equation 1 generates single-hit curves and is valid if a single Ab molecule is sufficient to produce neutralization; otherwise the curve would have an initial shoulder (i.e., it would be a multiple-hit curve).

†This equation is derived similarly to equation 1 above, except that the average number of neutralizing Ab molecules per virion is $c(A/V)$. Again, this equation requires that a single Ab molecule is sufficient for neutralization; otherwise the curve would have an initial shoulder.

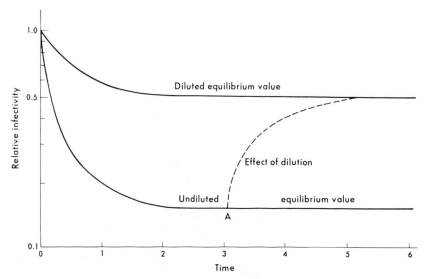

FIG. 52-4. The time course of neutralization in a virus–Ab mixture with readily reversible Ab–Ag combinations. One mixture is undiluted; the other is diluted fivefold. Note the different equilibrium values reached. If at time **A** the undiluted mixture is diluted fivefold the viral titer, corrected for dilution, increases, owing to dissociation of virus–Ab complexes, and reaches the same equilibrium value as the originally diluted mixture.

FIG. 52-5. Kinetic curves of neutralization of poliovirus with stable Ab–virion complexes and Ab excess. Note the linearity of the curves in the semilog plot **A**, with different slopes corresponding to different relative concentrations of Abs (given by the numbers near each line). In **B** the slopes of the curves of **A** are plotted versus the concentration of the antiserum (in relative values), yielding a straight line. The linearity of the two types of curves implies that a single Ab molecule is sufficient to neutralize a virion. (Dulbecco R et al: Virology 2:162, 1956)

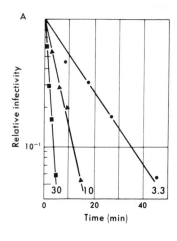

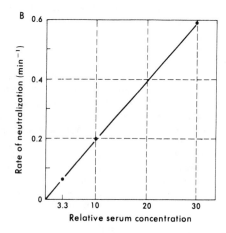

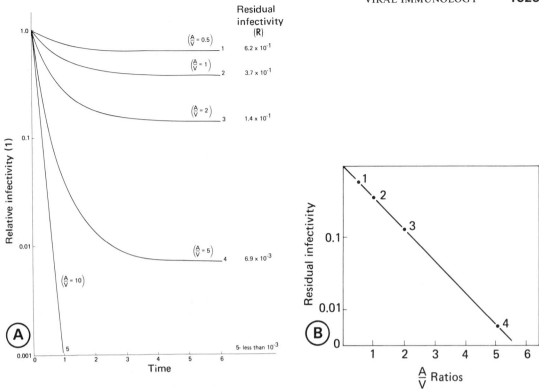

FIG. 52-6. A. Theoretical kinetic curves at different ratios of total Ab (A) to total virus (V) virion complexes. All the curves tend to plateaus, the relative infectivity of which is a function of the ratio A/V (in arbitrary units). When the residual infectivity values are plotted semilogarithmically versus the A/V ratio they generate a straight line (**B**). This result affords additional evidence that neutralization of a virion requires only the binding of a single Ab molecule.

3) **The persistent fraction.** The kinetic curves actually obtained with a large excess of Abs, however, show an abrupt plateau not justifiable on the basis of the Ab/virion ratio (Fig. 52-7). The fraction of virus that is not neutralized is unlike the residual infectivity of dissociable Ab–virion complexes at equilibrium, because the activity of this **persistent fraction** is not affected either by dilution or by the addition of a fresh charge of Ab. This phenomenon obscured the study of neutralization of animal viruses for a long time: since it was erroneously attributed to equilibrium, it prevented the recognition of the stable complexes. The cause was eventually shown to be either the formation of aggregates in which some virions remain unneutralized, or, more often, the interaction of virions with **Ab molecules that do not produce neutralization but prevent neutralization by the conventional interaction.** These inhibitory molecules appear to be of two classes: 1) readily dissociable molecules, whose action is duplicated by univalent Ab fragments; and 2) Ab causing conditional neutralization, which is ineffective with the cells used for determining the residual infectivity. The persistent fraction can be decreased by the further addition of Abs to the immunoglobulins causing neutralization.

MEASUREMENT OF NEUTRALIZING ANTIBODIES

This measurement rests on the relations derived above. If the serum–virus mixtures are assayed **without dilution** the relative concentration of Ab is derived from the proportion of neutralized virus, according to the **percentage law.** This method is simple to perform and measures both reversible and stable Ab–virion complexes; it is adequate, and is widely used, for diagnostic purposes, but it is unsuitable for accurate measurement.

If, alternatively, the serum–virus mixtures are assayed **with dilution** the relative concentration and combining power of the Ab can be calculated either from the **slope of kinetic curves,** according to equations 1 and 2, or from the **residual relative infectivity,** according to equation 3. Although these methods measure mostly stably bound Ab, they yield results of far greater precision than the first method.

The neutralizing power of the serum reflects the degree of protection against a virus. In the early period of immunization, however, with low-affinity Abs present, such protection is best estimated by carrying out **neutralization in the presence of complement** (see Sensitizing Antibodies, above).

ROLE OF COMPLEMENT IN ANTIVIRAL IMMUNITY

Complement (Ch. 20) has a considerable significance in immunity: it enhances the neutralization of viruses by

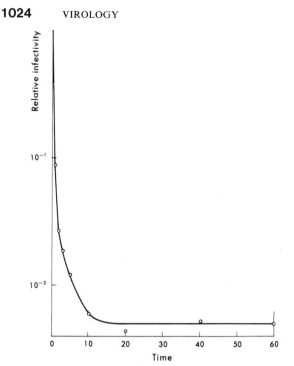

FIG. 52-7. Kinetic curve of neutralization of western equine encephalitis virus (a togavirus), showing the rather abrupt change into a plateau as the time of incubation increases. This plateau is not justified by the A/V ratio employed, which would have allowed a far greater neutralization. The plateau corresponds to the persistent fraction. (Dulbecco R et al: Virology 2:162, 1956)

specific Abs; it inactivates virions directly by blanketing them; in cooperation with Abs, it lyses enveloped virions in the same way that it lyses cells; and it lyses some types of virions even without Abs. In primary infections of experimental animals complement plays an important antiviral role before a strong Ab response is mounted; thus decomplementation by cobra venom factor (Ch. 20), or genetic deficiency of component C5 in mice, increases the duration and severity of primary influenza in mice.

HEMAGGLUTINATION INHIBITION

With virions that agglutinate RBCs, adding the appropriate Abs to the virus before adding the RBCs decreases the hemagglutinating titer by hindering adsorption of the virions to the RBCs. This is known as **hemagglutination inhibition** (HI). It differs from neutralization in several ways. In HI, Abs interfere with the adsorption of the virions, rather than with cell infection; and stable Ab–virion complexes do not appear to play a special role in HI, since it is effectively carried out by univalent Ab fragments. Finally, the number of Ab molecules per virion may have to be high enough to cover all the sites of the virion involved in adsorption.

DIAGNOSTIC USE OF IMMUNOLOGY

Immunologic tests can be used with standard antisera to identify and characterize a virus isolated from a patient and, with standard Ags, to detect antiviral Abs in the patient's serum. These two applications of viral immunology involve a number of general problems which will be dealt with before considering the diagnostic methodology.

TYPES OF VIRUS-SPECIFIC ANTIBODIES

Different kinds of viral preparations elicit the formation of different Abs. 1) Immunization with virions that cannot multiply in the host, e.g., with **killed virions,** produces Abs predominantly directed toward **surface components** of the virions; these Abs have neutralizing and hemagglutination-inhibiting (HI) activities against the virions, as well as complement-fixing (CF) and precipitating activities against Ags of the viral coat. 2) In contrast, **viruses that multiply in the host** and produce a cytopathic effect in some cells—as in natural infection or in vaccination with "live" vaccines—lead to the formation of Abs against **all the viral Ags,** including Abs for surface Ags, CF or precipitating Abs for both surface and internal

Ags, and Abs for nonvirion Ags. 3) Immunization with **internal components of the virions** produces CF and precipitating Abs active only toward the Ags of these components.

SENSITIVITY OF TEST METHODS

Neutralization is the most sensitive test method, followed by radioimmunoassay (RIA), HI, and complement fixation (CF). These differences in sensitivity parallel the amount of Ag required for each test: for example, about 10^3 influenza viral particles are required for neutralization tests, at least 10^7 for HI, and an even greater number for CF. For these reasons, in a given serum the neutralizing Abs usually have higher titers (i.e., are detected more easily) than HI or CF Abs specific for surface components of the virions.

After viral infection the titers of Abs to different components rise and fall with quite different time courses, as will be discussed in the chapters on specific viruses.

The Abs that react in the different tests may overlap though they may not be altogether identical. Neutralization may be primarily caused by Ab molecules specific for the sites of the virion that are involved in the release of nucleic acid into the cells, while CF may involve additional surface Ags. Neutralization probably requires molecules with a higher affinity for virions than do HI and CF. Only certain classes of immunoglobulins can participate in CF.

ANTIGENIC CLASSIFICATION OF VIRUSES

Because of their high specificity, immunologic methods can differentiate not only between viruses of different families but also between closely related viruses of the same family or subfamily. By these means **family Ags** may be identified; each family or subfamily may be subdivided into types **(species)** on the basis of **type-specific Ags;** some types can even be subdivided further **(intratypic differentiation).** The levels in this classification are obviously somewhat arbitrary. Usually Abs detected by neutralization tend to be less crossreactive, and thus are useful to define the immunologic type, whereas those detected by CF tend to be more crossreactive and are useful to define the family. By proper procedures, however, such as immunization with purified Ags, highly specific CF Abs can be prepared.

The resolving power of Abs is maximized by using **monoclonal Abs:** each is produced from a single clone of B cells and is therefore endowed with a single specificity. Monoclonal Abs to influenza or rabies virus can reveal antigenic differences that are not picked up by sera of immune animals, which usually recognize many specificities.

Monoclonal Abs were first produced from cultures of spleen fragments: spleen cells were transferred from a virus-immunized to an irradiated mouse (in which they reconstitute the lymphoid system) and 3 weeks later small fragments of the spleen of the recipient animal were separately cultivated in the presence of Ag. A fraction of the cultivated fragments release antiviral Abs into the culture supernatant; if this fraction is small most of the Abs are monoclonal. A more recent production technic employs **hybridomas:** spleen cells from immunized animals are fused to myeloma cells; the resulting hybrid cells are cloned to yield cultures (hybridomas) that release monoclonal Abs into the medium.

EVOLUTION OF VIRAL ANTIGENS

The great selectivity of neutralization can be understood on the basis of the structure, genetics, and evolution of viruses. Animal viruses that have evolved in the ecology of mammalian organisms have been opposed by the neutralizing Abs, which are able to block viral infection. Viral evolution must tend to select for mutations that change the antigenic determinants involved in neutralization. In contrast, other antigenic sites would tend to remain unchanged, because mutations affecting them would not be selected for. A virus would thus evolve from an original type to a variety of types, different in neutralization (and sometimes in HI) tests, but all retaining some of the original mosaic of antigenic determinants recognizable by CF.

These evolutionary arguments are consistent with the observation that the clearest differentiation of types within a family is present in viruses of rather complex architecture, in which the Ags involved in the attachment of the virion to the cell vary more than the other proteins. Thus enveloped viruses have a strain-specific envelope but a crossreactive internal capsid; adenoviruses have type-specific fibers and family-specific (but also type-specific) capsomers (Ch. 54). Moreover, the C Ag of polioviruses, which appears only after heating, crossreacts in all three viral types (Ch. 57). The heating evidently reveals antigenic sites that are normally hidden and hence are not affected by selective pressure.

The extent of antigenic variation differs widely among viruses. Variation is most extensive in influenza virus, an orthomyxovirus (see Ch. 58), in which the hemagglutinins are the main sites of neutralization (see The Envelope, Ch. 46). Epidemiologic studies show that this variation occurs by the appearance every decade or so of strains in which the hemagglutinins and the neuraminidase molecules are genetically unrelated to those of the previous years **(antigenic shift),** followed by the progressive change of the new strain in the following years **(antigenic drift)** (see Ch. 58). Antigenic shift is probably due to the appearance of recombinants of a human virus with a virus present in an animal reservoir (e.g., avian or equine); antigenic drift, in contrast, is due to the progressive accumulation of mutations, which partly overcome the immunity prevalent in the host population.

Habitat and Selection. Drift is probably favored by the habitat of influenza viruses, namely, cells lining the respiratory tract: here they are exposed only to IgA Abs, which tend to form reversible complexes with virions and hence are less likely to neutralize. The virus then can multiply (although at a reduced rate) even in an immune host, producing a large population as a source of mutants; these, being less sensitive to the Abs, can then be selected. In contrast, viruses that cause viremia are more effectively neutralized by IgG; hence mutants with a somewhat decreased binding of neutralizing Abs may still be eliminated. Thus viruses such as mumps and measles, which are structurally similar to influenza viruses, have persisted as single immunologic types with only minimal antigenic variation.

Also explainable in terms of selective evolution is the frequent **crossreactivity in neutralization of viruses** of different types within a family, the heterotypic neutralization titer of sera regularly being much lower than the homotypic titer. These crossreactions can be attributed to a residual antigenic specificity of the primordial type, persisting despite subsequent mutations.

METHODS FOR ANTIGENIC ANALYSIS OF VIRUSES

We shall consider here only those immunologic methods that are peculiar to viruses: neutralization, HI, and immuno-electron microscopy. A summary of the diagnostic serologic procedures used for a number of human viruses is given in Table 52-1.

NEUTRALIZATION TEST

Neutralization tests are mostly used in serologic diagnosis, in typing of viral isolates, and in characterizing related viruses. Three methods will be described.

Determining Neutralizing Antibody Titers by a 50% Endpoint Method. The titration is usually carried out with **constant virus and varying serum.**

Method 1. Endpoint Test. In order to carry out a constant-virus–varying-serum titration, the serum is inactivated at 56° C for 30 min to destroy labile substances that have antiviral activity or that affect neutralization (see below). The serum is then diluted serially, usually in twofold steps, and each dilution is added to a constant amount of virus (usually between 30 and 100 ID_{50}; see Appendix, Ch. 46). The amount of virus must be adequate to infect all the assay units in the subsequent titration and at the same time small enough to detect a low concentration of Abs. The mixtures are incubated for a selected time (at either 37° or 4° C, depending upon the virus employed); then a constant volume of each is assayed for infectivity by inoculation into 5–10 test units of a suitable assay system, such as mice, chick embryos, or tissue culture tubes. The neutralization titer of the serum is the dilution at which 50% of

TABLE 52-1. Summary of Serologic Tests for Viral Diseases

Virus	Neutralization			Other tests		
	Common virus source	Usual test host	Route of inoculation	Test	Common antigen source	Red cells
Enterovirus						
Growing in TC	TC fluid (rhesus monkey kidney)	TC (rhesus monkey kidney)		CF HI (some)	TC fluid (rhesus monkey kidney)	Human 0
Not growing in TC	Infected mouse torso suspensions	Suckling mice	SC, IP, or IC	CF (some) HI (some)	Mouse torso	Human 0
Rhinovirus	TC (diploid human fetal fibroblast strain; human lines; monkey kidney)	TC				
Influenza	Chick embryo	Chick embryo	Allantoic sac	HI	Allantoic fluid	Fowl, human 0, guinea pig
	TC (monkey or calf kidney)	TC (monkey or calf kidney)		CF	Allantoic fluid	
Parainfluenza	TC (monkey kidney)	TC (monkey kidney)		CF HI HAI	TC (monkey kidney)	Fowl, human 0, guinea pig
Mumps	Chick embryo (allantoic fluid, membrane, amniotic cavity)	Chick embryo TC (monkey kidney)		Allantoic cavity (production of HA)	Allantoic fluid and membrane	Fowl, monkey
Respiratory syncytial	TC (Hep-2)	TC (human lines)		CF	TC (Hep-2)	
Measles	TC (BSC-1)	TC (human, chick primary, human lines)		CF HI	TC (BSC-1) TC	Monkey
Rabiesvirus*	Mice TC (hamster kidney, human diploid, chick embryo fibroblasts)	Mice, 4–6 weeks	IC	Fluorescent Ab in brain smears CF	Mouse brain	
Togavirus	Mouse brain TC	Suckling or weanling mice TC	IC	HI Kaolin-absorbed or acetone-extracted sera CF	Mouse brain (fluorocarbon-extracted) TC fluids (chick embryo fibroblasts, hamster kidney, HeLa cells)	Day-old chick or goose cells

(continued)

TABLE 52-1. Summary of Serologic Tests for Viral Diseases (*continued*)

Virus	Neutralization			Other tests		
	Common virus source	Usual test host	Route of inoculation	Test	Common antigen source	Red cells
Reovirus	TC (rhesus monkey kidney)	TC (rhesus monkey kidney)		HI CF	TC fluids (rhesus monkey kidney)	Human 0, bovine
Herpes simplex	TC or CAM	TC (rabbit kidney) Embryonated chicken eggs	CAM	CF	TC (rabbit kidney or CAM)	
Varicella–herpes zoster	TC (human fibro-blasts)			CF	TC (primary human amnion)	
Human cytomegalic	TC (human fibro-blasts)			CF	TC (human fibro-blasts)	
Adenovirus	TC (HeLa or KB)	TC (HeLa or KB)	HI CF	TC (HeLa or KB)	Monkey or rat	
Variola; vaccinia	Chicken embryo (CAM)	Chicken egg (CAM)		HI	Chicken embryo	Chicken cells (selected)
	TC	TC (monkey kidney, HeLa, KB)		CF RIA	Chicken embryo (CAM)	
Rubella virus	TC (Monkey kidney)	TC (Monkey kidney)		HI CF RIA	TC (monkey kidney)	Human 0, fowl
Coronaviruses	TC (human primary)	TC (human primary)		CF HAI	Mouse brain (suckling)	Mouse, rat, chicken
	OC (human)	OC (human)				
Rotavirus				IEM CF	Fecal extracts Duodenal fluid	

*Rabiesvirus is usually identified by a mouse protection test. An unknown virus is inoculated intracerebrally in untreated mice and in mice previously vaccinated with an attenuated strain (HEP Flury vaccine virus). If the untreated animals die and those vaccinated survive, the identification of rabiesvirus is certain.

Abbreviations: TC, tissue cultures; OC, organ cultures; CAM, chorioallantoic membrane of chick embryo; IC, intracerebral; IP, intraperitoneal; SC, subcutaneous; CF, complement fixation; HI, hemagglutination inhibition; HAI, hemadsorption inhibition; HA, hemagglutinins; RIA, radioimmune assay; IEM, immuno-electron microscopy. HEP-2, BSC-1, HeLa, and KB are cell lines.

the units are protected (50% endpoint), calculated by using the method of Reed and Muench (see Appendix, Ch. 46).

The accuracy of the assay can be calculated by the same methods as used for viral assays (see Appendix, Ch. 46). (Since the serum dilutions are closely spaced the assay is more precise than the corresponding viral assays, which usually employ more widely spaced dilutions.) The constant-virus–varying-serum titration relies on the constancy of the virus employed; this requirement, however, is not critical because, according to the percentage law, the proportion of virus neutralized is (at least largely) independent of the titer of the virus.

This method is statistically more accurate than the **constant-serum-varying-virus** method because in virus–serum mixtures a small change in Ab titer usually produces a much larger change in the infectious titer of the virus. The reason is the exponential relation between residual infectivity and Ab concentration of equation 3 above.

Method 2. Plaque-Reduction Test. In this test an inoculum of about 100 plaque-forming units is incubated with serial dilutions of the serum; each mixture is then added to a monolayer culture, which is overlaid with agar and incubated. The endpoint is an 80% reduction in the number of plaques. The precision of the method depends on the number of plaques at the endpoint (see Appendix, Ch. 46).

These two methods are used either qualitatively, to demonstrate the presence of virus-specific Ab, or quantitatively, to compare Ab titers in different sera.

Method 3. Determining the Rate of Neutralization by the Plaque Method. This test measures the slope (k) of the kinetic curves described above. The values obtained are extremely reproducible; differences of about 20% are usually significant. Values obtained with the same serum provide a sensitive basis for distinguishing viral strains, including laboratory mutants (e.g., in work with vaccine strains of poliovirus).

In this test a virus sample of known titer is mixed with the antiserum; samples are taken at intervals, diluted, and assayed for plaques. The logarithm of the ratio of the titer of the sample to the original titer (the relative infectivity, I) is plotted against the time of sampling, yielding a straight line through or near the origin. The slope of that line is k, which is characteristic for the serum and the virus; it is determined from the relation $k = -(\ln I/tC)$, from equation 2 above, where I is the relative infectivity determined after t minutes of incubation of the neutralizing mixture and C is the concentration of the serum.

Example: At a serum dilution of 10^{-3} the relative infectivity is 3×10^{-2} after a 10-min incubation. Since $\ln 3 \times 10^{-2} = -3.5$, $k = 3.5/(10 \times 10^{-3}) = 350$.

HEMAGGLUTINATION-INHIBITION

For this test a serial twofold dilution of heat-inactivated serum is prepared in saline, and from each dilution 0.25 ml is mixed with 0.25 ml of a viral suspension containing 4 hemagglutinating units (defined in Ch. 46: see Hemagglutination Assay). (If hemagglutination-inhibitory substances are known to be present in the serum or the viral preparation they must be removed in advance; see

Complications, below.) To each mixture is added 0.25 ml of a 1% RBC suspension. The tubes are shaken and then incubated at the temperature and for the time required for optimal hemagglutination with the virus used. The agglutination pattern is read after incubation as discussed in Chapter 46. The **HI titer** is the reciprocal of the highest serum dilution that completely prevents hemagglutination. Example:

				Initial serum dilution					HI titer
	1:8	1:16	1:32	1:64	1:128	1:256	1:512	1:1024	
A	0	+	+	+	+	+	+	+	8
B	0	0	0	0	+	+	+	+	64

IMMUNO-ELECTRON MICROSCOPY

This technic is especially useful for human viruses (such as rotaviruses; Ch. 64) that cannot be grown in cell cultures or in convenient animal hosts, and do not produce hemagglutination, but have a high titer in the blood or excretions of infected individuals. When the virus-containing sample is mixed with appropriate Abs, electron microscopic examination reveals virions with an Ab halo.

TESTS ON PAIRED SERA

With a widespread virus a high Ab titer may have resulted from previous exposure, and the serologic diagnosis of acute infection is based on an increase of the Ab titer in **paired sera,** i.e., one obtained soon after the onset of disease **(acute serum)** and the other obtained 1 or 2 weeks later **(convalescent serum).** This comparison requires considerable precision and is carried out by the plaque-reduction method whenever possible, or by HI or CF tests. A fourfold increase in the titer is considered evidence of infection.

COMPLICATIONS OF IMMUNOLOGIC TESTS

Interfering substances occasionally obscure the significance of immunologic tests. Although they concern the specialist carrying out

the test, some knowledge of their nature is useful for evaluating test results.

Neutralization Test. Human and animal sera contain **nonspecific viral inhibitors,** especially active against influenza, mumps, herpes simplex, and togaviruses. They result from interactions with properdin and the alternate complement pathway. They are heat-labile and can be eliminated by incubation at 56°C for 30 min.

Hemagglutination-Inhibition Test. Most sera contain inhibitors of hemagglutination that are not Abs; they must be removed. The inhibitors for influenza viruses are inactivated by receptor-destroying enzyme (RDE; see Mechanism of Hemagglutination, Ch. 46), or by trypsin or periodate; those for togaviruses, by extraction with acetone–chloroform.

Complement Fixation Test. Performance of this test is frequently hampered by the presence of anticomplementary substances in crude tissue suspensions used as Ag. This is especially true of the infected brain suspension used with togaviruses; it can be freed of the anticomplementary factors by thorough extraction with acetone in the cold or by extraction with a fluorocarbon.

CELL-MEDIATED IMMUNITY

IMMUNITY BASED ON CYTOTOXIC THYMUS-DERIVED LYMPHOCYTES (CTLs)

As will be discussed in Chapter 53, this type of immunity appears to be very important not only in localizing viral infections and in ultimate recovery, but also in the pathogenesis of viral diseases.

The CTLs are usually found in the blood or the spleen, and sometimes in abundance in the lymph nodes draining local infection sites (e.g., in herpes simplex infection of mice and rabbits). CTLs are also present in exudates within affected organs: in mice they are found in the lung after infection with influenza virus, or in the cerebrospinal fluid after induction of meningitis by arenaviruses. CTLs kill virus-infected cells in vitro.

In experimental animals primary CTLs reach a maximal abundance about 6 days after a viral infection and

then disappear as infection subsides. However, the organism contains, for a long time, virus-specific memory T cells, which can be recognized by culturing spleen cells together with virus-infected target cells: within a few days secondary CTLs appear in the culture. Secondary CTLs are similarly produced in the body after a second infection.

Formation of CTLs is elicited by cell-associated Ags. Hence it occurs in all infections by enveloped viruses, whose glycoproteins are incorporated in the plasma membrane (see Enveloped Viruses, Ch. 50). Thus, with vesicular stomatitis virus the response is directed at the single envelope glycoprotein; under conditions in which temperature-sensitive (*ts*) mutants fail to introduce the protein into the membrane they do not elicit the cellular response. The viral glycoproteins not only are synthesized in infected cells, but they can derive exogenously from the infecting virions, whose envelopes fuse with the cell plasma membrane in the initial stage of viral penetration. Hence even noninfectious or inactivated viruses can elicit a cellular immune response. Moreover, the virions themselves may also be able to elicit the response after adsorbing to macrophages.

CTLs are also present in infections by some nonenveloped viruses (such as DNA-containing tumor viruses, Ch. 66, or coxsackie B viruses) but it is not clear which Ag they recognize.

The number and nature of the viral proteins present at the cell surface determines the specificity of the immune response. Thus studies with recombinants between different influenza virus A types show that the hemagglutinin elicits a type-specific immunity, and the matrix protein elicits an immunity that is crossreactive among several A types, but not between A and B types. Larger viruses (herpesviruses, poxviruses) elicit both type- and group-specific responses because the viral glycoproteins contain determinants of both types.

The spectrum of the secondary response depends on the relation between primary and secondary infection. For instance, with influenza virus the secondary response is type-specific if the second infection is caused by the same viral type as the first infection, whereas it is crossreacting between types if the second infection is of a different type. This mechanism can build a broad heterotypic cellular immunity to influenza viruses. The crossreactivity of cellular immunity appears to be greater than that of humoral immunity.

Major Histocompatibility (MHC) Restriction. As in the induction of CTLs to minor Ags of the target cells, the induction of CTLs specific for virus-infected cells and the expression of the killing action of the CTLs require that the immune cells and the target cells share an identical major histocompatibility (MHC) Ag. Thus in mice the virus-infected cells must have the same H-2 (D or K) Ag as the CTLs' precursors in order to elicit the formation of mature CTLs (either primary or secondary); and in turn the mature CTLs can only kill those target cells with the same antigenic specificity. In humans the HLA Ags must be matched. The restriction occurs because, during the maturation in the thymus, CTLs become determined to recognize the MHC Ag present on the thymus epithelia. Recognition of MHC Ag is essential in most cases: virus-infected cells of certain lines of cells that do not express MHC Ag on their surface are not lysed by any CTL specific for the infecting virus.

In some exceptional cases virus-specific CTLs also lyse infected cells with allogeneic MHC Ag. This occurs especially with influenza virus or herpesvirus, which, acting as nonspecific T cell mitogens, bypass the role of the thymus. Some of these CTLs may even lyse uninfected cells.

The important roles of immune CTLs in the pathogenesis of viral diseases will be reviewed in Ch. 53.

ANTIBODY-DEPENDENT COMPLEMENT-INDEPENDENT CELL-MEDIATED CYTOTOXICITY (ADCC)

In this response the effector cells are **killer (K) lymphocytes,** which lack T cell markers and surface Igs. In vitro these cells kill virus-infected cells sensitized by IgG from immune donors, but not unsensitized targets. The cytotoxic cells themselves are not virus-specific but acquire their specificity by reacting with the Fc region of the sensitizing Abs. The cytotoxic effect is inhibited by IgG F(ab')$_2$ fragments, which compete with the cell-bound IgG. This type of cytotoxic cell has been observed in humans and in animals infected by various enveloped viruses.

ADCC is very efficient in vitro against herpes simplex–infected cells, preventing the usual spread of the virus from infected to neighboring uninfected cells through cellular bridges; it may play a role in defense against human herpesvirus infections.

Herpesviruses induce Fc receptors on the surface of infected cells, apparently as part of virus-specified glycoproteins. After binding IgGs, these receptors cap and are internalized together with the IgG; it is not known whether these events play an antiviral role.

NATURAL KILLER (NK) CELL CYTOTOXICITY

The so-called natural killer (NK) cells, which are found in the peripheral blood of most humans, are distinguishable from T or B lymphocytes, macrophages, and polymorphonuclear cells. They do not contain Fc receptors, and are cytotoxic without requiring sensitizing Abs. They attack a variety of iso-, allo-, or xenogeneic cell types, either infected by viruses or uninfected. NK cells are especially abundant in individuals infected by enveloped viruses (e.g., mumps, herpes simplex, or lymphocytic choriomeningitis viruses) or after vaccination with vaccinia virus.

Probably NK cells are not directly virus-induced, because their presence is not related to a previous history of viral disease; however, they appear to be indirectly virus-induced by **interferon** produced by virus-infected cells. In fact, interferon confers a viral specificity on NK cells by enhancing their cytotoxic action (as well as that of Ab-dependent K cells) on infected target cells, while protecting uninfected cells from lysis.

IMMUNOSUPPRESSION

The immune response is frequently depressed in animals infected with oncogenic viruses (Ch. 66) containing either RNA (retroviruses) or DNA (Marek disease virus). This event favors oncogenesis by helping the spread of the virus and reducing the rejection of transformed cells. This type of suppression is manifested by an impaired responsiveness of lymphoid cells to T cell mitogens (phytohemagglutinin or concanavalin A) and sometimes by a reduced response of B cells to immunization with other Ags. It is probably mediated by thymus-derived suppressor cells, because early thymectomy reduces the severity of the disease produced by a later infection.

Cytomegalovirus also depresses both T and B cell functions, probably because it persistently infects lymphocytes and other immune cells. In the mixed lymphocyte reaction in cultures (Ch. 23), measles, herpes simplex, rubella, or poliovirus inhibit the blastogenesis of lymphocytes cultivated together with allogeneic lymphocytes, apparently by infecting monocytes and macrophages (which play an important role in that reaction).

SELECTED READING

BOOKS AND REVIEW ARTICLES

HERMANN EC: New concepts and developments in applied diagnostic virology. Progr Med Virol 17:222, 1974

KOPROWSKI C, KOPROWSKI H (eds): Viruses and Immunity. New York, Academic Press, 1975

MELNICK JL: Viral vaccines. Prog Med Virol 23:158, 1977

NOTKINS AL (ed): Viral Immunology and Immunopathology. New York, Academic Press, 1975

OLDSTONE MBA: Virus neutralizations and virus-induced immune complex disease. Progr Med Virol 19:85, 1975

SHVARTSMAN YA S, ZYKOV P: Secretory anti-influenza immunity. Adv Immunol 22:291, 1976

WOODRUFF JF, WOODRUFF JJ: T lymphocyte interaction with viruses and virus-infected tissues. Progr Med Virol 19:121, 1975

SPECIFIC ARTICLES

DANIELS CA, BORSOS T, RAPP HH et al: Neutralization of sensitized virus by purified components of complement. Proc Natl Acad Sci USA 65:528, 1970

DOHERTY PC, SOLTER D, KNOWLES BB: H-2 gene expression is required for T cell-mediated lysis of virus-infected target cells. Nature 226:361, 1977

GERHARD W, CROCE CM, LOPES D, KOPROWSKI H: Repertoire of antiviral antibodies expressed by somatic cell hybrids. Proc Natl Acad Sci USA 75:1510, 1978

GERHARD W, WEBSTER RG: Antigenic drift in influenza A viruses. I. Selection and characterization of antigenic variants of A/PR/8/34 (HON1) influenza virus with monoclonal antibody. J Exp Med 148:383, 1978

HALE AH, WITTE ON, BALTIMORE D, EISEN H: Vesicular stomatitis virus glycoprotein is necessary for H-2–restricted lysis of infected cells by cytotoxic T lymphocytes. Proc Natl Acad Sci USA 75:970, 1978

HAPEL AJ, BABLANIAN R, COLE GA: Inductive requirements for the generation of virus-specific T lymphocytes. J Immunol 121:736, 1978

JERNE NK, AVEGNO P: Development of the phage-inactivating properties of serum during the course of specific immunization of an animal: reversible and irreversible inactivation. J Immunol 76:200, 1956

KAHAN L, FENTON WA, MURAKAMI WT: Specificity and distribution of antigenic determinants on the polyoma virus capsid and nature of the reaction of immunoglobulin G antibody with the capsid surface. J Mol Biol 95:239, 1975

KELSEY DK, OVERALL JC, GLASGOW LA: Correlation of the suppression of mitogen responsiveness and the mixed lymphocyte reaction with the proliferative response to viral antigen of splenic lymphocytes from cytomegalovirus-infected mice. J Immunol 121:464, 1978

KOMATSU Y, NAWA Y, BELLAMY AR, MARBROOK J: Clones of cytotoxic lymphocytes can recognize uninfected cells in a primary response against influenza virus. Nature 274:802, 1978

MANDEL B: Neutralization of poliovirus: a hypothesis to explain the mechanism and the one-hit character of the neutralization reaction. Virology 69:500, 1976

SCHOLTISSEK C, ROHDE W, VON HOYNINGEN V, ROTT R: On the origin of the human influenza virus subtypes H2N2 and H3N2. Virology 87:13, 1978

WEBSTER RG, CAMPBELL CH: Studies on the origin of pandemic influenza. IV. Selection and transmission of "new" influenza viruses in vivo. Virology 62:404, 1974

WIKTOR TJ, DOHERTY PC, KOPROWSKI H: In vitro evidence of cell-mediated immunity after exposure of mice to both live and inactivated rabies virus. Proc Natl Acad Sci USA 75:334, 1977

YOUNG JF, PALESE P: Evolution of human influenza A viruses in nature: Recombination contributes to genetic variation of H1N1 strains. Proc Natl Acad Sci USA 76:6547, 1979

ZINKERNAGEL RM, OLDSTONE MBA: Cells that express viral antigens but lack H-2 determinants are not lysed by immune thymus-derived lymphocytes but are lysed by other antiviral immune attack mechanisms. Proc Natl Acad Sci USA 73:3666, 1976

ZWEERINK HJ, COURTNEIDGE SA, SJEHEL JJ, CRUMPTON MJ, ASKONAS BA: Cytotoxic T cells kill influenza virus infected cells but do not distinguish between serologically distinct type A viruses. Nature 267:354, 1977

chapter

PATHOGENESIS OF VIRAL INFECTIONS

The consequences of a viral infection depend upon a number of factors, both viral and host, that affect pathogenesis, including the number of infecting viral particles and their path to susceptible cells, the speed of viral multiplication and spread, the effect of the virus on cell functions, the host's secondary responses to cellular injury (edema, inflammation), and the host's defenses, both immunologic and nonspecific. Viral infection was long thought to produce only acute clinical disease, but other host responses are being increasingly recognized. These include asymptomatic infections, induction of various cancers, and chronic progressive neurologic disorders.

This chapter will deal with the general features of viral pathogenesis, and where possible it will describe the mechanisms involved. Subsequent chapters will discuss the particular characteristics and special effects of various viruses that infect man.

CELLULAR AND VIRAL FACTORS IN PATHOGENESIS

The effects of viral infection on cells depend on both the characteristics of the virus and susceptibility of the cells.

VIRAL CHARACTERISTICS

Viral virulence, like bacterial virulence, is under polygenic control and cannot be assigned to any single viral property: it is, however, frequently associated with several characteristics that promote viral multiplication and cell damage. Thus virulent viruses multiply well at the elevated temperatures that arise during illness (i.e., above 39° C), induce interferon poorly and resist its inhibitory action, block the biosynthesis of host macromolecules, and damage cell lysosomes or alter the cell membranes of the infected cells, producing damage in the infected animal or **cytopathic effects** in cell cultures. **Attenuated mutants,** altered in various of these functions, including **conditionally lethal temperature-sensitive (ts) mutants** (see Mutant Types, Ch. 50), produce less severe or no disease. However, some viruses that are attenuated in their behavior in animals (e.g., poliovirus vaccine strains) cause the same cytopathic effects as wild-type virus in cultures where their multiplication is not restrained.

CELL SUSCEPTIBILITY

The Role of Cell Receptors. The susceptibility of cells to viral infection is often determined by their early interactions with a virus: viral attachment or the release of its nucleic acid in the cells. With animal viruses, as was observed earlier with bacteriophages (Ch. 47), resistance of the cells is often caused by failure of viral adsorption: hence cells resistant to a virus may be susceptible to its extracted nucleic acid (see Productive Infection, Ch. 50). Indeed, differences in the adsorption of viruses to cells in culture have been correlated with differences in organ susceptibility, and also with changes in host susceptibility with age.

Both physiologic and genetic factors affect the presence or activation of receptors for viral adsorption, as well as other cellular properties that influence susceptibility.

Physiologic Factors. Cultivation may markedly alter the viral susceptibility of cells from that in the original organ. Hence many viruses can be propagated in cells that are readily cultured, obviating the need for cells that are hard to culture, or for intact animals. For instance, polioviruses, which multiply in the nervous tissue but not in the kidney of a living monkey, multiply well in cultures derived from the kidneys, since receptors develop in the cultivated kidney cells.

Marked changes in susceptibility accompany the **maturation** of animals. Many viruses are much more virulent in newborn animals (e.g., coxsackieviruses, herpes simplex virus) and others in adults (e.g., polioviruses, lymphocytic choriomeningitis [LCM] virus). There are several mechanisms: with coxsackieviruses in mice the change in susceptibility is correlated with receptor activity, although it may also depend on changes in interferon and Ab production; with foot-and-mouth-disease virus it involves the rate of viral multiplication; and with LCM virus it is due primarily to an immunologic mechanism (see below).

Genetic Factors. Genetic differences in susceptibility have been demonstrated in mice (with togaviruses, mouse hepatitis virus, and influenza virus) and in chickens (with oncogenic retroviruses). In crosses between resistant and susceptible animals the heterozygous first-generation (F_1) progeny are uniformly resistant or uniformly susceptible, depending on which allele is dominant. (Resistance is dominant with influenza and togaviruses, and susceptibility with mouse hepatitis and retroviruses.) Moreover, backcrosses of the F_1 individuals to the parent carrying the recessive allele yield 50% resistant animals. These results imply a difference in a single gene or a closely linked cluster. Some of these hereditary differences evidently involve the host cell–virus interactions, since they can also be demonstrated in vitro with macrophages and with cultured cells; but others could reflect control of the immune response.

CELLULAR RESPONSES TO VIRAL INFECTIONS

Cells can respond to viral infection in three different ways: 1) **no apparent change;** 2) **cytopathic effect and death;** 3) **hyperplasia,** which may be followed either by **death** (as in the pocks of poxviruses on the chorioallantoic membrane of the chick embryo: Ch. 56), or by **continued loss of growth control** (as in **viral transformation** of normal to cancer cells: Ch. 66). The development of inclusion bodies and chromosomal aberrations may be special features of these cellular responses.

Cytopathic Effects. Virus-induced cell injury has been most extensively studied with cultured cells since these are believed to reflect in vivo cell damage accurately, and their responses can be quantified. In vitro cell damage, termed the **cytopathic effect,** is recognized from various morphologic alterations, which are listed in Table 53-1; cell death usually follows.

The factors listed below appear to contribute to development of the various cytopathic effects.

1) **Effects on synthesis of cellular macromolecules.** Many virulent viruses cause an early depression of cellular syntheses. As noted in Chapter 50, DNA-containing viruses inhibit the synthesis of host cell DNA, but not of host cell RNA and protein until late in the multiplication cycle, whereas many RNA viruses inhibit host cell RNA and protein synthesis early in the multiplication cycle.

2) **Alteration of lysosomes.** Some viruses cause a **reversible** increase in lysosome permeability, without leakage of enzymes from the organelles. This change is shown by an increased binding of neutral red, the dye commonly used to stain the live cells in the plaque assay: the cells appear hyperstained and form "red plaques." Other viruses effect an **irreversible** damage resulting in disruption of the organelles and discharge of their hydrolytic enzymes into the cytoplasm. The cells lose their ability to be stained with neutral red and form the usual "white plaques." This profound effect appears to be due to proteins synthesized late in the viral multiplication cycle, possibly capsid subunits.

3) **Alterations of the cell membrane.** Many of the budding, enveloped viruses incorporate viral subunits, usually glycoproteins, into the infected cell membranes, as a preliminary to the formation of the viral envelope (see Enveloped Viruses, Ch. 50). Even some viruses that do not bud from the cell surface, such as herpes simplex and vaccinia viruses, insert novel Ags into the plasma membrane. These changes may be recognized by reaction with virus-specific Abs, or by **hemadsorption** (e.g.,

TABLE 53-1. Cellular Response to Viral Infection

Virus	Cell type*	Cytopathic response	Inclusion body
Adenoviruses	HeLa Rat embryo	Cell rounding and clumping Transformed	Nuclear
Herpesviruses (herpes simplex)	HeLa	Polykaryocytes (some strains); cell rounding	Nuclear
Poxviruses (variola)	HeLa	Slow rounding; hyperplastic foci	Cytoplasmic
Picornaviruses (poliovirus)	Monkey kidney	Cell lysis	None
Orthomyxoviruses (influenza virus)	Monkey kidney	Slow rounding	None
Paramyxoviruses (parainfluenza virus)	Monkey kidney	Fusion of cell membranes; syncytial formation	Cytoplasmic
Coronaviruses	Human diploid	Minimal; syncytia rarely	None
Togaviruses (eastern equine encephalitis virus)	Mouse L	Cell lysis	None
Rubella virus	Human amnion	Slow enlargement and rounding	Cytoplasmic
Reoviruses	Monkey kidney	Enlargement and vacuolation	Cytoplasmic
Rabiesvirus	Hamster kidney	Usually none	Cytoplasmic

*With many viruses several cell types can be used; in such instances, a commonly used type is listed.

orthomyxoviruses and paramyxoviruses), or by adsorption of increased quantities of plant **lectins** such as concanavallin A (e.g., poxviruses, paramyxoviruses). The inserted Ags make the cells targets for **immunologic destruction** by virus-specific Abs plus complement or by immune T lymphocytes. In addition, effects on their membranes probably play a large role in altering the shape and function of cells. In a striking effect, observed with paramyxoviruses and some herpesviruses, the infected cells fuse with adjacent cells (i.e., establish a continuity between the plasma membranes), forming **giant cells (polykaryocytes).**

4) **Cytopathic effect without synthesis of progeny virus.** A high concentration of virions or of viral coat proteins may initiate cytopathic changes without replication of infectious virus **(toxic effect).** For instance, vaccinia virus at high concentration can rapidly kill cultures of macrophages and other cells, mumps virus lyses RBCs and induces syncytia formation in cell cultures by a direct action on plasma membranes, and the cytopathic effect of adenovirus is reproduced by a purified capsid protein, the penton. Such effects of the viral coat (or some lytic agent associated with it) may be the equivalent of the "lysis from without" induced by bacteriophages (see Effect of Phage Attachment, Ch. 47).

5) **Abortive infection** may also cause cytopathic effects as a consequence of **incomplete viral synthesis:** for example, in cultured HeLa cells influenza viruses synthesize Ags and damage the cells, though they do not form infectious virions.

Viral toxic effects also occur in animals. In mice, for example, the intravenous injection of a concentrated preparation of influenza, mumps, or vaccinia virus causes hemorrhages and cellular necrosis in various organs, resulting in death within 24 h; a large intracerebral inoculum of influenza virus produces necrosis of brain cells; and a large intranasal inoculum of Newcastle disease virus (a paramyxovirus) produces extensive pneumonia. All these effects are produced without synthesis of viral components or with synthesis of only incomplete particles.

Development of Inclusion Bodies. Intracellular masses of new material may arise as accumulations either of virions or of unassembled viral components, in the nucleus (e.g., adenovirus), in the cytoplasm (e.g., rabiesvirus Negri bodies), or in both (e.g., measles). These inclusion bodies appear to disrupt the structure and function of the cells and to contribute to their death. Other inclusion bodies do not contain detectable virions or their components but are "scars" left by earlier viral multiplication (e.g., the eosinophilic, intranuclear inclusion bodies that eventually appear in cells infected by herpes simplex virus: Ch. 55).

Induction of Chromosomal Aberrations. In **primary cultures** chromosomal aberrations such as breaks or constrictions are commonly seen after infections with measles, rubella, several herpes, parainfluenza, mumps, adeno-, polyoma, simian 40 (SV40), or Rous sarcoma viruses. During **natural infections** measles virus produces similar chromosomal abnormalities in peripheral leukocytes. The alterations often appear to be an early expression of the cytopathic effect in cells that will die later. Some of these aberrations have characteristic features: for example, herpes simplex virus induces breaks only at certain sites of two specific chromosomes; and chromatid breaks may continue to occur during the multiplication of cells surviving infection by herpes simplex or polyoma virus, suggesting a persistent infection of the cell clones.

These chromosomal alterations may conceivably result from reaction with a viral protein or virus-specified enzymes, or even from interruption of DNA or protein synthesis. The latter possibilities are suggested by the occurrence of chromatid breaks in uninfected cells treated with inhibitors of DNA synthesis or deprived of essential amino acids.

Cell Transformation. Certain viruses which produce tumors or leukemia (Ch. 66) may have several effects on cultured cells: 1) stimulation of the synthesis of cellular DNA (e.g., polyoma virus); 2) surface alterations recognizable by the incorporation or uncovering of new antigenic specificities distinct from those of virion subunits, and by increased agglutinability by plant lectins; 3) chromosomal aberrations; and 4) alterations of the growth properties of the cells, resulting in cell hyperplasia because their division is no longer subject to **topoinhibition** (i.e., inhibition of growth in a dense culture). Moreover, growth is less dependent upon serum in the culture medium and does not require anchorage to the surface of a culture vessel—colonies of transformed cells grow in soft agar or methylcellulose. This conversion of a normal cultured cell to one resembling a malignant cell has been termed **transformation** (Ch. 66). DNA-containing viruses (adenoviruses, herpes simplex, polyoma, SV40) can transform only nonpermissive cells but at least a portion of the viral genome persists and continues to function. In contrast, cells transformed by RNA viruses (e.g., avian leukosis, murine leukemia) are permissive and usually continue to produce virions.

PATTERNS OF DISEASE

Viruses cause three basic patterns of infection: localized, disseminated, and inapparent.

Localized Infections. In these infections viral multiplication and cell damage remain localized near the site of entry (e.g., the skin or parts of the respiratory or gastrointestinal tract). When the virus spreads from the first infected cells to neighboring cells by diffusion across intercellular spaces, and by cell contact, the result is a single lesion or a group of lesions, as with warts. In a less strictly localized pattern transport of the virus by excretions or secretions within connected cavities causes diffuse involvement of an organ, as with influenza, the common cold, or viral gastroenteritis. Virus may in time spread to distant sites, but this dissemination is not essential for production of the characteristic illness.

Disseminated Infections. These infections develop through several sequential steps, as illustrated by Fenner's classic investigation of ectromelia (mousepox), summarized in Figure 53-1. Mousepox virus **enters** through an abrasion of the skin and **multiplies locally;** from there it spreads rapidly to **regional lymph nodes,** where it also multiplies. The virus then enters the lymphatics and the blood stream, and this **primary viremia** causes the dissemination of the virus to other susceptible organs, especially the liver and spleen. Viral multiplication results in **necrotic lesions** in these organs and a more intense **secondary viremia,** which disseminates the virus to the **target organ,** the skin. There the virus undergoes extensive multiplication, producing papules which eventually ulcerate. With the appearance of the papular rash the asymptomatic **incubation period** terminates and clinical disease begins.

The temporal relation between viral multiplication in the various organs, development of lesions, and formation of Abs should be noted (Fig. 53-1A). It is particularly striking that *overt disease begins only after virus becomes widely disseminated in the body and has attained maximum titers in the blood and the spleen.*

This model of dissemination is applicable not only to exanthematous diseases such as smallpox and measles but also to nonexanthematous diseases such as poliomyelitis and mumps. Thus the target organ for poliovirus is the central nervous system, and for mumps virus, the salivary and other glands. In some instances, such as poliovirus infections (Ch. 57) primary and secondary viremias are not distinguishable.

The dissemination of **neurotropic viruses** to the nervous system may occur by transmission **along nerves** as well as by viremia. For instance, in mice such centripetal transmission of herpes simplex virus after foot pad inoculation can be followed by assaying segments of nerves at various times after infection. The virus may conceivably travel either by diffusion within the axons or by multiplication in endoneural cells (Schwann's cells and fibro-

FIG. 53-1. Sequential events in pathogenesis of ectromelia (mousepox) in mice inoculated in foot pad. **A.** Relation between viral multiplication (in foot pad, spleen, blood, and skin), development of primary lesion and rash, and appearance of Abs **(E-AHA). B** Diagram of the dissemination of virus and pathogenesis of the rash in mousepox. (Fenner F: Lancet 2:915, 1948)

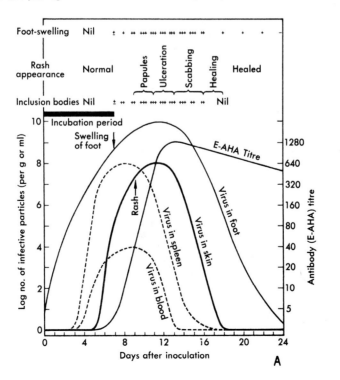

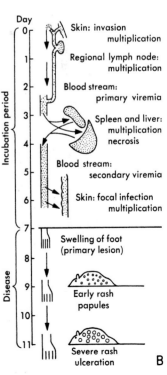

blasts), in which viral Ags can be localized by immunofluorescence.

For many years, before refined structural studies on virions became possible, animal viruses were classified primarily in terms of their viscerotropism, neurotropism, or dermotropism. The grouping of viruses on the basis of their target organs is presented in Table 53-2. However, the target organ for a given virus (where the susceptible cells are damaged) and the type of disease produced bear no relation to the taxonomic position of the virus, as defined in Chapters 46 (see The Viral Particles) and 50 (Table 50-1). In fact, such unrelated viruses as influenza and adenoviruses may produce diseases that cannot be clinically differentiated; and such related viruses as parainfluenza, mumps, and measles may produce completely different clinical syndromes.

Inapparent Infections. Transient viral infections without overt disease (inapparent infections) are very common. Moreover, they have great epidemiologic importance, for they represent an often unrecognized source for dissemination of a virus, and they also confer immunity. For example, for every paralytic case of poliomyelitis in the United States before the days of widespread immunization, 100–200 inapparent infections could be detected serologically or by viral isolation.

Several factors are involved in the production of inapparent infections.

1) **Moderate viruses** or **attenuated strains** (as in live vaccines) usually cause inapparent infections.

2) When the **host's defense mechanisms** are effective even viruses capable of causing acute disease may generate an inapparent infection. These defenses include

TABLE 53-2. Grouping of Viruses by Pathogenic Characteristics in Man

Classification by major target organs	Specific virus	Portal of entry	Other affected organs
Respiratory viruses	Influenza A, B, and C	Respiratory tract	
	Parainfluenza	Respiratory tract	
	Respiratory syncytial	Respiratory tract	
	Measles	Respiratory tract	Brain, skin, lung
	Mumps	Respiratory tract	CNS, testes, ovaries, pancreas
	Adenoviruses	Respiratory tract	
	Rhinoviruses	Respiratory tract	
	Coxsackieviruses (some)	Respiratory tract	CNS*
	Echoviruses (some)	Respiratory tract	CNS*, skin
	Reoviruses	Respiratory tract	?
	Lymphocytic choriomeningitis	Respiratory tract	CNS
	Coronavirus	Respiratory tract	
Enteric viruses	Polioviruses	Gastrointestinal tract	Muscles, CNS*
	Coxsackieviruses	Gastrointestinal tract	CNS*, skin
	Echoviruses	Gastrointestinal tract	CNS*, skin
	Rotavirus (reovirus)	Gastrointestinal tract	
	Hepatitis A and B; C (?)	Gastrointestinal tract and blood	Liver*
Neurotropic viruses	Polioviruses	Gastrointestinal tract	Upper respiratory tract
	Coxsackieviruses	Gastrointestinal tract	Upper respiratory tract
	Echoviruses	Gastrointestinal tract	Upper respiratory tract
	Rabies	Skin and blood	
	Mumps	Respiratory tract	Testes, ovaries, pancreas
	Arboviruses	Blood	
	Herpes simplex	Respiratory tract; genitalia	Skin
	Virus B	Respiratory tract; blood	Respiratory tract
	Varicella-zoster	Respiratory tract	Skin, cornea
	Lymphocytic choriomeningitis	Respiratory tract	Respiratory tract
Dermotropic viruses	Poxviruses	Respiratory tract	Respiratory tract, viscera, CNS
	Measles	Respiratory tract	Lung, brain
	Varicella-zoster	Respiratory tract	CNS, cornea
	Coxsackieviruses	Gastrointestinal tract	Upper respiratory tract
	Echoviruses	Gastrointestinal tract	CNS, gastrointestinal tract
	Herpes simplex	Skin	Peripheral ganglia, cornea, genitalia
	Rubella	Respiratory tract	Respiratory tract
	Human wart (papillomaviruses)	Skin	
	Molluscum contagiosum	Skin	

*Major involvement in clinical disease.

the host's **immunity,** especially the ability to produce a prompt secondary response (Abs or cytotoxic T lymphocytes), and the appearance of **viral interference** (Ch. 51).

3) **Failure of the virus to reach the target organ** is also an expression of host defense, but of a more obscure nature. Thus in experimental infections of adult mice with St. Louis or Japanese B encephalitis virus extraneural inoculation causes infection, but only intracerebral inoculation produces disease. A blood–brain barrier has been postulated, though this may be the capillary endothelium since passsage of viruses from the blood into the brain probably requires active multiplication of the virus in the endothelial cells.

EFFECTS OF VIRUSES ON EMBRYONIC DEVELOPMENT

The variation of susceptibility with age is especially striking for the embryonic period. Indeed, some viruses that produce mild disease in the adult produce extensive infection and severe malformations in the embryo. The role of viruses in the pathogenesis of congenital anomalies was not recognized until Gregg, in 1941, discovered that rubella virus may cause a variety of congenital anomalies if the mother is infected during the first 3 months of pregnancy (Ch. 63). Of all the viral infections that may occur during the first trimester of pregnancy, rubella is the major cause of fetal death and congenital malformations. But a few other viruses also have **teratogenic effects: cytomegalovirus** induces a low incidence of microcephaly, motor disability, and chorioretinitis; **group B coxsackieviruses** are responsible for some congenital heart lesions; and **type 2 herpes simplex virus** may cause microcephaly and other severe central nervous system malformations.

Malformations are readily produced in lower animals by various infections during embryonic development, such as reovirus in mice, influenza, mumps, and Newcastle disease viruses in chicks, and hog cholera virus in swine. Abortion may be produced by Japanese B virus infection in swine and by equine abortion virus (a herpesvirus) in horses, while rat virus (a parvovirus indigenous to rats) causes the birth of hamsters with malformations of the head resembling those of children with Down's syndrome (mongolism).

These infections have made it possible to study the pathways of invasion of the embryo. Some leukemia-inducing viruses of mice and chickens reach the embryo through the ovum or sperm, whereas LCM virus and rat virus reach it through the placenta.

Passage of a virus across the placenta apparently occurs only when the mother is viremic. Multiplication of the virus in the placenta probably favors transmission but is not strictly required since the small coliphage ϕX174, which is unable to multiply in animals, is transmitted (though with very low efficiency).

IMMUNOLOGIC AND OTHER SYSTEMIC FACTORS

CIRCULATING ANTIBODIES

Protection. Abs in serum and extracellular fluids provide the main protection against primary viral infections: i.e., at the site of viral entry into the host. For those infections in which viremia is an essential link in the pathogenesis of the disease (measles, poliomyelitis, yellow fever, smallpox) the degree of protection is directly related to the level of neutralizing Abs in the blood when virus enters it. The mechanism by which Abs neutralize viruses has been considered in Chapter 52.

Protection of the respiratory and gastrointestinal tracts is associated with **IgA Abs** (Ch. 17), which are secreted into the extracellular fluids. Hence by inducing the secretion of IgA Abs natural infections produce specific local as well as systemic immunity. Viral vaccines, particularly those containing live attenuated virus, also elicit the production of IgA Abs in the respiratory and gastrointestinal secretions. Although this feature theoretically affords a marked advantage to live viral vaccines, some vaccines produced with killed viruses, such as polioviruses and influenza viruses, have proved effective.

The protective role of Ab is also evident in the prophylactic effectiveness of **passive immunization.** Administration of immune serum or immune γ-**globulin,** before infection or early in the incubation period, can prevent or modify diseases with viremia and long incubation periods (greater than 12 days), such as measles, hepatitis A and B, poliomyelitis, and mumps. The striking protection of populations by some viral vaccines (Table 53-3) constitutes an additional demonstration of the prophylactic function of Abs. These specific vaccines will be discussed in the chapters that follow.

TABLE 53-3. Viral Diseases In Which Immunization Has Been Effective

Disease	Vaccine	
	Attenuated virus	Inactivated virus
Smallpox	+	
Yellow fever	+	
Poliomyelitis	+	+
Measles	+	
Influenza	+*	+
Mumps	+	+
Rabies	+†	+
Adenovirus infections‡	+	+
Rubella	+	

*Experimental.

† For veterinary use.

‡ Caused by types 3, 4, 7, and 21.

Recovery. Although Abs generally develop during recovery from a viral disease they appear to play a less prominent role in this process than in protection. Thus intracellular virus may continue to increase, and pathologic lesions to evolve, even while Abs are being elaborated. Moreover, in most patients with agammaglobulinemia recovery is normal: further evidence that factors other than circulating Abs act to limit the course of viral diseases. Furthermore, even patients with selective IgA deficiency do not develop more prolonged or more severe respiratory or enteric viral infections.

The limited effect of Abs on recovery is not surprising since they are ineffective against intracellular viral precursors and virions. Furthermore, many viruses can spread to contiguous uninfected cells through intercellular processes, thus remaining inaccessible to Abs. However, Abs do serve an important function in restricting the dissemination of some viruses (e.g., polioviruses and togaviruses) whose pathogenesis is dependent upon a viremic stage.

Persistence of Antibodies. The time course and the persistence of Ab production and immunity vary with 1) the virus, 2) the nature of the antigenic stimulus, and 3) the type of Ab. For example, 1) neutralizing Abs fall from their maximal level more rapidly, and to a lower titer, following influenza than following poliomyelitis infection; 2) immunity to measles persists for life following infection, but lasts only a few months following immunization with formalin-inactivated virus; 3) after infection complement-fixing Abs generally appear earlier but decrease much sooner than neutralizing Abs.

Long-lasting immunity, with persistence of circulating Abs, follows infection with a number of viruses, especially those causing viremia. Thus second attacks are extremely rare with measles, smallpox, yellow fever, or poliomyelitis, to mention only a few examples. In contrast, *second infections are common with most acute localized infections without viremia,* particularly respiratory diseases. Adenovirus infections (Ch. 54) are notable exceptions, perhaps because they frequently terminate in latent infections of lymphoid tissue in the respiratory and gastrointestinal tracts.

Latent infection with persistent synthesis of critical viral Ags offers the most reasonable explanation for the long-lasting immunity that follows many viral infections. **Repeated infections** by a prevalent virus, or **secondary Ab response** in a disease with a long incubation period, could also provide an explanation, but these possibilities seem much less plausible. Thus, neither of the latter mechanisms can account for the prolonged persistence of circulating Abs against smallpox or yellow fever in previously infected residents of the United States, where these diseases rarely, if ever, occur.

In Panum's classic observations on a measles epidemic in the Faroe Islands, those persons were immune who had been alive during the preceding epidemic, 67 years earlier, whereas the younger islanders were highly susceptible. Hence, not only did immunity persist in the absence of overt clinical reinfection, but the immune persons failed to infect their nonimmune contacts during all these years. This absence of transmission could be explained by incomplete viral multiplication, or by the continued neutralization of the virus produced in the immune persons.

CELL-MEDIATED IMMUNITY

Dysgammaglobulinemias and drug-mediated immune suppression in humans have provided the strongest evidence that humoral Abs do not play the determinant role in recovery from many viral infections. Patients who lack immunoglobulins but develop cell-mediated immunity (CMI) ordinarily recover from viral diseases without difficulty whereas patients with defective CMI but normal Abs recover poorly from certain viral infections. For example, in individuals with defective CMI smallpox immunization frequently leads to severe generalized vaccinia or to extensive necrosis of the skin and muscle of the affected extremity (**vaccinia gangrenosa;** Ch. 56). Moreover, these complications are unaffected by the administration of specific neutralizing Abs, but they are arrested by local injection of lymphoid cells from recently immunized donors, or by injection of **transfer factor** obtained from such cells (Ch. 22); and development of a delayed-type hypersensitivity reaction to heat-inactivated vaccinia virus accompanies this recovery. Finally, in experimental animals depression of CMI by **antilymphocytic serum** (which contains Abs to T cells) increases the severity or the duration of infection with a number of viruses, particularly those possessing envelopes (e.g., herpesviruses, poxviruses, and paramyxoviruses).

Cellular immunity also appears to play a critical role in **maintaining the suppressed state of latent viral infections:** activation of such infections has become common in patients with organ transplants or with malignancies whose CMI is suppressed by therapy. These latent infections include herpesviruses (varicella-zoster and cytomegalovirus), measles virus, human wart virus, and papovaviruses similar to simian virus 40. Herpesvirus infections are also often activated in patients with extensive burns (herpes simplex virus and cytomegalovirus) and in the aged (varicella-zoster virus) owing to diminished CMI.

Pseudotolerance. Some viruses that infect the embryo or the newborn without damaging host cells give rise to apparent immunologic tolerance. Thus chicks infected by the avion leukosis virus produce virus throughout life without detectably producing neutralizing Abs. However, this phenomenon is not due to true tolerance since

Abs are formed, but they are complexed with the large amount of virus produced. Similarly, humans infected in utero with rubella virus or cytomegalovirus, and fetal mice infected with influenza virus, produce virus-specific Abs for long periods after birth. And although mice with persistent **LCM** or **murine leukemia** virus infection give birth to offspring who are viremic and lack detectable virus-specific circulating Abs, the infants synthesize small amounts of Abs which complex with virus (see below).

The virus produced in large amounts throughout the life of such **pseudotolerant** animals, usually with viremia, is disseminated **vertically** to their offspring through the ovum, placenta, or milk, and **horizontally** to contacts, through excretions and secretions. In such animals infection is asymptomatic for most of their life span, but late in life a chronic disease may develop.

DISEASE BASED ON VIRUS-INDUCED IMMUNOLOGIC RESPONSE

The immune response, despite its protective and ameliorative effects, can also contribute to the production of disease, particularly with viruses that **antigenically alter the cell's surface membranes.** For instance, a severe, **hemorrhagic dengue,** often associated with a shock syndrome, occurs in those who have had prior infection with a different serotype; and children immunized with **inactivated measles** or **respiratory syncytial virus** develop unusually severe disease if subsequently infected by the same virus. This is probably due to cell-mediated hypersensitivity or to specific Ab-dependent, cell-mediated lysis of infected cells. In another mechanism, circulating virus-specific **Ag–Ab complexes** may lodge in organs, such as the brain or kidney, inducing inflammation and disease.

The central role of the immune response in the development of some viral diseases is dramatically demonstrated in mice infected with **LCM virus,*** an arenavirus. In adult mice severe, often fatal, disease follows about a week after intracerebral inoculation; but if the immune response has been suppressed (by neonatal thymectomy, chemicals, x-irradiation, or antilymphocytic serum) disease fails to develop although viral multiplication and spread are unrestrained. Moreover, after infection in utero or at birth circulating Abs and specific CMI are not detectable; the mice appear normal for 9–12 months in spite of widespread viral multiplication that produces persistent viremia and viruria, with viral Ags in most organs. Tissue injury can be initiated in such mice by transfer of spleen cells from a syngeneic immune donor

but not by immune spleen cells treated with anti-Θ serum (from animals immunized with Ag from T lymphocytes) or by large amounts of immune serum.

In adult mice viral replication and spread are relatively restricted in both neural and extraneural tissues, but the cell-mediated antiviral immune response is quick to develop and to elicit lethal disease by attacking a critical number of involved cells in the neural membranes. In contrast, in the fetus, neonate, or immunosuppressed adult mouse, with limited immunologic capabilities, infection proceeds unimpeded to eventual involvement of all tissues. The constant high level of virus that develops may conceivably depress the clonal expansion of virus-specific T lymphocytes, but eventually Abs are produced, and they complex with excess circulating viral Ags and complement. Filtration of these aggregates by the renal glomeruli initiates an inflammatory response, culminating in glomerulonephritis. In addition, necrotic lesions appear in the liver, brain, spleen, and other organs, apparently resulting from the reaction between virus-sensitized killer T lymphocytes and viral Ags present on the surface of many cells. After intracerebral infection of an adult animal, the brain damage is generated by a similar T-cell–dependent immunologic mechanism. In contrast to the devastating effects of natural LCM infection, the immunization of mice with inactivated LCM produces humoral Abs which upon subsequent infection restrict viral spread and prevent disease.

These observations lead to the conclusion that in the absence of effective immunity LCM virus multiplies harmlessly for a long time in mice, producing an inapparent infection similar to the endosymbiotic infection that it causes in cell cultures (see below). Thus every aspect of the disease, in both acute and chronic infection, can be shown to be immunologically mediated, and in this single animal model, LCM dramatically demonstrates both the benefits and the disadvantages derived from the immunologic response.

NONSPECIFIC SYSTEMIC FACTORS

Nonspecific factors that influence resistance to viral infection include various **hormones, temperature, inhibitors** other than Abs, and **phagocytes.** Nutrition may also affect the course of viral infections; e.g., measles has devastating consequences in malnourished children of West Africa. However, malnutrition influences so many aspects of the host defenses that specific analysis of cause and effect is difficult.

Infected cells may form and release an especially important factor, **interferon,** which inhibits viral multiplication by preventing the synthesis of viral proteins, and thus interferes with the infection of other cells by many viruses. Accordingly, interferon not only prevents the infection of cells but also limits viral spread and assists in

---

* This virus can also infect man, but it usually produces a mild respiratory infection, only occasionally followed by severe meningitis.

recovery. This agent has been discussed in Chapter 51 and will not be considered further here.

Phagocytosis does not appear to be as important a defense mechanism in viral as in bacterial infections. On the contrary, some viruses impair the antibacterial activity of **polymorphonuclear leukocytes** by producing leukopenia (e.g., measles virus) or by reducing phagocytic function (e.g., influenza virus). **Macrophages,** however, do appear to be important in viral infections: they rapidly take up certain viruses, and there is a correlation between host and macrophage susceptibility to viral infections (see Genetic Factors, above). Virus-infected macrophages can act as a source of infection for other cells, but with viruses that are unable to multiply in them macrophages appear to play their usual role as scavengers.

Hormones have a potential effect on viral infections, as can be illustrated by several examples. **Pregnancy** increases the severity of several viral diseases: paralytic poliomyelitis is more frequent and more extensive; smallpox has a more severe course and abortion is common; the complications of influenza, particularly pneumonia, are increased. **Cortisone** enhances the susceptibility of many animals to neurotropic viruses, and of chick embryos to influenza and mumps virus; in humans it causes enlargement and perforation of herpetic corneal ulcers, and induces extensive visceral spread of varicella virus that often terminates in severe pneumonia. These deleterious effects appear to result from the suppressive effects of cortisone on inflammatory reactions, cell-mediated immunity, and interferon production rather than from its action on the Ab response.

Temperature increases in the host may reduce viral replication 1) by suppressing a temperature-sensitive step; 2) by accelerating the inactivation of many heat-labile viruses (enveloped viruses); and 3) by increasing interferon production. Conversely, in mice held at 4° C, rather than 25° C, after infection with coxsackievirus B1, viral multiplication is excessive in many organs, little interferon is produced, and the mortality is strikingly increased. A rise in body temperature may therefore contribute to recovery from viral disease.

Virus-Neutralizing Inhibitors. Globulins and lipoproteins in the serum of uninfected animals can neutralize the infectivity of many enveloped viruses in vitro; they differ from Abs and complement. It is assumed that the host's resistance to infection is enhanced by these inhibitors.

STEADY-STATE VIRAL INFECTIONS

ENDOSYMBIOTIC INFECTIONS

In certain virus–cell systems the infected cells multiply for many generations while continuing to release virus. These cells generate populations in which almost all cells are infected (as shown by the release of virus or by the presence of viral nucleic acids or Ags), even when external reinfection is prevented by the presence of Abs. The cells may be functionally normal: for instance, chick

embryos carrying avian leukosis viruses (Ch. 66) develop normally.

Many viruses can cause such endosymbiotic infections. **Paramyxoviruses** (parainfluenza, measles, and mumps viruses) appear to be particularly suited to initiating persistent infections. Despite the superficial similarity to lysogeny in bacteria endosymbiotic infection is maintained by quite a different mechanism, although retroviruses do associate their viral nucleic acid with the cellular chromosome as a provirus. In other cases infection persists because the viruses do not have the pathogenic ability to disrupt their host cells or to prevent cell division; and there are enough viral nucleic acid molecules or viral particles in each cell to ensure infection of most daughter cells by random segregation at mitosis.

Endosymbiotic infections in cultured cells may be the in vitro counterpart of latent infections in animals (see below). Moreover, latency (for example, as occurs in subacute sclerosing panencephalitis; see below) may be assisted by the remarkable capacity of humoral Abs, in conjunction with the alternate complement pathway, to **redistribute ("cap") viral Ags on the plasma membrane.** This phenomenon (Fig. 53-2) appears to promote shedding of viral Ags from the cell surface, leaving the cell surface free of viral glycoproteins and the infected cell protected from either cell-mediated or Ab-mediated immunologic destruction.

Defective viral particles, whose genome is partially deleted, are commonly generated during replication of most viruses, particularly upon repeated passage at high viral concentration. These particles, termed **defective interfering** (or **DI**) **particles** (Ch. 51), produce effective interference with homotypic nondefective virus. They appear to play a central role in steady-state infections in cell cultures, and possibly in the establishment and maintenance of in vivo persistent infections.

CARRIER STATE INFECTIONS

Endosymbiotic infection must be distinguished from the **carrier state:** viral persistence based on the infection of a small proportion of the cell population in a culture. In these cells a regular viral multiplication cycle takes place, usually terminating in cell death, but the released virus infects only a small number of the other cells. This limited reinfection depends on special conditions of the culture, e.g., partial resistance of the cells to infection, or the presence in the medium of antiviral agents such as weak or highly dilute Ab or interferon, or cell-to-cell transmission of virus without release into the medium. If the continuing reinfection is prevented, as by adding antiviral Abs, the culture is often cured of the infection.

Table 53-4 summarizes the differences between endosymbiotic and carrier state infections.

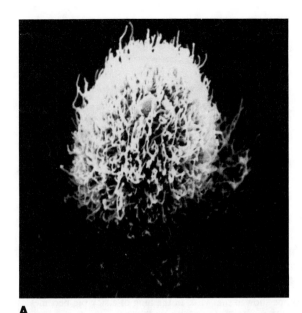

TABLE 53-4. Comparison of Steady-State Endosymbiotic and Carrier State Infections in Cultures

	Endosymbiotic infection	Carrier state infection
Proportion of cells infected	All or most	1% or less
Viral transmission	Intracellular: parent to progeny cells	Usually extracellular
Special properties of viruses	Viruses released by budding; do not kill cell or inhibit mitosis	Produced by any virus
Effect of antiviral Ab	Does not limit infection	Usually cures infection

LATENT INFECTIONS IN ANIMAL HOSTS

In **persistent in vivo inapparent infections,** i.e., **latent infections,** overt disease is not produced but the virus is not eradicated. This **equilibrium between host and parasite** is achieved in various ways by different viruses and hosts, as the following examples illustrate.

1) Some viruses **multiply profusely** without causing cell damage (i.e., they produce **endosymbiotic infections** in animals, as in cell cultures). As noted above, **LCM virus infection of mice** may persist without disease for long periods until an adequate immune response has developed. Some slowly developing viral diseases (see slow viral infections, below) may have this genesis.

2) **Herpes simplex virus** has a special pattern of **latency and recurrence.** This virus usually infects humans between 6 and 18 months of age, and the virus persists but cannot be found except during recurrent acute episodes, such as herpes labialis (fever blisters; Ch. 55). The form in which the latent (undetectable) virus persists between recurrent episodes has not been identified. Virus cannot be isolated from tissue homogenates; but by cultivating cells of sensory ganglia, virus has been detected in the human trigeminal (type 1 virus) and thoracic, lumbar, and sacral dorsal root (type 2 virus) ganglia, as well as in sensory ganglia of experimentally infected mice, rabbits, and monkeys. It may be that, as in virus-carrier cultures, infection is confined to only a small proportion (about 0.1%) of the ganglia cells by Abs, cellular immunity, viral interference, (by interferon or DI particles), or metabolic factors. Because Abs are present most of the extracellular virus is neutralized and goes undetected. Acute episodes probably depend on a transient change in the local level of immunity or changes in the susceptibility of the uninfected cells induced by a variety of physical and physiologic factors such as fever, intense sunlight, fatigue, or menstruation.

Experimental herpes simplex infections of rabbits illustrate the role of shifts in the immune system. Herpes encephalitis can be

FIG. 53-2. Distribution of measles virus Ags on the surface of infected HeLa cells before and after reaction with virus-specific Abs. **A.** Scanning electron microscopy demonstrates the abundant fine microvilli randomly distributed when infected cells were not reacted with Abs. (In contrast, uninfected cells display shorter, thicker, and less abundant villi). **B.** Marked redistribution (**"capping"**) of viral Ags seen after infected cells were mixed with virus-specific Abs at 37°C. Serum without measles virus Abs or serum with Abs directed against Ags of other viruses did not cap the measles virus Ags. (Lampert PW et al: J Virol 15:1248, 1975) (×4000)

reactivated 6 months after an acute encephalitis episode by inducing anaphylaxis with any Ag. Similarly, herpes keratitis can be provoked, after an acute corneal ulcer is healed, if the rabbit is made sensitive to horse serum and a corneal Arthus reaction is induced. Experimental herpes simplex virus infection can also be activated by nonspecific excitants such as ultraviolet light, histamine or epinephrine injection, corticosteroids, or surgical manipulation of sensory ganglia and nerves.

3) **Adenovirus infections** in humans are self-limited, but the virus frequently establishes a latent infection of tonsils and adenoids (see Pathogenesis, Ch. 54). Though these tissues fail to yield infectious virus when homogenized and tested in sensitive cell cultures, cultured fragments of about 85% of these "normal" tonsils and adenoids, after a variable time, show characteristic adenovirus-induced cytopathic changes and yield infectious virus.

Failure to recover infectious virus initially may be attributed to the paucity of virions, to their association with either Ab or receptor material, or to the absence of mature virions. The latent infection is evidently not the result of lysogeny since replication of infectious virus cannot be induced by ultraviolet irradiation, nor are the cells resistant to infection.

Similarly, cultured monkey kidney cells may yield unsuspected viruses that cannot be detected in homogenates of the extirpated organs. Thus, the oncogenic **simian virus 40 (SV40**, Ch. 66) commonly infects rhesus and cynomolgus monkeys in nature, and in cultures of cells from these species infectious virus appears without causing cytopathic changes. This virus also illustrates the role of host resistance in latency: in contrast to the findings with rhesus monkey cells, when the virus is added to cultures of kidney cells from the green monkey (which has never been found to carry the virus) extensive cell injury accompanies viral multiplication.

4) The **Shope rabbit papilloma virus** (Ch. 66) illustrates latency due to the **replication of viral nucleic acid without viral maturation.** This virus produces warts equally well in the skin of domestic or of wild cottontail rabbits, but infectious virus and viral Ags can usually be detected only in the tumors of the wild animals,* in the keratinized cells of the outer epidermal layer but not in the growing basal layer. However, the basal layer of warts from either wild or domestic rabbits yields infectious viral DNA when extracted with phenol. In man the most dramatic example of incomplete viral production is the persistence and continued replication of **measles virus nucleocapsids** in lesions of **subacute sclerosing panencephalitis (SSPE),** which may develop years after acute measles infection (see Slow Viral Infections, below).

5) **Swine influenza virus,** an orthomyxovirus related to influenza A virus, illustrates a complex ecologic situation

* Since the assay of this virus is extremely insensitive, it is not known whether failure to recover the virus signifies paucity or absence.

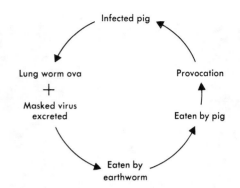

FIG. 53-3. Natural history of swine influenza infection.

in which the virus is **latent in two intermediate hosts** and requires **assistance from a bacterium,** *Hemophilus influenzae suis,* to induce the acute respiratory disease in pigs (Fig. 53-3).

Shope showed that virus in the lung of a sick pig becomes associated, in **occult** form, with ova of the lungworm (a common parasite of most pigs), which is coughed up with pulmonary secretions, swallowed, and eventually passed in the pig's feces. The contaminated feces are then eaten by earthworms, in which the lungworm ova develop into larvae, and where the virus can remain occult for at least 2 years. To complete the cycle the earthworm is eaten by a pig; and the lungworm larvae migrate to the pig's lungs, where they develop into mature lungworms. The virus remains noninfectious through this long odyssey; but when the parasitized pig is jolted by cold or by infection with *H. influenzae suis,* the occult virus is somehow induced to replicate; it may initiate an acute disease and induce viral spread. The nature of the viral occult form in the lungworm and earthworm, and the exact role of the bacterium are unknown.

EPIDEMIOLOGIC SIGNIFICANCE OF LATENT INFECTIONS

As these examples indicate, latent viral infections affect the incidence and the pathogenesis of **acute viral diseases** in several ways. 1) In herpes simplex infection **recurrent acute disease** is induced in the infected individual when the balance that maintains the latent infection is intermittently disturbed (see Herpes Simplex Viruses, Pathogenesis, Ch. 55). 2) Swine influenza mimics the way in which a virus widely seeded among humans may initiate, in response to **environmental changes,** an **explosive epidemic by viral reactivation,** in many areas at the same time, as noted in pandemics of influenza (see Epidemiology, Ch. 58). 3) A **reactivated latent (occult) virus** may spread and **initiate an epidemic** among susceptible contacts. Thus, a latent varicella virus, persisting after chickenpox, can be activated and produce the different clinical picture of herpes zoster (shingles); and the patient may then serve as a focus for initiating an epidemic of chickenpox (see Varicella–Herpes Zoster Virus, Epidemiology, Ch. 55). 4) Viral latency can also be seen in

TABLE 53-5. Examples of Slow Viral Infections

Disease	Host	Organ primarily affected	Virus
Kuru	Man	Brain	Unknown—?viroid
Creutzfeldt-Jakob disease	Man	Brain	Unknown—?viroid
Subacute sclerosing panencephalitis	Man	Brain	Paramyxovirus (measles)
Progressive encephalitis	Man	Brain	Togavirus (rubella)
Progressive multifocal leukoencephalopathy	Man	Brain	Papovavirus (SV40 and JC)
Scrapie	Sheep	Brain	Unknown—?viroid
Mink encephalopathy	Mink	Brain	Unknown—?viroid
Visna	Sheep	Brain	Retrovirus
Maedi	Sheep	Lung	Retrovirus
Progressive pneumonia	Sheep	Lung	Retrovirus
Lymphocytic choriomeningitis	Mouse	Kidney, brain, liver	Arenavirus
Canine demyelinating encephalomyelitis	Dog	Brain, spinal cord	Paramyxovirus (distemper)
Aleutian mink disease	Mink	Reticuloendothelial system	Parvovirus
Hard pad disease	Dog	Brain	Paramyxovirus (distemper)

the development of certain **chronic diseases** dependent on an **immunologic response,** as for LCM, noted above, and possibly for subacute sclerosing panencephalitis and some other so-called **slow viral infections.** 5) Some latent viral states produce **uncontrolled proliferation of cells,** i.e., tumors, as discussed in Chapter 66.

SLOW VIRAL INFECTIONS

The term **slow virus** has become associated with those viruses that require prolonged periods of infection (often years) before disease appears. The term, however, is somewhat misleading, for although the virus is patient and the disease process develops over a protracted period, viral multiplication may not be unusually slow. Moreover, diseases of this group may be caused not only by unusual viruses but also by others that ordinarily cause acute diseases. Table 53-5 lists some chronic degenerative diseases, particularly of the central nervous system, that belong to this group. Five that occur in man are discussed: subacute sclerosing panencephalitis, progressive encephalitis, kuru, Creutzfeldt-Jakob disease, and progressive multifocal leukoencephalopathy. Other chronic diseases may have similar origins (e.g., multiple sclerosis, systemic lupus erythematosus). Viral diseases that have a long incubation period but an acute course, e.g., type B hepatitis and rabies, are not considered to be slow viral infections, and are not discussed here.

Even some common human viruses appear to give rise occasionally to chronic degenerative diseases. **Subacute sclerosing panencephalitis (SSPE),** the best substantiated example, follows several years after measles, and perhaps even after measles immunization. A virus identical with or closely related to measles virus appears to be the etiologic agent: the lesions contain specific Ags closely related to measles virus and huge quantities of nucleocapsids characteristic of a paramyxovirus; and a virus similar to measles virus* has been isolated by co-cultivating affected brain tissue along with susceptible cells.

Progressive encephalitis, similar to but more rapidly progressive than SSPE, develops in a rare child who previously had congenital or early childhood rubella. Rubella virus rather than measles virus is recovered by culturing the affected brain in vitro with or without cocultivation with cells susceptible to rubella virus multiplication.

Progressive multifocal leukoencephalopathy is a rare subacute demyelinating disease. Two different species of papovaviruses (Ch. 66) have been isolated from the brains of victims, and also are seen in the intranuclear inclusion bodies of affected oligodendrocytes. All of the viruses isolated are of the SV40–polyoma subgroup: two are almost identical with SV40, but the remainder (e.g., JC virus) distinctly differ from SV40 immunologically, chemically, and biologically. This disease generally develops in patients with immunologic defects due to disorders of the reticuloendothelial system (such as Hodgkin's disease and leukemias) or to immunosuppressive therapy. Thus, the emergent viruses appear to be **opportunists** liberated from a latent infection in an immunologically compromised host; whether they are the etiologic agents of the disease or fellow-travelers is uncertain.

---

* The RNA of SSPE virus has about 10% more nucleotides than measles virus (Ch. 59).

Kuru, another slow viral disease of man, was first observed in 1957 in the Fore tribe of cannibals living in stone age conditions in New Guinea (*kuru* = shivering or trembling, in the Fore language); it is transmitted by consumption of the brains of deceased relatives, a tribal ceremonial ritual for children and young female adults. This degenerative disease of the cerebellum, manifested by ataxia, disturbed balance, clumsy gait, and tremor, progresses inexorably to death in less than a year after onset. A striking decrease in kuru was seen after the tribal chiefs prohibited this custom of cannibalism.

The pathologic findings do not include the customary inflammatory evidence of an infectious process but do resemble the findings in **scrapie,** a disease of sheep proved to be transmissible and caused by an unusual virus (i.e., difficult to inactivate and without detectable nucleic acid). This resemblance suggested a viral etiology for kuru, and Gajdusek and Gibbs, using brain material from kuru patients, transmitted the disease serially to chimpanzees. (Subsequently it was also transmitted to New and Old World monkeys, and even to mink and ferrets.) The degenerative process, which had the same clinical and pathologic characteristics as kuru in man (**subacute spongiform encephalopathy),** appeared 18–30 months after the initial inoculation of chimpanzees, and after 1 year in subsequent passages.

Creutzfeldt-Jacob disease, a fatal presenile dementia of midadult life, which is not geographically restricted and hence not so exotic, like kuru shows a **spongiform encephalopathy** and also appears to be a chronic viral disease. A similar disease has been serially transmitted from the brains of patients to chimpanzees, several species of monkeys, guinea pigs, and cats. Its epidemiology is not clear, but the disease has been accidentally transmitted to humans by a corneal transplant from a person who subsequently developed the fatal disease, and by electroencephalographic electrodes sterilized only with 70% ethanol and formaldehyde vapor after the electrodes had been used on a patient with the disease.

The Viruses. Owing in part to the long incubation periods and cumbersome assays, the etiologic agents of kuru and Creutzfeldt-Jakob disease, presumably viruses, have not been well characterized. The properties described, however, appear to place these viruses in a category quite separate from the well-known viruses of man and other animals.

The viruses are highly resistant to inactivation by the usual chemical and physical sterilizing agents: formaldehyde, β-propiolactone, proteases, nucleases (RNases and DNases), ultraviolet irradiation at 254 nm, and heat at $80\,^\circ$ C (they are only incompletely inactivated at $100\,^\circ$ C). Moreover, electron microscopic examinations of infected brain tissues (with as much as 10^{12} LD$_{50}$/gram) and virus concentrated in CsCl and sucrose gradients (10^7–10^8-LD$_{50}$/ml) did not reveal viral particles. The genome appears to be no more than 150,000 daltons in size, as determined by ionizing radiation target size. Neither infectious nucleic acids nor viral Ags have been detected.

Neither immunosuppression (e.g., from x-ray or cyclophosphamide) nor immunopotentiation (e.g., with adjuvants) affects the pathogenesis of either disease in experimental animals, and B and T cell function appears intact in the natural disease and in experimental infections. The characteristics of the agents of kuru and Creutzfeldt-Jakob disease are similar to those of scrapie and transmissible mink encephalopathy viruses. These are indeed unique viruses, if they are viruses. Further characterization may show them to be a new type of infectious agent, perhaps **viroids,** which thus far have been described only as pathogens of plants. A viroid consists of an infectious molecule of covalently closed circular single-stranded RNA (110,000–127,000 daltons) without associated protein (see Distinctive Properties, Ch. 46). Similar agents have not yet been isolated from animals, but the history of earlier studies on the distribution of other novel microbes suggests that viroids will also be found in organisms other than plants, and probably in animals.

The examples of slow viral infections of man are still few, and the evidence of causation is sparse. Nevertheless, suspicion of the role of viruses in chronic degenerative diseases is now high, and some conventional viruses (e.g., togaviruses, echoviruses) are being implicated.

SELECTED READING

BOOKS AND REVIEW ARTICLES

ALLISON AC: Lysosomes in virus-infected cells. Perspect Virol 5:29, 1967

COLE GA, NATHANSON N: Lymphocytic choriomeningitis pathogenesis. Prog Med Virol 18:94, 1974

DIENER TO: Viroids: the smallest known agents of infectious disease. Annu Rev Microbiol 28:23, 1974

FUCILLO DA, SEVER JL: Viral teratology. Bacteriol Rev 37:19, 1973

GAJDUSEK DC: Unconventional viruses and the origin and disappearance of kuru. Science 197:943, 1977

HORSTMANN DM: Viral infections in pregnancy. Yale J Biol Med 42:99, 1969

MERIGAN TC, STEVENS DA: Viral infections in man associated with acquired immunological deficiency states. Fed Proc 30:1858, 1971

NICHOLS WW: Virus-induced chromosome abnormalities. Annu Rev Microbiol 24:479, 1970

PORTER DD: Persistence of viral infection in the presence of immunity. In Notkins AB (ed): Viral Immunology and Immunopathology. New York, Academic Press, 1975, p 189

TER MEULEN V, HALL WW: Slow virus infections of the nervous system: virological, immunological and pathogenic considerations. J Gen Virol 41:1, 1978

SPECIFIC ARTICLES

BLANDEN RV: Mechanisms of recovery from a generalized viral infection. II. Passive transfer of recovery mechanisms with immune lymphoid cells. J Exp Med 133:1074, 1971

BORDEN EC, MURPHY FA, NATHANSON N, MONETH TPC: Effect of antilymphocyte serum on Tacaribe virus infection in infant mice. Infect Immunol 3:466, 1971

FULGINITI VA, KEMPE CH, HATHAWAY WE et al: Progressive vaccinia in immunologically deficient individuals. In Bergsiyia D, Good RA (eds): Birth Defects: Immunologic Deficiency Diseases in Man. New York, National Foundation, 1968, p 129

GALLOWAY DA, FENOGLIO C, SHEVCHUK M, MCDOUGALL JK: Detection of herpes simplex RNA in human sensory ganglia. Virology 95:265, 1979

NATHANSON N, COLE GA: Immunosuppression: a means to assess the role of the immune response in acute virus infection. Fed Proc 30:1822, 1971

OLDSTONE MBA, DIXON FJ: Pathogenesis of chronic disease associated with persistent lymphocytic choriomeningitis viral infection. II. Relationship of the antilymphocytic choriomeningitis immune response to tissue injury in chronic lymphocytic choriomeningitis disease. J Exp Med 131:1, 1970

VOLKERT M, LUNDSTEDT C: Tolerance and immunity to the lymphocytic choriomeningitis virus. Ann NY Acad Sci 181:183, 1971

chapter

ADENOVIRUSES

Because acute viral respiratory diseases impose huge clinical and economic burdens great efforts to isolate other major causative agents followed the isolation of influenza virus in 1933. The search was unsuccessful, however, until in 1953 two groups of investigators discovered **adenoviruses,** the first of several families now known to be etiologic agents of these acute infections. Rowe and colleagues, using cultures of human adenoids as a potentially favorable host for the elusive "common cold" virus, noted cytopathic changes in uninoculated cultures, after prolonged incubation, as well as in cells inoculated with respiratory secretions. The pathologic alterations were shown to be due to the emergence of previously unidentified viruses from latent infections of the adenoid tissues—hence the name adenoviruses. Hilleman and Werner, studying an epidemic of influenzalike disease in army recruits, isolated several similar cytopathic agents from respiratory secretions added to cultures of human tissues.

Adenoviruses cause acute respiratory and ocular infections. It soon became evident, however, that they are not the etiologic agents of the common cold and that they are responsible for only a small percent of acute viral respiratory infections.

GENERAL CHARACTERISTICS

Several characteristics of adenoviruses are of particular interest: 1) Adenoviruses are simple DNA-containing viruses (i.e., composed of only DNA and protein) that multiply in the cell nucleus. 2) They induce **latent infections** in tonsils, adenoids, and other lymphoid tissues of man, and they are readily activated. 3) Several adenoviruses are the first common viruses of humans shown to be **oncogenic** for lower animals under special experimental circumstances. 4) They serve as "helpers" for a group of small, defective DNA-containing viruses, the **adeno-associated viruses** (discussed at the end of this chapter), which cannot replicate in their absence. (Conversely, some adenoviruses cannot multiply in primary monkey cells unless the genetically unrelated simian virus 40 (SV40) is present as a helper; see Abortive Infections, Ch. 50.)

Adenoviruses are widespread in nature. The 89 accepted members of the adenovirus family have similar chemical and physical characteristics and a family crossreactive Ag (Table 54-1), but are distinguished by Abs to their individual type-specific Ags: 35 are from humans and the rest from various other animals. Comparative studies of the viruses from humans permit their classification into several groups (Tables 54-2 and 54-3).

PROPERTIES OF THE VIRUS

Structure. Electron microscopy shows the virions to be 60–90 nm in diameter. In sections the viral particles have a dense central **core** and an outer coat, the **capsid** (Fig. 54-1). Negative staining reveals icosahedral particles with capsids composed of 252 **capsomers** (Fig. 54-2): 240 **hexons** make up the faces and edges of the equilateral triangles, and 12 **pentons** comprise the vertices. The hexons are truncated triangular or polygonal prisms with a central hole (Figs. 54-3 and 54-4). The pentons are more complex, consisting of a polygonal **base** with an attached **fiber,** whose length varies with the viral type (Figs. 54-4 and 54-5). Enzymatic iodination of surface structures, chemical crosslinking of virion components, and controlled disruption of the virion reveal four additional minor capsid proteins (IIIa, VI, VIII, and IX) associated with the hexons or pentons in stoichiometric amounts (Fig. 54-6). These proteins appear to confer stability on the capsid, to form links with the core proteins, and to function in virion assembly.

Physical and Chemical Characteristics. Each virion contains one linear, double-stranded DNA molecule associated with proteins to form the **core.** The DNAs of different viral types vary in molecular weight and base composition (Table 54-2): viruses within each group share 70%–95% of their nucleotide sequences (as shown by DNA–DNA, and DNA–mRNA hybridization); and the DNAs of viruses of different groups have only 10%–25% homology.

The viral DNA has two novel features: 1) the terminal nucleotide sequences of each strand are inverted repetitions so that if the DNA is denatured both strands form single-stranded circles through "panhandles" produced between the complementary ends, and 2) a small protein of about 55,000 mol wt is covalently linked through the terminal deoxycytosine at the 5' end of each strand (Fig.

TABLE 54-1. Characteristics of Adenoviruses

1. Icosahedral symmetry
2. Diameter of 60–90 nm
3. Capsid contains 252 polygonal capsomers, 12 fibers, and 4 minor proteins
4. Double-stranded DNA genome
5. Resistance to lipid solvents (absence of lipids)
6. Related by family crossreacting soluble ags (except for the chicken adenoviruses)
7. Multiplication in cell nuclei

54-7). The functions of these unique terminal structures of the viral genome are unclear, but they may be important in replication.

The virion contains at least ten species of proteins associated with the capsid and the DNA–protein core (Fig. 54-6). Unlike most animal viruses, the capsid proteins have relatively strong noncovalent bonds between protomers in a capsomer, and weaker bonds between capsomers (see Ch. 46); hence the capsid can be artificially disrupted into intact capsomers (Figs. 54-3 and 54-4). Further dissociation of the hexons into their constituent polypeptide chains, in contrast, requires rigorous denaturing conditions (e.g., 6 M guanidine hydrochloride and a sulfhydryl reagent to block formation of disulfide bonds). The penton base is the least stable of the capsid's morphologic subunits. Noncovalent bonds associate the glycosylated fiber with the penton base.

The virion's core includes two basic proteins associated with the DNA to form a chromatin-like structure (see Ch. 49), and the covalently bonded 5′-terminal protein. One of these (designated protein VII, Fig. 54-6) is, like histones, rich in arginine (about 23 moles %).

Stability. Adenoviruses are relatively stable in homogenates of infected cells; they retain undiminished infectivity for several weeks at 4°C and for months at −25°C. Purified virions, however, are relatively unstable under all conditions of storage.

The pentons seem more weakly held in the capsid, and may be spontaneously released from purified viral particles, causing aggregation or disruption of the virions. Adenoviruses are resistant to lipid solvents.

Hemagglutination. Human adenoviruses differ in their ability to agglutinate rhesus monkey or rat RBCs. Hemagglutination occurs when the tips of fibers on virions or on aggregated pentons (commonly arranged as dodecagons) bind to the RBC surface and cause crosslinking. There is a remarkable correspondence between subgroups arranged according to their hemagglutination properties (Table 54-3), viral oncogenicity, and DNA characteristics (Table 54-2).

Dispersed purified fibers and single pentons cannot produce a lattice and therefore cause only incomplete hemagglutination unless they are crosslinked by heterologous Abs from a virus of the same subgroup (i.e., the crossreacting Abs recognize the penton base or shaft of the fiber but not the fiber tip). In contrast, agglutination is inhibited by type-specific Abs, which combine with the tip of the fiber. Viral Ags, as well as type-specific Abs, can be titrated by hemagglutination procedures.

The nature of the fiber receptor on susceptible RBCs is not known. The combination of adenoviruses with cell receptors is stable, and spontaneous elution of the hemagglutinin (as seen with influenza viruses) does not occur.

Immunologic Characteristics. The major immunologic reactivities of adenoviruses are expressed by the hexon and penton proteins (Table 54-4). The **hexons** contain **family-reactive determinants,** which crossreact with a similar Ag in all except the chicken adenoviruses. The hexons also possess a **type-specific reactive site,** which is the prevalent Ag exposed when hexons are assembled in virions (identified by neutralization titrations).

TABLE 54-2. Physical and Chemical Characteristics and Oncogenic Properties of Human Adenoviruses

Subgroup	Types*	Oncogenic potential†	Viral DNA		
			% of Virion	Mol Wt	% G + C
I	3, 7, 11, 14, 16, 21	Weak	12.5–13.7	~23 × 10⁶ (35 Kbp)	49–52
II	8–10, 13, 15, 17, 19, 20, 22–30	None	12.5–13.7	~23–25 × 10⁶ (35–38 Kbp)	57–61
III	1, 2, 4–6	None	12.5–13.7	~23 × 10⁶ (35 Kbp)	57–59
IV	12, 18, 31	High	11.6–12.5	~20 × 10⁶ (30 Kbp)	48–49

Kbp = kilobase pairs

*Types 32–35 have not been sufficiently characterized for inclusion in this table.

†Highly oncogenic adenoviruses induce tumors in newborn hamsters within 2 months after inoculation; weakly oncogenic viruses induce tumors in fewer animals in 4–18 months. Even those adenoviruses that are not oncogenic transform nonpermissive rodent cells in vitro.

TABLE 54-3. Subgroups of Human Adenoviruses According to Hemagglutination Reaction*

Subgroup	Viral types	Penton dodecagons	Agglutination of RBCs	
			Rhesus	Rat
I	3, 7, 11, 14, 16, 21	+	+	0
II	8–10, 13, 15, 17, 19, 20, 22–30	+	+† or 0	+
III	1, 2, 4–6	–	0	Partial‡
IV	12, 18, 31	–	0	Partial‡

*Classification suggested by Rosen L: Virology 5:574, 1958. The groups described by Rosen correspond precisely to subgroups ordered according to DNA structure and oncogenic properties (Table 54-2); hence they have been rearranged to employ a single classification.

†Some types (9, 13, and 15) agglutinate rhesus RBCs but to a lower titer than rat RBCs.

‡Complete hemagglutination occurs when heterologous Ab to virus of the same hemagglutination subgroup is added to the reaction mixture, producing groups of fibers or aggregation of pentons into regular groups of 12.

The **pentons** provide minor Ags of the virions and a **family-reactive soluble Ag** found in infected cells. The purified fibers contain a **major type-specific Ag** as well as a **minor subgroup Ag.** Although the fiber is the organ of attachment to the host cell, Abs to the fiber or to the intact penton only weakly neutralize viral infectivity.

Neutralizing, hemagglutination-inhibiting, and complement-fixing (CF) Abs appear about 7 days after the onset of illness and attain maximal titers after 2–3 weeks. IgA and IgG Abs appear in nasal secretions at the same time or within a week after the detection of serum Abs. The CF Abs begin to decline 2–3 months after infection but are usually still present 6–12 months after infection. Neutralizing and hemagglutination-inhibiting Abs persist longer, decreasing in titer only two- to threefold in 8–10 years. Minor rises in heterotypic neutralizing Abs may follow adenovirus infections, especially when Abs to several types are already present at the time of infection.

The same type of adenovirus rarely produces a second attack of disease. Such persistent type-specific immunity is unusual among viral respiratory diseases, resistance of relatively short duration being the rule (see Immunologic characteristics, Ch. 58 and Immunologic Characteristics under Parainfluenza Viruses in Ch. 59).

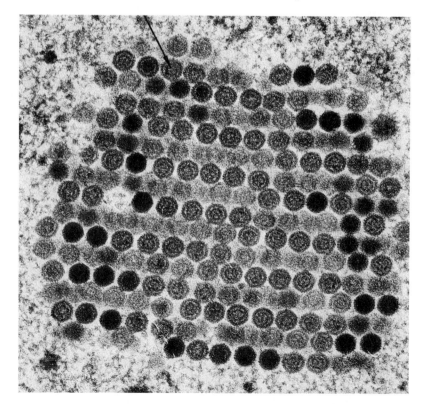

FIG. 54-1. Thin section of a crystalline mass of adenovirus particles in a cell infected with type 4. Two types of particles can be seen: dense particles with no discernible internal structure, and less dense particles showing the central body of the viral particles, the core **(arrows),** and the capsid. The polygonal shape of adenoviruses is apparent in many particles. Differences in appearance of particles are probably due to the relation of the center of the virion to the plane of section. (Courtesy of C. Morgan and H. M. Rose, Columbia University) (>110,000)

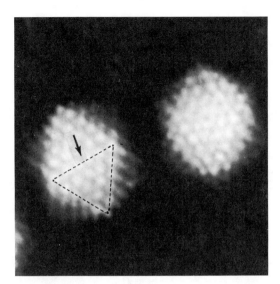

FIG. 54-2. Electron micrograph of purified particles of type 5 adenovirus embedded in sodium phosphotungstate. The icosahedral symmetry (Ch. 46) of the virion and subunit structure of the capsid are apparent. **Arrow** points to a virion's axis of twofold rotational symmetry. Capsomers at apices of **triangle** are centers of fivefold symmetry. The capsid parameters are $P = 1$, $f = 5$; hence the capsid consists of 252 capsomers (see Number of Capsomers in Icosahedral, in Ch. 46 Appendix). ($\times$440,-000, reduced)

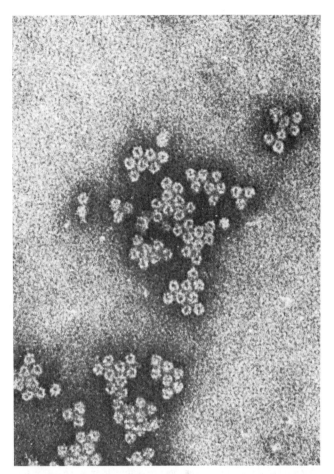

FIG. 54-3. Electron micrograph of purified type 5 hexons embedded in sodium silicotungstate. The polygonal shape of the capsomer with the central hole is apparent. Groups of nine hexons are seen in different orientations so that both tops (those with large holes) and bottom (small holes) are observed. The subunit structure can also be seen in many hexons. (Courtesy of M. V. Nermut, National Institute for Medical Research, London, England) ($\times$375,000)

Toxic Properties of the Penton. In addition to their immunologic reactivities, the penton and its individual components possess striking biologic activities. Thus, the intact penton causes rounding and clumping of cultured cells and detaches them from their support. Therefore the penton is also termed **toxin,** or **cell-detaching factor.** Hydrolysis of the penton's base by trypsin, leaving the fiber intact, destroys the cytopathic effect. The purified fiber, which is present in infected cells as a soluble protein as well as in pentons, has a different toxic action: in cultured cells it blocks biosynthesis of DNA, RNA, and protein; stops cell division; and inhibits the capacity of cells to support the multiplication of related or unrelated viruses (Table 54-4).

Host Range. Most adenoviruses of man do not produce recognizable disease in common laboratory animals, but **inapparent infections** follow intravenous inoculation in rabbits or intranasal instillation in hamsters, piglets, guinea pigs, and dogs. In rabbits type 5 virus persists for at least 6 months in the spleen, and it emerges when explants are cultured in vitro (similar to latent infection of human tonsils and adenoids; see Pathogenesis, below). Chick embryos are susceptible only to chicken adenoviruses. The nine members of subgroups I and IV (Table 54-2) produce tumors when inoculated in large amounts into newborn hamsters, rats, and mice (Ch. 66).

A variety of **cultured mammalian cells** support the multiplication of adenoviruses to a high titer and evince characteristic cytopathic changes, including pathognomonic nuclear alterations (see Effects of Viral Multiplication on Host cells, below). Epithelium-like human cell lines (such as HeLa cells) and primary cultures of human embryonic kidney are most satisfactory for human adenoviruses; primary cultures of various other types of human and animal cells also support viral multiplication but give much lower yields.

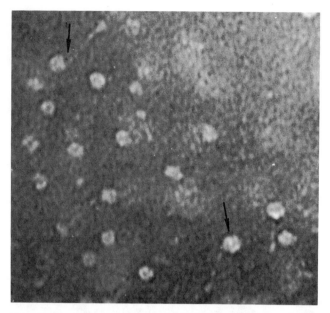

FIG. 54-4. Electron micrograph of the capsid components of purified type 5 adenovirus particles disrupted at pH 10.5. Note the polygonal, hollow capsomers—i.e., the **hexons,** which are 7–8.5 nm in diameter with a central hole about 2.5 nm across—and the fibers (1–2.5 nm in width and 20 nm in length), attached to polygonal bases—i.e., the **pentons (arrow).** By actual count there are 12 pentons per virion (480,000, reduced)

FIG. 54-5. Electron micrograph of purified type 5 adenovirus particles embedded in sodium silicotungstate. Micrographs obtained in areas where the silicotungstate was thin revealed the fiber components of the penton projecting from corners of the virion. Free pentons **(arrows)** and hexons are also present. (Valentine RC, Pereira, HG: J Mol Biol 13:13, 1965) (×350,000)

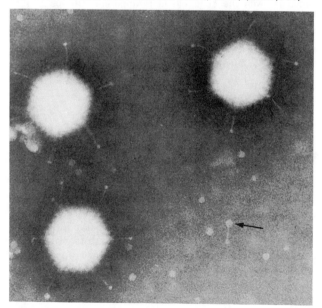

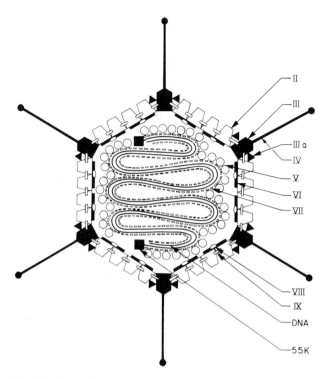

FIG. 54-6. Model of an adenovirus particle, showing the apparent architectural interrelationships of the structural proteins **(roman numerals)** and the nucleoprotein core in the virion. The hexon **(II)**, penton base **(III)**, and fiber **(IV)** and the hexon-associated proteins **(IIIa, VI, VIII,** and **IX)** make up the capsid. Proteins **V** and **VII** are core proteins associated with the viral DNA; the 55K protein is covalently linked to the 5′ end of the DNA.

FIG. 54-7. Diagram of the adenovirus genome, indicating the terminal inverted repetitions of nucleotide sequences and the protein covalently linked to a deoxycytosine at the 5′ terminus of each strand. **A.** The intact native linear viral DNA molecule. **B.** The configuration assumed by each strand after denaturation owing to hybridization of the complementary 3′ and 5′ ends of the strands to each other, forming a "panhandle."

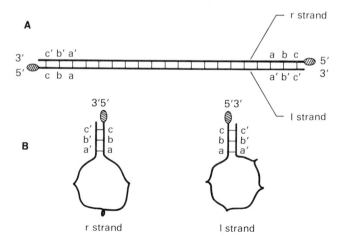

TABLE 54-4. Characteristics of Major Type 5 Adenovirus Proteins*

| Property | Hexon protein | Penton proteins | | | Internal proteins |
		Complete†	Base†	Fiber	
Immunologic reactivity	Type-specific and family cross-reactive	Family cross-reactive	Family cross-reactive	Type-specific	?
Biologic activity	None known	Cytopathic Attachment of virions to cells	Cytopathic	Blocks biosynthesis of macromolecules; inhibits viral multiplication	Probably aid in assembly of viral DNA
Hemagglutination	0	Partial	0	Partial	0
Mol wt of native protein	315,000‡	419,000§	236,000§	183,000§	
Polypeptide chains	105,000	—	80,000	61,000	55,000 48,000 (V) 19,000 (VII)

*Proteins of capsid, $\simeq 58\%$ of virion, consist of seven species; internal proteins, $\simeq 30\%$ of virion, consist of three species.

†A DNA endonuclease is closely associated with but physically separable from the penton base; its function during infection is unknown.

‡Molecular weights of hexons and fibers vary with type (e.g., type 2 hexon is 360,000 daltons).

§Estimated from sedimentation coefficient.

VIRAL MULTIPLICATION

The essential features of multiplication (Fig. 54-8) are similar for all adenovirus types. **Adsorption** to susceptible cultured cells is relatively slow, reaching a maximum after several hours. The viral particle then promptly penetrates the cell either directly through the plasma membrane, which re-forms rapidly (Fig. 54-9A), or by a process analogous to phagocytosis. In the latter case the virions must still find their way through the vacuole membrane into the cytoplasmic matrix for uncoating.

Uncoating of the viral DNA begins immediately after the virions have penetrated into the cytoplasm. It is detected biochemically when the nucleic acid becomes susceptible to DNase, and by electron microscopy when the virion appears spherical rather than polygonal (Fig. 54-9B). Initially, pentons and the immediate surrounding hexons are displaced, reducing the stability of the capsid. The other hexons and associated proteins then separate, and the naked **viral core** either enters the nucleus through nuclear pores, or releases the viral DNA into a nuclear pocket (Fig. 54-9C). The DNA thus gains access to the nucleus, where viral replication takes place (Fig. 54-10). Viral uncoating requires about 2 h.

Prior to and independent of viral DNA replication,

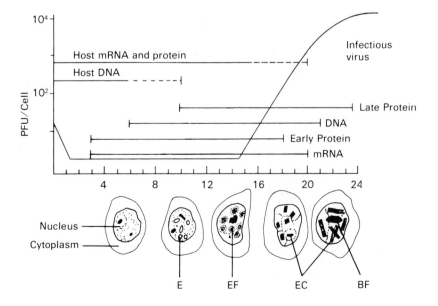

FIG. 54-8. Diagram of the sequential events in the biosynthesis of type 5 adenovirus, its effect on synthesis of host macromolecules, and the concomitant development of nuclear alterations. **E** = eosinophilic masses; **EF** = eosinophilic masses with basophilic Feulgen-positive border; **EC** = eosinophilic crystals; **BF** = basophilic Feulgen-positive masses; **PFU** = Plaque-forming units. Types 1, 2, and 6 have similar multiplication and nuclear changes.

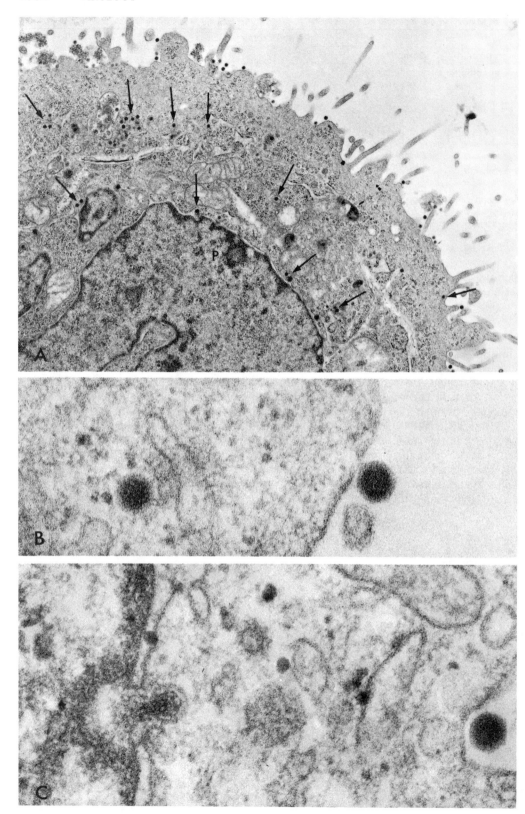

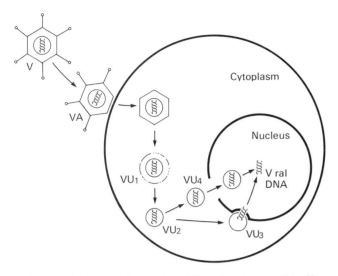

FIG. 54-10. Diagram of uncoating of the adenovirus particle. **V** = intact virion; **VA** = attachment of virion to cell, followed by penetration of virion into cell; **VU₁** = pentons are detached, leaving virion somewhat spherical and the viral DNA susceptible to DNase; **VU₂** = capsid disintegrates, leaving viral core which migrates to nucleus; **VU₃** = viral DNA is freed from core into nuclear pocket and free DNA enters nucleus; or **VU₄** = core enters nucleus through membrane pore, and DNA is then dissociated from proteins.

early messenger RNAs are transcribed from five separated regions of the genome (Fig. 54-11), corresponding to approximately 14% of the rightward (*r*) strand and 13% of the leftward (*l*) strand (designating the direction in which each strand is transcribed); these regions are identified by hybridization with restriction fragments of viral DNA and by electron microscopy of DNA–mRNA hybrids. As early as 2–3 h after infection early messengers are present on polyribosomes and are translated into several early proteins, which are detected by SDS–polyacrylamide gel electrophoresis. Studies using temperature-sensitive mutants reveal that at least two of these proteins are required for initiation of DNA replication: the **single-strand-specific DNA-binding protein** (mol wt 60,000–72,000 daltons, depending upon viral type) and

---

FIG. 54-9. Adsorption, penetration, and uncoating of adenovirus in HeLa cells. **A.** Numerous particles are adsorbed to the cell surface; others are present as free virions in the cytoplasm **(arrows).** Some particles in phagocytic vacuoles are also noted. A nuclear pocket **(P)** is also present. (×15,000) **B.** Higher magnification of a polygonal virion adsorbed to the plasma membrane and a virion that has assumed a spherical form in the cytoplasm. (×150,000) **C.** Partially uncoated viral particle releasing core material into a nuclear pocket. For comparison, an unaltered virion is shown on the cell surface. (×150,000) (Morgan C et al: J Virol 4:777, 1969)

an unidentified protein. Two of the early proteins can be detected immunologically: the single-strand-specific DNA-binding protein and the so-called **tumor (T) antigen,** which is present in transformed cells and adenovirus-induced tumors (see Ch. 66).

The activities of several enzymes involved in DNA synthesis may also increase prior to DNA replication. These early enzymes include aspartate transcarbamylase, thymidine kinase, deoxycytidylate deaminase, and DNA polymerase; but these enzymes are not unique to virus-infected cells, and it is uncertain whether they are products of the viral or the host cell genome. Inhibition of the synthesis of early mRNA (by dactinomycin [actinomycin D] or by pyrimidine analogs) or of early proteins (by cycloheximide or by amino acid analogs) prevents viral DNA synthesis.

Replication of viral DNA is semiconservative, is initiated at either end of the molecule, and occurs by an asymmetric displacement of either strand (Fig. 54-11). Replication begins in the nucleus 6–8 h after infection, attains its maximum rate by 18–20 h, and practically ceases by 22–24 h after infection.

Transcription of late mRNAs begins shortly after the initiation of DNA replication and cannot occur in its absence (e.g., when blocked by viral mutations or by metabolic inhibitors). Hybridization to separated DNA strands shows that late mRNAs are predominantly encoded in the *r* strand. Only one late mRNA appears to arise from the *l* strand (Fig. 54-11). The mRNAs are generated by **processing of long primary transcripts and by splicing of noncontiguous leader sequences** to sequences containing the message for a single protein (see Ch. 50). Thus at least 13 late messages are derived from primary transcripts stretching between map units 16 and 91. Processing probably regulates the relative proportions of late messengers, some of which are present in considerable abundance (e.g., those for the hexon and the 100K proteins). It follows that **late proteins,** which include the capsid proteins, are primarily products of the *r* strand; their formation depends upon prior replication of the viral DNA. Although early mRNAs continue to be transcribed late, only about 30% of them are expressed (e.g., translation of the DNA-binding protein mRNA is meager during the period of late mRNA and protein synthesis).

Although viral DNA and mRNAs are synthesized in the nucleus, and virions are assembled in the nucleus, viral proteins, like host proteins, are synthesized on polyribosomes in the cytoplasm. After their release from the polyribosomes, the polypeptide chains are immediately transported into the nucleus, where they assemble into the multimeric viral capsid proteins. **Assembly** of mature particles takes several steps that begin about 2–4 h after initiation of capsid protein production; this period is probably required to attain component pools of adequate size. The **eclipse period** lasts 13–17 h.

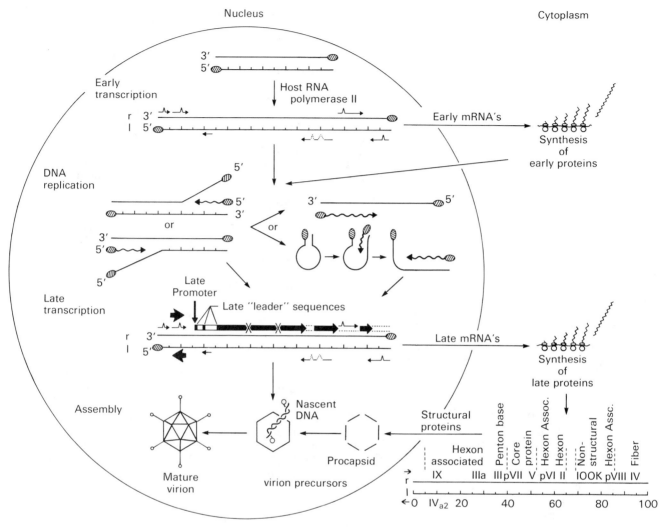

FIG. 54-11. Diagram of the biosynthetic events in the multiplication of type 2 adenovirus (used as a model since its transcription has been studied in greatest detail). Early mRNAs are transcribed from five separate regions of the genome; like late messages, they are processed from larger transcripts; and they have leader sequences transcribed from noncontiguous regions (the intervening absent sequences are indicated by a connecting caret [^]). The semiconservative, asymmetric DNA replication is shown. A mechanism for replication of the displaced single strands, using the inverted terminal repetition to form a circlelike intermediate, is also suggested. The function of the 5′ terminal 55K protein in DNA replication is unknown; the model presented, however, shows that it could serve as a primer for DNA replication. Late mRNAs, except that for protein IVa₂, have a single promoter at 16 map units on the r (rightward) strand. Note that each late "megatranscript" contains only one set of leader sequences, and therefore only a single mRNA can be derived from each transcript. Also indicated are the regions of the genome in which the late viral proteins are encoded, and their known functions.

Effects of Viral Multiplication on Host Cells. Adenovirus infection has a profound effect on the physiology of host cells. Production of host DNA stops abruptly 8–10 h after infection, and host biosynthesis of protein and RNA ceases 6–19 h later (Fig. 54-8). The division of infected cells also accordingly halts. This inhibition of mitosis is associated with production of chromosomal aberrations.

The hallmark of infection with adenoviruses is the development of characteristic **nuclear lesions** (Figs. 54-8

and 54-12), caused by the accumulation of unassembled viral components. In fact, the process of adenovirus assembly is quite inefficient: *only about 10%–15% of the new viral DNA and proteins is incorporated into virions.* Some cellular changes may be produced by the accumulation of fibers (which block the synthesis of macromolecules) and of DNA endonuclease. The basophilic inclusion bodies (Figs. 54-8 and 54-12) are composed of the excess viral DNA and structural protein, and the large baso-

FIG. 54-12. Sequential development of nuclear alterations in HeLa cells infected with type 7 adenovirus (types 3 and 4 produce similar effects). Earliest changes may be the formation of small eosinophilic inclusions (Feulgen-negative) **(1)** and clusters of granules **(G).** Clusters of granules gradually become larger and more prominent (Feulgen-positive) and form a large central mass **(CM).** In later stages the nucleus is enlarged and crystalline masses (Feulgen-positive) are apparent **(C).** (Boyer, GS et al: J Exp Med 110:327, 1959) (×1050)

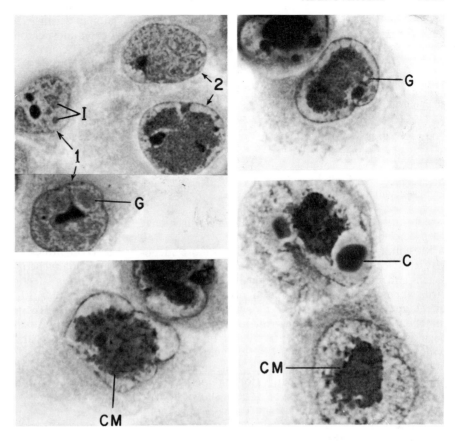

philic crystals in cells infected with type 3, 4, or 7 are made up of viral particles arranged in a crystalline lattice (Figs. 54-1 and 54-12). In cells infected with type 5 or 6 adenovirus prominent bar-shaped eosinophilic crystals are formed, mostly by the arginine-rich internal viral proteins.

Despite the extensive alterations, the infected cells remain intact and the nuclei do not release the newly synthesized virions. Less than 1% of the total virus is in the culture fluid when the maximal viral titer is attained (as measured after cell disruption). The infected cells also remain metabolically active, and even exhibit an increased utilization of glucose and production of organic acids. The culture medium therefore becomes acid, in contrast to the alkalinity that develops in cultures infected with agents that lyse the cells (e.g., polioviruses; see Laboratory Diagnosis, Ch. 57).

PATHOGENESIS

Progress in the study of pathogenesis has been impeded by the lack of satisfactory animal models. Accordingly, knowledge of human adenovirus infections is derived primarily from clinical observations and from experiments on volunteers. The recognized diseases (Table 54-5) predominantly involve the **respiratory tract** and the **eye.** Adenoviruses usually cause either inapparent infections or self-limited illnesses which are followed by complete recovery and persistent type-specific immunity.

Type 3 or 7 adenovirus has been isolated from a number of fatal cases of nonbacterial pneumonia in infants, and type 7 from rare cases of fatal pneumonia in military personnel. The pulmonary lesions observed were those of nonbacterial bronchopneumonia, but in addition there were numerous epithelial cells containing central basophilic masses in the nuclei, closely resembling those produced by the same adenovirus types in cell cultures (Fig. 54-12). Evidence is mounting that type 11 adenovirus, commonly isolated from the urine of children with **acute hemorrhagic cystitis,** is an etiologic agent of this disease.

Virus introduced by feeding, or swallowed in respiratory secretions, multiplies in cells of the gastrointestinal tract and is excreted in the feces (where it is still infectious, owing to its stability), but usually it does not produce **gastrointestinal disease.** However, some adenoviruses, most often type 1, 2, 5, or 6, have been isolated occasionally from cases of acute infectious diarrhea and intussusception, though an etiologic relation between virus and illness has not been clearly demonstrated. Viremia is not observed in human infections, and virus is not commonly transmitted to distant organs (although

TABLE 54-5. Clinical Syndromes Caused by Specific Types of Adenoviruses

Diseases syndrome	Adenovirus type	
	Most common	Less common
Acute respiratory disease of recruits	4, 7	3, 11, 14, 21
Pharyngoconjunctival fever; pharyngitis	3	5, 7, 21
Conjunctivitis	3, 7	2, 5, 6, 9, 10, 11
Epidemic keratoconjunctivitis	8	7, 19
Nonbacterial pneumonia of infants*	7	—
Acute hemorrhagic cystitis*	11	—

*Least common of diseases produced.

rare instances of adenovirus meningitis in children have been described).

Most individuals are infected with one or more adenoviruses before the age of 15. As a corollary, 50%–80% of tonsils and adenoids removed surgically yield an adenovirus when explants are cultured in vitro. The most frequent types are 1, 2, and 5, the types responsible for most infections in young children. Adenoviruses have also been isolated from cultured fragments of mesenteric lymph nodes. It thus appears that following an initial infection the virus frequently becomes **latent** in the lymphoid tissues, where it may persist for long periods. Recurrent illnesses, however, have not been shown to arise from these latent infections.

It appears likely that latent infections are readily established because the infected cells are not lysed and viral particles remain protected within their nuclei. In nature, the occult virus is confined to relatively few cells, probably by Abs. Culturing tissues in vitro, however, alters the cellular environment in ways (including dilution of viral inhibitors) that permit the virus to spread to uninfected cells, to multiply more rapidly, and to produce detectable cytopathic changes.

As noted above, some adenoviruses produce **tumors** in experimental animals after inoculation of large amounts. Ardent search, however, has not uncovered evidence that these viruses cause cancer in man.

LABORATORY DIAGNOSIS

Adenovirus infection can be diagnosed serologically and by isolation of the offending virus from respiratory and ocular secretions, urine, and feces. For isolation, infected material is inoculated into cultures of continuous lines of human cells (e.g., HeLa or KB) or into human embryo kidney cells. If virus is present, cytopathic changes (rounding and clumping of cells) develop after 2–14 days, the time depending upon the quantity of virus in the infected materials.

The virus is identified as an adenovirus by CF titration with a hyperimmune rabbit serum or a convalescent human serum. This procedure detects the crossreactive hexon and penton family Ags. The specific type of adenovirus can be ascertained most conveniently through hemagglutination-inhibition titrations; the required number of titrations can be reduced considerably by first determining the hemagglutination subgroup (Table 54-3). To establish a virus as a new serotype neutralization titration is the method of choice. A rapid presumptive diagnosis of adenovirus infection, but not of the specific viral type, can be made by examination of cells in respiratory or ocular secretions, or in urine, using the immunofluorescence technic with rabbit serum containing Abs to the family Ag.

The **serologic diagnosis** of an adenovirus infection is accomplished most conveniently by **CF titration,** which detects crossreactive family Ags. Unfortunately, this assay identifies fewer than 50% of new infections, because many people have a constant high level of Abs from prior infections. A more precise diagnosis requires **neutralization titrations** with acute- and convalescent-phase sera. **Hemagglutination-inhibition** assay* for Abs is practically as sensitive as neutralization, is simpler and less expensive, and is almost as accurate if nonspecific inhibitors are removed prior to use. However, owing to the type-specificity of the reaction all of the common viral types must be used if a virus has not been isolated from the patient.

EPIDEMIOLOGY

Man provides the only known reservoir for strains of adenoviruses that infect humans. Person-to-person spread in respiratory and ocular secretions is the most common mode of viral transmission, though dissemination in swimming pools has also been implicated in epidemics of **pharyngoconjunctival fever** and **conjunctivitis.** The spread of **epidemic keratoconjunctivitis** caused by type 8 adenovirus appears to be associated with conjunctival trauma produced by dust and dirt in shipyards and factories, or with improperly sterilized optical instruments. Adenoviruses are commonly present in the feces of infected persons, but there is no evidence that disease is transmitted by the fecal–oral route.

Despite the large number and the worldwide distribution of adenoviruses, their clinical importance is largely restricted to epidemics of **acute respiratory disease (ARD),** an influenzalike illness in military recruits, and to limited outbreaks among children (except for keratoconjunctivitis caused by type 8). Infections are observed throughout the year, but the greatest incidence and largest epidemics occur in late fall and winter. Types 7, 4, and 3 (in order of decreasing importance) are the viruses most

* Types 10 and 19 crossreact to such an extent that they cannot be distinguished by this procedure.

frequently responsible for epidemics of acute respiratory and ocular diseases (Table 54-5); types 11, 14, and 21 have been increasingly implicated in epidemics. Peculiarly, type 4 adenovirus commonly causes acute respiratory disease (ARD) in military recruits, but rarely produces infections in civilians. This epidemiologic phenomenon is without parallel; its explanation is unknown.

A relatively high proportion of adults have Abs to one or more types of adenoviruses (particularly types 1–3, 5, and 7), indicating previous infections. However, epidemiologic studies indicate that adenoviruses annually cause at most 4%–5% of viral respiratory illnesses in civilians.

PREVENTION AND CONTROL

Isolation of sick persons has little or no effect on the spread of adenoviruses since many healthy carriers exist. Immunization, however, offers an effective preventive measure, as would be expected from the lasting type-specific immunity produced by natural infections. A formalinized virus vaccine was experimentally successful, but its irregular antigenicity caused its abandonment in favor of a highly effective **live virus vaccine,** used primarily to protect military recruits. Type 4 virus was formerly the predominant agent responsible for acute respiratory disease of recruits in the United States, but type 7 replaced it as successful immunization suppressed ARD caused by type 4. Following continued immunization with vaccines containing live types 4 and 7, other types have appeared (e.g., type 21, which previously had been responsible for epidemics of ARD in European defense forces).

Vaccines for military use should contain at least types 3, 4, 7, and 21. In closed populations, such as chronic disease hospitals or homes for orphans, a vaccine containing types 1–7 may be useful for infants and young children.

Some common pathogenic adenoviruses that must be included in a vaccine (e.g., types 3 and 7) are oncogenic for animals other than man. Although there is no evidence that adenoviruses are oncogenic in humans, some consider it unwise to use these live adenoviruses to prevent mild, self-limiting diseases. A suitable vaccine, without the putative oncogenic danger, may be provided by using the **purified hexon and fiber capsid proteins,** which are present in abundance as soluble Ags in infected cells and can stimulate production of neutralizing Abs.

ADENO-ASSOCIATED VIRUSES

Small icosahedral virions, 20–25 nm in diameter, are often found growing with adenoviruses in cells of man and other animals. These **adeno-associated viruses (AAVs)** (members of the family Parvoviridae) are defective and depend totally upon the multiplication of the unrelated adenoviruses for their own replication. (Herpes simplex virus can assist in partial synthesis of AAV: some viral proteins are made, but infectious virus is not produced.) The viral genome consists of a linear molecule of **single-stranded DNA** with a mol wt of only 1.4×10^6 (4.3 Kb). The virions, however, can contain either a positive or a negative DNA strand which, when extracted from the viral particles, forms double-stranded molecules unless a reagent that blocks hydrogen bond formation (e.g., formalin) is present. The quantity of genetic information contained in this small DNA molecule is very limited and is largely utilized for specifying the viral capsid proteins. The precise functions that the helper supplies are unknown, but temperature-sensitive adenovirus mutants that are unable to replicate viral DNA at the restrictive temperature still serve as effective helpers. This implies that AAV utilizes a very early adenovirus protein or a cell protein induced by early adenovirus infection. It is striking that the complete multiplication of adenovirus is itself inhibited when it offers assistance to the defective AAV (and AAV also reduces adenovirus oncogenicity in hamsters). Although production of infectious AAV in cells coinfected with an adenovirus resembles complementation by two genetically related defective mutants (see Ch. 50), cross-hybridization of DNAs extracted from AAVs and several types of adenoviruses failed to detect homologous regions.

Four distinct immunologic types of AAVs have been identified as contaminants of human and simian adenoviruses, and AAV Abs are frequently found in humans and monkeys. About 70%–80% of infants acquire Abs to types 1, 2, and 3 within the first decade, and more than 50% of adults maintain detectable Abs. Adeno-associated viruses can occasionally be isolated from fecal, ocular, and respiratory specimens during acute adenovirus infections, but not during other illnesses. Thus, AAVs appear to be defective in man as well as in cell cultures. They have been suspected of playing a role in the pathogenesis of "slow" viral infections (Ch. 53), but so far this role is unconfirmed and they appear to be only silent, unobtrusive partners of adenovirus infection.

PARVOVIRUSES

Several other viruses (latent rat viruses, minute mouse virus, and porcine virus) have physical and chemical characteristics similar to AAV (although only the AAVs are defective in their natural hosts). They are called **parvoviruses** (L. *parvus,* small), and are grouped with the AAVs in the family Parvoviridae. The latent rat viruses, H_1 and H_3, are particularly intriguing: they are defective in human cells, and adenoviruses can serve as their helpers and are in turn inhibited. Moreover, the H_1 and H_3 viruses may serve as models for the study of human disease since they can produce dwarfism, mongoloid appearance, and a variety of other congenital lesions when pregnant hamsters and rats are infected (see Effects of Viruses on Embryonic Development, Ch. 53). However, the suggestion that AAV might have similar effects in man has not been confirmed.

SELECTED READING

BOOKS AND REVIEW ARTICLES

FLINT J: The topography and transcription of the adenovirus genome. Cell 10:153, 1977

GINSBERG HS: Adenovirus structural proteins. Compr Virol 13:409, 1979

GINSBERG HS, YOUNG CSH: Genetics of adenoviruses. Compr Virol 9:27, 1977

PHILIPSON L, LINDBERG U: Reproduction of adenoviruses. Compr Virol 3:143, 1974

ROSE JA: Parvovirus reproduction. Compr Virol 3:1 1974

WOLD SM, GREEN M, BÜTTNER W: Adenoviruses. In Nayak DP (ed): The Molecular Biology of Animal Viruses, Vol 2. New York, Marcel Dekker, 1978, p 673

SPECIFIC ARTICLES

BERGET SM, MOORE C, SHARP PA: Spliced segments at the 5′terminus of adenovirus 2 late mRNA. Proc Natl Acad Sci USA 74:3171, 1977

BERK AJ, SHARP PA: Structure of the adenovirus 2 early mRNAs. Cell 14:695, 1978

CARTER TH, GINSBERG HG: Viral transcription in KB cells infected by temperature sensitive "early" mutants of adenovirus type 5. J Virol 18:156, 1976

CHOW LT, GELINAS RE, BROKER TR, ROBERTS RJ: An amazing sequence arrangement at the 5′ ends of adenovirus 2 messenger RNA. Cell 12:1, 1977

EVANS R, FRASER N, ZIFF E et al: The initiation sites for RNA-transcription in AD2 DNA. Cell 12:733, 1977

HILLEMAN MR, WERNER JR: Recovery of a new agent from patients with acute respiratory illness. Proc Soc Exp Biol Med 85:183, 1954

HORNE RW, BRENNER S, WATERSON AP, WILDY P: The icosahedral form of an adenovirus. J Mol Biol I:84, 1959

HUEBNER RJ, ROWE WP, LANE WT: Oncogenic effects in hamsters of human adenovirus types 12 and 18. Proc Natl Acad Sci USA 48:2051, 1962

REKOSH DMK, RUSSELL WC, BELLET AJD, ROBINSON AJ: Identification of a protein linked to the ends of adenovirus DNA. Cell 11:283, 1977

ROWE WP, HUEBNER RJ, GILMORE LK et al: Isolation of a cytopathogenic agent from human adenoids undergoing spontaneous degeneration in tissue culture. Proc Soc Exp Biol Med 84:570, 1953

TRENTIN JJ, YABE Y, TAYLOR G: The quest for human cancer viruses. Science 137:835, 1962

VAN DER VLIET PC, LEVINE AJ: DNA-binding proteins specific for cells infected by adenovirus. Nature 246:170, 1973

chapter

55

HERPESVIRUSES

GENERAL CHARACTERISTICS

The herpesviruses, a family (Herpetoviridae)* of structurally similar viruses, are named (Gr. *herpein,* to creep) for those members responsible for two common diseases of man: **herpes simplex** (fever blisters) and **herpes zoster** (chickenpox or varicella, and shingles) (Gr. *zoster,* girdle). The skin lesions of the herpetic diseases illustrate the affinity of most herpesviruses for cells of ectodermal origin. Like the poxviruses, another group of DNA viruses, herpesviruses exhibit **focal cytopathogenicity,** producing vesicles or pocks in patients and in egg membranes. Another prominent characteristic is the production of **latent** and **recurrent** infections; although the mechanism is not clear, it is noteworthy that herpesviruses, like adenoviruses (which also initiate latent infection), are DNA viruses and replicate in the cell nucleus.

The unusual epidemiologic features of the common herpes infections long puzzled physicians until technics became available for identifying and characterizing the responsible viruses. Thus the multiple recurrence of fever blisters in certain individuals was bewildering until about 1950, when Burnet in Australia and Buddingh in the United States showed that herpes simplex virus often becomes latent after initiating a primary infection, usually in children, and is then repeatedly activated by subsequent provocations (see Latent Infections, Ch. 53). A similar mechanism was suspected in **chickenpox (varicella),** i.e., that it recurs as **herpes zoster (shingles),** since an outbreak of chickenpox was often observed to follow the sporadic appearance of zoster in an adult. However, because the initial and the subsequent syndromes are so different the relation was not certain until Weller in 1954 isolated in tissue cultures the same virus from patients with the two diseases. The virus closely resembles that of herpes simplex.

Besides the herpes simplex and varicella–zoster viruses, the major herpesviruses that infect man are cytomegalovirus (inclusion or salivary gland virus of man), and EB (Epstein–Barr) virus. These viruses are widely separated in evolution despite their structural similarities

* The genera are provisionally termed **Alphaherpesvirinae** (includes the herpes simplex types 1 and 2 viruses and the varicella–zoster virus), **Betaherpesvirinae** (cytomegalovirus), and **Gammaherpesvirinae** (Epstein–Barr Virus).

(Table 55-1). They differ strikingly in their DNA composition; herpes simplex viruses are immunologically related to each other but show no relatedness to other family members; and the other three human herpesviruses antigenically crossreact only slightly.

Herpesviruses have also been found in every other eukaryotic species examined, from fungi to monkeys. Examples include **B virus** of monkeys (which may infect man accidentally), **pseudorabies virus** of pigs, **virus III** of rabbits, **cytomegaloviruses of animals** (inclusion virus of guinea pigs, inclusion virus of mice, and the agent of inclusion body rhinitis in pigs), oncogenic viruses (Ch. 66) that produce lymphoproliferative malignancies in chickens (**Marek's disease virus**) and in monkeys (**herpesvirus saimirii**), and several viruses that may be associated with renal carcinoma in frogs. These animal viruses appear to have only minor immunologic relatedness to human herpesviruses except for the marked crossreactivity between herpes simplex virus and B virus. In this chapter only the viruses that infect man will be discussed.

Herpesvirus particles have a diameter from 180 nm (herpes simplex and varicella–zoster viruses) to 200 nm (cytomegalovirus). Within a population of virions many particles do not possess envelopes, and some are empty capsids (Fig. 55-1B and D). The virion components (Figs. 55-1 and 55-2) are arranged in: 1) a DNA-containing **toroidal core** about 75 nm in diameter, 2) an **icosahedral capsid** 95–105 nm in diameter, 3) a surrounding **granular zone (tegument)** composed of globular proteins, and 4) an encompassing **envelope** possessing periodic short projections. The capsid is composed of 162 elongated hexagonal prisms (**capsomers**), each 9.5 × 12.5 nm with a central hole approximately 4 nm in diameter.

The chemical composition of purified enveloped viral particles is consonant with their morphology. Herpes simplex virions contain 25–30 virus-specified proteins (70% of the virion); a large, linear, double-stranded DNA molecule (7% of the virion); envelope lipid (22% of the virion), which is chiefly host-specific phospholipid derived from the nuclear membrane; and small amounts of polyamines (spermine within the nucleocapsid, and spermidine within the envelope). Other herpesviruses examined are similar.

Like other enveloped viruses, herpesviruses are relatively unstable at room temperature and are readily inactivated by lipid solvents.

HERPES SIMPLEX VIRUSES

IMMUNOLOGIC CHARACTERISTICS

Two main immunologic variants, types 1 and 2, although they crossreact strongly, can be distinguished by neutralization titrations (they also differ in clinical patterns). Some additional minor antigenic variations have been observed, but do not warrant classification. Infected cells contain, along with virions, soluble Ags which elicit a delayed-type skin reaction in addition to the usual immunologic reactions; their molecular nature has not been characterized. Infected cells also are antigenically altered by the insertion of viral structural glycoproteins into their membranes, including the plasma membrane. Accordingly, infected cells may be damaged and lysed by reaction with virus-specific Abs plus complement, or with specifically activated T lymphocytes.

Specific Abs can be assayed by neutralization or complement-fixation (CF) tests; they reach maximum titers in about 14 days. During the early stages after a primary infection, neutralizing Abs can be detected only in the presence of complement. Thereafter, complement is not required for neutralization, but its presence raises Ab titers four- to eightfold. Neutralizing Abs are primarily directed to the envelope glycoproteins; CF Abs react with all the virion proteins.

Abs may drop to undetectable levels after the first infection, only to reappear with recurrent episodes. By adulthood, titers are generally high and persist indefinitely. Accordingly, an increase usually cannot be detected in recurrent adult disease, although infectious virus can be isolated readily from the lesion. Fetuses ac-

quire maternal Abs via placental transfer and these persist until about 4 months after birth, which probably explains the common occurrence of primary infection in babies from 6 to 18 months of age.

Cell-mediated immunity also develops after primary infection and probably is a major immunologic factor in maintaining a latent state. It appears that recurrences are initiated by episodic impairment of cellular immunity, as measured by diminished production of macrophage migration-inhibitory factor and diminished activity of sensitized T lymphocytes. Moreover, in immunosuppressed patients the virus is commonly activated and disseminated, leading to acute disease.

HOST RANGE

Man is the natural host for herpes simplex viruses, but a relatively wide range of animals are also susceptible, in-

FIG. 55-1. The four morphologic types of herpes simplex virus particles embedded in phosphotungstate. **A.** Enveloped full particle showing the thick envelope surrounding the nucleocapsid. **B.** Enveloped empty particle; the capsid does not contain viral DNA and therefore can be penetrated by the phosphotungstate. **C.** Naked full particle; the structure of the capsomers is plainly visible. **D.** Naked empty particle (Watson DH et al: Virology 19:250, 1963) (×200,000)

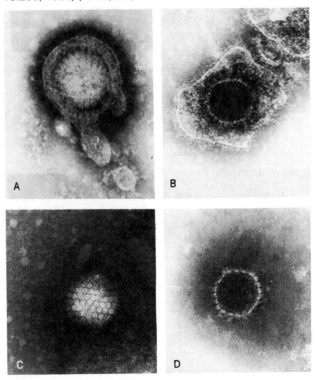

TABLE 55-1. Characteristics of Members of the Herpsevirus Family

Size	180–200 nm
Symmetry of capsid	Icosahedral
Number of capsomers	162
Lipid envelope	Present
Sensitivity to ether and chloroform	Inactivated
Nucleic acid	Double-stranded DNA*
Site of biosynthesis of viral DNA	Nucleus
Site of assembly of viral particles	Nucleus
Inclusion bodies	Intranuclear, eosinophilic†
Common family antigen	None

*Molecular weights (daltons): herpes simplex viruses, approximately 100×10^6 (154 Kbp); cytomegalovirus, 150×10^6 (231 Kbp); EB virus, 100×10^6 (154 Kbp). Guanine–cytosine content (moles %): cytomegalovirus, 58.5%; herpes simplex type 1, 68%; herpes simplex type 2, 70%; varicella-zoster, 46%; and EB virus, 59%. DNA–DNA hybridizations indicated that 40%–46% of the base sequences are homologous in types 1 and 2 herpes simplex viruses.

†Cytomegalovirus-infected cells may also contain basophilic cytoplasmic inclusion bodies.

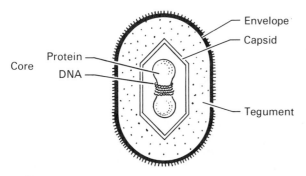

FIG. 55-2. Diagram of virion of herpes simplex virus. The DNA–protein core consists of DNA wrapped around an associated protein as on the spindle of a spool. The envelope contains a number of glycoproteins; the capsid is composed of at least four unique proteins; and the so-called tegument consists of about eight distinct polypeptides.

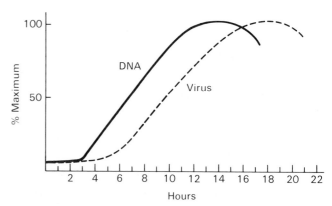

FIG. 55-3. Temporal relationship between viral DNA replication and the production of infectious virus.

cluding mice, guinea pigs, hamsters, and rabbits. The effects of infection depend upon the route of inoculation. For example, inoculation of the cornea in the rabbit results in keratoconjunctivitis or keratitis, whereas intracerebral inoculation produces fatal encephalitis. The chick embryo has been a convenient host: the production of pocks on the chorioallantoic membrane affords a reproducible method for detection and assay, similar to that employed with poxviruses.

Many **cultured cell types** support multiplication of herpes simplex virus, undergo extensive cytopathic changes, and develop intranuclear inclusion bodies; chromosomal breaks and aberrations are also observed. The response of the cells varies with the strain of virus employed: some strains cause marked clumping of cells, producing pocklike lesions; others produce multinucleated giant cells (polykaryocytes) by fusion of membranes and recruitment of the nuclei of adjoining cells; and some strains produce typical plaques with suitable cells.

VIRAL MULTIPLICATION

The viral envelope provides the normal **attachment** of the virions to susceptible cells. Following attachment, it appears to fuse with the cellular plasma membrane, permitting the nucleocapsid to enter directly into the cytoplasm (see Ch. 50). Intact virions may also enter via phagocytic vacuoles, from which they are released into the cytoplasm by similar viral envelope–membrane fusion. In the cytoplasm the capsid separates from the viral core, which enters the nucleus and initiates viral multiplication. The **eclipse period** is 5–6 h in monolayer cell cultures, and virus increases exponentially until approximately 17 h after infection (Fig. 55-3); each cell has then made 10^4–10^5 physical particles, of which about 100 are

infectious. Virions are **released** by slow leakage from infected cells.

As with other DNA-containing viruses (Ch. 50), the biochemical events are sequentially regulated (Fig. 55-4), but with herpes simplex virus (and probably with other herpesviruses as well) replication has unique features: **Replication of viral DNA** does not impose a strict control on transcription, and it is not essential for transcribing the "late" regions of the genome. Indeed, both early and late, a large fraction of the genome is transcribed, and the major control of gene expression seems to be effected by **posttranscriptional processing.** Thus, even in the absence of viral protein synthesis, initial transcripts from about 40% of the genome are processed to **immediate early (α) mRNAs,** which represent only about 10% of the genome (Fig. 55-5). Without translation, only these α mRNAs accumulate in the cytoplasm, and the larger, unprocessed transcripts remain in the nucleus. If protein synthesis is permitted, **delayed early (β) mRNAs,** representing about 40% of the genome, are formed. The β proteins block translation of α proteins and process transcripts into **late (γ) mRNAs,** derived from about 43% of the viral DNA sequences. Thus the synthesis and translation of the mRNAs are **coordinately regulated:** formation of the α proteins is necessary for synthesis of the β proteins, and both of these nonstructural and minor structural proteins are necessary for synthesis of the late, major structural γ proteins.

Viral DNA replication is carried out by both viral α and β proteins and host cellular enzymes. The reactions are not yet precisely understood, but they appear to involve complex replicative intermediates, including "head-to-tail" concatemeric circular and linear–circular forms (see Ch. 50) generated by the reiterated nucleotide sequences (Fig. 55-5).

Only about 25% of the viral DNA and protein made in cell cultures is assembled into virions. The production of host DNA, RNA, and protein declines, concomitant with

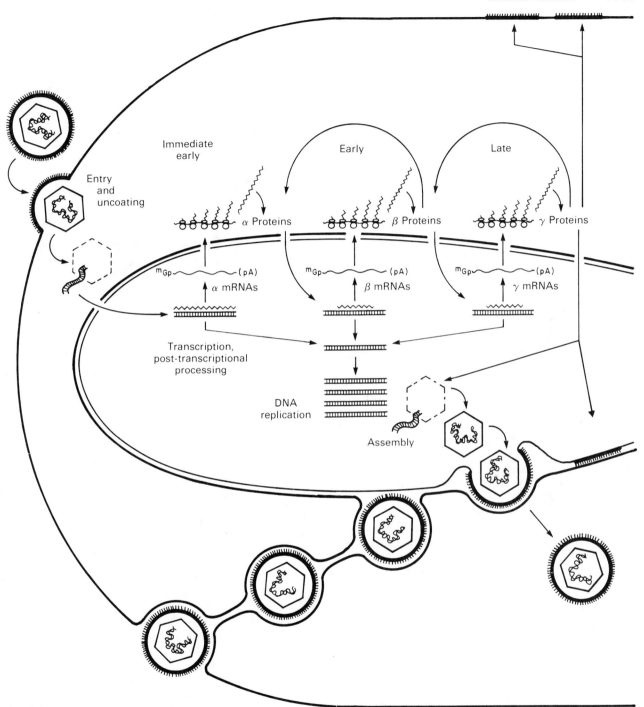

FIG. 55-4. Sequence of events in the multiplication of herpes simplex virus from entry of virus into the cell by fusion of the virion envelope with the cell plasma membrane to assembly of virions and their exit from the cell through the endoplasmic reticulum. Also illustrated are transcription and coordinated sequential processing of mRNAs and synthesis of sets of proteins ($\alpha \rightarrow \beta \rightarrow \gamma$) required for DNA replication and virion structures. (Modified from diagram kindly supplied by B. Roizman, University of Chicago.)

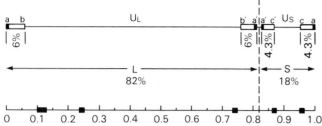

FIG. 55-5. Diagram of the DNA genome of type 1 herpes simplex virus, showing the arrangement of the unique and the reiterated nucleotide sequences. **Line 1. L** = large portion of the DNA, consisting of the unique regions, **U_L** (70% of the total DNA), bounded by the terminal sequence **ab** and its inverted repeated sequence **b'a'; S** = small portion of the genome, consisting of the unique region, **U_S** (9% of the total DNA), bounded by the terminal sequence **ca** and its inverted repeat sequence **a'c'; a** = the terminal redundant sequence bounding the genome. **Line 2.** Extent of the L and S regions of the genome. **Line 3.** Possible arrangement of the templates for the immediate early αmRNAs suggested by DNA–RNA hybridization analysis (relative sizes of the blocks correspond to the percent of the total viral DNA that hybridized to the viral mRNAs). (Modified from Jones PC et al: J Virol 21:268, 1977)

viral replication, and finally ceases within 3–5 h after infection.

The biosynthetic steps can be correlated with the development of **nuclear inclusion bodies** and the formation of viral particles. A basophilic, Feulgen-positive, granular mass of newly synthesized viral DNA develops centrally in the nucleus. The assembly of incomplete viral particles (Fig. 55-6) begins within this material. Initially only the capsid surrounds the electron-dense DNA-containing nucleoid; the particle is noninfectious and unstable. The envelope appears to be acquired from the inner nuclear membrane, which reduplicates (Fig. 55-7) to permit egress of viral particles into the cytoplasm and the endoplasmic reticulum. The envelope contains virus-specific subunits (particularly glycoproteins) as well as host cell materials. Mature infectious virus is slowly liberated from infected cells through the endoplasmic reticulum, but occasionally it also escapes by a process akin to reverse phagocytosis. Unlike the process with other enveloped viruses, envelopment and release of viral particles does not occur by budding from the plasma membrane. Rather, the plasma membrane is changed morphologically and contains virus-specific glycoprotein Ags.

The movement of viral particles from the nucleus into the cytoplasm is accompanied by the transport of soluble Ags into the cytoplasm; concomitantly the originally basophilic intranuclear inclusion body is converted into an eosinophilic, Feulgen-negative mass. Thus the eosinophilic inclusion body that is usually observed in infected cells does not contain viral particles or specific viral Ags (detectable by immunofluorescence) but actually is the burnt-out remnant of a viral factory.

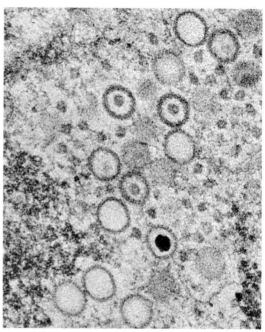

FIG. 55-6. Formation of herpes simplex virus particles within the cell nucleus. Some capsids are in the process of assembling, while others are complete. Cores of varying density are forming within the capsids. Indistinct particles probably represent virus sectioned at one margin, with loss of density and overlapping structure. (Nii S et al: J Virol 2:517, 1968) (×90,000)

PATHOGENESIS

The most striking characteristic of herpes simplex virus infection is its propensity for persisting in a quiescent or latent state in man, with recurrence of activity at irregular intervals. The initial infection occurs through a break in the mucous membranes (e.g., eye, mouth, throat, genitals) or skin, where local multiplication ensues. From this focus virus spreads to regional lymph nodes, where it multiplies further. On occasion virus disseminates, unrestricted by Ab, into the blood and to distant organs. Viremia can occasionally be detected by isolation of virus from the cells of the blood buffy coat.

The initial infection ordinarily occurs in children 6–18 months of age; serologic surveys have demonstrated that it is most often inapparent. But 10%–15% of those infected do develop a **primary disease,** usually **herpetic gingivostomatitis,** which is characterized by multiple vesicles in the oral mucous membranes and the mucocutaneous border. Similar lesions are less commonly seen in other regions, including the nostrils, the external genitalia or urethra, the cornea, and sites of trauma. More serious but rare complications include neonatal generalized infections, meningoencephalitis,* and diffuse skin

---

* Encephalitis occurs more frequently in adults, either as a primary infection or as a flare-up of a latent infection.

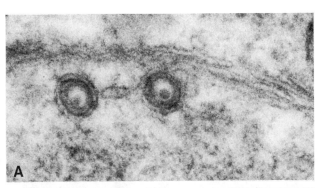

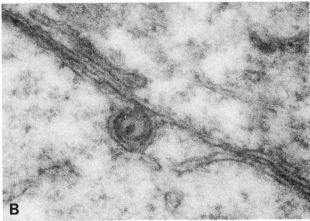

FIG. 55-7. A and **B.** Thin sections of mature particles of herpes simplex virus. Particles show the inner core and two or three membranes. In B the nuclear membrane partly surrounds a **viral** particle (apparently a step in the simultaneous assembly and egress of complete virions from the nucleus). (Morgan C et al: J Exp Med 110:643, 1959) (×87,000)

involvement in children with chronic eczema (Kaposi's varicelliform eruption, eczema herpeticum).

When the initial infection recedes the virus persists, despite the presence of a high Ab titer, producing a latent infection, probably in a sensory ganglion adjacent to the major site of the primary disease. The form of the occult virus remains unknown, but the balance may be readily upset, inducing viral replication in the nerve ganglion and thus provoking the second, **recurrent form** of herpes simplex disease. Since neutralizing Abs and cellular immunity are present the new virus cannot disseminate, but it does reach contiguous cells; recurrent herpes simplex therefore remains localized. In a given individual the clinical features are much the same with each episode. For example, if **gingivostomatitis** was the primary disease, the recurrent form is usually **herpes labialis** (fever blisters) in the same region of the lip; and if the primary disease was **herpetic keratitis,** the recurrent disease is also keratitis (which may lead to corneal scarring and is one of the major infectious causes of blindness).

Types 1 and 2 herpes simplex viruses differ signifi-cantly in their pathogenic potential. Type 1 virus is primarily associated with oral and ocular lesions and is transmitted in oral and respiratory secretions, whereas type 2 is isolated primarily from genital and anal lesions and is passed through sexual contact. Changes in sexual mores, however, have somewhat altered this common pattern: occasionally type 2 virus is isolated from oral lesions and type 1 virus from genital lesions. Mothers with **genital herpes** (a painful, persistent, recurrent infection) are the primary source of **neonatal infections** with type 2 virus, which are often severe and even fatal. It is also striking that in the United States more than 80% of women with cervical carcinoma have Abs to type 2 virus (Ch. 66). However, this correlation does not necessarily prove a causal relation but may only be a reflection of lower socioeconomic status and, possibly, of a more promiscuous sexual activity of many who become victims of cervical carcinoma.

Latency. In experimental models in mice, labial strains of type 1 are characteristically found latent in neurons of cervical ganglia, and type 2 genital strains are found in sacral ganglia. Infectious virus is not detectable as such in ganglia, but it emerges when ganglion fragments are cultured in vitro. In man the virus has similarly been detected in cultures of trigeminal ganglia (type 1) or sacral ganglia (type 2). Moreover, neurectomy of the facial branch of the trigeminal nerve (as therapy for persistent trigeminal neuralgia) characteristically activates latent herpes simplex virus, producing fever blisters. In experimental latent infections of mice and rabbits, trigeminal or sciatic nerve section likewise activates quiescent virus.

Viral multiplication and recurrent disease may also be induced in humans by naturally occurring factors such as heat, cold, or sunlight (ultraviolet light), and by immunologically unrelated hypersensitivity reactions, pituitary and adrenal hormones, and emotional disturbances. A few of these, such as epinephrine and hypersensitivity reactions, have evoked recurrences in experimental animals (see Latent Infections, Ch. 53).

LABORATORY DIAGNOSIS

The increasing occurrence of life-threatening herpesvirus infections in children with inherited T cell deficiencies, and in patients who are immunosuppressed because of transplantation or cancer therapy, has given great importance to rapid, accurate, and economic laboratory procedures for an etiologic diagnosis. The development of effective chemotherapeutic agents will make rapid diagnosis even more imperative. The following methods are available: 1) The quickest and most economic diagnostic test consists of demonstrating characteristic multinuclear giant cells, containing intranuclear eosinophilic **inclusion bodies,** in scrapings from the base of vesicles. This procedure cannot, however, distinguish a herpes simplex infection from one produced by varicella–zoster virus or cytomegalovirus. 2) Electron microscopic examination reveals cells that contain viral particles in various stages of maturation. 3) Specific Ags may be detected in cells from the lesions by immunofluorescent technics. 4) Virus can be isolated by inoculating material from lesions (especially vesicular fluid or scrapings) into susceptible tissue culture cells or newborn mice. 5) A **rise in Ab titer** is detected by neutralization, CF, indirect hemagglutination, or an enzyme-linked immunosorbent assay (ELISA) using serum obtained early in the primary disease and again 14–21 days after onset **(paired sera).** Following primary infections there is often also a small rise in neutralizing Abs for varicella–herpes zoster virus. Patients with recurrent disease have a high initial titer of circulating Abs and often do not show a significant increase.

EPIDEMIOLOGY, CONTROL, AND TREATMENT

Close person-to-person contact is the most common mechanism of viral transmission. Man is the only known natural host and source of virus. Secretions from lesions about the mouth (herpes labialis and stomatitis) and genitalia are the most frequent sources. Virus is also transmissible on eating and drinking utensils for a brief period after contamination. Occasionally virus is found in the saliva of healthy individuals, particularly children after the primary infection; the shedding may even continue for weeks or months.

Rare cases of primary infection have been reported in adults. Approximately 80% of adults have relatively high titers of neutralizing and CF Abs, as well as cellular immunity, and a substantial fraction of this population is subject to recurrent herpes.

Control is not feasible because of the large numbers of persons with inapparent infections and minor recurrent lesions from which virus is shed. However, it is important, whenever possible, to prevent contact between infants and persons who have herpetic lesions. A vaccine to prevent the primary infection would not have sufficient clinical application to warrant development, although the required knowledge is available.

Herpetic keratitis has been treated by repeated local application of 5-iodo-2′-deoxyuridine (idoxuridine), but recurrences often follow cessation of therapy, and drug-resistant herpesviruses may emerge. Cytosine arabinoside (1-β-D-arabinofuranosylcytosine; cytarabine) has been experimentally successful, but it is too toxic for clinical use. Adenine arabinoside (9-β-D-arabinofuranosyladenine; vidarabine) and its monophosphate have been employed locally for keratitis and herpes labialis, and intravenously in cases of encephalitis and disseminated herpes simplex infections. In early clinical trials, as well as experimentally in animals, adenine arabinoside monophosphate, in association with a deaminase inhibitor, appears to be the most promising therapeutic agent available (Ch. 51).

Cortisone, which is commonly employed in the treatment of inflammatory ocular lesions, is **contraindicated** in treating herpes simplex keratitis: the resulting greater viral multiplication leads to increased cellular damage, more extensive lesions, and often perforation of the cornea.

In **genital herpes,** unfortunately, the pyrimidine analogs described for herpetic keratitis have been of questionable value. Phototherapy, using dyes such as proflavine and exposure to light, has been proposed but since it has only limited effectiveness its use does not seem warranted in view of its potential carcinogenic effect.

B VIRUS (HERPESVIRUS SIMIAE)

In its natural host, the monkey, B virus produces latent infections like those of herpes simplex virus in man. However, in humans B virus can produce acute ascending myelitis and encephalomyelitis. The increased handling of monkeys and the widespread use of monkey kidney cells in the commercial production of poliovirus vaccines have augmented the transmission of this virus to humans: 24 cases have been reported in the past 25 years.

Recognition of the danger is critical, since the disease in humans is usually fatal.

B Virus closely resembles herpes simplex virus: the two are antigenically related but not identical. Antiserum from rabbits immunized with B virus neutralizes herpes simplex virus; however, Abs to herpes simplex virus neutralize B virus only slightly.

VARICELLA–HERPES ZOSTER VIRUS

The viruses isolated from patients with varicella **(chickenpox)** and from those with herpes zoster **(shingles)** are physically and immunologically indistinguishable (Fig. 55-8). Their identity has been further established by the production of typical varicella in children following inoculation of herpes zoster vesicle fluid.

PROPERTIES OF THE VIRUS

Varicella–zoster virus is relatively unstable. Even at −40° to −70°C, the infectivity of virus from tissue cultures cannot be maintained reliably for longer than 2 months. Vesicle fluid from a patient, however, remains

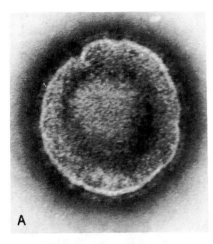

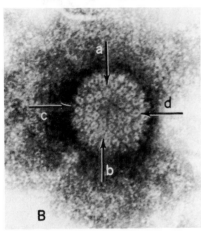

FIG. 55-8. Morphology of varicella–herpes zoster virus particles embedded in sodium phosphotungstate. **A.** Intact virion showing the envelope with surface projections and the centrally placed capsid. **B.** Viral particle in which the envelope has ruptured, revealing the structure of the capsid more clearly. **Arrows** point to capsomers situated on axes of fivefold symmetry. (Almeida JD et al: Virology 16:353, 1962) (×200,000, reduced)

infectious for many months at −70° C, perhaps because of the high titers of virus and the high concentrations of protein.

Immunologic Characteristics. Agar diffusion reveals five soluble Ags in both vesicle fluids and tissue culture preparations. In varicella, IgM Abs usually appear in the serum 2–3 days after onset of the exanthem; IgG Abs soon replace the IgM and continue to increase in titer for about 2 weeks. Titers of CF Abs decrease over a period of months, but neutralizing, immune adherence, and hemagglutinating Abs, as well as Abs to infected cell membrane Ags (detected by immunofluorescence), persist for many years after the primary infection. Cellular immunity, detected by lymphocyte responsiveness to the viral membrane Ag, also persists. Most herpes zoster patients

have a relatively high titer of viral IgG Abs at the onset of the disease, an indication that herpes zoster, like recurrent herpes simplex, occurs in previously infected, partially immune individuals.

Host Range. In contrast to herpes simplex virus, varicella virus does not cause reproducible disease in experimental animals or chick embryos, but it can be propagated in cultures of a variety of human and monkey cells. Little virus is found in the culture fluid, but stable virus can be obtained by sonic disruption of cells 24–36 h after infection. Serial propagation is also accomplished by transfer of infected cells. Characteristic cytopathic effects and eosinophilic intranuclear inclusion bodies develop (Fig. 55-9), and metaphase arrest and chromosomal aberrations have been observed.

Viral Multiplication. Multiplication of varicella–zoster virus is confined to the nucleus, and the developmental stages are similar to those of herpes simplex virus; the biochemical details have not yet been described. Infectious virus after assembly does not emerge readily from infected cells in culture. Natural infections, however, are highly contagious, and the virus is present extracellularly in high titer in the vesicle fluid of lesions.

FIG. 55-9. Typical eosinophilic inclusion bodies **(single arrows)** in nuclei of human embryonic cells infected with varicella–herpes zoster virus. Poorly differentiated, pale eosinophilic bodies are also present in the paranuclear area of several cells **(double arrows).** (Weller TH et al: J Exp Med 108:843, 1958) (×1260)

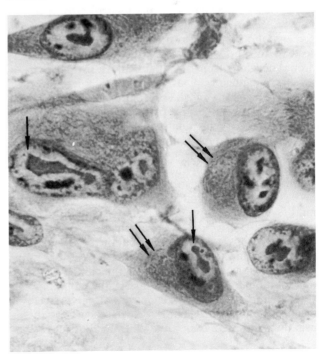

PATHOGENESIS

Varicella (Chickenpox). The primary disease produced in a host without immunity, is usually a mild, self-limited illness of young children. The clinical picture strongly suggests that the virus is spread by respiratory secretions, enters the respiratory tract, multiplies locally and possibly in regional lymph nodes, produces viremia, and is disseminated by the blood to the skin and internal organs. The virus prefers ectodermal tissues, particularly in children.

After an incubation period of 14–16 days, fever occurs, followed within a day by a papular rash of the skin and mucous membranes. The papules rapidly become vesicular and are accompanied by itching. The lesions, which are painless (in contrast to herpes zoster), occur in successive crops, and all stages can be observed simultaneously.

In the infrequent adult cases the disease is more severe, often with a diffuse nodular pneumonia; the mortality may be as high as 20%. Varicella is also usually diffuse and intense in children receiving adrenocortical steroids and in persons with immune deficiencies.

The vesicles evolve from a ballooning and degeneration of the prickle cells of the skin, along with formation of giant cells with intranuclear eosinophilic inclusion bodies (Fig. 55-9). In disseminated fatal varicella, lesions containing similar giant cells appear in liver, lungs, and nervous tissue. Basically identical lesions may occur in dorsal root ganglia of patients with herpes zoster.

Herpes Zoster (Shingles) is the recurrent form of the disease, occurring predominantly in adults. It affects persons who were previously infected with the varicella-zoster virus and who possess circulating Abs. The syndrome develops from an inflammatory involvement of sensory ganglia of spinal or cranial nerves, the virus having reached the ganglia earlier during acute varicella by travel along the nerves from the involved skin. The virus appears to remain latent in ganglionic nerve cells, and during activation it probably travels back along the nerve fibers to the skin.

Herpes zoster usually has a sudden onset of pain and tenderness along the distribution of the affected sensory nerve (frequently an intercostal nerve), accompanied by mild fever and malaise. A vesicular eruption, similar in pathology to varicella (except for its distribution), then occurs in crops along the distribution of the affected nerve; it is almost always unilateral. The vesicular eruption may last as long as 2–4 weeks; the pain may persist for additional weeks or months. If the inflammation spreads into the spinal cord or cranial nerve paralysis results. Meningoencephalitis, which occurs rarely, is usually manifested as an acute illness with severe symptoms (headache, ataxia, coma, convulsions), but most patients recover completely.

Clinically, herpes zoster may be activated by trauma, by injection of certain drugs (arsenic, antimony), or by tuberculosis, cancer, or leukemia. Moreover, in persons with immunodeficiency states, particularly when these are induced in the therapy of lymphoproliferative diseases, the activated virus may disseminate and cause serious, often fatal, illness.

LABORATORY DIAGNOSIS

A clinical diagnosis of either varicella or herpes zoster seldom offers serious difficulties, but rarely the identification of chickenpox may require laboratory tests. This can be done most rapidly and easily by preparing a smear from the base of a vesicle and staining (by the Giemsa method) to detect typical varicella giant cells and cells with characteristic inclusion bodies. Immunofluorescence, agar gel immunodiffusion, and electron microscopic examination of vesicular fluid provide rapid confirmation. In addition, virus may be isolated in cultured human or monkey cells and identified by serologic technics. Antibody determinations on paired sera from patients may also be useful. Indirect immunofluorescence, immune adherence, hemagglutination, and the enzyme-linked immunosorbent assay (ELISA) technics provide the most sensitive and rapid Ab assays. Antibody titrations are also useful to demonstrate susceptibility in seronegative adults or immunologically crippled children after exposure, as a basis for early prophylaxis with immune globulin.

EPIDEMIOLOGY AND CONTROL

Varicella virus is usually transmitted in respiratory secretions, producing a highly communicable disease with a high clinical attack rate. Epidemics are common among children, especially in the winter and spring. Second attacks of chickenpox apparently do not occur. Herpes zoster, in contrast, is of low incidence, is not seasonal, is recurrent, and is predominantly confined to persons over 20 years of age. As has been pointed out, a case of herpes zoster may initiate an outbreak of chickenpox, and contact with chickenpox is said to provoke attacks of shingles in partially immune individuals.

There are no widely applicable means for preventing and controlling varicella infection. An attenuated virus vaccine has been developed but not fully tested in controlled trials. In view of the seriousness of the disease in adults, however, one might question the wisdom of attempts to prevent infection in children unless the preventive procedure can offer as lasting protection as the natural disease. As in measles (see Measles, Prevention and Control, Ch. 59), chickenpox can be prevented or modified by administering high-Ab-titer IgG to contacts within 72 h of exposure. Prophylaxisis is of particular importance for susceptible adults and for children with impaired immunity.

Adenine arabinoside has been used with apparent success to treat disseminated disease in the seriously ill (particularly in immunologically suppressed patients), but additional control studies are necessary.

CYTOMEGALOVIRUS (SALIVARY GLAND VIRUS) GROUP

Salivary gland virus disease of newborns is a severe, often fatal illness, usually affecting the salivary glands, brain, kidneys, liver, and lungs. M. G. Smith in 1956 isolated the causative agent. The term cytomegalovirus was applied to the group because of the large size of the infected cells and their huge intranuclear inclusion bodies. Assignment to the herpesvirus family was based on the morphology of the viral particle (Fig. 55-10), the chemical composition of the virion (Table 55-1), and the characteristics of the intranuclear inclusion body present in infected cells (Figs. 55-11 and 55-12).

Although disease is rare, infections are widespread: congenital and neonatal subclinical infections (acquired before or at birth from the mother—see Pathogenesis) occur in 5%–7% of live births. The virus commonly produces latent infections which subsequently may be activated by pregnancy, multiple blood transfusions, or immunosuppression for organ transplantation.

PROPERTIES OF THE VIRUS

Immunologic Characteristics. CF titrations show a spectrum of crossreactive Ags relating all the human cyto-

megaloviruses. Although they are not antigenically homogeneous, data are not yet available to permit classification into distinct immunologic types. Moreover, DNA–DNA renaturation kinetics and restriction endonuclease maps indicate that the nucleotide sequences of different viral strains are largely homologous, indicating the genetic basis for the close immunologic relationships. Investigation of the virion's antigenic structure has been hindered by its low reactivity and poor immunogenicity.

Humoral Abs develop relatively early during infection (IgM Abs are present at birth in the congenitally infected newborn) and persist at high levels during viral excretion. Cellular immunity, however, appears to play the major role in suppression of viral multiplication, leading to latent infection or, less commonly, to viral eradication.

Host Range. Many species of animals are infected with their own specific cytomegaloviruses, but no laboratory

FIG. 55-11. Electron micrograph of a portion of an intranuclear inclusion in a cell infected with human cytomegalovirus. Inclusion is made up of viral particles in various stages of development. Particles are composed of a central core about 40 nm in diameter, surrounded by a pale zone and, externally, by a thin membranous shell. Only a few particles have a dark central core indicating the presence of nucleic acid. (Becker P et al: Exp Mol Pathol 4:11, 1965) (×40,000)

FIG. 55-10. Enveloped full particle of human cytomegalovirus. (Wright HT Jr et al: Virology 23:419, 1964) (×405,000)

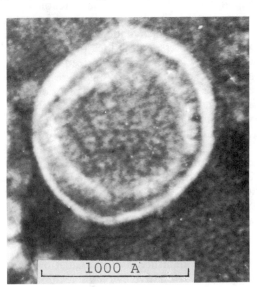

1000 A

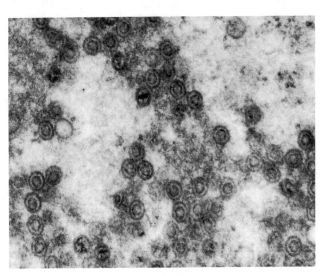

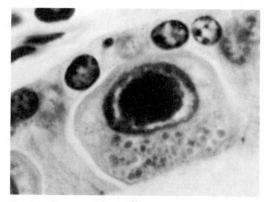

FIG. 55-12. Epithelial duct cell from a human submaxillary gland, showing typical eosinophilic nuclear and basophilic cytoplasmic inclusions produced by infection with human cytomegalovirus. (Nelson JS, Wyatt JP: Medicine 38: 223, 1959) (×1500)

animal has proved susceptible to infection with cytomegaloviruses of humans. Virus has been isolated and propagated only in cultured human fibroblasts. In vivo, however, virus appears to multiply in a variety of cell types, including many of epithelial morphology. In fact, in disseminated disease virus has been detected in cells of essentially every organ.

Viral Multiplication. Replication of the virus appears to be relatively slow, and newly made infectious particles are first detected 48–72 h after infection of cell cultures with a low virus-to-cell multiplicity. Under these conditions cytochemical and immunofluorescent technics best reveal the synthesis of viral proteins and the development of the cytologic lesions that accompany viral multiplication, namely, focal lesions, followed by generalized cytopathic changes including rounding of cells and the appearance of large intranuclear eosinophilic inclusion bodies (Fig. 55-12).

De novo biosynthesis of DNA and accumulation of early and late viral proteins are detected initially in the nucleus. Electron microscopic studies show that viral particles, like herpes simplex virions, are assembled in the nucleus (Fig. 55-11), attain their envelope at the nuclear membrane, and migrate through reduplications of the nuclear membrane into the cytoplasmic endoplasmic reticulum. The maturation of viral particles appears to be inefficient: only rare completely assembled virions can be detected among many incomplete particles (many of these are noninfectious dense bodies formed by enveloped viral proteins without DNA or assembled capsids). Hence the yield of infectious virus in cell cultures is low, and as many as 10^6 particles are needed to initiate infection of a new culture. Most infectious virus remains cell-associated, and therefore the addition of intact infected cells to a culture initiates viral propagation most efficiently.

PATHOGENESIS

A mother with a latent infection may transmit cytomegalovirus to the fetus, either by transplacental transfer during pregnancy or by excreting virus into the genital tract at the time of birth. Protracted viral shedding may follow (although neutralizing Abs develop), and the infant carrier may serve as a source for dissemination (in different studies 15%–60% of healthy children were shown to excrete virus during the first year of life). Although prolonged viral persistence is common following congenital infection, indefinite persistence is rare, and viral shedding is unusual in adults with Abs. Virus may also become latent in the newborn or during early childhood. When manifest illness does rarely occur in newborns and infants up to 4 months of age, it usually has a relentless progression, with hepatic and renal insufficiency, pneumonia, neurologic symptoms, and eventual death.

Patients with neoplastic diseases and recipients of organ transplants, subjected to corticosteroids or other immunosuppressive drugs, are particularly susceptible to viral activation or exogenous infection that results in localized or disseminated disease or inapparent infection. A syndrome resembling infectious mononucleosis (see below) may be observed in recipients of multiple transfusions of blood from latently infected donors. The syndrome has most frequently been reported in patients who have undergone open-heart surgery, probably because of the large volumes of blood they receive.

Like adenoviruses, cytomegalovirus can be isolated from explants of apparently normal adenoids and salivary glands cultured in vitro (see Pathogenesis, Ch. 54). Similar findings have been made with cytomegaloviruses of mice, rats, hamsters, and guinea pigs.

The pathologic lesion is characterized by necrosis and pathognomonic cellular alterations. The affected cells are greatly increased in size; the nucleus is enlarged and contains a brightly stained eosinophilic inclusion body up to 15μm in diameter (Fig. 55-12), larger than that produced by any other virus infecting man. In addition, the cytoplasm may be swollen and vacuolated and may show up to 20 minute basophilic and osmiophilic structures 2μm–4μm in diameter; these contain DNA and polysaccharide and therefore are positive with Feulgen and periodic acid–Schiff stains.

LABORATORY DIAGNOSIS

Infection can be identified by viral isolation, immunologic assays, and exfoliative cytologic technics:

1) **Isolation of virus** in cultures of human embryonic fibroblasts is the most sensitive method to detect infection in the newborn, but it is also the most expensive and cumbersome.

2) An **immunofluorescent assay** is also suitable for diagnosis in the newborn because it can distinguish the baby's IgM Abs from the maternal IgG Abs.

3) An **indirect hemagglutination test,** using tanned sheep RBCs to adsorb viral Ag, is sensitive and capable of detecting IgM and IgG Abs, and is rapid and inexpensive.

4) **Complement fixation** is the least expensive and most convenient method to assay Abs in older children and adults.

5) Identification of characteristic **cytomegalic cells** with intranuclear and cytoplasmic **inclusions** (particularly in urinary sediment and bronchial and gastric washings) offers an efficient and inexpensive diagnostic procedure. Detection of viral Ags by indirect immunofluorescence or the immunoperoxidase technic can rapidly confirm the cytologic diagnosis.

EPIDEMIOLOGY AND CONTROL

Infection with human cytomegalovirus appears to be worldwide and common despite the relative rarity of clinical disease. From 10% to 18% of all stillborns show characteristic lesions at autopsy. In adults above 16 years of age typical inclusions in salivary glands are rare; but in a sample taken in the United States, CF Abs were found in 53% of the population between 18 and 25 years old and in 81% of those over 35 years of age. Transplacental passage, infection at birth, and blood transfusion are the most apparent mechanisms for transmitting virus, but person-to-person spread in urine and respiratory secretions seems likely. Among pregnant women 10%–15% excrete virus during their third trimester, and at least 1% of newborns enter the world with viruria. Children and adults with immunologic deficiencies, naturally or iatrogenically acquired, are particularly susceptible to active disease. Vaccines consisting of attenuated strains are now being studied, but effective measures for prevention and control probably will not be available until more is known of viral characteristics and transmission.

EB (EPSTEIN-BARR) VIRUS AND INFECTIOUS MONONUCLEOSIS

In a search for the cause of Burkitt lymphoma, Epstein and Barr in 1964 observed herpeslike viral particles (termed **EB virus;** Fig. 55-13) in a small proportion (0.5%–10%) of the lymphoma cells repetitively cultured in vitro. The possible causal relation of EB virus to the malignancy is uncertain and will be discussed in Chapter 66.

Nevertheless, EB virus does appear to be the cause of infectious mononucleosis. The first indication of this relationship came 4 years after the discovery of EB virus, when G. and W. Henle discovered that lymphocytes from their technician, who had just recoverd from infectious mononucleosis, could be serially cultured in vitro, unlike normal lymphocytes; and a small number of these cultured cells were found to contain EB virus. A number of further observations supported the conclusion that EB virus is the etiologic agent of infectious mononucleosis: 1) The virus persists in lymphocytes cultured from patients long after recovery from infectious mononucleosis. 2) Transfusion of blood from such individuals produces the disease in recipients devoid of Abs. 3) Patients with infectious mononucleosis develop neutralizing and CF Abs, as well as Abs that react with EB-virus–infected cells in an immunofluorescence assay. 4) The Abs persist for years and their presence or absence is correlated with resistance or susceptibility to infectious mononucleosis. Finally, 5) EB virus has been isolated from pharyngeal secretions of patients with infectious mononucleosis.

PROPERTIES OF THE VIRUS

Virions purified from cultured Burkitt lymphoma cells (Fig. 55-14) are structurally similar to those of other herpesviruses (Fig. 55-1; Table 55-1). The virion contains about the same number and size proteins as herpes simplex virus. However, viral DNA from lymphoid cell lines derived from infectious mononucleosis patients, though of the same size, differs in about 35% of the sequences when compared to EB virus DNA derived from Burkitt lymphomas. Immunologically, EB virus is unrelated to other herpesviruses.

The virus multiplies in cultured lymphoreticular cells, but it is difficult to use cell-free virus for serial passage—possibly because almost all the particles are structurally defective, as judged by electron microscopic examination. Viral infectivity can be measured only by transformation of B lymphocytes from humans and a few other primate species, and even in these cells the viral yield is low. Hence it is difficult to study the biochemistry of viral propagation. The characteristic ordered sequence of biosynthetic events is apparent, however, when one measures the synthesis of viral proteins immunologically. Early Ags are made prior to and independently of DNA replication, and late Ags (i.e., viral capsid proteins) are synthesized after replication of DNA has begun. In addition, immunofluorescence detects both a virus-specific membrane Ag and a nuclear Ag which are also present in virus-free transformed cells (Ch. 66).

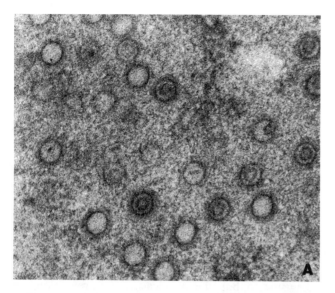

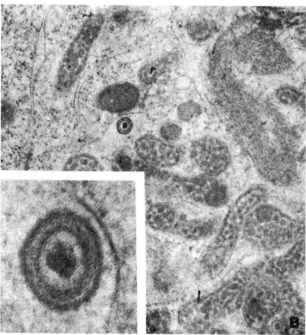

FIG. 55-13. Structure of EB virus in cells cultured from Burkitt lymphoma. **A.** Numerous developing immature particles in a thin section of a lymphoblast nucleus. (×76,500) **B.** Mature viral particle with envelope, capsid, and nucleoid in the cytoplasm. Epstein MA et al: J Exp Med 121:761, 1965) (×42,000; inset, ×213,500)

INFECTIOUS MONONUCLEOSIS

Pathogenesis. Infectious mononucleosis is an acute infectious disease, primarily affecting lymphoid tissue throughout the body. It is characterized by the appear-

ance of enlarged and often tender lymph nodes, an enlarged spleen, and abnormal lymphocytes in the blood (from which the disease derives its name). In addition, fever and sore throat are common. Occasionally other diverse manifestations are observed, including mild hepatitis, signs of meningitis or other central nervous system involvement, hematuria, proteinuria, thrombocytopenic purpura, and hemolytic anemia.

Epidemiologic and clinical evidence suggests that the virus enters the body in respiratory secretions, multiplies in local lymphatic tissues, and spreads into the blood stream, where it infects and multiplies in lymphocytes. Affected lymphocytes, unlike normal lymphocytes, can be readily cultured in vitro even after recovery from acute infections. This finding suggests that latent infections with EB virus are common and may be responsible for the persistence of specific Abs.

Laboratory Diagnosis. Increases in the titers of Abs to early Ags and to the viral capsid proteins of EB virus are required to establish the specific diagnosis.

The **indirect immunofluorescence assay** uses as capsid Ag a lymphoma-derived cell line producing EB virus, and as early Ag a nonproducer cell line superinfected with EB virus. **Complement-fixation** titrations use a homogenate of an infected continuous cell line as Ag. The technics are specific and sensitive, but they are highly specialized and not yet suitable for the general hospital laboratory.

The practical laboratory diagnosis rests upon two unique findings:

1) **Abnormal, large lymphocytes** with deeply basophilic, foamy cytoplasm and fenestrated nuclei appear, often accounting for 50%–90% of the circulating lymphocytes. Initially there is a leukopenia, but by the second week of disease the count may rise to 10,000–80,000 cells/mm^3.

2) **Heterophil Abs,** i.e., agglutinins for sheep RBCs, develop in 50%–90% of patients during the course of the disease.

The immunogen eliciting the heterophil Abs is unknown, but it appears to be distinct from the EB virus. Thus a small percentage of patients develop EB virus Abs but no heterophil Abs; heterophil Abs can be adsorbed from serum by beef RBCs without reducing the Ab titer for EB virus; and heterophil Abs are transient but Abs to EB virus persist for years, perhaps for life.

Heterophil Abs also appear during serum sickness following injection of horse serum, and they may be present in serum from healthy persons. However, a differential adsorption test of a patient's serum will distinguish these Abs. Thus heterophil Abs of infectious mononucleosis are adsorbed by beef RBCs but not by guinea pig kidney (which contains Forssman Ag); serum sickness agglutinins are adsorbed by both beef RBCs and guinea pig

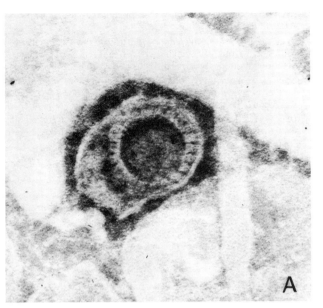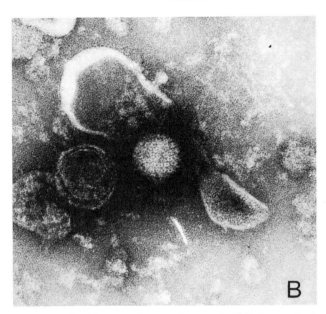

FIG. 55-14. Structure of EB virus obtained from cultured Burkitt lymphoma cells. Electron micrographs are of purified, negatively stained virus. **A.** Empty capsid enclosed in envelope. (×200,000) **B.** Viral particle with disrupted envelope. A capsid (about 75 nm in diameter) similar to that of herpesviruses is clearly visible. (Courtesy of Dr. Klaus Hummler) (×120,000)

kidney; and normal serum heterophil agglutinins are adsorbed by guinea pig kidney but not usually by beef RBCs.

Epidemiology and Control. Infectious mononucleosis appears to be primarily a disease of relatively affluent teenagers and young adults (such as college students), in whom it causes proved disease in about 15% of the susceptible population. (About 75% of entering college students in the United States are free of detectable Abs.) The peak incidence is at 15–20 years of age. The etiologic relation of EB virus to infectious mononucleosis furnished an explanation for these epidemiologic observations: members of lower socioeconomic groups develop EB virus Abs at an early age, when infections may be asymptomatic, whereas children at higher economic levels are not exposed to early infections and thus are more liable to develop disease from later infections.

Successful infection appears to require extensive exposure, or unknown cooperating factors, for multiple cases in families are infrequent despite the presence of virus in pharyngeal secretions of infected persons. In cases of infectious mononucleosis the source of the infecting virus is rarely obvious, perhaps because individuals with latent infections produce the virus in pharyngeal and oral secretions for prolonged periods after recovery. The common association of infectious mononucleosis with kissing may reflect a requirement for a large inoculum. No adequate control procedures are available.

SELECTED READING

BOOKS AND REVIEW ARTICLES

HANSHAW JB: Cytomegaloviruses. Virol Monogr 3:1, 1968

KAPLAN AS (ed): The Herpesviruses. New York, Academic Press, 1973

NAHMIAS AJ, ROIZMAN B: Infection with herpes-simplex virus 1 and 2. N Engl J Med 286:667, 719, 781, 1973

PAGANO JS: Diseases and mechanisms of persistent DNA virus infection: latency and cellular transformation. J Infect Dis 132:209, 1975

ROIZMAN B, FURLONG D: The replication of herpesviruses. In Fraenkel-Conrat H, Wagner RR (eds): Comprehensive Virology, Vol 3. New York, Plenum Press, 1974, p 229

STEWART JA, HERRMANN KL: Herpes simplex virus. In Rose NR, Friedman H (eds): Manual of Clinical Immunology. Washington DC, American Society for Microbiology, 1976

WELLER TH: The cytomegaloviruses: ubiquitous agents with protean clinical manifestations. N Engl J Med 285:203, 267, 1971

SPECIFIC ARTICLES

ALMEIDA JD, HOWATSON HF, WILLIAMS M: Morphology of varicella (chickenpox) virus. Virology 16:353, 1962

BEN-PORAT T, KAPLAN AS: Studies on the biogenesis of herpesvirus envelope. Nature 235:165, 1972

BUDDINGH GJ, SCHRUM DI, LANIER JC, GUIDRY DJ: Natural history of herpetic infections. Pediatrics 11:595, 1953

COOK MI, STEVENS JC: Replication of varicella-zoster virus in cell culture: an ultrastructural study. J Ultrastruct Res 32:334, 1971

EPSTEIN M, BOAV Y, ACHONG B: Studies with Burkitt's lymphoma. Wistar Inst Symp Monogr 4:69, 1965

HENLE G, HENLE W, DIEBL V: Relation of Burkitt's tumor-associated herpes-type virus to infectious mononucleosis. Proc Natl Acad Sci USA 59:94, 1968

OLDING LB, JENSEN FC, OLDSTONE MBA: Pathogenesis of cytomegalovirus infection. I. Activation of virus from bone marrow-derived lymphocytes by *in vitro* allogenic reaction. J Exp Med 141:561, 1975

SAWYER RN, EVANS AS, NIEDERMAN JC, MCCOLLUM RW: Prospective studies on a group of Yale University freshmen. I. Occurrence of infectious mononucleosis. J Infect Dis 123:263, 1971

SHELDRICK P, BERTHELOT N: Inverted repetitions in the chromosome of herpes simplex virus. Cold Spring Harbor Symp Quant Biol 39:667, 1974

WADSWORTH S, HAYWARD GS, ROIZMAN B: Anatomy of herpes simplex virus DNA. V. Terminally repetitive sequences. J Virol 17:503, 1975

chapter 56

POXVIRUSES

The smallpox was always present, filling the churchyards with corpses, tormenting with constant fears all whom it had striken, leaving on those whose lives it spared the hideous traces of its power, turning the babe into a changeling at which the mother shuddered, and making the eyes and cheeks of the bethrothed maiden objects of horror to the lover.

MACAULAY TB: The History of England from the Accession of James II, Vol. IV

Since the beginning of history, smallpox* has left its indelible mark on the medical, political, and cultural affairs of man. Records show severe epidemics from earliest times. Indeed, by the eighteenth century the disease had become endemic in the major cities of Europe. Terror of its presence was such that "no man dared to count his children as his own until they had had the disease."†

Because virulent smallpox strains appear to have no animal reservoir, and because vaccination is very effective, complete eradication of this scourge—long a dream of public health authorities—has now become a virtual reality as a result of the vigorous WHO program of confinement and immunization initiated in 1966. Whereas in 1945 most of the world's inhabitants lived in endemic areas, October 1977 saw the last reported case of natural infection, detected in Somalia. Immunization has eliminated smallpox from the rest of the world (including India and Pakistan, where severe epidemics still occurred in 1973), and no documented case has been reported in the United States since 1949. However, with the disappearance of the disease, and with the decline in enforced vaccination, susceptibility will slowly return, bringing a liability to massive epidemics if a source of infection should appear.

It is still too early to be certain that this virus is buried in the biologic graveyard of extinct organisms. Moreover, the undetected introduction of smallpox into a country, from a persistent, unrecognized focus, is facilitated by modern, rapid transportation and by physicians' unfamiliarity with the disease. Hence continued knowledge, constant vigilance, and effective emergency measures are still necessary until we can be sure that smallpox is truly eliminated from the earth. Indeed, to reduce the risk of escape of this virus (e.g. from laboratory accidents such as the one that occurred in England in 1978)‡ stocks of smallpox virus throughout the world have been destroyed. Thus, in the United States only a reference stock has been preserved, under stringent controls, at the Center for Disease Control in Atlanta.

GENERAL CHARACTERISTICS

The smallpox (variola) virus is representative of the **poxviruses,** a group of agents that infect both humans and lower animals and produce characteristic vesicular skin lesions, often called **pocks.** Poxviruses (family Poxviridae) are the largest of animal viruses (Table 56-1): they can be seen with phase optics or in stained preparations with the light microscope. The viral particles (originally called **elementary bodies**) are somewhat rounded, brick-shaped, or ovoid, and have a complex structure consisting of an internal central mass, the nucleoid, surrounded by two membrane layers (Figs. 56-1 through 56-3). The surface is covered with ridges which may be tubules or threads (Fig. 56-2). Poxviruses contain DNA, protein, and lipid. They are relatively resistant to inactivation by common disinfectants and by heat, drying, and cold.

All poxviruses studied are related immunologically by a common internal Ag extractable from viral particles by 1 N NaOH. They can be divided into genera on the basis of their more specific Ags, nucleic acid homology, morphology, and natural hosts.

The genus **Orthopoxvirus** consists of **viruses of certain mammals,** including variola, vaccinia, cowpox, ectromelia of mice, rabbitpox, and monkeypox. Other genera include viruses specific for birds (Avipoxvirus), ungulates (Capripoxvirus), and arthropods (Entomopoxvirus), and the tumor-producing (fibroma and myxoma) viruses of rabbits (Leporipoxvirus). Viruses of a sixth genus which resemble other poxviruses in structure but not immunologically (hence termed **Parapoxviruses**) include contagious pustular dermatitis (orf), paravaccinia (milker's nodules), and bovine papular stomatitis viruses. Some poxviruses, such as the molluscum contagiosum virus and Yaba monkey tumor virus cannot be classified immunologically in any of these genera.

‡ In August 1978 two cases of smallpox occurred in Birmingham, England, as a result of transmission of the virus to a room over a laboratory in which smallpox viruses were being investigated.

---

* A term initially employed to distinguish the disease from "large pox" (syphilis).
† The Comte de la Condamine, an eighteenth century French mathematician and scientist.

TABLE 56-1. Characteristics of the Poxviruses

Size	250–390 nm × 200–260 nm
Morphology	Brick-shaped to ovoid (Figs. 56-1, 56-11, 56-12)
Protein and lipid content	Present
Stability	Relatively resistant to inactivation by chemicals (disinfectants) or by heat, cold, or drying; inactivated by chloroform, variable inactivation by ether
Nucleic acid	Double-stranded DNA*
Mol wt and base ratio	Vaccinia DNA = 150 × 10⁶ daltons (231 Kbp); AT/GC = 1.67†
	Fowlpox DNA = 200 × 10⁶ daltons (307 Kbp); AT/GC = 1.84
Antigenicity	Common family and genus Ags
Multiplication‡	In cytoplasm of cells
Cytopathogenicity	Predilection for epidermal cells; eosinophilic inclusion bodies produced

Kbp = kilobase pairs.

*Complementary strands of vaccinia DNA are covalently crosslinked at or near the termini. Other poxviruses have not been examined for this structure.

†DNAs of cowpox, rabbitpox, and mousepox are similar.

‡A DNA-dependent RNA polymerase in virion.

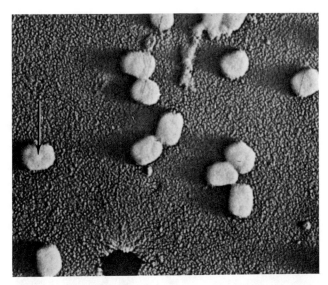

FIG. 56-1. Morphology of vaccinia virus as revealed by electron microscopic examination of shadowed preparation of purified viral particles. Note the central core surrounded by a depression **(arrow).** (Sharp DG et al: Proc Soc Exp Biol Med 61:259, 1946) (×28,000)

Poxviruses possess the unusual capacity to **reactivate** other members of the family previously inactivated by heat or 6 M urea* when both infectious and inactivated viruses infect the same cells. Reactivation is accomplished by the infectious viral particles producing the enzyme which uncoats the inactivated virus (see Ch. 50) and frees its intact DNA from the denatured viral coat, permitting it to express its genes and multiply. **Genetic recombination** may also occur, but only between two viruses of the same immunologic group **(homoreactivation).** The proportion of recombinants seems to be related to the degree of homology between the viral DNAs and parallels the degree of antigenic cross-reactivity.

Poxviruses vary widely in their ability to cause generalized infection, but they share a **predilection for epidermal cells,** in which they multiply in the cytoplasm and produce eosinophilic inclusion bodies (termed **Guarnieri bodies).** Fibroma and myxoma viruses have, in addition, a great affinity for subcutaneous connective tissues. Most poxviruses also multiply readily in epidermal cells of the chorioallantois of chick embryos, where they produce characteristic **nodular focal lesions,** termed **pocks** (Fig. 56-4). These lesions reflect a second characteristic of poxviruses: the propensity to cause cellular hyperplasia before cell necrosis. With myxoma, fibroma, and Yaba viruses the hyperplasia predominates, and tumors develop.

VARIOLA (SMALLPOX) AND VACCINIA

PROPERTIES OF THE VIRUSES

The virulence and contagiousness of smallpox virus have understandably limited its laboratory investigation. On the other hand the closely related and much less dangerous vaccinia virus is one of the most thoroughly investigated animal viruses. Since the two viruses are very similar they are discussed together.

* In virus capable of reactivation the protein capsid has been denatured by such treatment, but the viral DNA is unaffected.

MORPHOLOGY

Poxviruses have a complex architecture without obvious symmetry. Vaccinia virions are brick-shaped particles with rounded corners and a central dense region with crescentic dense areas on each side (Fig. 56-1). The viral surface layer, revealed by negatively stained and freeze-etched specimens (Fig. 56-2), is composed of threadlike, double-ridged, beaded structures that appear to curve around the particle. Thin sections (Fig. 56-3) disclose a central nucleoid with a dumbbell-shaped dense core

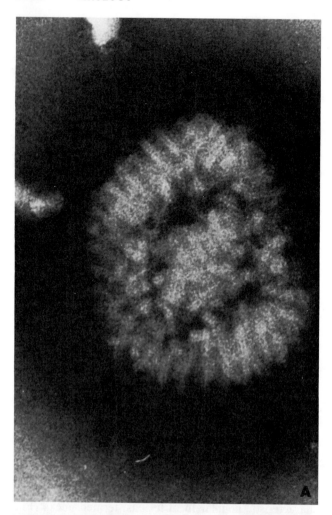

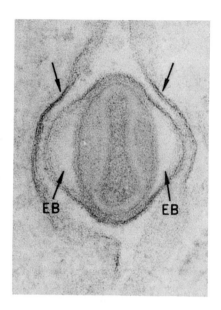

FIG. 56-3. Thin section of intact mature vaccinia virus particle showing the inner nucleic acid core and surrounding membranes. The elliptical body **(EB)** on either side of the nucleoid causes a prominent central bulging of the virion. Viral particle is lodged between two cells **(arrows).** (Dales S: J Cell Biol 18:51, 1963) (×120,000)

composed of regularly arranged, DNase-sensitive, electron-dense threadlike structures. The nucleoid is surrounded by lipoprotein membranes. Between the nucleoid and the outer viral coat is an ellipsoidal body which causes the central thickening of the virion. Viral particles from smallpox crusts and vesicle fluids are morphologically indistinguishable from vaccinia particles.

CHEMICAL AND PHYSICAL CHARACTERISTICS

Vaccinia virus was the first animal virus to be prepared in sufficient quantity and purity for detailed chemical and physical measurements. The viral particles are composed of about 3.2% DNA associated with spermidine, 91.6% protein, 5% lipid (including cholesterol, phospholipid, and neutral fat), and 0.2% non-DNA carbohydrate (present in the membrane glycoproteins).

FIG. 56-2. Fine structure of mature vaccinia virion from a purified suspension of viral particles. **A.** Virion negatively stained with phosphotungstate. The double tract of ridges and a suggestion of the subunit structure can be seen. The protrusion and sawtooth effect of the ridges is noticeable at the periphery. (×224,000) **B.** Freeze-etched virion, showing subunits, length, and random orientation of the double ridge. (Medzon EL, Bauer H: Virology 40:860, 1970) (×224,000)

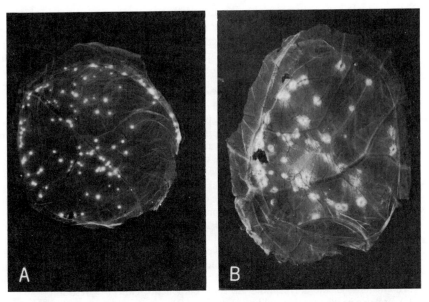

FIG. 56-4. Pocks produced by variola virus **(A)** and vaccinia virus **(B)** 3 days after viral inoculation of the chorioallantoic membranes of chick embryos. Note the small gray-white pocks produced by variola virus and the large pocks made by vaccinia virus. (Downie AW, MacDonald A: Br Med Bull 9:191, 1953)

The viral cores contain several enzymes, primarily for transcription of the **immediate early mRNA**: DNA-dependent RNA polymerase, polyadenylate polymerase, methyl- and guanylyltransferases, two nucleoside triphosphate phosphohydrolases, single-stranded DNA endonuclease and exonuclease, and a protein kinase.

Stability. Variola and vaccinia viruses are relatively stable; they resist drying and retain their infectivity for many months at $4°C$ and for years at $-20°C$ to $-70°C$. Thus exudate or crusts taken from smallpox patients may yield infectious virus after almost a year at room temperature, and diagnostic specimens need no refrigeration. The persistence of infectious variola virus on bedclothes causes a hazard not only for medical personnel but even for laundry workers. The relative resistance of the virus to dilute phenol and other common disinfectants complicates the decontamination of clothing, instruments, furniture, etc. However, variola and vaccinia viruses are inactivated by apolar lipophilic solvents (e.g., chloroform), by autoclaving, or by heating at $60°C$ for 10 min.

ANTIGENIC STRUCTURE AND IMMUNITY

The antigenic structure of poxviruses has been examined largely with vaccinia virus, but smallpox virus is very similar. Viral strains from cases of severe smallpox (variola major) and those from cases of variola minor (alastrim) are immunologically indistinguishable.

The Ags of poxviruses can be measured by all the usual immunologic technics, including hemagglutination inhibition and neutralization of infectivity. Viral neutralization is determined by inhibition of cytopathic effects on cells in tissue culture, by prevention of lesions in rabbit skin, or, most quantitatively, by reduction in the number of either pocks on the chorioallantoic membrane of the chick embryo or plaques on a monolayer of cultured cells.

Rivers and his colleagues, in the 1930s, began dissecting the complex antigenic structure of vaccinia viruses, as had been done earlier with bacteria (particularly the pneumococcus).* Their pioneering investigations revealed two multicomponent groups of major Ags, consisting of the DNA-containing core (termed the NP or nucleoprotein Ag) and coat material (called the LS Ag because it contains both heat-labile and stable components). The complex virion contains a minimum of 7

---

* The studies of this group were remarkable for the time. The investigators recognized the viral particle as the infectious unit; they developed precise assay technics; they applied statistical technics to their data; they prepared large quantities of virus in a highly purified state and determined its chemical composition; and they purified Ags from infected cells and studied their immunologic, chemical, and physical characteristics. Their data have withstood the test of time and the subsequent development of more elegant technics.

distinct major Ags revealed by immunodiffusion technics, and about 30 polypeptide chains identified by SDS–polyacrylamide gel electrophoresis, corresponding to no more than 25% of the coding capacity of the viral genome.

Stepwise dissection of the virion with trypsin or chymotrypsin, 2-mercaptoethanol, and a nonionic detergent (e.g., Nonidet P40) permits localization of some of the Ags and polypeptides. 1) Two Ags responsible for neutralizing Abs are present in the surface membrane; 2) one large and three small Ags are in the core; and 3) one large Ag has been located between the core and the surface structures. The neutralizing Abs induced by the two surface Ags neutralize only viruses from the homologous subgroup. A **family Ag,** which is a component of the core and one of the constituents of the NP fraction, can be extracted from virions with weak alkalai; it crossreacts, in complement-fixation (CF) and precipitin assays, with a similar Ag from all poxviruses. Although it is not yet possible to identify the Ags with specific polypeptide chains, most of the polypeptides have been topographically located in the virion.

The majority of the polypeptides are core components, either basic proteins needed for tight folding of the large DNA molecule or virion enzymes essential for uncoating and initiating viral multiplication.

Hemagglutinin. Extracts of variola- and vaccinia-infected cells, but not purified viral particles, hemagglutinate RBCs from about 50% of chickens (apparently a genetic trait of the bird). The hemagglutinin is a lipoprotein embedded in the membranes of infected cells and probably corresponds to a new cell surface Ag. It is detected by immunofluorescence shortly after vaccinia infection, and is also responsible for **hemadsorption** of susceptible chicken RBCs to infected cells. Hemagglutination and hemadsorption are blocked by virus-specific Abs, but these Abs do not neutralize infectious virus. In contrast to influenza viruses, vaccinia hemagglutinin does not elute spontaneously from the aggregated cells. The hemagglutination-inhibition titration may be used as a convenient and accurate procedure for diagnosis of infection.

Antibody response. Following infection or immunization, Abs develop to each of the viral Ags. The variations observed in their time of appearance and persistence depend on the nature and quantity of the Ag and not on the sensitivity of the titration technic employed. Thus, neutralizing and hemagglutination-inhibiting Abs are first detected about the sixth day after onset of illness in the unvaccinated person, whereas CF Abs ordinarily appear 2–3 days later. Neutralizing and hemagglutination-inhibition Abs persist at least 20 years after infection, but CF Abs remain less than 2 years. In those previously immunized, the various Abs generally appear 2–3 days sooner after onset of illness, reach higher titers, and persist longer.

Immunity to smallpox is long-lasting, if not persistent for life, following natural infection. In the rare reinfections that have been reported, the disease is usually atypical, very mild, and often without skin rash **(variola sine eruptione).** If infection occurs in artificially immunized individuals the clinical disease is also usually milder than in unimmunized neighbors.

The relative importance of **humoral** and **cellular immunity** remains largely unexplored. However, immunization with live vaccine has led to disease in persons with congenital defects of thymus-derived (T) cells (Ch. 51), as discussed under Complications, below.

Cell-mediated immunity may be critical for recovery from infection. For example, 80% of mice given antithymocyte serum succumbed to a subsequent mousepox virus infection, whereas only 5% of the control animals died; Abs and interferon levels were comparable in both groups. Transfer of spleen cells from immunized mice markedly reduced the viral titer, although only low levels of specific humoral Abs were attained and interferon was not detectable. The spleen cell immunity was virus-specific and appeared to be initiated by T cells: the antiviral activity was inhibited if, prior to transfer, the cells were irradiated or were reacted with antithymocyte or anti-Θ serum. Moreover, macrophages from vaccinia-immunized animals do not support multiplication of vaccinia virus, although unrelated poxviruses can replicate without restraint.

HOST RANGE

Variola virus has a much more limited host range than do vaccinia and other poxviruses. Monkeys are the only animals, other than man, known to be naturally infected; also, when variola virus is placed onto monkeys' scarified skin or inoculated intradermally, local lesions and fever follow.

FIG. 56-5. Eosinophilic cytoplasmic inclusion bodies (Guarnieri bodies) in corneal epithelial cells of rabbit eye infected with vaccinia virus. (Coriell LL et al: J Invest Dermatol 11:313, 1948) (×900)

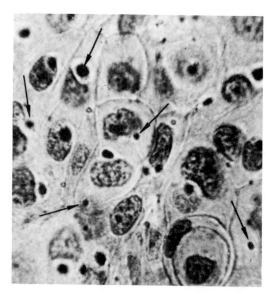

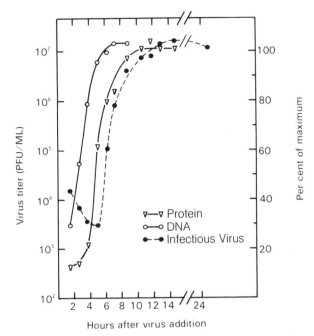

FIG. 56-6. Biosynthetic events in multiplication of vaccinia virus. (Modified from Salzman NP et al: Virology 19:542, 1963)

In rabbits variola virus can initiate keratitis and local skin lesions, but it cannot be propagated serially. On the chorioallantoic membrane of the chick embryo variola virus produces characteristic small, gray-white pocks that can be readily distinguished from the large pocks produced by vaccinia virus (Fig. 56-4). Variola minor (alastrim) and variola major viruses have the same host range but the former generally multiply to a lower titer in the chorioallantois of the chick embryo. Suckling and adult mice can also be infected.

Many animals, including the chick embryo, rabbit, calf, and sheep, are susceptible to **vaccinia virus** and have served as convenient hosts for investigation and for production of vaccine. Epidermal cells show the greatest sensitivity, but vaccinia also multiplies readily in the rabbit testis, and a number of strains have been adapted to the rabbit brain (**neurovaccinia**). Continued passage of vaccinia on the chick embryo chorioallantoic membrane tends to reduce its virulence for man and rabbits.

Cultured cells commonly used for propagation of variola virus are human embryonic kidney, monkey kidney, and continuous lines of human epitheliumlike cells such as HeLa cells. The slowly developing cytopathic effects become maximal in 5–8 days (depending on the size of the inoculum). The infected cells contain weakly eosinophilic cytoplasmic inclusion bodies. Infection may be detected even earlier by hemadsorption with susceptible chicken RBCs. Variola virus can be assayed in monolayer cultures by counting hyperplastic foci.

Vaccinia virus propagates in culture more rapidly, to a higher

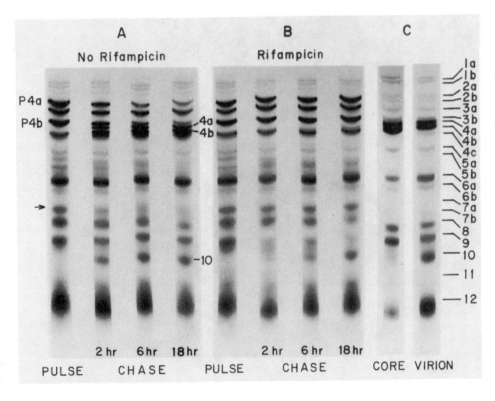

FIG. 56-7. Processing of precursor proteins during morphogenesis of vaccinia virions. Maturation is inhibited by the action rifampicin. The proteins were labeled for 20 min, the isotope was removed from the medium, and the incubation of cells continued **(A)** without rifampin, or **(B)** in the presence of rifampin (100μg/ml), which blocks the processing of two major proteins, **4a** and **4b,** from their precursors. Proteins labeled with ^{35}S-methionine are displayed by electrophoresis of infected cell extracts in polyacrylamide gels containing sodium dodecyl sulfate. **C.** Electrophoretic patterns of proteins from labeled purified cores and virions. (Modified from Moss B, Rosenblum EN: J Mol Biol 81:267, 1973).

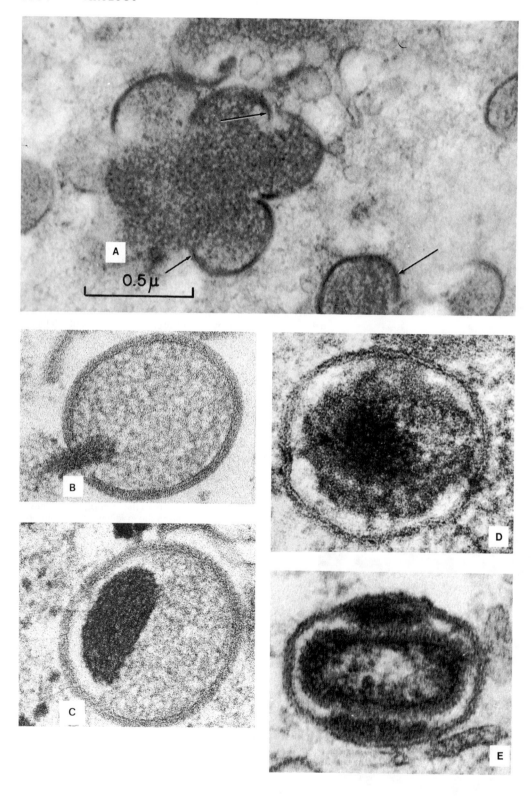

0.5µ

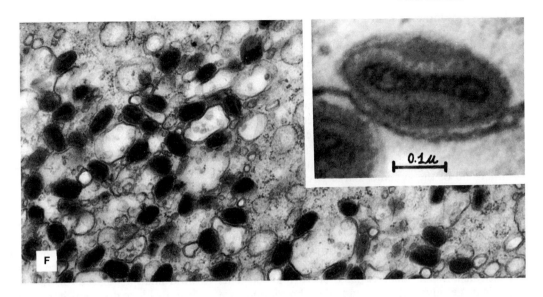

FIG. 56-8. Developmental stages in maturation and assembly of vaccinia virus. Electron micrographs of cells taken at intervals following infections. **A.** Dense trilaminar viral membranes forming within and around clumps of dense fibrillar material in the cytoplasm. (×48,000) **B.** Insertion of viral DNA and associated proteins in incomplete particles. (×150,000) **C.** Condensation of nucleoprotein within the immature particle. (×150,000) **D.** Maturation of a virion. The complete trilaminar lipoprotein envelope encloses the particle. Within is a dense nucleoid of fibrous nucleoprotein surrounded by a less dense homogeneous substance, thought to be material of the lateral bodies and core. (×170,000) **E.** A further stage in a maturing virion. The core and two lateral bodies are clearly differentiated within the viral envelope. (×170,000) **F.** Large group of mature vaccinia virions, smaller and denser than in their formative stages. (×24,000) **Inset** shows internal structure of a single particle at higher magnification. The dense dumbbell-shaped core is surrounded by a zone of lower density. (×150,000) (**A** and **F,** Dales S, Siminovich, L: J Biophys Biochem Cytol 10:475, 1961; **B** and **C,** Morgan, C: Virology 73:43, 1976, **D** and **E,** Dales S, Mosbach EG: Virology 35:564, 1968

titer, and in a wider variety of animal cells than does variola virus. Cytopathic changes are accompanied by the development of eosinophilic inclusion bodies similar to Guarnieri bodies (Fig. 56-5). Hemadsorption permits early detection of infection. Virus can be reliably quantitated by plaque assay, with or without an agar overlay,* and by pock counts on the chorioallantoic membrane.

VIRAL MULTIPLICATION

Although poxviruses contain DNA, *biosynthesis of the viral components and their assembly into viral particles take place entirely within the cytoplasm* of the cell. Moreover, the biosynthetic reactions can occur in enucleated cells, although infectious virus is not assembled.

Virus attaches to host cell receptors, whose chemical nature is still undefined, and it enters the cytoplasm by membrane-to-membrane fusion as well as by the process of engulfment (Ch. 50). After penetration, viral DNA is released by a **two-stage uncoating process** (Ch. 50). The **first stage** is initiated, almost immediately after penetration, by preexisting host cell enzymes, which break down

---

* Because infected cells are not lysed rapidly, and most of the newly synthesized virus spreads to cells in contact and is not released into the medium, an agar overlay is not necessary to permit formation of localized foci.

the viral membrane phospholipid and part of the protein coat of the viral particle to free the nucleoprotein core. The **second stage,** after a lag of 30–60 min. (depending upon viral multiplicity), results in breakdown of this core to liberate viral DNA. At the onset of this second stage, when the DNA within the intact core is still protected from DNase activity, a DNA-dependent RNA polymerase present in the core transcribes about 25% of the viral genome. The resulting transcript is processed within the core, and functional **immediate early mRNAs** emerge to code for proteins, including one, probably a proteolytic enzyme, required for the final uncoating events. The liberated viral DNA is stable and can transmit its genetic information for several hours. As the DNA is released from the core, the RNA for a second set of mRNAs, **delayed early,** is transcribed in the cytoplasm. Finally, after viral DNA replication begins, **late mRNAs** appear (derived from about 60% of the genome) and at least some early messengers continue to be made. Transcription continues until about 7 h after infection.

Synthesis of specific enzymes and of a few viral structural proteins begins early in the biosynthetic process, before replication of viral DNA. The products include the second-stage uncoating protein, three proteins associated with the nucleoprotein core, a protein essential for initiation of viral DNA replication, and enzymes related to DNA biosynthesis (thymidine kinase, DNA polymerase, nucleases, and polynucleotide ligase). Production of some of these enzymes is blocked 3–4 h later by translational control, though the mRNAs are not yet degraded (see Fig. 50-9).

Viral DNA begins to be synthesized 1.5–2 h after infection and attains its maximal concentration by the time newly made infectious virus is first detected (Fig. 56-6). The mode of **DNA replication** is imposed by the unique covalent crosslinking of the two strands (see Synthesis of DNA-Containing Viruses, Ch. 50). Synthesis is initiated at either end of the genome. Large circular and forked replicating forms are found, indicating that an endonuclease cleaves the single-stranded crosslinks at various times during replication. **Late viral proteins** are first detected about 4 h after infection, and infectious virus is formed about 1 h later by packaging viral DNA randomly selected from the preformed pool (Fig. 56-6).

Posttranslational modifications of several proteins (i.e., cleavage, glycosylation, and phosphorylation) are essential to virion maturation. Morphogenesis depends particularly upon cleavage of three major proteins, which together comprise about 35% of the virion's protein mass (Fig. 56-7A). Indeed, morphogenesis is completely blocked by rifampin, which prevents cleavage of the precursors of core proteins 4a and 4b (Fig. 56-7B).

Concomitant with the biosynthesis of virus-directed mRNA and viral DNA the following effects on **host cell macromolecules** are noted. Production of host proteins stops because initiation of polypeptide chain synthesis is blocked, and host cell polyribosomes are disrupted (liberating ribosomes for synthesis of viral proteins). Host DNA ceases replicating, but it is not degraded. Although the host mRNAs do not leave the nucleus, their synthesis continues unaltered for about 3 h, after which the rate of host transcription decreases and, by 7 h after infection, virtually stops. Processing of precursor mRNA molecules begins to decrease about 2 h after infection, and by 3 h synthesis of new ribosomes cannot be detected. How these controls of host cell biosynthesis are induced by the virus, and how they relate to cell injury, are still unknown.

The **morphologic counterparts** of the foregoing biochemical events have been observed in thin sections of infected cells (Fig. 56-8). As the viral DNA synthesis increases, regions of dense fibrous material appear in the cell cytoplasm. About 3 h after infection some of the early proteins form membrane-like structures, which begin to enclose patches of viral components and proceed to form immature particles into which DNA enters. (Fig. 56-8A and B). After the envelope is completed, the nucleoid begins to take shape within the immature particle; an additional membrane encloses the condensing DNA; the lateral bodies differentiate; and, finally, the outer coat structures are laid down on the previously

formed membrane, completing the assembly of mature virions (Fig. 56-9).

Lysis of infected cells is not a prerequisite for liberation of newly formed virions. The viral particles seem to be released through cell villi. This is apparently an inefficient process, at least in cell culture, since less than 10% of the formed virus is released. Radioautography and immunofluorescence reveal that viral materials may also be transmitted directly from cell to cell through villi.

SMALLPOX

PATHOGENESIS

Two basic forms of smallpox are recognized: **variola major,** which has a case fatality rate of approximately 25%, and **variola minor,** or **alastrim,** a less virulent form with a mortality rate below 1%. Although a variety of factors may influence the mortality rate in any epidemic, the epidemiologic evidence is convincing that severe and mild smallpox exist as distinct entities. Nevertheless, it is impossible to distinguish the viruses responsible for these two forms of the disease.

Virus multiplies first in the mucosa of the upper respiratory tract and then in the regional lymph nodes. A transient viremia then disseminates virus to internal organs (liver, spleen, lungs), where the virus propagates extensively. A second viral invasion of the blood stream terminates the **incubation period** (about 12 days) and initiates the **toxemic phase,** characterized by prodromal macular rashes, fever, generalized aching, headache, malaise, and prostration. Virus spreads to the skin and multiplies in the epidermal cells; the characteristic **skin eruption** follows in 3–4 days. Macular at onset, the rash progresses from papular to vesicular, finally becoming pustular in the second week of illness. In severe cases the rash may become hemorrhagic or confluent. The course of variola minor is similar but shorter, and the rash and other symptoms are less severe.

The inclusion bodies, or **Guarnieri bodies,** surrounded by a clear halo, characteristically develop in cells of the skin and mucous membranes infected with variola or vaccinia virus (Fig. 56-5). Each inclusion body consists of an accumulation of viral particles and viral Ags. (Similar masses are observed in cells infected with other poxviruses.) More than one may be present in a single infected cell.

A **hypersensitivity** response to the viral Ags may contribute to the eruptive lesions of smallpox, for when the Ab response and hypersensitivity are inhibited in infected rabbits by x-irradiation or cytotoxic drugs, the characteristic pustules do not form, although viral multiplication is unrestricted. The **toxinlike properties** of the viral particles may also play a role in the cell necrosis. It is clear, however, that pustule formation does not result from secondary bacterial infection, since bacterial cultures of the pustule fluid are ordinarily sterile.

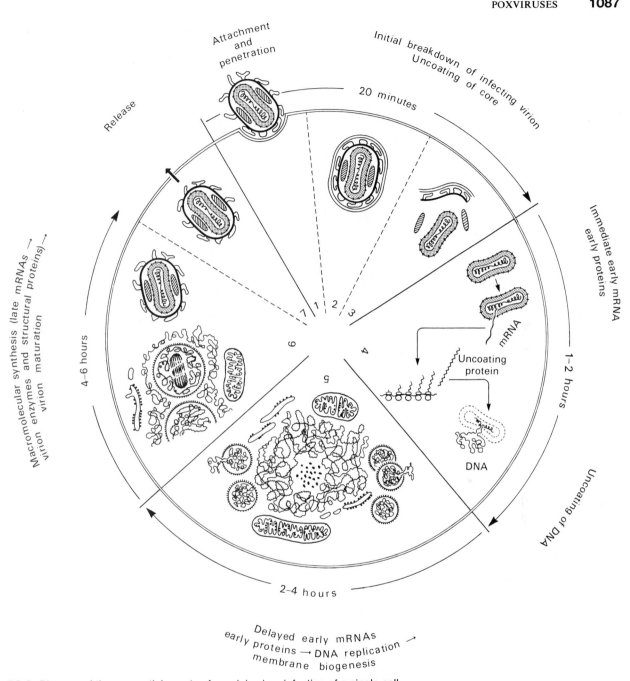

FIG. 56-9. Diagram of the sequential events of vaccinia virus infection of a single cell.

LABORATORY DIAGNOSIS

If smallpox has not been wholly eradicated, the recognition of sporadic cases will increasingly depend upon laboratory procedures as doctors become less familiar with the clinical manifestations. These procedures either identify or isolate the virus, detect viral Ag, or measure a rise in titer of specific Abs. The recommended procedures and the specimens to be collected depend upon the stage of the suspected disease (Table 56-2). The stability of the infectious virus and its Ags facilitates laboratory diagnosis because materials can be transported without danger of inactivation.

An early presumptive diagnosis can be made most rapidly by visualizing **viral particles** (elementary bodies) by electron microscopy, or by light microscopy in a Giemsa-stained smear of scrapings from the bases of skin lesions. Virus is abundant in lesions during the active disease; but it is rarely detected during the incubation period and prodromal stage before the rash appears. A positive smear is decisive, but a negative result calls for further tests. Identification can be confirmed by viral isolation or by immunologic technics.

Virus can be **isolated** most efficiently by inoculating blood or material from skin lesions onto the chorioallantoic membrane of the 12- to 14-day-old chick embryo: small grayish white pocks ap-

pear within 2–3 days (Fig. 56-4). The agent producing the pocks can be specifically identified by procedures summarized in Table 56-2. The development of pocks also affords a reliable quantitative viral assay.

The detection of soluble **viral Ags** (Table 56-2) is probably the most useful, economic, and efficient of the diagnostic procedures. Serologic diagnosis is generally reserved for atypical or mild cases occurring in partially immune individuals (e.g., **variola sine eruptione**). Since specific **Abs** measured by neutralization and by hemagglutination-inhibition titrations may persist for years following vaccination only a fourfold or greater increase is considered diagnostic in a previously immunized person. A positive CF reaction, however, rarely persists longer than 6–8 months after vaccination. Radioimmune assays correlate with viral neutralization but are more sensitive and precise.

EPIDEMIOLOGY

Smallpox is confined to man and is spread chiefly by person-to-person contact. Although smallpox is considered to be highly contagious, spread is slow, and the probability of infection from a single exposure appears to be low.

Initially the virus is transmitted from the lesions of the upper respiratory tract in droplet secretions or by con-

TABLE 56-2. Diagnosis of Variola by Laboratory Tests

| | Stage of variola and material tested | | | | | | | | | | |
| | Pre-eruptive | Maculopapular & papular | | | Vesicular | | Pustular | | Crusting | | Time required for test |
Detection of virus (method)	Blood	Blood	Skin lesions	Saliva	Blood	Skin lesions	Blood	Pustular fluid	Blood	Crusts	
Microscopic examination (Electron microscopy [most sensitive] or Giemsa-stained smears)			+			+		±		−	1 h
Viral isolation (Culture on chorioallantoic membrane of 12- to 14-day-old chick embryos, or in tissue culture)	±	±	+	+	±	+		+		+	1–3 days
Antigen detection* (Complement fixation, agar-gel precipitation, immunofluorescence, or radioimmune precipitation)	±	±	+			+		+		+	3–24 h
Detection of antibodies† (Hemagglutination inhibition, complement fixation, neutralization, or radioimmune precipitation)	−	±			+		+		+		3 h to 3 days

+ = Test usually positive; ± = test may or may not be positive; − = results usually negative; open spaces also indicate negative results.

*Probably the most useful, economic, and efficient of diagnostic procedures.

†Positive results indicate a rise in Ab titer.

(Modified from Downie AW, MacDonald A: Br Med Bull 9: 191, 1963)

tamination of drinking or eating utensils; later, when the vesicles or pustules rupture and crusts are formed, the skin lesions also become a source of contagion. Dissemination by fomites is important because the virus is resistant to ordinary temperatures and drying. Airborne transmission of variola virus is unusual but can occur, as has been demonstrated epidemiologically and experimentally.

Although any person infected with variola virus is potentially contagious, the most dangerous disseminators are persons with unrecognized disease, e.g., the partially immune patient who has relatively few lesions. Such cases, easily overlooked or misdiagnosed, have been primarily responsible for introducing smallpox into countries free of the disease.

PREVENTION AND CONTROL

From the systematic beginnings by Jenner in England at the close of the eighteenth century, artificial immunization has become increasingly effective.* The probable worldwide eradication of smallpox by the WHO demonstrates the ultimate effectiveness of vaccination. This remarkable achievement was based on the principle of surveillance and containment: i.e., isolation of cases and early immunization of all their contacts. The alternative, eradication by immunzing the entire population of a country, was discarded because it proved impossible to reach all who were well but susceptible.

Preparation of Vaccine. Since its original use for immunization against smallpox, cowpox virus has been propagated in many different laboratories under diverse conditions, and is now believed, on the basis of its antigenic structure, to have been inadvertently replaced with an attenuated smallpox virus. The vaccinia virus used today is distinctly different from the cowpox virus encountered in nature. There are now a number of different strains of varying virulence (e.g., dermovaccinia and neurovaccinia). The one accepted as the prototype for immunization is a dermal strain of uncertain origin. It infects the skin at the site of inoculation, and ordinarily does not produce viremia.

Successful immunization requires the use of infectious (atten-

* **Variolation** to protect against smallpox was practiced long before infectious agents and concepts of immunization were understood. The Chinese powdered old crusts and applied them to the nostrils; Brahmins in India preserved crusts and inoculated them into the skin of the unscarred; Persians ingested crusts from patients; and in Turkey fluid from pocks was inoculated. It was this last practice that Lady Mary Wortley Montague, wife of the British ambassador to Turkey, introduced into England in 1718. Crusts and vesicle fluids were selected from patients during epidemics of mild disease (alastrim). The practice spread to the colonies, where it was more widely used than in the British Isles, but it never became popular because of the risks involved. Jenner introduced the use of attenuated (cowpox) virus in 1776, prompted by the clinical observation that milkmaids who acquired cowpox usually escaped smallpox, even when the disease was rampant in the community.

uated) virus, because of the marked lability of the protective Ag. The vaccine most commonly employed is prepared from scrapings of vaccinial lesions on the skin of calves or sheep, with 1% phenol added to kill contaminating bacteria, and 40% glycerol to increase stability of the virus. To maintain its infectivity the vaccine must be stored frozen or refrigerated, a serious problem in areas where refrigerated transport and storage are difficult. The WHO successfully used lyophilized vaccines to overcome the problem of inactivation in hot climates.

Administration of Vaccine and Results. Vaccine is administered intradermally by gently breaking the epidermis under a drop of vaccine; air jet has been particularly effective for immunization of large numbers. Puncture or scarification permits infectious virus to enter the skin, where it multiplies in the deeper layers of the epidermis. The extent of multiplication and spread of virus, and thus the type of reaction that ensues, depend on the state of immunity (and hypersensitivity) of the host. One of three responses is seen (Fig. 56-10).

1) The **primary response** occurs in an individual who has no effective immunity. A small papule appears on the 4th day after vaccination, reaches its maximum with a pustule on the 8th–10th day, and eventually becomes crusted.

2) An **accelerated (vaccinoid) response** progresses through the same stages as in the primary reaction but does so more rapidly and with less intensity because the subject has partial immunity. The rate of development and the extent of the reaction reflect the balance between the degree of immunity and the extent of hypersensitivity.

3) The **early or immediate response** is primarily a **delayed hypersensitivity reaction** (Ch. 20). It does not require viral multiplication since it is due to the host reaction to viral protein; and in contrast to the primary and accelerated reactions, the early response can be elicited by noninfectious virus (inactivated by heat at 56° C for 30 min.) as well as by live virus. Hence, although an immune individual will show an early response, this reaction does not necessarily indicate immunity; it only implies previous infection by vaccinia virus.

Failure to elicit any dermal response is sometimes seen, but it is never the result of complete immunity; it simply indicates that the vaccination technic was faulty or the vaccine inadequate. Since immunity develops as a consequence of viral multiplication, which leads to a primary or accelerated reaction, to assure immunity the individual who has no reaction must be revaccinated until a reaction occurs.

Following the initial vaccination specific Abs, as measured by neutralization or hemagglutination-inhibition titrations, appear in most persons on about the 10th day; CF Abs are detected later, and in only about 50% of those immunized successfully. In revaccinated individuals Abs appear earlier and reach higher titers.

Protective immunity develops 7–10 days after vaccination, which is rapid enough to be effective if contacts of smallpox cases are vaccinated shortly after exposure (the incubation period is about 12 days). It has generally been stated that protection lasts for 3–7 years and that reimmunization every 3 years will afford uninterrupted immunity. However, mild smallpox has occurred only 1 year after known successful vaccination.

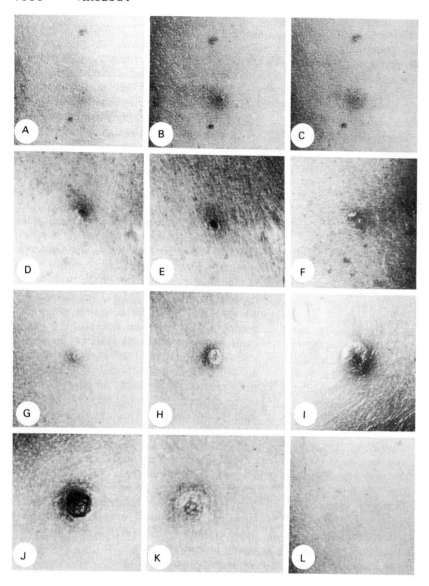

FIG. 56-10. Characteristic cutaneous responses to smallpox vaccine. The vaccine was administered by jet injection. **A, B, C. Immediate response. A:** day 1; **B,** day 3; **C,** day 5. **D, E, F. Accelerated Response: D,** day 4; **E,** day 6; **F,** day 11. **G–H. Primary response: G,** day 3, papule; **H,** day 6, vesicle; **I,** day 9, pustule with beginning crust formation; **J,** day 11, crust; **K,** day 17, crust shed leaving area of desquamation and discoloration; **L,** day 42, remaining scar. (Elisberg BL et al: J Immunol 77:340, 1956)

Complications. Though vaccination is relatively safe it gives rise to rare but occasionally fatal complications affecting the skin or central nervous system, especially with initial vaccinations (Table 56-3). One type results from widespread secondary implantation of virus on skin diseases, such as eczema **(eczema vaccinatum)** or even diaper rash. Viremia frequently occurs in vaccinees who have agammaglobulinemia or dysgammaglobulinemia, and results in progressive and often fatal disease **(generalized vaccinia).** A more frequent and less serious form of generalized vaccinia appears to be related to an allergic reaction, since viremia is apparently rare, virus is usually absent from the skin lesions, and vesicles do not form **(erythematous urticarial reactions).** Probably the most

alarming complication is progressive spread of a primary vaccination response with extensive necrosis of skin and muscle **(vaccinia gangrenosa)** in those rare persons with thymic dysplasias, who cannot develop cellular immunity (about 1.5 cases per million primary vaccinees). It is essential that physicians and public health officers be aware of these complications and not attempt to vaccinate persons with skin eruptions or defects in immunity. Those with similarly afflicted siblings also should not be vaccinated unless they can be conveniently separated from the family.

Postvaccinal encephalitis, a serious and often fatal form of demyelinating disease, has an incidence of about 2.9 per million primary vaccinations (Table 56-3). Since

TABLE 56-3. Incidence of Complications Associated with Smallpox Vaccination in the United States

Complication	Complications per 10^6 primary vaccinations (by age, in years, at vaccination)					Complications per 10^6 revaccinations, all ages
	<1	1-4	5-19	20+	All ages	
Death (from all complications)	5	0.5	0.5	Unknown	1.0	0.1
Postvaccinal encephalitis	6	2	2.5	4	2.9	0.0
Vaccinia gangrenosa	1	0.5	1	7	0.9	0.7
Eczema vaccinatum	14	44	35	30	38	3
Generalized vaccinia	394	233	140	212	242	9
Accidental vaccinia infection	507	577	371	606	529	42

(Modified from Center for Disease Control Morbidity and Mortality Weekly Report 20:340, 1971)

it almost never occurs in children less than 6 months old, or after revaccination, routine initial immunizations were given to children between 3 and 6 months of age whenever possible.

With smallpox essentially eradicated, the risk of a dangerous complication of vaccination outweighs the advantages of universal immunization. For example, smallpox has not been seen in the United States since 1949, but vaccination has caused an average of seven deaths per year. The US Public Health Service therefore recommended that immunization no longer be routine for children but that it be restricted to people at special risk. Since the risk of smallpox now appears to be non-existent, even selective immunization is now unnecessary.

TREATMENT

Limited but encouraging success in specific chemoprophylaxis and chemotherapy of poxvirus infections has been reported. Methisazone (N-methylisatin-β-thiosemicarbazone), which inhibits multiplication of poxviruses, has proved valuable in 1) treating cases of **vaccinia gangrenosa,** and 2) preventing smallpox in intimate contacts of proved cases (see Ch. 51). Treatment of smallpox patients with this drug during an epidemic, however, has not been effective. In vaccinia gangrenosa local injections of leukocytes or transfer factor from immune persons have effectively halted the necrotic process, and in generalized vaccinia γ-globulin from hyperimmune serum has been employed with variable success.

OTHER POXVIRUSES THAT INFECT MAN

Several diseases other than variola and vaccinia are caused by poxviruses: cowpox, molluscum contagiosum, contagious dermatitis, milker's nodules (paravaccinia), and tanapox.

COWPOX

Cowpox is a self-limiting occupational disease of man acquired from the udders and teats of infected cows. The vesicular inflammatory lesions are usually localized on the fingers, but the virus may be implanted on the face or other parts of the body.

Cowpox virus has properties similar to those of variola and vaccinia virus, but its antigenic structure differentiates it from the other agents in the subgroup. The host ranges of cowpox and vaccinia viruses are similar, but cowpox virus differs in several respects: 1) pocks appear more slowly on chorioallantoic membranes; 2) the virus has a tendency to invade mesodermal tissue, involving capillary endothelium and thus producing hemorrhagic ulcers in the pocks; 3) the inclusion bodies are larger and more eosinophilic than classic Guarnieri bodies; and 4) keratitis is produced slowly in rabbits, in comparison with the rapid development effected by vaccinia virus.

MOLLUSCUM CONTAGIOSUM

The molluscum contagiosum virus produces an uncommon skin disease affecting mainly children and young adults. The lesion is a chronic, proliferative process, restricted to the epithelium of the skin of the face, arms, legs, back, buttocks, and genitals. Electron microscopic observations reveal that the molluscum body (a large cytoplasmic inclusion body) is composed of virions indistinguishable from other poxviruses. Mature viral particles develop by a process resembling the formation of vaccinia virions. Molluscum contagiosum has been transmitted experimentally to man, and infections have been achieved in cultures of HeLa cells and primary human amnion or foreskin cells. Although virus could not be serially propagated, viral particles developed, as revealed by electron microscopic examination of thin sections, and cytopathic changes appeared in cultures of infected cells.

The tendency to induce proliferative lesions is even more striking with molluscum contagiosum virus than with other poxviruses. This virus thus appears to provide a link between the common pathogenic viruses of man and tumor-inducing viruses. However, it should be noted that infected cells do not continue to synthesize DNA; it is the neighboring uninfected cells whose rate of cell division is stimulated. The mechanism of this stimulation is unknown.

MILKER'S NODULES (PARAVACCINIA)

Jenner recognized the existence of two diseases affecting the udder and teats of cows: classic cowpox and a second condition consisting predominantly of vesicular lesions. The latter disease is also transmitted to man, producing painless smooth or warty "milker's nodules" on the hands and arms. The lesions rarely become pustular. The infected cells contain eosinophilic cytoplasmic inclusion bodies and elementary bodies characteristic of poxvirus infection. Disease is associated with only mild constitutional symptoms and enlargement of regional lymph nodes.

Infection does not confer immunity to either cowpox or vaccinia viruses. Paravaccinia virus cannot be propagated on the chorioallantoic membrane of chick embryos or in laboratory animals usually susceptible to cowpox. The virus, which was isolated from a milker's nodule of humans, has been serially cultured in fetal bovine kidney tissue cultures, as well as in diploid bovine conjunctival cells and human embryonic fibroblasts. In contrast to many poxviruses, it cannot be propagated in series in continuous human cell lines, such as HeLa cells. Its cytopathic effects resemble those produced by vaccinia virus, and infected cells prepared with Giemsa's stain show metachromatic cytoplasmic inclusion bodies.

Paravaccinia virus has stability characteristics similar to those of poxviruses. In thin sections the average viral particle measures 120 × 280 nm and has the typical morphology of a poxvirus. Electron microscopic observations of preparations stained with sodium phosphotungstate (Fig. 56-11) reveal ovoid virions whose size and surface structures are identical to those of contagious pustular dermatitis (Fig. 56-12) and bovine pustular stomatitis viruses.

CONTAGIOUS PUSTULAR DERMATITIS (ORF)

Although a natural affliction of sheep, mainly affecting lambs, contagious pustular dermatitis occurs rarely as an occupational disease of man. Sheep characteristically develop vesicles in the oral mucosa; these become encrusted and heal slowly after several weeks. In man, the infection usually causes a single lesion on a finger, beginning as a small, painless vesicle, which becomes pustular, encrusts, and finally heals. Transmission of infection from man to man has not been recorded.

The causative agent, orf virus, can be isolated in various animal cell cultures or on the chick chorioallantoic membrane. It produces characteristic cytopathic changes but rarely gives rise to eosinophilic inclusion bodies. Electron microscopic study of the viral particles (Fig. 56-12) reveals the prominent tubulelike structures characteristic of poxviruses; the virion, in contrast to vaccinia and variola virions, is ovoid (160 × 260 nm) and is encircled in a regular pattern by the surface tubules.

FIG. 56-12. Mature orf virus particle, negatively stained with phosphotungstic acid. The woven pattern of threads or tubules can be clearly seen. The apparent criss-crossing of the tubules results from the visualization of both the front and back faces of the particle. (Nagington J, Horne RW Virology 16:248, 1962) (×600,000, reduced)

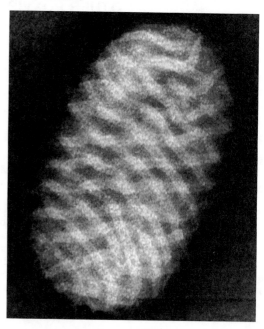

FIG. 56-11. Electron micrograph of viral particle isolated from lesion of milker's nodule. Negatively stained viral particle shows the characteristic morphology of a poxvirus. (Friedman-Kien, AE et al: Science 140:1335, 1963) (×114,000)

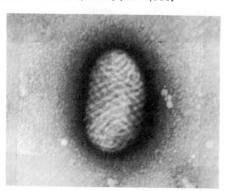

TANAPOX VIRUS

A newly identified virus with characteristic poxvirus morphology has been isolated from two epidemics affecting several hundred tribesmen along the Tana River in Kenya. The disease is characterized by one or two pocklike lesions on the exposed upper part of the body and a febrile illness accompanied by severe aching and prostration. The lesion begins with a papule that develops into a raised vesicle and then umbilicates. Initially the pock resembles that of smallpox, but pustulation never follows. The virus has a host range limited to human and monkey cell cultures. It is serologically distinguishable from other orthopoxviruses, including the Yaba poxvirus of monkeys. The tanapox virus resembles a monkeypox virus that has affected monkeys in captivity in the United States, but simian outbreaks in nature have not been detected. Tanapox virus is probably a monkeypox virus that has been transmitted to man in Africa where the natives use monkey meat and skins.

SELECTED READING

BOOKS AND REVIEW ARTICLES

BENENSON AS: Smallpox. In Evan AS (ed): Viral Infections of Humans: Epidemiology and Control. New York, Plenum Press, 1976, p 429.

FOEGE WH, EDDINS DL: Mass vaccination programs in developing countries. Prog Med Virol 15:205, 1973

LANE JM, MILLER D, NEFF JM: Smallpox and smallpox vaccination policy. Annu Rev Med 22:251, 1971

MACK TM: Smallpox in Europe, 1959–1971. J Infect Dis 125:161, 1972

MOSS B: Reproduction of poxvirus. In Fraenkel-Conrat H, Wagner RR (eds): Comprehensive Virology, Vol 3. New York, Plenum Press, 1975, p 405

SPECIFIC ARTICLES

ARITA I: Virological evidence for the success of the smallpox eradication programme. Nature 279:293, 1979

BLADEN RV: Mechanisms of recovery from a generalized viral infection: Mousepox. II. Passive transfer of recovery mechanisms with immune lymphoid cells. J Exp Med 133:1074, 1971

COUNCILMAN WT, MAGRATH GB, BRINCKERHOFF WR: The pathological anatomy and histology of variola. J Med Res 11:12, 1904

DOWNIE AW, TAYLOR-ROBINSON CH, COUNT AE et al: Tanapox: a new disease caused by a pox virus. Br Med J 1:363, 1971

GERHELIN P, BERNS KI: Characterization and localization of the naturally occurring cross-links in vaccinia virus DNA. J Mol Biol 88:785, 1974

KATES J, BEESON J: Ribonucleic acid synthesis in vaccinia virus. I. The mechanism of synthesis and release of RNA in vaccinia cores. J Mol Biol 50:1, 1970

MARSDEN JP: Variola minor: a personal analysis of 13,686 cases. Bull Hyg 23:735, 1948

MEIKELJOHN G: Smallpox: is the end in sight? J Infect Dis 133:347, 1976

PICORNAVIRUSES

HISTORY

Until the 1900s poliomyelitis (Gr. *poli,* gray, and *myelos,* spinal cord) was a disease primarily of infants (hence the name "infantile paralysis"), and this is still the pattern where sanitation is primitive. But with improved sanitation in many countries, in the 75 years prior to widespread immunization in the 1960s, epidemics increased, the age distribution advanced, and the disease showed increasing severity as it appeared in young adults. This paradoxic response to improved sanitation was eventually explained by the findings that 1) practically everyone became infected, though the paralytic disease was rare, and 2) the consequences were usually negligible if infection was acquired early in life but might be serious when infection was postponed.

Clinically severe poliomyelitis was never very prevalent. Thus, in 1953 there were 1450 deaths and about 7000 cases with residual paralyses in the United States (versus about 500 deaths from measles, which was considered hardly more than a nuisance). However, the visibility of the crippled survivors caused even small epidemics to be terrifying. The problem was dramatized by the severe handicap of Franklin D. Roosevelt, who acquired poliomyelitis as an adult. The public's generous financial support of research (through the March of Dimes) led within 20 years to essentially complete control by immunization. Though the poliovirus was one of the most difficult to work with at the start of this program, it became a model for investigation of many other animal viruses.

Landsteiner* and Popper, in 1909, transmitted poliomyelitis to monkeys by intracerebral inoculation of a spinal cord filtrate from a patient, and the responsible agent was shown to be a virus. However, progress depended on the development of improved technics for cultivating the virus. For example, as long as monkeys had to be used for experimentation, epidemiologic studies could only demonstrate that three antigenically distinct polioviruses exist. Adaptation of polioviruses to the cotton rat by Armstrong in 1939 was a substantial step forward. The turning point came, however, when Enders, Weller, and Robbins showed in 1945 that polioviruses can be isolated and readily propagated in cultures in **nonneural** human

---

* Karl Landsteiner was also distinguished for his profound contributions to the understanding of immunologic specificity and for the discovery of human blood groups.

or monkey tissue. The incisive investigations that followed soon led to control of the disease.

Many related viruses were discovered as accidental byproducts of the intensive pursuit of polioviruses. Thus, **coxsackieviruses†** were isolated in 1948 from the intestinal tract of children by intracerebral inoculation of newborn mice. The subsequent introduction of tissue culture technics revealed the **echoviruses** (enteric cytopathic human orphan viruses), a third group of viruses in the gastrointestinal tract of man. They were called "orphans" because they were not clearly associated with disease. A fourth group of related viruses, designated **rhinoviruses** (Gr. *rhino,* nose), was discovered in 1956 during studies of mild upper respiratory infections fitting the description of the common cold.

CLASSIFICATION AND GENERAL CHARACTERISTICS

Polioviruses, coxsackieviruses, and echoviruses are similar in epidemiologic pattern; in physical, chemical, and biologic characteristics; and in infecting the human gastrointestinal tract. They were originally given the name **enteroviruses,** but the inadequacy of this term became apparent when some coxsackieviruses and echoviruses were also found to produce acute respiratory infections. With the discovery of rhinoviruses, which have similar chemical and physical characteristics but produce primarily acute respiratory infection, the new term *Picornaviridae* (vernacular, **picornaviruses**) was coined as the family designation (*pico,* implying small, and RNA, the nucleic acid component). However, because of differences in certain physical, chemical, and biologic characteristics, human picornaviruses were classified into two genera: **Enteroviruses** (which occasionally cause respiratory rather than intestinal or neurologic disease) and **Rhinoviruses** (Tables 57-1 and 57-2).

The physical and chemical properties of picornaviruses are summarized in Table 57-2. They are small, contain RNA, and do not contain lipid. Polioviruses, described in detail in the following section, will serve as the prototype of the family.

---

† Named after Coxsackie, N.Y., the town from which the initial isolates were obtained.

TABLE 57-1. Classification of Picornaviruses Affecting Humans

Genus	Species	No. of types
Enterovirus	Poliovirus	3
	Coxsackievirus	
	Coxsackievirus A	23*
	Coxsackievirus B	6
	Echovirus	32†
	Enterovirus	4‡
Rhinovirus	Rhinovirus	113

*Type 23 was shown to be identical with echovirus type 9; A23 has been dropped and the number is unused.

†Type 10 has been reclassified as reovirus 1, and type 28 as rhinovirus 1; the numbers are now unused.

‡To simplify classification and to avoid confusion caused by overlap of host range characteristics, newly isolated enteroviruses are no longer divided into coxsackieviruses and echoviruses. From type 68 upward only the designation enterovirus is employed.

TABLE 57-2. Characteristics of Picornaviruses

Size	22–30 nm
Morphology	Icosahedral
Capsomers	Probably 32
Nucleic acid	Single-stranded RNA
Reaction to lipid solvents	Resistant
Stability at room temperature	Relatively stable
pH Stability	Enterovirus: stable at pH 3–9
	Rhinoviruses: unstable below pH 5–6
Stability at 50° C	Enteroviruses: relatively unstable
	Rhinoviruses; relatively stable
Density in CsCl	Enteroviruses: 1.33–1.34 g/cm³
	Rhinoviruses: 1.38–1.41 g/cm³

Viruses similar to human picornaviruses have been found in several species of lower animals: the agent of foot-and-mouth disease in cattle (a member of the genus Aphthovirus that is physically similar to rhinoviruses), Teschen disease viruses of pigs, and Mengo and encephalomyocarditis viruses of mice* (similar to enteroviruses).

POLIOVIRUSES

MORPHOLOGY

Electron micrographs of purified virus in thin sections of virus-infected cells reveal particles 27–30 nm in diameter, with a dense core of approximately 16 nm. Negative staining shows a capsid with a subunit arrangement consistent with icosahedral symmetry (Fig. 57-1). There appear to be 32 capsomers per virion. Polioviruses and viruses of the other three picornavirus groups present only minor differences in size and structure.

PHYSICAL AND CHEMICAL CHARACTERISTICS

The characteristics of the polioviruses are given in Table 57-3. Poliovirus was the first animal virus to be obtained in crystalline form (Fig. 57-2). A single molecule of single-stranded RNA constitutes about 30% of the virion; the remainder consists of four major and one minor species of proteins. There are three serotypes; infectious RNA can be extracted from all three. The virion RNA has the polarity of the viral mRNA (positive strand), and it can be translated in vitro. At its 3′ terminus is a poly(A) track of about 90 nucleotides, which is necessary for its infectivity. In addition, the virion RNA has two unusual features: 1) its 5′ end is not capped but terminates in pUp; and 2) a protein (**VPg**) of about 7000 daltons is covalently attached to the 5′ end of the genome. Although VPg is always covalently linked to the virion RNA, the RNA remains infectious if VPg is removed by pronase; its presence, however, seems essential for initiation of RNA replication. The physical properties of the three serotypes are identical and their base compositions are closely similar; their RNAs share 36%–52% of their nu-

cleotide sequences, as detected by cross-hybridization between RNA from virions and replicative forms of RNA from infected cells.

Polioviruses are more stable than many viruses (e.g., those with lipid envelopes). Hence their transmission is facilitated because they can remain infectious for relatively long periods in water, milk, and other foods. Polioviruses are readily inactivated by pasteurization (see Ch. 67) and by many other chemical and physical agents. Magnesium chloride (1 M) appears to stabilize their intercapsomeric bonds and hence markedly increases thermal stability.

IMMUNOLOGIC CHARACTERISTICS

The three distinct immunologic types of polioviruses can be recognized by neutralization, complement fixation (CF), or gel-diffusion precipitation reactions with type-specific sera.

Crude preparations of each type contain two type-specific Ags; each Ag is associated with a different physical form of the viral particle. The **D (dense) Ag** is associated with fully assembled infectious particles and induces an immunogenic response to all of the virion structural proteins. The **C (coreless) Ag** is associated with empty particles devoid of RNA; it is antigenically distinct because it is deficient in the proteins VPg, VP2, and VP4 but instead contains VP0, the precursor of the two miss-

* The latter two viruses, classified in the genus **Cardiovirus,** commonly infect mice and rodents, but rare infections of humans (clinically manifested as carditis and encephalitis) have been detected.

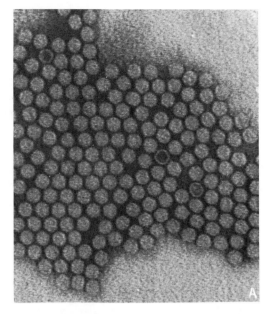

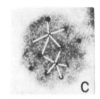

FIG. 57-1. Electron micrograph of a purified preparation of poliovirus negatively stained. **A.** Cubic symmetry of viral particles is evident (×150,000) **B.** Higher magnification of a viral particle printed in reverse contrast. Capsomers measure approximately 6 nm in diameter; their fine structure is not apparent (×600,000) **C.** Same particle as in **B,** marked to display two clear axes of fivefold symmetry **(white lines).** (Mayor HD: Virology 22:156, 1964)

TABLE 57-3. Characteristics of Polioviruses

Diameter of virion	27–30 nm
Diameter of internal core	16 nm
Diameter of capsomer	6 nm
Mol wt of RNA	2.5×10^6 daltons (7.7 Kb)
Base composition (G + C)	46 moles %†
Mole wt of virion proteins‡	
VP1	35×10^3 daltons
VP2	28×10^3 daltons
VP3	24×10^3 daltons
VP4	6×10^3 daltons
VPg	$\sim 7 \times 10^3$ daltons
Sedimentation coefficient of virion	157–160 S_{20}
Particle mass of virion	1.1×10^{-17}g
Mol wt of virion	$8–9 \times 10^6$ daltons

†Composition is closely similar for the three types.

‡Virion proteins 1–4 are present in equal molar amounts.

cross-absorption experiments and by the development of heterotypic Abs following natural infections or immunization. Slight crossreactivity between types 2 and 3 can be detected by neutralization, but not between types 1 and 3. The immunologic kinship between types 1 and 2 is substantiated epidemiologically: possession of type 2 Abs confers significant protection against the paralytic effects of subsequent type 1 infection.

Antigenic variants of types 1 and 2 viruses have been detected by precise plaque-reduction and CF studies with cross-absorbed sera. These antigenic differences, however, do not affect the capacity of Abs induced by one strain to protect against infection by all other strains of the same type. Despite these minor intratypic differences, **polioviruses** actually **show marked antigenic stability,** both in nature and in laboratory manipulations.

Neutralizing Abs appear early in the course of poliovirus infection, and they have usually reached a high titer by the time the patient is first seen by a physician (Fig.

FIG. 57-2. Crystals of purified type 1 poliovirus particles. (Schaffer FL, Schwerdt CE: Proc Natl Acad Sci USA 41:1020, 1955)

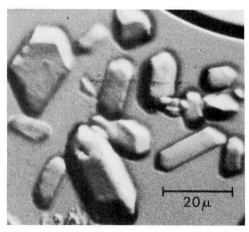

ing polypeptides (VP2 and VP4; see Fig. 50-14, in Ch. 50). Although C particles do not react in vitro with Abs to the D Ag, injection of C particles into rabbits induces production of Abs to D as well as to C, and immunization with either type of particle results in the formation of neutralizing Abs. Various mild denaturing procedures (e.g., exposure to 56° C) cause the D particle to lose VP4, VPg, and RNA and to convert to C Ag reactivity.

There is no common poliovirus group Ag, but antigenic relations between types do exist. The crossreactions are particularly prominent when heated virus is employed in CF titrations. The crossreactivity, however, can be demonstrated only when sera are obtained from humans who have been infected with more than one type of poliovirus; i.e., after an initial infection the Ab response is strictly type-specific, but upon infection with a second type, Abs develop to two or all three of the viruses. Immunologic crossreactivity between types 1 and 2 is also demonstrated in neutralization titrations by

57-3). They attain a maximum titer 2–6 weeks after the onset of disease, decrease to about one-fourth that level in 18–24 months, and then seem to persist indefinitely. Their presence confers clear protection against subsequent infection. Complement-fixing Abs (anti-D and anti-C) appear during the first 2–3 weeks after infection, reach maximum titers in about 2 months, and persist for an average of 2 years. Inapparent and nonparalytic infections result in Ab levels as high as those present after severe paralytic disease.

Second attacks of paralytic poliomyelitis are rare and invariably are due to a different viral type from that producing the first illness.

HOST RANGE

Man is the only known natural host for polioviruses. Abs are present in some monkeys and chimpanzees studied in captivity, but there is evidence that the infection is acquired only after capture.

Old World monkeys and chimpanzees are susceptible to infection (by the intracerebral, intraspinal, and oral routes) with fresh isolates as well as with laboratory strains. In contrast, nonprimates are relatively insusceptible. However, by serial passage, strains of poliovirus were adapted to cotton rats and mice. Strains of type 2 virus also have been adapted to suckling hamsters and the chick embryo.

An important development was the discovery that these viruses, hitherto considered purely neurotropic, can multiply and produce cytopathology in human extraneural tissues cultured in vitro. It then rapidly became evident that many tissues from primates can furnish cells susceptible to cytopathic changes (Fig. 57-4). Cultures of primate tissues are now widely used for isolation and identification, as well as for experimental studies with polioviruses and other picornaviruses.

VIRAL MULTIPLICATION

To initiate infection, polioviruses attach rapidly to specific host cell receptors (composed of lipid and glyco-

FIG. 57-3. Diagram of the times at which the clinical forms of poliomyelitis appear, correlated with times at which virus is present in various sites and with development and persistence of Abs. The high incidence of subclinical poliovirus infection is also noted. (Horstmann DM: Yale J Biol Med 36:5, 1963)

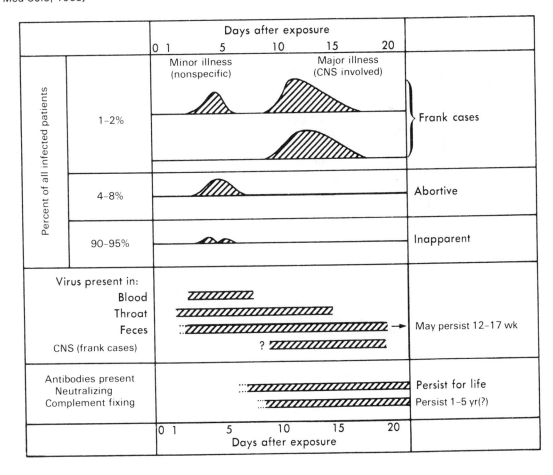

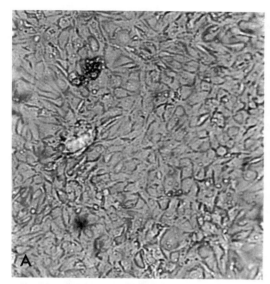
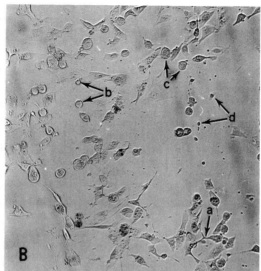

FIG. 57-4. Cytopathic changes in monkey kidney cell cultures infected with type 1 poliovirus. **A.** Monolayer of uninfected cells, unstained. (×200) **B.** Advanced cytopathic changes in infected cultures, unstained. Polioviruses, like most enteroviruses that infect monkey kidney cells, produce marked cell retraction **(a)**, rounding **(b)**, and occasionally ballooning of cells **(c)**, followed by rapid lysis leaving a granular debris **(d)**. (×200) (Ashkenazi A, Melnick JL: Am J Clin Pathol 38:209, 1962)

proteins), which are much more prevalent in susceptible than in nonsusceptible tissues. Such **adsorption** to susceptible cells is independent of temperature, but depends upon the concentration of electrolytes. Infectious D particles, but not C particles, can adsorb, an indication that either VP4 or the conformation of the capsid is critical for attachment. Very soon after attachment the viral capsids are altered, by loss of VP4, probably while still combined to receptors. About 50% of the altered particles also lose their RNA, become noninfectious, and elute from the cells. The remaining particles **penetrate** into the host cells and rapidly uncoat their RNA, which becomes susceptible to RNase within 30–60 min after infection.

The **replication** of infectious virus follows the general pattern of viruses with positive-strand RNA, which initially serves as mRNA for the synthesis of viral proteins, including an RNA replicase. The RNA replicative intermediates then develop (see Ch. 50; Synthesis of RNA Viruses), serving first for synthesis of complementary negative strands and then for positive strands. The small VPg is covalently linked to the 5′ terminal oligonucleotide of all nascent RNAs (positive and negative strands) during an early step in replication. The VPg, however, is cleaved from about one-half of the plus strands, and these are destined to become messengers. Hence VPg distinguishes virion RNAs from viral mRNAs of identical nucleotide sequences.

Replication of viral RNA is independent of biosynthesis of host cell DNA. The production of positive strands commences soon after viral uncoating is completed; but

all the early molecules, which are copied from the nascent complementary RNA templates, become messengers on very large cytoplasmic polyribosomes. Since internal initiation of protein synthesis on this message does not occur, this polygenic RNA of about 7000 nucleotides serves as a monocistronic message; it is *translated into a single long polypeptide,* which is subsequently cleaved into the four individual viral capsid proteins plus VPg and the nonvirion proteins (Ch. 50). Progeny RNA first appears in viral particles about 3 h after infection (Fig. 57-5A); once virion assembly has started, production of capsid proteins (Fig. 57-5B) and RNA replication are closely coupled, and newly made viral RNA is incorporated into virions within 5 min after synthesis. The final step in morphogenesis (Fig. 57-6) appears to be the combination of viral RNA with a shell of viral proteins termed the **procapsid** (Ch. 50), during which one of the procapsid proteins (VP0) is cleaved to yield two of the final capsid structures (VP2 and VP4). Since complete viral replication occurs in cells enucleated with cytochalasin B (see Ch. 49), host cell nuclear functions are not required.

Final **assembly** of infectious particles is accomplished rapidly. Approximately 500 virions per cell are produced. Initially, virions are released through vacuoles, but after several hours they escape in a burst, accompanied by death and lysis of the host cell.

Synthesis of host cell proteins is inhibited very shortly after viral infection (Fig. 57-5C), accompanied by disruption of the host cell polyribosomes. Synthesis of nor-

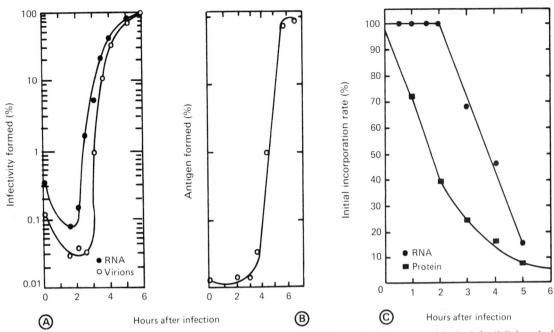

A ⓐ   Hours after infection   ⓑ   ⓒ   Hours after infection

FIG. 57-5. Biosynthetic events in poliovirus-infected cells. **A.** Time course of viral RNA synthesis (measured by its infectivity) and of maturation of virions. **B.** Biosynthesis of viral capsid proteins measured by incorporation of ^{14}C-labeled amino acids into Ab-precipitable material. **C.** Rate of total RNA and protein synthesis in poliovirus-infected cells measured by the incorporation of ^{14}C-uridine or ^{14}C-L-valine into acid-precipitable material at the indicated times after infection. (**A,** Darnell JE Jr. et al: Virology 13:271, 1961; **B,** Scharff MD, Levintow L: Virology 19:491, 1963; **C,** Zimmerman EF et al: Virology 19:400, 1963)

FIG. 57-6. Development of poliovirus particles in pieces of cytoplasmic matrix of artificially disrupted cells. Particles in various stages of assembly from empty shells **(s)** to complete virions **(v)** can be seen. (Horne RW, Nagington J: J Mol Biol 1:333, 1959. Copyright by Academic Press, Inc., [London] Ltd. (×200,000)

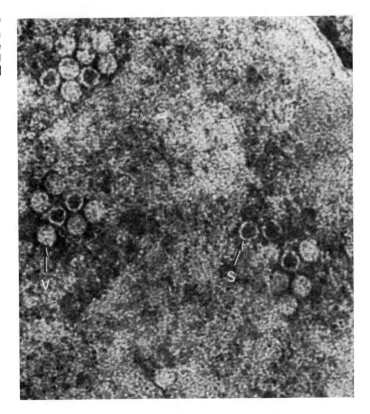

mal host cell RNA ceases about 2 h after infection, shortly after biosynthesis of viral RNA begins (Fig. 57-5).

The **cytopathologic changes** accompanying these biosynthetic events are diagrammed in Figure 57-7. Intranuclear alterations, consisting of rearrangement of chromatin material with condensation at the nuclear membrane, are the first changes detected. One or more small intranuclear eosinophilic inclusion bodies of unknown nature form, and the nucleus becomes distorted and wrinkled and gradually shrinks. These events are probably related to the inhibited synthesis of the host cell's protein and nuclear RNA (Fig. 57-5C). The cytoplasm then develops a large eosinophilic mass, which is the site of replication and assembly of viral subunits (Figs. 57-7 and 57-8), and knobs appear on the cell membrane as the result of cytoplasmic bubbling associated with the release of virus. Finally the nucleus becomes pyknotic, the nuclear chromatin becomes fragmented, and the cell becomes rounded and dies.

GENETIC CHARACTERISTICS

The demonstration in 1953 that poliovirus contains only RNA stimulated great interest because it identified RNA as genetic material, and it provided a theoretic basis for developing attenuated mutants for use as a live virus vaccine.

Polioviruses undergo mutations and recombination just as DNA viruses do. The number of mutant phenotypes observed (Table 57-4) is much larger than the number of viral genes, so that many phenotypes must arise from different mutations of the same gene.

Early studies revealed that recombination could occur between some of the mutant phenotypes. With conditionally lethal temperature-sensitive mutants several genes have been mapped: two genes for replication of viral RNA, one for synthesis of capsid proteins, and two for regulation of cell functions. Genetic maps have been obtained by recombination analyses and by the use of pactamycin to inhibit the initiation of protein synthesis;

FIG. 57-7. Diagram of the multiplication of poliovirus and the accompanying pathologic changes in infected cells. A perinuclear cytoplasmic eosinophilic mass (inclusion body) develops as viral multiplication reaches its maximum; the inclusion body impinges on the nucleus, which degenerates as the cell dies. (Adapted from Reissig M et al: J Exp Med 104:289, 1956)

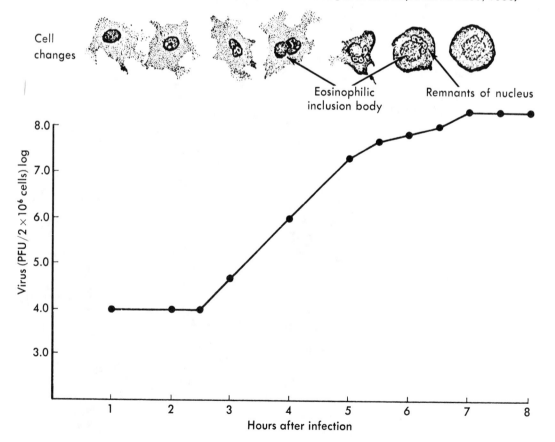

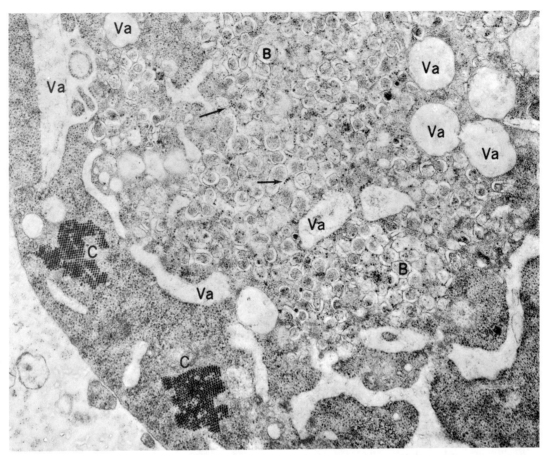

FIG. 57-8. Cytoplasmic changes and assembly of virions in a poliovirus-infected cell. Large numbers of membrane-enclosed pieces of cytoplasm **(B)** accumulate in the central region of the cell, pushing the nucleus to one side (nucleus is not shown in this picture). Large cytoplasmic vacuoles **(Va)** also develop. A large number of virions **(arrows)** are present in the cytoplasmic matrix, both between and within the membrane-enclosed bodies **(B).** Two large crystals of virions are present **(C).** (Dales S et al: Virology 26:379, 1965) (×25,000)

these maps locate the genes for virion proteins close to the 5′ end of the RNA genome and in the order (5′→3′) VP4-VP2-VP3-VP1. The nonstructural proteins are encoded toward the 3′ end of the genome.

Genetic studies have also revealed that many mutations are **pleiotropic,** i.e., two phenotypic traits are changed by the same mutation, although the two phenotypes can also be changed separately. The **d** and **e** phenotypes (Table 57-4), for example, can arise by a single mutation. Hence mutations affecting neurotropism can be found among mutant phenotypes that are easily detectable in vitro, whereas their direct detection in primates would be much more limited and costly.

Several of the mutant phenotypes that are frequently associated with **attenuation** (Table 57-4) affect the viral capsid, and others multiply preferentially at or below usual body temperatures. These results suggest that neurovirulence depends on the ability of the virus to in-

teract with certain cells and to replicate in febrile patients (*rct*/40 mutants). Furthermore, attenuated viruses induce the synthesis of more interferon than virulent viruses do, and are more readily inhibited by it (see Interferon, Ch. 49).

PATHOGENESIS

The major sequence of events in the multiplication and spread of polioviruses was revealed by studies in chimpanzees and man, as well as in cell cultures. In humans the progression of infection culminates in invasion of the target organs, the brain and spinal cord (Fig. 57-9).

Infection is initiated by the ingestion of virus and its **primary multiplication** in the oropharyngeal and the intestinal mucosa. It is not known, however, whether virus multiplies in epithelial or lymphoid cells of the alimentary tract. The tonsils and Peyer's patches of the ileum

TABLE 57-4. Examples of Poliovirus Mutants

Class of mutants	Characteristic	Marker name
Factors that affect cell–virus interaction and therefore viral multiplication	1. Ability to multiply at 40° C	rct/40
	*2. Inability to multiply at 40° C	rct/40−
	*3. Ability to multiply at 23° C	rct/23+
	*4. Heat defectiveness (inability to multiply at 40° C, but usual multiplication at 36° C)	hd
	5. Plaque size	s
	*6. Resistance to heating in $AlCl_3$	a
	7. Resistance to heat inactivation of virion	t
	8. Inability to grow in MS cells	ms
Variants distinguished by presence or absence of inhibitory substance in media	*1. Sensitivity to agar inhibitor at acid pH	d
	2. Sensitivity to agar inhibitor at neutral pH	m
	3. Cystine inhibition of multiplication	cy+
	4. Cystine dependence	cy^d
	5. Tryptophan dependence	
	6. Adenine resistance	
	7. Guanidine resistance	g^r
	8. Guanidine dependence	g^d
	9. Hydroxybenzylbenzimidazole (HBB) resistance	HBB^r
	10. Resistance to normal bovine serum inhibitor	bo
	11. Resistance to normal horse serum inhibitor	ho
Mutants whose markers are related to physical characteristics of the virus	*1. Poor elution from $Al(OH)_3$ gel	$Al(OH)_3$
	*2. Greater adsorbability to DEAE-cellulose	e
Immunologic variants	1. Intratypic antigenic variants	

*Phenotypes associated with attenuated viruses.

are invaded early in the course of infection, and extensive viral multiplication ensues in these loci, so that as much as 10^7–10^8 infectious doses (i.e., 10^7–10^8 times the mean tissue culture dose [TCD_{50}]) of virus per gram of tissue may accumulate. From the primary sites of propagation the virus drains into deep cervical and mesenteric lymph nodes, but since its titer there is relatively low, these nodes may not be important sites of viral replication for progressive infections. From the nodes the virus drains into the blood, resulting in a **transient viremia** which disseminates virus to other susceptible tissues, such as the brown fat (axillary, paravertebral, and suprasternal) and the viscera (probably in reticuloendothelial cells). In these extraneural sites the virus replicates, and it is continually fed back into the blood stream to establish and maintain a **persistent viremic stage.** In most natural infections, even in nonimmune individuals, only transient viremia occurs; the infection does not progress beyond the lymphatic stage, and clinical disease does not ensue.

Viral **spread to the central nervous system** requires persistent viremia, which implies that direct invasion through capillary walls is the major pathway of penetration into the CNS. Furthermore, the presence of specific Abs in the blood, even at the relatively low levels obtained by passive immunization, effectively halts viral spread and prevents invasion of the brain and spinal cord. However, transmission of virus along nerve fibers from peripheral ganglia may provide an additional route for entry into the CNS, because polioviruses are found in these ganglia during the progression of infection, and virus can spread along nerve fibers in both peripheral nerves and the CNS.

Poliomyelitis generally conjures up the picture of a severe, crippling, and occasionally fatal paralytic disease. However, probably no more than 1%–2% of infections culminate in that syndrome (Fig. 57-3). A moderate number of infections induce transient viremia, resulting in a mild febrile disease or so-called "summer grippe."

The **course of classic paralytic disease** is initiated by a **minor disease,** which is associated with the viremia and is characterized by constitutional and respiratory or gastrointestinal signs and symptoms (see Fig. 57-3). There follows, after 1–3 days or often without any interval, the **major disease,** characterized by headache, fever, muscle

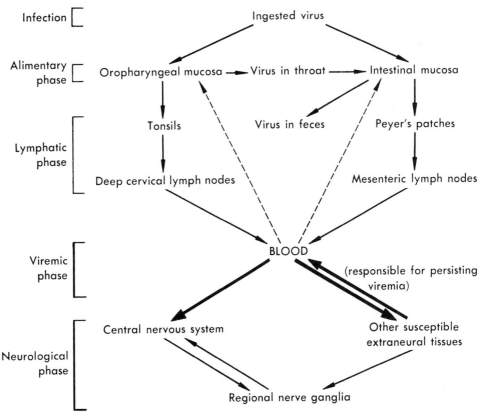

FIG. 57-9. Pathogenesis of poliomyelitis. Model is based on a synthesis of data obtained in man and chimpanzees. (Adapted from Sabin AB: Science 123:1151, 1956; Bodian D: Science 122:105, 1955)

stiffness, and paralysis associated with cell destruction in the CNS. Lesions causing **paralysis** occur most frequently in the anterior horn cells of the spiral cord **(spinal poliomyelitis)**; similar lesions may occur in the medulla and brain stem **(bulbar poliomyelitis)** and in the motor cortex **(encephalitic poliomyelitis)** (Fig. 57-10). Bulbar poliomyelitis is often fatal because of respiratory or cardiac failure; the other forms may result in survival with highly variable patterns of residual paralysis. Paralysis becomes maximal within a few days of onset, and is often followed by extensive recovery, which may be aided by physiotherapy. The recovery represents, in part, compensatory hypertrophy of muscles that have not lost their innervation. In addition, although the virus-infected cells in the lesions are irreversibly damaged, neighboring uninfected neurons may also contribute to the paralysis through edema and byproducts of necrosis but paralysis from this cause is reversible. The temporal relation between clinical manifestations, distribution of virus, and the appearance of Abs is summarized in Figure 57-3.

Several **host factors** may alter the course of infection: fatigue, trauma, injections of drugs and vaccines, tonsil-lectomy, pregnancy, and age. These factors do not increase the incidence of infection but do affect the frequency and severity of paralysis. Trauma, including hypodermic injections, tends to localize the paralysis to the traumatized muscles. Tonsillectomy, recent or of long standing, markedly increases the incidence of bulbar poliomyelitis. The mechanisms of action of these localizing host factors are not certain. They may increase infection of peripheral nerve ganglia or transmission of virus along peripheral nerves associated with the affected area, or they may increase the permeability of blood vessels in corresponding areas of the CNS. The greater severity of paralysis in adults, and even more in pregnant women, may be related to endocrine factors: steroids, for example, greatly heighten the severity of infection in experimental animals.

LABORATORY DIAGNOSIS

Laboratory methods for the diagnosis of poliomyelitis are simple and efficient, because of the ready availability of excellent tissue culture methods and immunologic

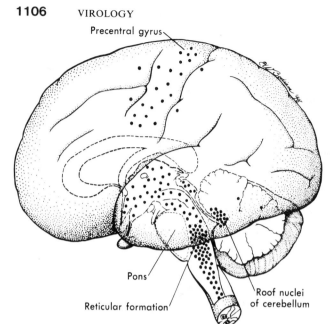

FIG. 57-10. Lateral view of the human brain and the midsagittal surface of the brain stem. **Black dots** = usual distribution of lesions. Spinal lesions usually occur in the anterior horn cells; lesions in the cerebral cortex are largely restricted to the precentral gyrus; those of the cerebellum are largely found in the roof nuclei; lesions of the brain stem centers are widespread. (Bodian D: Papers and Discussions Presented at the First International Poliomyelitis Conference. Philadelphia, Lippincott, 1949)

technics. Virus is isolated most readily from feces or rectal swabs for about 5 weeks after onset, and from pharyngeal secretions for the first 3–5 days of disease. Specimens with added antibiotics, are inoculated into several tissue culture tubes. Multiplication of virus is detected by the development of characteristic cytopathic changes (Fig. 57-4B), and by the failure of infected cultures to become acid in 1–4 days after inoculation, revealed by incorporating phenol red in the medium (since this indicator shifts from red to yellow on acidification). The latter criterion arises from the inability of the dying infected cells to produce organic acids from glucose. Final identification of the virus is accomplished by neutralization tests, using a standard serum for each of the three types.

For serologic diagnosis Ab levels are compared in sera obtained during the acute phase of the disease and 2–3 weeks after onset, using neutralization or, less often, CF titrations. Titration endpoints in neutralization are based on the inhibition of cytopathic effects or on continued acid production; the latter assay is especially convenient, as it can be carried out in disposable plastic trays, and the endpoints can be determined rapidly without microscopic observations. CF tests are usually less dependable, because both D and C Ags are present in viral prepara-

tions, and Abs directed against each Ag appear at different times.

EPIDEMIOLOGY

Serologic surveys show that polioviruses are globally disseminated. In densely populated countries with poor hygienic conditions practically 100% of the population over 5 years of age have Abs to all three types of poliovirus, epidemics do not occur, and paralytic disease is rare. In countries with improved sanitation, in contrast, the young are shielded from exposure, and prior to the effective use of vaccines many who reached adulthood had escaped infection and were therefore without protective Abs. Because the incidence and severity of paralytic disease increase with age, if infection is delayed until susceptible persons are above age 10–15 years more severe crippling disease occurs.

The widespread use of vaccines has strikingly altered the epidemiologic picture. In some countries (Sweden, Finland, Denmark, West Germany), paralytic poliomyelitis appears to have been eradicated. In the United States epidemics have been eliminated except in pockets of lower socioeconomic groups among whom immunization has not been widespread (although available without cost to the individual): it is estimated that in 1977 there were at least 20 million unimmunized children. Among those unprotected, small epidemics are again occurring in the very young (reviving the picture of "infantile paralysis") because, in contrast to the situation in countries with poor sanitation, polioviruses are not widely disseminated while babies are still protected by maternal Abs.

Transmission. Poliomyelitis occurs primarily in the summer, like the common summer diarrheal diseases. This finding first suggested transmission by the fecal route. Indeed, a large amount of virus is excreted in the feces for an average period of 5 weeks after infection, even in the presence of a high titer of circulating Abs (Fig. 57-3). A patient is maximally contagious, however, during the first week of illness, when pharyngeal excretion of virus also occurs. Multiple modes of infection probably account for the fact that infections can occur in any season of the year.

Person-to-person contact is the primary mode of spread, and transmission within families and schools appears to be the major mechanism of dispersion throughout a community. Flies may occasionally serve as accidental vectors, but they are not an important mode of distribution. Water- and milkborne epidemics caused by fecal contamination have also been reported. Dissemination of virus is rapid and extensive in nonimmune members of a family or in other contact groups, but the ratio of paralytic disease to inapparent infections is low—i.e., about 1:200 in temperate zones.

PREVENTION AND CONTROL

Until vaccine became available in 1954 the only approaches to the control of infection were passive immunization and nonspecific public health measures (isolation of patients; closing of such gathering-places as schools and swimming pools; widespread spraying of insecticides), none of which proved successful in preventing or stopping an epidemic.

The present era of successful control can be attributed to three major discoveries: 1) Protection is required against the three distinct antigenic types of poliovirus. 2) Multiplication of poliovirus to a high titer in cultures of nonnervous tissues affords a practicable procedure for preparing large quantities of virus which is free of the nervous tissue that may induce demyelinating encephalomyelitis. 3) As the role of viremia in pathogenesis suggested, the infection can be interrupted before the CNS is infected. In addition, the protection of monkeys, mice, and man by passive immunization with immune serum or pooled γ-globulin proved that even low titers of Abs can be effective in preventing paralytic poliomyelitis.

The development of poliomyelitis vaccine proceeded by two different approaches: the preparation of an **inactivated virus vaccine,** based on evidence that poliovirus inactivated with formalin could immunize monkeys; and the development of a **live attenuated virus vaccine,** modeled on the successful control of smallpox and yellow fever by such vaccines.

Inactivated Virus Vaccine. Salk demonstrated that all three types of polioviruses could be inactivated in about 1 week by 1:4000 formalin, pH 7, at 37° C, with retention of adequate antigenicity. When purified virus is used, the inactivation follows pseudo–first-order kinetics (Ch. 51). However, in crude viral preparations aggregation of the viral particles results in a complex inactivation curve: the exponential rate of viral inactivation is not constant, and the inactivation curve tails off markedly. Failure to recognize this complication led to some serious initial difficulties in vaccine production, exemplified by an incident in which residual infectious virus in several lots of commercial vaccine induced 260 cases of poliomyelitis with 10 deaths. Fortunately, the errors were soon rectified and a safe, highly effective vaccine was developed.* Extensive controlled studies showed an effective protection against paralytic poliomyelitis in 70%–90% of those immunized, and subsequent use of vaccine in the general population confirmed its protective ability.

* In the presence of 1 M MgCl$_2$ inactivation by formalin shows much less tailing off. Moreover, this procedure, which can be carried out at 50°, inactivates adventitious viruses present in monkey cell cultures. These viruses, which include SV40 virus (Ch. 66), are more resistant to formalin inactivation than poliovirus, but are not stabilized by MgCl$_2$ against heat inactivation. Because 1 M MgCl$_2$ also selectively reduced heat inactivation of infectious poliovirus, heating in its presence may similarly be used with infectious virus preparations to eliminate extraneous viruses such as SV40.

The inactivated vaccine has been supplanted by the live attenuated vaccine in the United States, but it is still popular elsewhere. Vaccine containing all three poliovirus types is administered in three intramuscular or subcutaneous injections over a 3- to 6-month period. Antibody levels for all three types appear to fall to approximately 20% of their maximum titer within 2 years, and thereafter decline at a slower rate. The actual persistence of Abs is difficult to evaluate, because of the uncontrolled occurrence of reinfections. Booster injections of vaccine every 5 years are therefore recommended.

Immunization does not prevent reinfection of the alimentary tract unless serum Ab levels are very high (which is unusual except shortly after booster doses). However, infection of the oropharyngeal mucosa and tonsils is generally prevented, eliminating transmission by pharyngeal secretions. This effect may explain the decreased incidence of infection observed after the widespread use of inactivated virus vaccine.

Live Attenuated Virus Vaccine. Infection of the alimentary canal with attenuated live viruses offers several hypothetical advantages: 1) long-lasting immunity, similar to that following natural infections; 2) prevention of reinfection of the gastrointestinal tract and therefore elimination of this route for transmission of the virus; and 3) inexpensive mass immunization without the need for sterile equipment.

Three different sets of attenuated viruses were independently selected by Cox, Koprowski, and Sabin by multiple passage in a foreign host, most frequently tissue culture. The strains developed by Sabin were chosen by the US Public Health Service for commercial production of vaccines. These strains lack neurovirulence for susceptible monkeys inoculated both intramuscularly and intracerebrally, but they occasionally cause paralysis following intraspinal inoculation. The types 1 and 2 strains are genetically stable, probably because they contain several mutations that decrease virulence (Table 57-4), although a minor increase in neurovirulence may occur during passage in man. Fortunately, however, the alimentary tract of man does not offer a marked selective advantage to mutants with increased neurovirulence. In contrast, the type 3 vaccine strain reverts more frequently; it is estimated to produce approximately one case of paralytic poliomyelitis for every 3×10^6 vaccinated people and the frequency is much greater in children with immunodeficiency diseases and in adult males. (The genetic markers in mutant strains used for immunization are of particular value in determining whether or not the vaccine was responsible for the rare postimmunization case of disease.)

In most individuals oral administration of a single type in a dose of 10^5–3.2×10^5 TCD$_{50}$ produces infection of the gastrointestinal canal, excretion of virus in high titer for 4–5 weeks, and development of Abs to a titer of approximately 1:128 in 3–4 weeks. Serologic conversion occurs in over 95% of those without Abs to any of the three types at the time the vaccine is administered. During the period of relatively high Ab titers natural reinfection of the alimentary tract is prevented, but the duration of this protection has not been clearly defined.

Since Abs and immunity follow natural infections a similar persistence was expected to follow immunization with infectious attenuated virus. In fact, however, Abs decrease at approximately the same rate as after immunization with inactivated viruses, i.e., a diminution in 2 years to about 20% of the maximum titer. Antibody levels are generally higher, however, following live than following inactivated vaccines.

At first the three viral types were fed sequentially because it was feared that simultaneous administration might result in interference with the multiplication of one or more types. However, this concern proved to be unfounded. The three viruses are now given together, and interference is minimized by adjusting the viral concentrations so that type 1 is present in the greatest quantity and type 2 in the least. For maximum Ab response immunization with three doses of the trivalent vaccine during the first year of life, preferably from 3 to 18 months of age, is recommended, a booster is also advised for all children at the time of entrance to elementary school. Further vaccine administration is believed unnecessary unless one is exposed to a known case of poliomyelitis or anticipates travel to a region where poliomyelitis is endemic.

Preexisting infection of the alimentary canal with other enteroviruses may interfere with successful implantation of the poliovirus vaccine strains. Hence community immunization programs are usually carried out in the winter or early spring, when enteroviruses are less prevalent.

Critique of Poliovirus Vaccines. Each class of the vaccines has advantages and disadvantages.

Inactivated virus vaccine, which is now of high potency and moderate purity, has the distinct **advantage** that it is safe and remarkably effective when properly employed. For example, the exclusive use of the inactivated vaccine in Finland and Sweden has apparently eliminated paralytic poliomyelitis in these countries. Indeed, there has not been a single case for over 10 years; and in Finland, despite constant surveillance, no poliovirus has been isolated during this period. However, the inactivated vaccine has the following **disadvantages:** 1) logistic problems of administration by sterile injection to large numbers of people, especially children; 2) greater cost, both for administration and for several doses of vaccine; 3) requirement for booster immunizations every 5 years; and 4) failure to eliminate intestinal reinfection and fecal excretion. This last feature, however, could also be an advantage if immunity is not long-lasting, since natural infection could then occur at the usual rate, inducing immunity without producing paralytic disease.

The **live attenuated virus vaccine** has clear **advantages:** 1) it is easily administered; 2) it is relatively inexpensive; 3) it results in synthesis and excretion of IgA Abs into the gastrointestinal tract, thus producing alimentary tract resistance, decreasing spread within the population, and, therefore, conferring **herd immunity** as well as **individual**

immunity; and 4) its effectiveness approaches 100%. The **disadvantages** of this vaccine as presently constituted are 1) reversion to increased virulence of the type 3 virus employed, and 2) dissemination of virus to unvaccinated contacts. The latter process might be advantageous by increasing the resistance of members of a group; it is, however, a potential hazard because the transmission is uncontrolled and the viruses may be mutants of increased virulence.

Despite the safe immunization of millions with live virus vaccine in many countries, its acceptance for general use was slow in the United States, owing to the earlier accident with the inactivated virus vaccine and the fear of reversion of the attenuated strains to neurovirulence. The initial hesitancy has been overcome, however, and at present only the live virus vaccine is used in the United States.

Whatever the advantages of either kind of vaccine may be, both have been used in the United States with remarkable effects (Fig. 57-11): In 1955, when the inactivated virus vaccine was approved for general use, 28,985 cases of poliomyelitis were reported in the United States; in the following year there were 15,140 cases, in 1964 only 122 cases, and in 1969 (after the shift to the live vaccine) a mere 20 cases of paralytic disease and no deaths. From 1969 to 1974, 110 cases of paralytic poliomyelitis occurred, in four general categories: 1) 33 cases in two well-defined **epidemics,** affecting 22 unimmunized preschool children of Mexican-American parentage in South Texas in 1970, and 11 unimmunized students in a Christian Science boarding school in Connecticut in 1972; 2) 6 cases **imported** into the United States, most commonly from Mexico; 3) 29 **endemic** cases, unassociated with travel or vaccine; and 4) 42 **vaccine-associated** cases, occurring in persons who had either received the live virus vaccine or were in contact with recipients; of these, 5 cases were in persons with immunodeficiency diseases. *Indeed, the majority of cases of paralytic disease in the United States now occur from vaccine administration,* in either recipients or contacts. This disadvantage of the live virus vaccine could be revealed only as a consequence of its impressive effectiveness.

Because there is no reservoir for polioviruses other than man, it is theoretically possible to eradicate the disease by global immunization. Owing to the practical problems of such a program, however, its fulfillment seems unlikely. Therefore, as sanitation improves and immunization decreases the prevalence of infection, as well as of disease, more and more people could reach adulthood, when disease is most dangerous, without ever encountering the virus, and therefore without protective Abs. Hence the constant threat that virus will be introduced into a poliovirus-free, nonimmune population will probably make vaccination, at least of children, necessary indefinitely. It is depressing, however, that social and economic factors hamper the attainment of even this

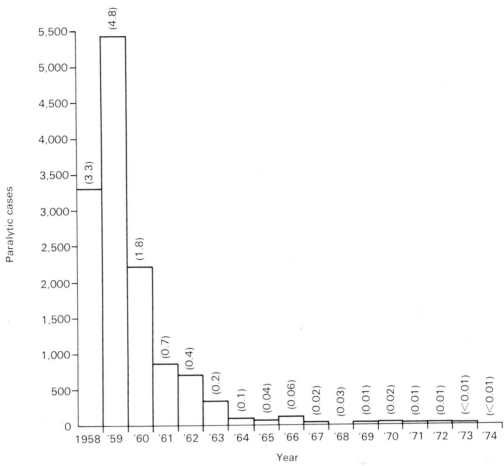

FIG. 57-11. Number of cases of paralytic poliomyelitis after administration of killed virus vaccine (1955–1965) and live virus vaccine (1961–1974). Figures in parentheses give the total case rate per 100,000 population. (Adapted from Center for Disease Control statistics)

limited goal: as noted above, many of the economically disadvantaged are not vaccinated, and if the number of susceptibles increases (for lack of early natural infection) the threat of devastating epidemics once again may arise.

COXSACKIEVIRUSES

Following the demonstration that yellow fever virus and other togaviruses (Ch. 62) are more infectious and pathogenic for newborn than for adult mice, Dalldorf and Sickles attempted to utilize this unique host for studies of poliomyelitis. Instead they isolated a new virus from the feces of two children in Coxsackie, N.Y., in 1948. This new development offered the first major clue that many viruses other than polioviruses infect the intestinal tract of man.

Coxsackieviruses are distinguished from other enteroviruses by their much greater pathogenicity for the suckling than for the adult mouse. They are divided into two groups on the basis of the lesions observed in suck-ling mice: **group A viruses** produce a diffuse myositis with acute inflammation and necrosis of fibers of voluntary muscles; **group B viruses** evoke focal areas of degeneration in the brain, focal necrosis in skeletal muscle, and inflammatory changes in the dorsal fat pads, pancreas, and occasionally the myocardium.

PROPERTIES OF THE VIRUSES

Physical and Chemical Characteristics. Those few coxsackieviruses that have been appropriately examined are similar to polioviruses in physical properties and chemical composition (Tables 57-2 and 57-3) but differ significantly in RNA base composition. Hybridization analyses also show only about 5% nucleotide sequence homology.

Immunologic Characteristics. Each of the 23 group A and 6 group B coxsackieviruses is identified by a type-specific Ag, measured by neutralization and CF. In addition, all from group B and one from group A (A9) share a group Ag which is detected by agar-gel diffusion. Crossreactivities have also been observed between several group A viruses, but no common group A Ag has been found. A type-specific virion Ag of a few types (see Laboratory Diagnosis) causes agglutination of group O human RBCs at 37°C (maximum titers are obtained with RBCs from newborns).

Type-specific Abs usually appear in the blood within a week after onset of infection in man, and they attain maximum titer by the third week. Neutralizing Abs persist for at least several years, but CF Abs decrease rapidly after 2–3 months. Resistance to reinfection, according to epidemiologic data, appears to be long-lasting. A purified viral protein (VP2), as well as whole virions, can induce production of neutralizing Abs. In patients infected with group B or A9 viruses the Ab directed against the so-called group-specific Ag appears earlier and persists longer than the type-specific Ab (suggesting a secondary response resulting from prior infection by a related virus).

Host Range. Suckling mice inoculated by the intracerebral, intraperitoneal, or subcutaneous route are employed for propagation and isolation of coxsackieviruses. Mice 4–5 days old are still susceptible to infection by group A viruses, but group B viruses multiply best in mice 1 day old or less. Even adult mice can be rendered susceptible by cortisone administration, x-irradiation, continuous exposure to cold (4°C) during the period of infection (group B viruses), or severe malnutrition (group B viruses). Denervation of the limb of an adult mouse increases susceptibility to group A viruses, with resulting myositis and muscle necrosis limited to the affected extremity.

The striking susceptibility of newborn mice may be partially explained by their failure to produce interferon when infected by coxsackieviruses. The increased susceptibility produced by cortisone may also be accounted for by inhibition of interferon synthesis (Ch. 51).

Newborn mice infected with **group A viruses** develop a total flaccid paralysis, resulting from severe and extensive degeneration of skeletal muscles; there are no significant lesions elsewhere. Muscle necrosis may be so extensive that a marked liberation of myoglobulin results, causing renal lesions similar to those developing in the crush syndrome.

Group B viruses produce quite different manifestations in suckling mice, including tremors, spasticity, and spastic paralysis. Degeneration of skeletal muscle is focal and limited. The most prominent pathologic lesions are necrosis of brown fat pads, encephalomalacia, pancreatitis, myocarditis, and hepatitis. Adult as well as suckling mice develop pancreatitis, but in adult mice most of the other lesions do not appear or are so minimal that the mice survive. Necrosis of the myocardium, however, is often noted and is markedly increased by cortisone. Cortisone or pregnancy in adult mice transforms an inapparent infection into a fatal one.

Intracerebral inoculation of rhesus monkeys with A7 and A14 viruses produces widespread degeneration of ganglionic cells of the CNS, followed by flaccid paralysis similar to that caused by poliovirus. (Because of this behavior A7 virus was initially mistaken for a new type of poliovirus.)

Tissue culture technics have become increasingly valuable for study and isolation of coxsackieviruses, and by repeated passage attenuated strains can be obtained. The group B and A9 viruses multiply readily in various cell cultures, but most group A viruses do not (Table 57-5). The similar tissue culture host range of group B and A9 viruses parallels their antigenic relation and further suggests that these viruses are very closely related despite their dissimilar pathologic effects in suckling mice.

Viral Multiplication. The multiplication cycle of coxsackieviruses is very similar to that of polioviruses. However, the assembled virions tend to remain within the cell rather than to be released rapidly into the culture medium. The cytopathic changes are also similar to those caused by other enteroviruses, but those produced by group A viruses develop much more slowly.

PATHOGENESIS

Most coxsackievirus infections in humans are mild; infections mimicking those of man have not been produced in laboratory animals. Hence we have very little knowledge of the pathogenesis of human infections or the pa-

TABLE 57-5. Multiplication of Coxsackieviruses in Cell Cultures

	Cell cultures		
	Monkey kidney	HeLa	Human amnion or embryonic kidney
Group A			
Types 1, 2, 4–6, 19, 22	–	–	–
Type 7	±*	±*	–
Type 9	+	±*	+
All others	±*	±†	±‡
Group B			
Types 1–6	+	+	+

*Not readily isolated in cell culture, but strains have been adapted to multiply in indicated cells.

†Types A13, 15, 18, and 21 multiply readily on first passage.

‡Types A11, 13, 15, 18, 20, and 21 grow in human embryonic kidney cells.

(Adapted from Wenner HA, Lenahan MF: Yale J Biol Med 34: 421, 1961)

TABLE 57-6. Clinical Syndromes Commonly Associated with Coxsackieviruses

Clinical syndrome	Coxsackieviruse	
	Group	Predominant types
Aseptic meningitis	A	2, 4–7, 9, 10, 12, 16
	B	All
Paralytic disease	A	2, 7, 9
	B	3–5
Herpangina	A	2, 4–6, 8–10, 16, 21
Fever, exanthema	A	2, 4, 9, 16
	B	4
Acute upper respiratory infection (cold)	A	10, 21, 24
	B	2–5
Epidemic pleurodynia or myalgia	B	1–5
Myocarditis of the newborn	B	2–5
Interstitial myocarditis and valvulitis in infants and children	B	2–5
Pericarditis	B	1–5
Undifferentiated febrile illness	All	All

thology of the lesions. The marked diversity of clinical syndromes, however, indicates that virus enters through either the mouth or the nose and follows a pathogenic course, from local multiplication through viremic spread, that is akin to that demonstrated in poliovirus infections (Fig. 57-9). In biopsies obtained from a few patients with coxsackievirus A infections, focal necrosis and myositis were noted, but the lesions were not distinctive. In children who died of **myocarditis of the newborn,** a highly fatal disease caused by group B coxsackieviruses, the myocardium showed edema, diffuse focal necrosis, and acute inflammation; focal necrosis with inflammatory reaction also occurred in the liver, adrenals, pancreas, and skeletal muscle; and occasionally there was diffuse meningoencephalitis. Group B coxsackieviruses also appear to cause mild interstitial focal myocarditis and occasionally valvulitis in infants and children.

The coxsackieviruses can produce a remarkable variety of illnesses (Table 57-6), and even the same virus may be responsible for quite divergent types of disease. Still, a number of group A viruses have not been definitely implicated as causative agents of any human disease. Some viruses in each group are associated with at least one distinctive syndrome, which can usually be diagnosed on clinical grounds alone. Thus herpangina* is caused by certain group A viruses, and epidemic pleurodynia† and

* **Herpangina** is an acute disease with sudden onset of fever, headache, sore throat, dysphagia, anorexia, and sometimes stiff neck. The diagnosis is dependent upon recognition of the pathognomonic lesions in the throat: at the onset small papules are present, but these soon become circular vesicles that ulcerate.

† **Epidemic pleurodynia (epidemic myalgia;** Bornholm disease) is an acute febrile disease with sudden onset of pain in the thorax (a "stitch in the side") which is aggravated by deep breathing (simulating pleurisy) and by movement. The pain may be chiefly abdominal or associated with other muscle groups, and may be accompanied by muscle tenderness.

myocarditis of the newborn by certain group B viruses. Other syndromes present no clinical features distinctive for coxsackieviruses: rarely, illnesses simulating paralytic poliomyelitis can be induced, particularly by A7 virus; a few group A and B viruses cause an acute upper respiratory illness; and pancreatitis, nephritis, and hepatitis have occasionally been associated with group A and B coxsackievirus infections.

Group B viruses may produce myocarditis of newborns in humans by intrauterine infection, as can certain group A viruses in mice. These findings suggest that coxsackieviruses may, like rubella virus, be responsible for some cases of congenital heart disease. Indeed, women with coxsackievirus infections during the first trimester of pregnancy have been shown to give birth to newborns with twice the normal incidence of congenital heart lesions.

LABORATORY DIAGNOSIS

The etiologic diagnosis of coxsackievirus infections depends upon isolation of the causative agent from feces, throat secretions, or cerebrospinal fluid by inoculating suckling mice. However, for the initial isolation of group B and A9 viruses inoculation of cell cultures is more suitable (Table 57-5). In autopsies of patients with myocarditis and valvulitis the antigen of group B viruses has been demonstrated by immunofluorescence.

A newly isolated virus is grouped as A or B on the basis of the lesions produced in suckling mice. Type identification is considerably more cumbersome, owing to the large number of types. One aid in identification is based on the fact that relatively few coxsackieviruses induce hemagglutination of human group O RBCs from newborns at 37° C; these types (B1, B3, B5, A20, A21, and A24) are rapidly distinguished from other coxsackieviruses and can be readily identified by hemagglutination-inhibition titrations. With group A viruses, because of the large number of types, identification is initiated by neutralization titrations using pools containing several type-specific sera, and final identification is accomplished with the individual type-specific sera.

While serologic diagnosis without viral isolation is not practicable, because of the large number of possible Ags, *identification of an isolated virus as the cause of a particular illness requires serologic confirmation of infection* (by neutralization, immunofluorescence, CF, or hemagglutination-inhibition titrations) because many enteroviruses appear to be present as harmless inhabitants of the intestinal tract rather than as etiologic agents of a current disease.

EPIDEMIOLOGY AND CONTROL

Coxsackieviruses are widely distributed throughout the world, as demonstrated by the occurrence of proved epidemics and by the results of serologic surveys. The type prevalent in any locality varies every few years, probably owing to the development of immunity in the population. For example, in 1947–1948 coxsackievirus B1 was predominant in epidemics observed in New York and New England, but by 1951 the B3 virus produced epidemics throughout the world, replacing the B1 virus.

Coxsackieviruses are highly infectious within a family or the closed population of an institution (about 75% of susceptibles are infected). However, the mechanism of spread may vary with the strain of virus and the clinical syndrome. Most clinical infections and epidemics occur in summer and fall, and the viruses are frequently present in the feces, suggesting a fecal–oral spread. However, viruses may also be isolated from nasal and pharyngeal secretions and produce acute respiratory disease, suggesting spread by the respiratory route as well.

No effective control measures are yet available. Immunization is not practical because of the large number of viruses that induce human disease and the relative infrequency of epidemics caused by any single virus.

ECHOVIRUSES

The first echoviruses were accidentally discovered in human feces, unassociated with human disease, during epidemiologic studies of poliomyelitis. Viruses are now termed **echoviruses** if they are found in the gastrointestinal tract, produce cytopathic changes in cell cultures, do not induce detectable pathologic lesions in suckling mice, and have the properties listed in Table 57-2. Most echoviruses, however, are no longer "orphans" in the world of human diseases, but have been associated with one or more clinical syndromes ranging from minor acute respiratory diseases to afflictions of the CNS (Table 57-7).

Initially, 33 viruses were assigned echovirus serotype designations. However, once they were characterized, echoviruses 10 and 28 were reclassified (Table 57-1).

PROPERTIES OF THE VIRUSES

Data on the characteristics of echoviruses are exceedingly fragmentary, and the viruses cannot be adequately compared.

Physical and Chemical Characteristics. The morphology and general chemical characteristics are similar to those of polioviruses and coxsackieviruses. Infectious RNA has been extracted from a number of echoviruses, but it has not been studied in detail.

Echoviruses are generally heat-stable, but there are marked variations, and some are considerably less stable than polioviruses. As with polioviruses, 1 M $MgCl_2$ stabilizes echoviruses to heat inactivation.

Hemagglutination. Of the 32 echoviruses, 12 show **hemagglutinating activity** with human group O erythrocytes.[*] Maximum titers are obtained with RBCs from newborn humans (as with some coxsackieviruses), but the optimum temperature for the reaction varies with the virus.[†] The hemagglutinin is an integral part of the viral particle. Some types (3, 11, 12, 20, and 25) elute spontaneously from agglutinated RBCs at 37°C, but unlike orthomyxoviruses and paramyxoviruses (Chs. 58 and 59) they do not remove the receptors from the cells, which are still agglutinable by the same or by other echoviruses.

Immunologic Characteristics. The type designation of each echovirus is dependent upon a specific Ag in the viral capsid, and neutralization titration is the most discriminating method for its identification. There is no group echovirus Ag, but heterotypic crossreactions occur between a few pairs,[‡] causing major difficulties in the identification of freshly isolated viruses and in the serologic diagnosis of infections.

Immunologic studies can also be carried out by CF and hemagglutination-inhibition titrations. The CF titrations have the advantage of simplicity but the disadvantage of increased crossreactivity among echoviruses; it is also difficult to obtain satisfactory Ag for this assay from some isolates.

Host Range. The original notion that echoviruses are not pathogenic for experimental animals has proved to be incorrect for at least 14 of the known viruses.[¶] Intraspinal or intracerebral inoculation of virus into rhesus and cynomolgus monkeys initiates viremia, neuronal lesions, and meningitis, occasionally associated with detectable muscle weakness. Some strains of types 6 and

---

[*] Types 3, 6, 7, 11–13, 19–21, 24, 29, and 30.

[†] Maximum titers are obtained at 4°C for types 3, 11, 13, and 19; and at 37° for types 6, 24, 29, and 30. Titers for types 7, 12, 20, and 21 are independent of temperature.

[‡] Types 1 and 8 show a major antigenic overlapping by neutralization titrations, and type 12 crossreacts to a lesser extent with type 29. Antibodies directed against type 23 neutralize type 22 virus, but the reciprocal reaction does not occur. Minor reciprocal cross-neutralization occurs between types 11 and 19, and between types 6 and 30.

[¶] Types 1–4, 6–9, 13, 14, 16–18, and 20.

TABLE 57-7. Clinical Syndromes Associated with Infection by Echoviruses

	Associated echovirus type			
	Common		Uncommon	
Clinical syndrome	Epidemic	Endemic	Epidemic	Sporadic
Aseptic meningitis	4, 6, 9, 30		3, 11, 16, 18, 19	1, 2, 5, 7, 13–15 17, 20–22, 25, 31
Neuronal injury				
Paralysis			4, 6, 30	1, 2, 9, 11, 16, 18
Encephalitis			3	2, 4, 6, 7, 9, 18, 19
Rash, fever	4, 9		16, 18	1–7, 14, 19
Acute upper respiratory infection		20?	19	4, 8, 11, 22, 25
Enteritis	6		14, 18	8, 11, 12, 19, 20, 22–24, 32
Pleurodynia				6
Myocarditis				6, 19

(Modified from Wenner H: Ann NY Acad Sci 101: 308, 1962)

9 produce lesions in newborn mice similar to those induced by group B and A coxsackieviruses, respectively.

Cultures of kidney cells from rhesus or cynomolgus monkeys are most suitable for isolation and propagation of all echoviruses (Table 57-8). The final cytopathic changes produced by most echoviruses are similar to those induced by polioviruses and coxsackieviruses (Fig. 57-4B).

Viral Multiplication. Judging from the fragmentary data available, the multiplication of echoviruses resembles that of polioviruses.

Echoviruses replicate in the cytoplasm of infected cells. The virions appear to assemble and become oriented in columns supported by a fine filamentous lattice distinct from the endoplasmic reticulum (Fig. 57-12), a procedure similar to the assembly process of coxsackieviruses. Crystalline viral arrays may form. Viral particles are subsequently dispersed in the cytoplasm and released from the host cell through small rents in the plasma membrane or in cytoplasmic protrusions that are shed from the cell. Eventually, infected cells disrupt. Types 22 and 23 echoviruses, in contrast, appear to have a different mode of replication: they cause characteristic nuclear changes, and unlike all other echoviruses they are not inhibited by 2(α-hydroxybenzyl)-benzimidazole or guanidine (Ch. 51).

TABLE 57-8. Susceptibility of Cell Cultures to Echoviruses

Cell culture	Echoviruses
Rhesus and cynomolgus monkey kidneys	All
Patas monkey kidney	Types 7, 12, 19, 22–25
Human amnion and kidney	All
Continuous human cell lines	Poor until adapted

PATHOGENESIS

Echoviruses usually enter humans by the oral route; a few probably infect through the respiratory tract. The majority of infections probably remain limited to the pri-

FIG. 57-12. Assembly of type 9 echovirus particles in the cytoplasm of infected cells. Viral particles appear to differentiate in columns upon a fine filamentous lattice at cytoplasmic template sites **(A)**. Another crystal-like array of virions associated with finely granular masses is shown **(B)**. (×50,800) **Inset.** A mass at higher magnification (×112,800). Transected fibrils lie between the particles **(arrows).** (Rifkind RA et al: J Exp Med 114:1, 1961)

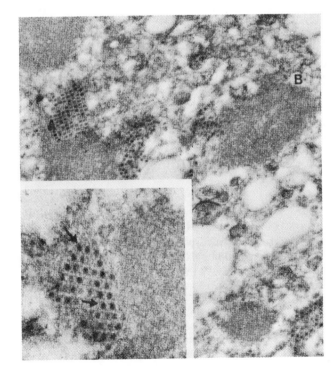

mary cells infected in the alimentary or respiratory tract. It is obvious, however, from the clinical manifestations elicited (Table 57-7), that virus occasionally disseminates beyond the initial organs infected, causing fever, rash, and symptoms of CNS infections. In fact, virus can be isolated from the blood in several of the syndromes listed, and from the cerebrospinal fluid in aseptic meningitis.

The pathologic effects of echovirus infection are still unknown, owing to the general mildness of the diseases. Cerebral edema and some focal destructive and infiltrative lesions have been noted in the CNS of the rare fatal cases examined, but the pathologic findings are not distinctive. Similar neurologic injury has been produced in monkeys and chimpanzees following intracerebral or intraspinal inoculation of the viruses.

LABORATORY DIAGNOSIS

Viral isolation in rhesus monkey kidney cells offers the most sensitive and reliable procedure for diagnosis of an echovirus infection. Feces and throat secretions are the most abundant sources of virus; infectious virus persists in feces longer than in any other body excretion or fluid. Use of kidney cell cultures from patas as well as rhesus monkeys is valuable for identification of the specific virus type isolated (Table 57-8). The differential host suscepti-

bility noted, and the limited number of viruses (12) possessing hemagglutinating activity, afford convenient tools for preliminary grouping of an unknown echovirus and reduce the expense of the immunologic identification of a freshly isolated agent. Neutralization titrations provide the final criterion for identification because of their greater specificity.

Diagnosis solely by serologic analysis of the patient's paired sera is cumbersome* and expensive and is usually employed only during an epidemic caused by a single virus type.

EPIDEMIOLOGY AND CONTROL

The epidemiologic features of echovirus infections resemble those of other enteroviruses, especially coxsackieviruses. But for those echoviruses that cause respiratory infections (particularly for those, such as type 9, which produce extensive waves of infection) respiratory secretions may be a more significant route of viral transmission than feces. This route is also suggested by the rapid and pervasive spread of virus within the family unit.

Immunization does not appear practicable or warranted because of the large number of viruses and the relative infrequency of epidemics produced by a single agent.

NEW ENTEROVIRUSES

Newly identified picornaviruses that are not polioviruses, but conform to the characteristics of enteroviruses, are no longer separated into the species coxsackieviruses and echoviruses because of ambiguities presented by overlapping host range variations. Of the four such enteroviruses isolated (enteroviruses 68–71; see Table 57-1), **enterovirus 70** merits special attention. This virus, which has the typical physical and chemical characteristics of other enteroviruses (Table 57-2), was isolated from many patients during epidemics of **acute hemorrhagic conjunctivitis** that swept through Africa, Asia, India, and Europe

in 1969–1974. The disease is characterized by sudden swelling, congestion, watering, and pain in the eyes accompanied by subconjunctival hemorrhages. Symptoms subside rapidly and recovery is usually complete within 1–2 weeks. The virus is readily isolated in diploid human embryonic lung fibroblasts and KB cells, but it can be easily adapted to propagation in primary monkey kidney cells. Enterovirus 70 multiplies best at 33° C but not at all at 38° C, usually a property of a rhinovirus rather than an enterovirus. This property correlates with its preferential infection of the conjunctivae.

RHINOVIRUSES

Acute afebrile upper respiratory diseases, grouped clinically as the **common cold,** are the most frequent afflictions of man. Although the diseases are not serious, they cause much discomfort as well as the loss of more than 200 million man-days of work and school each year in the United States alone.

There have been many attempts to discover the etiology of this syndrome. Kruse in 1914 showed that the common cold could be

transmitted to man by a filterable agent, but subsequent studies in every conceivable animal failed to isolate the virus; only man seemed susceptible! From extensive human transmission experiments by Andrewes and his colleagues in England, and by Dowling and Jackson in the United States, the notion emerged that the common cold was caused by a large number of viruses, rather than by a single agent, as had been commonly thought. (They also pre-

* Because of the virtual absence of shared Ags it would require that each serum be tested with 31 different viruses.

sented evidence that seems to explode the myth that cold, dampness, and thin clothes provoke the onset of a cold.) It is now known that viruses belonging to several different families can cause the common cold (see Epidemiology, below).

Since the initial isolations of viruses from patients with common cold in 1956, 113 immunologically distinct but biologically related viruses have been isolated. Their chemical and physical characteristics led to their classification into the genus designated **rhinoviruses,** of the picornavirus family.

It is not clear why rhinoviruses were not isolated sooner, despite many efforts, since suitable cells were available and the initial isolations were finally accomplished with methods employed unsuccessfully in earlier studies. The subsequent isolation of numerous types, however, was clearly facilitated by the important discovery that optimal propagation of rhinoviruses, in human embryo or monkey kidney cells, requires special conditions approximating those in the nasal cavities: an incubation temperature of 33° C and pH of 6.8–7.3. The viral multiplication and cytopathology are minimal—with some rhinoviruses, not even detectable—if the infected cell cultures are maintained under the more usual conditions of 37° C and pH 7.6

PROPERTIES OF THE VIRUSES

Structural Properties. Because of the relatively poor viral yield in cell cultures only a few rhinoviruses have been investigated in any detail. They closely resemble other picornaviruses (Table 57-2), but the subunit structures of the capsid seem to be more loosely bonded. The virion consists of a single molecule of RNA (2.3×10^6–2.8×10^6 daltons) and a capsid containing 60 copies of four different polypeptide chains with molecular weights similar to those of the four enterovirus proteins.

Rhinoviruses are sharply distinguished from other picornaviruses because they are **inactivated at low pH, maintain their infectivity at 50° C,** and **have higher buoyant density (1.38–1.41 g/cm^3) in CsCl.** When rhinoviruses are held at pH 3–5 for 1 h at 37° C the virions are disrupted, yielding RNA and denatured protein, and more than 99% of their infectivity is lost; the other picornaviruses are not affected by these conditions. Conversely, rhinoviruses are more stable than the other picornaviruses when heated at 50° C at neutral pH.

Not all rhinoviruses have yet been tested for heat stability, but the strains that multiply only in human cells (**H strains**) appear to be more stable than those that multiply also in monkey cells (**M strains**). Molar $MgCl_2$ partially stabilizes the M strains. The high buoyant density of rhinoviruses in CsCl is due to the permeability of their capsids to cesium ions (as many as 5000 cesium molecules may reversibly bind to an RNA molecule).

No hemagglutinating activity has been detected for rhinoviruses, but they have the novel ability to inhibit the spontaneous hemagglutination of trypsin-treated human RBCs.

Immunologic Characteristics. Each of the distinct rhinoviruses identified possesses a type-specific Ag. There is not a common group Ag but several consistent reciprocal crossreactions (e.g., types 2 and 49; types 13 and 41), and more frequent unilateral heterologous neutralization reactions, are noted with sera from immunized rabbits. Sequential immunization of rabbits with two or three different types further demonstrates immunologic relationships among rhinoviruses because acquired immunologic memory effects a secondary response to heterologous but related types. Thus several clusters of antigenically related rhinoviruses have been detected (e.g., types 13, 14, and 41; types 5, 17, and 42), which fact may be utilized for the development of an effective vaccine.

Natural human infections stimulate the production of type-specific neutralizing Abs which confer resistance to reinfection by virus of the same type (see Epidemiology, below). The Ab response appears to be greater to the M than to the H strains.

Host Range. Humans are the natural host for rhinoviruses. The chimpanzee is the only uniformly susceptible laboratory animal; after intranasal inoculation, virus multiplies in the nasal and pharyngeal mucosal cells and type-specific Abs appear, but disease does not develop.

Cell and organ cultures are the only practical experimental hosts available. The **H viruses** multiply and produce cytopathic changes in human embryo kidney cells, in certain human diploid cell lines, and in a specially selected HeLa cell line, termed HeLa "R". The **M viruses** multiply and produce cytopathic changes in primary cultures of rhesus monkey and human embryo kidney cells, human diploid cells, and continuous human cell lines (e.g., KB and HeLa cells). Organ cultures of human embryonic nasal and tracheal epithelium are particularly sensitive hosts for multiplication of rhinoviruses. A few recently isolated rhinoviruses multiplied only in these organ cultures, whereas others could be propagated in human cell cultures after preliminary isolation and several passages in organ cultures. Whether these rhinoviruses with limited host ranges to highly differentiated cells are new types or only uniquely fastidious strains of previously isolated types is unknown.

The **cytopathic changes** observed in cell cultures under optimal conditions qualitatively resemble those produced by other picornaviruses, but they are slower to develop and are usually incomplete. In infected organ cultures ciliary activity diminishes, and superficial epithelial cells begin to shed 18–22 h after infection. As in infections in humans, only the fully differentiated outer epithelial cells are injured; deeper cells appear unaffected.

Viral Multiplication. The biochemical events in the replication of rhinoviruses do not differ significantly from those of other picornaviruses. For maximum replication, as for viral isolation, a temperature of 33 °C is optimal. Higher temperatures restrict a temperature-sensitive step late in viral multiplication.

PATHOGENESIS

Human infections appear to be confined to the respiratory tract, and generally fit the syndrome termed the **common cold.** During the first 2–5 days of illness viruses can be isolated from nasopharyngeal secretions but not from other secretions or body fluids. Rhinoviruses have also been associated with some exacerbations of chronic bronchitis and a few cases of bronchopneumonia in children and young adults (i.e., so-called primary atypical pneumonia; see Ch. 42). The absence of a satisfactory experimental animal has hindered detailed studies of pathogenesis.

LABORATORY DIAGNOSIS

Isolation of the etiologic agent from nasopharyngeal secretions is the only practical method of establishing the diagnosis. Organ cultures of human embryonic nasal or tracheal epithelium are the most sensitive hosts and are required for isolation of some rhinoviruses. Such cultures, however, are inconvenient for routine diagnostic purposes; therefore, monolayer cultures of primary human embryo kidney cells, human diploid cells, or HeLa "R" cells are generally used for primary isolations. The isolated virus can be typed by neutralization titrations with standard sera. Because of the existence of at least 113 distinct immunologic types, the number of titrations required is first narrowed by preliminary neutralization using pools containing several type-specific sera.

Serologic diagnosis is not routinely practical, owing to the absence of a family crossreactive Ag, the existence of so many distinctive types, and the small amounts of virus obtained in cell cultures.

EPIDEMIOLOGY, PREVENTION, AND CONTROL

The epidemiology of specific rhinovirus infections is that of the common cold. Seroepidemiologic surveys demonstrate that Abs to several prototype viruses are prevalent in many parts of the world and that rhinovirus infections are geographically widespread. Antibodies are found in relatively few infants and children, whereas the majority of adolescents and adults have high titers of Abs to one or

more of the viruses studied. School children frequently introduce the virus into a family, where it spreads readily, particularly to those whose nasal secretions lack IgA Abs. The secondary attack rate may be as high as 70% in a family if the primary patient manifests symptoms of a common cold. The high potential infectivity of rhinoviruses is further evident from a study of military recruits during their initial 4 weeks of training: 90% became infected with one or more different viruses and 40% had two or more rhinovirus infections.

The incidence of isolation of rhinoviruses corresponds with the occurrence of minor respiratory infections, which is greatest in fall, winter, and early spring. Rhinoviruses play a significant role as causative agents of common colds, but clearly they are not solely responsible for production of these illnesses. For example, in one 2-year study of college students, rhinoviruses were isolated from 24% of patients with common colds, from 2% of those convalescing from a cold, and from 1.6% of well individuals; in a study of families with young children 20% of upper respiratory infections were associated with rhinoviruses; and in a study of military recruits, rhinoviruses were isolated from 31.5% of those with common colds. These studies have also demonstrated that inapparent infections occur.

It is common knowledge that the same individual may have repeated episodes of the common cold, even five or six times in a single year. One reason is the prevalence of a large number of immunologically unrelated viruses that cause this syndrome, including viruses other than rhinoviruses, e.g., A21 coxsackievirus, and coronaviruses (Ch. 60).

When volunteers were successively infected with four or five different rhinoviruses the infection conferred specific immunity for at least 2 years to homologous but not to heterologous rhinoviruses. The degree of protection appears to depend upon the Ab levels present at the time of reexposure, particularly the level of IgA Abs in nasal secretions. How long specific immunity persists is unknown.

Theoretically it should be possible to prepare a vaccine that could induce immunity for any single virus or for all rhinoviruses. In fact, an inactivated type 13 vaccine has proved effective when administered intranasally to volunteers. However, many rhinovirus types are widespread (in contrast to the few predominant special types seen with other organisms), and several may be prevalent concurrently. A vaccine containing 113 different rhinoviruses appears to be impractical, but if the antigenically related clusters of rhinoviruses are sufficiently broad, and encompass enough types, the development of an effective vaccine may yet be possible.

SELECTED READING

POLIOVIRUSES

BOOKS AND REVIEW ARTICLES

Biology of poliomyelitis. Ann NY Acad Sci 61:737, 1955

COOPER PD: Genetics of picornaviruses. Compr Virol 9:133, 1977

HOLLAND JH: Enterovirus entrance into specific host cells and subsequent alterations of cell protein and nucleic acid synthesis. Bacteriol Rev 28:3, 1964

KORANT DB: Regulation of animal virus replication by protein cleavage. In Reich E, Rifkin D, Shaw E (eds): Proteases and Biological Control. Cold Spring Harbor, NY, Cold Spring Harbor Laboratory, 1975, p 621

LEVINTOW L: Reproduction of picornaviruses. Compr Virol 2:109, 1974

MELNICK JL: Enteroviruses. In Evans AS (ed): Viral Infections of Humans. Epidemiology and Control. New York, Plenum Press, 1976, p 163

PAUL JR: Poliomyelitis (infantile paralysis). In Mudd S (ed): Infectious Agents and Host Reactions. Philadelphia, Saunders, 1970, p 519

RUECKERT RR: On the structure and morphogenesis of picornaviruses. Compr Virol 6:131, 1976

SPECIFIC ARTICLES

BODIAN D: Histopathologic basis of clinical findings in poliomyelitis. Am J Med 6:563, 1949

ENDERS JF, WELLER TH, ROBBINS FC: Cultivation of the Lansing strain of poliomyelitis virus in cultures of various human embryonic tissues. Science 109:85, 1945

GIRARD M: In vitro synthesis of poliovirus ribonucleic acid: role of the replicative intermediate. J Virol 3:376, 1969

HORSTMANN DM, MCCALLUM RW, MASCOLA AD: Viremia in human poliomyelitis. J Exp Med 99:355, 1954

JACOBSON MF, ASSO J, BALTIMORE D: Further evidence on the formation of poliovirus proteins. J Mol Biol 49:657, 1970

NOMOTO A, DETJEN B, POZZATTI R, WIMMER E: The location of the polio genome protein in viral RNAs and its implication for RNA synthesis. Nature 268:208, 1977

REKOSH D: Gene order of the poliovirus capsid proteins. J Virol 9:479, 1972

SABIN AB: Oral poliovirus vaccine: recent results and recommendations for optimum use. R Soc Health J 2:51, 1962

SALK JE: A concept of the mechanism of immunity for preventing poliomyelitis. Ann NY Acad Sci 61:1023, 1955

Special Advisory Committee on Oral Poliovirus Vaccine. Report to the Surgeon General, USPHS. JAMA 190:49, 1964

VILLA-KOMAROFF L, GUTTMAN N, BALTIMORE D, LODISH HF: Complete translations of poliovirus RNA in a eukaryotic cell-free system. Proc Natl Acad Sci USA 72:4157, 1975

COXSACKIEVIRUSES

BOOKS AND REVIEW ARTICLES

MELNICK JL, WENNER HA, PHILLIPS CA: The enteroviruses. In Lennette EH, Schmidt NJ (eds): Diagnostic Procedures for Viral and Rickettsial Disease, 5th ed. New York, American Public Health Association, 1980

SPECIFIC ARTICLES

BURCH GE, SUN S, CHU K et al: Interstitial and coxsackievirus B myocarditis in infants and children: a comparative histologic and immunofluorescent study of 50 autopsied hearts. JAMA 203:1, 1968

CROWELL RL, PHILIPSON L: Specific alterations of coxsackievirus B3 eluted from HeLa cells. J Virol 8:509, 1971

DALLDORF G, SICKLES GM: An unidentified, filtrable agent isolated from the feces of children with paralysis. Science 108:61, 1948

ECHOVIRUSES

BOOKS AND REVIEW ARTICLES

MIRKOVIC RR, KONO R, YIN-MURPHY M et al: Enterovirus type 70: the etiologic agent of pandemic acute hemorrhagic conjunctivitis. Bull WHO 49:341, 1973

WENNER HA, BEHBEHANI AM: Echoviruses. Virol Monogr 1:1, 1968

RHINOVIRUSES

BOOKS AND REVIEW ARTICLES

DINGLE JH: The curious case of the common cold. J Immunol 81:91, 1958

KAPIKIAN AZ: Rhinoviruses. In Lennette EH, Schmidt NJ (eds): Diagnostic Procedures for Viral and Rickettsial Diseases, 4th ed. New York, American Public Health Association, 1969, p 603

TYRRELL DAJ: Rhinoviruses. Virol Monogr 2:68, 1968

SPECIFIC ARTICLES

FOX JP, COONEY MK, HALL CE: The Seattle virus watch. V. Epidemiologic observations of rhinovirus infections, 1965–1969, in families with young children. Am J Epidemiol 101:122, 1975

PELON W, MOGABGAB WJ, PHILLIPS IA, PIERCE WE: A cytopathogenic agent isolated from naval recruits with mild respiratory illnesses. Proc Soc Exp Biol Med 94:262, 1957

PRICE WH: The isolation of a new virus associated with respiratory clinical disease in humans. Proc Natl Acad Sci USA 42:892, 1956

chapter

ORTHOMYXOVIRUSES

HISTORY AND CLASSIFICATION

In 1918–1919 one of the most devastating plagues in history swept the world, killing approximately 20 million persons and afflicting a huge part of the human population. The underlying disease, **influenza,*** had been known to occur in large epidemics for several centuries. Indeed, the pandemics of 1743 and 1889–1890 were only slightly less disastrous than that of World War I.

The influenza bacillus (*Hemophilus influenzae,* Ch. 34) was originally named as the primary cause of the disease by Pfeiffer in the great pandemic of 1889–1890. However, in 1933, Smith, Andrewes, and Laidlaw in England found that filtered, bacteria-free nasal washings from patients with influenza produced a characteristic febrile illness when inoculated intranasally into ferrets. The viral etiology was soon confirmed in other laboratories, and it eventually became clear that *H. influenzae* is only one of a number of bacterial pathogens (others include *Staphylococcus aureus* and *Streptococcus pneumoniae*) that may cause severe, often fatal secondary penumonia in patients with influenza.

Further progress in the investigation of the virus and the disease was accelerated by the fortunate findings that influenza viruses can multiply to high titer in the chick embryo, a convenient and inexpensive laboratory animal; and that they cause **hemagglutination** of chicken RBCs (see Hemagglutination, Ch. 46). This reaction, discovered by chance in 1941 by Hirst and also by McClelland and Hare, proved to be of great practical and theoretic importance: it provided a simple method for detecting and quantitating influenza

viruses; its specific inhibition by Abs to the virus provided a highly sensitive **hemagglutination-inhibition** test for measuring Ab; and its study revealed the mechanism of infection of host cells, since the receptor sites for the virus on the RBCs proved to be the same as those on the susceptible host cells. These receptors were shown to be mucoproteins possessing a terminal *N*-acetylneuraminic acid (NANA) group. As described earlier (see Hemagglutination, Ch. 46), adsorption of virus leads to release of NANA by a viral enzyme, neuraminidase; the RBCs thereby become inagglutinable, and soluble mucoproteins present in respiratory secretions become nonreactive with fresh virus.

Of major interest in this family of viruses is the frequent emergence of novel antigenic variants as the source of pandemics, and the analysis of the mechanism responsible for this unusual genetic instability.

The successful investigations of influenza viruses, and the general availability of tissue cultures, led to the discovery of additional viruses (e.g., parainfluenza viruses) that agglutinate RBCs and react with similar mucoproteins. These were originally classified together with influenza viruses and termed myxoviruses (Gr., *myxo,* mucus), but later discovery of major physical and chemical differences among the viruses led to their separation into two families (Table 58-1): **Orthomyxoviridae*** and **Paramyxoviridae** (vernacular, orthomyxoviruses and paramyxoviruses), whose distinguishing characteristics are listed in Table 58-2. Orthomyxoviruses (influenza viruses) will be described in this chapter and paramyxoviruses in Chapter 59.

INFLUENZA VIRUSES

After the isolation of the causative agent of influenza it soon became evident that a complex group of viruses was involved. The agents isolated from humans in England and the United States were found to be similar to, but not identical with, the swine influenza virus isolated by Shope in 1931; and many viral strains isolated proved to be antigenic variants when compared with the initial

isolates. In 1940 Francis and Magill, studying patients with influenza in the United States, independently isolated viruses that were immunologically distinct from the original strains. The agent isolated in 1933 was termed influenza A virus, and the second discovered was called influenza B virus. A third distinct antigenic type, influenza C virus, was subsequently isolated in 1949. In-

---

* Derived from an Italian form of Latin *influentia* (influence), reflecting the widespread supposition that epidemics resulted from an astrologic or other occult influence such as an unhappy conjunction of the stars.

* *Ortho*myxoviridae, to contrast with *Para*myxoviridae, is the designation assigned to this family by the International Committee on Viral Nomenclature.

TABLE 58-1. Classification of Orthomyxovirus and Paramyxovirus Families

Family	Genus (type)	Species (subtype)*
Orthomyxovirus	Influenzavirus A	H_0N_1 (A_0, human)
		H_1N_1 (A_1)
		H_2N_2 (A_2)
		H_3N_2 (A_{HK}, A_3)
		$H_{sw}N_1$ (Swine)
		$H_{eq}N$ (2 equine)
		$H_{av}N$ (8 avian)
	Influenzavirus B	B† (human)
	Influenzavirus C	(human)
Paramyxovirus	Paramyxovirus	Parainfluenza 1–4
		Simian SV5 parainfluenza
		Mumps
		Newcastle disease (NDV)
	Morbillivirus	Measles
		Rinderpest
		Canine distemper
	Pneumovirus	Respiratory syncytial

*Based on the immunologically distinct surface Ags, the hemagglutinin (H), and the neuraminidase (N), which undergo antigenic variation.

†Antigenic variations among strains are known but the information is inadequate to enable division into subtypes.

fluenza C virus rarely produces clinical disease and has not been responsible for epidemics.

Since the discovery of influenza viruses **major new antigenic variants** (i.e., **species,** or **subtypes;** see Table 58-1) of influenza A and B viruses have continually emerged; the new variants are only remotely related to the earlier viruses or to each other. The frequent recurrence of the epidemic disease reflects the genetic variability of influenza viruses.

Although bacterial types are narrow subgroups within a species, it should be noted that the original so-called **types** of influenza virus (Table 58-1) are broad groups, now called **genera.**

MORPHOLOGY

Influenza viruses are somewhat heterogeneous in size and shape, but generally are roughly spherical or ovoid (Fig. 58-1). Influenza A viruses have a mean diameter of 90–100 nm while influenza B viruses are somewhat larger, approximately 100 nm in diameter. Filamentous forms of similar diameter occur in fresh isolates (Fig. 58-2).

Influenza virus particles (like paramyxoviruses) are distinguished by spikes or rods that cover the entire surface (Fig. 58-1). These are evenly spaced and appear to be arranged in interlocking hexagons so that each rod has six neighbors. (Type C influenza virus particles display areas sparsely covered with spikes, revealing an underlying lattice of hexagonal and pentagonal units.) Beneath the outer zone is a continuous membrane (Fig. 58-3). Disruption of the particle by lipid solvents uncovers an inner helical component, the nucleocapsid (Fig. 58-3). The purified nucleocapsid (Fig. 58-4) is composed of filaments of variable length (averaging 60 nm), each containing one of the eight segments of the viral RNA.

Upon complete disruption of the virion with sodium dodecyl sulfate (SDS) the **hemagglutinin** and the **neuraminidase** subunits can be separated by electrophoresis or rate zonal centrifugation. Each of these viral surface projections has a distinct structure (Fig. 58-5). The dispersed hemagglutinins are univalent and therefore attach to RBCs but do not agglutinate them. When the SDS is removed the hemagglutins aggregate into clusters of radiating rods (Fig. 58-5C) which are multivalent and therefore produce hemagglutination. The neuraminidase subunits, after removal of SDS, aggregate into pinwheel-like structures (Fig. 58-5D); both the single and aggregated units have enzymatic activity. These arrangements originate from the mutual adherence of the terminal hydrophobic portions of the proteins, which are normally embedded in the lipid bilayer of the virion envelope.

The **filamentous forms** of virus (Fig. 58-2), often seen in freshly isolated strains, are very pleomorphic and frequently appear to be composed of spherical subunits; they may be as long as 1–2 μm and can be observed by darkfield microscopy. Evenly spaced spikes, similar to those of the more classic spherical particles, project from

TABLE 58-2. Characteristics of Orthomyxoviruses and Paramyxoviruses

Characteristics	Orthomyxoviruses	Paramyxoviruses
Particle size	Small (80–120 nm)	Large (125–250 nm)
Diameter of internal helical core (nucleocapsid)	9 nm	18 nm
Localization of nucleocapsid	Nucleus	Cytoplasm
Fragmentation of nucleocapsid	+	−
Virion RNA polymerase	+	+
Separate hemagglutinin and neuraminidase	+	0
Filamentous forms	Common	Observed
Hemolysin	0	+*
Prominent cytoplasmic inclusions	0	+
Syncytial formation	0	+

*All species except pneumoviruses (respiratory syncytial virus).

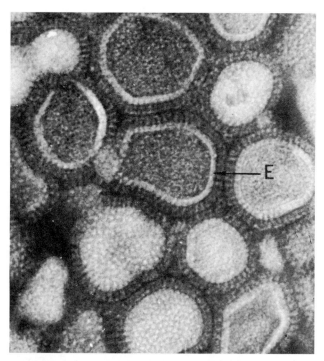

FIG. 58-1. Electron micrograph of intact influenza A virions embedded in phosphotungstate. The virions are of variable shape and size and show evenly spaced, short surface projections covering the entire surface of the particles. **E** = envelope. (Hoyle L et al: Virology 13:448, 1961) (×300,000)

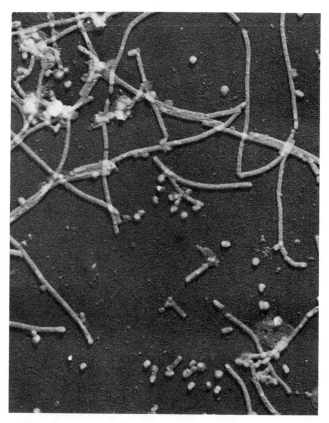

FIG. 58-2. Electron micrograph of influenza $H_2N_2(A_2)$ virus (third passage) showing filamentous viral particles as well as a few spherical ones. Chromium-shadowed preparation. (Choppin PW et al: J Exp Med 112:945, 1960). (×10,400)

the surfaces. Filaments, which are apparently infectious, appear to assemble as a result of some defect in the development of particles and their emergence from the cell membrane. The capacity of an influenza virus strain to produce a predominance of filamentous forms is a stable genetic attribute of the strain.

PHYSICAL AND CHEMICAL CHARACTERISTICS

The intact spherical particles contain approximately 0.8%–1.1% RNA, 70% protein, 6% carbohydrate, and 20%–24% lipids. A lower proportion of RNA is found in preparations containing numerous defective viral particles or filamentous forms.

Disruption of intact particles with lipid solvents, followed by removal of the hemagglutinin and neuraminidase by adsorption onto RBCs, reveals that the RNA is associated entirely with the inner helical core (the S Ag), which is 5% RNA by weight. In accord with electron micrographs of the nucleocapsid, gentle chemical extraction and physical separation of the nucleocapsid from other viral components yield nucleoproteins of three size classes. The genome extracted directly from the virion consists of eight separate pieces of single-stranded

RNA, corresponding to segments of the nucleocapsid (the influenza C genome, however, appears to consist of only 7 segments). All of the RNA pieces have almost identical 5' ends consisting of a sequence of about 13 nucleotides ending with an Appp. In addition, the 3' termini of all the genome RNAs have a high degree of conservation for the first 12 nucleotides (Fig. 58-6). It is striking that these conserved sequences at the 3' and 5' ends of the RNAs have extensive homology in all strains of types A, B, and C thus far examined. Since each RNA segment codes for a single viral protein (Fig. 58-7), this physical structure is consistent with the marked genetic lability and very high recombination rate of influenza viruses (see Immunologic Characteristics, and Genetic Chracteristics, below).

Influenza virus RNAs are single-stranded molecules (total mol wt $5.9–6.3 \times 10^6$ daltons, assuming one molecule of each RNA segment per virion). Influenza A, B, and C viruses are probably phylogenetically quite distant, for their genomes differ significantly in size and base

FIG. 58-3. Partially disrupted influenza virus particle. The disrupted outer membrane or envelope **(E)**, which is 6–10nm thick, appears to have collapsed and become distorted, revealing the nucleocapsid folded in parallel repeating bands. (Hoyle L et al: Virology 13:448, 1961) (×300,000)

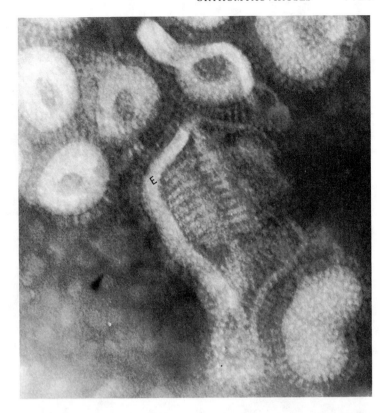

FIG. 58-4. Preparation of highly concentrated nucleocapsid prepared by ether disintegration of purified influenza A virus particles and embedding in phosphotungstate. The helical structure of the nucleocapsid is apparent, particularly in regions marked **H**. Particle at **A** is interpreted as part of an elongated structure viewed along the particle axis. (Hoyle L et al: Virology 13:448, 1961) (×270,000)

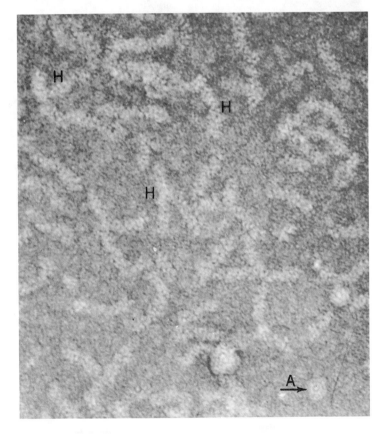

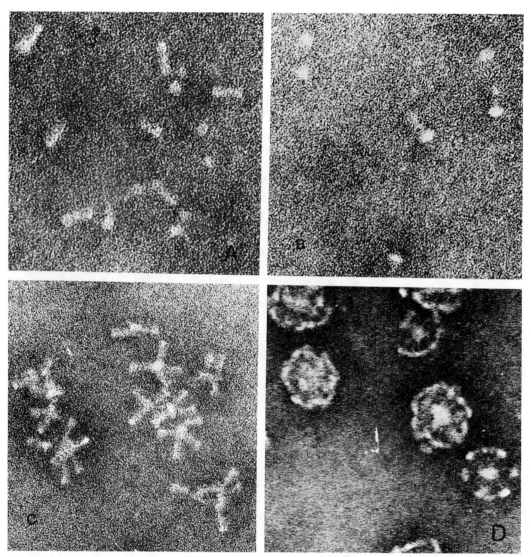

FIG. 58-5. Morphology of the hemagglutinin and neuraminidase subunits of influenza A virus. **A.** Single hemagglutinins appear as thick rods 14 × 4 nm in the presence of sodium dodecyl sulfate (SDS). **B.** Individual neuraminidase subunits dispersed in SDS are seen as oblong structures with a centrally located fiber possessing a terminal knob 40 nm in diameter. **C.** Clusters of hemagglutinins formed by removal of SDS. **D.** Neuraminidase subunits aggregated by the tips of their tails to form pinwheellike clusters when SDS was removed. (Laver WG, Valentine RC: Virology 38:105, 1969) (×500,000)

FIG. 58-6. Model illustrating the nucleotide sequences of the two conserved regions of all orthomyxovirus genome RNAs and the two classes of transcribed cRNAs: polyadenylated incomplete transcripts (i.e., the mRNAs); and nonpolyadenylated complete transcripts, which are the templates for virion RNAs. The sequences 14–16 [(- -)*] at the 5′ ends are variable. The mRNAs are copied by the virion transcriptase, whereas the complete cRNAs are transcribed by a polymerase modified by newly synthesized viral proteins. (Modified from Hay AJ, Skehel JJ: Br Med Bull 35:47, 1979)

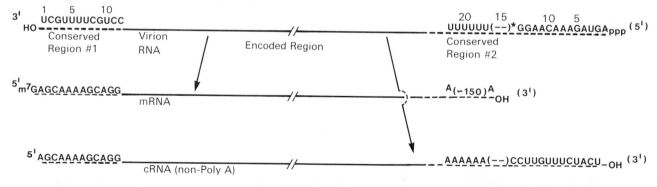

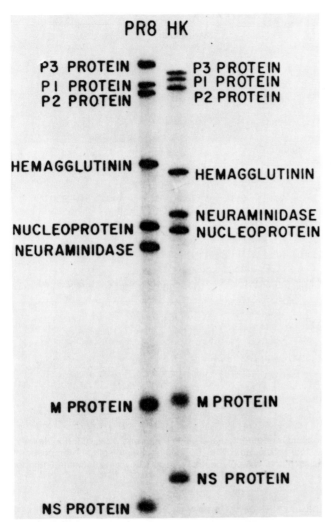

FIG. 58-7. Assignment of influenza A virus proteins to the RNA genome segment in which each is encoded. Virion RNAs from PR8 and HK type A viruses (of subtypes H_0N_1 and H_2N_2, respectively) were electrophoretically separated on polyacrylamide-urea gels. The assignments were developed from a series of antigenically characterized recombinant viruses, each of which showed an exchange of a single RNA segment and a difference in a single viral protein. Viral proteins were identified by analysis on SDS–polyacrylamide gels. (Ritchey MB et al: J Virol 20:307, 1976)

compositions (the $A + U/G + C$ ratio is about 1.25 for type A, 1.42 for type B, and 1.46 for type C).

Seven distinct **viral proteins** can be separated by polyacrylamide gel electrophoresis (fig. 58-7). Two other proteins, nonstructural components NS_1 and NS_2, which are derived by cleavage of a precursor, are only found in infected cells. The membrane **matrix (M) protein** (composed of identical small monomers) is the most abundant virion protein (Fig. 58-8); it is associated with the inner surface of the lipid bilayer and confers stability upon the viral envelope. The **hemagglutinin** and **neuraminidase** are glycoproteins (containing glucosamine, fucose, galactose, and mannose). The rod-shaped hemagglutinin is synthesized as a single glycoprotein, but for virions to attain infectivity its precursor must be proteolytically cleaved to two unique polypeptides (HA_1 and HA_2), which are held together by disulfide bonds. The neuraminidase consists of only one unique glycosylated polypeptide, and the nucleocapsid protein (NP) is a single phosphorylated polypeptide species. Three large internal polypeptides (P_1, P_2, and P_3), associated with RNA transcription and replication, are present in the nucleocapsid in relatively small numbers (Fig. 58-8).

Tryptic peptide maps show little difference between the nucleocapsid proteins from different strains of influenza A, but sharp differences between the hemagglutinins are noted. This result is consistent with evidence that within a given viral type (genus) the nucleocapsid proteins of different strains are antigenically similar, whereas the glycoprotein surface Ags are strain-specific.

The **lipid** of the viral particle is two-thirds phospholipid and one-third unesterified cholesterol. The kinds and concentrations of the individual lipids resemble those of the plasma membrane of the host cell. Thus when cells are labeled with ^{32}P before viral infection the lipids incorporated into viral particles, except for phosphatidic acid, are seen to be derived from the host cell. At least a portion of the viral polysaccharide is also of host origin. In contrast, the RNA and proteins are specified by the viral genome.

Stability. The high lipid content makes these viruses susceptible to rapid inactivation by lipid solvents and surface-active reagents. As with most other viruses with lipid envelopes, the infectivity of influenza viruses is relatively labile on storage at $-15°$ C or $4°$ C but is retained for long periods at $-70°$ C.

IMMUNOLOGIC CHARACTERISTICS

Viral Antigens. One of the surface glycoprotein Ags corresponds to some of the radiating spikes on the virion's surface, and is biologically and immunologically the same as the **hemagglutinin** (Figs. 58-5A and 58-8). This Ag is measured by inhibition of hemagglutination activity, neutralization of infectivity, or complement fixation (CF). The other surface glycoprotein of influenza A and B virions corresponds morphologically, immunologically, and biologically to the **neuraminidase** (Fig. 58-5B and 58-8). It is assayed by inhibition of the enzyme activity. Antibodies specific for neuraminidase do not neutralize infectivity, but they can inhibit the release of budding virus from infected cells and hence diminish the viral yield. Influenza C viruses, however, do not have neur-

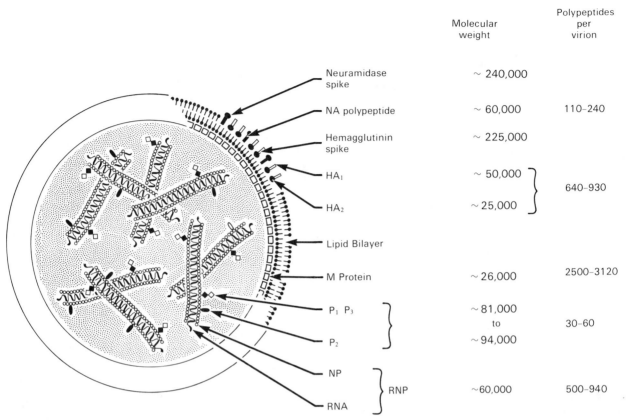

	Molecular weight	Polypeptides per virion
Neuramidase spike	~ 240,000	
NA polypeptide	~ 60,000	110–240
Hemagglutinin spike	~ 225,000	
HA₁	~ 50,000	640–930
HA₂	~ 25,000	
Lipid Bilayer		
M Protein	~ 26,000	2500–3120
P₁ P₃	~ 81,000 to ~ 94,000	30–60
P₂		
NP		
RNA } RNP	~60,000	500–940

FIG. 58-8. Schematic model for influenza virus particles. The nucleocapsid is segmented and gives the appearance of a double helix owing to association of internal proteins (**NP, P₁, P₂,** and **P₃**) with the single-stranded RNAs. The hemagglutinin spike is composed of three sets of **HA₁** and **HA₂** polypeptides. The neuraminidase spike consists of four **NA** polypeptides. **Inset** shows the estimated number of each polypeptide per virion.

aminidase activity, implying that virions can be released by another mechanism.

The major structural **matrix protein of the viral envelop (M protein)** is measured by CF and immunodiffusion; its specific Abs cannot neutralize infectivity or inhibit hemagglutinin and neuraminidase activities. The **internal** or **nucleocapsid Ag** corresponds morphologically to the internal helical component (the RNA–protein core, Fig. 58-4) and is immunologically identical with the **soluble Ag** that is present in infected cells. This Ag is assayed by CF titration.

Immunologic Grouping. On the basis of their nucleocapsid and M protein Ags the many influenza viruses are divided into three distinct **immunologic types (genera):** A, B, and C. The Ags of each type are unique and do not crossreact with those of the other two.

Within types A and B immunologic variants are distinguished by antigenic differences of the hemagglutinin and neuraminidase. The antigenic variations of these two proteins, however, are genetically independent. Over the past few decades a new major variant (i.e., **subtype**) of type A has emerged every 10–12 years, but only a single subtype is generally prevalent at any one time.

Antigenic Variation. Influenza A virus undergoes two distinct forms of antigenic variation: **Antigenic drift** reflects minor antigenic changes in either the hemagglutinin or the neuraminidase or in both. **Major antigenic shift** occurs infrequently and reflects the appearance of viral strains with surface Ags that are immunologically only distantly related to those on earlier strains. The antigenic shift may involve either the hemagglutinin alone or the neuraminidase as well. Influenza A viruses have undergone four major antigenic shifts since 1933, detected by immunologic studies on serum from persons of different ages and on the viruses isolated (Fig. 58-9). A virus similar to swine influenza virus ($H_{sw}N_1$)* presumably

* Isolated by Shope in 1931 from pigs with a severe respiratory infection.

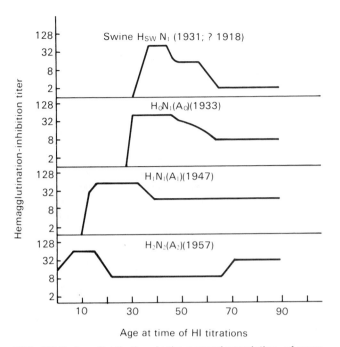

FIG. 58-9. Age distribution, in the general population, of mean hemagglutination-inhibition Abs to subtypes of influenza A viruses. Sera were obtained in 1952 and 1958 and each was assayed with the viruses noted. The highest Ab titers for each virus tested were in those persons who were probably infected when that subtype first became prevalent (dates in parentheses). (Modified from Davenport FM et al: J Exp Med 98:641, 1953, with addition of data for H_2N_2 (A_2) virus)

accounted for the human infections between 1918 and 1929, since serums from persons born during that period contains Abs to the swine agent. The first human influenza A virus [subtype H_0N_1 (A_0)]* was isolated in 1933; it was responsible for all influenza A infections until 1947; moreover, in serum from persons born during this period, regardless of their age when tested, the highest influenza Ab titers are to the H_0N_1 subtype.

In 1947 H_1N_1 (A_1) viruses emerged and supplanted all prior strains, as indicated by isolations and by the appearance of Abs. In 1957 the H_2N_2 (A_2, Asian) influenza virus became prevalent. In 1968 another relatively large antigenic shift occurred and the Hong Kong H_3N_2 (A_{HK}) virus emerged: the neuraminidase molecules are antigenically similar to those of the original A_2 virus but the hemagglutinin is chemically and immunologically unique.

* The genus is termed type A, and the subtype representing the first human influenza A virus to be isolated was originally called A_0. With the recognition that the hemagglutinin (H) and neuraminidase (N) glycoproteins vary independently and determine the antigenic characteristics of a viral strain the subtypes are now named accordingly; thus H_0N_1 is a synonym for A_0, etc. (see Table 58-1).

In 1976 swine influenza virus ($H_{sw}N_1$) unexpectedly appeared at a US Army post at Fort Dix, N.J., but it again disappeared after a brief encounter with about 200 soldiers. In the late fall of 1977 an H_1N_1 virus, another old acquaintance, emerged in the Soviet Union and Hong Kong, but this H_1N_1 is antigenically closer to the variants isolated in 1950 than to the original strains that prevailed in 1947.

Influenza B viruses also undergo antigenic variations, but these are neither so extreme nor so frequent as those of A viruses, and some immunogenic crossreactivity occurs among all the B variants. Hence, influenza B variants have not been classified into distinct subtypes. The originally isolated B virus was prevalent from 1936 to 1948, and the second variant appeared in 1954. Antigenic drift frequently occurred and eventually (in 1962) yielded an only distantly related antigenic variant.

The continual antigenic variation of influenza viruses is of considerable practical importance and theoretic interest. Each major shift has found a large proportion of the world population immunologically defenseless against the newly emerged virus. Furthermore, the neutralizing Abs induced by the vaccine current at the time did not react with the variant strain. The minor antigenic drifts that more frequently occur between pandemics, in contrast, may reduce the effectiveness of the vaccine but do not make it useless.

Basis for Antigenic Variations. The **minor antigenic variations** of antigenic drift result from mutations in the hemagglutinin and neuraminidase genes, and they emerge by selection of mutants that are less susceptible to Abs prevailing in the population. In fact, similar new variants have been selected experimentally by passage of viruses in the presence of small amounts of Ab in mice or chick embryos.

A **major antigenic shift**, in contrast, cannot be explained by a simple mutation because the peptide maps of the hemagglutinins from different viral subtypes differ greatly, indicating extensive diversity in amino acid sequences. The change probably results from **recombination** (i.e., **gene reassortment**) between a human and an animal strain, both influenza A viruses. Such recombinants between different viral species have been produced experimentally in animals and have been selected by passage in immunized animals.

Each new major variant results from the adding of a new Ag while some of the previous ones are retained. Hence, **primary immunization** with H_2N_2 (A_2) virus induces formation of neutralizing and hemagglutination-inhibiting Abs that react with A_2 virus itself and with H_1N_1 (A_1) and H_0N_1 (A_0) subtype viruses, although the Abs to the A_1 and A_0 Ags are of lower titer. However, primary immunization with H_1N_1 (A_1) virus does not elicit neutralizing or hemagglutination-inhibiting Abs able to react with the H_2N_2 (A_2) virus. The complexity of the immunologic reactivities reflects the independent changes of the hemagglutinin and the neuraminidase, which are encoded in different genes (Fig. 58-7). With a major antigenic change (a new subtype) the chemical changes of the hemagglutinin and neuraminidase, or of the hemagglutinin alone, are of such magnitude as to add antigenic reactivites that do not crossreact with the prior surface Ags. However, the antigenic specificities of the major internal NP and M proteins may not change even with major antigenic shifts.

On successive exposures to influenza viruses, whether by infection or by artificial immunization, *the Ab response is predominantly directed against the Ags of the viral strain with which one was initially infected.* Thus if a child were infected first with an H_0N_1 (A_0) virus in 1940 and then with an H_1N_1 (A_1) in 1947, his Ab response in 1947 would be greatest to the H_0N_1 (A_0) viral Ags, although he would also develop H_1N_1 (A_1) Abs. With advancing age and an increased number of infections the Ab response to infection becomes broader, but the titer of Abs against the Ag of the original infecting virus remains the highest. This phenomenon, termed **"the doctrine of original antigenic sin,"** is reflected in the Ab levels of persons in different age groups (Fig. 58-9); it suggests that the initial encounter with influenza virus elicits a primary response and that subsequent meetings induce higher Ab titers owing to stimulation of an enlarged population of crossreactive memory cells persisting since the primary antigenic response. It is striking, however, that the immunologic response to infection with H_2N_2 (A_2) viruses appears to ignore the previously observed generality of this doctrine.

The prominent antigenic shift, resulting in the appearance of the major antigenic variants described, has led to speculation about whether an almost limitless number of major antigenic changes can occur. However, continued studies of "serologic archeology," which led to the concept of original antigenic sin, imply that the number of subtypes is limited. Thus before the H_2N_2 (A_2) pandemic of 1957 only persons older than 70 possessed Abs for the new A_2 virus (Fig. 58-9); hence the 1889–1890 influenza pandemic was probably caused by a virus immunologically similar to the A_2 virus of 1957, and the H_2N_2 viral subtype was absent between 1890 and 1957. Moreover, the next major variant, which produced extensive influenza epidemics from 1968 to 1977 (H_3N_2 virus), was closely related immunologically to the virus prevalent about 1900, suggesting that in 1957 the virus had gone full cycle since 1889–1890. These observations, as well as the appearance first of $H_{sw}N_1$ (swine influenza virus) in 1976 and then of H_1N_1 (A_1) in 1977, strengthen the thesis that only a limited number of antigenic variations are possible. However, since $H_{sw}N_1$ did not flourish and

spread, and H_0N_1 (A_0) did not emerge after the H_3N_2 (A_3) virus, it appears that the subtypes are not compelled to reappear in a set order, although the immune status of the population must serve as a major selective mechanism.

Immunity. Either clinically evident or inapparent infection leads to immunity, and the immunity to viruses of the same antigenic structure appears to be long-lasting. However, reinfections can be caused by variants with minor antigenic differences. *Immunity is induced by the hemagglutin,* since it can be evoked by injection of purified hemagglutinin. Neuraminidase probably also plays a role, however, for antineuraminidase Abs effectively reduce viral spread from infected cells and therefore diminish the impact of infection. Thus immunization of mice with purified neuraminidase does not prevent infection on subsequent challenge but does markedly reduce the extent of lung lesions and prevents transmission of virus to unimmunized cohorts. Abs directed against the nucleocapsid Ag, in contrast, do not confer immunity.

Since immunity is highly type-specific, and generally subtype-specific, infection or artificial immunization with one influenza H_1N_1 (A_1) virus affords immunity against infection with other H_1N_1 viruses but not against H_2N_2 (A_2) viruses (nor, of course, against influenza B viruses). Circulating antihemagglutinin and antineuraminidase Abs are present for many years after infection, and disease is usually mild when due to reinfection by a subtype similar to one previously experienced.

Cell-Mediated Immunity. In mouse infections virus-specific T lymphocytes appear that are cytotoxic for infected cells, reacting with virus-specific glycoproteins as well as with the broader-reacting M protein. Thus, the immune cytotoxicity exhibits crossreactivity between influenza viruses of the same genus (e.g., influenza A) as well as subtype specificity. The reaction requires that the cytotoxic cells share the H-2K or H-2D regions of the host histocompatibility complex. The role of cell-mediated immunity, and its persistence in man, require intensive exploration.

Artificial Immunization. This is limited not only by the marked antigenic variation of the viruses but also by the restriction of the infections to the respiratory mucous membranes, where **secretory IgA** Abs are required, and where Ab concentrations are only approximately 10% of those in the blood. Hence minor antigenic modifications of the infecting virus permit it to escape neutralization more readily than it could if viremia were an essential part of the infectious process. The situation is analogous to the outgrowth of drug-resistant bacterial mutants when the drug concentration is borderline.

GENETIC CHARACTERISTICS

The remarkable genetic variability of influenza viruses involves not only antigenic subtypes but also other genetic markers. These include 1) reactions with RBCs from animals of different species, 2) avidity for Ab, 3) virulence, 4) reactions with soluble mucoprotein inhibitors, 5) heat resistance, 6) host range, and 7) morphology. Only a few of these mutations have an obvious bearing upon the behavior of influenza viruses in nature, but they illustrate the ease with which these agents vary and the potential types of selective pressures at work in nature.

Variation in the avidity of viruses for specific Ab (Ch. 52) is frequently noted. Thus, viruses isolated in the course of a single epidemic may vary in their susceptibility to neutralization by antiserum prepared with homologous virus (isolated during the same epidemic) or with heterologous strains. Strains isolated during the height of epidemics commonly react to high titer only with homologous Abs. Passage of such strains in the presence of increasing quantities of Ab selects for variants that are inhibited only minimally by either homologous or heterologous Abs, suggesting that these viruses have been modified so that the usual surface Ags are present in reduced amounts or are partially obscured. The emergence of similar variants in nature may permit the persistence of virus in the population during interepidemic periods, setting the stage for the appearance of new antigenic subtypes. The appearance of strains having a minimal crossreactivity with prior viruses may well be responsible for the initiation of another epidemic.

Numerous mutants with increased virulence for a given host or organ system have been isolated. Conversely, **temperature-sensitive (ts) mutants** unable to multiply effectively at temperatures above 37°C are less virulent and may prove valuable for live virus vaccines.

Genetic recombination has been extensively studied in influenza viruses because of its special epidemiologic and clinical implications for such a variable virus, the opportunity (which was initially unique) to investigate RNA as genetic material, and the numerous markers available.

The first evidence of recombination between animal viruses was obtained by Burnet in 1949, using influenza viruses: infections with mixtures of neurotropic and nonneurotropic strains of different antigenic identity yielded recombinants in which neuropathogenicity from one strain was combined with an antigenic character from the other. Subsequently, genetic recombination has been observed with many other naturally occurring strains carrying various markers, and also with conditionally lethal ts mutants. Extraordinarily high recombination frequencies have been reported (up to 50%). The new genotypes, however, are not the consequences of true recombination (Ch. 11) within an RNA molecule but rather they emerge as the result of an independent assortment and segregation of separate segments of the viral genomes. The high frequency of reassortment recombination between influenza A viruses of different subtypes has made it possible to identify the viral protein encoded in a particular RNA segment (Fig. 58-7) by associating a given segment with a given polypeptide in recombinant viruses. This approach is possible because each RNA segment represents a single gene, and because RNA and proteins from different subtypes have unique electrophoretic mobilities. In addition, ts mutants have been isolated with identifiable defects in each of the viral genes.

Recombination is detected only within a genus, and not between influenza A and B viruses. When a mixed infection is initiated with high multiplicities of influenza A and B viruses, however, viral particles appear that have surface Ags of both parent viruses and are therefore neutralizable by Abs to either. This property results from **phenotypic mixing** (Ch. 50) and is not passed on to the progeny.

In a process similar to **marker rescue of phage** (see Genetic Reactivation of Ultraviolet-Inactivated Phage, Ch. 47), recombination may be observed between ultraviolet-inactivated and infectious viruses with different genetic markers. The frequency is low (10^{-2}–10^{-3} with closely related strains, and 10^{-4}–10^{-5} with distantly related strains). **Multiplicity reactivation** (see Genetic Reactivation of Ultraviolet-Inactivated Phage, Ch. 47) is similarly seen with viral preparations partially inactivated by heat or radiation: under conditions of multiple infection of a single cell the yield of infectious virus is unexpectedly increased.

HOST RANGE

Strains of human influenza viruses are best propagated experimentally in the amniotic cavity of chick embryos (the most sensitive and the most convenient host) or in the respiratory tract of ferrets or mice. Many strains also multiply readily in cultures of monkey kidney, calf kidney, and chick embryo cells. Viruses can be readily adapted to propagation in the allantoic cavity of the chick embryo, as well as in the respiratory tracts of monkeys and many rodents.

VIRAL MULTIPLICATION

Entry. Infection is initiated with the attachment of virions to susceptible host cells by reactions between the hemagglutinin spikes and specific host cell mucoprotein receptors.* The host receptors are similar to or identical with those on RBCs and with the soluble mucoprotein inhibitors in human and animals secretions (see Hemagglutination, Ch. 46). After attachment the viral particles fuse with the cytoplasmic membrane (Fig. 58-10). Infectivity is rapidly lost **(viral eclipse)** as the virions are shorn of their envelope, and their nucleocapsid enters directly

---

* Although the neuraminidase-containing spikes also can react with cell surface receptors, virions remain infectious after these spikes are removed by trypsin. Antibodies to neuraminidase also cannot neutralize infectivity although they block enzyme activity.

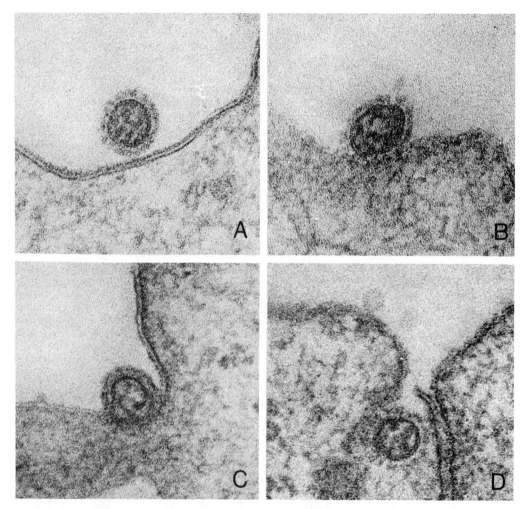

FIG. 58-10. Adsorption and penetration of influenza virus into cells of the chorioallantoic membrane of the chick embryo. **A.** Beginning attachment. **B.** Virion is attached to the cell and the viral envelope is fusing with the plasma membrane of the cell. **C.** Fusion of the viral envelope with the plasma membrane is more advanced. **D.** Nucleocapsid has penetrated into the cytoplasm. (Morgan C, Rose HM: J Virol 2:925, 1968) (×200,000)

into the cytoplasm through the rent in the fused virus–cell membranes.

Replication. The initial steps in viral replication following entrance of the viral genome into the cell are still unclear. Unlike other RNA-containing viruses (e.g., picornaviruses), influenza viruses cannot replicate in enucleated cells. Moreover, ultraviolet irradiation, dactinomycin, or mitomycin C blocks viral multiplication if administered during the first 2 h of infection but not thereafter (i.e., before synthesis of viral RNA is established). However, chemical inhibitors of DNA biosynthesis (e.g., arabinosylcytosine, hydroxyurea) do not reduce propagation of infectious virus. Hence, *functioning but not replicating host DNA is essential for early events* in

multiplication of influenza viruses. α-Amanitin, which inhibits RNA polymerase II (see Ch. 49), also blocks viral production. Thus the host continually supplies transcripts whose 5′ ends are cannibalized to provide caps for the 5′ termini of the viral mRNAs and to serve as primers for viral transcription.

Influenza viruses are negative-strand RNA viruses, and therefore contain an **RNA-dependent RNA polymerase (RNA transcriptase)** within the virion to transcribe the virion RNA segments into mRNAs. The assignment of viral proteins to RNA-dependent RNA polymerase activities is not precise, but temperature-sensitive mutants indicate that the virion transcriptase activity requires functional P_1 and P_3 proteins (Figs. 58-7 and 58-8), whereas the **replicase** activity also requires active P_2 and NP proteins for synthesis of new negative strands of virion RNA (Figs. 58-7 and 58-8). Two classes of complementary RNAs (cRNAs) are made in infected cells. 1) polyadenylated incomplete transcripts of the virion RNAs (terminated about 17 nucleotides from the conserved 5′ ends) which are associated with polysomes and serve as the mRNAs; and 2) non-polyadenylated complete transcripts, which are the templates for virion RNAs (Fig. 58-6).

Primary virion transcription, detected with radiolabeled probes within the first hour of infection, can occur in the absence of protein synthesis and yields only mRNAs. Replication of the nonpolyadenylated cRNAs requires synthesis of viral proteins, apparently to modify the RNA polymerase, permitting production of complete transcripts. The complementary RNAs are predominant during the first 2–3 h of infection, and virion RNAs predominate thereafter. Both classes of complementary RNAs and the virion RNAs are made separately for each viral RNA segment through the usual intermediary replicative forms (Ch. 50). The viral RNAs are probably synthesized in the nucleus.

The nucleocapsid protein (detected with fluorescein-labeled or ferritin-labeled Abs) is synthesized in the cytoplasm and is rapidly transported into the nucleus, where a significant proportion of the viral RNA is also found (Figs. 58–11 and 58–12). The nucleocapsid sub-

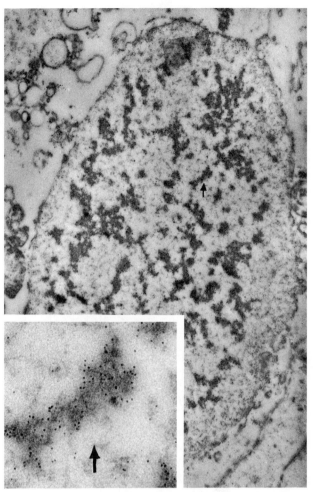

FIG. 58-11. Multiplication of influenza A virus. Diagram of the biosynthesis of virions and viral subunits measured by titrations and immunofluorescence.

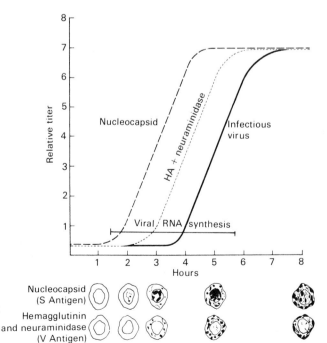

FIG. 58-12. Nucleus of influenza virus-infected cell showing aggregates of dense material labeled with ferritin-conjugated specific A **(arrow)**. The chromatin is sparse and the nuclear membranes are disrupted. (×26,000) **Inset.** Higher magnification of the portion of the nucleus marked by arrow. Ferritin-conjugated Ab is present within the aggregates of dense material. The intervening nuclear matrix is nearly devoid of ferritin, i.e., nucleocapsid Ag. (×97,000) (Morgan C et al: J Exp Med 114:833, 1961)

sequently moves into the cytoplasm and migrates to the cell membrane. The hemagglutinin and neuraminidase proteins remain in the cytoplasm throughout replication (Fig. 58–11).

Assembly. About 4 h after infection with influenza A virus the virion M protein becomes associated with the inner surface of the cell plasma membrane, and discrete patches of the membrane thicken and incorporate hemagglutinin and neuraminidase molecules, which gradually replace the host proteins in these segments (Figs. 58–13 and 58–14). As segments of the helical nucleocap-

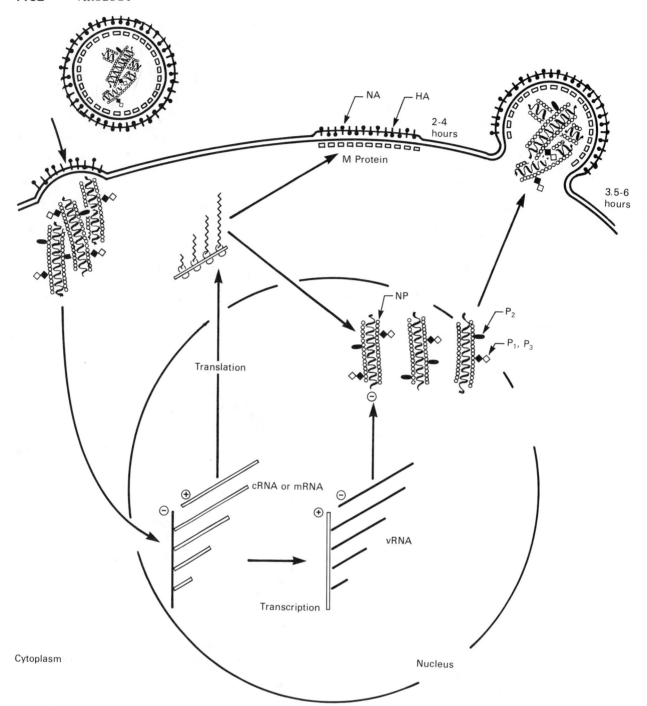

FIG. 58-13. Schematic representation of the steps in the biosynthesis of influenza viruses, from adsorption, fusion of the viral envelope, and penetration of the nucleocapsid into the cytoplasm to insertion of newly made viral proteins in the cell's plasma membrane, assembly, and budding of the virion. The site of viral RNA transcription and replication is unknown but probably occurs in the nucleus.

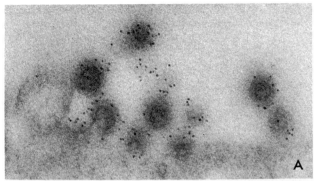

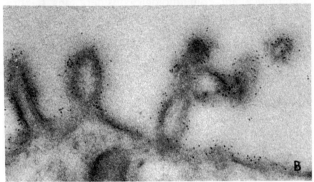

FIG. 58-14. Development of influenza virus particles at the surface of an infected cell. Viral particles and components labeled with ferritin-conjugated specific Ab. **A.** In addition to fully formed viral particles, ferritin–Ab complexes have tagged one particle (to the left of center) presumed to be in the process of budding and two others (below and to the right) probably at early stages of differentiation. (×140,000) **B.** Surface of an infected cell with several cytoplasmic protrusions to which ferritin–Ab complexes have combined as a result of the virus-specific antigenic change in the surface. No mature viral particles are evident. (×97,000) (Morgan, C et al: J Exp Med 114:825, 1961)

sid impinge on the altered membrane it buds and forms viral particles, which are released as they are completed; virions cannot be detected within the cell.

Although the assembly of virions is an imperfect process, yielding virions of considerable morphologic heterogeneity and many noninfectious particles, it must have an effective control for packaging the appropriate set of nucleocapsid segments. Thus recombinants derived from two subtypes containing distinguishable RNA segments never contain two copies of the same gene from different parents (e.g., a single virus does not contain the hemagglutinin genes from two parents).

Viral particles are released over many hours, without lysis of the infected cells, but eventually the cells die. The mechanism for releasing the budding virions is unclear. Neuraminidase may serve this function since specific Ab to neuraminidase blocks viral release though it cannot neutralize infectivity. However, univalent Fab fragments

of this Ab do not reduce viral release, though they neutralize neuraminidase activity in vitro. The bivalent Ab may thus block viral release by binding virions to the membrane rather than by inhibiting specific enzyme activity.

The sequence of events with influenza B virus is similar, but the latent period (identical with the eclipse period for viruses that are assembled at the cell membrane) is 1–2 h longer.

TOXIC PROPERTIES

Without production of virions, influenza A and B viruses in sufficient amounts can damage cells and induce a pyrogenic effect. Intravenous or intraperitoneal inoculation into mice causes hepatic necrosis and splenic enlargement in 12–48 h, while intracerebral inoculation causes brain damage accompanied by convulsions. Fever follows 1–2 h after intravenous inoculation of virus into rabbits, owing to the elaboration of an endogenous pyrogen by polymorphonuclear leukocytes (PMNs). In vitro the reaction of viral particles with PMNs inhibits the leukocytes from phagocytizing bacteria, which may be critical in producing secondary bacterial pneumonias (see below).

PATHOGENESIS

Influenza, an acute respiratory disease associated with constitutional symptoms, results from infection and destruction of cells lining the upper respiratory tract, trachea, and bronchi. Virus enters the nasopharynx and spreads to susceptible cells, whose membranes contains the specific mucoprotein receptors. The virus must first pass through respiratory secretions; and though these contain mucoproteins that can also combine with viral particles, infection is not blocked because the viral neuraminidase hydrolyzes the mucoproteins, rendering them ineffective as inhibitors.*

During acute illness the ciliated epithelial cells of the upper respiratory tract are primarily involved. Viral multiplication is followed by necrosis of infected cells and extensive desquamation of the respiratory epithelium, which is directly responsible for the respiratory signs and symptoms of the acute infection. In nude mice or in mice treated with anti-Θ serum experimental infections are mild, implying that most of the cell damage is caused by virion-activated cytotoxic T lymphocytes that recognize viral Ags on infected cells. (Such activated T cells are also cytotoxic to cells infected with other influenza viruses of the same genus.)

Early in the course of the uncomplicated disease constitutional symptoms—fever, chills, generalized aching

* The evolutionary selection of viruses containing neuraminidase is understandable. Were influenza virus devoid of its surface hydrolytic enzyme, the secretory mucoprotein would be as effective as Abs and infection would be difficult to establish.

(particularly muscular), headache, prostration, and anorexia—are more prominent than would be expected from a local infection of the respiratory tract. Viremia, however, is not an essential event in the pathogenesis of influenza infection, and although it has been detected on rare occasions, the constitutional symptoms are probably due to breakdown products of dying cells absorbed into the blood stream. The liberation of endogenous pyrogen from PMNs (see Toxic Properties, above) is probably not important in producing fever, since few such cells are found in the upper respiratory tract.

Normally, influenza is a self-limited disease lasting 3–7 days. About 10% of patients with clinical influenza have small areas of lobular pulmonary consolidation. **Secondary bacterial pneumonias** are the major cause of death. Although **fatal primary influenza virus pneumonia** without bacterial invasion is rare, it accounts for the unusual deaths from primary influenza. It occurs most frequently in persons with diminished respiratory function (e.g., those with mitral stenosis or chronic pulmonary disease, and women during the later stages of pregnancy). Viruses isolated from such fatal cases seem no more virulent in experimental animals than those causing minor illness.

Fatal nonbacterial pneumonia was more common in the 1918–1919 pandemic, and postmortem descriptions of the rare deaths in more recent epidemics are identical. The lungs appear unforgettably huge, distended with edema fluid and blood which pours out if the lung is cut. Microscopically, the tracheal and bronchial epithelium shows marked destruction and often denudation. The bronchioles and alveoli are distended with cell debris, blood, and edema fluid, but purulent exudate is absent. Fibrin-free hyaline-like membranes often line the alveoli. Apparently, the pathogenesis of influenza pneumonia evolves in two stages: from bronchial and bronchiolar epithelial necrosis to hemorrhage and edema.

The highly fatal pneumonia during the pandemic of 1918–1919 was predominantly characterized by **secondary bacterial complications.** The most prominent invaders were *Staphylococcus aureus, Hemophilus influenzae,* and β-hemolytic streptococci. In recent epidemics, coagulase-positive *S. aureus* and pneumococci have been most frequent. In addition to the usual severe epithelial injury of influenza, the secondary bacterial pneumonias show **deeper invasion** of the walls of the bronchi and bronchioles by bacteria, destruction of alveolar walls, purulent exudate, abscess formation, and vascular thrombi.

The influence of influenza virus infection on bacterial invasion has been studied experimentally. Introduction of pneumococci or streptococci into the lungs of healthy mice results in little if any inflammatory response, but extensive bacterial pneumonia develops in animals with pulmonary edema induced by influenza virus or irritant chemicals. This effect of influenza viruses is probably promoted by their capacity to inhibit phagocytosis of bacteria by PMNs.

Extrapulmonary lesions have also been observed in fatal cases of influenza pneumonia, including central nervous system involvement such as the Guillain-Barré syndrome (ascending myelitis) or encephalitis; Reye's syndrome (encephalopathy and fatty liver), which inexplicably has been particularly associated with influenza B epidemics; hemorrhage into the adrenals, pancreas, and ovaries; and renal tubular degeneration. Viruses have been isolated from such lesions only rarely, however, and the relation of viral infection to the lesions is obscure.

LABORATORY DIAGNOSIS

A presumptive diagnosis of influenza can often be made from clinical and epidemiologic considerations. Laboratory confirmation of a clinical diagnosis is generally too costly for individual or sporadic cases, but it is used to establish the presence of the agent in the community, to determine its specific type, and to carry out epidemiologic studies. Diagnosis may be established by: 1) viral isolation, 2) demonstration of specific Ab increase, and 3) immunofluorescent demonstration of specific Ags in epithelial cells present in nasal secretions or sputum.

Virus is usually **isolated** by inoculating nasal and throat washings or secretions into the amniotic sac of 11- to 13-day-old chick embryos or onto monolayers of monkey kidney cell cultures. The latter have the advantage of also supporting multiplication of many respiratory viruses other than influenza. Fresh isolates of influenza virus may fail to produce cytopathic changes in monkey kidney cells, and the presence of virus is best detected by the **hemadsorption** technic, employing guinea pig RBCs. With embryonated eggs, virus can be detected in the amniotic fluid, after 2–3 days' incubation, by **hemagglutination,** using guinea pig or human RBCs for influenza A and B viruses and chick RBCs for influenza C virus. The newly isolated virus from either chick embryos or cell cultures is usually typed by **hemagglutination inhibition,** using standard antisera.

Immunologic methods are used most frequently for diagnosis of influenza infection. Because the majority of persons already have influenza virus Abs at the time of infection, it is essential to demonstrate an **increase** by comparing titers in serum specimens obtained during both the acute and the convalescent phases of the disease (paired sera). **Hemagglutination-inhibition** technics are most often employed for this purpose, although they are handicapped by the troublesome presence in serum of **nonspecific** mucoprotein or protein **inhibitors.** These may be eliminated by treating the serum, after heat inactivation (56° C for 30 min.), with either the receptor-destroying enzyme (RDE) from *Vibrio cholerae* (a neuraminidase), or a mixture of trypsin and potassium periodate, or the adsorbent kaolin.

The **complement-fixation (CF)** assay is equally sensitive, and it circumvents the difficulties presented by nonspecific serum inhibitors. With crude preparations, which contain the nucleocapsid antigen, CF has broad specificity and can identify only viral type; but if one utilizes the hemagglutinin separated from virions the assay is just as strain-specific as hemagglutination inhibition. An increase in specific Abs can also be measured by **neutralization** titrations, but because of its greater expense and the time required this procedure is employed only for special purposes.

Immunofluorescent technics furnish a method for establishing the diagnosis of influenza while the patient is still acutely ill: fluorescein-conjugated Abs reveal the presence of virus-specific Ags in desquamated cells from the nasopharynx. This technic permits diagnosis of about 75% of the infections detected by viral isolation or by immunologic assays.

EPIDEMIOLOGY

Influenza occurs in recurrent epidemics which start abruptly, spread rapidly, and are frequently of worldwide distribution. An influenza A epidemic generally appears every 2–4 years and an influenza B epidemic every 3–6 years, but the patterns have not been completely predictable. Although epidemics occur periodically in any given geographic locality, outbreaks occur somewhere every year. Epidemics of influenza A viruses are usually more widespread and more severe than those of influenza B. Influenza C virus has not caused epidemics, and it usually produces inapparent infections.

The incidence is highest in the age group 5–9 years; above 35 it gradually declines with increasing age. The very young and the very old suffer the highest mortality, with about three-quarters of deaths occurring in those over 55 years of age. Indeed, even without virologic or serologic evidence an influenza epidemic is recognizable by the increased mortality due to pneumonia in the elderly. Other special groups showing elevated mortality include pregnant women and persons with chronic pulmonary disease or cardiac insufficiency. A striking exception to the usual age-related mortality rate, however, was noted in the severe 1918–1919 pandemic: the majority of the 20 million deaths occurred in young adults, probably because older persons had previous exposure to and hence some immunity against this unusually virulent influenza virus.

Epidemics are common from early fall to late spring. Outbreaks often develop in many places in a country at almost the same time and spread rapidly to neighboring communities and countries; with the common use of air travel, intercontinental spread has also become rapid. The rapid dissemination is not entirely due to the speed of modern transportation, however, for this characteristic was also noted when man could travel no faster than the speed of his horse. The pattern of epidemic spread may be related to the occurrence of sporadic cases and the probable seeding of the virus in the population several weeks prior to an explosive outbreak. Even during an epidemic, a **high ratio of infection to disease,** from 9:1 to 3:1, can be demonstrated serologically.

The pathway of widespread dissemination of the virus has now been exemplified by several well-studied episodes, such as the 1957 pandemic caused by a new variant, H_2N_2 (A_2, Asian) subtype, which apparently emerged from central China in February of 1957 (Fig. 58-15). The arrival of the virus in the United States was detected in Naval personnel in Newport, R.I., on June 2 and shortly thereafter in San Diego, Cal., without a traceable connection between the two episodes. The first civilian outbreak was observed at a conference in Davis, Cal., on June 20, followed by several similar small episodes elsewhere in California. From the conference the virus was carried directly by some of the more peripatetic members to a meeting of young people in Iowa, and from this location the virus was seeded throughout the country. This initial dissemination resulted in small, sporadic outbreaks until September, when epidemics occurred in almost all parts of the country. Similar spread along paths of travel occurred throughout the world.

In the summer of 1968, after an appropriate period of antigenic drift (11 years), another influenza A variant (H_3N_2) appeared in Hong Kong and produced a mild but widespread pandemic whose spread was strikingly similar to that of 1957. Although the 1968 Hong Kong virus had antigenic characteristics clearly different from the previously isolated H_2N_2 (A_2) viruses, the immunologic relatedness (owing to crossreacting neuraminidase) was sufficient so that the new isolates are not considered a new subtype. However, the H_3N_2 viruses isolated from epidemics in 1969 and 1970 showed further antigenic drift, suggesting that a new A subtype may soon emerge.

When swine influenza ($H_{sw}N_1$) virus was isolated from a fatal illness and from approximately 200 nonfatal infections at Fort Dix, N.J., in the spring of 1976, it was postulated that this virus would be the next pandemic subtype. This prediction was based on the concept that the genetic variation of influenza A virus is limited (see Basis for Antigenic Variations, above) and that the emergence of subtypes is cyclic. However, the feared $H_{sw}N_1$ virus did not spread, and the previously prevalent H_3N_2 subtype (A Victoria) remained epidemiologically viable. The emergence of an H_1N_1 (A_1) virus in China and Russia in 1977 and its rapid spread to other continents heralds this subtype as a prevalent species and again suggests the limited variations of influenza viruses.

Many questions concerning the epidemiology of influenza remain unanswered. Not the least puzzling among the unknowns are the following: 1) Why does the virus not spread rapidly at the time of the initial infections in a community? 2) Where is the virus during interepidemic intervals? 3) How does the virus become "masked" or "go underground" in the interval between its seeding and the occurrence of an epidemic? 4) What provocative factors induce the epidemic?

After seeding, or during the interval between epidemics, virus may conceivably 1) simply be transmitted slowly, producing inapparent infections or sporadic cases; 2) remain latent in the persons previously infected; or 3) reside, active or latent, in an animal reservoir. Influenza virus has rarely been isolated in nonepidemic periods, which speaks against the first possibility. The intriguing ecology of swine influenza virus, which is activated by cold weather in the presence of *H. influenzae suis* (Ch. 53, Latent Viral Infections),

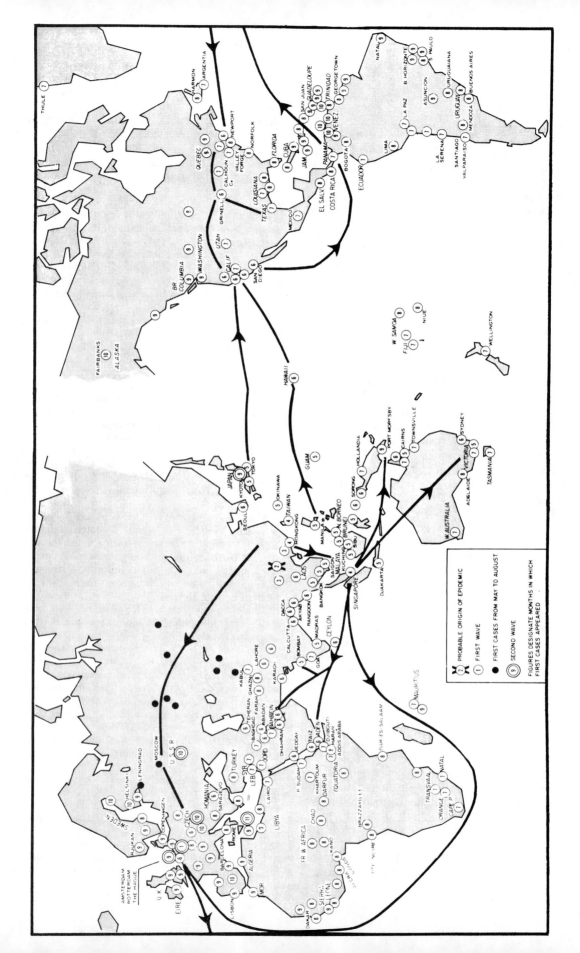

offers one example of the second mechanism. Finally, human strains of influenza A virus show immunologic and genetic relations to influenza viruses of horses, ducks, chickens, and pigs,* which supports the third mechanism and also suggests that animals may serve as a source of new variants by recombination with human strains.

The appearance of influenza viruses after a silent interval no doubt frequently depends on the development of a new antigenic variant that can escape an immunologic barrier existing in nature. However, "old" strains, having only minimal antigenic changes, also initiate epidemics, presumably because Abs in the population fall below the level necessary to prevent infections. The nature of the provoking factors that initiate an epidemic remains a mystery.

PREVENTION AND CONTROL

The high incidence of inapparent infections, the short incubation period, and the high infectivity preclude the successful use of isolation or quarantine procedures to control influenza. Quarantine of travelers entering a country can delay but not prevent the entrance of virus. In South Africa in 1957, for example, where ships were quarantined and the passengers and crew forbidden to land, infection did not enter through the ports; but virus finally entered from the north, probably being carried by immigrant laborers traveling overland.

Artificial immunization can prevent influenza to a significant extent (reducing the incidence up to 70%), but not completely. Viruses propagated in chick embryos, partially purified, and inactivated by formalin, can provide a highly effective vaccine if the viruses utilized include a strain whose hemagglutinin and neuraminidase glycoproteins are closely related immunologically to the currently prevalent strain. This requirement may be difficult to satisfy. For example, although influenza vaccines containing an H_0N_1 (A_0) virus were highly effective in 1943 and 1945, the vaccine employed in 1947 failed because it did not include the newly emerged antigenic variant, the H_1N_1 (A_1) virus. Hence, influenza must be under constant global surveillance, including accurate antigenic characterization of isolated strains.† Vaccines currently employed contain a mixture of several strains

---

* The N_1 in human viruses extant during the late 1930s is immunologically related to the neuraminidase present in at least four of the avian influenza viruses as well as to that in the swine virus; the N_2 present in the H_2N_2 and H_3N_2 viruses is similar to the neuraminidase of a turkey influenza virus isolated in 1966; and the hemagglutinin of the H_3N_2 virus is immunologically similar to the hemagglutinin of an equine influenza virus ($H_{eq2}N_{eq2}$).
† The World Health Organization has established centers throughout the world for this purpose.

FIG. 58-15. Progress of Asian influenza pandemic from its probable origin in Central China, Feb. 1957 to Jan. 1958. (Langmuir AD: Am Rev Resp Dis 83:1, 1961)

of influenza A and B viruses in order to cover the known antigenic spectrum. New major antigenic variants are added as they appear.

Despite the proved value of the available inactivated viral vaccines several factors have limited their use and possibly their effectiveness. 1) Pyrogenic reactions, accompanied by constitutional symptoms (not unlike the manifestations of mild influenza) and by local reactions, have been common, particularly in infants and young children; an incidence of 10%–20% is not unusual in children less than 6 years of age, even when partially purified whole virus is used. 2) Secretory IgA Abs in respiratory secretions are probably critical for successful protection, but subcutaneous injection of inactivated virus induces only low levels of such Abs in the respiratory tract. 3) Abs begin to decrease about 3 months after immunization and immunity is often lost within 6 months.

Generally, a single subcutaneous injection of 0.1–1 ml containing at least 300 chick cell agglutinating (CCA) units per inoculum will confer immunity in 2–4 weeks. Persons immunized with a new subtype, especially children, show a primary immunologic response, while those who have had previous exposure to the Ags in the vaccine exhibit a secondary response. Therefore, if the vaccine contains a new major antigenic variant to which most individuals have no detectable Abs, such as H_2N_2 (A_2) in 1957, a **second injection** is recommended a month after the first.

New methods for preparation and administration of vaccines are now being tested with encouraging results. 1) **Subunit vaccines,** consisting of the hemagglutinin and neuraminidase components of disrupted virions, are being used rather widely. This formulation permits administration of a greater antigenic mass, therefore effecting greater Ab responses, with fewer major toxic reactions. 2) **Attenuated infectious viruses** (selected by serial passage at low temperatures) have been widely used in the Soviet Union as well as in other countries, and those selected as temperature-sensitive mutants are being employed experimentally in the United States. The attenuated viruses have been shown to produce IgA Abs in respiratory secretions. With live vaccines, however, it is impossible to select and test rapidly a suitable derivative of the viral subtype that has recently emerged. 3) **Intranasal aerosol administration** of inactivated viral vaccine induces an adequate response of IgA Abs in nasal secretions as well as of specific IgA, IgM, and IgG Abs in serum, but this method has not yet been adequately evaluated. 4) **Vaccines prepared with adjuvants** (such as light mineral oil, peanut oil, or Arlacel A) result in a greater and more persistent Ab response. As yet, adjuvants have not been widely employed for fear of generalized reactions to the adjuvant and increased toxicity of the vaccine. The validity of the former concern is uncertain and vaccine in adjuvant may become more generally accepted. Use of purified viral vaccines should largely eliminate the latter problem.

A new approach to immunization has developed from the ability to "tailor-make" variants of influenza viruses by **genetic recombination** of a new antigenic variant with an avirulent temperature-sensitive mutant, or with an established strain, in order to ensure propagation of the newly emerged subtypes to high titers. (The swine A influenza $H_{sw}N_1$ vaccine virus used for nationwide immunization in 1976 was prepared by this method from the Fort Dix isolate.)

Although effective vaccines can now be developed, the implications of their widespread use merit consideration. It is undoubtedly desirable to protect persons at high risk from the severe complications of influenza. The advisability of mass immunization, however, may depend on whether influenza viruses have a virtually unlimited capacity for antigenic mutation, or whether the antigenic variations are very limited in their potential extent. If infinite antigenic variation is possible, mass immunization programs, which would increase herd immunity, might serve to accelerate the selection and appearance of new antigenic subtypes. But if the potential number of antigenic mutants is limited, which now seems likely, mass immunization with a broad antigenic spectrum of viruses would result in a population with broadly reactive Ab levels adequate to provide an effective barrier to the spread of influenza viruses.

Widespread immunization has revealed an additional source of concern, at least with inactivated intact virus vaccine. Following immunization of over 35 million persons with inactivated swine influenza ($H_{sw}N_1$) virus in the fall of 1976, the Guillain-Barré syndrome occurred in 354 recipients, with 28 deaths. Most instances occurred 2–4 weeks after immunization, and the incidence was almost 10 times greater than that in the nonimmunized population in the same period. These findings point to a causal relation between influenza immunization and this neurologic complication, but the etiology of the Guillain-Barré syndrome is not specifically related to influenza immunization. It has also been reported following other immunization procedures (e.g., rabies, smallpox), but the scale and the surveillance of the 1976 immunization program was never previously equaled so that an etiologic association could not be established with other immunogens. Such complications, in a mass program that retrospectively proved unnecessary, not only brought criticism on the use of inflenza vaccine but unfortunately also discouraged public acceptance of other immunization programs.

Chemotherapy. The discovery that **amantadine (1-adamantanamine)** can inhibit an early step in the multiplication (uncoating) of some influenza viruses (Ch. 51) reawakened hopes for the successful chemical control of viral diseases. This compound proved modestly successful in preventing influenza H_2N_2 (A_2) infections in limited clinical trials, providing about 50% protection against infection. Its clinical usefulness is limited, however, because its therapeutic value has been less striking than its prophylactic effect, its effectiveness is restricted to influenza A viruses (it does not affect influenza B), and it has neurologic toxic effects (particularly in the aged).

Ribavirin (1-β-D-ribofuranosyl-1,2,4-triazole-3-carboxamide, virazole), which inhibits synthesis of viral RNA by blocking guanine biosynthesis, has been more effective than amantadine in preventing experimental influenza A infection in cell cultures and animal models. Limited clinical trials suggest that it may be therapeutically effective.

SELECTED READING

BOOKS AND REVIEW ARTICLES

ANDREWES CH: Factors in virus evolution. Adv Virus Res 4:1, 1957

British Medical Bulletin: Schild GC (ed): Influenza. London, Medical Department, The British Council, 1979

COMPANS RW, CHOPPIN PW: Reproduction of myxoviruses. Compr Virol 4:179, 1975

DOWDLE WR, KENDAL AP, NOBLE GR: Influenza viruses. In Lennette EH, Schmidt NJ (eds): Diagnostic Procedures for Viral and Rickettsial Diseases, 5th ed. New York, American Public Health Association, 1980

KILBOURNE ED (ed): The Influenza Viruses and Influenza. New York, Academic Press, 1975

PALESE P: The genes of influenza virus. Cell 10:1, 1977

STUART-HARRIS CH, SHIELD GC: Influenza. The Virus and the Disease. Littleton, MA, Publishing Sciences Group, 1976

WEBSTER RG, LAVER WG: Antigenic variation in influenza virus. Prog Med Virol 13:271, 1971

SPECIFIC ARTICLES

DAVENPORT FM, MINUSE E, HENNESSY AV, FRANCIS T JR: Interpretations of influenza antibody patterns of man. Bull WHO 41:453, 1969

KILBOURNE ED: Future influenza vaccines and the use of genetic recombinants. Bull WHO 41:643, 1969

PLOTCH SJ, BOULOY M, KRUG RM: Transfer of 5′ terminal cap of globin mRNA to influenza viral complementary RNA during transcription in vitro. Proc Natl Acad Sci USA 76:1618, 1979

SCHULMAN JL, KILBOURNE ED: Independent variation in nature of hemagglutinin and neuraminidase antigens of influenza virus: distinctiveness of hemagglutinin antigen of Hong Kong/68 virus. Proc Natl Acad Sci USA 63:326, 1969

ZWIERINK HJ, COURTNEIDGE SA, SKEHEL JJ et al: Cytotoxic T cells kill influenza virus infected cells but do not distinguish between serologically distinct type A viruses. Nature 267:354, 1977

chapter

PARAMYXOVIRUSES

The paramyxoviruses (Paramyxoviridae) differ widely pathogenically. **Parainfluenza** and **respiratory syncytial viruses** produce **acute respiratory diseases, measles virus** causes a **generalized exanthematous disease,** and **mumps virus** initiates a **systemic disease** of which **parotitis** is a predominant feature. However, on the basis of chemical and several biologic properties the paramyxoviruses are relatively homogeneous (see Table 59-1). There are also sufficient differences in their characteristics to permit their classification into three genera: **Paramyxovirus; Morbillivirus;** and **Pneumovirus.** Respiratory syncytial virus, a pneumovirus, which cannot hemagglutinate or cause hemolysis of RBCs, differs most sharply from the other paramyxoviruses.

GENERAL PROPERTIES OF THE VIRUSES

The characteristics that are similar for all paramyxoviruses will be discussed in this section; the distinctive properties will be described in the following sections on the individual viruses.

Morphology. The virions are **roughly spherical enveloped particles** of heterogeneous sizes (Table 59-1), larger than influenza viruses (see Table 58-2). Electron microscopic examination of negatively stained virions discloses that they appear similar to orthomyxoviruses. The intact viral particle has a well-defined **outer envelope,** about 10 nm thick, covered with short (8–12 nm) **spikes** that are more or less regularly arranged (Fig. 59-1). Disruption of the envelope reveals an inner **helical nucleocapsid,** and serrations with a regular periodicity of about 5 nm (Figs. 59-2 through 59-4). The nucleocapsid is distinctly different from that of influenza viruses (see Table 58-2): its diameter is approximately twice as great; its characteristic periodic serrations are more discrete; and it can be isolated from the virion as a single long helical structure. Filamentous virions are observed in thin sections of infected cells but not in negatively stained preparations, because they are apparently disrupted during fixation.

Physical and Chemical Characteristics. The paramyxoviruses that have been purified and analyzed all have a similar composition: approximately 74% protein, 19% lipid, 6% carbohydrate, and 1.0% RNA. The nucleocap-

sid contributes 20%–25% of the mass of the virion; it consists of a single species of protein (NP) and **one large molecule of single-stranded RNA** having **negative polarity** (Table 59-1). The viral envelope contains three proteins. Two glycoproteins form the surface projections (Fig. 59-5): one (HN) has both **hemagglutinin and neuraminidase** activities; the other (F) is responsible for the virion's **hemolytic and cell fusion** functions. The third protein (M), which is nonglycosylated, forms the **inner layer of the envelope,** maintaining its structure and integrity.

Because the viral envelope has a high lipid content, organic solvents or surface-active agents rapidly inactivate the virions by dissolving their envelopes, thus liberating the envelope proteins and the nucleocapsids from the disrupted particles.

The virions are relatively unstable, losing 90%–99% of their infectivity in 2–4 h when suspended in a protein-free medium at room temperature or at 4° C.

Viral Multiplication. The multiplication cycles do not appear to differ for each paramyxovirus, except for the great variations in the length of various phases. For example, the eclipse period is 3–5 h for parainfluenza viruses, 16–18 h for mumps virus, and 9–12 h for measles virus. The basic biosynthetic events (predominantly derived from studies of parainfluenza and Newcastle disease viruses) are similar to those for other enveloped viruses that possess a negative single-stranded RNA and contain a RNA-dependent RNA polymerase within the virion (see Synthesis of RNA Viruses, Ch. 50), but there are several unique features. In contrast to orthomyxoviruses, paramyxoviruses multiply without restraint in the presence of actinomycin D. The unsegmented virion RNA has only a single promoter site for the RNA polymerase, and hence the monocistronic mRNAs are probably derived either by cleavage of a transcript of the entire genome or by precise sequential transcription, termination, and reinitiation, yielding a series of mRNAs of various lengths up to the size of the virion RNA. Cleavage and processing of a large transcript seems more likely, since excessive numbers of virion-length, single-stranded RNA molecules complementary to viral RNA are present in infected cells.

Virion infectivity, as well as hemolytic and cell fusion

TABLE 59-1. Characteristics of Human Paramyxoviruses

Common properties	
Average size	125–250 nm (range 100–800 nm)
Nucleocapsid diameter	18 nm (except RSV = 14 nm)
Viral genome	5–6 × 10^6 daltons (17–20 Kb); single negative-strand molecules
Virion RNA polymerase	+
Reaction with lipid solvents	Disrupts
Syncytial formation	+
Cytoplasmic inclusion bodies	+ (Measles virus also nuclear)
Site of multiplication	Cytoplasm

Distinguishing properties	Parainfluenza	Mumps	Measles	RSV
Hemagglutinin*	+	+	+	–
Hemadsorption†	+	+	+	–
Hemolysin	+	+	+	–
Neuraminidase	+	+	–	–
Antigenic types	4	1	1	1
Antigenic relationships	Mumps	Parainfluenza	–	–
Genus	Paramyxovirus	Paramyxovirus	Morbillivirus	Pneumovirus

*Chicken and guinea pig RBC.

†Infected cells adsorb guinea pig RBCs.

+ = Present; – = absent

FIG. 59-1. Electron micrograph of type 3 parainfluenza virus embedded in phosphotungstate. The envelope **(E)**, peripheral short projecting spikes **(H)**, and internal nucleocapsid **(NC)** are apparent. The helical nucleocapsid can also be seen escaping from a small break in the envelope. (Courtesy of A. P. Waterson, St. Thomas's Hospital Medical School, London) (×210,000)

FIG. 59-2. Partially disrupted mumps virus particle showing the envelope and the hollow helical strands forming the nucleocapsid. Several broken strands have been released, revealing periodic structures **(A)** which make up the helical nucleocapsid. Fine threads connecting separated pieces are visible at **B.** (Horne RW, Waterson AP: J Mol Biol 2:75, 1960. Copyright by Academic Press, Inc., (London) Ltd.) (×250,000)

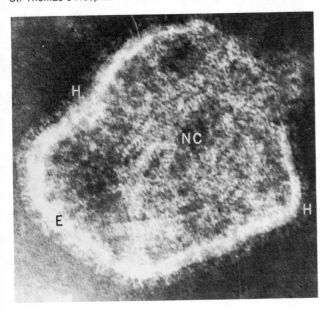

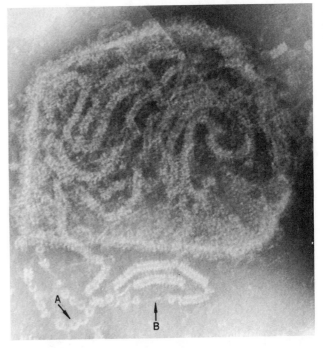

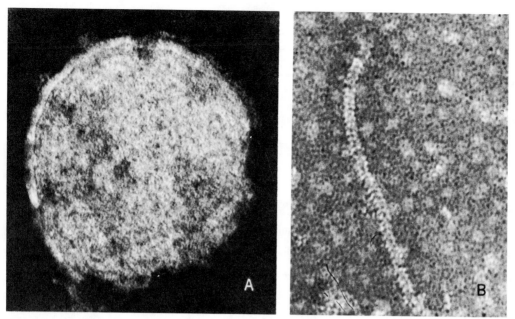

FIG. 59-3. Fine structure of measles virus, revealed by negative staining with sodium phosphotungstate. **A.** Particle showing the characteristic envelope and peripheral projections. The nucleocapsid is tightly packed and shows an appearance of concentric rings toward the periphery. (×280,000) **B.** Portion of the helical nucleocapsid released from a disrupted virion. (Horne RW, Waterson AP: J Mol Biol 2:75, 1960 Copyright by Academic Press, Inc., (London) Ltd.) (×240,000)

FIG. 59-4. Strand of helical nucleocapsid of type 1 (Sendai) parainfluenza virus embedded in phosphotungstate. (Examination by tilting through large angles reveals the sense of the helix to be left-handed.) (Horne RW, Waterson AP: J Mol Biol 2:75, 1960. Copyright by Academic Press, Inc., (London) Ltd.) (×200,000)

FIG. 59-5. Diagram of a paramyxovirus. (Different forms are shown for the **F** and **HN** glycoproteins to indicate their chemical differences, although they are not distinguishable in electron micrographs.) The F and HN glycoproteins appear to penetrate into the lipid bilayer and may traverse the entire envelope. The M protein forms the inner layer of the envelope and maintains its structure and integrity. The actual arrangement of the NP protein in the nucleocapsid is unknown.

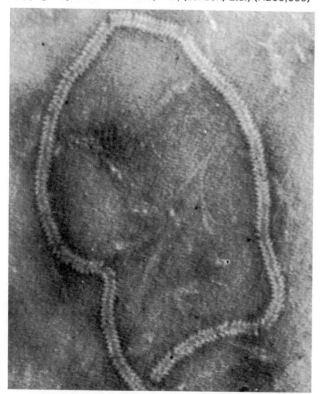

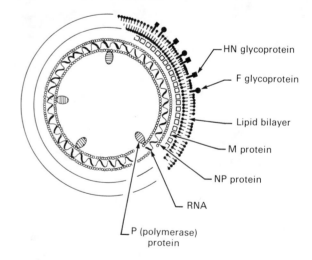

- HN glycoprotein
- F glycoprotein
- Lipid bilayer
- M protein
- NP protein
- RNA
- P (polymerase) protein

properties, requires **maturation of the F glycoprotein** by proteolytic cleavage of a larger precursor (F_0) by a cellular enzyme. With most paramyxoviruses the NP proteins of the nucleocapsid, as well as the HN (hemagglutinin–neuraminidase) protein, are detected only in the cytoplasm of infected cells, but the measles virus nucleocapsid is also present in the nucleus.

After viral RNA is synthesized in the cytoplasm, it is rapidly associated with the newly made nucleocapsid protein, but only a relatively small proportion of the nucleocapsids thus formed are assembled into virions. Occasionally, positive strands, owing to their excess, may be accidentally assembled in virions as well as in nucleocapsids. The cytoplasmic inclusion bodies (Fig. 59-6) are predominantly accumulations of the excess nucleocapsids.

Electron microscopic examination of infected cells reveals the remarkable assembly and maturation of paramyxoviruses at the plasma membrane (Fig. 59-7). Strands of viral nucleocapsid can be seen to associate with the viral M protein lining the regions of thickened cell membranes containing virus-specific proteins: these differentiated cell membranes are destined to become the viral envelopes. Intact viral particles are noted only at the cell membrane, where they are assembled and released

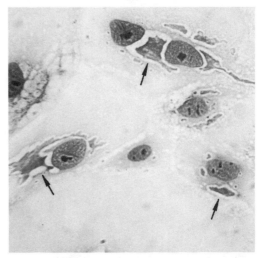

FIG. 59-6. Eosinophilic cytoplasmic inclusions **(arrows)** in dog kidney cells infected with type 2 parainfluenza virus. Nuclei appear unaffected. (Brandt CD: Virology 14:1, 1961) (×450)

from the cell by budding. The final assembly of the nucleocapsid and the specifically altered plasma membrane is especially prominent with these viruses (Fig. 59-7).

PARAINFLUENZA VIRUSES

Parainfluenza viruses were recognized in 1957 as important causes of acute respiratory infections of man, and were intially termed "hemadsorption virus," "croup-associated viruses," and "influenza D." There are at least four antigenic types that infect man and one that infects monkeys.

The major characteristics of parainfluenza viruses infecting man are listed in Table 59-1.

IMMUNOLOGIC CHARACTERISTICS

Each parainfluenza virus possesses three distinct Ags from which it derives its type specificity: the **HN (hemagglutinin–neuraminidase)** and the **F (fusion–hemolysin)** surface Ags, and an internal nucleocapsid Ag, the **NP protein** (Fig. 59-5). Parainfluenza viruses are immunologically unrelated to influenza viruses. Progressive major antigenic alterations have not been detected; only type 4 parainfluenza virus has subtypes, A and B.

Although there is not a single Ag common to all parainfluenza viruses, there are immunologic relations among the parainfluenza (at least types 1, 2, and 3) and with mumps viruses (Table 59-1). These are noted by heterotypic Ab responses to infection in those who have had prior infections with one or more of the other viruses.

The serologic response to the initial infection (the primary response) in humans, as well as in lower animal hosts, is strictly type-specific. Although crossreactions among parainfluenza viruses are not detected in antisera produced in animals, heterotypic Abs appear in humans, probably as a result of repeated infections with different members of the group; each additional infection broadens the Ab response. The human heterotypic responses to infections are sufficiently frequent to permit the following conclusions: 1) these agents constitute a group of viruses with crossreactive Ags, 2) the qualitative and quantitative characteristics of the heterotypic Ab response to infection depend upon prior immunologic experience, and 3) immunologic diagnosis of infection by a specific virus may be unreliable.

Most adults have a relatively high titer of circulating Abs to all antigenic types but usually lack the critical neutralizing IgA Abs in nasal secretions. These appear following acute infections, but they decrease substantially within 1–6 months although specific IgG serum Ab levels remain relatively high. Recurrent infections occur despite the presence of neutralizing Abs in the serum, although the severity of disease is usually reduced (perhaps owing to a secondary response of IgA Abs). The initial infection with a given type is the most severe and usually

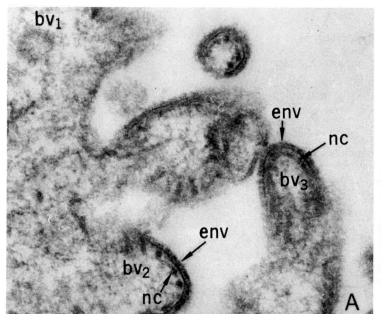

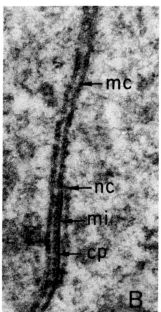

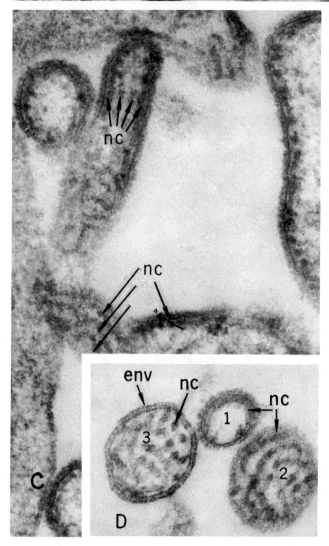

FIG. 59-7. Parainfluenza virions forming at the cell membrane of a chick embryo cell and being released by budding **A.** Viral buds **(bv)** at various stages of development. The altered cell membrane forming the viral envelope **(env)** and the nucleocapsid **(nc)** beneath the envelope can be clearly distinguished. The nucleocapsid is cut transversely in two of the buds (bv₁ and bv₂) and longitudinally in a third (bv₃). (×105,000) **B.** Differentiation of the viral envelope at the cell membrane. The cell membranes **(mc)** of two adjacent cells are shown. On contact with the nucleocapsid **(nc)** the cell membrane differentiates into the internal membrane of the envelope **(mi)** and the outer layer of short projections **(cp).** (×105,000) **C.** Arrangement of the nucleocapsid **(nc)** impinging on the altered cell membrane which is participating in the formation of the viral envelope. Notice the regularity of the arrangement of the nucleocapsid, seen in cross section and in longitudinal views. (×105,000) **D.** Thin section of virions. Note the structure of the envelope covered with short projections **(env)** and the various arrangements of the nucleocapsids **(nc).** In particle **1** the nucleocapsid is arranged parallel to the envelope. In particles **2** and **3** the nucleocapsid is arranged more irregularly. (Berkaloff A: J Microsc 2:633, 1963.) (×92,000)

occurs in children; infections in adults are commonly afebrile and minor.

REACTIONS WITH ERYTHROCYTES

The hemagglutinins and neuraminidases of parainfluenza viruses resemble those of influenza viruses in most biologic characteristics (see Hemagglutination, Ch. 46).

Maximum hemagglutination titers are obtained with chicken RBCs at 4° C (types 1 and 2) or with guinea pig RBCs at 25° (types 1 and 3). The **hemolysin (F) glycoprotein** is inhibited by type-specific antiserum and is similar to the hemolysins of mumps and Newcastle disease viruses. For hemolytic activity to occur, intact mucoprotein receptors on RBCs are required. A dialyzable material in allantoic and tissue culture fluids, possibly calcium, inhibits the hemolytic activity.

HOST RANGE

Primary cultures of cells from monkey and embryonic human kidneys are the hosts of choice for primary isolations, neutralization titrations, and investigations of the biologic properties of parainfluenza viruses. Organ cultures of tracheal and nasal epithelium have similarly proved to be sensitive for primary isolations. Types 2 and 3 also multiply well in human continuous epithelium-like cell lines (e.g., HeLa cells). Type 1 virus has been adapted to HeLa and human diploid cells, but neither cell line is satisfactory for initial viral isolation.

In cell cultures cytopathic changes develop very slowly (particularly with type 4); infection is most quickly detected by hemadsorption of guinea pig RBCs. The cytopathic changes that eventually appear consist of stringiness or rounding of cells (types 1 and 4) or formation of large **syncytia** containing eosinophilic **cytoplasmic inclusion bodies** (types 2 and 3; Fig. 59-6). Syncytia are produced by the viral surface F glycoprotein, which causes fusion and then dissolution of the fused membranes of infected cells (see Cytopathic Effects, Ch. 53). This remarkable capacity of parainfluenza viruses (even after ultraviolet or heat inactivation) to induce fusion of cells from many animal species has great general utility for studying the genetics of eukaryotic cells as well as for virology (see Cell Hybridization, Ch. 49).

Intranasal inoculation of very small amounts of virus into hamsters or guinea pigs at about 3 months of age results in infection which yields relatively high titers of virus in the lungs within 2–3 days, but pulmonary lesions do not form. The animals develop type-specific Abs and resist intranasal challenge with homologous virus for 1–3 months, but then susceptibility returns. This pattern mimics the relation of Abs and recurrent susceptibility in humans.

Type 3 virus is highly infectious for cattle. It appears to be harmless, however, except under conditions of stress, such as the herding of cattle together for transportation, when it may induce an acute febrile upper respiratory disease (hence the term "shipping fever").

Parainfluenza viruses have been adapted to propagation in the amniotic sacs of 7- to 9-day-old chick embryos, and thence to multiplication in the allantoic cavity.

PATHOGENESIS

For most parainfluenza viruses there is a lack of experimental animals suitable for studying the development of lesions. Infections in volunteers, however, and observations of natural infections, permit a few conclusions concerning pathogenesis. The virus enters by the respiratory route, and in most adults it multiplies and causes inflammation only in the upper segments of the tract. In infants and young children, however, the bronchi, bronchioles, and lungs are occasionally involved. *Viremia is neither an essential nor a common phase of infection.*

Parainfluenza viruses cause a spectrum of illnesses, primarily in infants and young children, ranging from mild upper respiratory infections to croup or pneumonia (Table 59-2). In infants type 1 and 2 viruses are particularly prone to produce laryngotracheobronchitis (croup) and type 3 virus to cause bronchiolitis and pneumonia. The occasional infections of adults usually evoke a subclinical illness or a mild "cold." Even in children the majority of infections with parainfluenza viruses appear to be clinically inapparent.

Parainfluenza viruses, like all other paramyxoviruses, can readily establish a persistent, endosymbiotic infection in vitro (Ch. 53). Scattered findings suggest that similar chronic infections may follow acute diseases in vivo. Thus, a parainfluenzalike virus has been isolated by cocultivation of susceptible cells and brain tissue from a patient with multiple sclerosis, and structures resembling paramyxovirus nucleocapsids have been observed in electron micrographs of tissues from patients with various collagen diseases (e.g., systemic lupus erythematosus).

LABORATORY DIAGNOSIS

An etiologic diagnosis of parainfluenza virus infection requires laboratory procedures; the lack of distinctive clinical features precludes etiologic diagnoses on this basis. Measurement of a **rise in serum Abs** by hemagglutination-inhibition or complement-fixation (CF) titration permits diagnosis conveniently and economic-

TABLE 59-2. Clinical Syndromes Associated with Parainfluenza Viruses

Diseases	Virus type
Minor upper respiratory disease	1, 3, 4*
Bronchitis	1, 3
Bronchopneumonia	1, 3
Croup	1, 2

*Clinical disease uncommon.

ally. However, serologic technics alone cannot reliably establish the specific type of virus that is responsible, because of the frequency and the degree of heterotypic Ab responses, as discussed under Immunologic Characteristics, above.

The specific parainfluenza virus responsible for an infection can be identified by **viral isolation,** although the marked instability of the virions makes isolation difficult. For this purpose nasopharyngeal secretions containing antibiotics are added to primary tissue cultures of monkey or embryonic human kidney. Although cytopathic changes may not be detectable except with type 2 virus, or may develop very slowly, viral infection can be recognized rapidly and conveniently by **immunofluorescence** or by **adsorption of guinea pig RBCs** to infected cells. Hemadsorption to cells infected with types 1 and 3 viruses can usually be detected within 5 days after inoculation of the patient's secretions, but types 2 and 4 often require 10 days or more. The specific type can be identified by hemadsorption-inhibition technics utilizing standard sera. Immunofluorescence can yield comparable diagnostic results in only 24–48 h. It should be remembered that the simian parainfluenza virus SV5 is a common latent agent in monkey kidney cultures, and its emergence from this tissue must not be confused with its primary isolation from man.

EPIDEMIOLOGY

Parainfluenza viruses produce disease throughout the year, but the peak incidence is noted during the "respiratory disease season" (late fall and winter). Most infections are endemic, but sharp, small epidemics occasionally occur with types 1 and 2. Parainfluenza virus infections are **primarily childhood diseases:** type 3 infec-

tions occur earliest and most frequently, so that 50% of children in the United States are infected during the first year of life, and almost all by 6 years of age; 80% of children are infected with types 1 and 2 by 10 years of age. Type 4 viruses induce few clinical illnesses but infections are common: by 10 years of age 70%–80% of children have Abs.

Parainfluenza viruses are **disseminated in respiratory secretions.** Type 3 shows the most effective spread, and during outbreaks in closed populations (e.g., in institutions or hospitals) all children who are free of neutralizing Abs become infected. Under similar circumstances only about 50% of children are infected with type 1 or type 2 virus.

The epidemiologic patterns and the clinical manifestations of parainfluenza virus infections, in children and adults, emphasize the protective effect of neutralizing Abs as well as the lack of complete or long-lasting immunity (probably owing to an inadequate level of IgA Abs in the respiratory secretions). Febrile and severe illness is observed only with the initial infection. Reinfection may be produced by the same virus within as little as 9–12 months, but it results in a much milder disease.

PREVENTION AND CONTROL

Reducing the attack rate of respiratory diseases is an important social and economic goal. However, prevention of parainfluenza virus infections would probably reduce the incidence of acute respiratory illnesses by only about 15% in children less than 10 years old, and by much less in adults. Nevertheless, an effective vaccine, evoking an Ab response in the respiratory tract, would be of value for young children, especially in hospitals and institutions.

MUMPS VIRUS

The unique clinical picture of mumps can be recognized in writings of Hippocrates from the fifth century B.C. The etiologic agent was not isolated and identified as a virus, however, until 1934, when Johnson and Goodpasture produced parotitis in monkeys by inoculating bacteria-free infectious material directly into Stensen's duct, but no further major progress was made until the virus was propagated in the chick embryo and was found (in 1945) to agglutinate chicken RBCs, like influenza viruses. The subsequently isolated parainfluenza viruses were found to have an even closer relation to mumps virus.

IMMUNOLOGIC CHARACTERISTICS

Mumps virus, like parainfluenza viruses, contains the surface **hemagglutinin–neuraminidase (HN) glycoprotein** (also termed the **V antigen**), which induces protective

Abs, **the hemolysis-cell fusion (F) glycoprotein Ag** and the internal **RNA-protein nucleocapsid** (NP), which is immunologically identical with the soluble Ag from cells (called the **S antigen**).

Antigenically mumps virus exists as a single type; no immunologic variants have been detected. It crossreacts significantly, however, with parainfluenza and Newcastle disease viruses. Therefore a rise in heterotypic Abs to parainfluenza viruses is seen in the serum from mumps patients.

Antibodies against the nucleocapsid Ag appear within 7 days, and high titers are attained within 2 weeks after the onset of clinical illness. The HN Abs appear later (2–3 weeks after onset), attain maximum titers in 3–4 weeks, and persist longer than the nucleocapsid Abs. The HN Abs are measured by neutralization, hemagglutination inhibition, or CF with purified HN glycoprotein and appear to reflect the degree of immunity. Nucleo-capsid Abs, which are conveniently assayed by CF titrations, do not afford protection against subsequent infection. Immunity develops after subclinical as well as clinical infections and usually persists for many years, although second infections have been reported.

Mumps infection induces **delayed hypersensitivity** which can be observed by a skin test with infectious or inactivated virus. A positive skin reaction correlates roughly with immunity, providing a useful epidemiologic tool, but false reactions (both positive and negative) occur.

REACTIONS WITH ERYTHROCYTES

The viral particle **agglutinates RBCs** from several animal species (see Hemagglutination, Ch. 46). Mumps virus, however, has only weak neuraminidase activity against soluble mucoproteins, and therefore hemagglutination is highly susceptible to these inhibitors in serum or culture fluids. The hemagglutinin present on the surface of infected cells can also be detected in tissue culture by **hemadsorption.**

Mumps virus, like parainfluenza viruses, hemolyzes susceptible RBCs at 37 °C by interacting with the same specific receptors as are involved in hemagglutination. Several viral particles per cell are required to produce hemolysis, and even under optimal conditions only about 50% of the hemoglobin is released. Calcium inhibits hemolysis. Viral hemagglutination, hemadsorption, and hemolysis are inhibited by Abs to the virion surface Ags (the HN and F glycoproteins).

HOST RANGE

Humans are the only natural hosts for mumps virus, but the virus can infect monkeys and 6- to 8-day-old chick embryos (amniotic cavity or yolk sac); no pathologic lesions of the embryos are noted. After adaptation by serial passage in the allantoic sac of the chick embryo the virus can infect guinea pigs, suckling mice, hamsters, and white rats but has lost its virulence for man and monkey.

In culture, chick embryo and many types of mammalian cells (including monkey and human kidney cells) support multiplication of mumps virus. Infected cultures can be recognized by the appearance of viral particles (hemadsorption or agglutination of RBCs) and soluble Ag (CF), as well as by the slow development of cytopathic effects (cytolysis and development of giant cells with cytoplasmic inclusions, Fig. 59-8).

TOXICITY

Mumps virus, like influenza viruses, has a lethal toxin-like action when inoculated intravenously in large amounts into mice, espe-

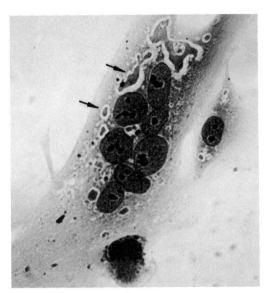

FIG. 59-8. Eosinophilic cytoplasmic inclusions and giant cell formations in monkey kidney cells infected with mumps virus. Inclusions **(arrows)** develop in close proximity to the nuclei, which are unaffected. (Brandt CD: Virology 14:1, 1961) (×450)

cially newborns. The toxic activity is not dissociable from the viral particle, but unlike the case with influenza virus, infectious virus is not required. The role of this toxic property in the pathogenesis of natural infections is not known.

PATHOGENESIS

Mumps typically has an acute onset of parotitis (with painful swelling of one or both glands) 16–18 days after exposure. The virus is **transmitted in saliva and respiratory secretions** and its **portal of entry** in man is the **respiratory tract.** It has been generally taught that virus enters Stensen's duct directly and multiplies initially in the duct and the parotid gland. The development of a sensitive viral assay, however, has provided evidence suggesting another pathway: after **primary multiplication** in the respiratory tract epithelium and cervical lymph nodes the salivary glands are infected via the blood stream (Fig. 59-9). Thus **viremia** begins several days before the development of mumps and before virus is present in the saliva. Virus is present in the blood and in saliva for 3–5 days after the onset of the disease and **virus in the urine** is common for 10 days or more after the onset. Moreover, the long incubation period reflects the time required for virus to establish a local infection, spread in the blood to the target organs, and multiply sufficiently to damage cells and induce inflammation. The occasional complications also suggest widespread dissemination of the virus.

The acute onset of fever and of salivary gland inflammation may be followed in 4–7 days by orchitis (in

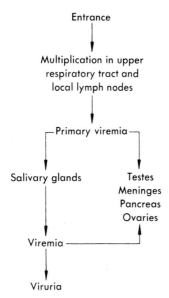

Entrance

↓

Multiplication in upper
respiratory tract and
local lymph nodes

↓

Primary viremia

Salivary glands   Testes
                        Meninges
                        Pancreas
                        Ovaries

Viremia

↓

Viruria

FIG. 59-9. Schematic representation of the pathogenesis of mumps.

20%–35% of males past puberty),* by meningitis or meningoencephalitis (0.2%–0.5% of cases), and occasionally by pancreatitis, oophoritis, or presternal edema. Even more rarely nephritis or paralytic manifestations may appear. The complications may develop at the same time as parotitis or even in the absence of salivary gland involvement.

The most complete pathologic descriptions of mumps virus lesions have been those of affected parotid glands and testes. The parotid lesions consist of interstitial inflammation of the gland and degeneration of the epithelium of the ducts. No characteristic inclusion bodies have been observed in patients' tissues, in contrast to tissue cultures (Fig. 59-8). Testes show diffuse degeneration, particularly of the epithelium of the seminiferous tubules, as well as edema, serofibrinous exudate, marked congestion, and punctate hemorrhages in the interstitial tissue. Meningoencephalitis or meningitis has been studied pathologically only in rare cases, and proof of etiology has often been lacking. The findings described are typical of postinfectious encephalitis, with perivascular demyelination.

LABORATORY DIAGNOSIS

Diagnosis can usually be made solely from clinical observations. However, to establish the diagnosis of mumps in atypical or subclinical infections (about one-third of infections) **viral isolation** and assays of **Ab response** are required. Virus can be isolated by inoculation of saliva, secretions from the parotid duct, or spinal fluid into the amniotic cavity of 8-day-old chick embryos or appropri-

* Despite the commonly told tales sterility rarely if ever results from mumps orchitis.

ate cell cultures; the latter appear to be more sensitive for primary isolation. Infection of test cells can be detected earliest by **immunofluorescence** or **hemadsorption** technics. Since the virion is unstable at 4° C or at room temperature, specimens must be used immediately or stored at −70° C.

Immunologic diagnosis is made by demonstrating an increase in Abs after infection. **Complement fixation** with the **nucleocapsid (soluble) Ag** can detect a rise within the first 7–10 days after onset of illness. After 14 days serologic diagnosis is accomplished most readily with the **viral particles (HN Ag)** in **hemagglutination-inhibition** or **CF** titrations. The antigenic crossreactions between the surface (HN) antigens of mumps and parainfluenza viruses complicate immunologic diagnosis, making it necessary to test all the related viral Ags; the highest Ab rise occurs in response to Ags of the infecting virus. When diagnosis is of critical importance the **plaque-neutralization** assay can detect Ab rises which may not be evident in other assays owing to the crossreactive CF Abs and the nonspecific inhibitors of hemagglutination.

Though the **skin test** may be useful in surveys for estimating immunity, it frequently leads to erroneous conclusions in individual cases and is not reliable for diagnosis. Indeed, because skin testing itself may induce a rise in Abs and confuse the interpretation of immunologic reactions, skin tests should not be used during the course of illness.

EPIDEMIOLOGY

Mumps is predominantly a **childhood disease** spread by droplets of saliva. Salivary secretions may contain infectious virus as early as 6 days before and as long as 9 days after the appearance of glandular swelling. Virus is also present in the saliva of patients with meningitis or orchitis, even in the absence of clinical involvement of the salivary glands. Although viruria develops, there is no evidence for transmission from this source.

Mumps is not nearly so contagious as the childhood exanthematous diseases (e.g., measles and chickenpox), and many children escape infection; hence **disease in adults** is not uncommon (about 50% of US Army recruits are without Abs). **Subclinical infections,** detectable immunologically, are also much more frequent than in other common childhood diseases.

Mumps appears most often during the winter and early spring months but it is endemic throughout the year. Large epidemics have been observed about every 3–5 years.

PREVENTION AND CONTROL

Because subclinical infections are common, control of infection by isolation is not effective. The disease can be prevented by immunization. Infectious attenuated virus, inoculated subcutaneously, induces development of Abs in greater than 95% of Ab-free subjects (both children and adults). Although vaccine induces an apparent infection, viremia and viruria are not detectable, clinical reactions do not occur, and virus does not spread to exposed contacts. The Ab response is not as great as that accompanying natural infections, but Ab levels persist for at least 8 years, and the vaccinees are almost uniformly protected upon exposure to mumps infections.

The live virus vaccine is of considerable value for susceptible adults in whom the disease is more severe and the complications more frequent. If widespread immunization of children is to be employed, the vaccine should induce persistent immunity, similar to that which follows the natural disease. The infectious attenuated virus vaccine may satisfy this requirement, but the data are not yet available to permit a confident recommendation.

Formalin-inactivated virus also induces Ab production after two subcutaneous injections, and its clinical effectiveness has been demonstrated in controlled studies. However, Abs decline 3–6 months after immunization, and neither the effectiveness nor the persistence of immunity appears to be as satisfactory as following the live virus vaccine.

Passive immunization with γ-globulin from convalescent serum has been employed to prevent infection after exposure, particularly in men, for whom orchitis is a relatively frequent and severely discomforting complication. Its effectiveness, however, has not been clearly demonstrated. Convalescent whole human serum was once used for similar purposes, but it is not recommended because of the considerable danger of contamination with hepatitis B or C virus.

MEASLES VIRUS

Measles **(morbilli)*** is one of the most infectious diseases known, and it is almost universally acquired in childhood; fortunately, immunity is essentially permanent. It was first described as an independent clinical entity by Sydenham in the seventeenth century. Although measles was demonstrated as early as 1758 to be transmissible in volunteers, and was transferred to monkeys in 1911, it was not until 1954 that the virus, through Enders' persistent and careful search, was isolated reproducibly from patients and was shown to produce cytopathic changes in tissue cultures. This achievement led to a rapid advance in knowledge of the virus and to the development of an effective vaccine.†

IMMUNOLOGIC CHARACTERISTICS

All measles strains studied belong to a **single antigenic type.**‡ Specific Abs are produced to each of the major viral Ags: the **hemagglutinin (H)**, the **hemolysin–cell fusion factor (F)**, and the **nucleocapsid (NP)**, all of which exist free in cell extracts as well as assembled in virions. Antibodies to the **virion surface glycoproteins** (hemagglutinin and hemolysin–cell fusion factor), in the presence of specific complement components (the alternate complement pathway), lyse infected cells, which contain these viral glycoproteins in their plasma membranes.

---

* **Rubeola** is often employed as a synonym; unfortunately, this term has also been used as a synonym for rubella (German measles).

† This brief account illustrates only in part the unique role Enders played in leading the way to the recent control of two important diseases, poliomyelitis and measles.

‡ Measles virus, however, is related to the viruses of canine distemper and rinderpest (of cattle) in antigenic, physical, and biologic properties.

These **cytotoxic Abs** probably play a role in certain aspects of the pathogenesis of the disease as well as in the elimination of infected cells during recovery. In the absence of complement, however, these same Abs cause the viral glycoproteins to accumulate ("cap") in a limited region of the infected cell surface, and finally to shed from the cell (Ch. 53). It is likely that this phenomenon permits infected cells to escape immunologic destruction and to persist, possibly causing chronic infections (Ch. 53).

Like most orthomyxoviruses and paramyxoviruses, measles virus (or its separated hemagglutinin) agglutinates monkey RBCs. In contrast to orthomyxoviruses and other paramyxoviruses, measles virus does not elute spontaneously from agglutinated cells, the RBC receptors are not destroyed by *Vibrio cholerae* neuraminidase, and the virions do not contain neuraminidase molecules.

Circulating Abs are detected 10–14 days after infection, i.e., when the rash appears or shortly thereafter, and reach maximal titer by the time the exanthem disappears. Antibody titers (neutralizing, CF, and hemagglutination-inhibiting) remain high following infection, and immunity persists for life (as shown by epidemiologic investigations; see Persistence of Abs, Ch. 53). In monkeys, reinfection can be produced 3–6 months after a primary infection, but clinical disease does not ensue.

Cell-mediated immunity is also demonstrable by the time the rash appears. Human T lymphocytes from normal as well as immune persons possess receptors for measles virus: the lymphocytes are agglutinated by purified virus, and they form rosettes upon reaction with infected cells. Following rosette formation, the virus-infected cells are killed. Although lymphocytes from both immune and susceptible persons are cytotoxic, the immune lymphocytes are more effectively so.

HOST RANGE

Measles virus is highly contagious for both humans and monkeys, its known natural hosts. Monkeys in captivity commonly develop spontaneous measles, humans probably serving as the source. Because of the resulting immunity only freshly captured, Ab-free monkeys are reliable for testing vaccines or studying pathogenesis.

Although unmodified measles virus from patients replicates poorly in vitro, multiplication has been achieved in a variety of mammalian as well as chick embryo cells. Both primary and continuous mammalian cell cultures are commonly employed.* Viruses adapted to propagation in vitro can also multiply in the amniotic sac or in the chorioallantoic membrane of chick embryos, and one strain has been adapted to propagation in brains of newborn mice.

The development of **large syncytial giant cells** is generally the major cytopathic effect produced by measles infection of cultured cells (Fig. 59-10), although some strains selected in vitro produce spindle-shaped rather than giant cells. Eosinophilic inclusion bodies develop in both the nuclei and the cytoplasm of syncytial cells (Fig. 59-10). The inclusions are composed of dense, highly ordered arrays of viral nucleocapsids (Figs. 59-11 and 59-12). Immunofluorescence studies reveal specific viral Ags in both the nuclear and cytoplasmic inclusion bodies.

PATHOGENESIS

Measles is a highly contagious, acute, febrile, exanthematous disease. The pathogenesis in man resembles the general pattern described for smallpox (Ch. 56) and mumps (Fig. 59-9), with **local multiplication** followed by **hematogenous dissemination.** Virus **transmitted in respiratory secretions** enters the upper respiratory tract, or perhaps the eye, and multiplies in the epithelium and regional lymphatic tissue. Virus may also disseminate to distant lymphoid tissue by a brief primary viremia. Viral multiplication in the upper respiratory tract and conjunctivae causes, after an incubation period of 10–12 days, the prodromal (i.e., prerash) symptoms of coryza, conjunctivitis, dry cough, sore throat, headache, low-grade fever, and Koplik spots (tiny red patches with central white specks on the buccal mucosa in which are noted characteristic giant cells containing viral nucleocapsids). Viremia occurs toward the end of the incubation period, permitting further widespread dissemination

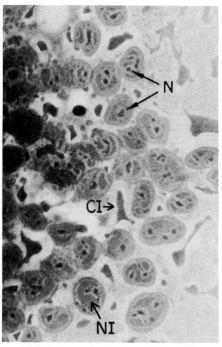

FIG. 59-10. Inclusion bodies of cells infected with measles virus. A giant cell consisting of a large syncytium of cells is illustrated; each large round body is a nucleus. Large eosinophilic cytoplasmic inclusions are indicated by **CI,** and numerous intranuclear eosinophilic inclusion bodies by **NI.** The nucleoli **(N)** are intact. (Kallman F et al: J Biophys Biochem Cytol 6:379, 1959) (×750)

of virus to the lymphoid tissue and skin. With the diffuse secondary multiplication of virus the prodromal symptoms are intensified and the typical red, maculopapular rash appears, first on the head and face and then on the body extremities.

Viral Ags and nucleocapsids are present in endothelial cells of the subcutaneous capillaries but usually infectious virus is not detectable in the affected superficial epidermal cells. The characteristic rash appears to result predominantly from interactions of sensitized lymphocytes, or of specific Abs and complement, with infected cells. It is also striking that when children with thymic dysplasia develop measles they do not display a rash but manifest extensive giant-cell pneumonia.

Virus is excreted in the secretions of the respiratory tract and eye, and in urine, during the prodromal phase and for about 2 days after the appearance of the rash. *This early shedding of virus, before the disease can be recognized, promotes its rapid epidemic spread.* The blood, lymph nodes, spleen, kidney, skin, and lungs also contain detectable virus during this period. Measles virus can multiply in and has been isolated from human macrophages and lymphocytes, suggesting that these cells may play a role in its dissemination in the body and in the

---

* 1) **Primary cultures:** human embryonic kidney; human amnion; monkey or dog kidney; chick embryo cells; bovine fetal tissue; 2) **continuous epithelium-like cell lines of human origin:** HeLa, KB, Hep-2, amnion, heart, nasal mucosa, bone marrow, kidney. Primary cultures of human embryonic or monkey kidneys are most susceptible to unadapted viruses.

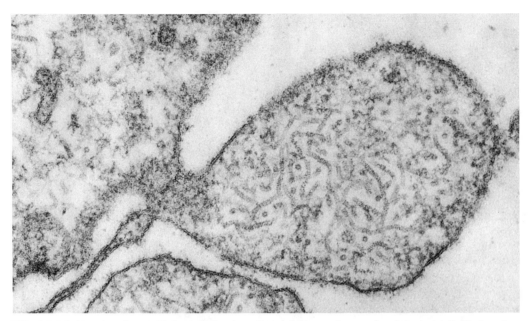

FIG. 59-11. Measles virus particles budding from the surface membrane of an infected cell. Viral nucleocapsids are seen within the forming particles; the fuzzy structures on the surfaces of the virions probably correspond to the surface projections seen by negative staining (Fig. 59-3). (Nakai T et al: Virology 38:50, 1969) (×98,000)

pathogenesis of the disease. The leukocytic involvement may also be responsible for the leukopenia observed during the prodromal stage as well as for depression of delayed-type hypersensitivity reactions (e.g., the tuberculin test). Measles virus can induce striking aberrations in the chromosomes of leukocytes during the acute disease. Although the chromosomal pulverization produced is probably lethal to the cell, a possible relation between some of the changes and the initiation of leukemia has been suggested.

The characteristic **viremia** in measles, in contrast to the more localized respiratory infections produced by influenza and parainfluenza viruses, probably contributes to the notably effective immunity conferred by the disease.

Complications. Bronchopneumonia and **otitis media,** with or without a bacterial component, are frequent complications of the disease. **Encephalomyelitis** is the most serious complication, appearing about 5–7 days after the rash. Its incidence in most epidemics is about 1:2000 cases (higher in children over 10 years); but in some outbreaks, particularly in the widespread infection of malnourished infants in Africa, the incidence has been much higher. The mortality rate of encephalomyelitis is

about 10%, and permanent mental and physical sequelae have been reported in 15%–65% of survivors.

There is no evidence that measles encephalomyelitis is due to increased virulence of the virus in certain epidemics. Indeed, measles virus cannot be isolated from the brain, but lymphocyte infiltration and demyelination are prominent pathologic features, reminiscent of allergic encephalitis. These findings suggest that measles encephalomyelitis may be a hypersensitivity response either to the measles virus or to virus-altered host tissue (i.e., an autoimmune phenomenon).

Giant cell pneumonia, a rare disease of debilitated children, or of those with immunodeficiency disease, was proved by isolation of virus to be due to measles.

Subacute Sclerosing Panencephalitis (SSPE). This progressive, degenerative neurologic disease of children and adolescents (causing mental and motor deterioration, myoclonic jerks, and electroencephalographic dysrhythmias) was unexpectedly discovered to be caused by measles virus. With this finding it became clear that a single virus could induce both an acute contagious disease and a chronic illness (see Slow Viral Infections, Ch. 53). The following data implicate measles virus in this severe chronic disease: 1) Almost all patients have had

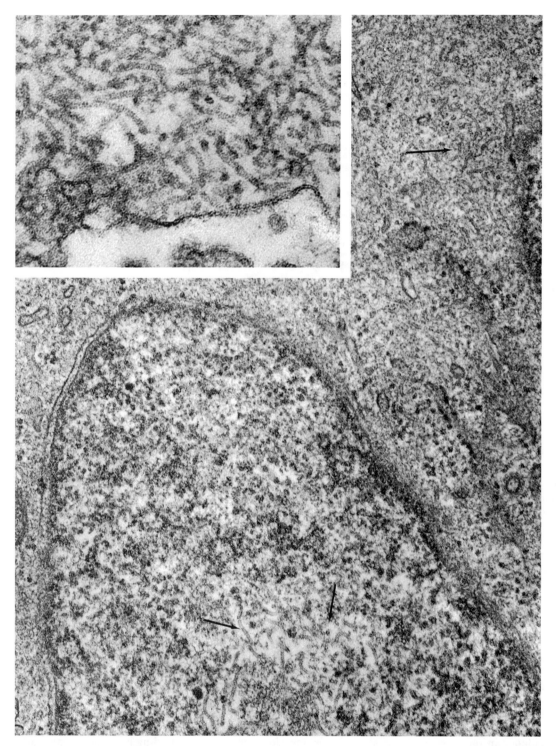

FIG. 59-12. Electron micrographs of cell infected with measles virus, showing intranuclear and intracytoplasmic matrices containing viral nucleocapsids **(arrows).** (×60,000) **Inset.** Higher magnification of an extensive accumulation of nucleocapsids. Where the tubules are favorably oriented cross-striations of the nucleocapsids can be seen. (Nakai T et al: Virology 38:50, 1969) (×140,000)

measles several years (up to 13) prior to onset of SSPE or have been immunized with live virus vaccine. 2) All have unusually high titers of measles Abs, even in the spinal fluid, and the Ab levels (IgM as well as IgG), often increase as the disease progresses. 3) Affected brain cells have nuclear and cytoplasmic inclusions similar to those seen in measles infections (Fig. 59-10); the inclusions consist of filamentous tubular structures indistinguishable from the nucleocapsids seen in cells infected with measles virus (Fig. 59-12). 4) Immunofluorescence study of the brain lesions reveals Ags that react with Abs to measles but not to distemper virus (a close relative). 5) A virus very similar to measles virus has been isolated in brains of ferrets and newborn mice, and also by cocultivation of affected brain cells with cells that readily support measles virus multiplication (e.g., African green monkey kidney cells, HeLa cells); and though virus has not been directly isolated from homogenate of brain cells, when the affected cells are cultured in vitro, they show typical inclusion bodies as well as the presence of viral Ag and viral nucleocapsid. The RNA of the SSPE virus contains all of the nucleotide sequences of measles virus RNA. About 10% consists of additional sequences, which may have been derived by recombination with the RNA of another virus.

The finding of measles Abs in the spinal fluid of patients with **multiple sclerosis** has raised the possibility that this chronic neurologic disorder is also a complication of prior measles.

Pathology. The development of very large multinuclear **giant cells** (Warthin–Finkeldey syncytial cells) is the predominant and characteristic feature of the pathology of measles. These distinctive cells are found in nasal secretions during the prodromal stage of the disease, as well as in lymphoid tissue of the gastrointestinal tract, particularly the appendix. Giant cells are also often observed in sputum from patients with bronchopneumonia and may contain eosinophilic nuclear and cytoplasmic inclusions, similar to those seen in infected cell cultures; and such cells are characteristic of the rare **giant cell pneumonia**. These giant cells are presumably produced by cell fusion, like syncytial cell formation in infected cell cultures.

In **measles encephalomyelitis** the brain shows perivascular hemorrhage and lymphocytic infiltration early in the disease; areas of demyelination later appear in the brain and spinal cord.

Brains from patients with **subacute sclerosing panencephalitis** display a degeneration of the cortex and especially the underlying white matter. Characteristically there are seen intranuclear and intracytoplasmic inclusion bodies not unlike those noted in acute measles, perivascular infiltration of plasma cells and lymphocytes, scattered degeneration of nerve cells, hypertrophy of astrocytes, microglial proliferation, and demyelination.

LABORATORY DIAGNOSIS

The epidemiologic and clinical features of measles are usually so characteristic that laboratory confirmation of the diagnosis is unnecessary except for investigative purposes. During the prodromal stage of the disease a rapid and simple presumptive diagnosis can be made by demonstrating specific viral immunofluorescence and characteristic giant cells in smears of the nasopharyngeal mucosa. Definitive diagnosis can be accomplished by isolation of virus from nasal or pharyngeal secretions, blood, or urine; viral isolations are best achieved in primary cultures of human embryonic or monkey kidneys. Serologic diagnosis can be made by comparing acute and convalescent sera using hemagglutination-inhibition or CF titrations, as described for other paramyxoviruses. Finding measles Abs in the cerebrospinal fluid is suggestive of severe neurologic disturbance such as acute encephalomyelitis or SSPE.

EPIDEMIOLOGY

Measles is a **highly contagious disease** in which virus is **spread in respiratory secretions.** It is predominantly a **childhood affliction** which occurs in epidemics during the winter and spring in rural areas. Since measles virus does not have a reservoir in nature other than humans, and since long-lasting immunity follows infection, persistence of the virus in a community depends upon endemic infections in a continuous supply of susceptible persons. It has been estimated that in urban areas 2500–5000 cases per year are required for continued transmission. Hence it is clear why in the more highly developed countries epidemics tend to appear in 2- to 3-year cycles, as a sufficient number of nonimmune children arises in the population, and why the disease disappears from small, isolated communities. However, even in a highly immunized population exogenous introduction of virus can initiate a limited epidemic: for example, epidemics are now appearing in high schools and universities in which over 95% of the students had been immunized as children.

In the United States the highest incidence is in children 5 to 7 years of age, and the disease is relatively mild. However, since 1973 the age-specific incidence has steadily increased to older children and teenagers. Moreover, the disease is more severe in adults and in young children (but transplacental immunity usually protects newborns and babies up to the age of 6 months). In communities having primitive and crowded living conditions (e.g., in West Africa) most cases occur in infants less than 2 years old; the illnesses tend to be severe and often fatal (probably owing to malnutrition and a high rate of secondary infections), and epidemics occur yearly because the close contacts probably decrease the proportion of susceptibles required. If the virus is introduced into isolated, unimmunized communities, where the exposure is rare and measles has not struck in many years, the incidence is very high, and the illness is severe and frequently is fatal to very young children and to the elderly (e.g., the Faroe Islands; see Persistence of Antibodies, Ch. 53).

PREVENTION AND CONTROL

Public health measures alone, such as isolation, have not successfully prevented or even limited measles epidemics. Prior to the development of vaccines for active immunization, passive immunization with pooled γ-globulin* was used to furnish temporary protection. Because of the long incubation period, large doses, even when administered shortly after exposure, prevent the disease, and small doses reduce its severity. This procedure is still effectively employed for exposed susceptibles, particularly adults.

After the successful cultivation of measles virus in tissue culture, vaccines were developed and effective control became possible. Following the methods that had proved so successful in the control of poliomyelitis both **attenuated live virus vaccine** and **formalin-inactivated virus vaccine** were prepared.

The initial "attenuated" virus induced mild measles with fever in about 80% of recipients, so that γ-globulin containing measles Abs was administered at a different site to reduce the reactions. Viruses of greater attenuation were eventually obtained, however, and they can now be used without an accompanying injection of γ-globulin. In children older than 1 year, the live attenuated virus selected for general immunization induces an Ab response in almost 100% of those who previously lacked Abs, but the vaccine still produces reactions in 15%–20% of children. The Ab titers are about 10%–25% of the levels observed after the natural disease. Neutralizing Abs endure without significant decline for at least 2 years after immunization but decrease two- to threefold by 5 years; CF Abs begin to decline in 6–8 months. Although effective immunity has been demonstrated at least 10 years after immunization, measles does occasionally occur in vaccinated children, particularly in those immunized during the first year of life. How long adequate protection persists after vaccination is still unclear, but the increasing number of measles outbreaks reported among adolescents who were previously immunized as young children suggests that the immunity may be less lasting than that following natural measles.

The vaccine does not cause acute neurologic complications and virus does not spread from vaccinees to susceptibles in the same family. Immunization with live virus vaccine has proved highly effective in large studies, and during epidemics it appears to prevent measles in more than 95% of children. In the United States, where widespread immunization has been employed, large epidemics have been eliminated and the incidence of measles has been reduced from about 500,000 reported cases per year in 1963 to 35,000 cases in 1976.

In spite of its proved effectiveness the vaccine must be used with caution until we know how likely it is to establish endosymbiotic infections that may lead to chronic diseases, such as SSPE. Of 350 cases of SSPE studied in the United States between 1960 and 1974, 40 were in children who had no history of measles but who had received live attenuated measles virus vaccine (on average, 3.3 years prior to onset of SSPE). Thus, the potential risk could be 0.48–1.13 cases of SSPE per million measles vaccine recipients, in contrast to 5.2–9.7 cases of SSPE per million cases of measles. Since without vaccination nearly everyone gets measles, the extensive use of measles immunization significantly reduces the actual number of cases of this dread complication.†

The attenuated virus vaccine has been lyophilized for general distribution. A melange of other attenuated viruses has also been added: for convenience the most popular combination is measles virus mixed with attenuated mumps and rubella viruses. It is comforting that the multiple-virus formulation has neither increased the reaction rate to measles attenuated virus nor decreased the Ab response to any of the viral immunogens present.

A **formalin-inactivated** virus vaccine has been shown to elicit Abs in about 75% of the recipients after a course of three intramuscular injections. This vaccine has not received acceptance, however, because 1) protection is only temporary, 2) neutralizing Abs as well as CF Abs begin to decline rapidly within 3–6 months, and 3) unanticipated severe disease has been reported in vaccinees who were subsequently infected naturally or reimmunized with live virus vaccine. (It should be noted, however, that this "atypical" measles is also observed in some recipients of live virus vaccine who subsequently develop natural measles.)

RESPIRATORY SYNCYTIAL VIRUS

From a chimpanzee with coryzal illness, and from a laboratory worker who had been in contact with the animal, Morris and coworkers isolated a new virus in 1956. The following year Chanock isolated similar viruses from two infants with pneumonia and croup. Subsequent studies have indicated that *the virus is a major cause of lower*

---

* Because the majority of adults in most countries have had measles, and because levels of circulating Abs remain high, pooled normal adult γ-globulin is effective in passive immunization.

† It is estimated that from 1963 to 1972 measles immunization prevented about 24 million cases of measles, saved approximately 2400 lives, and averted about 7900 cases of mental retardation and other neurologic complications.

respiratory tract disease during infancy and early childhood throughout the world. Since it characteristically induces formation of large syncytial masses in infected cell cultures, it was named **respiratory syncytial (RS) virus.**

PROPERTIES OF THE VIRUS

The RS virus appears to be related to measles and parainfluenza viruses, but several distinctions led to its classification in a separate genus, **Pneumovirus** (Table 59-1). In particular, hemagglutination, hemadsorption, hemolytic, and neuraminidase activities are not detectable despite the presence of regularly spaced clublike projections (peplomers) on the virion's surface (Fig. 59-13). Moreover, the virions and the nucleocapsids are extremely fragile, which makes preservation of infectivity difficult. In addition, filamentous virions are present in purified preparations (Fig. 59-13) and in thin sections (Fig. 59-14); the filaments, unlike those of influenza viruses (see Fig. 58-2, Ch. 58), are often much narrower than the spherical virions, and they have the appearance of an elongated nucleocapsid rather than a folded helix covered with an envelope. Finally, the spherical virions are somewhat less variable in size and are slightly smaller (Table 59-1) than the typical paramyxoviruses; the nucleocapsid has a slightly smaller diameter and the helix

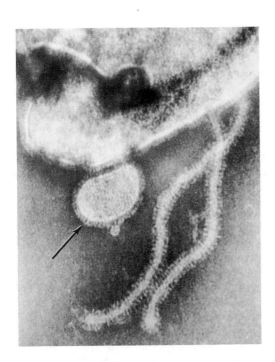

FIG. 59-13. Intact respiratory syncytial virus particle **(arrow)** and two filaments negatively stained with phosphotungstate; note regularly arranged clublike peripheral projections. (Bloth B et al: Arch Gesamte Virusforsch 13:582, 1963) (×110,000)

FIG. 59-14. Electron micrographs of thin sections of a cell culture infected with respiratory syncytial virus. Different stages are seen in the morphogenesis of the viral particles at the plasma membrane. **A.** Early stages of budding show thickening of the cell membrane with the appearance of fringelike projections. Submembranous accumulation of the nucleocapsid is also noted. (×75,000) **B.** Later stages of viral budding and a single free virion are seen. The circular arrangement of the nucleocapsid in the spherical particles suggests an organized packing of this component. (×76,000) **C.** Filamentous forms of viral particles are developing at the cell membrane. The linear arrangement of the nucleocapsid is visible within the filamentous particles. (Norrby E et al: J Virol 6:237, 1970) (×42,000)

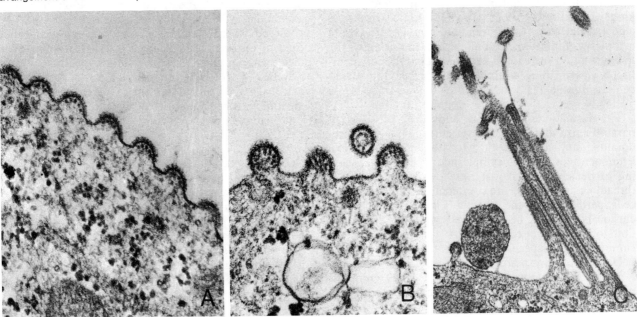

has a regular periodicity slightly larger than for other paramyxoviruses.

Immunologic Characteristics. At least **three antigenic variants** have been noted among the limited number of RS strains studied. The immunologic crossreactivity between variants is too great, however, to permit division into distinct types, and the antigenic differences do not appear to be progressive with successive epidemics. Each variant has a specific surface Ag which can be detected by plaque neutralization assays in cell cultures with standard sera. Purified virions and extracts of infected cells contain the specific viral surface Ag and also a soluble nucleocapsid Ag, detectable by CF titration, that is common to all RS viruses but not to other paramyxoviruses.

Protection from reinfection is brief following RS virus infections, probably because the titer of IgA Abs in nasal secretions declines rapidly. Serum Abs persist, but even a level of neutralizing Abs as high as 1:256 does not provide adults with effective protection against reinfection, and infants with neutralizing Abs may even develop bronchopneumonia on reinfection. Furthermore, babies during the first 6 months of life are most often infected and most seriously affected, despite the high level of serum IgG Abs derived from their mother.

Host Range. RS virus has been observed to cause illness only in humans and in the chimpanzee. In addition, intranasal inoculation of virus into ferrets, monkeys, and many other mammals produces inapparent infections, from which virus can be recovered.

RS virus multiplies and produces cytopathic changes in continuous human epithelium-like cell lines, in human diploid cell strains, and in primary simian and bovine kidney cell cultures. About 10 h after infection virus-specific Ags, detected by immunofluorescence, appear in the cytoplasm of infected cells but not in their nuclei. As the quantity of Ag increases it accumulates at the cell membrane, where virions assemble and bud from the altered cell membrane (as with other paramyxoviruses; Fig. 59-14).

The formation of syncytia and giant cells, as in parainfluenza and measles virus infections, is the predominant and characteristic cytopathology (Fig. 59-15). Prominent eosinophilic cytoplasmic inclusions, consisting of densely packed material having a granular or threadlike appearance, are commonly found in infected cells, particularly in the syncytia (Fig. 59-15). These inclusion bodies may be considered "scars" of the infection, for they do not contain RNA, viral particles, or accumulated virus-specific Ags.

Unlike orthomyxoviruses and most paramyxoviruses, RS virus has not been propagated in chick embryos, probably owing to its inability to combine with specific host mucoprotein receptors (which also reflects its failure to hemagglutinate RBCs).

PATHOGENESIS

The pathogenesis of RS virus infections has been largely surmised on the basis of clinical observations. The **initial infection** results from **viral multiplication in epithelial cells of the upper respiratory tract;** it most often ends at this stage in adults and older children. In about 50% of infected children less than 8 months old **virus spreads into the lower respiratory tract**—the bronchi, the bronchioli, and even the pulmonary parenchyma. Thus, in children RS virus may cause febrile upper respiratory disease, bronchitis, bronchiolitis, bronchopneumonia, and croup.

The dissemination of virus into the lower respiratory tract, with concomitant production of more serious disease, is probably due at least partially to the previously mentioned absence of IgA Abs in the respiratory secretions, and also to the slow development of Abs in the previously uninfected infant. With increasing age, and therefore repeated exposure and greater immunologic response to RS virus, infections are likely to be milder and confined to the upper respiratory tract, and most likely to be inapparent. In adults, because of the absence or limited quantity of virus-specific secretory (IgA) Abs in the respiratory tract, infections do occur despite the presence of circulating Abs; most frequently the disease is afebrile and clinically resembles the common cold.

Autopsy examination of infants dead of RS virus infection has shown severe necrotizing lesions of the epithelium of the bronchi and bronchioles, and interstitial pneumonia consisting of mononuclear infiltration, patchy atelectasis, and emphysema. Cytoplasmic inclusion bodies were noted, but syncytial formation was not observed. These changes represent one end of the spectrum of disease produced by RS virus. In infants who died with bronchiolitis the lungs contained small amounts of virus and intracellular Ags which could be detected by immunofluorescence, whereas in infants with pneumonia large quantities of infectious virus and viral Ags were present.

LABORATORY DIAGNOSIS

Detection of RS viral Ags in exfoliated cells of the respiratory tract by **specific immunofluorescence** offers the most rapid and economic means for diagnosing acute infections. **A precise diagnosis** of RS virus infection, however, requires either **isolation of the virus** or demonstration of a **rise in Abs.** Serologic diagnosis ordinarily is most reliable and easiest, using CF (the most convenient and economic) or neutralization titrations. Very young infants, however, may not produce a detectable Ab re-

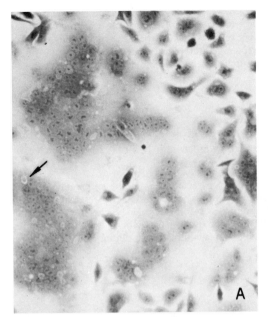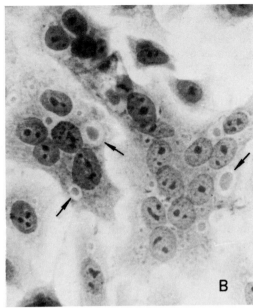

FIG. 59-15A, B. Syncytium formation and cytoplasmic inclusions **(arrows)** in Hep-2 cells infected with respiratory syncytial virus. Syncytia are characterized by the very large aggregates of intact nuclei and extensive cytoplasmic masses devoid of cell membranes. (Bennett CR Jr, Hamre D: J Infect Dis 110:8, 1962) **(A,** ×125; **B,** ×540)

sponse, in which case a laboratory diagnosis depends upon isolation of virus.

Because RS virus is highly labile, isolation is most efficient when nasal or pharyngeal secretions are inoculated directly from the patient into cultures of human continuous cell lines (Hep-2, HeLa, or KB). Characteristic giant cells develop within 2–14 days, but infection can be detected more rapidly by immunofluorescence. Newly isolated RS virus can be identified by CF or neutralization titrations with standard antisera.

Infectivity of RS virus is completely destroyed during storage at −15° to −25°C for only several days. Viral suspensions can be preserved without complete inactivation by adding protein (5%–10% normal serum or albumin), freezing rapidly, and maintaining at −70°C.

EPIDEMIOLOGY

RS virus is a **major cause of respiratory disease in young children.** Infections appear to have a worldwide distribution, and they occur in yearly epidemics of varying magnitude. Virus spreads rapidly through the susceptibles in a community, so that epidemics are sharply circumscribed and relatively brief. The outbreaks occur primarily in infants and children between late fall and early spring. Many adults may be infected during the episode, but they have mild disease or inapparent infections.

RS infections uniformly occur early in life. Indeed, RS virus is the only recognized virus that preferentially produces severe disease and has its maximum impact during the first 6 months of life. Although infants have a poor Ab response to infection, approximately one-third of infants in the United States develop Abs in the first year of life, and 95% by 5 years of age.

PREVENTION AND CONTROL

Because RS virus ranks so high on the list of causes of respiratory disease in young children, preventive methods are highly desirable. Isolation or general public health measures are not adequate to control the spread of infection: it is difficult to recognize the disease early, and inapparent cases are frequent.

Experience with an alum-precipitated, formalin-inactivated vaccine has been discouraging. Though the serum Ab response was good, the clinical response to subsequent RS virus infection was startling and paradoxic: both the **incidence and the severity** of disease were strikingly **increased,** particularly in infants. These results resembled the severe reactions in children immunized with inactive measles virus and then infected with measles virus after natural exposure. It is not certain whether this frightening response is due to a cell-mediated hypersensitivity, to interactions between Abs and virus-infected cells, or to Ab-virus complexes. However, it is clear that inactivated whole virus vaccines containing viruses with lipid envelopes must be used with caution.

Selection of attenuated viral variants, particularly temperature-sensitive mutants, is proceeding for the development of a live RS virus vaccine. But since the disease does not afford effective, lasting immunity, some view the utility of any vaccine with pessimism. Moreover, since enveloped viruses such as RS virus and other paramyxoviruses readily establish persistent infections in vitro, their use as live, temperature-sensitive mutants in a vaccine might greatly increase their potential for inducing chronic infections.

NEWCASTLE DISEASE VIRUS

Newcastle disease virus (NDV) is primarily a respiratory tract pathogen of birds, particularly chickens, but it occasionally produces accidental infections in man. Human infections are almost exclusively confined to poultry workers and laboratory personnel. The disease is characteristically mild and limited to conjunctivitis without corneal involvement. Although predominantly of veterinary interest, this virus merits brief mention because of the prominent role it has played in the investigation of paramyxoviruses.

NDV possesses the characteristic properties of paramyxoviruses listed in Table 59-1. This virus particularly resembles parainfluenza and mumps viruses in morphology, chemistry, and reactions with RBCs. Moreover, many patients with mumps virus infection develop hemagglutination-inhibiting and CF Abs to NDV.

SELECTED READING

PARAINFLUENZA VIRUSES

BOOKS AND REVIEW ARTICLES

CHANOCK RM: Parainfluenza viruses. In Lennette EH, Schmidt NJ (eds): Diagnostic Procedures for Viral, Rickettsial and Chlamydial Infections, 5th ed. Washington DC, American Public Health Association, 1980

CHOPPIN PW, COMPANS RW: Reproduction of paramyxoviruses. Compr Virol 4:95, 1975

GLEZEN WP, LODA FA, DENNY FW: The parainfluenza viruses. In Evans AS (ed): Viral Infections in Humans. Epidemiology and Control. New York, Plenum Medical Books, 1976, p 337

SPECIFIC ARTICLES

BLAIR CD, ROBINSON WS: Replication of Sendai virus. II. Steps in virus assembly. J Virol 5:639, 1970

CALIGUIRI LA, KLENK HD, CHOPPIN PW: The proteins of parainfluenza virus SV5. I. Separation of virion polypeptides by polyacrylamide electrophoresis. Virology 39:460, 1969

CHEN C, COMPANS RW, CHOPPIN PW: Parainfluenza virus surface projections: glycoproteins with hemagglutinin and neuraminidase activities. J Gen Virol 11:53, 1971

GLAZIER K, ROGHAW R, KINGSBURY DW: Regulation of Sendai virus transcription: evidence for a single promotor in vivo. J Virol 21:863, 1977

KLENK HD, CHOPPIN PW: Plasma membrane lipids and parainfluenza virus assembly. Virology 40:939, 1970

MUMPS VIRUS

BOOKS AND REVIEW ARTICLES

FELDMAN HA: Mumps. In Evans AS (ed): Viral Infections of Humans. Epidemiology and Control. New York, Plenum Medical Books, 1976, p 317

PARKMAN PD: Mumps. In Lennette EH, Schmidt NJ (eds): Diagnostic Procedures for Viral, Rickettsial and Chlamydial Infections, 5th ed. New York, American Public Health Association, 1980

SPECIFIC ARTICLES

LEVITT LP, MAHONEY DH, CASEY HL, BOND JO: Mumps in a general population: a seroepidemiologic study. Am J Dis Child 120:134, 1970

MEASLES VIRUS

BOOKS AND REVIEW ARTICLES

BLACK FL: Measles. In Evans AS (ed): Viral Infections of Humans. Epidemiology and Control. New York, Plenum Medical Books, 1976, p 297

GERSHON AA, KRUGMAN S: Measles virus. In Lennette EH, Schmidt NJ (eds): Diagnostic Procedures for Viral, Rickettsial and Chlamydial Infections, 5th ed. Washington DC, American Public Health Association, 1980

KRUGMAN S: Present status of measles and rubella immunization in the United States: a medical progress report. J Pediatr 90:1, 1977

MORGAN EM, RAPP F: Measles virus and its associated diseases. Bacteriol Rev 41:636, 1977

SPECIFIC ARTICLES

BLACK FL: Measles endemicity in insular populations: critical community size and its evolutionary implication. J Theor Biol 11:207, 1966

PANUM PL: Observations Made During the Epidemic of Measles on the Faroe Islands in the Year 1846. New York, American Publishing Association, 1940 (Reprint)

RESPIRATORY SYNCYTIAL VIRUS

BOOKS AND REVIEW ARTICLES

PARROTT RH: Respiratory syncytial virus. In Lennette EH, Schmidt NJ (eds): Diagnostic Procedures for Viral, Rickettsial and Chlamydial Infections, 5th ed. Washington DC, American Public Health Association, 1980

CHANOCK RM, KIM HW, BRANDT CD, PARROTT RH: Respiratory syncytial virus. In Evans AS (ed): Viral Infections of Humans. Epidemiology and Control. New York, Plenum Medical Books 1976, p 365

SPECIFIC ARTICLES

BENNETT CR, HAMRE D: Growth and serological characteristics of respiratory syncytial virus. J Infect Dis 110:8, 1962

JONCAS J, BERTHIAUME L, PAVILAVIS V: The structure of the respiratory syncytial virus. Virology 38:493, 1969

PARROTT RH, KIM HW, BRANDT CD, CHANOCK RM: Respiratory syncytial virus in infants and children. Prev Med 3:473, 1974

chapter

CORONAVIRUSES

After the discovery that rhinoviruses are major etiologic agents of the common cold (Ch. 57), more than 50% of illnesses still could not be associated with known causative agents. However, when Tyrrell and Bynoe in 1965 introduced the use of ciliated human embryonic tracheal and nasal organ cultures, they revealed a new group of viruses (and also improved the isolation of known agents). The unique properties of the new group (Table 60-1) included their distinctive club-shaped surface projections (Fig. 60-1), which give the appearance of a solar corona to the virion. Hence the family name **coronaviruses (Coronaviridae);** (L. *corona,* crown) was proposed to include the agents isolated initially as well as viruses isolated in human embryonic cell cultures by Hamre and Procknow from patients with acute respiratory diseases, and similar viruses from a variety of lower animals.*

PROPERTIES OF THE VIRUSES

Morphology. Electron microscopic examinations of negatively stained preparations reveal moderately pleomorphic spherical or elliptical virions. The surface is covered with distinctive pedunculated projections **(peplomers),** 20 nm long, with narrow bases and club-shaped ends (Fig. 60-1). Thin sections show virions with an outer membrane (envelope) 7–8 nm thick, a 4- to 8-nm electron-lucent intermediate zone, and an inner nucleocapsid consisting of an electron-dense shell 9–17 nm thick and a central zone containing amorphous material of variable density (Fig. 60-2). The nucleocapsid appears to be a loosely wound helix 14–16 nm in width.

Physical and Chemical Characteristics. Information on the structure of the virion is fragmentary. Two glycoproteins (mol wt 180,000 and 90,000) make up the surface projections, a third glycoprotein (mol wt 23,000) and a nonglycoprotein (mol wt 23,000) are constituents of the envelope membrane, and a nonglycosylated protein (mol wt 50,000) is a component of the nucleocapsid. The nu-

* Avian infectious bronchitis virus; calf neonatal diarrhea virus; murine hepatitis virus; porcine transmissible gastroenteritis virus; porcine hemagglutinating encephalitis virus; rat pneumotropic virus; rat sialodacryoadenitis virus; and turkey bluecomb disease virus.

cleocapsid contains one molecule of infectious single-stranded RNA (Table 60-1) and many molecules of the 50,000 daltons (50K) protein. The presence of lipid in the envelope makes the virion sensitive to lipid solvents. Deviations of as little as 0.5 u from pH 7.2 inactivate the virus, probably reflecting the lability of its envelope.

Immunologic Characteristics. Coronaviruses isolated from humans can be divided into at least three antigenic types by neutralization assays. The viruses initially isolated in organ cultures are immunologically distinct from those first detected in human embryo kidney cell cultures (Table 60-2), which is consistent with the strict host-range specificities (see Host Range, below). A third group of viruses obtained in human embryo tracheal organ cultures (termed **OC viruses**) crossreact weakly with the tissue culture viruses, but they are more closely related antigenically to the other organ culture isolates. A number of coronaviruses that are not immunologically related to the three defined antigenic types, or to each other, have also been isolated in organ cultures or in primary human embryonic kidney cells.

The viruses from humans appear antigenically unrelated to coronaviruses from other animals in neutralization tests, but complement-fixation (CF) assays indicate a partial immunologic relatedness of some human types with mouse hepatitis virus.

The virion contains at least three distinct antigenic structural units as detected by immunodiffusion studies; probably these same Ags react in CF assays as well. Two **organ culture (OC)** strains have been adapted to multiplication in the brains of suckling mice, and have attained the capacity to agglutinate RBCs from mice, rats, chickens, and humans—perhaps owing to a higher concentration of virus or to the selection of particular mutants. Unlike orthomyxoviruses, coronaviruses do not elute spontaneously from agglutinated RBCs, and treatment of susceptible RBCs with neuraminidase does not reduce hemagglutination. A rise in Abs (measured by CF, hemagglutination inhibition with adapted viruses, or neutralization) follows infection, but their duration and their protective value are still not known.

Host Range. Coronaviruses affecting man appear to be able to multiply in only a very limited range of host cells.

TABLE 60-1. Characteristics of Coronaviruses

Size	80–160 nm
Virion mol wt	$112 \pm 5 \times 10^6$ daltons
Morphology	Enveloped spherical; pleomorphic; pedunculated, club-shaped surface projections
Lipid content	Ether or chloroform inactivates
Nucleic acid	Single-stranded RNA, 6×10^6 daltons (18Kb)
Replication	Buds into cytoplasmic vacuoles

Viruses isolated in organ cultures of ciliated human embryonic tracheal or nasal tissues do not multiply well in monolayer cultures of human diploid or embryonic kidney cells, and vice versa; hence both culture systems must be employed for viral isolations.

A heteroploid human embryonic lung cell line and some human diploid cell strains have shown evidence of susceptibility to both types of viruses. Two of the OC strains have been adapted to growth in monkey kidney cells and in the brains of suckling mice, but other laboratory animals have not proved to be susceptible to infection.

FIG. 60-1. Coronaviruses. The negatively stained particles show the distinctive corona effect produced by the pedunculated surface projections, which are approximately 20 nm long and have a club-shaped end about 10 nm in width. The marked pleomorphism of the virions may be noted. (Kapikian AZ: In Diagnostic Procedures for Viral and Rickettsial Infections, 4th ed. New York, American Public Health Association, 1969) (×144,000)

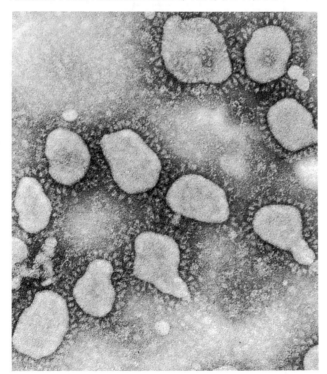

Viral Multiplication. Viral replication cannot yet be studied biochemically. Information derived from electron microscopy and fluorescent Ab investigations indicates that viral multiplication occurs solely in the cytoplasm and in close association with the smooth and the rough endoplasmic reticulum and the Golgi apparatus (Fig. 60-2). Nucleocapsids do not accumulate in masses visible in electron micrographs. Infrequently, thickened vesicle membranes and budding viral particles are observed; most commonly, however, formed virions are seen within vesicles or extracellularly. Viral multiplication is relatively slow compared with that of orthomyxoviruses or togaviruses: for example, the eclipse period lasts at least 6 h and maximum viral yield is not attained until about 24 h after infection. Viral multiplication is optimal at 32 ° C–33 ° C, as with rhinoviruses (Ch. 57), and infectivity is rapidly lost at higher temperatures or with prolonged incubation.

PATHOGENESIS

Coronaviruses constitute another group of viruses responsible for common colds and pharyngitis. Patients from whom the viruses were recovered, and volunteers who were inoculated intranasally to establish the etiologic relation of coronaviruses to disease, exhibited signs and symptoms of acute upper respiratory infections: coryza, nasal congestion, sneezing, and sore throat were common; less frequent symptoms were headache, cough, muscular or general aches, chills, and fever. Pulmonary involvement has not been noted. In volunteers inoculated intranasally, viral multiplication occurred in superficial cells of the respiratory tract, viral excretion was detectable at the time symptoms were first noted, about 3 days after infection, and the mean duration of illness was 6–7 days. Although coronaviruses have many distinctive viral characteristics the diseases they cause cannot be distinguished clinically from common colds produced by rhinoviruses or even from those occasionally initiated by influenza viruses.

Since the viruses have not been adapted to the respiratory tracts of lower animals, details of pathogenesis cannot be studied experimentally.

LABORATORY DIAGNOSIS

Coronaviruses are so fastidious in their host requirements that isolation is not practicable for routine diagnosis. In principle, the diagnostic serologic procedures of CF, neutralization, and hemagglutination-inhibition titrations can be used, but CF titration is the only practical procedure. Neutralization titrations are difficult and expensive, since they require human organ cultures and cell cultures; moreover, neutralizing Abs are present in

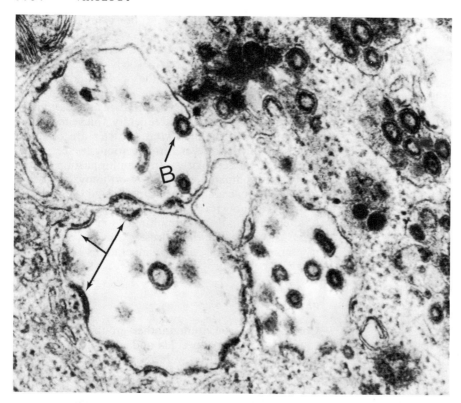

FIG. 60-2. Development of a coronavirus in a human diploid cell. Thickened membranes **(arrows)** indicate sites of early morphogenesis. A particle budding into a vacuole **(B),** and several mature particles within vacuoles, are also present. Mature particles show inner and outer shells with a translucent zone between them. (Becker WB et al: J Virol 1:1019, 1967) (×50,000)

50%–80% of the population, and although reinfection is common, a demonstrable increase in titer is difficult to measure. Hemagglutination has been demonstrated with only two viruses, both belonging to the same immunologic type of OC viruses.* For seroepidemiologic surveys, however, neutralization titrations are more informative than CF assays since CF Abs decline rapidly after infection.

EPIDEMIOLOGY

Acute upper respiratory infections generally have their greatest incidence in children, particularly those between 4 and 10 years of age. It is therefore especially striking that coronaviruses are frequently associated with common colds in adults, and it is a member aged 15 years or older who most frequently introduces infection into a family. Virus is not commonly isolated from children, and CF Abs are detected in only about 2% of them. Epidemiologic studies using neutralization titrations, however, have detected Abs in up to 50% of children 5–9 years old. Moreover, infection (with an increase in Ab

* Cultured human and monkey cells infected with OC 43 viruses adsorb rat and mouse RBCs and therefore the hemadsorption technic can be used for neutralization Ab studies with these viruses.

titer) is more common in older children and adults, although up to 70% of adults have Abs.

Coronavirus infections occur predominantly during the winter and early spring, and during a particular respiratory disease season only a single immunologic type appears to be the causative agent. In a seroepidemiologic study of adults covering a 4-year period in Washington, D.C., coronavirus infections occurred in 10%–24% of

TABLE 60-2. Neutralization Reactions of Representative Coronaviruses

Serum employed	Neutralization titer*		
	B814†	229E‡	OC 43§
B814	640	<5	<5
229E	<5	1600	<5
OC 43	<5	160	5120

*The initial isolates in each group serve as the group's prototype strain.

†Isolated by Tyrrell and Bynoe in human embryo tracheal organ culture.

‡Isolated by Hamre and Procknow in human embryo kidney cell culture.

§Isolated by McIntosh et al in human embryo tracheal organ culture.

(Modified from Bradburne AF, Tyrrell DAJ: Progr Med Virol 13:373, 1971)

those with upper respiratory illnesses. A study of Michigan families showed all age groups to be infected, from 0–4 years (29.2%) to over 40 years (22%), with the highest incidence at 15–19 years. These findings are in sharp contrast to the situation with other respiratory viruses (e.g., respiratory syncytial virus), where there is a marked decrease in disease incidence with increasing age. It is noteworthy that rhinovirus infections were uncommon during the periods of prevalent coronavirus infections. During epidemics all segments of the population are affected, but virus tends to spread preferentially within families. Unfortunately, because of the technical difficulties, comparatively few data on the epidemiologic incidence and spread of coronaviruses are available.

PREVENTION AND CONTROL

Additional data on duration of immunity following natural infection, on antigenic structure and immunogenic potential of the virions, and on incidence of infections are essential before the need for and the feasibility of a vaccine are established. Indeed, the high frequency of reinfection accompanied by disease suggests that successful immunization may not be possible.

SELECTED READING

BOOKS AND REVIEW ARTICLES

BRADBURNE AF, TYRRELL DAJ: Cornaviruses of man. Prog Med Virol 13:373, 1971

MCINTOSH K: Coronaviruses: a comparative review. Curr Top Microbiol Immunol 63:85, 1974

MONTO AS: Coronaviruses. Yale J Biol Med 47:234, 1974

SPECIFIC ARTICLES

HAMRE D, KINDIG DA, MANN J: Growth and intracellular development of a new respiratory virus. J Virol 1:810, 1967

HAMRE D, PROCKNOW JJ: A new virus isolated from the human respiratory tract. Proc Soc Exp Biol Med 121:190, 1966

HIERHOLZER JC, PALMER EL, WHITEFIELD SG, KAYE HS, DOWDLE WR: Protein composition of coronavirus OC 43. Virology 48:516, 1972

TYRRELL DAJ, BYNOE ML: Cultivation of a novel type of common-cold virus in organ cultures. Br Med J 1:1467, 1965

chapter **61**

RHABDOVIRUSES

Were the basis for evolutionary development not so indelibly imprinted on scientific thought a modern virologist might consider the emergence of the striking bulletlike morphology of rhabdoviruses (Gr. *rhabdos*, rod) a reflection of the violence of our times. This unique form, which was first described for vesicular stomatitis virus (VSV) of cattle and horses, is also associated with rabies virus (Fig. 61-1) and at least 25 other viruses that infect a variety of mammals, fish, insects, and plants. Properties of the agents in this group are summarized in Table 61-1. Several rhabdoviruses replicate in arthropods as well as in mammals (e.g., VSV, Hart Park virus, Flanders virus, Kern Canyon virus) and hence were previously considered to be arboviruses. Rabies virus is the only member of the group known to infect and produce disease in humans naturally; it will be considered in detail. The Marburg virus, a simian virus which only accidentally infects humans, has some similar features and will be discussed briefly.

RABIES VIRUS GROUP*

The terrifying change of a docile, friendly dog into a vicious rabid (L. *rabidus*, mad) beast, often with convulsions, struck terror in those in its vicinity and was long considered the work of supernatural causes. The infectious nature of rabies was recognized in 1804, but it was Pasteur in the 1880s who suggested that the responsible etiologic agent was not a bacterium. He used his knowledge of the properties of infectious agents, and his great intuition, to demonstrate for the first time that the pathogenicity of a virus (before viruses had actually been identified) could be modified by serial passage in an animal other than its natural host. Fifty serial intracerebral passages in rabbits yielded a modified† virus, **fixed virus** (as contrasted with the **wild-type** or **street virus**), which was used for immunization.

Upon discovery of the filterable causative agent in 1903, Negri described the presence of prominent cytoplasmic inclusion bodies **(Negri bodies)** in the nerve cells of infected human beings and animals. Their characteristic appearance and easy recognition made possible the rapid pathologic diagnosis of infection.

Although rabies was one of the first diseases of man to be recognized as caused by a virus, the agent was studied very little until the late 1960s, when methods for propagating attenuated viruses in cell cultures overcame the dangers encountered in handling the virus and the difficulties involved in growing it to high titer.

* Lyssavirus (Gr. *lyssa*, rage) is the official designation for this genus of the family Rhabdoviridae.
† Since the fixed virus strain is pathogenic for humans and other animals, it should be considered to be only partially attenuated.

PROPERTIES OF THE VIRUS

Morphology. The virions (which average 180–75 nm) are cylindrical, resembling a bullet (Fig. 61-1), with one rounded and one planar end, probably owing to collapse of the region where the budding particle is sealed. Regularly spaced projections, each with a knoblike structure at the distal end, cover the surface of the virion. Shorter bullet-shaped and cylindrical particles are also occasionally observed in electron micrographs. The helical nucleocapsid is symmetrically wound within the envelope along the axis of the virion, often giving the appearance of a series of transverse striations (Fig. 61-1A). The purified nucleoprotein is a ribbon-like helical strand, consisting of regular rod-like protein subunits attached to a thread of nucleic acid (Fig. 61-1B).

Physical and Chemical Properties. The viral envelope consists of a lipid bilayer covered by external surface projections composed of a glycoprotein **(G protein)** of 80,000 daltons, and two matrix **nonglycosylated proteins, M_1** (40,000 daltons) and M_2 (25,000 daltons), which reinforce the membrane internally. The glycoprotein surface projections, which act as a hemagglutinin of goose RBCs, serve for virion attachment to host cells. The nucleocapsid consists of one molecule of **single-stranded RNA of negative polarity,** many identical copies of an **N protein** of 69,000 daltons, and a few copies of the **RNA-dependent RNA transcriptase,** composed of a large **(L) protein** of 190,000 daltons and a smaller **(NS) protein.**

As with most enveloped viruses infectivity deteriorates rapidly at room or refrigerator temperatures in the ab-

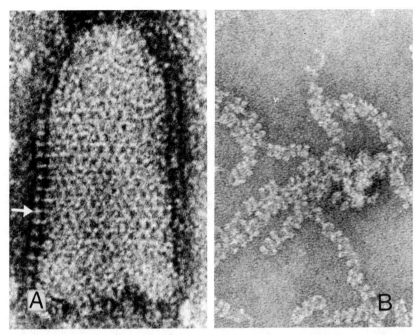

FIG. 61-1. Morphologic characteristics of rabies virus. **A.** Intact rabies virus particle embedded in phosphotungstate and viewed in negative contrast. On the left are well-resolved surface projections 6–7 nm long **(arrow)**. (×400,000) **B.** Helical nucleocapsid isolated from disrupted rabies virions. Note the tightly coiled and partially uncoiled regions of the single-stranded helix. (×212,000) (**A,** Hummeler K et al: J Virol 1:152, 1967; **B,** Sokol F et al: Virology 38:651, 1969)

sence of tissue proteins or added protein (normal serum or albumin). Inactivation is much slower, however, in crude tissue extracts or in infected tissues stored in neutral glycerol. Infectivity is quite stable in frozen or lyophilized tissue extracts.

Immunologic Characteristics. All the rabies viruses isolated from man and other animals, throughout the world, appear to be of a **single immunologic type.** Selected modified (fixed) and wild-type (street) viruses, prepared by many different methods and propagated in different tissues, are also immunologically similar. Several other viruses, immunologically related to rabies virus but dis-

tinguishable from it, appear to be limited geographically to regions of Africa and, in host range, primarily to lower animals and insects. However, some of these viruses have been isolated from human illnesses.*

The virion's surface structures are responsible for the production of neutralizing as well as hemagglutination-inhibiting Abs. Antibodies to the nucleocapsid, in contrast, are recognized by complement fixation (CF). A third class of virus-specific Abs appear after either infection or immunization; in the presence of complement they lyse infected cells whose plasma membranes have incorporated viral Ags. *These cytolytic Abs may play a deleterious rather than a protective role in pathogenesis.*

Owing to the long incubation period, circulating Abs may be present at the onset of illness. Neutralizing Abs also appear in high titer in the brain and cerebrospinal fluid of patients, and can serve as a valuable indicator of infection.

Host Range. Rabies virus can infect all mammals so far tested. Among domestic animals dogs, cats, and cattle are particularly susceptible. Skunks, bats, foxes, squirrels,

TABLE 61-1. Characteristics of Rhabdoviruses

Morphology	Bullet-shaped; 130–240 nm × 70–80 nm
Nucleic acid	Single-stranded RNA of negative polarity; 3.5–4.6×10^6 daltons (11–14 Kb); noninfectious
Virion enzyme	RNA transcriptase
Nucleocapsid	Helical; 18 nm in width
Effect of lipid solvents	Disrupt virions; inactivate infectivity
Maturation	Budding at cytoplasmic membranes
Hosts	Wide variety of mammals, fish, invertebrates, and plants
Common antigens	None

---

* Duvenhaga virus, isolated from the brain of a man who had been bitten by a bat in South Africa; Lagos bat virus (Nigeria); and Mokola shrew virus (Nigeria).

raccoons, coyotes, mongooses, and badgers are the principal wildlife hosts. Birds are also susceptible, but less so than mammals.

To establish laboratory infections hamsters, mice, guinea pigs, and rabbits (in order of decreasing susceptibility) are employed. Intracerebral inoculations are more reliable than subcutaneous or intramuscular inoculations.

Rabies virus can be propagated in chick or duck embryos. Attenuated strains developed by multiple passage in embryonated eggs now serve as important sources for vaccines.

Cultures of cells from many different animal species can also support viral multiplication. Hamster kidney, human diploid, and chick embryo cell cultures maintained at 31°C–33°C are most commonly used. Wild virus is propagated with greater difficulty than modified strains. Cytopathic changes are not usually observed in infected cultures, but intracytoplasmic Ag can be detected by immunofluorescence.

Viral Multiplication. Despite the long incubation period in natural infections, and in experimental animals, the characteristics of viral multiplication in cell cultures are not unusual: the eclipse period is 6–8 h, and the initial cycle of multiplication is completed in 19–24 h. Since cytopathic changes are absent or minimal, carrier cultures or endosymbiotic infections (Ch. 53) are established readily.

The mechanism of replication is probably the same as for VSV, which is morphologically and chemically similar. VSV virion RNA, a negative strand (i.e., noninfectious), is transcribed into two species of RNAs: 1) mRNAs transcribed by the **virion's RNA-dependent RNA polymerase (transcriptase);** and 2) 40S positive strand RNAs, which serve as templates for virion RNAs (probably synthesized by another polymerase, the **replicase**). The transcriptase has a single promoter on the virion genome, and therefore it starts at a single site and sequentially copies the genome: the mRNAs are then produced by either processing a large transcript of the entire genome or terminating and restarting transcription at each junction between messages. Viral RNA replication is accomplished through a characteristic replicative intermediate form (Ch. 50).

During viral replication cytoplasmic "factories" form prominent matrices of helical nucleocapsids. These masses of nucleocapsids appear as cytoplasmic inclusion bodies (i.e., Negri bodies) that can be identified by specific immunofluorescence. The virions then assemble by budding from cytoplasmic membranes (and from plasma membranes in many types of cultured cells).

PATHOGENESIS

A wound or abrasion of the skin, usually inflicted by a rabid animal, is the major portal of entry into man, virus entering with the animal's saliva. (A dense population of infected bats may also create an aerosol of infected secretions, by which virus appears to obtain entrance into the respiratory tract.) Virus multiplies in muscle and connective tissue but remains localized for periods that vary from days to months; it then **progresses along the axoplasm of peripheral nerves** to ganglia and eventually to the CNS, where it multiplies and produces severe and usually fatal encephalitis. Hematogenous spread of virus to the CNS has been claimed but not established. For transmission by a mammal the virus must reach the salivary glands, but it apparently does so via efferent nerves rather than through blood and lymph vessels.

The **incubation period** is usually 3–8 weeks but can be as short as 6 days or as long as 1 year. It depends upon the size of the viral inoculum, the severity of the wound, and the length of the neural path from the wound to the brain, i.e., it is shorter following bites on the face and head. Illness is ushered in by a **prodromal period,** with irritability, abnormal sensations about the wound site, and hyperesthesia of the skin. **Clinical disease** becomes apparent with the development of increased muscle tone and difficulty in swallowing, owing to painful and spasmodic contractions of the muscles of deglutition when fluid comes in contact with the fauces. Often the mere sight of liquids will induce such contractions; hence the common name **hydrophobia** (Gr., fear of water). The final stages of the disease result from the extensive damage in the CNS. A fatal outcome has been considered inevitable, but several patients with proved rabies have recovered after being given extensive care for sustaining vital functions. Epidemiologic data further suggest that recognizable rabies follows only 30%–50% of proven exposures: e.g., in an unintentional study about one-half of untreated persons developed clinical rabies and died following severe mutilation by a rabid wolf, which must certainly have effected a viral infection in all those attacked.

Pathologically, rabies is an encephalitis with neuronal degeneration of the spinal cord and brain. **Negri bodies** within affected neurons are the most characteristic and indeed are the only pathognomonic microscopic finding. These cytoplasmic inclusions are sharply defined, spherical or oval, eosinophilic, Feulgen-negative bodies, $2\mu m$–$10\mu m$ in diameter, containing a central mass of basophilic granules (Fig. 61-2). Several may be found in a single cell. Immunofluorescence studies have demonstrated specific viral Ags in the Negri body, and electron micrographs show (Fig. 61-3) that it consists of a large matrix of viral nucleocapsids (eosinophilic material) and budding virions (the basophilic granules).

Inclusion bodies are most abundant in Ammon's horn of the hippocampus but may also be found in large numbers in many other sites in the brain and in the posterior horn of the spinal cord. In the absence of identifiable Negri bodies a pathologic diagnosis of rabies cannot be made.

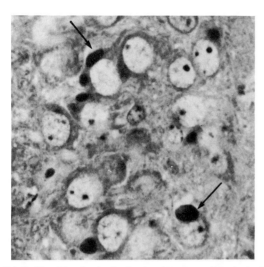

FIG. 61-2. Negri bodies in the brain of a mouse infected with rabies virus. Numerous large dark cytoplasmic inclusion bodies **(arrows)** are present. The matrix of the inclusion body is stained intensely; the internal granules appear as light vacuoles. Stained by the dinitrofluorobenzene method for protein-bound groups. (Courtesy of Dr. H. Koprowski, Wistar Institute, and Dr. R. Love, Jefferson University School of Medicine) (×2000, reduced)

LABORATORY DIAGNOSIS

Definitive diagnosis of infections in humans, and in suspected animal vectors, depends upon any one of the following findings: 1) identification of Negri bodies in brain tissue; 2) isolation of virus from brain or saliva; or 3) detection of viral Ags in specimens of brain, spinal cord, or skin by immunofluorescence (probably the method of choice considering speed, accuracy, and cost). Since the incubation period may be relatively short the need for an immediate definitive diagnosis makes serologic technics of little value early in the disease because circulating Abs appear slowly. Serum and cerebrospinal fluid Abs eventually reach high levels, however, and if clinical doubts exist during later stages, Ab titrations (neutralization, CF, immunofluorescence) can establish the diagnosis in an unimmunized patient.

Negri bodies are detected in impression smears prepared from the region of Ammon's horn. Seller's method, with a stain composed of a mixture of basic fuchsin and methylene blue, is commonly employed. Their distinctive stained appearance (**cherry red with deep blue granules**) differentiates them from other inclusion bodies, particularly those produced by distemper virus in dogs. Fluorescent Ab technics provide reliable confirmation of the specific nature of the inclusion body. The presence of Negri bodies is diagnostic, but *failure to detect them does not exclude rabies* and should be followed by attempts to isolate the virus.

Virus is preferably isolated by inoculating saliva, salivary gland tissue, or hippocampal brain tissue intracerebrally into infant mice. The mice develop paralysis after an incubation period of 6–21 days,

depending upon the quantity of virus present. The illness is not pathognomonic of rabies, and the virus must be identified by immunologic technics or by demonstrating Negri bodies in brain tissue from the inoculated animals.

EPIDEMIOLOGY

Although medical interest in rabies centers upon infections of man, epidemiologically this is a dead end to the infectious cycle, since humans contract rabies but do not normally transmit the disease. Dogs are generally the most dangerous source of infection for man, with cats next. In the United States, however, cattle are the most commonly infected domestic animals. The incidence of rabies in humans and dogs has decreased continuously in the United States since the institution of an effective immunization program and leash laws for dogs, and epizootic canine rabies (dog-to-dog transmission) has become rare. But rabies still remains enzootic in dogs, and in skunks, bats, and other wild animals, and control measures remain essential. For example, human rabies dropped from 33 cases in 1946 to only 1–3 cases per year since 1960, while rabies in animals (detected annually in almost every state) has continued a slight but uneven decline (Fig. 61-4).

Wild mammals serve as a large and uncontrollable reservoir of **sylvan rabies,** which is an increasing threat to people and domestic animals throughout the world. The most frequent wildlife sources in North America are skunks, foxes, and raccoons. Moreover, an epizootic in foxes has reintroduced rabies into Western Europe, which once was relatively free of the disease.

Vampire and insectivorous **bats** are important reservoirs and could be one of the most important links in the ecology of rabies: experimentally the virus can remain latent in these animals for long periods; and virus has been detected in the nasal mucosa, brown fat, and salivary glands of apparently healthy bats. Furthermore, the virus can be transmitted to man or other animals from bats without a direct bite, apparently by inhalation of infectious aerosols.

Worldwide, rabies in animals, including dogs, steadily increases. A worldwide total of about 1000 fatal human cases is reported annually to the WHO, and the actual number must be several times greater.

PREVENTION AND CONTROL

An effective program for rabies control must be directed toward preventing the disease in domestic animals as well as in man. However, control measures directed against wildlife, which is now the main source of human cases, are impracticable. In fact, the primary rabies problem in the United States and many other countries is no longer human cases, but rather the decision whether

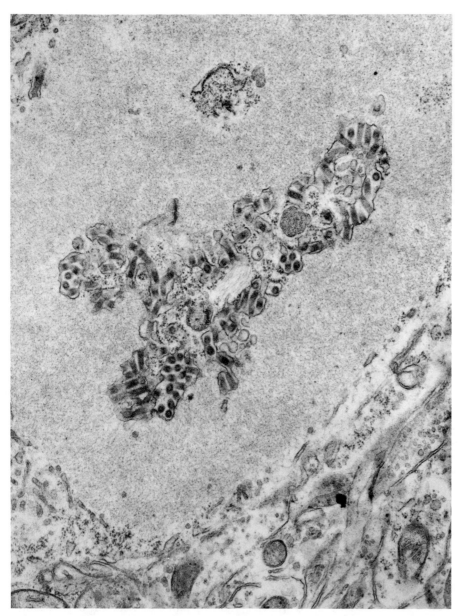

FIG. 61-3. Electron micrograph of a Negri body containing several inner bodies composed of developing and mature virions. (Matsumoto S: Adv Virus Res 16:257, 1970) (×19,500)

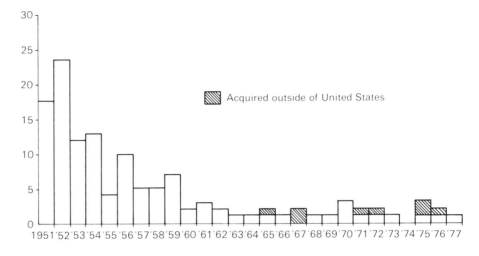

FIG. 61-4. Cases of rabies detected in the United States. **A.** In humans, 1951–1977. **B.** In wild and domestic animals, 1953–1977. (Center for Disease Control: Rabies Surveillance, Annual Summary, 1977)

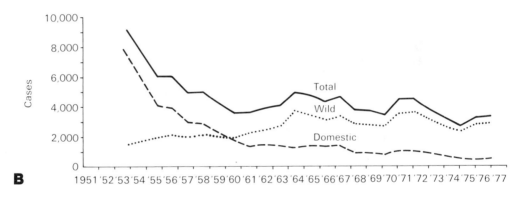

or not to undertake immunization, with its attendant dangers, after suspected exposure.

Vaccines. Pasteur's original vaccine was prepared by homogenization of partially dried* spinal cords from rabbits infected with modified (fixed) virus. Daily injections were given for 15–20 days with cords desiccated for progressively shorter periods. Frequent difficulties resulted, however, from the inexact method of viral inactivation and lack of quantitative controls.

A major advance was made with the development of **phenol-inactivated vaccines.** Semple vaccine, the one used in the United States until 1957, was a 5% suspension of infected nervous tissue (usually rabbit), treated with 0.25%–0.5% phenol. In attempts to reduce the demyelinating complications of immunization, nervous tissues of immature animals were also used, e.g., suckling mouse (or rat) brain. Daily subcutaneous injection for 7–14 days was recommended, depending upon the severity and location of the bite. Booster immunizations were given 10 and 20 days after the last

* Desiccation was employed to reduce infectivity since Pasteur had noted earlier that dried cultures of chicken cholera bacteria lost their pathogenicity, but not their immunogenicity, for chickens.

daily dose. Neutralizing Abs were usually detectable by 14 days after the first injection.

Despite the widespread acceptance and general use of phenolized vaccine, no valid control study had been carried out to prove its effectiveness. Moreover, the occurrence of demyelinating allergic encephalitis as a common complication of immunization made the continued search for an effective, safe vaccine imperative.

Rabies virus **propagated in embryonated duck eggs and inactivated** by β-propiolactone is the vaccine presently employed in the United States, owing to the lower risk of allergic encephalomyelitis. The course consists of 21 daily injections followed by booster inoculations 10 and 20 days later. Although the duck embryo vaccine is considerably less immunogenic than the nervous tissue vaccines, it nevertheless appears to be effective since only 13 cases of rabies occurred in 415,000 persons immunized with duck embryo vaccine from 1957 to 1975 in the United States. Fatalities have also been reported in several additional cases in which treatment was not started immediately or was not completed.

The low immunogenicity of the duck embryo vaccine

and the continuing occurrence of allergic encephalomyelitis have spurred efforts to replace it with a modified rabies virus propagated in cell cultures. Such a virus could be partially purified and concentrated in order to prepare highly immunogenic inactivated vaccines. A vaccine prepared from **virus grown in human diploid fibroblasts** has been successfully used in monkeys and rodents and shows great promise in human trials. Three injections during a 7-day period are equal in Ab levels to a full course of the phenolized or duck embryo vaccine. It reliably produces high titers of Abs, and it has been shown to protect humans bitten by rabid animals; it also appears to be free of encephalitogenic factors. This vaccine is being widely used in Europe and its use in the United States is steadily increasing. (In all probability it will become the worldwide vaccine of choice.) See addendum p. 1175.

Alternatively, the **purified virion surface glycoprotein** produces neutralizing Abs and may eventually be useful.

Several types of **live attenuated virus vaccines** have been prepared from strains of virus adapted to chick embryos and to various types of cultured cells. Failure of these adapted viruses to multiply in humans and the low antigenic mass injected have made them unsuitable for humans, but they have proved valuable for use in dogs, cats, and cattle. The low egg passage (LEP) Flury strain is the most frequently used for dogs. However, it is still weakly pathogenic for these animals when inoculated intramuscularly, and produces up to three cases per 10^6 doses of vaccine administered. Vaccination of dogs with LEP vaccine confers immunity for 3 years. The high egg passage (HEP) strain, which has had more than 180 egg transfers, is no longer virulent and is used for animals that cannot be immunized with the LEP vaccine (e.g., puppies and cats). But the HEP vaccine is not sufficiently immunogenic to control rabies in dogs, and therefore one with less virulence than the LEP strain is still necessary to avoid pathogenic effects but to afford good protection.

Public Health Measures. Primary control of rabies requires restriction of dogs and other domestic animals, as well as limitation of spread from wildlife to the greatest extent possible. The excellent WHO report (1973) recommends 1) compulsory prophylactic immunization of dogs with LEP and of cats with HEP Flury vaccines; 2) registration of all dogs, destruction of stray dogs, and isolation and observation of suspect dogs; and 3) attempted control of rabies in wildlife by trapping and other means. The danger of rabies is responsible for current severe restrictions on the transport of dogs across some national boundaries (e.g., England).

Prophylactic Treatment. Despite the low incidence of human rabies in many parts of the world, the question frequently arises as to the course to follow after an animal bite or scratch. Prophylactic immunization with phenolized rabbit vaccine is no longer recommended because of the relatively high reaction rate noted following its use. The incidence of demyelinating encephalomyelitis, caused by hypersensitivity to nervous tissue present in the vaccine, was 1:500 to 1:8,000, with a fatality rate of about 1:35,000. The duck vaccine lacks nervous tissue and is presently recommended. Unfortunately, it has not completely eliminated the occurrence of encephalomyelitis (neuroparalytic reactions about 1:25,000 immunized; deaths, 1:225,000 immunized). Systemic reactions (fever, myalgia, etc.) also occur in about one-third of the recipients after 5–8 doses.

In deciding on the management of a person bitten by an animal suspected of being rabid a number of factors must be considered, and it is not possible to present guidelines for all situations. However, the following is recommended: 1) vaccine should not be given to a person who has had only minimal contact with a questionable source of infection since allergic encephalitis may follow immunization; 2) if exposure to rabies appears definite, treatment with duck vaccine (or preferably with the human diploid cell vaccine) and with human immunoglobulin* should be instituted promptly; and 3) if the responsible animal can be captured alive it should be observed for at least 10 days for the development of symptoms, since Negri bodies may not be detectable in the brain early in the disease, and the virus can appear in the saliva several days before the onset of symptoms. Suspected bats, however, should be killed and examined at once.

Prophylactic treatment is directed toward confining the virus to the site of entry. Local treatment of the wound consists of thorough cleansing (cauterization is no longer suggested) and infiltration with antiserum. Because of the long incubation period active immunization usually produces an adequate Ab level before the virus has reached its target organs in the CNS. **Combined inoculation with hyperimmune serum* and vaccine** is the most effective regimen: serum Abs provide an immediate barrier to passage of virus, which lasts about 14 days, and meanwhile Abs are elicited by the vaccine. The use of hyperimmune serum, however, makes it mandatory to give the full course of vaccine injections in addition to two booster injections, since serum depresses Ab production. Antiserum alone should probably never be used.

Long-term Prophylaxis. Veterinarians, laboratory workers, dog handlers, wildlife workers, and certain hobbyists (e.g., spelunkers) may be sufficiently exposed to rabies to justify prophylactic immunization. Members of this high-risk group should receive a course of vaccine followed in several months by a booster injection. Subsequent booster doses should be given every 2–3 years or following a suspected exposure.

---

* A single injection of γ-globulin from hyperimmunized humans is recommended.

MARBURG VIRUS

In 1967 in Marburg and Frankfurt, Germany, and in Yugoslavia, an acute febrile illness appeared in laboratory workers handling tissues and cell cultures from recently imported African green monkeys. The 31 cases that occurred, including cases of nosocomial infections in hospital personnel, resulted in 7 fatalities. From blood and organ suspensions from fatal cases, a virus, termed the Marburg virus, was isolated by inoculation of guinea pigs.

The virus is inactivated by ether and sodium deoxycholate, and it appears to contain RNA since it replicates in the presence of inhibitors of DNA synthesis. Electron microscopic examinations of negatively stained preparations reveal particles that are somewhat rod-shaped, but with a variety of bizarre cylindrical and fishhook-like forms. The virions have a diameter of 75–80 nm but vary in length from 130 to 4000 nm. Most particles have one rounded end like rhabdoviruses; the other extremity is flat or occupied by a large bleb. Prominent cross-striations and an inner cylindrical structure add to its similarities with rhabdoviruses. Marburg virus is not antigenically related to any rhabdoviruses, and whether it should be considered a rhabdovirus is still unsettled.

Marburg virus disease has a sudden onset with high fever, gastrointestinal upset, constitutional symptoms, and marked prostration, followed by uremia, rash, hemorrhages, and CNS involvement. Fatal cases show necrotic foci in many organs, including the brain; the liver and lymphatic tissues are most severely affected. Medical personnel should take great precautions when handling saliva, urine, or blood from patients.

Serologic surveys in Uganda and Kenya show that the virus is present and produces **inapparent infections** in monkeys and humans. Although the disease has only reappeared once, in two tourists travelling in East Africa in 1975, the extensive use of primary monkey cell cultures makes it imperative that physicians and virologists be aware of this simian virus which on occasion also infects man.

SELECTED READING

BOOKS AND REVIEW ARTICLES

DEAN DJ, EVANS WM, MCCLURE RC: Pathogenesis of rabies. Bull WHO 29:803, 1963

Expert Committee Report on Rabies. Sixth Report, WHO Technical Report Series No. 523, World Health Organization, Geneva, 1973

HOWATSON AF: Vesicular stomatitis and related viruses. Adv Virus Res 16:196, 1970

MATSUMOTO S: Rabies virus. Adv Virus Res 16:257, 1970

SHOPE RE: Rabies. In Evans AS (ed): Viral Infections of Humans. Epidemiology and Control. New York, Plenum Medical Books, 1976

WAGNER RR: Reproduction of rhabdoviruses. Compr Virol 4:1, 1975

SPECIFIC ARTICLES

APPELBAUM E, GREENBERG M, NELSON J: Neurological complications following antirabies vaccination. JAMA 151:188, 1953

CAREY L, HATTWICK MAW: Treatment of persons exposed to rabies. JAMA 232:272, 1975

KABAT EA, WOLF A, BEZER AE: The rapid production of acute disseminated encephalomyelitis in rhesus monkeys by injections of heterologous and homologous brain tissue with adjuvants. J Exp Med 85:117, 1947

KISSLING RE: Marburg virus. Ann NY Acad Sci 174:932, 1970

PASTEUR L: Méthode pour prévenir la rage après morsure. CR Acad Sci [D] (Paris) 101:765, 1885

TIERKEL ES, SIKES RK: Preexposure prophylaxis against rabies. JAMA 201:911, 1967

ADDENDUM

The human diploid cell vaccine was approved by the Food and Drug Administration on June 9, 1980, for use in the United States. The licensing of the vaccine followed the findings that the Ab response to 5 injections of the human diploid cell vaccine is superior to that produced by 16–23 doses of the duck embryo vaccine; that of 153 persons who received the diploid cell vaccine plus immune globulin after being bitten by a rabid animal, none contracted rabies; and that serious adverse reactions are fewer than are attributed to the duck embryo vaccine.

chapter

TOGAVIRUSES, BUNYAVIRUSES, AND ARENAVIRUSES

The **arthropod-borne viruses (arboviruses)** multiply in both vertebrates and arthropods. In the cycle of transmission the former serve as reservoirs and the latter mostly as vectors, acquiring infection with a blood meal, but in some instances arthropods can also serve as reservoirs, maintaining the viruses by transovarian transmission. The virus is propagated in the arthropod's gut, attains a high titer in its salivary glands, and is transmitted when a fresh host is bitten. The viruses often cause disease in humans and other vertebrate hosts, but no ill effects are evident in the arthropods.

Most of the viruses described in this chapter were classified until recently, on epidemiologic grounds, as arboviruses. But increasing knowledge of their chemical and physical characteristics has revealed great heterogeneity among the arthropod-borne viruses. A number of these viruses with similar chemical and physical characteristics (Table 62-1), and of great medical significance to man, are now considered to be members of a single family, **Togaviridae** (vernacular, **togaviruses;** *toga,* coat), comprising four genera. The viruses infecting humans belong to: **Alphavirus** and **Flavivirus** (formerly called groups A and B arboviruses, respectively), and **Rubivirus** (rubella virus; German measles virus). The unique pathogenesis, epidemiology, and clinical problems of rubella, however, warrant its separate description (Ch. 63). This chapter will also consider two other medically significant groups of viruses whose characteristics distinguish them sharply from togaviruses. The family **Bunyaviridae** (previously called the Bunyamwera virus supergroup) includes the largest group of arthropod-borne viruses; it will undoubtedly be sudivided with further characterization. The family **Arenaviridae** (formerly included among the arboviruses), is also discussed in this chapter although arthropod transmission is not observed.

The frequent association of an enveloped structure with transmission by arthropods may be more than coincidental: enveloped viruses lose infectivity readily (e.g., on drying, on exposure to bile) and therefore must be spread either by intimate contact (e.g., ortho-myxoviruses, paramyxoviruses) or by insect bite, whereas naked viruses, such as picornaviruses, tend to be more stable and can survive a more circuitous fecal–oral spread.

Arthropod-borne diseases, because of their vectors, depend strongly on climatic conditions: they are endemic in areas of tropical rain forests, and epidemics in temperate areas usually appear after heavy rainfall has caused an increase in the vector population. Apart from agents known to cause human diseases, a large number of additional arboviruses have been isolated from mosquitoes and ticks trapped in forests and also from animals, especially monkeys, caged in the jungle as "sentinels" to permit insects to feed on them. These viruses are not known to cause prominent diseases of humans, but they are attracting a good deal of attention because of their threat as the world's expanding and increasingly mobile population impinges progressively on jungles. Over 350 arthropod-borne viruses have now been isolated (including some **rhabdoviruses,** discussed in Ch. 61, and **orbiviruses,** Ch. 64).

HISTORY

Yellow fever virus was the first arthropod-borne virus to be discovered, through the work of Major Walter Reed. He headed the US Army Yellow Fever Commission, established in 1901 to try to overcome the disastrous effect of yellow fever on American troops in Cuba during the Spanish-American War. Reed and the members of the Commission demonstrated transmission of this disease in bold experiments with human volunteers* built on the astute observations of Carlos Finlay, a Cuban physician, showing the association between yellow fever and mosquitoes. They also demonstrated the filterability of the agent. These studies established, for the first time, a virus as an agent of human disease, and an insect as the vector for a virus.

Their discoveries led to the eventual control of yellow fever, which for more than 200 years had intermittently been one of the world's major plagues, and, in fact, was a deciding factor in France's failure to complete the Panama Canal. This disease was by no means purely tropical: for example, an epidemic in the Mississippi Valley in 1878 caused 13,000 deaths, and substantial epidemics occurred in the nineteenth century as far north as Boston.

IMMUNOLOGIC CLASSIFICATION

The viruses isolated from arthropods may be divided into at least 34 distinct groups; many are still ungrouped.

---

* It should be noted that such experiments are not permitted today, and accordingly it is exceedingly difficult to establish whether a newly isolated virus is the cause of a disease or merely a fellow traveler with the undiscovered true etiologic agent.

TABLE 62-1. Properties of Togaviruses

Morphology	Icosahedral symmetry; lipid envelope Alphaviruses 45–75 nm Flaviviruses 37–50 nm
Nucleic acid	RNA; positive-strand, about 4×10^6 daltons (12Kb); infectious
Effect of lipid solvents	Inactivate
Stability	Unstable without added protein*
Hemagglutination	RBCs from newborn chicks or geese
Best animal host	Suckling mouse

*Survive several months at $-20°$ C, indefinitely at $-70°$ C or if lyophilized.

Table 62-2 lists the principal viruses that infect humans, the families and genera to which they have been assigned, and some of their clinical and epidemiologic characteristics.

Alphaviruses and flaviviruses are classified on the basis of hemagglutination-inhibition reactions: members of a genus crossreact with each other, but not with other togaviruses (Table 62-3). Bunyaviruses are classified into ten subgroups by complement fixation (CF), which shows greater crossreactivity among members of the family than does hemagglutination inhibition, but many

TABLE 62-2. Classification and Description of Togaviruses* and Clinically Related Viruses of Humans

Family	Genus (group)	Subgroup	Viral species	Vector	Clinical diseases in man	Geographic distribution
Togaviridae	Alphavirus	I	Eastern equine encephalitis (EEE)	Mosquito	Encephalitis	Eastern USA, Canada, Brazil, Cuba, Panama, Philippines, Dominican Republic, Trinidad
			Venezuelan equine encephalitis (VEE)	Mosquito	Encephalitis	Brazil, Colombia, Ecuador, Trinidad, Venezuela, Mexico, Florida, Texas
			Western equine encephalitis (WEE)	Mosquito	Encephalitis	Western USA, Canada, Mexico, Argentina, Brazil, British Guiana
			Sindbis	Mosquito	Subclinical	Egypt, India, South Africa, Australia
		II	Chikungunya	Mosquito	Headache, fever, rash, joint and muscle pains	East Africa, South Africa, Southeast Asia
			Semliki Forest	Mosquito	Fever or none	East Africa, West Africa
			Mayaro	Mosquito	Headache, fever, joint and muscle pains	Bolivia, Brazil, Colombia, Trinidad
			(13 others named)	Mosquito	Subclinical or none known	
	Flavivirus	I	St. Louis encephalitis	Mosquito	Encephalitis	USA, Trinidad, Panama
			Japanese B encephalitis	Mosquito	Encephalitis	Japan, Guam, Eastern Asian mainland, Malaya, India
			Murray Valley encephalitis	Mosquito	Encephalitis	Australia, New Guinea
			Ilheus	Mosquito	Encephalitis	Brazil, Guatemala, Trinidad, Honduras
			West Nile	Mosquito	Headache, fever, myalgia, rash, lymphadenopathy	Egypt, Israel, India, Uganda, South Africa
		II	Dengue (4 types)	Mosquito	Headache, fever, myalgia, prostration, rash (sometimes hemorrhagic)	Pacific islands, South and Southeast Asia, Northern Australia, New Guinea, Greece, Caribbean islands, Nigeria, Central and South America
		III	Yellow fever	Mosquito	Fever, prostration, hepatitis, nephritis	Central and South America, Africa, Trinidad
			(18 other viruses)	Mosquito		

(continued)

TABLE 62-2. Classification and Description of Togaviruses* and Clinically Related Viruses of Humans (*continued*)

Family	Genus (group)	Subgroup	Viral species	Vector	Clinical diseases in man	Geographic distribution
		IV	Tick-borne group (Russian spring-summer encephalitis group) 14 viruses	Tick	Encephalitis; meningoencephalitis, hemorrhagic fever	Russian spring-summer encephalitis; USSR; Canada, USA; others: Japan, Siberia, Central Europe, Finland, India, Malaya, Great Britain (louping ill)
			Rio Bravo (bat salivary gland) (15 others)	Unrecognized Unrecognized	Encephalitis	California, Texas
Bunyaviridae	Bunyavirus (Bunyamwera supergroup)	C group	Marituba and 10 others	Mosquito	Headache, fever	Brazil (Belem), Panama, Trinidad, Florida
		Bunyamwera group	Bunyamwera and 17 others	Mosquito	Headache, fever, myalgia, fever only; or none	Uganda, South Africa, India, Malaya, Columbia, Brazil, Trinidad, West Africa, Finland, USA
		California group	California encephalitis and 10 others	Mosquito	Encephalitis or none	USA, Trinidad, Brazil, Canada, Czechoslovakia, Mozambique
		7 other subgroups (7 ungrouped) members of genus)	46 viruses			
	(At least 2 others)	Phlebotomus fever group	20 viruses	Phlebotomus	Headache, fever, myalgia	Italy, Egypt
		Uukuniemi group	Uukuniemi and 6 others	Ticks		Finland
		(8 others)	21 viruses	Mosquitoes, ticks		
			8 unassigned viruses			
Arenaviridae	Arenavirus	Tacaribe,	Tacaribe, Junin Tamiami, Machupo, Pichinde, and 3 others Lymphocytic choriomeningitis Lassa		Headache, fever, myalgia, hemorrhagic signs	South and Central America
Ungrouped			Silverwater	Tick	None known	Canada
			Rift Valley fever	Mosquito	Headache, fever, myalgia, joint pains, hemorrhagic signs, rash	Africa
			Crimean hemorrhagic fever	Tick	Headache, fever, myalgia, hemorrhagic signs	Southern USSR
			36 others	Mosquito	None known	
Others			48 viruses (14 groups of 2-8 viruses)	Mosquito tick	None known in most cases	

*Rubivirus (rubella, or German measles, virus), also a togavirus, is discussed in Chapter 63.

TABLE 62-3. Results of Hemagglutination-Inhibition Titrations with Alpha- and Flavi-Togaviruses.

Immune serum	Genus	Virus (antigen)							
		EEE	VEE	WEE	Sindbis	Chikungunya	Mayaro	Semliki Forest	St. Louis encephalitis
EEE	Alphavirus	**10,240**	80	160	20	40	20	20	<10
VEE	Alphavirus	160	**640**	80	20	80	80	40	<10
WEE	Alphavirus	80	160	**10,240**	160	40	80	40	<10
Sindbis	Alphavirus	80	10	2560	**1280**	10	40	40	<10
Chikungunya	Alphavirus	20	20	40	<10	**1280+**	80	80	<10
Mayaro	Alphavirus	40	40	320	40	640	**1280+**	1280+	<10
Semliki Forest	Alphavirus	40	80	160	10	40	320	**2560**	<10
St. Louis encephalitis	Flavivirus	<10	<10	<10	<10	<10	<10	<10	**2560**

Note crossreactions among alphaviruses, but not between the alphaviruses and one flavivirus tested. Also note the crossreactivity among some viruses forming two subgroups; 1) EEE, VEE, WEE, Sindbis; and 2) Chikungunya, Mayaro, Semliki Forest. When neutralization assays are employed instead of hemagglutination inhibition, EEE appears to be antigenically unique, and VEE shows less crossreactivity with WEE and Sindbis viruses. At least four subgroups of flaviviruses have also been identified.

(Modified from Casals J: Ann NY Acad Sci 19:219, 1957)

bunyaviruses are still unclassified. Species within a subgroup are identified by neutralization with standardized antisera; there is less crossreaction with this test.

The immunologic crossreactions seen within the major groups of arthropod-borne viruses suggest close phylogenetic relations. Indeed, it is possible that many of these "species" may differ by only one or a few mutations. However, chemical analysis of the glycoprotein Ags and cross-hybridization of viral RNAs for characterizing genetic relatedness are just beginning to bring sharper taxonomic criteria into this field.

The immunologic crossreactions among arthropod-borne viruses are of practical as well as theoretic interest. Thus, with the viruses that show crossreactivity in hemagglutination-inhibition or CF tests, crossreacting neutralizing Abs may be evident after repeated immunization, though not after primary immunization. Moreover, infection by one virus may confer a demonstrable increase in resistance to subsequent challenge with another. Epidemiologic evidence suggests that such crossprotection may be important in nature; vaccines are being developed to take advantage of these findings.

TOGAVIRUSES

Among the major togaviruses pathogenic for man (Table 62-2), the alphaviruses produce severe encephalitis, particularly western and eastern equine encephalitis viruses (WEE and EEE, respectively), whereas the flaviviruses cause more varied illness (encephalitis, hemorrhagic diseases, and severe systemic illnesses).

PROPERTIES OF THE VIRUSES

Because of their wide distribution and the serious nature of many illnesses caused by togaviruses, particularly encephalitis, these viruses have been studied extensively. Most of them, however, are difficult and dangerous to work with, and the physical and chemical characterization is incomplete except for a few less pathogenic viruses that can be readily propagated in cell cultures. Sindbis and Semliki Forest viruses, which are not usually pathogenic for humans, have been most extensively used for biochemical studies.

Morphology. The virions are roughly spherical; in thin sections they have an outer membrane (lipoprotein en-

velope) and a core of electron-dense material (ribonucleoprotein), and they tend to pack in crystalline arrays (Figs. 62-1 and 62-2). Negative staining shows an outer membrane covered with fine projections, and a capsid consisting of 32 capsomers which appear to be arranged in icosahedral symmetry (Fig. 62-3). The alpha- and flaviviruses have the same general appearance, but alphaviruses are significantly larger (Table 62-1). Even within each genus the sizes of the virions are not identical; it is possible that these differences might be due to effects of the conditions of preparation on the size and shape of the envelope.

Physical and Chemical Characteristics. The purified virions that have been analyzed each contain one **positive-strand infectious RNA molecule** (Table 62-1), which amounts to 4%–8% of the particle weight. The RNA, like most viral and eukaryotic cell mRNAs, contains a polyadenylic acid–rich sequence at its 3' terminus and a cap at its 5' end. The proportions of protein and lipid reported vary considerably: for example, EEE virus has 54% lipid and 42% protein; Sindbis virus, 28% lipid and

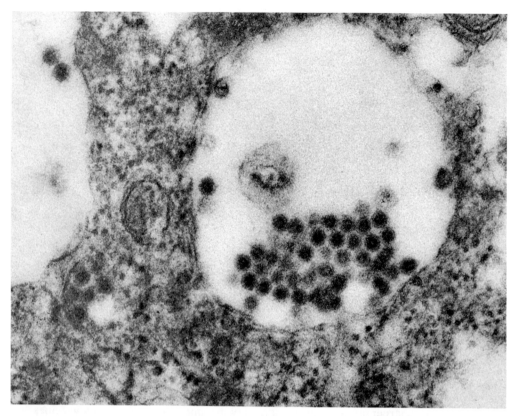

FIG. 62-1. Thin section of mature WEE virus within cytoplasmic vacuole of an infected cell. Viral particle has a dark central nucleocapsid 30 nm in diameter and a peripheral membrane about 2 nm thick. Note the cubical shape of the virions aggregated in a crystalline-like array. (Morgan C et al: J Exp Med 113:219, 1961) (×100,000)

66% protein. The lipids consist of phospholipid, neutral lipids (primarily unesterified cholesterol), and glycolipids. As with other enveloped viruses, the precise lipid composition reflects that of the host cells' plasma membranes.

Alphaviruses contain a basic nucleocapsid protein, and either two or three envelope glycoproteins (Sindbis and Venezuelan equine encephalitis [VEE] viruses, two; Semliki Forest virus, three) which form the surface spikes that span the lipid bilayer to establish association with the capsid (Fig. 62-4). Flaviviruses contain three proteins: a glycoprotein (E_1), the predominant species, is present in the spikes of the envelope; a nonglycosylated (E_2) small polypeptide is associated with the inner side of the envelope; and a small basic protein is associated with RNA in the nucleocapsid.

The surface spikes are the **hemagglutinins.** Virions agglutinate RBCs from newly hatched chicks or adult geese. Maximum hemagglutination (or **hemadsorption** on infected cell cultures) is effected within narrow ranges of pH and temperature which differ for alphaviruses and

for flaviviruses.* The virions become firmly bound to the RBC surface and do not elute spontaneously.

Cell lipids inhibit hemagglutination; hence, detection of hemagglutinating activity in lysates of cells (particularly brain or spinal cord) requires preliminary extraction with lipid solvents. Moreover, treatment of virions with such solvents or with nonionic detergents inactivates infectivity and liberates the glycoproteins that induce production of species-specific neutralizing Abs.

The infectivity of most togaviruses decreases rapidly at $35°–37°C$ in vitro; infectivity and hemagglutinating activity have maximum stability at about pH 8.5. These activities, in flaviviruses, unlike those of alphaviruses, are readily inactivated by small amounts of proteolytic enzymes (e.g., trypsin, chymotrypsin, and papain) and by reagents that attack sulfhydryl bonds.

Immunologic Characteristics. Alpha- and flavi-togaviruses also fall into immunologic subgroups whose members show crossreactivity of the hemagglutinin and nucleocapsid proteins (Table 62-3). Of practical concern for prolonged immunity and artificial immunization is

* Maximum agglutination is attained at pH 6.4 and $37°C$ for alphaviruses and at pH 6.5–7.0 and $4°–22°C$ for flaviviruses.

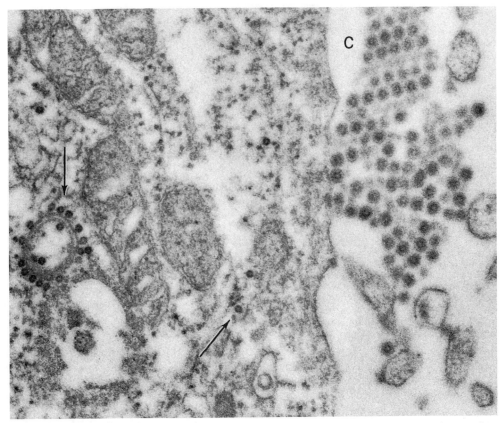

FIG. 62-2. Extracellular crystal (**C**) composed of WEE virus particles. Dense precursor particles are scattered in the cytoplasm and are present on opposite sides of two concentric lamellae near the left border (**arrows**). Mature virions are only seen within vacuoles (Fig. 62-1) or outside of cells. (Morgan C et al: J Exp Med 113:219, 1961) (× 86,000)

the finding that viruses isolated at different times (e.g., Murray Valley encephalitis viruses from 1956 and 1969) may be antigenically distinguishable.

Neutralizing Abs appear about 7 days after onset of disease and persist for many years, probably for life. Infection is followed by solid immunity which appears to be correlated with the development and persistence of neutralizing Abs. Hemagglutination-inhibiting Abs appear at the same time and are easier to assay; CF Abs also rise early but are not detectable after 12–14 months.

Most togaviruses maintain their immunogenicity and immunologic reactivity following destruction of infectivity by formalin, heat, or β-propiolactone. With yellow fever and dengue flaviviruses, however, formalin inactivation markedly impairs their capacity to elicit neutralizing Abs. Accordingly, preparation of successful vaccines required the development of attenuated viruses (see Prevention and Control, below). The existence of only a single known immunologic type of yellow fever virus accounts in part for the effectiveness of the vaccine.

Crossreactive hemagglutination-inhibiting Abs de-

velop after infection or artificial immunization; the degree of crossreactivity increases up to about 1 month after the immunogenic stimulus. For example, an animal infected with EEE virus develops homologous Abs about 7 days after infection, and 3–6 weeks later develops relatively high titers of hemagglutination-inhibiting Abs against all alphaviruses. If the same animal is subsequently inoculated with EEE virus or another alphavirus a rapid increase of hemagglutination-inhibiting Abs for all alphaviruses follows (a secondary group response); the Ab response is considerably greater than that following a single inoculation of either virus alone. Proposed immunization procedures, particularly with flaviviruses, take advantage of this broadened secondary response.

Host Range. Togaviruses multiply in a wide range of vertebrates and arthropods. Most togaviruses can also be propagated on a variety of primary and continuous cell cultures, including cultures of mosquito cells. They produce cytopathic effects (except in mosquito cells), and infected cells can also be detected by hemadsorption. A

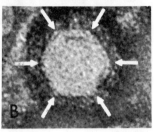

FIG. 62-3. Morphology of an alphatogavirus negatively stained. **A.** The spherical particles have a dense nucleocapsid and an envelope covered with fine projections. (×240,000) **B.** In an occasional virion a nucleocapsid is seen with a clear polygonal outline and the semblance of an ordered capsomeric structure. Arrows point to the axes of 5-fold symmetry of the icosahedron. (×360,000) (Simpson RW, Houser RE: Viology 34:358, 1968)

sensitive and reproducible plaque assay can be carried out with susceptible vertebrate cells. Because of their sensitivity and convenience, cell cultures have largely replaced the suckling mouse for experimental and diagnostic work.

Horses are readily and often fatally infected by the equine encephalitis viruses (whose initial isolation from horses, during epizootics, gave rise to their names). Monkeys are also useful hosts for studying the pathogenesis of infection, particularly with yellow fever virus.

Before its replacement by cell cultures the newborn mouse was the laboratory host of choice: it is highly susceptible to infection by all members of the family. Viruses multiply to high titer in brain, producing extensive pathologic changes. Some of the viruses also multiply in muscle, lymphoid, or vascular endothelium cells. Resistance increases with age: most mice by age 3–6 months are quite resistant to infection by peripheral routes.

Wild birds and domestic fowl, particularly when newly hatched, can be infected artificially or by the bite of an infected mosquito. Embryonated chicken eggs are sensitive, convenient hosts for many studies. Mosquitoes (*Culex, Anopheles,* and *Mansonia*) are the arthropod hosts in nature for alphaviruses.

In addition to their prolonged replication in mosquitoes and ticks, some flaviviruses can survive in ticks for many months by **transovarian transmission** (e.g., Russian spring-summer encephalitis viruses), and through periods of molting and metamorphosis, without apparent injury to the host; this survival during periods of poor transmission furnishes a possible mechanism for overwintering.

Viral Multiplication. Togaviruses, which multiply in the cytoplasm, show considerable variation in the lengths of their multiplication cycles although the temporal differences are relatively minor within each subgroup. Thus the duration of the cycle is relatively short for alphaviruses (e.g., WEE virus, Fig. 62-5), but longer for flaviviruses (e.g., the latent period of type 2 dengue virus is

FIG 62-4. Diagram of the Semliki Forest virus, showing the glycoprotein spikes spanning the envelope lipid bilayer and associating with the protein of the nucleocapsid. (Modified from Simons K et al. In Capaldi RA [ed]: Membrane Techniques. New York, Marcel Dekker, 1977)

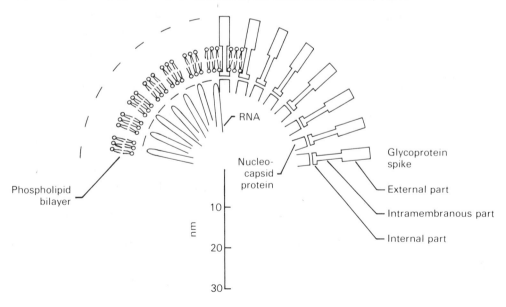

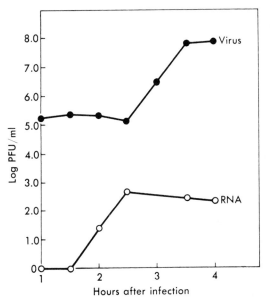

FIG 62-5. Temporal relation of the biosynthesis of infectious (viral) RNA and infectious WEE virus particles. (Wecker E, Richter A: Cold Spring Harbor Symp Quant Biol 27:137, 1962)

10–12 h, and the maximum yield is attained 20–30 h after infection).

Virus is adsorbed rapidly by susceptible cells in culture, and eclipse is evident within 1 h. The uncoated parental genome serves as mRNA for viral proteins, partic-

ularly the nonstructural proteins required for replicating RNA. The replication forms of viral RNA superficially resemble those in picornavirus-infected cells (Ch. 50); but in **alphavirus** infections two **RNA species** of positive polarity are synthesized: **42 S virion RNA and 26 S mRNA.** The 26S mRNA is identical with the 3′-OH terminal one-third of the 42S virion RNA and codes for the structural proteins, whereas the 42S RNA serves for translation of the nonstructural proteins (Fig. 62-6). Extensive replication of 26S RNAs amplifies the messenger for structural proteins when they are most needed. As with picornaviruses, the 26S RNA acts as a **monocistronic mRNA** to produce a polyprotein, which is subsequently cleaved to make the virion proteins (Fig. 62-6). In contrast, only genome RNAs of size 42S are found in **flavivirus-infected cells.**

After maturation by budding from membranes alphavirus particles (e.g., WEE and VEE) are rapidly released from the host cell (within 1 min.), suggesting that final **assembly of the virion** is accomplished at the cell surface and is concomitant with release (Fig. 62-7). Cells infected with Sindbis virus produce virions at the extraordinary rate of 10^4 per cell per hour for approximately 12 h. Flaviviruses are released more slowly, and the maximum yield appears to be significantly less.

Electron microscopy has contributed additional insight into the maturation and release of alphaviruses. RNA replication and the development of the viral nucleocapsids are associated with unique regions of cytoplasmic membranes close to newly formed cytoplasmic

FIG. 62-6. Model of translation and processing of alphavirus 42S and 26S mRNAs. The 42S and 26S mRNAs are translated into polyproteins: that from the 42S RNA is processed into the nonstructural proteins; that from the 26S RNA is sequentially processed into the viral structural proteins. (Recall that Sindbis virus has 2 glycoproteins, E_1 and E_2, whereas Semliki Forest virus has an additional (E_3) protein.) **P** indicates a precursor protein: e.g., P_{E2} signifies the precursor for protein E_2.

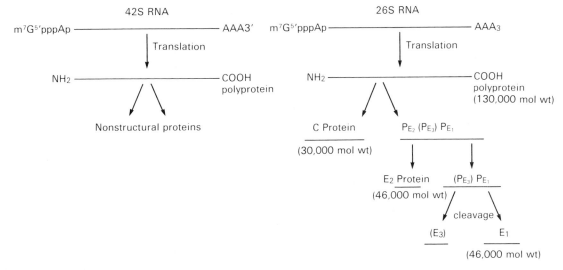

Viral structural proteins

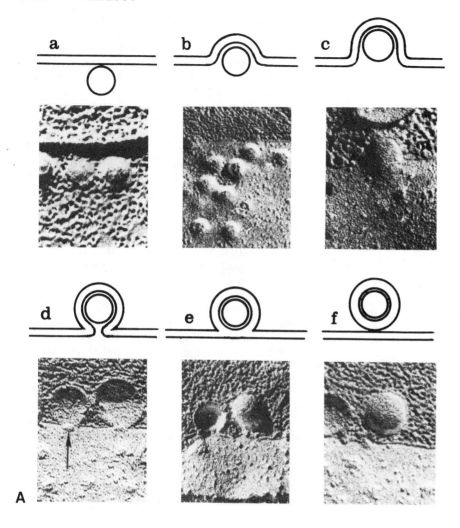

FIG 62-7. Morphogenesis of Sindbis virus (an alphavirus) at the plasma membrane of an infected cell as observed by freeze etching. **A.** Schematic representation of viral maturation with corresponding electron micrographs. **B.** Outer surface of the inner leaflet of a virus-infected cell, showing budding and mature virions (×15,000). (Brown DT, et al: J Virol 10:524, 1972)

vacuoles characteristic of the infections (Figs. 62-2 and 62-8). The nucleocapsids either pass into the lumen of the vacuoles, each acquiring its envelope in the process, or they are dispersed in the cytoplasm and subsequently receive their envelope as they are extruded through the cell membrane (Figs. 62-7 and 62-8). The particle becomes infectious when it acquires its final coat (i.e., the envelope). Flaviviruses are characteristically found in large numbers in prominent cytoplasmic vacuoles, and virions are released through narrow canaliculi connecting the endoplasmic reticulum with the plasma membrane.

PATHOGENESIS

When an infected mosquito bites a prospective host it injects virus from its salivary glands into the blood stream or lymph of its victim. Successful infection depends upon the presence of sufficient virus in the saliva of the mosquito and a paucity of neutralizing Abs in the host. Details of the pathogenesis of infections in humans are largely inferred from experimental studies in animals.

All togavirus infections are similar in the initial stages. Virus is removed from the blood by, and multiplies in,

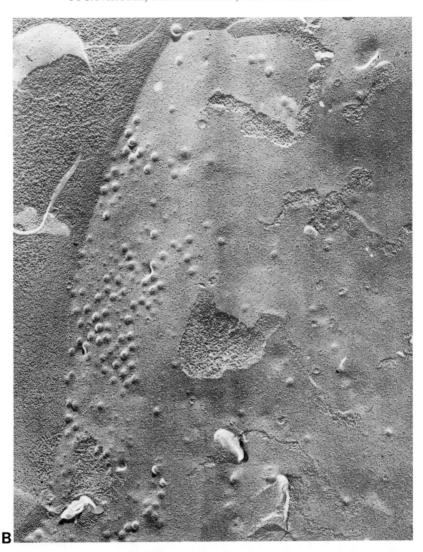

B

reticuloendothelial cells (mainly in the spleen and lymph nodes). Viremia follows, initiating the systemic phase of the clinical disease. Finally, virus invades various tissues—the central nervous system in the encephalitides; the skin and blood vessels in the hemorrhagic fevers; the skin, muscle, and viscera in dengue and yellow fevers. The mechanisms by which encephalitis viruses invade the central nervous system are not yet understood: entrance may be effected directly through the blood–brain barrier or, less likely, by transmission along nerves.

Alphaviruses. Despite similarities in certain pathogenic characteristics, considerable diversity exists in the diseases produced by viruses in this genus. Two types of **clinical syndromes** are seen (Table 62-2): 1) EEE, WEE, and VEE viruses produce a **systemic phase** of disease

(chills, fever, aches) resulting from the viremia, and an **encephalitic phase** may follow after a variable time; 2) infections by Chikungunya and Mayaro viruses are confined to the systemic phase, and the symptoms are solely constitutional.

EEE virus may produce severe illnesses, with high mortality, and often with severe residual neurologic damage among survivors. WEE virus, in contrast, usually causes a less severe disease (most often appearing in infants and children), and most patients recover completely. WEE virus also frequently initiates abortive disease (fever and headache) or clinically inapparent infections. VEE virus primarily infects horses; transmission to man usually results in a mild disease with variable systemic symptoms and only rarely causes severe encephalitis (for example, in the summer of 1971 a devastating

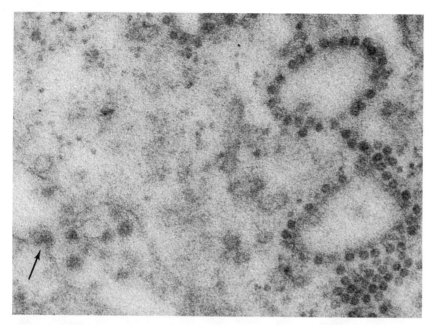

FIG. 62-8. Characteristic formation of particles of WEE virus at membrane of vacuoles in cytoplasm of infected cell. Extracellular virus is visible at lower left. One particle seems to be in the process of emerging from an adjacent cell membrane (**arrow**). (Morgan C et al: J Exp Med 113:219, 1961) (×96,000)

epizootic killed thousands of horses in Texas and Mexico, with mild disease occurring in over 100 humans as well).

Pathologically, severe alphavirus infections show gross involvement of viscera as well as of brain and spinal cord. The brain lesions caused by EEE virus are scattered in the white and gray matter, particularly in the brain stem and basal ganglia; the spinal cord shows milder changes. WEE infections, on the other hand, chiefly affect the brain, producing lymphocytic infiltration of the meninges and lesions of the parenchyma, predominantly in the gray matter. Lesions generally consist of necrosis of neurons, glial infiltration, and perivascular cuffing. Inflammatory reactions in walls of small blood vessels, and thrombi, may occur.

Flaviviruses. These produce three types of clinical syndromes (Table 62-2): 1) **central nervous system disease,** manifested mainly as encephalitis (St. Louis, Japanese B, Murray Valley, Ilheus, and Russian spring-summer encephalitis viruses), with pathologic features similar to those produced by alphaviruses; 2) **severe systemic disease** involving viscera such as liver and kidneys (yellow fever virus); and 3) **milder systemic disease** characterized by severe muscle pains and a rash which may be hemorrhagic (West Nile, dengue, and some of the tick-borne viruses). Despite the anxieties induced by the appearance of a flavivirus most of these viruses, except for yellow fever and dengue, predominantly produce subclinical or mild infections which can be recognized only by laboratory diagnostic procedures.

The pathogenesis of **yellow fever** differs from that of other flavivirus infections: after primary multiplication in lymph nodes extensive secondary viral multiplication, with cell destruction, occurs in many viscera, including the liver, spleen, kidneys, bone marrow, and lymph nodes. The organs involved and the severity of the lesions vary with the infecting strain. Yellow fever may be a grave disease, with a mortality of about 10%.

The pathology of yellow fever is characterized by degenerative lesions of the liver, kidney, and heart, accompanied by hemorrhages and bile staining of tissues. The most distinctive lesions occur in the liver, where a pathognomonic midzonal hyaline necrosis develops, with preservation of the basic liver architecture and without inflammatory reaction.

The severity and extent of the **encephalitides** due to flaviviruses vary with the etiologic agent. For example, **St. Louis encephalitis** virus produces mild lesions, low mortality, and few residua compared with **Japanese B encephalitis** virus, which has a mortality rate of about 8%, and causes neurologic residua in more than 30% and persistent mental disturbances in about 10% of clinically diagnosed infections.

In uncomplicated **dengue fever** deaths are rare; hence the pathologic lesions are not well known. Biopsies of the characteristic skin lesions reveal endothelial swelling, perivascular edema, and mononuclear infiltration in and about small vessels. In some epidemics, particularly in tropical Asia, hemorrhagic fever and shock syndrome may be prominent. **Hemorrhagic dengue** (characterized by high fever, hemorrhagic manifestations, shock, and a mortality as high as 15%) is most often seen in children sequentially infected with different immunologic types of virus during a limited period (about 2 years), suggesting a hypersensitivity or immune-complex type of disease. The pathologic findings and the marked depression of complement components (particularly C3) support this concept, suggesting that Ag–Ab complexes activate the complement system, with subsequent release of vasoactive peptides.

LABORATORY DIAGNOSIS

Togavirus infection is diagnosed by viral isolation or by serologic procedures. To isolate a virus from a patient, specimens are inoculated intraperitoneally and intracerebrally into newborn and suckling mice, or into susceptible cell cultures. The viremic phase of a togavirus encephalitis is usually completed by the time a patient seeks medical assistance. Hence isolation of virus from the blood and spinal fluid of patients during acute disease is unusual. With Chikungunya virus and many flaviviruses, however, the responsible virus may be isolated from blood during the initial stages of illness; in yellow fever and dengue fever, blood or serum may be a good source of virus during the first 2–4 days of disease. To obtain a virus from a fatal case of encephalitis, emulsions of brain and spinal cord are used.

A newly isolated virus is identified by 1) hemagglutination-inhibition titrations with standard antisera, to determine its immunologic group; and 2) neutralization titrations to establish its species. Because of the marked crossreactivity among flaviviruses and particularly within the same subgroup, considerable care is required to identify the agent.

Serologic diagnosis is made with serum drawn during the acute illness and during convalescence. Antigens are obtained from infected cell cultures or from brains of infected newborn mice extracted with acetone. Complement-fixation assays are preferred because of their simplicity, speed, economy, and comparative lack of crossreactivity. However, within an immunologic subgroup, rises in heterologous Ab may occur and obscure a precise etiologic diagnosis. Complement-fixation titrations are not adequate for epidemiologic surveys, moreover, because CF Abs do not persist as long as those measured by hemagglutination-inhibition or neutralization titrations.

EPIDEMIOLOGY

Alphaviruses. For most alphaviruses man is merely an incidental host. The mosquito is the common arthropod vector. The cycle of infection has been elucidated best for the equine encephalitis viruses, particularly **WEE** and **EEE,** and can be simply diagrammed as follows:

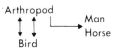

Despite the names of the diseases, the horse, like man, appears to be a dead end in the chain of infections (Fig. 62-9). In fact, horses are not significant reservoirs in nature for WEE and EEE viruses, probably because viremia does not usually reach sufficiently high levels to infect mosquitoes with regularity. It is noteworthy, however, that equine infections usually appear 2–3 weeks prior to the occurrence of disease in humans. Birds are the principal natural hosts of WEE and EEE viruses (Fig. 62-9). Birds likewise appear to be the most likely hosts in which viruses can persist in nonepidemic periods and during the seasons in which transmission by mosquitoes is not prominent (**overwintering**). Wild snakes and frogs, as well as some rodents, are probably secondary reservoirs for WEE. Hibernating mosquitoes are also a possible reservoir, for infectious virus can persist in them for at least 4 months.

WEE is generally found in the United States west of the Mississippi River, but increasingly the virus is also being isolated along the eastern seaboard. In contrast EEE is confined to the eastern part of the United States and Canada. Both viruses are also found in Caribbean islands, Central America, and South America.

The primary vector of WEE virus is the culicine mosquito, *Culex tarsalis* in central and western United States and *Culiseta melanura* in the northeast. The primary mosquito vector for EEE is not certain, but *Culiseta melanura* is susceptible to experimental infection and has been found infected in nature. Other mosquitoes also appear to be implicated in the epidemiologic cycles involving mammals.

Chikungunya (African for that which bends up) virus infections may be a notable exception to this general epidemiologic pattern: man is the only known vertebrate host, with *Anopheles* and *Aedes* mosquitoes the vectors.

VEE virus is distributed in the Everglades region of Florida (where it is endemic in rodents), in the southwestern United States, and in Central America, nothern South America, the Amazon Valley, and southern Mexico. The natural cycle of infections involves mammals and mosquitoes; the natural reservoir is in small mammals rather than birds. Several species of mosquitoes can probably transmit virus. Horses are invariably infected when human disease occurs and appear to be the major source of virus for mosquitoes. An example, the severe outbreak of 1971, has been mentioned under Pathogenesis.

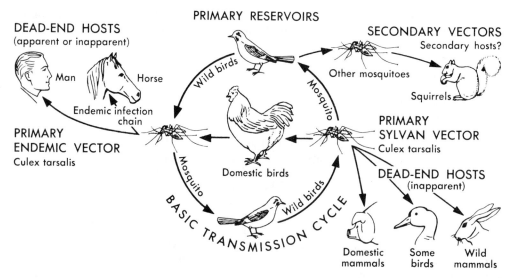

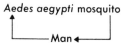

FIG. 62-9. Epidemiologic pattern for WEE virus infections. The chains for rural St. Louis encephalitis are similar, except that horses are inapparent, rather than apparent, hosts. EEE infections also have a similar summer infection chain, but a few significant differences exist: 1) the identity of the vector infecting man is unknown, 2) domestic birds do not appear to be a significant link in the chain, and 3) it has a bird-to-bird secondary cycle in pheasants whose role is unclear. (Hess AD, Holden P: Ann NY Acad Sci 70:294, 1958)

Flaviviruses. The epidemiologic patterns of flavivirus infections are more varied than those of alphaviruses. These patterns will be summarized for a few of the most important diseases.

St. Louis encephalitis* is the major flavivirus infection in the United States. The most severe epidemic to occur since reporting was initiated took place in 1975, affecting the central part of the country; 1815 cases with 140 deaths were recorded. The epidemiologic pattern is similar to that described for WEE virus infections. Wild birds are the major reservoir of the virus, and *Culex tarsalis* and the *C. pipiens* complex are the most common mosquito vectors. Man is an accidental, dead-end host (Fig. 62-9).

The epidemiology of **Murray Valley encephalitis, Japanese B encephalitis,** and **West Nile fever** is basically the same as that of St. Louis encephalitis except for the species of mosquito vectors and the avian reservoirs. Serologic surveys indicate that for each case of clinical disease produced by these viruses several hundred **inapparent infections** are also induced.

Yellow fever presents another complex ecologic situation.† Two distinct epidemiologic types exist, **urban** and **jungle yellow fever.** Each has a different cycle, but they may interact. In its simplest form the epidemiologic pattern of urban yellow fever simply involves man and the domestic mosquito, *Aedes aegypti:*

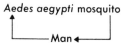

Viremia in man begins 1 or 2 days before, and persists for 2–4 days after, the onset of clinical illness. Viremia is greatest during this period and mosquitoes taking a blood meal may be infected. A 10- to 12-day period of viral multiplication in the cells lining the mosquito's intestinal tract is then required (the **extrinsic incubation period**) until sufficient virus accumulates in the salivary glands to permit transmission to man.

Jungle yellow fever is transmitted by various jungle mosquitoes, primarily to monkeys. Man becomes an accidental host when he enters the animals' domain.

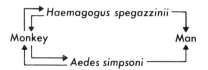

Infection of man may initiate a cycle of urban yellow fever.

Dengue fever resembles yellow fever epidemiologically. There is an urban cycle (man ⇌ *Aedes aegypti* mosquitoes), and probably a jungle cycle with the mon-

---

* The first epidemic due to this virus was recognized in St. Louis, Missouri, in 1933.

† The monograph entitled *Yellow Fever,* by G. K. Strode (ed), (McGraw-Hill, New York; 1951) reviews in exciting detail many of the important facts personally discovered by its authors.

key as the mammalian host. This disease is still prevalent in the Caribbean islands, as well as more distant subtropical areas (Table 62-2).

The **tick-borne** complex of viral infections introduces several unique features into the epidemiology of togavirus infections: ticks may serve as reservoirs by **transovarial transmission** of virus; and in addition to being transmitted by ticks (*Ixodes*), some of these viruses (e.g., **Russian spring-summer encephalitis virus**) may also be transmitted to man from the goat by milk instead of by an arthropod.

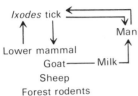

Omsk hemorrhagic fever virus is transmitted by *Dermacentor marginatus*, probably from muskrats, and **Powassan virus** (isolated in Canada and the United States) is transmitted by *Ixodes* ticks from small mammals, probably squirrels.

PREVENTION AND CONTROL

Control measures are aimed at 1) preventing transmission of the virus by eradicating, or at least reducing, the population of arthropod vectors, and 2) increasing the host resistance by artificial immunization. The former procedure has been relatively successful. Effective vaccines have been prepared against some of the viruses.

Yellow fever has played a special role in the development of concepts and methods for the control of insect-borne diseases. Reducing the population of the vector, *A. aegypti*, proved effective soon after Reed and his colleagues demonstrated the causative agent and the vector requirement. The use of modern insecticides has facilitated this control measure. It should be noted, however, that *A. aegypti* mosquitoes and other possible vectors are still present in many parts of the world, including the southeastern United States. Moreover, mosquito control measures cannot eliminate jungle yellow fever.

As noted above, the loss of immunogenicity of yellow fever virus upon inactivation made it necessary to seek a vaccine containing infectious virus. Theiler and his coworkers attenuated a mouse-adapted yellow fever virus by serial passage in tissue cultures, first of embryonic mouse tissues, then of embryonic chicks, and finally of embryonic chicks without brain or cord. For the most widely used vaccine, the attenuated virus (17D yellow fever virus) is propagated in embryonated chicken eggs. The French, chiefly in Africa, utilize a vaccine containing virus attenuated by serial passage in mouse brain. The 17D strain produces far fewer and less severe toxic reactions than the French vaccine, but the latter elicits a higher Ab response. The duration of protection afforded by the 17D vaccine is not known, but Ab has been detected at least 6 years after immunization.

Eradication of yellow fever in the United States, and substantial reduction of its incidence in South America and other parts of the world, initially suggested that this disease would be eliminated throughout the world. However, the reservoir of jungle yellow fever was later discovered, and yellow fever has increased in incidence in parts of Central and South America and has been creeping northward. With this nidus present the danger of epidemics is real.

Formalinized chick embryo **EEE and WEE vaccines** produce effective Ab responses in horses. But these vaccines have not been utilized in humans except for protection of laboratory workers; their effectiveness in man has not been established. A live, attenuated **VEE virus vaccine** was successfully used to immunize horses in Texas in 1971; this procedure, along with strict quarantine of the equine population, halted the epidemic. An attenuated VEE virus vaccine has also been used experimentally in humans, but its effectiveness has not yet been demonstrated. A **Japanese B encephalitis vaccine**, prepared by formalin inactivation of virus propagated in chick embryos, has been employed with apparent success in children in Japan. However, a similar vaccine was ineffective in US Army personnel stationed in Japan and the Far East. **Dengue vaccines** containing infectious attenuated viruses have been shown to have promise. However, until we explicitly understand the role of viral Ag–Ab complexes in the pathogenesis of dengue hemorrhagic fever and shock syndrome such vaccines should be used with caution.

The marked antigenic crossreactivity of flaviviruses may prove useful for immunization purposes. For example, in experimental animal studies immunization with an infectious attenuated virus, such as yellow fever virus, and subsequent injection of one or more inactivated or live attenuated heterologous viruses, resulted in a broad immunologic response that protected against a variety of flaviviruses.

BUNYAVIRIDAE

On the basis of immunologic, chemical, and morphologic features many so-called arboviruses have been grouped in a single family, Bunyaviridae (Table 62-2). Eighty-six members of this family have been classified into a single genus (Bunyavirus or Bunyamwera* supergroup) on the

* Named for Bunyamwera, Uganda, where the type species virus was isolated.

basis of their similarities. This large genus (Table 62-2) includes the California encephalitis, Bunyamwera, and C subgroups, which are the major ones affecting humans (additional members of the genus, the Bwamba, Capim, Guama, Koongol, Patois, Simbu, and Tete virus groups, have similar properties). More than 55 other viruses are candidates for Bunyaviridae family membership, but immunologic and chemical differences suggest that they may be divided into at least two additional genera (Table 62-2); these include viruses of the Uukuniemi, Phlebotomus fever, Crimean–Congo hemorrhagic fever, Anopheles A, Anopheles B, and Turlock groups.

PROPERTIES OF THE VIRUSES

The **structures** of all viruses studied show marked similarities to each other (Table 62-4) and clear distinctions from togaviruses. The virions are spherical particles, clearly larger than togaviruses. Their envelopes are covered with surface projections, which are indistinct and filamentous for most bunyaviruses but arranged in a regular lattice on the unclassified Uukuniemi virus. These spikes consist of two glycoproteins present in equimolar amounts; they have hemagglutinating activity and induce hemagglutination-inhibiting and neutralizing Abs. The nucleocapsid, when released from the virion, appears to have helical symmetry, is often seen in circular forms, and is present in three distinct segments (each consists of a common protein and a unique RNA molecule). The three species of RNA are linear, however, and have free 5′ ends. Perhaps the circles are formed by pairing of complementary segments at the 3′ and 5′ ends of the RNA molecules: in fact, free RNAs of the Uukuniemi virus are circular under nondenaturing conditions, and can re-form circles on annealing after denaturation. The viral RNAs are negative strands, their 3′ ends are not polyadenylated, their 5′ ends are not capped, and they are not infectious. The virions contain a RNA-dependent RNA polymerase, probably a transcriptase.

Immunologic Characteristics. Bunyaviridae, like togaviruses, can be studied by hemagglutination-inhibition, CF, and neutralization titrations. In contrast to togaviruses, however, the crossreactivity of bunyaviruses is maximal in CF rather than in hemagglutination-inhibition titrations. For example, CF titrations with one or two standard sera can identify an agent as a group C virus, and specific viral identification can be accomplished by hemaggluntination-inhibition and neutralization titrations. By these procedures, all bunyaviruses are shown to be antigenically related, probably through the crossreacting nucleocapsid protein. Nevertheless these viruses can be subdivided immunologically into ten subgroups (e.g., group C, Bwamba group, Bunyamwera group). The remaining viruses of the family, which belong to at least two genera, can be allocated into at least ten subgroups, each immunologically unrelated to viruses of the Bunyavirus genus.

Host Range. Suckling and newly weaned mice are the laboratory animals of choice for isolation of the viruses. These viruses also multiply and produce cytopathic changes and plaques in a variety of cultured cells; among the most useful are continuous human (e.g., HeLa) lines and baby hamster kidney (BHK21) cell lines.

Viral Multiplication. Bunyaviruses have not been studied in detail, but the negative polarity of the segmented virion RNA suggests that it replicates like orthomyxoviruses (Ch. 58). Viral multiplication takes place in the cytoplasm, and the virions first appear within small vesicles or cisternae in the region of the Golgi apparatus (Fig. 62-10). The particles form by budding into the vacuoles and consist of an electron-dense nucleocapsid core closely bound by an envelope (Fig. 62-10).

PATHOGENESIS AND EPIDEMIOLOGY

The **California encephalitis viruses,** which were initially isolated from mosquitoes in the San Joaquin Valley of

TABLE 62-4. Properties of Bunyaviridae and Arenaviridae

Property	Bunyaviridae	Arenaviridae
Morphology	Spherical; enveloped	Spherical, pleomorphic; enveloped
	Diameter 90–100 nm Helical nucleocapsid	Diameter 110–130 nm average
Nucleic acid*	RNA; single-stranded; 6–7×10^6 daltons; 3 segments: 4, 2, and 0.8×10^6 daltons, (12, 6 and 2.4 Kb)	RNA; single-stranded; 2 segments virus-specific: 3.6 and 1.6×10^6 daltons (11 and 4.8 Kb); (in 3 species host RNA also in virions)
Virion proteins*	2 glycoproteins: 115,000 and 38,000 daltons; 1 non-glycosylated protein: 19,000 daltons	2 glycoproteins: 72,000 and 34,000 daltons; 2 non-glycosylated proteins: 72,000 and 12,000 daltons.
Effect of lipid solvents	Inactivate	Inactivate
Stability	Unstable below pH 7.0	Unstable
Hemagglutination	1-day chick or goose RBCs	None
Best animal host	Suckling mice; chickens	Various rodents

*Molecular weights vary with different viral species.

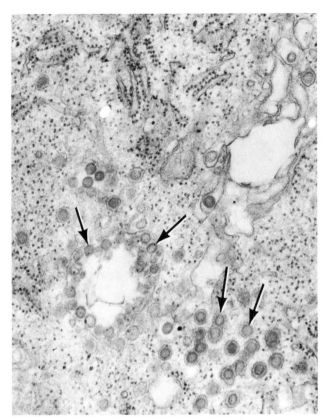

FIG. 62-10. Development of Bunyamwera virus in the cytoplasm of a neuron in a mouse brain. Viral particles are present in the cytoplasm and budding into Golgi vacuoles and cisternae of the endoplasmic reticulum. Virions have a mean diameter of 90–100 nm, a nucleocapsid core of 60–70 nm in diameter, and an envelope of 15–20 nm thick. (Murphy FA et al: J Virol 2:1315, 1968) (×39,500)

recorded. Clinical disease has been reported from 13 states in all regions of the United States as well as in the countries noted in Table 62-2. The natural reservoir of the virus in unknown, but the agent has been found in the blood of rabbits, squirrels, and field mice in titers adequate to infect mosquitoes. Although California encephalitis viruses have been isolated from several *Aedes* and culicine species, *Aedes triseriatus* appears to be the principal vector. Moreover, transovarial passage of the virus in this vector may furnish an important mechanism for its maintenance during the winter.

Among a large number of viruses isolated from experimental (sentinel) monkeys caged in the forested Belém area of Brazil seven different viruses were recognized as being immunologically related to each other but distinct from togaviruses. Several other immunologically distinguishable but related viruses have been isolated in the Florida Everglades and Central America. These viruses, previously called **group C arboviruses** but now recognized as a division of the Bunyavirus genus, produce mild disease in man, consisting of headache and fever; recovery is complete.

The natural reservoir of C viruses appears to be in monkeys and other forest mammals (e.g., opossums, rats, sloths). The specific mosquito vector has not been established, but culicine and sabethine mosquitoes are likely candidates.

Bunyamwera virus was first isolated from a mixed pool of *Aedes* mosquitoes trapped in Uganda. Eighteen immunologically related but distinct viruses have been recognized, including strains isolated in the United States (Florida, Virginia, Colorado, Illinois, and New Mexico). Disease attributed to the Bunyamwera group (Table 62-2) is rare and usually mild.

Among the many other viruses of this family only some species of the Phlebotomus fever group and Rift Valley fever virus are known to produce disease in man. Most of the viruses were isolated from insects (particularly mosquitoes and ticks) and animals captured in the wild.

California, are widely distributed. They produce prominent clinical illnesses, manifested by fever, headache, and mild or severe central nervous system involvement, particularly in children. Recovery is usually complete, although mild residua and even rare deaths have been

ARENAVIRUSES

On the basis of morphologic, immunologic, and clinical characterizations the seemingly disparate Lassa virus, lymphocytic choriomeningitis (LCM) virus, and Tacaribe group of viruses, previously considered to be arboviruses, are now classified into a single family called **Arenaviridae** (vernacular, **arenaviruses;** L. *arena,* sand), a term derived from the unique electron microscopic appearance of the virions (Fig. 62-11).

Complement fixation reveals the immunologic relatedness of the arenaviruses; the Tacaribe viruses, however, are more closely related to each other than to Lassa or LCM viruses. Neutralization titrations show the immunologic specificity of each arenavirus.

While arenaviruses were initially grouped together because of their immunologic relations and similar morphology, they also have comparable epidemiologic, ecologic, and pathogenic patterns. Tacaribe group, LCM, and Lassa viruses do not require arthropods for spread; and the natural hosts of all arenaviruses appear to be rodents, in which they often produce chronic infections.

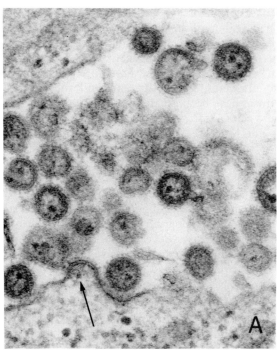
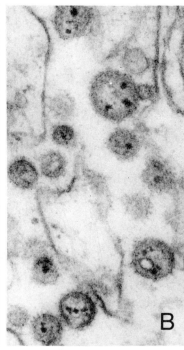

FIG. 62-11. A. Machupo virus (a Tacaribe group virus) particle budding (**arrow**) from the plasma membrane of an infected cell; the thickened membrane of the budding particle is prominent compared with the neighboring membrane. Many extracellular mature virions (mean diameter of 11–13 nm) are present: their prominent surface projections and internal, ribosome-like particles are readily seen.(×114,000). **B.** Lymphocytic choriomeningitis (LCM) virus particles in an infected culture of mouse macrophages. The morphology is strikingly similar to that of the Tacaribe group virus in **A.** (×82,000) (Murphy FA et al: J Virol 4: 535, 1969)

PROPERTIES OF THE VIRUSES

The viral particles (Table 62-4) are spherical or pleomorphic. As viewed in thin sections the virions consist of a dense, well-defined envelope with prominent, closely spaced projections and an unstructured interior containing a varying number of electron-dense granules, probably host ribosomes (Fig. 62-11), which cause the unique pebbly appearance from which the viruses gained their name. Negative-contrast electron micrographs also show spherical or pleomorphic virions with an envelope having pronounced and regularly spaced club-shaped surface projections.

Within the virion are several species of ribonucleoproteins which are both virus-specific and host cell–derived (Table 62-4). The host cell ribonucleoproteins have characteristics of ribosomes and contain 28S and 18S RNAs (a 4-6S RNA of host cell origin is also present). The viral RNA segments are single-stranded RNAs of "negative" polarity with the same characteristics as the RNAs of bunyaviruses. The virions contain two glycoproteins associated with the envelope, and two nonglycosylated proteins (a component of the nucleocapsid and a minor protein of unknown function) (Table 62-4).

The biochemical events of **viral multiplication** are largely unexplored, but it is assumed that arenavirus replication follows the patterns described for other negative-stranded RNA viruses (Ch. 50). It is striking that the appearance of ribosome-like RNAs in the virions, but neither the synthesis of viral RNAs nor production of infectious virus, is inhibited by Dactinomycin (Actinomycin D) (0.15μg/ml). Electron microscopic studies show that virions are formed by budding, chiefly from plasma membranes (Fig. 62-11). At the sites of budding, the host cell membrane becomes thickened, more clearly bilamellar, and covered with projections. The unique electron-dense particles, which also resemble ribosomes morphologically, are present within the budding particles before separation from the infected cell. Viral infection in vitro is not associated with extensive damage of the host cell.

PATHOGENESIS AND EPIDEMIOLOGY

Arenaviruses commonly produce chronic carrier states in their natural hosts, and virus may be isolated from the animals' urine, as well as from their blood and internal organs. The **Tacaribe group of viruses** (Tacaribe, Machupo, Junin, Tamiami, and four others) have been principally isolated from bats and cricetid rodents in the Western Hemisphere. Junin and Machupo viruses have been frequently isolated from cases of Argentinian and Bolivian hemorrhagic fevers, respectively, and appear to be the etiologic agents of these severe illnesses. Thus, the Tacaribe group viruses, similar to LCM virus of mice, appear to be **spread to humans in excretions of the naturally infected rodents.** Except for Lassa virus, there is no evidence of viral spread from patient to patient, and virus is rarely isolated from arthropods, even during epidemics.

LCM virus infects humans rarely, and usually under conditions in which the mouse population is very dense, or from contact with infected hamsters. The disease is generally mild, and is manifested most often as a lymphocytic form of meningitis or an influenza-like illness, but occasionally as a meningoencephalitis. Leukopenia and thrombocytopenia frequently develop. Very rarely, LCM virus produces severe and even fatal illnesses associated with hemorrhagic manifestations.

Lassa virus, first isolated in 1969 from an American missionary working in Nigeria, has attracted considerable interest because it is **highly contagious** and produces a **serious febrile illness.** The disease is characterized by severe generalized myalgia, marked malaise, and sore throat accompanied by patchy or ulcerative pharyngeal lesions. Fatal cases also develop myocarditis, pneumonia with pleural effusion, encephalopathy, and hemorrhagic lesions. Virus persists in the blood for 1–2 weeks, and during this period it can be isolated from urine, pleural fluid, and throat washings. The virus is more stable than togaviruses in body fluids, which probably permits its person-to-person contagion and accounts for the hazard it presents for laboratory isolation or study. Arthropods collected in Nigeria, the only known locale of natural infections, have not yielded virus, and insect cell cultures are insusceptible to viral propagation. The only cycle of Lassa virus transmission outside humans has been detected in the wild rodent *Mastomys natalensis,* which suggests that rodent control may limit viral transmission to man. Lassa virus produces an infection in mice similar to LCM virus infection, and chronic latent infections can be established.

GENERAL REMARKS

Only a few of the known viruses that multiply in arthropods and vertebrates have been discussed in this chapter. The properties of most of these agents, and even their clinical and epidemiologic behavior, are known in only a fragmentary fashion. Some, such as the phlebotomus (sandfly) fever virus, transmitted by the bite of the female sandfly *Phlebotomus papatasii,* assumed importance to the US Armed Forces during World War II, when the disease (which is not serious) appeared in military personnel in the Mediterranean area.

Many of the arthropod-borne viruses, including some that can cause serious disease, also produce a very much larger number of inapparent infections in endemic areas; hence the native human population carries a high level of immunity but the insect population is still highly infectious because of the viral reservoir in lower animals. Such diseases could increase dramatically in quantitative significance 1) when ecologic alterations cause development of a dense population of infected arthropods next to a nonimmune human population, or 2) when a large, immunologically virgin human population (e.g., military personnel) moves into an endemic area.

Because of their close antigenic relation to human pathogens, those arthropod-borne viruses that have not been associated with human disease cannot be ignored by medical investigators, however esoteric they may seem. Several such viruses have been isolated in the United States or Canada, for example, the Rio Bravo virus (flavivirus group, USA), California encephalitis and Trivittatus viruses (California group, USA), and Silverwater virus (ungrouped, Canada). Since a change in either the host reservoir, the vector, or the genetics of the viral population might permit these agents to infect man, they remain a potential hazard.

The comforting realization that as many as 150 arthropod-borne viruses of seemingly diverse immunologic groupings, or even without obvious relatives, may be segregated into one family, the Bunyaviridae, on the basis of physical and chemical characteristics, indicates that order is appearing in what previously seemed unmanageable. Thus, the ecologic–epidemiologic classification of otherwise disparate viruses as "arboviruses" is being replaced by classification on a broader base into several new families. In addition, a few viruses initially isolated from arthropods have been placed into well-established families (reo-, rhabdo-, and picornaviruses), and the arenaviruses have been shown not to be associated with arthropod vectors.

SELECTED READING

BOOKS AND REVIEW ARTICLES

CASALS J: Arenaviruses. Yale J Biol Med 48:115, 1975

DOWNS MG: Arboviruses. In Evans AS (ed): Viral Infections of Humans. Epidemiology and Control. New York, Plenum Medical Books, 1976, p 71

KÄÄRIÄINEN L, KERÄNEN S, LACHMI B, SÖDERLUND H, TUOMI K, ULMANEN I: Replication of Semliki Forest virus. Med Biol 53:342, 1975

PFAU CJ: Biochemical and biophysical properties of the arenaviruses. Prog Med Virol 18:64, 1974

PFEFFERKORN ER, SHAPIRO D: Reproduction of togaviruses. Compr Virol 2:171, 1974

PORTERFIELD JS: The basis of arbovirus classification. Med Biol 53:400, 1975

SCHLESINGER RW: Dengue Viruses. Virology Monograph, Vol 16. Vienna, Springer-Verlag, 1977

SHOPE RE: Arboviruses. In Lennette EH, Spaulding EH, Truant JP (eds): Manual of Clinical Microbiology, 2nd ed. Washington, DC, American Society for Microbiology, 1974, p 740

SPECIFIC ARTICLES

DALRYMPLE JM, SCHLESINGER S, RUSSELL PK: Antigenic characterization of two Sindbis envelope glycoproteins separated by isoelectric focusing. Virology 69:93, 1976

FRIEDMAN RM, LEVY HB, CARTER WB: Replication of Semliki forest virus: Three forms of viral RNA produced during infection. Proc Natl Acad Sci USA 56:440, 1966

GENTSCH J, BISHOP DHL, OBVISKI JF: The virus particle nucleic acids and proteins of four bunyaviruses. J Gen Virol 34:257, 1977

HEWLETT MJ, PETTERSSON RE, BALTIMORE D: Circular forms of Unkuniemi virion RNA: an electron microscopic study. J Virol 21:1085, 1977

HORZINEK MC: The structure of togaviruses and bunyaviruses. Med Biol 53:406, 1975

SIMMONS DT, STRAUSS JH: Translation of Sindbis virus 26S RNA and 49S RNA in lysate of rabbit reticulocytes. J Mol Biol 86:397, 1974

chapter

RUBELLA VIRUS

Rubella virus is the etiologic agent of German measles. This disease resembles measles but is milder and does not have the serious consequences often seen with measles in the very young. Because rubella seems to be such a harmless disease it did not receive much attention earlier, although its probable viral etiology was demonstrated in experiments with human volunteers in 1938. However, interest in rubella was much increased when Gregg, an Australian ophthalmologist, noted in 1941 that women contracting rubella during the first trimester of pregnancy frequently gave birth to babies with congenital defects. Nevertheless, the cultivation of rubella virus was not achieved until 1962 when Parkman, Buescher, and Artenstein detected the virus through its interference with type 11 echovirus in primary grivet monkey kidney cultures, and Weller and Neva demonstrated unique cytopathic changes (Fig. 63-1) in infected primary human amnion cultures.

Rubella virus is classified as a **togavirus** (genus Rubivirus) on the basis of its physical and chemical characteristics (Ch. 62). However, owing to the worldwide importance of German measles in humans, and its unique clinical features and pathogenesis, a separate chapter is devoted to this virus.

PROPERTIES OF THE VIRUS

Morphology. The virion is roughly spherical and has an average diameter of about 60 nm, in both thin sections and negatively stained preparations; it consists of a roughly isometric core of 30 nm, covered by a loose envelope (Fig. 63-2). Negative staining technics reveal small spikes projecting 5–6 nm from the envelopes of most particles. Gentle disruption of the envelope with sodium deoxycholate uncovers an angular core; definite symmetry of the nucleocapsid is obscure, but ringlike subunit structures are discernible.

The morphology and growth characteristics of rubella virus resemble those of the alpha-togaviruses (Ch. 62).

Chemical and Physical Properties. The virion contains one molecule of single-stranded RNA of 38–40S and a molecular weight of $2.5–3 \times 10^6$ daltons (7.7–9.2 Kb), significantly smaller than the RNAs of other togaviruses. Two glycoproteins are present in the envelope, and an arginine-rich protein is associated with the RNA. The virion RNA is infectious, like that of other togaviruses; therefore the RNA is a positive strand.

Like other enveloped viruses, rubella virus is rapidly inactivated by ether, chloroform, and sodium deoxycholate; it is relatively labile when stored at 4° C, and relatively stable at −60° to −70° C.

The **hemagglutinin** is an integral component of the virion, although it remains biologically active after gentle disruption of the viral particle. It is most effectively assayed with newborn chick, pigeon, goose, or human group O RBCs. As with other togaviruses the hemagglutinin does not elute spontaneously from the affected RBCs, and neuraminidase does not render the RBCs inagglutinable (see Mechanism of Hemagglutination, Ch. 46).

Immunologic Properties. Rubella virus is immunologically distinct from the other togaviruses; only a single antigenic type has been detected. Neutralizing and hemagglutination-inhibiting Abs develop during the incubation period of the disease and are commonly present when the rash appears (see Pathogenesis, below); they attain maximal titer during early convalescence and persist (along with immunity) for many years if not for life. It must be noted, however, that the hemagglutination-inhibiting Abs and the neutralizing Abs are responses to different antigens and can vary independently following German measles or immunization. Complement-fixing Abs appear about 6–10 days after onset, begin to diminish after 4–6 months, and disappear after a few years.

VIRAL MULTIPLICATION

Rubella virus can replicate in a variety of primary, continuous, and diploid cell cultures of monkey, rabbit, and human origin. When all cells are initially infected the eclipse period is 10–12 h. The viral titer reaches a maximum 30–40 h after infection and may remain high for weeks (indeed, carrier cultures are readily established; see Ch. 53).

As with other togaviruses, the infectious virion RNA initiates viral replication by serving as a mRNA for viral protein synthesis. Viral multiplication is confined to the cytoplasm, and its RNA synthesis proceeds through intermediate replicative forms, but it is not known whether,

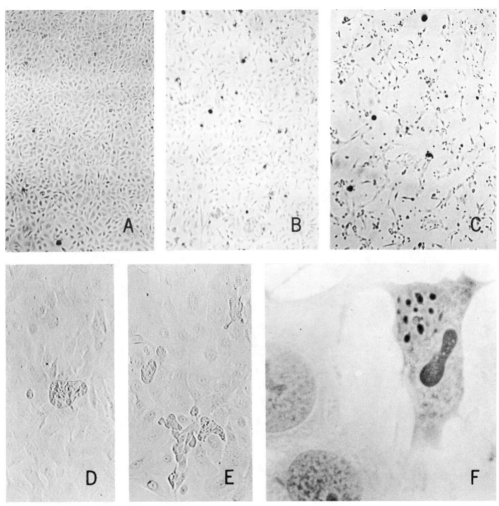

FIG. 63-1. Cytopathic effect of rubella virus in human amnion cultures. **A.** Appearance of normal amnion cell culture as viewed microscopically under low magnification. (× 33) **B.** Rubella-infected culture with estimated 20% destruction of cells on 14th day after inoculation. (× 33) **C.** Rubella-infected culture with 80% cell destruction on 28th day after inoculation. (× 33). **D.** Single affected cell with adjacent uninvolved cells on 10th day after inoculation. (× 132) **E.** Scattered infected cells showing ameboid distortion on 10th day after inoculation. (× 132) **F.** Infected cell with large eosinophilic cytoplasmic inclusion and basophilic aggregation of nuclear chromatin, as well as portions of two normal cells. (H & E stain; × 3500) (Neva FA et al: Bacteriol Rev 28:444, 1964)

as with alpha-togaviruses, 26S RNA as well as 42S RNA is made (see Synthesis of RNA Viruses, Ch. 50, and Viral Multiplication, Ch. 62). Morphogenesis of the virions occurs at cell membranes (particularly the plasma membrane), which differentiate by incorporating viral proteins. Nucleocapsids then bud through the thickened vacuolar and surface membranes to form mature viral particles (Fig. 63-3). Unlike the case with other togaviruses, however, nucleocapsids do not accumulate in the cytoplasm. Because viral components, including the hemagglutinin, are incorporated into the cell surface membrane during budding, the infected cells can be detected by hemadsorption.

In many infected cell cultures (e.g., grivet monkey kidney) an increase in viral titer is associated with **increased resistance** to infection with some challenge viruses (e.g, picornaviruses, orthomyxoviruses, measles virus), which provides another procedure for detecting infected cells and isolating viruses. The interference is induced by propagation of rubella virus, and it is not consistently associated with interferon production.

Cytopathic changes are not detectable in most rubella-infected cell cultures, but in cultured primary human cells distinctive cellular alterations appear slowly over 2–3 weeks. Affected cells are enlarged or rounded, and they often have ameboid pseudopods; staining re-

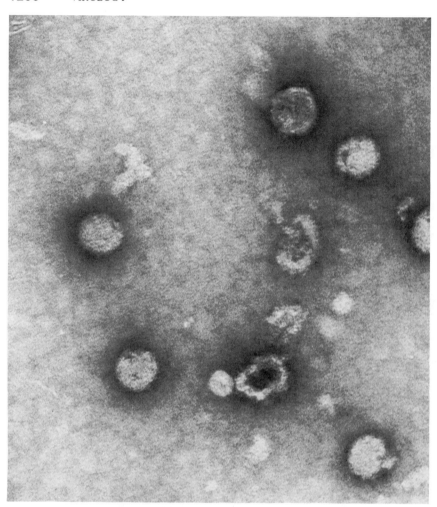

FIG. 63-2. Morphology of rubella virus. Note the nucleocapsids within and dissociated from the envelopes. The nucleocapsid cores are somewhat angular and when separate from the envelopes show a subunit structure. (Horzinek MC: Arch gesamte Virusforsch 33:306, 1971)

veals a disappearance of the nuclear membrane, prominent clumping of nuclear chromatin, and round or irregular eosinophilic cytoplasmic inclusion bodies (Fig. 63-1). The cytopathic effects are associated with inhibition of the biosynthesis of host macromolecules and the inability of infected cells to divide. However, the infected cells do not lyse.

In cultures of human diploid cells that are chronically infected **(carrier cultures)** many chromosomal breaks are evident. Such breaks may have a bearing on the pathogenesis of congenital lesions in the infected fetus.

PATHOGENESIS

The rash appears 14–25 days (average 18 days) after infection with rubella virus. During this prolonged incubation period viremia occurs and viral dissemination is widespread throughout the body, including the placenta during pregnancy. Virus multiplies in many organs but

few signs are manifested except for 1) a relatively common arthralgia and arthritis in women, accompanying infection of synovial membranes; 2) leukopenia from viral replication in lymphocytes; 3) occasional thrombocytopenia but uncommon purpuric manifestations; and 4) rare encephalitis (even a chronic progressive disease simulating measles-induced subacute sclerosing panencephalitis has been reported). Virus can be isolated from nasopharyngeal secretions (and occasionally from feces and urine) as early as 7 days before and as late as 7 days after the appearance of the exanthem. Respiratory secretions are probably the major vehicle for transmitting the virus.

The disease is not unlike measles, except that it is milder, is of shorter duration, and has fewer complications. It is initiated by a 1- to 2-day prodromal period of fever, malaise, mild coryza, and prominent cervical and occipital lymphadenopathy. During the prodromal illness, and for 1–2 days after the rash appears, virus can be

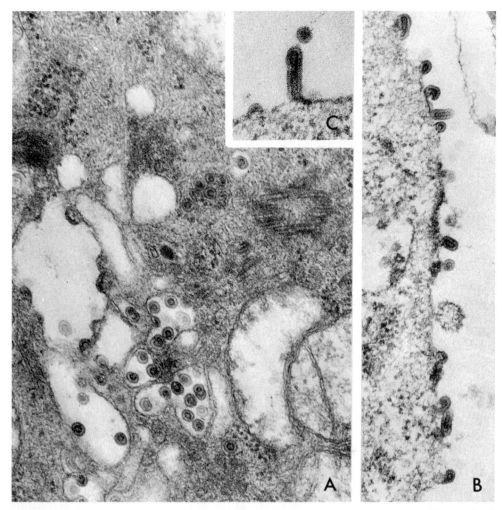

FIG. 63-3. Development of rubella virus in the surface and cytoplasmic membranes of infected cell cultures. **A.** Viral particles budding from cytoplasmic membranes into vacuoles and into the cytoplasm. Numerous mature virions are present within vacuoles. (× 60,000) **B.** Viral particles budding from the surface of an infected cell. (× 60,000) **C.** An elongated form in the budding process. (× 84,000) (Oshiro LS et al: J Gen Virol 5:205, 1969)

isolated from the blood. Virus can also be isolated from the skin lesions.

No characteristic pathologic lesions have been described in rubella, except for the **serious damage induced in infected fetuses.** This damage seems to involve tissues of all germ layers, and results from a combination of rapid death of some cells and persistent viral infection in others. The continued infection in turn frequently induces chromosomal aberrations and, finally, reduced cell division. The infant infected during the first trimester may be stillborn; if it survives, it may have deafness, cataracts, cardiac abnormalities, microcephaly, motor deficits, or other congenital anomalies in addition to throm-

bocytopenic purpura, hepatosplenomegaly, icterus, anemia, and low birth weight (the **rubella syndrome**). The greater susceptibility of the early embryo to damage is correlated with the greater placental transmission of virus at that stage: when infection occurs during the first 8 weeks of pregnancy virus can be isolated as often from an aborted fetus as from the placenta, whereas in later infections viral isolations are less frequent from the fetus than from the placenta.

Persistence. When a woman has clinical rubella during the first trimester of pregnancy the chance that the baby will have a structural abnormality is approximately 30%.

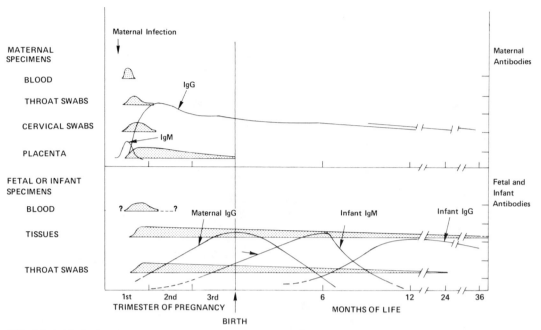

FIG. 63-4. Virologic events and antibody response in maternal–fetal rubella infection. (Modified from Meyer HM et al: Am J Clin Pathol 57:803, 1972)

Deformed infants born 6–8.5 months after intrauterine infection, and even those clinically normal, may still excrete virus in their nasopharyngeal secretions, although high titers of neutralizing IgM and IgG Abs and cell-mediated immunity are present (Fig. 63-4). Viral shedding persists until the clones of infected cells that are still able to divide eventually disappear. Indeed, one infant continued to shed virus in the presence of circulating Abs 2 years after birth, and in another the virus was isolated from cataract tissue at 3 years of age.

The Ab response in congenital rubella, like that noted in congenital cytomegalovirus infections (Ch. 55) and congenital syphilis (Ch. 39), indicates that intrauterine rubella infections *do not induce immunologic tolerance* to rubella virus Ags. However, immunity following congenital rubella shows differences from that following postnatal or childhood infections: some congenitally infected infants fail to develop typical Ab responses; others lose detectable Abs by 5 years of age; and others show depressed or absent cell-mediated immunity.

LABORATORY DIAGNOSIS

Rubella may be clinically confused with measles as well as with infections produced by a number of echo- and coxsackieviruses (Ch. 57). The diagnosis may be confirmed by inoculating infected materials (usually nasopharyngeal secretions) into susceptible cell cultures. The **interference assay,** in primary grivet monkey kidney cultures, is the quickest and most sensitive procedure for initial viral isolation. Infection may also be detected by the development of cytopathic changes in susceptible cell cultures, and by hemadsorption. Serologic diagnosis can be efficiently accomplished by hemagglutination-inhibition titrations, but care must be used to remove lipoprotein nonspecific inhibitors of hemagglutination (e.g., by adsorption of serum with dextran sulfate–$CaCl_2$). Complement-fixation and neutralization titrations and solid-phase immunoenzyme assays are also reliable.

Evaluation of the immune status is of particular importance for 1) women who are exposed to German measles during the first trimester of pregnancy (even a past history of rubella or previous immunization is not an absolute guarantee of immunity); 2) women of childbearing age who wish to determine whether they should be immunized; and 3) neonates born to a mother who was exposed to rubella during pregnancy or who have suggestive signs of the rubella syndrome.

EPIDEMIOLOGY

Rubella is a highly contagious disease, spread by nasal secretions. Unlike measles or chickenpox, however, *rubella infection is often inapparent,* thus fostering viral dissemination and rendering isolation of patients virtually useless. The ratio of inapparent infections to clinical cases is low (approximately 1:1) in children but as high as 9:1 in young adults. Patients are most infectious during the prodromal period, owing to the large amount of virus present in the nasopharynx, but communicability may persist for as long as 7 days after the appearance of the rash. Transmission usually occurs by direct contact with infected persons, mostly children 5–14 years of age. However, *the apparently normal infant excreting virus acquired in utero is perhaps the most dangerous carrier:* its infection is unrecognized and it comes in close contact with nurses, physicians, hospital visitors (including future mothers early in pregnancy), and later with other children at home.

In urban areas of Europe and the Americas, during the winter and spring months, minor outbreaks are noted every 1 or 2 years. Major epidemics recur every 6–9 years, and superepidemics erupt at intervals of up to 30 years. This epidemic pattern succeeds in immunizing 85% of the population up to 15 years of age (the age span of susceptibles is significantly more than in measles). Infection is almost always followed by long-lasting protection against clinical disease, although reinfections do occur. Most inapparent infections occur in those whose immunity has partially waned; the reinfection induces a secondary immune response (IgG Abs), which probably prevents or reduces the extent of viremia.

PREVENTION AND CONTROL

The extensive epidemic in the United States during 1964 resulted in congenital disabilities in approximately 20,-000 infants, causing enormous anguish and an economic loss of well over 1 billion dollars. To avoid these dire consequences, prevention of maternal infection is of utmost importance. Isolation procedures are rarely practical, and passive immunization is of questionable value. However, effective live **attenuated viral vaccines** have been developed and are presently being used to prevent the recurrence of such a devastating experience. These vaccines contain viruses either isolated in African green monkey cells and attenuated by further cell passage (in primary duck embryo cells) or isolated and passed in human embryo diploid cells. The vaccines require a single injection and are immunogenic in at least 95% of the recipients, but Abs appear later than those following natural infection, and at levels as much as tenfold lower. Nevertheless, immunization effectively protects the recipients from clinical rubella following exposure, even after extensive exposure during epidemics, and accordingly reduces the incidence of the congenital rubella syndrome (Fig. 63-5).

Though the vaccines employed are highly protective some drawbacks exist. 1) In 2–3 weeks after immunization small amounts of infectious virus appear in the nasopharynx. Viremia is unusual, however, and transmission to susceptible contacts has not been observed. 2) At the time of nasopharyngeal viral shedding mild arthralgias and occasional arthritis occur in 1%–2% of children and in 25%–40% of adult women; adults also occasionally experience mild rash, fever, and lymphadenopathy. 3) The relatively low Ab levels attained confer solid immunity only for 5–10 years rather than the long-lasting immunity that follows natural infection (unfortunately, the most attenuated viruses used in vaccines also are the least immunogenic). 4) A tenfold higher reinfection rate is observed in vaccinees as compared with those who have

FIG. 63-5. Cases of rubella by period of onset and of congenital rubella syndrome by period of birth, in the United States, 1969–1975. (Hayden GF et al: J Infect Dis 135:337, 1977)

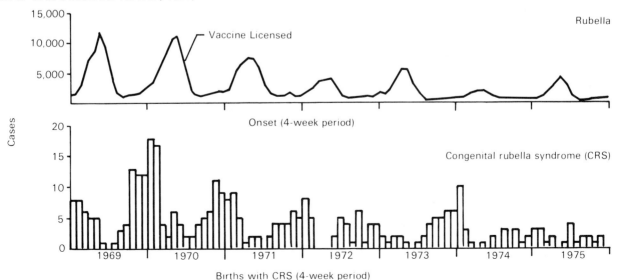

had natural infections. 5) An effective herd immunity is not produced, even when as many as 85% of 1- to 12-year-olds are immunized. 6) Despite the marked reduction in incidence of clinical disease, infection frequently is not prevented, even with exposure only 2–3 months after immunization. However, the inapparent infection could serve to induce a life-long immunity similar to that following the natural disease.

It should be emphasized that immunization has a unique goal: to protect an unborn fetus rather than the recipient of the vaccine. This goal has been approached by two almost opposing plans: 1) immunization of teenage girls, who are the prospective mothers; or 2) immunization of children 1–12 years of age, who are the major viral transmitters, thereby indirectly protecting pregnant women. In the United States the latter alternative has been chosen: i.e., routine immunization of all children. In addition, immunization is recommended for all women of childbearing age who are without protective Abs (birth control must be rigidly practiced for 2–3 months following immunization).

The immunization program pursued in the United States has elicited great concern, since the vaccine may not confer long-lasting immunity and therefore the women may become susceptible during their childbearing years. Moreover, the incidence of rubella is increasing in those 15 years old and older (in 1977, 70% of cases occurred in those aged 15 or older). In Great Britain immunization is given to all girls between 10 and 14 years old, and to women of childbearing age who do not have detectable hemagglutination-inhibiting Abs. It is not yet clear which approach is more successful in attaining the goal of preventing fetal damage.

If preventive measures fail many physicians recommend therapeutic abortion when rubella occurs during the first trimester of pregnancy.

SELECTED READING

BOOKS AND REVIEW ARTICLES

RUBELLA VIRUS

ALFORD CA JR: Rubella. In Remington JS, Klein JO (eds): Infectious Diseases of the Fetus and Newborn Infant. Philadelphia, Saunders, 1976, p 71

HORSTMAN DM: Rubella. In Evans AS (ed): Viral Infections of Humans. Epidemiology and Control. New York, Plenum Medical Books, 1976, p 409

International Conference on Rubella Immunization. Am J Dis Child 118:1, 1969

NORRBY E: Rubella virus. Virol Monogr 7:115, 1969

RAWLS WE: Congenital rubella: the significance of virus persistence. Progr Med Virol 10:238, 1968

RAWLS WE: Rubella virus. In Blair JE, Lennette EH, Truant JP (eds): Manual of Clinical Microbiology. Washington, DC, American Society for Microbiology, 1970, p 528

SPECIFIC ARTICLES

DAVIS WJ, LARSON HE, SIMSARIAN JP, PARKMAN PD, MEYER HM JR: A study of rubella immunity and resistance to infection. JAMA 215:600, 1971

GREGG NM: Congenital cataract following German measles in the mother. Trans Ophthalmol Soc NZ 3:35, 1941

HAYDEN GF. MÖDLIN JF, WITTE JJ: Current status of rubella in the United States 1969–1975. J Infect Dis 135:337, 1977

SEDWICK WD, SOKOL F: Nucleic acid of rubella virus and its replication in hamster kidney cells. J Virol 5:478, 1970

VAHERI A, HOVI T: Structural proteins and subunits of rubella virus. J Virol 9:10, 1972

chapter **64**

REOVIRUSES AND EPIDEMIC ACUTE GASTROENTERITIS VIRUS

CLASSIFICATION AND GENERAL CHARACTERISTICS

The term **reovirus** (respiratory enteric orphan virus) refers to a group of RNA viruses that infect both the respiratory and the intestinal tracts, usually without producing disease. Though originally considered members of the echovirus group (and classified as type 10 echovirus), reoviruses are larger and differ in producing characteristic cytoplasmic inclusion bodies. Moreover, these inclusion bodies, which contain specific viral Ags, stain green-yellow with acridine orange, like cellular DNA, rather than red, like the usual single-stranded RNA. This striking observation led to the discovery that *reovirus RNA is double-stranded* with a secondary structure similar to that of DNA. This finding was the first indication that such an unusual nucleic acid exists in nature. Viruses with similar chemical and physical properties have since been found to be widely disseminated in vertebrates, invertebrates, and plants. These viruses (more than 150) have been grouped into the family **Reoviridae** (vernacular, **reoviruses**).

All of these viruses have segmented, double-stranded RNA genomes. They are similar in morphology (Table 64-1), but differences in structure, antigenicity, stability, and preferred hosts are the bases for dividing Reoviridae into several genera: **Reovirus** includes species that infect humans, birds, dogs, and monkeys; **Orbivirus** (Latin *orbis,* ring) comprises members that multiply in insects, including Colorado tick fever virus, which infects humans; **Rotavirus** (Latin *rota,* wheel) includes some major etiologic agents of infectious infantile diarrhea in humans and several other animals. In addition, Reoviridae not yet classified include the **cytoplasmic polyhedrosis virus group,** which consists of viruses that infect Lepidoptera and Diptera, and the **plant reovirus group,** which contains viruses that infect many different types of plants (e.g., rice dwarf virus and clover wound tumor virus).

MORPHOLOGY

The virions of all Reoviridae have similar sizes, icosahedral symmetry, and not one but two icosahedral capsids. Fine-structure electron microscopy (Fig. 64-1) reveals that the outer capsid is probably constructed of 32 large capsomers (18 nm in diameter), and that neighboring capsomers share subunits (Fig. 64-1), which is apparently a unique feature of this family. The inner capsid (52 nm in diameter) has 12 prominent projections at its vertices and an undetermined number of intervening capsomers. Rotaviruses are slightly smaller than reoviruses (Table 64-1) but have similar double capsids.

Orbiviruses (e.g., blue-tongue virus of sheep, the type species) show another morphologic variation: the capsomer arrangement of the outer capsid is usually obscure, but exposure of the virions to CsCl below pH 8 removes a thin outer layer, revealing a capsid of 32 large, ring-shaped capsomers (hence its Latin derivation, *orbis*). Colorado tick fever virus has not been adequately studied.

Reoviridae of different genera do not display any immunologic relatedness.

CHEMICAL AND PHYSICAL CHARACTERISTICS

The virions of Reoviridae are composed of protein and about 15% RNA. The double-stranded nature of the RNA is shown by many properties: complementary base ratios (G=C = 20 moles percent); sharp thermal denaturation (T_M 90°–95°C); pronounced hyperchromicity on denaturation; resistance to a concentration of ribonuclease A that completely degrades single-stranded RNA; and characteristic density in $CsSO_4$. Electron micrography reveals stiff filaments, like those of DNA, and x-ray diffraction patterns are consistent with double-stranded molecules.

The **genomes** of Reoviridae have another surprising feature: the RNA extracted from purified virions or infected cells is found in **10 or 11 distinct pieces,** which are distributed in **three size classes** (Fig. 64-2). That these are distinct components of the virion rather than products of artificial fragmentation is shown by the following characteristics: the fragment sizes are reproducible within a genus; the different pieces do not cross-hybridize; each fragment contains a free 3′-OH terminal cytosine and a 5′ terminal guanosine-5-diphosphate; and electron micrographs of gently disrupted virions demonstrate molecules of viral nucleic acid of the expected lengths.

Reovirus virions (other genera have not been examined) also contain about 3.7×10^6 daltons of a heteroge-

TABLE 64-1. Characteristics of Reoviridae That Infect Man

	Reoviruses	Rotaviruses	Orbiviruses*
Morphology	Icosahedral; double capsid; no envelope	Icosahedral; double capsid; no envelope	Icosahedral; double capsid—outer is skin-like
Size (diameter)	75–80 nm	70 nm	70 nm
Nucleic acid	RNA; double-stranded; 10 segments, mol wt 0.5–3×10^6, total 15×10^6 daltons (46 Kb)	RNA; double-stranded; probably 11 segments, mol wt 0.23–2.04×10^6, total 10×10^6 daltons (31 Kb)	RNA; double-stranded; 10 segments, mol wt 0.3–2.7×10^6, total 12×10^6 daltons (37 Kb)
Effect of lipid solvents	Infectivity stable	Infectivity stable	Infectivity stable
Virion polypeptides	8	10	8

*Blue-tongue virus, the best characterized orbivirus, was used for this comparative analysis. The size range for the genus is 60–80 nm.

neous collection of small, single-stranded oligonucleotides, whose function is unknown.

The **reovirus** and **orbivirus capsids** contain eight species of polypeptides, and those of rotaviruses contain ten polypeptides; their molecular weights are distributed in three size classes. Since the structural proteins (Table 64-2) can be separated only after denaturation of the virion, it is impossible to identify directly their specific functions (i.e., hemagglutination, RNA polymerase, nu-cleoside phosphohydrolase, group antigenicity, and type-specific antigenicity). However, association of each polypeptide with a virion structure (or a nonstructural protein), the identification of the RNA segment (i.e., the gene) in which each polypeptide is encoded, and the correlation of some polypeptides with function have been possible using genetic and biochemical technics (Table 64-2).

REOVIRUSES

IMMUNOLOGIC CHARACTERISTICS

The σ_1 outer capsid protein induces specific neutralizing and hemagglutination-inhibiting Abs which distinguish **three immunologic types** of reoviruses. They are antigenically related, however, by three or four crossreacting Ags that can be measured by complement-fixation (CF) and immunoprecipitin tests. For example, heterotypic reovirus Abs appear in the serum of about 25% of persons who have primary infections with type 1 reovirus. Type-specific Abs to reoviruses appear 2–4 weeks after infection.

The σ_1 outer capsid protein is also responsible for hem-agglutination by all three immunologic types (types 1 and 2 agglutinate human RBCs; type 3 agglutinates bo-vine RBCs). Reoviruses agglutinate and elute from RBCs, but unlike orthomyxoviruses they do not destroy the receptor sites on the red cells, and the reovirus re-ceptors are not hydrolysed by neuraminidase. The RBC receptor is probably a glycoprotein, since either trypsin or periodate inactivates it and N-acetyl-D-glucosamine blocks hemagglutination by binding to the viral capsid.

HOST RANGE

Reoviruses appear to be ubiquitous in nature: specific viral inhibitors (presumably Abs) have been found in the serum of all mammals tested except the whale; and humans and many other species (including cattle, mice, and monkeys) are naturally susceptible to reoviruses. Newborn mice are particularly vulnerable to experi-mental infection, which is often fatal; when infected with type 3 reovirus they occasionally develop a chronic ill-ness similar to runt disease (Ch. 23).

Cell cultures, including primary cultures of epithelial cells from many animals and various continuous human cells lines, are used to isolate and study reoviruses. The infection causes gross cytopathic changes and permits **hemadsorption** of human group O RBCs. Distinctive eosinophilic **inclusion bodies** (Fig. 64-3) are seen in the cytoplasm of infected cells.

VIRAL MULTIPLICATION

The eclipse period is long (6–9 h, depending upon viral type and size of inoculum) compared with that of other RNA viruses with icosahedral symmetry (Fig. 64-4 and Ch. 50). Virus then increases exponentially, reaching a maximum titer (from 250 to 2500 plaque-forming units per cell) by approximately 15 h after infection. Infected cells are not rapidly lysed following viral replication, and release of infectious virus is incomplete.

Owing to the unique structure of the capsid and the novel nucleic acid, replication of reoviruses presents

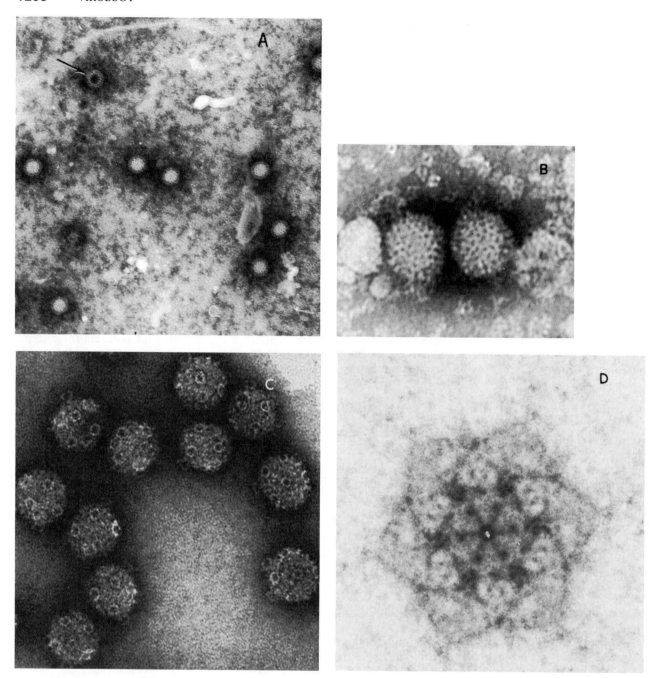

FIG. 64-1. Electron micrographs of negatively stained reovirus virions, cores, and capsomers. The virion is 75–80 nm in diameter. **A, B.** Cubic symmetry, structure of the capsid, and absence of an envelope illustrated by low (**A**) and high (**B**) magnification. Note the empty particle in **A** in which the inner coat that covers the core is apparent **(arrow).** (Gomatos PJ et al: Virology 17:441, 1962) (**A,** ×75,000; **B,** ×375,000) **C.** Cores prepared from purified virions by chymotrypsin degradation of the outer capsid followed by equilibrium centrifugation in a CsCl density gradient. Note the hollow spikes located at the 12 vertices of the icosahedral core. (White CK, Zweerink HJ: Virology 70:171, 1976) (×280,000) **D.** Cluster of reovirus capsid subunits in which the central capsomer was enhanced by an n = 6 rotation. Rotational enhancement of image detail makes evident the structure of the central capsomer, which is made of six wedge-shaped subunits that exhibit sharing with neighboring capsomers. Note also that each wedge-shaped subunit is also composed of subunits (i.e., polypeptide chains). (Palmer EL, Martin ML: Virology 76:109, 1977) (×1,200,000)

TABLE 64-2. Correlation Between Classes of Reovirus* Genomic Segments and Polypeptides in Virions and Infected Cells

Genome				Polypeptides				
Size class	Segment	Mol wt ($\times 10^{-6}$)	Time of expression	Size class	Component†	($\times 10^{-3}$)	origin‡	Function§
L	1(L_1)	2.7	Early¶	λ	λ_3	143	Core	RNA synthesis
	2(L_2)	2.6	Late		λ_2	148	Core	
	3(L_3)	2.5	Late		λ_1	153	Core	
M	4(M_1)	1.8	Late	μ	μ_2	74	Core	
	5(M_2)	1.7	Late		μ_{lc}#	78	Outer capsid	
	6(M_3)	1.6	Early¶		μ^N_s	72	Nonstructural	
S	7(S_1)	1.1	Late	σ	σ_1	54	Outer capsid	Cell association; hemagglutination; induction neutralizing Abs
	8(S_2)	0.85	Late		σ_2	52	Core	RNA synthesis
	9(S_3)	0.76	Early¶		σ^N_s	49	Nonstructural	RNA synthesis
	10(S_4)	0.71	Early¶		σ_3	43	Outer capsid	

*Type 3 (Dearing strain) is given as example. Sizes of the RNA segments and polypeptides are different for each type.

†Identification of polypeptide with double-stranded RNA segment was accomplished by in vitro translation of specific mRNAs and by intertypic genetic recombination, which takes advantage of the different sizes of RNA segments in different viral types.

‡Determined by controlled disruption of purified virions and SDS–polyacrylamide gel electrophoresis of infected cell cytoplasmic extracts.

§Identified by genetic studies using temperature-sensitive mutants.

¶Predominant messages transcribed during early stages of infection, and the mRNAs made in the presence of cyclohexamide added at time of infection.

#Processed from μ_1.

FIG. 64-2. Electrophoresis of the virion RNA of reovirus type 3, human rotavirus, calf rotavirus, and an orbivirus on 7.5% polyacrylamide gels (migration was from top to bottom). RNA segments for each virus are distributed in three size classes (L, M, S). (Modified from Schnagl RD, Holmes IH: J Virol 19:267, 1976)

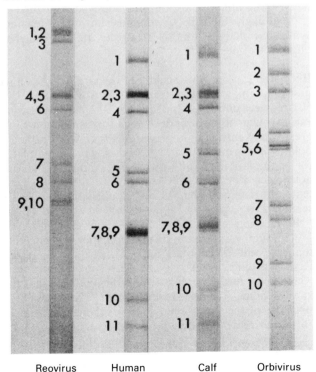

Reovirus   Human rotavirus   Calf rotavirus   Orbivirus

some unusual features. After **cell penetration** the virions become associated with lysosomes, whose proteolytic enzymes hydrolyze about one-half of the outer capsid (all of polypeptide σ_3 and an 8000-dalton piece of polypeptide μ_{lc}), forming a **subviral particle** that is smaller than the virion but larger than the viral core (inner capsid and RNAs). Thus, the core is only partially uncoated and the double-stranded RNA genome segments are not released free into the cytoplasm.

Transcription. The virion RNAs do not function as messengers; but, like DNA, they are first transcribed, and the transcripts are processed into single-stranded mRNAs. This transcription ensues entirely within the subviral particles, which leave the lysosomes after the uncoating process. Initially (beginning 2–4 h after infection) four segments (1, 6, 9, and 10 seen in Fig. 64-2 and Table 64-2) of the viral genome are predominantly transcribed, although all genome segments are copied. An RNA polymerase (transcriptase) contained within the core of the virion is responsible for this early transcription, which is accomplished without synthesis of new proteins or replication of the viral genome. Four additional enzymic activities within the virion (nucleotide phosphohydrolase, guanylyltransferase, and two methyltransferases) are responsible for synthesizing the caps at the 5′ termini of the mRNAs.

By 6 h after infection the ten segments of the viral genome are transcribed at comparable rates. The sizes of

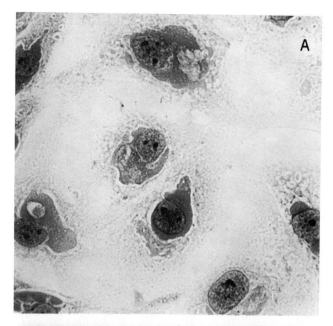

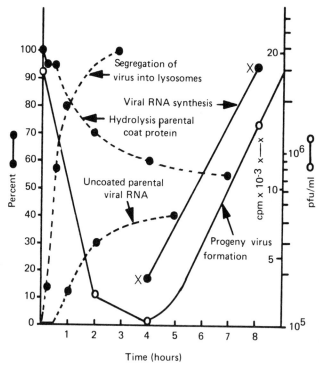

FIG. 64-4. Sequential events in the uncoating and replication of reovirus. (Silverstein SC, Dales S: J Cell Biol 36:197, 1968)

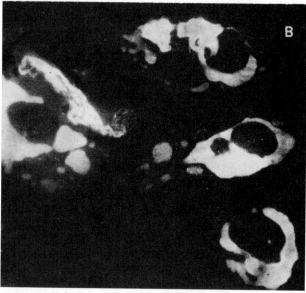

FIG. 64-3. Large eosinophilic cytoplasmic inclusion bodies in monkey kidney cells infected with type 1 reovirus. **A.** H & E stained and viewed with light microscope. (×1000) **B.** Stained with fluorescein-conjugated Ab and viewed with ultraviolet optics (×1500) (Rhim JS et al: Virology 17:342, 1962)

the mRNAs correspond to those of the virion RNAs (Fig. 64-2, Table 64-2), and each hybridizes specifically to one strand of the segment of the viral genome of corresponding size. Most of the newly synthesized single-stranded RNA molecules leave the subviral particles, rapidly become associated with polyribosomes, and ap-

pear to function as monocistronic mRNAs. Unexpectedly, these mRNAs are not polyadenylated at their 3′ ends though the virion contains an oligoadenylate synthetase activity and is relatively rich in poly(A) oligonucleotides.

RNA Replication. This activity of the viral RNA is also unusual. Unlike double-stranded DNA molecules (Ch. 50), reovirus double-stranded RNA is **replicated conservatively.** One of the strands of each segment is copied in great excess into plus single strands (i.e., with the same polarity as the mRNA), and these then serve as templates for synthesis of the minus strands. Free minus strands are not detected because they remain associated with their templates to form the various double-stranded RNA segments.

Replication of the viral RNA requires continuous protein synthesis, apparently to supply the virus-encoded replicase and transcriptase molecules. Newly replicated RNA molecules of all ten size classes are found in infected cells, further strengthening the evidence that the viral genome is indeed segmented. The segmented structure of the viral genome also explains the remarkably high recombination frequency (i.e., 3%–50%) detected with temperature-sensitive mutants (as with influenza viruses, Ch. 58).

Translation. About 75% of the newly synthesized plus strands become associated with cytoplasmic polyribosomes, which serve as the sites for synthesis of viral proteins. All eight virion structural proteins can be detected in infected cells as early as 3 h after infection, but only seven of these are primary gene products: μ_{1c} (Table 64-2) is derived from μ_1 by cleavage. In addition, two nonstructural viral polypeptides, μ_{Ns} and σ_{Ns}, whose functions are unknown, are also synthesized.

Assembly. Infectious virions begin to appear 6–7 h after infection (Fig. 64-4), but how the ten genomic segments (which may be likened to chromosomes) are segregated in the appropriate number remains unexplained. Excess viral Ags accumulate and viral particles assemble in close association with the spindle tubules (Figs. 64-5 and 64-6). However, the mitotic spindle is not essential for viral multiplication, since viral synthesis proceeds unhindered in nondividing cells arrested in metaphase by colchicine, which disaggregates the spindle. Unassembled, newly synthesized double-stranded RNAs and the excess viral Ags accumulate in large masses (Fig. 64-3), forming the characteristic cytoplasmic inclusion bodies which give green-yellow fluorescence with acridine orange staining.

Reovirus infection inhibits biosynthesis of host cell DNA and protein within 6 h, and cell division ceases. In such cells the mitotic index increases more than threefold, but the mitotic sequence is not completed and abnormal mitotic figures form.

PATHOGENESIS

Reoviruses have frequently been isolated from the feces and respiratory secretions of healthy persons, as well as from patients with a variety of clinical illnesses, particularly minor upper respiratory and gastrointestinal dis-

FIG. 64-5. A–D. Reovirus-infected cells stained with fluorescein-conjugated Ab. Viral Ag is closely associated with the mitotic spindle of virus-infected cells. Cells were examined by darkfield microscopy with ultraviolet illumination. (Spendlove RS et al: Immunol 90:554, 1963) (×750)

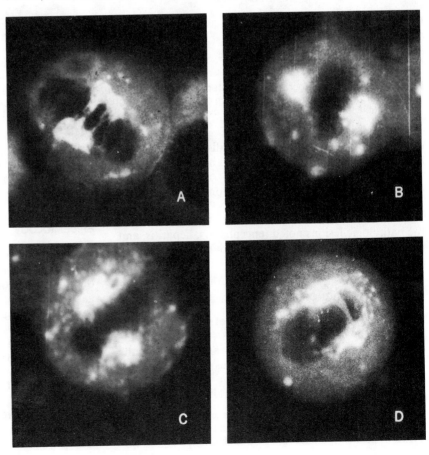

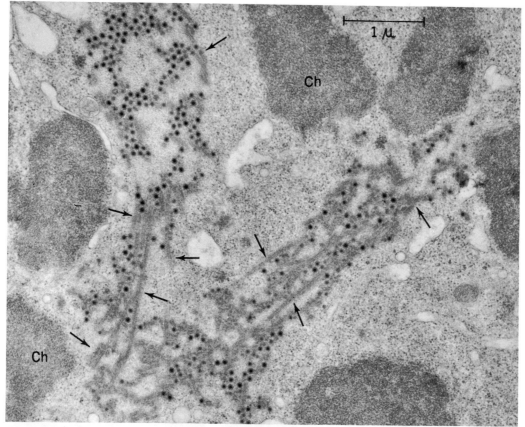

FIG. 64-6. Electron micrograph of a cell infected with type 3 reovirus. Section was made through the spindle and chromosomes **(Ch)** of an infected cell in mitosis, showing aggregates of viral particles closely associated with the tubules of the spindle (indicated by **arrows**). (Dales S: Proc Natl Acad Sci USA 50:268, 1963) (×15,000)

ease. The relation of these viruses to disease is not clear, and human transmission experiments have not been decisive: afebrile respiratory illnesses only occurred in one-third of the volunteers, the symptoms were irregular, and the illnesses were very mild.

Three deaths have been attributed to type 3 reoviruses; the pathologic lesions noted (encephalitis, hepatitis, and pneumonia) were similar to lesions found in experimental animals.

LABORATORY DIAGNOSIS

Reoviruses may be isolated from throat washings or fecal specimens by means of cell cultures (e.g., human embryonic or monkey kidney), and they are usually identified as belonging to the reovirus genus by CF tests. The specific immunologic type can then be identified by hemagglutination-inhibition or neutralization assays.

To permit recognition of Abs in a patient's serum by hemagglutination-inhibition titrations the serum should be pretreated with trypsin or periodate, or adsorbed with kaolin, to remove nonspe-

cific mucoprotein inhibitors. Since reoviruses are also frequently isolated from healthy persons, an increase in serum Ab titer must be demonstrated before an illness can be assumed to be caused by a reovirus.

EPIDEMIOLOGY AND CONTROL

Though reovirus infections do not seem to be of great clinical importance, further studies are needed to define their pathogenetic potential. Both the respiratory and the gastrointestinal tracts may well be sources of their spread. Unrecognized infections are common, for approximately 10% of children by 5 years of age and 65% of young adults in the United States have reovirus Abs in their serums. Antibodies are also frequently found in various wild and domestic animals, but it is not known whether the animals serve as reservoirs for human infections.

Since the meager data assign a limited pathogenicity to these viruses specific immunization procedures are not warranted.

ROTAVIRUSES

Acute **nonbacterial gastrointestinal infections** (epidemic gastroenteritis and infantile diarrhea) are second only to acute respiratory infections as the cause of illness in families with young children. Hence the recent discoveries of viruses that are major causes of these infections encourage the hope that these common illnesses will be better understood and controlled. A **rotavirus** (initially termed human reovirus-like agent) has been identified as a major cause of **sporadic acute enteritis** in infants and young children. In addition, two immunologically distinct viruses that resemble parvoviruses have been shown to be etiologic agents of **epidemic acute gastroenteritis** (discussed separately below).

These new viruses were discovered by direct electron microscopic observation of fecal extracts, duodenal fluid, and duodenal mucosa; none was detected by isolation in a cell culture or animal. A major diagnostic advance, **immune electron microscopy,** was of special utility: serums from convalescent patients were used as a source of specific Abs to agglutinate the virions and make them more easily recognizable, and to identify the virus as one that had actually produced an acute infection with an attendant Ab response (Figs. 64-7 and 64-8).

PROPERTIES OF THE VIRUS

All rotaviruses thus far isolated belong to a **single immunologic group.** Indeed, the human rotavirus and four rotaviruses isolated from animals* are antigenically similar, if not identical, as studied by CF and immune electron microscopy (by immunodiffusion the human and calf viruses are identical). The common rotavirus Ag appears to be a component of the inner capsid.

No appropriate cell culture system is available for study of the biochemical events in multiplication of a human rotavirus. However, virus has been serially propagated in human fetal intestinal organ cultures, and in primary embryonic human gut or in pig gut and kidney cells. In all cases only rare cells are infected; cytoplasmic multiplication of virus can be detected by immunofluorescence and electron microscopy, but cytopathic changes are minimal or absent. Human rotavirus can also infect, multiply in, and produce diarrhea in piglets (whether normally raised or germ-free), gnotobiotic

---

* Nebraska calf diarrhea virus, epizootic diarrhea of infant mice, virus simian agent, and the O virus (isolated from offal in compost).

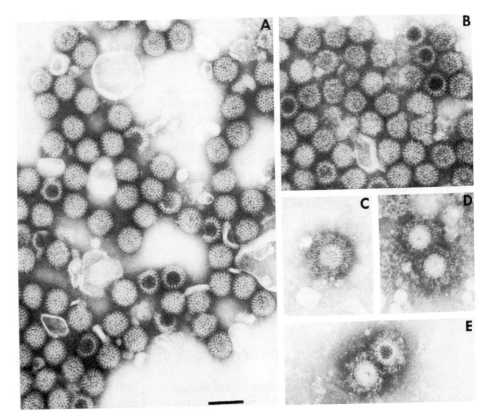

FIG. 64-7. Identification of a rotavirus from man by immune electron microscopy. **A.** Viral particles in stool filtrate. Note the capsomer structure and the appearance of a double capsid in the empty particles. **B.** Filtrate incubated with acute-phase serum. **C, D,** and **E.** Viral particles in the same stool preparation incubated with convalescent serum from the same patient. (Kapikian AZ et al: Science 185:1049, 1974. Copyright 1974 by the American Association for the Advancement of Science)

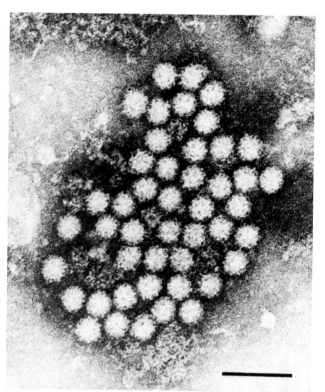

FIG. 64-8. Electron micrograph of an aggregate of acute non-bacterial gastroenteritis virus (Norwalk agent) particles in stool from an acutely ill patient. (Kapikian AZ et al: J Virol 10:1075, 1972) (×200,000)

newborn calves, and newborn rhesus monkeys. Virus can be detected in the duodenal mucosal cells, duodenal contents, and feces, and the affected animals show a brisk Ab response.

Thus the rotaviruses from different animals are both immunologically closely related and similar in their host responses. The four species of viruses also have the same number of genomic segments. However, the sizes of their RNAs (Fig. 64-2) and proteins differ.

PATHOGENESIS

Severe diarrhea and fever, occasionally accompanied by vomiting, is a common syndrome in children less than 2 years old, particularly infants. About half of the most severe illnesses of babies seen in hospitals are caused by the human rotavirus; during the fall and winter the inci-

dence is particularly high. Biopsies of the duodenal mucosa show it to be a principal site of viral multiplication, which may account for the symptoms observed (whether other parts of the gastrointestinal tract are also affected is uncertain). Viral excretion in feces is maximal during the first 4 days of illness, but it has been detected as long as 7 days after the onset of diarrhea. In experimental infections of rhesus monkeys and piglets, viral replication and pathologic lesions occur in mucosal and submucosal cells of the entire small intestine, but the large bowel and other organs do not appear to be involved.

LABORATORY DIAGNOSIS

Technics are not available to isolate rotaviruses by a conventional method, but they can be readily identified in stools by **immune electron microscopy** (Fig. 64-7). This same technic can be employed to recognize a patient's Ab response. The simplest and most economic diagnostic procedures are CF and microtiter solid-phase radioimmune assays of the patient's serums, using as Ag a known positive fecal extract or an extract of cell culture infected with the immunologically related calf diarrhea rotavirus. These Ab assays are just as sensitive and accurate for diagnosis of infection as identification of virus by immune electron microscopic examination of stools.

EPIDEMIOLOGY AND CONTROL

Acute gastroenteritis caused by a rotavirus is a worldwide, sporadic disease, primarily in young children 6–24 months of age, and is a leading cause of childhood deaths in developing countries. Its prevalence is demonstrated serologically: over 90% of children in the Washington, D.C., area were found to have CF Abs by age 3 years, like their Ab status to respiratory syncytial and parainfluenza viruses (Ch. 59). Although the virus is probably transmitted by the fecal–oral route, infection is most common during the cooler part of the year, unlike bacterial diarrheas and dysenteries.

In a small study at least one parent in a family had an inapparent infection at the time illness began in the baby, suggesting that the child's source of virus might literally have been the hand that fed it.

Control of this major disease in young children must await more extensive knowledge of the epidemiologic features of the infection, and the immunologic characteristics of the viruses, as well as better means for propagating the viruses in cell cultures.

COLORADO TICK FEVER VIRUS (ORBIVIRUS)

Colorado tick fever is the only tick-borne viral disease recognized in the United States, though Powassan virus,

a tick-borne flavi-togavirus (Ch. 62), has been isolated in Canada. The virions morphologically are typical orbiviruses (Table 64-1) having an outer capsid 75–80 nm in diameter and an inner capsid of 50 nm. The virus multiplies readily in hamsters, suckling and adult mice, and some continuous human cell lines. Virions replicate, in large numbers, free in the cytoplasm and unassociated with cell membranes or mitotic spindles.

Colorado tick fever virus is immunologically unrelated to reoviruses and rotaviruses. The virus is partially inactivated by ether, probably owing to its skinlike outer coat; but it is not inactivated by sodium deoxycholate (in contrast to viruses with classic envelopes, such as togaviruses, which are also transmitted by insects).

Colorado tick fever is an acute, febrile, nonexanthematous infection characterized by acute onset of fever, chills, headache, and severe pains in the muscles of the back and legs. The course of the disease is short and re-covery is complete. Infection induces long-lasting, probably lifelong, immunity.

The disease occurs in the western United States where its major vector, the tick *Dermacentor andersoni,* is found. The virus has also been isolated from the tick *D. variabilis,* collected on Long Island, N.Y., but no human infections have been reported from this locality. The golden ground squirrel appears to be the major animal reservoir; the virus has also been isolated from chipmunks, other squirrels, and a deer mouse. Infection of man is only incidental and is a dead end in the chain of transmission.

Prevention is directed primarily toward avoiding ticks, either by not entering infested areas or by wearing suitable clothing and using arthropod repellents.

EPIDEMIC ACUTE GASTROENTERITIS VIRUS

Epidemiologic studies in families and human volunteer experiments indicated that two immunologically distinct viruses cause acute nonbacterial gastroenteritis, one of the most common illnesses in the United States. Although the etiologic agents of epidemic acute viral gastroenteritis are unrelated to viruses of the family Reoviridae they will be described in this chapter, since their taxonomic niche is uncertain and their pathogenesis and epidemiology are so similar to rotavirus infections (which are generally more severe).

The viruses causing epidemic gastroenteritis have **not been successfully isolated** in animals or cell cultures. The etiologic agents* have been identified, however, by **immune electron microscopy** (Fig. 64-8) as described above for rotaviruses. The virions are 27 nm in diameter, spherical, probably of icosahedral symmetry, and without an envelope. On the basis of morphology, buoyant density of 1.40 gm/cm^3 in CsCl, relative resistance to acid and heat inactivation, and failure to be inactivated by ether, the viruses have been called **parvoviruslike** (Ch. 54). However, they are slightly larger than parvoviruses and, the nature of their nucleic acid has not been de-scribed because of the lack of adequate quantities of purified virions.

The disease occurs in outbreaks and is characterized by a combination of nausea, vomiting, diarrhea, low-grade fever, and abdominal pain. The illness is self-limited, usually lasting 1–2 days; it is commonly found in families or even in community-wide outbreaks. In contrast to the sporadic form of acute gastroenteritis caused in infants by rotaviruses, acute epidemic gastroenteritis mainly affects **school-age children and adults,** and often spreads to family contacts. Both forms of viral gastroenteritis occur most frequently from September to March.

Virions appear in stools with the onset of disease and are shed in greatest number during the first 24 h of illness. Virus has not been detected during the relatively short incubation period (about 48 h) or more than 72 h after onset. Antibodies develop following infection, but their persistence and effect on resistance to subsequent infection are still unknown. **Three distinct immunologic types** have been identified by immune electron microscopy. However, the frequent occurrence of disease in older children and adults may be due to the existence of many different immunologic types of viruses (like rhinoviruses) rather than to failure of Abs to persist or to produce resistance to infection.

---

* The first viruses were identified during an epidemic in Norwalk, Ohio and were termed the **Norwalk agents.**

SELECTED READING

BOOKS AND REVIEW ARTICLES

KAPIKIAN AZ, FEINSTONE SM, PURCELL RH, WYATT RG, THAMBILL TS, KALICA AR, CHANOCK RM: Detection and identification by immune electron microscopy of fastidious agents associated with respiratory illness, acute nonbacterial gastroenteritis, and hepati-tis A. In Pollard M (ed): Perspectives In Virology. New York, Academic Press, 1975, p 9

KAPIKIAN AZ, KIM HW, WYATT RG, CLINE WL, PARROTT RH, CHANOCK RM, ASROBIA JO, BRANDT CD, RODRIGUEZ WJ, KALICA AR, VANKIRK DH: Recent Advances in the aetiology of viral gastroenteritis. In Acute Diarrhea in Childhood. Ciba Foundation

Symposium 42, Amsterdam, Oxford, New York Elsevier, 1976, p 273

JOKLIK WK: Reproduction of Reoviridae. Compr Virol 3:231, 1974

SHATKIN AJ, BOTH GW: Reovirus mRNA: transcription and translation. Cell 7:305, 1976

SILVERSTEIN SC, CHRISTMAN JK, ACS G: The reovirus replicative cycle. Annu Rev of Biochem 45:375, 1976

STANLEY NF: The reovirus murine models. Prog Med Virol 18:257, 1974

SPECIFIC ARTICLES

BISHOP RF, DAVIDSON GP, HOLMES IH, RUCK BJ: Virus particles in epithelial cells of duodenal mucosa from children with acute nonbacterial gastroenteritis. Lancet 2:1281, 1973

FIELDS BH, JOKLIK WK: Isolation and preliminary genetic and biochemical characterization of temperature-sensitive mutants of reovirus. Virology 27:335, 1969

GOMATOS PJ, TAMM I: The secondary structure of reovirus RNA. Proc Natl Acad Sci USA 49:707, 1963

MCCRAE MA, JOKLIK WK: The nature of the polypeptide encoded by each of the 10 double-stranded segments of reovirus type 3. Virology 89:578, 1978

MUSTOE TA, RAMIG RF, SHARPE AH, FIELDS BN: Genetics of reovirus: identification of ds RNA segments encoding the polypeptides of the μ and σ size classes. Virology 89:594, 1978

SMITH RE, ZWEERINK HJ, JOKLIK WK: Polypeptide components of virions, top component and cores of reovirus type 3. Virology 39: 791, 1969

chapter **65**

HEPATITIS VIRUSES

The infectious nature of hepatitis was long unrecognized because the disease tends to occur sporadically and because jaundice, a prominent clinical sign, has many diverse causes. Since the time of Virchow the disease was believed to result from obstruction of the common bile duct by a plug of mucus, and it was known as **acute catarrhal jaundice.** In 1942 Voeght first transmitted hepatitis by feeding a patient's duodenal contents to volunteers. Subsequently it was found that the etiologic agents are filterable and that the disease may be transmitted in two ways: by the intestinal–oral route **(infectious hepatitis)** or by the injection of infected blood or its products **(serum hepatitis).** However, the differences in transmission are not absolute, since the virus of so-called infectious hepatitis can also produce disease when inoculated parenterally, and experimentally, the virus of serum hepatitis has been transmitted orally. The viruses causing these two types of hepatitis showed other differences, particularly the absence of crossimmunity in human transmission experiments. Accordingly, new names were assigned: **hepatitis A virus (HAV)** and **hepatitis B virus (HBV)** for infectious and serum hepatitis viruses, respectively. (That there is at least one additional hepatitis virus has recently been discovered, as will be discussed below.)

Additional evidence for the existence of more than one hepatitis virus evolved from the serendipitous discovery of the so-called **Australia antigen** (also called **hepatitis-associated Ag [HAA]** or **SH Ag),** which appears only in patients with hepatitis B (serum hepatitis) and is identified as the HBV surface Ag (HBsAg).

This Ag was first detected by Blumberg in the serum of an Australian aborigine. In a search for new serum alloantigens he happened to employ test serum from two hemophiliacs who had received multiple blood transfusions. The serum from these subjects, which contained Abs to the so-called Australia Ag, were then found to react with serum from a variety of patients who had received multiple transfusions or who resided in institutions (e.g., for mental defectives or for lepers) in which the inmates had a high incidence of hepatitis. The detection of Australia Ag was then recognized as signaling the presence of active or inactive serum hepatitis.

The discovery of this novel Ag has permitted further characterization of hepatitis B virus and has allowed diagnostic differentiation of the clinical disease. Although neither HAV nor HBV has yet been isolated in a cell culture, and only HAV has been propagated in vitro, both can produce infection in chimpanzees and monkeys and can be identified by electron microscopy. Moreover, the identification of hepatitis A and B viruses has led to the recognition that there is at least one other hepatitis virus, which is unrelated to HAV or HBV. This virus, termed **hepatitis C virus** (also called **non-A non-B hepatitis virus),** is at present the major etiologic agent of hepatitis following transfusions in the United States. The continued high incidence of hepatitis (e.g., in the United States approximately 53,000 cases were reported in 1978) and its great morbidity have stimulated increasing interest in these viruses.

PROPERTIES OF HEPATITIS A VIRUS

A 27-nm viral particle (Fig. 65-1) present in the stools of patients with hepatitis A is clearly the etiologic agent of the disease: 1) the particles (recognizable by electron microcopy) are found in the stools of patients with clinical hepatitis A during the peak of the disease and its associated biochemical changes; 2) these particles are not found in patients with serologically identified hepatitis B; 3) hepatitis A patients develop Abs that react with the viral particles (recognized by neutralization, immune adherence, and CF); and 4) the particles have been detected, by immune electron microscopy, in the serum, liver, and feces of marmosets and chimpanzees to whom the disease was transmitted with feces or blood from patients with hepatitis A.

The limited ability of hepatitis A virus to multiply in cell cultures (see below) and the inability to infect a convenient laboratory animal with HAV has made the characterization of this virus largely dependent upon virions isolated from the feces of infected humans and from the feces, blood, and liver of experimentally infected primates. This limited source of virus has sharply restricted investigations of the physical, chemical, and immunologic properties of the virus.

In **morphology** the virions resemble picornaviruses (Ch. 57), with an average diameter of 27 nm, icosahedral symmetry, and no envelope (Fig. 65-1). The **physical and chemical characteristics** of purified hepatitis A virions

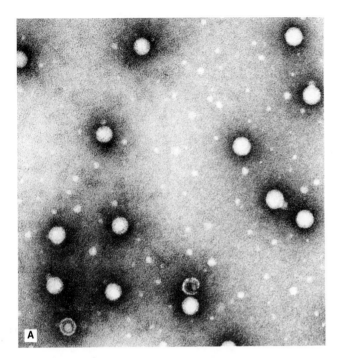
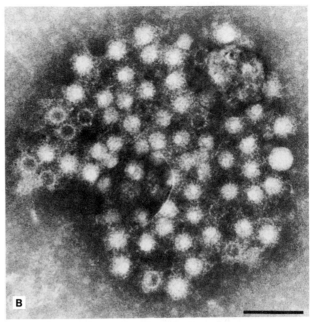

FIG. 65-1. Electron micrographs of hepatitis A virus (HAV) particles negatively stained with phosphotungstic acid. **A.** Virions purified from human stools. (×150,000) (Courtesy of Dr. John Gerin, Oak Ridge National Laboratory) **B.** Immune electron microscopy of HAV virions aggregated by serum from a late-convalescent patient with naturally acquired hepatitis A. Note the fuzzy margins of the virions and the lattice of Ab molecules connecting the virions. (×230,000) (Kapikian AZ et al: In Pollard M (ed.): Perspectives in Virology, vol. 9, p. 9. San Francisco, Academic Press, 1975)

(obtained from patients' feces) are also similar to those of enteroviruses: buoyant density of 1.32–1.35 g/cm³ in CsCl, a single-stranded RNA genome, contains three major polypeptides (mol wt 34,000, 25,000 and 23,000); and stable infectivity in ether and at pH 3.0. As with most viruses, including HBV, infectivity is inactivated by ultraviolet irradiation, formalin (1:400 dilution) for 3 days at 37°C, and heat for 5 min at 100°C (these properties, are important for vaccine development). The virions are relatively resistant to common disinfectants, however, and therefore special cleansing procedures must be used to prevent hospital spread.

Human transmission studies and extensive epidemiologic data suggest that the virus exists as a **single immunologic type** and that long-lasting immunity follows infection. Antibodies appear shortly after the onset of clinical disease (CF Abs being detectable earliest); Ab levels rise slowly, and persist for years (Fig. 65-2).

In addition to the replication of HAV in subhuman primates, **viral multiplication** occurs in vitro in a line of fetal rhesus monkey kidney cells and in hepatocytes growing in marmoset liver explant cultures. The virus multiplies in the cytoplasm of infected cells, but the quantity of virus produced is too small to permit biochemical studies of viral synthesis.

FIG. 65-2. Sequential relationship of viral shedding to clinical hepatitis, and development of virus-specific Abs during HAV infection. (Modified from Krugman S, Friedman H, Lattimer C: N Eng J Med 292:1141, 1975. Reprinted by permission from the New England Journal of Medicine)

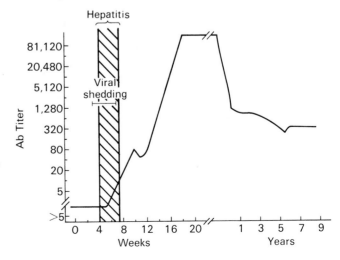

PROPERTIES OF HEPATITIS B VIRUS

Morphology. Serum from patients with clinical hepatitis B commonly contains three distinct structures (Fig. 65-3) that possess the hepatitis B surface antigen (HB$_s$Ag; see Antigenic Structure, below). The **Dane particle** (named for its first observer) is the least common form, but it alone has the structure attributed to viruses and appears to be infectious. This particle is a complex, double-layered sphere, 42 nm in diameter with an electron-dense core 28 nm in diameter (Fig. 65-3). Negative stain can penetrate the 42-nm particles, revealing a 7-nm outer layer or envelope surrounding the 28-nm core. Occasional particles lack the core and have a translucent center. The core structures can be isolated by reacting Dane particles with sodium dodecyl sulfate. **Spherical particles** with an average diameter of about 22 nm (range, 16–25 nm) are most numerous. These particles are unlike any known virus. Their surfaces appear to have a subunit structure similar to that of Dane particles, but no uniform symmetry is discernible. **Filamentous forms,** about 22 nm in diameter and ranging in length from 50 nm to greater than 230 nm, are also plentiful and have a similar surface structure. When the filaments are mixed with a nonionic detergent they form spheres that are morphologically and antigenically similar to the 22-nm particles, suggesting that the filaments are aggregates of the spherical particles.

The ease with which the above forms can be observed is a reflection of their amazing abundance in patients' serum. In one serum sample, for example, 3×10^{13} 22-nm particles/ml were counted, with 1/15 as many filaments, and 1/1500 as many 42-nm particles.

Physical and Chemical Characteristics. The abundance of viral structures has made physical and chemical analyses possible despite the inability to propagate the virus in cell cultures (Table 65-1). Purified **hepatitis B surface Ag(HB$_s$Ag),** which is present mostly in 22-nm spherical particles, consists of protein, lipid, and small quantities of carbohydrate; neither RNA nor DNA is detectable. The lipids are similar to those found in enveloped viruses. Several polypeptides, including two glycoproteins, have been identified. However, the immunologic and chemical similarities of some of the polypeptides indicate that they are not all primary viral gene products: some may be formed by proteolytic cleavage as well as by glycosylation.

Dane particles and isolated 28-nm cores contain a partially double-stranded circular DNA. The DNA molecules have a uniform length of 0.78 ± 0.09 nm, which corresponds to about 3600 nucleotides. Many Dane particles, and perhaps all, contain a DNA in which each strand has a gap of variable length and therefore each strand has a free 3' and 5' terminus and a single-stranded

region (Fig. 65-4). One of the strands is usually almost complete, whereas the shorter strand varies in different molecules, leaving a single-strand region of 10%–60% of the length of the DNA. This DNA is smaller than that of any other known virus containing double-stranded DNA. The DNA of the 42-nm Dane particle and of the isolated core is associated with a DNA polymerase which is dependent on the four deoxynucleoside triphosphates and Mg^{2+}, but its activity does not require an added DNA or RNA primer. The DNA polymerase of the particle can complete the single-stranded regions to form a complete double-stranded DNA molecule but the biologic utility of these findings is unknown (see Viral Multiplication). The DNA polymerase reaction increases the size of the DNA by about 30%.

Antigenic Structure. The 42-nm enveloped particle containing DNA, which is probably the hepatitis B virion, contains two distinct major Ags, the **surface Ag (HB$_s$Ag)** and the **core Ag (HB$_c$Ag).** The HB$_s$Ag is the only Ag on the two other hepatitis B particles circulating in the blood, i.e., the 22-nm spheres and the filamentous particles (Fig. 65-4).

The HB$_s$Ag contains several antigenic determinants: a **group-specific determinant, a,** which is present in all HB$_s$Ag-containing material, and two sets of generally mutually exclusive type-specific determinants, *d* or *y* and *w* or *r*, which represent sets of alleles of two independent genetic loci. In addition, four variants of the *w* allelic determinant have been observed. Thus, a number of phenotypic combinations or subtypes are possible (Table 65-2). A few sera containing surface antigenic reactivity of mixed subtype (*adyw* and *adyr*) have also been reported; whether these arise from mixed HBV infection or unusual phenotypic mixing is unclear.

The subtypes of HB$_s$Ag are not associated with different biologic activities of the viruses, and no clinical differences have been observed in the infections caused by the various subtypes. These subtypes are valuable as markers for studies of the epidemiologic behavior of the virus. For example, a study of several large epidemics in genetically unrelated populations has demonstrated that the subtype antigenic determinants are encoded in the viral genome and are not affected by the host.

The **HB$_c$Ag** is of a **single antigenic type** and is only found in the core of the 42-nm Dane particle and in free core particles present in the nuclei of hepatocytes from infected livers. Hence to detect the HB$_c$Ag in plasma one must disrupt the enveloped 42-nm particles with lipid solvents to expose the core Ag, whereas the HB$_s$Ag can be readily detected in serum during acute hepatitis B and in chronic carriers.

Some HB$_s$Ag-positive sera contain an additional specific Ag, designated **e (HB$_e$Ag),** which is considerably smaller than the HB$_s$Ag particle (about 12S); it is also distinct from the viral core Ag and the viral DNA polymerase. Nevertheless, HB$_e$Ag appears to be specific for hepatitis B virus infection. Two subdeterminants

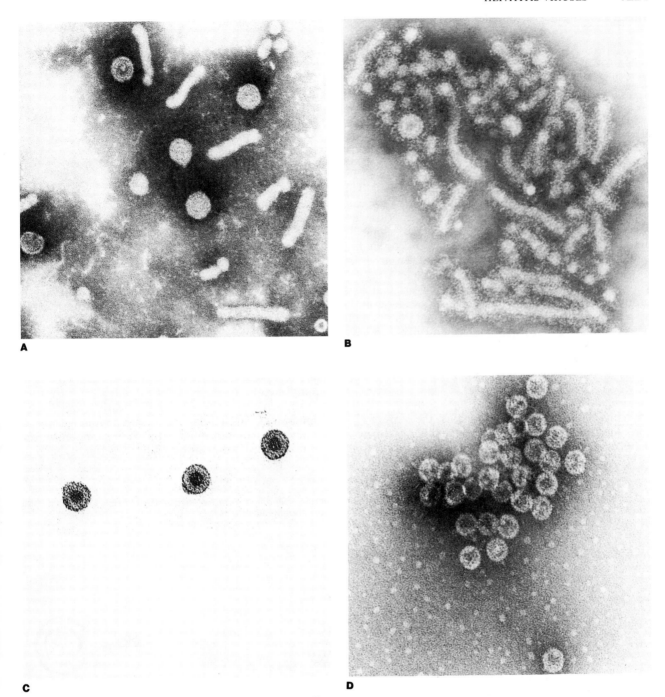

FIG. 65-3. Electron micrographs of negatively stained forms of hepatitis B virus (HBV). **A.** Dane particles, filaments, and 22-nm HB$_s$Ag particles present in the serum of a patient with active infection. (×150,000) **B.** Immunoelectron microscopy of Dane particles, filaments, and 22-nm HB$_s$Ag particles aggregated by serum from a late-convalescent patient. (×150,000). **C.** Purified Dane particles stained with uranyl acetate to show the DNA core. (×200,000). **D.** Cores of HBV isolated from the liver of a patient with active hepatitis. (×200,000). (**A–C,** courtesy of J Gerin, Oak Ridge National Laboratory; **D,** Kaplan PM et al: J Virol 17:885, 1976)

TABLE 65-1. Physical and Chemical Characteristics of Hepatitis B Viral Forms

	22-nm Spheres	Filamentous forms	Dane particles (42-nm spheres)	Core particles
Buoyant density (CsCl)	1.20 g/cm³	1.20 g/cm³	1.25 g/cm³	1.36 g/cm³
Sedimentation coefficient	54S		58.5S	110S
Protein	+		+	+
Glycoprotein	+		+	0
Lipid	+		+	0
Nucleic acid	0		Circular, interrupted double-stranded DNA	Circular, interrupted double-stranded DNA

(HB$_e$Ag/1 and HB$_e$Ag/2) have been detected. The HB$_e$Ag, in part bound to immunoglobulins, appears during the incubation period of acute B hepatitis, just after the appearance of HB$_s$Ag and prior to clinically apparent liver injury. Antibodies to HB$_e$Ag occur frequently in HB$_s$Ag-containing serum and even in serum having anti-HB$_s$ Abs. The function of the e Ag is unknown, but epidemiologic evidence suggests that it is probably a viral gene product whose presence in serum is associated with active viral replication and liver pathology. Patients with HB$_e$Ag are those most likely to have active disease and to be efficient transmitters of infection. But the presence of anti-HB$_e$Abs does not uniformly indicate a good prognosis, and serum from such patients may even be infectious.

Immunity. Antibodies directed against the surface and core Ags appear at different times and show different patterns of disappearance (Fig. 65-5). Antibodies to HB$_c$Ag increase rapidly during the early phase of clinical disease, and their appearance seems to be unrelated to recovery. These Abs are also consistently present in chronic carriers of HB$_s$Ag, and they may be a sensitive indicator of continued viral replication.

Cell-mediated immunity to HB$_s$Ag appears near the end of the acute phase of hepatitis, and its increase is correlated with the disappearance of the circulating Ag. In contrast, Abs to HB$_s$Ag do not appear until months after termination of the clinical illness (Fig. 65-5). It is also noteworthy that in chronic HB$_s$Ag carriers specific cell-mediated immunity is generally decreased and circulating anti-HB$_s$Abs are not demonstrable, but electron microscopic examination of serum detects HB$_s$Ag–Ab complexes.

An increase in Abs to HB$_s$Ag is clearly correlated with immunity. Thus human γ-globulin containing these Abs is of value in preventing hepatitis B; and immunization in humans, using heat-inactivated serum containing

FIG. 65-4. Schematic drawing of forms of HBV and the DNA structure. (Modified from Robinson WS, Lutwick LI: N Engl J Med 295:1232, 1976. Reprinted by permission from the New England Journal of Medicine)

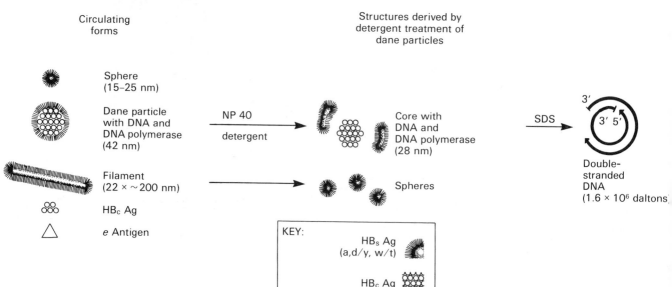

TABLE 65-2. Phenotypic Combinations or Subtypes of HBs Antigens Observed in Patients

Group determinant	Subtype determinant	
a	d/y	w/r
a	y	w_1
a	y	w_2
a	y	w_3
a	y	w_4
a	y	r
a	d	w_2
a	d	w_4

HB$_s$Ag, induces formation of Abs to HB$_s$Ag, accompanied by immunity to challenge. In addition, immunization of chimpanzees with purified HB$_s$Ag results in resistance to challenge with HBV-infected serum. (See Prevention and Control, p. 1227.)

Viral Multiplication. Without a susceptible cell for viral propagation, or even a convenient laboratory animal, the steps in viral replication are unexplored. The scanty data on viral multiplication come from examination of hepatocytes from infected humans or chimpanzees. Immunofluorescence detects HB$_c$Ag and electron microscopy reveals viral cores (nucleocapsids) in infected cell nuclei; HB$_s$Ag is abundant in the cytoplasm, particularly associated with membranes of the endoplasmic reticulum. This presence of the HBV surface Ag in the membranes of infected cells suggests that the putative virion, i.e., the Dane particle, acquires its envelope by budding through cytoplasmic membranes, like other enveloped nucleocapsids.

Viral DNA appears to replicate in the nucleus. The role of the core DNA polymerase, however, is unclear: it may be utilized to initiate DNA replication, which would be a unique, previously unknown strategy; the polymerase protein may also function as a structural nucleocapsid protein. Alternatively, because the polymerase is abundant in the nucleus, it may be fortuitously incorporated in the nucleocapsid during the assembly of incompletely replicated DNA molecules into defective particles. Overall, the process of viral multiplication must be highly inefficient since there is an excessive biosynthesis of the surface Ag in proportion to the number of 47-nm Dane particles made. This excess leads to the inordinate quantity of HB$_s$Ag particles in the blood (as much as 200μg of Ag/ml).

Although it has not been possible to study replication of HBV experimentally, a model for study may be possible since a virus, which appears to be in the same family as HBV, has been discovered to infect woodchucks in nature. The woodchuck hepatitis virus, which was found in the blood and livers of animals with chronic hepatitis, has the same morphologic forms as HBV, and like HBV the Dane-like particles contain DNA polymerase and partially double-stranded circular DNA (Fig. 65-4). However, the two hepatitis viruses clearly differ: they are immunologically unrelated; and their DNAs have only 3% to 5% nucleotide homology.

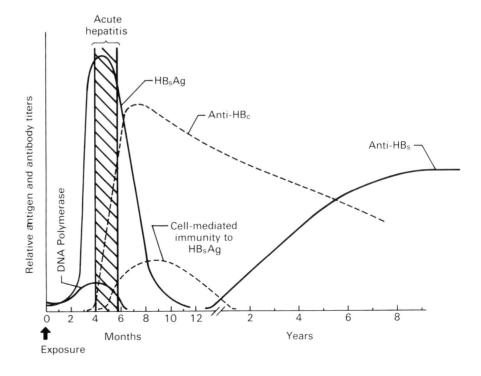

FIG. 65-5. Diagram of the time course of circulating HBV Ags and the immune response. Note that HB$_c$Ag is not found free in the plasma.

PROPERTIES OF HEPATITIS C VIRUS

The sensitive immunologic and electron microscopic methods used to detect the viruses and Ags of hepatitis A and B viruses have made it possible to identify at least two other viruses. One of these, hepatitis C has been transmitted to chimpanzees and the virus has been identified by immune electron microscopy. The virions are non-enveloped icosahedrons about 27 nm in diameter and like HAV, resemble picornaviruses in morphologic and some physical characteristics. The hepatitis C virions do not crossreact immunologically with Ags of hepatitis A or B viruses.

PATHOGENESIS

The response of humans to infection with hepatitis viruses ranges from inapparent infection and nonicteric hepatitis to severe jaundice, liver degeneration, and death; the disease is often debilitating and convalescence is prolonged. Though clinical differences in type A and type B hepatitis have been described, the acute diseases may be clinically indistinguishable, so that the two infections are often differentiated (Table 65-3) mainly by the routes of infection, the length of the incubation periods, and the laboratory identification of specific viruses, Ags, and Abs.

Hepatitis A virus (HAV) usually enters by the oral route and multiplies in the gastrointestinal tract (probably in the epithelium, although, as in poliovirus infection, mesenteric lymph nodes may also be involved). Viremia eventually occurs, and virus spreads to cells of the liver, kidney, and spleen. Virus can be detected in the feces and duodenal contents, and also in the blood and urine during the preicteric and the initial portion of the icteric phases (Fig. 65-6). When quantitative immune electron microscopy is used, HAV is first detected during the **preicteric period** and attains its highest concentration in the feces prior to the appearance of jaundice (Fig. 65-6). Indeed, the onset of jaundice usually heralds the approaching termination of viral shedding. More sensitive human transmission studies, however, indicate that virus may be shed in the feces for slightly longer periods. When virus is present in the feces it is also found in liver cells and in the bile. Antibodies to HAV appear as the viral titer decreases and liver damage becomes apparent (Fig. 65-6). This timing suggests that liver damage may be effected, at least in part, by immunologic mechanisms (although HAV–Ab complexes have not been detected in either chimpanzee or man).

Hepatitis B virus (HBV) enters predominantly by the parenteral route, but its primary site of replication is unknown. However, HB_cAg is present in the nuclei of hepatocytes as early as 2 weeks after experimental infection in chimpanzees, who develop a disease closely akin to that in man. During the latter half of the incubation period (Fig. 65-5), 5–8 weeks after infection, the blood becomes infectious (e.g., if used for transfusion), and it contains detectable HB_sAg and viral DNA polymerase. The HB_sAg has also been identified in bile, urine, semen,

TABLE 65-3. Differentiating Characteristics of Hepatitis Types A and B

Property	Type A	Type B
Usual transmission	Fecal-oral	Parenteral inoculation†
Characteristic incubation period*	15–40 days	60–160 days
Type of onset	Acute	Insidious
Fever >38°	Common	Uncommon
Seasonal incidence	Autumn and winter	Year-round
Age incidence	Commonest in children and young adults	All ages; commonest in adults
Size of virus	27 nm	42 nm
Viruses in feces	Incubation period and acute phase	Not demonstrated
Virus in blood	Incubation period and acute phase	Incubation period and acute phase; may persist for years
Appearance of HBsAg	Absent	30–50 days after infection
Detection of HBsAg		Blood (less often in feces, urine, semen, and bile)
Duration of HBsAg		60 days to years
Prophylactic value of γ-globulin	Good	Good if titer of anti-HBsAg is high

*Considerable overlapping in the duration of incubation periods for types A and B has been noted in volunteers as well as in patients during epidemics, i.e., as long as 85 days for type A hepatitis and as short as 20 days for type B hepatitis.

†Parental injection is probably not the predominant mode of transmission in developing countries where the means of spread is unknown.

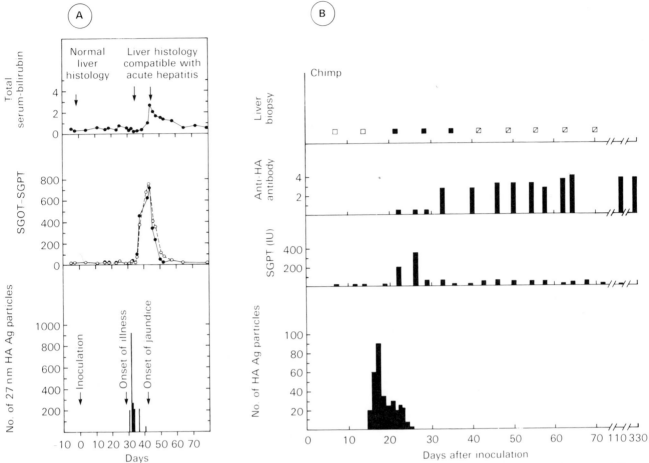

FIG. 65-6. Temporal relationship of clinical illness, liver pathology, serum enzyme alterations, fecal shedding of HAV Ag, and Ab response in experimental HAV infection (A) in man and (B) in chimpanzee. SGOT and SGPT = Serum glutamic oxaloacetic and pyruvic transaminases. Liver histology: □-normal; ■-acute hepatitis; ▨-resolving hepatitis. (Dienstag JL et al: Lancet 1:765, 1975; Dienstag JL et al: J Infect Dis 132:532, 1975)

feces, and nasopharyngeal secretions. It is striking that the HB$_s$Ag and DNA polymerase usually begin to decline in the blood before acute symptoms disappear and are not detectable by the time liver functions return to normal.

Whether immune reactions play a role in the pathogenesis of hepatitis B is still unclear but, as noted above, anti-HB$_c$ Abs appear at the onset of disease; HB$_s$Ag is present in large quantity weeks before anti-HB$_s$ Abs appear; and cell-mediated immunity to HB$_s$Ag becomes detectable during the active disease, and its increasing development correlates with the fall in HB$_s$Ag titer (Fig. 65-5).

Both HAV and HBV infections primarily affect the liver. When jaundice occurs it is usually preceded by anorexia, malaise, nausea, diarrhea, abdominal discomfort, fever, and chilliness. This preicteric phase may last from

2 days to 3 weeks. The icteric stage of hepatitis A (infectious) usually has an abrupt onset with a sharp rise in temperature, whereas that of hepatitis B (serum) characteristically appears more insidiously and with less fever. Nevertheless, *hepatitis B is ordinarily a more severe disease*, the fatality rate sometimes being as high as 50%, whereas in hepatitis A fatalities rarely exceed 1% (although convalescence is often prolonged). These differences, however, may not reflect only the characteristics of the viruses, for persons receiving blood or plasma transfusions are usually ill at the time of inoculation.

It has been proposed that HBV infection is causally related to hepatocarcinoma based on the evidence that: 1) there is a close correlation between the geographic distribution of hepatitis B and the incidence of hepatocarcinoma; and 2) HB$_s$Ag is detected in patients with hepatocarcinoma with an unusually high frequency. The striking association between the woodchuck hepatitis virus infections and hepatocarcinomas in these animals adds support to the hypothesis that HBV is an oncogenic virus.

Liver biopsies obtained during the course of the illness have revealed early cloudy swelling and fatty metamorphosis at the time that clinical symptoms begin, and diffuse parenchymal destruction by the time jaundice has developed. No inclusion bodies are seen in affected cells; intracellular HB$_s$Ag, however, can be detected by immunofluorescence early in the course of type B hepatitis, and occasionally 20- to 30-nm viruslike particles (probably the nucleocapsids or cores) are seen in liver biopsies. The degeneration of cells is not localized to any one part of the liver lobule, in contrast to the findings in yellow fever (Ch. 62) or chemical hepatitis. With recovery, hepatic cells regenerate; scar tissue (cirrhosis) develops only after extensive or long-standing destruction of cells, which is primarily associated with hepatitis B. In fatal infections, the liver parenchyma is often almost completely destroyed (**acute yellow atrophy**).

In 10%–20% of adult patients and about 35% of children HB$_s$Ag persists in the blood for extended periods, but less than half of these become **chronic Ag carriers.** More than half of the chronic carriers continue to manifest biochemical and pathologic evidence of chronic liver disease. Although the mechanism of persistent infection with antigenemia is unexplained, defects in cell-mediated immunity (e.g., that expressed to HB$_s$Ag) or other forms of immunodeficiency seem likely.

LABORATORY DIAGNOSIS

Immunologic technics provide the most sensitive and economic approach for detecting viral Ags and Abs. Immune electron microscopy can be used to identify HAV in feces and HBV in blood or liver biopsy tissue. If a liver biopsy is obtained, both viruses can be identified by immunofluorescent technics. The HAV Ag or Abs can be easily assayed by CF, immune adherence, or radioimmunoassay (RIA). The CF Abs are detectable earliest. The solid phase RIA is the most sensitive of these technics, but the immune adherence titration (which requires purified HAV Ag or standardized serum, guinea pig complement, and human group O RBCs) is simpler, less costly, and faster, as well as being quite sensitive and specific. The Abs in serum from clinical cases also neutralize HAV in chimpanzees and marmosets but this procedure is too expensive and time-consuming.

A number of tests have been devised to detect and measure HBV Ags, particularly the HB$_s$Ag. The most sensitive technics now available are RIAs with either a micro–solid-phase method or a double-Ab immunoprecipitation procedure. The passive hemagglutination assay (using RBCs coated with either HB$_s$Ag or anti-HB$_s$ Abs) and the enzyme immunoassay are also accurate and very sensitive methods for quantitating either Ag or Ab. Agar gel diffusion and countercurrent immunoelectro-

phoresis are of considerable value for identifying the subtype Ags and the e Ag. Since HB$_c$Ag does not circulate in the blood its measurement is not useful for clinical diagnosis. Detection of Abs to HB$_c$Ag is valuable, however, since their presence appears to reflect viral replication. Similarly, HB$_e$Ag has been associated with viral infectivity and therefore signifies specific HBV infection.

EPIDEMIOLOGY

Predominantly, the **type A hepatitis virus** (of the **short incubation disease** previously called infectious hepatitis) is spread by ingestion (particularly in epidemics), and **type B hepatitis virus** (of the **long incubation disease** formerly termed serum hepatitis) is disseminated by parenteral inoculation. However, the viruses of both diseases can be transmitted by either the oral or the parenteral route. The recently recognized **hepatitis C viruses** may have a relatively short incubation period, but their major spread is also by parenteral inoculation.

Hepatitis A. This form mimics poliomyelitis in many of its epidemiologic features. When environmental factors favor widespread intestinal–oral transmission the disease is **endemic,** and infection (usually mild or inapparent) occurs in the very young. Under these conditions the disease in adults is uncommon and epidemics are rare. On the other hand, under good sanitary conditions spread of the virus is restricted, and adulthood is frequently attained without immunity. In such nonimmune populations, especially in military groups, **epidemics** are likely, and the source of virus can usually be traced to **contamination of water or food** by infected humans.

The danger of HAV dissemination from an infected person is greatest during the latter part of the incubation period, when viral shedding in the feces is greatest but is unrecognized because jaundice is not yet present (Fig. 65-6). Because **subclinical infections,** particularly in children, often predominate, secondary person-to-person spread and contamination of food and drink are common. Such transmission is particularly favored by the unusual stability of hepatitis A virus and its notable resistance to disinfectants, such as chlorine at ordinary concentrations in water. Not only contaminated water but also **shellfish** that live in it (and concentrate the virus) may be sources of infection: for example, raw oysters and clams obtained from polluted waters have been the origin of numerous epidemics throughout the world.

Subhuman primates, particularly chimpanzees, are the only known natural nonhuman hosts. Hepatitis has developed among handlers 3–6 weeks after infected animals arrived in the United States from Africa. Apparently the young animals were infected by man after capture and subsequently excreted HAV in their feces, and hence infected their handlers (some animals showed clinical manifestations of disease).

Hepatitis B. This is readily distinguishable from hepatitis A not only by morphologic and immunologic properties but also on epidemiologic evidence, such as the long incubation period, parenteral transmission, increasing incidence with age, and nonepidemicity (Table 65-3).

Despite sensitive immunologic technics to detect viral carriers by screening the blood of donors for HB_sAg, blood and its products continue to be major sources of hepatitis in the United States. However, only a portion of these cases is caused by undetected hepatitis B virus. Indeed, about 75% of the cases now transmitted by **transfusion*** in the United States are caused by **type C** and other **non-A, non-B hepatitis viruses.** Hepatitis A virus is not a common cause of transfusion hepatitis because viremia is usually brief and chronic viremia occurs rarely, if ever.

The unrecognizable chronic carriers of hepatitis B or C virus are the pernicious sources of infection, spreading virus via transfusions of infected blood, plasma, or convalescent serum; via contaminated fibrinogen; and via inadequately sterilized syringes, needles, or instruments (medical and dental) containing traces of contaminated blood. The last source, which includes common stylets for blood counts and syringes for drawing blood, has been essentially eliminated in many parts of the world by the use of disposable instruments. However, needles used in tattooing, and communal equipment used by narcotics addicts, are still a common means of viral dissemination. Injection of as little as 10^{-6}–10^{-7} ml of contaminated blood may transmit infection, as predicted by the electron microscopic finding of as many as 10^7 Dane particles per ml of blood.

The large amount of virus and its Ags in blood, and the demonstration of HB_sAg (and the probable presence of virus) in other body fluids as well, also explain why type B hepatitis is an **occupational disease** of health professionals: dentists, physicians, nurses, and ward personnel who are frequently exposed to the unrecognized chronic carriers; laboratory workers who handle blood; and technicians who process human plasma and blood products.

Health personnel who work in hemodialysis units and in cancer therapy wards, where the chronic carrier rates among patients are high, appear to be particularly vulnerable. It is striking that infected professionals serving in renal dialysis units, for example, develop acute hepatitis, whereas the patients, who have various forms of immunodeficiencies, do not manifest clinical disease. The mechanism by which virus is transmitted from carriers to these healthy workers is not entirely clear. Many infected workers do not recall any accidental parenteral injections, although possible viral entrance through skin abrasions cannot be ignored. It should be recalled that HB_sAg is detectable in the saliva, urine, and feces of

* Overt hepatitis follows about 1% of blood transfusions in the United States. On the assumption that even more inapparent infections are produced, it has been estimated that 2%–3% of the adult population carries a hepatitis virus.

chronic carriers, as well as in blood and its products, so that entrance of the virus through membranes of the eye or mouth must be considered a plausible route of infection.

Additional epidemiologic observations provide further evidence that transmission of HBV by means other than parenteral injection is also likely. It is now clear that HBV has disseminated and persists throughout the world, even in remote and insular localities where medical care is primitive and blood and its products are not commonly used for therapeutic or prophylactic purposes. Hence nonparenteral transmission of virus must occur. Furthermore, family clusters of type B hepatitis are being observed with increasing frequency; these cases are grouped around an index case with whom family members have had close person-to-person contact but no known exchange of blood. As noted above, experimental oral transmission has been demonstrated, and sexual transmission is also probable. Neonates may be infected during gestation by placental transmission (cord blood often contains HB_sAg), or at the time of delivery, or in the postnatal period.

Although hepatitis B is usually sporadic, epidemics may occur when many samples of serum or plasma are pooled. For example, in the early 1940s more than 28,000 cases of serum hepatitis in American military personnel resulted from the use of yellow fever vaccine containing contaminated human serum to stabilize the live virus.

Hepatitis C. Sensitive immunologic technics used to screen blood donors for HBV Ags and to establish the diagnosis of viral hepatitis have sharply reduced the incidence of hepatitis B infections, but they have also revealed another agent (or agents) as a cause of posttransfusion hepatitis. **Non-A non-B hepatitis** (for convenience termed **hepatitis C**) is now the major form of posttransfusion hepatitis in the United States. The incubation period is usually 4 to 5 weeks; however, periods as short as 14 days and as long as 11 weeks have been recorded (which further implies that more than one virus is responsible for non-A non-B hepatitis). Transmission by means other than blood is still unknown since specific viral Ags have not yet been identified.

PREVENTION AND CONTROL

No specific therapy is available for any of the viral hepatitides. In all proved or suspected cases of viral hepatitis great care should be taken in the disposal of feces and of all syringes, needles, plastic tubing, and other equipment used for blood sampling and parenteral therapy. Whenever possible, disposable equipment (including needles and plastic syringes and even thermometers) should be used in hospital and office practice. A syringe, once used, should not be reused with a fresh needle, even merely to obtain a blood specimen. Nondisposable equipment and supplies (e.g., dishes and bed clothing) should be autoclaved at 15 lb pressure (121° C), boiled in water for at least 20 min, or heated to 180° for 1 h in a sterilizing oven.

Subjects giving a history of jaundice or with detectable HB_sAg in their blood should not be used as blood donors. Blood HB_sAg has been reported in about 0.3% of blood donors in New York and 0.2%–1.2% of different groups of blood donors in Tokyo; the carrier rate among commercial donors, particularly drug addicts, is 3–10 times higher than among volunteers. However, the incidence of hepatitis C Ags in the population is unknown. Because minute amounts of contaminated plasma can initiate infection, the practice of pooling plasmas should be avoided: then plasma from an infected individual, used unwittingly, will infect only one person.

The protection of individuals exposed to patients, carriers, or contaminated blood requires additional consideration. Since hepatitis A patients usually shed virus only briefly after jaundice appears (Fig. 65-2), and viremia is transient, these patients in hospitals and at home constitute a hazard for their contacts for only a short period. In contrast, hepatitis B (and probably hepatitis C) patients are potential sources of virus for transmission over a prolonged period, whether they are suffering from acute icteric hepatitis or are chronic carriers without clinical liver disease. Accordingly, all close contacts of carriers or possible carriers of HBV must take every precaution to prevent exposure to blood (and to objects potentially contaminated with blood—e.g., toothbrushes, razors) and also to body excreta. These precautions are of greatest significance for personnel working in hemodialysis units, intensive care units, and custodial mental institutions; for dentists; for technicians in clinical laboratories and blood processing facilities; and for close family contacts (particularly spouses).

Type A hepatitis may be prevented by **passive immunization.** Pooled human γ-globulin* reduces the incidence

---

* Infection with hepatitis A virus is so widespread that the serum of many adults contains anti-HAV Abs and a relatively high titer is present in concentrated γ-globulin pools. HAV and HBV have been eliminated from these pools, along with the fibrinogen, in the usual cold ethanol fractionation of plasma.

The recommended dose of γ-globulin is 0.06–0.15 ml/lb of body weight, administered by intramuscular injection as soon after exposure as possible.

of icteric disease (but not of infection) when given early in the incubation period. Initially, inconsistent results were obtained in preventing hepatitis B with pooled γ-globulin. However, the advent of assays for anti-HB_s Abs has made it apparent that immune γ-globulin containing a high titer of anti-HB_s Abs is partially effective; i.e., in different recipients it prevents disease, decreases the severity of hepatitis, or markedly prolongs the incubation period. Therefore the use of γ-globulin containing anti-HB_s Abs is recommended for prophylaxis in exposed persons and for those who are at a high risk of acquiring HBV infection owing to their occupation.

The identification of the probable etiologic agents of viral hepatitis and the ability to propagate these viruses in nonhuman primates have increased optimism concerning the production of effective vaccines. Moreover, the relatively large amounts of HB_sAg in the serum of chronic carriers make it possible to purify the Ags that are essential for inducing neutralizing Abs, and allow the use of this material for immunization after inactivation of any infectious virus present. Encouraging results have been obtained in humans and in chimpanzees using HB_sAg purified from human serum. However, despite purification, HB_sAg must be used with caution in humans since host cell glycoproteins and lipoproteins are probably present and may induce an immune response against the recipients' cells. This is of particular concern since cell-mediated immunity directed against hepatocytes may be present during acute infection (Fig. 65-5).

The continued prevalence of hepatitis and its serious consequences enhance the challenge to improve methods for replicating these recalcitrant viruses. The development of safe and effective vaccines for wide use may depend upon the progagation of these viruses in cell cultures. Indeed, prospects for an inactivated HAV vaccine are brightened by the finding that the virus can be propagated in a limited range of cell cultures. However, if a convenient method is not developed to produce abundant HB_sAg, either by viral propagation in cell culture or by recombinant DNA technology, we may eventually reach the unique position in which a successful vaccine had sufficiently reduced the source of Ag for humans to make continuation of widespread HBV immunization difficult.

SELECTED READING

BOOKS AND REVIEW ARTICLES

HOWARD CR, BURRELL CJ: Structure and nature of hepatitis B antigen. Prog Med Virol 22:36, 1976

LEBOUVIER GL, MCCOLLUM RW: Australia (hepatitis-associated) antigen: physicochemical and immunological characteristics. Adv Virus Res 16:357, 1970

MAYNARD JE: Hepatitis A. Yale J Biol Med 49:227, 1976

MELNICK JL, DREESMAN GR, HOLLINGER FB: Approaching the control of viral hepatitis type B. J Infect Dis 133:210, 1976

ROBINSON WS, LUTWICK LI: The virus of hepatitis, type B. N Engl J Med 295:1168, 1232, 1976

SZMUNESS W: Recent advances in the study of the epidemiology of hepatitis B. Am J Pathol 81:629, 1975

WATERSON AP: Infectious particles in hepatitis. Annu Rev Med 27:23, 1976

SPECIFIC ARTICLES

GROB PJ, JEMELKA H: Fecal SH (Australia) antigen in acute hepatitis. Lancet 1:206, 1971

HOLMES AW, WOLFE L, DEINHARDT F, CONRAD ME: Transmission of human hepatitis to marmosets: further coded studies. J Infect Dis 124:520, 1971

LEBOUVIER GL: The heterogeneity of Australia antigen. J Infect Dis 123:671, 1971

MAUPAS P, GOUDEAU A, COURSAGET P, DRUCKER J, BAGROS P: Immunisation against hepatitis B in man. Lancet 1:13647, 1976

PROVOST PJ, HILLEMAN MR: Propagation of human hepatitis A virus in cell culture in vitro. Proc Soc Exper Biol Med 160:213, 1979

SUMMERS J, SMOLEC JM, SNYDER R: A virus similar to hepatitis B virus associated with hepatitis and hepatoma in woodchucks. Proc Nat Acad Sci USA 75:4533, 1978

chapter

66

ONCOGENIC VIRUSES

The first tumor-producing (oncogenic) virus was discovered in 1908 by Ellerman and Bang, who demonstrated that seemingly spontaneous leukemias of chickens could be transmitted to other chickens by cell-free filtrates. Later (1911) Rous found that a chicken sarcoma, a solid tumor, can be similarly transmitted. These virus-induced tumors were then considered by many as a biologic curiosity, either not "true" cancers or perhaps a peculiarity of the avian species. These notions, however, were shaken when viral induction was also demonstrated with cutaneous fibroma and papilloma of wild rabbits (by Shope, in 1932) and the renal adenocarcinoma of the frog (Lucké, 1934).

The later discovery of virus-induced tumors in mice provided a particularly suitable system for experimental work. In 1936 Bittner demonstrated that a spontaneously occurring mouse adenocarcinoma is caused by a virus transmitted from the mother to the progeny through the milk, and in 1951 Gross discovered the first of many virus-induced murine leukemias. These studies revealed that the viral etiology of a cancer can easily go unrecognized for several reasons: 1) cancer can be caused by ubiquitous viruses, which may easily be considered as innocuous bystanders; 2) with some oncogenic viruses most viral particles can infect cells without inducing cancer; 3) the disease may not develop until long after infection; and 4) the cancers do not seem contagious, either because the efficiency of cancer production is low or because the method of transmission of the virus is inapparent (e.g., through the embryo or the milk).

The murine cancer viruses noted above were found to contain RNA. Further impetus to investigations of tumor viruses arose from the later discovery that several DNA-containing viruses also cause cancer in mice and other rodents: polyoma virus from leukemic mice, simian virus 40 (SV40) (as a passenger virus in cultures of rhesus monkey kidney cells), and finally **human** adenoviruses and herpesviruses. Moreover, by this time the study of bacterial lysogeny had clearly shown that the genetic material of viruses can become permanently integrated with that of the host. These realizations led to an explosive development of interest in viral carcinogenesis.

Soon the oncogenic effect of several viruses was demonstrated also in tissue cultures, in the form of **cell transformation** (Ch. 49). Studies in this model system led shortly to a shattering conclusion: *a virus that has induced cancer is often no longer recognizable in the culture by its infectivity,* or by its antigenicity. Traces could, however, be found in the form of viral DNA, RNA, and new Ags, in ways reminiscent of lysogeny. It thus became clear that time-tested technics and approaches for the identification of viral agents of disease may not be adequate in the search for viral agents of human cancer.

UNITY OF ONCOGENIC VIRUSES

The discovery of many new oncogenic viruses in the 1960s revealed a puzzling distribution: Oncogenic members were found in most classes of DNA-containing viruses, but in only one family of RNA-containing viruses, the group now called **retroviruses** (Table 66-1). The replication of retroviruses also displayed a sensitivity, peculiar for RNA viruses, to agents that interfere with DNA replication or transcription. This property was explained when Temin and Baltimore each discovered that the oncogenic retroviruses, unlike other RNA viruses, replicate through a DNA-containing intermediate, made by a **reverse transcriptase** (RNA-dependent DNA polymerase). This discovery brought unity into the field of oncogenic viruses, suggesting that *oncogenesis is an attribute of viral DNA.*

These developments, and the problems they raise, will be analyzed below by examining the characteristics of cell transformation induced by several viruses. Though none of the findings were obtained in humans, some were obtained in cultures of human cells, or with viruses that can cause human infections, raising important implications for human cancer.

DNA-CONTAINING TUMOR VIRUSES

These viruses are outstanding models for understanding the mechanisms of viral carcinogenesis. Viruses of the papova group (which includes polyoma and papilloma viruses) have been used most extensively in experimental work.

PAPOVA VIRUSES: POLYOMA AND RELATED VIRUSES

A virus isolated by Stewart and Eddy was found to produce neoplasia of different types when injected into

TABLE 66-1. Distribution of Oncogenic Viruses Among Animal Virus Families

Nucleic acid in virions	Viral group (or family)	Oncogenic viruses
RNA	Picornaviruses	None
	Togaviruses	None
	Orthomyxoviruses	None
	Paramyxoviruses	None
	Rhabdoviruses	None
	Coronaviruses	None
	Arenaviruses	None
	Retroviruses	Oncoviruses: leukosis viruses, sarcoma viruses*
	Reoviruses	None
DNA	Adenoviruses	Many types
	Papovaviruses	Polyoma virus, SV40, SV40-like human viruses, papilloma viruses
	Herpesviruses	Virus of neurolymphomatosis of chickens (Marek disease), Lucke's virus of frog renal adenocarcinoma, herpes simplex virus (cell transformation)†, Epstein–Barr virus (Burkitt lymphoma, nasopharyngeal carcinoma), cytomegalovirus†, primate herpesviruses
	Poxviruses	Fibroma virus
	Parvoviruses	None

*See Table 66-3.

†Transformation in vitro after ultraviolet irradiation.

newborn mice; it was therefore named **polyoma virus** (Gr. & L., agent of many tumors). The virus is widespread in mouse populations, both wild and in the laboratory; it is normally transmitted to animals after birth, through excretions and secretions. The structurally similar **SV40** was discovered by Sweet and Hilleman as an agent that multiplies silently in rhesus monkey kidney cultures (used for propagating poliovirus) but was found to produce cytopathic changes in similar cultures from African green monkeys (*Cercopithecus aethiops*). It was later shown to produce sarcomas after injection into newborn hamsters.

Subsequently human papova viruses were isolated: first from the brain of a patient with progressive multifocal leukoencephalopathy (PML);* next, others from patients with Wiskott-Aldrich syndrome (defects in cellular and humoral immunity, and reticulum cell sarcomas, due to an X-linked recessive allele); and then, frequently, from the urine of immunosuppressed individuals. These SV40-like viruses are designated by the initials of the persons from whom they were isolated (e.g., JC virus, from a PML patient; BK virus, from urine). These viruses, especially JC virus, produce tumors in hamsters

* It should be noted that viruses nearly identical to SV40 were also isolated from PML patients.

and transform hamster cells in vitro. Since 70% of humans have Abs to the SV40-like viruses, they may be the human equivalent of the widespread SV40 of rhesus cultures and polyoma virus of mice.

PROPERTIES OF THE VIRIONS

The virions of polyoma virus, SV40, and SV40-like human viruses are small naked icosahedrons with a diameter of 45 nm and with 72 capsomers (Ch. 46) containing a **cyclic double-stranded** DNA about 5 Kb long, associated with octamers of cellular histones (H2a, H2b, H3, H4) to form nucleosomes similar to those present in cellular chromatin (Ch. 49). The whole SV40 DNA and large parts of polyoma virus DNA have been sequenced.

Upon repeated serial passages at high multiplicity virions with **defective DNAs** accumulate. Some of these DNAs contain a **cellular DNA sequence** incorporated in the cyclic viral DNA; others are cyclic DNA molecules formed by **reiteration** (3- to 20-fold) of a small fragment of viral DNA, usually containing the origin of replication. Such cyclic molecules with several replication origins tend to outgrow the wild-type virus.

Hybridization experiments reveal a partial homology between the human viruses and SV40, but not with polyoma (however, SV40 and polyoma DNA sequences are related). Similarly, SV40 virions crossreact immunologically with BK virions but not with polyoma. Polyoma virus agglutinates guinea pig RBCs; BK and JC viruses agglutinate human group O RBCs.

All the viruses are very resistant to inactivation by heat or formalin; hence SV40 (from rhesus kidney cultures) survived in some early batches of formalin-killed poliovirus vaccine.

VIRAL MULTIPLICATION IN TISSUE CULTURE

These viruses can produce either a **productive** or a **nonproductive** infection. The outcome depends on the species of the cells and their physiologic state, because viruses with such small genomes depend heavily on cellular functions. Thus cells of certain **permissive** species (see Table 66-2) are killed by infection and yield virus, whereas those of **nonpermissive** species are **transformed** without virus production. In **semipermissive** cultures some cells are transformed while others are killed and yield virus (probably depending on the state of the cells). Permissive cells can be transformed by virus when viral multiplication is blocked by mutations or by DNA damage (see below).

Productive Infection. As with most DNA viruses, viral DNA replication and capsid assembly occur in the cell nucleus (capsid Ag is detected there by immunofluorescence—Fig. 66-1—and virions by electron microscopy); each nucleus can produce up to 10^7 viral particles. As with other naked viruses, release depends upon **dis-**

TABLE 66-2. Some Permissive and Nonpermissive Cells Used with Polyoma Virus and SV40

Cells	Virus	Type of Culture
Permissive	Polyoma virus	Secondary cultures of mouse embryo cells
		Primary cultures of mouse kidney cells
		3T3⎫ cell lines (mouse subcuta- 3T6⎭ neous tissue)
	SV40	Primary cultures of African green monkey kidney cells
		BSC-1⎫ CV-1 ⎬cell lines (African green VERO⎭ monkey kidney)
Nonpermissive or semipermissive	Polyoma virus	Secondary cultures of hamster embryo cells
		Secondary cultures of rat embryo cells
		BHK cell line (baby hamster kidney)
	SV40	Secondary cultures of hamster, rat, or mouse embryo cells
		3T3 cell line (mouse subcutaneous tissue)

FIG. 66-1. Fluorescence photomicrograph of cultured mouse kidney cells productively infected by polyoma virus. The accumulation of capsid Ag in the nuclei is revealed by its combination with fluorescein-conjugated Abs, which emit green fluorescence under ultraviolet light.

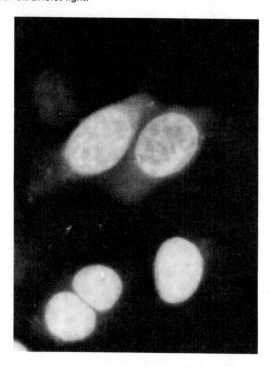

integration of the cells. Transcription of the supercoiled DNA occurs in **transcriptional complexes** and is carried out by the α-amanitin–sensitive host polymerase II (see Transcribing Enzymes, Ch. 49).

In the **early transcription,** genes for DNA replication are transcribed from the "early" DNA strand; in the **late transcription,** which begins after DNA replication has begun, genes for virion proteins are transcribed from the opposite, "late" DNA strand, at a much higher rate than the continuing early strand transcription. *Early transcription is autoregulated* because it is inhibited by a product of the early A gene (see Role of Viral Genes in Productive Infection, below).

Both early and late transcription appear to initiate near the origin of DNA replication (Fig. 66-2). Inefficient termination allows late transcription to go around the DNA circle even several times, generating long transcripts which are subsequently processed to mRNAs.

Viral Proteins (Fig. 66-3). The virions contain three late proteins: the main capsid protein (virion protein 1, VP1) and two minor proteins, VP2 and VP3 (VP3 being contained in VP2). In the infected cells three additional proteins are synthesized, collectively known as **T (tumor) antigens,** which are detected by immunofluorescence or immunoprecipitation with the serum of an animal bearing a large virus-induced tumor. Gel electrophoresis of the immunoprecipitate of infected cell extracts reveals three main bands: the nuclear **large T** protein (mol wt about 90,000) and **small T** (mol wt about 20,000) and, with polyoma virus only, the **middle T** (mol wt about

FIG. 66-2. Restriction enzyme site map of SV40 DNA (using *Hemophilus influenzae* I and II enzymes) and general transcription pattern. Letters **A** to **K** identify the fragments in order of decreasing size. The origin of divergent DNA replication is close to the C–A border. Both early and late transcriptions, on different DNA strands, begin close to the replication origin and continue beyond the boundaries of the early and late regions (**dashed lines**). The pattern of transcription is similar for polyoma virus.

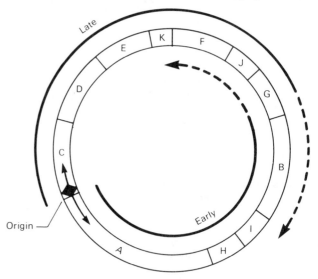

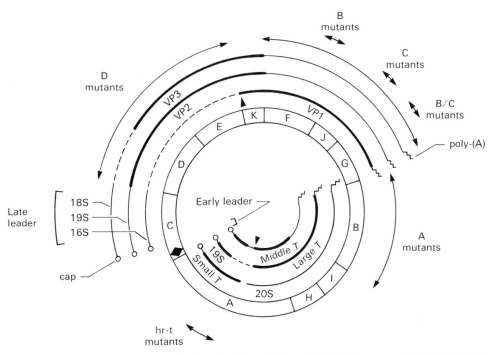

FIG. 66-3. SV40 mRNAs and proteins. **Heavy lines** indicate the translated parts of the mRNAs; **solid thin lines** are untranslated parts; **dashed lines** (introns) are parts removed by splicing. **Outer lines** indicate the location of various types of mutations. This map applies also to polyoma virus. The *hr-t* mutants located in the intron of the large T mRNA are known in polyoma virus; corresponding SV40 deletions are only defective in transforming abilities. The middle T is known only for polyoma virus. The two **black triangles** indicate shifts of the reading frame.

60,000) present in the plasma membrane. All three proteins share the same NH_2 end. Since T Ag is formed in the presence of arabinosyl cytosine (which inhibits the synthesis of viral DNA and of virion proteins) it is an "early" viral protein (see Ch. 50). The large T binds strongly to any double-stranded DNA, and especially to a segment of viral DNA near the origin of DNA replication and of early and late transcription. The middle T, like the transforming proteins of some retroviruses (see below) has protein kinase activity.

The infected cells also contain a perinuclear U Ag (see Fig. 66-8), a surface **tumor-specific transplantation Ag** (TSTA), and a **tumor-specific surface Ag** (TSSA), all of uncertain molecular nature (see Immunologic Relations, etc., below).

The polyoma virion proteins are synthesized on three late mRNAs (16S, 18S, 19S), all containing a common untranslated **(leader)** sequence derived from the 5' end of the late transcript, **spliced** to bodies of various length and location in the late region. The synthesis of VP2 (on the 19S mRNA) and VP3 (on the 18S mRNA) ends at a common terminator about halfway through the late region. VP1 is read on a different frame on the 16S mRNA and corresponds to the rest of the late region. The early proteins are again specified by two or three mRNAs with a common leader sequence: small T by a 20S mRNA containing a terminator, large T by a 19S mRNA from which the terminator has been spliced out,

and middle T by an mRNA with a different splicing, introducing a frame shift.

Thus, like small bacteriophages (Ch. 47), the small genome of these papovaviruses *expresses several distinct proteins from the same DNA segment.*

Viral **DNA replication** occurs in supercoiled replicative intermediates (Fig. 66-4) after an unusually long lag of 10–12 h. The lag is probably taken up by the **activation of cellular genes** essential for viral replication. Thus in crowded cultures infection stimulates the synthesis of cellular DNA and histones to a level comparable to that found in uninfected growing (i.e., uncrowded) cultures, and it also stimulates formation of enzymes involved in DNA synthesis: thymidine kinase, deoxycytidylate deaminase, and DNA polymerase α. The induced thymidine kinase is an enzyme normally found only in growing cells, showing that *the cells are converted to a growing state by the infection.*

The viral function that stimulates cellular DNA synthesis in crowded cultures may also be involved in transformation, because transformed cells have lost the regulatory mechanisms that inhibit DNA synthesis in uninfected crowded cultures (see Growth Regulation, Ch. 49).

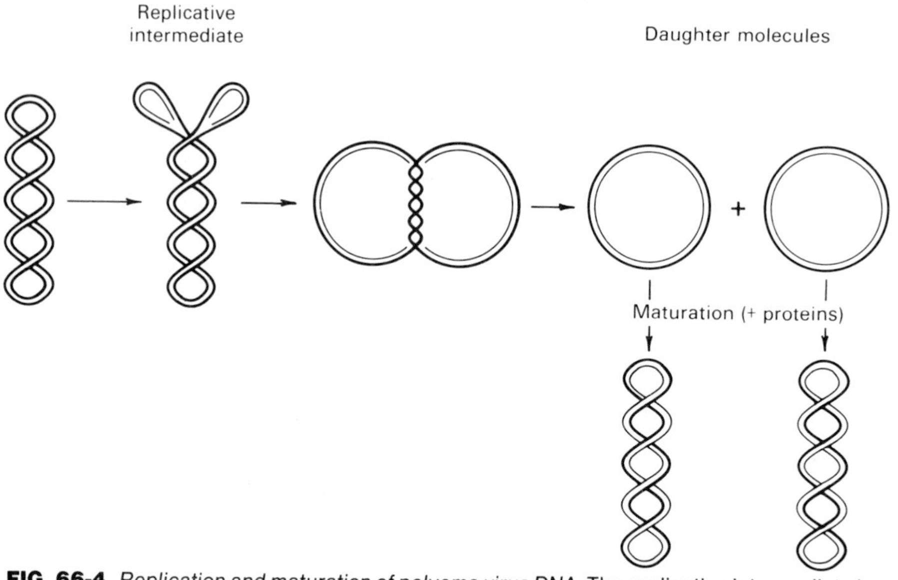

Replicative
intermediate   Daughter molecules

FIG. 66-4. *Replication and maturation of polyoma virus DNA.* The replicative intermediate is partly supercoiled, partly relaxed, and therefore has intermediate buoyant density in CsCl–ethidium bromide gradients. When replication is almost complete it yields two daughter molecules, each with a gap in the new strand at the replication terminus. These molecules are then sealed and, in conjunction with cellular histones, are converted to chromatin containing covalently closed DNA supercoils. Supercoiling is caused by unwinding of the helix after it binds the histones (see Chromatin Ch. 49).

Role of Viral Genes in Productive Infection. Several types of mutation are available for studying these functions: **temperature sensitivity** (*ts*), **host range, plaque size,** and **deletions.**

Small deletions can be produced, for instance, by using pancreatic DNase in the presence of Mn^{2+} to cause double-strand cuts at random locations in the viral DNA. When the cut (linear) DNA infects cells, its free ends often rejoin by recombination, eliminating a small sequence. Some deletions are viable; others are nonviable and require a helper virus for growth (usually a *ts* mutant is used).

The sites of mutations are **mapped** by physical methods. These methods utilize fragments produced by restriction endonucleases, which cut at characteristic sequences. First the order of the fragments is determined from overlaps of fragments produced by different endonucleases, producing a **restriction site map.** The position of individual mutations on this map is then established by **marker rescue,** as described in Fig. 66-5.

The analysis of mutants assigns **four genes** to polyoma virus or SV40 (see Fig. 66-3); mutations in different genes complement each other (i.e., cells with two mutant viruses yield wild-type virus under nonpermissive conditions).

Two genes are **late,** i.e., are expressed after DNA synthesis begins. Gene B/C specifies VP1: its mutations produce either heat-sensitive virions or small plaques. Gene D specifies VP2 and VP3; its mutations prevent penetration. Two other genes are **early.** Mutations in gene A render the large T Ag heat-labile, prevent viral DNA replication, and cause overproduction of early mRNAs by removing the negative regulation of the large T. Polyoma mutations in gene *hr-t* (for host range-transformation), which are not temperature-sensitive, restrict viral multiplication to special cell types.

CELL TRANSFORMATION

After cells are exposed to the virus, transformed clones are usually identified and isolated by one of the following characteristics: distinct morphology (Fig. 66-6), ability to form foci overgrowing a monolayer of untransformed cells, or ability to **grow in suspension** in soft agar (**anchorage independence**). The various procedures select for different subsets of transformed cells, with somewhat different properties.

Isolation of transformed clones in soft agar allows a distinction between cells undergoing **stable transformation,** which form rare large colonies, and those undergoing **abortive transformation,** which form small but much more frequent colonies. In abortive transformation the cells return to normality after four to six generations and stop growing in suspension; however, they continue to grow if transferred to a dish with liquid medium.

The number of stable transformed clones generated by a virus sample is much less (10^{-3}–10^{-5}, depending on the cell type) than the number of plaques on permissive cells. The low ratio is due to the difficulty of the virus in penetrating nonpermissive cells, because the ratio can be increased by infecting cells with DNA extracted from the virus (which presumably transfects the two cell

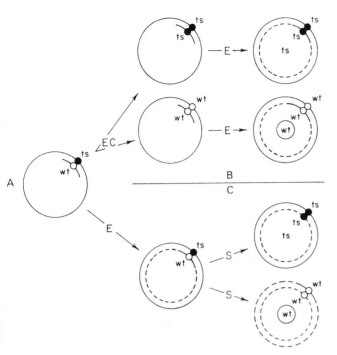

FIG. 66-5. Mapping of polyoma virus genes by marker rescue. The DNA of a *ts* mutant is nicked in one strand and the other strand is recovered as an intact circle and annealed to a fragment of wild-type (*wt*) DNA, obtained by a restriction endonuclease. When the fragment overlaps the mutated site (as in A) the progeny, formed by transfection, will include *wt* as well as *ts* genomes. The *wt* recombinants are formed either, as shown in B, by error correction (**EC**) within the heteroduplex DNA followed by elongation (**E**) of the fragment, or, as shown in C by elongation (**E**) of the fragment followed by segregation (**S**) at replication.

types equally). Transformation of a cell is evidently caused by a single virion, because its frequency is proportional to the virus titer (see The dose response curve of the plaque assay Ch. 46, Appendix).

Integration of the Viral DNA into the Cellular DNA. Mouse cells permanently transformed by SV40 contain one or a few viral genomes per cell, as shown by the kinetics of hybridization of their DNA with labeled viral DNA. During fractionation the viral DNA sequences are always found with the cellular DNA, even when completely denatured to single strands (by sedimentation in an alkaline sucrose gradient, Fig. 66-7). Hence the viral DNA is integrated, i.e., covalently bound to the cellular DNA as a **provirus,** equivalent to the prophage of lysogenic cells (Ch. 48). In contrast, there may be no integration in abortive transformation since the progeny of abortively transformed cells are virus-free. Integration probably occurs by a crossover between the cyclic viral DNA and the cellular DNA.

Most transformed cells contain more than one provirus. The provirus may be shorter than the viral DNA, but then a fraction of the early region, including the 5' end, is always retained. With polyoma virus, the fraction contains the *hr-t* gene. Transformed semipermissive cells also contain free viral DNA, either as a result of replication or as a plasmid.

To locate the viral and the cellular integration sites, fragments containing viral DNA generated by restriction endonucleases from the DNA of transformed cells or from the virus are compared to each other. Evidently only the hybrid linker fragments (containing both cellular and viral DNA), which derive from the two ends of the provirus, will differ from those obtained from the free viral DNA, showing in which fragment the viral DNA has been opened

FIG. 66-6. Colonies of the BHK line (hamster kidney) infected by polyoma virus. One colony is transformed (**arrows**) and is recognizable by its considerable thickness and the random orientation of its cells. The untransformed colonies are thin and contain cells that tend to orient parallel to each other. (Courtesy of M. Stoker)

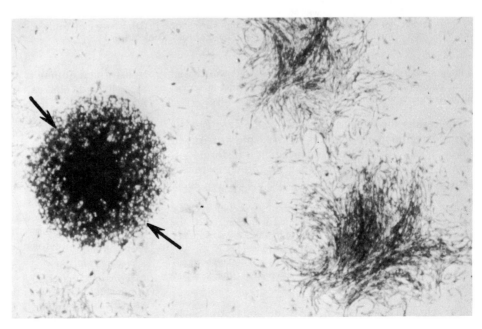

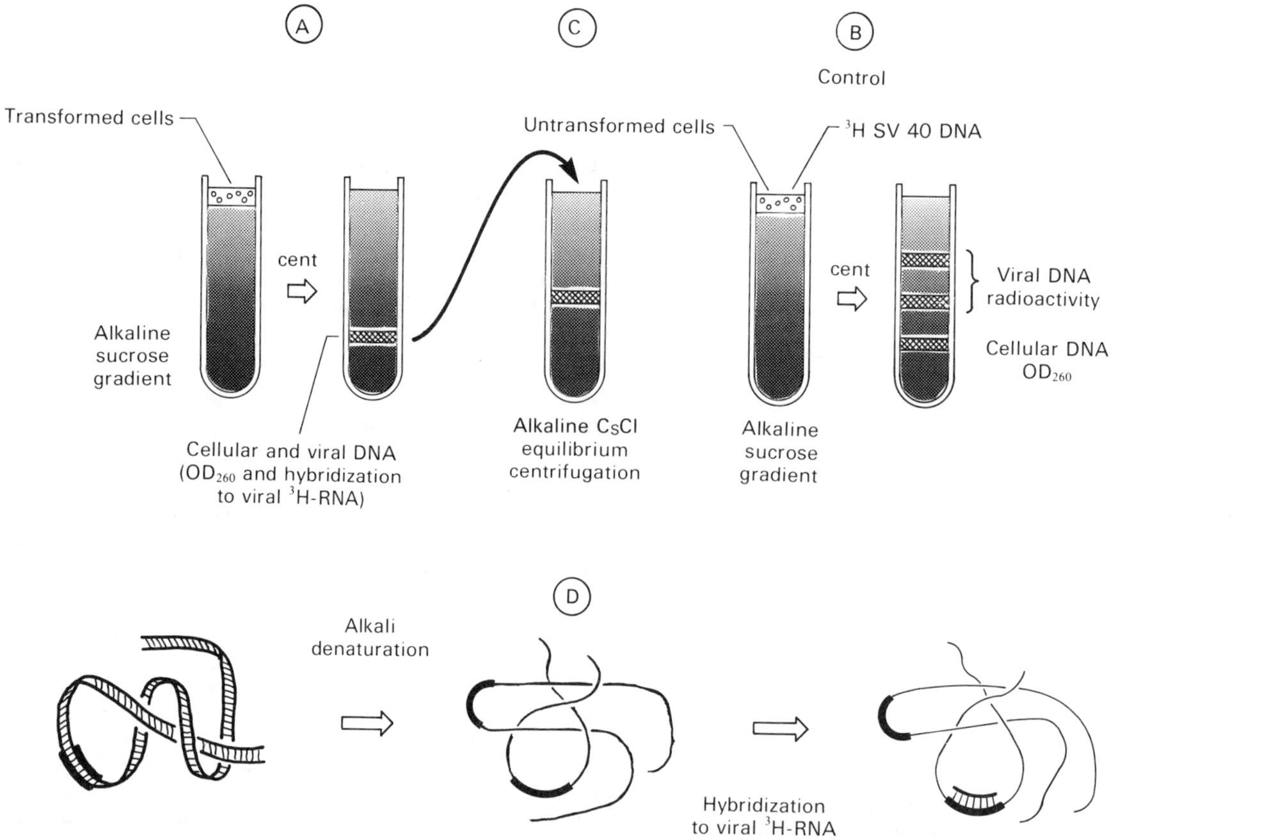

FIG. 66-7. SV40 DNA. Evidence for integration of SV40 DNA in the DNA of transformed SV3T3 cells. The main experiment (**A**) consisted of placing transformed cells on top of a preformed alkaline (pH 12.5) sucrose gradient, where they lysed; the gradient was then centrifuged to sediment the denatured cellular DNA toward the bottom of the tube. This technic, by avoiding the usual extraction of the cellular DNA, minimizes its breakage, so that (as shown in **B**) it sediments much faster than superhelical SV40 DNA, in spite of its fast sedimentation after denaturation (a characteristic of cyclic DNA). The band of cellular DNA contained all sequences hybridizable to viral RNA (made in vitro with *E. coli* RNA polymerase on SV40 DNA); hence these sequences were covalently bound to the cellular DNA (**C**): that free DNA was recovered from the gradient was shown by sedimenting it at equilibrium in CsCl, where it had the density characteristic of cellular DNA. The molecular species involved in the experiment are shown in (**D**), where viral sequences are represented by **heavy lines** and viral RNA (complementary only to one DNA strand) by **medium thick lines.**

at integration. The lengths of the hybrid fragments will depend mostly on the integration site in the cellular DNA; hence these sites can be compared in different clones of transformed cells.

The results show that *neither the viral nor the cellular sites of integration are constant* for the same provirus in cell clones that derive from independent transformation events; hence it is unlikely that the provirus causes transformation by its insertion in a specific cellular gene.

These experiments have also shown that in semipermissive cells many copies of viral DNA may be integrated in tandem at the same site (**oligomeric provirus**). They may arise by successive secondary integrations of the viral DNA that replicates in the cells; the secondary integrations, involving recombination between homologous sequences, would occur more often than the primary integration.

The **chromosomes** which have integrated an SV40 provirus can be determined in cell hybrids between

transformed human cells and normal mouse cells. Since these hybrid cells tend to lose human chromosomes (see Approaches to Gene Mapping, Ch. 49), the loss of transformation characters can be correlated with the loss of specific chromosomes. This approach has shown that *the SV40 provirus does not localize in a specific chromosome in transformed human lines.*

Transcription of the Provirus. Only the early strand of the proviral DNA is transcribed in nonpermissive cells, usually copying the whole early region but sometimes only part; hence not all of that region is essential for maintaining transformation.

The restriction of transcription is caused by chromatin proteins, as seems to be the case for cellular genes (see Transcription of Chromatin, Ch. 49), rather than by the specificity of the host transcriptase. In fact, with transcription in vitro by *Escherichia coli* transcriptase chromatin extracted from SV40-transformed cells yields the same RNA as is synthesized in vivo.

Induction of the Provirus. In some kinds of transformed cells the provirus can be induced, with production of infectious virus. Nonpermissive cells are usually induced by fusing them to permissive cells; virus replication occurs in the heterokaryons. Both nonpermissive and semipermissive cells are induced by exposure to mitomycin or radiations, which apparently cause the local replication of the provirus.

Permissive cells transformed by *ts* A mutants, although stable at high temperature, are induced to produce virus by lowering the temperature; presumably recovery of the A function causes provirus excision as well as viral DNA replication.

MECHANISM OF TRANSFORMATION

In permanently transformed cells the provirus maintains the phenotype changes: in cells containing a single provirus its loss causes the phenotype to revert to normal. However, the phenotype changes cannot be attributed to the integration of the provirus, because they are present in abortively transformed cells, which lack integration. Hence the provirus must alter the cell phenotype by the function of genes whose expression does not depend on integration.

Viral Gene Functions in Transformation. Only genes of the early region are required for transformation: viral DNA fragments lacking a large part of the late region can still transform. The A gene function may be involved in integration, because at the nonpermissive temperature *ts* A mutants can cause abortive but not stable transformation. The stable transformation obtained at the permissive temperature persists at the nonpermissive temperature, suggesting that the A gene plays a main role in the **initiation** of transformation. However, depending on the selection used for the isolation of the transformed cells, and the cell type, the transformed phenotype shows various degrees of reversion at the nonpermissive temperature, showing that the A gene also plays some role in the **maintenance** of transformation.

The *hr-t* gene appears to be mainly responsible for anchorage independence (i.e., growth in soft agar), which does not develop in cells infected by *hr-t* mutants. Cellular DNA replication in quiescent cultures may be induced by either the A or the *hr-t* gene, because neither mutation prevents it.

Role of Cellular Genes in Transformation. In addition to regulating transcription of the provirus cellular genes perform other functions in transformation, as shown by the appearance of **phenotypic revertants** of transformed cells, i.e., revertants in which the provirus (tested after induction by cell fusion) is not mutated.

That changes of cellular components may prevent the effect of the virus is understandable, because the transforming viral protein must interact with cellular proteins. Some types of phenotypic reversions are accompanied by gross changes in the number of chromosomes, suggesting that the **balance of cellular genes** is important for transformation. The required balance varies with the virus, since revertants of SV40-transformed 3T3 cells are retransformed by an RNA-containing sarcoma virus (see RNA-Containing Tumor Viruses, below) but not by SV40.

VIRUSES WITH HYBRID DNA; PSEUDOVIRIONS

Recombinants of adenovirus DNA and SV40 DNA enclosed in adenovirus capsids (**adeno–SV40 hybrids**) were observed in adenovirus grown in monkey cells, in which occult SV40 helps adenovirus multiplication. The SV40 **helper function** provides an initiation factor required for synthesis of adenovirus proteins. Some hybrids are **defective** in essential adenovirus functions and require an adenovirus helper for multiplication. Various **nondefective** hybrids may contain either a complete SV40 genome or parts of its early region. The partial hybrids expressing some SV40-specific Ags (T, U, TSTA) in the infected cells have been powerful tools for dissecting the function of the SV40 genome (Fig. 66-8). All hybrids express the helper function, which can be induced by injecting purified large T protein (see Viral Proteins, above) into the cells. Cells transformed by some of these hybrids are morphologically indistinguishable from cells transformed by SV40 alone, suggesting that such hybrids include SV40 genes important for transformation. The adenovirus and SV40 DNAs have very little homology; the recombination between them is analogous to the integration of SV40 DNA in the cellular DNA.

Pseudovirions. We have already noted the incorporation of cellular DNA (recognizable by hybridization) into covalently closed cyclic viral DNA during high multiplicity passages. This phenomenon resembles the formation of **specialized transducing phage** (Ch.

FIG. 66-8. SV40 functions expressed by nondefective adeno 2–SV40 hybrid viruses. The outer segments indicate the parts of SV40 DNA present and functional in different hybrids; the letters indicate the functions they express—U, tumor-specific transplantation Ag (TSTA), tumor (T) Ag, or the helper function for adenovirus replication in monkey cells.

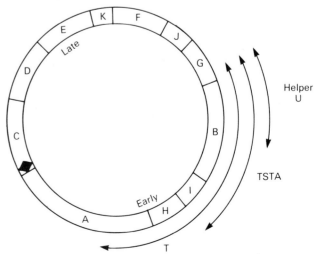

48). In addition, viral capsids can enclose cellular DNA fragments without viral sequences, just as in **generalized** transduction; the products are called **pseudovirions.**

The presence of cellular DNA in the virions might be useful for experimentally transducing genes from one animal cell to another. However, the efficiency would be extremely low, because the proportion of the animal cell genome carried in a virion is smaller, by a factor of at least 10^4, than the proportion of the smaller bacterial genome in a transducing phage.

ROLE OF SV40 AND RELATED VIRUSES IN HUMAN NEOPLASMS

The great diffusion of SV40-like viruses (e.g., BK virus) in humans, beginning in infancy, raises the question of whether they have an oncogenic role. However, such a role is not supported by studies of DNA homology, which show that cells of human malignancies drawn from a large sample lack BK virus sequences.

Suggestive evidence for an oncogenic role of SV40 has come from epidemiologic observations on children whose mothers received an SV40-contaminated batch of poliovaccine while pregnant. Among these children the incidence of brain malignancies was significantly higher than in a control population. This finding is in accord with the tropism displayed by SV40 and SV40-like viruses for human brain. Postnatal administration of the contaminated vaccine did not increase the incidence of malignancies.

PAPOVA VIRUSES: PAPILLOMA VIRUSES

The papilloma virions are structurally similar to polyoma virions, but somewhat larger (55 nm); they also contain a cyclic double-stranded DNA (mol wt about 5×10^6). Different members of the group produce **benign warts,** called **papillomas,** in several mammals, including man. In the rabbit and in cattle the tumors may become malignant if they persist for a sufficiently long time, or under the influence of small doses of chemical carcinogens; this malignant conversion occurs in the alimentary canal of cattle feeding on bracken, a carcinogen-producing fern.

The first of these viruses was discovered by Shope, who showed that extracts of warts of wild cottontail rabbits (*Sylvilagus floridanus*) produce warts when inoculated into the skin of either wild or domestic rabbits (*Oryctolagus cuniculus*). Papilloma viruses are readily recovered from extracts of human or other warts.

VIRAL MULTIPLICATION AND PATHOGENESIS

The papilloma viruses of man and rabbit induce papillomas in the skin and some mucous membranes in individuals of the same or related species. The cattle virus can also produce mesenchymal tumors in horses and hamsters.

Bioassay, based on wart formation in the skin, shows that the virus has extremely low efficiency. The ratio of physical particles to infectious units ranges between 10^5 and 10^8. In tissue cultures these viruses do not appear to propagate. Nonproductive infection, however, may occur: human skin cells can apparently be transformed in culture by the human wart virus, and fetal calf skin cells by the bovine virus.

Skin papillomas begin as a proliferation of the dermal connective tissue, followed by proliferation and hyperkeratinization of the epidermis. In warts of wild cottontail rabbits the nuclei of the keratohyaline and keratinized layer contain viral capsid Ag (recognizable by immunofluorescence) and viral particles (recognizable by electron microscopy and infectivity). The proliferating connective tissue and basal epidermis, in contrast, contain neither Ag nor virions (Fig. 66-9), but they do contain viral DNA, which appears to be responsible for the stimulus to proliferate. When rabbit papillomas progress to cancer the production of virus or viral DNA is markedly reduced or even **undetectable.**

Thus one such cancer (Vx7) after prolonged serial transplantation has retained the ability to produce these products only in very small amounts and in very few cells; another cancer (Vx2) has lost it completely. The continued production of virus by the Vx7 cancer could be due to the persistence of the viral genome in an integrated state, with occasional induction, or to a virus carrier state (Ch. 53).

Regression of Papillomas. Papillomas, including human warts, often regress spontaneously. If a rabbit has several separate papillomas they all regress simultaneously, suggesting an **immunologic mechanism** (see Immuno-

FIG. 66-9. Fluorescence photomicrograph of a frozen section of wild rabbit papilloma stained with fluorescent antiviral Abs. Capsid Ag is present in nuclei, and therefore appears as small, discrete, bright areas; it is restricted to the keratohyaline (**H**) and keratinized (**K**) layers of the epidermis. There is no capsid Ag in the cells of the proliferating basal layer (**P**). (Noyes WF, Mellors, RC: J Exp Med 104:555, 1957)

logic Relations, etc., below). In contrast, the papilloma-derived carcinomas in rabbits usually do not regress and can be serially transplanted. Presumably they are less immunogenic, or their high growth potential overcomes the allograft reaction (Ch. 21).

HUMAN ADENOVIRUSES

Human adenoviruses cause a productive infection of human cells and kill the cells. However, many of these viruses transform nonpermissive hamster or rat fibroblasts (in which infection is nonproductive). Moreover, some cause tumors when injected into newborn hamsters: at high frequency by viruses of group IV (**highly oncogenic**) and infrequently by those of group I (**weakly oncogenic**); viruses of groups II and III are **nononcogenic**. Viruses of all these classes transform hamster cells in vitro.

The virus-transformed cells contain integrated viral DNA sequences corresponding to a fraction of the genome or even to its entirety; usually many separate fragments are present, in nonequimolar amounts. Permissive human cells can be transformed by subgenomic viral DNA fragments, which lack cell-killing genes.

The genes essential for transformation are localized in a 6%–7% extreme left segment of the viral DNA, which is always present in transformed cells. Isolated DNA fragments of mol wt 10^6, if they contain that segment, can transform cells. Only early genes, including those of the transforming region, are transcribed in nonpermissive transformed cells; late genes, if present in the proviral DNA, are not transcribed, ensuring cell survival. Cells transformed by viruses of any group contain a 58,000-dalton tumor (T)Ag (revealed by Abs present in tumor-carrying animals), which is also made, as an early Ag, in the lytic infection of permissive cells. Like the transforming protein of avian sarcoma viruses (see below), it has protein kinase activity.

In spite of the wide distribution of adenoviruses, they *do not appear to be oncogenic in humans;* human neoplasms of various kinds were found to be free of human adenovirus DNA. Oncogenic activity is probably lacking, because human cells, being permissive, are killed rather than transformed.

ONCOGENIC HERPESVIRUSES

Infection of chickens or monkeys with certain herpesviruses causes the appearance of lymphomas. Indirect evidence suggests that herpesviruses may also be oncogenic for humans.

EPSTEIN–BARR VIRUS

The Epstein–Barr virus (EBV) is widespread in adult humans, in association with three types of disease: infectious mononucleosis, Burkitt lymphomas, and nasopharyngeal carcinoma. **Infectious mononucleosis** (Ch. 55), an extensive but self-limiting lymphoid proliferation, is the only disease certainly caused by EBV. Serologic studies show that EBV spreads horizontally in humans after birth, usually in childhood.

The Burkitt lymphoma, which occurs especially in children, is a B cell lymphoma. It is endemic in central Africa and New Guinea and sporadic throughout the world; even in endemic areas it is rather rare. Though the endemic and the sporadic lymphomas are clinically and pathologically similar, most endemic lymphomas contain EBV DNA, recognizable by hybridization, whereas most of the sporadic ones do not. Since the cells of sporadic lymphomas have a different pattern of B cell markers than do endemic lymphoma cells, they may originate from a different subset of B cells.

In vitro, lymphoma cells grow into **lymphoblastoid cell lines,** which retain the characteristics of the cells of origin. Similar lines are also generated by lymphocytes of infectious mononucleosis patients and, more rarely, of normal individuals with EBV Abs. All these lines contain 50–100 EBV genomes per cell, a few integrated but most as episomal DNA, i.e., free, covalently closed circles. Burkitt lymphomas that are EBV-negative by hybridization, or other malignant lymphoid tumors, yield lymphoblastoid lines without the EBV genome. All lines containing EBV DNA also contain the **EBV nuclear Ag (EBNA)** in their nuclei (see below).

In **nonproductive** lines there is no other expression of the EBV genome, except in rare cells; however, production of other viral Ags, and of virions, can be induced by idodeoxyuridine (IUdR), 5-bromodeoxyuridine (BUdR), mitomycin C, x-rays, or phorbol ester (a promoter of chemical carcinogenesis), and also by cocultivation with some permissive monkey cells. In **productive lines** a larger proportion of the cells synthesize virions, Ags, viral DNA, and infectious virus.

In Vitro Transformation. Virus from some productive lymphoblastoid lines transforms human or primate (marmoset or woolly monkey) B lymphocytes (which have EBV receptors), with single-hit kinetics. Transformation is recognized by morphologic changes in the cells, increased DNA synthesis, growth in soft agar over a fibroblastic feeder layer, and immortalization (i.e., lack of aging, Ch. 49). The transformed cells contain EBV nuclear Ag but no other viral Ags; hence transformation depends on the restriction of certain viral functions that would lead to viral multiplication and cell death. Inhibition by phosphonoacetic acid (which inhibits a herpesvirus DNA polymerase) suggests that transformation requires viral DNA replication.

Antigens. EBNA, a soluble complement-fixing Ag, usually recognized by anticomplement immunofluorescence, is present in most cell nuclei of lymphomas or of lines containing the viral genome, whether or not they are productive. The Ag is associated with chromosomes (or chromatin fibers in interphase). An **early Ag** complex is synthesized in productive or induced cells even in the absence of viral DNA replication; a **viral capsid Ag** complex and a **viral membrane Ag** (both late) require DNA replication. A surface Ag *not* found in virions is present in all cells containing the EBV genome. The appearance of the early and other Ags in a cell shows that a productive cycle has been initiated.

In individuals infected by EBV, Abs to the early complex and to the capsid Ag appear before Abs to EBNA. In Burkitt lymphoma patients the Ab titer is an indication of the state of the lymphoma: not only are titers higher in patients than in individuals without lymphoma, but they tend to be higher still in patients with recurrences after therapy.

Nasopharyngeal Carcinoma. This is the most frequent cancer in males in certain ethnic groups in China. The patients always have elevated titers to EBV-specific Abs. The cancer consists of anaplastic epithelial cells abundantly infiltrated with lymphocytes. EBV genomes (100–150/cell) are present in the epithelial cells but not in the infiltrating lymphocytes, as seen by cytologic hybridization of the cellular DNA with labeled probes of EBV DNA. Moreover, EBV genomes are detected in DNA extracted from lymphocyte-free cultures of the epithelial cells, and in cancer transplants in nude mice, where the lymphocytes are lost. The epithelial cells also contain the EBV nuclear Ag; early Ag is absent but can be induced by IUdR.

Etiologic Considerations. The association between EBV infection and endemic Burkitt lymphoma or nasopharyngeal carcinoma is not proof of an etiologic relationship. Thus most EBV-infected individuals (detected by Ab titer) do not develop lymphomas; and most sporadic Burkitt lymphomas do not contain demonstrable EBV (or its DNA or nuclear Ag). Hence in the EBV-positive lymphomas the virus might be a passenger that becomes established preferentially in tumor cells. However, EBV-negative lymphomas persist in seropositive (and therefore virus-infected) individuals, although their cells can be infected in vitro. Furthermore, even in areas where Burkitt lymphoma is endemic, lymphomas not of the Burkitt type are regularly EBV-negative. It thus appears that EBV-positive lymphomas probably do not acquire the virus by subsequent infection but contain it because they derive from a genome-carrying cell.

There is no doubt that EBV is potentially oncogenic, because it induces fatal lymphomas upon inoculation into marmosets or owl monkeys. The very low frequency of human Burkitt lymphoma or nasopharyngeal carcinoma, despite the wide distribution of the virus and the existence of endemic areas, could be explained by the existence of viral strains of different oncogenicity. However, though *noticeable DNA differences are observed between the DNAs of viruses isolated from infectious mononucleosis and from lymphomas,* they have not been detected among viral strains from different Burkitt lymphomas; but small differences may go unnoticed. Alternatively, **environmental or genetic factors** may be required to supplement the action of the virus. The etiologic need for environmental factors is supported by the clustering of the disease in space and time: indeed, the distribution of malaria corresponds to that of Burkitt lymphoma. Among possible genetic factors, a deletion of chromosome 14 is regularly observed in Burkitt lymphomas as well as in lymphoblastoid cell lines derived from them. The same deletion is frequently present in other types of lymphoma. Genetic factors may also be important in the genesis of nasopharyngeal carcinoma: in progeny of intermarriages between people from high and low cancer groups the frequency is intermediate; moreover, susceptibility seems higher for certain HLA types.

If a suitable EBV vaccine can be produced the role of the EBV in Burkitt lymphoma may be conclusively determined by vaccinating children in endemic areas before they become infected. Since the lymphoma usually occurs before the age of 10, a reduction in its incidence would be easy to observe and would establish an etiologic role, although not necessarily exclusive, of EBV.

HUMAN HERPES SIMPLEX VIRUS

Transformation of hamster embryo cells is obtained with either herpes simplex virus (HSV) type 1 or 2 if its ability to multiply and kill the cells is greatly reduced either by ultraviolet (UV) irradiation or by using *ts* mutants at the nonpermissive temperature. These transformed cells produce no infectious virus, and they contain one or a few copies of HSV DNA sequences, corresponding to 8%–32% of the whole viral genome.

Herpes simplex virus type 2 has been implicated in the etiology of **cervical cancer** in women by seroepidemiologic studies in several countries: women with this cancer generally have increased titers of HSV Abs. Some studies also show HSV sequences in DNA extracted from cancerous tissue; however, it is not clear whether the viral DNA was present *only* and in *all* cancer cells. The possible etiologic role of HSV in cervical cancer is even more difficult to establish than that of EBV in Burkitt lymphoma.

CYTOMEGALOVIRUS

This virus, especially if irradiated with UV light, also transforms hamster or human embryonic fibroblasts. The transformed human cells contain a nuclear Ag detected, like the EBV nuclear Ag, by anticomplement immunofluorescence; they also form tumors in nude mice.

Cytomegalovirus is frequently isolated from humans with malignancies, but here it is likely to be an opportunistic invader which becomes established when the immune system is depressed by debilitation or chemotherapy.

PRIMATE HERPESVIRUSES

Herpesvirus saimiri is indigenous but not pathogenic in squirrel monkeys, in which it spreads horizontally after birth. It induces lymphomas and reticulum cell sarcomas in owl or marmoset monkeys upon natural or laboratory infection. In contrast to EBV, it causes **T cell lymphomas.** The tumor cells release virus spontaneously or by cocultivation with African green monkey cells. Similarly, **herpesvirus ateles** is ubiquitous, but not pathogenic, in spider monkeys, and induces lymphomas after inoculation into some New World monkeys; it can also transform primate lymphocytes in vitro.

MAREK DISEASE VIRUS

The chicken herpesvirus, Marek disease virus (MDV), causes both a **productive** infection in the epithelium of feather follicles, and an **abortive** infection, with neoplastic transformation, in lymphoid T cells; the tumor cells infiltrate many visceral organs and the peripheral nerves. Artificial depletion of T cells (through neonatal thymectomy, administration of antithymocyte serum, or γ-radiation) reduces the incidence of lymphoma following infection. The involvement of T cells in the neoplasia, and the formation of suppressor cells, leads to severe **immunosuppression,** which apparently favors development of the disease by interfering with immunosurveillance.

Most MDV lymphoma cells contain little or no infectious virus, but their inoculation into healthy chickens transmits the disease. Serial transplantation leads to tumor strains with increased malignancy, apparently containing defective MDV DNA.

A successful **live vaccine** against MDV infection of chickens has been developed, using a virus adapted to the turkey. The vaccine is effective in markedly decreasing the incidence of lymphoma.

RNA-CONTAINING TUMOR VIRUSES: RETROVIRIDAE

The family of **Retroviridae** (L. *retro,* backward; vernacular, retroviruses) is characterized by the presence of a reverse transcriptase in the virions. Various **oncoviruses** (Gr. *onkos,* tumor) induce sarcomas, leukemias, lymphomas, and mammary carcinomas; in addition this family contains nononcogenic members.

The virions are enveloped and ether-sensitive, about 100 nm in diameter; the capsid, probably icosahedral, encloses the single-stranded RNA genome. These viruses are classified (Table 66-3) according to their appearance in electron micrographs of thin sections: mature virions of B particles have an eccentric core and C particles a central core; D particles have a morphology intermediate between B and C particles. A particles, found only within cells, have a double shell with an electronlucent center.

COMPONENTS OF VIRIONS

The **viral RNA** is a **single-stranded** molecule of about 10 Kb (sedimentation coefficient about 35S) but shorter in some viruses (Defective Leukosis Viruses of High Oncogenicity, Sarcoma Viruses, p. 1253), with a terminal repetition (21 nucleotides in avian viruses, about 50 in murine viruses) at the ends (Fig. 66-10). Like cellular mRNA, it has a poly(A) tail at the 3′ end, a cap at the 5′ end. Electron micrographs of the RNA extracted from several type C viruses show that *each virion contains two RNA molecules* held together by a dimer linkage structure near the 5′ end. The RNA dimer (70S) separates upon denaturation into the two identical molecules; hence the **virion is diploid.**

The identity of the two molecules was shown 1) by the kinetics of hybridization to DNA complementary to the viral RNA (C DNA) made in vitro (see Reverse Transcriptase, below), 2) by the analysis of oligonucleotides,* and 3) by genetic evidence (see Genetics, below).

Reverse Transcriptase. This enzyme is present in the amount of about 30 molecules per virion. It is specified by the viral genome, since it is heat-labile in virions carrying certain *ts* viral mutations. The enzyme synthesizes DNA complementary to the viral RNA, in the presence of the four deoxyriboside triphosphates and in a suitable ionic environment. Like other DNA polymerases, this **RNA-dependent DNA polymerase** elongates an oligonucleotide **primer** paired to a **template** strand, building the DNA complement of the template: *the primer is an associated tRNA.* The enzyme also has an exonuclease activ-

---

* **Oligonucleotide "fingerprints"** are obtained by digesting the ^{32}P-labeled RNA to completion with the G-specific RNase T1. The digest is fractionated by two-dimensional electrophoresis and chromatography, yielding a series of spots (detected by radioautography) whose coordinates depend on the size and composition of the oligonucleotides. In the digests of different RNAs the larger oligonucleotides are usually different and can be used as markers. Their relative positions in the viral RNA are determined by examining the fingerprints of poly(A)-ended RNA fragments of different lengths. Mutations or crossovers within an oligonucleotide usually change its coordinates. The fingerprint can be used to estimate the complexity of a viral RNA, i.e., the number of nucleotides contained in a nonrepeating sequence, because the expected distribution of oligonucleotide sizes is a function of such a number, assuming the nucleotide sequences to be random (as they are in practice).

TABLE 66-3. Retroviridae

Genus	Subgenus	Species

Cisternavirus A
  Mice, hamster, guinea
  pigs		

Oncovirus B
  Mammary carcinomas
  in mice
		Mouse mammary tumor viruses: MMTV-S (Bittner's
		  virus), MMTV-P (GR virus), MMTV-L

Oncovirus C	Avian	Rous sarcoma virus (RSV)
		Rous-associated virus (RAV)
		Other chicken sarcoma viruses
		Leukosis viruses (ALV)
		Reticuloendotheliosis viruses
		Pheasant viruses
	Mammalian	Murine sarcoma viruses (MSV)
		Murine leukosis virus G (Gross or AKR virus)
		Murine leukosis viruses (MLV)-F,M,R (Friend, Moloney, Rauscher viruses)
		Murine radiation leukemia virus
		Murine endogenous viruses
		Rat leukosis virus
		Feline leukosis viruses
		Feline sarcoma virus
		Feline endogenous virus (RD114)
		Hamster leukosis virus (HLV)
		Porcine leukosis virus
		Bovine leukosis virus
		Primate sarcoma viruses (woolly monkey; gibbon ape)
		Primate sarcoma-associated virus

		Primate endogenous viruses: baboon endogenous
		  virus (BaEV), stumptail monkey virus, (MAC-1),
		  owl monkey virus (OMC-1)

	Reptilian	Viper virus
Oncovirus D Primates		Mason–Pfizer monkey virus (MPMV)
		Langur virus
		Squirrel monkey virus
Lentivirus E		Visna virus of sheep
		Maedi virus
Spumavirus F		Foamy viruses of primates, cats, humans, and bovines

ity **(RNase H)** that degrades the RNA strand in DNA–RNA hybrids, apparently from the 5′ end of the RNA.

The avian enzyme contains two main subunits (Table 66-4). Subunit α, derived proteolytically from β, contributes mainly the polymerizing activity, whereas β is mainly responsible for recognizing the tRNA in the primer–template complex. The murine enzyme, in contrast, has a single subunit. This enzyme undergoes reversible phosphorylation, probably of regulatory significance, by a protein kinase and a phosphatase present in the virions. The polymerase of one virus can replicate the RNA of any other, showing **little specificity** for the primer–template complex.

VIRAL MULTIPLICATION

The adsorption of virions (through their glycoproteins to specific cell receptors) and penetration take place as with other enveloped viruses (see Ch. 50).

The outstanding feature of oncovirus multiplication is a **DNA intermediate** in the replication of the viral RNA. Such an intermediate was predicted by Temin from 1) the extraordinary sensitivity of oncornavirus multiplication to agents that inhibit DNA replication or transcription [BUdR, FUdR, or actinomycin D], and 2) the presence of sequences complementary to the viral RNA in the DNA of infected cells. The ability of this DNA to infect other cells, which then produce virus, later supplied direct evidence for a viral DNA.

Less than 1 h after infection of growing chicken cells with an avian oncovirus, viral DNA synthesis begins in the cytoplasm. A continuous **minus strand** (complementary to the viral RNA) is made by the viral reverse transcriptase, and even before it is completed the synthesis of the **plus DNA** strand (complementary to the minus strand) begins. When the minus strand is completed, the

Kilobases
from 3' end

```
10              8      5-6    2.5     1
Cap—TR——gag———pol———env——src——C—TR—poly (A)
```

Internal	Reverse	Envelope	Transformation
proteins	transcriptase	proteins	
precursor		precursor	

FIG. 66-10. Genetic map of a transforming, nondefective avian sarcoma virus. The 5' end is at left. **C** is a highly conserved sequence; **TR** is the terminal redundancy. Numbers at top indicate number of kilobases from the 3' end.

part of the viral RNA still paired to it is probably degraded by the RNase H.

Initially the viral DNA is a double-stranded **linear** molecule containing gaps, with a continuous minus strand and a discontinuous plus strand. Between 6 and 9 h after infection the gaps are filled, the DNA moves to the nucleus and becomes **cyclic;** by 24 h several DNA molecules have become integrated in the cellular DNA as **proviruses,** probably at a small number of specific sites present in each cell. Proviruses may sometimes integrate adjacent to a preexisting provirus (see Endogenous Oncoviruses, below). The **linear, cyclic, and proviral forms** of the viral DNA contain a complete genome because they are **all infectious.** The viral DNA does not seem to undergo independent replication either in linear or in cyclic form, probably because it is not a complete replicon. Progeny RNA is generated by regular transcription of the integrated provirus. This production implies that during integration the circular viral DNA opens at the end joint of the virion genome.

The evolution of this complex method of replication may be explained, as for temperate phage (Ch. 48), by the advantage of an integrated provirus, which is replicated by the cells. Reverse transcription perhaps also re-

flects a stage in the evolution from RNA to DNA genomes.

The main **molecular features of replication** emerge from in vitro and in vivo observations. The model of Fig. 66-11 is based on the nature and sizes of intermediates. The model includes two "jumps" of partially replicated molecules to other sites where replication can be completed. It is not known whether the DNA fragments jump to the same molecule on which synthesis began or to its partner (from the diploid virion). The latter event would account for the diploidy; template switching during chain elongation might account for the frequent recombination (see Genetics, p. 1248). The synthesis of the plus strand probably initiates at several places on the minus strand, using as primer oligonucleotides resulting from RNase digestion of the viral RNA (see Reverse Transcriptase, above). Multiple initiations explain the initial discontinuity of the plus strand.

The double-stranded DNA product has a **terminal redundancy** of about 600 base pairs, which is absent in the viral RNA; it allows the circularization of the molecule. The redundancy persists in the integrated provirus.

The **synthesis of viral proteins** begins at the same time as the synthesis of the viral DNA. The template for the synthesis of the *gag* polyprotein is the 35S viral RNA (either brought in by the infecting virus or a product of replication); the template for the *env* polyprotein is a 24S mRNA apparently derived from the 35S RNA by the splicing of a 500-base fragment derived from the 5' end to the 3' half of the molecule (see Transcription of Chromatin, Ch. 49). For some transforming viruses (see Sarcoma viruses, below) a shorter messenger is the template for the transforming protein. The transcriptase is apparently synthesized on the 35S RNA by occasional read-through beyond the *gag* gene, generating, 5% of the time, a large *gag-pol* polyprotein. The polyproteins are later cleaved into the final products (Table 66-4, Fig. 66-12). These mechanisms ensure that synthesis of the various proteins is commensurate with the requirements of viral replication.

Maturation (Fig. 66-13). The *env* polyprotein enters the plasma membrane (see Cellular Membranes, Ch. 49), probably during synthesis, and is then cleaved. About 8 h after infection the *gag* and *gag-pol* polyproteins, together with viral RNA, start to assemble under the cell plasma membrane, attracting envelope proteins already present

TABLE 66-4. Structural Components of Avian and Murine Oncovirus

| Gene | Precursor* | Virion polypeptide* | | Structural component | Virion substructure |
		Avian	Murine		
env	gp90	gp85-S	gp71-S	Knob	Envelope
		gp35-S	p15E-S	Spike	
		p10	p12E	Envelope-associated	
gag	p70–76	p19	p12	Inner coat†	Inner coat†
		p27	p30	Core shell	Core exterior
		p15	p15C	Core-associated	Nucleocapsid
		p12	p10	Nucleoprotein	
pol 1	p200	p91(β)	p70(α)	Reverse transcriptase	Ribonucleoprotein complex
		p64(α)			

*Numbers indicate sedimentation coefficients of the proteins; gp = glycoprotein; p = unglycosylated protein; S = disulfide bond.
†A small proportion of these phosphoproteins may bind to the viral RNA during maturation.
(Based on Data from Bolognesi DP et al: Science, 199:183, 1978)

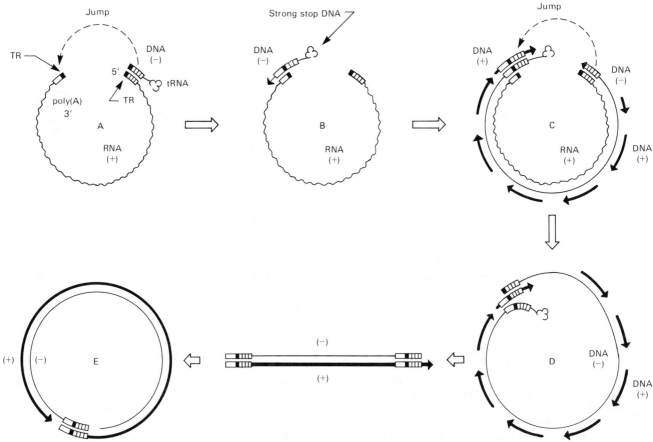

FIG. 66-11. Model of reverse transcription of a retrovirus DNA. **A.** Synthesis of the minus strand DNA initiates at the **tRNA** close to the 5′ end of the viral RNA and continues to its end, including the short terminal repetition (**TR**), then stops. This step yields the **strong stop DNA** attached to the tRNA primer. **B.** The strong stop (−) DNA "jumps" to the terminal repetition at the 3′ end, and then (**C**) is extended to a point in the proximity of the 5′ end of the RNA. At the same time the plus strand DNA initiates at several sites on the minus strand. The 3′ end of the minus strand "jumps" a second time to the other end, pairs with the end fragment of the plus strand, where it is completed (**D**). The various segments of the plus strand grow and are finally ligated together; removal of the tRNA primer results in a double-stranded DNA molecule longer than the viral RNA and with long terminal repeats (about 600 bases). This molecule can circularize (**E**). **Boxes** indicate terminal sequences. (Modified from Baltimore D et al: Cold Spring Harbor Symp Quantit. Biolo. 43:869, 1978)

FIG. 66-12. Mode of synthesis of leukosis virus proteins from the 35S and the 24S messengers.

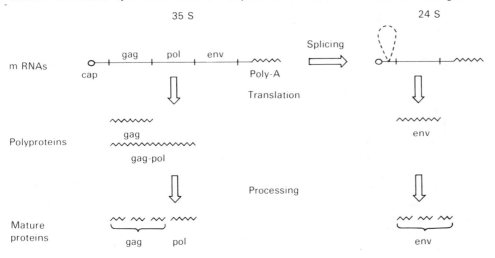

FIG. 66-13. Assembly and maturation of a leukosis virus. Incorporation of the viral glycoproteins (**VGP**) in the cellular membrane. B. Assembly of the core preproteins under the cellular membrane; interaction with VGP and viral RNA. C. Mature virion. (Modified from Bolognesi DP et al: Science 199:183, 1978. Copyright 1978 by the American Association for the Advancement of Science)

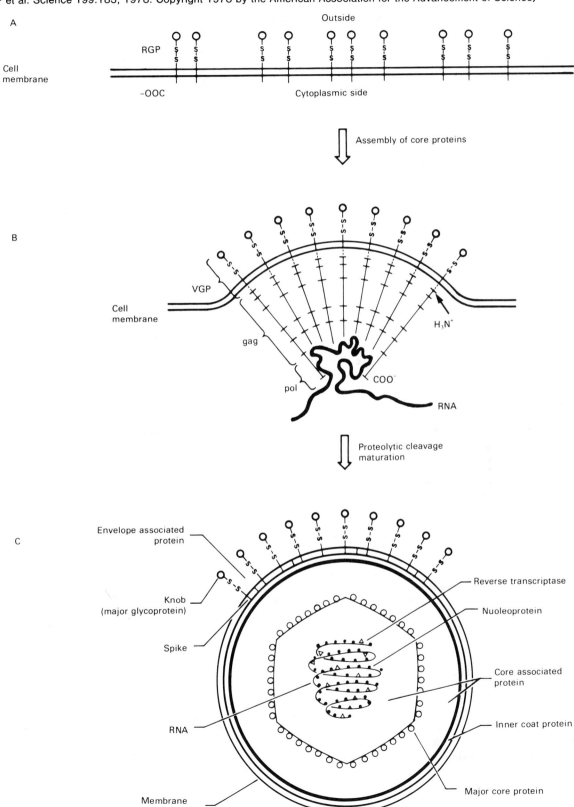

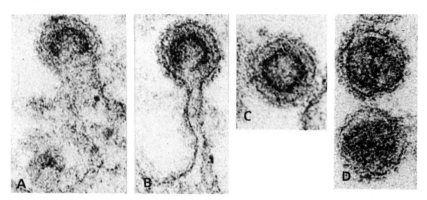

FIG. 66-14. Thin sections of type C particles from cells producing a mouse leukosis virus. **A** and **B.** Two phases of budding of the virions at the cell surface. **C.** Detached immature virion, with an electron-lucent nucleoid. **D.** Two mature particles with dense nucleoids. (Courtesy of L. Dmochowski)

in the membrane; the nucleocapsids assemble by a series of protein cleavages while budding of the virions takes place (Fig. 66-14).

Phenotypic Mixing (see Ch. 50). In cells infected by more than one kind of oncovirus some progeny virions contain glycoproteins not specified by the enclosed genome; the glycoproteins may even derive from an **unrelated virus,** such as vesicular stomatitis virus (VSV). Such **pseudotypes** have a **host range determined by the envelope,** which may be broader than the host range of the genotype. Thus a murine sarcoma virus genome in an envelope derived from a feline or a primate virus can induce tumors in the usual host of the latter.

Particles with a VSV genome and an avian leukosis virus envelope [VSV(ALV)] are readily selected because their production of VSV plaques is not neutralized by VSV antiserum. In the formation of these plaques the initial virus–cell interaction is determined by the ALV glycoproteins, whose functions can thus be dissociated from other events. Moreover, pseudotypes can reveal an **occult oncovirus infection** of a cell: e.g., infection with a VSV mutant producing a *ts* glycoprotein can yield heat-stable VSV(ALV) pseudotypes if the occult oncovirus can supply a heat-stable glycoprotein. This approach can detect oncoviruses with previously unknown characteristics (e.g., antigenic); therefore it can be useful in the search for oncogenic human viruses.

GENETICS

Many markers are available for genetic studies in both avian and murine viruses: **temperature-sensitive *(ts)* mutations** affecting the reverse transcriptase or the cell-transforming ability; differences in the **host range** or **antigenicity** determined by the glycoproteins; differences in the **electrophoretic mobility** of individual proteins; **oligonucleotide fingerprints** of different parts of the viral

RNA; and, for Rous sarcoma virus (RSV), morph^f mutants (see Fig. 66-15) producing **transformed cells** of **fusiform,** rather than round, morphology.

Oncoviruses have a **high mutation frequency,** suggesting frequent errors by the reverse transcriptase. For example, an avian sarcoma virus unable to infect duck cells generates mutants that are able to do so, with a frequency of 10^{-4}–10^{-5} per generation.

In genetic crosses with avian viruses *the frequency of recombination is very high,* 10%–20% between *env* and *pol* and between *env* and *src.* Indeed the progeny may have undergone 2–5 crossovers per molecule as judged from the new oligonucleotides present.

Vogt has proposed that this high frequency, far exceeding that of any other animal virus (DNA or RNA), is due to the diploidy of the virions. It can be shown that in crosses between two strains with different markers **heterozygous virions** are produced during the first multiplication cycle. In the next cycle they generate recombinants, possibly because the synthesis of the minus DNA strand frequently switches from one RNA molecule to its partner (after the initial switch required for replication; see above).

Recombination has permitted **mapping** of the viral genes: mutations are detected by changes (e.g., of electrophoretic mobility) in the viral proteins, and the sites of mutation in the genome are revealed by concomitant changes of oligonucleotides. The gene order for nondefective avian sarcoma viruses is shown in Fig. 66-10. It seems that the order is the same in murine viruses.

Recombination between an exogenous virus and an endogenous provirus (see Endogenous Oncoviruses, below) preexisting in the infected cell occasionally generates new virus types; it is of considerable significance in oncogenesis and a **powerful mechanism of oncovirus variation,** of great evolutionary significance (see Evolution of Endogenous Oncoviruses, below).

FIG. 66-15. Transformation of cultures of the iris of the chick embryo by either wild-type Rous sarcoma virus (**A**) or its morph^f mutant (**B**). Cells transformed by the wild-type virus are round, whereas those transformed by morph^f virus are fusiform and form bundles. The untransformed cells form an epithelial like sheet and can be recognized because they contain various amounts of black pigment, whereas both kinds of transformed cells are almost colorless (Courtesy of B. Ephrussi and H. Temin)

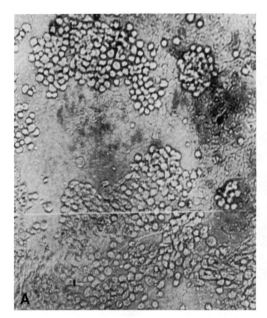

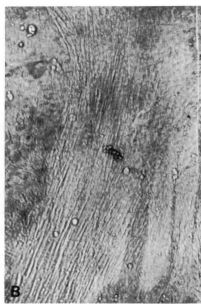

VIRAL ASSAY

In Vitro. Virus can be identified and assayed by several characteristics. In fibroblastic cultures sarcoma viruses transform cells which then form colonies (**foci**) over the normal resting cells. **Syncytial plaques** are produced by the fusion of XC cells (rat cells transformed by RSV) plated together with cells infected by murine leukosis viruses. **Immunofluorescent plaques,** revealed by fluorescent Abs to a viral Ag, are produced by many viruses. Virions in sufficient concentrations are assayed by measuring the **reverse transcriptase activity.**

In Vivo. Some assays use as endpoints the induction of leukemias or tumors; the production of spleen foci is used for Friend virus (Fig. 66-16) and an increase in spleen weight for Rauscher virus.

IMMUNOLOGIC REACTIONS OF VIRION PROTEINS

The virion proteins possess various kinds of antigenic sites which are useful for classifying the viruses. Antibodies elicited against the envelope glycoproteins react only with the same virus (**type-specific** Abs); those elicited against core proteins (**gs Ag**) react also with other viruses of the same viral species (**group-specific** Abs); some Abs react with viruses of other species as well (**interspecies** Abs). In the mammalian viruses two specificities can be distinguished in the same polypeptide chain of the major capsid protein: $gs1$ is common to all viruses of a given species but does not crossreact with viruses of other species, while $gs3$ is common to all mammalian (but not avian) oncoviruses, suggesting a **common origin** for all mammalian type C oncoviruses. The reverse transcriptase too is antigenic and contains type, group, and interspecies determinants.

Antibodies to various determinants may have different effects, e.g., immunoprecipitation or neutralization. Antibodies against envelope proteins are **cytotoxic** for infected or transformed cells carrying these proteins in their plasma membranes. In the absence of other signs of infection such surface Ags have in the past been mistaken for genuine cellular Ags. Thus GIX, originally recognized as a thymocyte differentiation Ag, was later shown to be a type-specific determinant of the main envelope glycoprotein (gp 70) of AKR virus (and is the only function expressed by an AKR provirus in those cells). Antibodies against glycoproteins are useful for **preventing viremia** in cats, and consequently preventing the horizontal spreading of infection. Heterologous Abs which are active especially toward interspecies Ags, can prevent tumor formation by murine or feline viruses even if administered several days after the virus.

Antibodies against viral glycoproteins in chronically infected mice (especially of the NZB strain) are correlated with an **autoimmune disease.** The same mechanism may be involved in the pathogenesis of human **lupus erythematosus,** because in this disorder the kidneys contain Ags that crossreact with the group-specific Ags of mammalian type C viruses.

CLASSIFICATION OF TYPE C ONCOVIRUSES

Type C oncoviruses are classified on the basis of serology, interference, and host range. These properties depend mostly on the envelope glycoprotein.

The subgroups, whether identified by serology, interference, or (for avian viruses) host range, are identical.

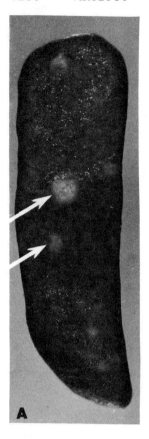

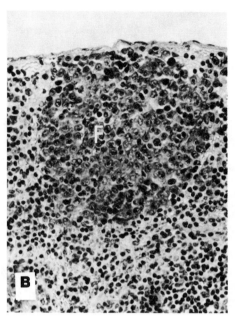

FIG. 66-16. Foci in the spleen of C3H mice inoculated with Friend leukemia virus. **A.** Macroscopic appearance of the foci (**arrows**) in whole spleen. **B.** Microscopic appearance at low magnification of a section of spleen with one focus (**F**). The focus contains large cells, characteristic of Friend's leukemia. (Axelrod AA, Steeves RA: Virology 24:513, 1964)

Avian type C viruses have seven subgroups (A to G), murine viruses three: G (AKR virus), FMR (Friend, Moloney, and Rauscher viruses), and NZB xenotropic.

Serology depends on antigenic differences of the glycoproteins (see Immunologic Reactions of Virion Proteins, above). **Interference** is the resistance to further infection of cells already harboring an identical or related virus whose envelope glycoproteins block the cell receptors for the superinfecting virus. The block is at the adsorption–penetration step.

Host Range. With **avian** viruses susceptibility of different breeds of chicken (and their cultured cells) is controlled by several dominant genes. Resistance can be bypassed by fusing the cells to virus-infected cells, or to the virus itself, using inactivated Sendai virus (see Cell Hybridization, Ch. 49); hence the block is at adsorption–penetration. The viruses are classified according to the genotype of the cells they can infect.

With **murine** viruses resistance of **mouse cells** is mainly controlled by the Fv-1 gene, which exists in the *n* or *b* alleles (so named from prototype mouse strains). Virus strains that grow in Fv-1^{nn} cells (e.g., Swiss NIH mice) are called N-tropic; those that grow in Fv-1^{bb} cells (e.g., BALB/c mice) are B-tropic.

Heterozygous strains of mice (Fv-1^{nb}) are resistant to both viruses; hence, in contrast to the avian case, resistance is dominant. Unlike interference, resistance affects a stage between penetration and integration. The tropism of the virus is determined by a *gag* protein present in the virions; restriction can be overcome by viral mutations generating NB-tropic viruses, or by treating cells with glucocorticoids. Another mouse gene, **Fv-2^r**, prevents focus formation by the Friend virus; susceptibility is dominant.

Murine viruses are also classified by their interaction with **cells of different species. Ecotropic** strains multiply only in murine cells, **xenotropic** strains only in cells of other species (e.g., mink), and rare **amphotropic** strains (from wild mice) in cells of both kinds. As in interference, infection is blocked at the adsorption–penetration step.

CATEGORIES OF TYPE C ONCOVIRUSES

There are two main categories.

1) **Nondefective leukosis viruses with low oncogenic potential** are often recovered from animals with lym-

phomas or leukemias. They can multiply in many kinds of cells in the body or in vitro without causing cell transformation, and they require no helper because they contain all the genes required for multiplication. These viruses transform only a very small proportion of specific **target cells** (mostly of the hemopoietic–lymphoid system) in certain stages of differentiation. Some viruses of this category are present in many animal species as silent proviruses, and can be activated to yield infectious virus which is usually not oncogenic. Moreover, as will be discussed below, some of these viruses generate oncogenic derivatives after recombining with viruses latent in the cells they infect.

2) **Leukosis or sarcoma viruses, usually defective, with high oncogenic activity** can be considered as derivatives of viruses of the first category containing in their genome *a fragment of cellular genetic material that confers a high oncogenic activity*. In most cases incorporation of the cellular fragment is accompanied by a loss of viral genes that are essential for reproduction; a nondefective virus is then required as helper. Some of the viruses, which produce leukemias, do not transform all kinds of cells in which they multiply but only specific target cells. Other viruses, which produce sarcomas in animals, transform every fibroblast that they infect in vitro, and apparently also in the animal because they are not found as silent proviruses.

NONDEFECTIVE LEUKOSIS VIRUSES OF LOW ONCOGENICITY

Murine leukemia viruses were recognized through observations in inbred mouse strains, some of which had been selected for a high incidence of leukemia at a young age (e.g., AKR, C58); others developed the disease infrequently and at an older age (e.g., BALB/c, C57BL, C3H/He), and some strains (e.g., NIH Swiss) were essentially leukemia-free. Gross found that filtered extracts of AKR leukemic cells could transmit leukemia to newborn low-leukemia (C3H) animals which, being immunologically immature, could not reject a virus or the tumor cells; a virus (called **AKR** or **Gross virus**) was identified as the agent. The different susceptibility of various mouse strains to spontaneous leukemia could be correlated with the frequency of spontaneous virus production in the spleen or thymus. *The target cells for the virus are in the thymus:* the earliest lesions are found in this organ: its removal at birth prevents development of the T cell leukemia, and this effect can be reversed by the subsequent grafting of an isologous thymus.

Gross also showed that animals of a high-leukemia strain did not need postnatal infection in order to develop leukemia, because their fertilized ova, when implanted into the uterus of a female of a low-leukemia strain, developed into mice with a high incidence of leukemia— evidently *the virus is transmitted congenitally* (**vertically**).

Subsequently other murine viruses of this category were isolated. One, isolated by Kaplan from x-irradiated mice, produces a low incidence of thymic lymphomas (**radiation leukemia virus**). Another, **Moloney leukemia virus,** produced a T cell lymphocytic leukemia much more efficiently than the Gross virus.

Several subgroups of **avian viruses** belong to this category. Most induce lymphatic leukemia, with B cells as targets. The leukemic cells produce IgM but not IgG, suggesting that transformation prevents the full differentiation of the cells. These viruses are transmitted genetically (transovarially), but can also infect virus-free animals, especially by laboratory infection. Some nondefective avian viruses that are genetically transmitted do not induce any kind of neoplasia.

Leukosis viruses similar to the avian viruses are present in other species. The **feline leukosis viruses** are transmitted only horizontally (i.e., by contagion from infected cats).

Endogenous Oncoviruses. The demonstration that some viruses are genetically transmitted shows that these viruses are not ordinary infectious agents. In fact, in crosses between high- and low-leukemia mouse strains transmission follows mendelian genetics, as expected of integrated **proviruses.** (AKR mice have two such proviruses Akv-1 and Akv-2, probably identical but at different locations.) In addition, AKR-like DNA sequences can be identified by nucleic acid hybridization in the DNA of cells of many mouse strains, including ones that *are not transformed and do not produce virus.* Similar observations were made in other species. Moreover, Rowe and Aaronson each showed that halogenated pyrimidines could induce a culture of seemingly virus-free BALB/c mouse cells to release hitherto unknown type C retroviruses. We shall call these **endogenous** proviruses, to distinguish them from the **exogenous** proviruses formed in somatic cells after infection.

Endogenous AKR proviruses were probably derived from infection of early mouse embryos at some distant time, because endogenous Moloney proviruses with similar properties have been produced by this mechanism in the laboratory. Hence the main difference between endogenous and exogenous proviruses is that the endogenous ones persist in the animal strain and are inherited like mendelian genes, whereas the exogenous ones exist only in the cell clones derived from an originally infected somatic cell.

Present evidence shows that *endogenous proviruses behaving like cellular genes are common in the DNA of many animal species.* DNA hybridization studies show that each provirus may be present in several copies (up to 100 or so per haploid genome), with some sequences more frequent than others, perhaps because some of the copies are defective (i.e., contain deletions). Because detection of endogenous proviruses depends on suitable hybridization probes, many may still be undiscovered.

The **induction** of endogenous proviruses to generate infectious virus can occur **spontaneously** in cell cultures. **Artificial induction** occurs in growing cultures after exposure to agents that also induce prophages (see Induction of a Lysogenic Cell, Ch. 48), such as halogenated pyrimidines, chemical carcinogens, radiations, or inhibitors of protein synthesis. Induction also occurs after infection of the cells with other oncogenic viruses, and, in lymphocytes, after mitogenic stimulation by lectins. Each procedure may induce some proviruses but not others in the same cell. Induction is probably the result of a change in the regulation of expression of proviruses. The viruses resulting from induction are usually infectious, sometimes defective; in mice, they may be ecotropic but, more frequently are xenotropic.

The **expression** of endogenous proviruses is controlled by cellular genes. Thus, depending on which alleles of three genes are present, chicken cells containing a RAV-0* provirus may: 1) release infectious RAV-0, 2) synthesize only its envelope proteins (known as **chicken cell associated helper factor, Chf),** which a superinfecting type C virus defective in envelope synthesis can use to form infectious pseudotypes, or 3) synthesize in addition some internal protein carrying the group-specific *(gs)* Ag. Synthesis of the major envelope glycoprotein generates interference against exogenous viruses of the same subgroup; it may also elicit an autoimmune disease.

Expression of endogenous proviruses is also affected by the age of the animal and the state of differentiation of the cells. Thus spontaneous induction frequently occurs in cells of the genital tract (ovary, placenta, testis, prostate); in some mouse strains different viruses are released into the blood and into the seminal fluid.

Some endogenous proviruses are not inducible (or are very poorly so), and are not infectious when the cellular DNA transfects other cells. Apparently they are closely associated to cellular regulatory elements which inhibit the expression of the provirus in both the donor and the recipient cells.

Basic of oncogenicity of nondefective leukosis viruses. The AKR and Moloney viruses are not themselves oncogenic: transformation is rare in comparison to the number of infected cells, and there is a long lag between the onset of virus multiplication in the thymus of mice and the appearance of neoplastic cells. Oncogenic activity is related to the formation of recombinants between the RNA of these ecotropic viruses and the RNA of an endogenous xenotropic virus which they probably induce. The recombinants appear in the preneoplastic period in the thymus, multiply like ecotropic viruses, in mouse cells and, like xenotropic viruses, in mink cells, forming cytopathic foci **(mink cell focus-forming virus, MCF).** Peptide analysis of the envelope glycoprotein, RNA fingerprinting, and analysis of heteroduplexes formed with the parental AKR or Moloney viral RNA show that *the MCF viruses arise by recombination within the* env *gene.* This finding explains the **dual host range** of MCF viruses, their interference by both ecotropic and xenotropic viruses, and their neutralization by both kinds of antisera. That this recombinant plays an important, possibly essential, role in oncogenesis is shown by its regular presence in all the lymphoma cells and by its ability to induce lymphomas upon injection into mice.

* RAV = Rous-associated virus. Such viruses are isolated from stocks of defective Rous sarcoma virus in which they act as helper (see Avian Sarcoma Viruses). Each one is identified by a number.

Possibly the recombinant envelope glycoprotein inserted in the cell membrane perturbs the function of the membrane, causing transformation.

DEFECTIVE LEUKOSIS VIRUSES OF HIGH ONCOGENICITY (ACUTE LEUKEMIA VIRUSES)

Among the **murine** viruses the **Friend virus** produces an erythroblastic leukemia with foci in the spleen (Fig. 66-16) and great enlargements of this organ and of the liver; the **Rauscher virus** produces a similar neoplasia but has a wider host range; the **Abelson virus,** isolated from Moloney virus–infected mice, produces B cell lymphomas, and also transforms fibroblasts in vitro. Among the **avian** strains, the **avian erythroblastosis virus (AEV), the avian myeloblastosis virus (AMV),** and the **myelocytomatosis virus (MC29)** produce the corresponding leukemias in chickens; some also induce sarcomas (AEV) or liver and kidney carcinomas (MC29).

These viruses have a broader range of target cells than the nondefective viruses, a much higher probability of transformation, and a much shorter latent period before the induction of neoplasia. Most of these viruses can be assayed in vitro by a **focus assay** on suitable cell cultures. Within the hemopoietic system, all of the viruses have a remarkable specificity for certain stages of differentiation.

Stocks of these viruses contain two types of particles (and two sizes of RNA): those of the defective transforming virus and those of the **helper,** a nondefective leukosis virus. The helper supplies functions essential for replication, and in some cases it may help also in transformation (for Abelson virus and probably also for the MCF viruses described in the previous section). **Nonproducer** transformed cells, which do not contain helper, can be obtained by transforming cells at low multiplicity. It is likely that the defective virus originally derived from the helper virus with which it is naturally associated.

Bases of High Oncogenicity. The association of defectiveness with high oncogenic activity, which is also present in some of the sarcoma viruses (see below), can be interpreted in two ways: either a segment of foreign DNA containing a transforming gene replaces part of the deleted viral genome, or a rearrangement of the deleted genome is transforming. Either mechanism may operate in different viruses.

A model for the first mechanism is the avian sarcoma viruses, which contain a transforming insertion (the *src* sequences, see below). Moreover, though the defective leukosis viruses do not contain *src* sequences some (Abelson, MC29) do contain foreign sequences in partial replacement of *gag-pol-env* sequences; and the Abelson extra sequences are, like the *src* sequences, cellular in origin and encode a protein kinase located at the cell membrane. However, the extra sequences of these acute leukemia viruses cannot be shown to be transforming for lack of mutations affecting transformation. The second mechanism, rearrangement, may operate in the Friend virus: here the defective component (**spleen focus-forming virus**, SFFV) is, like the MCF viruses, a recombinant whose *env* gene is homologous partly to the helper virus and partly to a xenotropic murine virus.

SARCOMA VIRUSES

These viruses induce sarcomas in animals with high efficiency and a short latent period, and they transform fibroblasts in vitro. When they are infecting cultures at low multiplicity every infected cell becomes transformed, producing characteristic **foci.** *Each of these viruses, contains a segment of foreign DNA.* Its origin and size, its location within the viral genome, and possibly its function may vary from virus to virus, yet each such segment confers high oncogenic activity.

Avian sarcoma viruses. These are generally known as **Rous sarcoma viruses** from F. P. Rous, who isolated the first strain. Some of the present strains derive from that original isolate but others have been independently isolated; they all behave similarly. The foreign sequences constitute the *src* gene (see Genetics, above). In most viruses the *src* sequences are merely inserted near the *env* gene at the 3' end of the viral RNA, making it 12% longer than that of nondefective avian leukosis viruses. These sarcoma viruses are **nondefective,** i.e., they replicate

without helper. Other strains in which the *src* sequences replace *env* sequences are **replication-defective** and require a leukosis virus as **helper** for replication but not for transformation. Defective viruses occur in stocks as pseudotypes with a helper envelope. Cells transformed by a defective virus release infectious virus only if superinfected by a helper able to replicate in the cells. This property allows the adaptation of a sarcoma virus to a new species (Fig. 66-17).

The *src* sequences constitute the transforming gene, i.e., are responsible for the transforming activity, because the ability to transform is lost in **nontransforming revertants** with a partial or complete deletion of *src* sequences. The deletions are located by genetic crosses with mutations that make transformation temperature sensitive.

Cellular *src* sequences. A cDNA hybridization probe specific for the viral *src* sequences hybridizes to a small number of similar sequences in the DNA of normal chicken or mammalian cells. The homology of the probe is highest, although not complete, to chicken cells: the melting temperature of the hybrids implies a 4% base difference. The homology to other species decreases like that of unique DNA sequences. Therefore the *src* sequences appear to be regular cellular sequences.

The functional equivalence of the cellular and viral *src* sequences is shown by the appearance of transforming sarcoma viruses when nontransforming revertants with a **partial *src* deletion** are inoculated into chickens. Sarcomas that occasionally develop after several months contain sarcoma viruses whose RNAs have reacquired

FIG. 66-17. Adaptation of murine sarcoma virus (**MSV**) to a new species (hamster) when a rare transformed hamster cell is superinfected by a hamster leukosis virus (**HLV**). The progeny contain particles with the MSV genome and the HLV envelope, which can infect hamster cells and transform them with high frequency.

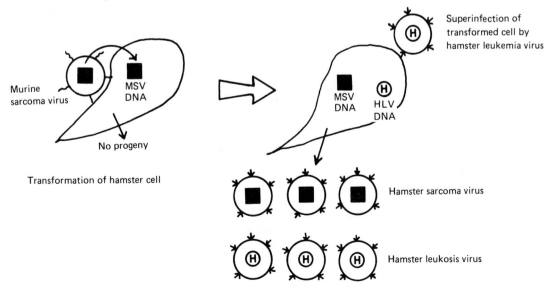

the characteristic longer length, while other genes carry markers of the infecting deleted virus; hence these new viruses are recombinants between the infecting virus and a cellular *sarc* sequence. Slight differences in sequences show that each recombinant arises by a separate event. Nontransforming revertants with a **complete *src* deletion** fail to generate transforming derivatives, pointing to the necessity for a region of homology for recombination.

In uninfected cells the *sarc* sequences are transcribed at a low rate in growing cells, resting cells, or cells transformed by chemicals; hence they do not seem to be unconditionally transforming.

The *src* product. In cells transformed (at the permissive temperature) with a *ts* mutant in the *src* gene, the transformed phenotype is lost within an hour after a shift to the nonpermissive temperature. Inhibition of protein synthesis at the permissive temperature also causes a reversion to the normal phenotype. These results show that some transformation characters are directly caused and are maintained by the continuous synthesis of the *src* product, a **viral transforming protein,** which, for ts mutants, becomes inactive at the nonpermissive temperature.

The *src* product has been identified by translating in vitro poly(A)-containing fragments of the viral RNA or a short mRNA extracted from the transformed cells (see Viral Multiplication, above). It is a phosphoprotein of 60,000 daltons (p60), which is also precipitated from extracts of both transformed and uninfected cells by Abs present in the serum of rabbits carrying a RSV-induced sarcoma; the p60 proteins of transformed and uninfected cells differ slightly in amino acid sequences. The transformed cells contain 50 times the amount of p60 present in uninfected cells, owing to the much faster expression of viral genes, which are transcribed from a viral promoter.

The viral p60, like the proteins implicated in the transformation by other oncoviruses and by polyoma virus (see Acute Leukemia Viruses and Papova Viruses) is a protein kinase: it causes phosphorylation of other proteins in the presence of ATP. Like these other proteins, it is also associated with the cell plasma membrane. Hence transformation might result from phosphorylation and the resulting changed activity of cellular proteins with growth-regulating functions (e.g., in the plasma membrane or in chromatin); this mechanism would explain the pleiotropism of transformation. However, the protein might have other functions more directly responsible for transformation. The isolated p60 is unstable; the half-life (about 30 min at 37° C) is adequate to explain the rapid reversion of the phenotype of cells transformed by *ts* mutants when the temperature of incubation is shifted.

There seems to be little doubt that the p60 is the viral

transforming protein. Why then does it not also transform uninfected cells? The reason may be qualitative, i.e., due to the small differences between the viral and the cellular protein, or quantitative, since the transformed cells contain so much more protein. That a large difference in concentration of a normal protein may shift the balance between the normal and the neoplastic state of the cells is not too surprising; it only implies that cellular growth control depends, among other things, on an accurate balance of all the participating functions. However, such a concept has profound implications for nonviral carcinogenesis, because it suggests that attention should be concentrated on alterations of regulatory cellular genes.

Murine Sarcoma Viruses. These viruses, which are all replication-defective, have leukosis virus genomes with large deletions that are partially replaced by the substitution of sequences derived from the cells in which each virus was isolated (Fig. 66-18). The RNAs of these viruses translated in vitro specify proteins not specified by the helper; they might be equivalent to the *src* protein and responsible for transformation in a similar way. However, lack of *ts* mutants for transformation has precluded their identification.

In the Kirsten and the Harvey sarcoma viruses the new sequences derive from a rat endogenous provirus; in Moloney sarcoma virus, from the mouse genome. In the Kirsten and Harvey viruses the rat substitutions leave only a small unaltered fraction of the natural helper virus from which they originated: 1.5 Kb at the 5′

FIG. 66-18. Heteroduplex of murine sarcoma virus (**MSV**) RNA with murine leukosis virus (**MLV**) cDNA. The MSV genome lacks three segments of the MLV genome (**A$_L$,B$_L$,C**) which are in part replaced by **A$_S$** and **B$_S$**. The MLV genome is about 10 Kb long, compared with 5.85 Kb for the MSV genome. The A$_S$ segment, close to the 3′ end of the MSV genome, probably corresponds to the *src* gene, which is characteristic of the sarcoma virus. The cyclic shape of the heteroduplex is related to the method of cDNA synthesis by reverse transcription (see Fig. 63-11). (Modified from Hu S, et al: Cell 10:469, 1977)

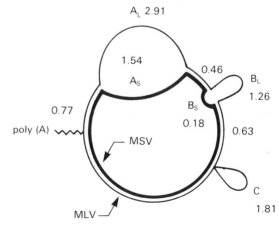

end and 1 Kb at the 3' end. Therefore none of the normal genes are entirely functional.

A possible factor in transformation is (as in the MCF viruses and in the SFFV) an *env* gene altered by recombination. The altered *env* preprotein in the cell plasma membrane may be related to the **tumor-specific surface antigen** (TSSA) observed in transformed cells (see Immunologic Relations, etc., below).

ROLE OF LEUKOSIS VIRUSES

ROLE IN CHEMICALLY AND PHYSICALLY INDUCED NEOPLASIA AND CELL TRANSFORMATION

Though mutagens and viruses have long appeared to provide alternative mechanisms for inducing cell transformation, they turn out not to be altogether independent. Thus the x-ray–induced mouse leukemia is caused by a virus (see Nondefective Leukosis Viruses, above). In another kind of interaction, many viruses can produce transformed foci in cell cultures in conjunction with carcinogens (such as 7,12-dimethylbenz[*a*] anthracene or 4-nitroquinoline-1-oxide), though neither agent alone does so. The chemical must act on the cells while they are infected, not before, suggesting that a virus-induced cell change promotes the occurrence and the stability of the primary alterations (i.e., DNA changes) produced by the chemical. Indeed, the infection seems to inhibit DNA repair in cells exposed to a carcinogen; however, it does not cause the expected enhancement of the induction of point mutations in the same cells.

These results have important implications for a possible participation of type C oncoviruses in oncogenesis by radiations and chemicals. In addition, the marked increase in sensitivity to transformation by chemicals affords an efficient method for detecting carcinogens and studying their mode of action.

ROLE OF TYPE C ONCOVIRUSES IN HUMAN NEOPLASIA

Since the viral etiology of leukemias and related tumors in many animal species has been recognized, a possible role of these viruses in human leukemias has been long suspected. This possibility is reinforced by some epidemiologic clustering of leukemia cases in contacts of leukemia patients, and by the apparent leukemic conversion of bone marrow cells from healthy donors transplanted into leukemic individuals.

Among known viruses those of cats and primates have been suspected as possible agents of human cancers. In fact, **feline leukemia and sarcoma viruses** are shed in the saliva of infected cats and can infect human cells in culture. However, in epidemiologic studies disease in pets does not correlate with human infection.

The **Type C viruses of primates** include both exogenous and endogenous members.

Exogenous Viruses. 1) The defective **simian sarcoma virus** (SSV), isolated with its associated helper, from a woolly monkey (a New World monkey), can produce tumors in newborn marmosets. 2) A group of **gibbon ape leukosis viruses** (GaLVs) were isolated in colonies of gibbons (Old World primates) with a high incidence of leukemia, usually from leukemic individuals. Most individuals in these colonies have Abs to GaLV. GaLVs are also oncogenic for some other primates. Both SSV and GaLV are *related by nucleotide sequences to each other and to type C viruses of rodents.*

Endogenous Viruses. 1) **Baboon endogenous xenotropic virus** (BaEV), isolated from placentas of normal animals, is related to an endogenous cat virus (RD114). Sequences with partial homology to BaEV are present in the DNA of all Old World monkeys (see Evolution of Endogenous Oncoviruses, below). 2) **MAC-1 virus,** from a stumptail monkey (it is not homologous to BaEV), has sequences that are present in multiple copies in the DNA of Old World primates. 3) **OMC-1 virus** came from an owl monkey. Its sequences are present in multiple copies in the DNA of New World monkeys.

There is evidence that type C viruses related to those of primates occur in human cells. This has been obtained in tissue culture, and also in fresh tissues, derived mostly from leukemic patients. 1) Electron microscopy reveals, in a small proportion of leukemic cells, **intracytoplasmic particles** similar to those often observed in animal leukemias (but no budding particles). 2) **High-molecular-weight RNA** and **reverse transcriptase** related to those of some primate or murine viruses have been extracted from intracellular particles of some patients. 3) **Ags** crossreacting with those of BaEV proteins have been observed in some human neoplastic tissue, as well as in kidney deposits of Ag–Ab complexes in patients with lupus erythematosus. (For a similar autoimmune disease of NZB mice, see Immunologic Relations of Virion Proteins, above.) 4) **Infectious virus** has been obtained in a few instances from human cell cultures: from a case of acute myeloblastic leukemia (HL23 virus), from one of histiocytic lymphoma, and from normal embryos. These isolates were all a mixture of two viruses, one closely related to SSV, the other to BaEV. The HL23 virus was not an accidental contaminant of the cultures, because RNA homologous to the cDNA of HL23 was obtained from particles of fresh, uncultured leukemic cells.

However, these observations do not prove that these type C viruses are the agents of human neoplasia, because the *neoplastic cells do not regularly contain in their DNA the corresponding proviruses.* The occasional presence in humans of viruses similar to those found in primates may well be due to accidental contact with viruses of murine origin related to the primate viruses GaLV or SSV, or from viruses of feline origin related to BaEV.

EVOLUTION OF ENDOGENOUS ONCOVIRUSES

The evolution of cellular DNA sequences related to type C viruses in primates has been investigated as an approach for determining whether these sequences have evolved at the rate of cellular genes, or at higher rates as viruses.

The RNA and the cDNA of the baboon endogenous virus, and the unique cellular sequences of the baboon, have been hybridized to the cellular DNA of the baboon and of other species. The difference in melting temperature (ΔTm) between a heterospecific hybrid and the homospecific hybrids was higher for viral than for cellular sequences; hence the *viral sequences have evolved more rapidly,* especially for certain species. Moreover, while some closely related species (langur and *Colobus,* mangabey and macaque) have distantly related viral sequences, some distant species (baboon and cat) have closely related viral sequences. These observations support a *proviral nature of the virus-related sequences in cellular DNAs.* In fact, a virus parasitizing a species would be expected both to participate in its evolution and to evolve independently, and it also might be able to cross species.

A related question is why *all* viral functions are conserved in those proviruses that are inducible. Clearly these functions are maintained by natural selection. Their maintenance is natural in viruses whose survival has been promoted by the infection of new cells from time to time. The alternative interpretation, that the viral genes are maintained because they perform functions valuable for the embryonic development of the host, is unlikely, because the number of genes expressed by endogenous proviruses is usually very limited. Furthermore, their expression in embryos varies enormously in different mouse strains, to the point of being undetectable in some; and in some species (e.g., the duck) no expression can be recognized. They thus do not appear to play an important part in development.

Therefore, the evidence strongly indicates that the virus-related sequences in the cellular DNA are proviruses derived from recurring infections, at intervals short enough to allow substantial evolution apart from that of the host.

OTHER ONCOVIRUSES

ONCOVIRUS B

Mouse Mammary Tumor Viruses (MMTVs). Some of these viruses are **exogenous,** while others are **endogenous** and genetically transmitted.

The prototype **exogenous** virus, **Bittner's virus,** was discovered as a result of questioning the seemingly obvious genetic basis of the high incidence of mammary cancers in some mouse strains such as C3H or RIII. In 1936 Bittner showed that mice of a low-cancer strain can be converted to a high incidence if newborns are nursed by females of a high-cancer strain. Conversely, the progeny of a high-cancer strain suckled by low-cancer strains had a low incidence of tumors. This simple experiment revealed that the cancer is induced by a transmissible agent present in the milk (the **milk agent**), later recognized as a virus.

Endogenous viruses were recognized, by both genetic studies and nucleic acid hybridization, in animals of many strains that develop spontaneous mammary neoplasms. The proviruses are present in all mouse strains, but not in other species.

Properties of the Viruses. The virions are generally similar to those of the C type described above, but differ in some details, such as the presence of spikes (Fig. 66-19) and the eccentricity of the cores (Fig. 66-20).

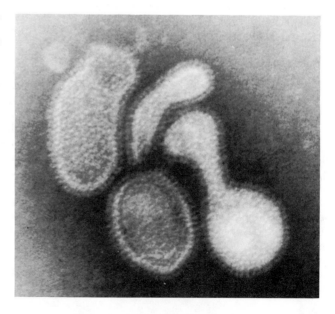

FIG. 66-19. Electron micrograph of the purified mouse mammary tumor virus (MMTV) with negative staining. Note the envelope, covered by spikes, which surrounds an internal component. **Marker** = 100 nm. (Lyons MJ, Moore DH: J Natl Cancer Inst 35:549, 1965)

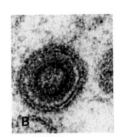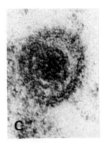

FIG. 66-20. Electron micrographs of thin sections of a mammary tumor producing mouse mammary tumor virus. **A.** Budding of the virions at the cell membrane. **B.** Immature virion with an electron-lucent core. **C.** Mature B-type particle with a dense, eccentric core. (Courtesy of L. Dmochowski)

Other differences include the molecular weights of the peptides, and a Mg^{2+} preference of the reverse transcriptase (containing two subunits) compared to a Mn^{2+} preference for type C viruses.

Viral replication in certain cell culture lines (such as the rat hepatoma line HTC, or lines of cat or mink cells) is similar to that of type C viruses. As with leukosis viruses, cells of many types are chronically infected but are not transformed. The **viral yield** in cultures is markedly **increased by glucocorticoids** (which probably increase transcription of the provirus). Halogenated pyrimidines are without effect, suggesting that the regulation of B and C oncoviruses is different.

Virus Types. MMTV-S (Bittner's virus) is exogenous, while MMTV-P of GR mice and MMTV-L of C3H mice are endogenous. All these viruses have grossly similar nucleic acid sequences but display minor antigenic differences in the coat proteins.

Assay. The bioassay uses as endpoint the development of breast cancer in susceptible, virus-free mice infected shortly after birth. It requires about 2 years and is only qualitative. Virus in sufficient amounts can be assayed by physical or biochemical methods, e.g., the number of virions, the amount of RNA hybridizable to cDNA, the activity of reverse transcriptase, the amount of Ag measured serologically.

Pathogenesis. MMTV-S viruses and -L viruses both induce **hyperplastic alveolar nodules,** but only the -S viruses cause mammary **adenocarcinomas,** which, once formed, are pregnancy-independent for growth.

The hyperplastic nodules, which are constituted of normal, milk-producing mammary tissue (Fig. 66-21),

contain large amounts of virus and appear to be the sites of cancer development. Thus pieces of infected mammary gland transplanted to gland-free mammary fat pads of isologous, virgin, virus-free females undergo neoplastic transformation more frequently if they contain nodules. On the other hand, MMTV-P virus induces **plaques** (not to be confused with those used for virus assay in cultures; see Ch. 46): these are benign tumors that appear during pregnancy and regress at its end. However, after several pregnancies they tend to generate pregnancy-independent carcinomas. These observations suggest that MMTVs of various kinds induce hyperplastic lesions (nodules, plaques) whose progression to cancer requires other factors.

Host and Hormonal Factors. The importance of hormonal factors is evident in the pregnancy-dependence of the plaques induced by the GR virus. Moreover, with the Bittner virus the cancer incidence is low in virgin infected females but high in infected females that are force-bred (i.e., made to bear litters in quick succession without nursing them) or given estrogen. It is even more striking that when virus-infected males are castrated and are injected with estradiol and deoxycorticosterone over a long period they frequently develop mammary cancer, whereas normal males have no mammary cancer at all.

Genetic Factors. The expression of an MMTV is strongly affected by the genetic constitution of the host, which determines the number of proviruses per haploid genome, and the time of tumor appearance.

Resistance of the C57BL strain to the Bittner virus (of which they are normally free) is determined by several genes that affect viral multiplication. Moreover, the development of cancers in the pres-

FIG. 66-21. Hyperplastic alveolar nodules in the mammary gland of a multiparous C3H female mouse bearing a mammary tumor. The nodules (some indicated by **arrows**) are filled with milk. (Nandi S et al: J Natl Cancer Inst 24:883, 1960) (Wholemount of gland, hematoxylin staining; ×6)

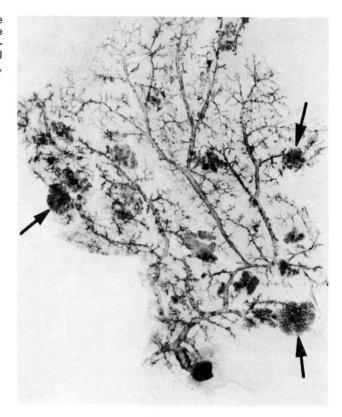

ence of the virus is controlled by two additional genes, for which the alleles for susceptibility are dominant. Various H2 alleles also affect the incidence of tumor formation.

Relevance in Human Breast Cancer.

It is not known whether viruses similar to MMTVs are involved in the genesis of human breast cancer. Particles containing 70s RNA and a reverse transcriptase have been isolated from neoplastic tissues, which sometimes contain Ags related to the viral core Ag of MMTV. Nucleic acid homology and serologic specificity show that the particles bear some relatedness to MMTV.

ONCOVIRUS D

The retroviruses of this group differ from others in virion morphology, nucleic acid sequences, and antigenicity. The virions resemble B oncovirus virions in the budding mechanism and in the preference of their reverse transcriptase for Mg^{2+}; they resemble type C virions in morphology and in the sizes of their peptides. Three members are known.

The exogenous **Mason–Pfizer monkey virus** (MPMV) was isolated from a mammary carcinoma of a rhesus monkey and also from normal rhesus tissues. Its glycoprotein contains interspecies

antigenic determinants that crossreact with the glycoprotein of the baboon endogenous virus (BaEV). The virus is propagated in rhesus or human cell lines, which continuously shed virus; as with MMTV, the yield is increased by dexamethasone. The syncytial plaque assay uses cells of a human glioma line transformed by RSV. The virus also transforms rhesus foreskin cultures, but inoculation in neonatal rhesus monkeys failed to induce neoplasia. This virus has been isolated from human cell cultures and from brain cells in Creutzfeldt–Jakob disease (a slow brain disease); its nucleic acid and Ags have also been found in other human tissues. Whether it is pathogenic or is an accidental invader is not clear.

Two endogenous type D viruses have been isolated, one from cells of an Asian primate (spectacled langur) and the other from a New World monkey (squirrel monkey). The langur virus sequences (but not those of the squirrel monkey virus) are related to MPMV.

CISTERNAVIRUS A

Intracytoplasmic A particles with a double shell and a clear core are observed by electron microscopy in the cisternae of the endoplasmic reticulum in oocytes, embryos, and neoplastic cells of murine origin, often associated with MMTV. Their structure, composition, and antigenic properties suggest that they are immature or defective MMTV nucleocapsids. The persistence of the particles in mouse hamster hybrids depends on the persistence of certain mouse chromosomes. They have no recognized biologic effects.

IMMUNOLOGIC RELATIONS OF VIRUS-TRANSFORMED CELLS TO ANIMAL HOSTS

Antigens that appear on the surface of transformed cells have great importance for the development of neoplasms, for they render the cells foreign to their own body, and the resulting **cell-mediated immune response** tends to destroy the transformed cells. Thus after neonatal thymectomy, or after administration of immunosuppressants, mice develop virus-induced leukemia more frequently and earlier. Conversely, mice **immunized with extracts of tumor** cells and later injected with virus develop leukemia less frequently and after a longer latent period. Moreover, **virus inoculation elicits immunity** to subsequently grafted transformed cells, presumably because the virus induces the appearance of its specific surface Ag in cells of the host, followed by an immune response.

In virus-producing cells transformed by herpesviruses or oncoviruses not only do the viral glycoproteins provide new cell surface Ags: a second type of **cell surface Ag, not present in the virions,** is found on the surface of cells transformed by **all oncogenic viruses.** It is not always known whether it is virus-specified or is a modified host Ag.

Cells transformed by papovaviruses have **tumor-specific transplantation Ags, TSTAs:** infection with SV40 induces resistance to transplantation of cells transformed by the same virus but not by polyoma virus, and vice versa (Table 66-5). The specificity suggests a viral origin, and adeno–SV40 hybrid viruses (see Fig. 66-8) show that SV40 TSTA is controlled by a central segment of the early region of the viral genome. In addition, serologic tests (in vitro cytolysis or immunofluorescence), with serum from animals immunized with transformed cells, reveal new **tumor-specific surface Ags, TSSAs.** These Ags are not necessarily related to TSTAs, and they may be less specific than TSTAs, because trypsin treatment of normal cells exposes an Ag that crossreacts with SV40 TSTA.

The TSSAs of cells transformed by avian or murine oncoviruses may be related to the recombinant *env* protein they contain (see Basis of Oncogenicity, above). In contrast, the **feline oncovirus-associated cell membrane Ag** (FOCMA) seems to be a polyprotein containing a part of the *gag* preprotein fused to a virus-specified nonvirion protein. Like TSTAs, FOCMA induces in cats an immunologic resistance to the induction of neoplasia by feline oncoviruses.

A third type of surface Ag on transformed cells, **oncofetal Ag,** crossreacts with Ags normally recognizable in the embryo but not in the adult.

Animals infected by oncovirus as embryos or at a young age, surprisingly, **do not develop immunologic tolerance:** although they contain much virus in their organs and blood, and can accept the transplantation of grafted tumors more readily than virus-free animals, they also produce Abs to viral Ags. As with the lymphocytic choriomeningitis virus (Ch. 62), the Abs are all bound to excess Ag; the Ag–Ab complexes are eliminated through the kidney, where they can be recognized.

In summary, various types of Ags in the surface of transformed cells induce in animals immunosurveillance against the tumor, probably because of the development of **killer cells** (see Cellular Immunity, Ch. 52). The **induction of Abs** also contributes to the resistance of the organism against tumor induction by Marek disease virus (see Oncogenic Herpesviruses) or by feline oncoviruses (see Immunologic Reactions of Virion Proteins); but formation of **blocking Abs** may favor tumor development.

"Slow" Infection. The presence of viral Ags on the surface of retrovirus-infected cells is also responsible for a different pathologic consequence—the "slow" infection produced in sheep brain by visna virus, a lentivirus (L. *lentus,* slow). The infection may not be directly harmful, but the strong cellular immune response elicited by the viral Ags (which are strong immunogens), reacting against similar Ags present in the cell surface, produces demyelinization and finally destruction of the nerve cells.

TABLE 66-5. Specificity of Resistance to Transformed Cells in Papovavirus-Immunized Hamsters*

Hamster immunized with	No. transformed cells inoculated per hamster			
	Polyoma-transformed		SV40-transformed	
	10^3		10^4	
	Fraction of recipients developing tumors			
Nothing	14/15	15/15	14/15	13/14
Polyoma virus	2/14	5/14	13/14	14/14
SV40	14/14	13/13	4/13	4/14

*Adult hamsters were immunized by intraperitoneal injection of about 2×10^6 PFU of either polyoma virus or SV40 and one month later they were inoculated subcutaneously with hamster cells transformed by either virus. The fractions give the number of animals later developing tumors, over the number receiving cells. The immunization with either virus is seen to be specific for the cells transformed by that virus.

(Data from Habel K, Ebby BE: Proc Soc Exp Biol Med 113:1, 1963)

COMPARISON OF ANIMAL CELL TRANSFORMATION WITH BACTERIAL LYSOGENIZATION

The mode of action of papovaviruses is akin to that of temperate phages (Ch. 48): both can elicit either a nonproductive transformation or a productive lytic infection, both viral DNAs are integrated in the cellular DNA, and both can induce cellular changes. However, no viral repressor has been identified in the animal system, whose stability appears to depend instead on the inability of the provirus to express all its genes—either because the viral genome is damaged or mutated, or because the cells are nonpermissive. Therefore the papovavirus system is less advantageous for viral spread than the lysogenic system, principally because it lacks an induction mechanism based on regulation.

The oncovirus system is even more advantageous than the lysogenic system, with which it shares a precise integration mechanism and the ability to convert cells: it lacks virulence and therefore does not need repression; and since it spreads as an RNA virus induction does not need provirus excision, only transcription.

SELECTED READING

BOOKS AND REVIEW ARTICLES

BISHOP JM: Retroviruses. Annu Rev Biochem 47:35, 1978

GRAF T, BEUG H: Avian leukemia viruses. Interaction with their target cells in vivo and in vitro. Biochim Biophys Acta 516:269, 1978

HIATT HH, WATSON JD, WINSTEN JA: Origins of human cancer. Book B. Mechanisms of Carcinogenesis. Cold Spring Harbor, NY, Cold Spring Harbor Laboratory, 1977

TOOZE J: The Molecular Biology of Tumor Viruses. Cold Spring Harbor, NY, Cold Spring Harbor Laboratory, 1979

WANG L-H: The gene order of avian RNA tumor viruses derived from biochemical analyses of deletion mutants and viral recombinants. Annu Rev Microbiol 32:561, 1978

WEISS R: Genetic transmission of RNA tumor viruses. Perspect Virol 9:165, 1975

SPECIFIC ARTICLES:

DNA VIRUSES

ASH JF, VOGT PK, SINGER SJ: Reversion from transformed to normal phenotype by inhibition of protein synthesis in rat kidney cells infected with a temperature-sensitive mutant of Rous sarcoma virus. Proc Natl Acad Sci USA 73:3603, 1976

BERK AJ, SHARP PA: Spliced early mRNA of Simian Virus 40. Proc Natl Acad Sci USA 75:1274, 1978

BLACK P, ROWE WP, TURNER HC, HUEBNER RJ: A specific complement fixing antigen present in SV40 tumor and transformed cells. Proc Natl Acad Sci USA 50:1148, 1963

BOTCHAN M, TOPP W, SAMBROOK J: The arrangement of Simian Virus 40 sequences in the DNA of transformed cells. Cell 9:269, 1976

CRAWFORD LV, COLE CN, SMITH AE, PAUCHA E, TEGTMEYER P, RUNDELL K, BERG P: Organization and expression of early genes of Simian Virus 40. Proc Natl Acad Sci USA 75:117, 1978

DULBECCO R, HARTWELL LH, VOGT M: Induction of cellular DNA synthesis by polyoma virus. Proc Natl Acad Sci USA 53:403, 1965

FIERS W, CONTRERAS R, HAEGEMAN T, ROGIERS R, VAN DE VOORDE A, VAN HEUVERSWYN H, VAN HERREWEGHE J, VOLCKAERT G, YSEBAERT M: Complete nucleotide sequence of SV40 DNA. Nature 273:113, 1978

FLUCK MM, STANELONI RJ, BENJAMIN T: Hr-t and ts-a: two early gene functions of polyoma virus. Virology 77:610, 1977

FRIED M: Cell transforming ability of a temperature-sensitive mutant of polyoma virus. Proc Natl Acad Sci USA 53:486, 1965

FRIED M, GRIFFIN BE, LUND E, ROBBERSON DL: Polyoma virus—A study of wild type, mutant, and defective DNA. Cold Spring Harbor Symp Quant Biol 34:45, 1975

HEINONEN OP, SHAPIRO S, MONSON RR, HARTZ SC, ROSENBERG L, SLONE D: Immunization during pregnancy against poliomyelitis and influenza in relation to childhood malignancy. Int J Epidemiol 2:229, 1973

HUTCHINSON MA, HUNTER T, ECKHART W: Characterization of T antigens in polyoma-infected and transformed cells. Cell 15:65, 1978

KAMEN R, LINDSTROM DM, SHURE H, OLD RW: Virus-specific RNA in cells productively infected or transformed by polyoma virus. Cold Spring Harbor Symp Quant Biol 34:187, 1975

KASCHKA-DIERICH C, ADAMS A, LINDHAL T, BARNKAMM GW, BJURSELL G, KLEIN G, GIOVANELLA BC, SINGH S: Intracellular forms of Epstein-Barr virus DNA in human tumor cells. Nature 260:302, 1976

KHOURY G, MARTIN M, LEE TNH, NATHANS D: A transcriptional map of the SV40 genome in transformed cell lines. Virology 63:263, 1975

KLEIN G, GIOVANELLA BC, LINDHAL T, FIALKOW PJ, SINGH S, STEHLIN JS: Direct evidence for the presence of Epstein-Barr virus DNA and nuclear antigen in malignant epithelial cells from patients with poorly differentiated carcinoma of the nasopharynx. Proc Natl Acad Sci USA 71:4737, 1974

MERZ JE, CARBON J, HERZBERG M, DAVIS RW, BERG P: Isolation and characterization of individual clones of Simian Virus 40 mutants containing deletions, duplications and insertions in their DNA. Cold Spring Harbor Symp Quant Biol 34:69, 1975

PRASAD I, ZOUZIAS D, BASILICO C: State of the viral DNA in rat cells transformed by polyoma virus. J Virol 18:436, 1976

REED SI, STARK GR, ALWINE JC: Autoregulation of Simian Virus 40 gene A by T antigen. Proc Natl Acad Sci USA 73:3083, 1976

RISSER R, RIFKIN D, POLLACK R: The stable classes of transformed cells induced by SV40 infection of established 3T3 cells and primary rat embryonic cells. Cold Spring Harbor Symp Quant Biol 34:317, 1975

SAMBROOK J, WESTPHAL H, SRINIVASAN P, DULBECCO R: The integrated state of viral DNA in SV40 transformed cells. Proc Natl Acad Sci USA 60:1288, 1968

SHOPE RE: Infectious papillomatosis of rabbits. J Exp Med 58:607, 1933

SMITH AE, SMITH R, GRIFFIN B, FRIED M: Protein kinase activity associated with polyoma virus middle T antigen in vitro. Cell 18:915, 1979

VOGT M, DULBECCO R: Virus-cell interaction with a tumor-producing virus. Proc Natl Acad Sci USA 46:365, 1960

WATKINS JF, DULBECCO R: Production of SV40 virus in heterokaryons of transformed and susceptible cells. Proc Natl Acad Sci USA 58:1396, 1967

ONCORNAVIRUS C

BALUDA MA: Widespread presence in chicken of DNA complementary to the RNA genome of avian leukosis viruses. Proc Natl Acad Sci USA 69:576, 1972

BENVENISTE RE, LIEBER MM, TODARO G: A distinct class of inducible murine type C viruses which replicate in the rabbit SIRC line. Proc Natl Acad Sci USA 71:602, 1974

ELDER JH, GAUTSCH JW, JENSEN FC, LERNER RA, HARTLEY JW, ROWE WP: Biochemical evidence that MCF murine leukemia viruses are envelope (env) gene recombinants. Proc Natl Acad Sci USA 74:4676, 1977

ERIKSON RL, COLLETT MS, ERIKSON E, PURCHIO AF: Evidence that avian sarcoma virus transforming gene product is a cyclic AMP—independent protein kinase. Proc Natl Acad Sci USA 76:6260, 1979

GROSS L: "Spontaneous" leukemia developing in C3H mice following inoculation, in infancy, with AK-leukemic extracts or AK-embryos. Proc Soc Exp Biol Med 76:27, 1951

HARADA F, SAWYER RC, DAHLBERG JE: A primer ribonucleic acid for initiation of in vitro Rous sarcoma virus deoxyribonucleic acid synthesis. J Biol Chem 250:3487, 1975

HARTLEY JW, WOLFORD NK, OLD LJ, ROWE WP: A new class of murine leukemia virus associated with development of spontaneous lymphomas. Proc Natl Acad Sci USA 74:789, 1977

HUGHES SH, SHANK PR, SPECTOR DH, KUNG HJ, BISHOP M, VARMUS HE, VOGT PK, BREITMAN ML: Proviruses of avian sarcoma virus are terminally redundant, co-extensive with unintegrated linear DNA and integrated at many sites. Cell 15:1397, 1978

KAPLAN HS: Influence of thymectomy, splenectomy and gonadectomy on incidence of radiation-induced lymphoid tumors in strain C-57 BL mice. J Natl Cancer Inst 11:83, 1950

KUNG HJ, HU S, BENDER W, BAILER JM, DAVIDSON N, NICOLSON MO, MCALLISTER RM: RD-114, Baboon, and Wooly monkey viral RNAs compared in size and structure. Cell 7:609, 1976

JAENISCH R: Germ line integration and Mendelian transmission of the exogenous Moloney leukemia virus. Proc Natl Acad Sci USA 73:1260, 1976

LAI CJ, NATHANS D: Mapping the genes of Simian Virus 40. Cold Spring Harbor Symp Quant Biol 34:53, 1975

LOWY DR, ROWE WP, TEICH N, HARTLEY JW: Murine leukemia virus: high frequency activation in vitro by 5-iododeoxyuridine and 5-bromodeoxyuridine. Science 174:155, 1971

MARTIN GS: Rous sarcoma virus: a function required for the maintenance of the transformed state. Nature 227:1021, 1970

MCCLAIN DA, MANESS PF, EDELMAN GM: Assay for early cytoplasmic effects of the src gene product of Rous sarcoma virus. Proc Natl Acad Sci USA 75:2750, 1978

PURCHIO AF, ERIKSON E, BRUGGE JS, ERIKSON RL: Identification of a polypeptide encoded by the avian src gene. Proc Natl Acad Sci USA 75:1567, 1978

ROUS P: Transmission of a malignant new growth by means of a cell-free filtrate. JAMA 56:198, 1911

STEHELIN D, GUNTAKA RV, VARMUS HE, BISHOP JM: Purification of DNA complementary to nucleotide sequences required for neoplastic transformation of fibroblasts by avian sarcoma viruses. J Mol Biol 101:349, 1976

TEMIN H: The RNA tumor viruses: Background and foreground. Proc Natl Acad Sci USA 69:1016, 1972

TROXLER DH, BOYARS JK, PARKS WP, SCOLNICK EM: Friend strain of spleen focus-forming virus: a recombinant between mouse type C ecotropic viral sequences and sequences related to xenotropic virus. J Virol 22:361, 1977

VARMUS HE, GUNTAKA RV, FAN WJW, HEASLEY S, BISHOP JM: Synthesis of viral DNA in the cytoplasm of duck embryo fibroblasts and in enucleated cells after infection by avian sarcoma virus. Proc Natl Acad Sci USA 71:3874, 1974

WANG LH, GALEHOUSE D, MELLON P, DUESBERG P, MASON WS, VOGT PK: Mapping oligonucleotides of Rous sarcoma virus RNA that segregate with polymerase and group-specific antigen markers in recombinants. Proc Natl Acad Sci USA 73:3952, 1976

WANG LH, HALPERN CC, NADEL M, HANAFUSA H: Recombination between viral and cellular sequences generates transforming sarcoma virus. Proc Natl Acad Sci USA 75:5812, 1978

WITTE ON, ROSENBERG N, PASKIND M, SHIELDS A, BALTIMORE D: Identification of an Abelson leukemia virus-encoded protein present in transformed fibroblasts and lymphoid cells. Proc Natl Acad Sci USA 75:2488, 1978

ONCORNAVIRUS B

BITTNER JJ: Some possible effects of nursing on the mammary gland tumor incidence in mice. Science 84:162, 1936

LYONS MJ, MOORE DH: Isolation of the mouse mammary tumor virus: chemical and morphological studies. J Natl Cancer Inst 35:549, 1965

NANDI S, HELMICH C: Transmission of mammary tumor virus by the GR mouse strain. Role of the virus in the production of lesions. J Natl Cancer Inst 52:1285, 1974

SARKAR NH, POMENTI AA, DION AS: Replication of mouse mammary tumor virus in tissue culture. Kinetics of virus production and the effect of RNA and protein inhibitors in viral synthesis. Virology 77:31, 1977

SCOLNICK EM, YOUNG HA, PARKS WP: Biochemical and physiological mechanisms in glucocorticoid hormone induction of mouse mammary tumor virus. Virology 69:148, 1976

WEISS DW, FAULKIN LJ JR, DEOME KB: Acquisition of heightened resistance and the susceptibility to spontaneous mouse mammary carcinomas in the original host. Cancer Res 24:732, 1964

chapter **67**

STERILIZATION AND DISINFECTION

BERNARD D. DAVIS
RENATO DULBECCO

Technics for sterilizing materials—i.e., for freeing them of contaminating viable organisms—were first developed as a prerequisite for the preparation of pure cultures in the laboratory. They were then rapidly adapted, in medicine, surgery, and public health, to prevent the spread of infectious disease (Ch. 1). This chapter will consider first the action of various lethal agents on bacteria and then special features of their action on viruses.

History. The early arts of civilization included practical means of preventing putrefaction and decay long before the role of microorganisms in these processes was appreciated. Perishable foods were preserved by drying, by salting, and by acid-producing fermentations. Embalming was practiced in ancient Egypt, but the essential oils used were probably less important than the dry climate. As was noted in Chapter 1, the canning of food was introduced 50 years before Pasteur's researches gave it a rational basis. Finally, chlorinated lime (calcium hypochlorite) and carbolic acid (phenol) were introduced in the early nineteenth century to deodorize sewage and garbage (and subsequently wounds), even before their germicidal action was recognized.

DEFINITIONS

Sterilization denotes the use of either physical or chemical agents to eliminate **all** viable microbes from a material, while **disinfection** generally refers to the use of germicidal chemical agents to destroy the potential **infectivity** of a material (which need not imply elimination of all viable microbes). **Sanitizing** refers to procedures used to lower the bacterial content of utensils used for food, without necessarily sterilizing them. **Antisepsis** usually refers to the topical application of chemicals to a body surface to kill or inhibit pathogenic microbes.

In contrast to chemotherapeutic agents, **disinfectants** must be effective against all kinds of microbes, must be relatively insensitive to their metabolic state, and need not be harmless to host cells. They are widely used for skin antisepsis in preparation for surgery, but for prophylactic application to open wounds, or for topical application to superficial infections, they have been largely replaced by various antibiotics which are painless and less damaging to the tissues.

CRITERIA OF VIABILITY

Unlike chemotherapeutic agents, which may be bactericidal or only bacteriostatic (Ch. 7), an effective disinfectant must be **bactericidal,** i.e., it must **destroy the ability of the organism to multiply** when placed in a suitable environment. Staining by dyes that ordinarily do not penetrate may be a useful indirect criterion of killing by heat or other methods that disrupt a permeability barrier, but such tests are not useful for methods, such as UV irradiation, that damage only DNA.

Tests for damaged cells' viability may give quite different cell counts in different media: the repair of damaged cells is evidently influenced by several factors, including osmotic tonicity and nutritional richness. For example, spores "sterilized" by Hg^{2+} can be **"resurrected"** by a wash in H_2S solution, which displaces Hg^{2+} from its inhibitory complexes with –SH groups in cell constituents.

The problem of defining viability assumes practical importance in the preparation of vaccines, which are often sterilized as gently as possible in order to retain maximal immunogenicity. Killing is ordinarily tested in culture media, but it is necessary to be sure that the organisms have also lost the ability to initiate infection in the animal body.

It should be emphasized that sterilization is not identical with **destruction** of bacteria or their products. For example, in solutions for intravenous administration it is necessary not only to ensure sterility but also to minimize prior bacterial contamination, since **pyrogenic** bacterial products (which can survive autoclaving or filtration) may subsequently produce a febrile, toxic response. Hence the water and the reagents used in the preparation of biologicals must satisfy criteria of purity quite different from those required for analytic chemical work.

Differential Susceptibility. The susceptibility of cells to disinfectants or to heat varies with their physiologic state. The cells in a young culture are generally more susceptible than those in a stationary-phase culture, in which significant changes in the cell wall and membrane have been noted (Ch. 6).

EXPONENTIAL KINETICS

Sterilization of bacteria by many agents exhibits the kinetics of a first-order reaction, in which the logarithm of the number of survivors decreases as a linear function of time of exposure:

$$\ln n = \ln n_o - kt$$

Such kinetics, observed for interactions with UV or ionizing radiation, are discussed in the Appendix of Chapter 11. Perhaps surprising is that similar kinetics are often observed for the action of heat (Fig. 67-1) or of various chemical disinfectants, such as phenol. Hence, even though heat and phenol no doubt damage a variety of protein molecules, in a cumulative manner, the lethal event in a given cell cannot be the denaturation of the last of many molecules of a given essential species of protein, for this process would give a multi-hit curve (see Fig.

67-8, below). The one-hit curve suggests, instead, damage to one or another single indispensable molecule, perhaps triggering irreversible damage to the cell membrane or irreversibly impairing DNA replication.

Heterogeneity of the population of cells, with respect to either composition or physiologic state, will of course distort first-order kinetics in the direction of multicomponent kinetics (Ch. 11; see also Fig. 67-8). Hence it is not surprising that the best exponential killing curves have been obtained with spores, which are more uniform than vegetative cells.

PHYSICAL AGENTS

TEMPERATURE

HEAT

Heat is generally preferred for sterilizing materials except those that it would damage. The process is rapid, all organisms are susceptible, and the agent penetrates clumps and reaches surfaces that might be protected

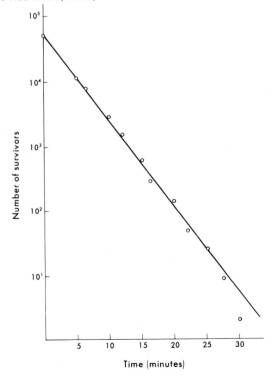

FIG. 67-1. Exponential killing, by heat, of spores of a thermophilic bacillus. Number of survivors are plotted on a logarithmic scale against time of exposure to 120°. (After Williams CC et al: Food Res 2:369, 1937)

from a chemical disinfectant. Fungi, most viruses, and the vegetative cells of various pathogenic bacteria are sterilized within a few minutes at 50°–70°C, and the spores of various pathogens at 100°. Consequently, it has been a common practice to sterilize syringes, needles, and instruments for minor surgery by heating for 10–15 min in boiling water or, even better, in a boiling dilute solution of alkali (e.g., washing soda).

The spores of some saprophytes, however, can survive boiling for hours. Since absolute sterility is essential for culture media, and for the instruments used in major surgical procedures, it has become standard practice to sterilize such materials by steam in an autoclave at a temperature of 121°C (250°F) for 15–20 min. At altitudes near sea level this temperature is attained by steam at a pressure of 15 lb per square inch (psi) in excess of atmospheric pressure. At high altitudes higher pressures are necessary (e.g., 3 psi higher in Denver, at an altitude of 5000 ft). The rapid action of steam depends in part on the large latent heat of water (540 cal/g): cold objects are thus rapidly heated by condensation on their surface.

In using an autoclave it is important that flowing steam be allowed to displace the air before building up pressure, for in steam mixed with air the temperature is determined by the partial pressure of the water vapor. Thus if air at 1 atm (15 psi) remains in the chamber and steam is added to provide an additional gauge pressure of 1 atm the average temperature will be only 100° (that of steam at 1 atm). Moreover, heating will be uneven since the air will tend to remain at the bottom of the chamber.

Vessels must be loosely plugged or capped and not completely filled with liquid, in order to permit free ebullition of the dissolved air during heating and free boiling of the superheated liquid when the steam pressure is lowered. With bulky porous objects (e.g., bundles of surgical dressings), or with large volumes of liquid, increased time must be allowed to ensure heating throughout. In modern autoclaves the steam pressure may be maintained in an outer jacket while the central chamber is decompressed, so that condensation water is rapidly evaporated.

Pasteurization, introduced to sterilize wine, is now used primarily for milk. It consists of heating at 62°C for 30 min or, in "flash" pasteurization, at a higher temperature for a fraction of a minute. Pasteurization is effective because the common milkborne pathogens (tubercle bacillus, *Salmonella, Streptococcus,* and *Brucella*) do not form spores and are reliably sterilized by this procedure; in addition, the total bacterial count is generally reduced by 97%–99%.

Kinetics. The sensitivity of an organism to heat is often expressed in practical work as the **thermal death point:** the lowest temperature at which a 10-min exposure of a given volume of a broth culture results in sterilization. The value is about 55°C for *E. coli,* 60° for the tubercle bacillus, and 120° for the most resistant spores.

For precise studies this qualitative endpoint has been replaced by quantitative determination of the numbers of survivors at different times. Since killing by heat turns out to have simple exponential kinetics the rate of killing can be expressed in terms of the **rate constant** k (see above) in the exponential decay curve (Fig. 67-1). Another convenient index is the **decimal reduction time,** D (the time required for a 10-fold reduction of viability), which is inversely proportional to the rate of killing. The logarithm of D varies linearly with temperature (Fig. 67-2); and from the slope of the curve it can be seen that the rate of killing (of the spores studied in this figure) increased about 10-fold with a rise of 10°. A suspension containing 10^5 cells would require about $5 \times D$ min to reduce the viable number to 1 cell.

Mechanism. Sterilization by heat involves **protein denaturation** (though the "melting" of membrane lipids may also be important): the temperature range of sterilization is one in which many proteins are denatured, with a high temperature coefficient. Moreover, both processes require a higher temperature when the material is thoroughly dried, or when the water activity of the medium is reduced by the presence of a high concentration of a neutral substance such as glycerol or glucose.

Moist Heat and Dry Heat. Denaturation is an unfolding of protein molecules from their native conformation as a consequence of the disruption of multiple bonds. Among these, hydrogen bonds (especially between a >C=O and an HN< group) are prominent, and they are more readily broken if they can be replaced with hydrogen bonds to water molecules (Fig. 67-3). Accordingly, bacteria and viruses, like isolated enzymes, require a higher temperature for irreversible damage in the dry state: reliable sterilization of spores by dry heat requires 160°C for 1–2 h. Moreover, hot air penetrates porous materials much more slowly than does condensing steam, so that after an hour in an oven at 160°C the center of a large package of surgical dressings may not even have reached 100°. Dry ovens are ordinarily used for sterilizing only glassware and metal objects. Intense dry heat is used in flaming contaminated surfaces in the process of aseptic transfer* of media or cultures and in disposing of infectious material by incineration.

---

* In the early days of developing a satisfactory ritual for aseptic transfer Pasteur recommended quick flaming of the bacteriologist's hands!

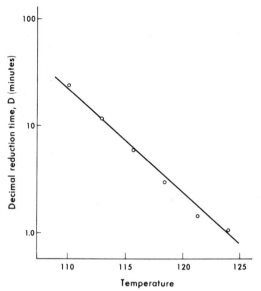

FIG. 67-2. Rate of killing of spores of a thermophilic bacillus at various temperatures, in terms of the time required for a ten-fold reduction in viability. (Schmidt CF: In Reddish CF [ed]: Antiseptics, Disinfectants, Fungicides, and Sterilization. Philadelphia, Lea & Febiger, 1957, p. 831)

The role of water in heat denaturation of proteins is illustrated by the usefulness of steam in pressing woolen fabrics (i.e., in shifting the multiple weak bonds between fibrous molecules of keratin).

FREEZING

When a suspension of bacteria is frozen the crystallization of the water results in the formation of tiny pockets of concentrated solutions of salts, which do not themselves crystallize unless the temperature is lowered below the eutectic point (about −20°C for NaCl); at this temperature the solution becomes saturated and the salt also crystallizes. The localized high concentrations of salt, and possibly the ice crystals, damage the bacteria, as shown by their increased sensitivity to lysozyme. Only some of the cells are killed, but repeated cycles of freezing and thawing result in a progressive decrease in the viable count.

Preservation. Once frozen, the surviving cells retain their viability indefinitely if the temperature is kept below the eutectic points of the various salts present; a household freezing unit (about −10°C) is not cold enough, but a satisfactory temperature is provided by solid CO_2-dry ice (−78°) or liquid N_2 (−180°). Freezing is therefore a useful means of preserving viable cultures.

In preserving bacteria, viruses, or animal cells by freezing it is helpful to add a relatively high concentration of glycerol, dimethylsulfoxide, or protein. These agents promote amorphous, vitreous solidification on cooling, instead of crystallization, thus avoiding local high concentrations of salt. Similarly, protein-rich materials (milk, serum) are added in the preservation of bacteria by **lyophilization** (desiccation from the frozen state).

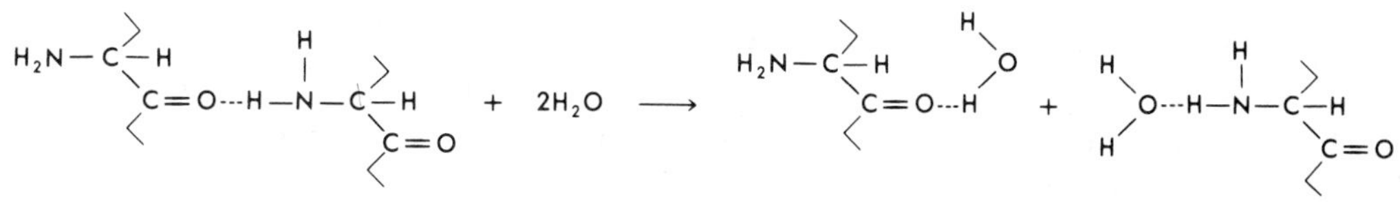

FIG. 67-3. Role of water in promoting the denaturation of protein by heat, by facilitating disruption of intramolecular hydrogen bonds between peptide groups.

RADIATION

ULTRAVIOLET RADIATION

With radiation of decreasing wave length the sterilization of bacteria first becomes appreciable at 330 nm and then increases rapidly (Fig. 67-4). The sterilizing effect of sunlight is due mainly to its content of UV light (300–400 nm). Most of the UV light approaching the earth from the sun, and all of that below 290 nm, is screened out by the ozone in the outer regions of the atmosphere; otherwise organisms could not survive on the earth's surface.

Photochemistry. The energy of UV light is absorbed in quanta by molecules of an appropriate structure. The absorbing molecule is thereby activated, resulting either in increased interatomic vibration or in excitation of an electron to a higher energy level. The activated molecule may undergo rupture of a chemical bond and formation of new bonds; or it may transfer most of its extra energy by collision to an adjacent molecule, which can then similarly undergo a chemical reaction; or the energy may be entirely dissipated by collision as increased translational energy (heat).

UV absorption by bacteria is due chiefly to the purines and pyrimidines of nucleic acids, with an average maximum at 260 nm. In addition, the aromatic rings of tryptophan, tyrosine, and phenylalanine in proteins absorb more moderately with an average maximum at 280 nm. The sterilization action spectrum (i.e., the efficiency of sterilization by radiation of various wave lengths: Fig. 67-4) parallels the absorption spectrum of the bacteria, suggesting that absorption by either nucleic acid or protein can have a lethal effect.

The **mechanism** of killing by UV light has been discussed in Chapter 11 (DNA Damages), where it was noted that lethal mutations make only a small contribution; the major mechanism is a series of alterations in the DNA that block its replication. Since most of these alterations can be repaired, by several different mechanisms, the quantum efficiency of UV sterilization is ordinarily very low.

The contribution of damage to other parts of the cell is negligible. Accordingly, in obtaining mRNA or proteins produced by a phage it is useful to infect cells that have been heavily irradiated, so that none of the cellular DNA is any longer transcribed: despite this overkill, the added infecting phage DNA is effectively transcribed and translated.

Practical Uses. Inexpensive low-pressure mercury vapor lamps, emitting 90% of their radiation at 254 nm, are widely used to decrease airborne infection, e.g., in places of public crowding, barracks, hospital wards, sur-

gical operating rooms, and rooms containing experimental animals. The effectiveness of UV treatment of air in public places seems uncertain, but it has been convincingly demonstrated in hospital wards and in animal houses, where the infected individuals cannot make close contact with other individuals.

In laboratory areas used for bacterial transfers UV lamps are similarly useful to decrease the contamination of cultures and infection of workers. It is important to protect the eyes by eyeglasses (glass is opaque to UV), since excessive exposure of the cornea to UV causes severe irritation, with a latent period of about 12 h.

In preparing killed bacterial vaccines UV irradiation has a theoretic advantage, since the genome is much more sensitive to UV damage than are the surface Ags. Nevertheless, this approach has not been particularly successful. One problem is the difficulty of avoiding clumps, in whose centers virulent organisms remain unexposed. Moreover, tissue extracts are quite opaque to UV light.

Ionizing Radiations. The lethal and mutagenic actions of x-ray and other ionizing radiations have been discussed in Chapter 11. Intense sources of radioactivity can be used to sterilize hospital goods, foods, etc. With foods, however, the large doses required (millions of rads) often have undesirable effects on flavor.

FIG. 67-4. Action spectrum of UV killing of *E. coli.* (After Gates, FL: J Gen Physiol 14:31, 1930)

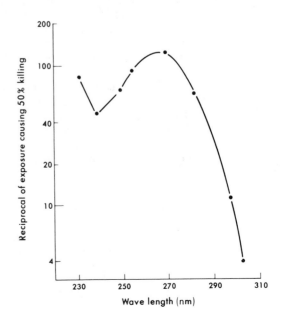

PHOTODYNAMIC SENSITIZATION (PHOTOOXIDATION)

In strong **visible light** certain fluorescent dyes (e.g., methylene blue, rose bengal, eosin) sterilize bacteria and viruses (and also lyse RBCs and denature proteins). These dyes retain an absorbed quantum for a comparatively long time (10^{-6}–10^{-8} sec), during which the energy sometimes is transferred to another molecule instead of being emitted as fluorescence. This transfer leads to oxidation of certain residues in proteins (especially histidine and tryptophan) and in nucleic acids.

Even in the absence of added dyes, intense visible light is capable of killing bacteria, presumably via physiologically occurring photosensitizing substances, such as **riboflavin** and **porphyrins.** It is therefore inadvisable to expose bacterial cultures to direct sunlight, even when protected by UV-absorbing glass. For example, BCG vaccine in glass ampules can lose all viability and effectiveness on exposure to bright sunlight in outdoor field stations.

Psoralen markedly increases sensitivity to killing by near UV: the radiation causes it to form cross-links between the two strands of DNA.

MECHANICAL AGENTS

ULTRASONIC AND SONIC WAVES

In the supersonic (ultrasonic) range, with a frequency of 15,000 to several hundred thousand per second, sound waves denature proteins, disperse a variety of materials, and sterilize and disintegrate bacteria. (Audible sonic waves of sufficient intensity are also bactericidal.) The effect has not been of practical value as a means of sterilization but it is useful for disrupting cells for experimental purposes **(sonication).**

FILTRATION

Bacteria-free filtrates may be obtained by the use of filters with a maximum pore size not exceeding 1 nm. This procedure is used for solutions that cannot tolerate sterilization by heat (e.g., sera, and media containing proteins or labile metabolites). The early, rather adsorptive filters of asbestos or diatomaceous earth were replaced by unglazed porcelain or sintered glass, and these in turn have been replaced by disposable membrane filters, of graded porosity, made of nitrocellulose.

Membrane filters can also be used to recover bacteria quantitatively for chemical or microbiologic analysis. Moreover, colonies can be grown from single cells on such a filter resting on a solid nutrient medium, and transfer of the filter permits convenient exposure of the colonies to a succession of different media.

CHEMICAL AGENTS

Among the many chemicals (including nutrients such as O_2 and fatty acids) that are bacteriostatic and even bactericidal at sufficiently high concentrations, the term **disinfectant** is restricted to those that are rapidly bactericidal at low concentrations. In contrast to most chemotherapeutic agents, which interact with various active metabolic systems, most disinfectants act either by dissolving lipids from the cell membrane (detergents, lipid solvents) or by damaging proteins or nucleic acids (denaturants, oxidants, alkylating agents, sulfhydryl reagents).

The rate of killing by disinfectants increases with concentration and with temperature. Anionic compounds are more active at low pH and cationic compounds at high pH. This effect results from the greater penetration of the undissociated form of the inhibitor, and possibly also from the increase in opposite charges in cell constituents.

DETERMINATION OF DISINFECTANT POTENCY

Phenol Coefficient. Ever since Lister began spraying surgical operating rooms with phenol this compound (see Specific Chemical Agents, below) has been considered the standard disinfectant, though it is required in a relatively high concentration: 0.9% to sterilize a suspension of *Salmonella typhosa* under ordinary conditions in 10 min. The phenol coefficient of a compound is the ratio of the minimal sterilizing concentration of phenol (under standard conditions) to that of the compound. In the official test of the US government a broth culture is diluted 1:10 with various concentrations of the test compound; the endpoint is the lowest concentration that yields, after incubation for 10 min at 20 °C, sterile loopful samples. The germicide is generally recommended for use at five times this concentration. Two organisms are ordinarily used: *S. typhosa* (as a representative enteric pathogen) and *Staphylococcus aureus* (the major source of wound infection).

The phenol coefficient provides a reasonable index for comparing various phenol derivatives, which exhibit similar kinetics and mode of action, but it is less satisfactory for other agents, which may differ in their concentration–action curves and in their susceptibility to neutralization by the environment. Thus the concentrations of a disinfectant, c, required to sterilize a bacterial population in varying time, t, generally correspond to a curve which may be fitted by the equation ($c^n t = k$); and while phenol has the remarkably high concentration coefficient *(n)* of 5 to 6, oxidants such as hypochlorite have a value of about 1. Hence the fivefold increase above the endpoint concentration provides a much wider margin of safety for phenols than for oxidants.

Problems in Evaluating Disinfectants. Some agents (e.g., mercurials, detergents) may adhere to the bacteria and thus exert a bacteriostatic action, in the subinoculated samples, that mimics bactericidal action. To test such materials for bactericidal action it is important to include a **neutralizing compound** in the test medium.

In addition, the effectiveness of a disinfecting procedure often depends strongly on the "cleanness" of the material. The presence of a large amount of organic matter (e.g., in excreta or discarded cultures) rapidly neutralizes the action of many agents, either by chemical reaction (e.g., with oxidants) or by adsorption. Moreover, the drying of a solution may encase bacteria in crystals and thus protect them from bactericidal gases.

Because of these considerations, different kinds of disinfectants are used for different purposes, such as skin antisepsis, sanitizing food containers, or rendering discarded cultures harmless. Efficacy must finally be tested under the conditions of use.

As with killing by heat (Fig. 67-1), the sensitivity of an organism to a disinfectant can be expressed more precisely as the slope of the semilogarithmic curve for killing as a function of time, rather than as an endpoint for complete sterilization. However, the curves for chemical disinfection are often imperfectly exponential, with the physiologically more resistant members of the population surviving longer than would be predicted by extrapolation. Hence the endpoint remains of practical value. The statistical problem of defining complete sterilization will be further considered below in connection with the preparation of viral vaccines.

SPECIFIC CHEMICAL AGENTS

Acids and Alkalis. Strongly acid and alkaline solutions are actively bactericidal. However, mycobacteria are relatively resistant, it being common practice to liquefy sputum, before plating, by exposure for 30 min to 1 N NaOH or H_2SO_4. This procedure depends on the survival of a fraction of the population, rather than on complete resistance of the bacteria. Gram-positive staphylococci and streptococci also frequently survive.

Weak organic acids exert a greater effect than can be accounted for by the pH: the presence of highly permeable undissociated molecules promotes penetration of the acid into the cells, and the increasing activity with chain length suggests a direct action of the organic compound itself. (Long-chain fatty acids will be considered under Surface-Active Agents.) Lactic acid is the natural preservative of many fermentation products; and salts of propionic acid (CH_3CH_2COOH) are now frequently added to bread and other foods to retard mold growth.

Salts. Pickling in brine, or treatment with solid NaCl, has been used for many centuries as a means of preserving perishable meats and fish. Bacteria vary widely in susceptibility.

Though physiologic saline (0.9% NaCl) is widely used as a diluent for bacteria, it is not very suitable: a balanced salt solution, containing Mg^{2+} and buffer, permits much better survival. Strains vary widely in their ability to survive in distilled water. Some of the killing, however, is due to traces of heavy metal ions, which are more bactericidal in the absence of competitive ions.

Heavy Metals. The various metallic ions can be arranged in a series of decreasing antibacterial activity. Hg^{2+} and Ag^+, at the head of the list, are effective at less than 1 part per million (ppm), because of their high affinity for –SH and other groups: bacteria killed by Ag^+ contain 10^5–10^7 Ag^+ ions per cell. The concentration required for killing is therefore markedly affected by the inoculum size. As was noted above, the antibacterial action of Hg^{2+} can be readily reversed by sulfhydryl compounds.

Various organic mercury compounds (e.g., Merthiolate, Mercurochrome, Metaphen), in which one of the valences of Hg is covalently combined, have been used as relatively nonirritating antiseptics, and also as preservatives for sera and vaccines. Silver has long been used in various forms as a mild antiseptic. Copper salts have great importance as fungicides in agriculture but not in medicine.

Inorganic Anions. Inorganic anions are much less toxic than some of the cations. Boric acid has found wide use as a mild antiseptic.

Halogens. **Iodine** combines irreversibly with proteins (e.g., by iodinating tyrosine residues), and it is an oxidant. **Tincture of iodine** (a 2%–7% solution of I_2 in aqueous alchohol containing KI) is a rapidly acting bactericide. It is a reliable antiseptic for skin and for minor wounds, but it has a painful and destructive effect on exposed tissue. I_2 complexes spontaneously with detergents to form **iodophors,** which provide a readily available reservoir of bound I_2 in equilibrium with free I_2 at an effective but nonirritating concentration.

Chlorine was the antiseptic introduced (as chlorinated lime) by O. W. Holmes in Boston in 1835, and by Semmelweis in Vienna in 1847, to prevent transmission of puerperal sepsis by the physician's hands. Chlorine combines with water to form hypochlorous acid (HOCl), a strong oxidizing agent:

$$Cl_2 + H_2O \rightleftharpoons HCl + HOCl$$
$$\text{or}$$
$$Cl_2 + 2NaOH \rightleftharpoons NaCl + NaOCl + H_2O$$

Hypochlorite solutions (200 ppm Cl_2) are used to sanitize clean surfaces in the food and the dairy industries and in restaurants; and Cl_2 gas, added at 1–3 ppm, is widely used to disinfect water supplies and swimming pools. Chlorine is a reliable, rapidly acting disinfectant for such "clean" materials, but it is less satisfactory for materials containing organic matter.

The "chlorine demand" of a water supply increases with its content of organic matter, and chlorination must be titrated to a definite level of free Cl_2. This reactivity of Cl_2 is a virtue in the sanitizing of food utensils: residual traces of chlorine will be rapidly destroyed on subsequent contact with food, leaving no flavor or odor.

Other Oxidants. **Hydrogen peroxide** (H_2O_2), in a 3% solution, was once widely used as an antiseptic, but bacteria vary greatly in their susceptibility, since some species possess catalase. **Potassium permanganate** ($KMnO_4$) is of value as a urethral antiseptic in concentrations around 1:1000. **Peracetic acid** ($CH_3CO-O-OH$), a strong oxidizing agent, is used as a vapor for the sterilization of chambers for germ-free animals; its reliable disinfection compensates for the inconveniently long flushing required because of its toxicity for animals. These compounds, as well as the halogens, presumably act by oxidizing –SH and S–S groups of enzymes and membrane components.

Alkylating Agents. Formaldehyde and ethylene oxide replace the labile H atoms on –NH₂ and –OH groups, which are abundant in proteins and nucleic acids, and also on –COOH and –SH groups of proteins (Fig. 67-5). The reactions of formaldehyde are in part reversible, but the high-energy epoxide bridge of ethylene oxide leads to irreversible reactions. These alkylating agents, in contrast to other disinfectants, are nearly as active against spores as against vegetative bacterial cells, presumably because they can penetrate easily (being small and uncharged) and do not require H_2O for their action.

Formaldehyde is a gas, usually marketed as a 37% aqueous solution (formalin). It has long been used at about 0.1% in preparing vaccines: i.e., for sterilizing bacteria, or inactivating toxins or viruses, without destroying their antigenicity.

Formaldehyde may be used as a gas for sterilizing dry surfaces. However, it is extensively absorbed on surfaces as a reversible polymer (paraformaldehyde), whose slow subsequent depolymerization provides an irritating residue.

Ethylene oxide, a highly water-soluble gas, has proved to be the most reliable substance available for gaseous disinfection of dry surfaces. However, its action is slower than that of steam, and its use is more expensive and presents some hazard of residual toxicity (vesicant action). Indeed, the potential hazards of mutagenicity and carcinogenicity for humans deserve careful investigation, since formaldehyde and ethylene oxide, like other alky-

lating agents (Ch. 11), have been shown to be mutagenic to bacteria, plant seeds, and *Drosophila*.

Ethylene oxide is widely used to sterilize heat-sensitive objects: plasticware; surgical equipment; hospital bedding; and books, leather, etc. handled by patients. Ethylene oxide is explosive, but this hazard is eliminated by using a mixture with 90% CO_2 or a fluorocarbon.

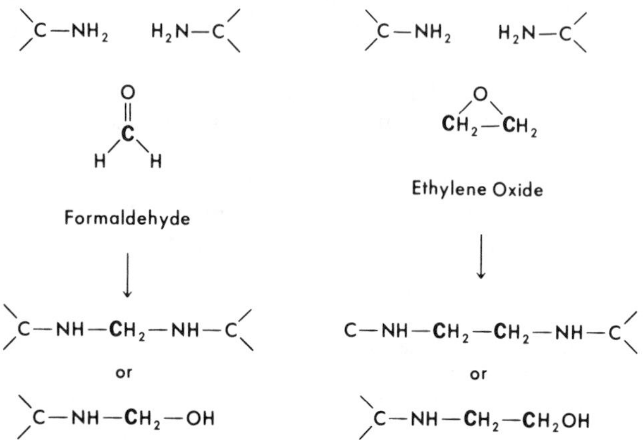

Surface-Active Agents. These compounds (**surfactants**) are generally called synthetic detergents. Such compounds, like fatty acids (soaps), contain both a hydrophobic and a hydrophilic portion; they therefore form micelles (large aggregates) in aqueous solution, in which only the hydrophilic portion is in contact with water; they can similarly form a layer that coats and solubilizes hydrophobic molecules or structures. Anionic detergents are only weakly bactericidal, perhaps because they are repelled by the net negative charge of the bacterial surface. Nonionic detergents are not bactericidal and may even serve as nutrients.

Cationic detergents, however, are active against all kinds of bacteria. The most effective types are the **quaternary** compounds, containing three short-chain alkyl groups as well as a long-chain alkyl group (e.g., benzalkonium chloride, Fig. 67-6). These compounds are widely used for skin antisepsis and for sanitizing food utensils. They act by disrupting the cell membrane, causing the release of metabolites; in addition, their detergent action provides the advantage of dissolving lipid films that may protect bacteria, and they leave a tenacious bactericidal surface film on the treated object.

A variety of cationic germicides are on the market. In the absence of adsorbing macromolecules or lipids these may be rapidly bactericidal at concentrations as low as 1 ppm; and unlike many other disinfectants they are not poisonous to man. Their activity is neutralized by soaps and phospholipids, since oppositely charged surfactants precipitate each other. With increasing chain length their bactericidal action (and their depression of surface tension) goes through a maximum because the tendency to form a surface layer increases but the solubility decreases.

Phenols. Phenol (C_6H_5OH) is both an effective denaturant of proteins and a detergent. Its bactericidal action involves cell lysis.

The antibacterial activity of phenol is increased by halogen or alkyl substituents on the ring, which increase the polarity of the

FIG. 67-5. Reactions of formaldehyde and ethylene oxide with amino groups. Similar condensations may take place with other nucleophilic groups. Bridges may be formed between groups on the same molecule or on different molecules.

FIG. 67-6. Benzalkonium chloride (benzyldimethyl alkonium chloride; Zephiran). A typical quaternary ammonium detergent; the long-chain alkyl group is a mixture obtained by the reduction and amination of the fatty acids of vegetable or animal fat.

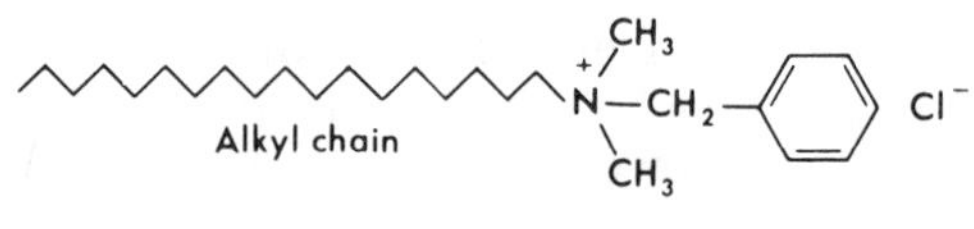

phenolic–OH group and also make the rest of the molecule more hydrophobic; the molecule becomes more surface-active and its antibacterial potency may be increased 100-fold or more. Phenols are more active when mixed with soaps, which increase their solubility and promote penetration. However, too high a proportion of soap impairs activity, presumably by dissolving the disinfectant too completely in soap micelles.

With increasing chain length the potency of phenols first increases and then decreases, presumably because of low solubility. With gram-negative organisms the maximum is reached at a relatively short length (Fig. 67-7); bulkier molecules are excluded, like certain drugs, by the outer membrane (Ch. 6).

A mixture of **tricresol** (mixed ortho-, meta-, and para-methylphenol) and soap is a widely used disinfectant for discarded bacteriologic materials. Its action is not impaired by the presence of organic matter, because it must be used in a relatively high concentration and it is not extensively destroyed or bound by organic molecules. **Hexylresorcinol** (4-hexyl-1,3-dihydroxybenzene) is used as a skin antiseptic.

Halogenated bis-phenols, such as **hexachlorophene**, are bac-

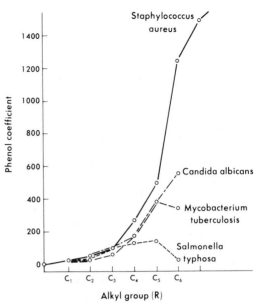

FIG. 67-7. Germicidal activity of homologous series of o-alkyl-p-chlorophenol derivatives against four organisms. The parent compound (p-chlorophenol) had a phenol coefficient (see text, above) of 4 against all the organisms. Note the strikingly increased activity of the longer-chain (C-6 and C-7) derivatives against the gram-positive Staphylococcus, and the opposite response of the gram-negative Salmonella. A similar divergence in response at greater chain lengths has also been observed with other homologous series of phenol derivatives. (Klarmann, EG, Wright ES: In Antiseptics, Disinfectants, Fungicides, and Sterilization. Reddish CF [ed] Philadelphia, Lea & Febiger, 1957, p. 506)

teriostatic in very high dilutions, and appear to be less inactivated by soaps or anionic detergents than are ordinary phenols. Hexachlorophene is also not very volatile and lacks the unpleasant odor of many phenols. It is widely used as a skin antiseptic (especially mixed with a detergent, as pHisoHex). However, its absorption through inflamed skin may cause serious systemic toxicity.

Alkyl esters of *p*-hydroxybenzoic acid are used as a preservative in foods and pharmaceuticals: they act on bacteria much like an alkyl-substituted phenol, but when taken by mouth they are nontoxic because they are rapidly hydrolyzed, yielding the harmless free *p*-hydroxybenzoate.

The **essential oils** of plants, which have been used since antiquity as preservatives and antiseptics, contain a variety of phenolic compounds, including thymol (5-methyl-2-isopropylphenol) and eugenol (4-allyl-2-methoxyphenol); the latter is used in dentistry as an antiseptic in cavities.

Alcohols. The disinfectant action of the aliphatic alcohols increases with chain length up to 8–10 carbon atoms, above which the water solubility becomes too low. Although **ethanol** (CH_3CH_2OH) has received widest use, **isopropyl alcohol** ($CH_3CHOHCH_3$) is less volatile and slightly more potent, and is not subject to legal restrictions.

The disinfectant action of alcohols, like their denaturing effect on proteins, involves the participation of water. Ethanol is most effective in 50%–70% aqueous solution: at 100% it is a poor disinfectant, in which anthrax spores have been reported to survive for as long as 50 days; and its bactericidal action is negligible at concentrations below 10%–20%. Some organic disinfectants (formaldehyde, phenol) are less effective in alcohol than in water because of the lowered affinity of the disinfectant for the bacteria relative to the solvent. On the other hand, alcohol removes lipid layers that may protect skin organisms from some other disinfectants.

Other organic solvents, such as ether, benzene, acetone, or chloroform, also kill bacteria but are not reliable disinfectants. However, the addition of a few drops of toluene or chloroform, which dissolve slightly in aqueous solutions, will prevent the growth of fungi or bacteria. **Glycerol** is bacteriostatic at concentrations exceeding 50% and is used as a preservative for vaccines and other biologicals, since it is not irritating to tissues.

INACTIVATION OF VIRUSES

Inactivation of viral particles is the permanent loss of infectivity. For some purposes (e.g., retention of immunogenic potency) it is important to know not only the degree of inactivation but also its consequences for various components of the virions.

THE DEGREE OF INACTIVATION

The exposure of a population of virions to a chemical or physical inactivating agent at a defined concentration for a limited time results in the inactivation of a proportion of the virions; the others retain infectivity. The proportion of virions that is inactivated is related to the **dose** (i.e., the product of time × concentration or intensity), as described for the action of radiations in Chapter 11, Appendix. The shape of the survival curve yields information about the mechanism of inactivation; and once the shape is known it is possible to calculate **the dose of the agent required** for a certain degree of inactivation.

What is the **required degree of inactivation?** As shown in Chapter 11, Appendix, in the usual semilog plot the survival curves may vary in shape at low doses of the inactivating agents and become straight lines at higher doses. In such a plot a survival equal to zero is never reached (since the log of zero is negative infinity). In practice, the sample exposed to inactivation contains a finite number of virions; therefore a dose can be reached at which less than a single surviving virion exists on the average in the whole sample. At such doses survival must be interpreted in terms of probability. For instance, if the theoretic survival is ½ a virion, there is a chance of ½ that a virion will survive in the sample; and if many similar samples are inactivated in the same way, half of them will contain an active virion. Although total inactivation thus cannot be reached with certainty it is possible to achieve a safe inactivation. The corresponding degree of theoretic inactivation varies for different purposes; it depends on the acceptable risk of retaining an active virion and on the total amount of the virus to be inactivated.

Example: A vaccine of inactivated virus is prepared, starting with a virus titer of 10^7 infectious units per ml, of which 0.1 ml is inoculated per individual; a 1% risk of infectious virus surviving in the inoculum is acceptable. A theoretic survival of 10^{-8} would be satisfactory for a single dose of vaccine. If a million individuals are to be inoculated, the same survival would cause 10^4 individuals to receive an infectious dose, and the risk would be unacceptable. For a million doses, the theoretic survival should become 10^{-14}, which would lower the risk to a 1% chance of one case for the total inoculated population.

Predictions Based on Survival Curves. Such low theoretic survivals cannot be determined experimentally, but can only be inferred by extrapolating the curves observed at much higher survivals. **Extrapolation always involves a risk** because one is never quite sure that the shape of the survival curve in the undetermined segment follows a predictable behavior. Barring unforeseeable events, however, extrapolation is justified if the survival curve is sufficiently well defined in the part that can be determined; faulty extrapolations usually derive from inadequate data in this region. Errors occur especially when the survival curve is of either the **multiple-hit** or the **multicomponent** type (Ch. 11, Appendix) and it is erroneously assumed that the slope of the lower part is known.

The **errors** that can arise are of **two types,** as shown in Figure 67-8. With a multiple-hit type of survival curve (case A) the dose of inactivating agent required for achieving an acceptable level of infectivity is overestimated, which may decrease the immunogenicity of the vaccine. With a multicomponent type of curve (case B) the dose is underestimated and a dangerously high level of infectious virus could remain. Errors of this type are more common, and their consequences may be more serious.

FIG. 67-8. Possible errors in predictions based on survival curves. In **A** the survival curve is multiple-hit and in **B** it is multicomponent; in both an inadequately determined curve leads to erroneous extrapolation. The estimated dose for acceptable inactivation is too large in **A**; in **B** it is only about 20% of that required, and the corresponding survival is about 10^4 times the acceptable level.

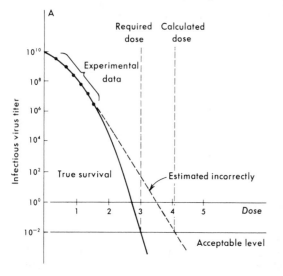
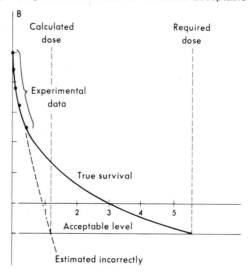

INACTIVATING AGENTS: MECHANISMS OF ACTION

Some agents are relatively selective, i.e., show a more pronounced action on one viral component than on another. Since this selectivity is important for practical applications, the agents are grouped according to their main target as nucleotropic, proteotropic, lipotropic, and universal (unselective) agents. However, the results vary with the composition of the virus. For instance, a nucleotropic agent will affect nucleic acid almost exclusively when acting on bacteriophages that contain 50% nucleic acid, but will also damage other components when acting on influenza virus, which contains only 1% nucleic acid. The various inactivating agents are as follows:

Nucleotropic agents: UV light of 260 nm wavelength; formaldehyde; nitrous acid; hydroxylamine; decay of radioactive elements that are incorporated in nucleic acid, such as ^{32}P or ^{3}H (incorporated as thymidine).

Proteotropic agents: UV light of 235 nm wavelength; heat; proteolytic enzymes; mild acid pH; sulfhydryl-containing or sulfhydryl-reactive compounds. Detergents and high concentrations of urea or guanidinium compounds disrupt inter- (and some intra-) capsomeric bonds, loosening the capsid of naked viruses.

Lipotropic agents: Lipolytic enzymes, lipid solvents.

Universal agents: X-rays; alkylating agents; photodynamic action.

The properties of some inactivating agents will be briefly analyzed in the following paragraphs.

Heat acts mostly by denaturation of capsid protein, since infectious nucleic acid can be extracted from inactivated virus. Often the effect is quite limited; for instance loss of the small capsid component VP4 in poliovirus (see Ch. 50) or of a minor component in the capsid of the helical phage f1 (see Ch. 47).

The half-life decreases rapidly with increasing temperature. Viruses vary greatly in their sensitivity: the very unstable Rous sarcoma virus has a half-life of only 1–2 h at 37°C. Viral half-lives are usually increased by the presence of various salts (especially Mg^{2+}), protein, cystine, or polysulfides, which stabilize the tertiary structure of the protein.

The curves for heat inactivation are usually of the multicomponent type. With poliovirus they clearly differentiate a major rapid inactivation (with a one-hit curve) and a minor slower inactivation. The minor component is probably **free viral RNA**, released from the virions by the heat and able to produce plaques with low efficiency (Ch. 50).

UV light at 260 nm acts mainly on nucleic acids, as noted above. This agent is not well suited for viral inactivation because its damage to nucleic acids can be repaired by a variety of enzymatic and genetic mechanisms (Ch. 11); the inactivation must be measured after allowing maximal reactivation. UV is also absorbed by many substances in biologic media, and shadows are projected by opaque particles; hence the risk of virions' escaping inactivation is high.

X-rays cause the production of highly reactive and short-lived chemicals within and around the virions; these unstable products interact with various viral components **(direct effect)**. The reactive chemicals also cause changes in components of the medium that produce long-lived inactivating poisons **(indirect effect)**; this effect is minimized by a medium rich in organic molecules, especially amino acids. The direct effect results in one-hit survival curves; the indirect effect results in multiple-hit curves.

Formaldehyde is widely used for the production of killed vaccines. It reacts mainly with amino groups in proteins and in single-stranded nucleic acids (Fig. 67-5). Thus virions with double-stranded nucleic acid are inactivated mostly by modification of the protein; but in the presence of formaldehyde the helical structure melts at relatively low temperatures, freeing the amino groups. The activation curve is of the multicomponent type, as shown in Figure 67-9; hence extrapolation to a given degree of inactivation is difficult, as discussed above.

Ethylene oxide (Fig. 67-5) is an effective agent for inactivating all viruses, with one-hit kinetics. Other alkylating agents, nitrous acid, and hydroxylamine act on nucleic acids as described in Chapter 11 (Action of Chemical Mutagens): the former two also act similarly on proteins.

Lipid Solvents. Ether (usually added at 20% to a suspension of virions) readily inactivates enveloped but not naked viruses. As we have seen in many preceding chapters, this simple test is widely used for distinguishing the two groups.

Enzymes. Most viruses are resistant to the action of trypsin, but some are sensitive (e.g., some arboviruses and, under special conditions, some orthomyxoviruses). Viruses are more readily inactivated by proteolytic enzymes of broader specificity, such as pronase. Enveloped viruses may be inactivated by some lipolytic enzymes: some arboviruses are inactivated by phospholipase A, and influenza virus treated with phospholipase C becomes susceptible to inactivation by a protease.

Photodynamic Inactivation. The presence of certain dyes (acridine orange, proflavin, methylene blue, neutral red) can render virions

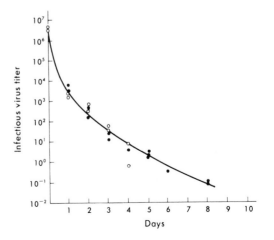

FIG. 67-9. Time course of the inactivation of type 2 poliovirus by formaldehyde, showing the marked multicomponent character of the survival curve. (Modified from Gard S: In Wolstenholme GEW, Millar ECP [eds]: The Nature of Viruses. CIBA Foundation Symposium; Boston, Little, Brown, 1957)

susceptible to inactivation by exposure to visible light, as described above for bacteria. Most viruses become photosensitive simply on being mixed with the dye, but others (e.g., poliovirus) must multi-ply in its presence, thus enclosing dye within the capsid. The latter group remain sensitive in the absence of external dye, but they lose photosensitivity on infecting a cell and releasing the nucleic acid.

SELECTED READING

BOOKS AND REVIEW ARTICLES

ALBERT A: Selective Toxicity, 5th ed. London, Methuen, 1978

BEUCHAT LR: Injury and repair of gram-negative bacteria, with special consideration of the involvement of the cytoplasmic membrane. Adv Applied Microbiol 23:219, 1978

BLOCK SS (eds): Disinfection, Sterilization, and Preservation. Philadelphia, Lea & Febiger, 1968

BRUCH CW: Gaseous sterilization. Annu Rev Microbiol 15:245, 1961

CHICK H, BROWNING CH: The theory of disinfection. Med Res Coun Syst Bacteriol [Lond] 1:179, 1930

GRAY TRG, POSTGATE JR (eds): The Survival of Vegetative Microbes. Cambridge University Press, 1976

HUGO WB (ed): Inhibition and Destruction of the Bacterial Cell. New York, Academic Press, 1971

LEA DE: Actions of Radiations on Living Cells, 2nd ed. London, Cambridge University Press, 1955

MOSELY BEB, WILLIAMS E: Repair of damaged DNA in bacteria. Adv Microb Physiol 16:99, 1977

PHILLIPS CB, WARSHOWSKY B: Chemical disinfectants. Annu Rev Microbiol 12:525, 1958

PIERSON MD, GOMEZ RF, MARTIN SE: The involvement of nucleic acids in bacterial injury. Adv Applied Microbiol 23:263, 1978

WELLS WF: Air-borne Contagion and Air Hygiene. Cambridge, Harvard University Press, 1955

INDEX

1320 / INDEX

Legionnaires' disease *(continued)*
  history, 798
  treatment, 799
Leishmania donovani, immune defense
    against, 503
Leishmaniasis, association with T cell
    deficiency, 509
Leistikow and Loeffler, 636
Lennette, 584
Lentivirus, 1259
Lentivirus E, 1244*t*
Leporipoxvirus, 1078
Lepromatous leprosy, association with T
    cell deficiency, 509
  tumor incidence with, 544
Lepromin, 740
Leprosy, 739–741
  clinical features, 740
  dapsone treatment, 117
  diagnosis, 741
  epidemiology, 741
  etiologic agent. See *Mycobacterium leprae.*
  history, 724
  lepromatous. See *Lepromatous leprosy.*
  phases, 740
  rat, 739
  treatment, 741
Leptospira biflexa, 760–761
Leptospira interrogans
  morphology, **753**
  serotype *autumnalis,* 761
  serotypes, 760
Leptospires. See also *Leptospirosis;*
    Spirochetes.
  antigenicity, 761
  classification, 760
  identification of pathogenic types, 761
  morphology, 760
Leptospirosis, 761
  incidence, 752
Leptotrichia, 668
Lesion
  Acinetobacter antitratum in, 674
  actinomycosal, 747
  edema zone, streptococci in, **610**
  nodular focal, 1079
  open and closed, 570
  pneumonic, **603**
  Pseudomonas aeruginosa in, 674
  suppurative, 575
  thrombotic, 621
  transmission of, 570
Leucine, biosynthesis, 52
Leuconostoc mesenteroides, fermentation yields,
    36
Leukemia
  acute lymphatic (ALL), 418
  bone marrow destruction, 538
  chronic lymphatic, increase in circulat-
    ing lymphocytes, 396*n*
  correlation to antibody titer, **545**
  lymphocytic, 396*n,* 541
  in mice, 396*n,* 541
Leukemia viruses, 1251
  acute, 1252
Leukoagglutination, 406
Leukocidin, 559

Leukocyte
  adhesion to endothelium, 562
  basophilic. See *Basophil.*
  chemotaxis, S-adenosyl methionine
    requirement, 101*n*
  eosinophilic. See *Eosinophil.*
  polymorphonuclear. See *Polymorpho-
    nuclear leukocyte.*
  role in defense, 568, **568**
Leukopenia, 515
Leukosis viruses. See also *Oncogenic viruses;
    oncoviruses.*
  as a helper, 1253
  assembly and maturation, **1247**
  type C, **1248**
  defective, 1251, 1252–1253
  nondefective, 1250–1252
  protein synthesis, **1246**
  role in neoplasia, 1255
Leukotriene C, 479*t,* 480
Levan, 55
Lewis blood groups, 528
Liebig, 4
Life
  defined in terms of self-replication, 12
  discussion of, 854–855
  minimal unit of, 12
Life cycle, 132
Ligand. See also *Antigen; Hapten; Immunogen.*
  association rates, 302
  binding sites, topography of, 305
  bonds to antibodies, 305
  definition, 298
  distal projections, 303
  equilibrium dialysis. See *Equilibrium
    dialysis.*
  fluorescence quenching by, 330
  incubation, 300
  intrinsic affinity of binding site, *298*
  ionic binding, 300
  monospecific system, 307, **307**
  non-ionic binding, 300
  partition of, 304
  and protein structure, 251
  and regulation, 262
  solubility and complex stability, 304
  specific binding, **301**
  terminal residues, 303
  thermodynamics of binding, 302
  univalent, 298
    binding by multivalent antibody, 334
Ligase, 174–175, 187
  T4, 176
Light (L) chains. See under *Immunoglobulin,
    light chains.*
Light microscope. See under *Microscope,
    light.*
Limiting dilutions, 6–7
Limulus polyphemus, use of blood in LPS
    assay, 655
Lincomycin, 120–121, **121,** 580
  action, 248
  induction of pseudomembranous entero-
    colitis, 721, 722
  resistance, 578
    mechanism, 125*t*
Linnaeus, Carolus, 2, 25*n*

Lipid A, 85, **85,** 654–655
  biosynthesis, 87, 90–91
  composition, variation in, 91
  interactions, 88
  mycobacterial, **727**
  in phages, 915
  structure, 90–91
Lipid solvents, inactivation of viruses,
    1273
Lipmann, 57, 241
Lipoamino acids, 91
Lipoic acid, 33
Lipopolysaccharide (LPS)
  antigenic effects, 394, 427
  attachment to outer membrane, 88
  biosynthesis, **86,** 87
  core, 85–86, **86**
  extraction, 85*n*
  gal mutant forms, 88
  gonococcal, 641
  in vitro assay, using limulus blood,
    655
  lipid A, **85**
  mitogenic activity, 395
  in outer membrane, 82–83
  and penicillin, 81
  polymyxins and, 92
  R mutants, 85–86
  side chains, 86–87
  structure, 85–87
  transfer to outer membrane, 87–88
  triggering surfaces on, 457
Lipoproteins, in outer membrane, 82–83
Liposomes, 459
Lipoteichoic acid, 78
Lipotropic agents, 1273
Listeria, characteristics, 799*t*
Listeria monocytogenes, 799–800. See also
    Listeriosis.
Listeriosis
  clinical manifestations, 800*t*
  history, 799
  pathogenesis and treatment, 800
Lister, Joseph, 6–7, 1268
Liver
  hepatitis of. See *Hepatitis.*
  immune disease of, 510*t*
Liver cell. See *Hepatocyte.*
Loeffler, 7–8, 586, 590
Loeffler and Frosch, 854
Loeffler's coagulated serum medium,
    587
Loeffler's slant, 591
Logarithms, 65
Lollipops, in electron micrographs, 198,
    220
Louse
  Borrelia in, 760
  carrier of typhus, 768
Lower intestinal tract, flora, 809*t*
LPS. See *Lippopolysaccharide.*
Luciferase, 38
Luciferin, oxidation of, 38
Lucke, 1232
Lucretius, 2
Lupus erythematosus, 461, 489, **490,**
    1255

W